Lippincott®

Illustrated Reviews

Biochemistry

Second South Asian Edition

As per the Indian Competency-Based Medical Curriculum

Lippincott® Illustrated Reviews

Biochemistry

Second South Asian Edition

As per the Indian Competency-Based Medical Curriculum

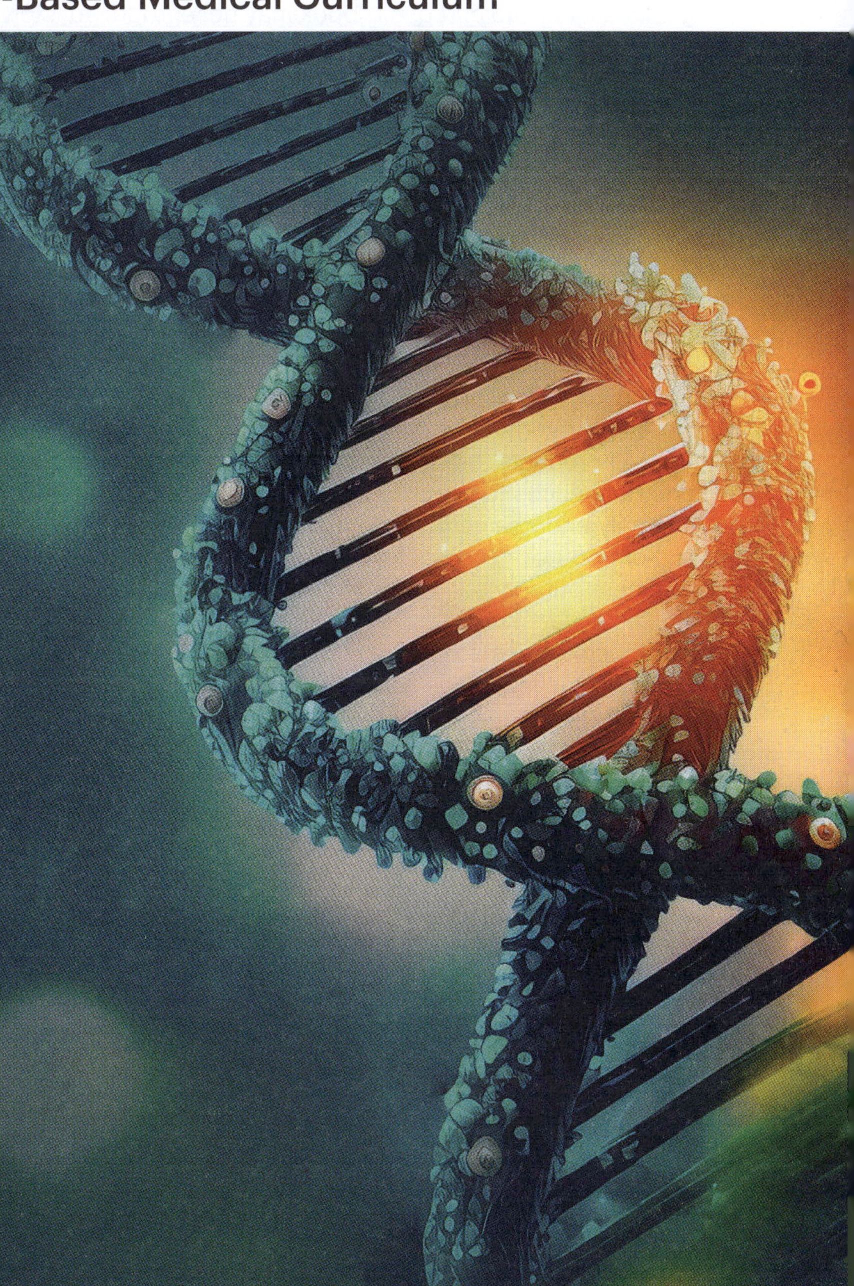

SAE Editors

Editor-in-Chief

Ritu Singh, MD
Director Professor and Head
Department of Biochemistry
Lady Hardinge Medical College and
Associated Smt. Sucheta Kriplani Hospital
and Kalawati Saran Children's Hospital
New Delhi

Co-Editor

Rajeev Goyal, MD
Associate Professor
Department of Biochemistry
Lady Hardinge Medical College and
Associated Smt. Sucheta Kriplani Hospital
and Kalawati Saran Children's Hospital
New Delhi

International Editor

Denise R. Ferrier, PhD
Professor Emerita
Department of Biochemistry and Molecular
Biology Drexel University College of Medicine
Philadelphia, Pennsylvania

Publishing Manager: Ravi Vishwakarma
Development Editor: Nayan Gogoi
Asst. Manager Manufacturing: Sumit Johry

3rd Floor, Tower B, Building No. 10, Phase II, DLF Cyber City, Gurgaon, Haryana - 122002

ISBN-13: 978-81-97055-97-3
E-ISBN: 978-81-97055-96-6
Second Impression, 2025

Published by Wolters Kluwer (India) Pvt. Ltd., New Delhi
Compositor: Design Modus, New Delhi (www.designmodus.in)
Printed at: Samrat Offset Pvt. Ltd.

For product enquiry, please contact – Marketing Department (marketing@wolterskluwerindia.co.in) or log on to our website www.wolterskluwerindia.co.in.

Preface

The first South Asian Edition of Lippincott® Illustrated Reviews: Biochemistry, was an adaptation of the international edition of Lippincott® Illustrated Reviews: Biochemistry, a long-standing book of international fame, to suit the requirements of South Asia, particularly of India. Fourteen new chapters were added to the previous edition to meet the requirements of the National Medical Commission's Competency-Based Curriculum, ensuring that primary healthcare physicians understand basic Biochemistry concepts and can apply them effectively in patient care.

The first South Asian edition of Lippincott® Illustrated Reviews: Biochemistry was well received by both students and teachers. It covered all the competencies in biochemistry for the MBBS first phase, as well as the topics required for vertical and horizontal integration with other subjects across all phases of the MBBS curriculum.

This **second South Asian edition of Lippincott® Illustrated Reviews: Biochemistry** has many important updates. This updated and revised edition emphasizes clinical application and as such several new clinical cases have been added to all the units. Each chapter now has 10 multiple choice questions of MBBS Ist professional level, with explanations provided online. Also, there are additional 5 MCQs of NEET PG/NExT level, which involve concepts covered across the three phases of MBBS. These additional 5 questions may be taken as self-directed learning (SDL) opportunities or they may be addressed as the student progresses to higher levels. New Clinical Application boxes have been added in some chapters to train students to confidently apply biochemistry in clinical scenarios.

Five clinical subject matter experts, from various institutions, have contributed to the Clinical Cases Questions given at the end of each Unit, which closely represent the actual case scenarios. As the knowledge required to interpret clinical cases is covered in several chapters, all the cases have been put together at the end of units. Further, some clinical cases may require knowledge from various units and such cases carry a pertinent note regarding this. This enables the student to have holistic knowledge applicable to real patients.

Biochemistry explores how our bodies use nutrients to create building blocks, fuels and cell communication molecules. It also covers chemical conversion and elimination processes. This book offers a concise review of these mechanisms, including illustrative examples and the use of concept maps for organization. The text has been modified to enhance understanding, with important topics highlighted in bold for easier recall by students.

In addition, there is a wide array of useful online resources—Clinical Cases ( ) and Multiple-Choice Questions with answers and detailed explanations, other than those present in the book.

Dr. Ritu Singh acknowledges the blessings of her parents, Dr. RV Singh and Mrs. Rama Singh, the encouragement of her husband, Dr. AV Sharma, who has been a pillar of support, and the understanding of her daughter, Dr. Ananya Sharma, son, Mr. Akash Sharma, and daughter-in-law, Mrs. Eepsa Malhotra.

Dr. Rajeev Goyal would like to acknowledge his mother late Smt. Saroj Goyal for her eternal blessings, father Sh. Suresh Chand, and wife Dr. Monil Singhai for their constant encouragement and support and children Advit and Maanya. The children have been patient with an understanding beyond their years.

The Editors would also like to express their gratitude to the publishing team at Wolters Kluwer (India), Ms Sheenam Aggarwal, Mr Ravi Vishwakarma and Mr Nayan Gogoi for their professionalism, expertise and commitment to excellence in bringing this edition to fruition.

Finally, and most importantly, we extend our gratitude to our students, who have been the driving force behind this endeavor. We believe that Biochemistry has often been portrayed as a dry subject focused on memorization and cycles. With this book, our aim is to cultivate a passion for comprehending biochemical reactions and their disruptions, which, once understood, are never forgotten. We wish our students all the best as they embark on the challenging and highly responsible journey of becoming medical doctors.

"To acquire Knowledge one must study; To acquire wisdom, one must observe"

Marilyn vos Savant

May the quest and curiosity for understanding the myriad reactions and their implications in our body—a complex soup of chemicals and charges—shape students into responsible learners. They must understand that today's biochemistry is tomorrow's medicine. This noble and compassionate profession of healing requires in-depth knowledge.

Best Wishes!

Ritu Singh

Rajeev Goyal

Contributors to
LIR Biochemistry Second South Asian Edition

Associate Editors

Ashikh Seethy, MD, PhD, Assistant Professor, Department of Biochemistry, All India Institute of Medical Sciences, Guwahati, Assam

Monica Kakkar, MD, Additional Professor, Department of Biochemistry & Associate Dean Academics, All India Institute of Medical Sciences, Bathinda, Punjab

Clinical Subject Matter Experts

Sharmila Banerjee Mukherjee, MD, Director Professor, Department of Paediatrics, Lady Hardinge Medical College & Associated Hospitals, New Delhi

Jagdish P. Goyal, MD, MAMS, Professor, Department of Pediatrics, All India Institute of Medical Sciences, Jodhpur, Rajasthan

Ritika Sud, MD, Professor, Department of Medicine, Lady Hardinge Medical College & SSK Hospital, New Delhi

Ankit Jain, MD, DM, Associate Professor, Department of Cardiology, Vardhman Mahavir Medical College and Safdarjung Hospital, New Delhi

Neha Taneja, MD, Department of Dermatology & Venereology, All India Institute of Medical Sciences, New Delhi

Ananya Sharma, MD, Department of Dermatology & Venereology, All India Institute of Medical Sciences, New Delhi

Publisher's Acknowledgments

The publisher would like to thank the following reviewers for providing their valuable suggestions.

A. Pullaiah, Professor and HOD, Department of Biochemistry, Kakitaya Medical College, Warangal

Alapaty Shailaja, Professor and HOD, Department of Biochemistry Vydehi Institute of Medical Sciences and Research Institute, Bangalore, Karnataka

Alka Singh, Professor and HOD, Department of Biochemistry, The Oxford Medical College, Hospital and Research Centre Bengaluru

Amareshwara M Maligi, Professor and HOD, Department of Biochemistry, Koppal Institute of Medical Sciences, Koppal, Karnataka

Amrut Dambal, Professor and HOD, Department of Biochemistry, KLE's JGMM Medical College, Hubballi, Karnataka

Anisa Atif, Professor and Head, Department of Biochemistry AIIMS, Rishikesh

Anita Verma, Professor and Head, Department of Biochemistry, Sardar Patel Medical College, Bikaner

Aparna Varma, Professor, Department of Biochemistry, All India Institute of Medical Sciences, Bibi Nagar, Telangana

Ashish Jain, Associate Professor, Department of Biochemistry, People's College of Medical Sciences (PCMC), Bhopal

Ashish Sharma, Professor and Head, Department of Biochemistry, Geetanjali Medical College and Hospital Udaipur

Asif, Professor and HOD, Department of Biochemistry, Shadan Medical College, Hyderabad

B. Aruna Kumari, Professor and HOD, Department of Biochemistry, Government Medical College, Nalgonda

B. Dharmendhar, Professor and HOD, Department of Biochemistry, Government Medical College, Khammam

Basavaraj Devaranavadagi, Professor and HOD, Department of Biochemistry, Shri B M Patil Medical College, Vijayapura, Karnataka

Basavaraj Savadi, Professor and Head Department of Biochemistry, Basaveshwara Medical College, Chitradurga, Karnataka

Bhagyashree K Bhuyar, Assistant Professor, Department of Biochemistry, Koppal Institute of Medical Sciences, Koppal, Karnataka

Bindu C M, Professor and HOD Department of Biochemistry RajaRajeshwari Medical College and Hospital, Bengaluru

Cariappa K V, Professor and HOD Department of Biochemistry Kodagu Institute of Medical Science, Madikeri

Desai Vidya Sripad, Professor and Head, Department of Biochemistry, AIIMS, Mangalagiri, Andhra Pradesh

Doddappa M. Bannigida, Assistant Professor, Department of Biochemistry, MVJ Medical College and Research Hospital, Bengaluru

Elvia Jamatia, Assistant Professor, Department of Biochemistry BSAMC, Delhi

G R Sreekanth, Professor and HOD, Department of Biochemistry, Raichur Institute of Medical Science, Raichur

G. Ravi Kumar, Professor and HOD, Department of Biochemistry, MNR Medical College, Sangareddy

G. Uma Ramani, Professor and HOD, Department of Biochemistry, Gitam Medical College, Visakhapatnam

Garima Sehgal, Professor and HOD Department of Biochemistry, The Oxford Medical College, Hospital and Research Centre Bengaluru

Geetha Meenakshi, Professor and HOD, Department of Biochemistry, Government Medical College, Mahabubnagar

Gitanjali Goyal, Professor and Head, AIIMS, Bhatinda

H Kishan Reddy, Professor and Head, Department of Biochemistry, Chalmeda Anand Rao Institution of Medical Sciences, Karimnagar, Telangana

Himanshu Madaan, Professor and Head, Department of Biochemistry KCGMC, Karnal

Isha Rani, Assistant Professor, MMU, Sadopur

J R Keshari, Additional Professor, IGIMS, Patna

Jagannadha Phaneendra DS, Professor and HOD, Department of Biochemistry, Neelima Medical College, Hyderabad

Jagat R Kanwar, Professor and Head, Department of Biochemistry AIIMS, Bhopal

Jairam Rawatani, Professor, Department of Biochemistry, Dr. Sampurnanand Medical College, Jodhpur

Jayanta De, Professor and HOD, RG Kar Medical College, Kolkata

Jusmita Dutta, Professor and Head, Department of Biochemistry LNMC, Bhopal

K. Pratibha, Professor and HOD Department of Biochemistry ESIC Medical College and PGIMSR, Bengaluru

K. Saradamba, Professor and HOD, Department of Biochemistry, RIMS Medical College, Srikakulam

Kalaivanam K N, Professor and HOD Department of Biochemistry Sapthagiri Institute of Medical Sciences and Research Center, Bengaluru

Kapil Gupta, Professor, Adesh Medical College, Bhatinda

Karthikeyan P, Assistant Professor, Department of Biochemistry AIIMS, Delhi

Kaushalraj Dandegaonker, Ex-Tutor, Department of Biochemistry, Rajiv Gandhi Medical College, Thane, Maharashtra

Kavita Rasalkar, Professor and HOD, Manipal Tata Medical College, Jamshedpur

Krishnaswamy Dyamaiah, Professor Department of Biochemistry, Vijayanagara Institute of Medical Sciences, Ballari, Karnataka

Laxmi Narayana, Professor and Head, Department of Biochemistry, TRR Institute of Medical Sciences, Inole-Hyderabad

M B Prathima, Professor and HOD, Department of Biochemistry, A J Institute of Medical Science, Mangaluru

M. Madhava Latha, Professor and HOD, Department of Biochemistry, Mamata Medical College, Khammam

M. Uma Devi, Professor and HOD, Department of Biochemistry, Bhaskar Medical College, Hyderabad

M.G Kalejar, Professor and Head, Department of Biochemistry RKDF Medical College, Bhopal

Mahadeva S K, Professor and HOD, Department of Biochemistry, Mandya Institute of Medical Science

Maitreyee D S, Professor and HOD, Department of Biochemistry, Father Muller Medical College, Mangaluru

Manish Kumar Sinha, Assistant Professor, Patna Medical College and Hospital, Patna

Manju Lata Kumawat, Professor, Department of Biochemistry, PGIMS, Rohtak

Manjula K S, Professor and HOD, Department of Biochemistry, Sri Chamundeshwari Medical College-SCMCH

Maria Thomas, Professor, Christian Medical College, Ludhiana

Minakshi, Associate Professor, Department of Biochemistry, Government Medical College, Dausa

Minni Verma, Professor, Gian Sagar Medical College, Banur

Mirza Sharif Ahmed Baig, Professor and HOD, Department of Biochemistry, Khaja Bandanawaz Institute of Medical Science, Kalaburagi

Mithu Banerjee, Professor and Head, Department of Biochemistry, All Indian Institute of Medical Science, Jodhpur

N. Krishnamurthy, Professor and HOD Department of Biochemistry BGS Global Institute of Medical Sciences, Bengaluru

Nagalakshmi C S, Professor and HOD Department of Biochemistry, Chikkaballapura Institute of Medical Sciences, Chikkaballapura, Karnataka

Neelam Aggarwal, Assistant Professor, Shaheed Nirmal Mahto Medical College, Dhanbad

Neeru Bhaskar, Professor and Head, Adesh Medical College, Shahbad

Nitin, Professor and HOD, Department of Biochemistry, KIMS Medical College, Narketpally

Pampa Reddy Kollur, Professor and HOD Department of Biochemistry Yadgir Institute of Medical Science, Yadgir

Prashant Paunipagar, Professor and HOD, Department of Biochemistry, ESIC Medical College and PGIMSR, Gulbarga

Preethi, Professor and HOD, Department of Biochemistry, NRI Medical College, Visakhapatnam

Purnima jindal, Professor, MMU Medical College, Sadopur

Raghavendra D S, Professor and HOD Department of Biochemistry, Sri Siddhartha Institute of Medical Sciences, Bangalore Rural, Karnataka

Raghavendra U, Professor and HOD Department of Biochemistry Yenepoya Medical College, Mangaluru

Raghunath H, Associate Professor, Department of Biochemistry, Mandya Institute of Medical Science

Rajinderjit Singh Ahi, Professor and Head, GGS Medical College, Faridkot

Rama Shouri, Professor and HOD, Department of Biochemistry, Gandhi Medical College, Hyderabad

Ramesh Bhatt, Professor and HOD, Mata Guzri Memorial Medical College, Kishanganj

Ravi Kishore Polepalli, Professor and HOD, Department of Biochemistry, MVJ Medical College and Research Hospital, Bengaluru

Ravirala Tagore, Professor and HOD, Department of Biochemistry, Mahaveer Medical College, Vikarabad

Ruqaya Aziz, Professor and Head, SKIMS, Srinagar

S M Awanti, Professor and HOD, Department of Biochemistry, Mahadevappa Rampure Medical College, Kalaburagi

Sahiba Kukreja, Professor and Head, SGRD Medical College, Amritsar

Sampath, Professor and HOD, Department of Biochemistry, ESI Medical College, Hyderabad

Sandhya Mishra, Professor and Head, Department of Biochemistry, SMS Medical College, Jaipur

Sangappa Kashinakunti, Professor and Head, Department of Biochemistry, S. Nijalingappa Medical College, Navangar, Bagalkot, Karnataka

Sapna Vyakaranam, Professor and Head, Department of Biochemistry, Mediciti Institution of Medical Sciences, Hyderabad

Sarla Mahawar, Professor, Department of Biochemistry, J.L.N. Medical College, Ajmer

Satish Kumar D, Professor, Department of Biochemistry, ESIC Medical College and PGIMSR, Kalaburgi

Shalini Gupta, Professor and Head, AIIMS, Mohali

Shilpa HD, Professor and HOD Department of Biochemistry, Dr. Chandramma Dayananda Sagar Institute of Medical Sciences and Education Research, Ramnagara District, Karnataka

Shishir Kumar, Assistant Professor, RIMS, Ranchi

Shivakumar, Professor and HOD Department of Biochemistry, Sri Siddartha Medical College, Tumkur, Karnataka

Siva Prabodh, Professor and Head, Department of Biochemistry, NRI Medical College, Mangalagiri, Andhra Pradesh

Smita Sonoli, Professor and HOD Department of Biochemistry, KAHER's Jawaharlal Nehru Medical College, Belagavi, Karnataka

Sridevi Vinit, Professor, Department of Biochemistry, Subbaiah Institute of Medical Sciences, Shivamogga, Karnataka

Suhasini, Professor and HOD, Department of Biochemistry, SVS Medical College, Mahabubnagar

Sukanya Shetty, Professor and HOD, Department of Biochemistry, K S Hegde Medical Academy, Mangaluru

Sukhraj Kaur Dhillon, Assistant Professor, GMC, Amritsar

Sumathi M E, Professor and HOD Department of Biochemistry, Shri Atal Bihari Vajpayee Medical College and Research Institute, Bengaluru

Sumit Jhajharia, Professor, KIMS, Bhubaneswar

Sunita Manhas, Professor and Head, Department of Biochemistry MM Medical College, Ambala

Sunitha, Professor and Head, Department of Biochemistry, Katuri Medical College, Guntur, Andhra Pradesh

Suresh Babu Ganji, Professor and Head, Department of Biochemistry, Government Medical College, Siddipet, Telangana

Suryabhan Lokhande, Associate Professor, Department of Biochemistry, AIIMS, Bhopal

(Maj Gen) T K Saha, Professor and Head, Department of Biochemistry, Amrita School of Medicine, Faridabad

Tarun Kumar Sharma, Assistant Professor, GMCH, Chandigarh

V Sunanda, Professor and Head, Department of Biochemistry, Malla Reddy Institution of Medical Sciences, Hyderabad, Telangana

V. Bhavani, Professor and Hod, Department of Biochemistry, Andhra Medical College, Visakhapatnam

Vanitha Gowda M N, Professor and HOD
Department of Biochemistry, Ramaiah Medical College and Hospital, Bangalore, Karnataka

Vasundhara Devi, Professor and HOD, Department of Biochemistry, Dr. Patnam Mahender reddy Medical College, Chevella

Veerabhadra Goud GK, Professor, Department of Biochemistry, MNR Medical College, Sangareddy

Vijaya Kumari, Professor and Head, Department of Biochemistry, ESIC Medical College, Alwar

Vijaya Lakshmi, Professor and Head, Department of Biochemistry, Malla Reddy Institution of Medical Sciences for Women, Hyderabad, Telangana

Vishal Kalasker, Professor and HOD, Department of Biochemistry, Navodaya Medical College Hospital and Research Center, Raichur

Viswa Kumar, Professor and Head, Department of Biochemistry, Narayana Medical College, Nellore, Andhra Pradesh

Vivian D Souza, Professor and HOD, Department of Biochemistry, Kanachur Institute of Medical Science, Mangaluru, Karnataka

Contents

How to Use the Book

Lippincott® Illustrated Reviews: Biochemistry

The second South Asian edition of Lippincott® Illustrated Reviews: Biochemistry comprehensively covers all competencies in biochemistry for undergraduate medical students. Its aim is to enable students to grasp concepts in a rational, logical, and practical manner, thereby preparing them for their future roles as primary healthcare physicians. As a student of medical biochemistry, it is important that clinical applications of the subject are understood. To facilitate this, numerous features have been integrated across all units.

Blue boxes, integrated throughout chapters, reflect relevance of important biochemical interactions or extra relevant knowledge.

> **One-carbon metabolism**
>
> One-carbon groups, collectively known as "**one carbon pool,**" are important donors of carbon atoms for synthesis of various compounds. These include **formyl, formimino, methenyl, hydroxymethyl, methylene and methyl groups**. Tetrahydrofolic acid (THFA) serves as carrier of these one-carbon units, accepting them during catabolic reactions and donating them during various synthetic reactions. (p. 402)

Throughout the chapter, essential texts have been emphasized in **bold** to facilitate easier comprehension and highlight key points.

> one-
> esis of
> ciency
> cause
> vomen
> vege-
>
> nents:
>), and
>
> cid, in
>
> **B. Folate deficiency**
>
> Inadequate serum levels of folate can be caused by:
>
> - **Increased demand:** Seen in pregnancy and lactation (see p. 514).
> - **Poor absorption:** Caused by **pathology of the small intestine, alcoholism**, or treatment with drugs (for example, methotrexate) that are *dihydrofolate reductase* inhibitors (see Fig. 32.2). A folate-free diet can cause a deficiency within a few weeks.
>
> Several folate antagonists, such as **methotrexate, aminopterin and amethopterin, are competitive inhibitors of enzyme DHFR, and are used as anti-cancer agents**. As THF is not available for DNA synthesis, DNA synthesis activity of tumor cells is inhibited.

> **Clinical Application 1.1: Slower, Longer-Acting Insulin Created by Substituting Amino Acids**
>
> Insulin glargine was first approved for use in the United States in the year 2000. It is a slower-acting form of insulin created in the laboratory by replacing the asparagine normally at position 21 on the A chain of insulin with glycine, and extending the carboxy terminus by two additional arginine residues. The result of these changes is a less water-soluble form of insulin with a net charge of +0.2, which is closer to 0, causing a slower absorption of insulin glargine from the site of injection. The glycine substitution prevents deamidation of the asparagine at acidic pH in the neutral, subcutaneous space. The additional arginine residues shift the isoelectric point from pH 5.4 to pH 6.7, making the molecule more soluble at acidic pH and less soluble at neutral pH. Insulin glargine is therefore a form of insulin that acts slowly, has longer activity, and requires less frequent injection. This form of insulin can be useful in the treatment of diabetes mellitus and help patients achieve better glycemic control. (See Chapter 23 for the structure of insulin.)

Clinical Application boxes train students to confidently apply biochemistry in clinical scenarios.

> **V. Medical Perspectives**
>
> - A resistance to methotrexate (MTX), an anticancer drug is often seen in patients receiving MTX for a long time. This is because of **gene amplification** of *dihydrofolate reductase (DHFR)* gene leading to increased number of *DHFR* genes, causing increased production of *DHFR*. Therefore, the clinician often needs to increase dosage of MTX in such patients
> - As the individual grows from the embryonic stage to an adult, his hemoglobin changes to a different type. This is achieved by **gene switching**. Embryonic hemoglobin has two alpha and two epsilon chains. This is replaced by fetal Hb at 6 months of life which has two alpha and two gamma chains. After birth, two alpha and two beta chains (HbA1) or two alpha and two delta chains are synthesized (HbA2). Defects in gene switching can give rise to severe hemoglobinopathies.
> - Gene switching is also responsible for class switching of immunoglobulins during transition from primary to secondary immune response. During **primary response IgM** is produced, but in **secondary response IgG** with same antigen specificity is synthesized.

At the end of each chapter, there is a crucial section titled "**Medical Perspectives,**" designed to assist aspiring doctors in grasping the practical application of the respective subject matter in common clinical conditions.

Clinical Case Questions

Chapter 31 Nutrition: Overview and Macronutrients

31.1. A 5-year-old boy is brought with complaints of failure to gain weight. His parents are illiterate laborers, with low income. There is a history of recurrent infections in this child. The child has not received any vaccination. On examination, child appears irritable. Both weight and height are lower than those expected for his age. The skin is dry, hair is brittle, and muscle appears wasted. There is no edema.

A. What is the diagnosis and management of this patient?

Answer—A: This is a case of Marasmus.

Marasmus is a form of severe malnutrition characterized by energy deficiency. A child with marasmus looks emaciated. Body weight is reduced to less than 60% of the expected body weight for the age.

Poor health conditions, insufficient maternal and child feeding and care, and inability to purchase nutritious food, compounded by high rates of illiteracy, play a complex and interrelated role as contributing factors to the undernutrition.

Clinical features include stunted growth, extreme muscle wasting, depletion of subcutaneous fat, weakness, and anemia. There is no edema.

Treatment aims to provide adequate nutrition, restore normal body composition, and cure the condition that caused the deficiency. Protein-rich

Furthermore, to reinforce the concepts, **clinical cases** commonly encountered in medical practice are included at the conclusion of each unit. These cases encompass topics discussed throughout the unit and sometimes extend across multiple units.

Appendix
Unit VII: Medical Nutrition

DIETS FOR SPECIFIC CONDITIONS

A. Diet for Diabetic Patients

The aim of diet management in Diabetes Mellitus is to maintain near normal blood glucose levels, control hyperlipidemia and hypertension, and prevent complications of diabetes.

Important features of diabetic diet are:

- Carbohydrate intake should be at least 130 mg/day, providing 45–65% of daily calories.
- Carbohydrates with low glycemic index should be included.

- Maintain calcium, phosphorus and vitamin D intake to prevent renal osteodystrophy

Diet Composition

- The total calorie requirement is 30–35 kcal/kg body wt/day.
- In predialysis patients, protein intake should be 0.5–0.6 g/kg body wt/day, while in patients on dialysis, 1.0–1.2 g/kg body wt/day protein of high biological values should be included.
- Fats should contribute 30% to 40% of energy and rest of non-protein calories should be provided by carbohydrates.

Appendices at the end of each unit offer valuable clinical and biochemical information pertinent to the subject matter.

Study Questions

Short-Answer Questions

Write short notes on:

32.1. Wernicke-Korsakoff syndrome
32.2. Role of folate in one-carbon metabolism
32.3. Biochemical significance of vitamin K
32.4. Biochemical basis of scurvy

Long-Answer Questions

32.1. Describe the biochemical basis and clinical manifestations of rickets.
32.2. Explain the role of vitamin A in health. Describe in the detail the biochemical mechanism of visual cycle.

Multiple Choice Questions

Choose the ONE best answer.

32.1. Which of the following vitamins and its deficiency manifestation is NOT correctly matched?
A. Folic acid – Neural tube defects
B. Niacin – Diarrhea and dermatitis
C. Vitamin B_{12} – Bleeding
D. Vitamin C – Sore, spongy gums, and loose teeth
Correct answer = C.

32.6. Which of the follo… co-enzyme functio…
A. Vitamin A
B. Vitamin D
C. Vitamin E
D. Vitamin K
Correct answer =

32.7. Accumulation of … soluble vitamins ca…
A. Niacin
B. Thiamine
C. Riboflavin
D. Folate
Correct answer =

32.8. Based on statem… correct option:

Multiple-Choice Questions (NExT/NEET PG)

Choose the ONE best answer.

32.1. A 52-year-old woman presents with fatigue of several months' duration. Blood studies reveal macrocytic anemia, reduced levels of hemoglobin, elevated levels of homocysteine, and normal levels of methylmalonic acid. Which of the following is most likely deficient in this woman?
A. Folic acid
B. Folic acid and vitamin B_{12}
C. Iron

Correct answer = C.

32.4. A 60-year-old male presents with… the skin and eyes, dark urine, an… History reveals significant weight los… Examination shows hepatomegaly a… gallbladder. Lab tests indicate ele… and alkaline phosphatase levels. Ima… biliary obstruction. Additionally, … reports experiencing easy bruising … bleeding from minor cuts. Impairme… the following is responsible for the… manifestations?

Each chapter provides important **multiple choice, long answer and short answer questions**, which are commonly asked in MBBS examination. Additionally, each chapter includes 5 **MCQs for NEET PG/NEXT** preparation. Since these exams are taken after completing the final professional year, these questions encompass concepts from all MBBS subjects, extending beyond the content covered in this book. Explanations for the multiple-choice questions will be available on the online learning platform.

CONCEPT MAPS

Understanding concepts is paramount in Biochemistry, as this subject cannot be memorized through rote learning alone. Retaining knowledge in Biochemistry necessitates comprehension, as mere memorization is insufficient.

Students sometimes view biochemistry as a list of facts or equations to be memorized, rather than a body of concepts to be understood. Details provided to enrich understanding of these concepts inadvertently turn into distractions. What seems to be missing is a road map—a guide that provides the student with an understanding of how various topics fit together to "tell a story." Therefore, in this text, a series of biochemical concept maps have been created to graphically illustrate relationships between ideas presented in a chapter and to show how the information can be grouped or organized. A concept map is, thus, a tool for visualizing the connections between concepts. Material is represented in a hierarchic fashion, with the most inclusive, most general concepts at the top of the map, and the more specific, less general concepts arranged beneath. The concept maps ideally function as templates or guides for organizing information, so the student can readily find the best ways to integrate new information into knowledge they already possess. Concept map construction is described below.

A. Concept boxes and links

Educators define concepts as "perceived regularities in events or objects." In the biochemical maps, concepts include abstractions (for example, free energy), processes (for example, oxidative phosphorylation), and compounds (for example, glucose 6-phosphate). These broadly defined concepts are prioritized with the central idea positioned at the top of the page. The concepts that follow from this central idea are then drawn in boxes (Fig. 1A). The size of the type indicates the relative importance of each idea. Lines are drawn between concept boxes to show which are related. The label on the line defines the relationship between two concepts, so that it reads as a valid statement (that is, the connection creates meaning). The lines with arrowheads indicate in which direction the connection should be read.

B. Cross-links

Unlike linear flow charts or outlines, concept maps may contain cross-links that allow the reader to visualize complex relationships between ideas represented in different parts of the map (Fig. 1B) or between the map and other chapters in this book (Fig. 1C). Cross-links can, thus, identify concepts that are central to more than one topic in biochemistry, empowering students to be effective in clinical situations and on the United States Medical Licensure Examination (USMLE) or other examinations that require integration of material. Students learn to visually perceive nonlinear relationships between facts, in contrast to cross-referencing within linear text.

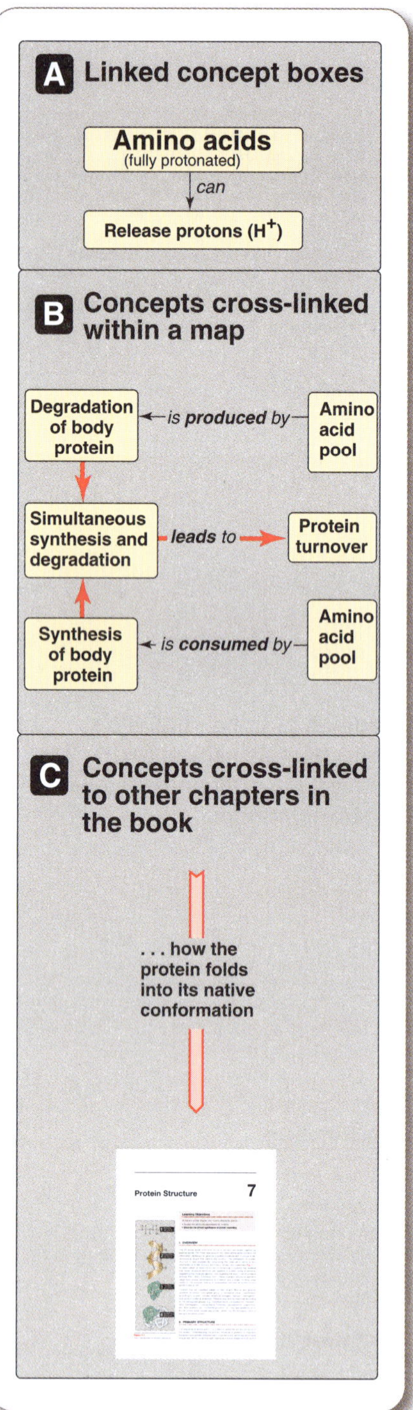

Figure 1 **A–C.** Symbols used in concept maps.

Unit I

Biochemistry—The Basis of Life

Competencies Addressed

- **BI1.1** Describe the molecular and functional organization of a cell and its subcellular components
- **BI6.7** Describe the processes involved in maintenance of normal pH, water and electrolyte balance of body fluids and the derangements associated with these
- **BI6.8** Discuss and interpret results of Arterial Blood Gas (ABG) analysis in various disorders

Unit I: Overview

Biochemistry is the branch of science concerned with the chemical basis of life. Derived from the Greek word "bio," meaning "life," it explores the intricate molecular interactions within living organisms. Cells, the fundamental units of life, contain a diverse array of biomolecules including proteins, carbohydrates, lipids, and nucleic acids. These biomolecules interact in complex, regulated pathways to facilitate essential metabolic processes, ensuring the functionality and vitality of the human body. Various chemical reactions are organized into metabolic pathways, with the majority catalyzed by enzymes. These enzymes play a pivotal role in regulating metabolic processes, ensuring they proceed efficiently. Their activity is finely tuned through regulatory mechanisms, which, in turn, govern the overall functioning of metabolic pathways. Any disruption to these interactions can lead to the onset of disease states, highlighting the delicate balance required for optimal cellular function.

Cell, the smallest functional and structural unit of an organism, can be prokaryotic (bacteria) or eukaryotic. Eukaryotic cells are composed of several membrane-enclosed organelles, each performing a unique function. Nucleus is the carrier of genetic information in the form of DNA and also coordinates activities of cell like growth, protein synthesis, metabolism and cell division. Mitochondria serve as the primary site for energy production, whereas the endoplasmic reticulum plays a crucial role in protein synthesis. Defect in the structure or functioning of these organelles leads to disease state (Chapter 2). Cells are specialized to perform specific functions within various tissues and organs. Diseases may arise from genetic or environmental factors or a combination of both.

Cell membrane is a lipid bilayer with integral and peripheral proteins. Cell membrane is selectively permeable, and the transport across the membrane, which can be active or passive, is mediated by various mechanisms. These transport mechanisms are important for various cell functions such as maintenance of potential difference, intake of essential substances or secretion of products (Chapter 3).

The pH of blood and body fluids is maintained in a narrow range, as it is important for activity of enzymes, ion channels and transport proteins. Several mechanisms, including buffers, operate to maintain pH in the optimal range. Changes in pH, caused due to metabolic, respiratory or renal derangements, can produce acidosis or alkalosis. Whenever this change in pH occurs, other systems try to compensate and restore it to normal (Chapter 4).

Water, which constitutes about 50%–60% of the body weight, is divided into intracellular and extracellular compartments. These compartments vary in their electrolyte and protein content. A balance is maintained between water intake and excretion, which can be disturbed in physiological and pathological conditions. Water can move from one compartment to another, movement being determined by that compartment's osmolality. Sodium (along with its associated anions), urea and glucose are the major determinants of plasma osmolality. There are several mechanisms to maintain the water and electrolyte equilibrium, and disturbance in these, as seen in nephrogenic diabetes insipidus and syndrome of inappropriate secretion of antidiuretic hormone (SIADH), can lead to alterations in volume or osmolality (Chapter 5).

For primary healthcare physicians, grasping the significance of organelles and their role in cellular and tissue pathology is crucial. Diseases stem from various sources, including biological infections (viruses, bacteria), chemical toxins such as drugs and pesticides, physical trauma, oxygen deprivation as seen in environmental pollution, genetic disorders, immunological responses, nutritional deficiencies, endocrine irregularities and many other derangements. Knowledge of membrane transporters is essential for understanding the pathology of diseases (e.g., cystic fibrosis) or for treatment modalities (e.g., oral rehydration solution). The pH of the body can be altered in many conditions. Difficulty in respiration, either due to nervous or pulmonary defects, can cause respiratory acidosis. Accumulation of metabolic acids, such as lactic acid or ketoacids, can cause metabolic acidosis. Several diseases and drugs can alter the water and electrolyte balance. Diuretics reduce the extracellular fluid volume by promoting renal excretion of Na^+ and water. Loss of water and electrolytes due to vomiting and diarrhea can cause dehydration. Hyperkalemia can cause arrhythmias. Acidosis and alkalosis can cause electrolyte derangement by shifting them from one compartment to another. The knowledge of biochemical basis will be helpful in identifying the underlying cause of a disorder and its management.

Biochemical Basis of Medicine

1

The subject of medical biochemistry involves studying metabolic processes in the human body, from clinical, diagnostic, prognostic and therapeutic perspectives.

Human body is composed of, and dependent on, a wide diversity of elements, compounds and biomolecules. The discipline of medical biochemistry involves the study of structure, properties and functions of biological molecules like nucleic acids, carbohydrates (Unit III), proteins (Unit II) and lipids (Unit IV) which influence health and disease. The ecosystem of life is orchestrated by complex metabolism regulated by various mechanisms such as enzyme activity, hormones and neurotransmitters, etc. In this, role of micro- and macronutrients, vitamins and minerals (Unit VII) also becomes significant because they facilitate optimal maintenance of different biochemical pathways. The compartmentalization of cells as part of various tissues as well as intracellular compartmentalization (Chapter 2) further adds to complexity as well as control of the metabolic processes. The body is essentially a "collection of chemicals" and has processes to maintain, repair and grow itself. It also has processes to recognize its own cells (self-tolerance) as well as obstruct invasion from outside, e.g., in the form of bacteria and viruses (Unit IX: Immunology). The metabolic processes are further controlled by gene-environmental interactions as the environment (inside body or outside body) decides the activation or suppression of gene expression. At a metabolic level, this translates to production or suppression of production of certain enzymes and proteins (Unit VIII: Storage and Expression of Genetic Information). Any derangement in the delicate balance of the myriad biochemical reactions in the body may pose a threat to health and begin the march of disease.

Countless chemical reactions that take place within the cells in our body, maintaining a homeostatic state, are collectively known as metabolism. These chemical reactions are organized into metabolic pathways, each step being catalyzed by a specific enzyme. Some reactions may even be non-enzymatic. Major metabolic pathways for several biological processes include carbohydrate and energy metabolisms, lipid and lipoprotein metabolisms, amino acid and protein metabolisms, and nucleic acid and protein biosynthesis. These processes are controlled by many factors including energy demand of cell (e.g., exercise, starvation and diabetic states), influence of hormones (e.g., insulin and glucagon), mental states (e.g., release of adrenaline and cortisol), diurnal rhythms (melatonin), food intake (e.g., obesity), genetics (e.g., thalassemias) and many more.

A primary healthcare physician is expected to take medical history of the patient, examine, elicit signs and symptoms, send test requisitions, interpret the data received, form an opinion about ongoing derangements in the body and organs or systems affected and treat the patient accordingly. The knowledge of medical biochemistry is essential for understanding the pathogenesis of disease and planning its management. Almost all patients admitted to a hospital have biochemical tests done to aid in the diagnosis and treatment; patient management requires understanding of the basic metabolic processes.

Certain cells and organs are specialized in metabolizing specific molecules. Changes in blood levels of these metabolites can provide clues to particular organ damage or genetic diseases (e.g., elevated ammonia levels indicating defective urea formation in the liver). Therefore, careful analysis of a patient's presentation, considered along with patient's ethnicity and age group, prevalent conditions in that geographical area and family history, can help in pinpointing the disease and preparing a differential diagnosis, which can then be confirmed by relevant investigations.

Measurement of number of specific markers gives useful information in the diagnosis and management of disease. Enzymes, such as those of the coagulation cascade, have a defined function in blood and their levels can be altered in disease. Other enzymes, such as *creatine kinase* appear in the blood incidentally and reflect tissue injury. Further, analysis of isoenzymes can indicate the source tissue (e.g., CK-MB indicates myocardial tissue injury, whereas CK-MM indicates skeletal muscle derangements). Enzymes are also the preferred markers in numerous pathological conditions such as hepatitis (*alanine transaminase*, *aspartate transaminase*, *alkaline phosphatase*), pancreatitis and many cancers. They provide vital insights into the disease process by way of diagnosis, progress and assessment of response to therapy. Instead of focusing on rare disorders, the more common diseases should be considered first. Generally, the investigations are ordered as a group, and their patterns of derangement are valuable in confirming the diagnosis. It is pertinent to note that the biological processes do not occur in isolation and perturbations in one process can have disturbing effects on many other processes.

To further orient the medical students to the vast and complicated world of biochemistry, and the application of which is the basis of medicine, a few examples are given below. They are intended solely for orientation purposes.

I. EDEMA

In a patient with swelling of the body due to accumulation of fluid, known as edema, several pathways could have been disturbed. Only if a physician knows about the affected pathways, he/she can work out the diagnosis. If edema is generalized, it can be due to hypoalbuminemia. In such a case, various factors affecting serum albumin concentration must be assessed, such as nutritional deficiency (Chapter 31) and hepatic disease (see Chapter 48). History, physical examination, urinalysis, and other investigations will help in identifying the probable cause (see Appendix Unit II, Table 1). Edema with hypoalbuminemia in a child in absence of nutritional deficiency is most likely due to nephrotic syndrome. If there is a history of chronic alcoholism, cirrhosis of liver causing decreased synthesis of albumin is the most probable cause.

II. ANEMIA

A patient may present with generalized weakness, loss of appetite and irritability. This is a general, nonspecific presentation. But as focused history is elicited and specific signs looked for, probable diagnosis comes up. If the patient has pallor, anemia is a probability, especially in children and women of poor socioeconomic status, due to iron deficiency or hookworm infestations. Diagnosis of iron-deficiency anemia can be confirmed by hemogram and peripheral smear (see Fig. 33.9, p. 558). Deficiency of Vitamin B_{12} or folic acid can lead to macrocytic anemia, also called megaloblastic anemia (see Chapter 32, p. 527). In case of megaloblastic anemia, it is important to understand that folic acid supplementation can partially reverse the hematologic abnormalities of vitamin B_{12} deficiency and, therefore, can mask a vitamin B_{12} deficiency, which can later result in CNS defects.

III. JAUNDICE

An increase in bilirubin, known as jaundice, may be a sign of problem with one or many of the processes involved in survival of red blood cells or functioning of liver and bile ducts. Jaundice may be physiological or pathological. It may further be prehepatic, hepatic or posthepatic. Blood tests for liver function give indication of the cause. Conjugated bilirubin, being polar and water soluble is excreted in urine. If urine bilirubin is present, it shows that the cause is hepatic or obstructive. The brown color of stool is due to stercobilin, which is produced by action of intestinal bacteria on bilirubin, therefore, light colored stool indicates obstruction to excretion of bile pigments.

IV. OBESITY

Fat cells in the adipose tissue store excess energy as triglyceride, to be released as free fatty acids for use at other sites, when required. This system allows survival during food deficit periods. However, with sedentary lifestyle and nutritional abundance, excessive fat stores produce obesity, which is linked to a large number of adverse health consequences, such as cardiovascular diseases, insulin resistance, diabetes mellitus (DM), obstructive sleep apnea, nonalcoholic fatty liver disease, osteoarthritis and increased risk of several types of cancer. For management of obesity, the knowledge of composition and energy content of common diets is required for diet therapy.

V. DIABETES MELLITUS

Diabetes mellitus which, biochemically, is just a persistent increase in blood glucose, has a significant economic cost, both due to treatment-related expenses and loss of productivity. It affects multiple metabolic pathways, almost all organs and the presentation can have both acute features as well as long-term complications. The diagnostic criteria of diabetes are a must-know as the Indian population has a very high incidence of DM. Further, the physician must do the tests like glycated hemoglobin to monitor the disease, regulate blood glucose levels by therapy and watch for long-term complications like nephropathy (test for albumin excretion rate), cataract, hyperlipidemias and neuropathy. Besides drug therapy, medical nutrition therapy, which includes optimal coordination of caloric intake, along with other aspects of diabetes therapy (insulin, exercise, weight loss) has significant role in preventing or delaying the onset and complications of type 2 DM in high-risk individuals. It is again based on basic knowledge of food components, their carbohydrate content and glycemic index. Regular monitoring of glucose and lipids and screening for complications are required in diabetic patients. The type of sample and timing of collection can affect the results and need to be taken care of.

VI. ELECTROLYTE BALANCE AND PH

Many conditions can affect the pH of blood. Though there are several mechanisms to ensure that blood pH is maintained in a narrow range, several disorders can overwhelm these mechanisms, producing acidosis or alkalosis. The differences in intra- and extracellular pH and composition of electrolytes can be disturbed in a wide variety of conditions and can have profound effects on critical body functions. Strenuous exercise can produce acidosis due to accumulation of lactic acid, produced during anaerobic glycolysis in skeletal muscles. Acidosis can cause hyperkalemia, which in turn, can lead to life-threatening arrhythmias. A patient of apparently common gastroenteritis, having severe vomiting may land in life-threatening alkalosis, due to loss of gastric HCl. Severe dehydration and fluid loss are major causes of diarrheal deaths, and knowledge of biochemistry of sodium and glucose absorption from intestine helps in preventing dehydration by using simple foods available at home. Further, IV fluids given in hospitals require an essential knowledge of electrolyte balance of body and normal sodium, potassium, and chloride levels which have to be maintained.

VII. DRUG INTERACTIONS

The biochemical properties of drugs can affect their absorption and action of drugs. Acidic drugs are absorbed from stomach, while basic drugs are absorbed from intestine. Acidification of urine promotes the excretion of basic drugs. Most of the drugs bind reversibly to plasma proteins, such as albumin. The bound drug remains restricted to vascular compartment, as protein-bound drugs do not cross membranes and also remain for longer time in the body as their metabolism and excretion are also delayed. More than one drug can bind to the same site within the protein, and a drug can displace another drug, changing its duration or intensity of action.

VIII. AGING

Aging is another condition which has multidimensional effects on normal metabolism, disease susceptibility, disease manifestations, response to treatment, and disease outcomes. In older individuals, normal homeostatic mechanisms also have reduced capabilities and can be overwhelmed by minor disturbances. Older individuals are also at increased risk for multiple coexisting diseases. Aging is also thought to be associated with increased oxidative stress damage. The primary healthcare physician while prescribing the type and dosage of medicines should take into account the reduced drug metabolizing and excretion capacities of the elderly. The values of several lab parameters also change with age and this should be considered when evaluating the lab results of aged patients. Both the energy requirement and availability also change with aging.

IX. PREANALYTICAL VARIATIONS IN SAMPLES

The primary healthcare physician must be very careful to do appropriate sampling such that preanalytical errors (i.e., errors that occur during a laboratory test, before the sample is measured by the analyzer) do not interfere with report authenticity. The type of sample, such as whole blood (venous, arterial or capillary), serum, plasma, urine or other body fluids (cerebrospinal fluid [CSF], pleural fluid, ascitic fluid, etc.) to be tested also needs to be decided as per suspected disease. Timing of sample collection (random or early morning, spot urine or 24-hour collection, a particular day of menstrual cycle for fertility-related hormones, etc.) and patient preparation, all affect the test results and have to be considered. Wrong sample collection techniques, inappropriate additives or preservatives and inappropriate sampling site can give erroneous results. Sample for blood glucose must be collected in a tube having sodium fluoride as it will inhibit in vitro further glycolysis in sample. For analysis of serum, a plain evacuated tube is required which will allow the sample to clot. If blood is required as for glycated hemoglobin, the sample collection tube must have ethylenediaminetetraacetic acid (EDTA) to prevent blood from clotting. If there is hemolysis during sample collection, red cell contents such as potassium are released, and their levels will be falsely elevated. If tourniquet is applied for prolonged time, water diffuses into the interstitial space and the serum or plasma sample will be concentrated and show falsely elevated levels of proteins and protein-bound components of plasma, such as calcium or thyroxine. Samples collected from a site of ongoing intravenous fluid infusion will give wrong results, as the fluid will either dilute the specimen, reducing the measured amount or give falsely elevated results (e.g., increased plasma glucose levels in a sample that was collected from a vein where dextrose intravenous drip was being administered).

X. GENETICS

The primary healthcare physician must be well aware of the genetic influences in disease (see Unit VIII: Storage and Expression of Genetic Information). This becomes very important in some areas like prenatal diagnosis of disease. Chorionic villi sampling and sampling by amniocentesis are used to diagnose diseases such as thalassemias, Duchenne muscular dystrophy and sickle cell anemia in unborn child. This is followed by genetic counseling of the antenatal mother so that she can make informed decisions. A positive quadruple screening test of maternal serum for trisomy 21 (Down syndrome) is further confirmed by karyotyping using an amniocentesis sample. Recombinant DNA technology is another area of biochemistry where the advances have led to quantum leaps in patient care. The production of human insulin, blood clotting factors VIII and IX, growth hormone, albumin, and thrombolytic agents are achievements of genetic engineering in the field of patient care. Infectious diseases like human immunodeficiency virus (HIV) infection and tuberculosis can be diagnosed by polymerase chain reaction (PCR) and specific probes, even in the window period. In the year 2020, the COVID-19 was declared a pandemic and the diagnosis can be made by real-time PCR by detection of specific genes of the SARS-CoV2 (severe acute respiratory syndrome-coronavirus 2). A doctor must be aware of the genetic diagnosis available and keep himself/herself updated on gene augmentation and therapy as it is being developed for patient care.

Molecular medicine has also altered the way diseases are classified. In cancers, the classification systems, which were earlier based on the extent of spread and histopathological features, are now expanding to include molecular features. The field is progressing from diagnosis of disease to prediction of occurrence of disease, and now to the choice of treatment based on individual genetic composition - the era of personalized medicine based on genetic analysis.

XI. Chapter Summary

- Much like the medical science itself, biochemistry is a vast area of research that yields profound discoveries each year. Without the ongoing flurry of scientific breakthroughs made by biochemists, we would not have had the precise chemical and molecular knowledge to create the vital drugs, therapies and diagnostic tools that are used every day.
- Biochemical techniques advance our understanding of the chemical structures and processes that underpin human health and disease, revealing the underlying transformations between these physiological and pathological states.
- Biochemistry and medicine enjoy a mutually stimulating relationship. Biochemistry has been offering insights and solutions to questions posed by physicians. It has been at the service of patients contributing to better understanding of the basis of disease at molecular level, offering appropriate tests for diagnosis of disease and explaining the impact particular drugs will have on molecular processes (including adverse effects).
- For physicians, a detailed knowledge of biochemistry is an essential prerequisite in implementing evidence-based medicine and improving patient outcomes.
- For all the superspecialized fields of medicine like diabetology, endocrinology, cardiology, gastroenterology, nephrology, neurology, oncology, etc., biochemistry is continuously fostering better disease understanding, and diagnostic and prognostic testing in interest of patient care. Notable contributions in the field of patient care include diagnosis and monitoring of diabetes with the discovery of glycated hemoglobin, understanding water, electrolyte and hydrogen ion homeostasis, contribution toward development and monitoring of intravenous nutrition, and in-depth understanding of role of lipids and lipoprotein transport mechanisms for management of cardiovascular diseases.
- A pathbreaking contribution was the development of point-of-care testing which provided an easy platform for bedside emergency testing (e.g., troponin in myocardial infarction) and self-testing by patients (e.g., blood glucose levels by glucometer).
- In recent years, significant progress has been made in laboratory automation and workflow management.
- Evidence-based clinical biochemistry is coming up in a big way in line with evidence-based medicine wherein clinical biochemists critically evaluate the existing evidence toward diagnostic tests.
- In recent years, substantial growth has been achieved in molecular diagnostics, genetic screening (prenatal and neonatal), screening for inborn errors of metabolism, immunochemistry, hormone analysis, mass spectrometric analysis, next-generation sequencing, PCR for infectious diseases, neurochemistry and many more.
- To conclude, the spectrum of biochemistry in medicine ranges from a simple observation that lemon juice, rich in vitamin C, can prevent life-threatening scurvy in sailors to the exciting possibility of gene therapy for treatment of genetic disorders such as thalassemias, cystic fibrosis, hemophilia and sickle cell disease.
- We must utilize the full scope of biochemistry and its vast knowledge to practice medicine effectively.

Cell and Subcellular Components

2

Learning Objectives

At the end of this chapter, the student will be able to:

- Explain the structural and functional organization of a cell
- Enlist different cell organelles, their functions and specific markers for each cell organelle
- Discuss diseases associated with cell organelles

I. OVERVIEW

Cells are the smallest functional and structural unit of an organism. **As an independent unit, a cell can metabolize its own nutrients, provide for its own energy and replicate itself to produce succeeding generations of cells.** It can be viewed as an enclosed compartment in which innumerable chemical reactions take place simultaneously, contributing to life and procreation of the cell. Collection of similar cells form tissues, and collection of tissues form organs, which carry out the functions necessary to sustain the life of an organism (Fig. 2.1). **An organelle or the subcellular components are different types of membrane-enclosed bodies in the cell, each performing a unique function.** Metabolism comprising of both, anabolism and catabolism, takes place in the cell—most of the body functions ultimately occur at the cellular level. Enzymes and various macromolecules inside the cell form the basis for cell growth, repair and carrying out of various functions. An understanding of components of cells and how cells work is fundamental to applications of diagnostic enzymes, metabolic disease processes, organelle specific diseases and treatments.

II. EUKARYOTIC CELL

There are two basic types of cells—prokaryotic and eukaryotic cells. Both share strong similarities in biochemical functions. **The prokaryotic cells do not have membrane-bound organelles,** are smaller in size (1–10 μm) and **have an ill-defined nucleus.**

The most characteristic feature of eukaryotic cell is the nucleus and membrane-enclosed organelles, i.e., mitochondria, endoplasmic reticulum (ER), Golgi apparatus, peroxisomes and lysosomes, etc. Most cells are microscopic and are typically 5–100 μm in diameter. Also present are nonmembranous structures such as starch and fat droplets (Fig. 2.2).

A. Nucleus

It is one of the largest organelles inside the cell. It is the **carrier of genetic information in the form of deoxyribonucleic acid (DNA**). The thread-like DNA is tightly packed with nuclear proteins (histones) to form chromatin. **Somatic cells in humans have 46 chromosomes compactly packed in the nucleus.** The nucleus coordinates various activities of cell like growth, protein synthesis, metabolism and cell division. It is separated from rest of the cell by double-layered nuclear membrane. **The nuclear membrane is continuous with endoplasmic reticulum and has pores which allow the movements of RNA, hormones and regulatory proteins to and fro from cytosol.** Some hormones interact at specific sites on the DNA to regulate transcription. Some cells like the human red blood cell (RBC) lose their nuclei on maturation. This is an adaptation—the amount of oxygen required by the cell for its own metabolism is low and thus it can effectively deliver oxygen to other tissues.

A dense, non-membranous mass called nucleolus is closely associated with inner nuclear membrane. It is the site for transcription and assembly of ribosomal RNA (rRNA).

B. Mitochondria

They are round or oval-shaped bodies whose **primary function is to generate large amount of energy in**

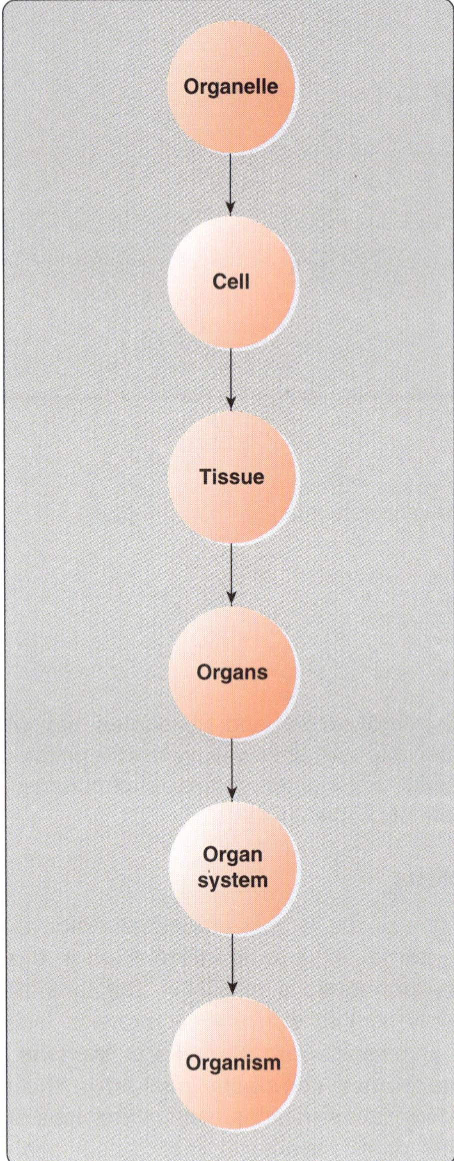

Figure 2.1 Levels of organization in human body.

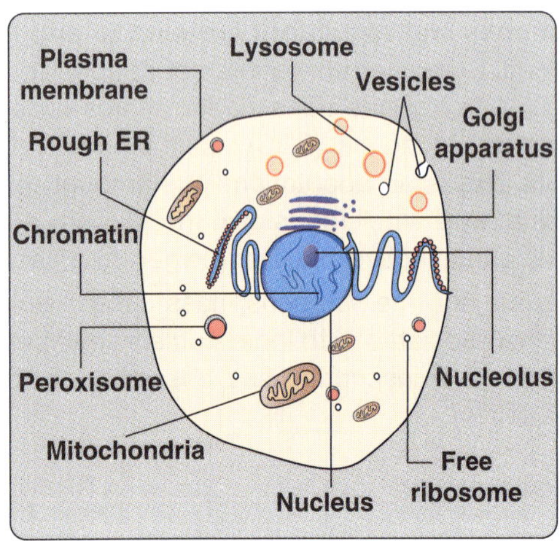

Figure 2.2 Eukaryotic cell.

form of ATP. Thus, they are nicknamed as the "powerhouses of the cell." A mammalian cell has about 800–2,500 mitochondria. The number per cell varies widely, as RBCs do not have mitochondria, whereas liver and muscle cells contain hundreds and thousands of them. **They are unlike other cell organelles in that they have two distinct membranes and a unique genome.** The organelle is composed of compartments that carry out specialized functions. These compartments or regions include the outer membrane, the intermembrane space, the inner membrane, and the cristae and matrix (Fig. 2.3).

The outer membrane is rich in lipids and is composed of phospholipids and cholesterol. It is freely permeable to smaller molecules through transmembrane proteins (porins) that make up the channels for free passage. It also contains enzymes for elongation of fatty acids, oxidation of epinephrine and degradation of tryptophan. The outer membrane can associate with the ER membrane, in a structure called MAM (mitochondria-associated ER membrane). This is involved in the transfer of lipids between the ER and mitochondria.

The inner mitochondrial membrane is rich in proteins (protein-lipid ratio of 3:1) and is not freely permeable as porins are absent in this membrane. It contains proteins that perform different functions:

- Proteins that perform redox reactions of oxidative phosphorylation.
- *ATP synthase* which generates ATP in the matrix.
- Transport proteins that regulate metabolite passage into and out of the mitochondrial matrix.
- Mitochondrial fission and fusion proteins. Mitochondria are dynamic organelles with the ability to fuse and divide (fission). This mitochondrial dynamics is important for the health of the cell and defects in dynamics lead to genetic disorders.

In addition, the inner membrane is rich in a phospholipid, cardiolipin. The inner membrane is folded, and the invaginations are called **cristae**. This enhances the surface area, thus enhancing the ability of the mitochondria to produce ATP. The density of crista varies and cells that have greater demand for ATP have more cristae. Also, there is a membrane potential across the inner membrane formed by the action of enzymes of the electron transport chain (ETC).

The mitochondrial matrix is the space enclosed by the inner membrane. The matrix contains a mixture of enzymes, ribosomes, tRNA and mitochondrial DNA. It also has enzymes for fatty acid oxidation, ketogenesis and the citric acid cycle.

Additional functions of mitochondria include apoptosis (programmed cell death), and hormonal signaling. Mitochondria of liver cells have enzymes that detoxify ammonia.

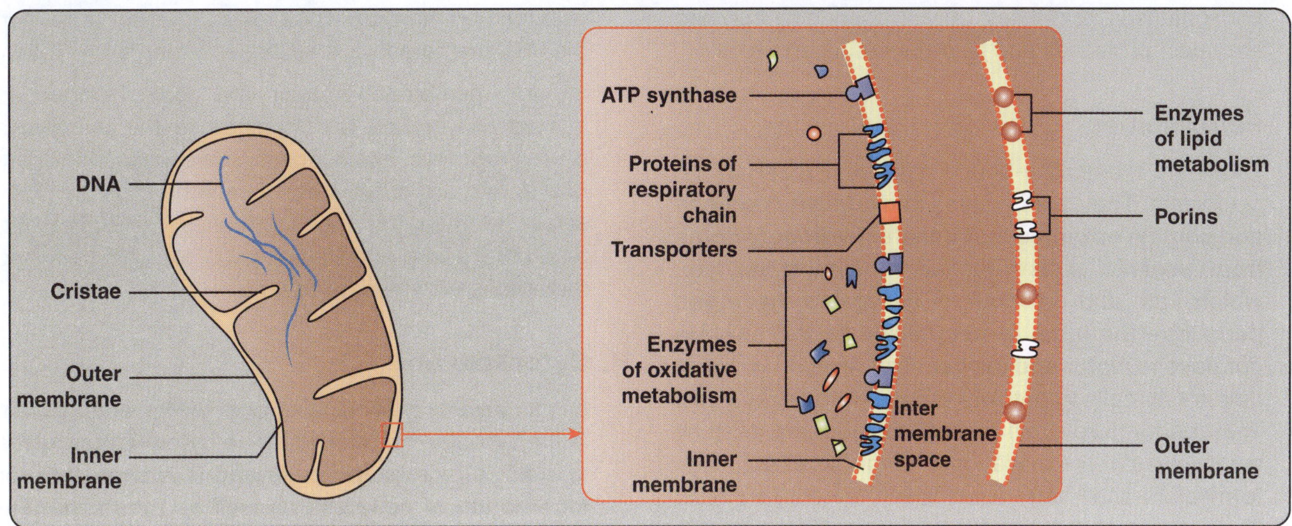

Figure 2.3 Structure of mitochondria.

C. Endoplasmic reticulum

It is a network of membranous tubules within the cytoplasm of eukaryotic cell. The membranes of ER are continuous with the outer nuclear membrane. There are two types of ER.

Rough endoplasmic reticulum (RER) has its cytosolic face studded with ribosomes—they are the **sites of protein synthesis**. It is especially prominent in cells such as hepatocytes. RER also synthesize secretory proteins, mostly glycoproteins, that are packed into transport vesicles, which move along the cytoskeleton toward their destination.

Smooth endoplasmic reticulum (SER) lacks ribosomes and is involved in **lipid synthesis and transport, production of steroid hormones, and detoxification of drugs and toxins** (hydroxylation and conjugation reactions). It is abundant in liver and gonadal cells.

D. Golgi apparatus

It is a complex made of series of flattened, stacked pouches called cisternae and vesicles. It is located next to ER and near the cell nucleus. **The Golgi apparatus is responsible for transporting, modifying, and packaging proteins and lipids into vesicles for delivery to targeted destinations.** The apparatus has three primary compartments, known as "cis" (cisternae nearest to the ER), "medial" (central layers of cisternae), and "trans" (cisternae farthest from ER).

Proximal or cis region receives the freshly synthesized proteins from ER. **The proteins must have the "signal peptide" to be accepted by Golgi apparatus for transport.**

At medial part, certain posttranslational modifications of proteins occur. These modifications involve addition of fatty acids to form lipidated proteins, or phosphate groups (phosphorylation) to form phosphoproteins, or oligosaccharides to the side chain to form glycoproteins.

In the final stage, modified proteins and lipids are sorted in **trans Golgi network** and are packaged into vesicles at the trans face. These vesicles then deliver the molecules to their target destination such as lysosomes or the cell membrane or for secretion outside the cells.

E. Lysosomes

They are **spherical vesicles that contain hydrolytic enzymes** (approximately 60 in number) that are involved in breakdown of biomolecules—carbohydrates, lipids, proteins and nucleic acids.

The enzymes of lysosomes are synthesized in RER where they receive an N-linked oligosaccharide. From there, they are transported to Golgi apparatus, where mannose residues of the oligosaccharide get phosphorylated; this helps in binding of these proteins to mannose-6-phosphate receptors present in the trans-Golgi; finally, they are directed toward lysosomes.

The lysosomal enzymes responsible for hydrolysis require an acidic environment for optimal activity (pH ~5.0). Lysosomes are capable of fusing with other organelles and digesting cellular debris, in a process known as **autophagy**. Lysosomes are abundantly present in macrophages. They can break down bacteria and other microorganisms following the phagocytosis of these particles by macrophages. So, they are aptly called as the "waste disposal system" of the cell. Lysosomes contain enzymes like phosphatases, *nucleases*, polysaccharide hydrolyzing enzymes, *proteases* and lipid-degrading enzymes.

F. Peroxisomes

These are membrane-bound organelles found in the cytoplasm. **They are involved in oxidative reactions and contain enzymes that transfer hydrogen atoms from organic substrates to molecular oxygen, which, as and in course generates hydrogen peroxide—the latter is neutralized by the enzyme catalase to water and oxygen.** This reaction is important in liver cells to detoxify various toxic substances. They perform key role in lipid metabolism, as they are involved in the oxidation of long-chain fatty acids to medium-chain fatty acids which are subsequently shuttled to mitochondria for β-oxidation. They are also involved in the metabolism of branched chain fatty acids.

G. Cell membranes

Cell membrane is a biological membrane that separates interior of all cells from the outside environment (the extracellular space). Some of the functions of the cell membrane are (i) selective transport of ions and organic molecules, (ii) cell adhesion, (iii) ion conductivity, and (iv) cell signaling.

Cell membrane consists of lipid bilayer in combination with membrane proteins such as integral proteins and peripheral proteins. The lipids are in the form of phospholipids and cholesterol; the latter assists in the maintenance of the fluidity of the membrane. Peripheral proteins are present on the surface of lipid bilayer, usually attached non-covalently with membrane lipids. The protein spectrin, present in the RBC membrane, is an example of one such protein.

Integral membrane proteins are deeply embedded in lipid bilayer and are attached to the membrane by hydrophobic interactions. Some span the entire bilayer and are called transmembrane proteins. They serve as receptors (for hormones, neurotransmitters, etc.), tissue-specific antigens, ion channels, and membrane enzymes (Chapter 3).

> **I-cell disease**
>
> Inclusion cell disease is a rare inherited metabolic disorder characterized by coarse facial features, skeletal abnormalities and mental retardation. It results when newly synthesized lysosomal enzymes fail to reach the proper intracellular locations, and are instead secreted outside the cell. Thus, the intracellular levels of these enzymes are reduced/absent, while the plasma levels are elevated. Mannose 6-phosphate is the marker to target the nascent enzymes to the lysosomes. In patients with I-cell disease, absence of a certain enzyme involved in the generation of mannose-6-phosphate residues prevents the proper targeting of the lysosomal *hydrolases* to the lysosomes. Since many macromolecules are degraded by these *hydrolases*, I-cell disease leads to accumulation of molecules like mucolipids and mucopolysaccharides within the cells of many tissues of the body, leading to the clinical manifestations.

H. Cytoskeleton

It is a network of interlinking protein filaments present in the cytoplasm of all cells. **The cytoskeleton gives specific shape to the cells and is responsible for locomotion of cell itself as well as the movement of various organelles within it.** It also facilitates the movement of chromosomes during cell division. The major type of filaments in cytoskeleton include microfilaments, intermediate filaments and microtubules.

Microfilaments are predominantly composed of actin—they help in muscle contraction, cell movement, maintenance of cell shape.

Intermediate filaments impart mechanical support to the cells. It is formed by different proteins in different tissues, but one of the most common proteins that comprise intermediate filaments is keratin.

Microtubules are hollow cylinders and are formed by polymerization of the protein, tubulin. The cilia and flagella are structural components of cell and they are maintained by microtubules which give them strength and movement. They also play a role in intracellular transport and in the formation of mitotic spindles during mitosis and meiosis.

I. Cytoplasm

Cytoplasm is a highly viscous solution with suspended particles and cell organelles that fills each cell and is enclosed by cell membrane. It makes up about 70% of the cell volume. It is site for various metabolic pathway (e.g., glycolysis, glycogenesis and glycogenolysis, hexose monophosphate [HMP] shunt, fatty acid synthesis, etc.). **The term cytosol is used to denote that portion of the cytoplasm from which the organelles are excluded**.

J. Ribosomes

They are nucleoprotein (RNA and proteins) assemblies which **bind messenger RNA and transfer RNA, to synthesize polypeptides and proteins.** Ribosomes, being entirely particulate, do not have a cell membrane and are often called **"non-membranous organelles." They have two subunits, the small subunits**

Cell Organelle	Function	Marker Enzymes/Molecules
Nucleus	DNA storage, DNA replication, RNA synthesis	DNA
Mitochondria	Energy production	*Cytochrome c oxidase, glutamate dehydrogenase*
Endoplasmic reticulum	Lipid production, detoxification Protein production	*Glucose 6-phosphatase*
Golgi apparatus	Protein modification and export	*Galactosyl transferase*
Peroxisome	Lipid metabolism	Catalase
Lysosome	Degradation of macromolecules, e.g., lipids	Acid phosphatase
Cytosol	Glycolysis, HMP shunt, glycogen metabolism, fatty acid synthesis	*Lactate dehydrogenase*
Cell membrane	Selective transport of ions and organic molecules, cell adhesion, ion conductivity, cell signaling	5′-Nucleotidase *Alkaline phosphatase*

DNA = deoxyribonucleic acid; SER = smooth endoplasmic reticulum; NADPH = reduced nicotinamide adenine dinucleotide phosphate; RER = rough endoplasmic reticulum; HMP shunt = hexose monophosphate shunt.

Figure 2.4 Overview of different cell organelles and marker enzymes.

read the mRNA, and the large subunits catalyze the formation of polypeptide chains from amino acids. Ribosomes can be free or membrane bound depending on their spatial distribution in the cell. They are membrane bound when attached to ER.

Marker enzymes are exclusively present in a particular organelle and can be used to characterize cell organelles. They are useful for in vivo disease diagnosis and in vitro cell fractionation. A brief description of different cell organelles, and their marker enzymes is given in Figure 2.4.

Cell Fractionation

Within the set-up of a research laboratory, many a times it is vital to separate an individual organelle from the rest of the cellular components. This holds true especially for identification and purification of proteins that are present in an organelle, and also while studying the effects that a drug or a disease exert on an organelle. Cell fractionation involves the separation of organelles into distinct fractions. **Cell fractionation is most commonly achieved by a process known as "differential centrifugation."** The cells that are being studied are lysed by disrupting the cell membrane with the help of detergents and/or hypotonic solutions. The resulting "lysate" is subjected to centrifugation, initially at lower speeds and durations to facilitate the sedimentation of denser organelles. The supernatant is then further subjected to centrifugations at higher speeds and prolonged durations, so that relatively less dense organelles can be separated. Centrifugation of the cell lysate at 500 g for 10 minutes leads to pelleting of the nuclear fraction; 10,000 g for 20 minutes results in sedimentation of mitochondria. Centrifugations at speeds of 100,000 g or more is termed as ultracentrifugation. Ultracentrifugation for 100,000 g and a duration of 1 hour leads to pelleting of fragments of endoplasmic reticulum (these are referred to as microsomes); the supernatant at this stage will contain soluble proteins of the cytoplasm. The purity of a fraction separated using differential centrifugation can be assessed by estimating the level or activity of certain molecules (usually enzymes) that are present exclusively in that organelle/cell fraction.

III. Medical Perspectives

- Protein synthesis occurs in ribosomes that are present free in the cytoplasm, or that are attached to the endoplasmic reticulum. From the site of their synthesis, the proteins are targeted to their destined organelles usually with the help of certain specific amino acid sequences known as "signal sequences." As already mentioned, lysosomal proteins are targeted with the help of a mannose-6-phosphate residue, rather than an amino acid sequence. Defects in the targeting of proteins to their respective intracellular sites can result in diseases. As an example, in the human body glyoxylate is converted to the amino acid glycine by the peroxisomal enzyme, *alanine glyoxylate aminotransferase*. A targeting defect of this enzyme resulting in its localization to mitochondria, rather than peroxisomes, leads to accumulation of glyoxylate; the latter is then converted to oxalate, which combines with calcium to form insoluble calcium oxalate, resulting in formation of renal stones—the condition is termed as type I hyperoxaluria (*hyper* – increased, *oxal* – oxalate, *uria* – presence in urine).

- Mitochondrial defects are caused by a mutation in mitochondrial genome and they are maternally inherited—some of the examples include Leber's hereditary optic neuropathy, myoclonic epilepsy, Kearns-Sayre syndrome, to name a few. Maternal inheritance occurs because the zygote derives its mitochondria from the ovum at the time of fertilization, rather than from the sperm. Incidence is 1 in 5,000 individuals. Since the primary function of mitochondria is energy production in the form of ATP, those tissues that have huge energy requirements (like neurons and muscles) are primarily affected in mitochondrial disorders.
- Lysosomal storage diseases (LSDs) are a group of genetically inherited diseases characterized by accumulation of undigested macromolecules inside the lysosomes, due to defects in various hydrolytic enzymes. The most common LSD is Gaucher's disease. Others include Tay-Sachs disease, Niemann-Pick disease and Hunter syndrome.

The accumulated macromolecules as such can cause cellular dysfunction. Further, the accumulated macromolecules within the lysosomes compromise other lysosomal functions like autophagy of old and worn-out organelles. Abnormal metabolites generated by these defective organelles also propagate the cellular dysfunction in lysosomal storage diseases.

- *Acid phosphatases* are marker enzymes for lysosomes. They are produced in many tissues/cells like osteoclasts, platelets, spleen, erythrocytes and prostate gland. Elevated levels of *acid phosphatase* in plasma results from release of the enzyme from these tissues in various pathologies affecting these tissues/cells.
- Zellweger syndrome is an inherited disorder characterized by complete absence of peroxisomes in the cells, due to defects in proteins required for the biogenesis of peroxisomes. It is an autosomal recessive disorder. Since peroxisomes are involved in the oxidation of long chain fatty acids and branched chain fatty acids, they accumulate in cells and tissues. Patients do not survive beyond 1 year.
- Hereditary spherocytosis is an inherited disorder in which there is defect of spectrin, a component of RBC cytoskeleton. The resulting alteration in the shape of RBCs leads to a loss of erythrocyte surface area. These cells are rapidly removed from the circulation by spleen resulting in anemia.
- Signaling through insulin receptor begins with binding of the insulin. In type 1 diabetes mellitus, there is insulin deficiency. Diabetes mellitus type II, on the other hand, is characterized by insulin receptor resistance.
- Anticancer drugs like vinblastine and vincristine inhibit the formation of mitotic spindle. Since this is essential for the proceeding of cell cycle, these drugs inhibit the cell division, thus abrogating cell proliferation in malignancies.

IV. Chapter Summary

- A eukaryotic cell has a plasma membrane, cytoplasm, ribosomes, nucleus and other membrane-bound organelles that allow compartmentalization of various biochemical functions.
- The cell membrane is a phospholipid bilayer embedded with proteins.
- The nucleolus within the nucleus is the site for ribosome assembly.
- Ribosomes are found in the cytoplasm or are attached to ER. They perform protein synthesis.
- Mitochondria perform cellular respiration and produce ATP.
- Peroxisomes break down fatty acids, amino acids, and some toxins.
- Vesicles are storage and transport compartments.
- Lysosomes are the digestive organelles of cells.
- The endomembrane system includes nuclear envelope, the ER, Golgi apparatus, lysosomes, and vesicles. These work together to modify, package, tag, and transport membrane lipids and proteins.
- The cytoskeleton provides rigidity and shape to cells.
- Microtubules pull replicated chromosomes to opposite ends of a dividing cell and are structural elements of cilia and flagella.
- The human body is complex and has levels of organization that build on each other. Cells make up tissues, tissues make up organs, and organs make up organ systems.

Study Questions

Short-Answer Questions

2.1. What are the functions of mitochondria? Which all organs are predominantly affected in diseases with mitochondrial inheritance?

2.2. Explain a technique by which various cell organelles can be isolated.

2.3. What are lysosomal storage diseases?

Long-Answer Question

2.1. Explain in detail how compartmentalization at the cellular level is important for carrying out various biochemical functions of the cell.

Multiple-Choice Questions

Choose the ONE best answer.

2.1. Which of the following is a function of rough endoplasmic reticulum?
A. Production of proteins
B. Synthesis of steroid hormones
C. Detoxification of certain substances
D. Lipid synthesis
Correct answer = A.

2.2. Which of the following statements is NOT true?
A. Lysosomes are rich in different hydrolytic enzymes.
B. pH inside lysosomes is around 7.
C. Acid phosphatase is a marker enzyme for lysosomes.
D. I-cell disease is a form of lysosomal storage disease.
Correct answer = B.

2.3. Which of the following statements about the nucleus is correct?
A. It is rich in both DNA and RNA.
B. Nuclear membrane is double layered.
C. It is present in prokaryotes.
D. RBCs have more than one nucleus for its effective functioning.
Correct answer = B.

2.4. Eukaryotic animal cells do not have one of the following organelles.
A. Mitochondria
B. Nucleus
C. Chloroplast
D. Peroxisome
Correct answer = C.

2.5. Which is the largest cell organelle of the eukaryotic cell having genetic information?
A. Mitochondria
B. Nucleus
C. Lysosome
D. Endoplasmic Reticulum
Correct answer = B.

2.6. Which of the following statements is NOT true?
A. RBCs have no mitochondria.
B. Prokaryotic cells have membrane bound organelle.
C. Cells of all eukaryotic tissues have well-defined nucleus.
D. Number of mitochondria per mammalian cell varies widely.
Correct answer = B.

2.7. Which is NOT true of mitochondria?
A. They have genetic information and are absolutely autonomous in nature.
B. They are called powerhouse of the cell.
C. Outer membrane of mitochondria is freely permeable.
D. Inner membrane of mitochondria lacks porins.
Correct answer = A.

2.8. Which of the following is NOT true about lysosomes?
A. They contain hydrolytic enzymes.
B. They are present abundantly in macrophages.
C. They are not capable of fusing with other organelles to digest their components.
D. They can break down bacteria and other microorganisms following their phagocytosis.
Correct answer = C.

2.9. Which organelle is not membrane bound in eukaryotic cell?
A. Nucleus
B. Mitochondria
C. Lysosomes
D. Ribosomes
Correct answer = D.

2.10. Which of the following pairs of organelles have oxidative systems?
A. Nucleus and mitochondria
B. Mitochondria and peroxisomes
C. Peroxisomes and lysosomes
D. Endoplasmic reticulum and ribosomes
Correct answer = B.

Multiple-Choice Questions (NExT/NEET PG)

Choose the ONE best answer.

2.1. Choose the correct statement
A. Catalase is present in lysosomes.
B. A typical cell has 800–2500 mitochondria.
C. Mitochondrial disorders are paternally inherited.
D. Zellweger syndrome is a lysosomal storage disease.
Correct answer = B.

2.2. Which of the following is NOT a function of cell membrane?

A. Ion conductivity
B. Cell signaling
C. Selective transport of molecules
D. Movement of chromosomes during cell division

Correct answer = D.

2.3. Based on the Statements given below, choose the correct option.

Statement 1: Refsum disease is associated with neurological manifestations.

Statement 2: In Refsum disease, there is accumulation of phytanic acid.

A. Both the Statements are correct, and Statement 2 is the correct explanation for Statement 1
B. Both the Statements are correct, but Statement 2 is NOT the correct explanation for Statement 1
C. Statement 1 is correct, but Statement 2 is NOT correct
D. Statement 1 is NOT correct, but Statement 2 is correct

Correct answer = A.

2.4. Mitochondrial genome:

A. contains genes that encode for all its proteins, transfer RNAs (tRNAs), and ribosomal RNAs (rRNAs).
B. encodes only 13 proteins, 22 transfer RNAs (tRNAs), and 2 ribosomal RNAs (rRNAs).
C. encodes all its proteins, but it does not encode transfer RNA (tRNA) or ribosomal RNA (rRNA).
D. encodes some of its own proteins only.

Correct answer = B.

2.5. Histones proteins found in the nuclei of eukaryotes are rich in:

A. Glycine and phenylalanine
B. Glycine and arginine
C. Lysine and arginine
D. Lysine and phenylalanine

Correct answer = C.

Cell Membranes

3

Learning Objectives

At the end of this chapter, the student will be able to:

- Describe the fluid mosaic model and the structure of membrane bilayer
- Discuss the different functions of various proteins in the cell membrane
- Explain the different types of membrane transport—passive and active transport, and endocytosis and exocytosis

I. OVERVIEW

The cell membrane is the physical barrier that separates interior of the cell from outside environment. Intracellularly, membranes also make different subcellular compartments which perform various functions. **Cell membrane and organelle membranes consist of a lipid bilayer in combination with membrane proteins. The membrane proteins and lipids carry out different functions—** (i) transport of materials into and out of the cell, (ii) cell sensing and signal transduction, (iii) cellular recognition of antigens, (iv) cell–cell communication via hormones and neurotransmitters, and (v) communication between cellular compartments. Membrane functions are therefore vital to health—specific defects in membrane proteins, and various toxins that act on the membrane, are associated with numerous known disease states.

II. BIOCHEMICAL COMPOSITION OF MEMBRANES

Cell membrane is composed of lipid bilayer with embedded (or integral) proteins and peripheral proteins. Carbohydrates are also present in the membranes as a part of glycoproteins and glycolipids. The proportions of lipids and proteins present in the cell membrane vary in different cells of the body; for example, nerve cell membrane is rich in lipids, whereas hepatocyte cell membrane is rich in proteins.

Among the intracellular organelles, the inner membrane of mitochondria, which is the location of a large number of enzymes involved in metabolism, is highly rich in proteins, compared to lipids. Further, the individual constituents of the membrane lipids and proteins greatly vary with different cells and different organelles. For instance, the phospholipid cardiolipin is present abundantly in the inner mitochondrial membrane, compared to other organelle membranes—it is required for the optimal functioning of various complexes of the electron transport system.

A. Lipid constituents

The major lipids present in the membranes are phospholipids, glycosphingolipids and cholesterol. The proportion of these lipids varies with the cells and the organelles. **The distribution of the lipids is also asymmetric between the two leaflets of lipid bilayer.** For example, phosphatidylserine is a glycerol containing phospholipid (glycerophospholipid) present in the inner lipid leaflet of the membrane bilayer. When a cell undergoes apoptosis (programmed cell death), phosphatidylserine is flipped to the outer leaflet of the cell membrane—there, it acts as a signal for macrophages to engulf and clear the dying cell.

1. **Phospholipids:** Glycerophospholipids are the most common amongst the two major phospholipids of the cell membrane (Fig. 3.1). The fatty acids of the glycerophospholipids usually have even numbered carbon atoms. They may be saturated or unsaturated (i.e., they contain one or more double bonds). **The presence of double bonds produces a "kink" in the structure of the fatty acid which leads to an increase in the membrane fluidity** (Fig. 3.1). The most abundant saturated and unsaturated fatty acids in the membranes are palmitic acid and oleic acid, respectively.

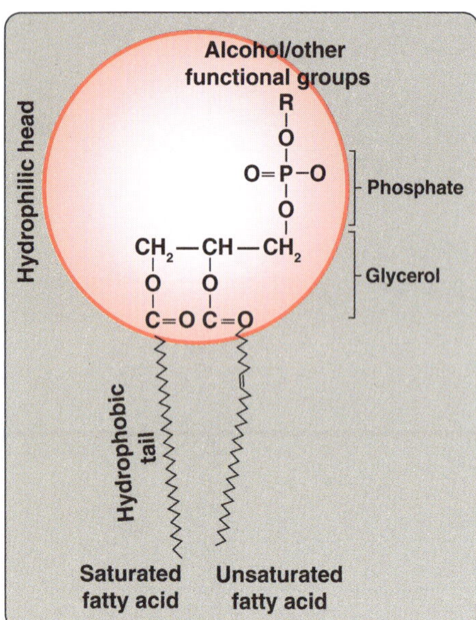

Figure 3.1 Structure of a glycerophospholipid molecule.

The second major type of phospholipid in the membrane is sphingomyelin. It is an example of a sphingophospholipid. Here, fatty acids and phosphate are attached to sphingosine, a long chain amino alcohol, rather than to glycerol. Fatty acid is attached by an amide linkage to the amino group of sphingosine. It also contains choline, or trimethyl ethanolamine. As the name suggests, sphingomyelin is a prominent component of myelin sheaths.

Phospholipids contain both hydrophobic and hydrophilic regions and therefore termed amphipathic. They contain one polar head and two nonpolar hydrophobic tails. The polar heads are formed by the phosphate group (and many a times the alcohol attached to it), whereas the hydrophobic tails are formed by the two fatty acids esterified to glycerol. **The amphipathic nature of phospholipids helps them to spontaneously arrange in such a way that the hydrophobic tails interact with each other, remaining away from the surrounding water molecules. The polar heads come in direct contact with the intra- and extracellular aqueous environment.** This whole arrangement allows the formation of various lipid enclosed structures in vivo and in vitro (Fig. 3.2).

Vesicles are formed by the shedding from the membranes; like the membranes from which they are derived, **they also have a lipid bilayer**. Vesicles are involved in transport of certain molecules between different organelles of the cell, and also in endocytosis and exocytosis of molecules, into and outside the cell, respectively.

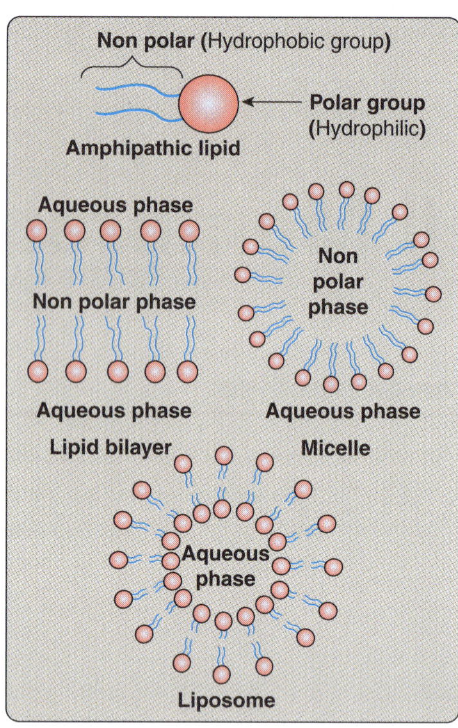

Figure 3.2 Formation of lipid bilayer, micelles and liposomes from amphipathic phospholipid.

Liposomes are prepared artificially by subjecting a solution of phospholipids to sound waves (sonication). They possess a lipid bilayer and an internal aqueous environment which can be loaded with a cargo, for example, a drug, while these particles are synthesized. **Liposomal preparations are currently in use for targeted delivery of drugs**, so as to reduce their unwanted effects in non-target tissues. Additionally, it is **also used for the delivery of nucleic acids in gene therapy trials**. The specificity of liposomes is enhanced by incorporating components like antibodies in the liposomal membrane so that they are targeted to specific tissues or tumors.

Micelles are composed of a single layer of lipids with the hydrophobic tails enclosed within—in the intestines, lipids are absorbed in the form of micelles.

2. **Cholesterol:** It is the major sterol present in the membranes. The steroid ring of cholesterol is embedded between the hydrophobic tails of phospholipids and therefore provides rigidity to the membrane. Only its hydrophilic hydroxyl group is at the aqueous interface. **Cholesterol acts as a fluidity buffer of the membranes—when the membrane fluidity increases, the rigid structure of cholesterol tends to restore it to a lower level.** On the other hand, cholesterol also prevents the reduction of membrane fluidity by preventing

the fully compact arrangement of the hydrophobic fatty acid chains.

3. **Lipid rafts, caveolae and tight junctions:** These are specialized structures in the cell membrane.

   Lipid rafts are dynamic areas on the outer leaflet of lipid bilayer and are enriched with cholesterol, sphingolipids and signaling proteins. They are involved in signal transduction and in response to external stimuli.

   Caveolae are small invaginations of the cell membrane into the cytoplasm; they are derived from the lipid rafts. They are flask-shaped structures containing the protein caveolin-1. Proteins of caveolae are components of signal transduction systems (e.g., insulin and some G proteins). They also mediate the transport of fatty acids and other molecules across the cell membrane.

B. Protein constituents

Proteins are the major functional molecules of membranes. They act as enzymes, pumps and transporters, channels, structural components, antigens and receptors for various molecules. There are two categories of membrane proteins—integral proteins and peripheral proteins.

1. **Integral or intrinsic proteins** span the lipid bilayer usually as a bundle of α-helical transmembrane segment. Some proteins like transporters, ion channels, various receptors and G proteins span the bilayer multiple times, while some proteins like glycophorin A in the RBCs span the membrane only once. Integral proteins can be separated from the membrane only with the help of membrane disruption using strong detergents.

2. **Peripheral proteins** are bound to hydrophilic regions of intrinsic proteins and/or polar head of phospholipids through electrostatic and van der Waals interactions. Unlike integral proteins, they can be easily removed from the membrane by treatment with ionic solutions. One such peripheral protein of the RBC membrane, ankyrin binds to a cytoskeletal protein, spectrin, and this interaction plays an important role in maintaining biconcave shape of the erythrocyte.

C. Carbohydrate constituents

Carbohydrates are located on the external surface of membranes as oligosaccharides attached predominantly to the proteins (termed as glycoproteins) or to the lipids (in which case they are called glycolipids). The monosaccharide moieties in oligosaccharides include glucose, mannose, galactose, fucose, etc. Carbohydrates of glycolipids of RBC cell membrane constitute the antigens of the ABO blood grouping system. **Glycocalyx** refers to the glycoprotein and glycolipid covering that surrounds the cell membrane of certain epithelial and endothelial cells. It protects the cell from digestion and restricts uptake of hydrophobic molecules. In addition, it contributes to cell-cell recognition, communication and intercellular adhesion.

III. FLUID MOSAIC MODEL OF CELL MEMBRANES

The fluid mosaic model was proposed by Singer and Nicolson in 1972. The model has evolved over the time, but it is still the best for explaining the function and structure of the cell membrane. **It describes the membrane structure as a mosaic of components—phospholipids, cholesterol, proteins and carbohydrates, that gives the membrane a fluid character.** It is likened to integral membrane proteins "icebergs" floating in a sea of fluid phospholipid molecules. Salient features explained by the model are (Fig. 3.3):

- **Phospholipids are the main structural components of the membrane.** They have a hydrophilic head and two hydrophobic tails. **Phospholipids form a bilayer** with heads facing the aqueous medium surrounding the membrane.

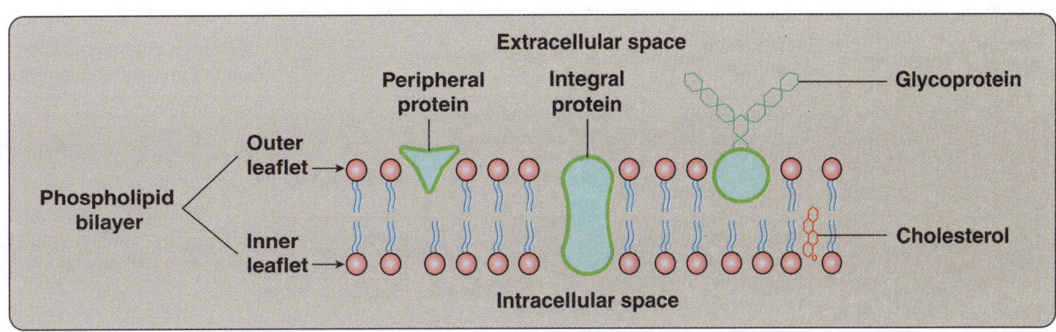

Figure 3.3 Fluid mosaic model of cell membrane.

- The membranes are **asymmetric**. Some proteins are embedded in the outer layer, some in the inner layer and some span the whole membrane. Carbohydrates are attached only on the outer surface forming glycoprotein and glycolipids.
- The **membranes are dynamic structures** because the position and orientations of the components keep changing. **Phospholipid molecules can undergo rapid lateral diffusion, can rotate around their axis, and can flip-flop**. Flip-flop is the migration of phospholipid from one side of the bilayer to the other, though it is not very common.
- The **fluidity of the membrane is dependent on lipid composition** of the membrane. In the bilayer, hydrophobic chains can be ordered to form a "liquid-ordered" state. As the temperature rises, the hydrophobic side chains become disordered taking on a more fluid arrangement (liquid-disordered state). This temperature is termed as the **transition temperature** (T_m). Presence of unsaturated fatty acids that exist in cis configuration also increase the fluidity of the bilayer (which means the T_m is reduced) by decreasing the compactness of the side chain packing.
- **Cholesterol** acts as a buffer to modify the fluidity of membrane. **At temperatures below T_m**, it interferes with the interaction of the tails of fatty acids and thus **increases fluidity**. **At temperatures above T_m**, it limits the disorder because it is more rigid than the hydrocarbon tails of the fatty acids and cannot move in the membrane to the same extent, and thus **decreases fluidity**.
- The fluidity of membrane significantly affects membrane function. **As the fluidity increases, the permeability of water and other small hydrophilic molecules also increases.**
- The **lateral mobility of integral protein increases as the fluidity increases**. Therefore, lipid phase effects can have a profound effect on a protein in which components span the membrane. The insulin receptor is an excellent example of a protein whose functioning is altered with changes in fluidity.
- At normal body temperature (37°C), the lipid bilayer is maintained in a state somewhere between the liquid-ordered and liquid-disordered states, due to the effect of unsaturated fatty acids and cholesterol present in the membranes.

IV. MEMBRANE TRANSPORT

A number of small non-polar molecules like oxygen, carbon dioxide, etc. can freely diffuse across the membrane, but since the membrane has a prominent hydrophobic core, most of the ions and polar molecules and large hydrophobic molecules are transported across the membranes with the help of specific proteins. **The cell membrane is selectively permeable**, in other words, permeable to certain substance and not to others. Membrane transport refers to the collective mechanisms that mediate the passage of ions and small molecules through lipid bilayers. **The transport across membrane can be broadly classified into passive transport and active transport** (Figs. 3.4 and 3.5).

A. Passive transport

Passive transport is a transport process in which molecules **move from a higher concentration to lower concentration without consumption of energy**.

1. **Simple diffusion:** Molecules move passively through the bilayer along the concentration gradient of the molecules on both sides of the

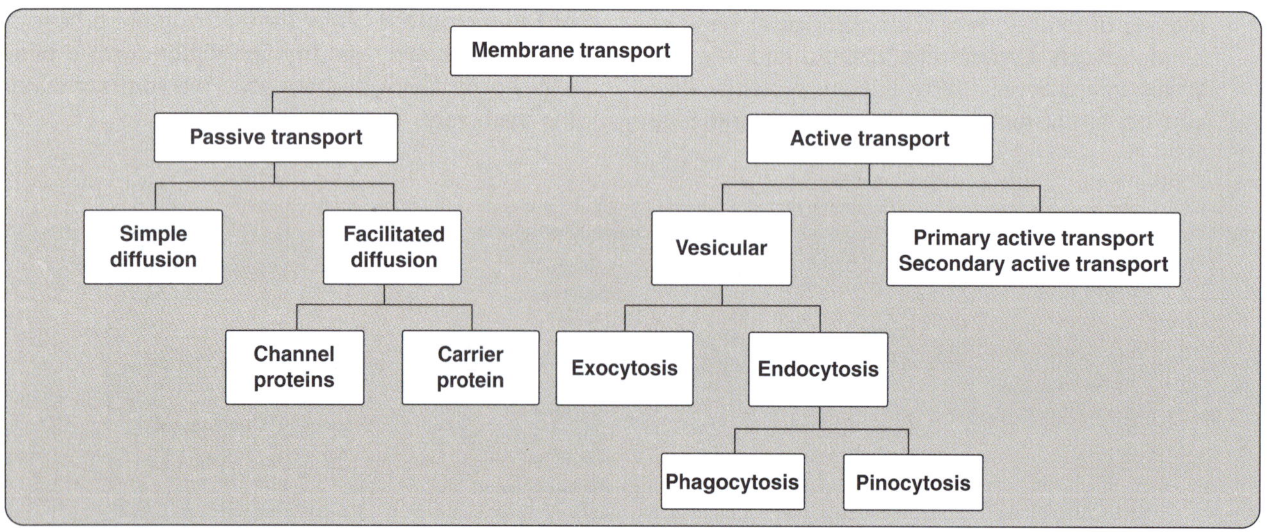

Figure 3.4 Membrane transport.

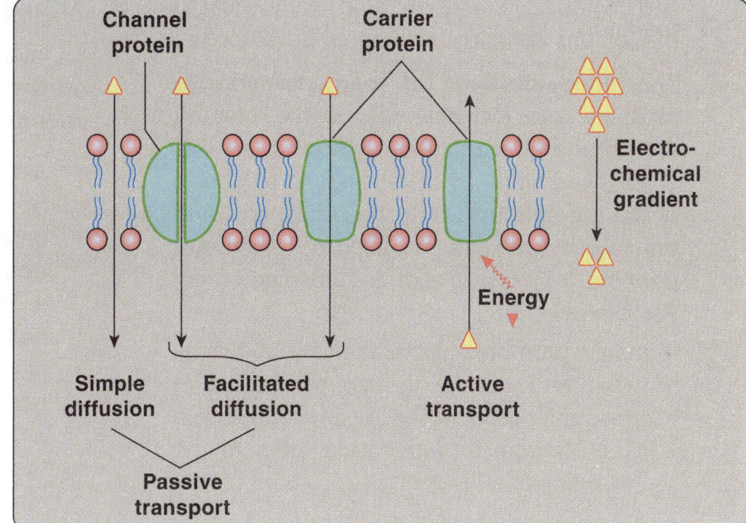

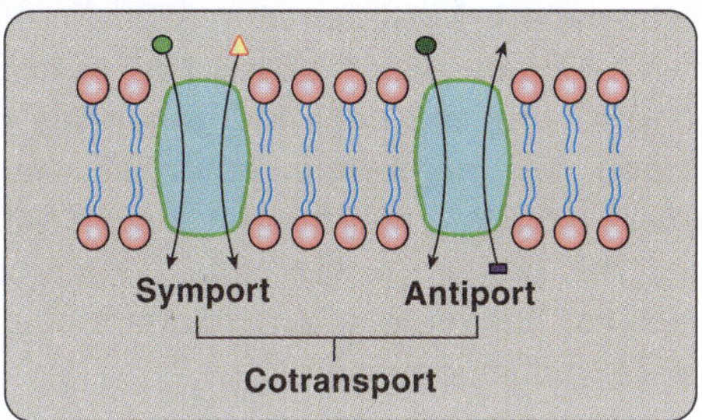

Figure 3.5 Types of transport—active and passive transport.

membrane, till equilibrium is reached. **Non-polar molecules such as gases and steroid hormones freely diffuse across the membrane through this process.** Exchange of oxygen and carbon dioxide in the lungs and at the tissue level is example of simple diffusion.

2. **Facilitated diffusion:** It is also a type of **transport of molecules from a higher to a lower concentration, but mediated by a specific protein transporter or ion channel. There is no energy expenditure**. Facilitated diffusion by transporter proteins is explained with the help of a **"ping-pong" mechanism**. The transporter protein exists in two forms—"ping" and "pong." In the "ping" state, the protein is exposed to a high concentration of the solute, the solute molecules bind to it and bring about a conformational change in the protein to convert it into "pong" state. The solute is then released on other side of the membrane. The process is reversible, based on the concentration gradient. The transporter proteins get saturated at higher concentrations of the solute molecules; thus, the rate of diffusion tends to flatten at higher concentrations of the solute.

   Ion channels are protein molecules that span across the cell membrane allowing the passage of ions. Most of them are highly selective for only one ion. Specific channels for Na^+, K^+, Ca^{2+} and Cl^- have been identified. They have an aqueous pore, which becomes accessible to ions after a conformational change in the protein structure. These channels can be of 3 types:

   a. **Ligand-gated channels** open or close on binding of ligands (specific molecule that bind to the receptor). The ligands can be hormones, drugs, or other intercellular messengers, which bind to their receptors to open or close the channels. Examples include:
      - **Calcium channels in sarcolemma** (endoplasmic reticulum in the muscle cells). These respond to a variety of metabolic and electrical stimuli and release Ca^{2+} into the cytosol.
      - **Acetylcholine receptors**. Acetylcholine binds to the receptor protein, which triggers the conformational change in the receptor, opening the channel for sodium ions. Rapid movement of sodium ions causes membrane depolarization and generation of action potential.

   b. **Voltage-gated channels** respond to the change in electrical potential across the membrane, which causes it to open. **Voltage-gated sodium channels and voltage-gated potassium channels are the examples of such channels**. These are involved in conduction of nerve impulses. Opening of sodium channel causes influx of Na^+ into the cells causing depolarization, whereas closing of sodium channel coupled with opening of potassium channel causes K^+ influx, result in repolarization.

   c. **Mechanically-gated channels or mechanosensitive channels** respond to various mechanical stimuli. They are important components of various sensory receptors that respond to touch, pressure, sound, vibration, etc.

   The kinetics of ions channels are different from that of transporters—they are not saturable, and can achieve diffusion rates nearly equal to that of ions diffusing unhindered in solutions.

The concentrations of many transporter proteins and ion channels in the plasma membrane are regulated by hormones; this is achieved either by increasing the synthesis of the proteins that form these channels and transporters, or by translocating the transporter proteins from cytoplasm to the membrane. For example,

- Insulin promotes glucose uptake by adipocytes and muscle tissues by facilitating the translocation of glucose transporters (GLUT4) from an intracellular pool to the cell membrane.
- Aldosterone is involved in renal sodium reabsorption. It increases the production of epithelial sodium (Na) channels (ENaCs) in the renal tubular cells that are involved in this process.

Aquaporins are tetrameric transmembrane proteins that form channels in the cell membranes for transport of molecules, especially water. **These water channels augment the movement of water by facilitated diffusion.** They selectively permit the passage of water and exclude the passage of ions, because the pores are too narrow to allow passage of larger ions. Aquaporins play a very important role in the water balance of the body. In case of a relative lowering of the body water content, antidiuretic hormone (ADH, or vasopressin) is secreted from the hypothalamus. ADH causes translocation of aquaporins from cytoplasm to the basolateral membrane of the collecting duct cells, promoting reabsorption of water from the renal tubules, thereby restoring the body water volume.

Mutations in cystic fibrosis transmembrane conductance regulator (CFTR) protein which acts as an ATP-dependent chloride channel, **results in cystic fibrosis, an autosomal recessive disease**, characterized by decreased transport of Cl^- across the membranes, along with a diminished transport of water. In most of the tissues, this results in decreased chloride and water content of the luminal contents—leading to manifestations like thick respiratory secretions predisposing to infections, stagnation of secretions of pancreatic enzymes causing pancreatitis and decreased levels of pancreatic enzymes in the intestines leading to impaired fat and protein absorption, plugging of bile ducts by thick biliary secretions, etc. In sweat glands, the reabsorption of chloride is compromised, leading to increased sweat chloride levels.

B. Active transport

It involves **transport against a concentration gradient**. This process **requires energy** and **involves specific transporter proteins known as pumps.**

1. **ATP-driven pumps:** The transport is directly coupled with hydrolysis of ATP to ADP and inorganic phosphate (Pi); this is also called **primary active transport**. Most common example of this is ***Na^+-K^+ ATPase* which drives Na^+ out of the cell, and K^+ into the cell, and generates what is termed as membrane potential**. Another example is multidrug resistance-1 protein (MDR-1 protein). This protein pumps certain drugs out of the cells, and therefore, is an important cause of resistance to chemotherapy.
2. **Coupled transporters or cotransporters:** The transport of one solute against a concentration gradient is coupled with transport of another along its electrochemical gradient. This is also termed as **secondary active transport**, called so because there is no direct hydrolysis of ATP. Instead it relies upon the electrochemical potential difference created by pumping of a different ion in/out of the cell, which is rather ATP dependent.

Note: Transporter proteins and channels can also be classified as uniporters, symporters and antiporters (Fig. 3.5).

Uniporters: Here, **only one type of molecule is transported** across by the transporter/channel. Examples include Glucose transporters (like GLUT4), epithelial sodium channel (ENaC), etc.

Symporters: Here, **two different molecules are transported in the same direction**, i.e., either into or outside the cells. An example is sodium-glucose symporters (SGLT1 and SGLT2), which cotransport one glucose into the cell for every two sodium ions they import into the cell. It is present in small intestine and kidneys, where they are involved in the absorption of sodium and glucose from the intestinal and tubular lumens, respectively.

Antiporters: Here, **two species of ions or solutes are pumped in opposite directions** across a membrane. For example, *Na^+-K^+ ATPase* pump transports three sodium ions out of the cell in exchange for two potassium ions into the cell.

C. Vesicular transport

It comprises of endocytosis and exocytosis. **Endocytosis** is the process by which cells take up large molecules; in **exocytosis**, molecules are released from the cells into the surrounding interstitial fluid.

Property	Passive Transport		Active Transport	
	Simple Diffusion	Facilitated Diffusion	Primary Active Transport	Coupled Transport
Requires specific protein	No	Yes	Yes	Yes
Solute transported against a gradient	No	No	Yes	Yes
Through direct ATP Hydrolysis	No	No	Yes	No
Through co-transported movement down the gradient	No	No	No	Yes
Examples	Gases: O_2, CO_2; Steroid hormones	Glucose, ions, H_2O	Ions, small water-soluble molecules	Glucose and amino acids

Figure 3.6 Differences between passive and active transport.

Both the processes involve formation of vesicles from the cell membrane.

Examples include release of insulin from the β-cells of pancreas which occurs via exocytosis, and uptake of cholesterol into the cells by endocytosis—the latter occurs following the binding of cholesterol carrying lipoproteins to certain extracellular receptors, by what is termed as "receptor-mediated endocytosis." Phagocytosis of microorganisms by macrophages is another example of endocytosis. The term pinocytosis refers to a type of endocytosis where liquid droplets containing small molecules are taken up by the cells.

A summary of differences between passive and active transport is given in Figure 3.6.

Ionophores

Ionophore means "ion carrier." They are **lipid-soluble molecules**; by getting inserted into the membranes, **they can shuttle the movement of ions across the membranes**. They can dissipate the electrochemical gradient across the membrane by facilitating such ion movements. Examples include valinomycin, an antibiotic which is a K^+ ion carrier, and uncoupling protein-1 (UCP-1) which is a proton ionophore. UCP-1 can insert into the inner mitochondrial membrane and dissipate the proton gradient which is required for the formation of ATP. However, the energy from the dissipation of proton gradient is released as heat, rather than as ATP. Thus, UCP-1 is also termed as thermogenin, and is important in generation of body heat, especially in the newborns.

V. Medical Perspectives

- Liposomes are artificial membranes produced outside the body by sonication. Such vesicles are surrounded by lipid bilayer with an aqueous interior. They are used clinically as carriers of drugs targeted to specific organs.
- Hereditary spherocytosis is caused by an inherited defect in the red cell cytoskeleton. Most commonly it is due to a deficiency of spectrin but it can also arise due to mutations in proteins like ankyrin, which is a peripheral membrane protein, and band 3 protein, which is an integral membrane protein in the RBC membrane, to which the cytoskeleton is linked via ankyrin. As a result of these defects, the cells are not able to maintain the biconcave shape and becomes more spherical in shape. Life span of RBCs is decreased because these are easily destroyed in the spleen, as their traverse through the splenic capillaries are slowed down—this leads to anemia.
- Myasthenia gravis is an autoimmune disorder characterized by antibodies against the acetylcholine receptors. As a result, the receptors cannot be activated by acetylcholine and the muscle cells containing these receptors cannot respond to the neurotransmitter-mediated stimulus. Thus, muscle contraction is impaired. A main symptom is muscle fatigue. The treatment consists of acetylcholinesterase inhibitors. These drugs prevent the breakdown of acetylcholine in the neuromuscular junctions, allowing acetylcholine to be present at the neuromuscular junction for an extended period and in higher amounts, stimulating the remaining receptors. Immunosuppressants are also used to reduce the generation of antibodies against the receptor.
- Many viruses and bacterial toxins gain their entry into the cell by binding to various components of the cell membrane. Vibrio cholerae produces cholera toxin which binds to membrane glycolipids, diphtheria toxin binds to a protein receptor at the outer leaflet of the plasma membrane, and SARS-CoV-2 (the causative of novel coronavirus disease—COVID-19) binds to another protein known as ACE2 receptor (ACE is angiotensin-converting enzyme).

- In the small intestine, glucose and Na^+ bind to different sites on Na^+-glucose symporter (also known as sodium glucose transporter or SGLT) at the apical surface of the cells. Na^+ moves into the cell down its electrochemical gradient and drags glucose with it. To maintain a steep Na^+ gradient, Na^+-glucose symporter is dependent upon *Na^+-K^+ ATPase* located at the basolateral membrane, which maintains a low intracellular concentration of Na^+ by pumping it out of the cell. The treatment of a patient suffering from diarrhea (e.g., in food poisoning and cholera) makes use of this information. Oral rehydration solution (ORS), consisting of NaCl and glucose, has been developed by World Health Organization (WHO) for administration during dehydration in diarrhea. The transport of glucose and Na^+ across the intestinal epithelial cells forces movement of water from lumen of the gut into intestinal cells, resulting in rehydration. Glucose alone or NaCl alone would not be effective.
- Fluidity of neuronal membranes is crucial in their functions, as many of the critical functions carried out by neurons, including neurotransmitter release and amino acid uptake for protein synthesis rely on membrane fluidity. Docosahexaenoic acid (*Docosa* – twenty-two, *hexa* – six, *ene* – double bond), commonly known as DHA, is a 22-carbon PUFA, widely used as a food additive. Neuronal membranes are rich in DHA, where it increases the fluidity of the membrane, due to the presence of six double bonds in its structure.
- In many membrane glycerophospholipids, the second carbon of glycerol is esterified to arachidonic acid, a 20-carbon polyunsaturated fatty acid (PUFA) that contains four double bonds. Arachidonic acid acts as a precursor for a large family of molecules known as eicosanoids (eicosa—twenty). These include inflammatory mediators like prostaglandins and thromboxanes. One mechanism by which corticosteroids, widely used anti-inflammatory molecules, exert their action is by inhibiting the release of arachidonic acid from the phospholipids, thus preventing the formation of eicosanoids.
- Sitosterols are a type of plant sterols or phytosterols. They are present in the diet, but in normal individuals only minimal quantities of sitosterols are absorbed, most of which are eliminated in the bile. Sitosterolemia refers to a rare genetic disease due to mutations in the transporters that promote the secretion of absorbed sitosterols from intestinal epithelial cells into the intestinal lumen, or biliary canalicular cells into biliary canaliculi. Accumulation of sitosterols in the membrane, instead of cholesterol, leads to membrane abnormalities in the RBCs. The patients are also at risk for lipid accumulation and atherosclerosis.
- The release of vesicles containing neurotransmitters is important in nerve conduction and muscle contraction. The exocytosis of these vesicles is inhibited by botulinum toxin, one of the most potent toxins in existence, produced by Clostridium botulinum, an anaerobic bacterium. Botulism is a rare, but potentially fatal disease caused due to infection by C. botulinum. It usually results from consumption of food contaminated by the bacteria, and can progress to muscle paralysis and death. Since the toxin prevents muscle contraction, botulinum toxin also finds a cosmetic use, where local injections of botulinum toxin to the facial muscles are used to remove wrinkling of face.
- Apart from cystic fibrosis, the CFTR protein is also involved in the pathogenesis of diarrhea due to cholera and certain toxin-producing strains of Escherichia coli. Cholera toxin and toxins from E. coli cause phosphorylation and persistent activation of the CFTR protein, leading to loss of chloride (which combines with Na^+ to form NaCl) and water from the intestinal cells—leading to diarrhea and dehydration. Indeed, mutations of CFTR protein in cystic fibrosis is thought to provide protection against cholera, by minimizing this water loss.

VI. Chapter Summary

- Membranes are dynamic structures composed of lipids, proteins and carbohydrates.
- The fluid mosaic model forms the basis for understanding membrane structure and function. Basically, all membranes are lipid bilayers formed by two layers of phospholipids. The polar heads of phospholipids are in the aqueous environment on the outer and inner surface of membrane. Fatty acid chains face the interior of the bilayer.
- The fluidity of the membrane is due to presence of cholesterol and unsaturated fatty acids.
- Two types of proteins are present in the cell, integral or intrinsic proteins and peripheral proteins.
- Various passive and active transport mechanisms are employed to maintain the gradient of different molecules across different membranes.
- Large molecules enter or leave the cell through endocytosis or exocytosis.
- Mutations that affect the structure of membrane proteins may cause different diseases.

Study Questions

Short-Answer Questions

3.1. What is the significance of fluidity in the functioning of the membranes? Name two factors that determine membrane fluidity.

3.2. Why is docosahexaenoic acid (DHA) used as a nutritional supplement, especially for children of growing age?

3.3. What is the rationale for the administration of ORS in cases of dehydration?

Long-Answer Question

3.1. Enumerate different types of transport process through cell membrane and explain the mechanisms involved in each.

3.2. Explain the fluid mosaic model of plasma membrane. What are the different types of proteins present in the membrane? Name a clinical condition that arises due to a defect in a membrane protein and explain its pathogenesis.

Multiple-Choice Questions

Choose the ONE best answer.

3.1. Which of the following is associated with ping-pong mechanism?
A. Simple diffusion
B. Facilitated diffusion
C. Active transport
D. Vesicular transport
Correct answer = B.

3.2. Which of the following statements about lipid rafts is FALSE?
A. They are rich in sphingolipids.
B. Their main function is signal transduction.
C. They are characterized by absence of receptor proteins.
D. They are present on the outer leaflet of phospholipid bilayer.
Correct answer = C.

3.3. Which of the following statements about aquaporins is FALSE?
A. They are predominantly channels.
B. They are abundant in erythrocytes and kidneys.
C. They are tetrameric proteins.
D. They mediate secondary active transport.
Correct answer = D.

3.4. Which of the following is an example of a symporter?
A. Na^+-K^+ ATPase
B. Glucose-Na^+ transporter
C. Epithelial sodium channel
D. CFTR chloride channel
Correct answer = B.

3.5. Increase in the membrane fluidity is due to the presence of:
A. Saturated fatty acids
B. Unsaturated fatty acids
C. Proteins
D. Glycoproteins
Correct answer = B.

3.6. Liposomes are prepared by sonication from a solution of:
A. Phospholipids
B. Glycosphingolipids
C. Cholesterol
D. Proteins
Correct answer = A.

3.7. Rigidity of the membrane is due to the presence of:
A. Cholesterol
B. Proteins
C. Phospholipids
D. Carbohydrates
Correct answer = A.

3.8. Which proteins span across the lipid bilayer:
A. Transporters only
B. Ion Channels only
C. Receptors and G Proteins only
D. All of the above proteins
Correct answer = D.

3.9. Which of the following get transported across membranes by coupled active transport?
A. Ions
B. Small water-soluble molecules
C. Glucose and amino acids
D. Steroid hormones
Correct answer = C.

3.10. Which of the following get transported across membranes by simple diffusion?
A. Oxygen, steroid hormones, carbon dioxide
B. Calcium, acetylcholine
C. Insulin, thermogenin
D. Aquaporin, chloride
Correct answer = A.

Multiple-Choice Questions (NExT/NEET PG)

Choose the ONE best answer.

3.1. Membrane fluidity is increased by:
A. Increasing the temperature
B. Presence of unsaturated fatty acids having kinks and cholesterol

C. Both A and B are correct
D. Neither A nor B is correct

Correct answer = E.

3.2. Based on statements given below, choose the correct option:

Statement 1: Valinomycin is used in potassium selective electrodes

Statement 2: Valinomycin is a potassium-specific transporter and facilitates the movement of potassium ions lipid membranes "down" the electrochemical gradient

A. Both the Statements are correct, and Statement 2 is the correct explanation for Statement 1
B. Both the Statements are correct, but Statement 2 is NOT the correct explanation for Statement 1
C. Statement 1 is correct, but Statement 2 is NOT correct
D. Statement 1 is NOT correct, but Statement 2 is correct

Correct answer = A.

3.3. Based on the statements given below, choose the correct option:

Statement 1: Cystic fibrosis patients are more prone to respiratory infections

Statement 2: In cystic fibrosis, transport of water across membranes is diminished due to mutation in a sodium channel

A. Both the Statements are correct, and Statement 2 is the correct explanation for Statement 1
B. Both the Statements are correct, but Statement 2 is NOT the correct explanation for Statement 1
C. Statement 1 is correct, but Statement 2 is NOT correct
D. Statement 1 is NOT correct, but Statement 2 is correct

Correct answer = C.

3.4. Which of the following organelles show similarity to a prokaryotic cell?

A. Mitochondria only
B. Chloroplast only
C. Both mitochondria and chloroplast
D. None of the above

Correct answer = C.

3.5. Cholera toxin affects chloride transporter that secretes chloride ion into the lumen of the small intestine. How does cholera toxin lead to dehydration?

A. The toxin turns off the transporter, so negative chloride ions pushes water into the lumen.
B. The toxin turns on the transporter, and water follows the ion due to osmosis.
C. Toxin turns off the transporter, so water is no longer exchanged for chloride ions.
D. Toxin turns on the transporter, and water is actively transported with the ion.

Correct answer = B.

Acid–Base Balance and Arterial Blood Gas (ABG) Analysis

4

Learning Objectives

At the end of this chapter, the student will be able to:

- Define acid, base, pH and buffer
- Discuss major buffer systems in body
- Explain how the pH of blood is maintained
- Describe acid–base disorders and compensatory mechanisms
- Describe the significance of ABG analysis

I. OVERVIEW

According to Bronsted-Lowry model, **acids are molecules that can donate H^+ (i.e., protons), which are accepted by bases**. The concentration of H^+ ions in the body fluids varies greatly—from ~10^{-6} mmol/l in the pancreatic secretions to ~10 mmol/l in the gastric juice; expressing the H^+ concentration in conventional units (mmol/l or mol/l) is hence inconvenient. Further, the H^+ concentration in the blood is physiologically maintained within a narrow range. The severity of the subtle changes in H^+ concentration of blood will not be reflected if expressed in mmol/l. To thwart these problems, pH scale is used to denote H^+ concentration in biological systems, similar to the way it is used in those chemical systems where H^+ concentrations are low.

pH is the negative logarithm (to the base 10) of the H^+ activity expressed in moles/liter (*Note: In this and all the subsequent discussions, we will consider "activity" same as "concentration", though "activity" is the "concentration" of a substance, after adjustments are made for intermolecular interactions.*)

Since the pH of the intracellular and the extracellular milieu is to be maintained within a narrow range, various short-term and long-term mechanisms are adopted by the body to ensure this. Alterations in the pH leading to values outside the physiological limits lead to cell dysfunction. **The pH of the arterial blood is 7.35–7.45. Acidemia refers to a pH <7.35 and alkalemia, a pH >7.45.** The processes that lead to acidemia and alkalemia are termed acidosis and alkalosis, respectively. Both acidosis and alkalosis, when severe, can be fatal.

II. MAINTENANCE OF PH IN THE BODY

The pH of intracellular fluid in most of the cells is maintained between 6.0 and 7.4, the exact values varying with different cell types. The maintenance of optimum pH is important for enzyme activity and cell metabolism, generation of membrane potential, functioning of ion channels and transporters, and in the functioning of organelles and cytoskeleton.

Oxidation of carbon fuels in the body for generation of energy results in production of acids such as carbonic acid during the citric acid cycle and generation of lactic acid during glycolysis. Because of the production of acids by the cells, **the pH of venous blood (7.32–7.38) is slightly lower than that of the arterial blood (7.35–7.45)**. Volatile acids (e.g., carbonic acid in the form of CO_2) are removed from the body via lungs, while non-volatile or fixed acids (e.g., lactic acid, sulfuric acid and phosphoric acid) are excreted in the urine. Approximately 80 mmol of non-volatile acids is produced by the body via metabolism.

A number of states, physiological and pathological, can lead to buildup or loss of H^+ from the body, but the body has inherent mechanisms operating to maintain the pH in the optimal range. This section describes the three mechanisms that operate at different levels to defend the alteration in the body pH—buffers, respiratory regulation, and renal regulation. The time frame and effects of their actions vary. **Buffers act immediately but for a short duration, while renal mechanisms set in after some hours to days and are sustained; respiratory regulation of pH occurs after a few minutes of deviation of pH from the**

normal. Aberrations in acid–base balance occur when these mechanisms dysfunction or are overwhelmed.

A. Buffers

A buffer is a mixture of a weak acid and its strong conjugate base (or a weak base and its strong conjugate acid) which resists changes in pH of the solution that contains it. Henderson-Hasselbalch equation for a weak acid and its strong conjugate base states that:

$$pH = pK_a + \log \frac{[Conjugate\ base]}{[Acid]}$$

where K_a is the dissociation constant of the acid.

The pH of a buffer is thus determined by two factors—the pK_a, and the relative concentration of conjugate base and acid. **Buffering capacity** is the ability of a buffer to resist the change in pH, when an acid or a base is added to the solution containing the buffer. Quantitatively, it can be expressed as the number of moles of a strong acid or a strong base required to alter the pH of 1 liter of a buffer solution by 1. **Buffering capacity is maximum when the pH of the solution is pK_a ± 1**; it also depends on the amount of buffer present in the system. Since different body fluid compartments differ in their composition (see Fig. 5.2, p. 40), the major buffers in body fluids also vary. Table 1 (see p. 57) gives the list of buffers in different body fluids. Their roles and significance are described below.

1. **Bicarbonate buffer:** The bicarbonate buffer system consists of H_2CO_3, and its conjugate base—HCO_3^-. When a strong acid (denoted as HA) is added to the system, H^+ combines with HCO_3^- to form H_2CO_3 and A^-, which being a weak acid and a weak base, respectively, will not affect the pH drastically: $HA + HCO_3^- \rightarrow H_2CO_3 + A^-$

   A strong base (denoted as AOH) when added, reacts with H_2CO_3 to form its bicarbonate salt and water—again resisting the change in the pH: $AOH + H_2CO_3 \rightarrow AHCO_3 + H_2O$

   Despite a pK_a of 6.1 which is outside the ± 1 range from the blood pH of 7.4, bicarbonate buffer is the major buffer of the extracellular fluid (ECF). There are two factors that facilitate the efficient functioning of bicarbonate buffer in the ECF:

   a. An unusually high ratio of 20 between the conjugate base and the acid (i.e., HCO_3^- and dissolved CO_2) seen in the bicarbonate buffer system draws its maximal buffering action towards the physiological pH, according to the Henderson-Hasselbalch equation.

   b. The levels of pCO_2 and HCO_3^- in the ECF are regulated by the lungs and the kidneys (see B. Respiratory Regulation of pH, p. 27 and C. Renal Regulation of pH, p. 27). This enables a fine control of the pH, since these components can be removed (and in case of HCO_3^-, added) to keep the pH within the optimal range.

An open system buffer is one in which both the components of the buffer (acid and its conjugate base/base and its conjugate acid) can be removed from the body. Since CO_2 is removed by the lungs, and HCO_3^- by the kidneys, the bicarbonate buffer system is an example of an open buffer system.

In contrast, **closed system buffers** do not have any gas equilibration, i.e., they do not have any removal of the components at the pulmonary level; at the best, one of the components may be removed by the kidneys. Closed system buffers in the human body include phosphate, protein, and hemoglobin buffers.

2. **Phosphate buffer:** Three pK_a values are possible for phosphoric acid (H_3PO_4), since it is a triprotic acid.

   $H_3PO_4 \rightleftharpoons H_2PO_4^- + H^+$ $pK_1 = 2.1$

   $H_2PO_4^- \rightleftharpoons HPO_4^{2-} + H^+$ $pK_2 = 6.8$

   $HPO_4^{2-} \rightleftharpoons PO_4^{3-} + H^+$ $pK_3 = 12.3$

   Since the pK_2 of 6.8 is close to the physiological pH, the components of phosphate buffer in the physiological systems are HPO_4^{2-} and $H_2PO_4^-$. They are present in a ratio of 4:1.

   Despite the optimal pH for buffering close to that of the ECF, phosphate buffer does not play a predominant role in defending the ECF pH. This is because of low levels of both HPO_4^{2-} and $H_2PO_4^-$ in the ECF. However, intracellular fluid is rich in organic and inorganic phosphates (see Fig. 5.2, p. 40); thus, **phosphate buffer is a major buffer system of the cell**. Similarly, it **also acts as a major buffer in the renal tubules**, where phosphate concentrations are high, and the pH is much closer to pK_2 of H_3PO_4 than the pH of the ECF.

3. **Protein and hemoglobin buffers:** Proteins form an important buffer system in the blood and the intracellular fluid due to their sheer abundance in these fluids. The buffering action of proteins (including hemoglobin) is conferred by the constituent amino acids. The **buffering action of proteins is mainly because of imidazole group of histidine, the sulfhydryl group of cysteine** [which have pK_a values near the physiological pH (6.0 and 8.3, respectively)], and **ionizable side chains** (or R-groups – see Fig. 6.1, p. 69).

Since histidine residues are much more common in proteins than cysteines (e.g., one molecule of hemoglobin contains 38 histidine residues and 6 cysteine residues), **histidine is the major contributor to the buffering action of proteins at the physiological pH** (Fig. 4.1).

Buffers act quickly to help maintain the pH of body fluids within a narrow range by absorbing excess acid or base. However, they do not eliminate the excess acid or base from the body; instead, they temporarily bind to it. The removal of excess acids or bases from the body is primarily managed by the respiratory system (lungs) and the renal system (kidneys), which can eliminate or excrete these substances to restore the body's acid–base balance.

B. Respiratory regulation of pH

In the body, gaseous exchange (between O_2 and CO_2) occurs at two sites—pulmonary capillaries and capillaries of peripheral tissues. CO_2, the product of tissue respiration, diffuses from the cell into the tissue fluid, and at the capillary interface, diffuses across the capillary membrane into the blood.

CO_2 is transported to the pulmonary capillaries from the tissue capillaries by three mechanisms (Fig. 4.2):

1. CO_2 enters the red blood cells (RBCs) where it reacts with water in presence of an enzyme, carbonic anhydrase, to form carbonic acid (H_2CO_3); the latter then dissociates to form H^+ and HCO_3^-. **H^+ reacts with oxyhemoglobin (HbO_2^-) to release oxygen and form deoxyhemoglobin (HHb), the so-called Bohr effect.** HCO_3^-, as more quantities are formed, is transported from RBCs into the plasma by an anion exchanger protein, in exchange for chloride from the plasma; this is called **Hamburger phenomenon**.

   Transport in the form of HCO_3^- accounts for about 70% of the total CO_2 transported.

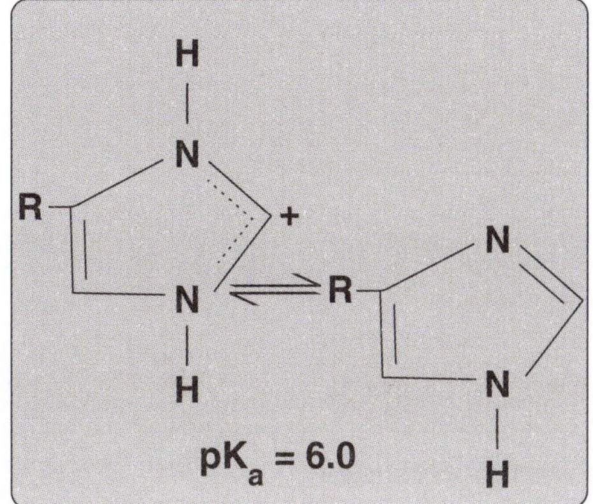

Figure 4.1 Buffering action by imidazole group of histidine.

2. CO_2 also directly reacts with the N-terminal amino groups of the globin chains of hemoglobin to form **carbaminohemoglobin**. This accounts for nearly **20% of the total CO_2 transported**.

3. Nearly **10% of CO_2 is transported in the dissolved state**, in the venous blood.

   Thus, CO_2 is transported in the venous blood without much of a change in the pH.

   In the pulmonary capillaries where O_2 tension is high and CO_2 tension is low, CO_2 is exchanged for O_2, by a reversal of the above mechanisms (Fig. 4.2). **Increased O_2 levels decrease the CO_2 and H^+ binding to hemoglobin; this is known as Haldane effect.** CO_2 thus released from carbaminohemoglobin, and that is formed as a reversal of the carbonic anhydrase reaction, diffuses from the pulmonary capillaries into the alveoli, to enter the expired air.

   Lungs regulate the pH by controlling the CO_2 levels in the blood. This is achieved by controlling the rate of respiration, which is ultimately determined by a set of receptors known as chemoreceptors.

   Central chemoreceptors in the respiratory center located in the medulla of the brain control the rate of ventilation and the tidal volume. **Accumulation of CO_2 leads to an involuntary increase in the rate and depth of ventilation, resulting in expiration of an increased amount of CO_2.** A converse happens when pH is increased or CO_2 levels in blood are low.

C. Renal regulation of pH

In acidemia and alkalemia, kidneys conserve or excrete the acids and the bases, as the case may be, to restore the normal pH. **Four mechanisms are employed by the kidneys to achieve this**:

1. **Bicarbonate reabsorption:** In thick ascending loop of Henle, HCO_3^- is totally reabsorbed, without net excretion of H^+ (Fig. 4.3).

2. **H^+ secretion:** In the collecting ducts, H^+ is actively secreted by *H^+-ATPase* and *K^+–H^+-ATPase*. This is the reason for the acidic pH of urine (Fig. 4.4).

3. **Combination of H^+ in the tubular fluid with phosphate and ammonia:** The **minimum pH that can be achieved for the urine by the kidneys is 4.5**. Below this pH, the gradients to overcome for the active secretion of H^+ becomes too high.

   The amount of H^+ that can be excreted in urine when pH is lowered from 7.4 to 4.5 = log $10^{-4.5}$ – log $10^{-7.4}$ = 3.2 x 10^{-5} mol/l. Thus, the amount of H^+ that can be excreted considering a urine

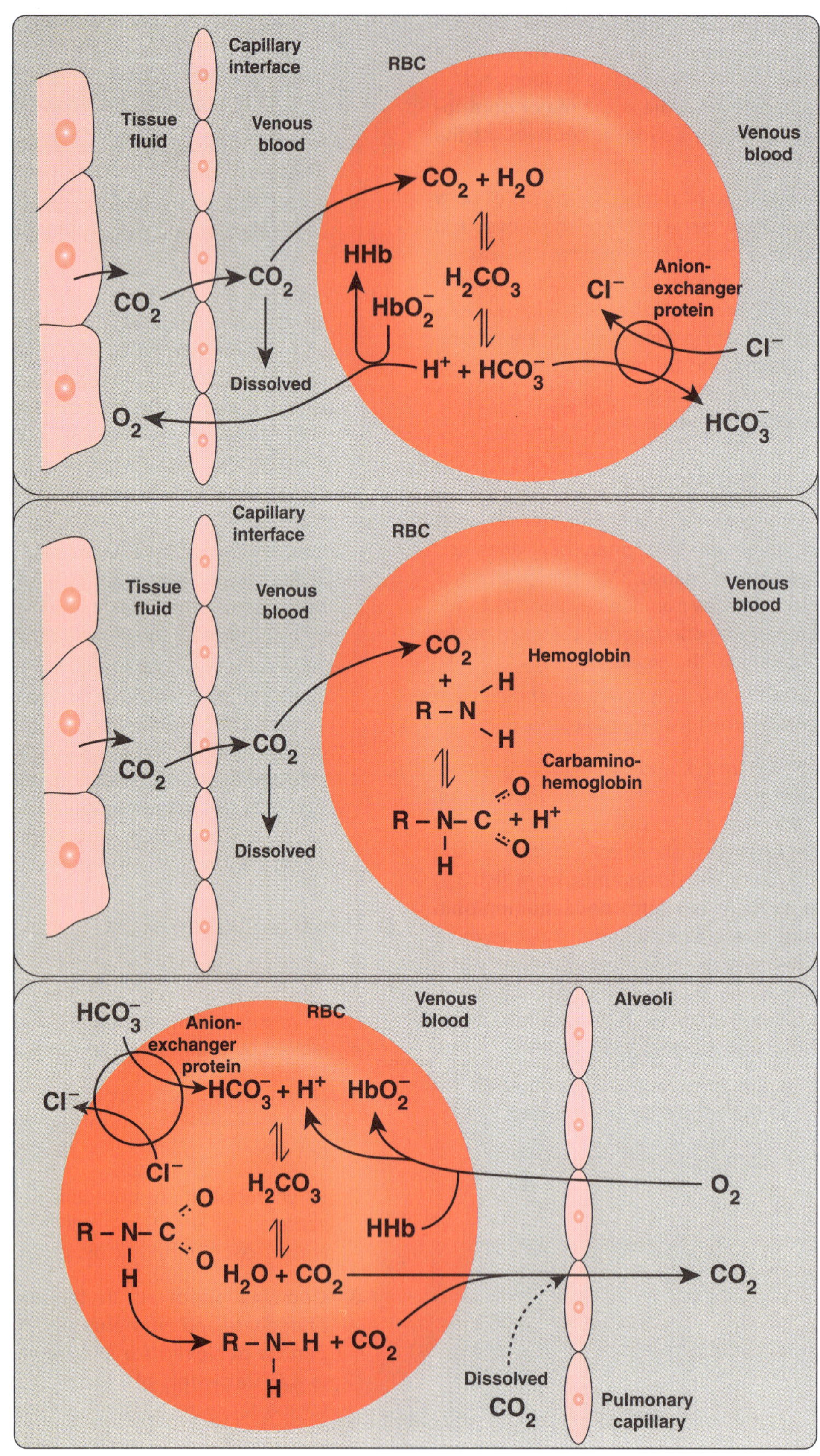

Figure 4.2 CO_2 transport from peripheral capillaries to pulmonary capillaries and alveoli.

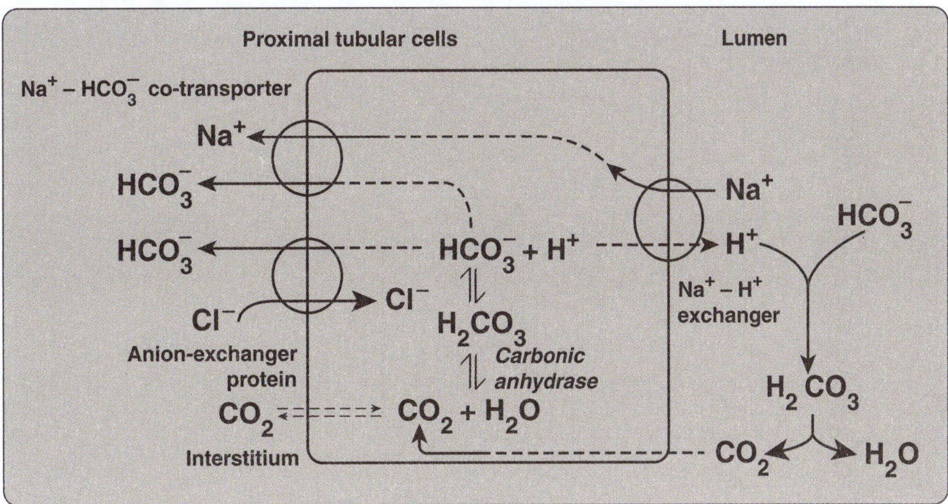

Figure 4.3 Bicarbonate reabsorption in the proximal tubule.

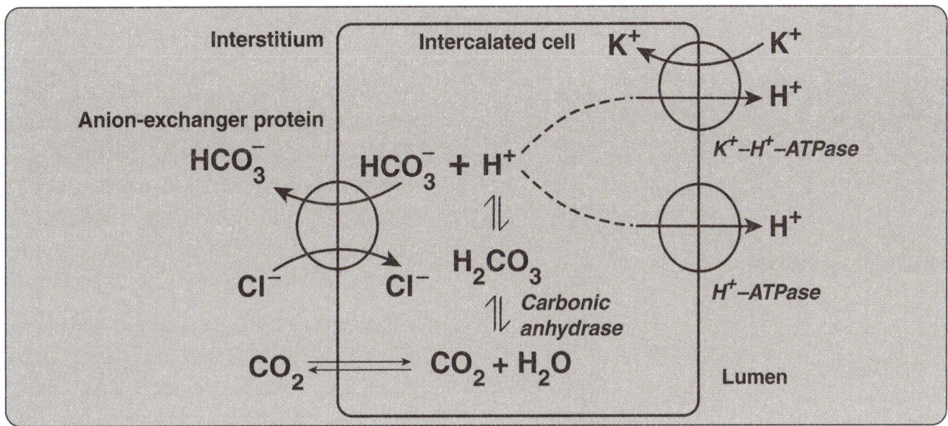

Figure 4.4 H^+ secretion by the intercalated cells of the collecting duct.

volume of 1.4 l/day is 44.8 µmol/day. This is negligible compared to the daily production of 80 mmol of H^+ as a result of metabolism. Thus, alternate means are required to excrete H^+ in the urine, that do not involve any subsequent reduction in pH. This is achieved by combining H^+ of the tubular fluid with phosphates and ammonia.

- H^+ generated by the intercalated cells combines with HPO_4^{2-} to generate $H_2PO_4^-$; the latter is excreted as NaH_2PO_4 (Fig. 4.5).
- Ammonium ion (NH_4^+) is produced mainly in the proximal tubules from glutamine by the action of the enzyme glutaminase. The product glutamate undergoes further metabolism to produce another NH_4^+ and two HCO_3^- molecules (Fig. 4.6).

In the thick ascending limb of loop of Henle, NH_4^+ re-enters the tubular cells via Na^+-K^+-$2Cl^-$ cotransporter (taking the place of K^+) and then enters the interstitium, where it dissociates into NH_3 and H^+ (Fig. 4.7).

In the collecting duct, NH_3 re-enters the lumen via specific transporters and combines with H^+ to form NH_4^+, which is excreted as NH_4Cl.

4. **New bicarbonate generation:** Apart from the reabsorption of filtered HCO_3^-, kidneys also play a role in replenishing the HCO_3^- reserve in the body. The glutaminase reaction will ultimately result in the generation of two HCO_3^- molecules that enter the blood. A closer look at Figures 4.4, 4.5, and 4.7 also reveals that the activity of *H⁺-ATPase*, and the buffering action of the phosphate buffer or NH_3—lead to one HCO_3^- molecule transported into the plasma, per proton that enters the luminal fluid.

In physiological states, renal tubules produce sufficient H^+ to re-absorb almost all of the HCO_3^- that is filtered by the glomeruli. An amount of H^+ produced in excess of this ensures the removal of the 80 mmol of non-volatile acids produced in the body. In acidosis, the H^+ secretion by the renal tubules is increased, so as to

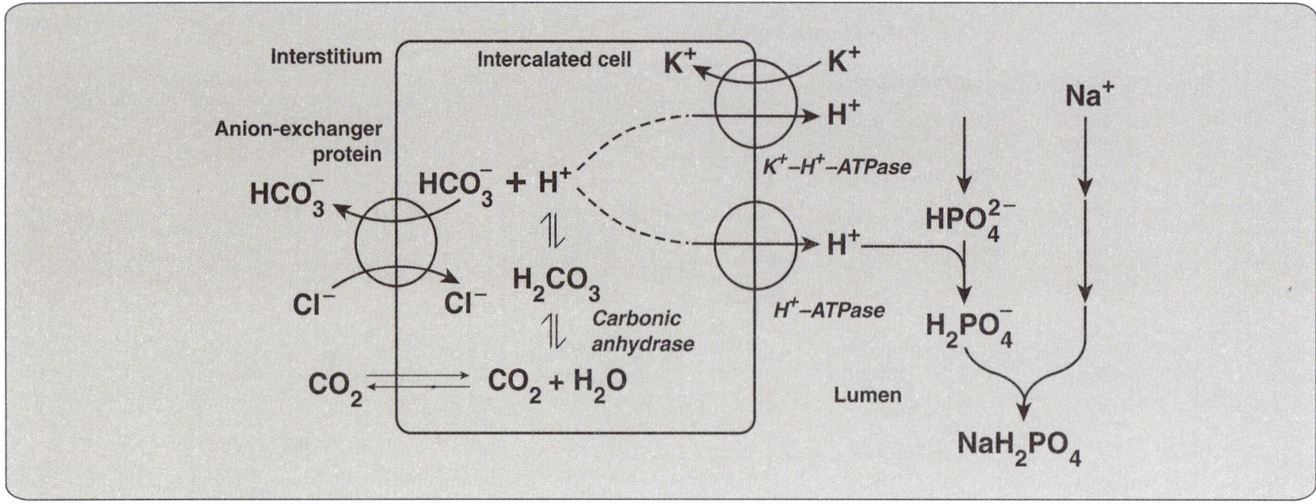

Figure 4.5 Phosphates form an important urine buffer by combining with H^+ in the lumen.

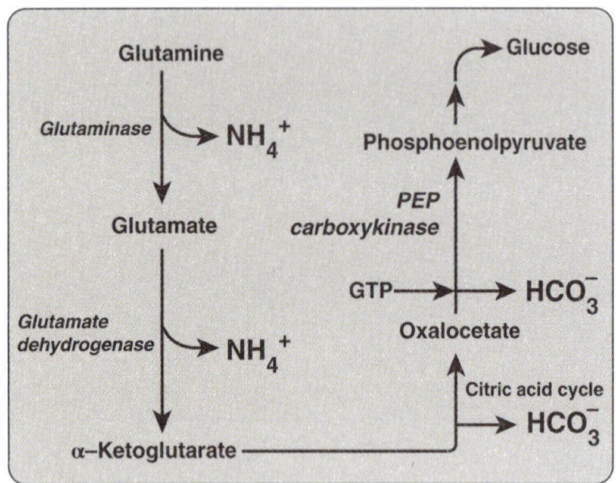

Figure 4.6 Metabolism of glutamine and its products in the proximal tubules generates $2NH_4^+$ and $2HCO_3^-$.

facilitate H^+ excretion; HCO_3^- is conserved—this brings about an increase in the pH of the blood. The opposite happens in case of alkalosis.

The signals to the kidneys to regulate the secretion of H^+ (and reabsorption of HCO_3^-) are summarized in Table 2 (see p. 57).

D. Acid–base disturbances and compensation

The pH of the arterial blood is 7.35–7.45. The body functions get impaired outside this range of pH. After an insult to the normal acid–base equilibrium, the defense mechanisms try to restore the pH to the normal.

Since buffers act as the first line of defense against the alterations in the pH, the aberrations in the pH of the body can be determined by looking at the status of the buffers.

Acidemia refers to a blood pH <7.35, and the conditions leading to it are referred to as **acidosis**. **This occurs when the ratio of HCO_3^- to pCO_2 is decreased**. When this occurs due to an increase in the pCO_2, the condition is referred to as **respiratory acidosis**; if the fall in the ratio is due to a decrease in the HCO_3^- levels, the term used is **metabolic acidosis**.

Alkalemia refers to a blood pH >7.45, and the conditions leading to it are referred to as alkalosis. The ratio of HCO_3^- to pCO_2 is increased. **Respiratory alkalosis** occurs when there is a decrease in the pCO_2; increase in the HCO_3^- levels lead to **metabolic alkalosis** (see p. 57).

Presence of any one of the above-mentioned disorders, along with appropriate compensation constitutes **"simple acid–base disorders."** It is also not uncommon to encounter situations when the acid–base disturbance is caused by more than one primary abnormality. Such disorders are known as **"mixed acid–base disorders."**

III. RESPIRATORY DISORDERS WITH AN ABNORMAL ACID–BASE STATUS

A. Respiratory acidosis

Respiratory acidosis is characterized by an increase in the pCO_2 of the blood and a decrease in the pH. **Respiratory acidosis almost always results from CO_2 retention in the body (hypercapnia).** Clinical features of respiratory acidosis will be predominantly that of the underlying disorder. However, CO_2 has a vasodilatory effect on cerebral vessels—CO_2 accumulation intrinsically can lead to neurological manifestations like headache, drowsiness, confusion, and even coma.

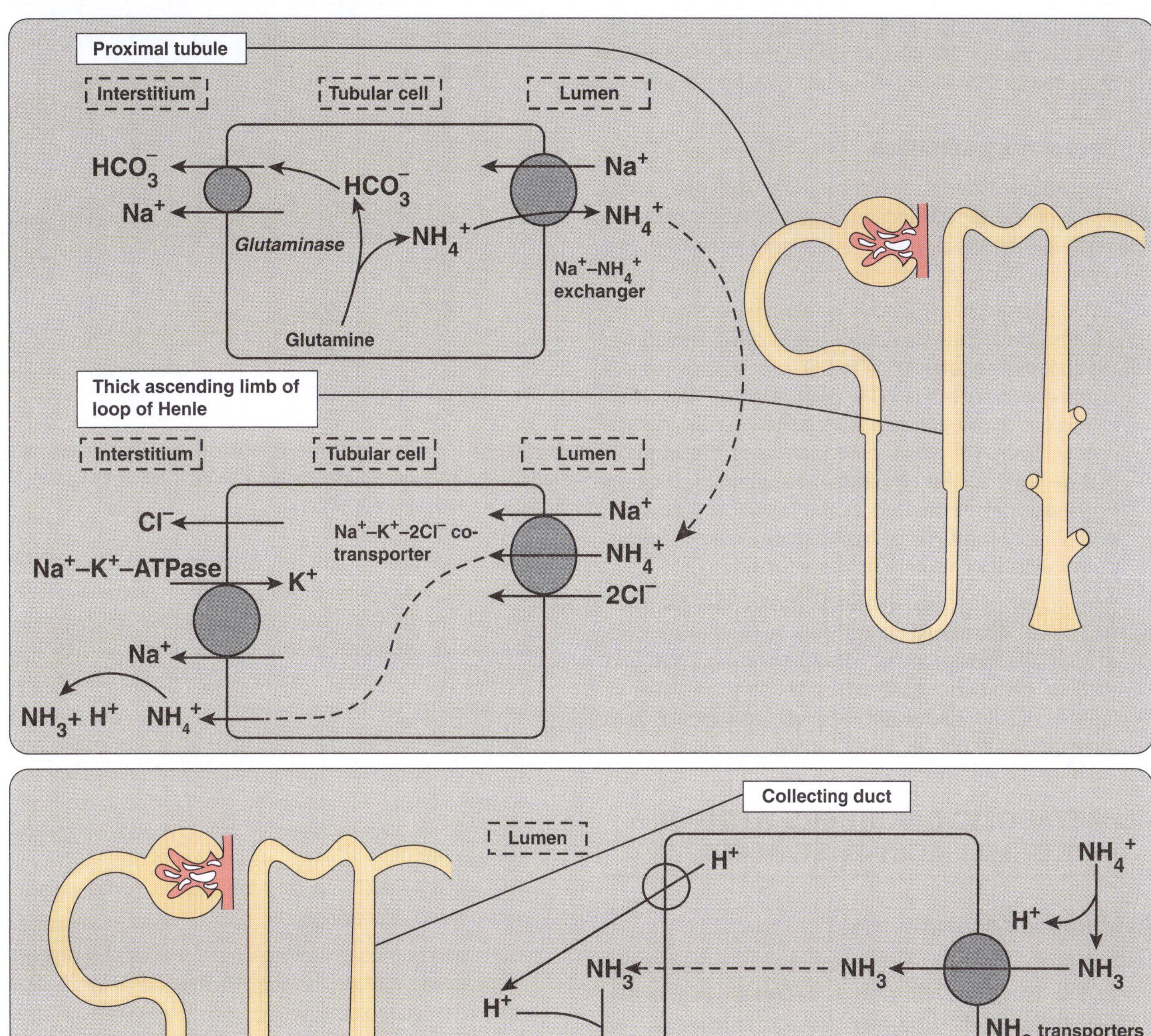

Figure 4.7 Renal generation and handling of NH_4^+.

The cause of CO_2 retention can be:

(i) An abnormality of the respiratory center (e.g., infections or tumors involving the central nervous system)

(ii) Respiratory muscle paralysis or fatigue (e.g., in poliomyelitis, muscular dystrophies, etc.)

(iii) Disease afflictions of airways and lungs (e.g., severe asthma, chronic bronchitis, etc.)

(iv) Conditions that restrict ventilation (e.g., rib fracture, skeletal deformities, etc.)

The compensation is by the kidneys, which results in an increase in the HCO_3^- in the blood. In disorders where the function of the respiratory center is intact, the increase in pCO_2 also increases the rate and depth of ventilation, to eliminate excess CO_2 from the body.

B. Respiratory alkalosis

Respiratory alkalosis results from a decrease in the pCO_2 of the blood and an increase in the pH. The renal compensation results in conservation of H^+ and excretion of HCO_3^-.

A decrease in the pCO_2 or hypocapnia is caused by an increase in the rate and/or depth of ventilation. As the vasodilatory effect due to CO_2 on cerebral vessels is abrogated, a decline in the cerebral blood flow leads to clinical manifestations like confusion, drowsiness and seizures. **Alkalosis also increases the binding of free Ca^{2+} in the circulation to albumin, leading on to sudden reduction in the levels of free Ca^{2+} and clinical features of hypocalcemia like tingling, numbness and muscle spasms (or tetany).**

Respiratory alkalosis generally arises due to stimulation of respiratory center. This is seen in hypoxia at high altitudes, hypoxia due to severe anemia and cardiac disorders, salicylate poisoning, in anxious individuals, etc. Respiratory alkalosis also results from overvigorous mechanical ventilation.

IV. METABOLIC DISORDERS WITH AN ABNORMAL ACID–BASE STATUS

A. Metabolic acidosis

Metabolic acidosis is associated with a **reduction in the HCO_3^- levels and a decrease in the pH.** The compensation by respiratory mechanisms, i.e., increase in the tidal volume and the rate of ventilation, leads to a reduction in pCO_2.

Metabolic acidosis can result broadly from four scenarios:

(i) Production of fixed acids which exceeds the body's capacity to eliminate them

(ii) Addition of fixed acids to the body by ingestion or infusion

(iii) Reduction of H^+ excretion by the kidneys

(iv) Loss of HCO_3^- from the body

However, for the ease of determining the etiology clinically, metabolic acidosis is classified based on the effect exerted by the underlying disorder on the anion gap.

Anion gap: The concentrations of anions and cations in the plasma are equal, so that the electrical neutrality is maintained. However, **since routine laboratory analysis of plasma electrolytes involves only selected electrolytes, there exists a gap between the measured cations and measured anions** (Fig. 4.8). Since Na^+ is the primarily measured cation and Cl^- and HCO_3^- are the anions that are primarily measured, **plasma anion gap = $[Na^+] - ([Cl^-] + [HCO_3^-])$**, all values in mmol/l. The reference interval for anion gap is 12 ± 4 mmol/l.

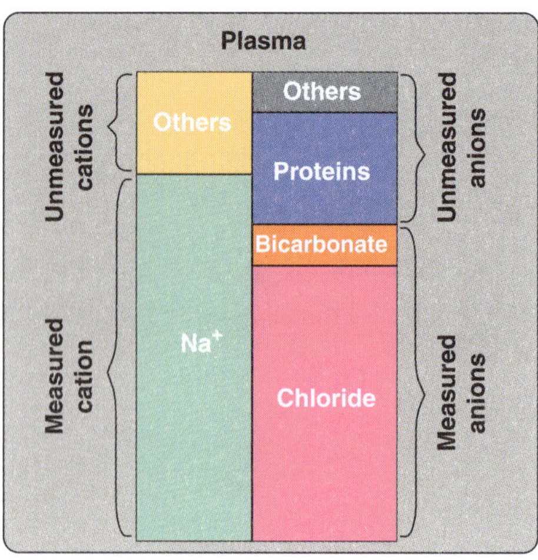

Figure 4.8 Anion gap is the difference between measured cation and measured anions (or alternatively unmeasured anions – unmeasured cations).

Note: The formula for calculation of anion gap given above is based on the convention followed in the United States. According to the European conventions, K^+ is also included in the calculation, so the formula gets modified to: plasma anion gap = $([Na^+] + [K^+]) - ([Cl^-] + [HCO_3^-])$, and the reference interval also gets altered accordingly.

Anion gap is thus, an indicator of unmeasured anions. Unmeasured cations include K^+, free Ca^{2+}, Mg^{2+}, and cationic immunoglobulins (e.g., IgG); unmeasured anions include anionic proteins like albumin and anionic immunoglobulins (e.g., IgA), phosphates, sulfates, lactates, and ketone bodies. Anion gap can also be expressed as (unmeasured anions – unmeasured cations).

A reduction in the anion gap is seen due to a reduction in the unmeasured anions (e.g., in hypoalbuminemia due to decrease in albumin) or an increase in the unmeasured cations (seen in lithium toxicity and in multiple myeloma when producing high levels of IgG). However, the most common cause of a reduced anion gap is laboratory error! So, the testing should be repeated, if the diagnosis does not fit. A high anion gap is seen when unmeasured anions are increased (e.g., hyperphosphatemia, or in multiple myeloma when producing high levels of IgA)—in these conditions, the pH of the blood is normal. **However, most commonly, an increase in the plasma anion gap is associated with metabolic acidosis.** Metabolic acidosis is thus classified into high anion gap metabolic acidosis and non-(or normal) anion gap metabolic acidosis.

1. **High anion gap metabolic acidosis:** Here non-measured anions accumulate, with a decrease in the blood pH. As H^+ is removed from the body, HCO_3^- is correspondingly decreased. Causes include:

   a. **Lactic acidosis**—when aerobic metabolism is compromised, pyruvate is converted to lactate, instead of getting converted to acetyl CoA (Chapter 13. V.K. Pyruvate reduction to lactate, p. 199). It is also seen in type 1a and 1b glycogen storage diseases, due to an increased flux through glycolysis (Fig. 16.8, p. 229).

   b. **Accumulation of acetoacetate and/or β-hydroxybutyrate**—seen in diabetic ketoacidosis (see Chapter 21, V.C. Excessive ketone body production in diabetes mellitus, p. 315), prolonged starvation (Fig. 28.12, p. 460) and excess ethanol consumption (Chapter 27. IV.C.4. Alcohol-related hypoglycemia, p. 447).

   c. **Renal failure**—due to impaired excretion of sulfuric, phosphoric and organic acids produced in the body.

   d. Toxicity due to accumulation of anions—seen in **poisoning with methanol, ethylene glycol, paraldehyde, salicylate**, etc.

> Salicylate alters the metabolism and produces different organic acids, leading onto metabolic acidosis. Salicylate also stimulates the respiratory center to increase the rate and depth of ventilation, leading to respiratory alkalosis. Since there is more than one primary disorder involved, salicylate poisoning is an example of mixed acid–base disorder.

2. **Non-anion gap metabolic acidosis:** This arises due to a direct loss of fluid rich in bicarbonate from the gastrointestinal tract or the renal tubule. To maintain the electrical neutrality of the blood, chloride is reabsorbed in excess leading onto hyperchloremia. Since a measured anion (HCO_3^-) is replaced by another measured anion (Cl^-), the anion gap remains more or less normal.

   a. **Gastrointestinal causes** include severe diarrhea, and intestinal fistula, where bicarbonate-rich fluid in the gastrointestinal tract is not properly reabsorbed.

   b. **Renal causes** include:

      i. Administration of drugs like acetazolamide which inhibit *carbonic anhydrase* in the proximal tubule, thereby decreasing the H^+ secretion into the lumen from the tubular cells, impeding the HCO_3^- reabsorption.

      ii. Renal tubular acidosis (RTA)—These are clinical conditions in which metabolic acidosis ensues because of a failure in the acidification of the urine (i.e., excretion of H^+ or reabsorption of HCO_3^-) by the kidneys, despite an intact glomerular filtration rate.

      Metabolic acidosis is clinically characterized by the features of the underlying disorder. Otherwise, a typical finding is a characteristic ventilation with increased depth rather than the rate, known as **Kussmaul respiration**.

B. Metabolic alkalosis

Metabolic alkalosis results in an increase in the HCO_3^- levels and an increase in the pH. Respiratory compensation leads to retention of CO_2 and an increase in pCO_2.

Metabolic alkalosis can result broadly from two scenarios—addition of exogenous HCO_3^- into the body, or reduced elimination of HCO_3^- from the body. Reduced elimination of HCO_3^- is from a renal retention of HCO_3^- which occurs in response to acid loss from the body, or when hypovolemia/hyponatremia leads to increased renal Na^+ (and hence HCO_3^-) absorption. Metabolic alkalosis can also result from hypokalemia, and also due to increased levels or action of mineralocorticoids. The causes and mechanisms are detailed below, and in Figure 4.9.

1. Exogenous HCO_3^- or base administration: Overtreatment with $NaHCO_3^-$ (e.g., in hyponatremia or in metabolic acidosis), or massive blood transfusions leading to citrate overload (citrate is used as an anticoagulant in stored blood).

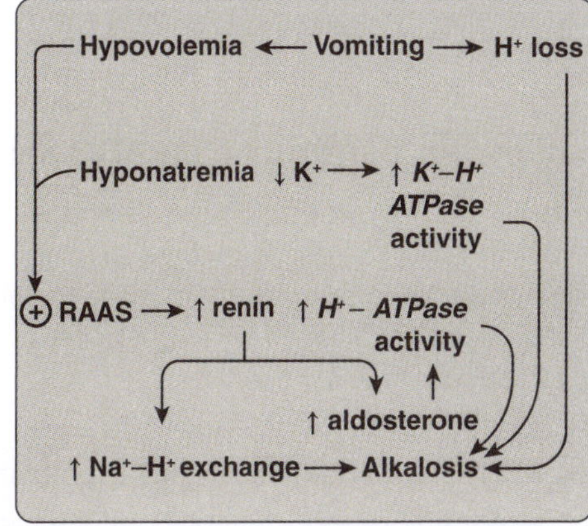

Figure 4.9 Causes of metabolic alkalosis and their mechanisms.

2. Hypovolemia leading to alkalosis is seen in vomiting where both fluid and H^+ loss occur. In addition, hypovolemia (and accompanying hyponatremia) as such can lead to alkalosis because of (1) increased renal Na^+ reabsorption, which is accompanied by HCO_3^- reabsorption and (2) stimulation of renin-angiotensin-aldosterone-system (RAAS) (see Chapter 5, Section III.C. Hormones regulate renal handling of Na^+ and K^+, p. 44)—renin increases Na^+-H^+-exchanger activity, aldosterone increases *H^+-ATPase* activity, and thus loss of H^+ in urine.

3. Hypokalemia leads to acidification of urine and increased blood pH, by increasing the activity of *K^+-H^+ ATPase*, and increasing the expression of NH_3 transporters in the collecting duct (Fig. 4.7).

4. Increased plasma aldosterone levels as seen in Conn's syndrome (aldosterone-secreting tumor of adrenal cortex) or in adrenal hyperplasia cause metabolic alkalosis by increasing the *H^+-ATPase* activity, and by causing hypokalemia.

   Clinical features of metabolic alkalosis, apart from that of the underlying disorder, include hypoventilation (as a compensation by the respiratory center) and features of hypocalcemia like tingling and numbness.

   Whatever be the cause of acid–base imbalance determination of the type and the severity is not possible solely on the basis of clinical features, because of the varied underlying causes and the mechanisms. Diagnosis is thus made with the help of a number of tests, collectively known as ABG analysis.

V. ARTERIAL BLOOD GAS (ABG) ANALYSIS

Arterial blood gas refers to **a battery of tests that includes estimation of pH, oxygen tension (pO_2), carbon dioxide tension (pCO_2), HCO_3^- concentration, and oxygen saturation of hemoglobin (sO_2) in arterial blood.** Specialized instruments are also available for measurement of methemoglobin and carboxyhemoglobin levels.

A. Indications for performing an ABG analysis

ABG analysis is performed to evaluate the acid–base status and the O_2-CO_2 gas exchange in an individual. ABG analysis is hence performed when respiratory/metabolic acidosis/alkalosis is suspected. Some other indications for ABG analysis are:

- For monitoring oxygen therapy
- In sleep disorder studies—here breathing gets interrupted briefly, but repeatedly during sleep
- For detection and quantification of the levels of abnormal hemoglobins (e.g. carbon monoxide poisoning [see Chapter 8, Section II.E.5. CO binding, p. 104] and methemoglobinemia [see Chapter 8, Section IV.D. Methemoglobinemias, p. 110])

B. Sample collection for ABG analysis

The arterial blood is obtained via percutaneous needle puncture or from an indwelling arterial catheter, the former being more commonly performed. **Heparin is the anticoagulant of choice for ABG sample**. The syringe is flushed with heparin solution, not allowing any liquid to remain behind.

Precautions while collecting the sample for ABG analysis:

- The samples should be collected under strict aseptic conditions.
- Contact of the sample with air should be prevented.
- Heparin must be completely removed from the syringe (when heparin solution is used).
- The sample should be transported to the analyzer without any delay. This is very important especially in patients with leukocytosis and thrombocytosis, as WBCs and platelets use up oxygen and generate carbon dioxide, as the analysis is delayed. Delay also will lead to diffusion of gases through the plastic syringes. To avoid the delay in transportation, ABG analysis is nowadays done as POCT (Point-of-Care Testing), placing the analyzers at the point of sample collection (like ICUs and wards).
- Samples should NOT be transported on ice when plastic syringes are used, since low temperatures will lead to contraction of polymer molecules, thus widening the pores in the material (in the atomic scales). This will enhance the diffusion of gases, leading to spurious results.
- Mixing of arterial and venous blood should be avoided since the reference intervals for parameters (especially pO_2) are different for arterial and venous samples.

C. Instrumentation

ABG analyses are performed by automated analyzers, which use electrochemical sensors, ion-selective electrodes, and the principles of potentiometry and amperometry. The analytes that are estimated include pH, pO_2 and pCO_2. The HCO_3^- concentration is calculated once the pH and pCO_2 values are obtained, using the Henderson-Hasselbalch equation.

D. Reference intervals

pH: 7.35–7.45 (mean: 7.40)

pCO_2: 35–45 mm of Hg (mean: 40 mm of Hg)

HCO_3^-: 22–30 mmol/l (mean: 24 mmol/l)

pO_2: 67–104 mm of Hg*

Carboxyhemoglobin and methemoglobin: ≤1%

*pO_2 should always be interpreted in the context of F_iO_2 or fraction of inspired oxygen. Natural air, which has 21% oxygen, has an F_iO_2 of 0.21. The value of F_iO_2 differs when patients are on oxygen supplementation. Generally, the expected pO_2 (in mm of Hg) is five times the F_iO_2. Further, these values should be interpreted considering the previous reports (if available).

VI. INTERPRETATION OF AN ABG ANALYSIS REPORT

Interpretation of an ABG analysis report involves addressing the following questions:

Is the ABG report valid?
The consistency of the ABG report is assessed with the help of the Henderson-Hasselbalch equation,

$$pH = 6.1 + \log \frac{[HCO_3^-]}{[0.03 \times pCO_2]}$$

which, on rearranging gives*:

$[H^+]$ in nanomoles/l = $24(pCO_2)/[HCO_3^-]$

If the values do not corroborate, the ABG report is not valid. This could arise due to an instrument error, or some laxity during the sample collection or transport.

Is there acidosis or alkalosis present?
Look at the pH. pH <7.35 denotes acidosis and pH >7.45 denotes alkalosis. It should be noted that acidosis or alkalosis may be present even if the pH is within the reference interval; this happens in cases where body mechanisms compensate for acidosis or alkalosis, and in mixed acid-base disorders.

Is the acid-base disturbance metabolic or respiratory?
In primary respiratory disorders, the pH and pCO_2 change in opposite directions. In primary metabolic disorders the pH and pCO_2 change in the same direction (see Table 3 given in the appendix of this Unit, p. 57)

Is there appropriate compensation?
Respiratory disorders are compensated by metabolic responses and vice versa. Compensatory mechanisms return the pH towards, but not always to, the normal value. However, in chronic respiratory alkalosis, pH may even reach the normal value due to compensation. See Table 4 given in the appendix of this Unit for a summary of the expected compensation in various acid-base disorders, p. 58.
It should be noted that no disorder of acid-base balance is over-compensated. Inappropriate (inadequate or excessive) compensatory responses should lead to a suspicion of a mixed acid-base disorder.

What is the plasma anion gap?
Plasma anion gap should be calculated when metabolic acidosis is present, using the formula $[Na^+]-([Cl^-] + [HCO_3^-])$, all values in mmol/l. Narrow down the causes based on high or non-anion gap metabolic acidosis. In case of high anion gap, it is also important to look at the osmolal gap (See Chapter 5, VIII. Medical Perspectives, p. 51), if available to calculate. High anion gap metabolic acidosis with increased osmolal gap suggests poisoning with methanol, ethanol or ethylene glycol.

Does the report correlate clinically?
Finally, ABG reports should always be interpreted in the context of the clinical history and examination findings and should correlate with the underlying cause of acid base imbalance. If there is no clinical correlation, a repeat sampling and report should be obtained; other underlying causes should be considered.

---

*The Henderson-Hasselbalch equation for bicarbonate buffer system is:

$$pH = 6.1 + \log \frac{[HCO_3^-]}{[H_2CO_3]}$$

Concentration of H_2CO_3 is determined by the partial pressure of dissolved CO_2. This is given by the partial pressure of CO_2 (pCO_2) and the solubility co-efficient of CO_2 in blood (α). Therefore,

$$pH = 6.1 + \log \frac{[HCO_3^-]}{[\alpha \times pCO_2]}$$

Substituting the physiological value for α which is 0.03 mmol/L per mm Hg:

$$pH = 6.1 + \log \frac{[HCO_3^-]}{[0.03 \times pCO_2]}$$

VII. Medical Perspectives

This section describes some additional medical perspectives in relation to the topics described in the previous sections. Most of the common conditions are already given in the text.

- ABG analysis is one of the most commonly done investigation in intensive care units. It is used to document respiratory failure and assess its severity, monitor patients on ventilators, assess acid–base balance in critical illness and assess response to therapeutic interventions like mechanical ventilation.
- Acute mountain sickness (AMS) develops in certain individuals at altitudes above 3,000–4,000 m; hypoxia leads to increased cerebral blood flow and increased fluid extravasation from the blood vessels into interstitial space leading to cerebral edema (known as high-altitude cerebral edema or HACE). The condition is potentially fatal, unless promptly managed immediately. Definite management is a rapid descent to lower attitudes. Pharmacological management for treatment and prevention of AMS involves administration of acetazolamide, an inhibitor of *carbonic anhydrase*, which leads to impaired HCO_3^- excretion and metabolic acidosis. Respiratory compensation by increased rate and depth of ventilation improves oxygenation to the brain. This, along with the diuretic action of acetazolamide relieves edema and the manifestations of AMS.
- Acidosis leads to hyperkalemia—what is the underlying mechanism? Na^+-H^+ exchanger transports Na^+ into the cells in exchange of H^+ transport into ECF. In acidosis, the H^+ concentration of ECF is already high, so the activity of this exchanger is diminished. Less intracellular sodium levels, which results, inhibit *Na^+-K^+ ATPase*, and thus retention of K^+ in the ECF (see Fig. 5.14, p. 50). The opposite happens in alkalosis, leading to hypokalemia.
- Mutations in an isoenzyme of *carbonic anhydrase*, CA-2 lead to impairment of acidification of bone by osteoclasts, a process which is important in bone resorption. This leads to formation of hard and dense bones—the condition is termed osteopetrosis (*osteo—bone, petra—rock*).

VIII. Chapter Summary

- pH of various body fluids has to be maintained within a narrow range, so as to ensure the proper functioning of various body processes. The pH is maintained within this range with the help of diverse mechanisms that act in a coordinated manner to resist the aberrations and revert any changes occurring to the pH.
- Buffers act quickly but only for short term. **The main physiological buffers in the body are bicarbonate buffer, phosphate buffer and protein (and hemoglobin) buffers.**
- In the ECF, bicarbonate buffer plays a prime role.
- Phosphate buffer is important in intracellular fluid and urine.
- Buffering action of proteins is mostly due to the histidine residues, as their pK_a value is close to the physiological pH. In RBCs, hemoglobin contributes to an additional mechanism in buffering.
- Lungs maintain the pH by regulating the levels of CO_2. Buildup of H^+ and CO_2 results in increased rate and depth of ventilation, and vice versa, thus regulating the CO_2 elimination and retention, respectively, regulating the pH.
- Excretion of non-volatile acid is the function of kidneys. Renal regulation of pH is slow acting—but sustained and comprehensive. **Kidneys reabsorb bicarbonate; kidneys also excrete H^+ directly and also in the form of ammonium ion and in combination with phosphates.** Kidneys also generate bicarbonate, to replenish base stores in the body.
- The physiological pH of arterial blood is 7.35–7.45; a decrease is termed **acidemia** and an increase is termed **alkalemia**.
- Respiratory causes of acid–base disturbances include primary alterations in the pCO_2; compensation is by renal mechanisms and involves HCO_3^- conservation or elimination as the case may be.
- Metabolic causes of acid–base disturbances include primary alterations in the HCO_3^-; compensation is by respiratory mechanisms and involves CO_2 conservation or elimination, depending on the underlying abnormality.
- Simple acid–base disorders never result in overcompensation; if overcompensation or long-term undercompensation is observed, mixed acid–base disorder is to be suspected, where there is more than one primary abnormality.
- Respiratory acidosis is caused by CO_2 retention due to depression of respiratory center, respiratory muscle fatigue or paralysis, diseases of airways and lungs, or conditions that restrict ventilation.
- Respiratory alkalosis results from CO_2 washout and occurs due to overvigorous mechanical ventilation or due to stimulation of respiratory center.
- Metabolic acidosis stems from HCO_3^- elimination and can be classified based on the anion gap into non-anion gap metabolic acidosis and high anion gap metabolic acidosis. The former is associated with hyperchloremia. Renal tubular acidosis, an example of non-anion gap metabolic acidosis, occurs due to failure in acidification of urine due to impaired excretion of H^+ or reabsorption of HCO_3^-. High anion gap metabolic acidosis is due to accumulation of a non-measured anion.

- Metabolic alkalosis is seen in hypovolemia, hypokalemia, and hyperaldosteronism and is associated with an increase in the HCO_3^- levels in the plasma (Fig. 4.10).
- Arterial blood gas analysis involves measurement of pH, pO_2, HCO_3^- levels and pCO_2 and is performed in a number of disorders, not limited to acid–base disturbances. Sample is most commonly collected from the radial artery and anticoagulation is achieved by addition of heparin. Interpretation of ABG report is done in a stepwise manner, and the report should always be correlated clinically.

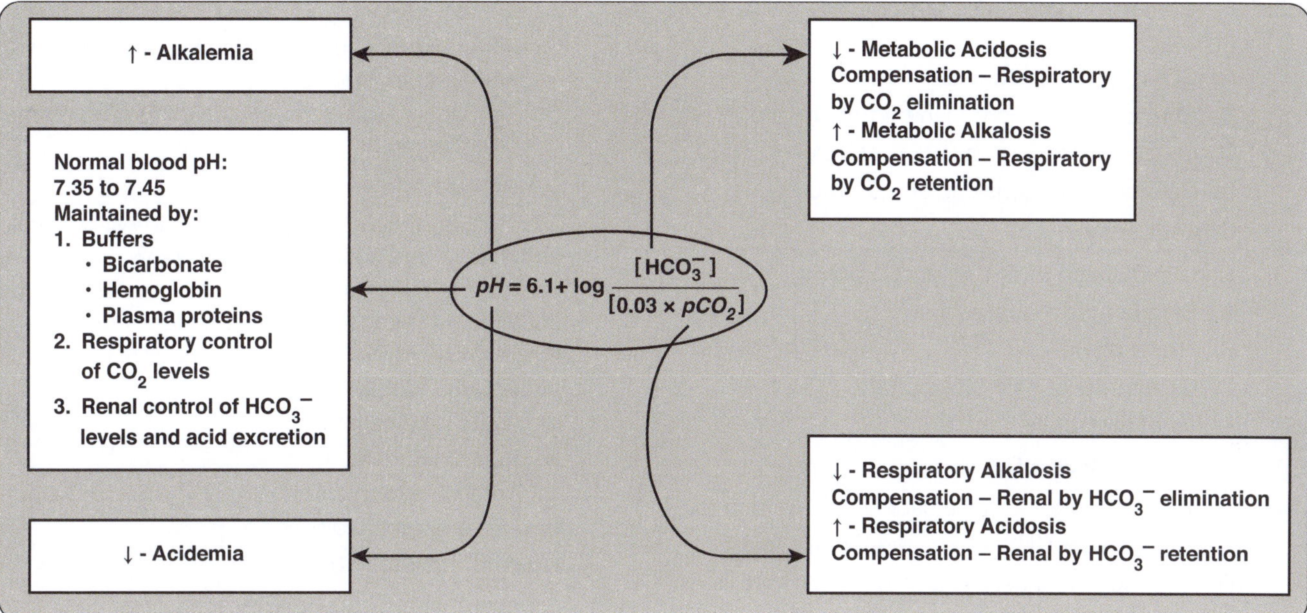

Figure 4.10 Key concept map of acid–base balance.

Study Questions

Short-Answer Questions

4.1. Chloride levels in venous plasma is lower than that of the arterial plasma. Why?

4.2. What is the limiting pH of urine? What is the reason the pH of urine cannot go lower than this value?

4.3. Explain why salicylate toxicity leads to the development of mixed acid–base disorder.

4.4. Name two conditions in which the plasma anion gap is decreased. What is the reason for the reduction of anion gap in these conditions?

Long-Answer Questions

4.1. What is the acid–base abnormality, the mechanism for its development, and the physiological compensation in the following conditions?
   A. Conn's syndrome
   B. Muscular dystrophy
   C. Uncontrolled type-I diabetes mellitus

4.2. Explain why bicarbonate buffer is the predominant buffer in the blood, despite a pK_a of 6.1 which is far from the normal pH of blood.

Multiple-Choice Questions

Choose the ONE best answer.

4.1. pH of the arterial blood is:
   A. Slightly less than that of the venous blood.
   B. Slightly more than that of the venous blood.
   C. Same as of that of the venous blood.
   D. None of the above statement is correct.

   Correct answer = B.

4.2. Bicarbonate buffer is an efficient buffer of the extracellular fluid because: in it the ratio between its conjugate base and the acid is:
   A. Same
   B. 5
   C. 10
   D. 20

   Correct answer = D.

4.3. The composition of the phosphate buffer under physiological condition is:
   A. H_3PO_4 and $H_2PO_4^{1-}$
   B. $H_2PO_4^{1-}$ and HPO_4^{2-}
   C. $H_2PO_4^{2-}$ and PO_4^{3-}
   D. H_3PO_4 and $H_2PO_4^{2-}$

   Correct answer = B.

4.4. Buffering action of proteins is primarily due to the presence of:

A. Glycine
B. Alanine
C. Arginine
D. Histidine

Correct answer = D.

4.5. In pulmonary capillaries and alveoli, CO_2 gets released because:

A. CO_2 and O_2 tensions are same.
B. CO_2 tension is more than that of O_2.
C. O_2 tension is more than that of CO_2.
D. Release of the tensions of CO_2 and O_2.

Correct answer = C.

4.6. Renal regulation of pH depends upon:

A. Bicarbonate reabsorption.
B. H^+ secretion.
C. Generation of new bicarbonate.
D. All of the above.

Correct answer = D.

4.7. Respiratory acidosis (hypercapnia) can be due to:

A. Infections of CNS and respiratory muscles.
B. Diseases of lungs and airways.
C. Both of A and B.
D. Salicylate poisoning only.

Correct answer = C.

4.8. Metabolic acidosis can result from:

A. Endogenous production and/or ingestion of more acids.
B. Reduction of H^+ excretion by the kidneys and/or loss of HCO_3^{1-}.
C. Both of A and B
D. Independent of A and B

Correct answer = C.

4.9. ABG analysis can be used for:

A. Monitoring O_2 therapy.
B. Evaluation of acid Base status.
C. Sleep disorders evaluation.
D. All of the above

Correct answer = D.

4.10. The kidneys contribute to acid base balance by:

A. Secretion of ammonia
B. Retention of bicarbonate
C. Increased ketogenesis
D. Decreased uptake of CO_2

Correct answer = B.

Multiple-Choice Questions (NExT/NEET PG)

Choose the ONE best answer.

4.1. Which of the following best describes the alterations observed in ethylene glycol poisoning?

A. Non-anion gap metabolic acidosis with increased osmolal gap
B. Non-anion gap metabolic acidosis with normal osmolal gap
C. High anion gap metabolic acidosis with increased osmolal gap
D. High anion gap metabolic acidosis with normal osmolal gap

Correct answer = C.

4.2. The results of investigations in a 30-year-old male admitted to the hospital is given below.
Blood pH: 7.24; bicarbonate: 20 mmol/l; plasma sodium: 140 mmol/l; plasma chloride: 112 mmol/l.
From which of the following is the most probable underlying disorder?

A. Acute renal failure
B. Type B lactic acidosis
C. Adrenal hyperplasia
D. Acute diarrheal disease

Correct answer = D.

4.3. A patient with history of ingestion of an unknown compound has the following ABG analysis report—pH: 7.43, pCO_2: 28 mm of Hg, HCO_3^-: 18 mmol/l. What is the acid–base disorder in this patient?

A. No acid–base disorder is evident
B. Metabolic acidosis
C. Respiratory alkalosis
D. Mixed acid–base disorder

Correct answer = D.

4.4. Which of the following statements best describe the Bohr's effect?

A. Decrease in H^+ in pulmonary capillaries leading to release of CO_2 from hemoglobin.
B. Increase in H^+ at tissue levels leading to release of O_2 from hemoglobin.
C. Increase in H^+ at tissue levels leading to relaxed state of hemoglobin.
D. Decrease in H^+ in pulmonary capillaries leading to taut state of hemoglobin.

Correct answer = B.

4.5. The major site for reabsorption of bicarbonate is:

A. Glomerulus
B. Proximal tubule
C. Loop of Henle
D. Distal tubule

Correct answer = B.

Water and Electrolyte Balance of Body Fluids

5

Learning Objectives

At the end of this chapter, the student will be able to:

- Describe the water and electrolyte distribution and homeostasis
- Describe disorders associated with imbalance of water and electrolytes in the body

I. OVERVIEW

Water is considered to be the **universal solvent** and acts as the medium for a large number of biochemical reactions that sustain life. Water gives a major contribution to the body weight in humans. Indeed, the human body can be considered as an organized solution of proteins, minerals, and fat in water! In multicellular organisms, water is present in distinct body compartments—the distribution and movement of water across these compartments are finely regulated. Numerous disease states lead to or result in aberrations of normal water distribution.

Closely associated with the distribution of body water is the concentration of electrolytes in various body compartments. Important electrolytes of the body include Na^+, K^+, Ca^{2+}, Mg^{2+}, Cl^- and HCO_3^-. Electrolytes serve diverse functions in the body, e.g., they maintain resting membrane potential, facilitate neurotransmission and muscle contraction, regulate myocardial electrical activity, and are involved in signal transduction. A number of mechanisms regulate the electrolyte and water distribution in the body, and not surprisingly, electrolyte imbalance is seen in several pathological conditions. A significant number of patients in most in-patient hospital settings have some form of electrolyte abnormality, and when left unaddressed is a major cause of morbidity and mortality.

This chapter describes the maintenance of water and electrolyte balance in the body. The discussion on electrolytes will be limited to sodium and potassium. Various causes and consequences of electrolyte abnormalities are also mentioned, along with the estimation of sodium and potassium in the laboratory.

> Water is considered to be universal solvent because it has the ability to dissolve more substances than any other liquid (it cannot dissolve everything). Water molecules have a polar arrangement of oxygen (negative electrical charge) on one side and hydrogen atoms (positive electrical charge) on the other side. Due to this, water molecules are attracted to many other molecules and dissolve them.

II. WATER HOMEOSTASIS

Water constitutes approximately 50% and 60% of the body weight in adult females and males, respectively, the difference arising due to a higher body fat in females. The values vary with age and malnourishment, as the body fat percentage deviates from that of an average adult. The rest of the body weight is contributed mostly by proteins, minerals, and fats—approximately 18%, 7%, and 15%, respectively, in adult males.

A. Body fluid compartments

A 70-kg normal adult male will have 42 L of total body water (considering that 60% of the body weight is due to water, and the density water is 1 kg/L). This water is distributed between intracellular and extracellular compartments (Fig. 5.1). **The major components of extracellular fluid (ECF) are interstitial fluid (ISF) and plasma.** ISF is the fluid that is present in between the cells and constitutes nearly three-fourths of the

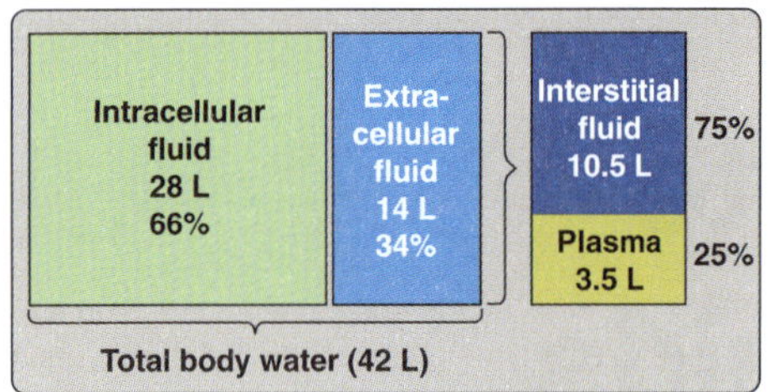

Figure 5.1 Body fluid compartments (the volumes shown are for a 70-kg normal male).

ECF volume, i.e., 10.5 L; it is also known as tissue fluid. Plasma is the acellular component of the blood and comprises 3.5 L of the total 5 L volume of blood. Other minor components the ECF include lymph, and transcellular fluids—fluids present within linings of epithelial cavities (e.g., cerebrospinal fluid, synovial fluid, intraocular fluid, pleural fluid, peritoneal fluid, and pericardial fluid).

The different fluid compartments vary in their electrolyte and protein content (Fig. 5.2). **Sodium is the major extracellular cation, and potassium is the major intracellular cation.** The composition of plasma and ISF are nearly identical, with respect to cations. Since ISF is formed by the filtration of plasma at capillary level, the protein content of ISF is less than that of the plasma (since proteins have a higher molecular weight and a lot of them hence resist filtration). To maintain the charge neutrality, there is increased level of small anions like chloride and bicarbonate in ISF, in lieu of the negatively charged proteins.

B. Water intake and removal

The total water volume in the body is determined by the addition of water to the body and its removal from it (see Table 5, appendix Unit V, p. 58). Daily intake of water through ingestion of fluids directly and as part of food amounts to 2,100 ml. Water added by the metabolic reactions that result in water generation—the so-called "metabolic water," accounts for an additional 200 ml. Thus, nearly 2,300 ml of water is added to the total body water per day.

The same volume of water gets removed daily from the body, maintaining the total water volume in the body relatively constant. Most of the water removal happens through excretion via the kidneys, resulting in a urine output of 1,400 ml/day or 1 ml/minute. Water is also lost from the skin and the respiratory tract as a result of evaporation. This "insensible water loss" removes 700 ml of water from the body every day. An additional 200 ml of water is lost through feces—the total daily water loss adding up to 2,300 ml.

This steady state in water balance is altered in physiological and pathological conditions. For instance, exertion, sweating, fever, and hyperventilation—all these lead to an enhanced "insensible water loss" through skin or respiratory tract or both. Diarrhea results in an increased water loss through feces. All these have to be compensated by an increase in the fluid intake and by the conservation of water by the kidneys. What acts as the signal for the body to deploy these mechanisms? The answer lies in certain receptors in the hypothalamus.

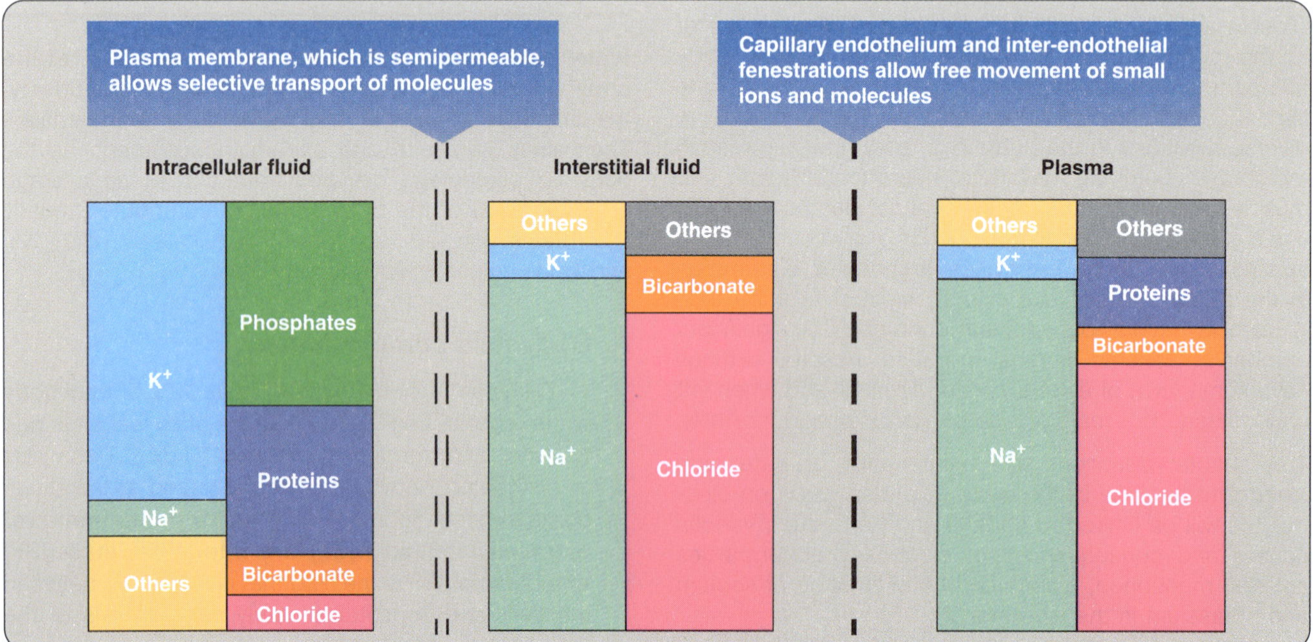

Figure 5.2 Composition of various body fluid compartments. Cations are shown on the left, and anions on the right. Note that the composition of interstitial fluid and plasma is mostly similar, except for the content of proteins.

C. Water balance is closely knit with osmolality of the ECF

Osmolality refers to the osmoles (or the number of osmotically active particles) per kilogram of a solvent, in the case of human body, water. Osmolality of plasma (which, for all the practical purposes, is equated with the osmolality of ECF) is determined by the cations, anions and other osmoles that are abundant in the plasma. **Plasma osmolality is calculated by the formula:**

Plasma osmolality = 2Na$^+$ + Glucose + Urea

where, the plasma concentrations of sodium, glucose and urea are expressed in mmol/l and the plasma osmolality is expressed in milliosmoles/kg (mOsmol/kg). Note the following:

- ***Osmolarity refers to the number of osmoles per liter of a solvent. Since 1 L of water weighs ~1 kg at body temperature, in biological systems, the terms osmolality and osmolarity are used interchangeably.***
- *In the formula, sodium concentration is multiplied by a factor of 2 to account for the anions that are associated with sodium (primarily, chloride and bicarbonate), instead of adding them separately to the equation, since these are also osmotically active.*
- *Urea and glucose are the other major osmoles. Proteins are not included in the formula despite their high concentration in the plasma (up to 8 g/dl), as the absolute number of protein molecules are negligible (in comparison to the other osmoles) due to the higher molecular weight of proteins.*
- *Reference interval for plasma osmolality is 275–295 mOsmol/kg.*

Conditions associated with an excess loss of water relative to sodium, from the ECF (e.g., sweating or exercise), is accompanied by an increase in the plasma osmolality. Hypothalamus contains certain special neurons with specific receptors that are known as **osmoreceptors**. Increased plasma osmolality by 10 mOsmol/kg or more leads to the activation and downstream signaling of these receptors. One effect of this signaling is the release of a peptide hormone vasopressin from the stored vesicles of the posterior pituitary (Fig. 5.3). Vasopressin brings about an increased water reabsorption from the renal collecting ducts as explained in the subheading **D**. Vasopressin causes solute-free water reabsorption by the nephrons. Osmoreceptor activation also leads to activation of thirst, so as to induce fluid intake, and restoration of the osmolality to the normal.

Apart from the alterations in the plasma osmolality, a reduction in the blood volume and blood pressure also acts as a stimulus for osmoreceptor activation. This is seen in conditions like diarrhea and blood loss. Hypervolemia (increase in the volume of ECF) and reductions in the plasma osmolality lead to diminished osmoreceptor signaling, reduced vasopressin in circulation, and suppression of thirst.

D. Vasopressin causes solute-free water reabsorption by the nephrons

Vasopressin (*vaso*: vessel, *press*: pressure), also known as antidiuretic hormone or ADH

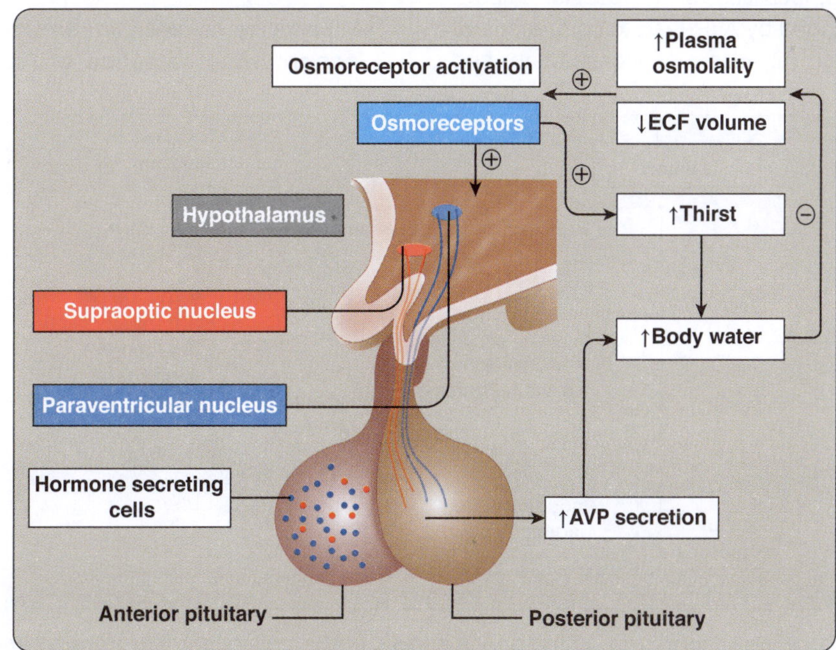

Figure 5.3 Vasopressin/arginine vasopressin (AVP)/antidiuretic hormone (ADH) is synthesized by hypothalamic nuclei and is stored and secreted from posterior pituitary, in response to a number of stimuli.

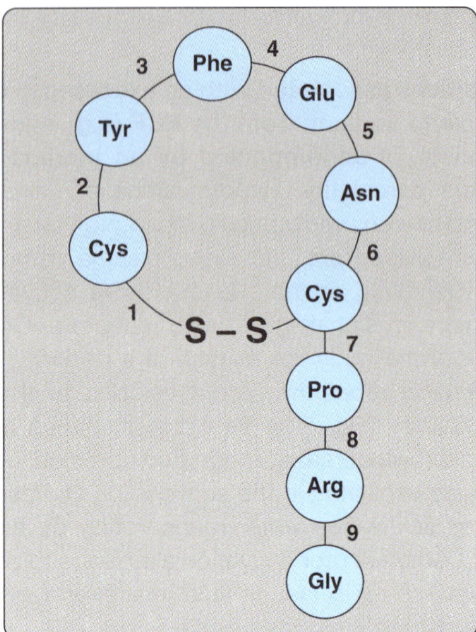

Figure 5.4 Structure of arginine vasopressin (AVP) – Note that the 8th amino acid in this nonapeptide is arginine.

(*anti*: against, *diuresis*: increased urination), is a nine amino-acid peptide (nonapeptide) hormone. In humans, the 8th amino acid is arginine, so the hormone is also called arginine vasopressin or AVP (Fig. 5.4). Vasopressin is produced by the supraoptic and paraventricular nuclei of the hypothalamus. It is then transported along the pituitary stalk by nerve fibers to the posterior pituitary for storage in vesicles, from where it is released on osmoreceptor stimulation.

AVP mediates its action principally through two receptors—**V_1 receptors**, localized to the vascular smooth muscles when engaged by AVP, lead to contraction of the vascular smooth muscles, vasoconstriction, and elevation of blood pressure. **V_2 receptors**, present in nephrons in the thick ascending limb of loop of Henle and the principal cells of the collecting duct, play a direct role in water homeostasis. Binding of AVP to V_2 receptors leads to insertion of the cytoplasmic aquaporins into the apical membrane of the cells (Fig. 5.5). The thick ascending limb of loop of Henle and the principal cells of the collecting duct are otherwise totally impermeable to water, and the transport of water across the apical membranes of these cells is entirely dependent on this insertion of aquaporin-2. At the basolateral surface, other aquaporins transport water into the interstitium, and then into the circulation (Fig. 5.5). V_2 receptor signaling also causes increased transcription of aquaporin gene. The overall effect is a reduction in the osmolality of the ECF, and an increase in the ECF volume.

An impairment in AVP production or function leads to impaired water reabsorption by the kidneys, resulting in increased urine volume (polyuria), decreased urine osmolality, hypernatremia, i.e., increased plasma concentration of sodium (as water is removed from the body, but sodium remains the same), increased plasma osmolality and increased thirst (polydipsia). The condition is known as **diabetes insipidus**, and based on the etiology is classified into central diabetes insipidus and nephrogenic diabetes insipidus. In the former, the production of AVP is impaired due to destruction of the hypothalamic cells due to autoimmunity, tumors or trauma, or due to mutations in the gene coding for AVP. Nephrogenic diabetes insipidus commonly occurs due to mutations in the genes encoding V_2 receptor or aquaporin-2.

A converse is seen in SIADH—the **Syndrome of Inappropriate secretion of AntiDiuretic Hormone,**

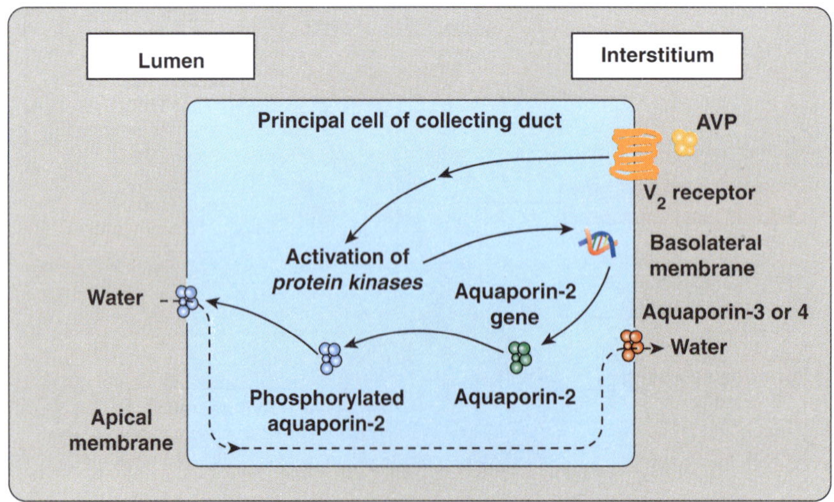

Figure 5.5 Action of arginine vasopressin on the principal cells of the collecting duct to bring about solute-free water reabsorption. Also note that the apical membrane of the cells is in contact with the lumen, and the basolateral membrane with the interstitium.

which is seen in disorders of central nervous system like stroke, infections, and trauma, and also in tumors (like certain lung cancers) that ectopically produce AVP. Increased AVP causes water retention and increased total body water, concentrated urine (increased urine osmolality), hyponatremia, and decreased plasma osmolality.

E. Blood volume is also determined by the plasma proteins

The functions of the ISF or tissue fluid include providing nutrition to the tissues, facilitating gaseous exchange to and from the tissues, and removal of waste from the tissues. To carry out these functions, ISF is in constant exchange with the fluid portion of the blood, i.e., the plasma, and this exchange occurs at the level of capillaries. However, since the diameter of the intercellular fenestrations is <7 nm, blood cells and larger molecules (e.g., most of the plasma proteins) cannot enter the ISF.

The filtration of fluid across the capillary membrane is determined by a balance of forces known as Starling forces (Fig. 5.6). The force that favors the fluid entry into the interstitial space is the hydrostatic pressure gradient (which is the difference between the hydrostatic pressures of the capillary and the ISF, the former being higher). The osmotic pressure difference between the plasma and the ISF also determines the fluid movement. The small molecules and ions are distributed equally between plasma and ISF—so the effective osmoles in this case are the molecules that do not diffuse—the plasma proteins. **The osmotic pressure contributed by the plasma proteins is known as oncotic pressure or colloid osmotic pressure, and this is thus important in the retention of fluid in the intravascular compartment and also in the resorption of fluids from ISF into the circulation.**

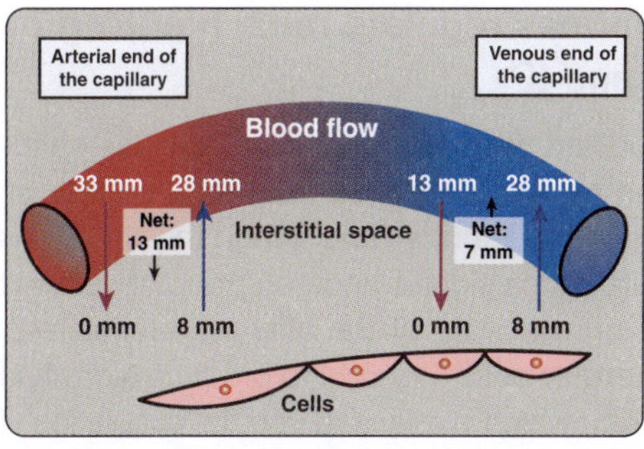

Figure 5.6 Starling forces that mediate fluid exchange between plasma and interstitial fluid at the capillaries. The hydrostatic pressure is shown in dark red, and oncotic pressure is shown in blue. All pressures are given in mm of Hg. There is a net pressure difference of 13 – 7 = 6 mm Hg between the arterial and the venous ends of the capillaries. The net filtration due to this pressure is compensated by a larger number of venous capillaries than arterial capillaries. Some fluid that is retained is returned to the circulation via lymph.

Overall, the hydrostatic pressure gradient is slightly higher than the opposing oncotic pressure gradient, resulting in some retention of fluid within the ISF. This is then returned to the intravascular compartment via lymphatics. **Accumulation of ISF in excess amounts is known as edema.** This is seen when:

(i) there is obstruction of lymphatics, the vessels that carry lymph.

(ii) the level of albumin, the major plasma protein, is low (hypoalbuminemia), causing a reduction in colloid osmotic pressure (or oncotic pressure).

> **Osmolality versus Tonicity**
>
> In biological systems, since certain ions can freely diffuse across the semipermeable membrane to reach an equilibrium, such ions will not contribute to the osmotic pressure, even though they will be included in the calculation of osmolality. Hence, a distinct term is required to express the effect of osmotically active molecules that are effective in bringing about a change in the osmotic pressure. **Tonicity is the concentration of effective osmoles in a solution.** Water thus moves into a hypertonic solution from a hypotonic solution, when separated by a semipermeable membrane.

III. REGULATION OF SODIUM AND POTASSIUM

The regulation of electrolytes of the ECF, especially sodium, is closely associated with the water balance, as seen in Section IIC. Water balance is closely knit with osmolality of the ECF (see p. 41), and is determined by AVP in response to plasma osmolality. Multiple other factors also determine the concentration of the important electrolytes in the body fluids. In this section, the additional mechanisms that keep the levels of these electrolytes within the physiological limits are described.

A. Sodium–Potassium ATPase (Na^+-K^+ ATPase) pump

The interior of the cells contains many negatively charged molecules like proteins and organic phosphates in higher amounts (Fig. 5.2). The plasma membrane is impermeable to these molecules. The presence of these molecules tends to attract positive ions from the ECF into the cells. Entry of these ions, predominantly sodium, which is the principal cation of the ECF, increases the intracellular osmolality. This unchecked presence of sodium within the cell will lead to osmosis of water into the cell, leading to cell swelling and lysis. This is abated by the sodium-potassium *ATPase* pump (Fig. 5.7). **For each molecule of ATP hydrolyzed, 3 Na^+ are extruded from the cell, while 2 K^+ are transported inside.** Because of the imbalance in the number of Na^+ and K^+ ions transported by the *Na^+-K^+ ATPase*, this protein acts as an electrogenic pump, and **is the prime reason for the existence of membrane potential.** Since more ions leave the cell due to the activity of *Na^+-K^+ ATPase*, water also moves out of the cell and prevents cell lysis. Indeed, in a resting individual, a major chunk of the total ATP consumption by the body is for the functioning of the *Na^+-K^+ ATPases*. Another implication of the activity of the *Na^+-K^+ ATPase* is the higher levels of potassium within the cell and sodium outside the cell, making these ions the principal cations of intracellular and extracellular fluids, respectively.

***Na^+-K^+ ATPase* is also involved in secondary active transport of molecules like glucose and amino acids in the intestines and the nephrons** (see Chapter 12, Section III·D. Intestinal absorption of monosaccharides, p. 180 and Chapter 24, Section III·D. Amino acid and small peptide intestinal absorption, p. 381, and Fig. 12.10 and see Fig. 5.17).

B. Renal handling of sodium and potassium

Electrolytes can be lost via multiple body secretions like sweat, mucosal secretions, and gastrointestinal fluids, but **kidneys are the only organs that have the ability to conserve or excrete electrolytes under a tight regulatory control**. Glomeruli freely filter Na^+ and K^+ into the renal tubules. More than 99% of the sodium and 87% of potassium that is filtered by the glomeruli are reabsorbed by the renal tubules. This reabsorption occurs at multiple sites in the nephrons, and is mediated by numerous transporters, and is under the influence of various regulatory factors (see Fig. 5.18).

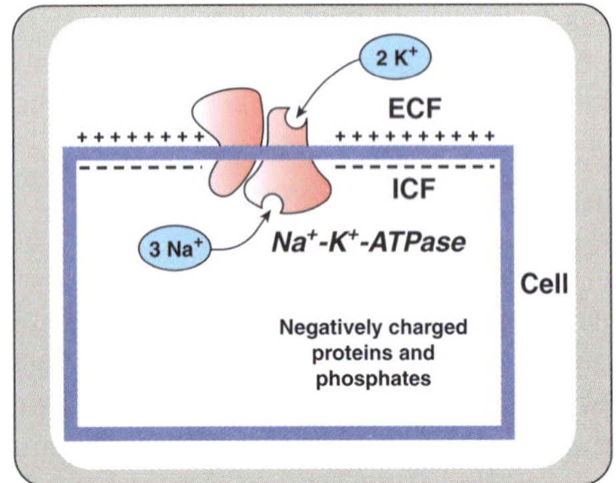

Figure 5.7 *Na^+-K^+ ATPase* pump.
ECF = extracellular fluid;
ICF = intracellular fluid.

In the proximal tubule, sodium is reabsorbed by the tubular epithelial cells from the lumen by at least three mechanisms:

(i) cotransport with glucose (via **S**odium-**GL**ucose Co**T**ransporter-**2**, SGLT2)

(ii) cotransport with amino acids

(iii) in exchange for H^+ (by a Na^+-H^+ exchanger).

An active transport mechanism that relies on the ATP hydrolysis by the *Na^+-K^+ ATPase* located on the basolateral membrane of the tubular epithelial cells, leads to pumping of sodium into the interstitium. Potassium which enters the cell is transported into the tubular fluid. In the thick ascending limb of loop of Henle, sodium is reabsorbed by two different transporters—Na^+-K^+-$2Cl^-$ cotransporter and Na^+-H^+ exchanger. A Na^+-Cl^- cotransporter in the distal tubule and epithelial sodium channels (ENaCs) in the principal cells of the collecting duct, further completes the sodium reabsorption.

Most of the K^+ reabsorption occurs in the proximal tubule predominantly by a paracellular route. This paracellular transport is mediated by a chemical gradient generated by reabsorption of Na^+ and an electrical gradient generated by Cl^- reabsorption. Paracellular transport also occurs in the thick ascending limb of loop of Henle, along with transcellular transport of K^+ via the Na^+-K^+-$2Cl^-$ cotransporter. Together, these account for reabsorption of about 87% of filtered K^+. In the intercalated cells of collecting duct, K^+ reabsorption is mediated by a *K^+-H^+ ATPase* in the apical surface and K^+ channels on the basolateral membrane. Potassium is also secreted by the principal cells of the collecting duct by a renal outer medullary K^+ channel (ROMK).

C. Hormones regulate renal handling of Na^+ and K^+

A modified region of the distal tubular epithelial cells known as **macula densa acts as a sensor of the Na^+ levels in the tubular fluid**. When the Na^+ levels are low, macula densa signals another group of cells in the proximity (on the walls of the afferent arteriole)—the **juxtaglomerular cells** (Fig. 5.8) to produce a *protease* known as **renin**, which is released into the circulation. The other stimulus for renin secretion is a low blood pressure in the afferent arteriole, which reflects a low ECF volume.

Renin cleaves the protein angiotensinogen, which is synthesized by the liver, to produce a decapeptide called angiotensin-I from the amino terminal of angiotensinogen. **Angiotensin-I undergoes further cleavage into an octapeptide, angiotensin-II.** This second cleavage is catalyzed by a peptidase that is present mostly in the pulmonary capillaries, known as angiotensin-converting enzyme (ACE) (Fig. 5.8).

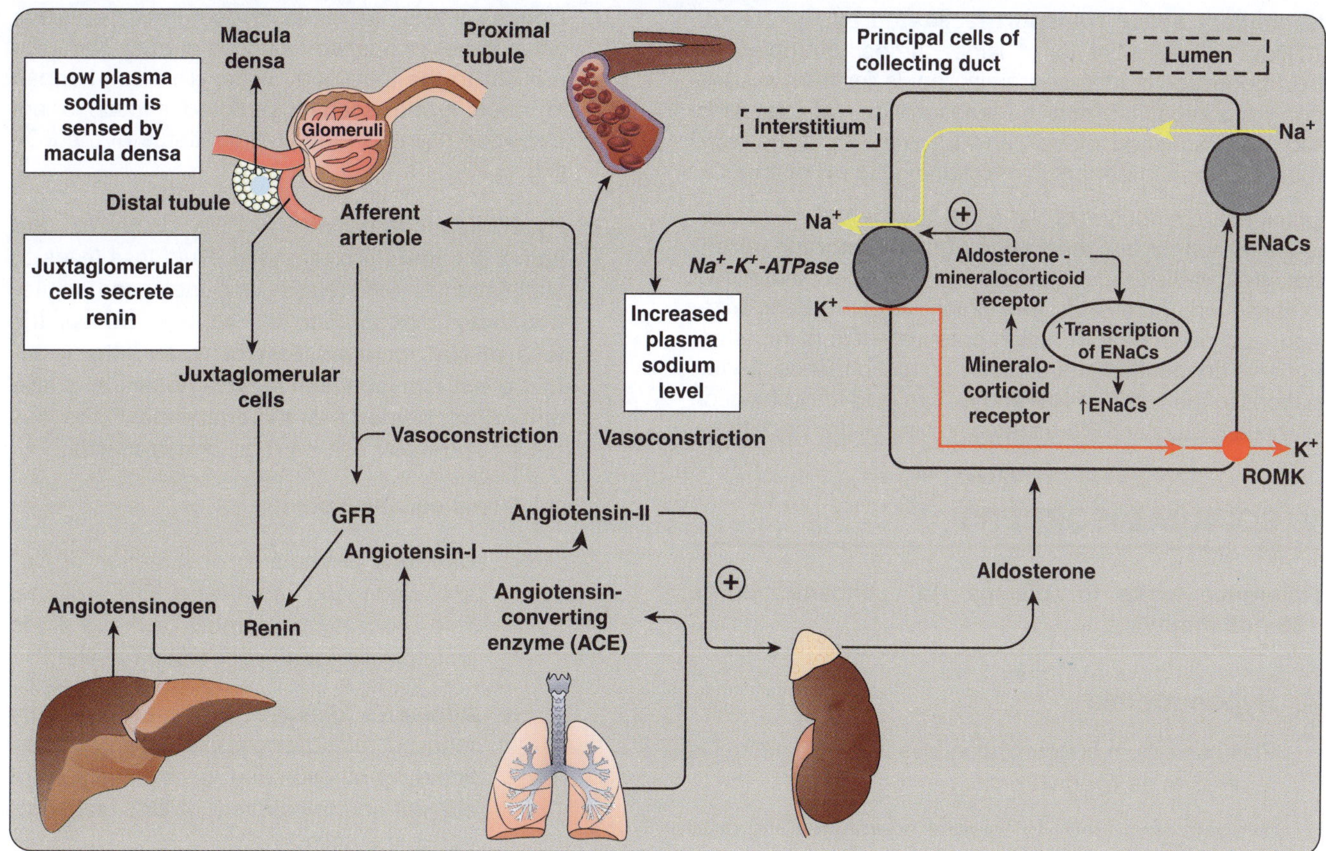

Figure 5.8 Renin-Angiotensin-Aldosterone System. GFR = glomerular filtration rate; ENaCs = epithelial sodium channels; ROMK = renal outer medullary K^+ channel.

Angiotensin-II binds to AT1 receptors, which are G-protein coupled receptors (GPCRs) on the plasma membrane. Downstream signaling of these receptors lead to an increase in the intracellular calcium levels. AT1 receptors are localized mainly to two sites in the body. In the vascular smooth muscle cells, increased Ca^{2+} leads to vasoconstriction (and elevation of blood pressure); vasoconstriction of the afferent arteriole reduces the rate of glomerular filtration, conserving Na^+ and water. Angiotensin-II also increases the activity of Na^+-H^+ exchanger. In the zona glomerulosa of the adrenal cortex, Ca^{2+} signaling leads to activation of aldosterone synthesis.

Aldosterone is a steroid hormone; it acts by binding to a mineralocorticoid receptor in the cytoplasm of the principal cells of the collecting duct. The hormone-receptor complex then translocates to the nucleus and increases the transcription of the ENaCs (Fig. 5.8). Aldosterone also promotes the insertion of ENaCs into the apical membrane of these cells, in addition to increasing the activity of the *Na^+-K^+ ATPase* on the basolateral surface. The effect is increased reabsorption of Na^+, and secretion of K^+. To summarize, in response to a low ECF sodium, or a low vascular volume, the **renin-angiotensin-aldosterone system (RAAS)** promotes sodium (and hence water) retention and potassium excretion.

In conditions that lead to hypervolemia, an increase in the ECF volume leads to stretching of the walls of atria and ventricles, in response to which two hormones are released from these structures—**atrial natriuretic peptide** (ANP; *natrium*: sodium *uresis:* excretion in urine) and **B-type natriuretic peptide** (BNP, previously known by a misnomer—brain natriuretic peptide). Their actions are similar and is mediated by cGMP signaling (see Chapter 18, Section IV·E·2. Nitric oxide and vascular endothelium, p. 253). **The natriuretic peptides inhibit the secretion of renin and reduce the sensitivity of zona glomerulosa to angiotensin-II—they eventually cause sodium and water excretion (natriuresis and diuresis)**.

IV. FUNCTIONS OF SODIUM AND POTASSIUM

Na^+ is the principal cation of the ECF and is the **prime determinant of the ECF volume and water balance**. Normal level of sodium in the ECF is required to prevent shifting of water to or from the intracellular compartment. Sodium thus prevents cell lysis and cell shrinkage, by maintaining the water level in the intracellular compartment. Sodium also maintains the intravascular volume and thus the integrity of circulation to the peripheral organs.

Secondary active transport facilitated by the *Na^+-K^+ ATPase* is essential for a large number of molecules like monosaccharides and amino acids for their absorption from intestines and the reabsorption from the renal tubules. Exchange of Na^+ and K^+ with H^+ in the renal tubules is also important in maintaining the pH of the ECF.

Differential distribution of Na^+ and K^+ by the *Na^+-K^+ ATPase* is the reason for the **generation of the membrane potential**. In excitable tissues like nerves and muscles, movement of these ions across the membranes leads to depolarization, and generation of action potential—this is important in nerve conduction and muscle contraction. Thus, perturbations in the levels of Na^+ and K^+ can lead to dysfunction of these tissues and life-threatening clinical manifestations.

V. SODIUM DISORDERS

Reference limits for plasma Na^+ concentration is 136–145 mmol/l.

A. Hyponatremia

Plasma sodium concentration less than 136 mmol/l is referred to as hyponatremia.

Hyponatremia leads to cerebral edema, leading onto neurological manifestations like headache, nausea and vomiting, and in severe cases, seizure, coma and death. Acute symptomatic hyponatremia is hence a medical emergency.

The effects of hyponatremia on a cell are shown in Figure 5.9. These effects are most pronounced in the nervous tissue.

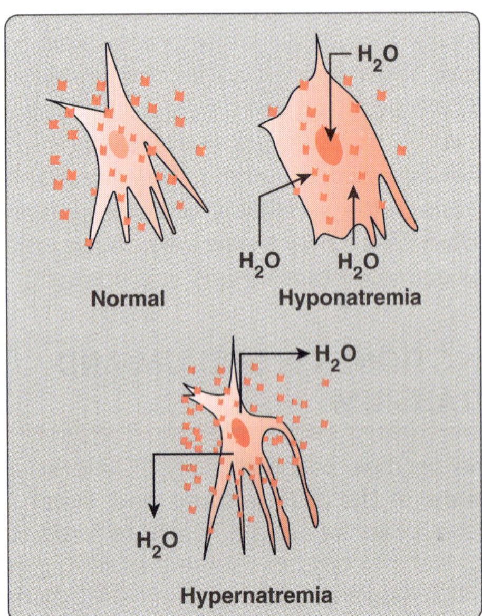

Figure 5.9 Effects of hypernatremia and hyponatremia on the cells. Na^+ ions are shown in red.

1. **Causes of hyponatremia:** Hyponatremia develops when there is a relative reduction of body Na^+, with respect to the total body water. This can happen in three broad scenarios—these situations are classified based on the volume status of the ECF (Fig. 5.10).

2. **Hypovolemic hyponatremia:** Here **both Na^+ and water are lost** from the body. Fluid loss leads to stimulation of AVP release and increased solute-free water reabsorption by kidneys. Hence, the **loss of Na^+ exceeds that of water**—the resultant effect is hyponatremia. This condition is also termed as **"depletional hyponatremia."** The loss of Na^+ and water can be renal or extrarenal.

   a. **Renal causes include:**

      i. **Diuretics** promote renal excretion of Na^+ and water and reduce the ECF volume. This is achieved by most of these drugs by inhibiting renal reabsorption of Na^+.

      ii. **Addison's disease** is characterized by reduced mineralocorticoid (and glucocorticoid) synthesis due to dysfunction of adrenal and impairment of Na^+ reabsorption in the collecting duct.

      iii. **In metabolic alkalosis** (see Chapter 4, Section IV·B. Metabolic alkalosis, p. 33), the body tries to conserve H^+, by promoting H^+ reabsorption via Na^+-H^+ exchangers resulting in increased sodium losses.

      iv. In **diabetic ketoacidosis** (DKA) (see Chapter 21, V.C. Excessive ketone body production in diabetes mellitus, p. 315) and starvation ketoacidosis, Na^+ gets excreted along with β-hydroxybutyrate in the urine. In DKA, increased plasma glucose levels also draw water from the intracellular compartment, leading to hyponatremia.

> Diarrhea can be due to multiple mechanisms—in secretory diarrhea (e.g., cholera), the bacterial toxins cause active sodium loss into the intestinal fluid, the fluid is isotonic to plasma; in osmotic diarrhea (seen in administration of certain laxatives) and nontoxigenic bacterial and viral diarrheas, the intestinal fluid is hypotonic to plasma. Vomitus and sweat are also hypotonic to plasma. Thus, vomiting, excess sweating, osmotic diarrhea, and diarrhea due to nontoxigenic strains are associated with loss of more water than Na^+, and thus hypernatremia. When these conditions are corrected by water alone, or hypotonic fluids, hyponatremia can set in.

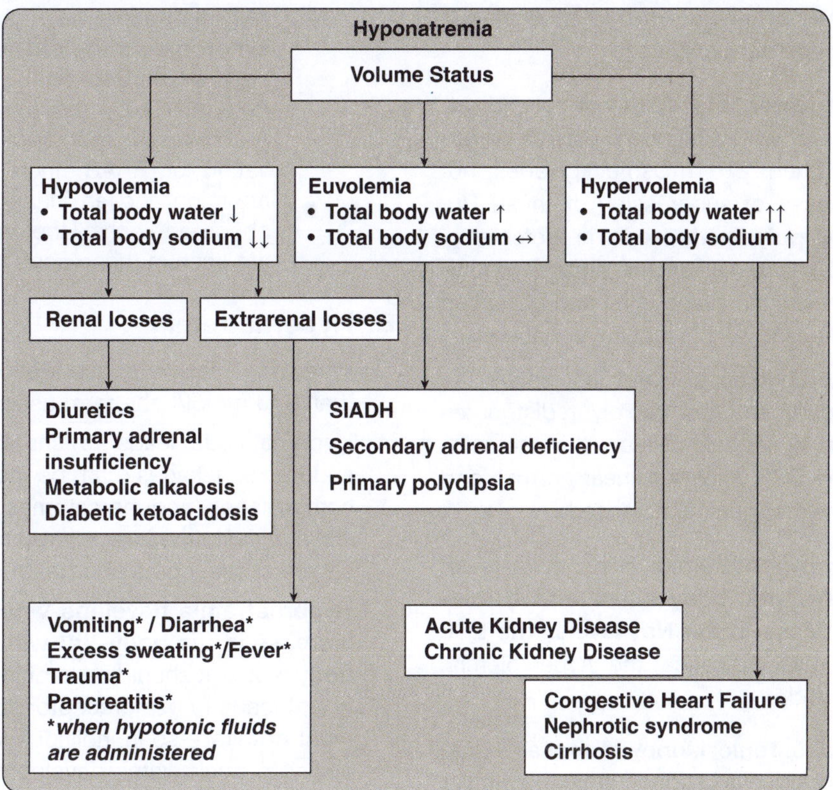

Figure 5.10 Classification and causes of hyponatremia.

b. Extrarenal causes include the following conditions, when associated with replacement of fluids that do not contain sufficient Na^+:

i. Gastrointestinal fluid and electrolyte loss which occurs in **vomiting and diarrhea**.

ii. Loss through skin in conditions like **fever, excessive sweating and burns**.

iii. Loss of fluid into a third space (which is not in equilibrium with intravascular compartment), which is seen in acute pancreatitis, in fractures causing internal bleeds, etc.

The difference between renal and extrarenal causes is also reflected in the urine sodium levels—a concentration of >20 mmol/l suggests renal losses, whereas in extrarenal causes the body tries to conserve Na^+ by increasing reabsorption of Na^+ by the renal tubules.

Clinically, hypovolemia or reduction in the volume of ECF manifests as increased thirst, dry skin and mucous membranes, diminished peripheral pulses, hypotension, tachycardia (due to baroreceptor reflex in response to hypotension), and loss of skin turgor.

> Primary adrenal insufficiency leads to hypovolemic hyponatremia, while secondary adrenal insufficiency causes euvolemic hyponatremia. This is because aldosterone secretion is not regulated by ACTH (but by renin), so in secondary adrenal insufficiency, aldosterone levels are normal, unlike that in primary adrenal insufficiency.

3. Euvolemic hyponatremia: Here, total body Na+ remains nearly same, with an increase in the total body water resulting in **"dilutional hyponatremia."** The accumulation of water is primarily in the intracellular compartment, the ECF volume remaining constant—hence the euvolemia.

Increase in the total body water occurs due to two mechanisms:

a. Increase in solute-free water absorption due to increase in AVP levels: This is classically seen in **SIADH** (see Section II·D Vasopressin causes solute-free water reabsorption by the nephrons, p. 41). The other reason for elevated AVP is glucocorticoid deficiency, due to an abnormality in the adrenocorticotropic hormone (ACTH) secretion by the pituitary (secondary adrenal insufficiency). Glucocorticoids have a negative feedback effect on AVP secretion, which is abrogated in secondary adrenal insufficiency.

b. **Increased water intake:** In normal individuals, intake of water beyond a plasma osmolality less than 275 mOsmol/kg does not occur because of suppression of thirst. But in **primary polydipsia** (seen in defects of thirst center in brain, and in certain psychiatric disorders), increased thirst and increased water intake lead to dilutional hyponatremia.

In both the conditions, water first enters the ECF compartment, but rapidly redistributes to the more hypertonic milieu within the cells, keeping the ECF volume to near normal, and total body water getting increased.

4. **Hypervolemic hyponatremia:** Here, there is **an increase in the total body water and a relatively less increase in the Na^+ levels**. The ECF volume is increased—hence the hypervolemia. The causes include:

a. **Acute and chronic kidney disease,** where the ability of kidneys to concentrate the tubular fluid is impaired, leading to an imbalance in water and sodium removal by the kidneys.

b. **Congestive heart failure** leads to hypotension and activation of RAAS (Fig. 5.11). Na^+ and water retention occur, but since the heart cannot maintain cardiac output, stasis of fluid occurs, which escapes into the interstitial space, leading to edema.

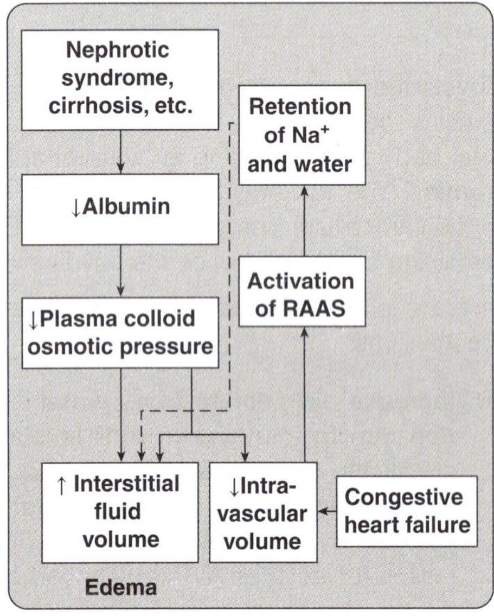

Figure 5.11 Mechanism of hyponatremia in congestive heart failure and hypoalbuminemia. RAAS = Renin-Angiotensin-Aldosterone System.

c. **Hypoalbuminemia, seen in nephrotic syndrome and cirrhosis,** also leads to edema (see Section II.E Blood volume is also determined by the plasma proteins, p. 43). A decrease in blood volume, followed by activation of RAAS, increased Na^+ and water retention and an inability to retain the water absorbed within the vascular compartment are seen in these conditions also.

B. Hypernatremia

When the plasma concentration of sodium is more than 145 mmol/l, the term used is hypernatremia.

Hypernatremia leads to shrinkage of brain cells—sudden shrinkages can be accompanied by brain hemorrhages and hematomas. Most frequent clinical manifestation is an altered mental status. In acute severe cases, coma and death ensue.

Hypernatremia develops when there is a relative increase in the body Na^+ with respect to the total body water. It should be kept in mind that an increase in Na^+ leads to osmoreceptor activation and development of thirst—so in individuals with intact thirst and ability to drink water, development of hypernatremia is rare. Similar to the classification of hyponatremia, hypernatremia is also classified based on the volume status of the ECF (Fig. 5.12).

1. **Hypernatremia with increased ECF volume:** Here, **both Na^+ and total body water is increased with a relatively larger increase in Na^+**. Volume overload leads to inhibition of AVP release and decreased solute-free water reabsorption by the kidneys. The causes include:

a. Administration of hypertonic saline (overcorrection of hyponatremia) or sodium bicarbonate (overtreatment of metabolic acidosis).

b. Increased plasma aldosterone levels as seen in Conn's syndrome (aldosterone-secreting tumor of adrenal cortex) or in adrenal hyperplasia. Increased Na^+ reabsorption in the collecting duct leads to hypernatremia.

2. **Hypernatremia with normal ECF volume:** Here, **total body water is decreased with a relatively normal levels of body Na^+**. Loss of water is predominantly from the intracellular compartment; ECF volume remains same, and hence the euvolemia. These findings are observed in central and nephrogenic diabetes insipidus, where AVP secretion and action, respectively, are hampered.

3. **Hypernatremia with decreased ECF volume:** Here, **both Na^+ and water are lost from the body with a relatively larger loss of water compared to Na^+**. The causes include:

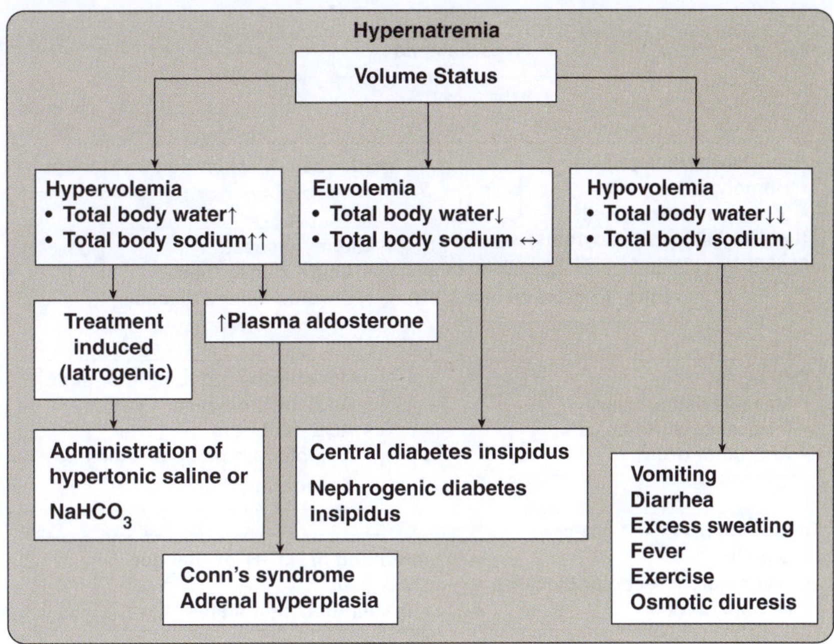

Figure 5.12 Classification and causes of hypernatremia.

a. Gastrointestinal fluid and electrolyte loss which occurs in vomiting and diarrhea (osmotic diarrhea, viral diarrhea and nontoxigenic bacterial infections of the intestine).

b. Loss through skin in conditions like fever, excessive sweating and burns.

c. Osmotic diuresis is seen following administration of certain drugs—like mannitol (used to lower raised intracranial pressure), and in conditions like hyperglycemia. Here, the increased levels of osmoles in plasma, and in urine, draw water from the cells into the ECF, which eventually gets excreted in urine.

VI. POTASSIUM DISORDERS

Reference limits for plasma K^+ concentration is 3.5–5.0 mmol/l. Since K^+ is predominantly intracellular, alterations in the plasma K^+ levels occur mostly due to a shift of K^+ between ECF and ICF compartments. Variations in the renal handling of K^+ is another major reason for the same. K^+ plays a very important role in maintaining the neuromuscular excitability, especially in the myocardium. Hence alterations in the ECF K^+ levels are associated with alterations in the heart rhythm—arrythmias, characteristic ECG changes, and cardiac arrest. Effects on the skeletal and intestinal smooth muscles are responsible for myopathy and muscle weakness, and paralysis of intestinal muscles, respectively.

A plasma concentration more than 5.0 mmol/l is referred to as hyperkalemia and less than 3.5 mmol/l is known as hypokalemia. Both the situations are medical emergencies and should be promptly managed.

A. Hyperkalemia

Hyperkalemia results either from shift of K^+ from ICF to ECF, or due to impaired excretion of K^+ by the kidneys (Fig. 5.13). Increased K^+ in plasma is also seen when cell lysis occurs in the samples collected—this occurs in hemolysis and in conditions that drastically increase WBCs (e.g., leukemias) or platelets (thrombocytosis) in blood; this is termed as pseudohyperkalemia.

Two other important determinants of K^+ shift are the *Na^+-K^+ ATPase* and Na^+-H^+ exchanger (Fig. 5.14)—both are located on the plasma membrane.

Aldosterone action is affected in patients receiving drugs like spironolactone (which is a blocker of mineralocorticoid receptor) or amiloride (which inhibits ENaCs); these are called potassium-sparing diuretics, as they promote Na^+ and water excretion, without lowering the plasma K^+ levels.

B. Hypokalemia

Hypokalemia results from shift of K^+ from ECF to ICF, or due to increased renal or extrarenal excretion of K^+, or due to reduced K^+ intake (seen in starvation) (Fig. 5.15).

1. **Causes of extracellular to intracellular K^+ shift:** The redistribution of K^+ between the body fluid compartments is determined by *Na^+-K^+ ATPase* and Na^+-H^+ exchanger (Fig. 5.14). Causes of an intracellular shift include:

   a. Alkalosis—due to increased activity of Na^+-H^+ exchanger to compensate for the increased

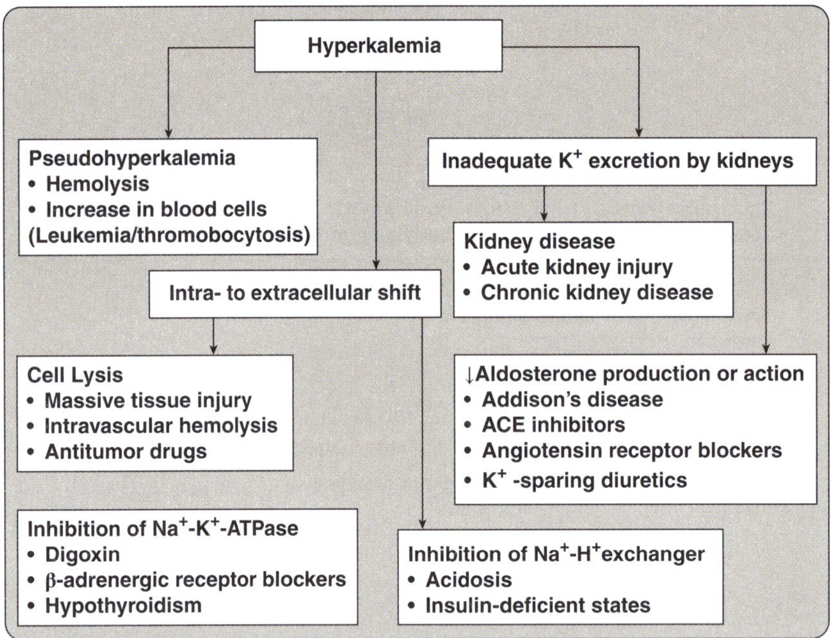

Figure 5.13 Classification and causes of hyperkalemia.

pH, causing increased intracellular Na^+, which gets exchanged for ECF K^+, by *Na^+-K^+ ATPase*.

b. Administration of drugs like insulin, which stimulates Na^+-H^+ exchanger activity, or drugs that stimulate the β-adrenergic receptors (thereby enhancing the activity of *Na^+-K^+ ATPase*).

c. Hyperthyroidism—as thyroxine is a stimulator of *Na^+-K^+ ATPase*.

2. **Causes of K^+ loss from the body:** Loss of K^+ can be renal or extrarenal.

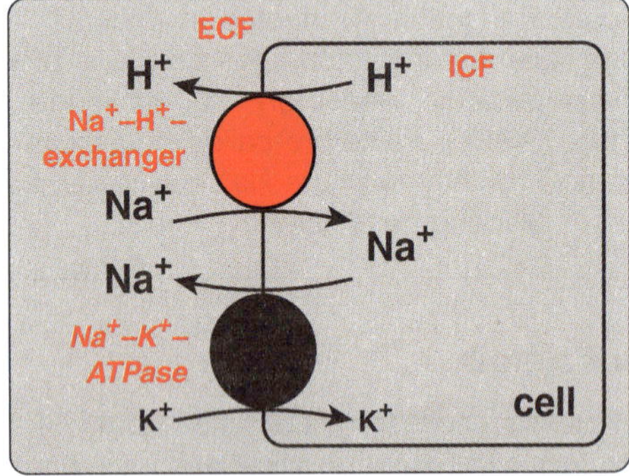

Figure 5.14 Na^+-H^+ exchanger and *Na^+-K^+ ATPase* are two important determinants of K^+ distribution between ECF and ICF. Na^+-H^+ exchanger transports Na^+ into the cell in exchange of H^+ transport into ECF. In acidosis, the H+ concentration of ECF is already high, so the activity of this exchanger is diminished. The resulting low intracellular sodium levels inhibit *Na^+-K^+ ATPase*. Na^+-H^+ exchanger activity is stimulated by insulin. So, insulin-deficient states (like DKA) are associated with hyperkalemia.

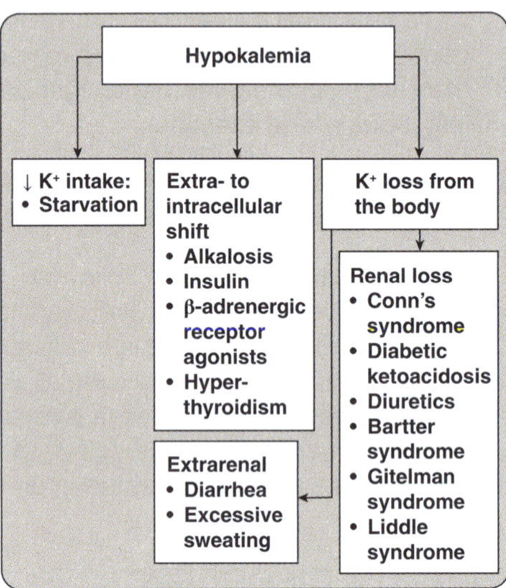

Figure 5.15 Classification and causes of hypokalemia.

a. Extrarenal loss occurs in diarrhea and excessive sweating.

b. Renal loss of K^+ is seen in:
   - Conn's syndrome—aldosterone excess leads to increased K^+ loss from the collecting ducts.
   - Diabetic ketoacidosis can lead to shift of K^+ into the ECF, increased plasma K^+ levels, and excretion of K^+ filtered in excess, which combines with the ketone bodies.
   - Diuretics (other than K^+-sparing diuretics) that inhibit the renal reabsorption of Na^+ leads to increased Na^+ levels in the tubular fluid. When this Na^+-rich fluid reaches the collecting duct, the Na^+ is reabsorbed by

ENaCs and K^+ is ultimately excreted. This leads to hypokalemia (see Fig. 5.18).

- A similar scenario is seen in certain inherited syndromes in which the tubular sodium transporters are defective. Bartter syndrome (defective Na^+-K^+-$2Cl^-$ cotransporter) and Gitelman syndrome (defective Na^+-Cl^- cotransporter) lead to increased Na^+ levels in the tubular fluid, while Liddle syndrome is associated with overexpression of ENaCs—all these lead to increased Na^+-K^+ exchange and hypokalemia (see Fig. 5.18).

Diabetic ketoacidosis can present as hyperkalemia or hypokalemia or with normal plasma K^+ levels, but the total K^+ stores in the body is always depleted in this condition. Treatment with insulin can cause rapid redistribution of K^+ into the cells and hypokalemia, even in DKA with normal plasma K^+ levels, leading onto cardiac complications. This should be considered while treating DKA.

VII. ESTIMATION OF SODIUM AND POTASSIUM

In the clinical laboratory, the widely used method for estimation of serum electrolytes employs ion-selective electrodes (ISEs) and is based on the principle of potentiometry (Fig. 5.16). Potentiometry involves the measurement of voltage difference between two half cells of an electrochemical cell.

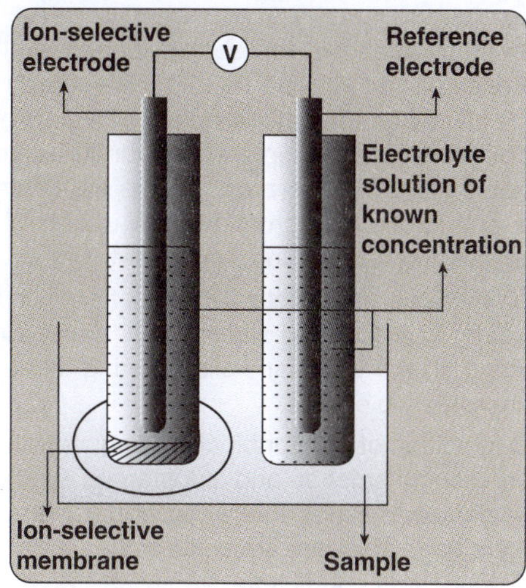

Figure 5.16 Ion selective electrode.

Ion-selective membranes, depending on their composition, provide selective permeability to one particular ion—this enables the ISEs to interact only with that ion. In case of Na^+ electrode, the ion-selective membrane is composed of glass (in a specific composition); ISE for K^+ uses valinomycin.

VIII. Medical Perspectives

Various disorders related to electrolyte imbalance have been extensively discussed throughout this chapter. This section describes some additional medical perspectives in relation to serum electrolytes.

- *Osmolal gap*: Plasma osmolality can be calculated by the formula, plasma osmolality = $2Na^+$ + Glucose + Urea. Plasma osmolality can also be estimated directly by measuring the depression of freezing point of the plasma. Normally, this correlates with the calculated plasma osmolality. When osmoles other than Na^+, glucose or urea are elevated in the plasma, the difference between the measured osmolality (by estimating the depression in freezing point of plasma) and the calculated osmolality [which is termed as osmolal gap] is increased. This is seen in poisoning with certain chemicals like ethylene glycol and methanol.
- In certain clinical conditions, there is increased levels of fluids within the transcellular fluid compartments—fluid accumulation within the pleural cavity is known as pleural effusion, and increased volume of peritoneal fluid is called ascites. Based on the etiology, the fluid can be a transudate or an exudate. Note that the transcellular fluids are ultrafiltrates of plasma, and the fluid exchange is in accordance with the Starling forces. Transudative pleural effusion and transudative ascites occur when the accumulation of fluid is due to an increase in the hydrostatic pressure or a reduction in the oncotic pressure of plasma—this is seen in ECF volume overload (e.g., heart failure) or in conditions leading to hypoalbuminemia. Transudates thus have a low protein content and a low cell number. On the contrary, exudative pleural effusion and exudative ascites result from an increase in the capillary permeability as a result of inflammation—this is mostly seen in infections and inflammatory disorders. As the capillary permeability is high, cells and proteins are present in increased amounts in exudates.
- The sodium-dependent transport of glucose in the proximal tubules of the kidneys is mediated by a transporter known as **S**odium **GL**ucose Co**T**ransporter-**2** (SGLT2). This cotransporter is present on the apical surface of the tubular epithelial cells. Sodium which enters the cell along with glucose is transported into the interstitium by *Na^+-K^+ ATPase*. Inhibitors of SGLT2 are known as glifozins. Since they prevent renal glucose reabsorption, they are used in the treatment of diabetes mellitus.

- In the small intestine, the sodium-dependent transport of glucose is mediated by a related cotransporter, **S**odium **GL**ucose co**T**ransporter-**1** (SGLT1). The absorption of glucose is dependent on the ATP hydrolysis by *Na^+-K^+ ATPase* (Fig. 5.17). Since the latter requires transport of Na^+ for functioning, in conditions like diarrhea where energy stores in the body has to be replenished by administration of carbohydrates, intake of carbohydrates alone will not be sufficient to enable glucose absorption. This is the basis of administering solutions containing both carbohydrates (glucose as such, or sucrose, or starch) and salt in oral rehydration therapy for the treatment of fluid and energy loss in diarrhea. Glucose is transported into the interstitium by a GLUT transporter.
- Recommended daily dietary intake of sodium is <2 g (<5 g NaCl) and that of potassium is >3.5 g. Increased intake of sodium is associated with risks of developing hypertension and heart diseases. Potassium lowers blood pressure by acting on the vascular walls, relaxing them. Potassium also promotes excretion of sodium—further augmenting its cardioprotective effects.
- Since inhibition of ACE or blocking angiotensin-II receptors can lead to diminished action of angiotensin-II—i.e., reduction in vasoconstriction and aldosterone synthesis—ultimately leading to decreased ECF Na^+ levels, ECF volume and blood pressure, drugs that bring about these effects like ACE inhibitors (e.g., enalapril) and angiotensin-II receptor blockers (e.g., losartan) are used in the treatment of hypertension.
- Angiotensin-II is inactivated via an enzymatic hydrolysis by angiotensin converting enzyme-2 (ACE-2). This enzyme is found bound to the membranes of lung, intestine, etc. SARS-CoV2, the etiological agent for the novel coronavirus disease—COVID-19, gains entry into the cell by binding to these proteins on the cell surface.
- *Chloride (Cl^-)*: Physiological functions of chloride are mostly similar to that of Na^+. Reference interval for plasma chloride concentration is 102–109 mmol/l. An increase or decrease in the chloride levels in the plasma as such does not lead to any clinical manifestations, and except for certain acid-base disorders (see p. 32), chloride levels in plasma increase or decrease in parallel to sodium levels.
- Estimation of urine osmolality can be of use in diagnosis of certain disorders. For instance, urine osmolality is low in diabetes insipidus and in chronic renal failure—in both these conditions, the concentrating capacity of the renal tubules is lost, and a diluted urine is excreted. On the other hand, urine osmolality is high in SIADH and in dehydration.
- The term Fractional Excretion of sodium (or FENa) refers to the percentage of sodium that is filtered by the kidneys, which is excreted in urine. Remember that most of the sodium that is filtered is reabsorbed. FENa is given by the formula:

$$FENa = \frac{Urine\ Sodium/Serum\ Sodium}{Urine\ Creatinine/Serum\ Creatinine} \times 100$$

Reference interval is 1%–2%. In conditions like hypovolemia, where the body tries to conserve sodium, the FeNa is <1%.

- Intravenous (i.v) fluid therapy can be for maintenance (in patients who are unable to consume food and water orally) or for replacement (for correction of water/electrolyte imbalance). The choice of i.v fluids highly depends on the underlying clinical condition. For example, normal saline (0.9% NaCl) is used in hypovolemic hyponatremia for sodium replacement, while half-normal saline with 5% dextrose and 20 mmol/l of potassium is usually used as a maintenance fluid.

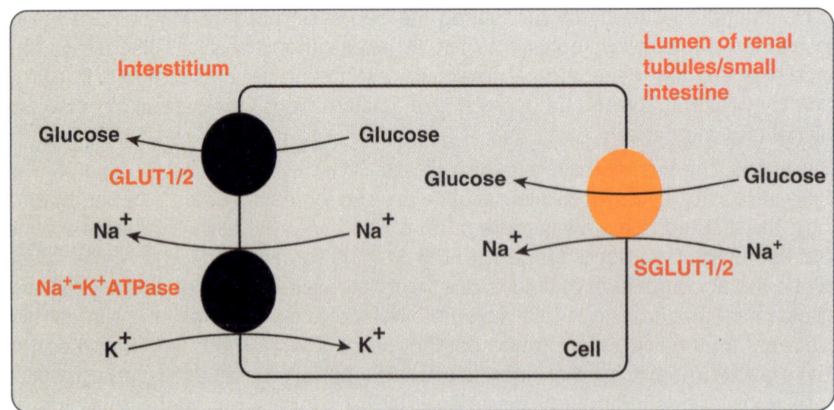

Figure 5.17 Secondary active transport of glucose from the lumen into the interstitium in small intestine and renal tubules.

IX. Chapter Summary

- Water gives a major contribution to the total body weight and is distributed in distinct compartments in the body.
- Most of the total body water is present intracellularly. Intracellular and extracellular compartments differ in the composition of electrolytes and proteins.
- Extracellular fluid (ECF) is mainly constituted by plasma and its filtrate—interstitial fluid (ISF). The movement of fluid between these two is primarily determined by the osmotic pressure contributed by the plasma proteins.
- Total body water is determined by the water intake and water removal from the body. Water balance is maintained by two mechanisms—stimulation of thirst and renal water reabsorption.
- The main regulator of body water is plasma osmolality. Increased plasma osmolality leads to stimulation of osmoreceptors in the hypothalamus, which stimulates thirst and also the release of AVP. AVP causes increased free water absorption from the nephrons. Water balance is altered in SIADH and diabetes insipidus.
- Sodium is the major extracellular cation, while potassium is the major intracellular cation. Sodium-potassium *ATPase* is the mechanism behind this differential distribution. Since it is an electrogenic pump, sodium and potassium are required for generation of various potentials and functioning of excitable tissues. The regulation of sodium and potassium is carried out by a number of mechanisms—most of them involving hormones like aldosterone and natriuretic peptides. Renin-angiotensin-aldosterone system is a key player in the maintenance of sodium and potassium balance in the body.
- Abnormalities in the level of electrolytes—both increase and decrease—can be fatal and should be promptly identified and treated.
- Sodium disorders present usually with neurological manifestations, while potassium disorders are associated with aberrations in the cardiac electrical activity. These disorders are classified, respectively, based on the ECF volume status or the etiology, so as to help make an easy diagnosis and initiate early treatment.
- Estimation of serum electrolytes, in modern clinical laboratories, is mostly done with the help of ISEs; the heart of such an electrode is the ion-selective membrane.

Location	Transporter/ Mechanism	Effect	Regulation	Comments
Glomeruli	Ultrafiltration of plasma	Na^+ and K^+ freely filtered into the tubular fluid	↓ by angiotensin-I; ↑ by ANP and BNP	-
Proximal tubule	SGLT2	Na^+ and Glucose transport into tubular cell	-	Inhibited by Glifozins
	Na^+-amino acid cotransporter	Na^+ and amino acid transport into tubular cell	-	-
	Na^+-H^+ exchanger	Na^+ into the tubular cell; H^+ into the lumen		↓ activity in metabolic alkalosis. Activity ↑ by angiotensin-II
	Na^+-K^+ ATPase	Na^+ from the tubular cells into the interstitium; K^+ from the interstitium into the tubular cell	-	Involved in active transport of Na^+, glucose and amino acids
	Paracellular transport	K^+ from lumen into the interstitium	-	-
Thick ascending limb of loop of Henle	Na^+-K^+-$2Cl^-$ cotransporter	Na^+, K^+, and Cl^- from the lumen into the tubular cells		Defective in Bartter syndrome; Inhibited by loop diuretics
	Na^+-H^+ exchanger	Na^+ into the tubular cell; H^+ into the lumen		↓ activity in metabolic alkalosis
	Na^+-K^+ ATPase	Na^+ from the tubular cells into the interstitium; K^+ from the interstitium into the tubular cell	-	Involved in active transport of Na^+
	Paracellular transport	K^+ from lumen into the interstitium	-	-
Distal tubule	Na^+-Cl^- **cotransporter**	Na^+ and Cl^- from the lumen into the tubular cells	-	Defective in Gitelman syndrome; Inhibited by thiazide diuretics
Principal cells of collecting duct	ENaC	Na^+ from the lumen into the tubular cells	↑ by aldosterone	↑ in Liddle syndrome; ↓ by spironolactone; activity inhibited by amiloride
	ROMK	K^+ from the tubular cell into the lumen		
	Na^+-K^+ ATPase	Na^+ from the tubular cells into the interstitium; K^+ from the interstitium into the tubular cell	-	Involved in active transport of Na^+
Intercalated cells of collecting duct	K^+ channels	K^+ from the tubular cell into the interstitium		
	K^+-H^+ ATPase	K^+ into the tubular cell; H^+ into the lumen		↓ activity in metabolic alkalosis

ANP = atrial natriuretic peptide; BNP = B-type natriuretic peptide; SGLT2 = Sodium GLucose Transporter-2; ENaC = epithelial sodium channel; ROMK = renal outer medullary K^+ channel

Figure 5.18 Summary of renal handling of Na^+ and K^+, and their regulation.

Study Questions

Short-Answer Questions

5.1. What are the causes of pseudohyperkalemia?

5.2. What is the reason for edema in chronic liver disease?

5.3. Discuss the mechanism of action of angiotensin-converting enzyme inhibitors.

Long-Answer Question

5.1. What is the effect of AVP on the renal tubules? Explain the biochemical mechanisms involved in bringing about this effect. Name a condition in which this action is defective and the electrolyte abnormality that is associated.

5.2. In context with Conn's syndrome, answer the following:

A. What is the biochemical defect?

B. What are the electrolyte abnormalities seen in this condition?
C. What is the biochemical basis for the above electrolyte abnormalities?

Multiple-Choice Questions

Choose the ONE best answer.

5.1. What is the normality of normal saline?
A. 2 Normal
B. 1 Normal
C. 154 milli-Normal
D. 1 milli-Normal

Correct answer = C.

5.2. Concentration of which of the following substance is different in plasma when compared to the interstitial fluid?
A. Sodium
B. Potassium
C. Proteins
D. Magnesium

Correct answer = C.

5.3. Abnormality of which of the following can lead to presence of glucose in urine despite a normal plasma glucose concentration?
A. SGLT1
B. SGLT2
C. GLUT4
D. GLUT5

Correct answer = B.

5.4. Approximately what percentage of body weight is due to water:
A. 10%
B. 20%
C. 40%
D. 60%

Correct answer = D.

5.5. The formula to calculate plasma osmolality is:
A. $2Na^+$ + glucose + urea + creatinine
B. $2Na^+$ + glucose + urea
C. $2Na^+ + 2\ K^+$ + glucose + urea
D. $2Na^+ + 2\ K^+$ + glucose

Correct answer = B.

5.6. Impairment of arginine vasopressin production leads to:
A. Decreased water reabsorption, decreased urine osmolality, hypernatremia with increased sodium concentration but same amount of sodium and increased plasma osmolality.
B. Decreased water reabsorption, decreased urine osmolality, hypernatremia with decreased sodium concentration and increased plasma osmolality.
C. Decreased water reabsorption, decreased urine osmolality, hypernatremia with increased sodium concentration but same amount of sodium and decreased plasma osmolality.
D. Decreased water reabsorption, decreased urine osmolality, hypernatremia with increased sodium concentration due to increase in its absolute amount and increased plasma osmolality.

Correct answer = A.

5.7. Which of the following is NOT a function of the interstitial fluid (ISF) or tissue fluid?
A. Providing nutrition to the tissues
B. Facilitating gaseous exchange
C. Blood clotting
D. Removal of waste from the tissues.

Correct answer = C.

5.8. Na^+ and K^+ ATPase regulates movement of the following in the cells:
A. Na^+ and K^+ ions only.
B. Movement of Na^+ and K^+ ions in addition to absorption of amino acids and glucose in intestines only.
C. Movement of Na^+ and K^+ ions in addition to absorption of amino acids and glucose in intestines and nephrons.
D. Movement of Na^+ and K^+ ions in addition to absorption of glucose in intestine and amino acids in nephrons only.

Correct answer = C.

5.9. In response to the low ECF sodium, the resulting sodium retention and potassium excretion is brought by:
A. Renin and angiotensin
B. Renin and aldosterone
C. Angiotensin and aldosterone
D. Renin-angiotensin-aldosterone system

Correct answer = D.

5.10. Which of the following mechanisms can lead to hyperkalemia by shifting potassium from the intracellular fluid (ICF) to the extracellular fluid (ECF)?
A. Acute kidney injury
B. Increased potassium intake
C. Cell lysis due to hemolysis
D. Addison's disease

Correct answer = C.

Multiple-Choice Questions (NExT/NEET PG)

Choose the ONE best answer.

5.1. What is plasma osmolality in a patient with the following lab report?
Plasma sodium, 138 mmol/l; serum urea, 120 mg/dl; plasma glucose, 90 mg/dl.
A. 163 mOsmol/kg
B. 301 mOsmol/kg
C. 348 mOsmol/kg
D. 486 mOsmol/kg

Correct answer = B.

5.2. Which of the following can lead to a spurious elevation in plasma K^+ levels?
A. Applying a tourniquet for prolonged duration while sample collection
B. Intravascular hemolysis
C. Tumor lysis due to treatment with anticancer drugs
D. Massive tissue injury due to a road traffic accident

Correct answer = A.

5.3. Which is a NOT a potential complication of massive blood transfusion?
A. Hyperkalemia
B. Hypercalcemia
C. Hypernatremia
D. Hypothermia

Correct answer = B.

5.4. Elevated blood pressure under resting conditions results in increased urine output because:
A. Baroreceptors signal the hypothalamus to increase ADH secretion
B. Baroreceptors signal the hypothalamus to decrease ADH secretion
C. Stimulation of renin secretion
D. Increase in aldosterone level

Correct answer = B.

5.5. A 65-year-old male patient with a history of chronic kidney disease presents to the emergency department with complaints of muscle weakness, fatigue, and irregular heartbeats over the past few days. On examination, his blood pressure is 160/90 mmHg, heart rate is irregular at 110 beats per minute, and he has decreased deep tendon reflexes. Which electrolyte imbalance is most likely responsible for the patient's symptoms?
A. Hypernatremia
B. Hyponatremia
C. Hyperkalemia
D. Hypokalemia

Correct answer = C.

Appendix

Unit I: Biochemistry—The Basis of Life

IMPORTANT TABLES

Table 1 Important physiological buffers in various body fluids

Body Fluid	Buffer	Components
Extracellular fluid	Bicarbonate buffer	HCO_3^- and H_2CO_3
	Proteins	Histidine and cysteine residues
	Phosphate buffer	$H_2PO_4^-$ and HPO_4^{2-}
Blood	Bicarbonate buffer	HCO_3^- and H_2CO_3
	Hemoglobin	HHb and HbO_2^-
	Plasma proteins	Histidine and cysteine residues
	Phosphate buffer	$H_2PO_4^-$ and HPO_4^{2-}
Intracellular fluid	Phosphate buffer	$H_2PO_4^-$ and HPO_4^{2-}
	Proteins	Histidine and cysteine residues
Urine	Ammonia	NH_4^+ and NH_3
	Phosphate	$H_2PO_4^-$ and HPO_4^{2-}

Table 2 Factors that affect renal secretion of H^+ (and reabsorption of HCO_3^-)

Signal	Change	Mechanism of Regulation	Effect of H^+ Secretion/ HCO_3^- Absorption
pCO_2	↑	↑ activity of carbonic anhydrase	↑ H^+ secretion; ↑ HCO_3^- absorption
H^+	↑	↑ secretion of H^+ into tubular fluid	
Na^+	↓	↑ activity of Na^+-H^+ exchanger; stimulation of RAAS	
ECF volume	↓	↑ activity of Na^+-H^+ exchanger; stimulation of RAAS	
Angiotensin	↑	↑ activity of Na^+-H^+ exchanger; ↑ aldosterone	
Aldosterone	↑	↑ activity of *H^+-ATPase*	↑ H^+ secretion
K^+	↓	↑ activity of *K^+-H^+ ATPase*	

RAAS = renin-angiotensin-aldosterone system; ECF = extracellular fluid

Table 3 Acid-base disorders and the compensation

Disorder	pH	Primary Change	Compensation
Respiratory acidosis	↓	pCO_2 ↑	HCO_3^- ↑
Metabolic acidosis	↓	HCO_3^- ↓	pCO_2 ↓
Respiratory alkalosis	↑	pCO_2 ↓	HCO_3^- ↓
Metabolic alkalosis	↑	HCO_3^- ↑	pCO_2 ↑

Table 4 Expected compensation in various acid-base disorders

Disorder		Expected Compensation
Metabolic acidosis		pCO_2 will ↓ 1.25 mmHg per 1 mmol/l ↓ in $[HCO_3^-]$
Metabolic alkalosis		pCO_2 will ↑ 6 mmHg per 10 mmol/l ↑ in $[HCO_3^-]$
Respiratory acidosis	Acute	$[HCO_3^-]$ will ↑ 0.1 mmol/l per 1 mmHg ↑ in pCO_2
	Chronic	$[HCO_3^-]$ will ↑ 0.4 mmol/l per 1 mmHg ↑ in pCO_2
Respiratory alkalosis	Acute	$[HCO_3^-]$ will ↓ 0.2 mmol/l per 1 mmHg ↓ in pCO_2
	Chronic	$[HCO_3^-]$ will ↓ 0.4 mmol/l per 1 mmHg ↓ in pCO_2

Table 5 Maintenance of total water volume in the body

Water Intake		Water Removal	
Ingestion of fluids (directly) and water in food	2,100 ml	Urine	1,400 ml
Metabolic water (from chemical reactions)	200 ml	Insensible water loss	700 ml
		Feces	200 ml
Total intake	*2,300 ml*	*Total removal*	*2,300 ml*

Clinical Case Questions

Chapter 2 Cell and Subcellular Components

2.1. A 12-year-old child with progressive hearing loss, visual disturbances, and abnormal gait was brought for consultation. History revealed that the child was normal at birth, but gradually developed loss of hearing and loss of smell. The mother revealed that developmental milestones were delayed and he has never been able to walk properly. Skin biopsy revealed variably sized vacuoles with lipid accumulations in the epidermal cells. Blood investigation showed elevated levels of phytanic acid—a branched chain fatty acid.

A. Which cell organelle could have been affected in this disease? Justify the answer.

Answer—A: Peroxisomes. The condition described is associated with elevated levels of a branched chain fatty acid known as phytanic acid in circulation. Normally, branched chain fatty acids are catabolized by enzymes located in peroxisomes. Defect in these enzymes or in the formation of peroxisome as such, results in accumulation of branched chain fatty acids in the cells and in circulation. The case given here is that of Refsum disease, which occurs most commonly due to mutations in the peroxisomal enzyme, phytanyl CoA hydroxylase. Refsum disease can also occur due to defective targeting of this enzyme to the peroxisomes

B. What is the reason for the various neurological manifestations?

Answer—B: The maintenance of normal cell membrane structure is very crucial in tissues like nervous tissue, where localized modifications of the cell membrane are involved in the formation of vesicles for release of neurotransmitters. Accumulation of phytanic acid leads to incorporation of this moiety in the lipid membranes, disrupting their normal functions, including neurotransmission—leading to neurological manifestations.

C. Give examples of diseases due to defective protein targeting to organelles.

Answer—C: In I-cell disease, lysosomal hydrolases are secreted outside the cell, instead of these enzymes being targeted to lysosomes. This results in accumulation of various macromolecules in the lysosomes, leading to cellular dysfunctions—causing coarse facial features, neurological abnormalities, etc.

In primary hyperoxaluria, the enzyme involved in conversion of glyoxylate to glycine is targeted to mitochondria instead of peroxisomes. Glyoxylate that accumulates is converted to oxalic acid—forming calcium oxalate renal stones.

D. What is the prevalence of Refsum disease?

Answer—D: Refsum disease is an extremely rare autosomal recessive disease with some cases documented in the United Kingdom and Norway. The exact prevalence of this condition worldwide and in India is not known.

2.2. A 14-year-old boy presented to the emergency with complaints of headache, vomiting, seizures, decreased vision, followed by hemiparesis (weakness of one-half of the body) and generalized weakness. On questioning, the mother revealed that the boy was normal at birth, but had delayed developmental milestones, learning difficulty, muscle weakness, and easy fatigability. He is the first child of a non-consanguineous marriage. His mother, maternal grandmother, maternal uncle and aunt were also suffering from varying degrees of muscle weakness and neurological manifestations, but his maternal cousins, the son and daughter of the maternal uncle, were normal. MRI scan of brain showed various areas of infarcted brain tissue. Muscle biopsy detected presence of abnormal muscle fibers. Based on these findings, a genetic disorder was suspected.

(**Note:** *Clinical Case 2.2 requires knowledge from Chapter 2, as well as an understanding of pedigree charting from Chapter 39.)*

A. Discuss the mode of inheritance of this disease.

Answer—A: The pedigree of the family is shown below:

The pedigree shows maternal inheritance, as the disease is transmitted to the next generation only by the affected females.

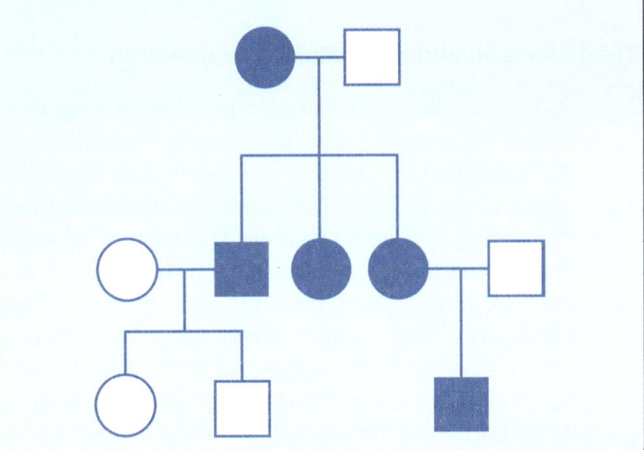

B. Which organelle is possibly affected?

Answer—B: Mitochondria. Maternal inheritance is a characteristic of mitochondrial disorders, as the mitochondria in the zygote is inherited exclusively from the egg. Not all the mitochondria that are transferred to the zygote are defective; also, not all the defective mitochondria are distributed to the daughter cells during the cell division of the zygote. Thus, mitochondrial diseases vary considerably in the clinical manifestations. However, the common manifestations are neuromuscular.

C. Why are the manifestations mostly in the neuromuscular system?

Answer—C: The proteins coded by the mitochondrial genome are involved in the electron transport chain and ATP synthesis. Since the primary function of mitochondria is energy production in the form of ATP, in mitochondrial disorders, the tissues that have huge energy requirements like neurons and muscles are primarily affected.

Chapter 3 Cell Membranes

3.1. A 1-year-old girl child of Caucasian background was brought to the clinic by her mother. She had been feverish for the past 24 hours and was coughing frequently. The mother stated that her daughter had experienced 3 attacks of bronchitis since birth. The mother had also noted that her daughter had been passing somewhat bulky, foul-smelling stools for the past several months and was not gaining weight as expected for age. The physician ordered for a sweat chloride test and sweat chloride was found to be elevated.

A. What is the probable diagnosis?

Answer—A: Cystic fibrosis. Frequent respiratory infections and increase in sweat chloride levels are suggestive. Caucasian populations show a high frequency of cystic fibrosis transmembrane conductance regulator (CFTR) mutations.

B. What is the reason for bulky, foul-smelling stools?

Answer—B: Stagnation of secretions in the pancreatic duct leads to decreased levels of these enzymes in the intestinal lumen, impairing fat and protein absorption. Presence of undigested fat causes bulky, foul-smelling stools.

C. What is the biochemical basis of this disease?

Answer—C: CFTR protein is a chloride channel. Mutations in CFTR leads to decreased transport of Cl^- across the membranes, along with a diminished transport of water. In most of the tissues, this results in decreased chloride and water in the luminal contents—leading to manifestations like thick respiratory secretions predisposing to infections, stagnation of secretions of pancreatic enzymes causing pancreatitis and decreased levels of pancreatic enzymes in the intestines leading to impaired fat and protein absorption, plugging of bile ducts by thick biliary secretions, etc. In sweat glands, the reabsorption of chloride is compromised, leading to increased sweat chloride levels.

3.2. A 30-year-old male presented to the emergency at 4 pm with history of 5–6 episodes of watery stools since morning. He also complained of severe abdominal cramps and mild fever. There was no history of blood in stools or vomiting. On further interrogation, he revealed that he has been traveling to the outskirts of the city since last 4 days and was consuming unhygienically made food from the roadside restaurants. On examination, the patient looked exhausted. His skin and oral mucosa were dry. The doctor admitted the patient and started him on oral rehydration solution (ORS).

(**Note:** *Clinical case 3.2 requires knowledge from Chapters 3, 20 and 37)*

A. Presuming that the etiology in this case is a toxigenic strain of Escherichia coli, ingested along with food, what could be the mechanism of diarrhea?

Answer—A: Toxins from E. coli cause phosphorylation and persistent activation of the CFTR protein, leading to loss of chloride (which combines with Na^+ to form NaCl) and water from the intestinal cells—leading to diarrhea and dehydration.

B. What is the rationale for the administration of ORS?

(**Note:** *The answer requires knowledge from Chapter 13, for glucose transporters)*

Answer—B: ORS, consists of NaCl and glucose, in addition to other components. In the intestinal epithelial cells, absorption of glucose is dependent on sodium, via the SGLT1 transporter. The transport of glucose and Na^+ across the intestinal epithelial cells forces the movement of water from lumen of the gut into intestinal cells, resulting in rehydration. Glucose alone or NaCl alone will not be effective.

Chapter 4 Acid-Base Balance and Arterial Blood Gas (ABG) Analysis

4.1. A 2-year-old boy was brought to the neurology OPD with complaints of progressive muscle weakness and delay in achieving developmental milestones. The child was born at full term and was normal at birth but started developing seizures at the age of 3 months. The child was unable to control his head, sit, or grasp objects.

Investigations showed blood pH of 7.24, pCO_2 38 mm of Hg, HCO_3^- 16 mmol/L, plasma chloride 95 mmol/l, plasma sodium 138 mmol/l, and elevated levels of serum lactate. Muscle biopsy and estimation of pyruvate dehydrogenase (PDH) complex activity revealed a decreased activity of the PDH enzyme complex.

Answer—A: pH of 7.24 is acidosis. Both pH and pCO_2 are decreased (hence the change in pH and pCO_2 are in the same direction), suggesting metabolic acidosis.

A. What is the acid-base disturbance in this patient?

B. What is the expected compensation?

Answer—B: Since this is a case of metabolic acidosis, the suggested compensation is respiratory, i.e., a reduction in the pCO_2. pCO_2 is expected to decrease by 1.25 mm of Hg per 1 mmol/l reduction in HCO_3^-.

Here, reduction in HCO_3^- = 24 – 16 = 8 mmol/l.

Therefore, expected reduction in pCO_2 = 8 × 1.25 = 10 mm of Hg.

But observed reduction in HCO_3^- = 40 – 38 = 2 mm of Hg.

Thus, this is a case of uncompensated metabolic acidosis.

C. What is the reason for the observed acid-base disturbance in this patient?

Answer—C: Since this is a case of metabolic acidosis, plasma anion gap has to be calculated.

Plasma anion gap = 138 – (95 + 16) = 27 mmol/l (which is elevated). Thus, this is a case of high anion gap metabolic acidosis.

Defect in PDH leads to decreased conversion of pyruvate produced by glucose metabolism into acetyl CoA. The pyruvate that accumulates is channeled into lactate (+ H^+) by the enzyme lactate dehydrogenase. Elevated lactate (+ H^+, which is secreted into the ECF) is the reason for the high anion gap metabolic acidosis.

D. What is the reason for the neurological manifestations?

Answer—D: Nervous tissue requires a lot of ATP for the normal functions like neurotransmitter synthesis, nerve conduction, and maintenance of osmotic balance in the cell via Na^+-K^+ ATPase. Nervous tissue depends on glucose metabolism for most of its energy requirements. Defective conversion of pyruvate to acetyl CoA causes a drastic reduction in the number of ATP molecules that can be generated from glucose—this leads to nerve dysfunction and neurological manifestations.

4.2. A 19-year-old male medical student was brought to the emergency by his friends with complaints of tingling and numbness of face and spasm of extremities for the past 30 minutes. On physical examination, he looked apprehensive. His respiratory rate was 26/min (normal: 12–16/min). Since his symptoms were consistent with hypocalcemia, investigations were ordered. The level of free calcium in the serum was 1.05 mmol/l (reference: 1.12–1.32 mmol/l). An arterial blood gas analysis gave the following results—blood pH, 7.51; pCO_2, 28 mm of Hg; HCO_3^-, 22 mmol/l.

On further questioning, it was revealed that the patient was supposed to give a Biochemistry theory examination on acid-base balance the next day, for which he had hardly prepared.

A. What is the acid-base abnormality in this case?

Answer—A: pH of 7.51 is alkalosis. Both pH is increased and pCO_2 is decreased (hence the change in pH and pCO_2 are in the opposite direction), suggesting respiratory alkalosis.

B. What is the reason for hypocalcemia in this patient?

Answer—B: Alkalosis increases the binding of free Ca^{2+} in the circulation to albumin, leading on to sudden reduction in the levels of free Ca^{2+} and clinical features of hypocalcemia like tingling, numbness and muscle spasms (or tetany).

C. How will you treat this patient?

Answer—C: Since this seems to be a case of hyperventilation due to anxiety, the patient is to be given reassurance. He can be asked to rebreathe into a bag, and this means the inspired air will have more CO_2 which helps in compensation of alkalosis.

Chapter 5 Water and Electrolyte Balance of Body Fluids

5.1. The emergency team in a hospital was directed to attend a 45-year-old male (body weight, 60 kg), who was undergoing treatment for bacterial sepsis and was on intravenous antibiotics, because he developed cardiac arrest. The intern incharge of the patient, who had just begun his clinical rotations, gave a history of administering a solution of 20% potassium chloride intravenously, mistaking it for an antibiotic. Before the administration of the drug, the plasma potassium of the patient was recorded to be 4.0 mmol/l. A blood sample sent to the lab soon after the cardiac event (about 5 minutes after the emergency team reaching the patient) showed a plasma potassium level of 5.0 mmol/l. Answer the questions given below (Presume that there was no intracellular shift of K^+ during the entire duration of these events).

A. What is the volume of plasma in this individual?

Answer—A: 60% of the body weight (in this case, 60 kg) is contributed by water. Hence, total body water here is 36 L. 33% of this is in the ECF, which is 12 L. Plasma constitutes 1/4th of the ECF, therefore plasma volume is 3 L.

B. How many millimoles of potassium was administered to this patient?

Answer—B: Increase in plasma K^+ concentration is 1 mmol/l (from 4.0 mmol/l to 5.0 mmol/l). Given that K^+ was distributed only extracellularly during the time of this increase, total K^+ that entered the ECF = 1 mmol/l x ECF volume (which is 12 L). Therefore, 12 mmol of K^+ was administered to this patient.

C. What is the reason for the cardiac arrest?

Answer—C: Even though hyperkalemic levels (>5.0 mmol/l) of plasma K^+ concentration was never documented in this patient, the actual K^+ level reached may not be so. Soon after administration of potassium chloride, 12 mmol of K^+ entered the plasma (to be distributed in a volume of 3 L). This would have contributed to a sudden increase in plasma K^+ by 12/3 = 4 mmol/l. Adding this to the initial plasma K^+ concentration of 4.0 mmol/l, the peak plasma K^+ following the intravenous administration of potassium chloride would have been 8 mmol/l. This could have incited the cardiac arrest.

5.2. A 30-year-old male presented to the emergency with a history of 5–6 episodes of watery stools. He also complained of severe abdominal cramps and mild fever. On further interrogation, he revealed that he was consuming food that was not properly stored, the previous day before the onset of the symptoms. He also told that he was drinking bottled water and packed fruit juices after the onset of loose stools.

On examination, he looked exhausted and drowsy. His pulse was 110/minute with low volume. His skin and oral mucosa were dry. The plasma sodium level was 125 mmol/l. As his condition seemed quite critical, the doctor on duty started a normal saline drip immediately. He was kept under observation and was switched to oral rehydration solution. He was discharged, after he showed significant improvement during the hospital stay.

A. What is the reason for hyponatremia in this patient?

Answer—A: Since the symptoms developed within 24 hours of consumption of non-properly stored food, the likely causative organism for diarrhea is a toxin-producing bacterium. Secretory diarrhea, seen in such cases, results in loss of fluid isotonic to that of plasma from the gastrointestinal tract. This, along with consumption of fluids that are devoid of sodium (bottled water and packed fruit juices in this case), leads to development of hypovolemic hyponatremia.

B. What are the clinical signs of low ECF volume?

Answer—B: Clinically, hypovolemia or reduction in the volume of ECF manifests as increased thirst, dry skin and mucous membranes, diminished peripheral pulses, hypotension, tachycardia (due to baroreceptor reflex in response to hypotension), and loss of skin turgor.

C. What is the rationale of Oral Rehydration Therapy?

Answer—C: In the small intestine, the sodium-dependent transport of glucose is mediated by a related cotransporter, SGLT1. The absorption of glucose is dependent on the ATP hydrolysis by Na^+-K^+ ATPase. Since the latter requires transport of Na^+ for functioning, in conditions like diarrhea where energy stores in the body have to be replenished by administration of carbohydrates, intake of carbohydrates alone will not be sufficient to enable glucose absorption. This is the basis of administering solutions containing both carbohydrates (glucose as such, or sucrose, or starch) and salt in oral rehydration therapy for the treatment of fluid and energy loss in diarrhea.

Unit II

Protein Structure and Function

Competencies Addressed

- **BI2.1** Explain fundamental concepts of enzyme, isoenzyme, alloenzyme, coenzyme and co-factors. Enumerate the main classes of IUBMB nomenclature
- **BI2.3** Describe and explain the basic principles of enzyme activity
- **BI2.4** Describe and discuss enzyme inhibitors as poisons and drugs and as therapeutic enzymes
- **BI2.5** Describe and discuss the clinical utility of various serum enzymes as markers of pathological conditions
- **BI2.7** Interpret laboratory results of enzyme activities and describe the clinical utility of various enzymes as markers of pathological conditions
- **BI5.1** Describe and discuss structural organization of proteins
- **BI5.2** Describe and discuss functions of proteins and structure-function relationships in relevant areas, e.g., hemoglobin and selected hemoglobinopathies
- **BI6.12** Describe the major types of hemoglobin and its derivatives found in the body and their physiological/pathological relevance
- **BI9.2** Discuss the involvement of ECM components in health and disease

Unit II: Overview

Proteins are made of amino acids joined by peptide bonds. Each amino acid, as the name implies, **has amino group and carboxyl group. The variation in side chains classifies the amino acid as those with polar side chains, nonpolar side chains, acidic or basic side chains**, etc. (Chapter 6). Proteins form enzymes, polypeptide hormones, contractile proteins, collagen, etc. The linear **primary structure** of the linked amino acids further folds into **secondary, tertiary and quaternary structures** to enable the protein to perform its various complicated functions (Chapter 7). Protein misfolding is associated with neurodegenerative disorders like **Parkinson's disease and Alzheimer's disease**. Chapter 8 deals with the globular proteins, hemoglobin and myoglobin. **Hemoglobin** is responsible for transfer of oxygen to tissues and any derangements in structure leads to diseases like sickle cell anemia and thalassemias. **Myoglobin,** on the other hand, works as an oxygen reservoir and carrier within the muscle. The fibrous proteins of the extracellular matrix are **collagen and elastin** (Chapter 9). They may be fibril forming (skin, cartilage and blood vessels), network forming (basement membrane and corneal epithelium) or fibril associated (cartilage and tendons). Chapter 10 discusses the function of **clinically important enzymes** which may be used in treatment (e.g., *streptokinase*) or diagnosis (e.g., *creatine kinase* isoenzyme [CK-MB] for diagnosis of myocardial infarction).

While understanding the basic concepts of biochemistry in medicine, the primary healthcare physician must also focus on common diseases associated with proteins like hypoalbuminemia, hypoproteinemia, hypergammaglobulinemia, thalassemia, sickle cell anemia, etc., and a clear understanding of therapeutic and diagnostic enzymes which are used in daily practice.

Amino Acids

6

Learning Objectives

At the end of this chapter, the student will be able to:

- Classify amino acids based on the variations in their side chains
- Describe the physiologically relevant properties of amino acids

I. OVERVIEW

Proteins are the most abundant and functionally diverse molecules in living systems. Virtually every life process depends on this class of macromolecules. For example, **enzymes and polypeptide hormones** direct and regulate metabolism in the body, whereas contractile proteins **in muscle** permit movement. In bone, the protein **collagen** forms a framework for the deposition of calcium phosphate crystals, acting like the steel cables in reinforced concrete. In the bloodstream, proteins, such as **hemoglobin and albumin**, transport molecules essential to life, whereas **immunoglobulin**s fight infectious bacteria and viruses. In short, proteins display an incredible diversity of functions, yet all share the common structural feature of being linear polymers of amino acids. This chapter describes the properties of amino acids. Chapter 7 explores how these simple building blocks are joined to form proteins that have unique three-dimensional structures, making them capable of performing specific biologic functions.

II. STRUCTURE

Although >300 different amino acids have been described in nature, **only 20 are commonly found as constituents of mammalian proteins.** [Note: These standard amino acids are the only amino acids that are encoded by DNA, the genetic material in the cell (see Chapter 37.II. Genetic Code, p. 635). **Nonstandard amino acids** are produced by chemical modification of standard amino acids (see p. 119).] Each amino acid has a carboxyl group, a primary amino group (except for proline, which has a secondary amino group), and a distinctive side chain (R group) bonded to the α-carbon atom. **At physiologic pH (~7.4), the carboxyl group is dissociated, forming the negatively charged carboxylate ion ($-COO^-$), and the amino group is protonated ($-NH_3^+$)** (Fig. 6.1A). In proteins, almost all of these carboxyl and amino groups

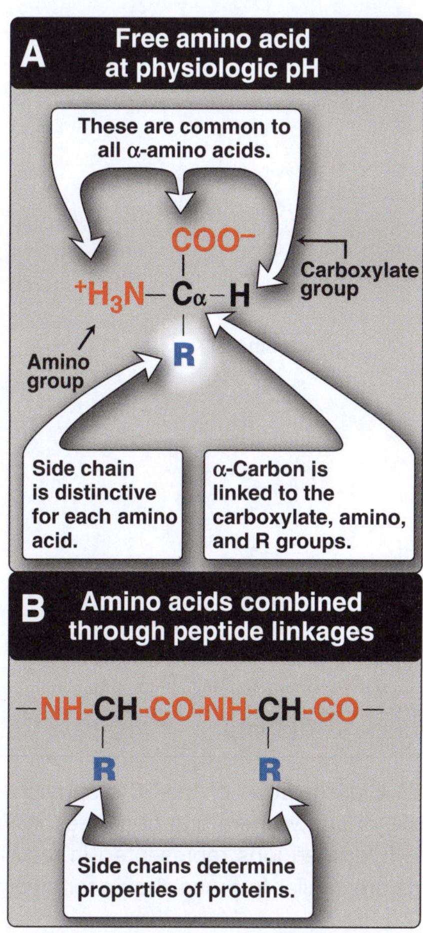

Figure 6.1 **A, B.** Structural features of amino acids.

are combined through peptide linkage and, in general, are not available for chemical reaction except for hydrogen bond formation (Fig. 6.1B). Thus, it is the nature of the side chains that ultimately dictates the role an amino acid plays in a protein. Therefore, it is useful to classify the amino acids according to the properties of their side chains, that is, whether they are nonpolar (have an even distribution of electrons) or polar (have an uneven distribution of electrons, such as acids and bases) as shown in Figures 6.2 and 6.3.

The 20 standard amino acids are discussed in detail below. **However, there is also a 21st and 22nd amino acid which is selenocysteine (Sec) and pyrrolysine (Pyl).** Sec and Pyl are rare amino acids coded by UGA and UAG, respectively. They are cotranslationally inserted into proteins and usually function as stop signals.

A. Amino acids with nonpolar side chains

Each of these amino acids has a nonpolar side chain that does not gain or lose protons or participate in hydrogen or ionic bonds (see Fig. 6.2). The side chains of these amino acids can be thought of as "oily" or lipid-like, a property that promotes hydrophobic interactions (see Fig. 7.10, pp. 88–89).

1. **Location in proteins:** In proteins found in aqueous solutions (a polar environment), the side chains of the nonpolar amino acids tend to cluster together in the interior of the protein (Fig. 6.4). This phenomenon, known as the **hydrophobic effect**, is the result of the hydrophobicity of the nonpolar R groups, which act much like droplets of oil that coalesce in an aqueous environment. By

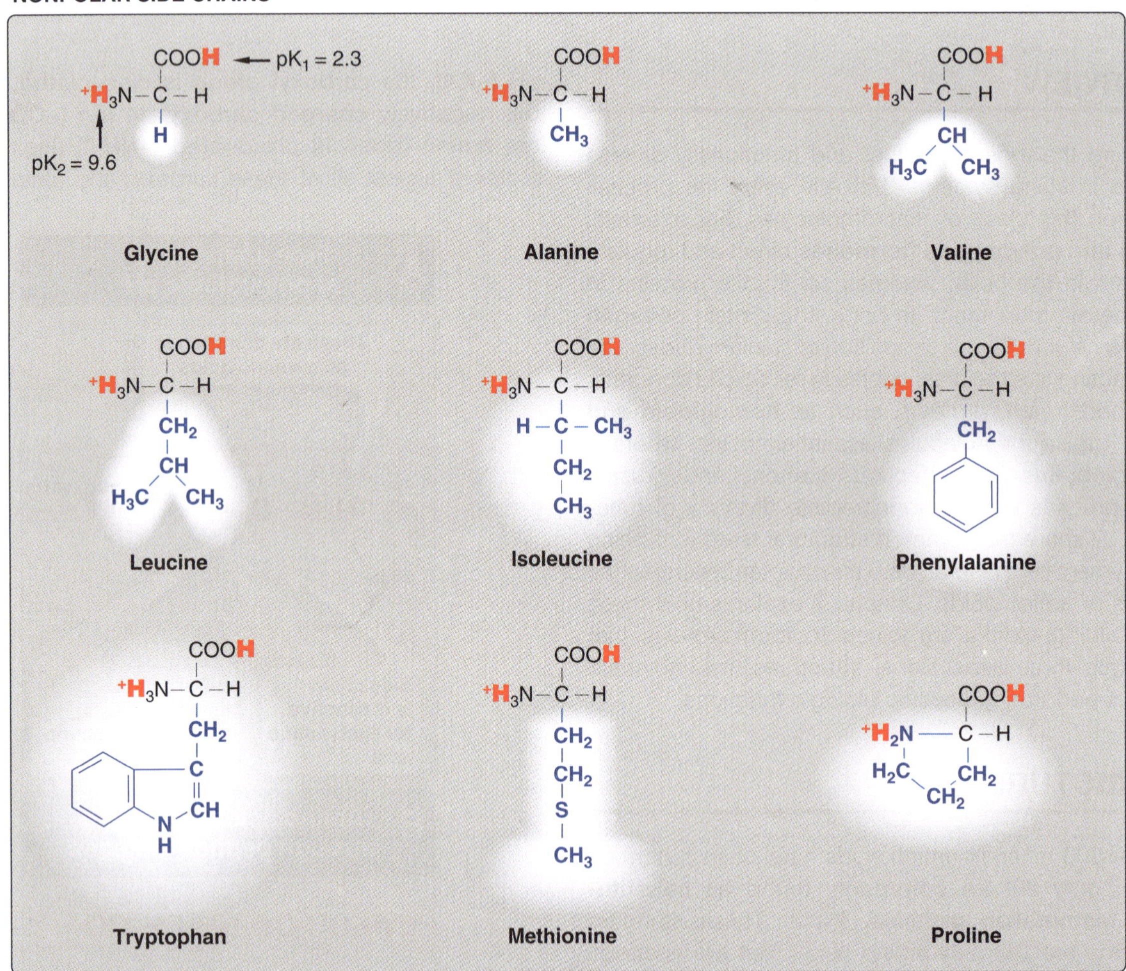

Figure 6.2 Classification of the 20 standard amino acids, according to the charge and polarity of their side chains at acidic pH, is shown here and continues in Figure 6.3. Each amino acid is shown in its fully protonated form, with dissociable hydrogen ions represented in red. The pK values for the α-carboxyl and α-amino groups of the nonpolar amino acids are similar to those shown for glycine.

UNCHARGED POLAR SIDE CHAINS

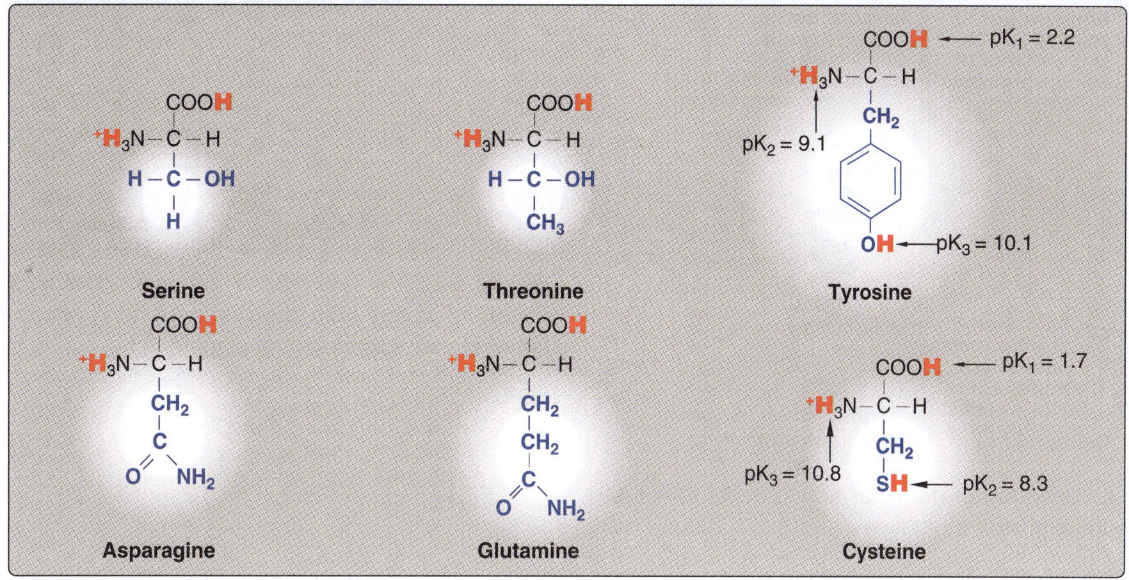

ACIDIC SIDE CHAINS

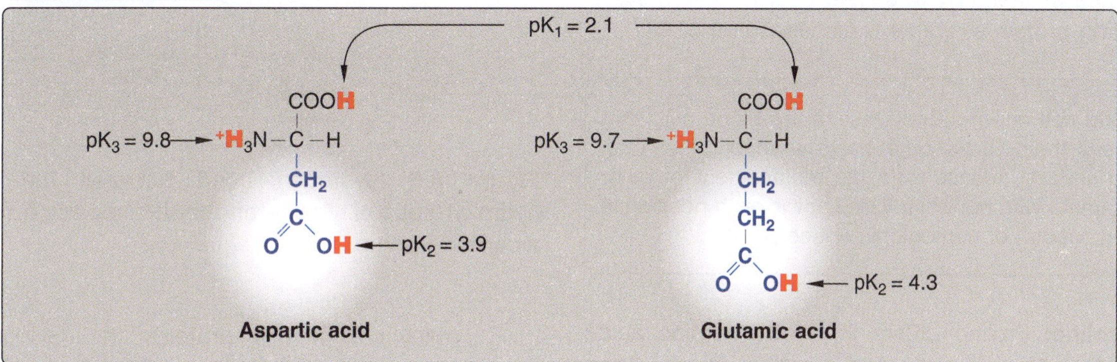

BASIC SIDE CHAINS

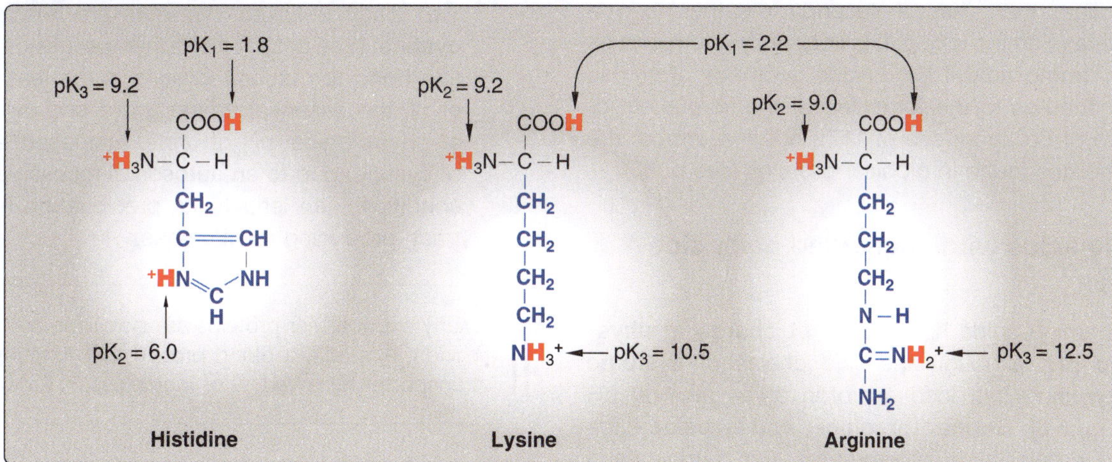

Figure 6.3 Classification of the 20 standard amino acids, according to the charge and polarity of their side chains at acidic pH (continued from Fig. 6.2). [Note: At physiologic pH (7.35 to 7.45), the α-carboxyl groups, the acidic side chains, and the side chain of free histidine are deprotonated.]

filling up the interior of the folded protein, these nonpolar R groups help give the protein its three-dimensional shape. However, for proteins that are located in a hydrophobic environment, such as a membrane, the nonpolar R groups are found on the outside surface of the protein, interacting with

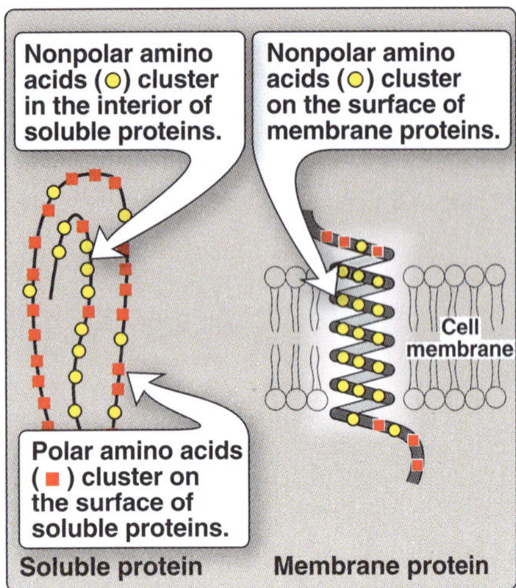

Figure 6.4 Location of nonpolar amino acids in soluble and membrane proteins.

the lipid environment (see Fig. 6.4). The importance of these hydrophobic interactions in stabilizing protein structure is discussed on p. 88.

> **Sickle cell anemia,** a disease of red blood cells that causes them to become sickle shaped rather than disc shaped, results from the replacement of polar glutamate with nonpolar valine at the sixth position in the β subunit of hemoglobin A (see p. 107).

2. **Proline:** Proline differs from other amino acids in that its side chain and α-amino nitrogen form a rigid, five-membered ring structure (Fig. 6.5). Proline, then, has a secondary (rather than a primary) amino group. It is frequently referred to as an **"imino acid."** The unique geometry of proline contributes to the formation of the fibrous structure of collagen (see p. 117), but it interrupts the α-helices found in globular proteins (see p. 85).

B. Amino acids with uncharged polar side chains

These amino acids have **zero net charge at physiologic pH**, although the side chains of **cysteine and tyrosine** can lose a proton at an alkaline pH (see Fig. 6.3**). Serine, threonine, and tyrosine** each contain a polar hydroxyl group that can participate in hydrogen bond formation (Fig. 6.6). The side chains of **asparagine and glutamine** each contain a carbonyl group and an amide group, both of which can also participate in hydrogen bonds.

1. **Disulfide bond:** The side chain of **cysteine** contains a sulfhydryl (thiol) group (–SH), which is an important component of the active site of many enzymes. In proteins, the –SH groups of two cysteines can be oxidized to form a covalent cross-link called a disulfide bond (–S–S–). Two disulfide-linked cysteines are referred to as **cystine**. (See p. 88 for a further discussion of disulfide bond formation.) Cystine is the least soluble of all the essential amino acids and is insoluble at physiological pH of urine. Increased excretion of cystine, due to an autosomal recessive genetic condition, can lead to its precipitation in urinary tract, producing renal stones.

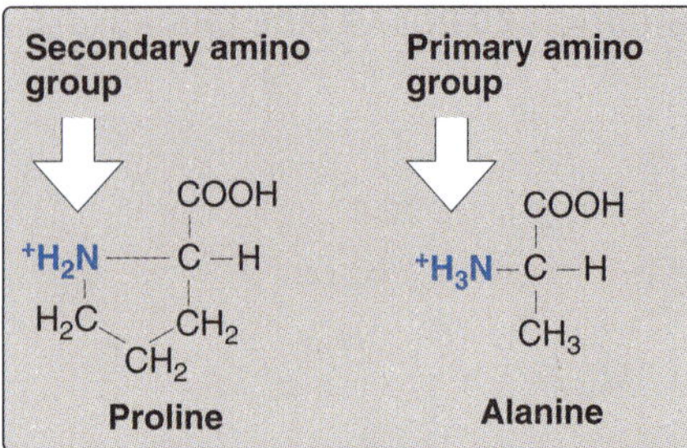

Figure 6.5 Comparison of the secondary amino group found in proline with the primary amino group found in other amino acids such as alanine.

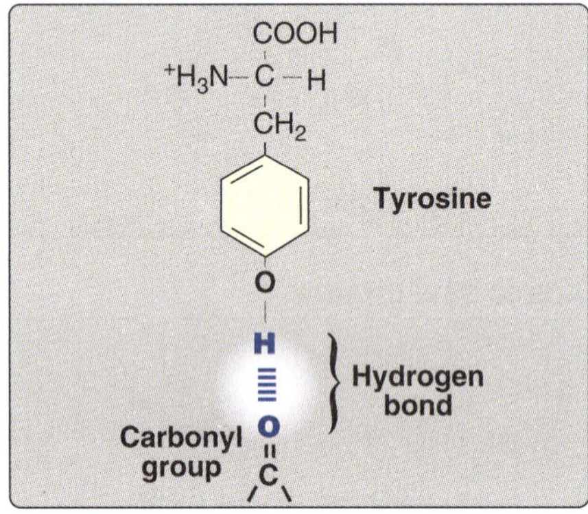

Figure 6.6 Hydrogen bond between the phenolic hydroxyl group of tyrosine and another molecule containing a carbonyl group.

> Many extracellular proteins are stabilized by disulfide bonds. Albumin, a blood protein that functions as a transporter for a variety of molecules, is an example.

2. **Side chains as attachment sites for other compounds:** The polar hydroxyl group of serine, threonine, and (rarely) tyrosine can serve as a site of attachment for structures such as a phosphate group. In addition, the amide group of asparagine, as well as the hydroxyl group of serine or threonine, can serve as a site of attachment for oligosaccharide chains in glycoproteins (see p. 269).

C. Amino acids with acidic side chains

The amino acids **aspartic acid and glutamic acid** are proton donors. At physiologic pH, the side chains of these amino acids are fully ionized, containing a negatively charged carboxylate group ($-COO^-$). The fully ionized forms are called aspartate and glutamate.

D. Amino acids with basic side chains

The side chains of the basic amino acids accept protons (see Fig. 6.3). At physiologic pH, the R groups of **lysine and arginine** are fully ionized and positively charged. In contrast, the free amino acid histidine is weakly basic and largely uncharged at physiologic pH. However, when histidine is incorporated into a protein, its R group can be either positively charged (protonated) or neutral, depending on the ionic environment provided by the protein. This important property of histidine contributes to the buffering role it plays in the functioning of such proteins as hemoglobin (see p. 26).

E. Abbreviations and symbols for commonly occurring amino acids

Each amino acid name has an associated three-letter abbreviation and a one-letter symbol (Fig. 6.7). The one-letter codes are determined by the following rules.

1. **Unique first letter:** If only one amino acid begins with a given letter, then that letter is used as its symbol. For example, V = valine.

2. **Most commonly occurring amino acids have priority:** If more than one amino acid begins with a particular letter, the most common of these amino acids receives this letter as its symbol. For example, glycine is more common than glutamate, so G = glycine.

3. **Similar sounding names:** Some one-letter symbols sound like the amino acid they represent. For example, F = phenylalanine, or W = tryptophan ("twyptophan" as Elmer Fudd would say).

4. **Letter close to initial letter:** For the remaining amino acids, a one-letter symbol is assigned that is as close in the alphabet as possible to the initial letter of the amino acid, for example, K = lysine. Furthermore, B is assigned to Asx, signifying either aspartic acid or asparagine; Z is assigned to Glx, signifying either glutamic acid or glutamine; and X is assigned to an unidentified amino acid.

F. Amino acid isomers

Because the α-carbon of an amino acid is attached to four different chemical groups, it is an asymmetric (chiral) atom. **Glycine is the exception** because its α-carbon has two hydrogen substituents Amino acids with a chiral α-carbon exist in two different isomeric forms, designated D **and** L, **which are enantiomers, or mirror images** (Fig. 6.8). [Note: Enantiomers **are optically active.** If an isomer, either

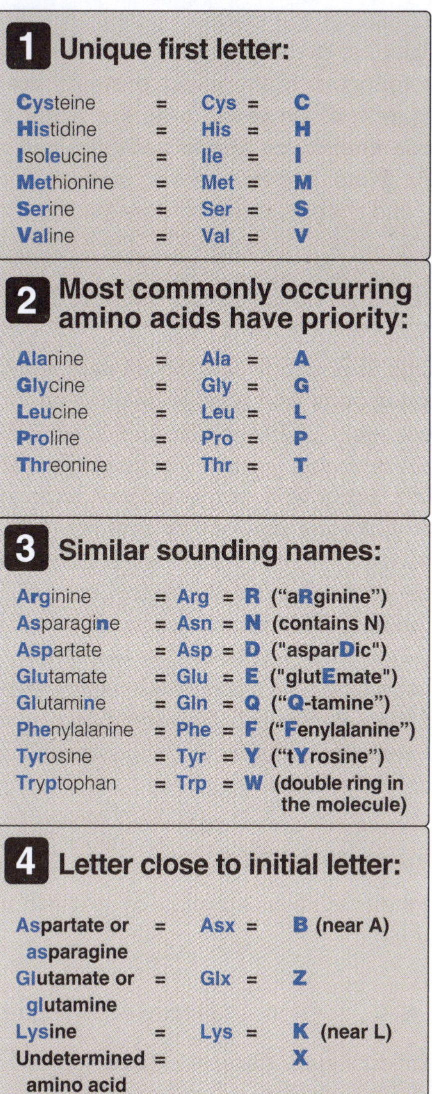

Figure 6.7 Abbreviations and symbols for the standard amino acids.

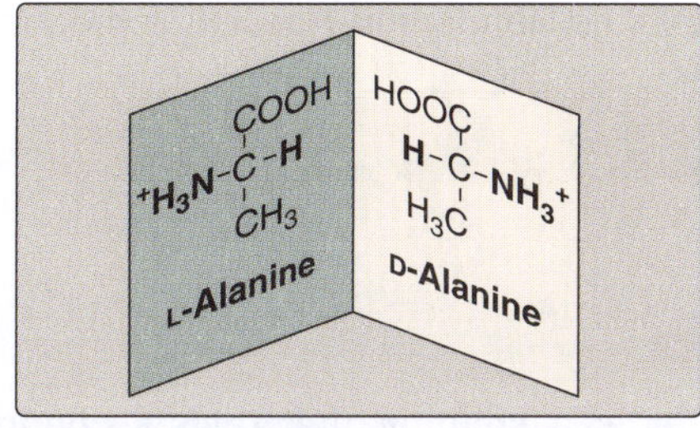

Figure 6.8 D and L forms of alanine are mirror images (enantiomers).

D or L, causes the plane of polarized light to rotate clockwise, it is designated the (+) form.] **All amino acids found in mammalian proteins are of the L configuration. However, D-amino acids are found in some antibiotics and in bacterial cell walls** (see p. 385). [Note: *Racemases* enzymatically interconvert the D- and L-isomers of free amino acids.]

III. ACIDIC AND BASIC PROPERTIES

Amino acids in aqueous solution contain **weakly acidic α-carboxyl groups and weakly basic α-amino groups**. In addition, each of the acidic and basic amino acids contains an ionizable group in its side chain. **Thus, both free amino acids and some amino acids combined in peptide linkages can act as buffers**. Acids may be defined as proton donors and bases as proton acceptors. Acids (or bases) described as weak ionize to only a limited extent. The concentration of protons ($[H^+]$) in aqueous solution is expressed as pH, where pH = log $1/[H^+]$ or $-\log [H^+]$. The quantitative relationship between the pH of the solution and concentration of a weak acid (HA) and its conjugate base (A^-) is described by the Henderson-Hasselbalch equation.

A. Equation derivation

Consider the release of a proton by a weak acid represented by HA:

$$\underset{\text{weak acid}}{HA} \rightleftarrows \underset{\text{proton}}{H^+} + \underset{\text{salt form or conjugate base}}{A^-}$$

The salt or conjugate base, A^-, is the ionized form of a weak acid. By definition, the dissociation constant of the acid, K_a, is:

$$K_a = \frac{[H^+]\,[A^-]}{[HA]}$$

[Note: **The larger the K_a, the stronger the acid, because most of the HA has dissociated into H^+ and A^-. Conversely, the smaller the K_a, the less acid has dissociated and, therefore, the weaker the acid.]** By solving for the $[H^+]$ in the above equation, taking the logarithm of both sides of the equation, multiplying both sides of the equation by –1, and substituting pH = $-\log [H^+]$ and $pK_a = -\log K_a$, we obtain the **Henderson-Hasselbalch equation**:

$$pH = pK_a + \log \frac{[A^-]}{[HA]}$$

B. Buffers

A buffer is a solution that resists change in pH following the addition of an acid or base. A buffer can be created by mixing a weak acid (HA) with its conjugate base (A^-). If an acid such as HCl is added to a buffer, A^- can neutralize it, being converted to HA in the process. If a base is added, HA can likewise neutralize it, being converted to A^- in the process. **Maximum buffering capacity occurs at a pH equal to the pK_a,** but a conjugate acid-base pair can still serve as an effective buffer when the pH of a solution is within approximately ±1 pH unit of the pK_a. **If the amounts of HA and A^- are equal, the pH is equal to the pK_a.** As shown in Figure 6.9, a solution containing acetic acid (HA = CH_3-COOH) and acetate ($A^- = CH_3-COO^-$) with a pK_a of 4.8 resists a change in pH from pH 3.8 to 5.8, with maximum buffering at pH 4.8. At pH values less than the pK_a, the protonated acid form (CH_3-COOH) is the predominant species in solution. At pH values greater than the pK_a, the deprotonated base form (CH_3-COO^-) is the predominant species.

C. Amino acid titration

The titration curve of an amino acid can be analyzed in the same way as described for acetic acid.

1. **Carboxyl group dissociation:** Consider alanine, for example, which contains an ionizable α-carboxyl and α-amino group. [Note: Its $-CH_3$ R group is nonionizable.] At a low (acidic) pH, both of these groups are protonated (Fig. 6.10). As the pH of the solution is raised, the –COOH group of form I can dissociate by donating a H^+ to the medium. The release of a H^+ results in the formation of the carboxylate group, $-COO^-$. This structure is shown as form II, which is the dipolar

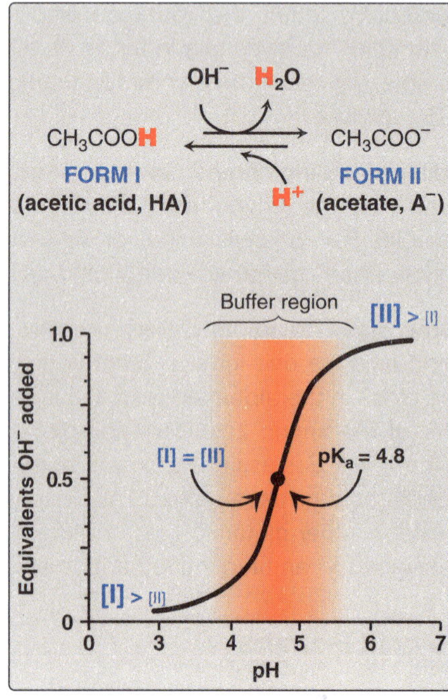

Figure 6.9 Titration curve of acetic acid.

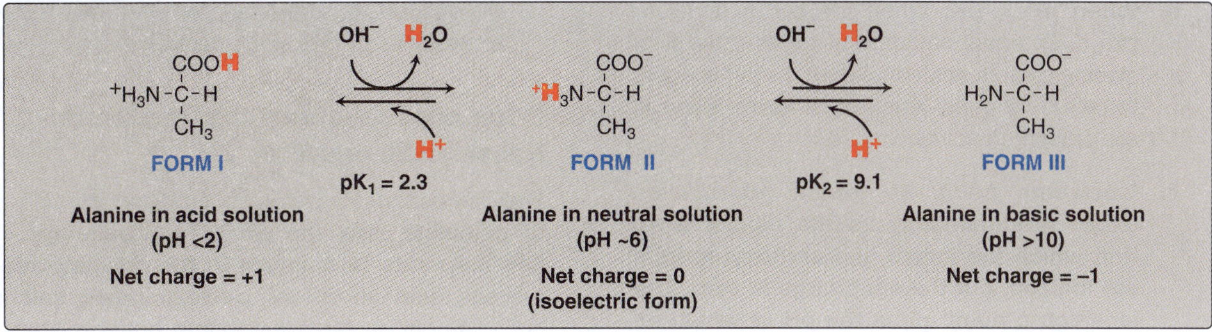

Figure 6.10 Ionic forms of alanine in acidic, neutral, and basic solutions.

form of the molecule (see Fig. 6.10). This form, also called a **zwitterion** (from the German word for "hybrid"), **is the isoelectric form of alanine, that is, it has an overall (net) charge of zero.**

2. **Application of the Henderson-Hasselbalch equation:** The dissociation constant of the carboxyl group of an amino acid is called K_1, rather than K_a, because the molecule contains a second titratable group. The Henderson-Hasselbalch equation **can be used to analyze the dissociation of the carboxyl group** of alanine in the same way as described for acetic acid:

$$K_1 = \frac{[H^+]\,[II]}{[I]}$$

where I is the fully protonated form of alanine and II is the isoelectric form of alanine (see Fig. 6.10). This equation can be rearranged and converted to its logarithmic form to yield:

$$pH = pK_1 + \log \frac{[II]}{[I]}$$

3. **Amino group dissociation: The second titratable group of alanine is the amino ($-NH_3^+$) group** shown in Figure 6.10. Because this is a much weaker acid than the –COOH group, it has a much smaller dissociation constant, K_2. [Note: Its pK_a is, therefore, larger.] Release of a H^+ from the protonated amino group of form II results in the fully deprotonated form of alanine, form III (see Fig. 6.10).

4. **Alanine pKs:** The sequential dissociation of H^+ from the carboxyl and amino groups of alanine is summarized in Figure 6.10. Each titratable group has a pK_a that is numerically equal to the pH at which exactly one half of the H^+ have been removed from that group. The pK_a for the most acidic group (–COOH) is pK_1, whereas the pK_a for the next most acidic group ($-NH_3^+$) is pK_2. [Note: The pK_a of the α-carboxyl group of amino acids is ~2, whereas that of the α-amino group is ~9.]

5. **Alanine titration curve:** By applying the Henderson-Hasselbalch equation to each dissociable acidic group, it is possible to calculate the complete titration curve of a weak acid. Figure 6.11 shows the change in pH that occurs during the addition of base to the fully protonated form of alanine (I) to produce the completely deprotonated form (III). Note the following:

   a. **Buffer pairs: The $-COOH/-COO^-$ pair can serve as a buffer in the pH region around pK_1, and the $-NH_3^+/-NH_2$ pair can buffer in the region around pK_2.**

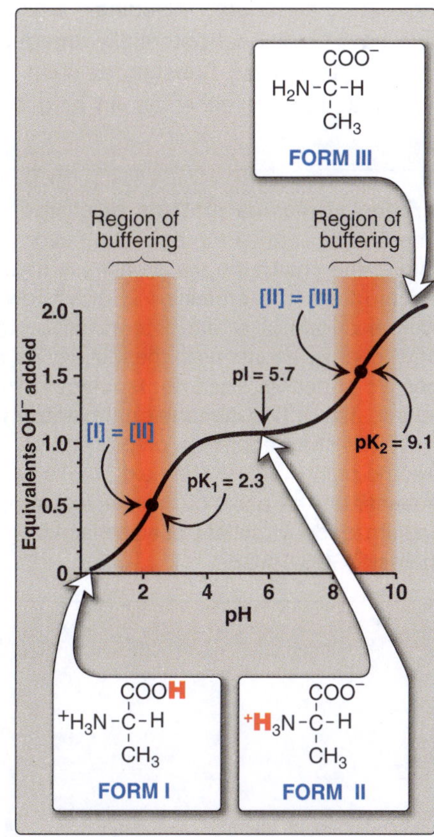

Figure 6.11 The titration curve of alanine.

b. **When pH = pK:** When the pH is equal to pK_1 (2.3), equal amounts of forms I and II of alanine exist in solution. When the pH is equal to pK_2 (9.1), equal amounts of forms II and III are present in solution.

c. **Isoelectric point: At neutral pH, alanine exists predominantly as the dipolar form II in which the amino and carboxyl groups are ionized, but the net charge is zero. The isoelectric point (pI) is the pH at which an amino acid is electrically neutral**, that is, in which the sum of the positive charges equals the sum of the negative charges. For an amino acid, such as alanine, that has only two dissociable hydrogens (one from the α-carboxyl and one from the α-amino group), the pI is the average of pK_1 and pK_2 (pI = [2.3 + 9.1]/2 = 5.7) as shown in Figure 6.11. The pI is, thus, midway between pK_1 (2.3) and pK_2 (9.1). pI corresponds to the pH at which the form II (with a net charge of zero) predominates and at which there are also equal amounts of forms I (net charge of +1) and III (net charge of −1).

6. **Net charge at neutral pH: At physiologic pH, amino acids have a negatively charged group ($-COO^-$) and a positively charged group ($-NH_3^+$), both attached to the α-carbon.** [Note: Glutamate, aspartate, histidine, arginine, and lysine have additional potentially charged groups in their side chains.] Substances such as amino acids that can act either as an acid or a base are defined as amphoteric and are referred to as ampholytes (amphoteric electrolytes).

> **Separation of plasma proteins** by charge typically is done at a pH above the pI of the major proteins. Therefore, the charge on the proteins is negative. In an electric field, the proteins will move toward the positive electrode at a rate determined by their net negative charge. Variations in the mobility pattern are suggestive of certain diseases. **Isoelectric focusing** is the first step of **two-dimensional electrophoresis** and involves the separation of proteins by their pI (isoelectric point) values. The next step is separation of proteins, on the basis of their molecular weights, by sodium dodecyl sulfate polyacrylamide gel electrophoresis **(SDS-PAGE)**.

D. Other applications of the Henderson-Hasselbalch equation

The Henderson-Hasselbalch equation can be used **to calculate how the pH of a physiologic solution responds to changes in the concentration of a weak acid and/or its corresponding salt** form. For example, in the bicarbonate buffer system, the Henderson-Hasselbalch equation predicts how shifts in the bicarbonate ion concentration, $[HCO_3^-]$, and the carbon dioxide concentration $[CO_2]$ influence pH (Fig. 6.12A). The equation is also useful for calculating the abundance of ionic forms of acidic and basic drugs. For example, most drugs are either weak acids or weak bases (Fig. 6.12B). Acidic drugs (HA) release a H^+, causing a charged anion (A^-) to form.

$$HA \rightleftarrows H^+ + A^-$$

Weak bases (BH^+) can also release a H^+. However, the protonated form of basic drugs is usually charged, and the loss of a proton produces the uncharged base (B).

$$BH^+ \rightleftarrows B + H^+$$

A drug passes through membranes more readily if it is uncharged. Thus, for a weak acid, such as **aspirin**, the uncharged HA can permeate through membranes, but A^- cannot. Likewise, for a weak base, such as **morphine**, the uncharged B form permeates through the cell membrane, but BH^+ does not. Therefore**, the effective concentration of the permeable form of each drug at its absorption site is determined by the relative concentrations of the charged (impermeant) and uncharged (permeant) forms.** The ratio between the two forms is determined by the pH at the site of absorption and by the strength of the weak acid or base, which is represented by the pK_a of the ionizable group. The Henderson-Hasselbalch equation is useful in determining how much drug is found on either side of a membrane that separates two compartments that differ in pH, for example, the stomach (pH 1.0–1.5) and blood plasma (pH 7.4).

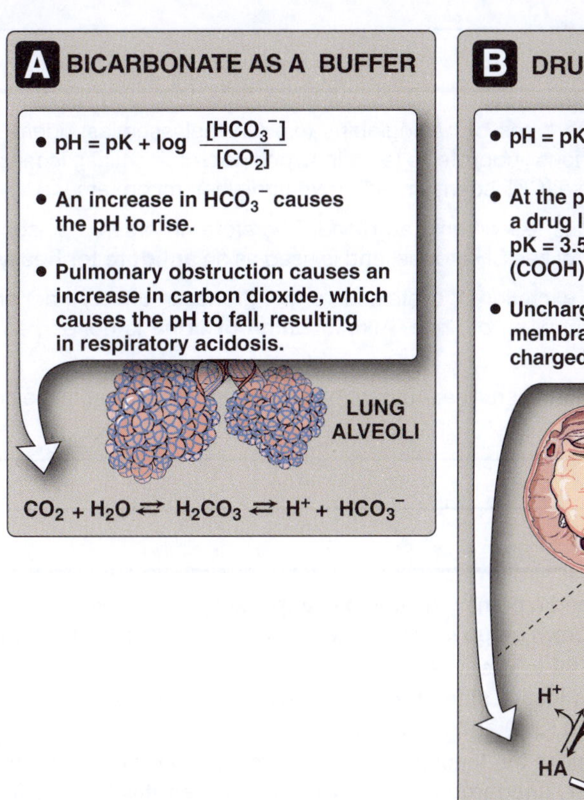

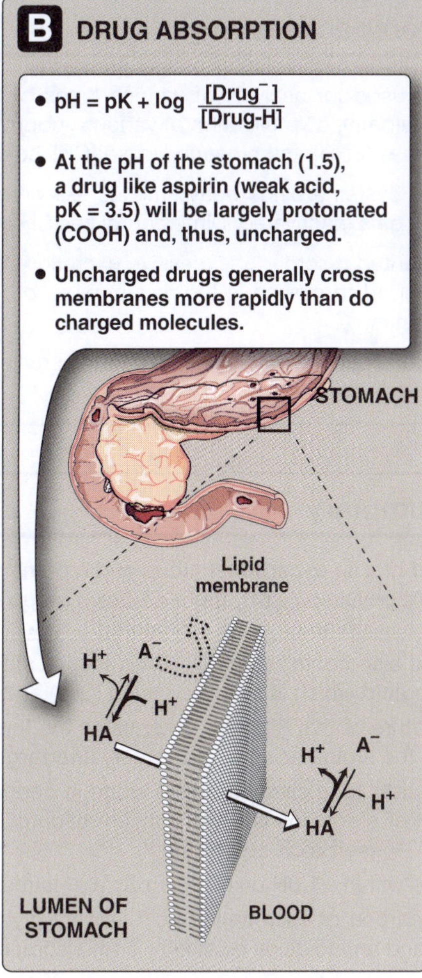

Figure 6.12 The Henderson-Hasselbalch equation is used to predict: (**A**) changes in pH as the concentrations of bicarbonate (HCO_3^-) or carbon dioxide (CO_2) are altered and (**B**) the ionic forms of drugs.

IV. Medical Perspectives

Electrophoresis is used for analysis of various types of hemoglobins (e.g., in thalassemias), identification of M Band (e.g., in multiple myeloma), quantification of various lipoproteins (e.g., in hyperlipoproteinemias), isoenzyme analysis (e.g., *lactate dehydrogenase* [LDH] and *creatine kinase* [CK] isoenzymes in myocardial infarction, etc.)

The sulfur group in cysteine binds heavy metals like arsenic and lead. Therefore, in arsenic poisoning, the SH group containing enzymes gets inactivated. Dimercaprol has SH groups, and is used as an **antidote for heavy metal poisoning**.

Cystinuria is an inherited disorder of excessive excretion of cysteine in urine. Stones in urinary tract are formed because cysteine gets converted to cystine in the acidic milieu of urine. **Alkalinization of urine** is done in an effort to decrease the formation of stones.

Posttranslational modifications of arginine may form citrulline. Autoantibodies may form against these citrullinated proteins leading to certain types of arthritis.

V. Chapter Summary

- Each amino acid has an **α-carboxyl** group and a primary **α-amino group** (except for **proline**, which has a **secondary amino group**). At physiologic pH, the α-carboxyl group is dissociated, forming the negatively charged carboxylate ion ($-COO^-$), and the α-amino group is protonated ($-NH_3^+$).
- Each amino acid also contains one of 20 distinctive **side chains** attached to the α-carbon atom. There is also a 21st and 22nd amino acid which is selenocysteine (Sec) and pyrrolysine (Pyl).
- The chemical nature of this **R group** determines the function of an amino acid in a protein and provides the basis for classification of the amino acids as **nonpolar**, **uncharged polar**, **acidic** (**polar negative**), or **basic** (**polar positive**).
- All free amino acids, plus charged amino acids in peptide chains, can serve as **buffers**. The quantitative relationship between the pH of a solution and the concentration of a weak acid (HA) and its conjugate base (A^-) is described by the **Henderson-Hasselbalch equation**.
- Buffering occurs within ±1 pH unit of the pK_a and is maximal when pH = pK_a, at which $[A^-] = [HA]$.
- Because the α-carbon of each amino acid (except glycine) is attached to four different chemical groups, it is asymmetric (**chiral**), and amino acids exist in D- and L-isomeric forms that are optically active mirror images (**enantiomers**).
- The L-form of amino acids is found in proteins synthesized by the human body.

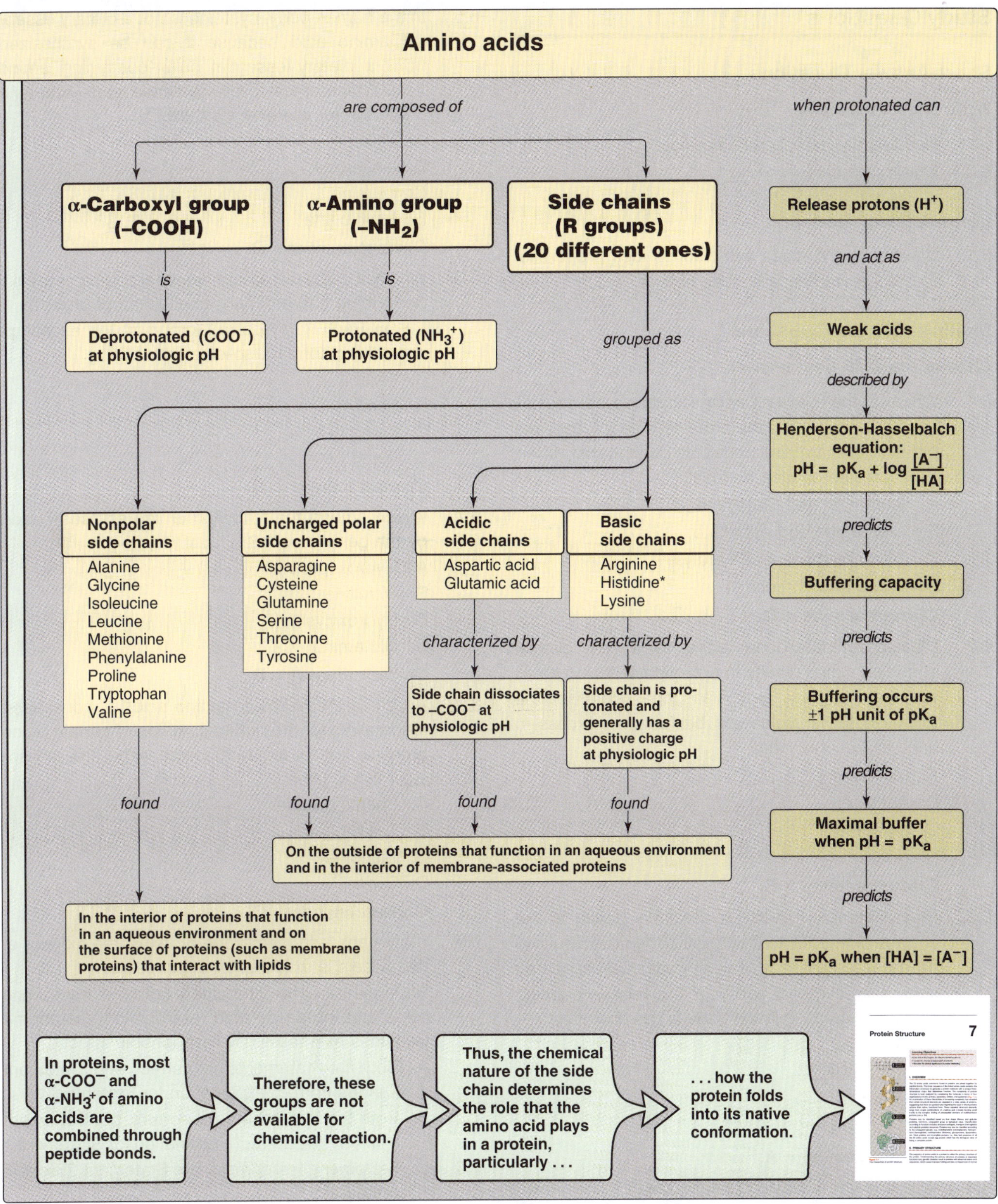

Figure 6.13 Key concept map for amino acids. [Note: *Free histidine is largely deprotonated at physiologic pH, but when incorporated into a protein, it can be protonated or deprotonated depending on the local environment.]

Study Questions

Short-Answer Questions

Write short notes on:

6.1. Henderson-Hasselbalch equation

6.2. Electrophoresis

Long-Answer Question

6.1. Classify amino acids and discuss their properties. Explain how they can act as buffers.

Multiple-Choice Questions

Choose the ONE best answer.

6.1. Which of the following amino-acids are co-translationally inserted into the proteins in which they are present, and are also coded by codons that otherwise function as stop codons?
A. Glutamine and Glutamate
B. Asparagine and Aspartate
C. Selenocysteine and Pyrrolysine
D. Serine and Threonine
Correct answer = C.

6.2. Dibasic aminoaciduria refers to a rare clinical condition which results in excretion of amino acids with two carboxyl groups in urine. Which of the following amino acids can be found in excess in urine in this condition?
A. Methionine
B. Arginine
C. Cysteine
D. Serine
Correct answer = B.

6.3. Phenylketonuria usually is due to a defect in the enzyme phenylalanine hydroxylase and this results in the reduced synthesis of an aromatic hydroxy-amino acid. The levels of which of the following amino acids is decreased in the body in this condition?
A. Tyrosine
B. Tryptophan
C. Histidine
D. Serine
Correct answer = A.

6.4. Which of the following techniques of protein fraction utilizes the existence of zwitterions?
A. Salt fractionation
B. Affinity chromatography
C. Iso-electric focusing
D. Ultracentrifugation
Correct answer = C.

6.5. In the human body, cysteine is not a dietary essential amino acid because it can be synthesized from a dietary essential sulfur-containing amino acid. Which of the following amino acids acts as a precursor for cysteine synthesis?
A. Valine
B. Isoleucine
C. Leucine
D. Methionine
Correct answer = D.

6.6. Which of the following techniques are employed while performing a two-dimensional electrophoresis?
I. SDS-PAGE II. Western blotting III. Ion exchange chromatography IV. Iso-electric focusing
A. I and II
B. I and IV
C. II and III
D. II and IV
Correct answer = B.

6.7. Which among the following amino acids is coded by the genetic code?
A. Hydroxyproline
B. Citrulline
C. Hydroxylysine
D. Glutamine
Correct answer = D.

6.8. Which of the following amino acids is considered responsible for the buffering action of certain blood proteins due its ability to ionize within the physiologic blood pH?
A. Phenyl alanine
B. Tryptophan
C. Histidine
D. Serine
Correct answer = C.

6.9. Statement 1: In membrane proteins, the exterior of the protein is made up of polar residues.
Statement 2: The interaction between membrane lipids and the amino acid residues in a membrane protein is maintained by hydrophobic effect.
A. Both the Statements are correct, and Statement 2 is the correct explanation for Statement 1
B. Both the Statements are correct, but Statement 2 is NOT the correct explanation for Statement 1
C. Statement 1 is correct, but Statement 2 is NOT correct
D. Statement 1 is NOT correct, but Statement 2 is correct
Correct answer = D.

6.10. Which of the following amino acids differs from the other in the nature of their amino group?
A. Proline

B. Glycine
C. Leucine
D. Alanine

Correct answer = A.

Multiple-Choice Questions (NExT/NEET PG)

Choose the ONE best answer.

6.1. Which one of the following statements concerning the titration curve for a nonpolar amino acid is correct? The letters A through D designate certain regions on the curve below.

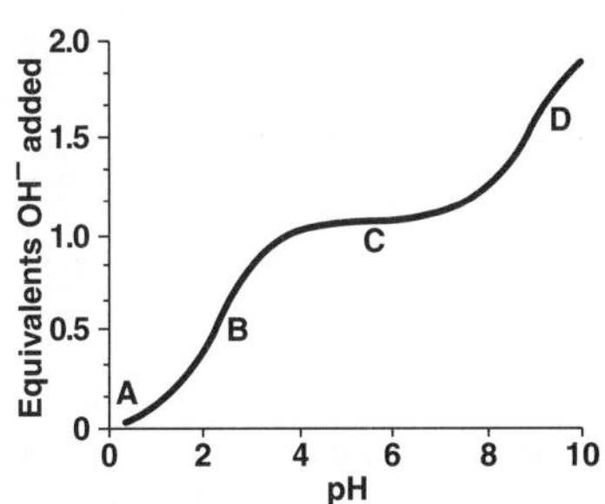

A. Point A represents the region where the amino acid is deprotonated.
B. Point B represents a region of minimal buffering.
C. Point C represents the region where the net charge on the amino acid is zero.
D. Point D represents the pK of the amino acid's carboxyl group.
E. The amino acid could be lysine.

Correct answer = C.

6.2. In Sickle Cell Anemia, the nucleotide alteration leads to substitution of a glutamate residue with valine at the 7th position in the β-globin chain. Which of the following best describes this substitution?

A. A hydroxy-amino acid is replaced by a simple amino acid
B. A polar amino acid is replaced by a non-polar amino acid
C. An aliphatic amino acid is replaced by an aromatic amino acid
D. A dietary essential amino acid is replaced by a non-essential amino acid

Correct answer = B.

6.3. Hartnup disease is due to defective transport of neutral amino acids in intestines and renal tubules. Which of the following amino acids is excreted in urine in this condition?

A. Arginine
B. Glutamate
C. Tryptophan
D. Histidine

Correct answer = C.

6.4. The bacterial transpeptidase that is involved in the formation of crosslinks between the peptide side chains of peptidoglycans in the cell wall is inhibited by penicillin and similar antibiotics. This is responsible for their bactericidal effect. Which of the following amino acids is involved in this cross-link formation?

A. D-Alanine
B. L-Alanine
C. Glycine
D. L-Arginine

Correct answer = A.

6.5. Many enzymes are regulated with the help of covalent modifications. Phosphorylation and dephosphorylation is an example of one such reversible covalent modification. A substitution or deletion of which among the following amino acids in the enzyme is most likely associated with an impairment of this regulatory mechanism?

A. Serine
B. Aspartate
C. Glutamine
D. Tryptophan

Correct answer = A.

Protein Structure

7

Learning Objectives

At the end of this chapter, the student will be able to:

- Explain the structural organization of proteins
- Describe the clinical significance of protein misfolding

I. OVERVIEW

The 20 amino acids commonly found in proteins are joined together by **peptide bonds**. The linear sequence of the linked amino acids contains the information necessary to generate a protein molecule with a unique three-dimensional shape that determines function. The complexity of protein structure is best analyzed by considering the molecule in terms of four organizational levels: **primary, secondary, tertiary, and quaternary** (Fig. 7.1). An examination of these hierarchies of increasing complexity has revealed that certain structural elements are repeated in a wide variety of proteins, suggesting that there are general rules regarding the ways in which proteins achieve their native, functional form. These repeated structural elements range from simple combinations of α-helices and β-sheets forming small motifs to the complex folding of polypeptide domains of multifunctional proteins (see p. 88).

Proteins may be **classified based on their shape (fibrous and globular proteins), functions, conjugated group or biological value**. Classification according to function includes **structural (collagen), transport (hemoglobin) and catalytic proteins (enzymes)**. Proteins may also be classified according to the **conjugated groups, e.g., metalloproteins (ceruloplasmin), hemoproteins (hemoglobin), nucleoproteins (histones), glycoproteins, lipoproteins, etc**. Most proteins are incomplete proteins, i.e., they lack presence of all the 20 amino acids except **egg protein which has the biological value of being a complete protein.**

II. PRIMARY STRUCTURE

The sequence of amino acids in a protein is called the primary structure of the protein. Understanding the primary structure of proteins is important because many genetic diseases result in proteins with abnormal amino acid sequences, which cause improper folding and loss or impairment of normal function. If the primary structures of the normal and the mutated proteins are known, this information may be used to diagnose or study the disease. **Protein sequencing by Sanger's/Edman's method or mass spectrometry may be used to study the primary structure of a protein.**

A. Peptide bond

In proteins, amino acids are joined covalently by **peptide bonds, which are amide linkages between the α-carboxyl group of one amino acid and the α-amino group of another**. For example, valine and alanine can form the dipeptide valylalanine through the formation of a peptide bond (Fig. 7.2). Peptide bonds are resistant to conditions that denature proteins, such as heating and high concentrations of urea (see p. 89). Prolonged exposure to a strong acid or base at elevated temperatures is required to break these bonds nonenzymically.

1. **Naming the peptide:** By convention, the free amino end (N-terminal) of the peptide chain is written to the left and the free carboxyl end (C-terminal) to the right. Therefore, **all amino acid sequences are read from the N- to the C-terminal end**. For example, in Figure 7.2A, the order of the amino acids in the dipeptide is valine, alanine. Linkage of ≥50 amino acids through peptide bonds results in an unbranched chain called a polypeptide, or protein. Each component amino acid is called a residue because it is the portion of the amino acid remaining after the atoms of water are lost in the formation

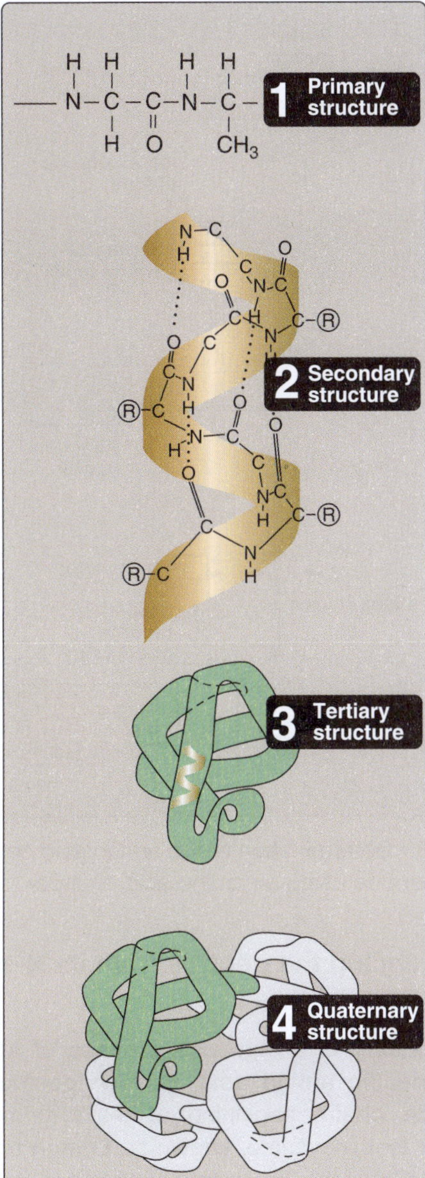

Figure 7.1 Four hierarchies of protein structure.

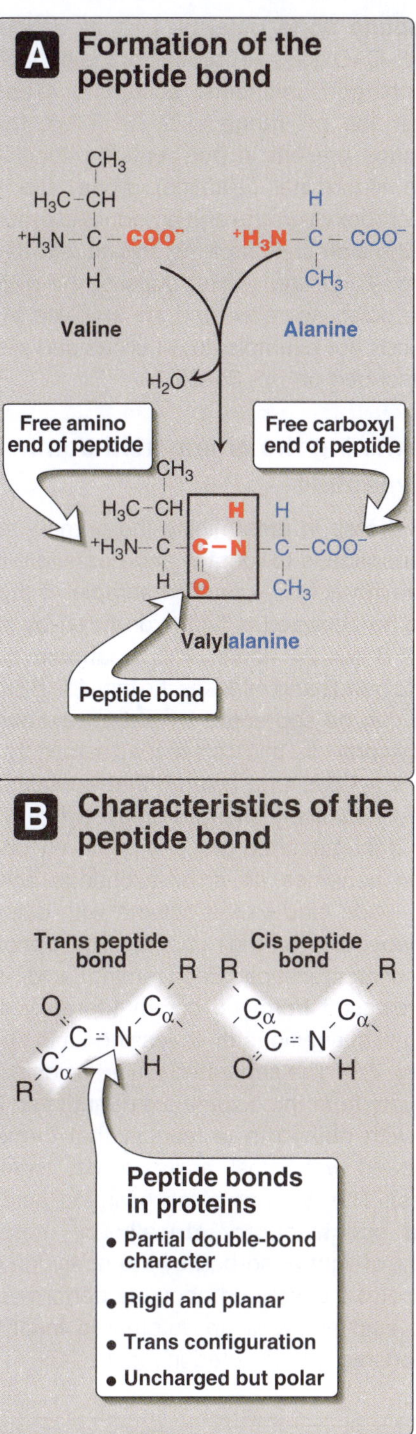

Figure 7.2 **A.** Formation of a peptide bond, showing the structure of the dipeptide valylalanine. **B.** Characteristics of the peptide bond. [Note: Peptide bonds involving proline may have a cis configuration.]

of the peptide bond. When a peptide is named, all amino acid residues have their suffixes (-ine, -an, -ic, or -ate) changed to -yl, with the exception of the C-terminal amino acid. For example, a tripeptide composed of an N-terminal valine, a glycine, and a C-terminal leucine is called valylglycylleucine.

2. **Peptide bond characteristics:** The peptide bond has a **partial double-bond** character, that is, it is shorter than a single bond and is rigid and planar (Fig. 7.2B). This prevents free rotation around the bond between the carbonyl carbon and the nitrogen of the peptide bond. However, the bonds between the α-carbons and the α-amino or α-carboxyl groups can be freely rotated (although they are limited by the size and character of the R groups). This allows the polypeptide chain to assume a variety of possible conformations. The peptide bond is **almost always in the trans configuration** (instead of the cis; see Fig. 7.2B), in large part because of steric interference of the R groups (side chains) when in the cis position.

3. **Peptide bond polarity:** Like all amide linkages, the –C=O and –NH groups of the peptide bond are uncharged, and neither accept nor release protons over the pH range of 2–12. Thus, the charged groups present in polypeptides consist solely of the N-terminal (α-amino) group, the C-terminal (α-carboxyl) group, and any ionized groups present in the side chains of the constituent amino acids. The –C=O and –NH groups of the peptide bond are polar, however, and are involved in hydrogen bonds (for example, in α-helices and β-sheets), as described on pp. 85–86.

B. Determining the amino acid composition of a polypeptide

The first step in determining the primary structure of a polypeptide is to identify and quantitate its constituent amino acids. A purified sample of the polypeptide to be analyzed is first hydrolyzed by strong acid at 110°C for 24 hours. This treatment cleaves the peptide bonds and releases the individual amino acids, which can be separated by **cation-exchange chromatography**. In this technique, a mixture of amino acids is applied to a column that contains a resin to which a negatively charged group is tightly attached. [Note: If the attached group is positively charged, the column becomes an anion-exchange column.] The amino acids bind to the column with different affinities, depending on their charges, hydrophobicity, and other characteristics. Each amino acid is sequentially released from the chromatography column by eluting with solutions of increasing ionic strength and pH (Fig. 7.3). The separated amino acids contained in the eluate from the column are quantitated by heating them with **ninhydrin** (a reagent that forms a purple compound with most amino acids, ammonia, and amines). The amount of each amino acid is **determined spectrophotometrically** by measuring the amount of light absorbed by the ninhydrin derivative. The analysis described above is performed using an amino acid analyzer, an automated machine whose components are depicted in Figure 7.3.

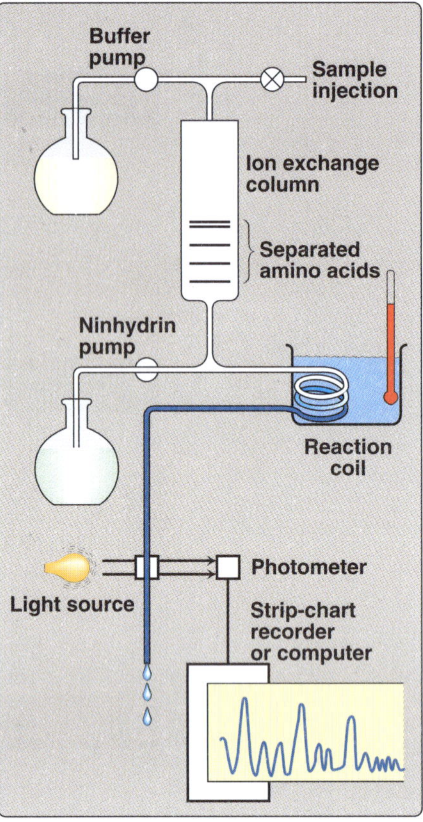

Figure 7.3 Determination of the amino acid composition of a polypeptide using an amino acid analyzer.

C. Sequencing the peptide from its N-terminal end

Sequencing is a stepwise process of identifying the specific amino acid at each position in the peptide chain, beginning at the N-terminal end. Phenyl isothiocyanate, known as **Edman reagent**, is used to label the amino-terminal residue under mildly alkaline conditions (Fig. 7.4). The resulting phenylthiohydantoin (PTH) derivative introduces an instability in the N-terminal peptide bond such that it can be hydrolyzed without cleaving the other peptide bonds. The identity of the amino acid derivative can

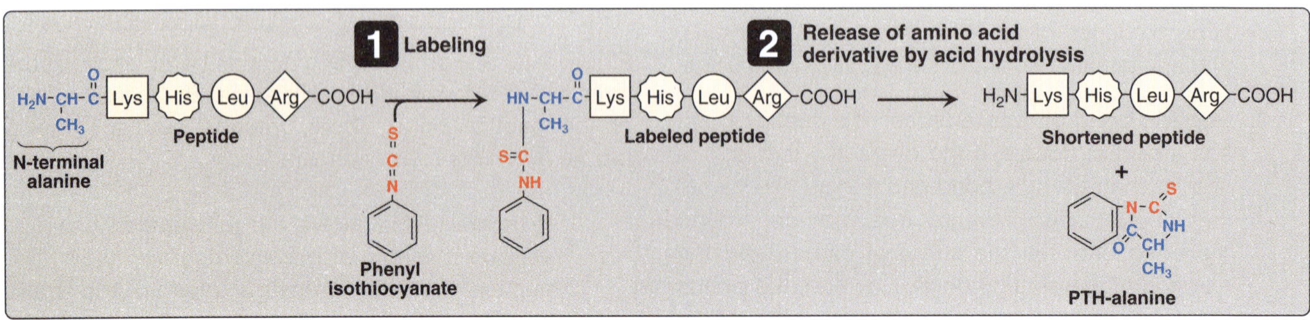

Figure 7.4 Determination of the amino (N)-terminal residue of a polypeptide by Edman degradation. PTH = phenylthiohydantoin.

then be determined. Edman reagent can be applied repeatedly to the shortened peptide obtained in each previous cycle. Automated sequencers are now used.

D. Cleaving the polypeptide into smaller fragments

Many polypeptides have a primary structure composed of >100 amino acids. Such molecules cannot be sequenced directly from end to end. However, these large molecules can be cleaved at specific sites and the resulting fragments sequenced. By using more than one cleaving agent (enzymes and/or chemicals) on separate samples of the purified polypeptide, overlapping fragments can be generated that permit the proper ordering of the sequenced fragments, thereby providing a complete amino acid sequence of the large polypeptide (Fig. 7.5). Enzymes that hydrolyze peptide bonds are termed ***peptidases (proteases).*** [Note: ***Exopeptidases* cut at the ends of proteins and are divided into *aminopeptidases* and *carboxypeptidases. Carboxypeptidases* are used in determining the C-terminal amino acid. *Endopeptidases* cleave within a protein.**]

E. Determining a protein's primary structure by DNA sequencing

The sequence of nucleotides in a protein-coding region of the DNA specifies the amino acid sequence of a polypeptide. Therefore, if the nucleotide sequence can be determined, knowledge of the genetic code (see p. 635) allows the sequence of nucleotides to be translated into the corresponding amino acid sequence of that polypeptide. This **indirect process**, although routinely used to obtain the amino acid sequences of proteins, has the **limitations of not being able to predict the positions of disulfide bonds in the folded chain and of not identifying any amino acids that are modified after their incorporation into the polypeptide** (posttranslational modification; see pp. 644–645). Therefore, **direct protein sequencing** is an extremely important tool for determining the true character of the primary sequence of many polypeptides.

> Circular proteins, such as cyclotides, are formed from linear protein precursors and retain their original functions. Their unique circular structure grants them increased stability and resistance to proteolytic degradation compared to linear proteins. Notably, they are more prevalent in plants than in bacteria or animals.

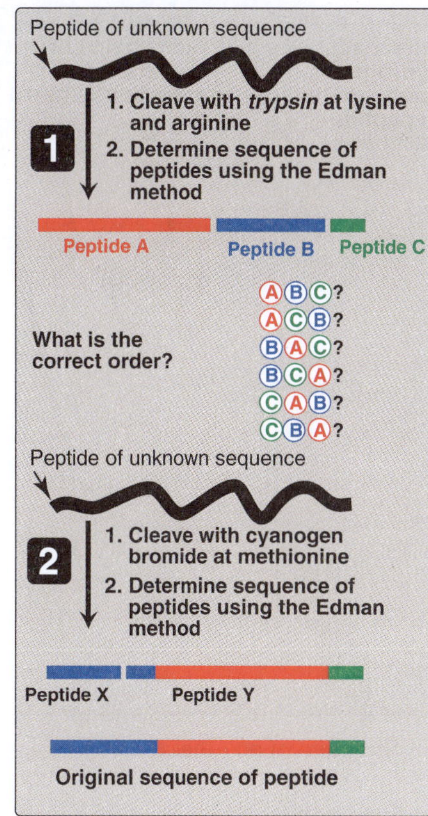

Figure 7.5 Overlapping of peptides produced by the cleavage action of *trypsin* and cyanogen bromide.

III. SECONDARY STRUCTURE

The polypeptide backbone does not assume a random three-dimensional structure but, instead, generally forms regular arrangements of amino acids that are located near each other in the linear sequence. These arrangements are termed the secondary structure of the polypeptide. **The α-helix, β-sheet, and β-bend (or, β-turn) are examples of secondary structures** commonly encountered in proteins. Each is stabilized by **hydrogen bonds** between atoms of the peptide backbone. [Note: **The collagen α-chain helix**, another example of secondary structure, is discussed on p. 117.] Secondary structure may be studied by circular dichroism spectroscopy.

A. α-Helix

Several different polypeptide helices are found in nature, but the α-helix is the most common. It is a rigid, right-handed spiral structure, consisting of a tightly packed, coiled polypeptide backbone core, with the side chains of the component L-amino acids extending outward from the central axis to avoid interfering sterically with each other (Fig. 7.6). A very diverse group of proteins contains α-helices. For example, the **keratins** are a family of closely related, rigid, fibrous proteins whose structure is nearly entirely α-helical. They are a major component of tissues such as hair and skin. In contrast to keratin, **myoglobin**, whose structure is also highly α-helical, is a globular, flexible molecule (see p. 98) found in muscles.

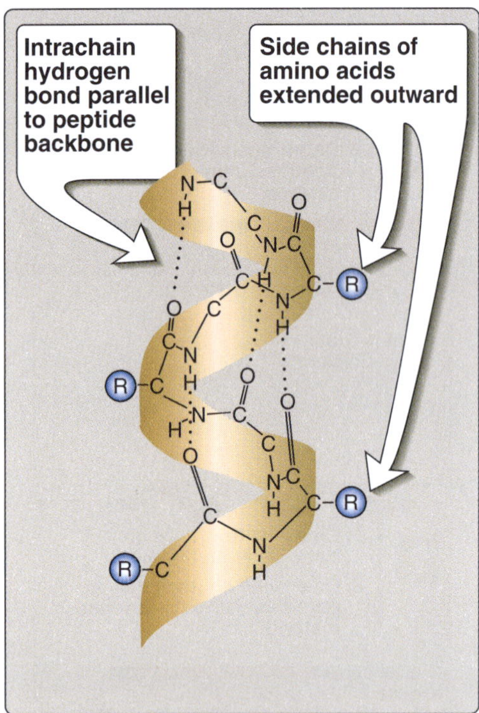

Figure 7.6 Structure of an α-helix.

1. **Hydrogen bonds:** An α-helix is stabilized by extensive hydrogen bonding between the peptide bond carbonyl oxygens and amide hydrogens that are part of the polypeptide backbone (see Fig. 7.6). The hydrogen bonds extend up and are parallel to the spiral from the carbonyl oxygen of one peptide bond to the –NH group of a peptide linkage four residues ahead in the polypeptide. This insures that all but the first and last peptide bond components are linked to each other through intrachain hydrogen bonds. **Hydrogen bonds are individually weak, but they collectively serve to stabilize the helix.**

2. **Amino acids per turn:** Each turn of an α-helix contains **3.6 amino acids**. Thus, amino acids spaced three or four residues apart in the primary sequence are spatially close together when folded in the α-helix.

3. **Amino acids that disrupt an α-helix:** The R group of an amino acid determines its propensity to be in an α-helix. **Proline disrupts an α-helix because its rigid secondary amino group is not geometrically compatible with the right-handed spiral of the α-helix**. Instead, it inserts a kink in the chain, which interferes with the smooth, helical structure. **Glycine is also a "helix breaker"** because its R group (a hydrogen) confers high flexibility. Additionally, amino acids with charged or bulky R groups (such as **glutamate and tryptophan**, respectively) and those with a branch at the β-carbon, the first carbon in the R group (for example, **valine)**, have low α-helix propensity.

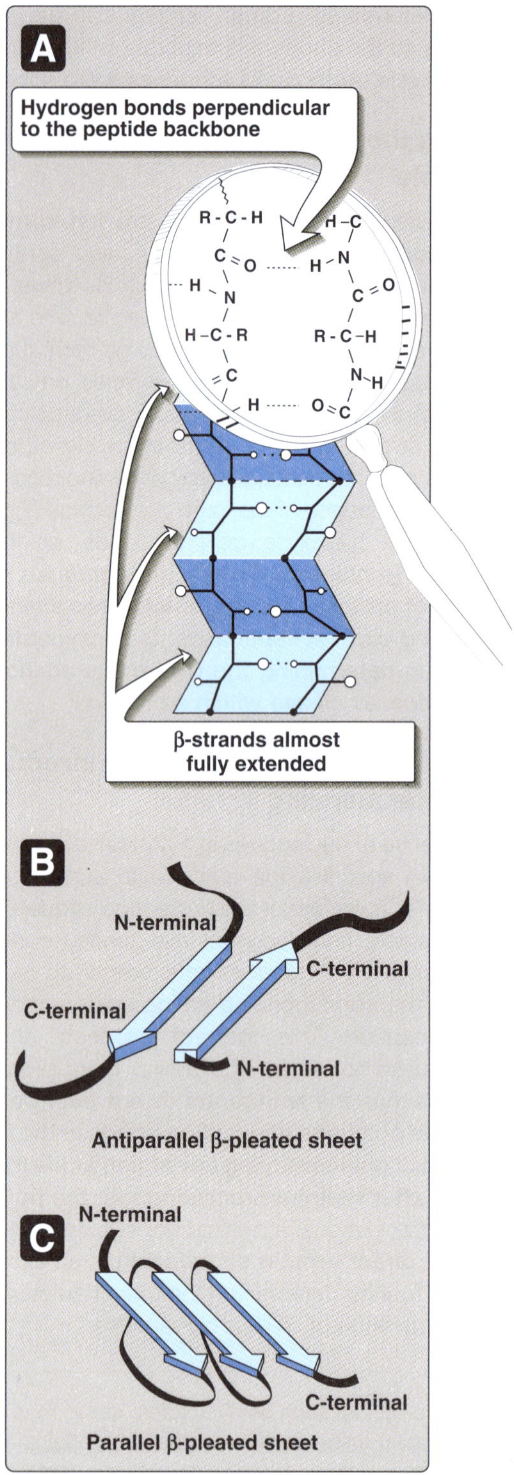

Figure 7.7 **A.** Structure of a β-sheet. **B.** An antiparallel β-sheet with the β-strands represented as broad arrows. **C.** A parallel β-sheet formed from a single polypeptide chain folding back on itself.

B. β-Sheet

The β-sheet is another form of secondary structure in which all of the peptide bond components are involved in **hydrogen bonding** (Fig. 7.7A). Because the surfaces of β-sheets appear "pleated," they are often called β-pleated sheets. [Note: Pleating results from successive α-carbons being slightly above or below the plane of the sheet.] Illustrations of protein structure often show β-strands as broad arrows (Fig. 7.7B).

1. **Formation:** A β-sheet is formed by two or more peptide segments (β-strands) aligned laterally and stabilized by hydrogen bonds between the carboxyl and amino groups of amino acids that either are far apart in a single polypeptide (intrachain bonds) or are in different polypeptide chains (interchain bonds). The adjacent β-strands are arranged either antiparallel to each other (with the N-termini alternating as shown in Fig. 7.7B) or parallel to each other (with the N-termini together as shown in Fig. 7.7C). On each β-strand, the R groups of adjacent amino acids extend in opposite directions, above and below the plane of the β-sheet. [Note: β-sheets are not flat and have a right-handed curl (twist) when viewed along the polypeptide backbone.]

2. **Comparing α-helices and β-sheets:** In β-sheets, the β-strands are almost fully extended and the hydrogen bonds between the strands are perpendicular to the polypeptide backbone (see Fig. 7.7A). In contrast, in α-helices, the polypeptide is coiled and the hydrogen bonds are parallel to the backbone (see Fig. 7.6).

> The orientation of the R groups of the amino acid residues in both the α-helix and the β-sheet can result in formation of polar and nonpolar sides in these secondary structures, thereby making them amphipathic.

C. β-Bends (reverse turns, β-turns)

β-Bends reverse the direction of a polypeptide chain, helping it form a compact, globular shape. They are usually found on the surface of protein molecules and often include charged residues. [Note: β-Bends were given this name because they often connect successive strands of antiparallel β-sheets.] β-Bends are generally composed of four amino acids, one of which may be proline, the amino acid that causes a kink in the polypeptide chain. **Glycine, the amino acid with the smallest R group, is also frequently found in β-bends**. β-Bends are stabilized by the formation of hydrogen bonds between the first and last residues in the bend.

D. Nonrepetitive secondary structure

Approximately one half of an average globular protein is organized into repetitive structures, such as the α-helix and β-sheet. The remainder of the polypeptide chain is described as having a loop or coil conformation. These **nonrepetitive secondary structures are not random** but rather simply have a less regular structure than those described above. [Note: **The term "random coil" refers to the disordered structure obtained when proteins are denatured** (see p. 89).]

E. Supersecondary structures (motifs)

Globular proteins are constructed by combining secondary structural elements (that is, α-helices, β-sheets, and coils), producing specific geometric patterns, or motifs. These form primarily the core (interior) region of the molecule. They are connected by loop regions (for example, β-bends) at the surface of the protein. Supersecondary structures are usually produced by the close packing of side chains from adjacent secondary structural elements. For example, α-helices and β-sheets that are adjacent in the amino acid sequence are also usually (but not always) adjacent in the final, folded protein. Some of the more common motifs are illustrated in Figure 7.8.

> Motifs may be associated with particular functions. Proteins that bind to DNA contain a limited number of motifs. The helix–loop–helix motif is an example found in a number of proteins that function as transcription factors (see p. 655). Motifs are discussed in detail in Chapter 38: Regulation of Gene Expression, p. 655.

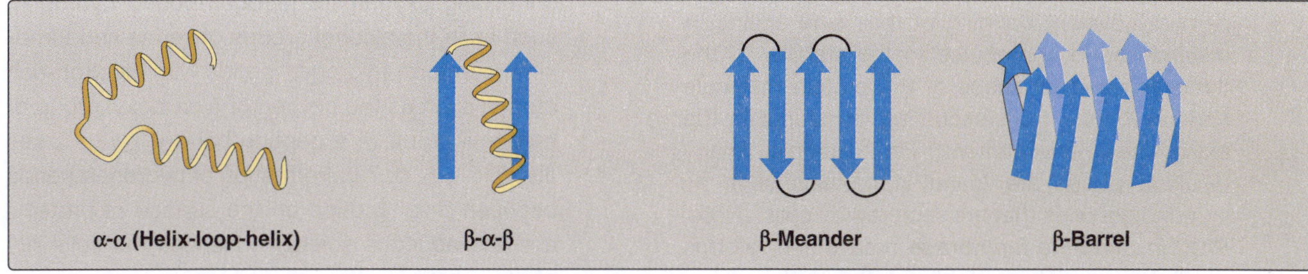

Figure 7.8 Common structural motifs involving α-helices and β-sheets. The names describe their schematic appearance.

IV. TERTIARY STRUCTURE

The primary structure of a polypeptide chain determines its tertiary structure. "Tertiary" refers both to the folding of domains (the basic units of structure and function; see A. below) and to the final arrangement of domains in the polypeptide. The tertiary structure of globular proteins in aqueous solution is compact, with a high density (close packing) of the atoms in the core of the molecule. Hydrophobic side chains are buried in the interior, whereas hydrophilic groups are generally found on the surface of the molecule. **Methods to study the tertiary structure include X-ray crystallography and nuclear magnetic resonance (NMR) spectroscopy**.

A. Domains

Domains are the fundamental functional and three-dimensional structural units of polypeptides. Polypeptide chains that are >200 amino acids in length generally consist of two or more domains. The core of a domain is built from combinations of supersecondary structural elements (motifs). Folding of the peptide chain within a domain usually occurs independently of folding in other domains. Therefore, each domain has the characteristics of a small, compact globular protein that is structurally independent of the other domains in the polypeptide chain.

B. Stabilizing interactions

The unique three-dimensional structure of each polypeptide is determined by its amino acid sequence. Interactions between the amino acid side chains guide the folding of the polypeptide to form a compact structure. The following four types of interactions cooperate in stabilizing the tertiary structures of globular proteins.

1. **Disulfide bonds: A disulfide bond (–S–S–) is a covalent linkage formed from the sulfhydryl group (–SH) of each of two cysteine residues to produce a cystine residue** (Fig. 7.9). The two cysteines may be separated from each other by many amino acids in the primary sequence of a polypeptide or may even be located on two different polypeptides. The folding of the polypeptide(s) brings the cysteine residues into proximity and permits covalent bonding of their side chains. A disulfide bond contributes to the stability of the three-dimensional shape of the protein molecule and prevents it from becoming denatured in the extracellular environment. For example, many disulfide bonds are found in proteins such as immunoglobulins that are secreted by cells. [Note: ***Protein disulfide isomerase*** breaks and reforms disulfide bonds during folding.]

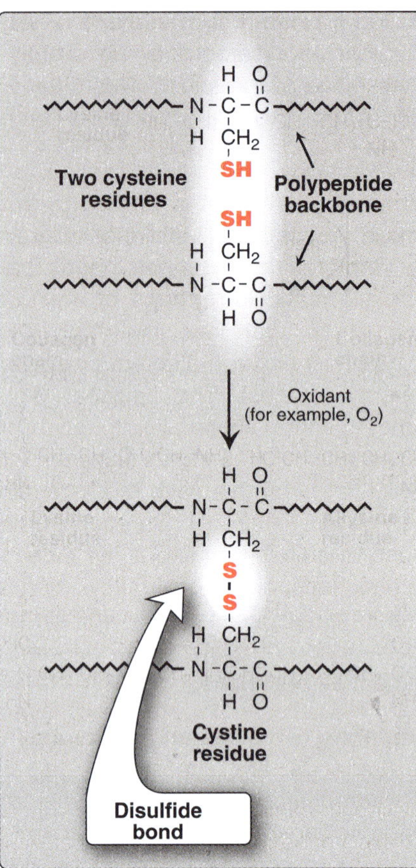

Figure 7.9 Formation of a disulfide bond by the oxidation of two cysteine residues, producing one cystine residue. O_2 = oxygen.

2. **Hydrophobic interactions: Amino acids with nonpolar side chains tend to be located in the interior of the polypeptide molecule**, where they associate with other hydrophobic amino acids (Fig. 7.10). In contrast, **amino acids with polar or charged side chains tend to be located on the surface of the molecule in contact with the polar solvent**. [Note: Recall that proteins located in nonpolar (lipid) environments, such as a membrane, exhibit the reverse arrangement (see Fig. 6.4, p. 72).] In each case, a segregation of R groups occurs that is energetically most favorable.

3. **Hydrogen bonds:** Amino acid side chains containing oxygen- or nitrogen-bound hydrogen, such as in the alcohol groups of serine and threonine, can form hydrogen bonds with electron-rich atoms, such as the oxygen of a carboxyl group or carbonyl group of a peptide bond (Fig. 7.11; see also Fig. 6.6, p. 72). Formation of hydrogen bonds between polar groups on the surface of proteins and the aqueous solvent enhances the solubility of the protein.

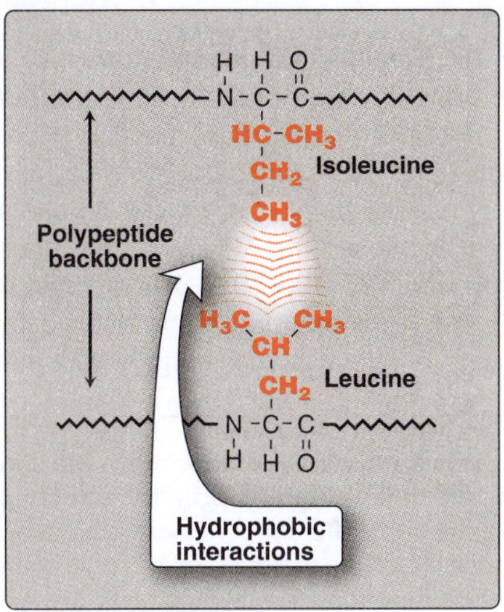

Figure 7.10 Hydrophobic interactions between amino acids with nonpolar side chains.

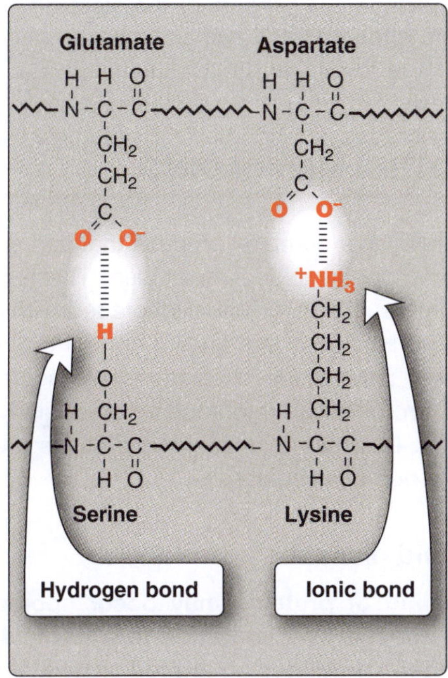

Figure 7.11 Interactions of side chains of amino acids through hydrogen bonds and ionic bonds (salt bridges).

4. **Ionic interactions:** Negatively charged groups, such as the carboxylate group ($-COO^-$) in the side chain of aspartate or glutamate, can interact with positively charged groups such as the amino group ($-NH_3^+$) in the side chain of lysine (see Fig. 7.11).

C. Protein folding

Interactions between the side chains of amino acids determine how a linear polypeptide chain folds into the intricate three-dimensional shape of the functional protein. Protein folding, which occurs within the cell in seconds to minutes, involves nonrandom, ordered pathways. As a peptide folds, secondary structures form, driven by the hydrophobic effect (that is, hydrophobic groups come together as water is released). These small structures combine to form larger structures. Additional events stabilize secondary structure and initiate formation of tertiary structure. In the last stage, the peptide achieves its fully folded, native (functional) form characterized by a low-energy state (Fig. 7.12). [Note: Some biologically active proteins or segments thereof lack a stable tertiary structure. They are referred to as intrinsically disordered proteins.]

D. Protein denaturation

Denaturation results in the **unfolding and disorganization of a protein's secondary and tertiary structures without the hydrolysis of peptide bonds**. Denaturing agents include **heat, urea, organic solvents, strong acids or bases, detergents, and ions of heavy metals such as lead.** Denaturation may, under ideal conditions, be reversible, such that the protein refolds into its original native structure when the denaturing agent is removed. However, most proteins remain permanently disordered once denatured. Denatured proteins are often insoluble and precipitate from solution.

E. Chaperones in protein folding

The information needed for correct protein folding is contained in the primary structure of the polypeptide. However, most denatured proteins do not resume their native conformations even under favorable environmental conditions. This is because, for many proteins, folding is a facilitated process that requires a specialized group of proteins, referred to as molecular chaperones, and ATP hydrolysis. The **chaperones, also known as heat shock proteins (HSP),** interact with a polypeptide at various stages during the folding process. Some chaperones bind hydrophobic regions of an extended polypeptide and are important in keeping the protein unfolded until its synthesis is completed (e.g., Hsp70). Others form cage-like macromolecular structures composed of two stacked rings. The partially folded protein enters the cage, binds the central cavity through hydrophobic interactions, folds, and is released (for example, mitochondrial Hsp60). [Note: Cage-like chaperones are sometimes referred to as chaperonins.] Chaperones, then, facilitate correct protein folding by binding to and stabilizing exposed, aggregation-prone hydrophobic regions in nascent (and denatured) polypeptides, preventing premature folding.

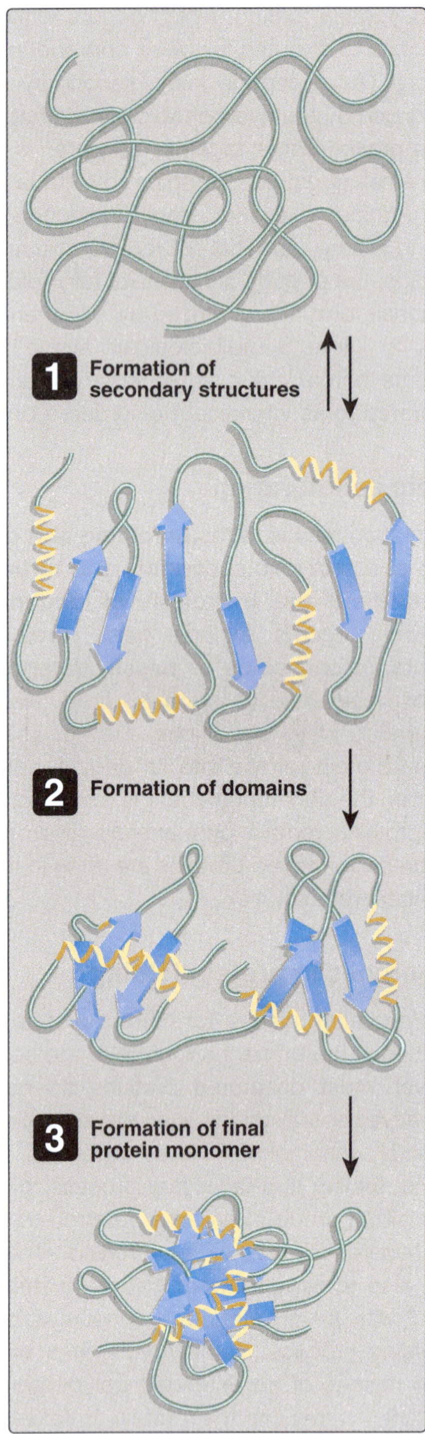

Figure 7.12 Steps in protein folding (simplified).

V. QUATERNARY STRUCTURE

Many proteins consist of a single polypeptide chain and are defined as monomeric proteins. However, others may consist of two or more polypeptide chains that may be structurally identical or totally unrelated. The arrangement of these polypeptide subunits is called the quaternary structure of the protein. **Subunits are held together primarily by noncovalent interactions (for example, hydrogen bonds, ionic bonds, and hydrophobic interactions**). Subunits either may function independently of each other or may work cooperatively, as in hemoglobin, in which the binding of oxygen to one subunit of the tetramer increases the affinity of the other subunits for oxygen (see pp. 100–101). **X-ray crystallography and NMR spectroscopy** may be used to detect spatial arrangements of subunits and to study their interactions.

> **Isoforms are proteins that perform the same function but have different primary structures.** They can arise from different genes or from tissue-specific processing of the product of a single gene. **If the proteins function as enzymes, they are referred to as isozymes** (see pp. 142–143).
>
> **Supramolecular assembly** is the molecular self-assembly of multimeric proteins to form a complex held together by noncovalent bonds. The examples include components of the **electron transport chain, ribosome and spliceosome.**

VI. PROTEIN MISFOLDING

Protein folding is a complex process that can sometimes result in improperly folded molecules. These misfolded proteins are usually tagged and degraded within the cell (see p. 379). However, this quality control system is not perfect, and intracellular or extracellular aggregates of misfolded proteins can accumulate, particularly as individuals age. Deposits of misfolded proteins are associated with a number of diseases.

A. Amyloid diseases

Misfolding of proteins may occur spontaneously or be caused by a mutation in a particular gene, which then produces an altered protein. In addition, some apparently normal proteins can, after abnormal proteolytic cleavage, take on a unique conformation that leads to the spontaneous formation of long, fibrillar protein assemblies consisting of β-pleated sheets. Accumulation of these insoluble fibrous protein aggregates, called amyloids, has been implicated in **neurodegenerative disorders such as Parkinson disease and Alzheimer disease** (AD). The dominant component of the amyloid plaque that accumulates in AD is **amyloid β (Aβ)**, an extracellular peptide containing 40–42 amino acid residues. X-ray crystallography and infrared spectroscopy demonstrate a characteristic β-pleated sheet secondary structure in nonbranching fibrils. This peptide, when aggregated in a β-pleated

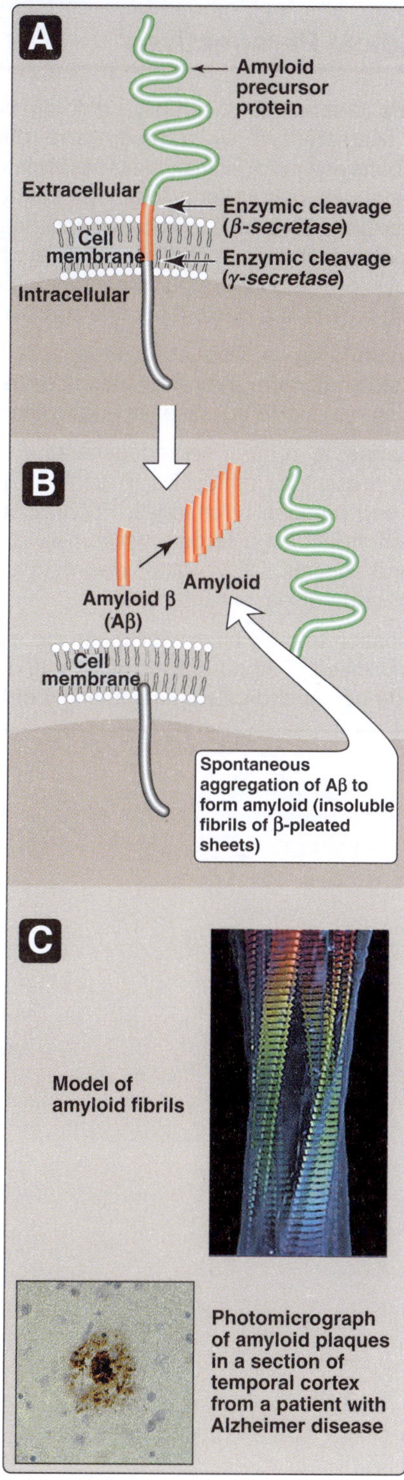

Figure 7.13 **A–C.** Formation of amyloid plaques found in Alzheimer disease (AD). [Note: Mutations to *presenilin*, the catalytic subunit of γ-*secretase*, are the most common cause of familial AD.]

sheet conformation, is neurotoxic and is the central pathogenic event leading to the cognitive impairment characteristic of the disease. The Aβ that is deposited in the brain in AD is derived by enzymic cleavages (by *secretases*) from the larger amyloid precursor protein, a single transmembrane protein expressed on the cell surface in the brain and other tissues (Fig. 7.13). The Aβ peptides aggregate, generating the amyloid that is found in the brain parenchyma and around blood vessels. Most cases of AD are not genetically based, although at least 5% of cases are familial. **A second biologic factor involved in the development of AD is the accumulation of neurofibrillary tangles inside neurons.** A key component of these tangled fibers is an abnormal form (hyperphosphorylated and insoluble) of the **tau (τ) protein**, which, in its healthy version, helps in the assembly of the microtubular structure. The defective τ appears to block the actions of its normal counterpart. [Note: **In Parkinson disease, amyloid is formed from α-synuclein protein.**]

B. Prion (proteinaceous infectious particle) diseases

The prion protein (PrP) is the causative agent of transmissible spongiform encephalopathies (TSE), including Creutzfeldt-Jakob disease (CJD) in humans, scrapie in sheep, and bovine spongiform encephalopathy in cattle (popularly called "mad cow" disease). After an extensive series of purification procedures, scientists were surprised to find that the infectivity of the agent causing scrapie in sheep was associated with a single protein species that was not complexed with detectable nucleic acid. This infectious protein is designated PrP^{Sc} (Sc = scrapie). It is highly resistant to proteolytic degradation and tends to form insoluble aggregates of fibrils, similar to the amyloid found in some other diseases of the brain. A noninfectious form of PrP^{C} (C = cellular), encoded by the same gene as the infectious agent, is present in normal mammalian brains on the surface of neurons and glial cells. Thus, PrP^{C} is a host protein. No primary structure differences or alternate post-translational modifications have been found between the normal and the infectious forms of the protein. The key to becoming infectious apparently lies in changes in the three-dimensional conformation of PrP^{C}. Research has demonstrated that a number of α-helices present in noninfectious PrP^{C} are replaced by β-sheets in the infectious form (Fig. 7.14). This conformational difference is presumably what confers relative resistance to proteolytic degradation of infectious prions and permits them to be distinguished from the normal PrP^{C} in infected tissue. The infective agent is, thus, an altered version of a normal protein, which acts as a template for converting the normal protein to the pathogenic conformation. The TSE are invariably fatal, and no treatment is currently available that can alter this outcome.

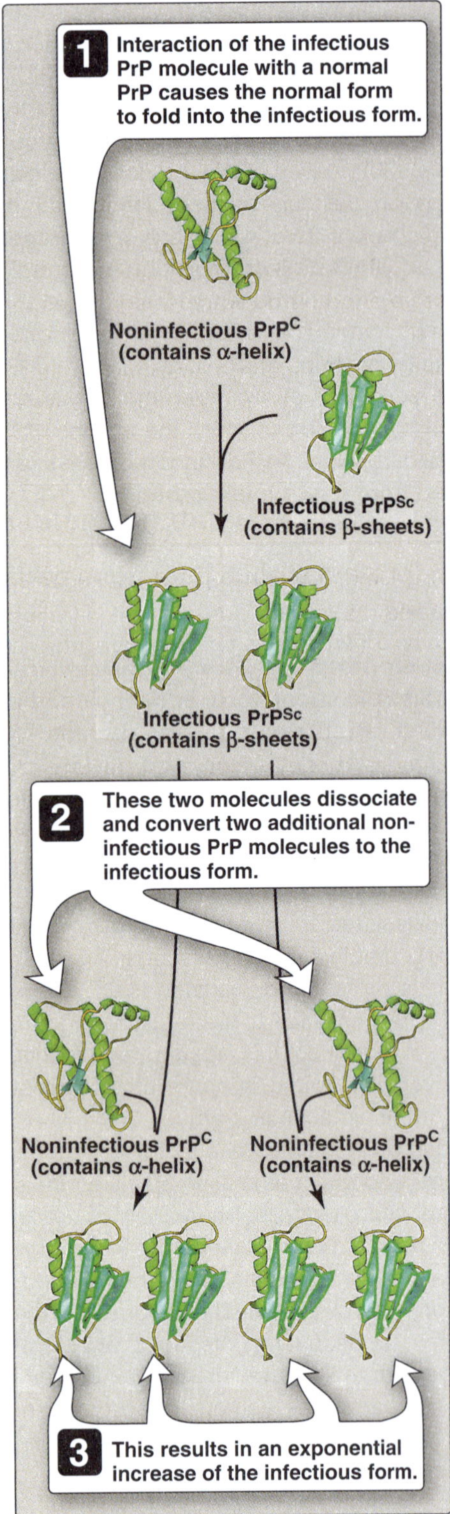

Figure 7.14 One proposed mechanism for multiplication of infectious prions. PrP = prion protein; PrP^{c} = prion protein cellular; PrP^{Sc} = prion protein scrapie.

VII. Medical Perspectives

Alzheimer's disease is the most common cause of dementia characterized by a progressive decline in thinking, behavior and social skills associated with brain degeneration and memory loss. Apparently normal proteins assume abnormal conformations, leading to amyloid plaques and neurofibrillary tangles which are mostly amyloid β peptides around neurons. The major aggregating protein is amyloid β peptide and tau protein.

Protein misfolding is also associated with cerebral beta-amyloid angiopathy, retinal ganglion cell degeneration in glaucoma, sickle cell disease and cystic fibrosis.

Prion disease is a rare neurodegenerative disorder which can present as difficulties in thinking, memory, judgment and personality changes. Prion diseases are protein conformational disorders which can infect both humans and animals. Unlike other disorders of protein misfolding, prion disease is infectious.

Heat application to curly hairs (which has more disulfide bonds crosslinked to alpha-helices of keratin) can break the disulfide bonds and cause straightening of hair.

VIII. Chapter Summary

- Central to understanding protein structure is the concept of the **native conformation** (Fig. 7.15), which is the functional, fully folded protein structure (for example, an active enzyme or structural protein).
- The unique three-dimensional structure of the native conformation is determined by its **primary structure**, that is, its amino acid sequence.
- Interactions between the amino acid side chains guide the folding of the polypeptide chain to form **secondary**, **tertiary**, and (sometimes) **quaternary** structures, which cooperate in stabilizing the native conformation of the protein.
- In addition, a specialized group of proteins named **chaperones** is required for the proper folding of many species of proteins.
- **Protein denaturation** results in the unfolding and disorganization of the protein's structure, which are not accompanied by hydrolysis of peptide bonds. Denaturation may be reversible or, more commonly, irreversible.
- Disease can occur when an apparently normal protein assumes a conformation that is cytotoxic, as in the case of **Alzheimer disease (AD)** and the **transmissible spongiform encephalopathies** (**TSE**), including **Creutzfeldt-Jakob disease**.
- In AD, normal proteins, after abnormal chemical processing, take on a unique conformational state that leads to the formation of neurotoxic **amyloid β peptide** (**Aβ**) assemblies consisting of β-pleated sheets.
- In TSE, the infective agent is an altered version of a normal **prion protein** that acts as a template for converting normal protein to the pathogenic conformation.

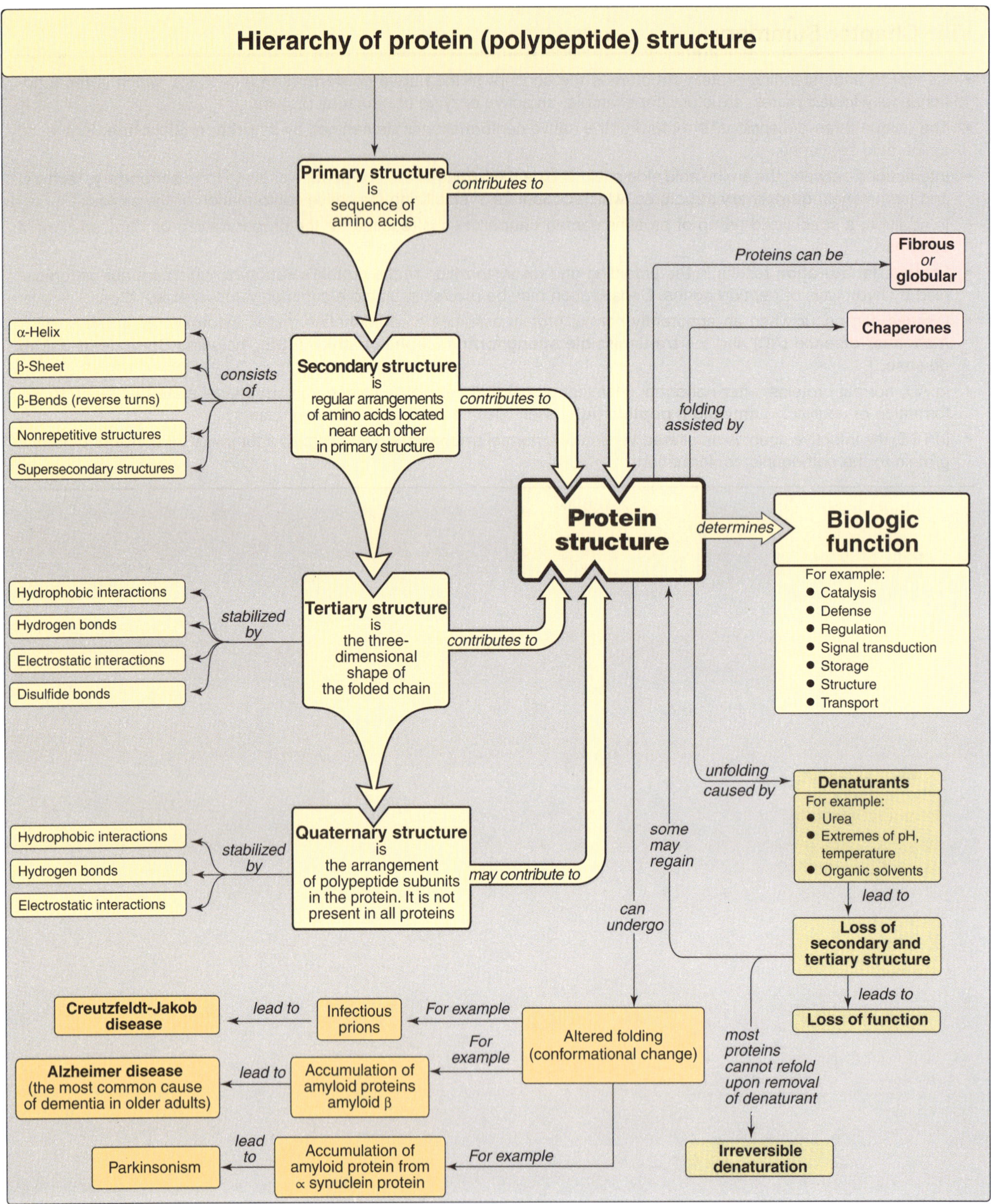

Figure 7.15 Key concept map for protein structure.

Study Questions

Short-Answer Questions

Write short notes on:

7.1. Supramolecular assembly
7.2. Classification of proteins

Long-Answer Question

7.1. Discuss the organization of protein structure with examples. Add a note on the diseases associated with protein misfolding.

Multiple-Choice Questions

Choose the ONE best answer.

7.1. Which one of the following statements concerning protein structure is correct?
 A. Proteins consisting of one polypeptide have quaternary structure that is stabilized by covalent bonds.
 B. The formation of a disulfide bond in a protein requires the participating cysteine residues to be adjacent in the primary structure.
 C. The denaturation of proteins leads to irreversible loss of secondary structural elements such as the α-helix.
 D. The primary driving force for protein folding is the hydrophobic effect.

 Correct answer = D.

7.2. A particular point mutation results in disruption of the α-helical structure in a segment of the mutant protein. The most likely change in the primary structure of the mutant protein is:
 A. glutamate to aspartate.
 B. lysine to arginine.
 C. methionine to proline.
 D. valine to alanine.

 Correct answer = C.

7.3. In comparing the α-helix to the β-sheet, which statement is correct only for the β-sheet?
 A. It may be found in typical globular proteins.
 B. It is stabilized by interchain hydrogen bonds.
 C. It is an example of secondary structure.
 D. It may be found in supersecondary structures.

 Correct answer = B.

7.4. An 80-year-old man presented with impairment of intellectual function and alterations in behavior. His family reported progressive disorientation and memory loss over the last 6 months. There is no family history of dementia. The patient was tentatively diagnosed with Alzheimer disease (AD). Which one of the following is true in this condition?
 A. It is associated with β-amyloid, an abnormal protein with an altered amino acid sequence.
 B. It results from accumulation of denatured proteins that have random conformations.
 C. It is associated with the accumulation of soluble amyloid precursor protein.
 D. It is associated with the deposition of neurotoxic amyloid β peptide aggregates.

 Correct answer = D.

7.5. Abnormal folding of proteins is the basis of pathogenesis of which of the following conditions?
 A. Creutzfeldt Jakob disease
 B. Hepatitis A
 C. Nephrotic syndrome
 D. Sickle cell disease

 Correct answer = A.

7.6. Which of the following pairs of proteins and their classification is NOT correctly matched?
 A. Ceruloplasmin - metalloprotein
 B. Hemoglobin - hemoprotein
 C. Histone - nucleoprotein
 D. Myoglobin – fibrous protein

 Correct answer = D.

7.7. Which of the following amino acids and the reason for their disrupting the formation of alpha-helix is NOT correctly matched?
 A. Glycine – high flexibility of R group
 B. Proline – rigid secondary amino group
 C. Tryptophan – bulky side chain
 D. Glutamate – branching at the β-carbon

 Correct answer = D.

7.8. Which of the following compounds can bind to amino acids, protein, and amines to form a purple colored complex, aiding in their detection during analytical processes like chromatography?
 A. Biuret reagent
 B. Nitric acid
 C. Edman's reagent
 D. Ninhydrin

 Correct answer = D.

7.9. Which of the following interactions is NOT involved in the stabilization of tertiary structure of proteins?
 A. Peptide bonds
 B. Hydrophobic interactions
 C. Hydrogen bonds
 D. Ionic interactions

 Correct answer = A.

7.10. Which of the following amino acids are frequently present in β-turns?
A. Lysine and Arginine
B. Valine and Leucine
C. Aspartate and Glutamate
D. Proline and Glycine
Correct answer = D.

Multiple-Choice Questions (NExT/NEET PG)

Choose the ONE best answer.

7.1. One among the most detrimental amino-acid changes that can occur in a protein, leading to pathogenic loss of function, occurs when a cysteine residue in a protein is substituted by some other amino acid. The formation of which among the following is affected in this situation, leading to protein instability?
A. Hydrogen bonds
B. Hydrophobic interactions
C. Disulfide linkages
D. Ionic bonds
Correct answer = C.

7.2. Which of the following techniques, respectively, are used for studying the primary structure, secondary structure, and tertiary structure of a protein?
A. Sanger sequencing, circular dichroism spectroscopy, and Edman's method.
B. Mass spectrometry, circular dichroism spectroscopy, and X-ray crystallography.
C. NMR spectroscopy, DNA sequencing, and Sanger sequencing.
D. Circular dichroism spectroscopy, X-ray crystallography, and DNA sequencing.
Correct answer = B.

7.3. G.N. Ramachandran, who first introduced it in 1963, described a plot to visualize the conformational flexibility of peptide chains in proteins, and the probable protein structures based on the flexibility. Rotation around which of the following bonds from the basis of this plot?
A. Peptide bond and C_{α}-C bond
B. Peptide bond and C_{α}-N bond
C. C_{α}-N bond and C_{α}-C bond
D. Hydrogen bond between the R-group and peptide bond
Correct answer = C.

7.4. The straightening of the curls of hair involves breakage of chemical bonds in keratin, and the alteration of its structure. Application of heat achieves this temporarily, while chemical treatment sustains the alteration for longer durations. Which of the following is a crucial component of the chemical hair straightening process?
A. Hydrogenation of peptide bonds
B. Reduction of disulfide bonds
C. Hydrolysis of ester linkages
D. Isomerization of double bonds
Correct answer = B.

7.5. Statement 1: Chaperones facilitate protein folding by binding and stabilizing hydrophobic regions in the nascent proteins.
Statement 2: Hydrophobic regions are aggregation prone and are prone for premature misfolding.
A. Both the Statements are correct, and Statement 2 is the correct explanation for Statement 1
B. Both the Statements are correct, but Statement 2 is NOT the correct explanation for Statement 1
C. Statement 1 is correct, but Statement 2 is NOT correct
D. Statement 1 is NOT correct, but Statement 2 is correct
Correct answer = A.

Globular Proteins

8

Learning Objectives

At the end of this chapter, the student will be able to:

- Describe the structure and functions of hemoglobin and myoglobin
- Describe the biochemical basis of hemoglobinopathies and their laboratory diagnoses
- Interpret the report of glycated hemoglobin

I. OVERVIEW

The previous chapter described the types of secondary and tertiary structures that are the bricks and mortar of protein architecture. By arranging these fundamental structural elements in different combinations, widely diverse proteins can be constructed that are capable of various specialized functions. This chapter examines the relationship between structure and function for the clinically important globular hemeproteins. Fibrous structural proteins are discussed in Chapter 9.

II. GLOBULAR HEMEPROTEINS

Hemeproteins are a group of specialized proteins that contain heme as a tightly bound prosthetic group. (See p. 131 for a discussion of prosthetic groups.) The role of the heme group is dictated by the environment created by the three-dimensional structure of the protein. For example, the heme group of a cytochrome functions as an electron carrier that is alternately oxidized and reduced (see p. 166). In contrast, the **heme group of the enzyme *catalase*** is part of the active site of the enzyme that catalyzes the breakdown of hydrogen peroxide (see p. 797). In **hemoglobin and myoglobin**, the two most abundant hemeproteins in humans, the heme group serves to reversibly bind oxygen (O_2).

A. Heme structure

Heme is a complex of protoporphyrin IX and ferrous iron (Fe^{2+}), as shown in Figure 8.1. The iron

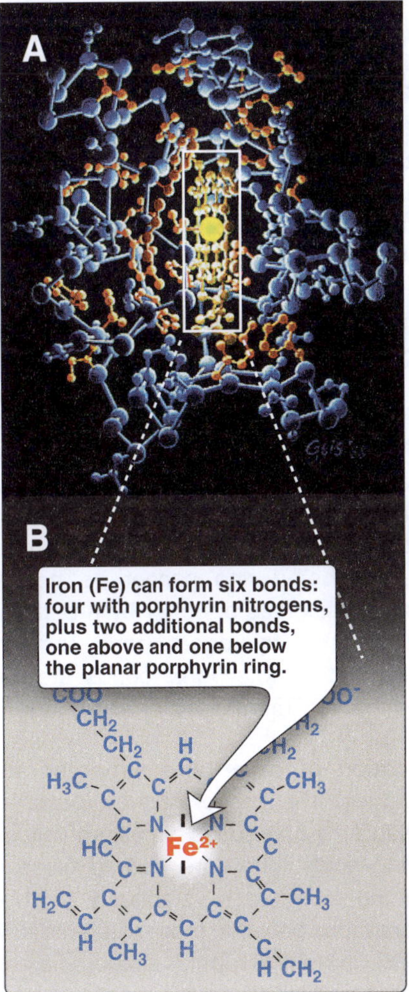

Figure 8.1 **A.** Hemeprotein (cytochrome c). **B.** Structure of heme.

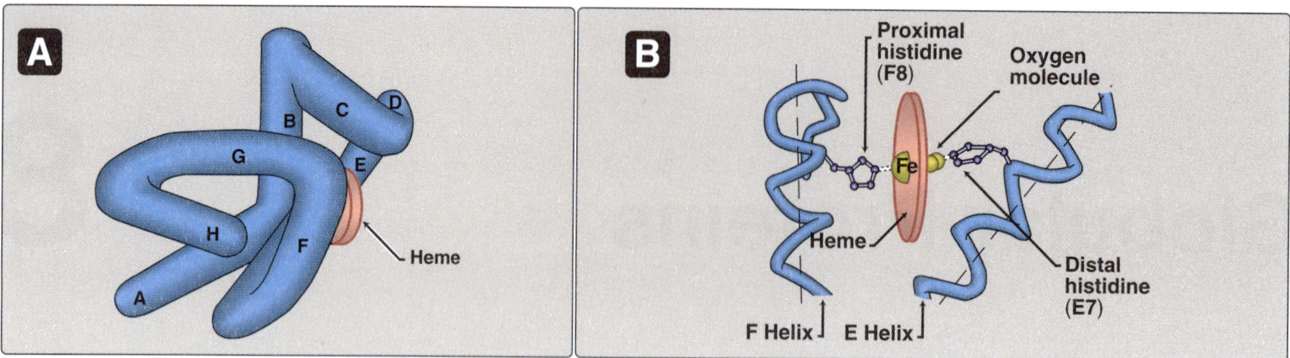

Figure 8.2 **A.** Model of myoglobin showing α-helices A to H. **B.** Schematic diagram of the oxygen-binding site of myoglobin.

is held in the center of the heme molecule by bonds to the four nitrogens of the porphyrin ring. The heme Fe^{2+} can form two additional bonds, one on each side of the planar porphyrin ring. In myoglobin and hemoglobin, one of these positions is coordinated to the side chain of a histidine residue of the globin molecule, whereas the other position is available to bind O_2 (Fig. 8.2). (See Chapter 26.)

B. Myoglobin structure

Myoglobin, a hemeprotein is present in heart and skeletal muscle. Myoglobin consists of a single polypeptide chain. It has 153 amino acids. **Myoglobin is structurally similar to the individual polypeptide chains of the tetrameric hemoglobin molecule**. This homology makes myoglobin a useful model for interpreting some of the more complex properties of hemoglobin.

1. **α-Helical content:** Myoglobin is a compact molecule, with ~80% of its polypeptide chain folded into eight stretches of α-helix. These α-helical regions, labeled A to H in Figure 8.2A, are terminated either by the presence of proline, whose five-membered ring cannot be accommodated in an α-helix (see p. 85) or by β-bends and loops stabilized by hydrogen bonds and ionic bonds (see p. 89). [Note: Ionic bonds are also termed electrostatic interactions or salt bridges.]

2. **Location of polar and nonpolar amino acid residues:** The interior of the globular myoglobin molecule is composed almost entirely of nonpolar amino acids. They are packed closely together, forming a structure stabilized by hydrophobic interactions between these clustered residues (see p. 88). In contrast, polar amino acids are located almost exclusively on the surface, where they can form hydrogen bonds, both with each other and with water.

3. **Binding of the heme group:** The heme group of the myoglobin molecule sits in a crevice, which is lined with nonpolar amino acids. Notable exceptions are two histidine residues (see Fig. 8.2B). **One, the proximal histidine (F8), binds directly to the Fe^{2+} of heme. The second, or distal histidine (E7), does not directly interact with the heme group but helps stabilize the binding of O_2 to Fe^{2+}.** Thus, the protein, or globin, portion of myoglobin creates a special microenvironment for the heme that permits the reversible binding of one oxygen molecule (oxygenation). The simultaneous loss of electrons by Fe^{2+} (oxidation to the ferric [Fe^{3+}] form) occurs only rarely.

4. **Functions of myoglobin**

   a. **Myoglobin functions both as an oxygen reservoir and as an oxygen carrier** that increases the rate of oxygen transport within the muscle cell. In aquatic mammals like seals that remain under water for long periods, PO_2 of muscles decreases to a very low level leading to release of significant amount of oxygen for the cellular respiration. In humans, the PO_2 of muscle cells hardly goes below a certain level. Hence, **even during strenuous exercise the release of oxygen from myoglobin is negligible as oxygen dissociation curve is hyperbolic** (Fig. 8.3). However, it has other functions.

   b. **Protein part of myoglobin prevents metmyoglobin formation**. Bound oxygen often forms superoxide with iron attached to heme of myoglobin. But because of H-bond that the iron forms with distal histidine of myoglobin, the superoxide is less likely to form metmyoglobin that causes metmyoglobinemia.

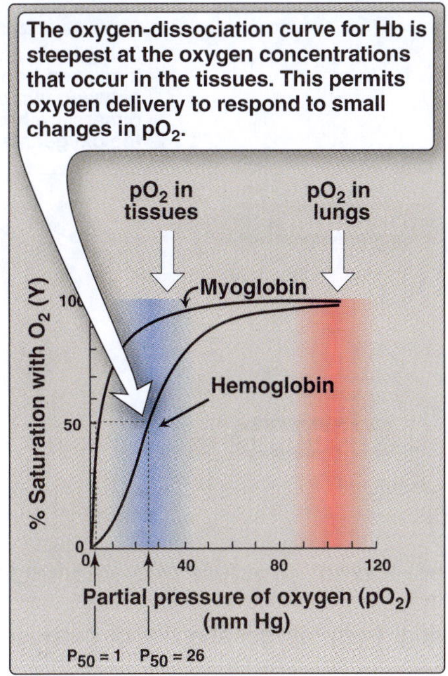

Figure 8.3 Oxygen-dissociation curves for myoglobin and hemoglobin (Hb).

c. **Oxymyoglobin detoxifies nitric oxide radicals by following reaction:**

NO + oxymyoglobin → NO_3^- + metmyoglobin

d. **Deoxymyoglobin prevents reperfusion injury**. It occurs on recanalization of blocked coronary arteries following therapy with clot busters in cases of acute myocardial infarction. Restart of oxygen supply makes electron transport chain (ETC) operational but leads to leaking of electrons and formation of free radicals which is harmful. Nitric oxide radical can inhibit ETC and thereby reduce reperfusion injury. **Deoxymyoglobin in infarcted myocardium can decrease reperfusion injury by forming nitric oxide:**

NO_2^- + H^+ + deoxymyoglobin →
NO + OH^- + metmyoglobin

C. Hemoglobin structure and function

Hemoglobin is found exclusively in red blood cells (RBC), where its main function is to transport O_2 from the lungs to the capillaries of the tissues. Hemoglobin A, the major hemoglobin in adults, is composed of four polypeptide chains (two α chains having 141 amino acids each and two β chains having 146 amino acids each) held together by noncovalent interactions (Fig. 8.4). Each chain (subunit) has stretches of α-helical structure and a hydrophobic heme-binding pocket similar to that described for myoglobin. However, the tetrameric hemoglobin molecule is structurally and functionally more complex than myoglobin. For example, hemoglobin can transport protons (H^+) and carbon dioxide (CO_2) from the tissues to the lungs and can carry four molecules of O_2 from the lungs to the cells of the body. Furthermore, the oxygen-binding properties of hemoglobin are regulated by interaction with allosteric effectors (see p. 101).

> Obtaining O_2 from the atmosphere solely by diffusion greatly limits the size of organisms. Circulatory systems overcome this, but transport molecules such as hemoglobin are also required because O_2 is only slightly soluble in aqueous solutions such as blood.

1. **Quaternary structure:** The hemoglobin tetramer can be envisioned as composed of **two identical**

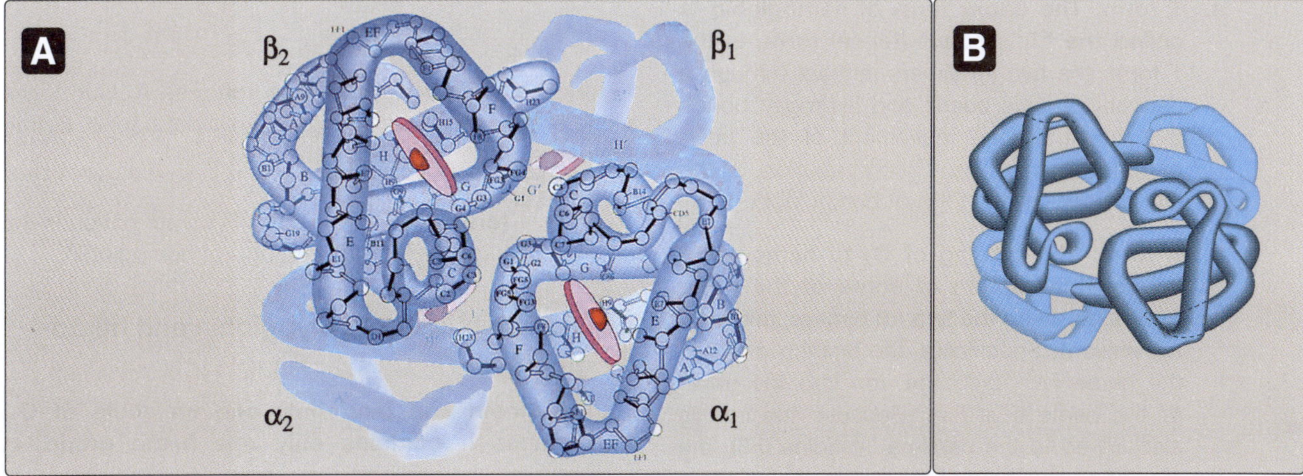

Figure 8.4 **A.** Structure of hemoglobin showing the polypeptide backbones. **B.** Simplified drawing showing the α-helices.

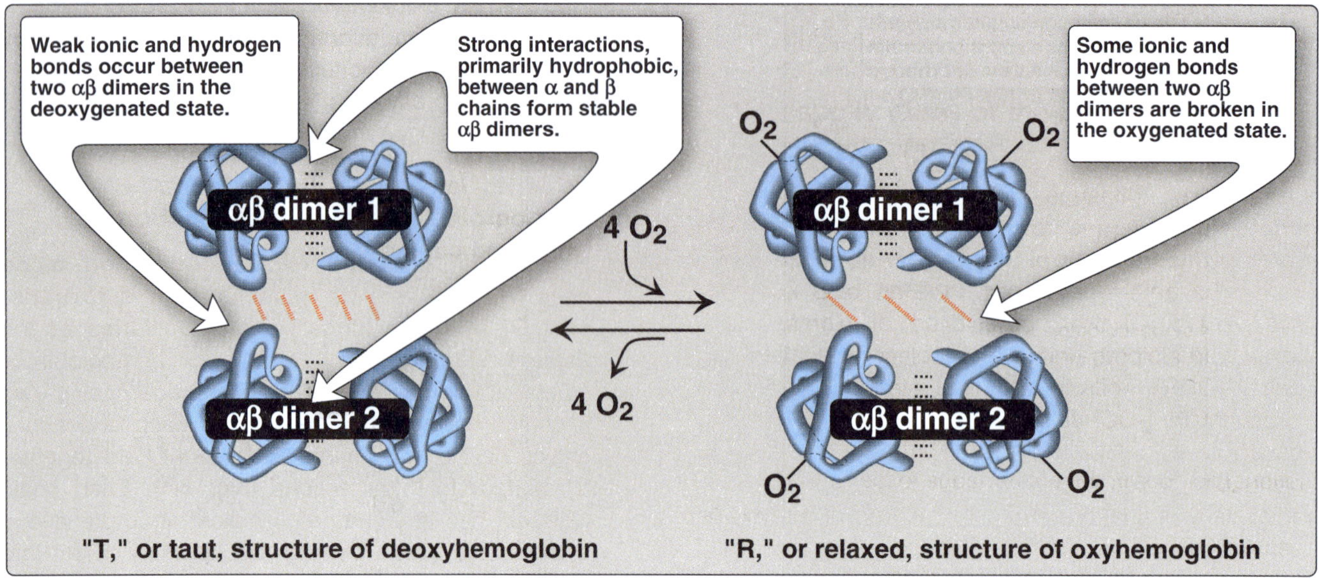

Figure 8.5 Schematic diagram showing structural changes resulting from oxygenation and deoxygenation of hemoglobin.

dimers, $(\alpha\beta)_1$ and $(\alpha\beta)_2$. The two polypeptide chains within each dimer are held tightly together primarily by hydrophobic interactions (Fig. 8.5). [Note: In this instance, hydrophobic amino acid residues are localized not only in the interior of the molecule but also in a region on the surface of each subunit. Multiple interchain hydrophobic interactions form strong associations between α-subunits and β-subunits in the dimers.] In contrast, the two dimers are held together primarily by polar bonds. **The weaker interactions between the dimers allow them to move with respect to one other. This movement results in the two dimers occupying different relative positions in deoxyhemoglobin as compared with oxyhemoglobin** (see Fig. 8.5).

a. **T form: The deoxy form of hemoglobin is called the "T," or taut (tense) form.** In the T form, the two αβ dimers interact through a network of ionic bonds and hydrogen bonds that constrain the movement of the polypeptide chains. The T conformation is the low-oxygen-affinity form of hemoglobin.

b. **R form: The binding of O_2 to hemoglobin causes the rupture of some of the polar bonds between the two αβ dimers, allowing movement.** Specifically, the binding of O_2 to the heme Fe^{2+} pulls the iron into the plane of the heme (Fig. 8.6). Because the iron is also linked to the proximal histidine (F8), the resulting movement of the globin chains alters the interface between the αβ dimers. This leads to a structure called the **"R," or relaxed form** (see Fig. 8.5). The R conformation is the high-oxygen-affinity form of hemoglobin.

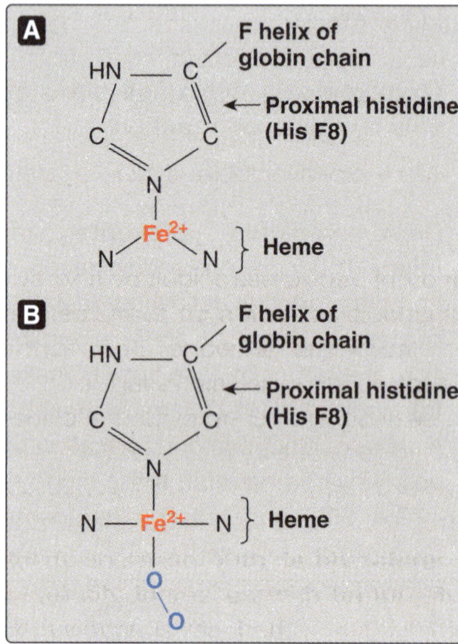

Figure 8.6 Movement of heme iron (Fe). **A.** Out of the plane of the heme when oxygen (O_2) is not bound. **B.** Into the plane of the heme upon O_2 binding.

D. Oxygen binding to myoglobin and hemoglobin

Myoglobin can bind only one molecule of O_2, because it contains only one heme group. In contrast, hemoglobin can bind four molecules of O_2, one at each of its four heme groups. The degree of saturation (Y) of these oxygen-binding sites on all myoglobin or hemoglobin molecules can vary between zero (all sites are empty) and 100% (all sites are full), as shown in Figure 8.3. [Note: **Pulse oximetry is a noninvasive, indirect method of measuring the oxygen saturation of arterial blood based on differences in light absorption by oxyhemoglobin and deoxyhemoglobin.]**

1. **Oxygen-dissociation curve:** A plot of Y measured at different partial pressures of oxygen (pO_2) is called the oxygen-dissociation curve. [Note: pO_2 may also be represented as PO_2.] The curves for myoglobin and hemoglobin show important differences (see Fig. 8.3). This graph illustrates that **myoglobin has a higher oxygen affinity at all pO_2 values than does hemoglobin.** The partial pressure of oxygen needed to achieve half saturation of the binding sites (P_{50}) is ~1 mm Hg for myoglobin and 26 mm Hg for hemoglobin. The higher the oxygen affinity (that is, the more tightly O_2 binds), the lower the P_{50}.

   a. **Myoglobin: The oxygen-dissociation curve for myoglobin has a hyperbolic shape** (see Fig. 8.3). This reflects the fact that **myoglobin reversibly binds a single molecule of O_2.** Thus, oxygenated (MbO_2) and deoxygenated (Mb) myoglobin exist in a simple equilibrium:

$$Mb + O_2 \rightleftarrows MbO_2$$

   The equilibrium is shifted to the right or to the left as O_2 is added to or removed from the system. [Note: Myoglobin is designed to bind O_2 released by hemoglobin at the low pO_2 found in muscle. Myoglobin, in turn, releases O_2 within the muscle cell in response to oxygen demand.]

   b. **Hemoglobin: The oxygen-dissociation curve for hemoglobin is sigmoidal** in shape (see Fig. 8.3), indicating that the subunits cooperate in binding O_2. **Cooperative binding** of O_2 by the four subunits of hemoglobin means that the binding of an oxygen molecule at one subunit increases the oxygen affinity of the remaining subunits in the same hemoglobin tetramer (Fig. 8.7). Although it is more difficult for the first oxygen molecule to bind to hemoglobin, the subsequent binding of oxygen molecules occurs with high affinity, as shown by the steep upward curve in the region near 20–30 mm Hg (see Fig. 8.3).

E. Allosteric effectors

The ability of hemoglobin to reversibly bind O_2 is affected by the **pO_2, the pH** of the environment, the partial pressure of carbon dioxide **(pCO_2),** and the availability of **2,3-bisphosphoglycerate (2,3-BPG).** These are collectively called **allosteric ("other site") effectors,** because their interaction at one site on the tetrameric hemoglobin molecule causes structural changes that affect the binding of O_2 to the heme iron at other sites on the molecule. [Note: The binding of O_2 to monomeric myoglobin is not influenced by allosteric effectors.]

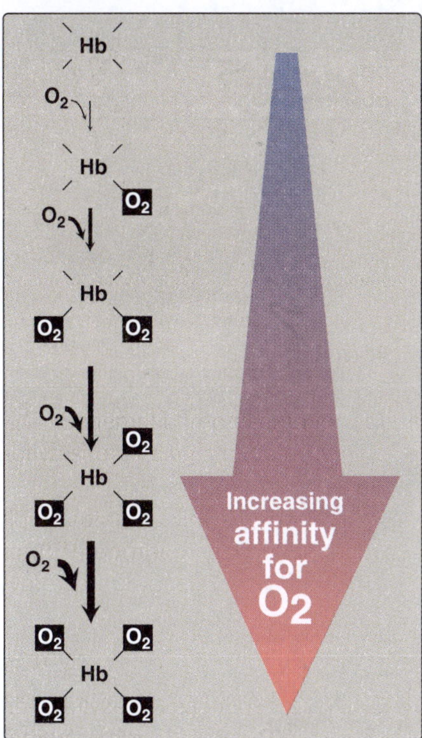

Figure 8.7 Hemoglobin (Hb) binds successive molecules of oxygen (O_2) with increasing affinity.

1. **Oxygen:** The sigmoidal oxygen-dissociation curve reflects specific structural changes that are initiated at one subunit and transmitted to other subunits in the hemoglobin tetramer. The net effect of this cooperativity is that the **affinity of hemoglobin for the last oxygen molecule bound is ~300 times greater than its affinity for the first oxygen molecule bound. Oxygen, then, is an allosteric effector of hemoglobin.** It stabilizes the R form.

   a. **Loading and unloading oxygen:** The **cooperative binding of O_2 allows hemoglobin to deliver more O_2 to the tissues in response to relatively small changes in the pO_2.** This can be seen in Figure 8.3, which indicates pO_2 in the alveoli of the lung and the capillaries of the tissues. For example, in the lung, oxygen concentration is high, and hemoglobin becomes virtually saturated (or "loaded") with O_2. In contrast, in the peripheral tissues,

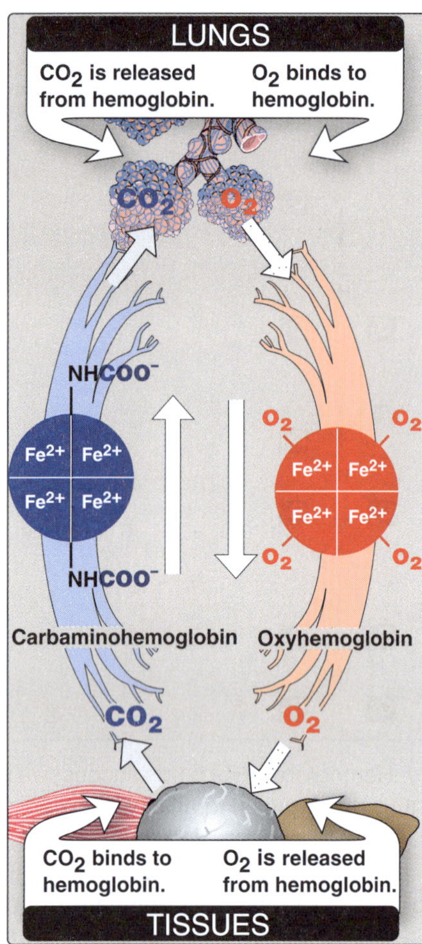

Figure 8.8 Transport of oxygen and carbon dioxide by hemoglobin. Fe = iron.

oxyhemoglobin releases (or "unloads") much of its O_2 for use in the oxidative metabolism of the tissues (Fig. 8.8).

b. **Significance of the sigmoidal oxygen-dissociation curve:** The steep slope of the oxygen-dissociation curve over the range of oxygen concentrations that occur between the lungs and the tissues **permits hemoglobin to carry and deliver O_2 efficiently from sites of high to sites of low pO_2**. A molecule with a hyperbolic oxygen-dissociation curve, such as myoglobin, could not achieve the same degree of O_2 release within this range of pO_2. Instead, it would have maximum affinity for O_2 throughout this oxygen pressure range and, therefore, would deliver no O_2 to the tissues.

2. **Bohr effect:** The release of O_2 from hemoglobin is enhanced when the pH is lowered (proton concentration $[H^+]$ is increased) or when the hemoglobin is in the presence of an increased pCO_2. Both result in decreased oxygen affinity of hemoglobin and, therefore, a shift to the right

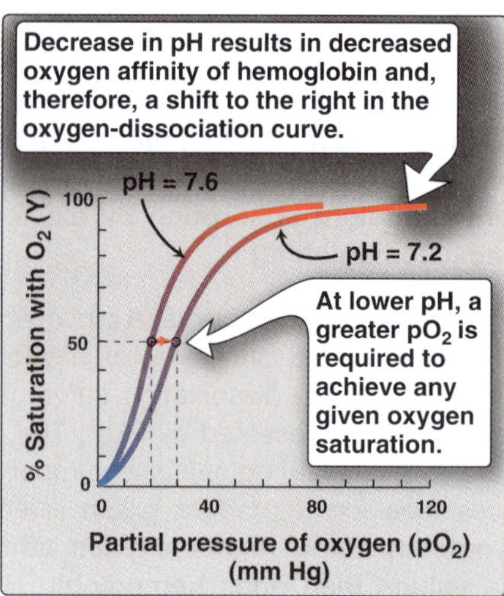

Figure 8.9 Effect of pH on the oxygen affinity of hemoglobin. Protons are allosteric effectors of hemoglobin.

in the oxygen-dissociation curve (Fig. 8.9). Both, then, stabilize the T (deoxy) form. **This change in oxygen binding is called the Bohr effect.** Conversely, raising the pH or lowering the concentration of CO_2 results in a greater oxygen affinity, a shift to the left in the oxygen-dissociation curve, and stabilization of the R (oxy) form.

a. **Source of the protons that lower pH:** The concentration of both H^+ and CO_2 in the capillaries of metabolically active tissues is higher than that observed in alveolar capillaries of the lungs, where CO_2 is released into the expired air. In the tissues, CO_2 is converted by zinc-containing ***carbonic anhydrase*** to carbonic acid:

$$CO_2 + H_2O \rightleftarrows H_2CO_3$$

which spontaneously loses a H^+, becoming bicarbonate (the major blood buffer):

$$H_2CO_3 \rightleftarrows HCO_3^- + H^+$$

The H^+ produced by this pair of reactions contributes to the lowering of pH. This differential pH gradient (that is, lungs having a higher pH and tissues a lower pH) favors the unloading of O_2 in the peripheral tissues and the loading of O_2 in the lung. Thus, the **oxygen affinity of the hemoglobin molecule responds to small shifts in pH** between the lungs and oxygen-consuming tissues, making hemoglobin a more efficient transporter of O_2.

b. **Mechanism of the Bohr effect:** The Bohr effect reflects the fact that the deoxy form of hemoglobin has a greater affinity for H^+ than does oxyhemoglobin. This is caused by ionizable groups such as specific histidine side chains that have a higher pK_a (see p. 74) in deoxyhemoglobin than in oxyhemoglobin. Therefore, an increase in the concentration of H^+ (resulting in a decrease in pH) causes these groups to become protonated (charged) and able to form ionic bonds (salt bridges). These bonds preferentially stabilize the deoxy form of hemoglobin, producing a decrease in oxygen affinity. [Note: Hemoglobin, then, is an important blood buffer.]

The Bohr effect can be represented schematically as:

$$\underset{\text{Oxyhemoglobin}}{HbO_2 + H^+} \rightleftarrows \underset{\text{Deoxyhemoglobin}}{HbH + O_2}$$

where **an increase in H^+ (or a lower pO_2) shifts the equilibrium to the right (favoring deoxyhemoglobin), whereas an increase in pO_2 (or a decrease in H^+) shifts the equilibrium to the left.**

3. **2,3-BPG effect on oxygen affinity:** 2,3-BPG is an important regulator of the binding of O_2 to hemoglobin. It is the most abundant organic phosphate in the RBC, where its concentration is approximately that of hemoglobin. 2,3-BPG is synthesized from an intermediate of the glycolytic pathway (Fig. 8.10; see p. 197 for a discussion of 2,3-BPG synthesis in glycolysis).

   a. **2,3-BPG binding to deoxyhemoglobin:** 2,3-BPG decreases the oxygen affinity of hemoglobin by binding to deoxyhemoglobin but not to oxyhemoglobin. This preferential binding stabilizes the T conformation of deoxyhemoglobin. The effect of binding 2,3-BPG can be represented schematically as:

$$\underset{\text{Oxyhemoglobin}}{HbO_2 + 2,3\text{-BPG}} \rightleftarrows \underset{\text{Deoxyhemoglobin}}{Hb-2,3\text{-BPG} + O_2}$$

   b. **2,3-BPG binding site:** One molecule of 2,3-BPG binds to a pocket, formed by the two β-globin chains, in the center of the deoxyhemoglobin tetramer (Fig. 8.11). This pocket contains several positively charged amino acids that form ionic bonds with the negatively charged phosphate groups of 2,3-BPG. [Note: Replacement of one of these amino acids can result in hemoglobin variants with abnormally high oxygen affinity that may be compensated for by increased RBC production (erythrocytosis).] Oxygenation of hemoglobin narrows the pocket and causes 2,3-BPG to be released.

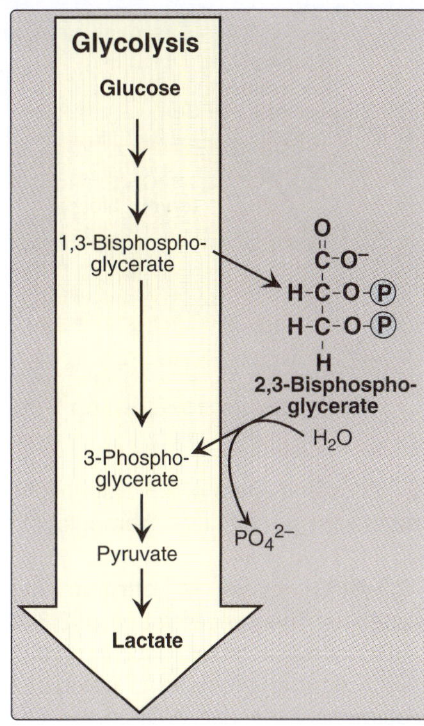

Figure 8.10 Synthesis of 2,3-bisphosphoglycerate. [Note: Ⓟ is a phosphoryl group, PO_3^{2-}.] In older literature, 2, 3-bisphosphoglycerate (2,3-BPG) may be referred to as 2,3-diphosphoglycerate (2,3-DPG).

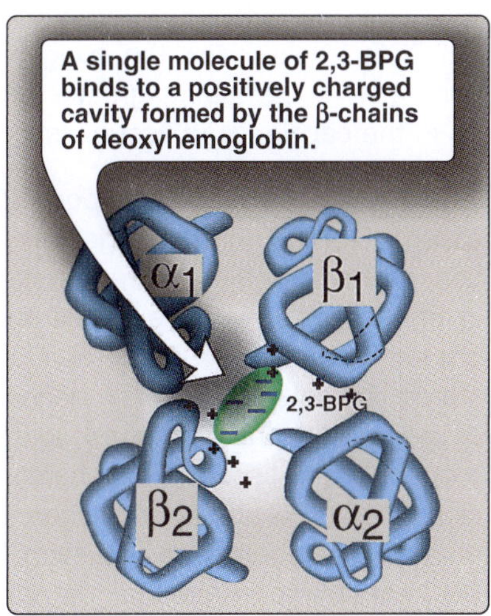

Figure 8.11 Binding of 2,3-bisphosphoglycerate (2,3-BPG) by deoxyhemoglobin.

   c. **Oxygen-dissociation curve shift:** Hemoglobin from which 2,3-BPG has been removed has high oxygen affinity. However, as seen in the RBC, the presence of 2,3-BPG significantly reduces the oxygen affinity of hemoglobin, shifting the oxygen-dissociation curve to the right (Fig. 8.12). This reduced affinity enables hemoglobin to release O_2 efficiently at the partial pressures found in the tissues.

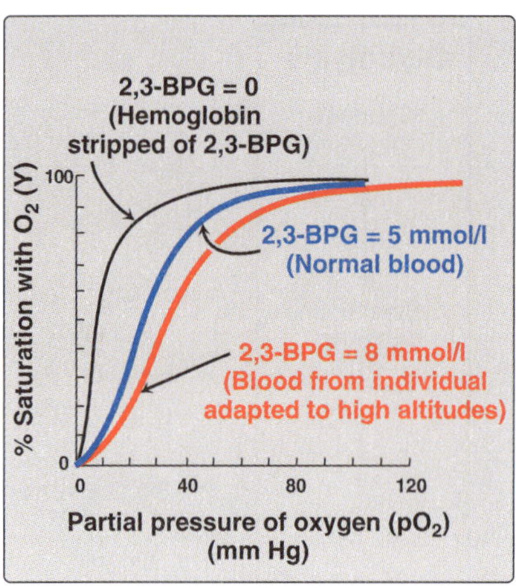

Figure 8.12 Allosteric effect of 2,3-bisphosphoglycerate (2,3-BPG) on the oxygen affinity of hemoglobin.

d. **2,3-BPG levels in chronic hypoxia or anemia:** The concentration of 2,3-BPG in the RBC increases in response to chronic hypoxia, such as that observed in chronic obstructive pulmonary disease (COPD) like emphysema, or at high altitudes during acclimatization, where circulating hemoglobin may have difficulty receiving sufficient O_2. Intracellular levels of 2,3-BPG are also elevated in chronic anemia, in which fewer than normal RBC are available to supply the body's oxygen needs. Elevated 2,3-BPG levels lower the oxygen affinity of hemoglobin, permitting greater unloading of O_2 in the capillaries of tissues (see Fig. 8.12).

e. **2,3-BPG in transfused blood:** 2,3-BPG is essential for the normal oxygen transport function of hemoglobin. However, storing blood in the currently available media results in the gradual depletion of 2,3-BPG. Consequently, stored blood displays an abnormally high oxygen affinity and fails to unload its bound O_2 properly in the tissues. Thus, hemoglobin deficient in 2,3-BPG acts as an oxygen "trap" rather than as an oxygen delivery system. Transfused RBC are able to restore their depleted supplies of 2,3-BPG in 6–24 hours. However, severely ill patients may be compromised if transfused with large quantities of such 2,3-BPG–depleted blood. Stored blood, therefore, is treated with a "**rejuvenation**" solution that rapidly restores 2,3-BPG. [Note: Rejuvenation also restores ATP lost during storage.]

4. **CO_2 binding:** Most of the CO_2 produced in metabolism is hydrated and transported as bicarbonate ion (see Fig. 6.12, p. 77). However, some CO_2 is carried as carbamate bound to the terminal amino groups of hemoglobin (forming carbaminohemoglobin as shown in Fig. 8.8), which can be represented schematically as follows:

$$\text{Hb} - \text{NH}_2 + \text{CO}_2 \rightleftarrows \text{Hb} - \text{NH}_2 - \text{COO}^- + \text{H}^+$$

The binding of CO_2 stabilizes the T, or deoxy, form of hemoglobin, resulting in a decrease in its oxygen affinity and a right shift in the oxygen-dissociation curve. In the lungs, CO_2 dissociates from the hemoglobin and is released in the breath.

5. **CO binding: Carbon monoxide (CO) binds tightly (but reversibly) to the hemoglobin iron, forming carboxyhemoglobin.** When CO binds to one or more of the four heme sites, hemoglobin shifts to the R conformation, causing the remaining heme sites to bind O_2 with high affinity. This shifts the oxygen-dissociation curve to the left and changes the normal sigmoidal shape toward a hyperbola. As a result, the affected hemoglobin is unable to release O_2 to the tissues (Fig. 8.13). **[Note: The affinity of hemoglobin for CO is 220 times greater than for O_2**. Consequently, even minute concentrations of CO in the environment can produce toxic concentrations of carboxyhemoglobin in the blood. For example, increased levels of CO are found in the blood of tobacco smokers**.** **CO toxicity appears to result from a combination of tissue hypoxia and direct CO-mediated damage at the cellular level.**] CO poisoning is treated with 100% O_2 at high pressure (hyperbaric oxygen therapy), which facilitates the dissociation of CO from the hemoglobin. [Note: **CO inhibits Complex IV of the electron transport chain** (see p. 166).] In addition to O_2, CO_2, and CO, nitric oxide gas (NO) also is carried by hemoglobin. NO is a potent vasodilator (see p. 252). It can be taken up (salvaged) or released from RBC,

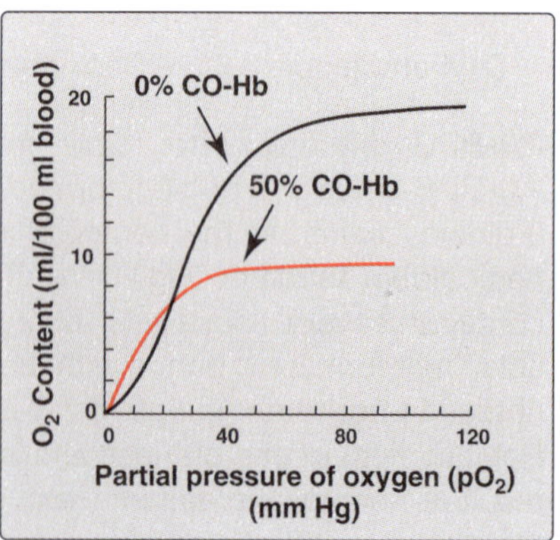

Figure 8.13 Effect of carbon monoxide (CO) on the oxygen affinity of hemoglobin. CO competes with O_2 for binding the heme iron. CO-Hb = carboxyhemoglobin (carbon monoxyhemoglobin).

thereby modulating NO availability and influencing vessel diameter.

F. Minor hemoglobins

It is important to remember that human hemoglobin A (**HbA**) is just one member of a functionally and structurally related family of proteins, the hemoglobins (Fig. 8.14). Each of these oxygen-carrying proteins is a tetramer, composed of two α-globin (or α-like) polypeptides and two β-globin (or β-like) polypeptides. Certain hemoglobins, such as **HbF,** are normally synthesized only during fetal development, whereas others, such as **HbA$_2$**, are synthesized in the adult, although at low levels compared with HbA. HbA can also become modified by the covalent addition of a hexose (see 3. below).

1. **Fetal hemoglobin: HbF is a tetramer consisting of two α chains identical to those found in HbA, plus two γ chains** ($\alpha_2\gamma_2$; see Fig. 8.14). The γ chains are members of the β-globin gene family (see pp. 106–107).

   a. **HbF synthesis during development: In the first month after conception, embryonic hemoglobins such as Hb Gower 1, composed of two α-like zeta (ζ) chains and two β-like epsilon (ε) chains ($\zeta_2\varepsilon_2$), are synthesized by the embryonic yolk sac**. In the fifth week of gestation, the site of globin synthesis shifts, first to the liver and then to the marrow, and the primary product is HbF. **HbF is the major hemoglobin found in the fetus and newborn,** accounting for ~60% of the total hemoglobin in the RBC during the last months of fetal life (Fig. 8.15). HbA synthesis starts in the bone marrow at about the eighth month of pregnancy and gradually replaces HbF. Figure 8.15 shows the relative production of each type of hemoglobin chain during fetal and postnatal life. [Note: **HbF represents <2% of the hemoglobin in most adults and is concentrated in RBC known as F cells.**]

Form	Chain composition	Fraction of total hemoglobin
HbA	$\alpha_2\beta_2$	90%
HbA$_2$	$\alpha_2\delta_2$	2%–3%
HbF	$\alpha_2\gamma_2$	<2%
HbA$_{1c}$	$\alpha_2\beta_2$-glucose	4%–6%

Figure 8.14 Normal adult human hemoglobins. HbA$_{1c}$ is a subtype of HbA (or, HbA$_1$). [Note: The α chains in these hemoglobins are identical.] Hb = hemoglobin.

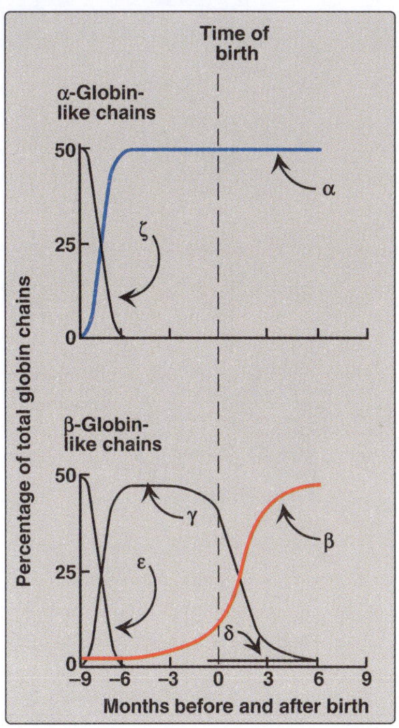

Figure 8.15 Developmental changes in globin production.

   b. **2,3-BPG binding to HbF:** Under physiologic conditions, HbF has a higher oxygen affinity than does HbA as a result of HbF only weakly binding 2,3-BPG. [Note: The γ-globin chains of HbF lack some of the positively charged amino acids that are responsible for binding 2,3-BPG in the β-globin chains.] **Because 2,3-BPG serves to reduce the oxygen affinity of hemoglobin, the weaker interaction between 2,3-BPG and HbF results in a higher oxygen affinity for HbF relative to HbA**. This is an important factor (besides the difference in PO_2 between maternal and fetal blood) for transfer of oxygen from maternal blood that contains HbA to fetal blood that contains HbF. In contrast, if both HbA and HbF are stripped of their 2,3-BPG, they then have a similar oxygen affinity. So, the higher oxygen affinity of HbF facilitates the transfer of O_2 from the maternal circulation across the placenta to the RBC of the fetus.

2. **Hemoglobin A$_2$: HbA$_2$ is a minor component of normal adult hemoglobin, first appearing shortly before birth and, ultimately, constituting ~2% of the total hemoglobin**. It is composed of **two α-globin chains and two δ-globin chains** ($\alpha_2\delta_2$; see Fig. 8.14).

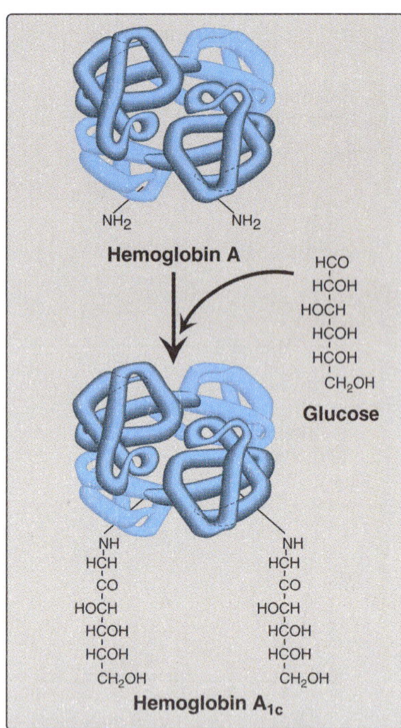

Figure 8.16 Nonenzymatic addition of glucose to hemoglobin. The nonenzymatic addition of a sugar to a protein is referred to as glycation.

Hemoglobin A_{1c}

Under physiologic conditions, HbA is slowly **glycated (nonenzymically condensed** with a hexose), the extent of glycation being dependent on the plasma concentration of the hexose. The most abundant form of glycated hemoglobin is HbA_{1c}. It has glucose residues attached predominantly to the amino groups of the N-terminal valines of the β-globin chains (Fig. 8.16). **Increased amounts of HbA_{1c} are found in RBC of patients with diabetes mellitus,** because their HbA has contact with higher glucose concentrations during the 120-day lifetime of these cells. (See p. 277 for a discussion of the use of HbA_{1c} levels in assessing average blood glucose levels in patients with diabetes.)

III. GLOBIN GENE ORGANIZATION

To understand diseases resulting from genetic alterations in the structure or synthesis of hemoglobin, it is necessary to grasp how the hemoglobin genes, which direct the synthesis of the different globin chains, are structurally organized into gene families and also how they are expressed.

A. α-Gene family

The genes coding for the α-globin and β-globin subunits of the hemoglobin chains occur in two separate gene clusters (or families) located on two different chromosomes (Fig. 8.17). The α-gene cluster on **chromosome 16 contains two genes for the α-globin chains**. It also contains the **ζ gene** that is expressed early in development as an α-globin-like component of embryonic hemoglobin. [Note: Globin gene families **also contain globin-like genes that are not expressed**, that is, their genetic information is not used to produce globin chains. These are called **pseudogenes**.]

B. β-Gene family

A single gene for **the β-globin chain is located on chromosome 11** (see Fig. 8.17). There are an additional four β-globin-like genes: the **ε gene** (which, like the ζ gene, is expressed early in embryonic development**), two γ genes** (G_γ and A_γ that are expressed in HbF), and the δ gene that codes for the globin chain found in the minor adult hemoglobin HbA_2.

C. Steps in globin chain synthesis

Expression of a globin gene begins in the nucleus of RBC precursors, where the DNA sequence encoding the gene is transcribed. The ribonucleic acid (RNA) produced by transcription is actually a precursor of the messenger RNA (mRNA) that is used as a template for the synthesis of a globin chain. Before it can serve this function, two noncoding stretches

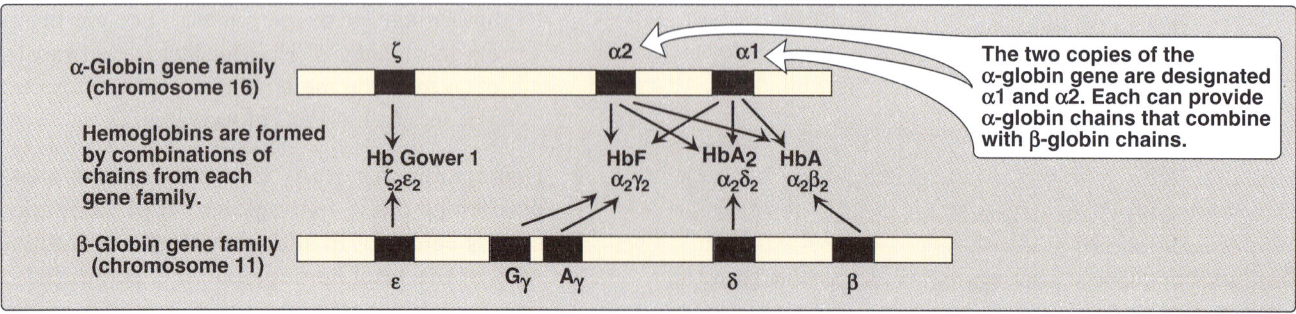

Figure 8.17 Organization of the globin gene families. Hb = hemoglobin.

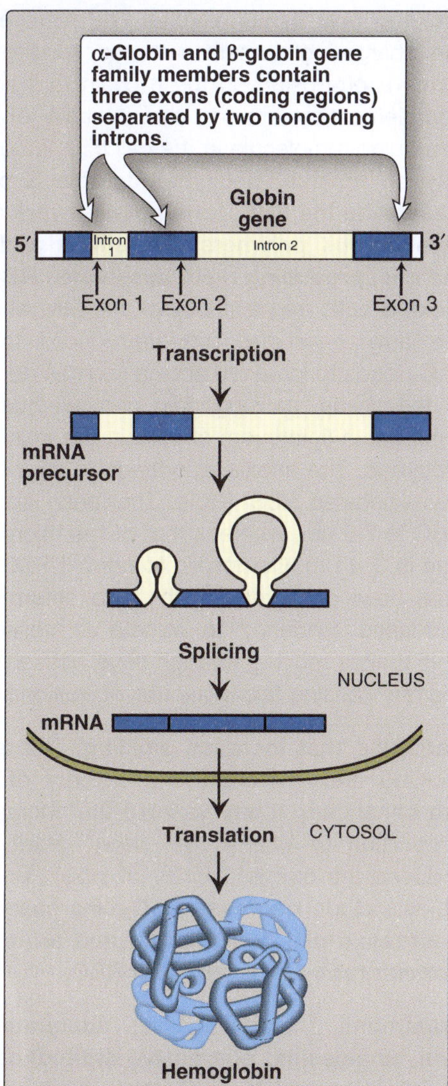

Figure 8.18 Synthesis of globin chains. mRNA = messenger ribonucleic acid.

of RNA (introns) must be removed from the mRNA precursor sequence and the remaining three fragments (exons) joined in a linear manner. The resulting mature mRNA enters the cytosol, where its genetic information is translated, producing a globin chain. (A summary of this process is shown in Figure 8.18. A more detailed description of gene expression is presented in Chapter 38.)

IV. HEMOGLOBINOPATHIES

Hemoglobinopathies are defined as a group of genetic disorders caused by production of a structurally abnormal hemoglobin molecule, synthesis of insufficient quantities of normal hemoglobin, or, rarely, both. Sickle cell anemia (HbS), **hemoglobin C disease** (HbC), **hemoglobin SC disease** (HbS + HbC = HbSC), and the **thalassemias** are representative hemoglobinopathies that can have severe clinical consequences. The first three conditions result from production of hemoglobin with an altered amino acid sequence (qualitative hemoglobinopathy), whereas the thalassemias are caused by decreased production of normal hemoglobin (quantitative hemoglobinopathy).

A. Sickle cell anemia (hemoglobin S disease)

Sickle cell anemia, the most common of the RBC sickling diseases, is a genetic disorder caused by a single nucleotide substitution (a point mutation, see p. 636) in the gene for β-globin. It is the most common inherited blood disorder in the United States, affecting 50,000 Americans. It occurs primarily in the African American population, affecting 1 in 500 newborn African American infants. Sickle cell anemia is **an autosomal-recessive disorder**. It occurs in individuals who have inherited two mutant genes (one from each parent) that code for synthesis of the β chains of the globin molecules. [Note: **The mutant β-globin chain is designated β^S, and the resulting hemoglobin, $\alpha_2\beta^S{}_2$, is referred to as HbS.**] An infant does not begin showing symptoms of the disease until sufficient HbF has been replaced by HbS so that sickling can occur (see p. 109). Sickle cell anemia is characterized **by lifelong episodes of pain ("crises"), chronic hemolytic anemia with associated hyperbilirubinemia** (see p. 422)**, and increased susceptibility to infections, usually beginning in infancy.** [Note: The lifetime of a RBC in sickle cell anemia is <20 days, compared with 120 days for normal RBC, hence, the anemia.] Other symptoms include **acute chest syndrome, stroke, splenic and renal dysfunction, and bone changes** due to marrow hyperplasia. Life expectancy is reduced. Heterozygotes, representing 1 in 12 African Americans, have one normal and one sickle cell gene. The blood cells of such heterozygotes contain both HbS and HbA, and these individuals have sickle cell trait. They usually do not show clinical symptoms (but may under conditions of extreme physical exertion with dehydration) and can have a normal life span.

1. **Amino acid substitution in HbS β chains:** A molecule of HbS contains two normal α-globin chains and two mutant β-globin chains (β^S), in which **glutamate at position six has been replaced with valine** (Fig. 8.19). Therefore, during electrophoresis at alkaline pH, HbS migrates more slowly toward the anode (positive electrode) than does HbA (Fig. 8.20). This altered mobility of HbS is a result of the absence of the negatively charged glutamate residues in the two β chains, thereby rendering HbS less negative than HbA. [Note: **Electrophoresis**

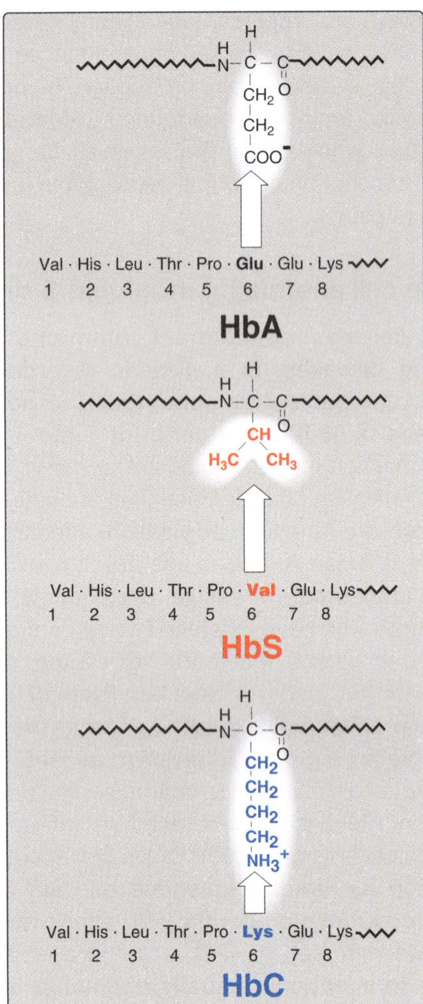

Figure 8.19 Amino acid substitutions in hemoglobin S (HbS) and hemoglobin C (HbC).

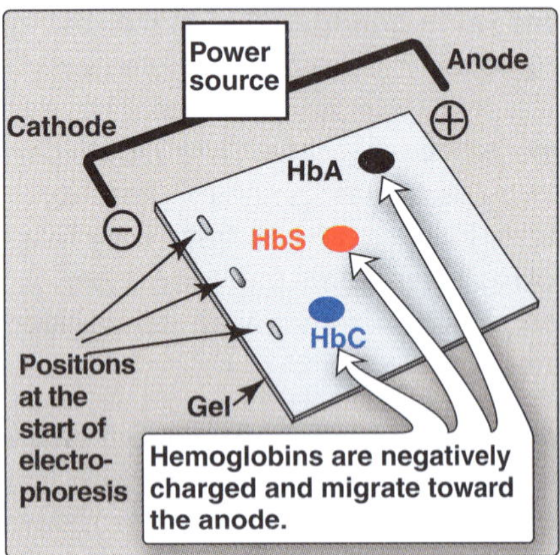

Figure 8.20 Diagram of hemoglobins (HbA), (HbS), and (HbC) after electrophoresis.

of hemoglobin obtained from lysed RBC is routinely used in the diagnosis of sickle cell trait and sickle cell anemia (or, sickle cell disease). DNA analysis also is used (see p. 678).]

2. **Sickling and tissue anoxia:** The replacement of the charged glutamate with the nonpolar valine forms a protrusion on the β chain that fits into a complementary site on the β chain of another hemoglobin molecule in the cell (Fig. 8.21). **At low oxygen tension, deoxyhemoglobin S polymerizes inside the RBC, forming a network of insoluble fibrous polymers that stiffen and distort the cell, producing rigid, misshapen RBC. Such sickled cells** frequently block the flow of blood in the narrow capillaries. This interruption in the supply of O_2 leads to localized anoxia (oxygen deprivation) in the tissue, causing pain and eventually ischemic death (infarction) of cells in the vicinity of the blockage. The anoxia also leads to an increase in deoxygenated HbS. [Note: The mean diameter of RBC is 7.5 μm, whereas that of the microvasculature is 3–4 μm. Compared to normal RBC, sickled cells have a decreased ability to deform and an increased tendency to adhere to vessel walls. This makes moving through small vessels difficult, thereby causing microvascular occlusion.]

3. **Variables that increase sickling:** The extent of sickling and, therefore, the severity of disease are enhanced by any variable that increases the proportion of HbS in the deoxy state (that is, reduces the oxygen affinity of HbS). These variables include **decreased pO_2, increased pCO_2, decreased pH, dehydration, and an increased concentration of 2,3-BPG in RBC.**

4. **Treatment:** Therapy involves **adequate hydration, analgesics, aggressive antibiotic therapy if infection is present, and transfusions** in patients at high risk for fatal occlusion of blood vessels. Intermittent transfusions with packed RBC reduce the risk of stroke, but the benefits must be weighed against the complications of transfusion, which include iron overload that can result in hemosiderosis (see p. 559), bloodborne infections, and immunologic complications. **Hydroxyurea (hydroxycarbamide), an antitumor drug, is therapeutically useful because it increases circulating levels of HbF, which decreases RBC sickling**. This leads to decreased frequency of painful crises and reduces mortality. Stem cell transplantation is possible. [Note: The morbidity and mortality associated with sickle cell anemia has led to its inclusion in newborn screening panels to allow prophylactic antibiotic therapy to begin soon after the birth of an affected child.]

5. **Possible selective advantage of the heterozygous state:** The high frequency of the β^S mutation among black Africans, despite its damaging effects in the homozygous state, suggests that a selective

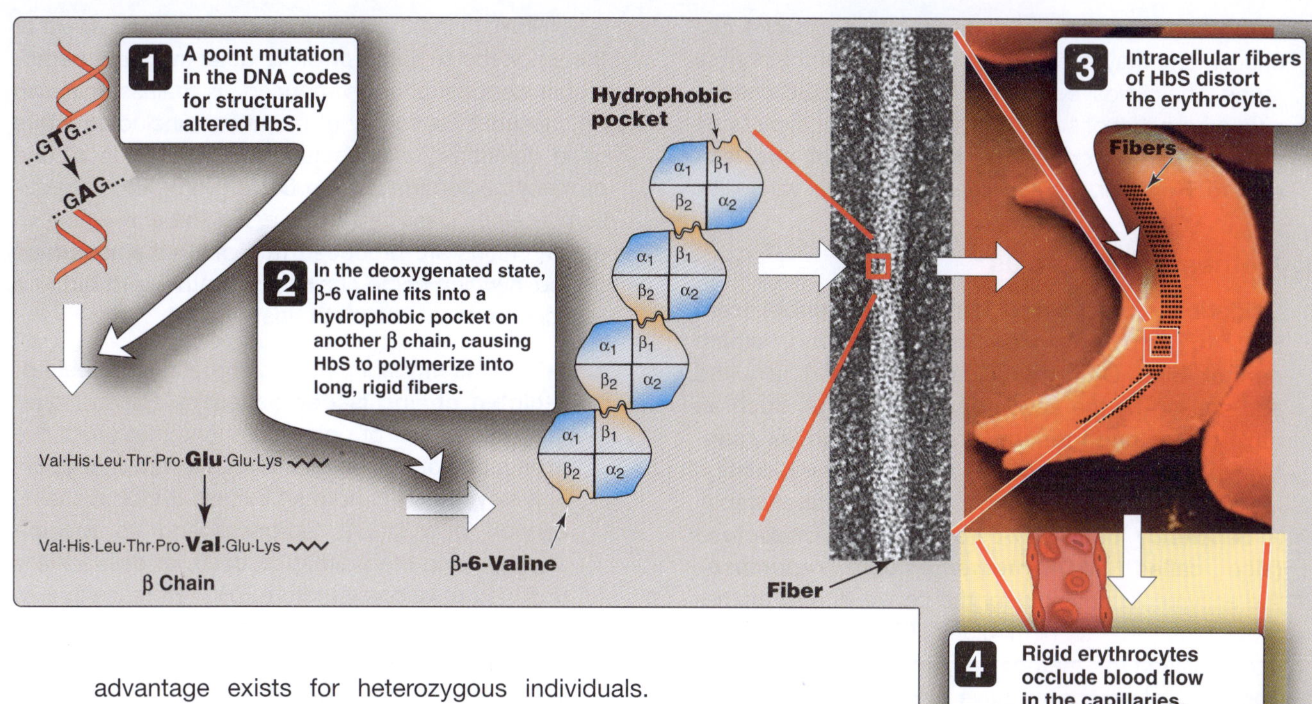

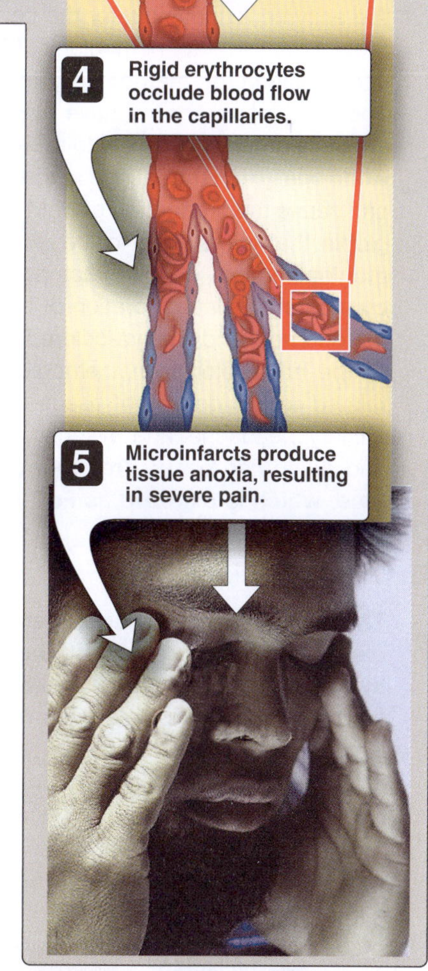

Figure 8.21 Molecular and cellular events leading to sickle cell crisis. HbS = hemoglobin S.

advantage exists for heterozygous individuals. For example, **heterozygotes for the sickle cell gene are less susceptible to the severe malaria caused by the parasite** Plasmodium falciparum. This organism spends an obligatory part of its life cycle in the RBC. One theory is that because these cells in individuals heterozygous for HbS, like those in homozygotes, have a **shorter life span than normal, the parasite cannot complete the intracellular stage of its development**. This may provide a selective advantage to heterozygotes living in regions where malaria is a major cause of death. For example, in Africa, the geographic distribution of sickle cell anemia is similar to that of malaria.

B. Hemoglobin C disease

Like HbS, HbC is a hemoglobin variant that has a single amino acid substitution in the sixth position of the β-globin chain (see Fig. 8.19). In HbC, however, a **lysine is substituted for the glutamate** (as compared with a valine substitution in HbS). [Note: This substitution causes HbC to move more slowly toward the anode than HbA or HbS does (see Fig. 8.20).] Rare patients homozygous for HbC generally have a **relatively mild, chronic hemolytic anemia**. They do not suffer from infarctive crises, and no specific therapy is required.

C. Hemoglobin SC disease

HbSC disease is another of the RBC sickling diseases. **In this disease, some β-globin chains have the sickle cell mutation, whereas other β-globin chains carry the mutation found in HbC disease.** [Note: Patients with HbSC disease are **doubly heterozygous**. They are called **compound heterozygotes** because both of their β-globin genes are abnormal, although different from each other.] Hemoglobin levels tend to be higher in HbSC disease than in sickle cell anemia and may even be at the low end of the normal range. The clinical course of adults with HbSC anemia differs from that of sickle cell anemia in that symptoms such as **painful crises are less frequent and less severe**. However, there is significant clinical variability.

D. Methemoglobinemias

Oxidation of the heme iron in hemoglobin from Fe^{2+} to Fe^{3+} produces methemoglobin, which cannot bind O_2. This oxidation may be acquired and caused by the action of certain **drugs, such as nitrates**, or **endogenous products such as reactive oxygen species** (see Chapter 47). The oxidation may also result from congenital defects, for example, a **deficiency of *NADH-cytochrome b_5 reductase* (also called *NADH-methemoglobin reductase*)**, the enzyme responsible for the conversion of methemoglobin (Fe^{3+}) to hemoglobin (Fe^{2+}), leads to the accumulation of methemoglobin (Fig. 8.22). [Note: The RBC of newborns have approximately half the capacity of those of adults to reduce methemoglobin.] Additionally, rare mutations in the α- or β-globin chain can cause the production of HbM, an abnormal hemoglobin that is resistant to the *reductase*. The methemoglobinemias are characterized by **"chocolate cyanosis"** (a blue coloration of the skin and mucous membranes and brown-colored blood) as a result of the dark-colored methemoglobin. Symptoms are related to the degree of tissue hypoxia and include anxiety, headache, and dyspnea. In rare cases, coma and death can occur. **Treatment is with methylene blue, which is oxidized as Fe^{3+} is reduced.**

E. Thalassemias

The thalassemias are hereditary hemolytic diseases in which an imbalance occurs in the synthesis of globin chains. As a group, they are the most common single-gene disorders in humans. Normally, synthesis of the α- and β-globin chains is coordinated, so that each α-globin chain has a β-globin chain partner. This leads to the formation of $\alpha_2\beta_2$ (HbA). In the thalassemias, the synthesis of either the α- or the β-globin chain is defective, and hemoglobin concentration is reduced. A thalassemia can be caused by a variety of mutations, including entire gene deletions, or substitutions or deletions of one of many nucleotides in the DNA. [Note: Each thalassemia can be classified as either a disorder in which no globin chains are produced **(α^0- or β^0-thalassemia)**, or one in which some chains are synthesized but at a reduced level **(α^+- or β^+-thalassemia).]**

1. **β-Thalassemias:** In these disorders, **synthesis of β-globin chains is decreased or absent**, typically as a result of point mutations that affect the production of functional mRNA. However, α-globin chain synthesis is normal. Excess α-globin chains cannot form stable tetramers and so precipitate, causing the premature death of cells initially destined to become mature RBC. Increase in $\alpha_2\delta_2$ (HbA_2) and $\alpha_2\gamma_2$ (HbF) also occurs. There are only two copies of the β-globin gene in each cell (one on each chromosome 11). Therefore, individuals with β-globin gene defects have either **β-thalassemia trait (β-thalassemia minor)** if they have only one defective β-globin gene or β-thalassemia major (Cooley anemia) if both genes are defective (Fig. 8.23). Because the β-globin gene is not expressed until late in prenatal development, the physical manifestations of β-thalassemias appear only several months after birth. Those individuals with β-thalassemia minor make some β chains and usually do not require specific treatment. However, those infants born with β-thalassemia major are seemingly healthy at birth but **become severely anemic, usually during the first or second year**

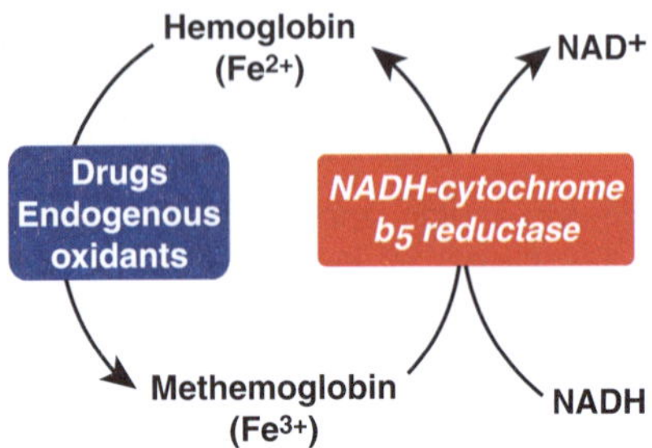

Figure 8.22 Formation of methemoglobin and its reduction to hemoglobin by *NADH-cytochrome b_5 reductase*.

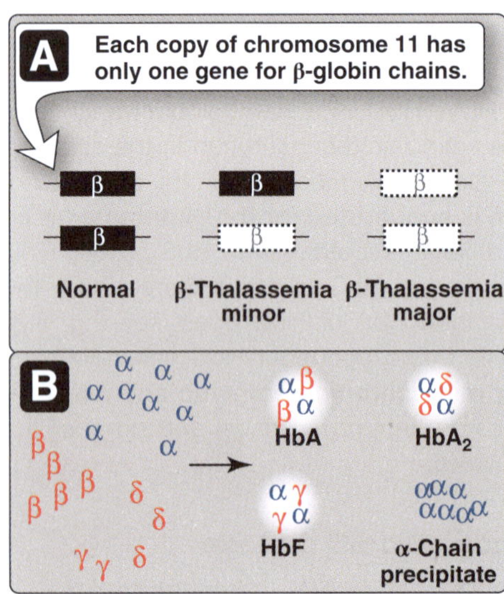

Figure 8.23 **A.** β-Globin gene mutations in the β-thalassemias. **B.** Hemoglobin (Hb) tetramers formed in β-thalassemias.

of life, due to ineffective erythropoiesis. **Skeletal changes as a result of extramedullary hematopoiesis** also are seen. These patients **require regular transfusions of blood.** [Note: Although this treatment is lifesaving, the cumulative effect of the transfusions **is iron overload**. Use of iron **chelation** therapy has improved morbidity and mortality.] The only curative option available is **hematopoietic stem cell transplantation.**

2. **α-Thalassemias:** In these disorders, **synthesis of α-globin chains is decreased or absent, typically as a result of deletional mutations**. Because each individual's genome contains four copies of the α-globin gene (two on each chromosome 16), there are several levels of α-globin chain deficiencies (Fig. 8.24). If one of the four genes is defective, the individual is termed **a "silent" carrier of α-thalassemia**, because no physical manifestations of the disease occur. If two α-globin genes are defective, the individual is designated as having **α-thalassemia trait.** If three α-globin genes are defective, the individual has **hemoglobin H (β_4) disease**, a hemolytic anemia of variable severity. If all four α-globin genes are defective, **hemoglobin Bart (γ_4) disease with hydrops fetalis and fetal death results,** because α-globin chains are required for the synthesis of HbF. [Note: Heterozygote advantage against malaria is seen in both α- and β-thalassemias.]

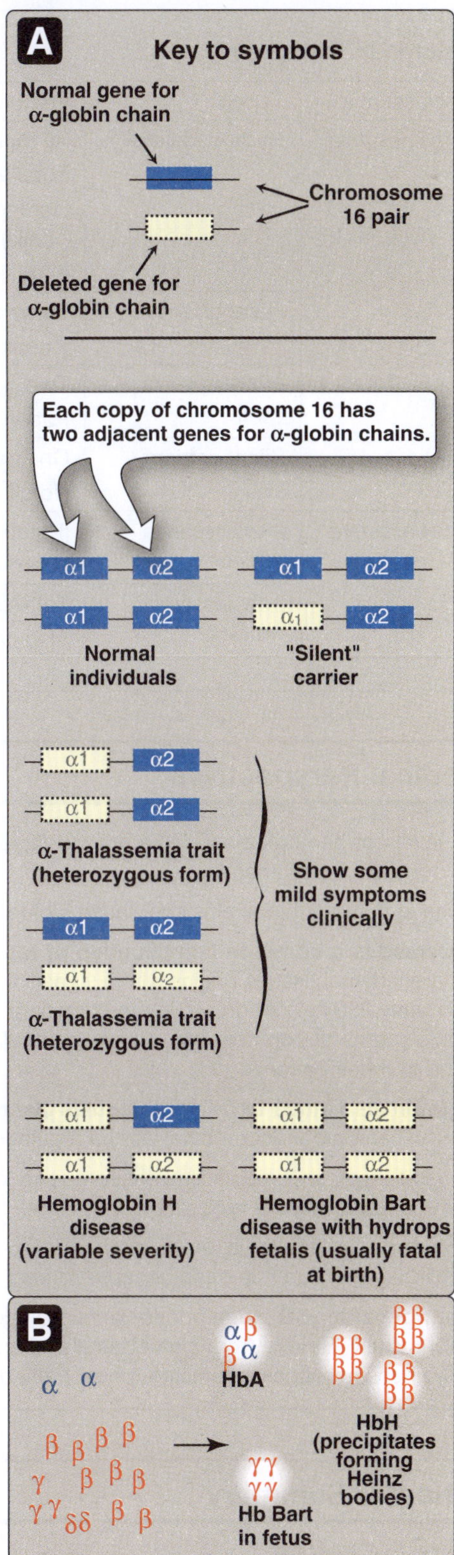

Figure 8.24 **A.** α-Globin gene deletions in the α-thalassemias. **B.** Hemoglobin (Hb) tetramers formed in α-thalassemias.

Features of thalassemias

Thalassemia	Types	Features
α-thalassemia	Hydrops fetalis	All four α genes deleted Total absence of α chain synthesis Hemoglobin Bart's (tetramer of γ chains, is the main hemoglobin in the red cells) Fatal (in utero or shortly after birth)
	Hemoglobin H disease	Three genes deleted β chains accumulate, and form tetramers known as hemoglobin H
	α-thalassemia trait	Two genes deleted Patients have a mild, microcytic, hypochromic anemia
	Silent carriers	One gene deleted No clinical manifestations
β-thalassemia	β-thalassemia minor	β-globin chain synthesis is significantly reduced There is a compensatory increase in the amount of hemoglobins A2 and F
	β-thalassemia major	Total absence of β chain synthesis Also called Cooley's anemia

V. Medical Perspectives

Normal levels of hemoglobin in India are 12–15 g/dl for women and 13–17 g/dl for men. Low levels of hemoglobin are seen in iron-deficiency anemia.

Mutations in hemoglobin chains can cause **sickle cell anemia** and various forms of **thalassemia.**

Iron overload is a common complication of repeated blood transfusions given to thalassemia patients. Iron deposition in parenchymal tissues begins within 1 year of starting the regular transfusions. One unit of transfused RBCs contains approximately 250 mg of iron, while the body cannot excrete more than 1 mg of iron per day. If a patient receives 25 units per year, 5 grams of iron per year will be accumulated in the absence of chelation. The intestinal iron absorption is also increased in these patients.

Hemoglobin also binds to other molecules like nitric oxide (NO). NO dilates blood vessels, which reduces blood pressure. Thus, hemoglobin contributes to the regulation of blood pressure by distributing nitric oxide through the blood.

After muscle injury, myoglobin is released quickly due to its low molecular weight and is an early marker of both myocardial infarction and rhabdomyolysis.

Myoglobinuria in adults is most often due to trauma or muscle necrosis because of prolonged immobilization. Myoglobinuria in children is often because of viral infections.

Carbon monoxide (CO) is an odorless and colorless gas that is produced when fuels such as gas, oil, coal, and wood do not burn fully. Smoke from cigarettes also produces CO. The most common symptoms of CO poisoning are headache, dizziness, weakness, vomiting, chest pain, and confusion.

VI. Chapter Summary

Hemoglobin A (HbA), the major hemoglobin in adults, is composed of four polypeptide chains (two α chains and two β chains, $\alpha_2\beta_2$) held together by noncovalent interactions (Fig. 8.25). The subunits occupy different relative positions in deoxyhemoglobin compared with oxyhemoglobin.

The **deoxy form** of Hb is called the **"T,"** or **taut (tense), conformation**. It has a constrained structure that limits the movement of the polypeptide chains. The T form is the **low-oxygen-affinity form** of Hb.

The binding of oxygen (O_2) to the heme iron causes rupture of some of the ionic and hydrogen bonds and movement of the dimers. This leads to a structure called the **"R,"** or **relaxed, conformation**. The R form is the **high-oxygen-affinity form** of Hb.

The **oxygen-dissociation curve** for Hb is **sigmoidal** in shape (in contrast to that of **myoglobin**, which is **hyperbolic**), indicating that the subunits cooperate in binding O_2. The binding of an oxygen molecule at one heme group increases the oxygen affinity of the remaining heme groups in the same Hb molecule (**cooperativity**).

Hb's ability to bind O_2 reversibly is affected by the partial pressure of oxygen (**pO_2**), the **pH** of the environment, the partial pressure of carbon dioxide (**pCO_2**), and the availability of **2,3-bisphosphoglycerate** (**2,3-BPG**). For example, the release of O_2 from Hb is enhanced when the pH is lowered or the pCO_2 is increased (the **Bohr effect**), such as in **exercising muscle**, and the oxygen-dissociation curve of Hb is shifted to the right.

To cope long-term with the effects of **chronic hypoxia** or **anemia**, the concentration of **2,3-BPG** in **red blood cells** increases. **2,3-BPG** binds to the Hb and decreases its oxygen affinity. It therefore also shifts the oxygen-dissociation curve to the right.

Fetal hemoglobin (HbF) binds 2,3-BPG less tightly than does HbA and has a higher oxygen affinity.

Carbon monoxide (**CO**) binds tightly (but reversibly) to the Hb iron, forming **carboxyhemoglobin**.

Hemoglobinopathies are disorders primarily caused either by production of a structurally abnormal Hb molecule as in **sickle cell anemia** or synthesis of insufficient quantities of normal Hb subunits as in the **thalassemias** (Fig. 8.26).

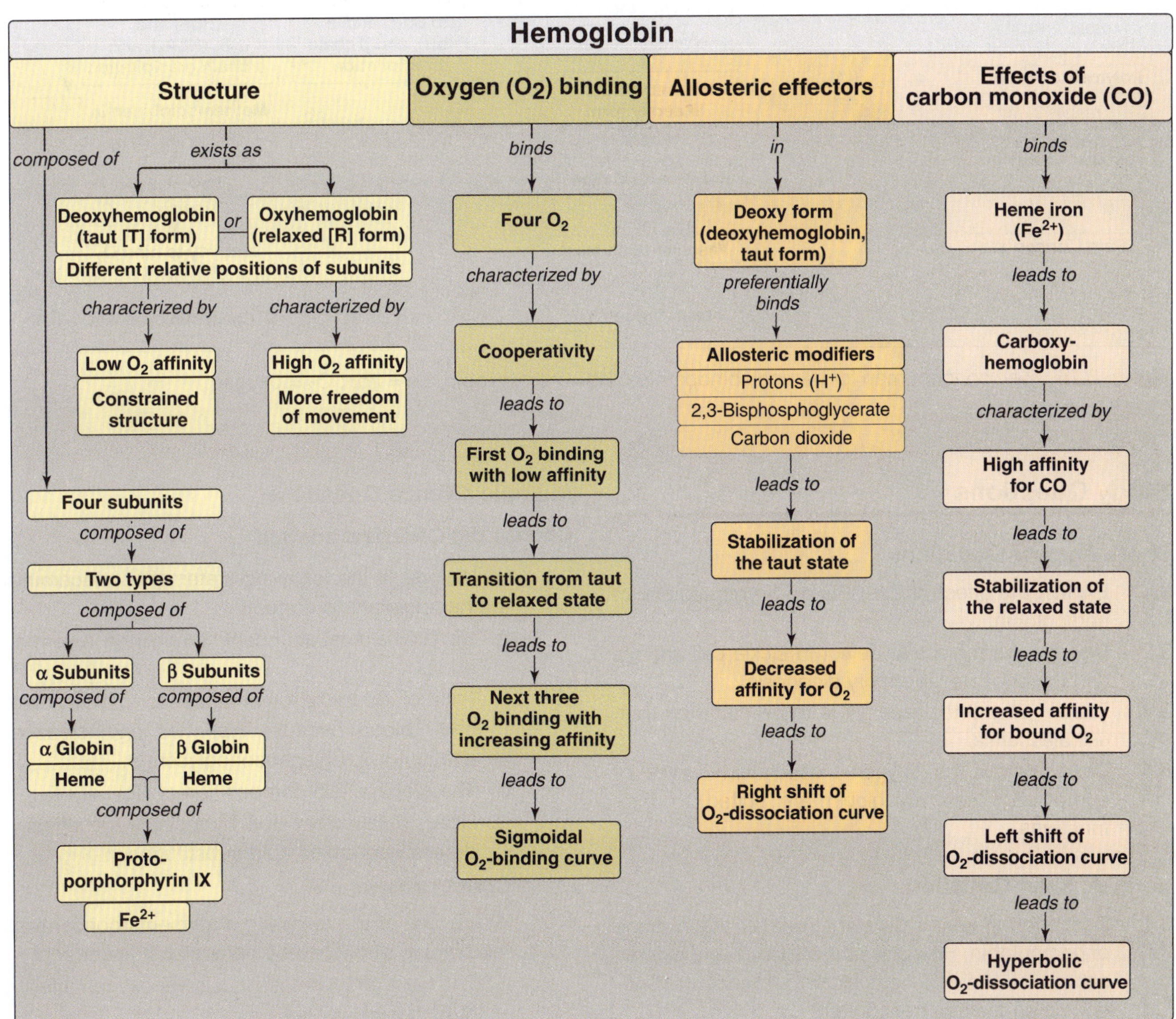

Figure 8.25 Key concept map for hemoglobin structure and function. Fe^{2+} = ferrous iron.

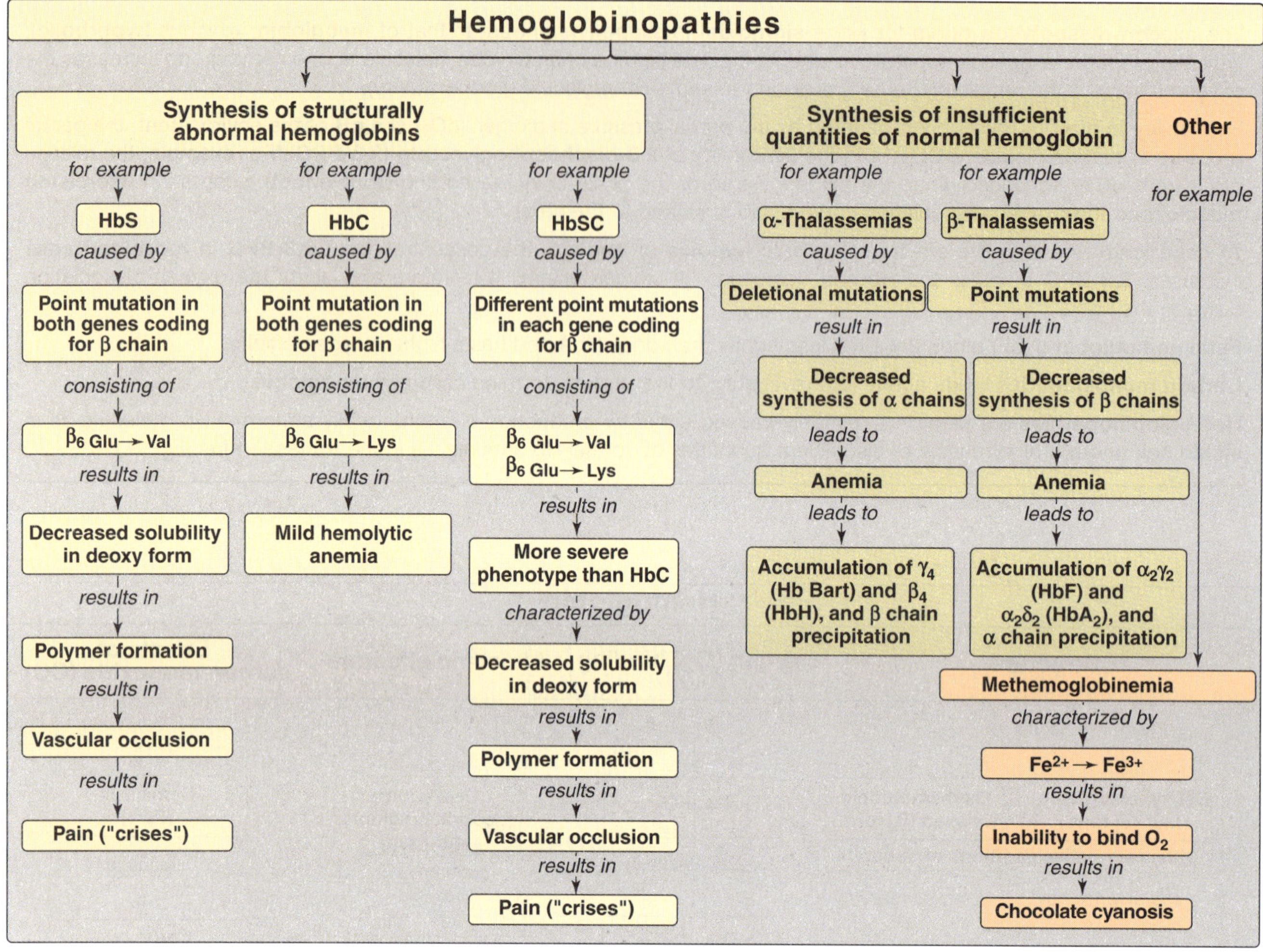

Figure 8.26 Key concept map for hemoglobinopathies. Hb = hemoglobin; Fe = iron; O_2 = oxygen.

Study Questions

Short-Answer Questions

8.1. Explain the effect of 2,3-BPG on oxygen affinity of hemoglobin.

8.2. Discuss the molecular defect in sickle cell anemia and describe its laboratory diagnosis.

8.3. Explain why β-thalassemia is more common than α-thalassemia.

8.4. Discuss how the oxygen dissociation curves of hemoglobin and myoglobin are suited to their functions.

Long-Answer Question

8.1. Describe different orders of structural organization of hemoglobin. How does the structure and oxygen saturation curve of hemoglobin influence its function as an oxygen transporter?

Multiple-Choice Questions

Choose the ONE best answer.

8.1. Which one of the following statements concerning the hemoglobins is correct?

A. HbA is the most abundant hemoglobin in normal adults.

B. Fetal blood has a lower affinity for oxygen than adult blood because HbF has an increased affinity for 2,3-bisphosphoglycerate.

C. The globin chain composition of HbF is $\alpha_2\delta_2$.

D. HbA_{1c} differs from HbA by a single, genetically determined amino acid substitution.

Correct answer = A.

8.2. Which one of the following statements concerning the binding of oxygen by hemoglobin is correct?

A. The Bohr effect results in a lower oxygen affinity at higher pH values.

B. Carbon dioxide increases the oxygen affinity of hemoglobin by binding to the C-terminal groups of the polypeptide chains.
C. The oxygen affinity of hemoglobin increases as the percentage saturation increases.
D. The hemoglobin tetramer binds four molecules of 2,3-bisphosphoglycerate.

Correct answer = C.

8.3. Statement 1: Homozygosity for Hemoglobin C results in sickle shaped RBCs.
Statement 2: In Hemoglobin C, polar glutamic acid residue is substituted by a polar lysine residue.
A. Both the Statements are correct, and Statement 2 is the correct explanation for Statement 1
B. Both the Statements are correct, but Statement 2 is NOT the correct explanation for Statement 1
C. Statement 1 is correct, but Statement 2 is NOT correct
D. Statement 1 is NOT correct, but Statement 2 is correct

Correct answer = D.

8.4. Which of the following is TRUE, regarding HbF?
A. It is produced in the embryonic life
B. It has a similar beta-chain compared to HbA
C. It has a p50 lower than that of HbA
D. It has a high net positive charge compared to HbA

Correct answer = C.

8.5. The effect of protons on oxygen binding to hemoglobin is referred to as:
A. Pasteur effect
B. Bohr effect
C. Haldane effect
D. Hamburger phenomenon

Correct answer = B.

8.6. Which of the following is the oxygen binding site of the hemoglobin?
A. Proximal histidines of both alpha and beta subunits
B. Proximal histidines of alpha subunits only
C. Proximal histidines of beta subunits only
D. None of these

Correct answer = D

8.7. Statement 1: Homozygous beta-thalassemia is much more common than homozygous alpha-thalassemia.
Statement 2: Alpha thalassemia usually arises due to gene mutations, rather than gene deletions.
A. Both the Statements are correct, and Statement 2 is the correct explanation for Statement 1
B. Both the Statements are correct, but Statement 2 is NOT the correct explanation for Statement 1
C. Statement 1 is correct, but Statement 2 is NOT correct
D. Statement 1 is NOT correct, but Statement 2 is correct

Correct answer = C.

8.8. Increase in the concentration of which of the following in the RBCs help in the delivery of oxygen to the tissues in high altitude acclimatization and in COPD?
A. 2,3-BPG
B. Nitric oxide
C. HbF
D. Bicarbonate

Correct answer = A.

8.9. Statement 1: β-thalassemia does not manifest at the time of birth.
Statement 2: Fetal hemoglobin is composed of alpha and delta globin chains, and these are not defective in β-thalassemia.
A. Both the Statements are correct, and Statement 2 is the correct explanation for Statement 1
B. Both the Statements are correct, but Statement 2 is NOT the correct explanation for Statement 1
C. Statement 1 is correct, but Statement 2 is NOT correct
D. Statement 1 is NOT correct, but Statement 2 is correct

Correct answer = C.

8.10. Blood that is stored for a longer period is less efficient in delivering oxygen to the peripheral tissues. Depletion of which of the following in RBCs is responsible for this?
A. Protons
B. Hemoglobin
C. Bicarbonate
D. 2,3-BPG

Correct answer = D.

Multiple-Choice Questions (NExT/NEET PG)

Choose the ONE best answer.

8.1. β-Lysine 82 in HbA is important for the binding of 2,3-bisphosphoglycerate. In Hb Helsinki, this amino acid has been replaced by methionine. Which of the following should be true concerning Hb Helsinki?
A. It should be stabilized in the taut, rather than the relaxed, form.

B. It should have increased oxygen affinity and, consequently, decreased oxygen delivery to tissues.
C. Its oxygen-dissociation curve should be shifted to the right relative to HbA.
D. It results in anemia.

Correct answer = B.

8.2. A 67-year-old man presented to the emergency department with a 1-week history of angina and shortness of breath. He complained that his face and extremities had taken on a blue color. His medical history included chronic stable angina treated with isosorbide dinitrate and nitroglycerin. Blood obtained for analysis was brown. Which one of the following is the most likely diagnosis?

A. Carboxyhemoglobinemia
B. Methemoglobinemia
C. Sickle cell anemia
D. β-Thalassemia

Correct answer = B.

8.3. Statement 1: The substitution of distal histidine by tyrosine in the alpha and beta globin chains, known respectively as Hb Saskatoon and Hb Boston, results in methemoglobinemia.

Statement 2: The distal histidine stabilizes the interaction between oxygen and heme, and prevents abnormal release of superoxide and oxidation of heme iron.

A. Both the Statements are correct, and Statement 2 is the correct explanation for Statement 1
B. Both the Statements are correct, but Statement 2 is NOT the correct explanation for Statement 1
C. Statement 1 is correct, but Statement 2 is NOT correct
D. Statement 1 is NOT correct, but Statement 2 is correct

Correct answer = A.

8.4. Rapaport-Leubering pathway is an off-shoot of glycolysis which is prominent in erythrocytes for the production of 2,3-bis phosphoglycerate. In a laboratory-generated mammal with a defect in this pathway, how would the p50 value of hemoglobin compare to that of the wild-type?

A. Lower than wild-type
B. Same as wild-type
C. Higher than wild-type
D. Information is insufficient to make a conclusion

Correct answer = A.

8.5. Binding of oxygen to iron in the heme pulls the iron into the plane of the heme. Because of this, there is an alteration in the magnetic properties of deoxyhemoglobin, compared to that of oxyhemoglobin. Blood flow to the localized regions of brain can hence be determined using this principle. Which of the following techniques is based on this?

A. PET scan
B. NMR spectroscopy
C. Functional MRI
D. 4D ultrasonography

Correct answer = C.

Fibrous Proteins and Extracellular Matrix

9

Learning Objectives

At the end of this chapter, the student will be able to:

- Describe collagen structure and function
- Discuss the synthesis and degradation of collagen and associated diseases
- Explain the characteristics of elastin and disorders of elastin degradation
- Discuss the structure and function of extracellular matrix

I. OVERVIEW

Collagen and elastin are examples of common, well-characterized fibrous proteins of the extracellular matrix (ECM) that serve structural functions in the body. For example, collagen and elastin are **found as components of skin, connective tissue, blood vessel walls, and the sclera and cornea of the eye**. Each fibrous protein exhibits special **mechanical properties**, resulting from its unique structure, which is obtained by **combining specific amino acids into regular, secondary structural elements**. This is in contrast to globular proteins (discussed in Chapter 8), whose shapes are the result of complex interactions between secondary, tertiary, and, sometimes, quaternary structural elements.

II. COLLAGEN

Collagen is the most abundant protein in the human body. A typical collagen molecule is a long, rigid structure in which three polypeptides (referred to as **α chains**) are wound around one another in a **rope-like triple helix** (Fig. 9.1). Although these molecules are found throughout the body, their types and organization are dictated by the structural role collagen plays in a particular organ. In some tissues, **collagen may be dispersed as a gel that gives support to the structure, as in the ECM or the vitreous humor of the eye.** In other tissues, **collagen may be bundled in tight, parallel fibers that provide great strength, as in tendons. In the cornea of the**

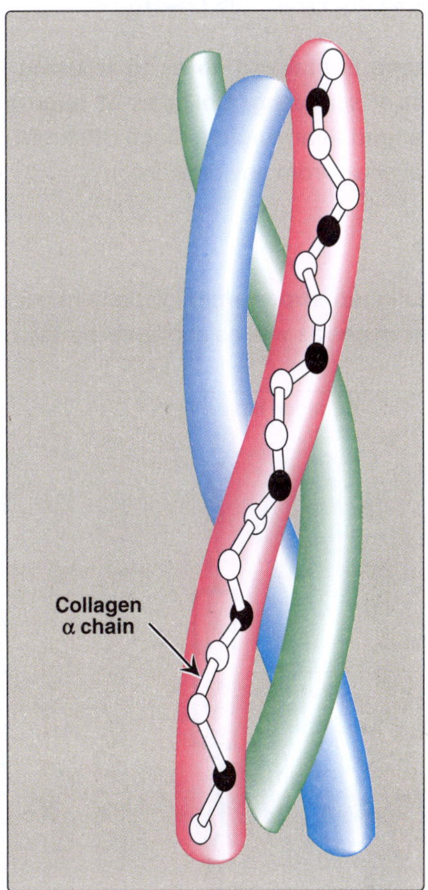

Figure 9.1 Triple-stranded helix of collagen formed from three α chains. [Note: The α chains themselves are helical in structure.]

TYPES OF COLLAGEN	TISSUE DISTRIBUTION
A. Fibril-forming	
I	Skin, bone, tendon, blood vessels, cornea
II	Cartilage, intervertebral disk, vitreous body
III	Blood vessels, skin, muscle
B. Network-forming	
IV	Basement membrane
VIII	Corneal and vascular endothelium
C. Fibril-associated*	
IX	Cartilage
XII	Tendon, ligaments, some other tissues

Figure 9.2 The most abundant types of collagen. [Note: ***Fibril-associated collagens with interrupted triple helices are known as FACIT.]**

eye, collagen is stacked so as to transmit light with a minimum of scattering. **Collagen of bone occurs as fibers arranged at an angle to each other** so as to resist mechanical shear from any direction.

A. Types

The collagen superfamily of proteins includes >25 collagen types as well as additional proteins that have collagen-like domains. The three polypeptide α chains are held together by interchain hydrogen bonds. Variations in the amino acid sequence of the α chains result in structural components that are about the same size ('1,000 amino acids long) but with slightly different properties. These α chains are combined to form the various types of collagen found in the tissues. For example, **the most common collagen, type I, contains two chains called α1 and one chain called α2 ($\alpha1_2\alpha2$), whereas type II collagen contains three α1 chains ($\alpha1_3$).**

The collagens can be organized into three groups, based on their location and functions in the body (Fig. 9.2).

1. **Fibril-forming collagens: Types I, II, and III are the fibrillar collagens and have the rope-like structure** described above for a typical collagen molecule. In the electron microscope, these linear polymers of fibrils have characteristic banding patterns, reflecting the regular staggered packing of the individual collagen molecules in the fibril (Fig. 9.3). **Type I collagen fibers (composed of collagen fibrils) are found in supporting elements of high tensile strength (for example, tendons and corneas)**, whereas fibers formed from **type II collagen molecules are restricted to cartilaginous structures**. The fibers derived **from type III collagen are prevalent in more distensible tissues such as blood vessels.**

2. **Network-forming collagens:** Types IV and VIII form a **three-dimensional mesh**, rather than distinct fibrils (Fig. 9.4). For example, type

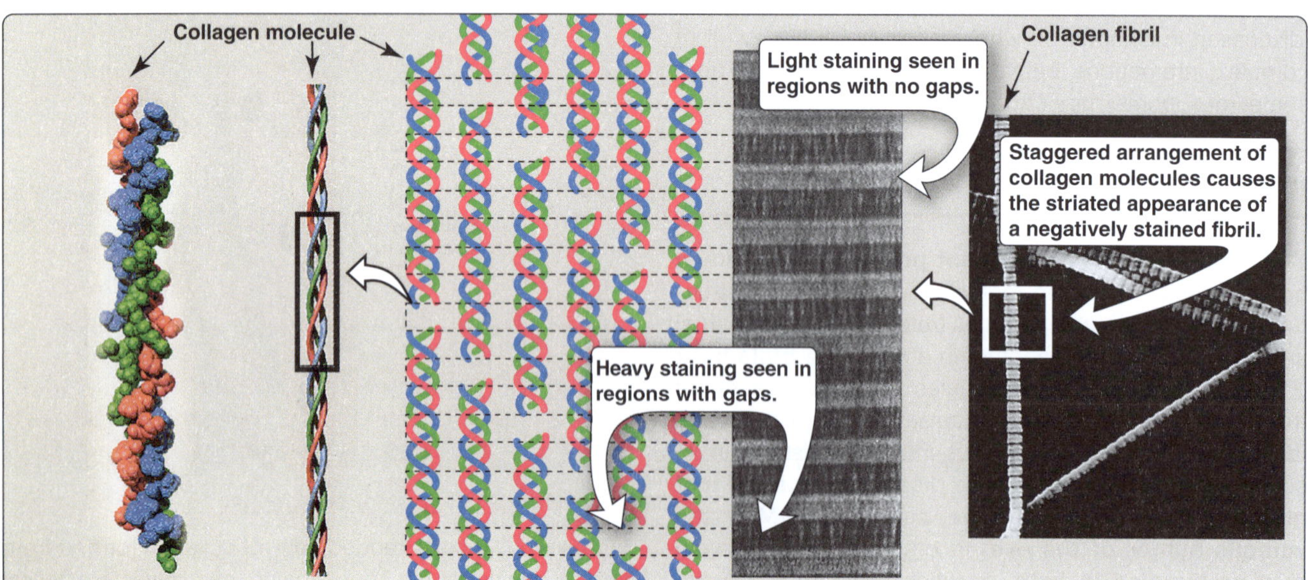

Figure 9.3 Collagen fibrils at right have a characteristic banding pattern, reflecting the regularly staggered packing of the individual collagen molecules in the fibril.

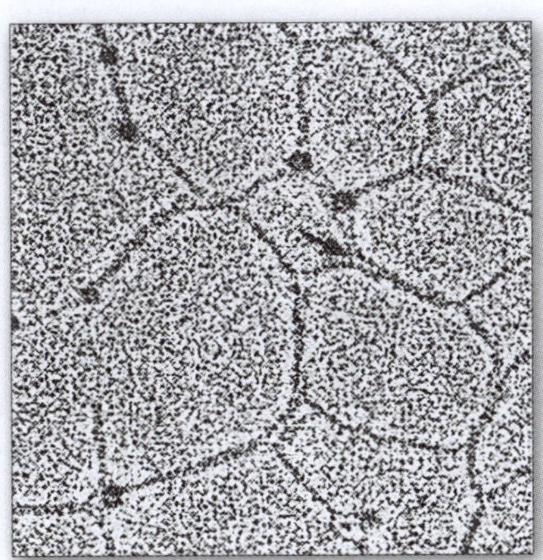

Figure 9.4 Electron micrograph of a polygonal network formed by association of collagen type IV monomers.

IV molecules assemble into a sheet or meshwork that constitutes a major part of **basement membranes.**

> Basement membranes are thin, sheet-like structures that provide mechanical support for adjacent cells and function as a semipermeable filtration barrier to macromolecules in organs such as the kidney and the lung.

3. **Fibril-associated collagens:** Types IX and XII bind to the surface of collagen fibrils, **linking these fibrils to one another and to other components in the ECM** (see Fig. 9.2).

B. Structure

Unlike most globular proteins that are folded into compact structures, collagen, a fibrous protein, has an elongated, triple-helical structure that is stabilized by interchain hydrogen bonds.

1. **Amino acid sequence: Collagen is rich in proline and glycine, both of which are important in the formation of the triple-stranded helix.** Proline facilitates the formation of the helical conformation of each α chain because its ring structure causes "kinks" in the peptide chain. [Note: The presence of proline dictates that the helical conformation of the α chain cannot be an α helix (see p. 85).] Glycine, the smallest amino acid, is found in every third position of the polypeptide chain. It fits into the restricted spaces where the three chains of the helix come together. The glycine residues are part of a repeating sequence, –Gly–X–Y–, where X is frequently proline, and Y is often hydroxyproline (but can be hydroxylysine, Fig. 9.5). **Thus, most of the α chain can be regarded as a polytripeptide whose sequence can be represented as (–Gly–Pro–Hyp–)$_{333}$.**

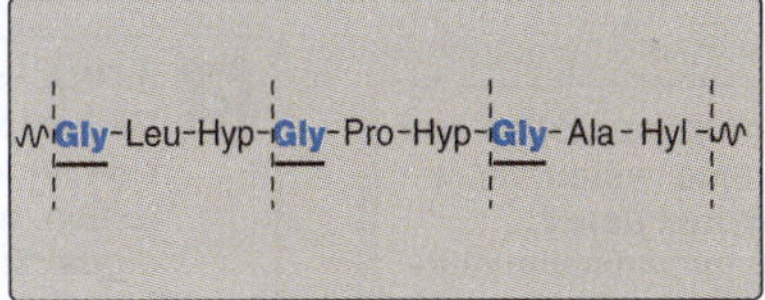

Figure 9.5 Amino acid sequence of a portion of the α1 chain of collagen. Hyp = hydroxyproline; Hyl = hydroxylysine.

2. **Hydroxyproline and hydroxylysine: Collagen contains hydroxyproline and hydroxylysine, which are nonstandard amino acids (see p. 119) not present in most other proteins**. They result from the hydroxylation of some of the proline and lysine residues after their incorporation into polypeptide chains (Fig. 9.6). Therefore, the hydroxylation is a posttranslational modification (see pp. 644–645). [Note: Generation of **hydroxyproline maximizes formation of interchain hydrogen bonds that stabilize the triple-helical structure**.]
3. **Glycosylation:** The hydroxyl group of the hydroxylysine residues of collagen may be enzymatically glycosylated. Most commonly, glucose and galactose are sequentially attached to the polypeptide chain prior to triple-helix formation (Fig. 9.7).

C. Biosynthesis

The polypeptide precursors of the collagen molecule are synthesized in fibroblasts (or in the related osteoblasts of bone and chondroblasts of

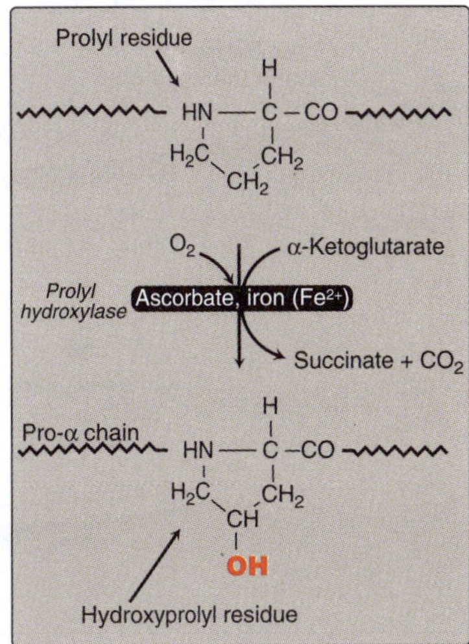

Figure 9.6 Hydroxylation of proline residues in pro-α chains of collagen by *prolyl hydroxylase*. [Note: Fe^{2+} (*hydroxylase* cofactor) is protected from oxidation to Fe^{3+} by ascorbate (vitamin C).]

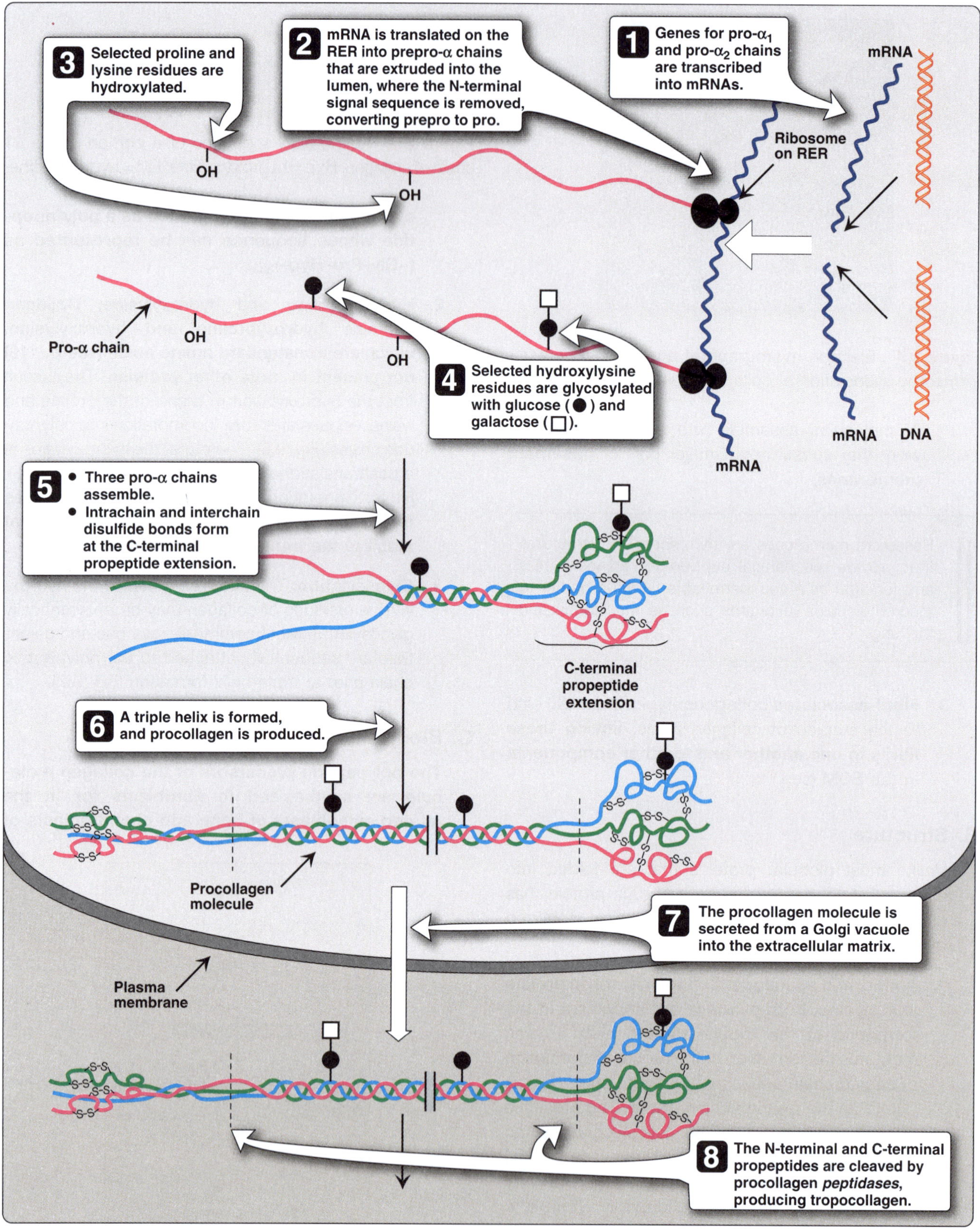

Figure 9.7 (continued on the next page) Synthesis of collagen. RER = rough endoplasmic reticulum; mRNA = messenger RNA. Synthesis of collagen.

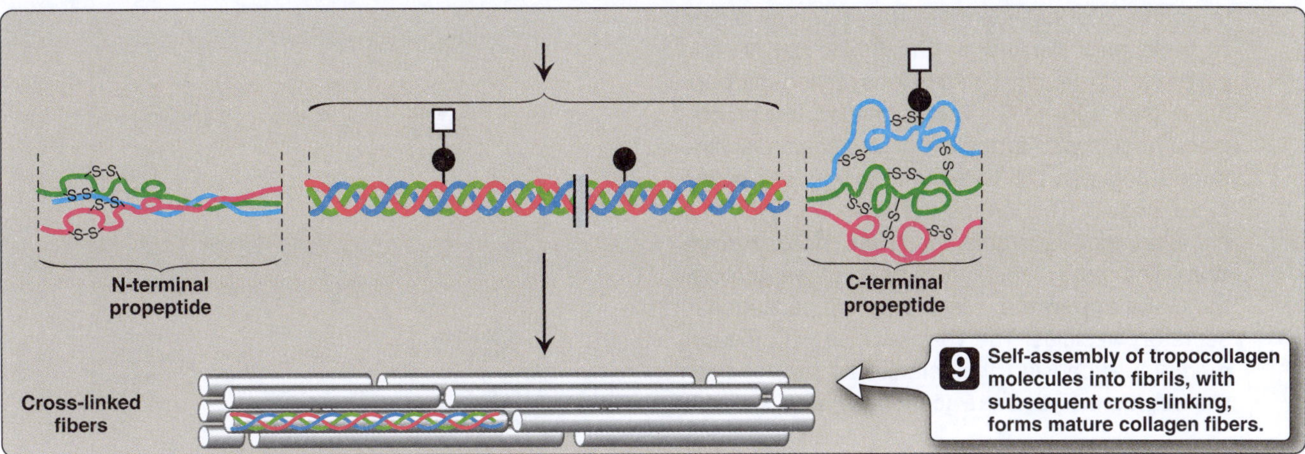

Figure 9.7 (continued from the previous page) Synthesis of collagen.

cartilage). They are enzymically modified and form the triple helix, which gets secreted into the ECM. After additional enzymic modification, the mature extracellular collagen fibrils aggregate and become cross-linked to form collagen fibers.

1. **Pro-α chain formation:** Collagen is one of many proteins that normally function outside of cells. Like most proteins produced for export, the **newly synthesized polypeptide precursors of α chains (prepro-α chains)** contain a special amino acid sequence at their amino (N)-terminal ends. This sequence acts as a signal that, in the absence of additional signals, **targets the polypeptide** (Note: See Chapter 37, p. 644, Fig. 37.15 for more information on protein targeting) being synthesized for secretion from the cell. The signal sequence facilitates the binding of ribosomes to the rough endoplasmic reticulum (RER) and directs the passage of the prepro-α chain into the lumen of the RER. The signal sequence is rapidly cleaved in the lumen to yield a precursor of collagen called a pro-α chain (see Fig. 9.7).

2. **Hydroxylation:** The pro-α chains are processed by a number of enzymic steps within the lumen of the RER while the polypeptides are still being synthesized (see Fig. 9.7). Proline and lysine residues found in the Y-position of the –Gly–X–Y– sequence can be hydroxylated to form hydroxyproline and hydroxylysine residues. These **hydroxylation reactions require molecular oxygen, ferrous iron (Fe^{2+}), and the reducing agent vitamin C** (ascorbic acid, see p. 530), without which the hydroxylating enzymes, ***prolyl hydroxylase*** **and** ***lysyl hydroxylase***, are unable to function (see Fig. 9.6). In the case of ascorbic acid deficiency (and, therefore, a lack of proline and lysine hydroxylation), interchain H-bond formation is impaired, as is formation of a stable triple helix. Additionally, collagen fibrils cannot be cross-linked (see 7. below), greatly decreasing the tensile strength of the assembled fiber. The resulting deficiency disease is known as **scurvy.** Patients with scurvy often show ecchymoses (bruise-like discolorations) on the limbs as a result of subcutaneous extravasation (leakage) of blood due to capillary fragility (Fig. 9.8).

3. **Glycosylation:** Some hydroxylysine residues are modified by glycosylation with glucose or glucosyl-galactose (see Fig. 9.7).

4. **Assembly and secretion:** After hydroxylation and glycosylation, three pro-α chains form procollagen,

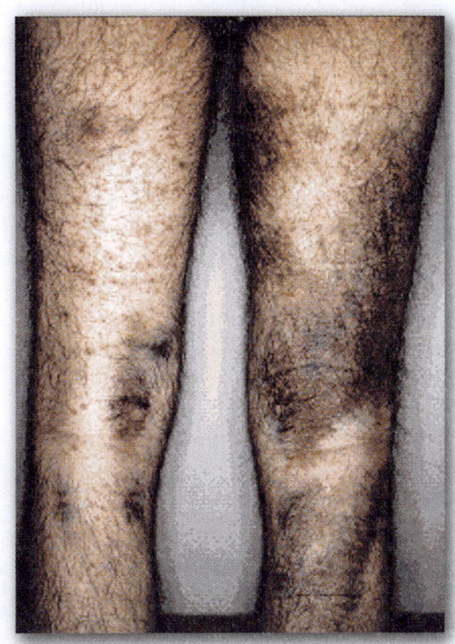

Figure 9.8 The legs of a 46-year-old man with scurvy.

a precursor of collagen that has a central region of triple helix flanked by the nonhelical N- and carboxyl (C)-terminal extensions called propeptides (see Fig. 9.7). The formation of procollagen begins with formation of interchain disulfide bonds between the C-terminal extensions of the pro-α chains. This brings the three α chains into an alignment favorable for **triple helix formation**. The procollagen molecules move through the **Golgi apparatus**, where they are packaged in secretory vesicles. The vesicles fuse with the cell membrane, causing the **release of procollagen molecules into the extracellular space.**

5. **Extracellular cleavage of procollagen molecules:** After their release, the triple-helical procollagen molecules are cleaved by ***N- and C-procollagen peptidases***, which remove the terminal propeptides, producing tropocollagen molecules.

6. **Collagen fibril formation:** Tropocollagen molecules spontaneously associate to form collagen fibrils. They form an ordered, parallel array, with adjacent collagen molecules arranged in a staggered pattern, each overlapping its neighbor by a length approximately three quarters of a molecule (see Fig. 9.7).

7. **Cross-link formation:** The fibrillar array of collagen molecules serves as a substrate for ***lysyl oxidase*.** **This copper-containing extracellular enzyme oxidatively deaminates some of the lysine and hydroxylysine residues** in collagen. The reactive aldehydes that result (allysine and hydroxyallysine) can spontaneously condense with lysine or hydroxylysine residues in neighboring collagen molecules **to form covalent cross-links** and, thus, mature collagen fibers (Fig. 9.9). [Note: Cross-links can form between two allysine residues.]

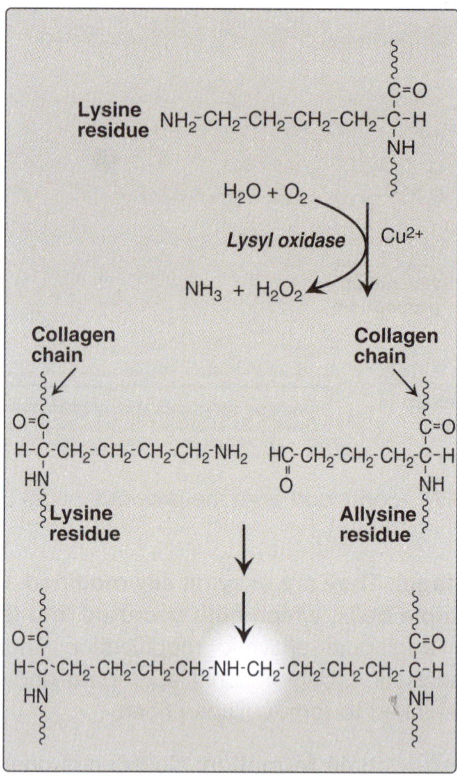

Figure 9.9 Formation of cross-links in collagen. [Note: *Lysyl oxidase* is irreversibly inhibited by a toxin present in seeds from Lathyrus odoratus (sweet pea), leading to a condition known as lathyrism that is characterized by skeletal and vascular problems.] Cu^{2+} = copper; NH_3 = ammonia; H_2O_2 = hydrogen peroxide.

> ***Lysyl oxidase* is one of several copper-containing enzymes.** Others include ***ceruloplasmin*** (see p. 556), *cytochrome c oxidase* (see p. 166), *superoxide dismutase* (see p. 793), and *tyrosinase* (see p. 407). Disruption in copper homeostasis causes copper deficiency (**X-linked Menkes syndrome**) or overload (**Wilson disease**, see p. 556).

D. Degradation

Normal collagens are highly stable molecules, having half-lives as long as several years. However, connective tissue is dynamic and is constantly being remodeled, often in response to growth or injury of the tissue. Breakdown of collagen fibers is dependent on the proteolytic action of ***collagenases***, which are part of a large family of matrix *metalloproteinases*. For type I collagen, the cleavage site is specific, generating three-quarter and one-quarter length fragments. These fragments are further degraded by other matrix *proteinases*.

E. Collagenopathies

Defects in any one of the many steps in collagen fiber synthesis can result in a genetic disease involving an inability of collagen to form fibers properly and, therefore, an inability to provide tissues with the needed tensile strength normally provided by collagen. More than 1,000 mutations have been identified in 23 genes coding for 13 of the collagen types. The following are examples of diseases (collagenopathies) that are the result of defective collagen synthesis.

1. **Ehlers-Danlos syndrome:** Ehlers-Danlos syndrome (EDS) is a heterogeneous group of connective tissue disorders that result from heritable defects in the metabolism of fibrillar collagen molecules. EDS can be caused by a **deficiency of collagen-processing enzymes (for example, *lysyl hydroxylase* or *N-procollagen peptidase*)**

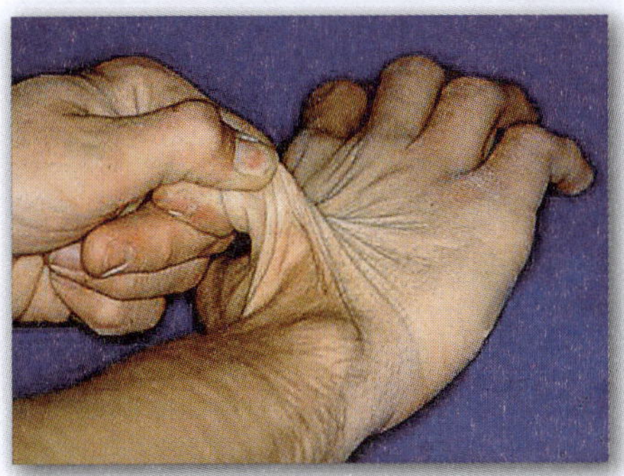

Figure 9.10 Stretchy skin of classic Ehlers-Danlos syndrome.

or from mutations in the amino acid sequences of collagen types I, III, and V. The classic form of EDS, caused by defects in type V collagen, is characterized by skin extensibility and fragility and joint hypermobility (Fig. 9.10). The vascular form, due to defects in type III collagen, is the most serious form of EDS because it is associated with potentially lethal arterial rupture. [Note: The classic and vascular forms show autosomal-dominant inheritance.] Collagen that contains mutant chains may have altered structure, secretion, or distribution, and it frequently is degraded. [Note: Incorporation of just one mutant chain may result in degradation of the triple helix. This is known as a dominant-negative effect.]

2. **Osteogenesis imperfecta:** This syndrome, known as **"brittle bone disease,"** is a genetic disorder of bone fragility characterized by bones that fracture easily, with minor or no trauma (Fig. 9.11). **Over 80% of cases of osteogenesis imperfecta (OI) are caused by dominant mutations to the genes that encode the α1 or α2 chains in type I collagen. The most common mutations cause the replacement of glycine (in –Gly–X–Y–) by amino acids with bulky side chains**. The resultant structurally abnormal α chains prevent the formation of the required triple-helical conformation. Phenotypic severity ranges from mild to lethal. Type I OI, the most common form, is characterized by mild bone fragility, hearing loss, and blue sclerae. Type II, the most severe form, is typically lethal in the perinatal period as a result of pulmonary complications. In utero fractures are seen (see Fig. 9.11). Type III is also a severe form and is characterized by multiple fractures at birth, short stature, spinal curvature leading to a humped-back (kyphotic) appearance, and blue sclerae. Dentinogenesis imperfecta, a disorder of tooth development, may be seen in OI.

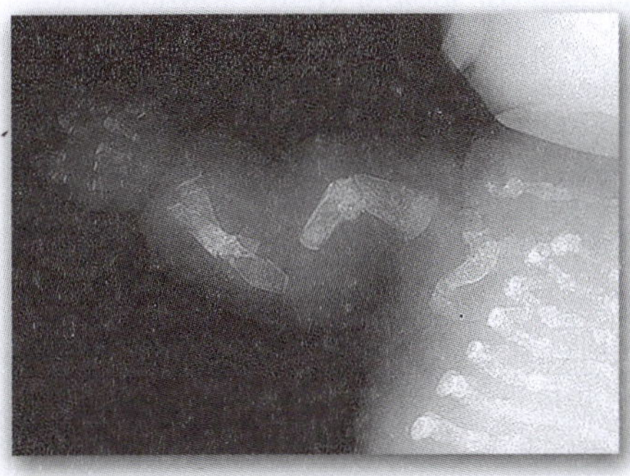

Figure 9.11 Lethal form (type II) of osteogenesis imperfecta in which the fractures appear in utero, as revealed by this radiograph of a stillborn fetus.

Extracellular matrix (ECM)

The extracellular matrix (ECM) is a complex of extracellular macromolecules and minerals secreted outside cells and tissues, with various structures and functions such as cell adhesion, cell-to-cell communication, support, and scaffolding. Sometimes referred to as "connective tissue," the multicellularity of ECM involves different types of cells and compounds, making a discussion on ECM difficult to confine to a single chapter.

- **Structural fibrous proteins like collagen and elastin:** Collagen and Elastin are discussed in detail in this Chapter.
- **Specialized proteins like fibronectin and laminin: Fibronectin** are high molecular weight glycoproteins of the extracellular matrix. They bind to integrins, collagen, fibrin, and proteoglycans, facilitating cell movement. Fibronectins also bind to platelets to facilitate movement to concerned areas in case of cell injury or blood clotting. **Laminins** are found in the basal lamina and form structures that resist tensile forces. They also bind to other ECM components like collagens and assist in cell adhesion.
- **Glycosaminoglycans, proteoglycans, and glycoproteins:** Examples include keratin sulfates, heparin sulfate, hyaluronic acid, chondroitin sulfates, dermatin sulfates, etc. They are discussed in detail in Chapter 19. Additionally, fibrillin-1 is a large glycoprotein secreted by fibroblasts and incorporated into insoluble microfibrils to form a scaffold for the deposition of elastin. A mutation in the Fibrillin-1 gene in chromosome 15 can cause Marfan syndrome, an inherited autosomal dominant trait, affecting the eye (ectopia lentis or dislocation of the eye) and skeletal muscle disorders like hyperextensibility of joints and long digits.

Specialized forms of ECM include bone and cartilage. Bone is a mineralized connective tissue that is dynamic and undergoes continuous remodeling, resorption, and deposition of new bone tissue.

It contains primarily collagen type 1 (approximately 90%) with some collagen type V. The non-collagen proteins in bone include proteoglycans, plasma proteins, Decorin, bone-specific CS-PGIII, osteonectin, osteocalcin, bone sialoprotein, and others. The inorganic (mineral) component of bone is hydroxyapatite, and minerals in bone include magnesium, carbonate, sodium, and fluoride. Osteoclasts and osteoblasts cause bone resorption and deposition, respectively. Osteoclasts are derived from pleuripotent hematopoietic stem cells. The proton-translocating ATPase in the apical membrane of osteoclasts expels protons, lowering pH and allowing demineralization to occur by increasing the solubility of hydroxyapatite.

Osteoblasts are derived from pleuripotent mesenchymal precursors and regulate the deposition of new bone matrix. The movement of calcium and phosphate across cell surfaces is regulated by osteoblasts.

Cartilage is an avascular tissue mainly consisting of collagen type II and non-collagen proteins like proteoglycans and chondronectin. Proteoglycans make the cartilage compressible, and chondronectin helps in the attachment of type II collagen to chondrocytes.

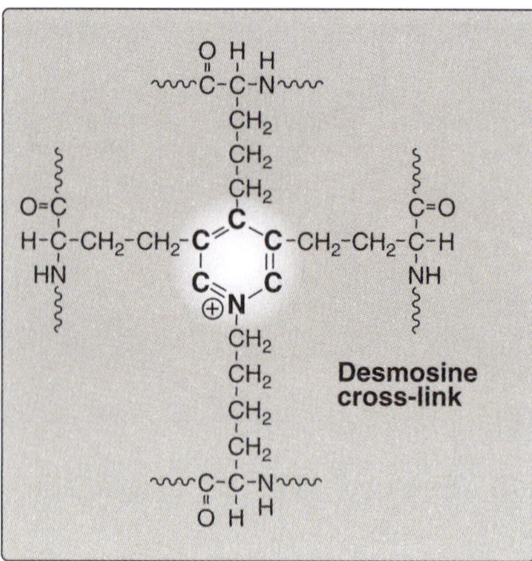

Figure 9.12 Desmosine cross-link unique to elastin.

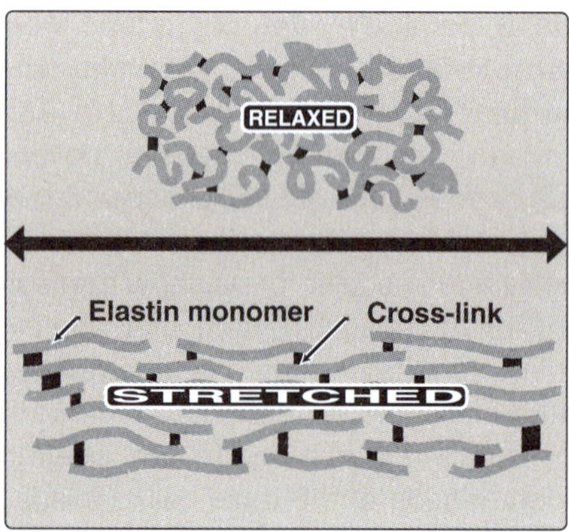

Figure 9.13 Elastin fibers in relaxed and stretched conformations.

III. ELASTIN

In contrast to collagen, which forms fibers that are tough and have high tensile strength, **elastin is a connective tissue fibrous protein with rubber-like properties. Elastic fibers composed of elastin and glycoprotein microfibrils are found in the lungs, the walls of large arteries, and elastic ligaments. They can be stretched to several times their normal length but recoil to their original shape when the stretching force is relaxed.**

A. Structure

Elastin is an insoluble protein polymer generated from a **precursor, tropoelastin**, which is a soluble polypeptide composed of '700 amino acids' that are primarily small and nonpolar (for example, glycine, alanine, and valine). **Elastin is also rich in proline and lysine but contains scant hydroxyproline and hydroxylysine**. Tropoelastin is secreted by the cell into the ECM. There, it **interacts with specific glycoprotein microfibrils, such as fibrillin**, which function as a scaffold onto which tropoelastin is deposited. Some of the lysyl side chains of the tropoelastin polypeptides are oxidatively deaminated by *lysyl oxidase*, forming allysine residues. Three of the allysyl side chains plus one unaltered lysyl side chain from the same or neighboring polypeptides form a desmosine cross-link (Fig. 9.12). This produces elastin, an extensively interconnected, rubbery network that can stretch and bend in any direction when stressed, giving connective tissue elasticity (Fig. 9.13). **Mutations in the fibrillin-1 protein** are responsible for **Marfan syndrome**, a connective tissue disorder characterized by impaired structural integrity in the skeleton, the eye, and the cardiovascular system. With this disease, abnormal fibrillin protein is incorporated into microfibrils along with normal fibrillin, inhibiting the formation of functional microfibrils. [Note: Patients with Marfan syndrome, OI, or EDS may have blue sclerae due to tissue thinning that allows underlying pigment to show through.]

B. α_1-Antitrypsin in elastin degradation

Blood and other body fluids contain a protein, **α_1-antitrypsin (AAT), which inhibits a number of proteolytic enzymes (called *peptidases*, *proteases*, or *proteinases*) that hydrolyze and destroy proteins**. [Note: The inhibitor was originally named AAT because it inhibits the activity of *trypsin*, a proteolytic enzyme synthesized as trypsinogen by the pancreas (see p. 380).] **AAT has the important physiologic role of inhibiting neutrophil *elastase*, a powerful *protease* that is released into the extracellular space and degrades elastin of alveolar walls** as well as other structural proteins in a variety of tissues (Fig. 9.14). Most of the AAT found in plasma is synthesized and secreted by the liver. Extrahepatic synthesis also occurs.

1. **α_1-Antitrypsin in the lungs:** In the normal lung, the alveoli are chronically exposed to low levels of neutrophil *elastase* released from activated and degenerating neutrophils. The proteolytic activity of *elastase* can destroy the elastin in alveolar walls if unopposed by the action of **AAT, the most important inhibitor of neutrophil *elastase*** (see Fig. 9.14). Because lung tissue cannot regenerate, the destruction of the connective tissue of alveolar walls caused by an imbalance between the *protease* and its inhibitor results in pulmonary disease.

2. **α_1-Antitrypsin deficiency and emphysema:** In the United States, '2%–5% of patients with emphysema are predisposed to the disease by **inherited defects in AAT.** A number of different mutations in the gene for AAT are known to cause a deficiency of the protein, but **one single purine base mutation (GAG to AAG, resulting in the substitution of lysine for glutamic acid at position 342 of the protein) is clinically the most widespread and severe**. [Note: The mutated protein is termed the **Z variant**.] The mutation causes the normally monomeric AAT to misfold, polymerize, and aggregate within the RER of hepatocytes, resulting in decreased secretion of AAT by the liver. **AAT deficiency is, therefore, a misfolded protein disease**. Consequently, blood levels of AAT are reduced, decreasing the amount that gets to the lung. The polymer that accumulates in the liver may result in cirrhosis (scarring of the liver). In the United States, the AAT mutation is most common in Caucasians of Northern European ancestry. An individual must inherit two abnormal AAT alleles to be at risk for the **development of emphysema**. In a heterozygote, with one normal and one defective gene, the levels of AAT are sufficient to protect the alveoli from damage. [Note: Methionine 358 in AAT is required for the binding of the inhibitor to its target *proteases*. Smoking causes the oxidation and subsequent inactivation of the methionine, thereby rendering the inhibitor powerless to neutralize *elastase*. **Smokers with AAT deficiency, therefore, have a considerably elevated rate of lung destruction** and a poorer survival rate than nonsmokers with the deficiency.] The deficiency of *elastase* inhibitor can be treated by weekly augmentation therapy, that is, intravenous administration of AAT. The AAT diffuses from the blood into the lung, where it reaches therapeutic levels in the fluid surrounding the lung epithelial cells.

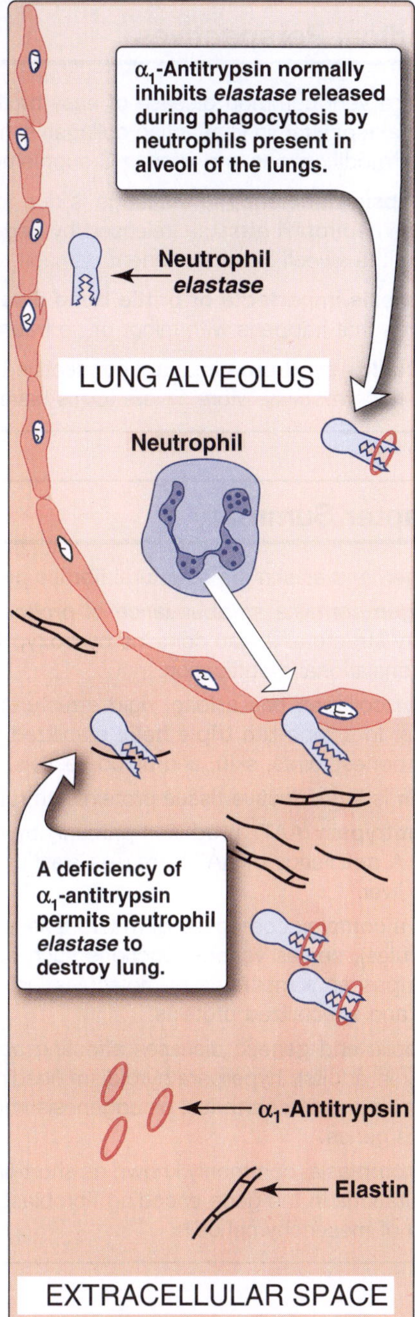

Figure 9.14 Destruction of alveolar tissue by *elastase* released from neutrophils activated as part of the immune response to airborne pathogens.

IV. Medical Perspectives

Scurvy is a very common disease of easy bruisability, hemorrhage, bleeding gums, weakness, etc. due to inadequate dietary intake of vitamin C affecting collagen synthesis. It is common in children and elderly population. It is easily treated by dietary modifications and vitamin C supplements.

α_1-antitrypsin deficiency in the lungs is due to inherited defects of α_1-antitrypsin leading to emphysema. Normally, in alveoli, the neutrophil *elastase* (released by degenerating neutrophils) is continuously inhibited by α_1-antitrypsin, thereby protecting the alveoli from permanent damage.

Osteogenesis imperfecta or brittle bone disease is a genetic disorder of bone fragility. Patient gives classical history of fractures that happens with minor or no trauma. Usually, it is diagnosed in infancy.

Ehlers-Danlos syndrome is a rare connective tissue disorder due to heritable genetic defects causing skin extensibility and joint hypermobility. More severe forms have vascular features also.

V. Chapter Summary

- Collagen and elastin are structural fibrous proteins of the extracellular matrix (Fig. 9.15).
- **Collagen** contains an abundance of **proline**, **lysine**, and **glycine**, the latter occurring at every third position in the primary structure. It also contains **hydroxyproline**, **hydroxylysine**, and **glycosylated hydroxylysine**, each formed by posttranslational modification.
- Fibrillar collagen has a long, rigid structure, in which three collagen polypeptide α chains are wound around one another in a rope-like **triple helix** stabilized by **interchain hydrogen bonds**. Diseases of fibrillar collagen synthesis affect bones, joints, skin, and blood vessels.
- **Elastin** is a connective tissue protein with rubber-like properties in tissues such as the lung.
- **α_1-Antitrypsin** (**AAT**), produced primarily by the liver, inhibits **elastase**-catalyzed degradation of elastin in the alveolar walls. A deficiency of AAT increases elastin degradation and can cause **emphysema** and, in some cases, **cirrhosis** of the liver.
- ECM, a complex composed of macromolecules and minerals secreted outside cells and tissues (extracellular macromolecules), serves various functions such as cell adhesion, cell-to-cell communication, support, and scaffolding. It encompasses a multitude of molecules including structural proteins, glycoproteins, proteoglycans, glycosaminoglycans, and specialized proteins.
- Metabolic and genetic diseases affecting bone and cartilage comprise osteomalacia (resulting from vitamin D deficiency in adults), hyperparathyroidism (leading to bone resorption), rickets (due to vitamin D deficiency in children), osteoporosis, osteoarthritis, osteogenesis imperfecta, and several chondrodysplasias caused by genetic mutations in various genes.
- Achondroplasia, commonly known as short-limbed dwarfism, may resemble a collagen disorder but is actually caused by mutations in the gene encoding fibroblast growth factor receptor 3. This mutation affects the growth and differentiation of mesenchymal cells.

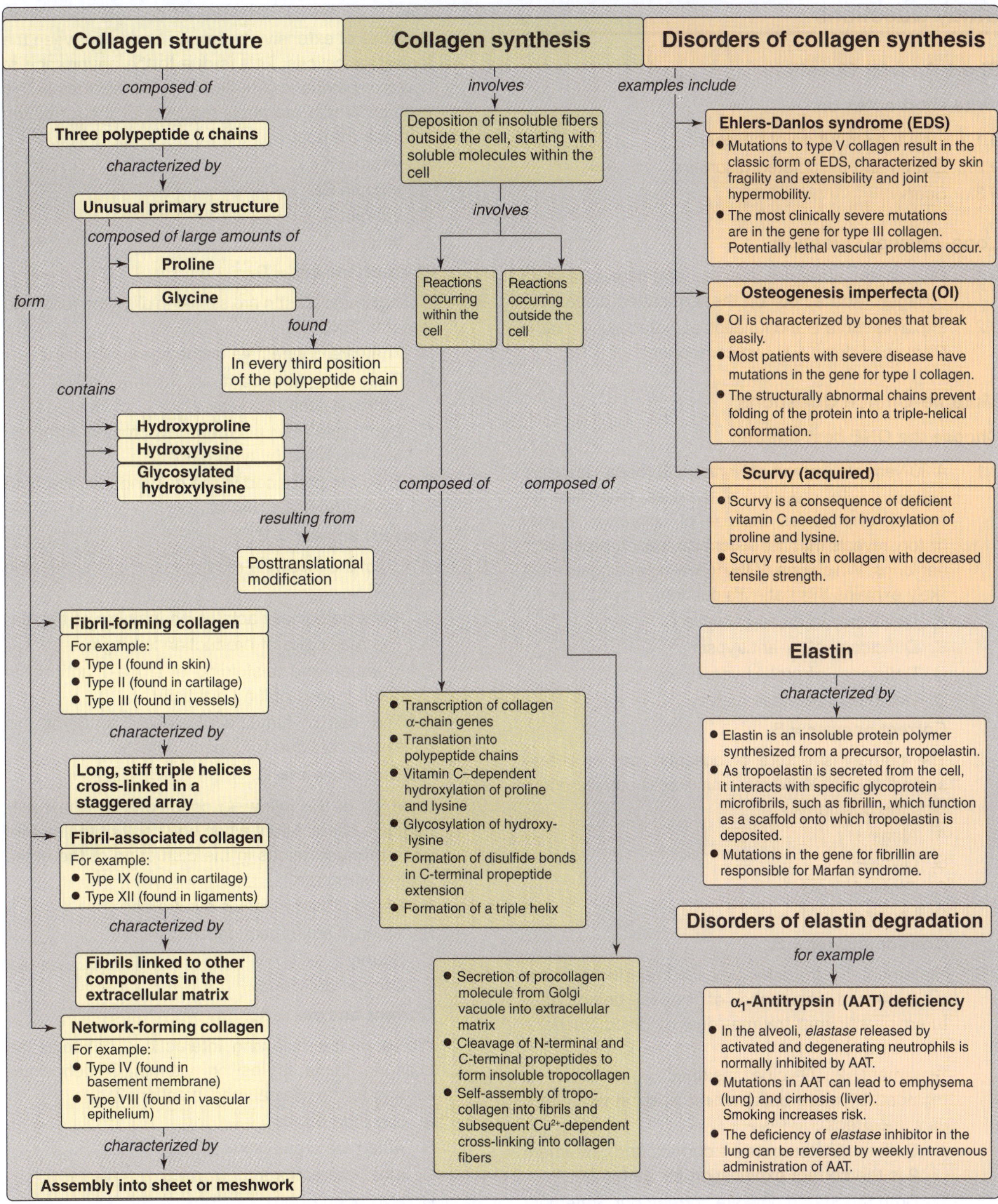

Figure 9.15 Key concept map for the fibrous proteins collagen and elastin. Cu^{2+} = copper.

Study Questions

Short-Answer Questions

Write short notes on:

9.1. Tissue distribution of collagen
9.2. Elastin and associated disorders
9.3. Scurvy

Long-Answer Question

9.1. Discuss the structure, function and biosynthesis of collagen. Add a note on the associated diseases.
9.2. Explain structure and function of extracellular matrix. Discuss in detail any two components of ECM.

Multiple-Choice Questions

Choose the ONE best answer.

9.1. A 30-year-old woman of Northern European ancestry presents with progressive dyspnea (shortness of breath). She denies the use of cigarettes. Family history reveals that her sister also has problems with her lungs. Which one of the following etiologies most likely explains this patient's pulmonary symptoms?
A. Deficiency in dietary vitamin C
B. Deficiency of α_1-antitrypsin
C. Deficiency of prolyl hydroxylase
D. Decreased elastase activity

Correct answer = B.

9.2. The primary structure of collagen can be represented as A-X-Y. Which amino acid usually occupies the position A?
A. Alanine
B. Glycine
C. Glutamic acid
D. Leucine

Correct answer = B.

9.3. Statement 1: In osteogenesis imperfecta, there is increased incidence of bones being fractured by minimal trauma, due to defective bone mineralization.
Statement 2: Glycine residues in collagen is replaced by more bulky amino acids in osteogenesis imperfecta.
A. Both the Statements are correct, and Statement 2 is the correct explanation for Statement 1.
B. Both the Statements are correct, but Statement 2 is NOT the correct explanation for Statement 1.
C. Statement 1 is correct, but Statement 2 is NOT correct
D. Statement 1 is NOT correct, but Statement 2 is correct

Correct answer = A.

9.4. The triple helical structure of collagen is formed because of extensive hydrogen bonds between the individual helices. This is due to the abundance of hydroxy-proline and hydroxy-lysine residues in this protein. Which vitamin is required for the formation of these residues?
A. Vitamin K
B. Vitamin B6
C. Vitamin A
D. Vitamin C

Correct answer = D.

9.5. Collagen and elastin are similar in all of the following aspects, EXCEPT:
A. They are connective tissue fibrous proteins
B. Both are abundant in hydroxyproline and hydroxylysine residues
C. Both have their polypeptides cross-link formation mediated by lysyl oxidase
D. They are produced in the cell and secreted into the extracellular matrix.

Correct answer = B.

9.6. Which of the following regarding alpha-1 antitrypsin is true?
A. It physiologically acts as an inhibitor of trypsin
B. The major site of production is lungs
C. Mutation and misfolding of alpha-1 antitrypsin result in loss of lung elasticity
D. The loss of function of alpha-1 antitrypsin is exclusively due to genetic defects

Correct answer = C.

9.7. In which of the following conditions is 'dominant-negative effect' seen, when one copy of the mutant gene product results in the disruption of the entire protein structure?
A. Classic Ehlers-Danlos syndrome
B. Alpha-1 antitrypsin deficiency
C. Scurvy
D. Copper deficiency

Correct answer = A.

9.8. Which of the following interactions between the collagen fibrils is lost in vitamin C deficiency, leading to the clinical manifestations?
A. Disulfide bonds
B. Aldehyde cross-links
C. Ionic interactions
D. Hydrogen bonds

Correct answer = D.

9.9. A 25-year-old male presented with a history of sudden visual problems and chest pain. Examination revealed a tall, slender body type with disproportionately long limbs and fingers, a high-arched palate, and lens dislocation. His cardiovascular assessment showed aortic root dilation. Family history uncovered a relative with similar features. Genetic testing confirmed defects in the fibrillin-1 gene. What is the most likely diagnosis?

A. Ehlers-Danlos syndrome
B. Osteogenesis imperfecta
C. Marfan syndrome
D. Alpha-1 antitrypsin deficiency

Correct answer = C.

9.10. Statement 1: Alpha-1 antitrypsin deficiency has cirrhosis as one of the clinical manifestations.

Statement 2: Defective protein gets misfolded in the liver, which is the site of alpha-1 antitrypsin synthesis.

A. Both the Statements are correct, and Statement 2 is the correct explanation for Statement 1.
B. Both the Statements are correct, but Statement 2 is NOT the correct explanation for Statement 1.
C. Statement 1 is correct, but Statement 2 is NOT correct
D. Statement 1 is NOT correct, but Statement 2 is correct

Correct answer = A.

Multiple-Choice Questions (NExT/NEET PG)

Choose the ONE best answer.

9.1. A 7-month-old child "fell over" while crawling and now presents with a swollen leg. Imaging reveals a fracture of a bowed femur, secondary to minor trauma, and thin bones (see x-ray below). Blue sclerae are also noted. At age 1 month, the infant had multiple fractures in various states of healing (right clavicle, right humerus, and right radius). A careful family history has ruled out nonaccidental trauma (child abuse) as a cause of the bone fractures. Which pairing of a defective (or deficient) molecule and the resulting pathology best fits this clinical description?

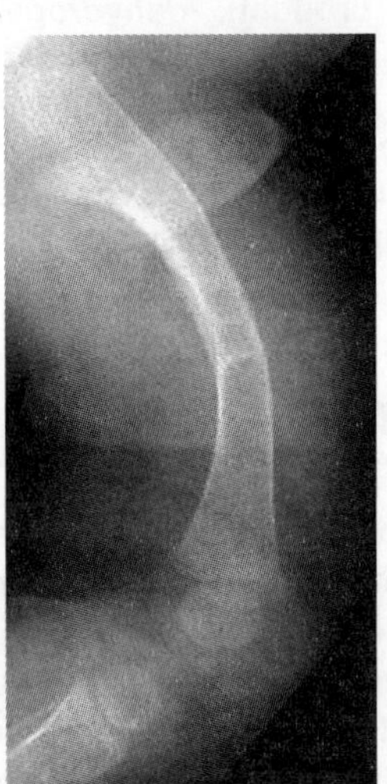

A. Elastin and emphysema
B. Fibrillin and Marfan disease
C. Type I collagen and osteogenesis imperfecta
D. Type V collagen and Ehlers-Danlos syndrome

Correct answer = C.

9.2. Menkes disease results from a decrease in the serum copper level dur to defective copper transport in the intestine. Certain clinical manifestations of this condition are due to defect in the collagen maturation. The functioning of which of the following enzymes is affected here?

A. Collagenases
B. Lysyl oxidase
C. Prolyl hydroxylase
D. Lysyl hydroxylase

Correct answer = B.

9.3. Alport syndrome is a genetic condition characterized by renal dysfunction and hearing loss. It arises due to a defect in the biosynthesis of collagen in the basement membrane. Which of the following collagens is affected in this condition?

A. Type I
B. Type II
C. Type III
D. Type IV

Correct answer = D.

9.4. Tumor metastasis involves migration of the cancer cells from the original (primary) tumor site to distant parts of the body, forming new (secondary) tumors in organs or tissues. This process involves cancer cells breaking away from the primary tumor, and traveling through the bloodstream or lymphatic system. Presence of which of the following in the tumor environment helps in this process?

A. Pro-collagen carboxy-proteinase
B. Alpha-1 antitrypsin
C. Matrix metalloproteinases
D. Prolyl hydroxylase

Correct answer = C.

9.5. Osteolathyrism is caused by chronic consumption of the *Lathyrus sativus* (grass pea) and exposure to its neurotoxic compound, β-aminopropionitrile (BAPN) which affects bone development. The pathogenic mechanism is considered as an inhibition of collagen and elastin cross-linking by this BAPN. Which of the following enzymes is inhibited by BAPN?

A. Pro-collagen amino-proteinase
B. Lysyl oxidase
C. Prolyl hydroxylase
D. Lysyl hydroxylase

Correct answer = B.

Enzymes

10

Learning Objectives

At the end of this chapter, the student will be able to:

- Classify enzymes
- Discuss the mechanism of action of enzymes and factors affecting enzyme activity
- Explain the types of enzyme inhibition with examples
- Discuss the enzymes used in clinical diagnosis

I. OVERVIEW

Virtually all reactions in the body are mediated by **enzymes, which are protein catalysts** that increase the rate of reactions without being changed in the overall process. Among the many biologic reactions that are energetically possible, enzymes selectively channel reactants (called substrates) into useful pathways. Thus, enzymes direct all metabolic events. This chapter examines the nature of these catalytic molecules and their mechanisms of action.

II. NOMENCLATURE

Each enzyme is assigned two names. The first is its short, recommended name, convenient for everyday use. The second is the more complete systematic name, which is used when an enzyme must be identified without ambiguity.

A. Recommended name

Most commonly used enzyme names have the suffix "*-ase*" attached to the substrate of the reaction (for example, *glucosidase* and *urease*) **or to a description of the action performed (for example, *lactate dehydrogenase* and *adenylyl cyclase*).** [Note: Some enzymes retain their original trivial names, which give no hint of the associated enzymic reaction, for example, *trypsin* and *pepsin*.]

B. Systematic name

In the systematic naming system, enzymes are divided into six major classes (Fig. 10.1), each with numerous subgroups. For a given enzyme, the suffix *-ase* is attached to a fairly complete description of the chemical reaction catalyzed, including the names of all the substrates, for example, *lactate:nicotinamide adenine dinucleotide (NAD^+) oxidoreductase*. [Note: Each enzyme is also assigned a classification number. *Lactate:NAD^+ oxidoreductase* is 1.1.1.27.] The systematic names are **unambiguous and informative** but are frequently too cumbersome to be of general use.

> Potentially confusing enzyme nomenclature includes ***synthetase*** **(requires ATP),** ***synthase*** **(no ATP required),** ***phosphatase*** **(uses water to remove a phosphate group),** ***phosphorylase*** **(uses inorganic phosphate to break a bond and generate a phosphorylated product),** ***dehydrogenase*** **(NAD^+ or flavin adenine dinucleotide [FAD] is an electron acceptor in a redox reaction),** ***oxidase*** **(oxygen is the acceptor, and oxygen atoms are not incorporated into substrate), and** ***oxygenase*** **(one or both oxygen atoms are incorporated).**

III. PROPERTIES

Enzymes are protein catalysts that increase the velocity of a chemical reaction and are not consumed during the reaction. [Note: Some ribonucleic acids (RNA) can catalyze reactions that affect phosphodiester and peptide bonds. **RNAs with catalytic activity are called ribozymes (see p. 620) and are much less common than protein catalysts.**]

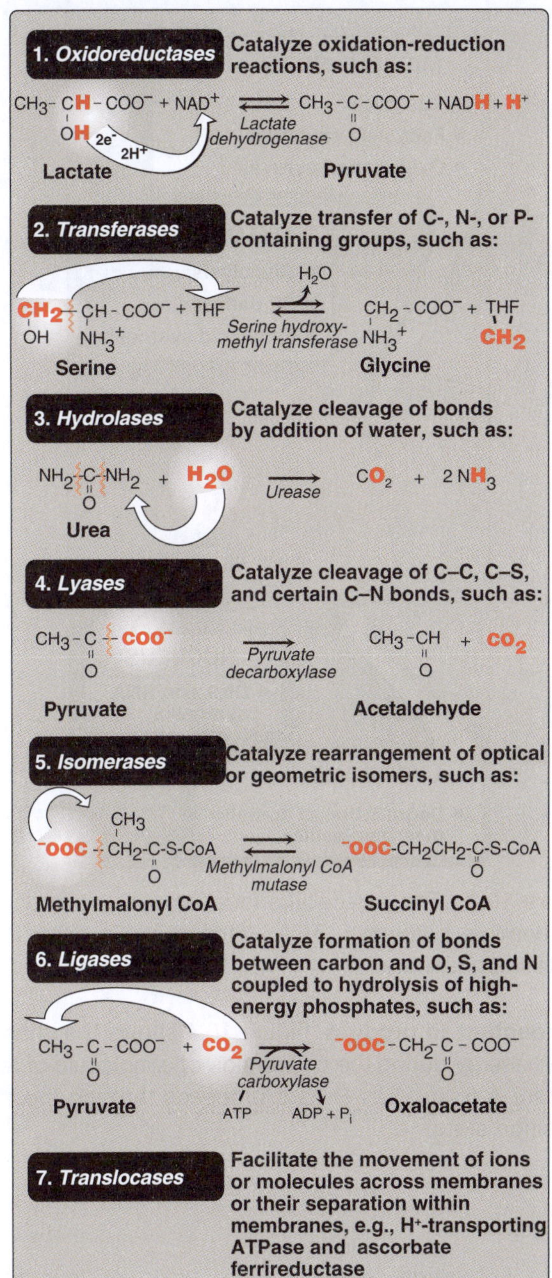

Figure 10.1 The **six major classes of enzymes with examples**. NAD(H) = nicotinamide adenine dinucleotide; THF = tetrahydrofolate; CoA = coenzyme A; CO_2 = carbon dioxide; NH_3 = ammonia; ADP = adenosine diphosphate; P_i = inorganic phosphate.

A. Active site

Enzyme molecules contain a special pocket or cleft called the active site. **The active site, formed by folding of the protein, contains amino acid side chains that participate in substrate binding and catalysis** (Fig. 10.2). The substrate binds the enzyme noncovalently, forming an **enzyme–substrate (ES) complex**. Binding is thought to cause a conformational change in the enzyme (**induced fit model**) that allows catalysis. ES is converted to an **enzyme–product (EP) complex** that subsequently dissociates to enzyme and product.

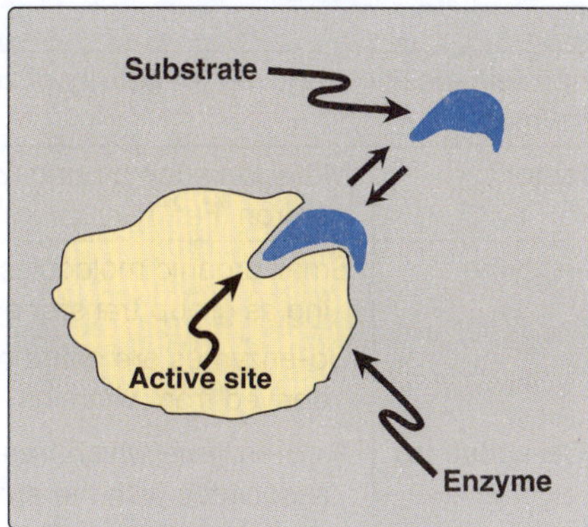

Figure 10.2 Schematic representation of an enzyme with one active site binding a substrate molecule.

B. Efficiency

Enzyme-catalyzed reactions are highly efficient, **proceeding from 10^3 to 10^8 times faster than uncatalyzed reactions**. The number of substrate molecules converted to product per enzyme molecule per second is called the turnover number, or k_{cat}, and typically is 10^2–10^4 s^{-1}. [Note: k_{cat} is the rate constant for the conversion of ES to E + P (see p. 135).]

C. Specificity

Enzymes **are highly specific**, interacting with one or a few substrates and catalyzing only one type of chemical reaction. The set of enzymes made in a cell determines which reactions occur in that cell.

D. Holoenzymes, apoenzymes, cofactors, and coenzymes

Some enzymes require nonproteins for enzymic activity. The term **holoenzyme** refers to the active enzyme with its nonprotein component, whereas the enzyme without its nonprotein moiety is termed an **apoenzyme** and is inactive. If the nonprotein moiety is a metal ion, such as zinc (Zn^{2+}) or iron (Fe^{2+}), it is called a **cofactor**. If it is a small organic molecule, it is termed a coenzyme. Coenzymes that only transiently associate with the enzyme are called **cosubstrates**. Cosubstrates dissociate from the enzyme in an altered state (NAD^+ is an example; see p. 199). If the coenzyme is permanently associated with the enzyme and returned to its original form, it is called a prosthetic group (FAD is an example; see p. 207). Coenzymes commonly are derived from vitamins. For example, NAD^+ contains niacin, and FAD contains riboflavin (see Chapter 32).

Nonprotein moieties required for activity of some enzymes	
Cofactor	Metal ion such as zinc (Zn^{2+}) or iron (Fe^{2+})
Co-enzyme	Small organic molecules serving as group transfer agents Co-enzymes are commonly derived from vitamins
a) Co-substrate	A co-enzyme which transiently associates with the enzyme and dissociates in its altered state. For e.g., NAD^+
b) Prosthetic group	A coenzyme which is tightly associated with the enzyme and returned to its original form. For e.g., FAD, pyridoxal phosphate, biotin

E. Regulation

Enzyme activity can be **regulated, that is, increased or decreased**, so that the rate of product formation responds to cellular need.

F. Location within the cell

Many enzymes are localized in **specific organelles within the cell** (Fig. 10.3). Such compartmentalization serves to isolate the reaction substrate or product from other competing reactions. This provides a favorable environment for the reaction and organizes the thousands of enzymes present in the cell into purposeful pathways.

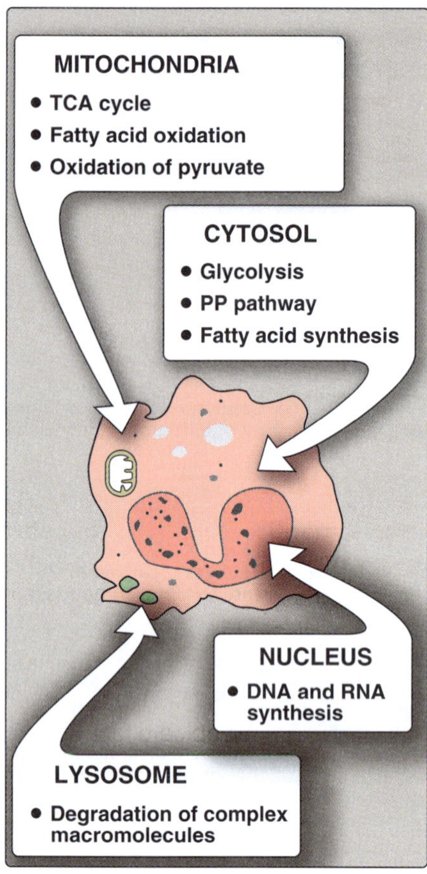

Figure 10.3 The intracellular location of some important biochemical pathways. TCA = tricarboxylic acid; PP = pentose phosphate.

IV. MECHANISM OF ENZYME ACTION

The mechanism of enzyme action can be viewed from two different perspectives. The first treats catalysis in terms of energy changes that occur during the reaction. That is, enzymes provide an alternate, energetically favorable reaction pathway different from the uncatalyzed reaction. The second perspective describes how the active site chemically facilitates catalysis.

A. Energy changes occurring during the reaction

Virtually all chemical reactions have an energy barrier separating the reactants and the products. **This barrier, called the activation energy (E_a), is the energy difference between that of the reactants and a high-energy intermediate, the transition state (T*), which is formed during the conversion of reactant to product.** Figure 10.4 shows the changes in energy during the conversion of a molecule of reactant A to product B as it proceeds through the transition state.

$$A \rightleftarrows T^* \rightleftarrows B$$

1. **Activation energy:** The peak of energy in Figure 10.4 is the **difference in free energy between the reactant and T*, in which the high-energy, short-lived intermediate is formed during the conversion of reactant to product**. Because of the high E_a, the rates of uncatalyzed chemical reactions are often slow.

2. **Rate of reaction:** For molecules to react, they must contain sufficient energy to overcome the energy barrier of the transition state. In the absence of an enzyme, only a small proportion of a population of molecules may possess enough energy to achieve the transition state between reactant and product. The rate of reaction is determined by the number of such energized molecules. **In general, the lower the E_a, the more molecules have**

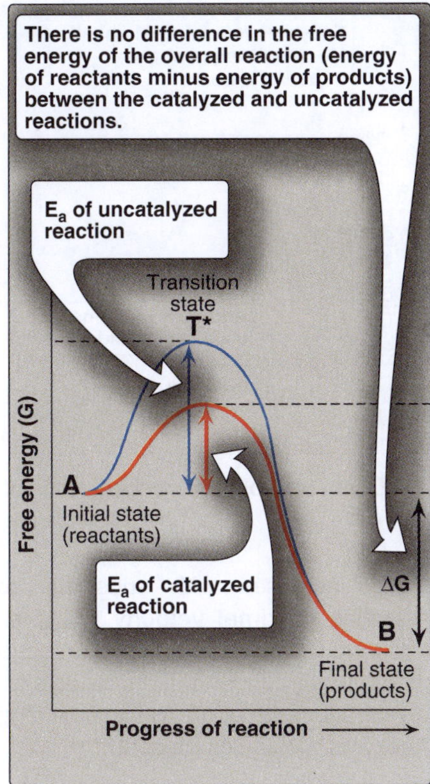

Figure 10.4 Effect of an enzyme on the activation energy (E_a) of a reaction. ΔG = change in free energy.

sufficient energy to pass through the transition state and, therefore, the faster the rate of the reaction.

3. **Alternate reaction pathway: An enzyme allows a reaction to proceed rapidly under conditions prevailing in the cell by providing an alternate reaction pathway with a lower E_a** (see Fig. 10.4). The enzyme does not change the free energies of the reactants (substrates) or products and, therefore, does not change the equilibrium of the reaction (see p. 161). It does, however, accelerate the rate by which equilibrium is reached.

B. Active site chemistry

The active site is not a passive receptacle for binding the substrate but, rather, is a complex molecular machine employing a diversity of chemical mechanisms to facilitate the conversion of substrate to product. **A number of factors are responsible for the catalytic efficiency of enzymes, including the following examples**.

1. **Transition-state stabilization:** The active site often acts as a flexible molecular template that binds the substrate and initiates its conversion to the transition state, a structure in which the bonds are not like those in the substrate or the product (see T* at the top of the curve in Fig. 10.4). By stabilizing the transition state, the enzyme greatly increases the concentration of the reactive intermediate that can be converted to product and, thus, accelerates the reaction. [Note: The transition state cannot be isolated.]

2. **Catalysis: The active site can provide catalytic groups that enhance the probability that the transition state is formed. In some enzymes, these groups can participate in general acid–base catalysis in which amino acid residues provide or accept proton**s. In other enzymes, catalysis may involve the transient formation of a covalent ES complex. [Note: The mechanism of action of ***chymotrypsin***, an enzyme of protein digestion in the intestine, includes general base, general acid and covalent catalysis. A histidine at the active site of the enzyme gains (general base) and loses (general acid) protons, mediated by the pK of histidine in proteins being close to physiologic pH. Serine at the active site forms a transient covalent bond with the substrate.]

3. **Transition-state visualization:** The enzyme-catalyzed conversion of substrate to product can be visualized as being similar to removing a sweater from an uncooperative infant (Fig. 10.5). The process has a high E_a because the only reasonable strategy for removing the garment (short of ripping it off) requires that the random flailing of the baby results in both arms being fully extended over the head, an unlikely posture. However, we can envision a parent acting as an enzyme, first coming in contact with the baby (forming ES) and then guiding the baby's arms into an extended, vertical position, analogous to the transition state. This posture (conformation) of the baby facilitates the removal of the sweater, forming the disrobed baby, which here represents product. [Note: The substrate bound to the enzyme (ES) is at a slightly lower energy than unbound substrate (S) and explains the small dip in the curve at ES.]

V. FACTORS AFFECTING REACTION VELOCITY

Enzymes can be isolated from cells and their properties studied in a test tube (that is, in vitro). Different enzymes show different responses to changes in substrate concentration, temperature, and pH. This section describes factors that influence the reaction velocity of enzymes. Enzymic responses to these factors give us valuable clues as to how enzymes function in living cells (that is, in vivo).

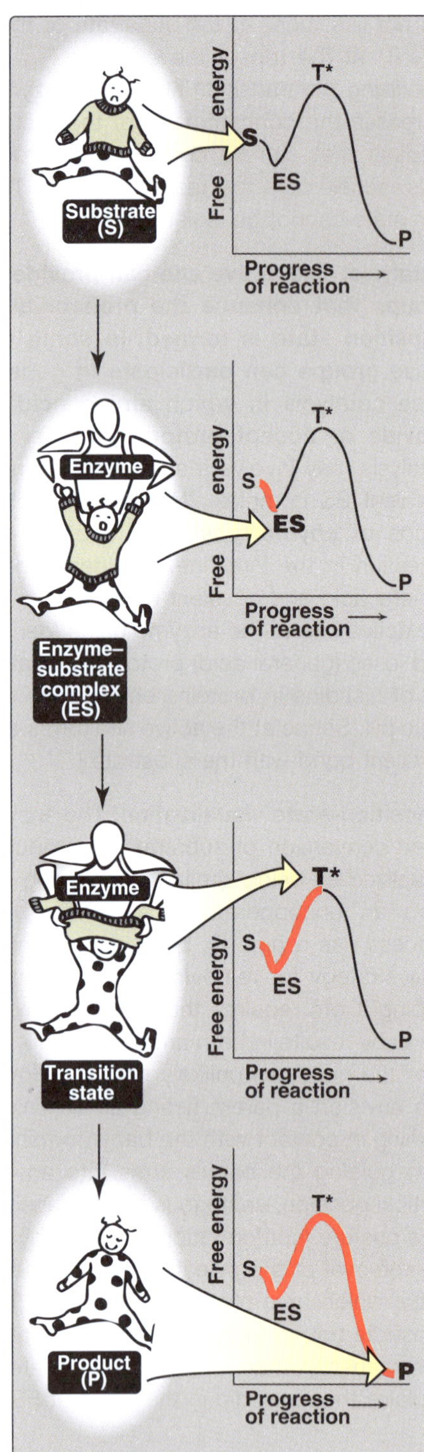

Figure 10.5 Schematic representation of energy changes accompanying formation of an enzyme–substrate complex and subsequent formation of a transition state.

A. Substrate concentration

1. **Maximal velocity: The rate or velocity of a reaction (v) is the number of substrate molecules converted to product per unit time.** Velocity is usually expressed as μmol of product formed per minute. The rate of an enzyme-catalyzed reaction increases with substrate concentration until a **maximal velocity (V_{max})** is reached (Fig. 10.6). The leveling off of the reaction rate at high substrate concentrations reflects the saturation with substrate of all available binding sites on the enzyme molecules present.

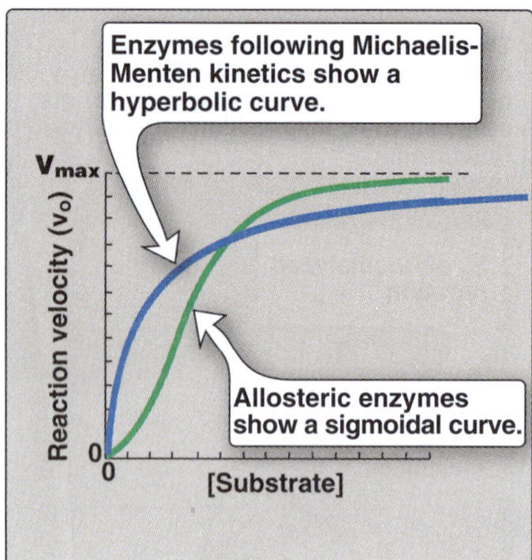

Figure 10.6 Effect of substrate concentration on reaction velocity.

2. **Shape of the enzyme kinetics curve:** Most enzymes show Michaelis-Menten kinetics (see p. 135), in which the plot of initial reaction velocity (v_o) against substrate concentration is **hyperbolic** (similar in shape to that of the oxygen-dissociation curve of myoglobin; see p. 99). In contrast, **allosteric enzymes do not follow Michaelis-Menten kinetics and show a sigmoidal curve** (see Fig. 10.6) that is similar in shape to the oxygen-dissociation curve of hemoglobin (see p. 99).

B. Temperature

1. **Velocity increase with temperature:** The reaction velocity **increases with temperature until a peak velocity is reached** (Fig. 10.7). This increase is the result of the increased number of substrate molecules having sufficient energy to pass over the energy barrier and form the products of the reaction.

2. **Velocity decrease with higher temperature:** Further elevation of the temperature causes a decrease in reaction velocity as a result of **temperature-induced denaturation of the enzyme** (see Fig. 10.7).

> The optimum temperature for most human enzymes is between 35°C and 40°C. Human enzymes start to denature (see p. 89) at temperatures above 40°C, but thermophilic bacteria found in hot springs have optimum temperatures of 70°C.

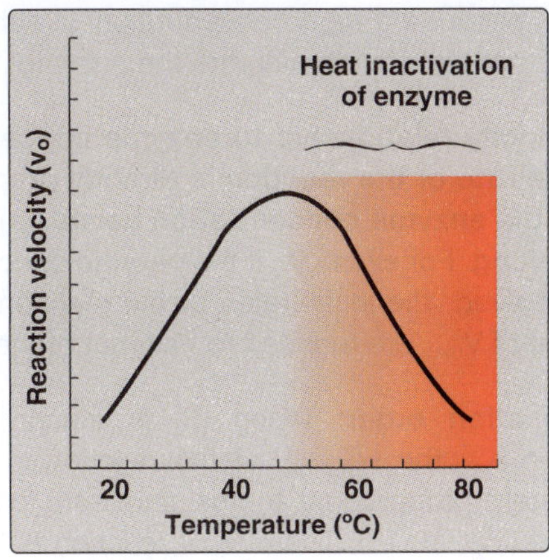

Figure 10.7 Effect of temperature on an enzyme-catalyzed reaction.

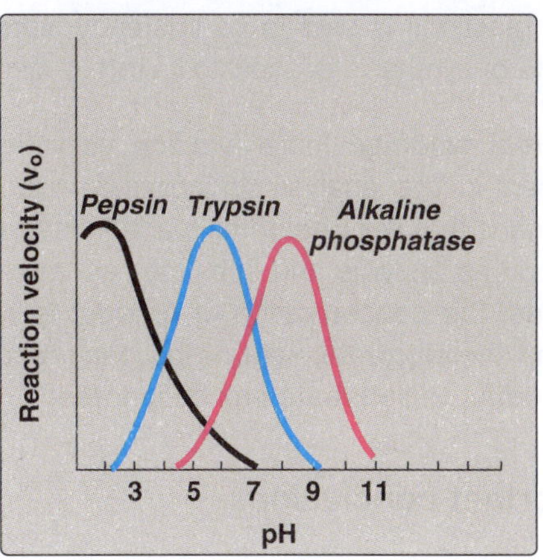

Figure 10.8 Effect of pH on enzyme-catalyzed reactions.

C. pH

1. **pH effect on active site ionization:** The concentration of protons ($[H^+]$) affects reaction velocity in several ways. First, the catalytic process usually requires that the enzyme and substrate have specific chemical groups in either an ionized or unionized state in order to interact. For example, catalytic activity may require that an amino group of the enzyme be in the protonated form ($-NH_3^+$). Because this group is deprotonated at alkaline pH, the rate of the reaction declines.

2. **pH effect on enzyme denaturation:** Extremes of pH can also lead to denaturation of the enzyme, because the structure of the catalytically active protein molecule depends on the ionic character of the amino acid side chains.

3. **Variable pH optimum: The pH at which maximal enzyme activity is achieved is different for different enzymes and often reflects the $[H^+]$ at which the enzyme functions** in the body. For example, ***pepsin,*** a digestive enzyme in the stomach, is maximally active at **pH 2,** whereas other enzymes, designed to work at neutral pH, are denatured by such an acidic environment (Fig. 10.8).

VI. MICHAELIS-MENTEN KINETICS

Leonor Michaelis and Maude Menten proposed a simple model that accounts for most of the features of many enzyme-catalyzed reactions. In this model, the enzyme reversibly combines with its substrate to form an ES complex that subsequently yields product, regenerating the free enzyme. The reaction model, involving one substrate molecule, is represented below:

$$E + S \underset{k_{-1}}{\overset{k_1}{\rightleftharpoons}} ES \xrightarrow{k_2} E + P$$

where S is the substrate.

E is the enzyme.

ES is the enzyme–substrate complex.

P is the product.

k_1, k_{-1}, and k_2 (or, k_{cat}) are rate constants.

A. Michaelis-Menten equation

The Michaelis-Menten equation describes how reaction velocity varies with substrate concentration:

$$v_o = \frac{V_{max}\,[S]}{K_m + [S]}$$

where v_o = initial reaction velocity

V_{max} = maximal velocity = $k_{cat}\,[E]_{Total}$

K_m = Michaelis constant = $(k_{-1} + k_2)/k_1$

$[S]$ = substrate concentration

The **following assumptions are made in deriving the Michaelis-Menten rate equation.**

1. **Enzyme and substrate relative concentrations:** The substrate concentration ([S]) is much greater than the concentration of enzyme so that the percentage of total substrate bound by the enzyme at any one time is small.

2. **Steady-state assumption:** The concentration of the ES complex does not change with time (the steady-state assumption), that is, the rate of formation of ES is equal to that of the breakdown of ES (to E + S and to E + P). In general, an intermediate in a series

of reactions is said to be in steady state when its rate of synthesis is equal to its rate of degradation.

3. **Initial velocity:** Initial reaction velocities (v_o) are used in the analysis of enzyme reactions. This means that the rate of the reaction is measured as soon as enzyme and substrate are mixed. At that time, the concentration of product is very small, and therefore, the rate of the back reaction from product to substrate can be ignored.

B. Important conclusions

1. **K_m characteristics:** K_m, the Michaelis constant, is characteristic of an enzyme and its particular substrate and reflects the affinity of the enzyme for that substrate. K_m is numerically equal to the substrate concentration at which the reaction velocity is equal to one half V_{max}. K_m does not vary with enzyme concentration.

    a. **Small K_m: A numerically small (low) K_m reflects a high affinity of the enzyme for substrate,** because a low concentration of substrate is needed to half-saturate the enzyme—that is, to reach a velocity that is one half V_{max} (Fig. 10.9).

    b. **Large K_m: A numerically large (high) K_m reflects a low affinity of enzyme for substrate** because a high concentration of substrate is needed to half-saturate the enzyme.

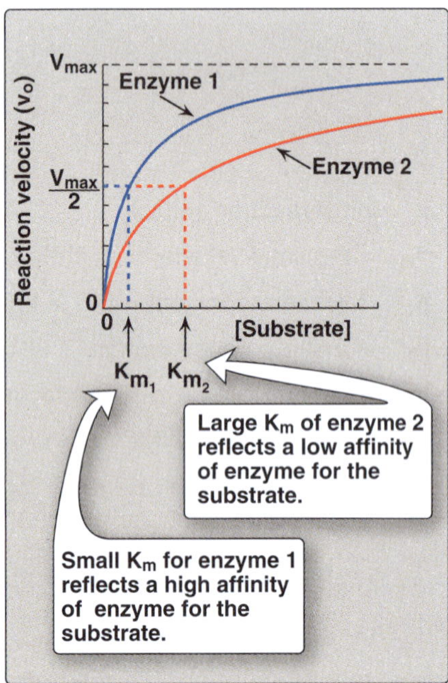

Figure 10.9 Effect of substrate concentration on reaction velocities for two enzymes: enzyme 1 with a small Michaelis constant (K_m) and enzyme 2 with a large K_m. V_{max} = maximal velocity.

2. **Velocity relationship to enzyme concentration: The rate of the reaction is directly proportional to the enzyme concentration because [S] is not limiting**. For example, if the enzyme concentration is halved, the initial rates of the reaction (v_o) and that of V_{max} are reduced to half that of the original.

3. **Reaction order:** When [S] is much less (<<) than K_m, the velocity of the reaction is approximately proportional to the substrate concentration (Fig. 10.10). The rate of reaction is then said to be **first order** with respect to substrate. When [S] is much greater (>>) than K_m, the velocity is constant and equal to V_{max}. The rate of reaction is then independent of substrate concentration (the enzyme is saturated with substrate) and is said to be **zero order** with respect to substrate concentration (see Fig. 10.10).

C. Lineweaver-Burk plot

When v_o is plotted against [S], it is not always possible to determine when V_{max} has been achieved because of the gradual upward slope of the hyperbolic curve

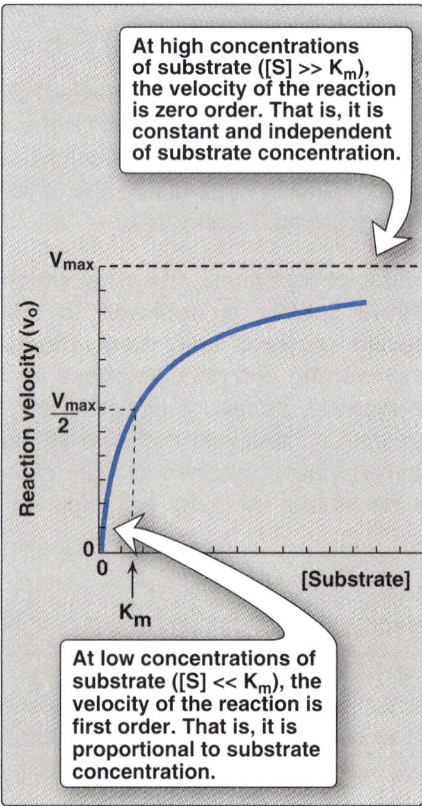

Figure 10.10 Effect of substrate concentration on reaction velocity for an enzyme-catalyzed reaction. V_{max} = maximal velocity; K_m = Michaelis constant.

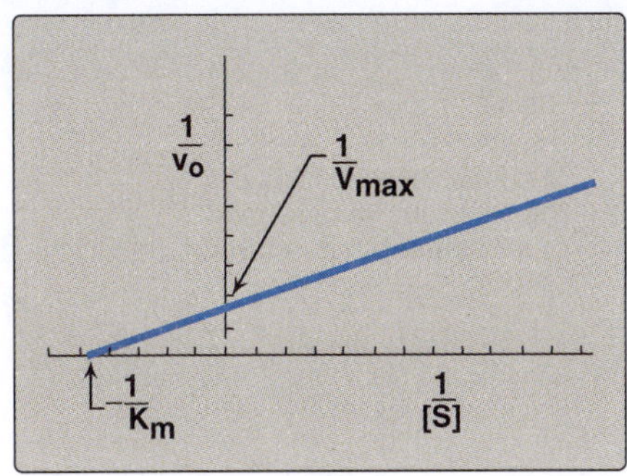

Figure 10.11 Lineweaver-Burk plot. v_o = initial reaction velocity; V_{max} = maximal velocity; K_m = Michaelis constant; [S] = substrate concentration.

at high substrate concentrations. **However, if $1/v_o$ is plotted versus 1/[S], a straight line is obtained** (Fig. 10.11). **This plot, the Lineweaver-Burk plot (also called a double-reciprocal plot) can be used to calculate K_m and V_{max} as well as to determine the mechanism of action of enzyme inhibitors.**

The equation describing the Lineweaver-Burk plot is:

$$\frac{1}{v_o} = \frac{K_m}{V_{max}\,[S]} + \frac{1}{V_{max}}$$

where the intercept on the x axis is equal to $-1/K_m$, and the intercept on the y axis is equal to $1/V_{max}$. [Note: The slope = K_m/V_{max}.]

VII. ENZYME INHIBITION

Any substance that can **decrease the velocity of an enzyme-catalyzed reaction** is called an inhibitor. Inhibitors can be reversible or irreversible. **Irreversible inhibitors** bind to enzymes through covalent bonds. Lead, for example, forms covalent bonds with the sulfhydryl side chain of cysteine in proteins. ***Ferrochelatase*, an enzyme involved in heme synthesis (see p. 417), is irreversibly inhibited by lead.** [Note: An important group of irreversible inhibitors are the mechanism-based inhibitors that are converted by the enzyme itself to a form that covalently links to the enzyme, thereby inhibiting it. They also are referred to as **"suicide" inhibitors.**] **Reversible inhibitors** bind to enzymes through noncovalent bonds and, thus, dilution of the enzyme–inhibitor complex results in dissociation of the reversibly bound inhibitor and recovery of enzyme activity. The two most commonly encountered types of reversible inhibition are competitive and noncompetitive.

A. Competitive inhibition

This type of inhibition occurs when the inhibitor binds reversibly to the same site that the substrate would normally occupy and, therefore, competes with the substrate for that site.

1. **Effect on V_{max}:** The effect of a competitive inhibitor is reversed by increasing the concentration of substrate**.** At a sufficiently high [S], the reaction velocity reaches the V_{max} observed in the absence of inhibitor, that is, **V_{max} is unchanged** (Fig. 10.12).

2. **Effect on K_m:** A competitive inhibitor **increases the apparent K_m** for a given substrate. This means that, in the presence of a competitive inhibitor, more substrate is needed to achieve one half V_{max}.

3. **Effect on the Lineweaver-Burk plot:** Competitive inhibition shows a characteristic Lineweaver-Burk plot in which the plots of the inhibited and uninhibited reactions intersect on the y axis at $1/V_{max}$ (V_{max} is unchanged). The inhibited and uninhibited reactions show different x-axis intercepts, indicating that the apparent K_m is increased in the presence of the competitive inhibitor because $-1/K_m$ moves closer to zero from a negative value (see Fig. 10.12). [Note: An important group of competitive inhibitors are the transition state analogs, stable molecules that approximate the structure of the transition state, and, therefore, bind the enzyme more tightly than does the substrate.]

4. **Statin drugs as examples of competitive inhibitors:** This group of antihyperlipidemic agents competitively inhibits the rate-limiting (slowest) step in cholesterol biosynthesis. This reaction is catalyzed by *hydroxymethylglutaryl coenzyme A reductase* (*HMG CoA reductase;* see p. 343). Statins, such as atorvastatin (Lipitor) and pravastatin (Pravachol), are structural analogs of the natural substrate for this enzyme and **compete effectively to inhibit *HMG CoA reductase*.** By doing so, they inhibit de novo cholesterol synthesis, thereby lowering plasma cholesterol levels (Fig. 10.13).

B. Noncompetitive inhibition

This type of inhibition is recognized by its characteristic effect on V_{max} (Fig. 10.14). Noncompetitive inhibition occurs when the inhibitor and substrate bind at different sites on the enzyme. The noncompetitive inhibitor can bind either free enzyme or the enzyme–substrate complex, thereby preventing the reaction from occurring (Fig. 10.15).

1. **Effect on V_{max}:** Noncompetitive inhibition cannot be overcome by increasing the concentration of substrate. Therefore, noncompetitive inhibitors **decrease the apparent V_{max}** of the reaction.

2. **Effect on K_m:** Noncompetitive inhibitors do not interfere with the binding of substrate to enzyme. Therefore, the enzyme shows the **same K_m** in the

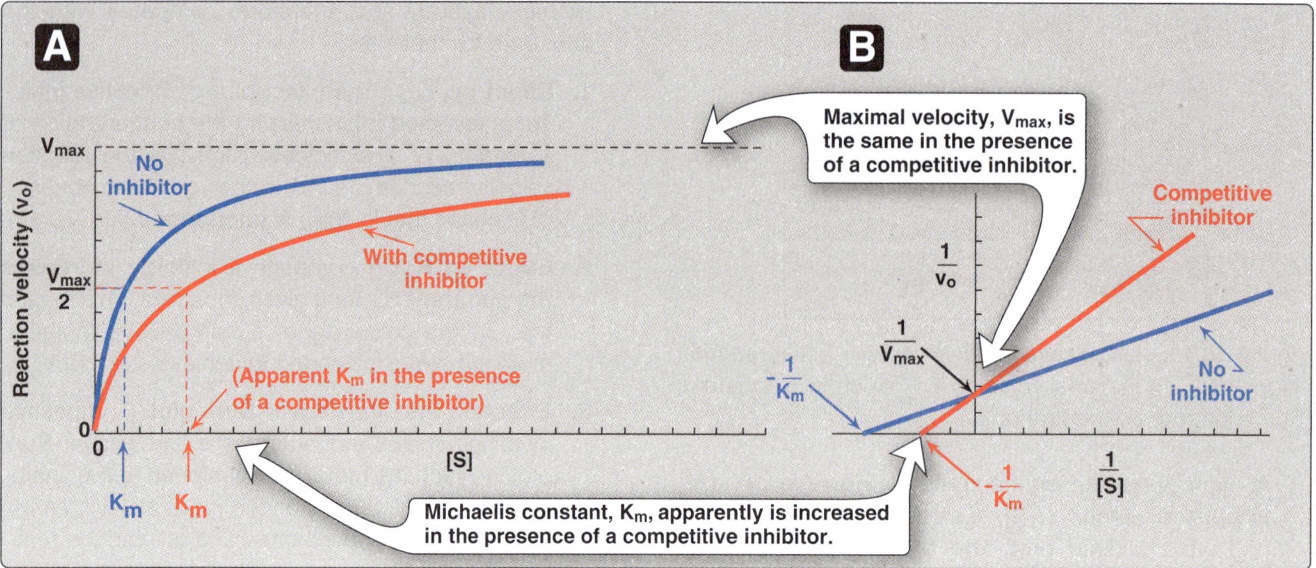

Figure 10.12 **A.** Effect of a competitive inhibitor on the reaction velocity versus substrate concentration ([S]) plot. **B.** Lineweaver-Burk plot of competitive inhibition of an enzyme. [Note: The slope increases if inhibitor concentration increases.]

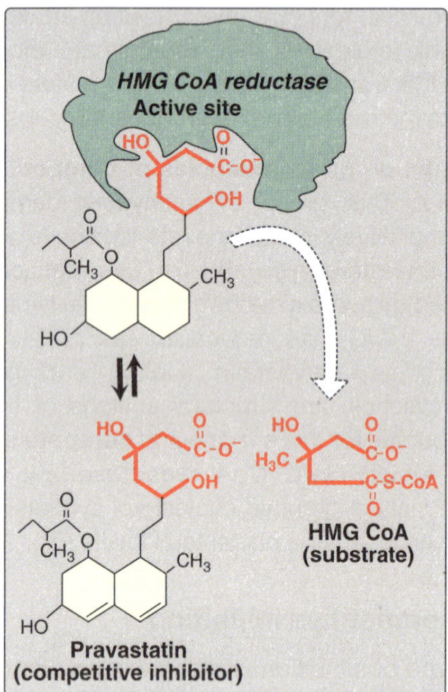

Figure 10.13 Pravastatin competes with hydroxymethylglutaryl coenzyme A (HMG CoA) for the active site of *HMG CoA reductase*.

presence or absence of the noncompetitive inhibitor, that is, K_m is unchanged.

3. **Effect on Lineweaver-Burk plot:** Noncompetitive inhibition is readily differentiated from competitive inhibition by plotting $1/v_o$ versus 1/[S] and noting that the apparent V_{max} decreases in the presence of a noncompetitive inhibitor, whereas **K_m is unchanged** (see Fig. 10.14). [Note: **Oxypurinol, a metabolite of the prodrug allopurinol, is a noncompetitive inhibitor of *xanthine oxidase***, an enzyme of purine degradation (see p. 585).]

Poisonous enzymes are often secreted by bacteria as a means to circumvent host defenses. These enzymes play various roles in undermining the immune response and aiding bacterial invasion and spread:

- **Hemolysins:** Produced by bacteria like streptococci hemolysins target and destroy red blood cells, aiding in tissue damage and facilitating bacterial survival.
- **Leucocidins:** These enzymes target and destroy white blood cells, which are essential components of the immune system, weakening the host's defense mechanisms.
- **Coagulase:** Coagulase enzymes promote blood clotting, forming a protective barrier around the bacteria that helps them evade phagocytosis by immune cells.
- **Streptokinase:** Streptokinase is capable of dissolving blood clots, enabling bacteria to invade host tissues. Interestingly, streptokinase is also used therapeutically to dissolve blood clots in medical settings.
- **Collagenase and hyaluronidase:** These enzymes break down collagen and hyaluronic acid, respectively. By degrading these tissues, collagenase and hyaluronidase facilitate the spread of bacterial infection within the host.

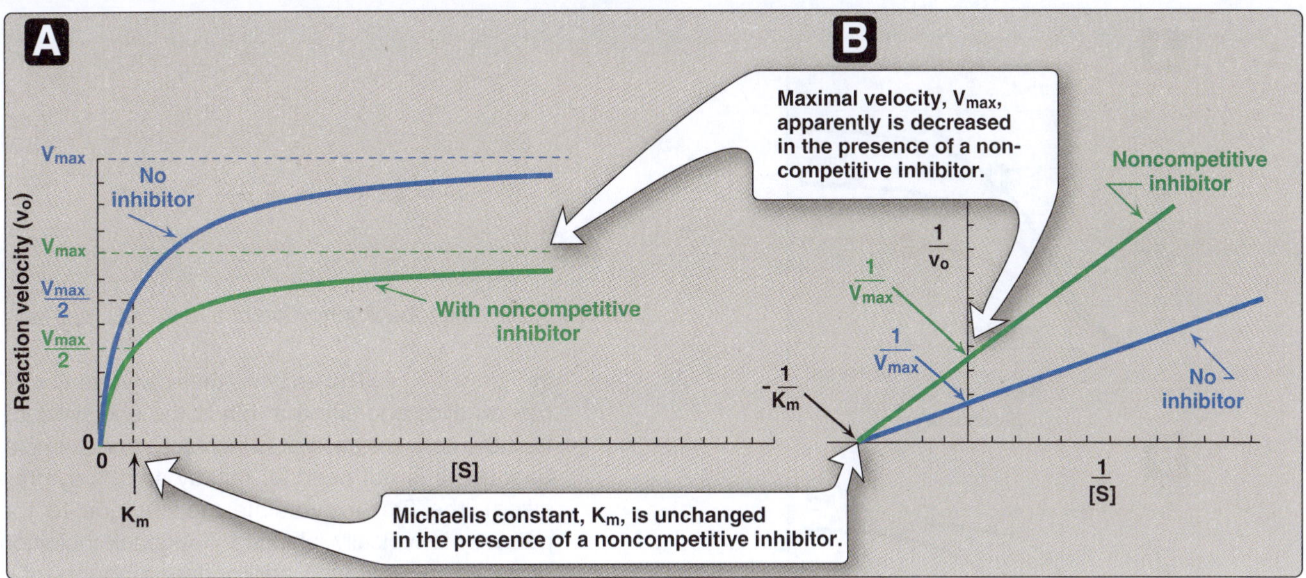

Figure 10.14 **A.** Effect of a noncompetitive inhibitor on the reaction velocity versus substrate concentration ([S]) plot. **B.** Lineweaver-Burk plot of noncompetitive inhibition of an enzyme. [Note: The slope increases if inhibitor concentration increases.]

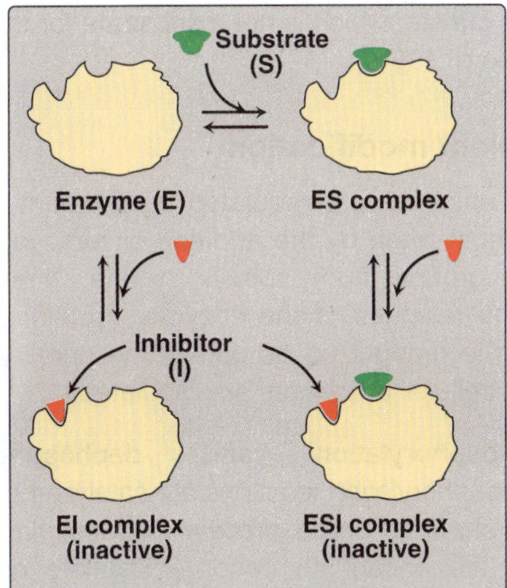

Figure 10.15 A noncompetitive inhibitor binding to both free enzyme and enzyme–substrate (ES) complex.

C. Enzyme inhibitors as drugs

At least half of the ten most commonly prescribed drugs in the United States act as enzyme inhibitors. For example, the widely prescribed **β-lactam antibiotics, such as penicillin and amoxicillin**, act by inhibiting enzymes involved in bacterial cell wall synthesis. Drugs may also act by inhibiting extracellular reactions. This is illustrated by ***angiotensin-converting enzyme (ACE)* inhibitors.** They lower blood pressure by blocking plasma *ACE* that cleaves angiotensin I to form the potent vasoconstrictor, angiotensin II. These drugs, which include **captopril, enalapril, and lisinopril**, cause vasodilation and, therefore, a reduction in blood pressure. **Aspirin**, a nonprescription drug, irreversibly inhibits prostaglandin and thromboxane synthesis by inhibiting *cyclooxygenase* (see p. 334).

VIII. ENZYME REGULATION

The regulation of the reaction velocity of enzymes is essential if an organism is to coordinate its numerous metabolic processes. The rates of most enzymes are responsive to changes in substrate concentration, because the intracellular level of many substrates is in the range of the K_m. Thus, an increase in substrate concentration prompts an increase in reaction rate, which tends to return the concentration of substrate toward normal. In addition, some enzymes with specialized regulatory functions respond to allosteric effectors and/or covalent modification or they show altered rates of enzyme synthesis (or degradation) when physiologic conditions are changed.

A. Allosteric enzymes

Allosteric enzymes are regulated by molecules called effectors that bind noncovalently at a site other than the active site. These enzymes are almost always composed of multiple subunits, and the regulatory (allosteric) site that binds the effector is distinct from the substrate-binding site and may be located on a subunit that is not itself catalytic. Effectors that inhibit enzyme activity are termed negative effectors, whereas those that increase enzyme activity are called positive effectors. **Positive and negative effectors** can affect the affinity of the enzyme for its

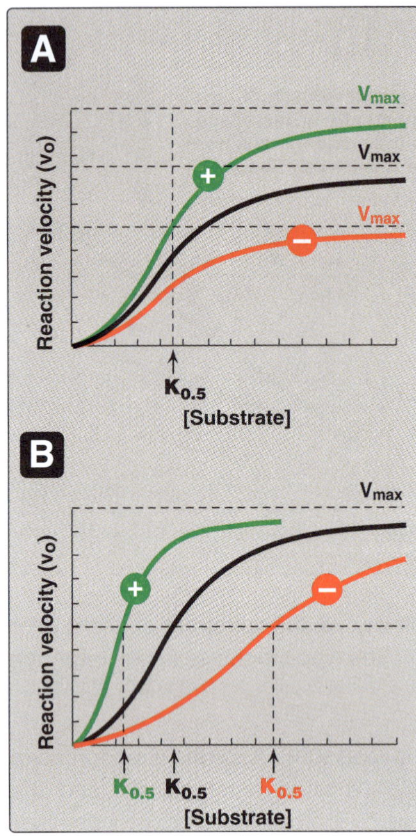

Figure 10.16 Effects of negative ⊖ or positive ⊕ effectors on an allosteric enzyme. **A.** Maximal velocity (V_{max}) is altered. **B.** The substrate concentration that gives half maximal velocity ($K_{0.5}$) is altered.

substrate ($K_{0.5}$), modify the maximal catalytic activity of the enzyme (V_{max}), or both (Fig. 10.16). [Note: Allosteric enzymes frequently catalyze the committed step, often the rate-limiting step, early in a pathway.]

1. **Homotropic effectors: When the substrate itself serves as an effector, the effect is said to be homotropic.** Most often, an allosteric substrate functions as a positive effector. In such a case, the presence of a substrate molecule at one site on the enzyme enhances the catalytic properties of the other substrate-binding sites. That is, their binding sites exhibit cooperativity. These enzymes show a sigmoidal curve when v_o is plotted against substrate concentration, as shown in Figure 10.16. This contrasts with the hyperbolic curve characteristic of enzymes following Michaelis-Menten kinetics, as previously discussed. [Note: **The concept of cooperativity of substrate binding is analogous to the binding of oxygen to hemoglobin** (see p. 101).]

2. **Heterotropic effectors: The effector may be different from the substrate, in which case the effect is said to be heterotropic.** For example, consider the feedback inhibition shown in Figure 10.17. The enzyme that converts D to E has an allosteric site that binds the end product, G. If the concentration of G increases (for example, because it is not used as rapidly as it is synthesized), the first irreversible step unique to the pathway is typically inhibited. Feedback inhibition provides the cell with appropriate amounts of a product it needs by regulating the flow of substrate molecules through the pathway that synthesizes that product. Heterotropic effectors are commonly encountered. **For example, the glycolytic enzyme *phosphofructokinase-1* is allosterically inhibited by citrate**, which is not a substrate for the enzyme (see p. 195).

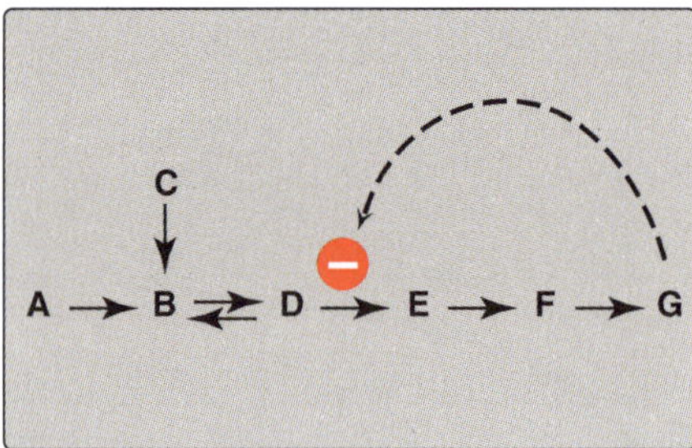

Figure 10.17 Feedback inhibition of a metabolic pathway.

B. Covalent modification

Many enzymes are regulated by covalent modification, most often by the addition or removal of phosphate groups from specific serine, threonine, or tyrosine residues of the enzyme. Protein phosphorylation is recognized as one of the primary ways in which cellular processes are regulated.

1. **Phosphorylation and dephosphorylation:** Phosphorylation reactions are catalyzed by a family of enzymes called ***protein kinases*** that use ATP as the phosphate donor. Phosphate groups are cleaved from phosphorylated enzymes by the action of ***phosphoprotein phosphatases*** (Fig. 10.18).

2. **Enzyme response to phosphorylation:** Depending on the specific enzyme, the phosphorylated form may be more or less active than the unphosphorylated enzyme. For example, hormone-mediated phosphorylation of ***glycogen phosphorylase*** (an enzyme that degrades glycogen) increases activity, whereas phosphorylation of ***glycogen synthase*** (an enzyme that synthesizes glycogen) decreases activity (see p. 232).

C. Enzyme synthesis

The regulatory mechanisms described above modify the activity of existing enzyme molecules. However, cells can also regulate the amount of enzyme present

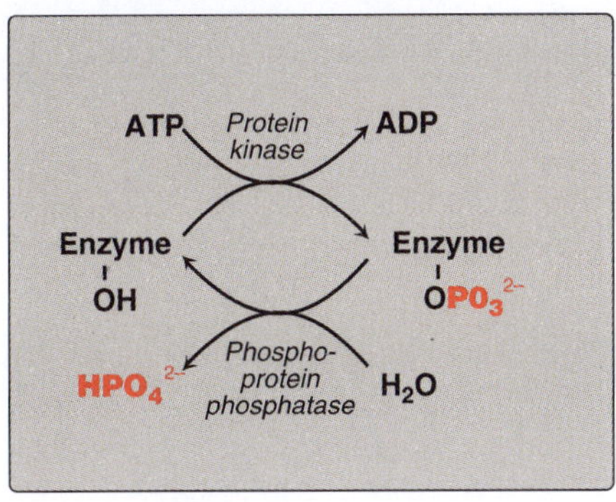

Figure 10.18 Covalent modification by the addition and removal of phosphate groups. [Note: HPO_4^{2-} may be represented as P_i and PO_3^{2-} as Ⓟ.] ADP = adenosine diphosphate.

by altering the rate of enzyme degradation or, more typically, the rate of enzyme synthesis. **The increase (induction) or decrease (repression) of enzyme synthesis** leads to an alteration in the total population of active sites. Enzymes subject to regulation of synthesis are often those that are needed at only one stage of development or under selected physiologic conditions. For example, elevated levels of insulin as a result of high blood glucose levels cause an increase in the synthesis of key enzymes involved in glucose metabolism (see p. 201). In contrast, enzymes that are in constant use are usually not regulated by altering the rate of enzyme synthesis. Alterations in enzyme levels as a result of induction or repression of protein synthesis are slow (hours to days), compared with allosterically or covalently regulated changes in enzyme activity, which occur in seconds to minutes. Figure 10.19 summarizes the common ways that enzyme activity is regulated.

IX. ENZYMES IN CLINICAL DIAGNOSIS

Plasma enzymes can be classified into two major groups. **First, a relatively small group of enzymes are actively secreted into the blood by certain cell types**. For example, the liver secretes zymogens (inactive precursors) of the enzymes involved in **blood coagulation**. **Second, a large number of enzyme species are released from cells during normal cell turnover.** These enzymes almost always function intracellularly and have no physiologic use in the plasma. In healthy individuals, the levels of these enzymes are fairly constant and represent a steady state in which the rate of release from damaged cells into the plasma is balanced by an equal rate of removal from the plasma. **Increased plasma levels of these enzymes may indicate tissue damage** (Fig. 10.20).

> Plasma is the fluid, noncellular part of blood. Laboratory assays of enzyme activity most often use serum, which is obtained by centrifugation of whole blood after it has been allowed to coagulate. Plasma is a physiologic fluid, whereas serum is prepared in the laboratory.

REGULATOR EVENT	TYPICAL EFFECTOR	RESULTS	TIME REQUIRED FOR CHANGE
Substrate availability	Substrate	Change in velocity (v_o)	Immediate
Product inhibition	Reaction product	Change in V_{max} and/or K_m	Immediate
Allosteric control	Pathway end product	Change in V_{max} and/or $K_{0.5}$	Immediate
Covalent modification	Another enzyme	Change in V_{max} and/or K_m	Immediate to minutes
Synthesis or degradation of enzyme	Hormone or metabolite	Change in the amount of enzyme	Hours to days

Figure 10.19 Mechanisms for regulating enzyme activity. [Note: Inhibition by pathway end product is also referred to as feedback inhibition.]

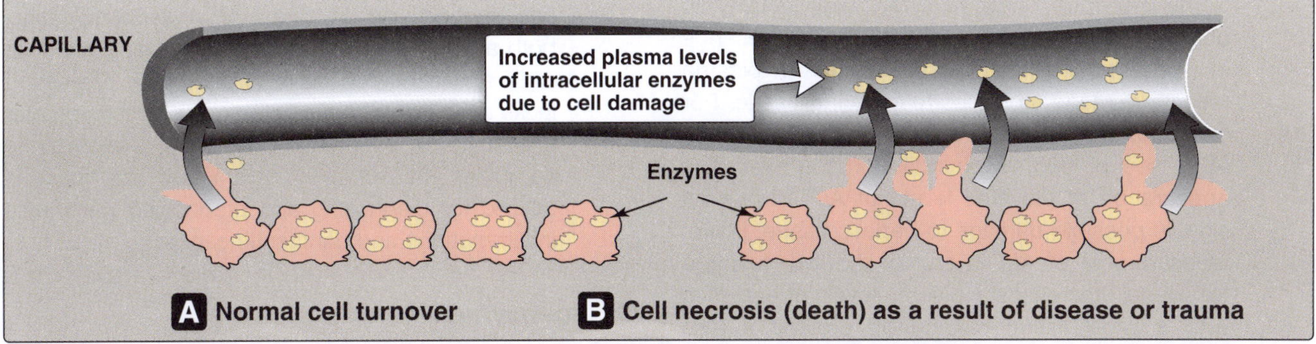

Figure 10.20 Release of enzymes from normal (**A**) and diseased or traumatized (**B**) cells.

A. Plasma enzyme levels in disease states

Many diseases that cause tissue damage result in an increased release of intracellular enzymes into the plasma. The activities of many of these enzymes are routinely determined for diagnostic purposes in diseases of the heart, liver, skeletal muscle, and other tissues. **The level of specific enzyme activity in the plasma frequently correlates with the extent of tissue damage**. Therefore, determining the degree of elevation of a particular enzyme activity in the plasma is often useful in evaluating the prognosis for the patient.

B. Plasma enzymes as diagnostic tools

(**See Appendix Unit II Table 2: Enzymes of clinical significance.** Refer to Chapter 48, page 807, for information on the diagnostic significance of ALP, GGT, and 5'nucleotidase. Additionally, Chapter 48, Liver function tests, page 805, and Figure 48.1 provide insights into diagnostic enzymes of hepatocellular damage, namely AST, ALP, and ALT.)

1. **Diagnosis of myocardial infarction:** Measurement of blood levels of proteins with cardiac specificity (biomarkers) is used in the diagnosis of MI. Myocardial muscle is the only tissue that contains >5% of the total ***CK* activity as the *CK2* (MB)** isoenzyme. Appearance of this hybrid isoenzyme in plasma is virtually specific for infarction of the myocardium. Following an acute MI, *CK2* appears in plasma within 4–8 hours following onset of chest pain, reaches a peak of activity at ~24 hours, and returns to baseline after 48–72 hours (Fig. 10.21). **Troponins T (TnT) and I (TnI)** are regulatory proteins involved in muscle contractility. **Cardiac-specific isoforms (cTn)** are released into the plasma in response to cardiac damage. They are highly sensitive and specific for damage to cardiac tissue. cTn appear in plasma within 4–6 hours after an MI, peak in 24–36 hours, and remain elevated for 3–10 days. Elevated cTn, in combination with the clinical presentation and characteristic changes in the ECG, are currently considered the "gold standard" in the diagnosis of an MI.

   Some enzymes show relatively high activity in only one or a few tissues. Therefore, the presence of increased levels of these enzymes in plasma reflects damage to the corresponding tissue. For example, the enzyme *alanine aminotransferase* (*ALT*; see p. 805) is abundant in the liver. The appearance of elevated levels of *ALT* in plasma signals possible damage to hepatic tissue. [Note: Measurement of *ALT* is part of the liver function test panel.] Increases in plasma levels of enzymes with a wide tissue distribution provide a less specific indication of the site of cellular injury and limits their diagnostic value.

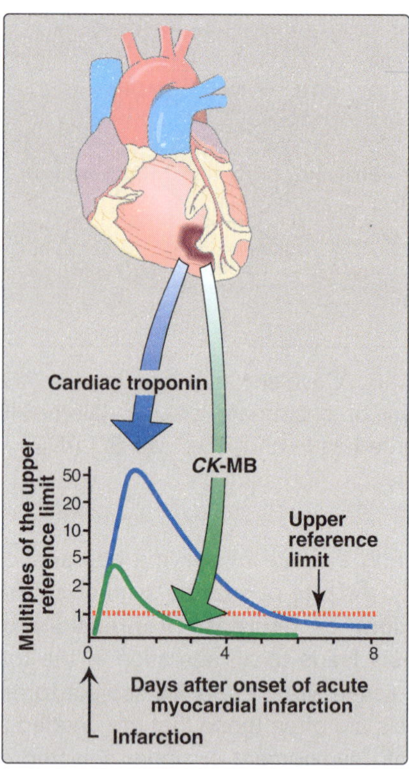

Figure 10.21 Appearance of *creatine kinase* isozyme *CK*-MB and cardiac troponin in plasma after a myocardial infarction. [Note: Either cardiac troponin T or I may be measured.]

2. **Diagnosis of prostatic diseases:** Prostate-specific antigen (PSA), secreted from the secretary epithelium of the prostate gland, serves as a significant biomarker for both benign prostatic hyperplasia (BPH) and prostate cancers. In normal conditions, its level ranges between 1 to 4 mg/L. However, in cases of cancer, PSA levels typically elevate, often exceeding 10 mg/L, though levels lower than 10 mg/L can also be indicative of cancer. Conversely, BPH generally presents with PSA values below 10 mg/L.

3. **Diagnosis of pancreatic diseases:** Amylase and lipase are enzymes found in pancreatic secretions, aiding in the digestion of carbohydrates and fats, respectively. In cases of acute pancreatitis, both amylase and lipase levels increase. However, while amylase is sensitive to pancreatic disease, it lacks specificity. Conversely, lipase exhibits greater specificity for pancreatic conditions. Clinically, the combined measurement of amylase and lipase is utilized for diagnosing and prognosing pancreatic inflammations and pancreatic cancers.

C. Isoenzymes

Isoenzymes (also called isozymes) are enzymes that catalyze the same reaction. However, they do not necessarily have the same physical properties because of genetically determined differences in amino acid sequence. For this reason, isoenzymes may contain different numbers of charged amino acids, which allows electrophoresis (the movement of charged particles in an electric field) to separate them (Fig. 10.22). Different organs commonly contain characteristic proportions of different isoenzymes. The pattern of isoenzymes found in the plasma may, therefore, serve as a means of identifying the site of tissue damage. For example, the plasma levels of *creatine kinase* (*CK*) are commonly determined in the diagnosis of myocardial infarction (MI). They are particularly useful when the electrocardiogram (ECG) is difficult to interpret such as when there have been previous episodes of heart disease.

1. **Isoenzyme of creatinine kinase:** Many isoenzymes contain different subunits in various combinations. For example, ***CK* occurs as three isoenzymes**. Each isoenzyme is a dimer composed of two polypeptides (called B and M subunits) associated in one of three combinations: *CK1* = BB, *CK2* = MB, and *CK3* = MM. Each *CK* isoenzyme shows a characteristic electrophoretic mobility (see Fig. 10.21). [Note: **Virtually all *CK* in the brain is the BB isoform, whereas it is MM in skeletal muscle. In cardiac muscle, about one third is MB with the rest as MM.**]

2. **Isoenzymes of alkaline phosphatase:** Alkaline phosphatase (ALP) Isoenzymes of alkaline phosphatase (ALP) play crucial roles in various physiological and pathological processes. The total ALP levels typically range from 40 to 130 U/L. ALPI, the intestinal ALP isoenzyme secreted by enterocytes, exhibits altered expression in conditions such as inflammatory bowel disease. Liver ALP is primarily expressed in the liver, with markedly elevated levels observed in cases of extrahepatic cholestasis. Additionally, increased liver ALP levels may occur in intrahepatic cholestasis due to viral hepatitis or as a side effect of certain medications. The bone isoenzyme of ALP typically increases during periods of skeletal growth in children and can be elevated in various bone disorders, including rickets, Paget's disease, osteomalacia, osteoblastoma, and bone cancers. Furthermore, the placental isoenzyme of ALP is detected in the blood during normal pregnancy. These distinct isoenzymes of ALP serve as valuable diagnostic markers in different clinical scenarios, aiding in the assessment and management of various diseases and conditions affecting the liver, intestines, bones, and pregnancy.

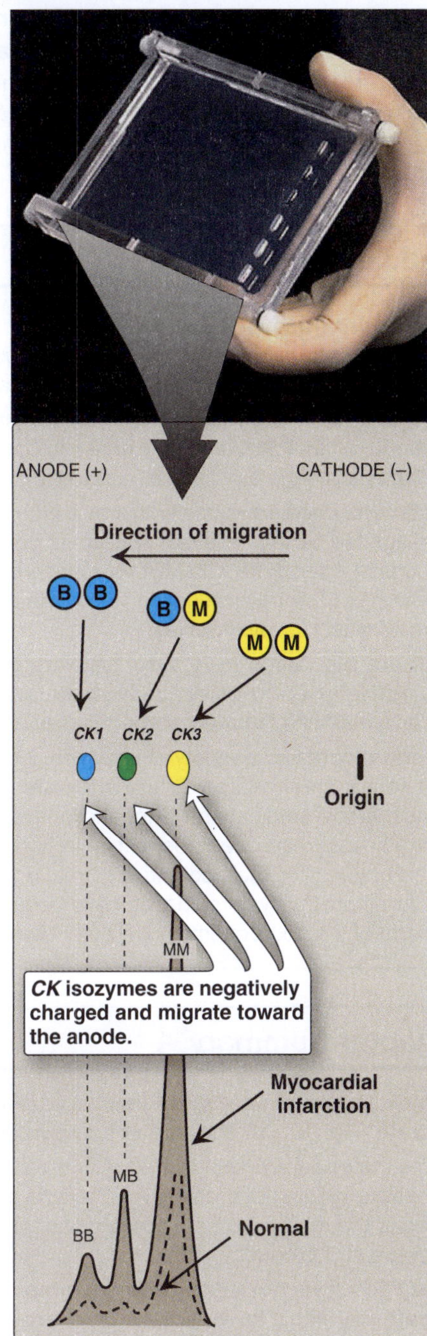

Figure 10.22 Subunit composition, electrophoretic mobility, and enzyme activity of *creatine kinase* (*CK*) isoenzymes.

3. **Isoenzymes of LDH:** LDH plays a critical role in converting lactate to pyruvate and serves as a valuable diagnostic marker. LDH enzyme levels in serum are commonly utilized in clinical settings for diagnostic purposes (see Appendix Unit 2, p. 149). LDH is a tetramer composed of four subunits, which can consist of either heart (H) or muscle (M) polypeptide chains. The LDH enzyme has five isoenzymes, each associated with specific organs affected by disease. LDH1 (H4) predominates

in heart muscle, while LDH2 (H3M1) and LDH3 (H2M2) are prominent in red blood cells and brain tissue, respectively. LDH4 (H1M3) and LDH5 (M4) are primarily found in the liver and skeletal muscle, respectively. Elevations in LDH levels are observed in various conditions, including myocardial infarction, cancers, hemolytic anemia, hepatocellular damage, and muscular dystrophies. Monitoring LDH levels provides valuable insight into tissue damage and disease progression, aiding in the diagnosis and management of these disorders.

X. Medical Perspectives

Therapeutic enzymes are those which are given to patients for treatment of diseases, e.g., *streptokinase* (lyses intravascular clots in acute myocardial infarction), *asparaginase* (used in acute lymphoblastic leukemia), pancreatin (*lipase* and *trypsin* for pancreatic insufficiency) and *papain* (anti-inflammatory).

Many drugs have components to inhibit enzymes of body and include statins (HMG CoA inhibitors to reduce cholesterol), allopurinol (inhibits *xanthine oxidase* in gout), methotrexate (inhibition of *dihydrofolate reductase* in cancer), sulfonamides (inhibits *pteroid synthetase* in bacteria), acyclovir (antiviral DNA polymerase) and many more.

Diagnostic enzymes are those enzymes which are released into blood when there is disease, and their blood levels can help us diagnose certain diseases. Enzymes commonly used to diagnose diseases include *lactate dehydrogenase*, *AST*, *creatine kinase* isoenzyme (CK-MB) for myocardial infarction, and *AST*, *ALT*, *alkaline phosphatase (ALP)*, *gamma glutamyltransferase (GGT)* for liver function tests (see Appendix Unit II). *Creatine kinase* isoenzymes (CK-MM) are also used for diagnosis of muscular dystrophies.

Enzymes are also used in in vitro test kits in labs for analysis and include *glucose oxidase* (used for estimation of glucose), *lipase* (used to detect triglyceride levels), *urease* (used to determine urea levels) and *horseradish peroxidase* (used in enzyme-linked immunosorbent assay [ELISA]).

Snake venom contains a variety of enzymes, including serine proteinases, metalloproteinases, phospholipase A2, acetylcholinesterases, and hyaluronidases, which are responsible for its toxic effects. However, these venom components have also found medical applications. Phospholipases A2 and hyaluronidases have shown promise in managing certain cancers. Serine proteinases have been beneficial in the treatment of acute ischemic stroke and thrombosis. Metalloproteinases have been studied for their potential in treating neurodegenerative conditions like Parkinson's disease and Alzheimer's disease. Additionally, hyaluronidases have been utilized in aesthetic cosmetology procedures. Thus, while snake venom can be harmful, its components have been harnessed for therapeutic purposes.

XI. Chapter Summary

- Enzymes are **protein catalysts** that increase the velocity of a chemical reaction by lowering the energy of the **transition state** (Fig. 10.23). They are not consumed during the reaction.
- Enzyme molecules contain a special cleft called the **active site**, which contains amino acid side chains that participate in substrate binding and catalysis. The active site binds the substrate, forming an **enzyme–substrate** (**ES**) **complex**. Binding is thought to cause a conformational change in the enzyme (induced fit) that allows catalysis. ES is converted to enzyme and product.
- An enzyme allows a reaction to proceed rapidly under conditions prevailing in the cell by providing an **alternate reaction pathway** with a **lower activation energy** (**E_a**). Because the enzyme does not change the free energies of the reactants or products, it does not change the equilibrium of the reaction.
- Most enzymes show **Michaelis-Menten kinetics**, and a plot of the **initial reaction velocity** (v_o) against **substrate concentration** (**[S]**) has a **hyperbolic** shape similar to the oxygen-dissociation curve of myoglobin. A **Lineweaver-Burk** plot of 1/v and 1/[S] allows determination of **V_{max}** (maximal velocity) and **K_m** (Michaelis constant, which reflects affinity for substrate).
- Any substance that can decrease the velocity of an enzyme-catalyzed reaction is called an **inhibitor**. The two most common types of reversible inhibition are **competitive** (which **increases** the **apparent K_m**) and **noncompetitive** (which **decreases** the **apparent V_{max}**).
- In contrast, the **multisubunit allosteric enzymes** show a **sigmoidal curve** similar in shape to the oxygen-dissociation curve of hemoglobin. They typically catalyze the **committed step** of a pathway.
- **Allosteric enzymes** are regulated by molecules called **effectors** that bind noncovalently at a site other than the active site. Effectors can be either **positive** (increase enzyme activity) or **negative** (decrease enzyme activity). An allosteric effector can alter the affinity of the enzyme for its substrate ($K_{0.5}$), the maximal catalytic activity of the enzyme (V_{max}), or both.
- Enzymes can also be regulated by **covalent modification** and by changes in the rate of synthesis or degradation. Enzymes have diagnostic and therapeutic value in medicine.

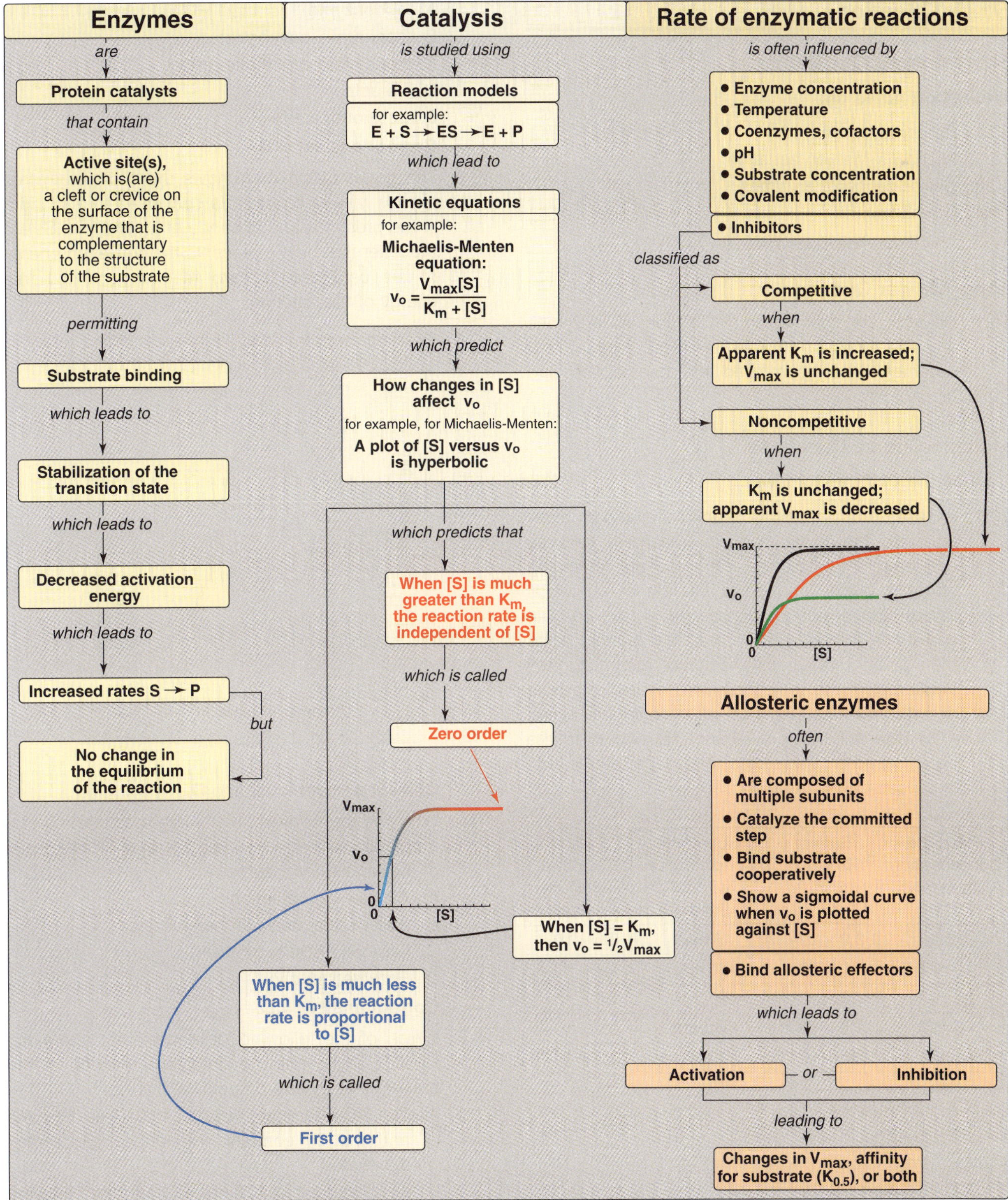

Figure 10.23 Key concept map for the enzymes. S = substrate; [S] = substrate concentration; P = product; E = enzyme; v_o = initial velocity; V_{max} = maximal velocity; K_m = Michaelis constant; $K_{0.5}$ = substrate concentration that gives half maximal velocity.

Study Questions

Short-Answer Questions

Write short notes on:

10.1. Properties of enzymes
10.2. Michaelis-Menten equation
10.3. Classification of enzymes
10.4. Isoenzymes
10.5. Lab diagnosis of myocardial infarction

Long-Answer Question

10.1. Discuss the mechanism of action of enzymes and the types of enzyme inhibition. Explain how enzymes have been used for therapeutics and diagnostics.

Multiple-Choice Questions

Choose the ONE best answer.

10.1. In cases of ethylene glycol poisoning and its characteristic metabolic acidosis, treatment involves correction of the acidosis, removal of any remaining ethylene glycol, and administration of an inhibitor of alcohol dehydrogenase (ADH), the enzyme that oxidizes ethylene glycol to the organic acids that cause the acidosis. Ethanol (grain alcohol) frequently is the inhibitor given to treat ethylene glycol poisoning. Results of experiments using ADH with and without ethanol are shown to the right. Based on these data, what type of inhibition is caused by the ethanol?

Substrate Concentration with Ethanol	Rate of Reaction (mol/L/s)	Substrate Concentration without Ethanol	Rate of Reaction (mol/L/s)
5 mM	3.0×10^{-7}	5 mM	8.0×10^{-7}
10 mM	5.0×10^{-7}	10 mM	1.2×10^{-6}
20 mM	1.0×10^{-6}	20 mM	1.8×10^{-6}
40 mM	1.6×10^{-6}	40 mM	1.9×10^{-6}
80 mM	2.0×10^{-6}	80 mM	2.0×10^{-6}

A. Competitive
B. Feedback
C. Irreversible
D. Noncompetitive

Correct answer = A.

10.2. Alcohol dehydrogenase (ADH) requires oxidized nicotinamide adenine dinucleotide (NAD^+) for catalytic activity. In the reaction catalyzed by ADH, an alcohol is oxidized to an aldehyde as NAD^+ is reduced to NADH and dissociates from the enzyme. The NAD^+ is functioning as a/an:

A. apoenzyme.
B. coenzyme–cosubstrate.
C. coenzyme–prosthetic group.
D. cofactor.
E. heterotropic effector.

Correct answer = B.

10.3. The graph below that shows the changes in free energy when a reactant is converted to a product in the presence and absence of an enzyme. Select the letter that best represents the activation energy of the catalyzed forward reaction and the free energy of the reaction.

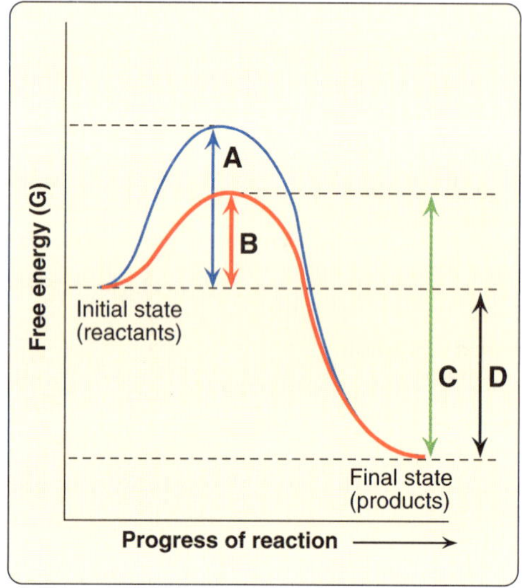

Correct answers = B and **D.**

10.4. Which of the following regulatory mechanisms of a metabolic pathway involves alteration in the levels of the regulatory enzymes?

A. Allosteric modulation
B. Sub-cellular compartmentalization
C. Enzyme phosphorylation
D. Enzyme induction

Correct answer = D.

10.5. Which of the following describes best about the kinetics of an enzyme catalyzed reaction in the presence of a non-competitive inhibitor?

A. The inhibitor can bind to both free Enzyme and the ES complex; Km unchanged; Vmax decreased
B. The inhibitor can bind to both free Enzyme and the ES complex; Km decreased; Vmax unchanged
C. The inhibitor can bind only to free Enzyme; Km unchanged; Vmax decreased
D. The inhibitor can bind only to free Enzyme; Km decreased; Vmax unchanged

Correct answer = A.

10.6. All the following statements about isoenzymes are correct, EXCEPT:
A. They are transcribed from the same gene
B. They can be identified using electrophoresis
C. They show different enzyme kinetics
D. They act on similar substrates
Correct answer = A.

10.7. Which of the following is the newly added 7th major class of enzymes?
A. Oxidoreductases
B. Translocases
C. Hydrolases
D. Lyases
Correct answer = B.

10.8. Which one of the following is an example of a 'mechanism-based inhibitor' of an enzyme?
A. Acetazolamide
B. Atorvastatin
C. Allopurinol
D. Methotrexate
Correct answer = C.

10.9. All the following enzymes can be used as therapeutic agents, EXCEPT:
A. Asparaginase
B. Alkaline phosphatase
C. Trypsin
D. Streptokinase
Correct answer = B.

10.10. Which of the following best describes the role of heme in hemoglobin and other hemoproteins?
A. Co-factor
B. Co-enzyme
C. Co-substrate
D. Prosthetic group
Correct answer = D.

Multiple-Choice Questions (NExT/NEET PG)

Choose the ONE best answer.

10.1. Which of the following statements about the active site of an enzyme is NOT correct?
A. The active site constitutes only a small part of the enzyme
B. The substrate binds to the active site using weak non-covalent interactions
C. It is constituted by residues that are far apart in the primary structure of the enzyme
D. The active site is in perfect complementarity with the substrate
Correct answer = D.

10.2. Statement 1: ATP acts a homotropic allosteric modulator of phosphofructokinase-1.
Statement 2: A homotropic allosteric modulator acts a substrate, as well as a modulator, of the enzyme.
A. Both the Statements are correct, and Statement 2 is the correct explanation for Statement 1
B. Both the Statements are correct, but Statement 2 is NOT the correct explanation for Statement 1
C. Statement 1 is correct, but Statement 2 is NOT correct
D. Statement 1 is NOT correct, but Statement 2 is correct
Correct answer = A.

10.3. Statement 1: All enzymes are proteins in nature.
Statement 2: The active site of an enzyme has critical residues in three dimensional space, which is possible only if the molecule has a properly folded 3-dimensional structure.
A. Both the Statements are correct, and Statement 2 is the correct explanation for Statement 1
B. Both the Statements are correct, but Statement 2 is NOT the correct explanation for Statement 1
C. Statement 1 is correct, but Statement 2 is NOT correct
D. Statement 1 is NOT correct, but Statement 2 is correct
Correct answer = D.

10.4. A drug used for anti-cancer treatment is 5-fluro-uracil. In the body, it is converted into 5-fluro-UMP and this molecule binds to the active site of the enzyme thymidylate synthase to form a stable complex to inactivate the enzyme. Which of the following enzyme inhibitions is this an example of?
A. Competitive inhibition
B. Non-competitive inhibition
C. Uncompetitive inhibition
D. Irreversible inhibition
Correct answer = D.

10.5. One form of monogenic diabetes arises due to genetic variants in the glucokinase gene. Glucokinase is involved in the eventual formation of ATP though glycolysis and subsequent insulin secretion in the pancreatic β-cells, in response to a high glucose concentration after a meal. Which of the following is true regarding the kinetics of the defective enzyme in this form of monogenic diabetes?
A. Maximal velocity is increased
B. K_m is unaltered
C. Turn-over number is decreased
D. K_m and turn-over number are increased
Correct answer = C.

Appendix
Unit II: Protein Structure and Function

CLINICAL RELEVANCE FOR PRIMARY HEALTHCARE PHYSICIANS

Blood levels of total protein, albumin, globulin and albumin/globulin ratio (A:G ratio) are commonly performed laboratory tests.

Table 1 Reference values and causes of derangements for commonly assessed plasma proteins

Analyte	Reference Range	Abnormalities
Total protein	6.0–8.0 g/dl	*Hypoproteinemia:* Increased loss (renal, gastrointestinal, blood loss), reduced intake of amino acids *Hyperproteinemia:* Dehydration, excessive production of gamma globulins, multiple myeloma, Waldenström's macroglobulinemia, chronic diseases
Albumin	3.5–5.5 g/dl	*Hyperalbuminemia:* Dehydration *Hypoalbuminemia:* Malnutrition, malabsorption, liver disease, protein-losing enteropathy, gastrointestinal loss in diarrhea, renal loss such as nephrotic syndrome, skin loss such as in burns, hemodilution
Globulin	2.0–3.5 g/dl	*Hyperglobulinemia:* Chronic infections, multiple myeloma
A:G ratio	1.2–1.6	*Decreased:* Nephrotic syndrome *Reversal of A:G ratio:* Liver disease, malnutrition, chronic infections, multiple myeloma

Glycated albumin is a sensitive indicator of short-term hyperglycemic control than glycated hemoglobin (HbA_{1c}) in diabetes because glycated albumin has a shorter half-life.

High-Sensitivity C-Reactive Protein

C-reactive protein (CRP) is an acute-phase protein, which is synthesized in the liver and is one of the first acute-phase proteins to rise in response to inflammation. High-sensitivity C-reactive protein (hsCRP) is the same protein but can be detected at lower levels by using monoclonal antibody-based tests. High levels of hsCRP predict recurrent coronary events and lower survival in patients with unstable angina and acute myocardial infarction.

Cardiac Troponin (cTn)

Cardiac troponin (cTn) is considered a "gold standard" in the diagnosis of acute coronary syndrome. Cardiac troponin T (cTnT) and troponin I (cTnI) are cardiac regulatory proteins that control the calcium-mediated interaction between actin and myosin. They are more specific for cardiac injury and are elevated for a prolonged time after acute coronary event, making them preferable over *creatine kinase-MB* (CK-MB) measurement.

Urinary Proteins

Proteins in urine can be plasma proteins (filtered and not reabsorbed by kidneys) or from kidney, urinary tract, vagina and prostate. Urinary protein tests can be done qualitatively (commonly done using a reagent strip) or quantitatively. Most quantitative assays are performed on 24-hour urine specimens, which allow for circadian rhythmic changes in excretion at certain times of the day.

For spot urine assay, creatinine is also measured, and results are expressed as protein/creatinine ratio.

Cerebrospinal Fluid Proteins

The reference range for cerebrospinal fluid (CSF) protein is 15–45 mg/dl. Increased total CSF proteins may be found in bacterial, viral, and fungal meningitis, multiple sclerosis, obstruction, neoplasm and cerebral infarction.

Patients of multiple sclerosis have multiple, distinct oligoclonal bands in the γ region on CSF electrophoresis.

Amino Acids

Amino acids, in addition to synthesis of body proteins, are also required for synthesis of nonprotein nitrogen-containing compounds; they also regulate several metabolic processes.

Phenylalanine promotes alertness, elevates mood, decreases pain, and is used to treat arthritis and depression. It is found naturally in eggs, milk and cheese and often added to soft drinks. Phenylalanine must not be taken if a person suffers from phenylketonuria (PKU) in which case a phenylalanine-free diet early in childhood can protect from mental disorders.

Aspartame, a common sweetener used as sugar replacement, contains phenylalanine. Tryptophan, a precursor of serotonin is a natural relaxant which reduces anxiety and induces sleep. It is used in the treatment of migraine headaches and controls hyperactivity in children. It is also useful in weight control as it reduces appetite. Tryptophan occurs naturally in chocolates, eggs, oats, milk and cheese.

Cysteine is used in the food industry for production of flavors and as a processing aid for baking. Cysteine is a popular target for site-directed labeling experiments for biomolecular structure. Monosodium glutamate (MSG), the sodium salt of glutamic acid is used as a food additive and flavor enhancer. Selenocysteine, in which a selenium atom replaces sulfur, is present in several enzymes, such as *formate dehydrogenases*, glycine reductases, and some hydrogenases. HIV-1 encodes a functional selenoprotein, and selenium levels in patients with human immunodeficiency virus (HIV) infection are low. Studies have shown that selenium supplementation can delay CD4 decline in HIV-infected patients, thus delaying the onset of acquired immunodeficiency syndrome (AIDS).

Screening at birth is now available for several aminoacidopathies. Newborn screening helps in early identification and treatment. For diagnosis of PKU, several tests based on tandem mass spectrometry (MS/MS) as well as DNA analysis are available. Sapropterin dihydrochloride, available as Kuvan, is approved for treatment of PKU. It is an analogue of tetrahydrobiopterin (BH4) that stimulates *phenylalanine hydroxylase* (PAH) enzyme to process phenylalanine in people with PKU. For tyrosinemia, the drug nitisinone (NTBC), is available, which prevents the formation of toxic metabolites.

Elevated homocysteine is associated with increased cardiovascular risk, with a large proportion of individuals with untreated homocystinuria experience a thromboembolic event at a younger age. Patients with cystinuria suffer from recurrent renal stones. Intake of high volumes of fluid prevents precipitation of cystine and urinary stone formation. For uncontrolled cases, drug penicillamine, which forms a more soluble complex with cystine can be used.

Collagen and Aging

With increasing age, collagen production declines. This decline, along with muscle weakness, leads to wrinkles and sagging skin. Soft tissue fillers, such as hyaluronic acid, are injected to restore the skin tightness. Botulinum toxin, which is a neurotoxin, is also used. It acts by decreasing the activity of the muscles of expression responsible for the creation of skin folds, lines and wrinkles around the mouth and in the neck.

Table 2 Enzymes of clinical significance

Enzyme	Normal Range	Clinical Significance
Aspartate transaminase (AST)	5–30 U/l	Hepatic disorder, myocardial infarction (MI). In MI, *AST* levels start rising within 6–8 hours, peak at 24 hours, and return to normal within 5 days
Alanine transaminase (ALT)	6–37 U/l	Hepatic disorders
Alkaline phosphatase (ALP)	30–90 U/l	Obstructive hepatobiliary and bone disorders
Creatine kinase (CK)	Male: 15–160 U/l Female: 15–130 U/l	Acute myocardial infarction, muscular dystrophy In MI, the CK-MB levels start rising within 4–8 hours, peak at 12–24 hours, and return to normal levels within 48–72 hours
Lactate dehydrogenase (LDH)	100–225 U/l	Cardiac, hepatic, skeletal muscles, and renal diseases Hematologic and neoplastic disorders In MI, *LDH* levels start rising in 12–24 hours, reach peak levels in 48–72 hours, and may remain elevated for 10 days
Amylase	Serum: 25–130 U/l Urine: 1–15 U/h	Acute pancreatitis Salivary gland lesions, such as mumps and parotitis Perforated peptic ulcer Intestinal obstruction Ruptured ectopic pregnancy
Lipase	0–1.0 U/ml	Acute pancreatitis
Gamma-glutamyl transpeptidase	Male: 6–45 U/l Female: 5–30 U/l	Hepatobiliary disorders, chronic alcoholism
Acid phosphatase (ACP)	0–3.5 ng/ml	Prostatic carcinoma

Clinical Case Questions

Chapter 6 Amino Acids

6.1. A 5-year-old boy was brought to the pediatric OPD. His mother reported that he complains of pain during micturition. On examination, the boy exhibited tenderness in the flank region. Urine examination revealed hematuria and hexagonal crystals. CT scan and X-rays further confirmed the diagnosis of cystine stones.

A. What is the biochemical basis of cystine stones?

B. How can their formation be minimized?

Answer—A: Cystine, being the least soluble among essential amino acids, can precipitate and form stones. An increase in cysteine excretion may occur in certain autosomal recessive genetic conditions. This heightened cysteine excretion leads to the formation of cystine in the acidic pH of urine, consequently contributing to the development of urinary stones.

Answer—B: The formation of stones can be minimized by increasing the volume of urine (by increasing water intake) and alkalinizing the urine as cystine is soluble in alkaline pH.

Chapter 7 Protein Structure

7.1. A 72-year-old male, residing in Delhi, was brought to the OPD by his daughter. She expressed concerns about her father's progressive memory loss and confusion over the past 1 year. He struggles with recalling recent events, managing finances and occasionally becomes disoriented even in familiar surroundings. Additionally, he experiences difficulty locating his belongings and has trouble articulating himself during conversations. Mood and personality changes are evident, with increased withdrawal and occasional irritability. He abstains from smoking and alcohol consumption.

He has hypertension, managed with antihypertensive medication, and well-controlled type 2 diabetes managed with oral hypoglycemics. There is no history of stroke or transient ischemic attacks.

Family history: His mother had dementia in her late 80s.

Physical examination: Well-groomed and cooperative but anxious.

General physical examination: CVS, CNS and respiratory system within normal limits.

Neurological examination shows no focal deficits.

Cognitive assessment shows mild to moderate cognitive impairment.

Investigations: Complete blood count, liver and kidney function tests, thyroid function tests and vitamin B_{12} levels are normal.

Brain MRI shows mild atrophy, more pronounced in the hippocampal region, with no signs of stroke or other focal lesions.

A. What is the likely diagnosis?

Answer—A: Alzheimer's dementia is the most likely diagnosis in view of no other apparent secondary cause of dementia and history of memory loss for recent events and organizational tasks. It is the most common cause of dementia characterized by a progressive decline in thinking, behavior and social skills associated with brain degeneration and memory loss.

B. What is the underlying biochemical defect in this condition?

Answer—B: Alzheimer disease is caused by abnormal folding of proteins characterized by senile neuritic plaques, neurofibrillary triangles and cerebral amyloid deposits. Apparently normal proteins undergo a transformation into abnormal conformations, giving rise to amyloid plaques and neurofibrillary tangles, primarily composed of amyloid β peptides surrounding neurons. The major aggregating protein is amyloid β peptide and tau protein.

C. Is it hereditary in nature?

Answer—C: The presence of Alzheimer disease in a first degree relative increases the relative risk of its occurrence. However, individuals without a family history of the disease can still develop Alzheimer. Researchers have identified several genes that contribute to an increased risk of Alzheimer disease.

Chapter 8 Globular Proteins

8.1. A 19-year-old college girl from northeastern part of India studying in Delhi was admitted to a hospital with high-grade fever with mild anemia, no cyanosis but had a just palpable spleen. Blood investigation showed RBCs with Plasmodium vivax. Her Hb level was 9g%, RBCs were microcytic and hypochromic, and target cells were found on examination of peripheral blood. High-performance liquid chromatography (HPLC) of Hb showed 75% HbA, 15% HbE and 10% HbF. After 2 days, her Hb level dropped to 6g%, but oxygen saturation was 98% and blood transfusion was done. Her fever subsided on 4th day, and on 6th day she was discharged with an advice to take 5 mg of folic acid daily for rest of her life.

A. What is HbE? Write about its molecular defect.

Answer—A: HbE is due to a mutation in Hb β chain gene that leads to replacement of glutamine with lysine at 26th residue of β chain of Hb. HbE is unstable at oxidative stress situation and at high temperature during fever. A cryptic splice site is formed in RNA formed from this mutated gene causing abnormal RNA processing leading to decreased normally spliced β^E-mRNA. This leads to features of mild β-thalassemia. HbE has a right shift of oxygen dissociation curve indicating its higher capacity to deliver oxygen to tissues that compensates the reduced oxygenation of tissues due to anemia.

B. How does the patient having HbE usually present?

Answer—B: Heterozygotes are normal and go mostly unnoticed. Homozygotes have mild anemia and sometimes have delayed puberty associated with less secondary sex characters. Hypochromic microcytic anemia with target cells are seen. Anemia is severe when associated with HbC or β-thalassemia.

C. Write about its laboratory diagnosis.

Answer—C: In suspected cases of hemoglobinopathies, the diagnostic tests are Hb HPLC or capillary electrophoresis of Hb. In HbE disease, usually the anemia is mild and blood picture is as described above.

8.2. A 56-year-old male subject presented to a doctor's clinic with acute pain of a carbuncle on his right axilla and revealed that he had carbuncles on three occasions during last 6 months. He had his breakfast before attending the hospital. His glycated hemoglobin was measured to be 12%. After incision and drainage of the carbuncle, the patient was advised to take oral hypoglycemic drugs for 2 months, analgesic as and when required and antibiotics for 5 days. After 2 months when the patient attended, his glycated hemoglobin was 9%. The dose of his oral hypoglycemic agent was increased. He was advised to take a diabetic diet and to engage in physical exercise regularly.

A. Define glycated hemoglobin. How is it produced?

B. Write the role of estimation of glycated hemoglobin in diagnosis and monitoring of diabetes mellitus.

Answer—A: Under physiologic conditions, HbA is slowly glycated (nonenzymically condensed with a hexose), the extent of glycation being dependent on the plasma concentration of the hexose. The most abundant form of glycated hemoglobin is HbA_{1c}. It has glucose residues attached predominantly to the amino groups of the N-terminal valines of the β-globin chains. Increased amounts of HbA_{1c} are found in RBC of patients with diabetes mellitus, because their HbA has contact with higher glucose concentrations during the 120-day lifetime of these cells.

Answer—B: Glycated Hb (particularly HbA_{1c}) is used for both diagnosis and long-term monitoring of diabetes mellitus. Although fasting plasma glucose estimation is the preferred test for screening and diagnosis of diabetes mellitus and self-monitoring of glucose for its day-to-day monitoring, HbA_{1c} ≥6.5% is also diagnostic of diabetes. The normal level is <5.7%. The levels between 5.7% and 6.4% is considered as prediabetic. It is used to monitor the glycemic control of diabetic patients for last 6–8 weeks. It is indicated at least twice in year for well-controlled diabetics and quarterly when therapy is changed or when patient fails to meet the glycemic goal. The therapeutic target for nonpregnant adult diabetic patient is <7%. More stringent control targeting HbA_{1c} <6.0% is recommended in selective cases (patients without significant hypoglycemia, having long-life expectance, no cardiovascular disease and no polypharmacy), less stringent criteria of HbA_{1c} <8% may be appropriate in patients with a history of severe hypoglycemia, limited life expectancy, advanced microvascular or macrovascular complications, extensive comorbid conditions, etc.

8.3. A 10-year-old male child from a village in Madhya Pradesh, India, presents to the emergency department with severe abdominal pain and jaundice for the past two days. The patient's past history reveals multiple acute painful episodes of chest pain and abdominal pain. The patient also has a history of hospital admission for acute osteomyelitis of the tibia, caused by Salmonella typhi, for which the patient was treated with antibiotics and then discharged.

On examination, the patient is found to have pallor, icterus, and splenomegaly. Investigations reveal severe anemia with an increased reticulocyte count and elevated unconjugated bilirubin. USG shows massive splenomegaly. Peripheral smear demonstrates sickled RBCs.

(**Note:** *Clinical Case 8.3 requires knowledge from Chapters 8 and 39 [Fig. 39.12]*)

A. What is the likely diagnosis?

Answer—A: Sickle cell disease.

B. What is the diagnostic test for sickle cell disease? What is the best test to confirm the diagnosis but not done due to nonavailability of resources?

Answer—B: The diagnostic test for sickle cell disease is hemoglobin electrophoresis (HPLC). However, the definitive test to confirm the diagnosis is genetic testing using polymerase chain reaction (PCR) and restriction fragment length polymorphism analysis (RFLP), or by using allele-specific oligonucleotide probes, although this test is often unavailable in rural areas due to resource constraints.

C. What is the molecular basis for sickle cell disease?

Answer—C: In sickle cell anemia, the hydrophobic glutamic acid is replaced by hydrophilic valine in the β chain of hemoglobin due to a point mutation at position 6 of the globin gene. This change in the primary structure causes hemoglobin to form long fibers that distort the disk-shaped red blood cells into a sickle shape which clogs the arteries. It is interesting to note that a single base mutation (primary structure) in the approximately 600 amino acids (1,800 bases) of the hemoglobin molecule leads to secondary structural changes, resulting in disease.

8.4. An 8-year-old child presents to the OPD clinic with complaints of easy fatigability and recurrent episodes of jaundice. Further history reveals that the patient's elder sister also had similar complaints and a history of regular 3-weekly blood transfusions. On examination, the patient has chipmunk facies, pallor, icterus and splenomegaly. Investigations reveal severe anemia with unconjugated hyperbilirubinemia and a raised reticulocyte count. Peripheral blood smear shows microcytic hypochromic cells with target RBCs. HPLC shows an increase in HbA2 to 40% and an increase in HbF to 15%, while HbA levels were found to be 45%.

Answer—A: This is a case of thalassemia major and mode of inheritance is autosomal recessive.

A. What is the diagnosis and the mode of inheritance of this disease?

B. What is the cause of thalassemia?

Answer—B: Thalassemia can be caused by a variety of mutations, including entire gene deletions, substitutions, or deletions of one or more nucleotides in the DNA. In beta thalassemia, the synthesis of β-globin chains is decreased or absent, usually due to point mutations that affect the production of functional mRNA. However, α-globin chain synthesis remains normal. The excess α-globin chains cannot form stable tetramers and precipitate, leading to the premature death of cells that were meant to become mature red blood cells. This process also causes an increase in $\alpha 2\delta 2$ (HbA2) and $\alpha 2\gamma 2$ (HbF) levels. Since there are only two copies of the β-globin gene in each cell (one on each chromosome 11), individuals with β-globin gene defects have either β-thalassemia trait (β-thalassemia minor) if they have only one defective β-globin gene or β-thalassemia major (Cooley anemia) if both genes are defective.

C. What are the complications of this disease and of regular blood transfusion?

Answer—C: Complications of thalassemia include hemochromatosis, which can lead to cardiomyopathy, liver cirrhosis, and diabetes due to iron overload from regular blood transfusions. Additionally, there is a high risk of HIV, hepatitis B, and hepatitis C infections due to the need for blood transfusions. Thalassemia can also cause delayed growth and sexual development.

Chapter 9 Fibrous Proteins

9.1. A 24-year-old girl presented to the OPD with bruising, rash, bleeding gums and severe weakness. She gave history of an abnormal fad diet for weight loss in which she only took lean protein and fat for the past many months. The diet restrictions did not allow her any fruits and vegetables. She had also lost many kilograms on this diet. On examination, she had multiple bruises and rashes all over her body, but systemic examination was normal. Physical violence was ruled out.

(**Note:** *Clinical Case 9.1 requires knowledge from Chapters 9 and 32)*

A. What is the probable diagnosis? What advice would you give to the patient?

Answer—A: The history and clinical signs and symptoms point to a diagnosis of scurvy due to lack of vitamin C. Vitamin C is mostly found in fruits and vegetables. It is essential for the hydroxylation and cross-linking of proline and lysine during biosynthesis of collagen. Deficiency decreases synthesis of collagen type IV which is found in blood vessels as also the basement membrane separating the epidermis from the dermis. Scurvy results in decreased tensile strength of collagen leading to easy bruisability. Vitamin C deficiency is a product of inadequate dietary intake as humans lack L-gulonolactone oxidase, the enzyme required for its synthesis. The body store of vitamin C is only 1,500 mg (mainly in leukocytes and adrenal glands), and normal levels are dependent on recent uptake. In scurvy, hemorrhage can occur in almost any organ. Further, bone formation is altered and bones become brittle.

The patient is advised to go back to a normal diet. Initially, ascorbic acid supplements should be given.

Chapter 10 Enzymes

10.1. A group of five poor laborers decided to rejoice by drinking as they were just paid their salary after the current project was over. They purchased alcohol from the local village shop. However, within few minutes to hours of drinking the alcohol, they had nausea, vomiting, diarrhea, amnesia, drowsiness, dizziness, and blurred vision. They were brought to the local doctor who suspected they had consumed fake spurious liquor. He immediately administered ethanol to them after which four of them started feeling better. They were then sent to the district hospital.

A. Why did the doctor administer ethanol?

Answer—A: The doctor suspects methanol poisoning, which is a common adulterant in alcohol. Ethanol therapy is recommended when methanol more than 30 ml is ingested. Ethanol and methanol both are substrates for alcohol dehydrogenase with ethanol having a much lower K_m for the enzyme compared to methanol. This prevents the methanol from being converted to formaldehyde which is highly toxic and not metabolized further. On the other hand, ethanol is converted to the nontoxic acetic acid.

B. What further treatment is expected to be done at the district hospital?

Answer—B: Further treatment, for which the laborers are being sent to district hospital, may involve hemodialysis. They also have to be assessed for retinal damage by formic acid.

Advanced Cases/Self-Directed Learning

II.1. A 3-year-old girl presented with acute onset generalized tonic-clonic convulsions and altered sensorium for a few hours. There was no history of preceding fever, diarrhea, vomiting, jaundice, decreased urine output, trauma or accidental drug ingestion. However, there was a past history of recurrent abdominal pain, decreased appetite and lassitude for the past 2 months. On further inquiry, regression of developmental milestones was found, attributed to the poor appetite. Both parents were illiterate and belonged to the lower socio-economic strata. The family had migrated from Rajasthan a year before and worked as unskilled laborers in a paint factory. They resided in an illegally constructed shack on the premises.

The child was hemodynamically stable and was started on maintenance intravenous fluid therapy. Both pupils were of normal size and had a normal reaction. The Glasgow Coma Scale was E2M2V2, warranting elective intubation and ventilatory support. Random blood sugar, serum electrolytes, arterial blood gas, renal function and kidney function tests were normal. Biomarkers for sepsis were negative.

A. Which heavy metal poisoning would you suspect in the given circumstances?

> **Answer—A:** Lead poisoning, due to the family's association with the paint industry, where exposure to lead is common.

B. What is the mechanism of lead toxicity?

> **Answer—B:** Lead toxicity interferes with calcium-dependent processes, including cellular functions, muscle contraction, and neurotransmitter release. It also interacts with proteins containing sulfhydryl groups and enzymes, impairing protein folding. Lead interferes with signal transduction pathways, induces oxidative stress, causes cellular damage, and leads to dysfunction of the vascular endothelium.

C. How will you confirm the diagnosis?

> **Answer—C:** Lead levels in the blood can be estimated by using Atomic Absorption Spectrophotometry or Inductive Coupled Plasma Mass Spectrometry (ICP-MS). In this case, lead levels are likely to be elevated due to the family's association with the paint industry. Normal lead levels are typically considered to be less than 5 µg/dL in adults and less than 3.5 µg/dL in children, although ideally, there should be no lead in the blood.

D. At what blood lead level will you initiate chelation in children and what is the preferred drug?

> **Answer—D:** Chelation is recommended at blood lead levels > 45 µg/dL and the chelator of choice is Dimercaprol. It binds lead and excretes it through urine.

II.2. A 55-year-old male presented to the emergency department with a history of crushing chest pain associated with diaphoresis for the past 3 hours. His past medical history is significant for type 2 diabetes mellitus, hypertension, and a history of smoking. On examination his BP was 100/60, Spo_2 was 94%. His ECG was performed and was reported as normal. Troponin T levels were also not elevated.

A. What should be the next logical step?

Answer—A: Serial ECG and troponin levels should be performed as it may take time to show derangements.

B. After how much time the second troponin test should be done.

Answer—B: It is advisable to repeat after 6–12 hours after presentation to the emergency department.

C. The patient was diagnosed with acute coronary syndrome and underwent myocardial revascularization. However, on the 4th day post-procedure, he again developed chest pain with dynamic ECG changes. Which cardiac biomarker testing can indicate re-infarction?

Answer—C: CK-MB is the marker of choice in this situation. The reason is that troponins appear 4–12 hours after an myocardial infarction (MI), peak at 12–48 hrs and remain elevated for 4–10 days. Hence troponin levels cannot be diagnostic for a recurrent MI on 4th day. On the other hand, CK-MB is elevated 3–6 hours after a MI, peaks at 16–30 hours but returns to normal in 72 hours. Hence, a chest pain on 4th day with dynamic ECG changes is best evaluated by levels of CK-MB.

Unit III

Carbohydrates

Competencies Addressed

- **BI3.1** Discuss and differentiate monosaccharides, disaccharides and polysaccharides giving examples of main carbohydrates as energy fuel, structural element and storage in the human body
- **BI3.2** Describe the processes involved in digestion and assimilation of carbohydrates and storage
- **BI3.3** Describe and discuss the digestion and assimilation of carbohydrates from food
- **BI3.4** Define and differentiate the pathways of carbohydrate metabolism (glycolysis, gluconeogenesis, glycogen metabolism, HMP shunt)
- **BI3.6** Describe and discuss the concept of TCA cycle as an amphibolic pathway and its regulation
- **BI3.7** Describe the common poisons that inhibit crucial enzymes of carbohydrate metabolism (e.g., fluoride, arsenate)
- **BI3.8** Discuss and interpret laboratory results of analytes associated with metabolism of carbohydrates
- **BI3.9** Discuss the mechanism and significance of blood glucose regulation in health and disease
- **BI3.10** Interpret the results of blood glucose levels and other laboratory investigations related to disorders of carbohydrate metabolism
- **BI6.6** Describe the biochemical processes involved in generation of energy in cells
- **BI9.1** List the functions and components of the extracellular matrix (ECM)
- **BI9.2** Discuss the involvement of ECM components in health and disease

Unit III: Overview

Carbohydrates are a source of energy, structural components, intracellular communications, and fuel storage in body. They are divided into simple and complex sugars, and dietary malabsorption and enzymatic deficiencies lead to diseases like lactose intolerance (Chapter 12). Bioenergetics decides whether a particular reaction is possible given the energy shifts, and the electron transport chain participates in trapping the energy released from carbohydrates like glucose into high-energy compounds like ATP (Chapter 11). These high-energy compounds give their energy wherever required to keep the metabolic processes going.

Metabolic pathways can be **anabolic** (synthetic) or **catabolic** (degradative). Carbohydrate metabolism has the anabolic processes of glycogenesis (storage of excess carbohydrates by forming glycogen [Chapter 16]) and gluconeogenesis (making glucose when there is not enough ingested, like in starvation [Chapter 15). The catabolic processes to release energy from compounds include glycolysis (breakdown of glucose to acetyl-CoA with the release of ATP [Chapter13]), further breakdown of acetyl-CoA via TCA cycle to release more energy (Chapter 14) and glycogenolysis (breakdown of stored glycogen to release energy and glucose [Chapter 16]).

Further, many special compounds are made in these metabolic processes, e.g., NADPH and ribose-P by PPP pathway (Chapter 18). Glycosaminoglycans have ability to bind water and form gel-like substance which is the basis of the body's matrix (Chapter 19). Besides glucose, fructose and galactose also contribute to the energy metabolism and complex molecules like glycosylated proteins (Chapter 17).

All pathways interact with protein, lipid and other metabolisms in some way or other. The interactions are covered in the Unit VI: Integration of Metabolism. Since these pathways are interconnected, derangement in one invariably leads to other consequences (Unit VI).

Most pathways are under hormonal and energy balance controls with feedback mechanisms and rate-limiting steps to fine tune the process. The **rate-limiting step is the slowest step of a pathway with maximum control** in form of hormones, coenzymes, cofactors, feedback mechanisms acting at this step. Crossing this step usually means that the pathway will go to completion. A **pathway is called a cycle when the initial substrates are regenerated**, e.g., acetyl-CoA in TCA cycle.

The primary healthcare physician must know the normal levels of common analytes like blood glucose and be able to interpret derangements (Appendix Unit III) and understand the hormonal control of blood glucose levels. There are few conditions in which gluconeogenesis must happen at an accelerated rate and ketone bodies will appear in urine (Unit VI). The diagnosis of common diseases like Diabetes Mellitus (also see Unit VI) and an understanding of the biochemical basis of acute and chronic complications of Diabetes Mellitus are a must for patient care. Other topics to be understood include G6PD deficiency which can lead to hemolytic anemia (Chapter 18) and glycogen storage disorders (Chapter 16).

Bioenergetics and Oxidative Phosphorylation 11

Learning Objectives

At the end of this chapter, the student will be able to:

- Describe the factors that makes biochemical reaction spontaneous and thermodynamically favorable
- Describe the process of ATP biosynthesis by oxidative phosphorylation
- List the inhibitors of ATP biosynthesis by oxidative phosphorylation
- Explain the role of mitochondria in apoptosis and mutation of mitochondrial DNA in pathogenesis of myopathies

I. OVERVIEW

Bioenergetics describes the transfer and utilization of energy in biologic systems. It concerns the **initial and final energy states of the reaction components**, not the reaction mechanism or how much time it takes for the chemical change to occur. Bioenergetics makes use of a few basic ideas from the field of thermodynamics, particularly the concept of free energy. Because changes in free energy provide a measure of the energetic feasibility of a chemical reaction, they allow prediction of whether a reaction or process can take place. In short, bioenergetics predicts if a process is possible, whereas kinetics measures the reaction rate (see p. 133).

II. FREE ENERGY

The direction and extent to which a chemical reaction proceeds are determined by the degree to which two factors change during the reaction. These are **enthalpy** (**ΔH**, a measure of the change [Δ] in heat content of the reactants and products) and **entropy** (**ΔS**, a measure of the change in randomness or disorder of the reactants and products), as shown in Figure 11.1. Neither of these thermodynamic quantities by itself is sufficient to determine whether a chemical reaction will proceed spontaneously in the direction it is written. However, when combined mathematically (see Fig. 11.1), enthalpy and entropy can be used to define a third quantity, **free energy (G)**, which predicts the direction in which a reaction will spontaneously proceed.

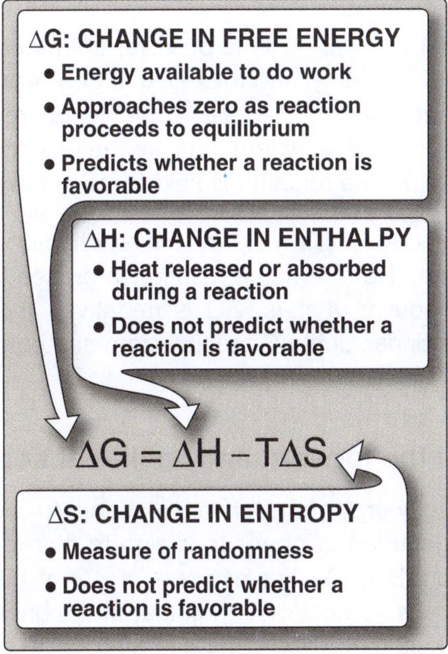

Figure 11.1 Relationship between changes in free energy (G), enthalpy (H), and entropy (S). T is the absolute temperature in Kelvin (K), where K = °C + 273.

III. FREE ENERGY CHANGE

The change in free energy is represented in two ways, ΔG and $ΔG^0$. The first, **ΔG** (without the superscript "0"), represents the change in free energy and, thus, the direction of a reaction at any specified concentration of products and reactants. ΔG, then, is a variable. This contrasts with the standard free energy change, $\mathbf{ΔG^0}$ (with the superscript "0"), which is the energy change when reactants and products are at a concentration of 1 mol/l. [Note: The concentration of protons (H^+) is assumed to be 10^{-7} mol/l (that is, pH = 7). This may be shown by a prime sign (′), for example, $\Delta G^{0\prime}$.] Although ΔG^0, a constant, represents energy changes at these nonphysiologic concentrations of reactants and products, it is nonetheless useful in comparing the energy changes of different reactions. Furthermore, ΔG^0 can readily be determined from measurement of the equilibrium constant (see p. 161). [Note: This section outlines the uses of ΔG, and ΔG^0 is described in D. below.]

A. ΔG and reaction direction

The sign of ΔG can be used to predict the direction of a reaction at constant temperature and pressure. Consider the reaction:

$$A \rightleftarrows B$$

1. **Negative ΔG:** If ΔG is negative, then there is a net loss of energy, and the reaction goes spontaneously as written (that is, A is converted into B) as shown in Figure 11.2A. The reaction is said to be **exergonic**.
2. **Positive ΔG:** If ΔG is positive, then there is a net gain of energy, and the reaction does not go spontaneously from B to A (Fig. 11.2B). Energy must be added to the system to make the reaction go from B to A. The reaction is said to be **endergonic**.
3. **Zero ΔG:** If ΔG = 0, then the reaction is in **equilibrium**. [Note: When a reaction is proceeding spontaneously (that is, ΔG is negative), the reaction continues until ΔG reaches zero and equilibrium is established.]

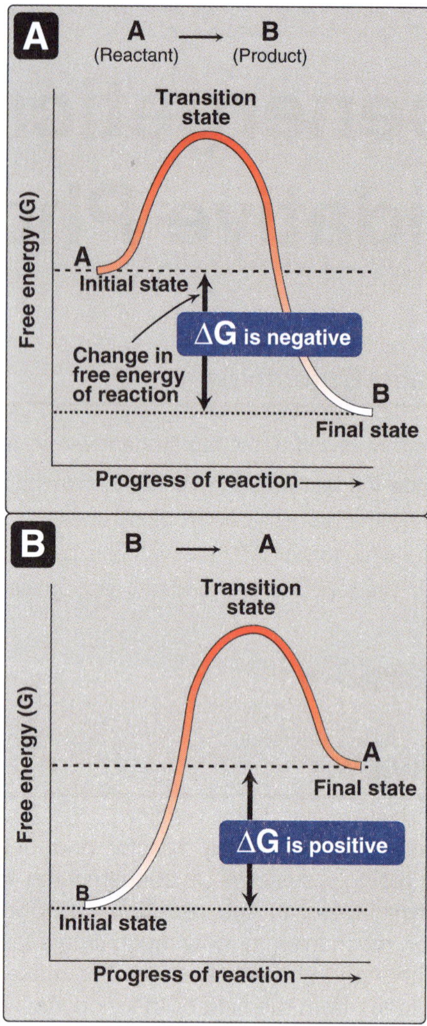

Figure 11.2 Change in free energy (ΔG) during a reaction. **A.** The product has a lower free energy (G) than the reactant. **B.** The product has a higher free energy than the reactant.

B. ΔG of the forward and back reactions

The free energy of the forward reaction (A → B) is equal in magnitude but opposite in sign to that of the back reaction (B → A). For example, if ΔG of the forward reaction is –5 kcal/mol, then that of the back reaction is +5 kcal/mol. [Note: ΔG can also be expressed in kilojoules per mole or kJ/mol (1 kcal = 4.2 kJ).]

C. ΔG and reactant and product concentrations

The ΔG of the reaction A → B depends on the concentration of the reactant and product. At constant temperature and pressure, the following relationship can be derived:

$$\Delta G = \Delta G^0 + RT \ln \frac{[B]}{[A]}$$

where ΔG^0 is the standard free energy change (see D. below)

R is the gas constant (1.987 cal/mol K)

T is the absolute temperature (K)

[A] and [B] are the actual concentrations of the reactant and product

ln represents the natural logarithm.

A reaction with a positive ΔG^0 can proceed in the forward direction if the ratio of products to reactants ([B]/[A]) is sufficiently small (that is, the ratio of reactants to products is large) to make ΔG negative. For example, consider the reaction:

Glucose 6-phosphate ⇄ fructose 6-phosphate

Figure 11.3A shows reaction conditions in which the concentration of reactant, glucose 6-phosphate, is

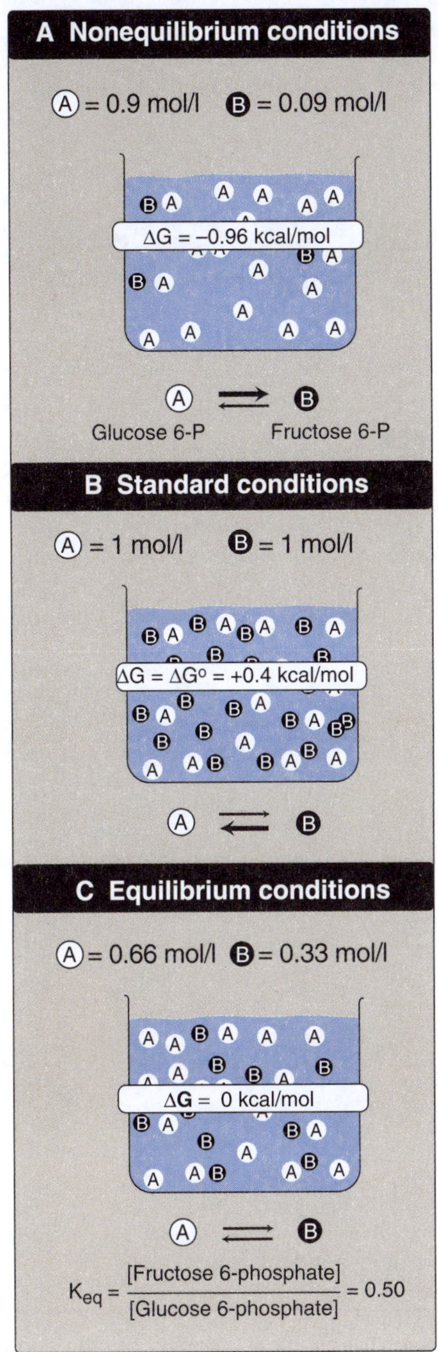

Figure 11.3 Free energy change (ΔG) of a reaction depends on the concentration of reactant Ⓐ and product **B**. For the conversion of glucose 6-phosphate to fructose 6-phosphate, ΔG is negative when the ratio of reactant Ⓐ to product **B** is large (top, panel **A**), is positive under standard conditions (middle, panel **B**), and is zero at equilibrium (bottom, panel **C**). ΔG^0 = standard free energy change.

high compared with the concentration of product, fructose 6-phosphate. This means that the ratio of the product to reactant is small, and RT ln([fructose 6-phosphate]/[glucose 6-phosphate]) is large and negative, causing ΔG to be negative despite ΔG^0 being positive. Thus, the reaction can proceed in the forward direction.

D. Standard free energy change

The standard free energy change, ΔG^0, is so called because it is equal to the free energy change, ΔG, under standard conditions (that is, **when reactants and products are at 1 mol/l concentrations**; Fig. 11.3B). Under these conditions, the natural logarithm of the ratio of products to reactants is zero (ln1 = 0), and, therefore, the equation shown at the bottom of the previous page becomes:

$$\Delta G = \Delta G^0 + 0$$

1. **ΔG^0 and reaction direction:** Under standard conditions, ΔG^0 can be used to predict the direction a reaction proceeds because, under these conditions, ΔG^0 is equal to ΔG. However, **ΔG^0 cannot predict the direction of a reaction under physiologic conditions** because it is composed solely of constants (R, T, and K_{eq} [see 2. below]) and is not, therefore, altered by changes in product or substrate concentrations. If ΔG^0 is negative, the probability of ΔG being negative is high. Hence such reactions are called thermodynamically favorable reactions. But many thermodynamically unfavorable reactions where ΔG is positive do occur in living organisms. In a metabolic pathway, if A and B are metabolic intermediates and ΔG^0 is positive for reaction A $\rightarrow$ B, substance A comes to the system from the previous reaction and B is used up in the subsequent reaction and that make [B] to [A] ratio so low that RT ln[B]/[A] becomes a greater negative integer than ΔG^0 making ΔG of the reaction negative. One example of such reaction in glycolytic pathway is:

   DHAP $\rightarrow$ Glyceraldehyde 3-phosphate [ΔG^0 is positive]

2. **Relationship between ΔG^0 and K_{eq}:** In a reaction A $\rightleftarrows$ B, a point of equilibrium is reached at which no further net chemical change takes place (that is, when A is being converted to B as fast as B is being converted to A). In this state, the ratio of [B] to [A] is constant, regardless of the actual concentrations of the two compounds:

$$K_{eq} = \frac{[B]_{eq}}{[A]_{eq}}$$

   where K_{eq} is the equilibrium constant, and $[A]_{eq}$ and $[B]_{eq}$ are the concentrations of A and B at equilibrium. If the reaction A $\rightleftarrows$ B is allowed to go to equilibrium at constant temperature and pressure, then, at equilibrium, the overall ΔG is zero (Fig. 11.3C). Therefore,

$$\Delta G = 0 = \Delta G^0 + RT \ln \frac{[B]_{eq}}{[A]_{eq}}$$

where the actual concentrations of A and B are equal to the equilibrium concentrations of reactant and product ($[A]_{eq}$ and $[B]_{eq}$), and their ratio is equal to the K_{eq}. Thus,

$$\Delta G^0 = -RT \ln K_{eq}$$

This equation allows some simple predictions:

If $K_{eq} = 1$, then $\Delta G^0 = 0$ $A \rightleftarrows B$

If $K_{eq} > 1$, then $\Delta G^0 < 0$ $A \leftarrow B$

If $K_{eq} < 1$, then $\Delta G^0 > 0$ $A \rightarrow B$

3. **ΔG^0s of two consecutive reactions:** The ΔG^0s are **additive** in any sequence of consecutive reactions, as are the ΔGs. For example:

Glucose + ATP	→ glucose 6-phosphate + ADP	$\Delta G^0 = -4{,}000$ cal/mol
Glucose 6-phosphate	→ fructose 6-phosphate	$\Delta G^0 = +400$ cal/mol
Glucose + ATP	→ fructose 6-phosphate + ADP	$\Delta G^0 = -3{,}600$ cal/mol

By this way, a thermodynamically unfavorable reaction does occur in living system by being coupled with a thermodynamically highly favorable reaction.

4. **ΔGs of a pathway:** The additive property of ΔG is very important in biochemical pathways through which substrates (reactants) must pass in a particular direction (for example, A → B → C → D → ...). As long as the sum of the ΔGs of the individual reactions is negative, the pathway can proceed as written, even if some of the individual reactions of the pathway have a positive ΔG. However, the actual rates of the reactions depend on the lowering of activation energies (E_a) by the enzymes that catalyze the reactions (see p. 132).

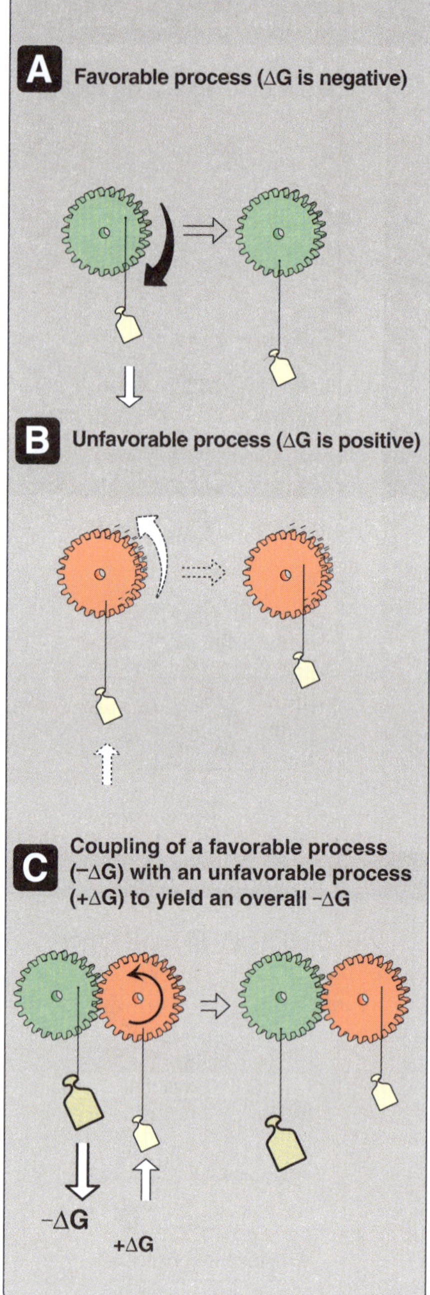

Figure 11.4 Mechanical model of the coupling of favorable and unfavorable processes. **A.** Gear with weight attached spontaneously turns in the direction that achieves the lowest energy state. **B.** The reverse movement is energetically unfavorable (not spontaneous). **C.** The energetically favorable movement can drive the unfavorable one. ΔG = change in free energy.

IV. ATP: AN ENERGY CARRIER

Reactions or processes that have a large positive ΔG, such as moving ions against a concentration gradient across a cell membrane, are made possible by coupling the endergonic movement of ions with a second, spontaneous process with a large negative ΔG such as the exergonic hydrolysis of ATP (see p. 180). [Note: In the absence of enzymes, ATP is a stable molecule because its hydrolysis has a high E_a.] Figure 11.4 shows a mechanical model of energy coupling. The simplest example of energy coupling in biologic reactions occurs when the energy-requiring and the energy-yielding reactions share a common intermediate.

A. Common intermediates

Two chemical reactions have a common intermediate when they occur sequentially in that the product of the first reaction is a substrate for the second. For example, given the reactions

$$A + B \rightarrow C + D$$

$$D + X \rightarrow Y + Z$$

D is the common intermediate and can serve as a carrier of chemical energy between the two reactions. [Note: The intermediate may be linked to an enzyme.] Many coupled reactions use ATP to generate a common intermediate. These reactions may involve the transfer of a phosphate group from ATP to another molecule. Other reactions involve the transfer of phosphate from an energy-rich intermediate to adenosine diphosphate (ADP), forming ATP. Thus ATP-ADP cycle is operative in our body that fulfills the need of approximate 80 kg of ATP from approximately 20 g of adenine nucleotide present in a person having 60 kg body weight.

B. Energy carried by ATP

ATP consists of a molecule of adenosine (adenine + ribose) to which three phosphate groups are attached (Fig. 11.5). Removal of one phosphate produces ADP, and removal of two phosphates produces adenosine monophosphate (AMP). **For ATP, the ΔG^0 of hydrolysis is approximately –7.3 kcal/mol for each of the two terminal phosphate groups**. Because of this large negative ΔG^0 of hydrolysis, **ATP is called a high-energy phosphate compound**. Any phosphorylated compound whose ΔG^0 of hydrolysis is less than or equal to that of ATP is considered as high-energy phosphates, because of their high phosphoryl group transfer potential. As ATP is abundant in cells among all high-energy phosphates, it is selected by nature as currency of energy.

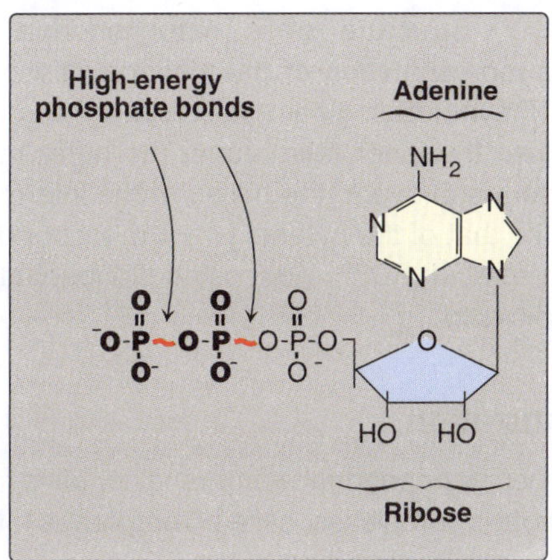

Figure 11.5 Adenosine triphosphate (ATP).

Adenine nucleotides are interconverted (2 ADP ⇄ ATP + AMP) by ***adenylate kinase***. Adenylate kinase plays a crucial role in energy homeostasis of cells by three different ways:

(a) By producing AMP, it stimulates ATP-producing metabolic pathways as AMP act as allosteric stimulator of those pathways

(b) In severely ischemic tissues where ATP production by oxidative phosphorylation stops, the cells try to produce one ATP from two ADP molecules by using above reaction

(c) AMP forms ADP by this reaction to take part in ATP-ADP cycle. **AMP if not converted to ADP undergoes catabolism to form uric acid** and is lost through urine. AMP loss is equivalent to an energy loss as *de novo* AMP synthesis is an energy-consuming process.

V. ELECTRON TRANSPORT CHAIN

Energy-rich molecules, such as glucose, are metabolized by a series of oxidation reactions ultimately yielding carbon dioxide and water (H_2O), as shown in Figure 11.6. The **metabolic intermediates of these reactions donate electrons**

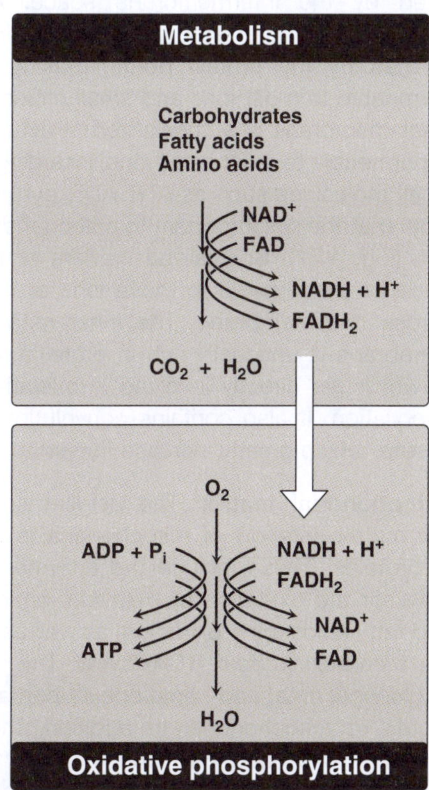

Figure 11.6 The metabolic breakdown of energy-yielding molecules. NAD(H) = nicotinamide adenine dinucleotide; $FAD(H_2)$ = flavin adenine dinucleotide; ADP = adenosine diphosphate; P_i = inorganic phosphate; CO_2 = carbon dioxide.

to specific coenzymes, nicotinamide adenine dinucleotide (NAD^+) and flavin adenine dinucleotide (FAD), to form the energy-rich reduced forms, NADH and $FADH_2$. These reduced coenzymes can, in turn, each donate a pair of electrons to a **specialized set of electron carriers, collectively called the electron transport chain** (ETC), described in this section. As electrons are passed down the ETC, they lose much of their free energy. This energy is used to move H^+ across the inner mitochondrial membrane, creating a H^+ gradient that drives the production of ATP from ADP and inorganic phosphate (P_i), described on p. 167. The **coupling of electron transport with ATP synthesis is called oxidative phosphorylation, sometimes denoted as OXPHOS**. It proceeds continuously in all tissues that contain mitochondria. [Note: The free energy not trapped as ATP is used to drive ancillary reactions such as transport of calcium ions into mitochondria and to generate heat.]

A. Mitochondrial electron transport chain

The ETC (except for cytochrome c, see p. 165) is located in the **inner mitochondrial membrane** and is the **final common pathway** by which electrons derived from different fuels of the body flow to oxygen (O_2), reducing it to H_2O (see Fig. 11.6).

1. **Mitochondrial membranes:** The mitochondrion contains an outer and an inner membrane separated by the intermembrane space. Although the outer membrane contains special channels (formed by the protein porin), making it freely permeable to most ions and small molecules, the inner membrane is a specialized structure that is impermeable to most small ions, including H^+, and small molecules such as ATP, ADP, pyruvate, and other metabolites important to mitochondrial function (Fig. 11.7). Specialized carriers or transport systems are required to move ions or molecules across this membrane. The inner mitochondrial membrane is unusually rich in proteins, over half of which are directly involved in oxidative phosphorylation. It also contains convolutions, called cristae, which greatly increase its surface area.

2. **Mitochondrial matrix:** The gel-like solution of the matrix (interior) of mitochondria is also rich in proteins. These include the enzymes responsible for the oxidation of pyruvate, amino acids, and fatty acids (by β-oxidation) as well as those of the tricarboxylic acid (TCA) cycle. The synthesis of glucose, urea, and heme occurs partially in the matrix of mitochondria. In addition, the matrix contains NAD^+ and FAD (the oxidized forms of the two coenzymes that are required as electron acceptors), and ADP and P_i, which are used to produce ATP. [Note: The matrix also contains mitochondrial deoxyribonucleic acid (mtDNA), ribonucleic acid (mtRNA), and ribosomes.]

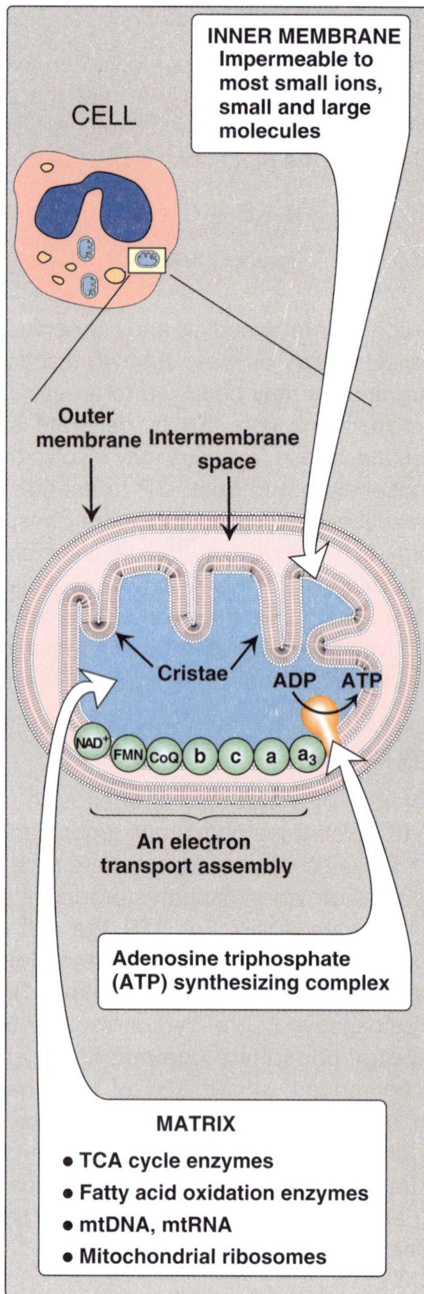

Figure 11.7 Structure of a mitochondrion showing schematic representation of the electron transport chain and the ATP synthesizing complex on the inner membrane. [Note: Unlike the inner membrane, the outer membrane is highly permeable, and the milieu of the intermembrane space is like that of the cytosol.] mt = mitochondrial; RNA = ribonucleic acid; ADP = adenosine diphosphate; TCA = tricarboxylic acid.

B. Organization

The inner mitochondrial membrane contains four separate protein complexes, called **Complexes I, II, III, and IV** that each contain part of the ETC (Fig. 11.8). These complexes accept or donate electrons to the relatively

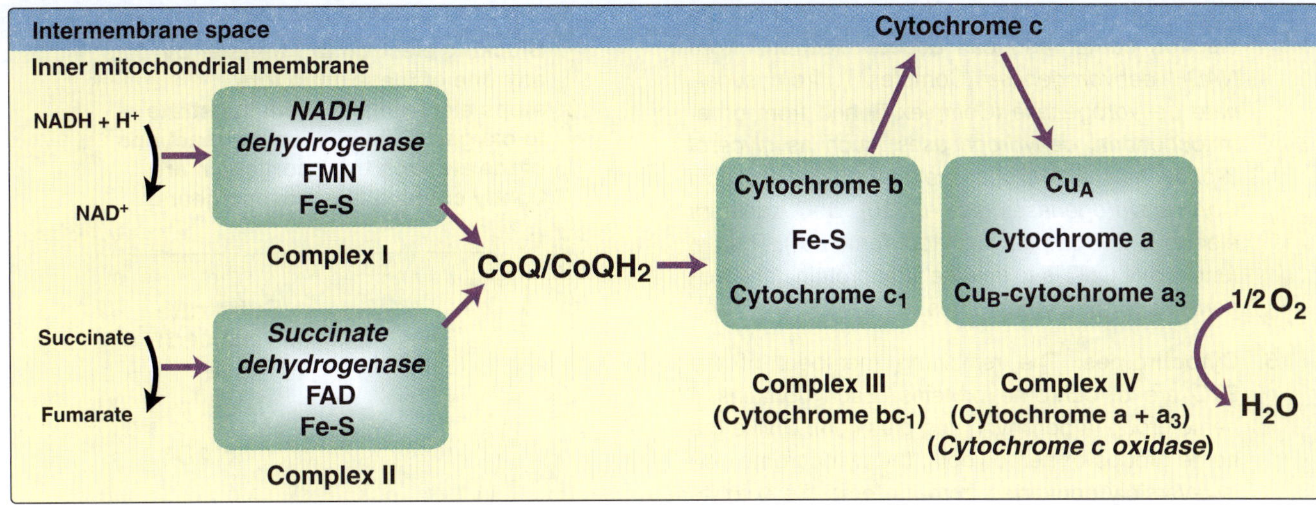

Figure 11.8 Electron transport chain. Electron flow is shown by magenta arrows. NAD(H) = nicotinamide adenine dinucleotide; FMN = flavin mononucleotide; FAD = flavin adenine dinucleotide; Fe-S = iron-sulfur; CoQ = coenzyme Q; Cu = copper.

mobile electron carrier coenzyme Q (CoQ) and cytochrome c. Each carrier in the ETC can receive electrons from an electron donor and can subsequently donate electrons to the next acceptor in the chain. The electrons ultimately combine with O_2 and H^+ to form H_2O. This requirement for O_2 makes the electron transport process the respiratory chain, which accounts for the greatest portion of the body's use of O_2.

C. Reactions

With the exception of CoQ, which is a lipid-soluble quinone, all members of the ETC are proteins. These may function as enzymes as is the case with the flavin-containing *dehydrogenases*, may contain iron as part of an iron-sulfur (Fe-S) center, may contain iron as part of the porphyrin prosthetic group of heme as in the cytochromes, or may contain copper (Cu) as does the cytochrome a + a_3 complex.

1. **NADH formation:** NAD^+ is reduced to NADH by *dehydrogenases* that remove two hydrogen atoms from their substrate. [Note: For examples of these reactions, see the discussion of the *dehydrogenases* of the TCA cycle, p. 206.] **Both electrons but only one H^+ (that is, a hydride ion [:H^-]) are transferred to the NAD^+, forming NADH plus a free H^+.**

2. **NADH dehydrogenase:** The free H^+ plus the hydride ion carried by NADH are transferred to *NADH dehydrogenase*, a protein complex (Complex I) embedded in the inner mitochondrial membrane. Complex I has a tightly bound molecule of flavin mononucleotide (FMN), a coenzyme structurally related to FAD (see Fig. 32.15, p. 535) that accepts the two hydrogen atoms (2 electrons + 2 H^+), becoming $FMNH_2$. *NADH dehydrogenase* also contains peptide subunits with Fe-S centers (Fig. 11.9). At Complex I, electrons move from NADH to FMN to the iron of the Fe-S centers and then to CoQ. **As electrons flow, they lose energy. This energy is used to pump four H^+ across the inner mitochondrial membrane, from the matrix to the intermembrane space.**

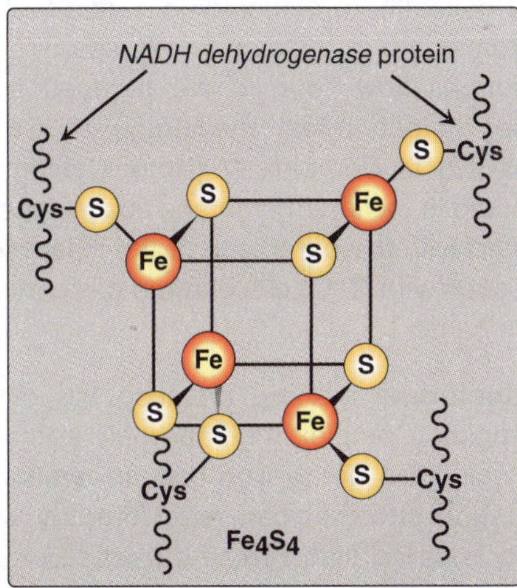

Figure 11.9 Iron-sulfur (Fe-S) center of Complex I. [Note: Complexes II and III also contain Fe-S centers.] NADH = nicotinamide adenine dinucleotide; Cys = cysteine.

3. **Succinate dehydrogenase:** At Complex II, electrons from the *succinate dehydrogenase*–catalyzed oxidation of succinate to fumarate move from the coenzyme, $FADH_2$, to an Fe-S protein, and then to CoQ. [Note: Because no energy is lost in this process, **no H^+ are pumped at Complex II.**]

4. **Coenzyme Q:** CoQ is a quinone derivative with a long, hydrophobic isoprenoid tail. It is made from an intermediate of cholesterol synthesis (see p. 343). [Note: It is also called **ubiquinone** because it is

ubiquitous in biologic systems.] CoQ is a mobile electron carrier and can accept electrons from *NADH dehydrogenase* (Complex I), from *succinate dehydrogenase* (Complex II) and from other mitochondrial *dehydrogenases*, such as *glycerol 3-phosphate dehydrogenase* (see p. 171) and *acyl CoA dehydrogenases* (see p. 310). CoQ transfers electrons to Complex III (cytochrome bc_1). Thus, a function of CoQ is to link the flavoprotein *dehydrogenases* to the cytochromes.

5. **Cytochromes:** The remaining members of the ETC are cytochrome proteins. Each contains a heme group (a porphyrin ring plus iron). Unlike the heme groups of hemoglobin, the cytochrome iron is reversibly converted from its ferric (Fe^{3+}) to its ferrous (Fe^{2+}) form as a normal part of its function as an acceptor and donor of electrons. Electrons are passed along the chain from cytochrome bc_1 (Complex III), to cytochrome c, and then to cytochromes a + a_3 ([Complex IV], see Fig. 11.8). As electrons flow, four H^+ are pumped across the inner mitochondrial membrane at Complex III and two at Complex IV. [Note: Cytochrome c is located in the intermembrane space, loosely associated with the outer face of the inner membrane. As seen with CoQ, cytochrome c is a mobile electron carrier.]

6. **Cytochrome a + a_3:** Because this cytochrome complex (Complex IV) is the only electron carrier in which the **heme iron has an available coordination site that can react directly with O_2**, it also is called *cytochrome c oxidase*. At Complex IV, the transported electrons, O_2, and free H^+ are brought together, and **O_2 is reduced to H_2O** (see Fig. 11.8). [Note: Four electrons are required to reduce one molecule of O_2 to two molecules of H_2O.] *Cytochrome c oxidase* contains Cu atoms that are required for this complicated reaction to occur. Electrons move from Cu_A to cytochrome a to cytochrome a_3 (in association with Cu_B) to O_2.

7. **Site-specific inhibitors:** Inhibitors of specific sites in the ETC have been identified and are illustrated in Figure 11.10. These respiratory inhibitors prevent the passage of electrons by binding to a component of the chain, blocking the oxidation-reduction reaction. Other than these inhibitors shown in the Figure 11.10, carboxin and thenoyltrifluoroacetone (TTFA) are known to block electron flow from complex II. Therefore, all electron carriers before the block are fully reduced, whereas those located after the block are oxidized. [Note: Inhibition of the ETC inhibits ATP synthesis because these processes are tightly coupled (see p. 169).]

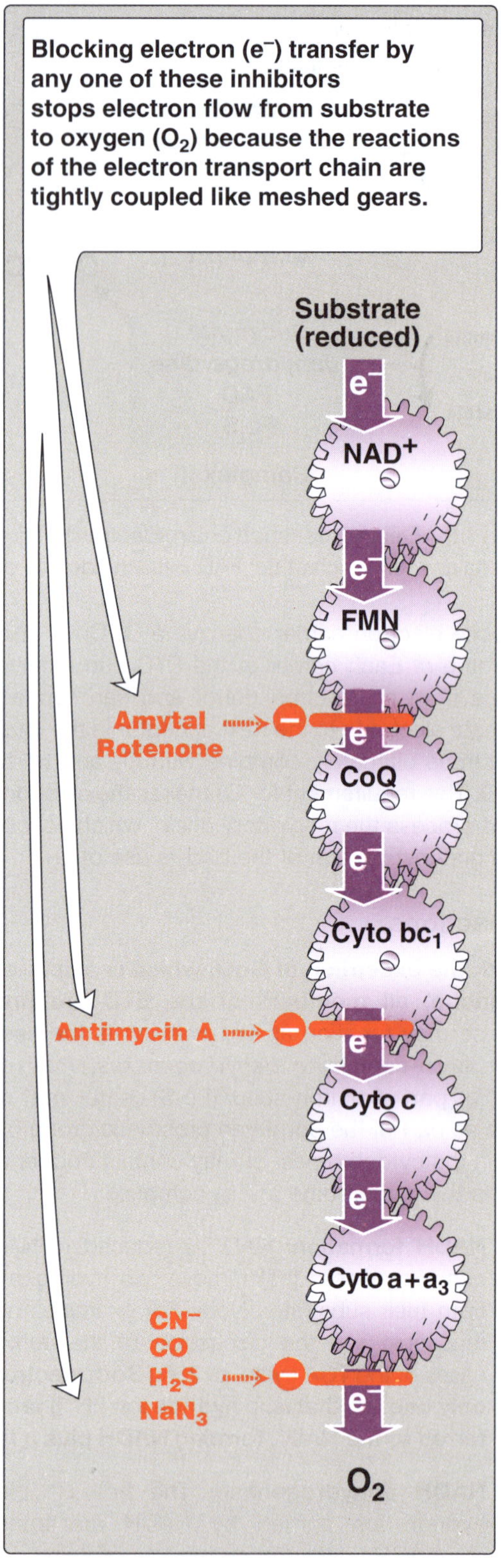

Figure 11.10 Site-specific inhibitors of electron transport shown using a mechanical model for the coupling of oxidation-reduction reactions. [Note: Normal direction of electron flow is illustrated.] NAD^+ = nicotinamide adenine dinucleotide; FMN = flavin mononucleotide; CoQ = coenzyme Q; Cyto = cytochrome; CN^- = cyanide; CO = carbon monoxide; H_2S = hydrogen sulfide; NaN_3 = sodium azide.

Leakage of electrons from the ETC produces reactive oxygen species (ROS), such as superoxide ($O_2^{\bullet-}$), hydrogen peroxide (H_2O_2), and hydroxyl radicals ($OH\bullet$). ROS damage DNA and proteins and cause lipid peroxidation. Enzymes such as *superoxide dismutase (SOD)*, *catalase*, and *glutathione peroxidase* are cellular defenses against ROS (see p. 797).

D. Free energy release during electron transport

The free energy released as electrons are transferred along the ETC from an electron donor (reducing agent or reductant) to an electron acceptor (oxidizing agent or oxidant) is used to pump H^+ at Complexes I, III, and IV. [Note: The electrons can be transferred as hydride ions to NAD^+; as hydrogen atoms to FMN, CoQ, and FAD; or as electrons to cytochromes.]

1. **Redox pairs:** Oxidation (loss of electrons) of one substance is always accompanied by reduction (gain of electrons) of a second. For example, Figure 11.11 shows the oxidation of NADH to NAD^+ by *NADH dehydrogenase* at Complex I, accompanied by the reduction of FMN, the prosthetic group, to $FMNH_2$. Such redox reactions can be written as the sum of two separate half reactions, one an oxidation and the other a reduction (see Fig. 11.11). NAD^+ and NADH form a redox pair, as do FMN and $FMNH_2$. Redox pairs differ in their tendency to lose electrons. This tendency is a characteristic of a particular redox pair and can be quantitatively specified by a constant, E_0 (the standard reduction potential), with units in volts.

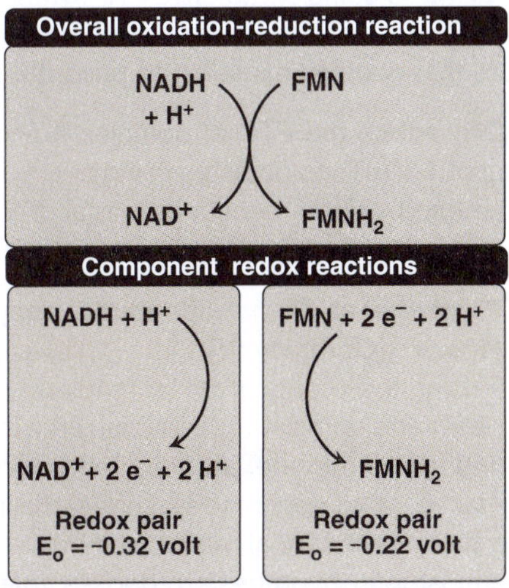

Figure 11.11 Oxidation of NADH by FMN, separated into two component half reactions. NAD(H) = nicotinamide adenine dinucleotide; FMN(H_2) = flavin mononucleotide; e^- = electron; H^+ = proton; E_0 = standard reduction potential.

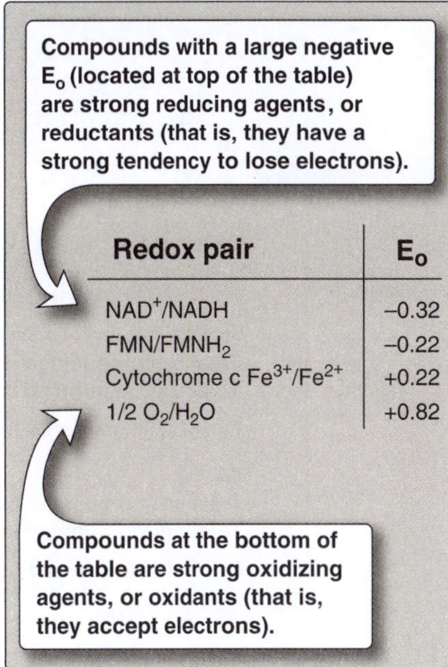

Redox pair	E_o
NAD^+/NADH	−0.32
FMN/$FMNH_2$	−0.22
Cytochrome c Fe^{3+}/Fe^{2+}	+0.22
1/2 O_2/H_2O	+0.82

Figure 11.12 Standard reduction potentials (E_0) of some reactions. NAD(H) = nicotinamide adenine dinucleotide; FMN(H_2) = flavin mononucleotide; Fe = iron.

2. **Standard reduction potential:** The E_0 of various redox pairs can be ordered from the most negative E_0 to the most positive. The more negative the E_0 of a redox pair, the greater the tendency of the reductant member of that pair to lose electrons. The more positive the E_0, the greater the tendency of the oxidant member of that pair to accept electrons. Therefore, **electrons flow from the pair with the more negative E_0 to that with the more positive E_0.** The E_0 values for some members of the ETC are shown in Figure 11.12. **The components of the electron transport chain are arranged in order of increasingly positive E_0 values, indicating the progression of electron transfer in a thermodynamically favorable direction.**

3. **Relationship of ΔG^0 to ΔE_0:** The ΔG^0 is related directly to the magnitude of the change in E_0:

$$\Delta G^0 = -nF\ \Delta E_0,$$

where n = number of electrons transferred (1 for a cytochrome, 2 for NADH, $FADH_2$, and CoQ)

F = Faraday constant (23.1 kcal/volt mol)

ΔE_0 = E_0 of the electron-accepting pair minus the E_0 of the electron-donating pair

ΔG^0 = change in the standard free energy

4. **ΔG^0 of ATP:** The ΔG^0 for the phosphorylation of ADP to ATP is +7.3 kcal/mol. The transport of a pair of

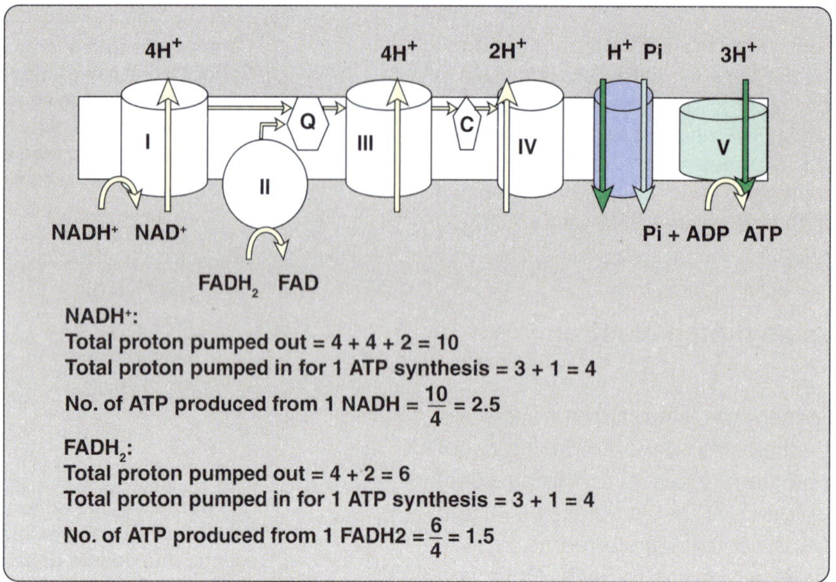

Figure 11.13 Synthesis of ATP in oxidative phosphorylation.

electrons from NADH to O_2 through the ETC releases 52.6 kcal ($nF\Delta E_0$ = 2*23.1*1.14). Total number of protons pumped out of inner mitochondrial membrane is 10. For synthesis of one ATP, average number of protons that need to enter through FoF1 complex is 3. For transport of Pi to mitochondrial matrix, one hydroxyl ion goes out of matrix by the Pi transporter. This is equivalent to entry of one proton. So for synthesis of one ATP, four protons enter into the matrix. **Thus, one NADH produces 2.5 (10/4 = 2.5) ATP**, sometimes expressed as a **P/O ratio** (ATP made per O atom reduced) of 2.5 (Fig. 11.13). Energy captured as currency of energy is 18.25 (2.5 * 7.3 = 18.25) kcal. Efficiency of an energy-capturing system is calculated by using following equation:

Efficiency = (Energy captured ÷ energy released) * 100

So, the **efficiency of oxidative phosphorylation when one NADH forms 2.5 ATP is 34.7%** [(18.25/52.6)*100]. The remaining calories are used for ancillary reactions or released as heat. [Note: The P:O for $FADH_2$ is 1.5:1 because Complex I is bypassed.]

VI. PHOSPHORYLATION OF ADP TO ATP

The transfer of electrons down the ETC is energetically favored because NADH is a strong electron donor and O_2 is an avid electron acceptor. However, the flow of electrons does not directly result in ATP synthesis.

A. Chemiosmotic hypothesis

The chemiosmotic hypothesis (also known as the Mitchell hypothesis) explains how the free energy generated by the transport of electrons by the ETC is used to produce ATP from ADP + P_i.

1. **Proton pump:** Electron transport is coupled to ADP phosphorylation by the pumping of H^+ across the inner mitochondrial membrane, from the matrix to the intermembrane space, at Complexes I, III, and IV. **For each pair of electrons transferred from NADH to O_2, 10 H^+ are pumped.** This creates an electrical gradient (with more positive charges on the cytosolic side of the membrane than on the matrix side) and a pH (chemical) gradient (the cytosolic side of the membrane is at a lower pH than the matrix side), as shown in Figure 11.14. The energy (**proton-motive force**) generated by these gradients is sufficient to drive ATP synthesis. Thus, the H^+ gradient serves as the common intermediate that couples oxidation to phosphorylation.

   $FADH_2$ enters the ETC at complex II and yields a total of 1.5 ATP, as depicted in Figure 11.13. On the other hand, $NADH^+$ generates a total of 2.5 ATP (as shown in Fig. 11.13) when it enters at complex I.

2. **ATP synthase:** The **multisubunit enzyme *ATP synthase* ([Complex V]** Fig. 11.15) synthesizes ATP using the energy of the H^+ gradient. It contains a membrane domain (F_o) that spans the inner mitochondrial membrane and an extramembranous domain (F_1) that appears as a sphere that protrudes into the mitochondrial matrix (see Fig. 11.14). The chemiosmotic hypothesis proposes that after H^+ have been pumped to the cytosolic side of the inner mitochondrial membrane, they reenter the matrix by passing through a H^+ channel in the F_o domain, driving rotation of the c ring of F_o and, at

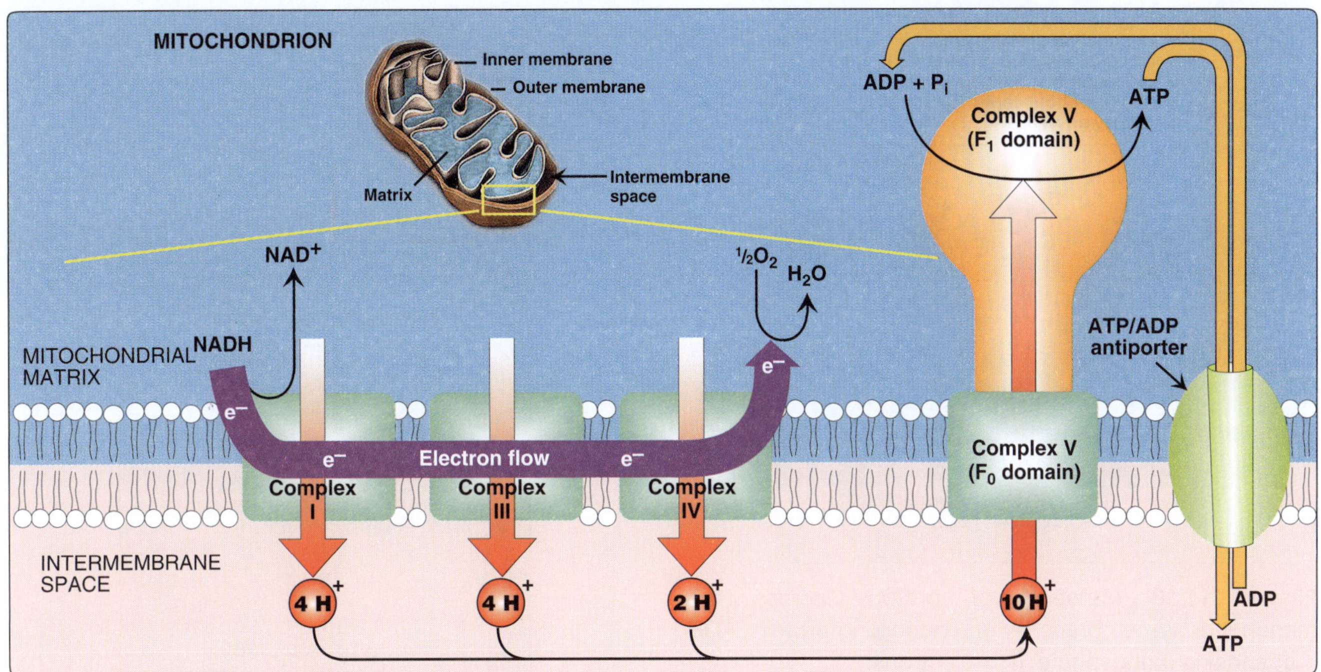

Figure 11.14 Electron transport chain shown in association with proton (H^+) pumping. Ten H^+ are pumped for each nicotinamide adenine dinucleotide (NADH) oxidized. [Note: H^+ are not pumped at Complex II.] e^- = electron; Complex V = *ATP synthase*.

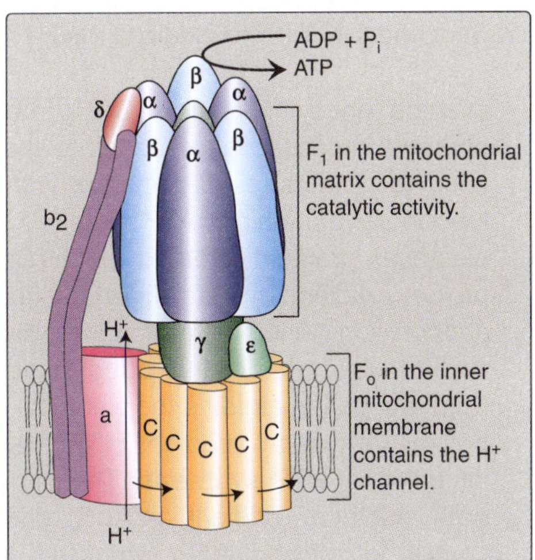

Figure 11.15 *ATP synthase* (*F_1F_o-ATPase*). [Note: The c ring of vertebrates contains eight subunits. **One complete turn of the ring is driven by eight H^+ (protons) moving through the F_o domain.** The resulting conformational changes in the three β subunits of the F_1 domain allow phosphorylation of three adenosine diphosphates (ADP) to three ATP.] P_i = inorganic phosphate.

the same time, dissipating the pH and electrical gradients. Rotation in F_o causes conformational changes in the three β subunits of F_1 that allow them to bind ADP + P_i, phosphorylate ADP to ATP, and release ATP. **One complete rotation of the c ring produces three ATP**. [Note: *ATP synthase* is also called *F_1/F_o-ATPase* because the enzyme can also catalyze the hydrolysis of ATP to ADP and P_i.]

a. **Coupling in oxidative phosphorylation:** In normal mitochondria, ATP synthesis is coupled to electron transport through the H^+ gradient. Increasing (or decreasing) one process has the same effect on the other. For example, hydrolysis of ATP to ADP and P_i in energy-requiring reactions increases the availability of substrates for *ATP synthase* and, thus, increases H^+ flow through the enzyme. Electron transport and H^+ pumping by the ETC increase to maintain the H^+ gradient and allow ATP synthesis.

b. **Oligomycin:** This drug binds to the $\mathbf{F_o}$ (hence the letter "o") domain of *ATP synthase*, closing the H^+ channel and preventing reentry of H^+ into the matrix, thereby inhibiting phosphorylation of ADP to ATP. Because the pH and electrical gradients cannot be dissipated in the presence of this phosphorylation inhibitor, electron transport stops because of the difficulty of pumping any more H^+ against the steep gradient. This dependency of cellular respiration on the ability to phosphorylate ADP to ATP is known as **respiratory control** and is the consequence of the tight coupling of these processes.

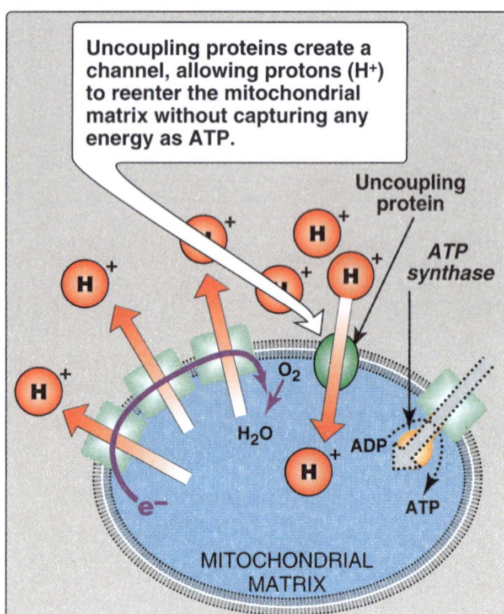

Figure 11.16 Transport of protons across the mitochondrial membrane by an uncoupling protein. ADP = adenosine diphosphate; e^- = electrons.

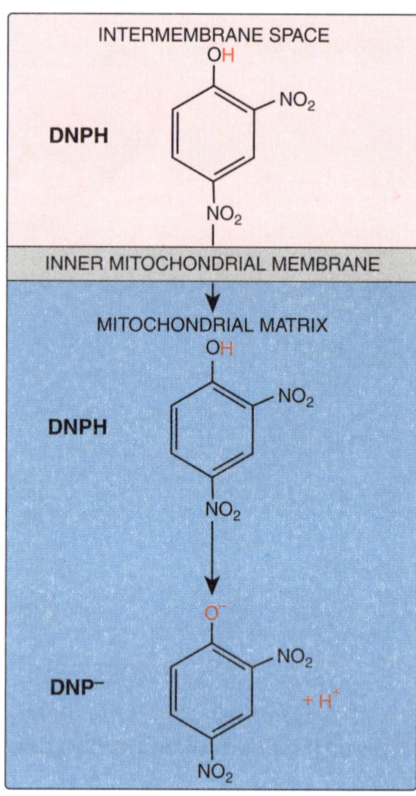

Figure 11.17 2,4-Dinitrophenol (DNP), a proton (H^+) carrier, shown in its reduced (DNPH) and oxidized (DNP^-) forms.

c. **Uncoupling proteins:** Uncoupling proteins (UCP) occur in the inner mitochondrial membrane of mammals, including humans. These proteins form channels that allow H^+ to reenter the mitochondrial matrix without energy being captured as ATP (Fig. 11.16). The energy is released as heat, and the process is called **nonshivering thermogenesis**. UCP1, also called **thermogenin**, is responsible for heat production in the mitochondria-rich brown adipocytes of mammals. [Note: Cold causes catecholamine-dependent activation of UCP1 expression.] In brown fat, unlike the more abundant white fat, ~90% of its respiratory energy is used for thermogenesis in infants in response to cold. Thus, **brown fat is involved in energy expenditure, whereas white fat is involved in energy storage**. [Note: Brown fat depots have recently been shown to be present in adults.]

d. **Synthetic uncouplers:** Electron transport and phosphorylation of ADP can also be uncoupled by compounds that shuttle H^+ across the inner mitochondrial membrane, dissipating the gradient. The classic example is **2,4-dinitrophenol**, a lipophilic H^+ carrier (ionophore) that readily diffuses through the mitochondrial membrane (Fig. 11.17). This uncoupler causes electron transport to proceed at a rapid rate without establishing a H^+ gradient, much as do the UCP. There are many examples of uncouplers e.g., carbonyl cyanide m-chlorophenyl hydrazone (CCCP), carbonyl cyanide phenylhydrazone (CCP), etc. Again, energy is released as heat rather than being used to synthesize ATP. **Valinomycin** is a K^+ ionophore that inhibits ATP synthesis because electrical gradient created by ETC due to proton gradient is lost. **Nigericin** creates an antiport on inner mitochondrial membrane and brings in protons into but extrudes out K^+ from matrix. It maintains the electrical gradient but pH (or proton) gradient is lost, leading to inhibition of ATP synthesis. This proves that both electrical gradient and pH gradient are crucial for driving *ATP synthase*. [Note: **In high doses, aspirin and other salicylates uncouple oxidative phosphorylation.** This explains the fever that accompanies toxic overdoses of these drugs.]

> Substrate level phosphorylation is different from oxidative phosphorylation and is a process in which the oxidation and phosphorylation is NOT coupled. In substrate level phosphorylation, ATP and GTP are produced from ADP and GDP respectively using released chemical energy, the Gibbs free energy, to transfer a phosphoryl group to ADP or GDP. Examples include specific reactions in glycolysis (Chapter 13) and citric acid cycle (Chapter 14)

B. Membrane transport systems

The inner mitochondrial membrane is impermeable to most charged or hydrophilic substances. However, it contains numerous transport proteins that permit passage of certain molecules from the cytosol to the mitochondrial matrix.

1. **ATP and ADP transport:** The inner membrane requires specialized carriers to transport ADP and P_i from the cytosol (where ATP is hydrolyzed to ADP in many energy-requiring reactions) into mitochondria, where ATP can be resynthesized. An adenine nucleotide antiporter imports one ADP from the cytosol into the matrix, while exporting one ATP from the matrix into the cytosol (see Fig. 11.14). Atractyloside and bongkrekic acid are inhibitors of adenine nucleotide antiporter. A symporter cotransports P_i and H^+ from the cytosol into the matrix.

2. **Reducing equivalent transport:** The inner mitochondrial membrane lacks an NADH transporter, and NADH produced in the cytosol (for example, in glycolysis; see p. 197) cannot directly enter the mitochondrial matrix. However, reducing equivalents of NADH are transported from the cytosol into the matrix using substrate shuttles. In the **glycerol 3-phosphate shuttle** (Fig. 11.18A), two electrons are transferred from NADH to dihydroxyacetone phosphate by cytosolic *glycerol 3-phosphate dehydrogenase*. The glycerol 3-phosphate produced is oxidized by the mitochondrial isozyme as FAD is reduced to $FADH_2$. CoQ of the ETC oxidizes the $FADH_2$. Therefore, the glycerol 3-phosphate shuttle results in the synthesis of 1.5 ATP for each cytosolic NADH oxidized. This contrasts with the **malate-aspartate shuttle** (Fig. 11.18B), which produces NADH (rather than $FADH_2$) in the mitochondrial matrix, thereby yielding 2.5 ATP for each cytosolic NADH oxidized by *malate dehydrogenase* as oxaloacetate is reduced to malate. A transport protein moves malate into the mitochondrial matrix. The glycerol 3-phosphate shuttles are abundant in skeletal muscles but hepatocytes and cardiac myocytes contain mostly malate-aspartate shuttles.

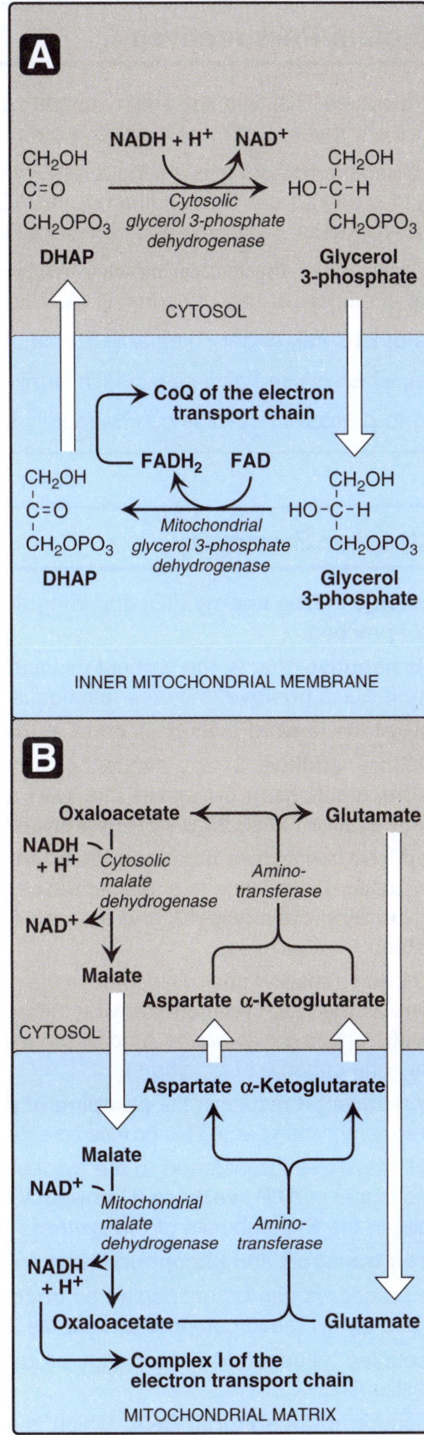

Figure 11.18 Substrate shuttles for the transport of reducing equivalents across the inner mitochondrial membrane. **A.** Glycerol 3-phosphate shuttle. **B.** Malate-aspartate shuttle. DHAP = dihydroxyacetone phosphate; NAD(H) = nicotinamide adenine dinucleotide; H^+ = proton; $FAD(H_2)$ = flavin adenine dinucleotide; CoQ = coenzyme Q.

C. Mitochondria and apoptosis

The process of apoptosis (programmed cell death) may be initiated through the intrinsic (mitochondrial-mediated) pathway by the formation of pores in the outer mitochondrial membrane. These pores allow cytochrome c to leave the intermembrane space and enter the cytosol. There, cytochrome c, in association with proapoptotic factors, activates a family of proteolytic enzymes (the *caspases*), causing cleavage of key proteins and resulting in the morphologic and biochemical changes characteristic of apoptosis.

VII. Medical Perspectives

Carbon monoxide (CO), cyanide and hydrogen sulfide are inhibitors of electron transfer from cyt C to O_2. Carbon monoxide inhibits the enzyme *cytochrome c oxidase* of human mitochondrial respiratory chain.

Carbon monoxide poisoning is a serious risk whenever CO is released in a closed space like a closed car or closed room. Sleeping in a parked car with engine on and AC running has caused deaths due to CO poisoning. Symptoms start as dizziness, weakness, confusion and can progress to arrhythmias, seizures and death.

OXPHOS disease is the clinical disease involving components of oxidative phosphorylation. They may be caused by mutations in nuclear or mitochondrial DNA. They are described in the Clinical Case Question below.

Inhibitors of ETC can target complexes I, II, III, IV, *ATP synthase* and ATP-ADP antiporter.

Rotenone, which is used as a pesticide in farming, inhibits aerobic cellular respiration by blocking *NADH dehydrogenase*.

Exposure to potassium cyanide can be rapidly fatal.

VIII. Chapter Summary

- The **change** in **free energy** (ΔG) occurring during a reaction predicts the **direction** in which that reaction will spontaneously proceed.
- If ΔG is **negative** (that is, the product has a lower free energy than the substrate), then the reaction is **spontaneous** as written. If ΔG is **positive**, then the reaction is **not spontaneous**. If ΔG **= 0**, then the reaction is in **equilibrium**.
- The ΔG of the forward reaction is equal in magnitude but opposite in sign to that of the back reaction.
- The ΔG are **additive** in any sequence of consecutive reactions, as are the standard free energy changes (ΔG^0). Therefore, reactions or processes that have a large, positive ΔG are made possible by **coupling** with those that have a large, negative ΔG such as **ATP hydrolysis**.
- The reduced coenzymes **nicotinamide adenine dinucleotide** (**NADH**) and **flavin adenine dinucleotide** (**$FADH_2$**) each donate a pair of electrons to a specialized set of **electron carriers**, consisting of **flavin mononucleotide** (**FMN**), **iron-sulfur centers**, **coenzyme Q**, and a series of heme-containing **cytochromes**, collectively called the **electron transport chain**.
- The electron transport chain is located in the inner mitochondrial membrane, which is impermeable to most substances. It serves as the final common pathway through which electrons, derived from various body fuels, flow to oxygen (O_2), ultimately reducing it to water. Among the cytochromes involved, cytochrome c oxidase, the terminal cytochrome, is the only one capable of binding O_2.
- Electron transport results in the **pumping of protons** (**H^+**) across the inner mitochondrial membrane from the matrix to the intermembrane space. This process creates **electrical** and **pH gradients** across the inner mitochondrial membrane.
- After H^+ have been transferred to the cytosolic side of the membrane, they reenter the matrix by passing through the **F_o** H^+ channel in ***ATP synthase*** (**Complex V**), dissipating the pH and electrical gradients and causing conformational changes in the **F_1** β subunits of the *synthase* that result in the synthesis of ATP from ADP + inorganic phosphate.
- **Electron transport** and **phosphorylation** are tightly coupled in **oxidative phosphorylation** ([**OXPHOS**], Fig. 11.19).
- These processes can be uncoupled by **uncoupling protein-1** of the inner mitochondrial membrane of brown adipocytes and by synthetic compounds such as **2,4-dinitrophenol** and **aspirin**, all of which dissipate the H^+ gradient.
- In uncoupled mitochondria, the energy produced by electron transport is released as **heat** rather than being used to synthesize ATP.
- Mutations in **mitochondrial DNA**, which is **maternally inherited**, are responsible for some cases of mitochondrial diseases such as **Leber hereditary optic neuropathy**.
- The release of **cytochrome c** into the cytoplasm and subsequent activation of proteolytic *caspases* results in **apoptotic cell death**.

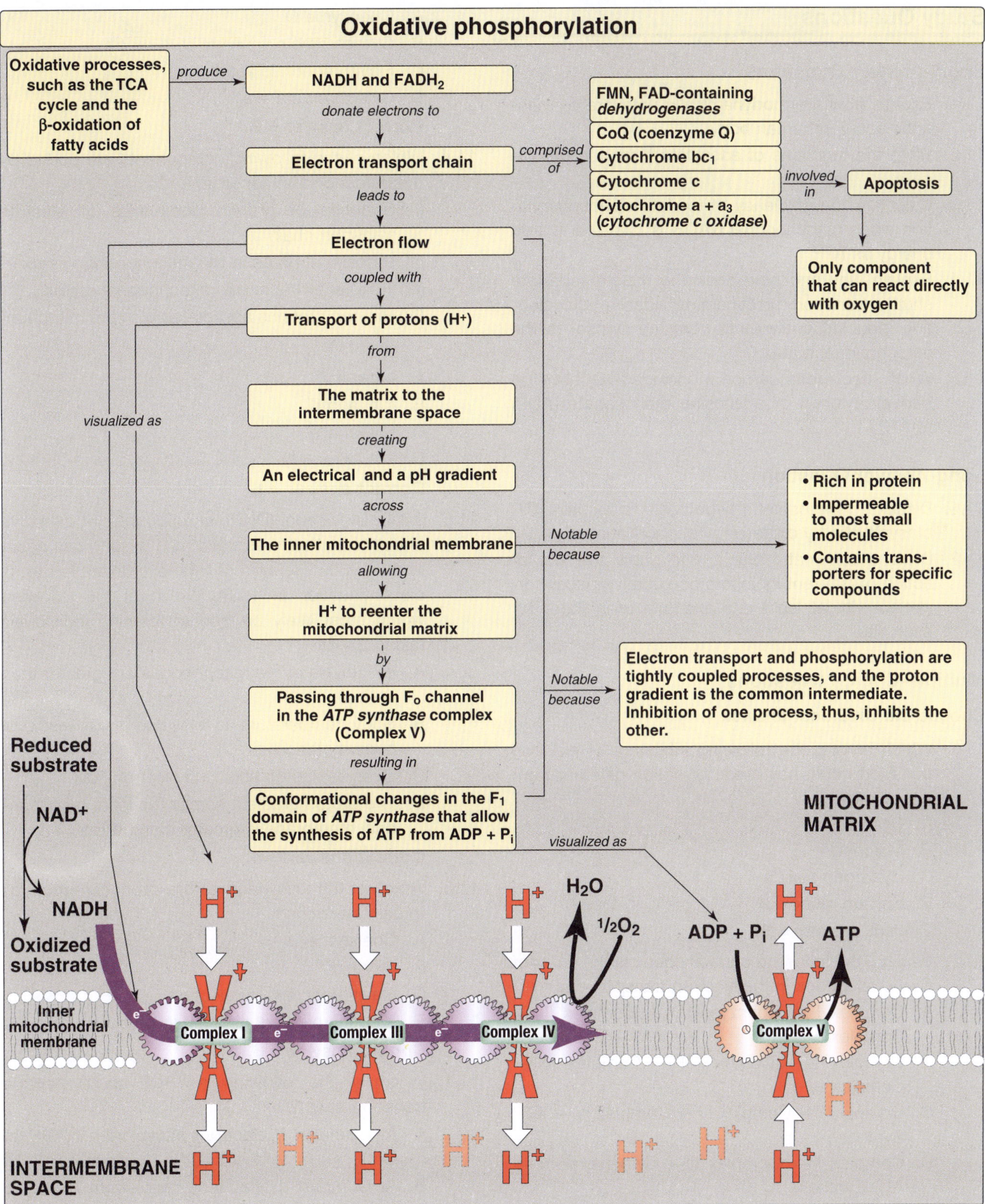

Figure 11.19 Key concept map for oxidative phosphorylation (OXPHOS). [Note: Electron (e^-) flow and ATP synthesis are shown as sets of interlocking gears to emphasize coupling.] TCA = tricarboxylic acid; NAD(H) = nicotinamide adenine dinucleotide; FAD(H_2) = flavin adenine dinucleotide; FMN = flavin mononucleotide; ADP = adenosine diphosphate.

Study Questions

Short-Answer Questions

11.1. Explain how thermodynamically unfavorable reactions occur in human body.

11.2. Write the functions of adenylate kinase in muscle cells.

11.3. Calculate the efficiency of oxidative phosphorylation when mitochondrial NADH is oxidized to trap energy as ATP.

11.4. Explain why and how does the malate-aspartate shuttle transport nicotinamide adenine dinucleotide reducing equivalents from the cytosol to the mitochondrial matrix.

11.5. What effect does carbon monoxide (CO) have on phosphorylation of adenosine diphosphate (ADP) to ATP?

Long-Answer Question

11.1. Discuss chemiosmotic hypothesis to explain ATP biosynthesis by oxidative phosphorylation.

11.2. Draw a schematic diagram to show the site of action of different inhibitors of oxidative phosphorylation with at least one example of inhibitor for each site.

Multiple-Choice Questions

Choose the ONE best answer.

11.1. Which among the following inhibitors of electron transport chain has a site of action different from the others?

A. Cyanide
B. Malonate
C. Hydrogen sulfide
D. Carbon monoxide

Correct answer = B.

11.2. Which if the following correctly depicts the order of flow of electrons in the electron transport chain?

A. Complex I → Coenzyme Q → Complex III → Complex IV
B. Complex I → Complex II → Complex III → Complex IV
C. Complex I → Complex III → Complex IV → ATP Synthase
D. Complex I → Complex III→ Cytochrome c → ATP Synthase

Correct answer = A.

11.3. Which of the following proteins, a mobile electron carrier in the electron transport chain located in the inter-membrane space, can mediate apoptosis (a form of programmed cell-death) when it is released into the cytoplasm?

A. Coenzyme Q
B. Cytochrome c
C. Flavin mononucleotide
D. Cytochrome a + a_3

Correct answer = B.

11.4. Which of the following components are coded by the mitochondrial genome?

(I) Subunits of protein complexes of electron transport chain
(II) Ribosomal RNAs in the mitochondrial matrix
(III) Transfer RNAs in the mitochondrial matrix
(IV) Histone proteins associated with the mitochondrial DNA

A. I and II
B. I, II, and III
C. I, II, and IV
D. I, II, III, and IV

Correct answer = B.

11.5. 2,4-Dinitrophenol (DNP), an uncoupler of oxidative phosphorylation, was used as a weight-loss agent in the 1930s. Reports of fatal overdoses led to its discontinuation in 1939. Which of the following would most likely be true concerning individuals taking 2,4-DNP?

A. ATP levels in the mitochondria are greater than normal.
B. Body temperature is elevated as a result of hypermetabolism.
C. Cyanide has no effect on electron flow.
D. The proton gradient across the inner mitochondrial membrane is greater than normal.

Correct answer = B.

11.6. Which of the following has the strongest tendency to gain electrons?

A. Coenzyme Q
B. Oxygen
C. Flavin adenine dinucleotide
D. Nicotinamide adenine dinucleotide

Correct answer = B.

11.7. Which of the following is TRUE about electron transport chain?

A. At Complex I, electrons move from NADH to CoQ
B. At Complex II, electrons move from $FADH_2$ to Cytochrome c
C. At Complex III, electrons move from $FADH_2$ to CoQ
D. At Complex IV, electrons move from CoQ to oxygen

Correct answer = A.

11.8. In which complex of electron transport chain does iron has a free coordination site?

A. Complex I
B. Complex II
C. Complex III
D. Complex IV

Correct answer = D.

11.9. Which of the following is TRUE about Complex II of electron transport chain?

A. Electrons move from fumarate to succinate
B. No H^+ are pumped at Complex II
C. Complex II has a tightly bound molecule of flavin mononucleotide (FMN)
D. It also is called cytochrome c oxidase

Correct answer = B.

11.10. Carbon monoxide (CO) inhibits transfer of electrons in electron transport chain at:

A. Complex I
B. Complex II
C. Complex III
D. Complex IV

Correct answer = D.

Multiple-Choice Questions (NExT/NEET PG)

Choose the ONE best answer.

11.1. One of the strategies in the treatment of cyanide toxicity is to administer a hydroxylated form of a vitamin; this combines with cyanide to form a much less toxic product, which can be excreted from the body. Identify the vitamin.

A. Thiamine
B. Niacin
C. Cobalamin
D. Folate

Correct answer = C.

11.2. Clinical manifestations of hyperthyroidism include weight loss despite a normal or increased appetite and intolerance to heat. Which of the following effects of thyroxine on oxidative phosphorylation gives the best explanation for these clinical features?

A. Thyroxine activates complex IV of the electron transport system leading to increased oxidation of glucose and fats.
B. Thyroxine stimulates the Fo-F1 ATP synthase and increases the production of ATP.
C. Thyroxine facilitates the proton pumping across complexes I, III, and IV, thereby enhancing the proton gradient across the inner mitochondrial membrane.
D. Thyroxine uncouples oxidation from phosphorylation by allowing entry of protons into mitochondrial matrix, bypassing the Fo-F1 ATP synthase.

Correct answer = D.

11.3. "Statin-associated muscle symptoms" is a group of documented side effects of use of statin. A probable explanation for these symptoms is a decrease in the ATP production in the muscles, as statins inhibit the production of isoprenes which are derived from an intermediate of cholesterol synthesis. Which component of the electron transport chain is affected in this situation?

A. NADH dehydrogenase
B. Succinate dehydrogenase
C. Coenzyme Q
D. Cytochrome c

Correct answer = C.

11.4. Pathogenesis in a certain form of familial amyotropic lateral sclerosis, a progressive neurodegenerative disorder, is due to the accumulation of protein aggregates as a result of reactive oxygen species-induced damage. Variants in the gene coding for which of the following proteins can lead to this condition?

A. Aldose reductase
B. Superoxide dismutase
C. L-amino acid oxidase
D. ATP synthase

Correct answer = B.

11.5. A patient suspected of having an inherited disorder affecting oxidative phosphorylation is undergoing DNA sequencing to identify the implicated gene. In which of the following locations is the defective gene likely to be found?

A. Nuclear DNA
B. Mitochondrial DNA
C. Either nuclear DNA or mitochondrial DNA
D. Neither nuclear DNA nor mitochondrial DNA

Correct answer = C.

Introduction to Carbohydrates

12

Learning Objectives

At the end of this chapter, the student will be able to:

- Discuss and differentiate monosaccharides, disaccharides and polysaccharides
- Describe the processes involved in digestion and assimilation of carbohydrates.

I. OVERVIEW

Carbohydrates (saccharides) are the most abundant organic molecules in nature. They have a wide range of functions, including providing a significant fraction of the dietary calories for most organisms, acting as a storage form of energy in the body, and serving as cell membrane components that mediate some forms of intercellular communication. Carbohydrates also serve as a structural component of many organisms, including the cell walls of bacteria, the exoskeleton of insects, and the fibrous cellulose of plants. [Note: The full set of carbohydrates produced by an organism is its glycome.] The empiric formula for many of the simpler carbohydrates is $(CH_2O)_n$, where $n \geq 3$, hence the name "hydrate of carbon."

GENERIC NAMES		EXAMPLES
3 Carbons:	trioses	Glyceraldehyde
4 Carbons:	tetroses	Erythrose
5 Carbons:	pentoses	Ribose
6 Carbons:	hexoses	Glucose
7 Carbons:	heptoses	Sedoheptulose
9 Carbons:	nonoses	Neuraminic acid

Figure 12.1 Examples of monosaccharides found in humans, classified according to the number of carbons they contain.

II. CLASSIFICATION AND STRUCTURE

Monosaccharides (simple sugars) can be classified according to the number of carbon atoms they contain. Examples of some monosaccharides commonly found in humans are listed in Figure 12.1. They can also be classified by the type of carbonyl group they contain. Carbohydrates with an aldehyde as their carbonyl group are called aldoses, whereas those with a keto as their carbonyl group are called ketoses (Fig. 12.2). For example, glyceraldehyde is an aldose, whereas dihydroxyacetone is a ketose. Carbohydrates that have a free carbonyl group have the suffix -ose. [Note: Ketoses have an additional "ul" in their suffix such as xylulose. There are exceptions, such as fructose, to this rule.] Monosaccharides can be linked by glycosidic bonds to create larger structures (Fig. 12.3). Disaccharides contain two monosaccharide units, oligosaccharides contain three to ten monosaccharide units, and polysaccharides contain more than ten monosaccharide units and can be hundreds of sugar units in length.

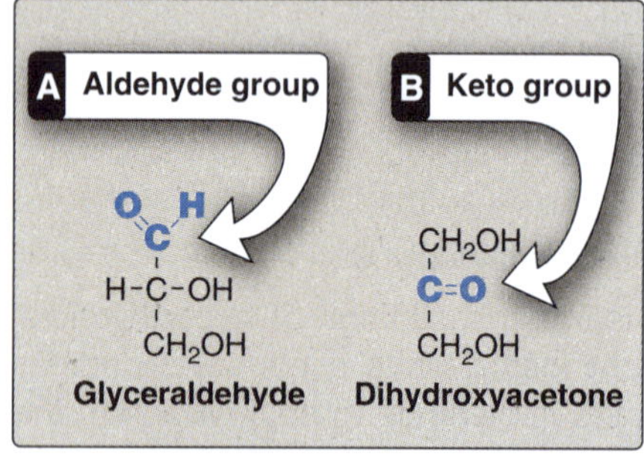

Figure 12.2 Examples of an aldose (**A**) and a ketose (**B**) sugar.

A. Isomers and epimers

Compounds that have the **same chemical formula but have different structures are called isomers**. For example, fructose, glucose, mannose, and galactose

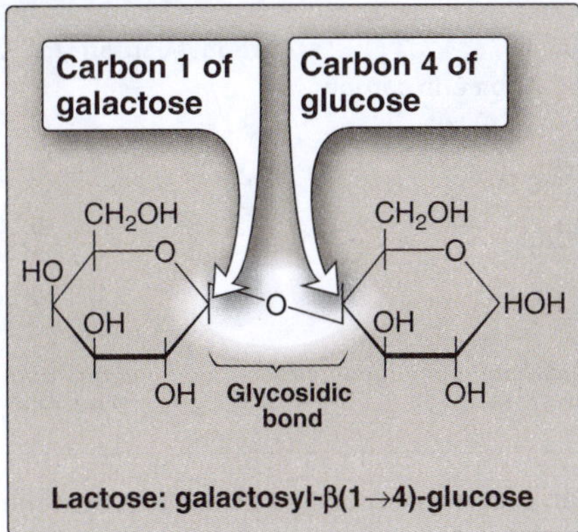

Figure 12.3 A glycosidic bond between two hexoses producing a disaccharide.

are all isomers of each other, having the same chemical formula, $C_6H_{12}O_6$. **Carbohydrate isomers that differ in configuration around only one specific carbon atom (with the exception of the carbonyl carbon, see C. 1. below) are defined as epimers of each other**. For example, glucose and galactose are C-4 epimers because their structures differ only in the position of the –OH (hydroxyl) group at carbon 4. [Note: The carbons in sugars are numbered beginning at the end that contains the carbonyl carbon (that is, the aldehyde or keto group), as shown in Fig. 12.4.] Glucose and mannose are C-2 epimers. However, because galactose and mannose differ in the position of –OH groups at two carbons (carbons 2 and 4), they are isomers rather than epimers (see Fig. 12.4).

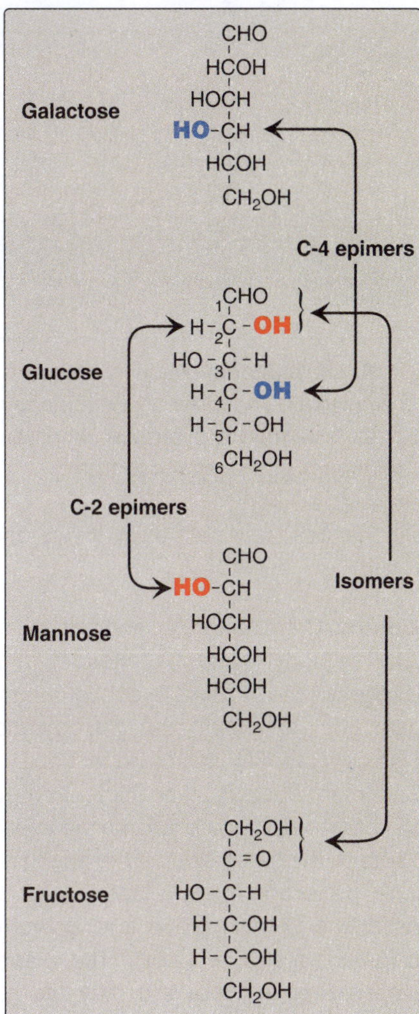

Figure 12.4 Carbon-2 (C-2) and C-4 epimers and an isomer of glucose.

B. Enantiomers

Compounds which have same chemical and structural formula but differ in spatial arrangement of groups are known as **stereoisomers**. This type of isomerism is found in the pairs of structures that are mirror images of each other. **These mirror images are called enantiomers**, and the two members of the pair are designated as a D- and an L-sugar (Fig. 12.5). The vast majority of the sugars in humans are D-isomers. In the D-isomeric form, the –OH group on the **asymmetric carbon** (a carbon linked to four different atoms or groups) farthest from the carbonyl carbon is on the right, whereas in the L-isomer, it is on the left. Most enzymes are specific for either the D or the l form, but enzymes known as *isomerases* are able to interconvert D- and L-isomers.

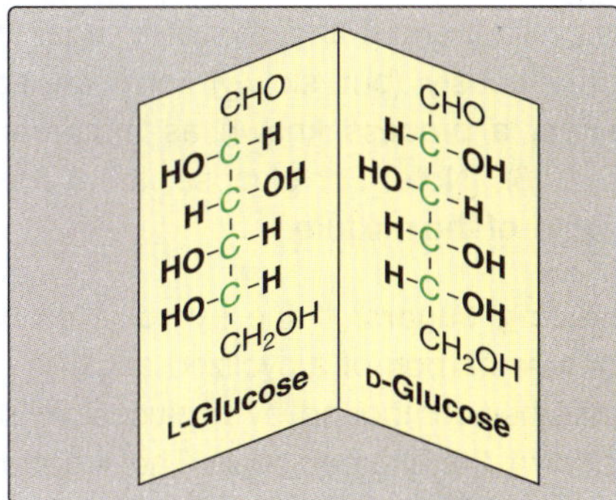

Figure 12.5 Enantiomers (mirror images) of glucose. Designation of D and L is by comparison to a triose, glyceraldehyde. [Note: The asymmetric carbons are shown in green.]

C. Monosaccharide cyclization

Monosaccharides with five or more carbons commonly exist in a cyclic form, where the carbonyl group (aldehyde or ketone) reacts with a hydroxyl group on the same molecule, forming a ring structure. This creates an asymmetric carbon, known as the **anomeric carbon**, which can exist in two different configurations: alpha and beta. However, in rare cases,

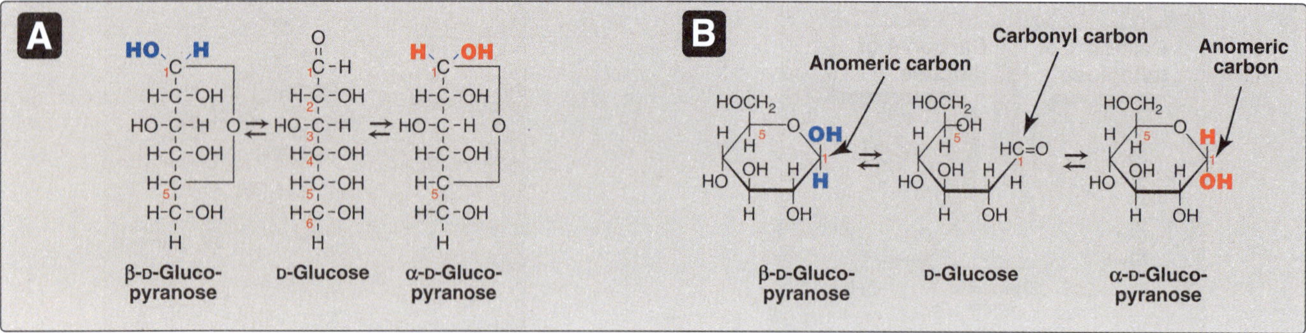

Figure 12.6 **A.** The interconversion (**mutarotation**) of the α and β anomeric forms of glucose shown as modified Fischer projection formulas. **B.** The interconversion shown as Haworth projection formulas. [Note: A sugar with a six-membered ring (5 C + 1 O) is termed a pyranose, whereas one with a five-membered ring (4 C + 1 O) is a furanose. Virtually all glucose in solution is in the pyranose form.]

monosaccharides can also exist in an open-chain, acyclic form.

1. **Anomers:** Creation of an anomeric carbon (the former carbonyl carbon) generates a new pair of isomers, the α and β configurations of the sugar (for example, α-D-glucopyranose and β-D-glucopyranose), as shown in Figure 12.6, that are anomers of each other. [Note: In the α configuration, the –OH group on the anomeric carbon projects to the same side as the ring in a modified Fischer projection formula (see Fig. 12.6A) and is trans to the CH_2OH group in a Haworth projection formula (see Fig. 12.6B). The α and β forms are not mirror images, and they are referred to as **diastereomers**.] Enzymes are able to distinguish between these two structures and use one or the other preferentially. For example, glycogen is synthesized from α-D-glucopyranose, whereas cellulose is synthesized from β-D-glucopyranose. The cyclic α and β anomers of a sugar in solution spontaneously (but slowly) form an equilibrium mixture, a process known as mutarotation (see Fig. 12.6). [Note: For glucose, the α form makes up 36% of the mixture.]

2. **Reducing sugars:** If the hydroxyl group on the anomeric carbon of a cyclized sugar is not linked to another compound by a glycosidic bond (see E. below), the ring can open. The sugar can act as a reducing agent and is termed a reducing sugar. Such sugars can react with chromogenic agents (for example, the Benedict reagent) causing the reagent to be reduced and colored as the aldehyde group of the acyclic sugar is oxidized to a carboxyl group. **All monosaccharides, but not all disaccharides, are reducing sugars.** [Note: Fructose, a ketose, is a reducing sugar because it can be isomerized to an aldose.]

> A colorimetric test can detect a reducing sugar in urine. A positive result is indicative of an underlying pathology (because sugars are not normally present in urine) and can be followed up by more specific tests to identify the reducing sugar.

D. Monosaccharide joining

Monosaccharides can be joined to form disaccharides, oligosaccharides, and polysaccharides. See Box below for important polysaccharides which include branched glycogen (from animal sources) and starch (plant sources) and unbranched cellulose (plant sources). Each is a polymer of glucose.

> **Important Disaccharides**
> - lactose = galactose + glucose
> - sucrose = glucose + fructose
> - maltose = glucose + glucose

> **Important Polysaccharides**
>
> **From animal sources**
> - Glycogen : Branched polymer of Glucose
>
> **From plant sources**
> - Starch : Branched polymer of glucose
> - Cellulose : Unbranched polymer of glucose

E. Glycosidic bonds

The bonds that link sugars are called glycosidic bonds. They are formed by enzymes known as *glycosyltransferases* that **use nucleotide sugars (activated sugars) such as uridine diphosphate glucose as substrates.** Glycosidic bonds between sugars are named according to the numbers of the connected carbons and with regard to the position of the anomeric hydroxyl group of the first sugar involved in the bond. If this anomeric hydroxyl is in the α configuration, then the linkage is an α-bond. If it is in the β configuration, then the linkage is a β-bond.

For example, Lactose is synthesized by forming a glycosidic bond between carbon 1 of β-galactose and carbon 4 of glucose. Therefore, the linkage is a β(1→4) glycosidic bond (see Fig. 12.3). Because the anomeric end of the glucose residue is not involved in the glycosidic linkage, lactose remains a reducing sugar.

F. Carbohydrate linkage to noncarbohydrates

Carbohydrates can be attached by glycosidic bonds to noncarbohydrate structures, including purine and pyrimidine bases (found in nucleic acids), aromatic rings (such as those found in steroids and bilirubin), proteins (found in glycoproteins and proteoglycans), and lipids (found in glycolipids). If the group on the noncarbohydrate molecule to which the sugar is attached is an $-NH_2$ group, then the bond is called an **N-glycosidic link.** If the group is an –OH, then the bond is an **O-glycosidic link** (Fig. 12.7). [Note: All sugar-sugar glycosidic bonds are O-type linkages.]

III. DIETARY CARBOHYDRATE DIGESTION

The principal sites of dietary carbohydrate digestion are the mouth and intestinal lumen. This digestion is rapid and

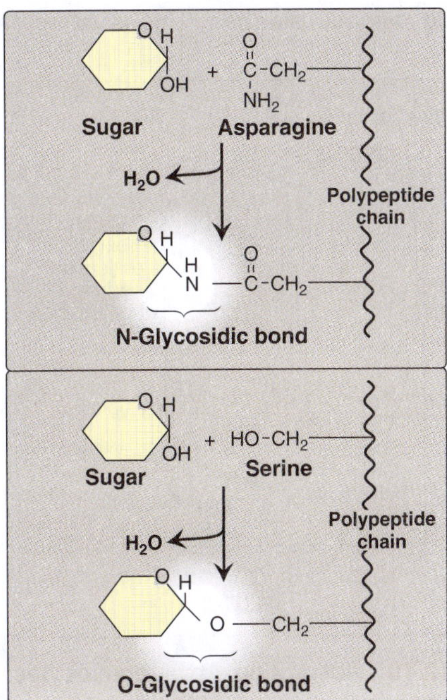

Figure 12.7 Examples of N- and O-glycosidic bonds in glycoproteins.

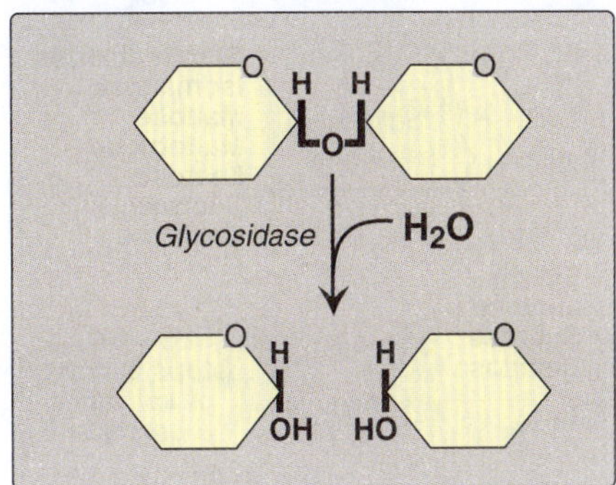

Figure 12.8 Hydrolysis of a glycosidic bond.

is catalyzed by enzymes known as *glycoside hydrolases* (*glycosidases*) that hydrolyze glycosidic bonds (Fig. 12.8). Because little monosaccharide is present in diets of mixed animal and plant origin, the enzymes are primarily *endoglycosidases* that hydrolyze polysaccharides and oligosaccharides and *disaccharidases* that hydrolyze tri- and disaccharides into their reducing sugar components. *Glycosidases* are usually specific for the structure and configuration of the glycosyl residue to be removed as well as for the type of bond to be broken. The final products of carbohydrate digestion are the monosaccharides glucose, galactose, and fructose that are absorbed by cells (enterocytes) of the small intestine.

A. Salivary α-amylase

The major dietary polysaccharides are of plant (starch, composed of amylose and amylopectin) and animal (glycogen) origin. During mastication (chewing), *salivary α-amylase* acts briefly on dietary starch and glycogen, hydrolyzing random α(1→4) bonds. Because branched amylopectin and glycogen also contain α(1→6) bonds, which *α-amylase* cannot hydrolyze, the digest resulting from its action contains a mixture of short, branched and unbranched oligosaccharides known as **dextrins** (Fig. 12.9). [Note: Disaccharides are also present as they, too, are resistant to *amylase*.] Carbohydrate digestion halts temporarily in the stomach, because the high acidity inactivates *salivary α-amylase*.

> Cellulose is a polysaccharide of plant origin, containing β(1→4) glycosidic bonds between glucose residues. Humans do not produce *β(1→4)-endoglucosidases, and* therefore, are unable to digest cellulose, Cellulose acts as dietary fiber (p. 510)

B. Pancreatic α-amylase

When the acidic stomach contents reach the small intestine, they are neutralized by bicarbonate secreted by the pancreas, and *pancreatic α-amylase* continues the process of starch digestion.

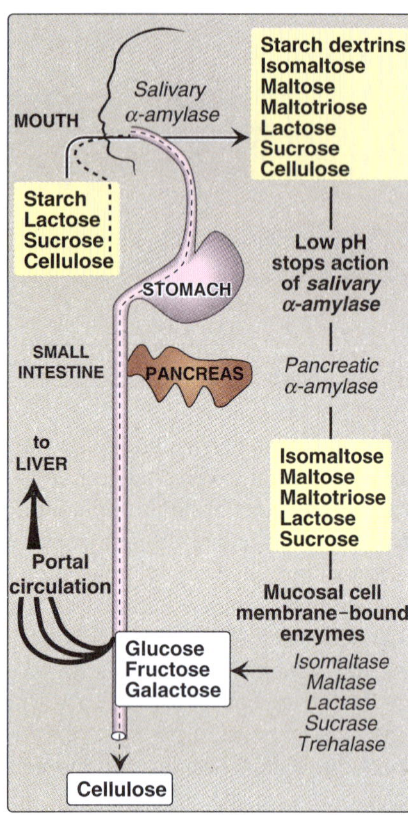

Figure 12.9 Digestion of carbohydrates. [Note: Indigestible cellulose enters the colon and is excreted.]

C. Intestinal disaccharidases

The final digestive processes occur primarily at the mucosal lining of the duodenum and upper jejunum and include the action of several *disaccharidases* (see Fig. 12.9). For example, *isomaltase* cleaves the $\alpha(1\rightarrow6)$ bond in isomaltose, and *maltase* cleaves the $\alpha(1\rightarrow4)$ bond in maltose and maltotriose, each producing glucose. *Sucrase* cleaves the $\alpha(1\rightarrow2)$ bond in sucrose, producing glucose and fructose, and *lactase* (*β-galactosidase*) cleaves the $\beta(1\rightarrow4)$ bond in lactose, producing galactose and glucose. [Note: The substrates for *isomaltase* are broader than its name suggests, and it hydrolyzes the majority of maltose.] Trehalose, an $\alpha(1\rightarrow1)$ disaccharide of glucose found in mushrooms and other fungi, is cleaved by *trehalase*. These enzymes are transmembrane proteins of the brush border on the luminal (apical) surface of the enterocytes.

> *Sucrase* and *isomaltase* are enzymic activities of a single protein that is cleaved into two functional subunits, which remain associated in the cell membrane and form the *sucrase-isomaltase* (*SI*) complex. In contrast, *maltase* is one of two enzymic activities of the single membrane protein *maltase-glucoamylase* (*MGA*) that does not get cleaved. Its second enzymic activity, *glucoamylase*, cleaves $\alpha(1\rightarrow4)$ glycosidic bonds in dextrins.

D. Intestinal absorption of monosaccharides

The upper jejunum absorbs the bulk of the monosaccharide products of digestion. However, different sugars have different mechanisms of absorption (Fig. 12.10). For example, galactose and glucose are taken into enterocytes by secondary active transport that requires a concurrent uptake (**symport**) of sodium (Na^+) ions. The transport protein is the sodium-dependent glucose cotransporter 1 (**SGLT-1**). [Note: Sugar transport is driven by the Na^+ gradient created by the *Na^+-potassium (K^+) ATPase* that moves Na^+ out of the enterocyte and K^+ in (see Fig. 12.10).] Fructose absorption utilizes an energy- and Na^+-independent monosaccharide transporter (GLUT-5). All three monosaccharides are transported from the enterocytes into the portal circulation by yet another transporter, GLUT-2. [Note: See p. 192 for a discussion of these transporters.]

E. Abnormal degradation of disaccharides

The overall process of carbohydrate digestion and absorption is so efficient in healthy individuals that ordinarily all digestible dietary carbohydrate is absorbed by the time the ingested material reaches the lower jejunum. However, because only monosaccharides are absorbed, any deficiency (genetic or acquired) in a specific *disaccharidase* activity of the intestinal mucosa causes the passage of undigested carbohydrate into the large intestine. As a consequence of the presence of this osmotically active material, water is drawn from the mucosa into the large intestine, causing osmotic diarrhea. This is reinforced by the

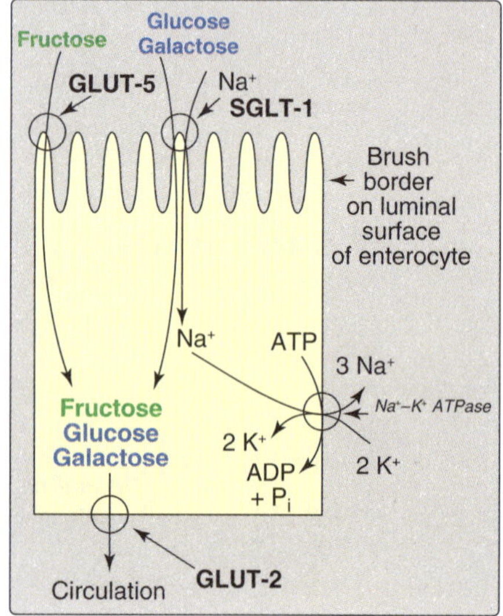

Figure 12.10 Absorption by enterocytes of the monosaccharide products of carbohydrate digestion. GLUT = glucose transporter; SGLT-1 = sodium (Na^+)-dependent glucose cotransporter. K^+ = potassium.

bacterial fermentation of the remaining carbohydrate to two- and three-carbon compounds (which are also osmotically active) plus large volumes of carbon dioxide and hydrogen gas (H_2), causing abdominal cramps, diarrhea, and flatulence.

1. **Digestive enzyme deficiencies:** Genetic deficiencies of the individual *disaccharidases* result in disaccharide intolerance. Alterations in disaccharide degradation can also be caused by a variety of intestinal diseases, malnutrition, and drugs that injure the mucosa of the small intestine. For example, brush border enzymes are rapidly lost in normal individuals with severe diarrhea, causing a temporary, acquired enzyme deficiency. Therefore, patients suffering or recovering from such a disorder cannot drink or eat significant amounts of dairy products or sucrose without exacerbating the diarrhea.

2. **Lactose intolerance:** The digestion of lactose requires intestinal brush-border enzyme lactase. The deficiency of lactase, either primary or secondary, results in an inability to digest lactose. Over 60% of the world's adults are lactose intolerant (Fig. 12.11). This is particularly manifested in certain populations. For example, up to 90% of adults of African or Asian descent are *lactase* deficient. In **primary lactase deficiency**, a genetically caused decrease or absence of lactase is seen. **Secondary lactase deficiency** occurs in association with small-intestinal mucosal diseases, such as celiac disease, which affects the structure and function of brush-border enzymes. The age-dependent loss of *lactase* activity starting at approximately 2 years of age represents a reduction in the amount of enzyme produced. It is thought to be caused by small variations in the DNA sequence of a region on chromosome 2 that controls expression of the gene for *lactase*, also on chromosome 2.

   The undigested lactose undergoes bacterial fermentation in colon, where large quantities of CO_2 and organic acids are produced and patients have abdominal symptoms such as diarrhea, abdominal pain, cramps, and flatus.

   Treatment for this disorder is to reduce consumption of milk; eat yogurts and some cheeses (bacterial action and aging process decrease lactose content) as well as green vegetables, such as broccoli, to ensure adequate calcium intake; use *lactase*-treated products; or take *lactase* in pill form prior to eating.

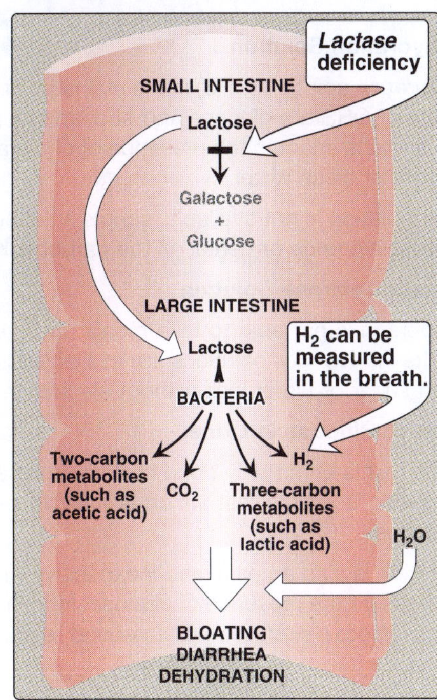

Figure 12.11 Abnormal lactose metabolism. CO_2 = carbon dioxide; H_2 = hydrogen gas.

3. **Congenital sucrase-isomaltase deficiency:** This autosomal-recessive disorder results in an intolerance of ingested sucrose. Congenital *SI* deficiency has a prevalence of 1:5,000 in individuals of European descent and appears to be much more common (up to 1:20) in the Inuit people of Greenland and Canada. Treatment includes the dietary restriction of sucrose and enzyme replacement therapy.

4. **Diagnosis:** Identification of a specific enzyme deficiency can be obtained by performing oral tolerance tests with the individual disaccharides. Measurement of H_2 in the breath is a reliable test for determining the amount of ingested carbohydrate not absorbed by the body, but which is metabolized instead by the intestinal flora (see Fig. 12.11).

IV. Medical Perspectives

Lactose Intolerance

Lactose intolerance is a common clinically seen disorder of carbohydrate malabsorption. Lactose which remains undigested due to deficiency of *lactase* is fermented in colon, producing gases and organic acids, which cause osmotic diarrhea. Affected persons show improvement with lactose-free diet. Another alternative is using curd because lactobacilli in curd produce lactase.

Oral Rehydration Solution

Oral rehydration solution (ORS) is the mainstay of treatment in cholera and other dehydrating diarrheal diseases. The main ingredients of ORS are glucose and sodium chloride, as glucose-facilitated absorption of sodium and water in the small intestine remains intact in the presence of cholera toxin. The ORS is available as dry powder, to be reconstituted in specified amount of clean water.

If ORS formulation is not available, simple solution of sugar and salt can also effectively prevent dehydration. ORS formulations containing rice or cereal as the carbohydrate source can also be effectively used.

Intravenous Dextrose Solution

When a person is not able to take meals orally, intravenous solution of dextrose is given. Dextrose is a sugar similar to glucose. Its intravenous solutions are available is different concentration, such as 5% or 10%. It can be also used to treat low-blood sugar or water loss without electrolyte loss.

Detection of Glucose in Urine

Benedict's test is commonly done to detect presence of glucose in urine. This test is positive for several reducing sugars also but being a simple test which is easy to perform, it is commonly used in primary health care centers for testing urinary glucose.

Urine dipstick is also an easy and inexpensive method for urinary glucose analysis. It uses two coupled enzymatic reactions to measure the presence of glucose. In the first reaction, glucose is oxidized to gluconic acid and hydrogen peroxide by enzyme *glucose oxidase*. In the second reaction, hydrogen peroxide oxidizes a dye to form a colored product.

Mutarotation

In medical laboratories, standard solution of glucose is commonly prepared. In a freshly prepared solution, most of the glucose molecules are in α-form. *Glucose oxidase*, which is used to estimate the amount of glucose, acts on β-form. Therefore, the value obtained for a fresh glucose solution will be incorrect. Keeping the glucose solution for 12 hours allows mutarotation to occur, and then we get correct value of glucose in the solution.

V. Chapter Summary

- **Monosaccharides** (Fig. 12.12) containing an aldehyde group are called **aldoses**, and those with a keto group are called **ketoses**.
- **Disaccharides**, **oligosaccharides**, and **polysaccharides** consist of monosaccharides linked by **glycosidic bonds**.
- **Isomers**: Compounds with the same chemical formula but different structures.
- **Epimers**: Two monosaccharide isomers differing in configuration around one specific carbon atom (not the carbonyl carbon).
- In **enantiomers** (mirror images), the members of the sugar pair are designated as **d**-and **l-isomers**.
- When the aldehyde group on an acyclic sugar gets oxidized as a chromogenic agent gets reduced, that sugar is a **reducing sugar**.
- When a sugar cyclizes, an **anomeric carbon** is created from the carbonyl carbon of the aldehyde or keto group.
- A sugar can have its anomeric carbon linked to an $-NH_2$ or an –OH group on another structure through **N-** and **O-glycosidic bonds**, respectively.
- ***Salivary α-amylase*** initiates digestion of **dietary polysaccharides** producing oligosaccharides. ***Pancreatic α-amylase*** continues the process.
- The final digestive processes occur at the **mucosal lining** of the **small intestine**.
- *Disaccharidases* (e.g., ***lactase***, ***sucrase***, ***isomaltase***, and ***maltase***) produce monosaccharides.
- Absorption of the monosaccharides requires specific **transporters**.
- If carbohydrate degradation is deficient (as a result of heredity, disease, or drugs that injure the intestinal mucosa), undigested carbohydrate will pass into the large intestine, where it can cause **osmotic diarrhea**. Bacterial fermentation of the material produces large volumes of carbon dioxide and hydrogen gas, causing abdominal cramps, diarrhea, and flatulence.
- **Lactose intolerance**, primarily caused by the age-dependent loss of ***lactase*** (**adult-type hypolactasia**), is by far the most common of these deficiencies.

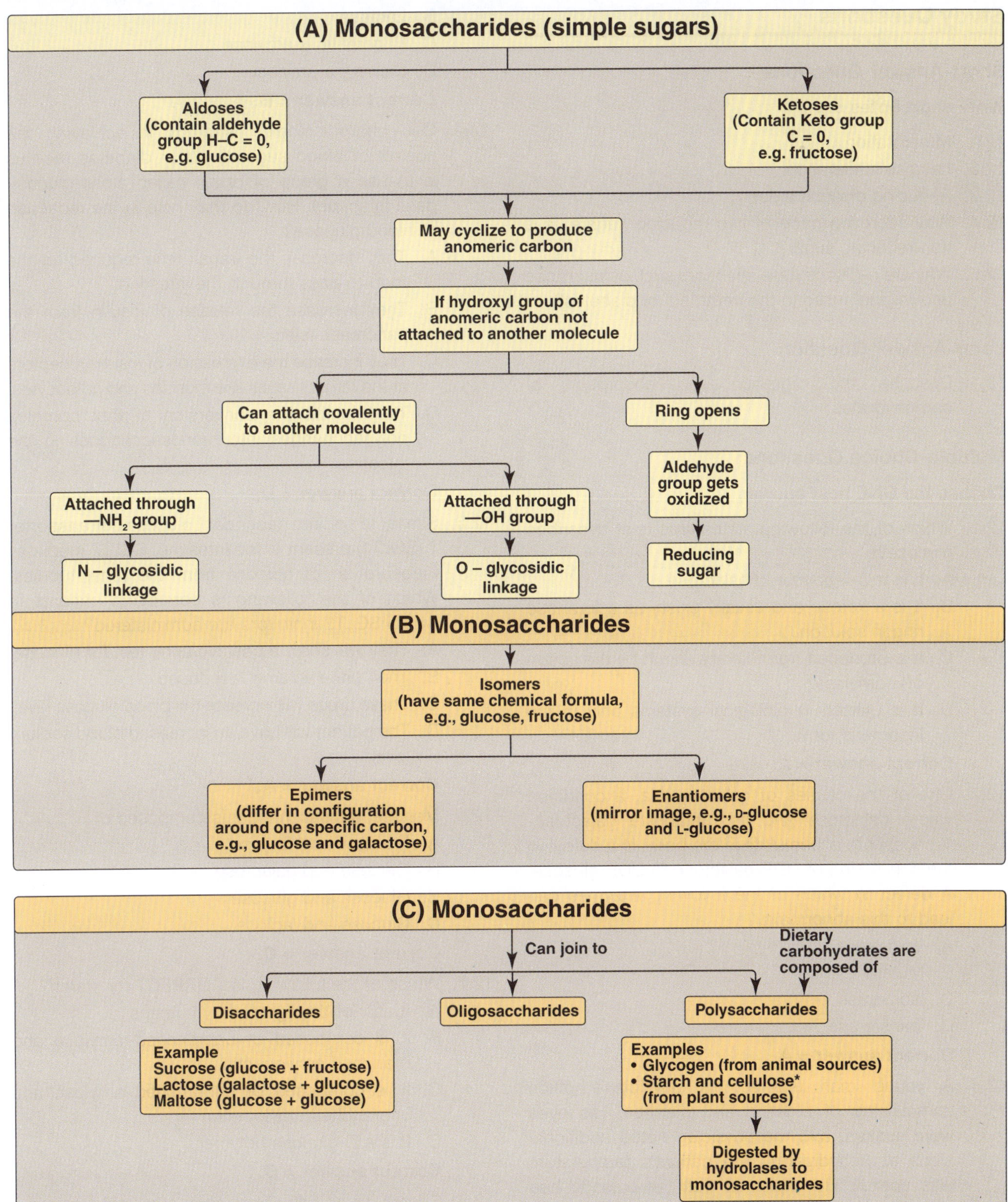

Figure 12.12 Key concept map for the classification and structure of monosaccharides and the digestion of dietary carbohydrates.

Study Questions

Short-Answer Questions

Write short notes on:

12.1. Mutarotation
12.2. Lactose intolerance
12.3. Reducing disaccharides
12.4. Why sucrose, made of two reducing sugars, is a non-reducing sugar?
12.5. Why are α-glucosidase inhibitors such as acarbose and miglitol, used in the treatment of diabetes?

Long-Answer Question

12.1. Describe the digestion and absorption of carbohydrates.

Multiple-Choice Questions

Choose the ONE best answer.

12.1. Which of the following statements best describes glucose?
A. It is a C-4 epimer of galactose.
B. It is a ketose and usually exists as a furanose ring in solution.
C. It is produced from dietary starch by the action of α-amylase.
D. It is utilized in biological systems only in the l-isomeric form.
Correct answer = A.

12.2. One of the causes of galactosemia, a condition where galactose and its metabolite galactose-1-phosphate accumulate in the body, is a defect in the conversion of UDP-galactose to UDP-glucose. A defect in which of the following enzymes can lead to this abnormality?
A. Epimerase
B. Mutase
C. Isomerase
D. Transferase
Correct answer = A.

12.3. A young man entered his physician's office complaining of bloating and diarrhea. His eyes were sunken, and the physician noted additional signs of dehydration. The patient's temperature was normal. He explained that the episode had occurred following a birthday party at which he had participated in an ice cream–eating contest. The patient reported prior episodes of a similar nature following ingestion of a significant amount of dairy products. This clinical picture is most probably due to a deficiency in the activity of:
A. Isomaltase
B. Lactase
C. Pancreatic α-amylase
D. Salivary α-amylase.
Correct answer = B.

12.4. One pharmacological strategy in achieving the control of blood sugar levels in diabetes mellitus is to use a group of drugs called alpha-glucosidase inhibitors. How do they help in the reduction of blood glucose?
A. They decrease the transit time required for the food to pass through the intestine.
B. They increase the release of insulin from the pancreatic islets.
C. They increase the expression of insulin receptors at the target tissues like muscles and adipocytes.
D. They inhibit the conversion of that complex polysaccharides into monosaccharides in the intestine.
Correct answer = D.

12.5. Similar to sodium dependent glucose cotransporter 1 (SGLT1) present in the intestine, SGLT2 mediates reabsorption of glucose from the renal tubules. Which of the following is correct, in patients in whom SGLT2 inhibitors are administered?
A. They will show a positive urine test for glucose.
B. Their urine volume is reduced.
C. These drugs will increase the blood glucose level.
D. The patient will have an increased blood sodium level.
Correct answer = A.

12.6. Maltose, a disaccharide, is composed of
A. Fructose and lactose
B. Glucose and galactose
C. Glucose and glucose
D. Glucose and lactose
Correct answer = C.

12.7. Which of the following is CORRECT for starch?
A. It cannot be digested by humans
B. It is composed of unbranched amylose and branched amylopectin
C. It is composed of unbranched amylose and unbranched amylopectin
D. It is a D-glucose polymer
Correct answer = B.

12.8. Sugars which differ from each other only around a single carbon atom are called:
A. Epimers
B. Enantiomers
C. Optical isomers
D. None of the above
Correct answer = A.

12.9. Which of the following is a homopolysaccharide made of fructose?
   A. Cellulose
   B. Starch
   C. Inulin
   D. None of the above
   Correct answer = C.

12.10. Which of the following is a pentose sugar?
   A. Dihydroxyacetone
   B. Ribulose
   C. Erythrose
   D. Glucose
   Correct answer = B.

Multiple-Choice Questions (NExT/NEET PG)

12.1. A 30-year-old male presented to the emergency department with sudden onset severe upper abdominal pain and two episodes of vomiting. The pain radiated to the back and improved on bending forward. He gave a history of binge alcohol consumption for the past 2 days. On examination the patient was feverish with a heart rate of 98 bpm and BP of 96/60 mm Hg. Tenderness was present in the epigastric region. Which of the following enzymes is most likely elevated in the serum of this patient?
   A. Creatine kinase
   B. Alkaline phosphatase
   C. Amylase
   D. Alanine transaminase
   Correct answer = C.

12.2. A 34-year-old woman experiences bloating, abdominal pain, and diarrhea, worsened by consuming mushrooms, dried fruit, and nutrition bars. Despite trying gluten-free and lactose-free diets, her symptoms persist. Laboratory tests are normal. Which condition is most likely responsible for her symptoms?
   A. Gluten sensitivity
   B. Lactose intolerance
   C. Trehalose intolerance
   D. Fructose malabsorption
   Correct answer = C.

12.3. Statement 1: In astrobiology, one method of determining the potential existence of life in extraterrestrial materials, such as meteorites, involves analyzing the ratio of D- and L-isomers of certain organic compounds.
   Statement 2: Biological systems, which rely on enzymatic activity, typically produce exclusively one isomer form (either D or L) for specific compounds, whereas non-biological chemical synthesis tends to produce a racemic mixture, meaning an equal proportion of D- and L-isomers.
   A. Both the Statements are correct, and Statement 2 is the correct explanation for Statement 1
   B. Both the Statements are correct, but Statement 2 is NOT the correct explanation for Statement 1
   A. Statement 1 is correct, but Statement 2 is NOT correct
   D. Statement 1 is NOT correct, but Statement 2 is correct
   Correct answer = A.

12.4. The only carbohydrate which does not have any chiral carbon atoms is:
   A. Glyceraldehyde
   B. Erythrose
   C. Dihydroxyacetone
   D. Erythrulose
   Correct answer = C.

12.5. Examination of the urine of an asymptomatic pediatric patient showed a positive reaction with Benedict's test but a negative reaction with the glucose oxidase test. Which of the following statements will be TRUE in this case:
   A. Glucose and fructose can be present in urine
   B. Sucrose and glucose can be present in urine
   C. Glucose and fructose cannot be present in urine
   D. Sucrose and glucose cannot be present in urine
   Correct answer = D.

Introduction to Metabolism and Glycolysis 13

Learning Objectives

At the end of this chapter, the student will be able to:

- Describe concepts of metabolic pathways and mechanisms of their regulation
- Discuss second-messenger systems
- Explain the role of glucose transporters
- Explain the pathway of glycolysis, its regulation and energetics

I. METABOLISM OVERVIEW

In Chapter 10, individual enzymic reactions were analyzed in an effort to explain the mechanisms of catalysis. However, in cells, these reactions rarely occur in isolation. Instead, they are organized into multistep sequences called pathways, such as that of glycolysis (Fig. 13.1). In a pathway, the product of one reaction serves as the substrate of the subsequent reaction. Most pathways can be classified as either catabolic (degradative) or anabolic (synthetic). Catabolic pathways break down complex molecules, such as proteins, polysaccharides, and lipids, to a few simple molecules (for example, carbon dioxide, ammonia, and water). Anabolic pathways form complex end products from simple precursors, for example, the synthesis of the polysaccharide glycogen from glucose. [Note: Pathways that regenerate a component are called cycles.] Different pathways can intersect, forming an integrated and purposeful network of chemical reactions. Metabolism is the sum of all the chemical changes occurring in a cell, a tissue, or the body. The next several chapters focus on the central metabolic pathways that are involved in synthesizing and degrading carbohydrates, lipids, and amino acids.

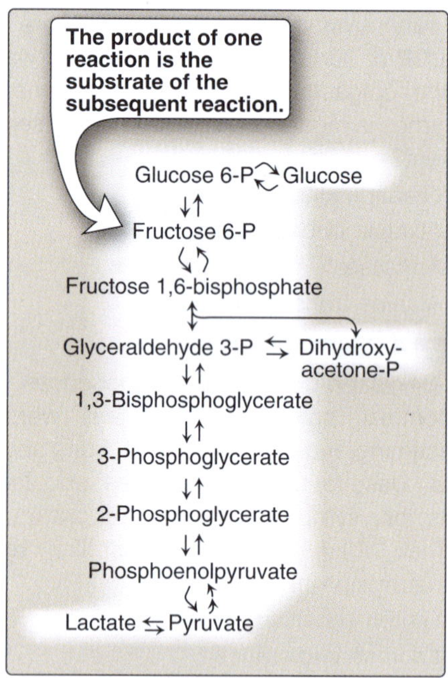

Figure 13.1 Glycolysis, an example of a metabolic pathway. [Note: Pyruvate to phosphoenolpyruvate requires two reactions.] Curved reaction arrows (⥂) indicate forward and reverse reactions that are catalyzed by different enzymes. P = phosphate.

A. Metabolic map

Metabolism is best understood by examining its component pathways. Each pathway is composed of multienzyme sequences, and each enzyme, in turn, may exhibit important catalytic or regulatory features. A metabolic map containing the important central pathways of energy metabolism is presented in Figure 13.2. This "big picture" view of metabolism is useful in tracing connections between pathways, visualizing the purposeful movement of metabolic intermediates (metabolites), and depicting the effect on the flow of intermediates if a pathway is blocked (for example, by a drug or an inherited deficiency of an

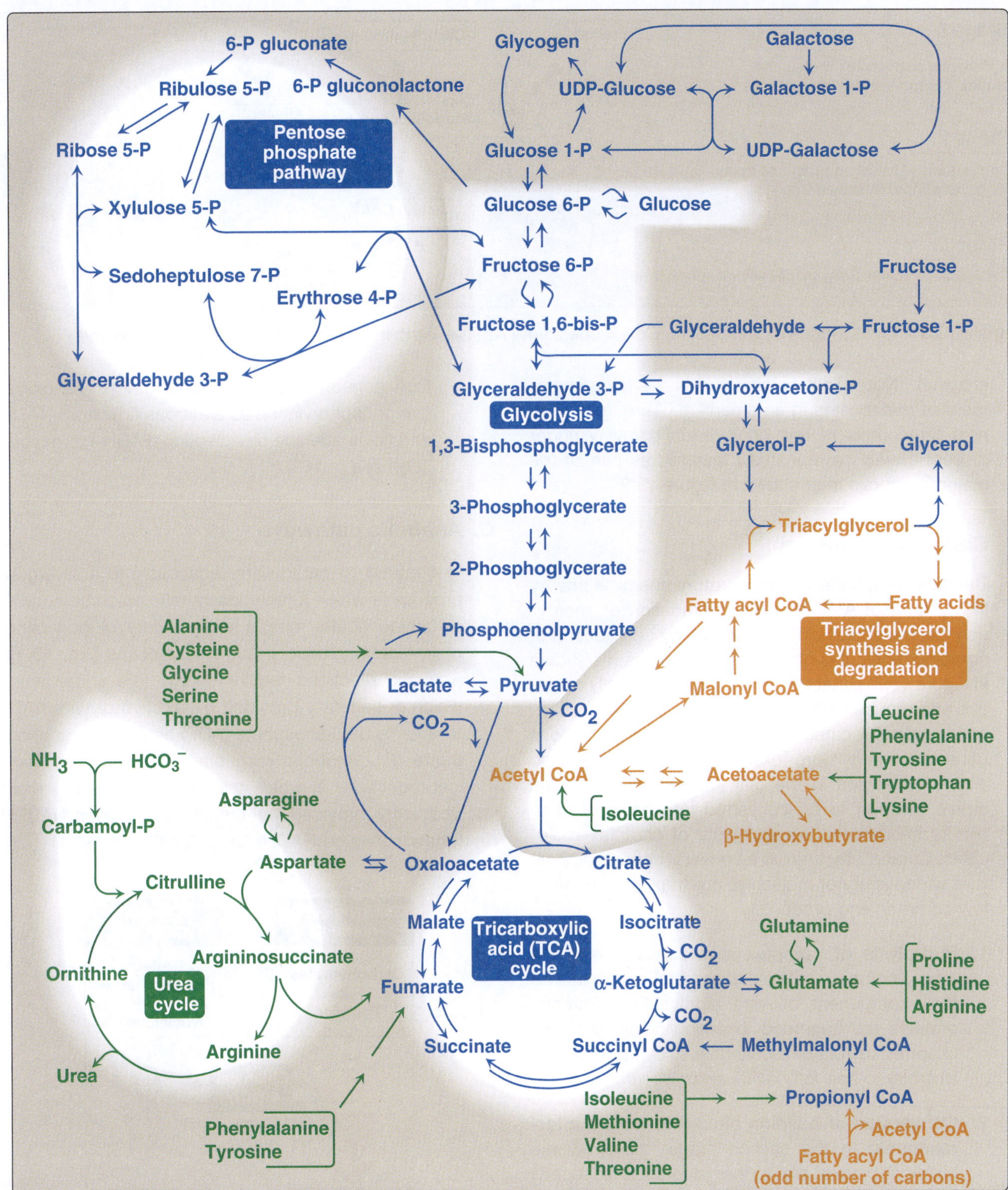

Figure 13.2 Important reactions of intermediary metabolism. Several important pathways to be discussed in later chapters are highlighted. Curved reaction arrows (⇄ curved) indicate forward and reverse reactions that are catalyzed by different enzymes. The straight arrows (⇆) indicate forward and reverse reactions that are catalyzed by the same enzyme. **Blue text** = intermediates of carbohydrate metabolism; **brown text** = intermediates of lipid metabolism; **green text** = intermediates of protein metabolism. UDP = uridine diphosphate; P = phosphate; CoA = coenzyme A; CO_2 = carbon dioxide; HCO_3^- = bicarbonate; NH_3 = ammonia.

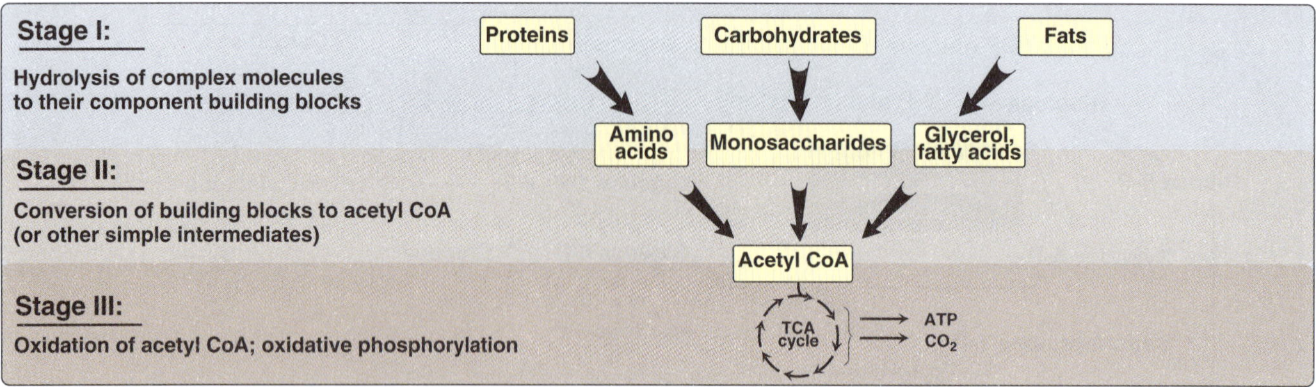

Figure 13.3 Three stages of catabolism. CoA = coenzyme A; TCA = tricarboxylic acid; CO_2 = carbon dioxide.

enzyme). [Note: The metabolome is the full complement of metabolites in an organism.] Throughout the next three units of this book, each pathway under discussion will be repeatedly featured as part of the major metabolic map shown in Figure 13.2.

B. Catabolic pathways

Catabolic reactions serve to capture chemical energy in the form of ATP from the degradation of energy-rich fuel molecules. ATP generation by degradation of complex molecules occurs in three stages, as shown in Figure 13.3. Catabolic pathways are typically oxidative and require oxidized coenzymes such as nicotinamide adenine dinucleotide (NAD^+). These reactions generate energy (are exergonic). Catabolism also allows molecules in the diet (or nutrient molecules stored in cells) to be converted into basic building blocks needed for the synthesis of complex molecules. Catabolism, then, is a convergent process (that is, a wide variety of molecules are transformed into a few common end products).

1. **Hydrolysis of complex molecules:** In the first stage, complex molecules are broken down into their component building blocks. For example, proteins are degraded to amino acids, polysaccharides to monosaccharides, and fats (triacylglycerols) to free fatty acids and glycerol.

2. **Conversion of building blocks to simple intermediates:** In the second stage, these diverse building blocks are further degraded to acetyl coenzyme A (CoA) and a few other simple molecules. Some energy is captured as ATP, but the amount is small compared with the energy produced during the third stage of catabolism.

3. **Oxidation of acetyl coenzyme A:** The tricarboxylic acid (TCA) cycle (see p. 206) is the final common pathway in the oxidation of fuel molecules that produce acetyl CoA. Oxidation of acetyl CoA generates large amounts of ATP via oxidative phosphorylation as electrons flow from NADH and flavin adenine dinucleotide ($FADH_2$) to oxygen ([O_2], see p. 163).

C. Anabolic pathways

In contrast to catabolism, anabolism is a divergent process in which a few biosynthetic precursors (such as amino acids) form a wide variety of polymeric, or complex, products (such as proteins [Fig. 13.4]). Anabolic reactions require energy (are endergonic), which is generally provided by the hydrolysis of ATP to adenosine diphosphate (ADP) and inorganic phosphate (P_i). Anabolic reactions often involve chemical reductions in which the reducing power is most frequently provided by the electron donor NADPH (phosphorylated NADH, see p. 249).

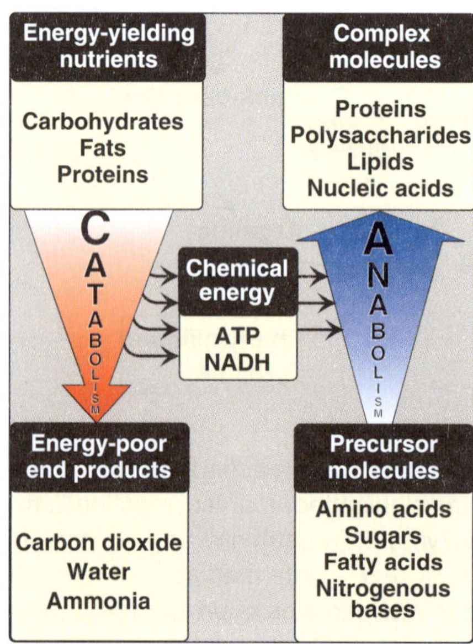

Figure 13.4 Comparison of catabolic and anabolic pathways. NADH = nicotinamide adenine dinucleotide.

II. METABOLISM REGULATION

The pathways of metabolism must be coordinated so that the production of energy or the synthesis of end products meets the needs of the cell. Furthermore, individual cells function as part of a community of interacting tissues, not in isolation. Thus, a sophisticated communication system has evolved to coordinate the functions of the body. Regulatory signals that inform an individual cell of the metabolic state of the body as a whole include hormones, neurotransmitters, and the availability of nutrients. These, in turn, influence signals generated within the cell (Fig. 13.5).

A. Intracellular communication

The rate of a metabolic pathway can respond to regulatory signals that arise from within the cell. For example, the rate may be influenced by the availability of substrates, product inhibition, or alterations in the levels of allosteric activators or inhibitors. These intracellular signals typically elicit rapid responses and are important for the moment-to-moment regulation of metabolism.

B. Intercellular communication

The ability to respond to intercellular signals is essential for the development and survival of organisms. **Signaling between cells provides for long-range integration of metabolism** and **usually results in a response, such as a change in gene expression, that is slower** than is seen with intracellular signals. Communication between cells can be mediated, for example, by surface-to-surface contact and, in some tissues, by formation of gap junctions, allowing direct communication between the cytoplasms of adjacent cells. However, for energy metabolism, the most important route of communication is chemical signaling between cells by blood-borne hormones or by neurotransmitters.

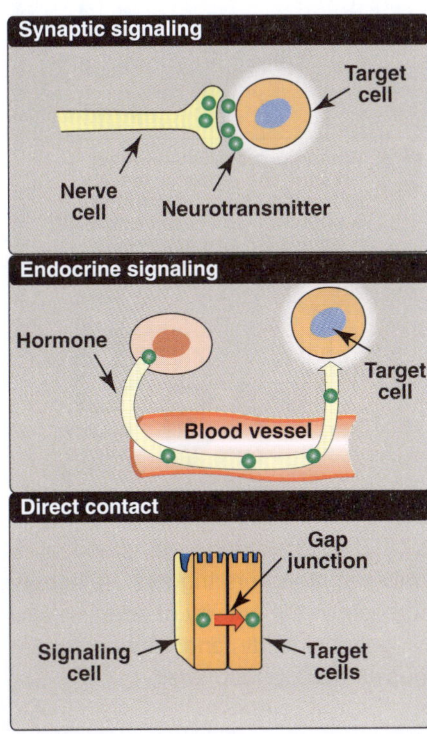

Figure 13.5 Some commonly used mechanisms for transmission of regulatory signals between cells.

C. Second messenger systems

Hormones and neurotransmitters can be thought of as signals and their receptors as signal detectors. Receptors respond to a bound ligand by initiating a series of reactions that ultimately result in specific intracellular responses. Second messenger molecules, so named because they intervene between the original extracellular messenger (the neurotransmitter or hormone) and the ultimate intracellular effect, are part of the cascade of events that converts (transduces) ligand binding into a response. Two of the most widely recognized second messenger systems are the **calcium/phosphatidylinositol system** (see p. 325) and the ***adenylyl cyclase (adenylate cyclase)*** system, which is particularly important in regulating the pathways of intermediary metabolism. Both involve the binding of ligands, such as epinephrine or glucagon, to specific **G protein–coupled receptors** (GPCR) on the cell (plasma) membrane. GPCR are characterized by an extracellular ligand-binding domain, seven transmembrane α helices, and an intracellular domain that interacts with trimeric G proteins (Fig. 13.6). [Note: Insulin, another key regulator of metabolism, binds a membrane tyrosine kinase receptor (see p. 442) and not a GPCR.]

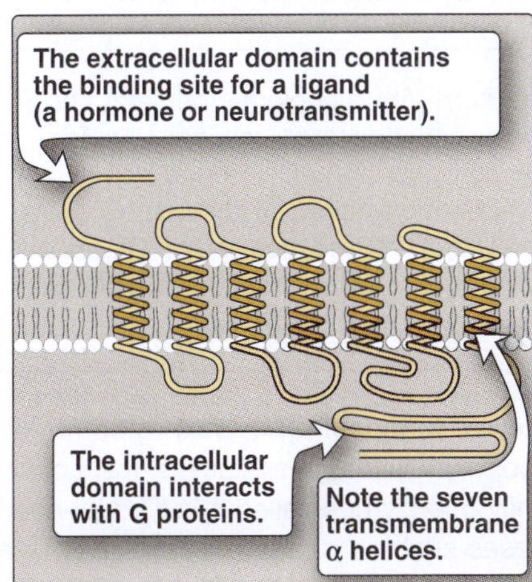

Figure 13.6 Structure of a typical G protein–coupled receptor of the plasma membrane.

D. Adenylyl cyclase

Adenylyl cyclase (AC) is a membrane-bound enzyme that **converts ATP to 3′,5′-adenosine monophosphate (cyclic AMP, or cAMP)**. The recognition of a chemical signal, such as hormones or neurotransmitter, by some GPCR, such as the β- and α_2-adrenergic receptors, triggers either an increase or a decrease in the activity of AC. Each chemical signal binds to a unique type of GPCR. Therefore, tissues that respond to more than one signal must have several different GPCR, each of which can be linked to *AC*.

1. **Guanosine triphosphate–dependent regulatory proteins:** The effect of the activated, occupied GPCR on second messenger formation is indirect, mediated by specialized trimeric proteins (α, β, and γ subunits) of the cell membrane. These proteins, referred to as **G proteins** because the α subunit binds guanosine di- or triphosphates (GDP or GTP), form a link in the chain of communication between the receptor and *AC*. In the inactive form of a G protein, the α subunit is bound to GDP (Fig. 13.7). Ligand binding causes a conformational change in the receptor, triggering replacement of this GDP with GTP. The GTP-bound form of the α subunit dissociates from the βγ subunits and moves to *AC*, affecting enzyme activity. Many molecules of active Gα protein are formed by one activated receptor. [Note: The ability of a hormone or neurotransmitter to stimulate or inhibit *AC* depends on the type of Gα protein that is linked to the receptor. One type, designated G_s, stimulates *AC* (see Fig. 13.7), whereas another type, designated G_i, inhibits the enzyme (not shown).] **The actions of the Gα–GTP complex are short-lived because Gα has an inherent *GTPase* activity**, resulting in the rapid hydrolysis of GTP to GDP. This causes inactivation of Gα, its dissociation from *AC*, and its reassociation with the βγ dimer.

> Toxins from <u>Vibrio cholerae</u> (cholera) and <u>Bordetella pertussis</u> (whooping cough) cause inappropriate activation of *AC* through covalent modification (ADP-ribosylation) of different G proteins. With cholera, the *GTPase* activity of $G\alpha_s$ is inhibited in intestinal cells. With whooping cough, $G\alpha_i$ is inactivated in respiratory tract cells.

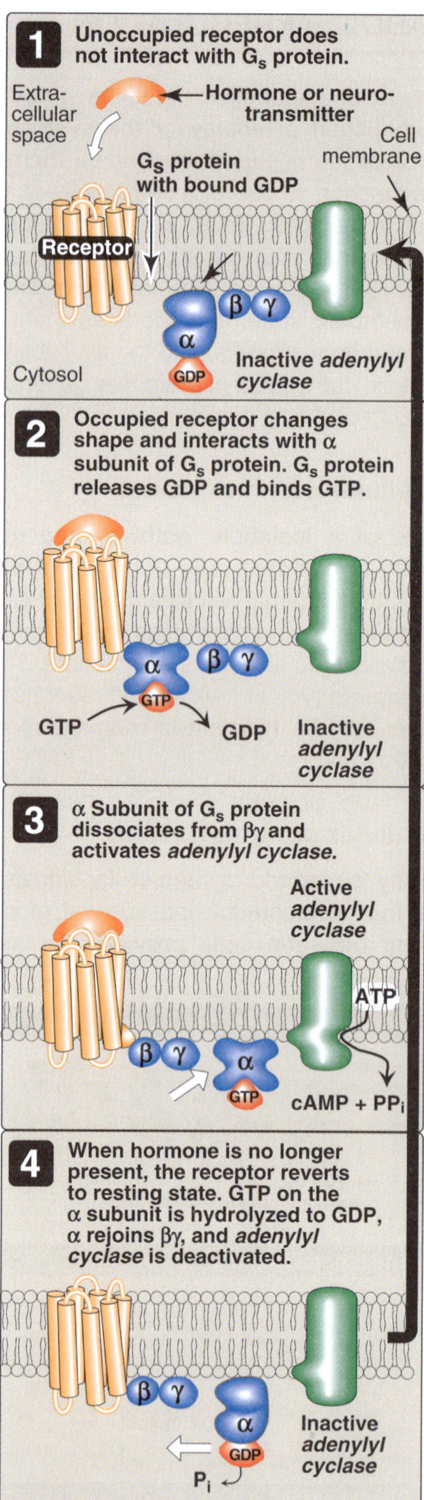

Figure 13.7 The recognition of chemical signals by certain membrane receptors triggers an increase (or, less often, a decrease) in the activity of *adenylyl cyclase*. GDP and GTP = guanosine di- and triphosphates; cAMP = cyclic adenosine monophosphate.

2. **Protein kinases:** The next step in the cAMP second messenger system is the activation of a family of enzymes called *cAMP-dependent protein kinases* such as *protein kinase A (PKA)*, as shown in Figure 13.8. **cAMP activates *PKA*** by binding to its two regulatory subunits, causing the release of its two catalytically active subunits. These subunits transfer phosphate from ATP to specific serine or threonine residues of protein substrates.

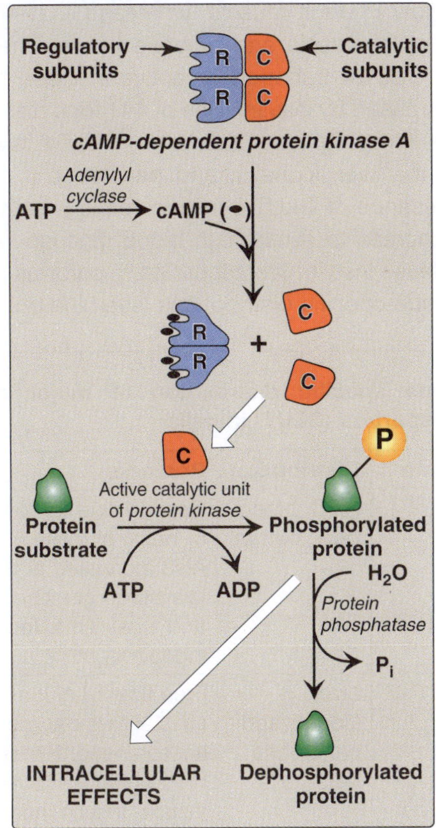

Figure 13.8 Actions of cyclic adenosine monophosphate (cAMP). (P) = phosphate; ADP = adenosine diphosphate; P_i = inorganic phosphate.

The phosphorylated proteins may act directly on the cell's ion channels or, if enzymes, may become activated or inhibited. [Note: Several types of *protein kinases* are not cAMP dependent, for example, *protein kinase C*, described on p. 325.]

3. **Protein phosphatases:** The phosphate groups added to proteins by *protein kinases* are removed by *protein phosphatases*, enzymes that hydrolytically cleave phosphate esters (see Fig. 13.8). This insures that changes in protein activity induced by phosphorylation are not permanent.

4. **cAMP hydrolysis:** cAMP is rapidly hydrolyzed to 5′-AMP by ***cAMP phosphodiesterase*** that cleaves the cyclic 3′,5′-phosphodiester bond. 5′-AMP is not an intracellular signaling molecule. Therefore, the effects of neurotransmitter- or hormone-mediated increases of cAMP are rapidly terminated if the extracellular signal is removed. [Note: *cAMP phosphodiesterase* is inhibited by caffeine, a methylxanthine derivative.]

III. GLYCOLYSIS OVERVIEW

The glycolytic pathway is used by all tissues for the oxidation of glucose to provide energy (as ATP) and intermediates for other metabolic pathways. Glycolysis is at the hub of carbohydrate metabolism because virtually all sugars, whether arising from the diet or from catabolic reactions in the body, can ultimately be converted to glucose (Fig. 13.9A). **Pyruvate is the end product of glycolysis in cells with mitochondria and an adequate supply of O_2.** This series of ten reactions is called aerobic glycolysis because O_2 is required to reoxidize the NADH formed during the oxidation of glyceraldehyde 3-phosphate (Fig. 13.9B). **Aerobic glycolysis** sets the stage

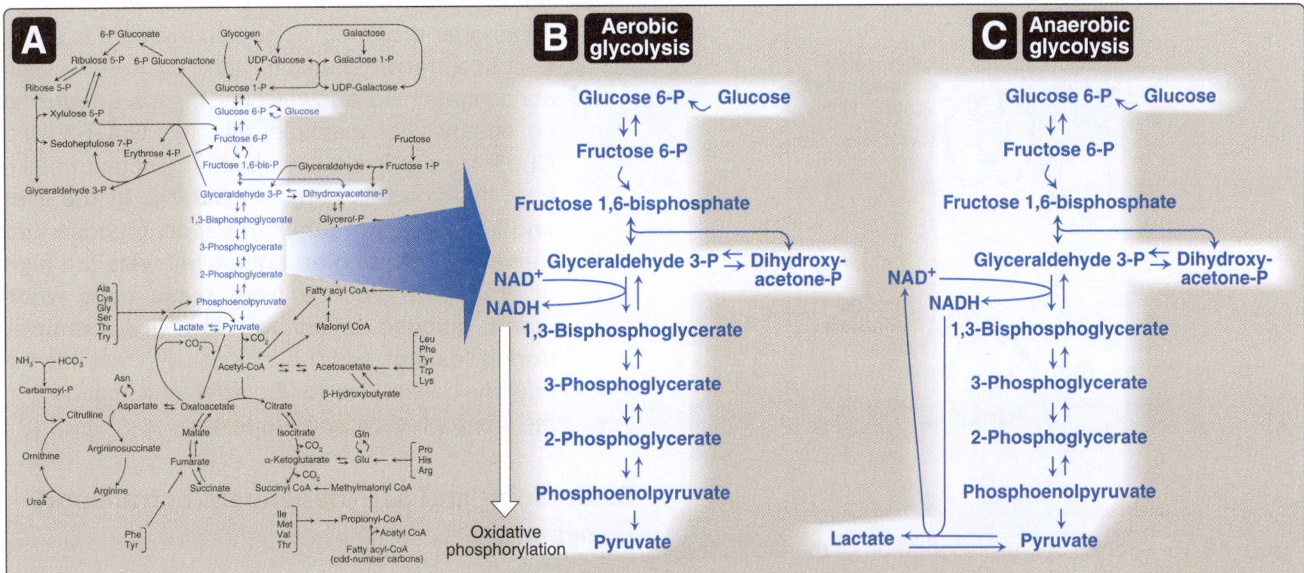

Figure 13.9 **A.** Glycolysis shown as one of the essential pathways of energy metabolism. **B.** Reactions of aerobic glycolysis. **C.** Reactions of anaerobic glycolysis. NAD(H) = nicotinamide adenine dinucleotide; P = phosphate.

for the oxidative decarboxylation of pyruvate to acetyl CoA, a major fuel of the TCA cycle. Alternatively, pyruvate is reduced to lactate as NADH is oxidized to NAD^+ (Fig. 13.9C). This conversion, called anaerobic glycolysis, can occur without the participation of O_2. **Anaerobic glycolysis** allows the production of ATP in tissues that lack mitochondria (for example, red blood cells [RBC] and parts of the eye) or in cells deprived of sufficient O_2 (hypoxia).

IV. GLUCOSE TRANSPORT INTO CELLS

Glucose cannot diffuse directly into cells but enters by one of two transport systems: a sodium (Na^+)- and ATP-independent transport system or a Na^+- and ATP-dependent cotransport system.

A. Sodium- and ATP-independent transport system

This **passive** system is mediated by a family of 14 glucose transporter (GLUT) isoforms found in cell membranes. They are designated GLUT-1 to GLUT-14. These monomeric protein transporters exist in the membrane in two conformational states (Fig. 13.10). Extracellular glucose binds to the transporter, which then alters its conformation, transporting glucose across the cell membrane via facilitated diffusion. Because GLUT transport one molecule at a time, they are **uniporters**.

1. **Tissue specificity:** GLUT display a tissue-specific pattern of expression. For example, GLUT-1 is abundant in RBC and the blood–brain barrier but is low in adult muscle, whereas GLUT-4 is abundant in muscle and adipose tissue. The number of GLUT-4 transporters active in these tissues is increased by insulin. (See p. 441 for a discussion of insulin and glucose transport.) GLUT-2 is abundant in the liver, kidneys, and pancreatic β cells. The presence of GLUT-2 in pancreatic β cells enables pancreas to sense high blood glucose levels and release insulin. The other GLUT isoforms also have tissue-specific distributions (see blue box below).

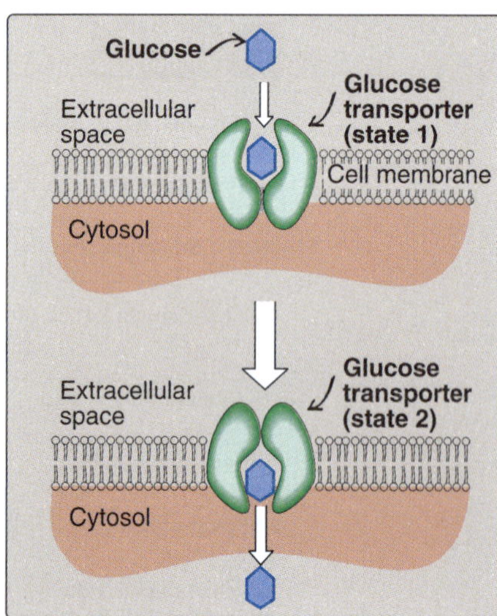

Figure 13.10 Schematic representation of the facilitated transport of glucose through a cell membrane. [Note: Glucose transporter proteins are monomeric and contain 12 transmembrane α helices.]

Tissue specific distribution of major glucose transporters (GLUT) in cells

Name	Distribution	Remarks
GLUT-1	RBC, blood-brain barrier	Responsible for low levels of basal glucose intake. Also increased activity if low blood glucose levels and vice-versa. Insulin independent
GLUT-2	Liver, kidney, and pancreatic beta-cells	Free flow of glucose into cells allows pancreas to assess need: Senses high blood glucose to release insulin. Insulin independent
GLUT-3	Neurons	High affinity allows glucose uptake even in low blood glucose concentrations
GLUT-4 (insulin dependent)	Muscle and adipose tissue	Insulin-regulated glucose transporter for glucose storage: Activity increased by insulin

2. **Specialized functions:** In facilitated diffusion, transporter-mediated glucose movement is down a concentration gradient (that is, from a high concentration to a lower one, therefore requiring no energy). For example, GLUT-1, GLUT-3, and GLUT-4 are primarily involved in glucose uptake from the blood. In contrast, **GLUT-2, in the liver and kidneys, can either transport glucose into these cells when blood glucose levels are high or transport glucose from these cells when blood glucose levels are low (for example, during fasting).** GLUT-5 is unusual in that it is the primary transporter for fructose (not glucose) in the small intestine and the testes.

B. Sodium- and ATP-dependent cotransport system

This **energy-requiring process** transports glucose against (up) its concentration gradient (that is, from low extracellular concentrations to higher intracellular concentrations) as Na^+ is transported down its electrochemical gradient. [Note: The gradient is created by the *Na^+-potassium (K^+) ATPase* (see Fig. 12.10, p. 180).] Because this secondary active transport process requires the concurrent uptake (**symport**) of Na^+, the transporter is a sodium-dependent glucose cotransporter (SGLT). This type of cotransport occurs in the epithelial cells of the intestine (see p. 180), renal tubules, and choroid plexus. [Note: The choroid plexus, part of the blood–brain barrier, also contains GLUT-1.] **This transport system is the biochemical basis of adding sodium chloride to glucose in oral rehydration fluids to ensure uptake of glucose and water.**

V. GLYCOLYSIS REACTIONS

The conversion of glucose to pyruvate occurs in two stages (Fig. 13.11). **The first five reactions of glycolysis correspond to an energy-investment phase** in which the phosphorylated forms of intermediates are synthesized at the expense of ATP. The **subsequent reactions of glycolysis constitute an energy-generation phase** in which a net of two molecules of ATP are formed by substrate-level phosphorylation (see p. 197) per glucose molecule metabolized.

A. Glucose phosphorylation

Phosphorylated sugar molecules do not readily penetrate cell membranes because there are no specific transmembrane carriers for these compounds and because they are too polar to diffuse through the lipid core of membranes. Therefore, the **irreversible phosphorylation** of glucose (Fig. 13.12) effectively traps the sugar as cytosolic glucose 6-phosphate and commits it to further metabolism in the cell. Mammals have four isozymes (I–IV) of the enzyme *hexokinase* that catalyze the phosphorylation of glucose to glucose 6-phosphate.

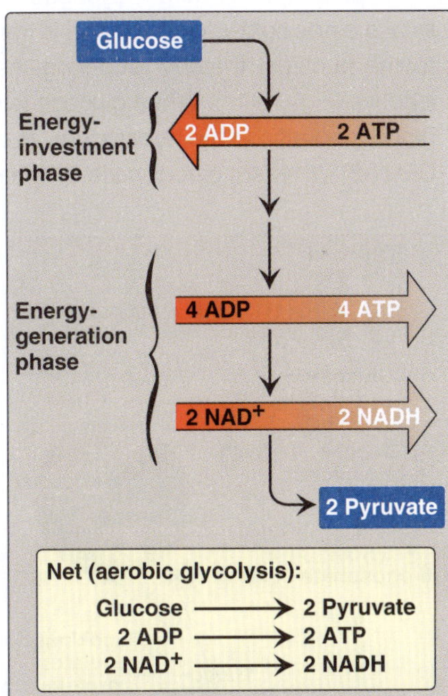

Figure 13.11 Two phases of aerobic glycolysis. NAD(H) = nicotinamide adenine dinucleotide; ADP = adenosine diphosphate.

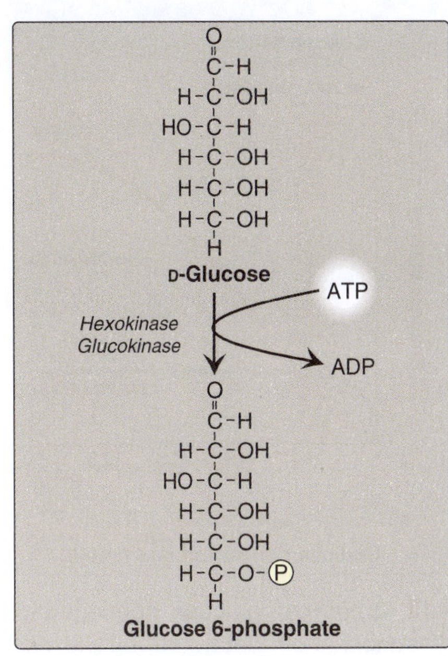

Figure 13.12 Energy-investment phase: phosphorylation of glucose. [Note: *Kinases* utilize ATP complexed with a divalent metal ion, most typically magnesium.] ADP = adenosine diphosphate; Ⓟ = phosphate.

1. **Hexokinases I–III:** In most tissues, glucose phosphorylation is catalyzed by one of these isozymes of *hexokinase*, which is one of three regulatory enzymes of glycolysis (along with *phosphofructokinase* and *pyruvate kinase*). They are inhibited by the reaction product glucose 6-phosphate, which accumulates when further metabolism of this hexose phosphate is reduced. *Hexokinases I–III* have a low Michaelis constant (K_m) and, therefore, a high affinity (see p. 135) for glucose. This permits the efficient phosphorylation and subsequent metabolism of glucose even when tissue concentrations of glucose are low (Fig. 13.13). However, because these isozymes have a low maximal velocity ([V_{max}], see p. 134) for glucose, they do not sequester (trap) cellular phosphate in the form of phosphorylated glucose or phosphorylate more glucose than the cell can use. [Note: These isozymes have broad substrate specificity and are able to phosphorylate several hexoses in addition to glucose.]

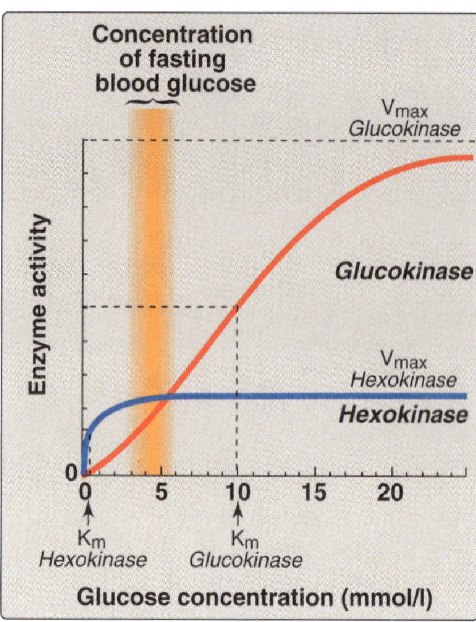

Figure 13.13 Effect of glucose concentration on the rate of phosphorylation catalyzed by *hexokinase* and *glucokinase*. K_m = Michaelis constant; V_{max} = maximal velocity.

2. **Hexokinase IV:** In liver parenchymal cells and pancreatic β cells, *glucokinase* (the *hexokinase IV* isozyme) is the predominant enzyme responsible for glucose phosphorylation. In β cells, *glucokinase* functions as a glucose sensor, determining the threshold for insulin secretion (see p. 439). [Note: *Hexokinase IV* also serves as a glucose sensor in hypothalamic neurons, playing a key role in the adrenergic response to hypoglycemia (see p. 445).] In the liver, the enzyme facilitates glucose phosphorylation during hyperglycemia. Despite the popular but misleading name *glucokinase*, the sugar specificity of the enzyme is similar to that of other *hexokinase* isozymes.

   a. **Kinetics:** *Glucokinase* differs from *hexokinases I–III* in several important properties. For example, it has a much higher K_m, requiring a higher glucose concentration for half-saturation (see Fig. 13.13). Thus, *glucokinase* functions only when the intracellular concentration of glucose in the hepatocyte is elevated such as during the brief period following consumption of a carbohydrate-rich meal, when high levels of glucose are delivered to the liver via the portal vein. *Glucokinase* has a high V_{max}, allowing the liver to effectively remove the flood of glucose delivered by the portal blood. This prevents large amounts of glucose from entering the systemic circulation following such a meal, thereby minimizing hyperglycemia during the absorptive period (see blue box below). [Note: GLUT-2 ensures that blood glucose equilibrates rapidly across the hepatocyte membrane.]

Comparison of *glucokinase* and *hexokinase*

Characteristics	*Glucokinase*	*Hexokinase*
K_m value	High	Low
Affinity	Low	High
Induced by	Insulin and glucose	Not induced
Function	Active at levels of glucose higher than 100 mg/dl to allow glucose to be taken up by liver for conversion to glycogen Active during high postprandial blood glucose levels	Utilization of blood glucose even at low levels
Occurrence	Liver, pancreatic β cells	In all tissues

   b. **Regulation:** *Glucokinase* activity is not directly inhibited by glucose 6-phosphate as are the other *hexokinases*. Instead, it is indirectly inhibited by fructose 6-phosphate (which is in equilibrium with glucose 6-phosphate, a product of *glucokinase*) and is indirectly stimulated by glucose (a substrate of *glucokinase*). Regulation is achieved by reversible binding to the hepatic protein glucokinase regulatory protein (GKRP). In the presence of fructose 6-phosphate, *glucokinase* binds tightly to GKRP and is translocated to the nucleus, thereby rendering the enzyme inactive (Fig. 13.14). When glucose levels in the blood (and also in the hepatocyte, as a result of GLUT-2) increase, *glucokinase* is released from

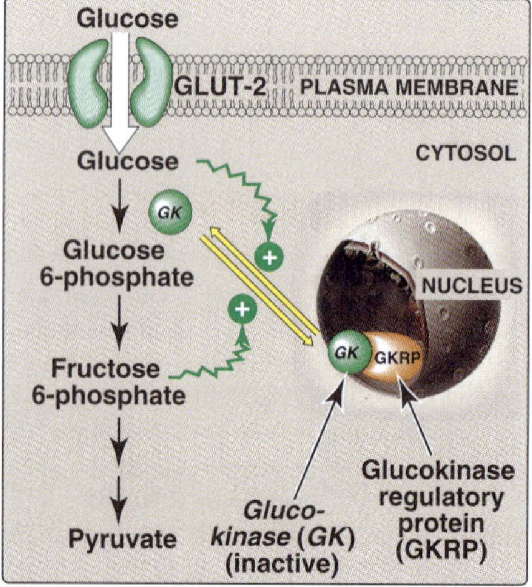

Figure 13.14 Regulation of *glucokinase* activity by glucokinase regulatory protein. GLUT = glucose transporter.

GKRP, and the enzyme reenters the cytosol where it phosphorylates glucose to glucose 6-phosphate. [Note: GKRP is a competitive inhibitor of glucose use by *glucokinase*.]

> *Glucokinase* functions as a glucose sensor in blood glucose homeostasis. Inactivating mutations of *glucokinase* are the cause of a rare form of diabetes, maturity onset diabetes of the young type 2 (MODY 2) that is characterized by impaired insulin secretion and hyperglycemia.

B. Glucose 6-phosphate isomerization

The isomerization of glucose 6-phosphate to fructose 6-phosphate is catalyzed by *phosphoglucose isomerase* (Fig. 13.15). The reaction is readily reversible and is not a rate-limiting or regulated step.

C. Fructose 6-phosphate phosphorylation

The **irreversible phosphorylation** reaction catalyzed by *phosphofructokinase-1 (PFK-1)* is the **most important control point** and the **rate-limiting** and **committed step** of glycolysis (Fig. 13.16). *PFK-1* is controlled by the available concentrations of the substrates ATP and fructose 6-phosphate as well as by other regulatory molecules.

1. **Regulation by intracellular energy levels:** *PFK-1* is inhibited allosterically by elevated levels of ATP, which act as an energy-rich signal indicating an abundance of high-energy compounds. Elevated levels of citrate, an intermediate in the TCA cycle (see p. 208), also inhibit *PFK-1*. [Note: Inhibition by citrate favors the use of glucose for glycogen synthesis (see p. 225).] Conversely, *PFK-1* is activated allosterically by high concentrations of AMP, which signal that the cell's energy stores are depleted.

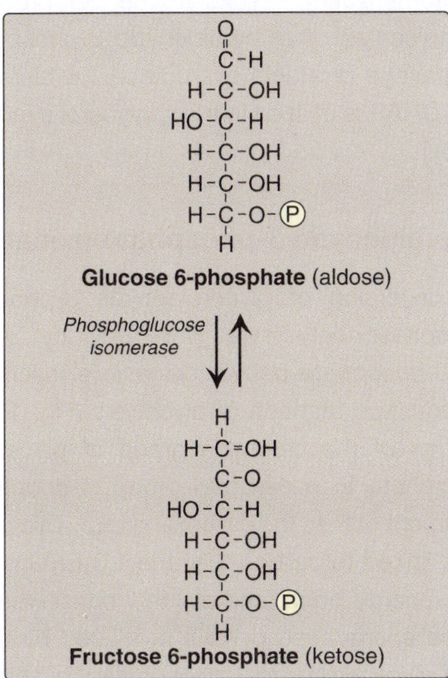

Figure 13.15 Aldose-ketose isomerization of glucose 6-phosphate to fructose 6-phosphate. Ⓟ = phosphate.

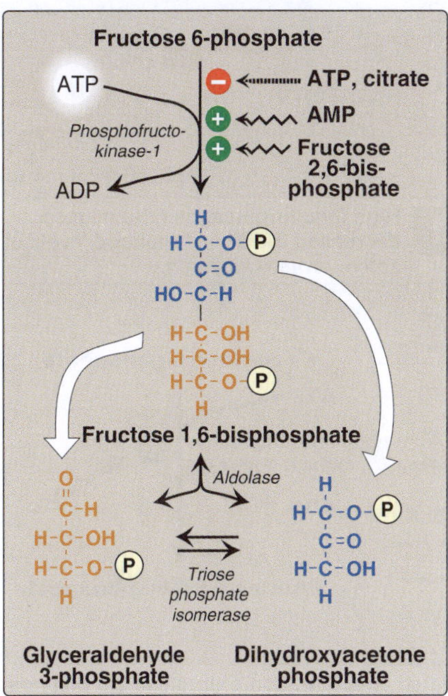

Figure 13.16 Energy-investment phase (continued): conversion of fructose 6-phosphate to triose phosphates. Ⓟ = phosphate; AMP and ADP = adenosine mono- and diphosphates.

2. **Regulation by fructose 2,6-bisphosphate:** Fructose 2,6-bisphosphate is the most potent activator of *PFK-1* (see Fig. 13.16) and is able to activate the enzyme even when ATP levels are high. It is formed from fructose 6-phosphate by *phosphofructokinase-2 (PFK-2)*. Unlike *PFK-1*, ***PFK-2* is a bifunctional protein** that has both the *kinase* activity that produces fructose 2,6-bisphosphate and the *phosphatase* activity that dephosphorylates fructose 2,6-bisphosphate to fructose 6-phosphate. In the liver isozyme, phosphorylation of *PFK-2* inactivates the *kinase* domain and activates the *phosphatase* domain (Fig. 13.17). The opposite is seen in the cardiac isozyme. Skeletal *PFK-2* is not covalently regulated. [Note: Fructose 2,6-bisphosphate is an inhibitor of *fructose 1,6-bisphosphatase*, an enzyme of gluconeogenesis (see p. 218). The reciprocal actions of fructose 2,6-bisphosphate on glycolysis (activation) and gluconeogenesis (inhibition) insure that both pathways are not fully active at the same time, preventing a futile cycle of glucose oxidation

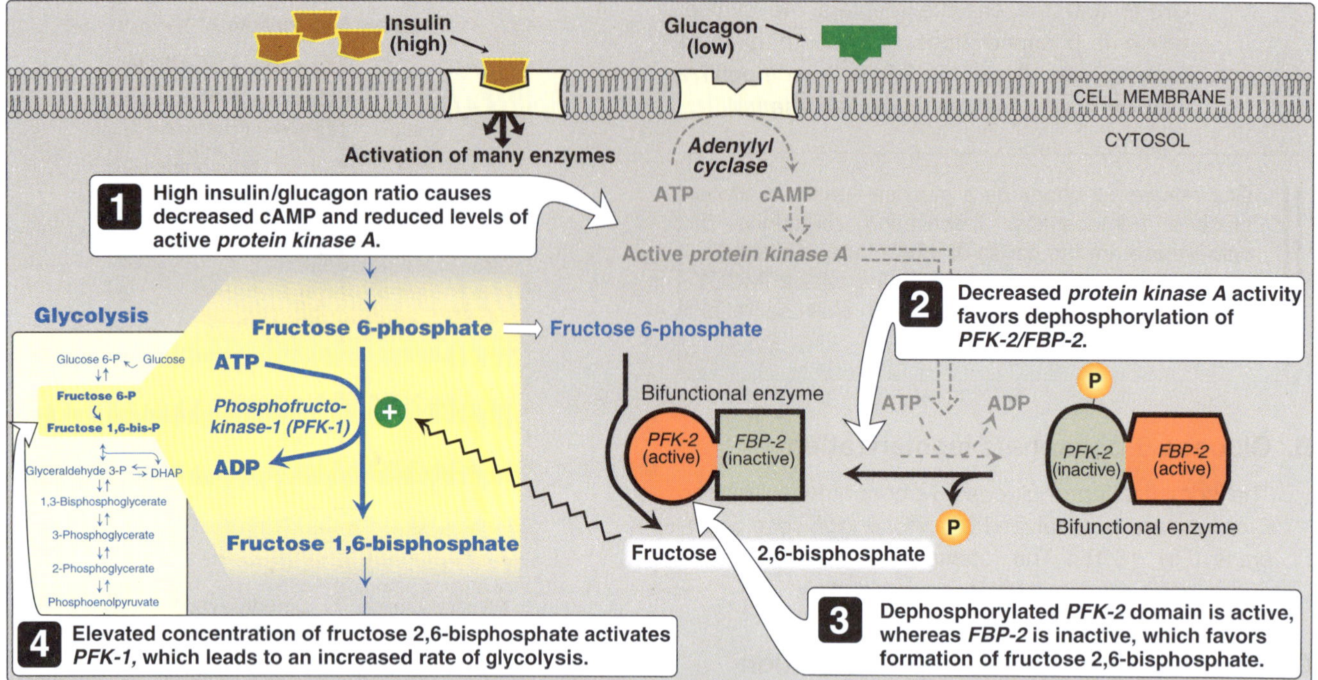

Figure 13.17 Effect of elevated insulin concentration on the intracellular concentration of fructose 2,6-bisphosphate in the liver. *PFK-2 = phosphofructokinase-2*; *FBP-2 = fructose 2,6-bisphosphatase*; AMP and ADP = adenosine mono- and diphosphates; cAMP = cyclic AMP; (P) = phosphate.

to pyruvate followed by glucose resynthesis from pyruvate.]

a. **During the well-fed state:** Decreased levels of glucagon and elevated levels of insulin (such as occur following a carbohydrate-rich meal) cause an increase in hepatic fructose 2,6-bisphosphate (*PFK-2* is dephosphorylated) and, thus, in the rate of glycolysis (see Fig. 13.17). Therefore, fructose 2,6-bisphosphate acts as an intracellular signal of glucose abundance.

b. **During fasting:** By contrast, the elevated levels of glucagon and low levels of insulin that occur during fasting (see p. 458) cause a decrease in hepatic fructose 2,6-bisphosphate (*PFK-2* is phosphorylated). This results in inhibition of glycolysis and activation of gluconeogenesis.

D. Fructose 1,6-bisphosphate cleavage

Aldolase cleaves fructose 1,6-bisphosphate to dihydroxyacetone phosphate (DHAP) and glyceraldehyde 3-phosphate (see Fig. 13.16). The reaction is reversible and not regulated. [Note: *Aldolase B*, the hepatic isoform, also cleaves fructose 1-phosphate and functions in dietary fructose metabolism (see p. 239).]

E. Dihydroxyacetone phosphate isomerization

Triose phosphate isomerase interconverts DHAP and glyceraldehyde 3-phosphate (see Fig. 13.16). DHAP must be isomerized to glyceraldehyde 3-phosphate for further metabolism by the glycolytic pathway. This isomerization results in the net production of two molecules of glyceraldehyde 3-phosphate from the cleavage products of fructose 1,6-bisphosphate. [Note: DHAP is utilized in triacylglycerol synthesis (see p. 306).]

F. Glyceraldehyde 3-phosphate oxidation

The conversion of glyceraldehyde 3-phosphate to 1,3-bisphosphoglycerate (1,3-BPG) by *glyceraldehyde 3-phosphate dehydrogenase* is the first oxidation-reduction reaction of glycolysis (Fig. 13.18). The oxidation of the aldehyde group of glyceraldehyde 3-phosphate to a carboxyl group is coupled to the attachment of P_i to the carboxyl group. This phosphate group, linked to carbon 1 of the 1,3-BPG product by a high-energy bond (see p. 163), conserves much of the free energy (see p. 159) produced by the oxidation of glyceraldehyde 3-phosphate. This high-energy phosphate drives ATP synthesis in the next reaction of glycolysis.

Because there is a limited amount of NAD^+ in the cell, the NADH formed by the *dehydrogenase* reaction must be oxidized for glycolysis to continue. Two major mechanisms for oxidizing NADH to NAD^+ are the reduction of pyruvate to lactate by *lactate dehydrogenase (LDH)* (anaerobic, see p. 199) and the electron transport chain ([ETC] aerobic, see p. 164). Because NADH cannot cross the inner mitochondrial membrane, the ETC requires the malate-aspartate and glycerol 3-phosphate substrate shuttles to move NADH reducing equivalents into the mitochondrial matrix (see p. 171).

1. **Arsenic poisoning:** The toxicity of arsenic is due primarily to the inhibition by **trivalent arsenic (arsenite)** of enzymes such as the *pyruvate dehydrogenase complex (PDHC)*, which require lipoic acid as a coenzyme (see p. 207). However, **pentavalent arsenic (arsenate)** competes with P_i as a substrate for *glyceraldehyde 3-phosphate dehydrogenase*, forming a complex that spontaneously hydrolyzes to form 3-phosphoglycerate (see Fig. 13.18). This prevents net ATP and NADH production by glycolysis, without inhibiting the pathway itself, thus depriving the cell of energy usually obtained from the glycolytic pathway.

   Arsenate also competes with P_i binding to the F_1 domain of *ATP synthase* (see p. 168), resulting in formation of ADP-arsenate that is rapidly hydrolyzed.

2. **2,3-Bisphosphoglycerate synthesis in RBC:** Some of the 1,3-BPG is converted to 2,3-BPG by the action of *bisphosphoglycerate mutase* (see Fig. 13.18). 2,3-BPG, which is found in only trace amounts in most cells, is **present at high concentration in RBC and serves to increase O_2 delivery** (see p. 103). 2,3-BPG is hydrolyzed by a *phosphatase* to 3-phosphoglycerate, which is also an intermediate in glycolysis (see Fig. 13.18). In the RBC, glycolysis is modified by inclusion of these shunt reactions, known as **Rapaport-Luebering Cycle (BPG Shunt)**.

G. 3-Phosphoglycerate synthesis and ATP production

When 1,3-BPG is converted to 3-phosphoglycerate, the high-energy phosphate group of 1,3-BPG is used to synthesize ATP from ADP (see Fig. 13.18). This reaction is catalyzed by phosphoglycerate kinase, which, unlike most other kinases, is physiologically reversible. Because two molecules of 1,3-BPG are formed from each glucose molecule, this *kinase* reaction replaces the two ATP molecules consumed by the earlier formation of glucose 6-phosphate and fructose 1,6-bisphosphate. [Note: This reaction is an example of **substrate-level phosphorylation**, in which the

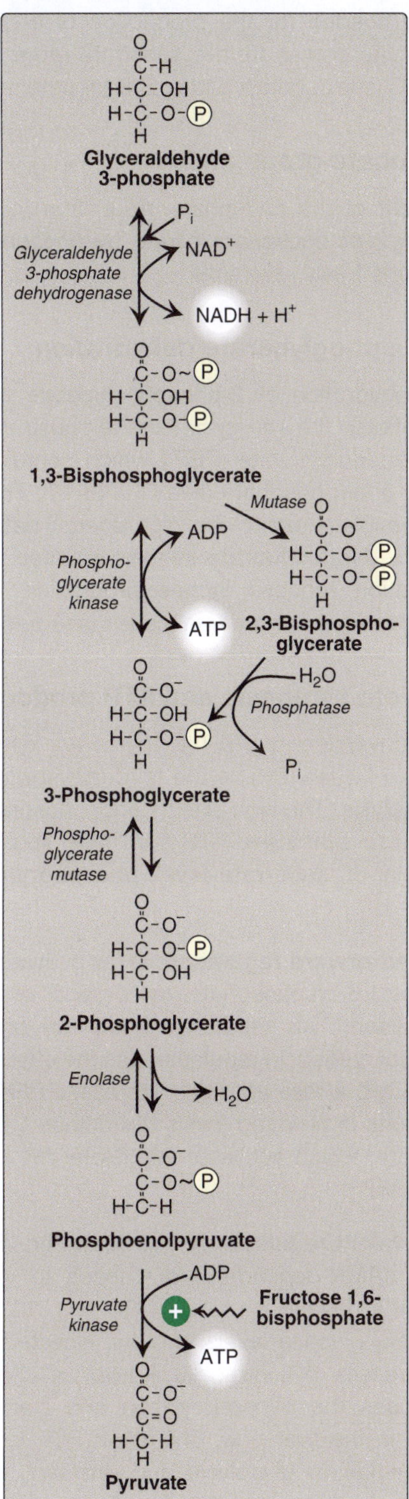

Figure 13.18 Energy-generating phase: conversion of glyceraldehyde 3-phosphate to pyruvate. NAD(H) = nicotinamide adenine dinucleotide; Ⓟ = phosphate; P_i = inorganic phosphate; ~ = high-energy bond; ADP = adenosine diphosphate.

energy needed for the production of a high-energy phosphate comes from a substrate rather than from the ETC (see J. below and p. 209 for other examples).]

H. Phosphate group shift

The shift of the phosphate group from carbon 3 to carbon 2 of phosphoglycerate by *phosphoglycerate mutase* is freely reversible.

I. 2-Phosphoglycerate dehydration

The dehydration of 2-phosphoglycerate by *enolase* redistributes the energy within the substrate, forming phosphoenolpyruvate (PEP), which contains a high-energy enol phosphate (see Fig. 13.18). The reaction is reversible, despite the high-energy nature of the product. [Note: **Fluoride inhibits *enolase***, and water fluoridation reduces lactate production by mouth bacteria, decreasing dental caries (see p. 560).]

J. Pyruvate synthesis and ATP production

The conversion of PEP to pyruvate, catalyzed by *pyruvate kinase* (*PK*), is the third irreversible reaction of glycolysis. The high-energy enol phosphate in PEP is used to synthesize ATP from ADP and is another example of **substrate-level phosphorylation** (see Fig. 13.18).

1. **Feedforward regulation:** *PK* is activated by fructose 1,6-bisphosphate, the product of the *PFK-1* reaction. This feedforward (instead of the more usual feedback) regulation has the effect of linking the two *kinase* activities: increased *PFK-1* activity results in elevated levels of fructose 1,6-bisphosphate, which activates *PK*. [Note: *PK* is inhibited by ATP.]

2. **Covalent regulation in the liver:** Phosphorylation by *cAMP-dependent PKA* leads to inactivation of the hepatic isozyme of *PK* (Fig. 13.19). When blood glucose levels are low, elevated glucagon increases the intracellular level of cAMP, which causes the phosphorylation and inactivation of *PK* in the liver only. Therefore, PEP is unable to continue in glycolysis and, instead, enters the gluconeogenesis pathway. This partly explains the observed inhibition of hepatic glycolysis and stimulation of gluconeogenesis by glucagon. Dephosphorylation of *PK* by a *phosphatase* results in reactivation of the enzyme.

3. **Pyruvate kinase deficiency causes anemia:** Because mature RBC lack mitochondria, they are completely dependent on glycolysis for ATP production. ATP is required to meet the metabolic needs of RBC and to fuel the ion pumps necessary for the maintenance of the flexible, biconcave shape that allows them to squeeze through narrow capillaries. The anemia observed in glycolytic enzyme deficiencies is a consequence of the reduced rate of glycolysis, leading to decreased ATP production by substrate-level phosphorylation. The resulting alterations in the RBC membrane lead to changes in cell shape and, ultimately, to phagocytosis by cells of the mononuclear phagocyte system, particularly splenic macrophages. The premature death and lysis of RBC result in mild-to-severe nonspherocytic hemolytic anemia, with the severe form requiring regular transfusions. Among patients with rare genetic defects of glycolytic enzymes, the majority has a deficiency in *PK*. Severity depends both on the degree of enzyme deficiency (generally 5%–35% of normal levels) and on the extent to which RBC compensate by synthesizing increased levels of 2,3-BPG (see p. 103). Almost all individuals with *PK* deficiency have a mutant enzyme that shows altered kinetics or decreased stability (Fig. 13.20). Individuals heterozygous for *PK* deficiency have resistance to the most severe forms of malaria.

   Liver *PK* is encoded by the same gene as the RBC isozyme. However, liver cells show no effect because they can synthesize more *PK* and can also generate ATP by oxidative phosphorylation.

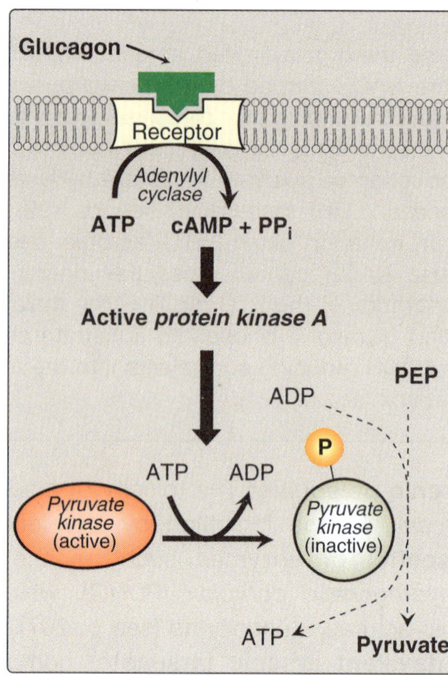

Figure 13.19 Covalent modification of hepatic *pyruvate kinase* results in inactivation of the enzyme. cAMP = cyclic adenosine monophosphate; PEP = phosphoenolpyruvate; (P) = phosphate; PP_i = pyrophosphate; ADP = adenosine diphosphate.

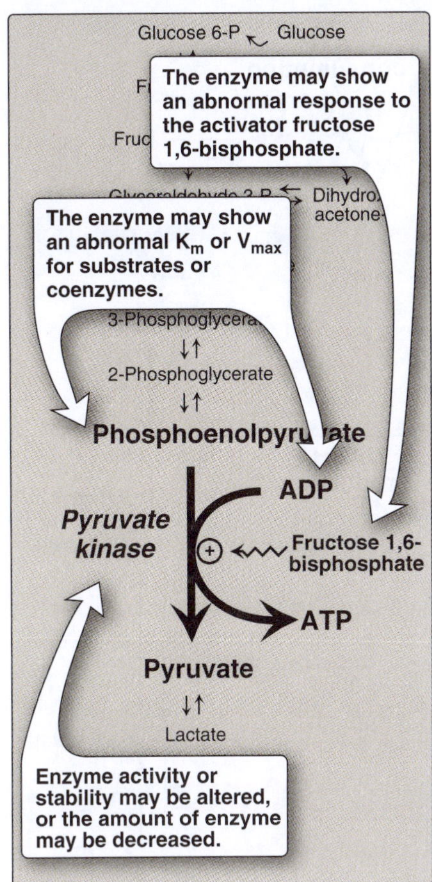

Figure 13.20 Alterations observed with various mutant forms of *pyruvate kinase*. K_m = Michaelis constant; V_{max} = maximal velocity; ADP = adenosine diphosphate.

> The tissue-specific expression of *PK* in RBC and the liver results from the use of different start sites in transcription (see p. 662) of the gene that encodes the enzyme.

K. Pyruvate reduction to lactate

Lactate, formed from pyruvate by *LDH*, is the final product of anaerobic glycolysis in eukaryotic cells (Fig. 13.21). Reduction to lactate is the major fate for pyruvate in tissues that are poorly vascularized (for example, the lens and cornea of the eye and the kidney medulla) or in RBC that lack mitochondria.

1. **Lactate formation in muscle:** In exercising skeletal muscle, NADH production (by *glyceraldehyde 3-phosphate dehydrogenase* and by the three NAD^+-linked *dehydrogenases* of the TCA cycle, see p. 210) exceeds the oxidative capacity of the ETC. This results in an elevated NADH/NAD^+ ratio, favoring reduction of pyruvate to lactate by *LDH*. **Therefore, during intense exercise, lactate accumulates in muscle, causing a drop in the intracellular pH, potentially resulting in cramps.** Much of this lactate eventually diffuses into the bloodstream and can be used by the liver to make glucose (see p. 215). The cyclical process of conversion of glucose to lactate in muscle (by anaerobic glycolysis) and conversion of lactate to glucose in liver (by gluconeogenesis) is known as **Cori's cycle.**

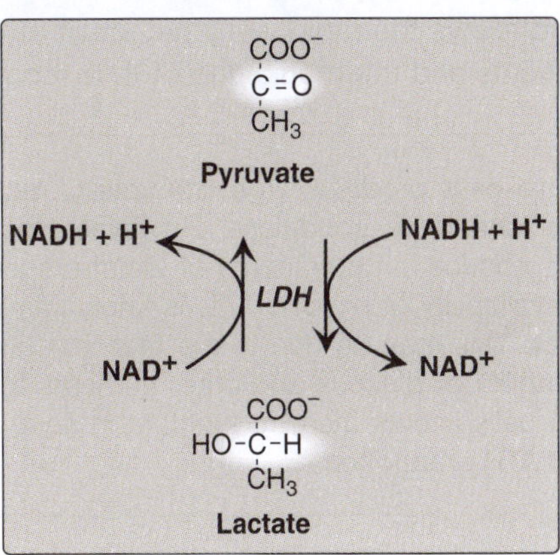

Figure 13.21 Interconversion of pyruvate and lactate by *lactate dehydrogenase* (*LDH*). NAD(H) = nicotinamide adenine dinucleotide.

2. **Lactate utilization:** The direction of the *LDH* reaction depends on the relative intracellular concentrations of pyruvate and lactate and on the ratio of NADH/NAD^+. For example, in the liver and heart, this ratio is lower than in exercising muscle. Consequently, the liver and heart oxidize lactate (obtained from the blood) to pyruvate. In the liver, pyruvate is either converted to glucose by gluconeogenesis or converted to acetyl CoA that is oxidized in the TCA cycle. Heart muscle exclusively oxidizes lactate to carbon dioxide and water via the TCA cycle.

3. **Lactic acidosis:** Elevated concentrations of lactate in the plasma, termed lactic acidosis (a type of metabolic acidosis), occur when there is a collapse of the circulatory system, such as with myocardial infarction, pulmonary embolism, and uncontrolled hemorrhage, or when an individual is in shock. The failure to bring adequate amounts of O_2 to the tissues results in impaired oxidative phosphorylation and decreased ATP synthesis. To survive, the cells rely on anaerobic glycolysis for generating ATP, producing lactic acid as the end product. [Note: Production of even meager amounts of ATP may be lifesaving during the period required to reestablish adequate blood flow to the tissues.] The additional O_2 required to recover from a period when O_2 availability has been inadequate is termed the **O_2 debt.** [Note: The O_2 debt is often related to patient morbidity or mortality. In many clinical situations, measuring the blood levels of lactic

acid allows the rapid, early detection of O_2 debt in patients and the monitoring of their recovery.]

Normal cells produce large amounts of lactate only under anaerobic conditions. However, most cancer cells produce large amounts of lactate regardless of the availability of oxygen. This is known as **Warburg effect**. This may be due to the fact that continuous availability of glucose and other nutrients to proliferating cells allows them to function in spite of inefficient ATP production.

L. Energy yield from glycolysis

Despite the production of some ATP by substrate-level phosphorylation during glycolysis, the end product, pyruvate or lactate, still contains most of the energy originally contained in glucose. The TCA cycle is required to release that energy completely (see p. 206).

1. **Anaerobic glycolysis:** A net of two molecules of ATP are generated for each molecule of glucose converted to two molecules of lactate (Fig. 13.22). There is no net production or consumption of NADH.

2. **Aerobic glycolysis:** The generation of ATP is the same as in anaerobic glycolysis (that is, a net gain of two ATP per molecule of glucose). Two molecules of NADH are also produced per molecule of glucose. Ongoing aerobic glycolysis requires the oxidation of most of this NADH by the ETC, producing 2.5 ATP for each NADH molecule entering the chain (see p. 164). [Note: NADH cannot cross the inner mitochondrial membrane, and substrate shuttles are required (see p. 171).]

VI. HORMONAL REGULATION

Regulation of the activity of the irreversible glycolytic enzymes by allosteric activation/inhibition or covalent phosphorylation/dephosphorylation is short term (that is, the effects occur over minutes or hours). Superimposed on these effects on the activity of preexisting enzyme molecules are the long-term hormonal effects on the number of new enzyme molecules. These hormonal effects can result in 10- to 20-fold increases in enzyme synthesis that typically occur over hours to days. Regular consumption of meals rich in carbohydrate or administration of insulin initiates an increase in the amount of *glucokinase*, *PFK-1*, and *PK* in the liver (Fig. 13.23). The change reflects an increase in gene transcription, resulting in increased enzyme synthesis. Increased availability of these three enzymes favors the conversion of glucose to pyruvate, a characteristic of the absorptive state (see p. 452). [Note: The transcriptional effects of insulin and carbohydrate

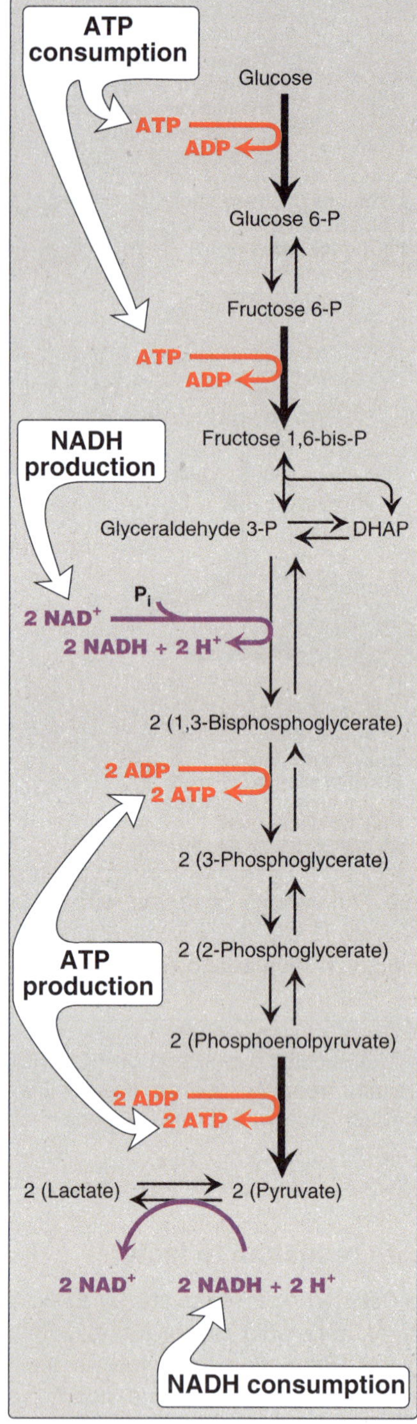

Figure 13.22 Summary of anaerobic glycolysis. Reactions involving the production or consumption of ATP or nicotinamide adenine dinucleotide (NADH) are indicated. The three irreversible reactions of glycolysis are shown with thick arrows. DHAP = dihydroxyacetone phosphate; ADP = adenosine diphosphate; P = phosphate.

(specifically glucose) are mediated by the transcription factors sterol regulatory element–binding protein-1c and carbohydrate response element–binding protein, respectively. These factors also regulate transcription of genes

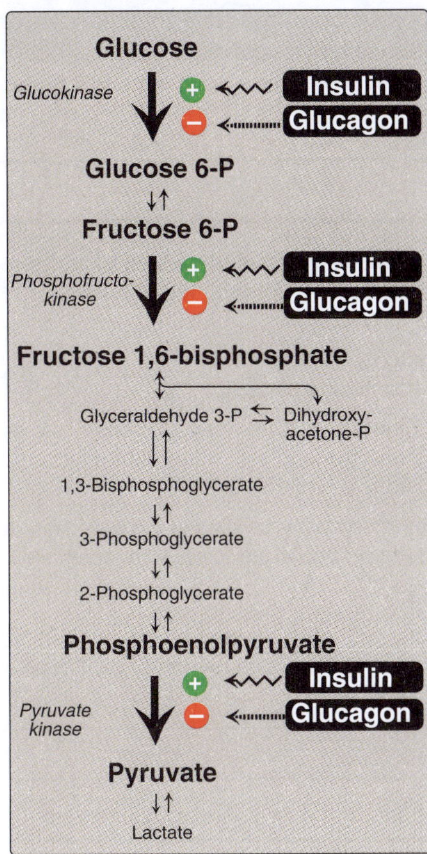

Figure 13.23 Effect of insulin and glucagon on the expression of key enzymes of glycolysis in the liver. P = phosphate.

involved in fatty acid synthesis (see p. 302).] Conversely, gene expression of the three enzymes is decreased when plasma glucagon is high and insulin is low (for example, as seen in fasting or diabetes).

VII. ALTERNATE FATES OF PYRUVATE

Pyruvate can be metabolized to products other than lactate.

A. Oxidative decarboxylation to acetyl CoA

Oxidative decarboxylation of pyruvate by the *PDHC* is an important pathway in tissues with a high oxidative capacity such as cardiac muscle (Fig. 13.24). *PDHC* irreversibly converts pyruvate, the end product of aerobic glycolysis, into acetyl CoA, a TCA cycle substrate (see p. 206) and the carbon source for fatty acid synthesis (see p. 301).

B. Carboxylation to oxaloacetate

Carboxylation of pyruvate to oxaloacetate by *pyruvate carboxylase* is a biotin-dependent reaction (see Fig. 13.24). This irreversible reaction is important

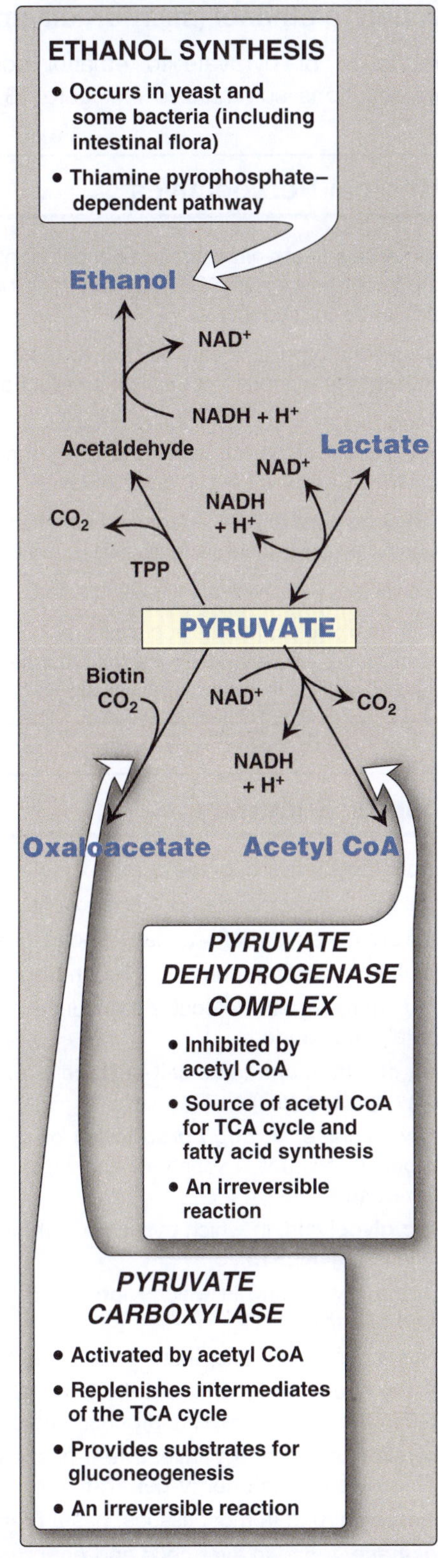

Figure 13.24 Summary of the metabolic fates of pyruvate. TPP = thiamine pyrophosphate. TCA = tricarboxylic acid; NAD(H) = nicotinamide adenine dinucleotide; CoA = coenzyme A; CO_2 = carbon dioxide.

because it replenishes the TCA cycle intermediate and provides substrate for gluconeogenesis (see p. 215).

C. Reduction to ethanol (microorganisms)

The reduction of pyruvate to ethanol occurs by the two reactions summarized in Figure 13.24. The decarboxylation of pyruvate to acetaldehyde by thiamine-requiring *pyruvate decarboxylase* occurs in yeast and certain other microorganisms but not in humans.

VIII. Medical Perspectives

Renal glomerulus filters glucose, and almost all of it is reabsorbed in the proximal convoluted tubules by sodium-dependent glucose cotransporter (SGLT). Glucose in glomerular filtrate beyond the saturation level of 180 mg/dl (known as **renal threshold**) leads to glucosuria.

Diabetes mellitus type II is caused due to insulin resistance, in which glucose absorption in skeletal muscles through GLUT-4 (under the influence of insulin) is reduced. This is discussed in detail in Chapter 29.

When blood is taken for glucose estimation, blood cells keep utilizing glucose and this can decrease the estimated amount of glucose. Fluoride, which removes magnesium required by the enzyme enolase, can inhibit glycolysis and is added to blood samples so that glucose levels do not decrease in vitro before estimation is done.

The 2,3-BPG content of RBCs decreases in stored blood, which decreases their ability to release oxygen. This reduction can be significant for critical patients and to prevent this, additives are added to blood storage bags to replenish 2,3-BPG.

Inherited deficiency of *pyruvate kinase* is a rare cause of hemolytic anemia.

During intense exercise, lactate accumulates in muscle, reducing intracellular pH, which causes cramps. When lactic acid increases in blood, as seen in myocardial infarction, pulmonary embolism, and uncontrolled hemorrhage, the pH of blood decreases causing a life-threatening condition called metabolic acidosis.

IX. Chapter Summary

- **Catabolic pathways** degrade complex molecules to a few simple products with **ATP production**.
- **Anabolic** pathways synthesize complex end products from simple precursors with **ATP hydrolysis**.
- The rate of a metabolic pathway can respond to **regulatory signals** such as **intracellular allosteric activators** or **inhibitors**.
- **Intercellular** signaling, primarily through **hormones** or **neurotransmitters,** provides for the integration of metabolism.
- **Second messenger molecules** transduce a chemical signal (hormone or neurotransmitter binding) to appropriate intracellular responders.
- Binding of a hormone to its **cell-surface G protein–coupled receptor**, ***activates*** **G protein**, which in turn, activates ***adenylyl cyclase (AC***).
- ***Adenylyl cyclase, through production of*** **cyclic adenosine monophosphate (*cAMP*)**, activates ***protein kinase A***, which phosphorylates a variety of enzymes, causing their activation or deactivation. Phosphorylation is reversed by ***phosphatases***.
- **Aerobic glycolysis**, in which **pyruvate** is the end product, occurs in cells with mitochondria and an adequate supply of oxygen ([O_2], Fig. 13.25).
- **Anaerobic glycolysis**, in which **lactic acid** is the end product, occurs in cells that lack mitochondria and in cells deprived of sufficient O_2.
- Glucose is passively transported across membranes by 1 of 14 **glucose transporter (GLUT) isoforms**.
- **GLUT-1** is abundant in RBC and the brain, **GLUT-4** (which is **insulin dependent**) in muscle and adipose tissue, and **GLUT-2** in the liver, kidneys, and pancreatic β cells.
- **Glycolysis** has an **energy-investment** phase in which phosphorylated intermediates are synthesized at the expense of ATP, followed by an **energy-generation** phase in which ATP is produced through **substrate-level phosphorylation**.
- ***Hexokinase,*** found in most tissues, has a **high affinity (low K_m)** and a **low maximal velocity (V_{max})** for glucose.
- ***Glucokinase,*** found in **liver cells** and **pancreatic β cells,** has a high K_m and a high V_{max} for glucose.
- *Phosphofructokinase-1* (*PFK-1*) mediated phosphorylation of fructose 6-phosphate is the most important control point and rate limiting step of glycolysis.
- **Fructose 2,6-bisphosphate**, whose synthesis by bifunctional ***phosphofructokinase-2*** **(*PFK-2*)** is increased in the liver by insulin and decreased by glucagon, is the most potent allosteric activator of *PFK-1*.
- ***PK*** **deficiency** accounts for the majority of all inherited defects in glycolytic enzymes. Effects are restricted to **RBC** and present as mild-to-severe **chronic**, **nonspherocytic hemolytic anemia**.

- In cells such as RBC that lack mitochondria and in tissues such as exercising muscle, **anaerobic glycolysis** occurs, where lactate is produced.
- Elevated concentrations of lactate in the plasma (**lactic acidosis**) occur with circulatory system collapse or shock.
- Pyruvate also can be 1) **oxidatively decarboxylated** to acetyl CoA by ***pyruvate dehydrogenase***, 2) **carboxylated** to oxaloacetate (a TCA cycle intermediate) by ***pyruvate carboxylase***, or 3) **reduced** to ethanol by microbial ***pyruvate decarboxylase***.

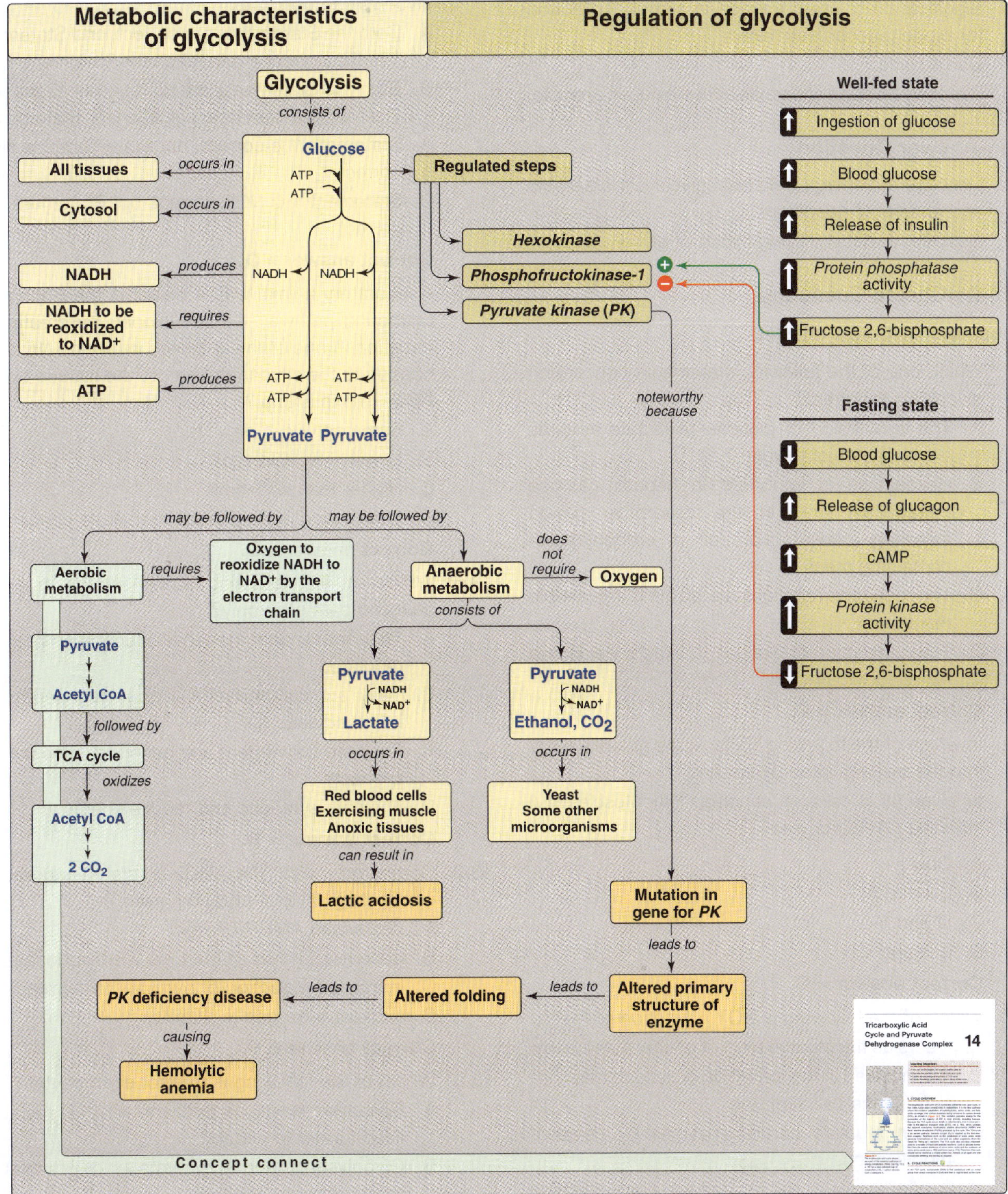

Figure 13.25 Key concept map for glycolysis. NAD(H) = nicotinamide adenine dinucleotide; cAMP = cyclic adenosine monophosphate; CoA = coenzyme A; TCA = tricarboxylic acid; CO_2 = carbon dioxide.

Study Questions

Short-Answer Questions

Write short notes on:

13.1. Differences between glucokinase and hexokinase.
13.2. Significance of BPG shunt.
13.3. Significance of using fluoride in sample collection for blood glucose estimation.
13.4. Cori's cycle.
13.5. Biochemical basis of cramps after strenuous exercise.

Long-Answer Question

13.1. Describe the energy yield from glycolysis in aerobic and anaerobic conditions.
13.2. Describe in detail the regulation of glycolysis.

Multiple-Choice Questions

Choose the ONE best answer.

13.1. Which one of the following statements concerning glycolysis is correct?

A. The conversion of glucose to lactate requires the presence of oxygen.
B. Hexokinase is important in hepatic glucose metabolism only in the absorptive period following consumption of a carbohydrate-containing meal.
C. The regulated reactions are also the irreversible reactions.
D. The conversion of glucose to lactate yields two ATP and two NADH.

Correct answer = C.

13.2. In which of the following site(s) is/are glucose entry into the cell regulated by insulin?

(I) Liver (II) β-cells of pancreas (III) Muscles (IV) Intestine (V) Adipocytes

A. Only I
B. I, II and IV
C. III and V
D. I, III and V

Correct answer = C.

13.3. Which of the following is NOT a function of ATP?

A. It acts as the storage form of energy in the body.
B. It is involved in the formation of second messengers during cell signaling.
C. It can regulate certain enzymes by covalent modification.
D. It can regulate certain enzymes by allosteric modification.

Correct answer = A.

13.4. Statement 1: Arsenate, a compound which inhibits glyceraldehyde 3-phosphate dehydrogenase, can be used as an inhibitor of glycolysis while collecting a blood sample for estimation of glucose.

Statement 2: The product that is formed after incorporation of arsenate undergoes spontaneous hydrolysis to form 3-phosphoglycerate, allowing glycolysis to continue.

A. Both the Statements are correct, and Statement 2 is the correct explanation for Statement 1
B. Both the Statements are correct, but Statement 2 is NOT the correct explanation for Statement 1
A. Statement 1 is correct, but Statement 2 is NOT correct
D. Statement 1 is NOT correct, but Statement 2 is correct

Correct answer = D.

13.5. A laboratory animal with a defect in the Rapaport-Leubering pathway was developed by creating a mutation in one of the enzymes involved. What will happen to the oxygen delivery to the tissues by the RBCs in this animal?

A. Same as wild-type
B. Lower than wild-type
C. Higher than wild-type
D. Information is insufficient to make a conclusion

Correct answer = B.

13.6. Which of the following statements is true for anabolic pathways only?

A. Their irreversible (nonequilibrium) reactions are regulated.
B. They are called cycles if they regenerate an intermediate.
C. They are convergent and generate a few simple products.
D. They are synthetic and require energy.

Correct answer = D.

13.7. Compared with the resting state, vigorously contracting skeletal muscle shows:

A. decreased AMP/ATP ratio.
B. decreased levels of fructose 2,6-bisphosphate.
C. increased reduction of pyruvate to lactate.
D. increased oxygen availability.

Correct answer = C.

13.8. Which of the following is true for erythrocytes?

A. Glycolysis produces pyruvate which is metabolized further to lactate
B. Glycolysis produces pyruvate which is metabolized further to ethanol
C. Glycolysis produces lactate which is metabolized further to hemoglobin

D. Glycolysis produces pyruvate which is metabolized further to CO_2

Correct answer = A.

13.9. Which of the following statements is correct?

A. Glucose uptake by brain cells is through energy-requiring (active) transport.

B. Glucose uptake by intestinal mucosal cells requires insulin.

C. Glucose uptake by liver cells is through facilitated diffusion involving a glucose transporter.

D. Glucose uptake by most cells is through simple diffusion up a concentration gradient.

Correct answer = C.

13.10. Energy released by oxidation of glucose is stored as:

A. ATP.

B. NADH.

C. both ATP and NADH.

D. neither ATP nor NADH.

Correct answer = C.

Multiple-Choice Questions (NExT/NEET PG)

Choose the ONE best answer.

13.1. A 8-month-old child is being evaluated for anemia. There is a history of dark colored urine. On examination, the spleen and liver are enlarged. The oxygen dissociation curve for hemoglobin in his erythrocytes is shifted to right, as compared with the curve obtained with normal red cells. A deficiency of which enzyme is most likely to account for the anemia in this patient?

A. Pyruvate kinase

B. Glucose 6-P dehydrogenase

C. Pyruvate carboxylase

D. Glutathione reductase

Correct answer = A.

13.2. Patients with certain forms of epilepsy are found to respond to a ketogenic diet. A probable mechanism here is that the neurons are unable to utilize glucose for energy production. Which of the following statements can be TRUE in this situation?

A. Glucose transporter 1, which does not respond to insulin, is most likely defective in these patients

B. Glucose transporter 1, which responds to insulin, is most likely defective in these patients

C. Glucose transporter 4, which does not respond to insulin, is most likely defective in these patients

D. Glucose transporter 4, which responds to insulin, is most likely defective in these patients

Correct answer = A.

13.3. Which of the following best describes the activity level and phosphorylation state of the listed hepatic enzymes in an individual who consumed a carbohydrate-rich meal about an hour ago? PFK-1 = phosphofructokinase-1; PFK-2 = phosphofructokinase-2; Ⓟ = phosphorylated.

Choice	PFK-1		PFK-2		Pyruvate Kinase	
	Activity	Ⓟ	Activity	Ⓟ	Activity	Ⓟ
A.	Low	No	Low	No	Low	No
B.	High	Yes	Low	Yes	Low	Yes
C.	High	No	High	No	High	No
D.	High	Yes	High	Yes	High	Yes

Correct answer = C.

13.4. A patient with severe hyperglycemia was administered insulin. Which of the following statements is true in this case?

A. Glucose uptake will be stimulated in liver through GLUT2

B. Glucose uptake will be stimulated in muscle through GLUT2

C. Glucose uptake will be stimulated in liver through GLUT4

D. Glucose uptake will be stimulated in muscle through GLUT4

Correct answer = D.

13.5. A 30-year-old man presented to the emergency department with a 1-day history of multiple episodes of watery diarrhea, severe abdominal cramps, and mild fever. No blood in the stools or vomiting was reported. He had eaten local food from roadside restaurants during recent travel. Physical signs suggested dehydration, indicated by dry skin and mucosa. The patient appeared fatigued and was diagnosed with likely infectious gastroenteritis, associated with contaminated food intake. Which of the following is the cause of intestinal fluid loss in this individual?

A. Persistent activation of $G_{\alpha s}$ subunit of G-protein coupled receptor

B. Persistent activation of $G_{\alpha i}$ subunit of G-protein coupled receptor

C. Mutation in sodium dependent glucose cotransporter 1 (SGLT1)

D. Inhibition of cAMP dependent protein kinase

Correct answer = A.

Tricarboxylic Acid Cycle and Pyruvate Dehydrogenase Complex

14

Learning Objectives

At the end of this chapter, the student will be able to:

- Describe the reactions of the tricarboxylic acid cycle
- Explain the physiological regulation of TCA cycle
- Explain the energy generation in various steps of this cycle
- Discuss how acetyl CoA is at the crossroads of metabolism

I. CYCLE OVERVIEW

The tricarboxylic acid cycle ([TCA cycle] also called the citric acid cycle, or the Krebs cycle) plays several roles in metabolism. **It is the final pathway where the oxidative catabolism of carbohydrates, amino acids, and fatty acids converge, their carbon skeletons being converted to carbon dioxide (CO_2),** as shown in Figure 14.1. This oxidation provides energy for the production of the majority of ATP in most animals, including humans. Because the TCA cycle occurs totally in mitochondria, it is in close proximity to the electron transport chain ([ETC], see p. 163), which oxidizes the reduced coenzymes nicotinamide adenine dinucleotide (NADH) and flavin adenine dinucleotide ($FADH_2$) produced by the cycle. The TCA cycle is an **aerobic pathway**, because oxygen (O_2) is required as the final electron acceptor. Reactions such as the catabolism of some amino acids generate intermediates of the cycle and are called **anaplerotic** (from the Greek for "filling up") reactions. The TCA cycle also provides intermediates for a number of important anabolic reactions, such as glucose formation from the carbon skeletons of some amino acids and the synthesis of some amino acids (see p. 403) and heme (see p. 415). Therefore, this cycle should not be viewed as a closed system but, instead, as an open one with compounds entering and leaving as required.

II. CYCLE REACTIONS

In the TCA cycle, oxaloacetate (OAA) is first condensed with an acetyl group from acetyl coenzyme A (CoA) and then is regenerated as the cycle is completed (see Fig. 14.1). **Two carbons enter the cycle as acetyl CoA and two leave as CO_2.** Therefore, the entry of one acetyl CoA into one round of the TCA cycle does not lead to the net production or consumption of intermediates.

A. Acetyl CoA production

The major sources of acetyl CoA are glucose, pyruvate, fatty acids, and ketogenic amino acids. The major utilization of acetyl CoA is for oxidation, fatty acid formation, and formation of cholesterol and ketone bodies.

The major source of acetyl CoA for the TCA cycle is the oxidative decarboxylation of pyruvate by the multienzyme *pyruvate dehydrogenase complex* (*PDH complex*, or *PDHC*). However, the *PDHC* (described below) is not a component of the TCA cycle. Pyruvate, the end product of aerobic glycolysis, is transported from the cytosol into the mitochondrial matrix by the pyruvate mitochondrial carrier of the inner mitochondrial membrane. In the matrix, the *PDHC* converts pyruvate to acetyl CoA. [Note: Fatty acid oxidation is another source of acetyl CoA (see p. 311).]

1. **PDHC component enzymes:** The *PDHC* is a **protein aggregate of multiple copies of three enzymes**, *pyruvate decarboxylase* ([*E1*] sometimes called *pyruvate dehydrogenase*), *dihydrolipoyl transacetylase* (*E2*), and *dihydrolipoyl dehydrogenase* (*E3*). Each catalyzes a part of the overall reaction (Fig. 14.2). Their physical

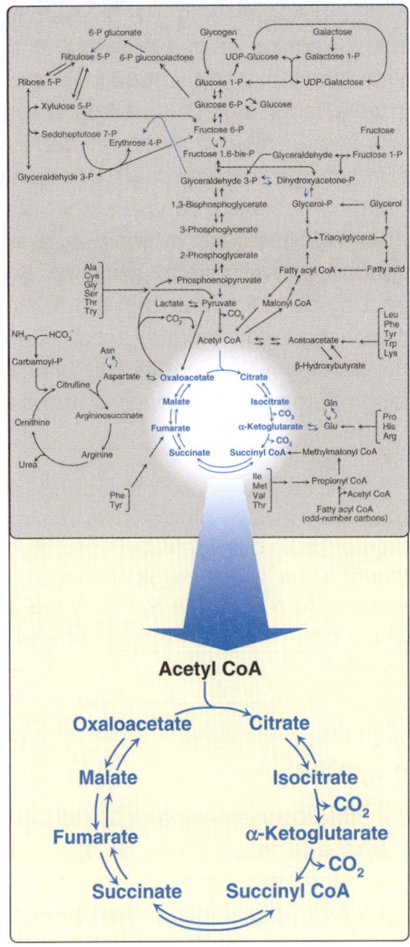

Figure 14.1 The tricarboxylic acid cycle shown as a part of the essential pathways of energy metabolism. [Note: See Fig. 13.2, p. 187 for a more detailed map of metabolism.] CO_2 = carbon dioxide; CoA = coenzyme A.

association links the reactions in proper sequence without the release of intermediates. In addition to the enzymes participating in the conversion of pyruvate to acetyl CoA, the *PDHC* also contains two regulatory enzymes, *pyruvate dehydrogenase kinase* (*PDH kinase*) and *pyruvate dehydrogenase phosphatase* (*PDH phosphatase*).

2. **Coenzymes:** The *PDHC* contains **five coenzymes** that act as carriers or oxidants for the intermediates of the reactions shown in Figure 14.2. *E1* requires thiamine pyrophosphate (TPP), *E2* requires lipoic acid and CoA, and *E3* requires FAD and NAD^+.

   TPP, lipoic acid, and FAD are tightly bound to the enzymes and function as coenzymes–prosthetic groups (see p. 131).

Deficiencies of thiamine or niacin can cause serious central nervous system problems. This is because brain cells are unable to produce sufficient ATP (via the TCA cycle) if the *PDHC* is inactive. Wernicke-Korsakoff, an encephalopathy-psychosis syndrome due to thiamine deficiency, may be seen with alcohol abuse (see p. 533).

3. **Regulation:** Covalent modifications by the two regulatory enzymes of the *PDHC* alternately activate and inactivate *E1*. *PDH kinase* phosphorylates and inactivates *E1*, whereas *PDH phosphatase* dephosphorylates and activates *E1* (Fig. 14.3). The *kinase* itself is allosterically activated by ATP, acetyl CoA, and NADH. Therefore, **in the presence of these high-energy products, the *PDHC* is turned off**. [Note:

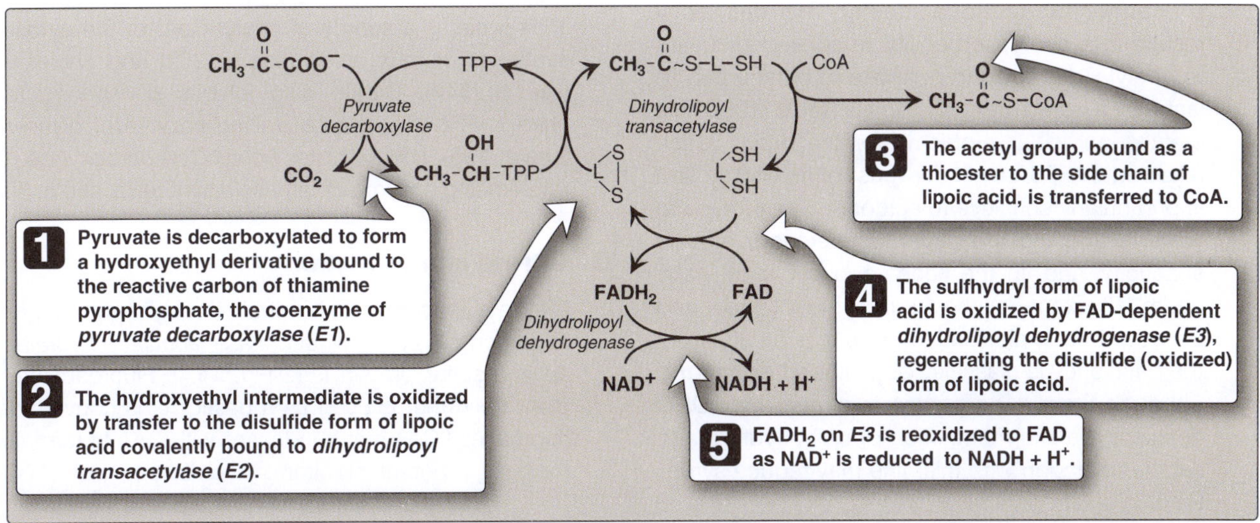

Figure 14.2 Mechanism of action of the enzymes (*E*) of the *pyruvate dehydrogenase complex*. [Note: All the coenzymes of the complex, except for lipoic acid, are derived from vitamins. TPP is from thiamine, FAD from riboflavin, NAD from niacin, and CoA from pantothenic acid.] CO_2 = carbon dioxide; TPP = thiamine pyrophosphate; L = lipoic acid; CoA = coenzyme A; $FAD(H_2)$ and NAD(H) = flavin and nicotinamide adenine dinucleotides; ~ = high-energy bond.

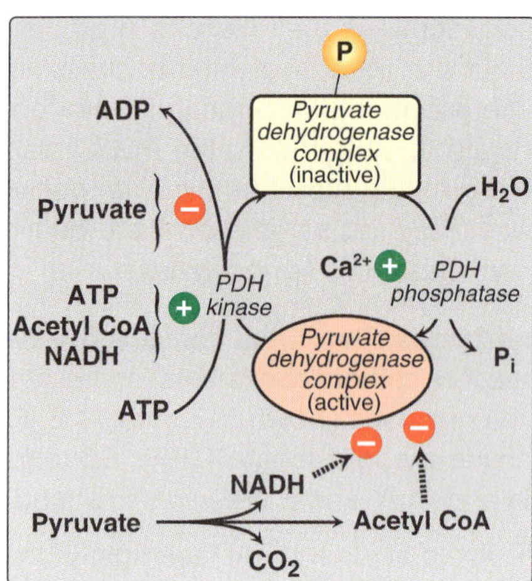

Figure 14.3 Regulation of *pyruvate dehydrogenase (PDH) complex*. (P) = phosphate [⇢ denotes product inhibition.]

It is actually the rise in the ATP/ADP (adenosine diphosphate), $NADH/NAD^+$, or acetyl CoA/CoA ratios that affects enzymic activity.] Pyruvate is a potent inhibitor of *PDH kinase*. Therefore, if pyruvate concentrations are elevated, *E1* will be maximally active. Calcium (Ca^{2+}) is a strong activator of *PDH phosphatase*, stimulating *E1* activity. This is particularly important **in skeletal muscle, where Ca^{2+} release during contraction stimulates the *PDHC*** and, thus, energy production. [Note: Although covalent regulation by the *kinase* and *phosphatase* is primary, the *PDHC* is also subject to product (NADH and acetyl CoA) inhibition.]

4. **Deficiency:** A deficiency of the α subunits of the tetrameric *E1* component of the *PDHC*, although very rare, is the most common biochemical cause of **congenital lactic acidosis**. The deficiency results in a decreased ability to convert pyruvate to acetyl CoA, causing pyruvate to be shunted to lactate via *lactate dehydrogenase* (see p. 199). This creates particular problems for the brain, which relies on the TCA cycle for most of its energy and is particularly sensitive to acidosis. Symptoms are variable and include neurodegeneration, muscle spasticity, and, in the neonatal-onset form, early death. The gene for the α subunit is X linked, and because both males and females may be affected, the deficiency is classified as **X-linked dominant**. Although there is no proven treatment for *PDHC* deficiency, dietary restriction of carbohydrate and supplementation with thiamine may reduce symptoms in select patients.

5. **Arsenic poisoning:** As previously described (see p. 197), pentavalent arsenic (arsenate) can interfere with glycolysis at the *glyceraldehyde 3-phosphate* step, thereby decreasing ATP production. However, arsenic poisoning is due primarily to inhibition of enzyme complexes that require lipoic acid as a coenzyme, including *PDH*, *α-ketoglutarate dehydrogenase* (see E. below), and *branched-chain α-keto acid dehydrogenase* (see p. 401). Arsenite (the trivalent form of arsenic) forms a stable complex with the thiol (–SH) groups of lipoic acid, making that compound unavailable to serve as a coenzyme. When it binds to lipoic acid in the *PDHC*, pyruvate (and, consequently, lactate) accumulates. As with *PDHC* deficiency, this particularly affects the brain, causing neurologic disturbances and death.

Leigh syndrome (subacute necrotizing encephalomyelopathy) is a rare, progressive, neurodegenerative disorder caused by defects in mitochondrial ATP production, primarily as a result of mutations in genes that encode proteins of the *PDHC*, the ETC, or *ATP synthase*. Both nuclear and mitochondrial DNA can be affected.

B. Citrate synthesis

The irreversible condensation of acetyl CoA and OAA to form citrate (a tricarboxylic acid) is catalyzed by *citrate synthase*, the initiating enzyme of the TCA cycle (Fig. 14.4). This aldol condensation has a highly negative change in standard free energy ([ΔG^0], see p. 159), which strongly favors citrate formation. The enzyme is inhibited by citrate (product inhibition). Substrate availability is another means of regulation for *citrate synthase*. The binding of OAA greatly increases the enzyme's affinity for acetyl CoA.

Citrate, in addition to being an intermediate in the TCA cycle, is a source of acetyl CoA for the cytosolic synthesis of fatty acids (see p. 302) and cholesterol (see p. 342). Citrate also inhibits *phosphofructokinase-1 (PFK-1)*, the rate-limiting enzyme of glycolysis (see p. 195), and activates *acetyl CoA carboxylase* (the rate-limiting enzyme of fatty acid synthesis, see p. 302).

C. Citrate isomerization

Citrate is isomerized to isocitrate through hydroxyl group migration catalyzed by *aconitase* (*aconitate hydratase*), an iron-sulfur protein (see Fig. 14.4). Fluoroacetate, a plant toxin that is used as a pesticide, is converted to fluoroacetyl CoA that condenses with OAA to form fluorocitrate, a potent inhibitor of *aconitase*.]

D. Oxidative decarboxylation of isocitrate

Isocitrate dehydrogenase catalyzes the irreversible oxidative decarboxylation of isocitrate to α-ketoglutarate,

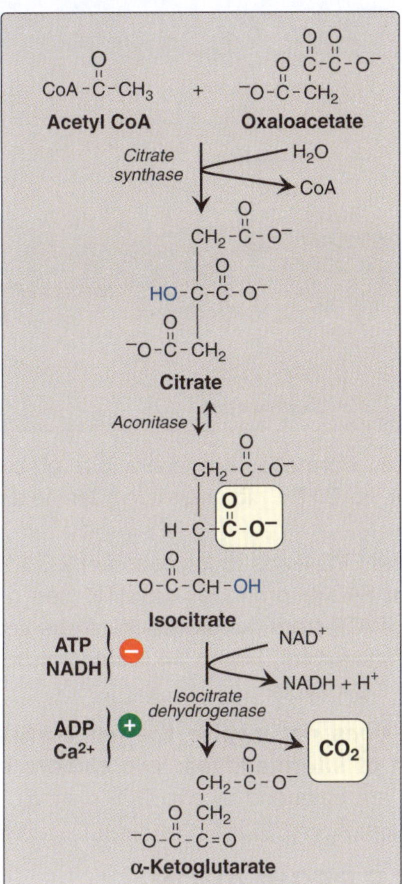

Figure 14.4 Formation of α-ketoglutarate from acetyl coenzyme A (CoA) and oxaloacetate. NAD(H) = nicotinamide adenine dinucleotide; CO_2 = carbon dioxide.

yielding the first of three NADH molecules produced by the cycle and the first release of CO_2 (see Fig. 14.4). This is one of the rate-limiting steps of the TCA cycle. The enzyme is allosterically activated by ADP (a low-energy signal) and Ca^{2+} and is inhibited by ATP and NADH, levels of which are elevated when the cell has abundant energy stores.

E. Oxidative decarboxylation of α-ketoglutarate

The irreversible conversion of α-ketoglutarate to succinyl CoA is catalyzed by the *α-ketoglutarate dehydrogenase complex*, a protein aggregate of multiple copies of three enzymes (Fig. 14.5). The mechanism of this oxidative decarboxylation is very similar to that used for the conversion of pyruvate to acetyl CoA by the *PDHC*. **The reaction releases the second CO_2 and produces the second NADH of the cycle.** The coenzymes required are TPP, lipoic acid, FAD, NAD^+, and CoA. Each functions as part of the catalytic mechanism in a way analogous to that described for the *PDHC* (see p. 206). The large negative ΔG^0 of the reaction favors formation of succinyl CoA, a high-energy thioester similar to acetyl CoA. The *α-ketoglutarate dehydrogenase complex* is inhibited by its products, NADH and succinyl CoA, and activated by Ca^{2+}. However, it is not regulated by phosphorylation/dephosphorylation reactions as described for the *PDHC*. [Note: α-Ketoglutarate is also produced by the oxidative deamination (see p. 384) and transamination of the amino acid glutamate (see p. 382).]

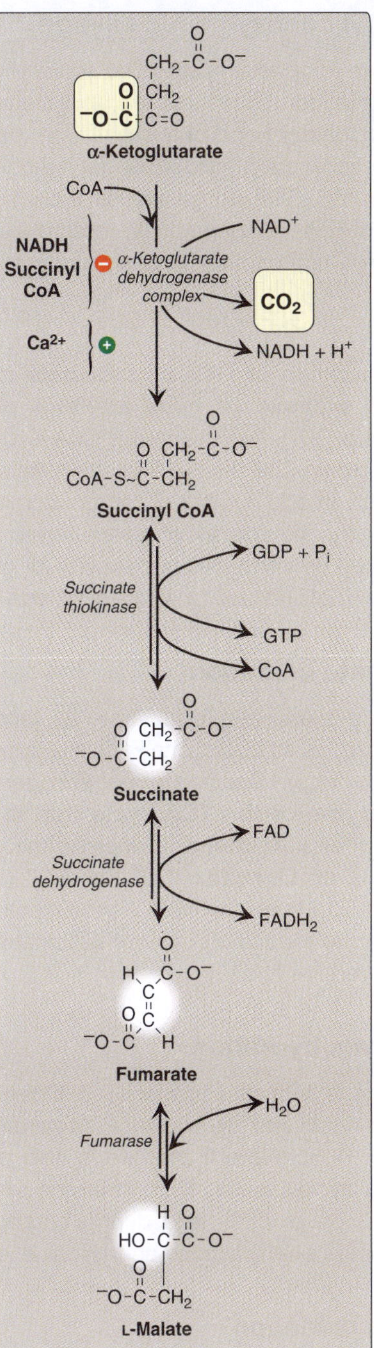

Figure 14.5 Formation of malate from α-ketoglutarate. $FAD(H_2)$ and NAD(H) = flavin and nicotinamide adenine dinucleotides; GDP and GTP = guanosine di- and triphosphates; ~ = high-energy bond; CoA = coenzyme A.

F. Succinyl coenzyme A cleavage

Succinate thiokinase (also called *succinyl CoA synthetase*, named for the reverse reaction) cleaves the high-energy thioester bond of succinyl CoA (see Fig. 14.5). This reaction is coupled to phosphorylation of guanosine diphosphate (GDP) to guanosine triphosphate (GTP). GTP and ATP are energetically interconvertible by the *nucleoside diphosphate kinase* reaction:

$$GTP + ADP \rightleftarrows GDP + ATP$$

The generation of GTP by *succinate thiokinase* is another example of substrate-level phosphorylation (see p. 197). [Note: Succinyl CoA is also produced from propionyl CoA derived from the metabolism of fatty acids with an odd number of carbon atoms (see p. 312) and from the metabolism of several amino acids (see p. 401). It can be converted to pyruvate for gluconeogenesis (see p. 215) or used in heme synthesis (see p. 415).]

G. Succinate oxidation

Succinate is oxidized to fumarate by *succinate dehydrogenase*, as its coenzyme FAD is reduced to $FADH_2$ (see Fig. 14.5). ***Succinate dehydrogenase* is the only enzyme of the TCA cycle that is embedded in the inner mitochondrial membrane. As such, it functions as Complex II of the ETC** (see p. 165). [Note: FAD, rather than NAD^+, is the electron acceptor because the reducing power of succinate is not sufficient to reduce NAD^+.]

H. Fumarate hydration

Fumarate is hydrated to malate in a freely reversible reaction catalyzed by *fumarase* (*fumarate hydratase*, see Fig. 14.5). [Note: Fumarate is also produced by the urea cycle (see p. 385), in purine synthesis (see Fig. 34.7 on p. 580), and during catabolism of the amino acids phenylalanine and tyrosine (see p. 398).]

I. Malate oxidation

Malate is oxidized to OAA by *malate dehydrogenase* (Fig. 14.6). **This reaction produces the third and final NADH of the cycle.** The ΔG^0 of the reaction is positive, but the reaction is driven in the direction of OAA by the highly exergonic *citrate synthase* reaction. [Note: OAA is also produced by the transamination of the amino acid aspartic acid (see p. 382).]

III. ENERGY PRODUCED BY THE CYCLE

Four pairs of electrons are transferred during one turn of the TCA cycle: three pairs reducing three NAD^+ to NADH and one pair reducing FAD to $FADH_2$. Oxidation of one NADH by the ETC leads to formation of 2.5 ATP, whereas oxidation of $FADH_2$ produces 1.5 ATP (see p. 168). The total yield of ATP from the oxidation of one acetyl CoA is shown in Figure 14.7. Figure 14.8 summarizes the reactions of the TCA cycle.

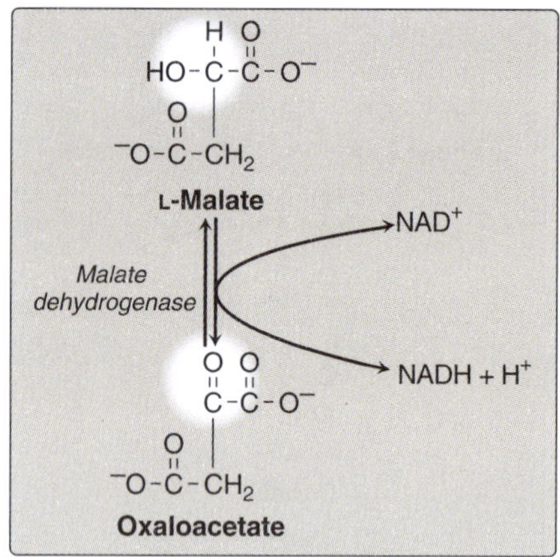

Figure 14.6 Formation (regeneration) of oxaloacetate from malate. NAD(H) = nicotinamide adenine dinucleotide.

The cycle does not involve the net consumption or production of intermediates. Two carbons entering as acetyl CoA are balanced by two CO_2 exiting.

IV. CYCLE REGULATION

In contrast to glycolysis, which is regulated primarily by *PFK-1*, the TCA cycle is controlled by the regulation of several enzymes (see Fig. 14.8). The most important of these regulated enzymes are those that catalyze reactions with highly negative ΔG^0: *citrate synthase*, *isocitrate dehydrogenase*, and the *α-ketoglutarate dehydrogenase complex*. Reducing equivalents needed for oxidative phosphorylation are generated by the *PDHC* and the TCA cycle, and both processes are upregulated in response to a decrease in the ATP/ADP ratio.

Energy-producing reaction	Number of ATP produced
3 NADH → 3 NAD^+	7.5
$FADH_2$ → FAD	1.5
GDP + P_i → GTP	1
	10 ATP/acetyl CoA oxidized

Figure 14.7 Number of ATP molecules produced from the oxidation of one molecule of acetyl coenzyme A (CoA) using both substrate-level and oxidative phosphorylation. NAD(H) and FAD(H_2) = nicotinamide and flavin adenine dinucleotides; GDP and GTP = guanosine di- and triphosphates; P_i = inorganic phosphate.

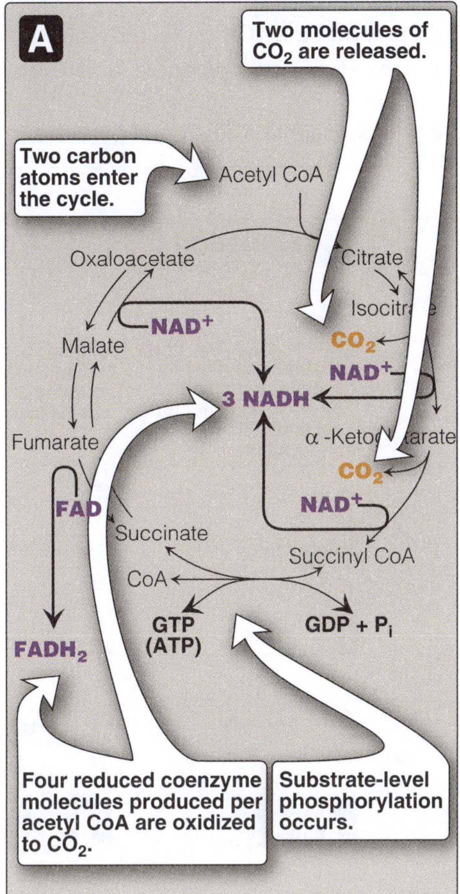

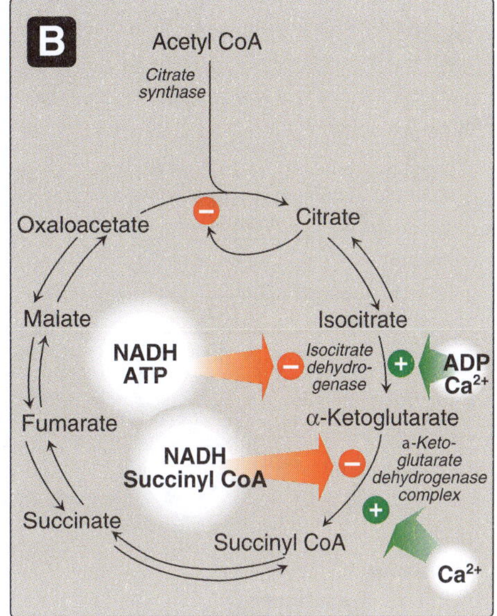

Figure 14.8 **A.** Production of reduced coenzymes, ATP, and carbon dioxide (CO_2) in the tricarboxylic acid cycle. [Note: Guanosine triphosphate (GTP) and ATP are interconverted by *nucleoside diphosphate kinase*.] **B.** Inhibitors and activators of the cycle.

V. Medical Perspectives

As TCA cycle is an important metabolic cycle, its defects are usually not compatible with life, therefore such patients are rarely seen. However, some recessive deficiencies of *succinate dehydrogenase* and *isocitrate dehydrogenase* leading to severe multisystem syndromes have been documented.

Toxic or poisonous inhibitors of TCA include fluoroacetate (inhibitor of acotinase) and heavy metal ions like arsenite and mercury. Arsenite inhibits the reactions that use inorganic phosphate like *succinate thiokinase* and *pyruvate dehydrogenase*.

VI. Chapter Summary

- **Pyruvate** is oxidatively decarboxylated by the ***pyruvate dehydrogenase complex*** (***PDHC***), producing **acetyl coenzyme A** (**CoA**), which is the major fuel for the **tricarboxylic acid** (**TCA**) **cycle** (Fig. 14.9).
- The multienzyme *PDHC* requires five coenzymes: **thiamine pyrophosphate**, **lipoic acid**, **flavin adenine dinucleotide** (**FAD**), **nicotinamide adenine dinucleotide** (**NAD**+), and **CoA**.
- **Deficiency of *E1 of PDHC*** is the most common biochemical cause of **congenital lactic acidosis**. The brain is particularly affected in this **X-linked dominant** disorder.
- **Arsenic poisoning** causes inactivation of the *PDHC* by binding to lipoic acid.
- In the TCA cycle, **citrate** is synthesized from **oxaloacetate** (**OAA**) and **acetyl CoA** by ***citrate synthase.***
- **The two carbons of acetyl CoA are released as carbon dioxide** (**CO_2**), in reactions catalyzed by ***isocitrate dehydrogenase and α-ketoglutarate dehydrogenase.***

- **Three NADH**, **one FADH$_2$**, and **one GTP** are produced by one round of the TCA cycle, providing 10 ATP.
- The generation of acetyl CoA by the oxidation of pyruvate via the *PDHC* also produces an NADH, which produces another 2.5 ATP in ETC.
- Therefore, a total of 12.5 ATP are produced from the complete mitochondrial oxidation of pyruvate to CO_2.

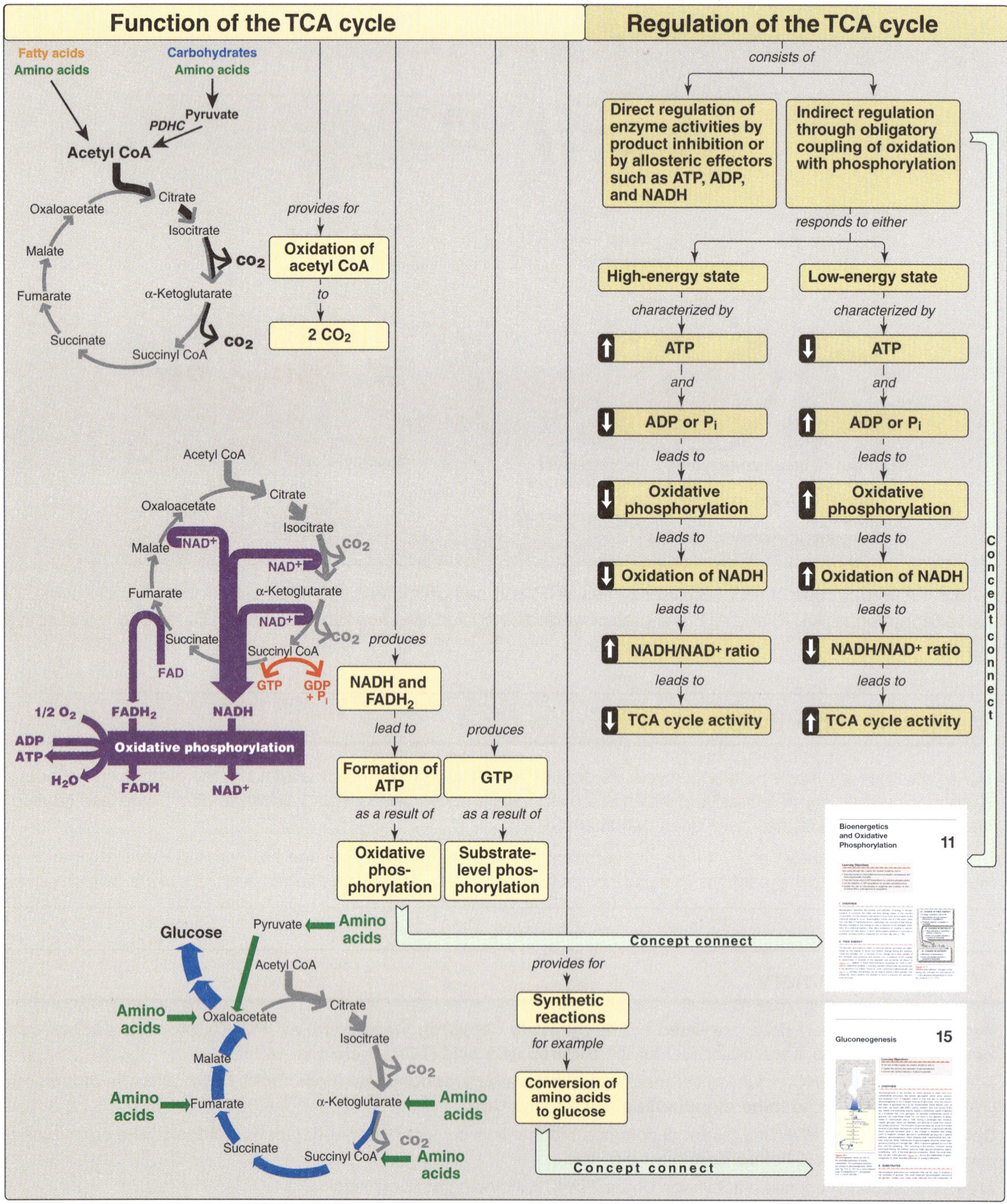

Figure 14.9 Key concept map for the tricarboxylic acid (TCA) cycle. *PDHC = pyruvate dehydrogenase complex*; CoA = coenzyme A; CO_2 = carbon dioxide; NAD(H) = nicotinamide adenine dinucleotide; FAD(H_2) = flavin adenine dinucleotide; GDP and GTP = guanosine di- and triphosphates; ADP = adenosine diphosphate; P_i = inorganic phosphate.

Study Questions

Short-Answer Questions

Write short notes on:

14.1. Arsenic poisoning
14.2. Pyruvate dehydrogenase complex
14.3. Energy produced by the TCA cycle
14.4. Non-Physiological inhibition of TCA cycle

Long-Answer Question

14.1. Discuss the amphibolic role of TCA cycle in the integration of metabolism.
14.2. Highlight the sources and utilization of acetyl CoA.
14.3. Explain the physiological regulation of TCA cycle. Add a note on energy generation in this cycle.

Multiple-Choice Questions

Choose the ONE best answer.

14.1. The conversion of pyruvate to acetyl coenzyme A and carbon dioxide:
- A. involves the participation of lipoic acid.
- B. is activated when pyruvate decarboxylase of the pyruvate dehydrogenase complex (PDHC) is phosphorylated by PDH kinase in the presence of ATP.
- C. is reversible.
- D. occurs in the cytosol.

Correct answer = A.

14.2. Which one of the following conditions decreases the oxidation of acetyl coenzyme A by the citric acid cycle?
- A. A high availability of calcium
- B. A high acetyl CoA/CoA ratio
- C. A low ATP/ADP ratio
- D. A low NAD^+/NADH ratio

Correct answer = D.

14.3. Which coenzyme–cosubstrate is used by dehydrogenases in both glycolysis and the tricarboxylic acid cycle?
- A. NAD^+
- B. FAD
- C. Both A and B
- D. Neither A nor B

Correct answer = C.

14.4. Which of the following enzymes involved in catalyzing the reactions of the citric acid cycle in eukaryotes is located at a site different from the others?
- A. Succinate dehydrogenase
- B. α-Ketoglutarate dehydrogenase
- C. Citrate synthase
- D. Isocitrate dehydrogenase

Correct answer = A.

14.5. Which of the following components of pyruvate dehydrogenase complex is NOT derived from a vitamin?
- A. Flavin adenine dinucleotide
- B. Thiamine pyrophosphate
- C. Lipoic acid
- D. Coenzyme A

Correct answer = C.

14.6. Krebs cycle takes place in:
- A. Cytosol in all organs of the body
- B. Mitochondria in all organs of the body
- C. Cytosol of liver cells only
- D. Mitochondria of liver cells only

Correct answer = B.

14.7. Which of the following is TRUE about Krebs cycle?
- A. Pyruvate condenses with oxaloacetate to form citrate
- B. Acetyl CoA condenses with oxaloacetate to form citrate
- C. Succinyl CoA is decarboxylated to fumarate
- D. Succinate is decarboxylated to fumarate

Correct answer = B.

14.8. Which of the following is TRUE about aconitase?
- A. Aconitase is an iron sulfur protein
- B. Aconitase is stimulated by fluoroacetate
- C. Aconitase is allosterically activated by ATP
- D. None of the above

Correct answer = A.

14.9. At which point does phenylalanine enter the TCA cycle
- A. Succinyl CoA
- B. Acetyl CoA
- C. Fumarate
- D. Malate

Correct answer = C.

14.10. Choose the correct option:
- A. GTP is produced during conversion of succinyl CoA to succinate in TCA cycle
- B. NADH is produced during conversion of succinyl CoA to succinate in TCA cycle
- C. GTP is produced during conversion of succinate to fumarate in TCA cycle
- D. NADH is produced during conversion of succinate to fumarate in TCA cycle

Correct answer = A.

Multiple-Choice Questions (NExT/NEET PG)

Choose the ONE best answer.

14.1. The following is the sum of three steps in the citric acid cycle.

$A + B + FAD + H_2O \rightarrow C + FADH_2 + NADH$

Choose the lettered answer that corresponds to the missing "A," "B," and "C" in the equation.

Reactant A	Reactant B	Product C
A. Succinyl CoA	GDP	Succinate
B. Succinate	NAD^+	Oxaloacetate
C. Fumarate	NAD^+	Oxaloacetate
D. Succinate	NAD^+	Malate
E. Fumarate	GTP	Malate

Correct answer = B.

14.2. A 1-month-old male presents with neurological problems and lactic acidosis. An enzyme assay for pyruvate dehydrogenase complex (PDHC) activity on extracts of cultured skin fibroblasts showed 5% of normal activity with a low concentration of thiamine pyrophosphate (TPP), but 80% of normal activity when the assay contained a thousand-fold higher concentration of TPP. Which one of the following statements concerning this patient is correct?

A. Administration of thiamine is expected to reduce his serum lactate level and improve his clinical symptoms.
B. A high-carbohydrate diet would be expected to be beneficial for this patient.
C. Citrate production from aerobic glycolysis is expected to be increased.
D. PDH kinase, a regulatory enzyme of the PDHC, is expected to be active.

Correct answer = A.

14.3. The dysfunction of which of the following enzymes gives rise to the accumulation of lactate and a high-output cardiac failure in wet beriberi?

A. Pyruvate dehydrogenase
B. Pyruvate kinase
C. Transketolase
D. Lactate dehydrogenase

Correct answer = A.

14.4. Based on Statements 1 and 2, choose the correct option.

Statement 1: The citric acid cycle enzyme aconitase (which converts citrate to isocitrate) is inhibited by the pesticide fluroacetate.

Statement 2: Fluoroacetate is converted to fluoroacetyl CoA that condenses with oxaloacetate to form fluorocitrate, a potent inhibitor of aconitase.

A. Both the statements are correct, and Statement 2 is the correct explanation for Statement 1
B. Both the statements are correct, but Statement 2 is NOT the correct explanation for Statement 1
C. Statement 1 is correct, but Statement 2 is NOT correct
D. Statement 1 is NOT correct, but Statement 2 is correct

Correct answer = A.

14.5. Based on Statements 1 and 2, choose the correct option.

Statement 1: One potential therapeutic strategy to treat patients with defects in pyruvate dehydrogenase (PDH) complex is to alter the activity of pyruvate dehydrogenase kinase.

Statement 2: PDH kinase phosphorylates and activates the PDH complex.

A. Both the Statements are correct, and Statement 2 is the correct explanation for Statement 1
B. Both the Statements are correct, but Statement 2 is NOT the correct explanation for Statement 1
C. Statement 1 is correct, but Statement 2 is NOT correct
D. Statement 1 is NOT correct, but Statement 2 is correct

Correct answer = C.

Gluconeogenesis

15

Learning Objectives

At the end of this chapter, the student will be able to:

- Explain the process and regulation of gluconeogenesis
- Discuss the medical relevance of gluconeogenesis

I. OVERVIEW

Gluconeogenesis is the process by which glucose is made from non-carbohydrate precursors like lactate, glucogenic amino acids, glycerol, and propionyl CoA. It happens mainly in liver but also in renal cortex. Gluconeogenesis is not a simple reversal of glycolysis, since the irreversible steps of glycolysis have to be circumvented. Some tissues, such as the brain, red blood cells (RBC), kidney medulla, lens and cornea of the eye, testes, and exercising muscle, require a continuous supply of glucose as a metabolic fuel. Liver glycogen, an essential postprandial source of glucose, can meet these needs for <24 hours in the absence of dietary intake of carbohydrate (see p. 224). During a prolonged fast, however, hepatic glycogen stores are depleted, and glucose is made from noncarbohydrate precursors. The formation of glucose does not occur by a simple reversal of glycolysis, because the overall equilibrium of glycolysis strongly favors pyruvate formation (that is, the change in standard free energy [ΔG^0] is negative). Instead, glucose is synthesized de novo by a special pathway, gluconeogenesis, which requires both mitochondrial and cytosolic enzymes. [Note: Deficiencies of gluconeogenic enzymes cause hypoglycemia.] During an overnight fast, ~90% of gluconeogenesis occurs in the liver, with the remaining ~10% occurring in the kidneys. However, during prolonged fasting, the kidneys become major glucose-producing organs, contributing ~40% of the total glucose production. [Note: The small intestine can also make glucose.] Figure 15.1 shows the relationship of gluconeogenesis to other essential pathways of energy metabolism.

II. SUBSTRATES

Gluconeogenic precursors are molecules that can be used to produce a net synthesis of glucose. The most important gluconeogenic precursors are **glycerol, lactate, and α-keto acids obtained from the metabolism of glucogenic amino acids**. [Note: All but two amino acids (leucine and lysine) are glucogenic (see p. 397).]

A. Glycerol

Glycerol is released during the hydrolysis of triacylglycerols (TAG) in adipose tissue (see p. 309) and is delivered by the blood to the liver. Glycerol is phosphorylated by *glycerol kinase* to glycerol 3-phosphate, which is oxidized by *glycerol 3-phosphate dehydrogenase* to dihydroxyacetone phosphate, an intermediate of glycolysis and gluconeogenesis.

B. Lactate

Lactate from anaerobic glycolysis is released into the blood by exercising skeletal muscle and by cells that lack mitochondria such as RBC. In the **Cori cycle**, this lactate is taken up by the liver and oxidized to pyruvate that is converted to glucose, which is released back into the circulation (Fig. 15.2).

C. Amino acids

Amino acids produced by hydrolysis of tissue proteins are the major sources of glucose during a fast. Their metabolism generates α-keto acids, such as pyruvate that is converted to glucose, or α-ketoglutarate that can enter the tricarboxylic acid (TCA) cycle and form oxaloacetate (OAA), a direct precursor of

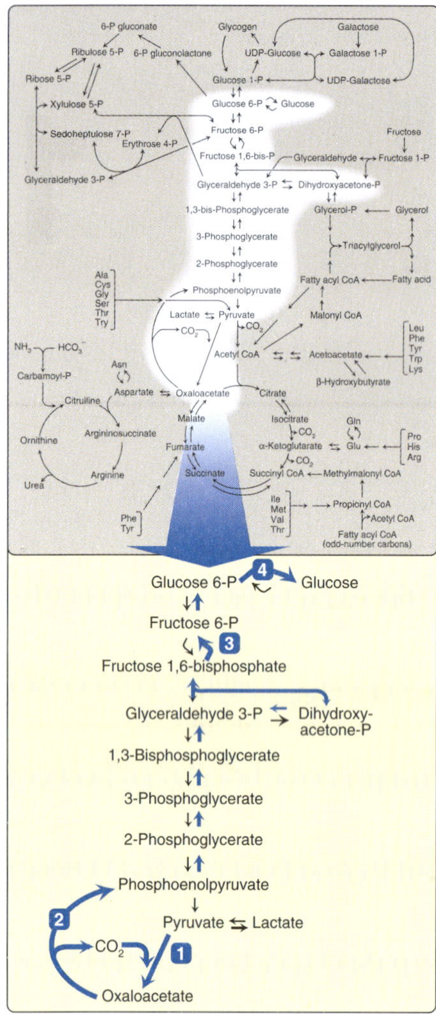

Figure 15.1 Gluconeogenesis shown as one of the essential pathways of energy metabolism. The numbered reactions are unique to gluconeogenesis. [Note: See Fig. 13.2, p. 187, for a more detailed map of metabolism.] P = phosphate; CO_2 = carbon dioxide.

phosphoenolpyruvate (PEP). The nitrogen groups remaining after amino acid degradation are transferred to pyruvate by the enzyme alanine transaminase, resulting in the production of alanine. In the liver, alanine is utilized in the urea cycle for nitrogen disposal and in gluconeogenesis to generate glucose from pyruvate. This metabolic pathway is commonly referred to as the **Cahill cycle** or **glucose-alanine cycle**.

> **Acetyl CoA and compounds that give rise only to acetyl CoA (for example, acetoacetate, lysine, and leucine) cannot give rise to a net synthesis of glucose.** This is because of the irreversible nature of the *pyruvate dehydrogenase complex* (*PDHC*), which converts pyruvate to acetyl CoA (see p. 206). These compounds give rise instead to ketone bodies (see p. 314) and are termed ketogenic.

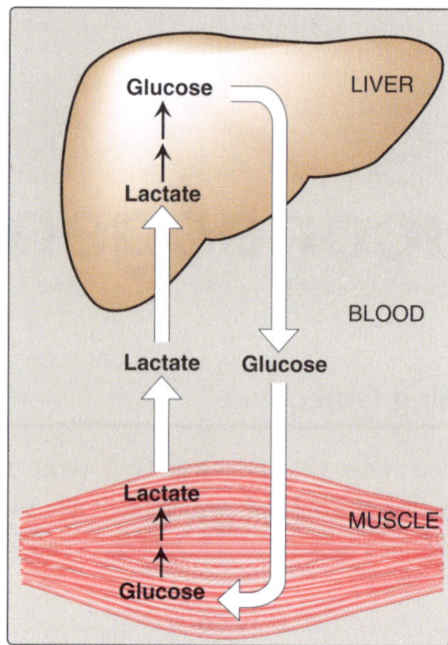

Figure 15.2 The intertissue Cori cycle links gluconeogenesis with glycolysis. [Note: Diffusion of lactate and glucose across membranes is facilitated by transport proteins.]

III. REACTIONS

Seven glycolytic reactions are reversible and are used in the synthesis of glucose from lactate or pyruvate. However, **three glycolytic reactions are irreversible** and must be **circumvented by four alternate reactions** that energetically favor the synthesis of glucose. These irreversible reactions, which together are unique to gluconeogenesis, are described below.

A. Pyruvate carboxylation

The first roadblock to overcome in the synthesis of glucose from pyruvate is the irreversible conversion in glycolysis of PEP to pyruvate by *pyruvate kinase (PK).* In gluconeogenesis, pyruvate is carboxylated by *pyruvate carboxylase (PC)* to OAA, which is converted to PEP by *PEP-carboxykinase* (*PEPCK*) (Fig. 15.3).

1. **Biotin:** *PC* requires the **coenzyme biotin** (see p. 534) covalently bound to the ε-amino group of a lysine residue in the enzyme (see Fig. 15.3). ATP hydrolysis drives formation of an enzyme–biotin–carbon dioxide (CO_2) intermediate, which then carboxylates pyruvate to form OAA. [Note: HCO_3^- provides the CO_2.] **The *PC* reaction occurs in the mitochondria of liver and kidney cells and has two purposes: to allow production of PEP, an important substrate for gluconeogenesis, and to provide OAA that can replenish the TCA cycle intermediates** that may become depleted. Muscle cells also contain *PC* but use the OAA

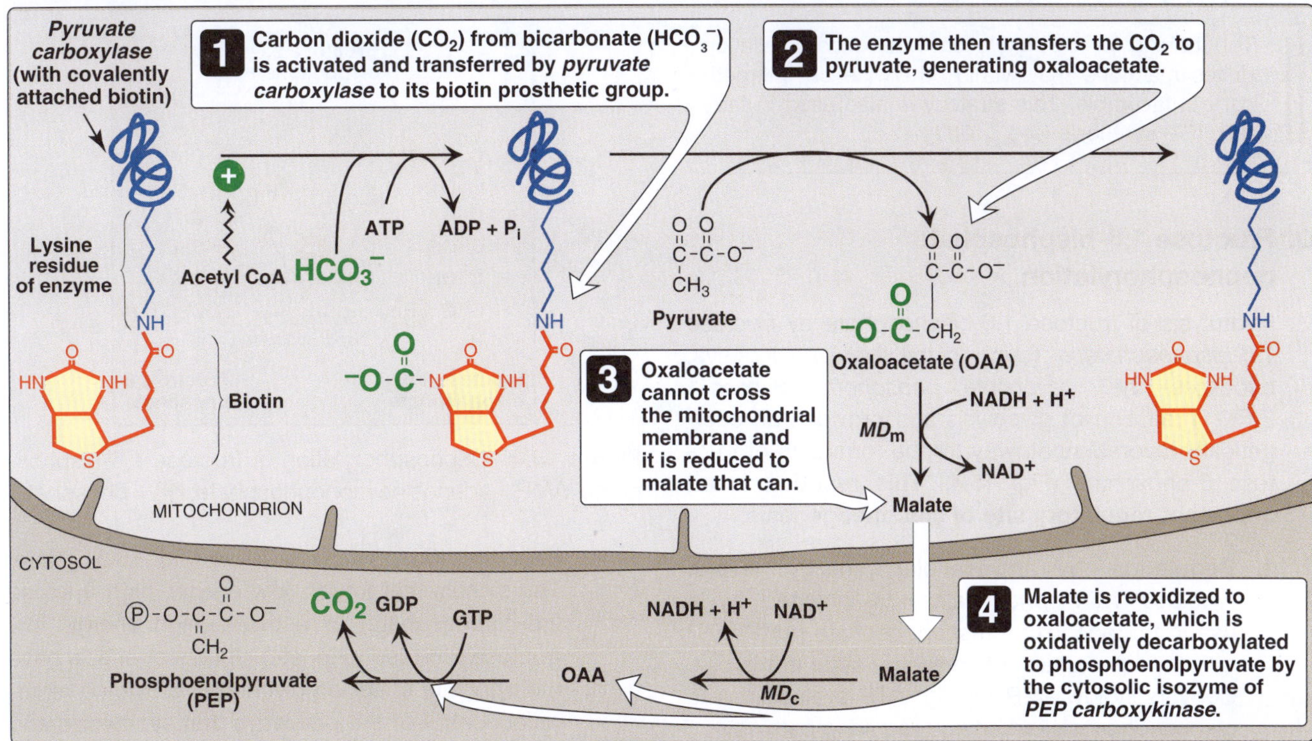

Figure 15.3 PEP synthesis in the cytosol. [Note: The process moves nicotinamide adenine dinucleotide (NADH) reducing equivalents required for gluconeogenesis out of mitochondria into the cytosol.] MD_m and MD_c = mitochondrial and cytosolic isozymes of *malate dehydrogenase*; GTP and GDP = guanosine tri- and diphosphates; ADP = adenosine diphosphate.

product only for the replenishment (anaplerotic) purpose and do not synthesize glucose. [Note: Pyruvate carrier protein moves pyruvate from the cytosol into mitochondria.]

> *PC* is one of several *carboxylases* that require biotin. Others include *acetyl CoA carboxylase* (p. 302), *propionyl CoA carboxylase* (p. 312).

2. **Allosteric regulation:** *PC* is allosterically activated by acetyl CoA. Elevated levels of acetyl CoA in mitochondria signal a metabolic state in which increased synthesis of OAA is required. For example, this occurs during fasting, when OAA is used for gluconeogenesis in the liver and kidneys. Conversely, at low levels of acetyl CoA, *PC* is largely inactive, and pyruvate is primarily oxidized by the *PDHC* to acetyl CoA that can be further oxidized by the TCA cycle (see p. 206).

B. Oxaloacetate transport to the cytosol

For gluconeogenesis to continue, OAA must be converted to PEP by *PEPCK*. PEP production in the cytosol requires transport of OAA out of mitochondria. However, there is no OAA transporter in the inner mitochondrial membrane. A series of reactions, known as **malate shuttle**, helps in this transport. OAA is first reduced to malate by mitochondrial *malate dehydrogenase* (*MD*). Malate is transported into the cytosol and reoxidized to OAA by cytosolic *MD* as nicotinamide adenine dinucleotide (NAD^+) is reduced to NADH (see Fig. 15.3). The NADH is used in the reduction of 1,3-bisphosphoglycerate to glyceraldehyde 3-phosphate by *glyceraldehyde 3-phosphate dehydrogenase* (see p. 197), a reaction common to glycolysis and gluconeogenesis. [Note: When abundant, lactate is oxidized to pyruvate as NAD^+ is reduced. The pyruvate is transported into mitochondria and carboxylated by *PC* to OAA, which can be converted to PEP by the mitochondrial isozyme of *PEPCK*. PEP is transported to the cytosol. OAA can also be converted to aspartate that is transported into the cytosol.]

C. Cytosolic oxaloacetate decarboxylation

OAA is decarboxylated and phosphorylated to PEP in the cytosol by *PEPCK*. The reaction is driven by hydrolysis of guanosine triphosphate ([GTP], see Fig. 15.3). The combined actions of *PC* and *PEPCK* provide an energetically favorable pathway from pyruvate to PEP. PEP is then acted on by the reactions of glycolysis running in the reverse direction until it becomes fructose 1,6-bisphosphate.

> The pairing of carboxylation with decarboxylation drives reactions that would otherwise be energetically unfavorable. This strategy is also used in fatty acid (FA) synthesis (see p. 303).

D. Fructose 1,6-bisphosphate dephosphorylation

Hydrolysis of fructose 1,6-bisphosphate by *fructose 1,6-bisphosphatase*, found in the liver and kidneys, bypasses the irreversible *phosphofructokinase-1* (*PFK-1*) reaction of glycolysis and provides an energetically favorable pathway for the formation of fructose 6-phosphate (Fig. 15.4). **This reaction is an important regulatory site of gluconeogenesis.**

1. **Regulation by intracellular energy levels:** *Fructose 1,6-bisphosphatase* is inhibited by a rise in the adenosine monophosphate (AMP)/ATP ratio, which signals a low-energy state in the cell. Conversely, low AMP and high ATP levels stimulate gluconeogenesis, an energy-requiring pathway.

2. **Regulation by fructose 2,6-bisphosphate:** *Fructose 1,6- bisphosphatase* is inhibited by fructose 2,6-bisphosphate, an allosteric effector whose concentration is influenced by the insulin/glucagon ratio. When glucagon is high, the effector is not made by hepatic *PFK-2* (see p. 195), and thus, the *phosphatase* is active (Fig. 15.5). [Note: The signals that inhibit (low energy, high fructose 2,6-bisphosphate) or activate (high energy, low fructose 2,6-bisphosphate) gluconeogenesis have the opposite effect on glycolysis, providing reciprocal control of the pathways that synthesize and oxidize glucose (see p. 196).]

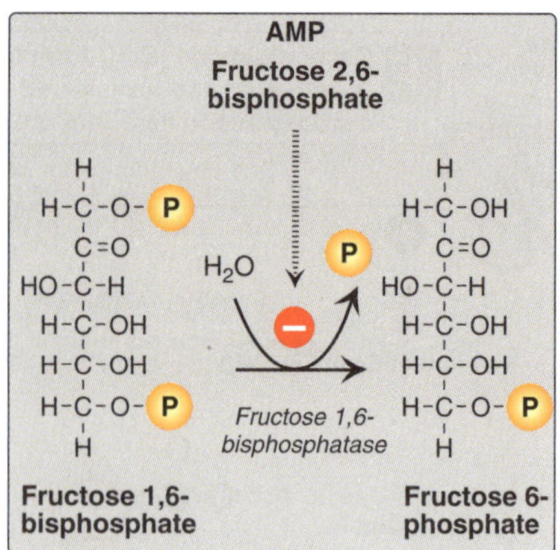

Figure 15.4 Dephosphorylation of fructose 1,6-bisphosphate. AMP = adenosine monophosphate; P = phosphate.

E. Glucose 6-phosphate dephosphorylation

Glucose 6-phosphate hydrolysis by *glucose 6-phosphatase* bypasses the irreversible *hexokinase/glucokinase* reaction and provides an energetically favorable

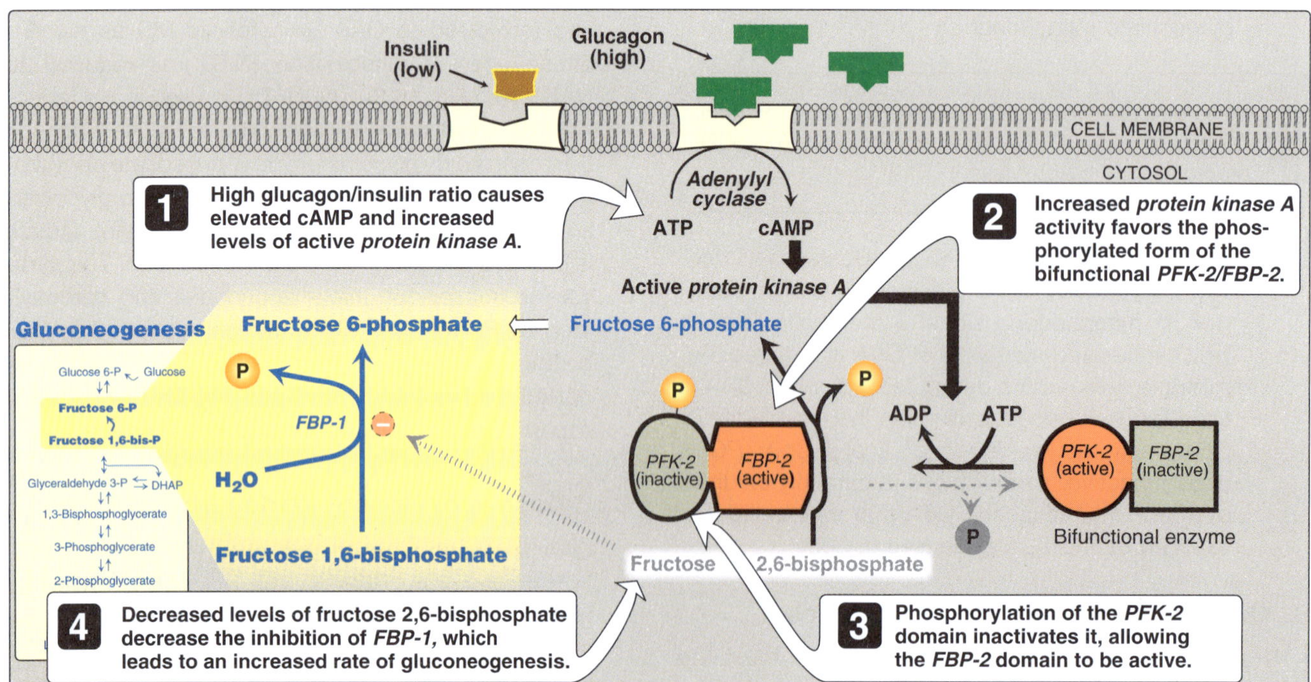

Figure 15.5 Effect of elevated glucagon on the intracellular concentration of fructose 2,6-bisphosphate in the liver. AMP and ADP = adenosine mono- and diphosphates; cAMP = cyclic AMP; *PFK-2* = *phosphofructokinase-2*; *FBP-2* = *fructose 2,6-bisphosphatase*; *FBP-1* = *fructose 1,6-bisphosphatase*; P and P = phosphate.

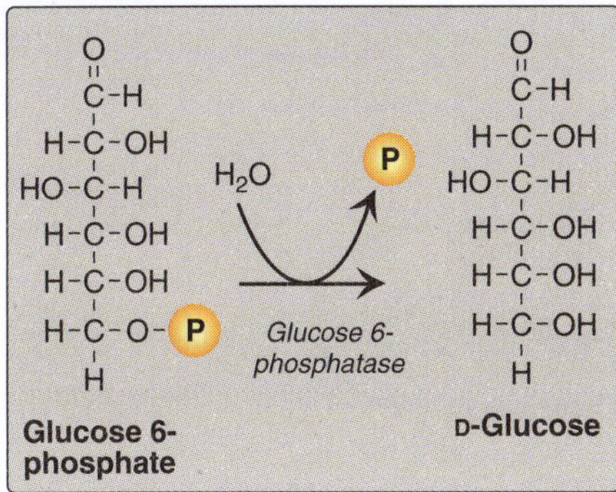

Figure 15.6 Dephosphorylation of glucose 6-phosphate allows release of free glucose from gluconeogenic tissues (primarily the liver) into blood. P = phosphate.

pathway for the formation of free glucose (Fig. 15.6). **The liver is the primary organ that produces free glucose from glucose 6-phosphate.** This process requires a complex of two proteins found only in gluconeogenic tissue: glucose 6-phosphate translocase, which transports glucose 6-phosphate across the endoplasmic reticular (ER) membrane, and *glucose 6-phosphatase*, which removes the phosphate, producing free glucose (see Fig. 15.6). [Note: These ER membrane proteins are also required for the final step of glycogen degradation (see p. 229).] Specific transporters are responsible for moving the free glucose into the cytosol and then into blood.

Glycogen storage diseases Ia and Ib, caused by deficiencies in the *phosphatase* and the translocase, respectively, are characterized by severe fasting hypoglycemia, because free glucose is unable to be produced from either gluconeogenesis or glycogenolysis.

F. Summary of the reactions of glycolysis and gluconeogenesis

Of the 11 reactions required to convert pyruvate to free glucose, 7 are catalyzed by reversible glycolytic enzymes (Fig. 15.7).

The 3 irreversible reactions (catalyzed by *hexokinase/glucokinase*, *PFK-1*, and *PK*) are circumvented by reactions catalyzed by *glucose 6-phosphatase*, *fructose 1,6-bisphosphatase*, *PC, and PEPCK.*

In gluconeogenesis, the equilibria of the reversible glycolytic reactions are pushed toward glucose synthesis as a result of the essentially irreversible formation of PEP, fructose 6-phosphate, and glucose by the gluconeogenic enzymes. The energy from 4 ATP, 2 GTP and 2 NADH is utilized in formation of 1 molecule of glucose.

IV. REGULATION

The moment-to-moment regulation of gluconeogenesis is determined primarily by the circulating level of glucagon

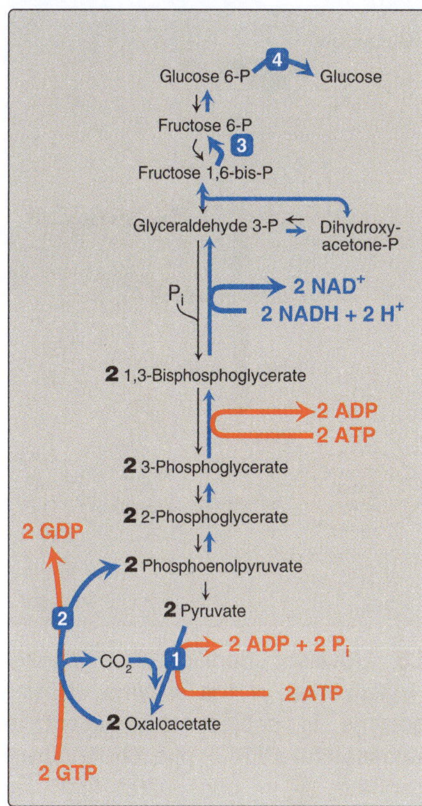

Figure 15.7 Summary of the reactions of glycolysis and gluconeogenesis, showing the energy requirements of gluconeogenesis. The numbered reactions are unique to gluconeogenesis. P = phosphate; GDP and GTP = guanosine di- and triphosphates; NAD(H) = nicotinamide adenine dinucleotide; ADP = adenosine diphosphate.

and by the availability of gluconeogenic substrates. In addition, slow adaptive changes in enzyme amount result from an alteration in the rate of enzyme synthesis or degradation or both. [Note: Hormonal control of the glucoregulatory system is presented in Chapter 27.]

A. Glucagon

This peptide hormone from pancreatic islet α cells (see p. 443) stimulates gluconeogenesis by three mechanisms.

1. **Changes in allosteric effectors:** Glucagon lowers hepatic fructose 2,6-bisphosphate, resulting in *fructose 1,6-bisphosphatase* activation and *PFK-1* inhibition, thereby favoring gluconeogenesis over glycolysis (see Fig. 15.5). [Note: See p. 195 for the role of fructose 2,6-bisphosphate in glycolysis regulation.]

2. **Covalent modification of enzyme activity:** Glucagon binds its G protein–coupled receptor (see p. 189) and, via an elevation in cyclic AMP (cAMP) level and *cAMP-dependent protein kinase A* activity, stimulates the conversion of hepatic *PK* to its inactive (phosphorylated) form. This decreases

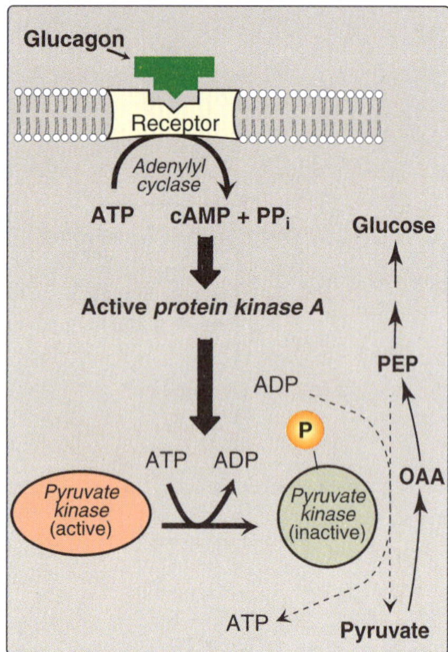

Figure 15.8 Covalent modification of *pyruvate kinase* results in inactivation of the enzyme. [Note: Only the hepatic isozyme is subject to covalent regulation.] OAA = oxaloacetate; PEP = phosphoenolpyruvate; PP_i = pyrophosphate; (P) = phosphate; AMP and ADP = adenosine mono- and diphosphates; cAMP = cyclic AMP.

PEP conversion to pyruvate, which has the effect of diverting PEP to gluconeogenesis (Fig. 15.8).

3. **Induction of enzyme synthesis:** Glucagon increases transcription of the gene for *PEPCK* via the transcription factor cAMP response element–binding (CREB) protein, thereby increasing the availability of this enzyme as levels of its substrate rise during fasting. [Note: Cortisol (a glucocorticoid) also increases expression of the gene, whereas insulin decreases expression.]

B. Substrate availability

The availability of gluconeogenic precursors, particularly glucogenic amino acids, significantly influences the rate of glucose synthesis. Decreased insulin levels favor mobilization of amino acids from muscle protein to provide the carbon skeletons for gluconeogenesis. The ATP and NADH coenzymes required for gluconeogenesis are primarily provided by FA oxidation.

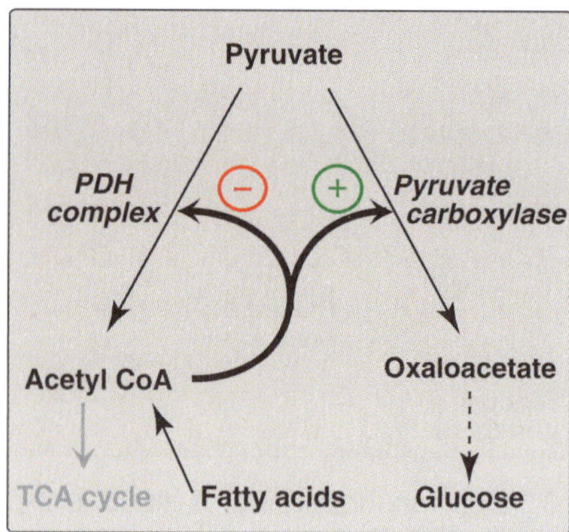

Figure 15.9 Acetyl coenzyme A (CoA) diverts pyruvate away from oxidation and toward gluconeogenesis. *PDH* = *pyruvate dehydrogenase*; TCA = tricarboxylic acid.

C. Allosteric activation by acetyl CoA

Allosteric activation of hepatic *PC* by acetyl CoA occurs during fasting. As a result of increased TAG hydrolysis in adipose tissue, the liver is flooded with FA (see p. 461). The rate of formation of acetyl CoA by β-oxidation of these FA exceeds the capacity of the liver to oxidize it to CO_2 and water. As a result, acetyl CoA accumulates and activates *PC*. [Note: Acetyl CoA inhibits the *PDHC* (by activating *PDH kinase*; see pp. 207–208). Thus, this single compound can divert pyruvate toward gluconeogenesis and away from the TCA cycle (Fig. 15.9).]

D. Allosteric inhibition by AMP

Fructose 1,6-bisphosphatase is inhibited by AMP, a compound that activates *PFK-1*. This results in reciprocal regulation of glycolysis and gluconeogenesis seen previously with fructose 2,6-bisphosphate (see p. 218). [Note: Thus, elevated AMP stimulates energy-producing pathways and inhibits energy-requiring ones.]

V. Medical Perspectives

Gluconeogenesis is activated in fasting or starvation to give a constant supply of glucose to vital organs and cells particularly RBCs, brain, kidney medulla, lens and cornea of eye which need a constant supply of glucose.

Excessive intake of alcohol causes hypoglycemia by elevating cytoplasmic NADH. Ethanol inhibits gluconeogenesis by preventing the oxidation of lactate to pyruvate. Hypoglycemia is associated with blurring of vision, confusion and inability to complete normal tasks.

Very rare inborn genetic disorder leading to deficiency of *pyruvate carboxylase* has been documented. It presents with seizures and neurological symptoms because of hypoglycemia and lactic acidosis within first few hours or days of life. Symptoms can be controlled if diagnosed early.

VI. Chapter Summary

- **Gluconeogenic precursors** include **glycerol**, **lactate**, and **α-keto acids** (e.g., **α-ketoglutarate** and **pyruvate**) (Fig. 15.10).
- Seven of the reactions of glycolysis are reversible and are used for gluconeogenesis in the liver and kidneys.
- Three reactions, catalyzed by ***pyruvate kinase***, ***phosphofructokinase-1***, and *glucokinase/hexokinase*, are **physiologically irreversible** and must be circumvented.
- **Pyruvate** is converted to **oxaloacetate** and then to **phosphoenolpyruvate** (**PEP**) by ***pyruvate carboxylase*** (***PC***) and ***PEP-carboxykinase*** (***PEPCK***).
- **Fructose 1,6-bisphosphate** is converted to **fructose 6-phosphate** by ***fructose 1,6-bisphosphatase***.
- **Glucose 6-phosphate** is dephosphorylated to **glucose** by ***glucose 6-phosphatase***. Its deficiency results in severe, fasting hypoglycemia.

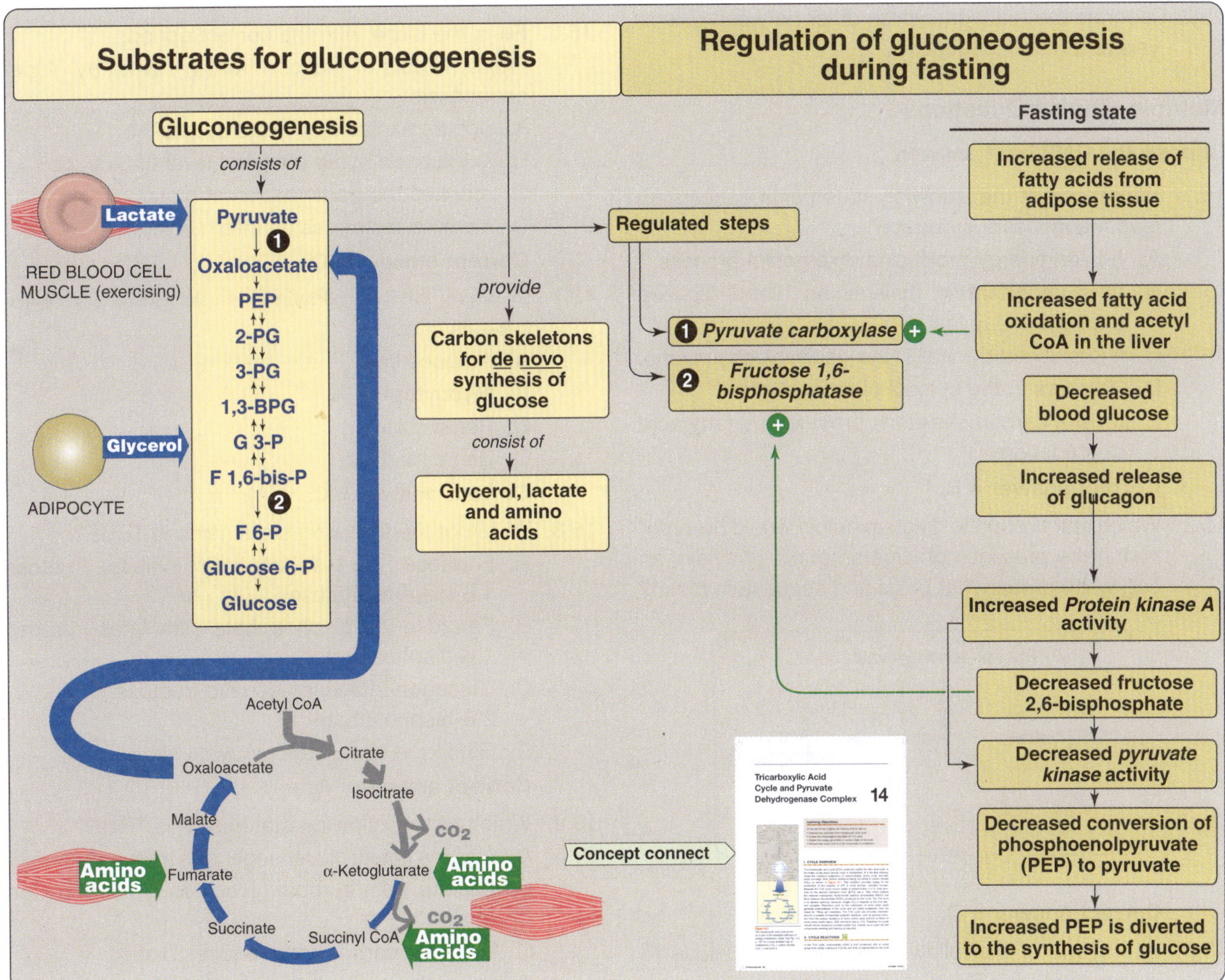

Figure 15.10 Key concept map for gluconeogenesis. TCA = tricarboxylic acid. CoA = coenzyme A; cAMP = cyclic adenosine monophosphate; P = phosphate; (B)PG = (bis)phosphoglycerate; G = glyceraldehyde; F = fructose; CO_2 = carbon dioxide.

Study Questions

Short-Answer Questions

Write short notes on:

15.1. Gluconeogenic precursors
15.2. Regulation of gluconeogenesis
15.3. Why alcoholics are susceptible to hypoglycemia?

Long-Answer Question

15.1. Discuss the process of gluconeogenesis. Add a note on the tissues which are dependent on a constant supply of glucose.
15.2. Describe the reactions which differ between glycolysis and gluconeogenesis.

Multiple-Choice Questions

Choose the ONE best answer.

15.1. Which one of the following statements concerning gluconeogenesis is correct?
A. It is an energy-producing (exergonic) process.
B. It is important in maintaining blood glucose during a 2-day fast.
C. It is inhibited by a fall in the insulin/glucagon ratio.
D. It occurs in the cytosol of muscle cells.
E. It uses carbon skeletons provided by fatty acid degradation.

Correct answer = B.

15.2. Which reaction in the diagram below would be inhibited in the presence of large amounts of avidin, an egg white protein that binds and sequesters biotin?

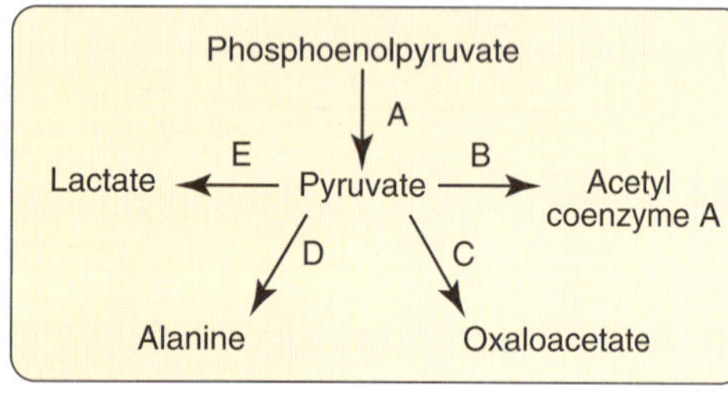

Correct answer = C.

15.3. Which one of the following reactions is unique to gluconeogenesis?
A. 1,3-Bisphosphoglycerate → 3-phosphoglycerate
B. Lactate → pyruvate
C. Oxaloacetate → phosphoenolpyruvate
D. Phosphoenolpyruvate → pyruvate

Correct answer = C.

15.4. Which of the following molecules cannot give rise to glucose via gluconeogenesis?
A. Odd-chain fatty acids
B. Glycerol
C. Lactate
D. Even-chain fatty acids

Correct answer = D.

15.5. Which of the following intracellular locations are involved in the conversion of lactate to glucose by gluconeogenesis?
A. Cytosol and mitochondria
B. Cytosol only
C. Mitochondria only
D. Cytosol, mitochondria and endoplasmic reticulum

Correct answer = D.

15.6. Fill in the blank with the correct option:
The synthesis of glucose from pyruvate by gluconeogenesis ________________.
A. occurs exclusively in the cytosol.
B. is inhibited by an elevated level of glucagon.
C. requires the participation of biotin.
D. involves lactate as an intermediate.

Correct answer = C.

15.7. In which of the following tissues does gluconeogenesis occur?
A. Adipocytes
B. Myocytes
C. Renal cortex
D. Renal medulla

Correct answer = C.

15.8. Which of the following statements is TRUE?
A. Fructose 2,6-bisphosphate inhibits fructose 1,6-bisphosphatase
B. Fructose 2,6-bisphosphate stimulates fructose 1,6-bisphosphatase
C. Glucagon increases hepatic fructose 2,6-bisphosphate
D. Glucagon stimulates pyruvate kinase

Correct answer = A.

15.9. Which of the following statements is TRUE?
A. Leucine enters gluconeogenesis through fumarate
B. Lactate from aerobic glycolysis is used in Cori cycle
C. Pyruvate carboxylase requires biotin
D. Pyruvate carboxylase is allosterically inactivated by acetyl CoA

Correct answer = C.

15.10. Malate shuttle is used in:
A. Glycolysis and gluconeogenesis
B. Glycolysis and glycogenesis
C. Gluconeogenesis and glycogenesis
D. Gluconeogenesis and glycogenolysis

Correct answer = A.

Multiple-Choice Questions (NExT/NEET PG)

Choose the ONE best answer.

15.1. The figure below shows the entry of various amino acid metabolites into the TCA cycle.
Based on the figure above, which of the following amino acids are purely glucogenic?

A. Alanine and serine
B. Leucine and lysine
C. Phenyl alanine and glycine
D. Tryptophan and aspartate

Correct answer = A.

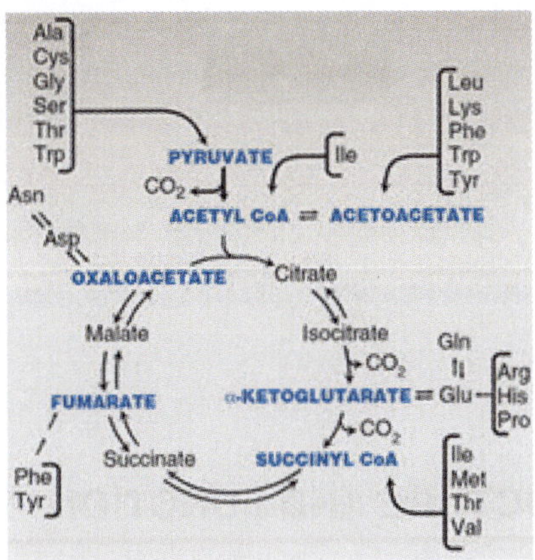

15.2. A 28-year-old gym enthusiast reports persistent fatigue and muscle cramps on waking up from sleep. He also experiences confusion and sweating—signs of hypoglycemia. His diet includes a significant intake of raw egg whites, which contain a substance that binds to and inhibits absorption of a vitamin. Which among the following enzymes is not functioning properly in this individual?

A. Pyruvate kinase
B. Pyruvate carboxylase
C. Pyruvate decarboxylase
D. Pyruvate dehydrogenase complex

Correct answer = B.

15.3. Based on the statements given below, choose the correct option
Statement 1: Patients with a genetic defect in glucose-6-phosphatase experience manifestations of hypoglycemia.
Statement 2: Glucose-6-phosphatase catalyzes the formation of glucose during gluconeogenesis and glycogen breakdown.

A. Both the Statements are correct, and Statement 2 is the correct explanation for Statement 1
B. Both the Statements are correct, but Statement 2 is NOT the correct explanation for Statement 1
C. Statement 1 is correct, but Statement 2 is NOT correct
D. Statement 1 is NOT correct, but Statement 2 is correct

Correct answer = A.

15.4. Why is the production of acetyl CoA in fatty acid oxidation essential for gluconeogenesis?

A. Acetyl CoA inhibits the pyruvate dehydrogenase complex and activates pyruvate carboxylase
B. Acetyl CoA activates the pyruvate dehydrogenase complex and inhibits pyruvate carboxylase
C. Acetyl CoA inhibits the pyruvate dehydrogenase complex and pyruvate carboxylase
D. Acetyl CoA activates the pyruvate dehydrogenase complex and pyruvate carboxylase

Correct answer = A.

15.5. A person has consumed large amount of alcohol. Which of the following will be true in this case?

A. The metabolism of ethanol by alcohol dehydrogenase produces reduced nicotinamide adenine dinucleotide (NADH)
B. The increase in NADH as ethanol is oxidized increases the availability of oxaloacetate (OAA)
C. Larger amount of lactate will be converted to pyruvate
D. Increase in pyruvate and oxaloacetate promotes gluconeogenesis

Correct answer = A.

Glycogen Metabolism and Glycogen Storage Disorders

16

Learning Objectives

At the end of this chapter, the student will be able to:

- Describe the structure and function of glycogen
- Explain the steps of glycogen synthesis and degradation
- Discuss regulation of glycogenesis and glycogenolysis
- Explain glycogen storage diseases

I. OVERVIEW

A constant source of blood glucose is an absolute requirement for human life. Glucose is the greatly preferred energy source for the brain and the required energy source for cells with few or no mitochondria such as mature red blood cells. Glucose is also essential as an energy source for exercising muscle, where it is the substrate for anaerobic glycolysis. Blood glucose can be obtained from three primary sources: the diet, glycogen degradation, and gluconeogenesis. Dietary intake of glucose and glucose precursors, such as starch (a polysaccharide), disaccharides, and monosaccharides, is sporadic and, depending on the diet, is not always a reliable source of blood glucose. In contrast, gluconeogenesis (see p. 215) can provide sustained synthesis of glucose, but it is somewhat slow in responding to a falling blood glucose level. **Therefore, the body has developed mechanisms for storing a supply of glucose in a rapidly mobilized form, namely, glycogen.** In the absence of a dietary source of glucose, this sugar is rapidly released into the blood from liver glycogen. Similarly, muscle glycogen is extensively degraded in exercising muscle to provide that tissue with an important energy source. When glycogen stores are depleted, specific tissues synthesize glucose de novo, using glycerol, lactate, pyruvate, and amino acids as carbon sources for gluconeogenesis (see Chapter 15). Figure 16.1 shows the reactions of glycogen synthesis and degradation as part of the essential pathways of energy metabolism.

II. STRUCTURE AND FUNCTION

The main stores of glycogen are found in **skeletal muscle and liver**, although most other cells store small amounts of glycogen for their own use. **The function of muscle glycogen is to serve as a fuel reserve for the synthesis of ATP during muscle contraction. That of liver glycogen is to maintain the blood glucose concentration**, particularly during the early stages of a fast (Fig. 16.2; also see p. 460). [Note: Liver glycogen can maintain blood glucose for <24 hours.]

A. Amounts in liver and muscle

Approximately 400 g of glycogen make up 1%–2% of the fresh weight of resting muscle, and ~100 g of glycogen make up to 10% of the fresh weight of a well-fed adult liver. What limits the production of glycogen at these levels is not clear. However, in some glycogen storage diseases (GSD) (see Fig. 16.8), the amount of glycogen in the liver and/or muscle can be significantly higher. [Note: In the body, muscle mass is greater than liver mass. Consequently, most of the body's glycogen is found in skeletal muscle.]

B. Structure

Glycogen is a branched-chain polysaccharide made exclusively from α-d-glucose. The primary glycosidic bond is an α(1→4) linkage. After an average of 8–14 glucosyl residues, there is a branch containing an

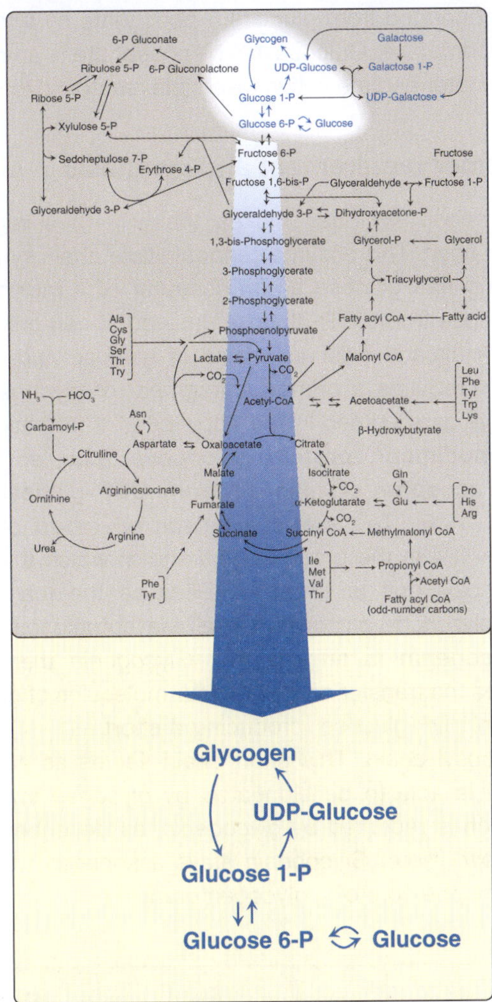

Figure 16.1 Glycogen synthesis and degradation shown as a part of the essential pathways of energy metabolism. [Note: See Fig. 13.2, p. 187, for a more detailed map of metabolism.] P = phosphate; UDP = uridine diphosphate.

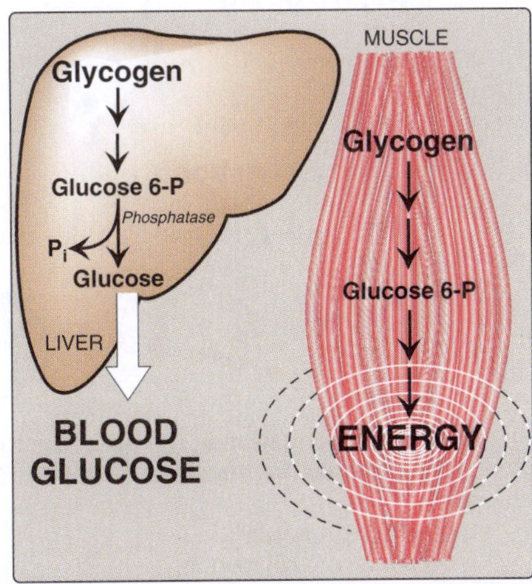

Figure 16.2 Functions of muscle and liver glycogen. [Note: The presence of *glucose 6-phosphatase* in liver allows release of glucose into blood.] P = phosphate; P_i = inorganic phosphate.

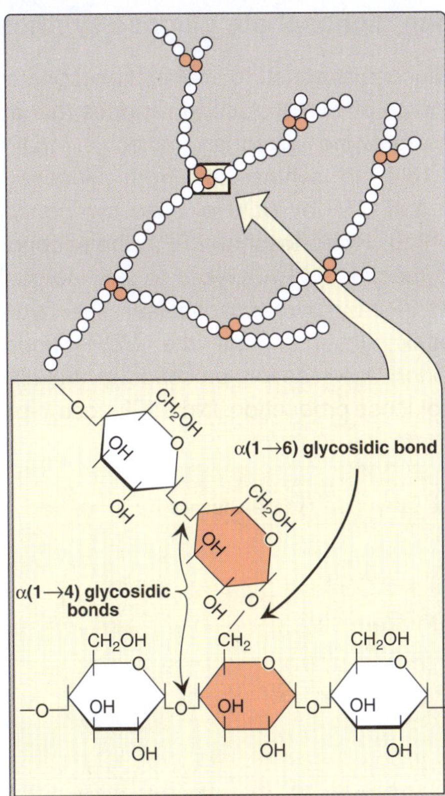

Figure 16.3 Branched structure of glycogen, showing α(1→4) and α(1→6) glycosidic bonds.

α(1→6) linkage (Fig. 16.3). A single glycogen molecule can contain up to 55,000 glucosyl residues. These polymers of glucose exist as large, spherical, cytoplasmic granules (particles) that also contain most of the enzymes necessary for glycogen synthesis and degradation.

C. Glycogen store fluctuation

Liver glycogen stores increase during the well-fed state (see p. 454) and are depleted during a fast (see p. 460). Muscle glycogen is not affected by short periods of fasting (a few days) and is only moderately decreased in prolonged fasting (weeks). Muscle glycogen is synthesized to replenish muscle stores after they have been depleted following strenuous exercise. [Note: Glycogen synthesis and degradation go on continuously. The difference between the rates of these two processes determines the levels of stored glycogen during specific physiologic states.]

III. SYNTHESIS (GLYCOGENESIS)

Glycogen is synthesized from molecules of α-d-glucose. The process occurs in the cytosol and requires energy supplied by ATP (for the phosphorylation of glucose) and uridine triphosphate (UTP).

A. Uridine diphosphate glucose synthesis

α-d-Glucose attached to uridine diphosphate (UDP) is the source of all the glucosyl residues that are added to the growing glycogen molecule. UDP-glucose (Fig. 16.4) is synthesized from glucose 1-phosphate and UTP by *UDP–glucose pyrophosphorylase* (Fig. 16.5). Pyrophosphate (PP_i), the second product of the reaction, is hydrolyzed to two inorganic phosphates (P_i) by *pyrophosphatase*. The hydrolysis is exergonic, insuring that the *UDP–glucose pyrophosphorylase* reaction proceeds in the direction of UDP-glucose production. [Note: Glucose 1-phosphate is generated from glucose 6-phosphate by *phosphoglucomutase*. Glucose 1,6-bisphosphate is an obligatory intermediate in this reversible reaction (Fig. 16.6).]

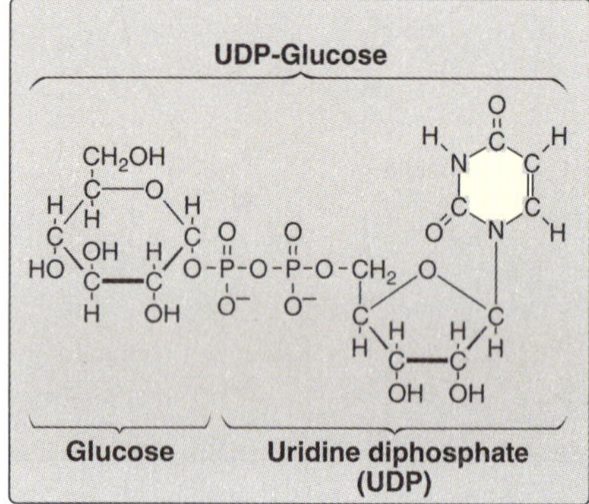

Figure 16.4 The structure of UDP-glucose, a nucleotide sugar.

B. Primer requirement and synthesis

Glycogen synthase makes the α(1→4) linkages in glycogen. This enzyme cannot initiate chain synthesis using free glucose as an acceptor of a molecule of glucose from UDP-glucose. Instead, it can only elongate already existing chains of glucose and, therefore, requires a primer. A fragment of glycogen can serve as a primer. In the absence of a fragment, **the homodimeric protein *glycogenin* can serve as an acceptor of glucose from UDP-glucose** (see Fig. 16.5). The side-chain hydroxyl group of tyrosine-194 in the protein is the site at which the initial glucosyl unit is attached. Because the reaction is catalyzed by *glycogenin* itself via autoglucosylation, ***glycogenin* is an enzyme**. *Glycogenin* then catalyzes the transfer of at least four molecules of glucose from UDP-glucose, producing a short, α(1→4)-linked glucosyl chain. This short chain serves as a primer that is able to be elongated by *glycogen synthase*, which is recruited by *glycogenin*, as described in C. below. [Note: *Glycogenin* stays associated with and forms the core of a glycogen granule.]

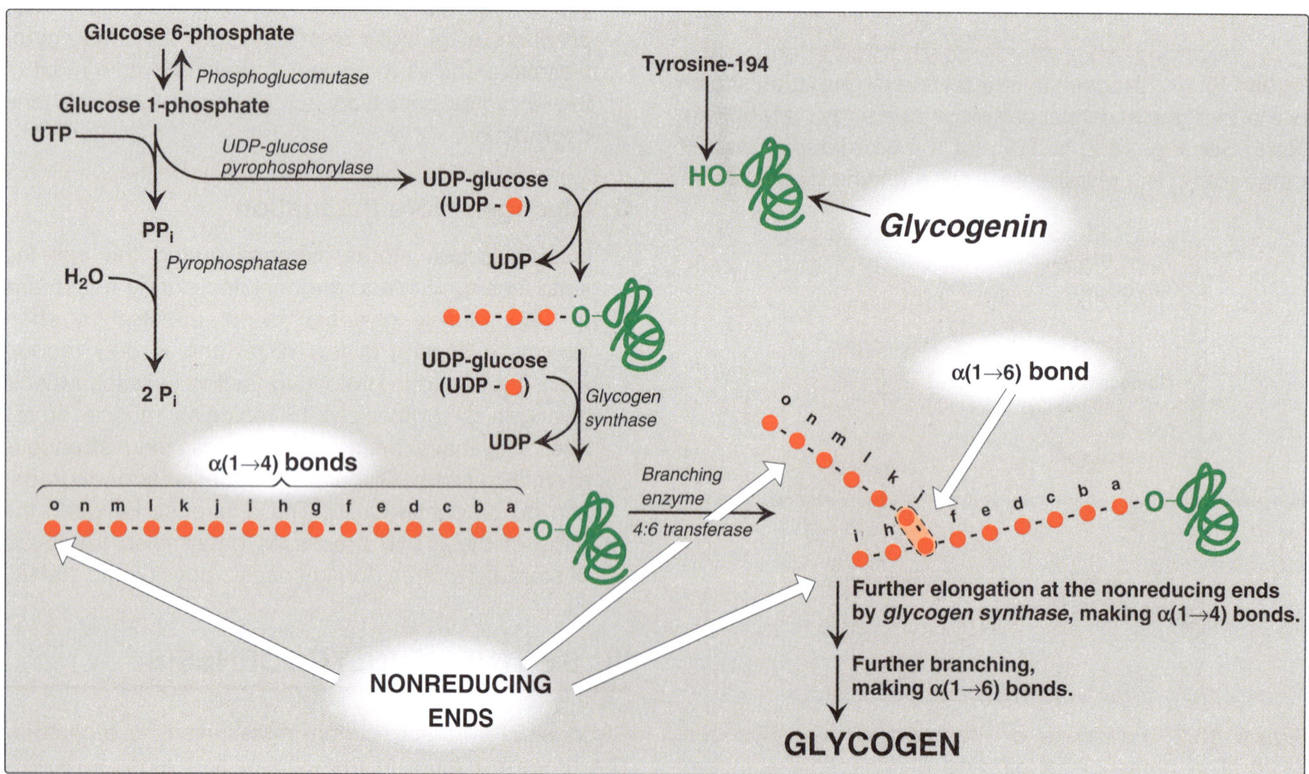

Figure 16.5 Glycogen synthesis. UDP and UTP = uridine di- and triphosphates; PP_i = pyrophosphate; P_i = inorganic phosphate.

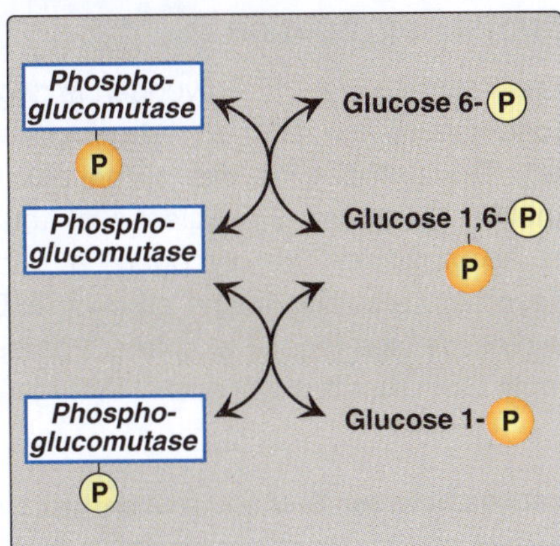

Figure 16.6 Interconversion of glucose 6-phosphate and glucose 1-phosphate by *phosphoglucomutase*. (P) and (P) = phosphate.

C. Elongation by glycogen synthase

Elongation of a glycogen chain involves the **transfer of glucose from UDP-glucose to the nonreducing end of the growing chain, forming a new glycosidic bond** between the anomeric hydroxyl group of carbon 1 of the activated glucose and carbon 4 of the accepting glucosyl residue (see Fig. 16.5). [Note: The nonreducing end of a carbohydrate chain is one in which the anomeric carbon of the terminal sugar is linked by a glycosidic bond to another molecule, making the terminal sugar nonreducing (see p. 178).] The enzyme responsible for making the α(1→4) linkages in glycogen is *glycogen synthase*. [Note: The UDP released when the new α(1→4) glycosidic bond is made can be phosphorylated to UTP by *nucleoside diphosphate kinase* (UDP + ATP ⇄ UTP + ADP; see p. 582).]

D. Branch formation

If no other synthetic enzyme acted on the chain, the resulting structure would be a linear (unbranched) chain of glucosyl residues attached by α(1→4) linkages. Such a compound is found in plant tissues and is called amylose. In contrast, glycogen has branches located, on average, eight glucosyl residues apart, resulting in a highly branched, tree-like structure (see Fig. 16.3) that is far more soluble than the unbranched amylose. Branching also increases the number of nonreducing ends to which new glucosyl residues can be added (and also, as described in IV. below, from which these residues can be removed), thereby greatly accelerating the rate at which glycogen synthesis can occur and dramatically increasing the size of the glycogen molecule.

1. **Branch synthesis:** Branches are made by the action of the branching enzyme, *amylo-α(1→4)→α(1→6)-transglycosylase*. This enzyme removes a set of six to eight glucosyl residues from the nonreducing end of the glycogen chain, breaking an α(1→4) bond to another residue on the chain, and attaches it to a nonterminal glucosyl residue by an α(1→6) linkage, thus functioning as a *4:6 transferase*. The resulting new, nonreducing end (see "i" in Fig. 16.5), as well as the old nonreducing end from which the six to eight residues were removed (see "o" in Fig. 16.5), can now be further elongated by *glycogen synthase*.

   Deficient activity of the glycogen-branching enzyme (***4:6 transferase***), results in accumulation of abnormal glycogen, with longer-than-normal outer chains in the liver, muscle, and/or other tissues. This is known as **Andersen disease** (glycogen storage disease type IV).

2. **Additional branch synthesis:** After elongation of these two ends has been accomplished, their terminal six to eight glucosyl residues can be removed and used to make additional branches.

IV. DEGRADATION (GLYCOGENOLYSIS)

The degradative pathway that mobilizes stored glycogen in liver and skeletal muscle is not a reversal of the synthetic reactions. Instead, a separate set of cytosolic enzymes is required. When glycogen is degraded, the primary product is glucose 1-phosphate, obtained by breaking α(1→4) glycosidic bonds. In addition, free glucose is released from each α(1→6)–linked glucosyl residue (branch point).

A. Chain shortening

Glycogen phosphorylase sequentially cleaves the α(1→4) glycosidic bonds between the glucosyl residues at the nonreducing ends of the glycogen chains by simple phosphorolysis (producing glucose 1-phosphate) until four glucosyl units remain on each chain at a branch point (Fig. 16.7). The resulting structure is called a **limit dextrin**, and *phosphorylase* cannot degrade it any further (Fig. 16.8). *Phosphorylase* requires pyridoxal phosphate (a derivative of vitamin B_6; see p. 531) as a coenzyme.

B. Branch removal

Branches are removed by the two enzymic activities of a single bifunctional protein, the debranching enzyme (see Fig. 16.8). First, *oligo-α(1→4)→α(1→4)-glucantransferase* activity removes the outer three of the four glucosyl residues remaining at a branch. It next transfers them to the nonreducing end of another chain, lengthening it accordingly. Thus, an α(1→4) bond is broken and an α(1→4) bond is made, and the enzyme functions as a *4:4 transferase*. Next,

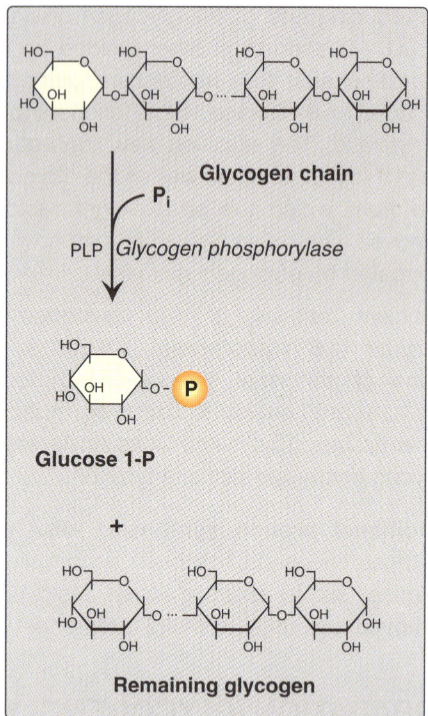

Figure 16.7 Cleavage of an α(1→4)-glycosidic bond. PLP = pyridoxal phosphate; P_i = inorganic phosphate; (P) = phosphate.

the remaining glucose residue attached in an α(1→6) linkage is removed hydrolytically by *amylo-α(1→6)-glucosidase* activity, releasing free (nonphosphorylated) glucose. The glucosyl chain is now available again for degradation by *glycogen phosphorylase* until four glucosyl units in the next branch are reached.

C. Glucose 1-phosphate isomerization to glucose 6-phosphate

Glucose 1-phosphate, produced by *glycogen phosphorylase*, is isomerized in the cytosol to glucose 6-phosphate by *phosphoglucomutase* (see Fig. 16.6). In the liver, glucose 6-phosphate is transported into the endoplasmic reticulum (ER) by glucose 6-phosphate translocase. There, it is dephosphorylated to glucose by *glucose 6-phosphatase* (the same enzyme used in the last step of gluconeogenesis; see p. 218). The glucose is then transported from the ER to the cytosol. Hepatocytes release glycogen-derived glucose into the blood to help maintain blood glucose levels until the gluconeogenic pathway is actively producing glucose. [Note: Muscle lacks *glucose 6-phosphatase*. Consequently, glucose 6-phosphate cannot be dephosphorylated and sent into the blood. Instead, it enters glycolysis, providing energy needed for muscle contraction. As no ATP is required for phosphorylation in first step of glycolysis, 3 ATPs are produced, as compared to 2 ATP/glucose in liver, where free glucose is produced.]

D. Lysosomal degradation

A small amount (1%–3%) of glycogen is degraded by the lysosomal enzyme, *acid α(1→4)-glucosidase* (*acid maltase*). The purpose of this autophagic pathway is unknown. However, a deficiency of this enzyme causes accumulation of glycogen in vacuoles in the lysosomes, resulting in the serious GSD type II: Pompe disease (see Fig. 16.8). [Note: Pompe disease is the only GSD that is a lysosomal storage disease.]

Differences between liver and muscle glycogenolysis

Liver Glycogenolysis	Muscle Glycogenolysis
Glucose is formed from *glucose-6-phosphatase*	*Glucose-6-phosphatase* is absent
Glycogenolysis stimulated by Glucagon	Glycogenolysis stimulated by Adrenaline
Glucose inhibits hepatic *glycogen phosphorylase*	Glycogenolysis stimulated by AMP

Lysosomal storage diseases are genetic disorders characterized by the accumulation of abnormal amounts of carbohydrates or lipids primarily due to their decreased lysosomal degradation resulting from decreased activity or amount of lysosomal *acid hydrolases*.

V. GLYCOGENESIS AND GLYCOGENOLYSIS REGULATION

Because of the importance of maintaining blood glucose levels, the synthesis and degradation of its glycogen storage form are tightly regulated. In the liver, glycogenesis accelerates during periods when the body has been well fed, whereas glycogenolysis accelerates during periods of fasting. In skeletal muscle, glycogenolysis occurs during active exercise, and glycogenesis begins as soon as the muscle is again at rest. Regulation of synthesis and degradation is accomplished on two levels. First, *glycogen synthase* and *glycogen phosphorylase* are hormonally regulated (by covalent phosphorylation/dephosphorylation) to meet the needs of the body as a whole. Second, these same enzymes are allosterically regulated (by effector molecules) to meet the needs of a particular tissue.

A. Covalent activation of glycogenolysis

The binding of hormones, such as glucagon or epinephrine, to plasma membrane G protein–coupled receptors ([GPCR], see p. 189) signals the need for glycogen to be degraded, either to elevate blood glucose levels or to provide energy for exercising muscle.

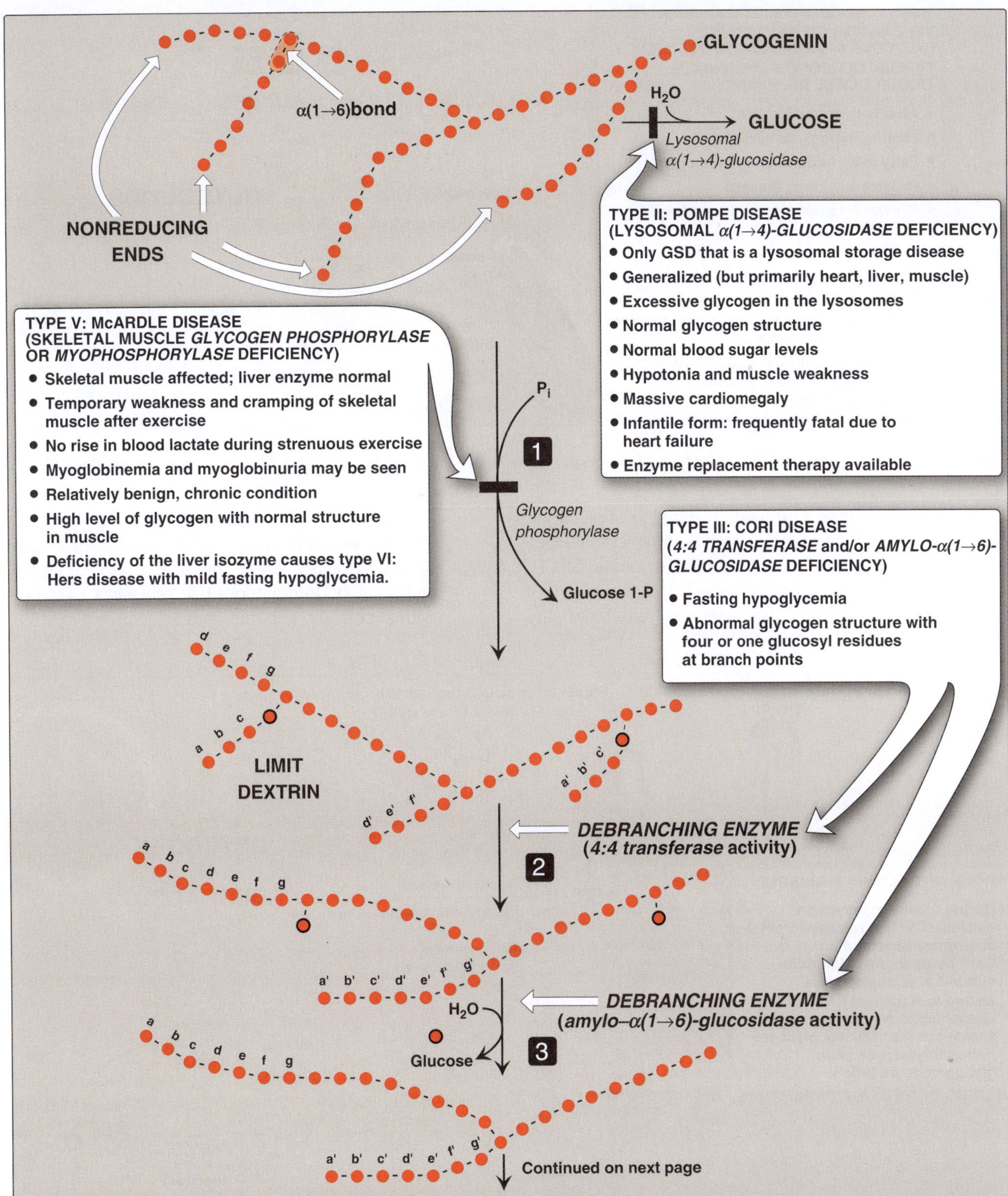

Figure 16.8 (continued on next page) Glycogen degradation, showing some of the glycogen storage diseases (GSD). [Note: GSD type IV: Andersen disease is caused by defects in *branching enzyme*, an enzyme of synthesis, resulting in liver cirrhosis that can be fatal in early childhood.] P_i = inorganic phosphate; P = phosphate.

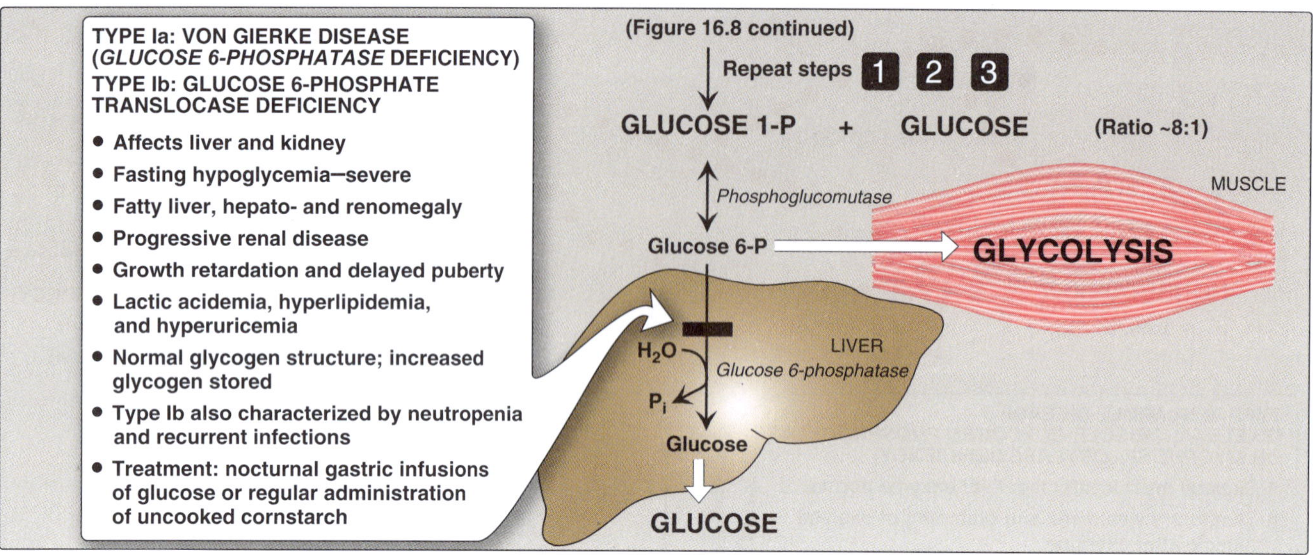

Figure 16.8 (continued from previous page) Glycogen degradation, showing some of the glycogen storage diseases (GSD).

1. **Protein kinase A activation:** Binding of glucagon or epinephrine to their specific hepatocyte GPCR, or of epinephrine to a specific myocyte GPCR, results in the G protein–mediated activation of *adenylyl cyclase*. This enzyme catalyzes the synthesis of cyclic adenosine monophosphate (cAMP), which activates *cAMP-dependent protein kinase A* (*PKA*). cAMP binds the two regulatory subunits of tetrameric *PKA*, releasing two individual catalytic subunits that are active (Fig. 16.9;

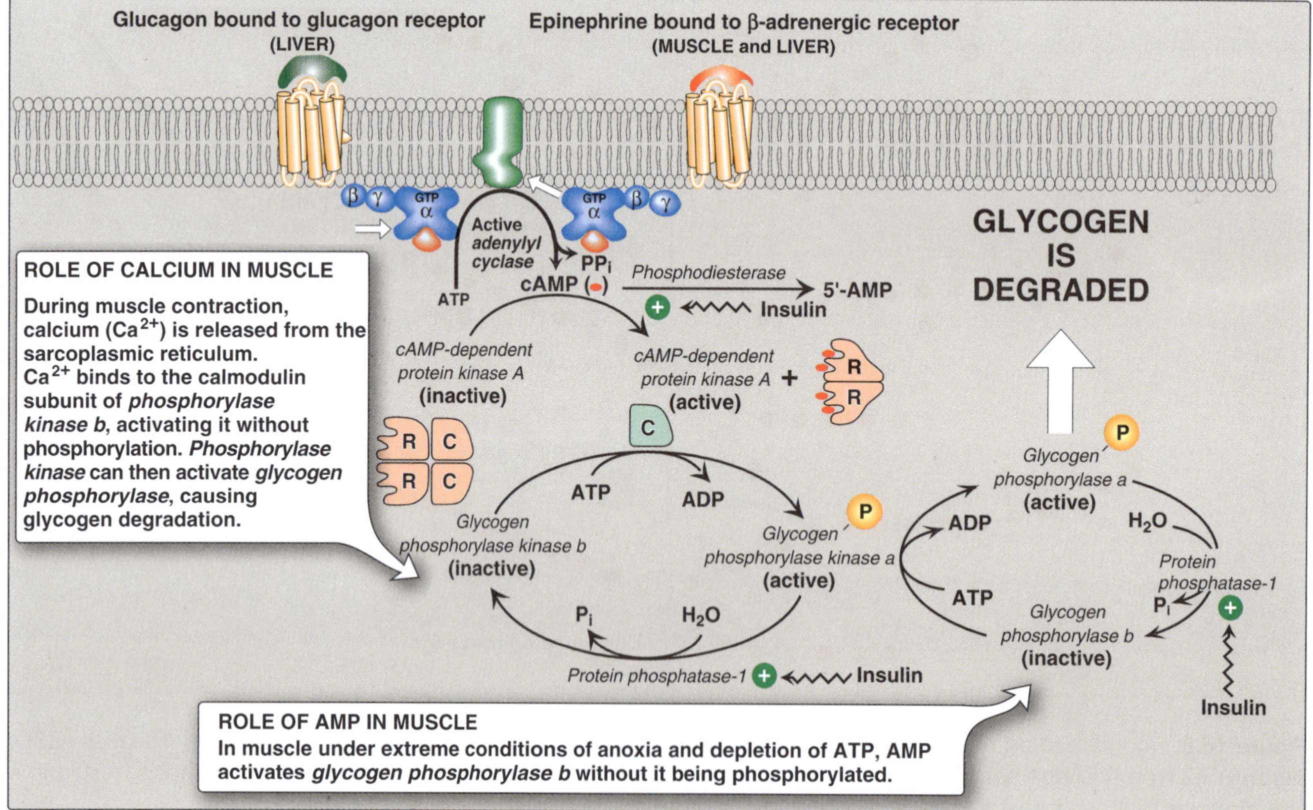

Figure 16.9 Stimulation and inhibition of glycogen degradation. AMP = adenosine monophosphate; cAMP = cyclic AMP; GTP = guanosine triphosphate; (P) = phosphate; PP_i = pyrophosphate; R = regulatory subunit; C = catalytic subunit.

also see p. 190). *PKA* then phosphorylates several enzymes of glycogen metabolism, affecting their activity. [Note: When cAMP is removed, the inactive tetramer reforms.]

2. **Phosphorylase kinase activation:** *Phosphorylase kinase* exists in two forms: an inactive "b" form and an active "a" form. Active *PKA* phosphorylates the inactive "b" form of *phosphorylase kinase*, producing the active "a" form (see Fig. 16.9).

3. **Glycogen phosphorylase activation:** *Glycogen phosphorylase* also exists in a dephosphorylated, inactive "b" form and a phosphorylated, active "a" form. *Phosphorylase kinase a* is the only enzyme that phosphorylates *glycogen phosphorylase b* to its active "a" form, which then begins glycogenolysis (see Fig. 16.9).

4. **Signal amplification:** The cascade of reactions described above activates glycogenolysis. The large number of sequential steps serves to amplify the effect of the hormonal signal (that is, a few hormone molecules binding to their GPCR result in a number of *PKA* molecules being activated that can each activate many *phosphorylase kinase* molecules). This causes the production of many active *glycogen phosphorylase a* molecule that can degrade glycogen.

5. **Phosphorylated state maintenance:** The phosphate groups added to *phosphorylase kinase* and *phosphorylase* in response to cAMP are maintained because the enzyme that hydrolytically removes the phosphate, *protein phosphatase-1* (*PP1*), is inactivated by inhibitor proteins that are also phosphorylated and activated in response to cAMP (see Fig. 16.9). [Note: *PP1* is activated by a signal cascade initiated by insulin (see Fig. 27.7, p. 441). Because insulin also activates the *phosphodiesterase* that degrades cAMP, it opposes the effects of glucagon and epinephrine.]

B. Covalent inhibition of glycogenesis

The regulated enzyme in glycogenesis, *glycogen synthase*, also exists in two forms, the active "a" form and the inactive "b" form. However, in contrast to *phosphorylase kinase* and *phosphorylase*, the active form of *glycogen synthase* is dephosphorylated, whereas the inactive form is phosphorylated at several sites on the enzyme, with the level of inactivation proportional to the degree of phosphorylation (Fig. 16.10). Phosphorylation is catalyzed by several different *protein kinases* in response to cAMP (for example, *PKA* and *phosphorylase kinase*) or other signaling mechanisms (see C. below). *Glycogen synthase b* can be reconverted to the "a" form by *PP1*. Figure 16.11 summarizes the covalent regulation of glycogen metabolism.

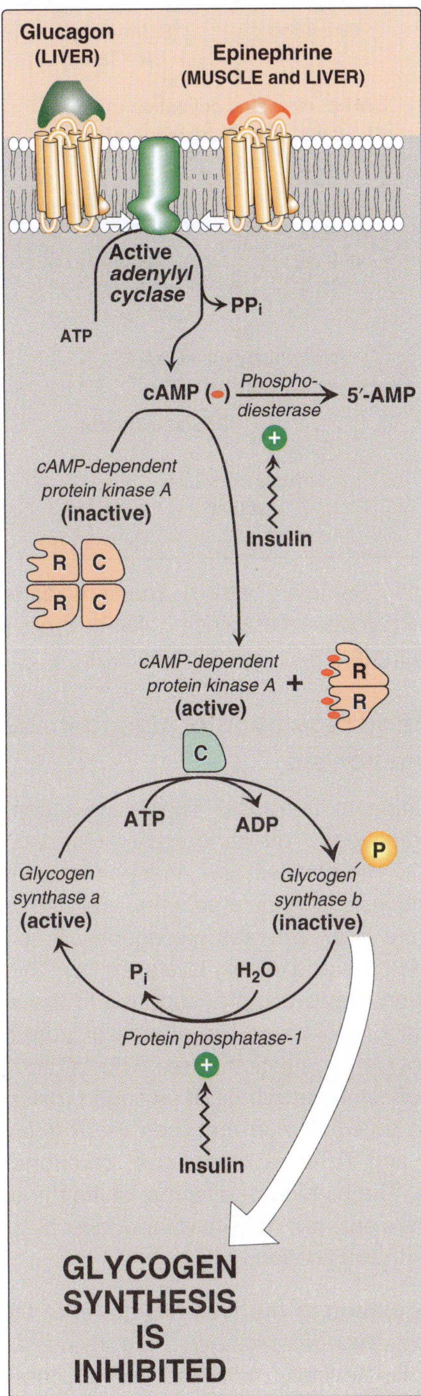

Figure 16.10 Hormonal regulation of glycogen synthesis. [Note: In contrast to *glycogen phosphorylase, glycogen synthase* is inactivated by phosphorylation.] cAMP = cyclic adenosine monophosphate; (P) = phosphate; PP_i = pyrophosphate; R = regulatory subunit; C = catalytic subunit; ADP = adenosine diphosphate.

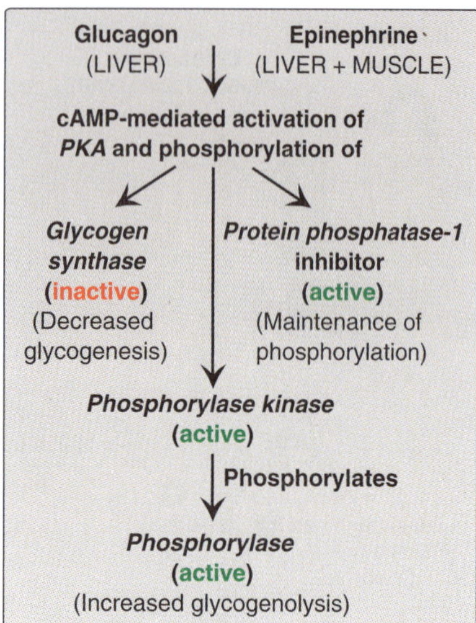

Figure 16.11 Summary of the hormone-mediated covalent regulation of glycogen metabolism. cAMP = cyclic adenosine monophosphate; *PKA = protein kinase A.*

C. Allosteric regulation of glycogenesis and glycogenolysis

In addition to hormonal signals, *glycogen synthase* and *glycogen phosphorylase* respond to the levels of metabolites and energy needs of the cell. Glycogenesis is stimulated when glucose and energy levels are high, whereas glycogenolysis is increased when glucose and energy levels are low. This allosteric regulation allows a rapid response to the needs of a cell and can override the effects of hormone-mediated covalent regulation. [Note: The "a" and "b" forms of the allosteric enzymes of glycogen metabolism are each in an equilibrium between the R (relaxed, more active) and T (tense, less active) conformations (see p. 100). The binding of effectors shifts the equilibrium and alters enzymic activity without directly altering the covalent modification.]

1. **Regulation in the well-fed state:** In the well-fed state, *glycogen synthase b* in both liver and muscle is allosterically activated by glucose 6-phosphate, which is present in elevated concentrations (Fig. 16.12). In contrast, *glycogen phosphorylase a* is allosterically inhibited by glucose 6-phosphate, as well as by ATP, a high-energy signal. [Note: In liver, but not muscle, free glucose is also an allosteric inhibitor of *glycogen phosphorylase a*.]

2. **Glycogenolysis activation by AMP:** Muscle *glycogen phosphorylase* (*myophosphorylase*), but not the liver isozyme, is active in the presence of the high AMP concentrations that occur under extreme conditions of anoxia and ATP depletion. AMP binds to *glycogen phosphorylase b*, causing its activation without phosphorylation (see Fig. 16.9). [Note: Recall that AMP also activates *phosphofructokinase-1* of glycolysis (see p. 195), allowing glucose from glycogenolysis to be oxidized.]

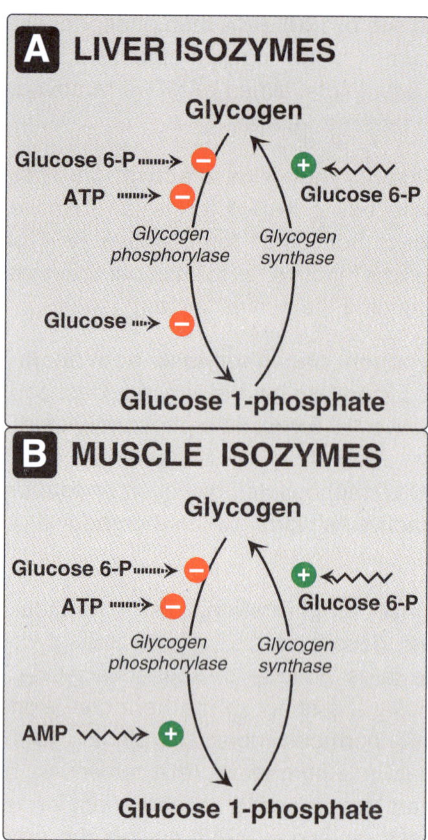

Figure 16.12 Allosteric regulation of glycogenesis and glycogenolysis in liver (**A**) and muscle (**B**). P = phosphate; AMP = adenosine monophosphate.

3. **Glycogenolysis activation by calcium:** Calcium (Ca^{2+}) is released into the sarcoplasm in muscle cells (myocytes) in response to neural stimulation and in the liver in response to epinephrine binding to α_1-adrenergic receptors. The Ca^{2+} binds to calmodulin (CaM), the most widely distributed member of a family of small, Ca^{2+}-binding proteins. The binding of four molecules of Ca^{2+} to CaM triggers a conformational change such that the activated Ca^{2+}–CaM complex binds to and activates protein molecules, often enzymes, that are inactive in the absence of this complex (Fig. 16.13). Thus, CaM functions as an essential subunit of many complex proteins. One such protein is the tetrameric *phosphorylase kinase*, whose "b" form is activated by the binding of Ca^{2+} to its δ subunit (CaM) without the need for the *kinase* to be phosphorylated by *PKA*. [Note: Epinephrine at β-adrenergic receptors signals through a rise in cAMP, not Ca^{2+} (see p. 230).]

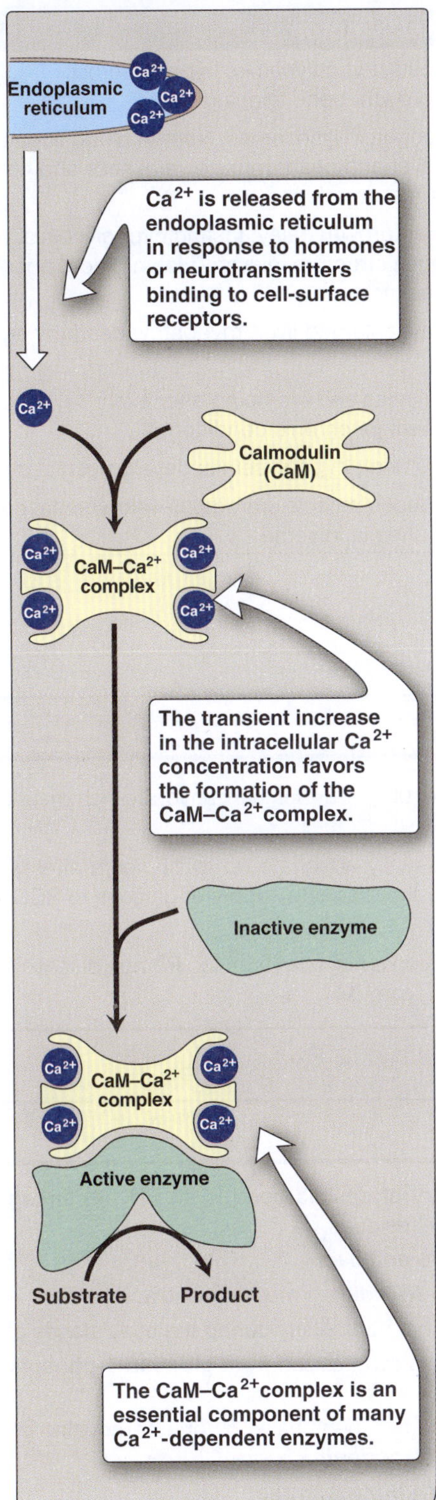

Figure 16.13 Calmodulin mediates many effects of intracellular calcium (Ca^{2+}). [Note: Ca^{2+} activates *phosphorylase kinase* in liver and muscles.]

a. **Muscle phosphorylase kinase activation:** During muscle contraction, there is a rapid and urgent need for ATP. It is supplied by the degradation of muscle glycogen to glucose 6-phosphate, which enters glycolysis. Nerve impulses cause membrane depolarization, which promotes Ca^{2+} release from the sarcoplasmic reticulum into the sarcoplasm of myocytes. The Ca^{2+} binds the CaM subunit, and the complex activates muscle *phosphorylase kinase b* (see Fig. 16.9).

b. **Liver phosphorylase kinase activation:** During physiologic stress, epinephrine is released from the adrenal medulla and signals the need for blood glucose. This glucose initially comes from hepatic glycogenolysis. Binding of epinephrine to hepatocyte α_1-adrenergic GPCR activates a phospholipid-dependent cascade (see p. 325) that results in movement of Ca^{2+} from the ER into the cytoplasm. A Ca^{2+}–CaM complex forms and activates hepatic *phosphorylase kinase b*. [Note: The released Ca^{2+} also helps to activate *protein kinase C* that can phosphorylate (therefore, inactivate) *glycogen synthase a*]

VI. GLYCOGEN STORAGE DISEASES

GSD are a group of genetic diseases caused by defects in enzymes required for glycogen degradation or, more rarely, glycogen synthesis. They result either in formation of glycogen that has an abnormal structure or in the accumulation of excessive amounts of normal glycogen in specific tissues as a result of impaired degradation. A particular enzyme may be defective in a single tissue, such as the liver (resulting in hypoglycemia) or muscle (causing muscle weakness), or the defect may be more generalized, affecting a variety of tissues, such as the heart and kidneys. Severity ranges from fatal in early childhood to mild disorders that are not life threatening. Some of the more prevalent GSD are illustrated in Figure 16.8 and tabulated in Figure 16.14.

Type[a]	Deficient Enzyme	Main Signs/Symptoms
I – Von Gierke disease	Glucose-6-phosphatase	Lactic acidosis, hypoglycemia, hyperuricemia, Impaired growth, bone thinning
II – Pompe disease[a]	Acid α-glucosidase (acid maltase)	Excess glycogen in lysosomes. Normal blood sugar. Enlarged liver and heart; muscle weakness and heart problems in severe forms
III – Cori disease[a]	Glycogen debranching enzyme (4:4 transferase)	Enlarged liver, growth delay, fasting hypoglycemia, abnormal glycogen structure, elevated fat in blood, possible muscle weakness
IV – Andersen disease	Glycogen branching enzyme (4:6 transferase)	Growth delay, enlarged liver, myopathy; death by age 5 usually
V – McArdle disease[a]	Muscle glycogen phosphorylase (myophosphorylase)	Muscle weakness and cramping after exercise; usually a relatively benign, chronic condition
VI – Hers disease	Liver glycogen phosphorylase	Liver enlargement; hypoglycemia; developmental delay
VII – Tarui disease	Muscle phosphofructokinase	Exercise-induced muscle cramps, developmental delay, hemolytic anemia in some

[a]This describes 7 of the 15 types of GSDs. See also Figure 16.8.

Figure 16.14 Descriptions of glycogen storage diseases.

VII. Medical Perspectives

During overeating, the liver cannot store glycogen beyond approximately 100 g. Further, carbohydrate-rich meal will lead to deposition of fat via de novo synthesis.

Fasting releases glucose by breakdown of glycogen from liver. This is usually adequate up to 17 hours after which fat and protein breakdown starts as well as gluconeogenesis happens to ensure adequate supply of glucose to RBCs, brain, etc. (please see chapter 28 for details).

Glycogen storage diseases are explained in Figure 16.8 and include Von Gierke disease (Type Ia), Pompe disease (Type II), Cori disease (Type III), Andersen disease (Type IV) and McArdle Disease (Type V).

VIII. Chapter Summary

- Glycogen is a highly **branched polymer** of **α-d-glucose**. The primary glycosidic bond is an **α(1→4) linkage, with** branch points containing an **α(1→6) linkage a**fter about 8–14 glucosyl residues,
- The main stores of **glycogen** in the body are found in **skeletal muscle and liver**
- Skeletal muscle glycogen serves as a fuel reserve for the synthesis of **ATP** during muscle **contraction**
- The **liver glycogen is** used to maintain the **blood glucose** concentration, particularly during the early stages of a **fast**.
- **Uridine diphosphate** (**UDP**)**-glucose**, the building block of glycogen, is synthesized from **glucose 1-phosphate** and **UTP** by ***UDP–glucose pyrophosphorylase*** (Fig. 16.15).
- **Glucose** from UDP-glucose is transferred to the **nonreducing ends** of glycogen chains by primer-requiring ***glycogen synthase***, which makes the α(1→4) linkages. The **primer** is made by ***glycogenin***.
- Branches are formed by ***amylo-α(1→4)→α(1→6)-transglycosylase*** (a ***4:6 transferase***).
- Pyridoxal phosphate–requiring ***glycogen phosphorylase*** cleaves the α(1→4) bonds between glucosyl residues at the nonreducing ends of the glycogen chains, producing **glucose 1-phosphate**.
- The branch point is removed by the bifunctional **debranching** enzyme, having ***α(1→4)→α(1→4)-glucantransferase*** and ***α(1→6) glucosidase*** activity.
- Glucose 1-phosphate is converted to **glucose 6-phosphate** by ***phosphoglucomutase***.
- In **muscle**, glucose 6-phosphate enters glycolysis.
- In **liver**, the phosphate is removed by ***glucose 6-phosphatase***, releasing free glucose that can be used to maintain blood glucose levels at the beginning of a fast.

- A deficiency of the *phosphatase* causes **glycogen storage disease Ia (von Gierke disease**) and results in fasting hypoglycemia.
- Glycogen synthesis and degradation are **reciprocally regulated** to meet whole-body needs by the same hormonal signals (namely, an **elevated insulin** level results in overall **increased glycogenesis** and **decreased glycogenolysis**, whereas an **elevated glucagon**, or **epinephrine**, level causes the opposite effects).
- In the well-fed state, *glycogen synthase* is activated by glucose 6-phosphate, but *glycogen phosphorylase* is inhibited by glucose 6-phosphate as well as by ATP.
- In the liver, free glucose also serves as an allosteric inhibitor of *glycogen phosphorylase*.
- The rise in **calcium** in muscle during exercise and in liver in response to epinephrine activates *phosphorylase kinase* by binding to the enzyme's **calmodulin** subunit. This allows the enzyme to activate *glycogen phosphorylase*, thereby causing glycogen degradation. **AMP** activates *glycogen phosphorylase* (*myophosphorylase*) in muscle.

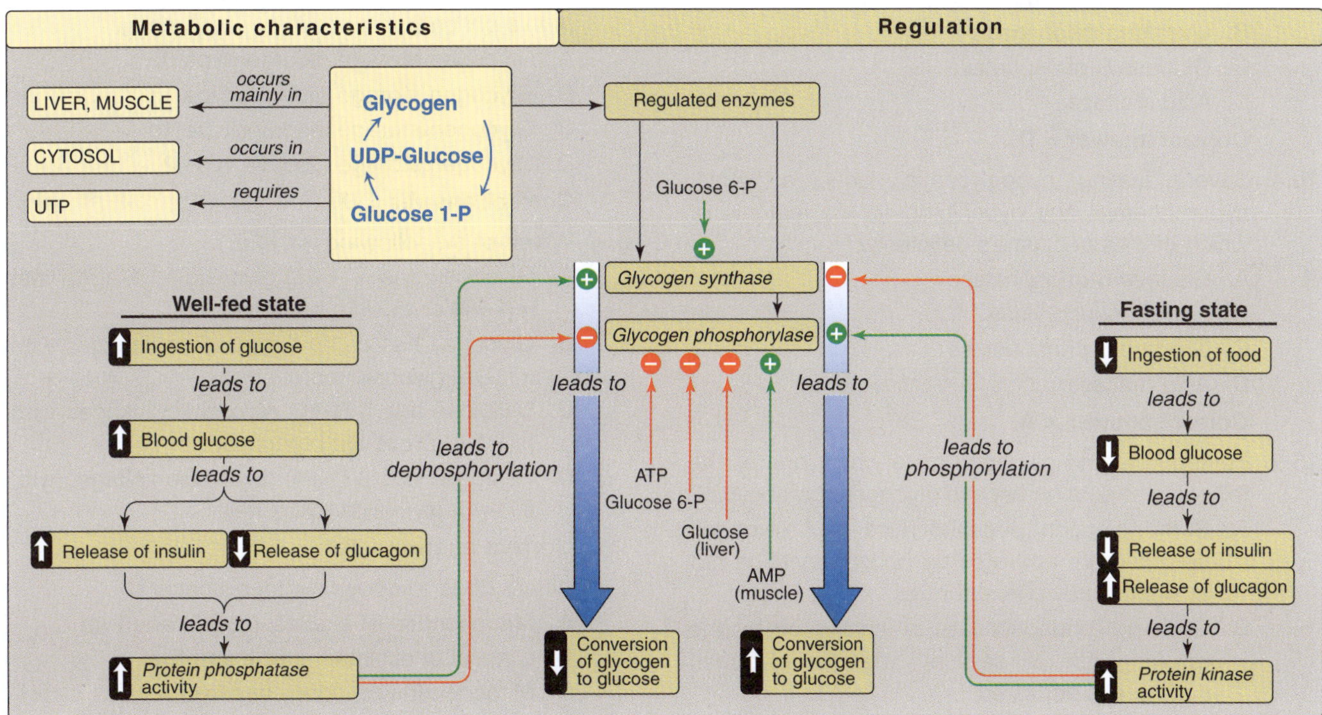

Figure 16.15 Key concept map for glycogen metabolism in the liver. [Note: *Glycogen phosphorylase* is phosphorylated by *phosphorylase kinase*, the "b" form of which can be activated by calcium.] UDP and UTP = uridine di- and triphosphates; P = phosphate; AMP = adenosine monophosphate.

Study Questions

Short-Answer Questions

Write short notes on:

16.1. Von Gierke disease

16.2. Structure and function of glycogen

16.3. Calmodulin-calcium complex in regulation of phosphorylase kinase

16.4. Glycogen in starvation and well-fed state

Long-Answer Questions

16.1. Discuss the process of glycogenolysis and glycogen synthesis. Add a note on their reciprocal regulation.

16.2. Explain how muscle and liver differ in their storage of glycogen and the release of glucose from these stores. Discuss how this is relevant to their functions.

Multiple-Choice Questions

Choose the ONE best answer.

16.1. A patient complains of exercise intolerance. However there is no rise in blood lactate during exercise. These features suggest deficiency of which of these enzymes?

A. Myophosphorylase

B. Liver phosphorylase

C. Glucose 6-phosphatase

D. Acid maltase

Correct answer = A.

16.2. Fatal, progressive cirrhosis and glycogen with longer-than-normal outer chains are features of which of the following?
A. Type I Glycogen storage disease
B. Type II Glycogen storage disease
C. Type III Glycogen storage disease
D. Type IV Glycogen storage disease
Correct answer = D.

16.3. Generalized accumulation of glycogen, severe hypotonia, and death from heart failure is seen in deficiency of which of these enzymes?
A. Myophosphorylase
B. Liver phosphorylase
C. Glucose 6-phosphatase
D. Acid maltase
Correct answer = D.

16.4. Severe fasting hypoglycemia, lactic acidemia, hyperuricemia, and hyperlipidemia are features of which of these enzyme's deficiency?
A. Glucose 6-phosphatase
B. Liver phosphorylase
C. Glucose 6-phosphatase
D. Acid maltase
Correct answer = A.

16.5. Epinephrine and glucagon have which one of the following effects on hepatic glycogen metabolism?
A. Both glycogen phosphorylase and glycogen synthase are activated by phosphorylation but at significantly different rates.
B. Glycogen phosphorylase is inactivated by the resulting rise in calcium, whereas glycogen synthase is activated.
C. Glycogen phosphorylase is phosphorylated and active, whereas glycogen synthase is phosphorylated and inactive.
D. The net synthesis of glycogen is increased.
Correct answer = C.

16.6. In contracting skeletal muscle, a sudden elevation of the sarcoplasmic calcium concentration will result in:
A. activation of cyclic adenosine monophosphate (cAMP)-dependent protein kinase A.
B. conversion of cAMP to AMP by phosphodiesterase.
C. direct activation of glycogen synthase b.
D. direct activation of phosphorylase kinase b.
Correct answer = D.

16.7. Which among the following proteins catalyze the initial steps of glycogen synthesis when glucose molecules are added directly to glycogenin from UDP-glucose?
A. Glycogen synthase
B. Glycogen phosphorylase
C. Branching enzyme
D. Glycogenin
Correct answer = D.

16.8. After a normal term delivery and unremarkable nursery stay, a 2-month-old female infant was admitted to a hospital because of decreased activity and feeding and breathing difficulties. On examination, the child was underweight and showed signs of respiratory distress. Liver was enlarged and muscle tone was reduced. Estimation of acid alpha-glucosidase in the leucocytes revealed a decreased activity. What is the most likely disorder in this patient?
A. Glycogen storage disorder type 0
B. Glycogen storage disorder type 1a
C. Glycogen storage disorder type 1b
D. Glycogen storage disorder type 2
Correct answer = D.

16.9. Which of the following is TRUE?
A. Glycogen has α (1→6) glycosidic linkages with α (1→4) glycosidic branches.
B. Glycogen has α (1→4) glycosidic linkages with α (1→6) glycosidic branches.
C. Cellulose has β (1→6) glycosidic linkages with β (1→4) glycosidic branches.
D. Glycogen has β (1→4) glycosidic linkages with β (1→6) glycosidic branches.
Correct answer = B.

16.10. Where does glycogen synthesis occur?
A. Mitochondria of skeletal muscle and liver
B. Cytosol of skeletal muscle and liver
C. Mitochondria of kidney and liver
D. Cytosol of kidney and liver
Correct answer = B.

Multiple-Choice Questions (NExT/NEET* PG)

***(Proposed National Exit Test for MBBS Final professional Completed Students in India)**

16.1. Based on the statements given below, choose the correct option.
Statement 1: Muscle glycogenolysis does not produce any free glucose.
Statement 2: Glucose-6-phosphatase is absent in muscles.
A. Both the Statements are correct, and Statement 2 is the correct explanation for Statement 1
B. Both the Statements are correct, but Statement 2 is NOT the correct explanation for Statement 1
C. Statement 1 is correct, but Statement 2 is NOT correct

D. Statement 1 is NOT correct, but Statement 2 is correct

Correct answer = D.

16.2. A 17-year-old male presented to the OPD with complaints of painful muscle cramps and weakness after performing strenuous exercise. He was able to resume exercising if a brief rest was taken at the appearance of premature fatigue during exercise. He indicated that mild to moderate exercise was not associated with any symptoms. Blood examination revealed elevated creatine kinase (CK); plasma glucose was within the reference limits. A deficiency in which of the following enzymes is most likely the cause of the patient's muscle cramps?

A. Glucose-6-phosphatase
B. Liver phosphorylase
C. Muscle phosphorylase
D. Debranching enzyme

Correct answer = C.

16.3. Which of the following statements is correct?

(I) Epinephrine promotes glycogenolysis in both skeletal muscles and liver.
(II) Skeletal muscles when exercising depend on muscle glycogen stores and eventually liver glycogen stores for energy.
(III) Evolutionarily, epinephrine release and the subsequent activation of glycogenolysis helps in the 'flight' response.

A. I and II
B. I, II, and III
C. I and III
D. II and III

Correct answer = B.

16.4. Patients with a certain form of glycogen storage disease have a deficiency of phosphorylase kinase. However, their muscle glycogen stores can be depleted if they perform strenuous exercise. Effect of which of the following mechanisms (and the corresponding mediator) is responsible for the utilization of muscle glycogen stores in these patients?

A. Glucagon-cAMP
B. Epinephrine-cAMP
C. Calcium-calmodulin
D. Allosteric activation-AMP

Correct answer = D.

16.5. A 4-month-old infant was brought by his parents with the complaints of multiple episodes of tremor and seizure. On examination, the child was found to have a doll like face. Abdomen was enlarged. Investigations revealed low plasma glucose and elevated serum triglycerides, low density lipoprotein (LDL), lactate, and urate. Ultrasound scan of the abdomen revealed enlarged liver and kidneys. Which of the following enzymes is most likely deficient in this patient?

A. Glucose-6-phosphatase
B. Liver phosphorylase
C. Muscle phosphorylase
D. Debranching enzyme

Correct answer = A.

Monosaccharide and Disaccharide Metabolism

17

Learning Objectives

At the end of this chapter, the student will be able to:

- Describe the metabolism of fructose and its associated diseases
- Explain the sorbitol pathway and its clinical importance
- Describe the metabolism of galactose and its associated diseases

I. OVERVIEW

Glucose is the most common monosaccharide consumed by humans, and its metabolism has already been discussed. Two other monosaccharides, fructose and galactose, also occur in significant amounts in the diet (primarily in disaccharides) and make important contributions to energy metabolism. In addition, galactose is an important component of glycosylated proteins. Figure 17.1 shows the metabolism of fructose and galactose as part of the essential pathways of energy metabolism.

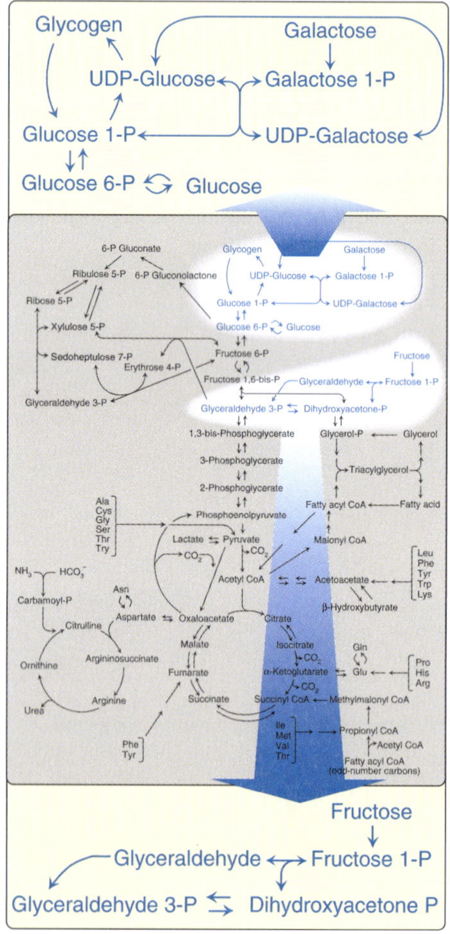

Figure 17.1 Galactose and fructose metabolism as part of the essential pathways of energy metabolism. [Note: See Fig. 13.2, p. 187, for a more detailed map of metabolism.] UDP = uridine diphosphate; P = phosphate.

II. FRUCTOSE METABOLISM

About 10% of the calories in the Western diet are supplied by fructose (~55 g/day). **The major source of fructose is the disaccharide sucrose**, which, when cleaved in the intestine, releases equimolar amounts of fructose and glucose. **Fructose is also found as a free monosaccharide in many fruits, in honey, and in high-fructose corn syrup** (typically, 55% fructose and 45% glucose), which is used to sweeten soft drinks and many foods (see p. 510). **Fructose transport into cells is not insulin dependent** (unlike that of glucose into certain tissues; see p. 191), and, in contrast to glucose, fructose does not promote the secretion of insulin. Fructose is seen in large quantities in seminal plasma, where it provides energy for mobility of spermatozoa.

A. Phosphorylation

For fructose to enter the pathways of intermediary metabolism, it must first be phosphorylated (Fig. 17.2). This can be accomplished by either *hexokinase* or

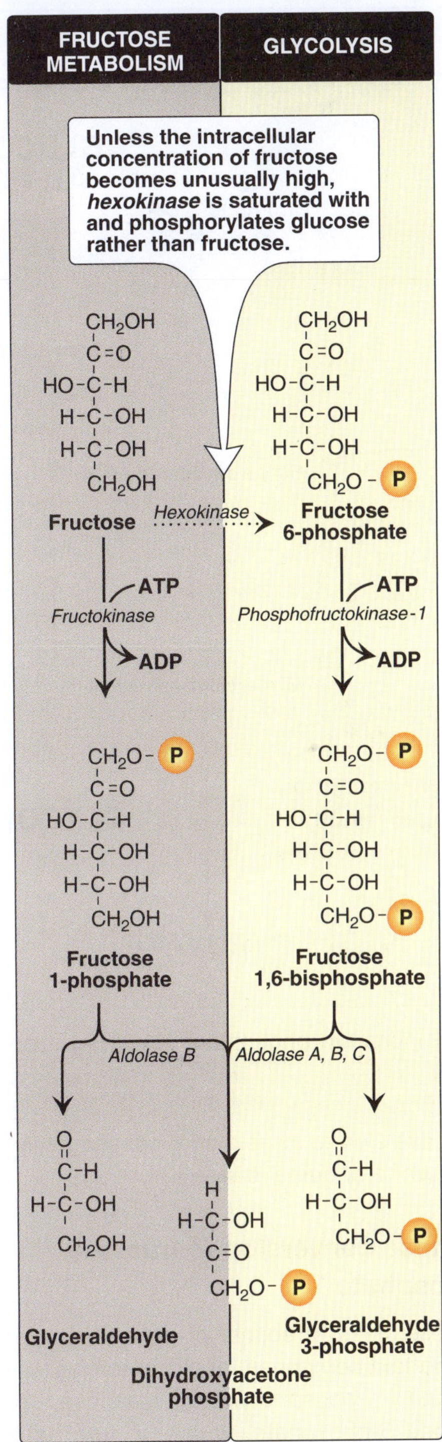

Figure 17.2 Fructose phosphorylation products and their cleavage. P = phosphate; ADP = adenosine diphosphate.

fructokinase. *Hexokinase* phosphorylates glucose in most cells of the body (see p. 193), and several additional hexoses can serve as substrates for this enzyme. However, it has a low affinity (that is, a high Michaelis constant [K_m]; see p. 135) for fructose. Therefore, unless the intracellular concentration of fructose becomes unusually high, the normal presence of saturating concentrations of glucose means that little fructose is phosphorylated by *hexokinase*. ***Fructokinase* provides the primary mechanism for fructose phosphorylation** (see Fig. 17.2). The enzyme has a low K_m for fructose and a high V_{max} ([maximal velocity], see p. 134). It is found in the liver (which processes most of the dietary fructose), kidneys, and the small intestine and converts fructose to fructose 1-phosphate, using ATP as the phosphate donor. [Note: These three tissues also contain *aldolase B*, discussed in section B.] *Fructokinase* phosphorylates at first position, while *hexokinase* phosphorylates at sixth position.

B. Fructose 1-phosphate cleavage

Fructose 1-phosphate is not phosphorylated to fructose 1,6-bisphosphate as is fructose 6-phosphate (see p. 195) but is cleaved by *aldolase B* (also called *fructose 1-phosphate aldolase*) to two trioses, dihydroxyacetone phosphate (DHAP) and glyceraldehyde. DHAP can be used in glycolysis or gluconeogenesis, whereas glyceraldehyde can be metabolized by a number of pathways, as illustrated in Figure 17.3.

Humans express three *aldolase* isozymes (the products of three different genes): *aldolase A* in most tissues; *aldolase B* in the liver, kidneys, and small intestine; and *aldolase C* in the brain. All cleave fructose 1,6-bisphosphate produced during glycolysis to DHAP and glyceraldehyde 3-phosphate (see p. 196), but only *aldolase B* cleaves fructose 1-phosphate.

C. Kinetics

The **rate of fructose metabolism is more rapid than that of glucose** because triose production from fructose 1-phosphate bypasses *phosphofructokinase-1*, the major rate-limiting step in glycolysis (see p. 195).

D. Disorders

A deficiency of one of the key enzymes required for the entry of fructose into metabolic pathways can result in either a benign condition as a result of ***fructokinase* deficiency (essential fructosuria)** or a severe disturbance of liver and kidney metabolism as a result of ***aldolase B* deficiency (hereditary fructose intolerance [HFI])**, which occurs in ~1:20,000 live births (see Fig. 17.3).

1. **Hereditary fructose intolerance:** It is an autosomal recessive disorder caused due to deficiency of enzyme ***aldolase B***. The first symptoms of HFI appear when a baby is weaned from lactose-containing milk and begins ingesting food containing sucrose or fructose. Fructose

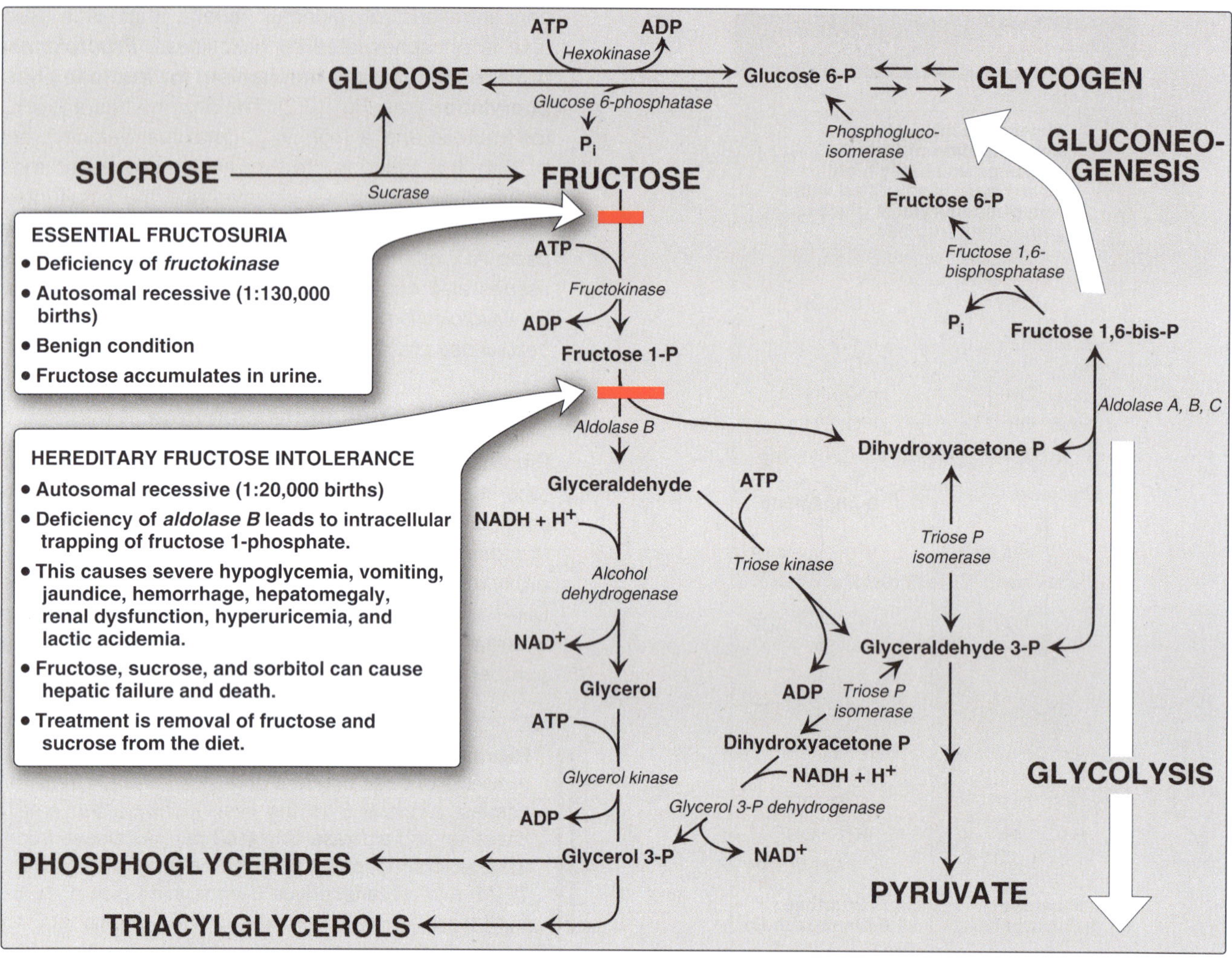

Figure 17.3 Summary of fructose metabolism. P = phosphate; P_i = inorganic phosphate; NAD(H) = nicotinamide adenine dinucleotide; ADP = adenosine diphosphate.

1-phosphate accumulates, resulting in a drop in the level of inorganic phosphate (P_i) and, therefore, of ATP production. Fructose 1-phosphate also inhibits *glycogen phosphorylase*, which causes accumulation of glycogen in liver and **hypoglycemia**. As ATP falls, adenosine monophosphate (AMP) rises. The AMP is degraded, causing **hyperuricemia** (and lactic acidemia). The decreased availability of hepatic ATP decreases gluconeogenesis (causing **hypoglycemia** with vomiting) and protein synthesis (causing a decrease in blood-clotting factors and other essential proteins). Renal reabsorption of P_i is also decreased. [Note: The drop in P_i also inhibits glycogenolysis (see p. 227).] **Diagnosis of HFI can be made on the basis of fructose in the urine, enzyme assay using liver cells, or by DNA-based testing** (see Chapter 44). With HFI, sucrose, as well as fructose, must be removed from the diet to prevent liver failure and possible death. [Note: Individuals with HFI display an aversion to sweets and, consequently, have an absence of dental caries.]

E. Mannose conversion to fructose 6-phosphate

Mannose, the C-2 epimer of glucose (see p. 177), is an important component of glycoproteins (see p. 268). *Hexokinase* phosphorylates mannose, producing mannose 6-phosphate, which, in turn, is reversibly isomerized to fructose 6-phosphate by *phosphomannose isomerase*. [Note: Most intracellular mannose is synthesized from fructose or is preexisting mannose produced by the degradation of glycoproteins and salvaged by *hexokinase*. Dietary carbohydrates contain little mannose.]

F. Glucose conversion to fructose via sorbitol

Most sugars are rapidly phosphorylated following their entry into cells. Therefore, they are trapped within the cells, because organic phosphates cannot freely cross membranes without specific transporters. An alternate mechanism for metabolizing a monosaccharide is to convert it to a polyol (sugar alcohol) by the reduction of an aldehyde group, thereby producing an additional hydroxyl group. This pathway is known as **sorbitol (or polyol) pathway**.

1. **Sorbitol synthesis:** *Aldose reductase* reduces glucose, producing sorbitol (or, glucitol; Fig. 17.4), but the K_m is high. ***Aldose reductase* is found in many tissues, including the retina, lens, kidneys, peripheral nerves, ovaries, and seminal vesicles**. **A second enzyme, *sorbitol dehydrogenase*, can oxidize sorbitol to fructose in cells of the liver, ovaries, and seminal vesicles** (see Fig. 17.4). The two-reaction pathway from glucose to fructose in the seminal vesicles benefits sperm cells, which use fructose as a major carbohydrate energy source. The pathway from sorbitol to fructose in the liver provides a mechanism by which any available sorbitol is converted into a substrate that can enter glycolysis.

2. **Hyperglycemia and sorbitol metabolism:** Because insulin is not required for the entry of glucose into cells of the retina, lens, kidneys, and peripheral nerves, large amounts of glucose may enter these cells during times of hyperglycemia (for example, in uncontrolled diabetes). Elevated intracellular glucose concentrations and an adequate supply of reduced nicotinamide adenine dinucleotide phosphate (NADPH) cause *aldose reductase* to produce a significant increase in the amount of sorbitol, which cannot pass efficiently through cell membranes and, therefore, remains trapped inside the cell (see Fig. 17.4). This is exacerbated **when *sorbitol dehydrogenase* is low or absent** (for example, in cells of the retina, lens, kidneys, and peripheral nerves). As a result, **sorbitol accumulates in these cells, causing strong osmotic effects and cell swelling due to water influx and retention.** Some of the pathologic alterations associated with diabetes can be partly attributed to this osmotic stress, including **cataract formation, peripheral neuropathy, and microvascular problems leading to nephropathy and retinopathy** (see p. 478). [Note: Use of NADPH in the *aldose reductase* reaction decreases the generation of reduced glutathione, an important antioxidant (see p. 797), and may be related to diabetic complications.]

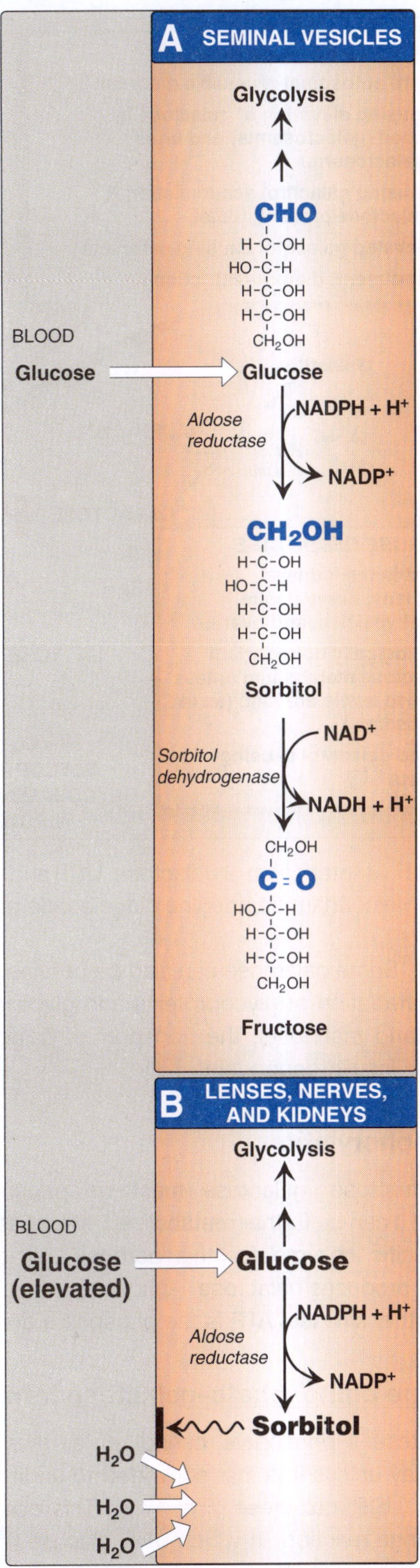

Figure 17.4 Sorbitol metabolism. NAD(H) = nicotinamide adenine dinucleotide; NADP(H) = nicotinamide adenine dinucleotide phosphate.

III. GALACTOSE METABOLISM

The major dietary source of galactose is lactose (galactosyl β-1,4-glucose) obtained from milk and milk products. [Note: The digestion of lactose by *β-galactosidase* (*lactase*) of the intestinal mucosal cell membrane was discussed

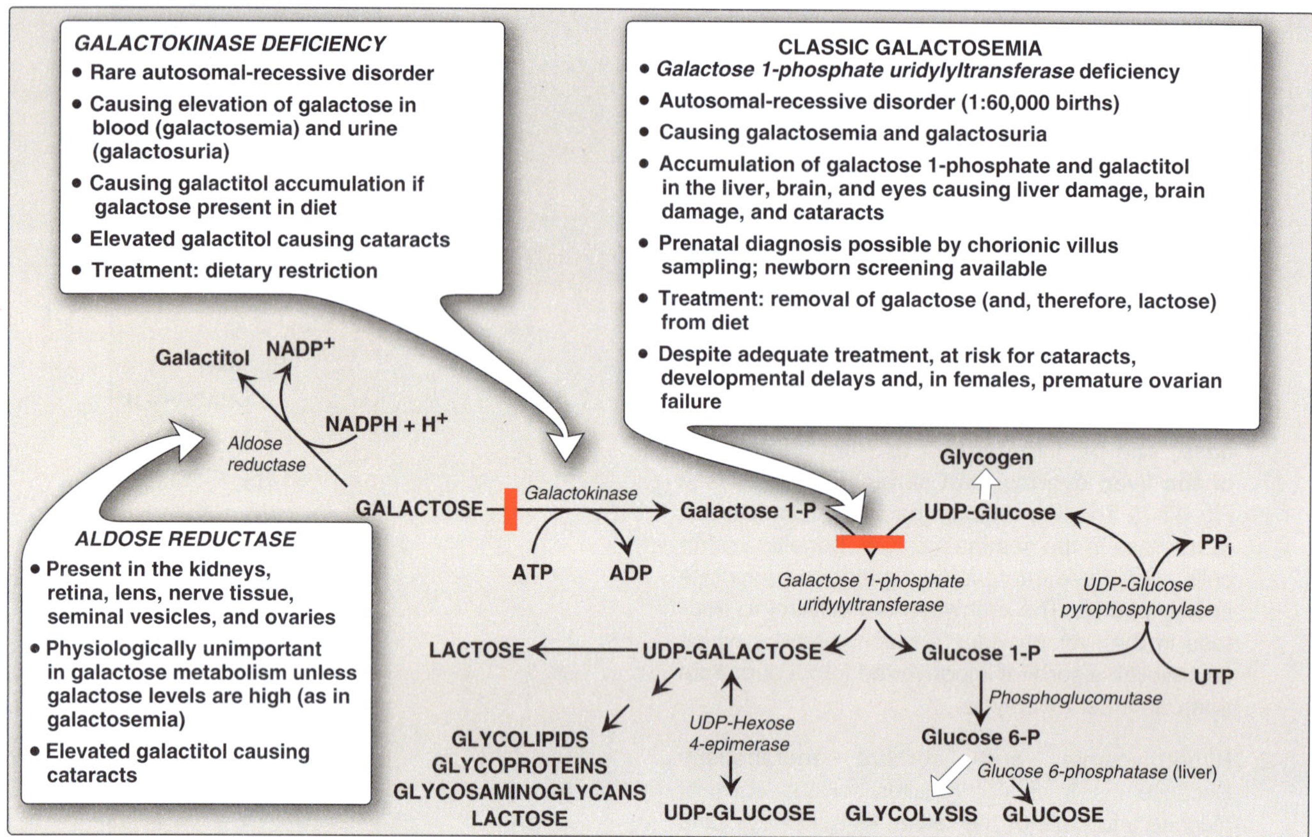

Figure 17.5 Metabolism of galactose. UDP and UTP = uridine di- and triphosphates; P = phosphate; PP_i = pyrophosphate; NADP(H) = nicotinamide adenine dinucleotide phosphate; ADP = adenosine diphosphate.

on p. 180.] Some galactose can also be obtained by lysosomal degradation of glycoproteins and glycolipids. Like fructose (and mannose), the transport of galactose into cells is not insulin dependent.

A. Phosphorylation

Like fructose, galactose must be phosphorylated before it can be further metabolized. Most tissues have a specific enzyme for this purpose, *galactokinase*, which produces galactose 1-phosphate (Fig. 17.5). As with other *kinases*, ATP is the phosphate donor.

B. Uridine diphosphate–galactose formation

Galactose 1-phosphate cannot enter the glycolytic pathway unless it is first converted to uridine diphosphate (UDP)-galactose (Fig. 17.6). This occurs in an exchange reaction, in which UDP-glucose reacts with galactose 1-phosphate, producing UDP-galactose and glucose 1-phosphate (see Fig. 17.5). **The reaction is catalyzed by *galactose 1-phosphate uridylyltransferase*** (*GALT*). This is the **rate-limiting enzyme** for galactose metabolism. [Note: The glucose 1-phosphate product can be isomerized to glucose 6-phosphate, which can enter glycolysis or be dephosphorylated.]

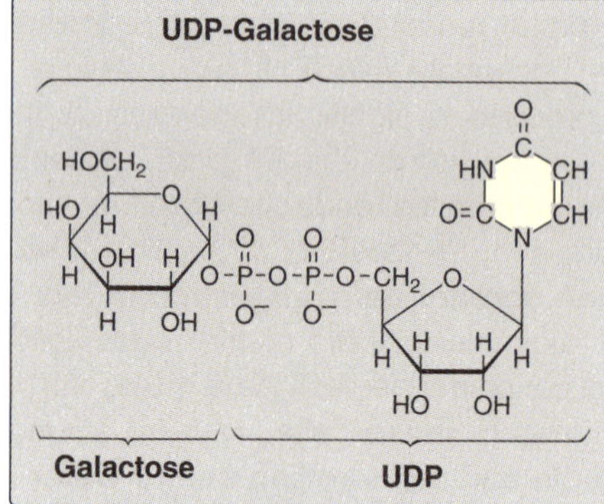

Figure 17.6 Structure of UDP-galactose. UDP = uridine diphosphate.

C. UDP-galactose conversion to UDP-glucose

For UDP-galactose to enter the mainstream of glucose metabolism, it must first be isomerized to its C-4 epimer, UDP-glucose, by *UDP-hexose 4-epimerase*. This "new" UDP-glucose (produced from the original UDP-galactose) can participate in biosynthetic reactions (for example, glycogenesis) as well as in the *GALT* reaction. [Note: See Fig. 17.5 for a summary of the interconversions.]

D. UDP-galactose in biosynthetic reactions

UDP-galactose can serve as the donor of galactose units in a number of synthetic pathways, including synthesis of lactose (see IV. below), glycoproteins (see p. 269), glycolipids (see p. 327), and glycosaminoglycans (see p. 260). [Note: If galactose is not provided by the diet (for example, when it cannot be released from lactose owing to a lack of *β-galactosidase* in people who are lactose intolerant), all tissue requirements for UDP-galactose can be met by the action of *UDP-hexose 4-epimerase* on UDP-glucose, which is efficiently produced from glucose 1-phosphate and uridine triphosphate (see Fig. 17.5).]

E. Disorders

GALT is severely deficient in individuals with **classic galactosemia** (see Fig. 17.5). In this disorder, galactose 1-phosphate and, therefore, galactose accumulate. Physiologic consequences are similar to those found in HFI (see p. 239), but a broader spectrum of tissues is affected. The accumulated galactose is shunted into side pathways such as that of galactitol production. This reaction is catalyzed by *aldose reductase*, the same enzyme that reduces glucose to sorbitol (see p. 240). ***GALT*** **deficiency is part of the newborn screening panel.** Treatment of galactosemia requires removal of galactose and lactose from the diet. [Note: Deficiencies in *galactokinase* and the *epimerase* result in less severe disorders of galactose metabolism, although cataracts are common (see Fig. 17.5).]

IV. LACTOSE SYNTHESIS

Lactose is a disaccharide that consists of a molecule of β-galactose attached by a β(1→4) linkage to glucose. Therefore, **lactose is galactosyl β(1→4)-glucose**. Because lactose (milk sugar) is made by lactating (milk-producing) mammary glands, milk and other dairy products are the dietary sources of lactose. Lactose is synthesized in the Golgi by *lactose synthase* (*UDP-galactose:glucose galactosyltransferase*), which transfers galactose from UDP-galactose to glucose, releasing UDP (Fig. 17.7).

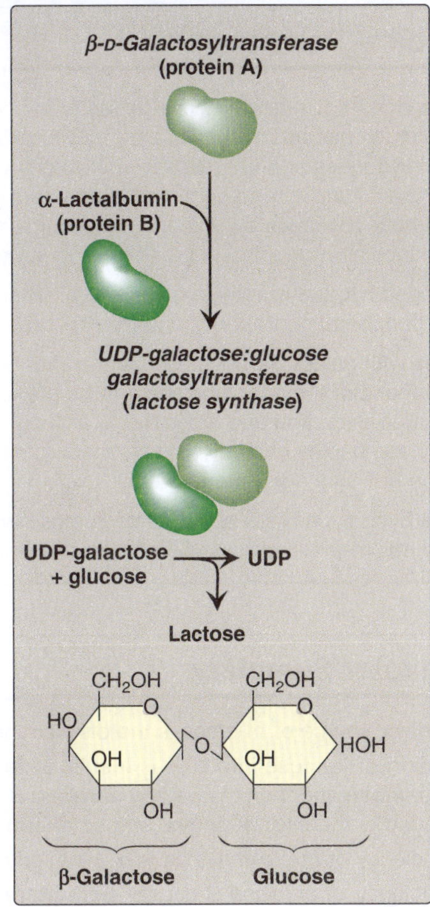

Figure 17.7 Lactose synthesis. UDP = uridine diphosphate.

Lactose synthase is composed of two proteins, A and B. Protein A is a *β-D-galactosyltransferase* and is found in a number of body tissues. In tissues other than the lactating mammary gland, this enzyme transfers galactose from UDP-galactose to N-acetyl-D-glucosamine, forming the same β(1→4) linkage found in lactose, and producing N-acetyllactosamine, a component of the structurally important N-linked glycoproteins (see p. 269).

In contrast, protein B is found only in lactating mammary glands. It is α-lactalbumin, and its synthesis is stimulated by the peptide hormone prolactin. Protein B forms a complex with the enzyme, protein A, changing the specificity of that *transferase* (by decreasing the K_m for glucose) so that lactose, rather than N-acetyllactosamine, is produced (see Fig. 17.7).

V. Medical Perspectives

Fructose does not require insulin for entry into cells and initial metabolic reactions. Due to this, fructose was considered good for patients with diabetes mellitus. But due to rapid entry in cells, it increases fatty acid synthesis, raises serum triglycerides and low-density lipoprotein (LDL) cholesterol level. It also increases triacylglycerol (TAG) synthesis by increasing availability of glycerol phosphate from glycolysis. Accumulation of fructose-1-phosphate lowers Pi levels in cell, which remove the inhibitory effect on adenosine deaminase. There is increased conversion of adenosine monophosphate (AMP) to inosine monophosphate (IMP) and subsequent increase in uric acid production.

In prolonged hyperglycemia, large quantities of sorbitol accumulate in cells of retina, lens, kidneys, and peripheral nerves, causing peripheral neuropathy, and microvascular problems leading to nephropathy and retinopathy.

Neonates with galactosemia present early after birth with vomiting and jaundice. Even with early diagnosis and treatment, manifestations of the condition seem to be present in most individuals with galactosemia, including some mental delays, learning disorders, fine and gross motor skill complications. The treatment involves a galactose-restricted diet and counseling. These babies should to be weaned immediately from breast milk or formula containing lactose or galactose and should be put on a soya-metabolism.

Neonatal hypoglycemia is a feature of fructose intolerance and galactosemia. It is also seen in glycogen storage disease type 1, medium-chain *fatty acyl-CoA dehydrogenase* deficiency and long-chain *fatty acyl-CoA dehydrogenase* deficiency.

VI. Chapter Summary

- The major source of fructose is the disaccharide **sucrose**
- Transport of fructose into cells is **insulin independent**. Fructose is first phosphorylated to **fructose 1-phosphate** by ***fructokinase*** and then cleaved by ***aldolase B*** to **dihydroxyacetone phosphate** and **glyceraldehyde**. These enzymes are found in the **liver**, **kidneys**, and **small intestine**.
- A deficiency of *fructokinase* causes a benign condition (**essential fructosuria**).
- A deficiency of *aldolase B* causes **hereditary fructose intolerance (HFI)**, in which **severe hypoglycemia** and **liver failure** lead to **death** if fructose (and sucrose) is not removed from the diet.
- **Mannose**, an important component of **glycoproteins**, is phosphorylated by ***hexokinase*** to **mannose 6-phosphate**, which is reversibly isomerized to **fructose 6-phosphate** by ***phosphomannose isomerase***.
- Glucose can be reduced to **sorbitol** (**glucitol**) by ***aldose reductase*** in many tissues, including the **lens**, **retina**, **peripheral nerves**, **kidneys**, **ovaries**, and **seminal vesicles**. In the liver, ovaries, and seminal vesicles, a second enzyme, ***sorbitol dehydrogenase***, can oxidize sorbitol to produce **fructose**.
- **Hyperglycemia** results in the accumulation of sorbitol in those cells lacking *sorbitol dehydrogenase*. The resulting **osmotic events** cause cell swelling and may contribute to the **cataract formation**, **peripheral neuropathy**, **nephropathy**, and **retinopathy** seen in **diabetes**.
- The major dietary source of **galactose** is **lactose**. The transport of galactose into cells is insulin independent.
- A deficiency of ***galactokinase*** results in cataracts.
- A deficiency of ***galactose 1-phosphate uridylyltransferase*** (***GALT***) causes **classic galactosemia**. Galactose 1-phosphate accumulates, and excess galactose is converted to **galactitol** by ***aldose reductase***. This causes **liver damage**, **brain damage**, and **cataracts**. Treatment requires removal of galactose (and lactose) from the diet.
- **Lactose** is a disaccharide of **galactose** and **glucose**. Milk and other **dairy products** are the dietary sources of lactose.
- Lactose is synthesized by ***lactose synthase*** from **UDP-galactose** and **glucose** in the **lactating mammary gland**.

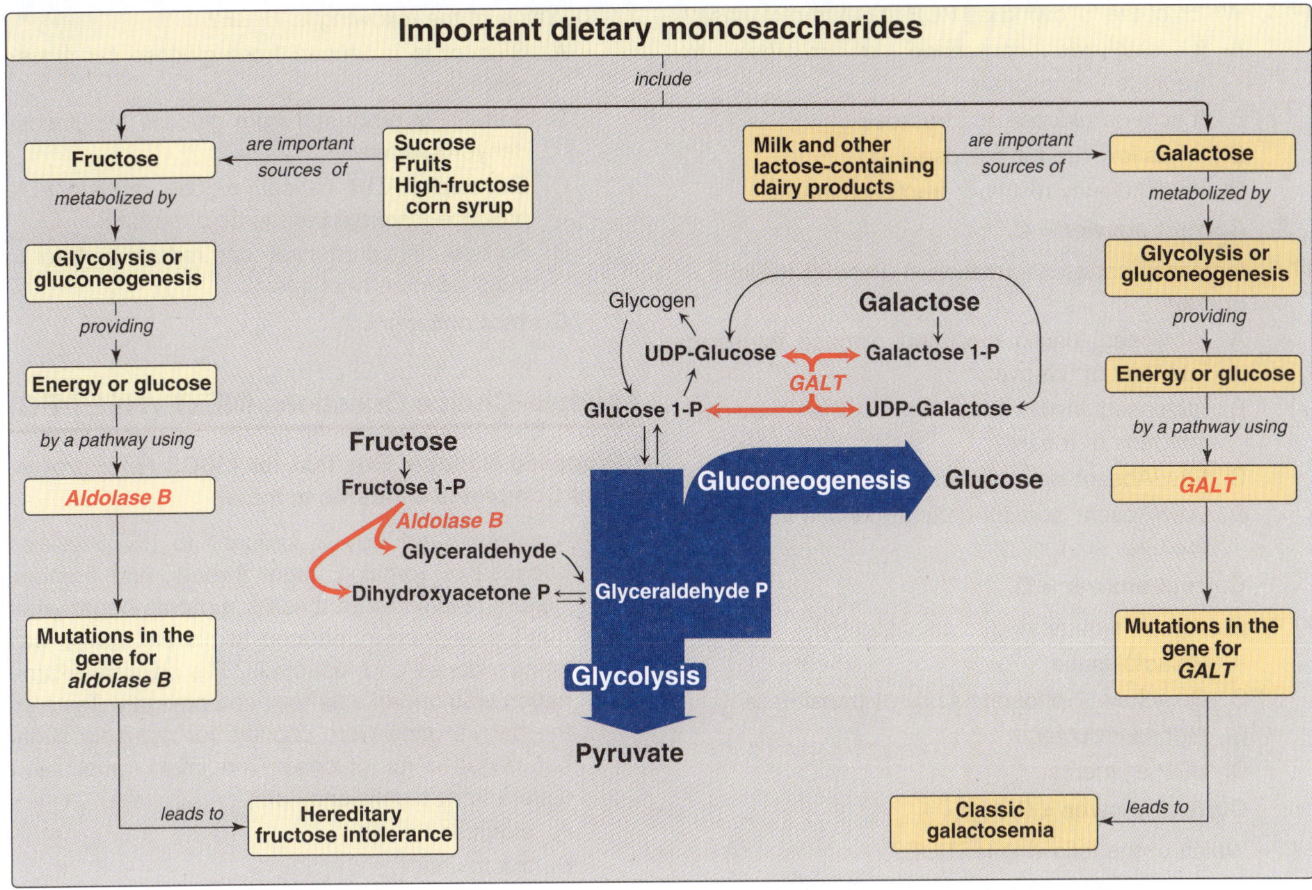

Figure 17.8 Key concept map for metabolism of fructose and galactose. *GALT* = *galactose 1-phosphate uridylyltransferase*; UDP = uridine diphosphate; P = phosphate.

Study Questions

Short-Answer Questions

Write short notes on:

17.1. Congenital galactosemia
17.2. Biochemical basis of cataract in diabetes mellitus
17.3. Hereditary fructose intolerance

Long-Answer Question

17.1. Describe in detail the metabolism of galactose. Add a note on associated disease.

Multiple-Choice Questions

Choose the ONE best answer.

17.1. A nursing female with classic galactosemia is on a galactose-free diet. She is able to produce lactose in breast milk because:

A. galactose can be produced from fructose by isomerization.
B. galactose can be produced from a glucose metabolite by epimerization.
C. hexokinase can efficiently phosphorylate galactose to galactose 1-phosphate.
D. the enzyme affected in galactosemia is activated by a hormone produced in the mammary gland.

Correct answer = B.

17.2. Which among the following is a disaccharide that can be synthesized from monosaccharides in humans?

A. Lactose
B. Maltose
C. Sucrose
D. Trehalose

Correct answer = A.

17.3. Lactose synthesis is essential in the production of milk by mammary glands. In lactose synthesis:

A. galactose from galactose 1-phosphate is transferred to glucose by galactosyltransferase (protein A), generating lactose.
B. protein A is used exclusively in lactose synthesis.
C. α-lactalbumin (protein B) regulates the specificity of protein A by decreasing its affinity for glucose.
D. protein B expression is stimulated by prolactin.

Correct answer = D.

17.4. Which of the following is TRUE about fructokinase?
- A. It catalyzes conversion of fructose to fructose-6-phosphate
- B. It acts on glucose and fructose
- C. It has low Km for fructose
- D. Its deficiency results in fructosemia

Correct answer = C.

17.5. Increased cataract formation in diabetes mellitus is due to:
- A. Increased insulin mediated glucose transport into lens of the eye
- B. Increased insulin mediated fructose transport into lens of the eye
- C. Low/Absent aldose reductase in lens of the eye
- D. Low/Absent sorbitol dehydrogenase in lens of the eye

Correct answer = D.

17.6. Galactose is converted to galactitol by:
- A. Galactokinase
- B. Galactose 1-phosphate uridylyltransferase
- C. Aldose reductase
- D. UDP epimerase

Correct answer = C.

17.7. Which of the following is TRUE?
- A. Lactose is α -galactose attached to glucose by a α (1-4) linkage
- B. Lactose is α -galactose attached to glucose by a β (1-4) linkage
- C. Lactose is β-galactose attached to glucose by a α (1-4) linkage
- D. Lactose is β-galactose attached to glucose by a β (1-4) linkage

Correct answer = D.

17.8. Lactose is synthesized in:
- A. Cytosol
- B. Golgi
- C. Endoplasmic reticulum
- D. Mitochondria

Correct answer = B.

17.9. Rate of fructose metabolism is more rapid than that of glucose because:
- A. Fructose Insulin stimulates more rapid fructose intake into cells
- B. Triose production from fructose 1-phosphate bypasses *phosphofructokinase-1*
- C. Fructose 1-phosphate is rapidly phosphorylated to fructose 1,6- bisphosphate
- D. Fructose can enter intermediary metabolism without phosphorylation

Correct answer = B.

17.10. Which of the following is TRUE?
- A. Glucitol is produced from glucose by aldose reductase
- B. Sorbitol is produced from glucose by glucose dehydrogenase
- C. Elevated NADH causes aldose reductase to produce increased amounts of sorbitol
- D. Sorbitol dehydrogenase can reduce sorbitol to fructose in seminal vesicles

Correct answer = A.

Multiple-Choice Questions (NExT*/NEET PG)

***(Proposed National Exit Test for MBBS Final professional Completed Students in India)**

17.1. A 5-month-old boy is brought to his physician because of vomiting, night sweats, and tremors. History revealed that these symptoms began after fruit juices were introduced to his diet as he was being weaned off breast milk. The physical examination was remarkable for hepatomegaly. Tests on the baby's urine were positive for reducing sugar but negative for glucose. The infant most likely suffers from a deficiency of:
- A. aldolase B.
- B. fructokinase.
- C. galactokinase.
- D. β-galactosidase.

Correct answer = A.

17.2. A 3-month-old girl is developing cataracts. Other than not having a social smile or being able to track objects visually, all other aspects of the girl's examination are normal. Tests on the baby's urine are positive for reducing sugar but negative for glucose. Which enzyme is most likely deficient in this girl?
- A. Aldolase B
- B. Fructokinase
- C. Galactokinase
- D. Galactose 1-phosphate uridylyltransferase

Correct answer = C.

17.3. Based on statements given, choose the correct option.

Statement 1: Activation of aldose reductase pathway as a result of cellular hyperglycemia is one of the mechanisms of development of complications in diabetes mellitus.

Statement 2: This pathway results in depletion of NADPH, which is required for anti-oxidant defense mechanisms in the cells.
- A. Both the Statements are correct, and Statement 2 is the correct explanation for Statement 1
- B. Both the Statements are correct, but Statement 2 is NOT the correct explanation for Statement 1

C. Statement 1 is correct, but Statement 2 is NOT correct

D. Statement 1 is NOT correct, but Statement 2 is correct

Correct answer = A.

17.4. A three-week-old, breastfed infant presented with jaundice, an enlarged liver, and cataract. Laboratory tests indicated presence of reducing substances in the urine and a reduced activity of the enzyme galactose-1-phosphate uridyltransferase in red blood cells. What is the most important long-term treatment of this condition?

A. Liver transplantation

B. Gene therapy

C. Dietary galactose restriction

D. Enzyme replacement therapy

Correct answer = C.

17.5. Which of the following is TRUE about classical galactosemia?

A. Urinary reducing sugar tests are negative

B. Hypoglycemia occurs due to reduced gluconeogenesis

C. It is caused due to deficiency of galactokinase

D. Accumulation of sorbitol causes cataracts

Correct answer = B.

Pentose Phosphate Pathway and Nicotinamide Adenine Dinucleotide Phosphate

18

Learning Objectives

At the end of this chapter, the student will be able to:

- Describe the pentose phosphate pathway
- Describe disorders associated with deficiencies in pentose phosphate pathway
- Enumerate uses of NADPH
- Describe the biochemical role of *Glucose 6-phosphate dehydrogenase*
- Describe disorders of *Glucose 6-phosphate dehydrogenase* deficiency

I. OVERVIEW

The pentose phosphate pathway, also known as **hexose monophosphate [HMP] shunt** or, phosphogluconate pathway, occurs in the cytosol. **It includes an irreversible oxidative phase, followed by a series of reversible sugar–phosphate interconversions** (Fig. 18.1). In the oxidative phase, carbon 1 of a glucose 6-phosphate molecule is released as carbon dioxide (CO_2), and one pentose sugar-phosphate plus two reduced nicotinamide adenine dinucleotide phosphates (NADPH) are produced. The rate and direction of the reversible reactions are determined by the supply of and demand for intermediates of the pathway. The pentose phosphate pathway **provides a major portion of the body's NADPH**, which functions as a biochemical reductant. It **also produces ribose 5-phosphate**, required for nucleotide biosynthesis (see p. 579), and provides a mechanism for the conversion of pentose sugars to triose and hexose intermediates of glycolysis. **No ATP is directly consumed or produced in the pathway**.

II. IRREVERSIBLE OXIDATIVE REACTIONS

The oxidative portion of the pentose phosphate pathway consists of three irreversible reactions that lead to the formation of ribulose 5-phosphate, CO_2, and two molecules of NADPH for each molecule of glucose 6-phosphate oxidized (Fig. 18.2).

This portion of the pathway is particularly important in:

- The liver, lactating mammary glands, and adipose tissue for the NADPH-dependent **biosynthesis of fatty acids** (see p. 301)
- The testes, ovaries, placenta, and adrenal cortex for the NADPH-dependent **biosynthesis of steroid hormones** (see p. 359)
- Red blood cells (RBC) for the NADPH-dependent **reduction of glutathione** (see p. 797).

A. Glucose 6-phosphate dehydrogenation

***Glucose 6-phosphate dehydrogenase* (*G6PD*)** catalyzes the oxidation of glucose 6-phosphate to 6-phosphogluconolactone as the coenzyme $NADP^+$ gets reduced to NADPH. This initial reaction is the **committed, rate-limiting, and regulated step** of the pathway. NADPH is a potent competitive inhibitor of *G6PD*, and the ratio of NADPH/$NADP^+$ is sufficiently high to substantially inhibit the enzyme under most metabolic conditions. However, with increased demand for NADPH, the ratio of NADPH/$NADP^+$ decreases, and flux through the pathway increases

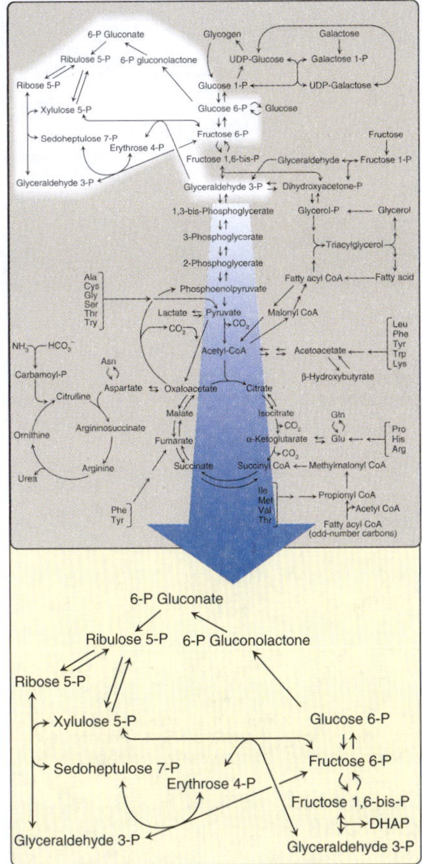

Figure 18.1 Pentose phosphate pathway shown as a component of the metabolic map. [Note: See Fig. 13.2, p. 187 for a more detailed map of metabolism.] P = phosphate; DHAP = dihydroxyacetone phosphate.

in response to the enhanced activity of *G6PD*. [Note: Insulin upregulates expression of the gene for *G6PD*, and flux through the pathway increases in the absorptive state (see p. 453).]

B. Ribulose 5-phosphate formation

6-Phosphogluconolactone is hydrolyzed by *6-phosphogluconolactone hydrolase* in the second step. The oxidative decarboxylation of the product, 6-phosphogluconate, is catalyzed by *6-phosphogluconate dehydrogenase.* This third irreversible step produces ribulose 5-phosphate (a pentose sugar–phosphate), CO_2 (from carbon 1 of glucose), and a second molecule of NADPH (see Fig. 18.2).

III. REVERSIBLE NONOXIDATIVE REACTIONS

The nonoxidative reactions of the pentose phosphate pathway occur in all cell types synthesizing nucleotides and nucleic acids. These reactions catalyze the interconversion of sugars containing three to seven carbons (see Fig. 18.2). **These reversible reactions permit ribulose 5-phosphate (produced by the oxidative portion of the pathway) to be converted either to ribose 5-phosphate (needed for nucleotide synthesis; see pp. 574–575) or to intermediates of glycolysis (that is, fructose 6-phosphate and glyceraldehyde 3-phosphate).** For example, many cells that carry out reductive biosynthetic reactions have a greater need for NADPH than for ribose 5-phosphate. In this case, ***transketolase*** (which transfers two-carbon units in a thiamine pyrophosphate [TPP]-requiring reaction) and ***transaldolase*** (which transfers three-carbon units) convert the ribulose 5-phosphate produced as an end product of the oxidative phase to glyceraldehyde 3-phosphate and fructose 6-phosphate, which are glycolytic intermediates. In contrast, when the demand for ribose for nucleotides and nucleic acids is greater than the need for NADPH, the nonoxidative reactions can provide the ribose 5-phosphate from glyceraldehyde 3-phosphate and fructose 6-phosphate in the absence of the oxidative steps (Fig. 18.3).

In addition to *transketolase*, TPP is required by the multienzyme complexes *pyruvate dehydrogenase* (see p. 206), *α-ketoglutarate dehydrogenase* of the tricarboxylic acid cycle (see p. 209), and *branched-chain α-keto acid dehydrogenase* of branched-chain amino acid catabolism (see p. 402).

A. Regulation

Glucose 6-phosphate dehydrogenase and *6-phosphogluconate dehydrogenase* are the regulatory enzymes of HMP shunt. Synthesis of both these enzymes is induced by insulin. G6PD activity is competitively inhibited by NADPH.

IV. NADPH USES

The coenzyme NADPH differs from nicotinamide adenine dinucleotide (NADH) only by the presence of a phosphate group on one of the ribose units (Fig. 18.4). This seemingly small change in structure allows NADPH to interact with NADPH-specific enzymes that have unique roles in the cell. For example, in the cytosol of hepatocytes, the steady-state $NADP^+$/ NADPH ratio is ~0.1, which favors the use of NADPH in reductive biosynthetic reactions. This contrasts with the high NAD^+/NADH ratio (~1,000), which favors an oxidative role for NAD^+. This section summarizes some important NADPH-specific functions in reductive biosynthesis and detoxification reactions.

A. Reductive biosynthesis

Like NADH, NADPH can be thought of as a high-energy molecule. However, the electrons of NADPH

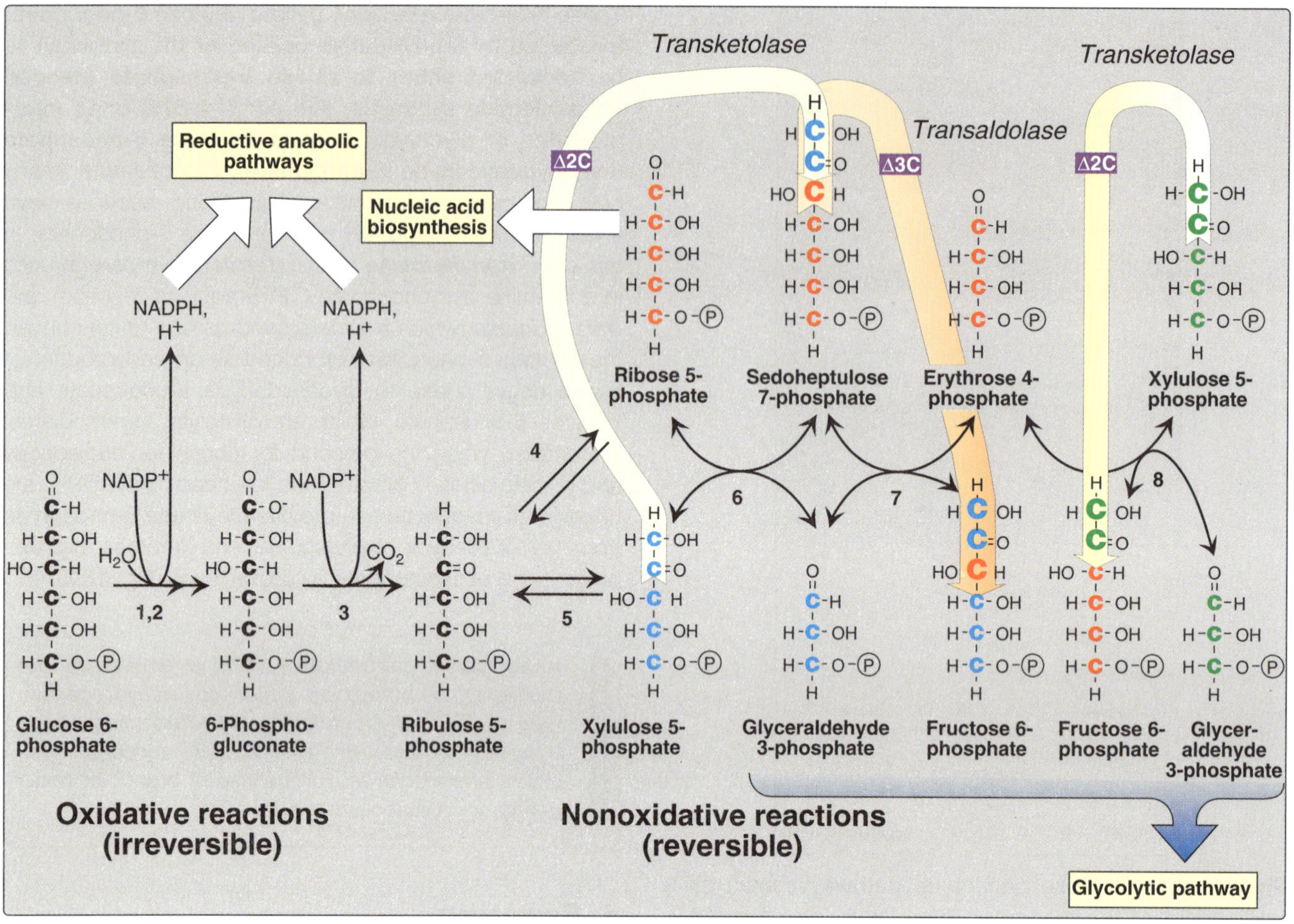

Figure 18.2 Reactions of the pentose phosphate pathway. Enzymes numbered above are: (1, 2) *glucose 6-phosphate dehydrogenase* and *6-phosphogluconolactone hydrolase*, (3) *6-phosphogluconate dehydrogenase*, (4) *ribose 5-phosphate isomerase*, (5) *phosphopentose epimerase*, (6, 8) *transketolase* (coenzyme: thiamine pyrophosphate), and (7) *transaldolase*. $\Delta 2C$ = two carbons are transferred from a ketose donor to an aldose acceptor in *transketolase* reactions; $\Delta 3C$ = three carbons are transferred in the *transaldolase* reaction. This can be represented as: 5C sugar + 5C sugar $\xrightarrow{\Delta 2C}$ 7C sugar + 3C sugar $\xrightarrow{\Delta 3C}$ 4C sugar + 6C sugar. NADP(H) = nicotinamide adenine dinucleotide phosphate; Ⓟ = phosphate; CO_2 = carbon dioxide.

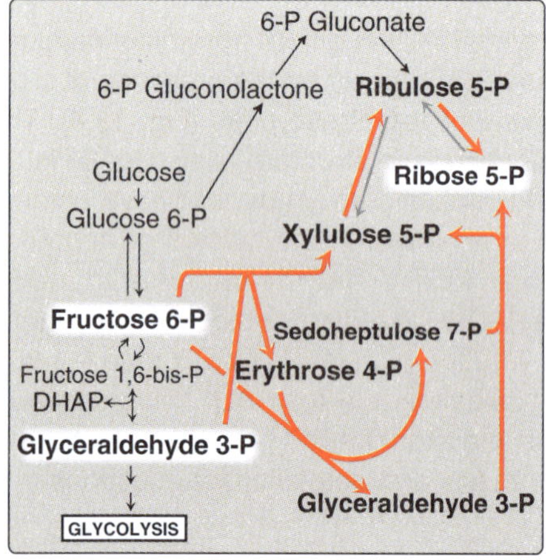

Figure 18.3 Formation of ribose 5-phosphate from intermediates of glycolysis. P = phosphate; DHAP = dihydroxyacetone phosphate.

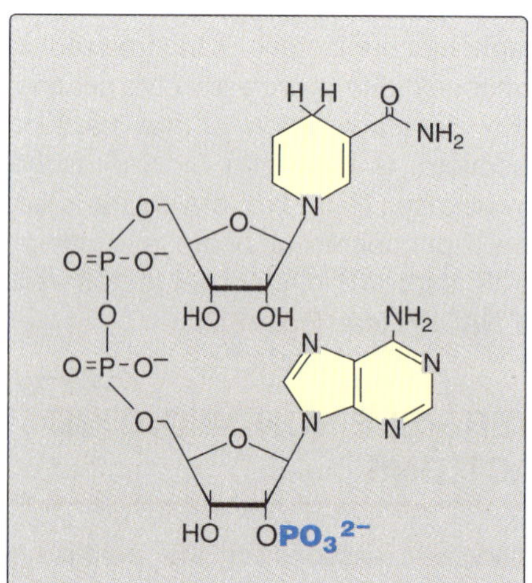

Figure 18.4 Structure of reduced nicotinamide adenine dinucleotide phosphate (NADPH).

are used for reductive biosynthesis, rather than for transfer to the electron transport chain as is seen with NADH (see p. 163). Thus, in the metabolic transformations of the pentose phosphate pathway, part of the energy of glucose 6-phosphate is conserved in NADPH, a molecule with a negative reduction potential (see p. 167), that, therefore, can be used in reactions requiring an electron donor, such as **fatty acid (see p. 301), cholesterol (see p. 343), and steroid hormone (see p. 359) synthesis.**

B. Hydrogen peroxide reduction

Hydrogen peroxide (H_2O_2) is one of a family of **reactive oxygen species** (ROS) that are formed from the partial reduction of molecular oxygen (O_2). These compounds are formed continuously as byproducts of aerobic metabolism, through reactions with drugs and environmental toxins, or when the level of antioxidants is diminished, all creating the condition of oxidative stress. These highly reactive oxygen intermediates can cause serious chemical damage to DNA, proteins, and unsaturated lipids and can lead to cell death. ROS have been implicated in a number of pathologic processes, including reperfusion injury, cancer, inflammatory disease and aging (see p. 795). The cell has several protective mechanisms that minimize the toxic potential of these compounds. NADPH provides electrons for the reduction of H_2O_2 by providing reducing equivalents for regeneration of **reduced glutathione in reaction catalyzed by *glutathione reductase***.

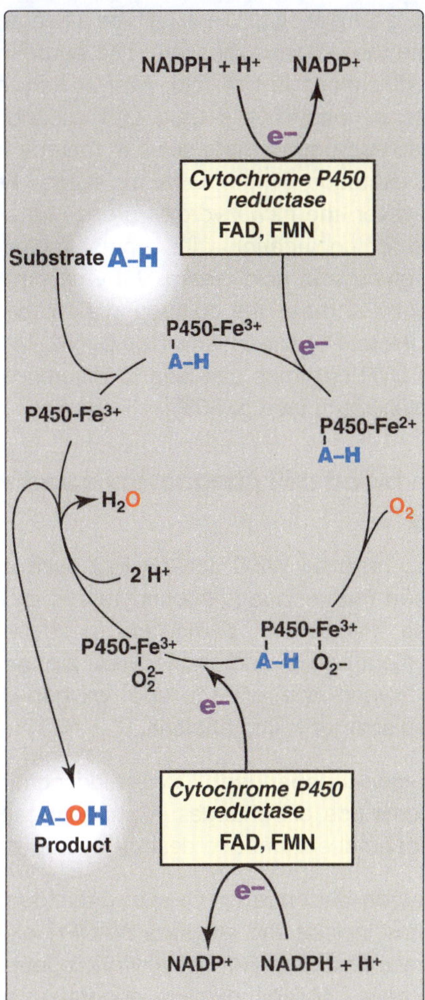

Figure 18.5 *Cytochrome P450* (*CYP*) *monooxygenase* catalytic cycle (simplified). Electrons (e^-) move from nicotinamide adenine dinucleotide phosphate (NADPH) to flavin adenine dinucleotide (FAD) to flavin adenine mononucleotide (FMN) of the *reductase* and then to the heme iron (Fe) of the microsomal *CYP* enzyme. [Note: In the mitochondrial system, e^- move from FAD to an iron-sulfur protein and then to the *CYP* enzyme.]

C. Cytochrome P450 monooxygenase system

Monooxygenases (*mixed-function oxidases*) incorporate one atom from O_2 into a substrate (creating a hydroxyl group), with the other atom being reduced to water (H_2O). In the *cytochrome P450* (*CYP*) *monooxygenase* system, NADPH provides the reducing equivalents required by this series of reactions (Fig. 18.5). The overall reaction catalyzed by a *CYP* enzyme is

$$R\text{–}H + O_2 + NADPH + H^+ \rightarrow R\text{–}OH + H_2O + NADP^+$$

where R may be a steroid, drug, or other chemical. [Note: *CYP* enzymes are actually a superfamily of related, heme-containing *monooxygenases* that participate in a broad variety of reactions. The P450 in the name reflects the absorbance at 450 nm by the protein.]

1. **Mitochondrial system:** An important function of the *CYP monooxygenase* system found associated with the inner mitochondrial membrane is the biosynthesis of steroid hormones. In steroidogenic tissues, such as the placenta, ovaries, testes, and adrenal cortex, it is used to hydroxylate intermediates in the conversion of cholesterol to steroid hormones, a process that makes these hydrophobic compounds more water soluble (see p. 359). The liver uses this same system in bile acid synthesis (see p. 346) and the hydroxylation of cholecalciferol to 25-hydroxycholecalciferol ([vitamin D_3], see p. 540), and the kidney uses it to hydroxylate vitamin D_3 to its biologically active 1,25-dihydroxylated form.

2. **Microsomal system:** The microsomal *CYP monooxygenase* system found associated with the membrane of the smooth endoplasmic reticulum (particularly in the liver) functions primarily in the detoxification of foreign compounds (xenobiotics). These include numerous drugs and such varied pollutants as petroleum products and pesticides. *CYP* enzymes of the microsomal system

(for example, *CYP3A4*) can be used to hydroxylate these toxins (phase I). The purpose of these modifications is two-fold. First, it may itself activate or inactivate a drug and second, make a toxic compound more soluble, thereby facilitating its excretion in the urine or feces. Frequently, however, the new hydroxyl group will serve as a site for conjugation with a polar molecule, such as glucuronic acid (see p. 268), which will significantly increase the compound's solubility (phase II). [Note: Polymorphisms (see p. 682) in the genes for *CYP* enzymes can lead to differences in drug metabolism] (see p. 783).]

D. White blood cell phagocytosis and microbe killing

Phagocytosis by WBC (leukocytes) such as neutrophils and macrophages (monocytes) is an important defense mechanism, particularly in bacterial infections. Neutrophils and monocytes are armed with both oxygen-independent and oxygen-dependent mechanisms for killing bacteria.

1. **Oxygen-independent:** Oxygen-independent mechanisms use pH changes in phagolysosomes and lysosomal enzymes to destroy pathogens.

2. **Oxygen-dependent:** Oxygen-dependent mechanisms include the enzymes *NADPH oxidase* and *myeloperoxidase* (*MPO*) that work together in killing bacteria. NADPH oxidase generates superoxide by oxidizing NADPH, and myeloperoxidase uses this superoxide to produce other reactive oxygen species that help destroy bacteria (see p. 793).

E. Nitric oxide synthesis

Nitric oxide (NO) is recognized as a mediator in a broad array of biologic systems. NO is the **endothelium-derived relaxing factor that causes vasodilation** by relaxing vascular smooth muscle. It also acts as a **neurotransmitter**, **prevents platelet aggregation**, and plays an essential role in macrophage function. It has a very short half-life in tissues (3–10 seconds) because it reacts with O_2 and $O_2^{\bar{\bullet}}$ and is converted into nitrates and nitrites including peroxynitrite ($O{=}NOO^-$), a reactive nitrogen species (RNS). [Note: NO is a free radical gas that is often confused with nitrous oxide (N_2O), the "laughing gas" that is used as an anesthetic and is chemically stable.]

1. **Nitric oxide synthase:** Arginine, O_2, and NADPH are substrates for cytosolic *NO synthase* ([*NOS*], Fig. 18.6). Flavin mononucleotide (FMN), FAD, heme, and tetrahydrobiopterin (see p. 404) are coenzymes, and NO and citrulline are products of the reaction. **Three *NOS* isozymes**, each the product of a different gene, have been identified. Two are constitutive (synthesized at a constant rate), calcium (Ca^{2+})–calmodulin (CaM)-dependent

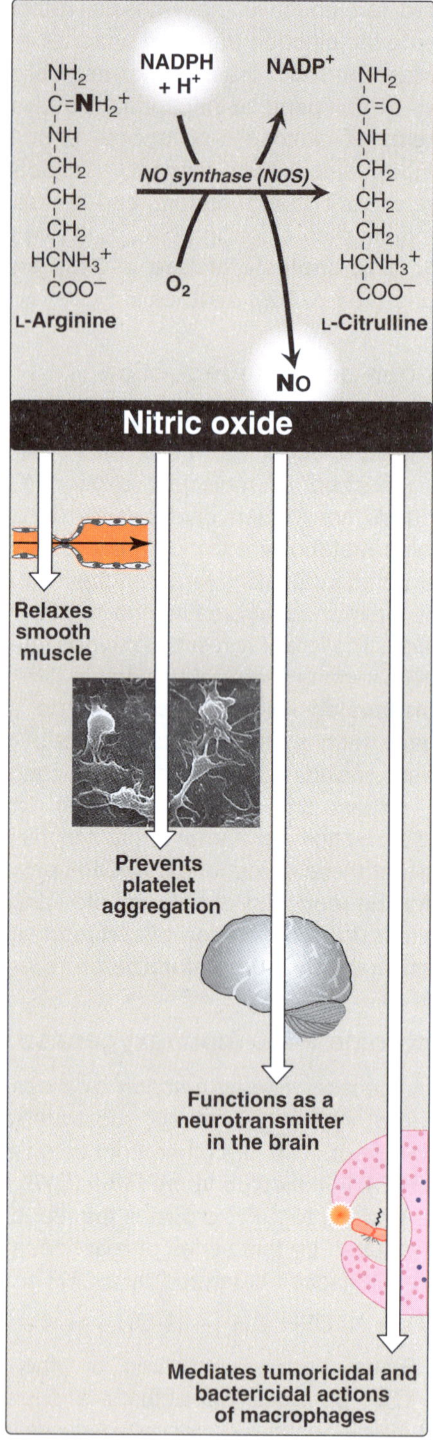

Figure 18.6 Synthesis and some actions of nitric oxide (NO). [Note: Flavin mononucleotide, flavin adenine dinucleotide, heme, and tetrahydrobiopterin are additional coenzymes required by *NOS*.] NADP(H) = nicotinamide adenine dinucleotide phosphate.

enzymes (see p. 232). They are found primarily in endothelium (*eNOS*) and neural tissue (*nNOS*) and constantly produce very low levels of NO for vasodilation and neurotransmission. An inducible, Ca^{2+}-independent enzyme (*iNOS*) can be expressed in many cells, including macrophages and neutrophils, as an early defense against pathogens. The specific inducers for *iNOS* vary with cell type and include proinflammatory cytokines, such as tumor necrosis factor-α (TNF-α) and interferon-γ (IFN-γ), and bacterial endotoxins such as lipopolysaccharide (LPS). These compounds promote synthesis of *iNOS*, which can result in large amounts of NO being produced over hours or even days.

2. **Nitric oxide and vascular endothelium:** NO is an important mediator in the control of vascular smooth muscle tone. NO is synthesized by *eNOS* in endothelial cells and diffuses to vascular smooth muscle, where it activates the cytosolic form of *guanylyl cyclase* (or, *guanylate cyclase*) to form cyclic guanosine monophosphate (cGMP). [Note: This reaction is analogous to the formation of cyclic adenosine monophosphate (cAMP) by *adenylyl cyclase* (see p. 190).] The resultant rise in cGMP causes activation of *protein kinase G*, which phosphorylates Ca^{2+} channels, causing decreased entry of Ca^{2+} into smooth muscle cells. This decreases the Ca^{2+}–CaM activation of *myosin light-chain kinase*, thereby decreasing smooth muscle contraction and favoring relaxation. Vasodilator nitrates, such as nitroglycerin, are metabolized to NO, which causes relaxation of vascular smooth muscle and, therefore, lowers blood pressure. Thus, NO can be envisioned as an endogenous nitrovasodilator. [Note: Under hypoxic conditions, nitrite (NO_2^-) can be reduced to NO, which binds to deoxyhemoglobin. The NO is released into the blood, causing vasodilation and increasing blood flow.]

3. **Nitric oxide and macrophage bactericidal activity:** In macrophages, *iNOS* activity is normally low, but synthesis of the enzyme is significantly stimulated by bacterial LPS and by release of IFN-γ and TNF-α in response to infection. Activated macrophages form $O_2^{\bullet-}$ radicals that combine with NO to form intermediates that decompose, producing the highly bactericidal OH· radical.

4. **Additional functions:** NO is a potent inhibitor of platelet adhesion and aggregation (by activating the cGMP pathway). It is also characterized as a neurotransmitter in the central and peripheral nervous systems.

V. G6PD DEFICIENCY

G6PD deficiency is a hereditary condition characterized by **hemolytic anemia** caused by the inability to detoxify oxidizing agents. *G6PD* deficiency is the most common disease-producing enzyme abnormality in humans, affecting >400 million individuals worldwide. This deficiency has the highest prevalence in the Middle East, tropical Africa and Asia, and parts of the Mediterranean. *G6PD* deficiency is **X linked** and is, in fact, a family of deficiencies caused by a number of different mutations in the gene encoding *G6PD*. Only some of the resulting protein variants cause clinical symptoms. The life span of individuals with a severe form of *G6PD* deficiency may be somewhat shortened as a result of complications arising from chronic hemolysis. This negative effect of *G6PD* deficiency has been balanced in evolution by an advantage in survival, an increased resistance to Plasmodium falciparum malaria. [Note: Sickle cell trait and the thalassemias also confer resistance to malaria.]

In addition to hemolytic anemia, a clinical manifestation of *G6PD* deficiency is neonatal jaundice appearing 1–4 days after birth. The jaundice, which may be severe, typically results from increased production of unconjugated bilirubin (see p. 422).

A. G6PD role in RBCs

Diminished *G6PD* activity **impairs the ability of the cell to form the NADPH that is essential for the maintenance of the G-SH pool**. This results in (i) decrease in the detoxification of free radicals and peroxides formed within the cell (Fig. 18.7), (ii)oxidation of sulfhydryl groups in proteins, leading to the formation of denatured proteins that form insoluble masses (called Heinz bodies) that attach to RBC membranes (Fig. 18.8), and (iii)additional oxidation of membrane proteins causing RBC to be rigid (less deformable), and they are removed from the circulation by macrophages in the spleen and liver.

Although *G6PD* deficiency occurs in all cells of the affected individual, it is most severe in RBC, where the pentose phosphate pathway provides the only means of generating NADPH. Additionally, the RBC has no nucleus or ribosomes and cannot renew its supply of the enzyme. Thus, **RBC are particularly vulnerable** to enzyme variants with diminished stability. [Note: Other tissues have an alternative source of NADPH (*NADP⁺-dependent malate dehydrogenase* [*malic enzyme*]; see p. 305) that can keep G-SH reduced.]

B. Precipitating factors in G6PD deficiency

Most individuals who have inherited one of the *G6PD* mutations do not show clinical manifestations (that

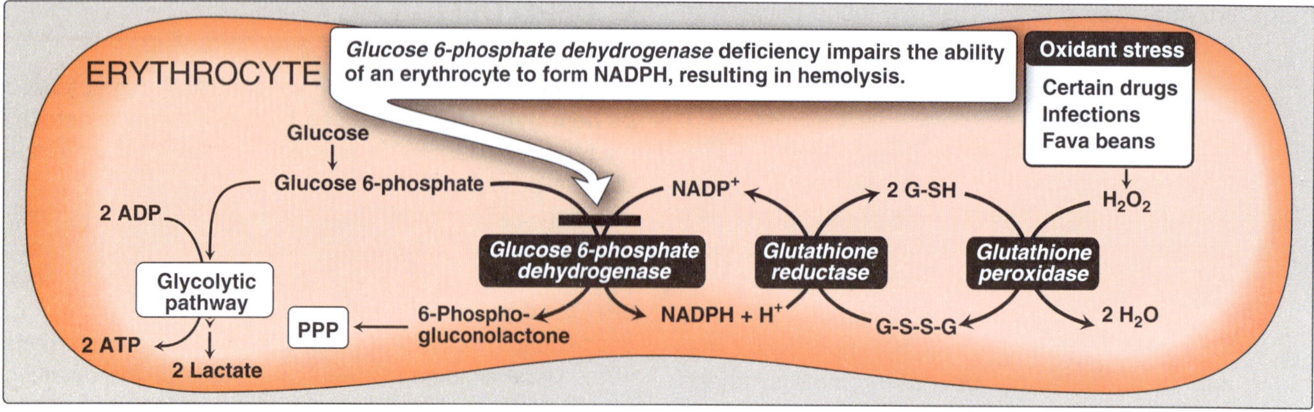

Figure 18.7 Pathways of glucose 6-phosphate metabolism in the erythrocyte. NADP(H) = nicotinamide adenine dinucleotide phosphate; G-SH = reduced glutathione; G-S-S-G = oxidized glutathione; H_2O_2 = hydrogen peroxide; PPP = pentose phosphate pathway.

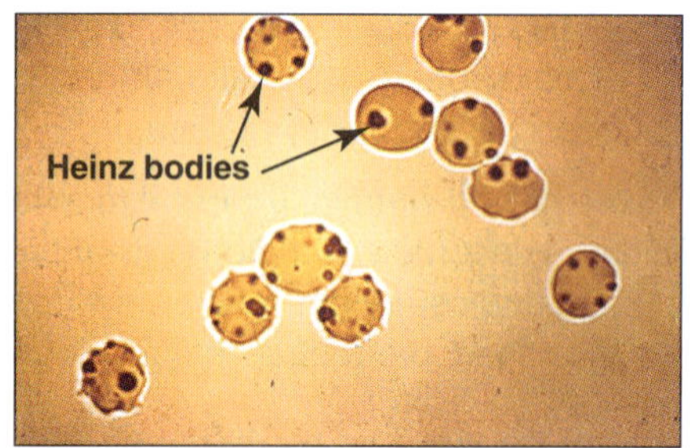

Figure 18.8 Heinz bodies in the erythrocytes of a patient with *glucose 6-phosphate dehydrogenase* deficiency.

is, they are asymptomatic). However, **some patients with *G6PD* deficiency develop hemolytic anemia if they are treated with an oxidant drug, ingest fava beans, or contract a severe infection**.

1. **Oxidant drugs:** Commonly used drugs that produce hemolytic anemia in patients with *G6PD* deficiency are best remembered from the mnemonic AAA: **antibiotics** (for example, sulfamethoxazole and chloramphenicol), **antimalarials** (for example, primaquine but not chloroquine or quinine), and **antipyretics** (for example, acetanilide but not acetaminophen).

2. **Favism:** Some forms of *G6PD* deficiency, for example, the Mediterranean variant, are particularly susceptible to the hemolytic effect of the fava (broad) bean, a dietary staple in the Mediterranean region. Favism, the hemolytic effect of ingesting fava beans, is not observed in all individuals with *G6PD* deficiency, but all patients with favism have *G6PD* deficiency.

3. **Infection:** Infection is the **most common precipitating factor** of hemolysis in *G6PD* deficiency. The inflammatory response to infection results in the generation of free radicals in macrophages. The radicals can diffuse into the RBC and cause oxidative damage.

C. Variant G6PD properties

Almost all *G6PD* variants are caused by **point mutations** (see p. 636) in the gene for *G6PD*. Some mutations do not affect enzymic activity. However, other mutations result in decreased catalytic activity, decreased stability, or an alteration of binding affinity for $NADP^+$ or glucose 6-phosphate. [Note: Active *G6PD* exists as a homodimer or tetramer. Mutations at the interface between subunits can affect stability.] The severity of the disease usually correlates with the amount of residual enzyme activity in the patient's RBC. For example, variants can be classified as shown in Figure 18.9. *G6PD* A^- is the prototype of the moderate (class III) form of the disease. The RBC contain an unstable but kinetically normal *G6PD*, with most of the enzyme activity present in the reticulocytes and younger RBC (Fig. 18.10). Therefore, the **oldest RBC have the lowest level of enzyme**

Class	Clinical symptoms	Residual enzyme activity
I	Very severe (chronic, nonspherocytic hemolytic anemia)	<10%
*II	Severe (acute hemolytic anemia)	<10%
*III	Moderate	10%–60%
IV	None	>60%

Figure 18.9 Classification of *glucose 6-phosphate dehydrogenase* (*G6PD*) deficiency variants. [Note: Class V variants (not shown) result in overproduction of *G6PD*.] * = most common.

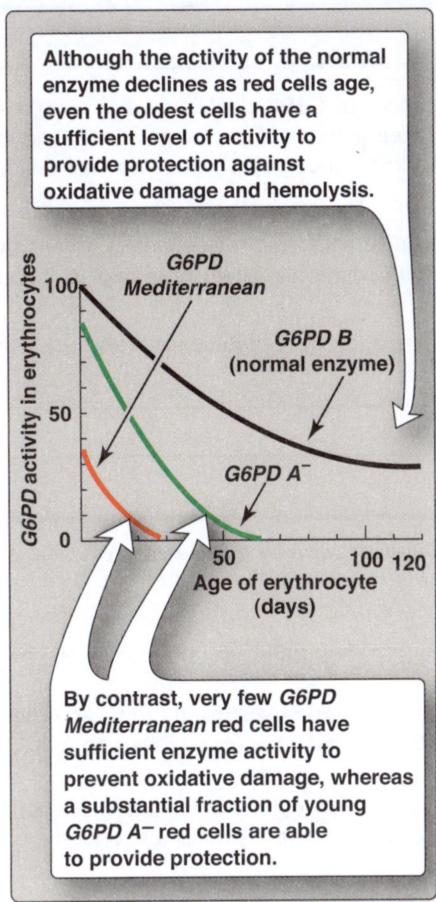

Figure 18.10 Decline of erythrocyte *glucose 6-phosphate dehydrogenase* (*G6PD*) activity with cell age for the three most commonly encountered forms of the enzyme.

activity and are preferentially removed in a hemolytic episode. Because hemolysis does not affect younger cells, the episodes are self-limiting. *G6PD Mediterranean* is the prototype of a more severe (class II) deficiency. Class I mutations (rare) are the most severe and are associated with chronic nonspherocytic hemolytic anemia, which occurs even in the absence of oxidative stress.

D. G6PD molecular biology

The cloning of the gene for *G6PD* and the sequencing of its DNA (see Chapter 44) have permitted the identification of mutations that cause *G6PD* deficiency. More than 400 *G6PD* variants have been identified, a finding that explains the numerous biochemical and clinical phenotypes that have been described. Most mutations that result in enzymic deficiency are missense mutations (see p. 636) in the coding region. Both *G6PD A⁻* and *G6PD Mediterranean* represent mutant enzymes that differ from the respective normal variants by a single amino acid. Large deletions or frameshift mutations have not been identified, suggesting that complete absence of *G6PD* activity is likely lethal.

VI. Medical Perspectives

Defects in the HMP shunt have major health consequences. The consequences of G6PD deficiency are mostly seen in RBCs, which can no longer defend against oxidative stress. The deficiency leads to hemolytic anemia, with aggregates of globin chains forming the characteristic Heinz bodies. G6PD deficiency confers a degree of resistance to malaria.

Persons with G6PD deficiency also show increased methemoglobin in circulation.

HMP shunt is used for diagnosing thiamine (vitamin B_1) deficiency. Thiamine is administered to suspected patient and the activity of the *transketolase* enzyme of HMP shunt within RBCs is measured. If *transketolase* activity increases, it confirms the diagnosis of thiamine deficiency.

Rare genetic defects in *NADPH oxidase* cause **chronic granulomatous disease**, an immunodeficient state characterized by recurrent bacterial and fungal infections.

VII. Chapter Summary

- The **pentose phosphate pathway** includes an irreversible oxidative phase followed by a series of reversible sugar–phosphate interconversions (Fig. 18.11).
- No ATP is directly consumed or produced in the pathway.
- The NADPH-producing **oxidative portion** of the pathway is important in providing reducing equivalents for reductive biosynthesis and detoxification reactions.
- **Glucose 6-phosphate** is irreversibly converted to **ribulose 5-phosphate**, and **two NADPH** are produced. The regulated step is catalyzed by ***glucose 6-phosphate dehydrogenase*** (***G6PD***), which is strongly inhibited by a rise in the **NADPH/NADP$^+$ ratio**.
- **Reversible nonoxidative reactions** interconvert sugars. This part of the pathway converts ribulose 5-phosphate to **ribose 5-phosphate**, required for nucleotide and nucleic acid synthesis, or to **fructose 6-phosphate** and **glyceraldehyde 3-phosphate** (glycolytic intermediates).
- NADPH is a source of **reducing equivalents** in **reductive biosynthesis**, such as the production of fatty acids in liver, adipose tissue, and the mammary gland; cholesterol in the liver; and steroid hormones in the placenta, ovaries, testes, and adrenal cortex.
- NADPH is also required by **red blood cells** (**RBC**) for **hydrogen peroxide reduction. Reduced glutathione** (**G-SH**) is used by ***glutathione peroxidase*** to reduce the peroxide to water. The **oxidized glutathione** (**G-S-S-G**) produced is reduced by ***glutathione reductase***, using NADPH as the source of electrons.
- NADPH provides reducing equivalents for the **mitochondrial *cytochrome P450 monooxygenase* system**, which is used in **steroid hormone synthesis** in steroidogenic tissue, **bile acid synthesis** in the liver, and **vitamin D activation** in the liver and kidneys.
- The **microsomal system** uses NADPH to **detoxify** foreign compounds (xenobiotics), such as drugs and a variety of pollutants.
- NADPH provides the reducing equivalents for phagocytes in the process of eliminating invading microorganisms. ***NADPH oxidase*** uses molecular oxygen (O_2) and electrons from NADPH to produce **superoxide radicals**, which, in turn, can be converted to peroxide by ***superoxide dismutase***. ***Myeloperoxidase*** catalyzes the formation of bactericidal hypochlorous acid from peroxide and chloride ions.
- ***G6PD* deficiency** impairs the ability of the cell to form the NADPH that is essential for the maintenance of the G-SH pool. The cells most affected are **RBC** because they do not have additional sources of NADPH.
- *G6PD* deficiency is an **X-linked disease** characterized by **hemolytic anemia** caused by the production of free radicals and peroxides following administration of **oxidant drugs**, ingestion of **fava beans**, or severe **infections**. The extent of the anemia depends on the amount of residual enzyme. Babies with *G6PD* deficiency may experience **neonatal jaundice**.

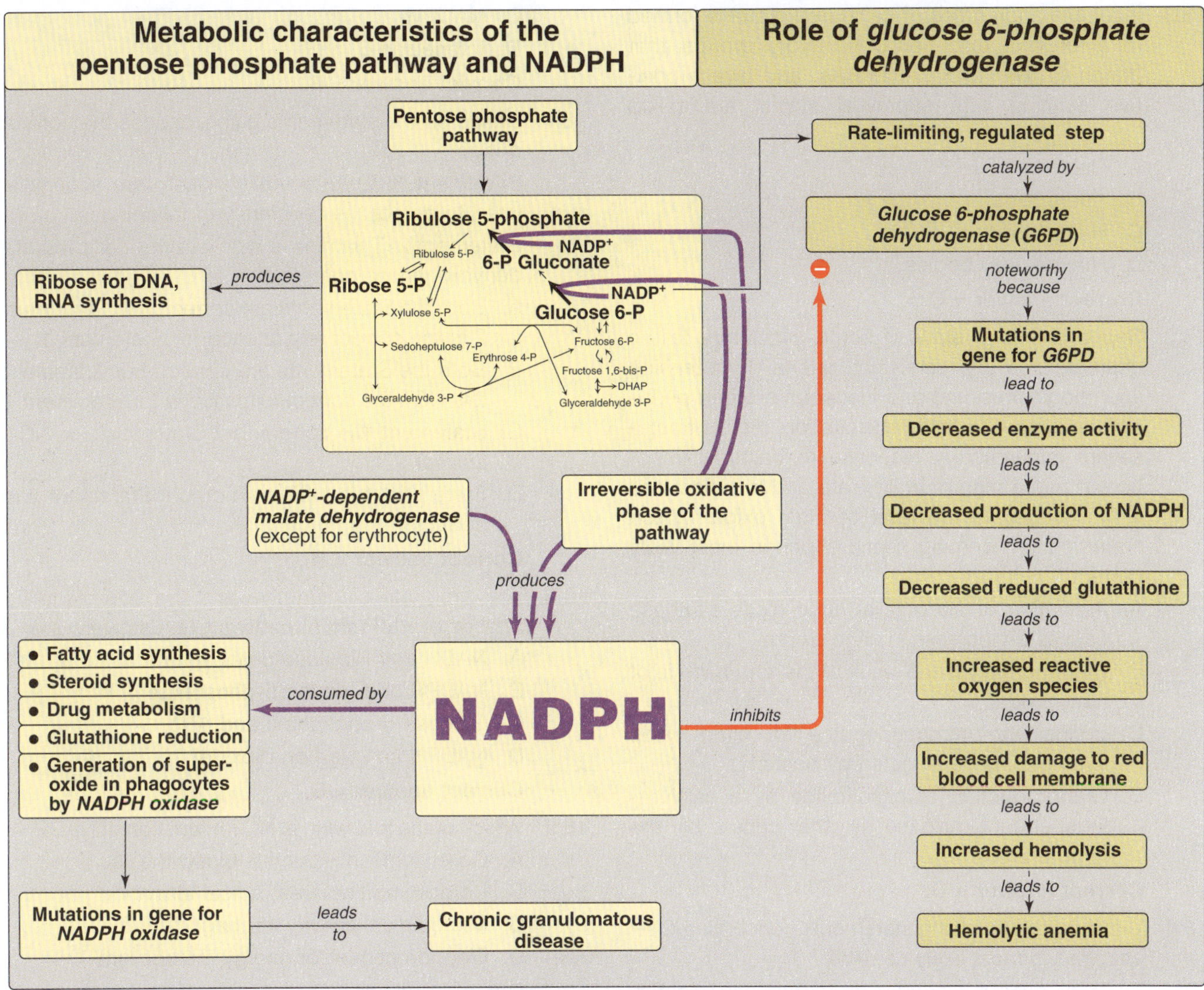

Figure 18.11 Key concept map for the pentose phosphate pathway and nicotinamide adenine dinucleotide phosphate (NADPH).

Study Questions

Short-Answer Questions

Write short notes on:

18.1. Glucose 6-phosphate dehydrogenase deficiency
18.2. Effect of nitric oxide on vascular endothelium
18.3. Cytochrome P450 monooxygenase system
18.4. Biological role of nitric oxide
18.5. Measurement of the activity of which enzyme in red blood cells could be used to determine thiamine status in the body? Explain its biochemical basis.

Long-Answer Questions

18.1. Describe the phases and regulation of pentose phosphate pathway.
18.2. Describe the uses of NADPH.

Multiple-Choice Questions

Choose the ONE best answer.

18.1. In preparation for a trip to an area of India where chloroquine-resistant malaria is endemic, a young man is given primaquine prophylactically. Soon thereafter, he develops a hemolytic condition due to a deficiency in glucose 6-phosphate dehydrogenase. A less-than-normal level of which of the following is a consequence of the enzyme deficiency and the underlying cause of the hemolysis?

A. Glucose 6-phosphate
B. Oxidized form of nicotinamide adenine dinucleotide
C. Reduced form of glutathione
D. Ribose 5-phosphate

Correct answer = C.

18.2. Insoluble aggregates of denatured proteins formed as a result of oxidation of sulfhydryl groups, that attach to the RBC membranes, and give a positive staining with supravital stains, are called __________.

A. Dohle bodies
B. Auer rods
C. Heinz bodies
D. Howell–Jolly bodies

Correct answer = C.

18.3. Septic shock, a state of acute circulatory failure characterized by persistent arterial hypotension (low blood pressure) and inadequate organ perfusion refractory to fluid resuscitation, results from a severe inflammatory response to bacterial infection. It has a high mortality rate and is associated with changes in the level of nitric oxide. Which statement concerning septic shock is most likely correct?

A. Activation of endothelial nitric oxide synthase causes an increase in nitric oxide.
B. High mortality is the result of the long half-life of nitric oxide.
C. Lysine, the nitrogen source for nitric oxide synthesis, is deaminated by bacteria.
D. Overproduction of nitric oxide by a calcium-independent enzyme is the cause of the hypotension.

Correct answer = D.

18.4. Which of the following statements concerning CYP enzymes is most likely correct?

A. They accept electrons from reduced nicotinamide adenine dinucleotide.
B. They catalyze the hydroxylation of hydrophobic molecules.
C. They differ from nitric oxide synthase in that they contain heme.
D. They function in association with an oxidase.

Correct answer = B.

18.5. Which of the following is the most common reaction catalyzed by cytochrome P450 enzymes?

A. Dehydrogenation
B. Mono-oxygenation
C. Di-oxygenation
D. Hydro-peroxidation

Correct answer = B.

18.6. Sequestration of which of the following molecules by intravascular heme is thought to be responsible for vasoconstriction and pain manifestations in patients with sickle cell disease?

A. Arginine
B. Nitric oxide
C. NADPH
D. Interleukin-6

Correct answer = B.

18.7. Based on the statements given, choose the correct option.

Statement 1: Erythrocyte transketolase activity is used to detect the deficiency of thiamine.

Statement 2: Thiamine is a co-enzyme for pyruvate dehydrogenase complex.

A. Both the Statements are correct, and Statement 2 is the correct explanation for Statement 1
B. Both the Statements are correct, but Statement 2 is NOT the correct explanation for Statement 1
C. Statement 1 is correct, but Statement 2 is NOT correct
D. Statement 1 is NOT correct, but Statement 2 is correct

Correct answer = B.

18.8. The products of oxidative and the non-oxidative phases of HMP shunt pathway, respectively, are:

A. NADH and Ribulose-5-phosphate
B. NADPH and Ribose-5-phosphate
C. Glucose-6-phosphate and ATP
D. NADPH and Carbon dioxide

Correct answer = B.

18.9. Which of the following is NOT a function of NADPH?

A. Co-enzyme in reductive biosynthesis
B. Defense against free radical formation
C. Microbicidal activity in phagocytosis
D. Electron carrier for energy metabolism

Correct answer = D.

18.10. In which of the following cells/tissues is the production of nitric oxide increased by the action of tumor necrosis factor-α and interferon-γ?

A. Macrophages
B. Neurons
C. Endothelium
D. Skeletal muscles

Correct answer = A.

Multiple-Choice Questions (NExT/NEET PG)

Choose the ONE best answer.

18.1. During evolution, humans and certain primates have lost their ability to synthesize a vitamin. This change is attributed to the availability and consumption of citrus fruits, which relieved the selection pressure on the gene required for that vitamin synthesis. Gene for which among the following enzymes was lost in this process?

A. Aldonolactonase

B. Glucuronate reductase

C. Gulonolactone oxidase

D. UDP-glucose dehydrogenase

Correct answer = C.

18.2. In male patients who are hemizygous for glucose 6-phosphate dehydrogenase (G6PD) deficiency, pathophysiologic consequences are more apparent in erythrocytes (RBC) than in other cells, such as in the liver. Which one of the following provides the most reasonable explanation for this different response by these individual tissue types?

A. Excess glucose 6-phosphate in the liver, but not in RBCs, can be channeled to glycogen, thus averting cellular damage.

B. Liver cells, in contrast to RBCs, have alternative mechanisms for supplying the NADPH required for keeping metabolic and cellular integrity.

C. Glucose 6-phosphatase activity in RBCs decreases the level of glucose 6-phosphate, thus resulting in cell damage. This does not happen in the hepatocyte.

D. Because RBCs do not have mitochondria, production of ATP required to keep cell integrity depends exclusively on the routing of glucose 6-phosphate to the pentose phosphate pathway.

Correct answer = B.

18.3. A G6PD A^- individual was treated with primaquine from day 0 to day 120. Hemolysis occurred immediately after initiation of drug therapy, as indicated by progressive anemia, hemoglobinuria, and reticulocytosis. However, despite continued administration of the drug, the hemolysis spontaneously decreased and red cell survival improved with time. Red cell G6PD activity measured 2 months after termination of therapy was 10% of normal. Which one of the following statements about this patient is correct?

A. The patient shows increased bone marrow erythropoiesis.

B. Erythrocytes in this patient exhibit a longer lifetime than in normal individuals.

C. During the period of peak hemolysis, erythrocytes from the patient will show no G6PD activity.

D. The intracellular concentration of NADPH in the patient's erythrocytes is greater than normal.

Correct answer = A.

18.4. A 6-year-old boy was brought to the pediatric clinic with a history of recurrent bacterial and fungal infections, including abscesses and pneumonia. His mother reported his symptoms began in infancy with frequent skin and lung infections that were difficult to treat. On examination, the boy appeared pale with multiple healed and active skin lesions. Nitroblue tetrazolium (NBT) reduction test was positive. What is the most likely diagnosis in this patient?

A. G-6-PD deficiency

B. Chronic granulomatous disease

C. Hyperimmunoglobulin E syndrome

D. Severe combined immunodeficiency

Correct answer = B.

18.5. Beutler test, also known as the fluorescent spot test, is based on the fluorescence shown by NADPH on exposure to ultraviolet light. In this test, the contents of the lysed RBCs from the blood of the subject are incubated with the substrate and this is then applied on a filter paper. The filter paper is subsequently exposed to ultraviolet light to monitor the NADPH levels. Which of the following conditions can be screened using this test?

A. G-6-PD deficiency

B. Alkaptonuria

C. Hereditary fructose intolerance

D. Essential fructosuria

Correct answer = A.

Glycosaminoglycans, Proteoglycans, and Glycoproteins

19

Learning Objectives

At the end of this chapter, the student will be able to:

- Describe the structure and classification of glycosaminoglycans
- Explain the synthesis and degradation of glycosaminoglycans
- Discuss the diseases associated with glycosaminoglycan metabolism
- Describe the structure and synthesis of glycoproteins

I. GLYCOSAMINOGLYCAN OVERVIEW

Glycosaminoglycans (GAG) are large complexes of negatively charged heteropolysaccharide chains. They are generally associated with a small amount of protein (core protein), forming proteoglycans, which typically consist of up to 95% carbohydrate. GAG have the special ability to bind large amounts of water, thereby producing the gel-like matrix. This matrix forms the basis of the body's ground substance, which, along with fibrous structural proteins such as collagen, elastin, and fibrillin-1, and adhesive proteins such as fibronectin, makes up the extracellular matrix (ECM). **The hydrated GAG serve as a flexible support for the ECM**, interacting with the structural and adhesive proteins, and as a molecular sieve, influencing movement of materials through the ECM. The viscous, lubricating properties of mucous secretions also result from the presence of GAG, which led to the original naming of these compounds as mucopolysaccharides.

II. STRUCTURE

GAG are **long, unbranched, heteropolysaccharide chains composed of a repeating disaccharide unit** [acidic sugar–amino sugar]$_n$ (Fig. 19.1). [Note: A single exception is keratan sulfate, which contains galactose rather than an acidic sugar.]

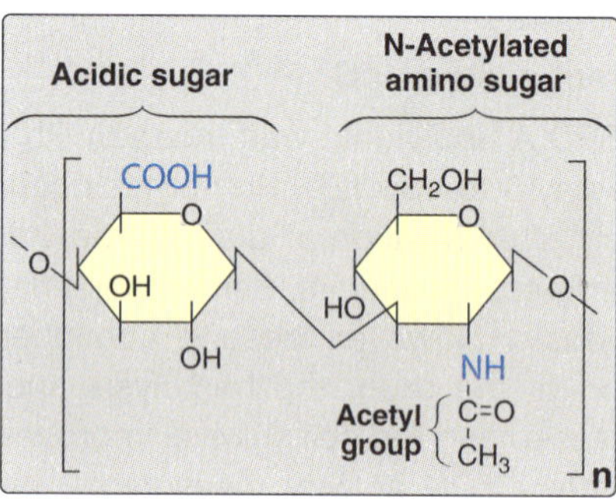

Figure 19.1 Repeating disaccharide unit of glycosaminoglycans.

The **amino sugar is either D-glucosamine or D-galactosamine**, in which the amino group is usually acetylated, thus eliminating its positive charge. The amino sugar may also be sulfated on carbon 4 or 6 or on a nonacetylated nitrogen.

The **acidic sugar is either D-glucuronic acid or its C-5 epimer L-iduronic acid** (Fig. 19.2). These uronic sugars contain carboxyl groups that are negatively charged at physiologic pH and, together with the sulfate groups ($-SO_4^{2-}$), give GAG their strongly negative nature.

A. Structure–function relationship

Because of the **high concentration of negative charges**, these heteropolysaccharide chains tend to be extended in solution. They **repel each other** and

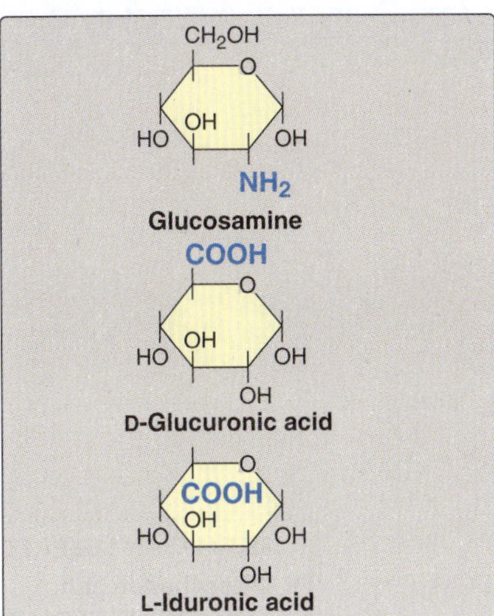

Figure 19.2 Some monosaccharide units found in glycosaminoglycans.

are **surrounded by a shell of water molecules**. When brought together, they slide past each other, much as two magnets with the same polarity seem to slide past each other. This produces the **slippery consistency of mucous secretions and synovial fluid.** When a solution of GAG is compressed, the water is squeezed out, and the GAG are forced to occupy a smaller volume. When the compression is released, the GAG spring back to their original, hydrated volume because of the repulsion of their negative charges. This property contributes to the **resilience of synovial fluid and the vitreous humor of the eye** (Fig. 19.3).

B. Classification

The six major types of GAG are divided according to monomeric composition, type of glycosidic linkages, and degree and location of sulfate units. The structure of the GAG and their distribution in the body is illustrated in Figure 19.4. All GAG, except for hyaluronic acid (HA), are sulfated and are found covalently attached to protein, forming proteoglycan monomers.

C. Proteoglycans

Proteoglycans are found in the ECM and on the outer surface of cells.

1. **Monomer structure:** A proteoglycan monomer found in cartilage consists of a core protein to which up to 100 linear chains of GAG are covalently attached. These chains, which may each be composed of up to 200 disaccharide units, extend out from the core protein and remain separated from each other because of charge repulsion. The resulting structure resembles a bottle brush (Fig. 19.5). In cartilage proteoglycans, the

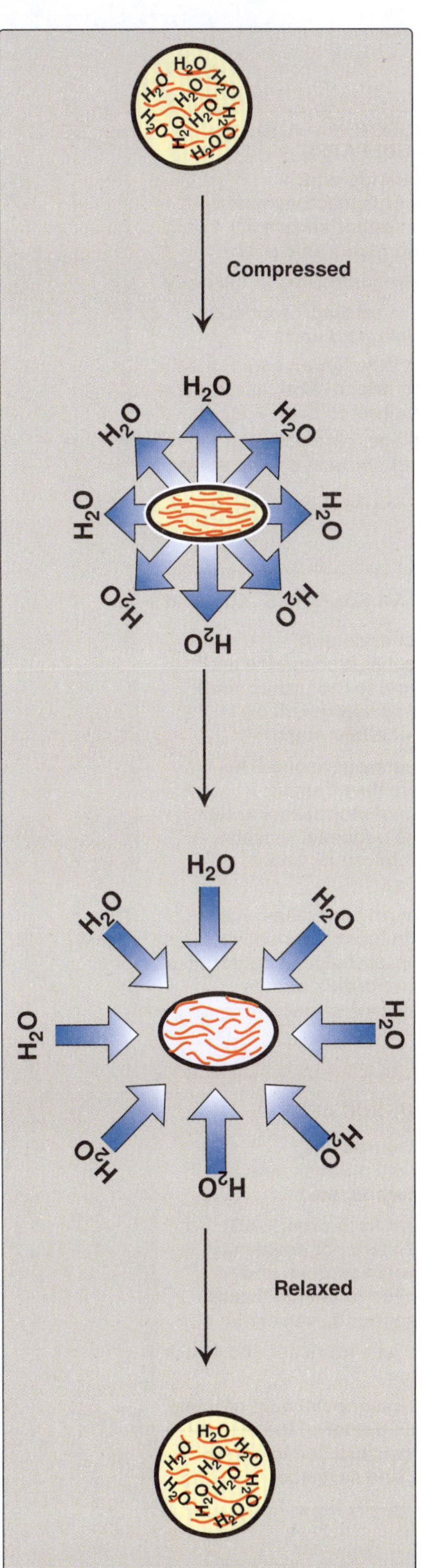

Figure 19.3 Resilience of glycosaminoglycans.

species of GAG include chondroitin sulfate and keratan sulfate. [Note: Proteoglycans are grouped into gene families that encode core proteins with common structural features. The aggrecan family

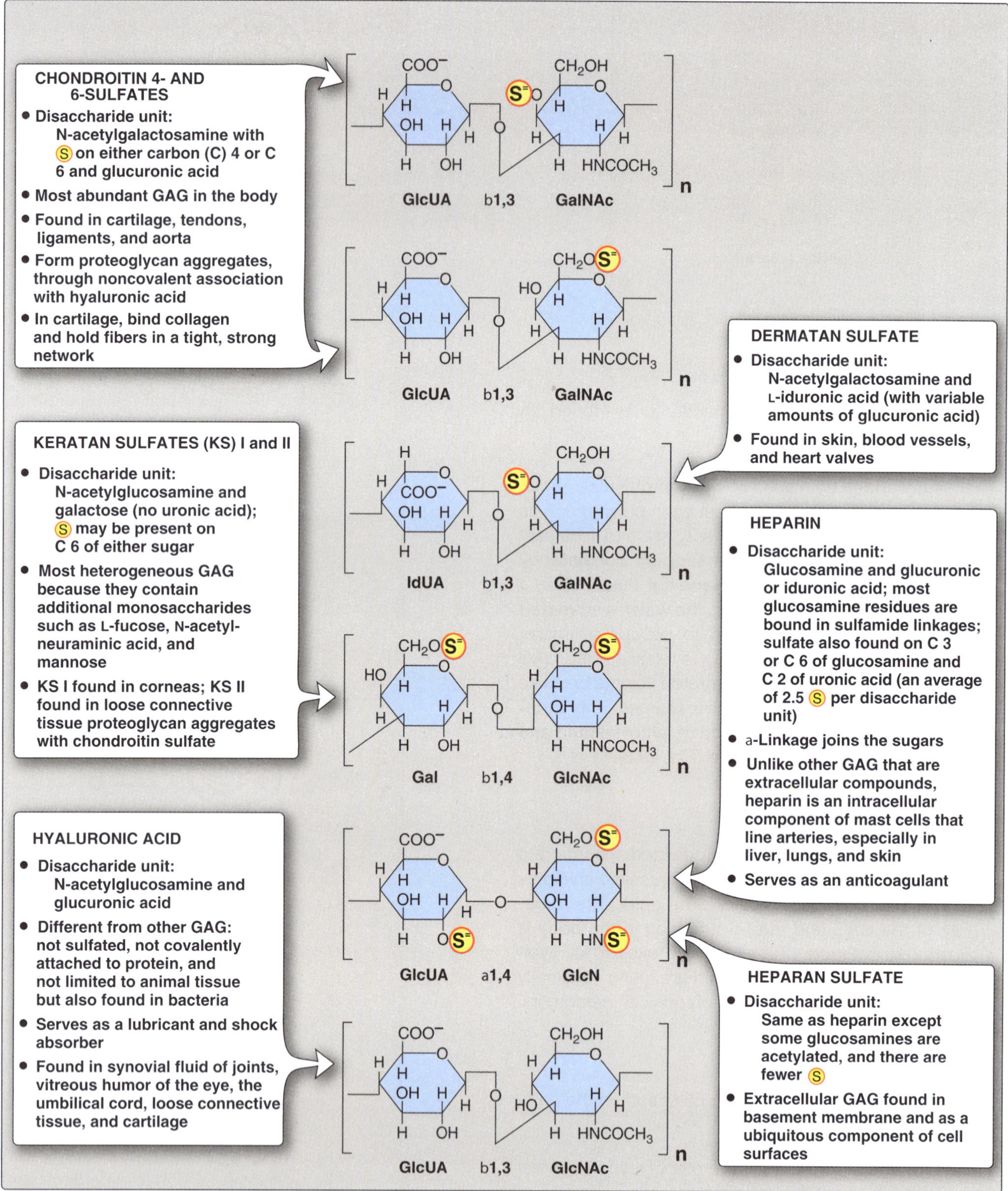

Figure 19.4 Structure of repeating units in and distribution of glycosaminoglycans (GAG). Sulfate groups (S) are shown in all possible positions. GlcUA and IdUA = glucuronic and iduronic acids; GalNAc = N-acetylgalactosamine; GlcNAc = N-acetylglucosamine; GlcN = glucosamine; Gal = galactose.

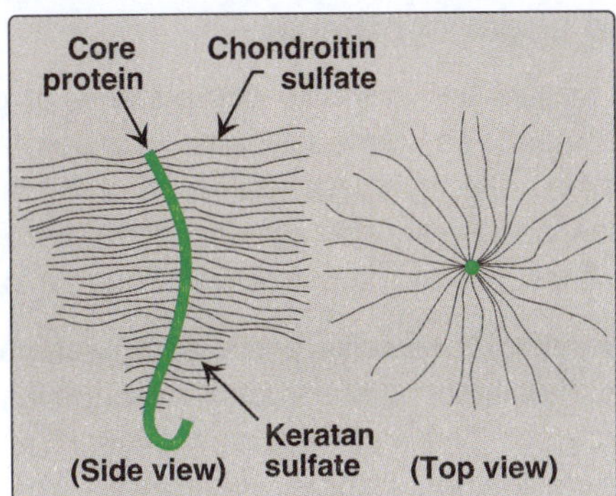

Figure 19.5 Bottle brush model of a cartilage proteoglycan monomer.

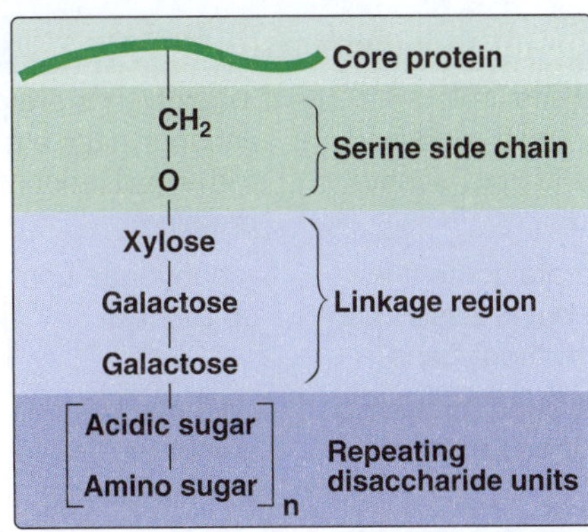

Figure 19.6 Glycosaminoglycan linkage regions.

(aggrecan, versican, neurocan, and brevican), abundant in cartilage, is an example.]

2. **GAG–protein linkage:** This covalent linkage is most commonly through a trihexoside (galactose-galactose-xylose) and a serine residue in the protein. An O-glycosidic bond (see p. 178) is formed between the xylose and the hydroxyl group of the serine (Fig. 19.6).

3. **Aggregate formation:** Many proteoglycan monomers can associate with one molecule of HA to form proteoglycan aggregates. The association is not covalent and occurs primarily through ionic interactions between the core protein and the HA. The association is stabilized by additional small proteins called link proteins (Fig. 19.7).

4. **Difference between proteoglycans and glycoproteins:** Glycoproteins and proteoglycans are two types of molecules that contain both proteins and carbohydrate units. The carbohydrate units are covalently bound to protein molecules, varying in size from monosaccharides to polysaccharides.

The key difference between proteoglycans and glycoproteins is that proteoglycans have long, unbranched chains with disaccharide units as repeating structures, while glycoproteins have short, highly branched glycan chains with no repeating units. The differences between proteoglycans and glycoproteins are shown in Figure 19.8.

Clinical Application 19.1: Proteoglycans, Cartilage, and Osteoarthritis

Osteoarthritis affects millions of individuals worldwide. In this disease, joint cartilage is degraded and proteoglycans that normally help provide a cushion for the joint are lost. Without the resilience of the cartilage protecting the joint, there is pain, stiffness, and swelling, with progressive worsening of signs and symptoms. Glucosamine and chondroitin have been reported both to relieve pain and to stop progression of osteoarthritis. Based on several well-controlled clinical studies, it appears that glucosamine sulfate (but not glucosamine hydrochloride) and chondroitin sulfate may have a small to moderate effect in relieving symptoms of osteoarthritis.

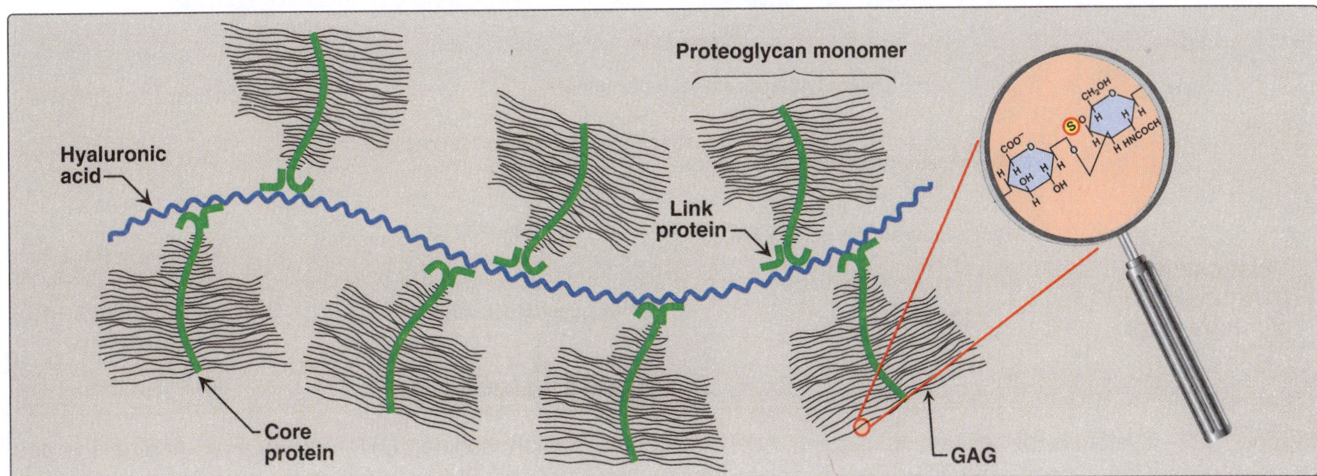

Figure 19.7 Proteoglycan aggregate. GAG = glycosaminoglycan.

Glycoproteins	Proteoglycans
Glycoproteins are proteins to which oligosaccharides are covalently attached	Proteoglycans are glycosaminoglycans associated with small amount of protein
Carbohydrate content is variable but is lesser than that of proteoglycans	Carbohydrate content is up to 95%
They have short, branched heterooligosaccharide chains	They have long, unbranched, heteropolysaccharide chains
They do not have repeating disaccharide units	Heteropolysaccharide chains are composed of a repeating disaccharide unit
Carbohydrate chains may or may not be negatively charged	Carbohydrate chains are negatively charged
They combine with collagen to form cartilage, modulation of cellular development	Cell-to-cell recognition and signaling

Figure 19.8 Differences between glycoproteins and proteoglycans

III. SYNTHESIS

The heteropolysaccharide chains are elongated by the sequential addition of alternating acidic and amino sugars donated primarily by their uridine diphosphate (UDP) derivatives. The reactions are catalyzed by a family of specific *glycosyltransferases*. Because GAG are produced for export from the cell, their synthesis occurs primarily in the Golgi and not in the cytosol.

A. Amino sugar synthesis

Amino sugars are essential components of glycoconjugates such as proteoglycans, glycoproteins, and glycolipids. The synthetic pathway of amino sugars (hexosamines) is very active in connective tissues, where as much as 20% of glucose flows through this pathway.

1. **N-Acetylglucosamine and N-acetylgalactosamine:** The monosaccharide fructose 6-phosphate is the precursor of N-acetylglucosamine (GlcNAc) and N-acetylgalactosamine (GalNAc). A hydroxyl group on the fructose is replaced by the amide nitrogen of a glutamine, and the glucosamine 6-phosphate product gets acetylated, isomerized, and activated, producing the nucleotide sugar UDP-GlcNAc (Fig. 19.9). UDP-GalNAc is generated by the epimerization of UDP-GlcNAc. It is these nucleotide sugar forms of the amino sugars that are used to elongate the carbohydrate chains.

2. **N-Acetylneuraminic acid:** NANA, a nine-carbon, acidic monosaccharide (see Fig. 22.15, p. 330), is a member of the family of sialic acids, each of which is acylated at a different site. These compounds are usually found as terminal carbohydrate residues of oligosaccharide side chains of glycoproteins, of glycolipids, or, less frequently, of GAG. N-Acetylmannosamine 6-phosphate (derived from fructose 6-phosphate) and phosphoenolpyruvate (an intermediate in glycolysis; see p. 198) are the immediate sources of the carbons and nitrogens for NANA synthesis (see Fig. 19.9).

Before NANA can be added to a growing oligosaccharide, it must be activated to cytidine monophosphate (CMP)-NANA by reacting with cytidine

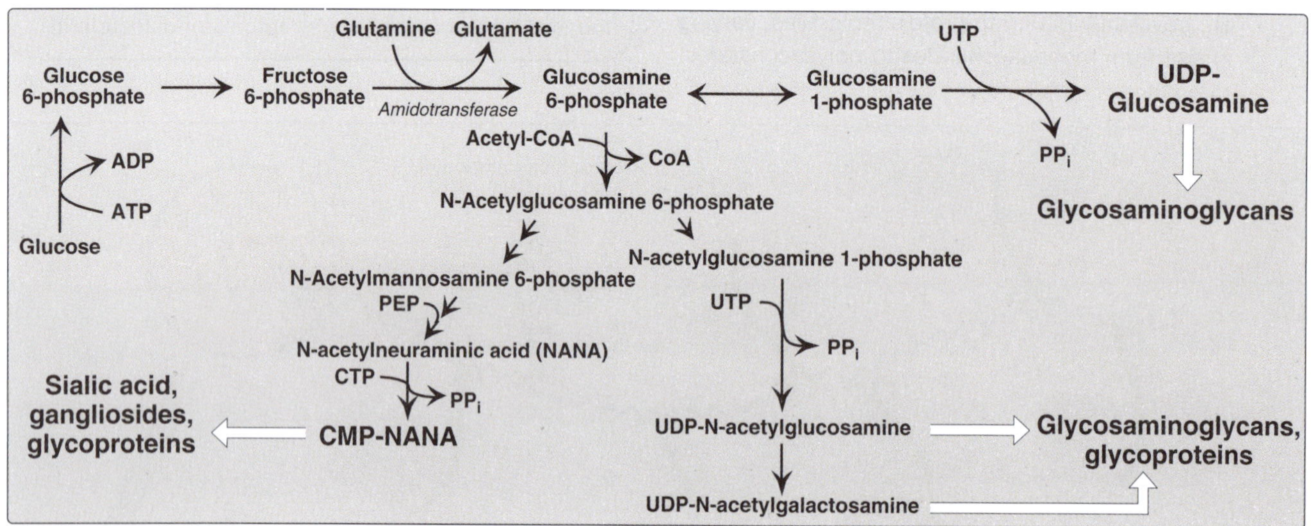

Figure 19.9 Synthesis of the amino sugars. ADP = adenosine diphosphate; UTP and UDP = uridine tri- and diphosphates; CoA = coenzyme A; PEP = phosphoenolpyruvate; CTP and CMP = cytidine tri- and monophosphates; PP$_i$ = pyrophosphate.

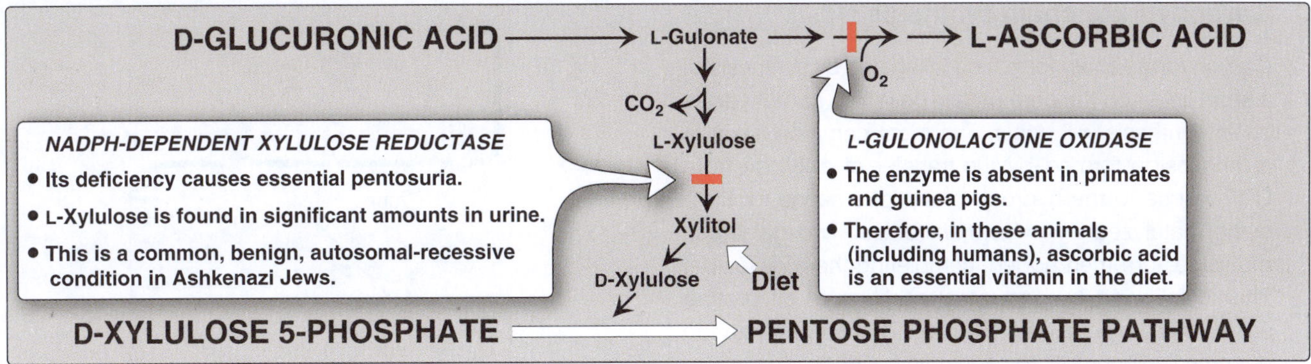

Figure 19.10 Metabolism of glucuronic acid. NADPH = reduced nicotinamide adenine dinucleotide phosphate; CO_2 = carbon dioxide.

triphosphate (CTP). *CMP-NANA synthetase* catalyzes the reaction. [Note: CMP-NANA is the only nucleotide sugar in human metabolism in which the carrier nucleotide is a monophosphate rather than a diphosphate.]

B. Acidic sugar synthesis

D-Glucuronic acid and its C-5 epimer, L-iduronic acid, are essential components of GAG. Glucuronic acid is also required for the detoxification of lipophilic compounds, such as bilirubin, steroids (see p. 786), and many drugs, including the statins (see p. 345), because conjugation with glucuronate (glucuronidation) increases water solubility. In plants and mammals (other than guinea pigs and primates, including humans), glucuronic acid is a precursor of ascorbic acid (vitamin C) as shown in Figure 19.10. This uronic acid pathway also provides a mechanism by which dietary D-xylulose can enter the central metabolic pathways.

1. **Glucuronic acid:** Glucuronic acid can be obtained in small amounts from the diet and from the lysosomal degradation of GAG. It also can be synthesized by the uronic acid pathway, in which glucose 1-phosphate reacts with uridine triphosphate (UTP) and is converted to UDP-glucose. Oxidation of UDP-glucose produces UDP-glucuronic acid, the form that supplies glucuronic acid for GAG synthesis and glucuronidation (Fig. 19.11). The end product of glucuronic acid metabolism in humans is D-xylulose 5-phosphate, which can enter the pentose phosphate pathway and produce the glycolytic intermediates glyceraldehyde 3-phosphate and fructose 6-phosphate (see Fig. 19.10; see also Fig. 18.2, p. 250).

2. **l-Iduronic acid:** Synthesis of L-iduronic acid occurs after D-glucuronic acid has been incorporated into the carbohydrate chain. *Uronosyl 5-epimerase* causes epimerization of the D- to the L-sugar.

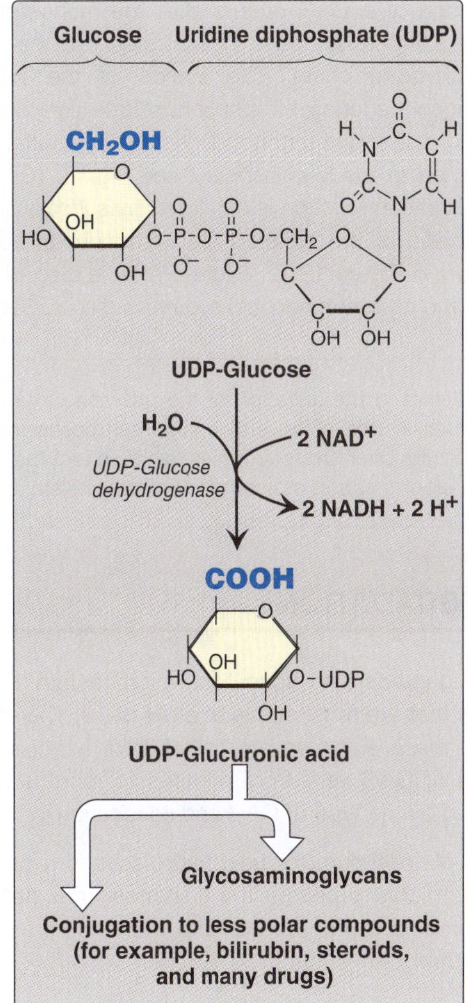

Figure 19.11 Oxidation of UDP-glucose to UDP-glucuronic acid. NAD(H) = nicotinamide adenine dinucleotide.

C. Core protein synthesis

The core protein is made by ribosomes on the rough endoplasmic reticulum (RER), enters the RER lumen, and then moves to the Golgi, where it is glycosylated by membrane-bound *glycosyltransferases*.

D. Carbohydrate chain synthesis

Carbohydrate chain formation is initiated by synthesis of a short linker on the core protein on which carbohydrate chain synthesis will occur. The most common linker is a trihexoside formed by the transfer of a xylose from UDP-xylose to the hydroxyl group of a serine (or threonine) catalyzed by *xylosyltransferase*. Two galactose molecules are then added, completing the trihexoside. This is followed by sequential addition of alternating acidic and amino sugars (Fig. 19.12) and epimerization of some D-glucuronyl to L-iduronyl residues.

E. Sulfate group addition

Sulfation of a GAG occurs after the monosaccharide to be sulfated has been incorporated into the growing carbohydrate chain. The source of the sulfate is 3′-phosphoadenosyl-5′-phosphosulfate ([PAPS] a molecule of adenosine monophosphate with a sulfate group attached to the 5′-phosphate; see Fig. 22.16, p. 330). The sulfation reaction is catalyzed by *sulfotransferases*. Synthesis of the sulfated GAG chondroitin sulfate is shown in Figure 19.12. [Note: PAPS is also the sulfur donor in glycosphingolipid synthesis (see p. 330).]

> A defect in the sulfation of the growing GAG chains results in one of several autosomal-recessive disorders, the chondrodystrophies, which affect the proper development and maintenance of the skeletal system.

IV. DEGRADATION

GAG are degraded in lysosomes, which contain hydrolytic enzymes that are most active at a pH of ~5. Therefore, as a group, these enzymes are called *acid hydrolases*. The half-lives of GAG vary from minutes to months and are influenced by the type of GAG and its location in the body.

The low pH optimum for acid hydrolases is a protective mechanism that prevents the enzymes from destroying the cell should leakage occur into the cytosol where the pH is neutral.

A. GAG phagocytosis

Because GAG are extracellular or cell-surface compounds, they must first be engulfed by invagination of the cell membrane (phagocytosis), forming a vesicle inside of which are the GAG to be degraded. This vesicle then fuses with a lysosome, forming a single digestive vesicle in which the GAG are efficiently degraded (see p. 252 for a discussion of phagocytosis).

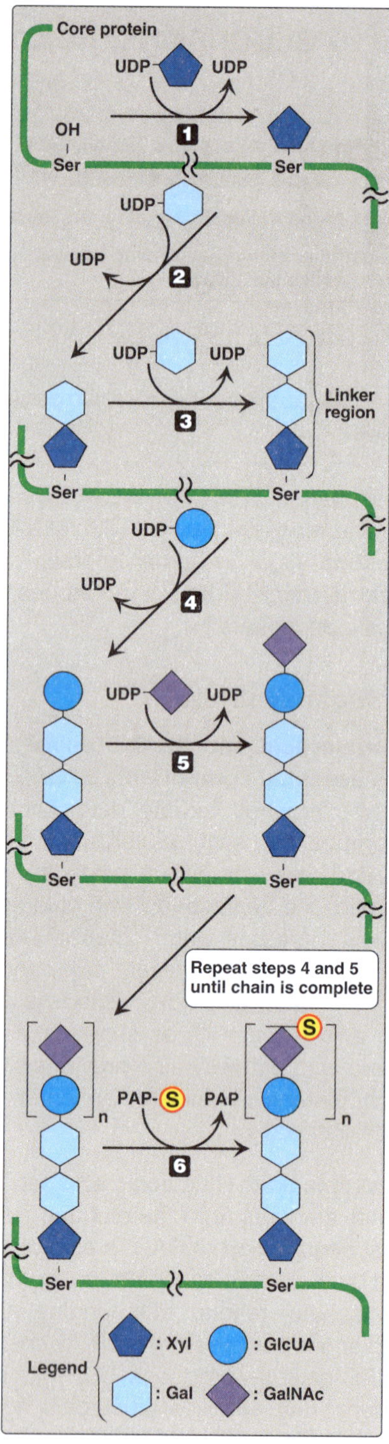

Figure 19.12 Synthesis of chondroitin sulfate. PAP-Ⓢ = 3′-phosphoadenosyl-5′- phosphosulfate; Ser = serine.

B. Lysosomal degradation

The lysosomal degradation of GAG requires a large number of *acid hydrolases* for complete digestion. First, the polysaccharide chains are cleaved by *endoglycosidases*, producing oligosaccharides. Further degradation of the oligosaccharides occurs sequentially from the nonreducing end (see p. 227) of each

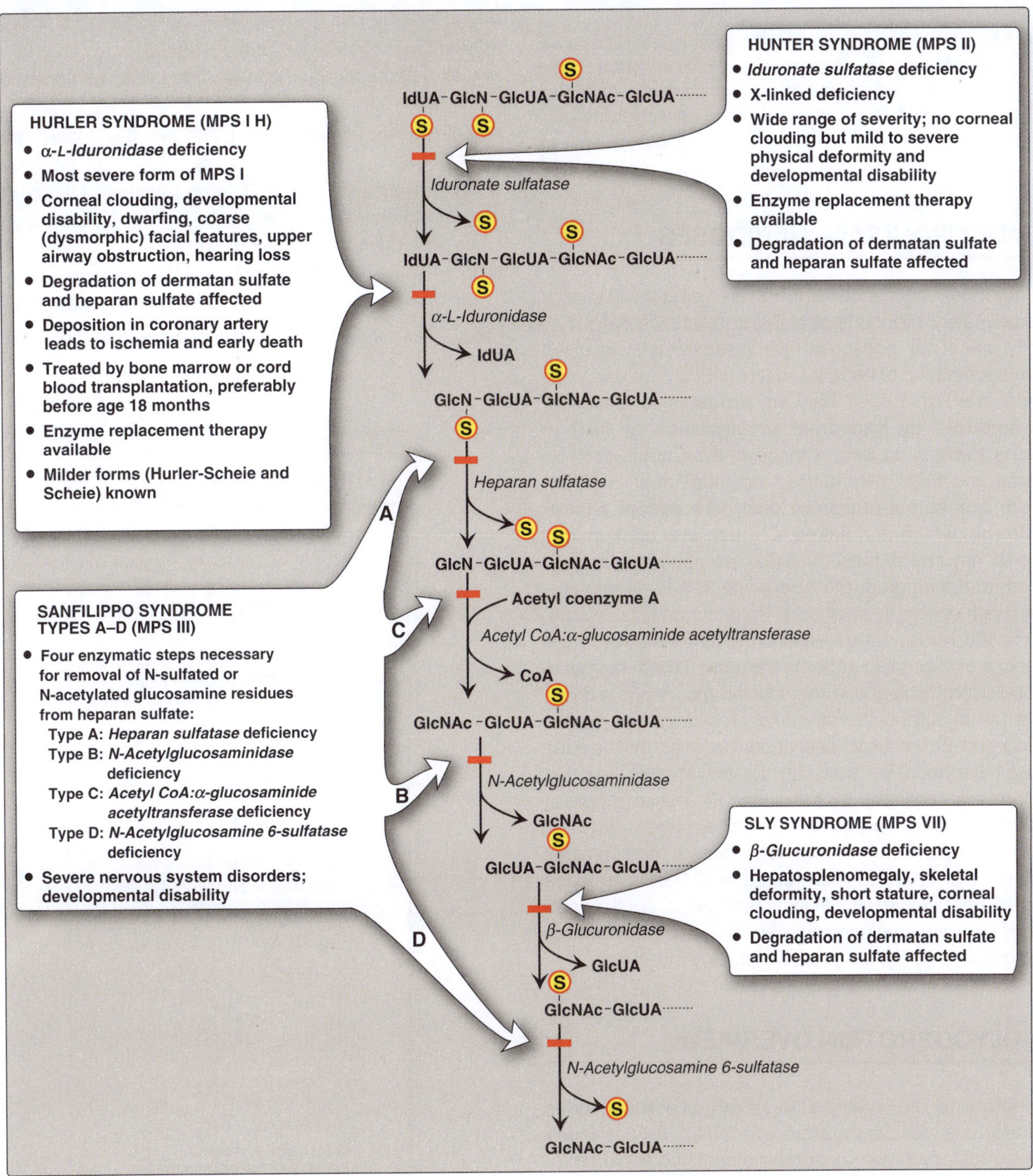

Figure 19.13 Degradation of the glycosaminoglycan heparan sulfate by lysosomal enzymes, indicating sites of enzyme deficiencies in some representative mucopolysaccharidoses (MPS). [Note: Deficiencies in *galactosamine 6-sulfatase* and *β-galactosidase* that degrade keratan sulfate result in Morquio syndrome (MPS IV), A and B, respectively. Deficiencies in *arylsulfatase B* that degrades dermatan sulfate result in Maroteaux-Lamy syndrome (MPS VI).] GlcUA and IdUA = glucuronic and iduronic acids; GalNAc = N-acetylgalactosamine; GlcNAc = N-acetylglucosamine; GlcN = glucosamine; Ⓢ = sulfate.

chain, the last group (sulfate or sugar) added during synthesis being the first group removed (by *sulfatases* or *exoglycosidases*).

Endo- and *exoglycosidases* are also involved in the lysosomal degradation of glycoproteins (see p. 272) and glycolipids (see p. 327). Deficiencies in these enzymes result in the accumulation of partially degraded carbohydrates, causing tissue damage. Examples of some of these enzymes and the bonds they hydrolyze are shown in Figure 19.13.

Multiple *sulfatase* deficiency (Austin disease) is a rare lysosomal storage disease in which all *sulfatases* are nonfunctional because of a defect in the formation of formylglycine, an amino acid derivative required at the active site for enzymic activity to occur.

V. MUCOPOLYSACCHARIDOSES

The mucopolysaccharidoses (MPS) are hereditary diseases (approximately 1:25,000 live births) caused by a deficiency of any one of the lysosomal *hydrolases* normally involved in the degradation of heparan sulfate (HS) and/or dermatan sulfate (see Fig. 19.13). They are **progressive disorders characterized by lysosomal accumulation of GAG in various tissues**, causing a range of symptoms, such as skeletal and ECM deformities, and intellectual disability. **All are autosomal-recessive disorders except Hunter syndrome, which is X linked**. Children who are homozygous for any one of these diseases are apparently normal at birth and then gradually deteriorate. In severe deficiencies, death occurs in childhood. There currently is no cure. Incomplete lysosomal degradation of GAG results in the presence of oligosaccharides in the urine. These fragments can be used to diagnose the specific mucopolysaccharidosis by identifying the structure present on the nonreducing end of the oligosaccharide, because that residue would have been the substrate for the missing enzyme. Diagnosis is confirmed by measuring the patient's cellular level of the lysosomal *hydrolases*. Bone marrow and cord blood transplants, in which transplanted macrophages produce the enzymes that degrade GAG, have been used to treat Hurler and Hunter syndromes, with limited success. Enzyme replacement therapy is available for both syndromes but does not prevent neurologic damage.

VI. GLYCOPROTEIN OVERVIEW

Glycoproteins are proteins to which oligosaccharides (glycans) are covalently attached. [Note: Glycosylation is the most common posttranslational modification of proteins.] They differ from the proteoglycans in several important ways. Glycoproteins contain highly variable amounts of carbohydrate but typically less than that of proteoglycans. For example, the glycoprotein immunoglobulin G (IgG) contains <4% of its mass as carbohydrate, whereas the proteoglycan aggrecan contains >80%. In glycoproteins, the glycan is relatively short (usually two to ten sugar residues in length, although it can be longer); does not contain repeating disaccharide units and, consequently, is structurally diverse; is often branched instead of linear; and may or may not be negatively charged.

Membrane-bound glycoproteins participate in a broad range of cellular phenomena, including:

- Cell-surface recognition (by other cells, hormones, and viruses)
- Cell-surface antigenicity (such as the blood group antigens)
- As components of the ECM and of the mucins of the gastrointestinal and urogenital tracts, where they act as protective biologic lubricants
- In addition, all of the globular proteins present in human plasma are glycoproteins (albumin is an exception)

Figure 19.14 summarizes some glycoprotein functions.

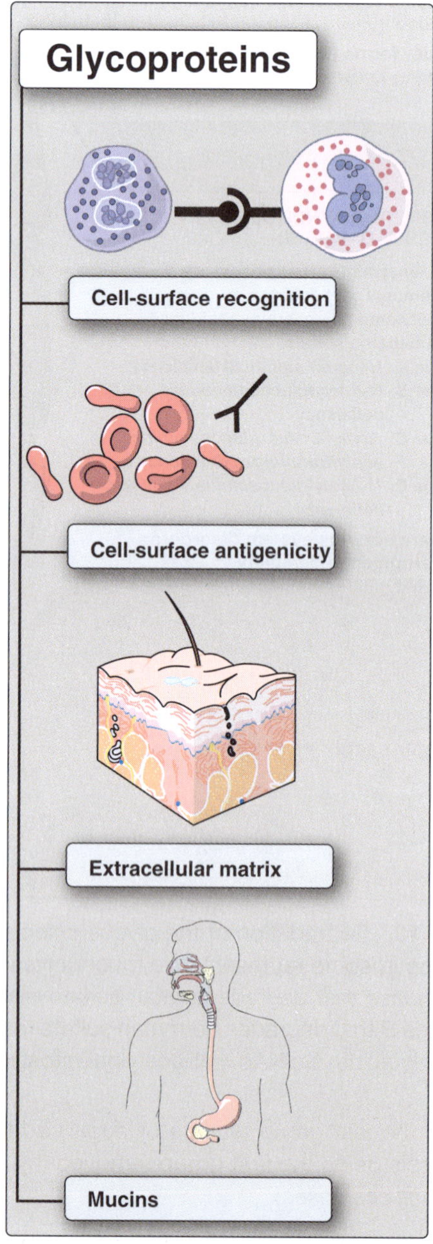

Figure 19.14 Functions of glycoproteins.

VII. OLIGOSACCHARIDE STRUCTURE

The oligosaccharide (glycan) components of glycoproteins are generally branched heteropolymers composed primarily of D-hexoses, with the addition in some cases of neuraminic acid (a nonose) and of L-fucose, a 6-deoxyhexose.

A. Carbohydrate–protein linkage

The glycan may be attached to the protein through an N- or an O-glycosidic link (see p. 179). In the former case, the sugar chain is attached to the amide group of an asparagine side chain and, in the latter case, to the hydroxyl group of either a serine or threonine side chain. [Note: In the case of collagen, there is an O-glycosidic linkage between galactose or glucose and the hydroxyl group of hydroxylysine (see p. 119).]

B. N- and O-Linked oligosaccharides

A glycoprotein may contain only one type of glycosidic linkage (N or O linked) or may have both types within the same molecule.

1. **O-Linked:** The O-linked glycans may have one or more of a wide variety of sugars arranged in either a linear or a branched pattern. Many are found in extracellular glycoproteins or as membrane glycoprotein components. For example, O-linked oligosaccharides on the surface of red blood cells help provide the **ABO blood group** determinants. If the terminal sugar on the glycan is GalNAc, the blood group is A. If it is galactose, the blood group is B. If neither GalNAc nor galactose is present, the blood group is O.

2. **N-Linked:** The N-linked glycans fall into two broad classes: complex oligosaccharides and high-mannose oligosaccharides. Both contain the same pentasaccharide core shown in Figure 19.15, but the complex oligosaccharides contain a diverse group of additional sugars, for example, GlcNAc, GalNAc, L-fucose, and NANA, whereas the high-mannose oligosaccharides contain primarily mannose.

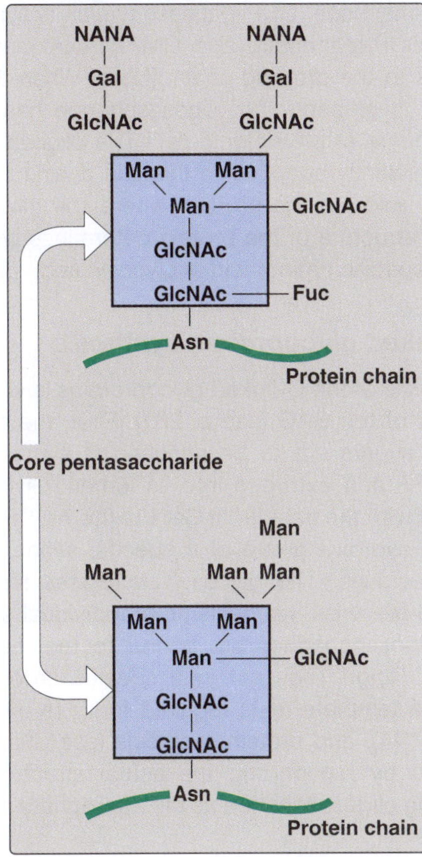

Figure 19.15 Complex (top) and high-mannose (bottom) N-linked oligosaccharides. [Note: Members of each class contain the same pentasaccharide core (shown inside the box).] NANA = N-acetylneuraminic acid; Gal = galactose; GlcNAc = N-acetylglucosamine; Man = mannose; Fuc = fucose; Asn = asparagine.

VIII. GLYCOPROTEIN SYNTHESIS AND SORTING

Proteins destined to function in the cytoplasm are synthesized on free cytosolic ribosomes. However, proteins, including glycoproteins, that are destined for cellular membranes, lysosomes, or to be exported from the cell, are synthesized on ribosomes attached to the RER. These proteins contain **specific signal sequences** that act as molecular addresses, **targeting the proteins to their proper destinations**. An N-terminal hydrophobic sequence initially directs these proteins to the RER, allowing the growing polypeptide to be extruded into the lumen (see p. 644). The proteins are then transported via secretory vesicles to the **Golgi, which acts as a sorting center** (Fig. 19.16). In the Golgi, those glycoproteins that are to be secreted from the cell (or are targeted for lysosomes) are packaged into vesicles that fuse with the cell (or lysosomal) membrane and release their contents. Those that are destined to become components of the cell membrane are integrated into the Golgi membrane, which buds off, forming vesicles that add their membrane-bound glycoproteins to the cell membrane. [Note: Therefore, the membrane glycoproteins are oriented with the carbohydrate portion on the outside of the cell (see Fig. 19.16).]

A. Carbohydrate components

The precursors of the carbohydrate components of glycoproteins are nucleotide sugars, which include UDP-glucose, UDP-galactose, UDP-GlcNAc, and UDP-GalNAc. In addition, guanosine diphosphate

(GDP)-mannose, GDP-L-fucose (which is synthesized from GDP-mannose), and CMP-NANA may donate sugars to the growing chain. [Note: When the acidic NANA is present, the oligosaccharide has a negative charge at physiologic pH.] The oligosaccharides are covalently attached to the side chains of specific amino acids in the protein, where the three-dimensional structure of the protein determines whether or not a specific amino acid is glycosylated.

B. O-Linked glycoprotein synthesis

Synthesis of the O-linked glycoproteins is very similar to that of the GAG (see p. 261). First, the protein to which sugars are to be attached is synthesized on the RER and extruded into its lumen. Glycosylation begins with the transfer of GalNAc (from UDP-GalNAc) to the hydroxyl group of a specific serine or threonine residues. The *glycosyltransferases* responsible for the stepwise synthesis (from individual sugars) of the oligosaccharides are bound to the membranes of the Golgi. They act in a specific order, without using a template as is required for DNA, ribonucleic acid (RNA), and protein synthesis (see Unit VIII), but instead by recognizing the actual structure of the growing oligosaccharide as the appropriate substrate.

C. N-Linked glycoprotein synthesis

Synthesis of N-linked glycoproteins occurs in the lumen of the RER and requires the participation of the phosphorylated form of dolichol (dolichol pyrophosphate),

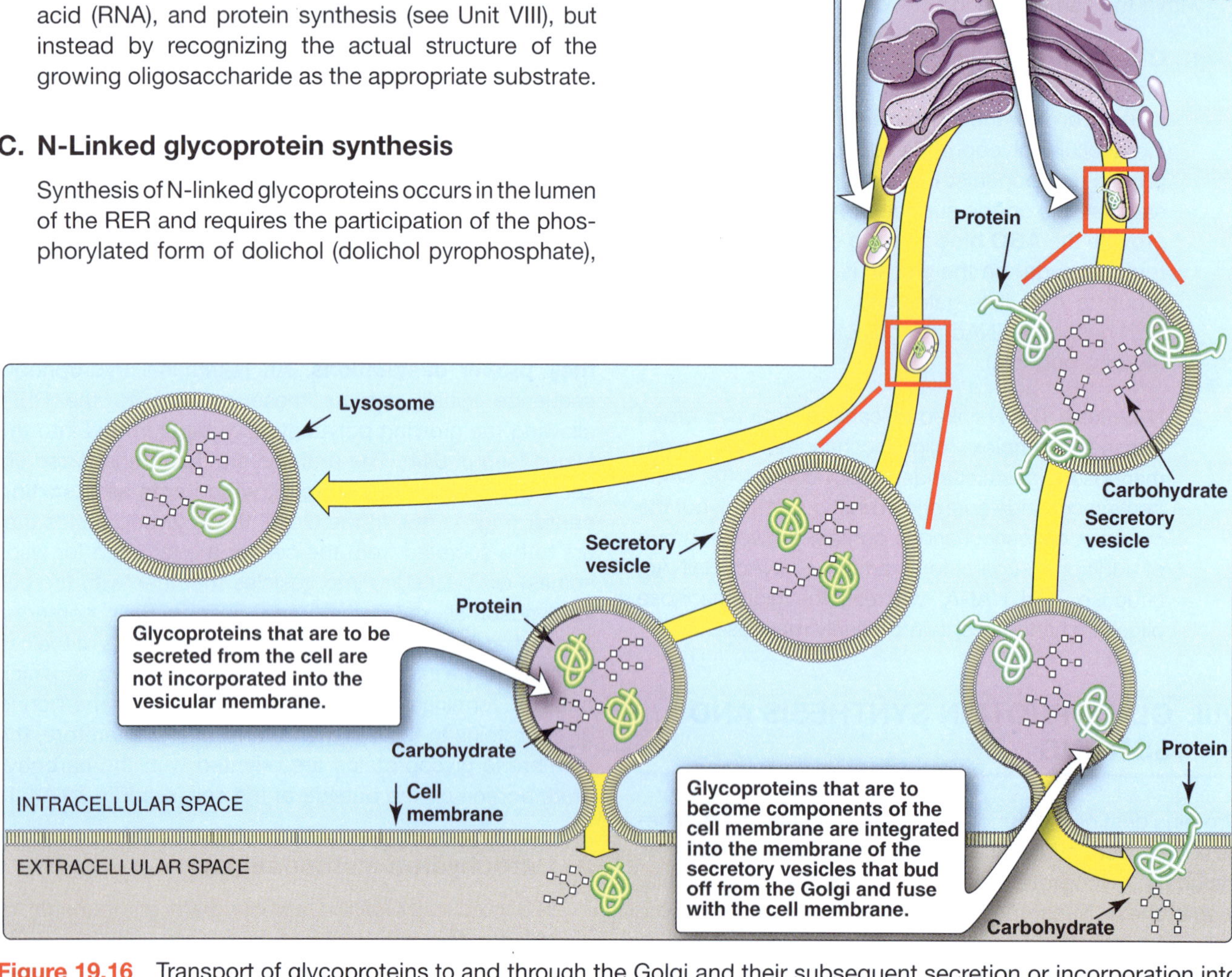

Figure 19.16 Transport of glycoproteins to and through the Golgi and their subsequent secretion or incorporation into a lysosome or the cell membrane.

a lipid of the RER membrane (Fig. 19.17). The initial product is processed in the RER and Golgi.

1. **Dolichol-linked oligosaccharide synthesis:** As with the O-linked glycoproteins, the protein is synthesized on the RER and enters its lumen. However, it does not become glycosylated with individual sugars. Instead, a lipid-linked oligosaccharide is first constructed. This consists of dolichol (an RER membrane lipid made from an intermediate of cholesterol synthesis; see p. 343) attached through a pyrophosphate linkage to an oligosaccharide containing GlcNAc, mannose, and glucose. The sugars to be added sequentially to the dolichol by membrane-bound *glycosyltransferases* are first GlcNAc, followed by mannose and glucose (see Fig. 19.17). The entire 14-sugar oligosaccharide is then transferred from dolichol to the amide nitrogen of an asparagine residue in the protein to be glycosylated by a *protein–oligosaccharide transferase* present in the RER. [Note: Tunicamycin inhibits N-linked glycosylation.]

> Congenital disorders of glycosylation (CDG) are syndromes caused primarily by defects in the N-linked glycosylation of proteins, either oligosaccharide assembly (type I) or processing (type II).

2. **N-Linked oligosaccharide processing:** After addition to the protein, the N-linked oligosaccharide is processed by the removal of specific mannosyl and glucosyl residues as the glycoprotein moves through the RER. Finally, the oligosaccharide chains are completed in the Golgi by addition of a variety of sugars (for example, GlcNAc, GalNAc, and additional mannoses and then fucose or NANA as terminal groups) to produce a complex glycoprotein. Alternatively, they are not processed further, leaving branched, mannose-containing chains in a high-mannose glycoprotein (see Fig. 19.17). The ultimate fate of N-linked glycoproteins is the same as that of the O-linked glycoproteins (for example, they can be released by the cell or become part of a cell membrane). In addition, N-linked glycoproteins can be targeted to

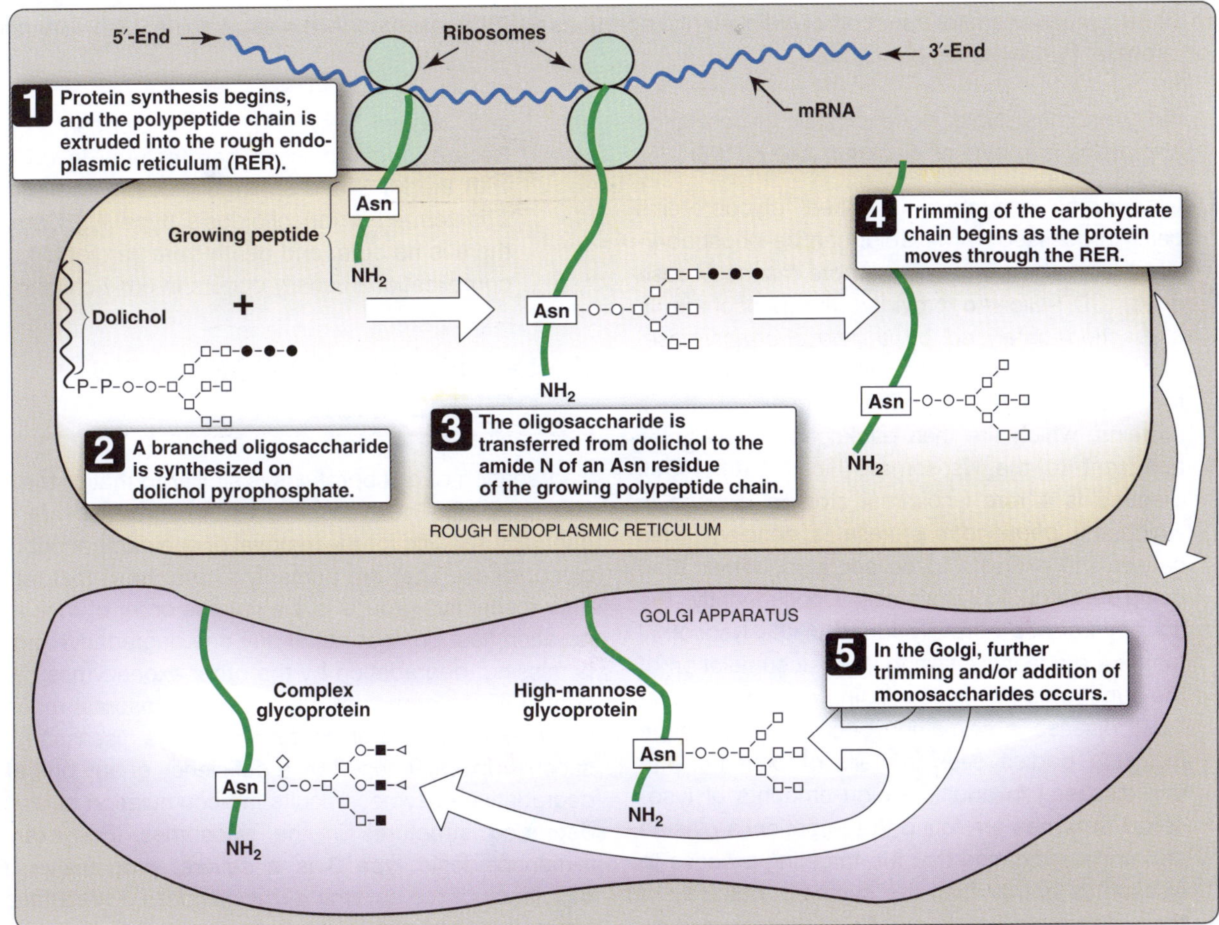

Figure 19.17 Synthesis of N-linked glycoproteins. ○ = N-acetylglucosamine; □ = mannose; • = glucose; ■ = galactose; ◇ or ◁ = terminal group (fucose or N-acetylneuraminic acid); mRNA = messenger RNA; Asn = asparagine.

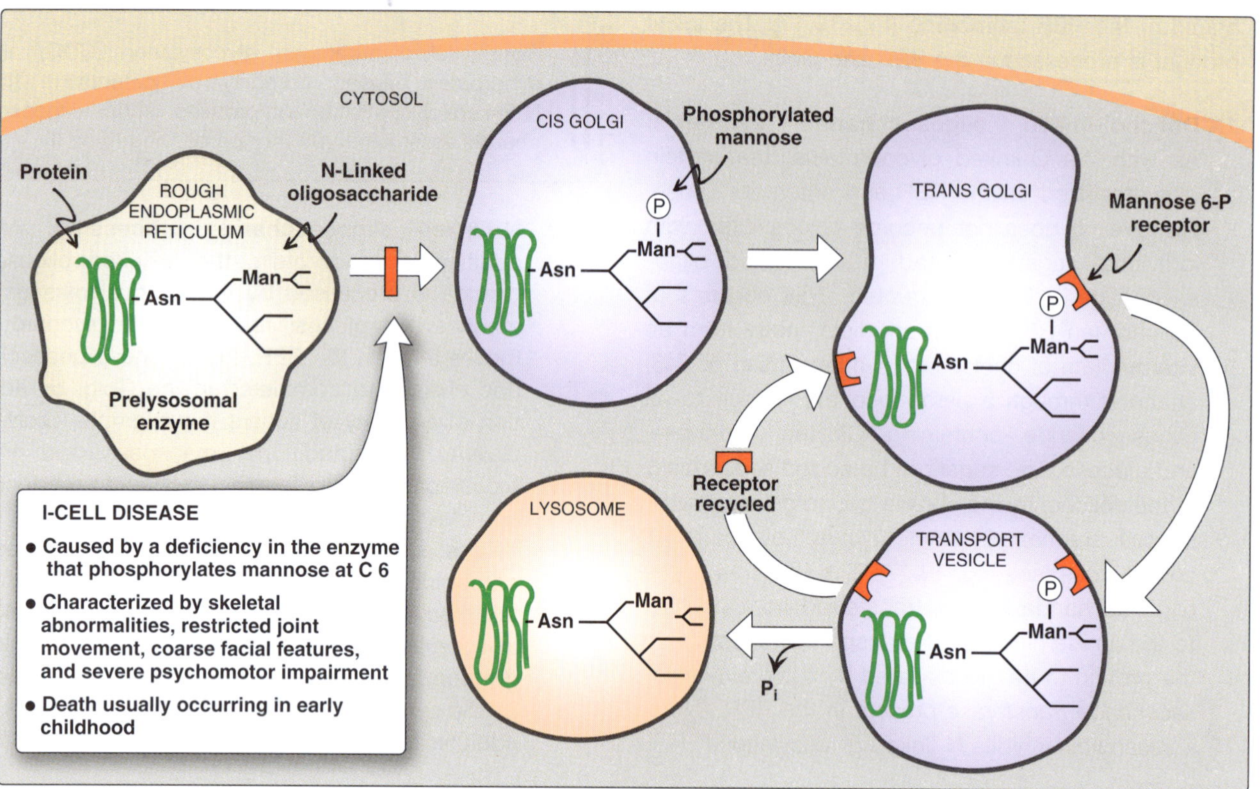

Figure 19.18 Mechanism for transport of N-linked glycoproteins to the lysosomes. Asn = asparagine; Man = mannose; Ⓟ = phosphate; P_i = inorganic phosphate.

the lysosomes. [Note: Nonenzymatic glycosylation of proteins is known as glycation (see p. 106).]

3. **Lysosomal enzymes:** N-Linked glycoproteins being processed in the Golgi can be phosphorylated on carbon 6 of one or more mannosyl residues. UDP-GlcNAc provides the phosphate in a reaction catalyzed by a *phosphotransferase*. Receptors, located in the Golgi membrane, bind the mannose 6-phosphate residues of these proteins, which are then packaged into vesicles and sent to the lysosomes (Fig. 19.18). **I-Cell disease** is a rare lysosomal storage disease in which the *phosphotransferase* is deficient. This causes the proteins to be secreted, rather than being targeted to lysosomes. Consequently, the *acid hydrolases* normally found in the lysosomal matrix are absent, resulting in an accumulation of the substrates for these missing enzymes. [Note: I-Cell disease is so named because of the large inclusion bodies seen in cells of patients with this disease.] In addition, high amounts of lysosomal enzymes are found in the patient's plasma and urine, indicating that the targeting process to lysosomes (rather than the synthetic pathway of these enzymes) is deficient. I-Cell disease is characterized by skeletal abnormalities, restricted joint movement, coarse (dysmorphic) facial features, and severe psychomotor impairment. [Note: Because I-cell disease has features in common with the MPS and sphingolipidoses (see p. 331), it is termed a mucolipidosis (type II).] Currently, there is no cure, and death from cardiopulmonary complications usually occurs in early childhood.

IX. LYSOSOMAL GLYCOPROTEIN DEGRADATION

Degradation of glycoproteins is similar to that of the GAG (see p. 266). The lysosomal *acid hydrolases* are each generally specific for the removal of one component of the glycoprotein. They are primarily exoenzymes that remove their respective groups in the reverse order of their incorporation (last on, first off). If any one degradative enzyme is missing, degradation by the other exoenzymes cannot continue. A group of very rare autosomal-recessive diseases called the glycoprotein storage diseases (oligosaccharidoses), caused by a deficiency of any one of the degradative enzymes, results in accumulation of partially degraded structures in the lysosomes. For example, α-mannosidosis type 3 is a severe, progressive, fatal deficiency of the enzyme *α-mannosidase*. Presentation is similar to Hurler syndrome, but immune deficiency is also seen. Mannose-rich oligosaccharide fragments appear in the urine. Diagnosis is by enzyme assay.

X. Medical Perspectives

Mucopolysaccharidoses are a rare group of inherited conditions. Signs and symptoms and the severity of the condition vary significantly by the type of disorder. In general, most affected people appear healthy at birth and have normal development initially. However, as condition progresses, there is a decline in physical and/or mental function. The two main treatment options for MPS include hematopoietic stem cell transplant (HSCT) and enzyme replacement therapy (ERT). For MPS I, Aldurazyme is approved as ERT. However, as it does not cross the blood-brain barrier, problems with mental development persist. Galsulfase is approved for the treatment of MPS VI. Sialic acid supplementation is being tried in *N-Acetylneuraminic Acid Synthase* (NANS) deficiency. As sialic acid is available in breast milk and common foodstuffs, its supplementation is considered a safe option.

Heparin acts as an anticoagulant. Heparin is clinically used in patients for anticoagulation in several conditions such as acute coronary syndrome, deep vein thrombosis, and pulmonary embolism. It is also used as anticoagulant in vitro for blood sample collection when non-clotted whole blood or plasma is required for analysis.

Heparan sulfate regulates the interactions between cells, and between cells and extracellular matrix (ECM), by binding themselves to matrix macromolecules such as fibronectin or laminin. HS and heparin also have a preventive role in metastases and wound healing.

Hyaluronic acid is added to ophthalmic formulations to enhance hydration and lubrication of the eye, promote wound healing, or extend drug residence time. HA has also been found to be beneficial for the treatment of joint and skin diseases, and for wound healing. HA and its synthetic derivatives can be used as 3D scaffold structures, providing physical support for in vitro cell culture efforts. HA-based formulations are also used for cosmetic uses, for antiwrinkle effect. HA is being studied as a potential drug delivery system.

XI. Chapter Summary

- **Glycosaminoglycans** (**GAG**) are **long**, **negatively charged**, **unbranched**, **heteropolysaccharide chains** generally composed of a **repeating disaccharide unit** [acidic sugar–amino sugar]$_n$ (Fig. 19.19).
- The **amino sugar** is either **D-glucosamine** or **D-galactosamine**. The amino sugar may also be **sulfated** on carbon 4 or 6 or on a nonacetylated nitrogen.
- The **acidic sugar** is either **D-glucuronic acid** or its C-5 epimer **L-iduronic acid**.
- GAG bind large amounts of water, thereby producing the gel-like matrix that forms the basis of the body's **ground substance**.
- The viscous, lubricating properties of **mucous secretions** are also caused by the presence of GAG, which led to the original naming of these compounds as **mucopolysaccharides**.
- There are six major types of GAG, including **chondroitin 4-** and **6-sulfates**, **keratan sulfate**, **dermatan sulfate**, **heparin**, **heparan sulfate**, and **hyaluronic acid (HA)**.
- All GAG, except HA, are found covalently attached to a **core protein**, forming **proteoglycan monomers**.
- Many proteoglycan monomers associate with a molecule of **HA** to form **proteoglycan aggregates**. GAG are synthesized in the **Golgi**. The polysaccharide chains are elongated by the sequential addition of alternating acidic and amino sugars, donated by their **UDP derivatives**. D-Glucuronate may be epimerized to **L-iduronate**.
- The completed proteoglycans are secreted into the **extracellular matrix** (**ECM**) or remain associated with the outer surface of cells.
- GAG are degraded by **lysosomal *acid hydrolases***. A deficiency of any one of the *hydrolases* results in a **mucopolysaccharidosis**. These are hereditary disorders in which GAG accumulate in tissues, causing symptoms such as **skeletal** and **ECM deformities** and **intellectual disability**. Examples of these genetic diseases include **Hunter** (X-linked) and **Hurler syndromes**.
- **Glycoproteins** are proteins to which **oligosaccharides** (**glycans**) are covalently attached. They differ from the proteoglycans in that the length of the glycoprotein's carbohydrate chain is relatively short, may be branched, and does not contain serial disaccharide units.
- **Membrane-bound** glycoproteins participate in a broad range of cellular phenomena, including **cell-surface recognition**, **cell-surface antigenicity** (such as the blood group antigens), and as components of the ECM and of the **mucins** of the gastrointestinal and urogenital tracts. In addition, almost all of the globular proteins present in human plasma are glycoproteins.
- Glycoproteins are synthesized in the **rough endoplasmic reticulum** (**RER**) and the **Golgi**. The precursors of the carbohydrate components of glycoproteins are **nucleotide sugars**.
- A deficiency in the ***phosphotransferase*** that phosphorylates mannose residues at carbon 6 in N-linked glycoprotein enzymes destined for the lysosomes results in **I-cell disease**.
- Glycoproteins are degraded in lysosomes by ***acid hydrolases***. A deficiency of any one of these enzymes results in a **lysosomal glycoprotein storage disease** (**oligosaccharidosis**), resulting in accumulation of partially degraded structures in the lysosome.

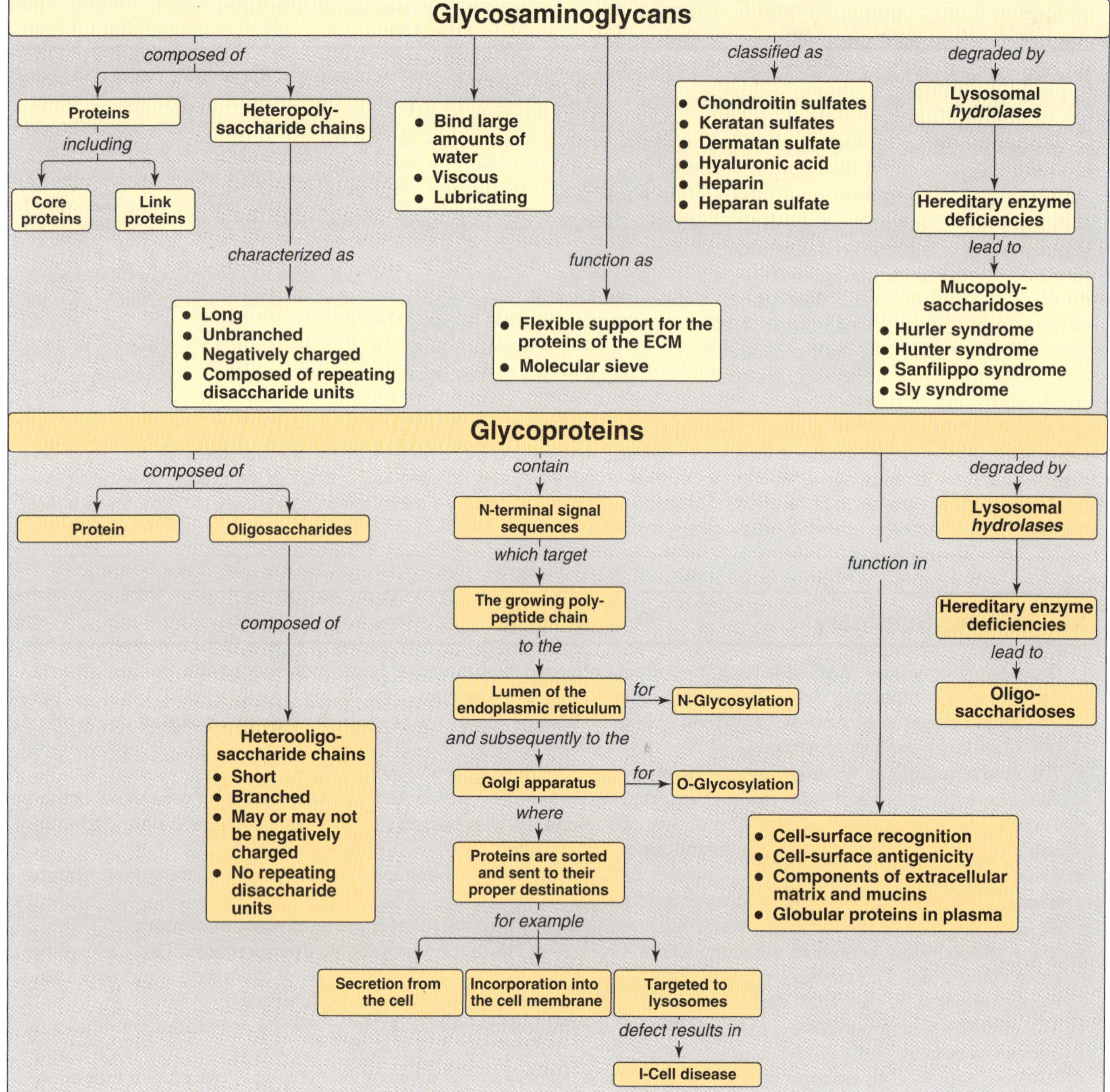

Figure 19.19 Key concept map for glycosaminoglycans and glycoproteins. ECM = extracellular matrix.

Study Questions

Short-Answer Questions

Write short notes on:

19.1. Blood group antigens

19.2. I-cell disease

19.3. Biological basis of slippery consistency of mucous secretions

19.4. Differences between glycoproteins and proteoglycans.

Long-Answer Questions

19.1. What are mucopolysaccharides? Describe their basic structure and important functions.

19.2. Describe the structure and synthesis of glycoproteins.

Multiple-Choice Questions

Choose the ONE best answer.

19.1. Among the glycosaminoglycans, chondroitin sulfate (which contains glucuronic acid) is present in the bones while dermatan sulfate (which contains iduronic acid) is present in the cartilages. The absence of which among the following enzymes in osteoblasts is responsible for this tissue distribution of glycosaminoglycans?

A. Epimerase
B. Mutase
C. Isomerase
D. Transferase

Correct answer = A.

19.2. The presence of the following compound in the urine of a patient suggests a deficiency in which one of the enzymes listed below?

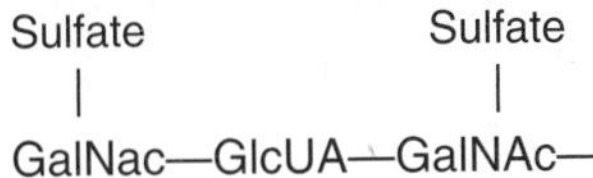

A. Galactosidase
B. Glucuronidase
C. Sulfatase
D. Mannosidase

Correct answer = C.

19.3. Hunter syndrome is caused by an accumulation of glycosaminoglycans due to a defective catabolism. In which of the following intracellular locations is the defective enzyme located in this condition?

A. Endoplasmic reticulum
B. Golgi apparatus
C. Lysosomes
D. Peroxisomes

Correct answer = C.

19.4. Which of the following is CORRECT about proteoglycans?

A. Proteoglycans are found in the ECM and on the outer surface of cells.
B. Many proteoglycan monomers can associate covalently with one molecule of HA to form proteoglycan aggregates
C. The core protein is made by Golgi
D. None of the above

Correct answer = A.

19.5. Based on the statements given below, choose the correct option.

Statement 1: Proteoglycans are involved in providing a cushioning effect and resilience effect to the tissues where they are present.

Statement 2: Glycosaminoglycans, the carbohydrate components of the proteoglycans, are rich in negative charge, causing them to absorb water; this water can be squeezed out when the molecule is compressed.

A. Both the Statements are correct, and Statement 2 is the correct explanation for Statement 1
B. Both the Statements are correct, but Statement 2 is NOT the correct explanation for Statement 1
C. Statement 1 is correct, but Statement 2 is NOT correct
D. Statement 1 is NOT correct, but Statement 2 is correct

Correct answer = A.

19.6. Which of the following mucopolysaccharidoses is inherited in a manner different from that of the others?

A. Hurler syndrome
B. Hunter syndrome
C. Sanfilippo syndrome
D. Sly syndrome

Correct answer = B.

19.7. Tunicamycin is a natural product antibiotic that prevents the addition of N-acetylglucosamine (GlcNAc) to dolichol phosphate in the endoplasmic reticulum. Synthesis of which of the following is inhibited by tunicamycin?

A. N-linked glycoproteins
B. O-linked glycoproteins
C. Proteoglycans
D. Glycosaminoglycans

Correct answer = A.

19.8. Which of the following is the predominant intracellular location of glycosyltransferases?

A. Nucleus
B. Mitochondria
C. Peroxisomes
D. Golgi apparatus

Correct answer = D.

19.9. Which of the following best describes the functions of glucuronic acid?

I. Detoxification of drugs II. Excretion of bilirubin III. Formation of glycosaminoglycans IV. Glucose donor in glycogen synthesis

A. I and II
B. I, II, and III
C. I, II, and IV
D. I, II, III, and IV

Correct answer = B.

19.10. Which of the following is found in high amounts in the serum and urine of patients with I-cell disease?

A. Heparan sulfate
B. Chondroitin sulfate
C. Dermatan sulfate
D. Lysosomal hydrolases

Correct answer = D.

Multiple-Choice Questions (NExT/NEET PG)

Choose the ONE best answer.

19.1. Which among the following best differentiates a glycoprotein from a proteoglycan?

A. Percentage carbohydrate content which is higher in proteoglycans

B. Presence of sulfated residues exclusive to proteoglycans

C. Presence of unbranched 2-unit repeats in proteoglycans, but not glycoproteins

D. Presence of N-acetyl galactosamine in proteoglycans

Correct answer = C.

19.2. Mucopolysaccharidoses are inherited lysosomal storage diseases. They are caused by:

A. An increased rate of synthesis of the carbohydrate component of proteoglycans.

B. The synthesis of polysaccharides with an altered structure.

C. Defects in the degradation of the GAGs in proteoglycans.

D. The synthesis of abnormally small amounts of protein cores.

Correct answer = C.

19.3. An 8-month-old boy with coarse facial features, skeletal abnormalities, and delays in both growth and development is diagnosed with I-cell disease based on his presentation, and on histologic and biochemical testing. Which of the following statements is correct about this condition?

A. There is a decreased production of cell-surface glycoproteins

B. The ability to ubiquitinate proteins is compromised

C. The levels of glycosylates proteins are reduced

D. There is incorrect targeting of lysosomal proteins

Correct answer = D.

19.4. A way of assessing alcohol use disorder is to estimate the level of carbohydrate deficient transferrin in the serum. The probable mechanism here is the impairment of the addition of the carbohydrate (sialic acid or N-acetylneuraminic acid) to the plasma protein transferrin, as it is synthesized in the liver. Functioning of which of the following enzymes is impaired in this condition?

A. Cytochrome P450

B. Aldehyde dehydrogenase

C. Glycosyltransferases

D. Acid hydrolases

Correct answer = C.

19.5. An infant with corneal clouding has dermatan sulfate and heparan sulfate in his urine and is suspected to have Hurler syndrome. Decreased activity of which of the following enzymes can confirm the diagnosis in this child?

A. α-L-Iduronidase

B. α-Glucuronidase

C. Glycosyltransferase

D. Iduronate sulfatase

Correct answer = A.

Appendix

Unit III: Carbohydrates

APPLICATIONS OF CARBOHYDRATE METABOLISM FOR THE PRIMARY HEALTHCARE PHYSICIAN

Diabetes mellitus is characterized by increased blood glucose levels, either due to insulin deficiency or insulin resistance. It can be diagnosed by measuring plasma glucose or glycated hemoglobin levels. For details, see Chapter 29.

Glycated Hemoglobin (HbA_{1c})

It is the hemoglobin in which glucose is non-enzymatically attached to N-terminal valine residue of a beta-globin chain. This glycation is irreversible. As the life span of hemoglobin in RBCs is 90–120 days, HbA_{1c} reflects the glucose concentrations over past 90–120 days.

As values of HbA_{1c} can differ with the method used, it should be measured using the National Glycohemoglobin Standardization Program (NGSP) certified method. The results of HbA_{1c} assay can be affected by numerous conditions such as sickle cell disease, pregnancy, hemodialysis, blood loss or transfusion, or erythropoietin therapy.

Fructosamine

Fructosamine includes all ketoamine linkages that result from glycation of serum proteins. As albumin is the most abundant of serum proteins, fructosamine is predominantly a measure of **Glycated Albumin**.

The life span of hemoglobin in red blood cells is 90–120 days, serum proteins have a lower half-life of 14–21 days. Therefore, while HbA_{1c} provides a long-term record of glycemic control (i.e., over a period of 2–3 months), the measurement of fructosamine reflects glucose control over previous 2 weeks.

Microalbuminuria

Excretion of amounts of albumin too small to be detected by urinary dipstick or conventional measures of urine protein is known as microalbuminuria. It refers to presence of albumin in urine in range of 30–300 mg/24 hours. It is also known **moderately increased albuminuria**.

It is a good screening test for early detection of microvascular renal disease.

As the test for microalbumin requires collection of urine for 24 hours, it can be cumbersome. For spot urine samples, **albumin creatinine ratio (ACR)** is used. Values less than 30 μg/mg creatinine are normal, 30–300 μg/mg creatinine is microalbuminuria and more than 300 μg/mg creatinine is clinical albuminuria.

Glucose Intolerance

Glucose intolerance (GI) can be defined as dysglycemia that comprises both prediabetes and diabetes. It includes the conditions of impaired fasting glucose (IFG) and impaired glucose tolerance (IGT) and diabetes mellitus (DM).

IFG and IGT are collectively also known as prediabetes.

Two-Hour Oral Glucose Tolerance Test (OGTT)

In this test, plasma glucose level is measured before and 2 hours after ingestion of 75 g of anhydrous glucose in 250–300 mL of water. Patients need to consume a diet with at least 150 g per day of carbohydrates for 3–5 days and not take any medications that can impact glucose tolerance such as steroids and thiazide diuretics.

OGTT is performed:

- When fasting plasma glucose is between 100 and 126 mg/dl (to confirm IGT)
- For diagnosis of gestational diabetes mellitus
- For confirmation of renal glycosuria

IGT is diagnosed if plasma glucose (PG) level in the 2-hour sample is between 140 to 199 mg/dl (7.8 to 11.0 mmol/L). DM is diagnosed if the PG is greater than or equal to 200 mg/dl.

Gestational Diabetes Mellitus

Gestational Diabetes Mellitus (GDM) is defined as hyperglycemia with onset or first recognition during pregnancy. Undiagnosed or inadequately treated GDM can lead to significant maternal and fetal complications. Women with GDM and their offspring are at increased risk of developing type 2 diabetes later in life.

Diagnosis of GDM

For diagnosis of GDM, WHO recommends OGTT using 75 g glucose, irrespective of the last meal, with a threshold value of 2-hour blood glucose >140 mg/dl.

Universal screening of all pregnant women at first antenatal contact is recommended. If the first test is negative, second test should be done at 24–28 weeks of gestation.

In the postpartum period, OGTT should be repeated at 6 weeks after delivery.

Glucose Challenge Test (GCT)

It is used as a screening test for GDM. 50 g glucose is given orally, irrespective of the last meal. If plasma glucose at 1-hour after the load is greater than or equal to 140 mg/dl (7.8 mmol/L), OGTT is performed. GCT has now been replaced by OGTT with 75 g of glucose.

Renal Glycosuria

Normal renal threshold for glucose is 180 mg/dl. Up to this level, glucose is completely reabsorbed from the glomerular filtrate and there is no excretion of glucose in urine. If blood glucose level is higher than 180 mg/dl, glucose is excreted in urine. However, if the renal transporters for glucose are defective, glucose will be excreted in urine at a lower blood glucose level. This condition is known as renal glycosuria.

Renal glycosuria is physiologically seen in pregnancy and old age.

Estimated Average Glucose

Estimated average glucose (eAG) is a value calculated from HbA_{1c}, providing values in mg/dl, the same units that patients see routinely in blood glucose measurements. eAG can be used to help improve the discussion of glucose control with patients.

The formula for conversion of HbA_{1c} to eAG is:

$$28.7 \times HbA_{1c} - 46.7 = eAG$$

Table 1 Estimated average glucose concentration in blood derived from measured HbA_{1c} values

HbA_{1c}	eAG (mg/dl)*
6	126
6.5	140
7	154
7.5	169
8	183
8.5	197
9	212
9.5	226
10	240

Complications of Diabetes Mellitus

The complications of Diabetes Mellitus may be acute (sudden onset) or chronic (after prolonged disease)

Common acute complications include:

(a) Diabetic ketoacidosis

(b) Hyperosmolar hyperglycemic nonketotic coma

(c) Hypoglycemia due to lack of food or insulin/medication overdose

Chronic complications may be microvascular or macrovascular.

Microvascular

1. Retinopathy
2. Neuropathy (sensory, motor, autonomic)
3. Nephropathy

Macrovascular

1. Coronary heart disease
2. Peripheral arterial disease
3. Cerebrovascular disease

Therefore, a patient of DM should be periodically evaluated by the physician or referred to specialists for the above complications.

Hormones Which Increase Blood Glucose Level (Hyperglycemic hormones)

1. Glucagon
2. Epinephrine
3. Adrenocorticotropic hormone (ACTH)
4. Glucocorticoids
5. Growth hormone
6. Thyroxine

Reducing Substances in Urine

Benedict's test is commonly used to detect the presence of reducing substances in urine. The most common reducing sugar in urine is glucose. However, other reducing substances can also be excreted in urine. They are:

1. Fructose
2. Lactose
3. Galactose
4. Pentoses
5. Ascorbic acid
6. Salicylates
7. Homogentisic acid

Clinical Case Questions

Chapter 11 Bioenergetics and Oxidative Phosphorylation

11.1. A 20-year-old male subject developed sudden loss of vision of his right eye. Color vision was impaired and scotoma (partial loss of vision) in visual field examination was observed on the other eye. Fundus examination revealed edema in nerve fiber layer and microangiopathy. After 8 weeks, acute vision loss and similar findings on fundal examination were observed in the other eye. Mitochondrial DNA analysis revealed a mutation in NADH dehydrogenase and a diagnosis of Leber hereditary optic neuropathy (LHON) was confirmed.

A. Explain the biochemical basis of LHON.

Answer—A: Thirteen of the ~90 polypeptides required for oxidative phosphorylation are encoded by mtDNA and synthesized in mitochondria, whereas the remaining proteins are encoded by nuclear DNA synthesized in the cytosol, and then transported into mitochondria. Defects in oxidative phosphorylation are more likely a result of alterations in mtDNA, which has a mutation rate about 10 times greater than that of nuclear DNA. Tissues with the greatest ATP requirement (e.g., the central nervous system, skeletal and heart muscles, and the liver) are most affected by defects in oxidative phosphorylation. Mutations in mtDNA are responsible for several diseases, including some cases of mitochondrial myopathies, and LHON, a disease in which bilateral loss of central vision occurs as a result of neuroretinal degeneration, including damage to the optic nerve. Mutations in NADH Q reductase (MT-ND1, MT-ND4, MT-ND4L, and MT-ND6) cause this disease.

B. List four diseases resulting out of mutation in mitochondrial DNA and mention their biochemical defects.

Answer—B:

- LHON (Leber hereditary optic neuropathy): Mutation of NADH dehydrogenase as mentioned above.
- MERRF syndrome (Myoclonic Epilepsy with Ragged Red Fibers): Point mutation disrupts the mitochondrial gene for tRNA-Lys
- MELAS (Mitochondrial encephalomyopathy, lactic acidosis, and stroke-like episodes): NADH dehydrogenase defect
- NARP (Neuropathy, ataxia, and retinitis pigmentosa) syndrome: Defective ATP synthase

C. How are these diseases inherited?

Answer—C: mtDNA is maternally inherited and hence, so are these diseases because mitochondria from the sperm cell do not enter the fertilized egg.

Chapter 12 Introduction to Carbohydrates

12.1. A 30-year-old male presents with abdominal cramps and diarrhea. He had taken milkshake few hours ago. He had experienced discomfort in abdomen several times earlier also. His pain is associated with lot of flatulence. He reveals that he regularly takes curd with food but has not experienced such episodes at those times.

A. What is the most probable diagnosis?

Answer—A: Abdominal cramps, flatulence and diarrhea after taking milk product is suggestive of lactose intolerance.

B. What advice will you give to this patient?

Answer—B: Patient should be advised to reduce consumption of milk and eat yogurts or take *lactase* in pill form prior to eating.

12.2. A 2-year-old girl presents with diarrhea persisting for 3 weeks. She has received two courses of appropriate oral antibiotics, each lasting a week, along with zinc syrup. Although there was a transient decrease in frequency after the first week, the diarrhea continues. There is no passage of blood in the stools, and she has not experienced fever, vomiting, abdominal distension, or lethargy. The child continues to be fed a predominantly milk-based diet, with a slightly decreased appetite. She appears irritable, especially during urination or defecation. Examination reveals normal vitals, underweight, mild pallor, and an absence of dehydration, but the perianal region is raw and inflamed. There are no significant systemic findings.

A. What is the clinical diagnosis?

Answer—A: Since the duration of diarrhea is more than 14 days the clinical diagnosis is persistent diarrhea with no dehydration.

B. What is the most common cause of persistent diarrhea in young children?

Answer—B: Secondary lactose intolerance caused by the rapid purge rate during the initial phase of infectious diarrhea.

C. How will you investigate a child with suspected secondary lactose intolerance?

Answer—C: Analysis of the watery part of the stool will reveal presence of reducing substances and the stool pH will be acidic (< 6.0) on two occasions.

Chapter 13 Introduction to Metabolism and Glycolysis

13.1. A 63-year-old female presented to the clinic with the complaint of frequent urination and burning during micturition. She revealed that she had been repeatedly suffering from these problems. Urine analysis revealed glycosuria. Fasting blood glucose and HbA_{1c} were within normal range.

A. What is the probable diagnosis?

Answer—A: Urinary glucose with normal fasting blood glucose and normal HbA_{1c} indicates that this is a case of renal glycosuria. Urinary tract infection is a possibility in this case and must be investigated with urine culture and routine tests.

B. What is the biochemical basis of this condition?

Answer—B: Renal glycosuria is caused by a decreased ability of the kidneys to reabsorb filtered glucose in the proximal tubules due to a defect in the sodium-dependent glucose cotransporter (SGLT). It is often considered a benign condition that can occur in individuals of any age, not specifically limited to old age. The high glucose in the urine can predispose the individual to frequent urinary tract infections.

Chapter 14 Tricarboxylic Acid Cycle and Pyruvate Dehydrogenase Complex

14.1. A man presented to the primary healthcare center with diarrhea, abdominal pain, numbness, and history of cancer and heart disease. His relatives informed that this illness was very common in their village. The villagers were mainly miners and lived in the vicinity of the mines. Besides giving symptomatic treatment, the doctor advised the *Tahsildar* of the district to have the water purity in the village supply checked. Why?

Answer: The symptoms and medical history of the man, along with the fact that this illness was common in village, indicate typical arsenic poisoning. Long-term arsenic poisoning is usually because of contaminated groundwater and may be due to mining or agriculture. Arsenic causes cellular apoptosis by allosteric inhibition of pyruvate dehydrogenase complex. Arsenic also prevents use of thiamine, resulting in clinical picture of thiamine deficiency. Low levels of potassium in cells also interferes with heart rhythm. Poisoning with arsenic can cause lactic acidosis and low potassium levels in cells.

Chapter 15 Gluconeogenesis

15.1. A 45-year-old healthy male completed a 20 km long marathon in record time. He then quickly consumed alcohol to celebrate his success. He was already having muscle cramps and pain but now he suddenly fainted.

A. What could be the reason for this?

Answer—A: Marathon runners will build up lactic acid in muscle which causes muscle cramps and pain. This lactate also increases in blood and is taken up by liver to convert it to pyruvate via the Cori cycle. This process of conversion of lactate to pyruvate requires NAD^+. However, in this case the sudden alcohol intake compromised this conversion as NAD^+ is being used now for alcohol metabolism. The high amount of cytoplasmic NADH formed by alcohol dehydrogenase further inhibits gluconeogenesis leading to hypoglycemia.

Chapter 16 Glycogen Metabolism and Glycogen Storage Disorders

16.1. A medical student was posted in emergency ward last evening and did not get time to have dinner. Subsequently, she could hardly sleep for 3 hours and she also missed her breakfast. She had an exam at 12 noon. During the exam, she felt very dizzy, weak and could not remember anything at all.

A. Discuss the biochemical state leading to these symptoms.

Answer—A: Assuming the student had a normal lunch the day before, she was fasting for 24 hours by the time she came to give her exam. The glycogen stores of liver would release glucose for a maximum of 17–20 hours (under influence of glucagon) and gluconeogenesis should be induced in the meantime. Further, stress hormones like cortisol and adrenaline released by lack of sleep and examination stress would further cause heart rate variability and increased blood pressure. The symptoms are a net result of all above.

16.2. A 35-year-old lady was an avid gym enthusiast. However, she always had episodes of cramping somewhat lessened by a glucose-rich drink. Therefore, she exercised less and carefully. One day however when she exercised too much, she had extreme weakness in both legs, muscle cramps and later excreted red-brown urine.

A. What is the possible diagnosis?

Answer—A: The lady could be suffering from McArdle disease (myophosphorylase deficiency), thereby having an inability to efficiently breakdown muscle glycogen to glucose-6-phosphate. The situation is improved by having her customary glucose-rich drink, which provides her dietary glucose. The red-brown urine could be because of myoglobinuria.

16.3. A 1-year-old boy presents with generalized tonic-clonic seizures. There is no history of fever. Vitals are stable. Past history is not contributory. The child has short stature, a doll-like face (round, full cheeks, short nose and small chin), thin limbs and a protuberant abdomen that reveals firm hepatomegaly with sharp margins. The remaining systemic examination is normal. Preliminary investigations reveal hypoglycemia and the absence of hypocalcemia, hypo- or hypernatremia, uremia and deranged liver function tests. Metabolic acidosis and elevated lactate levels are found on the arterial blood gas analysis.

(**Note:** *Clinical Case 16.3 requires knowledge from Chapters 4, 16 and 48)*

A. What clinical diagnosis will you consider in this child?

Answer—A: A glycogen storage disorder that primarily involves the liver.

B. What other biochemical tests will you order and what abnormalities would you expect to find?

Answer—B: Detailed kidney function tests, uric acid levels and a lipid profile will be requested. It is expected that hyperuricemia and hyperlipidemia will be seen.

C. How will you confirm the diagnosis in type 1 glycogen storage disorder?

Answer—C: Enzyme assays will be required to analyze deficiency of the enzyme glucose 6 phosphatase.

Chapter 17 Monosaccharide and Disaccharide Metabolism

17.1. A 5-day-old baby boy, born full term normal delivery, is brought to your clinic with complaint of vomiting after feeds. The baby is breastfed and had been taking all feeds normally since birth. There was no adverse history at the time of birth. On examination, you notice yellowing of sclera. The baby is dehydrated, but on per abdomen examination there is no apparent organomegaly. Urine examination revealed presence of reducing sugar.

A. What is the most probable diagnosis?

Answer—A: This may be a case of galactosemia. Newborns with classic galactosemia present very soon after birth with symptoms, which include jaundice and vomiting. Enlarged liver may also be palpated on examination.

B. Which other investigation will you do to confirm the diagnosis?

Answer—B: Reduced or absence of activity of galactose-1-phosphate uridyltransferase in RBC will confirm the diagnosis.

17.2. A 35-year-old man presented with a two-year history of chest pain and dyspnea on exertion. He also reported a history of neuropathic pain in bilateral limbs and hearing loss for the past 5 years. His blood pressure was normal. An ECG revealed left ventricular hypertrophy with T-wave inversion in the lateral and inferior leads and a short PR interval. Echocardiography showed concentric severe left ventricular hypertrophy. Cardiac MRI revealed late gadolinium enhancement at the inferolateral left ventricle. Proteinuria was present on urine examination. A provisional diagnosis of Fabry's disease was made.

A. What is the cause of Fabry's disease?

Answer—A: It is due to a deficiency of alpha galactosidase A.

B. How will you confirm the diagnosis?

Answer—B: Blood galactosidase A activity or genetic analysis for mutations in the galactosidase gene may be performed.

C. What is the prognosis?

Answer—C: The prognosis is good if enzyme replacement therapy is initiated early in the course of the disease. Since the disease is curable, it should always be suspected in cases of unexplained left ventricular hypertrophy.

17.3. A 20-day-old boy presents with progressively increasing jaundice noted from the second week of life. The baby is irritable, not breastfeeding well and experiencing vomiting, but is passing pigmented stools of normal consistency and frequency. The baby is the third born of a consanguineous union, with a significant family history of previous infant deaths at 3 months of age with similar symptoms (cause unknown). He was born at term gestation with a birth weight of 2.2 kg. Anthropometry reveals failure to gain weight, with a current weight of 2 kg, but age-appropriate length and head circumference. The only salient clinical findings are an enlarged liver and spleen, both firm in consistency. The liver function test is as follows: S. bilirubin (total) 12 mg/dL, S. bilirubin (direct) 11.2 mg/dL, S. bilirubin (indirect) 0.8 mg/dL; SGOT 280 IU/L; SGPT 560 IU/L; total proteins 5.2 g/L with S. albumin 2.5 g/L. The abdominal ultrasonogram reveals no radiological signs suggestive of biliary atresia.

A. What clinical diagnosis can be made based on the clinical and biochemical profile?

Answer—A: Infantile hepatitis.

B. The treating physician suspects an inborn error of metabolism and orders a urinary test for reducing substances. Which disorder is probably being suspected?

Answer—B: Galactosemia is the probable disorder, given the clinical and diagnostic picture.

C. The urinary test for reducing substances is strongly positive. What confirmatory tests should be done to establish the diagnosis?

Answer–C: Elevated erythrocyte galactose-1-phosphate concentration, reduced erythrocyte galactose-1-phosphate uridylyl transferase (GALT) enzyme activity and/or biallelic pathogenic variants found in *GALT* will be useful for establishing a diagnosis.

Chapter 18 Pentose Phosphate Pathway and Nicotinamide Adenine Dinucleotide Phosphate

18.1. A 2-month-old girl, born full term, is brought with complaint of persistent pneumonia. The girl suffered from several episodes of gastroenteritis, since few days after birth. She has received three recent courses of broad-spectrum antibiotics due to persistent pneumonia. On admission, physical examination revealed mild chest retraction and rales in both lung fields. She was alert and responsive, with no neurological deficit. Her height and weight were low for her age. Immunologic studies showed normal concentrations of immunoglobulins. Neutrophil nitroblue tetrazolium (NBT) slide test was abnormal.

Answer–A: The history of recurrent infections, normal immunoglobulin levels and abnormal neutrophil nitroblue tetrazolium (NBT) test indicate that this is a case of Chronic Granulomatous Disease (CGD). In the NBT test, neutrophils from peripheral blood are activated to produce superoxide which is detected by the NBT, which is converted from a yellow water-soluble compound to a dark-blue insoluble formazan, which can be seen microscopically.

A. What is the most probable diagnosis?

B. What is the biochemical basis of this disease?

Answer–B: CGD is caused due to genetic defects in NADPH oxidase.

18.2. A 5-year-old boy presents with fever, increased frequency, painful stream of urination, and urgency, suggestive of a urinary tract infection. He is started on standard dosage of Nalidixic acid. When he returns after three days, the clinician notes new clinical pallor and very light jaundice. The child is passing very dark urine, attributed to the medication. A hemogram and liver function test are ordered. The hemogram shows a drop in hemoglobin by 5 g/dL compared to the previous one. Total and indirect serum bilirubin are elevated, while direct serum bilirubin, SGOT, SGPT, and serum albumin levels are normal. The treating physician suspects acute hemolysis and orders a peripheral smear and other tests.

Answer–A: Nalidixic acid induced hemolysis in a possible setting of Glucose 6 phosphate dehydrogenase (G6PD) deficiency.

A. What do you think caused this acute episode of hemolysis?

B. What will the peripheral smear show?

Answer–B: Anisopoikilocytosis, blister/bite cells and Heinz bodies (on supravital staining).

C. How will you confirm the diagnosis?

Answer–C: Enzyme assay of erythrocytic G6PD levels should be performed. G6PD will show levels <10% of normal in case of G6PD deficiency.

Chapter 19 Glycosaminoglycans, Proteoglycans, and Glycoproteins

19.1. A 10-year-old boy presented with a chief complaint of irregularly placed front teeth. Dental history did not reveal delay or early eruption of milk teeth. Patient's parents gave history of frequent episodes of seizures from the age of 8 years and a surgery for inguinal hernia on right side last year. His brother had similar complaint of inguinal hernia. Family history revealed early death of his mother's two siblings at the age of 13 years.

On general physical examination, patient had growth retardation, macrocephaly with bilateral temporal and frontal bossing, coarse facial features and depressed nasal bridge. He had claw like hands and stiff joints. Skin of extremities was thickened and inelastic. Physical examination also revealed protruded abdomen and hepatosplenomegaly. Corneal clouding was not seen. Urine analysis showed the presence of GAGs. Serum alkaline phosphatase level was increased.

A. What is the most probable diagnosis?

B. What is the biochemical basis of this disease?

Answer—A: This is a case of Hunter syndrome. Coarsening of facial features, macrocephaly, broad noses with flared nostrils, prominent supraorbital ridges, thick lips and large tongue are common features of this disease. The skin of these patients is inelastic and thickened. Mobility of joints is restricted. GAG storage leads to hepatomegaly and splenomegaly. Umbilical hernia is commonly observed. Analysis of urine GAG levels can be used to confirm MPS.

It may be type A classic severe form or type B milder form. In type A, clinical features are apparent in the first year of life; somatic and neurologic development is slow with severe mental retardation. In type B, there is mild mental retardation, but intelligence is usually normal, and progression of the disease is very slow.

Answer—B: In Hunter syndrome, there is deficiency in the activity of lysosomal enzyme, iduronate-2-sulfatase. GAG are partially degraded or remains undegraded and get accumulated within the lysosomes and is excreted in urine.

Unit IV

Lipid Metabolism

Competencies Addressed

- **BI4.1** Describe and discuss main classes of lipids (essential/non-essential fatty acids, cholesterol and hormonal steroids, triglycerides, major phospholipids and sphingolipids) relevant to human system and their major functions
- **BI4.2** Describe the processes involved in digestion and absorption of dietary lipids and also the key features of their metabolism
- **BI4.3** Explain the regulation of lipoprotein metabolism and associated disorders
- **BI4.4** Describe the structure and functions of lipoproteins, their functions, interrelations and relations with atherosclerosis
- **BI4.5** Interpret laboratory results of analytes associated with metabolism of lipids
- **BI4.6** Describe the therapeutic uses of prostaglandins and inhibitors of eicosanoid synthesis
- **BI4.7** Interpret laboratory results of analytes associated with metabolism of lipids.

Unit IV: Overview

Lipids are a **heterogenous group of water-insoluble molecules**, whereas the body is mostly having aqueous solutions like blood and cell cytosol. The lipids help to **compartmentalize** the cell's internal milieu (see p. 15), are **stored for energy in adipocytes** (see p. 307) and are **transported as lipoproteins** (see Fig. 23.17, p. 352) in blood. Lipids are stored as triacylglycerol for energy (see p. 305), **make steroid hormones** (see Fig. 23.24, p. 359) and important compounds like **eicosanoids and prostaglandins** (see p. 333) and are involved in metabolism of fat-soluble vitamins. Chapter 20 deals with the digestion and absorption of lipids.

Lipids are **synthesized from acetyl CoA** (which occupies a central role in intermediary metabolism) whenever there is excess production of acetyl CoA. This situation of excess acetyl CoA favors **synthesis of fatty acids (promoted by insulin),** which are stored as triacylglycerol (TAG) (Chapter 21). Increased storage of TAG leads to obesity.

When body needs energy and glycogen stores have been depleted (starvation, diabetes), **this stored triacylglycerol is broken down to acetyl CoA in presence of hormones like glucagon**. However, all of this acetyl CoA cannot be sent to tricarboxylic acid (TCA) cycle as the oxaloacetate (first step of TCA cycle) is being used for gluconeogenesis in these conditions. So now the excess acetyl CoA (generated from breaking down triacylglycerol) will start forming ketone bodies to get whatever energy it can (chapter 21). Further, lipids make certain special compounds like phospholipids, prostaglandins, eicosanoids and sphingomyelins (chapter 22) for various designated roles in membranes, surfactants, myelin, etc. The student will do well to understand that the synthesis of TAG from acetyl CoA and breakdown of TAG to acetyl CoA is happening inside cells, whereas the lipoprotein assisted transfer of cholesterol and triglycerides is happening in blood.

Lipoproteins transfer lipids from intestine to liver as well as from liver to all parts of body (chapter 23). They (low-density lipoprotein) give cholesterol to tissues so that steroid hormones can be made, and various components of cell membrane and cell structure can be maintained. The excess cholesterol is brought back to liver by high-density lipoprotein to break cholesterol to bile for excretion. However, in certain disease conditions like **atherosclerosis** (chapter 23) the excess lipoproteins get modified and deposited in arterial vessel walls to form plaques. These atherosclerotic plaques may be stable or vulnerable to rupture and thrombus formation. If thrombi are released from plaques, it may block end arteries like the ones in myocardium and cause angina or myocardial infarction which is an emergency situation. Such thrombi also cause ischemia in brain leading to stroke and paralytic events.

Hence it is very important to control lipid levels. The suggested normal ranges to be achieved by lifestyle modifications and drugs (if required) are given in the appendix for this unit. The primary healthcare physician must understand the central role of lipids and integrated metabolism in controlling the epidemic of **coronary artery disease** that is surging in India and advise patients accordingly.

Lipids: Classification, Digestion, Absorption, and Physiological Functions 20

Learning Objectives

At the end of this chapter, the student will be able to:

- Discuss the processes involved in digestion and absorption of dietary lipids and the key features in their metabolism
- Classify lipids and discuss their functions
- Explain the importance of polyunsaturated fatty acids (PUFAs)

I. OVERVIEW

Lipids are a heterogeneous group of water-insoluble (hydrophobic) organic molecules (Fig. 20.1). **Because of their insolubility in aqueous solutions, body lipids are generally found compartmentalized, as in the case of membrane-associated lipids or droplets of triacylglycerol in adipocytes, or transported in blood in association with protein, as in lipoprotein particles** (see p. 349) or on albumin. Lipids may be simple, compound, derived or complexed to other compounds (see p. 291, blue box: Classification of Lipids). Lipids are a major source of energy for the body, and they also provide the hydrophobic barrier that permits partitioning of the aqueous contents of cells and subcellular structures. Lipids serve additional functions in the body (for example, some fat-soluble vitamins have regulatory or coenzyme functions, and the prostaglandins and steroid hormones play major roles in the control of the body's homeostasis, see p. 290, blue box: Functions of Lipids). Deficiencies or imbalances of lipid metabolism can lead to some of the major clinical problems encountered by physicians, such as atherosclerosis, diabetes, and obesity.

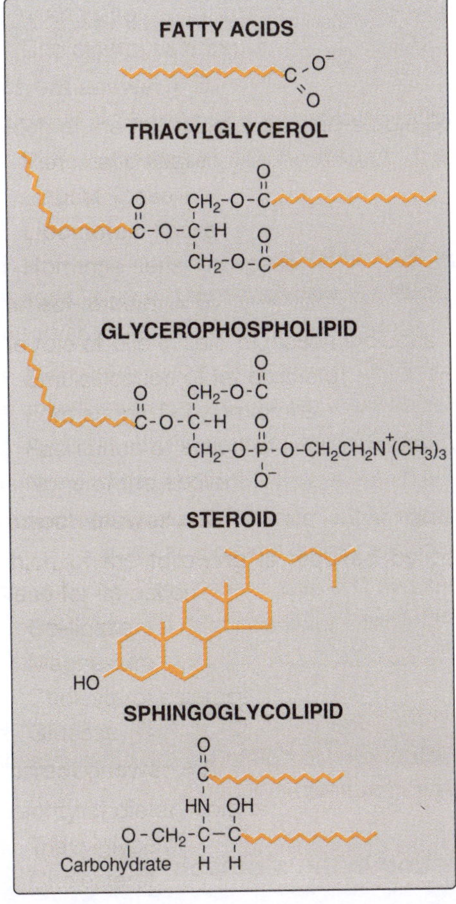

Figure 20.1 Structures of some common classes of lipids. Hydrophobic portions of the molecules are shown in orange.

II. DIGESTION, ABSORPTION, SECRETION, AND UTILIZATION

The average daily intake of lipids by U.S. adults is ~78 g, of which >90% is triacylglycerol ([TAG], formerly called triglyceride [TG]), that consists of three fatty acids (FA) esterified to a glycerol backbone (Fig. 20.1). The remainder of the dietary lipids consists primarily of cholesterol, cholesteryl esters, phospholipids, and nonesterified

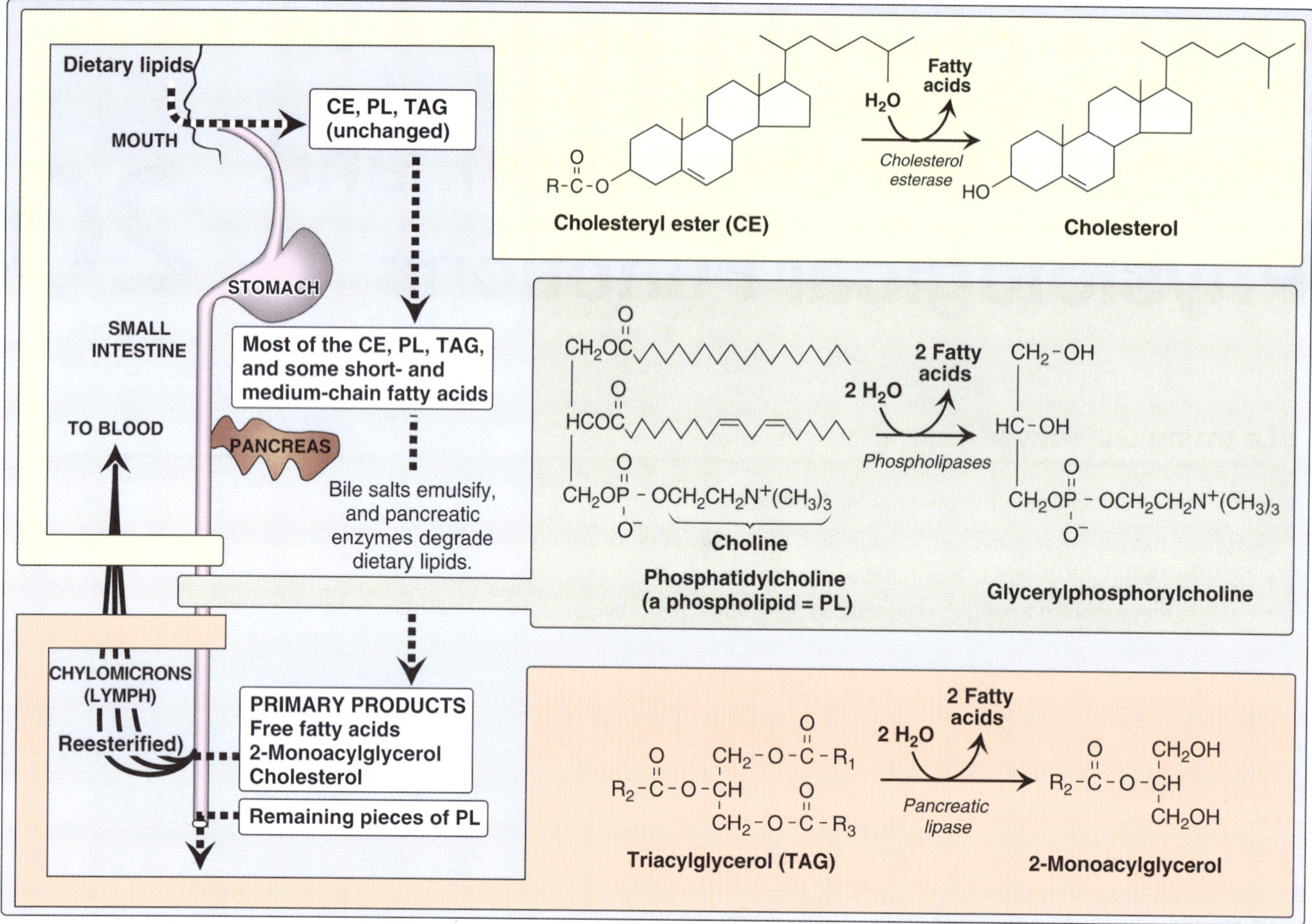

Figure 20.2 Overview of lipid digestion.

> **Functions of lipids**
>
> - Lipids like triacylglycerols are the storage form of energy. These are stored in adipocytes.
> - Phospholipids and cholesterol are part of the structure of biomembranes.
> - Lipid-derived steroid hormones like cortisol, estrogen and androgens influence many metabolic processes.
> - Lipids like dipalmitoylphosphatidylcholine act as surfactants in lung alveoli.
> - Lipids such as galactoceramide and cholesterol function as electrical insulators between neurons.
> - Lipids are essential for the absorption of fat-soluble vitamins A, D, E and K.
> - Lipids in the form of lipoproteins transport triacylglycerol and cholesterol to and fro from intestine, liver and tissues to facilitate various functions of lipids.
> - Vitamin D is itself made from 7-dehydrocholesterol.
> - Physically, lipids also form thermal barriers (subcutaneous fat) and cushion internal organs.

(free) FA (FFA). The digestion of dietary lipids begins in the stomach and is completed in the small intestine. The process is summarized in Figure 20.2.

A. Digestion in the stomach

Lipid digestion in the stomach is limited. It is catalyzed by ***lingual lipase*** that originates from glands at the back of the tongue and ***gastric lipase*** that is secreted by the gastric mucosa. Both enzymes are relatively acid stable, with optimal pH values of 4 to 6. **These *acid lipases*** hydrolyze FA from TAG molecules, particularly those containing short- or medium-chain-length (≤12 carbons) FA such as are found in milk fat. Consequently, these ***lipases* play a particularly important role in lipid digestion in infants for whom milk fat is the primary source of calories**. They also become important digestive enzymes in individuals with pancreatic insufficiency

Classification of lipids

I. Simple lipids	Triacylglycerol
II. Compound lipids	
(A) Phospholipids	1. Nitrogen glycerophosphatides e.g., lecithin, cephalin and phosphatidylserine 2. Non-nitrogen glycerophosphatides e.g., phosphatidylglycerol, phosphatidylinositol, cardiolipin 3. Plasmalogens e.g., choline and ethanolamine plasmalogens 4. Phosphosphingosides e.g., sphingomyelin
(B) Nonphosphorylated lipids	1. Glycerosphingolipids e.g., cerebrosides, globosides and gangliosides 2. Sulfatides e.g., sulfated cerebroside, sulfated gangliosides and sulfated globosides
III. Derived lipids	Fatty acids, leukotrienes, steroids, prostaglandins, etc.
IV. Lipids complexed to other compounds	Lipoproteins

such as those with cystic fibrosis (CF). *Lingual* and *gastric lipases* aid these patients in degrading TAG molecules (especially those with short- to medium-chain FA) despite a near or complete absence of *pancreatic lipase* (see Section D.1).

B. Cystic fibrosis

CF is the most common lethal genetic disease in Caucasians of Northern European ancestry and has a prevalence of ~1:3,300 births in the United States. CF is an **autosomal-recessive** disorder caused by **mutations to the gene for the CF transmembrane conductance regulator (CFTR)** protein that functions as a chloride channel on epithelium in the pancreas, lungs, testes, and sweat glands. Defective CFTR results in decreased secretion of chloride and increased uptake of sodium and water. In the pancreas, the depletion of water on the cell surface results in thickened mucus that clogs the pancreatic ducts, **preventing pancreatic enzymes from reaching the intestine, thereby leading to pancreatic insufficiency**. In CF, abnormally thick, sticky mucus traps germs in the airways, leading to chronic lung infections and progressive pulmonary disease. Male infertility may also be seen. Treatment includes replacement of these enzymes and supplementation with fat-soluble vitamins.

C. Emulsification in the small intestine

The critical process of dietary lipid emulsification occurs in the duodenum. **Emulsification increases the surface area of the hydrophobic lipid droplets** so that the digestive enzymes, which work at the interface of the droplet and the surrounding aqueous solution, can act effectively. **Emulsification is accomplished by two complementary mechanisms, namely, use of the detergent properties of the conjugated bile salts and mechanical mixing due to peristalsis**. Bile salts, made in the liver and stored in the gallbladder, are amphipathic derivatives of cholesterol (see Fig. 23.11, p. 348). Conjugated bile salts consist of a hydroxylated sterol ring structure with a side chain to which a molecule of glycine or taurine is covalently attached by an amide linkage (Fig. 20.3). These emulsifying agents interact with the dietary lipid droplets and the aqueous duodenal contents, thereby stabilizing the droplets as they become smaller from peristalsis and preventing them from coalescing. [Note: See p. 346 for a more complete discussion of bile salt metabolism.]

D. Degradation by pancreatic enzymes

The dietary TAG, cholesteryl esters, and phospholipids are enzymatically degraded (digested) in the

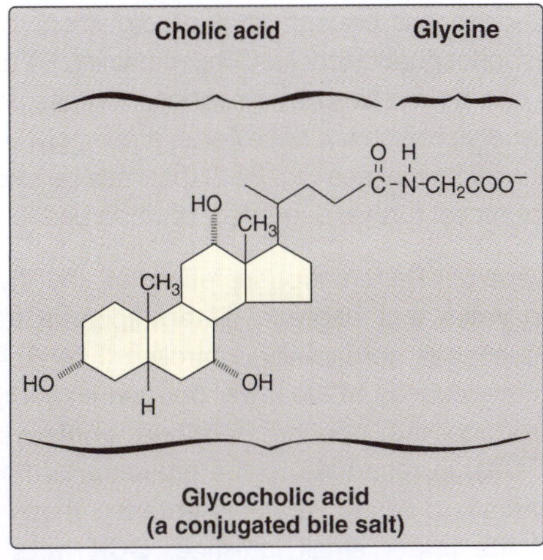

Figure 20.3 Structure of glycocholic acid.

small intestine by pancreatic enzymes, whose secretion is hormonally controlled.

1. **Triacylglycerol degradation:** TAG molecules are too large to be taken up efficiently by the mucosal cells (enterocytes) of the intestinal villi. Therefore, they are hydrolyzed by an ***esterase, pancreatic lipase***, which preferentially removes the FA at carbons 1 and 3. The primary products of hydrolysis are, thus, a mixture of **2-monoacylglycerol (2-MAG) and FFA (see Fig. 20.2**). A second protein, ***colipase,*** also secreted by the pancreas as a zymogen, binds the *lipase* at a ratio of 1:1 and anchors it at the lipid–aqueous interface. *Colipase* restores activity to *lipase* in the presence of inhibitory substances like bile salts that bind the micelles. **Orlistat, an antiobesity drug, inhibits *gastric* and *pancreatic lipases*, thereby decreasing fat absorption, resulting in weight loss.**

   Pancreatic lipase is found in high concentrations in pancreatic secretions, and it is highly efficient catalytically, thus ensuring that only severe pancreatic deficiency, such as that seen in CF, results in significant malabsorption of fat.

2. **Cholesteryl ester degradation:** Most dietary cholesterol is present in the free (nonesterified) form, with 10%–15% present in the esterified form. Cholesteryl esters are hydrolyzed by pancreatic ***cholesteryl ester hydrolase*** **(*cholesterol esterase*),** which produces cholesterol plus FFA (see Fig. 20.2). Activity of this enzyme is greatly increased in the presence of bile salts.

3. **Phospholipid degradation:** Pancreatic juice is rich in the proenzyme of ***phospholipase*** $\boldsymbol{A_2}$ that, like procolipase, is activated by *trypsin* and, like *cholesteryl ester hydrolase*, requires bile salts for optimum activity. ***Phospholipase*** $\boldsymbol{A_2}$ **removes one FA from carbon 2 of a phospholipid, leaving a lysophospholipid**. For example, phosphatidylcholine (the predominant phospholipid of digestion) becomes lysophosphatidylcholine. The remaining FA at carbon 1 can be removed by *lysophospholipase*, leaving a glycerylphosphoryl base (for example, glycerylphosphorylcholine, see Fig. 20.2) that may be excreted in the feces, further degraded, or absorbed.

4. **Control:** Pancreatic secretion of the hydrolytic enzymes that degrade dietary lipids in the small intestine is hormonally controlled (Fig. 20.4). Cells in the mucosa of the lower duodenum and jejunum produce the peptide hormone **cholecystokinin (CCK),** in response to the presence of lipids and partially digested proteins entering these regions of the upper small intestine. **CCK acts on the gallbladder (causing it to contract and release bile**, a mixture of bile salts, phospholipids, and free cholesterol) and **on the exocrine cells of the pancreas (causing them to release digestive enzymes**). It also **decreases gastric motility**, resulting in a slower release of gastric contents into the small intestine (see Fig. 23.4). Other intestinal cells produce another peptide hormone, **secretin,** in response to the low pH of the chyme entering the intestine from the stomach. Secretin **causes the pancreas to release a solution rich in bicarbonate** that helps neutralize the pH of the intestinal contents, bringing them to the appropriate pH for digestive activity by pancreatic enzymes.

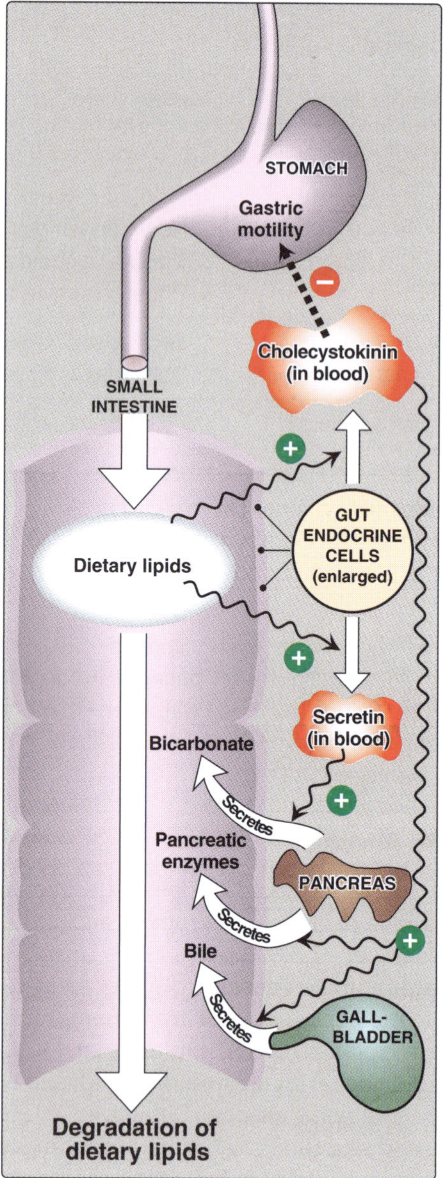

Figure 20.4 Hormonal control of lipid digestion in the small intestine. [Note: The small intestine is divided into three parts: the duodenum (upper 5%), the jejunum, and the ileum (lower 55%).]

E. Absorption by enterocytes

FFA, free cholesterol, and 2-MAG are the primary products of lipid digestion in the jejunum. These, plus bile salts and fat-soluble vitamins (A, D, E, and K), **form mixed micelles** (that is, disc-shaped clusters of a mixture of amphipathic lipids that coalesce with their hydrophobic groups on the inside and their hydrophilic groups on the outside). Therefore, mixed micelles are soluble in the aqueous environment of the intestinal lumen (Fig. 20.5). These particles approach the primary site of lipid absorption, the brush border membrane of the enterocytes. This microvilli-rich apical membrane is separated from the liquid contents of the intestinal lumen by an unstirred water layer that mixes poorly with the bulk fluid. The hydrophilic surface of the micelles facilitates the transport of the hydrophobic lipids through the unstirred water layer to the brush border membrane where they are absorbed. Bile salts are absorbed in the terminal ileum, with <5% being lost in the feces. [Note: Relative to other dietary lipids, cholesterol is only poorly absorbed by the enterocytes. Drug therapy (for example, with **ezetimibe**) can further reduce cholesterol absorption in the small intestine.]

Because short- and medium-chain FA are water soluble, they do not require the assistance of mixed micelles for absorption by the intestinal mucosa.

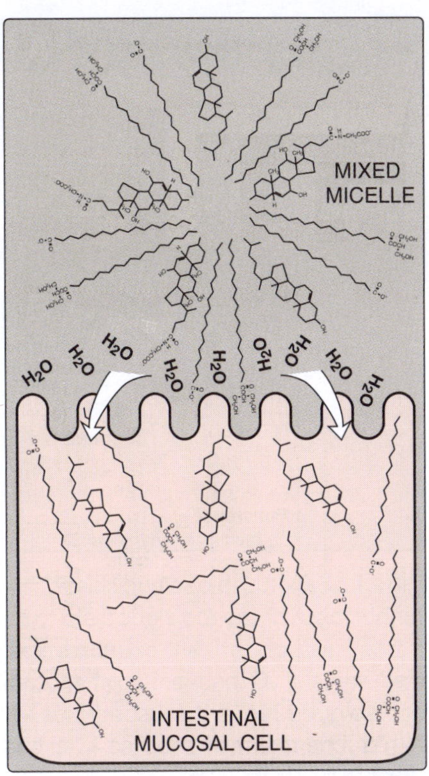

Figure 20.5 Absorption of lipids contained in a mixed micelle by an intestinal mucosal cell. The micelle itself is not absorbed. [Note: Short- and medium-chain-length fatty acids do not require incorporation into micelles.]

F. Triacylglycerol and cholesteryl ester resynthesis

The mixture of lipids absorbed by the enterocytes migrates to the smooth endoplasmic reticulum (SER) where biosynthesis of complex lipids takes place. The long-chain FA are first converted into their activated form by *fatty acyl coenzyme A (CoA) synthetase (thiokinase)*, as shown in Figure 20.6. Using the fatty acyl CoA derivatives, the 2-MAG absorbed by the enterocytes are converted to TAG through sequential reacylations by two *acyltransferases*, *acyl CoA:monoacylglycerol acyltransferase* and *acyl CoA:diacylglycerol acyltransferase*. Lysophospholipids are reacylated to form phospholipids by a family of *acyltransferases*, and cholesterol is acylated primarily by *acyl CoA:cholesterol acyltransferase*

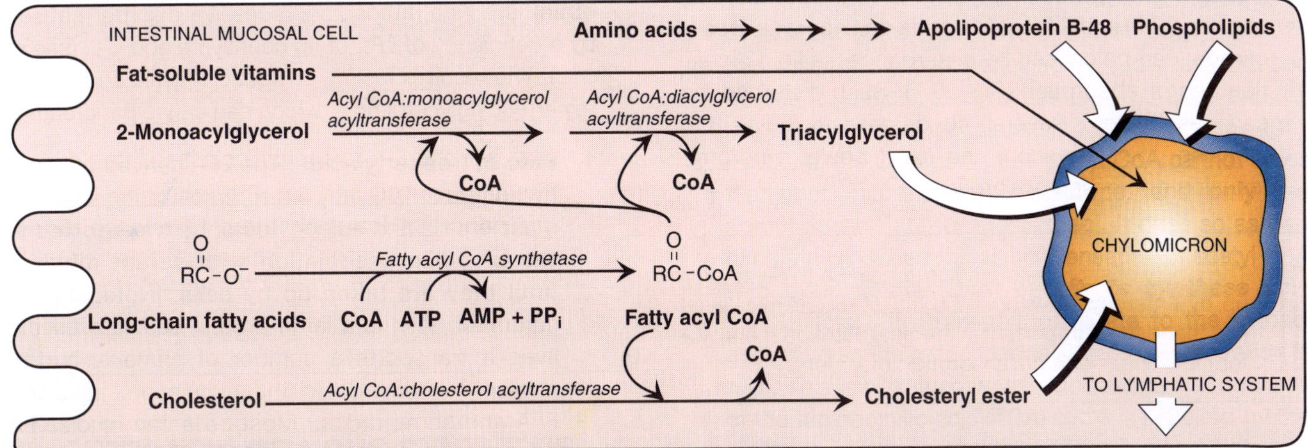

Figure 20.6 Assembly and secretion of chylomicrons by intestinal mucosal cells. [Note: Short- and medium-chain-length fatty acids do not require incorporation into chylomicrons and directly enter into the blood.] CoA = coenzyme A; AMP = adenosine monophosphate; PP_i = pyrophosphate.

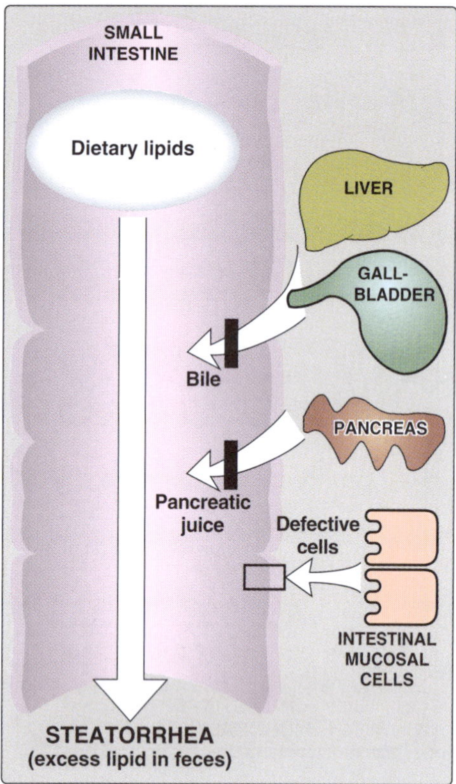

Figure 20.7 Possible causes of steatorrhea.

(see Fig. 23.21). [Note: Virtually all long-chain FA entering the enterocytes are used in this fashion to form TAG, phospholipids, and cholesteryl esters. **Short- and medium-chain FA are not converted to their CoA derivatives and are not reesterified to 2-MAG.** Instead, they are released into the portal circulation, where they are carried by serum albumin to the liver.]

G. Lipid malabsorption

Lipid malabsorption, resulting in increased lipid (including the fat-soluble vitamins and essential FA, see Fig. 20.7) in the feces, a condition known as **steatorrhea**, can be caused by disturbances in lipid digestion and/or absorption (Fig. 20.7). Such disturbances can result from several conditions, including CF (causing poor digestion) and short bowel syndrome (causing decreased absorption).

> The ability of short- and medium-chain FA to be taken up by enterocytes without the aid of mixed micelles has made them important in medical nutrition therapy for individuals with malabsorption disorders.

H. Secretion from enterocytes

The newly resynthesized TAG and cholesteryl esters are very hydrophobic and aggregate in an aqueous environment. Therefore, they must be packaged as particles of lipid droplets surrounded by a thin layer composed of phospholipids, nonesterified cholesterol, and a molecule of the protein apolipoprotein (apo) B-48 (see p. 350). This layer stabilizes the particle and increases its solubility, thereby preventing multiple particles from coalescing. [Note: Microsomal triglyceride transfer protein is essential for the assembly of all TAG-rich apo B–containing particles in the ER (see p. 350).] The lipoprotein particles are released by exocytosis from enterocytes into the lacteals (lymphatic vessels in the villi of the small intestine). The presence of these particles in the lymph after a lipid-rich meal gives it a milky appearance. This lymph is called **chyle** (as opposed to chyme, the name given to the semifluid mass of partially digested food that passes from the stomach to the duodenum), and the particles are named chylomicrons. **Chylomicrons** follow the lymphatic system to the thoracic duct and are then conveyed to the left subclavian vein, where they enter the blood. **The steps in the production of chylomicrons are summarized in Figure 20.6.** [Note: Once released into blood, the nascent (immature) chylomicrons pick up apolipoproteins E and C-II from high-density lipoproteins and mature. (For a more detailed description of chylomicron structure and metabolism, see Fig. 23.16, p. 351.)]

I. Use by the tissues

Most of the TAG contained in chylomicrons is broken down in the capillary beds of skeletal and cardiac muscle and adipose tissue. The TAG is degraded to FFA and glycerol by ***lipoprotein lipase*** **(*LPL*)**. This enzyme is synthesized and secreted primarily by adipocytes and muscle cells. **Secreted *LPL* is anchored to the luminal surface of endothelial cells in the capillaries of muscle and adipose tissues.**

Familial chylomicronemia (type I hyperlipoproteinemia) is a rare, autosomal-recessive disorder caused by a deficiency of *LPL* or its coenzyme apo C-II (see p. 351). The result is fasting chylomicronemia and severe hypertriacylglycerolemia, which can cause pancreatitis.

1. **Fate of free fatty acids:** The FFA derived from the hydrolysis of TAG may either directly enter adjacent muscle cells and adipocytes or be **transported in the blood in association with serum albumin until they are taken up by cells**. [Note: Human serum albumin is a large protein secreted by the liver. It transports a number of primarily hydrophobic compounds in the circulation, including FFA and some drugs.] Most cells can oxidize FA to produce energy (see p. 309). **Adipocytes can also reesterify FFA to produce TAG molecules,** which are stored until the FA are needed by the body (see p. 305).

2. **Fate of glycerol:** Glycerol released from TAG is taken up from the blood and phosphorylated by hepatic **glycerol kinase** to produce glycerol 3-phosphate, which can enter **either glycolysis or gluconeogenesis** by oxidation to dihydroxyacetone phosphate (see p. 196) or be used in TAG synthesis (see Fig. 21.14, p. 308).

3. **Fate of chylomicron remnants:** After most of the TAG has been removed, the chylomicron remnants (which contain cholesteryl esters, phospholipids, apolipoproteins, fat-soluble vitamins, and a small amount of TAG) **bind to receptors on the liver** (apo E is the ligand; see Fig. 23.16, p. 351) and **are endocytosed**. The intracellular remnants are hydrolyzed to their component parts. Cholesterol and the nitrogenous bases of phospholipids (for example, choline) can be recycled by the body. [Note: If removal of remnants by the liver is decreased because of impaired binding to their receptor, they accumulate in the plasma. This is seen in the rare **type III hyperlipoproteinemia** (also called familial dysbetalipoproteinemia or broad beta disease, see p. 353).]

III. Medical Perspectives

Disorders of lipid digestion and absorption could be due to intraluminal defects of impaired emulsification (deficiency of bile), deficiency of pancreatic enzymes, and impaired pH regulation. Diseases of mucosa can be disease related (tropical sprue) or surgery related (intestinal bypass, intestinal resection).

Fat maldigestion or malabsorption causes excretion of undigested lipids in stools. This is called **steatorrhea**. Loss of fat-soluble vitamins will also be seen in lipid malabsorption. Steatorrhea may be due to tropical sprue, nontropical sprue, celiac sprue, intestinal lipodystrophy (Whipple's disease), etc. Absence or deficiency of bile also leads to steatorrhea.

Lipid malabsorption also leads to defective absorption of fat-soluble vitamins like vitamin K. This may cause prolongation of prothrombin time. Lipid malabsorption further leads to deficiency of essential fatty acids and vitamin D. It also causes secondary hypolipoproteinemia.

Gallbladder removal is a common surgery. If the gallbladder is removed, the liver secretes bile directly into the small intestine. This allows for digestion of most foods, but high-volume, fatty, and greasy foods may be difficult to digest, leading to flatulence and discomfort.

Cystic fibrosis is an autosomal recessive disorder caused by mutations to the gene for the cystic fibrosis transmembrane conductance regulator (CFTR) protein that functions as a chloride channel on epithelium in the pancreas, lungs, testes, and sweat glands. Defective CFTR results in decreased secretion of chloride ions, and increased uptake of sodium ions and water. In the pancreas, the depletion of water on the cell surface results in thickened mucus that clogs the pancreatic ducts, preventing pancreatic enzymes from reaching the intestine, thereby leading to pancreatic insufficiency.

Pancreatic lipase inhibitors are used in the treatment of hyperlipidemia and obesity. Orlistat inhibits pancreatic and gastric lipases and is used as an antiobesity drug. The drug is not to be taken without medical supervision as it has many side effects.

IV. Chapter Summary

- **Dietary lipid digestion** begins in the stomach and continues in the small intestine (Fig. 20.8).
- **Cholesteryl esters**, **phospholipids**, and **triacylglycerols** (**TAG**) containing **long-chain-length fatty acids** (**FA**) are degraded in the **small intestine** by **pancreatic enzymes**. The most important of these enzymes are ***cholesterol esterase***, ***phospholipase A_2***, and ***pancreatic lipase***. In **cystic fibrosis**, thickened mucus prevents these enzymes reaching the intestine.
- In contrast, TAG in **milk fat** contain **short-** to **medium-chain-length FA** and are degraded in the **stomach** by ***acid lipases*** (***lingual lipase*** and ***gastric lipase***).
- The **hydrophobic** nature of lipids requires that dietary lipids be **emulsified** for efficient degradation. Emulsification occurs in the small intestine using **peristaltic action** (mechanical mixing) and **bile salts** (detergents).
- The primary products of dietary lipid degradation are **2-monoacylglycerol**, nonesterified (free) **cholesterol**, and **free FA**.
- These compounds, plus the **fat-soluble vitamins**, form **mixed micelles** that facilitate dietary lipid absorption by **intestinal mucosal cells** (**enterocytes**).
- Enterocytes use activated long-chain FA to regenerate TAG and cholesteryl esters and also synthesize protein (**apolipoprotein** [**apo**] **B-48**), all of which are then assembled with the fat-soluble vitamins into **lipoprotein particles** called **chylomicrons**.
- Short- and medium-chain FA enter blood directly.

- Chylomicrons are first released into the **lymph** and then enter the **blood**, where their lipid core is degraded by ***lipoprotein lipase*** (with **apo C-II** as the coenzyme) in the **capillaries** of **muscle** and **adipose** tissues. Thus, dietary lipids are made available to the peripheral tissues.
- **Fat maldigestion** or malabsorption causes **steatorrhea** (lipid in the feces). A deficiency in the ability to degrade chylomicron components, or remove chylomicron remnants after TAG has been degraded, results in accumulation of these particles in blood.

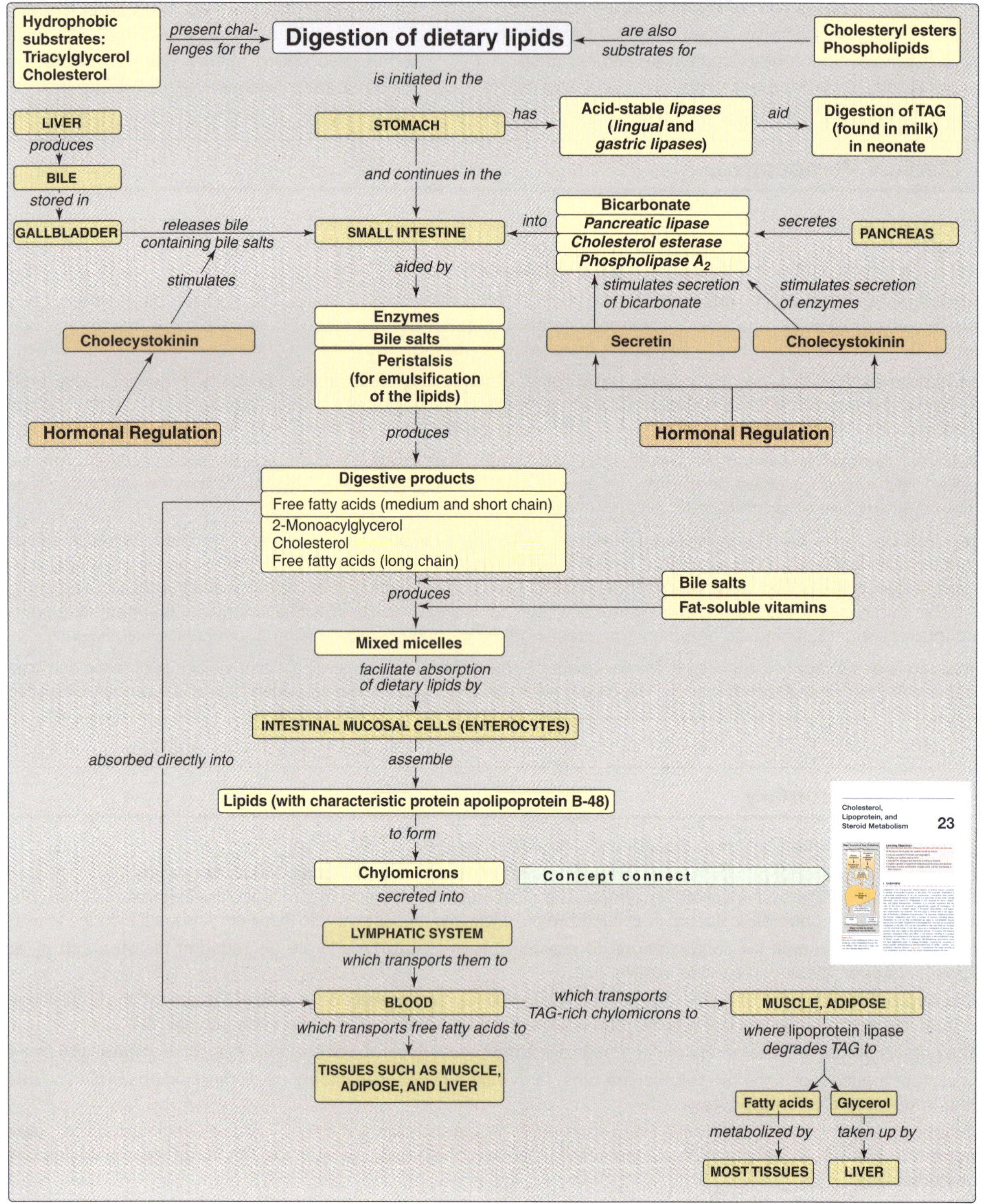

Figure 20.8 Key concept map for metabolism of dietary lipids.

Study Questions

Short-Answer Questions

Write short notes on:

20.1. Causes of steatorrhea
20.2. Hormonal control of lipid digestion
20.3. Cystic fibrosis
20.4. Classification of lipids
20.5. Chylomicrons
20.6. Role of pancreas in lipid digestion
20.7. Role of bile in lipid digestion

Long-Answer Question

20.1. Discuss digestion and absorption of dietary lipids. Add a note on fate of end products of lipid digestion.

Multiple-Choice Questions

Choose the ONE best answer.

20.1. Which one of the following statements about lipid digestion is correct?

A. Large lipid droplets are emulsified in the mouth through the act of chewing
B. The enzyme colipase facilitates the binding of bile salts to mixed micelles, maximizing the activity of pancreatic lipase.
C. The peptide hormone secretin causes the gall-bladder to contract and release bile.
D. Patients with cystic fibrosis have difficulties with digestion because their pancreatic secretions are less able to reach the small intestine, the primary site of lipid digestion.
E. Formation of triacylglycerol-rich chylomicrons is independent of protein synthesis in the intestinal mucosa.

Correct answer = D.

20.2. Which one of the following statements about lipid absorption from the intestine is correct?

A. Dietary triacylglycerol must be completely hydrolyzed to free fatty acids and glycerol before absorption.
B. The triacylglycerol carried by chylomicrons is degraded by lipoprotein lipase, producing fatty acids that are taken up by muscle and adipose tissues and glycerol that is taken up by the liver.
C. Fatty acids that contain ≤12 carbon atoms are absorbed and enter the circulation primarily via the lymphatic system.
D. Deficiencies in the ability to absorb fat result in excessive amounts of chylomicrons in the blood.

Correct answer = B.

20.3. Which of the following can be classified as a sphingolipid as well as a phospholipid?

A. Gangliosides
B. Plasmalogens
C. Sphingomyelins
D. Globosides

Correct answer = C.

20.4. Biological membranes are composed of molecules like cholesterol and phospholipids. Which among the following properties are common to these molecules that enable them to be present in the membranes?

A. They have a polar and a non-polar component
B. Both can associate with fatty acids
C. They are both lipids
D. They both can give rise to second messengers

Correct answer = A.

20.5. Orlistat is a medication prescribed to assist with weight loss in cases of obesity. It acts by inhibiting the absorption of fats from the intestine. Which of the following enzymes is inhibited by Orlistat?

A. Chymotrypsin
B. Pancreatic lipase
C. α-Amylase
D. Cholesterol esterase

Correct answer = B.

20.6. Which of the following is an acid-stable lipase?

A. Pancreatic lipase
B. Lingual lipase
C. Lipoprotein lipase
D. Hormone sensitive lipase

Correct answer = B.

20.7. The role of bile in lipid digestion includes

A. Emulsification of fat droplets
B. Formation of chylomicrons
C. Facilitation of action of lingual lipase
D. None of the above

Correct answer = A.

20.8 Which of the following is required by pancreatic lipase for its action?

A. Co-lipase
B. Magnesium
C. Cholesterol esterase
D. Glucose

Correct answer = A.

20.9. Majority of dietary lipid is

A. Triacylglycerol
B. Cholesterol
C. Cholesterol ester
D. Phospholipids

Correct answer = A.

20.10. Absorbed dietary lipids are transported as
A. VLDL
B. Chylomicrons
C. Mixed micelles
D. Monoacylglycerol
Correct answer = B.

Multiple-Choice Questions (NExT/NEET PG)

***(Proposed National Exit Test for MBBS Final professional Completed Students in India)**

Choose the ONE best answer.

20.1. Which of the following is NOT a correct statement about cystic fibrosis?
A. Treatment involves administration of certain enzymes
B. Sweat chloride levels are elevated
C. Genetic testing is not necessary before initiating the treatment
D. Inherited in an autosomal recessive manner
Correct answer = C.

20.2. Steatorrhea refers to increased lipid in the feces due to malabsorption. Which of the following conditions can lead to steatorrhea?
A. Chronic pancreatitis, leading to insufficient pancreatic enzyme production
B. Decrease in the intake of dietary fibers
C. Lactose intolerance, causing difficulty in digesting dairy products
D. Achlorhydria, that is decrease in the production of gastric acid
Correct answer = A.

20.3. Which one of the following statements about the absorption of lipids from the intestine is correct?
A. Dietary triacylglycerol must be completely hydrolyzed to free fatty acids and glycerol before absorption.
B. The TAG carried by chylomicrons is degraded to free fatty acids and glycerol by lipoprotein lipase.
C. Fatty acids that contain 10 carbons or less are absorbed and enter the circulation primarily via the lymphatic system.
D. Deficiencies in the ability to absorb fat result in excessive amounts of chylomicrons in the blood.
Correct answer = B.

20.4. Statement 1: Sitosterolemia is a condition in which plant sterols (sitosterols) are absorbed into the circulation from the intestine but cannot get excreted.
Statement 2: The plant sterols cannot get excreted because of an autosomal recessive disorder, leading to clinical manifestations including xanthomas, joint pains, and hemolysis.
A. Both the Statements are correct, and Statement 2 is the correct explanation for Statement 1
B. Statement 1 is correct, but Statement 2 is NOT correct
C. Statement 1 is NOT correct, but Statement 2 is correct
D. Both the Statements are incorrect
Correct answer = A.

20.5. Chyluria is a condition in which
A. Lymphatic fluid containing lipid droplets leaks into the kidney
B. Urine becomes milky white within few hours of taking fatty meal
C. Filariasis is often associated with chyluria
D. All of the above
Correct answer = D.

Fatty Acid, Triacylglycerol, and Ketone Body Metabolism

21

Learning Objectives

At the end of this chapter, the student will be able to:

- Describe essential and non-essential fatty acids
- Describe synthesis and degradation of fatty acids and triacylglycerols
- Describe regulation of fat mobilization and fatty acid oxidation
- Describe ketone body formation

I. OVERVIEW

Fatty acids exist free in the body (that is, they are nonesterified) **and as fatty acyl esters in more complex molecules such as triacylglycerols** (TAG). **Actually free fatty acids are never free**—they are always bound to albumin. Low levels of free fatty acids (FFA) occur in all tissues, but substantial amounts can sometimes be found in the plasma, particularly during fasting. Plasma FFA (transported bound to serum albumin) are in route from their point of origin (TAG of adipose tissue or circulating lipoproteins) to their site of consumption (most tissues). FFA can be oxidized by many tissues, particularly liver and muscle, to provide energy and, in the liver, to provide the substrate for ketone body synthesis. Fatty acids are also structural components of membrane lipids, such as phospholipids and glycolipids (see p. 321). Fatty acids attached to certain proteins enhance the ability of those proteins to associate with membranes (see p. 327). Fatty acids are also precursors of the hormone-like prostaglandins (see p. 333). Esterified fatty acids, in the form of TAG stored in white adipose tissue (WAT), serve as the major energy reserve of the body. Alterations in fatty acid metabolism are associated with obesity and diabetes. Figure 21.1 illustrates the metabolic pathways of fatty acid synthesis and degradation and their relationship to carbohydrate metabolism.

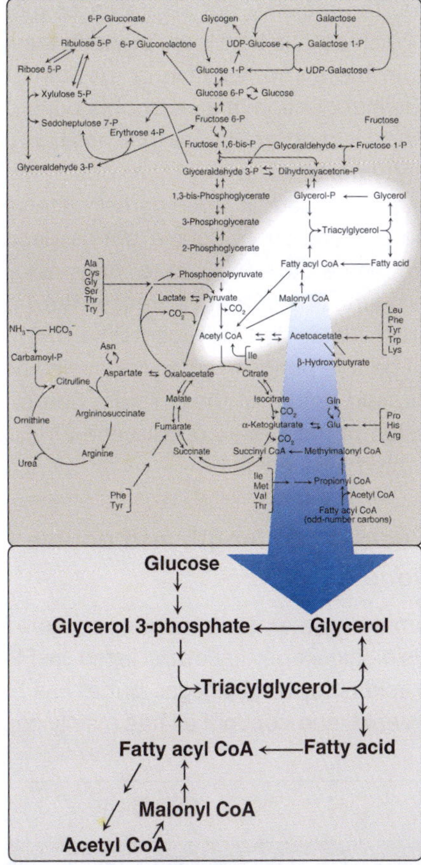

Figure 21.1 Triacylglycerol synthesis and degradation. CoA = coenzyme A.

II. FATTY ACID STRUCTURE

A fatty acid consists of a hydrophobic hydrocarbon chain with a terminal carboxyl group that has a pK_a (see p. 74) of ~4.8 (Fig. 21.2). At physiologic pH, the terminal carboxyl group (–COOH) ionizes, becoming –COO^-. [Note: When the pH is above the pK, the deprotonated form predominates (see p. 74).] This anionic group has an affinity for water, giving the fatty acid its amphipathic nature (having both a hydrophilic and a hydrophobic region). However, for long-chain-length fatty acids (LCFA), the hydrophobic portion is predominant. These molecules are highly water insoluble and must be transported in the circulation in association with protein. More than 90% of the fatty acids found in plasma are in the form of fatty acid esters (primarily TAG, cholesteryl esters, and phospholipids) contained in circulating lipoprotein particles (see p. 349). FFA are transported in the circulation in association with albumin, the most abundant protein in serum.

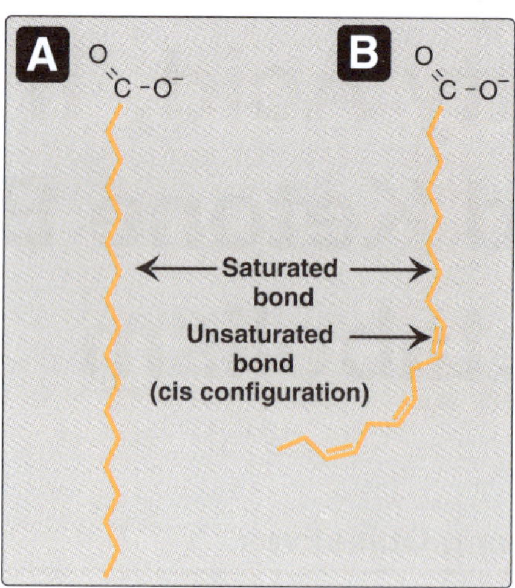

Figure 21.3 A saturated (**A**) and an unsaturated (**B**) fatty acid. Orange denotes hydrophobic portions of the molecules. [Note: Cis double bonds cause a fatty acid to kink.]

A. Fatty acid saturation

Fatty acid chains may contain no double bonds (that is, be **saturated**) or contain one or more double bonds (that is, be **mono- or polyunsaturated**). In humans, the majority are saturated or monounsaturated. When double bonds are present, they are nearly always in the cis rather than in the trans configuration. The introduction of a cis double bond causes the fatty acid to bend or kink at that position (Fig. 21.3). If the fatty acid has two or more double bonds, they are always spaced at three-carbon intervals. [Note: In general, addition of double bonds decreases the melting temperature (T_m) of a fatty acid, whereas increasing the chain length increases the T_m. Because membrane lipids typically contain LCFA, the presence of double bonds in some fatty acids helps maintain the fluid nature of those lipids. See p. 508 for a discussion of the dietary occurrence of cis and trans unsaturated fatty acids.]

B. Fatty acid chain length and double bond positions

The common names and structures of some fatty acids of physiologic importance are listed in Figure 21.4. Acetic acid is found in vinegar, butter has butyric and caproic acid, and coconut oil has mainly capric, lauric and myristic acid. Mustard oil has erucic acid (22:1, ω-9) and vegetable oils have linolenic, linoleic and arachidonic acid. Brain lipids have cerebronic acid (24) and nervonic (24:1, ω-9) acids.

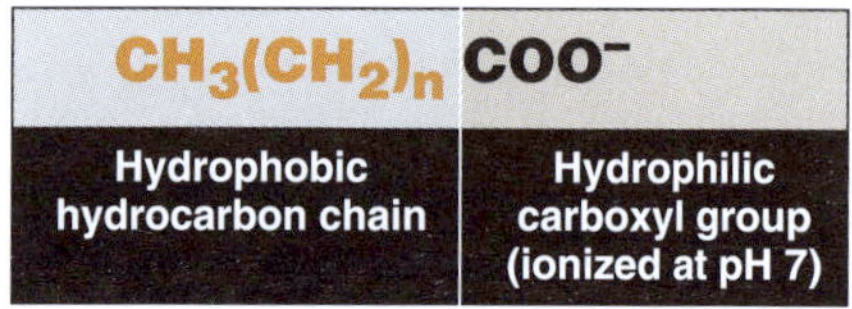

Figure 21.2 Structure of a fatty acid.

In humans, fatty acids with an even number of carbon atoms (16, 18, or 20) predominate, with longer fatty acids (>22 carbons) being found in the brain. The carbon atoms are numbered, beginning with the carbonyl carbon as carbon 1. The number before the colon indicates the number of carbons in the chain, and those after the colon indicate the numbers and positions (relative to the carboxyl end) of double bonds. For example, as denoted in Figure 21.4, arachidonic acid, 20:4(5,8,11,14), is 20 carbons long and has four double bonds (between carbons 5–6, 8–9, 11–12, and 14–15). [Note: Carbon 2, the carbon to which the carboxyl group is attached, is also called the α-carbon, carbon 3 is the β-carbon, and carbon 4 is the γ-carbon. The carbon of the terminal methyl group is called the ω-carbon regardless of the chain length.] The double bonds in a fatty acid can also be referenced relative to the ω (methyl) end of the chain. Arachidonic acid is referred to as an ω-6 fatty acid because the terminal double bond is six bonds from the ω end (Fig. 21.5A). [Note: The equivalent designation of n-6 may also be used (Fig. 21.5B).] Another ω-6 fatty acid is the essential linoleic acid 18:2(9,12). In contrast, α-linolenic acid, 18:3(9,12,15), is an essential ω-3 fatty acid.

Polyunsaturated fatty acids (PUFA) are the fatty acids having two or more double bonds, e.g., arachidonic acid. All essential fatty acids are PUFA. They are also responsible for fluidity of membranes and are also precursors for synthesis of other important compounds like eicosanoids.

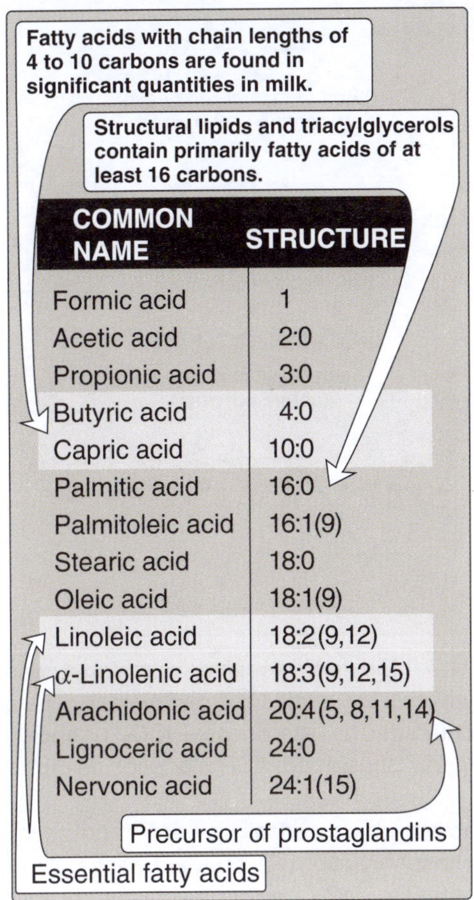

COMMON NAME	STRUCTURE
Formic acid	1
Acetic acid	2:0
Propionic acid	3:0
Butyric acid	4:0
Capric acid	10:0
Palmitic acid	16:0
Palmitoleic acid	16:1(9)
Stearic acid	18:0
Oleic acid	18:1(9)
Linoleic acid	18:2(9,12)
α-Linolenic acid	18:3(9,12,15)
Arachidonic acid	20:4(5, 8,11,14)
Lignoceric acid	24:0
Nervonic acid	24:1(15)

Figure 21.4 Some fatty acids of physiologic importance. [Note: A fatty acid containing 2–4 carbons is considered short; 6–12, medium; 14–20, long; and ≥22, very long.]

C. Essential fatty acids

Linoleic acid, the precursor of ω-6 arachidonic acid that is the substrate for prostaglandin synthesis (see p. 333), and **α-linolenic acid**, the precursor of ω-3 fatty acids that are important for growth and development, are dietary essentials in humans because we lack the enzymes needed to synthesize them. Plants provide us with these essential fatty acids. [Note: **Arachidonic acid becomes essential if linoleic acid is deficient in the diet**. See p. 509 for a discussion of the nutritional significance of ω-3 and ω-6 fatty acids.]

> Essential fatty acid deficiency (rare) can result in a dry, scaly dermatitis as a result of an inability to synthesize molecules that provide the water barrier in skin (see p. 326).

III. FATTY ACID DE NOVO SYNTHESIS

Carbohydrates and proteins obtained from the diet in excess of the body's needs for these nutrients can be

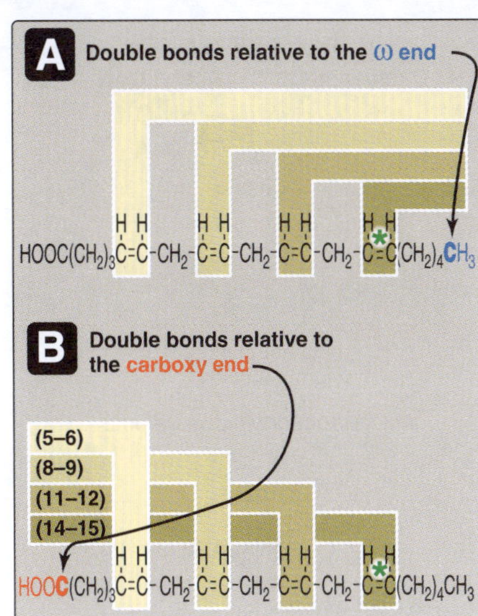

Figure 21.5 Arachidonic acid, 20:4(5,8,11,14), illustrating the position of the double bonds. **A.** Arachidonic acid is an ω-6 fatty acid because the first double bond from the ω end is 6 carbons from that end. **B.** It is also referred to as an n-6 fatty acid because the last double bond from the carboxyl end is 14 carbons from that end: 20 – 14 = 6 = n. Thus, the "ω" and "n" designations are equivalent (see *).

converted to fatty acids. In adults, de novo fatty acid synthesis occurs **primarily in the liver** and lactating mammary glands and, to a lesser extent, in adipose tissue. This cytosolic process is endergonic (see p. 160) and reductive. It incorporates carbons from acetyl coenzyme A (CoA) into the growing fatty acid chain, using ATP and reduced nicotinamide adenine dinucleotide phosphate (NADPH). [Note: Dietary TAG also supply fatty acids. See p. 452 for a discussion of the metabolism of dietary nutrients in the well-fed state.]

A. Cytosolic acetyl CoA production

The first step in fatty acid synthesis is the transfer of acetate units from mitochondrial acetyl CoA to the cytosol. Mitochondrial acetyl CoA is produced by the oxidation of pyruvate (see p. 206) and by the catabolism of certain amino acids (see p. 401). However, the CoA portion of acetyl CoA cannot cross the inner mitochondrial membrane, and **only the acetyl portion enters the cytosol.** It does so as part of citrate produced by the condensation of acetyl CoA with oxaloacetate (OAA) by ***citrate synthase*** (Fig. 21.6). [Note: The transport of citrate to the cytosol occurs when the mitochondrial citrate concentration is high. This is observed when *isocitrate dehydrogenase* of the tricarboxylic acid (TCA) cycle is inhibited by the presence of large amounts of ATP, causing citrate and isocitrate to accumulate (see pp. 208–209). Therefore, cytosolic citrate may be viewed as a high-energy signal. Because a large amount of ATP is needed for

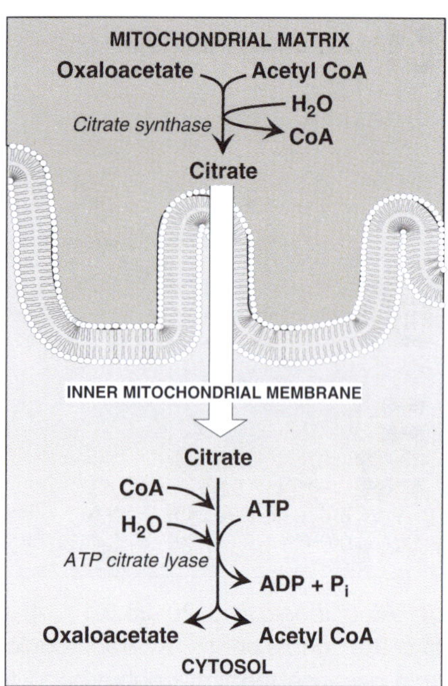

Figure 21.6 Production of cytosolic acetyl coenzyme A (CoA). [Note: Citrate is transported by the tricarboxylate transporter system.] ADP = adenosine monophosphate; P_i = inorganic phosphate.

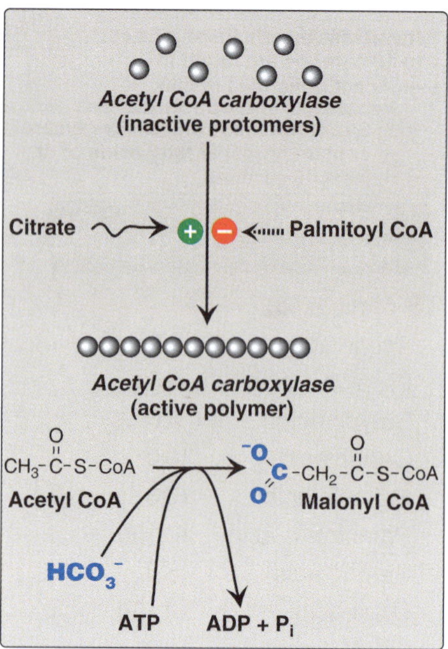

Figure 21.7 Allosteric regulation of malonyl coenzyme A (CoA) synthesis by *acetyl CoA carboxylase*. The carboxyl group contributed by bicarbonate (HCO_3^-) is shown in blue. P_i = inorganic phosphate; ADP = adenosine diphosphate.

fatty acid synthesis, the increase in both ATP and citrate enhances this pathway.] In the cytosol, citrate is cleaved to OAA and acetyl CoA by *ATP citrate lyase*.

B. Acetyl CoA carboxylation to malonyl CoA

The energy for the carbon-to-carbon condensations in fatty acid synthesis is supplied by the carboxylation and then decarboxylation of acyl groups in the cytosol. The carboxylation of acetyl CoA to malonyl CoA is catalyzed by ***acetyl CoA carboxylase***(*ACC*) (Fig. 21.7). *ACC* transfers carbon dioxide (CO_2) from bicarbonate (HCO_3^-) in an **ATP-requiring reaction**. The coenzyme is biotin (vitamin B_7), which is covalently bound to a lysyl residue of the *carboxylase* (see Fig. 32.16, p. 535). *ACC* carboxylates the bound biotin, which transfers the activated carboxyl group to acetyl CoA.

1. **Acetyl CoA carboxylase short-term regulation: This carboxylation is both the rate-limiting and the regulated step in fatty acid synthesis** (see Fig. 21.7). The inactive form of *ACC* is a protomer (complex of ≥2 polypeptides). The enzyme is **allosterically activated** by citrate, which causes protomers to polymerize, and allosterically inactivated by palmitoyl CoA (the end product of the pathway), which causes depolymerization. A second mechanism of short-term regulation is by **reversible phosphorylation**. *Adenosine monophosphate–activated protein kinase* (*AMPK*) phosphorylates and inactivates *ACC*. *AMPK* itself is activated allosterically by AMP and covalently by phosphorylation via several *kinases*. At least one of these *AMPK kinases* is activated by *cyclic AMP (cAMP)–dependent protein kinase A* (*PKA*). Thus, in the presence of counterregulatory hormones, such as epinephrine and glucagon, *ACC* is phosphorylated and inactive (Fig. 21.8). **In the presence of insulin, *ACC* is dephosphorylated and active.** [Note: This is analogous to the regulation of *glycogen synthase* (see p. 231).]

2. **Acetyl CoA carboxylase long-term regulation:** Prolonged consumption of a diet containing excess calories (particularly high-carbohydrate, low-fat diets) causes an **increase in *ACC* synthesis**, thereby increasing fatty acid synthesis. A low-calorie or a high-fat, low-carbohydrate diet has the opposite effect. [Note: *ACC* synthesis is upregulated by carbohydrate (specifically glucose) via the transcription factor carbohydrate response element–binding protein (ChREBP) and by insulin via the transcription factor sterol regulatory element–binding protein-1c (SREBP-1c). *Fatty acid synthase* (see C. below) is similarly regulated. The function and regulation of SREBP are described on p. 345.] Metformin, used in the treatment of type 2 diabetes, lowers plasma TAG through activation of *AMPK*, resulting in inhibition of *ACC* activity (by phosphorylation) and inhibition of *ACC* and *fatty acid synthase* expression (by decreasing SREBP-1c). **Metformin lowers**

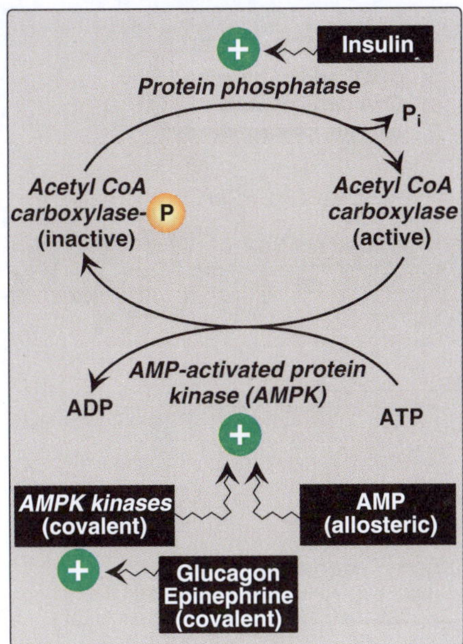

Figure 21.8 Covalent regulation of *acetyl CoA carboxylase* by *AMPK*, which itself is regulated both covalently and allosterically. CoA = coenzyme A; ADP and AMP = adenosine di- and monophosphates; (P) = phosphate; P_i = inorganic phosphate.

blood glucose by increasing *AMPK*-mediated glucose uptake by muscle.

C. Eukaryotic fatty acid synthase

The remaining series of reactions of fatty acid synthesis in eukaryotes is catalyzed by the multifunctional, homodimeric enzyme ***fatty acid synthase*** (*FAS*). The process involves the addition of two carbons from malonyl CoA to the carboxyl end of a series of acyl acceptors. Each *FAS* monomer is a multicatalytic polypeptide with six different enzymic domains plus a 4′-phosphopantetheine-containing acyl carrier protein (ACP) domain. 4′-Phosphopantetheine, a derivative of pantothenic acid (vitamin B_5, see p. 535), carries acyl units on its terminal thiol (–SH) group and presents them to the catalytic domains of *FAS* during fatty acid synthesis. It also is a component of CoA. [Note: In prokaryotes, *FAS* is a multienzyme complex.] The reaction numbers in brackets below refer to Figure 21.9.

1. An acetyl group is transferred from acetyl CoA to the –SH group of the ACP. Domain: ***Malonyl/acetyl CoA–ACP transacylase***.
2. Next, this two-carbon fragment is transferred to a temporary holding site, the –SH group of a cysteine residue on the **condensing enzyme domain** (see [4] below).
3. The now-vacant ACP accepts a three-carbon malonyl group from malonyl CoA. Domain: ***Malonyl/acetyl CoA–ACP transacylase***.
4. The acetyl group on the cysteine residue condenses with the malonyl group on ACP as the CO_2 originally added by *ACC* is released. The result is a four-carbon unit attached to the ACP domain. The loss of free energy from the decarboxylation drives the reaction. Domain: ***3-Ketoacyl–ACP synthase*, also known as condensing enzyme**.

   The next three reactions convert the 3-ketoacyl group to the corresponding saturated acyl group by a pair of NADPH-requiring reductions and a dehydration step.

5. The keto group is reduced to an alcohol. Domain: ***3-Ketoacyl–ACP reductase.***
6. A molecule of water is removed, creating a trans double bond between carbons 2 and 3 (the α- and β-carbons). Domain: *3-Hydroxyacyl–ACP dehydratase.*
7. The double bond is reduced. Domain: ***Enoyl–ACP reductase.***

   This sequence of steps results in the production of a four-carbon group (butyryl) whose three terminal carbons are fully saturated and which remains attached to the ACP domain. The steps are repeated (indicated by an asterisk), beginning with the transfer of the butyryl unit from the ACP to the cysteine residue [2*], the attachment of a malonyl group to the ACP [3*], and the condensation of the two groups liberating CO_2 [4*]. The carbonyl group at the β-carbon (carbon 3, the third carbon from the sulfur) is then reduced [5*], dehydrated [6*], and reduced [7*], generating hexanoyl-ACP. **This cycle of reactions is repeated five more times**, each time incorporating a two-carbon unit (derived from malonyl CoA) into the growing fatty acid chain at the carboxyl end. When the fatty acid reaches a length of 16 carbons, the synthetic process is terminated with **palmitoyl-S-ACP**. [Note: Shorter-length fatty acids are produced in the lactating mammary gland.] ***Palmitoyl thioesterase***, the final catalytic activity of *FAS*, cleaves the thioester bond, releasing a fully saturated molecule of palmitate (16:0). [Note: All the carbons in palmitic acid have passed through malonyl CoA except the two donated by the original acetyl CoA (the first acyl acceptor), which are found at the methyl (ω) end of the fatty acid. This underscores the rate-limiting nature of the *ACC* reaction.]

D. Reductant sources

The synthesis of one palmitate requires 14 NADPH, a reductant (reducing agent). **The pentose phosphate pathway (see p. 249) is a major supplier of the NADPH.** Two NADPH are produced for each molecule of glucose 6-phosphate that enters this pathway. The

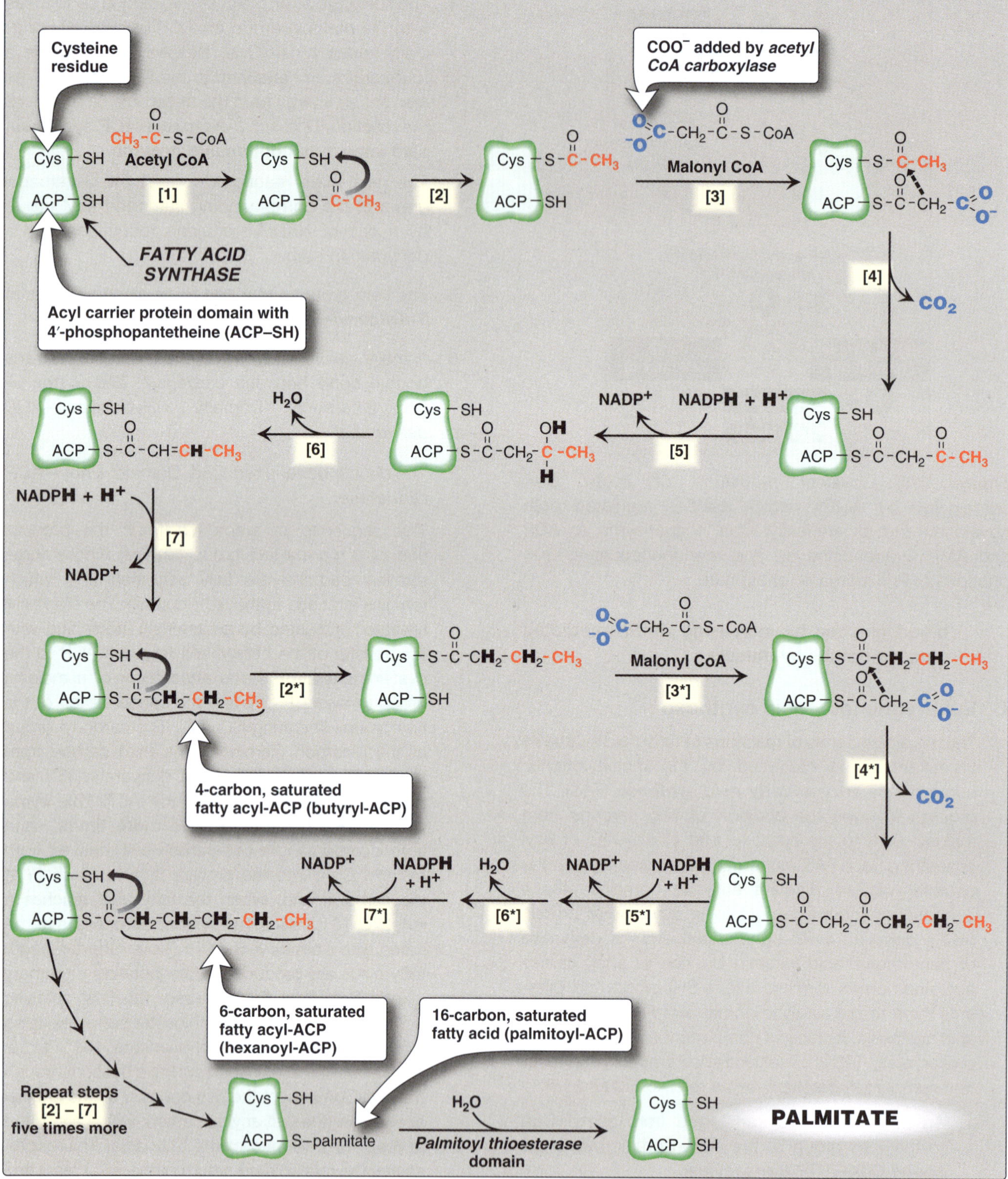

Figure 21.9 Synthesis of palmitate (16:0) by multifunctional *fatty acid synthase*. [Note: Numbers in brackets correspond to bracketed numbers in the text. A second repetition of the steps is indicated by numbers with an asterisk (*). Carbons provided directly by acetyl coenzyme A (CoA) are shown in red.] ACP = acyl carrier protein domain; CO_2 = carbon dioxide; NADP(H) = nicotinamide adenine dinucleotide phosphate.

cytosolic conversion of malate to pyruvate, in which malate is oxidized and decarboxylated by ***cytosolic malic enzyme*** (*NADP⁺-dependent malate dehydrogenase*), also produces cytosolic NADPH (and CO_2), as shown in Figure 21.10. [Note: Malate can arise from the reduction of OAA by cytosolic *NADH-dependent malate dehydrogenase* (see Fig. 21.10). One source of the cytosolic NADH required for this reaction is glycolysis (see p. 196). OAA, in turn, can arise from citrate cleavage by *ATP citrate lyase*.] A summary of the interrelationship between glucose metabolism and palmitate synthesis is shown in Figure 21.11.

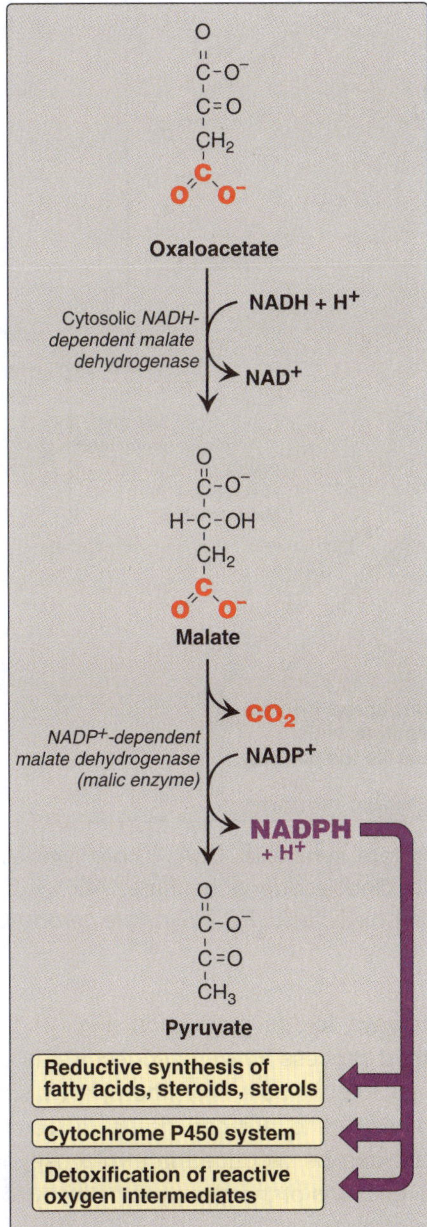

Figure 21.10 Cytosolic conversion of oxaloacetate to pyruvate with the generation of nicotinamide adenine dinucleotide phosphate (NADPH). [Note: The pentose phosphate pathway is also a source of NADPH.] NAD(H) = nicotinamide adenine dinucleotide; CO_2 = carbon dioxide.

E. Further elongation

Although palmitate, a 16-carbon, fully saturated LCFA (16:0), is the primary end product of *FAS* activity, it can be further elongated by the addition of two-carbon units to the carboxylate end primarily in the **smooth endoplasmic reticulum (SER)**. Elongation requires a system of separate enzymes rather than a multifunctional enzyme. Malonyl CoA is the two-carbon donor, and NADPH supplies the electrons. **The brain** has additional elongation capabilities, allowing it to produce **the very-long-chain fatty acids ([VLCFA] over 22 carbons) that are required for synthesis of brain lipids.**

F. Chain desaturation

Enzymes (***fatty acyl CoA desaturases***) also present in the SER are responsible for desaturating LCFA (that is, adding cis double bonds). The desaturation reactions require oxygen (O_2), NADH, cytochrome b_5, and its flavin adenine dinucleotide (FAD)-linked *reductase*. The fatty acid and the NADH get oxidized as the O_2 gets reduced to H_2O. The first double bond is typically inserted between carbons 9 and 10, producing primarily oleic acid, 18:1(9), and small amounts of palmitoleic acid, 16:1(9). A variety of polyunsaturated fatty acids can be made through additional desaturation combined with elongation.

> Humans have carbon 9, 6, 5, and 4 *desaturases* but lack the ability to introduce double bonds from carbon 10 to the ω end of the chain. This is the basis for the nutritional essentiality of the polyunsaturated ω-6 linoleic acid and ω-3 linolenic acid.

G. Storage as triacylglycerol components

Mono-, di-, and triacylglycerols consist of one, two, or three molecules of fatty acid esterified to a molecule of glycerol. Fatty acids are esterified through their carboxyl groups, resulting in a loss of negative charge and formation of neutral fat. [Note: **An acylglycerol that is solid at room temperature is called a fat. If liquid, it is an oil.**]

1. **Arrangement:** The three fatty acids esterified to a glycerol molecule to form a TAG are usually not of the same type. The fatty acid on carbon 1 is typically saturated, that on carbon 2 is typically unsaturated, and that on carbon 3 can be either. Recall that the presence of the unsaturated fatty acid(s) decrease(s) the T_m of the lipid. An example of a TAG molecule is shown in Figure 21.12.

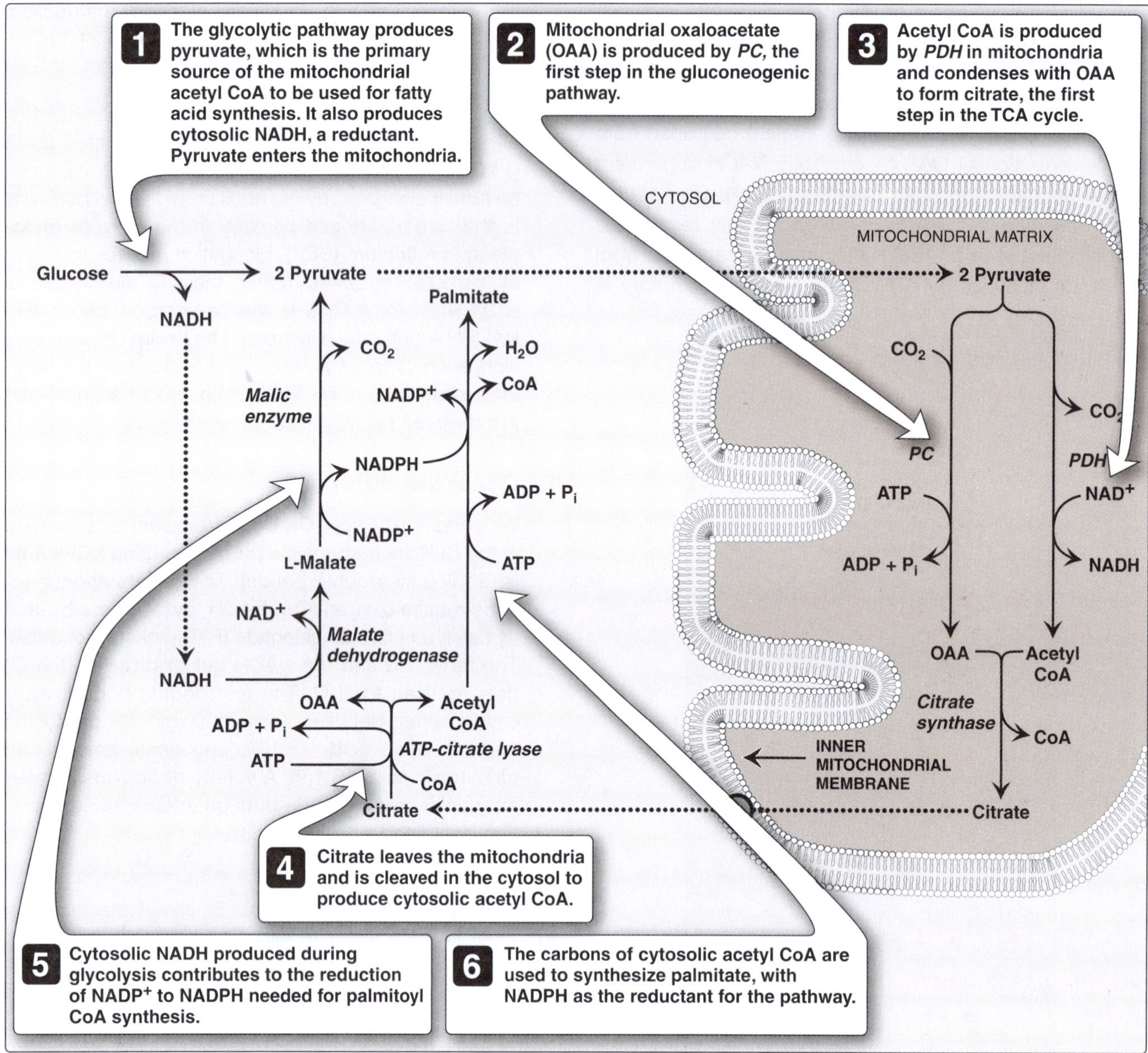

Figure 21.11 Interrelationship between glucose metabolism and palmitate synthesis. CoA = coenzyme A; NAD(H) = nicotinamide adenine nucleotide; NADP(H) = nicotinamide adenine dinucleotide phosphate; ADP = adenosine diphosphate; P_i = inorganic phosphate; CO_2 = carbon dioxide; TCA = tricarboxylic acid; *PC* = *pyruvate carboxylase*; *PDH* = *pyruvate dehydrogenase*.

2. **Triacylglycerol storage and function:** Because TAG are only slightly soluble in water and cannot form stable micelles by themselves, they **coalesce within white adipocytes to form large oily droplets that are nearly anhydrous. These cytosolic lipid droplets are the major energy reserve of the body.** [Note: TAG stored in brown adipocytes serve as a source of heat through nonshivering thermogenesis (see p. 170).]

3. **Glycerol 3-phosphate synthesis:** Glycerol 3-phosphate is the initial acceptor of fatty acids during TAG synthesis. There are two major pathways for its production (Fig. 21.13). [Note: A third process (glyceroneogenesis) is described on p. 309.] In both liver (the primary site of TAG synthesis) and adipose tissue, glycerol 3-phosphate can be produced from glucose, first using the reactions of the glycolytic pathway to produce dihydroxyacetone phosphate ([DHAP], see p. 196). DHAP is reduced by *glycerol 3-phosphate dehydrogenase* to glycerol 3-phosphate. A second pathway found in the liver, but not in adipose tissue, uses *glycerol kinase* to convert free glycerol to glycerol 3-phosphate (see Fig. 21.13).

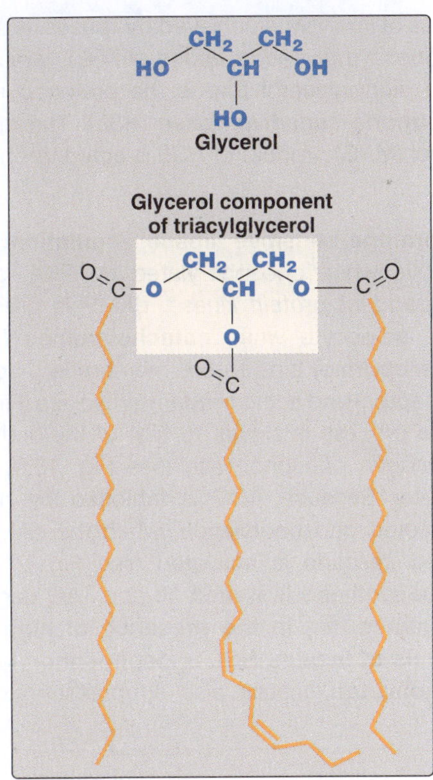

Figure 21.12 A triacylglycerol with an unsaturated fatty acid on carbon 2. Orange denotes the hydrophobic portions of the molecule.

[Note: **The glucose transporter in adipocytes (GLUT-4) is insulin dependent** (see p. 442). Thus, when plasma glucose levels are low, adipocytes have only a limited ability to synthesize glycerol phosphate and cannot produce TAG de novo.]

4. **Fatty acid activation:** A free fatty acid must be converted to its activated form (bound to CoA through a thioester link) before it can participate in metabolic processes such as TAG synthesis. This reaction, illustrated in Figure 20.6 on p. 293, is catalyzed by a family of *fatty acyl CoA synthetases* (***thiokinases***).

5. **Triacylglycerol synthesis:** This pathway from glycerol 3-phosphate involves four reactions, shown in Figure 21.14. These include the sequential addition of two fatty acids from fatty acyl CoA, the removal of phosphate, and the addition of the third fatty acid.

H. Triacylglycerol fate in liver and adipose tissue

In WAT, TAG is stored in a nearly anhydrous form as fat droplets in the cytosol of the cells. It serves as "depot fat," ready for mobilization when the body requires it for fuel. Little TAG is stored in healthy liver. Instead, most is exported, packaged with other lipids and apolipoproteins to form lipoprotein particles called very-low-density lipoproteins (VLDL). Nascent VLDL are secreted directly into the blood where they mature and function to deliver the endogenously derived lipids to the peripheral tissues. [Note: Recall from Chapter 20 that chylomicrons carry dietary (exogenously derived) lipids. Plasma lipoproteins are discussed in Chapter 23.]

IV. FAT MOBILIZATION AND FATTY ACID OXIDATION

Fatty acids stored in WAT, in the form of neutral TAG, serve as the body's major fuel storage reserve. TAG provide concentrated stores of metabolic energy because they are highly reduced and largely anhydrous. The yield from the complete oxidation of fatty acids to CO_2 and H_2O is 9 kcal/g fat (as compared to 4 kcal/g protein or carbohydrate, see Fig. 31.5 on p. 504).

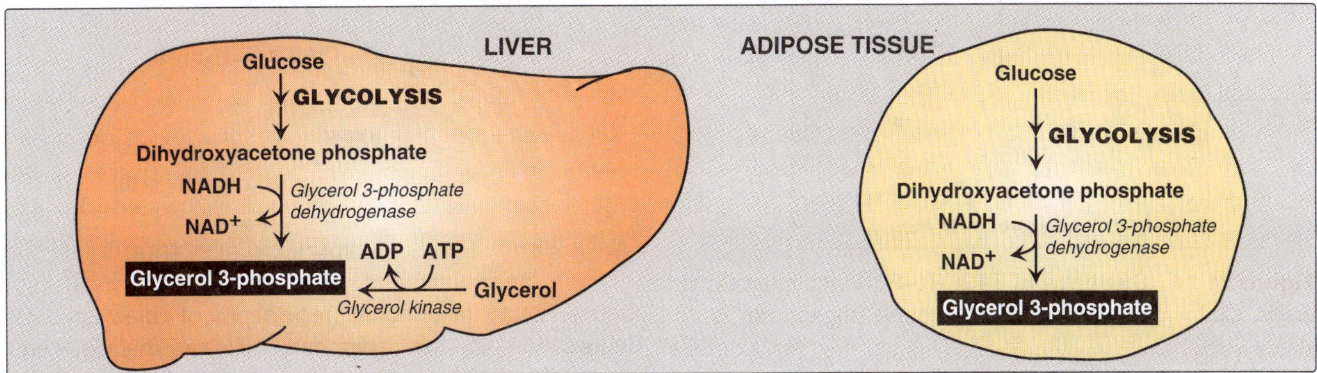

Figure 21.13 Pathways for production of glycerol 3-phosphate in liver and adipose tissue. [Note: Glycerol 3-phosphate can also be generated by glyceroneogenesis.] NAD(H) = nicotinamide adenine dinucleotide; ADP = adenosine diphosphate.

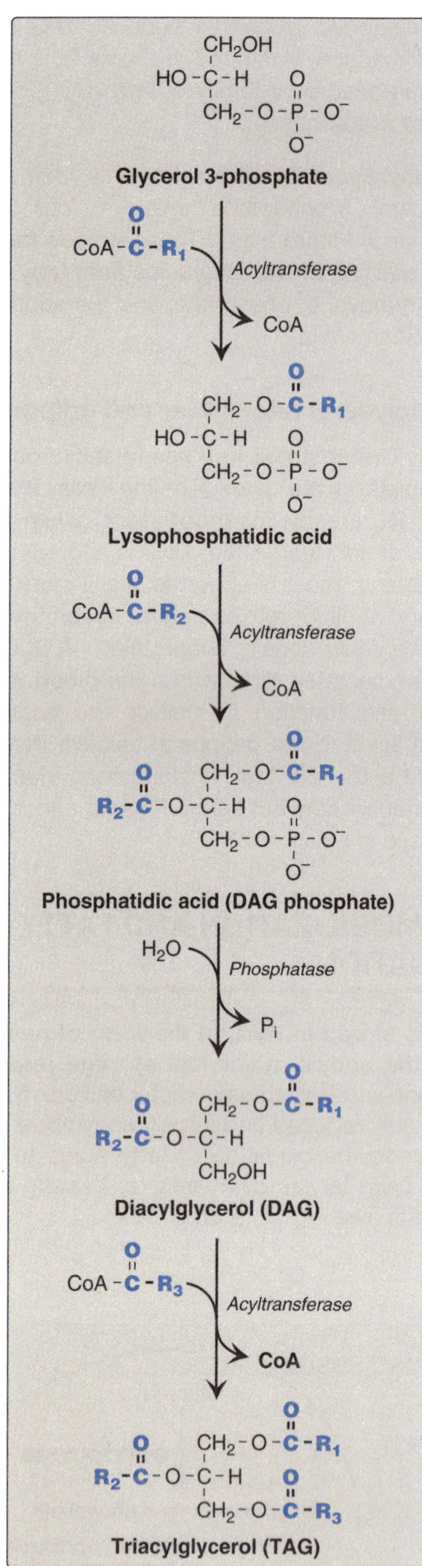

Figure 21.14 Synthesis of TAG. R1–R3 = activated fatty acids. CoA = coenzyme A; P_i = inorganic phosphate.

A. Fatty acid release from fat

The mobilization of stored fat requires the hydrolytic release of FFA and glycerol from their TAG form. This process of lipolysis is achieved by *lipases*. It is initiated by *adipose triglyceride* ***lipase*** (*ATGL*), which generates a diacylglycerol that is the preferred substrate for ***hormone-sensitive lipase*** (*HSL*). The monoacylglycerol (MAG) product of *HSL* is acted upon by *MAG lipase*.

1. **Hormone-sensitive lipase regulation:** *HSL* is active when phosphorylated by *PKA*, a *cAMP-dependent protein kinase*. cAMP is produced in the adipocyte when **catecholamines (such as epinephrine)** bind to cell membrane β-adrenergic receptors and activate *adenylyl cyclase* (Fig. 21.15). The process is similar to that of the activation of *glycogen phosphorylase* (see Fig. 16.9, p. 230). [Note: Because *ACC* is inhibited by hormone-directed phosphorylation, when the cAMP-mediated cascade is activated (see Fig. 21.8), fatty acid synthesis is turned off and TAG degradation is turned on.] **In the presence of high plasma levels of insulin, *HSL* is dephosphorylated and inactivated.** Insulin also suppresses expression

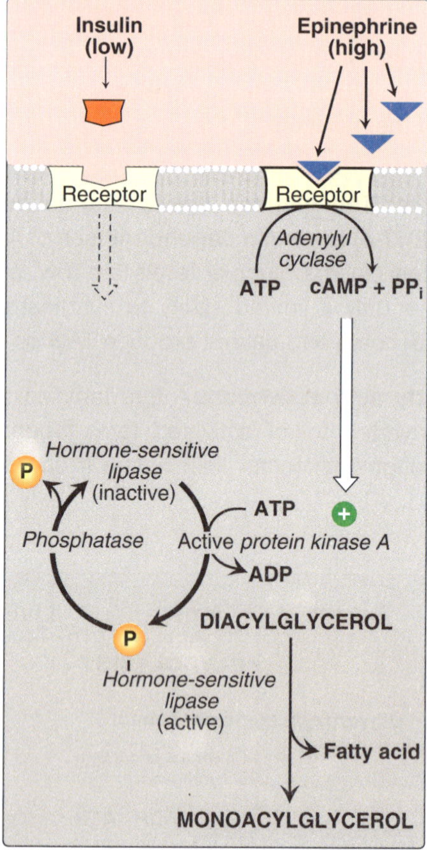

Figure 21.15 Hormonal regulation of diacylglycerol degradation in the adipocyte. [Note: Triacylglycerol is degraded to diacylglycerol by *adipose triglyceride lipase*.] cAMP = cyclic adenosine monophosphate; PP_i = pyrophosphate; ADP = adenosine diphosphate; (P) = phosphate.

of *ATGL*. [Note: **Fat droplets are coated by a protein (perilipin) that limits access of *HSL*.** Phosphorylation of perilipin by *PKA* allows translocation and binding of phosphorylated *HSL* to the droplet.]

2. **Fate of glycerol:** The **glycerol released during TAG degradation cannot be metabolized by adipocytes because they lack *glycerol kinase*.** Rather, glycerol is transported through the blood to the liver, which has the *kinase*. The resulting glycerol 3-phosphate can be used to form TAG in the liver or can be converted to DHAP by reversal of the *glycerol 3-phosphate dehydrogenase* reaction illustrated in Figure 21.13. DHAP can participate in glycolysis or gluconeogenesis.

3. **Fate of fatty acids:** The FFA move through the cell membrane of the adipocyte and bind to serum albumin. They **are transported to tissues such as muscle**, enter cells, get activated to their CoA derivatives, and are **oxidized for energy** in mitochondria. Regardless of their levels, plasma FFA cannot be used for fuel by red blood cells (RBC), which have no mitochondria. **The brain does not use fatty acids for energy to any appreciable extent**, but the reasons are less clear. [Note: Over 50% of the fatty acids released from adipose TAG are reesterified to glycerol 3-phosphate. WAT does not express *glycerol kinase*, and the glycerol 3-phosphate is produced by glyceroneogenesis, an incomplete version of gluconeogenesis: pyruvate to OAA via *pyruvate carboxylase* and OAA to phosphoenolpyruvate (PEP) via *phosphoenolpyruvate carboxykinase*. The PEP is converted (by reactions common to glycolysis and gluconeogenesis) to DHAP, which is reduced to glycerol 3-phosphate. The process decreases plasma FFA, molecules associated with insulin resistance in type 2 diabetes and obesity (see p. 476).]

B. Fatty acid β-oxidation

The major pathway for catabolism of fatty acids is a **mitochondrial pathway called β-oxidation**, in which two-carbon fragments are successively removed from the carboxyl end of the fatty acyl CoA, producing acetyl CoA, NADH, and $FADH_2$.

1. **Long-chain fatty acid transport into mitochondria:** After a LCFA enters a cell, it is converted in the cytosol to its CoA derivative by *long-chain fatty acyl CoA synthetase* (*thiokinase*), an enzyme of the outer mitochondrial membrane. Because β-oxidation occurs in the mitochondrial matrix, **the fatty acid must be transported across the inner mitochondrial membrane that is impermeable to CoA.** Therefore, a specialized carrier transports the long-chain acyl group from the cytosol into the mitochondrial matrix. This **carrier is carnitine, and this rate-limiting transport process is called the carnitine shuttle** (Fig. 21.16).

   a. **Translocation steps:** First, the acyl group is transferred from CoA to carnitine by ***carnitine-palmitoyltransferase I* (*CPT-I*), an enzyme of the outer mitochondrial membrane.** [Note: *CPT-I* is also known as *CAT-I* for *carnitine acyltransferase I*.] This reaction forms an acylcarnitine and regenerates free CoA. Second, the acylcarnitine is transported into the mitochondrial matrix in exchange for free carnitine by carnitine–acylcarnitine translocase. ***Carnitine palmitoyltransferase 2* (*CPT-II*, or *CAT-II*), an enzyme of the inner mitochondrial membrane**, catalyzes the transfer of the acyl group from carnitine to CoA in the mitochondrial matrix, thus regenerating free carnitine.

   b. **Carnitine shuttle inhibitor: Malonyl CoA inhibits *CPT-I*,** thus preventing the entry of

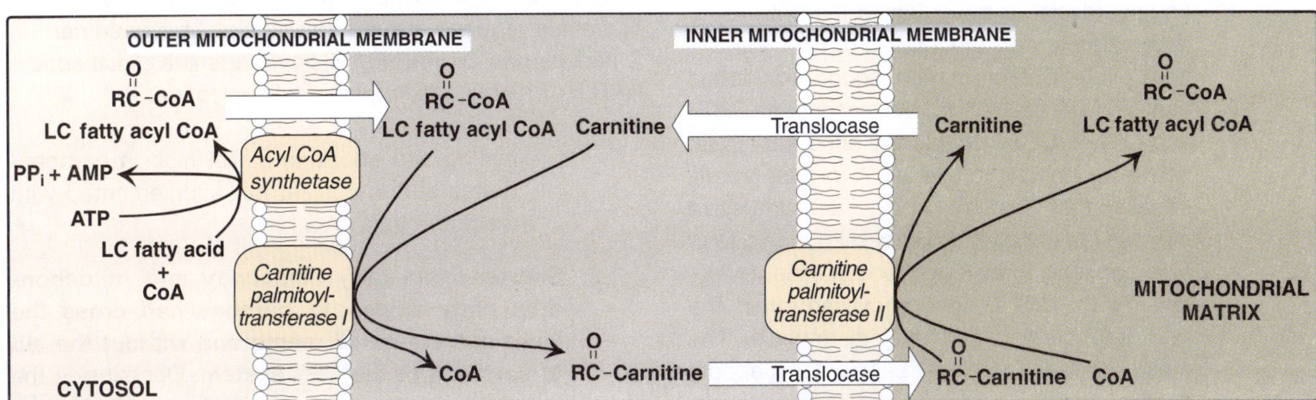

Figure 21.16 Carnitine shuttle. The net effect is that a long-chain (LC) fatty acyl coenzyme A (CoA) is transported from the outside to the inside of mitochondria. AMP = adenosine monophosphate; PP_i = pyrophosphate.

long-chain acyl groups into the mitochondrial matrix. Therefore, when fatty acid synthesis is occurring in the cytosol (as indicated by the presence of malonyl CoA), the newly made palmitate cannot be transferred into mitochondria and degraded. [Note: Muscle tissue, although it does not synthesize fatty acids, contains the mitochondrial isozyme of *ACC* (*ACC2*), allowing regulation of β-oxidation. The liver contains both isozymes.] Fatty acid oxidation is also regulated by the acetyl CoA/CoA ratio: As the ratio increases, the CoA-requiring *thiolase* reaction decreases (Fig. 21.17).

c. **Carnitine sources:** Carnitine can be obtained from the diet, where it is found primarily in meat products. It can also be synthesized from the amino acids lysine and methionine by an enzymatic pathway found in the liver and kidneys but not in skeletal or cardiac muscle. Therefore, these latter tissues are totally dependent on uptake of carnitine provided by endogenous synthesis or the diet and distributed by the blood. [Note: Skeletal muscle contains ~97% of all carnitine in the body.]

d. **Carnitine deficiencies:** Such deficiencies result in decreased ability of tissues to use LCFA as a fuel. **Primary carnitine deficiency** is caused by defects in a membrane transporter that prevent uptake of carnitine by cardiac and skeletal muscle and the kidneys, causing carnitine to be excreted. Treatment includes carnitine supplementation. **Secondary carnitine deficiency** occurs primarily as a result of defects in fatty acid oxidation leading to the accumulation of acylcarnitines that are excreted in the urine, decreasing carnitine availability. **Acquired secondary carnitine deficiency** can be seen, for example, in patients with liver disease (decreased carnitine synthesis) or those taking the antiseizure drug valproic acid (decreased renal reabsorption). [Note: Defects in mitochondrial oxidation can also be caused by deficiencies in *CPT-I* and *CPT-II*. ***CPT-I* deficiency** affects the liver, where an inability to use LCFA for fuel greatly impairs that tissue's ability to synthesize glucose (an endergonic process) during a fast. This can lead to severe hypoglycemia, coma, and death. ***CPT-II* deficiency** can affect the liver and cardiac and skeletal muscle. The most common (and least severe) form affects skeletal muscle. It presents as muscle weakness with myoglobinemia following prolonged exercise. Treatment includes avoidance of fasting and adopting a diet high in carbohydrates and low in fat but supplemented with medium-chain TAG.]

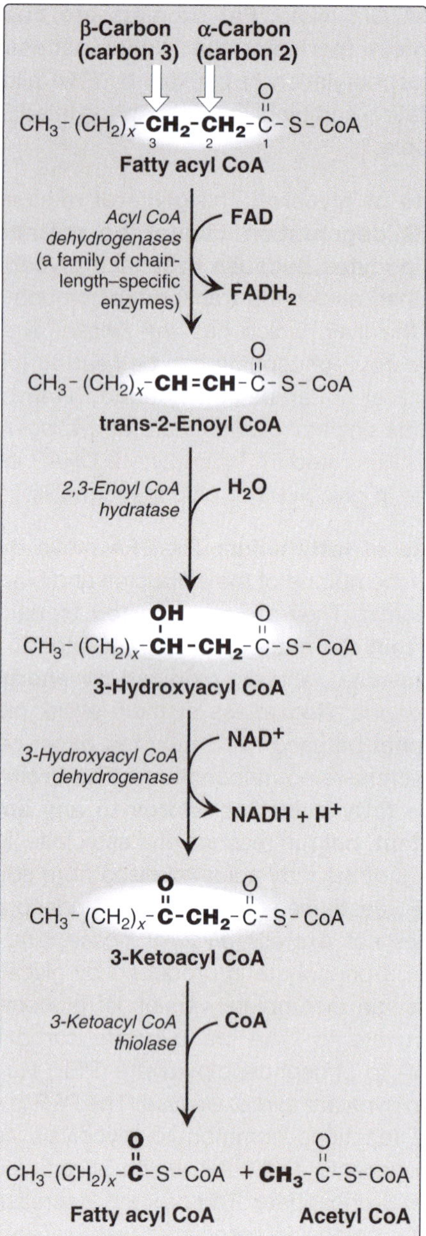

Figure 21.17 Enzymes involved in the β-oxidation of fatty acyl coenzyme A (CoA). [Note: *2,3-Enoyl CoA hydratase* requires a trans double bond between carbon 2 and carbon 3.] $FAD(H_2)$ = flavin adenine dinucleotide; NAD(H) = nicotinamide adenine dinucleotide.

2. **Shorter-chain fatty acid entry into mitochondria: Fatty acids ≤12 carbons can cross the inner mitochondrial membrane without the aid of carnitine or the *CPT* system.** Once inside the mitochondria, they are activated to their CoA derivatives by matrix enzymes and are oxidized. [Note: Medium-chain fatty acids are plentiful in human

milk. Because their oxidation is not dependent on *CPT-I*, malonyl CoA is not inhibitory.]

3. **β-Oxidation reactions:** The first cycle of β-oxidation is shown in Figure 21.17. It consists of a sequence of **four reactions involving the β-carbon (carbon 3) that results in shortening the fatty acid by two carbons at the carboxylate end.** The steps include an oxidation that produces $FADH_2$, a hydration, a second oxidation that produces NADH, and a CoA-dependent thiolytic cleavage that releases a molecule of acetyl CoA. Each step is catalyzed by enzymes with chain-length specificity. [Note: For LCFA, the last three steps are catalyzed by a trifunctional protein.] These four steps are repeated for saturated fatty acids of even-numbered carbon chains (n/2) – 1 times (where n is the number of carbons), each cycle producing one acetyl CoA plus one NADH and one $FADH_2$. The final cycle produces two acetyl CoA. The acetyl CoA can be oxidized or used in hepatic ketogenesis (see V. below). The reduced coenzymes are oxidized by the electron transport chain, NADH by Complex I, and $FADH_2$ by coenzyme Q (see p. 165). [Note: **Acetyl CoA is a positive allosteric effector of *pyruvate carboxylase* (see p. 216), thus linking fatty acid oxidation and gluconeogenesis.]**

4. **β-Oxidation energy yield:** The energy yield from fatty acid β-oxidation is high. For example, the oxidation of a molecule of palmitoyl CoA to CO_2 and H_2O produces 8 acetyl CoA, 7 NADH, and 7 $FADH_2$, from which 108 ATP can be generated. However, activation of the fatty acid requires two ATP. Therefore, the net yield from palmitate is 106 ATP (Fig. 21.18). A comparison of the processes of synthesis and degradation of long-chain saturated fatty acids with an even number of carbon atoms is provided in Figure 21.19.

5. **Medium-chain fatty acyl CoA dehydrogenase deficiency:** In mitochondria, there are four *fatty acyl CoA dehydrogenase* species, each with distinct but overlapping specificity for either short-, medium-, long-, or very-long-chain fatty acids. ***Medium-chain***

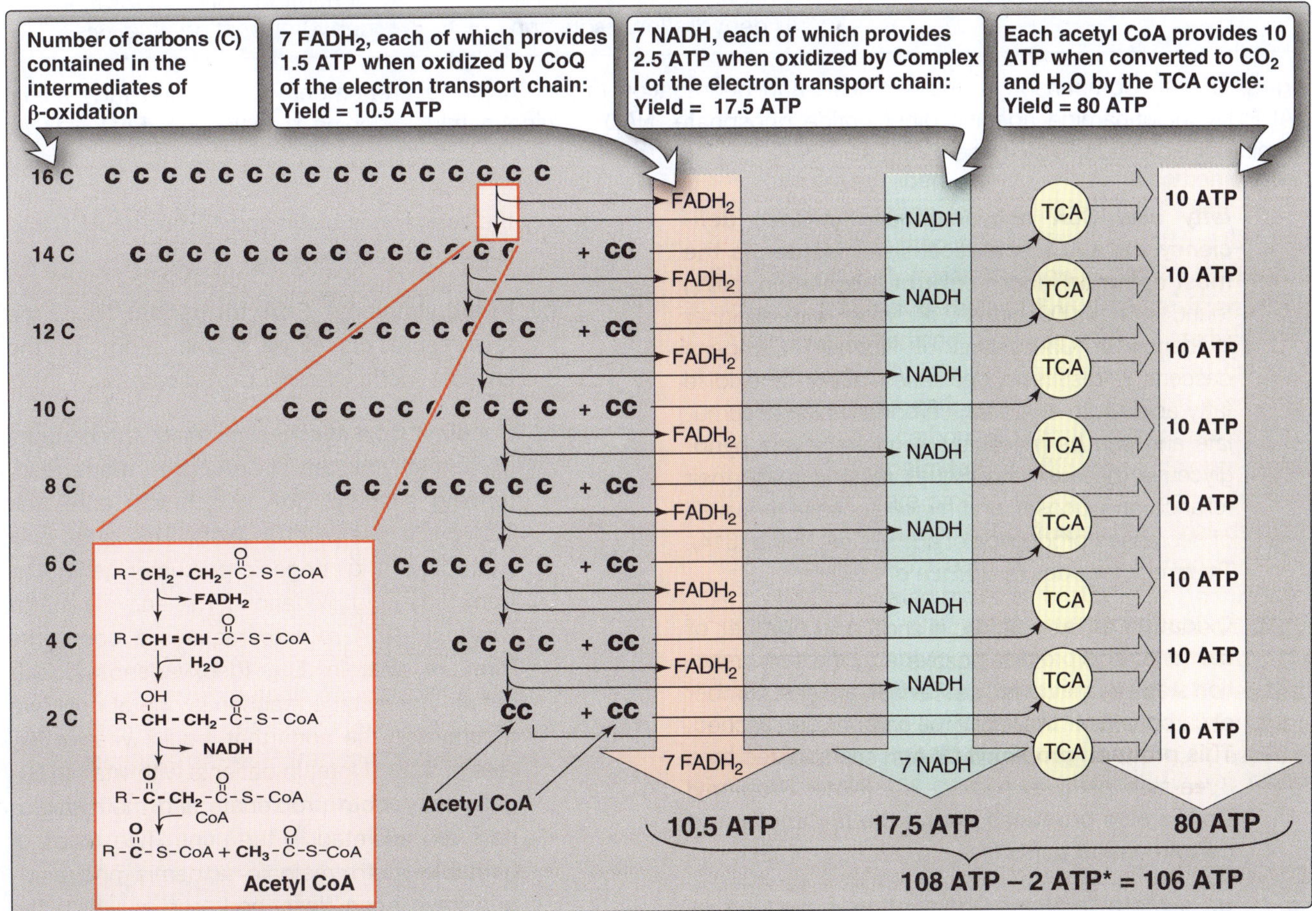

Figure 21.18 Summary of the energy yield from the oxidation of palmitoyl coenzyme A (CoA) (16 carbons). [Note: *Activation of palmitate to palmitoyl CoA requires the equivalent of 2 ATP (ATP → AMP + PP_i).] $FADH_2$ = flavin adenine dinucleotide; NADH = nicotinamide adenine dinucleotide; TCA = tricarboxylic acid; CoQ = coenzyme Q; CO_2 = carbon dioxide.

Variable	Synthesis	Degradation
Greatest flux through pathway	After carbohydrate-rich meal	In starvation
Hormonal state favoring pathway	High insulin/glucagon ratio	Low insulin/glucagon ratio
Major tissue site	Primarily liver	Muscle, liver
Subcellular location	Cytosol	Primarily mitochondria
Carriers of acyl/acetyl groups between mitochondria and cytosol	Citrate (mitochondria to cytosol)	Carnitine (cytosol to mitochondria)
Phosphopantetheine-containing active carriers	Acyl carrier protein domain, coenzyme A	Coenzyme A
Oxidation/reduction coenzymes	NADPH (reduction)	NAD^+, FAD (oxidation)
Two-carbon donor/product	Malonyl CoA: donor of one acetyl group	Acetyl CoA: product of β-oxidation
Activator	Citrate	—
Inhibitor	Palmitoyl CoA (inhibits *acetyl CoA carboxylase*)	Malonyl CoA (inhibits *carnitine palmitoyltransferase-I*)
Product of pathway	Palmitate	Acetyl CoA
Repetitive four-step process	Condensation, reduction dehydration, reduction	Dehydrogenation, hydration dehydrogenation, thiolysis

Figure 21.19 Comparison of the synthesis and degradation of long-chain, even-numbered, saturated fatty acids. NADPH = nicotinamide adenine dinucleotide phosphate; NAD^+ = nicotinamide adenine dinucleotide; FAD = flavin adenine dinucleotide; CoA = coenzyme A.

***fatty acyl CoA dehydrogenase* (*MCAD*) deficiency, an autosomal-recessive disorder, is the most common inborn error of β-oxidation**, being found in 1:14,000 births worldwide, with a higher incidence in Caucasians of Northern European descent. It results in decreased ability to oxidize fatty acids with six to ten carbons (which accumulate and can be measured in urine), severe hypoglycemia (because the tissues must increase their reliance on glucose), and hypoketonemia (because of decreased production of acetyl CoA; see p. 314). Treatment includes avoidance of fasting.

6. **Oxidation of fatty acids with an odd number of carbons:** This process proceeds by the same reaction steps as that of fatty acids with an even number of carbons, until the final three carbons are reached. **This product, propionyl CoA**, is metabolized by a three-step pathway (Fig. 21.20). [**Note: Propionyl CoA is also produced during the metabolism of certain amino acids** (see Fig. 25.11, p. 402).]

   a. **D-Methylmalonyl CoA synthesis:** First, propionyl CoA is carboxylated, forming D-methylmalonyl CoA. The enzyme *propionyl CoA carboxylase* has an absolute requirement for the coenzymes biotin and ATP, as do *ACC* and most other *carboxylases*.

   b. **l-Methylmalonyl CoA formation:** Next, the D-isomer is converted to the L-form by the enzyme *methylmalonyl CoA racemase*.

   c. **Succinyl CoA synthesis:** Finally, the carbons of L-methylmalonyl CoA are rearranged, forming succinyl CoA, which can enter the TCA cycle (see p. 210). [Note: This is the only example of a glucogenic precursor generated from fatty acid oxidation.] The enzyme *methylmalonyl CoA mutase* requires a coenzyme form of vitamin B_{12} (deoxyadenosylcobalamin). The *mutase* reaction is one of only two reactions in the body that require vitamin B_{12} (see p. 528). [Note: In patients with vitamin B_{12} deficiency, both propionate and methylmalonate are excreted in the urine. Two types of heritable methylmalonic acidemia and aciduria have been described: one in which the *mutase* is missing or deficient (or has reduced affinity for the coenzyme) and one in which the patient is unable to convert vitamin B_{12} into its

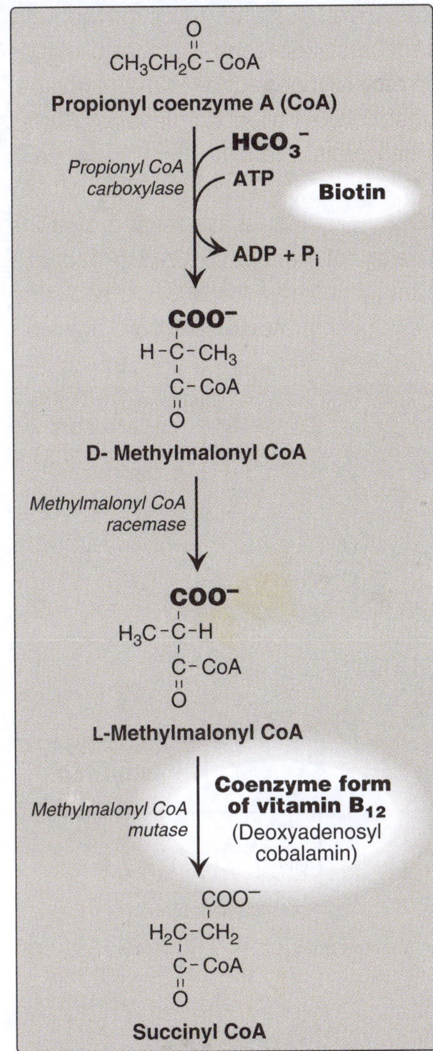

Figure 21.20 Metabolism of propionyl CoA. ADP = adenosine diphosphate; HCO_3^- = bicarbonate; P_i = inorganic phosphate.

coenzyme form. Either type results in metabolic acidosis and neurologic manifestations.]

7. **Unsaturated fatty acid β-oxidation:** The oxidation of unsaturated fatty acids generates intermediates that cannot serve as substrates for *2,3-enoyl CoA hydratase* (see Fig. 21.17). Consequently, additional enzymes are required. Oxidation of a double bond at an odd-numbered carbon, such as 18:1(9) (oleic acid), requires one additional enzyme, *3,2-enoyl CoA isomerase*, which converts the 3-cis derivative obtained after three rounds of β-oxidation to the 2-trans derivative required by the *hydratase*. Oxidation of a double bond at an even-numbered carbon, such as 18:2(9,12) (linoleic acid), requires an *NADPH-dependent 2,4-dienoyl CoA reductase* in addition to the *isomerase*. [Note: Because unsaturated fatty acids are less reduced than saturated fatty acids, fewer reducing equivalents are produced by their oxidation.]

8. **Peroxisomal β-oxidation: VLCFA ≥22 carbons in length undergo a preliminary β-oxidation in peroxisomes, because peroxisomes and not mitochondria are the primary site of the *synthetase* that activates fatty acids of this length**. The shortened fatty acid (linked to carnitine) diffuses to a mitochondrion for further oxidation. In contrast to mitochondrial β-oxidation, the initial dehydrogenation in peroxisomes is catalyzed by a FAD-containing *acyl CoA oxidase*. The $FADH_2$ produced is oxidized by O_2, which is reduced to hydrogen peroxide (H_2O_2). Therefore, no ATP is generated from this step. The H_2O_2 is reduced to H_2O by *catalase* (see p. 793). [Note: **Genetic defects in the ability either to target matrix proteins to peroxisomes (resulting in Zellweger syndrome, a peroxisomal biogenesis disorder**) or to transport VLCFA across the peroxisomal membrane (resulting in **X-linked adrenoleukodystrophy**) lead to accumulation of VLCFA in the blood and tissues.]

C. Peroxisomal α-oxidation

Branched-chain phytanic acid, a product of chlorophyll metabolism, is not a substrate for *acyl CoA dehydrogenase* because of the methyl group on its β-carbon (Fig. 21.21). Instead, it is hydroxylated at the α-carbon by *phytanoyl CoA α-hydroxylase* (*PhyH*); carbon 1 is released as CO_2; and the product, 15-carbon-long pristanal, is oxidized to **pristanic acid**, which is activated to its CoA derivative and undergoes β-oxidation. **Refsum disease is a rare, autosomal-recessive disorder caused by a deficiency of peroxisomal *PhyH***. This results in the accumulation of phytanic acid in the plasma and tissues. The symptoms are primarily

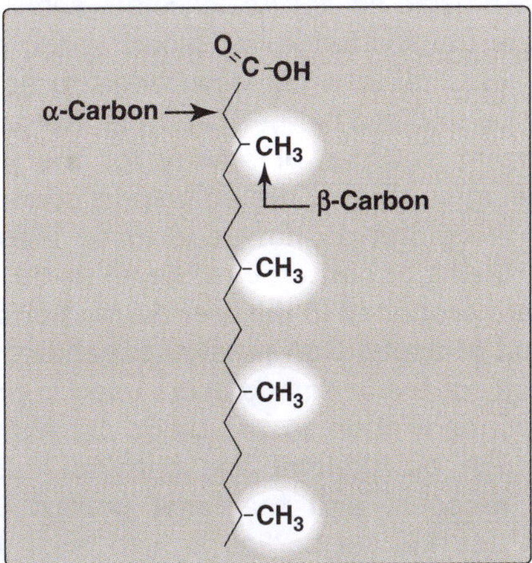

Figure 21.21 Phytanic acid, a branched-chain fatty acid 16 carbons in length.

neurologic, and the treatment involves dietary restriction to halt disease progression. [Note: ω-Oxidation (at the methyl terminus) also is known and generates dicarboxylic acids. Normally a minor pathway of the SER, its upregulation is seen with conditions such as *MCAD* deficiency that limit fatty acid β-oxidation.]

Physiologically important lipases

Lipase	Site of Action	Substrate
Lingual lipase	Mouth and stomach	Triglycerides with short- and medium-chain fatty acids
Pancreatic lipase	Small intestine	Triacylglycerol with long-chain fatty acids
Phospholipase A_2	Small intestine	Phospholipids with unsaturated fatty acids
Intestinal lipase	Small intestine	Acts on triacylglycerol with medium-chain fatty acids
Hormone-sensitive lipase	Adipose tissue cells	Stored triacylglycerol
Lipoprotein lipase	Capillary walls	Triacylglycerols in chylomicron or VLDL

V. KETONE BODIES: ALTERNATIVE FUEL FOR CELLS

Liver mitochondria have the capacity to convert acetyl CoA derived from fatty acid oxidation into ketone bodies. The compounds categorized as **ketone bodies are acetoacetate, 3-hydroxybutyrate (also called β-hydroxybutyrate), and acetone** (a nonmetabolized side product, Fig. 21.22). [Note: The two functional ketone bodies are organic acids.] Acetoacetate and 3-hydroxybutyrate are transported in the blood to the peripheral tissues. There they can be reconverted to acetyl CoA, which can be oxidized by the TCA cycle. **Ketone bodies are important sources of energy for the peripheral tissues because they 1) are soluble in aqueous solution and, therefore, do not need to be incorporated into lipoproteins or carried by albumin as do the other lipids; 2) are produced in the liver during periods when the amount of acetyl CoA present exceeds the oxidative capacity of the liver; and 3) are used in proportion to their concentration in the blood by extrahepatic tissues, such as skeletal and cardiac muscle, the intestinal mucosa, and the renal cortex. Even the brain can use ketone bodies** to help meet its energy needs if the blood levels rise sufficiently. **Thus, ketone bodies spare glucose, which is particularly important during prolonged periods of fasting** (see pp. 462–463).

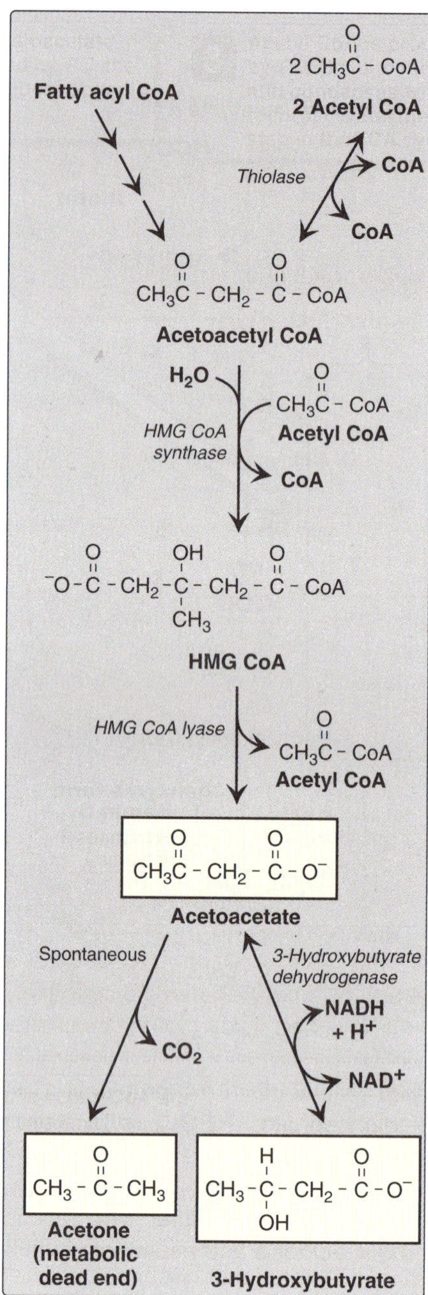

Figure 21.22 Synthesis of ketone bodies. [Note: The release of CoA in ketogenesis supports continued fatty acid oxidation.] CoA = coenzyme A; HMG = hydroxymethylglutarate; NAD(H) = nicotinamide adenine dinucleotide; CO_2 = carbon dioxide.

[Note: Disorders of fatty acid oxidation present with the general picture of hypoketosis (because of decreased availability of acetyl CoA) and hypoglycemia (because of increased reliance on glucose for energy).]

A. Ketone body synthesis by the liver: Ketogenesis

During a fast, the liver is flooded with fatty acids mobilized from adipose tissue. The resulting **elevated**

hepatic acetyl CoA produced by fatty acid oxidation inhibits *pyruvate dehydrogenase* (see p. 206) and **activates *pyruvate carboxylase*** ([*PC*], see p. 217). The OAA produced by *PC* is used by the liver for gluconeogenesis rather than for the TCA cycle. Additionally, fatty acid oxidation decreases the NAD^+/NADH ratio, and the rise in NADH shifts OAA to malate (see p. 210). The decreased availability of OAA for condensation with acetyl CoA results in the increased use of acetyl CoA for ketone body synthesis. [Note: Acetyl CoA for ketogenesis is also generated by the catabolism of ketogenic amino acids (see p. 396).]

1. **3-Hydroxy-3-methylglutaryl CoA synthesis:** The first step, formation of acetoacetyl CoA, occurs by reversal of the final *thiolase* reaction of fatty acid oxidation (see Fig. 21.17). Mitochondrial *3-hydroxy-3-methylglutaryl* (*HMG*) *CoA synthase* combines a third molecule of acetyl CoA with acetoacetyl CoA to produce HMG CoA. ***HMG CoA synthase* is the rate-limiting step in the synthesis of ketone bodies** and is present in significant quantities only in the liver. [Note: HMG CoA is also an intermediate in cytosolic cholesterol synthesis (see p. 342). The two pathways are separated by location in, and conditions of, the cell.]

2. **Ketone body synthesis:** HMG CoA is cleaved by ***HMG CoA lyase*** to produce acetoacetate and acetyl CoA, as shown in Figure 21.22. Acetoacetate can be reduced to form 3-hydroxybutyrate with NADH as the electron donor. [Note: Because ketone bodies are not linked to CoA, they can cross the inner mitochondrial membrane.] Acetoacetate can also spontaneously decarboxylate in the blood to form acetone, a volatile, biologically nonmetabolized compound that can be detected in the breath. The equilibrium between acetoacetate and 3-hydroxybutyrate is determined by the NAD^+/NADH ratio. Because this ratio is low during fatty acid oxidation, 3-hydroxybutyrate synthesis is favored.

B. Ketone body use by the peripheral tissues: Ketolysis

Although the liver constantly synthesizes low levels of ketone bodies, their production increases during fasting when ketone bodies are needed to provide energy to the peripheral tissues. 3-Hydroxybutyrate is oxidized to acetoacetate by *3-hydroxybutyrate dehydrogenase*, producing NADH (Fig. 21.23). Acetoacetate is then provided with a CoA molecule taken from succinyl CoA by *succinyl CoA:acetoacetate CoA transferase* (*thiophorase*). This reaction is reversible, but the product, acetoacetyl CoA, is actively removed by its cleavage to two acetyl CoA by *thiolase*. This pulls the reaction forward. Extrahepatic tissues, including the brain but excluding cells lacking mitochondria (for example, RBC), efficiently oxidize acetoacetate and 3-hydroxybutyrate in this manner. In contrast**, although the liver actively produces ketone bodies, it lacks *thiophorase* and, therefore, is unable to use ketone bodies as fuel.**

C. Excessive ketone body production in diabetes mellitus

When the rate of formation of ketone bodies is greater than the rate of their use, their levels begin

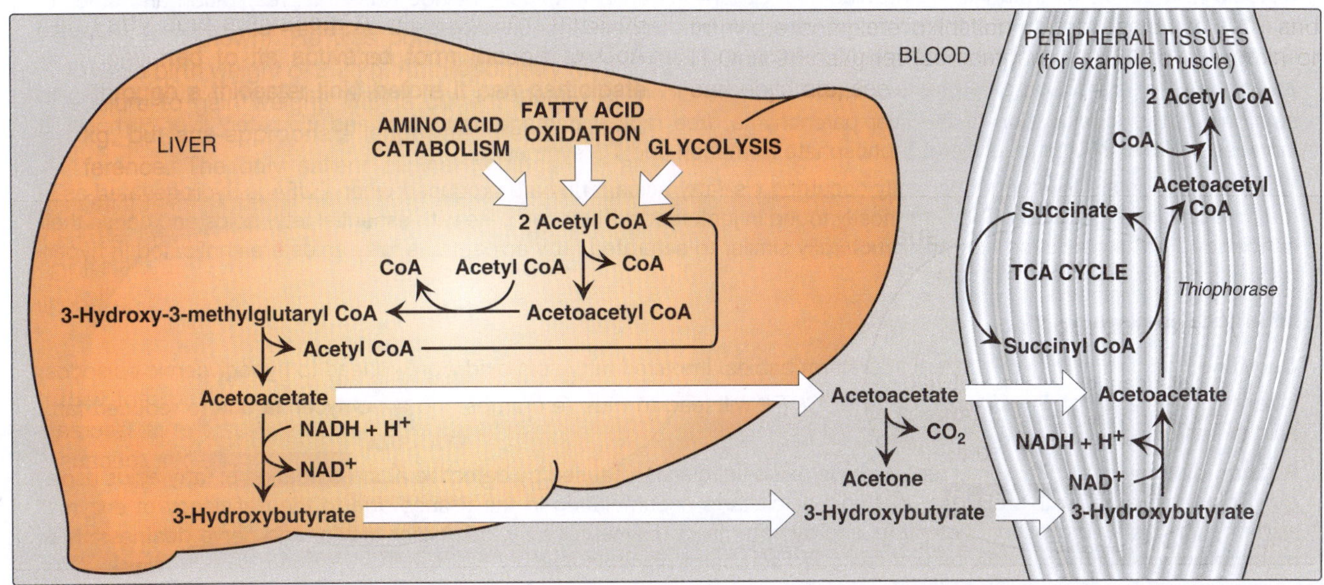

Figure 21.23 Ketone body synthesis in the liver and use in peripheral tissues. The liver and red blood cells cannot use ketone bodies. [Note: *Thiophorase* is also known as *succinyl CoA:acetoacetate CoA transferase*.] CoA = coenzyme A; NAD(H) = nicotinamide adenine dinucleotide; TCA = tricarboxylic acid; CO_2 = carbon dioxide.

to rise in the blood (ketonemia) and, eventually, in the urine (ketonuria). This is seen most often in cases of uncontrolled **type 1 diabetes mellitus (T1D)**, where the blood concentration of ketone bodies may reach 90 mg/dl (versus <3 mg/dl in normal individuals), and urinary excretion of ketone bodies may be as high as 5,000 mg/24 hour. The elevation of the ketone body concentration in the blood can result in **acidemia.** [Note: The carboxyl group of a ketone body has a pK_a of ~4. Therefore, each ketone body loses a proton (H^+) as it circulates in the blood, which lowers the pH.] Also, in uncontrolled T1D, urinary loss of glucose and ketone bodies results in dehydration. Therefore, the increased number of H^+ circulating in a decreased volume of plasma can cause a **severe acidosis (ketoacidosis**, Fig. 21.24) known as **diabetic ketoacidosis (DKA).]** A frequent symptom of DKA is a **fruity odor on the breath**, which results from increased production of acetone. Ketoacidosis may also be seen in cases of **prolonged fasting** (see p. 461) and **excessive ethanol consumption** (see p. 447).

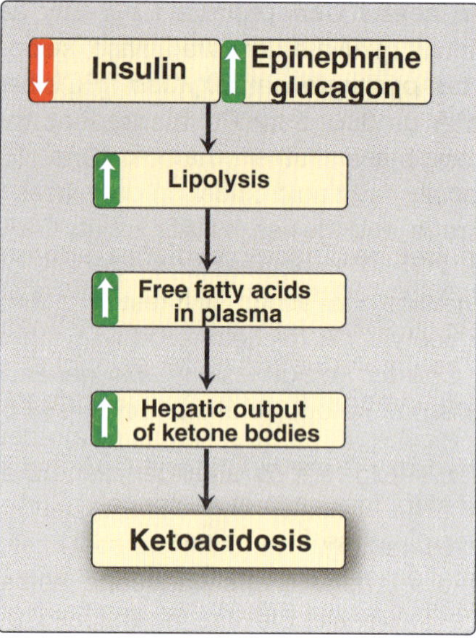

Figure 21.24 Mechanism of diabetic ketoacidosis seen in uncontrolled type 1 diabetes.

VI. Medical Perspectives

Fatty Acids and Diet

Linoleic and linolenic acids are **essential fatty acids** and must be supplied in diet as we cannot synthesize them. Arachidonic acid is the precursor of prostaglandins and can be synthesized in body if essential fatty acids are present.

Omega-3 fatty acids in diet appear to decrease cardiovascular risk factors and they decrease triglyceride levels. They are also associated with decrease in thromboxane levels and platelet aggregation. They are found in fish, walnuts and flax seeds.

PUFA, though recommended as beneficial in preventing cardiovascular disease, is also prone to being the target of free radicals generated by lipid peroxidation.

Common Diseases

Ketosis may be seen in starvation and diabetes. Salient features of ketosis include Kussmaul respiration and smell of acetone in patient's breath. Ketone bodies in urine can be diagnosed by urine dipstick test and Rothera's test. Diabetes is one of the most common health problems in India and is discussed in detail in chapter 29.

Alcoholism is also a health hazard worldwide. Chronic alcoholism interferes with lipid metabolism in liver leading to steatosis, fatty degeneration of the liver parenchyma, free radical damage, fatty liver and ultimately liver cirrhosis. It increases NADH/NAD ratio, glycerol 3-phosphate formation and TG synthesis.

Trans fatty acids are isomers of normally occurring cis-fatty acids. They are produced when PUFA is hydrogenated as in the production of margarine. They are mostly found in junk foods. They compete with essential fatty acids and cause their deficiency. Further, because they are structurally similar to saturated fatty acids, trans fatty acids are implicated in hypercholesterolemia and atherosclerosis.

Inherited Diseases

Carnitine deficiency may be present in preterm babies. Impaired fatty acid oxidation will lead to hypoglycemic episodes.

Inherited Deficiency of *Translocase*, *CPT-I* or *CPT-II* will lead to muscle cramps and hypoglycemia due to reduced fatty acid oxidation.

Refsum disease is an autosomal recessive peroxisomal disease caused by defective alpha-oxidation of fatty acids especially phytanic acid. In this disease, phytanic acid starts accumulating in the tissues due to the deficiency of enzyme, *alpha hydroxylase*. The patient has signs and symptoms of polyneuropathy, cerebellar ataxia, and nerve deafness. Milk must be avoided in these patients as it is a good source of phytanic acid.

Medium chain acyl CoA dehydrogenase deficiency is the most common mitochondrial beta-oxidation defect. In periods of fasting, there is a deficiency of glucose. Further, the infant is unable to maintain a supply of energy by breaking down medium-chain fatty acids. Therefore, hypoketotic hypoglycemia may occur and lead to sudden infant death syndrome. The primary goal of treatment is to provide continuous carbohydrates and avoid hypoglycemia.

Toxicity

The unripe ackee fruit causes **Jamaican vomiting sickness** because of hypoglycin present in it. Hypoglycin inhibits beta-oxidation by inhibiting medium- and short-chain *acyl-CoA dehydrogenase*. Unripe lychee fruit is postulated to act in a similar manner and cause acute toxic encephalopathy in Indian pediatric population.

Triclosan is an inhibitor of bacterial fatty acid synthesis and is used in cosmetics, toothpaste and toys. However, it is linked to abnormal endocrine signaling, weakening of the immune system, asthma and allergies.

VII. Chapter Summary

- A **fatty acid**, generally a linear hydrocarbon chain with a terminal carboxyl group, can be **saturated** or **unsaturated.**
- Two unsaturated fatty acids are dietary essentials: **linoleic** and **α-linolenic acids.**
- Fatty acids are synthesized in the **liver cytosol** following a meal containing excess carbohydrate and protein.
- Carbons used to synthesize fatty acids are provided by **acetyl coenzyme A** (**CoA**), energy by **ATP**, and reducing equivalents by **nicotinamide adenine dinucleotide phosphate** ([**NADPH**], Fig. 21.25) provided by the **pentose phosphate pathway** and ***malic enzyme***.
- **Citrate** carries two-carbon acetyl units from the mitochondrial matrix to the cytosol.
- The regulated step in fatty acid synthesis is the carboxylation of acetyl CoA to **malonyl CoA** by **biotin**- and **ATP**-requiring ***acetyl CoA carboxylase*** (***ACC***).
- **Citrate** allosterically activates ACC, and **palmitoyl CoA** inhibits it. ***ACC*** can also be activated by **insulin** and inactivated by ***adenosine monophosphate–activated protein kinase*** (***AMPK***) in response to **epinephrine**, **glucagon**, or a rise in **AMP.**
- The remaining steps in fatty acid synthesis are catalyzed by the multifunctional enzyme, ***fatty acid synthase***, which produces **palmitoyl CoA** by adding two-carbon units from malonyl CoA to a series of acyl acceptors.
- Fatty acids can be **elongated** and **desaturated** in the **smooth endoplasmic reticulum** (**SER**).
- When fatty acids are required for energy, ***hormone-sensitive lipase*** (**activated** by **epinephrine**, and **inhibited** by **insulin**), along with other *lipases*, degrades **triacylglycerol** (**TAG**) stored in **adipocytes.**
- The fatty acid products are carried by **serum albumin** to the liver and peripheral tissues, where their oxidation provides energy. The **glycerol** backbone of the degraded TAG is carried by the blood to the **liver**, where it serves as a **gluconeogenic precursor**. Fatty acid degradation (**β-oxidation**) occurs in **mitochondria.**
- The **carnitine shuttle** is required to transport long-chain fatty acids from the cytosol to the mitochondrial matrix. A translocase and the enzymes ***carnitine palmitoyltransferases*** (***CPT***) ***I*** and ***II*** are required. *CPT-I* is inhibited by **malonyl CoA**, thereby preventing simultaneous synthesis and degradation of fatty acids.
- Mitochondrial fatty acid β-oxidation produces **acetyl CoA, nicotinamide adenine dinucleotide** (**NADH**), and **flavin adenine dinucleotide** (**$FADH_2$**). The first step in β-oxidation is catalyzed by one of four *acyl CoA dehydrogenases*, each with chain-length specificity.
- ***Medium-chain fatty acyl CoA dehydrogenase*** (***MCAD***) **deficiency** causes a decrease in fatty acid oxidation (process stops once a medium-chain fatty acid is produced), resulting in **hypoketonemia** and severe **hypoglycemia.**
- Oxidation of fatty acids with an **odd number** of carbons proceeds two carbons at a time (producing acetyl CoA) until three-carbon **propionyl CoA** remains. This compound is carboxylated to **methylmalonyl CoA** (by **biotin-** and **ATP-requiring *propionyl CoA carboxylase***), which is then converted to **succinyl CoA** (a gluconeogenic precursor) by **vitamin B_{12}-requiring *methylmalonyl CoA mutase***.
- A genetic error in the *mutase* or vitamin B_{12} deficiency causes **methylmalonic acidemia** and **aciduria**. β-Oxidation of **unsaturated** fatty acids requires additional enzymes.
- β-Oxidation of **very-long-chain** fatty acids and α-oxidation of **branched-chain** fatty acids occur in the **peroxisome.** Deficiencies result in **X-linked adrenoleukodystrophy** and **Refsum disease**, respectively.
- **ω-Oxidation**, normally a minor pathway, occurs in the SER.
- Liver mitochondria can convert acetyl CoA derived from fatty acid oxidation into **acetoacetate** and **3-hydroxybutyrate** (**ketone bodies**). Peripheral tissues possessing mitochondria can oxidize 3-hydroxybutyrate to acetoacetate, which can be cleaved to two acetyl CoA, thereby producing energy for the cell. Unlike fatty acids, ketone bodies are utilized by the **brain** and, therefore, are important fuels during a fast. Because the liver lacks ***thiophorase*** required to degrade ketone bodies, it synthesizes them specifically for the peripheral tissues.
- **Ketoacidosis** occurs when the rate of ketone body formation is greater than the rate of use, as is seen in cases of uncontrolled **type 1 diabetes mellitus.**

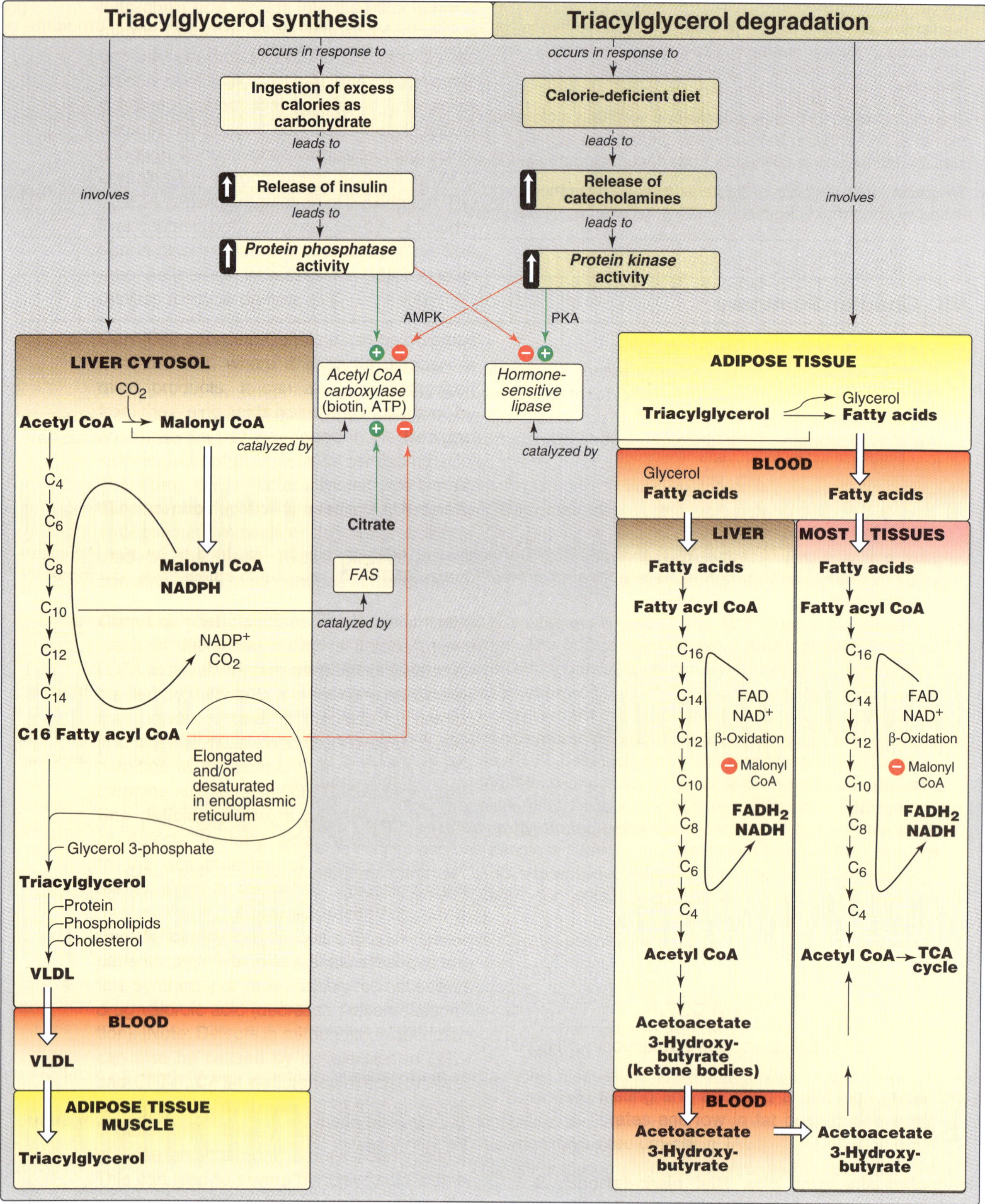

Figure 21.25 Key concept map for fatty acid and triacylglycerol metabolism. *AMPK = adenosine monophosphate–activated protein kinase*; *PKA = protein kinase A*; CoA = coenzyme A; NADP(H) = nicotinamide adenine dinucleotide phosphate; FAD(H_2) = flavin adenine dinucleotide; *FAS= fatty acid synthase*; CO_2 = carbon dioxide; NAD(H) = nicotinamide adenine dinucleotide; TCA = tricarboxylic acid; VLDL = very-low-density lipoprotein.

Study Questions

Short-Answer Questions

Write short notes on:

21.1. Saturated and unsaturated fatty acids
21.2. Trans-fatty acids
21.3. Lipoprotein lipase and hormone-sensitive lipase
21.4. Carnitine shuttle
21.5. Ketogenesis
21.6. Rate-limiting step of fatty acid synthesis

Long-Answer Question

21.1. How and when is the triacylglycerol stored in adipocytes broken down? Explain the process and end-product utilization in detail. Add a note on the hormones involved in this process.

Multiple-Choice Questions

Choose the ONE best answer.

21.1. When oleic acid, 18:1(9), is desaturated at carbon 6 and then elongated, what is the correct representation of the product?
A. 19:2(7,9)
B. 20:2 (ω-6)
C. 20:2(6,9)
D. 20:2(8,11)
Correct answer = D.

21.2. A 4-month-old child is being evaluated for fasting hypoglycemia. Laboratory tests at admission reveal low levels of ketone bodies (hypoketonemia), free carnitine, and long-chain acylcarnitines in the blood. Free fatty acid levels in the blood were elevated. Deficiency of which of the following would best explain these findings?
A. Adipose triglyceride lipase
B. Carnitine transporter
C. Carnitine palmitoyltransferase-I
D. Long-chain fatty acid dehydrogenase
Correct answer = B.

21.3. A teenager, concerned about his weight, attempts to maintain a fat-free diet for a period of several weeks. If his ability to synthesize various lipids were examined, he would be found to be most deficient in his ability to synthesize:
A. cholesterol
B. glycolipids
C. phospholipids
D. prostaglandins
E. triacylglycerol
Correct answer = D.

21.4 Which of the following represents the organs that can both synthesize and utilize ketone bodies?
A. Liver and brain
B. Kidney and brain
C. Kidney and liver
D. None of the above
Correct answer = D.

21.5 Which among the following pairs is correctly matched?
A. α-oxidation: Endoplasmic reticulum
B. β-oxidation: Peroxisome
C. ω-oxidation: Cytosol
D. Odd-chain fatty acid oxidation: Mitochondria
Correct answer = D.

21.6 The rate limiting step in beta oxidation of fatty acids is catalyzed by
A. Thiokinase
B. Carnitine palmitoyl transferase-1
C. Acyl CoA dehydrogenase
D. Thiolase
Correct answer = B.

21.7 The net ATP yield on beta oxidation of palmitic acid (C_{16}) is
A. 131 ATP
B. 129 ATP
C. 108 ATP
D. 106 ATP
Correct answer = D.

21.8 Which of the following is false regarding alpha oxidation?
A. One carbon is lost in form of CO_2 per cycle
B. It yields energy as ATP
C. It occurs in the peroxisomes
D. It requires phytanoyl CoA alpha hydroxylase
Correct answer = B.

21.9 Long-chain fatty acids penetrate the inner mitochondrial membrane as
A. Carnitine derivatives
B. Acetyl CoA
C. Malonyl CoA
D. None of the above
Correct answer = A.

21.10 Energy requiring step of fatty acid beta oxidation is catalyzed by
A. Acyl CoA dehydrogenase
B. Thiokinase
C. Thiolase
D. 3-Hydroxyacyl CoA dehydrogenase
Correct answer = B.

Multiple-Choice Questions (NExT/NEET PG)

Choose the ONE best answer.

21.1. A 6-month-old boy was hospitalized following a seizure. History revealed that for several days prior, his appetite was decreased. At admission, his blood glucose was 24 mg/dl (age-referenced normal is 60–100). His urine was negative for ketone bodies and positive for a variety of dicarboxylic acids. Blood carnitine levels (free and acyl bound) were normal. A tentative diagnosis of medium-chain fatty acyl coenzyme A dehydrogenase (MCAD) deficiency is made. In patients with MCAD deficiency, the fasting hypoglycemia is a consequence of:

A. decreased acetyl coenzyme A production.

B. decreased ability to convert acetyl coenzyme A to glucose.

C. increased conversion of acetyl coenzyme A to acetoacetate.

D. increased production of ATP and nicotinamide adenine dinucleotide.

Correct answer = A.

21.2. Which of the following is correct ?

A. In Zellweger syndrome, both very-long-chain fatty acids (VLCFA) and long-chain phytanic acid accumulate, whereas with X-linked adrenoleukodystrophy, only VLCFA accumulate.

B. In Zellweger syndrome, α-oxidation in peroxisomes is normal whereas in X-linked adrenoleukodystrophy α-oxidation is not normal

C. In Zellweger syndrome the peroxisomes are functional whereas in X-linked adrenoleukodystrophy it is not functional

D. In Zellweger syndrome very-long-chain fatty acids (VLCFA) accumulate but long-chain phytanic acid does not accumulate

Correct answer = A.

21.3. Triclosan is a component of cosmetics and toothpastes, where it inhibits the bacterial growth. This chemical specifically inhibits membrane synthesis that is required for the proliferation of the bacteria. Which of the following pathways is inhibited by this chemical?

A. Glycolysis

B. Citric acid cycle

C. Fatty acid synthesis

D. Electron transport chain

Correct answer = C.

21.4. All the following vitamins are required for the subsequent metabolism of propionyl CoA, after it has been formed from amino acids or fatty acids, EXCEPT:

A. Biotin

B. Niacin

C. Pantothenic acid

D. Cobalamin

Correct answer = A.

21.5. Bempedoic acid is a new drug that is used to manage increased serum cholesterol levels. It acts by inhibiting the export of 2-carbon units from the mitochondria to the cytoplasm which is the site of cholesterol synthesis. Inhibition of which of the following enzymes is brought about by Bempedoic acid?

A. Citrate synthase

B. ATP citrate lyase

C. Acetyl CoA carboxylase

D. HMG CoA synthase

Correct answer = B.

Phospholipid, Glycosphingolipid, and Eicosanoid Metabolism

22

Learning Objectives

At the end of this chapter, the student will be able to:

- Describe structure, metabolism and function of major phospholipids, eicosanoids and sphingolipids
- Describe therapeutic use of prostaglandins and inhibitors of eicosanoid synthesis
- Explain the genetic defects in sphingolipidosis

I. PHOSPHOLIPID OVERVIEW

Phospholipids are polar, ionic compounds composed of an alcohol that is attached by a phosphodiester bond to either diacylglycerol (DAG) or sphingosine. Like fatty acids (FA), phospholipids are amphipathic in nature. That is, each has a hydrophilic head, which is the phosphate group plus whatever alcohol is attached to it (for example, serine, ethanolamine, and choline; highlighted in blue in Fig. 22.1A), and a long, hydrophobic tail containing FA or FA-derived hydrocarbons (shown in orange in Fig. 22.1A). **Phospholipids are the predominant lipids of cell membranes.** In membranes, the hydrophobic portion of a phospholipid molecule is associated with the nonpolar portions of other membrane constituents, such as glycolipids, proteins, and cholesterol. The hydrophilic (polar) head of the phospholipid extends outward, interacting with the intracellular or extracellular aqueous environment (see Fig. 22.1A). Membrane phospholipids also function as a reservoir for intracellular messengers, and, for some proteins, phospholipids serve as anchors to cell membranes. Nonmembrane phospholipids serve additional functions in the body, for example, as **components of lung surfactant and essential components of bile**, where their detergent properties aid cholesterol solubilization.

II. PHOSPHOLIPID STRUCTURE

There are **two classes of phospholipids**: those that have glycerol (from glucose) as a backbone and those that have sphingosine (from serine and palmitate). Both classes are found as structural components of membranes, and both play a role in the generation of lipid signaling molecules.

A. Glycerophospholipids

Phospholipids that contain glycerol are called glycerophospholipids (or phosphoglycerides). Glycerophospholipids constitute the major class of phospholipids and are the predominant lipids in membranes. All contain (or are derivatives of) phosphatidic acid (PA), which is DAG with a phosphate group on carbon 3 (Fig. 22.1B). PA is the simplest phosphoglyceride and is the precursor of the other members of this group.

1. **From phosphatidic acid and an alcohol:** The phosphate group on PA can be esterified to a compound containing an alcohol group (see Fig. 22.1). For example:

Serine	+ PA	→	phosphatidylserine (PS)
Ethanolamine	+ PA	→	phosphatidylethanolamine (PE)
Choline	+ PA	→	phosphatidylcholine (PC) (lecithin)
Inositol	+ PA	→	phosphatidylinositol (PI)
Glycerol	+ PA	→	phosphatidylglycerol (PG)

2. **Cardiolipin: Two molecules of PA esterified through their phosphate groups to an additional molecule of glycerol form cardiolipin, or**

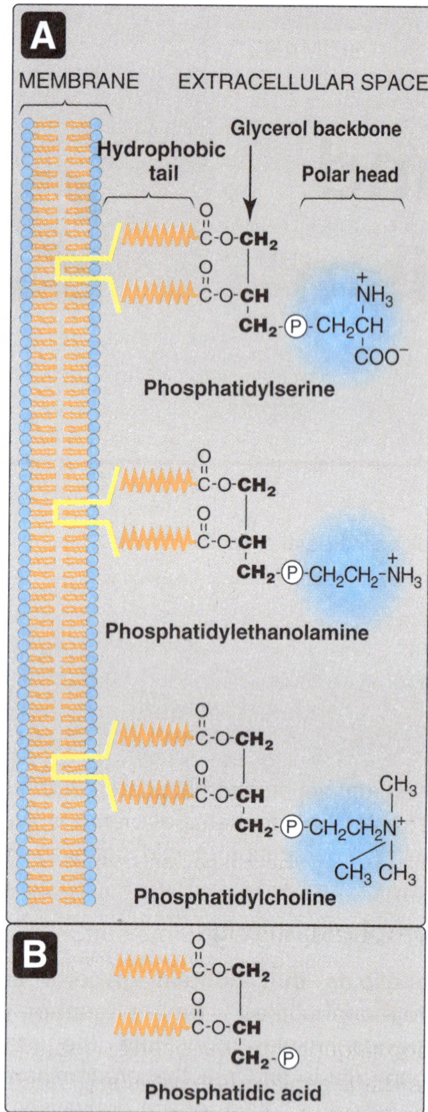

Figure 22.1 **A.** Structures of some glycerophospholipids. **B.** Phosphatidic acid. Ⓟ = phosphate (an anion).

diphosphatidylglycerol (Fig. 22.2). Cardiolipin is found in membranes in bacteria and eukaryotes. In eukaryotes, cardiolipin is virtually exclusive to the **inner mitochondrial membrane**, where it maintains the structure and function of certain respiratory complexes of the electron transport chain. [Note: Cardiolipin is antigenic and is recognized by antibodies (Ab) raised against Treponema pallidum, the bacterium that causes syphilis. The Wasserman test for syphilis detects these Ab.]

3. **Plasmalogens: When the FA at carbon 1 of a glycerophospholipid is replaced by an unsaturated alkyl group attached by an ether (rather than by an ester) linkage to the core glycerol molecule, an ether phosphoglyceride known as a plasmalogen** is produced. For example, phosphatidylethanolamine, which is abundant in nerve

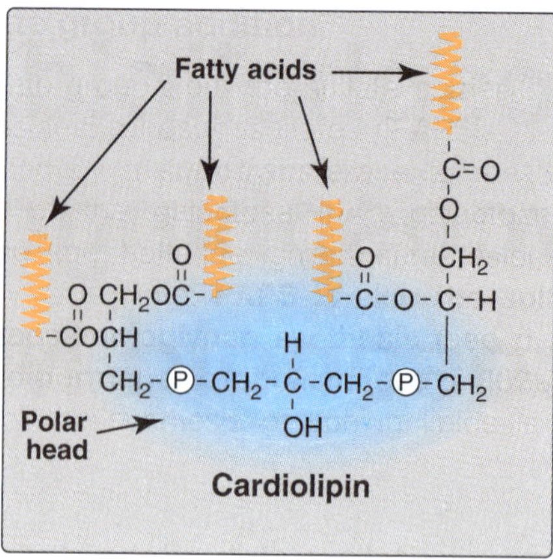

Figure 22.2 Structure of cardiolipin (diphosphatidylglycerol). Ⓟ = phosphate.

tissue (Fig. 22.3A), is the plasmalogen that is similar in structure to phosphatidylethanolamine. Phosphatidylcholine (abundant in heart muscle) is the other quantitatively significant ether lipid in mammals. [Note: Plasmalogens have "al" rather than "yl" in their names.]

4. **Platelet-activating factor:** A second example of an **ether glycerophospholipid** is platelet-activating factor (PAF), which has a saturated alkyl group in an ether link to carbon 1 and an acetyl

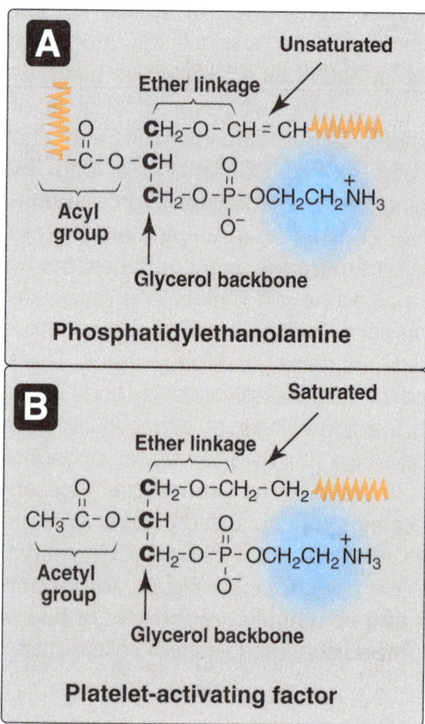

Figure 22.3 The ether glycerophospholipids. **A.** The plasmalogen phosphatidylethanolamine. **B.** Platelet-activating factor. (⁄\/\/\/\ is a long, hydrophobic hydrocarbon chain.

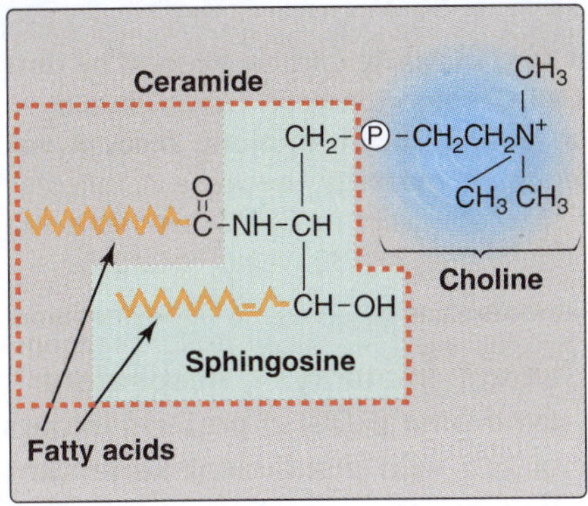

Figure 22.4 Structure of sphingomyelin, showing sphingosine (in green box) and ceramide components (in dashed box). Ⓟ = phosphate.

residue (rather than a FA) at carbon 2 of the glycerol backbone (Fig. 22.3B). PAF is synthesized and released by a variety of cell types. It binds to surface receptors, triggering potent thrombotic and acute inflammatory events. For example, **PAF activates inflammatory cells and mediates hypersensitivity, acute inflammatory, and anaphylactic reactions**. It causes platelets to aggregate and activate and neutrophils and alveolar macrophages to generate superoxide radicals to kill bacteria (see p. 252). It also lowers blood pressure. [Note: PAF is one of the most potent bioactive molecules known, causing effects at concentrations as low as 10^{-11} mol/l.]

B. Sphingophospholipids: Sphingomyelin

The backbone of sphingomyelin is the amino alcohol sphingosine, rather than glycerol (Fig. 22.4). A long-chain-length FA (LCFA) is attached to the amino group of sphingosine through an amide linkage, producing a ceramide, which can also serve as a precursor of glycolipids (see p. 329). The alcohol group at carbon 1 of sphingosine is esterified to phosphorylcholine, producing sphingomyelin, the only significant sphingophospholipid in humans. **Sphingomyelin is an important constituent of the myelin sheath of nerve fibers.** [Note: The myelin sheath is a layered, membranous structure that insulates and protects neuronal axons of the central nervous system (CNS).]

III. PHOSPHOLIPID SYNTHESIS

Glycerophospholipid synthesis involves either the donation of PA from cytidine diphosphate (CDP)-DAG to an alcohol or the donation of the phosphomonoester of the alcohol from CDP-alcohol to DAG (Fig. 22.5). In both cases, the CDP-bound structure is considered an activated intermediate, and cytidine monophosphate

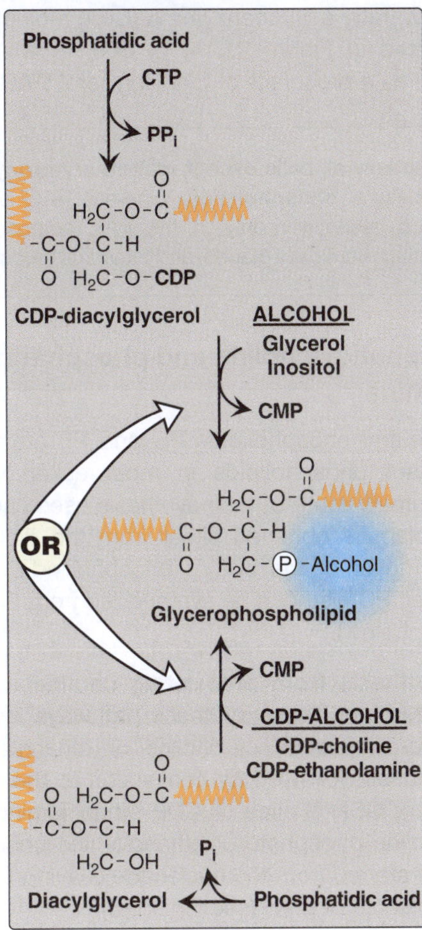

Figure 22.5 Glycerophospholipid synthesis requires activation of either diacylglycerol or an alcohol by linkage to cytidine diphosphate (CDP). CMP and CTP = cytidine mono- and triphosphates; P_i = inorganic phosphate; PP_i = pyrophosphate. (〰〰 is a fatty acid hydrocarbon chain.)

(CMP) is released as a side product. Therefore, a key concept in glycerophospholipid synthesis is activation, of either DAG or the alcohol to be added, by linkage with CDP. [Note: This is similar in principle to the activation of sugars by their attachment to uridine diphosphate (UDP) (see p. 226).] The FA esterified to the glycerol alcohol groups can vary widely, contributing to the heterogeneity of this group of compounds, with saturated FA typically found at carbon 1 and unsaturated ones at carbon 2. Most phospholipids are **synthesized in the smooth endoplasmic reticulum** (SER). From there, they are transported to the Golgi and then to membranes of organelles or the plasma membrane or are secreted from the cell by exocytosis. [Note: Ether lipid synthesis from dihydroxyacetone phosphate begins in peroxisomes.]

A. Phosphatidic acid

PA is the **precursor of other glycerophospholipids**. The steps in its synthesis from glycerol 3-phosphate and two fatty acyl coenzyme A (CoA) molecules were illustrated in Figure 21.14, p. 308, in which PA is shown as a precursor of triacylglycerol (TAG).

> **Essentially all cells except mature erythrocytes can synthesize phospholipids**, whereas TAG synthesis occurs essentially only in the liver, adipose tissue, lactating mammary glands, and intestinal mucosal cells.

B. Phosphatidylcholine and phosphatidylethanolamine

The neutral phospholipids PC and PE are the most abundant phospholipids in most eukaryotic cells. The primary route of their synthesis uses choline and ethanolamine obtained either from the diet or from the turnover of the body's phospholipids. [Note: In the liver, PC also can be synthesized from PS and PE (see 2. below).]

1. **Synthesis from preexisting choline and ethanolamine:** These synthetic pathways involve the phosphorylation of choline or ethanolamine by ***kinases***, followed by conversion to the activated form, CDP-choline or CDP-ethanolamine. Finally, choline phosphate or ethanolamine phosphate is transferred from the nucleotide (leaving CMP) to a molecule of DAG (see Fig. 22.5).

    a. **Significance of choline reutilization:** The reutilization of choline is important because, **although humans can synthesize choline de novo, the amount made is insufficient for our needs. Thus, choline is an essential dietary nutrient** with an adequate intake (see p. 502) of 550 mg for men and 425 mg for women. [Note: Choline is also used for the synthesis of acetylcholine, a neurotransmitter.]

    b. **Phosphatidylcholine in lung surfactant:** The pathway described above is the principal pathway for the synthesis of dipalmitoylphosphatidylcholine (DPPC or, dipalmitoyl lecithin). In DPPC, positions 1 and 2 on the glycerol are occupied by palmitate, a saturated LCFA. DPPC, made and secreted by type II pneumocytes, is a major lipid component of lung surfactant, which is the extracellular fluid layer lining the alveoli. **Surfactant serves to decrease the surface tension of this fluid layer, reducing the pressure needed to reinflate alveoli, thereby preventing alveolar collapse (atelectasis).** [Note: Surfactant is a complex mixture of lipids (90%) and proteins (10%), with DPPC being the major component for reducing surface tension.]

> **Fetal lung maturity can be gauged by determining the DPPC/sphingomyelin ratio, usually written as L (for lecithin)/S, in amniotic fluid. A value ≥2 is evidence of maturity**, because it reflects the shift from sphingomyelin to DPPC synthesis that occurs in pneumocytes at ~32 weeks' gestation.

    c. **Lung maturity: Respiratory distress syndrome (RDS) in preterm infants is associated with insufficient surfactant production and/or secretion and is a significant cause of all neonatal deaths.** Lung maturation can be accelerated by giving the mother **glucocorticoids** shortly before delivery to induce expression of specific genes. Postnatal administration of natural or synthetic surfactant (by intratracheal instillation) is also used. [Note: Acute RDS, seen in all age groups, is the result of alveolar damage (due to infection, injury, or aspiration) that causes fluid to accumulate in the alveoli, impeding the exchange of oxygen (O_2) and carbon dioxide (CO_2).]

2. **Phosphatidylcholine synthesis from phosphatidylserine:** The liver requires a mechanism for producing PC, even when free choline levels are low, because it exports significant amounts of PC in the bile and as a component of plasma lipoproteins. To provide the needed PC, PS is decarboxylated to PE by *PS decarboxylase*. PE then undergoes three methylation steps to produce PC, as illustrated in Figure 22.6. S-adenosylmethionine is the methyl group donor (see p. 400).

C. Phosphatidylserine

PS synthesis in mammalian tissues is provided by the base exchange reaction, in which the ethanolamine of PE is exchanged for free serine (see Fig. 22.6). This reaction, although reversible, is used primarily to produce the PS required for membrane synthesis. PS has a net negative charge.

D. Phosphatidylinositol

PI is synthesized from free inositol and CDP-DAG, as shown in Figure 22.5. PI is an unusual phospholipid in that it most frequently contains stearic acid on carbon 1 and arachidonic acid on carbon 2 of the glycerol. Therefore, PI serves as a **reservoir of arachidonic acid in membranes and, thus, provides the substrate for prostaglandin (see p. 333) synthesis when required.** Like PS, PI has a net negative charge. [Note: There is asymmetry in the phospholipid composition of the cell membrane. PS and PI, for example,

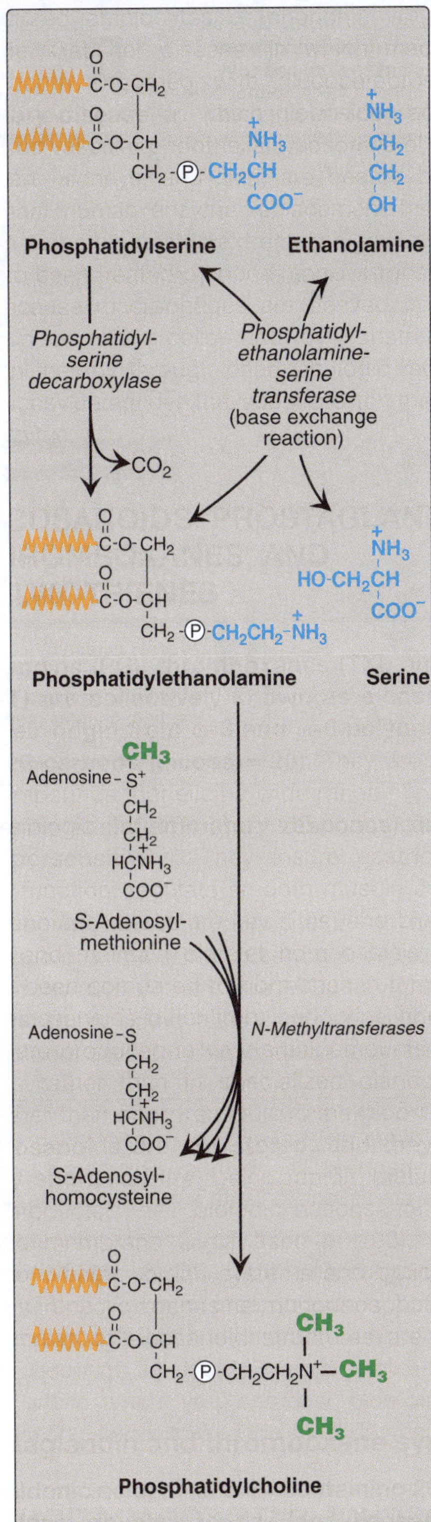

Figure 22.6 Synthesis of phosphatidylcholine from phosphatidylserine in the liver. (WWWW is a fatty acid hydrocarbon chain.) Ⓟ = phosphate; CO_2 = carbon dioxide.

are found primarily on the inner leaflet. Asymmetry is achieved by ATP-dependent enzymes known as "flippases" and "floppases."]

1. **Role in signal transduction across membranes: The phosphorylation of membrane-bound PI produces polyphosphoinositides such as phosphatidylinositol 4,5-bisphosphate ([PIP_2];** Fig. 22.7). The cleavage of PIP_2 by *phospholipase C* occurs in response to the binding of various neurotransmitters, hormones, and growth factors to G protein–coupled receptors (GPCR), such as the α_1 adrenergic receptor, on the cell membrane and activation of the G_q α-subunit (Fig. 22.8). The products of this cleavage, inositol 1,4,5-trisphosphate (IP_3) and DAG, mediate the mobilization of intracellular calcium and the activation of *protein kinase C*, which act synergistically to evoke specific cellular responses. **Signal transduction** across the membrane is, thus, accomplished.

2. **Role in membrane protein anchoring: Specific proteins can be covalently attached through a carbohydrate bridge to membrane-bound PI (**Fig. 22.9**). For example, *lipoprotein lipase*, an enzyme that degrades triacylglycerol in lipoprotein particles (see p. 351), is attached to capillary endothelial cells by a glycosyl phosphatidylinositol (GPI) anchor**. [Note: GPI-linked proteins are also found in a variety of parasitic protozoans, such as trypanosomes and *Leishmania*.] Being attached to a membrane lipid (rather than being an integral part of the membrane) allows GPI-anchored proteins increased lateral mobility on the extracellular surface of the plasma membrane. The protein can be cleaved from its anchor by the action of *phospholipase C* (see Fig. 22.9). [Note: **A deficiency in the synthesis of GPI in hematopoietic cells results in the hemolytic disease**

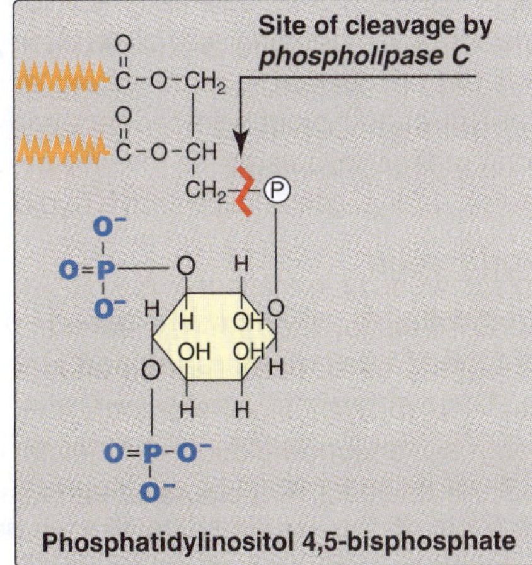

Figure 22.7 Structure of phosphatidylinositol 4,5-bisphosphate (PIP_2). Cleavage by *phospholipase C* produces inositol 1,4,5-trisphosphate (IP_3) and diacylglycerol. (WWWW is a fatty acid hydrocarbon chain.) Ⓟ = phosphate.

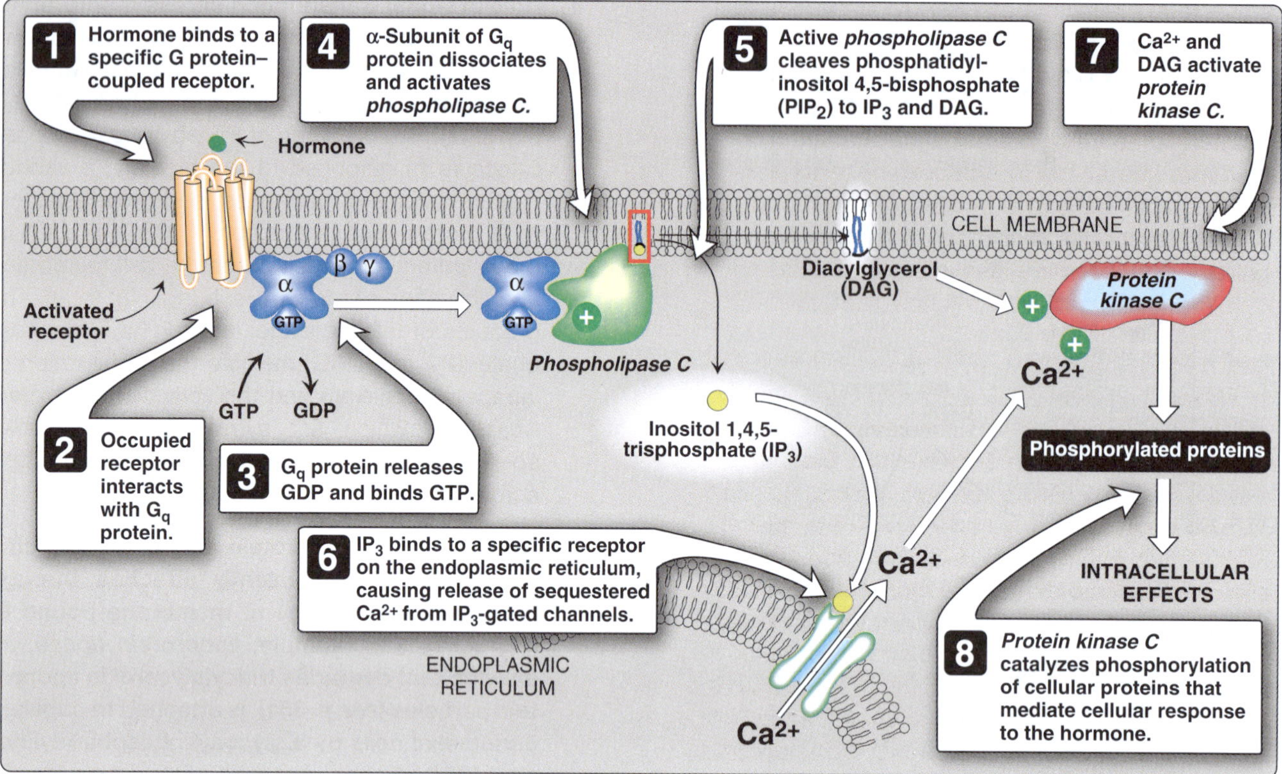

Figure 22.8 Role of inositol triphosphate and diacylglycerol in cell signaling. GDP and GTP = guanosine di- and triphosphates; Ca^{2+} = calcium.

paroxysmal nocturnal hemoglobinuria, because GPI-anchored proteins protect blood cells from complement-mediated lysis.]

E. Phosphatidylglycerol and cardiolipin

Phosphatidylglycerol occurs in relatively large amounts in mitochondrial membranes and is a precursor of cardiolipin (diphosphatidy glycerol). It is synthesized from CDP-DAG and glycerol 3-phosphate. Cardiolipin (see Fig. 22.2) is synthesized by the transfer of DAG 3-phosphate from CDP-DAG to a pre-existing molecule of phosphatidylglycerol.

F. Sphingomyelin

Sphingomyelin, a sphingosine-based phospholipid, is found in cell membranes and in the myelin sheath. The synthesis of sphingomyelin is shown in Figure 22.10. Briefly, palmitoyl CoA condenses with serine, as CoA and the carboxyl group (as CO_2) of serine are lost. [Note: This reaction, like the decarboxylation reactions involved in the synthesis of PE from PS and of regulators from amino acids (for example, the catecholamines from tyrosine; see p. 423), requires pyridoxal phosphate (a derivative of vitamin B_6) as a coenzyme.] The product is reduced in a nicotinamide adenine dinucleotide phosphate (NADPH)-requiring reaction to sphinganine (dihydrosphingosine). The sphinganine is acylated at the amino group with one of a variety of LCFA and then desaturated to produce a ceramide, the immediate precursor of sphingomyelin (and other sphingolipids, as described on p. 327).

> Ceramides play a key role in maintaining the skin's water-permeability barrier. Decreased ceramide levels are associated with a number of skin diseases.

Phosphorylcholine from PC is transferred to the ceramide, producing sphingomyelin and DAG. [Note: Sphingomyelin of the myelin sheath contains predominantly longer-chain FA such as lignoceric acid and nervonic acid, whereas gray matter of the brain has sphingomyelin that contains primarily stearic acid.]

IV. PHOSPHOLIPID DEGRADATION

The degradation of phosphoglycerides is performed by *phospholipases* found in all tissues and pancreatic juice. [Note: For a discussion of phospholipid digestion, see p. 291.] A number of toxins and venoms have *phospholipase* activity, and **several pathogenic bacteria produce *phospholipases* that dissolve cell membranes and allow the spread of infection.** Sphingomyelin is

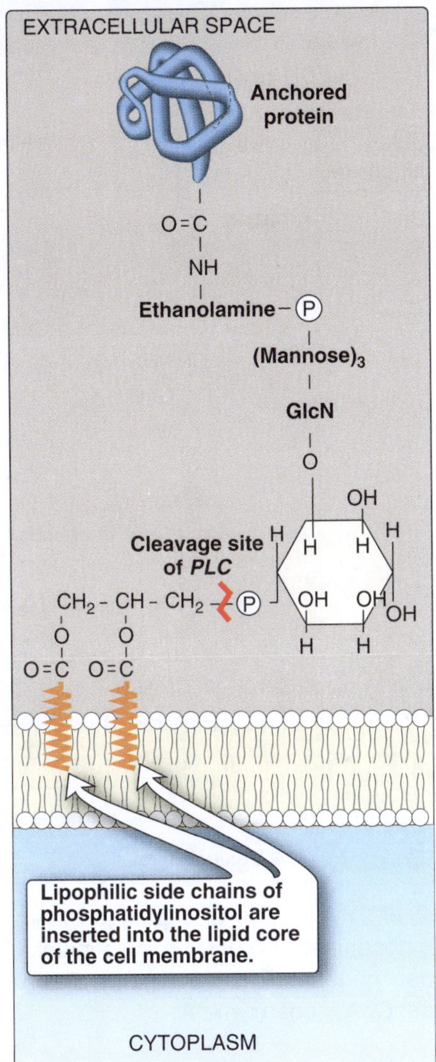

Figure 22.9 Example of a glycosyl phosphatidylinositol (GPI) membrane protein anchor. GlcN = glucosamine; Ⓟ = phosphate; *PLC* = *phospholipase C*.

degraded by the lysosomal *phospholipase*, *sphingomyelinase* (see B. below).

A. Phosphoglycerides

Phospholipases hydrolyze the phosphodiester bonds of phosphoglycerides, with each enzyme cleaving the phospholipid at a specific site. The major *phospholipases* are shown in Figure 22.11. [Note: Removal of the FA from carbon 1 or 2 of a phosphoglyceride produces a lysophosphoglyceride, which is the substrate for *lysophospholipases*.] ***Phospholipases* release molecules that can serve as second messengers (for example, DAG and IP_3)** or that are the substrates for synthesis of messengers (for example, arachidonic acid). *Phospholipases* are responsible not only for degrading phospholipids but also for remodeling them. For example, *phospholipases A_1* and *A_2* remove specific FA from membrane-bound phospholipids, which can be replaced with different FA using *fatty acyl CoA transferase*. This mechanism is used as one way to create the unique lung surfactant DPCC (see p. 324) and to insure that carbon 2 of PI (and sometimes of PC) is bound to arachidonic acid. [Note: Barth syndrome, a rare X-linked disorder characterized by cardiomyopathy, muscle weakness, and neutropenia, is the result of defects in cardiolipin remodeling.]

B. Sphingomyelin

Sphingomyelin is degraded by *sphingomyelinase*, a lysosomal enzyme that removes phosphorylcholine, leaving a ceramide. The ceramide is, in turn, cleaved by *ceramidase* into sphingosine and a free FA (Fig. 22.12). [Note: **The released ceramide and sphingosine regulate signal transduction pathways,** in part by influencing the activity of *protein kinase C* and, thus, the phosphorylation of its protein substrates. They also promote apoptosis.] **Niemann-Pick disease (types A and B) is an autosomal-recessive disorder caused by the inability to degrade sphingomyelin due to a deficiency of *sphingomyelinase*, a type of *phospholipase C*.** In the severe infantile form (type A, which shows <1% of normal enzymic activity), the liver and spleen are the primary sites of lipid deposits and are, therefore, greatly enlarged. The lipid consists primarily of the sphingomyelin that cannot be degraded (Fig. 22.13). Infants with this lysosomal storage disease experience rapid and progressive neurodegeneration as a result of deposition of sphingomyelin in the CNS, and they die in early childhood. A less severe variant (type B, which shows up to 10% of normal activity) with a later age of onset and a longer survival time causes little to no damage to neural tissue, but lungs, spleen, liver, and bone marrow are affected, resulting in a chronic form of the disease. Although Niemann-Pick disease occurs in all ethnic groups, type A occurs with greater frequency in the Ashkenazi Jewish population.

V. GLYCOLIPID OVERVIEW

Glycolipids are molecules that contain both carbohydrate and lipid components. Like the phospholipid sphingomyelin, glycolipids are derivatives of ceramides in which a LCFA is attached to the amino alcohol sphingosine. Therefore, they are more precisely called glycosphingolipids. [Note: Thus, ceramides are the precursors of both phosphorylated and glycosylated sphingolipids.] Like the phospholipids, glycosphingolipids are essential components of all membranes in the body, but they are found in greatest amounts in nerve tissue. They are located in the outer leaflet of the plasma membrane, where they interact with the extracellular environment. As such, they play a role in the regulation of cellular interactions (for example, adhesion and recognition), growth, and development.

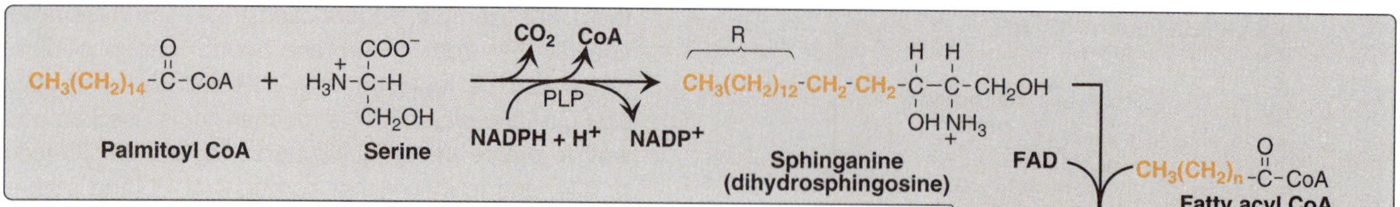

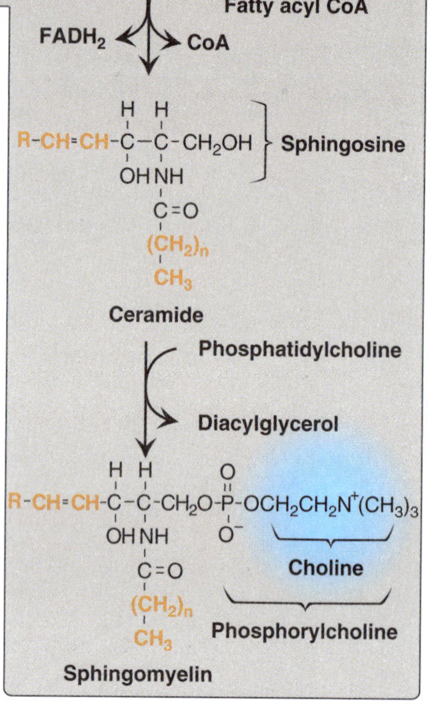

> Membrane glycosphingolipids associate with cholesterol and GPI-anchored proteins to form **lipid rafts, laterally mobile microdomains of the plasma membrane** that function to organize and regulate membrane signaling and trafficking functions.

Glycosphingolipids are antigenic and are the source of ABO blood group antigens (see p. 269), various embryonic antigens specific for particular stages of fetal development, and some tumor antigens. [Note: **The carbohydrate portion of a glycolipid is the antigenic determinant**.] They have been coopted for use as cell surface receptors for cholera and tetanus toxins as well as for certain viruses and microbes. Genetic disorders associated with an inability to properly degrade the glycosphingolipids result in lysosomal accumulation of these compounds. [Note: Changes in the carbohydrate portion of glycosphingolipids (and glycoproteins) are characteristic of transformed cells (cells with dysregulated growth).]

Figure 22.10 Synthesis of sphingomyelin. PLP = pyridoxal phosphate; NADP(H) = nicotinamide adenine dinucleotide phosphate; FAD(H_2) = flavin adenine dinucleotide; CoA = coenzyme A.

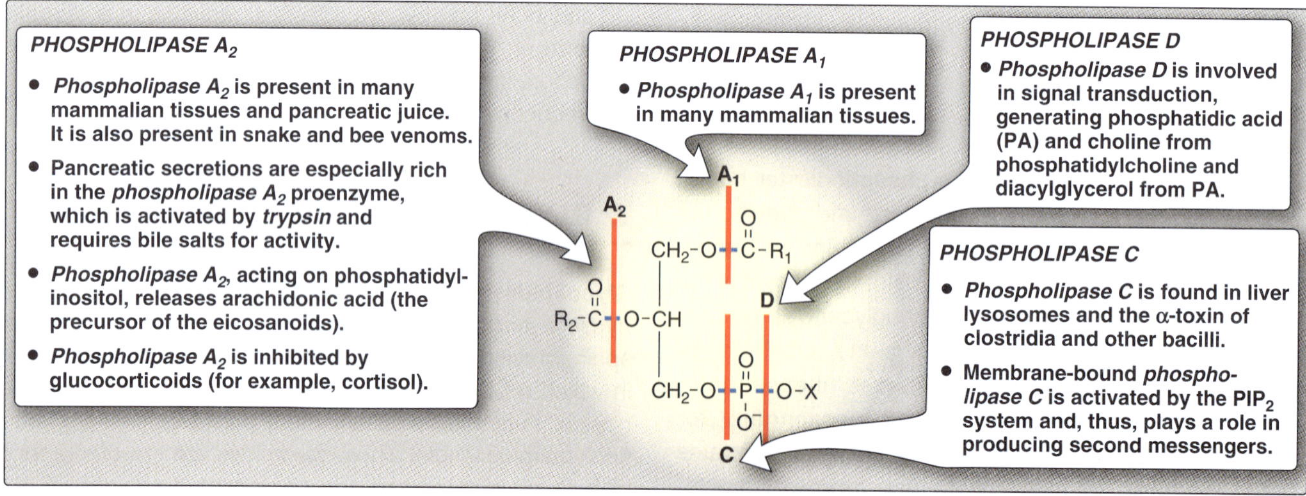

Figure 22.11 Degradation of glycerophospholipids by *phospholipases*. PIP_2 = phosphatidylinositol 4,5-bisphosphate; R_1 and R_2 = fatty acids; X = an alcohol.

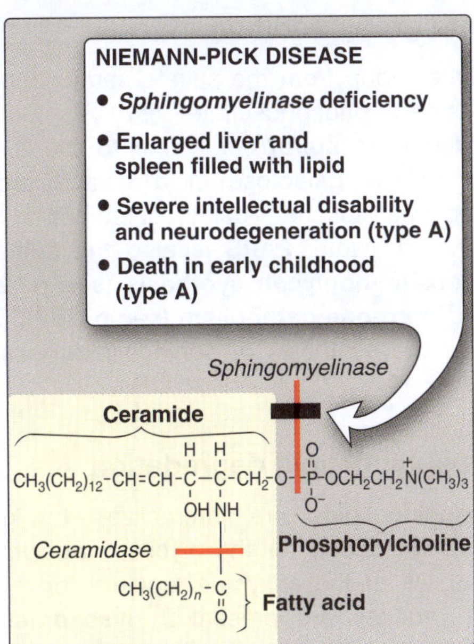

Figure 22.12 Degradation of sphingomyelin. [Note: Type B is the nonneuropathic form. It has a later age of onset and a longer survival time than type A.]

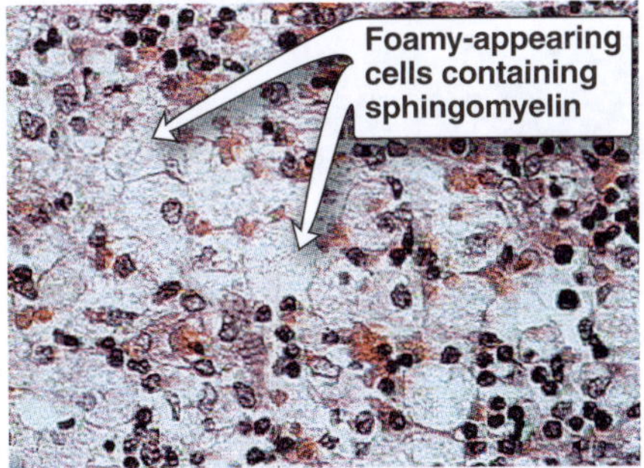

Figure 22.13 Accumulation of lipids in spleen cells from a patient with Niemann-Pick disease.

VI. GLYCOSPHINGOLIPID STRUCTURE

The glycosphingolipids differ from sphingomyelin in that they do not contain phosphate, and the polar head function is provided by a monosaccharide or oligosaccharide attached directly to the ceramide by an O-glycosidic bond (Fig. 22.14). The number and type of carbohydrate moieties present determine the type of glycosphingolipid.

A. Neutral glycosphingolipids

The simplest neutral glycosphingolipids are the **cerebrosides. These are ceramide monosaccharides that contain either a molecule of galactose (forming ceramide-galactose or galactocerebroside, the most common cerebroside found in myelin,** as shown in Fig. 22.14**) or glucose (forming ceramide-glucose or glucocerebroside, an intermediate in the synthesis and degradation of the**

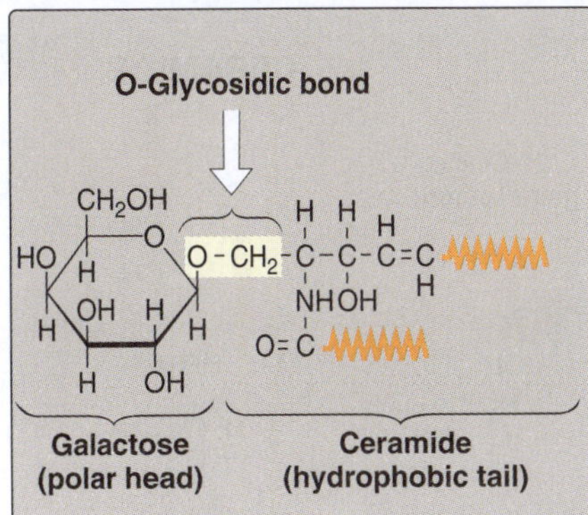

Figure 22.14 Structure of a neutral glycosphingolipid, galactocerebroside. (∧∧∧∧∧ is a hydrophobic hydrocarbon chain.)

more complex glycosphingolipids). [Note: Members of a group of galacto- or glucocerebrosides may also differ from each other in the type of FA attached to the sphingosine.] As their name implies, cerebrosides are found predominantly in the brain and peripheral nerves, with high concentrations in the myelin sheath. Ceramide oligosaccharides (or globosides) are produced by attaching additional monosaccharides to a glucocerebroside, for example, ceramide-glucose-galactose (also known as lactosylceramide). The additional monosaccharides can include substituted sugars such as *N*-acetylgalactosamine.

B. Acidic glycosphingolipids

Acidic glycosphingolipids are negatively charged at physiologic pH. The negative charge is provided by N-acetylneuraminic acid ([NANA], a sialic acid, as shown in Fig. 22.15) in gangliosides or by sulfate groups in sulfatides.

1. **Gangliosides: These are the most complex glycosphingolipids and are found primarily in the ganglion cells of the CNS, particularly at the nerve endings. They are derivatives of ceramide oligosaccharides and contain one or more molecules of NANA** (from CMP-NANA). The notation for these compounds is G (for ganglioside) plus a subscript M, D, T, or Q to indicate whether there is one (mono), two (di), three (tri), or four (quatro) molecules of NANA in the ganglioside, respectively. Additional numbers and letters in the subscript designate the monomeric sequence of the carbohydrate attached to the ceramide. (See Fig. 22.15 for the structure of G_{M2}.) Gangliosides are of medical interest because several lipid storage disorders involve the accumulation of NANA-containing glycosphingolipids in cells (see Fig. 22.20, p. 332).

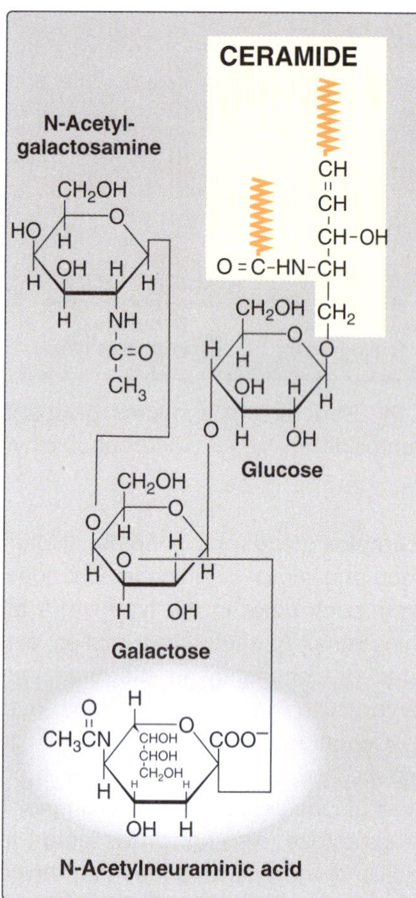

Figure 22.15 Structure of the ganglioside G_{M2}. (〰 is a hydrophobic hydrocarbon chain.)

2. **Sulfatides: These sulfoglycosphingolipids are sulfated galactocerebrosides that are negatively charged at physiologic pH.** Sulfatides are found **predominantly in the brain and kidneys.**

VII. GLYCOSPHINGOLIPID SYNTHESIS AND DEGRADATION

Synthesis of glycosphingolipids occurs primarily **in the Golgi by sequential addition of glycosyl monomers transferred from UDP-sugar donors to the acceptor molecule.** The mechanism is similar to that used in glycoprotein synthesis (see p. 269).

A. Enzymes involved in synthesis

The enzymes involved in the synthesis of glycosphingolipids are ***glycosyltransferases*** that are specific for the type and location of the glycosidic bond formed. [Note: These enzymes can recognize both glycosphingolipids and glycoproteins as substrates.]

B. Sulfate group addition

A sulfate group from the sulfate carrier 3′-phosphoadenosine-5′-phosphosulfate ([PAPS], Fig. 22.16) is added by a ***sulfotransferase*** to the 3′-hydroxyl group of the galactose in a galactocerebroside, forming the sulfatide galactocerebroside 3-sulfate (Fig. 22.17). [Note: **PAPS is also the sulfur donor in glycosaminoglycan synthesis (see p. 266) and steroid hormone catabolism (see p. 362).]** An overview of the synthesis of sphingolipids is shown in Figure 22.18.

C. Glycosphingolipid degradation

Glycosphingolipids are internalized by phagocytosis as described for the glycosaminoglycans (see p. 266). All of the enzymes required for the degradative process are present in **lysosomes,** which fuse with the phagosomes. The lysosomal enzymes

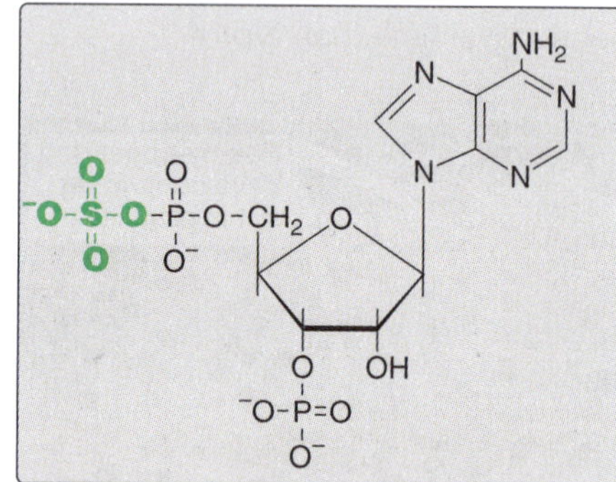

Figure 22.16 Structure of 3′-phosphoadenosine-5′-phosphosulfate.

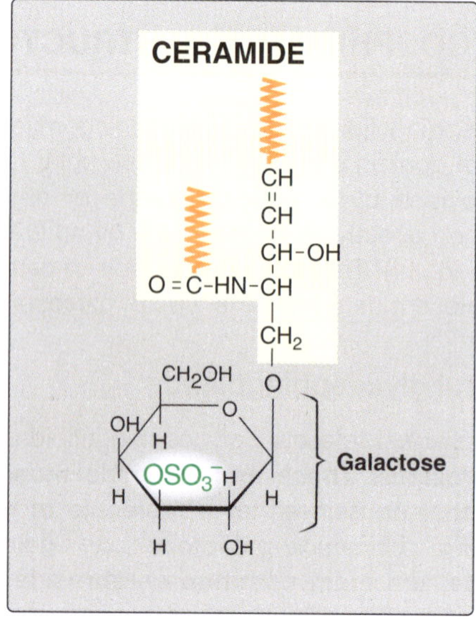

Figure 22.17 Structure of galactocerebroside 3-sulfate. (〰 is a hydrophobic hydrocarbon chain.)

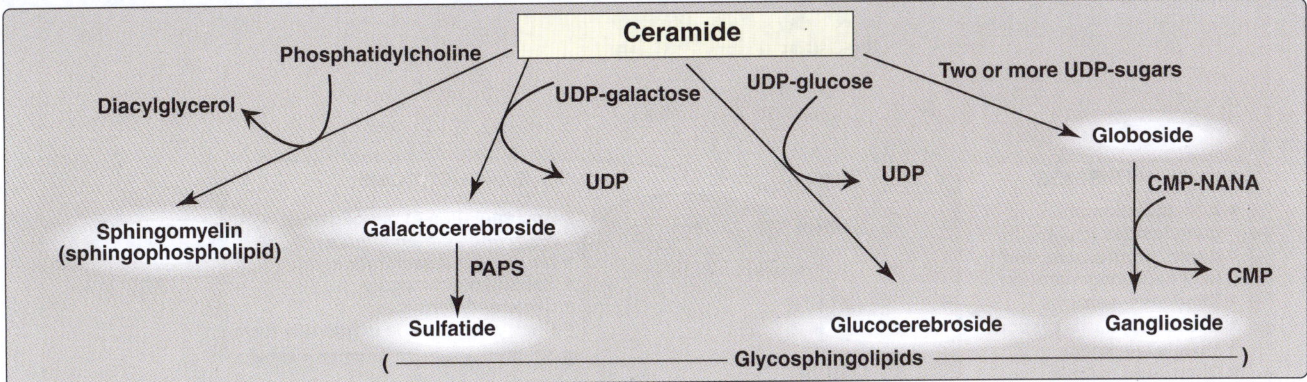

Figure 22.18 Overview of sphingolipid synthesis. UDP = uridine diphosphate; CMP = cytidine monophosphate; NANA = N-acetylneuraminic acid; PAPS = 3′-phosphoadenosine-5′-phosphosulfate.

hydrolytically and irreversibly cleave specific bonds in the glycosphingolipid. As seen with the glycosaminoglycans and glycoproteins (see p. 269), degradation is a sequential process following the rule "last on, first off," in which the last group added during synthesis is the first group removed in degradation. Therefore, defects in the degradation of the polysaccharide chains in these three glycoconjugates result in lysosomal storage diseases.

D. Sphingolipidoses

In a normal individual, synthesis and degradation of glycosphingolipids are balanced, so that the amount of these compounds present in membranes is constant. **If a specific lysosomal *acid hydrolase* required for degradation is partially or totally missing, a sphingolipid accumulates. Lysosomal lipid storage diseases caused by these deficiencies are called sphingolipidoses**. The result of a specific *acid hydrolase* deficiency may be seen dramatically in nerve tissue, where neurologic deterioration can lead to early death. Figure 22.20 provides an outline of the pathway of sphingolipid degradation and descriptions of some sphingolipidoses. [Note: Some sphingolipidoses can also result from defects in lysosomal activator proteins (for example, the **saposins**) that facilitate access of the *hydrolases* to short carbohydrate chains as degradation proceeds.]

1. **Common properties: A specific lysosomal hydrolytic enzyme is deficient in the classic form of each disorder. Therefore, usually, only a single sphingolipid (the substrate for the deficient enzyme) accumulates in the involved organs in each disease.** [Note: The rate of biosynthesis of the accumulating lipid is normal.] The disorders are progressive and, although many are fatal in childhood, extensive phenotypic variability is seen leading to the designation of different clinical types, such as types A and B in Niemann-Pick disease. Genetic variability is also seen because a given disorder can be caused by any one of a variety of mutations within a single gene. **The sphingolipidoses are autosomal-recessive disorders, except for Fabry disease, which is X linked**. The incidence of the sphingolipidoses is low in most populations, except for Gaucher and Tay-Sachs diseases, which, like Niemann-Pick disease, show a high frequency in the Ashkenazi Jewish population. [Note: Tay-Sachs also has a high frequency in Irish American, French Canadian, and Louisiana Cajun populations.]

2. **Diagnosis and treatment: A specific sphingolipidosis can be diagnosed by measuring enzyme activity in cultured fibroblasts or peripheral leukocytes or by analyzing DNA** (see Chapter 39). Histologic examination of the affected tissue is also useful. [Note: Shell-like inclusion bodies are seen in **Tay-Sachs**, and a crumpled tissue paper appearance of the cytosol is seen in **Gaucher disease** (Fig. 22.19).] Prenatal diagnosis, using

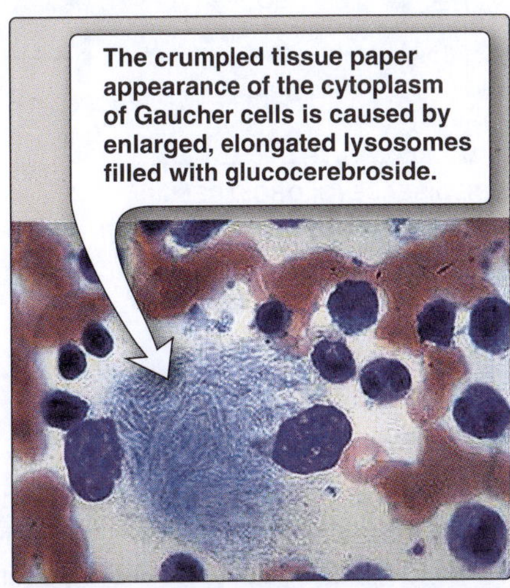

Figure 22.19 Aspirated bone marrow cells from a patient with Gaucher disease.

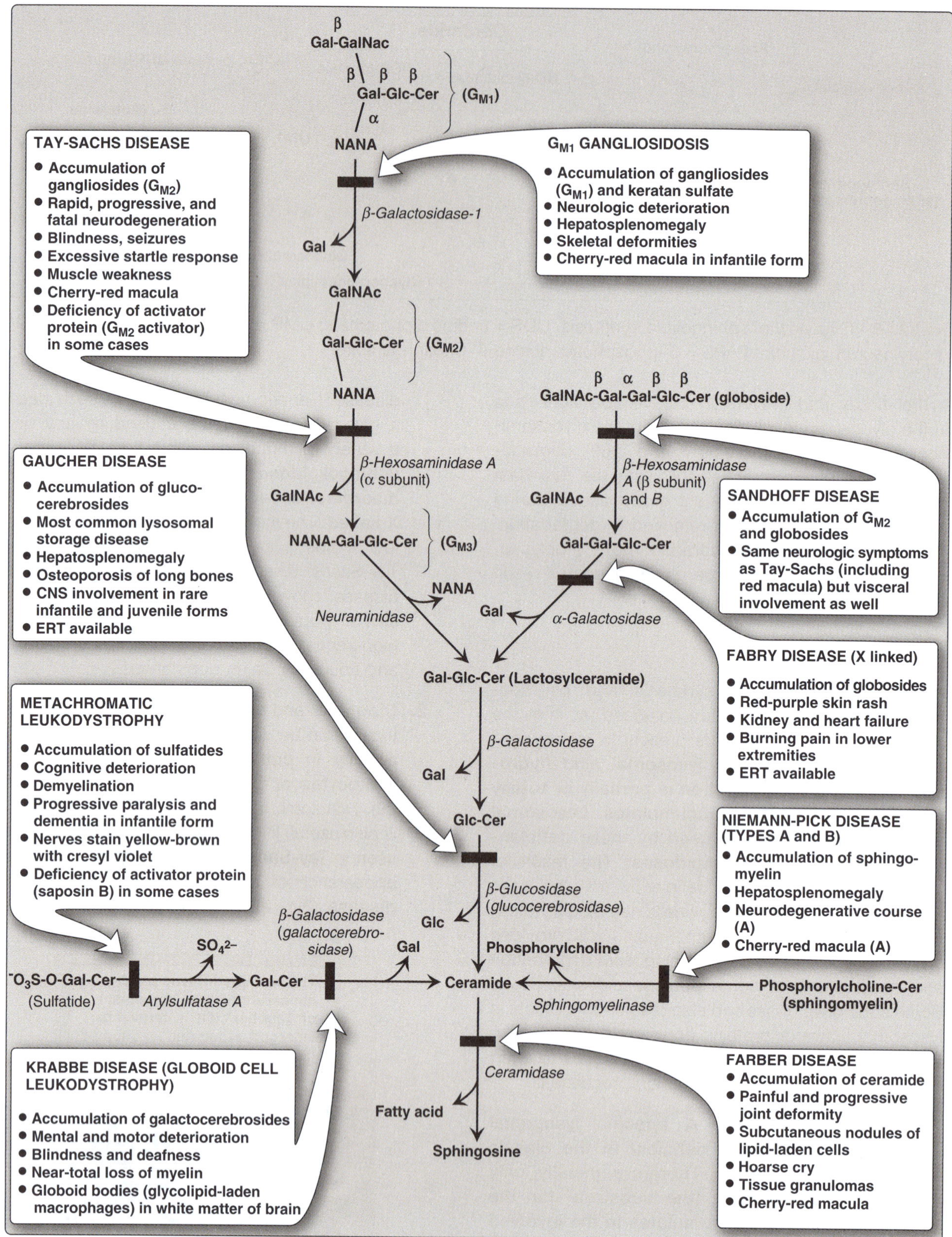

Figure 22.20 Degradation of sphingolipids showing the lysosomal enzymes affected in related genetic diseases, the sphingolipidoses. All are autosomal-recessive diseases except Fabry, which is X linked, and all can be fatal in early life. Cer = ceramide; Gal = galactose; Glc = glucose; GalNAc = N-acetylgalactosamine; NANA = N-acetylneuraminic acid; CNS = central nervous system. SO_4^{2-} = sulfate; ERT = enzyme replacement therapy.

cultured amniocytes or chorionic villi, is available. Gaucher disease, in which macrophages become engorged with glucocerebroside, and **Fabry disease**, in which globosides accumulate in the vascular endothelial lysosomes of the brain, heart, kidneys, and skin, are treated by recombinant human enzyme replacement therapy, but the monetary cost is extremely high. Gaucher has also been treated by bone marrow transplantation (because macrophages are derived from hematopoietic stem cells) and by substrate reduction therapy through pharmacologic reduction of glucosylceramide, the substrate for the deficient enzyme.

VIII. EICOSANOIDS: PROSTAGLANDINS, THROMBOXANES, AND LEUKOTRIENES

Prostaglandins (PG), thromboxanes (TX), and leukotrienes (LT) are collectively known as eicosanoids to reflect their origin from ω-3 and ω-6 polyunsaturated FA with 20 carbons (eicosa = 20). They are extremely potent compounds that elicit a wide range of responses, both physiologic (**inflammatory response**) and pathologic (**hypersensitivity**). They insure gastric integrity and renal function, regulate smooth muscle contraction (the intestine and uterus are key sites) and blood vessel diameter, and maintain platelet homeostasis. Although they have been compared to hormones in terms of their actions, eicosanoids differ from endocrine hormones in that they are produced in very small amounts in almost all tissues rather than in specialized glands and act locally rather than after transport in the blood to distant sites. **Eicosanoids are not stored, and they have an extremely short half-life,** being rapidly metabolized to inactive products. Their biologic actions are mediated by plasma membrane GPCR (see p. 190), which are different in different organ systems and typically result in changes in cyclic adenosine monophosphate production. Examples of eicosanoid structures are shown in Figure 22.21.

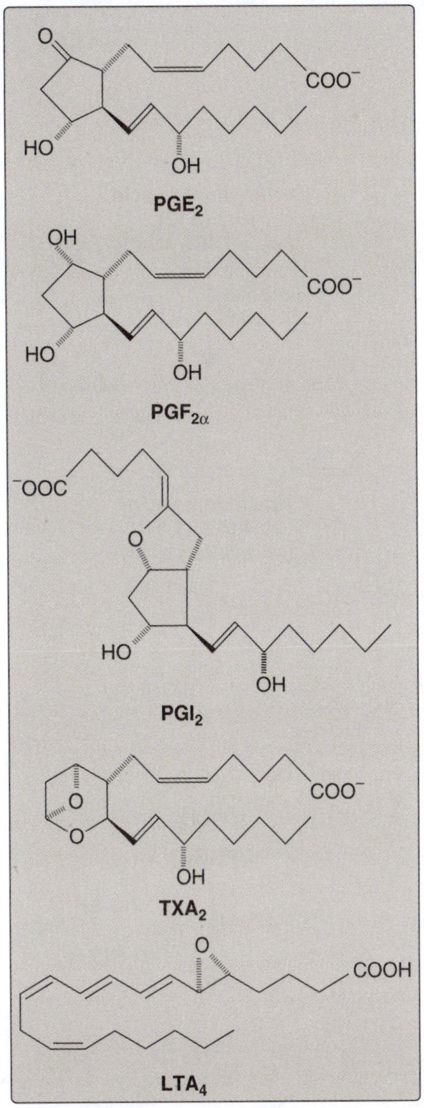

PGE_2

$PGF_{2\alpha}$

PGI_2

TXA_2

LTA_4

Figure 22.21 Examples of eicosanoid structures. [Note: Prostaglandins are named as follows: PG plus a third letter (for example, E), which designates the type and arrangement of functional groups in the molecule. The subscript number indicates the number of double bonds in the molecule. PGI_2 is also known as prostacyclin. Thromboxanes are designated by TX and leukotrienes by LT.]

A. Prostaglandin and thromboxane synthesis

Arachidonic acid, an ω-6 FA containing 20 carbons and four double bonds (an eicosatetraenoic FA), is the immediate precursor of the predominant type of human PG (series 2 or those with two double bonds, as shown in Fig. 22.22). It is derived by the elongation and desaturation of the essential FA linoleic acid, also an ω-6 FA. Arachidonic acid is incorporated into membrane phospholipids (typically PI) at carbon 2, from which it is released by *phospholipase A_2* (Fig. 22.23) in response to a variety of signals, such as a rise in calcium. [Note: Series 1 PG contain one double bond and are derived from an ω-6 eicosatrienoic FA, dihomo-γ-linolenic acid, whereas series 3 PG contain three double bonds and are derived from eicosapentaenoic acid (EPA), an ω-3 FA. See p. 507.]

1. **Prostaglandin H2 synthase:** The first step in PG and TX synthesis is the oxidative cyclization of free arachidonic acid to yield PGH_2 by ***PGH_2 synthase*** (or, *prostaglandin endoperoxide synthase*). This enzyme is an ER membrane-bound protein that

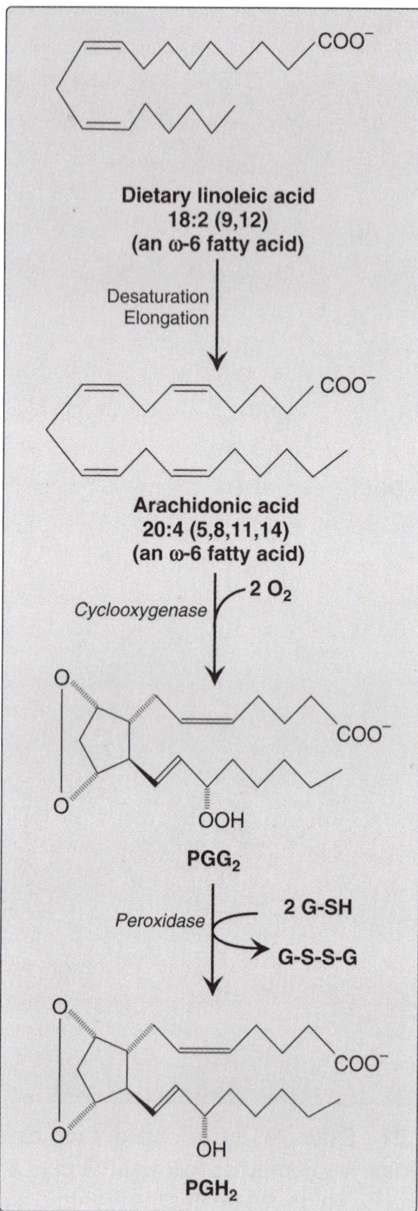

Figure 22.22 Oxidation and cyclization of arachidonic acid by the two catalytic activities (*cyclooxygenase* and *peroxidase*) of *PGH_2 synthase* (*prostaglandin endoperoxide synthase*). G-SH = reduced glutathione; G-S-S-G = oxidized glutathione; PG = prostaglandin.

has two catalytic activities: *fatty acid cyclooxygenase* (*COX*), which requires two molecules of O_2, and *peroxidase*, which requires reduced glutathione (see p. 250). PGH_2 is converted to a variety of PG and TX, as shown in Figure 22.23, by cell-specific *synthases*. [Note: PG contain a five-carbon ring, whereas TX contain a heterocyclic six-membered oxane ring (see Fig. 22.21).] Two **isozymes of *PGH_2 synthase***, usually denoted as *COX-1* and *COX-2*, are known. ***COX-1*** **is made constitutively in most tissues and is required for maintenance of healthy gastric tissue, renal homeostasis, and platelet aggregation. *COX-2* is inducible in a limited number of tissues in response to products of activated immune and inflammatory cell**s. [Note: The increase in PG synthesis subsequent to the induction of *COX-2* mediates the pain, heat, redness, and swelling of inflammation and the fever of infection.]

2. **Synthesis inhibition:** The synthesis of PG and TX can be inhibited by unrelated compounds. For example**, cortisol (a steroidal anti-inflammatory agent) inhibits *phospholipase A_2* activity** (see Fig. 22.23) and, therefore, arachidonic acid, the substrate for PG and TX synthesis, is not released from membrane phospholipids. **Aspirin, indomethacin, and phenylbutazone (all nonsteroidal anti-inflammatory drugs [NSAID]) inhibit both *COX-1* and *COX-2* and, thus, prevent the synthesis of the parent molecule, PGH_2.** [Note: Systemic inhibition of *COX-1*, with subsequent damage to the stomach and the kidneys and impaired clotting of blood, is the basis of aspirin's toxicity.] **Aspirin (but not other NSAID) also induces synthesis of lipoxins (anti-inflammatory mediators made from arachidonic acid) and resolvins and protectins (inflammation-resolving mediators made from EPA).** Inhibitors specific for *COX-2* (the coxibs, for example, **celecoxib**) were designed to reduce pathologic inflammatory processes mediated by *COX-2* while maintaining the physiologic functions of *COX-1*. However, their use has been associated with increased risk of heart attacks, likely as a result of decreased PGI_2 synthesis (see B. below), and some have been withdrawn from the market.

B. Thromboxanes and prostaglandins in platelet homeostasis

Thromboxane A_2 (TXA_2) is produced by *COX-1* in activated platelets. It **promotes platelet adhesion and aggregation and contraction of vascular smooth muscle, thereby promoting formation of blood clots (thrombi**). Prostacyclin (PGI_2), produced by *COX-2* in vascular endothelial cells, inhibits platelet aggregation and stimulates vasodilation and, so, impedes thrombogenesis. **The opposing effects of TXA_2 and PGI_2 limit thrombi formation to sites of vascular injury.** [Note: Aspirin has an antithrombogenic effect. It inhibits TXA_2 synthesis by *COX-1* in

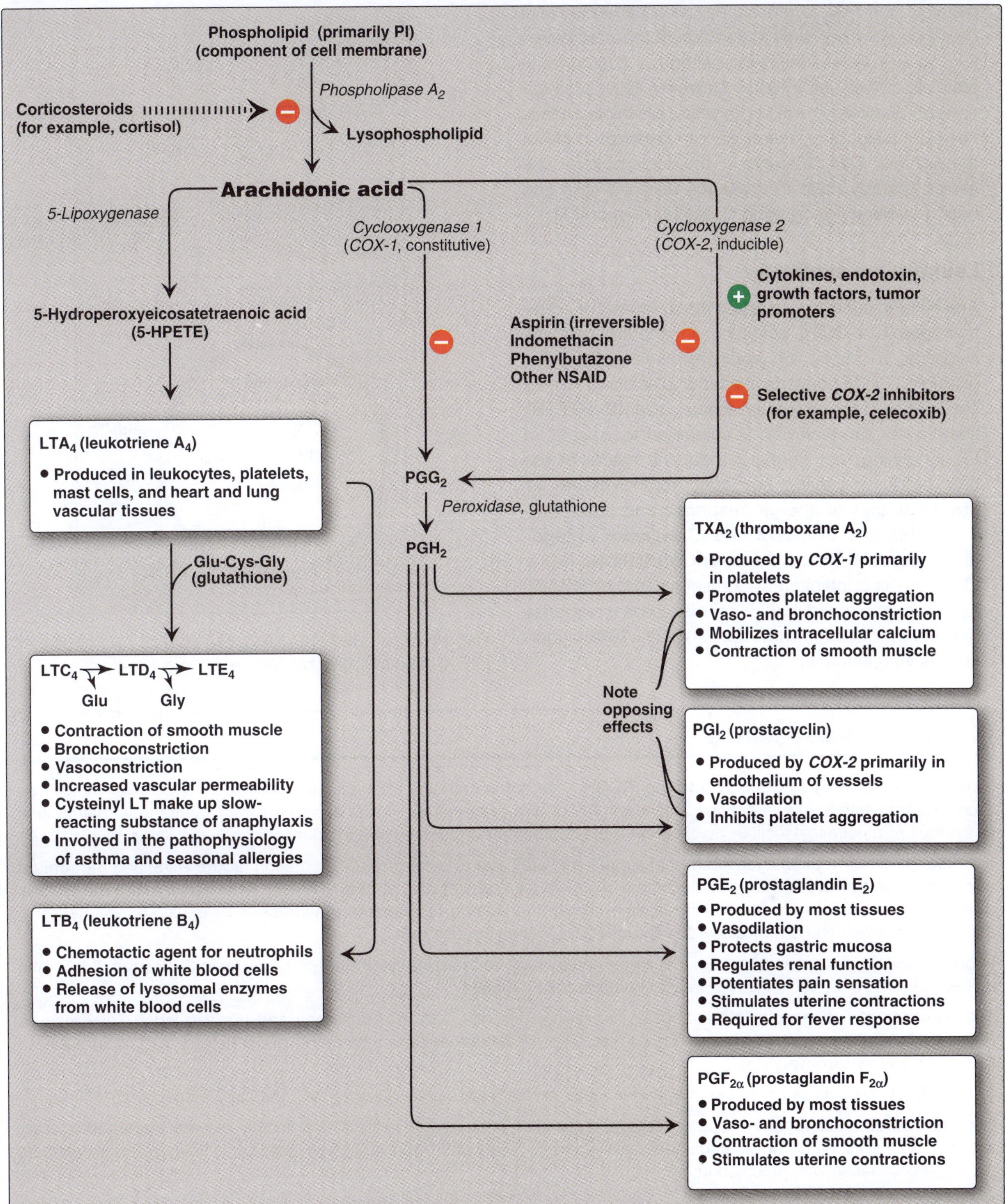

Figure 22.23 Overview of the biosynthesis and function of some important prostaglandins (PG), leukotrienes (LT), and a thromboxane (TX) from arachidonic acid. [Note: The arachidonic acid in the membrane phospholipid was derived from the ω-6 essential fatty acid (FA), linoleic, also an ω-6 FA.] PI = phosphatidylinositol; NSAID = nonsteroidal anti-inflammatory drugs; Glu = glutamate; Cys = cysteine; Gly = glycine.

platelets and PGI_2 synthesis by *COX-2* in endothelial cells through irreversible acetylation of these isozymes (Fig. 22.24). *COX-1* inhibition cannot be overcome in platelets, which lack nuclei. However, *COX-2* inhibition can be overcome in endothelial cells because they have a nucleus and, therefore, can generate more of the enzyme. This difference is the basis of low-dose aspirin therapy used to lower the risk of stroke and heart attacks by decreasing formation of thrombi.]

C. Leukotriene synthesis

Arachidonic acid is converted to a variety of linear hydroperoxy (–OOH) acids by a separate pathway involving a family of ***lipoxygenases*** (*LOX*). For example, *5-LOX* converts arachidonic acid to 5-hydroperoxy-6,8,11,14 eicosatetraenoic acid ([5-HPETE]; see Fig. 22.23). 5-HPETE is converted to a series of LT containing four double bonds, the nature of the final products varying according to the tissue. **LT are mediators of allergic response and inflammation. Inhibitors of *5-LOX* and LT-receptor antagonists are used in the treatment of asthma.** [Note: LT synthesis is inhibited by cortisol and not by NSAID. Aspirin-exacerbated respiratory disease is a response to LT overproduction with NSAID use in ~10% of individuals with asthma.]

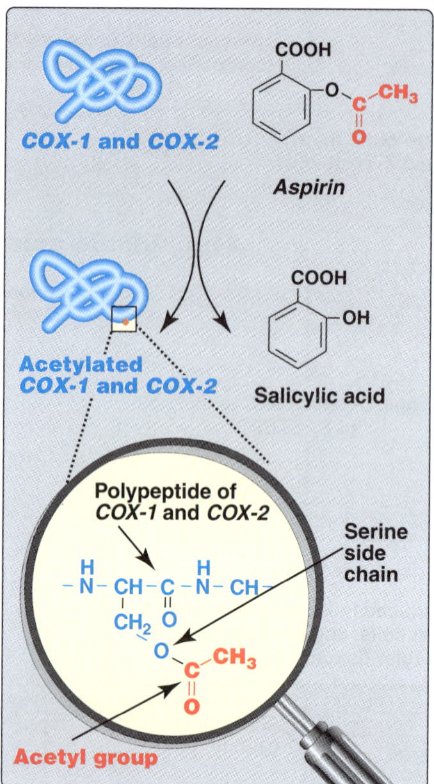

Figure 22.24 Irreversible acetylation of *cyclooxygenase* (*COX*)-*1* and *COX-2* by aspirin.

IX. Medical Perspectives

Neonatal Respiratory Distress Syndrome (RDS) is one of the most common cause of morbidity in preterm infants. The baby presents with apnea, cyanosis, inspiratory stridor and tachycardia. This is due to the inability of the immature lung to synthesize dipalmitoyl lecithin which is the main component of pulmonary surfactant.

Neonatal RDS has to be differentiated from acute lung injury due to pulmonary inflammation **(acute RDS)** and **meconium aspiration syndrome** (aspiration of meconium in utero or during birth due to fetal distress). **Neonatal hyaline membrane disease** is due to lining of alveoli with hyaline membrane leading to respiratory distress. This is also associated with surfactant deficiency and may also have an underlying genetic cause to it.

In pregnancies greater than 30 weeks, the lung maturity may be tested by **lecithin/sphingomyelin (L/S) ratio**. If L/S ratio is less than 2:1, the fetal lungs are likely to be surfactant deficient.

The presence of antiphospholipid antibodies in pregnancies can lead to miscarriage and preterm labor. They may be developed in patients with autoimmune diseases. The antibodies against membrane phospholipid cardiolipin leads to **APL antibody syndrome.**

Anticardiolipin antibodies are found in systemic lupus erythematosus (see Chapter 41, Medical Perspectives).

Nonsteroidal anti-inflammatory drugs **(NSAIDs)** have anti-inflammatory effects and inhibit platelet aggregation, e.g., aspirin acts by inhibition of prostaglandin synthetase by acetylation and ibuprofen acts by inhibiting *cyclooxygenase (COX)* by competing with arachidonate.

Prostaglandins and prostacyclins cause vasodilation and inhibit platelet aggregation. PGF2 is used for medical termination of pregnancy and in inducing labor. PGE is a bronchodilator and is used in aerosols for relieving bronchospasm. PGE2 and PGD2 produce inflammation by increasing capillary permeability. This produces erythema and wheal at site of injury. Prostaglandins inhibit gastric secretions and are therefore useful in treatment of acid peptic disease (see Fig. 22.23).

Inhibitors of eicosanoid synthesis include aspirin, indomethacin, phenylbutazone and selective COX-2 inhibitors like celecoxib (see Fig. 22.25).

X. Chapter Summary

- **Phospholipids** are polar, ionic compounds composed of an **alcohol** (for example, **choline** or **ethanolamine**) attached by a phosphodiester bond either to **diacylglycerol** (**DAG**), producing **phosphatidylcholine** or **phosphatidylethanolamine**, or to the amino alcohol **sphingosine** (Fig. 22.25). Addition of a long-chain fatty acid to sphingosine produces a **ceramide**. Addition of **phosphorylcholine** produces the phospholipid **sphingomyelin**.
- Phospholipids are the predominant lipids of **cell membranes.** Nonmembrane phospholipids serve as components of **lung surfactant** and **bile. Dipalmitoylphosphatidylcholine**, also called **dipalmitoyl lecithin**, is the major lipid component of **lung surfactant**. Insufficient surfactant production causes **respiratory distress syndrome**.
- **Phosphatidylinositol** (**PI**) serves as a reservoir for **arachidonic acid** in membranes. The phosphorylation of membrane-bound PI produces **phosphatidylinositol 4,5-bisphosphate** (**PIP**$_2$). This compound is degraded by ***phospholipase C*** in response to the binding of various neurotransmitters, hormones, and growth factors to membrane **G protein-coupled receptors.** The products of this degradation, **inositol 1,4,5-trisphosphate** (**IP**$_3$) and **DAG,** mediate the mobilization of intracellular **calcium** and the activation of ***protein kinase C***, which act synergistically to evoke cellular responses.
- Specific proteins can be covalently attached via a carbohydrate bridge to membrane-bound PI, forming a **glycosyl phosphatidylinositol** (**GPI**) **anchor**. A deficiency in GPI synthesis in hematopoietic cells results in the hemolytic disease **paroxysmal nocturnal hemoglobinuria.**
- The degradation of phosphoglycerides is performed by ***phospholipases*** found in all tissues and pancreatic juice.
- **Sphingomyelin** is degraded to a ceramide plus phosphorylcholine by the lysosomal enzyme ***sphingomyelinase***, a deficiency of which causes **Niemann-Pick** (**A and B**) **disease**.
- **Glycosphingolipids** are derivatives of **ceramides** to which carbohydrates have been attached. Adding one sugar molecule to the ceramide produces a **cerebroside**, adding an oligosaccharide produces a **globoside**, and adding an acidic **N-acetylneuraminic acid** molecule produces a **ganglioside**. Glycosphingolipids are found predominantly in cell membranes of the **brain** and **peripheral nervous tissue**, with high concentrations in the **myelin sheath**. They are **antigenic**. Glycolipids are degraded in the **lysosomes** by ***acid hydrolases***. A deficiency of any one of these enzymes causes a **sphingolipidosis**, in which a characteristic sphingolipid accumulates.
- **Prostaglandins** (**PG**), **thromboxanes** (**TX**), and **leukotrienes** (**LT**), the **eicosanoids**, are produced in very small amounts in almost all tissues, act locally, and have an extremely short half-life. They serve as mediators of the **inflammatory response**.
- **Arachidonic acid** is the immediate precursor of the predominant class of human PG (those with two double bonds). It is derived by the elongation and desaturation of the essential fatty acid **linoleic acid** and is stored in the membrane as a component of a phospholipid, generally PI. Arachidonic acid is released from the phospholipid by ***phospholipase A***$_2$ (inhibited by **cortisol**).
- Synthesis of the **PG** and **TX** begins with the oxidative cyclization of free arachidonic acid to yield PGH_2 by ***PGH***$_2$ ***synthase*** (or, ***prostaglandin endoperoxide synthase***), an endoplasmic reticular membrane protein that has two catalytic activities: ***fatty acid cyclooxygenase*** (***COX***) and ***peroxidase***. There are two isozymes of *PGH*$_2$ *synthase*: ***COX-1*** (constitutive) and ***COX-2*** (inducible). **Aspirin** irreversibly inhibits both. Opposing effects of PGI_2 and TXA_2 limit clot formation.
- **LT** are linear molecules produced from arachidonic acid by the ***5-lipoxygenase*** (***5-LOX***) pathway. They mediate allergic response. Their synthesis is inhibited by cortisol and not by aspirin.

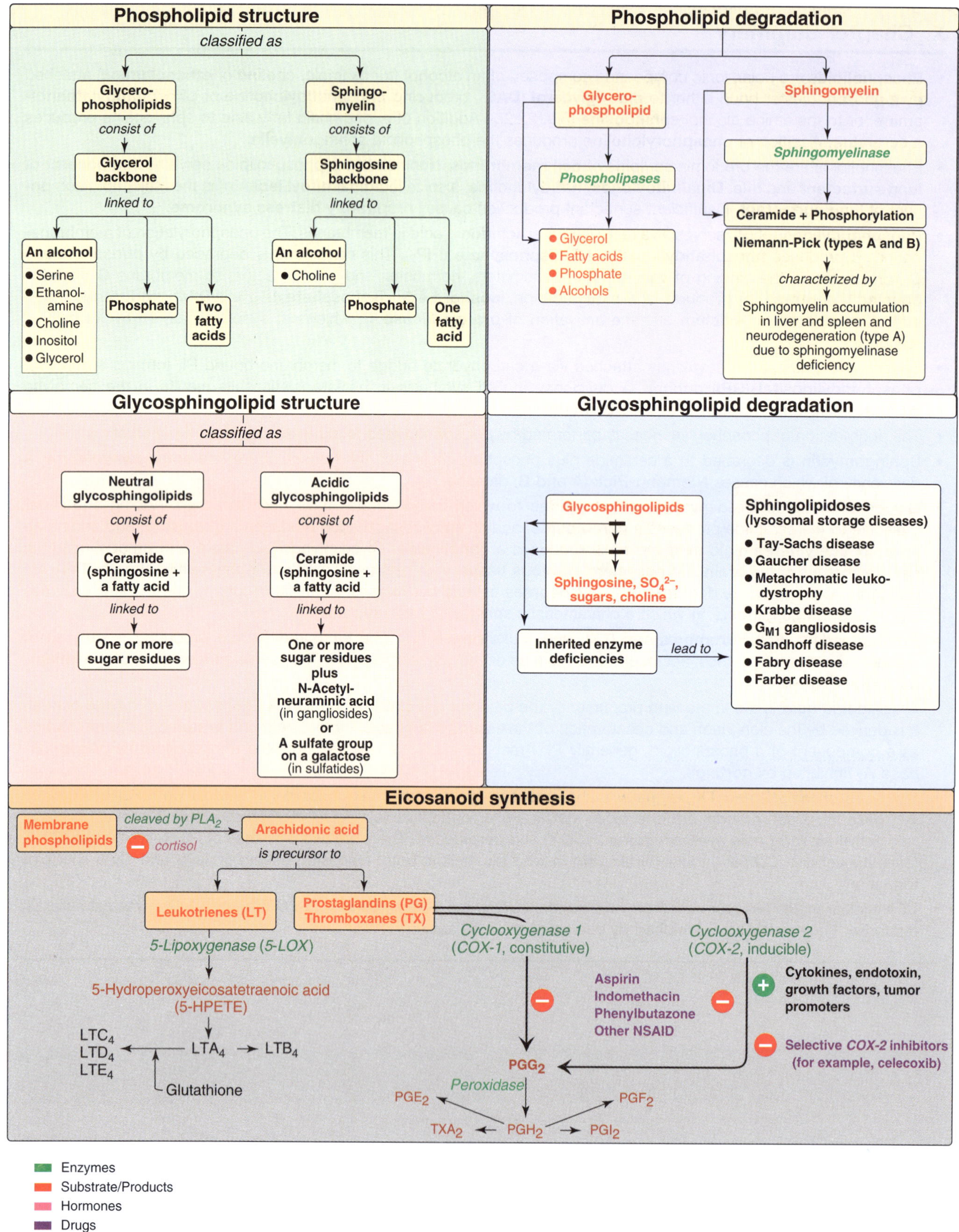

Figure 22.25 Key concept map for phospholipids, glycosphingolipids, and eicosanoids. *PLA_2 = phospholipase A_2*; SO_4^{2-} = sulfate ion; NSAID = nonsteroidal anti-inflammatory drugs.

Study Questions

Short-Answer Questions

Write short notes on:

22.1. *Phospholipases*
22.2. Lecithin/sphingomyelin ratio
22.3. Sphingolipidosis
22.4. Synthesis of Eicosanoids
22.5. Mechanism of action of aspirin
22.6. Lipid rafts
22.7. Sphingomyelin
22.8. Therapeutic uses of prostaglandins
22.9. Inhibitors of eicosanoid synthesis

Long-Answer Questions

22.1. Discuss the synthesis and functions of important prostaglandins, leukotrienes and thromboxanes. Add a note on the drugs acting on these compounds.
22.2. Explain the role of inositol triphosphate and diacylglycerol in cell signaling.

Multiple-Choice Questions

Choose the ONE best answer.

22.1. Which of the following about thromboxane A_2 is incorrect?
A. Produced by COX-1
B. Promotes platelet aggregation
C. Promotes vasodilatation
D. Mobilizes intracellular calcium
Correct answer = C.

22.2. Which of the following is required for synthesis of arachidonic acid from phospholipids?
A. Phospholipase A2
B. Cyclooxygenase 1
C. Cyclooxygenase 2
D. 5-Lipoxygenase
Correct answer = A.

22.3. Which of the following is not correct about cyclooxygenase 2?
A. Made constitutively in most tissues
B. Inducible in response to products of activated inflammatory cells
C. Mediates heat, redness, and swelling of inflammation by increasing prostaglandin synthesis
D. Inhibited by celecoxib
Correct answer = A.

22.4. Thromboxanes are made primarily in
A. Platelets
B. Liver
C. Spleen
D. Lung
Correct answer = A.

22.5. The most common glycosphingolipid found in myelin is
A. Galactocerebrosides
B. Gangliosides
C. Sulfoglycosphingolipids
D. None of the above
Correct answer = A.

22.6. Which of the following is true regarding cerebrosides?
A. Are neutral glycosphingolipids
B. Found primarily in ganglion cells of CNS
C. Have negative charge provided by N-acetylneuraminic acid
D. Contain phosphate
Correct answer = A.

22.7. Arachidonic acid is
A. an ω-6 fatty acid containing 20 carbons and four double bonds fatty acid
B. derived by the desaturation of the linoleic acid, an ω-3 fatty acid
C. an ω-3 fatty acid containing 20 carbons and four double bonds fatty acid
D. an ω-6 fatty acid containing 16 carbons and four double bonds fatty acid
Correct answer = A.

22.8. Which of the following is correct?
A. Tay-Sachs disease is due to deficiency of alpha galactosidase
B. Gaucher disease is due to deficiency of ceramidase
C. Niemann-Pick disease is due to deficiency of sphingomyelinase
D. Fabry's disease is due to deficiency of arylsulfatase A
Correct answer = C.

22.9. Which of the following regarding phospholipids is correct?
A. Phospholipids are polar, ionic, and amphipathic compounds
B. Non-membrane phospholipids are components of lung surfactants
C. Contain sphingosine or diacylglycerol
D. All of the above
Correct answer = D.

22.10. Leukotrienes
A. are made by action of cyclooxygenase 2 on arachidonic acid
B. increase vascular permeability
C. synthesis is inhibited by NSAID
D. are cyclic hydroperoxy acids
Correct answer = B.

Multiple-Choice Questions (NExT/NEET PG)

Choose the ONE best answer.

22.1. Aspirin-exacerbated respiratory disease (AERD) is a severe reaction to nonsteroidal anti-inflammatory drugs (NSAID) characterized by bronchoconstriction 30 minutes to several hours after ingestion. Which of the following statements best explains the symptoms seen in patients with AERD? NSAID:

A. inhibit the activity of the cystic fibrosis transmembrane conductance regulator protein, resulting in thickened mucus that block airways.

B. inhibit *cyclooxygenase* but not *lipoxygenase*, resulting in the flow of arachidonic acid to leukotriene synthesis.

C. activate the *cyclooxygenase* activity of prostaglandin H_2 synthase, resulting in increased synthesis of prostaglandins that promote vasodilation.

D. activate *phospholipases*, resulting in decreased amounts of dipalmitoylphosphatidylcholine and alveolar collapse (atelectasis).

Correct answer = B.

22.2. An infant, born at 28 weeks' gestation, rapidly gave evidence of respiratory distress. Clinical laboratory and imaging results supported the diagnosis of infant respiratory distress syndrome. Which of the following statements about this syndrome is true?

A. It is unrelated to the baby's premature birth.

B. It is a consequence of too few type II pneumocytes.

C. The lecithin/sphingomyelin ratio in the amniotic fluid is likely to be high (>2).

D. The concentration of dipalmitoylphosphatidylcholine in the amniotic fluid would be expected to be lower than that of a full-term baby.

E. It is an easily treated disorder with low mortality.

F. It is treated by administering surfactant to the mother just before she gives birth.

Correct answer = D.

22.3. A 10-year-old boy was evaluated for burning sensations in his feet and clusters of small, red-purple spots on his skin. Laboratory studies revealed protein in his urine. Enzymatic analysis revealed a deficiency of α-galactosidase, and enzyme replacement therapy was recommended. The most likely diagnosis is:

A. Fabry disease

B. Farber disease

C. Gaucher disease

D. Krabbe disease

E. Niemann-Pick disease

Correct answer = A.

22.4. Current medical advice for individuals experiencing chest pain is to call emergency medical services and chew a regular strength, noncoated aspirin. What is the basis for recommending aspirin?

A. Aspirin has an antithrombogenic effect: It prevents formation of blood clots that could occlude heart vessels.

B. Aspirin inhibits thromboxane A_2 synthesis by cyclooxygenase-1 in platelets through irreversible acetylation

C. Inhibits platelet activation and vasoconstriction.

D. Chewing a noncoated aspirin increases the rate of its absorption.

E. All of the above

Correct answer = E.

22.5. Which of the following intermediate products of the cyclooxygenase pathway are unstable?

A. PGG_2 and PGH_2

B. PGG_2 and PGE_2

C. PGE_2 and PGF_{2alpha}

D. PGD_2 and PGH_2

Correct answer = A.

Cholesterol, Lipoprotein, and Steroid Metabolism

23

Learning Objectives

At the end of this chapter, the student will be able to:

- Discuss cholesterol synthesis and degradation
- Explain role of cholesterol in body
- Describe the structure and functions of hormonal steroids
- Explain regulation of lipoprotein metabolism and its associated disorders
- Describe structure and function of lipoproteins and their relationship to atherosclerosis

I. OVERVIEW

Cholesterol, the characteristic steroid alcohol of animal tissues, performs a number of essential functions in the body. For example, **cholesterol is a structural component of all cell membranes, modulating their fluidity, and, in specialized tissues, cholesterol is a precursor of bile acids, steroid hormones, and vitamin D.** Cholesterol is also required for **nerve conduction and signal transduction**. Therefore, it is critically important that the cells of the body be assured an appropriate supply of cholesterol. To meet this need, a complex series of transport, biosynthetic, and regulatory mechanisms has evolved. The liver plays a central role in the regulation of the body's cholesterol homeostasis. For example, cholesterol enters the hepatic cholesterol pool from a number of sources including dietary cholesterol as well as that synthesized de novo by extrahepatic tissues and by the liver itself. Cholesterol is **eliminated from the liver as unmodified cholesterol in the bile, or it can be converted to bile salts that are secreted into the intestinal lumen**. It can also serve as a component of plasma lipoproteins that carry lipids to the peripheral tissues. In humans, the balance between cholesterol influx and efflux is not precise, resulting in a gradual deposition of cholesterol in the tissues, particularly in the endothelial linings of blood vessels. This is a potentially life-threatening occurrence when the lipid deposition leads to plaque formation, causing the narrowing of blood vessels **(atherosclerosis)** and increased risk of cardio-, cerebro-, and peripheral vascular disease. Figure 23.1 summarizes the major sources of liver cholesterol and the routes by which cholesterol leaves the liver.

Summary of the role of cholesterol in the body:

- Cholesterol serves as a structural component of all cell membranes, regulating their fluidity (Chapter 3).
- It acts as a precursor for the synthesis of all classes of steroid hormones, including glucocorticoids (e.g., cortisol), mineralocorticoids (e.g., aldosterone) and sex hormones (e.g., estrogens and androgens).
- Cholesterol is necessary for the synthesis of vitamin D.
- It serves as a precursor for bile acids and bile salts, essential for digestion and absorption of dietary fats.
- Cholesterol plays a role in signal transduction, facilitating communication between cells.
- It is a component of plasma lipoproteins, which transport lipids to extrahepatic tissues and back to the liver.
- Cholesterol is distributed in all body tissues, with particularly high concentrations in the brain and spinal cord.

II. CHOLESTEROL STRUCTURE

Cholesterol is a **very hydrophobic compound**. It consists of four fused hydrocarbon rings (A–D) called the steroid nucleus, and it has an eight-carbon, branched

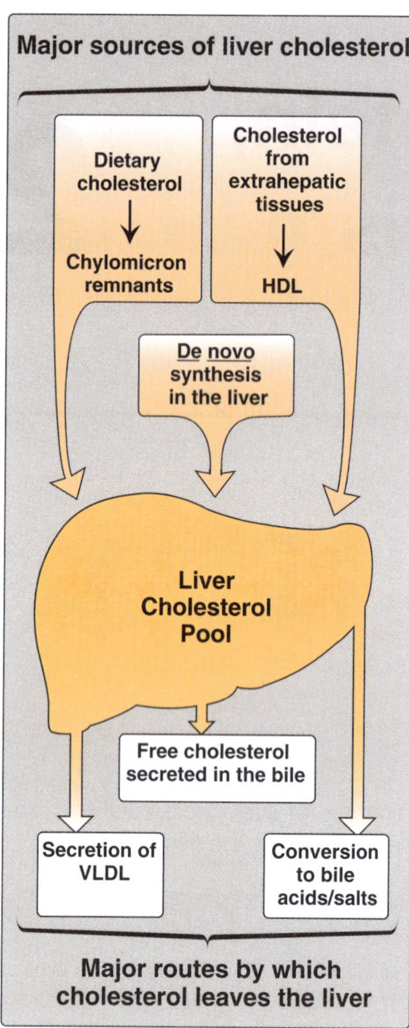

Figure 23.1 Sources of liver cholesterol (influx) and routes by which cholesterol leaves the liver (efflux). HDL and VLDL = high- and very-low-density lipoproteins.

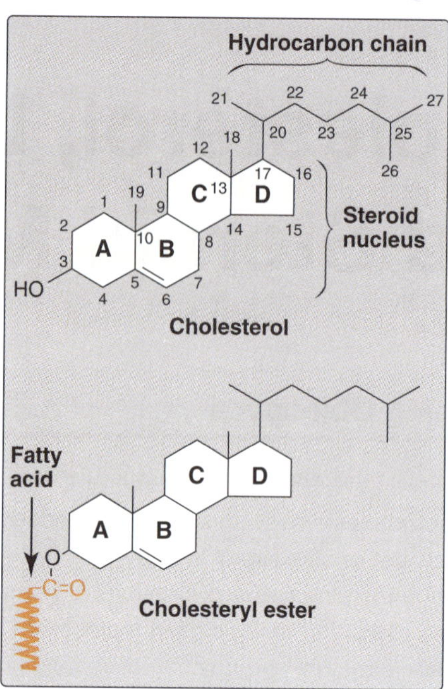

Figure 23.2 Structure of cholesterol and its ester.

hydrocarbon chain attached to carbon 17 of the D ring. Ring A has a hydroxyl group at carbon 3, and ring B has a double bond between carbon 5 and carbon 6 (Fig. 23.2).

A. Sterols

Steroids with 8 to 10 carbon atoms in the side chain at carbon 17 and a hydroxyl group at carbon 3 are classified as sterols. Cholesterol is the major sterol in animal tissues. It arises from de novo synthesis and absorption of dietary cholesterol. Intestinal uptake of cholesterol is mediated by the **Niemann-Pick C1-like 1 protein, the target of the drug ezetimibe** that reduces absorption of dietary cholesterol (see p. 293). [Note: Plant sterols (phytosterols), such as β-sitosterol, are poorly absorbed by humans (5% absorbed as compared to 40% for cholesterol). After entering the enterocytes, they are actively transported back into the intestinal lumen. **Defects in the efflux transporter (ABCG5/8) result in the rare condition of sitosterolemia.** Because some cholesterol is transported back as well, plant sterols reduce the absorption of dietary cholesterol. Daily ingestion of plant sterol esters supplied, for example, in spreads, is one of a number of dietary strategies to reduce plasma cholesterol levels (see p. 507).]

B. Cholesteryl esters

Most plasma cholesterol is in an esterified form (with a fatty acid [FA] attached at carbon 3, as shown in Fig. 23.2), which makes the structure even more hydrophobic than free (nonesterified) cholesterol. Cholesteryl esters are not found in membranes and are normally present only in low levels in most cells. Because of their hydrophobicity, cholesterol and its esters must be transported in association with protein as a component of a lipoprotein particle (see p. 349) or be solubilized by phospholipids and bile salts in the bile (see p. 349).

III. CHOLESTEROL SYNTHESIS

Cholesterol is synthesized by **virtually all tissues in humans**, although liver, intestine, adrenal cortex, and reproductive tissues, including ovaries, testes, and placenta, make the largest contributions to the cholesterol pool. As with FA, **all the carbon atoms in cholesterol are provided by acetyl coenzyme A (CoA), and nicotinamide adenine dinucleotide phosphate (NADPH) provides the reducing equivalents**. The pathway is **endergonic,** being driven by hydrolysis of the high-energy thioester bond of acetyl

CoA and the terminal phosphate bond of ATP. Synthesis requires enzymes in the cytosol, the membrane of the smooth endoplasmic reticulum (SER), and the peroxisome. The pathway is responsive to changes in cholesterol concentration, and regulatory mechanisms exist to balance the rate of cholesterol synthesis against the rate of cholesterol excretion. An imbalance in this regulation can lead to an elevation in circulating levels of plasma cholesterol, with the potential for vascular disease.

A. 3-Hydroxy-3-methylglutaryl coenzyme A synthesis

The first two reactions in the cholesterol biosynthetic pathway are similar to those in the pathway that produces ketone bodies (see Fig. 21.22, p. 314). They result in the production of 3-hydroxy-3-methylglutaryl CoA ([HMG CoA], Fig. 23.3). First, two acetyl CoA molecules condense to form acetoacetyl CoA. Next, a third molecule of acetyl CoA is added by *HMG CoA synthase*, producing HMG CoA, a six-carbon compound. [Note: Liver parenchymal cells contain two isoenzymes of the *synthase*. The cytosolic enzyme participates in cholesterol synthesis, whereas the mitochondrial enzyme functions in the pathway for ketone body synthesis.]

B. Mevalonate synthesis

HMG CoA is reduced to mevalonate by ***HMG CoA reductase*. This is the rate-limiting and key regulated step in cholesterol synthesis.** It occurs in the cytosol, uses two molecules of NADPH as the reducing agent, and releases CoA, making the reaction irreversible (Fig. 23.4). [Note: *HMG CoA reductase* is an integral membrane protein of the SER, with its catalytic domain projecting into the cytosol. Regulation of *reductase* activity is discussed in D. below.]

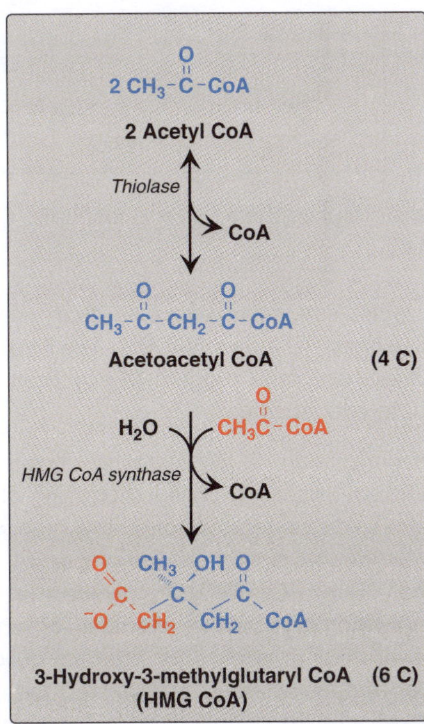

Figure 23.3 Synthesis of HMG CoA. CoA = coenzyme A.

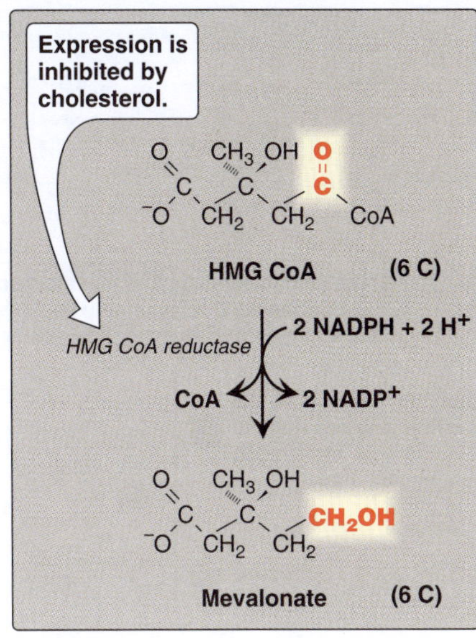

Figure 23.4 Synthesis of mevalonate. HMG CoA = hydroxymethylglutaryl coenzyme A; NADP(H) = nicotinamide adenine dinucleotide phosphate.

C. Cholesterol synthesis from mevalonate

The reactions and enzymes involved in the synthesis of cholesterol from mevalonate are illustrated in Figure 23.5. [Note: The numbers shown in brackets below correspond to numbered reactions shown in this figure.]

1. Mevalonate is converted to 5-pyrophosphomevalonate in two steps, each of which transfers a phosphate group from ATP.

2. A five-carbon isoprene unit, isopentenyl pyrophosphate (IPP), is formed by the decarboxylation of 5-pyrophosphomevalonate. The reaction requires ATP. [Note: IPP is the precursor of a family of molecules with diverse functions, the isoprenoids. Cholesterol is a sterol isoprenoid. Nonsterol isoprenoids include dolichol (see p. 271) and ubiquinone (or, coenzyme Q; see p. 165).]

3. IPP is isomerized to 3,3-dimethylallyl pyrophosphate (DPP).

4. IPP and DPP condense to form 10-carbon geranyl pyrophosphate (GPP).

5. A second molecule of IPP then condenses with GPP to form 15-carbon farnesyl pyrophosphate

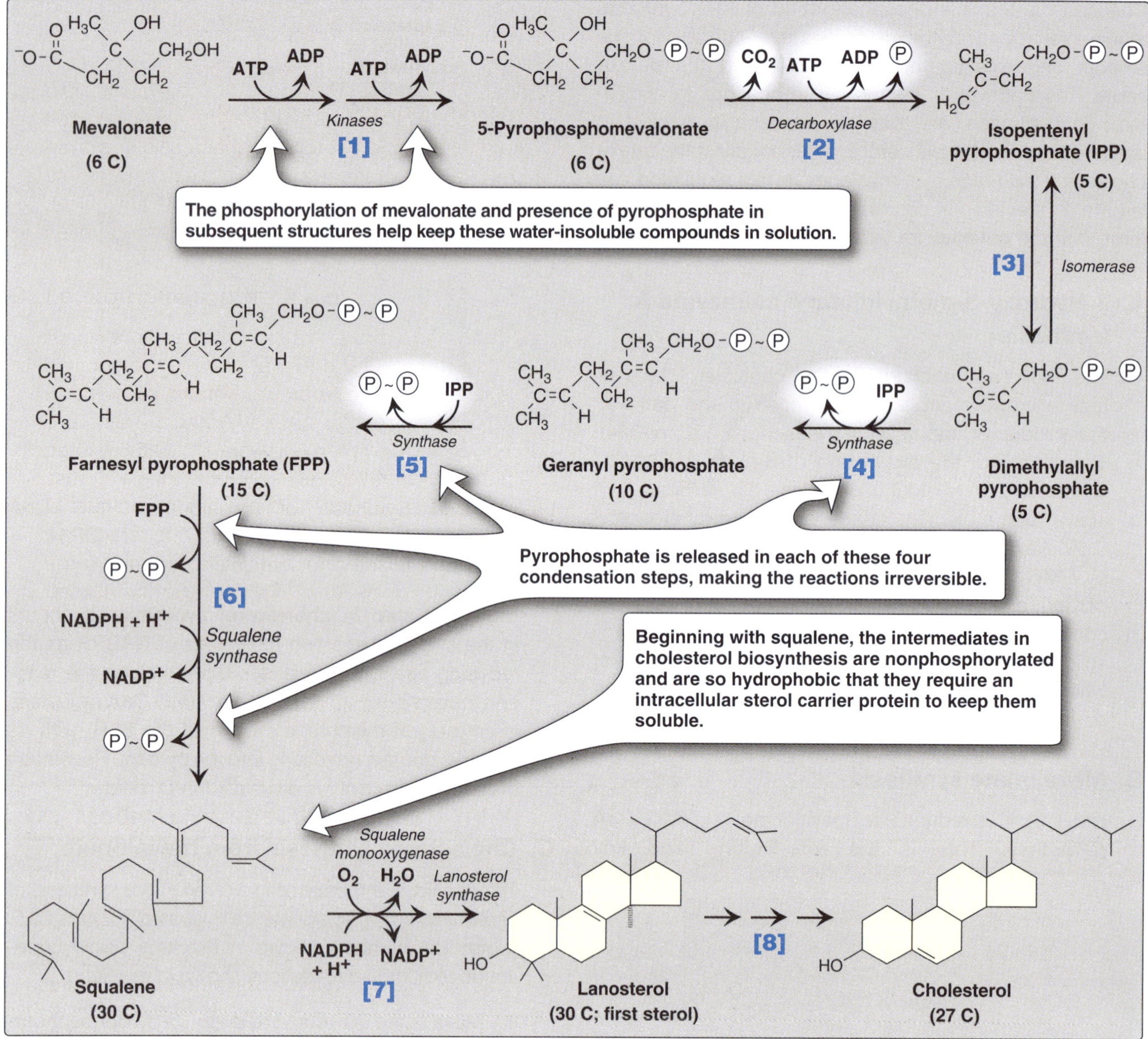

Figure 23.5 Synthesis of cholesterol from mevalonate. ADP = adenosine diphosphate; Ⓟ = phosphate; Ⓟ ~ Ⓟ = pyrophosphate; NADP(H) = nicotinamide adenine dinucleotide phosphate.

(FPP). [Note: Covalent attachment of farnesyl to proteins, a process known as **prenylation,** is one mechanism for anchoring proteins (for example, ras) to the inner face of plasma membranes.]

6. Two molecules of FPP combine, releasing pyrophosphate, and are reduced, forming the 30-carbon compound squalene. [Note: Squalene is formed from six isoprenoid units. Because 3 ATP are hydrolyzed per mevalonate residue converted to IPP, a total of 18 ATP are required to make the polyisoprenoid squalene.]

7. Squalene is converted to the sterol lanosterol by a sequence of two reactions catalyzed by SER-associated enzymes that use molecular oxygen (O_2) and NADPH. The hydroxylation of linear squalene triggers the cyclization of the structure to lanosterol.

8. The conversion of lanosterol to cholesterol is a multistep process involving shortening of the side chain, oxidative removal of methyl groups, reduction of double bonds, and migration of a double bond. **Smith-Lemli-Opitz syndrome (SLOS)**, an autosomal-recessive disorder of cholesterol biosynthesis, is caused by a partial deficiency in *7-dehydrocholesterol-7-reductase*, the enzyme that reduces the double bond in 7-dehydrocholesterol (7-DHC), thereby converting it to cholesterol.

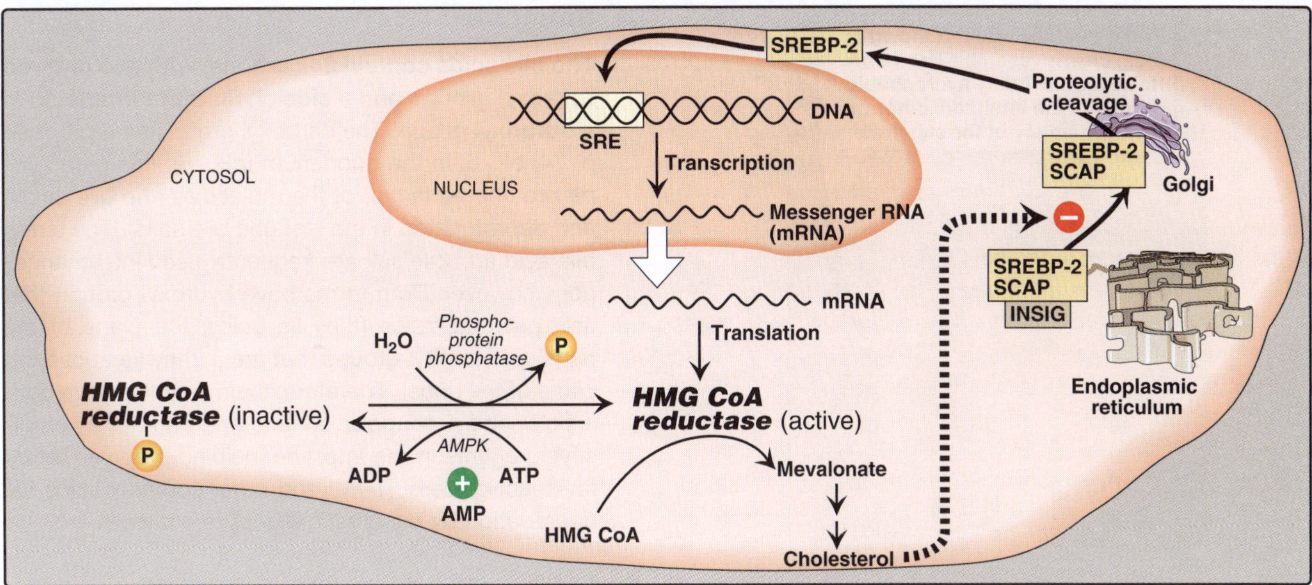

Figure 23.6 Regulation of *hydroxymethylglutaryl coenzyme A* (*HMG CoA*) *reductase*. SRE = sterol regulatory element; SREBP = SRE-binding protein; SCAP = SREBP cleavage–activating protein; *AMPK* = *adenosine monophosphate–activated protein kinase*; ADP = adenosine diphosphate; (P) = phosphate; INSIG = insulin-induced gene protein.

SLOS is one of several multisystem, embryonic malformation syndromes associated with impaired cholesterol synthesis. [Note: 7-DHC is converted to vitamin D_3 in the skin (see p. 540).]

D. Cholesterol synthesis regulation

***HMG CoA reductase* is the major control point for cholesterol biosynthesis and is subject to different kinds of metabolic control.**

1. **Sterol-dependent regulation of gene expression: Expression of the gene for *HMG CoA reductase* is controlled by the trans-acting factor, sterol regulatory element–binding protein-2 (SREBP-2),** which binds DNA at the cis-acting sterol regulatory element (SRE) upstream of the *reductase* gene. Inactive SREBP-2 is an integral protein of the SER membrane and associates with a second SER membrane protein, SREBP cleavage–activating protein (SCAP). When sterol levels in the SER are low, the SREBP-2–SCAP complex moves from the ER to the Golgi. In the Golgi membrane, SREBP-2 is sequentially acted upon by two *proteases*, which generate a soluble fragment that enters the nucleus, binds the SRE, and functions as a transcription factor. This results in increased synthesis of *HMG CoA reductase* and, therefore, increased cholesterol synthesis (Fig. 23.6). However, if sterols are abundant, they bind SCAP at its sterol-sensing domain and induce the binding of SCAP to yet other ER membrane proteins, the insulin-induced gene proteins (INSIG). This results in the retention of the SCAP–SREBP complex in the SER, thereby preventing the activation of SREBP-2 and leading to downregulation of cholesterol synthesis. [Note: SREBP-1c upregulates expression of enzymes involved in FA synthesis in response to insulin (see p. 301).]

2. **Sterol-accelerated enzyme degradation:** The *reductase* itself is a sterol-sensing integral protein of the SER membrane. When sterol levels in the SER are high, the enzyme binds to INSIG proteins. Binding leads to cytosolic transfer, ubiquitination, and proteasomal degradation of the *reductase* (see p. 379).

3. **Sterol-independent phosphorylation/dephosphorylation:** *HMG CoA reductase* activity is controlled covalently **through the actions of *adenosine monophosphate* (*AMP*)*–activated protein kinase* ([*AMPK*], see p. 301) and a *phosphoprotein phosphatase*** (see Fig. 23.6). The phosphorylated form of the enzyme is inactive, whereas the dephosphorylated form is active. [Note: Because *AMPK* is activated by AMP, cholesterol synthesis, like FA synthesis, is decreased when ATP availability is decreased.]

4. **Hormonal regulation:** The activity of *HMG CoA reductase* is controlled hormonally. An **increase in insulin favors dephosphorylation (activation) of the *reductase*,** whereas an increase in glucagon and epinephrine has the opposite effect.

5. **Drug inhibition: The statin drugs (atorvastatin, fluvastatin, lovastatin, pravastatin, rosuvastatin,**

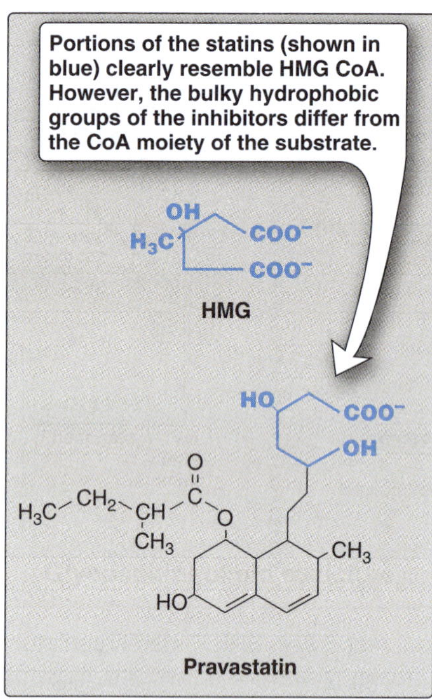

Figure 23.7 Structural similarity of hydroxymethylglutaric acid (HMG) and pravastatin, a clinically useful cholesterol-lowering drug of the statin family. CoA = coenzyme A.

and simvastatin) are structural analogs of HMG CoA and are (or are metabolized to) **reversible, competitive inhibitors of *HMG CoA reductase*** (Fig. 23.7). They are used to decrease plasma cholesterol levels in patients with hypercholesterolemia.

IV. CHOLESTEROL DEGRADATION

Humans cannot metabolize the cholesterol ring structure to carbon dioxide and water. Rather, the **intact steroid nucleus is eliminated from the body by conversion to bile acids and bile salts**, a small percentage of which is excreted in the feces, and by secretion of cholesterol into the bile, which transports it to the intestine for elimination. [Note: Some of the cholesterol in the intestine is modified by bacteria before excretion. The primary compounds made are the isomers coprostanol and cholestanol, which are reduced derivatives of cholesterol. Together with cholesterol, these compounds make up the bulk of neutral fecal sterols.]

V. BILE ACIDS AND BILE SALTS

Bile consists of a watery mixture of organic and inorganic compounds. **Phosphatidylcholine** (PC), or **lecithin** (see p. 321), and **conjugated bile salts** are quantitatively the most important organic components of bile. Bile can either pass directly from the liver, where it is synthesized, into the duodenum through the common bile duct, or be stored in the gallbladder when not immediately needed for digestion.

A. Structure

The bile acids contain 24 carbons, with two or three hydroxyl groups and a side chain that terminates in a carboxyl group. The carboxyl group has a pK_a (see p. 74) of ~6. In the duodenum (pH ~6), this group will be protonated in half of the molecules (the bile acids) and deprotonated in the rest (the bile salts). The terms bile acid and bile salt are frequently used interchangeably, however. Both forms have hydroxyl groups that are α in orientation (they lie below the plane of the rings) and methyl groups that are β (they lie above the plane of the rings). Therefore, the molecules have both a polar and a nonpolar surface and can act as emulsifying agents in the intestine, helping prepare dietary fat (triacylglycerol [TAG]) and other complex lipids for degradation by pancreatic digestive enzymes.

B. Synthesis

Bile acids are synthesized in the liver by a multistep, multiorganelle pathway in which hydroxyl groups are inserted at specific positions on the steroid structure; the double bond of the cholesterol B ring is reduced; and the hydrocarbon chain is shortened by three carbons, introducing a carboxyl group at the end of the chain. The most common resulting compounds, cholic acid (a triol) and chenodeoxycholic acid (a diol), as shown in Figure 23.8, are called primary bile acids. [Note: The **rate-limiting step in bile acid synthesis is the introduction of a hydroxyl group at carbon 7 of the steroid nucleus by *7-α-hydroxylase*, a SER-associated *cytochrome P450 monooxygenase***

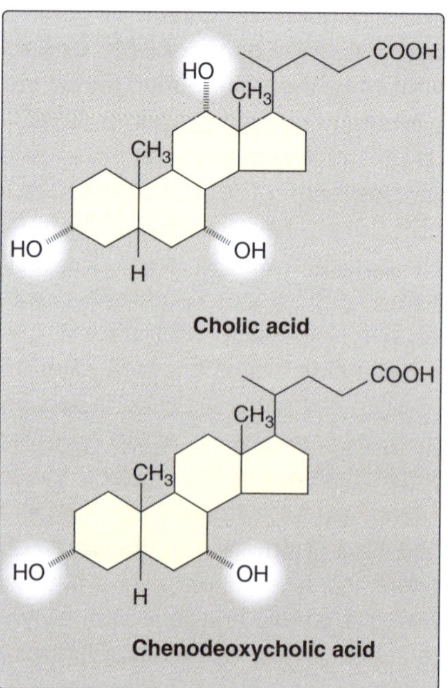

Figure 23.8 Bile acids. [Note: The ionized forms are bile salts.]

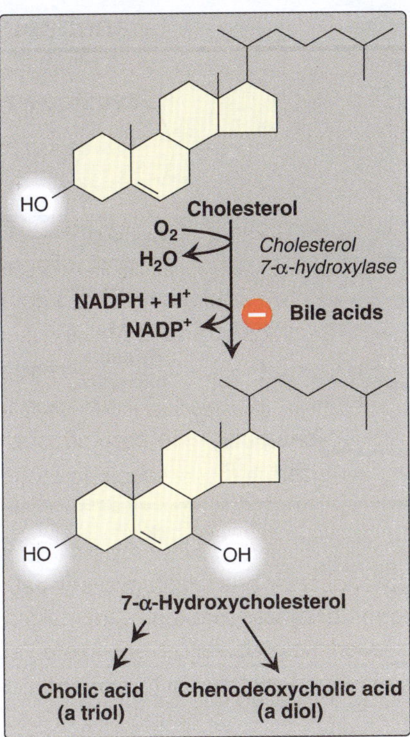

Figure 23.9 Synthesis of the bile acids cholic acid and chenodeoxycholic acid from cholesterol.

found only in liver. Expression of the enzyme is downregulated by bile acids (Fig. 23.9)].

C. Conjugation

Before the bile acids leave the liver, they are **conjugated to a molecule of either glycine or taurine** (an end product of cysteine metabolism) by an amide bond between the carboxyl group of the bile acid and the amino group of the added compound. These new structures include **glycocholic and glycochenodeoxycholic acids and taurocholic and taurochenodeoxycholic acids** (Fig. 23.10). The ratio of glycine to taurine forms in the bile is ~3/1. Addition of glycine or taurine results in the presence of a carboxyl group with a lower pK_a (from glycine) or a sulfonate group (from taurine), both of which are fully ionized (negatively charged) at the alkaline pH of bile. The conjugated, ionized bile salts are more effective detergents than the unconjugated ones because of their enhanced amphipathic nature. Therefore, only the conjugated forms are found in the bile. Individuals with genetic deficiencies in the conversion of cholesterol to bile acids are treated with exogenously supplied chenodeoxycholic acid.

Bile salts provide the only significant mechanism for cholesterol excretion, both as a metabolic product of cholesterol and as a solubilizer of cholesterol in bile.

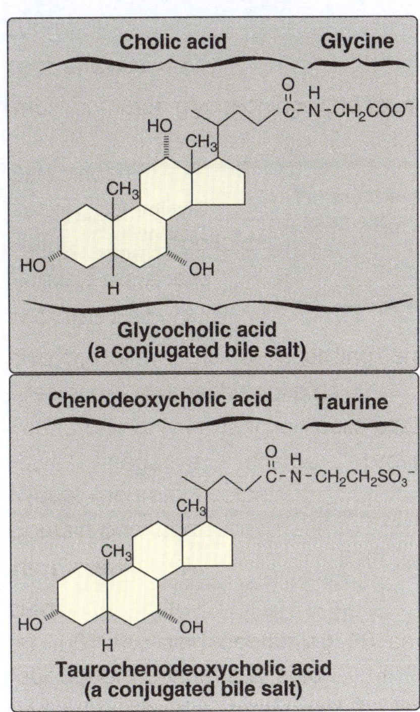

Figure 23.10 Conjugated bile salts. Note "cholic" in the names.

D. Bacterial action on bile salts

Bacteria of the intestinal microbiota (see p. 519) can deconjugate (remove glycine and taurine) bile salts. They can also **dehydroxylate carbon 7, producing secondary bile salts such as deoxycholic acid from cholic acid and lithocholic acid from chenodeoxycholic acid.**

E. Enterohepatic circulation

Bile salts secreted into the intestine are **efficiently reabsorbed (>95%) and reused**. The liver actively secretes bile salts via the bile salt export pump. In the intestine, they are reabsorbed in the terminal ileum via the apical sodium (Na^+)–bile salt cotransporter and returned to the blood via a separate transport system. [Note: Lithocholic acid is only poorly absorbed.] They are efficiently taken up from blood by the hepatocytes via an isoform of the cotransporter and reused. [Note: Albumin binds bile salts and transports them through the blood as was seen with FA (see p. 300).] The continuous process of secretion of bile salts into the bile, their passage through the duodenum where some are deconjugated then dehydroxylated to secondary bile salts, their uptake in the ileum, and their subsequent return to the liver as a mixture of primary and secondary forms is termed the enterohepatic circulation (Fig. 23.11). Between 15 and 30 g of bile salts are secreted from the liver into the duodenum each day, yet only ~0.5 g (<3%) is lost daily in the feces.

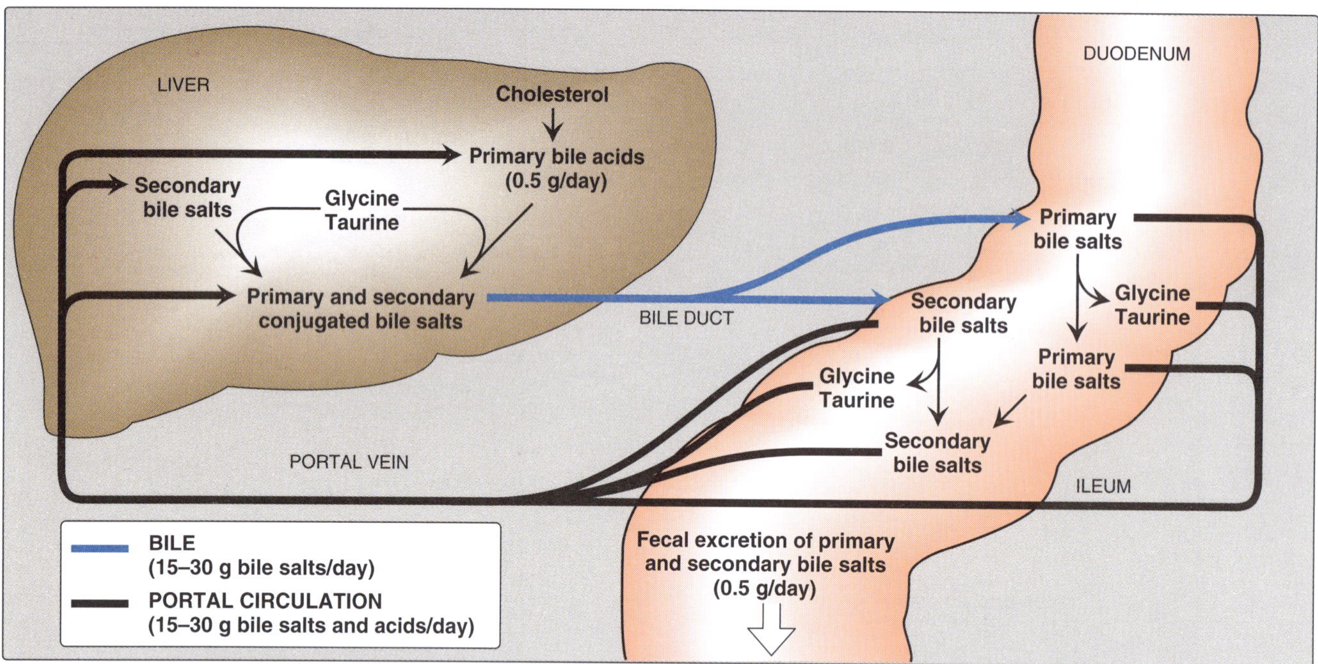

Figure 23.11 Enterohepatic circulation of bile salts. [Note: Ionized bile acids are called bile salts.]

Approximately 0.5 g/day is synthesized from cholesterol in the liver to replace the amount lost. Bile acid sequestrants, such as cholestyramine, bind bile salts in the gut and prevent their reabsorption, thereby promoting their excretion. They are used in the treatment of hypercholesterolemia because the removal of bile salts relieves the inhibition on bile acid synthesis in the liver, thereby diverting additional cholesterol into that pathway. [Note: Dietary fiber also binds bile salts and increases their excretion (see p. 510).]

F. Bile salt deficiency: Cholelithiasis

The movement of cholesterol from the liver into the bile must be accompanied by the simultaneous secretion of phospholipid and bile salts. If this dual process is disrupted and more cholesterol is present than can be solubilized by the bile salts and PC present, the cholesterol may precipitate in the gallbladder, leading to cholesterol gallstone disease or cholelithiasis (Fig. 23.12). This disorder is typically caused by a decrease of bile acids in the bile. Cholelithiasis also may result from increased secretion of cholesterol into bile, as seen with the use of **fibrates (for example, gemfibrozil)** to reduce cholesterol (and TAG) in the blood. **Laparoscopic cholecystectomy** (surgical removal of the gallbladder through a small incision) is currently the treatment of choice. However, for patients who are unable to undergo surgery**, oral administration of chenodeoxycholic acid** to supplement the body's supply of bile acids results in a gradual (months to years) dissolution of the gallstones. [Note: Cholesterol stones account for >85% of cases of cholelithiasis, with bilirubin and mixed stones accounting for the rest].

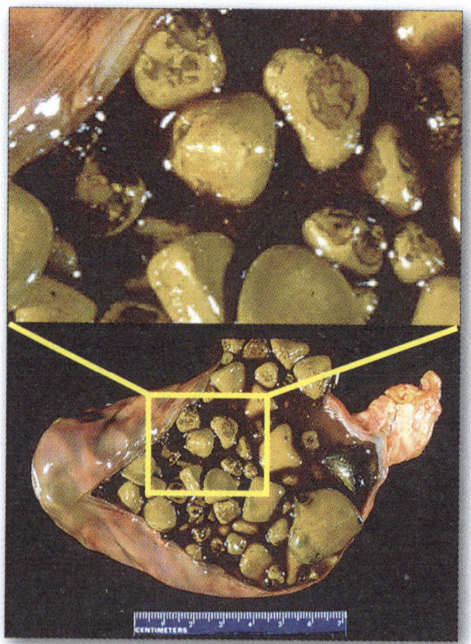

Figure 23.12 Gallbladder with gallstones.

VI. PLASMA LIPOPROTEINS

The plasma lipoproteins are spherical macromolecular complexes of lipids and proteins (apolipoproteins). The lipoprotein particles include chylomicrons, very-low-density lipoproteins (VLDL), low-density lipoproteins (LDL), and high-density lipoproteins (HDL). They differ in

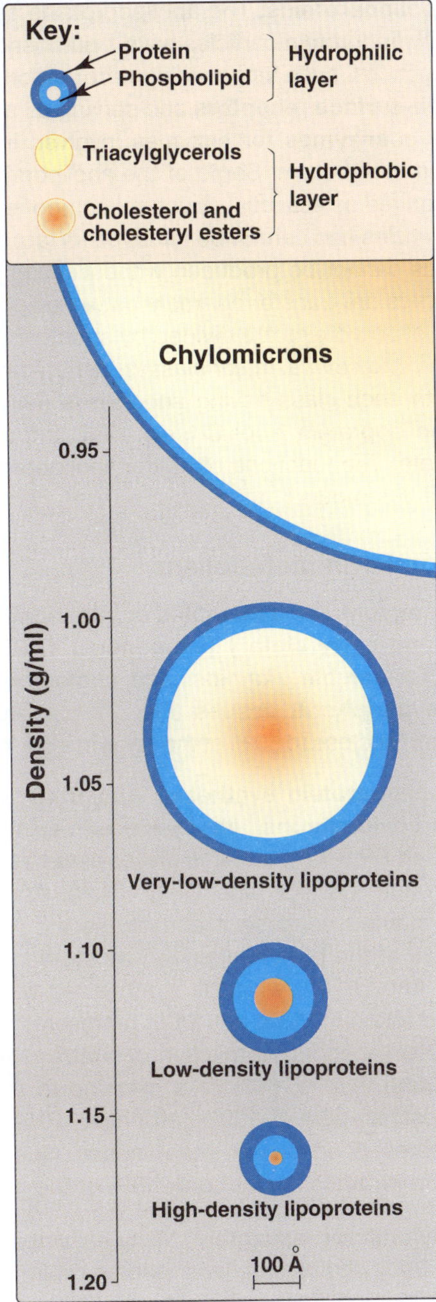

Figure 23.13 Plasma lipoprotein particles exhibit a range of sizes and densities, and typical values are shown. Ring widths approximate the amount of each component. [Note: Although cholesterol and its esters are shown as one component in the center of each particle, physically, cholesterol is on the surface, whereas cholesteryl esters are in the interior.]

lipid and protein composition, size, density (Fig. 23.13), and site of origin. [Note: Because lipoprotein particles constantly interchange lipids and apolipoproteins, the actual apolipoprotein and lipid content of each class of particles is somewhat variable.] **Lipoproteins function both to keep their component lipids soluble as they transport them in the plasma and to provide an efficient mechanism for transporting their lipid contents to (and from) the tissues.** In humans, the transport system is less perfect than in other animals and, as a result, humans experience a gradual deposition of lipid (especially cholesterol) in tissues.

A. Composition

Lipoproteins are composed of a neutral lipid core (containing TAG and cholesteryl esters) surrounded by a shell of amphipathic apolipoproteins, phospholipid, and nonesterified (free) cholesterol (Fig. 23.14). These amphipathic compounds are oriented such that their polar portions are exposed on the surface of the lipoprotein, thereby rendering the particle soluble in aqueous solution. The TAG and cholesterol carried by the lipoproteins are obtained either from the diet (exogenous source) or from de novo synthesis (endogenous source). [Note: The cholesterol (C) content of plasma lipoproteins is now routinely measured in fasting blood. Total C = LDL-C + HDL-C + VLDL-C, where VLDL-C is calculated by dividing TAG by 5 because the TAG/cholesterol ratio is 5/1 in VLDL. The goal value for total cholesterol is <200 mg/dl.]

1. **Size and density:** Chylomicrons are the lipoprotein particles lowest in density and largest in size and that contain the highest percentage of lipid (as TAG) and the lowest percentage of protein. VLDL and LDL are successively denser, having

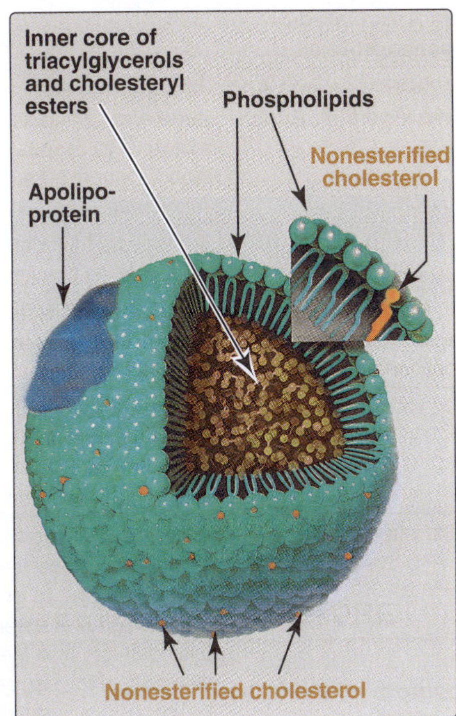

Figure 23.14 Structure of a typical lipoprotein particle.

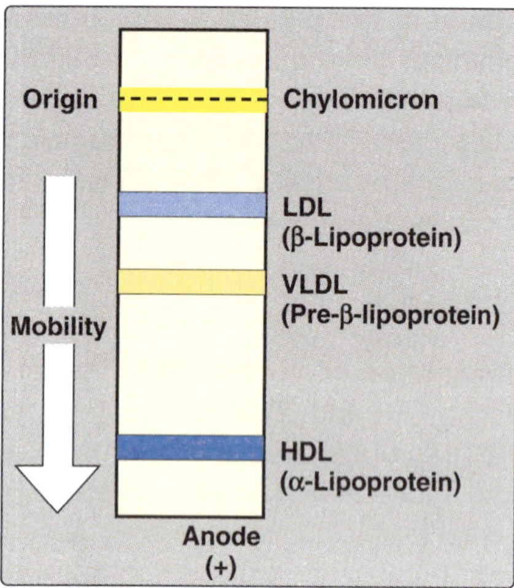

Figure 23.15 Electrophoretic mobility of plasma lipoprotein particles. [Note: The order of low-density lipoprotein (LDL) and very-low-density lipoprotein (VLDL) is reversed if ultracentrifugation is used as the separation technique.] HDL = high-density lipoprotein.

higher ratios of protein to lipid. HDL particles are the smallest and densest. Plasma lipoproteins can be separated on the basis of their electrophoretic mobility, as shown in Figure 23.15, or on the basis of their density by ultracentrifugation.

Functions of lipoproteins and associated apolipoproteins

Lipoprotein and Associated Apolipoproteins	Functions
Chylomicrons with Apo B-48, Apo C-II, ApoE	Transports dietary triglyceride and cholesterol from intestine to tissues Apo C-II activates *lipoprotein lipase*
VLDL with Apo B-100, Apo C-II, ApoE	Transports triglyceride from liver to tissues
IDL (intermediate-density lipoprotein) with ApoE and Apo B-100	VLDL remnant which takes up cholesterol from HDL to become LDL
LDL with Apo B-100	Delivers cholesterol into cells and then goes to liver via Apo B-100 recognizing LDL receptor in liver
HDL with Apo A-1	Picks up cholesterol accumulating in blood vessels, activates LCAT to convert it to cholesterol esters and then delivers it to liver and steroidogenic tissues via scavenger receptor SR-B1

2. **Apolipoproteins:** The apolipoproteins associated with lipoprotein particles have a number of diverse functions, such as providing **recognition sites for cell-surface receptors** and serving as **activators or coenzymes** for enzymes involved in lipoprotein metabolism. Some of the apolipoproteins are required as essential structural components of the particles and cannot be removed (in fact, the particles cannot be produced without them), whereas others are transferred freely between lipoproteins. Apolipoproteins are divided by structure and function into several major classes, denoted by letters, with each class having subclasses (for example, apolipoprotein [apo] C-I, apo C-II, and apo C-III). [Note: The functions of all the apolipoproteins are not yet known.]

B. Chylomicron metabolism

Chylomicrons are assembled in intestinal mucosal cells and carry dietary (exogenous) TAG, cholesterol, fat-soluble vitamins, and cholesteryl esters to the peripheral tissues (Fig. 23.16). [Note: TAG account for close to 90% of the lipids in a chylomicron.]

1. **Apolipoprotein synthesis: Apo B-48 is unique to chylomicrons.** Its synthesis begins on the rough ER (RER), and it is glycosylated as it moves through the RER and Golgi. [Note: Apo B-48 is so named because it constitutes the N-terminal 48% of the protein encoded by the gene for apo B. Apo B-100, which is synthesized by the liver and found in VLDL and LDL, represents the entire protein encoded by this gene. **Posttranscriptional editing** (see p. 662) of a cytosine to a uracil in intestinal apo B-100 messenger RNA (mRNA) creates a nonsense (stop) codon (see p. 636), allowing translation of only 48% of the mRNA.]

2. **Chylomicron assembly:** Many enzymes involved in TAG, cholesterol, and phospholipid synthesis are located in the SER. Assembly of the apolipoprotein and lipid into chylomicrons requires microsomal triglyceride transfer protein ([MTP], see p. 294), which loads apo B-48 with lipid. This occurs before transition from the ER to the Golgi, where the particles are packaged in secretory vesicles. These fuse with the plasma membrane releasing the lipoproteins, which then enter the lymphatic system and, ultimately, the blood. [Note: Chylomicrons leave the lymphatic system via the thoracic duct that empties into the left subclavian vein.]

3. **Nascent chylomicron modification:** The particle released by the intestinal mucosal cell is called a nascent chylomicron because it is functionally

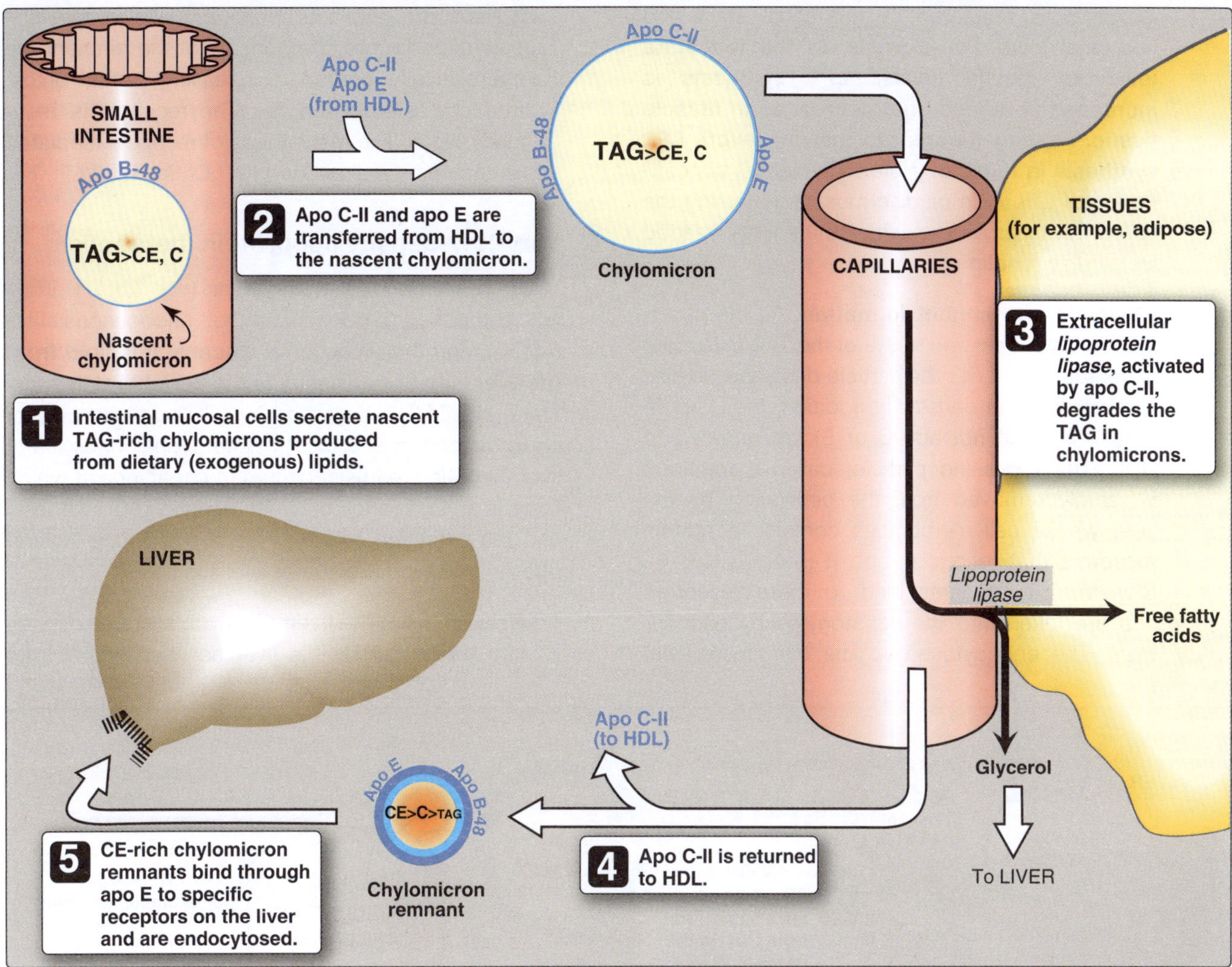

Figure 23.16 Metabolism of chylomicrons. Apo B-48, apo C-II, and apo E are apolipoproteins found as components of plasma lipoprotein particles. The particles are not drawn to scale (see Fig. 23.13 for details of their size and density). TAG = triacylglycerol; C = cholesterol; CE = cholesteryl ester; HDL = high-density lipoprotein.

incomplete. When it reaches the plasma, the particle is rapidly modified, receiving apo E (which is recognized by hepatic receptors) and apo C. The latter includes **apo C-II, which is necessary for the activation of *lipoprotein lipase* (*LPL*),** the enzyme that degrades the TAG contained in the chylomicron. The source of these apolipoproteins is circulating HDL (see Fig. 23.16). [Note: Apo C-III on TAG-rich lipoproteins inhibits *LPL*.]

4. **Triacylglycerol degradation by lipoprotein lipase: *LPL* is an extracellular enzyme that is anchored to the capillary walls of most tissues but predominantly those of adipose tissue and cardiac and skeletal muscle**. The adult liver does not express this enzyme. [Note: A *hepatic lipase* is found on the surface of endothelial cells of the liver. It plays a role in TAG degradation in chylomicrons and VLDL and is important in HDL metabolism (see p. 358).] *LPL*, activated by apo C-II on circulating chylomicrons, hydrolyzes the TAG in these particles to FA and glycerol. The FA are stored (in adipose) or used for energy (in muscle). The glycerol is taken up by the liver, converted to dihydroxyacetone phosphate (an intermediate of glycolysis), and used in lipid synthesis or gluconeogenesis. [Note: **Patients with a deficiency of *LPL* or apo C-II (type I hyperlipoproteinemia or familial chylomicronemia) show a dramatic accumulation (≥1,000 mg/dl) of chylomicron-TAG in the plasma (hypertriacylglycerolemia)** even in the fasted state. They are at increased risk for acute pancreatitis. Treatment is the reduction of dietary fat.]

5. **Lipoprotein lipase expression:** *LPL* is synthesized by adipose tissue and by cardiac and skeletal muscle. Expression of the tissue-specific

isozymes is regulated by nutritional state and hormonal level. For example, **in the fed state (elevated insulin levels), *LPL* synthesis is increased in adipose but decreased in muscle tissue. Fasting (decreased insulin) favors *LPL* synthesis in muscle**. [Note: The highest concentration of *LPL* is in cardiac muscle, reflecting the use of FA to provide much of the energy needed for cardiac function.]

6. **Chylomicron remnant formation:** As the chylomicron circulates, and >90% of the TAG in its core is degraded by *LPL*, the particle decreases in size and increases in density. In addition, the C apolipoproteins (but not apo B or E) are returned to HDL. The remaining particle, called a remnant, is rapidly removed from the circulation by the liver, whose cell membranes contain lipoprotein receptors that recognize apo E (see Fig. 23.16). Chylomicron remnants bind to these receptors and are taken into the hepatocytes by endocytosis. The endocytosed vesicle then fuses with a lysosome, and the apolipoproteins, cholesteryl esters, and other components of the remnant are hydrolytically degraded, releasing amino acids, free cholesterol, and FA. The receptor is recycled. [Note: The mechanism of receptor-mediated endocytosis is illustrated for LDL in Fig. 23.20.]

C. Very-low-density lipoprotein metabolism

VLDL are produced in the **liver** (Fig. 23.17). They are composed predominantly of endogenous TAG (~60%), and their function is to **carry this lipid from the liver (site of synthesis) to the peripheral tissues**. There, the TAG is degraded by *LPL*, as discussed for chylomicrons (see p. 351). [Note: Nonalcoholic fatty liver (hepatic steatosis) occurs in conditions in which there is an imbalance between hepatic TAG synthesis and the secretion of VLDL. Such conditions include obesity and type 2 diabetes mellitus (see p. 475).]

1. **Release from the liver:** VLDL are secreted directly into the blood by the liver as nascent particles

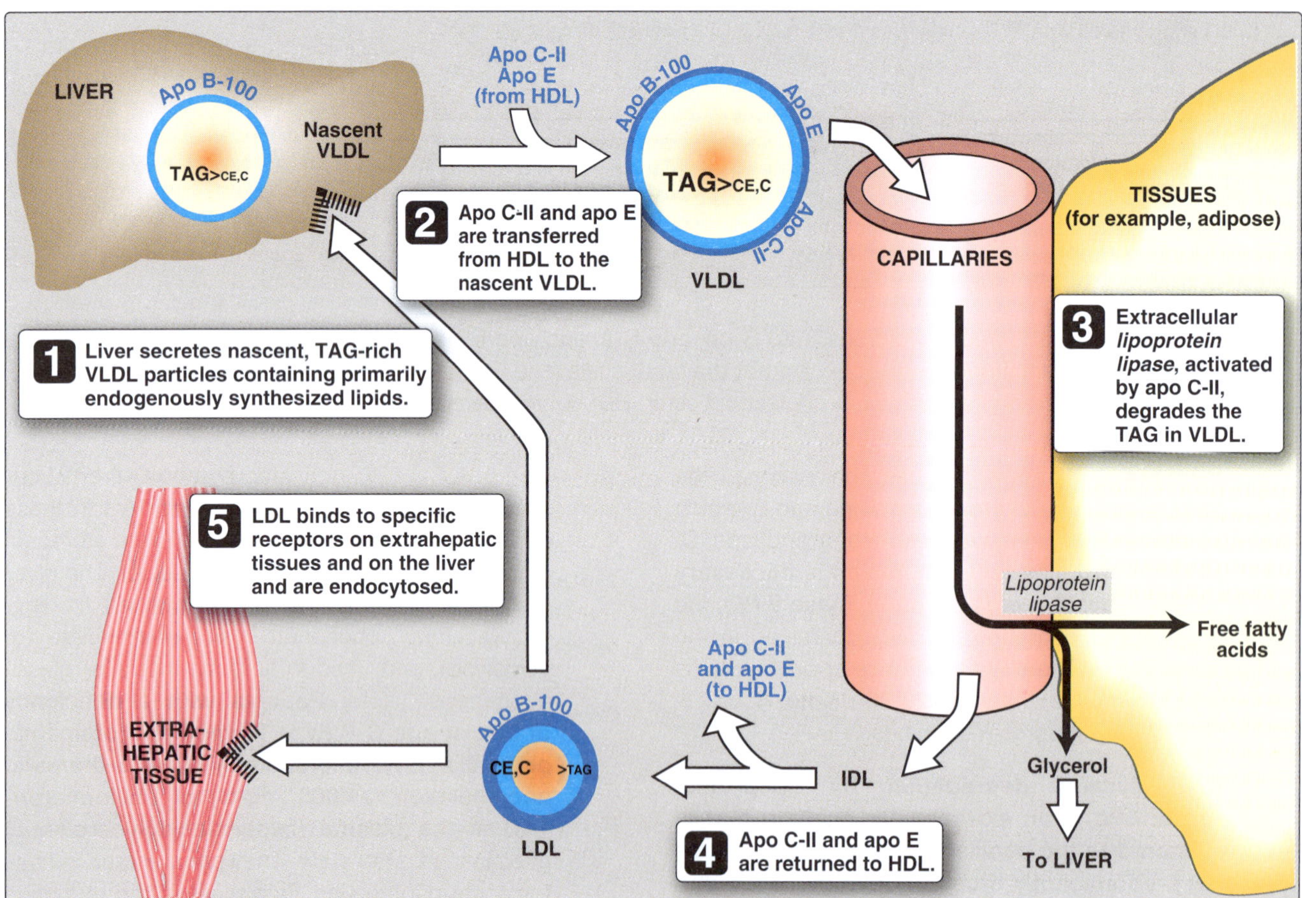

Figure 23.17 Metabolism of very-low-density lipoprotein (VLDL) and low-density lipoprotein (LDL) particles. Apo B-100, C-II, and E are apolipoproteins found as components of plasma lipoprotein particles. The particles are not drawn to scale (see Fig. 23.13 for details of their size and density). [Note: IDL can also be taken up by liver.] TAG = triacylglycerol; HDL and IDL = high- and intermediate-density lipoproteins; C = cholesterol; CE = cholesteryl ester.

containing apo B-100. They must obtain apo C-II and apo E from circulating HDL (see Fig. 23.17). As with chylomicrons, apo C-II is required for activation of *LPL*. [Note: Abetalipoproteinemia is a rare hypolipoproteinemia caused by a defect in MTP, leading to an inability to load apo B with lipid. Consequently, few VLDL or chylomicrons are formed, and TAG accumulates in the liver and intestine. Absorption of fat-soluble vitamins is decreased. LDL are low.]

2. **Modification in the circulation:** As VLDL pass through the circulation, TAG is degraded by *LPL*, causing the VLDL to decrease in size and become denser. Surface components, including the C and E apolipoproteins, are returned to HDL, but the particles retain apo B-100. Additionally, some TAG are transferred from VLDL to HDL in an exchange reaction that concomitantly transfers cholesteryl esters from HDL to VLDL. This exchange is accomplished by ***cholesteryl ester transfer protein*** **(*CETP*),** as shown in Figure 23.18.

3. **Conversion to low-density lipoproteins:** With these modifications, the VLDL is converted in the plasma to LDL. **Intermediate-density lipoproteins (IDL) of varying sizes are formed during this transition.** IDL can also be taken up by liver cells through receptor-mediated endocytosis that uses apo E as the ligand. Apo E is normally present in three isoforms, E-2 (the least common), E-3 (the most common), and E-4. Apo E-2 binds poorly to receptors, and patients who are homozygotic for apo E-2 are deficient in the clearance of IDL and chylomicron remnants. These individuals have familial type III hyperlipoproteinemia (familial dysbetalipoproteinemia or broad beta disease), with hypercholesterolemia and premature atherosclerosis. [Note: The apo E-4 isoform confers increased susceptibility to an earlier age of onset of the late-onset form of Alzheimer disease. The effect is dose dependent, with homozygotes being at greatest risk. Estimates of the risk vary.]

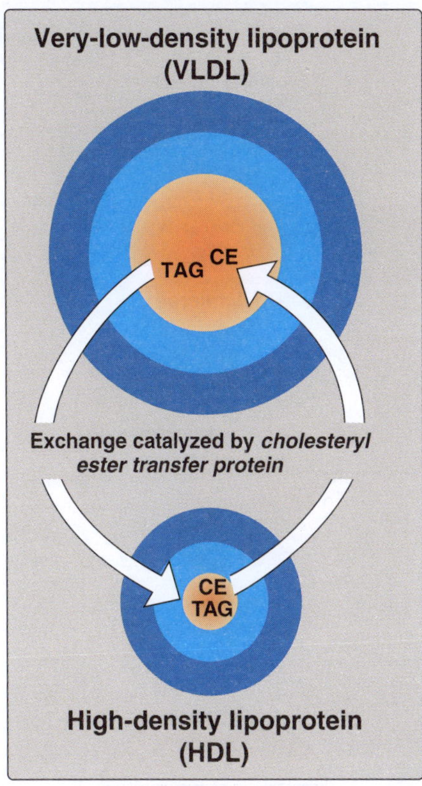

Figure 23.18 Transfer of cholesteryl ester (CE) from HDL to VLDL in exchange for triacylglycerol (TAG).

D. Low-density lipoprotein metabolism

LDL particles contain much less TAG than their VLDL predecessors and have a high concentration of cholesterol and cholesteryl esters (Fig. 23.19). About 70% of plasma cholesterol is in LDL.

1. **Receptor-mediated endocytosis:** The primary function of LDL particles is to provide cholesterol to the peripheral tissues (or return it to the liver).

Hypolipoproteinemias

Name	Laboratory Findings	Remarks
Familial hypobetalipoproteinemia	Decreased LDL, Apo B-100, and cholesterol; normal triacylglycerol	Decreased risk of CAD
Abetalipoproteinemia	Decrease in VLDL, LDL, Apo B-48, Apo B-100, cholesterol and triglycerides	Malabsorption states Mental retardation Physical milestones delayed
Hypoalphalipoproteinemia	Decrease in HDL and Apo A-1 with normal cholesterol and triglycerides	Increased risk of CAD
Familial alpha lipoprotein deficiency	Decrease in HDL and Apo A-1 with normal cholesterol and triglycerides	Increased risk of CAD

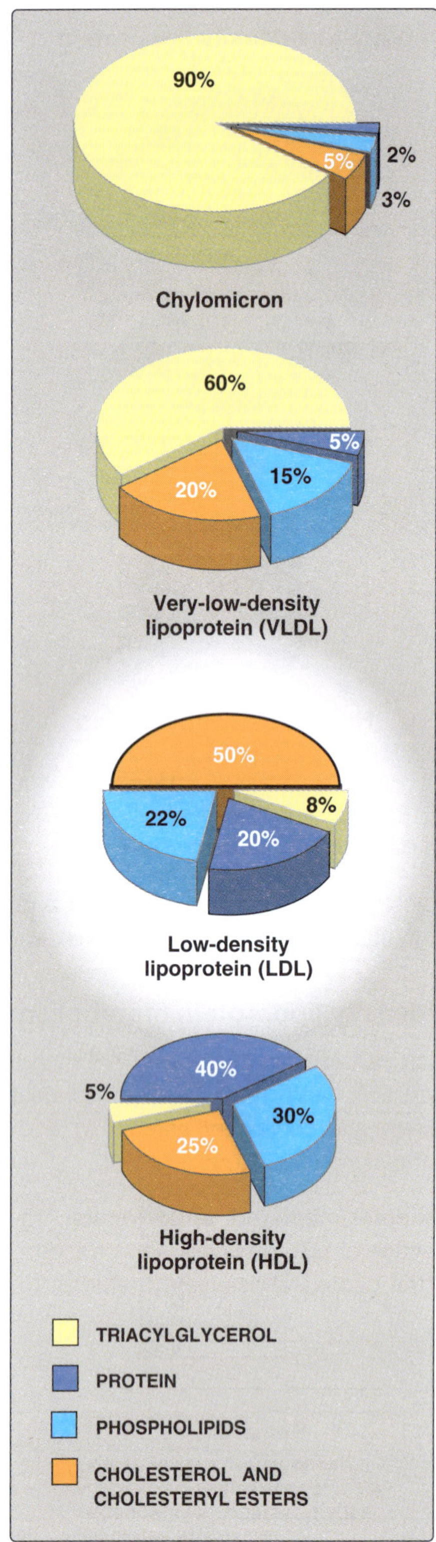

Figure 23.19 Composition of the plasma lipoprotein particles. Note the high concentration of cholesterol and cholesteryl esters in LDL.

They do so by binding to plasma membrane LDL receptors that recognize apo B-100 (but not apo B-48). Because these LDL receptors can also bind apo E, they are known as apo B-100/apo E receptors. A summary of the uptake and degradation of LDL particles is presented in Figure 23.20. [Note: The numbers in brackets below refer to corresponding numbers on that figure.] A similar mechanism of receptor-mediated endocytosis is used for the uptake and degradation of chylomicron remnants and IDL by the liver.

a. LDL receptors are negatively charged glycoproteins that are clustered in pits on cell membranes. The cytosolic side of the pit is coated with the protein clathrin, which stabilizes the pit.

b. After binding, the LDL–receptor complex is endocytosed. [Note: **Defects in the synthesis of functional LDL receptors causes a significant elevation in plasma LDL-C. Patients with such deficiencies have type IIa hyperlipidemia (familial hypercholesterolemia [FH]) and premature atherosclerosis.** Autosomal dominant hypercholesterolemia can also be caused by defects in apo B-100 that reduce its binding to the receptor and by **increased activity of a *protease, proprotein convertase subtilisin/kexin type 9* (*PCSK9*)**, which promotes internalization and lysosomal degradation of the receptor. *PCSK9* inhibitors are now available for the treatment of hypercholesterolemia.]

c. The vesicle containing LDL loses its clathrin coat and fuses with other similar vesicles, forming larger vesicles called endosomes.

d. The pH of the endosome falls (due to the proton-pumping activity of endosomal *ATPase*), which allows separation of the LDL from its receptor. The receptors then migrate to one side of the endosome, whereas the LDL stay free within the lumen of the vesicle.

e. The receptors can be recycled, whereas the lipoprotein remnants in the vesicle are transferred to lysosomes and degraded by lysosomal *acid hydrolases*, releasing free cholesterol, amino acids, FA, and phospholipids. These compounds can be reutilized by the cell. [Note: Lysosomal storage diseases result from rare autosomal-recessive deficiencies in the ability to hydrolyze lysosomal cholesteryl esters (Wolman disease) or to transport free cholesterol out of the lysosome (Niemann-Pick disease, type C).]

2. **Endocytosed cholesterol and cholesterol homeostasis:** The chylomicron remnant–, IDL-, and LDL-derived cholesterol affects cellular cholesterol content in several ways (see Fig. 23.20). First,

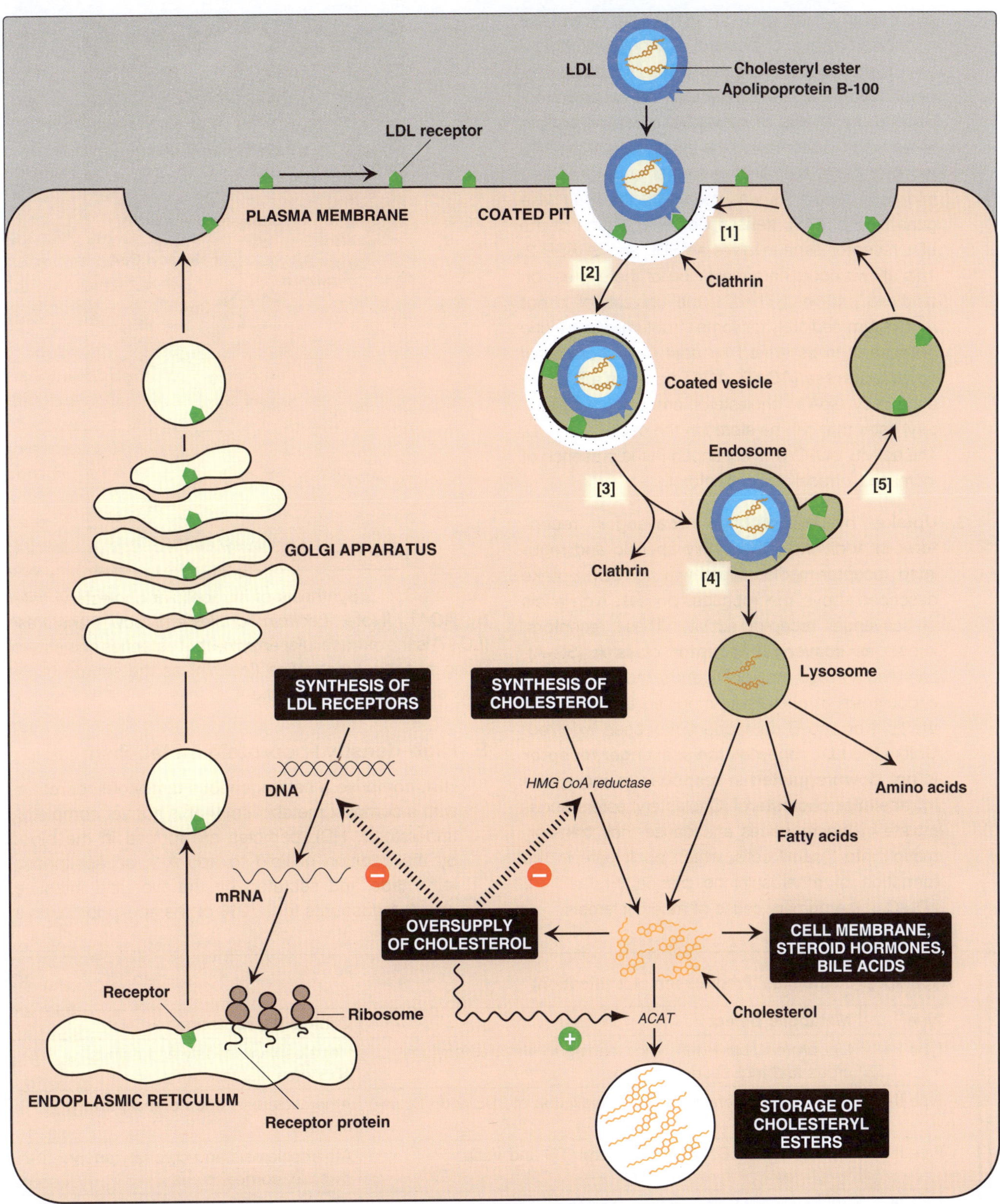

Figure 23.20 Cellular uptake and degradation of low-density lipoprotein (LDL) particles. [Note: Oversupply of cholesterol accelerates the degradation of *HMG CoA reductase*. It also decreases transcription of its gene as seen with the LDL receptor.] *ACAT* = *acyl CoA:cholesterol acyltransferase*; HMG CoA = hydroxymethylglutaryl coenzyme A; mRNA = messenger RNA.

expression of the gene for *HMG CoA reductase* is inhibited by high cholesterol, and de novo cholesterol synthesis decreases as a result. Additionally, degradation of the *reductase* is accelerated. Second, **synthesis of new LDL receptor protein** is reduced by decreasing the expression of the LDL receptor gene, thus limiting further entry of LDL-C into cells. [Note: As was seen with the *reductase* gene (see p. 345), transcriptional regulation of the LDL receptor gene involves an SRE and SREBP-2. This allows coordinate regulation of the expression of these proteins.] Third, if the cholesterol is not required immediately for some structural or synthetic purpose, it is **esterified by *acyl CoA:cholesterol acyltransferase* (*ACAT*).** *ACAT* transfers a FA from a fatty acyl CoA to cholesterol, producing a cholesteryl ester that can be stored in the cell (Fig. 23.21). The activity of *ACAT* is enhanced in the presence of increased intracellular cholesterol.

3. **Uptake by macrophage scavenger receptors:** In addition to the highly specific and regulated receptor-mediated pathway for LDL uptake described above, macrophages possess high levels of scavenger receptor activity. These receptors, known as **scavenger receptor class A (SR-A)**, can bind a broad range of ligands and mediate the endocytosis of chemically modified LDL in which the lipid or apo B component has been oxidized. **Unlike the LDL receptor, the scavenger receptor is not downregulated in response to increased intracellular cholesterol**. Cholesteryl esters accumulate in macrophages and cause their transformation into "foam" cells, which participate in the formation of atherosclerotic plaque (Fig. 23.22). LDL-C is the primary cause of atherosclerosis.

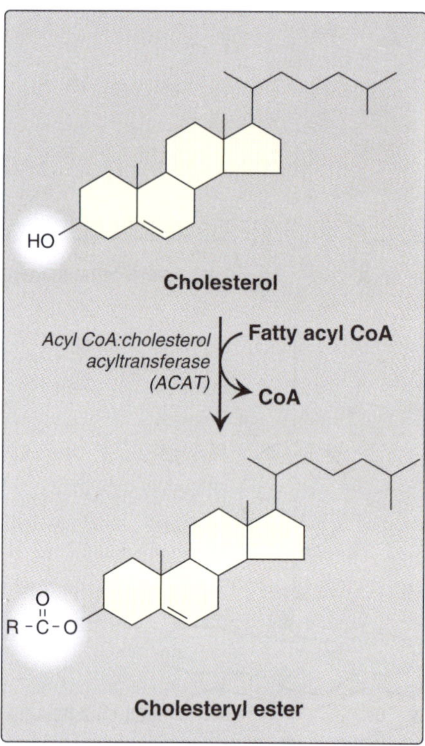

Figure 23.21 Synthesis of intracellular cholesteryl ester by *ACAT*. [Note: *Lecithin: cholesterol acyl transferase* (*LCAT*) is the extracellular enzyme that esterifies cholesterol using phosphatidylcholine (lecithin) as the source of the fatty acid.] CoA = coenzyme A.

E. High-density lipoprotein metabolism

HDL comprise a heterogeneous family of lipoproteins with a complex metabolism that is not yet completely understood. HDL particles are formed in the blood by the addition of lipid to apo A-1, an apolipoprotein made and secreted by the liver and intestine. Apo A-1 accounts for ~70% of the apolipoproteins in

Hyperlipoproteinemias (Fredrickson classification)

Type	Metabolic Defect	Clinical Presentations
Type I	*Lipoprotein lipase* deficiency leading to elevated chylomicrons and triglycerides in blood	Hepatosplenomegaly and pain in abdomen, xanthomas
Type IIA	LDL receptor defect leading to elevation of LDL and cholesterol in blood	Atherosclerosis and coronary artery disease
Type IIB	Increase in ApoB, LDL, cholesterol, TG and VLDL	Atherosclerosis and coronary artery disease, corneal arcus
Type III	Abnormal ApoE and increase in Apo C-II leading to elevation in chylomicrons, cholesterol, TG and VLDL	Vascular diseases
Type IV	Increase in Apo C-II and excess production of VLDL, increase in cholesterol and TG	Coronary artery disease
Type V	Increase in chylomicrons, TG and VLDL because of secondary causes	Heart diseases

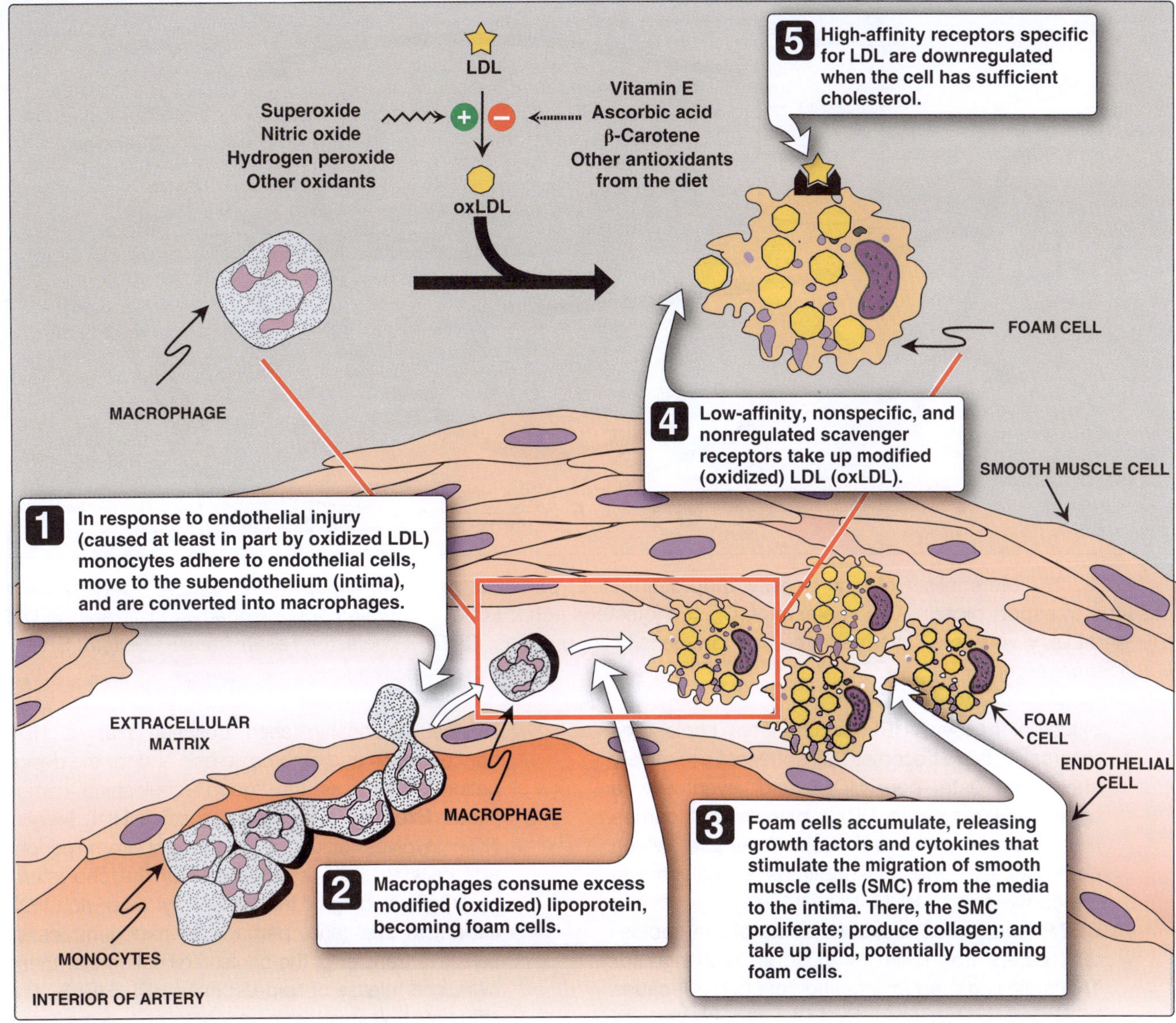

Figure 23.22 Role of oxidized low-density lipoprotein (LDL) particles in plaque formation in an arterial wall.

HDL. HDL perform a number of important functions, including the following.

1. **Apolipoprotein supply:** HDL particles serve as a circulating reservoir of **apo C-II** (the apolipoprotein that is transferred to VLDL and chylomicrons and is an activator of *LPL*) and **apo E** (the apolipoprotein required for the receptor-mediated endocytosis of IDL and chylomicron remnants).

2. **Nonesterified cholesterol uptake:** Nascent HDL are disc-shaped particles containing primarily phospholipid (largely PC) and apo A, C, and E. They take up cholesterol from nonhepatic (peripheral) tissues and return it to the liver as cholesteryl esters (Fig. 23.23). [Note: HDL particles are excellent acceptors of nonesterified cholesterol as a result of their high concentration of phospholipids, which are important solubilizers of cholesterol.]

3. **Cholesterol esterification:** The cholesterol taken up by HDL is immediately esterified by the plasma enzyme ***lecithin:cholesterol acyltransferase* (*LCAT*, also known as *PCAT*, in which P stands for phosphatidylcholine, the source of the FA).** This enzyme is synthesized and secreted by the liver. *LCAT* binds to nascent HDL and is activated by apo A-I. *LCAT* transfers the FA from carbon 2 of PC to cholesterol. This produces a hydrophobic cholesteryl ester, which is sequestered in the core of the HDL, and lysophosphatidylcholine, which binds to albumin. [Note: Esterification maintains the cholesterol concentration gradient, allowing continued efflux of cholesterol to HDL.] As the

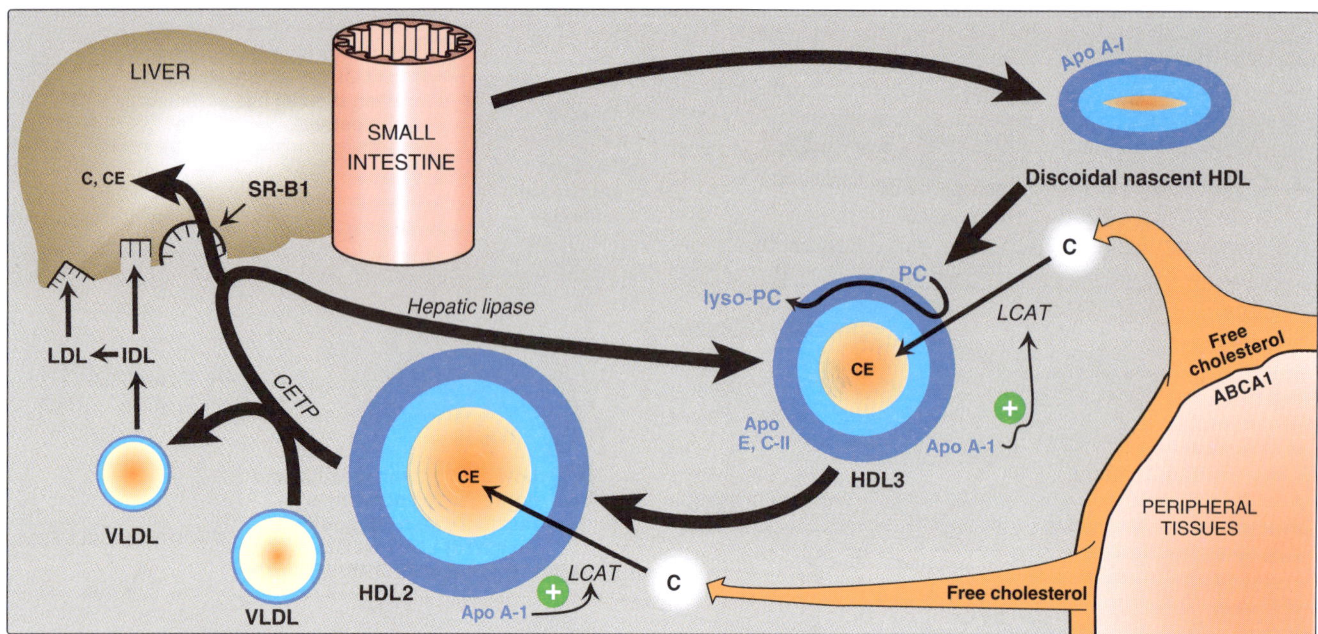

Figure 23.23 Metabolism of high-density lipoprotein (HDL) particles. Apo = apolipoprotein; ABCA1 = ATP-binding cassette transport protein; C = cholesterol; CE = cholesteryl ester; *LCAT* = *lecithin:cholesterol acyltransferase*; VLDL, IDL, and LDL = very-low-, intermediate-, and low-density lipoproteins; *CETP* = *cholesteryl ester transfer protein*; SR-B1 = scavenger receptor B1.

discoidal nascent HDL accumulates cholesteryl esters, it first becomes a spherical, relatively cholesteryl ester–poor HDL3 and, eventually, a cholesteryl ester–rich HDL2 particle that carries these esters to the liver. ***Hepatic lipase*,** which degrades TAG and phospholipids, participates in the conversion of HDL2 to HDL3 (see Fig. 23.23). ***CETP*** (see p. 358) transfers some of the cholesteryl esters from HDL to VLDL in exchange for TAG, relieving product inhibition of *LCAT*. Because VLDL are catabolized to LDL, the cholesteryl esters transferred by *CETP* are ultimately taken up by the liver (see p. 358).

> ABCA1 is an ATP-binding cassette (ABC) protein. ABC proteins use energy from ATP hydrolysis to transport materials, including lipids, in and out of cells and across intracellular compartments. In addition to Tangier disease, defects in specific ABC proteins result in sitosterolemia, cystic fibrosis, X-linked adrenoleukodystrophy, respiratory distress syndrome due to decreased surfactant secretion, and liver disease due to decreased bile salt secretion.

4. **Reverse cholesterol transport:** The selective transfer of cholesterol from peripheral cells to HDL and from HDL to the liver for bile acid synthesis or disposal via the bile is a key component of cholesterol homeostasis. This process of reverse cholesterol transport (RCT) is, in part, the basis for the inverse relationship seen between plasma HDL concentration and atherosclerosis and for the designation of HDL as the "good" cholesterol carrier. [Note**: Exercise and estrogen raise HDL levels.**] RCT involves efflux of cholesterol from peripheral cells to HDL, esterification of the cholesterol by *LCAT*, binding of the cholesteryl ester–rich HDL (HDL2) to liver (and, perhaps, steroidogenic cells), selective transfer of the cholesteryl esters into these cells, and release of lipid-depleted HDL (HDL3). The efflux of cholesterol from peripheral cells is mediated primarily by the transport protein ABCA1. [**Note: Tangier disease is a very rare deficiency of ABCA1 and is characterized by the virtual absence of HDL particles due to degradation of lipid-poor apo A-1.**] Cholesteryl ester uptake by the liver is mediated by the cell-surface receptor SR-B1 (scavenger receptor class B type 1) that binds HDL (see p. 357 for SR-A receptors). The HDL particle itself is not taken up. Instead, there is selective uptake of the cholesteryl ester from the HDL particle. [Note: Low HDL-C is a risk factor for atherosclerosis.]

F. Lipoprotein (a) and heart disease

Lipoprotein (a), or Lp(a), is nearly identical in structure to an LDL particle. Its distinguishing feature is the presence of an additional apolipoprotein molecule, apo(a), which is covalently linked at a single site to apo B-100. Circulating levels of Lp(a) are determined primarily by genetics. However, factors such as diet may play some role, as trans FA have been reported to increase it. The physiologic function of Lp(a) is unknown. When present in large quantities in the plasma, Lp(a) is associated with an increased risk of coronary heart disease. [Note: Apo(a) is structurally homologous to plasminogen, the precursor of a blood *protease* whose target is fibrin, the main protein component of blood clots. It is hypothesized that elevated Lp(a) slows the breakdown of blood clots that trigger heart attacks because it competes with plasminogen for binding to fibrin.] Niacin reduces Lp(a), as well as LDL-C and TAG, and raises HDL-C.

VII. STEROID HORMONES

Cholesterol is the precursor of all classes of steroid hormones: glucocorticoids (for example, cortisol), mineralocorticoids (for example, aldosterone), and the sex hormones (that is, androgens, estrogens, and progestins), as shown in Figure 23.24. [Note: Glucocorticoids and mineralocorticoids are collectively called corticosteroids.] Synthesis and secretion occur in the adrenal cortex (cortisol, aldosterone, and androgens), ovaries and placenta (estrogens and progestins), and testes (testosterone). Steroid hormones are transported by the blood from their sites of synthesis to their target organs. Because of their hydrophobicity, they must be complexed with a plasma protein. Albumin can act as a nonspecific carrier and does carry aldosterone. However, specific steroid-carrier plasma proteins bind the steroid hormones more tightly than does albumin (for example, **corticosteroid-binding globulin, or transcortin, is responsible for transporting cortisol**). A number of genetic diseases are caused by deficiencies in specific steps in the biosynthesis of steroid hormones. Some representative diseases are described in Figure 23.25.

A. Synthesis

Synthesis involves shortening the hydrocarbon chain of cholesterol and hydroxylating the steroid nucleus. The initial and rate-limiting reaction converts cholesterol to the 21-carbon pregnenolone. It is catalyzed by the *cholesterol side-chain cleavage enzyme*, a *cytochrome P450* (*CYP*) *mixed function oxidase* of the inner mitochondrial membrane (see p. 251) that is also known as $P450_{scc}$ and *desmolase*. NADPH and O_2 are required for the reaction. The cholesterol substrate can be newly synthesized, taken up from lipoproteins, or released by an *esterase* from cholesteryl esters stored in the cytosol of steroidogenic tissues. The cholesterol moves to the outer mitochondrial membrane. An important control point is the subsequent movement from the outer to the inner mitochondrial membrane. This process is mediated by StAR (steroidogenic acute regulatory) protein.

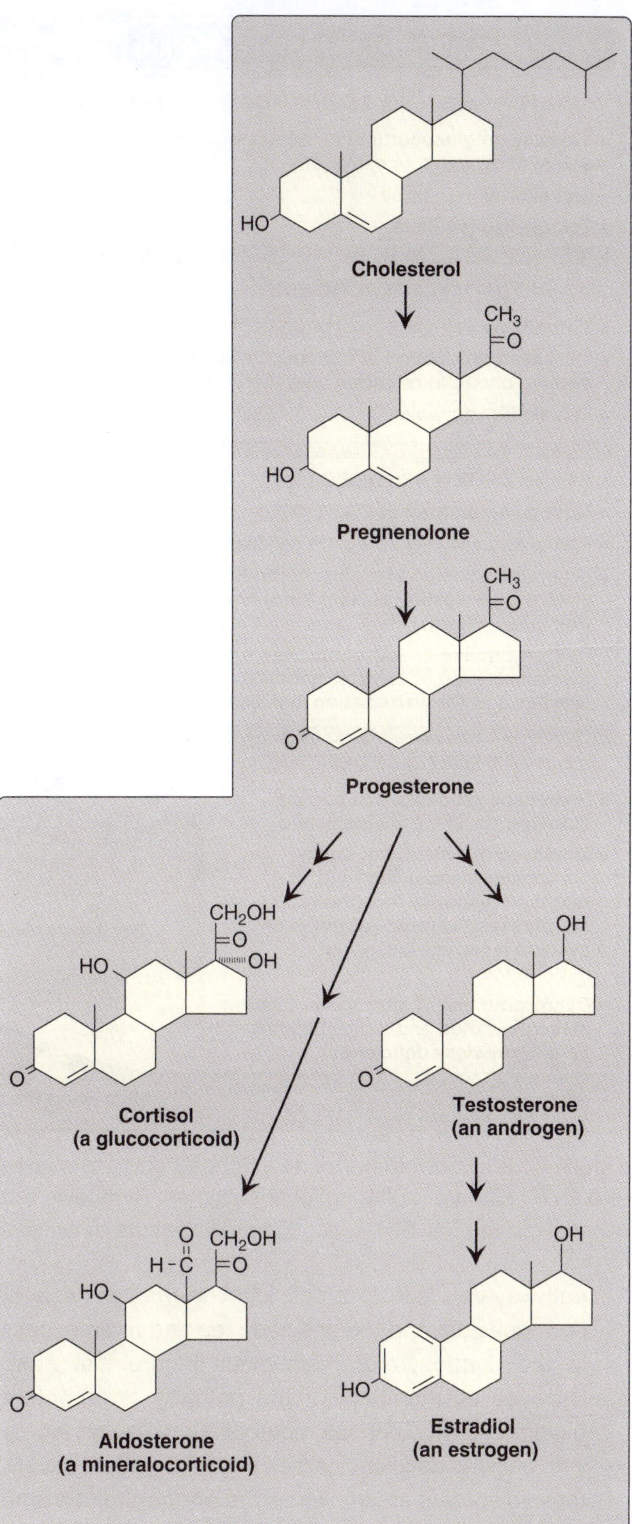

Figure 23.24 Key steroid hormones.

Pregnenolone is the parent compound for all steroid hormones (see Fig. 23.25). It is oxidized and then isomerized to progesterone, which is further modified to the other steroid hormones by CYP protein–catalyzed hydroxylation reactions in the SER and mitochondria. A defect in the activity or amount of an enzyme in this

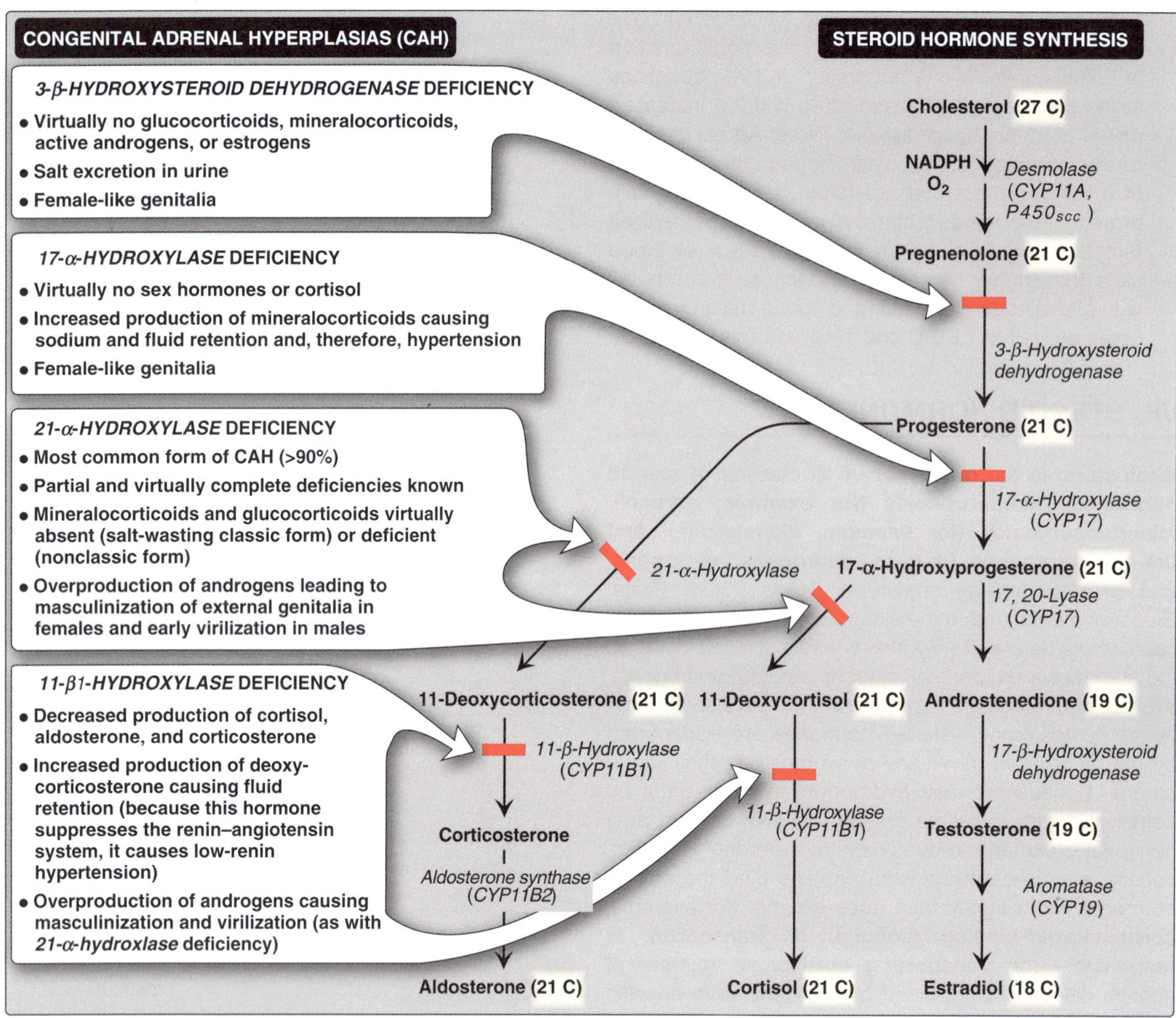

Figure 23.25 Steroid hormone synthesis and associated diseases. [Note: *3-β-Hydroxysteroid dehydrogenase*, *CYP17*, and *CYP11B2* are multifunctional enzymes. Synthesis of testosterone and the estrogens occurs primarily outside of the adrenal gland.] NADPH = nicotinamide adenine dinucleotide phosphate; CYP = cytochrome P450.

pathway can lead to a deficiency in the synthesis of hormones beyond the affected step and to an excess in the hormones or metabolites before that step. Because all members of the pathway have potent biologic activity, serious metabolic imbalances occur with enzyme deficiencies (see Fig. 23.25). Collectively, these disorders are known as the **congenital adrenal hyperplasias** (CAH), because they result in enlarged adrenals. [Note: Addison disease, due to autoimmune destruction of the adrenal cortex, is characterized by adrenocortical insufficiency.]

B. Adrenal cortical steroid hormones

Steroid hormones are synthesized and secreted in response to hormonal signals. The corticosteroids and androgens are made in different regions of the adrenal cortex and are secreted into blood in response to different signals. [Note: The adrenal medulla makes catecholamines (see p. 422).]

1. **Cortisol:** Its production in the middle layer (zona fasciculata) of the adrenal cortex is controlled by the hypothalamus, to which the pituitary gland is attached (Fig. 23.26). In response to severe stress (for example, infection), corticotropin-releasing hormone (CRH), produced by the hypothalamus, travels through capillaries to the anterior lobe of the pituitary, where it induces the production and secretion of adrenocorticotropic hormone (ACTH), a peptide. ACTH stimulates the adrenal cortex to synthesize and secrete the glucocorticoid cortisol,

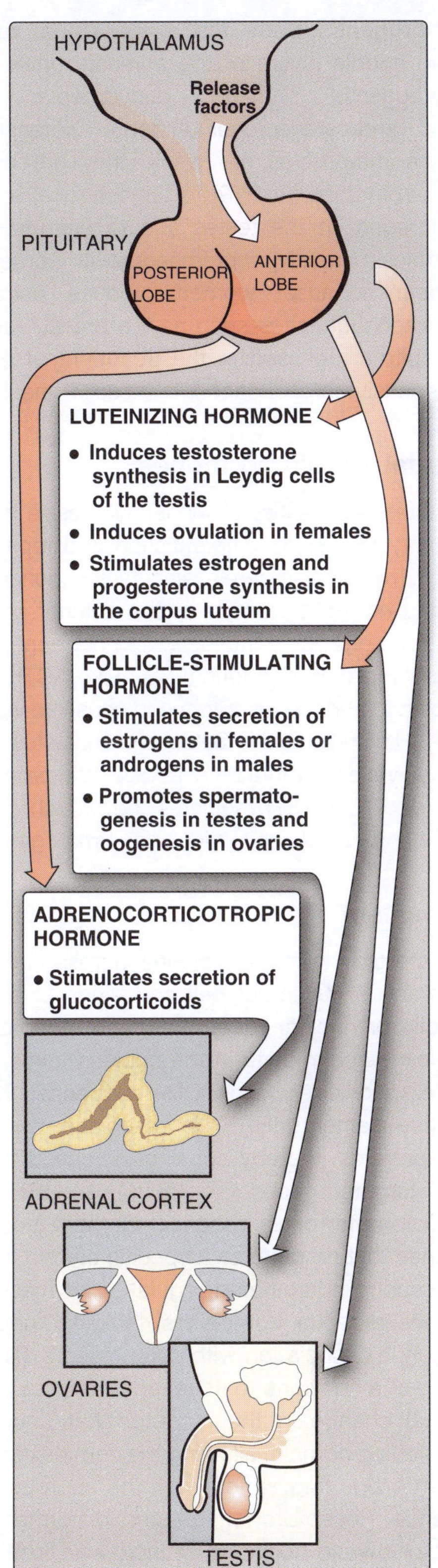

Figure 23.26 Pituitary hormone stimulation of steroid hormone synthesis and secretion.

the stress hormone. [Note: ACTH binds to a membrane G protein–coupled receptor, resulting in cyclic AMP (cAMP) production and activation of *protein kinase A* ([*PKA*], see p. 190). *PKA* phosphorylates and activates both the *esterase* that converts cholesteryl ester to free cholesterol and StAR protein.] Cortisol allows the body to respond to stress through its effects on intermediary metabolism (for example, increased gluconeogenesis) and the inflammatory and immune responses (which are decreased). As cortisol levels rise, the release of CRH and ACTH is inhibited. [Note: The reduction of cortisol in CAH results in a rise in ACTH that causes adrenal hyperplasia.]

Apolipoproteins

Apolipoprotein	Component of	Normal Plasma Level (mg/dl)
Apo A-I	HDL	100–200
Apo A-II	HDL	20–50
Apo B-100	LDL, VLDL	70–125
Apo B-48	Chylomicrons	< 5
Apo C-I	Chylomicrons, VLDL	5–8
Apo C-II	Chylomicrons, VLDL	3–7
Apo C-III	Chylomicrons, VLDL	10–12
Apo E	Chylomicrons, VLDL, IDL	3–15
Apo Lp(a)	Lp(a)	<30

2. **Aldosterone:** Its production in the outer layer (zona glomerulosa) of the adrenal cortex is induced by a decrease in the plasma Na^+/potassium (K^+) ratio and by the hormone angiotensin II (Ang-II). Ang-II (an octapeptide) is produced from angiotensin I ([Ang-I] a decapeptide) by *angiotensin-converting enzyme* (*ACE*), an enzyme found predominantly in the lungs but also distributed widely in the body. [Note: Ang-I is produced in the blood by cleavage of an inactive precursor, angiotensinogen, secreted by the liver. Cleavage is catalyzed by *renin*, made and secreted by the kidneys.] Ang-II binds to cell surface receptors. However, in contrast to ACTH, its effects are mediated through the phosphatidylinositol 4,5-bisphosphate pathway (see p. 325) and not by cAMP. Aldosterone's primary effect is on the kidney tubules, where it stimulates Na^+ and water uptake and K^+ excretion (Fig. 23.27). [Note: An effect of aldosterone is an increase in blood pressure. Competitive inhibitors of *ACE* are used to treat *renin*-dependent hypertension.]

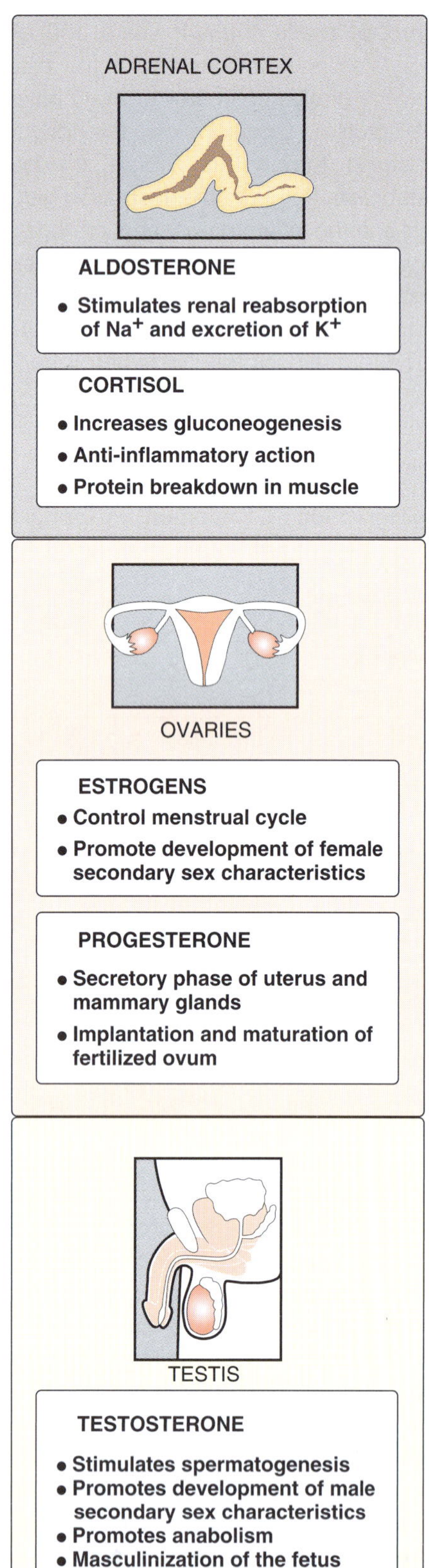

Figure 23.27 Actions of steroid hormones. Na^+ = sodium; K^+ = potassium.

3. **Androgens:** Both the inner (zona reticularis) and middle layers of the adrenal cortex produce androgens, primarily dehydroepiandrosterone and androstenedione. Although adrenal androgens themselves are weak, they are converted by *aromatase* (*CYP19*) to testosterone, a stronger androgen, in the testes and to estrogens in the ovaries (primarily) of premenopausal women. [Note: Postmenopausal women produce estrogen at extragonadal sites such as the breast. *Aromatase* inhibitors are used in the treatment of estrogen-responsive breast cancer in these women.]

C. Gonadal steroid hormones

The testes and ovaries (gonads) synthesize hormones necessary for sexual differentiation and reproduction. A single hypothalamic-releasing factor, gonadotropin-releasing hormone, stimulates the anterior pituitary to release the glycoproteins luteinizing hormone (LH) and follicle-stimulating hormone (FSH). Like ACTH, LH and FSH bind to surface receptors and cause an increase in cAMP. LH stimulates the testes to produce testosterone and the ovaries to produce estrogens and progesterone (see Fig. 23.27). FSH regulates the growth of ovarian follicles and stimulates testicular spermatogenesis.

D. Mechanism

Each steroid hormone diffuses across the plasma membrane of its target cell and binds to a specific cytosolic or nuclear receptor. These receptor–ligand complexes accumulate in the nucleus, dimerize, and bind to specific regulatory DNA sequences (hormone response elements [HRE]) in association with coactivator proteins, thereby causing increased transcription of targeted genes (Fig. 23.28). An HRE is found in the promoter or an enhancer element (see p. 627) for genes that respond to a specific steroid hormone, thus insuring coordinated regulation of these genes. Hormone–receptor complexes can also inhibit transcription in association with corepressors. [Note: The binding of a hormone to its receptor causes a conformational change in the receptor that uncovers its DNA-binding domain, allowing the complex to interact through a zinc finger motif with the appropriate DNA sequence. Receptors for the steroid hormones, plus those for thyroid hormone, retinoic acid (see p. 535), and 1,25-dihydroxycholecalciferol (vitamin D; see p. 540), are members of a superfamily of structurally related gene regulators that function in a similar way.]

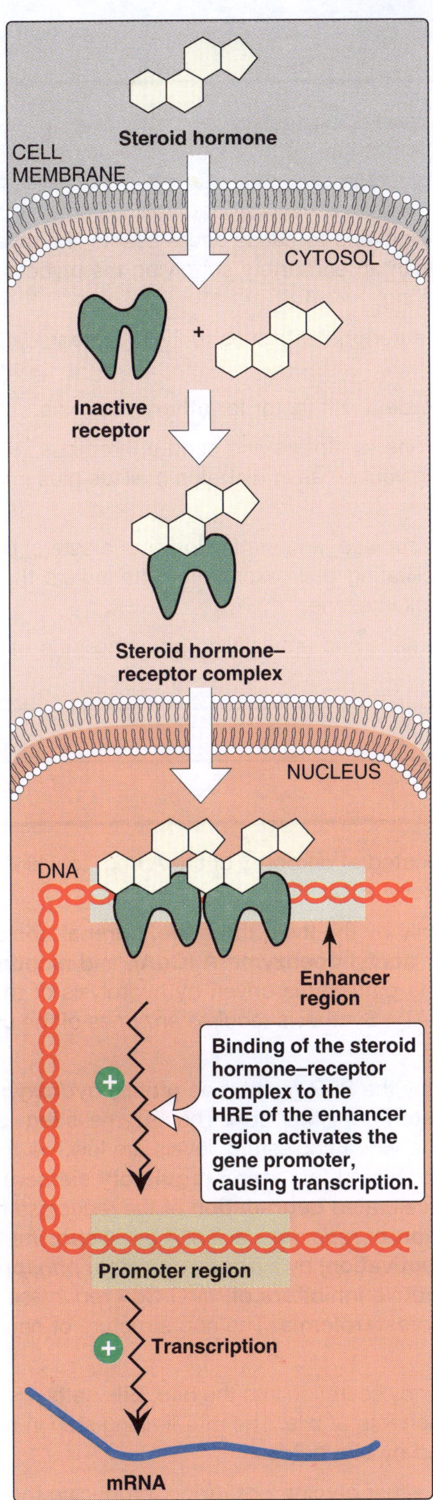

Figure 23.28 Activation of transcription by interaction of steroid hormone–receptor complex with hormone response element (HRE). The receptor contains domains that bind the hormone, DNA, and coactivating proteins. mRNA = messenger RNA.

E. Further metabolism

Steroid hormones are generally converted into inactive metabolic excretion products in the liver. Reactions include reduction of unsaturated bonds and the introduction of additional hydroxyl groups. The resulting structures are made more soluble by conjugation with glucuronic acid or sulfate (from 3′-phosphoadenosyl-5′-phosphosulfate; see p. 266). These conjugated metabolites are fairly water soluble and do not need protein carriers. They are eliminated in feces and urine.

VIII. Medical Perspectives

Low-density lipoprotein is considered a "bad" cholesterol as it transports cholesterol from the liver to peripheral tissues. This is essential for various utilizations of cholesterol like formation of steroid hormones, bile acids and vitamin D. It gets its "bad" name from the fact that modified lipids (oxidized and glycated LDL) may deposit in blood vessel walls and lead to phagocytosis by macrophages, foam cell formation, immune response, release of cytokines and metalloproteinases and ultimately rupture of atherosclerotic plaque and release of embolus to cause acute events like myocardial infarction or stroke. This happens because the ruptured thrombus lodges in an end artery supplying the myocardium or other vulnerable tissues leading to ischemia of the area that it was supplying.

The HDL is considered "good" because it transfers cholesterol from the peripheral tissues to liver for degradation and elimination. Therefore, HDL is considered anti-atherogenic.

A high LDL/HDL ratio as well as a high ApoB/ApoA ratio is considered to be a risk factor for **atherosclerosis.**

The atherosclerotic plaque may be stable or unstable (vulnerable) and prone to rupture and send a thrombus into circulation. However, it is now clear that oxidative stress, like in smoking and glycation as in diabetes mellitus plus inflammation, has a major contributory role in vulnerability of atherosclerotic plaque.

Statins are inhibitors of *hydroxymethylglutaryl coenzyme A (HMG CoA) reductase*, and statins lower cholesterol by inhibiting *HMG CoA reductase*. **Cholestyramine** binds bile salts, hence facilitating their excretion. This lowers the serum cholesterol. Ezetimibe selectively inhibits cholesterol absorption from small intestine.

Bile adequacy is important for cholesterol excretion. Low content of bile salts and emulsifying phospholipids as well as increased cholesterol favor formation of gallstones in gallbladder.

IX. Chapter Summary

Cholesterol is a hydrophobic compound, with a single hydroxyl group located at carbon 3 of the A ring, to which a fatty acid (FA) can be attached, producing an even more hydrophobic **cholesteryl ester**.

Cholesterol is synthesized by virtually all human tissues, although primarily by the **liver**, **intestine**, **adrenal cortex**, and **reproductive tissues** (Fig. 23.29). All the carbon atoms are provided by **acetyl coenzyme A** (**CoA**), and **nicotinamide adenine dinucleotide phosphate** provides the reducing equivalents. The pathway is driven by hydrolysis of the high-energy thioester bond of acetyl CoA and the terminal phosphate bond of ATP. Synthesis requires enzymes of the **cytosol**, **smooth endoplasmic reticulum** (**SER**), and **peroxisomes**.

The rate-limiting and regulated step in cholesterol synthesis is catalyzed by the SER-membrane protein ***hydroxymethylglutaryl coenzyme A*** (***HMG CoA***) ***reductase***, which produces **mevalonate** from HMG CoA. The enzyme is regulated by a number of mechanisms: 1) increased **expression** of the *reductase* **gene** when cholesterol levels are low, via the transcription factor, **sterol regulatory element–binding protein-2** (**SREBP-2**), bound to a **sterol regulatory element** (**SRE**), resulting in increased enzyme and, therefore, cholesterol, synthesis; 2) accelerated **degradation** of the *reductase* **protein** when cholesterol levels are high; 3) **phosphorylation** (causing **inactivation** of *reductase* activity) by ***adenosine monophosphate–activated protein kinase*** [***AMPK***] and **dephosphorylation** (**activation**) by a ***phosphoprotein phosphatase***; and 4) hormonal regulation by **insulin** and **glucagon**. **Statins** are **competitive inhibitors** of *HMG CoA reductase*. These drugs are used to decrease plasma cholesterol in patients with **hypercholesterolemia**. The ring structure of cholesterol cannot be degraded in humans.

Cholesterol is eliminated from the body either by conversion to bile salts or by secretion into the **bile**. **Bile salts** and **phosphatidylcholine** (**PC**) are quantitatively the most important organic components of bile. The rate-limiting step in bile acid synthesis is catalyzed by ***cholesterol-7-α-hydroxylase***, which is inhibited by **bile acids**.

Before the bile acids leave the liver, they are conjugated to a molecule of either **glycine** or **taurine**, producing the **conjugated bile salts glycocholic** or **taurocholic acid** and **glycochenodeoxycholic** or **taurochenodeoxycholic acid**.

Bile salts (deprotonated) are more amphipathic than bile acids (protonated) and, therefore, are more effective **emulsifiers** of dietary fat. Intestinal **bacteria** can remove the glycine and taurine as well as a hydroxyl group from the steroid nucleus, producing the **secondary bile salts**, **deoxycholic** and **lithocholic acids**.

Bile salts are efficiently reabsorbed (>95%) in the intestinal ileum by a **sodium–bile salt cotransporter**, returned to the blood, and carried by **albumin** back to the liver where they are taken up by the hepatic isoform of the cotransporter and reused (**enterohepatic circulation,** which **bile acid sequestrants** reduce).

If more cholesterol enters the bile than can be solubilized by the available bile salts and PC, **cholesterol gallstone disease** (**cholelithiasis**) can occur.

The plasma lipoproteins (see Fig. 23.29) include **chylomicrons**, **very-low-density lipoproteins** (**VLDL**), **intermediate-density lipoproteins** (**IDL**), **low-density lipoproteins** (**LDL**), and **high-density lipoproteins** (**HDL**). They function to keep lipids (primarily **triacylglycerol** [**TAG**] and **cholesteryl esters**) soluble as they transport them between tissues. Lipoproteins are composed of a **neutral lipid** (**TAG**, **cholesteryl esters**, or both) core surrounded by a shell of **amphipathic apolipoproteins**, **phospholipid**, and **nonesterified cholesterol**.

Chylomicrons are assembled in **intestinal mucosal cells** from **dietary lipids** (primarily TAG). Each nascent chylomicron particle has one molecule of **apolipoprotein** (**apo**) **B-48**. They are released from the cells into the lymphatic system and travel to the blood, where they receive **apo C-II** and **apo E** from HDL. Apo C-II activates endothelial ***lipoprotein lipase*** (***LPL***), which degrades the TAG in chylomicrons to FA and glycerol. The **FA** that are released are stored (in **adipose tissue**) or used for energy (in **muscle**). The **glycerol** is metabolized by the **liver**. Patients with a **deficiency** of *LPL* or apo C-II show a dramatic accumulation of chylomicrons in the plasma (**type I hyperlipoproteinemia** or **familial chylomicronemia**) even if fasted. After most of the TAG is removed, apo C-II is returned to HDL, and the **chylomicron remnant**, carrying most of the **dietary cholesterol**, binds to a liver receptor that recognizes apo E. The particle is **endocytosed**, and its contents degraded by **lysosomal enzymes**. Defective uptake of these remnants (and IDL) causes **type III hyperlipoproteinemia** or **dysbetalipoproteinemia**.

Nascent VLDL are produced in the liver and are composed predominantly of TAG. They contain a single molecule of **apo B-100**. Like chylomicrons, VLDL receive apo C-II and apo E from HDL in the plasma. VLDL carry hepatic TAG to the peripheral tissues where *LPL* degrades the lipid. Additionally, the VLDL particle receives **cholesteryl esters** from HDL in exchange for TAG. This process is accomplished by ***cholesteryl ester transfer protein*** (***CETP***).

VLDL in the plasma is first converted to IDL and then to LDL, a much smaller, denser particle. Apo C-II and apo E are returned to HDL, but the LDL retains apo B-100, which is recognized by receptors on peripheral tissues and the liver. LDL undergo **receptor-mediated endocytosis**, and their contents are degraded in the **lysosomes**. The *protease* ***proprotein convertase subtilisin/kexin type 9 (PCSK9)*** prevents receptor recycling. Defects in the synthesis of functional LDL receptors causes **type IIa hyperlipoproteinemia** (**familial hypercholesterolemia** [**FH**]). The endocytosed cholesterol decreases expression of *HMG CoA reductase* (and LDL receptors) through prevention of SREBP-2 binding to the SRE. Some of it can be esterified by ***acyl CoA:cholesterol acyltransferase*** (***ACAT***) and stored.

HDL are created by **lipidation** of **apo A-1** synthesized in the liver and intestine. They have a number of functions, including 1) serving as a circulating **reservoir** of apo C-II and apo E for chylomicrons and VLDL; 2) removing **cholesterol** from peripheral tissues via ABCA1 and esterifying it using ***lecithin:cholesterol acyl transferase*** (***LCAT***), a liver-synthesized plasma enzyme that is activated by **apo A-1**; and 3) delivering these cholesteryl esters to the liver (**reverse cholesterol transport**) for uptake via **scavenger receptor-B1** (**SR-B1**).

Cholesterol is the precursor of all classes of **steroid hormones**, which include **glucocorticoids**, **mineralocorticoids**, and the **sex hormones** (**androgens**, **estrogens**, and **progestins**). Synthesis, using primarily **cytochrome P450** ***mixed function oxidases***, occurs in the **adrenal cortex** (**cortisol** in the **zona fasciculata**, **aldosterone** in the **zona glomerulosa**, and **androgens** in the **zona reticularis**), **ovaries** and **placenta** (**estrogens** and **progestins**), and **testes** (**testosterone**). The initial and rate-limiting step is the conversion of cholesterol to **pregnenolone** by the side-chain cleavage enzyme $\boldsymbol{P450_{scc}}$. Deficiencies in synthesis lead to **congenital adrenal hyperplasia** (**CAH**). Each steroid hormone diffuses across the plasma membrane of its target cell and binds to a specific intracellular receptor. These **receptor–hormone complexes** accumulate in the nucleus, dimerize, and bind to specific regulatory DNA sequences (**hormone response elements**) in association with coactivator proteins, thereby causing increased **transcription** of targeted genes. In association with corepressors, transcription is decreased.

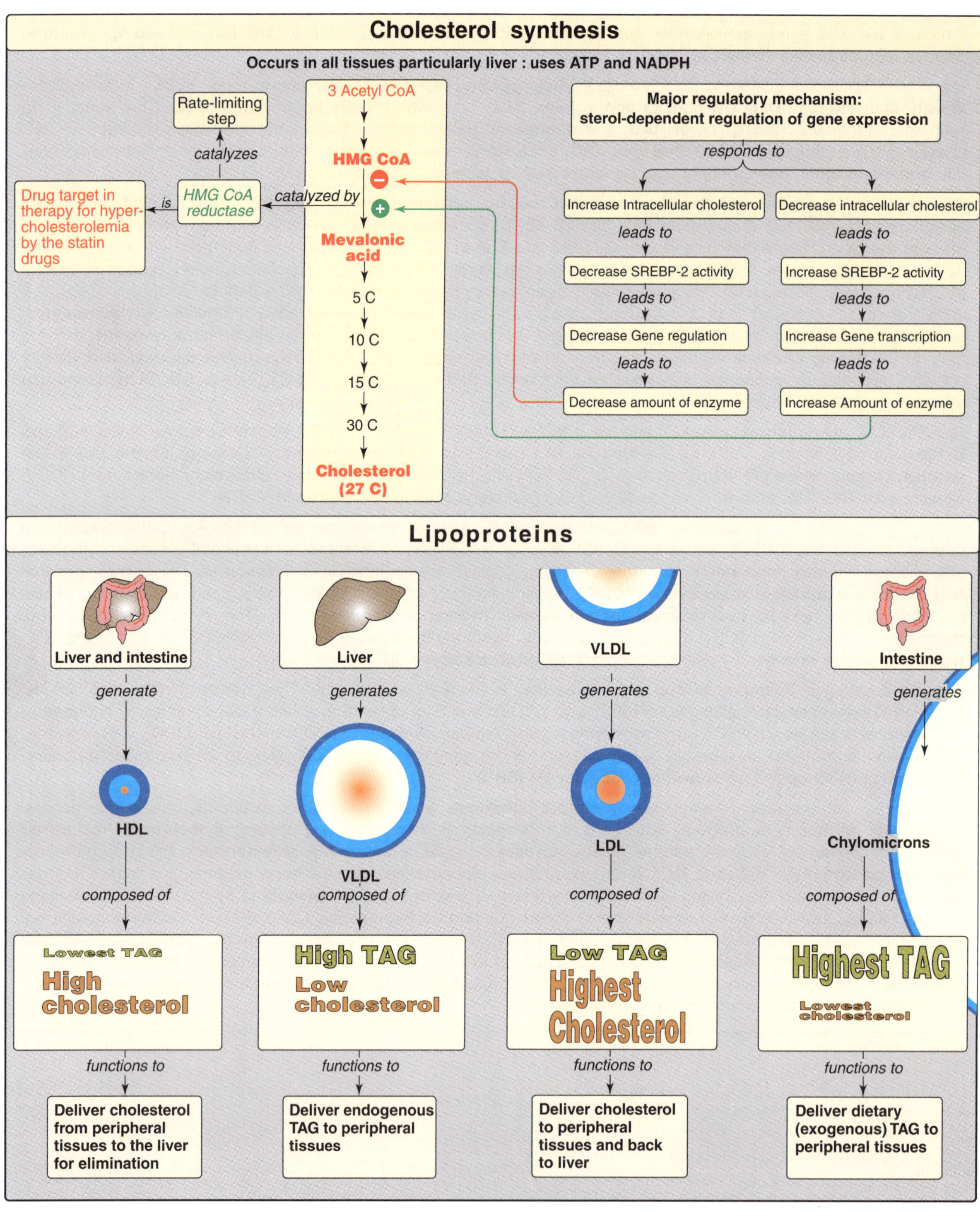

Figure 23.29 Concept map for cholesterol and the lipoproteins. HMG CoA = hydroxymethylglutaryl coenzyme A; SREBP = sterol regulatory element–binding protein; HDL, LDL, and VLDL = high-, low-, and very-low-density lipoproteins; TAG = triacylglycerol; NADPH = nicotinamide adenine dinucleotide phosphate; C = carbon.

Study Questions

Short-Answer Questions

Write short notes on:

23.1. Bile acids and bile salts
23.2. Statins
23.3. Apolipoproteins
23.4. Good and bad cholesterols
23.5. Biological role of cholesterol

Long-Answer Questions

23.1. Discuss the steps of cholesterol synthesis. Add a note on the rate-limiting step and its regulation.

23.2. What is the normal cholesterol level in blood? Discuss how lipoproteins are responsible for efficient transport of cholesterol and triglycerides to and from tissues to liver. Add a note on deposition of lipoproteins in blood vessels.

23.3. Explain steroid hormone synthesis and associated diseases.

Multiple-Choice Questions

Choose the ONE best answer.

23.1. Cholesterol degradation and excretion does not include
A. Degradation to carbon dioxide and water
B. Conversion to bile salts and bile acids
C. Conversion to coprostanol
D. Conversion to cholestanol

Correct answer = A.

23.2. Which is the first sterol to be produced during the synthesis of cholesterol?
A. Squalene
B. Mevalonate
C. Lanosterol
D. Acetoacetyl CoA

Correct answer = C.

23.3. Which of the following is true for secondary bile acids?
A. They are synthesized by bacteria
B. They are synthesized by hepatocytes
C. They include cholic acid and chenodeoxycholic acid
D. Produced by action of 12 alpha hydroxylase on primary bile acids

Correct answer = A.

23.4. The rate-limiting step in bile acid synthesis
A. Involves the introduction of a hydroxyl group at carbon 7 of the steroid nucleus
B. Is catalyzed by *7-α-hydroxylase*, a SER-associated *cytochrome P450 monooxygenase*
C. Occurs only in liver
D. All of the above

Correct answer = D.

23.5. The macrophage scavenger receptor pathway of LDL uptake
A. Is highly specific and regulated
B. Mediates endocytosis of modified LDL
C. Is downregulated in response to increased intracellular cholesterol
D. Is not involved in formation of foam cells

Correct answer = B.

23.6. The maturation of HDL involves the conversion of polar lipids to non-polar lipids, facilitating their movement into the core of the lipoprotein. Which of the following enzymes is involved in this process?
A. Paraoxonase
B. Cholesterol ester transfer protein
C. Lecithin cholesterol acyl transferase
D. Endothelial lipase

Correct answer = C.

23.7. Familial chylomicronemia syndrome is characterized by an elevation in the serum triacyl glycerols due to a defect in the lipolysis of triacyl glycerols present in chylomicrons. A defect in which of the following proteins can lead to this condition?
I. Apo CII
II. Apo B-100
III. LDL-receptor
IV. Lipoprotein lipase

A. I and III
B. II and IV
C. I and IV
D. II and III

Correct answer = C.

23.8. A blood sample was collected from a 25-year-old male to estimate the fasting lipid profile. On centrifugation, the serum appeared to be milky white in color. Which of the following is the most likely disorder associated with this finding?
A. Abetalipoproteinemia
B. Familial hypercholesterolemia
C. Familial chylomicronemia syndrome
D. Tangier's disease

Correct answer = C.

23.9. Effects of cortisol include
A. Increased gluconeogenesis and decreased inflammatory and immune responses
B. Increased gluconeogenesis and increased protein breakdown in muscles

C. Inhibition of ACTH release and decreased inflammatory responses
D. All of above

Correct answer = D.

23.10. Apo B 48 is
A. Unique to chylomicrons
B. Formed by post translational editing
C. Synthesized in smooth endoplasmic reticulum
D. Helps in binding of chylomicron remnant to the liver

Correct answer = A.

Multiple-Choice Questions (NExT/NEET PG)

Choose the ONE best answer.

23.1. Mice were genetically engineered to contain hydroxymethylglutaryl coenzyme A reductase in which serine 871, a phosphorylation site, was replaced by alanine. Which of the following statements concerning the modified form of the enzyme is most likely to be correct?
A. The enzyme is nonresponsive to ATP depletion.
B. The enzyme is nonresponsive to statin drugs.
C. The enzyme is nonresponsive to the sterol response element–sterol response element–binding protein system.
D. The enzyme is unable to be degraded by the ubiquitin–proteasome system.

Correct answer = A.

23.2. Calculate the amount of cholesterol in the low-density lipoproteins in an individual whose fasting blood gave the following lipid-panel test results: total cholesterol = 300 mg/dl, high-density lipoprotein cholesterol = 25 mg/dl, triglycerides = 150 mg/dl.
A. 55 mg/dl
B. 95 mg/dl
C. 125 mg/dl
D. 245 mg/dl

Correct answer = D.

For Questions NExT/NEET PG 23.3 and 23.4, use the following scenario.
A young girl with a history of severe abdominal pain was taken to her local hospital at 5 a.m. in severe distress. Blood was drawn, and the plasma appeared milky, with the triacylglycerol level >2,000 mg/dl (normal = 4–150 mg/dl). The patient was placed on a diet extremely limited in fat but supplemented with medium-chain triglycerides.

23.3. Which of the following lipoprotein particles are most likely responsible for the appearance of the patient's plasma?
A. Chylomicrons
B. High-density lipoproteins
C. Intermediate-density lipoproteins
D. Low-density lipoproteins
E. Very-low-density lipoproteins

Correct answer = A.

23.4. Which one of the following proteins is most likely to be deficient in this patient?
A. Apolipoprotein A-I
B. Apolipoprotein B-48
C. Apolipoprotein C-II
D. Cholesteryl ester transfer protein
E. Microsomal triglyceride transfer protein

Correct answer = C.

23.5. Complete the table below for an individual with classic 21-α-hydroxylase deficiency relative to a normal individual.

Variable	Increased	Decreased
Aldosterone		
Androstenedione		
Cortisol		
Blood glucose		
Adrenocorticotropic hormone		
Blood pressure		

How might the results be changed if this individual were deficient in 17-α-hydroxylase, rather than 21-α-hydroxylase?

Answer: Classic 21-α-hydroxylase deficiency causes mineralocorticoids (aldosterone) and glucocorticoids (cortisol) to be virtually absent. Because aldosterone increases blood pressure, and cortisol increases blood glucose, their deficiencies result in a decrease in blood pressure and blood glucose, respectively. Cortisol normally feeds back to inhibit adrenocorticotropic hormone (ACTH) release by the pituitary, and, so, its absence results in an elevation in ACTH. The loss of 21-α-hydroxylase pushes progesterone and pregnenolone to androgen synthesis and, therefore, causes androstenedione levels to rise. With 17-α-hydroxylase deficiency, sex hormone synthesis would be decreased. Mineralocorticoid production would be increased, leading to hypertension.

Appendix
Unit IV: Lipid Metabolism

CLINICAL RELEVANCE FOR PRIMARY HEALTHCARE PHYSICIANS

Table 1 Reference ranges for lipids

Lipids and Lipoproteins	Reference Range	Low Risk	Borderline Risk	High Risk
Total cholesterol	140–200 mg/dl	< 200 mg/dl	200–239 mg/dl	≥ 240 mg/dl
HDL cholesterol	40–75 mg/dl	> 60 mg/dl	–	< 40 mg/dl
LDL cholesterol	50–130 mg/dl	< 130 mg/dl	130–159 mg/dl	> 160 mg/dl
Triglyceride	60–150 mg/dl	< 150 mg/dl	200–499 mg/dl	≥ 500 mg/dl

The normal level of lipids and lipoproteins varies with age. The given values are of a normal adult aged 20–40 years. They also vary slightly in males and females.

Table 2 Reference ranges for lipids in children

Lipids and Lipoproteins	Reference Range	Borderline	Abnormal
Total cholesterol	< 170 mg/dl	170–199 mg/dl	≥ 200 mg/dl
HDL cholesterol	> 45 mg/dl	40–45 mg/dl	< 40 mg/dl
LDL cholesterol	< 110 mg/dl	110–129 mg/dl	≥ 130 mg/dl
Triglyceride (0–9 years)	< 75 mg/dl	75–99 mg/dl	≥ 100 mg/dl
Triglyceride (10–19 years)	< 90 mg/dl	90–129 mg/dl	≥ 130 mg/dl

Measurement of blood lipids is done from blood collected in a vacutainer with no additive and after 12 hours of fasting. The method is based on spectrophotometry and may be manual or automated.

Table 3 Target lipid ratios for reduced atherosclerotic risk

	Target Lipid Ratios	
	Men	Women
Total Cholesterol/HDL	< 4.5	< 4.0
LDL-C/HDL-C	< 3.0	< 2.5
ApoB/ApoA-I	< 0.9	< 0.8

The most important way to prevent atherosclerotic vascular disease is to follow a healthy lifestyle throughout life. A healthy lifestyle reduces atherosclerotic cardiovascular disease (ASCVD) risk at all ages. Unfavorable lipid profile is recognized as an important risk factor in the development and progression of cardiovascular disease (CVD).

Blood cholesterol level is a major modifiable risk factor, along with other health and lifestyle factors, for ASCVD risk. ASCVD risk calculator combines all major risk factors, including age, smoking, high blood pressure, abnormal cholesterol, and diabetes to estimate a patient's probability for developing ASCVD. Statins are the first-line agents used to decrease cholesterol and reduce the risk of ASCVD events. The 2018 American Heart Association (AHA) Guideline recommends that low-density lipoprotein cholesterol (LDL-C) should be measured and baseline LDL-C documented for adults 20 years or older who are free from ASCVD (and not on lipid-lowering therapy).

For statin therapy, ASCVD risk, patient's age, comorbid conditions such as diabetes and LDL-C levels are taken into account. There is no ideal target blood level for LDL-C. The 2018 AHA Guidelines recognize that "lower is better." Studies suggest that an optimal total cholesterol level is about 150 mg/dl, with LDL-C at or below 100 mg/dl. Lowering LDL-C levels by 1% generally equals about 1% reduction in heart disease and stroke risk.

Statins can be used as different regimes (high, moderate or low intensity), depending on percentage reduction in LDL-C levels. If a patient is unable to tolerate statins or statin alone is insufficient, other drugs such as bile acid sequestrants/ezetimibe or a PCSK9 inhibitor can be added.

DRUGS USED FOR LOWERING LDL-C LEVELS

- *Statins*: These drugs block intracellular cholesterol synthesis by inhibiting *3-hydroxy-3-methylglutaryl CoA (HMG-CoA) reductase*. As cholesterol level in hepatocytes is reduced, expression of the LDL receptor increases, which removes LDL from the circulation, thus reducing its entry into vessels and the formation of plaques. Examples include atorvastatin, rosuvastatin, lovastatin, simvastatin, pravastatin, fluvastatin.
- *Ezetimibe*: This is a cholesterol absorption inhibitor, which interacts with cholesterol transporters at the brush border of the small intestine and inhibits the intestinal absorption of cholesterol and related phytosterols. It helps in reduction of hepatic cholesterol stores and a reduction of total cholesterol, LDL-C, and other triglycerides in the blood.
- *Bile acid sequestrants*: These drugs work by sequestering cholesterol in the gut so that it is not absorbed. The adverse effects include bloating and constipation and these drugs are poorly tolerated. Example includes cholestyramine.
- *Proprotein convertase subtilisin/kexin-like9 (PCSK9) inhibitors*: *PCSK9* interacts with LDL receptor and enhances its degradation, causing higher circulating LDL concentrations. Monoclonal antibodies against *PCSK9* lower LDL levels and subsequently, cardiovascular risk.
- *Niacin*: Niacin increases high-density lipoprotein cholesterol (HDL-C) and also significantly decreases concentrations of all apoB containing lipoproteins, i.e., very low-density lipoprotein (VLDL), intermediate-density lipoprotein (IDL), low-density lipoprotein (LDL), and lipoprotein a [Lp(a)]. Its adverse effects include gastrointestinal discomfort, pruritus and flushing.
- *Fibrates*: Fibric acid derivatives or fibrates decrease triglyceride levels. They also reduce LDL-C levels and help to raise HDL-C. Fibrates act through a protein called peroxisome proliferator-activated receptor alpha (PPAR-alpha), which activates the enzyme *lipoprotein lipase*.

Lipid Malabsorption

Fat maldigestion or malabsorption causes excretion of undigested lipids in stools (steatorrhea).

Common causes are gallbladder removal and obstructive jaundice. This allows for digestion of most foods but high-volume, fatty greasy food and high-fiber foods may be difficult to digest leading to flatulence and discomfort.

Lipid malabsorption also leads to defective absorption and therefore, deficiency of fat-soluble vitamins like vitamin K and vitamin D.

Familial Hypercholesterolemia

Familial hypercholesterolemia (FH) is an inherited condition (autosomal dominant) characterized by high levels of LDL-C and increased risk of premature cardiovascular disease. There is decreased LDL receptor function (LDL receptor is not synthesized, or it is not properly transported from the endoplasmic reticulum to the Golgi apparatus, or the receptor does not properly bind LDL or it is not recycled back to the cell surface). All patients with LDL-C levels ≥ 190 mg/dl are prescribed statin therapy.

ATHEROSCLEROTIC CARDIOVASCULAR DISEASE (ASCVD) RISK FACTORS

- Family history of premature ASCVD
- Persistently elevated LDL (≥ 160 mg/dl)
- Persistently elevated triglycerides (≥ 175 mg/dl)
- Lp(a) levels ≥ 50 mg/dl
- ApoB levels ≥ 130 mg/dl
- Metabolic syndrome
- Chronic kidney disease

Besides using lipid profile as predictors of cardiovascular disease, several lipoprotein ratios or atherogenic indices are also used.

Clinical Case Questions

Chapter 20 Lipids: Classification, Digestion, Absorption and Physiological Functions

20.1. A 40-year-old obese lady presented to the OPD with complaints of bloating and discomfort after having heavy fatty meal and occasional severe abdominal pain relieved by antispasmodics. She was advised ultrasound abdomen which showed multiple small gallstones.

A. What could be the reason for her inability to digest fatty meals?

Answer A: The lady probably had intermittent obstruction of common bile duct by her multiple small gallstones leading to severe abdominal pain. A high fatty meal would cause the gallbladder to contract to excrete bile. A bloating, belching and flatulent response to fatty meal could be because of insufficient bile release due to obstruction. Bile is critical to digestion and absorption of fat. Bile breaks down fat to small droplets for providing a larger surface area for enzymes to act. Bile also is alkaline and neutralizes the stomach acid for further enzymatic activity.

20.2. A 45-year-old male, who works as a security guard in a school, presents with complaints of altered bowel habits (increased frequency and bulk of stools), excessive bloating and significant weight loss despite a normal appetite over the past 10–12 months. He describes his stools as bulky, foul-smelling and difficult to flush. He also experiences abdominal cramps and flatulence after consuming a fatty meal. The patient is not diabetic and has no known comorbidities or addictions. He underwent a cholecystectomy a year ago. Despite being advised a course of antibiotics and given a trial of lactose-free and gluten-free diets, his symptoms did not improve.

Physical examination: Dry scaly skin with loss of subcutaneous fat. A cholecystectomy scar present in the right hypochondrium.

A. What is the likely cause of the patient's condition and its biochemical basis?

Answer A: Removal of gall bladder leads to secretion of bile directly into the small intestine. This direct secretion of bile is unable to digest a large bulky fatty meal causing flatulence, bloating and steatorrhea.

B. What nutritional deficiencies is the patient likely to be suffering from?

Answer B: The patient is likely to be deficient in fat soluble vitamins A, D, E, and K.

Chapter 21 Fatty Acid, Triacylglycerol, and Ketone Body Metabolism

21.1. AK, a 59-year-old male with slurred speech, ataxia (loss of skeletal muscle coordination), and abdominal pain, was dropped off at the Emergency Department (ED).

He was binge drinking last night and did not eat anything since yesterday. He has a 6-year history of chronic, excessive alcohol consumption.

On examination, his liver was slightly enlarged. Bedside tests revealed hypoglycemia and hyperketonemia. His fasting blood glucose was 58 mg/dl (normal reference 70–100 mg/dl) and 3-hydroxybutyrate was 50 mg/dl (normal reference range 0–3 mg/dl). His liver enzymes were raised by 3–5 times of normal.

(**Note:** *Clinical Case 21.1 requires knowledge from Chapters 21, 28, and 29)*

A. Why were ketone bodies elevated in this patient?

Answer—A: The patient had consumed alcohol and was fasting. The stored triacylglycerol was broken down for energy. However, acetyl CoA formed from breakdown of triacylglycerol could not enter the TCA cycle. The reason is that oxidation of ethanol and fatty acids generates a large amount of NADH. The high NADH/NAD+ ratio shifts all the oxaloacetate out of the TCA cycle to malate (see Chapter 14). The acetyl CoA therefore has to move to ketone synthesis as it cannot synthesize citrate in the absence of oxaloacetate.

B. Mention other conditions in which you may see:
   (i) Hypoglycemia with ketonemia
   (ii) Hyperglycemia with ketonemia

Answer—B: Ketonemia will be seen whenever there is excess breakdown of stored triacylglycerol for energy.

(i) Starvation will cause hypoglycemia (because of lack of food intake) and ketonemia (because of excess acetyl CoA from breakdown of stored fats which cannot go to TCA cycle) (see Chapter 28).

(ii) Diabetes mellitus (see Chapter 29 for details).

Chapter 22 Phospholipid, Glycosphingolipid, and Eicosanoid Metabolism

22.1. A young primigravida gave birth to a baby at 32 weeks of gestation in a remote Indian village. The baby was not having normal respiration because of which the parents got the child to the nearby healthcare center. The doctor on duty put the baby on oxygen and sent them to the city hospital immediately. At city hospital, a chest X-ray was done which showed pronounced hypoaeration and bilateral fine granular opacities in pulmonary parenchyma. The baby was given surfactant replacement therapy as well as put on ventilator. He survived the crisis and was discharged after a few weeks.

A. What was the diagnosis?

Answer—A: Respiratory distress syndrome.

B. Why is it common in premature neonates?

Answer—B: Respiratory distress syndrome is one of the most common cause of morbidity in preterm infants. The baby presents with apnea, cyanosis, inspiratory stridor and tachycardia. This is due to inadequate synthesis of dipalmitoyl lecithin which is the main component of pulmonary surfactant. Surfactant reduces the surface tension in the alveoli by forming a thin layer film on the alveolar wall. It maintains the shape of the alveoli and prevents its collapse. Within the fetus, the surfactant is not required. However, at birth it must be present in adequate levels to prevent lung collapse. It starts getting synthesized at approximately 30 weeks of gestation. Hence premature births, as in this case, are prone to respiratory distress syndrome.

Chapter 23 Cholesterol, Lipoprotein, and Steroid Metabolism

23.1. An obese (BMI 32 kg/m^2), 45-year-old male presented to the emergency with acute chest pain radiating to the left arm. His ECG showed changes typical of an acute myocardial infarction. His troponin levels were also raised. He was given oxygen, vasodilators, thrombolytics, and pain medication. Angioplasty was done promptly. Before discharge, the resident on duty got his lipid profile done. His cholesterol levels were 300 mg/dl, LDL was 200 mg/dl and HDL was 31 mg/dl. His glycated Hb was 7.8% and he was not a known diabetic. He was put on statins and medications for diabetes mellitus.

A. How is BMI calculated and what level signifies obesity?

Answer—A: Body mass index (BMI) is the weight in kg divided by square of height (in meter). A BMI of over 30 implies obesity.

B. What are the normal levels of cholesterol, LDL, and HDL?

Answer—B: Cholesterol levels of a 45-year-old male should be less than 200 mg/dl and greater than 240 mg/dl is high risk

LDL cholesterol 130–160 mg/dl is borderline high and greater than 160 mg/dl is high risk.

HDL cholesterol of 40 mg/dl or less is considered a risk factor for heart disease.

C. What is the mechanism of action of statins?

Answer—C: Statins are HMG CoA reductase inhibitors and therefore inhibit the rate-limiting step of cholesterol synthesis and decrease blood cholesterol.

D. Why glycated Hb test was done for him and did the result matter for his ailment?

Answer—D: The normal glycated Hb levels should be less than 5.7%. A high value shows that the patient is also diabetic and therefore more prone to myocardial infarction given the lipid modifications, oxidative stress and inflammatory responses in diabetics.

23.2. A 22-year-old male, during routine screening, was found to have an LDL cholesterol of 200 mg/dL and a total cholesterol of 240 mg/dL. The patient does not report any symptoms and is active, consuming a balanced diet. A detailed examination revealed tendon xanthomas of the left Achilles tendon. His father was diagnosed with coronary artery disease at 45 years old, and his paternal grandfather suffered a fatal heart attack at the age of 50. His father's sister has been on medication for high cholesterol since her 30s.

Examination: Weight: 70 kg, Height: 180 cm, BMI: 21.6 kg/m^2 (normal), Blood pressure: 120/80 mmHg

Physical examination reveals tendon xanthomas on the Achilles tendons. No corneal arcus or xanthelasma noted.

A. What does the young man's clinical picture and lipid profile indicate? What investigation will confirm the diagnosis?

Answer—A: An elevated serum LDL of >190 mg/dL, along with tendon xanthomas at a young age and a family history of premature ischemic heart disease in a first-degree relative <55 years of age, is suggestive of familial hypercholesterolemia. Genetic testing to identify mutations should be performed.

B. What further advice would you give to this family?

Answer—B: Cascade screening involves screening close relatives of a person diagnosed with a specific genetic condition to identify individuals with the mutation. This can be followed by genetic testing to confirm the diagnosis.

23.3. A 52-year-old male software engineer visits the OPD for his annual health checkup. He has no specific complaints and has not undergone any health evaluation previously. The man leads a sedentary lifestyle with minimal physical activity, spending most of his day sitting at a computer with no regular exercise routine. His diet is high in fats and refined carbohydrates, with frequent consumption of fast food, sweets and sugary beverages. He smokes very rarely and consumes alcohol on social occasions. His father is diabetic and had a myocardial infarction at 55 years of age.

Physical examination: Weight: 90 kg, Height: 175 cm, BMI: 29.4 kg/m^2, Blood pressure: 145/90 mmHg. No visible xanthomas or xanthelasmas. Abdominal examination reveals central obesity (waist circumference 108 cm).

A. Enumerate modifiable and non-modifiable risk factors for coronary heart disease.

Answer—A: Modifiable risk factors are smoking, alcohol, obesity, lifestyle and exercise.

Non modifiable risk factors are family history with dyslipidemia, diabetes and hypertension,

B. What are the components of metabolic syndrome?

Answer—B: The five recognized components of metabolic syndrome are: waist circumference based on ethnicity to identify abdominal obesity, high blood pressure, impaired fasting blood sugar levels, higher levels of triglycerides and a reduction in the amount of HDL. See Appendix of Unit VI for diagnostic criteria of metabolic syndrome.

Unit V

Nitrogen Metabolism: Amino Acids

Competencies Addressed

- **BI5.3** Describe the digestion and absorption of dietary proteins
- **BI5.4** Describe common disorders associated with protein metabolism
- **BI5.5** Interpret laboratory results of analytes associated with metabolism of proteins
- **BI6.11** Describe the functions of heme in the body and describe the processes involved in heme and porphyrin metabolism

Unit V: Overview

Amino acid metabolism is focused on digestion and assimilation, *de novo* synthesis, or degradation of body protein. As such, **there is no storage form of amino acids** (unlike carbohydrates which are stored as glycogen and fats which are stored as triacylglycerol**). If ingested in excess, the amino acids are degraded into amino groups and carbon skeletons**. The amino groups form **urea** (Chapter 24) and the carbon skeletons form many intermediate compounds as well as specialized products (Chapters 25 and 26).

The amino groups of the amino acids are first channelized to **glutamate (through transamination reaction**) and then given into liver urea cycle to be made into urea. This urea, made in liver, is excreted out through the kidney. Therefore, liver disorders raise ammonia in blood (**hyperammonemia** because of non-detoxified amino groups), whereas kidney disorders raise blood urea because the kidney is unable to excrete the urea formed in liver (Chapter 24).

The student will do well to differentiate urea (disposal form of amino groups from amino acids) from uric acid (degradation product of purine catabolism) as they have no relation to each other but often uremia is confused with hyperuricemia in the learning stages.

The **carbon skeleton of amino acids can be glucogenic or ketogenic** and forms intermediaries of metabolism. Many genetic defects of amino acid metabolism have been discovered and lead to rare diseases like phenylketonuria and alkaptonuria. Special products like **tyrosine, catecholamine and serotonin** are synthesized from amino acids (Chapter 25). Amino acids are also the precursors of many important nitrogen-containing compounds like **hemeproteins** (hemoglobin, myoglobin, cytochromes, cytochrome P450, etc.), neurotransmitters, hormones, purines and pyrimidines (Chapter 26).

The primary healthcare physician often sees high blood urea in his patients which point towards kidney diseases or urinary tract obstruction. Detailed kidney function tests must be done for these patients (see Chapter 48: Organ function tests). Hyperammonemia symptoms are commonly seen in liver diseases like cirrhosis, often due to alcoholism. Liver function tests, ultrasound and CT scans will be required.

Digestion and Absorption of Proteins and Nitrogen Disposal from Amino Acids

24

Learning Objectives

At the end of this chapter, the student will be able to:

- Describe dietary protein digestion
- Discuss nitrogen removal from amino acids and the urea cycle
- Explain hyperammonemia and its causes

I. OVERVIEW

Unlike fats and carbohydrates, amino acids are not stored by the body. That is, no protein exists whose sole function is to maintain a supply of amino acids for future use. Therefore, amino acids must be obtained from the diet, synthesized de novo, or produced from the degradation of body protein. **Any amino acids in excess of the biosynthetic needs of the cell are rapidly degraded**. The first phase of catabolism involves the removal of the α-amino groups (usually by **transamination and subsequent oxidative deamination**), forming ammonia and the corresponding α-keto acids, the carbon skeletons of amino acids. A portion of the free ammonia is excreted in the urine, but most is used in the **synthesis of urea** (Fig. 24.1), which is quantitatively the most important route for disposing of nitrogen from the body. In the second phase of amino acid catabolism, described in Chapter 25, the carbon skeletons of the α-keto acids are converted to common intermediates of energy-producing metabolic pathways. These compounds can be metabolized to carbon dioxide (CO_2) and water (H_2O), glucose, fatty acids, or ketone bodies by the central pathways of metabolism described in Units III and IV.

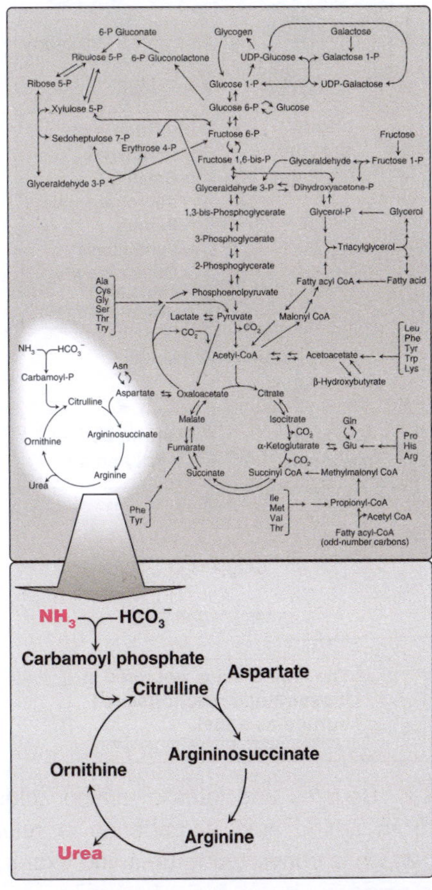

Figure 24.1 Urea cycle shown as part of the essential pathways of energy metabolism. [Note: See Fig. 13.2, p. 187, for a more detailed map of metabolism.] NH_3 = ammonia; CO_2 = carbon dioxide.

II. OVERALL NITROGEN METABOLISM

Amino acid catabolism is part of the larger process of the metabolism of nitrogen-containing molecules. Nitrogen enters the body in a variety of compounds present in food, the most important being amino acids contained in dietary

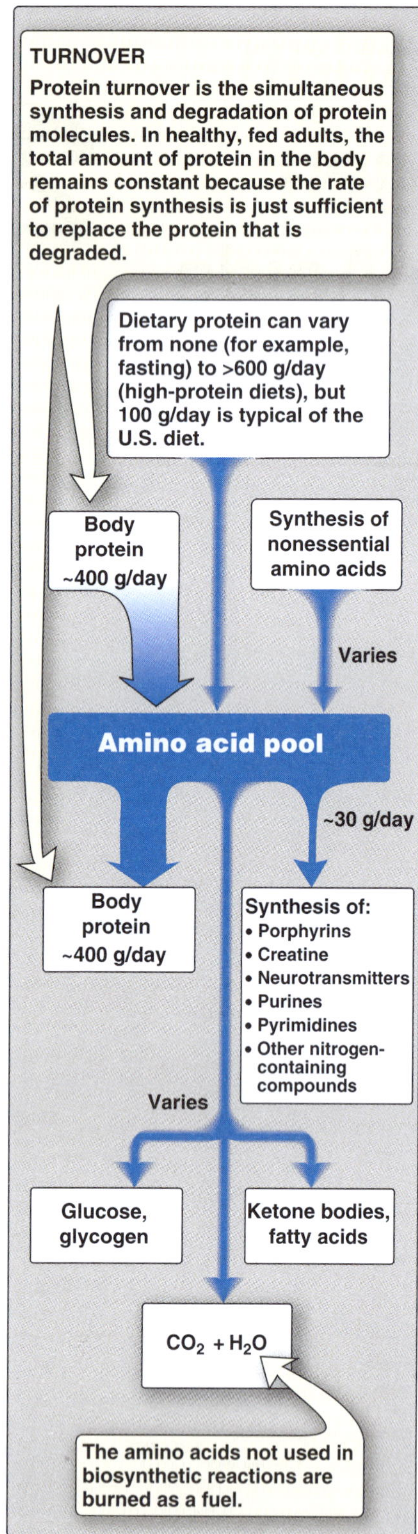

Figure 24.2 Sources and fates of amino acids. [Note: Nitrogen from amino acid degradation is released as ammonia, which is converted to urea and excreted.] CO_2 = carbon dioxide.

protein. Nitrogen leaves the body as urea, ammonia, and other products derived from amino acid metabolism (such as creatinine, see p. 424). The role of body proteins in these transformations involves two important concepts: the amino acid pool and protein turnover.

A. Amino acid pool

Free amino acids are present throughout the body, such as in cells, blood, and the extracellular fluids. For the purpose of this discussion, envision all of these amino acids as if they belonged to a single entity, called the amino acid pool. **This pool is supplied by three sources:** 1) amino acids provided by the degradation of **endogenous (body) proteins**, most of which are reutilized; 2) amino acids derived from **exogenous (dietary) protein**; and 3) nonessential amino acids **synthesized** from simple intermediates of metabolism (Fig. 24.2). Conversely, the **amino acid pool is depleted by three routes:** 1) **synthesis of body protein**, 2) consumption of amino acids as precursors of essential nitrogen-containing small molecules (**specialized products**), and 3) conversion of amino acids to **glucose, glycogen, fatty acids, and ketone bodies or oxidation to CO_2 + H_2O** (see Fig. 24.2). Although the amino acid pool is small (comprising ~90–100 g of amino acids) in comparison with the amount of protein in the body (~12 kg in a 70-kg man), it is conceptually at the center of whole-body nitrogen metabolism.

> In healthy, well-fed individuals, the input to the amino acid pool is balanced by the output. That is, the amount of amino acids contained in the pool is constant. The amino acid pool is said to be in a steady state, and the individual is said to be in nitrogen balance (see p. 513).

B. Protein turnover

Most proteins in the body are constantly being synthesized and then degraded (turned over), permitting the removal of abnormal or unneeded proteins. For many proteins, regulation of synthesis determines the concentration of protein in the cell, with protein degradation assuming a minor role. For other proteins, the rate of synthesis is constitutive (that is, essentially constant), and cellular levels of the protein are controlled by selective degradation.

1. **Rate:** In healthy adults, the total amount of protein in the body remains constant because the rate of protein synthesis is just sufficient to replace the protein that is degraded. This process, called protein turnover, leads to the **hydrolysis and resynthesis of 300–400 g of body protein each day**. The rate of protein turnover varies widely

for individual proteins. Short-lived proteins (for example, many regulatory proteins and misfolded proteins) are rapidly degraded, having half-lives measured in minutes or hours. Long-lived proteins, with half-lives of days to weeks, constitute the majority of proteins in the cell. Structural proteins, such as collagen, are metabolically stable and have half-lives measured in months or years.

2. **Protein degradation:** There are two major enzyme systems responsible for degrading proteins: the **ATP-dependent ubiquitin (Ub)–proteasome system** of the cytosol and the **ATP-independent degradative enzyme system of the lysosomes**. Proteasomes selectively degrade damaged or short-lived proteins. Lysosomes use ***acid hydrolases*** (see p. 266) to nonselectively degrade intracellular proteins (autophagy) and extracellular proteins (heterophagy), such as plasma proteins, that are taken into the cell by endocytosis. Liver cells can take up desialylated plasma proteins by asialoglycoprotein receptor-mediated endocytosis.

   a. **Ubiquitin–proteasome system:** Proteins selected for degradation by the cytosolic ubiquitin–proteasome system are first modified by the covalent attachment of Ub, a small, globular, nonenzymic protein that is highly conserved across eukaryotic species. **Ubiquitination** of the target substrate occurs through isopeptide linkage of the α-carboxyl group of the C-terminal glycine of Ub to the ε-amino group of a lysine in the protein substrate by a three-step, enzyme-catalyzed, ATP-dependent process. [Note: *Enzyme 1* (*E1*, an activating enzyme) activates Ub, which is then transferred to *E2* (a conjugating enzyme). *E3* (a *ligase*) identifies the protein to be degraded and interacts with *E2*-Ub. There are many more *E3* proteins than there are *E1* or *E2*.] The consecutive addition of four or more Ub molecules to the target protein generates a polyubiquitin chain. Proteins tagged with Ub chains are recognized by a large, barrel-shaped, macromolecular, proteolytic complex called a **proteasome** (Fig. 24.3). The proteasome unfolds, deubiquitinates, and cuts the target protein into fragments that are then further degraded by cytosolic *proteases* to amino acids, which enter the amino acid pool. The Ub is recycled. It is noteworthy that the selective degradation of proteins by the ubiquitin–proteosome complex (unlike simple hydrolysis by proteolytic enzymes) requires ATP hydrolysis.

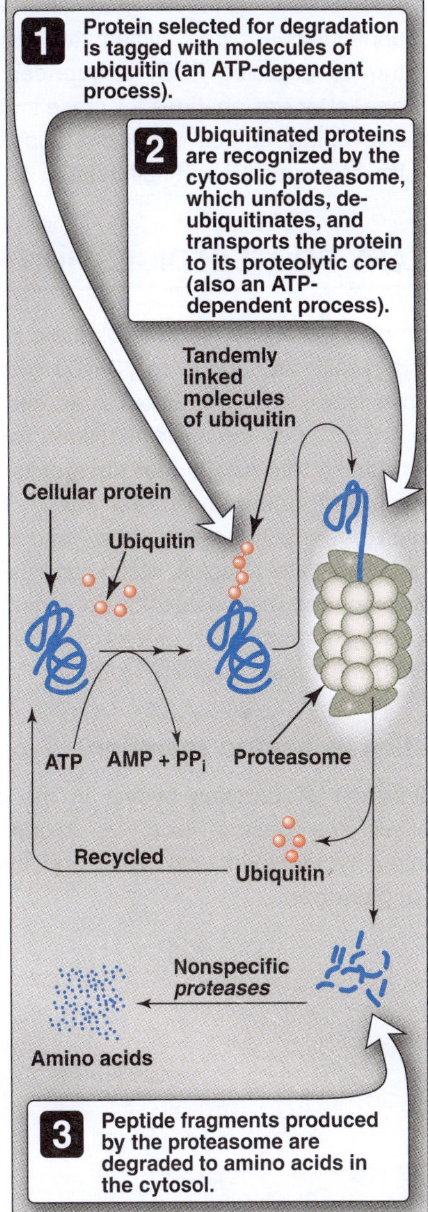

Figure 24.3 The ubiquitin–proteasome degradation pathway of proteins. AMP = adenosine monophosphate; PP_i = pyrophosphate.

   b. **Degradation signals:** Because proteins have **different half-lives**, it is clear that protein degradation cannot be random but, rather, is influenced by some structural aspect of the protein that serves as a **degradation signal**, which is recognized and bound by an *E3*. The half-life of a protein is also influenced by the amino (N)-terminal residue, the so-called N-end rule, and ranges from minutes to hours. Destabilizing N-terminal amino acids include **arginine** and posttranslationally modified amino acids such as **acetylated alanine**. In contrast, serine is a stabilizing amino acid.

Additionally, proteins rich in sequences containing proline, glutamate, serine, and threonine (called **PEST sequences** after the one-letter designations for these amino acids) are rapidly ubiquitinated and degraded and, therefore, have short half-lives.

III. DIETARY PROTEIN DIGESTION

Most of the nitrogen in the diet is consumed in the form of protein, typically amounting to 70–100 g/day in the American diet (see Fig. 24.2). Proteins are generally too large to be absorbed by the intestine. [Note: An example of **an exception to this rule is that newborns can take up maternal antibodies in breast milk**.] Therefore, proteins must be hydrolyzed to yield di- and tripeptides as well as individual amino acids, which can be absorbed. **Proteolytic enzymes** responsible for degrading proteins are produced by three different organs: the stomach, the pancreas, and the small intestine (Fig. 24.4).

A. Digestion by gastric secretion

The digestion of proteins begins in the stomach, which secretes gastric juice, a unique solution containing hydrochloric **acid (HCl)** and the **proenzyme pepsinogen**.

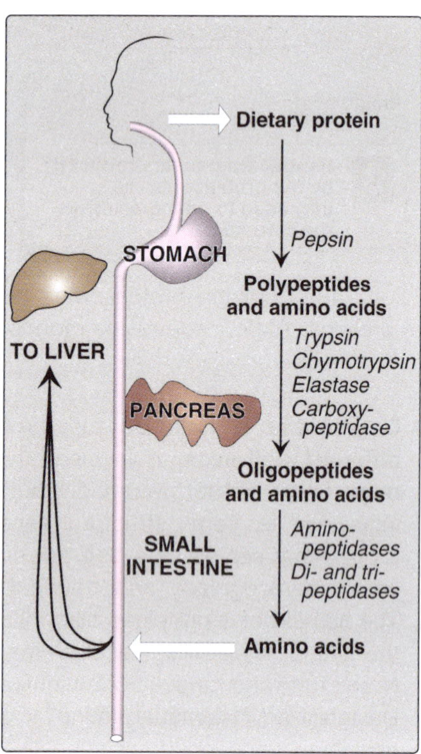

Figure 24.4 Digestion of dietary proteins by the proteolytic enzymes of the gastrointestinal tract.

1. **Hydrochloric acid:** Stomach HCl is too dilute (pH 2–3) to hydrolyze proteins. The acid, secreted by the parietal cells of the stomach, functions instead to kill some bacteria and to denature proteins, thereby making them more susceptible to subsequent hydrolysis by *proteases*.

2. **Pepsin:** This **acid-stable *endopeptidase*** is secreted by the chief cells of the stomach as an **inactive zymogen (or proenzyme), pepsinogen**. [Note: In general, zymogens contain extra amino acids in their sequences that prevent them from being catalytically active. Removal of these amino acids permits the proper folding required for an active enzyme.] In the presence of HCl, pepsinogen undergoes a conformational change that allows it to cleave itself (autocatalysis) to the **active form, *pepsin***, which releases polypeptides and a few free amino acids from dietary proteins and cuts a peptide bond that involves a tyrosine or a phenylalanine.

B. Digestion by pancreatic enzymes

On entering the small intestine, the polypeptides produced in the stomach by the action of *pepsin* are further cleaved to oligopeptides and amino acids by a group of pancreatic ***proteases*** that include both ***endopeptidases*** (that cleave within) and ***exopeptidases*** (that cut at an end). [Note: **Bicarbonate (HCO_3^-),** secreted by the pancreas in response to the intestinal hormone secretin, raises the intestinal pH.]

1. **Specificity:** Each of these enzymes has a different specificity for the amino acid R-groups adjacent to the susceptible peptide bond (Fig. 24.5). For example, ***trypsin*** cleaves only when the carbonyl group of the peptide bond is contributed by arginine or lysine. These enzymes, like *pepsin* described above, are synthesized and secreted as inactive zymogens.

2. **Zymogen release:** The release and activation of the pancreatic zymogens are mediated by the secretion of **cholecystokinin**, a polypeptide hormone of the small intestine (see p. 292).

3. **Zymogen activation: *Enteropeptidase*** (also called *enterokinase*), a *serine protease* synthesized by and present on the luminal (apical) surface of intestinal mucosal cells (enterocytes) of the brush border, converts the pancreatic zymogen trypsinogen to *trypsin* by removal of a hexapeptide from the N-terminus of trypsinogen. *Trypsin* subsequently converts other trypsinogen molecules to *trypsin* by cleaving a limited number of specific peptide bonds in the zymogen. Thus,

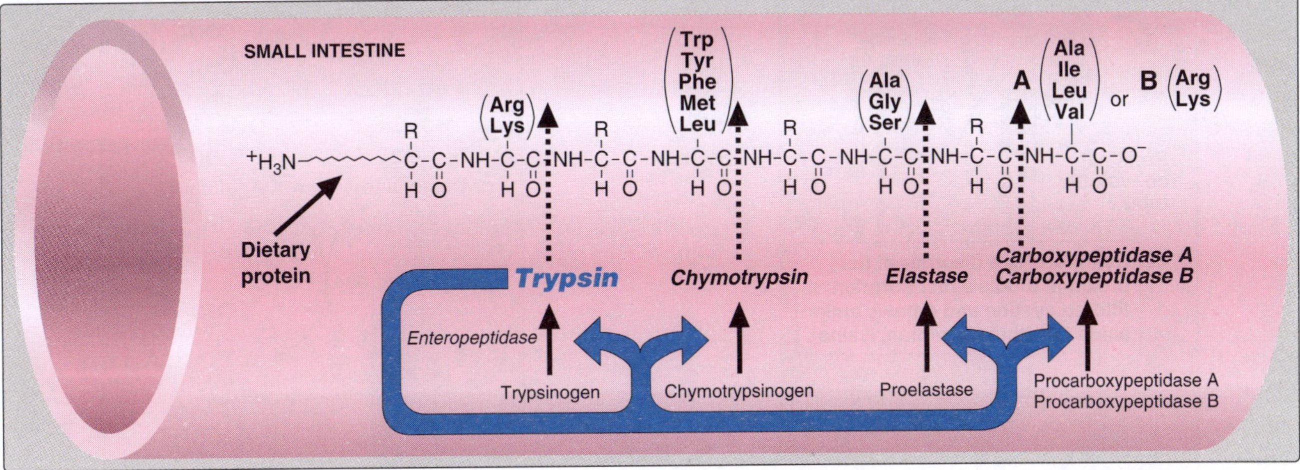

Figure 24.5 Cleavage of dietary protein in the small intestine by pancreatic *proteases*. The peptide bonds susceptible to hydrolysis are shown for each of the five major pancreatic *proteases*. [Note: The first three are serine *endopeptidases*, whereas the last two are *exopeptidases*. Each is produced from an inactive zymogen.]

enteropeptidase unleashes a cascade of proteolytic activity because *trypsin* is the common activator of all the pancreatic zymogens (see Fig. 24.5).

4. **Digestion abnormalities:** In individuals with a deficiency in pancreatic secretion (for example, because of **chronic pancreatitis, cystic fibrosis, or surgical removal** of the pancreas), the digestion and absorption of fat and protein are incomplete. This results in the abnormal appearance of lipids in the feces (a condition called **steatorrhea;** see p. 292) as well as undigested protein.

> **Celiac disease (celiac sprue)** is a disease of malabsorption resulting from immune-mediated damage to the small intestine in response to ingestion of gluten (or gliadin produced from gluten), a protein found in wheat, barley, and rye.

C. Digestion of oligopeptides by small intestine enzymes

The luminal surface of the enterocytes contains *aminopeptidase*, an *exopeptidase* that repeatedly cleaves the N-terminal residue from oligopeptides to produce even smaller peptides and free amino acids.

D. Amino acid and small peptide intestinal absorption

Most free amino acids are taken into enterocytes via sodium-dependent secondary active transport by **solute carrier (SLC) proteins of the apical membrane**. At least seven different transport systems with overlapping amino acid specificities are known. Di- and tripeptides, however, are taken up by a **proton-linked peptide transporter (PepT1)**. The peptides are then hydrolyzed to free amino acids. Regardless of their source, free amino acids are released from enterocytes into the portal system by sodium-independent transporters of the basolateral membrane. Therefore, only free amino acids are found in the portal vein after a meal containing protein. ORS fortified with amino acid **(super ORS)** enhances water and sodium uptake from intestine and rapidly corrects fluid-electrolyte imbalance in diarrhea. These amino acids are either metabolized by the liver or released into the general circulation. [Note: **Branched-chain amino acids (BCAA) are not metabolized by the liver but, instead, are sent from the liver to muscle via the blood.]**

E. Absorption abnormalities

The small intestine and the proximal tubules of the kidneys have common transport systems for amino acid uptake. Consequently, a defect in any one of these systems results in an inability to absorb particular amino acids into the intestine and into the kidney tubules. For example, one system is responsible for the uptake of cystine and the dibasic amino acids ornithine, arginine, and lysine (represented as **COAL**). In the **inherited disorder cystinuria**, this carrier system is defective, and all four amino acids appear in the urine (Fig. 24.6). **Cystinuria** occurs at a frequency of 1 in 7,000 individuals, making it one of the most common inherited diseases and the most common genetic error of amino acid transport. The disease expresses itself clinically by the precipitation of cystine to form kidney stones (calculi), which can block the urinary tract. Oral hydration is an important part of treatment for this disorder. [Note: **Defects in the uptake of tryptophan by a neutral amino acid**

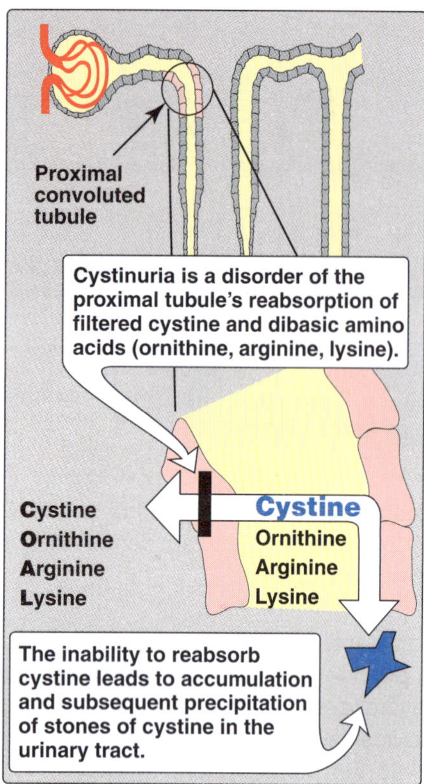

Figure 24.6 Genetic defect seen in cystinuria. [Note: Cystinuria is distinct from cystinosis, a rare defect in the transport of cystine out of lysosomes that results in the formation of cystine crystals within the lysosome and widespread tissue damage.]

transporter can result in Hartnup disorder and pellagra-like (see p. 534) dermatologic and neurologic symptoms as niacin is synthesized from tryptophan in our body.]

IV. NITROGEN REMOVAL FROM AMINO ACIDS

The presence of the α-amino group keeps amino acids safely locked away from oxidative breakdown. Removing the α-amino group is essential for producing energy from any amino acid and is an obligatory step in the catabolism of all amino acids. Once removed, this nitrogen can be incorporated into other compounds or excreted as urea, with the carbon skeletons being metabolized. This section describes transamination and oxidative deamination, reactions that ultimately provide ammonia and aspartate, the two sources of urea nitrogen (see p. 386).

A. Transamination: Funneling amino groups to glutamate

The first step in the catabolism of most amino acids is the transfer of their α-amino group to α-ketoglutarate

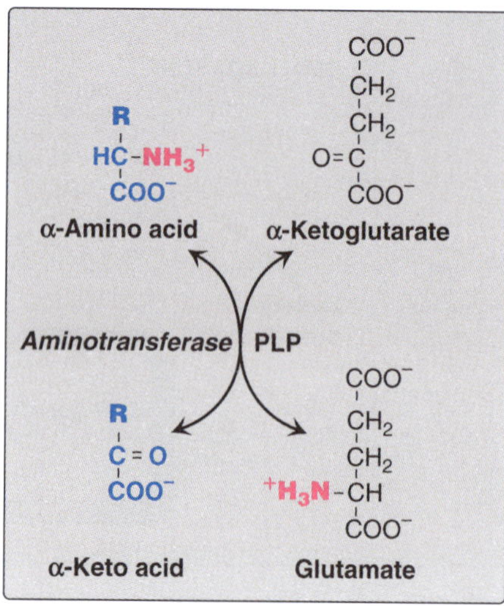

Figure 24.7 *Aminotransferase* reaction using α-ketoglutarate as the amino group acceptor. PLP = pyridoxal phosphate.

(Fig. 24.7), producing an α-keto acid (derived from the original amino acid) and glutamate. α-Ketoglutarate plays a pivotal role in amino acid metabolism by accepting the amino groups from most amino acids, thereby becoming glutamate. Glutamate produced by transamination can be oxidatively deaminated (see B. below) or used as an amino group donor in the synthesis of nonessential amino acids. This transfer of amino groups from one carbon skeleton to another is catalyzed by a family of enzymes called ***aminotransferases*** **(also called *transaminases*).** These enzymes are found in the cytosol and mitochondria of cells throughout the body. All amino acids, with the exception of lysine and threonine, participate in transamination at some point in their catabolism. [Note: These two amino acids lose their α-amino groups by deamination (see p. 401).]

1. **Substrate specificity:** Each *aminotransferase* is specific for one or, at most, a few amino group donors. *Aminotransferases* are named after the specific amino group donor, because the acceptor of the amino group is almost always α-ketoglutarate. Two important *aminotransferase* reactions are catalyzed by ***alanine aminotransferase*** **(*ALT*) and *aspartate aminotransferase* (*AST*)**, as shown in Figure 24.8.

   a. **Alanine aminotransferase:** *ALT* is present in many tissues. The enzyme catalyzes the transfer of the amino group of alanine to α-ketoglutarate, resulting in the formation of pyruvate and glutamate. The reaction is readily reversible. However, during amino acid catabolism, this enzyme (like most *aminotransferases*) functions in the direction of glutamate

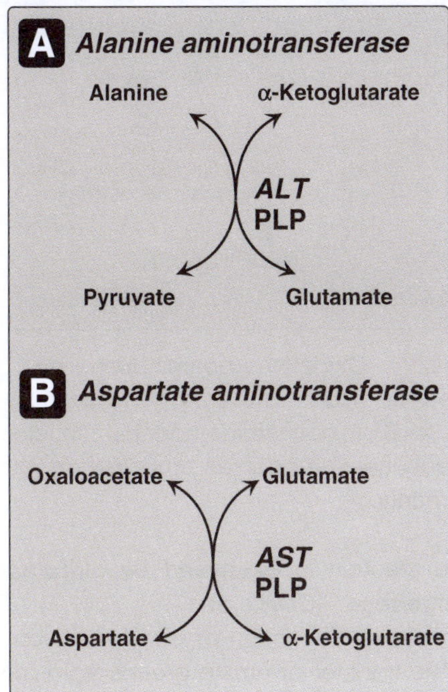

Figure 24.8 Reactions catalyzed during amino acid catabolism. **A.** *Alanine aminotransferase* (*ALT*). **B.** *Aspartate aminotransferase* (*AST*). PLP = pyridoxal phosphate.

synthesis. [Note: In effect, glutamate acts as a collector of nitrogen from most amino acids.]

b. **Aspartate aminotransferase:** *AST* is an exception to the rule that *aminotransferases* funnel amino groups to form glutamate. During amino acid catabolism, *AST* primarily transfers amino groups from glutamate to oxaloacetate, forming aspartate, which is used as a source of nitrogen in the urea cycle (see p. 385). Like other transaminations, the *AST* reaction is reversible.

2. **Mechanism: All *aminotransferases* require the coenzyme pyridoxal phosphate (a derivative of vitamin B_6;** see p. 531), which is covalently linked to the ε-amino group of a specific lysine residue at the active site of the enzyme. *Aminotransferases* act by transferring the amino group of an amino acid to the pyridoxal part of the coenzyme to generate pyridoxamine phosphate. The pyridoxamine form of the coenzyme then reacts with an α-keto acid to form an amino acid, at the same time regenerating the original aldehyde form of the coenzyme. Figure 24.9 shows these two component reactions for the transamination catalyzed by *AST*.

3. **Equilibrium:** For most transamination reactions, the equilibrium constant is near 1. This allows the reaction to function in both amino acid degradation through removal of α-amino groups (for example, after consumption of a protein-rich meal) and biosynthesis of nonessential amino acids through addition of amino groups to the carbon skeletons of α-keto acids (for example, when the supply of amino acids from the diet is not adequate to meet the synthetic needs of cells).

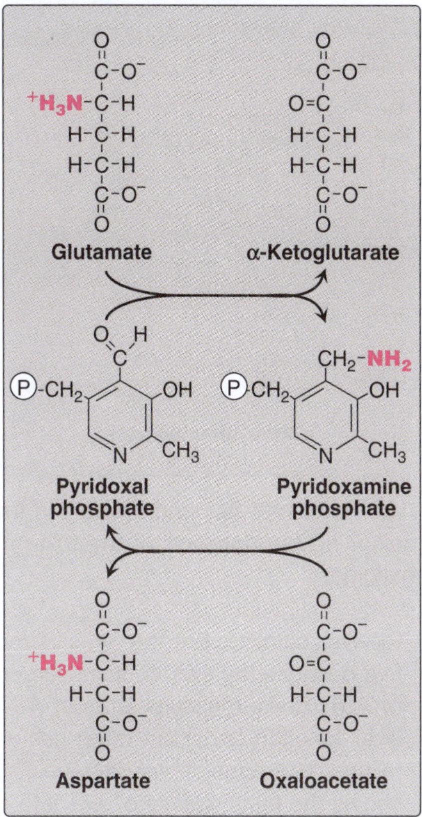

Figure 24.9 Cyclic interconversion of pyridoxal phosphate and pyridoxamine phosphate during the *aspartate aminotransferase* reaction. Ⓟ = phosphate group.

4. **Diagnostic value:** *Aminotransferases* are normally intracellular enzymes, with the low levels found in the plasma representing the release of cellular contents during normal cell turnover. Elevated plasma levels of *aminotransferases* indicate damage to cells rich in these enzymes. For example, physical trauma or a disease process can cause cell lysis, resulting in release of intracellular enzymes into the blood. Two *aminotransferases*, *AST* and *ALT*, are of particular diagnostic value when they are found in the plasma.

a. **Hepatic disease: Plasma *AST* and *ALT*** are elevated in nearly all hepatic diseases but are particularly high in conditions that cause extensive cell necrosis, such as severe viral hepatitis, toxic injury, and prolonged circulatory collapse. *ALT* is more specific than *AST*

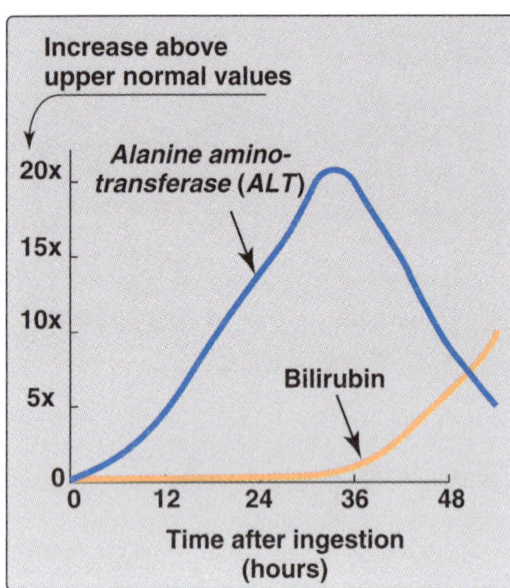

Figure 24.10 Pattern of *ALT* and bilirubin in the plasma, following poisoning by ingestion of the toxic mushroom Amanita phalloides.

for liver disease, but the latter is more sensitive because the liver contains larger amounts of *AST*. Serial measurements of *AST* and *ALT* (liver function tests) are often useful in determining the course of liver damage. Figure 24.10 shows the early release of *ALT* into the blood, following ingestion of a liver toxin. [Note: The elevation in bilirubin results from hepatocellular damage that decreases the hepatic conjugation and excretion of bilirubin (see p. 420).]

b. **Nonhepatic disease:** *Aminotransferases* may be elevated in nonhepatic diseases such as those that cause damage to cardiac or skeletal muscle. However, these disorders can usually be distinguished clinically from liver disease.

B. Oxidative deamination: Amino group removal

In contrast to transamination reactions that transfer amino groups, **oxidative deamination reactions result in the liberation of the amino group as free ammonia** (Fig. 24.11). These reactions occur primarily in the **liver and kidney**. They provide α-keto acids that can enter the central pathways of energy metabolism and ammonia, which is a source of nitrogen in hepatic urea synthesis. [Note: Ammonia exists primarily as ammonium (NH_4^+) in aqueous solution, but it is the unionized form (NH_3) that crosses membranes.]

1. **Glutamate dehydrogenase:** As described above, the amino groups of most amino acids are ultimately funneled to glutamate by means of transamination with α-ketoglutarate. **Glutamate is unique in that it is the only amino acid that undergoes oxidative deamination** at an appreciable rate.

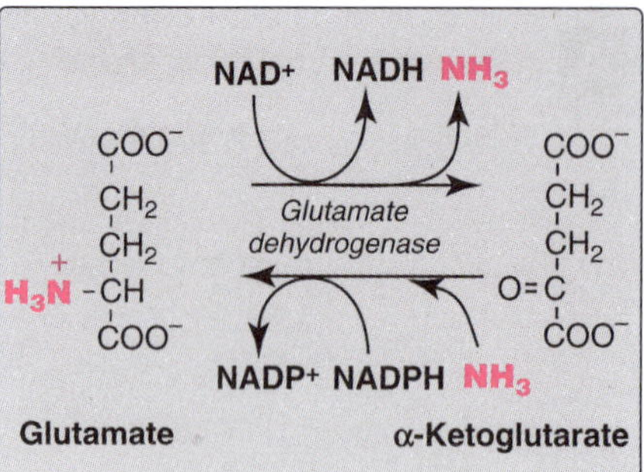

Figure 24.11 Oxidative deamination by *glutamate dehydrogenase*. [Note: The enzyme is unusual in that it uses both NAD^+ (nicotinamide adenine dinucleotide) and NADPH (nicotinamide adenine dinucleotide phosphate).] NH_3 = ammonia.

The reaction is **catalyzed by *glutamate dehydrogenase*** ([*GDH*], see Fig. 24.11). Therefore, the sequential action of transamination (resulting in the transfer of amino groups from most amino acids to α-ketoglutarate to produce glutamate) and the oxidative deamination of that glutamate (regenerating α-ketoglutarate) provide a pathway whereby the amino groups of most amino acids can be released as ammonia.

a. **Coenzymes: *GDH*, a mitochondrial enzyme**, is unusual in that it can use either nicotinamide adenine dinucleotide (NAD^+) or its phosphorylated reduced form (NADPH) as a coenzyme (see Fig. 24.11**). NAD^+ is used primarily in oxidative deamination** (the simultaneous loss of ammonia coupled with the oxidation of the carbon skeleton, as shown in Fig. 24.12A), whereas **NADPH is used in reductive amination** (the simultaneous gain of ammonia coupled with the reduction of the carbon skeleton, as shown in Fig. 24.12B).

b. **Reaction direction:** The direction of the reaction depends on the relative concentrations of glutamate, α-ketoglutarate, and ammonia and the ratio of oxidized to reduced coenzymes. For example, after ingestion of a meal containing protein, glutamate levels in the liver are elevated, and the reaction proceeds in the direction of amino acid degradation and the formation of ammonia (see Fig. 24.12A). High ammonia levels are required to drive the reaction to glutamate synthesis.

c. **Allosteric regulators: Guanosine triphosphate is an allosteric inhibitor of *GDH*, whereas adenosine diphosphate is an activator**. Therefore, when energy levels are low in the cell, amino acid degradation by *GDH*

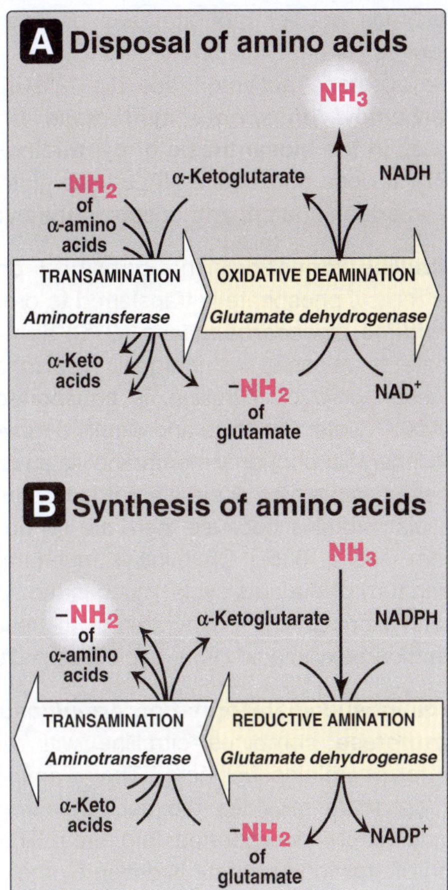

Figure 24.12 **A, B.** Combined actions of *aminotransferase* and *glutamate dehydrogenase* reactions. [Note: Reductive amination occurs only when ammonia (NH_3) level is high.] NAD(H) = nicotinamide adenine dinucleotide; NADP(H) = nicotinamide adenine dinucleotide phosphate.

is high, facilitating energy production from the carbon skeletons derived from amino acids.

2. **d-Amino acid oxidase: D-Amino acids (see p. 74) are supplied by the diet but are not used in the synthesis of mammalian proteins.** They are, however, efficiently **metabolized to α-keto acids, ammonia, and hydrogen peroxide** in the peroxisomes of liver and kidney cells by flavin adenine dinucleotide–dependent ***D-amino acid oxidase*** **(*DAO*).** The α-keto acids can enter the general pathways of amino acid metabolism and be reaminated to L-isomers or catabolized for energy. [Note: *DAO* degrades D-serine, the isomeric form of serine that modulates *N*-methyl-D-aspartate (NMDA)-type glutamate receptors. Increased *DAO* activity has been linked to increased susceptibility to schizophrenia. *DAO* also converts glycine to glyoxylate (see p. 398).] ***L-Amino acid oxidases*** are found in snake venom and in liver and kidney peroxisomes in low concentrations.

C. Ammonia transport to the liver

Two mechanisms are available in humans for the transport of ammonia from peripheral tissues to the liver for conversion to urea. Both are important in, but not exclusive to, skeletal muscle. The first uses ***glutamine synthetase*** to combine ammonia with glutamate to form glutamine, a nontoxic transport form of ammonia (Fig. 24.13). The glutamine is transported in the blood to the liver where it is cleaved by ***glutaminase*** to glutamate and ammonia (see p. 389). The glutamate is oxidatively deaminated to ammonia and α-ketoglutarate by *GDH*. The ammonia is converted to urea. The second transport mechanism involves the formation of alanine by the transamination of pyruvate produced from both aerobic glycolysis and metabolism of the succinyl coenzyme A (CoA) generated by the catabolism of the BCAA isoleucine and valine. Alanine is transported in the blood to the liver, where it is transaminated by *ALT* to pyruvate. The pyruvate is used to synthesize glucose, which can enter the blood and be used by muscle, a pathway called the glucose–alanine cycle. The glutamate product of *ALT* can be deaminated by *GDH*, generating ammonia. Thus, both alanine and glutamine carry ammonia to the liver.

V. UREA CYCLE

Urea ($H_2N{-}\overset{\displaystyle O}{\overset{\|}{C}}{-}NH_2$) is the major disposal form of amino groups derived from amino acids and accounts for ~90% of the nitrogen-containing components of urine. One nitrogen of the urea molecule is supplied by free ammonia and the other nitrogen by aspartate. [Note: **Glutamate is the immediate precursor of both ammonia** (through oxidative deamination by *GDH*) **and aspartate nitrogen** (through transamination of oxaloacetate by *AST*).] The carbon and oxygen of urea are derived from CO_2 (as HCO_3^-). Urea is produced by the liver and then is transported in the blood to the kidneys for excretion in the urine.

A. Reactions

The first two reactions leading to the synthesis of urea occur in the mitochondrial matrix, whereas the remaining cycle enzymes are located in the cytosol (Fig. 24.14). [Note: Gluconeogenesis (see p. 215) and heme synthesis (see p. 416) also involve both the mitochondrial matrix and the cytosol.]

1. **Carbamoyl phosphate formation:** Formation of carbamoyl phosphate by ***carbamoyl phosphate synthetase I (CPS I)*** is driven by cleavage of two molecules of ATP. Ammonia incorporated into carbamoyl phosphate is provided primarily by the oxidative deamination of glutamate by

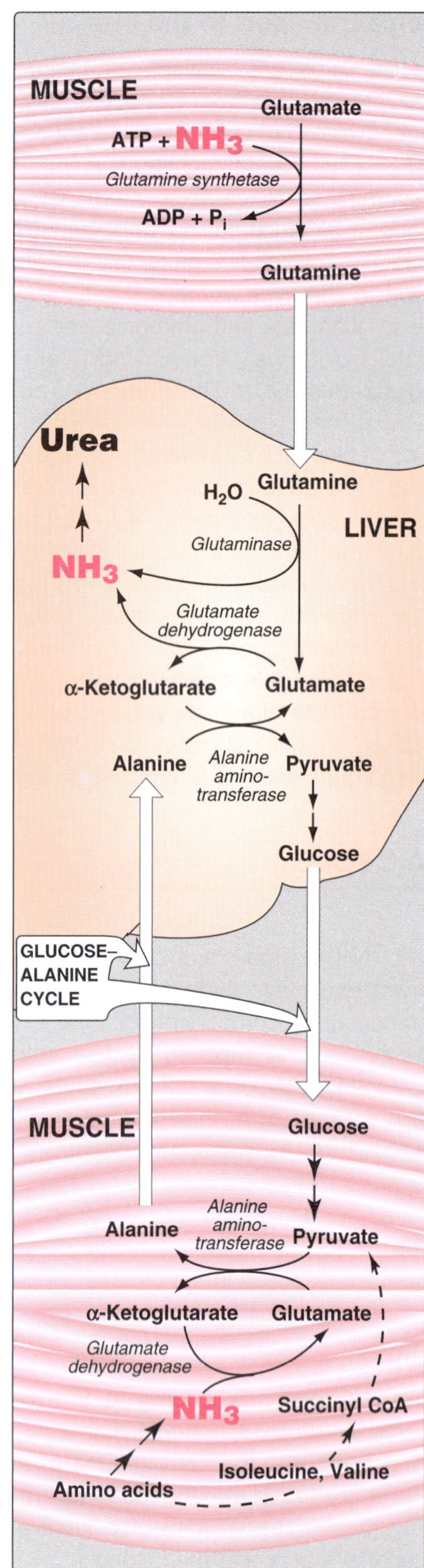

Figure 24.13 Transport of ammonia (NH_3) from muscle to the liver. ADP = adenosine diphosphate; P_i = inorganic phosphate; CoA = coenzyme A.

mitochondrial *GDH* (see Fig. 24.11). Ultimately, the nitrogen atom derived from this ammonia becomes one of the nitrogens of urea. *CPS I* requires N-acetylglutamate (NAG) as a positive allosteric activator (see Fig. 24.14). [**Note: *Carbamoyl phosphate synthetase II* participates in the biosynthesis of pyrimidines** (see p. 587). It does not require NAG, uses glutamine as the nitrogen source, and occurs in the cytosol.]

2. **Citrulline formation:** The carbamoyl portion of carbamoyl phosphate is transferred to ornithine by ***ornithine transcarbamylase* (*OTC*)** as the phosphate is released as inorganic phosphate. The reaction product, citrulline, is transported to the cytosol. [Note: Ornithine and citrulline move across the inner mitochondrial membrane via an antiporter. These basic amino acids are not incorporated into cellular proteins because there are no codons for them (see p. 635).] Ornithine is regenerated with each turn of the urea cycle, much in the same way that oxaloacetate is regenerated by the reactions of the tricarboxylic acid (TCA) cycle (see p. 206).

3. **Argininosuccinate formation:** ***Argininosuccinate synthetase*** combines citrulline with aspartate to form argininosuccinate. The α-amino group of aspartate provides the second nitrogen that is ultimately incorporated into urea. The formation of argininosuccinate is driven by the cleavage of ATP to adenosine monophosphate and pyrophosphate. This is the third and final molecule of ATP consumed in the formation of urea. As it is converted to AMP, this consumption is equivalent to two molecules of ATP. So total consumption of ATP for one molecule of urea is four. Hence, detoxification of ammonia is a costly affair energetically.

4. **Argininosuccinate cleavage:** Argininosuccinate is cleaved by ***argininosuccinate lyase*** to yield arginine and fumarate. The arginine serves as the immediate precursor of urea. The fumarate is hydrated to malate, providing a link with several metabolic pathways. Malate can be oxidized by *malate dehydrogenase* to oxaloacetate, which can be transaminated to aspartate (see Fig. 24.8) and enter the urea cycle (see Fig. 24.14). Alternatively, malate can be transported into mitochondria via the malate–aspartate shuttle (see p. 171), reenter the TCA cycle, and get oxidized to oxaloacetate, which can be used for gluconeogenesis (see p. 217). Thus, **urea cycle is linked to TCA cycle and hence it is called urea bicycle or Krebs-Henseleit cycle**. [Note: Malate oxidation generates NADH for oxidative phosphorylation (see p. 54), thereby providing ATP for urea cycle.]

5. **Arginine cleavage to ornithine and urea:** ***Arginase-I*** hydrolyzes arginine to ornithine and

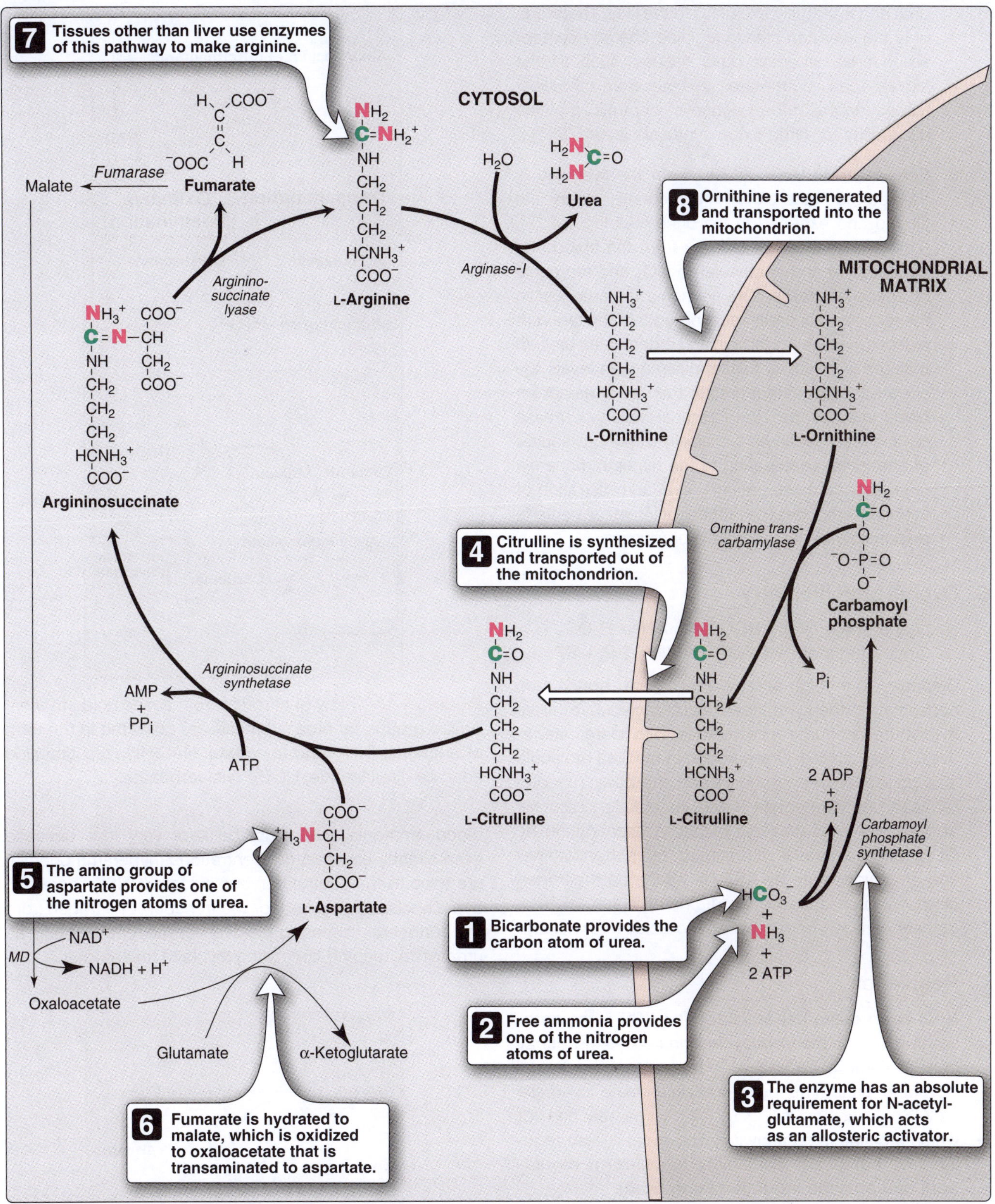

Figure 24.14 Reactions of the urea cycle. [Note: An antiporter transports citrulline and ornithine across the inner mitochondrial membrane.] ADP = adenosine diphosphate; AMP = adenosine monophosphate; PP_i = pyrophosphate; P_i = inorganic phosphate; NAD(H) = nicotinamide adenine dinucleotide; *MD = malate dehydrogenase.*

urea and is virtually exclusive to the liver. Therefore, only the liver can cleave arginine, thereby synthesizing urea, whereas other tissues, such as the kidney, can synthesize arginine from citrulline. [Note: *Arginase-II* in kidneys controls arginine availability for nitric oxide synthesis (see p. 251).]

6. **Fate of urea:** Urea diffuses from the liver and is transported in the blood to the kidneys, where it is filtered and excreted in the urine (see Fig. 24.19). A portion of the urea diffuses from the blood into the intestine and is cleaved to CO_2 and ammonia by bacterial *urease*. The ammonia is partly lost in the feces and is partly reabsorbed into the blood. It reaches liver via portal vein and regenerates urea. In patients with kidney failure, plasma urea levels are elevated, promoting a greater transfer of urea from blood into the gut. The intestinal action of *urease* on this urea becomes a clinically important source of ammonia, contributing to the hyperammonemia often seen in these patients. Oral administration of antibiotics reduces the number of intestinal bacteria responsible for this ammonia production.

B. Overall stoichiometry

$$\text{Aspartate} + NH_3 + HCO_3^- + 3\text{ ATP} + H_2O \rightarrow \text{urea} + \text{fumarate} + 2\text{ ADP} + \text{AMP} + 2\text{ P}_i + \text{PP}_i$$

Because four high-energy phosphate bonds are consumed in the synthesis of each molecule of urea, the synthesis of urea is irreversible, with a large, negative ΔG (see p. 159). One nitrogen of the urea molecule is supplied by free ammonia and the other nitrogen by aspartate. Glutamate is the immediate precursor of both ammonia (through oxidative deamination by *GDH*) and aspartate nitrogen (through transamination of oxaloacetate by *AST*). In effect, both nitrogen atoms of urea arise from glutamate, which, in turn, gathers nitrogen from other amino acids (Fig. 24.15).

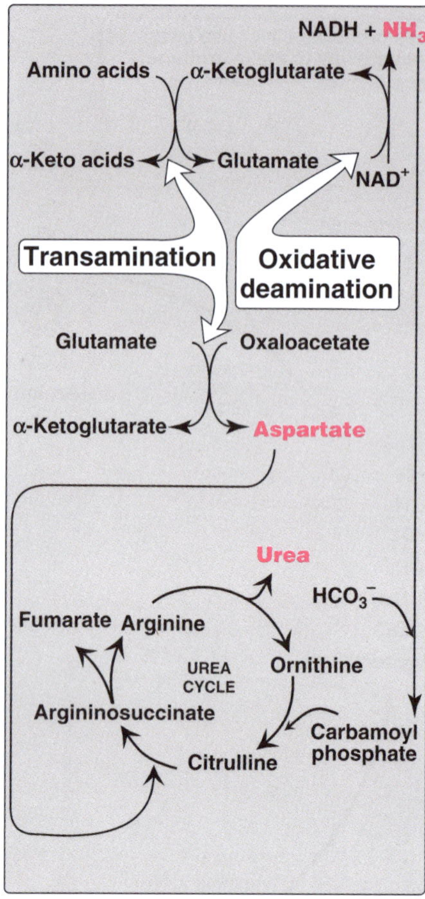

Figure 24.15 Flow of nitrogen from amino acids to urea. Amino groups for urea synthesis are collected in the form of ammonia (NH_3) and aspartate. NAD(H) = nicotinamide adenine dinucleotide; HCO_3^- = bicarbonate.

C. Regulation

NAG is an essential activator for *CPS I*, the rate-limiting step in the urea cycle. It increases the affinity of *CPS I* for ATP. NAG is synthesized from acetyl CoA and glutamate by *N-acetylglutamate synthase* (*NAGS*), as shown in Figure 24.16, in a reaction for which arginine is an activator. The cycle is also regulated by **substrate availability (short-term regulation) and enzyme induction (long term).**

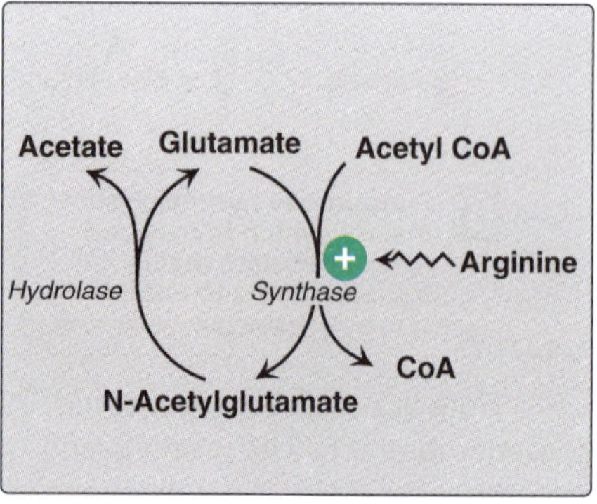

Figure 24.16 Formation and degradation of *N*-acetylglutamate, an allosteric activator of *carbamoyl phosphate synthetase I*. CoA = coenzyme A.

VI. AMMONIA METABOLISM

Ammonia is produced by all tissues during the metabolism of a variety of compounds, and it is disposed of primarily by formation of urea in the liver. However, the blood ammonia level must be kept very low, because even slightly elevated concentrations (hyperammonemia) are toxic to the central nervous system (CNS). Therefore, a mechanism is required for the transport of nitrogen from the peripheral tissues to the liver for ultimate disposal as urea while keeping circulating levels of free ammonia low.

A. Sources

Amino acids are quantitatively the most important source of ammonia because most Western diets are high in protein and provide excess amino acids, which travel to the liver and undergo transdeamination (that is, the linking of the *aminotransferase* and *GDH* reactions), producing ammonia. [Note: The liver catabolizes straight-chain amino acids, primarily.] However, substantial amounts of ammonia can be obtained from other sources.

1. **Glutamine:** An important source of plasma glutamine is from the catabolism of BCAA in skeletal muscle. This glutamine is taken up by cells of the intestine, the liver, and the kidneys. The liver and kidneys generate ammonia from glutamine by the actions of *glutaminase* (Fig. 24.17) and *GDH*. In the kidneys, most of this ammonia is excreted into the urine as NH_4^+, which provides an important mechanism for maintaining the body's acid–base balance through the excretion of protons. In the liver, the ammonia is detoxified to urea and excreted. [Note: α-Ketoglutarate, the second product of *GDH*, is a glucogenic precursor in the liver and kidneys.] Ammonia is also generated by intestinal *glutaminase*. Enterocytes obtain glutamine either from the blood or from digestion of dietary protein. [Note: Intestinal glutamine metabolism also produces alanine, which is used by the liver for gluconeogenesis, and citrulline, which is used by the kidneys to synthesize arginine.]

2. **Intestinal bacteria:** Ammonia is formed from urea by the action of bacterial *urease* in the lumen of the intestine. This ammonia is absorbed from the intestine by way of the portal vein, and virtually all is removed by the liver via conversion to urea.

3. **Amines:** Amines obtained from the diet and monoamines that serve as hormones or neurotransmitters give rise to ammonia by the action of *monoamine oxidase* (see p. 423).

4. **Purines and pyrimidines:** In the catabolism of purines and pyrimidines, amino groups attached to the ring atoms are released as ammonia (see Fig. 34.15, p. 585).

B. Transport in the circulation

Although ammonia is constantly produced in the tissues, it is present at very low levels in blood. This is due both to the rapid removal of blood ammonia by the liver and to the fact that several tissues, particularly muscle, release amino acid nitrogen in the form of glutamine and alanine, rather than as free ammonia (see Fig. 24.13).

1. **Urea:** Formation of urea in the liver is quantitatively the most important disposal route for ammonia. Urea travels in the blood from the liver to the kidneys, where it passes into the glomerular filtrate. See Chapter 48 p. 811 for renal function tests.

2. **Glutamine:** This amide of glutamate provides a nontoxic storage and transport form of ammonia (Fig. 24.18). The ATP-requiring formation of glutamine from glutamate and ammonia by *glutamine synthetase* occurs primarily in skeletal muscle and the liver but is also important in the CNS, where it is the major mechanism for the removal of ammonia in the brain. Glutamine is found in plasma

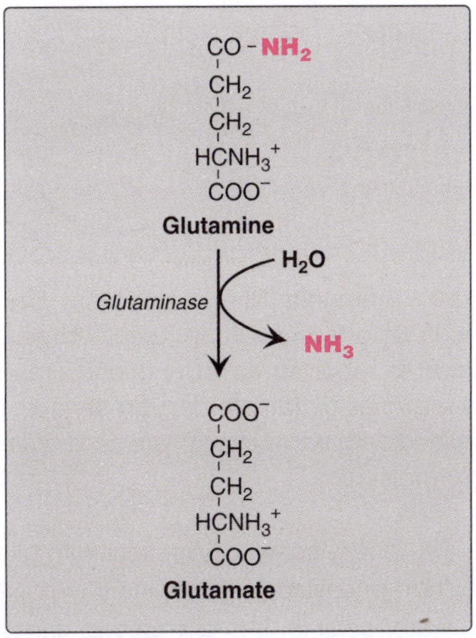

$CO-NH_2$
CH_2
CH_2
$HCNH_3^+$
COO^-
Glutamine
H_2O
Glutaminase
NH_3
COO^-
CH_2
CH_2
$HCNH_3^+$
COO^-
Glutamate

Figure 24.17 Hydrolysis of glutamine to form ammonia (NH_3).

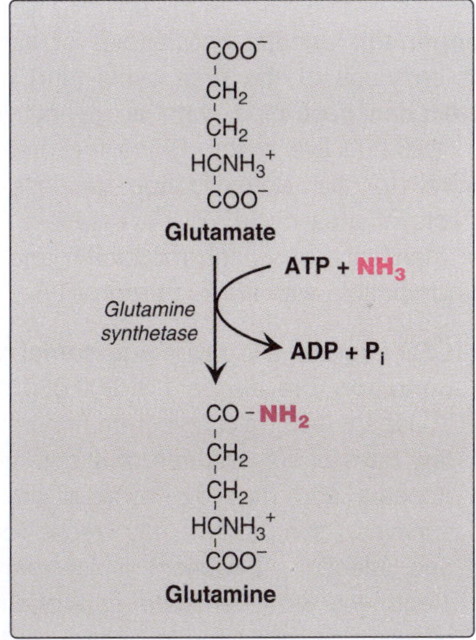

Figure 24.18 Synthesis of glutamine. ADP = adenosine diphosphate; P_i = inorganic phosphate; NH_3 = ammonia.

at concentrations higher than other amino acids, a finding consistent with its transport function. [Note: The liver keeps blood ammonia levels low through *glutaminase*, *GDH*, and the urea cycle in periportal (close to inflow of blood) hepatocytes and through *glutamine synthetase* as an ammonia scavenger in the perivenous hepatocytes.] Ammonia metabolism is summarized in Figure 24.19.

C. Hyperammonemia

The capacity of the hepatic urea cycle exceeds the normal rates of ammonia generation, and the levels of **blood ammonia are normally low (5–35 µmol/l).** However, when liver function is compromised, due either to genetic defects of the urea cycle or liver disease, blood levels can be >1,000 µmol/l. Such hyperammonemia is a medical emergency, because ammonia has a direct neurotoxic effect on the CNS. For example, elevated concentrations of ammonia in the blood cause the symptoms of ammonia intoxication, which include **tremors, slurring of speech, somnolence (drowsiness), vomiting, cerebral edema, and blurring of vision**. At high concentrations, ammonia can cause coma and death. There are two major types of hyperammonemia.

1. **Acquired: Liver disease** is a common cause of acquired hyperammonemia in adults.

   Diagnosis of liver diseases is done by blood tests (bilirubin, ALT, AST, *alkaline phosphatase* [ALP], albumin, total protein, gamma-glutamyl transferase [GTT], *lactate dehydrogenase* [LDH], ammonia and prothrombin time). Chapter 48 on Organ Function Tests may be seen for further details. Radiologically, ultrasound, MRI and CT scan aid the diagnosis.

2. **Congenital:** Genetic deficiencies of each of the five enzymes of the urea cycle (and of *NAGS*) have been described, with an overall incidence of ~1:25,000 live births. Since the urea cycle is central to the detoxification process, genetic defects of urea cycle will have severe neurological manifestations (high morbidity) and may not be compatible with life (high mortality)

   a. ***CPS I* defect:** It is a **rare autosomal recessive condition** (prevalence 1 in 800,000 newborns in Japan) leading to hyperammonemia. In first few days of life, features of hyperammonemia appears and may die soon. If the neonate survives, symptoms may recur on stress and infection. Treatment is low-protein diet. Treatment with benzoate and phenylacetate gives some benefits. As urea is not formed, excess nitrogen is accumulated in the form of glutamine and glycine. Activated phenylacetate, i.e., phenylacetyl CoA reacts with glutamine to form phenylacetylglutamine (Fig. 24.20). Being water soluble, it is excreted in urine. Similarly, benzyl CoA reacts with glycine to synthesize hippuric acid, that is also excreted in urine and

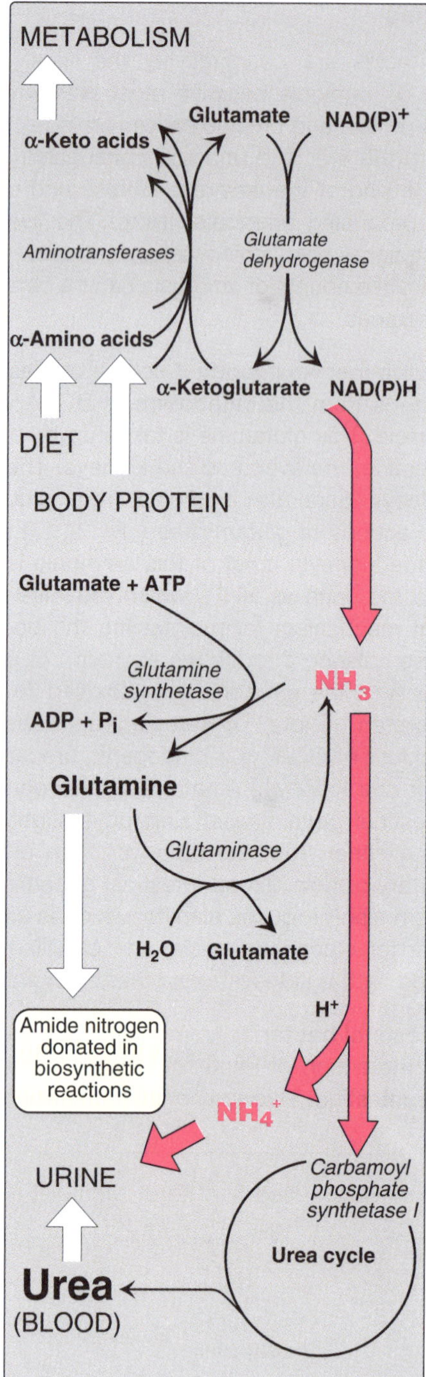

Figure 24.19 Ammonia (NH_3) metabolism. Urea content in the urine is reported as urinary urea nitrogen, or UUN. Urea in blood is reported as BUN (blood urea nitrogen). [Note: The enzymes *glutamate dehydrogenase*, *glutamine synthetase*, and *carbamoyl phosphate synthetase I* fix NH_3 into organic molecules.]

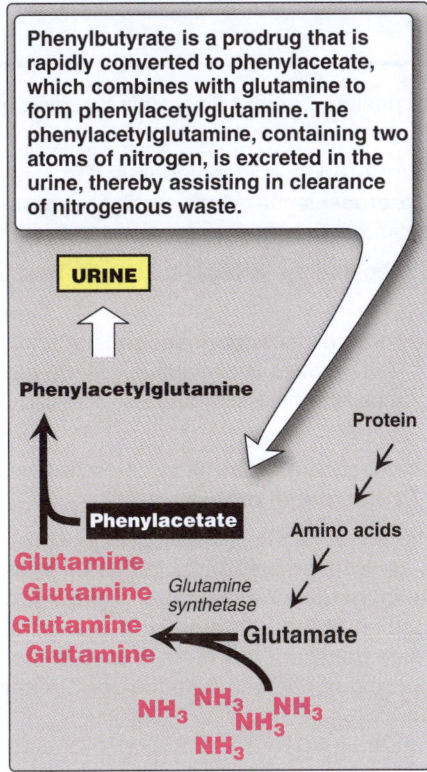

Figure 24.20 Treatment of patients with urea cycle defects by administration of phenylbutyrate to aid in excretion of ammonia (NH_3).

thus body can dispose of some amount of nitrogen. Rarely, dialysis is also done.

b. **NAGS deficiency:** NAGS deficiency mimics hyperammonemia due to CPS-I deficiency except that it responds to the treatment of NAG.

c. **Ornithine permease defect:** Ornithine is not transported to mitochondria from cytoplasm leading to hyperornithinemia and hyperammonemia. In absence of its normal substrate (ornithine), mitochondria act on lysine and carbamoylates to homocitrulline leading to homocitrullinuria. Treatment is similar to that of *CPS I* defect.

d. **X-linked *OTC* deficiency:** It is the most common of these disorders, predominantly affecting males, although female carriers may become symptomatic. Treatment is similar to that of *CPS I* defect. All of the other urea cycle disorders follow an autosomal-recessive inheritance pattern. In each case, the failure to synthesize urea leads to hyperammonemia during the first weeks following birth.

e. **Arginosuccinate synthase deficiency:** Blood, CSF and urine level of citrulline is high. The treatment options do not differ much.

f. **Arginosuccinate lyase defect:** Blood, CSF and urine level of arginosuccinate is high. The patients have low serum arginine, hepatomegaly, mental retardation and congenital trichorrhexis nodosa (friable and tufted hair). The treatment is complicated and out of scope of this book

g. **Arginase defect:** The hyperammonemia seen with *arginase* deficiency is less severe because arginine contains two waste nitrogens and can be excreted in the urine. Diagnosis of urea cycle defects require estimation of arterial blood gas, blood ammonia, and blood glucose.

Plasma amino acids and urine organic acids are done by tandem mass spectroscopy. Genetic tests and liver biopsy are additional tests that may be done.

VII. Medical Perspectives

Dietary protein digestion is inadequate in deficiency of pancreatic secretion as in **chronic pancreatitis and cystic fibrosis**. There is weight loss, abdominal pain and undigested protein and fat in stools.

Kwashiorkor is protein malnutrition with adequate energy intake. This presents with edema, inability to gain height, swelling of hands and feet, and bulging of stomach.

Hyperammonemia is when blood ammonia concentration is greater than normal. It is seen when ammonia is not converted to urea because of acquired diseases or genetic disorders. The acquired diseases are much more common and include **alcoholism** leading to damaged liver, fatty liver and liver cirrhosis. The acquired causes usually seen in adults also include **drugs (hepatotoxins)**, overwhelming **hepatitis B** infections and chronic liver diseases. Hyperammonemia can cause brain injury and death.

The congenital reasons include defects in urea cycle and certain organic acidurias.

Infants with urea cycle defects appear normal at birth but rapidly develop lethargy, seizures, neurological symptoms, cerebral edema and coma.

On the other hand, **uremia (high blood urea)** is a common abnormality seen in patients who are unable to excrete urea, mostly because of **chronic renal disease** (see Chapter 48: Organ Function Tests). Note that raised blood urea is not because of liver diseases. Liver diseases may cause hyperammonemia.

VIII. Chapter Summary

Nitrogen enters the body in a variety of compounds present in food, the most important being **amino acids** contained in **dietary protein**.

Nitrogen leaves the body as **urea**, **ammonia**, and other products derived from amino acid metabolism (Fig. 24.21). Free amino acids in the body are produced by hydrolysis of dietary protein by ***proteases*** activated from their **zymogen** form in the stomach and intestine, degradation of tissue proteins, and de novo synthesis.

This **amino acid pool** is consumed in the synthesis of body protein, metabolized for energy, or its members used as precursors for other nitrogen-containing compounds.

Free amino acids from digestion are taken up by intestinal **enterocytes** via **sodium-dependent secondary active transport**. Small peptides are taken up via **proton-linked transport**. Note that body protein is simultaneously degraded and resynthesized, a process known as **protein turnover**. The concentration of a cellular protein may be determined by regulation of its synthesis or degradation.

The ATP-dependent, cytosolic, selective **ubiquitin–proteasome** and ATP-independent, relatively nonselective **lysosomal *acid hydrolases*** are the two major enzyme systems that are responsible for **degrading proteins**.

Nitrogen cannot be stored, and amino acids in excess of the biosynthetic needs of the cell are quickly degraded. The first phase of **catabolism** involves the transfer of the α-amino groups through **transamination** by **pyridoxal phosphate–** dependent ***aminotransferases*** (***transaminases***), followed by **oxidative deamination of glutamate** by ***glutamate dehydrogenase***, forming **ammonia** and the corresponding **α-keto acids**. A portion of the **free ammonia** is excreted in the **urine**. Some ammonia is used in converting glutamate to **glutamine** for safe transport, but most is used in the hepatic synthesis of **urea**, which is quantitatively the most important route for disposing of nitrogen from the body. **Alanine** also carries nitrogen to the liver for disposal as urea.

The two major causes of **hyperammonemia** (with its neurologic effects) are acquired liver disease and congenital deficiencies of urea cycle enzymes such as X-linked ***ornithine transcarbamylase***.

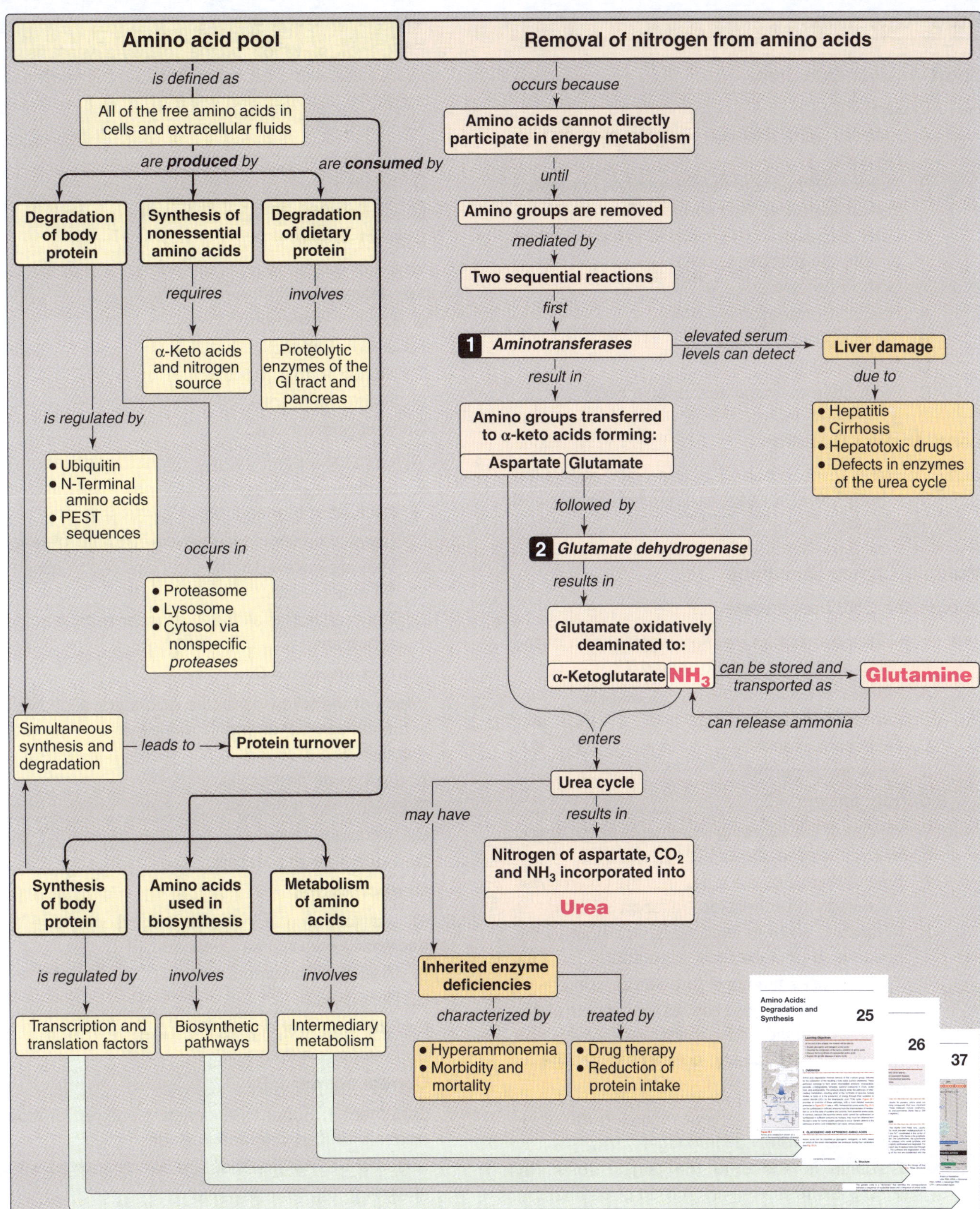

Figure 24.21 Key concept map for nitrogen metabolism. GI = gastrointestinal; PEST = proline, glutamate, serine, threonine; NH_3 = ammonia; CO_2 = carbon dioxide.

Study Questions

Short-Answer Questions

24.1. Explain why:

A. Patients with Hartnup disorder present with pellagra.

B. Super ORS corrects fluid-electrolyte imbalance in diarrhea faster than normal ORS.

C. GDH is claimed to be involved in incorporation of both the nitrogen in urea.

24.2. Write short notes on:

A. Ubiquitin-proteasome system

B. Oxidative deamination

C. Transamination

D. Production of amino acid pool in body

Long-Answer Question

24.1. How are amino group of amino acids eliminated from body? Add a note on ammonia toxicity and uremia.

Multiple-Choice Questions

Choose the ONE best answer.

24.1. In this transamination reaction (right), which of the following are the products of X and Y?

A. Alanine, α-ketoglutarate

B. Aspartate, α-ketoglutarate

C. Glutamate, alanine

D. Pyruvate, aspartate

Oxaloacetate X

Glutamate Y

Correct answer = B.

24.2. Which one of the following statements about amino acids and their metabolism is correct?

A. Free amino acids are taken into the enterocytes by a single proton-linked transport system.

B. In healthy, well-fed individuals, the input to the amino acid pool exceeds the output.

C. The liver uses ammonia to buffer protons.

D. The lungs and adipose tissues exhibit both glutamine synthetase and glutaminase activity and hence can produce and consume glutamine.

Correct answer = D.

24.3. Identify the correct statement regarding the ubiquitin-mediated protein degradation.

A. Ubiquitin is added to the target protein by linking it with the δ-amino group of ornithine.

B. There are more ubiquitin ligases than ubiquitin activating and conjugating enzymes.

C. The minimum number of ubiquitins to be added to the target protein for its degradation is five.

D. Ubiquitin-mediated protein degradation is an energy independent process.

Correct answer = B.

24.4. Deficiency of which of the following vitamins is associated with a decrease in serum transaminase activity?

A. Vitamin B_{12}

B. Niacin

C. Biotin

D. Pyridoxine

Correct answer = D.

24.5. Which of the following is the primary site of oxidative deamination in the human body?

A. Liver — mitochondria

B. Liver — cytosol

C. Kidney — mitochondria

D. Brain — cytosol

Correct answer = A.

24.6. Which of the following is true about transaminases?

A. Glutamate-α-ketoglutarate pair is always involved in the reaction.

B. They are present at high levels in normal plasma.

C. They are involved in the synthesis of non-essential amino-acids.

D. They can act on all the 21 standard amino acids in humans.

Correct answer = C.

24.7. Which of the following amino acid pairs is involved in the transport of ammonia to the liver from peripheral tissues?

A. Glutamate and serine.

B. Serine and threonine.

C. Valine and leucine.

D. Glutamine and alanine.

Correct answer = D.

24.8. All the following conditions are associated with decreased serum urea levels, EXCEPT:

A. Urea cycle disorders

B. Dehydration

C. Extensive liver damage

D. Low protein intake

Correct answer = B.

24.9. Statement 1: Super ORS contains oral rehydration solution fortified with amino acids.

Statement 2: Amino acids are co-transported with glucose in the small intestine.

A. Both the Statements are correct, and statement 2 is the correct explanation for statement 1

B. Both the statements are correct, but statement 2 is NOT the correct explanation for statement 1

C. Statement 1 is correct, but statement 2 is NOT correct

D. Statement 1 is NOT correct, but statement 2 is correct

Correct answer = A.

24.10. Which of the following enzymes is supplemented in a patient with chronic pancreatitis?

A. Alanine transaminase
B. Glutamate dehydrogenase
C. Enterokinase
D. Trypsin

Correct answer = D.

Multiple-Choice Questions (NExT/NEET PG)

Choose the ONE best answer.

24.1. A female neonate appeared healthy until age ~24 hours, when she became lethargic. A sepsis workup proved negative. At 56 hours, she started showing focal seizure activity. The plasma ammonia level was found to be 887 **μmol**/L (normal 5–35 **μmol**/L). Quantitative plasma amino acid levels revealed a marked elevation of citrulline but not argininosuccinate. Which one of the following enzymic activities is most likely to be deficient in this patient, and which corresponding analyte is elevated in the blood of this patient?

A. Arginase and asparagine
B. Argininosuccinate lyase and glutamine
C. Argininosuccinate synthetase and glutamine
D. Carbamoyl phosphate synthetase I and urea

Correct answer = C.

24.2. Hair proteins have a significant arginine content, and a reduction in arginine levels is one of the causes of easily breakable hair. Which of the following urea cycle disorders can lead to similar abnormalities in the structure of the hair?

A. Hyperammonemia type 2
B. Citrullinemia type 1
C. Hyperammonemia type 1
D. Argininosuccinic acidemia

Correct answer = D.

24.3. All the following are probable reasons why ammonia is toxic to the neurons, EXCEPT:

A. Ammonia can activate ion channels that can cause osmotic imbalance in the neurons
B. Ammonia leads to accumulation of alpha-ketoglutarate, thus affecting the TCA cycle
C. Ammonia favors glutamine synthesis from glutamate, which affects synthesis of GABA from glutamate
D. Glutamine is osmotic and causes entry of water into astrocytes, leading to cerebral edema.

Correct answer = B.

24.4. A 2-year-old female child presented with lethargy, intellectual disability, and aversion to protein diet. Laboratory investigations showed hyperammonemia. With a suspicion of hyperammonemia type I, sequencing of carbamoyl phosphate synthetase-1 was performed; however, no pathogenic variants were found in this gene. Defect in which other protein can give rise to a similar presentation?

A. N-acetyl glutamate synthase
B. Ornithine transcarbamylase
C. Arginosuccinate lyase
D. Arginase

Correct answer = A.

24.5. A strategy of treatment in hyperammonemia is the administration of phenyl butyrate (which is metabolized to phenyl acetate) and benzoate. What is the mechanism of action of these compounds?

A. They activate the enzymes of urea cycle to convert ammonia to urea.
B. They inhibit the catabolism of amino acids.
C. They provide an alternative mechanism for ammonia excretion.
D. They inhibit the diffusion of ammonium ions into the neurons.

Correct answer = C.

Amino Acids: Degradation and Synthesis

25

Learning Objectives

At the end of this chapter, the student will be able to:

- Explain glucogenic and ketogenic amino acids
- Describe the metabolism of the carbon skeleton of amino acids
- Discuss the biosynthesis of nonessential amino acids
- Explain the genetic diseases of amino acids

I. OVERVIEW

Amino acid degradation involves removal of the **α-amino group**, followed by the catabolism of the resulting **α-keto acids** (carbon skeletons). These pathways converge to form seven intermediate products: **oxaloacetate, pyruvate, α-ketoglutarate, fumarate, succinyl coenzyme A (CoA), acetyl CoA, and acetoacetate**. The products directly enter the pathways of intermediary metabolism, resulting either in the synthesis of glucose, ketone bodies, or lipids or in the production of energy through their oxidation to carbon dioxide (CO_2) by the tricarboxylic acid (TCA) cycle. Figure 25.1 provides an overview of these pathways, with a more detailed summary presented in Figure 25.15 (see p. 405). Nonessential amino acids (Fig. 25.2) can be synthesized in sufficient amounts from the intermediates of metabolism or, as in the case of cysteine and tyrosine, from essential amino acids. In contrast, because the essential amino acids cannot be synthesized (or synthesized in sufficient amounts) by humans, they must be obtained from the diet in order for normal protein synthesis to occur. Genetic defects in the pathways of amino acid metabolism can cause serious disease.

II. GLUCOGENIC AND KETOGENIC AMINO ACIDS

Amino acids can be classified as glucogenic, ketogenic, or both, based on which of the seven intermediates are produced during their catabolism (see Fig. 25.2).

A. Glucogenic amino acids

Amino acids whose catabolism yields pyruvate or one of the intermediates of the TCA cycle are termed glucogenic. Because these intermediates are substrates for gluconeogenesis (see p. 215), they can give rise to the net synthesis of glucose in the liver and kidney.

Color-coding used in this chapter:	• **BLUE CAPS TEXT** = names of seven products of amino acid metabolism • **Red text** = names of glucogenic amino acids • **Brown text** = names of glucogenic and ketogenic amino acids • **Green text** = names of ketogenic amino acids • **Cyan text** = one-carbon compounds

B. Ketogenic amino acids

Amino acids whose catabolism yields either acetoacetate or one of its precursors (acetyl CoA or acetoacetyl CoA) are termed ketogenic (see Fig. 25.2). Acetoacetate is one of the ketone bodies, which also include 3-hydroxybutyrate and acetone (see p. 195). Leucine and lysine are the only exclusively ketogenic amino acids found in proteins. Their carbon skeletons are not substrates for gluconeogenesis and, therefore, cannot give rise to the net synthesis of glucose.

III. AMINO ACID CARBON SKELETON CATABOLISM

The pathways by which amino acids are catabolized are conveniently organized according to which one (or more)

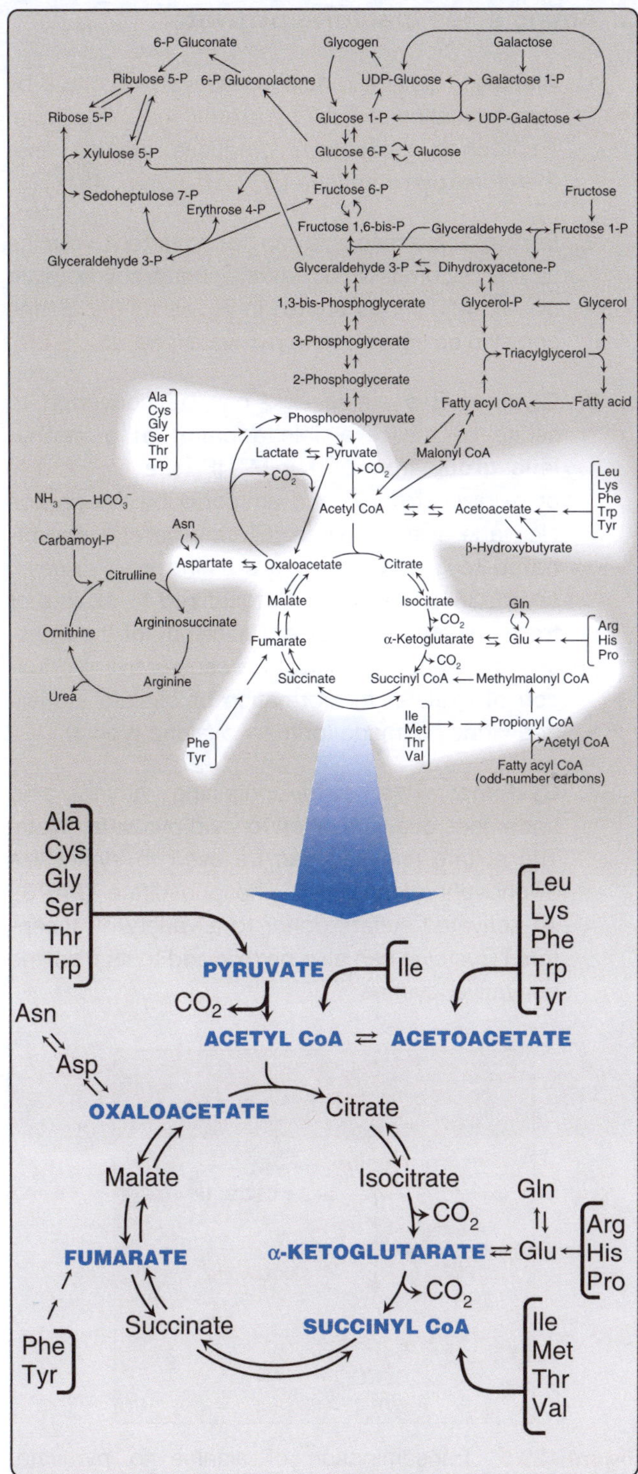

Figure 25.1 Amino acid metabolism shown as a part of the essential pathways of energy metabolism. (See Fig. 13.2, p. 187, for a more detailed map of metabolism.) CoA = coenzyme A; CO_2 = carbon dioxide.

of the seven intermediates listed above is produced from a particular amino acid.

	Glucogenic	Glucogenic and Ketogenic	Ketogenic
Nonessential	Alanine Arginine Asparagine Aspartate Cysteine Glutamate Glutamine Glycine Proline Serine	Tyrosine	
Essential	Histidine Methionine Threonine Valine	Isoleucine Phenyl-alanine Tryptophan	Leucine Lysine

Figure 25.2 Classification of amino acids. [Note: Some amino acids can become conditionally essential. For example, supplementation with glutamine and arginine has been shown to improve outcomes in patients with trauma, postoperative infections, and immunosuppression.]

A. Amino acids that form oxaloacetate

Asparagine is hydrolyzed by *asparaginase*, liberating ammonia and aspartate (Fig. 25.3). Aspartate loses its amino group by transamination to form oxaloacetate (see Fig. 25.3).

> Some rapidly dividing leukemic cells are unable to synthesize sufficient asparagine to support their growth. This makes asparagine an essential amino acid for these cells, which, therefore, require asparagine from the blood. *Asparaginase*, which hydrolyzes asparagine to aspartate, can be administered systemically to treat leukemia. *Asparaginase* lowers the level of asparagine in the plasma, thereby depriving cancer cells of a required nutrient.

B. Amino acids that form α-ketoglutarate via glutamate

1. **Glutamine:** This amino acid is hydrolyzed to glutamate and ammonia by the enzyme ***glutaminase*** (see p. 389). Glutamate is converted to α-ketoglutarate by transamination or through oxidative deamination by ***glutamate dehydrogenase*** (see p. 384).

2. **Proline:** This amino acid is oxidized to glutamate. Glutamate is transaminated or oxidatively deaminated to form α-ketoglutarate.

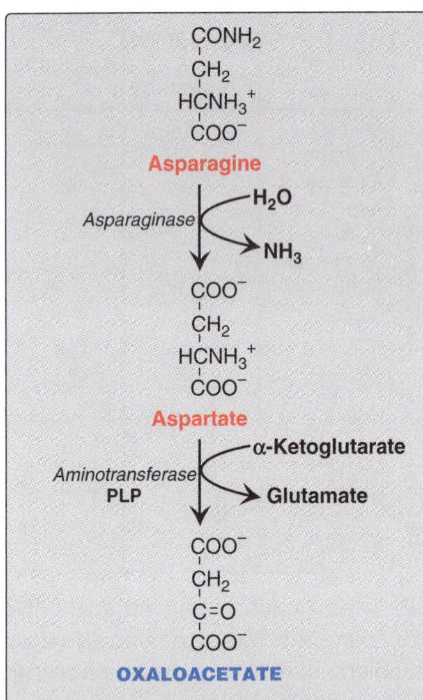

Figure 25.3 Metabolism of asparagine and aspartate. PLP = pyridoxal phosphate; NH_3 = ammonia.

3. **Arginine:** This amino acid is hydrolyzed by ***arginase*** to produce ornithine (and urea). [Note: The reaction occurs primarily in the liver as part of the urea cycle (see p. 388).] Ornithine is subsequently converted to α-ketoglutarate, with glutamate semialdehyde as an intermediate.

4. **Histidine:** This amino acid is oxidatively deaminated by ***histidase*** to urocanic acid, which subsequently forms **N-formiminoglutamate ([FIGlu],** Fig. 25.4). FIGlu donates its formimino group to tetrahydrofolate (THF), leaving glutamate, which is degraded as described above.

> Individuals deficient in folic acid excrete increased amounts of FIGlu in the urine, particularly after ingestion of a large dose of histidine. The FIGlu excretion test has been used in diagnosing a deficiency of folic acid. (See p. 402 for a discussion of folic acid, THF, and one-carbon metabolism.)

C. Amino acids that form pyruvate

1. **Alanine:** This amino acid loses its amino group by transamination to form pyruvate (Fig. 25.5). [Note: Tryptophan catabolism produces alanine and, therefore, pyruvate (see Fig. 25.10 on p. 401).]

2. **Serine:** This amino acid can be converted to glycine as THF becomes N^5,N^{10}-methylenetetrahydrofolate (N^5,N^{10}-MTHF), as shown in Figure 25.6A. Serine can also be converted to pyruvate (see Fig. 25.6B).

3. **Glycine:** This amino acid can be converted to serine by the reversible **addition of a methylene group** from N^5,N^{10}-MTHF (see Fig. 25.6A) or oxidized to CO_2 and ammonia by the glycine cleavage system. [Note: Glycine can be **deaminated** to glyoxylate (by a *d-amino acid oxidase*; see p. 386), which can be **oxidized** to oxalate or transaminated to glycine. Deficiency of the *transaminase* in liver peroxisomes causes overproduction of oxalate, the formation of oxalate stones, and kidney damage (primary oxaluria type 1).]

4. **Cysteine:** This sulfur-containing amino acid undergoes desulfurization to yield pyruvate. [Note: The sulfate released can be used to synthesize 3′-phosphoadenosine-5′-phosphosulfate **(PAPS)**, an activated sulfate donor to a variety of acceptors.] Cysteine can also be oxidized to its disulfide derivative, cystine.

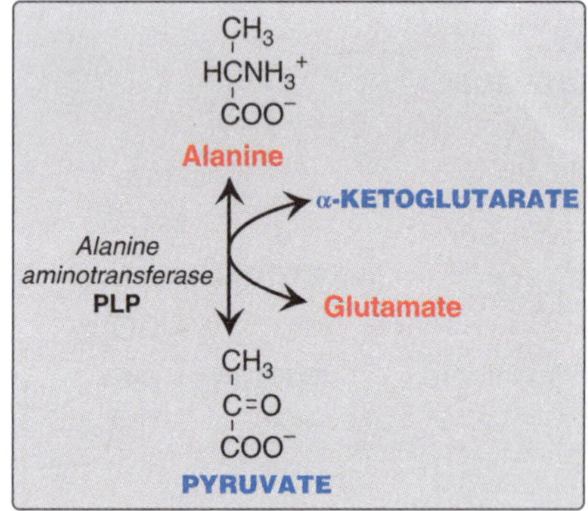

Figure 25.5 Transamination of alanine to pyruvate. PLP = pyridoxal phosphate.

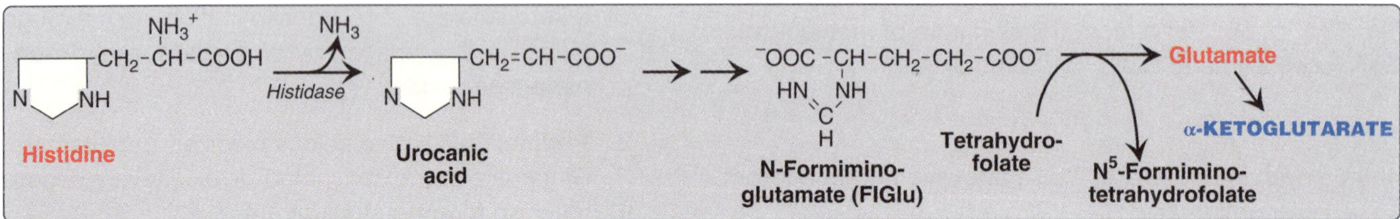

Figure 25.4 Degradation of histidine. NH_3 = ammonia.

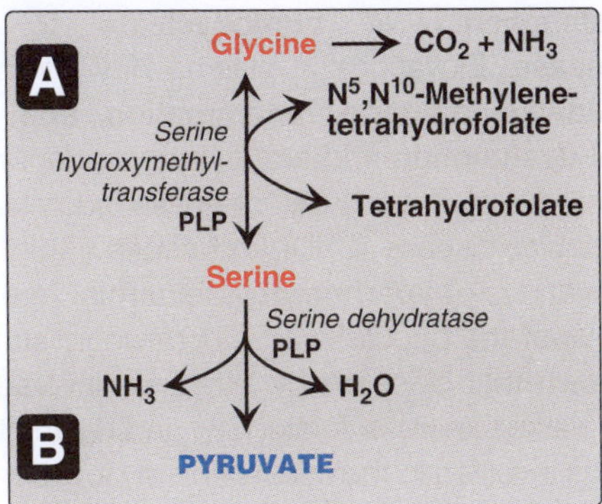

Figure 25.6 **A.** Interconversion of serine and glycine and oxidation of glycine. **B.** Dehydration of serine to pyruvate. PLP = pyridoxal phosphate; NH_3 = ammonia.

5. **Threonine:** This amino acid is converted to pyruvate in most organisms but is a minor pathway (at best) in humans.

D. Amino acids that form fumarate

1. **Phenylalanine and tyrosine:** Hydroxylation of phenylalanine produces tyrosine (Fig. 25.7). This irreversible reaction, catalyzed by tetrahydrobiopterin-requiring ***phenylalanine hydroxylase*** (***PAH***), initiates the catabolism of phenylalanine. Thus, phenylalanine metabolism and tyrosine metabolism merge, leading ultimately to fumarate and acetoacetate formation. Therefore, phenylalanine and tyrosine are both glucogenic and ketogenic.

2. **Inherited deficiencies:** Inherited deficiencies in the enzymes of phenylalanine and tyrosine metabolism lead to the diseases phenylketonuria (PKU) (see p. 406), tyrosinemia (see p. 271), and alkaptonuria (see p. 405) as well as the condition of albinism (see p. 408).

E. Amino acids that form succinyl CoA: Methionine

Methionine is one of four amino acids that form succinyl CoA. This sulfur-containing amino acid deserves special attention because it is converted to ***S*-adenosylmethionine (SAM),** the major methyl group donor in one-carbon metabolism (Fig. 25.8). Methionine is also the source of **homocysteine (Hcy),** a metabolite associated with atherosclerotic vascular disease and thrombosis (see p. 401).

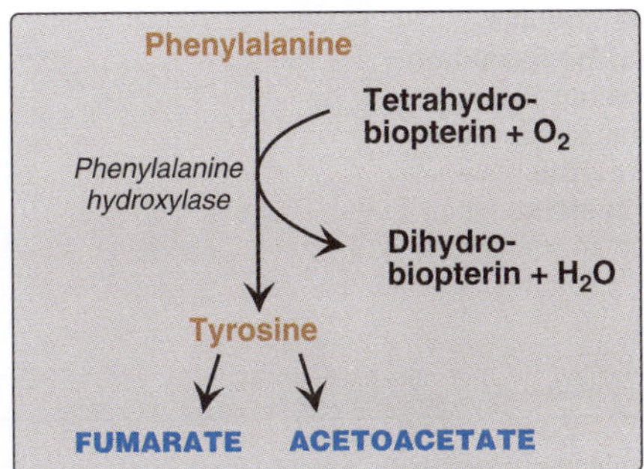

Figure 25.7 Degradation of phenylalanine.

1. **S-Adenosylmethionine synthesis:** Methionine condenses with ATP, forming SAM, a high-energy compound that is unusual in that it contains no phosphate. The formation of SAM is driven by hydrolysis of all three phosphate bonds in ATP (see Fig. 25.8).

2. **Activated methyl group:** The methyl group attached to the sulfur in SAM is activated and can be transferred by ***methyltransferases*** to a variety of acceptors such as norepinephrine in the synthesis of epinephrine. The methyl group is usually transferred to nitrogen or oxygen atoms (as with epinephrine synthesis and degradation, respectively; see p. 424) and sometimes to carbon atoms (as with cytosine). The reaction product, *S*-adenosylhomocysteine (SAH), is a simple thioether, analogous to methionine. The resulting loss of free energy makes methyl transfer essentially irreversible.

3. **S-Adenosylhomocysteine hydrolysis:** After donation of the methyl group, SAH is hydrolyzed to Hcy and adenosine. Hcy has two fates. If there is a deficiency of methionine, Hcy may be remethylated to methionine (see Fig. 25.8). If methionine stores are adequate, Hcy may enter the transsulfuration pathway, where it is converted to cysteine.

   a. **Methionine resynthesis:** Hcy accepts a methyl group from N^5-methyltetrahydrofolate (N^5-methyl-THF) in a reaction requiring methylcobalamin, a coenzyme derived from vitamin B_{12} (see p. 528). [Note: The methyl group is transferred by ***methionine synthase*** from the B_{12} derivative to Hcy, regenerating methionine. Cobalamin is remethylated from N^5-methyl-THF.]

   b. **Cysteine synthesis:** Hcy condenses with serine, forming cystathionine, which is hydrolyzed to α-ketobutyrate and cysteine (see Fig. 25.8). This vitamin B_6–requiring sequence has the net effect of converting serine to cysteine and Hcy to α-ketobutyrate, which is oxidatively decarboxylated to form propionyl CoA. Propionyl CoA is converted to succinyl CoA (see Fig. 21.20, p. 313). Because Hcy is synthesized from the essential amino acid methionine, cysteine is not an essential amino acid as long as sufficient methionine is available.

4. **Relationship of homocysteine to vascular disease:** Elevations in plasma Hcy levels promote **oxidative damage, inflammation, and endothelial dysfunction** and are an independent risk factor for occlusive vascular diseases such as cardiovascular disease (CVD) and stroke (Fig. 25.9). Mild elevations (**hyperhomocysteinemia**) are seen in ~7% of the population. Epidemiologic studies have shown that plasma Hcy levels are inversely related to plasma levels of folate, B_{12}, and B_6, the three vitamins involved in the conversion of Hcy to methionine and cysteine. Supplementation with these vitamins has been shown to reduce circulating levels of Hcy. However, in patients with established CVD, vitamin therapy does not decrease cardiovascular events or death. This raises the question as to whether Hcy is a cause of the vascular damage or merely a marker of such damage. [Note: Large elevations in plasma Hcy as a result of rare deficiencies in *cystathionine β-synthase* of the transsulfuration pathway are seen in patients with **classic homocystinuria** (resulting from severe hyperhomocysteinemia [>100 μmol/l], see p. 408).] Deficiencies in the remethylation reaction also result in a rise in Hcy.

> Elevated homocysteine and decreased folic acid levels in pregnant women are associated with increased incidence of neural tube defects (improper closure, as in spina bifida) in the fetus. Periconceptual supplementation with folate reduces the risk of such defects.

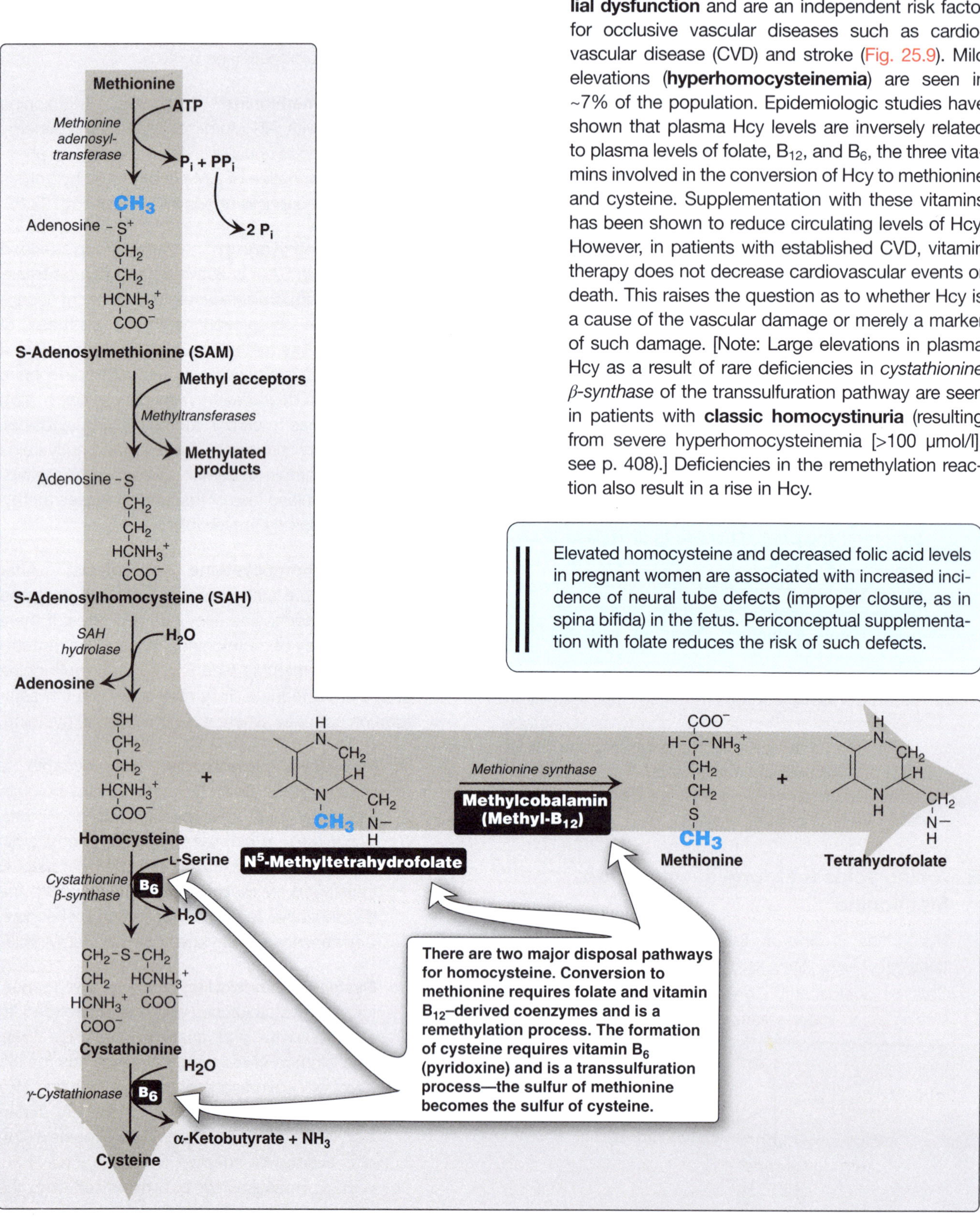

Figure 25.8 Degradation and resynthesis of methionine. [Note: The resynthesis of methionine from homocysteine is the only reaction in which tetrahydrofolate both carries and donates a methyl (–CH_3) group. In all other reactions, SAM is the methyl group carrier and donor.] PP_i = pyrophosphate; P_i = inorganic phosphate; NH_3 = ammonia.

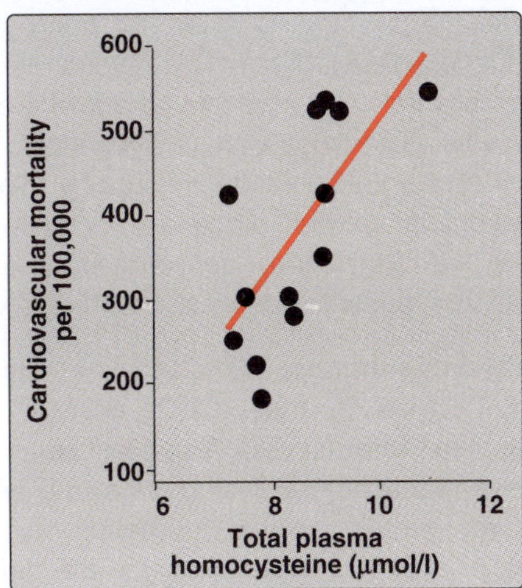

Figure 25.9 Association between cardiovascular disease mortality and total plasma homocysteine.

F. Other amino acids that form succinyl CoA

Degradation of valine, isoleucine, and threonine also results in the production of succinyl CoA, a TCA cycle intermediate and gluconeogenic compound. [Note: It is metabolized to pyruvate.]

1. **Valine and isoleucine:** These amino acids are branched-chain amino acids (BCAA) that generate propionyl CoA, which is converted to methylmalonyl CoA and then succinyl CoA by biotin- and vitamin B_{12}–requiring reactions.

2. **Threonine:** This amino acid is dehydrated to α-ketobutyrate, which is converted to propionyl CoA and then to succinyl CoA. Propionyl CoA, then, is generated by the catabolism of the amino acids methionine, valine, isoleucine, and threonine. [Note: Propionyl CoA also is generated by the oxidation of odd-numbered fatty acids (see p. 312).]

G. Amino acids that form acetyl CoA or acetoacetyl CoA

Tryptophan, leucine, isoleucine, and lysine form acetyl CoA or acetoacetyl CoA directly, without pyruvate serving as an intermediate. As noted earlier, phenylalanine and tyrosine also give rise to acetoacetate during their catabolism (see Fig. 25.7). Therefore, there are a total of six partly or wholly ketogenic amino acids.

1. **Tryptophan:** This amino acid is both glucogenic and ketogenic, because its catabolism yields alanine and acetoacetyl CoA (Fig. 25.10). [Note: Quinolinate from tryptophan catabolism is used in the synthesis of nicotinamide adenine dinucleotide ([NAD], see p. 533).]

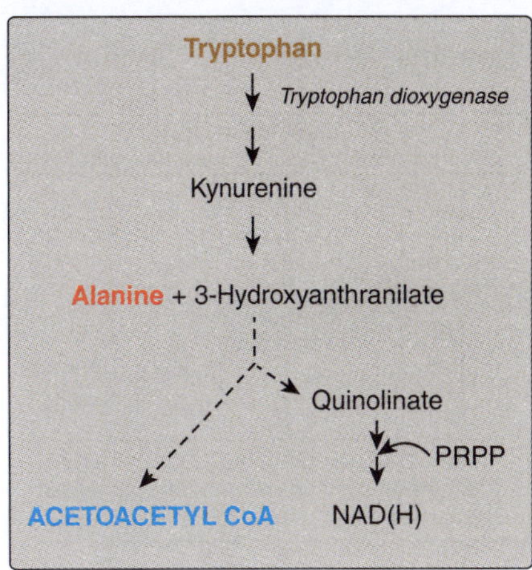

Figure 25.10 Metabolism of tryptophan by the kynurenine pathway (abbreviated). CoA = coenzyme A; PRPP = phosphoribosyl pyrophosphate; NAD(H) = nicotinamide adenine dinucleotide.

2. **Leucine:** This amino acid is **exclusively ketogenic**, because its catabolism yields acetyl CoA and acetoacetate (Fig. 25.11). The first two reactions in the catabolism of leucine and the other BCAA, isoleucine and valine, are catalyzed by enzymes that use all three BCAA (or their derivatives) as substrates (see H. below).

3. **Isoleucine:** This amino acid is **both ketogenic and glucogenic**, because its metabolism yields acetyl CoA and propionyl CoA.

4. **Lysine:** This amino acid is **exclusively ketogenic** and is unusual in that neither of its amino groups undergoes transamination as the first step in catabolism. Lysine is ultimately converted to acetoacetyl CoA.

H. Branched-chain amino acid degradation

The BCAA isoleucine, leucine, and valine are essential amino acids. In contrast to other amino acids, they are catabolized primarily by the peripheral tissues (particularly muscle), rather than by the liver. Because these three amino acids have a similar route of degradation, it is convenient to describe them as a group (see Fig. 25.11).

1. **Transamination:** Transfer of the amino groups of all three BCAA to α-ketoglutarate is catalyzed by a single, vitamin B_6–requiring enzyme, *branched-chain amino acid aminotransferase*, that is expressed primarily in skeletal muscle.

2. **Oxidative decarboxylation:** Removal of the carboxyl group of the α-keto acids derived from leucine, valine, and isoleucine is catalyzed by a single multienzyme complex, ***branched-chain***

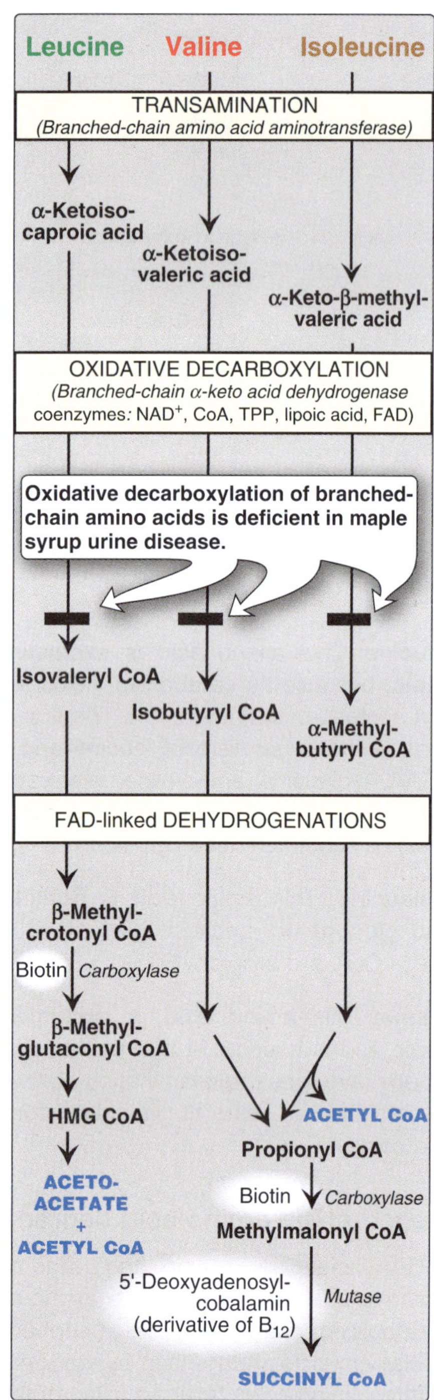

Figure 25.11 Degradation of leucine, valine, and isoleucine. [Note: *β-Methylcrotonyl CoA carboxylase* is one of four biotin-requiring *carboxylases* discussed in this book. The other three are *pyruvate carboxylase*, *acetyl CoA carboxylase*, and *propionyl CoA carboxylase*.] TPP = thiamine pyrophosphate; FAD = flavin adenine dinucleotide; CoA = coenzyme A; NAD = nicotinamide adenine dinucleotide; HMG = hydroxymethylglutarate.

α-keto acid dehydrogenase (BCKD) complex. This complex uses thiamine pyrophosphate, lipoic acid, oxidized flavin adenine dinucleotide (FAD), NAD^+, and CoA as its coenzymes and produces NADH. [Note: This reaction is similar to the conversion of pyruvate to acetyl CoA by the *pyruvate dehydrogenase* (*PDH*) *complex* (see p. 207) and α-ketoglutarate to succinyl CoA by the *α-ketoglutarate dehydrogenase* complex (see p. 209). The *dihydrolipoyl dehydrogenase* (*Enzyme 3*, or *E3*) component is identical in all three complexes.]

3. **Dehydrogenations:** Oxidation of the products formed in the *BCKD* reaction produces α-β-unsaturated acyl CoA derivatives and $FADH_2$. These reactions are analogous to the FAD-linked dehydrogenation in the β-oxidation of fatty acids (see p. 310). [Note: Deficiency in the ***dehydrogenase*** specific for isovaleryl CoA causes neurologic problems and is associated with a "sweaty feet" odor in body fluids.]

4. **End products:** The catabolism of isoleucine ultimately yields acetyl CoA and succinyl CoA, rendering it both ketogenic and glucogenic. Valine yields succinyl CoA and is glucogenic. Leucine is ketogenic, being metabolized to acetoacetate and acetyl CoA. In addition, NADH and $FADH_2$ are produced in the decarboxylation and dehydrogenation reactions, respectively. [Note: BCAA catabolism also results in glutamine and alanine being synthesized and sent out into the blood from muscle (see p. 386).]

IV. FOLIC ACID AND AMINO ACID METABOLISM

Some synthetic pathways require the addition of single-carbon groups that exist in a variety of oxidation states, including **formyl, methenyl, methylene, and methyl**. **These single-carbon groups can be transferred from carrier compounds such as THF and SAM** to specific structures that are being synthesized or modified. The **"one-carbon pool"** refers to the single-carbon units attached to this group of carriers. [Note: CO_2, coming from bicarbonate (HCO_3^-), is carried by the vitamin biotin (see p. 534), which is a prosthetic group for most carboxylation reactions but is not considered a member of the one-carbon pool. Defects in the ability to add or remove biotin from *carboxylases* result in multiple *carboxylase* deficiency. Treatment is supplementation with biotin.]

A. Folic acid and one-carbon metabolism

The active form of folic acid, THF, is produced from folate by ***dihydrofolate reductase*** in a two-step reaction requiring two nicotinamide adenine dinucleotide phosphate (NADPH). The one-carbon unit carried by THF is bound to N^5 or N^{10} or to both N^5 and N^{10}. Figure 25.12 shows the structures of the various

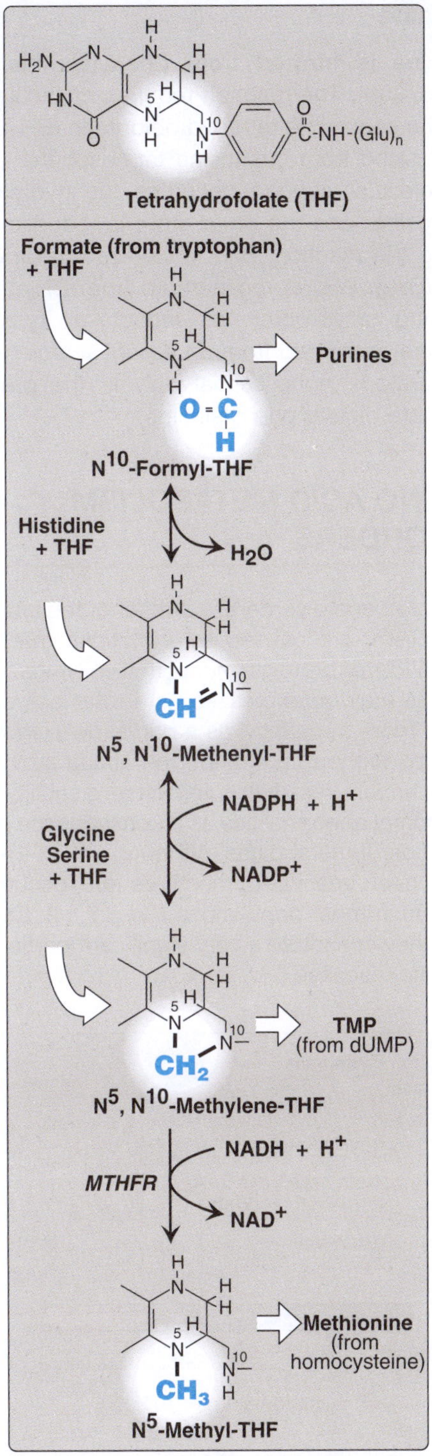

Figure 25.12 Summary of the interconversions and uses of THF. [Note: N^5,N^{10}-Methenyl-THF also arises from N^5-formimino-THF (see Fig. 20.4).] NADP(H) = nicotinamide adenine dinucleotide phosphate; NAD(H) = nicotinamide adenine dinucleotide; TMP = thymidine monophosphate; dUMP = deoxyuridine monophosphate; *MTHFR* = *N^5,N^{10}-methylene-THF reductase*.

members of the THF family and their interconversions and indicates the sources of the one-carbon units and the synthetic reactions in which the specific members participate. [Note: Folate deficiency presents as a megaloblastic anemia because of decreased availability of the purines and of the thymidine monophosphate needed for DNA synthesis (see p. 587).]

V. BIOSYNTHESIS OF NONESSENTIAL AMINO ACIDS

Nonessential amino acids are synthesized from intermediates of metabolism or, as in the case of tyrosine and cysteine, from the essential amino acids phenylalanine and methionine, respectively. The synthetic reactions for the nonessential amino acids are described below and are summarized in Figure 25.15. [Note: Some amino acids found in proteins, such as hydroxyproline and hydroxylysine (see p. 119), are produced by posttranslational modification (after incorporation into a protein) of their precursor (parent) amino acids.]

A. Synthesis from α-keto acids

Alanine, aspartate, and glutamate are synthesized by transfer of an amino group to the α-keto acids pyruvate, oxaloacetate, and α-ketoglutarate, respectively. These transamination reactions (Fig. 25.13; also see p. 248) are the most direct of the biosynthetic pathways. Glutamate is unusual in that it can also be synthesized by reversal of oxidative deamination, catalyzed by *glutamate dehydrogenase*, when ammonia levels are high (see p. 382).

B. Synthesis by amidation

1. **Glutamine:** This amino acid, which contains an amide linkage with ammonia at the γ-carboxyl, is formed from glutamate by ***glutamine synthetase***

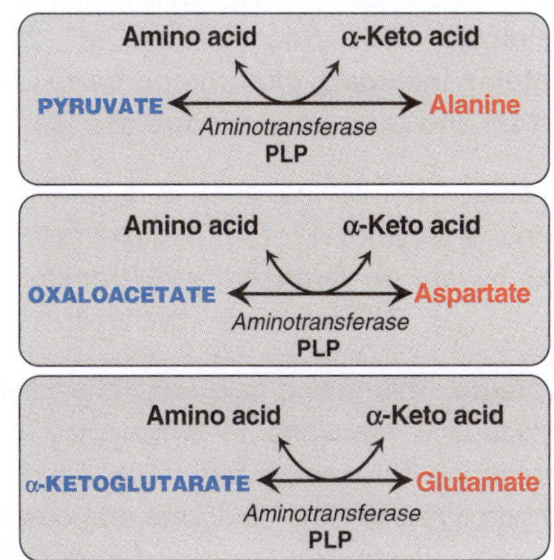

Figure 25.13 Formation of alanine, aspartate, and glutamate from the corresponding α-keto acids by transamination. PLP = pyridoxal phosphate.

(see Fig. 24.18, p. 389). The reaction is driven by the hydrolysis of ATP. In addition to producing glutamine for protein synthesis, the reaction also serves as a major mechanism for the transport of ammonia in a nontoxic form. (See p. 388 for a discussion of ammonia metabolism.)

2. **Asparagine:** This amino acid, which contains an amide linkage with ammonia at the β-carboxyl, is formed from aspartate by ***asparagine synthetase,*** using glutamine as the amide donor. Like the synthesis of glutamine, the reaction requires ATP and has an equilibrium far in the direction of amide synthesis.

C. Proline

Glutamate via glutamate semialdehyde is converted to proline by cyclization and reduction reactions. [Note: The semialdehyde can also be transaminated to ornithine.]

D. Serine, glycine, and cysteine

The pathways of synthesis for these amino acids are interconnected.

1. **Serine:** This amino acid arises from 3-phosphoglycerate, a glycolytic intermediate (see Fig. 13.18, p. 197), which is first oxidized to 3-phosphopyruvate and then transaminated to 3-phosphoserine. Serine is formed by hydrolysis of the phosphate ester. Serine can also be formed from glycine through transfer of a hydroxymethyl group by ***serine hydroxymethyltransferase* using N^5,N^{10}-MTHF as the one-carbon donor** (see Fig. 25.6A). [Note: Selenocysteine (Sec), the 21st genetically encoded amino acid, is synthesized from serine and selenium (see p. 561), while serine is attached to transfer RNA. **Sec is found in ~25 human proteins including *glutathione peroxidase* (see p. 142) and *thioredoxin reductase*** (see p. 583).]

2. **Glycine:** This amino acid is synthesized from serine by removal of a hydroxymethyl group, also by ***serine hydroxymethyltransferase*** (see Fig. 25.6A). THF is the one-carbon acceptor.

3. **Cysteine:** This amino acid is synthesized by two consecutive reactions in which Hcy combines with serine, forming cystathionine, which, in turn, is hydrolyzed to α-ketobutyrate and cysteine (see Fig. 25.8). [Note: Hcy is derived from methionine, as described on p. 400. Because methionine is an essential amino acid, cysteine synthesis requires adequate dietary intake of methionine.]

E. Tyrosine

Tyrosine is formed from phenylalanine by *PAH* (see p. 399). The reaction requires molecular oxygen and the coenzyme tetrahydrobiopterin (BH_4), which is synthesized from guanosine triphosphate. One atom of molecular oxygen becomes the hydroxyl group of tyrosine, and the other atom is reduced to water. During the reaction, BH_4 is oxidized to dihydrobiopterin (BH_2). BH_4 is regenerated from BH_2 by NADH-requiring *dihydropteridine reductase.* Tyrosine, like cysteine, is formed from an essential amino acid and is, therefore, nonessential only in the presence of adequate dietary phenylalanine.

VI. AMINO ACID METABOLISM DISORDERS

These single gene disorders, a subset of the inborn errors of metabolism, are caused by mutations that generally result in abnormal proteins, most often enzymes. The inherited defects may be expressed as a total loss of enzyme activity or, more frequently, as a partial deficiency in catalytic activity. Without treatment, the amino acid disorders almost invariably result in intellectual disability or other developmental abnormalities as a consequence of harmful accumulation of metabolites. Although >50 of these disorders have been described, many are rare, occurring in <1 per 250,000 in most populations (Fig. 25.14). Collectively, however, they constitute a very significant portion of pediatric genetic diseases (Fig. 25.15).

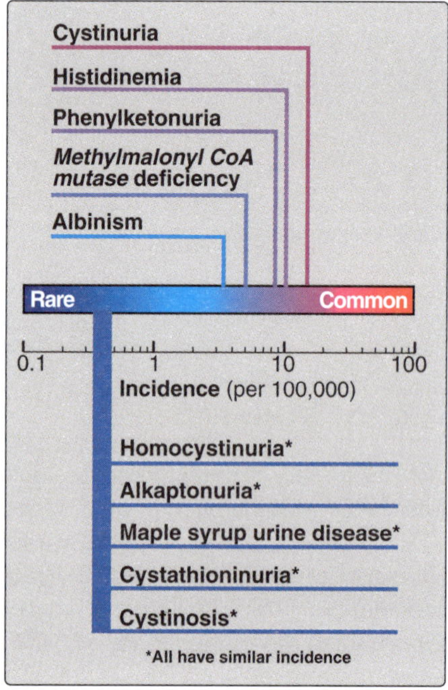

Figure 25.14 Incidence of inherited diseases of amino acid metabolism. [Note: Cystinuria is the most common inborn error of amino acid transport.]

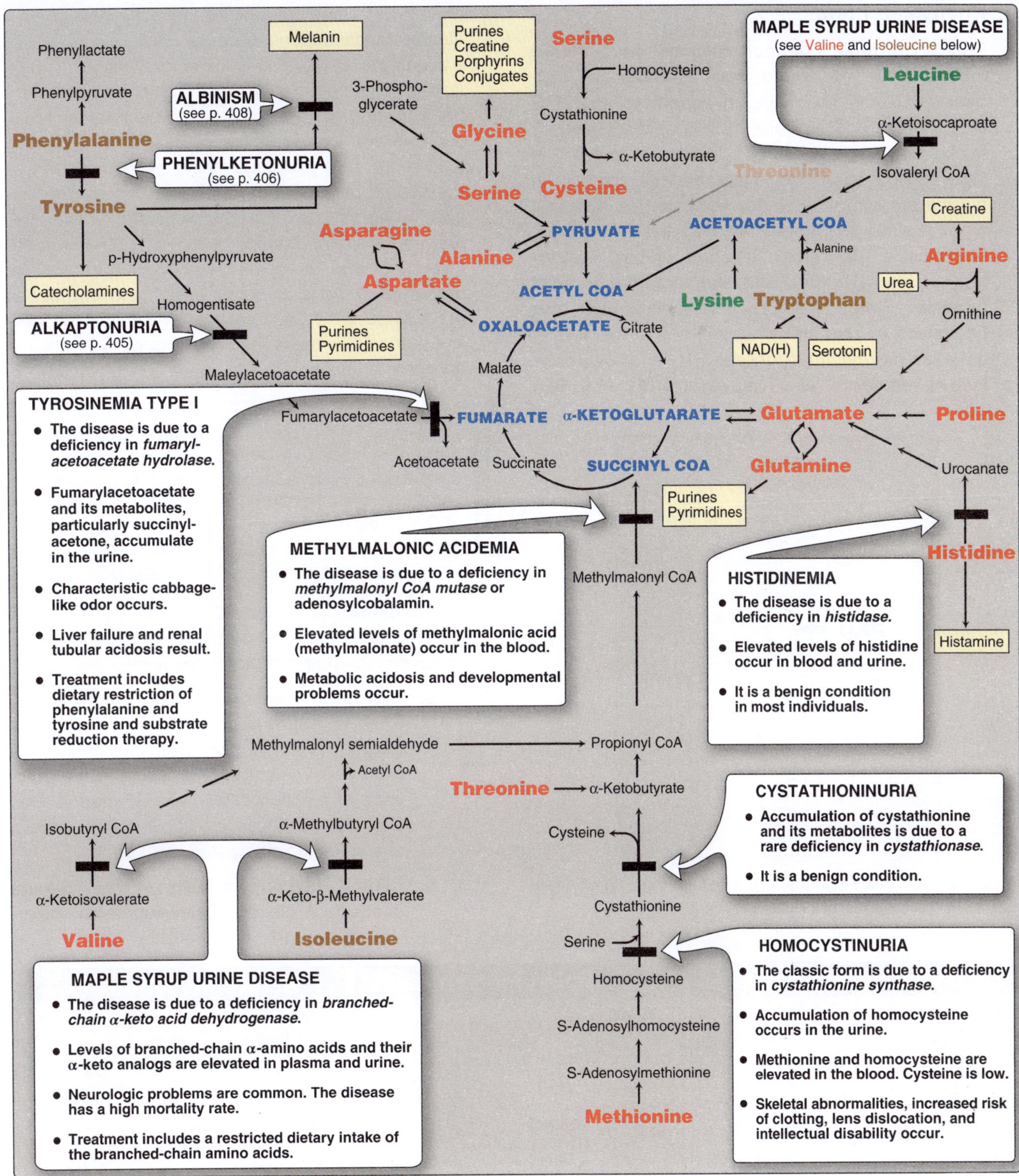

Figure 25.15 Summary of the metabolism of amino acids in humans. Genetically determined enzyme deficiencies are summarized in white boxes. Nitrogen-containing compounds derived from amino acids are shown in small, yellow boxes. Classification of amino acids is color coded: Red = glucogenic; brown = glucogenic and ketogenic; green = ketogenic. Compounds in BLUE ALL CAPS are the seven metabolites to which all amino acid metabolism converges. CoA = coenzyme A; NAD(H) = nicotinamide adenine dinucleotide.

A. Phenylketonuria

PKU is the **most common clinically encountered inborn error of amino acid metabolism** (incidence 1:15,000). It is caused by a **deficiency of *PAH*** (Fig. 25.16). Biochemically, PKU is characterized by hyperphenylalaninemia. Phenylalanine is present in high concentrations (ten times normal) not only in plasma but also in urine and body tissues. Tyrosine, which normally is formed from phenylalanine by *PAH*, is deficient. **Treatment includes dietary restriction of phenylalanine and supplementation with tyrosine.** [Note: Hyperphenylalaninemia may also be caused by rare deficiencies in any of the several enzymes required to synthesize BH_4 or in *dihydropteridine reductase*, which regenerates BH_4 from BH_2 (Fig. 25.17). Such deficiencies indirectly raise phenylalanine concentrations, because *PAH* requires BH_4 as a coenzyme. BH_4 is also required for *tyrosine hydroxylase* and *tryptophan hydroxylase*, which catalyze reactions leading to the synthesis of neurotransmitters, such as serotonin and the catecholamines. Simply restricting dietary phenylalanine does not reverse the central nervous system effects due to deficiencies in neurotransmitters. Supplementation with BH_4 and replacement therapy with l-3,4-dihydroxyphenylalanine and 5-hydroxytryptophan (products of the affected *tyrosine hydroxylase*– and *tryptophan hydroxylase*–catalyzed reactions) improves the clinical outcome in these variant forms of hyperphenylalaninemia, although the response is unpredictable.]

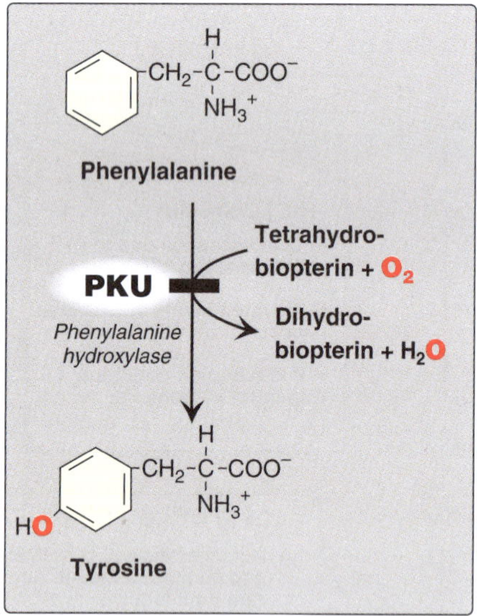

Figure 25.16 A deficiency in *phenylalanine hydroxylase* results in the disease phenylketonuria (PKU).

> Screening of newborns for a number of treatable disorders, including inborn errors of amino acid metabolism, is done by fluorimetry and tandem mass spectrometry of blood obtained from a heel prick. It is advisable that newborn screening for inborn errors of metabolism be done routinely.

1. **Additional characteristics:** As the name suggests, PKU is also characterized by elevated levels of a phenylketone in the urine.

   a. **Elevated phenylalanine metabolites:** Phenylpyruvate (a phenylketone), phenylacetate, and phenyllactate, which are not normally produced in significant amounts in the presence of functional *PAH*, are elevated in PKU (Fig. 25.18). These metabolites give urine a characteristic musty ("mousy") odor.

   b. **Central nervous system effects: Severe intellectual disability, developmental delay,**

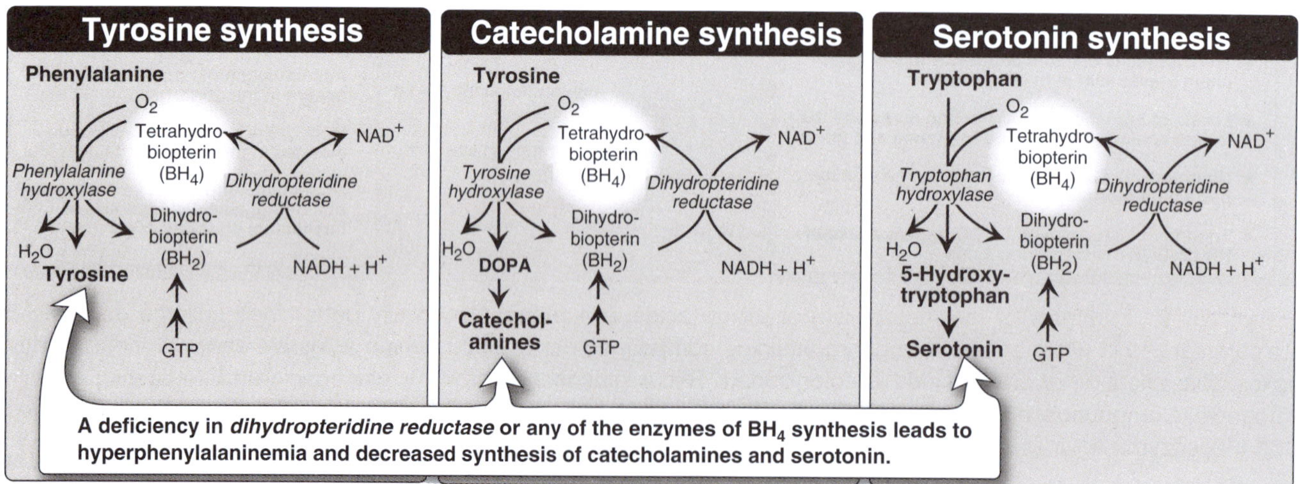

Figure 25.17 Biosynthetic reactions involving amino acids and tetrahydrobiopterin. [Note: Aromatic amino acid *hydroxylases* use BH_4 and not PLP (pyridoxal phosphate).] NAD(H) = nicotinamide adenine dinucleotide; GTP = guanosine triphosphate; DOPA = l-3,4-dihydroxyphenylalanine; O_2 = oxygen.

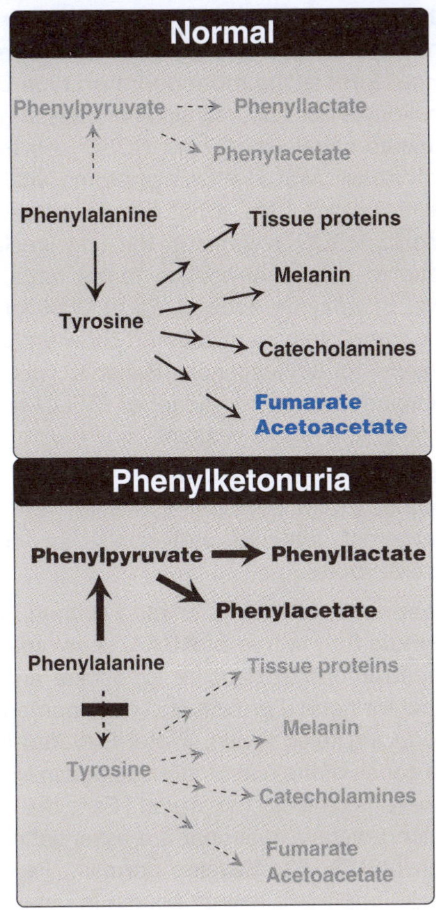

Figure 25.18 Pathways of phenylalanine metabolism in normal individuals and in patients with phenylketonuria.

microcephaly, and seizures are characteristic findings in untreated PKU. The affected individual typically shows symptoms of intellectual disability by age 1 year and rarely achieves an intelligence quotient (IQ) >50 (Fig. 25.19). [Note: These clinical manifestations are now rarely seen as a result of newborn screening programs, which allow early diagnosis and treatment.]

c. **Hypopigmentation:** Patients with untreated PKU may show a deficiency of pigmentation (fair hair, light skin color, and blue eyes). The hydroxylation of tyrosine by copper-requiring *tyrosinase*, which is the first step in the formation of the pigment melanin, is decreased in PKU because tyrosine is decreased.

2. **Newborn screening and diagnosis: Early diagnosis of PKU is important because the disease is treatable by dietary means.** Because of the lack of neonatal symptoms, laboratory testing for elevated blood levels of phenylalanine is mandatory for detection. However, the infant with PKU frequently has normal blood levels of phenylalanine at birth because the mother clears increased blood phenylalanine in her affected fetus through

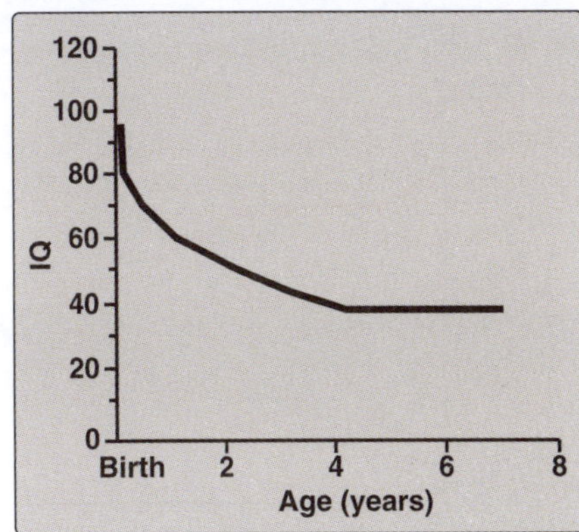

Figure 25.19 Typical intellectual ability in untreated patients of different ages with phenylketonuria. IQ = intelligence quotient.

the placenta. Normal levels of phenylalanine may persist until the newborn is exposed to 24–48 hours of protein feeding. Thus, screening tests are typically done after this time to avoid false negatives. For newborns with a positive screening test, diagnosis is confirmed through quantitative determination of phenylalanine levels.

3. **Prenatal diagnosis:** Classic PKU is caused by any of 100 or more different mutations in the gene that encodes *PAH*. The frequency of any given mutation varies among populations, and the disease is often doubly heterozygous (that is, the *PAH* gene has a different mutation in each allele). Despite this complexity, prenatal diagnosis is possible (see p. 682).

4. **Treatment:** Because most natural protein contains phenylalanine, an essential amino acid, it is impossible to satisfy the body's protein requirement without exceeding the phenylalanine limit when ingesting a normal diet. Therefore, in PKU, blood phenylalanine level is maintained close to the normal range by **feeding synthetic amino acid preparations free of phenylalanine, supplemented with some natural foods** (such as fruits, vegetables, and certain cereals) selected for their low phenylalanine content. The amount is adjusted according to the tolerance of the individual as measured by blood phenylalanine levels. The earlier treatment is started, the more completely neurologic damage can be prevented. Individuals who are appropriately treated can have normal intelligence. [Note: **Treatment must begin during the first 7–10 days of life to prevent cognitive impairment**.] Because phenylalanine is an essential amino acid, overzealous treatment that results in blood phenylalanine levels below normal is avoided. In patients with PKU, tyrosine cannot be synthesized from phenylalanine, and, therefore, it becomes an essential amino acid and

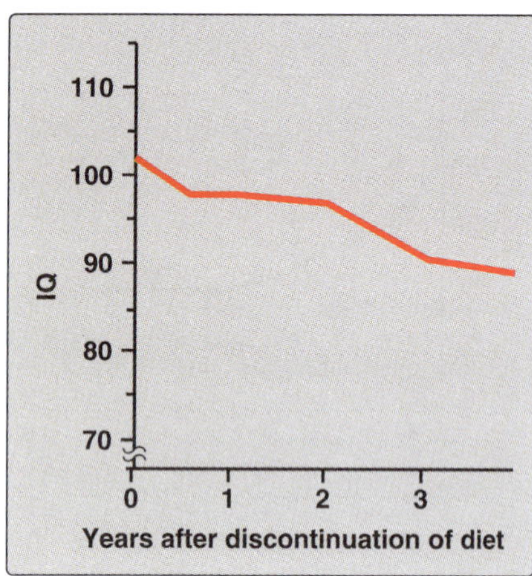

Figure 25.20 Changes in intelligence quotient (IQ) scores after discontinuation of low-phenylalanine diet in patients with phenylketonuria.

must be supplied in the diet. Discontinuance of the phenylalanine-restricted diet in early childhood is associated with poor performance on IQ tests. Adult PKU patients show deterioration of IQ scores after discontinuation of the diet (Fig. 25.20). Therefore, lifelong restriction of dietary phenylalanine is recommended. [Note: Individuals with PKU are advised to avoid aspartame, an artificial sweetener that contains phenylalanine.]

5. **Maternal phenylketonuria:** If women with PKU who are not on a low-phenylalanine diet become pregnant, the offspring can be affected with maternal PKU syndrome. High blood phenylalanine in the mother has a teratogenic effect, causing microcephaly and congenital heart abnormalities in the fetus. Because these developmental responses to high phenylalanine occur during the first months of pregnancy, dietary control of blood phenylalanine must begin prior to conception and be maintained throughout the pregnancy.

B. Maple syrup urine disease

Maple syrup urine disease (MSUD) is a rare (1:185,000), **autosomal-recessive disorder** in which there is a **partial or complete deficiency in *BCKD***, the mitochondrial enzyme complex that oxidatively decarboxylates leucine, isoleucine, and valine (see Fig. 25.11). These BCAA and their corresponding α-keto acids accumulate in the blood, causing a toxic effect that interferes with brain functions. The disease is characterized by feeding problems, vomiting, ketoacidosis, changes in muscle tone, neurologic problems that can result in coma (primarily because of the rise in leucine), and a characteristic maple syrup–like odor of the urine because of the rise in isoleucine. If untreated, the disease is fatal. If treatment is delayed, intellectual disability results.

1. **Classification:** MSUD includes a classic type and several variant forms. The **classic, neonatal-onset form is the most common type of MSUD**. Leukocytes or cultured skin fibroblasts from these patients show little or no *BCKD* activity. Infants with classic MSUD show symptoms within the first several days of life. If not diagnosed and treated, classic MSUD is lethal in the first weeks of life. Patients with intermediate forms have a higher level of enzyme activity (up to 30% of normal). The symptoms are milder and show an onset from infancy to adolescence. Patients with the rare thiamine-dependent variant of MSUD respond to large doses of this vitamin.
2. **Screening and diagnosis:** As with PKU, prenatal diagnosis and newborn screening are available, and most affected individuals are compound heterozygotes.
3. **Treatment:** MSUD is treated with a **synthetic formula that is free of BCAA**, supplemented with limited amounts of leucine, isoleucine, and valine to allow for normal growth and development without producing toxic levels. [Note: Elevated leucine is the cause of the neurologic damage in MSUD, and its level is carefully monitored.] Early diagnosis and lifelong dietary treatment are essential if the child with MSUD is to develop normally. [Note: BCAA are an important energy source in times of metabolic need, and individuals with MSUD are at risk of decompensation during periods of increased protein catabolism.]

C. Albinism

Albinism refers to a group of conditions in which a **defect in tyrosine metabolism results in a deficiency in the production of melanin.** These defects result in the partial or full absence of pigment from the skin, hair, and eyes. Albinism appears in different forms, and it may be inherited by one of several modes: **autosomal recessive (primary mode), autosomal dominant, or X linked.** Total absence of pigment from the hair, eyes, and skin (Fig. 25.21), *tyrosinase*-negative oculocutaneous albinism (type 1 albinism), results from an absent or defective copper-requiring *tyrosinase*. It is the most severe form of the condition. In addition to hypopigmentation, affected individuals have vision defects and photophobia (sunlight hurts their eyes). They are at increased risk for skin cancer.

D. Homocystinuria

The homocystinurias are a group of disorders involving defects in the metabolism of Hcy. These autosomal-recessive diseases are characterized by high urinary levels of Hcy, high plasma levels of Hcy and methionine, and low plasma levels of cysteine. The most

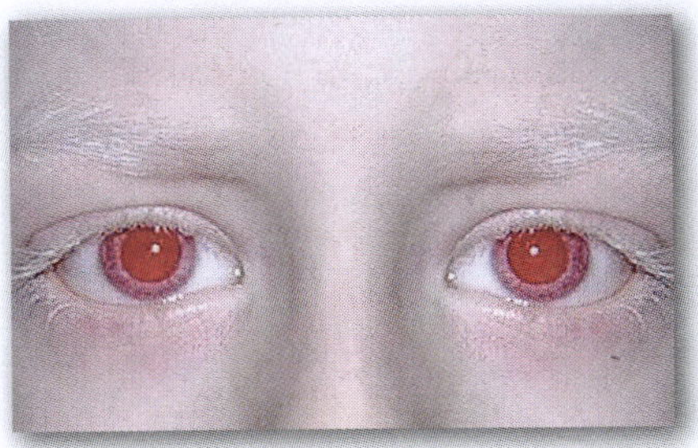

Figure 25.21 Patient with oculocutaneous albinism, showing white eyebrows and lashes and eyes that appear red in color.

common cause of homocystinuria is a **defect in the enzyme *cystathionine β-synthase***, which converts Hcy to cystathionine (Fig. 25.22). Individuals homozygous for *cystathionine β-synthase* deficiency exhibit dislocation of the lens (ectopia lentis), skeletal anomalies (long limbs and fingers), intellectual disability, and an increased risk for developing thrombi (blood clots). Thrombosis is the major cause of early death in these individuals. Treatment includes restriction of methionine and supplementation with vitamin B_{12} and folate. Additionally, some patients are responsive to oral administration of pyridoxine (vitamin B_6), which is converted to pyridoxal phosphate, the coenzyme of *cystathionine β-synthase.* These patients usually have a milder and later onset of clinical symptoms compared with B_6-nonresponsive patients. [Note: Deficiencies in methylcobalamin (see Fig. 25.8) or *N^5,N^{10}-MTHF reductase* ([*MTHFR*]; see Fig. 25.12) also result in elevated Hcy.]

E. Alkaptonuria

Alkaptonuria is a rare organic aciduria involving **a deficiency in *homogentisic acid oxidase***, resulting in the accumulation of homogentisic acid (HA), an intermediate in the degradative pathway of tyrosine (see Fig. 25.15 on p. 405). The condition has three characteristic symptoms: homogentisic aciduria (the urine contains elevated levels of HA, which is oxidized to a dark pigment on standing, as shown in Fig. 25.23A),

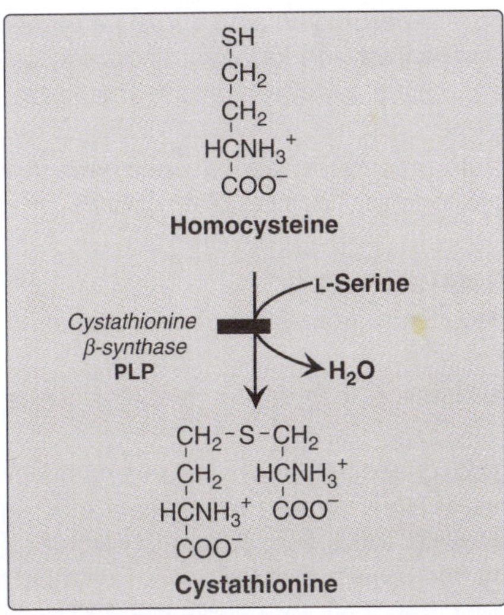

Figure 25.22 Enzyme deficiency in homocystinuria. PLP = pyridoxal phosphate.

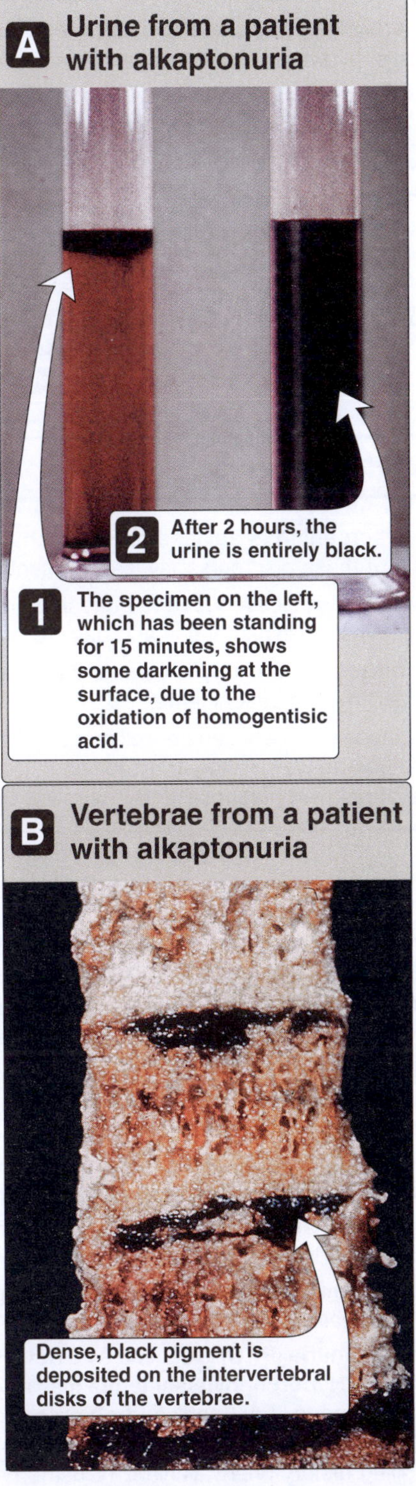

Figure 25.23 Specimens from a patient with alkaptonuria. **A.** Urine. **B.** Vertebrae.

early onset of arthritis in the large joints, and deposition of black pigment (**ochronosis**) in cartilage and collagenous tissue (see Fig. 25.23B). Dark staining of diapers can indicate the disease in infants, but usually no symptoms are present until about age 40 years. Treatment includes dietary restriction of phenylalanine and tyrosine to reduce HA levels. Although alkaptonuria is not life threatening, the associated arthritis may be severely crippling. [Note: Deficiencies in *fumarylacetoacetate hydrolase*, the terminal enzyme of tyrosine metabolism, result in tyrosinemia type I (see Fig. 25.15) and a characteristic cabbage-like odor to urine.]

F. Methylmalonic acidemia

Methylmalonic acidemia (MMA) is a rare (1:100,000) autosomal recessive disorder caused by a deficiency in methylmalonyl CoA mutase, which converts L-methylmalonyl CoA to succinyl CoA. Since the mutase requires vitamin B_{12}, the disease can also result from a severe B_{12} deficiency. The breakdown of odd-chain length fatty acids, valine, isoleucine, methionine, and threonine can all result in MMA, due to this enzyme deficiency. Elevation of methylmalonate in the blood and urine can result in a metabolic acidosis. There may also be an increase in propionyl-CoA, exacerbating the aciduria with an accumulation of additional propionic acid. Symptoms appear in early infancy, varying due to the degree of the enzyme deficiency, including failure to thrive, vomiting, dehydration, hypotonia, developmental delay, seizures, hepatomegaly, hyperammonemia, and a progressive encephalopathy. If severe and left untreated, it can lead to intellectual disability, chronic renal or hepatic damage, pancreatitis, and coma or death. Treatment includes a low-protein, high-calorie diet, and vitamin B_{12} supplementation. The diet limits the intake of isoleucine, threonine, methionine, and valine, as these amino acids can lead to the buildup of methylmalonic acid by the mutase deficiency.

VII. Medical Perspectives

Newborn screening is advised for all neonates to check for inborn errors of metabolism at the very early stage. Late diagnosis of these disorders may lead to mental retardation and other irreparable damages.

The common diseases required to be tested are hypothyroidism, congenital adrenal hypoplasia, G6PD deficiency, hemoglobinopathies, phenylketonuria and galactosemia.

Tandem mass spectroscopy is used to detect amino acid disorders like phenylketonuria, tyrosinemia, maple syrup urine disease and hypermethioninemia.

Some of these disorders can be well managed if diagnosed early before irreparable damage has taken place, e.g., phenylketonuria can be well managed by a lifelong low-protein phenylalanine restricted diet. But if mental retardation has set in because of delayed diagnosis and treatment, that cannot be reversed.

VIII. Chapter Summary

- **Amino acids** whose catabolism yields **pyruvate** or an **intermediate** of the **tricarboxylic acid cycle** are termed **glucogenic** (Fig. 25.24). They can give rise to the net formation of **glucose** in the **liver** and **kidneys**. The solely glucogenic amino acids are glutamine, glutamate, proline, arginine, histidine, alanine, serine, glycine, cysteine, methionine, valine, threonine, aspartate, and asparagine.
- Amino acids whose catabolism yields either **acetoacetate** or one of its precursors, **acetyl coenzyme A** (**CoA**) or **acetoacetyl CoA**, are termed **ketogenic**. Leucine and lysine are solely ketogenic. Tyrosine, phenylalanine, tryptophan, and isoleucine are both ketogenic and glucogenic.
- Tyrosine, phenylalanine, tryptophan, and isoleucine are both ketogenic and glucogenic.
- **Nonessential amino acids** can be synthesized from metabolic intermediates or from the carbon skeletons of essential amino acids.
- **Essential amino acids** need to be obtained from the **diet**. They include histidine, methionine, threonine, valine, isoleucine, phenylalanine, tryptophan, leucine, and lysine.
- **Phenylketonuria** (**PKU**) is caused by a **deficiency** of ***phenylalanine hydroxylase*** (***PAH***), which converts phenylalanine to tyrosine. **Hyperphenylalaninemia** may also be caused by deficiencies in the enzymes that synthesize or regenerate the coenzyme for *PAH*, **tetrahydrobiopterin**. Untreated individuals with PKU suffer from severe intellectual disability, developmental delay, microcephaly, seizures, and a characteristic musty (mousy) smell of the urine. Treatment involves controlling dietary phenylalanine. **Tyrosine** becomes an essential dietary component for people with PKU.
- **Maple syrup urine disease** (**MSUD**) is caused by a partial or complete deficiency in ***branched-chain a-keto acid dehydrogenase***, the enzyme that decarboxylates the **branched-chain amino acids** (**BCAA**) **leucine**, **isoleucine**, and

valine. Symptoms include feeding problems, vomiting, ketoacidosis, changes in muscle tone, and a characteristic sweet smell of the urine. If untreated, the disease leads to neurologic problems that result in death. Treatment involves controlling BCAA intake.

- Other important genetic diseases associated with amino acid metabolism include **albinism**, **homocystinuria**, **methylmalonic acidemia**, **alkaptonuria**, **histidinemia**, **tyrosinemia**, and **cystathioninuria**.

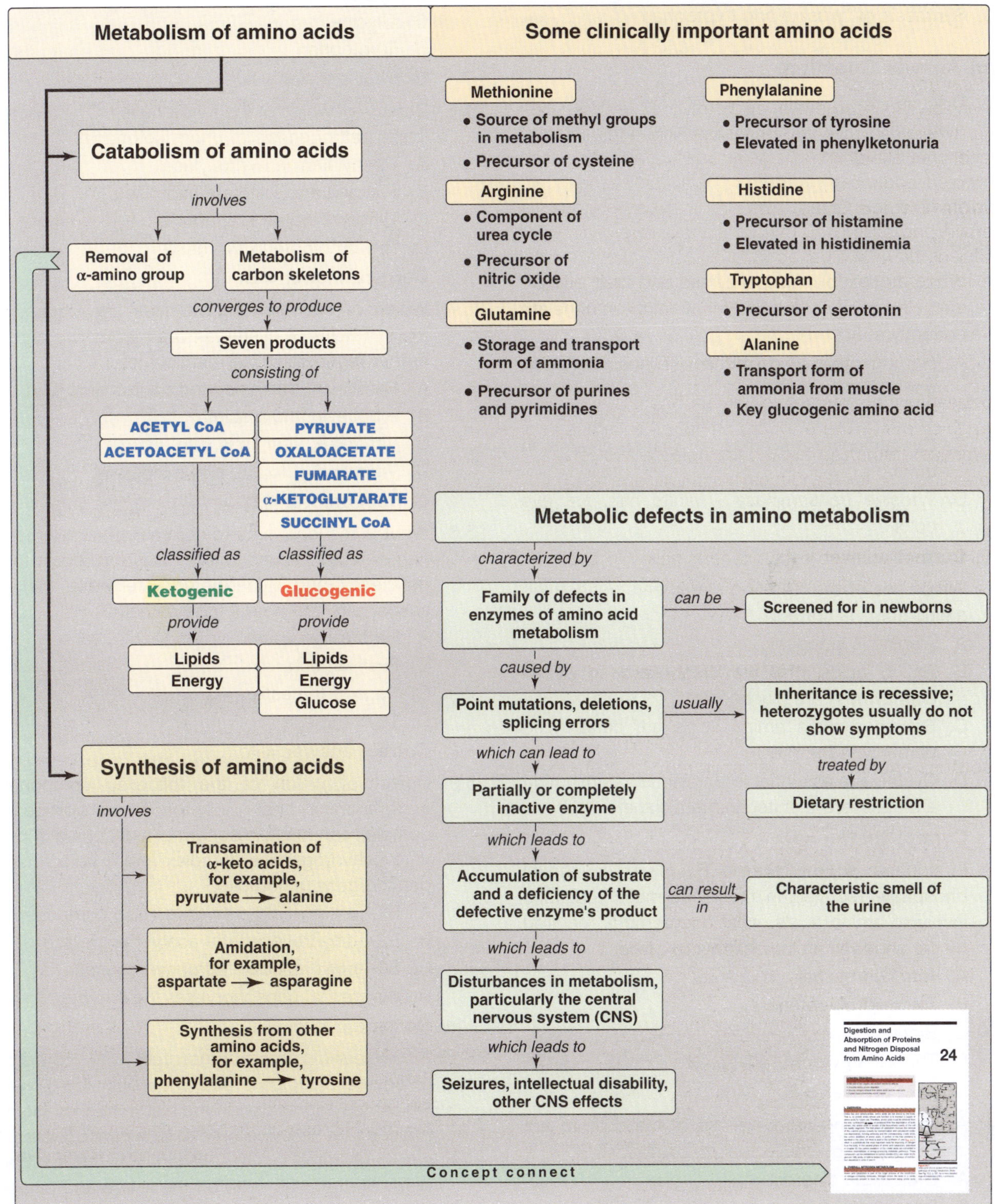

Figure 25.24 Key concept map for amino acid metabolism. CoA = coenzyme A.

Study Questions

Short-Answer Questions

Write short notes on:

25.1. Glucogenic and ketogenic amino acids

25.2. Methionine metabolism

25.3. Synthesis of tyrosine and tryptophan

Long-Answer Question

25.1. Discuss the genetic disorders of amino acid metabolism. Add a note on neonatal screening and prenatal diagnosis.

Multiple-Choice Questions

Choose the ONE best answer.

25.1. Which of the following enzymes and their associated clinical signs or laboratory findings in urine are correctly matched?

A. Homogentisic acid oxidase – Black pigmentation of cartilage

B. Tyrosinase – Sweaty feet–like odor of fluids

C. Cystathionine β-synthase – Cystine crystals in urine

D. Tyrosine transaminase – White hair, red eye color

Correct answer = A.

25.2. Which one of the following statements concerning amino acids is correct?

A. Alanine is ketogenic.

B. Amino acids that are catabolized to acetyl coenzyme A are glucogenic.

C. Branched-chain amino acids are catabolized primarily in the liver.

D. Cysteine is essential for individuals consuming a diet severely limited in methionine.

Correct answer = D.

25.3. In contrast to the vitamin B_6-derived pyridoxal phosphate required in most enzymatic reactions involving amino acids, what coenzyme is required by the aromatic amino acid hydroxylases?

A. Tetrahydrofolate

B. Tetrahydrobiopterin

C. Biotin

D. Thiamine pyrophosphate

Correct answer = B.

25.4. All the following inborn errors of metabolism are characterized by intellectual disability, EXCEPT:

A. Urea cycle disorders

B. Phenylketonuria

C. Maple syrup urine disease

D. Alkaptonuria

Correct answer = D.

25.5. All the following amino acids or their derivatives serve as neurotransmitters or their precursors, EXCEPT:

A. Threonine

B. Tyrosine

C. Glycine

D. Tryptophan

Correct answer = A.

25.6. Defects of which of the following is NOT implicated in the pathogenesis of phenylketonuria?

A. Phenylalanine hydroxylase

B. Fumaryl acetoacetate hydrolase

C. Dihydrobiopterin reductase

D. Dihydrobiopterin synthesis

Correct answer = B.

25.7. Which of the following vitamins are required for the synthesis of cysteine from methionine and the further regeneration of methionine?

A. Riboflavin, thiamine, and pantothenic acid

B. Riboflavin and niacin

C. Vitamin E, vitamin K, and vitamin C

D. Vitamin B_{12}, folate, and pyridoxine

Correct answer = D.

25.8. What is the ideal time to commence supplementation of folate in pregnant females to prevent folate deficiency and hyper-homocysteinemia that are associated with neural tube defects?

A. Before conception

B. First trimester

C. Second trimester

D. Third trimester

Correct answer = A.

25.9. Defect of which of the following components of branched chain α-ketoacid dehydrogenase complex can result in abnormal functioning of pyruvate dehydrogenase complex and α-ketoglutarate dehydrogenase complex?

A. E1—α-keto-acid dehydrogenase component

B. E2—dihydrolipoamide acetyl transferase

C. E3—dihydrolipoamide dehydrogenase

D. Kinase that phosphorylates the enzyme complex

Correct answer = C.

25.10. A 39-year-old male presented with complains of severe back pain and knee pain for 5 years, which has aggravated in the last 3 months. He had noticed that his urine used to turn dark on standing since childhood. On examination bluish black pigmentation was seen on the sclera and pinna of the ear. Routine blood investigations were normal. His urine sample turned black on alkalinization with few drops of 10% ammonia. Radiological examination of spine showed reduction of disc spaces and fusion of vertebral bodies. What is the most likely cause of the clinical manifestations here?

A. Deficiency of synthesis of a neurotransmitter.

B. Deposition of a pigment in the cartilages.

C. Accumulation of a metabolite that can lead to metabolic alkalosis.

D. Disruption of extracellular matrix due to chemical modifications by an accumulated intermediate.

Correct answer = B.

Multiple-Choice Questions (NExT/NEET PG)

Choose the ONE best answer.

25.1. A 1-week-old infant, who was born at home in a rural, medically-underserved area, has undetected classic phenylketonuria. Which statement about this baby and/or her treatment is correct?

A. A diet completely devoid of phenylalanine should be initiated immediately.

B. Dietary treatment will be discontinued in adulthood.

C. Supplementation with vitamin B_6 is required.

D. Tyrosine is an essential amino acid.

Correct answer = D.

25.2. Many neutral amino acids are transported by a common transporter. An increase in the plasma level of any one neutral amino acid can impact the entry of other neutral amino acids into the cell. The clinical manifestations in which of the following disorders are due to this mechanism?

A. Cystinuria and maple syrup urine disease

B. Phenylketonuria and maple syrup urine disease

C. Phenylketonuria and alkaptonuria

D. Cystinuria and tyrosinemia

Correct answer = B.

25.3. In the human body, certain amino acids are not dietary essential because these can be synthesized in the human body from other precursor molecules, including certain other amino acids. However, in certain metabolic disorders, this process is disrupted and amino acids that are otherwise non-essential in healthy individuals become dietarily essential. Which of the following diseases exemplify this?

A. Cystinuria and maple syrup urine disease

B. Primary hyperoxaluria and tyrosinemia

C. Phenylketonuria and alkaptonuria

D. Phenylketonuria and homocystinuria

Correct answer = D.

25.4. An 8-day-old male neonate was brought by his parents to the hospital with the complaints of lethargy and poor feeding. The baby was born at full term with uneventful antenatal and perinatal history. The parents of the baby were consanguineously married, and he has one healthy male sibling. On examination, the child was found to be hypotonic and minimally responsive, with diminished reflexes in all the four limbs. Blood investigations revealed metabolic acidosis. Plasma and CSF amino acid analysis by HPLC revealed elevated leucine, isoleucine, and valine levels. What is/are the most likely defective enzyme(s) in this child?

A. Pyruvate dehydrogenase complex

B. Branched chain α-ketoacid dehydrogenase complex

C. Isovaleryl CoA dehydrogenase

D. Glycine cleavage system

Correct answer = B.

25.5. Which of the following inborn errors of amino acid metabolism and the coenzyme supplementation that can improve the clinical manifestations is NOT correctly matched?

A. Homocystinuria—Pyridoxine

B. Maple syrup urine disease—Thiamine

C. Phenylketonuria—Biopterin

D. Albinism—Biotin

Correct answer = D.

Amino Acids: Conversion to Specialized Products Including Porphyrins

26

Learning Objectives

At the end of this chapter, the student will be able to:

- Discuss porphyrin metabolism and associated diseases
- Explain the types of jaundice with biochemical reasoning
- Describe synthesis of catecholamines

I. OVERVIEW

In addition to serving as building blocks for proteins, amino acids are precursors of many nitrogen-containing compounds that have important physiologic functions (Fig. 26.1). These molecules include **porphyrins, neurotransmitters, hormones, purines, and pyrimidines**. [Note: See p. 252 for the synthesis of nitric oxide from arginine.]

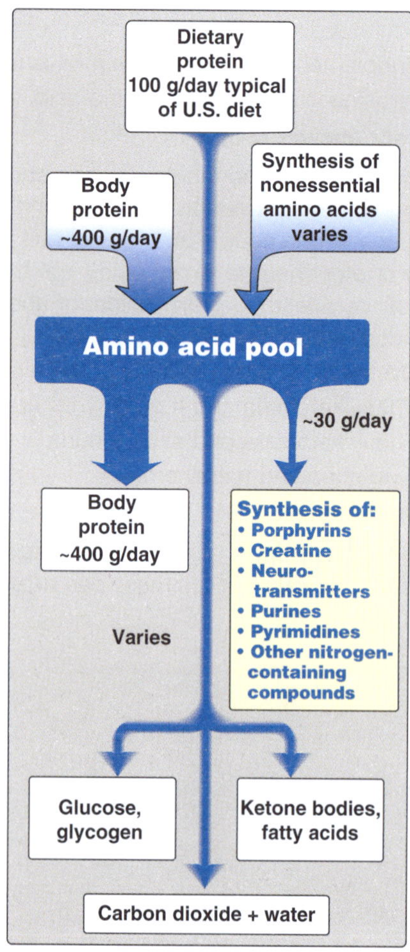

Figure 26.1 Amino acids as precursors of nitrogen-containing compounds.

II. PORPHYRIN METABOLISM

Porphyrins are cyclic compounds that readily bind metal ions, usually ferrous (Fe^{2+}) or ferric (Fe^{3+}) iron. The most prevalent metalloporphyrin in humans is heme, which consists of one Fe^{2+} coordinated in the center of the tetrapyrrole ring of protoporphyrin IX (see p. 115). Heme is the prosthetic group for hemoglobin (Hb), **myoglobin, the cytochromes, the *cytochrome P450* (*CYP*) *monooxygenase* system, *catalase*, *nitric oxide synthase*, and *peroxidase*.** These hemeproteins are rapidly synthesized and degraded. For example, 6–7 g of Hb is synthesized each day to replace heme lost through the normal turnover of erythrocytes. The synthesis and degradation of the associated porphyrins and recycling of the iron are coordinated with the turnover of hemeproteins.

A. Structure

Porphyrins are cyclic planar molecules formed by the linkage of four pyrrole rings through methenyl

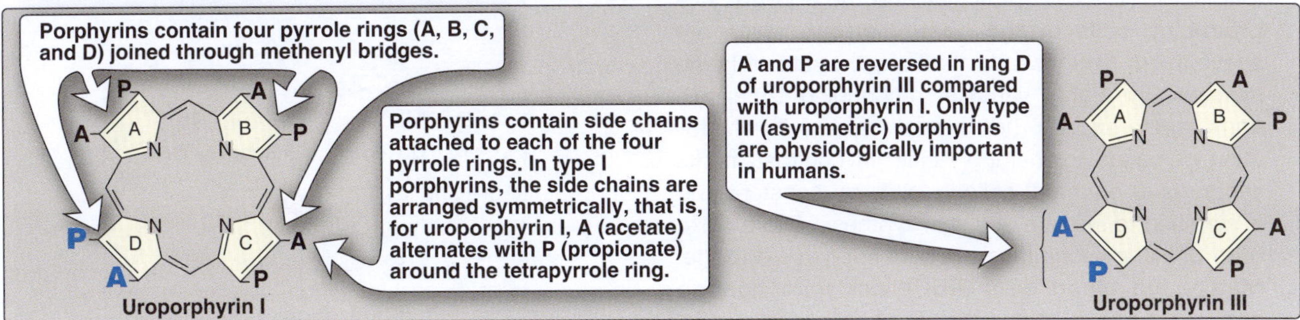

Figure 26.2 Structures of uroporphyrin I and uroporphyrin III.

bridges (Fig. 26.2). Three structural features of these molecules are relevant to understanding their medical significance.

Hemoproteins contain heme as the prosthetic group and serve diverse functions in the body.

- **Hemoproteins that reversibly bind to oxygen**:
  - Hemoglobin – oxygen and carbon dioxide transport; acts as a buffer
  - Myoglobin – oxygen reserve in muscles
  - Neuroglobin – oxygen reserve in neurons
  - Cytoglobin – oxygen reserve in multiple tissues
- **Hemoproteins with enzymatic activity**:
  - *Catalase* – conversion of H_2O_2 into water
  - *Peroxidases* – neutralization of harmful effects of peroxides
  - *Cytochrome P_{450} monooxygenases* – in drug/chemical detoxification and steroid metabolism
  - *Nitric oxide synthase* – generation of nitric oxide (NO)
  - *Cyclooxygenase* – in prostaglandin synthesis
- **Hemoproteins that form complexes and mobile carriers of electron transport system**
  - *Q-cytochrome c oxidoreductase* – in oxidative phosphorylation
  - *Cytochrome c oxidase* – in oxidative phosphorylation
  - Cytochrome c – in oxidative phosphorylation

Note: Apart from these roles of hemoproteins, dietary heme acts as an important source for iron for the body.

1. **Side chains:** Different porphyrins vary in the nature of the side chains attached to each of the four pyrrole rings. Uroporphyrin contains acetate ($-CH_2-COO-$) and propionate ($-CH_2-CH_2-COO-$) side chains; coproporphyrin contains methyl ($-CH_3$) and propionate groups; and protoporphyrin IX (and heme b, the most common heme) contains vinyl ($-CH=CH_2$), methyl, and propionate groups. [Note: The methyl and vinyl groups are produced by decarboxylation of acetate and propionate side chains, respectively.]

2. **Side chain distribution:** The side chains of porphyrins can be ordered around the tetrapyrrole nucleus in four different ways, designated by Roman numerals I to IV. Only type III porphyrins, which contain an asymmetric substitution on ring D (see Fig. 26.2), are physiologically important in humans. [Note: Protoporphyrin IX is a member of the type III series.]

3. **Laboratory diagnosis of porphyrias:** The samples for PBG should be collected preferably during an acute attack.

   In hereditary coproporphyria, urine and fecal coproporphyrin-III are elevated. **In variegate porphyria**, urine and fecal coproporphyrin-III and fecal protoporphyrin-IX are elevated.

   Acidified urine or fecal extracts containing porphyrins absorb light at 400 nm, which can be detected spectrophotometrically – the so-called **Soret band or Soret peak**. Individual porphyrins also show their own characteristic absorption spectra, which can be detected on a spectrometric scan.

   Genetic testing for detection of mutations in the genes involved in heme synthesis is also currently available.

4. **Porphyrinogens:** These porphyrin precursors (for example, uroporphyrinogen) exist in a chemically reduced, colorless form and serve as intermediates between porphobilinogen (PBG) and the oxidized, colored protoporphyrins in heme biosynthesis.

B. Heme biosynthesis

The major sites of heme biosynthesis are the **liver,** which synthesizes a number of heme proteins (particularly the *CYP* proteins), and the **erythrocyte-producing cells of the bone marrow**, which are active in Hb synthesis. In the liver, the rate of heme synthesis is highly variable, responding to alterations in the cellular heme pool caused by fluctuating demands for hemeproteins. In contrast, heme synthesis in erythroid cells is relatively constant and is matched to the rate of globin synthesis. [Note: Over 85% of all heme synthesis occurs in erythroid tissue. **Mature red blood cells (RBC) lack mitochondria and are unable to synthesize heme.]** The initial reaction and the last three steps in the formation of porphyrins occur in mitochondria, whereas the intermediate steps of the biosynthetic pathway occur in the cytosol. [Note: Fig. 26.8 summarizes heme synthesis.]

1. **δ-Aminolevulinic acid formation:** All the carbon and nitrogen atoms of the porphyrin molecule are provided by glycine (a nonessential amino acid) and succinyl coenzyme A (a tricarboxylic acid cycle intermediate) that condense to form δ-aminolevulinic acid (ALA) in a reaction catalyzed by ***ALA synthase*** ([*ALAS*], Fig. 26.3). This reaction requires pyridoxal phosphate ([PLP], see p. 531) as a coenzyme and is the committed and **rate-limiting step** in porphyrin biosynthesis. [Note: There are two *ALAS* isoforms, each produced by different genes and controlled by different mechanisms. *ALAS1* is found in all tissues, whereas *ALAS2* is erythroid specific. Loss-of-function mutations in *ALAS2* result in X-linked sideroblastic anemia and iron overload.]

   a. **Heme (hemin) effects:** When porphyrin production exceeds the availability of the apoproteins that require it, heme accumulates and is converted to **hemin by the oxidation of Fe^{2+} to Fe^{3+}.** Hemin decreases the amount (and, thus, the activity) of *ALAS1* by repressing transcription of its gene, increasing degradation of its messenger RNA, and decreasing import of the enzyme into mitochondria. [Note: In erythroid cells, *ALAS2* is controlled by the availability of intracellular iron (see p. 664).]

   b. **Drug effects:** Administration of any of a large number of drugs results in a significant increase in hepatic *ALAS1* activity. These drugs are metabolized by the ***microsomal CYP monooxygenase*** **system, a hemeprotein *oxidase*** system found in the liver (see p. 251). In response to these drugs, the synthesis of *CYP* proteins increases, leading to an enhanced consumption of heme, a component of these proteins. This, in turn, causes a decrease in the concentration of heme in liver cells. The lower intracellular heme concentration leads to an increase in the synthesis of *ALAS1* and prompts a corresponding increase in the synthesis of ALA.

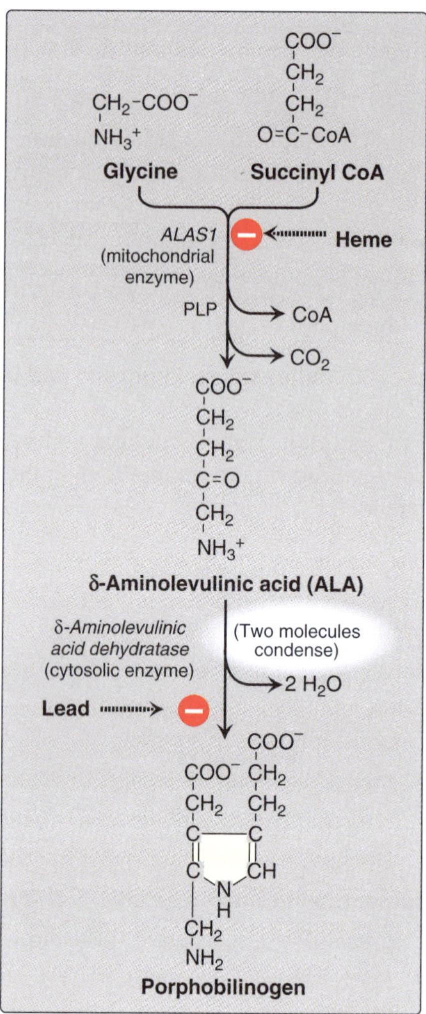

Figure 26.3 Pathway of porphyrin synthesis: Formation of porphobilinogen. [Note: *ALAS2* is regulated by iron.] *ALAS* = *δ-aminolevulinic acid synthase*; CoA = coenzyme A; CO_2 = carbon dioxide; PLP = pyridoxal phosphate. (Continued in Figs. 26.4 and 26.5.)

2. **Porphobilinogen formation:** The cytosolic condensation of two ALA to form PBG by zinc-containing ***ALA dehydratase (PBG synthase)*** is extremely sensitive to inhibition by heavy metal ions (for example, lead) that replace the zinc (see Fig. 26.3). This inhibition is, in part, responsible for the elevation in ALA and the anemia seen in lead poisoning.

3. **Uroporphyrinogen formation:** The condensation of four PBG produces the linear tetrapyrrole hydroxymethylbilane, which is cyclized and isomerized by ***uroporphyrinogen III synthase*** to produce the asymmetric uroporphyrinogen III. This cyclic tetrapyrrole undergoes decarboxylation of

its acetate groups by ***uroporphyrinogen III decarboxylase* (*UROD*),** generating coproporphyrinogen III (Fig. 26.4). The reactions occur in the cytosol.

4. **Heme formation:** Coproporphyrinogen III enters the mitochondrion, and two propionate side chains are decarboxylated by ***coproporphyrinogen III oxidase*** to vinyl groups generating **protoporphyrinogen IX**, which is oxidized to protoporphyrin IX. The introduction of iron (as Fe^{2+}) into protoporphyrin IX produces heme. This step can occur spontaneously, but the **rate is enhanced by *ferrochelatase***, an enzyme that, like *ALA dehydratase*, is inhibited by lead (Fig. 26.5).

C. Porphyrias

Porphyrias are rare, inherited (or sometimes acquired) defects in heme synthesis, resulting in the accumulation and increased excretion of porphyrins or porphyrin precursors (see Fig. 26.8). [Note: Inherited porphyrias are autosomal-dominant (AD) or autosomal-recessive (AR) disorders.] Each porphyria results in the accumulation of a unique pattern of intermediates caused by the deficiency of an enzyme in the heme synthetic pathway. Acquired porphyrias can occur due to factors such as drugs like sulfonamides, barbiturates, anti-cancer drugs, hemochromatosis, hepatitis C infections, lead poisoning, alcoholism and smoking. [Note: Porphyria, derived from the Greek for purple, refers to the red-blue color caused by pigment-like porphyrins in the urine of some patients with defects in heme synthesis.]

1. **Clinical manifestations:** The porphyrias are classified as **erythropoietic or hepatic,** depending on whether the enzyme deficiency occurs in the erythropoietic cells of the bone marrow or in the liver. Hepatic porphyrias can be further classified as chronic or acute. In general, individuals with an enzyme defect prior to the synthesis of the tetrapyrroles manifest abdominal and neuropsychiatric signs, whereas those with enzyme defects leading to the accumulation of tetrapyrrole intermediates show photosensitivity (that is, their skin itches and burns [pruritus] when exposed to sunlight). [Note: **Photosensitivity** is a result of the oxidation of colorless porphyrinogens to colored porphyrins, which are photosensitizing molecules thought to participate in the formation of superoxide radicals from oxygen. These radicals can oxidatively damage membranes and cause the release of destructive enzymes from lysosomes.]

   a. **Chronic hepatic porphyria: Porphyria cutanea tarda, the most common porphyria, is a chronic disease of the liver.** The disease is associated with severe deficiency of *UROD*, but clinical expression of the deficiency is influenced by various factors, such as hepatic

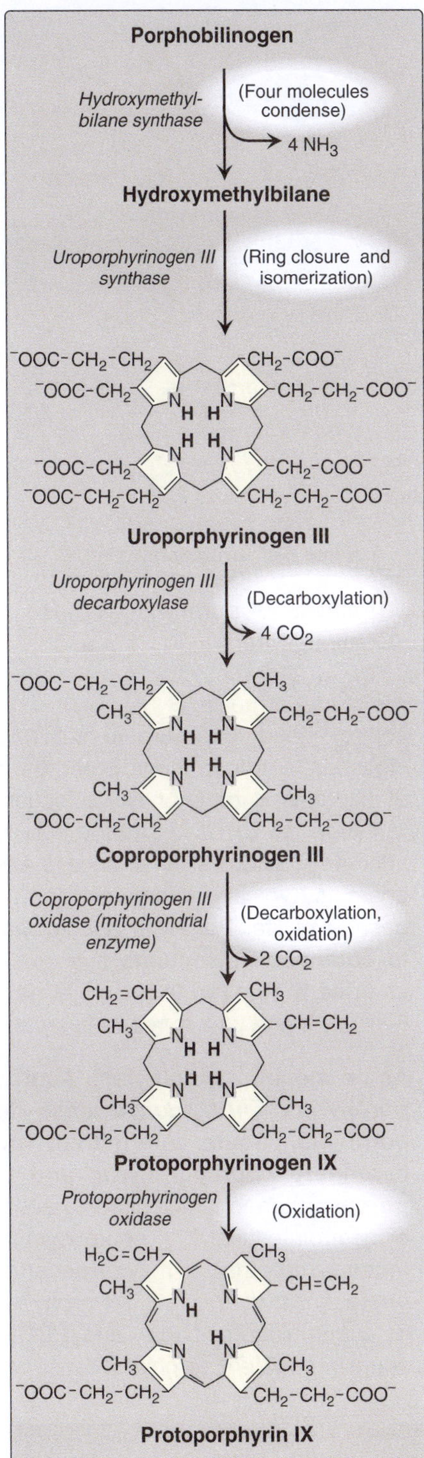

Figure 26.4 Pathway of porphyrin synthesis: formation of protoporphyrin IX. (Continued from Fig. 26.3.) The prefixes -uro (urine) and -copro (feces) reflect initial sites of discovery. [Note: Deficiency in *uroporphyrinogen III synthase* prevents isomerization, resulting in production of type I porphyrins.]

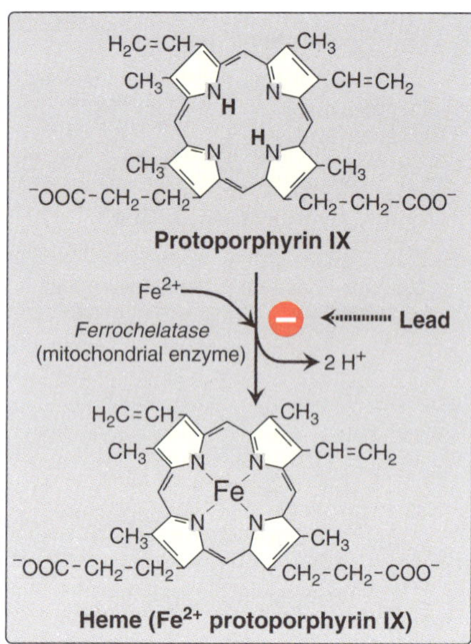

Figure 26.5 Pathway of porphyrin synthesis: formation of heme b. (Continued from Figs. 26.3 and 26.4.) Fe^{2+} = ferrous iron.

iron overload, exposure to sunlight, alcohol ingestion, estrogen therapy, and the presence of hepatitis B or C or HIV infections. [Note: Mutations to *UROD* are found in only 20% of affected individuals. Inheritance is AD.] Clinical onset is typically during the fourth or fifth decade of life. Porphyrin accumulation leads to **cutaneous symptoms** (Fig. 26.6) as well as **urine that is red to brown in natural light** (Fig. 26.7) and pink to red in fluorescent light.

b. **Acute hepatic porphyrias: Acute hepatic porphyrias (*ALA dehydratase*–deficiency porphyria, acute intermittent porphyria, hereditary coproporphyria, and variegate porphyria**) are characterized by acute attacks of gastrointestinal (GI), neuropsychiatric, and motor symptoms that may be accompanied by photosensitivity (Fig. 26.8). Porphyrias leading to accumulation of ALA and PBG, such as acute intermittent porphyria, cause **abdominal**

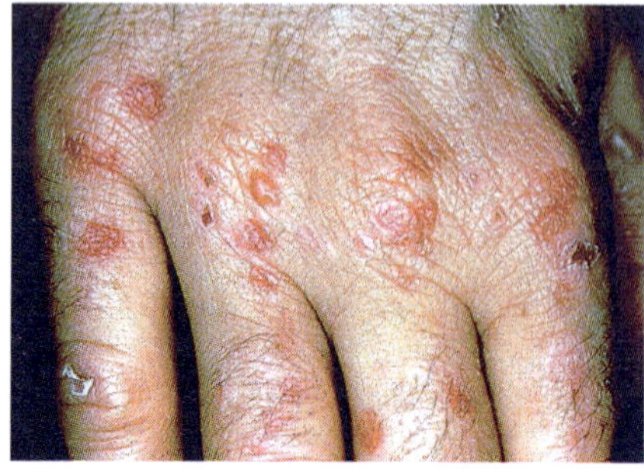

Figure 26.6 Skin eruptions in a patient with porphyria cutanea tarda.

Figure 26.7 Urine from a patient with porphyria cutanea tarda (right) and from a patient with normal porphyrin excretion (left).

pain and **neuropsychiatric disturbances**, ranging from anxiety to delirium. Symptoms of the acute hepatic porphyrias are **often precipitated by use of drugs, such as barbiturates and ethanol,** which induce the synthesis of the heme-containing *CYP* microsomal drug-oxidation system. This further decreases the amount of available heme, which, in turn, promotes increased synthesis of *ALAS1*.

c. **Erythropoietic porphyrias:** The **chronic erythropoietic porphyrias** (congenital erythropoietic porphyria and erythropoietic protoporphyria) cause photosensitivity characterized by skin rashes and blisters that appear in early childhood (see Fig. 26.8).

2. **Increased δ-aminolevulinic acid synthase activity:** One common feature of the hepatic porphyrias is decreased synthesis of heme. In the liver, heme normally functions as a repressor of the *ALAS1* gene. Therefore, the absence of this end product results in an increase in the synthesis of *ALAS1* (derepression). This causes an increased synthesis of intermediates that occur prior to the genetic block. The accumulation of these toxic intermediates is the major pathophysiology of the porphyrias.

3. **Laboratory diagnosis of porphyrias:** The samples for PBG should be collected preferably during an acute attack.

   In **hereditary coproporphyria, urine and fecal coproporphyrin-III** are elevated. **In variegate porphyria**, urine and fecal coproporphyrin-III and fecal protoporphyrin-IX are elevated.

   Acidified urine or fecal extracts containing porphyrins absorb light at 400 nm, which can be detected spectrophotometrically – the so-called **Soret band or Soret peak**. Individual porphyrins also show their own characteristic absorption spectra, which can be detected on a spectrometric scan.

   Genetic testing for detection of mutations in the genes involved in heme synthesis is also currently available.

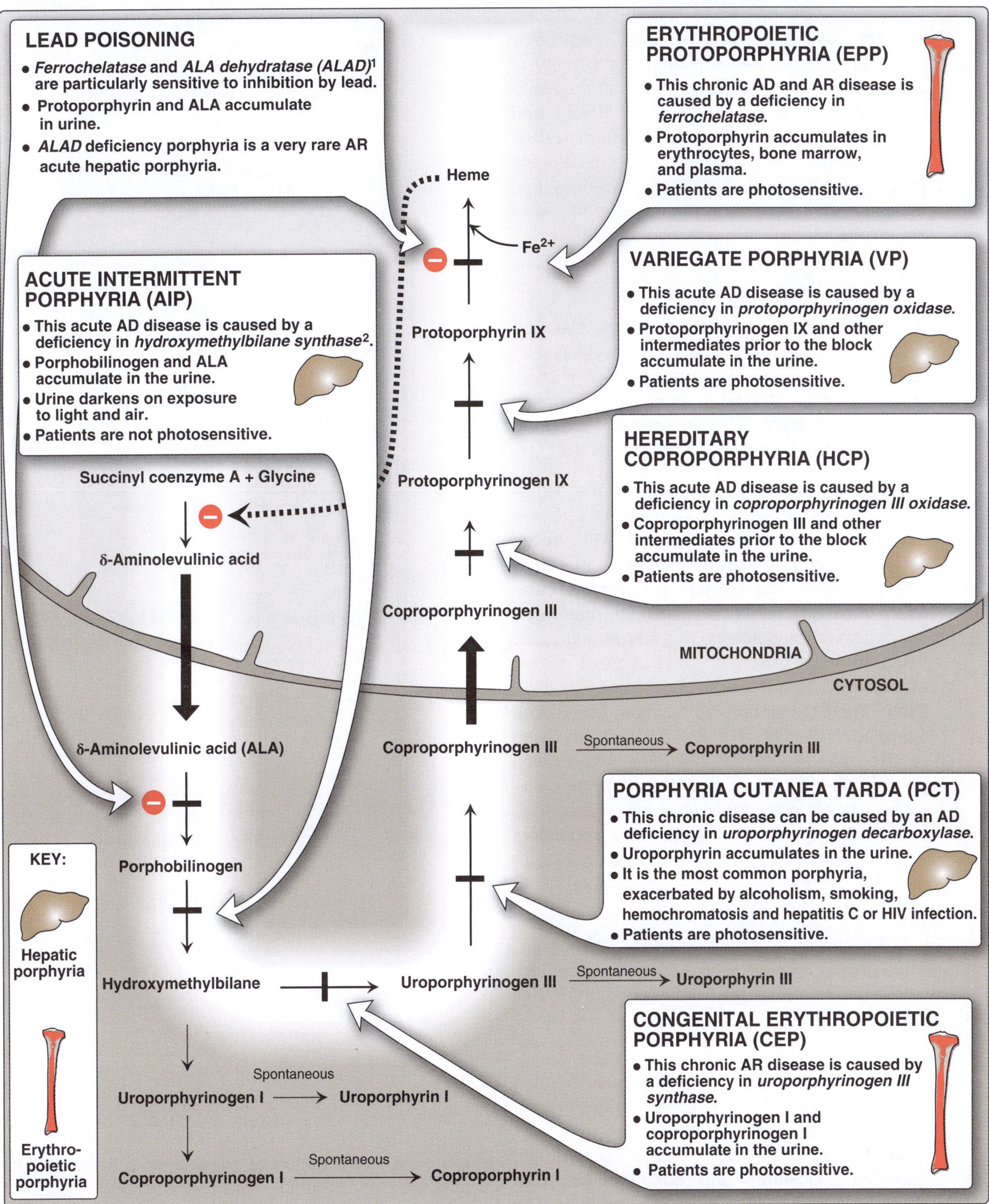

Figure 26.8 Summary of heme synthesis. [1]Also referred to as *porphobilinogen synthase*. [2]Also referred to as *porphobilinogen deaminase*. [Note: Symptomatic deficiencies in *ALA synthase-1* (*ALAS1*) are unknown. Deficiencies in X-linked *ALAS2* result in an anemia.] ALA = δ-aminolevulinic acid; AD = autosomal dominant; AR = autosomal recessive; Fe = iron.

4. **Treatment:** During acute porphyria attacks, patients require medical support, particularly treatment for pain and vomiting. The severity of acute symptoms of the porphyrias can be diminished by **intravenous injection of hemin and glucose**, which decreases the synthesis of *ALAS1*. Protection from sunlight, ingestion of β-carotene (provitamin A; see p. 536) that scavenges free radicals, and phlebotomy (removes porphyrins) are helpful in porphyrias with photosensitivity.

D. Heme degradation

After ~120 days in the circulation, RBC are taken up and degraded by the mononuclear phagocyte system (MPS), particularly in the liver and spleen (Fig. 26.9). Approximately 85% of heme destined for degradation comes from senescent RBC. The remainder is from the degradation of hemeproteins other than Hb.

1. **Bilirubin formation:** The first step in the degradation of heme is catalyzed by microsomal ***heme oxygenase*** in macrophages of the MPS. In the presence of nicotinamide adenine dinucleotide phosphate and oxygen, the enzyme catalyzes three successive oxygenations that result in opening of the porphyrin ring (converting cyclic heme to linear biliverdin), production of carbon monoxide (CO), and release of Fe^{2+} (see Fig. 26.9). [Note: The CO has biologic function, acting as a signaling molecule and anti-inflammatory. Iron is discussed in Chapter 33.] Biliverdin, a green pigment, is reduced, forming the red-orange bilirubin. Bilirubin and its derivatives are collectively termed bile pigments. [Note: The changing colors of a bruise reflect the varying pattern of intermediates that occurs during heme degradation.]

> Bilirubin, unique to mammals, appears to function at low levels as an antioxidant. In this role, it is oxidized to biliverdin, which is then reduced by *biliverdin reductase*, regenerating bilirubin.

2. **Bilirubin uptake by the liver:** Because bilirubin is only slightly soluble in plasma, it is transported through blood to the liver by **binding noncovalently to albumin**. [Note: Certain anionic drugs, such as **salicylates and sulfonamides, can displace bilirubin** from albumin, permitting bilirubin to enter the central nervous system (CNS). This causes the potential for neural damage in infants (see p. 422).] Bilirubin dissociates from the carrier albumin molecule, enters a hepatocyte via facilitated diffusion, and binds to intracellular proteins, particularly the protein ligandin.

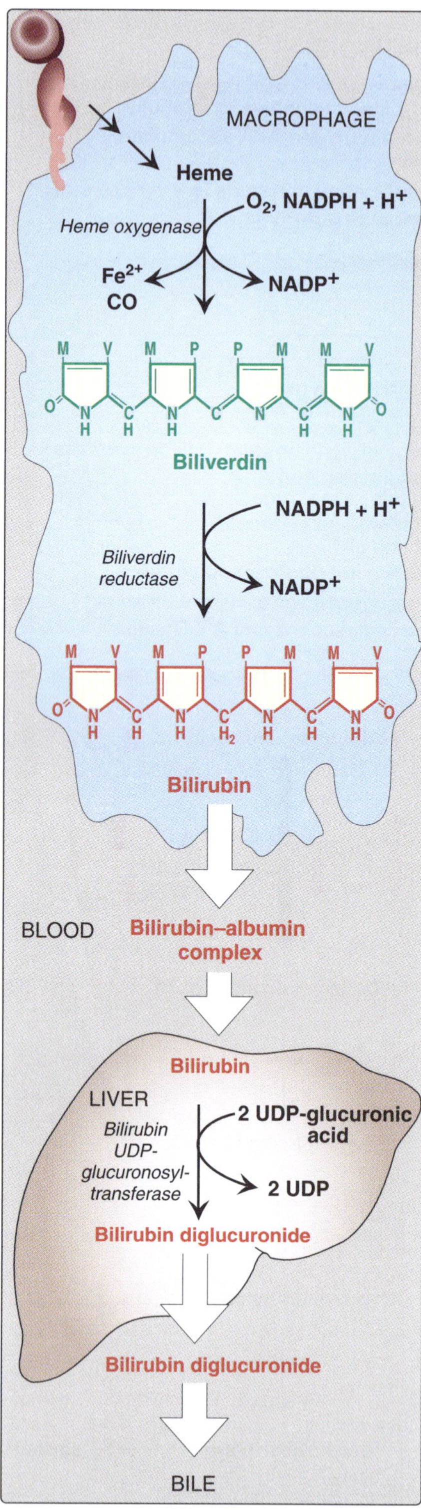

Figure 26.9 Formation of bilirubin from heme and its conversion to bilirubin diglucuronide. UDP = uridine diphosphate; Fe = iron; CO = carbon monoxide; NADP(H) = nicotinamide adenine dinucleotide phosphate.

3. **Bilirubin diglucuronide formation:** In the hepatocyte, bilirubin solubility is increased by the sequential addition of two molecules of glucuronic acid in a process called **conjugation**. The

reactions are catalyzed by **microsomal *bilirubin UDP-glucuronosyltransferase* (*bilirubin UGT*) using uridine diphosphate (UDP)-glucuronic acid** as the glucuronate donor. The bilirubin diglucuronide product is referred to as conjugated bilirubin (CB). [Note: Varying degrees of deficiency of *bilirubin UGT* result in **Crigler-Najjar I and II and Gilbert syndrome**, with Crigler-Najjar I being the most severe.]

4. **Bilirubin secretion into bile:** CB is actively transported against a concentration gradient into the bile canaliculi and then into the bile. This energy-dependent, rate-limiting step is susceptible to impairment in liver disease. [Note: A rare deficiency in the protein required for transport of CB out of the liver results in **Dubin-Johnson syndrome**.] Unconjugated bilirubin (UCB) is normally not secreted into bile.

5. **Urobilin formation in the intestine:** CB is hydrolyzed and reduced by gut bacteria to yield urobilinogen, a colorless compound. Most of the urobilinogen is further oxidized by bacteria to stercobilin, which gives feces the characteristic brown color. However, some is reabsorbed from the gut and enters the portal blood. A portion of this urobilinogen participates in the enterohepatic urobilinogen cycle in which it is taken up by the liver and then resecreted into the bile. The remainder of the urobilinogen is transported by the blood to the kidney, where it is converted to yellow urobilin and excreted, giving urine its characteristic color. The metabolism of bilirubin is summarized in Figure 26.10.

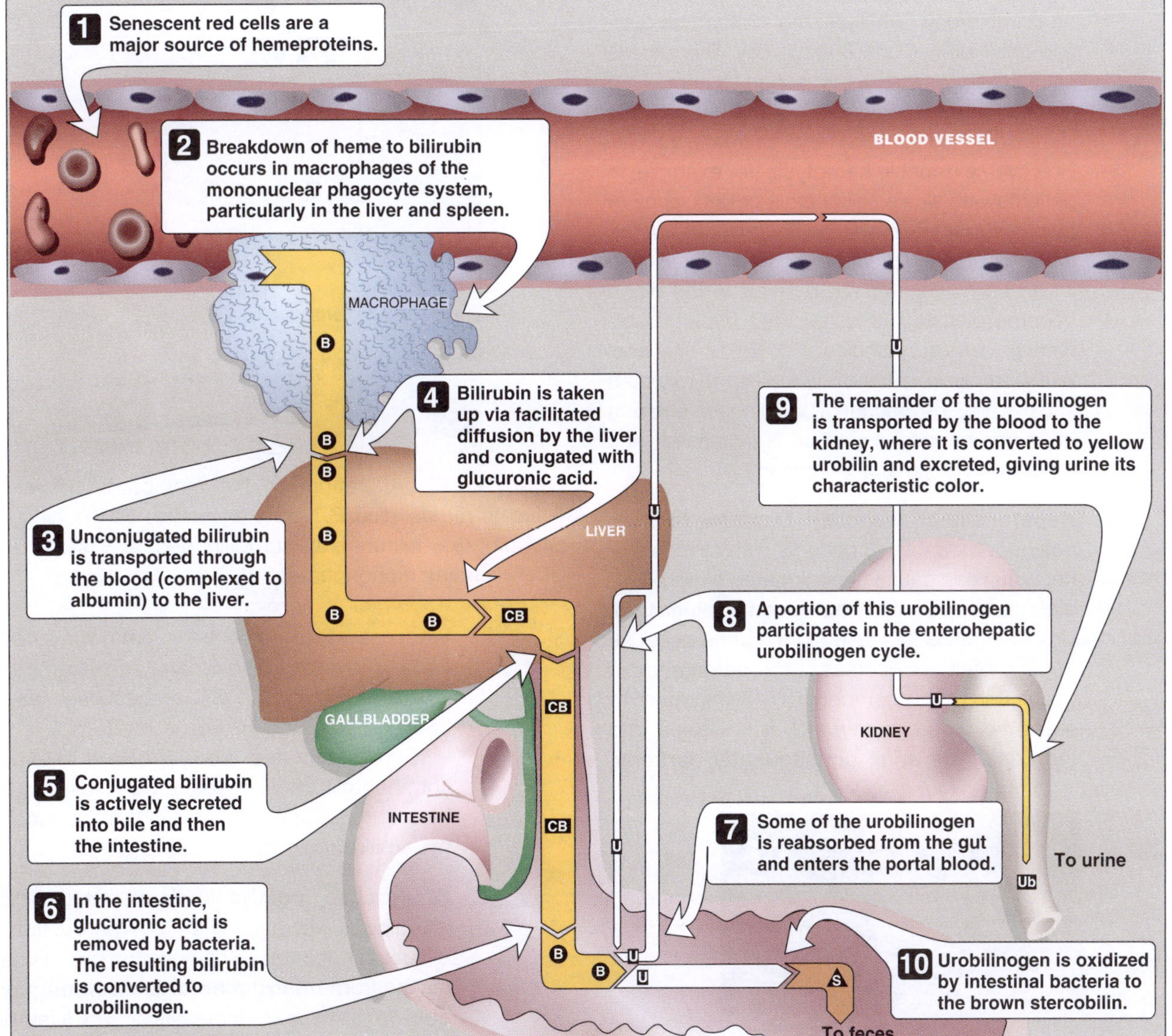

Figure 26.10 Catabolism of heme. B = bilirubin; CB = conjugated bilirubin; U = urobilinogen; Ub = urobilin; S = stercobilin.

E. Jaundice

Jaundice (or, icterus) refers to the **yellow color of skin, nail beds, and sclerae** (whites of the eyes) caused by bilirubin deposition, secondary to increased bilirubin levels in the blood (hyperbilirubinemia) as shown in Figure 26.11. Although not a disease, jaundice is usually a symptom of an underlying disorder. [Note: Blood bilirubin levels are normally ≤1 mg/dl. Jaundice is seen at 2–3 mg/dl.]

1. **Types:** Jaundice can be classified into three major types described below (also see Chapter 48). However, in clinical practice, jaundice is often more complex than indicated in this simple classification. For example, the accumulation of bilirubin may be a result of defects at more than one step in its metabolism.

   a. **Hemolytic (prehepatic):** The liver has the capacity to conjugate and excrete >3,000 mg of bilirubin/day, whereas the normal production of bilirubin is only 300 mg/day. This excess capacity allows the liver to respond to increased heme degradation with a corresponding increase in conjugation and secretion of CB. However, **extensive hemolysis** (for example, in patients with sickle cell anemia or deficiency of *pyruvate kinase* or *glucose 6-phosphate dehydrogenase*) may produce bilirubin faster than it can be conjugated. UCB levels in the blood become elevated (unconjugated hyperbilirubinemia), causing jaundice (Fig. 26.12A). [Note: With hemolysis, more CB is made and excreted into the bile, the amount of urobilinogen entering the enterohepatic circulation is increased, and urinary urobilinogen is increased.]

   b. **Hepatocellular (hepatic): Damage to liver cells** (for example, in patients with cirrhosis or hepatitis) can cause unconjugated hyperbilirubinemia as a result of **decreased conjugation**. Urobilinogen is increased in the urine because hepatic damage decreases the enterohepatic circulation of this compound, allowing more to enter the blood, from which it is filtered into the urine. The **urine consequently darkens, whereas stools may be a pale, clay color.** Plasma levels of *alanine* and *aspartate transaminases* (*ALT* and *AST*, respectively; see p. 383) are elevated. If CB is made but is not efficiently secreted from the liver into bile (intrahepatic cholestasis), it can leak into the blood (regurgitation), causing a conjugated hyperbilirubinemia.

   c. **Obstructive (posthepatic):** In this instance, jaundice is not caused by overproduction of bilirubin or decreased conjugation but, instead, results from **obstruction of the common bile duct** (extrahepatic cholestasis). For example, the presence of a tumor or bile stones may block the duct, preventing passage of CB

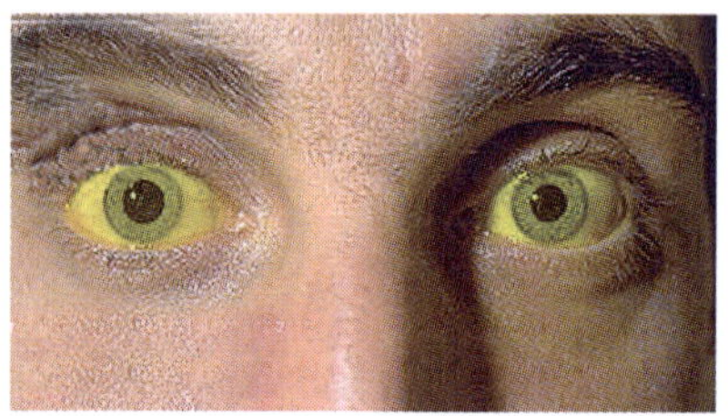

Figure 26.11 Jaundiced patient with the sclerae of his eyes appearing yellow.

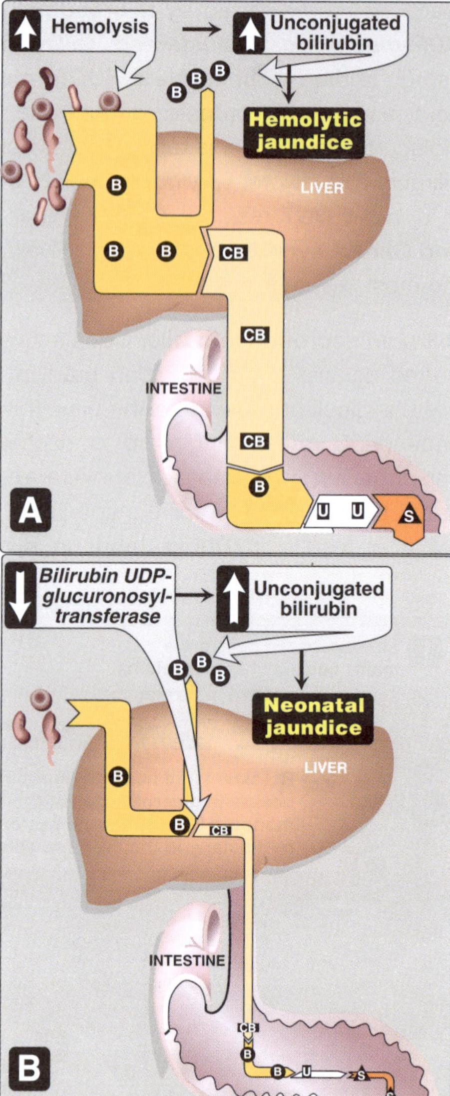

Figure 26.12 Alterations in the metabolism of heme. **A.** Hemolytic jaundice. **B.** Neonatal jaundice. CB = conjugated bilirubin; B = bilirubin; U = urobilinogen; S = stercobilin; UDP = uridine diphosphate.

into the intestine. Patients with obstructive jaundice experience GI pain and nausea and produce stools that are a pale, clay color. The CB regurgitates into the blood (conjugated hyperbilirubinemia). The CB is eventually excreted in the urine (which darkens over time) and is referred to as urinary bilirubin. Urinary urobilinogen is absent.

2. **Jaundice in newborns:** Most newborn infants (60% of full term and 80% of preterm) show a rise in UCB in the first postnatal week (and a **transient, physiologic jaundice) because the activity of hepatic *bilirubin UGT* is low at birth** (it reaches adult levels in about 4 weeks), as shown in Figures 26.12B and 26.13. Elevated UCB, in excess of the binding capacity of albumin (20–25 mg/dl), can diffuse into the basal ganglia, causing toxic encephalopathy (kernicterus) and a pathologic jaundice. Therefore, newborns with significantly elevated bilirubin levels are treated with **blue fluorescent light (phototherapy)**, as shown in Figure 26.14, which converts bilirubin to more polar and, therefore, water-soluble isomers. These photoisomers can be excreted into the bile without conjugation to glucuronic acid. [Note: Because of solubility differences, only UCB crosses the blood–brain barrier, and only CB appears in urine.]

3. **Bilirubin measurement:** Bilirubin is commonly measured by the **van den Bergh reaction**, in which diazotized sulfanilic acid reacts with bilirubin to form red azodipyrroles that are measured colorimetrically. In aqueous solution, the water-soluble CB reacts rapidly with the reagent (within 1 minute) and is said to be direct reacting. The UCB, which is much less soluble in aqueous solution, reacts more slowly. However, when the reaction is carried out in methanol, both CB and UCB are soluble and react with the reagent, providing the total bilirubin value. The indirect-reacting bilirubin, which corresponds to the UCB, is obtained by subtracting the direct-reacting bilirubin from the total bilirubin. [Note: In normal plasma, only ~4% of the total bilirubin is conjugated, or direct reacting, because most is secreted into bile.]

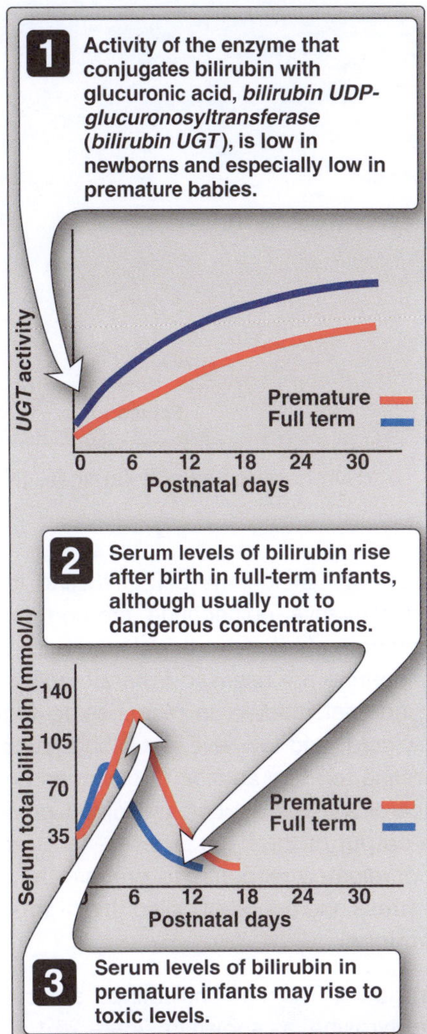

Figure 26.13 Neonatal jaundice. UDP = uridine diphosphate.

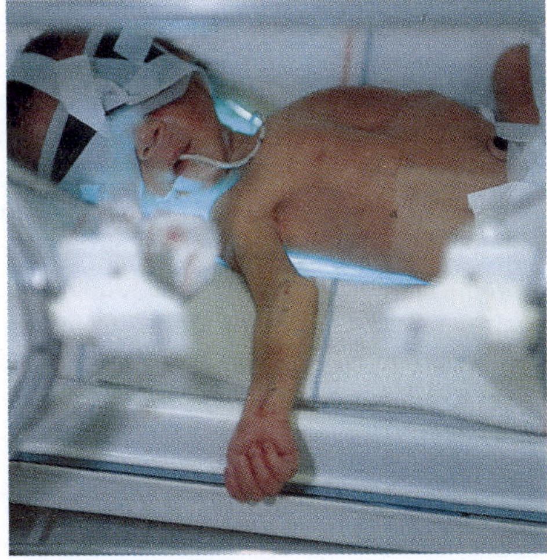

Figure 26.14 Phototherapy in neonatal jaundice.

III. OTHER NITROGEN-CONTAINING COMPOUNDS

A. Catecholamines

Dopamine, norepinephrine (NE), and epinephrine (or, adrenaline) are biologically active (biogenic) amines that are collectively termed catecholamines. Dopamine and NE are synthesized in the brain and function as neurotransmitters. Epinephrine is synthesized from NE in the adrenal medulla.

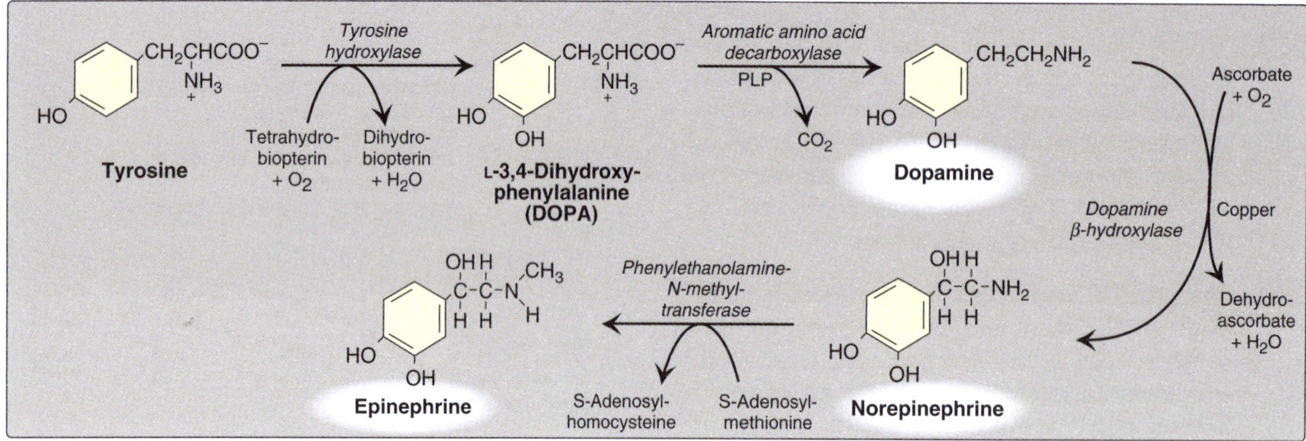

Figure 26.15 Synthesis of catecholamines. [Note: Catechols have two adjacent hydroxyl groups.] PLP = pyridoxal phosphate.

1. **Function:** Outside the CNS, NE and its methylated derivative, epinephrine, are hormone regulators of carbohydrate and lipid metabolism. NE and epinephrine are released from storage vesicles in the adrenal medulla in response to fright, exercise, cold, and low levels of blood glucose. They increase the degradation of glycogen and triacylglycerol as well as increase blood pressure and the output of the heart. These effects are part of a coordinated response to prepare the individual for stress and are often called the **"fight-or-flight" reactions.**

2. **Synthesis:** The catecholamines are synthesized from tyrosine, as shown in Figure 26.15. **Tyrosine** is first hydroxylated by ***tyrosine hydroxylase*** to form l-3,4-dihydroxyphenylalanine **(DOPA)** in a reaction analogous to that described for the hydroxylation of phenylalanine (see p. 398). The tetrahydrobiopterin (BH_4)-requiring enzyme is abundant in the CNS, the sympathetic ganglia, and the adrenal medulla, and it catalyzes the rate-limiting step of the pathway. DOPA is decarboxylated in a reaction requiring PLP to form dopamine, which is hydroxylated by *dopamine β-hydroxylase* to yield NE in a reaction that requires ascorbic acid (vitamin C) and copper. **Epinephrine** is formed from NE by an N-methylation reaction using *S*-adenosylmethionine (SAM) as the methyl donor (see p. 399).

> **Parkinson disease**, a neurodegenerative movement disorder, is due to insufficient dopamine production as a result of the idiopathic loss of dopamine-producing cells in the brain. Administration of L-DOPA (levodopa) is the most common treatment, because dopamine cannot cross the blood–brain barrier.

3. **Degradation:** The catecholamines are inactivated by oxidative deamination catalyzed by ***monoamine oxidase*** **(*MAO*)** and by O-methylation catalyzed by ***catechol-O-methyltransferase*** **(*COMT*)** using SAM as the methyl donor (Fig. 26.16). The reactions can occur in either order. The aldehyde products of the *MAO* reaction are oxidized to the corresponding acids. The products of these reactions are **excreted in the urine as**

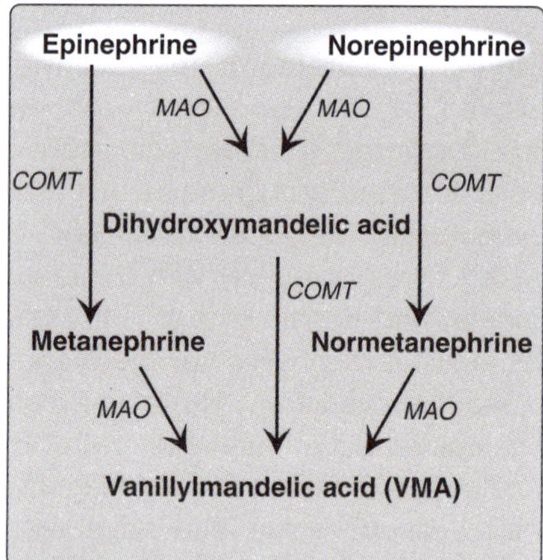

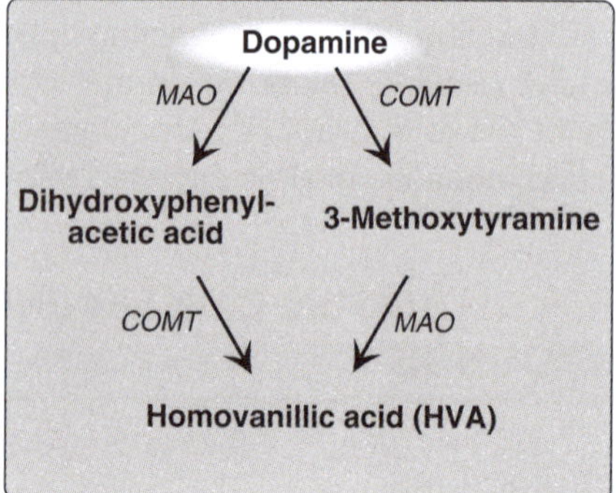

Figure 26.16 Metabolism of the catecholamines by *catechol-O-methyltranferase (COMT)* and *monoamine oxidase (MAO)*. [Note: *COMT* requires *S*-adenosylmethionine.]

vanillylmandelic acid (VMA) from epinephrine and NE and homovanillic acid (HVA) from dopamine. [Note: VMA and the metanephrines are increased with **pheochromocytomas,** rare tumors of the adrenal gland characterized by excessive production of catecholamines.]

4. **Monoamine oxidase inhibitors:** *MAO* is found in neural and other tissues, such as the intestine and liver. In the neuron, this enzyme **oxidatively deaminates and inactivates any excess neurotransmitter molecules** (NE, dopamine, or serotonin) that may leak out of synaptic vesicles when the neuron is at rest. ***MAO* inhibitors (MAOI)** may irreversibly or reversibly inactivate the enzyme, permitting neurotransmitter molecules to escape degradation and, therefore, both to accumulate within the presynaptic neuron and to leak into the synaptic space. This causes activation of NE and serotonin receptors and may be responsible for the **antidepressant action of MAOI.** [Note: The interaction of MAOI with tyramine-containing foods is discussed on p. 519.]

B. Histamine

Histamine is a chemical messenger that mediates a wide range of cellular responses, including **allergic and inflammatory reactions** and gastric acid secretion. A powerful **vasodilator**, histamine is formed by decarboxylation of histidine in a reaction requiring PLP (Fig. 26.17). It is secreted by mast cells as a result of allergic reactions or trauma. Histamine has no clinical applications, but agents that interfere with the action of histamine have important therapeutic applications.

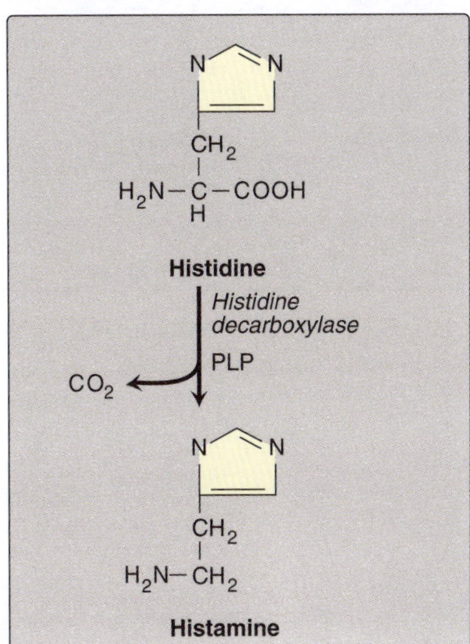

Figure 26.17 Biosynthesis of histamine. PLP = pyridoxal phosphate.

C. Serotonin

Serotonin, also called 5-hydroxytryptamine (5-HT), is synthesized and/or stored at several sites in the body (Fig. 26.18). The largest amount by far is found in the intestinal mucosa. Smaller amounts occur in the CNS, where it functions as a neurotransmitter, and in platelets. Serotonin is **synthesized from tryptophan**, which is hydroxylated in a BH_4-requiring reaction analogous to that catalyzed by *phenylalanine hydroxylase*. The product, 5-hydroxytryptophan, is decarboxylated to 5-HT. Serotonin has multiple physiologic roles including pain perception and regulation of sleep, appetite, temperature, blood pressure, cognitive functions, and mood (causes a feeling of well-being). [Note: **Selective serotonin reuptake inhibitors (SSRI)** maintain serotonin levels, thereby functioning as **antidepressants**.] Serotonin is degraded by *MAO* to 5-hydroxy-3-indoleacetic acid (5-HIAA).

D. Creatine

Creatine phosphate (also called phosphocreatine), the phosphorylated derivative of creatine found in muscle, is a high-energy compound that provides a

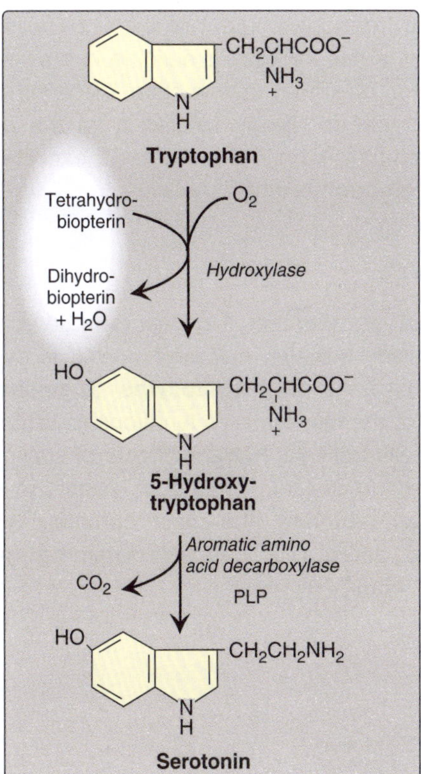

Figure 26.18 Synthesis of serotonin. [Note: Serotonin is converted to melatonin, a regulator of circadian rhythm, in the pineal gland.] PLP = pyridoxal phosphate; CO_2 = carbon dioxide.

small but rapidly mobilized reserve of high-energy phosphates that can be reversibly transferred to adenosine diphosphate (Fig. 26.19) to maintain the intracellular level of ATP during the first few minutes of intense muscular contraction. [Note: The amount of creatine phosphate in the body is **proportional to the muscle mass.]**

1. **Synthesis:** Creatine is **synthesized in the liver and kidneys from glycine and the guanidino group of arginine, plus a methyl group from SAM** (see Fig. 26.19). Animal products are dietary sources. Creatine is reversibly phosphorylated to creatine phosphate by *creatine kinase*, using ATP as the phosphate donor. [Note: The presence of ***creatine kinase*** **(*MB* isozyme)** in the plasma is indicative of heart damage and is used in the diagnosis of **myocardial infarction** (see p. 142).]

2. **Degradation:** Creatine and creatine phosphate spontaneously cyclize at a slow but constant rate to form creatinine, which is excreted in the urine. The amount excreted is proportional to the total creatine phosphate content of the body and, therefore, can be used to estimate muscle mass. When muscle mass decreases for any reason (for example, from paralysis or muscular dystrophy), the creatinine content of the urine falls. In addition, a rise in blood creatinine is a sensitive indicator of kidney malfunction, because creatinine normally is rapidly cleared from the blood and excreted. A **typical adult male excretes ~1–2 g of creatinine/day.**

E. Melanin

Melanin is a pigment that occurs in several tissues, particularly the eye, hair, and skin. It is synthesized **from tyrosine in melanocytes (pigment-forming cells**) of the epidermis. It functions to protect underlying cells from the harmful effects of sunlight. [Note: A defect in melanin production results in **oculocutaneous albinism**, the most common type being due to **defects in copper-containing *tyrosinase*** (see p. 408).]

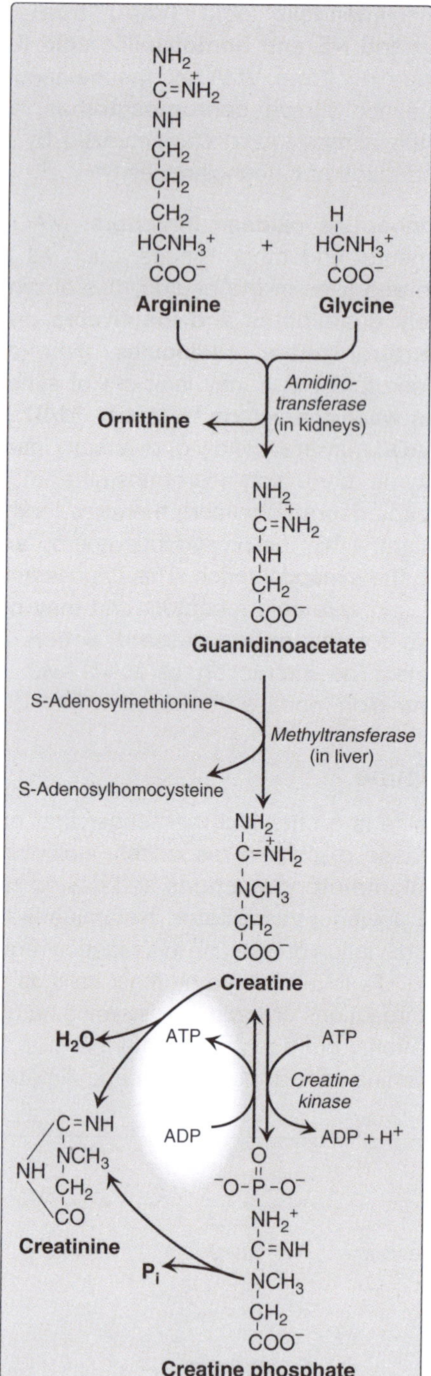

Figure 26.19 Synthesis of creatine. ADP = adenosine diphosphate; P_i = inorganic phosphate.

IV. Medical Perspectives

Jaundice is commonly seen in clinical practice and is basically a symptom of an underlying disease. It is characterized by yellow sclera, nail bed and yellowish color of skin due to deposition of bilirubin. It may be hemolytic (prehepatic), hepatocellular (damage to liver cells) or obstructive (extrahepatic cholestasis). The type of jaundice is confirmed by testing for levels of conjugated, unconjugated and total blood bilirubin as well as levels of liver enzymes like *ALT*, *AST* and *alkaline phosphatase (ALP)*, (see Liver function tests in chapter 48 for diagnostic details).

Parkinsonism is a relatively common neurogenerative movement disorder because of insufficient production of dopamine in the brain. Replacement by levodopa is a common treatment.

Porphyrias are very rare inherited disorders of heme synthesis resulting in accumulation of porphyrin or porphyrin precursors. The signs and symptoms range from photosensitivity to neuropsychiatric disorders, depending on the enzymatic defect.

Porphyrias may be exacerbated by consumption of alcohol, fasting, certain drugs (e.g., barbiturates and antifungals like griseofulvin), and increased progesterone during pregnancy and during the luteal phase of menstrual cycle.

Lead poisoning is a buildup of lead in the body over a period of months to years. Common sources for lead include occupations such as battery manufacturing, automobile repairs, exposure to lead-based paints, cosmetics, toys and paint dust. Lead in household plumbing pipes may also contaminate drinking water. Exposure can occur through inhalation, contact with the skin or mucous membranes, or ingestion. Lead has a sweet taste, and ingestion exposure is of special concern for infants and toddlers. Symptoms of lead poisoning may include developmental delays, learning disabilities and low IQ, abdominal pain, constipation, neurological changes and irritability. Very high lead levels can be fatal. Lead inhibits ALA dehydratase and ferrochelatase, both enzymes involved in the synthesis of heme, and therefore causes a decrease in heme synthesis. Further, high levels of lead impair iron utilization. This results in increased use of zinc (instead of iron) as substrate for chelation to protoporphyrin IX by ferrochelatase. Consequently, patients with lead poisoning may present with anemia and elevated levels of zinc protoporphyrin. The increase in ALA can be toxic to neurons. Lead can also cross the blood–brain barrier and is neurotoxic. The usual treatment is to remove the source of exposure to the lead contaminant, but in cases of severe lead poisoning (greater than 45 pg/dl measured in the serum), divalent chelators such as succimer (DMSA, 2,3-dimercaptosuccinic acid), calcium disodium ethylenediaminetetraacetic acid (EDTA) or others may be used to remove excess lead ions from the blood.

V. Chapter Summary

- **Amino acids** are **precursors** of many nitrogen (N)-containing compounds including **porphyrins**, which, in combination with **ferrous** (**Fe^{2+}**) **iron**, form **heme** (Fig. 26.20). The major sites of **heme biosynthesis** are the **liver**, which synthesizes a number of hemeproteins (particularly ***cytochrome P450*** **enzymes**), and the **erythrocyte-producing cells** of the bone marrow, which are active in **hemoglobin** synthesis. In the liver, the rate of heme synthesis is highly variable, responding to alterations in the cellular heme pool caused by fluctuating demands for hemeproteins.
- In contrast, heme synthesis in erythroid cells is relatively constant and is matched to the rate of globin synthesis. Heme synthesis starts with **glycine** and **succinyl coenzyme A**. The **committed step** is the formation of **δ-aminolevulinic acid** (**ALA**). This mitochondrial reaction is catalyzed by ***ALA synthase-1*** (***ALAS1***) in the liver (inhibited by **hemin**, the oxidized form of heme that accumulates when heme is being underutilized) and ***ALAS2*** in erythroid tissues (regulated by iron).
- **Porphyrias** are caused by inherited or acquired (**lead poisoning**) defects in heme synthesis, resulting in the accumulation and increased excretion of porphyrins or porphyrin precursors. Enzymic defects early in the pathway cause **abdominal pain** and **neuropsychiatric symptoms**, whereas later defects cause **photosensitivity**.
- **Degradation** of heme occurs in the **mononuclear phagocyte system**, particularly in the **liver** and **spleen**. The first step is the production by ***heme oxygenase*** of **biliverdin**, which is subsequently reduced to **bilirubin**.
- Bilirubin is transported by **albumin** to the liver, where its solubility is increased by the addition of two molecules of **glucuronic acid** by ***bilirubin uridine diphosphate-glucuronosyltransferase*** (***bilirubin UGT***). **Bilirubin diglucuronide** (**conjugated bilirubin**) is transported into the **bile canaliculi**, where it is first hydrolyzed and reduced by gut bacteria to yield **urobilinogen**, which is further oxidized by bacteria to **stercobilin**.
- **Jaundice** (**icterus**) refers to the yellow color of the skin and sclerae that is caused by deposition of bilirubin, secondary to increased bilirubin levels in the blood. Three commonly encountered types of jaundice are **hemolytic** (**prehepatic**), **obstructive** (**posthepatic**), and **hepatocellular** (**hepatic**) (see Fig. 26.20).
- Other important N-containing compounds derived from amino acids include the **catecholamines** (**dopamine**, **norepinephrine**, and **epinephrine**), **creatine**, **histamine**, **serotonin**, **melanin**, and **nitric oxide**.

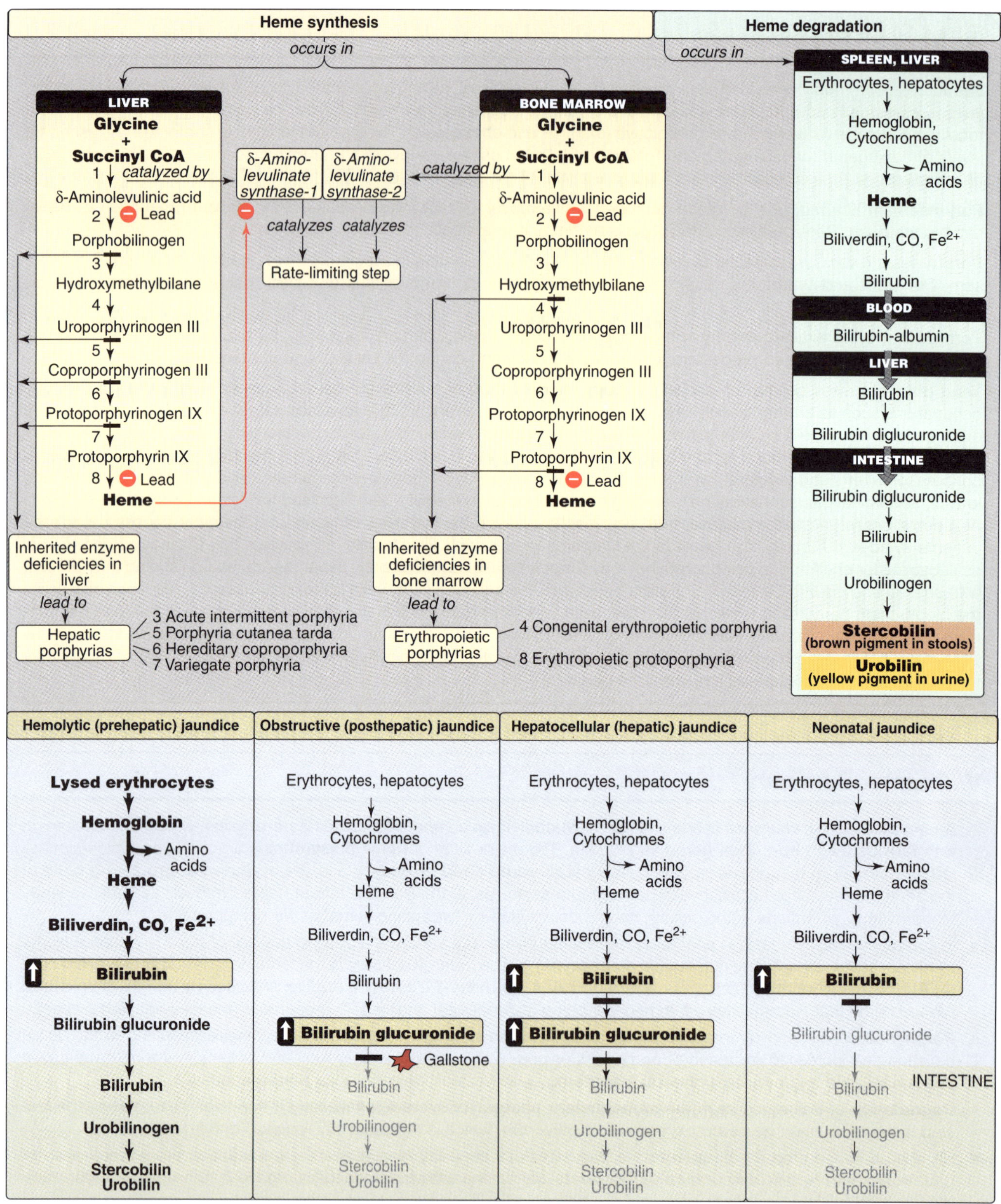

Figure 26.20 Key concept map for heme metabolism. ▬ = Block in the pathway. [Note: Hepatocellular jaundice can be caused by decreased conjugation of bilirubin or decreased secretion of conjugated bilirubin from the liver into bile.] CoA = coenzyme A; CO = carbon monoxide; Fe = iron.

Study Questions

Short-Answer Questions

Write short notes on:

26.1. Acute intermittent porphyrias
26.2. Synthesis of catecholamines
26.3. Lead poisoning
26.4. Physiological jaundice of newborn

Long-Answer Questions

26.1. Describe the synthesis of heme in liver and bone marrow. Add a note on the associated diseases.
26.2. What is jaundice? Describe the causes and laboratory diagnosis of different types of jaundice.

Multiple-Choice Questions

Choose the ONE best answer.

26.1. δ-Aminolevulinic acid synthase activity:
A. catalyzes the committed step in porphyrin biosynthesis.
B. is decreased by iron in erythrocytes.
C. is decreased in the liver in individuals treated with certain drugs such as the barbiturate phenobarbital.
D. requires tetrahydrobiopterin as a coenzyme.

Correct answer = A.

26.2. A 50-year-old man presented with painful blisters on the back of his hands. He was a golf instructor and indicated that the blisters had erupted shortly after the golfing season began. He did not have recent exposure to common skin irritants. He had partial complex seizure disorder that had begun ~3 years earlier after a head injury. The patient had been taking phenytoin (his only medication) since the onset of the seizure disorder. He admitted to an average weekly ethanol intake of ~18 12-oz cans of beer. The patient's urine was reddish orange. Cultures obtained from skin lesions failed to grow organisms. A 24-hour urine collection showed elevated uroporphyrin (1,000 mg; normal, <27 mg). The most likely diagnosis is:
A. acute intermittent porphyria.
B. erythropoietic protoporphyria.
C. hereditary coproporphyria.
D. porphyria cutanea tarda.

Correct answer = D.

26.3. A patient presents with jaundice, abdominal pain, and nausea. Clinical laboratory results are shown below.

Plasma bilirubin	Serum transaminases	Urinary bilirubin
Increase in conjugated bilirubin; unconjugated bilirubin is not elevated	AST and ALT are not elevated	Present

What is the most likely cause of the jaundice?
A. Decreased hepatic conjugation of bilirubin
B. Decreased hepatic uptake of bilirubin
C. Decreased secretion of bile into the intestine
D. Increased hemolysis

Correct answer = C.

26.4. Which of the following is a TRUE statement regarding ALA synthase?
A. ALA synthase is a cytoplasmic enzyme that is regulated by heme and iron levels.
B. ALA synthase requires vitamin B_6 as a cofactor.
C. ALA synthase is found exclusively in the liver and the bone marrow.
D. ALA synthase is inhibited by lead and other heavy metals.

Correct answer = B.

26.5. Heme synthesis occurs in all the following cells, EXCEPT:
A. RBCs
B. Hepatocytes
C. Neurons
D. Cardiomyocytes

Correct answer = A.

26.6. In the human body, the formation of creatinine from creatine is catalyzed by:
A. Creatine kinase
B. Creatine dehydratase
C. Creatine dehydrogenase
D. None of the above

Correct answer = D.

26.7. A 60-year-old male presented to the surgery department with complaints of gradually progressive yellowish discoloration of sclera and urine, fullness of abdomen, loss of appetite, and pale greasy stools. He reported a weight loss of 8 kg within the last 2 months. On physical examination, he was jaundiced. Radiological investigations suggested a diagnosis of carcinoma of the head of the pancreas. Which fraction of bilirubin is most likely elevated in the serum of this patient?
A. Conjugated bilirubin
B. Unconjugated bilirubin
C. Both conjugated bilirubin and unconjugated bilirubin
D. Neither conjugated bilirubin nor unconjugated bilirubin

Correct answer = A.

26.8. Accumulation of porphyrins in skin leads to tissue destruction because porphyrins:

A. Inhibit Na+-K+ ATPase
B. Cause phototoxicity
C. Lead to autoimmunity
D. Cause hypoxic injury

Correct answer = B.

26.9. Statement 1: Lead poisoning can mimic the symptoms of porphyria.

Statement 2: Lead can bind to sulfhydryl groups of ALA synthase, the regulatory enzyme of heme synthesis, and inhibit the activity of the enzyme.

A. Both the Statements are correct, and Statement 2 is the correct explanation for Statement 1
B. Both the statements are correct, but Statement 2 is NOT the correct explanation for Statement 1
C. Statement 1 is correct, but Statement 2 is NOT correct
D. Statement 1 is NOT correct, but Statement 2 is correct

Correct answer = C.

26.10. Which of the following conditions results in increased urinary excretion of VMA/metanephrines?

A. Gastrinoma
B. Pheochromocytoma
C. Hartnup disease
D. Lymphoma

Correct answer = B.

Multiple-Choice Questions (NExT/NEET PG)

Choose the ONE best answer.

26.1. A 2-year-old child was brought to his pediatrician for evaluation of gastrointestinal problems. The parents report that the boy has been listless for the last few weeks. Lab tests reveal a microcytic, hypochromic anemia. Blood lead levels are elevated. Which of the enzymes listed below is most likely to have higher-than-normal activity in the liver of this child?

A. δ-Aminolevulinic acid synthase
B. Bilirubin UDP glucuronosyltransferase
C. Ferrochelatase
D. Heme oxygenase

Correct answer = A.

26.2. A sample for porphyrin/precursors is to be collected for the laboratory diagnosis in a patient suspected to have porphyria. Which of the following statements is NOT correct regarding the sampling?

A. In acute porphyrias, the samples should be collected during an acute attack.
B. The sample should be protected from light, and should be analyzed as soon as possible.
C. In cutaneous porphyrias, the preferred sample for porphyrin estimation is the skin tissue at the exposed site.
D. The choice of sample to be analyzed—urine or stool—depends on the solubility of the porphyrin suspected to be elevated.

Correct answer = C.

26.3. Carcinoid syndrome occurs in the setting of a neuroendocrine tumor and results from the secretion of vasoactive peptides like serotonin into the circulation. An elevated level of which of the following in urine or plasma is suggestive of carcinoid syndrome?

A. Xanthurenic acid
B. Urocanic acid
C. Formiminoglutamic acid
D. 5-hydroxyindoleacetic acid

Correct answer = D.

26.4 Which of the following is also called plumboporphyria?

A. Doss porphyria
B. Congenital erythropoietic porphyria
C. Acute intermittent porphyria
D. Porphyria cutanea tarda

Correct answer = A.

26.5. Which of the statements regarding the genetic defects in *ALAS2* is NOT correct?

A. They are inherited in an X-linked manner, as *ALAS2* is located on the X chromosome.
B. Loss of function of *ALAS2* results in sideroblastic anemia, as porphyrin synthesis is affected, and excess iron unbound to heme deposits in the bone marrow.
C. Gain of function of *ALAS2* results in X-linked protoporphyria, due to unregulated synthesis of protoporphyrin IX in RBC precursors.
D. All the nucleated cells in the body are affected, since *ALAS2* is the constitutive enzyme for synthesis of heme that is incorporated in cytochromes.

Correct answer = D.

Appendix
Unit V: Nitrogen Metabolism: Amino Acids

ADDITIONAL CLINICAL PERSPECTIVES FOR PRIMARY HEALTHCARE PHYSICIAN

Various disorders have already been extensively discussed in Unit V. This section describes some additional clinical perspectives.

- Vitamin B_6 (pyridoxine) deficiency (e.g., in individuals receiving isoniazid [see Chapter 32.V.A]) can lead to anemia because δ-aminolevulinic acid (ALA) synthase requires pyridoxal phosphate (PLP), the active form of pyridoxine, as coenzyme.
- **Photodynamic therapy** refers to the use of light and a photosensitive chemical to bring about toxicity to the cells, in conditions like cancer. Because of the inherent photosensitive nature of porphyrins, porphyrin derivatives are among the chemicals used in photodynamic therapy. An example is porfimer sodium which is used in the treatment of esophageal cancer and lung cancer.
- In hemolytic anemia, excess bilirubin in bile precipitates with calcium as insoluble calcium bilirubinate, which results in formation of gallstones.
- Unconjugated bilirubin, when bound to albumin cannot cross the blood-brain barrier. Two molecules of unconjugated bilirubin bind noncovalently with one molecule of albumin. Certain drugs (like salicylates and sulfonamides) can displace unconjugated bilirubin from albumin, increasing the risk of bilirubin encephalopathy in individuals with unconjugated hyperbilirubinemias.
- Conjugated bilirubin in circulation (e.g., in hepatitis and obstructive jaundice) can bind covalently with albumin to form δ-bilirubin, also known as biliprotein. This leads to persistence of bilirubin in circulation in cholestatic conditions, since the half-life of bilirubin (originally 4 hours) will now be same as that of albumin (half-life, 14 days). This might mislead the clinicians who encounter otherwise significant clinical improvement in their patients who are being treated for hepatitis or cholestasis.
- Blue to blue-green light (460–490 nm) converts insoluble form of unconjugated bilirubin to a water-soluble structural isomer known as lumirubin, in which the moieties involved in internal hydrogen bond formation are exposed. This forms the basis of phototherapy in Crigler-Najjar syndrome and other unconjugated hyperbilirubinemias.
- The color changes that is observed in a bruise or a contusion is due to the pigments that predominate at different stages of healing. The initial red color is because of oxygenated hemoglobin, followed by blue/purple from deoxygenated hemoglobin, which then degrades to biliverdin (green color), and then to bilirubin (yellow-orange), which finally undergoes resorption. Based on the time duration of each stage, this is also useful in calculating the time of bruising, which has important applications in forensics.

Table 1 Classification of jaundice

Level of Defect	Causes	Fraction of Bilirubin Elevated
Prehepatic	*Hemolytic anemias* Defective erythroid maturation Massive hematoma	Unconjugated bilirubin
Hepatic	Crigler-Najjar syndrome—I & II Gilbert's syndrome *Benign neonatal hyperbilirubinemia*	Unconjugated bilirubin
	Dubin-Johnson syndrome Rotor syndrome	Conjugated bilirubin
	Common conditions like *Viral hepatitis* *Drug-induced liver injury* *Alcoholic liver disease*	Both unconjugated bilirubin and conjugated bilirubin
Posthepatic	*Stones in the common bile duct* *Parasitic infections of the bile duct* *Tumors of bile duct* *Carcinoma head of pancreas* *Malignancy around ampulla of Vater*	Conjugated bilirubin

(Note: Common conditions are in *italics*)

Liver function tests and renal function tests are given in Unit X, Chapter 48: Organ Function Tests.

Clinical Case Questions

Chapter 24 Digestion and Absorption of Proteins and Nitrogen Disposal from Amino Acids

24.1. A 68-year-old man presented to the medicine OPD with complaint of progressive weakness, disorientation, fatigue, nausea, vomiting, loss of appetite, and shallow respiration. The doctor after examining him asked for routine blood tests. His Hb was 13 g/dl, blood urea 98 mg/dl, fasting glucose 100 mg/dl and blood assay showed metabolic acidosis. His liver function tests were normal.

A. Discuss whether the blood urea and glucose are in normal range for this patient. Explain the expected disease.

Answer—A: The patient is suffering from kidney disorder as his blood urea levels are high. Normal blood urea levels are 20–40 mg/dl. Normal fasting glucose is 70–100 mg/dl. Urea is synthesized in the liver and excreted through the kidney.

B. What other tests would you like to do for this patient?

Answer—B: Further kidney function tests like creatinine and creatinine clearance tests must be done. High uremia usually signifies extreme damage to the kidney such that the kidneys are no longer able to filter the waste from the body. See Chapter 48 for kidney function tests.

24.2. A 50-year-old chronic alcoholic presented to the emergency department with a bout of hematemesis. He was already on treatment for cirrhosis and has severe jaundice and ascitis. But this time he had additional altered sensorium, tremors in hand, very musty breath and was extremely drowsy. His blood ammonia levels were 140 μmol/l (normal reference range, 6–47 μmol/l).

A. What is the probable diagnosis?

Answer—A: Hepatic encephalopathy is the probable diagnosis. The signs of altered sensorium, tremors in hand and extreme drowsiness signify a deterioration in condition of the chronic alcoholic patient such that the ammonia toxicity is now affecting the brain.

B. What is the biochemical basis of hyperammonemia in this case?

Answer—B: Chronic alcoholism and liver cirrhosis lead to a high ammonia level because liver is unable to detoxify ammonia to urea. The episode of gastrointestinal bleeding leading to hematemesis has further aggravated the condition. Blood is high in protein and is reabsorbed from bowel adding to the protein and ammonia overload. Ammonia crosses the blood-brain barrier easily, and here it has caused hepatic encephalopathy which has the potential of leading to coma.

Chapter 25 Amino Acids: Degradation and Synthesis

25.1. An 8-month-old baby was brought to the pediatrician by a mother of three children. The mother is a known case of hypothyroidism and diabetes mellitus. The doctor observed that the child cannot sit without support. The mother also reports poor feeding and choking episodes. He was born at home and not shown to any doctor. On examination, delayed milestones were observed. The child had puffy face and thick tongue. The pediatrician advised admission as well as neonatal screening panel.

A. Why did the pediatrician ask for neonatal screening?

(**Note:** *Clinical Case 25.1 requires knowledge from Chapters 25 and 48)*

Answer: Newborn screening is done for genetic or inborn metabolic diseases. One of the commonest signs is delayed milestones and mental retardation. Most babies with metabolic disorders look normal at birth. Newborn screening is commonly done for hypothyroidism, congenital adrenal hypoplasia, G6PD deficiency, phenylketonuria and galactosemia. Delayed milestones, poor feeding, choking episodes, puffy face and thick protruding tongue are features suggestive of neonatal hypothyroidism. A diagnosis in time can save the baby from mental retardation, morbidity and mortality. Hence the pediatrician chose to get this screening done while looking for infective and other causes of delayed milestones.

Chapter 26 Amino Acids: Conversion to Specialized Products Including Porphyrins

26.1. An 8-year-old boy presented with fever, pallor, fatigue and breathlessness. On examination, his sclera was yellow in color. History revealed that he is often jaundiced since birth. His Hb was 8.5 g/dl (normal Hb for his age is 10–15 g/dl), bilirubin was 9 g/dl (normal bilirubin <1.0 mg/dl), conjugated bilirubin 1.3 g/dl (normal conjugated bilirubin <0.2 mg/dl), unconjugated bilirubin was 7.7 g/dl (normal unconjugated bilirubin 0.2–0.8 mg/dl), AST was 30 mg/dl (normal AST 5–30 U/l), ALT was 34 mg/dl (normal ALT 6–37 U/l), ALP was 50 mg/dl (normal ALP 30–90 U/l), and LDH was 920 IU/L (normal LDH 100–225 U/l).

A. What is the likely diagnosis?

Answer—A: The patient appears to have G6PD deficiency since the history and tests are suggestive of anemia and hemolytic jaundice. Liver enzymes are normal, but LDH is very high. His history is suggestive of repeated triggering of hemolytic episodes by recurrent infections.

B. What further tests can be done to confirm the diagnosis?

Answer—B: G6PD activity in red blood cells can be analyzed by spectrophotometry.

Advanced Cases/Self-Directed Learning

V.1. A 7-year-old girl presented with a history of fragility of the skin and spontaneous blistering over the face and extremities. She had severe pallor and a history of transfusion-dependent anemia. Abdominal examination revealed gross distension with massive hepatosplenomegaly on ultrasonography. Closer examination revealed crusted erosions, scarring, and dyspigmentation, along with a few milia over photo-exposed sites including the face and distal extremities. This was associated with periorbital edema and sclerodermoid changes over acral areas. There was diffuse hyperpigmentation but no hypertrichosis. The teeth showed a reddish hue (Fig. A), and gross urine examination showed reddish discoloration (Fig. B). On probing, there was a history of photosensitivity and pink discoloration of the diaper since birth. She was the first-order child of a second-degree consanguineous marriage, and no other family member was affected.

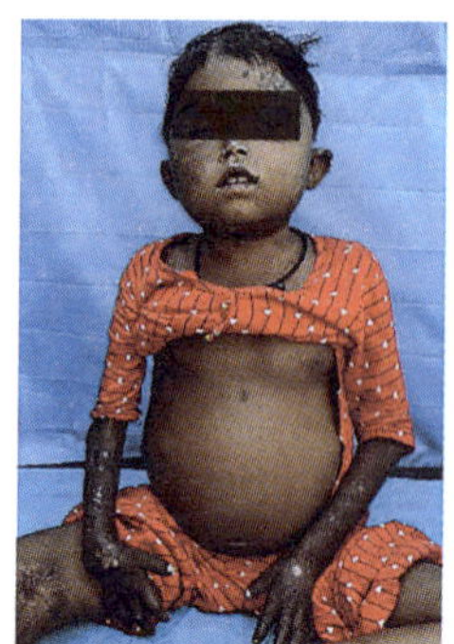

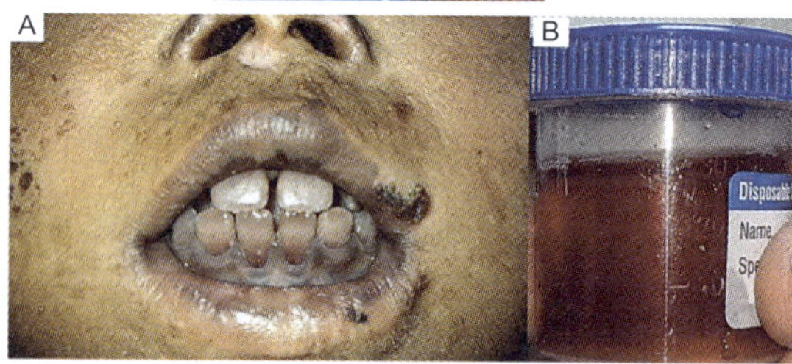

Photo courtesy: *Dr. Ananya Sharma, Dr. Neha Taneja* and *Dr. Rajat Choudhary*, All India Institute of Medical Sciences, New Delhi.

A. What is your diagnosis?

Answer—A: This is a case of congenital erythropoietic porphyria.

B. What is the cause of this condition?

Answer—B: This condition is caused by a deficiency in uroporphyrinogen III synthase. As a result, uroporphyrinogen I and coproporphyrinogen I accumulate in the urine.

Unit VI

Integration of Metabolism

Competencies Addressed

- **BI3.5** Describe and discuss the regulation, functions and integration of carbohydrate along with associated diseases/disorders
- **BI3.9** Discuss the mechanism and significance of blood glucose regulation in health and disease, also see Unit III: Carbohydrates
- **BI3.10** Interpret the results of blood glucose levels and other laboratory investigations related to disorders of carbohydrate metabolism, also see Unit III: Carbohydrates
- **BI6.1** Discuss the metabolic processes that take place in specific organs in the body in the fed and fasting states
- **BI8.4** Describe the causes (including dietary habits), effects and health risks associated with being overweight/obesity

Unit VI: Overview

All cells and tissues are capable of performing various catabolic and anabolic processes. It is necessary that both these processes are regulated and coordinated to each other. **This coordination of intricate network of reactions is known as integration of metabolism**. Insulin and glucagon are two major hormones which control this integration. Insulin, the hormone of abundance, ensures that body conserves energy and proteins when food is readily available. Insulin increases glucose uptake in muscles and adipose tissues and the synthesis of glycogen, protein, and triacylglycerol. When food supplies are low, glucagon, along with catecholamines, ensures that a minimal level of blood glucose is available to carry out body functions. Glucagon increases glycogenolysis, gluconeogenesis, fatty acid oxidation, ketogenesis, and amino acid uptake (Chapter 27).

Under the hormonal influences, **different organs have different metabolic profile in fed and fasting states**. The flow of intermediates through metabolic cycles is controlled by availability of substrates, allosteric regulation and covalent modification of enzymes as well as induction and repression of enzyme synthesis. Depending on availability, some organs can also change the metabolic fuel they use (Chapter 28).

Diabetes mellitus is a condition when blood glucose is persistently elevated, either due to relative or absolute deficiency of insulin. It is the leading cause of several diseases, including renal failure, nerve damage, heart attacks, and stroke. Diabetes can be classified into two groups, type 1 (T1D) and type 2 (T2D), T1D being caused by absolute deficiency of insulin due to autoimmune destruction of the pancreatic β cells (Chapter 29).

When energy intake exceeds energy expenditure, excess body fat accumulates, causing obesity. It has a major influence on several health risks. Appetite is influenced by neural signals as well as circulating hormones such as leptin, and metabolites. Lifestyle changes, pharmacotherapy and surgical procedure can help in weight management (Chapter 30).

The primary healthcare physician must be aware that **all metabolic pathways are interrelated.** Excess carbohydrate can be converted and stored as fat. During energy deficit, fat can be mobilized to provide acetyl CoA and ketone bodies for energy. Excessive ketone bodies, seen in diabetic ketoacidosis or starvation, can alter the blood pH, causing acidosis. Diabetes mellitus has several long-term complications and its early identification helps in reducing the morbidity and mortality. Obesity is a contributing factor to several health problems. Primary healthcare physician must know the assessment methods for obesity and its basic management.

The importance of understanding the biochemical basis of these diseases can be understood by the fact that the Government of India launched the National Programme for Prevention and Control of Cancer, Diabetes, CVD and Stroke (NPCDCS) in October 2010. India is home to nearly 50 million people with diabetes and has 15% of the global disease burden. Atherosclerosis is another disease where integration of metabolism is to be understood. Atherosclerosis is one of the main reasons for myocardial infarction and stroke (Unit IV: Chapter 23). Obesity is an emerging health problem in India with more than 21% females and 19% males being obese.

The flow of information is such that Unit VI must be read after Unit III on Carbohydrates. Further, appendix of Unit VI can be considered as an extension of Unit III.

Metabolic Effects of Insulin and Glucagon

27

Learning Objectives

At the end of this chapter, the student will be able to:

- Describe the structure and secretion of insulin and glucagon
- Describe the metabolic effects of insulin and glucagon
- Briefly describe the receptors and signal transduction of insulin and glucagon
- Enumerate the clinical features and types of hypoglycemia
- Describe the regulation of blood glucose level

I. OVERVIEW

Four major tissues play a dominant role in fuel metabolism: liver, adipose, muscle, and brain. These tissues contain unique sets of enzymes, such that each tissue is specialized for the storage, use, or generation of specific fuels. **These tissues do not function in isolation but rather form part of a network** in which one tissue may provide substrates to another or process compounds produced by other tissues. Communication between tissues is mediated by the nervous system, by the availability of circulating substrates, and by variation in the levels of plasma hormones (Fig. 27.1). **The integration of energy metabolism is controlled primarily by the actions of two peptide hormones, insulin and glucagon** (secreted in response to changing substrate levels in the blood), with the catecholamines epinephrine and norepinephrine (secreted in response to neural signals) playing a supporting role. Changes in the circulating levels of these hormones allow the body to store energy when food is abundant or to make stored energy available such as during survival crises (for example, famine, severe injury, and "fight-or-flight" situations). This chapter describes the structure, secretion, and metabolic effects of the two hormones that most profoundly affect energy metabolism.

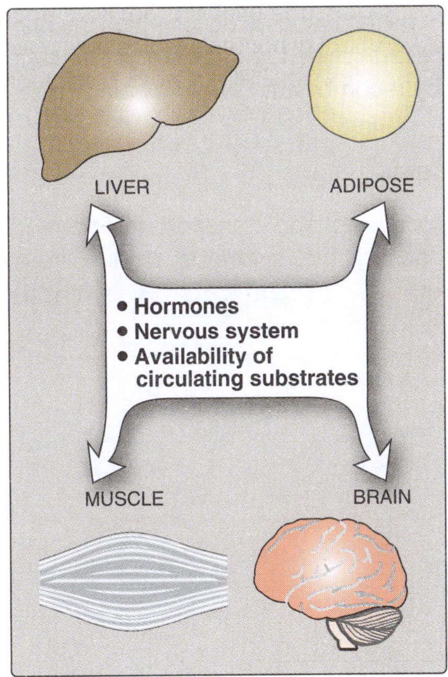

Figure 27.1 Mechanisms of communication between four major tissues.

II. INSULIN

Insulin is a peptide hormone produced by the β cells of the islets of Langerhans, which are clusters of cells embedded in the endocrine portion of the pancreas (Fig. 27.2). [Note: "Insulin" is from the Latin for island.] The islets make up only about 1%–2% of the total cells of the pancreas. Insulin is the **most important hormone coordinating the use of fuels** by tissues. Its metabolic effects are **anabolic**, favoring, for example, synthesis of glycogen, triacylglycerol (TAG), and protein.

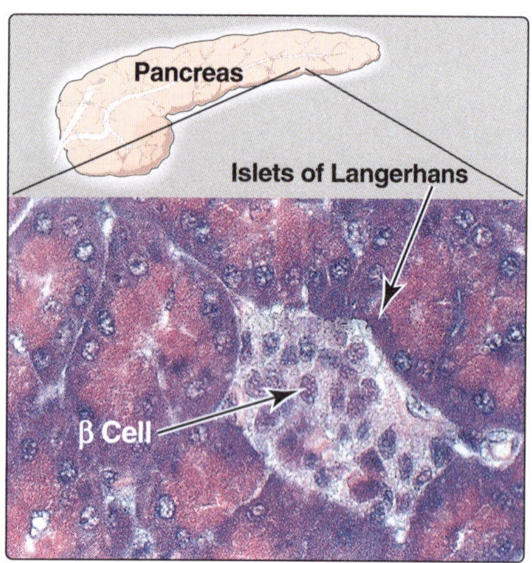

Figure 27.2 Islets of Langerhans.

A. Structure

Insulin is composed of 51 amino acids arranged in two polypeptide chains, designated A (21 amino acids) and B, which are linked together by two disulfide bonds (Fig. 27.3A). The insulin molecule also contains an intramolecular disulfide bond between amino acid residues of the A chain.

Insulin was the first peptide for which the primary structure was determined and the first therapeutic molecule made by recombinant DNA technology (see p. 677).

B. Synthesis

The processing and transport of intermediates that occur during the synthesis of insulin are shown in Figures 27.3B and 27.4. Biosynthesis involves production of two inactive precursors, **preproinsulin** (109 amino acids) and **proinsulin** (86 amino acids), which are sequentially cleaved to form the active hormone plus the connecting or **C-peptide** in a 1:1 ratio (see Fig. 27.4). Insulin is stored in cytosolic granules that, given the proper stimulus (see C.1. below), are released by exocytosis. (See p. 644 for a discussion of the synthesis of secreted proteins.) Insulin is degraded by *insulin-degrading enzyme*, which is present in the liver and, to a lesser extent, in the kidneys. Insulin has a plasma half-life of ~6 minutes. This short duration of action permits rapid changes in circulating levels of the hormone.

The C-peptide is essential for proper insulin folding. Also, because its half-life in plasma is longer than that of insulin, the C-peptide level is a good indicator of insulin production and secretion.

C. Secretion regulation

Secretion of insulin is regulated by bloodborne fuels and hormones.

1. **Increased secretion:** Insulin secretion by the pancreatic β cells is closely coordinated with the secretion of glucagon by pancreatic α cells (Fig. 27.5). **The relative amounts of glucagon and insulin** released are normally regulated such that the rate of hepatic glucose production is kept equal to the use of glucose by peripheral tissues. This **maintains blood glucose between 70 and 140 mg/dl**. In view of its coordinating role, it is not surprising that the β cell responds to a variety of stimuli. In particular, insulin

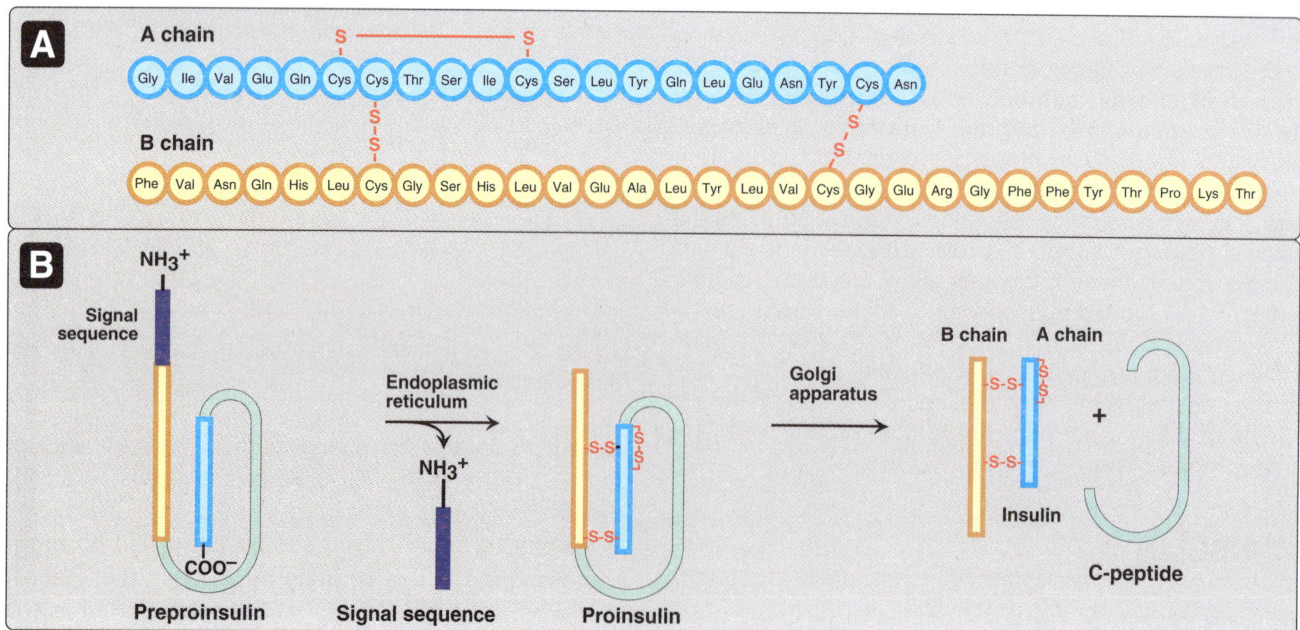

Figure 27.3 **A.** Structure of insulin. **B.** Formation of human insulin from preproinsulin. S-S = disulfide bond.

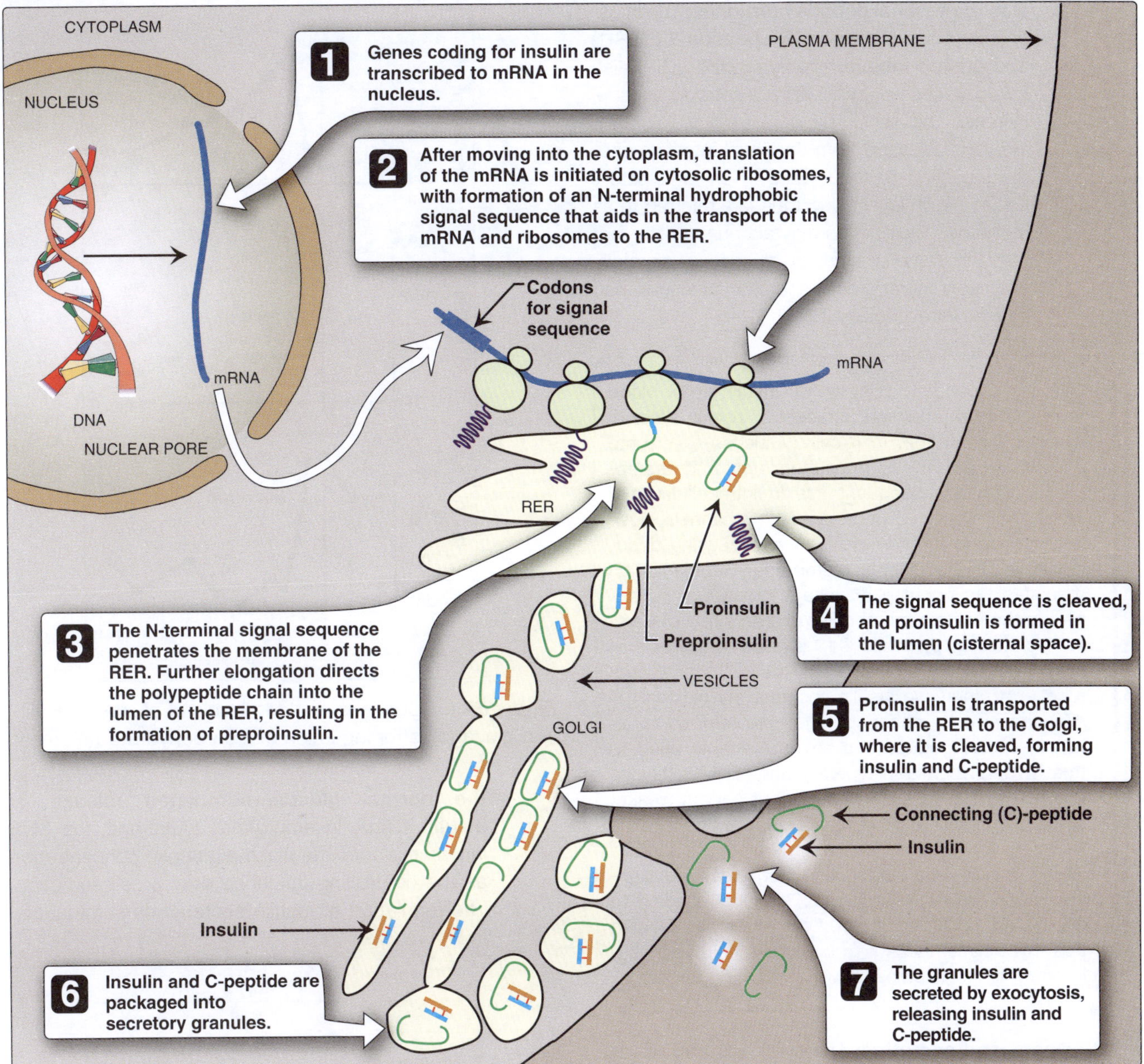

Figure 27.4 Intracellular movements of insulin and its precursors. mRNA = messenger RNA; RER = rough endoplasmic reticulum.

secretion is increased by glucose, amino acids, and gastrointestinal peptide hormones.

a. **Glucose:** Ingestion of a carbohydrate-rich meal leads to a rise in blood glucose, the primary stimulus for insulin secretion (see Fig. 27.5). The β cells are the most important glucose-sensing cells in the body. Like the liver, **β cells contain GLUT-2 transporters** and express *glucokinase* (*hexokinase IV*; see p. 193). **At blood glucose levels >45 mg/dl, *glucokinase* phosphorylates glucose in amounts proportional to the glucose concentration.** Proportionality results from the lack of direct inhibition of *glucokinase* by glucose 6-phosphate, its product. Additionally, the sigmoidal relationship between the velocity of the reaction and substrate concentration (see p. 193) maximizes the enzyme's responsiveness to changes in blood glucose level. Metabolism of glucose 6-phosphate generates ATP, leading to insulin secretion (see blue box below).

b. **Amino acids:** Ingestion of protein causes a transient rise in plasma amino acid levels (for example, arginine) that enhances the glucose-stimulated secretion of insulin. [Note: Fatty acids have a similar effect.]

c. **Gastrointestinal peptide hormones:** The intestinal peptides **glucagon-like peptide-1** (GLP-1) and **gastric inhibitory polypeptide** ([GIP] also called glucose-dependent insulinotropic peptide) increase the sensitivity of β cells to glucose. They are released from the small intestine after the ingestion of food, causing an anticipatory rise in insulin levels and, thus, are referred to as **incretins**. Their action may account for the fact that the same amount of glucose given orally induces a much greater secretion of insulin than if given intravenously (IV).

Glucose-dependent release of insulin into blood is mediated through an intracellular rise in calcium (Ca^{2+}) concentration in β cells. Glucose taken into β cells by GLUT-2 is phosphorylated and metabolized, with a subsequent rise in intracellular ATP levels. ATP-sensitive potassium (K^+) channels close in response to the rise in ATP levels, causing depolarization of the plasma membrane. Depolarization mediates the opening of voltage-gated Ca^{2+} channels in the plasma membrane, and an influx of Ca^{2+}. An increase in cytosolic Ca^{2+} signals exocytosis of insulin-containing vesicles from β cells. Sulfonylureas function to increase insulin secretion by closing the ATP-sensitive K^+ channels. They are referred to as insulin secretagogues and are oral agents used to treat hyperglycemia in type 2 diabetes. Meglitinides function similar to sulfonylureas, but have a weaker binding affinity and higher dissociation for the ATP-sensitive K^+ channels, and therefore are shorter-acting insulin secretagogues. Conversely, diazoxides open the ATP-sensitive K^+ channel, leading to decreased insulin secretion. Diazoxides are used to treat hypoglycemia caused by congenital hyperinsulinism or in insulinomas (insulin-producing tumors).

2. **Decreased secretion:** The synthesis and release of insulin are decreased when there is a scarcity of dietary fuels and also during periods of physiologic stress (for example, infection, hypoxia, and vigorous exercise), thereby preventing hypoglycemia. These effects are mediated primarily by the catecholamines norepinephrine and epinephrine, which are made from tyrosine in the sympathetic nervous system (SNS) and the adrenal medulla and then secreted. Secretion is largely controlled by neural signals. **The catecholamines (primarily epinephrine) have a direct effect on energy metabolism**, causing a rapid mobilization of energy-yielding fuels, including glucose from the liver (produced by glycogenolysis or gluconeogenesis; see p. 218) and fatty acids (FA) from adipose tissue (produced by lipolysis; see p. 307). In addition, **these biogenic amines can override the normal glucose-stimulated release of insulin**. Thus, in emergency situations, the SNS largely replaces the plasma glucose concentration as the controlling influence over β-cell secretion. The regulation of insulin secretion is summarized in Figure 27.6.

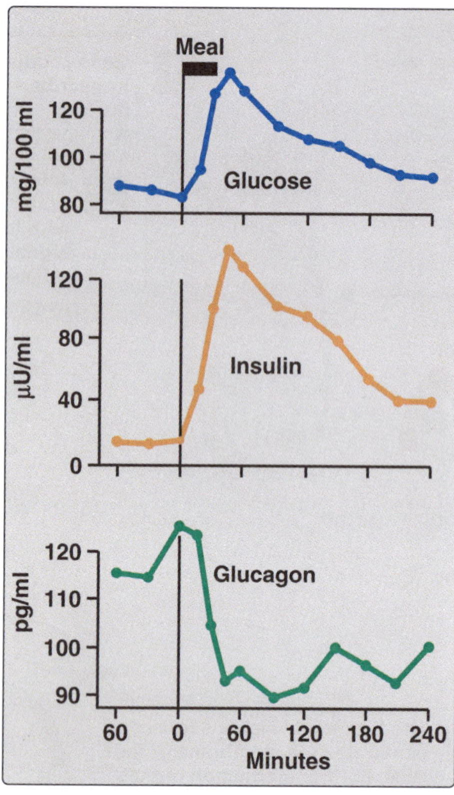

Figure 27.5 Changes in blood levels of glucose, insulin, and glucagon after ingestion of a carbohydrate-rich meal.

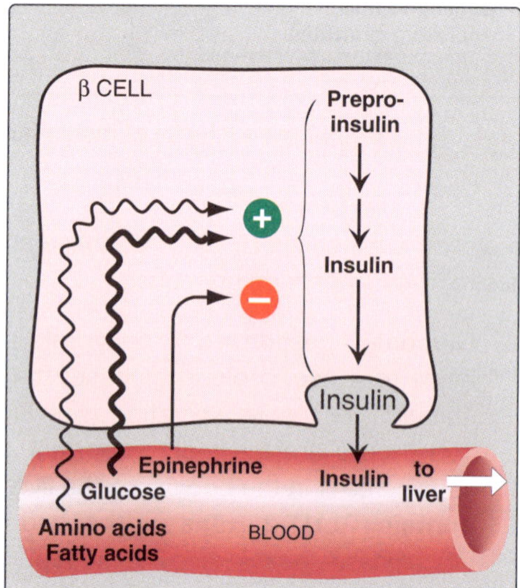

Figure 27.6 Regulation of insulin release from pancreatic β cells. [Note: Gastrointestinal peptide hormones also stimulate insulin release.]

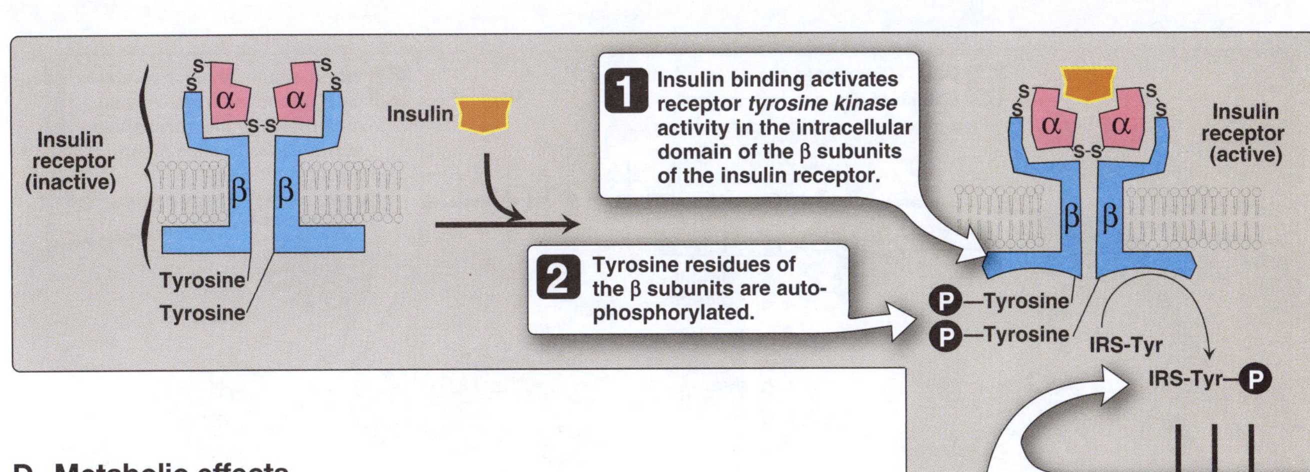

D. Metabolic effects

Insulin promotes the storage of nutrients as glycogen, TAG, and protein and inhibits their mobilization.

1. **Effects on carbohydrate metabolism:** The effects of insulin on glucose metabolism promote its storage and are **most prominent in three tissues: liver, muscle, and adipose**. In liver and muscle, insulin increases glycogen synthesis. In muscle and adipose, insulin increases glucose uptake by increasing the number of glucose transporters (GLUT-4; see p. 192) in the cell membrane. Thus, the IV administration of insulin causes an immediate decrease in blood glucose level. In the liver, insulin decreases the production of glucose through the inhibition of glycogenolysis and gluconeogenesis.

   The effects of insulin are due not just to changes in enzyme activity but also in enzyme amount insofar as insulin alters gene transcription.

2. **Effects on lipid metabolism:** A rise in insulin rapidly causes a significant reduction in the release of FA from adipose tissue by **inhibiting the activity of *hormone-sensitive lipase***, a key enzyme of TAG degradation in adipocytes. Insulin acts by promoting the dephosphorylation and, hence, inactivation of the enzyme (see p. 308). Insulin also increases the transport and metabolism of glucose into adipocytes, providing the glycerol 3-phosphate substrate for TAG synthesis (see p. 306). Expression of the gene for *lipoprotein lipase*, which degrades TAG in circulating chylomicrons and very-low-density lipoproteins ([VLDL], see p. 352), is increased by insulin in adipose, thereby providing FA for esterification to the glycerol. [Note: Insulin also promotes the conversion of glucose to TAG in the liver. The TAG are secreted in VLDL.]

3. **Effects on protein synthesis:** In most tissues, insulin stimulates both the entry of amino acids into cells and protein synthesis (translation). [Note: Insulin stimulates protein synthesis through covalent activation of factors required for translation initiation.]

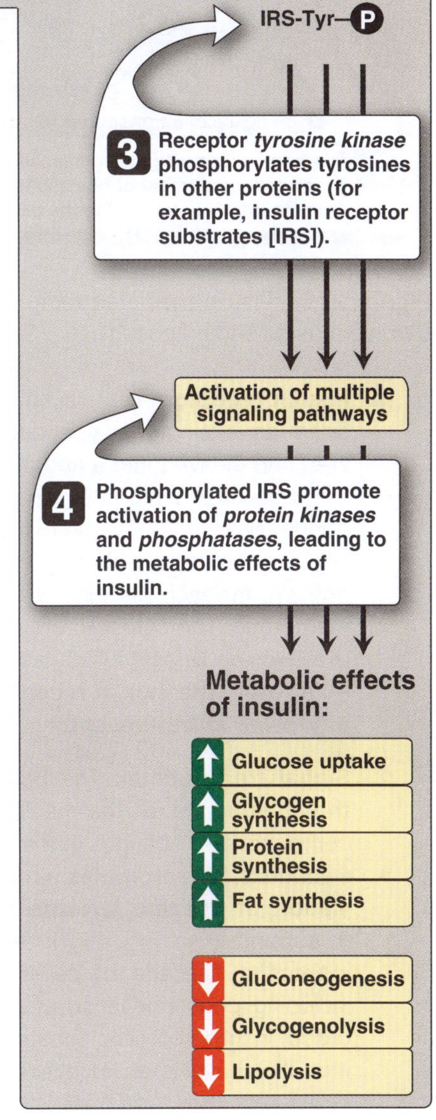

Figure 27.7 Mechanism of action of insulin. P = phosphate; Tyr = tyrosine; S-S = disulfide bond.

E. Mechanism

Insulin binds to specific, high-affinity receptors in the cell membrane of most tissues, including liver, muscle, and adipose. This is the first step in a cascade of reactions ultimately leading to a diverse array of biologic actions (Fig. 27.7).

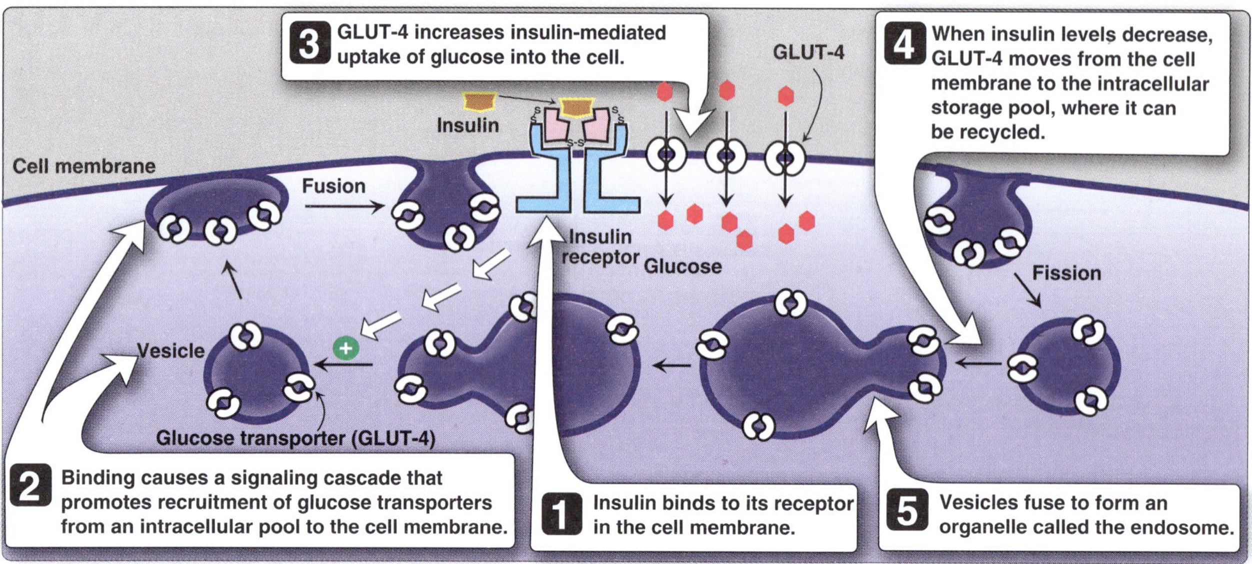

Figure 27.8 Insulin-mediated recruitment of GLUT-4 from intracellular stores to the cell membrane in skeletal and cardiac muscle and adipose tissue. S-S = disulfide bond.

1. **Insulin receptor:** The insulin receptor is synthesized as a single polypeptide that is glycosylated and cleaved into α and β subunits, which are then assembled into a tetramer linked by disulfide bonds (see Fig. 27.7). The extracellular α subunits contain the insulin-binding site. A hydrophobic domain in each β subunit spans the plasma membrane. The cytosolic domain of the β subunit is a *tyrosine kinase*, which is activated by insulin. As a result, **the insulin receptor is classified as a *tyrosine kinase* receptor**.

2. **Signal transduction:** The **binding of insulin to the α subunits** of the insulin receptor induces conformational changes that are transmitted to the β subunits. This **promotes a rapid autophosphorylation of specific tyrosine residues on each β subunit** (see Fig. 27.7). Autophosphorylation initiates a cascade of cell-signaling responses, including phosphorylation of a family of proteins called insulin receptor substrates (IRS). At least four IRS have been identified that show similar structures but different tissue distributions. Phosphorylated IRS proteins interact with other signaling molecules through specific domains (known as SH2), activating a number of pathways that affect gene expression, cell metabolism, and growth. The actions of insulin are terminated by dephosphorylation of the receptor.

3. **Membrane effects:** Glucose transport in some tissues, such as muscle and adipose, increases in the presence of insulin (Fig. 27.8). **Insulin promotes movement of insulin-sensitive glucose transporters (GLUT-4) from a pool located in intracellular vesicles to the cell membrane.** This movement is the result of a signaling cascade in which an IRS binds to and activates a *kinase* (*phosphoinositide 3-kinase*), leading to phosphorylation of the membrane phospholipid phosphatidylinositol 4,5-bisphosphate (PIP_2) to the 3,4,5-trisphosphate form (PIP_3) that binds to and activates *phosphoinositide-dependent kinase 1*. This *kinase*, in turn, activates *Akt* (or *protein kinase B*), resulting in GLUT-4 movement. In contrast, other tissues have insulin-insensitive systems for glucose transport (Fig. 27.9). For example, hepatocytes, erythrocytes, and cells of the nervous system, intestinal mucosa, renal tubules, and cornea do not require insulin for glucose uptake.

4. **Receptor regulation:** Binding of insulin is followed by internalization of the hormone–receptor complex. Once inside the cell, insulin is degraded in the lysosomes. The receptors may be degraded, but most are recycled to the cell surface. [Note: Elevated levels of insulin promote the degradation of receptors, thereby decreasing the number of surface receptors. This is one type of downregulation.]

5. **Time course:** The binding of insulin provokes a wide range of actions. The most immediate response is an increase in glucose transport into adipocytes and skeletal and cardiac muscle cells that occurs within seconds of insulin binding to its membrane receptor. **Insulin-induced changes in enzymic activity in many cell types occur**

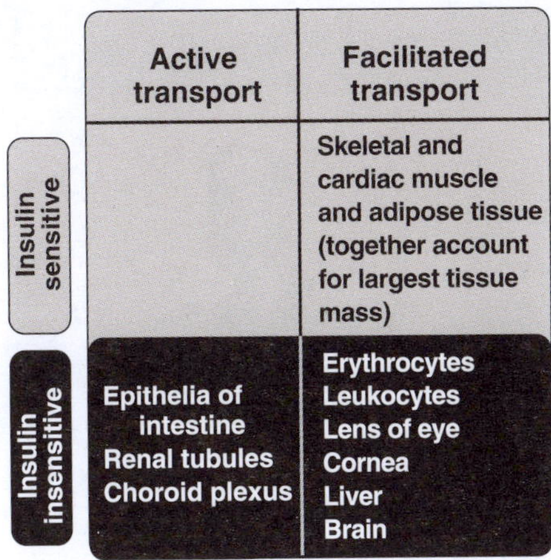

Figure 27.9 Characteristics of glucose transport in various tissues.

over minutes to hours and reflect changes in the phosphorylation states of existing proteins. **Insulin-induced increase in the amount of many enzymes, such as *glucokinase*, liver *pyruvate kinase*, *acetyl coenzyme A* (*CoA*) *carboxylase* (*ACC*), and *fatty acid synthase*, requires hours to days**. These changes reflect an increase in gene expression through increased transcription (mediated by sterol regulatory element–binding protein-1c; see p. 302) and translation.

III. GLUCAGON

Glucagon is a peptide hormone secreted by the α cells of the pancreatic islets of Langerhans. **Glucagon, along with epinephrine, norepinephrine, cortisol, and growth hormone (the counterregulatory hormones), opposes many of the actions of insulin** (Fig. 27.10). Most importantly, glucagon acts to maintain blood glucose levels by activation of hepatic glycogenolysis and gluconeogenesis. Glucagon is composed of 29 amino acids arranged in a single polypeptide chain. Glucagon is synthesized as a large precursor molecule (preproglucagon) that is converted to glucagon through a series of selective proteolytic cleavages, similar to those described for insulin biosynthesis (see Fig. 27.3). In contrast to insulin, preproglucagon is processed to different products in different tissues, for example, GLP-1 in intestinal L cells. Like insulin, glucagon has a short half-life.

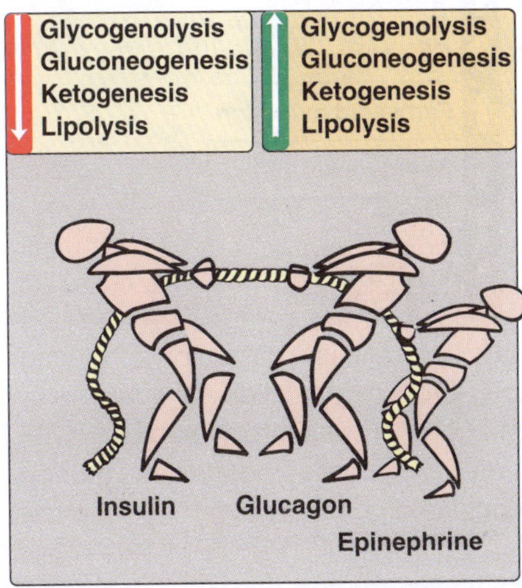

Figure 27.10 Opposing actions of insulin and glucagon plus epinephrine.

Unlike insulin, the amino acid sequence of glucagon is the same in all mammalian species examined to date.

A. Increased secretion

The α cell is responsive to a variety of stimuli that signal actual or potential hypoglycemia (Fig. 27.11). Specifically, **glucagon secretion is increased by low blood glucose, amino acids, and catecholamines**.

1. **Low blood glucose:** A decrease in plasma glucose concentration is the **primary stimulus for glucagon release**. During an overnight or prolonged fast, elevated glucagon levels prevent hypoglycemia (see Section IV below for a discussion of hypoglycemia).

2. **Amino acids:** Amino acids (for example, arginine) derived from a meal containing protein stimulate the release of glucagon. The **glucagon effectively prevents the hypoglycemia that would otherwise occur as a result of the increased insulin secretion that also occurs after a protein meal**.

3. **Catecholamines:** Elevated levels of circulating epinephrine (from the adrenal medulla), norepinephrine (from sympathetic innervation of the pancreas),

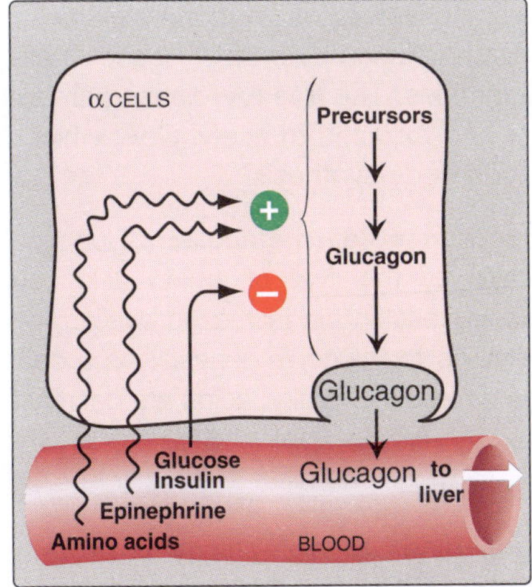

Figure 27.11 Regulation of glucagon release from pancreatic α cells. [Note: Amino acids increase release of insulin and glucagon, whereas glucose increases release of insulin and decreases release of glucagon.]

or both stimulate the release of glucagon. Thus, during periods of physiologic stress, the elevated catecholamine levels can override the effect on the α cell of circulating substrates. In these situations, regardless of the concentration of blood glucose, **glucagon levels are elevated in anticipation of increased glucose use**. In contrast, insulin levels are depressed.

B. Decreased secretion

Glucagon secretion is significantly decreased by elevated blood glucose and by insulin. Both substances are increased following ingestion of glucose or a carbohydrate-rich meal (see Fig. 27.5). The regulation of glucagon secretion is summarized in Figure 27.11.

C. Metabolic effects

Glucagon is a catabolic hormone that promotes the maintenance of blood glucose levels. Its primary target is the liver.

1. **Effects on carbohydrate metabolism:** The IV administration of glucagon leads to an immediate rise in blood glucose. This results from an increase in the degradation of liver glycogen and an increase in hepatic gluconeogenesis.

2. **Effects on lipid metabolism:** The primary effect of glucagon on lipid metabolism is inhibition of FA synthesis through phosphorylation and subsequent inactivation of *ACC* by *adenosine monophosphate (AMP)–activated protein kinase* (see p. 302). The resulting decrease in malonyl CoA production removes the inhibition on long-chain FA β-oxidation (see p. 309). Glucagon also plays a role in lipolysis in adipocytes, but the major activators of *hormone-sensitive lipase* (via phosphorylation by *protein kinase A*) are the catecholamines. The free FA released are taken up by liver and oxidized to acetyl CoA, which is used in ketone body synthesis.

3. **Effects on protein metabolism:** Glucagon increases uptake by the liver of amino acids supplied by muscle, resulting in increased availability of carbon skeletons for gluconeogenesis. As a consequence, plasma levels of amino acids are decreased.

D. Mechanism

Glucagon binds to high-affinity **G protein–coupled receptors** (GPCR) on the cell membrane of hepatocytes. The GPCR for glucagon is distinct from the GPCR that bind epinephrine. [Note: Glucagon receptors are not found on skeletal muscle.] Glucagon binding results in activation of *adenylyl cyclase* in the plasma membrane (Fig. 27.12; also see p. 190). This causes a rise in cyclic AMP (cAMP), which, in turn, activates *cAMP-dependent protein kinase A* and

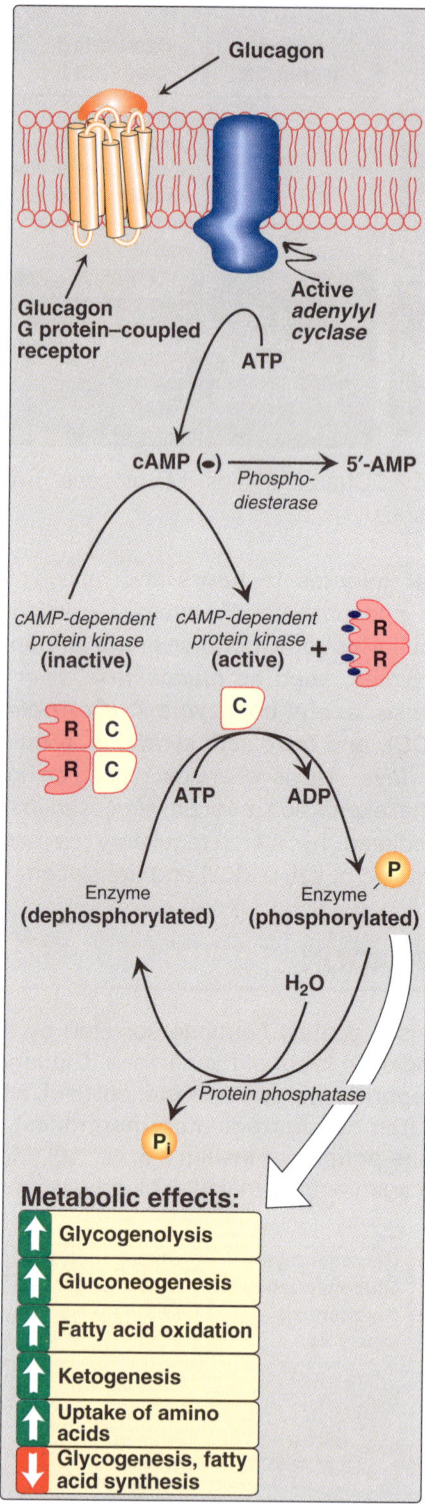

Figure 27.12 Mechanism of action of glucagon. [Note: For clarity, G-protein activation of *adenylyl cyclase* has been omitted.] R = regulatory subunit; C = catalytic subunit; cAMP = cyclic adenosine monophosphate; ADP = adenosine diphosphate; P = phosphate.

increases the phosphorylation of specific enzymes or other proteins. This cascade of increasing enzymic activities results in the phosphorylation-mediated activation or inhibition of key regulatory enzymes involved in carbohydrate and lipid metabolism. An example of such a cascade in glycogen degradation is shown in Figure 16.9 on p. 230. [Note: Glucagon, like insulin, affects gene transcription. For example, glucagon induces expression of *phosphoenolpyruvate carboxykinase* (see p. 221).]

IV. HYPOGLYCEMIA

Hypoglycemia is characterized by 1) **central nervous system (CNS) symptoms**, including confusion, aberrant behavior, or coma; 2) **a simultaneous blood glucose level ≤50 mg/dl**; and 3) **symptoms being resolved within minutes following glucose administration** (Fig. 27.13). **Hypoglycemia is a medical emergency** because the CNS has an absolute requirement for a continuous supply of bloodborne glucose to serve as a metabolic fuel. Transient hypoglycemia can cause cerebral dysfunction, whereas severe, prolonged hypoglycemia causes brain damage. Therefore, it is not surprising that the body has multiple overlapping mechanisms to prevent or correct hypoglycemia. The **most important hormone changes in combating hypoglycemia are increased secretion of glucagon and the catecholamines, combined with decreased insulin secretion**.

A. Symptoms

The symptoms of hypoglycemia can be divided into two categories. **Adrenergic** (neurogenic, autonomic) symptoms, such as anxiety, palpitation, tremor, and sweating, are mediated by catecholamine release (primarily epinephrine) regulated by the hypothalamus in response to hypoglycemia. Adrenergic symptoms typically occur when blood glucose levels fall abruptly. The second category of hypoglycemic symptoms is **neuroglycopenic**. The impaired delivery of glucose to the brain (neuroglycopenia) results in impairment of brain function, causing headache, confusion, slurred speech, seizures, coma, and death. Neuroglycopenic symptoms often result from a gradual decline in blood glucose, often to levels <50 mg/dl. The slow decline

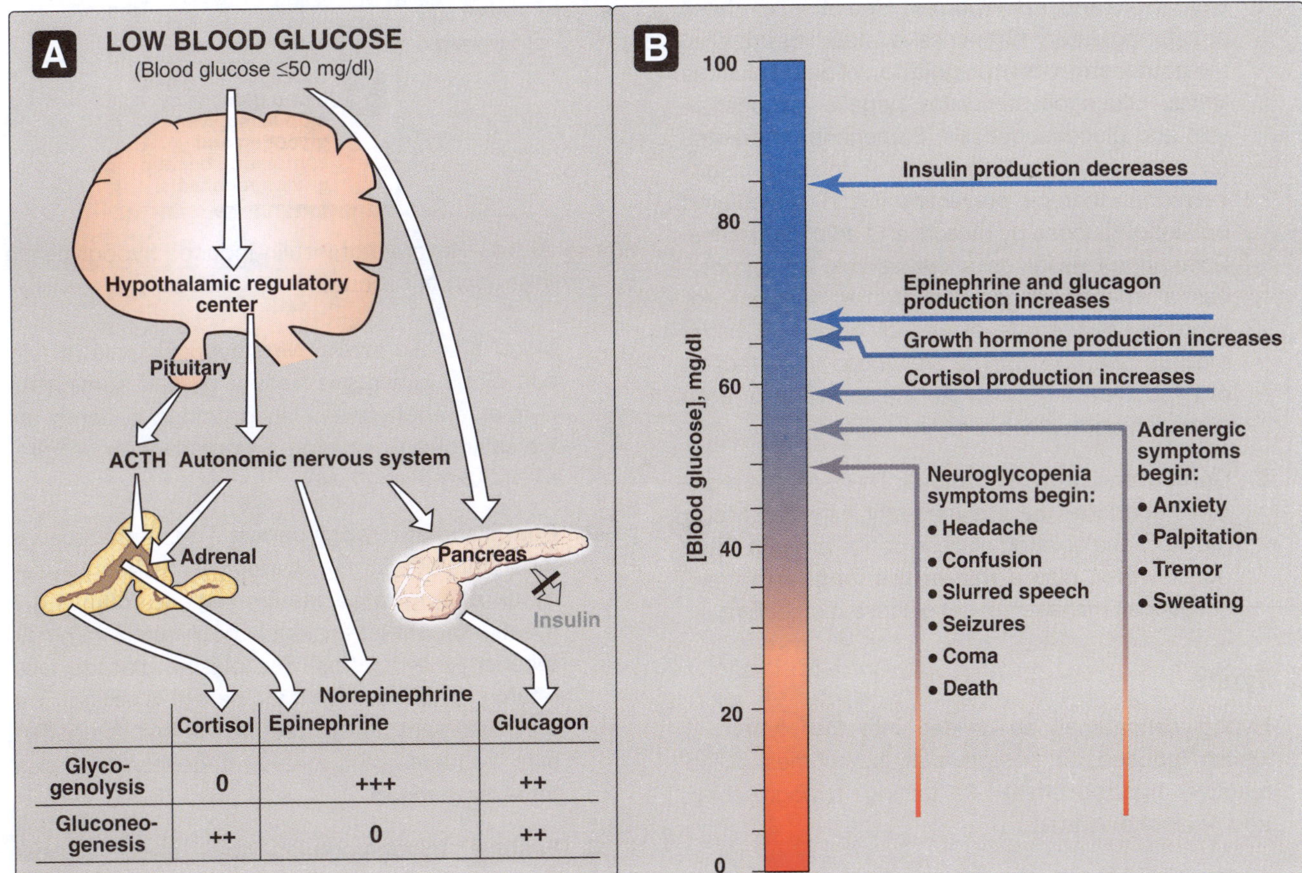

	Cortisol	Epinephrine	Glucagon
Glycogenolysis	0	+++	++
Gluconeogenesis	++	0	++

Figure 27.13 **A.** Actions of some of the glucoregulatory hormones in response to low blood glucose. **B.** Glycemic thresholds for the various responses to hypoglycemia. [Note: Normal fasted blood glucose is 70–99 mg/dl.] + = weak stimulation; ++ = moderate stimulation; +++ = strong stimulation; 0 = no effect; ACTH = adrenocorticotropic hormone.

in glucose deprives the CNS of fuel but fails to trigger an adequate adrenergic response.

B. Glucoregulatory systems

Humans have **two overlapping glucose-regulating systems that are activated by hypoglycemia**: 1) the pancreatic α cells, which release glucagon, and 2) receptors in the hypothalamus, which respond to abnormally low concentrations of blood glucose.

The hypothalamic glucoreceptors can trigger both:

(i) the secretion of catecholamines (mediated by the sympathetic division of the autonomic nervous system)

(ii) release of adrenocorticotropic hormone (ACTH) and growth hormone by the anterior pituitary (see Fig. 27.13).

[Note: ACTH increases cortisol synthesis and secretion in the adrenal cortex (see p. 360).] Glucagon, the catecholamines, cortisol, and growth hormone are sometimes called the **counterregulatory hormones** because each opposes the action of insulin on glucose use.

1. **Glucagon and epinephrine:** Secretion of these counterregulatory hormones is most important in the **acute, short-term regulation** of blood glucose levels. Glucagon stimulates hepatic glycogenolysis and gluconeogenesis. Epinephrine promotes glycogenolysis and lipolysis. It inhibits insulin secretion, thereby preventing GLUT-4–mediated uptake of glucose by muscle and adipose tissues. Epinephrine assumes a critical role in hypoglycemia when glucagon secretion is deficient, for example, in the late stages of type 1 diabetes mellitus (see p. 473). The prevention or correction of hypoglycemia fails when the secretion of both glucagon and epinephrine is deficient.

2. **Cortisol and growth hormone:** These counterregulatory hormones are less important in the short-term maintenance of blood glucose concentrations. They do, however, play a **role in the long-term** (transcriptional) management of glucose metabolism.

C. Types

Hypoglycemia may be divided into four types: 1) insulin induced, 2) postprandial (sometimes called reactive hypoglycemia), 3) fasting hypoglycemia, and 4) alcohol related.

1. **Insulin-induced hypoglycemia:** Hypoglycemia occurs frequently **in patients with diabetes who are receiving insulin treatment**, particularly those striving to achieve tight control of blood glucose levels. Mild hypoglycemia in fully conscious patients is treated by oral administration of carbohydrate. Unconscious patients are typically given glucagon subcutaneously or intramuscularly (Fig. 27.14).

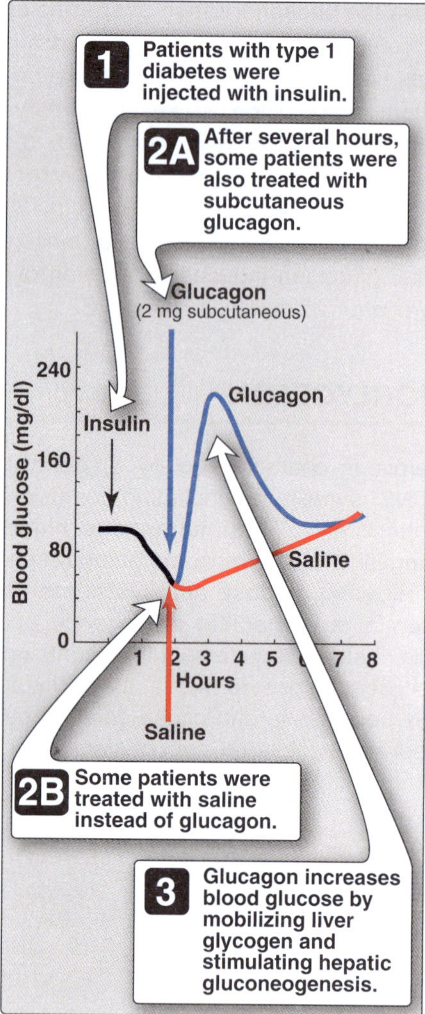

Figure 27.14 Reversal of insulin-induced hypoglycemia by administration of subcutaneous glucagon.

2. **Postprandial hypoglycemia:** This is the second most common form of hypoglycemia. It is caused by an **exaggerated insulin release following a meal**, prompting transient hypoglycemia with mild adrenergic symptoms. The plasma glucose level returns to normal even if the patient is not fed. The only treatment usually required is that the patient eats frequent small meals rather than the usual three large meals.

3. **Fasting hypoglycemia:** Low blood glucose during fasting is rare but is more likely to present as a serious medical problem. Fasting hypoglycemia, which tends to produce neuroglycopenic symptoms, may result from a reduction in the

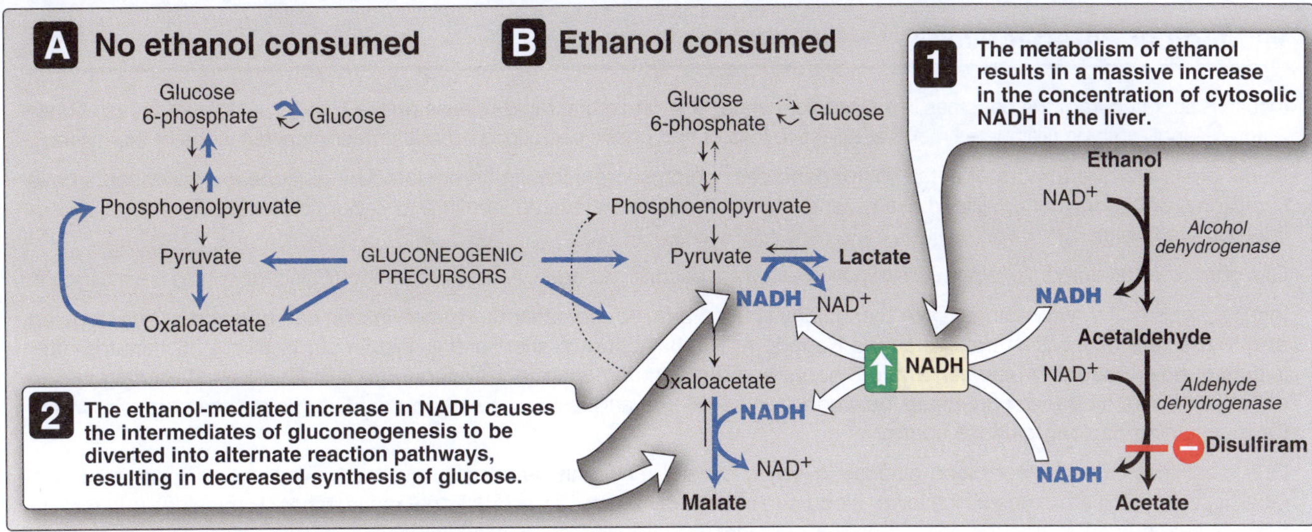

Figure 27.15 **A.** Normal gluconeogenesis in the absence of ethanol consumption. **B.** Inhibition of gluconeogenesis resulting from hepatic metabolism of ethanol. NAD(H) = nicotinamide adenine dinucleotide.

rate of glucose production by hepatic glycogenolysis or gluconeogenesis. Thus, low blood glucose levels are often seen in patients with hepatocellular damage or adrenal insufficiency or in fasting individuals who have consumed large quantities of ethanol (see 4. below). Alternately, fasting hypoglycemia may be the result of an increased rate of glucose use by the peripheral tissues because of overproduction of insulin by rare pancreatic tumors. If left untreated, a patient with fasting hypoglycemia may lose consciousness and experience convulsions and coma. [Note: Certain inborn errors of metabolism, for example, defects in FA oxidation, result in fasting hypoglycemia.]

4. **Alcohol-related hypoglycemia:** Alcohol (ethanol) is metabolized in the liver by two oxidation reactions (Fig. 27.15). Ethanol is first converted to acetaldehyde by zinc-containing *alcohol dehydrogenase*. Acetaldehyde is subsequently oxidized to acetate by *aldehyde dehydrogenase* (*ALDH*). In each reaction, electrons are transferred to oxidized nicotinamide adenine dinucleotide (NAD^+), resulting in an increase in the ratio of the reduced form (NADH) to NAD^+. **The abundance of NADH favors the reduction of pyruvate to lactate and of oxaloacetate (OAA) to malate.** Recall from p. 113 that pyruvate and OAA are substrates in the synthesis of glucose. Thus, the **ethanol-mediated increase in NADH causes these gluconeogenic precursors to be diverted into alternate pathways, resulting in the decreased synthesis of glucose**. This can precipitate hypoglycemia, particularly in individuals who have depleted their stores of liver glycogen. Decreased availability of OAA allows acetyl CoA to be diverted to ketone body synthesis in the liver (see p. 314) and can result in alcoholic ketosis that may result in ketoacidosis.

Hypoglycemia can produce many of the **behaviors associated with alcohol intoxication**, such as agitation, impaired judgment, and combativeness. Therefore, alcohol consumption in vulnerable individuals (such as those who are fasted or have engaged in prolonged, strenuous exercise) can produce hypoglycemia that may contribute to the behavioral effects of alcohol. Because alcohol consumption can also increase the risk for hypoglycemia in patients using insulin, those in an intensive insulin treatment protocol (see p. 473) are counseled about the increased risk of hypoglycemia that generally occurs many hours after alcohol ingestion.

Aldehyde dehydrogenase (*ALDH*) is inhibited by disulfiram, a drug that is used in the treatment of chronic alcoholism. The resulting rise in acetaldehyde results in flushing, tachycardia, hyperventilation, and nausea

Chronic alcohol consumption can also result in alcoholic fatty liver because of increased hepatic synthesis of TAG coupled with impaired formation or release of VLDL. This occurs as a result of decreased FA oxidation because of a fall in the NAD^+/NADH ratio and increased lipogenesis because of the increased availability of FA (decreased catabolism) and of glyceraldehyde 3-phosphate (the *dehydrogenase* is inhibited by the low NAD^+/NADH ratio; see p. 196). With continued alcohol consumption, alcoholic fatty liver can progress first to alcoholic hepatitis and then to alcoholic cirrhosis.

V. Medical Perspectives

Insulin is secreted from the pancreas in a biphasic manner, a short-lasting rapid-release phase 1 and a sustained-release phase 2. Interestingly, in type II diabetes mellitus, selective loss of first phase secretion precedes other manifestations of the disease.

Insulinomas are rare tumors of beta cells of pancreas, which secrete insulin unregulated by glucose levels. Patients have symptoms of hypoglycemia. Blood assays show low glucose and elevated insulin and C-peptide levels. Most of the insulinomas are benign.

Glucagonomas are tumors of alpha cells of pancreas, which secrete glucagon. A large proportion of glucagonomas are malignant.

Insulin prepared by recombinant DNA technology is used by diabetic patients. Human insulin can be short acting (regular) or intermediate acting (Neutral Protamine Hagedorn, NPH). Action of short-acting insulin starts about 30 minutes after injecting, peak occurring between 2 and 3 hours and the duration is up to 10 hours. Effect of intermediate-acting insulin starts 2–4 hours after injection, peaks between 4 and 10 hours and lasts up to 18 hours. Mixtures of these are prescribed for maintaining daylong glucose control.

Oral medicines, which lower blood glucose levels by several mechanisms are also available. *Alpha-glucosidase* inhibitors (acarbose, miglitol) are drugs which lower blood sugar levels by blocking the breakdown of starches. Biguanides (metformin) act primarily by decreasing the hepatic output of glucose. Sulfonylureas, thiazolidinediones, sodium-glucose transporter 2 (SGLT2) inhibitors, meglitinide and *dipeptidyl peptidase 4 (DPP-4)* inhibitors are other antihyperglycemic drugs.

As C-peptide is secreted in equimolar concentration to insulin, its serum levels can be used to assess pancreatic beta cell function. Its levels reduce in type 1 diabetes mellitus and exogenous insulin administration.

VI. Chapter Summary

- The integration of **energy metabolism** is controlled primarily by **insulin** and the opposing actions of **glucagon** and the catecholamines, particularly **epinephrine** (Fig. 27.16).
- Changes in the circulating levels of these hormones allow the body to store energy when food is abundant or to make stored energy available in times of **physiologic stress** (for example, during survival crises, such as famine).
- **Insulin** is a peptide hormone produced by the **β cells** of the **islets of Langerhans** of the **pancreas**. It consists of disulfide-linked A and B chains.
- A rise in blood glucose is the most important signal for insulin **secretion**. The **catecholamines**, secreted in response to stress, trauma, or extreme exercise, inhibit insulin secretion.
- Insulin increases glucose uptake (by glucose transporters (**GLUT-4**) in muscle and adipose tissue) and the synthesis of **glycogen**, **protein**, and **triacylglycerol**: It is an **anabolic** hormone. These actions are mediated by binding to its membrane ***tyrosine kinase* receptor**.
- **Glucagon** is a monomeric peptide hormone produced by the **α cells** of the pancreatic islets.
- Glucagon, along with epinephrine, norepinephrine, cortisol, and growth hormone (the **counterregulatory hormones**), opposes many of the actions of insulin. Glucagon acts to maintain blood glucose during periods of potential hypoglycemia. Glucagon increases **glycogenolysis**, **gluconeogenesis**, **fatty acid oxidation**, **ketogenesis**, and **amino acid uptake**: It is a **catabolic** hormone.
- Glucagon secretion is stimulated by low blood glucose, amino acids, and the catecholamines. Its secretion is inhibited by elevated blood glucose and by insulin.
- Glucagon binds to high-affinity **G protein–coupled receptors** on the cell membrane of hepatocytes. Binding results in the activation of ***adenylyl cyclase***, which produces the second messenger **cyclic adenosine monophosphate** (**cAMP**), leading to activation of ***cAMP-dependent protein kinase A.***
- Both insulin and glucagon affect **gene transcription**.
- **Hypoglycemia** is characterized by low blood glucose accompanied by **adrenergic** and **neuroglycopenic symptoms** that are rapidly resolved by the administration of glucose.
- Insulin-induced, postprandial, and fasting hypoglycemia result in release of glucagon and epinephrine.
- The rise in **nicotinamide adenine dinucleotide** (**NADH**) that accompanies **ethanol** metabolism inhibits gluconeogenesis, leading to hypoglycemia in individuals with depleted stores. Alcohol consumption also increases the risk for hypoglycemia in patients using insulin. Chronic alcohol consumption can cause **fatty liver disease**.

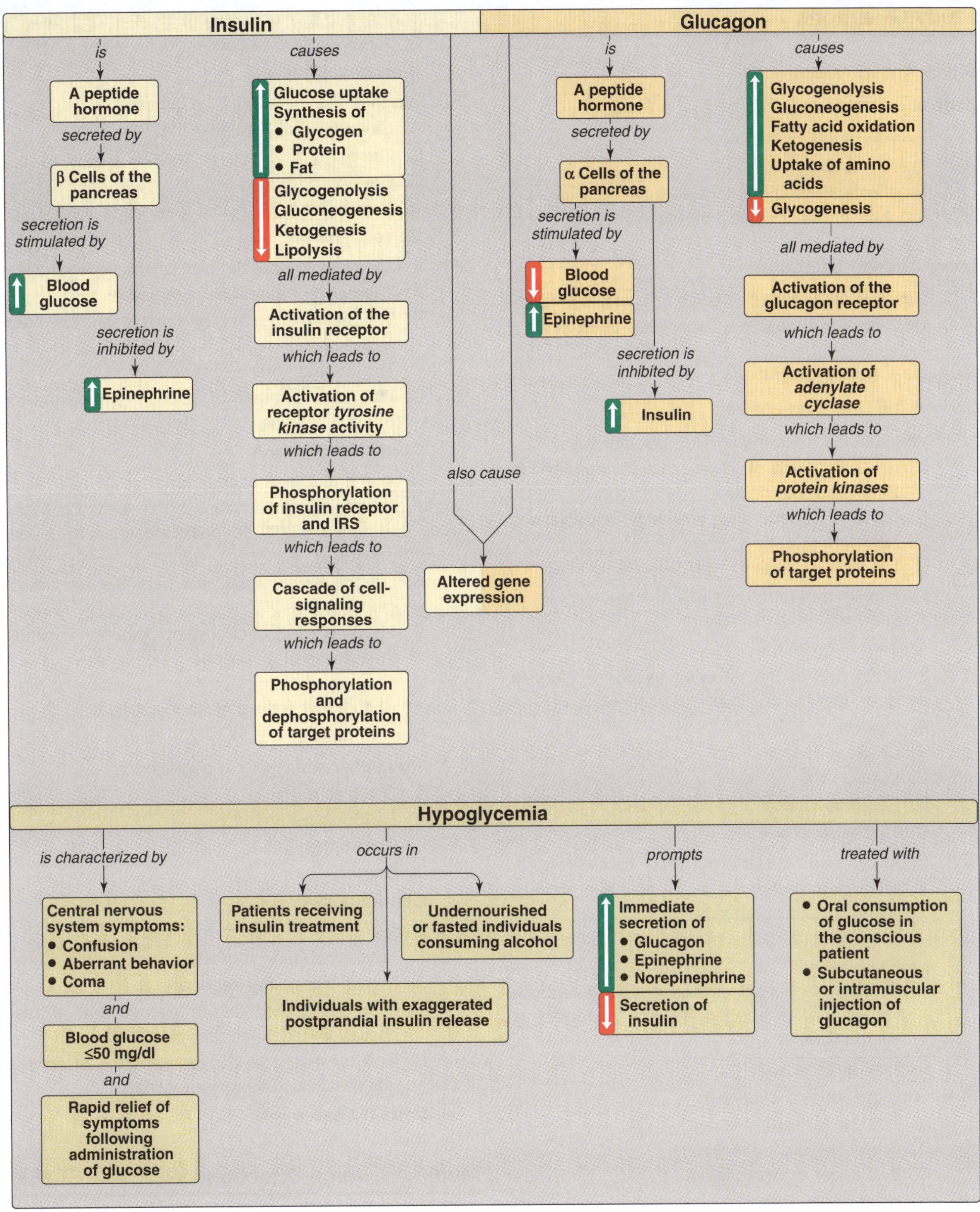

Figure 27.16 Key concept map for the metabolic effects of insulin and glucagon as well as hypoglycemia. IRS = insulin receptor substrates.

Study Questions

Short-Answer Questions

Write short notes on:

27.1. Counterregulatory hormones
27.2. Insulin receptor
27.3. Biochemical effects of glucagon
27.4. Biochemical basis of alcohol induced hypoglycemia

Long-Answer Questions

27.1. Describe the regulation of blood glucose by insulin.
27.2. Describe the structure and functions of insulin.

Multiple-Choice Questions

Choose the ONE best answer.

27.1. Which of the following is TRUE about insulin?
- A. It is composed of 53 amino acids arranged in 2 polypeptide chains
- B. Insulin is stored in granules in endoplasmic reticulum
- C. Insulin has a plasma half-life of ~6 minutes.
- D. Ingestion of protein inhibits the glucose-stimulated secretion of insulin.

Correct answer = C.

27.2. In which one of the following tissues is glucose transport into the cell insulin dependent?
- A. Adipose
- B. Brain
- C. Liver
- D. Red blood cells

Correct answer = A.

27.3. Choose the correct statement about insulin receptor.
- A. The insulin receptor is assembled into a dimer linked by disulfide bonds
- B. The α subunit has tyrosine kinase activity
- C. The β subunit has tyrosine kinase activity
- D. Binding of insulin promotes a rapid autophosphorylation of specific tyrosine residues on each α subunit

Correct answer = C.

27.4. Glucagon acts by binding to:
- A. Tyrosine kinase receptors
- B. G protein coupled receptors
- C. Zinc finger motif receptors
- D. None of the above

Correct answer = B.

27.5. Glucose-dependent release of insulin into blood is mediated through:
- A. A rise in sodium concentration in the β cell
- B. A rise in potassium concentration in the β cell
- C. A rise in calcium concentration in the β cell
- D. A rise in chloride concentration in the β cell

Correct answer = C.

27.6. Which of the following is an incretin?
- A. Glucagon-like peptide-1 (GLP-1)
- B. Gastric inhibitory polypeptide
- C. Both A and B
- D. Neither A nor B

Correct answer = C.

27.7. Insulin binds to specific, high-affinity receptors in:
- A. The cell membrane of most tissues
- B. The cell membrane of muscle and adipose tissue only.
- C. The cell membrane of liver only
- D. The cell membrane of adipose tissue and liver only

Correct answer = A.

27.8. Choose the correct statement:
- A. Insulin-induced increase in the amount of many enzymes, such as *glucokinase* requires few minutes.
- B. Dietary amino acids stimulate the release of glucagon.
- C. Elevated levels of circulating epinephrine inhibit the release of glucagon.
- D. Glucagon binding results in activation of tyrosine kinase in the plasma membrane

Correct answer = B.

27.9. Inside the cells, insulin is degraded in:
- A. Cytoplasm
- B. Endoplasmic reticulum
- C. Lysosomes
- D. Golgi

Correct answer = C.

27.10. Ethanol is metabolized in the liver:
- A. By alcohol dehydrogenase
- B. By aldehyde dehydrogenase
- C. First by alcohol dehydrogenase and subsequently by aldehyde dehydrogenase.
- D. First by aldehyde dehydrogenase and subsequently by alcohol dehydrogenase

Correct answer = C.

Multiple-Choice Questions (NExT/NEET PG)

Choose the ONE best answer.

27.1. Which of the following statements is true for insulin but not for glucagon?
- A. It is a peptide hormone secreted by pancreatic cells.

B. Its actions are mediated by binding to a receptor found on the cell membrane of liver cells.
C. Its secretion is increased by amino acids.
D. Its secretion is decreased by the catecholamines.

Correct answer = D.

27.2. A 39-year-old woman is brought to the emergency room complaining of weakness and dizziness. She recalls getting up early that morning to do her weekly errands and had skipped breakfast. She drank a cup of coffee for lunch and had nothing to eat during the day. She met with friends at 8 p.m. and had a few drinks. As the evening progressed, she soon became weak and dizzy and was taken to the hospital. Laboratory tests revealed her blood glucose to be 45 mg/dl (normal = 70–99). She was given orange juice and immediately felt better. The biochemical basis of her alcohol-induced hypoglycemia is an increase in:
A. fatty acid oxidation.
B. the ratio of the reduced and oxidized forms of nicotinamide adenine dinucleotide
C. oxaloacetate and pyruvate.
D. use of acetyl coenzyme A in fatty acid synthesis.

Correct answer = B.

27.3. Insulinomas are characterized by:
A. Increased blood insulin and increased blood glucose
B. Increased blood insulin and decreased blood glucose
C. Decreased blood insulin and increased blood glucose
D. Decreased blood insulin and decreased blood glucose

Correct answer = B.

27.4. Sulfonylureas, oral agents used to treat type 2 diabetes, increase insulin secretion by:
A. Closing ATP-sensitive K^+ channels
B. Closing ATP-sensitive Ca^+ channels
C. Opening ATP-sensitive K^+ channels
D. Opening ATP-sensitive Ca^+ channels

Correct answer = A.

27.5. Based on statements given below, choose the correct option:
Statement 1: The intestinal peptide glucagon-like peptide-1 (GLP-1) increase the sensitivity of β cells to glucose.
Statement 2: Preproglucagon is processed to GLP-1 in intestinal L cells.
A. Statement 1 is true but statement 2 is false.
B. Statement 2 is true but statement 2 is false.
C. Both statements are true
D. Both statements are false

Correct answer = C.

The Feed–Fast Cycle 28

Learning Objectives

At the end of this chapter, the student will be able to:

- Describe the metabolism in liver, adipose tissue, skeletal muscle and brain in fed state
- Describe the metabolism in kidney, adipose tissue, skeletal muscle and brain in fasting state

I. OVERVIEW OF THE ABSORPTIVE STATE

The **absorptive (well-fed) state is the 2- to 4-hour period after ingestion of a normal meal**. During this interval, transient increases in plasma glucose, amino acids, and triacylglycerols (TAG) occur, the latter primarily as components of chylomicrons synthesized and secreted by the intestinal mucosal cells (see p. 350). Islet tissue of the pancreas responds to the elevated level of glucose with increased secretion of insulin and decreased secretion of glucagon. **The elevated insulin/glucagon ratio and the ready availability of circulating substrates make the absorptive state an anabolic period** characterized by increased synthesis of TAG and glycogen to replenish fuel stores as well as increased synthesis of protein. During this absorptive period, virtually all tissues use glucose as a fuel, and the metabolic response of the body is dominated by alterations in the metabolism of liver, adipose tissue, skeletal muscle, and brain. In this chapter, an "organ map" is introduced that traces the movement of metabolites between tissues. The goal is to create an expanded and clinically useful vision of whole-body metabolism.

II. REGULATORY MECHANISMS

The flow of intermediates through metabolic pathways is controlled by four mechanisms:

1) the availability of substrates
2) allosteric regulation of enzymes
3) covalent modification of enzymes,
4) induction-repression of enzyme synthesis, primarily through regulation of transcription.

Although this scheme may at first seem redundant, each mechanism operates on a different timescale (Fig. 28.1)

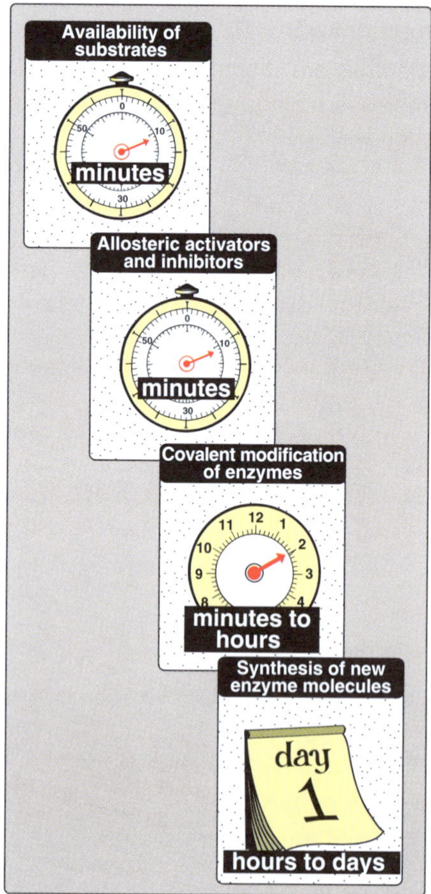

Figure 28.1 Control mechanisms of metabolism and some typical response times. [Note: Response times may vary according to the nature of the stimulus and from tissue to tissue.]

and allows the body to adapt to a wide variety of physiologic situations. In the absorptive state, these regulatory mechanisms insure that available nutrients are captured as glycogen, TAG, and protein.

A. Allosteric effectors

Allosteric changes **usually involve rate-determining reactions**. For example, glycolysis in the liver is stimulated following a meal by an increase in fructose 2,6-bisphosphate, an allosteric activator of *phosphofructokinase-1* ([*PFK-1*], see p. 195). In contrast, gluconeogenesis is decreased by fructose 2,6-bisphosphate, an allosteric inhibitor of *fructose 1,6-bisphosphatase* (see p. 220).

B. Covalent modification

The activity of many enzymes is regulated by the **addition** (via *kinases*, such as *cyclic adenosine monophosphate [cAMP]–dependent protein kinase A* [*PKA*] and *adenosine monophosphate–activated protein kinase* [*AMPK*]) **or removal** (via *phosphatases*) **of phosphate groups** from specific serine, threonine, or tyrosine residues of the protein.

In the absorptive state, most of the covalently regulated enzymes are in the dephosphorylated form and are active (Fig. 28.2). Three exceptions are *glycogen phosphorylase kinase* (see p. 228), *glycogen phosphorylase* (see p. 228), and *hormone-sensitive lipase* (*HSL*) (see p. 307), which are inactive in their dephosphorylated form. [Note: In the liver, the *phosphatase* domain of bifunctional *phosphofructokinase-2* (*PFK-2*) is inactive when the protein is dephosphorylated (see p. 196).]

C. Induction and repression of enzyme synthesis

Increased (induction of) or decreased (repression of) enzyme synthesis leads to **changes in the number of enzyme molecules**, rather than changing the activity of existing enzyme molecules. Enzymes subject to synthesis regulation are often those that are needed under specific physiologic conditions.

For example, **in the well-fed state**, elevated insulin levels result in an increase in the synthesis of key enzymes, such as *acetyl coenzyme A (CoA) carboxylase* (*ACC*) and *fatty acid synthase* (see p. 443), involved in anabolic metabolism. **In the fasted state**, glucagon induces expression of *phosphoenolpyruvate carboxykinase* (*PEPCK*) of gluconeogenesis (see p. 444). [Note: Both hormones affect gene transcription.]

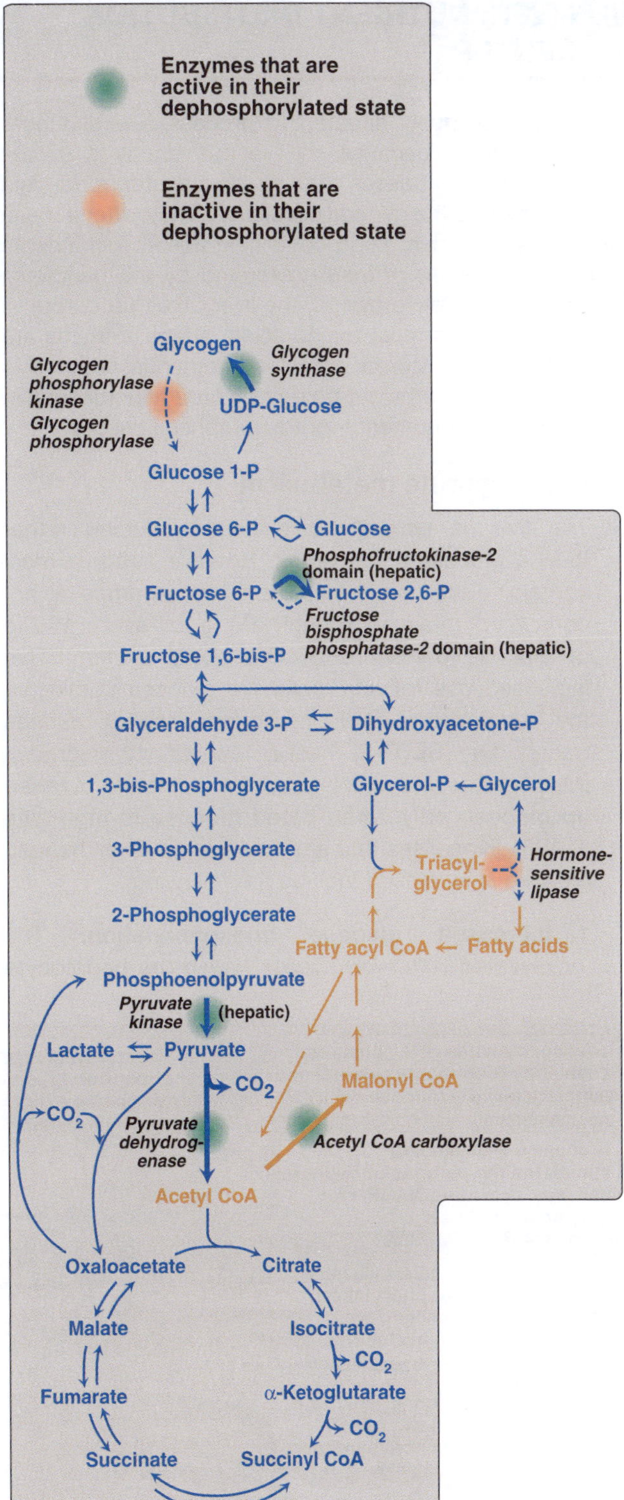

Figure 28.2 Important reactions of intermediary metabolism regulated by enzyme phosphorylation. **Blue text** = intermediates of carbohydrate metabolism; **brown text** = intermediates of lipid metabolism; P = phosphate; CoA = coenzyme A; CO_2 = carbon dioxide.

III. LIVER: NUTRIENT DISTRIBUTION CENTER

The liver is uniquely situated to process and distribute dietary nutrients because the venous drainage of the gut and pancreas passes through the hepatic portal vein before entry into the general circulation. Thus, after a meal, the liver is bathed in blood containing absorbed nutrients and elevated levels of insulin secreted by the pancreas. During the absorptive period, the liver takes up carbohydrates, lipids, and most amino acids. These nutrients are then metabolized, stored, or routed to other tissues. In this way, the liver smooths out potentially broad fluctuations in the availability of nutrients for the peripheral tissues.

A. Carbohydrate metabolism

The liver is normally a glucose-producing rather than a glucose-using organ. However, after a meal containing carbohydrate, the liver becomes a net consumer, retaining roughly 60 g of every 100 g of glucose presented by the portal system. This increased use reflects increased glucose uptake by the hepatocytes. Their insulin-independent glucose transporter (**GLUT-2**) has a low affinity (high K_m [Michaelis constant]) for glucose and, therefore, **takes up glucose only when blood glucose is high** (see p. 193). Processes that are upregulated when hepatic glucose is increased include the following:

1. **Increased glucose phosphorylation:** The elevated levels of glucose within the hepatocyte (as a result of elevated extracellular levels) allow *glucokinase* to phosphorylate glucose to glucose 6-phosphate (Fig. 28.3, ❶). [Note: *Glucokinase* has a high K_m for glucose, is not subject to direct product inhibition, and has a sigmoidal reaction curve (see p. 193).]

2. **Increased glycogenesis:** The conversion of glucose 6-phosphate to glycogen is favored by the activation of *glycogen synthase*, both by dephosphorylation and by increased availability of glucose 6-phosphate, its positive allosteric effector (see Fig. 28.3, ❷).

3. **Increased pentose phosphate pathway activity:** The increased availability of glucose 6-phosphate, combined with the active use of nicotinamide adenine dinucleotide phosphate (NADPH) in hepatic lipogenesis, stimulates the pentose phosphate pathway (see p. 248). This pathway typically accounts for 5%–10% of the glucose metabolized by the liver (see Fig. 28.3, ❸).

4. **Increased glycolysis:** In the liver, glycolysis is significant only during the absorptive period following a carbohydrate-rich meal. The conversion of glucose to pyruvate is stimulated by the elevated insulin/glucagon ratio that results in increased amounts of the regulated enzymes of glycolysis: *glucokinase*, *PFK-1*, and *pyruvate kinase* ([*PK*], see p. 202).

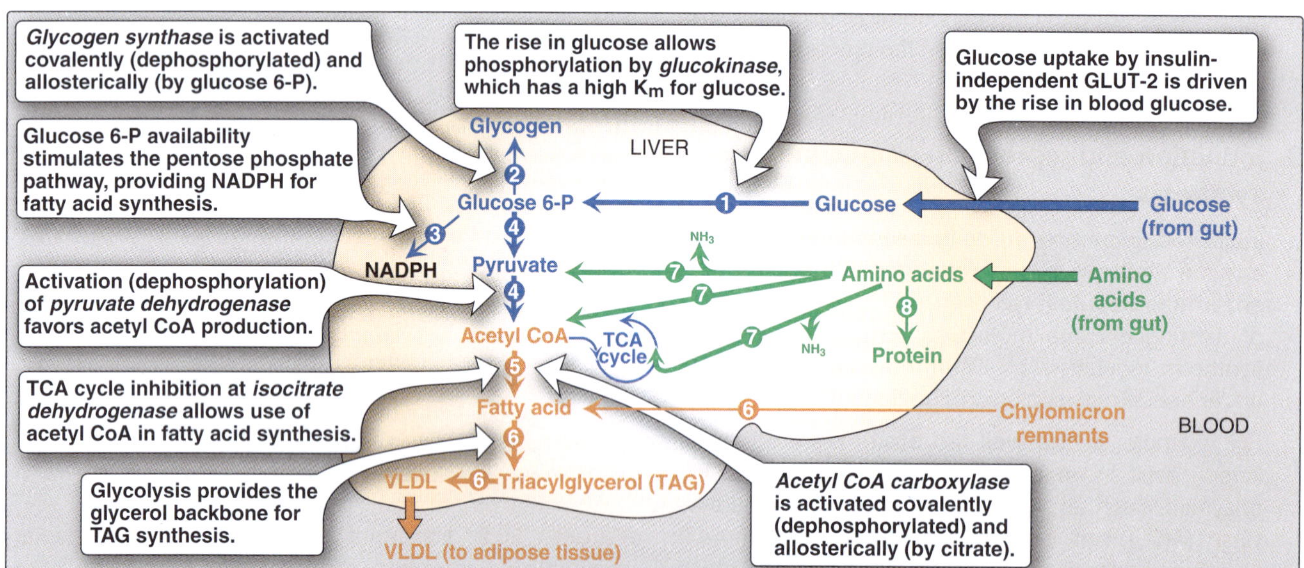

Figure 28.3 Major metabolic pathways in the liver in the absorptive state. [Note: The acetyl coenzyme A (CoA) is also used for cholesterol synthesis.] The numbers in circles, which appear both in the figure and in the text, indicate important pathways for carbohydrate, fat, or protein metabolism. **Blue text** = intermediates of carbohydrate metabolism; **brown text** = intermediates of lipid metabolism; **green text** = intermediates of protein metabolism; P = phosphate; TCA = tricarboxylic acid; VLDL = very-low-density lipoprotein; GLUT = glucose transporter; NADPH = nicotinamide adenine dinucleotide phosphate; NH_3 = ammonia.

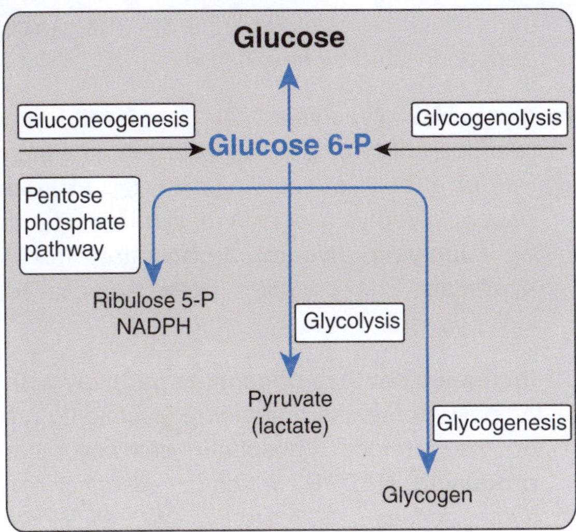

Figure 28.4 Central role of glucose 6-phosphate in metabolism. [Note: The presence of *glucose 6-phosphatase* in the liver allows the production of free glucose from the glucose 6-phosphate produced in glycogenolysis and gluconeogenesis.] NADPH = nicotinamide adenine dinucleotide phosphate; P = phosphate.

Additionally, *PFK-1* is allosterically activated by fructose 2,6-bisphosphate generated by the active (dephosphorylated) *kinase* domain of bifunctional *PFK-2*. *PK* is dephosphorylated and active. *Pyruvate dehydrogenase* (*PDH*), which converts pyruvate to acetyl CoA, is active (dephosphorylated) because pyruvate inhibits *PDH kinase* (see Fig. 28.3, ❹). The acetyl CoA either is used as a substrate for fatty acid (FA) synthesis or is oxidized for energy in the tricarboxylic acid (TCA) cycle. (See Fig. 28.4 for the central role of glucose 6-phosphate.)

5. **Decreased glucose production:** While pathways that promote glucose storage, such as glycolysis and glycogenesis, are being stimulated in the liver in the absorptive state, pathways that generate glucose, including gluconeogenesis and glycogenolysis, are being inhibited. *Pyruvate carboxylase* (*PC*), which catalyzes the first step in gluconeogenesis, is largely inactive because of low levels of its allosteric activator, acetyl CoA, which is being used for FA synthesis (see p. 217).

The high insulin/glucagon ratio also favors inactivation of other gluconeogenic enzymes such as *fructose 1,6-bisphosphatase* (see Fig. 13.17, p. 196). Glycogenolysis is inhibited by dephosphorylation of *glycogen phosphorylase* and *phosphorylase kinase*.

The increased uptake and decreased production of blood glucose in the absorptive period prevents hyperglycemia.

B. Fat metabolism

1. **Increased fatty acid synthesis:** Liver is the primary site of de novo synthesis of FA (see Fig. 28.3, ❺). **FA synthesis, a cytosolic process, is favored in the absorptive period by:**

(i) availability of the substrates acetyl CoA (from glucose and amino acid metabolism) and NADPH (from glucose metabolism in the pentose phosphate pathway) and,

(ii) by the activation of *ACC*, both by dephosphorylation and by the presence of its allosteric activator, citrate. [Note: Inactivity of *AMPK* favors dephosphorylation.]

ACC catalyzes the formation of malonyl CoA from acetyl CoA, the rate-limiting reaction for FA synthesis (see p. 301). Malonyl CoA inhibits *carnitine palmitoyltransferase-I* (*CPT-I*) of FA oxidation (see p. 309). **Thus, citrate directly activates FA synthesis and indirectly inhibits FA degradation**.

a. **Source of cytosolic acetyl coenzyme A:** Pyruvate from aerobic glycolysis enters mitochondria and is decarboxylated by *PDH*. The acetyl CoA product is combined with oxaloacetate (OAA) to form citrate via *citrate synthase* of the TCA cycle. Citrate leaves the mitochondria (as a result of the inhibition of *isocitrate dehydrogenase* by ATP) and enters the cytosol. Citrate is cleaved by *ATP citrate lyase* (induced by insulin), producing the acetyl CoA substrate of *ACC* plus OAA.

b. **Additional source of NADPH:** The OAA is reduced to malate, which is oxidatively decarboxylated to pyruvate by *malic enzyme* as NADPH is formed (see Fig. 21.11, p. 306).

2. **Increased triacylglycerol synthesis:** TAG synthesis is favored because fatty acyl CoA are available both from de novo synthesis and from hydrolysis of the TAG component of chylomicron remnants removed from the blood by hepatocytes (see p. 295). Glycerol 3-phosphate, the backbone for TAG synthesis, is provided by glycolysis (see p. 307). The liver packages these endogenous TAG into very-low-density lipoprotein (VLDL) particles that are secreted into the blood for use by extrahepatic tissues, particularly adipose and muscle tissues (see Fig. 28.3, ❻).

C. Amino acid metabolism

1. **Increased amino acid degradation:** In the absorptive period, more amino acids are present than the liver can use in the synthesis of proteins and other

nitrogen-containing molecules. The **surplus amino acids are not stored but are either released into the blood for other tissues to use in protein synthesis or deaminated, with the resulting carbon skeletons being degraded by the liver to pyruvate, acetyl CoA, or TCA cycle intermediates**. These metabolites can be oxidized for energy or used in FA synthesis (see Fig. 28.3, ❼). The **liver has limited capacity to initiate degradation of the branched-chain amino acids** (BCAA) leucine, isoleucine, and valine. They pass through the liver essentially unchanged and are metabolized in muscle (see p. 401).

2. **Increased protein synthesis:** The body does not store protein for energy in the same way that it maintains glycogen or TAG reserves. However, a transient increase in the synthesis of hepatic proteins does occur in the absorptive state, resulting in replacement of any proteins that may have been degraded during the previous period of fasting (see Fig. 28.3, ❽).

IV. ADIPOSE TISSUE: ENERGY STORAGE DEPOT

Adipose tissue is second only to the liver in its ability to distribute fuel molecules. In a 70-kg man, white adipose tissue (WAT) weighs ~14 kg, or about half as much as the total muscle mass. Nearly the entire volume of each adipocyte in WAT can be occupied by a droplet of anhydrous, calorically dense TAG (Fig. 28.5).

A. Carbohydrate metabolism

1. **Increased glucose transport:** Circulating insulin levels are elevated in the absorptive state, resulting in an influx of glucose into adipocytes via insulin-sensitive GLUT-4 recruited to the cell surface from intracellular vesicles (Fig. 28.6, ❶). The glucose is phosphorylated by *hexokinase*.

2. **Increased glycolysis:** The increased intracellular availability of glucose results in an enhanced rate of glycolysis (see Fig. 28.6, ❷). **In adipose tissue, glycolysis serves a synthetic function by supplying glycerol 3-phosphate for TAG synthesis** (see p. 306). [Note: Adipose tissue lacks *glycerol kinase*.]

3. **Increased pentose phosphate pathway activity:** Adipose tissue can metabolize glucose by means of the pentose phosphate pathway, thereby **producing NADPH**, which is essential for FA synthesis (see p. 307 and Fig. 28.6, ❸). However, in humans, de novo synthesis is not a major source

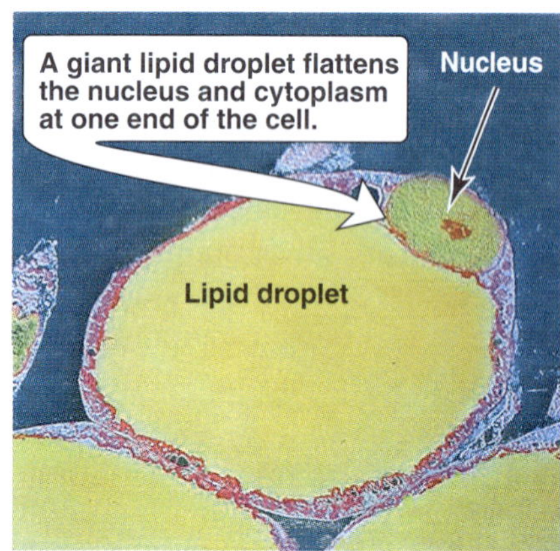

Figure 28.5 Colorized transmission electron micrograph of adipocytes.

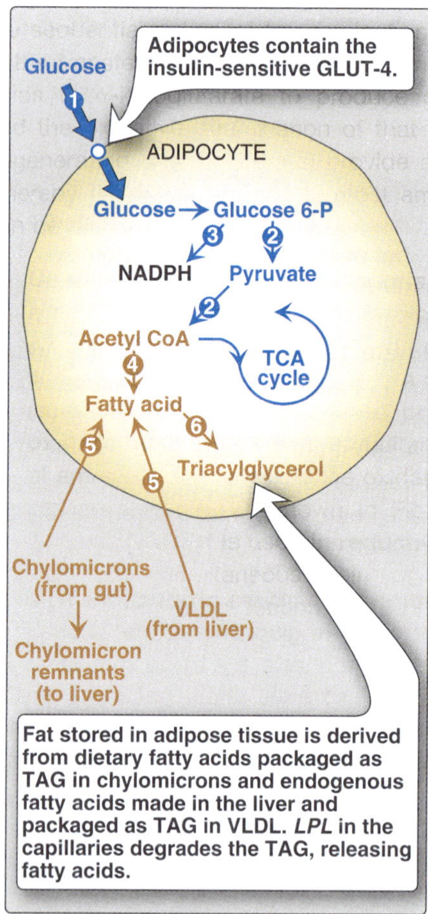

Figure 28.6 Major metabolic pathways in adipose tissue in the absorptive state. [Note: The numbers in the circles, which appear both in the figure and in the corresponding text, indicate important pathways for adipose tissue metabolism.] GLUT = glucose transporter; P = phosphate; NADPH = nicotinamide adenine dinucleotide; CoA = coenzyme A; TCA = tricarboxylic acid; TAG = triacylglycerol; VLDL = very-low-density lipoprotein; LPL = *lipoprotein lipase*.

of FA in adipose tissue, except when refeeding a previously fasted individual (see Fig. 28.6, ❹).

B. Fat metabolism

Most of the FA added to the TAG stores of adipocytes after consumption of a lipid-containing meal are provided by the degradation of **exogenous (dietary) TAG in chylomicrons** sent out by the intestine and **endogenous TAG in VLDL** sent out by the liver (see Fig. 28.6, ❺).

The FA are released from the lipoproteins by *lipoprotein lipase* (*LPL*), an extracellular enzyme attached to the endothelial cells of capillary walls in many tissues, particularly adipose and muscle (see p. 350). **In adipose tissue, *LPL* is upregulated by insulin.** Thus, in the fed state, elevated levels of glucose and insulin favor storage of TAG (see Fig. 28.6, ❻), all the carbons of which are supplied by glucose. [Note: Elevated insulin favors the dephosphorylated (inactive) form of *HSL* (see p. 307), thereby inhibiting lipolysis in the well-fed state.]

V. RESTING SKELETAL MUSCLE

Skeletal muscle accounts for ~40% of the body mass in individuals of healthy weight, and **it can use glucose, amino acids, FA, and ketone bodies as fuel**. In the well-fed state, muscle takes up glucose via GLUT-4 (for energy and glycogen synthesis) and amino acids (for energy and protein synthesis). In contrast to liver, there is no covalent regulation of *PFK-2* in skeletal muscle. However, in the cardiac isozyme, the *kinase* domain is activated by epinephrine-mediated phosphorylation (see p. 197).

> Skeletal muscle is unique in being able to respond to substantial changes in the demand for ATP that accompanies contraction. At rest, muscle accounts for ~25% of the oxygen (O_2) consumption of the body, whereas during vigorous exercise, it is responsible for up to 90%. This underscores the fact that skeletal muscle, despite its potential for transient periods of anaerobic glycolysis, is an oxidative tissue.

A. Carbohydrate metabolism

1. **Increased glucose transport:** The transient increase in plasma glucose and insulin after a carbohydrate-rich meal leads to an increase in glucose transport into muscle cells (myocytes) by GLUT-4 (see p. 192 and Fig. 28.7, ❶), thereby reducing blood glucose. Glucose is phosphorylated to glucose 6-phosphate by *hexokinase* and metabolized to meet the energy needs of myocytes.

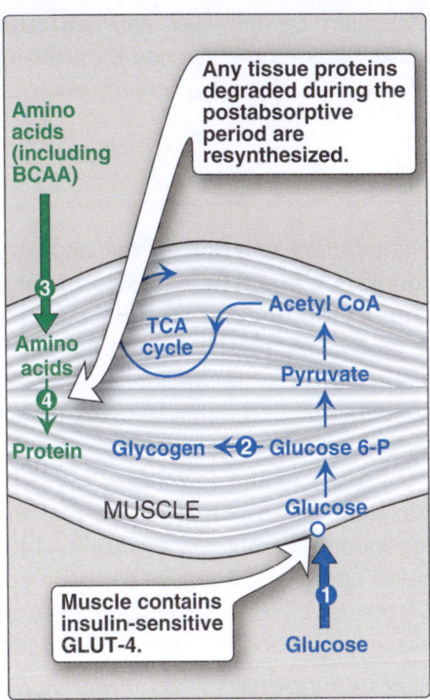

Figure 28.7 Major metabolic pathways in skeletal muscle in the absorptive state. [Note: The numbers in circles, which appear both in the figure and in the text, indicate important pathways for carbohydrate or protein metabolism.] CoA = coenzyme A; P = phosphate; GLUT = glucose transporter; BCAA = branched-chain amino acids; TCA = tricarboxylic acid.

2. **Increased glycogenesis:** The increased insulin/glucagon ratio and the availability of glucose 6-phosphate favor glycogen synthesis, particularly if glycogen stores have been depleted as a result of exercise (see p. 225 and Fig. 28.7, ❷).

B. Fat metabolism

FA are released from chylomicrons and VLDL by the action of *LPL* (see pp. 350 and 353). However, FA are of secondary importance as a fuel for resting muscle during the well-fed state, in which glucose is the primary source of energy.

C. Amino acid metabolism

1. **Increased protein synthesis:** An increase in amino acid uptake and protein synthesis occurs in the absorptive period after ingestion of a meal containing protein (see Fig. 28.7, ❸ and ❹). This synthesis replaces protein degraded since the previous meal.

2. **Increased branched-chain amino acid uptake:** Muscle is the **principal site for degradation of the BCAA** because it contains the required *transaminase* (see p. 401). The dietary BCAA escape

metabolism by the liver and are taken up by muscle, where they are used for protein synthesis (see Fig. 28.7, ❸) and as energy sources.

VI. BRAIN

Although contributing only 2% of the adult weight, the **brain accounts for a consistent 20% of the basal O_2 consumption of the body at rest**. Because the brain is vital to the proper functioning of all organs of the body, special priority is given to its fuel needs. To provide energy, substrates must be able to cross the endothelial cells that line the blood vessels in the brain (the blood–brain barrier [BBB]). **In the fed state, the brain exclusively uses glucose as a fuel** (GLUT-1 of the BBB is insulin independent), completely oxidizing ~140 g/day to carbon dioxide and water. Because the brain contains no significant stores of glycogen, it is completely dependent on the availability of blood glucose (Fig. 28.8, ❶).

If blood glucose levels fall to <50 mg/dl (normal fasted blood glucose is 70–99 mg/dl), cerebral function is impaired (see p. 445). The brain also lacks significant stores of TAG, and the FA circulating in the blood make little contribution to energy production for reasons that are unclear. The intertissue exchanges characteristic of the absorptive period are summarized in Figure 28.9.

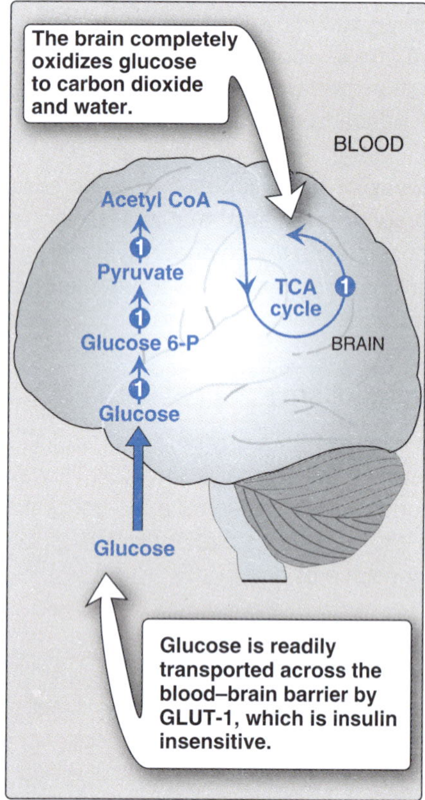

Figure 28.8 Major metabolic pathways in the brain in the absorptive state. [Note: The numbers in circles, which appear both in the figure and in the text, indicate important pathways for carbohydrate metabolism.] CoA = coenzyme A; TCA = tricarboxylic acid; P = phosphate; GLUT = glucose transporter.

VII. OVERVIEW OF THE FASTED STATE

Fasting begins if no food is ingested after the absorptive period. It may result from an inability to obtain food, the desire to lose weight rapidly, or clinical situations in which an individual cannot eat (for example, because of trauma, surgery, cancer, or burns). In the absence of food, plasma levels of glucose, amino acids, and TAG fall, triggering a decline in insulin secretion and an increase in glucagon, epinephrine, and cortisol secretion.

The decreased insulin/counterregulatory hormone ratio and the decreased availability of circulating substrates make the postabsorptive period of nutrient deprivation a catabolic period characterized by degradation of TAG, glycogen, and protein. This sets into motion an exchange of substrates among the liver, adipose tissue, skeletal muscle, and brain that is guided by two priorities: 1) the **need to maintain adequate plasma levels of glucose** to sustain energy metabolism in the brain, red blood cells, and other glucose-requiring tissues and 2) the **need to mobilize FA from TAG in WAT** for the synthesis and release of ketone bodies by the liver to supply energy to other tissues and spare body protein. As a result, blood glucose levels are maintained within a narrow range in fasting, while FA and ketone body levels increase. [Note: Maintaining glucose requires that the substrates for gluconeogenesis (such as pyruvate, alanine, and glycerol) be available.]

A. Fuel stores

The metabolic fuels available in a normal 70-kg man at the beginning of a fast are shown in Figure 28.10. Observe the enormous caloric stores available in the form of TAG compared with those contained in glycogen.

Although protein is listed as an energy source, each protein also has a function unrelated to energy metabolism (for example, as a structural component of the body or as an enzyme). Therefore, only about one third of the body's protein can be used for energy production without fatally compromising vital functions.

B. Enzymic changes

In fasting (as in the well-fed state), the flow of intermediates through the pathways of energy metabolism is controlled by four mechanisms: 1) the availability of substrates, 2) allosteric regulation of enzymes, 3) covalent modification of enzymes, and 4) induction-repression of enzyme synthesis. The

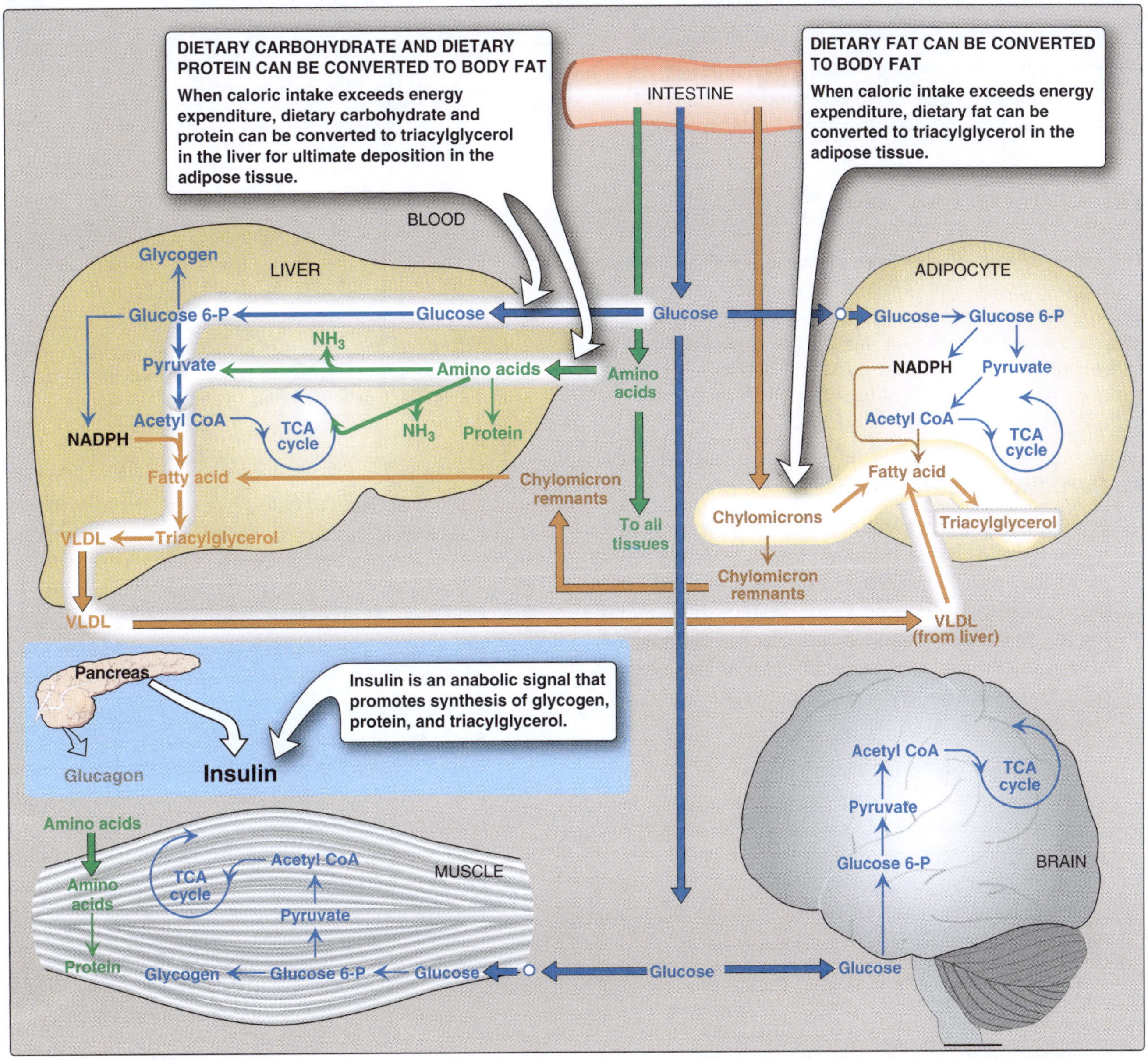

Figure 28.9 Intertissue relationships in the absorptive state and the hormonal signals that promote them. [Note: Small circles on the perimeter of muscle and the adipocyte indicate insulin-dependent glucose transporters.] P = phosphate; CoA = coenzyme A; NADPH = nicotinamide adenine dinucleotide phosphate; TCA = tricarboxylic acid; VLDL = very-low-density lipoprotein.

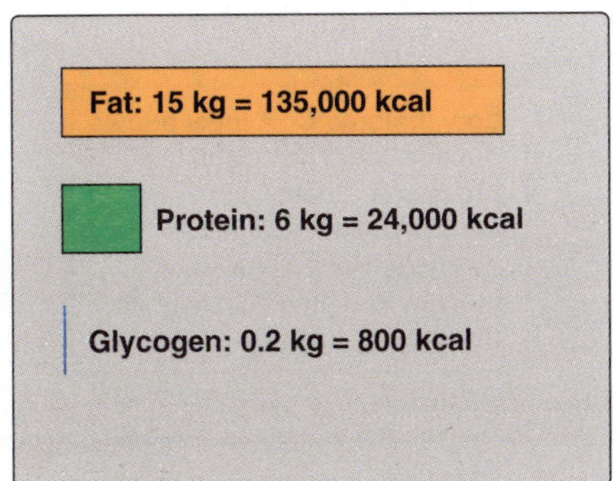

Figure 28.10 Metabolic fuels present in a 70-kg man at the beginning of a fast.

metabolic changes observed in fasting are generally opposite those described for the absorptive state (see Fig. 28.9). For example, **although most of the enzymes regulated by covalent modification are dephosphorylated and active in the well-fed state, they are phosphorylated and inactive in the fasted state.** Three exceptions are *glycogen phosphorylase* (see p. 231), *glycogen phosphorylase kinase* (see p. 231), and *HSL* (see p. 307), which are active in their phosphorylated states. In fasting, substrates are not provided by the diet but are available from the breakdown of stores and/or tissues, such as glycogenolysis with release of glucose from the liver, lipolysis with release of FA and glycerol from TAG in adipose tissue, and proteolysis with release of amino acids from muscle. Recognition that the changes in fasting are the reciprocal of those in the fed state is helpful in understanding the ebb and flow of metabolism.

VIII. LIVER IN FASTING

The primary role of the liver in fasting is maintenance of blood glucose through the **production of glucose** (from glycogenolysis and gluconeogenesis) for glucose-requiring tissues and the **synthesis and distribution of ketone bodies** for use by other tissues. Therefore, hepatic metabolism is distinguished from peripheral (or extrahepatic) metabolism.

A. Carbohydrate metabolism

The **liver first uses glycogen degradation and then gluconeogenesis to maintain blood glucose** levels to sustain energy metabolism of the brain and other glucose-requiring tissues in the fasted state. [Note: Recall that the presence of *glucose-6-phosphatase* in the liver allows the production of free glucose both from glycogenolysis and from gluconeogenesis (see Fig. 28.4).]

1. **Increased glycogenolysis:** Figure 28.11 shows the sources of blood glucose after ingestion of 100 g of glucose. During the brief absorptive period, ingested glucose is the major source of blood glucose. Several hours later, blood glucose levels have declined sufficiently to cause increased secretion of glucagon and decreased secretion of insulin. The increased glucagon/insulin ratio causes a rapid mobilization of liver glycogen stores (which contain ~80 g of glycogen in the fed state) because of *PKA*-mediated phosphorylation (and activation) of *glycogen phosphorylase kinase* that phosphorylates (and activates) *glycogen phosphorylase* (see p. 228). Figure 28.11 shows that because **liver glycogen is exhausted by 24 hours of fasting**, hepatic glycogenolysis is a transient response to early fasting. Figure 28.12, ❶,

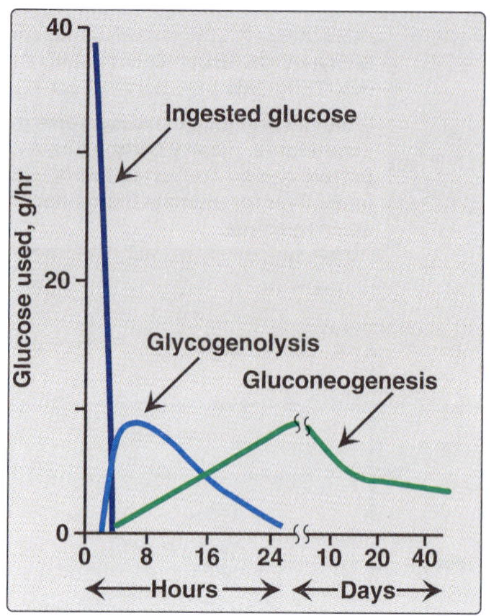

Figure 28.11 Sources of blood glucose after ingestion of 100 g of glucose. [Note: See Section B.2 for an explanation of the decline in gluconeogenesis.]

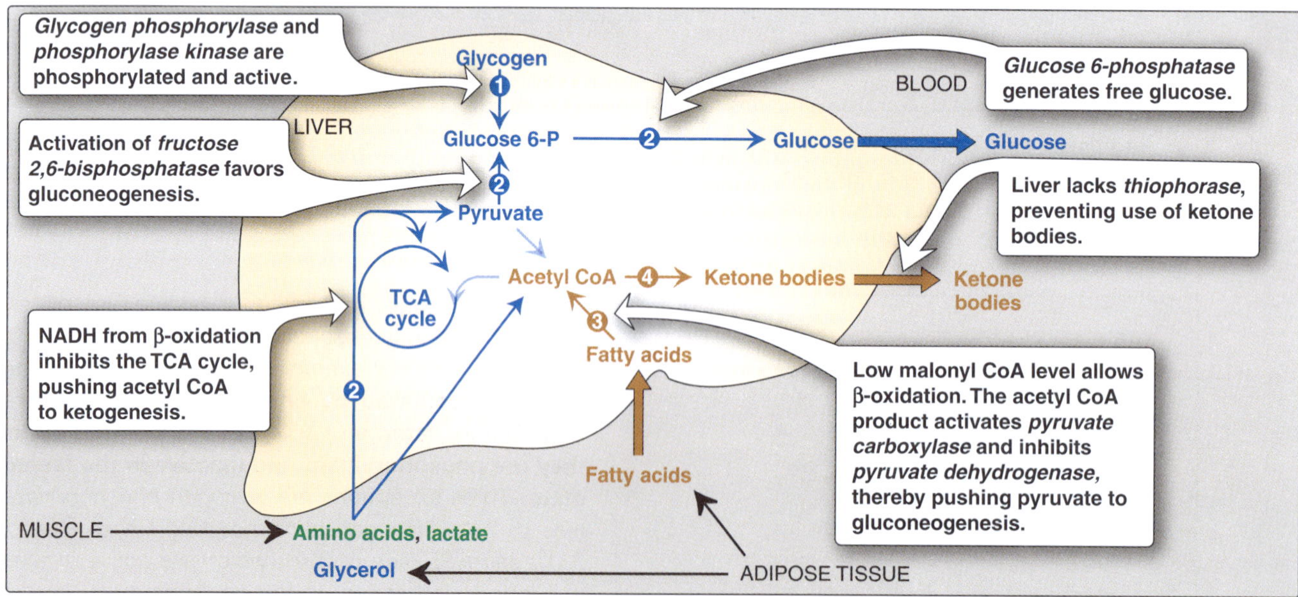

Figure 28.12 Major metabolic pathways in the liver during fasting. [Note: The numbers in circles, which appear both in the figure and in the corresponding citation in the text, indicate important metabolic pathways for carbohydrate or fat.] P = phosphate; CoA = coenzyme A; TCA = tricarboxylic acid; NADH = nicotinamide adenine dinucleotide.

shows glycogen degradation as part of the overall metabolic response of the liver during fasting. [Note: Phosphorylation of *glycogen synthase* simultaneously inhibits glycogenesis.]

2. **Increased gluconeogenesis:** The synthesis of glucose and its release into the circulation are vital hepatic functions during short- and long-term fasting (see Fig. 28.12, ❷). The **carbon skeletons for gluconeogenesis are derived primarily from glucogenic amino acids and lactate from muscle and glycerol from adipose tissue**. Gluconeogenesis, favored by activation of *fructose 1,6-bisphosphatase* (because of decreased availability of its inhibitor fructose 2,6-bisphosphate; see p. 217) and by induction of *PEPCK* by glucagon (see p. 221), begins 4–6 hours after the last meal and becomes fully active as stores of liver glycogen are depleted (see Fig. 28.11). [Note: The decrease in fructose 2,6-bisphosphate simultaneously inhibits glycolysis at *PFK-1* (see p. 195).]

B. Fat metabolism

1. **Increased fatty acid oxidation:** The oxidation of FA obtained from TAG hydrolysis in adipose tissue is the major source of energy in hepatic tissue in the fasted state (see Fig. 28.12, ❸). The fall in malonyl CoA because of phosphorylation (inactivation) of *ACC* by *AMPK* removes the brake on *CPT-I*, allowing β-oxidation to occur (see p. 309).

   FA oxidation generates NADH, flavin adenine dinucleotide ($FADH_2$), and acetyl CoA.

   - The NADH inhibits the TCA cycle and shifts OAA to malate. This results in acetyl CoA being available for ketogenesis.
   - The acetyl CoA is also an allosteric activator of *PC* and an allosteric inhibitor of *PDH*, thereby favoring use of pyruvate in gluconeogenesis (see Fig. 15.9, p. 220). [Note: **Acetyl CoA cannot be used as a substrate for gluconeogenesis**, in part because the *PDH* reaction is irreversible.]

   Oxidation of NADH and $FADH_2$ coupled with oxidative phosphorylation supplies the energy required by the *PC* and *PEPCK* reactions of gluconeogenesis.

2. **Increased ketogenesis:** The liver is unique in being able to synthesize and release ketone bodies, primarily 3-hydroxybutyrate but also acetoacetate, for **use as fuel by peripheral tissues but not by the liver itself because liver lacks *thiophorase*** (see p. 315). Ketogenesis, which starts during the first days of fasting (Fig. 28.13), is favored when

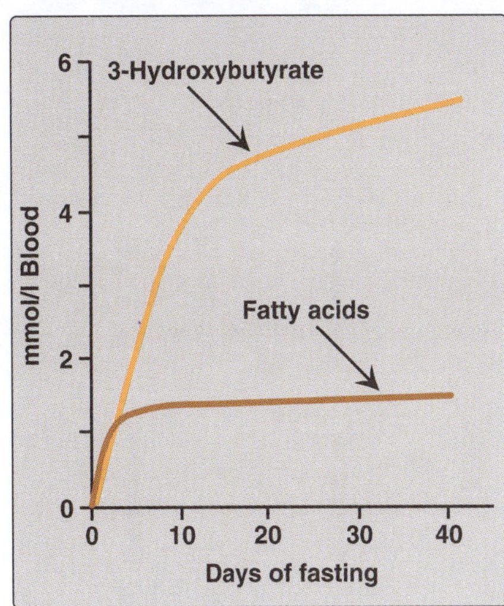

Figure 28.13 Concentrations of fatty acids and 3-hydroxybutyrate in the blood during fasting. [Note: 3-Hydroxybutyrate is made from the NADH-requiring reduction of acetoacetate.]

the concentration of acetyl CoA from FA oxidation exceeds the oxidative capacity of the TCA cycle. [Note: Ketogenesis releases CoA, insuring its availability for continued FA oxidation.] The availability of circulating **water-soluble ketone bodies is important in fasting because they can be used for fuel by most tissues**, including the brain, once their blood level is high enough. Ketone body concentration in blood increases from ~50 μM to ~6 mM in fasting. **This reduces the need for gluconeogenesis from amino acid carbon skeletons, thus preserving essential protein** (see Fig. 28.11). Ketogenesis as part of the overall hepatic response to fasting is shown in Figure 28.12, ❹. [Note: Ketone bodies are organic acids and, when present at high concentrations, can cause ketoacidosis.]

IX. ADIPOSE TISSUE IN FASTING

A. Carbohydrate metabolism

Glucose transport by insulin-sensitive GLUT-4 into the adipocyte and its subsequent metabolism are decreased because of low levels of circulating insulin. This results in decreased TAG synthesis.

B. Fat metabolism

1. **Increased fat degradation:** The *PKA*-mediated phosphorylation and activation of *HSL* (see p. 307) and subsequent hydrolysis of stored fat (TAG) are enhanced by the elevated catecholamines norepinephrine and epinephrine. These hormones, which

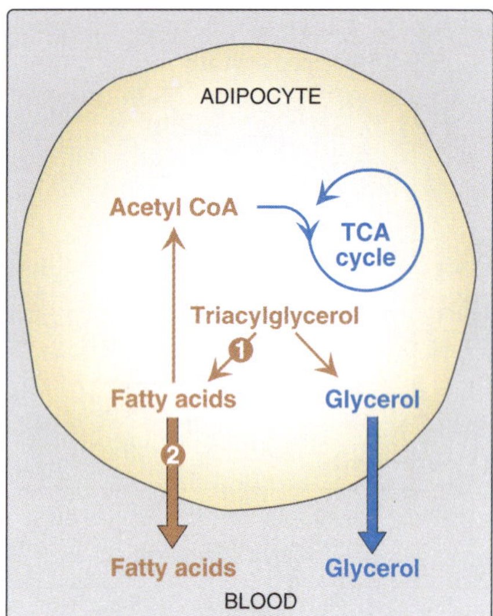

Figure 28.14 Major metabolic pathways in adipose tissue during fasting. [Note: The numbers in the circles, which appear both in the figure and in the corresponding citation in the text, indicate important pathways for fat metabolism.] CoA = coenzyme A; TCA = tricarboxylic acid.

are secreted from the sympathetic nerve endings in adipose tissue and/or from the adrenal medulla, are physiologically important activators of *HSL* (Fig. 28.14, ❶).

2. **Increased fatty acid release:** FA obtained from hydrolysis of TAG stored in adipocytes are primarily released into the blood (see Fig. 28.14, ❷). **Bound to albumin, they are transported to a variety of tissues for use as fuel.** The glycerol produced from TAG degradation is used as a gluconeogenic precursor by the liver, which contains *glycerol kinase*. [Note: FA can also be oxidized to acetyl CoA, which can enter the TCA cycle, thereby producing energy for the adipocyte. They also can be reesterified to glycerol 3-phosphate (from glyceroneogenesis, see p. 308), generating TAG and reducing plasma FA concentration.]

> Thiazolidinedione antihyperglycemic drugs function to increase transcription of PEPCK in adipose tissues, increasing glyceroneogenesis and TAG reesterification. The reduction of plasma FA concentration improves insulin sensitivity, as muscle and other tissue types become more dependent on glucose oxidation for energy.

3. **Decreased fatty acid uptake:** In fasting, *LPL* activity of adipose tissue is low. Consequently, FA in circulating TAG of lipoproteins are less available to adipose tissue than to muscle.

X. RESTING SKELETAL MUSCLE IN FASTING

Resting muscle **switches from glucose to FA** as its major fuel source in fasting.

By contrast, **exercising muscle initially uses creatine phosphate and its glycogen stores**. During intense exercise, glucose 6-phosphate from glycogenolysis is converted to lactate by anaerobic glycolysis (see p. 215). The lactate is used by the liver for gluconeogenesis (Cori cycle; see p. 215). **As these glycogen reserves are depleted, free FA provided by the degradation of TAG in adipose tissue become the dominant energy source**. The contraction-based rise in AMP activates *AMPK* that phosphorylates and inactivates the muscle isozyme of *ACC*, decreasing malonyl CoA and allowing FA oxidation (see p. 301).

A. Carbohydrate metabolism

Glucose transport into skeletal myocytes via insulin-sensitive GLUT-4 (see p. 192) and subsequent glucose metabolism are decreased because circulating insulin levels are low. Therefore, the glucose from hepatic gluconeogenesis is unavailable to muscle and adipose.

B. Lipid metabolism

Early in fasting, muscle uses FA from adipose tissue and ketone bodies from the liver as fuels (Fig. 28.15, ❶ and ❷). The utilization of ketone bodies by cardiac and skeletal muscles in the early stages of fasting conserves glucose for the central nervous system. **In prolonged fasting, muscle decreases its use of**

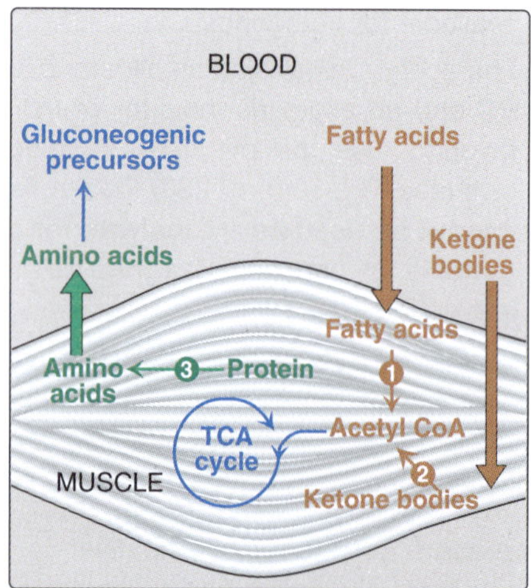

Figure 28.15 Major metabolic pathways in skeletal muscle during fasting. [Note: The numbers in the circles, which appear both in the figure and in the corresponding citation in the text, indicate important pathways for fat or protein metabolism.] CoA = coenzyme A; TCA = tricarboxylic acid.

ketone bodies (thus sparing them for the brain) and oxidizes FA almost exclusively.

The acetyl CoA from FA oxidation indirectly inhibits *PDH* (by activation of *PDH kinase*) and spares pyruvate, which is transaminated to alanine and used by the liver for gluconeogenesis (glucose–alanine cycle; see p. 386).

C. Protein metabolism

During the first few days of fasting, there is a **rapid breakdown of muscle protein** (for example, glycolytic enzymes), **providing amino acids that are used by the liver for gluconeogenesis** (see Fig. 28.15, ❸). Because muscle does not have glucagon receptors, muscle proteolysis is initiated by a fall in insulin and sustained by a rise in glucocorticoids. [Note: **Alanine and glutamine are quantitatively the most important** glucogenic amino acids released from muscle. They are produced by the catabolism of BCAA (see p. 402). The glutamine is used as a fuel by enterocytes, for example, which send out alanine that is used in hepatic gluconeogenesis (glucose–alanine cycle)]. **In the second week of fasting, the rate of muscle proteolysis decreases**, paralleling a decline in the need for glucose as a fuel for the brain, which has begun using ketone bodies as a source of energy.

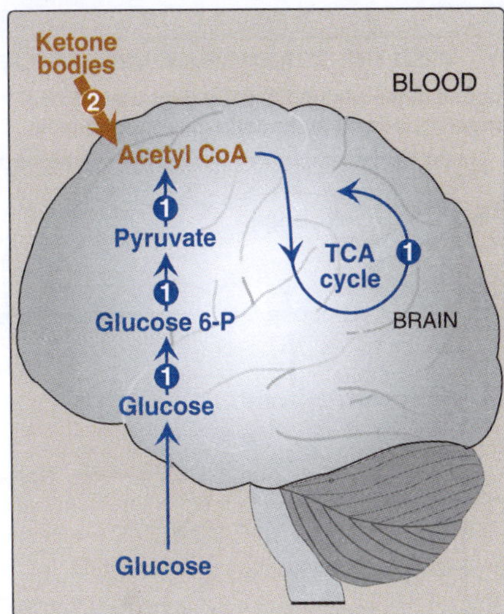

Figure 28.16 Major metabolic pathways in the brain during fasting. [Note: The numbers in the circles, which appear both in the figure and in the corresponding citation in the text, indicate important pathways for metabolism of fat or carbohydrates.] CoA = coenzyme A; TCA = tricarboxylic acid; P = phosphate.

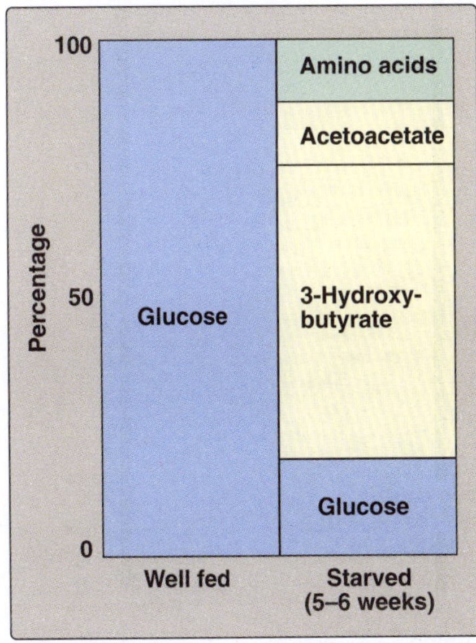

Figure 28.17 Fuel sources used by the brain to meet energy needs in the well-fed and starved states.

XI. BRAIN IN FASTING

During the early days of fasting, the brain continues to use only glucose as a fuel (Fig. 28.16, ❶). Blood glucose is maintained by hepatic gluconeogenesis from glucogenic precursors, such as amino acids from proteolysis and glycerol from lipolysis. **In prolonged fasting (beyond 2–3 weeks), plasma ketone bodies (see Fig. 28.12) reach significantly elevated levels and replace glucose as the primary fuel for the brain** (see Figs. 28.16, ❷ and 28.17). This reduces the need for protein catabolism for gluconeogenesis: Ketone bodies spare glucose and, thus, muscle protein.

As the duration of a fast extends from overnight to days to weeks, blood glucose levels initially drop and then are maintained at the lower level (65–70 mg/dl). The metabolic changes that occur during fasting insure that all tissues have an adequate supply of fuel molecules. The response of the major tissues involved in energy metabolism during fasting is summarized in Figure 28.18.

XII. KIDNEY IN LONG-TERM FASTING

As fasting continues into early starvation and beyond, the kidney plays important roles. The **renal cortex expresses the enzymes of gluconeogenesis**, including *glucose 6-phosphatase*, and, **in late fasting, ~50% of gluconeogenesis occurs here**. [Note: A portion of this glucose is used by the kidney itself.]

The **kidney also provides compensation for the acidosis that accompanies the increased production of ketone bodies** (organic acids). The glutamine released from the muscle's metabolism of BCAA is taken up by the kidney and acted upon by renal *glutaminase* and *glutamate dehydrogenase* (see p. 389), producing α-ketoglutarate, which can be used as a substrate for gluconeogenesis, plus ammonia (NH_3). The NH_3 picks up protons from ketone body dissociation and is excreted in

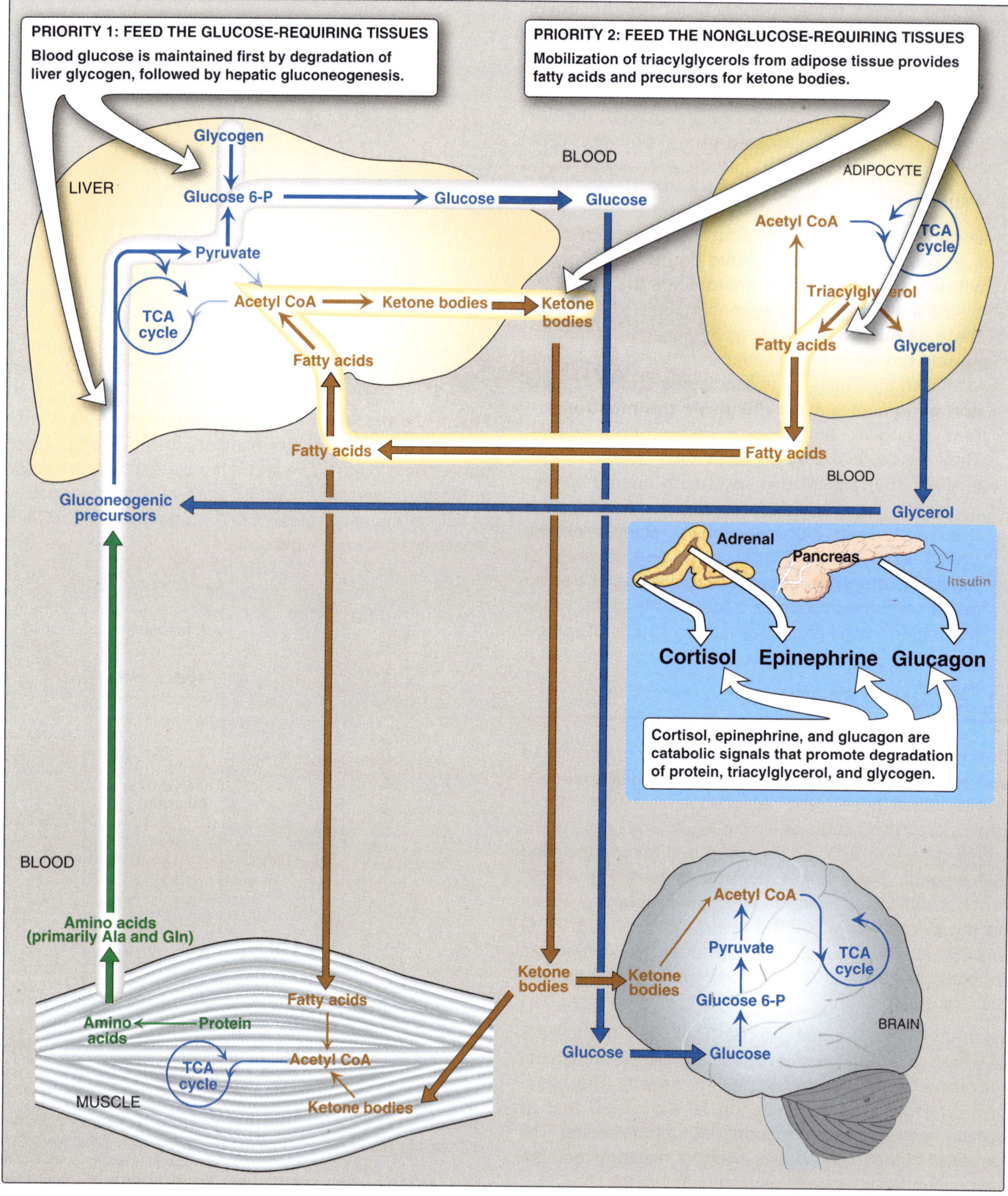

Figure 28.18 Intertissue relationships during fasting and the hormonal signals that promote them. P = phosphate; TCA = tricarboxylic acid; CoA = coenzyme A; Ala = alanine; Gln = glutamine.

the urine as ammonium (NH_4^+), thereby decreasing the acid load in the body (Fig. 28.19). Therefore, in long-term fasting, there is a switch from nitrogen disposal in the form of urea to disposal in the form of NH_4^+. [Note: As ketone body concentration rises, enterocytes, typically consumers of glutamine, become consumers of ketone bodies. This allows more glutamine to be available to the kidney.]

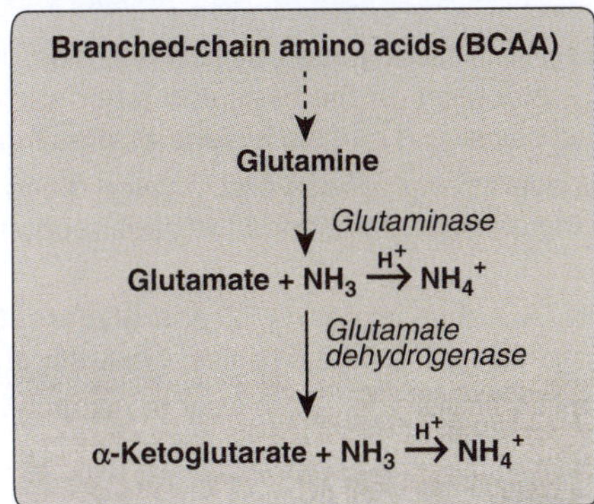

Figure 28.19 Use of glutamine from BCAA catabolism in muscle to generate ammonia (NH_3) used for the excretion of protons (H^+) as ammonium (NH_4^+) in the kidneys.

XIII. MEDICAL PERSPECTIVES

For estimation of serum lipoproteins, a fasting sample is preferred, as triglycerides (TGs) are elevated for several hours after a meal. Total cholesterol (TC), low-density lipoprotein cholesterol (LDL-C), TG and high-density lipoprotein cholesterol (HDL-C) can be measured directly by spectrophotometry. However, very low-density lipoprotein (VLDL) is usually calculated as TG/5. LDL may also be calculated using Friedewald equation.

Friedewald equation says that the TC = LDL-C + HDL-C + VLDL-C. Therefore, a raised TG will lead to an erroneously decreased LDL-C in calculations. LDL-C is an important target of lipid lowering therapy.

Intermittent fasting includes diet patterns with little or no energy intake for extended periods. It is expected to promote weight loss by utilizing fats for energy, after muscle glycogen stores have been depleted.

The branched-chain amino acids (BCAAs) leucine, valine and isoleucine are used by sportspersons as dietary supplements to produce an anabolic response for muscle protein synthesis. However, this belief is scientifically unsubstantiated as for synthesis of new muscle protein, all the essential amino acids along with non-essential amino acids must be present in adequate amounts. Increasing intake of just BCAAs, without a concurrent increased consumption of other essential amino acids doesn't provide any additional benefit.

XIV. CHAPTER SUMMARY

- The flow of intermediates through metabolic pathways is controlled by **four regulatory mechanisms**: 1) the availability of substrates, 2) allosteric activation and inhibition of enzymes, 3) covalent modification of enzymes, and 4) induction-repression of enzyme synthesis.
- In the **absorptive state**, the 2- to 4-hour period after ingestion of a meal, these mechanisms insure that available nutrients are captured as **glycogen**, **triacylglycerol (TAG)**, and **protein** (Fig. 28.20).
  - During this interval, transient increases in plasma glucose, amino acids, and TAG occur, the last primarily as components of **chylomicrons** synthesized by the intestinal mucosal cells.
  - The **pancreas** responds to the elevated levels of glucose with an increased secretion of insulin and a decreased secretion of glucagon.
  - The elevated **insulin/glucagon ratio** and the ready availability of circulating substrates make the absorptive state an **anabolic period** during which virtually all tissues use **glucose** as a fuel.
  - In addition, the **liver** replenishes its glycogen stores, replaces any needed hepatic proteins, and increases TAG synthesis. The latter are packaged in **very-low-density lipoproteins**, which are exported to the peripheral tissues.
  - **Adipose tissue** increases TAG synthesis and storage.
  - **Muscle** increases protein synthesis to replace protein degraded since the previous meal.
  - In the fed state, the **brain** uses glucose exclusively as a fuel.
- In **fasting**, plasma levels of glucose, amino acids, and TAG fall, triggering a decline in insulin secretion and an increase in glucagon and **epinephrine** secretion.
  - The decreased **insulin/counterregulatory hormone ratio** and the decreased availability of circulating substrates make the fasting state a **catabolic period**. This sets into motion an **exchange of substrates** among the liver,

adipose tissue, skeletal muscle, and brain that is guided by two priorities: 1) the need to maintain adequate plasma levels of glucose to sustain energy metabolism of the brain and other glucose-requiring tissues and 2) the need to mobilize fatty acids (FA) from adipose tissue and release **ketone bodies** from liver to supply energy to other tissues.

- To accomplish these goals, the liver degrades glycogen and initiates **gluconeogenesis**, using increased **FA oxidation** to supply the energy and reducing equivalents needed for gluconeogenesis and the acetyl coenzyme A building blocks for **ketogenesis**.
- The adipose tissue degrades stored TAG, thus providing FA and **glycerol** to the liver. The muscle can also use FA as fuel as well as ketone bodies supplied by the liver. The liver uses the glycerol for gluconeogenesis.
- **Muscle protein** is degraded to supply amino acids for the liver to use in gluconeogenesis but decreases as ketone bodies increase.
- The brain can use both glucose and ketone bodies as fuels.
- From late fasting into starvation, the **kidneys** play important roles by synthesizing glucose and excreting the **protons** from ketone body dissociation as **ammonium** (**NH_4^+**).

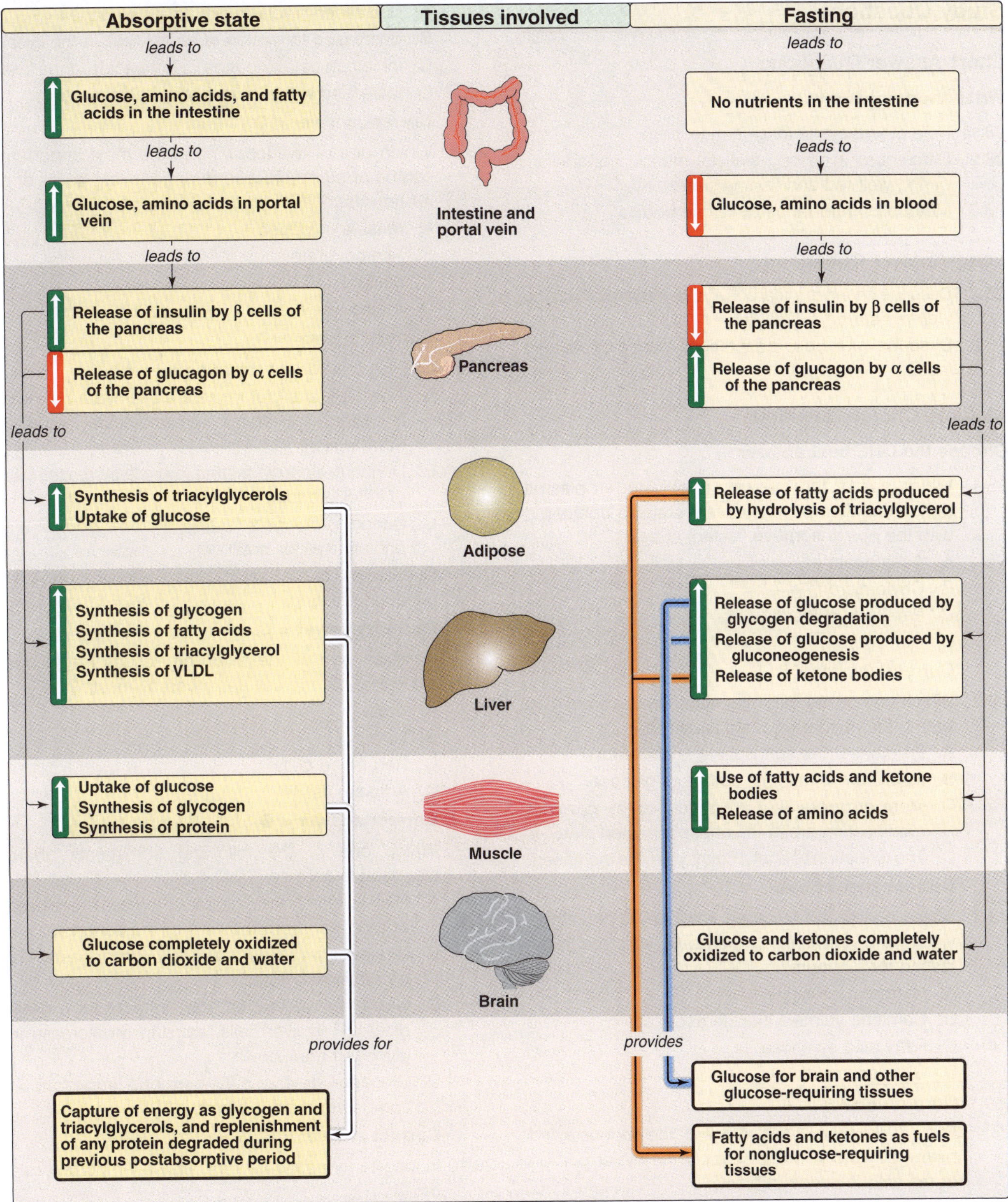

Figure 28.20 Key concept map for the feed–fast cycle. VLDL = very-low-density lipoprotein.

Study Questions

Short-Answer Questions

Write short notes on:

28.1. Role of kidneys in long-term fasting
28.2. Differences in resting skeletal muscle metabolism during well-fed and fasting states
28.3. Metabolic importance of ketone bodies

Long-Answer Questions

28.1. Describe changes in carbohydrate metabolism during well-fed state.
28.2. Describe metabolic adaptations during prolonged fasting.

Multiple-Choice Questions

Choose the ONE best answer.

28.1. Which one of the following is elevated in plasma during the absorptive (well-fed) state as compared with the postabsorptive (fasted) state?
A. Acetoacetate
B. Chylomicrons
C. Free fatty acids
D. Glucagon

Correct answer = B.

28.2. Which one of the following statements concerning liver in the absorptive state is correct?
A. Fructose 2,6-bisphosphate is elevated.
B. Insulin stimulates the uptake of glucose.
C. Most enzymes that are regulated by covalent modification are in the phosphorylated state.
D. The oxidation of acetyl coenzyme A is increased.

Correct answer = A.

28.3. Which one of the following enzymes is phosphorylated and active in an individual who has been fasting for 12 hours?
A. Hormone-sensitive lipase
B. Carnitine palmitoyltransferase-I
C. Fatty acid synthase
D. Glycogen synthase

Correct answer = A.

28.4. For a 70-kg man, in which one of the periods listed below do ketone bodies supply the major portion of the caloric needs of brain?
A. Absorptive period
B. Overnight fast
C. Three-day fast
D. Four-week fast

Correct answer = D.

28.5. Ketone bodies formation is increased during fasting because of:
A. decreased levels of circulating glucagon
B. decreased formation of acetyl CoA in the liver
C. inhibition of β-oxidation of fatty acids in the liver
D. increased levels of free fatty acids in serum

Correct answer = D.

28.6. Which one of the following is the most important source of blood glucose during the last hours of a 48-hour fast?
A. Muscle glycogen
B. Acetoacetate
C. Lactate
D. Amino acids

Correct answer = D.

28.7. Choose the correct statement:
A. Aspartate and glutamate are quantitatively most important glucogenic amino acids released from muscle.
B. During prolonged fasting, muscle increases use of ketone bodies.
C. During prolonged fasting, ketone bodies are primary fuel for brain.
D. Glucose is the major source of energy for liver during fasting.

Correct answer = C.

28.8. In which one of the following tissues is glucose transport into the cell enhanced by insulin?
A. Brain
B. Liver
C. Red blood cells
D. Adipose tissue

Correct answer = D.

28.9. Which one of the following statements about glucagon is correct?
A. High levels of blood glucose increase the release of glucagon from the cells of the pancreas.
B. Glucagon levels decrease following ingestion of a protein-rich meal.
C. Glucagon increases the intracellular levels of cAMP in liver cells, causing an increase in glycogen breakdown.
D. Glucagon is the only hormone important in combating hypoglycemia.

Correct answer = C.

28.10. In long-term fasting, kidneys play an important role by
A. decreasing the expression of glucose 6-phosphatase
B. providing compensation for the alkalosis
C. producing α-ketoglutarate that can be used as a substrate for gluconeogenesis
D. producing more of glutamine

Correct answer = C.

Multiple-Choice Questions (NExT/NEET PG)

Choose the ONE best answer.

28.1. The diagram below shows inputs to and outputs from pyruvate, a central molecule in energy metabolism.

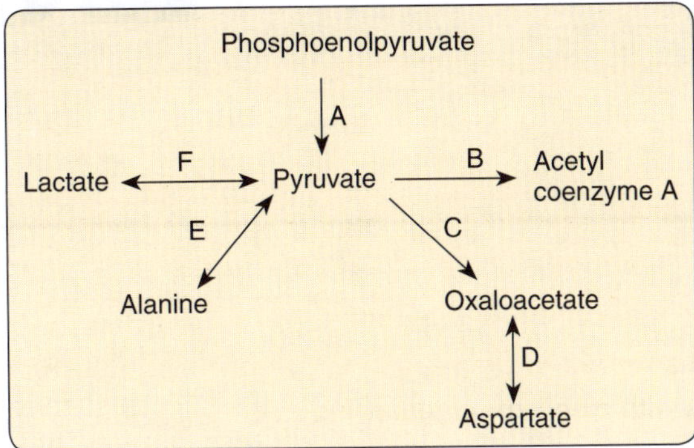

Which letter on the diagram represents a reaction that requires biotin and is activated by acetyl coenzyme A?

Correct answer = C.

28.2. A 39-year-old woman is brought to the emergency room complaining of dizziness. She recalls getting up early that morning to do as much shopping as possible and had skipped breakfast. She drank a cup of coffee for lunch and had nothing to eat during the day. She met with friends at 8 p.m. and had a drink at the bar. She soon became weak and dizzy and was transported to the hospital. Following examination, the patient was given orange juice and immediately felt better. The possible explanation in this case is

A. Blood glucose greater than 70 mg/dl
B. Elevated insulin
C. Elevated glucagon
D. Elevated liver glycogen.

Correct answer = C.

28.3 Which of the GLUT receptors has highest affinity for glucose?

A. GLUT 1
B. GLUT 2
C. GLUT 3
D. GLUT 4

Correct answer = C.

28.4. A 50-year-old man is brought to the emergency department after a motor vehicle accident. He is unconscious upon arrival, and his blood glucose level is found to be 30 mg/dL. Initial management includes administration of intravenous dextrose. Which hormone is responsible for the counter-regulatory response to hypoglycemia in this patient?

A. Insulin
B. Glucagon
C. Epinephrine
D. Growth hormone

Correct answer = C.

28.5. Gluconeogenesis decreases with fasting due to depletion of

A. pyruvate
B. oxaloacetate
C. acetyl CoA
D. alpha ketoglutarate

Correct answer = B.

Diabetes Mellitus

29

Learning Objectives

At the end of this chapter, the student will be able to:

- Describe the cause of type 1 and type 2 diabetes mellitus
- Describe the metabolic changes associated with diabetes mellitus
- Explain the acute and chronic complications of diabetes mellitus with biochemical reasoning

I. OVERVIEW

Diabetes mellitus (diabetes) is not one disease but rather is a heterogeneous group of multifactorial, primarily polygenic syndromes characterized by an **elevated fasting blood glucose (FBG) caused by a relative or absolute deficiency in insulin**. Over 100 million people in India have diabetes. Diabetes is the leading cause of adult blindness and amputation and a major cause of renal failure, nerve damage, heart attacks, and strokes. Most cases of diabetes mellitus can be separated into two groups (Fig. 29.1), type 1 ([T1D] formerly called insulin-dependent

Characteristics	Type 1 Diabetes	Type 2 Diabetes
AGE OF ONSET	Usually during childhood or puberty; symptoms develop rapidly	Frequently after age 35 years; symptoms develop gradually
NUTRITIONAL STATUS AT TIME OF DISEASE ONSET	Frequently undernourished	Obesity usually present
PREVALENCE	<10% of diagnosed diabetics	>90% of diagnosed diabetics
GENETIC PREDISPOSITION	Moderate	Very strong
DEFECT OR DEFICIENCY	β-Cell destruction, eliminating production of insulin	Insulin resistance combined with inability of β cells to produce appropriate quantities of insulin
FREQUENCY OF KETOSIS	Common	Rare
PLASMA INSULIN	Low to absent	High early in disease; low to absent in disease of long duration
ACUTE COMPLICATIONS	Ketoacidosis	Hyperosmolar hyperglycemic state
RESPONSE TO ORAL HYPOGLYCEMIC DRUGS	Unresponsive	Responsive
TREATMENT	Insulin always necessary	Diet, exercise, oral hypoglycemic drugs, insulin (may or may not be necessary); reduction of risk factors (weight reduction, smoking cessation, blood pressure control, treatment of dyslipidemia) essential to therapy

Figure 29.1 Comparison of type 1 and type 2 diabetes mellitus. [Note: The name of the disease reflects the clinical presentation of copious amounts of glucose-containing urine and is derived from the Greek word for siphon (diabetes) and the Latin word for honey-sweet (mellitus).]

diabetes mellitus) and type 2 ([T2D] formerly called non–insulin-dependent diabetes mellitus). The incidence and prevalence of T2D is increasing because of the aging of the population and the increasing prevalence of obesity and sedentary lifestyles (see p. 483).

II. TYPE 1

T1D is characterized by an **absolute deficiency of insulin** caused by an autoimmune attack on the islet β cells of the pancreas. In T1D, the islets of Langerhans become infiltrated with activated T lymphocytes, leading to a condition called insulitis. Over a period of years, this autoimmune attack on the β cells leads to gradual depletion of the β-cell population (Fig. 29.2). However, **symptoms appear abruptly when 80%–90% of the β cells have been destroyed**. At this point, the pancreas fails to respond adequately to ingestion of glucose, and insulin therapy is required to restore metabolic control and prevent life-threatening ketoacidosis. β-Cell destruction requires both a stimulus from the environment (such as a viral infection) and a genetic determinant that causes the β cells to be mistakenly identified as "nonself." [Note: Among monozygotic (identical) twins, if one sibling develops T1D, the other twin has only a 30%–50% chance of developing the disease. This contrasts with T2D (see p. 474), in which the genetic influence is stronger and, in virtually all monozygotic twinships, the disease eventually develops in both individuals.]

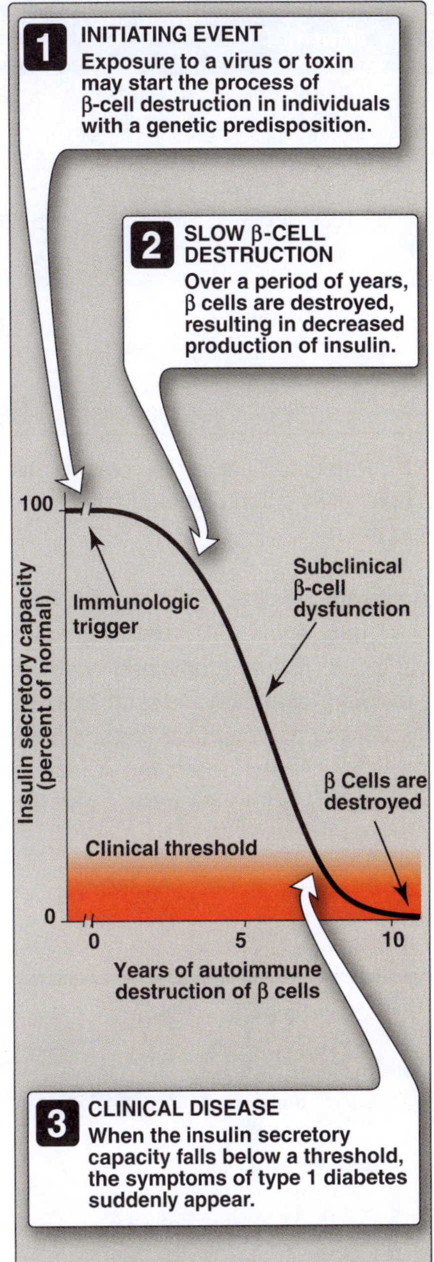

Figure 29.2 Insulin secretory capacity during onset of type 1 diabetes. [Note: Rate of autoimmune destruction of β cells may be faster or slower than shown.]

A. Diagnosis

The **onset of T1D is typically during childhood or puberty**, and symptoms develop suddenly. Individuals with T1D can usually be recognized by the abrupt appearance of **polyuria** (frequent urination), **polydipsia** (excessive thirst), and **polyphagia** (excessive hunger), often triggered by physiologic stress such as an infection. These symptoms are usually accompanied by fatigue and weight loss. The **diagnosis is confirmed by blood glucose or glycated hemoglobin levels**. Reference values for plasma glucose and HbA_{1c} and diagnostic criteria for prediabetes and diabetes mellitus Type 2 are given in Figure 29.3.

> When blood glucose is >180 mg/dl, the ability of renal sodium-dependent glucose transporters (SGLT) to reclaim glucose is impaired, and glucose "spills" into urine. The loss of glucose is accompanied by the loss of water, resulting in the characteristic polyuria (with dehydration) and polydipsia of diabetes.

B. Metabolic changes

The metabolic abnormalities of T1D result from a deficiency of insulin that profoundly affects metabolism in three tissues: liver, skeletal muscle, and white adipose (Fig. 29.4).

1. **Hyperglycemia and ketonemia:** Elevated levels of blood glucose and ketone bodies are the hallmarks of untreated T1D (see Fig. 29.4). 

   Hyperglycemia is caused by increased hepatic production of glucose via gluconeogenesis, combined with diminished peripheral utilization (muscle and adipose tissue have the insulin-regulated glucose transporter GLUT-4; see p. 192).

Test	Normal	Prediabetes	Diabetes Mellitus
Fasting plasma glucose (FPG)	70–99 mg/dl	100–125 mg/dl	≥ 126 mg/dl
Post-prandial glucose	100–140 mg/dl	140–199 mg/dl	≥ 200 mg/dl
Random plasma glucose	< 140 mg/dl	-	≥ 200 mg/dl
HbA_{1c}	< 5.7%	5.7%–6.4%	≥ 6.5%

Figure 29.3 Reference values for plasma glucose and HbA_{1c} and diagnostic criteria for prediabetes and Diabetes Mellitus Type 2.

Ketonemia results from increased mobilization of fatty acids (FA) from triacylglycerol (TAG) in adipose tissue, combined with accelerated hepatic FA β-oxidation and synthesis of 3-hydroxybutyrate and acetoacetate (ketone bodies; see p. 314). [Note: Acetyl coenzyme A from β-oxidation is the substrate for ketogenesis and the allosteric activator of *pyruvate carboxylase*, a gluconeogenic enzyme.]

Diabetic ketoacidosis (DKA), a type of metabolic acidosis caused by an imbalance between ketone body production and use, occurs in 25%–40% of those newly diagnosed with T1D and may recur if the patient becomes ill (most commonly with an infection) or does not comply with therapy. DKA is treated by replacing fluid and electrolytes and administering short-acting insulin to gradually correct hyperglycemia without precipitating hypoglycemia.

2. **Hypertriacylglycerolemia:** Not all of the FA flooding the liver can be disposed of through oxidation and ketone body synthesis. These excess FA are converted to TAG, which are packaged and secreted in very-low-density lipoproteins ([VLDL], see p. 352). Chylomicrons rich in dietary TAG are secreted by the intestinal mucosal cells following a meal (see p. 351). Because lipoprotein **TAG degradation catalyzed by *lipoprotein lipase* in the capillary beds of adipose tissue (see p. 352) is low in diabetes** (synthesis of the enzyme is decreased when insulin levels are

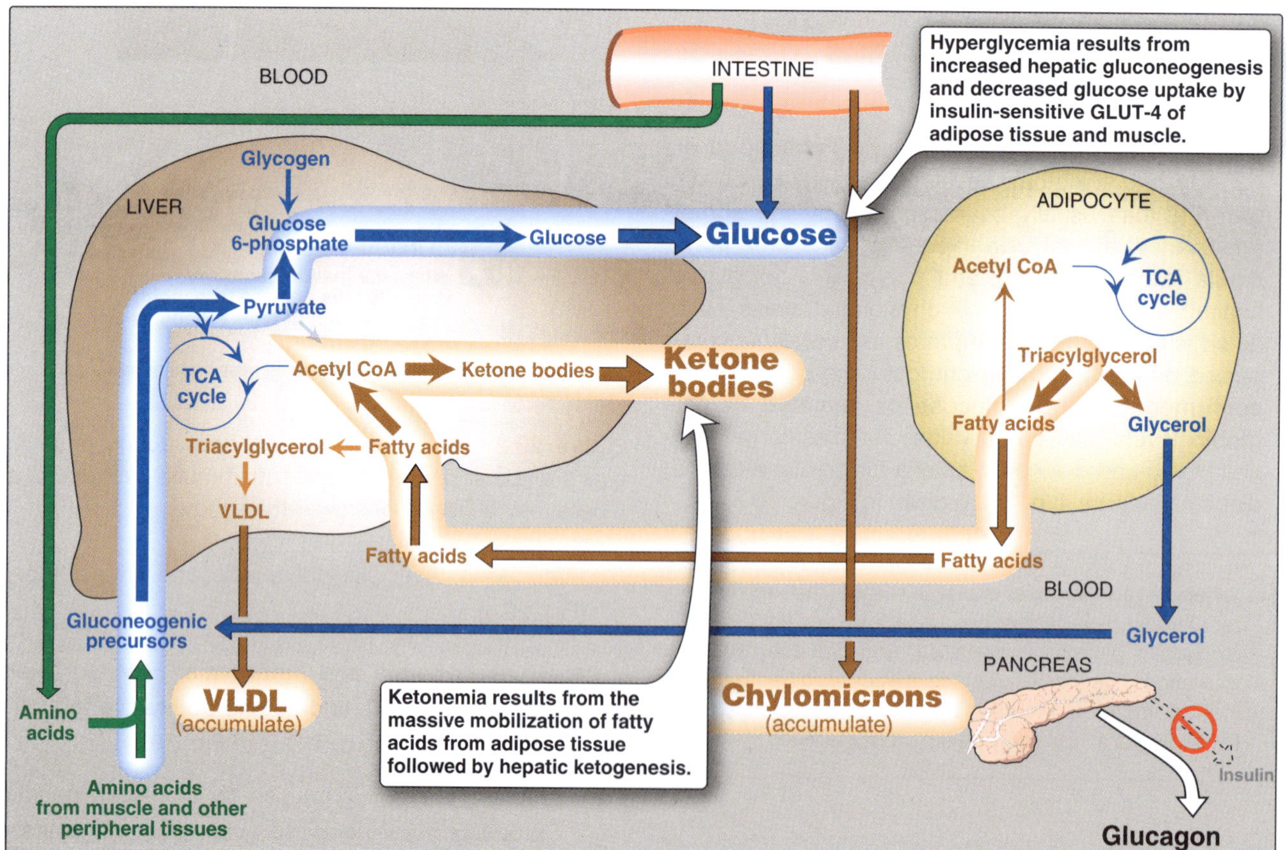

Figure 29.4 Intertissue relationships in type 1 diabetes. TCA = tricarboxylic acid; CoA = coenzyme A; VLDL = very-low-density lipoproteins; GLUT = glucose transporter.

low), the **plasma chylomicron and VLDL levels are elevated,** resulting in hypertriacylglycerolemia (see Fig. 29.4).

C. Treatment

Individuals with T1D must rely on **exogenous insulin** delivered subcutaneously (subq) either by periodic injection or by continuous pump-assisted infusion to control the hyperglycemia and ketonemia. Two types of therapeutic injection regimens are currently used, standard and intensive. [Note: Pump delivery is also considered intensive therapy.]

1. **Standard versus intensive treatment:** Standard treatment is typically two to three daily injections of recombinant human insulin. Mean blood glucose levels obtained are typically 225–275 mg/dl, with a glycated hemoglobin (HbA_{1c}) level (see p. 105) of 8%–9% of the total hemoglobin (blue arrow in Fig. 29.5). In contrast to standard therapy, intensive treatment seeks to more closely normalize blood glucose through more frequent monitoring and subsequent injections of insulin, typically ≥4 times a day. Mean blood glucose levels of 150 mg/dl can be achieved, with HbA_{1c} ~7% of the total hemoglobin (see red arrow in Fig. 29.5). [Note: Normal mean blood glucose is ~100 mg/dl, and HbA_{1c} is ≤ 5.7% (see black arrow in Fig. 29.5).] Therefore, normalization of glucose values (euglycemia) is not achieved even in intensively treated patients. Nonetheless, patients on intensive therapy show a ≥50% reduction in the long-term microvascular complications of diabetes (that is, retinopathy, nephropathy, and neuropathy) compared with patients receiving standard care. This confirms that the complications of diabetes are related to an elevation of plasma glucose.

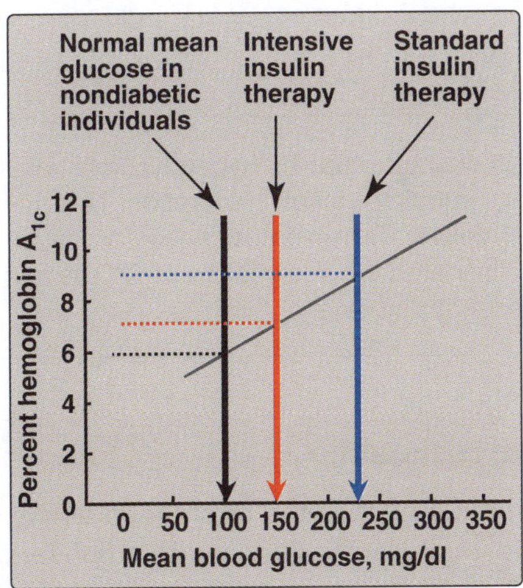

Figure 29.5 Correlation between mean blood glucose and percent hemoglobin A_{1c} in patients with type 1 diabetes receiving intensive or standard insulin therapy. [Note: Nondiabetic individuals are included for comparison.]

2. **Hypoglycemia:** One of the therapeutic goals in cases of diabetes is to decrease blood glucose levels in an effort to minimize the development of long-term complications of the disease (see p. 476 for a discussion of the chronic complications of diabetes). However, appropriate dosage of insulin is difficult to achieve. **Hypoglycemia caused by excess insulin is the most common complication of insulin therapy**, occurring in >90% of patients. The frequency of hypoglycemic episodes, seizures, and coma is particularly high with intensive treatment regimens designed to achieve tight control of blood glucose (Fig. 29.6). **In normal individuals, hypoglycemia triggers a compensatory secretion of counterregulatory hormones**, most notably glucagon and epinephrine, which promote hepatic production of glucose (see p. 445). **However, patients with T1D also develop a deficiency of glucagon secretion**. This defect occurs early in the disease and is almost universally present 4 years after diagnosis. Therefore, **these patients rely on epinephrine secretion to prevent severe hypoglycemia**. However, as the disease progresses, **T1D patients show diabetic autonomic neuropathy and impaired ability to secrete epinephrine in response to hypoglycemia**. The **combined deficiency of glucagon**

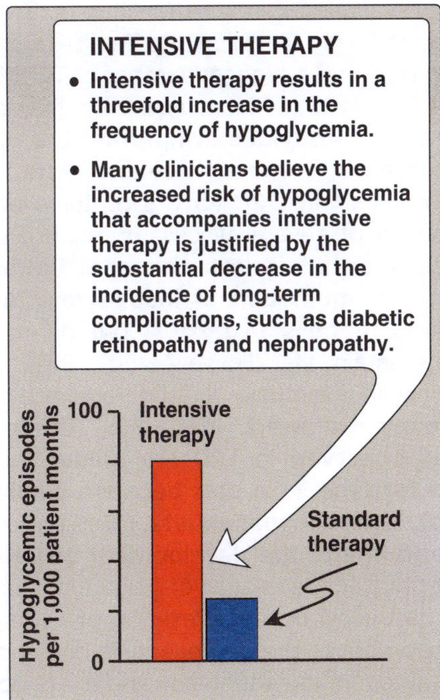

Figure 29.6 Effect of tight glucose control on hypoglycemic episodes in a population of patients on intensive therapy or standard therapy.

and epinephrine secretion creates a symptom-free condition sometimes called "hypoglycemia unawareness." Thus, patients with long-standing T1D are particularly vulnerable to hypoglycemia. Hypoglycemia can also be caused by strenuous exercise. Because exercise promotes glucose uptake into muscle and decreases the need for exogenous insulin, patients are advised to check blood glucose levels before or after intensive exercise to prevent or abort hypoglycemia.

3. **Contraindications for tight control:** Children are not put on a program of tight control of blood glucose before age 8 years because of the risk that episodes of hypoglycemia may adversely affect brain development. Elderly people typically do not go on tight control because hypoglycemia can cause strokes and heart attacks in this population. Also, the major goal of tight control is to prevent complications many years later. Tight control, then, is most worthwhile for otherwise healthy people who can expect to live at least 10 more years. [Note: For most nonpregnant adults with diabetes, the individual treatment strategies and goals are based on the duration of diabetes, age/life expectancy, and known comorbid conditions.]

III. TYPE 2

The burden of Type 2 Diabetes Mellitus (T2DM) in India is massive, and with 77 million people, the country ranks second for having the highest number of T2DM patients in the world, following China. **Typically, T2D develops gradually without obvious symptoms. The disease is often detected by routine screening tests**. However, many individuals with T2D have symptoms of polyuria and polydipsia of several weeks' duration. Polyphagia may be present but is less common. **Patients with T2D have a combination of insulin resistance and dysfunctional β cells** (Fig. 29.7) but do not require insulin to sustain life. However, in >90% of these patients, insulin eventually will be required to control hyperglycemia and keep HbA_{1c}<7%. **The metabolic alterations observed in T2D are milder than those described for type 1, in part because insulin secretion in T2D, although inadequate, does restrain ketogenesis and blunts the development of DKA.** [Note: Insulin suppresses the release of glucagon (see p. 444).] Diagnosis is based on the presence of hyperglycemia as described above. The pathogenesis does not involve viruses or autoimmune antibodies and is not completely understood. [Note: An acute complication of T2D in the elderly is a hyperosmolar hyperglycemic state characterized by severe hyperglycemia and dehydration and altered mental status.]

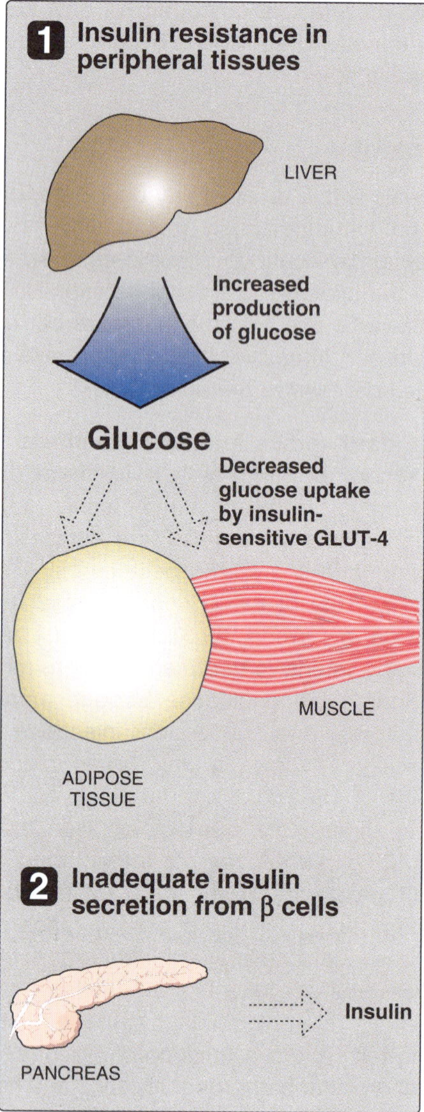

Figure 29.7 Major factors contributing to hyperglycemia observed in type 2 diabetes. GLUT = glucose transporter.

> T2D is characterized by hyperglycemia, insulin resistance, impaired insulin secretion, and, ultimately, β-cell failure. The eventual need for insulin therapy has eliminated the designation of T2D as non–insulin-dependent diabetes.

A. Insulin resistance

Insulin resistance is the decreased ability of target tissues, such as the liver, white adipose, and skeletal muscle, to respond properly to normal (or elevated) circulating concentrations of insulin. For example, insulin resistance is characterized by increased hepatic glucose production, decreased glucose uptake by muscle and adipose tissue, and

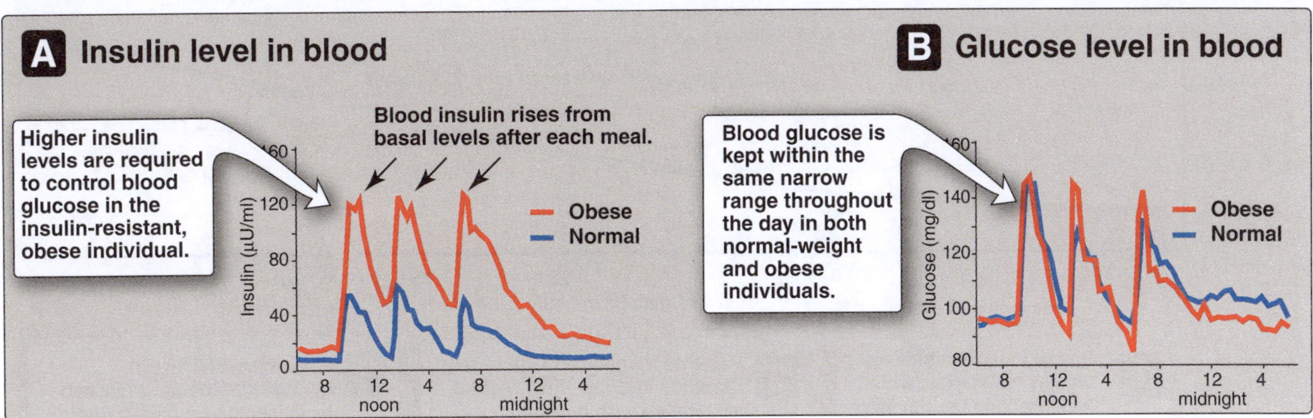

Figure 29.8 Blood insulin **(A)** and blood glucose levels **(B)** in normal-weight and obese subjects.

increased adipose lipolysis with production of free fatty acids (FFA).

1. **Insulin resistance and obesity:** Although obesity is the most common cause of insulin resistance and increases the risk of T2D, most people with obesity and insulin resistance do not develop diabetes. In the absence of a defect in β-cell function, obese individuals can compensate for insulin resistance with elevated levels of insulin. For example, Figure 29.8A shows that insulin secretion is two to three times higher in obese subjects than it is in lean individuals. This higher insulin concentration compensates for the diminished effect of the hormone (as a result of insulin resistance) and produces blood glucose levels similar to those observed in lean individuals (Fig. 29.8B).

2. **Insulin resistance and type 2 diabetes:** Insulin resistance alone will not lead to T2D. Rather, T2D develops in insulin-resistant individuals who also show impaired β-cell function. Insulin resistance and subsequent risk for the development of **T2D is commonly observed in individuals who are obese, physically inactive, or elderly** and in the 3%–5% of pregnant women who develop **gestational diabetes**. These patients are unable to sufficiently compensate for insulin resistance with increased insulin release. Figure 29.9 shows the time course for the development of hyperglycemia and the loss of β-cell function.

3. **Causes of insulin resistance:** Insulin resistance increases with weight gain and decreases with weight loss, and excess adipose tissue (particularly in the abdomen, known as central obesity) is key in the development of insulin resistance. Adipose is not simply an energy storage tissue, but also a secretory tissue. **With obesity, there are changes in adipose secretions that result in insulin resistance** (Fig. 29.10). These include secretion of **proinflammatory cytokines such as interleukin 6 and tumor necrosis factor-α** by activated macrophages (inflammation is associated with insulin resistance); **increased synthesis of leptin**, a protein with proinflammatory effects (see p. 486 for additional effects of leptin); and **decreased secretion of adiponectin** (see p. 483), a protein with anti-inflammatory effects. The net result is chronic, low-grade inflammation. **One effect of insulin resistance is increased lipolysis and production of FFA** (see Fig. 29.10). FFA availability decreases use of glucose, contributing to hyperglycemia, and increases ectopic deposition of TAG in liver (**hepatic steatosis**).

Steatosis results in nonalcoholic fatty liver disease (NAFLD). If accompanied by inflammation, a more serious condition, nonalcoholic steatohepatitis (NASH), can develop.] FFA also have a proinflammatory effect. In the long term, FFA impair insulin signaling. [Note: Adiponectin increases FA β-oxidation (see p. 485). Consequently, a decrease in this adipocyte protein contributes to FFA availability.

B. Dysfunctional β cells

In T2D, the pancreas initially retains β-cell capacity, resulting in insulin levels that vary from above normal to below normal. **However, with time, the β cell becomes increasingly dysfunctional and fails to secrete enough insulin** to correct the prevailing hyperglycemia. For example, insulin levels are high in typical, obese, T2D patients but not as high as in similarly obese individuals who do not have diabetes. Thus, the natural progression of the disease results in a declining ability to control hyperglycemia with endogenous secretion of insulin (Fig. 29.11). Deterioration of β-cell function may be accelerated by the toxic effects of sustained hyperglycemia and elevated FFA and a proinflammatory environment.

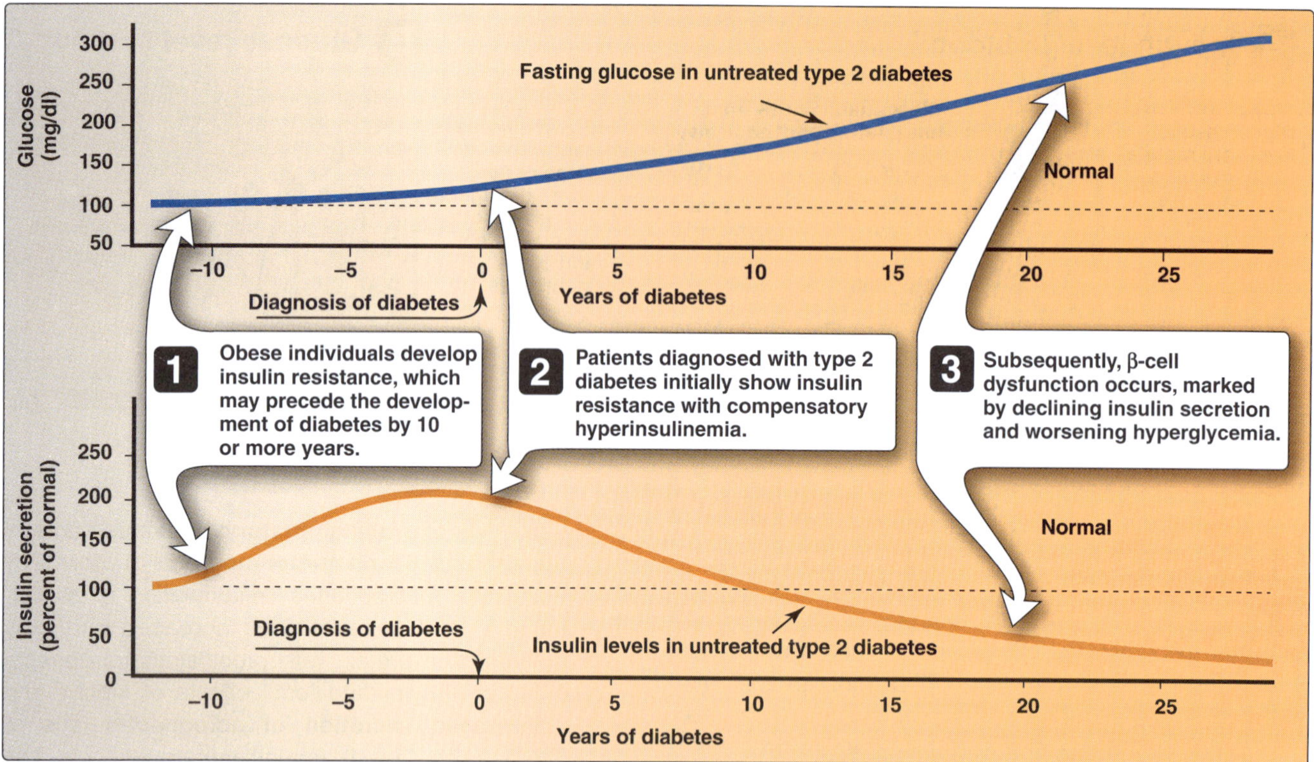

Figure 29.9 Progression of blood glucose and insulin levels in patients with type 2 diabetes.

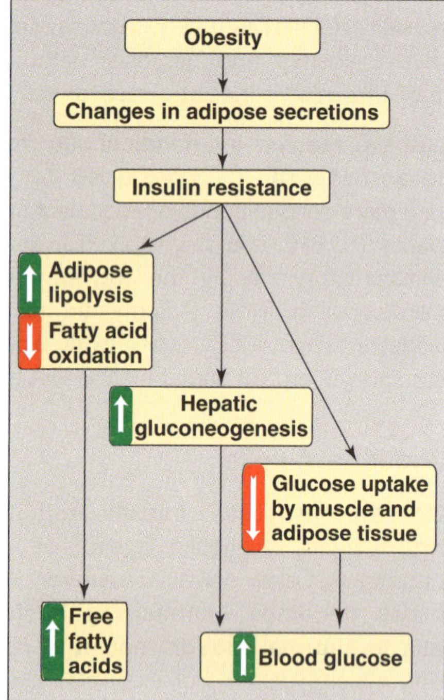

Figure 29.10 Obesity, insulin resistance and hyperglycemia. [Note: Inflammation also is associated with insulin resistance.]

C. Metabolic changes

The abnormalities of glucose and TAG metabolism in T2D are the result of insulin resistance expressed primarily in liver, skeletal muscle, and white adipose tissue (Fig. 29.12).

1. **Hyperglycemia:** Hyperglycemia is caused by increased hepatic production of glucose, combined with diminished use of glucose by muscle and adipose tissues. **Ketonemia is usually minimal or absent in patients with T2D because the presence of insulin**, even in the presence of insulin resistance, restrains hepatic ketogenesis.

2. **Dyslipidemia:** In the liver, FFA are converted to TAG, which are packaged and secreted in VLDL. Dietary TAG–rich chylomicrons are synthesized and secreted by the intestinal mucosal cells following a meal. Because lipoprotein TAG degradation catalyzed by *lipoprotein lipase* in adipose tissue is low in diabetes, the plasma chylomicron and VLDL levels are elevated, resulting in **hypertriacylglycerolemia** (see Fig. 29.11). Low levels of high-density lipoproteins are also associated with T2D, likely as a result of increased degradation.

D. Treatment

Weight reduction, exercise, and medical nutrition therapy (dietary modifications) often correct the hyperglycemia of newly diagnosed T2D.

Oral hypoglycemic agents, or subq insulin therapy may be required to achieve satisfactory plasma

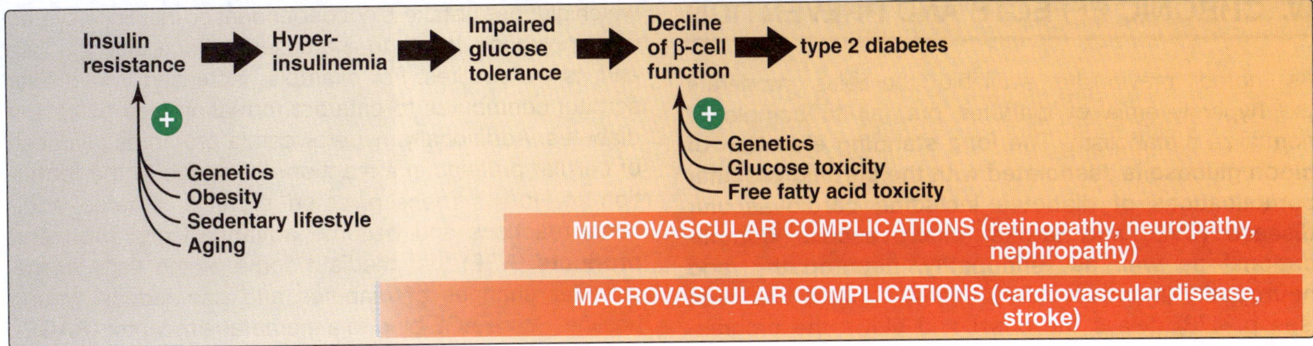

Figure 29.11 Typical progression of type 2 diabetes.

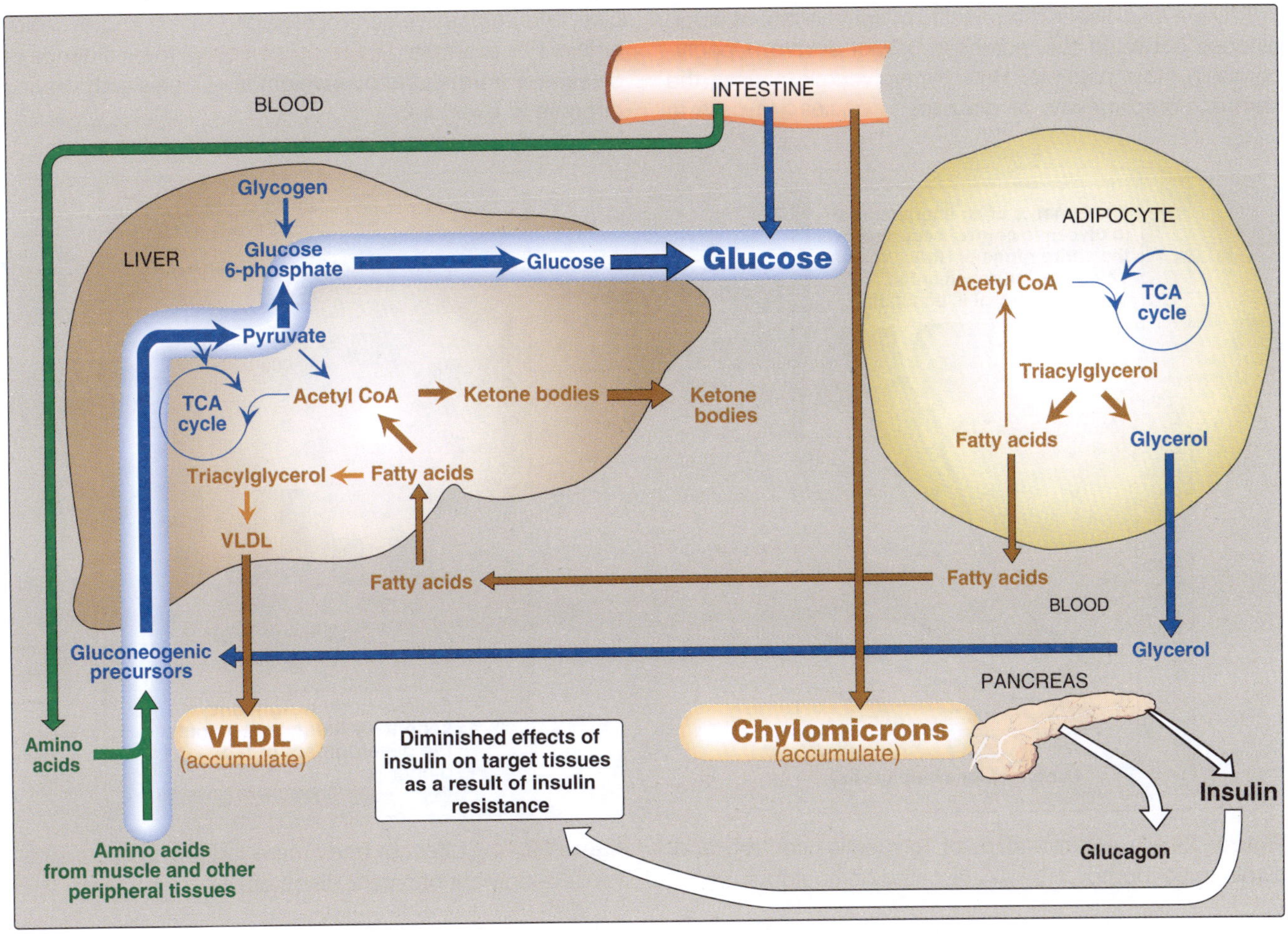

Figure 29.12 Intertissue relationships in type 2 diabetes. [Note: Ketogenesis is restrained as long as insulin action is adequate.] TCA = tricarboxylic acid; CoA = coenzyme A; VLDL = very-low-density lipoproteins.

glucose levels. [Note: Bariatric surgery in morbidly obese individuals with T2D has been shown to result in disease remission in most patients. Remission may not be permanent.]

> Hyperglycemic Hyperosmolar Nonketotic Coma (HONK) is a complication of diabetes mellitus, typically observed in individuals with type 2 diabetes. Patients exhibit profound hyperglycemia and glycosuria without ketosis due to adequate insulin production by the pancreas, which effectively inhibits ketogenesis.

IV. CHRONIC EFFECTS AND PREVENTION

As noted previously, available therapies moderate the hyperglycemia of diabetes but fail to completely normalize metabolism. **The long-standing elevation of blood glucose is associated with the chronic vascular complications of diabetes including cardiovascular disease (CVD) and stroke (macrovascular complications) as well as retinopathy, nephropathy, and neuropathy (microvascular).** Intensive insulin treatment (see p. 473) delays the onset and slows the progression of some long-term complications. For example, the incidence of retinopathy decreases as control of blood glucose improves and HbA_{1c} levels decrease (Fig. 29.13). [Note: Data concerning the effect of tight control on CVD in T2D are less clear.] The benefits of tight control of blood glucose outweigh the increased risk of severe hypoglycemia in most patients. How hyperglycemia causes the chronic complications of diabetes is unclear. In cells in which glucose uptake is not dependent on insulin, elevated blood glucose leads to increased intracellular glucose and its metabolites. For example, increased intracellular sorbitol contributes to cataract formation (see p. 241) in diabetes. Additionally, hyperglycemia promotes glycation of cellular proteins in a reaction analogous to the formation of HbA_{1c}. These glycated proteins undergo additional reactions and become **advanced glycation end products** (AGE) that mediate some of the early microvascular changes of diabetes and can reduce wound healing. Some AGE bind to a membrane receptor (**RAGE**), causing the release of proinflammatory molecules. There is currently no preventative treatment for T1D. The risk for T2D can be significantly decreased by a combined regimen of medical nutrition therapy, weight loss, exercise, and aggressive control of hypertension and dyslipidemias. For example, Figure 29.14 shows the incidence of disease in normal and overweight individuals with varying degrees of exercise.

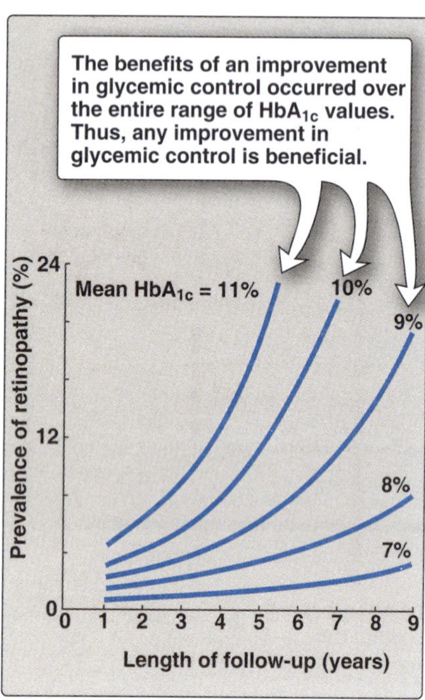

Figure 29.13 Relationship of glycemic control and diabetic retinopathy. HbA_{1c} = glycated hemoglobin.

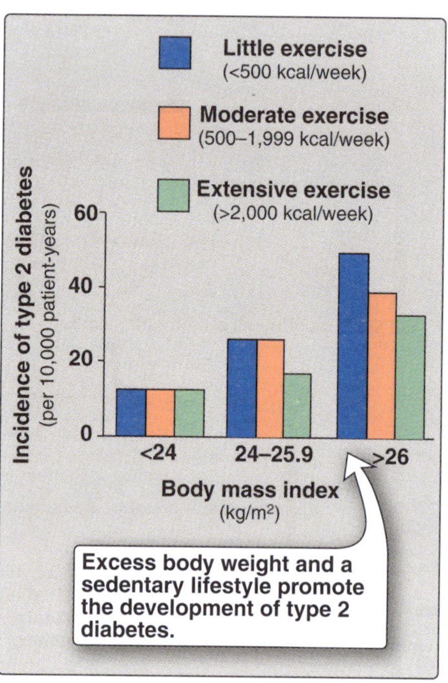

Figure 29.14 Effect of body mass index and exercise on the development of type 2 diabetes.

V. Medical Perspectives

Diabetes mellitus (DM) is a major global health problem. The global prevalence of diabetes among adults is nearly 8.5%, with DM being a leading cause of death. Asians are at higher risk to develop insulin resistance, which may be linked to central obesity, genetic factors, dietary and lifestyle patterns. Diabetes is a major cause of blindness, kidney failure, heart attacks, stroke and lower limb amputation. Besides the micro- and macrovascular complications, diabetic patients are also more prone to infections, especially of skin and urinary tract.

Ketosis may be seen in starvation and diabetes. Salient features of ketosis include Kussmaul respiration and smell of acetone in patient's breath. Excessive ketone bodies in blood can cause metabolic acidosis, electrolyte imbalance and dehydration. Ketone bodies in urine can be diagnosed by urine dipstick test and Rothera's test. Normal level for ketones in blood is <1 mg/dl. The urine ketone dipstick test is a simple, semi-quantitative method for detection (based on nitroprusside reaction) of acetoacetate. It can even be done by patient himself. It does not detect β-hydroxybutyrate.

Diabetic patients, due to neuropathy, may not manifest symptoms of myocardial infarction (MI), a condition known as **silent MI**. Lack of pain in silent MI increases morbidity and mortality, since patients do not seek medical treatment in time.

Homeostatic Model Assessment of Insulin Resistance (HOMA-IR) is a method for assessing β-cell function and insulin resistance (IR) from fasting glucose and insulin or C-peptide concentrations. The relationship between glucose and insulin in the fasting (basal) state reflects the balance between hepatic glucose output and insulin secretion.

Diabetes can be treated and its consequences avoided or delayed with lifestyle modifications (diet, physical activity), medication and regular screening and treatment for complications. Acute complications of diabetics are mainly ketoacidosis and hypoglycemia (because of treatment overdose or lack of food after insulin medication).

Long-term diabetics should be routinely assessed for retinopathy, cataract, peripheral neuropathy, nephropathy and macrovascular complications like cardiovascular risk.

Blood glucose values can be expressed in mg/dl or mmol/l. Values in mmol/l can be multiplied with 18 to get the value in mg/dl.

World Diabetes Day is observed on 14 November every year.

VI. Chapter Summary

- **Diabetes mellitus** is a heterogeneous group of syndromes characterized by an **elevation** of **fasting blood glucose** that is caused by a relative or absolute deficiency of insulin (Fig. 29.15).
- Diabetes is the leading cause of **adult blindness** and **amputation** and a major cause of **renal failure**, **nerve damage**, **heart attacks**, and **stroke**.
- Diabetes can be classified into two groups, **type 1** (**T1D**) and **type 2** (**T2D**).
- T1D constitutes about 10% of total diabetes cases in India. The disease is characterized by an **absolute deficiency** of **insulin** caused by an **autoimmune attack** on the **pancreatic β cells**. This destruction requires an **environmental stimulus** (such as a viral infection) and a **genetic determinant** that causes the β cell to be mistakenly identified as "nonself."
- The **metabolic abnormalities** of T1D include **hyperglycemia**, **diabetic ketoacidosis** (**DKA**), and **hypertriacylglycerolemia** that result from a deficiency of insulin. Those with T1D must rely on **exogenous insulin** delivered subcutaneously to control hyperglycemia and ketoacidosis.
- **T2D** has a strong **genetic** component. It results from a combination of **insulin resistance** and **dysfunctional β cells**.
- Insulin resistance is the decreased ability of target tissues, such as liver, white adipose, and skeletal muscle, to respond properly to normal (or elevated) circulating concentrations of insulin.
- **Obesity** is the most common cause of insulin resistance. However, most people with obesity and insulin resistance do not develop diabetes. In the absence of a defect in β-cell function, obese individuals without diabetes can compensate for insulin resistance with elevated levels of insulin. Insulin resistance alone will not lead to T2D. Rather, T2D develops in insulin-resistant individuals who also show impaired β-cell function.
- The acute **metabolic alterations** observed in T2D are **milder** than those described for the insulin-dependent form of the disease, in part because insulin secretion in T2D, although inadequate, does restrain **ketogenesis** and blunts the development of DKA.
- Available treatments for diabetes moderate the hyperglycemia but fail to completely normalize metabolism.
- The long-standing elevation of blood glucose is associated with the **chronic complications** of diabetes including **cardiovascular disease** and **stroke** (**macrovascular**) as well as **retinopathy**, **nephropathy**, and **neuropathy** (**microvascular**).

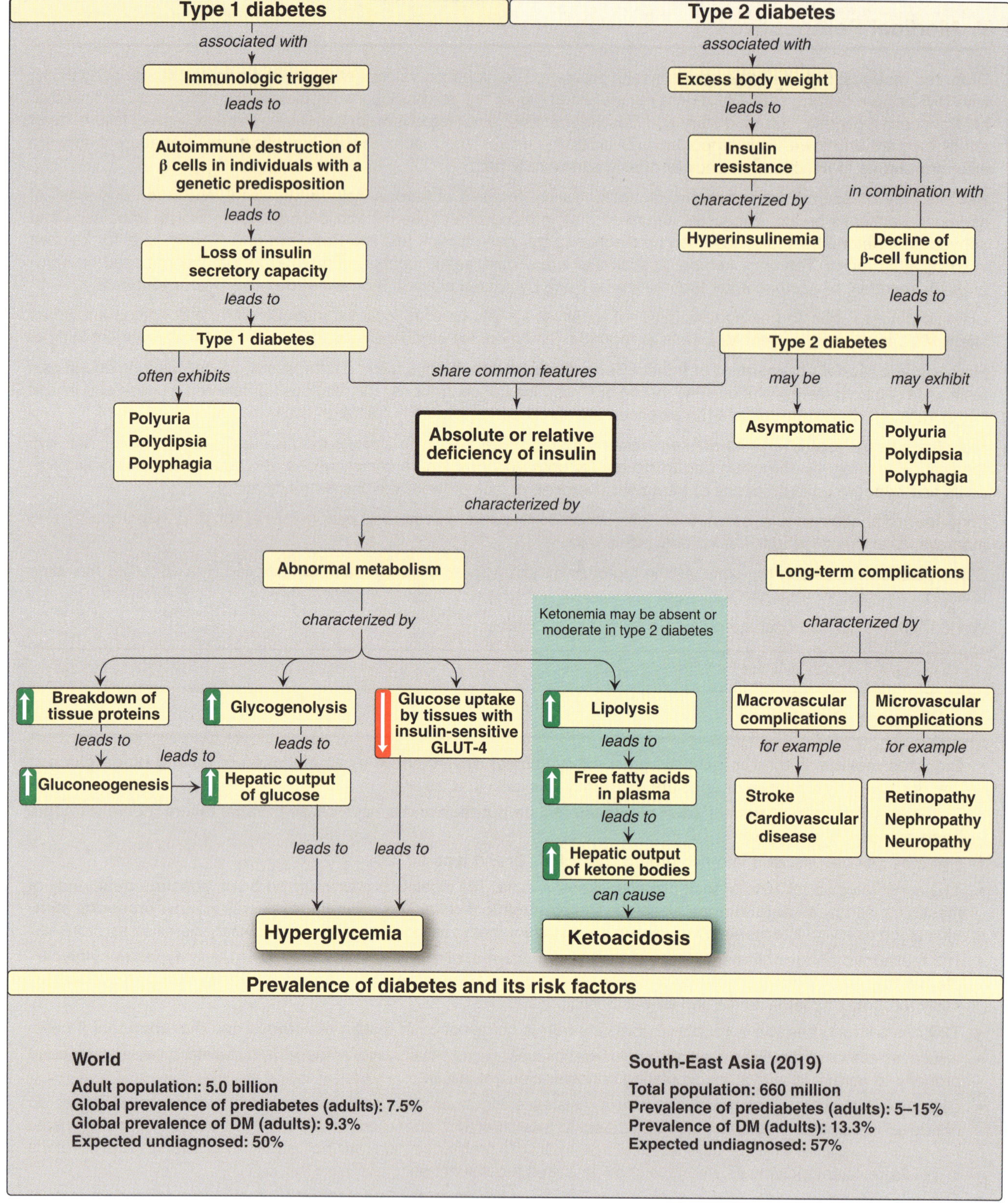

Figure 29.15 Key concept map for diabetes. [Note: Data are from 2019.] GLUT = glucose transporter.

Study Questions

Short-Answer Questions

Write short notes on:

29.1. Important differences between type 1 and type 2 diabetes mellitus
29.2. Insulin resistance
29.3. Biochemical basis of dyslipidemia in diabetes mellitus

Long-Answer Question

29.1. Describe the acute and long-term complications of diabetes mellitus.

Multiple-Choice Questions

Choose the ONE best answer.

29.1. Three patients being evaluated for gestational diabetes are given an oral glucose tolerance test. Based on the data shown below, which patient is prediabetic?

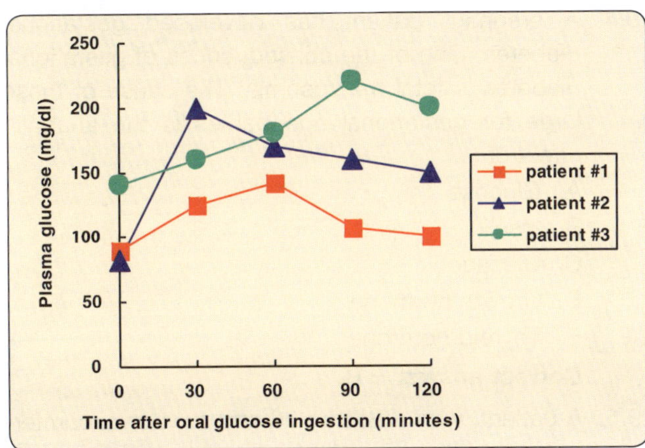

A. Patient #1
B. Patient #2
C. Patient #3
D. None

Correct answer = B.

29.2. Relative or absolute lack of insulin in humans would result in which one of the following reactions in the liver?

A. Decreased activity of hormone-sensitive lipase
B. Decreased gluconeogenesis from lactate
C. Decreased glycogenolysis
D. Increased formation of 3-hydroxybutyrate

Correct answer = D.

29.3. Which one of the following is characteristic of untreated diabetes regardless of the type?

A. Hyperglycemia
B. Ketoacidosis
C. Low levels of hemoglobin A_{1c}
D. Normal levels of C-peptide

Correct answer = A.

29.4. An obese individual with type 2 diabetes typically:

A. benefits from receiving insulin about 6 hours after a meal.
B. has a lower plasma level of glucagon than does a normal individual.
C. has a lower plasma level of insulin than does a normal individual early in the disease process.
D. shows improvement in glucose tolerance if body weight is reduced.

Correct answer = D.

29.5. What is the hallmark of untreated type 1 diabetes mellitus?

A. Insulin levels are high initially and fall later in the disease
B. Hyperglycemia and hyperosmolar state
C. Hyperglycemia and ketoacidosis
D. Hypertriacylglycerolemia and obesity

Correct answer = C.

29.6. Which of the following can help to distinguish between type 1 and type 2 diabetes mellitus?

A. Oral glucose tolerance test
B. Circulating islet-cell antibodies
C. HbA1c levels
D. Hyperglycemia

Correct answer = B.

29.7. Which of the following is an advantage of intensive treatment over standard treatment of Type 1 diabetes?

A. Decrease in the frequency of hypoglycemic episodes, coma, and seizures
B. Decrease in the incidence of long-term complications, such as diabetic retinopathy and nephropathy
C. Insulin injections are given only once or twice daily
D. Mean blood glucose levels are not close to that in nondiabetics

Correct answer = B.

29.8. Which of the following is the major defining difference between a type 1 diabetic and a type 2 diabetic?

A. Weight
B. LDL levels
C. Blood glucose levels
D. Ability to produce insulin

Correct answer = D.

29.9. Type 1 diabetics, prior to diagnosis, display polydipsia, polyuria, and polyphagia. Polyuria is due to which of the following?

A. Insulin stimulation of urea production
B. Osmotic imbalance due to elevated ketones in the blood
C. Osmotic imbalance due to reduced glucose levels in the urine
D. Osmotic imbalance due to increased glucose levels in the urine

Correct answer = D.

29.10. The frequency of hypoglycemic episodes, coma, and seizures is particularly high with intensive treatment regimens used in the management of type 1 diabetes because of
A. deficiency of glucagon and epinephrine secretion
B. increased insulin secretion
C. increased glucagon and epinephrine secretion
D. decreased exercise and sedentary lifestyle

Correct answer = A.

Multiple-Choice Questions (NExT/NEET PG)

Choose the ONE best answer.

29.1. A 45-year-old man presents to his primary care physician for a routine health checkup. He has a history of obesity and a family history of type 2 diabetes mellitus. His fasting blood glucose level is measured at 110 mg/dL, and his fasting insulin level is 20 mU/L.
Based on the provided information, what can be inferred about the patient's insulin resistance status using Homeostatic Model Assessment of Insulin Resistance (HOMA-IR)?
A. The patient is likely insulin sensitive.
B. The patient is likely insulin resistant.
C. The patient's insulin resistance status cannot be determined from the provided data.
D. The patient is likely in a state of hyperinsulinemia.

Correct answer = B.

29.2. A 56-year-old female patient visits the hospital for follow-up after discharge. She was treated for ketoacidosis and hyperglycemia and now is on basal and rapid acting insulins. To differentiate whether she has type 1 diabetes mellitus and was in ketoacidosis or has type 2 diabetes mellitus and had a hyperosmolar state with lactic acidosis, which of the following lab tests should be done?
A. Insulin levels
B. C-peptide levels
C. Fasting blood glucose levels
D. Random blood glucose levels
E. HbA1c levels

Correct answer = B.

29.3. Sequelae of insulin resistance in type 2 diabetes mellitus and metabolic syndrome is reduced secretion of insulin in response to increases in blood glucose. Insulin release from the pancreas appears to be dependent upon increase in the concentration of which pair of metabolites?
A. ATP and CO_2
B. ATP and NADH + H^+
C. ATP and NADPH + H^+
D. Glucose-6-phosphate and CO_2
E. Glucose-6-phosphate and NADH + H^+

Correct answer = C.

29.4. A pregnant patient has developed gestational diabetes. One of the consequences of gestational diabetes is fetal macrosomia. The cause of these large for gestational age babies is the anabolic effect of
A. Glucose
B. Insulin
C. Glucagon
D. Growth hormone
E. Thyroid hormone

Correct answer = B.

29.5. A patient, with a BMI of 36 and a waist circumference of 44 inches, has a fasting blood glucose level of 145 mg/dL. One reason for the elevated blood glucose is
A. Enhanced release of glucose from the intestinal epithelial cells
B. Stimulation of GLUT4 transporters in muscle
C. Activation of pyruvate carboxylase
D. Activation of protein kinase B
E. Inhibition of liver GLUT4 transporters

Correct answer = C.

Obesity

30

Learning Objectives

At the end of this chapter, the student will be able to:

- Describe the causes (including dietary habits) of obesity
- Discuss the metabolic effects of obesity
- Describe the health risks associated with being overweight/obese

I. OVERVIEW

Obesity is a disorder of body weight regulatory systems characterized by an accumulation of excess body fat. In primitive societies, in which daily life required a high level of physical activity and food was only available intermittently, a genetic tendency favoring storage of excess calories as fat may have had a survival value. Today, however, the sedentary lifestyle and abundance and wide variety of palatable, inexpensive foods in industrialized societies has undoubtedly contributed to an obesity epidemic. As adiposity has increased, so has the risk of developing associated diseases, such as type 2 diabetes (T2D), cardiovascular disease (CVD), hypertension, cancer, and arthritis. Particularly alarming is the explosion of obesity in children and adolescents, which has shown a threefold increase in prevalence over the last four decades. [Note: Approximately 17% of those age 2–19 years are obese.] A new global analysis, published by The Lancet, found that 12.5 million children (7.3 million boys and 5.2 million girls) in the country, aged between 5 and 19, were grossly overweight in 2022. Obesity has increased globally, and, by some estimates, there are more obese than undernourished individuals worldwide.

II. ASSESSMENT

Because the amount of body fat is difficult to measure directly, it is usually determined from an indirect measure, the body mass index (BMI), which has been shown to correlate with the amount of body fat in most individuals. [Note: Exceptions are athletes who have large amounts of lean muscle mass.] Measuring the waist size with a tape measure is also used to screen for obesity, because this measurement reflects the amount of fat in the central abdominal area of the body. The presence of excess central fat is associated with an increased risk for morbidity and mortality, independent of the BMI. [Note: A waist size ≥40 in (men) and ≥35 in (women) is considered a risk factor.]

A. Body mass index

The **BMI (defined as weight in kg/[height in m]^2)** provides a measure of relative weight, adjusted for height. This allows comparisons within and between populations. The healthy range for the BMI is between 18.5 and 24.9. Individuals with a BMI between 25 and 29.9 are considered overweight, those with a BMI ≥30 are defined as obese, and a BMI >40 is considered severely (morbidly) obese. These cutoffs are based on studies examining the relationship of BMI to premature death and are similar in men and women. Nearly two thirds of U.S. adults are overweight, and more than one third of those are obese. In Asians, the cutoffs for overweight (≥23.0 kg/m^2) and obesity (≥25.0 kg/m^2) are lower due to other risk factors and morbidities. Children with a BMI-for-age above the 95th percentile are considered obese.

B. Anatomic differences in fat deposition

The anatomic distribution of body fat has a major influence on associated health risks. **A waist/hip ratio (WHR) >0.8 for women and >1.0 for men is defined as android, apple-shaped, or upper-body obesity** and is associated with more fat deposition in the trunk (Fig. 30.1A). In contrast, **a WHR of**

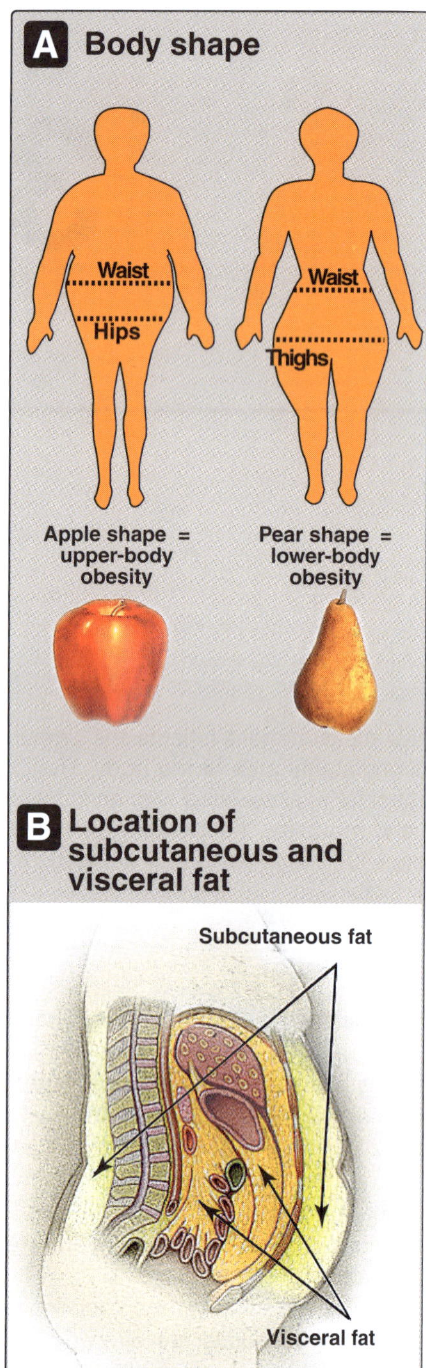

Figure 30.1 **A.** Individuals with upper-body obesity (left) have greater health risks than individuals with lower-body obesity (right). **B. Visceral fat** is located inside the abdominal cavity, packed in between the internal organs. **Subcutaneous fat** is found underneath the skin.

<0.8 for women and <1.0 for men is called gynoid, pear-shaped, or lower-body obesity. A lower WHR reflects a preponderance of fat distributed in the hips and thighs. **The pear shape, more commonly found in women, presents a much lower risk of metabolic disease**, and some studies indicate that it may actually be protective. Thus, the clinician can use simple indices of body shape to identify those who may be at higher risk for metabolic diseases associated with obesity.

> About 80%–90% of human body fat is stored in subcutaneous (subq) depots in the abdominal (upper body) and the gluteal-femoral (lower body) regions. The remaining 10%–20% is in visceral depots located deep within the abdominal cavity (Fig. 30.1B). Excess fat in visceral and abdominal subq stores increases health risks associated with obesity.

C. Biochemical differences in regional fat depots

The regional types of fat described above are biochemically different.

Subq adipocytes from the lower body, particularly in women, are larger, very efficient at fat (triacylglycerol [TAG]) deposition, and tend to mobilize fatty acids (FA) more slowly than subq adipocytes from the upper body. **Visceral adipocytes are the most metabolically active**.

In obese individuals, both abdominal subcutaneous and visceral depots have high rates of lipolysis and contribute to increased availability of free fatty acids (FFA). These metabolic differences may contribute to the higher health risk found in individuals with upper body (abdominal) obesity. [Note: FFA impair insulin signaling and are proinflammatory (see p. 475).]

1. **Endocrine function:** White adipose tissue, once thought to be a passive reservoir of TAG, is now known to play an active role in body weight regulatory systems. For example, the **adipocyte is an endocrine cell that secretes a number of protein regulators (adipokines), such as the hormones leptin and adiponectin.** Leptin regulates appetite as well as metabolism (see p. 486). Adiponectin reduces FFA levels in the blood (by increasing FA oxidation in muscles) and has been associated with improved lipid profiles, increased insulin sensitivity resulting in better glycemic control, and reduced inflammation in patients with diabetes.

   Adiponectin levels decrease as body weight increases, whereas leptin levels increase.

2. **Importance of portal circulation: With obesity, there is increased release of FFA and secretion of proinflammatory cytokines**, such as interleukin 6 (IL-6) and tumor necrosis factor-α (TNF-α), from adipose tissue. [Note: Cytokines are small proteins that regulate the immune system.]

One hypothesis for why abdominal adipose depots have such a large influence on metabolic dysfunction in obesity is that the FFA and cytokines released from these depots enter the portal vein and, therefore, have direct access to the liver. **In the liver, they may lead to insulin resistance (see p. 475) and increased hepatic synthesis of TAG,** which are released as components of very-low-density lipoprotein particles and contribute to the hypertriacylglycerolemia associated with obesity. By contrast, FFA from lower body subq adipose depots enter the general circulation, where they can be oxidized in muscle and, therefore, reach the liver in lower concentration.

D. Adipocyte size and number

As TAG are stored, adipocytes can expand to an average of two to three times their normal volume (Fig. 30.2). However, the ability of fat cells to expand is limited. With prolonged overnutrition, preadipocytes within adipose tissue are stimulated to proliferate and differentiate into mature fat cells, increasing the number of adipocytes. **Thus, most obesity is due to a combination of increased fat cell size (hypertrophy) and number (hyperplasia).** Obese individuals can have up to five times the normal number of adipocytes. [Note: Like other tissues, the adipose tissue undergoes continuous remodeling. Contrary to early dogma, we now know that adipocytes can die. The estimated average lifespan of an adipocyte is 10 years.] If excess calories cannot be accommodated within adipose tissue, the excess FA "spill over" into other tissues, such as muscle and the liver. The amount of this ectopic fat is strongly associated with insulin resistance. **With weight loss in an obese individual, the size of the fat cells is reduced, but the number is not usually affected.** Thus, a normal amount of body fat is achieved by decreasing the size of the fat cell below normal. However, small fat cells are very efficient at reaccumulating fat, and this may drive appetite and weight regain.

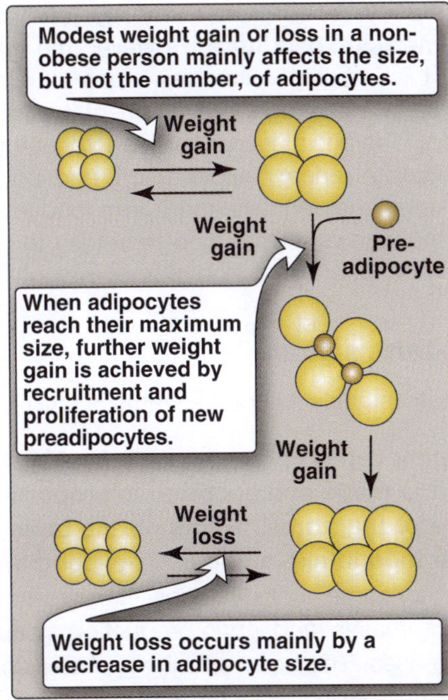

Figure 30.2 Hypertrophic (increased size) and hyperplastic (increased number) changes to adipocytes are thought to occur in severe obesity.

III. BODY WEIGHT REGULATION

The body weight of most individuals tends to be relatively stable over time. This observation prompted the hypothesis that each individual has a biologically predetermined "set point" for body weight. The body attempts to add to adipose stores when the body weight falls below the set point and to lose adipose from stores when the body weight rises above the set point. Thus, the body defends the set point. For example, with weight loss, appetite increases and energy expenditure falls, whereas with overfeeding, appetite falls and energy expenditure may slightly increase (Fig. 30.3). However, a strict set point model explains neither why some individuals fail to revert to their starting weight after a period of overeating nor the current epidemic of obesity.

A. Genetic contributions

It is now evident that genetic mechanisms play a major role in determining body weight.

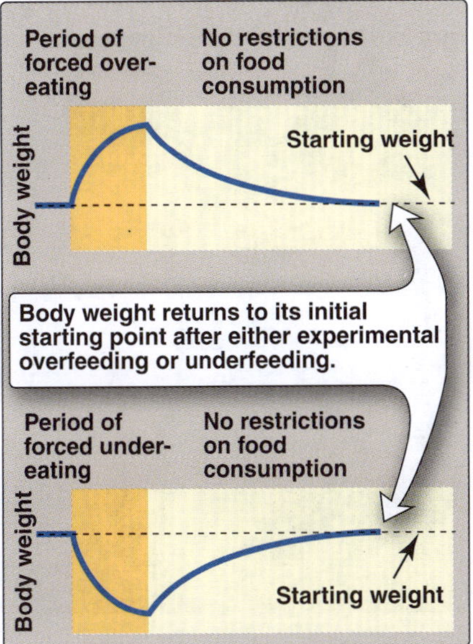

Figure 30.3 Weight changes following episodes of overfeeding or underfeeding followed by feeding with no restrictions.

Figure 30.4 Identical twins with combined weight of 1,300 lb. Note similarity in body shape.

1. **Biologic origin:** The importance of genetics as a determinant of obesity is indicated by the observation that children who are adopted usually show a body weight that correlates with their biologic rather than adoptive parents. Furthermore, identical twins have very similar BMI (Fig. 30.4), whether reared together or apart, and their BMI are more similar than those of nonidentical, dizygotic twins.

2. **Mutations:** Rare, single gene mutations can cause human obesity. For example, **mutations in the gene for leptin (causing decreased production) or its receptor (decreased function) result in hyperphagia (increased appetite for and consumption of food) and severe obesity** (Fig. 30.5), underscoring the importance of the leptin system in regulating human body weight (see IV below). [Note: Most obese humans have elevated leptin levels but are resistant to the appetite-regulating effects of this hormone.]

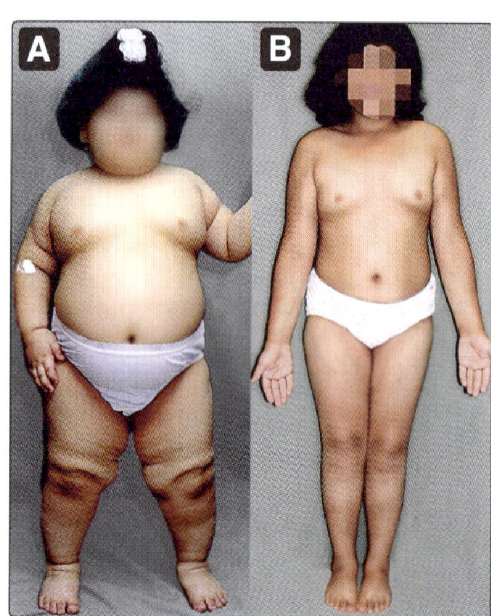

Figure 30.5 **A.** Patient with leptin deficiency before initiation of therapy at age 5 years. **B.** Patient at age 9 years after 48 months of therapy with subcutaneous injections of recombinant leptin.

B. Environmental and behavioral contributions

The epidemic of obesity occurring over the last several decades cannot be simply explained by changes in **genetic factors**, which are stable on this short time scale. Clearly, environmental factors, such as the **ready availability of palatable, energy-dense foods**, play a role. Furthermore, **sedentary lifestyles** decrease physical activity and enhance the tendency to gain weight. Eating behaviors, such as portion size, variety of foods consumed, an individual's food preferences, and the number of people present during eating, also influence food consumption. However, it is important to note that many individuals in this same environment do not become obese. The susceptibility to obesity appears to be explained, at least in part, by an interaction of an individual's genes and his or her environment and can be influenced by additional factors such as maternal under- or overnutrition that may "set" the body regulatory systems to defend a higher or lower level of body fat. Thus, epigenetic changes (see p. 665) likely influence the risk for obesity.

IV. MOLECULAR INFLUENCES

The cause of obesity can be summarized in a deceptively simple application of the first law of thermodynamics: **Obesity results when energy (caloric) intake exceeds energy expenditure**. However, the mechanism underlying this imbalance involves a complex interaction of biochemical, neurologic, environmental, and psychologic factors. The basic neural and humoral pathways that regulate appetite, energy expenditure, and body weight involve systems that regulate short-term food intake (meal to meal), and signals for the long-term (day to day, week to week, year to year) regulation of body weight (Fig. 30.6).

A. Long-term signals

Long-term signals reflect the status of fat (TAG) stores.

1. **Leptin:** Leptin is an adipocyte **peptide hormone that is made and secreted in proportion to the size of fat stores. It acts on the brain to regulate food intake and energy expenditure.** When we consume more calories than we need, body fat increases, and leptin production by adipocytes increases. The body adapts by increasing energy use (increasing activity) and decreasing appetite (an anorexigenic effect). When body fat decreases, the opposite effects occur. Unfortunately, **most obese individuals are leptin resistant**, and the

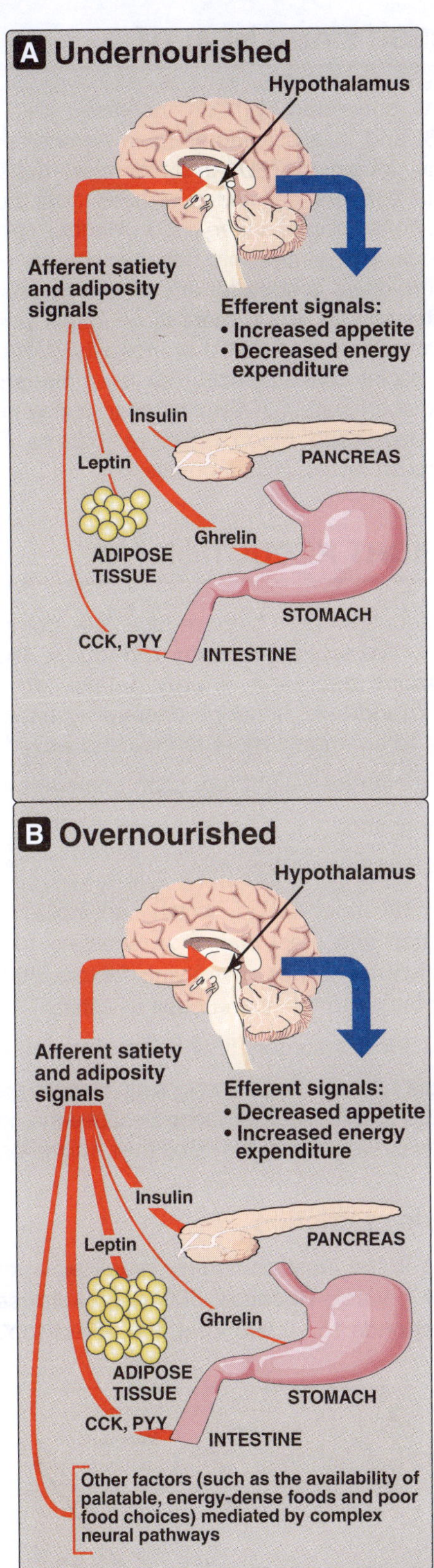

Figure 30.6 Some signals that influence appetite and satiety in undernourished **(A)** and overnourished **(B)** states. CCK = cholecystokinin; PYY = peptide YY.

leptin system may be better at preventing weight loss than preventing weight gain. [Note: Leptin's effects are mediated through binding to receptors in the arcuate nucleus of the hypothalamus.]

2. **Insulin:** Obese individuals are also hyperinsulinemic. Like leptin, insulin acts on hypothalamic neurons to dampen appetite. (See Chapter 27 for the effects of insulin on metabolism.) [Note: Obesity is associated with insulin resistance (see p. 475).]

Adiponectin is an adipokine secreted by adipocytes. Its levels are lower in obese people compared to the non-obese, as well as in individuals with insulin resistance and T2DM. It is a homeostatic factor for regulating glucose levels, lipid metabolism, and insulin sensitivity through its anti-inflammatory, anti-fibrotic, and antioxidant effects. It has an inverse relationship with obesity, its circulating levels increase with weight loss, including a reduction in body fat.

Resistin is also an adipokine that regulates glucose metabolism in mammalians. High resistin levels induce insulin resistance and exert proinflammatory effects.

Visfatin is an adipocytokine that is abundantly produced in visceral adipose tissue. It is also expressed in the bone marrow, liver, muscles, heart, placenta, lung, and kidney. Visfatin causes hypoglycemia by reducing glucose release from liver and increasing its utilization in adipocytes and myocytes.

B. Short-term signals

Short-term signals from the gastrointestinal (GI) tract control hunger and satiety, which affect the size and number of meals over a time course of minutes to hours. In the absence of food intake (between meals), the stomach produces **ghrelin**, an orexigenic (appetite-stimulating) hormone that drives hunger. As food is consumed, GI hormones, including **cholecystokinin** and **peptide YY**, among others, induce satiety (an anorexigenic effect), thereby terminating eating, through actions on gastric emptying and neural signals to the hypothalamus. Within the hypothalamus, neuropeptides (such as orexigenic **neuropeptide Y** [NPY] and anorexigenic **α-melanocyte–stimulating hormone** [α-MSH]) and neurotransmitters (such as anorexigenic **serotonin** and **dopamine**) are important in regulating hunger and satiety. Long-term and short-term signals interact, insofar as leptin increases secretion of α-MSH and decreases secretion of NPY. Thus, there are many complex regulatory loops that control the size and number of meals in relationship to the status of body fat stores. [Note: α-MSH, a cleavage product of proopiomelanocortin, binds to the melanocortin-4 receptor (MC4R). Loss-of-function mutations to MC4R are associated with early-onset obesity.]

V. METABOLIC EFFECTS

The primary metabolic effects of obesity include:

(i) dyslipidemias

(ii) glucose intolerance

(iii) insulin resistance expressed primarily in the liver, skeletal muscle, and adipose tissue.

These metabolic abnormalities reflect molecular signals originating from the increased mass of adipocytes (see Fig. 29.9, p. 476, and Fig. 30.6). [Note: About 30% of obese individuals do not show these metabolic abnormalities.]

A. Metabolic syndrome

Abdominal obesity is associated with a **cluster of metabolic abnormalities (hyperglycemia, insulin resistance, hyperinsulinemia, dyslipidemia [low levels of high-density lipoprotein (HDL) and elevated TAG], and hypertension) that is referred to as the metabolic syndrome** (Fig. 30.7). It is a risk factor for CVD and T2D. The low-grade, chronic, systemic inflammation seen with obesity contributes to the pathogenesis of insulin resistance and T2D and likely plays a role in metabolic syndrome. In obesity, adipocytes release proinflammatory mediators such as IL-6 and TNF-α. Additionally, levels of adiponectin, which normally dampens inflammation and sensitizes tissues to insulin, are low.

B. Nonalcoholic liver disease

Obesity is associated with **ectopic deposition of TAG in the liver (hepatic steatosis)** and results in increased risk for nonalcoholic fatty liver disease ([NAFLD], see p. 475).

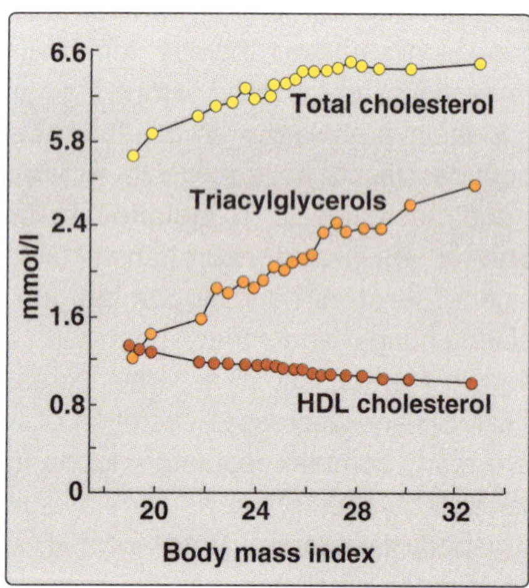

Figure 30.7 Body mass index and changes in blood lipids. HDL = high-density lipoprotein.

VI. OBESITY AND HEALTH

Obesity is correlated with an increased risk of death (Fig. 30.8) and is a risk factor for a number of chronic conditions, including T2D, dyslipidemias, hypertension, CVD, some cancers, gallstones, arthritis, gout, pelvic floor disorders (for example, urinary incontinence), NAFLD, and sleep apnea. The relationship between obesity and associated morbidities is stronger among individuals age <55 years. After age 74 years, there is no longer an association between increased BMI and mortality. [Note: Obesity also has social consequences (for example, stigmatization and discrimination).] Weight loss in obese individuals leads to decreased blood pressure, plasma TAG, and blood glucose levels. HDL increase.

VII. WEIGHT REDUCTION

Weight reduction can help reduce the complications of obesity. **To achieve weight reduction, the obese patient must decrease energy intake or increase energy expenditure**, although decreasing energy intake is thought to contribute more to inducing weight loss.

Typically, a plan for weight reduction combines:

(i) dietary change

(ii) increased physical activity

(iii) behavioral modification, which can include nutrition education and meal planning, recording food intake through food diaries, modifying factors that lead to overeating, and relearning cues to satiety.

(iv) Medications or surgery may be recommended.

Once weight loss is achieved, weight maintenance is a separate process that requires vigilance because the majority of patients regain weight after they stop their weight-loss efforts.

A. Caloric restriction

Dieting is the most commonly practiced approach to weight control. **Because 500 g of adipose tissue corresponds to ~3,900 kcal, the effect that caloric**

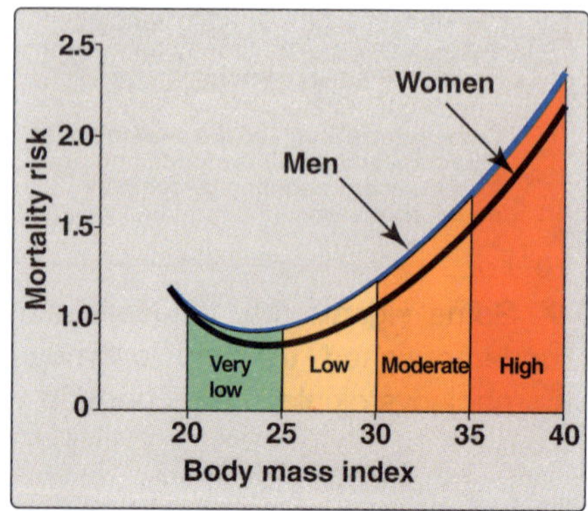

Figure 30.8 Body mass index and the relative risk of death.

restriction will have on the amount of adipose tissue can be estimated. Weight loss on calorie-restricted diets is determined primarily by caloric intake and not nutrient composition. [Note: However, compositional aspects can affect glycemic control and the blood lipid profile.] Caloric restriction is ineffective over the long term for many individuals. Over 90% of people who attempt to lose weight regain the lost weight when dietary intervention is suspended. Nonetheless, although few individuals will reach their ideal weight with treatment, weight losses of 10% of body weight over a 6-month period often reduce blood pressure and lipid levels and enhance control of T2D.

B. Physical activity

An increase in physical activity can create an energy deficit. Although adding exercise to a hypocaloric regimen may not produce a greater weight loss initially, exercise is a key component of programs directed at maintaining weight loss. **In addition, physical activity increases cardiopulmonary fitness and reduces the risk of CVD, independent of weight loss**. Persons who combine caloric restriction and exercise with behavioral treatment may expect to lose ~5%–10% of initial body weight over a period of 4–6 months. Studies show that individuals who maintain their exercise program regain less weight after their initial weight loss.

C. Pharmacologic treatment

The U.S. Food and Drug Administration has approved several weight-loss medications for use in adults. They include:

- **Orlistat** (decreases absorption of dietary fat)
- **Lorcaserin** and **phentermine** in combination with **topiramate** (promote satiety through serotonin signaling)
- **Liraglutide** (decreases appetite by activating the glucagon-like peptide 1 receptor)
- **Buproprion** in combination with **naltrexone** (increase metabolism by increasing norepinephrine).
- **Setmelanotide** (melanocortin-4 receptor agonist used for chronic weight management in patients with obesity resulting from proopiomelanocortin [POMC]), proprotein convertase subtilisin/kexin type 1 (PCSK1) or leptin receptor (LEPR) deficiency, confirmed by genetic testing.
- **Metreleptin** (a leptin analog approved as a replacement therapy for leptin deficiency in patients with congenital or acquired lipodystrophy and associated co-morbidities).

Their effects on weight reduction tend to be modest. [Note: Pharmacologic activation of brown adipocytes (see p. 170) is being explored.]

D. Surgical treatment

Gastric bypass and restriction surgeries are effective in causing weight loss in severely obese individuals. Through mechanisms that remain poorly understood, these operations greatly improve glycemic control in morbidly obese diabetic individuals. [Note: Implantation of a device that electrically stimulates the vagus nerve to decrease food intake has been approved.]

VIII. Medical Perspectives

Obesity rates are increasing worldwide, and it is recognized by the WHO as one of the “greatest public health challenges of the 21st century.” As it is a complex problem with numerous interconnected variables, reducing obesity poses a significant challenge.

The **cause of obesity is an imbalance between the energy intake and the energy expenditure.** The excess energy is stored in fat cells that enlarge and/or increase in number. The consequences arise from the increased mass of fat itself (stigma of obesity, behavioral changes, osteoarthritis, sleep apnea) or from the metabolic changes associated with excess fat (diabetes mellitus, gallbladder disease, hypertension, cardiovascular disease).

Hormones and cytokines secreted by adipose tissues are known as adipokines. They can have both pro- and anti-inflammatory effects and include TNF-α and IL-6 (pro-inflammatory), leptin, and adiponectin. Plasma levels of adiponectin are inversely related to fat stores and low adiponectin levels are associated with increased insulin resistance, whereas high adiponectin levels indicate improved insulin sensitivity.

To assess the body fat composition, surrogate markers such as BMI and waist: hip ratio are commonly used. For better composition analysis, dual-energy X-ray absorptiometry (DEXA) can be used. DEXA uses X-rays to measure whole body bone mass and soft tissue composition. It can measure regional body composition by subdividing the body using specific well-defined cut lines.

The foremost strategy to manage weight is dietary and lifestyle interventions. WHO recommends at least 150 minutes of moderate intensity exercises every week to maintain good health. Obesity due to other medical disorders like hypothyroidism, Cushing’s syndrome, etc. require the management of underlying diseases.

IX. Chapter Summary

- **Obesity**, the accumulation of excess body fat, results when **energy (caloric) intake** exceeds **energy expenditure** (Fig. 30.9).
- Obesity is increasing in industrialized countries because of a reduction in daily energy expenditure and an increase in energy intake resulting from the increasing availability of palatable, inexpensive foods.
- The **body mass index (BMI)** is easy to determine and highly correlated to body fat.
- The anatomic distribution of body fat has a major influence on associated health risks. Excess fat located in the **abdomen** (upper body, apple shape), as reflected in **waist size**, is associated with greater risk for **hypertension**, **insulin resistance**, **diabetes**, **dyslipidemia**, and **coronary heart disease** as compared to fat located in the hips and thighs (lower body, pear shape).
- A person's weight is determined by genetic and environmental factors.
- **Appetite** is influenced by afferent, or incoming, signals (that is, neural signals, circulating hormones such as **leptin**, and metabolites) that are integrated by the **hypothalamus**. These diverse signals prompt release of hypothalamic peptides (such as **neuropeptide Y** and **α-melanocyte–stimulating hormone**) and activate outgoing, efferent neural signals.
- Obesity is correlated with an increased risk of **death** and is also a risk factor for a number of **chronic conditions**.
- **Weight reduction** is achieved best with **negative energy balance**, that is, by decreasing caloric intake and increasing physical activity. Virtually all diets that limit particular groups of foods or macronutrients lead to short-term weight loss.
- Long-term maintenance of weight loss is difficult to achieve. Modest reduction in food intake occurs with **pharmacologic treatment**. **Surgical procedures**, such as gastric bypass, designed to limit food intake are an option for the severely obese patient who has not responded to other treatments.

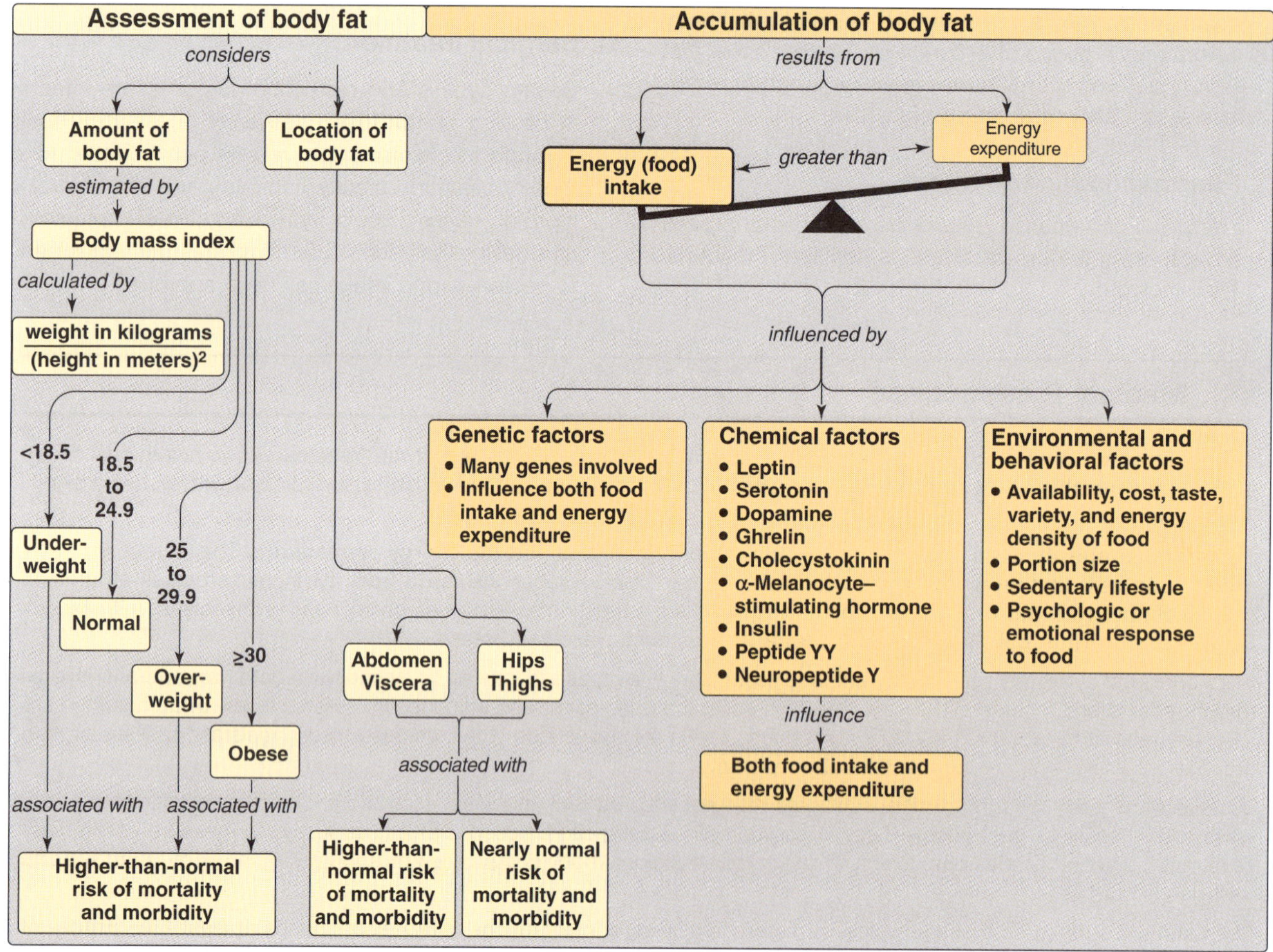

Figure 30.9 Key concept map for obesity. [Note: Body mass index may also be calculated by weight in pounds/(height in inches)^2 × 703.]

Study Questions

Short-Answer Questions

Write short notes on:

30.1. Body mass index

30.2. Leptin

Long-Answer Questions

30.1. Describe the effects of obesity on health

30.2. Describe strategies for body weight reduction

Multiple-Choice Questions

Choose the ONE best answer.

30.1. BMI (defined as weight in kg/[height in m]^2) provides a measure of relative weight, adjusted for height. Which of the following statements is INCORRECT regarding BMI?

A. BMI between 18.5 and 24.9 is considered healthy range.

B. BMI between 25 and 29.9 is considered overweight, those with a BMI ≥30 are defined as obese.

C. In Asians, the cut-offs for overweight (≥23.0 kg/m^2) and obesity (≥25.0 kg/m^2) are lower due to other risk factors and morbidities.

D. Children with a BMI-for-age above the 85th percentile are considered obese.

Correct answer = D.

30.2. Waist/hip ratio (WHR) >0.8 for women and >1.0 for men is

A. android, pear-shaped, or lower-body obesity

B. gynoid, apple-shaped, or upper-body obesity

C. android, apple-shaped, or upper-body obesity

D. gynoid, pear-shaped, or lower-body obesity

Correct answer = C.

30.3. Visceral adipose depots have larger influence on metabolic dysfunction in obesity than subcutaneous body adipose depots because they

A. release free fatty acids that enter the general circulation

B. secrete cytokines and release free fatty acids that enter the liver

C. tend to mobilize fatty acids more slowly

D. exhibit low rates of lipolysis

Correct answer = B.

30.4. Most obesity is due to a combination of increased fat cell size (hypertrophy) and number (hyperplasia). Which of the following statements is INCORRECT regarding size and number of fat cells?

A. Modest weight gain or loss in a nonobese person mainly affects the size, but not the number of adipocytes.

B. When adipocytes reach their maximum size, further weight gain is achieved by recruitment and proliferation of new pre-adipocytes.

C. Weight loss occurs mainly by a decrease in fat cell size.

D. Small fat cells are less efficient at reaccumulating fat.

Correct answer = D.

30.5. Which of the following hormones is primarily responsible for regulating appetite and energy balance?

A. Insulin

B. Leptin

C. Cortisol

D. Thyroxine

Correct answer = B.

30.6. Which of the following health conditions is NOT commonly associated with obesity?

A. Type 2 diabetes mellitus

B. Hypertension

C. Osteoporosis

D. Cardiovascular disease

Correct answer = C.

30.7. Which of the following dietary factors is most strongly linked to the development of obesity?

A. High protein intake

B. High carbohydrate intake

C. High fat intake

D. High fiber intake

Correct answer = B.

30.8. Which of the following is INCORRECT?

A. Ghrelin is an orexigenic hormone

B. Cholecystokinin induces satiety

C. peptide YY (PYY) produced by gut drives hunger

D. α-Melanocyte–stimulating hormone (α-MSH) is anorexigenic

Correct answer = C.

30.9. Which of the following criteria is NOT included in the diagnosis of metabolic syndrome according to the International Diabetes Federation (IDF) criteria?

A. Elevated blood pressure

B. Elevated fasting plasma glucose

C. Elevated triglyceride levels

D. Elevated high-density lipoprotein (HDL) cholesterol levels

Correct answer = D.

30.10. What is the primary goal of obesity management?
A. Achieving a specific target weight
B. Eliminating all dietary fat
C. Improving overall health and reducing obesity-related complications
D. Achieving a specific body mass index (BMI)
Correct answer = C.

Multiple-Choice Questions (NExT/NEET PG)

Choose the ONE best answer.

30.1. A 40-year-old woman, 5 ft, 1 in (155 cm) tall and weighing 188 lb (85.5 kg), seeks your advice on how to lose weight. Her waist measured 41 in and her hips 39 in. The remainder of the physical examination and the blood laboratory data were all within the normal range. Her only child (14 years of age), her sister, and both of her parents are overweight. The patient recalls being overweight throughout her childhood and adolescence. Over the past 15 years, she had been on seven different diets for periods of 2 weeks to 3 months, losing from 5 to 25 lb each time. On discontinuation of the diets, she regained weight, returning to 188 lb. Calculate and interpret the body mass index for the patient.
A. BMI = 35.6, patient is classified as obese
B. BMI = 38.5, patient is classified as obese
C. BMI = 35.6, patient is classified as overweight
D. BMI = 38.5, patient is classified as obese
Correct answer = A.

30.2. Which one of the following statements best describes the patient mentioned in question 30.1? She would be expected to have
A. the same number of adipocytes as an individual of normal weight, but each adipocyte is larger.
B. an apple pattern of fat distribution.
C. higher-than-normal levels of adiponectin.
D. lower-than-normal levels of circulating leptin.
E. lower-than-normal levels of circulating triacylglycerols.
Correct answer = B.

30.3. A 40-year-old man, with a BMI of 37, has a fasting blood glucose level of 168 mg/dL, a value that has been steadily increasing over the past 5 years. His HbA1c level is 8.9. Analysis of his blood, using Western blot technology, demonstrates the presence of antibodies against the insulin receptor. These antibodies can lead to
A. Downregulation of the insulin receptors
B. Upregulation of the insulin receptors
C. Enhanced release of pancreatic glucagon
D. Enhanced glucose uptake into muscle cells
Correct answer = A.

30.4. Considering the patient in the previous question, which of the following medical disorders would you expect this patient to also have?
A. Low blood pressure
B. Polycystic ovary syndrome (PCOS)
C. Fatty liver-nonalcoholic hepatitis (NASH–nonalcoholic steatohepatitis)
D. Hypouricemia
E. High HDL cholesterol
Correct answer = C.

30.5. Considering the patient in question 30.3, which of the following body types would you expect this patient to have?
A. Body builder
B. Central obesity
C. Obesity mostly centered in the hips and breasts
D. Tall and slender
E. Ideal body weight
Correct answer = B.

Appendix
Unit VI: Integration of Metabolism

CLINICAL RELEVANCE FOR PRIMARY HEALTHCARE PHYSICIANS

The prevalence of diabetes is about 9% in Indian adults. Diabetes may remain undiagnosed for long, with continuous disease progression leading to complications. Screening of non-communicable diseases (NCDs) can lead to early diagnosis and management with lifestyle modifications and treatment. Screening can be done by plasma glucose or HbA_{1c} level (Reference: Appendix Unit III—Carbohydrates).

In India, the National Programme for the Prevention and Control of Cancer, Diabetes, Cardiovascular Diseases and Stroke (NPCDCS) was initiated in 2010 with the focus on screening for common NCDs, at District and Community Health Center (CHC) levels, through NCD clinics.

Diabetic patients on antihyperglycemic medications can develop hypoglycemia. Symptoms of hypoglycemia include anxiety, sweating, confusion, feeling lightheaded or dizzy, nausea, blurred vision, etc. In such patients, blood glucose should be checked, and if it is low, patient should be given carbohydrate such as glucose or table sugar.

Self-monitoring of blood glucose (SMBG) is blood glucose testing at home for people with diabetes. It can be easily done using a glucometer. SMBG helps patients in monitoring their own health. However, there can be significant difference between capillary glucose estimated with glucometers and venous glucose estimated in laboratories. Laboratory estimation is nearer the true value, but glucometer readings have easy monitoring advantage. Chronic diabetic patients should also be monitored for cataract, dyslipidemia and renal function. Urine microalbumin estimation provides an early indication of diabetic nephropathy. Laboratory diagnostic details for diabetes may be seen from Appendix of Unit III—Carbohydrates.

Atherosclerosis is another epidemic leading to high prevalence of cardiovascular diseases. This is discussed in detail in chapter 23 of Unit IV. Atherosclerosis can lead to a stable ischemic plaque or a vulnerable ischemic plaque. The release of thrombus from a vulnerable ischemic plaque in blood vessels supplying end arteries is the main cause of an acute coronary event like myocardial infarction and also stroke. Dyslipidemia is discussed in Unit IV on Lipid Metabolism.

While it has been emphasized that obesity can cause several diseases, obesity itself can be due to several diseases, such as hypothyroidism, polycystic ovary syndrome, and Cushing's syndrome. Several medications, including certain antidepressants, anticonvulsants, insulin, sulfonylureas, and thiazolidinediones, oral contraceptives, and most corticosteroids also cause weight gain.

Excess carbohydrates taken as food are converted to fat and stored in adipose tissue. High fructose corn syrup containing foods promote weight gain and are associated with metabolic syndrome, diabetes mellitus and cardiovascular risk (Chapter 31—Nutrition: Overview and Macronutrients).

Ketone bodies are indicative of excess breakdown of fatty acids and absence of cellular glucose, as seen in uncontrolled diabetes mellitus and starvation. They can be easily assessed by the ketone urine dipstick test.

Table 1 Normal serum values

Analyte	Level (SI Unit)	Level (Conventional Units)
Insulin	14.35–143.5 pmol/l	2–20 μU/ml
Glucagon	40-130 ng/l	40–130 pg/ml
C-peptide	0.27–1.19 nmol/l	(0.8–3.5 ng/ml)
Ketone bodies	Nil	Nil

Metabolic Syndrome

Metabolic syndrome (MS) is cluster of conditions that increases the risk of heart disease, stroke and diabetes. Insulin resistance and obesity are significant factors for causation of MS.

Table 2 Diagnostic criteria for metabolic syndrome

According to the new International Diabetes Federation (IDF) definition, for a person to be defined as having the metabolic syndrome they must have:	
Central obesity (defined as waist circumference with ethnic specific values, for South Asians, male ≥90 cm and female ≥80 cm)	
Plus, any two of the following four factors:	
Raised triglycerides	≥ 150 mg/dl (1.7 mmol/l) or specific treatment for this lipid abnormality
Reduced HDL cholesterol	< 40 mg/dl (1.03 mmol/l) in males, < 50 mg/dl (1.29 mmol/l) in females or specific treatment for this lipid abnormality
Raised blood pressure	Systolic BP ≥ 130 or diastolic BP ≥ 85 mm Hg or treatment of previously diagnosed hypertension
Raised fasting plasma glucose	(FPG) ≥ 100 mg/dl (5.6 mmol/l), or previously diagnosed type 2 diabetes

Clinical Case Questions

Chapter 27 Metabolic Effects of Insulin and Glucagon

27.1. A 60-year-old male is brought to emergency in an unconscious state. History reveals that the patient has diabetes since last 5 years and is regularly taking insulin as per prescription. He did not take breakfast today and was feeling dizzy, sweaty and confused before losing consciousness. Blood test revealed plasma glucose was 48 mg/dl.

A. What is the most probable diagnosis?

B. What will be the immediate management in this patient?

Answer—A: This is a case of insulin-induced hypoglycemia. As the patient had taken insulin but skipped breakfast, he developed hypoglycemia.

Answer—B: As the patient is unconscious, IV glucose should be immediately given. In case of conscious patient, glucose or other simple sugar-containing foods can be given orally.

Chapter 28 The Feed–Fast Cycle

28.1. A 35-year-old male patient was brought to the emergency room with behavioral disturbances, including aggressiveness. History revealed voluntary complete fasting for over a week. The patient appeared ill with psychomotor impairment and incoherent speech. On physical examination, he was hemodynamically stable, and he had a body mass index (BMI) of 16. Laboratory studies showed blood glucose was 57 mg/dl; venous pH 7.08; total proteins: 4.3 g/dl; albumin 2.3 g/dl; and lactate 2.61 mM/L (0.45–1.90); urinalysis was positive for ketone bodies.

A. What is the probable diagnosis?

B. What is the biochemical significance of ketone body production in starvation?

Answer—A: As ketone bodies are present in urine, it may be due to diabetic ketoacidosis or starvation. Low blood glucose (57 mg/dl) rules out diabetic ketoacidosis. Also, total protein and albumin are decreased in this patient. Therefore, this is a case of starvation.

Answer—B: During starvation, the concentration of acetyl CoA is increased because of excessive fatty acid oxidation. However, because of a paucity of oxaloacetate (which is diverted to gluconeogenesis), citrate cannot be synthesized. Therefore, excess acetyl CoA is now diverted to ketone body formation to generate some energy.

Chapter 29 Diabetes Mellitus

29.1. A 14-year-old girl is brought to the emergency room with a history of the first episode of altered sensorium (disorientation and confusion) for a few hours. The child has not had any seizures. There is no history of any preceding fever, head injury, loose stools, jaundice, or accidental poisoning. The mother gives a history of progressive weight loss noticed over a few weeks but was not concerned since her daughter's appetite had increased substantially. On probing, it is found that the teenager had complained that her classes and sleep were getting interrupted due to frequent urges to urinate.

A. What differential diagnosis will you consider on the basis of history?

Answer—A: Diabetic ketoacidosis due to diabetes mellitus type I.

B. Which first line investigations will you order to confirm your suspicion and why?

Answer—B: Blood sugar (random) level for hyperglycemia; urine sugar and ketone levels for glycosuria and ketonuria, respectively; and arterial blood gas analysis for metabolic acidosis.

C. The investigations confirm your clinical suspicion. What is the underlying cause for the altered sensorium in this case?

Answer—C: There is an interplay of multiple factors that lead to mild levels of altered sensorium in diabetic ketoacidosis, and the mechanism is not very clearly understood. These include cerebral hypoperfusion, metabolic acidosis, hyperglycemia and resultant hyperosmolarity, direct toxicity of the ketone bodies and cerebral edema.

29.2. A 54-year-old female visits the medical OPD with complaints of increased thirst, frequent urination and unexplained weight loss over the past 4–6 months. The patient has been feeling fatigued and has noticed that minor cuts and bruises are taking longer to heal. She also mentions experiencing numbness and tingling in the toes. She consumes a vegetarian diet and her lifestyle is sedentary. Her mother and elder brother have type 2 diabetes mellitus.

Physical examination reveals Weight: 78 kg, Height: 160 cm, BMI: 30.5 kg/m^2 (classified as obese), Blood pressure: 145/85 mmHg, and a healing wound on her shin.

Laboratory investigation—Fasting blood glucose: 180 mg/dL.

A. What is the likely diagnosis?

Answer—A: Polyuria, polydipsia and unexplained weight loss in a 54-year-old obese female, who has a strong family history of type 2 diabetes mellitus, point to a likely diagnosis of diabetes mellitus. A fasting blood sugar of >126 mg/dl confirms the diagnosis of diabetes mellitus.

B. Name chronic complications of type 2 DM.

Answer—B: Retinopathy, neuropathy and nephropathy are common chronic complications of prolonged uncontrolled hyperglycemia.

Chapter 30 Obesity

30.1. A 65-year-old man with type 2 diabetes was referred by his primary care physician to the weight center for an evaluation of his obesity and recommendations for treatment options, including weight-loss surgery. His weight is 102 kg and height 152 cm. He has sleep apnea. He is taking his insulin regularly. His fasting blood glucose levels are 130 mg/dl and HbA_{1c} level is 6.1%. He has participated in both commercial and medical weight-loss programs but has regained any weight lost within months of discontinuing the programs.

A. Calculate the BMI of this patient.

B. Is weight loss surgery indicated in this patient?

Answer—A: BMI = Weight (in kg)/Height (in meters)^2 = 102 /(1.52)^2 = 44. He is a case of morbid obesity and requires medical intervention.

Answer—B: Indications for weight-loss surgery are

(i) clinically severe obesity (BMI ≥ 40 kg/m^2 or ≥ 35 kg/m^2 with comorbid conditions),

(ii) when less invasive methods of weight loss have failed, and

(iii) the patient is at high risk for obesity-associated morbidity or mortality.

This patient meets the criteria for surgery, with a BMI of 44 kg/m^2, numerous previous attempts at weight loss, and he has obesity-related comorbidities, including diabetes and sleep apnea. Also, he has no contraindications for surgery (end-stage lung disease, unstable cardiovascular disease, multiorgan failure, gastric varices, uncontrolled psychiatric disorders, ongoing substance abuse, and noncompliance with current regimens).

30.2. A 38-year-old, male shopkeeper by occupation, visits the General Medicine outpatient department for a general health checkup due to concerns about his weight. He reports increasing fatigue and difficulty in managing daily activities. He has experienced a gradual weight gain over the past 5 years. He attributes this to his sedentary job, lack of physical activity and poor eating habits, including frequent snacking and consumption of high-calorie food.

His mother has type 2 diabetes and father is overweight and has hypertension. His examination reveals—Weight: 105 kg, Height: 172 cm, BMI: 35.5 kg/m^2, Blood pressure: 140/94 mmHg. Waist circumference: 112 cm. Has acanthosis nigricans around the neck.

Laboratory investigation—Fasting blood glucose: 120 mg/dl, HbA1c: 6.3%, lipid profile shows elevated triglycerides and low HDL cholesterol.

A. Calculate the BMI and mention the class of obesity.

B. Name a surrogate marker of body fat composition for use in clinical practice.

C. Interpret the glycemic status of the patient.

Answer—A: Using the weight in kg/height in m^2 (105 ÷ 1.72 × 1 = 35.5 kg/m^2). This BMI shows that the patient is very obese.

Answer—B: BMI and waist-hip ratio are easy-to-use surrogate markers of body fat composition in the OPD setting.

Answer—C: Fasting blood sugar (FBS) of 120 mg/dl and HbA1c of 6.3% are suggestive of prediabetes or a state of impaired glucose tolerance, as per the American Diabetes Association (ADA) guidelines.

Unit VII

Medical Nutrition

Competencies Addressed

- **BI6.5** Describe the biochemical role of vitamins in the body and explain the manifestations of their deficiency
- **BI6.9** Describe the functions of various minerals in the body, their metabolism and homeostasis
- **BI6.10** Enumerate and describe the disorders associated with mineral metabolism
- **BI8.1** Discuss the importance of various dietary components and explain importance of dietary fiber
- **BI8.2** Describe the types and causes of protein energy malnutrition and its effects
- **BI8.3** Provide dietary advice for optimal health in childhood and adult, in disease conditions like diabetes mellitus, coronary artery disease and in pregnancy
- **BI8.5** Summarize the nutritional importance of commonly used items of food including fruits and vegetables (macro-molecules and its importance)

Unit VII: Overview

Nutrition deals with food and nutrients, their uses in body and their relation to health and disease. All energy (calories) is provided by three classes of nutrients: **fats**, **carbohydrates**, and **proteins**. Proteins also help in growth and are precursors of specialized products. Besides these three macronutrients, **micronutrients** (vitamins and minerals) are also required for optimal health and growth.

There is a specific requirement of these nutrients, and a diet which provides all of them in appropriate quantities is known as **balanced diet**. Failure to adhere to balanced diet can lead to either **undernutrition** or **overnutrition**. Earlier, undernutrition was a major health concern, but overnutrition, leading to **overweight** and **obesity** is also a health concern now. Changes in food habits, with more intake of processed food, have led to situation where obesity (due to overnutrition) and micronutrient deficiency coexist.

Vitamins are chemically unrelated organic compounds that cannot be synthesized in adequate quantities by humans and, therefore, must be supplied by the diet or obtained from intestinal flora. Vitamins can broadly be classified as **water soluble** or **fat soluble**. Vitamins are required to perform specific cellular functions. For example, many of the water-soluble vitamins are precursors of coenzymes for the enzymes of intermediary metabolism. Water-soluble vitamins can be excreted in urine, so their excess intake is not associated with toxicity. However, fat-soluble vitamins are stored in the body in larger amounts and excess can lead to toxicity.

Minerals are inorganic substances (elements) required in small amounts by the body. They function in a number of processes including formation of bones and teeth, fluid balance, nerve conduction, muscle contraction, signaling, and catalysis. Several minerals are essential enzyme cofactors. Deficiency of minerals can affect several critical processes and produce deficiency disease, e.g., iron deficiency is a very common treatable cause of anemia, calcium deficiency is common and can lead to bone deformities, etc.

This unit will cover various macro- and micronutrients, their biochemical role, daily requirement, sources and diseases associated with deficiency/excess.

Nutrition: Overview and Macronutrients

31

Learning Objectives

At the end of this chapter, the student will be able to:

- Describe the importance of various dietary components and dietary fiber
- Describe types and causes of protein energy malnutrition and its effects

I. OVERVIEW

Nutrients are the constituents of food necessary to sustain the normal functions of the body. **All energy (calories) is provided by three classes of nutrients: fats, carbohydrates, and protein** (Fig. 31.1). Because the intake of these energy-rich molecules is larger (g amounts) than that of the other dietary nutrients, they are called **macronutrients**. Energy requirements of an individual depend on many variables, including age, gender, state of growth, disease, etc. Deficiency or excess of any nutrient can cause disease. **A balanced diet provides the energy as well as macro- and micronutrients in proper proportions to maintain health**. This chapter focuses on the kinds and amounts of macronutrients that are needed to maintain optimal health and prevent chronic disease. Those nutrients needed in lesser amounts (mg or μg), vitamins and minerals, are called **micronutrients** and are considered in Chapters 32 and 33.

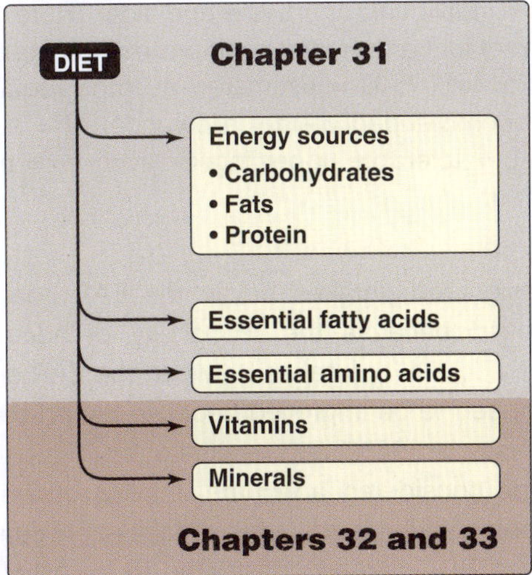

Figure 31.1 Essential nutrients obtained from the diet. [Note: Ethanol may provide a significant contribution to the daily caloric intake of some individuals.]

II. DIETARY REFERENCE INTAKES

Committees of U.S. and Canadian experts organized by the Food and Nutrition Board of the Institute of Medicine of the National Academy of Sciences have compiled **Dietary Reference Intakes (DRI), which are estimates of the amounts of nutrients required to prevent deficiencies and maintain optimal health and growth**. The DRI expands on the **Recommended Dietary Allowances (RDA)**, which have been published with periodic revisions since 1941. Unlike the RDA, the DRI establishes upper limits on the consumption of some nutrients and incorporates the role of nutrients in lifelong health, going beyond deficiency diseases. **Both the DRI and the RDA refer to long-term average daily nutrient intakes, because it is not necessary to consume the full RDA every day.**

> The Indian Council of Medical Research published recommendations in 2020 for DRI based on recent data on energy expenditure, protein metabolism, as well as available data on mineral and vitamin losses and absorption, to estimate nutrient requirements for Indians.

A. Definition

The **DRI consists of four dietary reference standards** for the intake of nutrients designated for specific life stage (age) groups, physiologic states, and gender (Fig. 31.2).

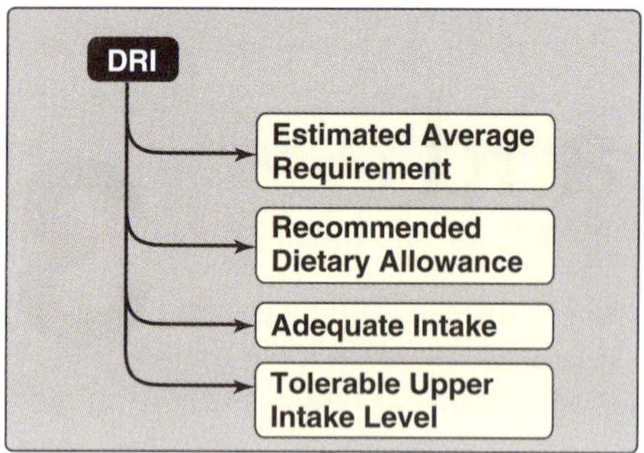

Figure 31.2 Components of the Dietary Reference Intakes (DRI).

1. **Estimated average requirement (EAR):** The average daily nutrient level estimated to **fulfil the requirement of half of the healthy individuals** (50%) of a particular age, gender, and life stage is the EAR.

2. **Recommended dietary allowance:** The average daily nutrient level estimated to **fulfil the requirement of 97%–98% individuals** of a particular age, gender, and life stage is RDA. It is 2 standard deviations (SDs) above the EAR. (that is, RDA = EAR + 2 SD_{EAR}).

3. **Adequate intake (AI):** This is based on nutrient intake by a group of apparently healthy individuals. AI is used when sufficient scientific data are not available to calculate and EAR or RDA.

   For example, the AI for young infants, for whom human milk is the recommended sole source of food for the first 6 months, is based on the estimated daily mean nutrient intake supplied by human milk for healthy, full-term infants who are exclusively breast-fed.

4. **Tolerable upper intake level (UL, or TUL):** It is the highest average daily intake level with which there is no risk of any adverse effects in almost all individuals in the general population.

   As intake increases above the UL, the potential risk of adverse effects may increase. The UL is useful because of the increased availability of fortified foods and the increased use of dietary supplements. For some nutrients, there may be insufficient data on which to develop a UL.

MICRO-NUTRIENT	EAR, RDA, or AI	UL
Thiamine	EAR, RDA	—
Riboflavin	EAR, RDA	—
Niacin	EAR, RDA	UL
Vitamin B_6	EAR, RDA	UL
Folate	EAR, RDA	UL
Vitamin B_{12}	EAR, RDA	—
Pantothenic acid	AI	—
Biotin	AI	—
Choline	AI	UL
Vitamin C	EAR, RDA	UL
Vitamin A	EAR, RDA	UL
Vitamin D	EAR, RDA	UL
Vitamin E	EAR, RDA	UL
Vitamin K	AI	—
Boron	—	UL
Calcium	EAR, RDA	UL
Chromium	AI	—
Copper	EAR, RDA	UL
Fluoride	AI	UL
Iodine	EAR, RDA	UL
Iron	EAR, RDA	UL
Magnesium	EAR, RDA	UL
Manganese	AI	UL
Molybdenum	EAR, RDA	UL
Nickel	—	UL
Phosphorus	EAR, RDA	UL
Selenium	EAR, RDA	UL
Vanadium	—	UL
Zinc	EAR, RDA	UL

Figure 31.3 Dietary Reference Intakes for vitamins and minerals in individuals age 1 year and older. [Note: An RDA has been set for carbohydrate and protein (macronutrients) but not for fat.] EAR = estimated average requirement; RDA = recommended dietary allowance; AI = adequate intake; UL = tolerable upper intake level; — = no value established.

B. Using the dietary reference intakes

Most nutrients have a set of DRI (Fig. 31.3). Usually a nutrient has an EAR and a corresponding RDA. Most are set by age and gender and may be influenced by special factors, such as pregnancy and lactation in women (see p. 519). When the data are not sufficient to estimate an EAR (or an RDA), an AI is designated. Intakes below the EAR need to be improved because the probability of adequacy is ≤50% (Fig. 31.4). Intakes between the EAR and RDA likely need to be improved because the probability of adequacy is <98%, and intakes at or above the RDA can be considered adequate. Intakes above the AI can be considered adequate. Intakes between the UL and the RDA can be considered to have no risk for adverse effects. [Note: Because the DRI is designed to meet the nutritional needs of the healthy, it does not include any special needs of the sick.]

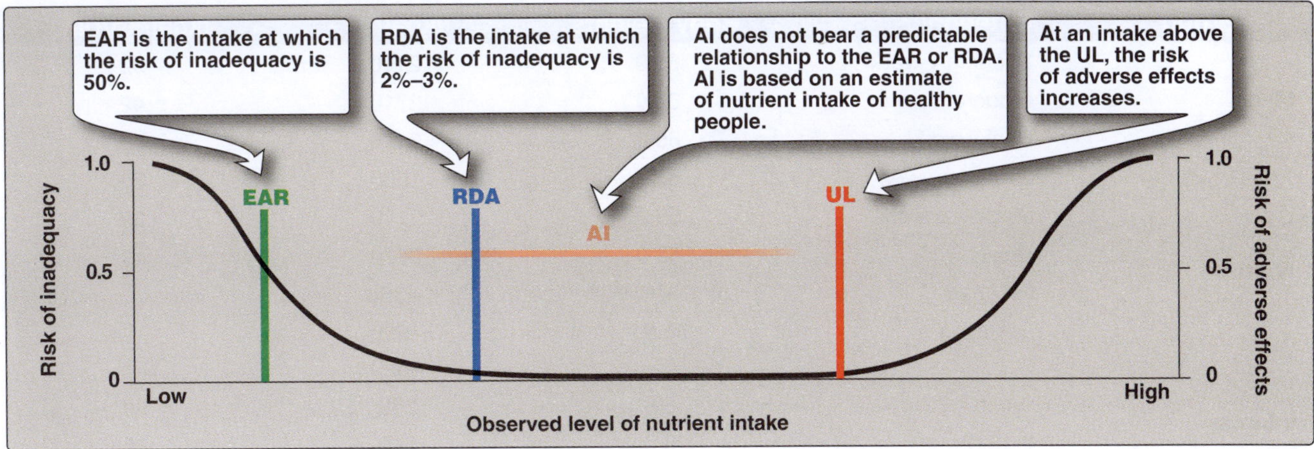

Figure 31.4 Comparison of the components of the Dietary Reference Intakes. EAR = estimated average requirement; RDA = recommended dietary allowance; AI = adequate intake; UL = tolerable upper intake level.

III. ENERGY REQUIREMENT IN HUMANS

The **Estimated Energy Requirement (EER)** is the average dietary energy intake predicted to maintain an energy balance (that is, the calories consumed are equal to the energy expended) in a healthy adult of a defined age, gender, and height whose weight and level of physical activity are consistent with good health. Differences in the genetics, body composition, metabolism, and behavior of individuals make it difficult to accurately predict a person's caloric requirements. However, some simple approximations can provide useful estimates.

- Sedentary adults require 30 kcal/kg/day
- Moderately active adults require 35 kcal/kg/day
- Very active adults require 40 kcal/kg/day

Studies conducted in India indicate that the energy expended for different physical activities is lower compared to other populations. This difference could be attributed to the fact that the energy expended for a given activity (physical activity ratio, PAR) is not constant across the range of body weights, and extrapolating from a heavier individual may overestimate the energy spent for a lighter individual. To address this, a recalculation was performed using PAR values specific to the Indian population, which showed that the PAR for sedentary individuals was lower, decreasing from 1.53 (used in the previous recommendation, RDA-2010) to 1.40. This trend was also observed for other levels of physical activity, indicating that the energy spent for each physical activity is lower for Indians (Fig. 31.5).

A. Energy content of food

To plan a proper diet, it is important to know the energy content of food. **The energy content of food is calculated from the heat released by the total combustion of food in a calorimeter.** It is expressed in kilocalories (kcal, or Cal). The standard conversion factors for determining the metabolic caloric value of fat, protein, and carbohydrate are shown in Figure 31.6. Note that the energy content of fat is more than twice that of carbohydrate or protein, whereas the energy content of ethanol is intermediate between those of fat and carbohydrate.

[Note: The joule (J) is a unit of energy widely used in countries other than the United States. One cal = 4.2 J; 1 kcal = 1,000 cal = 4.2 kJ. For uniformity, many scientists are promoting the use of joules rather than calories in the United States. However, kcal still predominates and is used throughout this text.]

> Calorimeter provides the physical energy content of food, as food undergoes complete combustion. However, in the human body, food is broken down in several stages, during which a comparably very small amount of energy is released. This energy, known as physiological energy, is lesser than the physical energy of food.

B. Use of food energy in the body

The energy generated by metabolism of the macronutrients is used for three energy-requiring processes that occur in the body:

- Resting metabolic rate (RMR)
- Physical activity
- The thermic effect of food

The number of kcal expended by these processes in a 24-hour period is the total energy expenditure (TEE).

1. **Resting metabolic rate:** RMR is defined as the **energy required by the body in a resting, post-absorptive condition**. It represents the energy required to carry out the normal body functions,

Age Group	Category	Body weights	(kcal/d)[a]	(kcal/kg/day)
Men	Sedentary work	65.0	2110	**32**
	Moderate work	65.0	2710	**42**
	Heavy work	65.0	3470	**53**
Women	Sedentary work	55.0	1660	**30**
	Moderate work	55.0	2130	**39**
	Heavy work	55.0	2720	**49**
	Pregnant	55.0 + GWG[b]	+ 350	
	Lactating	55.0+	+600 +520	
Infants	0-6 m	5.8	530	**90**
	6-12m	8.5	660	**80**
Children[c]	1-3y	12.9	1110	**83**
	4-6y	18.3	1360	**74**
	7-9 y	25.3	1700	**67**
Boys	10-12y	34.9	2220	**64**
Girls	10-12y	36.4	2060	**57**
Boys	13-15y	50.5	2860	**57**
Girls	13-15y	49.6	2400	**49**
Boys	16-18y	64.4	3320	**52**
Girls	16-18y	55.7	2500	**45**

[a]Rounded off to the nearest 10 kcal/d

[b]GWG: Energy need in pregnancy should be adjusted for actual bodyweight, observed weight gain and activity pattern for the population

[c]Energy needs of children and adolescents have been computed for reference children and adolescents; with a moderate daily physical activity level.

*The actual requirement in specific population groups should be adjusted for the actual weight and physical activity of that population.

Figure 31.5 Energy Requirement*

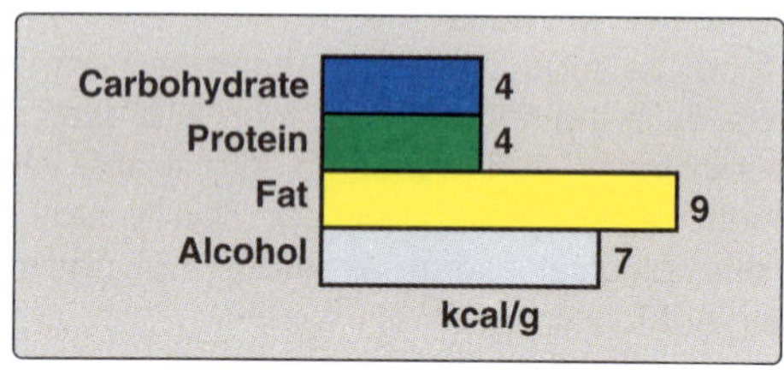

Figure 31.6 Average energy available from the macronutrients and alcohol.

such as respiration, blood flow, and ion transport. **When RMR is estimated for 24 hours, it is called resting energy expenditure (REE).**

RMR can be determined by measuring oxygen consumed or carbon dioxide produced. RMR reflects lean muscle mass, and can be estimated factoring in sex, age, height, and weight. A commonly used rough estimate for RMR is 1 kcal/kg/hour for men and 0.9 kcal/kg/hour for women. **RMR is about 10% higher than BMR.**

Basal metabolic rate (BMR) is **measured in more stringent conditions**, such as in the morning after an overnight fast, no exercise for the previous 24 hours, free from emotional stress, familiar with the apparatus, and the subject completely rested. It is not measured routinely.

For adult men weighing 70 kg, RMR is 1,800 kcal. For adult women weighing 50 kg, RMR is 1,300 kcal. REE contributes to 60%–75% of TEE in sedentary individuals. As disease conditions are hypercatabolic states, RMR is multiplied by an injury factor that ranges from 1.0 (mild infection) to 2.0 (severe burns) in calculating their TEE.

Respiratory quotient: Respiratory quotient, (RQ), is defined as the **volume of carbon dioxide released over the volume of oxygen absorbed during respiration**. It is a dimensionless number used in calculation for BMR.

As the food is metabolized for energy, oxygen is consumed, and carbon dioxide is produced. This uptake of oxygen is measured by a process known as indirect calorimetry.

Carbohydrates are oxidized through aerobic respiration resulting in an equal ratio of CO_2 release and oxygen consumption, therefore RQ for carbohydrates is 1. Fats undergo beta-oxidation to

SUBSTRATE	RQ
Carbohydrate	1.00
Protein	0.84
Fat	0.71

Figure 31.7 The respiratory quotient (RQ). [Note: For protein, the nitrogen is removed and excreted, and the α-keto acids are oxidized.]

form acetyl CoA which does not generate carbon dioxide. Therefore, when using fats over carbohydrates as a form of fuel, less CO_2 is generated for every oxygen consumed. The RQ for fats and proteins is 0.71 and 0.84, respectively (Fig. 31.7).

2. **Physical activity:** Muscular activity provides the greatest variation in the TEE. The amount of energy consumed depends on the duration and intensity of the exercise. This energy cost is expressed as a multiple of the RMR (range is 1.1 to >8.0) that is referred to as the **physical activity ratio (PAR) or the metabolic equivalent of the task (MET)**. In general, a lightly active person requires ~30%–50% more calories than the RMR (see Fig. 31.8), whereas a highly active individual may require ≥100% calories above the RMR.

3. **Thermic effect of food:** The amount of energy expenditure above the RMR due to the ingestion and digestion of food for use as energy or conversion to a storage form is called the thermic effect of food. It is also known as **specific dynamic action** (SDA) or **dietary induced thermogenesis**. The thermic effect of food may account for 5%–10% of the TEE.

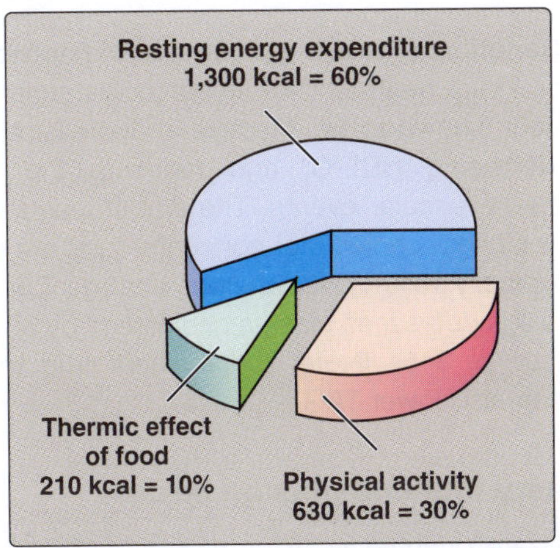

Figure 31.8 Estimated total energy expenditure in a healthy 20-year-old woman, 5 ft, 4 in (165 cm) tall, weighing 110 lb (50 kg), and engaged in light activity.

IV. PROXIMATE PRINCIPLES OF FOOD

Carbohydrates, fats and proteins are the three proximate principles of food. These are also known as macronutrients as they provide energy. Besides providing energy, these components also have many other structural and functional roles.

A. Acceptable macronutrient distribution ranges

Acceptable Macronutrient Distribution Ranges (AMDR) are defined as a range of intakes for a particular macronutrient that is associated with reduced risk of chronic disease while providing adequate amounts of essential nutrients.

The AMDR for adults is (Fig. 31.9):

- 45%–65% of total calories from carbohydrates
- 20%–35% of total calories from fats
- 10%–35% of total calories from proteins

The biologic properties of dietary fat, carbohydrate, and protein are described below.

MACRONUTRIENT	AMDR (percent of energy)
Fat	20–35
ω-6 Polyunsaturated fatty acids	5–10
ω-3 Polyunsaturated fatty acids	0.6–1.2*
Approximately 10% of the total fat can come from longer-chain, ω-3 or ω-6 fatty acids.	
Carbohydrate	45–65
• RDA: Men and women: 130 g/day	
No more than 10% of total calories should come from added sugars.	
Fiber	
• AI: Men: 38 g/day; women: 25 g/day	
Protein	10–35
• RDA: Men: 56 g/day; women: 46 g/day	

Figure 31.9 Acceptable Macronutrient Distribution Ranges (AMDR) in adults. [Note: *A growing body of evidence suggests that higher levels of ω-3 polyunsaturated fatty acids provide protection against coronary heart disease.] RDA = recommended dietary allowance; AI = adequate intake.

V. DIETARY FATS

Fats are a concentrated source of energy. They provide texture and palatability to the food. They are also required for intestinal absorption of fat-soluble vitamins. **Triacylglycerols (TAGs) constitute about 90% of dietary fat**, with smaller amounts of cholesterol, phospholipids, glycolipids, etc.

The type of fatty acid in TAG with respect to **number of double bonds** (saturated vs mono and polyunsaturated), **position of double bonds** (ω-3 vs ω-6) and **configuration** (trans vs cis) are important features which affect the influence of TAG on blood lipids.

The incidence of a number of chronic diseases is significantly influenced by the kinds and amounts of nutrients consumed (Fig. 31.10). **Dietary fats most strongly influence the incidence of coronary heart disease (CHD)**, but evidence linking dietary fat and the risk for cancer or obesity is much weaker.

> Earlier recommendations emphasized decreasing the total amount of dietary fat. Unfortunately, this resulted in increased consumption of refined grains and added sugars. Data now show that the type of fat is a more important risk factor than the total amount of fat.

A. Plasma lipids and coronary heart disease

Plasma cholesterol may arise from the diet or from endogenous biosynthesis. In either case, cholesterol is transported between the tissues in combination with protein and phospholipids as lipoproteins.

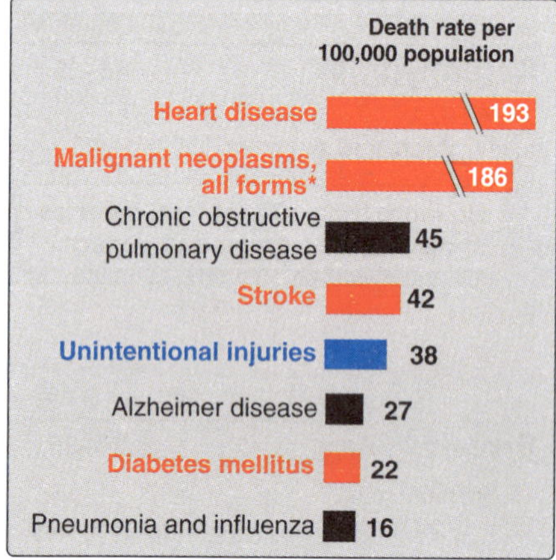

Figure 31.10 Influence of nutrition on some common causes of death in the United States in the year 2010. Red indicates causes of death in which the diet plays a significant role. Blue indicates causes of death in which excessive alcohol consumption plays a part. [Note: *Diet plays a role in only some forms of cancer.]

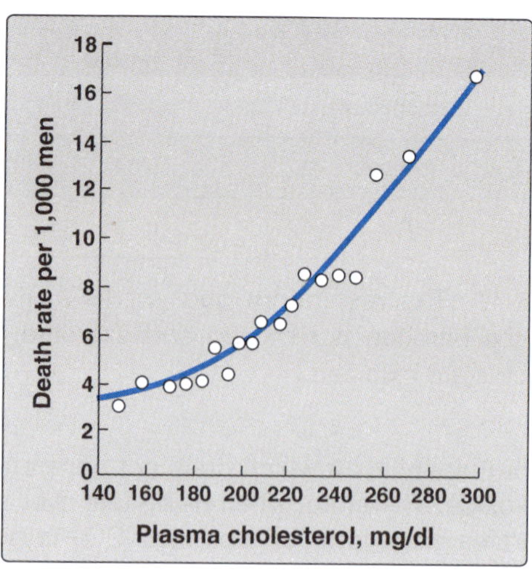

Figure 31.11 Correlation of the death rate from coronary heart disease with the concentration of plasma cholesterol. [Note: The data were obtained from a multiyear study of men with the death rate adjusted for age.]

1. **Low-density and high-density lipoproteins:** The level of plasma cholesterol is not precisely regulated but, rather, varies in response to diet. Elevated levels of total cholesterol (hypercholesterolemia) result in an increased risk for CHD (Fig. 31.11). A much stronger correlation exists between CHD and the level of cholesterol in low-density lipoproteins ([LDL-C], see p. 357). **As LDL-C increases, CHD increases**. In contrast, **elevated levels of high-density lipoprotein cholesterol (HDL-C) have been associated with a decreased risk for heart disease** (see p. 358). [Note: Elevated plasma triacylglycerol (TAG) is associated with CHD, but a causative relationship has yet to be demonstrated.] Abnormal levels of plasma lipids (dyslipidemias) act in combination with smoking, obesity, sedentary lifestyle, insulin resistance, and other risk factors to increase the risk of CHD.

2. **Benefits of lowering plasma cholesterol:** Dietary or drug treatment of hypercholesterolemia has been shown to be effective in decreasing LDL-C, increasing HDL-C, and reducing the risk for cardiovascular events. The diet-induced changes in plasma cholesterol concentrations are modest, typically 10%–20%, whereas treatment with statin drugs decreases plasma cholesterol by 30%–60% (see p. 345). [Note: Dietary and drug treatment can also lower TAG.]

B. Dietary fats and plasma lipids

TAGs are quantitatively the most important class of dietary fats. The influence of TAG on blood lipids is determined by the chemical nature of their constituent fatty acids. The absence or presence and number of double bonds (saturated versus mono- and polyunsaturated), the location of the double bonds (ω-6 versus ω-3), and the cis versus trans configuration of the unsaturated fatty acids are the most important structural features that influence blood lipids.

1. **Saturated fats:** TAG composed primarily of fatty acids whose hydrocarbon chains do not contain any double bonds are referred to as saturated fats. **Consumption of saturated fats is positively associated with high levels of total plasma cholesterol and LDL-C and an increased risk of CHD.** The main sources of saturated fatty acids are dairy and meat products and some vegetable oils, such as coconut and palm oils (a major source of fat in Latin America and Asia, although not in the United States). Many experts strongly advise limiting intake of saturated fats to <10% of total caloric intake and replacing them with unsaturated fats (and whole grains).

> Saturated fatty acids with carbon chain lengths of 14 (myristic) and 16 (palmitic) are most potent in increasing the plasma cholesterol level. Stearic acid (18 carbons, found in many foods including chocolate) has little effect on blood cholesterol.

2. **Monounsaturated fats:** TAG containing primarily fatty acids with one double bond are referred to as monounsaturated fats. These are **important components of membrane structural lipids, particularly in nervous tissue myelin.** Monounsaturated fatty acids (MUFA) are generally obtained from plant-based oils. When substituted for saturated fatty acids in the diet, **MUFA lower both total plasma cholesterol and LDL-C and maintain or increase HDL-C.** This ability of MUFA to favorably modify lipoprotein levels may explain, in part, the observation that Mediterranean cultures, with diets rich in olive oil (high in monounsaturated oleic acid), show a low incidence of CHD. [Note: Although there is no AMDR for MUFA, a common recommendation is 10%–20% of caloric intake.]

   a. **The Mediterranean diet:** The Mediterranean diet is an example of a diet rich in MUFA (from olive oil) and polyunsaturated fatty acids or PUFA (from fish oils, plant oils, and some nuts) but low in saturated fat. For example, Figure 31.12 shows the composition of the Mediterranean diet in comparison with both a Western diet similar to that consumed in the United States and a typical low-fat diet. The Mediterranean diet contains seasonally fresh food, with an abundance of plant material, low amounts of red meat, and olive oil as the principal source of fat. The Mediterranean diet is associated with decreased plasma total cholesterol and LDL-C, decreased TAG, and increased HDL-C when compared with a typical Western diet higher in saturated fats.

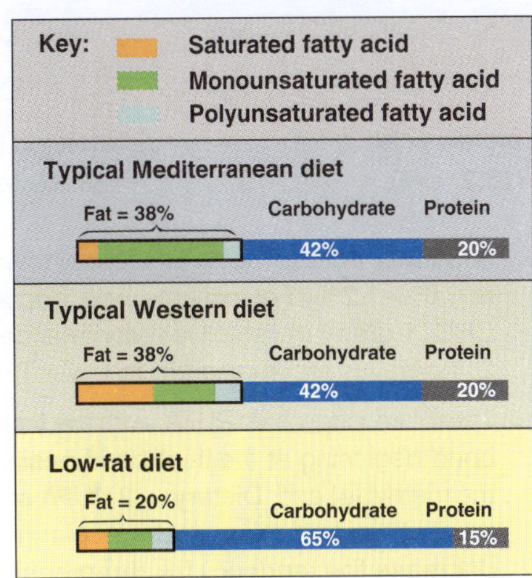

Figure 31.12 Composition of typical Mediterranean, Western, and low-fat diets.

3. **Polyunsaturated fats:** TAG containing primarily fatty acids with more than one double bond are referred to as polyunsaturated fats. Depending on the location of the double bond, PUFA can be ω-3 or ω-6. The effects of PUFA on cardiovascular disease are influenced by the location of the double bonds within the molecule.

   a. **ω-6 Fatty acids:** These are long-chain PUFA, with the first double bond beginning at the sixth bond position when starting from the methyl (ω) end of the fatty acid molecule. [Note: They are also called n-6 fatty acids (see p. 300).] Consumption of fats containing ω-6 PUFA, principally linoleic acid (18:2 [9,12]), obtained from vegetable oils, lowers plasma cholesterol when substituted for saturated fats. **Plasma LDL-C is lowered, but HDL-C, which protects against CHD, is also lowered, partially offsetting the benefits of lowering LDL-C.**

   Nuts, avocados, olives, soybeans, and various oils, including sunflower and corn oil, are common sources of these fatty acids. The AMDR for linoleic acid is 5%–10%. [Note: The lower recommendation for intake of PUFA relative to MUFA is because of concern that free radical–mediated oxidation (peroxidation) of PUFA may lead to deleterious products.]

Linoleic acid (18:2, ω-6) **α-Linolenic acid (18:3, ω-3)**

b. **ω-3 Fatty acids:** The AMDR for α-linolenic acid is 0.6%–1.2%. For patients with documented CHD, 1 g/day of fish oil is recommended while 2–4 g/day is recommended to lower TAG.

These are long-chain PUFA, with the first double bond beginning at the third bond position from the methyl (ω) end. **Dietary ω-3 PUFA suppress cardiac arrhythmias, reduce plasma TAG, decrease the tendency for thrombosis, lower blood pressure, and substantially reduce risk of cardiovascular mortality (Fig. 31.13), but they have little effect on LDL-C or HDL-C levels.** Evidence suggests that they have anti-inflammatory effects. The ω-3 PUFA, principally α-linolenic acid, 18:3 (9, 12, 15), are found in plant oils, such as flaxseed and canola, and some nuts, such as walnuts. The AMDR for α-linolenic acid is 0.6%–1.2%.

Fish oil contains the long-chain ω-3 **docosahexaenoic acid (DHA, 22:6)** and **eicosapentaenoic acid (EPA, 20:5)**. Two fatty fish (for example, salmon) meals per week are recommended. For patients with documented CHD, 1 g/day of fish oils is recommended, while 2–4 g/day is prescribed to lower TAG. DHA is a structural constituent of membranes specifically in the central nervous system. **DHA is known to promote brain development.** Since endogenous production of DHA is relatively low, it is included in infant formulas. Linoleic and α-linolenic acids are essential fatty acids (EFA) required for membrane fluidity and synthesis of eicosanoids (see p. 333). EFA deficiency, caused primarily by fat malabsorption, is characterized by scaly dermatitis as a result of the depletion of skin ceramides with long-chain fatty acids (see p. 326).

4. **Trans fatty acids:** Trans fatty acids (Fig. 31.14) are unsaturated fatty acids which contain at least one double bond in the trans configuration. These are **chemically classified as unsaturated fatty acids but behave more like saturated fatty acids in the body because they elevate LDL-C and lower HDL-C, thereby increasing the risk of CHD.** Trans fatty acids do not occur naturally in plants but occur in small amounts in animals. However, **trans fatty acids are formed during the hydrogenation of vegetable oils** (for example, in the manufacture of margarine and partially hydrogenated vegetable oil). Trans fatty acids are a major component of many commercial baked goods, such as cookies, and most deep-fried foods. Many manufacturers have reformulated their products to be free of trans fats. In 2006, the U.S. Food and Drug Administration required that Nutrition Facts labels (see p. 515) portray the trans fat content of packaged food. By 2018, virtually no industrial trans fatty acids will be permitted in food.

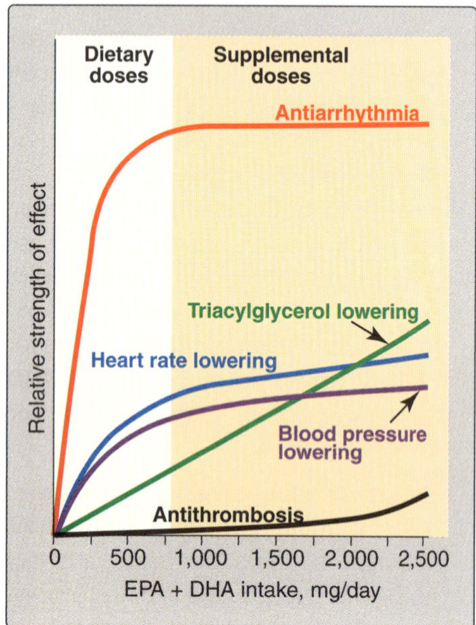

Figure 31.13 Dose responses of physiologic effects of fish oil (ω-3) intake. EPA = eicosapentaenoic acid (20:5); DHA = docosahexaenoic acid (22:6).

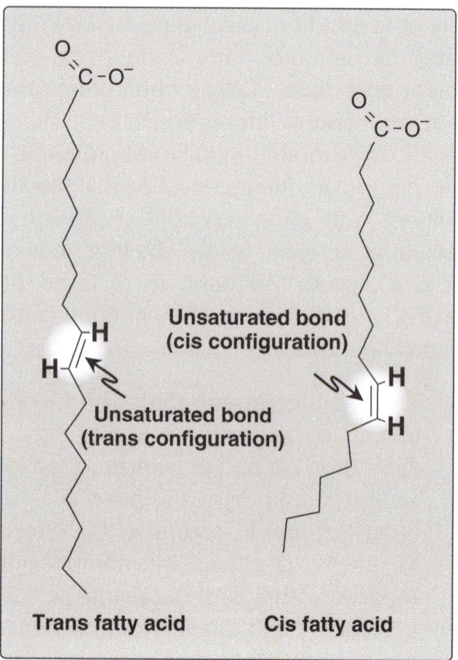

Figure 31.14 Structure of cis and trans fatty acids.

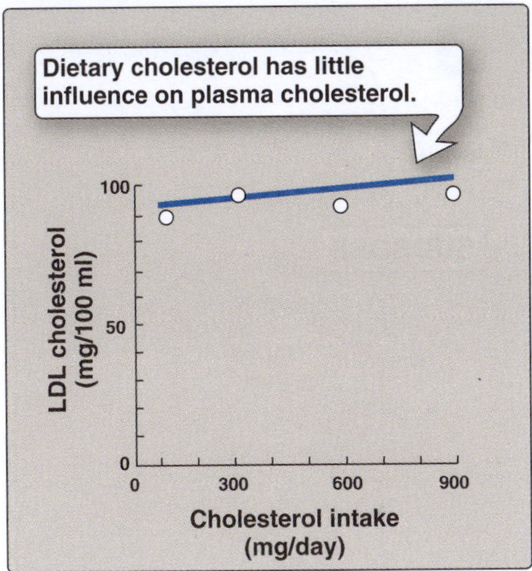

Figure 31.15 Response of plasma low-density lipoprotein (LDL) cholesterol concentrations to an increase in dietary cholesterol intake.

5. **Dietary cholesterol:** Cholesterol is found only in animal products. The effect of dietary cholesterol on plasma cholesterol (Fig. 31.15) is less important than the amount and types of fatty acids consumed. Many experts recommend ≤300 mg/day. However, having an upper limit has become controversial.

C. Other dietary factors affecting coronary heart disease

Moderate consumption of alcohol (up to 1 drink/day for women and up to 2 drinks/day for men) decreases the risk of CHD, because there is a positive correlation between moderate alcohol (ethanol) consumption and the plasma concentration of HDL-C. However, because of the potential dangers of alcohol abuse, health professionals are reluctant to recommend increased alcohol consumption to their patients. Red wine may provide cardioprotective benefits in addition to those resulting from its alcohol content (for example, red wine contains phenolic compounds that inhibit lipoprotein oxidation; see p. 357). [Note: These antioxidants are also present in raisins and grape juice.] Figure 31.16 summarizes the effects of dietary fats. [Note: Recent studies (including meta-analyses) have raised questions concerning the current guidelines for dietary fat in the prevention of CHD.]

VI. DIETARY CARBOHYDRATES

The carbohydrates constitute a major portion of the diet and provide 40–65% of daily energy requirements. The rising incidence of obesity has coincided with increased carbohydrate consumption (as fat consumption decreased), leading some observers to link obesity with carbohydrate consumption. However, **obesity has also been related to increasingly inactive lifestyles and to calorie-dense foods served in expanded portion size.** Carbohydrates are not inherently fattening.

A. Dietary carbohydrates can be of two types.

1. **Digestible carbohydrates:** These can be digested and utilized for energy production. Examples are starch, sucrose, etc.

TYPE OF FAT	METABOLIC EFFECTS	EFFECTS ON DISEASE PREVENTION
Trans fatty acid	↑ LDL ↓ HDL	↑ Incidence of coronary heart disease
Saturated fatty acid	↑ LDL Little effect on HDL	↑ Incidence of coronary heart disease; may increase risk of prostate, colon cancer
Monounsaturated fatty acid	↓ LDL Maintain or increase HDL	↓ Incidence of coronary heart disease
Polyunsaturated ω–6 fatty acids such as linoleic acid	↓ LDL ↓ HDL Provide arachidonic acid, which is an important precursor of prostaglandins and leukotrienes	↓ Incidence of coronary heart disease
Polyunsaturated ω–3 fatty acids such as DHA	Little effect on LDL Little effect on HDL Suppress cardiac arrhythmias, reduce serum triacylglycerols, decrease the tendency for thrombosis, lower blood pressure, reduce inflammation	↓ Incidence of coronary heart disease ↓ Risk of sudden cardiac death

Figure 31.16 Effects of dietary fats. LDL = low-density lipoprotein; HDL = high-density lipoprotein; DHA = docosahexaenoic acid.

2. **Indigestible carbohydrates:** These cannot be digested, hence do not provide energy. These are **useful as dietary fibers**. Examples are cellulose, hemicellulose, pectins, and lignin.

   a. **High-fructose corn syrup:** High-fructose corn syrups (HFCS) are corn syrups that **have undergone enzymatic processing to convert their glucose into fructose and have then been mixed with pure corn syrup (100% glucose) to produce a desired sweetness.** In the United States, HFCS 55 (containing 55% fructose and 42% glucose) is commonly used as a substitute for sucrose in beverages, including soft drinks, with HFCS 42 used in processed foods. **The composition and metabolism of HFCS and sucrose are similar, the major difference being that HFCS is ingested as a mixture of monosaccharides** (Fig. 31.17). Most studies have shown no significant difference between sucrose and HFCS meals in either postprandial glucose or insulin responses. [Note: The rise in the use of HFCS parallels the rise in obesity, but a causal relationship has not been demonstrated.]

   b. **Fiber**

      i. **Dietary fibers:** Dietary fibers are carbohydrates that are not digested in the small intestine and do not provide energy. Fiber can absorb 10–15 times its own weight in water, drawing fluid into the lumen of the intestine and promoting bowel motility.

      These include soluble fibers (cellulose and lignin) and insoluble fibers (pectins). Soluble fibers are completely or partially fermented by bacteria to short-chain fatty acids in the large intestine. Insoluble fibers pass through the large intestine unchanged. Fruits, vegetables, whole grains, and legumes are rich in fibers.

      Dietary fibers are an important component of diet.

      ii. **Advantages of fiber**

      - Dietary fiber retains water, absorbing 10–15 times its own weight in water, **adding bulk to the stool and softening it**, making it easier to pass (Fig. 31.18).
      - A high-fiber diet **lowers risk of developing constipation**, hemorrhoids and diverticulosis.
      - Fiber delays gastric emptying, resulting in a **sensation of satiety**.

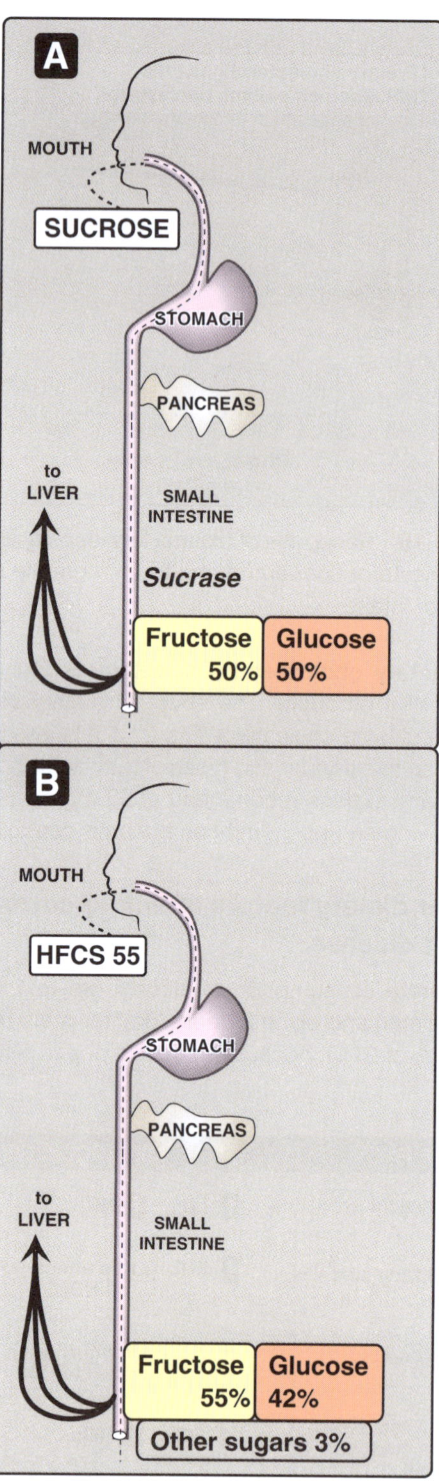

Figure 31.17 Digestion of sucrose (**A**) or high-fructose corn syrup (HFCS) 55 (**B**) leads to absorption of glucose plus fructose.

      - Fiber can also reduce spikes in post-meal blood glucose, thereby reducing the glycemic response.
      - Dietary fiber **interferes with intestinal bile acid reabsorption, reducing cholesterol and LDL levels**. This

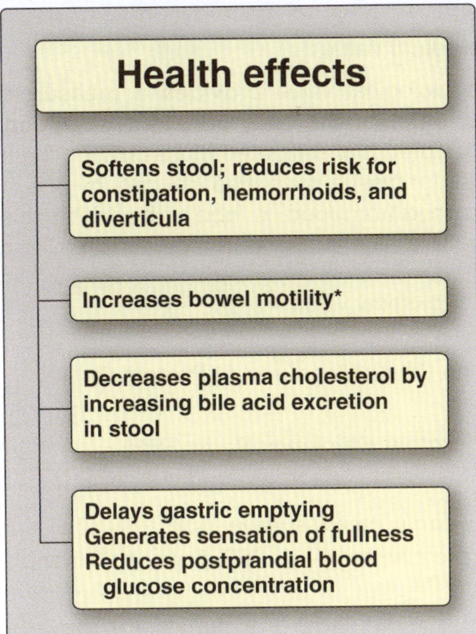

Figure 31.18 Actions of dietary fiber. [Note: *Increasing bowel motility decreases exposure time of the intestines to carcinogens.]

reduces the risk of coronary heart disease (see p. 347). For example, diets rich (25–50 g/day) in the soluble fibers such as oat bran are associated with a modest, but significant reduction in risk for CHD by lowering total cholesterol and LDL-C levels.

- A high-fiber diet likely **lowers the risk of colorectal cancer**. Increasing bowel motility reduces the exposure time of the intestine to carcinogens.

The AI for dietary fiber is 25 g/day for women and 38 g/day for men. However, traditional Indian diets are high-fiber diets. However, nowadays urban diets are rich in processed food and have far lower fiber content. [Note: "Functional fiber" is the term used for isolated fiber that has proven health benefits such as commercially available fiber supplements.]

B. Dietary carbohydrate and blood glucose

Some carbohydrate-containing foods produce a rapid rise followed by a steep fall in blood glucose concentration, whereas others result in a gradual rise followed by a slow decline (Fig. 31.19). Thus, they differ in their glycemic response (GR). [Note: Fiber blunts the GR.]

Glycemic index (GI) is a measure of the effects of carbohydrates on blood glucose levels. It ranks carbohydrate-rich foods on a scale of 0–100. **Glycemic load is calculated as the area under the 2-hour blood glucose response curve after ingestion of 50 g of carbohydrate.** Glycemic load of a food, compared with glycemic load of standard (glucose/white bread) gives the GI.

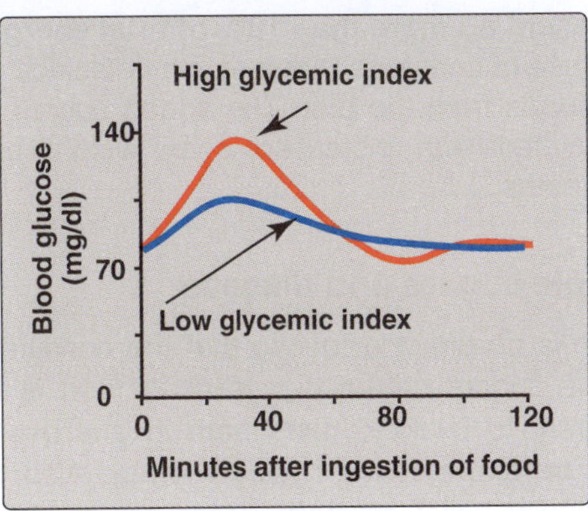

Figure 31.19 Blood glucose concentrations following ingestion of food with a low or high glycemic index (GI). [Note: The GI is defined as the area under the blood glucose curve.]

A low GI is <55, whereas a high GI is ≥70. Evidence suggests that a **low-GI diet improves glycemic control in diabetic individuals**. Food with a low GI tends to create a sense of satiety over a longer period of time and may be helpful in limiting caloric intake. [Note: How much a typical serving size of a food raises blood glucose is referred to as the glycemic load. A food (for example, carrots) can have a high GI and a low glycemic load.]

1. **Foods with high GI:** White wheat bread, whole wheat/whole meal bread, corn flakes, rice porridge, oats porridge, boiled potatoes, boiled white rice.

2. **Foods with low GI:** Barley, boiled carrot, kidney beans, chickpeas, soybeans, milk (full fat/skimmed).

C. Carbohydrate requirements

Carbohydrates are **not essential nutrients**, because the carbon skeletons of most amino acids can be converted into glucose (see p. 397). **However, the absence of dietary carbohydrate leads to ketogenesis (see pp. 314–315) and degradation of body protein whose constituent amino acids provide carbon skeletons for gluconeogenesis** (see p. 215).

The RDA **for carbohydrate is set at 130 g/day for adults and children**, based on the amount of glucose used by carbohydrate-dependent tissues, such as the brain and erythrocytes. However, this level of intake is usually exceeded. **Adults should consume 45%–65% of their total calories from carbohydrates. It is now recommended that added sugars**

represent no more than 10% of total energy intake because of concerns that they may displace nutrient-rich foods from the diet. The added sugars are also associated with increased body weight and type 2 diabetes.

D. Simple sugars and disease

There is no direct evidence that the consumption of simple sugars naturally present in food is harmful. Contrary to folklore, **diets high in sucrose do not lead to diabetes or hypoglycemia**. Also contrary to popular belief, **carbohydrates are not inherently fattening**. They yield 4 kcal/g (the same as protein and less than one half that of fat; see Fig. 31.6) and **result in fat synthesis only when consumed in excess of the body's energy needs**. However, there is an association between sucrose consumption and dental caries, particularly in the absence of fluoride treatment (see p. 560).

VII. DIETARY PROTEIN

The AMDR for protein is 10%–35%. Dietary protein provides the essential amino acids (see Fig. 25.2, p. 397). Nine of the 20 amino acids needed for the synthesis of body proteins are essential (that is, they cannot be synthesized in humans).

A. Protein quality

The quality of a dietary protein is a measure of its ability to provide the essential amino acids (EAA) required for tissue maintenance. Most government agencies have adopted the **Protein Digestibility–Corrected Amino Acid Score** (PDCAAS) as the standard by which to evaluate protein quality. PDCAAS is based on the profile of EAA after correcting for the digestibility of the protein. The highest possible score under these guidelines is 1.00. This amino acid score provides a method to balance intakes of poorer-quality proteins with high-quality dietary proteins.

1. **Proteins from animal sources:** Proteins from animal sources (meat, poultry, milk, and fish) have a high quality because they contain all the EAA in proportions similar to those required for synthesis of human tissue proteins (Fig. 31.20), and they are more readily digested. [Note: Gelatin prepared from animal collagen is an exception. It has a low biologic value as a result of deficiencies in several EAA.]

2. **Proteins from plant sources:** Plant proteins have a lower quality than do animal proteins. However, proteins from different plant sources may be combined in such a way that the result is equivalent in nutritional value to animal protein. This is known as mutual supplementation (see box). Animal proteins can also complement the biologic value of plant proteins.

Biological value

The biological value **provides a measurement of how efficient the body utilizes protein consumed in the diet.** It is estimated by calculating the nitrogen used for tissue formation divided by the nitrogen absorbed from food. A food with a high value correlates to a high supply of the essential amino acids.

$$\text{Biological value} = \frac{\text{Nitrogen retained}}{\text{Nitrogen absorbed}} \times 100$$

Some Other Measures of Protein Quality

(a) ***Protein Efficiency Ratio*** *(PER)* $= \frac{\text{Gain in body weight (g)}}{\text{Protein fed (g)}}$

Animal proteins have high PER (2.5) and plant proteins have low PER (between 0.5 and 2.5).

(b) ***Chemical Score*** $= \frac{\text{mg of limiting amino acid in 1 gram test protein}}{\text{mg of same amino acid in 1 gram of reference protein}} \times 100$

Egg protein has a chemical score of 100 and is considered a reference protein.

(c) ***Net Protein Utilization*** *(NPU)* $= \frac{\text{Nitrogen retained}}{\text{Intake of nitrogen}} \times 100$

Source	PDCAAS value
Animal proteins	
Egg	1.00
Milk protein	1.00
Beef/poultry/fish	0.82–0.92
Gelatin	0.08
Plant proteins	
Soybean protein	1.00
Kidney beans	0.68
Whole wheat bread	0.40

Figure 31.20 Relative quality of some common dietary proteins. PDCAAS = Protein Digestibility–Corrected Amino Acid Score.

Limiting amino acids

In any food, the essential amino acid which is in smallest quantity is known as limiting amino acid. Most plant-based proteins have a limiting amino acid. Protein synthesis is limited by the rate at which this limiting amino acid is available. For example, for wheat and rice, lysine is the limiting amino acid; for maize, lysine and tryptophan are limiting amino acids.

To overcome this limit, two incomplete proteins can be combined to produce a complete protein. This is known as **mutual supplementation**. For example, **wheat (lysine deficient but methionine rich) can be combined with kidney beans (lysine rich but methionine deficient)** to provide all essential amino acids (Fig 31.21). For vegetables, methionine is a limiting amino acid but combining them with grains or nuts can provide complete protein.

B. Nitrogen balance

Nitrogen balance occurs when the amount of nitrogen consumed equals that of the nitrogen excreted in the urine (primarily as urinary urea nitrogen, or UUN), sweat, and feces. Most healthy adults are normally in nitrogen balance. [Note: There is, on average, 1 g nitrogen in 6.25 g protein.]

1. **Positive nitrogen balance:** This occurs when nitrogen intake exceeds nitrogen excretion. It is observed during situations in which tissue growth occurs, for example, in childhood, pregnancy, or during recovery from an emaciating illness.

2. **Negative nitrogen balance:** This occurs when nitrogen loss is greater than nitrogen intake. It is associated with inadequate dietary protein; lack of an essential amino acid; or during physiologic stresses, such as trauma, burns, illness, or surgery.

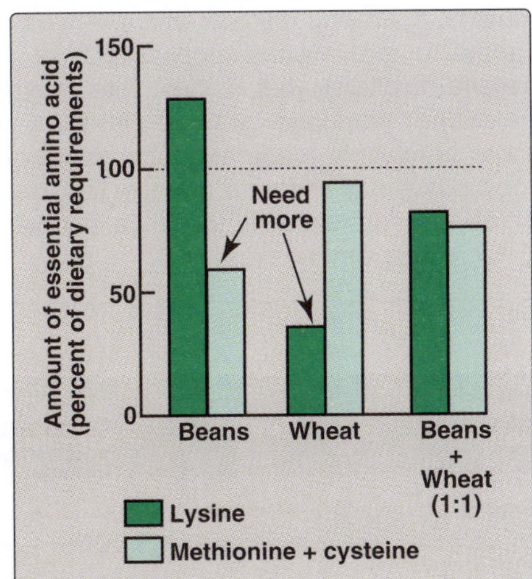

Figure 31.21 Combining two incomplete proteins that have complementary amino acid deficiencies results in a mixture with a higher biologic value.

Nitrogen (N) balance (g N_{in} − g N_{out}) in a 24-hour period can be determined by the formula, N balance = protein intake in g/6.25 − (UUN + 4 g), where 4 g accounts for urinary loss in forms other than UUN plus loss in skin and feces.

C. Protein requirements

The amount of dietary protein required in the diet varies with its biologic value. **The greater the proportion of animal protein in the diet, the less protein is required.**

- The RDA for protein is computed for proteins of mixed biologic value at 0.8 g/kg of body weight for adults, or ~56 g of protein for a 70-kg individual.
- People who exercise strenuously on a regular basis may benefit from extra protein to maintain muscle mass, and a daily intake of ~1 g/kg has been recommended for athletes.
- Women who are pregnant or lactating require up to 30 g/day in addition to their basal requirements.
- To support growth, infants should consume 2 g/kg/day.
- Disease states influence protein needs. Protein restriction may be needed in kidney disease, whereas burns require increased protein intake.

1. **Consumption of excess protein:** There is no physiologic advantage to the consumption of more protein than the RDA. Protein consumed in excess of the body's needs is deaminated, and the resulting carbon skeletons are metabolized to provide energy or acetyl coenzyme A for fatty acid synthesis. When excess protein is eliminated from the body as urinary nitrogen, it is often accompanied by increased urinary calcium, thereby increasing the risk of nephrolithiasis (kidney stones) and osteoporosis.

2. **The protein-sparing effect of carbohydrates:** The dietary protein requirement is influenced by the carbohydrate content of the diet. When the intake of carbohydrates is low, amino acids are deaminated to provide carbon skeletons for the synthesis of glucose that is needed as a fuel by the central nervous system. If carbohydrate intake is <130 g/day, substantial amounts of protein are metabolized to provide precursors for gluconeogenesis. Therefore, carbohydrate is considered to be "protein-sparing," because it allows amino

acids to be used for repair and maintenance of tissue protein rather than for gluconeogenesis.

D. Protein-energy (calorie) malnutrition

Protein-energy malnutrition (PEM) is a spectrum of conditions caused by protein and/or energy deficiencies. In developed countries, protein-energy malnutrition, also known as protein-energy undernutrition (PEU), is most commonly seen in patients with medical conditions that decrease appetite or alter how nutrients are digested or absorbed or in hospitalized patients with major trauma or infections. [Note: Such highly catabolic patients frequently require intravenous (IV, or parenteral) or tube-based (enteral) administration of nutrients.] PEM may also be seen in children or the elderly who are malnourished. **In developing countries, an inadequate intake of protein and/or calories is the primary cause of PEM.** Affected individuals show a variety of symptoms, including a depressed immune system with a reduced ability to resist infection. Death from secondary infection is common. **PEM is a spectrum of degrees of malnutrition, and two extreme forms are kwashiorkor and marasmus** (Fig. 31.22). [Note: Marasmic kwashiorkor has features of both forms.]

> PEM is caused by deficient intake or absorption of food (anorexia, oral intake restriction, malabsorption, etc) or due to conditions associated with increased energy requirements (growth, infection, trauma, cancer, etc.). **In children, undernutrition is often termed failure to thrive, in which a child fails to gain weight or linear growth or head circumference appropriately.**

1. **Kwashiorkor:** Kwashiorkor occurs **when protein deprivation is relatively greater than the reduction in total calories.** Protein deprivation is associated with severely decreased synthesis of visceral protein. Kwashiorkor is commonly seen in developing countries **in children after weaning at about age 1 year**, when their diet consists predominantly of carbohydrates or when there is very less gap in successive pregnancies. Typical symptoms include stunted growth, skin lesions, depigmented hair, anorexia, fatty liver, bilateral pitting edema, and decreased serum albumin concentration. **Edema results from the lack of adequate blood proteins**, primarily albumin, to maintain the distribution of water between blood and tissues. It may mask muscle and fat loss. The **child has a plump appearance** due to edema. Therefore, chronic malnutrition is reflected in the level of serum albumin. Nutritional repletion restores hair color, leaving a band of hair with altered pigmentation followed by a band with normal pigmentation (**flag sign**).

   [Note: Because caloric intake from carbohydrates may be adequate, insulin levels suppress lipolysis and proteolysis. Kwashiorkor is, therefore, nonadapted malnutrition.]

2. **Marasmus:** Marasmus occurs **when calorie deprivation is relatively greater than the reduction in protein.** It usually occurs in developing countries in children younger than age 1 year when breast milk is supplemented or replaced with watery gruels of native cereals that are usually deficient in both protein and calories. Typical symptoms include **arrested growth, extreme muscle wasting and depletion of subcutaneous fat** (emaciation), weakness, and anemia (Fig. 31.23). Individuals with marasmus **do not show the edema** observed in kwashiorkor. [Note: Refeeding severely malnourished individuals can result in hypophosphatemia (see p. 553), because any available phosphate is used to phosphorylate carbohydrate intermediates. Milk is frequently given because it is rich in phosphate.]

> **Cachexia**, a wasting disorder characterized by loss of appetite and muscle atrophy (with or without increased lipolysis) that cannot be reversed by conventional nutritional support, is seen with a number of chronic diseases, such as cancer and chronic pulmonary and renal disease. It is associated with decreased treatment tolerance and response and decreased survival time.

Type of PEM	Weight for age (% expected)	Weight for height	Edema	Muscle and fat content
Kwashiorkor	60–80	Normal or ↓	Present	↓
Marasmus	<60	Markedly ↓	Absent	Markedly ↓

Figure 31.22 Physical features of extreme protein-energy malnutrition (PEM) in children. [Note: The fatty liver and skin and hair changes of kwashiorkor are not seen in marasmus.]

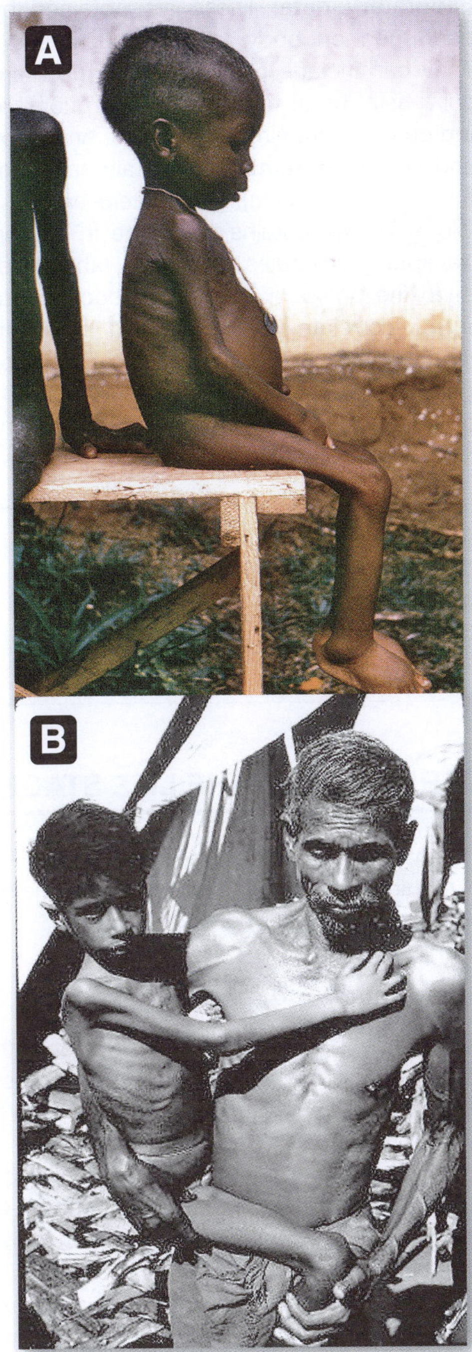

Figure 31.23 **A.** Listless child with kwashiorkor. Note the swollen belly and lower legs. **B.** Child suffering with marasmus.

VIII. NUTRITION TOOLS

A set of tools has been developed that gives consumers information about what (and how much) they should eat as well as the nutritional content of the foods they do eat. Additional tools allow medical professionals to assess whether or not the nutritional needs of an individual are being met.

A. MyPlate

MyPlate was designed by the U.S. Department of Agriculture (USDA) to graphically illustrate its recommendations as to what food groups and how much of each should be consumed daily. In MyPlate, the relative amounts of each of five food groups (vegetables, grains, protein, fruit, and dairy) are represented by the relative size of their section on the plate (Fig. 31.24). The number of servings depends on variables that include age and sex. [Note: MyPlate replaced MyPyramid in 2011.]

B. Nutrition facts label

Most types of packaged goods are required to have a Nutrition Facts label, or "food label" (Fig. 31.25),

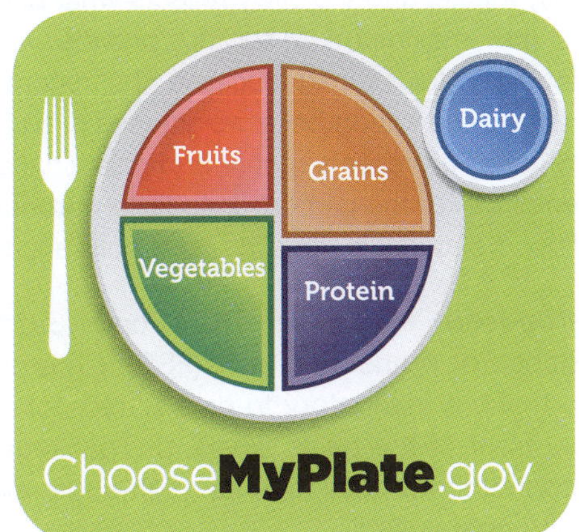

Figure 31.24 MyPlate.

Nutrition Facts			
Serving size 45*	Qty per 100g	Qty per 45g	% RDA# per serve
Energy (kcal)	344	155	8
Protein (g)	22	10	22
Carbohydrate (g)	66	30	23
Total sugars (g)	0	0.0	0.0
Added sugars (g)	0	0.0	0.0
Total dietary fiber (g)	11	5	12
Total fat (g)	1.5	0.7	1
Saturated fat (g)	0.1	0.06	0.3
Trans fat (g)	0	0.0	0.0
Cholesterol (mg)	0	0.0	-
Sodium (mg)	54	24	1.2

*Pack contains approximately 5 servings.

#Percent recommended dietary allowance based on a 2000 calorie diet (sedentary woman, ICMR RDA 2020). Your daily value may be higher or lower depending on your calorie needs.

Figure 31.25 Sample nutrition fact label.

that includes the size of a single serving, the Cal it provides, and the number of servings per container. In addition, a percent daily value (%DV) is shown for most nutrients listed. [Note: The %DV is based on a 2,000-Cal diet for healthy adults.]

1. **Percent daily value:** The %DV compares the amount of a given nutrient in a single serving of a product to the recommended daily intake for that nutrient. For example, the %DV for the micronutrients listed, as well as for total carbohydrates and fiber, are based on their recommended minimum daily intake. Thus, if the label lists 20% for calcium, one serving provides 20% of the minimum recommended amount of calcium needed each day. In contrast, the %DV for saturated fat, cholesterol, and sodium are based on their recommended maximum daily intake, and the %DV reflects what percentage of this maximum a serving provides. There is no %DV for protein because the recommended intake depends on body weight (see p. 513). [Note: "Sugars" represents mono- and disaccharides. The remainder of the carbohydrate (total carbohydrate – [fiber + sugars]) is the oligo- and polysaccharides.]

2. **Proposed revisions:** The Food Safety and Standards Authority of India (FSSAI), has proposed to make it mandatory to display red color-coding on front-of-the-pack labels on packaged food products that have high-fat, high-sugar and high-salt content levels.

C. Nutrition assessment

Evaluation of nutritional status on the basis of clinical information is nutritional assessment. This assessment can help in identifying patients who can benefit from nutritional therapy, that is the change in diet for treatment of disease conditions (for example, replacement of long-chain TAG with medium-chain TAG in malabsorption disorders) and/or the method of intake (for example, enteral [tube] or parenteral [IV] feeding). The assessment includes:

1. **Dietary history:** This is the record of food intake over a period of time. Food diary can be used to record specific types and amount of food eaten for a period of 3–7 days. Food frequency questionnaire and 24-hour recall of specific foods can also be used.

2. **Anthropometric measures:** These are the physical measurements of the body. These include weight, height, body mass index (BMI), skin fold thickness of triceps or subscapular (indicator of subcutaneous fat), and waist circumference (an indicator of abdominal fat).

3. **Laboratory data:** These are obtained by tests performed on body fluids, tissues, and waste. They can include **plasma LDL-C** (for cardiovascular risk), **fecal fat** (for malabsorption), **red cell indices** (for vitamin deficiencies), and **N balance** and **serum proteins** (such as albumin and transthyretin [prealbumin]) for protein–energy status. [Note: These proteins are made in the liver and transport molecules such as fatty acids and thyroxine (see p. 560) through blood. Low albumin levels correlate with increased morbidity and mortality in hospitalized patients. The short half-life (2–3 days) of transthyretin as compared to that of albumin (20 days) has led to its use in monitoring the progress of hospitalized patients.]

> Nutritional insufficiency can be the result of inadequate nutrient intake (caused, for example, by an inability to eat, loss of appetite, or decreased availability), inadequate absorption, decreased utilization, increased excretion, or increased requirements.

IX. NUTRITION AND THE LIFE STAGES

Macronutrient energy sources, micronutrients, EFA, and EAA are required at every life stage. Additionally, each stage has specific nutrition needs.

A. Infancy, childhood, and adolescence

The rapid growth and development in infancy (birth to age 1 year) and childhood (age 1 year to adolescence) necessitate higher energy and protein needs relative to body size than are required in subsequent life stages. In adolescence, the marked increases in height and weight that occur increase nutritional needs. Growth charts (Fig. 31.26) are used to compare an individual's stature (height) and/or weight to the expected values for others of the same age (≤20 years) and sex. They are based on data from large numbers of normal individuals over time. [Note: Deviations from the expected growth curve, as reflected in the crossing of two or more percentile lines, raise concern.]

1. **Infants:** Breast milk is ideal infant nutrition because it contains calories as well as most micronutrients in appropriate quantities. Breast milk contains carbohydrates (lactose and nearly 200 unique oligosaccharides), fats and proteins in a 7:3:1 ratio. [Note: About 90% of the microbiota (the population of microbes) in the breast-fed infant's intestine is represented by one type, Bifidobacterium infantis, which expresses all the enzymes needed to degrade these complex sugars. The sugars, in

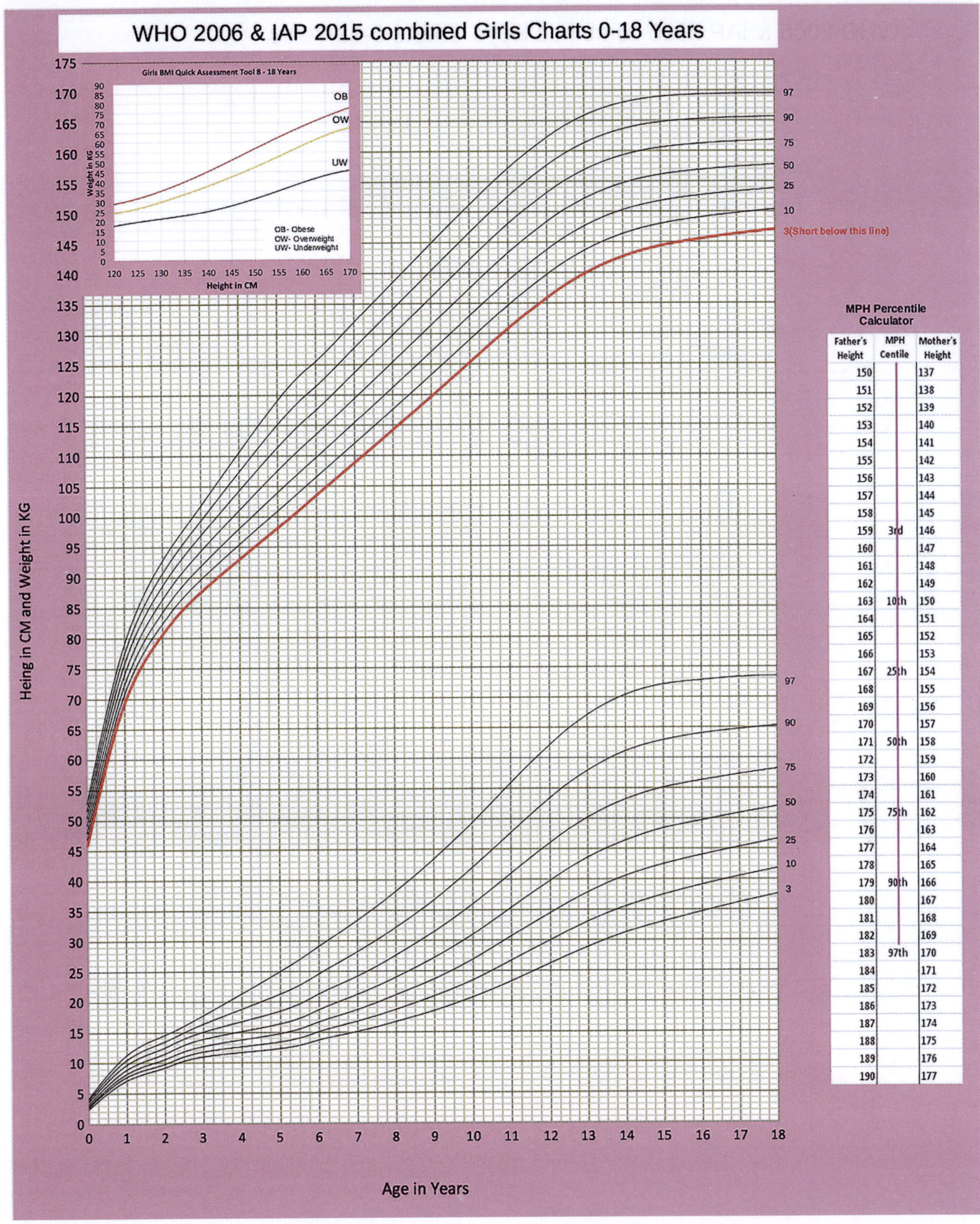

Father's Height	MPH Centile	Mother's Height
150		137
151		138
152		139
153		140
154		141
155		142
156		143
157		144
158		145
159	3rd	146
160		147
161		148
162		149
163	10th	150
164		151
165		152
166		153
167	25th	154
168		155
169		156
170		157
171	50th	158
172		159
173		160
174		161
175	75th	162
176		163
177		164
178		165
179	90th	166
180		167
181		168
182		169
183	97th	170
184		171
185		172
186		173
187		174
188		175
189		176
190		177

Figure 31.26 (continued on next page)
Growth chart of girls and boys (from 0 to 18 years) (see https://iapindia.org/iap-growth-charts/).

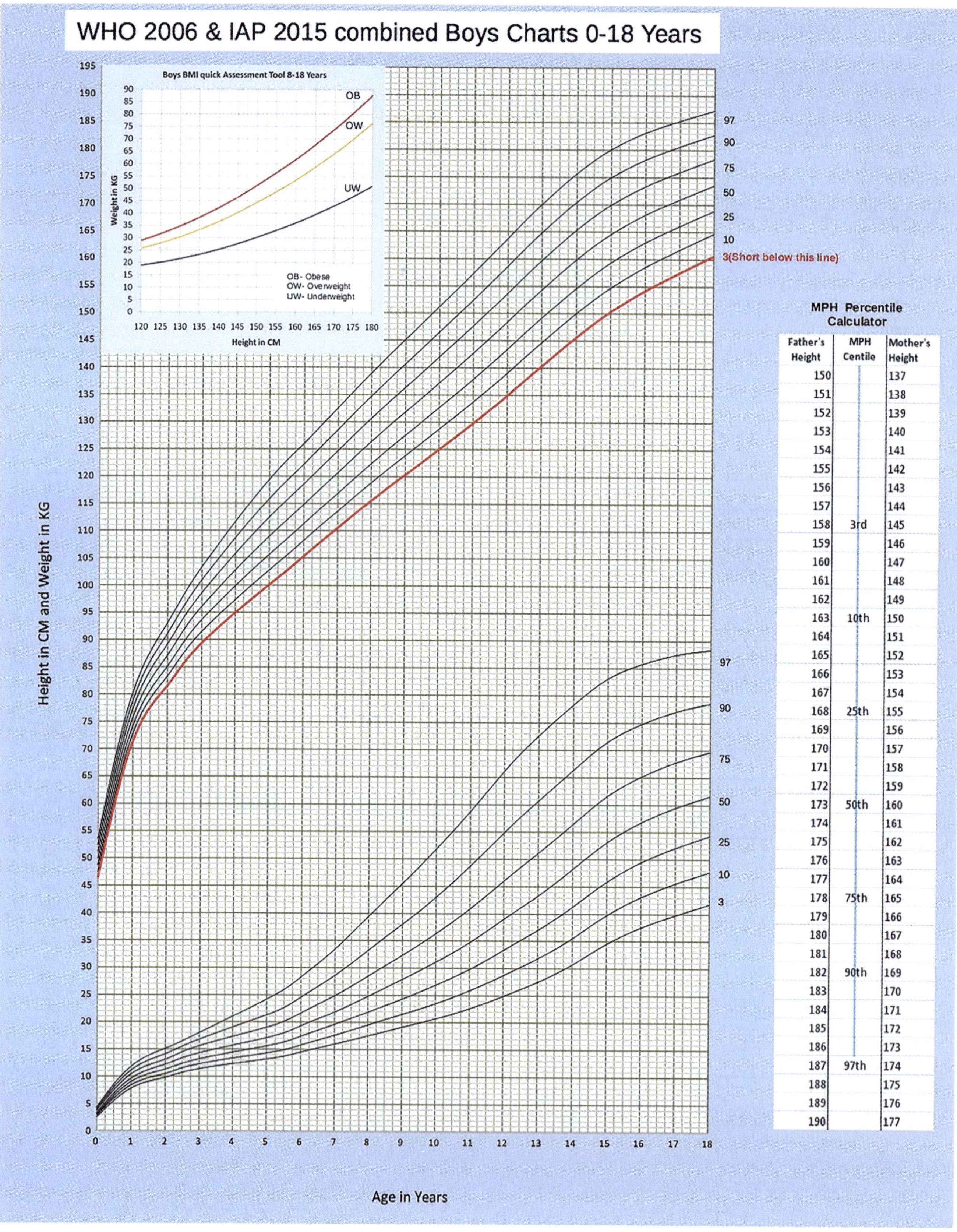

Father's Height	MPH Centile	Mother's Height
150		137
151		138
152		139
153		140
154		141
155		142
156		143
157		144
158	3rd	145
159		146
160		147
161		148
162		149
163	10th	150
164		151
165		152
166		153
167		154
168	25th	155
169		156
170		157
171		158
172		159
173	50th	160
174		161
175		162
176		163
177		164
178	75th	165
179		166
180		167
181		168
182	90th	169
183		170
184		171
185		172
186		173
187	97th	174
188		175
189		176
190		177

Figure 31.26 (continued on previous page)
Growth chart of girls and boys (from 0 to 18 years) (see https://iapindia.org/iap-growth-charts/).

turn, act as prebiotics that support the growth of B. infantis, a probiotic (helpful bacteria).] Breast milk is low in vitamin D, however, and exclusively breast-fed babies require vitamin D supplementation. [Note: Human milk provides antibodies and other proteins that reduce the risk of infection.]

The microbiota in and on the human body plus their genomes are referred to as the microbiome. It is acquired at birth from the environment and changes with the life stages. The gut microbiome influences host nutrition by facilitating processing of food consumed and is itself influenced by that food. Its relationship with undernutrition, obesity, and diabetes is under investigation.

2. **Children:** As with infants, children have increased need for calories and nutrients. In children, larger proportion of protein is required for maintenance of nitrogen balance and for meeting the requirements of growth and development of the body. Additional allowances for micronutrients are also required as deficiencies of iron and calcium are a concern at this stage.

3. **Adolescents:** In the teen years, the increases in height and weight increase the need for calories, protein, calcium, iron, and phosphorus. Eating patterns in this stage can result in overconsumption of fat, sodium, and sugar and underconsumption of vitamin A, thiamine, and folic acid. [Note: Eating disorders and obesity are concerns in this age group.]

B. Adulthood

Overnutrition is a concern in young adults, whereas malnutrition is a concern in older adults.

1. **Young adults:** Nutrition in young adults focuses on the maintenance of good health and the prevention of disease. The goal is a diet rich in plant-based foods (with a focus on fiber and whole grains), limited intake of saturated fat and trans fatty acids, and balanced intake of ω-3 and ω-6 PUFA.

2. **Pregnant or lactating women:** The requirements for calories, protein, and virtually all micronutrients increase in pregnancy and lactation.

   Nutrition during pregnancy affects the growth and development of the fetus and also affects the subsequent health of the offspring. Maternal weight gain during pregnancy is a good indicator of adequate nutrition. Pregnant women should increase their energy intake by 360 kcal/day in the second trimester and by 475 kcal/day in third trimester. Extra protein (1 g, 9 g, and 31 g per day in the first, second and third trimester, respectively) and micronutrients are also required. Supplementation with folic acid (to prevent neural tube defects), vitamin D, calcium, iron, iodine, and DHA are recommended.

   Similarly, during lactation for extra energy, proteins and micronutrients need to be included in the diet.

3. **Older adults:** Aging increases the risk of malnutrition. Nutrient intake may be reduced, due to reduced sense of taste (dysgeusia) and smell (hyposmia), along with physical limitations such as dental problems. Inadequate intake of protein, calcium, and vitamins D and B_{12} is common. B_{12} deficiency can result from decreased absorption caused by achlorhydria (reduced stomach acid, see p. 529). In aging, lean muscle mass decreases and fat increases, resulting in decreased RMR. [Note: Drug–nutrient interactions can occur at any life stage but are more common as the number of medications increases as in aging.]

Monoamine oxidase inhibitors (MAOI), used to treat depression (see p. 424) and early Parkinson disease, can interact with tyramine-containing foods. Tyramine is a monoamine derived from the decarboxylation of tyrosine during the curing, aging, or fermentation of food (Fig. 31.27). It causes the release of norepinephrine, increasing blood pressure and heart rate. Patients who take an MAOI and consume such foods are at risk for a hypertensive crisis.

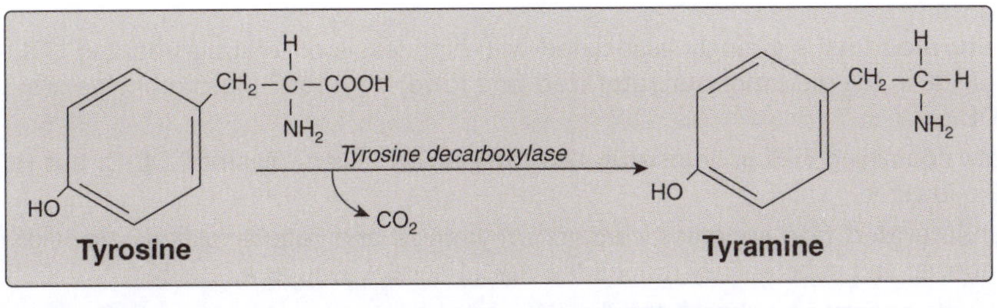

Figure 31.27 Decarboxylation of tyrosine to tyramine. CO_2 = carbon dioxide.

X. Medical Perspectives

Access to nutrition is essential for development. However, many populations face a challenge in providing optimal nutrition, leading to malnutrition, stunting and mortality, with children, pregnant and lactating women being most vulnerable. More than a third of the world's children are affected by protein energy malnutrition. In Asia, most countries have high or very high prevalence.

In India, children below 6 years, and pregnant and lactating mothers are given supplementary nutrition in **Integrated Child Development Services (ICDS) Scheme**. The school children are also provided nutritious one meal under the **Mid-Day Meal scheme**.

Along with undernutrition, overnutrition, causing obesity, also coexists. According to World Health Organization (WHO), in 2016, 39% of world's adult population were overweight and about 13% were obese.

Globally, excessive trans fat intake is responsible for more than 500,000 deaths of people from cardiovascular disease every year. Larger amount of trans fat in our food is contributed by **industrially-produced trans-fatty acids (iTFA)**, which are used in baked and fried foods, pre-packaged snacks and food, and cooking oils and spreads. WHO has launched a comprehensive plan, **REPLACE action package**, to eliminate industrially-produced trans-fats from the global food supply by 2023.

XI. Chapter Summary

- The **Dietary Reference Intakes** (**DRI**) provide estimates of the amounts of nutrients required to prevent deficiencies and maintain optimal health and growth. They consist of:
  - **Estimated Average Requirement** (**EAR**), the average daily nutrient intake level estimated to meet the requirement of 50% of the healthy individuals in a particular life stage (age) and gender group
  - **Recommended Dietary Allowance** (**RDA**), the average daily dietary intake level that is sufficient to meet the nutrient requirements of nearly all (97%–98%) individuals in a life stage and gender group
  - **Adequate Intake** (**AI**), which is set instead of an RDA if sufficient scientific evidence is not available to calculate the RDA
  - **Tolerable Upper Intake Level** (**UL**), the highest average daily nutrient intake level that is likely to pose no risk of adverse health effects to almost all individuals in the general population.
- Macronutrients provide energy (9 kcal/g of fat, 4 kcal/g of proteins and carbohydrates) for **Total Energy Expenditure (TEE)**, which comprises of three components: **resting metabolic rate**, **physical activity,** and **thermic effect of food.**
- **Acceptable Macronutrient Distribution Ranges** (**AMDR**) are defined as the ranges of intake for a particular macronutrient that are associated with reduced risk of chronic disease while providing adequate amounts of essential nutrients.
- Adults should consume **45%–65%** of their **total calories** from **carbohydrates**, **20%–35%** from **fat**, and **10%–35%** from **protein** (Fig. 31.28).
- **Dietary fiber** does not provide energy but is important as it is protective against several diseases.
- **Glycemic index (GI)** is a measure of the effects of carbohydrates on blood glucose levels. Diets with lower GI help in the management of blood glucose and cholesterol levels.
- Elevated levels of cholesterol in low-density lipoproteins (LDL-C) result in increased risk for **coronary heart disease** (**CHD**). In contrast, elevated levels of cholesterol in high-density lipoproteins (HDL-C) have been associated with a decreased risk for CHD.
- Dietary or drug treatment of **hypercholesterolemia** is effective in decreasing LDL-C, increasing HDL-C, and reducing the risk for CHD.
- Consumption of **saturated fats** is strongly associated with high levels of total plasma and LDL-C. When substituted for saturated fatty acids in the diet, **monounsaturated fats** lower both total plasma cholesterol and LDL-C but maintain or increase HDL-C.
- Consumption of fats containing **ω-6 polyunsaturated fatty acids** lowers plasma LDL-C, but HDL-C, which protects against CHD, is also lowered.
- Dietary **ω-3 polyunsaturated fats** suppress cardiac arrhythmias and reduce plasma triacylglycerols, decrease the tendency for thrombosis, and substantially reduce the risk of cardiovascular mortality.
- **Carbohydrates** provide **energy** and **fiber** to the diet. When they are consumed as part of a diet in which caloric intake is equal to energy expenditure, they do not promote obesity.

- Dietary **protein** provides **essential amino acids**. **Protein quality** is a measure of its ability to provide the essential amino acids required for tissue maintenance. Proteins from animal sources, in general, have a higher-quality protein than that derived from plants. However, proteins from different plant sources may be combined in such a way that the result is equivalent in nutritional value to animal protein.
- **Positive nitrogen** (**N**) **balance** occurs when N intake exceeds N excretion. It is observed in situations in which tissue growth occurs, for example, in childhood, pregnancy, or during recovery from an emaciating illness. **Negative N balance** occurs when N losses are greater than N intake. It is associated with inadequate dietary protein; lack of an essential amino acid; or during physiologic stresses such as trauma, burns, illness, or surgery.
- **Kwashiorkor** occurs when protein deprivation is relatively greater than the reduction in total calories. It is characterized by edema. **Marasmus** occurs when calorie deprivation is relatively greater than the reduction in protein. No edema is seen. Both are extreme forms of **protein-energy malnutrition** (**PEM**).
- **Nutrition Facts labels** give consumers information about the nutritional content of packaged foods.
- Medical assessment of nutritional status includes **dietary history**, **anthropometric measures**, and **laboratory data**.
- Each life stage has specific nutrition needs. **Growth charts** are used to monitor the growth pattern of an individual from birth through adolescence.
- Balanced diet meets the daily energy, protein, vitamin and mineral requirements. Different stages of life have their special dietary requirements. Specific diets are also required for different disease conditions.

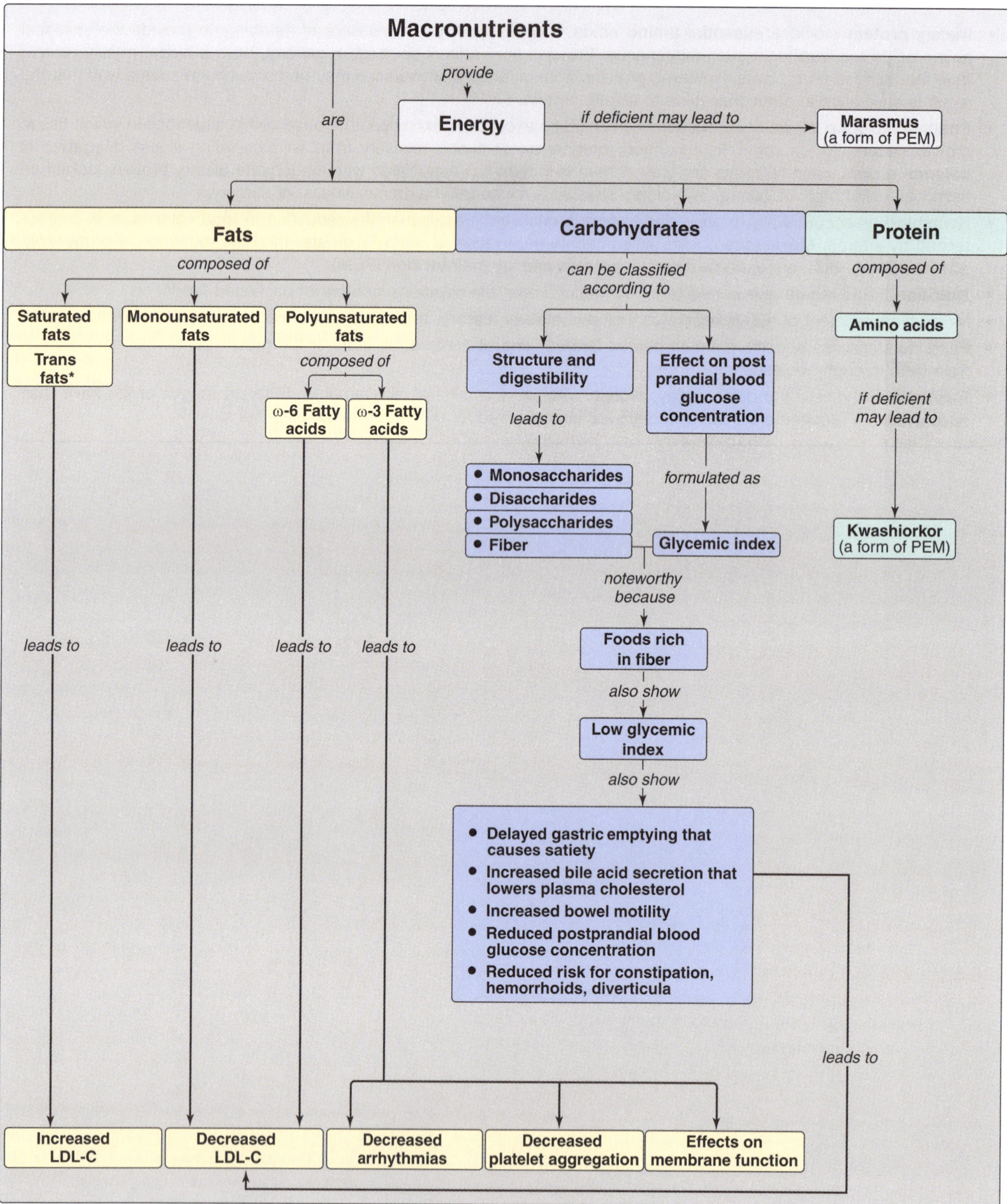

Figure 31.28 Key concept map for the macronutrients. [Note: *Trans fatty acids are chemically classified as unsaturated.] PEM = protein-energy malnutrition; LDL = low-density lipoprotein; C = cholesterol.

Study Questions

Short-Answer Questions

31.1. Describe the importance of dietary fibers.

31.2. What is the significance of the glycemic index of food?

31.3. Describe the basic components of a balanced diet.

31.4. How will you assess the nutritional status of a person?

31.5. How will you assess an individual's total energy expenditure?

31.6. Enumerate the differences between Kwashiorkor and marasmus.

Long-Answer Questions

31.1. What are the proximate principles of food? Describe various types of dietary fats and their significance in health.

31.2. Describe the types, causes and clinical features of Protein Energy Malnutrition

Multiple-Choice Questions

Choose the ONE best answer.

31.1. A 4-year-old girl presents with stunted growth. On examination, the skin has patchy lesions, hair is brittle and depigmented. The body appears swollen. What is the likely cause of this condition?

A. Protein-calorie malnutrition
B. Vitamin A deficiency
C. Protein deficiency, regardless of calories
D. Vitamin C deficiency

Correct answer = C.

31.2. What is the daily calorie requirement, percentage carbohydrate need, and the amount of carbohydrate to be included in the daily diet, in a 35-year-old lightly active, healthy man?

A. 2100 kcal, 45%–65%, 300 g
B. 1800 kcal, 30%–50%, 200 g
C. 3000 kcal, 60%–80%, 400 g
D. 1500 kcal, 35%–50%, 150 g

Correct answer = A.

31.3. The average daily nutrient level estimated to fulfil the requirement of 97%–98% individuals of a particular age, gender and life stage is known as:

A. Estimated Average Requirement
B. Recommended Dietary Allowance
C. Adequate Intake
D. Tolerable upper intake level

Correct answer = B.

31.4. Which one of the following statements concerning dietary fat is correct?

A. Coconut oil is rich in monounsaturated fats, and olive oil is rich in saturated fats.
B. Fatty acids containing trans double bonds, unlike the naturally occurring cis isomers, raise high-density lipoprotein cholesterol levels.
C. The polyunsaturated fatty acids linoleic and linolenic acids are required components.
D. Triacylglycerols obtained from plants generally contain less unsaturated fatty acids than do those from animals.

Correct answer = C.

31.5. Given the information that a 70-kg man is consuming a daily average of 275 g of carbohydrate, 75 g of protein, and 65 g of fat, which one of the following conclusions can reasonably be drawn?

A. About 20% of calories are derived from fats.
B. The individual is in nitrogen balance.
C. The proportions of carbohydrate, protein, and fat in the diet conform to current recommendations.
D. The total energy intake per day is about 3,000 kcal.

Correct answer = C.

31.6. Which of the following is the best explanation for the beneficial effects of dietary fibers?

A. They accelerate glucose and cholesterol absorption from the gut
B. They increase the transit time to food through the intestine
C. They produce growth promoting molecules that help in cell proliferation
D. They destroy the microbes present in the intestine

Correct answer = B.

31.7. A sedentary 50-year-old man weighing 176 lb (80 kg) requests a health checkup. He denies any health problems. Routine blood analysis is unremarkable except for plasma total cholesterol of 295 mg/dl. (Reference value is <200 mg.) The man refuses drug therapy for his hypercholesterolemia. Analysis of a 1-day dietary recall showed the following:

Kilocalories	3,475 kcal	Cholesterol	822 mg
Protein	102 g	Saturated fat	69 g
Carbohydrate	383 g	Total fat	165 g
Fiber	6 g	Polyunsaturated fats	20 g

Decreasing which one of the following dietary components would have the greatest effect in lowering the patient's plasma cholesterol?

A. Polyunsaturated fats
B. Cholesterol
C. Fiber
D. Saturated fat

Correct answer = D.

31.8. Statement 1: Deficiency of essential fatty acids (EFA) results in dry skin and scaly dermatitis.

Statement 2: Skin ceramides containing linoleic acid require EFA for their synthesis.

A. Both the Statements are correct, and Statement 2 is the correct explanation for Statement 1
B. Both the Statements are correct, but Statement 2 is NOT the correct explanation for Statement 1
C. Statement 1 is correct, but Statement 2 is NOT correct
D. Statement 1 is NOT correct, but Statement 2 is correct

Correct answer = A.

31.9. Which of the following non-nutritive sweeteners should be avoided in patients with phenylketonuria?
A. Acesulfame
B. Sucralose
C. Saccharin
D. Aspartame

Correct answer = D.

31.10. Which of the following foods and the limiting amino acids are correctly matched?
I. Wheat – Lysine, II. Rice – Leucine, III. Maize – Tryptophan, IV. Kidney beans – Methionine.
A. I and II only
B. I, II, and IV only.
C. I, III, and IV only.
D. I, II, III, and IV.

Correct answer = C.

Multiple-Choice Questions (NExT/NEET PG)

Choose the ONE best answer.

31.1. For the child shown below, which of the statements would support a diagnosis of kwashiorkor? The child:
A. appears plump due to increased deposition of fat in adipose tissue.
B. displays abdominal and peripheral edema.
C. has a serum albumin level above normal.
D. has markedly decreased weight for height.

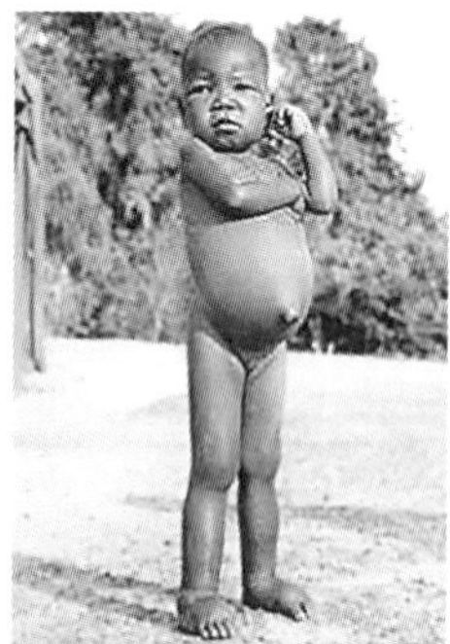

Correct answer = B.

31.2. Which of the following statements regarding dietary reference intakes (DRI) is NOT correct?
A. Estimated Average Requirement is the average daily nutrient level estimated to fulfill the requirement of 50% of the healthy individuals.
B. Recommended Dietary Allowance is the average daily nutrient level estimated to fulfill the requirement of 97%–98% of the individuals.
C. Nutrient requirements as per the DRI should be consumed every day to prevent deficiency manifestations.
D. DRI includes a tolerable upper intake level, which limits the intake above a certain level to avoid adverse effects.

Correct answer = C.

31.3. In chronic bronchitis, excessive mucus production causes airway obstruction that results in hypoxemia (low blood oxygen level), impaired expiration, and hypercapnia (carbon dioxide retention). Why might a high-fat, low-carbohydrate diet be recommended for a patient with chronic obstructive pulmonary disease caused by chronic bronchitis?
A. Fat contains more oxygen atoms relative to carbon or hydrogen atoms than do carbohydrates.
B. Fat is calorically less dense than carbohydrates.
C. Fat metabolism generates less carbon dioxide.
D. The respiratory quotient (RQ) for fat is higher than the RQ for carbohydrates.

Correct answer = C.

31.4. A certain myth exists about "negative caloric balance" food which suggests that the body uses more calories to digest and process these foods than the food contains. This idea is often associated with foods like celery, lettuce, or cucumbers, which are very low in calories and high in water content. A high value of which of the following is attributed to the mythical "negative caloric balance" in such foods?
A. Thermic effect of food
B. Fiber content
C. Carbohydrate content
D. Respiratory quotient

Correct answer = A.

31.5. Based on statements given below, choose the correct option:
Statement 1: Adding long-chain polyunsaturated fatty acids (LCPUFAs) to the enteral nutrition of infants is suggested to enhance their development, especially in terms of neurodevelopmental achievements.
Statement 2: Docosahexaenoic acid (DHA) is integral to the structure of cell membranes, particularly in the brain and eyes, contributing to the fluidity of these membranes and neurotransmitter signaling.
A. Both the Statements are correct, and Statement 2 is the correct explanation for Statement 1
B. Both the Statements are correct, but Statement 2 is NOT the correct explanation for Statement 1
C. Statement 1 is correct, but Statement 2 is NOT correct
D. Statement 1 is NOT correct, but Statement 2 is correct

Correct answer = A.

Micronutrients: Vitamins 32

Learning Objectives

At the end of this chapter, the student will be able to:

- Describe the classification and enumerate the fat-soluble and water-soluble vitamins and their major dietary sources and recommended daily allowances.
- Describe the functions and biochemical role of vitamins A, D, E and K.
- Discuss biochemical role of vitamins of the B-complex group and vitamin C.
- Describe in brief the diseases associated with the deficiency of each of these vitamins.

I. OVERVIEW

Vitamins are chemically unrelated organic compounds that cannot be synthesized in adequate quantities by humans and, therefore, must be supplied by the diet. Chemically similar substances with similar vitamin activity are termed vitamers. For instance, pyridoxine, pyridoxal, and pyridoxamine are vitamers of vitamin B6, while ergocalciferol (D2) and cholecalciferol (D3) are vitamers of vitamin D. Vitamins can broadly be classified as water soluble or fat soluble. Nine vitamins (folic acid, cobalamin, ascorbic acid, pyridoxine, thiamine, niacin, riboflavin, biotin, and pantothenic acid) are classified as **water soluble**. Because they are readily excreted in the urine, toxicity is rare. However, deficiencies can occur quickly. Four vitamins (A, D, K, and E) are termed **fat soluble** (Fig. 32.1). They are released, absorbed, and transported (in chylomicrons, see p. 350) with dietary fat. They are not readily excreted, and significant quantities are stored in the liver and adipose tissue. In fact, consumption of

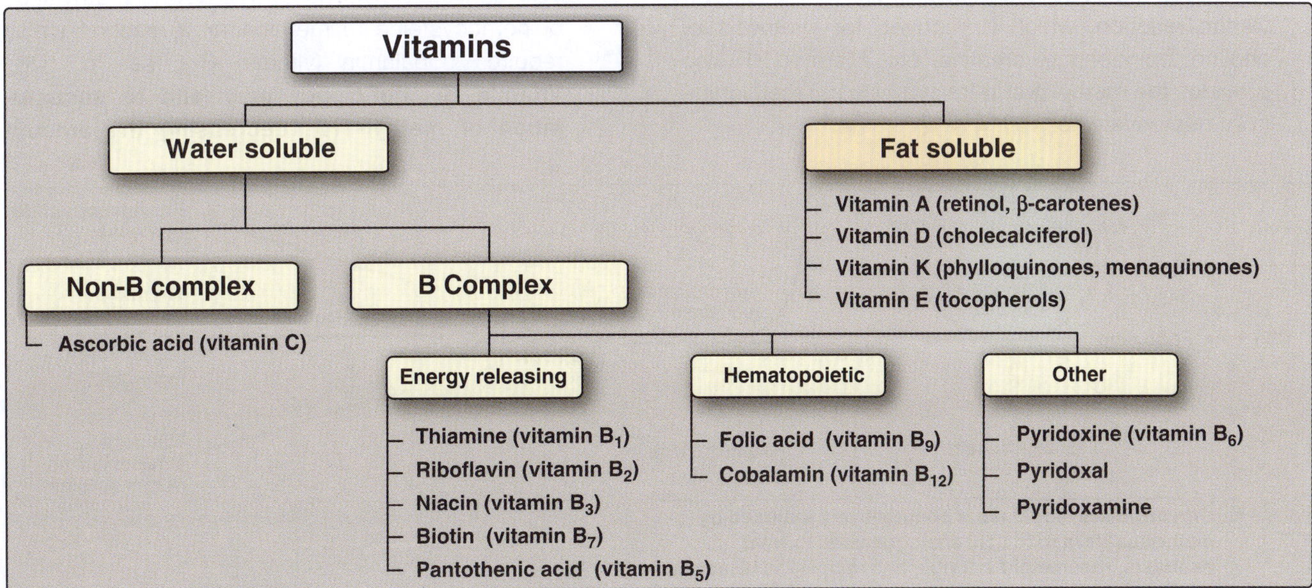

Figure 32.1 Classification of the vitamins. Because they are required in lesser amounts than the macronutrients (carbohydrate, protein, and lipid), vitamins are termed micronutrients.

vitamins A and D in excess of the Dietary Reference Intakes (see Chapter 31) can lead to accumulation of toxic quantities of these compounds. Vitamins are required to perform specific cellular functions. For example, many of the water-soluble vitamins are precursors of coenzymes for the enzymes of intermediary metabolism. In contrast to the water-soluble vitamins, only one fat-soluble vitamin (vitamin K) has a coenzyme function.

II. FOLIC ACID (VITAMIN B_9)

Folic acid (or, folate), which plays a **key role in one-carbon metabolism**, is essential for the biosynthesis of several compounds. Prevalence of folic acid deficiency varies from 10% to 60% in different Indian states because of dietary differences, particularly among pregnant women and individuals with alcoholism. **Leafy, dark-green vegetables are a good source of folic acid.**

Chemically, folic acid is composed of three components: pteridine group, para-aminobenzoic acid (PABA), and glutamic acid.

A. Function

Folic acid is absorbed in upper part of jejunum.

In liver, it is reduced to 5,6,7,8-tetrahydrofolic acid, in two steps, both catalyzed by NADPH-dependent *dihydrofolate reductase*. **Tetrahydrofolate (THF), receives one-carbon fragments from donors such as serine, glycine, and histidine and transfers them to intermediates in the synthesis of amino acids, purine nucleotides, and thymidine monophosphate (TMP)**, a pyrimidine nucleotide incorporated into DNA (Fig. 32.2).

N5-methyl THFA carries methyl group for transmethylation reaction, which is required for synthesis of choline, epinephrine, creatine, etc. Methyl-THF also provides the methyl group for synthesis of methionine from homocysteine.

> **One-carbon metabolism**
>
> One-carbon groups, collectively known as "**one carbon pool**," are important donors of carbon atoms for synthesis of various compounds. These include **formyl, formimino, methenyl, hydroxymethyl, methylene and methyl groups**. Tetrahydrofolic acid (THFA) serves as carrier of these one-carbon units, accepting them during catabolic reactions and donating them during various synthetic reactions. (p. 402)

B. Folate deficiency

Inadequate serum levels of folate can be caused by:

- **Increased demand:** Seen in pregnancy and lactation (see p. 519).
- **Poor absorption:** Caused by **pathology of the small intestine, alcoholism**, or treatment with drugs (for example, methotrexate) that are *dihydrofolate reductase* inhibitors (see Fig. 32.2). A folate-free diet can cause a deficiency within a few weeks.

  Several folate antagonists, such as **methotrexate, aminopterin and amethopterin, are competitive inhibitors of enzyme DHFR, and are used as anti-cancer agents**. As THF is not available for DNA synthesis, DNA synthesis activity of tumor cells is inhibited.

  Many microorganisms can synthesize folic acid from para-aminobenzoic acid (PABA). PABA analogues, such as **sulfonamides, can inhibit the synthesis of folic acid**, therefore inhibiting bacterial growth. These are used as antibacterial agents.
- **Folate trap:** The only reaction by which methyl-THF can be converted to THF is the methylation of homocysteine to methionine, a reaction which requires cobalamin (vitamin B_{12}, see p. 528). **Vitamin B_{12} deficiency may lead to accumulation of methyl-THF, decreasing the amount**

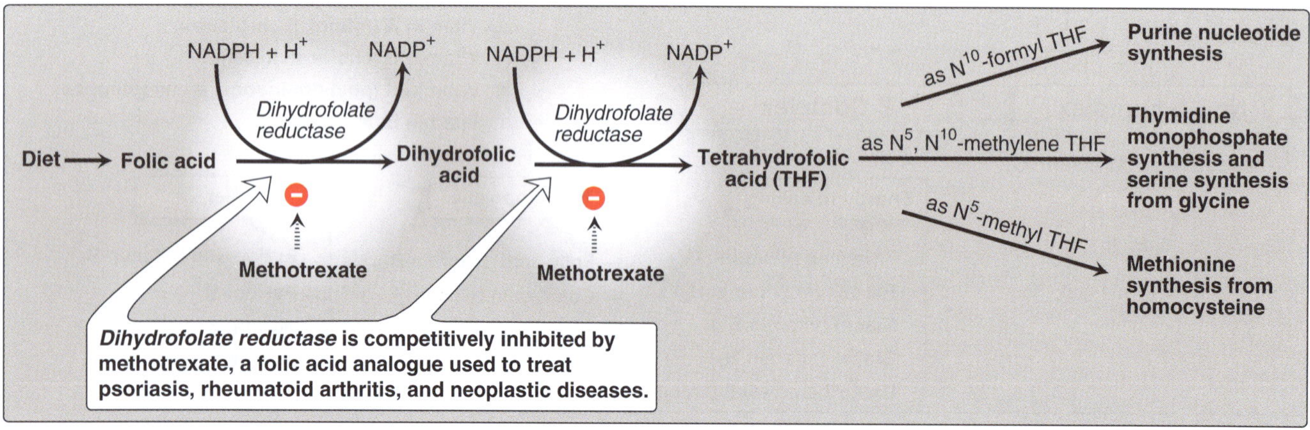

Figure 32.2 Production and use of tetrahydrofolate. NADP(H) = nicotinamide adenine dinucleotide phosphate.

of available free THF. This condition, known as folate trap, can mimic folate deficiency.

> **Nutritional anemias**
>
> Anemia is a condition in which the blood has a lower than normal concentration of hemoglobin, which results in a reduced ability to transport oxygen (O_2). Nutritional anemias are caused by inadequate intake of one or more essential nutrients. These can be classified according to the size of the red blood cells (RBCs), or mean corpuscular volume (MCV), observed in the blood (Fig. 32.3).
>
> **Microcytic anemia** (MCV below normal), caused by lack of iron, is the most common form of nutritional anemia.
>
> **Macrocytic anemia** (MCV above normal), results from a deficiency in folic acid or vitamin B_{12}. Macrocytic anemias are commonly called megaloblastic anemias because a deficiency of either vitamin (or both) causes accumulation of large, immature RBC precursors, known as megaloblasts, in the bone marrow and the blood (Fig. 32.4). Hypersegmented neutrophils are also seen.

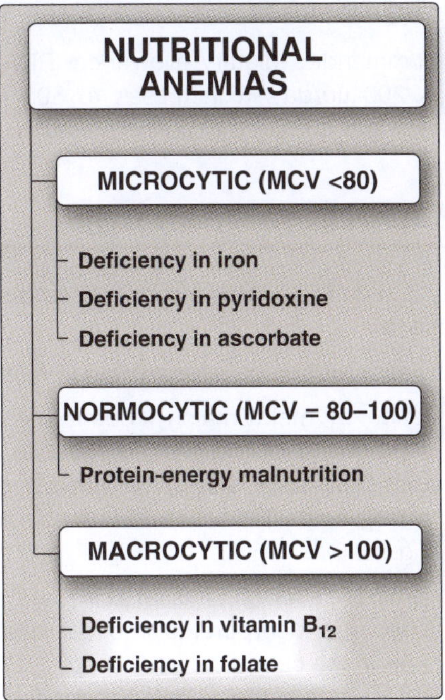

Figure 32.3 Classification of nutritional anemias by red cell size. The normal mean corpuscular volume (MCV) for people older than age 18 years is 80–100 μm^3. [Note: Microcytic anemia is also seen with heavy metal (for example, lead) poisoning.]

C. Folate deficiency diseases

1. **Folate and anemia:** A primary result of folic acid deficiency is **megaloblastic anemia** (see Fig. 32.4), caused by diminished synthesis of purine nucleotides and TMP, which leads to an inability of cells (including RBC precursors) to make DNA and, therefore, an inability to divide. As protein synthesis continues, hemoglobin accumulates in RBC precursors. The abnormal RBCs are rapidly destroyed in the spleen, causing anemia.

2. **Folate and neural tube defects:** Spina bifida and anencephaly, the most common neural tube defects (NTD), affect ~3,000 pregnancies in the United States annually. **Folic acid supplementation before conception and during the first trimester has been shown to significantly reduce NTD**. Therefore, all women of childbearing age are advised to consume 0.4 mg/day (400 μg/day) of folic acid to reduce the risk of having a pregnancy affected by NTD and ten times that amount if a previous pregnancy was affected.

   Adequate folate nutrition must occur **at the time of conception** because critical folate-dependent development occurs in the first weeks of fetal life, at a time when many women are not yet aware of their pregnancy. In 1998, the U.S. Food and Drug Administration authorized the addition of folic acid to wheat flour and enriched grain products, resulting in a dietary supplementation of ~0.1 mg/day. This supplementation allows ~50% of all reproductive-aged women to receive 0.4 mg of folate from all sources.

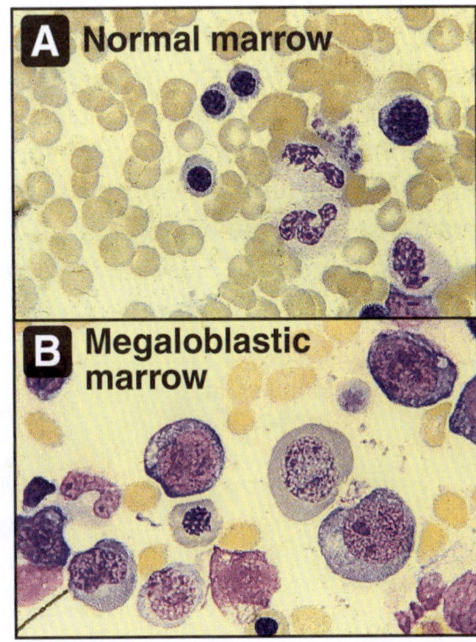

Figure 32.4 Bone marrow histology in normal (**A**) and folate-deficient (**B**) individuals.

3. **Folate and homocysteinemia:** As folate is required for conversion of homocysteine to methionine, its deficiency results in hyperhomocysteinemia.

D. Daily requirements

The recommended dietary allowance (RDA) for free folate is 200 μg/day. It increases to 500 μg/day in pregnancy and 300 μg/day during lactation.

III. COBALAMIN (VITAMIN B_{12})

Vitamin B_{12} is required in humans for two essential enzymatic reactions:

(i) Remethylation of homocysteine (Hcy) to methionine.

(ii) Isomerization of methylmalonyl coenzyme A (CoA), which is produced during the degradation of some amino acids (isoleucine, valine, threonine, and methionine) and fatty acids (FA) with odd numbers of carbon atoms (Fig. 32.5).

When cobalamin is deficient, unusual (branched) FA accumulate and become incorporated into cell membranes, including those of the central nervous system (CNS). This may account for some of the neurologic manifestations of vitamin B_{12} deficiency. [Note: Folic acid (as N^5-methyl THF) is also required in the remethylation of Hcy. Therefore, deficiency of B_{12} or folate results in elevated Hcy levels.]

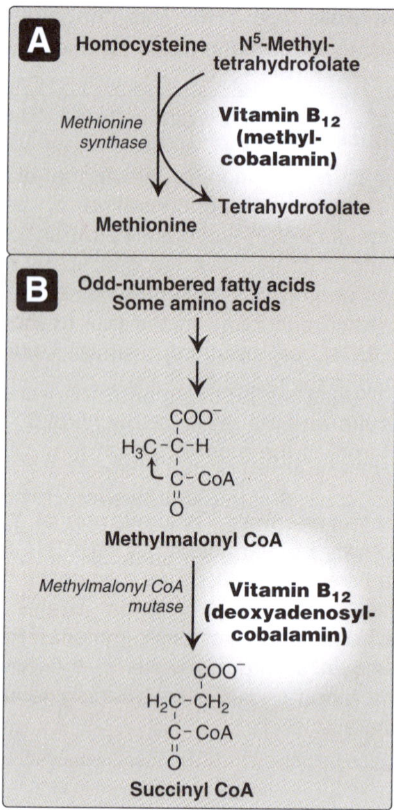

Figure 32.5 **A, B.** Reactions requiring coenzyme forms of vitamin B_{12}. CoA = coenzyme A.

A. Structure and coenzyme forms

Cobalamin contains a corrin ring system that resembles the porphyrin ring of heme (see p. 414), but differs in that two of the pyrrole rings are linked directly rather than through a methane bridge. Cobalt (see p. 562) is held in the center of the corrin ring by four coordination bonds with the nitrogens of the pyrrole groups. The remaining coordination bonds of the cobalt are with the nitrogen of 5,6-dimethylbenzimidazole and with cyanide in commercial preparations of the vitamin in the form of cyanocobalamin (Fig. 32.6). **The physiologic coenzyme forms of cobalamin are 5′-deoxyadenosylcobalamin and methylcobalamin**, in which cyanide is replaced with 5′-deoxyadenosine or a methyl group, respectively (see Fig. 32.6).

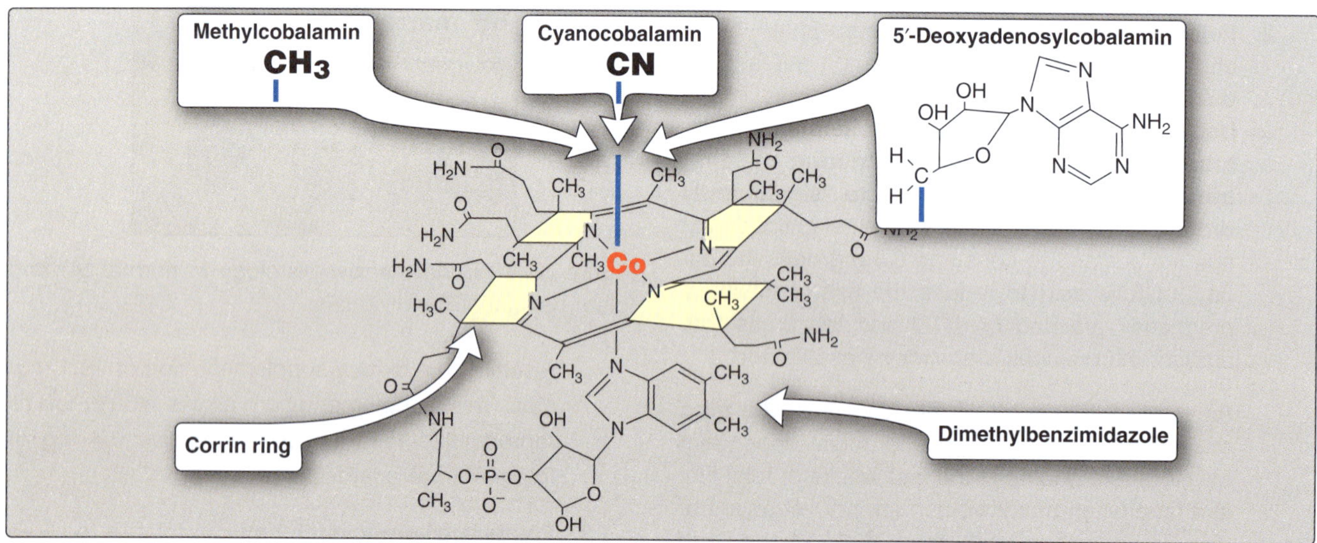

Figure 32.6 Structure of vitamin B_{12} (cyanocobalamin) and its coenzyme forms (methylcobalamin and 5′-deoxyadenosylcobalamin).

B. Distribution

Vitamin B_{12} is **synthesized only by microorganisms, and it is not present in plants.** Animals obtain the vitamin preformed from their intestinal microbiota (see p. 519) or by eating foods derived from other animals. Cobalamin is present in appreciable amounts in liver, red meat, fish, eggs, dairy products, and fortified cereals.

C. Causes of deficiency

- **Nutritional deficiency**: Nutritional deficiency of vitamin B_{12} is **common in vegetarians who do not take milk or milk products.**
- **Decreased absorption**: Due to gastrectomy, gastric atrophy or in pernicious anemia (see below).

D. Folate trap hypothesis

The **effects of cobalamin deficiency are most pronounced in rapidly dividing cells**, such as the erythropoietic tissue of bone marrow and the mucosal cells of the intestine. Such tissues need both the N^5,N^{10}-methylene and N^{10}-formyl forms of THF for the synthesis of nucleotides required for DNA replication (see pp. 579 and 589). However, **in vitamin B_{12} deficiency, the utilization of the N^5-methyl form of THF in the B_{12}-dependent methylation of Hcy to methionine is impaired.** Because the methylated form cannot be converted directly to other forms of THF, folate is trapped in the N^5-methyl form, which accumulates. The levels of the other forms decrease. Thus, cobalamin deficiency leads to a deficiency of the THF forms needed in purine and TMP synthesis, resulting in the symptoms of megaloblastic anemia.

E. Clinical indications for cobalamin

In contrast to other water-soluble vitamins, significant amounts (2–5 mg) of vitamin B_{12} are stored in the body. As a result, it may take several years for the clinical symptoms of B_{12} deficiency to develop as a result of decreased intake of the vitamin. [Note: Deficiency happens much more quickly (in months) if absorption is impaired (see below).

1. **Pernicious anemia:** Vitamin B_{12} deficiency is most commonly seen **in patients who fail to absorb the vitamin from the intestine** (Fig. 32.7). Vitamin B_{12} is released from food in the acidic environment of the stomach. [Note: Malabsorption of cobalamin in the elderly is most often due to reduced secretion of gastric acid (achlorhydria).] Free B_{12} then binds a glycoprotein (R-protein or haptocorrin), and the complex moves into the intestine. B_{12} is released from the R-protein by pancreatic enzymes and binds another glycoprotein, **intrinsic factor** (IF).

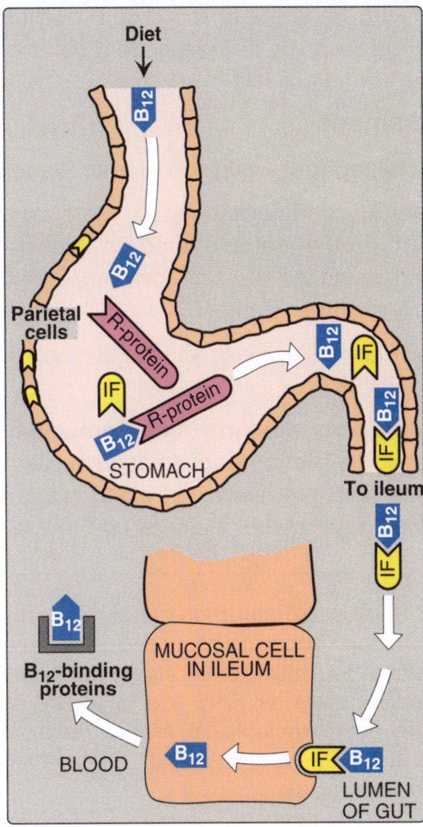

Figure 32.7 Absorption of vitamin B_{12}. [Note: Acid-dependent release of B_{12} from food is not shown.] IF = intrinsic factor.

The cobalamin–IF complex travels through the intestine and binds to a receptor (cubilin) on the surface of mucosal cells in the ileum. The cobalamin is transported into the mucosal cell and, subsequently, into the general circulation, where it is carried by its binding protein (transcobalamin). B_{12} is taken up and stored in the liver, primarily. It is released into bile and efficiently reabsorbed in the ileum. Severe malabsorption of vitamin B_{12} leads to pernicious anemia.

This disease is **most commonly caused due to an autoimmune destruction of the gastric parietal cells** that are responsible for the synthesis of IF (lack of IF prevents B_{12} absorption).

Patients who have had a partial or total gastrectomy become IF deficient and, therefore, B_{12} deficient.

Individuals with cobalamin deficiency are usually **anemic** (folate recycling is impaired), and they show **neuropsychiatric symptoms** as the disease develops. **The CNS effects are irreversible**. Pernicious anemia requires lifelong treatment with either high-dose oral B_{12} or intramuscular injection of cyanocobalamin. [Note: Supplementation works even in the absence of IF because ~1% of B_{12} uptake is by IF-independent diffusion.]

F. Assessment

The Schilling test evaluates vitamin B_{12} absorption.

Vitamin B_{12} deficiency can be determined by the **level of methylmalonic acid in blood**, which is elevated in individuals with low intake or decreased absorption of the vitamin.

ELISA: Plasma levels of vitamin B_{12} can be quantified by ELISA.

Methylmalonic aciduria and homocystinuria: As cobalamin deficiency leads to accumulation of methylmalonic acid (methylmalonic aciduria)and homocysteine (homocystinuria), these can be measured.

> Folic acid supplementation can partially reverse the hematologic abnormalities of B_{12} deficiency and, therefore, can mask a cobalamin deficiency. Thus, to prevent the later CNS effects of B_{12} deficiency, therapy for megaloblastic anemia is initiated with both vitamin B_{12} and folic acid until the cause of the anemia can be determined.

G. Daily requirements

The RDA of vitamin B_{12} is 1 μg for adults.

IV. ASCORBIC ACID (VITAMIN C)

The active form of vitamin C is ascorbic acid (Fig. 32.8). **Its main function is as a reducing agent**. Only the L-ascorbic acid and dehydroascorbic acid have reducing activity, the D-ascorbic acid does not have this activity. The structure of vitamin C resembles carbohydrates. Most animals and plants can synthesize vitamin C from glucose; however, **humans lack the functional enzyme, gulonolactone oxidase, for its synthesis**.

A. Biochemical functions of vitamin C

- Vitamin C is a coenzyme in hydroxylation reactions (for example, **hydroxylation of prolyl and lysyl residues in collagen**; see p. 119), where its role is to keep the iron (Fe) of *hydroxylases* in the reduced, ferrous (Fe^{+2}) form. Thus, vitamin C is required for the maintenance of normal connective tissue as well as for wound healing.
- Vitamin C also **facilitates the absorption of dietary nonheme iron** from the intestine by reduction of the ferric form (Fe^{+3}) to Fe^{+2} (see p. 557).
- Ascorbate is required in **tyrosine catabolism** as a coenzyme for 4-hydroxyphenylpyruvate dioxygenase.
- It is also required for hydroxylation of tryptophan to 5-hydroxytryptophan, **in synthesis of serotonin**.
- 7-alpha-hydroxylase requires ascorbic acid **for synthesis of bile acids**.
- It is also required for **synthesis of norepinephrine from dopamine** by the enzyme dopamine *beta-monooxygenase*.
- **Carnitine synthesis** requires ascorbate-dependent dioxygenases.
- Vitamin C is a **strong antioxidant**. The role of Vitamin C in prevention of cancer, cardiovascular diseases and cataract is under research.

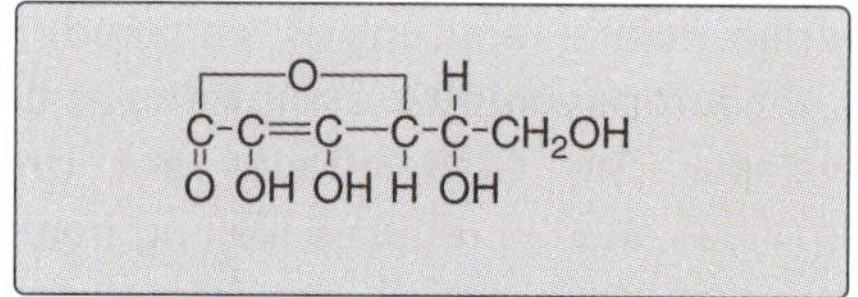

Figure 32.8 Structure of ascorbic acid.

B. Deficiency

Ascorbic acid deficiency results in **scurvy**, a disease characterized by **sore and spongy gums, loose teeth, fragile blood vessels, hemorrhage, swollen joints, bone changes, and fatigue** (Fig. 32.9). Many of the deficiency symptoms can be explained **by the decreased hydroxylation of collagen**, resulting in defective connective tissue. A **microcytic anemia** caused by decreased absorption of iron may also be seen.

C. Chronic disease prevention

Vitamin C is one of a group of nutrients that includes vitamin E (see p. 546) and β-carotene (see p. 536), which are known as **antioxidants**. [Note: Vitamin C regenerates the functional, reduced form of vitamin E.] Consumption of diets rich in these compounds is associated with a decreased incidence of some

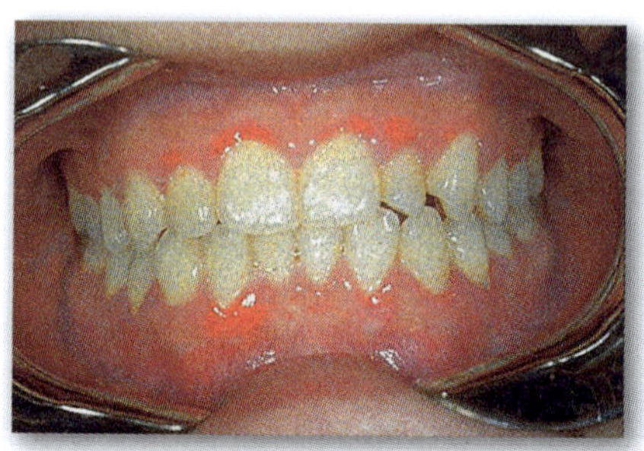

Figure 32.9 Oral manifestations in a patient with scurvy.

chronic diseases, such as cardiovascular disease (CVD) and certain cancers. However, clinical trials involving supplementation with the isolated antioxidants have failed to demonstrate any convincing preventive effects.

D. Dietary sources of vitamin C

Vitamin C is abundant in **citrus fruits**, such as lime, lemon, *amla* (Indian gooseberry) and broccoli, strawberries and tomatoes.

E. Daily requirements

An adult requires about 40 mg/day of vitamin C.

V. PYRIDOXINE (VITAMIN B_6)

Vitamin B_6 is a collective term for pyridoxine, pyridoxal, and pyridoxamine, all derivatives of pyridine. They differ only in the nature of the functional group attached to the ring (Fig. 32.10). Pyridoxine occurs primarily in plants, whereas pyridoxal and pyridoxamine are found in foods obtained from animals. Yeast, wheat germs, cereals, eggs, milk, tuna, and salmon are rich sources of vitamin B_6. **All three compounds can serve as precursors of the biologically active coenzyme, pyridoxal phosphate (PLP).** PLP functions as a coenzyme for a large number of enzymes, particularly those that catalyze reactions involving amino acids, for example, in the transsulfuration of Hcy to cysteine (see p. 400). [Note: PLP is also required by *glycogen phosphorylase* (see p. 227).]

Reaction type	Example
Transamination	Oxaloacetate + glutamate ⇄ aspartate + α-ketoglutarate
Deamination	Serine → pyruvate + NH_3
Decarboxylation	Histidine → histamine + CO_2
Condensation	Glycine + succinyl CoA → δ-aminolevulinic acid
Transsulfuration	Homocysteine + Serine → Cystathionine
	Cystathionine → Homoserine + Cysteine

Pyridoxal phosphate is a cofactor for the enzyme *kynureninase*, in catabolism of tryptophan. In deficiency of PLP, alternate metabolite, xanthurenic acid is excreted in urine. **Increased urinary excretion of xanthurenic acid is an indicator of pyridoxine deficiency.**

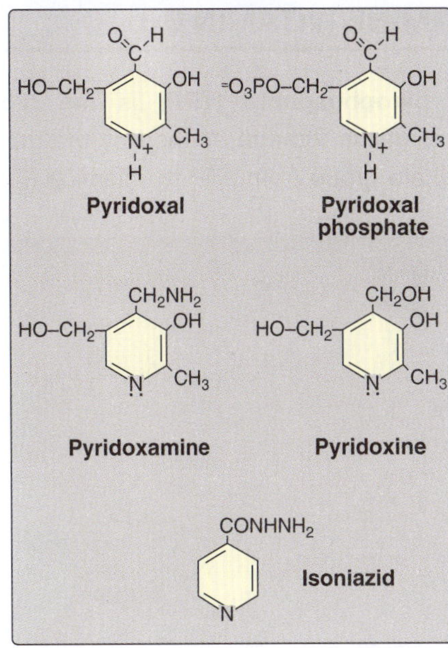

Figure 32.10 Structures of vitamin B_6 and the antituberculosis drug isoniazid.

A. Clinical indications for pyridoxine

Isoniazid, a drug commonly used to treat tuberculosis, **can induce a vitamin B_6 deficiency** by forming an inactive derivative with PLP. Thus, dietary supplementation with B_6 is an adjunct to isoniazid treatment. Cycloserine, another antitubercular drug, also acts as B_6 antagonist.

Otherwise, dietary deficiencies in pyridoxine are rare but have been observed in newborn infants fed formulas low in B_6, in women taking oral contraceptives, and in those with alcoholism.

The major manifestations of B_6 deficiency include **neurological dysfunction and anemia.** Due to B_6 deficiency, synthesis of serotonin, epinephrine and GABA is impaired. Decreased synthesis of sphingolipids can cause demyelination of nerves and peripheral neuritis. Decreased GABA formation can produce convulsions. Anemia is caused due to impaired heme synthesis.

B. Toxicity

Vitamin B_6 is the **only water-soluble vitamin with significant toxicity.** Neurologic symptoms (sensory neuropathy) occur at intakes above 500 mg/day, an amount nearly 400 times the RDA and over 5 times the tolerable upper limit (UL). (See Chapter 31 for a discussion of RDA and UL.) Substantial improvement, but not complete recovery, occurs when the vitamin is discontinued.

C. Daily requirements

The RDA for vitamin B_6 is about 2 mg/day.

VI. THIAMINE (VITAMIN B_1)

Thiamine pyrophosphate (TPP) is the biologically active form of the vitamin, formed by the transfer of a pyrophosphate group from ATP to thiamine (Fig. 32.11).

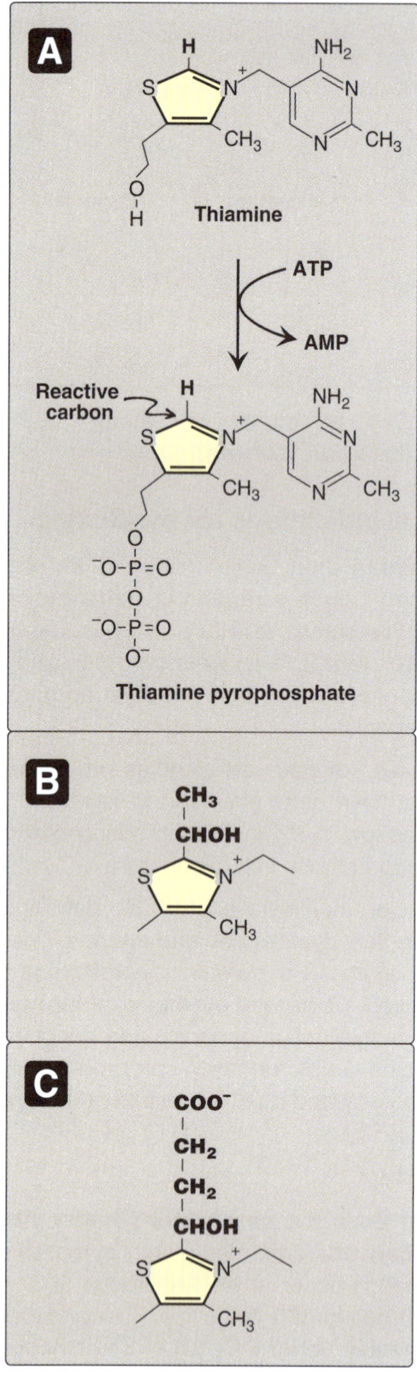

Figure 32.11 A. Structure of thiamine and its coenzyme form, thiamine pyrophosphate. **B.** Structure of intermediate formed in the reaction catalyzed by *pyruvate dehydrogenase*. **C.** Structure of intermediate formed in the reaction catalyzed by *α-ketoglutarate dehydrogenase*. AMP = adenosine monophosphate.

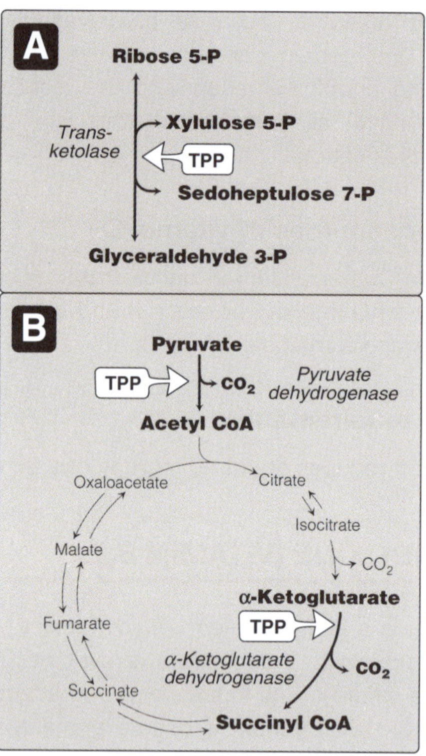

Figure 32.12 Reactions that use thiamine pyrophosphate (TPP) as coenzyme. **A.** *Transketolase.* **B.** *Pyruvate dehydrogenase* and *α-ketoglutarate dehydrogenase.* [Note: TPP is also used by *branched-chain α-keto acid dehydrogenase.*] P = phosphate; CoA = coenzyme A; CO_2 = carbon dioxide.

TPP serves as a coenzyme in the formation or degradation of α-ketols by *transketolase* (Fig. 32.12A) and in the oxidative decarboxylation of α-keto acids (Fig. 32.12B).

A. Distribution

Germs of cereals, legumes, liver, nuts, yeast and outer layers of grains. **When the grains are polished, the outer layer (aleurone), which contains thiamine, is removed**.

B. Clinical indications for thiamine

The oxidative decarboxylation of pyruvate and α-ketoglutarate, which plays a key role in energy metabolism of most cells, is particularly important in tissues of the CNS. In thiamine deficiency, the activity of these two *dehydrogenase*-catalyzed reactions is decreased, resulting in decreased production of ATP and, therefore, impaired cellular function. TPP is also required by *branched-chain α-keto acid dehydrogenase* of muscle (see p. 401). [Note: It is the *decarboxylase* of each of these *α-keto acid dehydrogenase* multienzyme complexes that requires TPP.]

1. **Beriberi:** This severe thiamine-deficiency syndrome is found in areas where polished rice is the major component of the diet. Adult beriberi is classified as **dry** (characterized by peripheral neuropathy with sensory disturbances, especially in the legs) or **wet** (characterized by edema because of dilated cardiomyopathy, palpitations and breathlessness). **Infantile beriberi** is seen in breastfed infants, born to mothers having thiamine deficiency.

2. **Wernicke-Korsakoff syndrome:** Some individuals with alcoholism develop Wernicke-Korsakoff syndrome, a thiamine-deficiency state characterized by mental confusion, gait ataxia, nystagmus (a to-and-fro motion of the eyeballs), and ophthalmoplegia (weakness of eye muscles) with Wernicke encephalopathy as well as memory problems and hallucinations with Korsakoff dementia. The syndrome is treatable with thiamine supplementation, but recovery of memory is typically incomplete.

C. Assessment of thiamine level

Thiamine deficiency is diagnosed by an increase in **erythrocyte *transketolase* activity** observed with addition of TPP.

D. Daily requirements

The RDA for vitamin B_1 is 1.0–1.7 mg/day.

> **Vitamin deficiencies in chronic alcoholics**
>
> Both acute and chronic alcohol consumption can cause malnutrition by multiple mechanisms, which include decreasing dietary caloric intake, impairing nutrient digestion and absorption, and increasing breakdown and excretion of nutrients.
>
> Both fat- and water-soluble vitamin deficiencies occur with moderate to severe alcohol use disorder. Vitamin A and K deficiencies are rare in the absence of liver disease or chronic pancreatitis.
>
> Thiamine deficiency is found in up to 80%, pyridoxine deficiency in over 50% and folate deficiency in about one-third of alcoholic patients.

VII. NIACIN (VITAMIN B_3)

Niacin, or nicotinic acid, is a substituted pyridine derivative. **The biologically active coenzyme forms are nicotinamide adenine dinucleotide (NAD^+) and its phosphorylated derivative, nicotinamide adenine dinucleotide phosphate ($NADP^+$)**, as shown in Figure 32.13. Nicotinamide, a derivative of nicotinic acid that contains an amide instead of a carboxyl group, also occurs in the diet. Nicotinamide is readily deaminated in the body and, therefore, is nutritionally equivalent to nicotinic acid. NAD^+ and $NADP^+$ serve as coenzymes in oxidation–reduction reactions in which the coenzyme undergoes reduction of the pyridine ring by accepting two electrons from a hydride ion, as shown in Figure 32.14. The reduced forms of NAD^+ and $NADP^+$ are NADH and NADPH, respectively. NAD^+ is used in the catabolic redox reactions, while NADPH is used in reductive biosynthesis. NAD^+ is also a donor of ADP-ribose in ADP-ribosylation reactions required for DNA repair and post-translational modifications in eukaryotes.

A metabolite of tryptophan, quinolinate, can be converted to NAD(P). In comparison, **60 mg of tryptophan = 1 mg of niacin**. Corn is low in both niacin and tryptophan. Corn-based diets can cause niacin deficiency.

A. Distribution

Niacin is found in unrefined and enriched grains and cereal, milk, legumes, and lean meats (especially liver).

B. Clinical indications for niacin

1. **Deficiency:** A deficiency of niacin causes **pellagra**, a disease involving the skin, gastrointestinal tract, and CNS. **The symptoms of pellagra progress through the three Ds: dermatitis (photosensitive), diarrhea, and dementia. If untreated, death (a fourth D) occurs**. Hartnup disorder, characterized by defective absorption of tryptophan, can result in pellagra-like symptoms.]

2. **Hyperlipidemia treatment:** Niacin at doses of 1.5 g/day, or 100 times the RDA, strongly **inhibits lipolysis in adipose tissue**, the primary producer of circulating free fatty acids (FFA). The liver normally uses these circulating FFA as a major precursor for triacylglycerol (TAG) synthesis. Thus, niacin causes a decrease in liver TAG synthesis, which is required for very-low-density lipoprotein ([VLDL], see p. 352) production. Low-density lipoprotein (LDL, the cholesterol-rich lipoprotein) is derived from VLDL in the plasma. Thus, **both plasma TAG (in VLDL) and cholesterol (in LDL) are lowered**. Therefore, niacin is particularly useful in the treatment of type IIb hyperlipoproteinemia, in which both VLDL and LDL are elevated. Niacin raises high-density lipoprotein and lowers Lp(a) levels (see p. 358).

However, the high doses of niacin required can cause acute, prostaglandin-mediated flushing. Aspirin can reduce this side effect by inhibiting

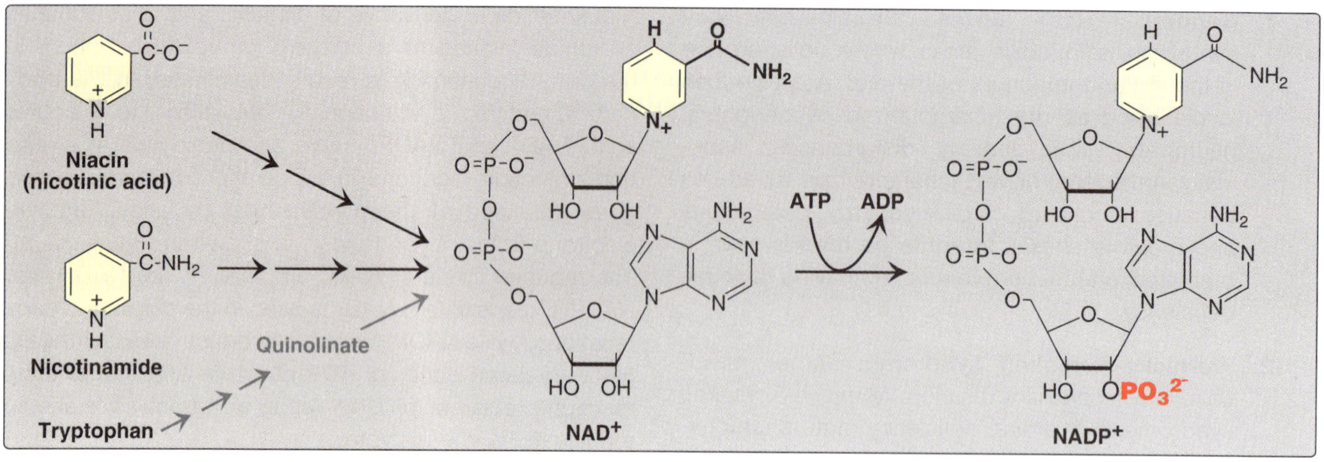

Figure 32.13 Structure and biosynthesis of oxidized nicotinamide adenine dinucleotide (NAD^+) and nicotinamide adenine dinucleotide phosphate ($NADP^+$). ADP = adenosine diphosphate.

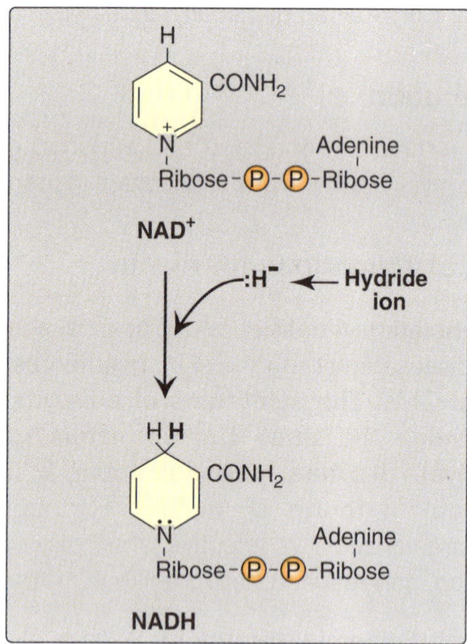

Figure 32.14 Reduction of oxidized nicotinamide adenine dinucleotide (NAD^+) to NADH. [Note: The hydride ion consists of a hydrogen (H) atom plus an electron.] = phosphate.

prostaglandin synthesis (see p. 334). Itching may also occur.

C. Daily requirements

The RDA for niacin is 12–21 mg/day.

VIII. RIBOFLAVIN (VITAMIN B_2)

Riboflavin consists of an isoalloxazine ring with a ribitol (alcohol of ribose sugar) side chain. **The two biologically active forms of B_2 are flavin mononucleotide (FMN) and flavin adenine dinucleotide (FAD)**, formed by the transfer of an adenosine monophosphate moiety from ATP to FMN (Fig. 32.15). FMN and FAD are each capable of reversibly accepting two hydrogen atoms, forming $FMNH_2$ or $FADH_2$, respectively. FMN and FAD are bound tightly, sometimes covalently, to flavoenzymes (for example, *NADH dehydrogenase* [FMN] and *succinate dehydrogenase* [FAD]) that catalyze the oxidation or reduction of a substrate.

Riboflavin deficiency is uncommon as it is synthesized by the intestinal bacteria. Riboflavin deficiency is not associated with a major human disease, although it frequently accompanies other vitamin deficiencies. Deficiency symptoms include dermatitis, cheilosis (fissuring at the corners of the mouth), and glossitis (the tongue appearing smooth and dark).

Riboflavin is heat stable but sensitive to light. Due to light sensitivity, phototherapy for hyperbilirubinemia (see p. 423) may require supplementation with the vitamin.

A. Distribution

Riboflavin is present in eggs, milk, yeast, legumes, whole cereals and green leafy vegetables.

B. Daily requirements

The RDA for riboflavin is about 1.3–2.1 mg/day.

IX. BIOTIN (VITAMIN B_7)

Biotin is a coenzyme in carboxylation reactions, in which it **serves as a carrier of activated carbon dioxide (CO_2)** (see Fig. 15.3, p. 217, for the mechanism of biotin-dependent carboxylations). Biotin is covalently bound to the ε-amino group of lysine residues in biotin-dependent enzymes (Fig. 32.16). Biotin deficiency does not occur naturally because the vitamin is widely distributed in

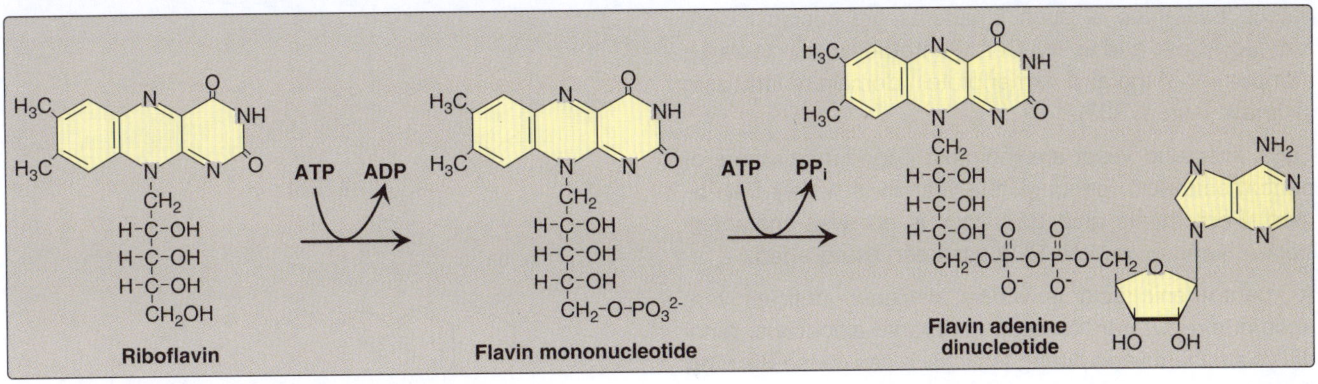

Figure 32.15 Structure and biosynthesis of the oxidized forms of flavin mononucleotide and flavin adenine dinucleotide. ADP = adenosine diphosphate; PP_i = pyrophosphate.

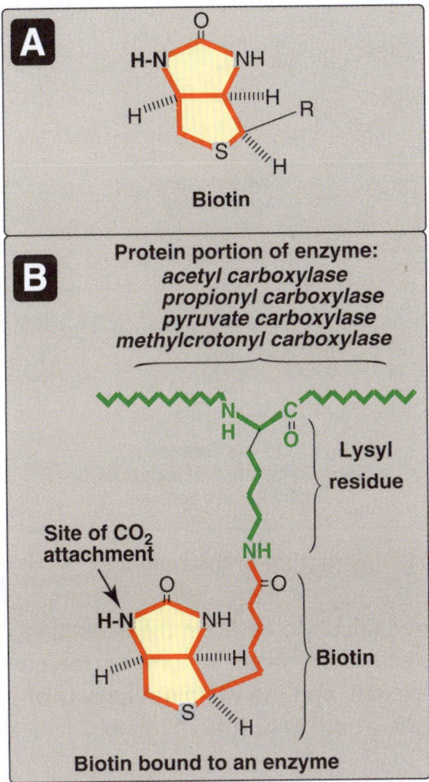

Figure 32.16 **A.** Structure of biotin. **B.** Biotin covalently bound to a lysyl residue of a biotin-dependent enzyme. CO_2 = carbon dioxide.

food. Also, a large percentage of the biotin requirement in humans is supplied by intestinal bacteria. However, the addition of raw egg white to the diet as a source of protein can cause biotin deficiency. Raw egg white contains the glycoprotein avidin, which tightly binds biotin and prevents its absorption from the intestine. Symptoms of biotin deficiency include dermatitis, hair loss, loss of appetite, and nausea. With a normal diet, however, it has been estimated that 20 eggs/day would be required to induce a deficiency syndrome. [Note: Inclusion of raw eggs in the diet is not recommended because of the possibility of salmonellosis caused by infection with Salmonella enterica.]

> Multiple carboxylase deficiency results from decreased ability to add biotin to *carboxylases* during their synthesis or to remove it during their degradation. Treatment is biotin supplementation.

A. Daily requirements

The RDA for biotin is 30 µg/day.

X. PANTOTHENIC ACID (VITAMIN B_5)

Pantothenic acid is **a component of CoA**, which functions in the transfer of acyl groups (Fig. 32.17). CoA contains a thiol group that carries acyl compounds as activated thiol

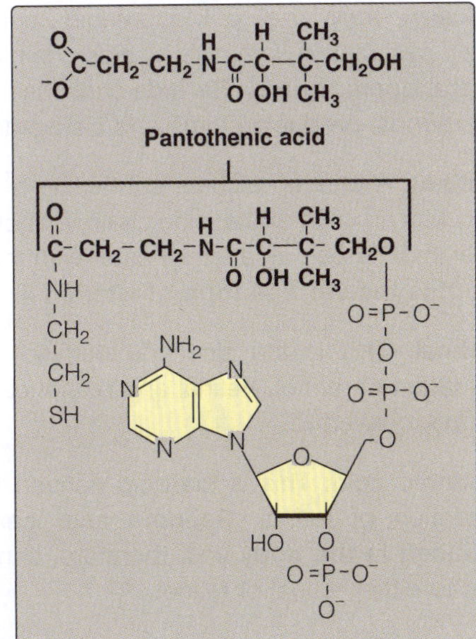

Figure 32.17 Structure of coenzyme A.

esters. Examples of such structures are succinyl CoA, fatty acyl CoA, and acetyl CoA. Pantothenic acid is also a component of the acyl carrier protein domain of *fatty acid synthase* (see p. 303).

Eggs, liver, and yeast are the most important sources of pantothenic acid, although the vitamin is widely distributed. Pantothenic acid deficiency is not well characterized in humans, and no RDA has been established.

As pantothenic acid is widely available, deficiency is uncommon but can be seen in chronic alcoholics, renal dialysis patients and during famines. It can cause **Burning Feet Syndrome**, characterized by burning pain in lower limbs, gait disturbances and sleep disturbances.

A. Daily requirements

RDA of pantothenic acid is 5 mg/day.

XI. VITAMIN A

Vitamin A is a fat-soluble vitamin that comes primarily from animal sources as retinol (preformed vitamin A), a retinoid. The retinoids, a family of structurally related molecules, are essential for vision, reproduction, growth, and maintenance of epithelial tissues. They also play a role in immune function. **Retinoic acid, derived from oxidation of retinol, mediates most of the actions of the retinoids, except for vision, which depends on retinal, the aldehyde derivative of retinol.**

A. Structure

The retinoids include the natural forms of vitamin A, retinol and its metabolites (Fig. 32.18), and synthetic forms (drugs). Retinoids are found only in animals. Chemically, they are polyisoprenoid compounds having a cyclohexenyl ring with a side chain made up of several isoprene units. The side chain has alternate double bonds, producing trans and cis isomers.

1. **Retinol:** A primary alcohol containing a β-ionone ring with an unsaturated side chain, retinol is found in animal tissues as a retinyl ester with long-chain FA. **It is the storage form of vitamin A.**

2. **Retinal:** This is the aldehyde derived from the oxidation of retinol. Retinal and retinol can readily be interconverted.

3. **Retinoic acid:** This is the acid derived from the oxidation of retinal. **Retinoic acid cannot be reduced in the body** and, therefore, cannot give rise to either retinal or retinol.

4. **β-Carotene: Plant foods contain β-carotene (provitamin A)**, which can be oxidatively and symmetrically cleaved in the intestine to yield two molecules of retinal. In humans, the conversion is inefficient, and the vitamin A activity of β-carotene is only about 1/12 that of retinol.

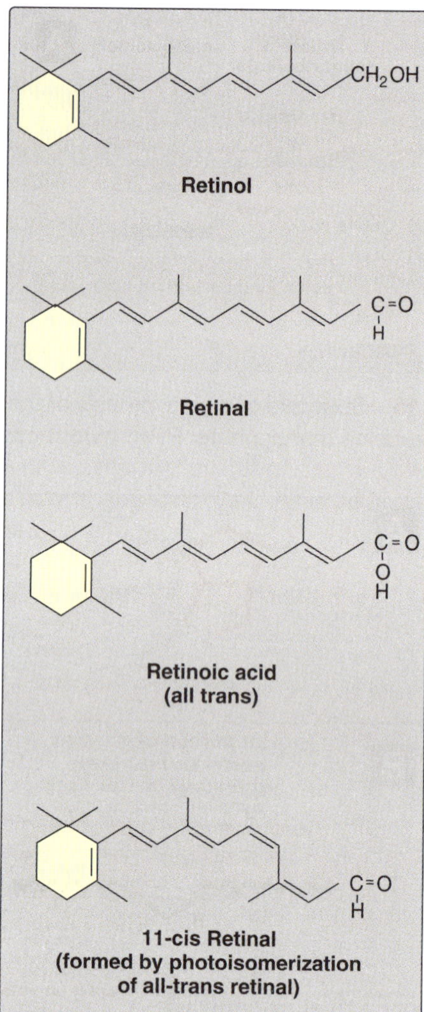

Figure 32.18 Structure of the retinoids.

B. Absorption and transport to the liver

Retinyl esters from the diet are hydrolyzed in the intestinal mucosa, releasing retinol and FFA (Fig. 32.19). Retinol derived from esters and from the reduction of retinal from β-carotene cleavage is reesterified to long-chain FA within the enterocytes and secreted as a component of chylomicrons into the lymphatic system. Retinyl esters contained in chylomicron remnants are taken up by, and stored in, the liver. Liver stores 90% of body's vitamin A. [Note: All fat-soluble vitamins are carried in chylomicrons.]

C. Release from the liver

When needed, retinol is released from the liver and **transported through the blood to extrahepatic**

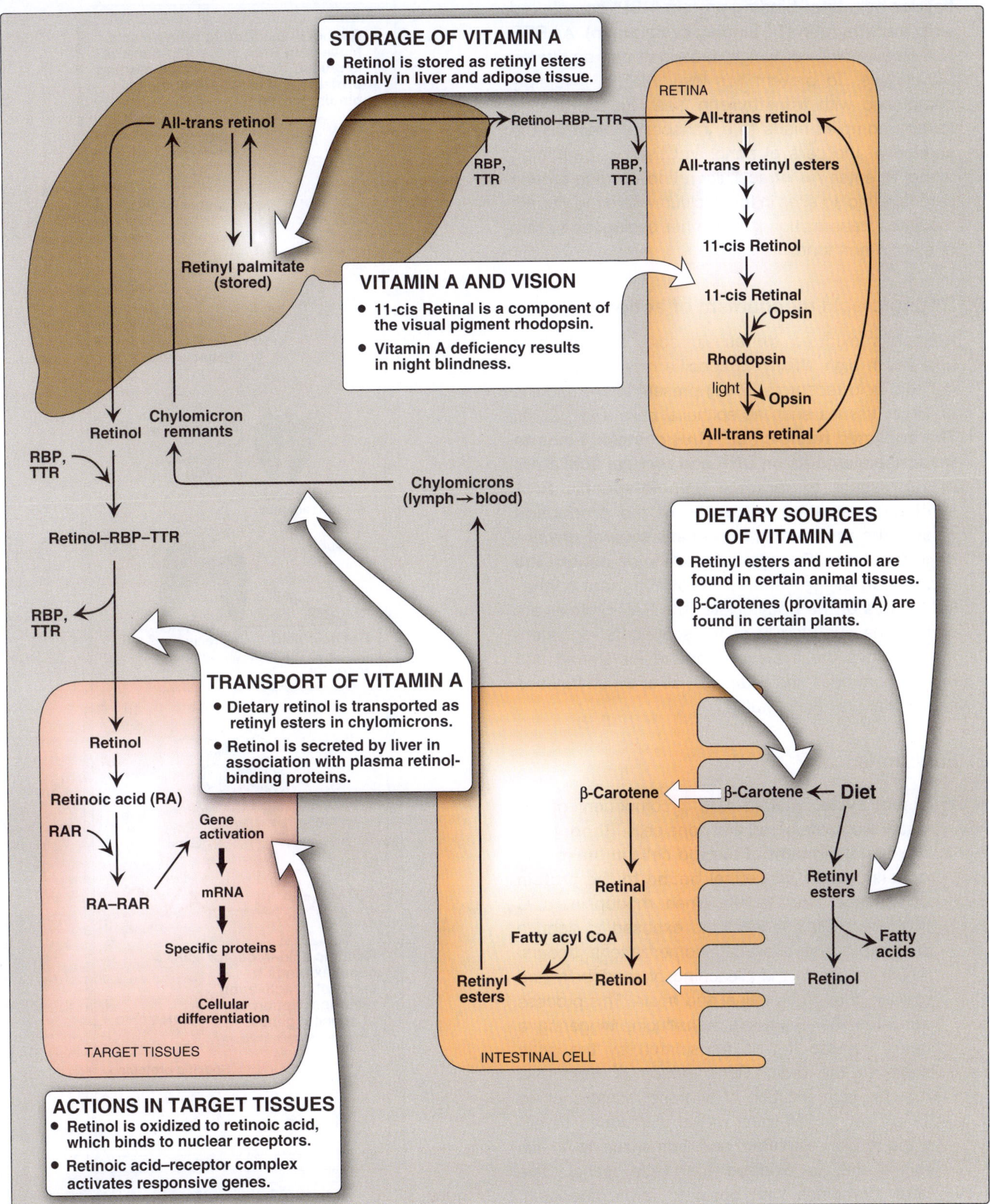

Figure 32.19 Absorption, transport, and storage of vitamin A and its derivatives. [Note: β-Carotene is a carotenoid, a plant pigment with antioxidant activity.] RBP = retinol binding protein; TTR = transthyretin; RAR = retinoic acid receptor; CoA = coenzyme A; mRNA = messenger RNA.

tissues by retinol-binding protein (RBP) complexed with transthyretin (T4 binding prealbumin). As RBP is a small molecular weight protein, it can be filtered by kidneys. To prevent this loss, RBP is reversibly complexed with **transthyretin** (see Fig. 32.19). The ternary complex binds to a transport protein on the surface of the cells of peripheral tissues, permitting retinol to enter. An intracellular retinol-binding protein carries retinol to sites in the nucleus where the vitamin regulates transcription in a manner analogous to that of steroid hormones.

D. Retinoic acid mechanism of action

Retinol is oxidized to retinoic acid. Retinoic acid binds with high affinity to specific receptor proteins (retinoic acid receptors [RAR]) present in the nucleus of target tissues such as epithelial cells (Fig. 32.20). **The activated retinoic acid–RAR complex binds to response elements on DNA and recruits activators or repressors to regulate retinoid-specific RNA synthesis, resulting in control of the production of specific proteins that mediate several physiologic functions. For example, retinoids control the expression of the gene for keratin in most epithelial tissues of the body.** [Note: The RAR proteins are part of the superfamily of transcriptional regulators that includes the nuclear receptors for steroid and thyroid hormones and vitamin D, all of which function in a similar way (see p. 362).]

E. Functions

1. **Visual cycle:** Vitamin A is a component of the visual pigments of rod and cone cells. **Rhodopsin, the visual pigment of the rod cells in the retina, consists of 11-cis retinal bound to the protein opsin** (see Fig. 32.19). When rhodopsin, a G protein–coupled receptor, is exposed to light, a series of photochemical isomerizations occurs, which results in the bleaching of rhodopsin and release of all-trans retinal and opsin. This process activates the G protein transducin, triggering a nerve impulse that is transmitted by the optic nerve to the brain. Regeneration of rhodopsin requires isomerization of all-trans retinal back to 11-cis retinal. All-trans retinal is reduced to all-trans retinol, esterified, and isomerized to 11-cis retinol that is oxidized to 11-cis retinal. The latter combines with opsin to form rhodopsin, thus completing the cycle. **This whole cycle of changes in rhodopsin on exposure to light and its regeneration is known as Wald's visual cycle**. Similar reactions, performed by conopsin (similar to rhodopsin), are responsible for color vision in the cone cells. There are three types of conopsin: cyanopsin (for blue color), iodopsin (for green color) and porphyropsin (for red color).

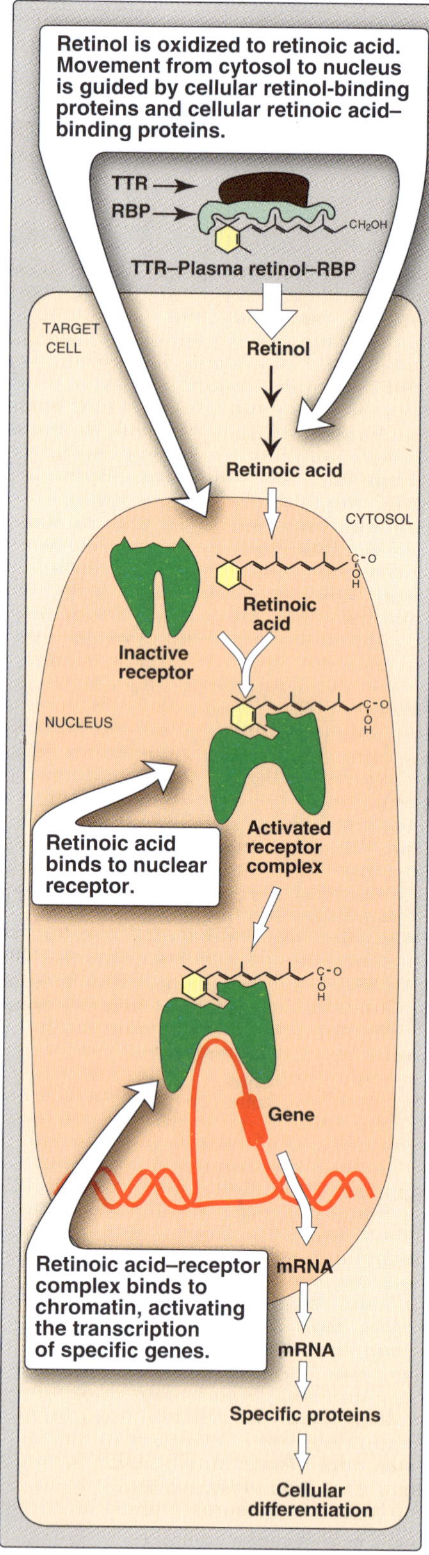

Figure 32.20 Action of the retinoids. [Note: Retinoic acid–receptor complex forms a dimer, but is shown as monomer for simplicity.] TTR = transthyretin; RBP = retinol-binding protein; mRNA = messenger RNA.

> **Dark adaptation**
>
> When a person shifts from bright light to dark, there is difficulty in seeing but the vision improves after a few minutes. This delay is due to time taken for regeneration of rhodopsin after its stores are depleted by bright light. This time is known as dark adaptation time. **It increases in vitamin A deficiency.**

2. **Epithelial cell maintenance:** Vitamin A is essential for normal differentiation of epithelial tissues and mucus secretion and, thus, supports the body's barrier-based defense against pathogens.

3. **Reproduction:** Retinol and retinal are essential for normal reproduction, supporting spermatogenesis in the male and preventing fetal resorption in the female. Retinoic acid is inactive in maintaining reproduction and in the visual cycle but promotes growth and differentiation of epithelial cells.

4. **Antioxidant property:** Carotenoids have been reported to have antioxidant property, with preventive role in cancer and cardiovascular diseases.

F. Distribution

Liver, kidney, cream, butter, and egg yolk are good sources of preformed vitamin A. Yellow, orange, and dark-green vegetables and fruits are good sources of the carotenes (provitamin A).

G. Daily requirements

The RDA of retinol for adults is 600 μg/day. In comparison, 1 RAE = 1 μg of retinol, 12 μg of β-carotene, or 24 μg of other carotenoids.

H. Clinical indications for vitamin A

Although chemically related, retinoic acid and retinol have distinctly different therapeutic applications. **Retinol and its carotenoid precursor are used as dietary supplements, whereas various forms of retinoic acid are useful in dermatology** (Fig. 32.21).

1. **Deficiency:** Vitamin A deficiency can occur due to:
   - Inadequate dietary intake
   - Fat malabsorption which decreases absorption of vitamin A
   - Impaired absorption due to biliary obstruction

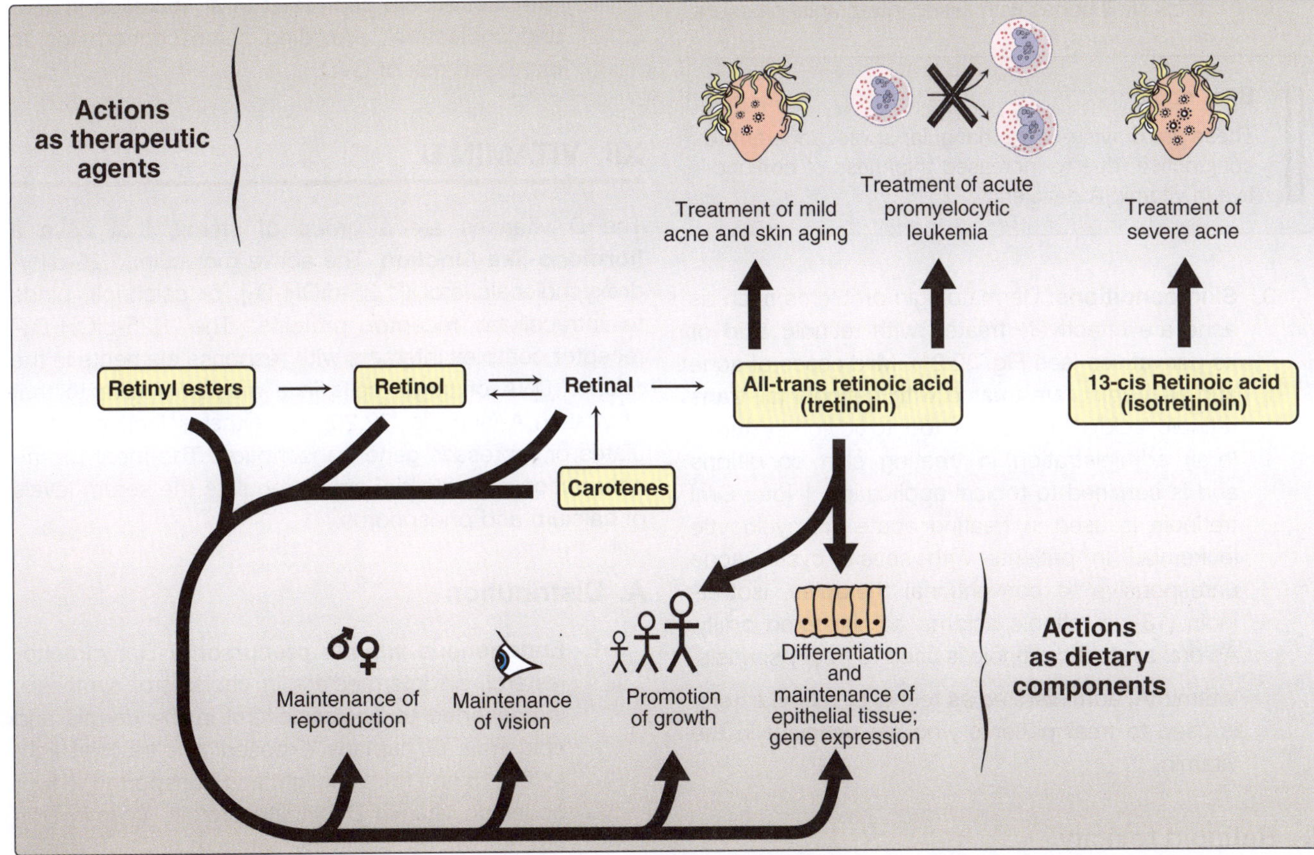

Figure 32.21 Summary of actions of retinoids. Compounds in **boxes** are available as dietary components or as pharmacologic agents.

- Impaired storage due to liver disease

2. **Deficiency manifestations**
   - Ocular manifestations

   Night blindness (nyctalopia) is one of the earliest signs of vitamin A deficiency. The visual threshold is increased, making it difficult to see in dim light. Prolonged deficiency leads to an irreversible loss in the number of visual cells.

   Severe deficiency leads to xerophthalmia, a pathologic dryness of the conjunctiva and cornea, caused, in part, by increased keratin synthesis. **If untreated, xerophthalmia results in corneal ulceration and, ultimately, in blindness** because of the formation of opaque scar tissue. The condition is most commonly seen in children in developing tropical countries. Over 500,000 children worldwide are blinded each year by xerophthalmia caused by insufficient vitamin A in the diet.

   - Gonadal dysfunctions: Testicular degeneration in males and an increased incidence of miscarriage.
   - Epithelial changes: Vitamin A deficiency is associated with epithelial changes such as increased keratinization and squamous metaplasia.
   - Bone remodeling defects: These can produce thick skull bones with raised intracranial pressure.

Bitot spots

These are grayish-white triangular spots, adherent to conjunctiva, due to increased thickness of conjunctiva in vitamin A deficiency.

3. **Skin conditions:** Dermatologic problems such as acne are effectively treated with retinoic acid or its derivatives (see Fig. 32.21). Mild cases of acne and skin aging are treated with tretinoin (all-trans retinoic acid). Tretinoin is too toxic for systemic (oral) administration in treating skin conditions and is confined to topical application. [Note: Oral tretinoin is used in treating acute promyelocytic leukemia.] In patients with severe cystic acne unresponsive to conventional therapies, isotretinoin (13-cis retinoic acid) is administered orally. An oral synthetic retinoid is used to treat psoriasis.

   Vitamin A, administered as retinol or retinyl esters, is used to treat patients who are deficient in the vitamin.

I. Retinoid toxicity

1. **Vitamin A:** Excessive intake of vitamin A (but not carotene) produces a toxic syndrome called hypervitaminosis A. Amounts exceeding 7.5 mg/day of retinol should be avoided. **Early signs of chronic hypervitaminosis A are reflected in the:**
   - Skin, which becomes dry and pruritic (because of decreased keratin synthesis)
   - Liver, which becomes enlarged and can become cirrhotic
   - CNS, where a rise in intracranial pressure may mimic the symptoms of a brain tumor

   Pregnant women, in particular, should not ingest excessive quantities of vitamin A because of its **potential for teratogenesis** (causing congenital malformations in the developing fetus). UL is 3,000 μg preformed vitamin A/day.

   [Note: Vitamin A promotes bone growth. In excess, however, it is associated with decreased bone mineral density and increased risk of fractures.]

2. **Isotretinoin:** The drug, an isomer of retinoic acid, **is teratogenic** and absolutely contraindicated in women with childbearing potential unless they have severe, disfiguring cystic acne that is unresponsive to standard therapies. Pregnancy must be excluded before treatment begins, and birth control must be used. Prolonged treatment with isotretinoin can result in an increase in TAG and cholesterol, providing some concern for an increased risk of CVD.

XII. VITAMIN D

The D vitamins are a group of sterols that have a **hormone-like function**. The active molecule, 1,25-dihydroxycholecalciferol ([1,25-diOH-D_3], or calcitriol), binds to intracellular receptor proteins. The 1,25-diOH-D_3–receptor complex interacts with response elements in the nuclear DNA of target cells in a manner similar to that of vitamin A (see Fig. 32.20) and either selectively stimulates or represses gene transcription. The most prominent actions of calcitriol are to regulate the serum levels of calcium and phosphorus.

A. Distribution

1. **Endogenous vitamin precursor:** 7-Dehydrocholesterol, an intermediate in cholesterol synthesis, is converted to cholecalciferol in the dermis and epidermis of humans exposed to ultraviolet light (290–315 nm) from sunlight and transported to liver bound to vitamin D–binding protein. Vitamin D is also known as "sunshine" vitamin.

2. **Diet:** Ergocalciferol (vitamin D_2), found in plants, and cholecalciferol (vitamin D_3), found in animal

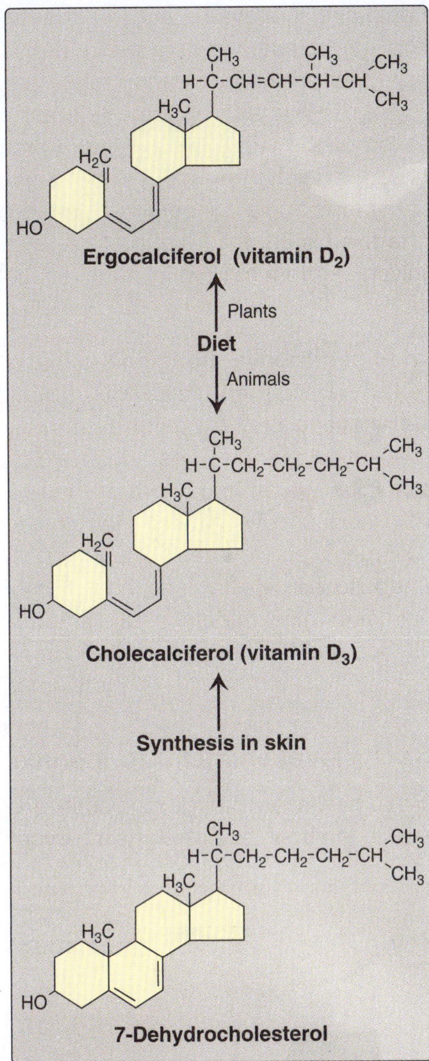

Figure 32.22 Sources of vitamin D. Vitamins D_2 and D_3 are first converted to calcidiol and then to calcitriol (active vitamin D). [Note: 7-Dehydrocholesterol (provitamin D_3) is decreased in the skin of older adults.]

tissues, are sources of preformed vitamin D activity (Fig. 32.22). Vitamin D_2 and vitamin D_3 differ chemically only in the presence of an additional double-bond and methyl group in the plant sterol. Dietary vitamin D is packaged into chylomicrons. [Note: Preformed vitamin D is a dietary requirement only in individuals with limited exposure to sunlight.]

B. Metabolism

1. **1,25-Dihydroxycholecalciferol formation:** Vitamins D_2 and D_3 are not biologically active but are **converted <u>in vivo</u> to calcitriol, the active form of the D vitamin, by two sequential hydroxylation reactions** (Fig. 32.23). The first hydroxylation occurs at the 25 position and is catalyzed by a specific *25-hydroxylase* in the liver. The product of the reaction, 25-hydroxycholecalciferol ([25-OH-D_3], calcidiol), is the predominant form of vitamin D in the serum and the major storage form. In the kidney, 25-OH-D_3 is further hydroxylated at the 1 position by *25-hydroxycholecalciferol 1-hydroxylase*, resulting in the formation of 1,25-diOH-D_3 (calcitriol). [Note: Both *hydroxylases* are cytochrome P450 proteins (see p. 251).]

2. **Hydroxylation regulation:** Calcitriol is the most potent vitamin D metabolite. Its formation is **tightly regulated by the level of serum phosphate (PO_4^{3-}) and calcium ions (Ca^{2+})** as shown in Figure 32.24. ***25-Hydroxycholecalciferol 1-hydroxylase*** **activity is increased directly by low serum PO_4^{3-}. Low serum Ca^{2+} triggers the secretion of parathyroid hormone (PTH) from the chief cells of the parathyroid gland, which upregulates the *1-hydroxylase*.** Thus, hypocalcemia caused by insufficient dietary Ca^{2+} results in elevated levels of serum 1,25-diOH-D_3. [Note: 1,25-diOH-D_3 inhibits expression of PTH, forming a negative feedback loop. It also inhibits activity of the *1-hydroxylase*.]

C. Function

The overall function of calcitriol is **to maintain adequate serum levels of Ca^{2+}**. It performs this function by 1) increasing uptake of Ca^{2+} by the intestine, 2) minimizing loss of Ca^{2+} by the kidney by increasing reabsorption, and 3) stimulating resorption (demineralization) of bone when blood Ca^{2+} is low (see Fig. 32.23).

1. **Effect on the intestine: Calbindin** is a calcium-binding protein, which is required for absorption of calcium from the intestinal cell to blood. Calcitriol stimulates intestinal absorption of Ca^{2+} by first entering the intestinal cell and binding to a cytosolic receptor. The 1,25-diOH-D_3–receptor complex then moves to the nucleus where it selectively interacts with response elements on the DNA. As a result, Ca^{2+} uptake is enhanced by increased expression of calbindin. Thus, the **mechanism of action of 1,25-diOH-D_3 is typical of steroid hormones** (see p. 362).

2. **Effect on bone:** Bone is composed of collagen and crystals of $Ca_5(PO_4)_3OH$ (hydroxyapatite). **When blood Ca^{2+} is low, 1,25-diOH-D_3 stimulates bone resorption by a process that is enhanced by PTH**. The result is an increase in serum Ca^{2+}. Therefore, bone is an important reservoir of Ca^{2+} that can be mobilized to maintain

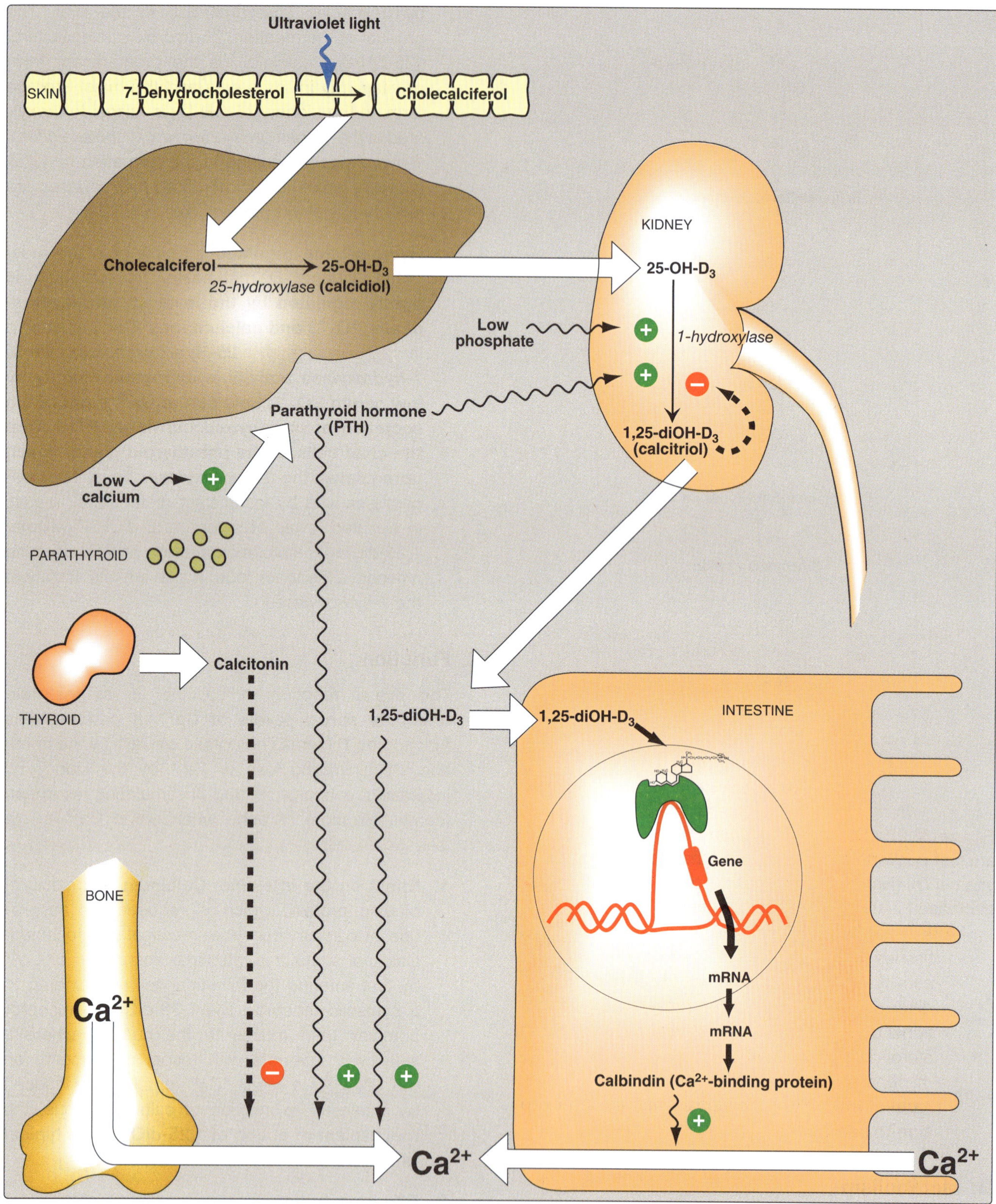

Figure 32.23 Metabolism and actions of vitamin D. [Note: Calcitonin, a thyroid hormone, decreases blood calcium (Ca^{2+}) by inhibiting mobilization from bone, absorption from the intestine, and reabsorption by the kidney. It opposes the actions of PTH.] mRNA = messenger RNA; 25-OH-D_3 = 25-hydroxycholecalciferol; 1,25-diOH-D_3 = 1,25-dihydroxycholecalciferol.

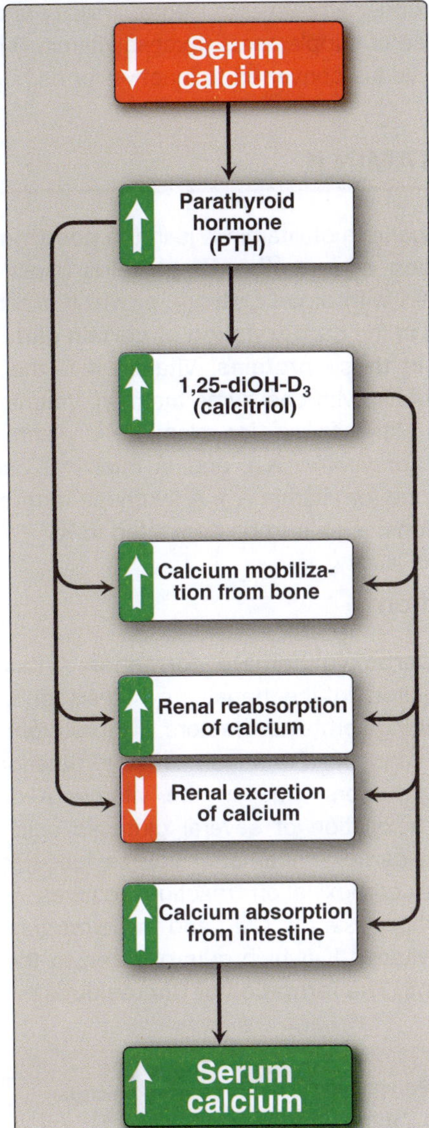

Figure 32.24 Response to low serum calcium. 1,25-diOH-D_3 = 1,25-dihydroxycholecalciferol. [Note: Calcitriol also increases intestinal absorption and renal reabsorption of phosphate. In contrast, PTH decreases renal reabsorption of phosphate.]

serum levels. [Note: PTH and calcitriol also work together to prevent renal loss of Ca^{2+}.]

Vitamin D also increases the mineralization of bone. This effect is mediated by multiple mechanisms, including increased production of calbindin-9 and increased local concentration of phosphate in bone which promotes bone mineralization.

3. **Effect on kidneys:** Calcitriol increases the reabsorption of both calcium and phosphorus by renal tubules.

D. Distribution and daily requirements

Vitamin D occurs naturally in fatty fish, liver, and egg yolk. Milk, unless it is artificially fortified, is not a good source. The RDA for individuals ages 1–70 years is 10 μg/day. [Note: 1 μg vitamin D = 40 international units (IU).]

Because breast milk is a poor source of vitamin D, supplementation is recommended for breastfed babies.

E. Causes of vitamin D deficiency

- Inadequate dietary intake
- Malabsorption of vitamin D: Seen in obstructive jaundice and steatorrhea.
- Impaired activation of vitamin D: In liver or kidney disorders, vitamin D may not be activated.
- Inadequate exposure to sunlight: This is a common cause of vitamin D deficiency, especially in people living in congested areas with poor exposure to sunlight or in people who spend very little time outdoors.

F. Clinical indications for vitamin D

1. **Nutritional rickets:** Vitamin D deficiency causes **a net demineralization of bone, resulting in rickets in children and osteomalacia in adults** (Fig. 32.25). Rickets is characterized by the continued formation of the collagen matrix of bone, but incomplete mineralization results in soft, pliable bones. Weight-bearing bones are bent.

   Features of rickets include (i) bowed legs, (ii) knock-knee, (iii) Harrison's sulcus (transverse depression extending from costal cartilage to axilla), (iv) rickety rosary (beading of ribs at costochondral junctions), (v) bossing of frontal bones, and (vi) pigeon chest.

   In osteomalacia, demineralization of preexisting bones increases their susceptibility to fracture. Insufficient exposure to daylight and/or deficiencies in vitamin D consumption occur predominantly in infants and the elderly. **In both rickets and osteomalacia, plasma calcium and phosphorus levels are low and *alkaline phosphatase* level (bone fraction) is markedly elevated.** Vitamin D deficiency is more common in the northern latitudes, because less vitamin D synthesis occurs in the skin as a result of reduced exposure to ultraviolet light. [Note: Loss-of-function mutations in the vitamin D receptor result in hereditary vitamin D–deficient rickets.]

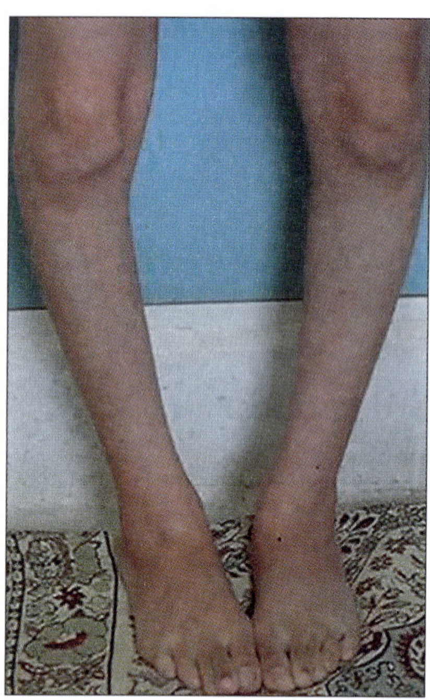

Figure 32.25 Bowed legs of middle-aged man with osteomalacia, a nutritional vitamin D deficiency that results in demineralization of the skeleton.

2. **Renal osteodystrophy:** Chronic kidney disease causes **decreased ability to form active vitamin D as well as increased retention of PO_4^{3-}, resulting in hyperphosphatemia and hypocalcemia.** The low blood Ca^{2+} causes a rise in PTH and associated bone demineralization with release of Ca^{2+} and PO_4^{3-}. Supplementation with vitamin D is an effective therapy. However, supplementation must be accompanied by PO_4^{3-} reduction therapy to prevent further bone loss and precipitation of calcium phosphate crystals.

3. **Hypoparathyroidism:** Parathyroid hormone increases blood calcium level (by increased resorption of bone) and decreases blood phosphate level (by increasing its renal excretion). **Lack of PTH causes hypocalcemia and hyperphosphatemia**. Patients may be treated with vitamin D and calcium supplementation.

G. Toxicity

Like all fat-soluble vitamins, vitamin D can be stored in the body and is only slowly metabolized. High doses (100,000 IU for weeks or months) can cause **toxicity, manifested by loss of appetite, nausea, thirst, and weakness.** Enhanced Ca^{2+} absorption and bone resorption results in hypercalcemia, which can lead to deposition of calcium salts in soft tissue (**metastatic calcification**). The UL is 100 μg/day (4,000 IU/day) for individuals ages 9 years or older, with a lower level for those under age 9 years. [Note: Toxicity is only seen with use of supplements. Excess vitamin D produced in the skin is converted to inactive forms.]

XIII. VITAMIN K

The principal role of vitamin K is in the **posttranslational modification of a number of proteins** (most of which are involved with blood clotting), in which it serves as a coenzyme in the **carboxylation of certain glutamic acid residues in these proteins**. Vitamin K is the only fat-soluble vitamin with coenzyme function. Vitamin K exists in several active forms, for example, in plants as phylloquinone (or vitamin K_1), and in intestinal bacteria as menaquinone (or vitamin K_2). A synthetic form of vitamin K, menadione, is able to be converted to K_2.

A. Function

1. **γ-Carboxyglutamate formation:** Vitamin K is required in the hepatic synthesis of the blood clotting proteins, factors II (prothrombin), VII, IX, and X. Formation of the functional clotting factors requires the vitamin K–dependent carboxylation of several glutamic acid residues to γ-carboxyglutamate (Gla) residues (Fig. 32.26). The carboxylation reaction requires *γ-glutamyl carboxylase*, O_2, CO_2, and the hydroquinone form of vitamin K (which gets oxidized to the epoxide form). **The formation of Gla residues is sensitive**

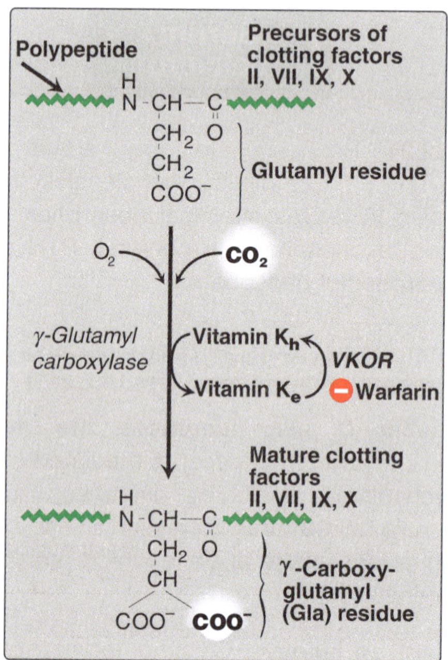

Figure 32.26 Carboxylation of glutamate to form γ-carboxyglutamate. h = hydroquinone; e = epoxide; *VKOR = vitamin K epoxide reductase.*

to inhibition by warfarin, a synthetic analog of vitamin K that inhibits *vitamin K epoxide reductase* (*VKOR*), the enzyme required to regenerate the functional hydroquinone form of vitamin K.

2. **Prothrombin interaction with membranes**

The Gla residues are good chelators of positively charged calcium ions, because of their two adjacent, negatively charged carboxylate groups. With prothrombin, for example, the prothrombin–calcium complex is able to bind to negatively charged membrane phospholipids on the surface of damaged endothelium and platelets. Attachment to membrane increases the rate at which the proteolytic conversion of prothrombin to thrombin can occur (Fig. 32.27).

3. **γ-Carboxyglutamate residues in other proteins:** Gla residues are also present in proteins other than those involved in forming a blood clot. For example, **osteocalcin** and matrix Gla protein of bone and proteins C and S (involved in limiting the formation of blood clots) also undergo γ-carboxylation. Osteocalcin is a protein synthesized by osteoblasts which binds to hydroxyapatite crystals and retains calcium.

B. Distribution and daily requirements

Vitamin K is found in cabbage, kale, spinach, egg yolk, and liver. The adequate intake for vitamin K is 55 μg/day. There is also synthesis of the vitamin by the gut microbiota.

C. Clinical indications for vitamin K

1. **Deficiency:** A true vitamin K deficiency is unusual because adequate amounts are generally obtained from the diet and produced by intestinal bacteria. **Increased bleeding tendency confirmed by increased prothrombin time (PT) is an important manifestation** of vitamin K deficiency. Deficiency may be caused due to:

a. **Treatment with antibiotics:** If the bacterial population in the gut is decreased, the amount of endogenously formed vitamin is decreased, and this can lead to hypoprothrombinemia in the marginally malnourished individual (for example, a debilitated geriatric patient). This condition may require supplementation with vitamin K to correct the bleeding tendency.

b. **Fat malabsorption syndromes**

c. **Vitamin K antagonists:** Agents, such as dicoumarin and warfarin are competitive inhibitors of vitamin K. In addition, certain cephalosporin antibiotics (for example, cefamandole) cause hypoprothrombinemia, apparently by a warfarin-like mechanism that inhibits *VKOR*. Consequently, their use in treatment is usually supplemented with vitamin K. Deficiency can also affect bone health.

2. **Deficiency in the newborn:** Because newborns have sterile intestines, they initially lack the bacteria that synthesize vitamin K. Because human milk provides only about one fifth of the daily requirement for vitamin K, it is recommended that all newborns receive a single intramuscular dose of vitamin K as prophylaxis against hemorrhagic disease of the newborn.

D. Toxicity

Prolonged administration of **large doses of menadione can produce hemolytic anemia and jaundice in the infant, because of toxic effects on the RBC**

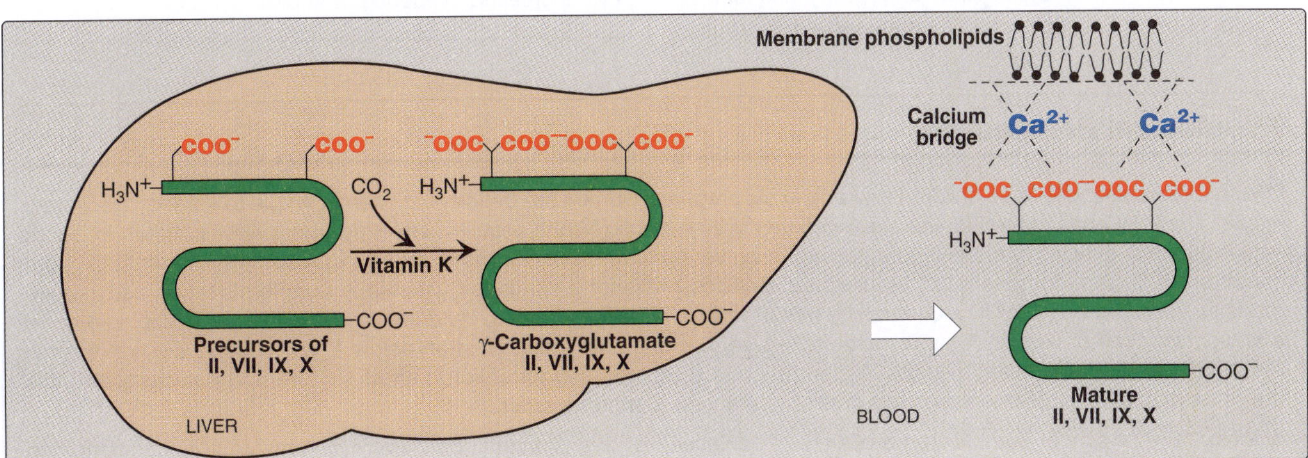

Figure 32.27 Role of vitamin K in blood coagulation. CO_2 = carbon dioxide.

membrane. Therefore, it is no longer used to treat vitamin K deficiency. No UL for the natural form has been set.

XIV. VITAMIN E

The E vitamins **consist of eight naturally occurring tocopherols**, of which α-tocopherol is the most active (Fig. 32.28). Vitamin E **functions as an antioxidant** in prevention of nonenzymic oxidations (for example, oxidation of LDL (see p. 356) and peroxidation of polyunsaturated FA by O_2 and free radicals). Vitamin E is a prominent membrane antioxidant, where it prevents membrane lipids from oxidation by endogenous as well as exogenous free radicals. [Note: Vitamin C regenerates active vitamin E.]

Due to its antioxidant activity, vitamin E is also considered to be anti-atherogenic, preventing the oxidation of LDL.

> **Relationship with Selenium**
>
> Selenium is present in glutathione peroxidase, an important antioxidant enzyme. Selenium decreases the requirement of vitamin E and vice versa.

A. Distribution and daily requirements

Vegetable oils are rich sources of vitamin E, whereas liver and eggs contain moderate amounts. The RDA for α-tocopherol is 7.5–10 mg/day for adults. The vitamin E requirement increases as the intake of polyunsaturated FA increases to limit FA peroxidation.

B. Deficiency

Newborns have low reserves of vitamin E, but breast milk (and formulas) contain the vitamin. Very-low-birth-weight infants may be given supplements to prevent the hemolysis and retinopathy associated with vitamin E deficiency. When observed in adults, deficiency is usually associated with defective lipid absorption or transport. [Note: Abetalipoproteinemia, caused by a defect in the formation of chylomicrons (and VLDL), results in vitamin E deficiency (see p. 352).]

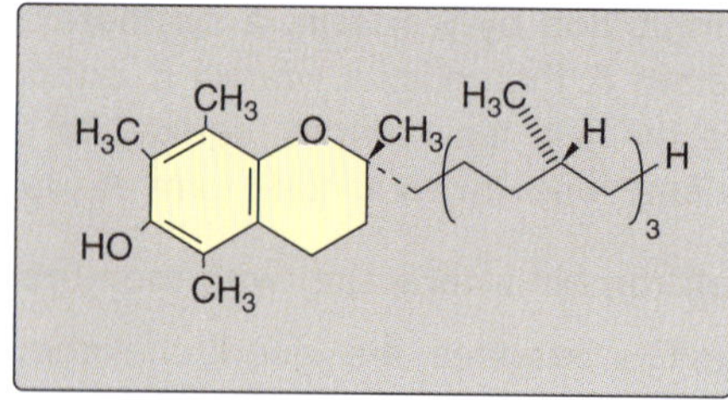

Figure 32.28 Structure of vitamin E (α-tocopherol).

C. Clinical indications for vitamin E

Vitamin E is not recommended for the prevention of chronic disease, such as CVD or cancer. Clinical trials using vitamin E supplementation have been uniformly disappointing. [Note: Vitamins E and C are used to slow the progression of age-related macular degeneration.]

D. Toxicity

Vitamin E is the least toxic of the fat-soluble vitamins, and no toxicity has been observed at doses of 300 mg/day (UL = 1,000 mg/day).

> Populations consuming diets high in fruits and vegetables show decreased incidence of some chronic diseases. However, clinical trials have failed to show a definitive benefit from supplements of folic acid; vitamins A, C, or E; or antioxidant combinations for the prevention of cancer or CVD. Subjects in the Alpha-Tocopherol, Beta-Carotene Cancer Prevention Study trial who received high doses of vitamin E not only lacked cardiovascular benefit but also had an increased incidence of stroke.

XV. Medical Perspectives

One in three preschool-aged children and one in six pregnant women are deficient in vitamin A due to inadequate dietary intake. The WHO has classified vitamin A deficiency as a public health problem affecting about 44% children in South Asia. Vitamin A deficiency is the leading cause of preventable childhood blindness and increases the risk of death from common childhood illnesses such as diarrhea. Besides preventing blindness, vitamin A supplementation can reduce mortality in children. The WHO recommends that in vitamin A-deficient areas, children 6 months to 5 years should receive a preventive dose of vitamin A supplementation every 6 months. In India, children below 1 year receive the first vitamin A supplementation dose with measles immunization at 9 months and for children aged 1–5 years, the subsequent nine doses of vitamin A supplementation are given twice a year, 6 months apart.

In India, 18% of school-age children and 24% of adolescents have vitamin D deficiency (Serum 25[OH]D concentration < 12 ng/mL).

Besides supplementation, edible oils and milk are also fortified with vitamin A and D.

Nearly 17% of school-age children and 31% of adolescents of age 10–19 years have vitamin B_{12} deficiency while almost 25% children have folate deficiency.

Folic acid supplementation is also given to the population between the ages of 6–60 months and all women in reproductive age group.

XVI. Chapter Summary

The vitamins are summarized in Figure 32.29 on pp. 547–549.

Active Form	Functions	Daily Requirement	Sources	Deficiency Disease	Toxicity	Notes
Water-Soluble						
Vitamin B_9 (Folic acid)						
Tetrahydrofolic acid	Transfer one carbon units, synthesis of methionine, serine, purine nucleotides, and thymidine monophosphate	200 μg 500 μg in pregnancy 300 μg during lactation	Leafy, dark-green vegetables	Megaloblastic anemia, neural tube defects	None	Administration of high levels of folate can mask vitamin B_{12} deficiency
Vitamin B_{12} (Cobalamin)						
Methylcobalamin, Deoxyadenosyl-cobalamin	Coenzyme for reactions: homocysteine to methionine, Methylmalonyl CoA to succinyl CoA	1 μg	Liver, red meat, fish, eggs, dairy products, and fortified cereals. It is produced by intestinal bacteria in animals	Pernicious anemia, dementia, spinal degeneration	None	Pernicious anemia is treated with intramuscular or high-dose oral vitamin B_{12}
Vitamin C (Ascorbic acid)						
Ascorbic acid	Antioxidant coenzyme for hydroxylation reactions, for example: In procollagen: Proline to hydroxyproline, Lysine to Hydroxylysine Facilitates the absorption of dietary nonheme iron For synthesis of carnitine, serotonin, norepinephrine, bile acids	40 mg	Citrus fruits, such as lime, lemon, *Amla* (Indian gooseberry) and broccoli, strawberries and tomatoes	Scurvy (Sore, spongy gums, loose teeth, poor wound healing, bleeding)	None	Benefits of supplementation not established in controlled trials

Figure 32.29 (continued on next page)
Summary of vitamins. [Note: Choline, like vitamin D, is considered an essential micronutrient in humans even though we are able to synthesize it.] P = phosphate; NAD(P) = nicotinamide adenine dinucleotide (phosphate); FMN = flavin mononucleotide; FAD = flavin adenine dinucleotide; CoA = coenzyme A.

Active Form	Functions	Daily Requirement	Sources	Deficiency Disease	Toxicity	Notes
Vitamin B_6 (Pyridoxine, Pyridoxamine, Pyridoxal)						
Pyridoxal phosphate	Coenzyme for enzymes, particularly in amino acid metabolism	2 mg	Yeast, wheat germs, cereals, eggs, milk, tuna, and salmon	Glossitis, neuropathy	Yes (sensory neuropathy)	Deficiency can be induced by isoniazid Sensory neuropathy occurs at high doses
Vitamin B_1 (Thiamine)						
Thiamine pyrophosphate (TPP)	Coenzyme for enzymes: pyruvate to acetyl CoA α-ketoglutarate to Succinyl CoA	1.0–1.7 mg	Germ of cereals, legumes, liver, nuts, yeast and outer layers of grains	Beriberi (peripheral neuropathy in dry form, edema and cardiomyopathy in wet form)	None	-
	Ribose 5-P + xylulose 5-P to Sedoheptulose 7-P + Glyceraldehyde 3-P Branched chain α-keto acid oxidation			Wernicke Korsakoff syndrome (most common in alcoholism, confusion, ataxia, memory loss, hallucinations, dysregulated eye movements		
Vitamin B_3 (Niacin)						
NAD^+ $NADP^+$	Electron transfer	12–21 mg	Unrefined and enriched grains and cereal, milk, legumes and lean meats	Pellagra (dermatitis, diarrhea, and dementia)	None	High doses of niacin used to treat hyperlipidemia
Vitamin B_2 (Riboflavin)						
FMN, FAD	Electron transfer	1.1–2.1 mg	Egg, milk, yeast, legumes, whole cereals and green leafy vegetables	Dermatitis, cheilosis, and glossitis	None	-
Vitamin B_7 (Biotin)						
Enzyme-bound biotin	Carboxylation reactions	30 μg	Fruits, Liver, Pork, Salmon, Whole grains	Dermatitis	None	Raw egg white contains the glycoprotein avidin, which tightly binds biotin and prevents its absorption from the intestine

Figure 32.29 (continued from previous page)
Summary of vitamins.

Active Form	Functions	Daily Requirement	Sources	Deficiency Disease	Toxicity	Notes
Vitamin B_5 (Pantothenic acid)						
Coenzyme A	Acyl carrier	5 mg	Eggs, liver, and yeast	Rare	None	-
Fat-Soluble						
Vitamin A (Retinol, Retinal, Retinoic acid, β-Carotene)						
Retinol Retinal Retinoic acid	Vision Maintenance of reproduction Differentiation and maintenance of epithelial tissues Promotion of growth Gene expression	600 retinol activity equivalents (RAE) Pregnant females: 800 RAE	Preformed vitamin A: Liver, kidney, cream, butter, and egg yolk Carotenes: Yellow, orange, and dark-green vegetables and fruits	Night blindness (increased visual threshold) Xerophthalmia (dryness of cornea) Infertility Growth retardation	Yes	β-Carotene not acutely toxic, but supplementation is not recommended Excess vitamin A can increase incidence of fractures
Vitamin D (Cholecalciferol, ergocalciferol)						
1,25-Dihydroxy-cholecalciferol	Calcium uptake Gene expression	400 IU (10 μg)	Fatty fish, liver, and egg yolk Fortified milk and edible oils	Rickets (in children) Osteomalacia (in adults)	Yes	Vitamin D is not a true vitamin because it can be synthesized in skin; application of sunscreen lotions or presence of dark skin color decreases this synthesis
Vitamin K (phylloquinone, menaquinone, menadione)						
phylloquinone menaquinone menadione	γ-carboxylation of glutamate residues in clotting and other proteins	55 μg	Cabbage, kale, spinach, egg yolk, and liver	Bleeding	Rare	Vitamin K is produced by intestinal bacteria Its deficiency is common in newborn Intramuscular treatment with vitamin K is recommended at birth
Vitamin E (α-tocopherol)						
Any of several tocopherol derivatives	Antioxidant	7.5–10 mg	Vegetable oils, liver and eggs	Rare, red blood cell fragility leads to hemolytic anemia	None	Benefits of supplementation for disease prevention not established in controlled trials

Figure 32.29 (continued from previous page)
Summary of vitamins.
Reference: Food Safety and Standards Authority of India, 2019.

Study Questions

Short-Answer Questions

Write short notes on:

32.1. Wernicke-Korsakoff syndrome
32.2. Role of folate in one-carbon metabolism
32.3. Biochemical significance of vitamin K
32.4. Biochemical basis of scurvy

Long-Answer Questions

32.1. Describe the biochemical basis and clinical manifestations of rickets.
32.2. Explain the role of vitamin A in health. Describe in the detail the biochemical mechanism of visual cycle.

Multiple Choice Questions

Choose the ONE best answer.

32.1. Which of the following vitamins and its deficiency manifestation is NOT correctly matched?
A. Folic acid – Neural tube defects
B. Niacin – Diarrhea and dermatitis
C. Vitamin B_{12} – Bleeding
D. Vitamin C – Sore, spongy gums, and loose teeth
Correct answer = C.

32.2. Which one of the following statements concerning vitamin D is correct?
A. A deficiency results in an increased secretion of calbindin.
B. Chronic kidney disease results in overproduction of 1,25-dihydroxycholecalciferol (calcitriol).
C. 25-Hydroxycholecalciferol (calcidiol) is the active form of the vitamin.
D. It is required in the diet of individuals with limited exposure to sunlight.
Correct answer = D.

32.3. Which of the following vitamins are involved in carboxylation reactions?
A. Thiamine and biotin
B. Thiamine and pyridoxine
C. Vitamin K and biotin
D. Vitamin K and pyridoxine
Correct answer = C.

32.4. Which of the following molecules can be used as precursors for synthesis of vitamins by human cells?
A. 7-dehydrocholesterol and tryptophan
B. Para-amino benzoic acid and threonine
C. 7-dehydrocholesterol and para-amino benzoic acid
D. Tyrosine and β-carotene
Correct answer = A.

32.5. Which of the following vitamin deficiencies and the molecule accumulating due to the resultant impaired metabolism is NOT correctly matched?
A. Thiamine – Lactate
B. Vitamin B_{12} – Methyl malonate
C. Pyridoxine – Xanthurenic acid
D. Vitamin C – Hydroxyproline
Correct answer = D.

32.6. Which of the following fat soluble vitamins has a co-enzyme function?
A. Vitamin A
B. Vitamin D
C. Vitamin E
D. Vitamin K
Correct answer = D.

32.7. Accumulation of which of the following water-soluble vitamins can lead to toxicity manifestations?
A. Niacin
B. Thiamine
C. Riboflavin
D. Folate
Correct answer = A.

32.8. Based on statements given below, choose the correct option:
Statement 1: Pyridoxine deficiency can lead to clinical manifestations of pellagra.
Statement 2: Kynureninase, a key enzyme of oxidative pathway of tryptophan metabolism, requires pyridoxal phosphate as a co-enzyme.
A. Both the Statements are correct, and Statement 2 is the correct explanation for Statement 1
B. Both the Statements are correct, but Statement 2 is NOT the correct explanation for Statement 1
C. Statement 1 is correct, but Statement 2 is NOT correct
D. Statement 1 is NOT correct, but Statement 2 is correct
Correct answer = A.

32.9. Which of the following vitamins are involved in maintenance of blood glucose levels due to their involvement in glycogenolysis and gluconeogenesis, respectively?
A. Niacin and thiamine
B. Thiamine and folate
C. Pyridoxine and biotin
D. Folate and vitamin B_{12}
Correct answer = C.

32.10. Deficiency of which of the following vitamins can lead to folate trap?

A. Folate

B. Thiamine

C. Biotin

D. Vitamin B_{12}

Correct answer = D.

Multiple-Choice Questions (NExT/NEET PG)

Choose the ONE best answer.

32.1. A 52-year-old woman presents with fatigue of several months' duration. Blood studies reveal macrocytic anemia, reduced levels of hemoglobin, elevated levels of homocysteine, and normal levels of methylmalonic acid. Which of the following is most likely deficient in this woman?

A. Folic acid

B. Folic acid and vitamin B_{12}

C. Iron

D. Vitamin C

Correct answer = A.

32.2. Abetalipoproteinemia is a rare autosomal recessive disorder resulting from pathogenic variants in the microsomal triglyceride transfer protein. This protein is required for the assembly of apo-B containing lipoproteins in liver and intestine. Abetalipoproteinemia is characterized by symptoms like peripheral neuropathy, delayed intellectual development, muscle weakness, and loss of deep tendon reflexes. The neurologic manifestations in this condition result from the inability to absorb which of the following vitamins from the intestine and the resulting deficiency thereof?

A. Folic acid

B. Vitamin B_{12}

C. Biotin

D. Vitamin E

Correct answer = D.

32.3. GABA or γ-amino butyrate is an inhibitory neurotransmitter that is synthesized from glutamic acid after decarboxylation. A reduced level of GABA is considered responsible for neurological manifestations, including seizures, in which of the following vitamin deficiencies?

A. Vitamin B_{12}

B. Folate

C. Pyridoxine

D. Thiamine

Correct answer = C.

32.4. A 60-year-old male presents with yellowing of the skin and eyes, dark urine, and pale stools. History reveals significant weight loss and pruritus. Examination shows hepatomegaly and a palpable gallbladder. Lab tests indicate elevated bilirubin and alkaline phosphatase levels. Imaging suggests biliary obstruction. Additionally, the patient reports experiencing easy bruising and prolonged bleeding from minor cuts. Impairment of which of the following is responsible for the hemorrhagic manifestations?

A. Synthesis of clotting factors

B. Absorption of fat-soluble vitamins from the intestine

C. Platelet plug formation

D. Leucocyte migration to the site of the injury

Correct answer = B.

32.5. A 55-year-old female with a history of atrial fibrillation was initiated on warfarin for stroke prevention. Despite adherence to prescribed doses, her International Normalized Ratio (INR) values (which is used to monitor the anticoagulation achieved by warfarin treatment) fluctuated significantly, posing challenges in achieving therapeutic anticoagulation. Testing of genetic variants that can alter the activity of which of the following enzymes is suggested in this patient?

A. γ-glutamyl carboxylase

B. γ-glutamyl transpeptidase

C. Vitamin K epoxide reductase

D. Prolyl hydroxylase.

Correct answer = C.

Micronutrients: Minerals 33

Learning Objectives

At the end of this chapter, the student will be able to:

- Describe the functions of various minerals in the body, their metabolism, and homeostasis.
- Enumerate and describe the disorders associated with mineral metabolism

I. OVERVIEW

Minerals are inorganic substances (elements) required in small amounts by the body. They function in a number of processes including formation of bones and teeth, fluid balance, nerve conduction, muscle contraction, signaling, and catalysis. Several minerals are essential enzyme cofactors. **Like the organic vitamins (see Chapter 32), minerals are micronutrients required in mg or µg amounts.** Those required by adults in the largest amounts (>100 mg/day) are referred to as the **macrominerals**. Minerals required in amounts between 1 and 100 mg/day are the **microminerals** (trace minerals). **Ultratrace minerals** are required in amounts <1 mg/day (Fig. 33.1). [Note: The classification of specific minerals into these categories can vary among sources.] Mineral concentrations in the body are influenced by their rates of absorption and excretion.

MINERAL CLASSIFICATIONS	RDA (OR AI*) FOR ADULTS
MACROMINERALS	
Calcium (Ca)	600 mg
Chloride (Cl)	1,800–2,300 mg*
Magnesium (Mg)	310–340 mg
Phosphorus (P)	600–1,200 mg
Potassium (K)	3,200–3,750 mg*
Sodium (Na)	1,900–2,100 mg*
MICROMINERALS (TRACE)	
Copper (Cu)	1.7 mg
Fluorine (as fluoride [F^-])	3–4 mg
Iron (Fe)	17–21 mg
Manganese (Mn)	4 mg*
Zinc (Zn)	10–12 mg
MICROMINERALS (ULTRATRACE)	
Chromium (Cr)	50 µg
Iodine (I)	150–250 µg
Molybdenum (Mo)	45 µg
Selenium (Se)	40 µg

Figure 33.1 Classification of minerals and recommended amounts to be consumed/day by adults. [Note: *An adequate intake (AI) is set if insufficient scientific evidence is available to calculate a Recommended Dietary Allowance (RDA).]

II. MACROMINERALS

The macrominerals include calcium (Ca^{2+}), phosphorus ([P] as inorganic phosphate [P_i, or PO_4^{3-}]), magnesium (Mg^{2+}), sodium (Na^+), chloride (Cl^-), and potassium (K^+). [Note: The free ionic forms are electrolytes.]

A. Calcium and phosphorus

These macrominerals are considered together because they are components of hydroxylapatite ($Ca_5[PO_4]_3OH$), which makes up bones and teeth.

1. **Calcium:** Ca^{2+} is the **most abundant mineral in the body**, with ~98% being found in bones and

teeth. The remainder is involved in a number of processes such as

- ***Activation of enzymes***: Calcium activates several enzymes, either directly (*succinate dehydrogenase*, *pancreatic lipase*) or through intracellular calcium binding protein, calmodulin (see p. 232), (adenylate cyclase, *glycogen synthase*, *pyruvate dehydrogenase*).
- ***Signaling***: Calcium acts as an important intracellular signal (second messenger) for several extracellular agents.
- ***Muscle contraction***: Calcium binding to troponin is required for interaction of actin and myosin filaments for muscle contraction.
- ***Blood clotting***: Several coagulation factors, including prothrombin, contain gamma carboxyglutamate residues, which bind to calcium during coagulation.
- **Calcium binds to a variety of proteins** including *phospholipase* A_2 (see p. 333), and *protein kinase C* (see p. 325) **and alters their activity**.
- Adequate calcium intake is essential to offset obligatory losses and prevent reduction in bone mass, which is one of the causes of osteoporosis. It is particularly important to ensure optimal calcium intake during pregnancy.

a. **Sources of calcium:** Dairy products and many green vegetables (for example, broccoli, but not spinach) are good dietary sources.

Absorption of dietary calcium is affected by several factors. **Calbindin** is a vitamin D-induced intracellular Ca^{2+}-binding protein involved in Ca^{2+} absorption in the intestine (see p. 541). **Calcitriol and parathyroid hormone (PTH) increase the absorption of calcium, while oxalates, phytates and dietary fiber decrease its absorption** (by forming insoluble complexes with calcium).

Toxicity is seen only with supplements (tolerable upper limit [UL] = 2,500 mg/day for adults).

RDA for calcium is 600 mg in adults and 1200 mg in pregnant and lactating women.

Normal serum calcium levels are 9–11 mg/dl, of which 50% is ionized, 40% is bound to proteins and 10% is complexed with other anions, such as phosphate and bicarbonate.

Hypercalcemia (elevated serum Ca^{2+}) can result from overproduction of PTH. This may cause constipation and kidney stones. **Hypercalcemia of malignancy** caused due to PTH-related protein (PTHrP) is another important cause of hypercalcemia seen in advanced malignancy.

Symptoms of hypercalcemia include abdominal pain, vomiting and constipation, muscle weakness and tiredness, psychiatric disturbances, and renal stones.

Hypocalcemia (low serum Ca^{2+}) can result from a deficiency of PTH or vitamin D. Chronic renal disease can also lead to hypocalcemia due to calcitriol deficiency. It can lead to bone demineralization (resorption). [Note: The hormonal regulation of serum Ca^{2+} levels was presented in the vitamin D section of Chapter 32 and is reviewed in 3. below.]

> Bone mass increases from infancy through the early reproductive years and then shows an age-related loss in both men and women that increases the risk for fracture. This process of bone resorption is known as osteoporosis. This loss is greatest in postmenopausal Caucasian women. Postmenopausal osteoporosis is managed by estrogen replacement along with calcium and vitamin D supplementation.

2. **Phosphorus:** Free phosphate (P_i) is the **most abundant intracellular anion**. However, 85% of the body's phosphorus is in the form of inorganic hydroxylapatite, with most of the remainder in intracellular organic compounds such as phospholipids, nucleic acids, ATP, and creatine phosphate.

a. **Functions of phosphate**

- **High-energy compounds:** Phosphate is a component of high-energy compounds such as ATP, GTP, CTP.
- **Phosphorylated compounds:** Phosphate is supplied as ATP for *kinases* and as P_i for *phosphorylases* (for example, *glycogen phosphorylase*, see p. 227) for various biochemical reactions.
- **Regulation of enzymes:** Its addition (by *kinases*) or removal (by *phosphatases*) is an important means of covalent regulation of enzymes (see Chapter 27).]
- Phosphate is an **important intracellular buffer**.
- Phosphorus is required **for calcification of bones**.

b. **Sources:** Phosphorus is widely distributed in food (milk, meat, eggs, and cereals are good sources), and dietary deficiency is rare.

RDA for phosphorus is 600 mg in adults and 1200 mg in pregnant and lactating women.

The **normal level of phosphorus is 2.5–4.5 mg/dl.**

Hypophosphatemia can be caused by refeeding carbohydrates to malnourished patients (refeeding syndrome, see p. 514), overuse of aluminum-containing antacids (aluminum chelates P_i), and increased urinary loss in response to increased production of PTH (see below). Muscle weakness is a common symptom. Rickets-like features may be seen due to decreased mineralization of bones.

Hyperphosphatemia is caused primarily by decreased PTH levels. Acute as well as chronic renal failure can also cause hyperphosphatemia. The excess P_i can combine with Ca^{2+} and form crystals that deposit in soft tissue (**metastatic calcification**). [Note: The Ca^{2+}/P_i ratio is important for bone formation (the ratio is ~2/1 in bone), and some experts are concerned that replacement of Ca^{2+}-rich milk by Ca^{2+}-poor, P_i-rich soft drinks can affect bone health.]

3. **Hormonal regulation:** Serum levels of Ca^{2+} and P_i are primarily controlled by calcitriol (1,25-dihydroxycholecalciferol, the active form of vitamin D) and PTH, both of which respond to a decrease in serum Ca^{2+}.

   a. **Calcitriol,** produced by the kidneys, increases serum Ca^{2+} and P_i by increasing bone resorption, intestinal absorption, and renal reabsorption of Ca^{2+} and P_i (Fig. 33.2).

   b. **PTH** (from the parathyroid glands) increases serum Ca^{2+} by increasing bone resorption, increasing renal reabsorption of Ca^{2+}, and activating the renal *1-hydroxylase* that produces calcitriol from calcidiol (see p. 540) (Fig. 33.3).

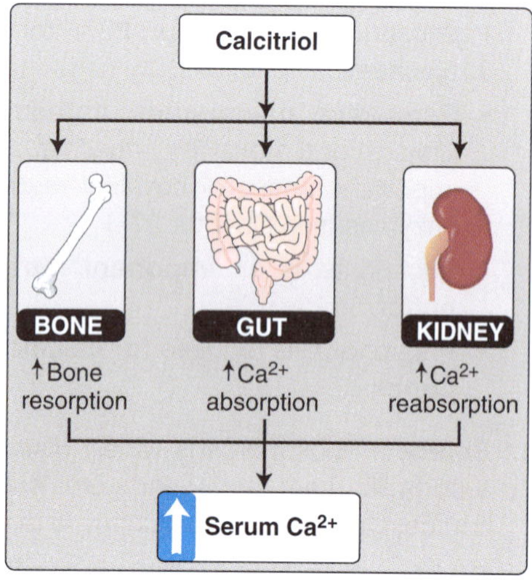

Figure 33.2 Effect of calcitriol on serum calcium (Ca^{2+}).

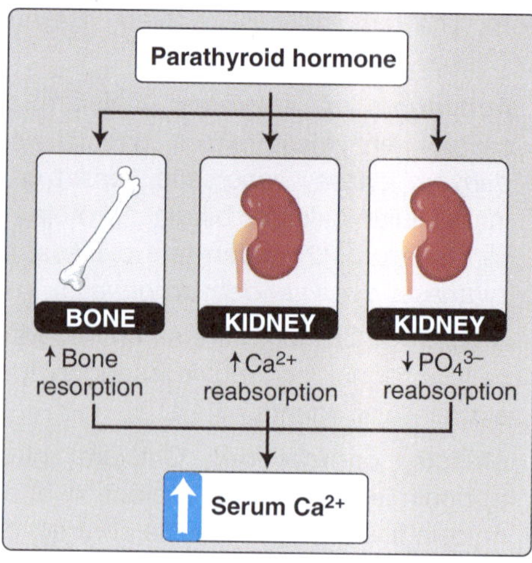

Figure 33.3 Effect of parathyroid hormone on serum calcium (Ca^{2+}). PO_4^{3-} = phosphate.

In contrast to calcitriol, PTH decreases P_i reabsorption in the kidneys, lowering serum P_i. High serum P_i increases PTH and decreases calcitriol.

   c. A third hormone, **calcitonin** (from the C cells of the thyroid gland), responds to elevated serum Ca^{2+} levels by promoting bone mineralization and increasing renal excretion of Ca^{2+} (and P_i).

B. Magnesium

About 60% of the body's Mg^{2+} is in bone, but it accounts for just 1% of the bone mass. The mineral is required by a variety of enzymatic reactions, including phosphorylation by *kinases* (Mg^{2+} binds the ATP cosubstrate) and phosphodiester bond formation by *DNA* and *RNA polymerases*.

The daily requirement of magnesium is about 310–340 mg. Mg^{2+} is widely distributed in foods, but the average intake in the United States is below the recommended level. Nuts, whole grains, beans, and cereals are major sources.

Normal serum concentration of magnesium is 1.6–2.3 mg/dl. **Hypomagnesemia** can result from decreased absorption or increased excretion of Mg^{2+}. Symptoms include hyperexcitability of skeletal muscles and nerves and cardiac arrhythmias. With **hypermagnesemia**, hypotension is seen. [Note: Magnesium sulfate is used in the treatment of preeclampsia, a hypertensive disorder of pregnancy.]

C. Sodium, chloride, and potassium

These macrominerals are considered together because they play important roles in several physiologic processes. For example, they maintain water balance, osmotic equilibrium, acid–base balance (pH), and the electrical gradients across cell membranes (membrane potential) that are essential for the functioning of neurons and myocytes. Sodium and potassium are discussed in detail in Chapter 5.

1. **Sodium and chloride:** Na^+ and Cl^- are primarily extracellular electrolytes. **Sodium is the most abundant cation in the extracellular fluid** (ECF) and important determinant of the osmolality of the plasma. They are readily absorbed from foods containing salt (NaCl), much of which comes from processed foods.

   Na^+ is required for the intestinal absorption (and renal reabsorption) of glucose and galactose (see p. 180) and free amino acids (see p. 381) by Na^+-linked transporters. Cl^- is used to form hydrochloric acid required for digestion (see p. 380).

   RDA for sodium is 1900–2100 mg in adults.

   Normal plasma level of sodium is 135–145 mEq/L. The kidneys can conserve or excrete large amounts of sodium, depending on its concentration in the ECF and the blood volume. Aldosterone and antidiuretic hormone (ADH, vasopressin) increase sodium reabsorption in kidneys.

   WHO recommends a reduction in sodium intake to reduce blood pressure and risk of cardiovascular disease, stroke and coronary heart disease in adults to <2 g/day sodium (5 g/day salt) in adults.

   a. **Hypertension:** Na^+ intake is related to blood pressure (BP).

      Ingestion of Na+ stimulates thirst centers in the brain and secretion of antidiuretic hormone from the pituitary, leading to water retention. This results in an increase in plasma volume and, consequently, an increase in BP. Chronic hypertension can damage the heart, kidneys, and blood vessels. Modest reductions in Na^+ intake have been shown to result in modest reductions in BP.

   b. **Hyper- and hyponatremia:** Hypernatremia, typically caused by excess loss of water relative to sodium loss, decreased water intake, or increased sodium intake or retention, and hyponatremia, typically caused by decreased ability to excrete water, can result in severe brain damage.

      [Note: Chronic hyponatremia increases Ca^{2+} excretion and can result in osteoporosis (low bone mass).]

      Causes of hypernatremia are:

      - Cushing's disease
      - Primary hyperaldosteronism
      - Prolonged steroid administration
      - Excessive intake of salt

      Hyponatremia is defined as a serum sodium level less than 135 mmol/L. It is one of the most common electrolyte disorders in hospitalized patients.

      Causes of hyponatremia are:

      - Addison's disease
      - Syndrome of inappropriate antidiuretic hormone (SIADH)
      - Vomiting, diarrhea, burns
      - Potassium deficiency

2. **Potassium:** In contrast to Na^+, **K^+ is primarily an intracellular electrolyte.** There is a narrow range for normal serum K^+ levels (3.5 – 5.5 mEq/L). [Note: The concentration differential of Na^+ and K^+ across the cell membrane is maintained by the *Na^+/K^+ATPase* (Fig. 33.4).] Even modest changes (up or down, resulting in hyper- or hypokalemia) can result in cardiac arrhythmias and skeletal muscle weakness. ECG shows inverted T waves in hypokalemia and elevated T waves in hyperkalemia.

   Causes of hyperkalemia are:

   - Decreased renal excretion (acute or chronic renal failure, hypoaldosteronism, Addison's disease)
   - Cellular shift (acidosis, cellular injury, hemolysis, cell lysis following chemotherapy)
   - Increased intake through food or supplements

   Causes of hypokalemia are:

   - Gastrointestinal loss (vomiting, diarrhea)
   - Renal loss (hyperaldosteronism, Cushing's syndrome, diuretics such as thiazides)
   - Cellular shift (alkalosis, insulin)
   - Decreased dietary intake, such as in chronic starvation
   - Inappropriate use of laxatives to lose weight

   RDA for potassium is 3225–3750 mg in adults.

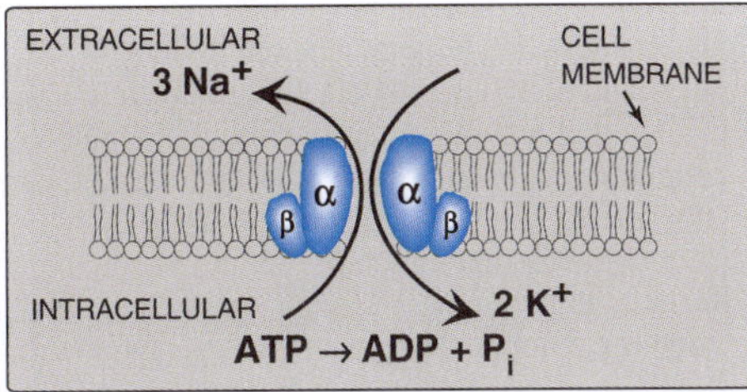

Figure 33.4 *Na^+/K^+ATPase*. Na^+ = sodium; K^+ = potassium; ADP = adenosine diphosphate; P_i = phosphate.

III. MICROMINERALS (TRACE MINERALS)

The trace minerals include copper (Cu), iron (Fe), manganese (Mn), and zinc (Zn). They are required by adults in amounts between 1 and 100 mg/day.

A. Copper

Cu is a key component of several enzymes that play critical functions in the body (Fig. 33.5). These include *ferroxidases* such as the *ceruloplasmin* and *hephaestin* involved in the oxidation of ferrous iron (Fe^{2+}) to the ferric form (Fe^{3+}) that is required for its intracellular storage or transport through blood (see B.1. below).

1.7 mg of Cu is required per day. Normal plasma concentration is 100–200 mg/dl. Meat, shellfish, nuts, and whole grains are good dietary sources of Cu. Dietary deficiency is uncommon. **If a deficiency does develop, anemia may be seen because of the effect on Fe metabolism**. Toxicity from dietary sources is rare (UL = 10 mg/day). **Menkes syndrome and Wilson disease are genetic causes of Cu deficiency and Cu overload, respectively**.

1. **Menkes syndrome:** In Menkes syndrome ("kinky hair" disease), a rare **X-linked** (1:140,000 males) disorder, **efflux of dietary Cu out of intestinal enterocytes into the circulation by a Cu-transporting *ATPase* (*ATP7A*) is impaired**. This results in systemic Cu deficiency. Consequently, **urinary and serum free (unbound) Cu are low, as is the concentration of *ceruloplasmin***, which carries over 90% of the Cu in the circulation (Fig. 33.6). **Progressive neurologic degeneration and connective tissue disorders are seen**, as are changes to hair. Parenteral administration of Cu has been used as a treatment with varying success. [Note: The mildest form of Menkes syndrome is called occipital horn syndrome.]

2. **Wilson disease:** In Wilson disease, an **autosomal-recessive** (AR) disorder affecting 1:35,000 live births, **efflux of excess Cu from the liver by *ATP7B* is impaired**. Cu accumulates in the liver; leaks into the blood; and is deposited in the brain, eyes, kidneys, and skin. In contrast to Menkes syndrome, **urinary and serum free Cu are high** (see Fig. 33.6). Hepatic dysfunction and neurologic and psychiatric symptoms are seen. **Kayser-Fleischer rings** (corneal deposits of Cu) may be present (Fig. 33.7). Renal damage and hemolytic anemia (due to copper deposits in bone marrow) may also be seen. Life-long use of Cu-chelating agents, such as penicillamine, is the treatment.

Cu-REQUIRING ENZYME	FUNCTION
Cytochrome c oxidase	Transfers electrons from cytochrome c to oxygen in the ETC (see p. 165)
Dopamine β-hydroxylase	Hydroxylates dopamine to norepinephrine (see p. 424)
Ferroxidases	Oxidize iron (see. p. 557)
Lysyl oxidase	Forms cross-links in collagen and elastin (see p. 122)
Tyrosinase	Synthesizes melanin (see p. 426)
Superoxide dismutase (nonmitochondrial form; also requires zinc)	Converts superoxide to hydrogen peroxide (see p. 793)

Figure 33.5 Examples of enzymes that require copper (Cu). ETC = electron transport chain.

VARIABLE	MENKES	WILSON
Whole-body Cu	Low	High
Free serum Cu	Low	High
Urinary Cu	Low	High
Inheritance	X-linked	AR
Cu-transporting *ATPase* affected	*ATP7A*	*ATP7B*

Figure 33.6 Comparison of Menkes syndrome and Wilson disease. Cu = copper; AR = autosomal recessive.

Ceruloplasmin

Ceruloplasmin is a protein with *ferroxidase* activity. It is produced in liver by incorporation of copper into apoceruloplasmin (an alpha 2 globulin). It has eight copper atoms, four covalently bound and four loosely bound. Ceruloplasmin is also as an acute-phase reactant protein.

The bioavailability (percent of the amount ingested that is able to be absorbed) of a mineral can be influenced by other minerals. For example, excess Zn decreases the absorption of Cu, and Cu is needed for the absorption of Fe.

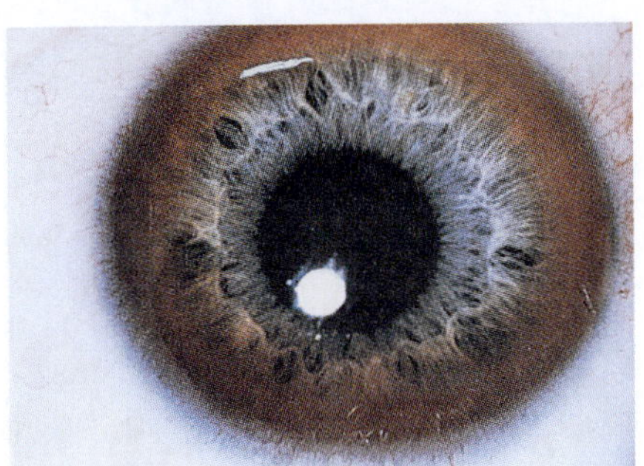

Figure 33.7 Kaiser-Fleischer rings.

B. Iron

Iron has several biochemical functions, such as:

- It is a **component of many proteins**, both catalytic (for example, *hydroxylases* such as *prolyl hydroxylase*, see p. 119) and noncatalytic (such as transferrin, hemosiderin).
- Iron can be linked to sulfur (S) as seen in the **Fe–S proteins of the electron transport chain** (see p. 165).
- It is **part of the heme prosthetic group** (see p. 97) in proteins such as hemoglobin (~70% of all Fe), myoglobin, and the cytochromes.

Free ionic (Fe) is toxic because it can cause production of the hydroxyl radical, a reactive oxygen species (ROS). Dietary Fe is available as Fe^{2+} in heme (animal sources) and Fe^{3+} in nonheme sources (plants). **Heme iron is less abundant, but it is better absorbed**. Also, plant sources contain inorganic molecules, such as phytates, which form complexes with iron and slow down its absorption.

The adult body typically contains 3–4 g of Fe. Meat, poultry, some shellfish, ready-to-eat cereals, lentils, and molasses are good dietary sources of Fe. Daily requirement is 17 mg for men and 21 mg for women, of which about 10% is absorbed. This amount, ~1–2 mg/day, is sufficient to replace Fe lost from the body primarily by the sloughing of cells.

The requirement of iron is higher in women due to extra losses in menstruation. Pregnancy and lactation also increase the iron requirement due to uptake by developing fetus and for breastfeeding (breastfed babies are dependent on mother for their iron requirements). They require about 40 mg/day of dietary iron.

1. **Absorption, storage, and transport:** Intestinal uptake of heme is by a heme carrier protein (Fig. 33.8). Within the enterocytes, ***heme oxygenase*** releases Fe^{2+} from heme (see p. 420). Nonheme Fe is taken up via the apical membrane protein **divalent metal ion transporter-1 (DMT-1)**.

   Absorbed Fe^{2+} from heme and nonheme sources has two possible fates: It can be

   (1) oxidized to Fe^{3+} and stored by the intracellular protein ferritin or

   (2) transported out of the enterocyte by the basolateral membrane protein ferroportin, oxidized by the Cu-containing membrane protein *hephaestin*, and taken up by the plasma transport protein transferrin (2 Fe^{3+}/transferrin), as shown in Figure 33.8.

   [Note: Cells other than enterocytes use the Cu-containing plasma protein *ceruloplasmin* in place of *hephaestin*.]

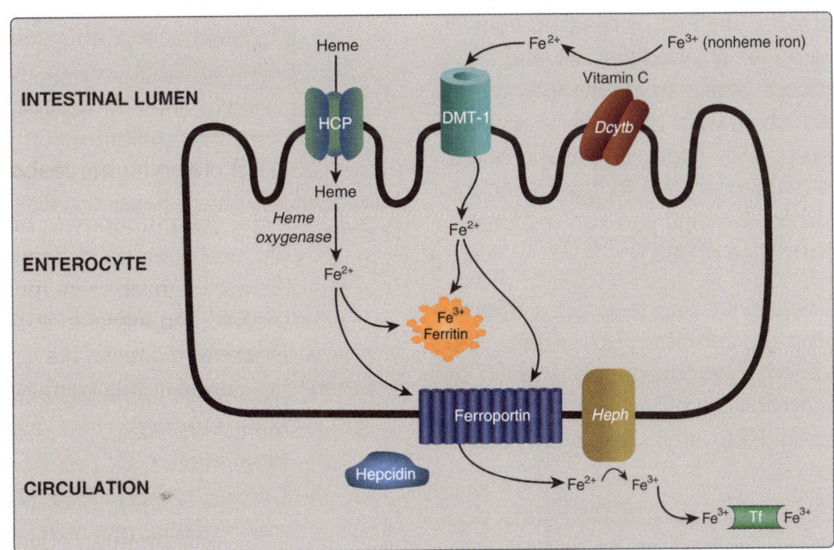

Figure 33.8 Absorption, storage, and transport of dietary iron (Fe). HCP = heme carrier protein; DMT = divalent metal ion transporter; *Dcytb* = *duodenal cytochrome b* (a *ferrireductase*); *Heph* = *hephaestin*; Tf = transferrin.

In normal individuals, transferrin (Tf) is about one third saturated with Fe^{3+}. **Ferroportin, the only known exporter of Fe from cells to the blood in humans, is regulated by the hepatic peptide hepcidin** that induces internalization and lysosomal degradation of ferroportin. Therefore, **hepcidin is the central molecule in Fe homeostasis.** [Note: Transcription of hepcidin is suppressed when Fe is deficient.]

Vitamin C enhances absorption of nonheme Fe because it is the coenzyme for *duodenal cytochrome b* (*Dcytb*), a *ferrireductase* that reduces Fe^{3+} to Fe^{2+}.

Mucosal block theory

Intestinal absorption of iron is important regulator of iron stores in the body. As iron cannot be excreted but can be lost only through sloughing of cells or by blood loss, this regulation is important to prevent toxicity. Iron stored in enterocytes (as ferritin) is lost on shedding of mucosal cells, if not required by the body.

2. **Recycling:** Macrophages phagocytose old and/or damaged red blood cells (RBC), freeing heme Fe that is sent out of the cells via ferroportin, oxidized by *ceruloplasmin*, and transported by Tf as described above. **This recycled Fe meets ~90% of our daily need, which is predominantly for erythropoiesis. Haptoglobin, an alpha-2 globulin, binds to free Hb and prevents its loss in urine, thus conserving iron.**

3. **Uptake:** Tf-bound Fe^{3+} from enterocytes and macrophages binds to receptors (TfR) on erythroblasts and other Fe-requiring cells and is taken up by receptor-mediated endocytosis. The Fe^{3+} is released from Tf for use (or stored on ferritin), and the TfR (and Tf) is recycled in a process similar to the receptor-mediated endocytosis seen with low-density lipoprotein particles (see p. 353). [Note: Regulation of the translation of the messenger RNA for ferritin and the TfR by iron regulatory proteins and iron-responsive elements is discussed on p. 663.]

4. **Deficiency:** Fe deficiency can result in a **microcytic, hypochromic anemia** (Fig. 33.9), the most common anemia worldwide, as a result of decreased hemoglobin synthesis and, consequently, decreased RBC size. Treatment is the administration of Fe.

   Iron deficiency can also cause achlorhydria due to atrophy of gastric epithelium. As achlorhydria reduces iron absorption, this further aggravates iron deficiency.

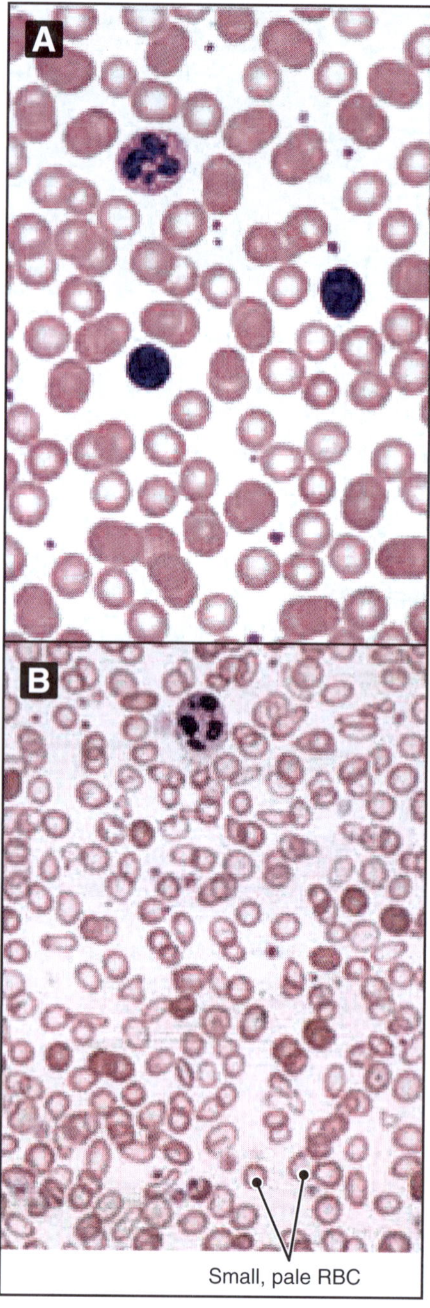

Figure 33.9 **A.** Normal red blood cells (RBC). **B.** Small (microcytic), pale (hypochromic) RBC in microcytic anemia.

   Causes of iron deficiency are:

   - Poor nutritional intake
   - Decreased absorption: Achlorhydria (gastric acid promotes iron absorption by reducing ferric ions)
   - Chronic blood loss: Hookworm infections, menorrhagia, hemorrhoids, etc.

5. **Excess:** Iron overload can occur with accidental ingestion (UL = 40 mg/day for children, 45 mg/

day for adults). People receiving repeated blood transfusions, such as those having thalassemia, are at risk for iron overload causing hemochromatosis. Treatment is use of an Fe chelator such as deferoxamine.

a. **Hereditary hemochromatosis:** Overload can also occur with genetic defects. An example is **hereditary hemochromatosis** (HH), an **autosomal recessive disorder of iron overload found primarily in those of Northern European** ancestry. It is most commonly caused by mutations to the HFE (high iron) gene. Hyperpigmentation with hyperglycemia **("bronze diabetes")** and damage to the liver (a major storage site for Fe), pancreas, and heart may be seen. In HH, serum Fe and Tf saturation are elevated. Treatment is phlebotomy or use of Fe chelators. [Note: Fe overload is seen with mutations to proteins of Fe metabolism that result in inappropriately low levels of hepcidin. It can result in hemosiderosis.]

> **Hemosiderosis** is a condition of iron overload as demonstrated by an increased serum iron and total iron binding capacity (TIBC) or transferrin, but without demonstrable tissue damage. When functional impairment of organs occurs because of iron overload, condition is now called as **hemochromatosis.**

6. **Laboratory evaluation:** Iron status is evaluated by packed cell volume, hemoglobin, red blood cell count and indices, total iron and total iron binding capacity (TIBC), percent saturation, transferrin, and ferritin.

In iron deficiency:

- Early stage of iron deficiency is characterized by normal levels of hemoglobin and serum iron. However, bone marrow iron stores are decreased and the serum ferritin level falls to < 30 ng/mL. Due to compensatory increase in iron absorption, transferrin and TIBC are increased.
- If iron deficiency continues, total iron and transferrin saturation are reduced. This is followed by anemia.

In iron overload and hemochromatosis:

- Transferrin and TIBC are decreased.

C. Manganese

Mn is important for the function of several enzymes (Fig. 33.10). Whole grains, legumes (for example, beans and peas), nuts, and tea (especially green tea) are good sources of the mineral. Consequently, Mn deficiency in humans is rare. Toxicity from foods and/or supplements is also rare (UL = 11 mg/day for adults).

Mn-REQUIRING ENZYME	FUNCTION
Arginase-I	Hydrolyzes arginine to urea plus ornithine in the urea cycle (see p. 385)
Glycosyltransferases	Transfer sugars in proteoglycan synthesis (see p. 264)
Pyruvate carboxylase	Carboxylates pyruvate to OAA in gluconeogenesis (see p. 216)
Superoxide dismutase (mitochondrial form)	Converts superoxide to hydrogen peroxide (see p. 797)

Figure 33.10 Examples of enzymes that require manganese (Mn). OAA = oxaloacetate.

D. Zinc

Zn plays important structural and catalytic functions in the body.

- Zinc fingers are supersecondary structures (motifs, see p. 655) in proteins (for example, transcription factors) that bind to DNA and regulate gene expression (Fig. 33.11).
- Hundreds of enzymes require Zn for activity. Examples include *alcohol dehydrogenase*, which oxidizes ethanol to acetaldehyde (see p. 784); *carbonic anhydrase*, which is important in the bicarbonate buffer system (see p. 26); *porphobilinogen synthase* of heme synthesis, which is inhibited by

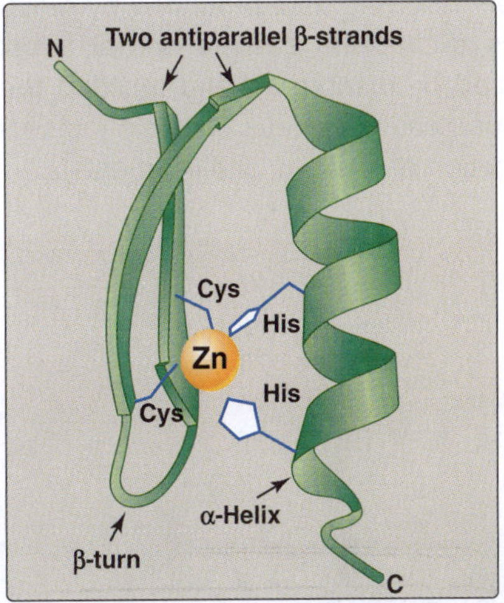

Figure 33.11 Zinc (Zn) finger is a common motif in proteins that bind DNA. Cys = cysteine; His = histidine.

lead (lead replaces the zinc; see p. 416); and the nonmitochondrial isoform of *superoxide dismutase* (*SOD*), which also requires Cu (see Fig. 33.5).

- Zinc is also required for DNA and RNA polymerase and reverse transcriptase.

Dietary sources of Zn include meat, fish, eggs, and dairy products. Phytates (phosphate storage molecules in some plant products) irreversibly bind Zn in the intestine, decreasing its absorption, and can result in a deficiency. [Note: Phytates may also bind Ca^{2+} and nonheme Fe.] Several drugs (for example, penicillamine) chelate metals, and their use may cause Zn deficiency. [Note: Severe deficiency is seen with a defect in the intestinal transporter for Zn that results in the malabsorption disorder acrodermatitis enteropathica. Symptoms include rashes, slowed growth and development, diarrhea, and immune deficiencies. Vision problems may also occur because Zn is needed in the metabolism of vitamin A.]

> Eukaryotic cells infected with bacteria can restrict availability of the essential micronutrients Fe, Mn, and Zn to the pathogens. This decreases the intracellular survival of the pathogen and is known as "**nutritional immunity**."

E. Other microminerals

Chromium (Cr) and fluorine (F) also play roles in the body. Cr potentiates the action of insulin by an unknown mechanism. It is found in fruits, vegetables, dairy products, and meat. **F (as fluoride [F^-]) is added to water in many parts of the world to reduce the incidence of dental caries** (Fig. 33.12). F^- replaces the hydroxyl group of hydroxylapatite, forming fluorapatite that is more resistant to the enamel-dissolving acid produced by mouth bacteria. **Excess of fluorine causes dental fluorosis characterized by discolored and mottled teeth**. Higher levels can cause skeletal fluorosis (osteoporosis and calcification of tendons and ligaments).

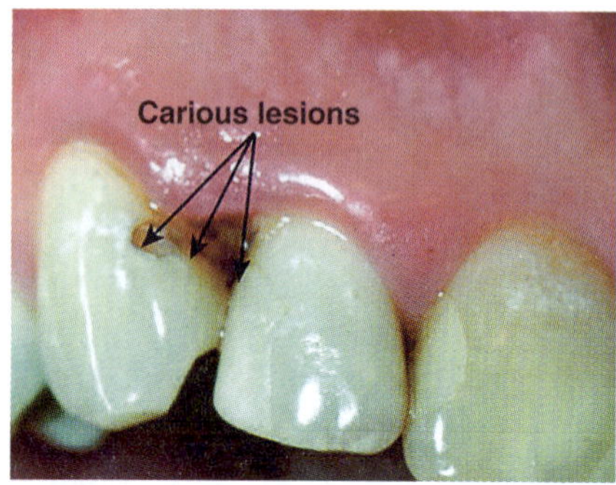

Figure 33.12 Dental caries (cavities).

IV. ULTRATRACE MINERALS

The ultratrace minerals include iodine (I), selenium (Se), and molybdenum (Mo). They are required by adults in amounts <1 mg/day.

A. Iodine

I is utilized in the synthesis of the thyroid hormones triiodothyronine (T_3) and thyroxine (T_4) that are required for development, growth, and metabolism. Dietary sources of iodine include dairy products, breads, seafood, and iodized salt. Upper regions of Himalayas are deficient in iodine and are called **goitrous belt**. The RDA is 150 μg for adults, and 250 μg for pregnant and lactating women.

1. **Synthesis of thyroid hormones**

   Circulating iodide (I^-) is taken up ("trapped") and concentrated in the epithelial follicular cells of the thyroid gland. It then is sent into the colloid of the follicular lumen where it is oxidized to iodine (I_2) by ***thyroperoxidase*** (*TPO*), as shown in Figure 33.13. *TPO* then uses I_2 to iodinate selected tyrosine residues in thyroglobulin (Tg), forming **monoiodinated tyrosine** (MIT) and **diiodinated tyrosine** (DIT), as shown in Figure 33.14. [Note: Tg is synthesized and secreted into colloid by follicular cells.] **The coupling of two DIT on Tg gives T_4, whereas coupling one MIT and one DIT gives T_3.** The iodinated Tg is endocytosed and stored in follicular cells until needed, at which time it is proteolytically digested to release T_3 and T_4, which are secreted into the circulation (see Fig. 33.13).

   Under normal conditions, ~90% of secreted thyroid hormone is T_4 that is carried by transthyretin.

   In target tissues (for example, the liver and developing brain), T_4 is converted to T_3 (the more active form) by Se-containing *deiodinases*. **T_3 binds to a nuclear receptor that binds DNA at thyroid response elements and functions as a transcription factor.** [Note: Thyroid hormone production is controlled by thyrotropin (thyroid-stimulating hormone ([TSH]) from the anterior pituitary. TSH secretion is itself controlled by thyrotropin-releasing hormone (TRH) from the hypothalamus.]

> Iodine is used in the **contrast medium for computed tomography (CT) scans** and to visualize arteries during heart catheterization. It is also present in amiodarone, used to treat cardiac arrhythmias.

2. **Hypothyroidism:** Underingestion of I can result in **goiter** (enlargement of the thyroid in response

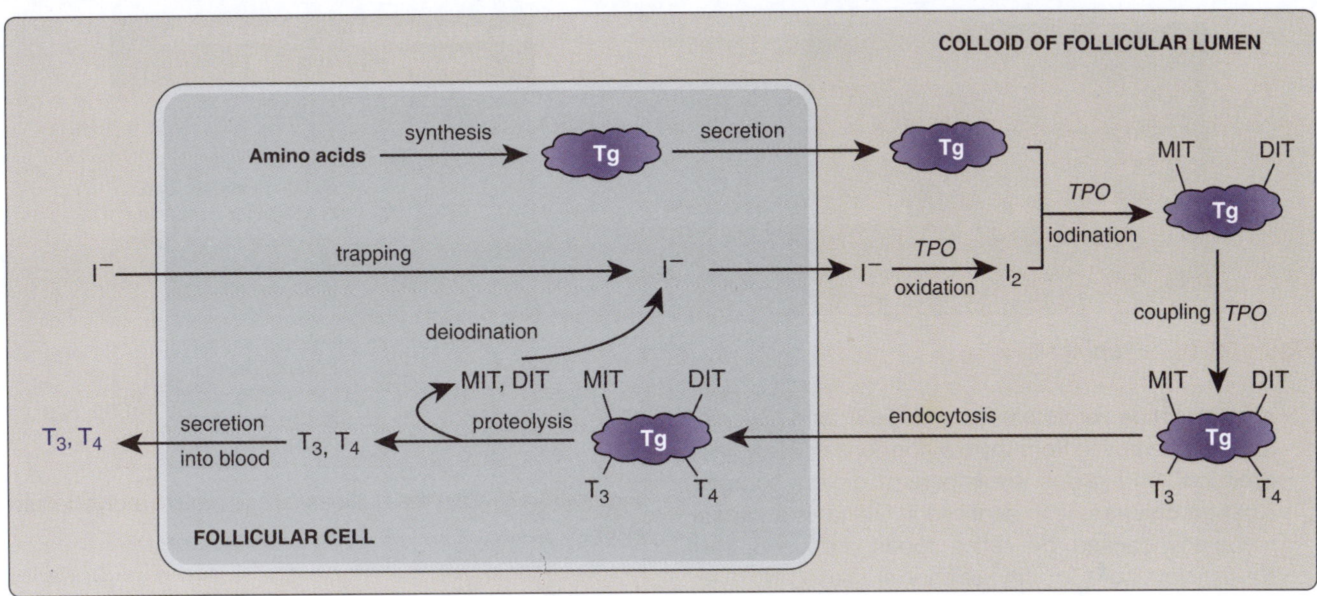

Figure 33.13 Thyroid hormone synthesis. *Tg* = thyroglobulin; *I* = iodide; I_2 = iodine; *TPO* = *thyroperoxidase*; *MIT* = monoiodinated tyrosine; *DIT* = diiodinated tyrosine; T_3 = triiodothyronine; T_4 = thyroxine.

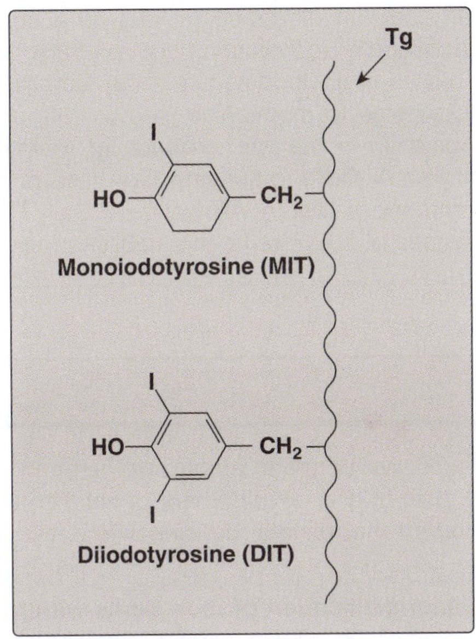

Figure 33.14 Iodination of thyroglobulin (Tg) with production of MIT and DIT.

to excessive stimulation by TSH), as shown in Figure 33.15. More severe deficiency results in **hypothyroidism that is characterized by fatigue, weight gain, decreased thermogenesis, and decreased metabolic rate** (see p. 503). If hormone deficiency occurs during fetal and infant development (**congenital hypothyroidism**), irreversible intellectual disability (formerly called "**cretinism**"), hearing loss, spasticity, and short stature can result.

In Hashimoto thyroiditis, primary hypothyroidism occurs due to autoimmune destruction of *TPO*.

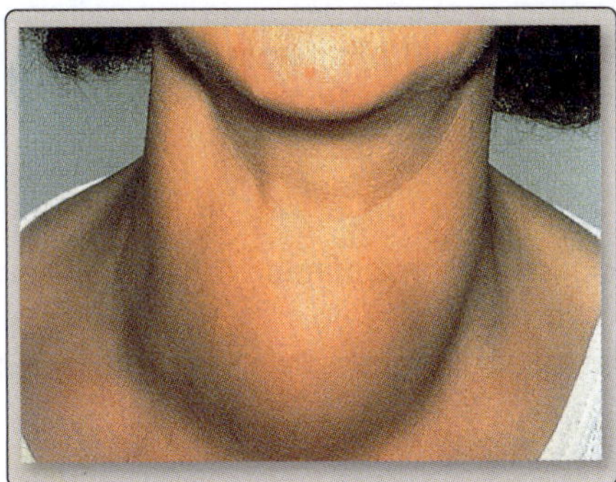

Figure 33.15 Goiter.

3. **Hyperthyroidism:** This condition is the result of overproduction of thyroid hormone. Although it can be caused by overingestion of I-containing supplements (UL = 1.1 g/day for adults), the **most common cause of hyperthyroidism is Graves' disease, in which an antibody that mimics the effect of TSH is produced**, resulting in dysregulated production of thyroid hormone. This can cause **nervousness, weight loss, increased perspiration and heart rate, protruding eyes (exophthalmos, Fig. 33.16), and goiter**.

B. Selenium

Se is present in ~25 human proteins (selenoproteins) **as a constituent of the amino acid selenocysteine, which is derived from serine** (see p. 404). Selenoproteins include *glutathione peroxidase* that oxidizes glutathione in the reduction of hydrogen peroxide, a ROS, to water (see p. 793); *thioredoxin reductase* that reduces thioredoxin, a coenzyme of

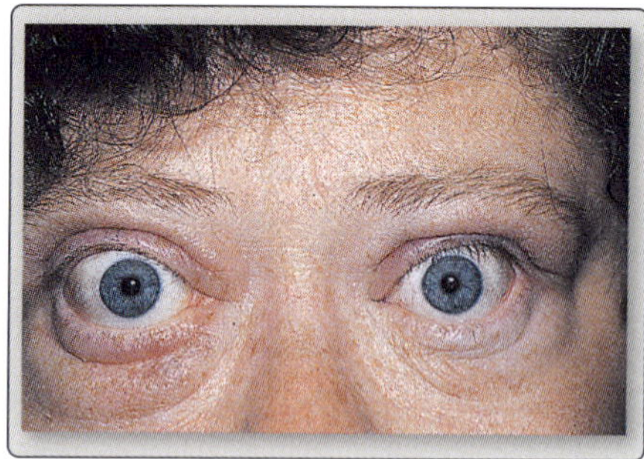

Figure 33.16 Exophthalmos.

ribonucleotide reductase (see p. 583); and *deiodinases* that remove I from thyroid hormones. Meat, dairy products, and grains are important dietary sources. **Keshan disease**, first identified in China, is a cardiomyopathy caused by eating foods produced from Se-deficient soil. Toxicity (selenosis) caused by overingestion of supplements causes brittle nails and hair. Cutaneous and neurologic effects may also be seen (UL = 400 μg in adults).

As selenium prevents peroxidation of lipids, a function similar to that of vitamin E, they both complement each other's action.

C. Molybdenum

Mo functions as a cofactor for a small number of mammalian *oxidases* (Fig. 33.17). Legumes are important dietary sources. No dietary deficiency syndromes are known. Mo has low toxicity in humans (UL = 2 mg/day in adults).

Mo-REQUIRING ENZYME	FUNCTION
Aldehyde oxidase	Metabolizes drugs
Sulfite oxidase	Converts sulfite to sulfate in metabolism of the sulfur-containing amino acids methionine and cysteine (see pp. 398–400)
Xanthine oxidase	Oxidizes hypoxanthine to xanthine and xanthine to uric acid in purine degradation (see p. 584)

Figure 33.17 Enzymes (*oxidases*) that require molybdenum (Mo).

Cobalt (Co), an ultratrace mineral, is a component of vitamin B_{12} (cobalamin, see p. 528), which is required as methylcobalamin in the remethylation of homocysteine to methionine (see p. 400) or adenosylcobalamin in the isomerization of methylmalonyl coenzyme A (CoA) to succinyl CoA (see p. 313). No Recommended Dietary Allowance or Daily Reference Intake (see p. 502) has been established for Co.

V. Medical Perspectives

Mineral deficiencies, including deficiencies of iron, iodine, zinc and folic acid, are common among women and children in low- and middle-income countries. These deficiencies restrict survival, growth and development of children and also contribute to morbidity and mortality in vulnerable populations. It is important that mineral deficiency is detected early and managed.

While dietary diversification is the most desirable way, **supplementation and fortification of food items with minerals is also done**. Several governments run health programs to ensure that women of reproductive age, pregnant women and children have sufficient essential micronutrients.

Iron deficiency is the most significant cause of anemia worldwide. The prevalence of iron-deficiency anemia was 53% (in 2015–16) for Indian women aged 15–49 years. In India, under the **National Iron+ Initiative**, the population between the ages of 6 to 60 months and all women in reproductive age group, are covered for supplementation of iron and folic acid. In populations with low calcium intake, calcium supplementation is recommended as part of the antenatal care for the prevention of preeclampsia in pregnant women.

Almost 20% of children suffer from zinc deficiency, which can lead to growth retardation, loss of appetite, and impaired immune function. WHO recommends zinc supplementation for 10–14 days in children suffering from diarrhea, as zinc reduces the duration and severity of diarrhea and also prevents subsequent episodes.

In India, wheat and rice flour are fortified with iron, folic acid and vitamin B_{12} and salt is fortified with iron and iodine. It is estimated that with iodine fortification, about 240 million people, including 5.5 million newborns, are protected every year from the brain damage and learning ability losses associated with iodine deficiency.

VI. Chapter Summary

The minerals are summarized in Figure 33.18 on pp. 563–564.

Mineral	Function	Daily Requirement	Sources	Notes
Macrominerals (> 100 mg/day for adults)				
Calcium	Component of hydroxylapatite ($Ca_5[PO_4]_3OH$) of bones and teeth, muscle contraction, signaling, blood clotting	600 mg	Dairy products, many green vegetables, fortified juice	Toxicity from supplements, hypocalcemia with PTH or vitamin D deficiency causes kidney stones, hypercalcemia with increased PTH causes bone resorption. Calcium deficiency is more commonly due to vitamin D deficiency.
Chloride	Digestion Fluid balance (along with sodium and potassium)	1,800–2,300 mg	Table salt, olives, lettuce, olives	Dietary deficiency rare, overingested as NaCl
Magnesium	Component (minor) of bones, regulates enzyme activity (binds substrate or enzyme)	310–340 mg	Nuts, whole grains, beans and cereals.	Hyperexcitability and arrhythmias seen with hypomagnesemia, hypotension with hypermagnesemia
Phosphorus	Component of hydroxylapatite ($Ca_5[PO_4]_3OH$) of bones and teeth, energy storage, membrane structure, regulation	600–1,200 mg	Milk, meat, eggs, and cereals	Dietary deficiency rare Hypophosphatemia with muscle weakness in refeeding syndrome, increased PTH and use of aluminum-containing antacids, hyperphosphatemia with metastatic calcification in PTH deficiency
Potassium	Membrane potential, blood pressure	3,200–3,750 mg	Bananas, fruit juices, milk, potatoes, spinach, tomatoes and tomato products, yogurt	Modest increase or decrease in serum level result in arrhythmia and muscle weakness
Sodium	Membrane potential, blood volume and pressure, uptake of glucose, galactose and amino acids	1,900–2,100 mg	It is widely present. Processed foods are rich in sodium. Other sources are: Cheese, meat, savory snacks (e.g., chips), soups, table salt	Dietary deficiency rare, overingested as NaCl, hyponatremia seen with excess water loss, hypernatremia with water retention
Microminerals (trace, 1–100 mg/day)				
Copper	Enzyme cofactor	1.7 mg	Chocolate and cocoa, shellfish, lentils, nuts and seeds	Dietary deficiency rare; Menkes disease (genetic systemic Cu deficiency) and Wilson disease (genetic systemic Cu overload)

Figure 33.18 (continued on next page)
Summary of minerals. PTH = parathyroid hormone; Cl^- = chloride; S = sulfur; T_3 = triiodothyronine; T_4 = thyroxine.

Mineral	Function	Daily Requirement	Sources	Notes
Fluorine (as fluoride)	Increases resistance to enamel-dissolving acid of mouth bacteria	-	Present in all foodstuffs, water and non-dairy beverages are the main sources	Deficiency results in dental caries Excess can cause dental or skeletal fluorosis
Iron (Fe)	Enzyme co-factor, oxygen binding, Fe-S protein	17–21 mg Pregnant women: 35 mg Lactating women: 21 mg	Meat, poultry, some shellfish, ready-to-eat cereals, lentils, and molasses	Dietary deficiency causes microcytic anemia; hereditary hemochromatosis, a genetic disease of Fe overload, with "bronze diabetes" (hyperglycemia, hyperpigmentation)
Manganese (Mn)	Enzyme cofactor	4 mg	Beans, nuts, spinach, whole grains, green tea	Dietary deficiency rare
Zinc (Zn)	Enzyme cofactor, protein structure (Zn finger)	10–12 mg	Meat, fish, eggs, and dairy products	Phytates and some drugs decrease absorption, severe deficiency (acrodermatitis enteropathica) with transported defect
Microminerals (ultratrace, < 1 mg/day)				
Chromium	Potentiates insulin action	50 μg	Fruits (e.g., apple and banana), meats, spices (e.g., garlic and basil), whole grains	Mechanism unknown
Iodine (I)	Thyroid hormone (T3 and T4) synthesis	150–250 μg	Dairy products, breads, seafood and iodized salt	Underdigestion causes goiter, hypothyroidism with fatigue, weight gain, and decreased metabolic rate; neurologic damage in congenital deficiency; hyperthyroidism in Graves' disease
Molybdenum (Mo)	Enzyme cofactor	45 μg	Beans and peas, nuts, whole grains	Dietary deficiency unknown
Selenium (Se)	Found (as selenocysteine) in selenoproteins	40 μg	Eggs, meats, whole grains nuts and seeds, seafood	Vitamin E-sparing effect Dietary deficiency rare (Keshan disease with Se-deficient soil), toxicity from supplements

Figure 33.18 (continued from previous page)
Summary of minerals. *Reference*: Food Safety and Standards Authority of India, 2019.

Study Questions

Short-Answer Questions

Write short notes on:

33.1. Wilson's disease
33.2. Hemochromatosis
33.3. Biochemical significance of iodine
33.4. Hyperkalemia

Long Answer Questions

33.1. Describe the functions of calcium in body. How is serum calcium level regulated?
33.2. Describe the absorption, transport and storage of iron in human body.

Multiple-Choice Questions

Choose the ONE best answer.

33.1. Which of the following minerals is required for the formation of a super-secondary protein structure that allows binding to DNA?

A. Sodium
B. Potassium
C. Zinc
D. Manganese

Correct answer = C.

33.2. A decrease in the levels of which mineral is seen in refeeding syndrome and with overuse of aluminum-containing antacids?

A. Calcium
B. Phosphate
C. Zinc
D. Magnesium

Correct answer = B.

33.3. Deficiency of which mineral can cause bone pain, tetany (intermittent muscle spasms), paresthesia (a "pins and needles" sensation), and an increased tendency to bleed?

A. Calcium
B. Selenium
C. Copper
D. Magnesium

Correct answer = A.

33.4. Which mineral is a constituent of some amino acids found in proteins involved in antioxidant defense, thyroid hormone metabolism, and redox reactions?

A. Iron
B. Selenium
C. Cobalt
D. Molybdenum

Correct answer = B.

33.5. The storage and release of insulin from β-cells of pancreas and its action on the insulin receptor require all the following micronutrients, EXCEPT:

A. Copper
B. Calcium
C. Zinc
D. Chromium

Correct answer = A.

33.6. Based on statements given below, choose the correct option:

Statement 1: A deficiency of selenium in the body can lead to hypothyroidism.

Statement 2: Selenium is required for the functioning of thioredoxin reductase and glutathione peroxidase.

A. Both the Statements are correct, and Statement 2 is the correct explanation for Statement 1
B. Both the Statements are correct, but Statement 2 is NOT the correct explanation for Statement 1
C. Statement 1 is correct, but Statement 2 is NOT correct
D. Statement 1 is NOT correct, but Statement 2 is correct

Correct answer = B.

33.7. Which of the following is affected the earliest in a patient with iron deficiency anemia?

A. Serum transferrin saturation
B. Serum ferritin
C. Mean corpuscular volume
D. Serum iron levels

Correct answer = B.

33.8. Which of the following enzymes and its metal co-factor is NOT correctly matched?

A. Xanthine oxidase – Molybdenum
B. RNA polymerase – Zinc
C. Ceruloplasmin – Copper
D. 5'-deiodinase – Chromium

Correct answer = D.

33.9. Based on the statements given below, choose the correct option:

Statement 1: Consumption of tea for 1 hour before and after taking the iron supplement should be avoided.

Statement 2: The absorption of iron is interfered by tannins and oxalates.

A. Both the Statements are correct, and Statement 2 is the correct explanation for Statement 1
B. Both the Statements are correct, but Statement 2 is NOT the correct explanation for Statement 1
C. Statement 1 is correct, but Statement 2 is NOT correct
D. Statement 1 is NOT correct, but Statement 2 is correct

Correct answer = A.

33.10. Which of the following is used in the treatment of pre-eclampsia due to its ability to decrease neuro-excitability and to induce vasodilation?

A. Copper
B. Magnesium
C. Zinc
D. Chromium

Correct answer = B.

Multiple-Choice Questions (NExT/NEET PG)

Choose the ONE best answer.

33.1. DiGeorge syndrome is a congenital condition that results in structural anomalies and failure of the thymus and parathyroid glands to develop. Clinical manifestations include recurrent infections as a consequence of a deficiency in T cells. Which one of the following is another expected clinical consequence in this condition?

A. Increased bone resorption
B. Increased calcium reabsorption in the kidney
C. Increased serum calcitriol
D. Increased serum phosphate

Correct answer = D.

33.2. A 52-year-old female is presented to the hospital because of unplanned changes in the pigmentation of her skin that give her a tanned appearance. Physical examination shows hyperpigmentation, hepatomegaly, and mild scleral icterus. Laboratory tests are remarkable for elevated serum transaminases (liver function tests) and fasting blood glucose. What are the anticipated lab finding(s) in this individual?

A. Increase in serum copper
B. Increase in serum iron and transferrin saturation
C. Decrease in serum copper and ceruloplasmin
D. Increased serum phosphate

Correct answer = B.

33.3. A 28-year-old male presents with recent, severe, upper-right-quadrant pain, along with some difficulty with fine motor tasks. Physical examination shows no jaundice. Laboratory tests reveal elevated liver function tests (serum aspartate and alanine aminotransferases) and elevated urinary calcium and phosphate. Ophthalmology consult reveals Kayser-Fleischer rings in the cornea. Which of the following is the least likely finding in this individual?

A. Increased 24-hour urinary copper excretion
B. Decreased serum ceruloplasmin
C. Decreased hepatic copper concentration
D. Pathogenic variants in *ATP7B* gene

Correct answer = C.

33.4. Which of the following combination of lab findings is consistent with iron deficiency anemia?

A. Low serum iron, low serum ferritin, and increased TIBC
B. Low serum iron, low serum ferritin, and decreased TIBC
C. Low serum iron, low serum ferritin, and increased transferrin saturation
D. Low serum iron, normal serum ferritin, and normal TIBC

Correct answer = A.

33.5. Based on statements given below, choose the correct option:

Statement 1: Oral zinc is used in the treatment of Wilson's disease because it induces metallothionein production in the intestine, metallothionein being an endogenous chelator of metals.

Statement 2: Metallothionein binds luminal copper, preventing its entry into the circulation, and the bound copper is then excreted in the stool with the normal turnover of intestinal cells.

A. Both the Statements are correct, and Statement 2 is the correct explanation for Statement 1
B. Both the Statements are correct, but Statement 2 is NOT the correct explanation for Statement 1
C. Statement 1 is correct, but Statement 2 is NOT correct
D. Statement 1 is NOT correct, but Statement 2 is correct

Correct answer = A.

Appendix
Unit VII: Medical Nutrition

DIETS FOR SPECIFIC CONDITIONS

A. Diet for Diabetic Patients

The aim of diet management in Diabetes Mellitus is to maintain near normal blood glucose levels, control hyperlipidemia and hypertension, and prevent complications of diabetes.

Important features of diabetic diet are:

- Carbohydrate intake should be at least 130 mg/day, providing 45–65% of daily calories.
- Carbohydrates with low glycemic index should be included.
- Fructose-rich foods, such as fruits and vegetables, have a lower postprandial glucose response than sucrose or starch. They should be included in diet. But as fructose can adversely affect plasma lipids, their intake should be limited.
- 30–40 g/day of natural dietary fiber should also be included.
- Fats should be included to provide 20–35% of total daily calories, with omega-3 PUFA, MUFA being preferred fats. Saturated fatty acids should provide <7% of total calories and cholesterol intake should be <200 mg/day, with minimal trans fatty acids intake.
- Proteins should provide 10–20% of total calories.

B. Diet Plan for Chronic Kidney Disease

Many patients with chronic kidney disease tend to develop protein-energy malnutrition. Due to rise in blood urea, there is a loss of appetite. Besides altered protein metabolism, these patients also develop hyperparathyroidism and insulin resistance. There is also loss of nutrients in dialysate in patients undergoing dialysis.

Goals

- Ensure adequate nutrition
- Control protein intake to minimize uremia
- Control sodium, potassium and fluid intake to prevent fluid overload and electrolyte imbalance
- Maintain calcium, phosphorus and vitamin D intake to prevent renal osteodystrophy

Diet Composition

- The total calorie requirement is 30–35 kcal/kg body wt/day.
- In predialysis patients, protein intake should be 0.5–0.6 g/kg body wt/day, while in patients on dialysis, 1.0–1.2 g/kg body wt/day protein of high biological values should be included.
- Fats should contribute 30% to 40% of energy and rest of non-protein calories should be provided by carbohydrates.

C. Diet Plan for Coronary Heart Disease

The level of plasma cholesterol varies in response to diet. Elevated levels of total cholesterol result in an increased risk for coronary heart disease (CHD). Increased LDL-C levels and decreased HDL-C levels are strongly associated with CHD.

Correcting the lipid abnormalities and maintaining optimal lipid levels through dietary and other measures can significantly reduce the burden of CHD. This can also reduce the mortality, coronary artery procedures and stroke in persons with established CHD. The main goal is to reduce LDL-cholesterol down to LDL goal. The goals vary with risk profile of each individual.

Table 1 LDL goal according to CHD risk

Risk Category	LDL Goal (mg/dl)
Known CHD or carrying CHD risk equivalents	< 100
More than two CHD risk factors	< 130
Zero to two CHD risk factors	< 160

Diet Composition

- Carbohydrates: 50–60% of calories from foods rich in complex carbohydrates.
- Protein: 15% of total calories

- Fats: Total fats should provide 25–30% of calories, with limits on various types of fats that can be included.
  - Saturated fatty acids: Less than 7% of total calories
  - Trans fatty acids: Less than 1% of total calories
  - Dietary cholesterol: Less than 200 mg/day
  - Polyunsaturated fatty acids (PUFAs): Up to 10% of total calories
  - Monounsaturated fatty acids (MUFAs): Up to 20% of total calories

Table 2 Nutritional importance of commonly used food items

Food Group	Nutritional Significance
Grains (Wheat, rice, corn, oats, etc.)	A major source of carbohydrates, proteins, minerals and fiber. They also provide the bulk of energy
Vegetables and fruits	Rich in vitamins, minerals and fiber
Legumes (Lentils, beans, pulses)	Good sources of protein, carbohydrate, vitamins, minerals and fiber
Meat, fish, eggs	Rich sources of protein, vitamins and minerals
Oils and fats	Rich sources of essential fatty acids, vitamins A and E
Milk and dairy products	Major sources of protein, calcium and vitamins

Clinical Case Questions

Chapter 31 Nutrition: Overview and Macronutrients

31.1. A 5-year-old boy is brought with complaints of failure to gain weight. His parents are illiterate laborers, with low income. There is a history of recurrent infections in this child. The child has not received any vaccination. On examination, child appears irritable. Both weight and height are lower than those expected for his age. The skin is dry, hair is brittle, and muscle appears wasted. There is no edema.

A. What is the diagnosis and management of this patient?

Answer—A: This is a case of Marasmus.

Marasmus is a form of severe malnutrition characterized by energy deficiency. A child with marasmus looks emaciated. Body weight is reduced to less than 60% of the expected body weight for the age.

Poor health conditions, insufficient maternal and child feeding and care, and inability to purchase nutritious food, compounded by high rates of illiteracy, play a complex and interrelated role as contributing factors to the undernutrition.

Clinical features include stunted growth, extreme muscle wasting, depletion of subcutaneous fat, weakness, and anemia. There is no edema.

Treatment aims to provide adequate nutrition, restore normal body composition, and cure the condition that caused the deficiency. Protein-rich food along with vitamin supplementation is given. Complications of the disorder must be treated.

31.2. A 40-year-old woman is admitted to the hospital with a history of prolonged chest pain. She had such episodes earlier also but did not seek medical attention. The blood pressure was 170/90 mm of Hg, random blood glucose 340 mg/dl, total cholesterol 300 mg/dl, triglycerides 280 mg/dl, LDL-C 290 mg/dl and HDL 28 mg/dl. An ECG was performed which revealed ischemic changes. A diagnosis of myocardial infarction was made.

A. Prepare a diet plan for this patient.

Answer—A: The patient has Coronary Heart Disease along with hypertension, dyslipidemia, and Type 2 Diabetes mellitus.

Maintaining optimal lipid levels through dietary measures can significantly reduce further complications. As the patient also has hypertension, total daily salt intake should be restricted.

Carbohydrates: 50%–60% of calories from foods rich in complex carbohydrates. As the patient is diabetic, carbohydrates with a low glycemic index should be included.

Protein: 15% of total calories

Fats: Total fats should provide 25%–30% of calories, with limits on various types of fats that can be included.

- Saturated fatty acids: Less than 7% of total calories
- Trans fatty acids: Less than 1% of total calories
- Dietary cholesterol: Less than 200 mg/day
- Polyunsaturated fatty acids (PUFAs): up to 10% of total calories
- Monounsaturated fatty acids (MUFAs): up to 20% of total calories

Chapter 32 Micronutrients: Vitamins

32.1. A 60-year-old man presents with confusion, disorientation, and gait imbalance. Other than disorientation, bilateral paresis of gaze is noted. He has history of heavy alcohol intake for 15 years. He has progressive dementia. Physical examination shows weight loss, decreased body mass index, and muscle wasting. There is no hepatomegaly or splenomegaly. CT head is negative for any tumor/injury.

A. What is the probable diagnosis?

Answer—A: The most likely diagnosis is Wernicke-Korsakoff syndrome. This presents with ataxia, confusion, and ocular disturbances.

B. What is the cause of this condition?

Answer—B: It is caused due to thiamine deficiency in chronic alcoholics. As the oral absorption of thiamine is impaired, thiamine is given parenterally as treatment.

C. Why thiamine should be given before or with glucose in this condition?

Answer—C: If glucose is administered before or without thiamine, Wernicke-Korsakoff syndrome can be worsened or even precipitated. Since aerobic metabolism and the Krebs cycle is dependent on thiamine, if it is not present, anaerobic respiration will occur, increasing the levels of lactate.

32.2. A 7-year-old girl was brought to her pediatrician's office by her parents due to concerns about her feeling tired all the time and a decreased appetite over the past few months. There was a history of Pica. She follows a strict vegan diet and does not consume any animal products or fortified foods. Upon examination, the pediatrician notes that she appears pale, and her tongue looks reddish with angular stomatitis. Hyperpigmentation of the knuckles was noted, and deep tendon reflexes are diminished. Blood tests are ordered to investigate further.

A. How can a strict vegan diet contribute to vitamin B_{12} deficiency?

Answer—A: Vitamin B_{12} is primarily found in animal-based foods such as meat, dairy and eggs. Without supplementation or consumption of fortified foods, individuals following a strict vegan diet may not get enough vitamin B_{12}.

B. How does vitamin B_{12} deficiency impact growth and development in children, and what are the potential long-term consequences if left untreated?

Answer—B: Vitamin B_{12} deficiency can negatively impact growth and development in children by causing stunted growth, delayed milestones, impaired cognitive function and neurological damage. Long-term consequences of untreated vitamin B_{12} deficiency may include permanent neurological damage, learning disabilities and developmental disorders.

C. What specific laboratory tests are useful in diagnosing vitamin B_{12} deficiency in pediatric patients, and what are the reference ranges for serum vitamin B_{12} levels in this age group?

Answer—C: Laboratory tests useful in diagnosing vitamin B_{12} deficiency in pediatric patients include serum vitamin B_{12} levels and a complete blood count (CBC) to assess for macrocytic anemia. Serum vitamin B_{12} levels below 200 pg/mL are considered deficient.

32.3. A 2-month-old male child presented to the emergency department with complaints of swelling over the body for 3 days, breathing difficulty for 1 day, and oliguria for 1 day. There was a history of hospital admission for loose stools multiple times previously. There was no history of fever, cough, cyanosis, or changes in feeding patterns. The child was exclusively breastfed. Dietary history of the mother revealed a predominant intake of polished rice. On examination, the child is lethargic and cachexic, with mild pallor and anasarca.

A. What differential diagnosis will you consider based on history?

Answer—A: The probable diagnosis is wet beriberi.

B. What are the various forms of beriberi?

Answer—B: The different types of beriberi are dry beriberi, wet beriberi and infantile beriberi.

C. What investigation confirms your clinical suspicion?

Answer—C: The investigations required are erythrocyte transketolase activity and 24-hour urinary thiamine excretion.

32.4. A 4-year-old male child presented with complaints of inadequate height gain since 1 year of age and pain during walking/climbing stairs. In the further history, the mother mentioned that the child only engages in indoor games. There is a history of two previous hospitalizations for cough and breathing difficulty.

O/E: Mild pallor

Widening of wrist and ankle joints

Beaded appearance of anterior costochondral junction present

Protuberant abdomen +

Investigations:

Calcium = 6.5 mg/dl

Phosphorous = 2.6

Vit. D = 8 ng/Ml

Sodium = 140 mEq/L

Potassium = 3.95 mEq/L

A. What is the most likely diagnosis of the case scenario?

Answer—A: Vitamin d deficiency rickets.

B. Mention characteristic physical features of the above-mentioned condition.

Answer—B: The features include frontal bossing, delayed dentition, rachitic rosary, Harrison's sulcus, bowing of legs, protuberant abdomen and increased frequency of fractures.

C. What preventive measures can be taken?

Answer—C: There should be adequate sun exposure, proper nutritional intake and vitamin D_3 supplementation for infants if required.

32.5. A 5-year-old boy is brought to the pediatrician by his parents due to a 2-month history of irritability, fatigue and difficulty to walk. His parents also report that he has been complaining of joint pain, especially in his legs. They have noticed that his gums bleed easily, and he has developed small, red bleeding spots on his skin. The boy has a limited diet and prefers to eat only a few types of food, such as bread and cheese. On examination, the child appears pale, and his gums are swollen and bleed upon slight pressure. His skin shows multiple petechiae.

A. What is the most likely diagnosis for this patient?

Answers—A: This patient is likely to have scurvy.

B. What is the primary cause of this condition?

Answer—B: Primary cause is deficiency of vitamin C in diet.

C. How can you prevent this condition?

Answer—C: This condition can be prevented by foods rich in vitamin C including citrus fruits (oranges, lemons, limes), strawberries, kiwi and broccoli.

32.6. A 10-year-old girl presented to your OPD with complaints of difficulty seeing at night, trouble walking through dark rooms, and dry eyes for the past 2 months. She is a pure vegetarian, but her diet lacks green leafy vegetables. On examination, white, fleshy lesions were noted on the temporal side of the conjunctiva.

A. What is your probable diagnosis?

Answer—A: The diagnosis is of vitamin A deficiency.

B. What are the white, fleshy lesions on conjuctiva?

Answer—B: These are bitot spots, caused due to keratin buildup. They are seen in vitamin A deficiency.

C. What is the treatment of the above micronutrient deficiency?

Answer—C: Treatment required is the administration of vitamin A as per the recommended dosage.

Chapter 33 Micronutrients: Minerals

33.1. A 55-year-old man presents with complaints of weakness, loss of appetite and weight loss. Recently, he has noticed yellow discoloration of his eyes. The patient has history of repeated blood transfusions due to a red cell disorder. Physical examination reveals increased skin pigmentation and enlarged liver. Lab investigations showed increased total bilirubin, alkaline phosphatase, AST, ALT, serum iron, total iron binding capacity and transferrin.

A. What is the probable diagnosis?

Answer–A: Clinical picture of weakness along with diabetes and liver function derangement and serum iron panel values (increased serum iron and total iron binding capacity [TIBC] and transferrin) indicate that this is a case of hemochromatosis. It can be primary or secondary.

B. What is the biochemical cause of this disease?

Answer–B: Primary hemochromatosis is due to homozygous recessive mutation leading to abnormally high iron absorption. Secondary hemochromatosis may result from excessive dietary, medicinal, or transfusional iron intake (as in thalassemias). Treatment includes therapeutic phlebotomy or administration of chelators, such as deferoxamine.

33.2. A 72-year-old man presents to emergency department with fatigue, lethargy and chest pain. He has a history of renal dysfunction and diabetes. An acute myocardial infarction was ruled out. His ECG shows tall, peaked t-waves.

A. What is the probable diagnosis?

Answer—A: Clinical presentation of fatigue, weakness and chest pain, which is not myocardial infarction, is suggestive of hyperkalemia. ECG finding of tall, peaked t-waves further strengthens the diagnosis of hyperkalemia. Arrythmia is the most lethal complication of hyperkalemia.

B. What is the likely cause of this condition in this patient?

Answer—B: Acute or chronic kidney disease is a common cause of hyperkalemia. This patient has a history of renal dysfunction. Increased intake of potassium, in dried fruits and nuts can cause hyperkalemia in renal disease patients.

Unit VIII

Storage and Expression of Genetic Information

Competencies Addressed

- **BI6.2** Describe and discuss the metabolic processes in which nucleotides are involved
- **BI6.3** Describe the common disorders associated with nucleotide metabolism
- **BI6.4** Describe the laboratory results of analytes associated with gout and Lesch-Nyhan syndrome
- **BI7.1** Describe the structure and functions of DNA and RNA and outline the cell cycle
- **BI7.2** Describe the processes involved in replication and repair of DNA and the transcription and translation mechanisms
- **BI7.3** Describe gene mutations and basic mechanism of regulation of gene expression
- **BI7.4** Describe applications of molecular technologies like recombinant DNA technology, PCR in the diagnosis and treatment of diseases with genetic basis
- **BI9.3** Describe protein targeting and sorting along with its associated disorders

Unit VIII: Overview

DNA, found in the nucleus of cells**, is the chemical basis of heredity.** DNA is made of nucleotides which consist of a sugar (deoxyribose in DNA), a phosphate group, and a nitrogenous base. The five nucleobases—adenine (A), cytosine (C), guanine (G), thymine (T), and uracil (U)—serve as the fundamental units of the genetic code, with the bases A, G, C, and T being found in DNA while A, G, C, and U are found in RNA (Chapter 34). The nucleotides, purines and pyrimidines are synthesized de novo or reused.

Genes are the functional basis of hereditary. They are the units of inheritance and their expression leads to RNA synthesis (Transcription Chapter 36) and further to protein formation (Translation Chapter 37). However, gene expression is still an enigma as different cells express different genes (that is, make different proteins). The control of gene expression (Chapter 38) is understood in principle but not in entirety. Recombinant DNA technology (Chapter 39) deals with the ways in which human genome has been manipulated in the laboratory to assist in medical diagnostics, therapeutics, and production of medically required molecules like insulin.

As a primary healthcare physician, you will need to understand medical perspectives of genetics.

- **Single gene disorders** which follow mendelian laws—the disease may then be autosomal dominant (e.g., neurofibromatosis), autosomal recessive (e.g., albinism, cystic fibrosis) or X-linked dominant (e.g., vitamin D-resistant rickets), and X-linked recessive (e.g., hemophilia, color blindness).
- **Multifactorial disorders** are the most common disorders, like diabetes mellitus type 2 and hypertension wherein the disease occurrence depends on the environment and multiple gene interactions. Physiologically, phenotypes like height and weight are also a result of gene-environment interactions.
- **Chromosomal anomalies** like Down syndrome (trisomy 21) and Turner's syndrome (45XO) wherein a chromosomal error is present.
- **Somatic diseases** may also be acquired by mutations affecting replications, e.g., some cancers.

Genetics offers the possibility of gene therapy but currently it is in research stages.

The main uses of genetics as of now is in recombinant DNA technology, diagnosis of some bacterial and viral infections and in prenatal diagnosis of common diseases like thalassemia. Stem cell therapy and gene augmentation are other promising areas.

Nucleotide Metabolism

34

Learning Objectives

At the end of this chapter, the student will be able to:

- Describe the structure and function of purine and pyrimidine nucleotides
- Describe the synthesis and degradation of nucleotides and the diseases associated with them

I. OVERVIEW

Ribonucleoside and deoxyribonucleoside phosphates **(nucleotides)** are essential for all cells. Without them, neither ribonucleic acid (RNA) nor deoxyribonucleic acid (DNA) can be produced, and, therefore, proteins cannot be synthesized or cells proliferate. Nucleotides also serve as **carriers of activated intermediates** in the synthesis of some carbohydrates, lipids, and conjugated proteins (for example, uridine diphosphate [UDP]-glucose and cytidine diphosphate **[CDP]-choline**) and are structural components of several essential coenzymes, such as **coenzyme** A, flavin adenine dinucleotide **(FAD[H_2]),** nicotinamide adenine dinucleotide (NAD[H]), and nicotinamide adenine dinucleotide phosphate **(NADP[H]).** Nucleotides, such as cyclic adenosine monophosphate **(cAMP)** and cyclic guanosine monophosphate **(cGMP),** serve as second messengers in signal transduction pathways. In addition, nucleotides play an important role as energy sources in the cell. Finally, nucleotides are important regulatory compounds for many of the pathways of intermediary metabolism, inhibiting or activating key enzymes. The purine and pyrimidine bases found in nucleotides can be synthesized de novo or can be obtained through salvage pathways that allow the reuse of the preformed bases resulting from normal cell turnover. [Note: Little of the purines and pyrimidines supplied by diet is utilized and is degraded instead.]

II. STRUCTURE

Nucleotides are composed of a nitrogenous base; a pentose monosaccharide; and one, two, or three phosphate groups. The nitrogen-containing bases belong to two families of compounds: the purines and the pyrimidines.

A. Purine and pyrimidine bases

Both DNA and RNA contain the same **purine bases: adenine (A) and guanine (G).** Both DNA and RNA contain the pyrimidine **cytosine (C),** but they differ in their second pyrimidine base: **DNA contains thymine (T), whereas RNA contains uracil (U).** T and U differ in that only T has a methyl group (Fig. 34.1). Unusual (modified) bases are occasionally found in some species of DNA (for example, in some viral DNA) and RNA (for example, in transfer RNA [tRNA]). Base modifications include methylation, glycosylation, acetylation, and reduction. Some examples of unusual bases are shown in Figure 34.2. [Note: The presence of an

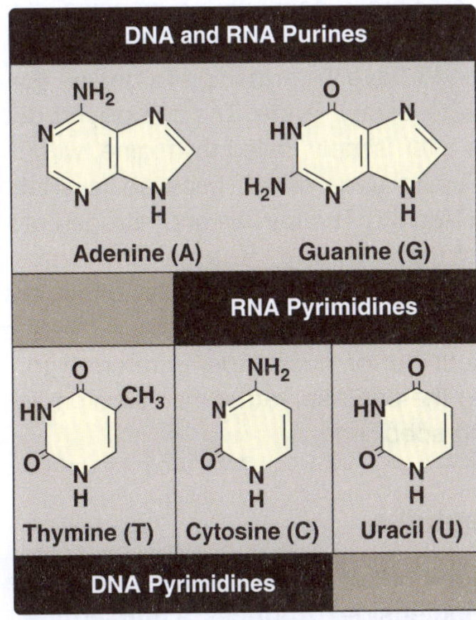

Figure 34.1 Purines and pyrimidines commonly found in DNA and RNA.

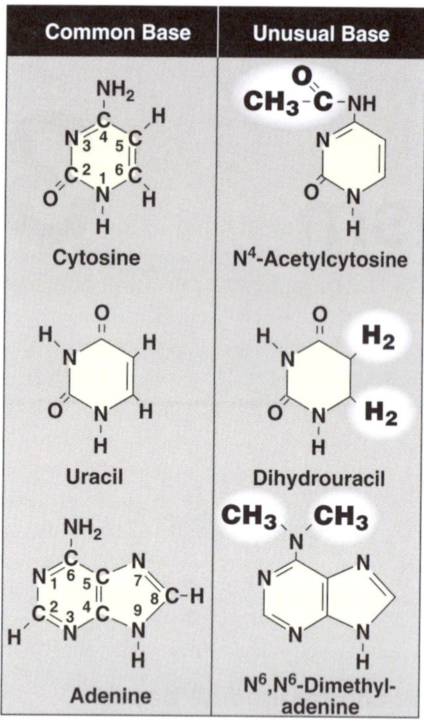

Figure 34.2 Examples of unusual bases.

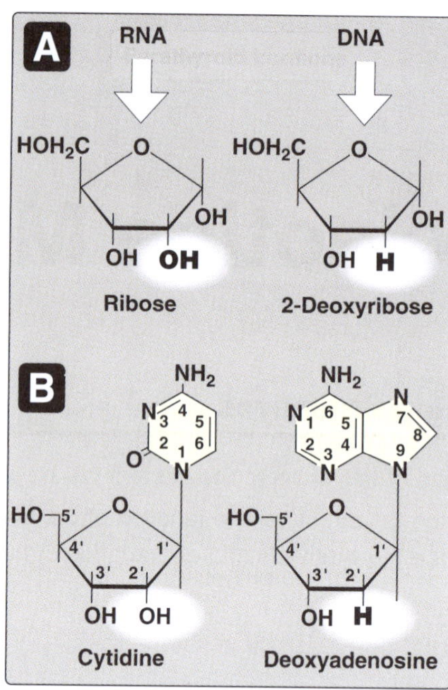

Figure 34.3 **A.** Pentoses found in nucleic acids. **B.** Examples of the numbering systems for purine- and pyrimidine-containing nucleosides.

unusual base in a nucleotide sequence may aid in its recognition by specific enzymes or protect it from being degraded by *nucleases*.]

B. Nucleosides

The addition of a pentose sugar to a base through an N-glycosidic bond produces a nucleoside. If the sugar is ribose, a ribonucleoside is produced, and if the sugar is 2-deoxyribose, a deoxyribonucleoside is produced (Fig. 34.3A). The ribonucleosides of A, G, C, and U are named adenosine, guanosine, cytidine, and uridine, respectively. The deoxyribonucleosides of A, G, C, and T have the added prefix deoxy- (for example, deoxyadenosine). [Note: The compound deoxythymidine is often simply called thymidine, with the deoxy-prefix being understood, because it is incorporated into DNA only.] The carbon and nitrogen atoms in the rings of the base and the sugar are numbered separately (see Fig. 34.3B). [Note: Carbons in the pentose are numbered 1′ to 5′. Thus, when the 5′-carbon of a nucleoside (or nucleotide) is referred to, a carbon atom in the pentose, rather than an atom in the base, is being specified.]

C. Nucleotides

The addition of one or more phosphate groups to a nucleoside produces a nucleotide. The first phosphate group is attached by an ester linkage to the 5′-OH of the pentose, forming a nucleoside 5′-phosphate or a 5′-nucleotide. The type of pentose is denoted by the prefix in the names 5′-ribonucleotide and 5′-deoxyribonucleotide. If one phosphate group is attached to the 5′-carbon of the pentose, the structure is a nucleoside monophosphate, like adenosine monophosphate (AMP, or adenylate). If a second or third phosphate is added to the nucleoside, a nucleoside diphosphate (for example, adenosine diphosphate [ADP] or triphosphate, for example, ATP) results (Fig. 34.4). The second and third phosphates are each connected to the nucleotide by a "high-energy bond" (a bond with a large, negative change in free energy [$-\Delta G$, see p. 160] of hydrolysis). [Note: The phosphate

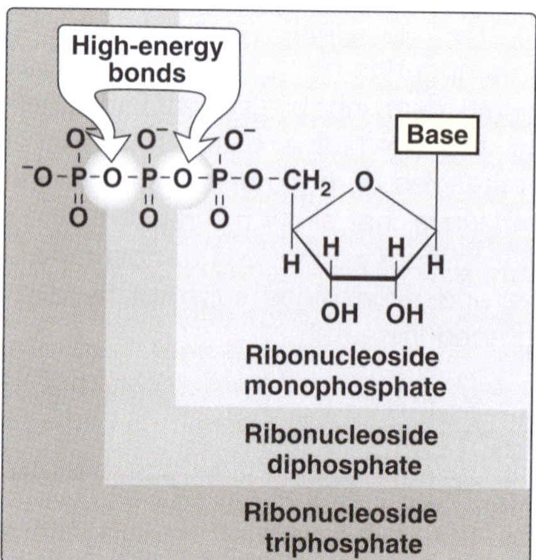

Figure 34.4 Ribonucleoside monophosphate, diphosphate, and triphosphate.

groups are responsible for the negative charges associated with nucleotides and cause DNA and RNA to be referred to as nucleic acids.]

III. PURINE NUCLEOTIDE SYNTHESIS

The **atoms of the purine ring** are contributed by a number of compounds, including amino acids (**aspartate, glycine, and glutamine**), carbon dioxide (**CO_2**), and N^{10}-formyltetrahydrofolate (**N^{10}-formyl-THF**), as shown in Figure 34.5. The purine ring is constructed primarily in the liver by a series of reactions that add the donated carbons and nitrogens to a preformed ribose 5-phosphate. [Note: Synthesis of ribose 5-phosphate from glucose 6-phosphate by the pentose phosphate pathway is discussed on p. 250.]

A. 5-Phosphoribosyl-1-pyrophosphate synthesis

5-Phosphoribosyl-1-pyrophosphate (PRPP) is an activated pentose that participates in the synthesis and salvage of purines and pyrimidines. Synthesis of PRPP from ATP and ribose 5-phosphate is catalyzed by *PRPP synthetase* (Fig. 34.6). This X-linked enzyme is activated by inorganic phosphate and inhibited by purine nucleotides (end-product inhibition). [Note: Because the sugar moiety of PRPP is ribose, ribonucleotides are the end products of de novo purine synthesis. When deoxyribonucleotides are required for DNA synthesis, the ribose sugar moiety is reduced (see p. 583).]

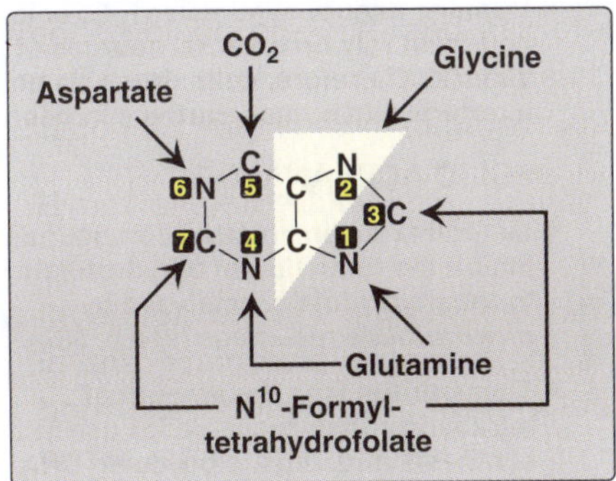

Figure 34.5 Sources of the individual atoms in the purine ring. The order in which the atoms are added is shown by the numbers in the black boxes (see Fig. 34.7). CO_2 = carbon dioxide.

B. 5-Phosphoribosylamine synthesis

Synthesis of 5-phosphoribosylamine from PRPP and glutamine is shown in Figure 34.7. The amide group of glutamine replaces the pyrophosphate group attached to carbon 1 of PRPP. This is the committed step in purine nucleotide biosynthesis. The enzyme that catalyzes the reaction, ***glutamine:phosphoribosylpyrophosphate amidotransferase* (*GPAT*),** is inhibited by the purine 5′-nucleotides AMP and guanosine monophosphate (GMP, or guanylate), the end products of the pathway. The rate of the reaction is also controlled by the intracellular concentration of PRPP. [Note: The concentration of PRPP is normally far below the Michaelis constant (K_m) for the *GPAT*. Therefore, any small change in the PRPP concentration causes a proportional change in rate of the reaction (see p. 136).]

C. Inosine monophosphate synthesis

The next nine steps in purine nucleotide biosynthesis leading to the synthesis of inosine monophosphate ([IMP] whose base is hypoxanthine) are illustrated in Figure 34.7. IMP is the parent purine nucleotide for AMP and GMP. Four steps in this pathway require ATP as an energy source, and two steps in the pathway require N^{10}-formyl-THF as a one-carbon donor (see p. 404). [Note: Hypoxanthine is found in tRNA (see Fig. 37.9 on p. 641).]

D. Synthetic inhibitors

Some **synthetic inhibitors of purine synthesis (for example, the sulfonamides)** are designed to inhibit

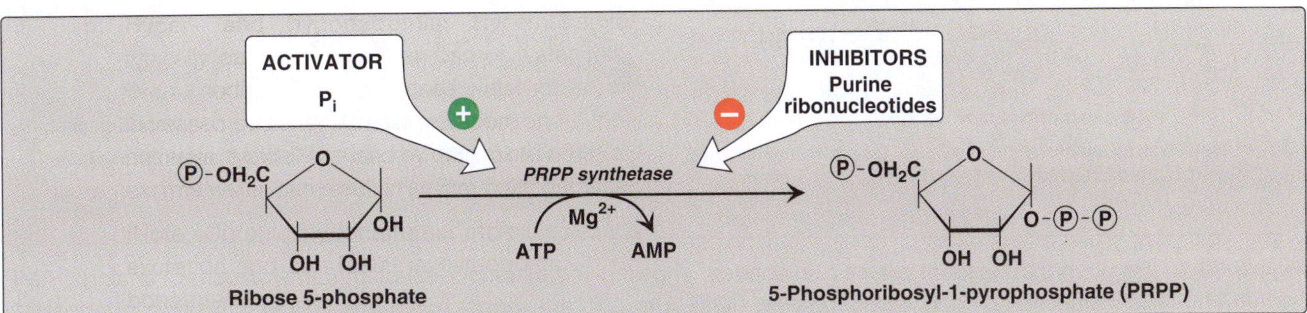

Figure 34.6 Synthesis of PRPP, showing the activator and inhibitors of the reaction. [Note: This is not the committed step of purine synthesis because PRPP is used in other pathways such as salvage (see p. 582).] = phosphate; P_i = inorganic phosphate; AMP = adenosine monophosphate; Mg = magnesium.

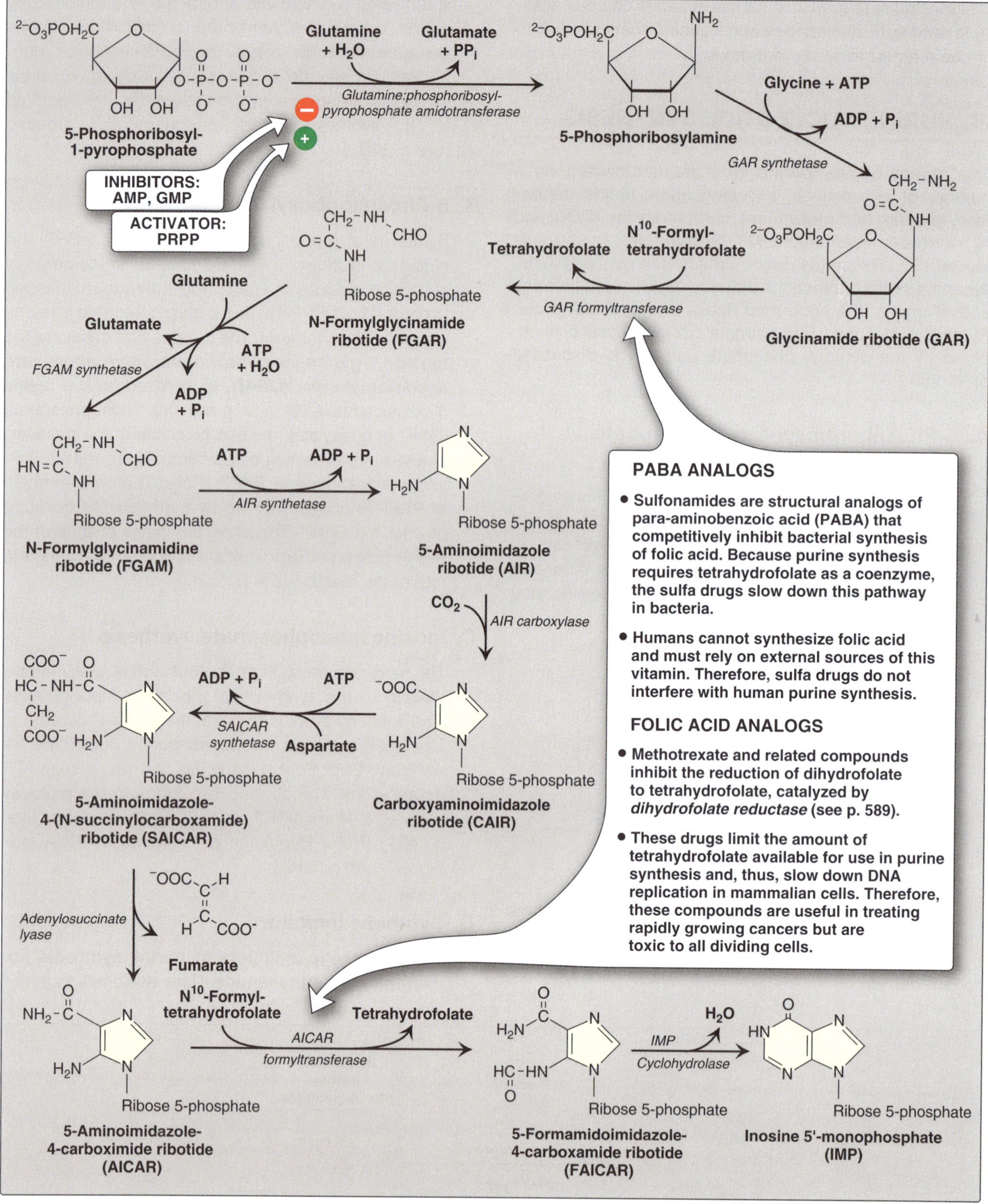

Figure 34.7 De novo synthesis of purine nucleotides, showing the inhibitory effect of some structural analogs. AMP and ADP = adenosine mono- and diphosphates; GMP = guanosine monophosphate; PRPP = 5-phosphoribosyl-1-pyrophosphate; P_i = inorganic phosphate; PP_i = pyrophosphate; CO_2 = carbon dioxide.

the growth of rapidly dividing microorganisms without interfering with human cell functions (see Fig. 34.7). Other **purine synthesis inhibitors, such as structural analogs of folic acid (for example, methotrexate),** are used pharmacologically to control the spread of cancer by interfering with the synthesis of nucleotides and, therefore, of DNA and RNA (see Fig. 34.7).

> Inhibitors of human purine synthesis are extremely toxic to tissues, especially to developing structures such as in a fetus, or to cell types that normally replicate rapidly, including those of bone marrow, skin, gastrointestinal (GI) tract, immune system, or hair follicles. As a result, individuals taking such anticancer drugs can experience adverse effects, including anemia, scaly skin, GI tract disturbance, immunodeficiency, and hair loss.

E. Adenosine and guanosine monophosphate synthesis

The conversion of IMP to either AMP or GMP uses a two-step, energy- and nitrogen-requiring pathway (Fig. 34.8). [Note: AMP synthesis requires guanosine triphosphate (GTP) as an energy source and aspartate as a nitrogen source, whereas GMP synthesis requires ATP and glutamine.] Also, the first reaction in each pathway is inhibited by the end product of that pathway. This provides a mechanism for diverting IMP to the synthesis of the purine present in lesser amounts. If both AMP and GMP are present in adequate amounts, the de novo pathway of purine nucleotide synthesis is inhibited at the *GPAT* step.

F. Nucleoside di- and triphosphate synthesis

Nucleoside diphosphates are synthesized from the corresponding nucleoside monophosphates by base-specific *nucleoside monophosphate kinases* (Fig. 34.9). [Note: These *kinases* do not discriminate between ribose or deoxyribose in the substrate.] ATP is generally the source of the transferred phosphate because it is present in higher concentrations than the other nucleoside triphosphates. *Adenylate kinase* is particularly active in the liver and in muscle, where the turnover of energy from ATP is high. Its function is to maintain equilibrium among the adenine nucleotides (AMP, ADP, and ATP). Nucleoside diphosphates and triphosphates are interconverted by *nucleoside*

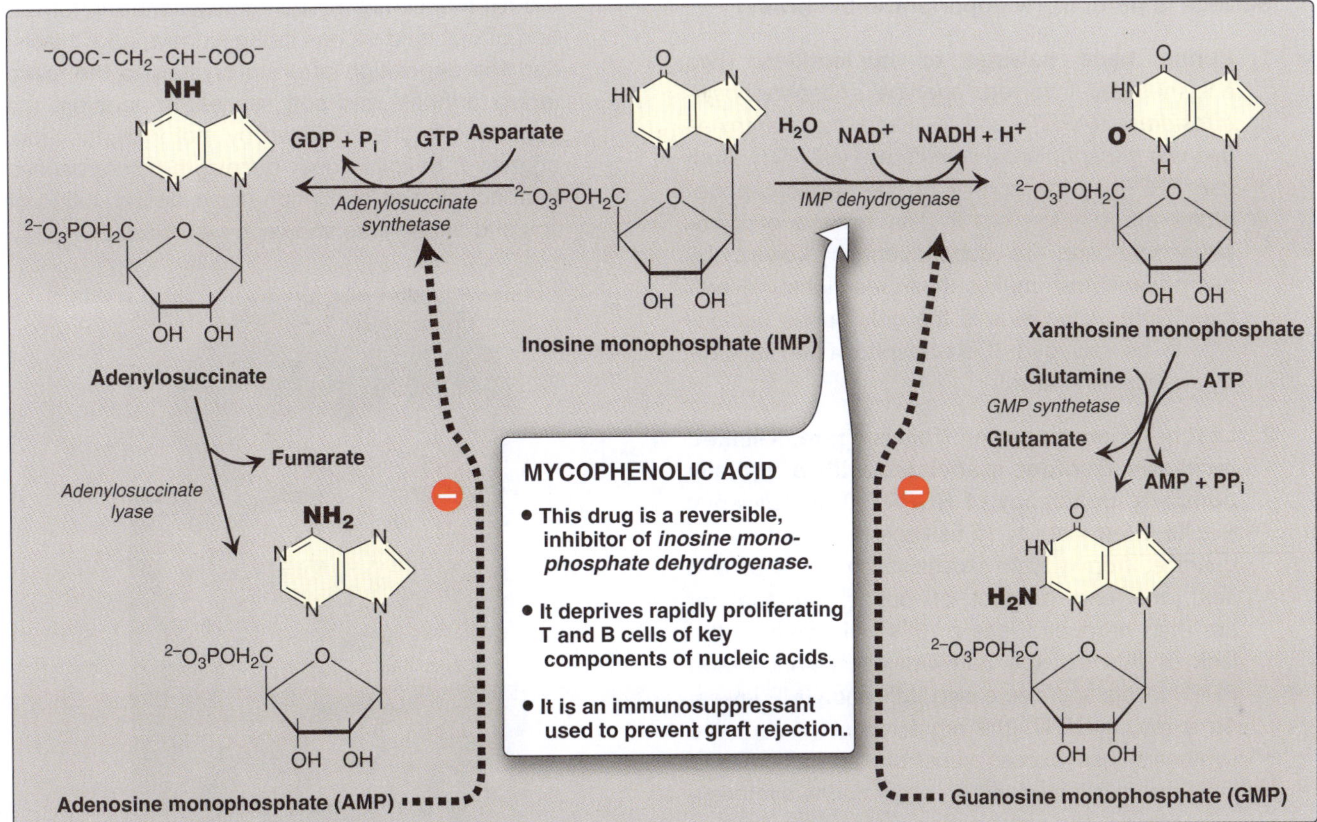

Figure 34.8 Conversion of IMP to AMP (or, adenylate) and GMP (or, guanylate) showing feedback inhibition. NAD(H) = nicotinamide adenine dinucleotide; GDP and GTP = guanosine di- and triphosphates; P_i = inorganic phosphate; PP_i = pyrophosphate.

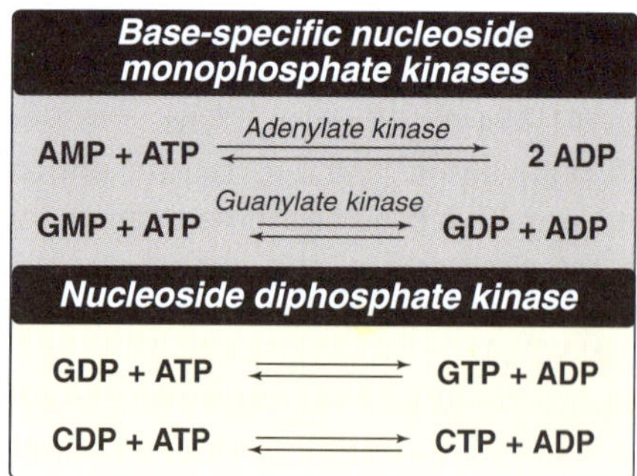

Figure 34.9 Conversion of nucleoside monophosphates to di- and triphosphates. AMP and ADP = adenosine mono- and diphosphates; GMP, GDP, and GTP = guanosine mono-, di-, and triphosphates; CDP and CTP = cytidine di- and triphosphates.

diphosphate kinase, an enzyme that, unlike the *monophosphate kinases*, has broad substrate specificity.

G. Purine salvage pathway

Purines that result from the normal turnover of cellular nucleic acids, or the small amount that is obtained from the diet and not degraded, can be converted to nucleoside triphosphates and used by the body. This is referred to as the salvage pathway for purines. [Note: **Salvage is particularly important in the brain.]**

1. **Purine base salvage to nucleotides:** Two enzymes are involved: *adenine phosphoribosyltransferase* (*APRT*) and X-linked *hypoxanthine–guanine phosphoribosyltransferase* (*HGPRT*). Both use PRPP as the source of the ribose 5-phosphate group (Fig. 34.10). The release of pyrophosphate and its subsequent hydrolysis by *pyrophosphatase* makes these reactions irreversible. [Note: Adenosine is the only purine nucleoside to be salvaged. It is phosphorylated to AMP by *adenosine kinase*.]

2. **Lesch-Nyhan syndrome: This is a rare, X-linked recessive disorder associated with a virtually complete deficiency of *HGPRT*.** The deficiency results in an inability to salvage hypoxanthine or guanine, from which excessive amounts of uric acid, the end product of purine degradation, are then produced (see p. 584). In addition, the lack of this salvage pathway causes increased PRPP levels and decreased IMP and GMP levels. As a result, *GPAT* (the regulated step in purine synthesis) has excess substrate and decreased inhibitors available, and de novo purine synthesis is increased. The combination of decreased purine reutilization and increased purine synthesis results in increased degradation of purines and the production of large amounts of uric acid, making *HGPRT* deficiency an inherited cause of hyperuricemia. In patients with Lesch-Nyhan syndrome, the hyperuricemia frequently results in the formation of uric acid stones in the kidneys (urolithiasis) and the deposition of urate crystals in the joints (gouty arthritis) and soft tissues. In addition, the syndrome is characterized by motor dysfunction, cognitive deficits, and behavioral disturbances that include self-mutilation (for example, biting of lips and fingers), as shown in Figure 34.11.

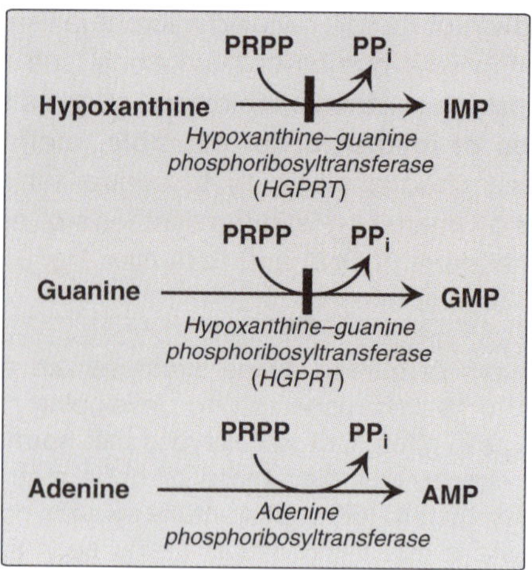

Figure 34.10 Salvage pathways of purine nucleotide synthesis. [Note: Virtually complete deficiency of ***HGPRT*** results in Lesch-Nyhan syndrome. Partial deficiencies of *HGPRT* are known. As the amount of functional enzyme increases, the severity of the symptoms decreases.] IMP, GMP, and AMP = inosine, guanosine, and adenosine monophosphates; PRPP = 5-phosphoribosyl-1-pyrophosphate; PP_i = pyrophosphate.

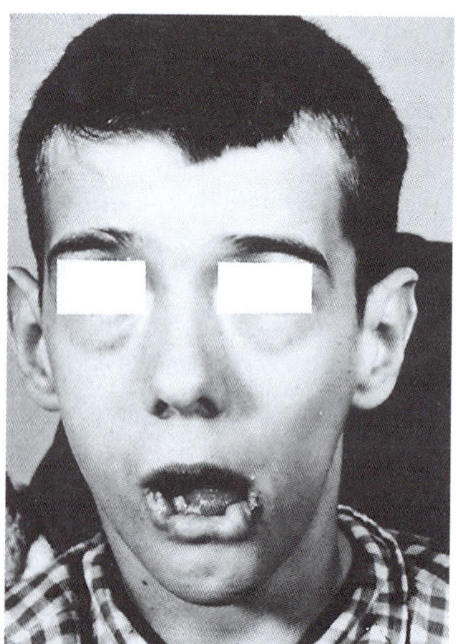

Figure 34.11 Lesions on the lips of a patient with Lesch-Nyhan syndrome.

IV. DEOXYRIBONUCLEOTIDE SYNTHESIS

The nucleotides described thus far all contain ribose (ribonucleotides). DNA synthesis, however, requires 2′-deoxyribonucleotides, which are produced from ribonucleoside diphosphates by the enzyme *ribonucleotide reductase* during the S-phase of the cell cycle (see p. 583). [Note: The same enzyme acts on pyrimidine ribonucleotides.]

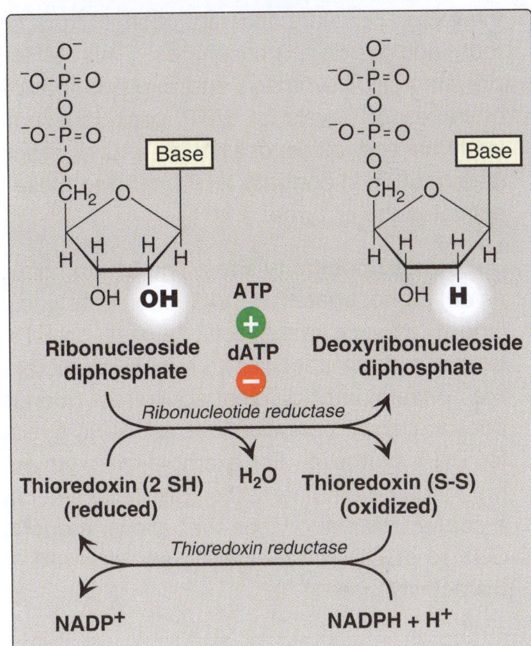

Figure 34.12 Conversion of ribonucleotides to deoxyribonucleotides. NADP(H) = nicotinamide adenine dinucleotide phosphate; dATP = deoxyadenosine triphosphate.

A. Ribonucleotide reductase

Ribonucleotide reductase (*ribonucleoside diphosphate reductase*) is a dimer composed of two nonidentical subunits, R1 (or, α) and the smaller R2 (or, β), and is specific for the reduction of purine nucleoside diphosphates (ADP and GDP) and pyrimidine nucleoside diphosphates (CDP and UDP) to their deoxy forms (dADP, dGDP, dCDP, and dUDP). The immediate donors of the hydrogen atoms needed for the reduction of the 2′-hydroxyl group are two sulfhydryl (–SH) groups on the enzyme itself (R1 subunit), which form a disulfide bond during the reaction (see p. 88). [Note: R2 contains the stable tyrosyl radical required for catalysis at R1.]

1. **Reduced enzyme regeneration:** In order for *ribonucleotide reductase* to continue to produce deoxyribonucleotides at R1, the disulfide bond created during the production of the 2′-deoxy carbon must be reduced. The source of the reducing equivalents is thioredoxin, a protein coenzyme of *ribonucleotide reductase*. Thioredoxin contains two cysteine residues separated by two amino acids in the peptide chain. The two –SH groups of thioredoxin donate their hydrogen atoms to *ribonucleotide reductase*, forming a disulfide bond in the process (Fig. 34.12).

2. **Reduced thioredoxin regeneration:** Thioredoxin must be converted back to its reduced form in order to continue performing its function. The reducing equivalents are provided by NADPH + H^+, and the reaction is catalyzed by *thioredoxin reductase*, a selenoprotein.

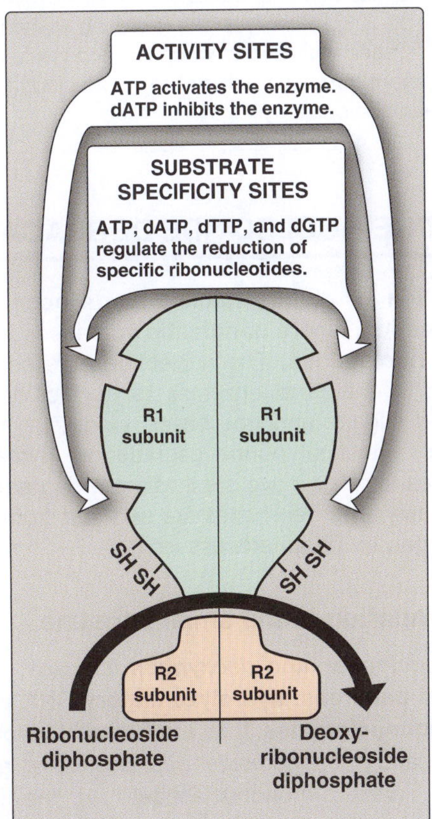

Figure 34.13 Regulation of *ribonucleotide reductase*. dATP, dTTP, and dGTP = deoxyadenosine, deoxythymidine, and deoxyguanosine triphosphates. [Note: The R1 subunit is also referred to as α and the R2 as β.]

B. Deoxyribonucleotide synthesis regulation

Ribonucleotide reductase is responsible for maintaining a balanced supply of the deoxyribonucleotides required for DNA synthesis. Consequently, the regulation of the enzyme is complex. In addition to the catalytic site, R1 contains two distinct allosteric sites involved in regulating enzymic activity (Fig. 34.13).

1. **Activity sites:** The binding of dATP to allosteric sites (known as activity sites) on R1 inhibits the overall catalytic activity of the enzyme and,

therefore, prevents the reduction of any of the four nucleoside diphosphates. This effectively prevents DNA synthesis and explains the toxicity of increased levels of dATP seen in conditions such as *adenosine deaminase (ADA)* deficiency (see p. 587). In contrast, ATP bound to these sites activates the enzyme.

2. **Substrate specificity sites:** The binding of nucleoside triphosphates to additional allosteric sites (known as substrate specificity sites) on R1 regulates substrate specificity, causing an increase in the conversion of different species of ribonucleotides to deoxyribonucleotides as they are required for DNA synthesis. For example, deoxythymidine triphosphate binding at the specificity site causes a conformational change that allows reduction of GDP to dGDP at the catalytic site when ATP is at the activity site.

The drug hydroxyurea (hydroxycarbamide) inhibits *ribonucleotide reductase*, thereby inhibiting the generation of substrates for DNA synthesis. The drug is an antineoplastic agent and is used in the treatment of cancers such as melanoma. Hydroxyurea is also used in the treatment of sickle cell anemia (see p. 108). However, the increase in fetal hemoglobin seen with hydroxyurea is because of changes in gene expression and not to *ribonucleotide reductase* inhibition.

V. PURINE NUCLEOTIDE DEGRADATION

Degradation of dietary nucleic acids occurs in the small intestine, where pancreatic *nucleases* hydrolyze them to nucleotides. The nucleotides are sequentially degraded by intestinal enzymes to nucleosides, phosphorylated sugars, and free bases. Uric acid is the end product of intestinal purine degradation. [Note: Purine nucleotides from de novo synthesis are degraded in the liver primarily. The free bases are sent out from the liver and salvaged by peripheral tissues.]

A. Degradation in the small intestine

***Ribonucleases* and *deoxyribonucleases*, secreted by the pancreas, hydrolyze dietary RNA and DNA to oligonucleotides** that are further hydrolyzed by pancreatic *phosphodiesterases*, producing a mixture of 3′- and 5′-mononucleotides. At the intestinal mucosal surface, *nucleotidases* remove the phosphate groups hydrolytically, releasing nucleosides that are taken into enterocytes by sodium-dependent transporters and degraded by *nucleosidases* (*nucleoside phosphorylases*) to free bases plus (deoxy) ribose 1-phosphate. Dietary purine bases are not used to any appreciable extent for the synthesis of tissue nucleic acids. Instead, they are degraded to uric acid in the enterocytes. Most of the uric acid enters the blood and is eventually excreted in the urine. A summary of this pathway is shown in Figure 34.14. [Note: Mammals other than primates express *urate oxidase* (*uricase*), which cleaves the purine ring, generating allantoin. Modified recombinant *urate oxidase* is now used clinically to lower urate levels.]

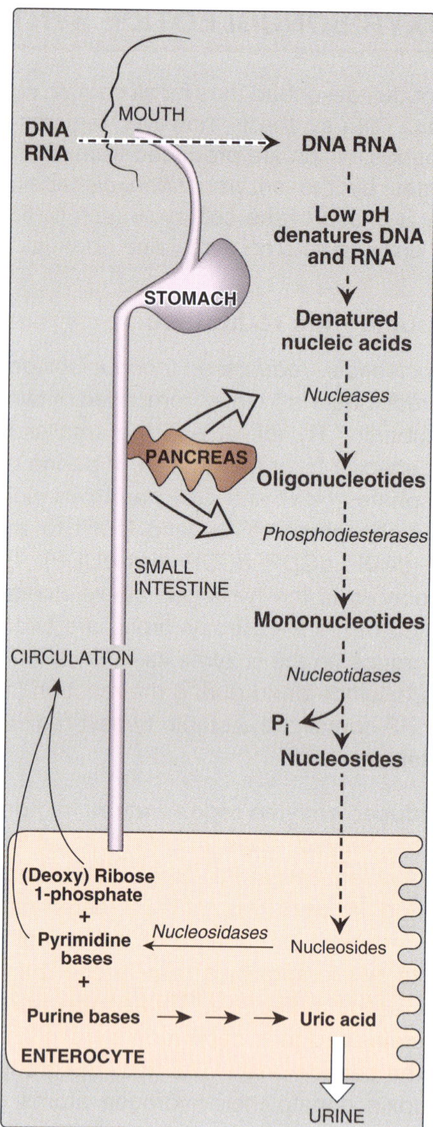

Figure 34.14 Digestion of dietary nucleic acids. P_i = inorganic phosphate.

B. Uric acid formation

A summary of the steps in the production of uric acid and the genetic diseases associated with deficiencies of specific degradative enzymes are shown in Figure 34.15. [Note: The bracketed numbers refer to specific reactions in the figure.]

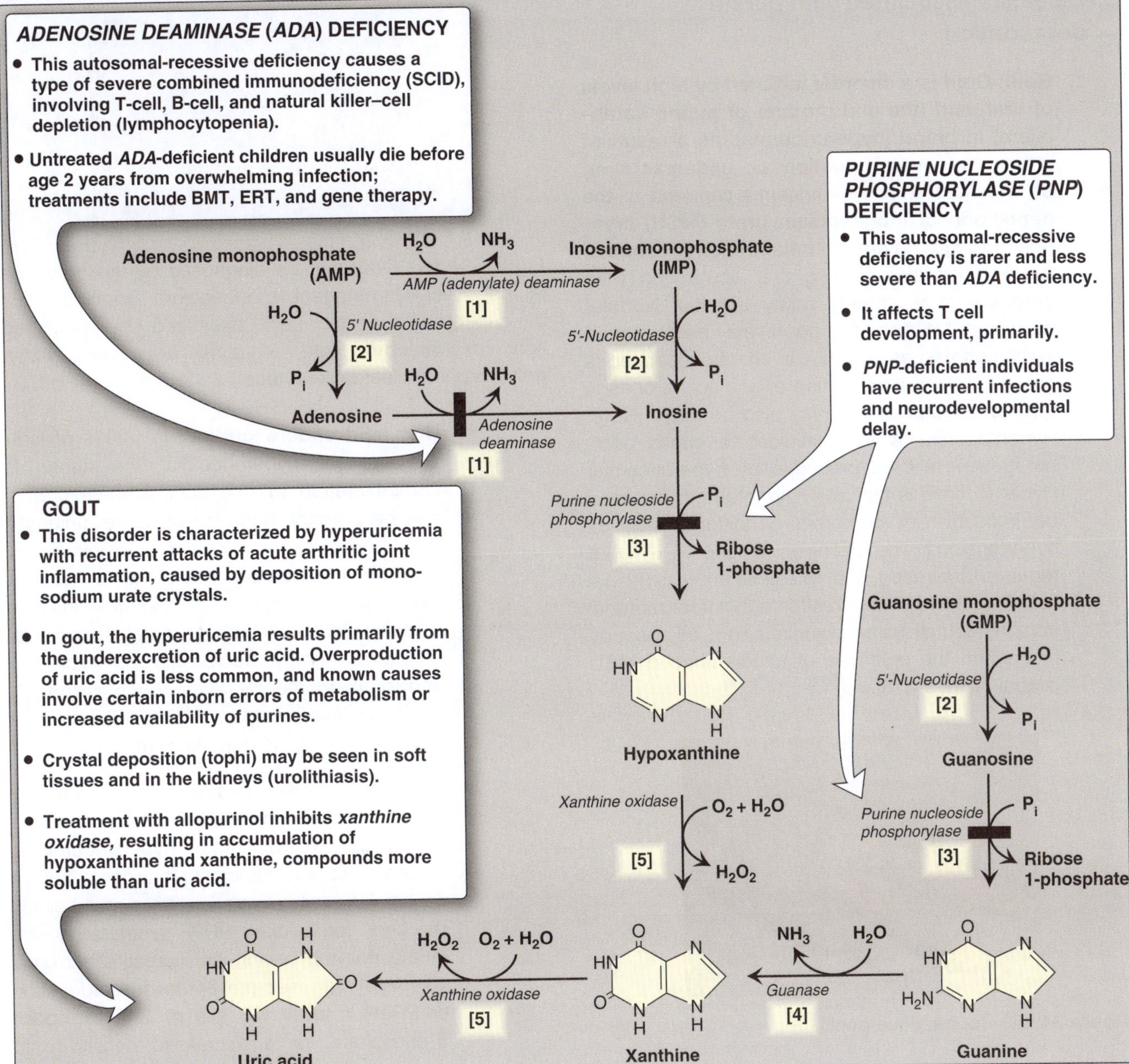

Figure 34.15 The degradation of purine nucleotides to uric acid, illustrating some of the genetic diseases associated with this pathway. [Note: The numbers in brackets refer to the corresponding numbered citations in the text.] BMT = bone marrow transplantation; ERT = enzyme replacement therapy; P_i = inorganic phosphate; H_2O_2 = hydrogen peroxide; NH_3 = ammonia.

1. An amino group is removed from AMP to produce IMP by *AMP (adenylate) deaminase* or from adenosine to produce inosine (hypoxanthine-ribose) by *adenosine deaminase*.

2. IMP and GMP are converted into their respective nucleoside forms, inosine and guanosine, by the action of *5′-nucleotidase*.

3. *Purine nucleoside phosphorylase* converts inosine and guanosine into their respective purine bases, hypoxanthine and guanine. [Note: A *mutase* interconverts ribose 1- and ribose 5-phosphate.]

4. Guanine is deaminated to form xanthine.

5. Hypoxanthine is oxidized by molybdenum-containing *xanthine oxidase* (*XO*) to xanthine, which is further oxidized by *XO* to uric acid, the final product of human purine degradation. Uric acid is excreted primarily in the urine.

C. Diseases associated with purine degradation

1. **Gout: Gout is a disorder initiated by high levels of uric acid (the end product of purine catabolism) in blood (hyperuricemia), as a result of either the overproduction or underexcretion of uric acid.** The hyperuricemia can lead to the **deposition of monosodium urate (MSU) crystals in the joints and an inflammatory response to the crystals**, causing first acute and then progressing to chronic gouty arthritis. Nodular masses of MSU crystals (tophi) may be deposited in the soft tissues, resulting in chronic tophaceous gout (Fig. 34.16). Formation of uric acid stones in the kidney (urolithiasis) may also be seen. [Note: Hyperuricemia is not sufficient to cause gout, but gout is always preceded by hyperuricemia. Hyperuricemia is typically asymptomatic but may be indicative of comorbid conditions such as hypertension.] The definitive diagnosis of gout requires aspiration and examination of synovial fluid (Fig. 34.17) from an affected joint (or material from a tophus) using polarized light microscopy to confirm the presence of needle-shaped MSU crystals (Fig. 34.18).

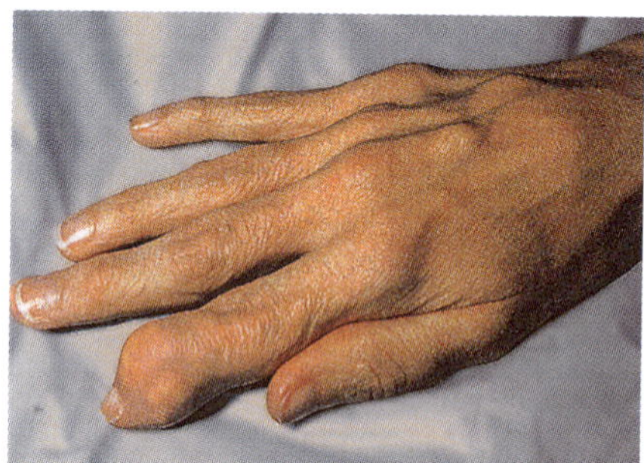

Figure 34.16 Tophaceous gout.

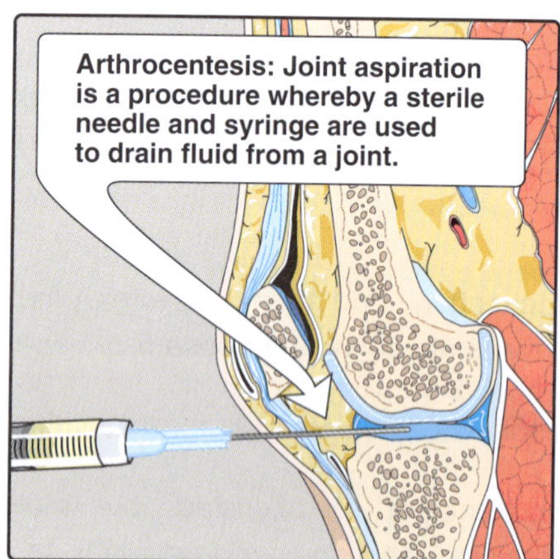

Figure 34.17 Analysis of joint fluid can help to define causes of joint swelling and arthritis, such as infection, gout, and rheumatoid disease.

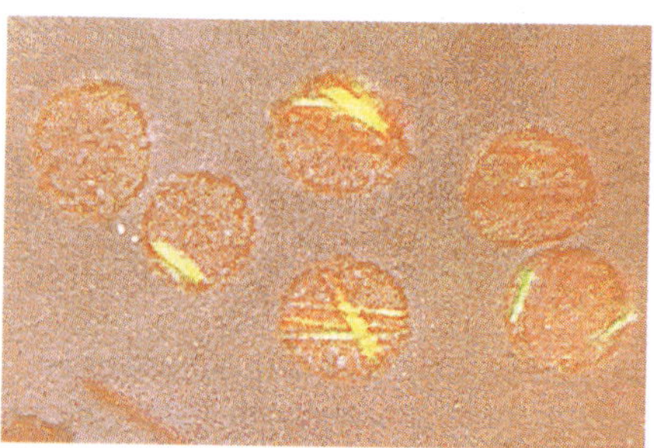

Figure 34.18 Gout can be diagnosed by the presence of negatively birefringent monosodium urate crystals in aspirated synovial fluid examined by polarized light microscopy. Here, crystals are seen within polymorphonuclear leukocytes.

a. **Uric acid underexcretion:** In >90% of individuals with hyperuricemia, the cause is underexcretion of uric acid. Underexcretion can be primary, because of as-yet-unidentified inherent excretory defects, or secondary to known disease processes that affect how the kidney handles urate (for example, in lactic acidosis, lactate increases renal urate reabsorption, thereby decreasing its excretion) and to environmental factors such as the use of drugs (for example, thiazide diuretics) or exposure to lead (saturnine gout).

b. **Uric acid overproduction:** A less common cause of hyperuricemia is from the overproduction of uric acid. **Primary hyperuricemia** is, for the most part, idiopathic (having no known cause). However, several identified mutations in the gene for X-linked *PRPP synthetase* result in the enzyme having an increased maximal velocity ([V_{max}], see p. 134) for the production of PRPP, a lower K_m (see p. 136) for ribose 5-phosphate, or a decreased sensitivity to purine nucleotides, its allosteric inhibitors (see p. 139). In each case, increased availability of PRPP increases purine production, resulting in elevated levels of plasma uric acid. Lesch-Nyhan syndrome (see p. 582) also causes hyperuricemia as a result of the decreased salvage of hypoxanthine and guanine and the subsequent increased availability of PRPP. Secondary hyperuricemia is typically the consequence of increased availability of purines (for example, in patients with myeloproliferative disorders or who are undergoing chemotherapy and so have a high rate of cell turnover). Hyperuricemia can also be the result of seemingly unrelated metabolic diseases, such as von Gierke disease (see Fig. 16.8, p. 230) or hereditary fructose intolerance (see p. 239).

A diet rich in meat, seafood (particularly shellfish), and ethanol is associated with increased risk of gout, whereas a diet rich in low-fat dairy products is associated with a decreased risk.

c. **Treatment:** Acute attacks of gout are treated with anti-inflammatory agents. Colchicine, steroidal drugs such as prednisone, and nonsteroidal drugs such as indomethacin are used. [Note: **Colchicine** prevents formation of microtubules, thereby decreasing the movement of neutrophils into the affected area. **Like the other anti-inflammatory drugs, it has no effect on uric acid levels.**] **Long-term therapeutic strategies for gout involve lowering the uric acid level below its saturation point (6.5 mg/dl), thereby preventing the deposition of MSU crystals. Uricosuric agents, such as probenecid or sulfinpyrazone**, that increase renal excretion of uric acid, are used in patients who are underexcretors of uric acid. Allopurinol, a structural analog of hypoxanthine, inhibits uric acid synthesis and is used in patients who are overproducers of uric acid. Allopurinol is oxidized to oxypurinol, a long-lived inhibitor of *XO*. This results in an accumulation of hypoxanthine and xanthine (see Fig. 34.15), compounds more soluble than uric acid and, therefore, less likely to initiate an inflammatory response. In patients with normal levels of *HGPRT*, the hypoxanthine can be salvaged, reducing the levels of PRPP and, therefore, de novo purine synthesis. Febuxostat, a nonpurine inhibitor of *XO*, is also available. [Note: Uric acid levels in the blood normally are close to the saturation point. One reason for this may be the strong antioxidant effects of uric acid.]

2. **Adenosine deaminase deficiency:** *ADA* is expressed in a variety of tissues, but, in humans, lymphocytes have the highest activity of this cytoplasmic enzyme. A deficiency of *ADA* results in an accumulation of adenosine, which is converted to its ribonucleotide or deoxyribonucleotide forms by cellular *kinases*. As dATP levels rise, *ribonucleotide reductase* is inhibited, thereby preventing the production of all deoxyribose-containing nucleotides (see p. 583). Consequently, cells cannot make DNA and divide. [Note: The dATP and adenosine that accumulate in *ADA* deficiency lead to developmental arrest and apoptosis of lymphocytes.] In its most severe form, this autosomal-recessive disorder causes a type of **severe combined immunodeficiency disease (SCID)**, involving a decrease in T cells, B cells, and natural killer cells. *ADA* deficiency accounts for ~14% of cases of SCID in the United States. Treatments include bone marrow transplantation, enzyme replacement therapy, and gene therapy (see p. 691). Without appropriate treatment, children with this disorder usually die from infection by age 2 years. [Note: *Purine nucleoside phosphorylase* deficiency results in a less severe immunodeficiency primarily involving T cells.]

VI. PYRIMIDINE SYNTHESIS AND DEGRADATION

Unlike the synthesis of the purine ring, which is constructed on a pre-existing ribose 5-phosphate, the pyrimidine ring is synthesized before being attached to ribose 5-phosphate, which is donated by PRPP. The sources of the atoms in the pyrimidine ring are glutamine, CO_2, and aspartate (Fig. 34.19).

A. Carbamoyl phosphate synthesis

The regulated step of this pathway in mammalian cells is the synthesis of carbamoyl phosphate from glutamine and CO_2, catalyzed by ***carbamoyl phosphate synthetase (CPS) II***. *CPS II* is inhibited by uridine triphosphate (the end product of this pathway, which can be converted into the other pyrimidine nucleotides) and is activated by PRPP. [**Note: Carbamoyl phosphate, synthesized by *CPS I*, is also a precursor of urea (see p. 385)**. Defects in *ornithine transcarbamylase* of the urea cycle promote pyrimidine synthesis because of increased availability of carbamoyl phosphate. A comparison of the two enzymes is presented in Figure 34.20.]

B. Orotic acid synthesis

The second step in pyrimidine synthesis is the formation of carbamoylaspartate, catalyzed by *aspartate transcarbamoylase*. The pyrimidine ring is then closed by *dihydroorotase*. The resulting dihydroorotate is oxidized to produce orotic acid (orotate), as shown

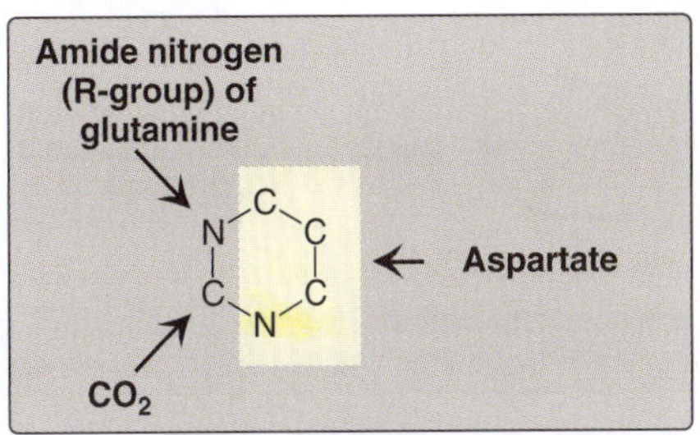

Figure 34.19 Sources of the individual atoms in the pyrimidine ring. CO_2 = carbon dioxide.

Variable	*CPS I*	*CPS II*
Cellular location	Mitochondria	Cytosol
Pathway involved	Urea cycle	Pyrimidine synthesis
Source of nitrogen	Ammonia	γ-Amide group of glutamine
Regulators	Activator: N-acetyl-glutamate	Activator: PRPP Inhibitor: UTP

Figure 34.20 Summary of the differences between *carbamoyl phosphate synthetase* (*CPS*) *I* and *II*. PRPP = 5-phosphoribosyl-1-pyrophosphate; UTP = uridine triphosphate.

in Figure 34.21. The human enzyme that produces orotate, *dihydroorotate dehydrogenase*, is a flavin mononucleotide-containing protein of the inner mitochondrial membrane. All other enzymes in pyrimidine biosynthesis are cytosolic. [Note: The first three enzymic activities in this pathway (*CPS II*, *aspartate transcarbamoylase*, and *dihydroorotase*) are actually three different catalytic domains of a single polypeptide known as *CAD* from the first letter in the name of each domain. (See p. 88 for a discussion of domains.) This is an example of a multifunctional or multicatalytic polypeptide that facilitates the ordered synthesis of an important compound. Synthesis of the purine nucleotide IMP also involves multifunctional proteins.]

C. Pyrimidine nucleotide synthesis

The completed pyrimidine ring is converted to the nucleotide orotidine monophosphate (OMP) in the second stage of pyrimidine nucleotide synthesis (see Fig. 34.21). As seen with the purines, PRPP is the ribose 5-phosphate donor. The enzyme *orotate phosphoribosyltransferase* produces OMP and releases pyrophosphate, thereby making the reaction biologically irreversible. [Note: Both purine and pyrimidine synthesis require glutamine, aspartic acid, and PRPP as essential precursors.] OMP (orotidylate) is decarboxylated to uridine monophosphate (UMP) by *orotidylate decarboxylase*. The *phosphoribosyltransferase* and *decarboxylase* activities are separate catalytic domains of a single polypeptide called *UMP synthase*.

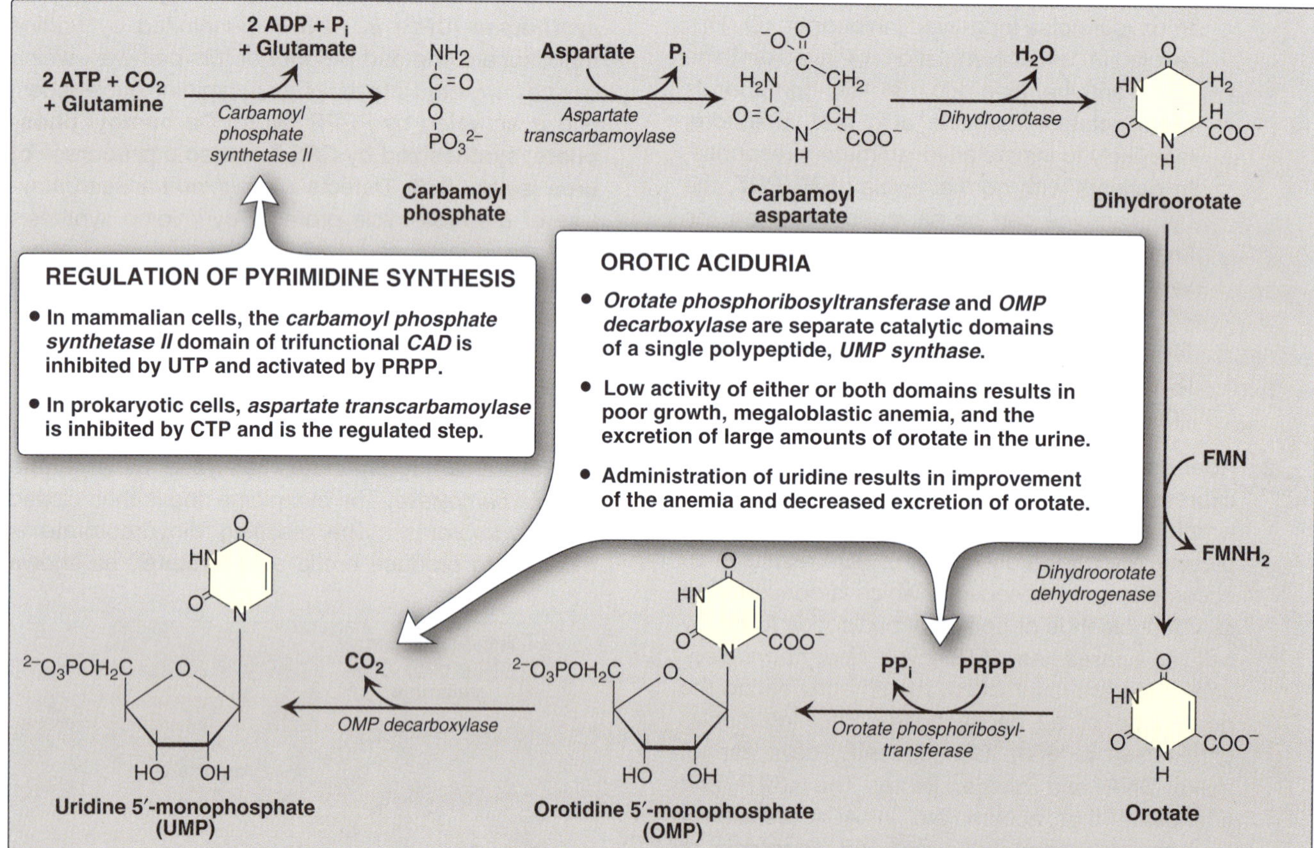

Figure 34.21 De novo pyrimidine synthesis. ADP = adenosine diphosphate; P_i = inorganic phosphate; FMN(H_2) = flavin mononucleotide; CTP = cytidine triphosphate; PRPP = 5-phosphoribosyl-1-pyrophosphate; PP_i = pyrophosphate.

Hereditary orotic aciduria (a very rare disorder) may be caused by a deficiency of one or both activities of this bifunctional enzyme, resulting in orotic acid in the urine (see Fig. 34.21). UMP is sequentially phosphorylated to UDP and UTP. [Note: The UDP is a substrate for *ribonucleotide reductase*, which generates dUDP. The dUDP is phosphorylated to dUTP, which is rapidly hydrolyzed to dUMP by *UTP diphosphatase* (*dUTPase*). Thus, *dUTPase* plays an important role in reducing availability of dUTP as a substrate for DNA synthesis, thereby preventing erroneous incorporation of uracil into DNA.]

D. Cytidine triphosphate synthesis

Cytidine triphosphate (CTP) is produced by amination of UTP by *CTP synthetase* (Fig. 34.22), with glutamine providing the nitrogen. Some of this CTP is dephosphorylated to CDP, which is a substrate for *ribonucleotide reductase*. The dCDP product can be phosphorylated to dCTP for DNA synthesis or dephosphorylated to dCMP that is deaminated to dUMP.

E. Deoxythymidine monophosphate synthesis

dUMP is converted to deoxythymidine monophosphate (dTMP) by *thymidylate synthase*, which **uses N^5,N^{10}-methylene-THF** as the source of the methyl group (see p. 402). This is an unusual reaction in that THF contributes not only a one-carbon unit but also two hydrogen atoms from the pteridine ring, resulting in the oxidation of THF to dihydrofolate ([DHF], Fig. 34.23). Inhibitors of *thymidylate synthase* include thymine analogs such as 5-fluorouracil, which serve as antitumor agents. 5-Fluorouracil is metabolically converted to 5-fluorodeoxyuridine monophosphate (5-FdUMP), which becomes permanently bound to the inactivated *thymidylate synthase*, making the drug a suicide inhibitor (see p. 137). DHF can be reduced to THF by *dihydrofolate reductase* (see Fig. 32.2, p. 526), an enzyme that is inhibited by folate analogs such as methotrexate. By decreasing the supply of THF, these drugs not only inhibit purine synthesis (see Fig. 34.7), but, by preventing methylation of dUMP to dTMP, they also decrease the availability of this essential component of DNA. DNA synthesis is inhibited and cell growth slowed. Thus, these drugs are used to treat cancer. [Note: Acyclovir (a purine analog) and AZT (3′-azido-3′-deoxythymidine, a pyrimidine analog) are used to treat infections of herpes simplex virus and human immunodeficiency virus, respectively. Each inhibits the viral *DNA polymerase*.]

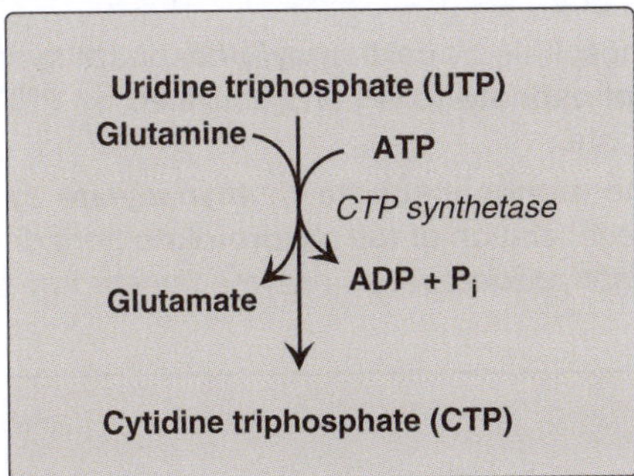

Figure 34.22 Synthesis of CTP from UTP. [Note: CTP, required for RNA synthesis, is converted to dCTP for DNA synthesis.] ADP = adenosine diphosphate; P_i = inorganic phosphate.

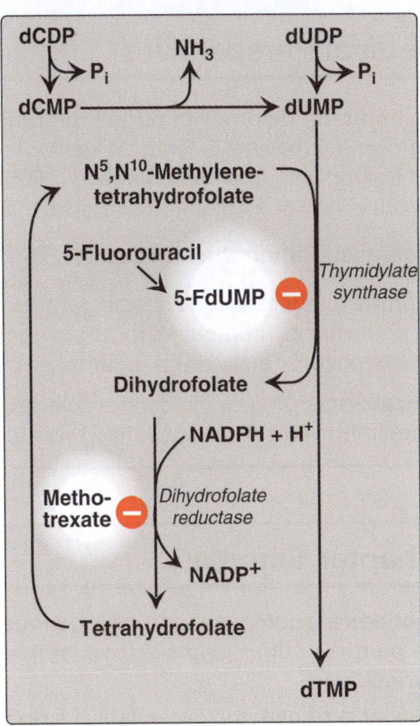

Figure 34.23 Synthesis of dTMP from dUMP, illustrating sites of action of antineoplastic drugs.

F. Pyrimidine salvage and degradation

Unlike the purine ring, which is not cleaved in humans and is excreted as poorly soluble uric acid, the pyrimidine ring is opened and degraded to highly soluble products, β-alanine (from the degradation of CMP and UMP) and β-aminoisobutyrate (from TMP degradation), with the production of ammonia and CO_2. Pyrimidine bases can be salvaged to nucleosides, which are phosphorylated to nucleotides. However, their high solubility makes pyrimidine salvage less significant clinically than purine salvage. [Note: The salvage of pyrimidine nucleosides is the basis for using uridine in the treatment of hereditary orotic aciduria (see p. 588).]

VII. Medical Perspectives

Gout is a **common disorder of purine degradation**. It is characterized by a high uric acid in blood (normal levels: males 3.4–7 mg/dl and females 2.5–6 mg/dl) leading to symptoms of gouty arthritis (pain and swelling in joints of big toe, ankles, feet). It is treated with **anti-inflammatory medication** and allopurinol. **Allopurinol inhibits *xanthine oxidase*** and therefore the conversion of xanthine to uric acid.

Drugs Acting on Purine Nucleotides

Sulfonamides interfere with folic acid synthesis of bacteria by preventing addition of para-aminobenzoic acid into the pteridine precursor by competing for the enzyme, *dihydropteroate synthetase*. Humans cannot synthesize folic acid and hence interference by sulfa drugs is not possible here.

Methotrexate acts on growing cancer cells as well as normal cells by inhibiting the conversion of dihydrofolate to tetrahydrofolate in human cells. Hence, this treatment of cancers comes with side effects.

VIII. Chapter Summary

- **Nucleotides** are composed of a **nitrogenous base** (**adenine** = **A**, **guanine** = **G**, **cytosine** = **C**, **uracil** = **U**, and **thymine** = **T**); a **pentose sugar**; and one, two, or three **phosphate groups** (Fig. 34.24). A and G are **purines**, and C, U, and T are **pyrimidines**.
- If the sugar is **ribose**, the nucleotide is a **ribonucleoside phosphate** (for example, adenosine monophosphate [AMP]), and it can have several functions in the cell, including being a component of **RNA**. If the sugar is **deoxyribose**, the nucleotide is a **deoxyribonucleoside phosphate** (for example, deoxyAMP) and will be found almost exclusively as a component of **DNA**.
- The **committed step** in **purine synthesis** uses **5-phosphoribosyl-1-pyrophosphate** ([**PRPP**], an activated pentose that provides the **ribose 5-phosphate** for de novo purine and pyrimidine synthesis and salvage) and nitrogen from **glutamine** to produce phosphoribosylamine. The enzyme is ***glutamine:phosphoribosylpyrophosphate amidotransferase*** and is inhibited by AMP and guanosine monophosphate (the end products of the pathway) and activated by PRPP.
- Purine nucleotides can also be produced from preformed purine bases by using **salvage reactions** catalyzed by ***adenine phosphoribosyltransferase*** (***APRT***) and ***hypoxanthine–guanine phosphoribosyltransferase*** (***HGPRT***). A near-total deficiency of *HGPRT* causes **Lesch-Nyhan syndrome**, a severe, inherited form of hyperuricemia accompanied by compulsive self-mutilation.
- All deoxyribonucleotides are synthesized from ribonucleotides by the enzyme ***ribonucleotide reductase***. This enzyme is highly regulated (for example, it is strongly inhibited by **deoxyadenosine triphosphate** [**dATP**], a compound that is overproduced in bone marrow cells in individuals with ***adenosine deaminase*** [***ADA***] **deficiency**). *ADA* deficiency causes **severe combined immunodeficiency disease**.
- The end product of purine degradation is **uric acid**, a compound of low solubility whose overproduction or undersecretion causes **hyperuricemia** that, if accompanied by the deposition of **monosodium urate crystals** in joints and soft tissues and an inflammatory response to those crystals, results in **gout**.
- The first step in **pyrimidine synthesis**, the production of carbamoyl phosphate by ***carbamoyl phosphate synthetase II***, is the **regulated** step in this pathway (it is inhibited by **uridine triphosphate** [**UTP**] and activated by PRPP). The UTP produced by this pathway can be converted to cytidine triphosphate.
- **Deoxyuridine monophosphate** can be converted to **deoxythymidine monophosphate** by ***thymidylate synthase***, an enzyme targeted by anticancer drugs such as **5-fluorouracil**. The regeneration of **tetrahydrofolate** from dihydrofolate produced in the *thymidylate synthase* reaction requires ***dihydrofolate reductase***, an enzyme targeted by the drug **methotrexate**. Pyrimidine degradation results in soluble products.

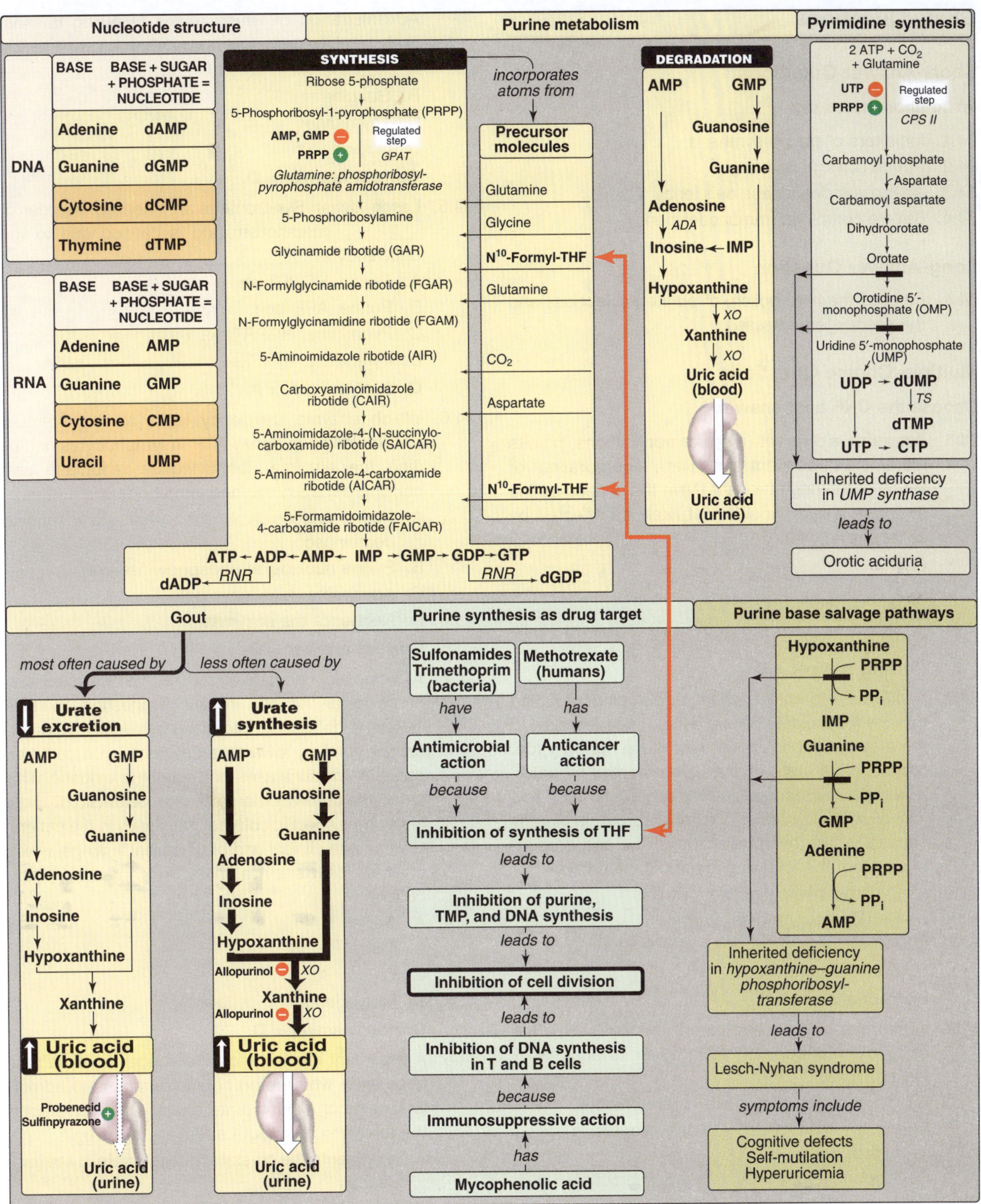

Figure 34.24 Key concept map for nucleotide metabolism. THF = tetrahydrofolate; *GPAT = glutamine: phosphoribosylpyrophosphate amidotransferase; ADA = adenosine deaminase*; *XO = xanthine oxidase*; *TS = thymidylate synthase*; *RNR = ribonucleotide reductase*; *CPS II = carbamoyl phosphate synthetase II*; AMP, GMP, CMP, TMP, and IMP = adenosine, guanosine, cytidine, thymidine, and inosine monophosphates; d = deoxy; PP_i = pyrophosphate; PRPP = 5-phosphoribosyl-1-pyrophosphate.

Study Questions

Short-Answer Questions

Write short notes on:

34.1. Inhibitors of purine synthesis
34.2. Purine salvage pathway
34.3. Adenosine deaminase deficiency
34.4. Severe combined immunodeficiency

Long-Answer Question

34.1. Discuss the metabolism of purine nucleotides and the associated diseases.

Multiple-Choice Questions

Choose the ONE best answer.

34.1. Azaserine, a drug with research applications, inhibits glutamine-dependent enzymes. Incorporation of which of the ring nitrogens (N) in the generic purine structure shown would most likely be affected by azaserine?

A. 1
B. 3
C. 7
D. 9

Correct answer = D.

34.2. A 42-year-old male patient undergoing radiation therapy for prostate cancer develops severe pain in the metatarsal phalangeal joint of his right big toe. Monosodium urate crystals are detected by polarized light microscopy in fluid obtained from this joint by arthrocentesis. This patient's pain is directly caused by the overproduction of the end product of which of the following metabolic pathways?

A. De novo pyrimidine biosynthesis
B. De novo purine biosynthesis
C. Purine salvage
D. Purine degradation

Correct answer = D.

34.3. Which one of the following enzymes of nucleotide metabolism is correctly paired with its pharmacologic inhibitor?

A. Dihydrofolate reductase—methotrexate
B. Inosine monophosphate dehydrogenase—hydroxyurea
C. Ribonucleotide reductase—5-fluorouracil
D. Thymidylate synthase—allopurinol

Correct answer = A.

34.4. A 1-year-old female patient is lethargic, weak, and anemic. Her height and weight are low for her age. Her urine contains an elevated level of orotic acid. Activity of uridine monophosphate synthase is low. Administration of which of the following is most likely to alleviate her symptoms?

A. Adenine
B. Guanine
C. Thymidine
D. Uridine

Correct answer = D.

34.5. Lesch-Nyhan Syndrome is an inherited disorder of ________ metabolism and is caused due to the deficiency of __________.

A. Purine, HGPRTase
B. Purine, APRTase
C. Pyrimidine, HGPRTase
D. Pyrimidine, APRTase

Correct answer = A.

34.6. Which enzyme deficiency leads to the eventual accumulation of deoxy-ATP in lymphocytes, inhibiting the enzyme ribonucleotide reductase and causing severe combined immunodeficiency?

A. Purine nucleoside phosphorylase and adenosine deaminase
B. Purine nucleoside phosphorylase only
C. Adenosine deaminase only
D. Neither of the enzymes

Correct answer = C.

34.7. A 3-day-old neonate presented with seizures and neurological issues, initially suggesting multiple sulfatase deficiency. Blood tests revealed hypouricemia, pointing towards a deficiency in a common cofactor for sulfatases and xanthine oxidase. This led to the consideration of a rare metabolic disorder involving a specific cofactor deficiency in the patient. Given these clinical and biochemical findings, which cofactor deficiency is most likely present in this patient?

A. Cobalt
B. Nickel
C. Chromium
D. Molybdenum

Correct answer = D.

34.8. A treatment modality for hyperuricemia, especially in patients who are on chemotherapy is to administer a recombinant urate oxidase. Which among the following is brought about by this drug?

A. It prevents the crystallization of urate in the joint spaces
B. It inhibits leukocyte migration to the joint spaces
C. It converts urate to allantoin to facilitate its excretion
D. It inhibits the degradation of purines that are released from the damaged DNA

Correct answer = C.

34.9. Formation of which among the following is inhibited during the administration of anti-cancer drugs methotrexate and 5-flurouracil?

A. dUMP
B. dTMP
C. dCMP
D. dAMP

Correct answer = B.

34.10. Both overproduction and a decreased excretion of urate is attributed to the manifestation of hyperuricemia in which among the following conditions?

A. Glycogen storage disorder type 1a
B. Lesch-Nyhan syndrome
C. Adenosine deaminase deficiency
D. Chronic kidney disease

Correct answer = A.

Multiple-Choice Questions (NExT/NEET PG)

34.1. Mycophenolate is used as an immunosuppressant to prevent graft rejection and to manage various autoimmune disorders. The mechanism of its action is the inhibition of the proliferation of lymphocytes, by reversible inhibition of the enzyme IMP dehydrogenase. Which of the following pathways and the reactions is inhibited by this drug?

A. De novo purine synthesis – IMP to GMP
B. Purine salvage pathway – Hypoxanthine to IMP
C. Purine degradation – IMP to inosine
D. Purine salvage pathway – Hypoxanthine to AMP

Correct answer = A.

34.2. A 52-year-old male presented to the clinic with severe pain in his right big toe. He stated that he drank about four cans of beer the previous day. On examination, the metatarsophalangeal joint of the right big toe was warm and red, and the patient winced with pain on pressing the first toe. Biochemical investigations revealed a serum urate level of 11 mg/dL. Which of the following is true in this patient?

A. This condition arises due to the deposition of uric acid in the synovial spaces.
B. Allopurinol, a competitive inhibitor of xanthine oxidase, can be prescribed for the long term management of this condition.
C. Lactic acidosis could have interfered with the renal handling of urate to cause the increased serum levels of urate.
D. Alcohol favors the conversion of uric acid to urate, and urate has a very low solubility compared to uric acid.

Correct answer = C.

34.3. A 4-year-old girl is presented with megaloblastic anemia, failure to thrive, recurrent diarrhea, and pulmonary infections, despite being born in a normal condition. Her anemia persists despite treatments with vitamin B12, folic acid, or pyridoxine. Over the last three years, she has experienced two similar infection episodes treated with antibiotics and blood transfusions, yet her condition worsened. A notable finding in her urine is crystalline sediments of orotic acid, adhering to the sides of glass containers, suggesting a metabolic disorder. What is the biochemical basis underlying her clinical manifestations?

A. Defective pyrimidine salvage pathway and impaired conversion of uridine to UMP.
B. Defective conversion of dUMP to dTMP and hence DNA formation.
C. Impaired pyrimidine degradation, leading to pyrimidine accumulation and orotic acid formation.
D. Impaired pyrimidine synthesis, leading to impaired DNA replication.

Correct answer = D.

34.4. Statement 1: Deficiency of ornithine transcarbamylase (OTC), an X-linked urea cycle disorder, has orotic aciduria, apart from hyperammonemia, as a clinical manifestation.

Statement 2: De novo pyrimidine synthesis has carbamoyl phosphate as one of the intermediate molecules, and carbamoyl phosphate accumulates in OTC deficiency.

A. Both the Statements are correct, and Statement 2 is the correct explanation for Statement 1
B. Both the Statements are correct, but Statement 2 is NOT the correct explanation for Statement 1
C. Statement 1 is correct, but Statement 2 is NOT correct
D. Statement 1 is NOT correct, but Statement 2 is correct

Correct answer = A.

34.5. One of the rare causes of primary gout is the overactivity of the enzyme 5-phosphoribosyl-1-pyrophosphate (PRPP) synthetase. The mechanism here is the increased levels of PRPP and hence purine production, and a subsequent purine degradation, eventually leading to increase urate levels. PRPP is also involved in de novo pyrimidine synthesis; however, the increased flux in this pathway does not lead to any clinical manifestations. Why?

A. Pyrimidines are catabolized to water-soluble products that are easily excreted by the kidneys.
B. Pyrimidine are less salvageable than purines; hence their content in the body do not increase to significantly high levels.

C. Unlike urate which induces inflammation, crystals formed during pyrimidine catabolism do not activate lymphocytes.

D. Only two purines are involved in nucleic acid synthesis, but three pyrimidines are required for DNA or RNA formation. Hence the pyrimidine levels are always lower than purine levels in the body.

Correct answer = A.

DNA Structure, Replication, and Repair

35

Learning Objectives

At the end of this chapter, the student will be able to:

- Describe the structure, organization and function of prokaryotic and eukaryotic DNA
- Describe the processes involved in replication of DNA
- Explain the DNA repair mechanisms
- Discuss how this knowledge may be used for diagnosis and treatment

I. OVERVIEW

Nucleic acids are required for the storage and expression of genetic information. There are two chemically distinct types of nucleic acids: deoxyribonucleic acid (DNA) and ribonucleic acid ([RNA], see Chapter 36). DNA, the repository of genetic information (or, genome), is present not only in chromosomes in the nucleus of eukaryotic organisms, but also in prokaryotic cells which lack nuclei and have a single chromosome. Further, prokaryotic cells may contain non-chromosomal DNA in form of plasmids.

The genetic information found in DNA is copied and transmitted to daughter cells through **DNA replication.** The DNA contained in a fertilized egg encodes the information that directs the development of an organism. This development may involve the production of billions of cells. Each cell is specialized, expressing only those functions that are required for it to perform its role in maintaining the organism. Therefore, DNA must be able not only to replicate precisely each time a cell divides, but also to have the information that it contains be selectively expressed. Further, any error made in the new copy during DNA replication must be zealously corrected as altering of the genome is a serious issue. Hence there are **DNA repair mechanisms**.

> The flow of information from DNA to RNA to protein is termed the "**central dogma**" of molecular biology (Fig. 35.1) and is descriptive of all organisms, with the exception of some viruses that have RNA as the repository of their genetic information.

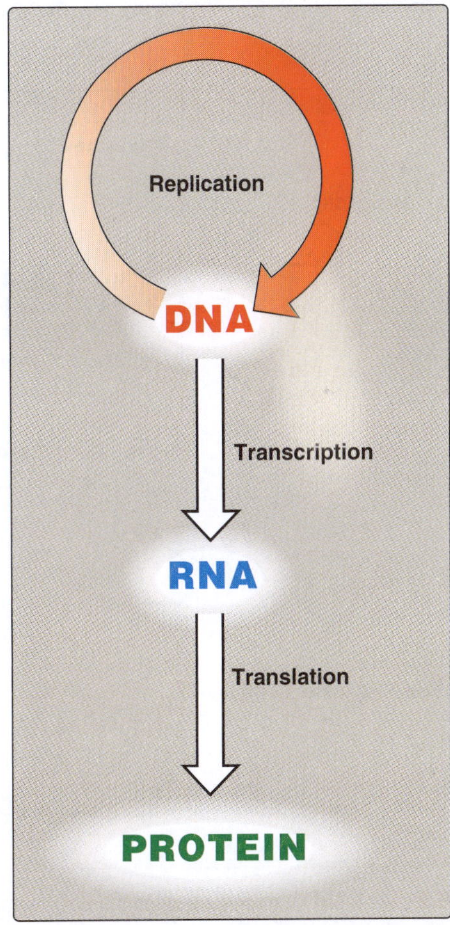

Figure 35.1 The "central dogma" of molecular biology.

II. DNA STRUCTURE

DNA is the chemical basis of heredity, whereas a gene is the fundamental unit of genetic information.

DNA consists of

- Bases
  - Purines: Adenine and Guanine
  - Pyrimidines: Cytosine and Thymine
- Deoxyribose sugar

Nucleosides are derivatives of purines and pyrimidines that are linked to a sugar through a ring nitrogen. **Nucleotides** are phosphorylated nucleotides, i.e., adenosine monophosphate (AMP).

DNA is a polymer of deoxyribonucleoside monophosphates (dNMP) covalently linked by 3′→5′-phosphodiester bonds. With the exception of a few viruses that contain single-stranded DNA (ssDNA), DNA exists as a double-stranded molecule (dsDNA), in which the two strands wind around each other, forming a double helix. [Note: The sequence of the linked dNMP is primary structure, whereas the double helix is secondary structure.] In eukaryotic cells, DNA is found associated with various types of proteins (known collectively as nucleoprotein) present in the nucleus, whereas the protein–DNA complex is present in a non–membrane-bound region known as the nucleoid in prokaryotes.

A. 3′→5′-Phosphodiester bonds

Phosphodiester bonds join the 3′-hydroxyl group of the deoxypentose of one nucleotide to the 5′-hydroxyl group of the deoxypentose of an adjacent nucleotide through a phosphoryl group (Fig. 35.2). The resulting long, unbranched chain has polarity, with both a 5′-end (the end with the free phosphate) and a 3′-end (the end with the free hydroxyl) that are not attached to other nucleotides. By convention, the bases located along the resulting deoxyribose-phosphate backbone are always written in sequence from the 5′-end of the chain to the 3′-end. For example, the sequence of bases in the DNA shown in Figure 35.2D (5′-TACG-3′) is read "thymine, adenine, cytosine, guanine." Phosphodiester linkages between nucleotides can be hydrolyzed enzymatically by a family of *nucleases*, *deoxyribonucleases* for DNA and *ribonucleases* for RNA, or cleaved hydrolytically by chemicals. [Note: Only RNA is cleaved by alkali.]

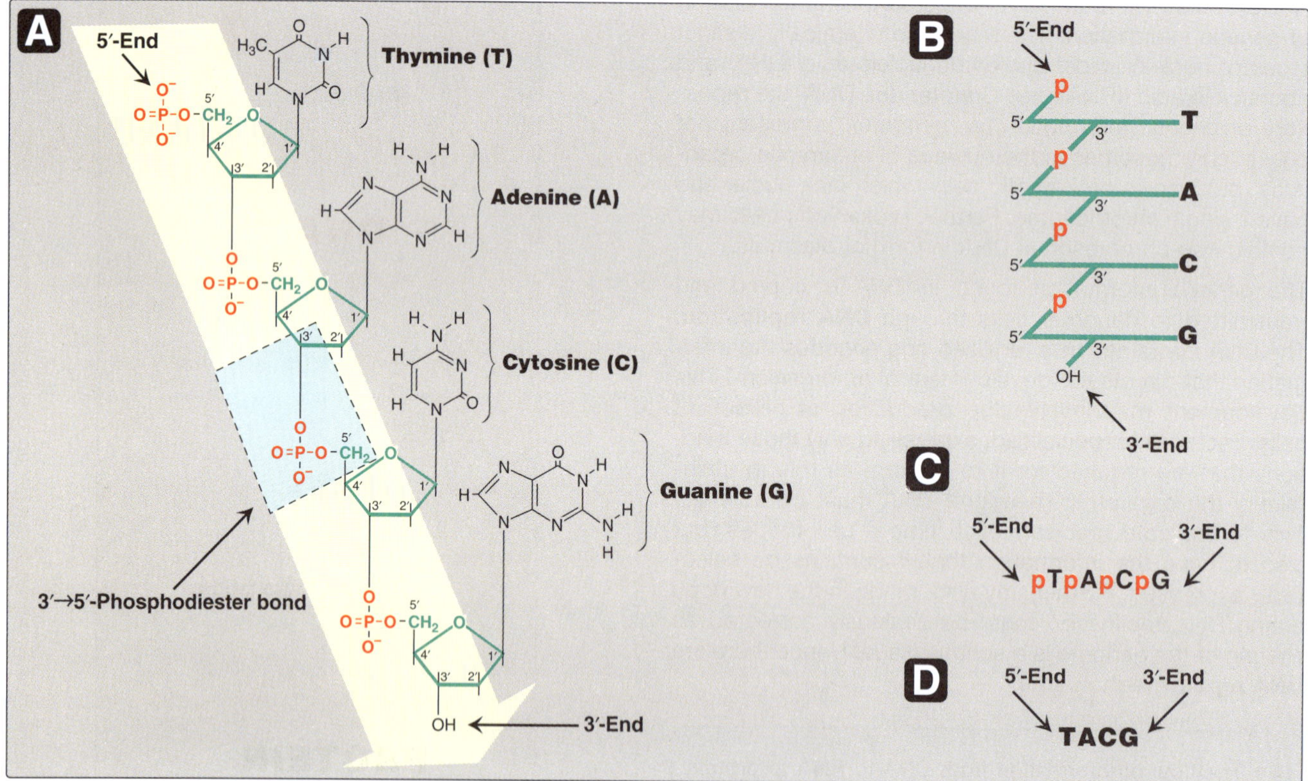

Figure 35.2 **A.** DNA with the nucleotide sequence shown written in the 5′→3′ direction. A 3′→5′-phosphodiester bond is shown highlighted in the blue box, and the deoxyribose-phosphate backbone is shaded in yellow. **B.** DNA written in a more stylized form, emphasizing the deoxyribose-phosphate (p) backbone. **C.** A simpler representation of the nucleotide sequence. **D.** The simplest (and most common) representation. [Note: The nucleotide base sequence is assumed to be written in the 5′→3′ direction unless otherwise indicated.]

B. Double helix

In the double helix, the two chains are coiled around a common axis called the helical axis. The chains are paired in an antiparallel manner (that is, the 5′-end of one strand is paired with the 3′-end of the other strand), as shown in Figure 35.3. In the DNA helix, the hydrophilic deoxyribose-phosphate backbone of each chain is on the outside of the molecule, whereas the hydrophobic bases are stacked inside. The overall structure resembles a twisted ladder. The spatial relationship between the two strands in the helix creates a major (wide) groove and a minor (narrow) groove. These grooves provide access for the binding of regulatory proteins to their specific recognition sequences along the DNA chain.

1. **Base-pairing:** The bases of one strand of DNA are paired with the bases of the second strand, so that an adenine (A) is always paired with a thymine (T), and a cytosine (C) is always paired with a guanine (G). [Note: The base pairs are perpendicular to the helical axis (see Fig. 35.3).] Therefore, one polynucleotide chain of the DNA double helix is always the complement of the other. Given the sequence of bases on one chain, the sequence of bases on the complementary chain can be determined (Fig. 35.4). [Note: The specific base-pairing in DNA leads to the Chargaff rule, which states that in any sample of dsDNA, the amount of A equals the amount of T, the amount of G equals the amount of C, and the total amount of purines (A + G) equals the total amount of pyrimidines (T + C).] The base pairs are held together by hydrogen bonds: two between A and T and three between G and C (Fig. 35.5). These hydrogen bonds, plus the hydrophobic interactions between the stacked bases, stabilize the structure of the double helix.

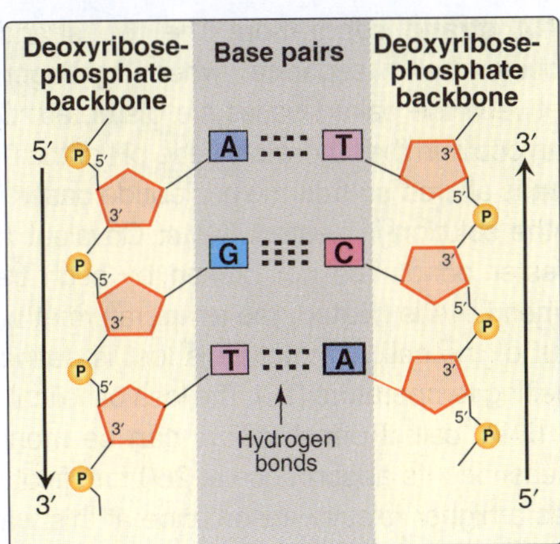

Figure 35.4 Two complementary DNA sequences. T = thymine; A = adenine; C = cytosine; G = guanine.

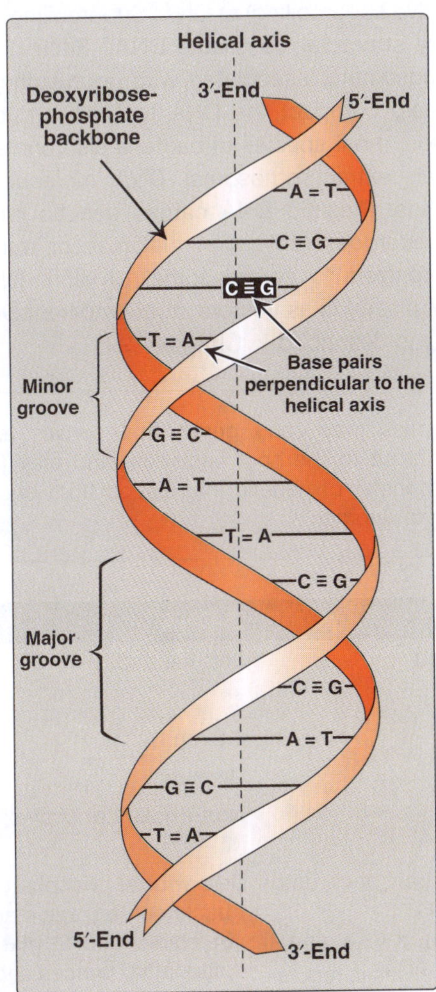

Figure 35.3 DNA double helix, illustrating some of its major structural features.

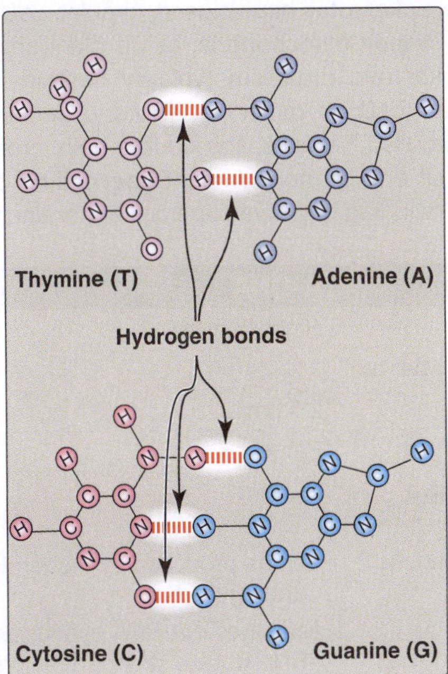

Figure 35.5 Hydrogen bonds between complementary bases.

2. **DNA strand separation:** The two strands of the double helix separate when hydrogen bonds between the paired bases are disrupted. Disruption can occur in the laboratory if the pH of the DNA solution is altered so that the nucleotide bases ionize, or if the solution is heated. [Note: Covalent phosphodiester bonds are not broken by such treatment.] When DNA is heated, the temperature at which one half of the helical structure is lost is defined as the melting temperature (T_m). The loss of helical structure in DNA, called denaturation, can be monitored by measuring its absorbance at 260 nm. [Note: ssDNA has a higher relative absorbance at this wavelength than does dsDNA.] Because there are three hydrogen bonds between G and C but only two between A and T, DNA that contains high concentrations of A and T denatures at a lower temperature than does G- and C-rich DNA (Fig. 35.6). Under appropriate conditions, complementary DNA strands can reform the double helix by the process called renaturation (or, reannealing). [Note: Separation of the two strands over short regions occurs during both DNA and RNA synthesis.]

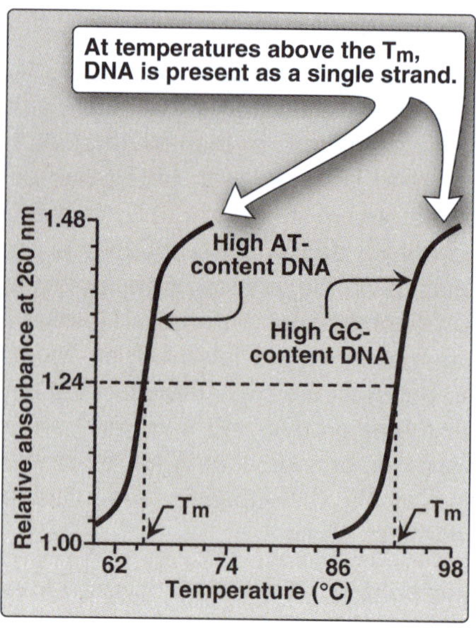

Figure 35.6 Melting temperatures (T_m) of DNA molecules with different nucleotide compositions. A = adenine; T = thymine; G = guanine; C = cytosine.

3. **Structural forms:** The structural forms of DNA are A-DNA, B-DNA and Z-DNA. Their salient features are given in Figure 35.7.

C. Linear and circular DNA molecules

Each chromosome in the nucleus of a eukaryote consists of one long, linear molecule of dsDNA, which is bound by a complex mixture of proteins (histone and nonhistone, see p. 608) to form chromatin. Eukaryotes have closed, circular, dsDNA molecules in their mitochondria, as do plant chloroplasts. A prokaryotic organism typically contains a single, circular, dsDNA molecule. [Note: Circular DNA is "supercoiled," that is, the double helix crosses over on itself one or more times. **Supercoiling** can result in overwinding (positive supercoiling) or underwinding (negative supercoiling) of DNA. Supercoiling, a type of tertiary structure, compacts DNA.] Each prokaryotic chromosome is associated with nonhistone proteins that help compact the DNA to form a nucleoid. In addition, most species of bacteria also contain small, circular, extrachromosomal DNA molecules called plasmids. Plasmid DNA carries genetic information and undergoes replication that may or may not be synchronized to chromosomal division. [Note: The use of plasmids as vectors in recombinant DNA technology is described in Chapter 39.]

Plasmids may carry genes that convey antibiotic resistance to the host bacterium and may facilitate the transfer of genetic information from one bacterium to another.

	A-DNA	B-DNA	Z-DNA
Helix	Right-handed	Right-handed	Left-handed
Base pairs per turn	11	10	12
Diameter	2.5 nm	2.3 nm	1.8 nm
Minor groove	Present	Present	Deep left
Major groove	Present	Present	Non-existent or convex shaped
Special features	Can be produced by moderately dehydrating the B form. Also, the DNA-RNA hybrids and RNA-RNA double stranded regions are similar to A form	Main constituent of chromosomal DNA May transit to Z form during some processes in regulation of gene expression	Deoxyribose phosphate backbone zigzags mainly in regions of DNA that have alternating purines and pyrimidines

Figure 35.7 Salient features of structural forms of DNA.

III. STEPS IN PROKARYOTIC DNA REPLICATION

When the two strands of dsDNA are separated, each can serve as a template for the replication (synthesis) of a new complementary strand. This produces two daughter molecules, each of which contains two DNA strands (one old, one new) in an antiparallel orientation (see Fig. 35.3). This process is called semiconservative replication because, although the parental duplex is separated into two halves (and, therefore, is not conserved as an entity), each of the parental strands remains intact in one of the two new duplexes (Fig. 35.8). The enzymes involved in DNA replication are template-directed, magnesium (Mg^{2+})-requiring *polymerases* that can synthesize the complementary sequence of each strand with extraordinary fidelity. The reactions described in this section were first known from studies of the bacterium Escherichia coli (E. coli), and the description given below refers to the process in prokaryotes. DNA synthesis in higher organisms is more complex but involves the same types of mechanisms. In either case, initiation of DNA replication commits the cell to continue the process until the entire genome has been replicated.

The enzymes and proteins involved in DNA replication are mainly,

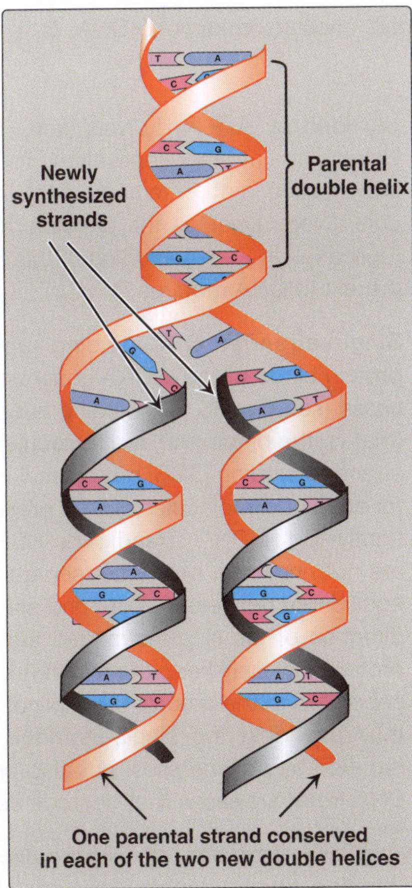

Figure 35.8 Semiconservative replication of DNA. T = thymine; A = adenine; C = cytosine; G = guanine.

- ***DNA helicases*** which unwind the double helix
- ***Topoisomerase*** which relieves the torsional strain produced by *helicase* activity
- *DNA primase* which cause formation of RNA primers
- **Single-stranded binding proteins** which prevent premature re-annealing of the separated dsDNA
- ***DNA polymerase*** which facilitates the formation of new strands
- ***DNA ligase*** which connects the newly formed DNA including the Okazaki fragments

A. Complementary strand separation

In order for the two complementary strands of the parental dsDNA to be replicated, they must first **separate (or "melt")** over a small region, because the *polymerases* use only ssDNA as a template. In prokaryotic organisms, DNA replication begins at a single, unique nucleotide sequence, a site called the origin of replication, or ori (oriC in E. coli), as shown in Figure 35.9A. [Note: This sequence is referred to as a **consensus sequence**, because the order of nucleotides is essentially the same at each site.] The ori includes short, AT-rich segments that facilitate melting. In eukaryotes, replication begins at multiple sites along the DNA helix (Fig. 35.9B). Having multiple origins of replication provides a mechanism for rapidly replicating the great length of eukaryotic DNA molecules.

B. Replication fork formation

As the two strands unwind and separate, synthesis occurs at two replication forks that move away from the origin in opposite directions (bidirectionally), generating a replication bubble (see Fig. 35.9). [Note: The term **"replication fork"** derives from the Y-shaped structure in which the tines of the fork represent the separated strands (Fig. 35.10).]

1. **Required proteins:** Initiation of DNA replication requires the recognition of the origin (start site) by a group of proteins that form the prepriming complex. These proteins are responsible for melting at the ori, maintaining the separation of the parental strands, and unwinding the double helix ahead of the advancing replication fork. In E. coli, these proteins include the following.

   a. **DnaA protein:** *DnaA* protein initiates replication by binding to specific nucleotide sequences (*DnaA* boxes) within oriC. Binding causes an AT-rich region (the DNA unwinding element) in the origin to melt. Melting (strand separation) results in a short, localized region of ssDNA.

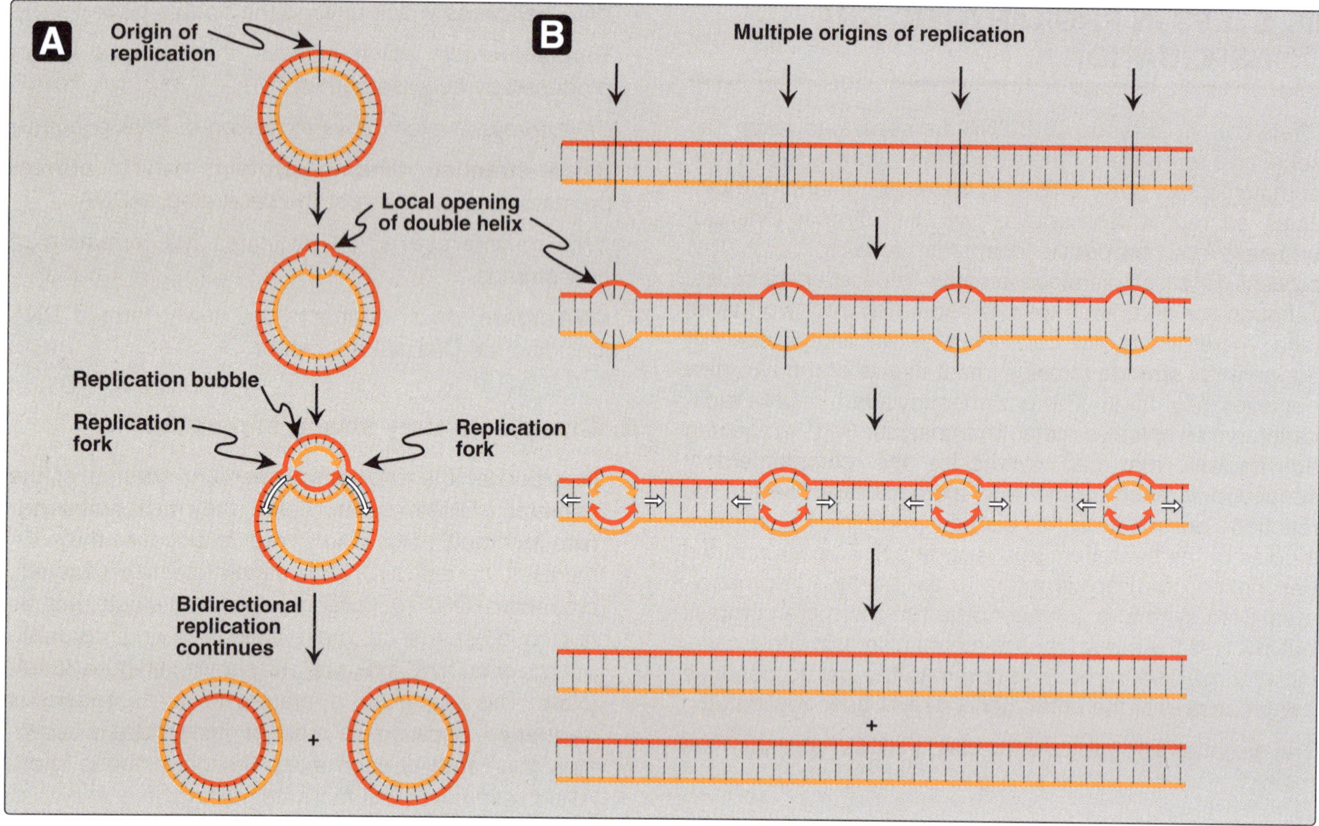

Figure 35.9 Replication of DNA: origins and replication forks. **A.** Small, circular prokaryotic DNA. **B.** Long, linear eukaryotic DNA.

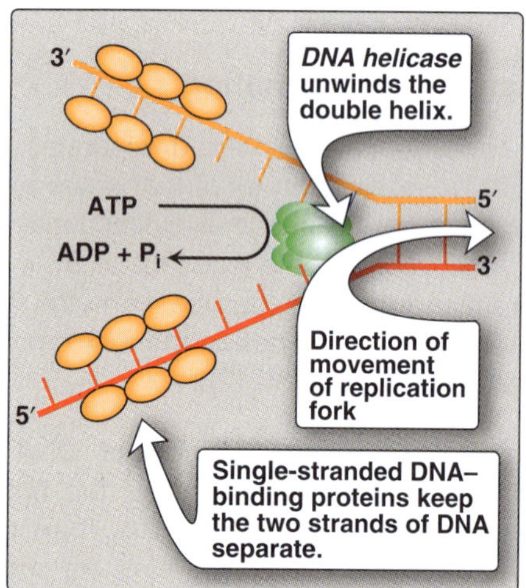

Figure 35.10 Proteins responsible for maintaining the separation of the parental strands and unwinding the double helix ahead of the advancing replication fork (≻). ADP = adenosine diphosphate; P_i = inorganic phosphate.

b. **DNA helicases:** These enzymes bind to ssDNA near the replication fork and then move into the neighboring double-stranded region, forcing the strands apart (in effect, unwinding the double helix). *Helicases* require energy provided by ATP hydrolysis (see Fig. 35.10). Unwinding at the replication fork causes supercoiling in other regions of the DNA molecule. [Note: *DnaB* is the principal *helicase* of replication in E. coli. Binding of this hexameric protein to DNA requires *DnaC*.]

c. **Single-stranded DNA–binding protein:** This protein binds to the ssDNA generated by *helicases* (see Fig. 35.10). Binding is cooperative (that is, the binding of one molecule of single-stranded binding [SSB] protein makes it easier for additional molecules of SSB protein to bind tightly to the DNA strand). The SSB proteins are not enzymes, but rather serve to shift the equilibrium between dsDNA and ssDNA in the direction of the single-stranded forms. These proteins not only keep the two strands of DNA separated in the area of the replication origin, thus providing the single-stranded template required by *polymerases*, but also protect the DNA from *nucleases* that degrade ssDNA.

2. **Solving the problem of supercoils:** As the two strands of the double helix are separated, a problem is encountered, namely, the appearance of positive supercoils in the region of DNA

ahead of the replication fork as a result of overwinding (Fig. 35.11) and negative supercoils in the region behind the fork. The accumulating positive supercoils interfere with further unwinding of the double helix. [Note: Supercoiling can be demonstrated by tightly grasping one end of a helical telephone cord while twisting the other end. If the cord is twisted in the direction of tightening the coils, the cord will wrap around itself in space to form positive supercoils. If the cord is twisted in the direction of loosening the coils, the cord will wrap around itself in the opposite direction to form negative supercoils.] To solve this problem, there is a group of enzymes called ***DNA topoisomerases,*** which are responsible for removing supercoils in the helix by transiently cleaving one or both of the DNA strands.

a. **Type I DNA topoisomerases:** These enzymes reversibly cleave one strand of the double helix. They have both strand-cutting and strand-resealing activities. They do not require ATP, but rather appear to store the energy from the phosphodiester bond they cleave, reusing the energy to reseal the strand (Fig. 35.12). Each time a transient nick is created in one DNA strand, the intact DNA strand is passed through the break before it is resealed, thus relieving (relaxing) accumulated supercoils.

Type I topoisomerases **relax negative supercoils** (that is, those that contain fewer turns of the helix than does relaxed DNA) in E. coli and both negative and positive supercoils (that is, those that contain fewer or more turns of the helix than does relaxed DNA) in many prokaryotic cells (but not E. coli) and in eukaryotic cells.

b. **Type II DNA topoisomerases:** These enzymes bind tightly to the DNA double helix and make transient breaks in both strands. The enzyme then causes a second stretch of the DNA double helix to pass through the break and, finally, reseals the break (Fig. 35.13). As a result, both negative and positive supercoils can be relieved by this ATP-requiring process. ***DNA gyrase,* a *type II topoisomerase* found in bacteria** and plants, has the unusual property of being able to introduce negative supercoils into circular DNA using energy from the hydrolysis of ATP. This facilitates the replication of DNA because the negative supercoils neutralize the positive supercoils introduced during opening of the double helix. It also aids in the transient strand separation required during transcription (see p. 598).

C. Direction of DNA replication

The *DNA polymerases* (*DNA pols*) responsible for copying the DNA templates are only able to read the parental nucleotide sequences in the 3′→5′ direction, and they synthesize the new DNA strands only in the

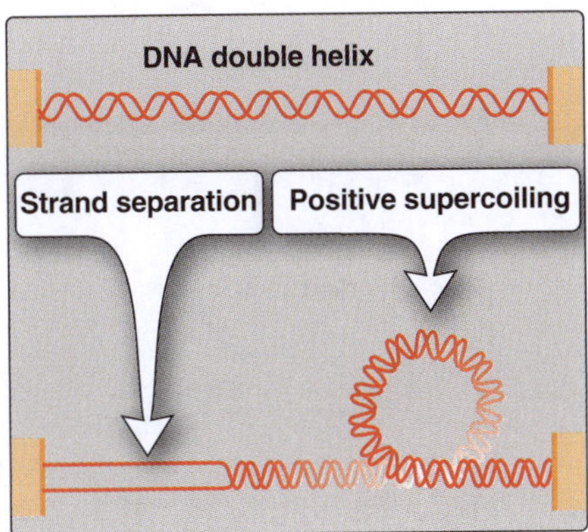

Figure 35.11 Positive supercoiling resulting from DNA strand separation.

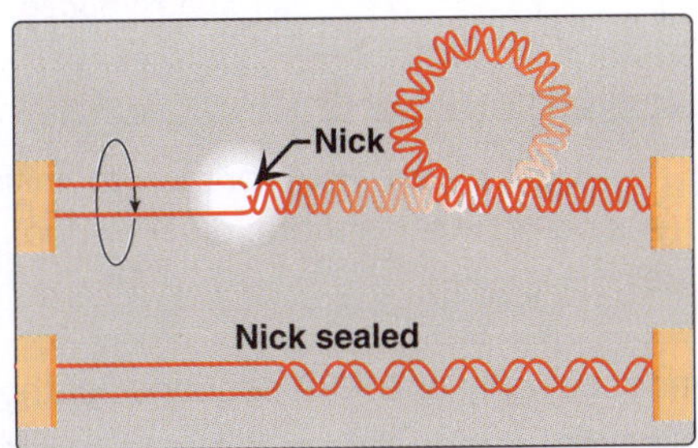

Figure 35.12 Action of *type I DNA topoisomerases*.

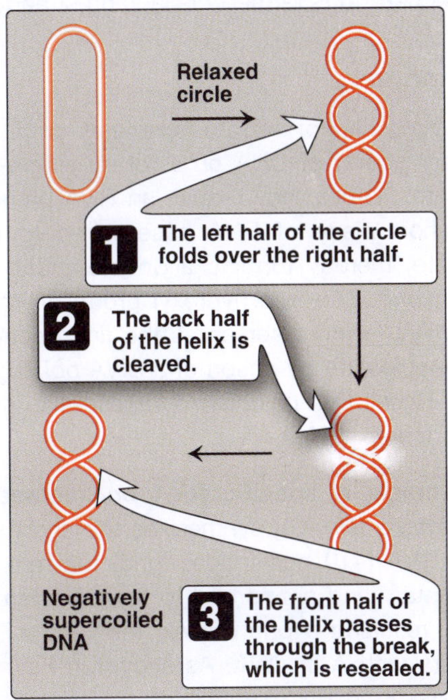

Figure 35.13 Action of *type II DNA topoisomerase*.

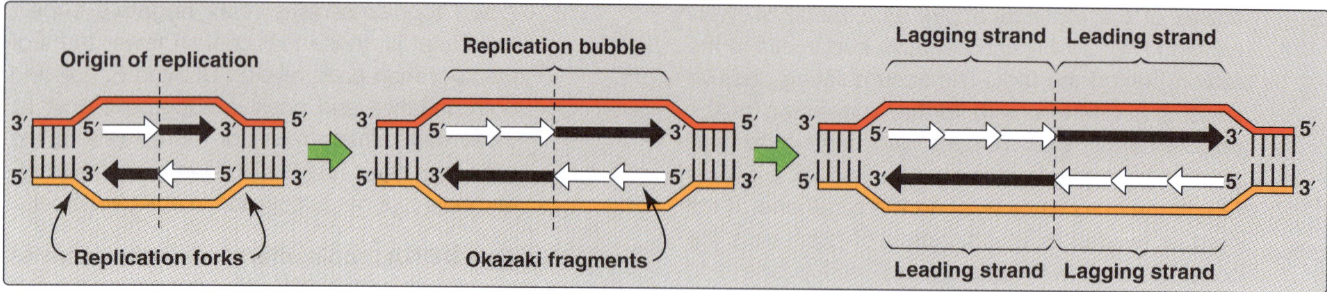

Figure 35.14 Semidiscontinuous synthesis of DNA. Black arrows = continuous synthesis; white arrows = discontinuous.

5′→3′ (antiparallel) direction. Therefore, beginning with one parental double helix, the two newly synthesized stretches of nucleotide chains must grow in opposite directions, one in the 5′→3′ direction toward the replication fork and one in the 5′→3′ direction away from the replication fork (Fig. 35.14). This feat is accomplished by a slightly different mechanism on each strand.

1. **Leading strand:** The strand that is being copied in the direction of the advancing replication fork is synthesized continuously and is called the leading strand.

2. **Lagging strand:** The strand that is being copied in the direction away from the replication fork is synthesized discontinuously, with small fragments of DNA being copied near the replication fork. These short stretches of discontinuous DNA, termed **Okazaki fragments,** are eventually joined (ligated) by *ligase* to become a single, continuous strand. The new strand of DNA produced by this mechanism is termed the lagging strand.

D. RNA primer

DNA pols cannot initiate synthesis of a complementary strand of DNA on a totally single-stranded template. Rather, they require an RNA primer, which is a short piece of RNA base-paired to the DNA template, thereby forming a double-stranded DNA–RNA hybrid. The free hydroxyl group on the 3′-end of the RNA primer serves as the first acceptor of a deoxynucleotide by action of a *DNA pol* (Fig. 35.15). [Note: Recall that *glycogen synthase* also requires a primer (see p. 225).]

1. **Primase:** A **specific *RNA polymerase*, called *primase* (DnaG),** synthesizes the short stretches of RNA (~10 nucleotides long) that are complementary and antiparallel to the DNA template. In the resulting hybrid duplex, the U (uracil) in RNA pairs with A in DNA. As shown in Figure 35.16, these short RNA sequences are constantly being synthesized at the replication fork on the lagging strand, but only one RNA sequence at the origin of replication is required on the leading strand. The substrates for this process are 5′-ribonucleoside

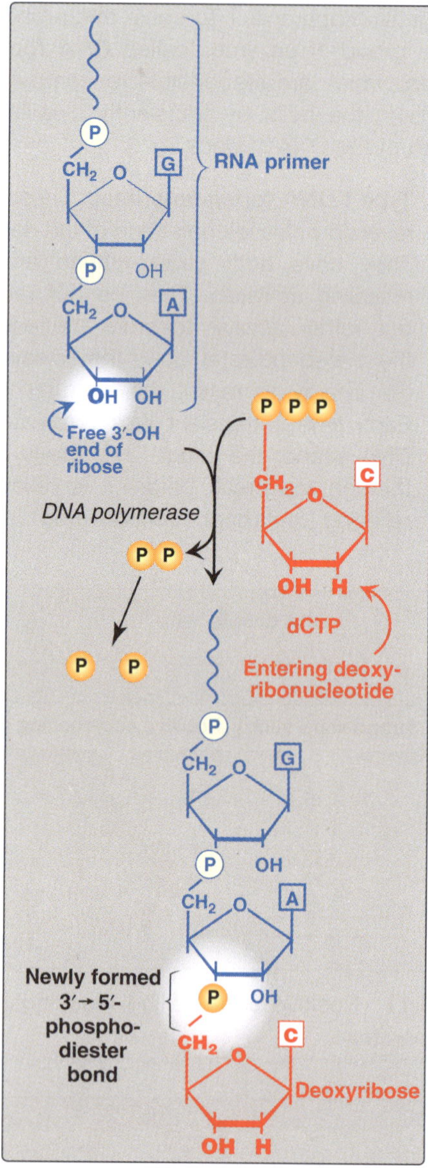

Figure 35.15 Use of an RNA primer to initiate DNA synthesis. (P) and (P) = phosphate; dCTP = deoxycytidine triphosphate.

triphosphates, and pyrophosphate is released as each ribonucleoside monophosphate is added through formation of a 3′→5′-phosphodiester bond. [Note: The RNA primer is later removed, as described in F. below.]

2. **Primosome:** The addition of *primase* converts the prepriming complex of proteins required for DNA strand separation (see p. 599) to a primosome. The primosome makes the RNA primer required for leading-strand synthesis and initiates Okazaki fragment formation in discontinuous lagging-strand synthesis. As with DNA synthesis, the direction of synthesis of the primer is 5′→3′.

E. Chain elongation

Prokaryotic (and eukaryotic) *DNA pols* elongate a new DNA strand by adding deoxyribonucleotides, one at a time, to the 3′-end of the growing chain (see Fig. 35.16). The sequence of nucleotides that are added is dictated by the base sequence of the template strand with which the incoming nucleotides are paired.

1. **DNA polymerase III:** DNA chain elongation is catalyzed by the multisubunit enzyme, *DNA pol III*. Using the 3′-hydroxyl group of the RNA primer as the acceptor of the first deoxyribonucleotide, *DNA pol III* begins to add nucleotides along the single-stranded template that specifies the sequence of bases in the newly synthesized chain. *DNA pol III* is a highly processive enzyme (that is, it remains bound to the template strand as it moves along and does not diffuse away and then rebind before adding each new nucleotide). The processivity of *DNA pol III* is the result of the β subunits of the holoenzyme forming a ring that encircles and moves along the template strand of the DNA, thus serving as **a sliding DNA clamp**. [Note: Clamp formation is facilitated by a protein complex, the clamp loader, and ATP hydrolysis.] The new (daughter) strand grows in the 5′→3′ direction, antiparallel to the parental strand (see Fig. 35.16). The nucleotide substrates are 5′-deoxyribonucleoside triphosphates. Pyrophosphate (PP_i) is released when each new deoxynucleoside monophosphate is added to the free 3′-hydroxyl group of the growing chain through a 3′→5′-phosphodiester bond (see Fig. 35.15). Hydrolysis of PP_i to 2 P_i by *pyrophosphatase* means that a total of two high-energy bonds are used to drive the addition of each deoxynucleotide.

> The production of PP_i with subsequent hydrolysis to 2 P_i is a common theme in biochemistry. Removal of the PP_i product drives a reaction in the forward direction, making it essentially irreversible.

All four substrates (deoxyadenosine triphosphate [dATP], deoxythymidine triphosphate [dTTP], deoxycytidine triphosphate [dCTP], and deoxyguanosine triphosphate [dGTP]) must be present for DNA elongation to occur. If one of the four is in short supply, DNA synthesis stops when that nucleotide is depleted.

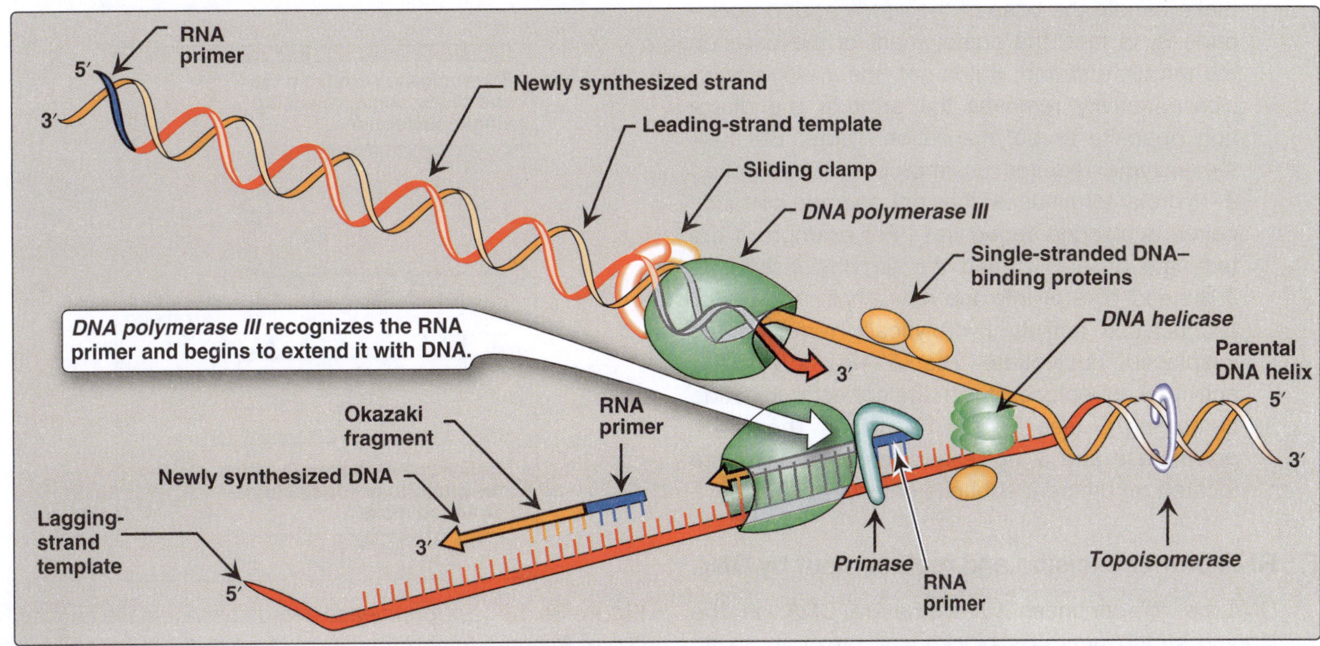

Figure 35.16 Elongation of the leading and lagging strands. [Note: The DNA sliding clamp is not shown for the lagging strand.]

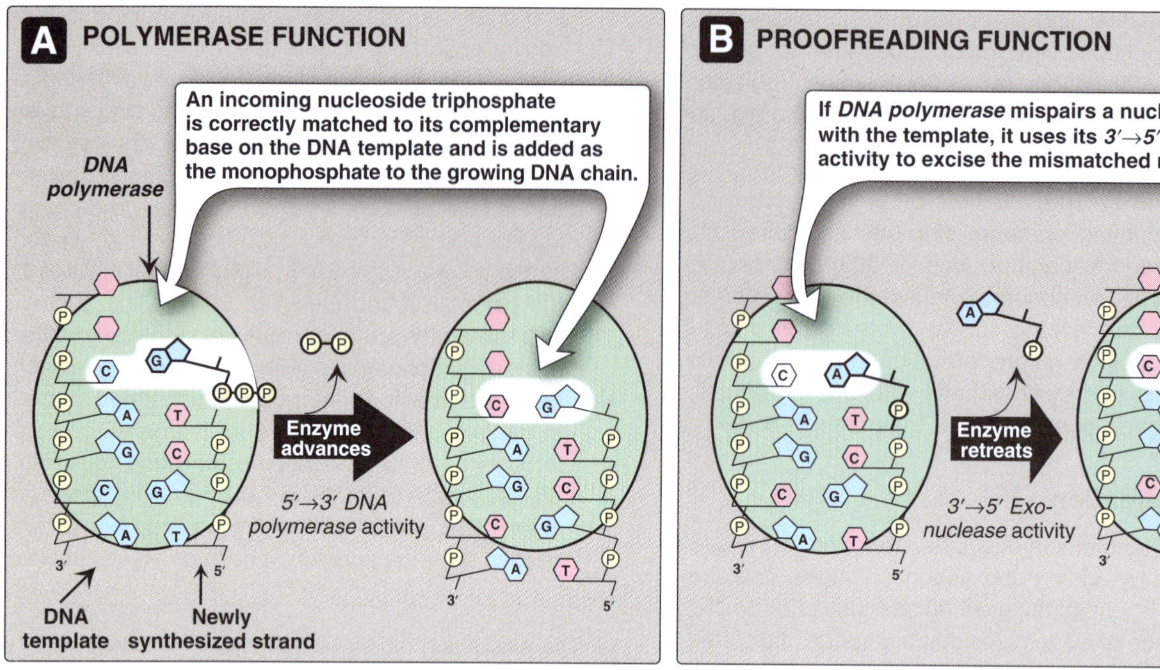

Figure 35.17 *3′→5′ Exonuclease* activity enables *DNA polymerase III* to proofread the newly synthesized DNA strand.

2. **Proofreading newly synthesized DNA:** It is highly important for the survival of an organism that the nucleotide sequence of DNA be replicated with as few errors as possible. Misreading of the template sequence could result in deleterious, perhaps lethal, mutations. To insure replication fidelity, ***DNA pol III* has a proofreading activity (*3′→5′ exonuclease*, Fig. 35.17) in addition to its *5′→3′ polymerase* activity**. As each nucleotide is added to the chain, *DNA pol III* checks to make certain the base of the newly added nucleotide is, in fact, the complement of the base on the template strand. If it is not, the *3′→5′ exonuclease* activity removes the error in the direction opposite to polymerization. [Note: Because the enzyme requires an improperly base-paired 3′-hydroxy terminus, it does not degrade correctly paired nucleotide sequences.] For example, if the template base is C and the enzyme inserts an A instead of a G into the new chain, the *3′→5′ exonuclease* activity hydrolytically removes the misplaced nucleotide. The *5′→3′ polymerase* activity then replaces it with the correct nucleotide containing G (see Fig. 35.17). [Note: The *5′→3′ polymerase* and *3′→5′ exonuclease* domains are located on different subunits of *DNA pol III*.]

F. RNA primer excision and replacement by DNA

DNA pol III continues to synthesize DNA on the lagging strand until it is blocked by proximity to an RNA primer. When this occurs, the RNA is excised and the gap filled by *DNA pol I*.

1. ***5′→3′* Exonuclease activity:** In addition to having the *5′→3′ polymerase* activity that synthesizes DNA and the *3′→5′ exonuclease* activity that proofreads the newly synthesized DNA like *DNA pol III*, monomeric *DNA pol I* also has a *5′→3′ exonuclease* activity that is able to hydrolytically remove the RNA primer. [Note: *Exonucleases* remove nucleotides from the end of the DNA chain, rather than cleaving the chain internally as do *endonucleases* (Fig. 35.18).] First, *DNA pol I* locates the space

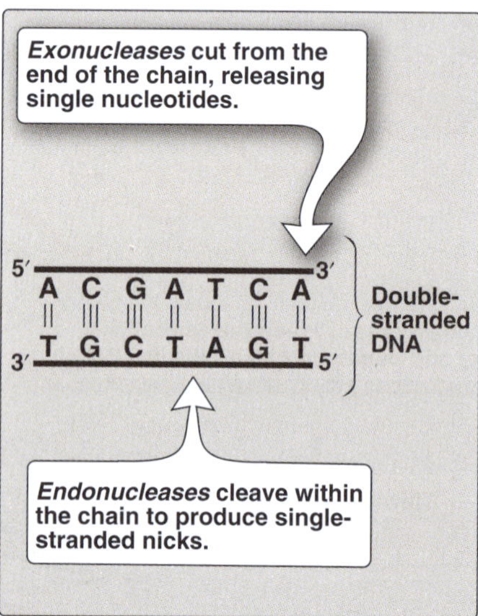

Figure 35.18 *Endonuclease* versus *exonuclease* activity. [Note: *Restriction endonucleases* (see p. 672) cleave both strands.] T = thymine; A = adenine; C = cytosine; G = guanine.

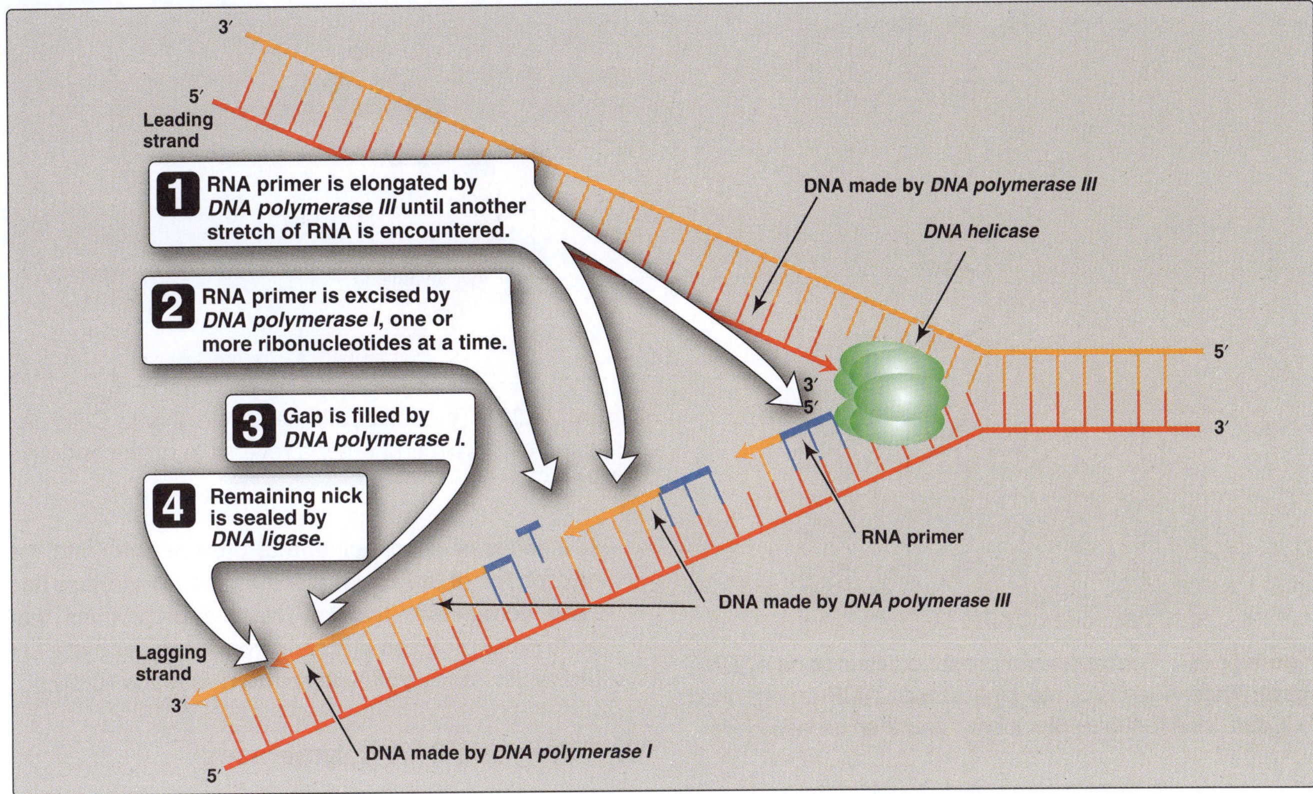

Figure 35.19 Removal of RNA primer and filling of the resulting gaps by *DNA polymerase I*.

(nick) between the 3′-end of the DNA newly synthesized by *DNA pol III* and the 5′-end of the adjacent RNA primer. Next, *DNA pol I* hydrolytically removes the RNA nucleotides ahead of itself, moving in the 5′→3′ direction (*5′→3′ exonuclease* activity). As it removes ribonucleotides, *DNA pol I* replaces them with deoxyribonucleotides, synthesizing DNA in the 5′→3′ direction (*5′→3′ polymerase* activity). As it synthesizes the DNA, it also proofreads using its *3′→5′ exonuclease* activity to remove errors. This removal/synthesis/proofreading continues until the RNA primer is totally degraded, and the gap is filled with DNA (Fig. 35.19). [Note: *DNA pol I* uses its *5′→3′ polymerase* activity to fill in gaps generated during most types of DNA repair (see p. 604).]

2. **Comparison of *5′→3′* and *3′→5′ exonuclease* activities:** The *5′→3′ exonuclease* activity of *DNA pol I* allows the *polymerase*, moving 5′→3′, to hydrolytically remove one or more nucleotides at a time from the 5′-end of the ~10 nucleotide–long RNA primer. In contrast, the *3′→5′ exonuclease* activity of *DNA pol I* and *pol III* allows these *polymerases*, moving 3′→5′, to hydrolytically remove one misplaced nucleotide at a time from the 3′-end of a growing DNA strand, increasing the fidelity of replication such that newly replicated DNA has one error per 10^7 nucleotides.

G. DNA ligase

The final phosphodiester linkage between the 5′-phosphate group on the DNA synthesized by *DNA pol III* and the 3′-hydroxyl group on the DNA made by *DNA pol I* is catalyzed by *DNA ligase* (Fig. 35.20). The joining of these two stretches of DNA requires energy, which in most organisms is provided by the cleavage of ATP to adenosine monophosphate + PP_i.

H. Termination

Replication termination in E. coli is mediated by sequence-specific binding of the protein Tus (terminus utilization substance) to replication termination (ter) sites on the DNA, stopping the movement of the replication fork.

IV. EUKARYOTIC DNA REPLICATION

The process of eukaryotic DNA replication closely follows that of prokaryotic DNA synthesis. Some differences, such as the multiple origins of replication in eukaryotic cells versus single origins of replication in prokaryotes, have already been noted. Eukaryotic origin recognition proteins, ssDNA-binding proteins, and ATP-dependent *DNA helicases* have been identified, and their functions are analogous to those of the prokaryotic proteins previously discussed. In contrast, RNA primers are removed

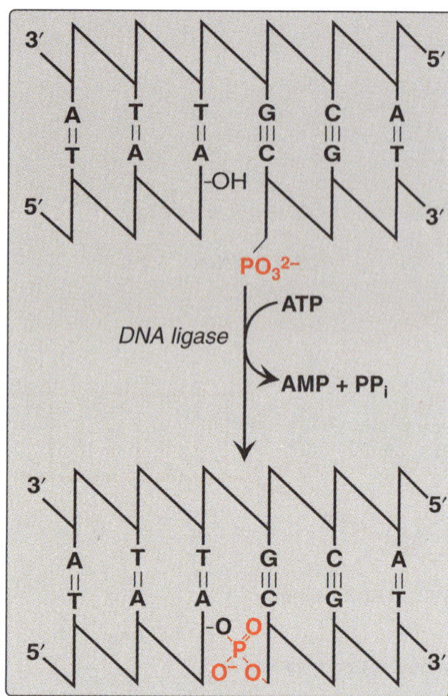

Figure 35.20 Formation of a phosphodiester bond by *DNA ligase*. [Note: Adenosine monophosphate (AMP) is first linked to *ligase*, then to the 5′-phosphate, and then released.]

FUNCTION	PROTEIN (S)
Origin recognition	ORC
Helicase activity	*MCM*
ssDNA protection	RPA
Primer synthesis	*Pol a/primase*
Sliding clamp	PCNA
Primer removal	*RNase H, FEN1*

Figure 35.21 Proteins and their function in eukaryotic replication. ORC = origin recognition complex; *MCM* = *minichromosome maintenance* (*complex*); RPA = replication protein A; PCNA = proliferating cell nuclear antigen; *FEN* = *flap endonuclease*.

by ***RNase H*** **and *flap endonuclease 1* (*FEN1*)** rather than by a *DNA pol* (Fig. 35.21).

A. Eukaryotic cell cycle

The events surrounding eukaryotic DNA replication and cell division (mitosis) are coordinated to produce the cell cycle (Fig. 35.22). The period preceding replication is called the G_1 phase (Gap 1). DNA replication occurs during **the S (synthesis) phase**. Following DNA synthesis, there is another phase **(G_2, or Gap 2)** before mitosis (M). Cells that have stopped dividing, such as mature T lymphocytes, are said to have gone out of the cell cycle into the **G_0 phase**. Such quiescent cells can be stimulated to reenter the G_1 phase to resume division. [Note: The cell cycle is controlled

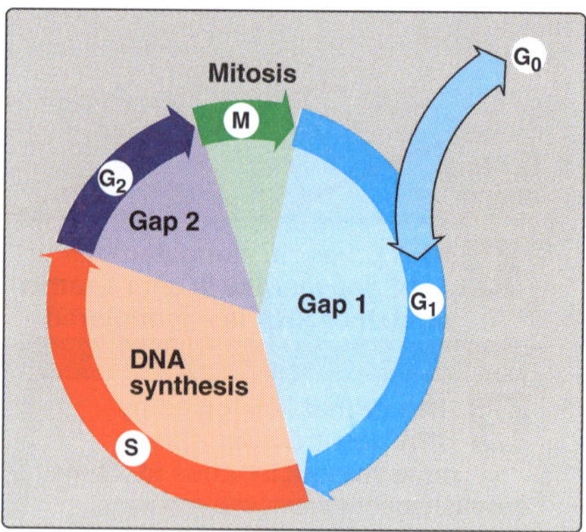

Figure 35.22 The eukaryotic cell cycle. [Note: Cells can leave the cell cycle and enter a reversible quiescent state called G_0.]

at a series of checkpoints that prevent entry into the next phase of the cycle until the preceding phase has been completed. Two key classes of proteins that control the progress of a cell through the cell cycle are the cyclins and ***cyclin-dependent kinases* (*Cdk*).**]

B. Eukaryotic DNA polymerases

At least five high-fidelity eukaryotic *DNA pols* have been identified and categorized on the basis of molecular weight, cellular location, sensitivity to inhibitors, and the templates or substrates on which they act. They are designated by Greek letters rather than by Roman numerals (Fig. 35.23).

1. **Pol α:** *Pol α* is a multisubunit enzyme. One subunit has *primase* activity, which initiates strand synthesis on the leading strand and at the beginning of each Okazaki fragment on the lagging strand. The *primase* subunit synthesizes a short RNA primer that is extended by the *5′→3′*

POLY-MERASE	FUNCTION	PROOF-READING*
Pol α (alpha)	• Contains *primase* • Initiates DNA synthesis	–
Pol β (beta)	• Repair	–
Pol δ (delta)	• Elongates Okazaki fragments of the lagging strand	+
Pol ε (epsilon)	• Elongates the leading strand	+
Pol γ (gamma)	• Replicates mitochondrial DNA	+

Figure 35.23 Activities of eukaryotic *DNA polymerases* (*pol*). [Note: The asterisk (*) denotes *3′→5′ exonuclease* activity.]

polymerase activity of *pol α*, generating a short piece of DNA. [Note: *Pol α* is also referred to as *pol α/primase*.]

2. **Pol ε and pol δ:** *Pol ε* is recruited to complete DNA synthesis on the leading strand, whereas *pol δ* elongates the Okazaki fragments of the lagging strand, each using *3′→5′ exonuclease* activity to proofread the newly synthesized DNA. [Note: *DNA pol ε* associates with proliferating cell nuclear antigen (PCNA), a protein that serves as a sliding DNA clamp in much the same way the β subunits of *DNA pol III* do in E. coli, thus insuring high processivity.]

3. **Pol β and pol γ:** *Pol β* is involved in gap filling in DNA repair. *Pol γ* replicates mitochondrial DNA.

C. Telomeres

Telomeres are complexes of DNA plus proteins (collectively known as shelterin) located at the ends of linear chromosomes. They maintain the structural integrity of the chromosome, preventing attack by *nucleases*, and allow repair systems to distinguish a true end from a break in dsDNA. In humans, telomeric DNA consists of several thousand tandem repeats of a noncoding hexameric sequence, AGGGTT, base-paired to a complementary region containing C and A. The G-rich strand is longer than its C-rich complement, leaving ssDNA a few hundred nucleotides in length at the 3′-end. The single-stranded region is thought to fold back on itself, forming a loop structure that is stabilized by protein.

1. **Telomere shortening:** Eukaryotic cells face a special problem in replicating the ends of their linear DNA molecules. Following removal of the RNA primer from the extreme 5′-end of the lagging strand, there is no way to fill in the remaining gap with DNA. Consequently, in most normal human somatic cells, telomeres shorten with each successive cell division. Once telomeres are shortened beyond some critical length, the cell is no longer able to divide and is said to be senescent. **In germ cells and stem cells, as well as in cancer cells, telomeres do not shorten** and the cells do not senesce. This is a result of the ribonucleoprotein *telomerase*, which maintains telomeric length in these cells.

2. **Telomerase:** This complex contains a **protein (Tert) that acts as a *reverse transcriptase*** and a short piece of RNA (Terc) that acts as a template. The C-rich RNA template base-pairs with the G-rich, single-stranded 3′-end of telomeric DNA (Fig. 35.24). The *reverse transcriptase* uses the RNA template to synthesize DNA in the usual 5′→3′ direction, extending the already longer 3′-end.

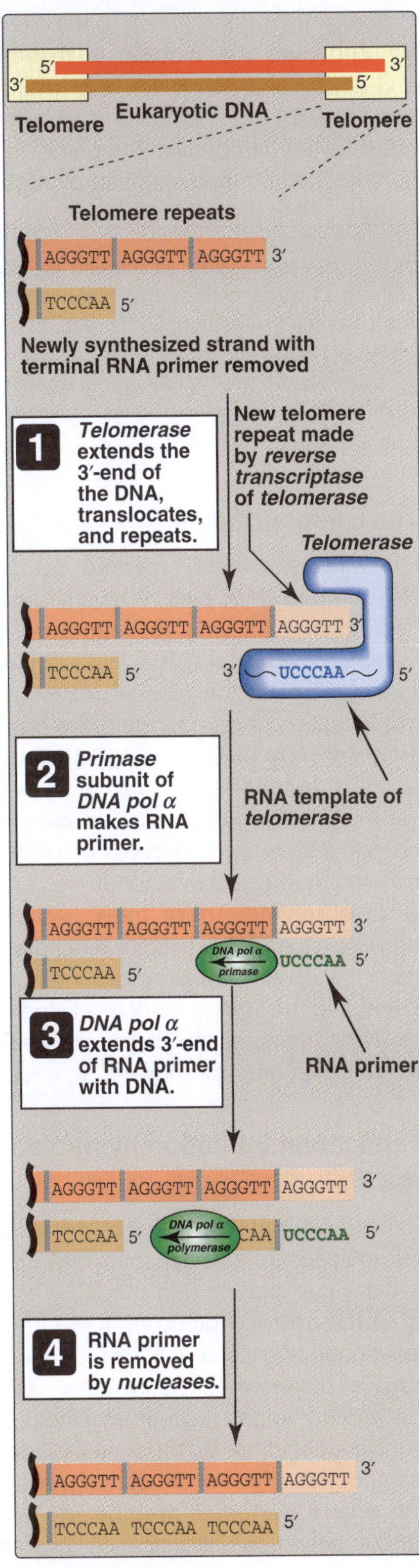

Figure 35.24 Mechanism of action of *telomerase*, a ribonucleoprotein. T = thymine; A = adenine; C = cytosine; G = guanine; *pol* = *polymerase*.

Telomerase then translocates to the newly synthesized end, and the process is repeated. Once the G-rich strand has been lengthened, *primase* activity of *DNA pol α* can use it as a template to synthesize an RNA primer. The primer is extended by *DNA pol α* and then removed by *nucleases*.

Telomeres may be viewed **as mitotic clocks** in that their length in most cells is inversely related to the number of times the cells have divided. The study of telomeres provides insight into the biology of normal aging, diseases of premature aging (the progerias), and cancer.

D. Reverse transcriptases

As seen with *telomerase*, ***reverse transcriptases* are RNA-directed *DNA pols***. A *reverse transcriptase* is involved in the replication of retroviruses, such as human immunodeficiency virus (HIV). These viruses carry their genome in the form of ssRNA molecules. Following infection of a host cell, the viral enzyme *reverse transcriptase* uses the viral RNA as a template for the 5′→3′ synthesis of viral DNA, which then becomes integrated into host chromosomes. *Reverse transcriptase* activity is also seen with transposons, DNA elements that can move about the genome (see p. 665). In eukaryotes, most transposons are transcribed to RNA, the RNA is used as a template for DNA synthesis by a *reverse transcriptase* encoded by the transposon, and the DNA is randomly inserted into the genome. [Note: Transposons that involve an RNA intermediate are called retrotransposons or retroposons.]

E. DNA replication inhibition by nucleoside analogs

DNA chain growth can be blocked by the incorporation of certain nucleoside analogs that have been modified on the sugar portion (Fig. 35.25). For example, removal of the hydroxyl group from the 3′-carbon of the deoxyribose ring as in 2′,3′-dideoxyinosine ([ddI] also known as didanosine), or conversion of the deoxyribose to another sugar, such as arabinose, prevents further chain elongation. By blocking DNA replication, these compounds slow the division of rapidly growing cells and viruses. **Cytosine arabinoside** (cytarabine, or araC) has been used in anticancer chemotherapy, whereas **adenine arabinoside** (vidarabine, or araA) is an antiviral agent. Substitution on the sugar moiety, as seen in **azidothymidine (AZT)**, also called zidovudine (ZDV), also terminates DNA chain elongation. [Note: These drugs are generally supplied as nucleosides, which are then converted to nucleotides by cellular *kinases*.]

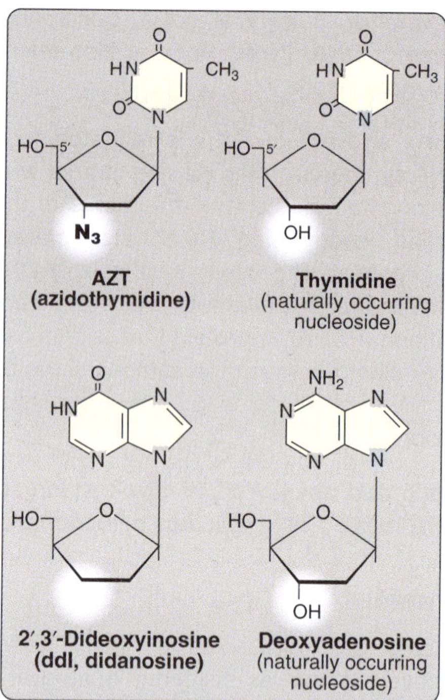

Figure 35.25 Examples of nucleoside analogs that lack a 3′-hydroxyl group. [Note: The ddI is converted to its active form (dideoxy ATP).]

V. EUKARYOTIC DNA ORGANIZATION

A typical (diploid) human somatic cell contains 46 chromosomes, whose total DNA is ~2 m long! It is difficult to imagine how such a large amount of genetic material can be effectively packaged into a volume the size of a cell nucleus so that it can be efficiently replicated and its genetic information expressed. To do so requires the interaction of DNA with a large number of proteins, each of which performs a specific function in the ordered packaging of these long molecules of DNA. Eukaryotic DNA is associated with tightly bound basic proteins, called histones. These serve to order the DNA into fundamental structural units, called nucleosomes, which **resemble beads on a string**. Nucleosomes are further arranged into increasingly more complex structures that organize and condense the long DNA molecules into chromosomes that can be segregated during cell division. [Note: The complex of DNA and protein found inside the nuclei of eukaryotic cells is called **chromatin**.]

A. Histones and nucleosome formation

There are five classes of histones, designated H1, H2A, H2B, H3, and H4. These small, evolutionally conserved proteins are positively charged at physiologic pH as a result of their high content of lysine and arginine. Because of their positive charge, they form ionic bonds with negatively charged DNA. Histones, along with ions such as Mg^{2+}, help neutralize the negatively charged DNA phosphate groups.

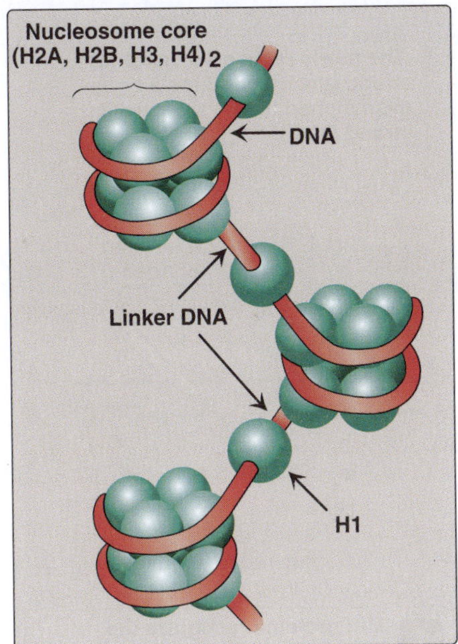

Figure 35.26 Organization of human DNA, illustrating the structure of nucleosomes. H = histone.

1. **Nucleosomes:** Two molecules each of H2A, H2B, H3, and H4 form the octameric core of the individual **nucleosome "beads**." Around this structural core, a segment of dsDNA is wound nearly twice (Fig. 35.26). Winding eliminates a helical turn, causing negative supercoiling. [Note: The N-terminal ends of these histones can be acetylated, methylated, or phosphorylated. These reversible covalent modifications influence how tightly the histones bind to the DNA, thereby affecting the expression of specific genes. Histone modification is an example of **epigenetics, or heritable changes in gene expression caused without alteration of the nucleotide sequence**.] Neighboring nucleosomes are joined by linker DNA ~50 bp long. H1 is not found in the nucleosome core, but instead binds to the linker DNA chain between the nucleosome beads. H1 is the most tissue specific and species specific of the histones. It facilitates the packing of nucleosomes into more compact structures.

2. **Higher levels of organization:** Nucleosomes can be packed more tightly (stacked) to form a nucleofilament. This structure assumes the shape of a coil, often referred to as a 30-nm fiber. The fiber is organized into loops that are anchored by a nuclear scaffold containing several proteins. Additional levels of organization lead to the final chromosomal structure (Fig. 35.27).

B. Nucleosome fate during DNA replication

Parental nucleosomes are disassembled to allow access to DNA during replication. Once DNA is synthesized, nucleosomes form rapidly. Their histone proteins come both from de novo synthesis and from the transfer of parental histones.

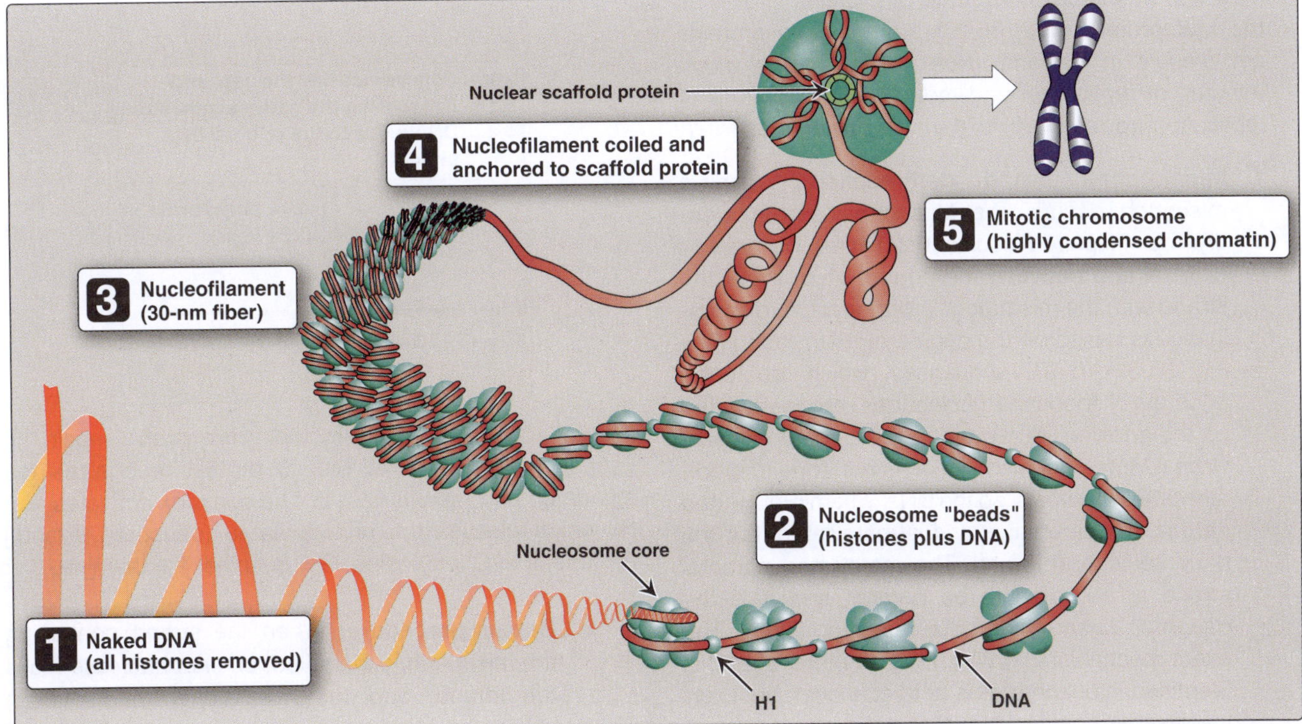

Figure 35.27 Structural organization of eukaryotic DNA. [Note: A 104 linear compaction is seen from 1–5.] H = histone.

VI. DNA REPAIR

Despite the elaborate proofreading system employed during DNA synthesis, errors (including incorrect base-pairing or insertion of one to a few extra nucleotides) can occur. In addition, DNA is constantly being subjected to environmental insults that cause the alteration or removal of nucleotide bases. The damaging agents can be either chemicals (for example, nitrous acid, which can deaminate bases) or radiation (for example, nonionizing ultraviolet [UV] radiation, which can fuse two pyrimidines adjacent to each other in the DNA, and high-energy ionizing radiation, which can cause double-strand breaks). Bases are also altered or lost spontaneously from mammalian DNA at a rate of many thousands per cell per day. If the damage is not repaired, a permanent change (mutation) is introduced that can result in any of a number of deleterious effects, including loss of control over the proliferation of the mutated cell, leading to cancer. Luckily, cells are remarkably efficient at repairing damage done to their DNA. Most of the repair systems involve recognition of the damage (lesion) on the DNA, removal or excision of the damage, replacement or filling the gap left by excision using the sister strand as a template for DNA synthesis, and ligation. These excision repair systems remove one to tens of nucleotides. [Note: Repair synthesis of DNA can occur outside of the S phase.]

A. Mismatch repair

Sometimes replication errors escape the proofreading activity during DNA synthesis, causing a mismatch of one to several bases. In E. coli, mismatch repair (MMR) is mediated by a group of proteins known as the Mut proteins (Fig. 35.28). Homologous proteins are present in humans. [Note: MMR occurs within minutes of replication and reduces the error rate of replication from 1 in 10^7 to 1 in 10^9 nucleotides.]

1. **Mismatched strand identification:** When a mismatch occurs, the Mut proteins that identify the mispaired nucleotide(s) must be able to discriminate between the correct strand and the strand with the mismatch. In prokaryotes, discrimination is based on the degree of methylation (see Fig. 35.32). GATC sequences, which are found once every thousand nucleotides, are methylated on the adenine (A) residue by *DNA adenine methylase* (*DAM*). This methylation is not done immediately after synthesis, so the DNA is hemimethylated (that is, the parental strand is methylated, but the daughter strand is not). The methylated parental strand is assumed to be correct, and it is the daughter strand that gets repaired. [Note: The exact mechanism by which the daughter strand is identified in eukaryotes is not yet known, but likely involves recognition of nicks in the newly synthesized strand.]

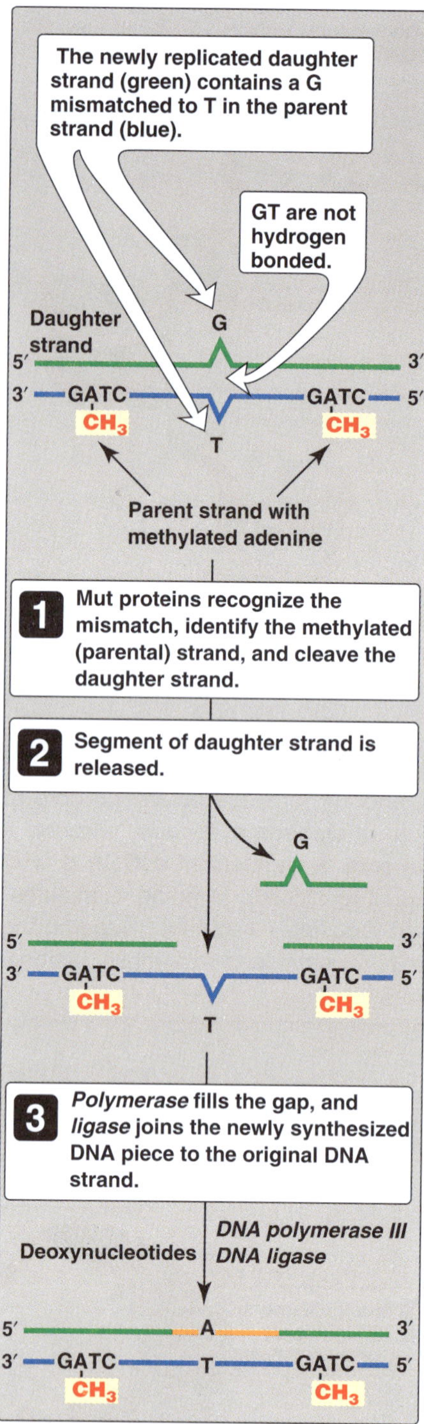

Figure 35.28 Methyl-directed mismatch repair in Escherichia coli. [Note: Mut S protein recognizes the mismatch and recruits Mut L. The complex activates Mut H, which cleaves the unmethylated (daughter) strand.] A = adenine; C = cytosine; G = guanine; T = thymine.

2. **Repair procedure:** When the strand containing the mismatch is identified, an *endonuclease* nicks the strand, and the mismatched nucleotide(s) is/are removed by an *exonuclease*. Additional nucleotides at the 5′- and 3′-ends of the mismatch

are also removed. The gap left by removal of the nucleotides is filled, using the sister strand as a template, by a *DNA pol*, typically *DNA pol III*. The 3′-hydroxyl of the newly synthesized DNA is joined to the 5′-phosphate of the remaining stretch of the original DNA strand by *DNA ligase*.

Mutation to the proteins involved in MMR in humans is associated with hereditary nonpolyposis colorectal cancer (HNPCC), also known as Lynch syndrome. Although HNPCC confers an increased risk for developing colon cancer (as well as other cancers), only about 5% of all colon cancer is the result of mutations in MMR.

B. Nucleotide excision repair

Exposure of a cell to UV radiation can result in the covalent joining of two adjacent pyrimidines (usually thymines), producing a dimer. These intrastrand cross-links prevent *DNA pol* from replicating the DNA strand beyond the site of dimer formation. Thymine dimers are excised in bacteria by UvrABC proteins in a process known as nucleotide excision repair (NER), as illustrated in Figure 35.29. A related pathway is present in humans (see 2. below). [Note: Transcription-coupled repair, a type of NER, fixes DNA lesions encountered during RNA synthesis.]

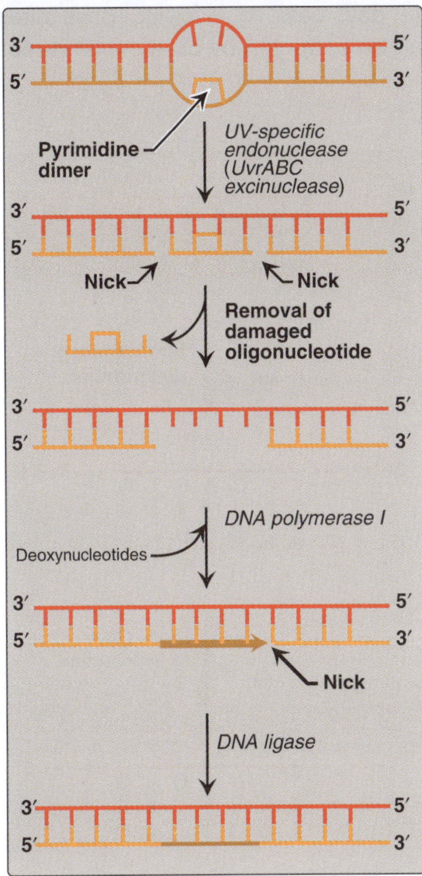

Figure 35.29 Nucleotide excision repair of pyrimidine dimers in Escherichia coli DNA. UV = ultraviolet.

1. **Recognition and excision of UV-induced dimers:** A *UV-specific endonuclease* (called *UvrABC excinuclease*) recognizes the bulky dimer and cleaves the damaged strand on both the 5′-side and 3′-side of the lesion. A short oligonucleotide containing the dimer is excised, leaving a gap in the DNA strand. This gap is filled in using a *DNA pol I* and *DNA ligase*. NER occurs throughout the cell cycle.

2. **UV radiation and cancer:** Pyrimidine dimers can be formed in the skin cells of humans exposed to UV radiation in unfiltered sunlight (see Fig. 35.32). In the rare genetic disease xeroderma pigmentosum (XP), the cells cannot repair the damaged DNA, resulting in extensive accumulation of mutations and, consequently, early and numerous skin cancers (Fig. 35.30). XP can be caused by defects in any of the several genes that code for the XP proteins required for NER of UV damage in humans.

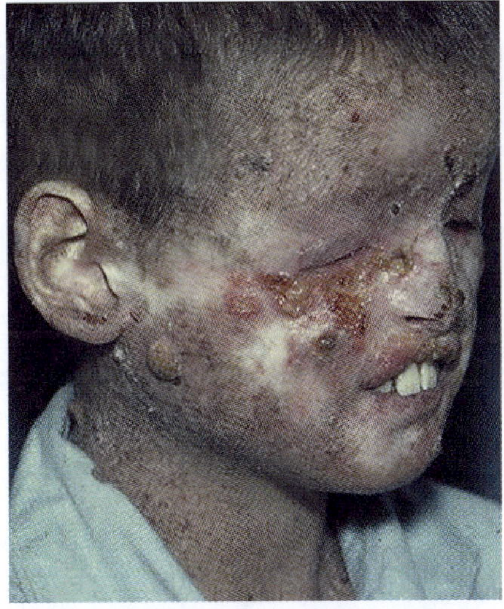

Figure 35.30 Patient with xeroderma pigmentosum.

C. Base excision repair

DNA bases can be altered, either spontaneously, as is the case with cytosine, which slowly undergoes deamination (the loss of its amino group) to form uracil, or by the action of deaminating or alkylating compounds. For example, nitrous acid, which is formed by the cell from precursors such as the nitrates, deaminates cytosine, adenine (to hypoxanthine), and guanine (to xanthine). Dimethyl sulfate can alkylate (methylate) adenine. Bases can also be lost spontaneously. For example, ~10,000 purine bases are lost this way per cell per day. Lesions involving base alterations or loss can be corrected by base excision repair ([BER], Fig. 35.31).

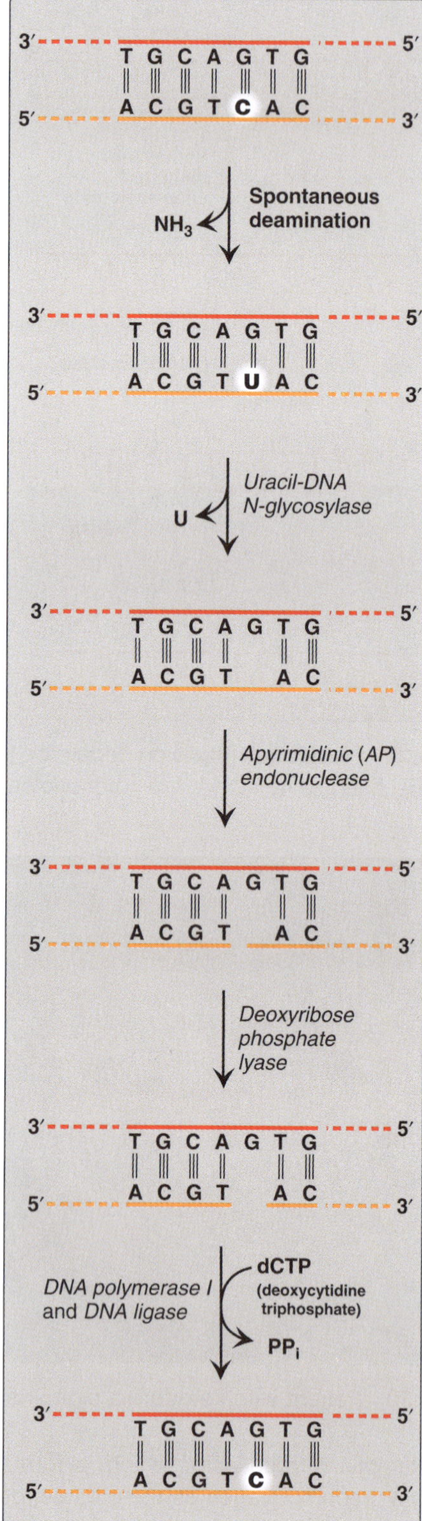

Figure 35.31 Correction of base alterations by base excision repair. C = cytosine; U = uracil; NH_3 = ammonia; PP_i = pyrophosphate.

1. **Abnormal base removal:** In BER, abnormal bases, such as uracil, which can occur in DNA by either deamination of cytosine or improper use of dUTP instead of dTTP during DNA synthesis, are recognized by specific *DNA glycosylases* (see Fig. 35.32) that hydrolytically cleave them from the deoxyribose-phosphate backbone of the strand. This leaves an apyrimidinic site, or apurinic if a purine was removed, both referred to as AP sites.
2. **AP site recognition and repair:** Specific *AP endonucleases* recognize that a base is missing and initiate the process of excision and gap filling by making an endonucleolytic cut just to the 5′-side of the AP site. A *deoxyribose phosphate lyase* removes the single, base-free, sugar phosphate residue. *DNA pol I* and *DNA ligase* complete the repair process.

D. Double-strand break repair

Ionizing radiation, chemotherapeutic agents such as doxorubicin, and oxidative free radicals (see p. 795) can cause double-strand breaks in DNA that can be lethal to the cell. [Note: Such breaks also occur naturally during genetic recombination.] dsDNA breaks cannot be corrected by the previously described strategy of excising the damage on one strand and using the undamaged strand as a template for replacing

Mechanism	Error	Repaired by
Mismatch repair	Error in copying 1-5 bases	Methylated DNA identified, digestion by *exonucleases* and replacement
Nucleotide excision repair	Damage to a segment of DNA because of UV radiation, chemicals, etc.	Approx. 30 bp nucleotide removed by UvrABC *excinuclease* and then replaced
Base excision repair	Single base damage because of radiation, chemicals, etc.	Base excision by *DNA-glycosylase*, recognition by AP *endonucleases* and repair
Double strands break	Both strands of DNA break due to ionizing radiations, chemotherapy, oxidative-free radicals, etc.	Approximation, unwinding, ligation by NHEJ or homologous recombination

Figure 35.32 Salient features of DNA repair mechanisms.

the missing nucleotide(s). Instead, they are repaired by one of two systems. The first is **nonhomologous end joining (NHEJ),** in which a group of proteins **(Ku and DNA-dependent *protein kinases*)** mediates the recognition, processing, and ligation of the ends of two DNA fragments. However, some DNA is lost in the process. Consequently, NHEJ is error prone and mutagenic. Defects in NHEJ are associated with a predisposition to cancer and immunodeficiency syndromes. The second repair system, homologous recombination (HR), uses the enzymes that normally perform genetic recombination between homologous chromosomes during meiosis. This system is much less error prone ("error-free") than NHEJ because any DNA that was lost is replaced using homologous DNA as a template. HR occurs in late S and G_2 of the cell cycle, whereas NHEJ can occur anytime (Fig. 35.32). [Note: Mutations to the proteins BRCA1 or BRCA2 (breast cancer 1 or 2), which are involved in HR, increase the risk for developing breast and ovarian cancer.]

VII. Medical Perspectives

Based on the knowledge gained in this chapter, the following medical perspectives must be understood.

Medications

- **Antibiotics**: *DNA gyrase* is a bacterial *topoisomerase* which acts on closed circular DNA of bacteria. Therefore, when drugs are given specifically against bacterial *DNA gyrase* it does not harm the human DNA enzymes. Some examples are the common antibiotics **ciprofloxacin, norfloxacin and nalidixic acid.**
- **Anti-cancer agents**: Anti-cancer drugs, when acting against eukaryotic human DNA, cannot differentiate between normal and abnormal cells. Therefore, chemotherapeutic drugs have many side effects.
- Etoposide and Adriamycin both inhibit human *topoisomerase* and are used as anti-cancer drugs. 5-mercaptopurine inhibits human *DNA polymerase* and 5-fluorouracil inhibits *thymidylate synthase*. Anti-cancer agents, such as the camptothecins, target human *type I topoisomerases*, whereas etoposide targets human *type II topoisomerases*.

Rare Diseases

Diseases associated with DNA repair mechanisms are rare but may be encountered in clinical practice.

- **Hereditary non-polyposis colon cancer** is a defect in mismatch repair of hMSH 1 and 2 genes. This cancer accounts for only 5% of all colon cancers
- **Xeroderma pigmentosum** is the increased sensitivity to sunlight and hyperpigmentation, and it occurs due to defective nucleotide excision repair. Though its prevalence is only one in a million, it is strongly associated with skin cancers.
- **Double strand break repair** happens in genetic recombination of immunoglobulins normally. However, abnormal double strand breaks happen due to chemotherapeutic drugs like doxorubicin and ionizing radiation, etc.
- **Fanconi anemia** is a very rare genetic condition with defect in DNA cross-link repair mechanisms. This leads to cancers and ultimately to bone marrow failure.
- **Ataxia telangiectasia** is due to a defective ATM gene which plays a role in regulating cell division in DNA damage. Symptoms include lymphoreticular neoplasms and coordinated difficulty in movement.

VIII. Chapter Summary

- DNA is a polymer of deoxynucleoside monophosphates covalently linked by **3′→5′-phosphodiester bonds** (Fig. 35.33). The resulting long, unbranched chain has **polarity**, with both a 5′-end (free phosphate) and a 3′-end (free hydroxyl). The sequence of nucleotides is read 5′→3′.
- DNA exists as a **double-stranded** molecule, in which the two chains are paired in an **antiparallel** manner and wind around each other, forming a **double helix**. **Adenine** pairs with **thymine**, and **cytosine** pairs with **guanine**. Each strand of the double helix serves as a **template** for constructing a **complementary** daughter strand (**semiconservative replication**).
- DNA replication occurs in the **S phase** of the **cell cycle** and begins at an **origin of replication**. As the two strands unwind and separate, synthesis occurs at two **replication forks** that move away from the origin in opposite directions (bidirectionally).
- ***Helicase*** unwinds the double helix. As the two strands of the double helix are separated, positive **supercoils** are produced in the region of DNA ahead of the replication fork and negative supercoils behind the fork. ***DNA topoisomerases* types I** and **II** remove supercoils.

- ***DNA polymerases*** (***pols***) synthesize new DNA strands only in the **5′→3′** direction. Therefore, one of the newly synthesized stretches of nucleotide chains must grow in the 5′→3′ direction toward the replication fork (**leading strand**) and one in the 5′→3′ direction away from the replication fork (**lagging strand**).
- *DNA pols* require a **primer**, a short stretch of RNA synthesized by ***primase***. Leading-strand synthesis needs only one RNA primer (**continuous** synthesis), whereas the lagging strand needs many (**discontinuous** synthesis involving **Okazaki fragments**).
- In Escherichia coli (E. coli), DNA chain elongation is catalyzed by ***DNA pol III***, using **5′-deoxyribonucleoside triphosphates** as substrates. The enzyme **proofreads** the newly synthesized DNA, removing terminal mismatched nucleotides with its ***3′→5′ exonuclease*** activity. RNA primers are removed by ***DNA pol I***, using its ***5′→3′ exonuclease*** activity. This enzyme fills the gaps with DNA, proofreading as it synthesizes.
- The final phosphodiester linkage is catalyzed by ***DNA ligase***. There are at least five high-fidelity **eukaryotic *DNA pols***. ***Pol α*** is a multisubunit enzyme, one subunit of which is a ***primase***. *Pol α 5′→3′ polymerase* activity adds a short piece of DNA to the RNA primer. ***Pol ε*** completes DNA synthesis on the leading strand, whereas ***pol δ*** elongates each lagging strand fragment. ***Pol β*** is involved with DNA repair, and ***pol γ*** replicates mitochondrial DNA. *Pols ε, δ,* and *γ* use *3′→5′ exonuclease* activity to proofread.
- **Nucleoside analogs** containing modified sugars can be used to block DNA chain growth. They are useful in anticancer and antiviral chemotherapy.
- **Telomeres** are stretches of **highly repetitive DNA** complexed with protein that protect the **ends** of linear chromosomes. As most cells divide and age, these sequences are shortened, contributing to senescence. In cells that do not senesce (for example, germline and cancer cells), the ribonucleoprotein **telomerase** employs its protein component ***reverse transcriptase*** to extend the telomeres, using its **RNA** component as a **template**.
- There are five classes of **positively charged histone** (**H**) **proteins**. Two of each of histones H2A, H2B, H3, and H4 form an octameric structural core around which DNA is wrapped, creating a **nucleosome**.
- The DNA connecting the nucleosomes, called **linker DNA**, is bound to H1. Nucleosomes can be packed more tightly to form a nucleofilament. Additional levels of organization create a chromosome.
- Most DNA damage can be corrected by **excision repair** involving recognition and removal of the damage by repair proteins, followed by replacement by *DNA pols* and joining by *ligase*. **Ultraviolet radiation** can cause **thymine dimers** that are recognized and removed in E. coli by **uvrABC proteins of nucleotide excision repair**.
- Defects in the **XP proteins** needed for nucleotide excision repair of thymine dimers in humans result in **xeroderma pigmentosum**.
- **Mismatched** bases are repaired by a similar process of recognition and removal by **Mut proteins** in E. coli. The extent of **methylation** is used for strand identification in prokaryotes. Defective mismatch repair by homologous proteins in humans is associated with **hereditary nonpolyposis colorectal cancer**.
- Abnormal bases (such as uracil) are removed by ***DNA N-glycosylases*** in **base excision repair**, and the sugar phosphate at the apyrimidinic or apurinic site is cut out.
- Double-strand breaks in DNA are repaired by **nonhomologous end joining** (error prone) and template-requiring **homologous recombination** ("error-free").

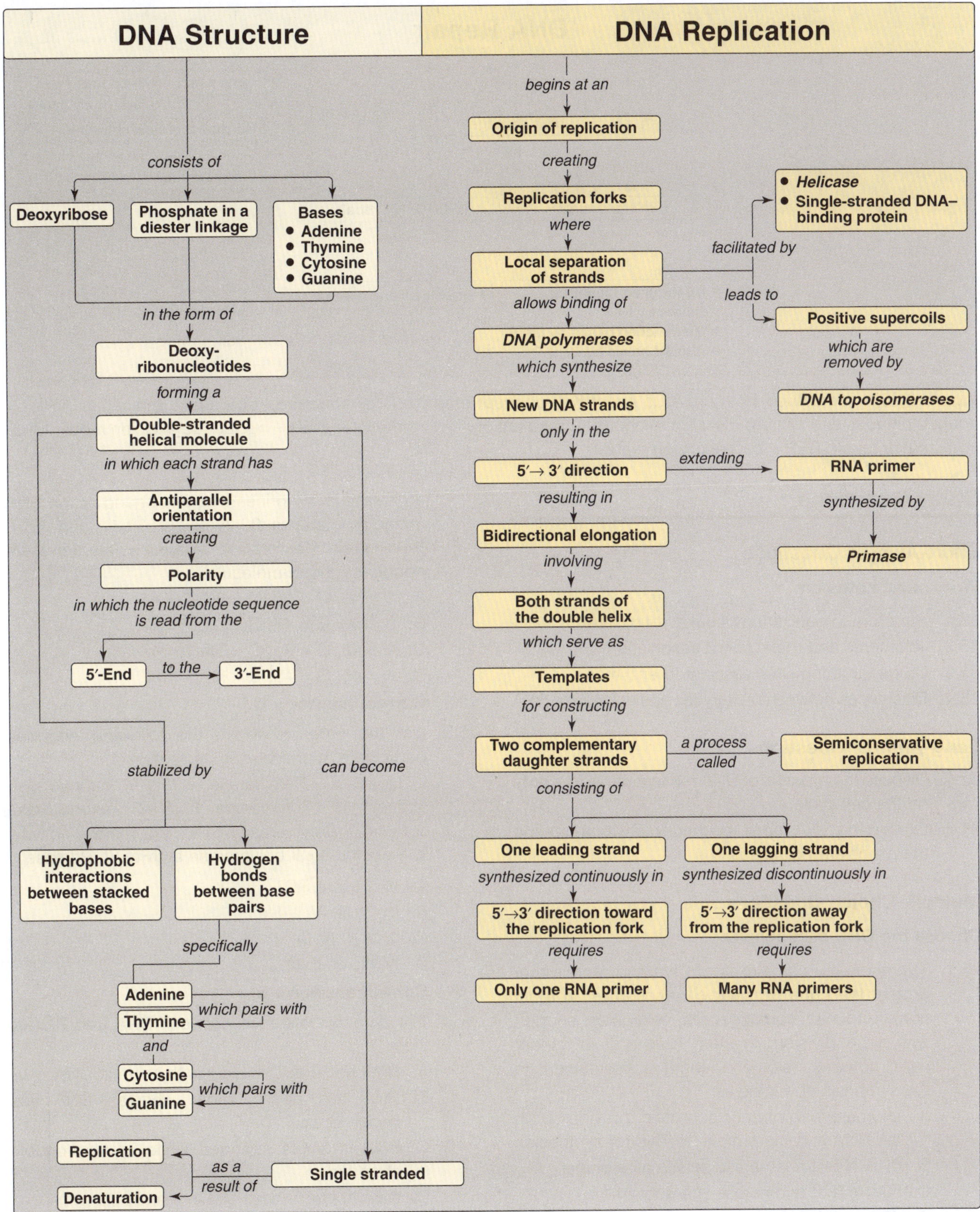

Figure 35.33 (continued on next page) Key concept map for DNA structure, replication, and repair. Key concept map for DNA structure, replication, and repair. NHEJ = nonhomologous end joining; HR = homologous recombination; UV = ultraviolet; NER = nucleotide excision repair; MMR = mismatch repair; BER = base excision repair. Key concept map for DNA structure, replication, and repair.

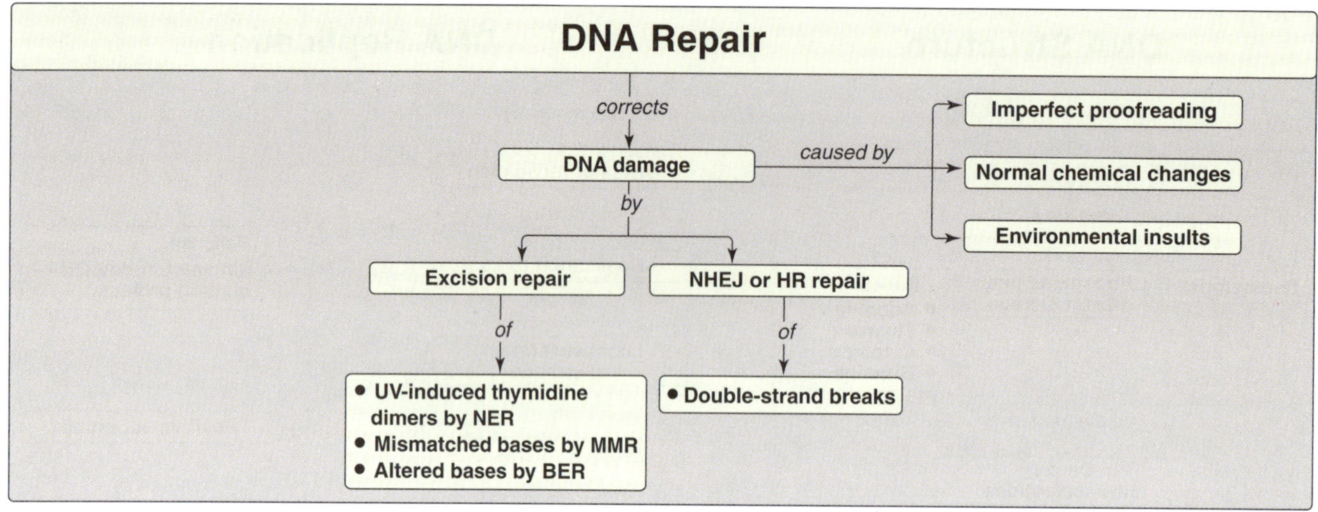

Figure 35.33 (continued from previous page) Key concept map for DNA structure, replication, and repair. NHEJ = nonhomologous end joining; HR = homologous recombination; UV = ultraviolet; NER = nucleotide excision repair; MMR = mismatch repair; BER = base excision repair.

Study Questions

Short-Answer Questions

Write short notes on:

35.1. Structural organization of eukaryotic DNA

35.2. Telomeres and their clinical association

35.3. Drugs targeting DNA associated enzymes

35.4. Differences between prokaryotic and eukaryotic DNA

Long-Answer Questions

35.1. Discuss the process of DNA replication. Add a note on the cell cycle.

35.2. Explain the DNA repair mechanisms with a note on the associated diseases.

Multiple-Choice Questions

Choose the ONE best answer.

35.1. Telomeres are complexes of DNA and protein that protect the ends of linear chromosomes. In most normal human somatic cells, telomeres shorten with each division. In stem cells and in cancer cells, however, telomeric length is maintained. In the synthesis of telomeres:

A. telomerase, a ribonucleoprotein, provides both the RNA and the protein needed for synthesis.
B. the RNA of telomerase serves as a primer.
C. the RNA of telomerase is a ribozyme.
D. the protein of telomerase is a DNA-directed DNA polymerase.

Correct answer = A.

35.2. While studying the structure of a small gene that was sequenced during the Human Genome Project, an investigator notices that one strand of the DNA molecule contains 20 A, 25 G, 30 C, and 22 T. How many of each base is found in the complete double-stranded molecule?

A. A = 40, G = 50, C = 60, T = 44
B. A = 50, G = 47, C = 50, T = 47
C. A = 42, G = 55, C = 55, T = 42
D. A = 45, G = 45, C = 52, T = 52

Correct answer = C.

35.3. List the order in which the following enzymes participate in prokaryotic replication.

I. Ligase II. Polymerase I (3′→5′ exonuclease activity) III. Polymerase I (5′→3′ exonuclease activity) IV. Polymerase I (5′→3′ polymerase activity) V. Polymerase III polymerase activity VI. Primase

A. VI, IV, V, II, III, I
B. VI, V, III, IV, II, I
C. II, VI, I, IV, III, V
D. IV, III, VI, II, VI, I

Correct answer = B.

35.4. The primary difference between DNA and RNA is that:

A. RNA has uracil instead of thymine in DNA.
B. RNA has ribose sugar whereas DNA has deoxyribose.
C. RNA is single stranded while DNA is double stranded.
D. DNA is much more stable than RNA.

Correct answer = B.

35.5. In the typical set-up for DNA electrophoresis, the direction of the migration of the DNA is toward the anode. Which of the following confers this particular electrophoretic mobility to the DNA?

A. Deoxyribose sugar
B. Pyrimidine bases
C. Purine bases
D. Phosphodiester linkages

Correct answer = D.

35.6. Fluoroquinolones like ciprofloxacin exhibit their antibiotic activity by inhibiting the bacterial topoisomerases. Which of the following processes is inhibited in a bacterial cell that is exposed to ciprofloxacin?

A. Addition of the deoxynucleotide to the growing DNA strand
B. Unwinding of the double helix
C. Relaxing of supercoiled DNA
D. Joining of the Okazaki fragments

Correct answer = C.

35.7. Which of the following DNA repair mechanism and the clinical condition arising due to its defect is NOT correctly matched?

A. Mismatch repair – Lynch syndrome
B. Nucleotide excision repair – Xeroderma pigmentosa
C. Base excision repair – Ataxia telangiectasia
D. Homologous recombination – Hereditary breast and ovarian cancer

Correct answer = C.

35.8. Pathogenic alterations in the gene coding for the subunits of DNA polymerase γ is known to cause various disorders with a wide range of clinical manifestations like neuropathy, ataxia, encephalopathy, and seizures. Which of the following cellular processes is affected in these conditions?

A. Replication of mitochondrial DNA
B. Mis-match repair in nuclear DNA
C. Synthesis of ribosomal RNA
D. Removal of pyrimidine dimers in DNA

Correct answer = A.

35.9. Which of the following phases of cell-cycle is inhibited by human topoisomerase inhibitors?

A. G1
B. S
C. M
D. G2

Correct answer = B.

35.10. Base excision repair is primarily involved in repairing DNA alterations caused by:

A. Ultraviolet radiation
B. Chemical agents and oxidative stress
C. Replication errors during DNA replication
D. Spontaneous deamination of bases

Correct answer = D.

Multiple-Choice Questions (NExT/NEET PG)

Choose the ONE best answer.

35.1. A 10-year-old girl is brought by her parents to the family doctor. She has many freckles on her face, neck, arms, and hands, and the parents report that she is unusually sensitive to sunlight. She is referred to a tertiary care hospital as the doctor has not seen such a case before. Here, two basal cell carcinomas are also identified on her face. Based on the clinical picture, which of the following processes is most likely to be defective in this patient?

A. Repair of double-strand breaks by error-prone homologous recombination
B. Removal of mismatched bases from the 3′-end of Okazaki fragments by a methyl-directed process
C. Removal of uracil from DNA by base excision repair
D. Removal of pyrimidine dimers from DNA by nucleotide excision repair

Correct answer = D.

35.2. When a cell is grown in a medium that includes radioactively tagged thymidine and it then undergoes four replication cycles, what percentage of the resulting cells will display radioactive labeling on both strands of their DNA?

A. 50%
B. 75%
C. 87.5%
D. 100%

Correct answer = C.

35.3. Hayflick limit refers to the number of times a normal cell population can divide until cell division stops. This limit is attributed to the shortening of telomeres, the protective sequences at the ends of chromosomes, which occurs with each cell division. Germ cells are an exception to this limit, as they are considered 'immortal'. Which of the following enzyme activities in these cells is responsible for this 'immortality'?

A. RNA dependent DNA polymerase activity
B. DNA dependent DNA polymerase activity
C. RNA dependent RNA polymerase activity
D. DNA dependent RNA polymerase activity

Correct answer = A.

35.4. A method to decipher the sequence of DNA is based on the synthesis of DNA strands, where the addition of nucleotides results in a detectable light signal. The light signal is generated with the help of the enzyme luciferase, which is ATP-dependent. The ATP is formed by the reaction between adenosine 5'-phosphosulfate (APS) and the molecule X released during the specific nucleotide addition to the growing DNA strand. Identify X.

A. Proton
B. Pyrophosphate
C. Magnesium
D. Guanosine triphosphate

Correct answer = B.

35.5. Proliferating Cell Nuclear Antigen (PCNA) is highly expressed in rapidly dividing cells, making it a valuable marker for assessing tumor aggressiveness and growth rates. Elevated levels of PCNA in tumor tissues compared to normal tissues can indicate a high proliferation rate, often correlating with a more aggressive tumor phenotype and poorer prognosis. It helps in differentiating between benign and malignant lesions in various cancers. Which of the following processes during cell proliferation is mediated by PCNA?

A. Maintenance of processivity of DNA polymerases
B. Facilitation of the action of ATP-dependent topoisomerases
C. Maintenance of the telomere length in cancer cells
D. Loading of single strand DNA binding proteins to DNA during replication

Correct answer = A.

RNA Structure, Synthesis, and Processing

36

Learning Objectives

At the end of this chapter, the student will be able to:

- Describe the structure and function of RNA
- Describe the process of transcription
- Explain expression of a gene in prokaryotes and eukaryotes

I. OVERVIEW

The genetic master plan of an organism is contained in the sequence of deoxyribonucleotides in its DNA. However, it is through ribonucleic acid (RNA), the "working copies" of deoxyribonucleic acid (DNA), that the master plan is expressed (Fig. 36.1). **The copying process, during which a DNA strand serves as a template for the synthesis of RNA, is called transcription.** Transcription produces messenger RNA (mRNA), which are translated into sequences of amino acids (proteins), and ribosomal RNA (rRNA), transfer RNA (tRNA), and additional RNA molecules that perform specialized structural, catalytic, and regulatory functions and are not translated. That is, they are noncoding RNA (ncRNA). Therefore, the final product of gene expression can be RNA or protein, depending upon the gene. [Note: Only ~2% of the genome encodes proteins.] A central feature of transcription is that it is highly selective. For example, many transcripts are made of some regions of the DNA. In other regions, few or no transcripts are made. This selectivity is due, at least in part, to signals embedded in the nucleotide sequence of the DNA. These signals instruct the *RNA polymerase* where to start, how often to start, and where to stop transcription. Several regulatory proteins are also involved in this selection process. The biochemical differentiation of an organism's tissues is ultimately a result of the selectivity of the transcription process. [Note: This selectivity of transcription is in contrast to the "all-or-none" nature of genomic replication.] Another important feature of transcription is that many RNA transcripts that initially are faithful copies of one of the two DNA strands may undergo various modifications, such as terminal additions, base modifications, trimming, and internal segment removal,

Comparison of DNA and RNA

	DNA	RNA
Sugar	Deoxyribose	Ribose
Bases	ATGC	AUGC
Synthesis	Genomic replication is in totality	RNA synthesis is selective depending on transcription required
Modifications	In replication, DNA is copied as it is	RNA synthesis is usually followed by post-transcriptional modifications
Proofreading	Major proofreading mechanisms in place	Minimal proofreading
Location in cell	Mainly inside nucleus	Mainly inside cytoplasm
Strands	Double stranded which become partly single stranded during processes like replication and transcription	Single strands but can fold into complex structures

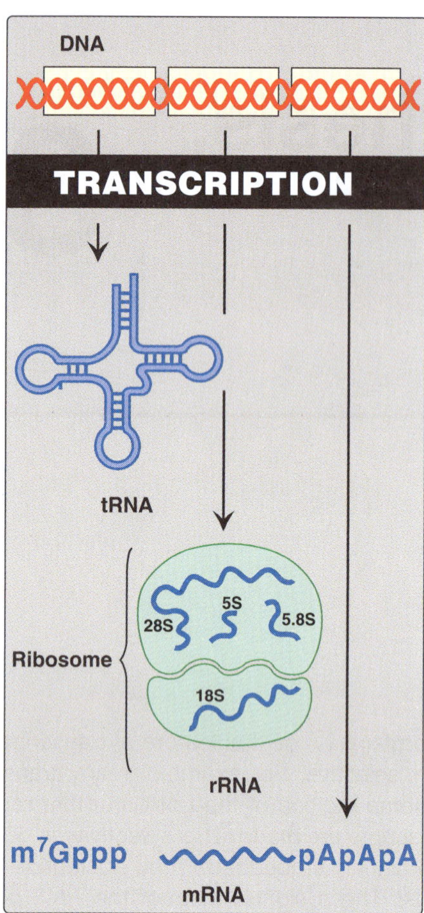

Figure 36.1 Expression of genetic information by transcription. [Note: RNA shown are eukaryotic.] tRNA = transfer RNA; rRNA = ribosomal RNA; mRNA = messenger RNA; m7Gppp = 7-methylguanosine-triphosphate cap; pApApA = poly-A tail; p = phosphate.

which convert the inactive primary transcript into a functional molecule. The transcriptome is the complete set of RNA transcripts expressed by a genome.

II. RNA STRUCTURE

There are three major types of RNA that participate in the process of protein synthesis: rRNA, tRNA, and mRNA. Like DNA, these RNA are unbranched polymeric molecules composed of nucleoside monophosphates joined together by 3′→5′-phosphodiester bonds (see p. 579). However, they differ from DNA in several ways. For example, they are considerably smaller than DNA, contain ribose instead of deoxyribose and uracil instead of thymine, and exist as single strands that are capable of folding into complex structures. The three major types of RNA also differ from each other in size, function, and special structural modifications. **Non coding RNA is a RNA molecule that is not translating to a protein. They therefore include ribosomal RNA, tRNA and many small RNA like miRNA.** [Note: In eukaryotes, additional small ncRNA molecules found in the nucleolus (snoRNA), nucleus (snRNA), and cytoplasm (microRNA [miRNA]) perform specialized functions as described on p. 664.]

A. Ribosomal RNA

rRNA are found in association with several proteins as components of the ribosomes, the complex structures that serve as the sites for protein synthesis (see p. 642). Prokaryotic cells contain three distinct size species of rRNA (23S, 16S, and 5S, where S is the Svedberg unit for sedimentation rate that is determined by the size and shape of the particle), as shown in Figure 36.2. Eukaryotic cells contain four rRNA species (28S, 18S, 5.8S, and 5S). Together, rRNA make up ~80% of the total RNA in the cell.

Catalytic RNA molecules are known as ribozymes. For example, RNA like snRNA possesses catalytic activity in spliceosomes. *RNAase-P* cuts the 5′end of pre-tRNA and is also a ribozyme. Further *peptidyl transferase* present in ribosomes catalyzes peptide bond formation qualifying it to be a ribozyme.

B. Transfer RNA

tRNA are the smallest (4S) of the three major types of RNA molecules. There is at least one specific type of tRNA molecule for each of the 20 amino acids commonly found in proteins. Together, tRNA make up ~15% of the total RNA in the cell. The tRNA molecules contain a high percentage of unusual (modified) bases, for example, dihydrouracil (see Fig. 34.2, p. 578), and have extensive intrachain base-pairing (Fig. 36.3) that leads to characteristic secondary and tertiary structure. Each tRNA serves as an adaptor molecule that carries its specific amino acid, covalently attached to its 3′-end, to the site of protein synthesis. There, it recognizes the genetic code sequence on an mRNA, which specifies the addition of that amino acid to the growing peptide chain (see p. 635).

C. Messenger RNA

mRNA comprises only ~5% of the RNA in a cell, yet is by far the most heterogeneous type of RNA in size

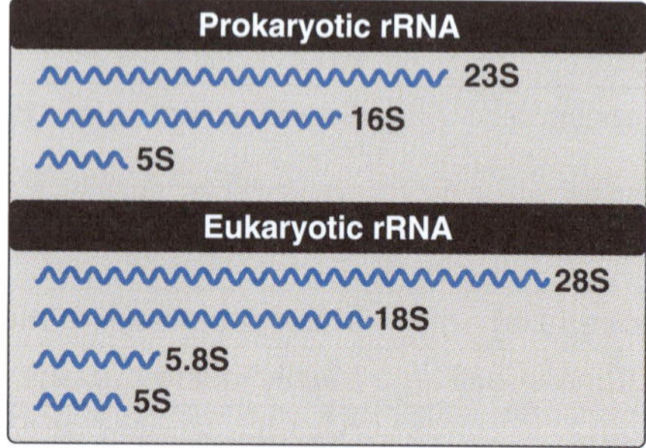

Figure 36.2 Prokaryotic and eukaryotic ribosomal RNA (rRNA). S = Svedberg unit.

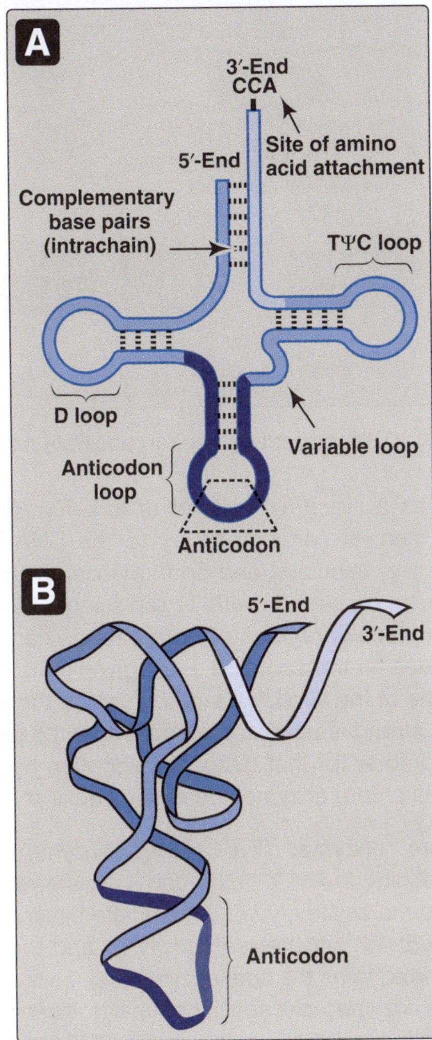

Figure 36.3 **A.** Characteristic transfer RNA (tRNA) secondary structure (cloverleaf). **B.** Folded (tertiary) tRNA structure found in cells. D = dihydrouracil; Ψ = pseudouracil; T = thymine; C = cytosine; A = adenine.

and base sequence. mRNA is coding RNA in that it carries genetic information from DNA for use in protein synthesis. In eukaryotes, this involves transport of mRNA out of the nucleus and into the cytosol. An mRNA carrying information from more than one gene is polycistronic. **Polycistronic mRNA** is characteristic of prokaryotes. An mRNA carrying information from only one gene is monocistronic and is characteristic of eukaryotes. **A cistron** is the smallest unit of genetic expression and therefore includes the gene as well as its promoter elements. In addition to the protein-coding regions that can be translated, mRNA contains untranslated regions at its 5′- and 3′-ends (Fig. 36.4). Special structural characteristics of eukaryotic (but not prokaryotic) mRNA include a long sequence of adenine nucleotides (a poly-A tail) on the 3′-end of the RNA, plus a cap on the 5′-end consisting of a molecule of 7-methylguanosine attached through an unusual (5′→5′) triphosphate linkage. The mechanisms for modifying mRNA to create these special structural characteristics are discussed on pp. 628–629.

RNA editing is the term given to changes in RNA after its formation. For example, a 100 kDa protein

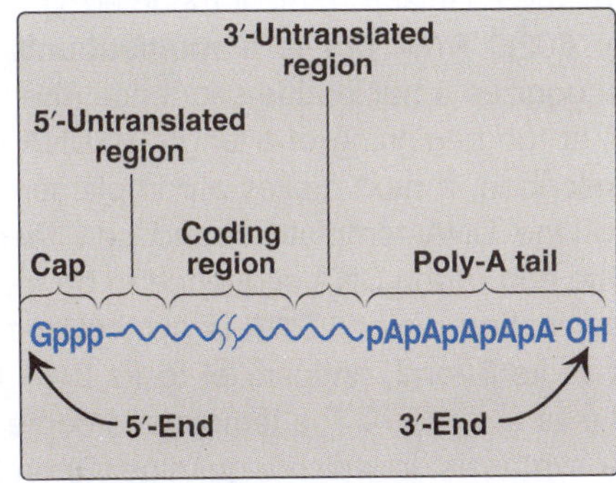

Figure 36.4 Structure of eukaryotic messenger RNA. G = guanine; A = adenine.

Types of RNA

Name	Details	Function
Messenger RNA (mRNA)	About 5% of total RNA of a cell. Quickly degraded	Carriers of genetic information from the gene for protein synthesis
Ribosomal RNA (rRNA)	About 80% of all RNA in a cell. Very stable	Involved in protein synthesis
Transfer RNA (tRNA)	About 15% of total RNA of a cell, very stable. More than 20 different tRNA. Contain high number of modified bases	Serves for transfer of specific amino acid to the mRNA during protein synthesis
Small RNA	About 1–2% of total RNA. There are about 30 different types of small RNA including snRNA (small nuclear RNA)	SnRNA like U1-U5 are catalytic RNA for mRNA splicing
Micro RNA or mi-RNA	Less than 1% of total RNA	Modulates the function of mRNA

(ApoB100) is made in liver from ApoB gene. But in the intestine, though the same primary transcript is formed, a *cytidine deaminase* converts a CAA codon in the mRNA into a UAA codon. This causes a change from the glutamine codon to a termination codon. Now the protein formed in intestine has a premature termination of chain and is ApoB48.

III. PROKARYOTIC GENE TRANSCRIPTION

The structure of magnesium-requiring *RNA polymerase* (*RNA pol*), the signals that control transcription, and the varieties of modification that RNA transcripts can undergo differ among organisms, particularly from prokaryotes to eukaryotes. Therefore, the discussions of prokaryotic and eukaryotic transcription are presented separately.

A. Prokaryotic RNA polymerase

In bacteria, one species of *RNA pol* synthesizes all of the RNA except for the short RNA primers needed for DNA replication [Note: RNA primers are synthesized by the specialized, monomeric enzyme *primase* (see p. 602).] ***RNA pol* is a multisubunit enzyme** that recognizes a nucleotide sequence (the promoter region) at the beginning of a length of DNA that is to be transcribed. It next makes a complementary RNA copy of the DNA template strand and then recognizes the end of the DNA sequence to be transcribed (the termination region). RNA is synthesized from its 5′-end to its 3′-end, antiparallel to its DNA template strand (see Fig. 36.5). The template is copied as it is in DNA synthesis, in which a guanine (G) on the DNA specifies a cytosine (C) in the RNA, a C specifies a G, a thymine (T) specifies an adenine (A), but an A

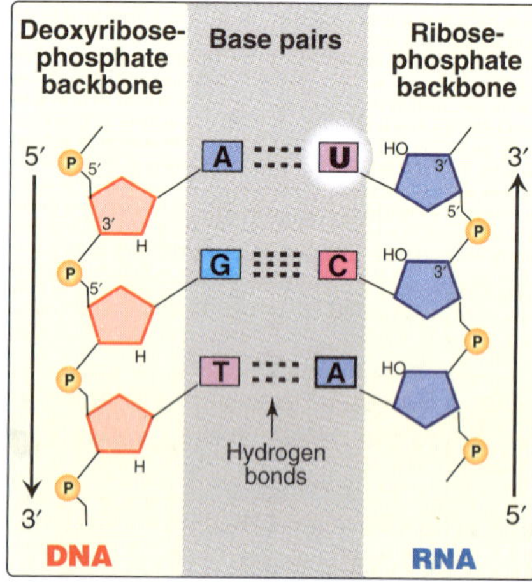

Figure 36.5 Antiparallel, complementary base pairs between DNA and RNA. T = thymine; A = adenine; C = cytosine; G = guanine; U = uracil.

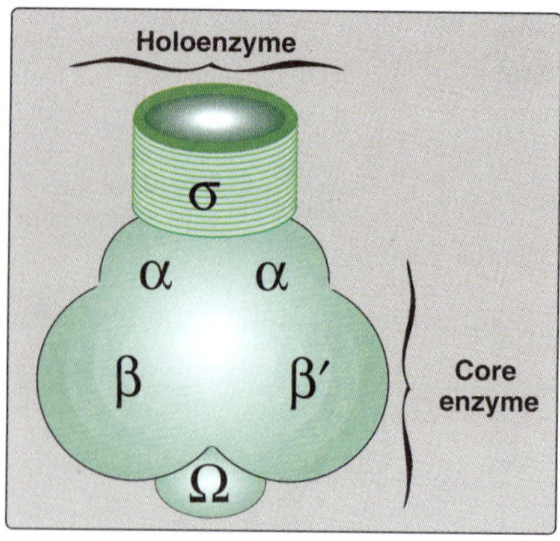

Figure 36.6 Components of prokaryotic *RNA polymerase*.

specifies a uracil (U) instead of a T (Fig. 36.5). The RNA, then, is complementary to the DNA template (antisense, minus) strand and identical to the coding (sense, plus) strand, with U replacing T. Within the DNA molecule, regions of both strands can serve as templates for transcription. For a given gene, however, only one of the two DNA strands can be the template. Which strand is used is determined by the location of the promoter for that gene. Transcription by *RNA pol* involves a core enzyme and several auxiliary proteins.

1. **Core enzyme:** Five of the enzyme's peptide subunits, 2 α, 1 β, 1 β′, and 1 Ω, are required for enzyme assembly (α, Ω), template binding (β′), and the *5′→3′ polymerase* activity (β) and together are referred to as the core enzyme (Fig. 36.6). However, this enzyme lacks specificity (that is, it cannot recognize the promoter region on the DNA template).

2. **Holoenzyme:** The **σ subunit (sigma factor)** enables *RNA pol* to recognize promoter regions on the DNA. The σ subunit plus the core enzyme make up the holoenzyme. [Note: Different σ factors recognize different groups of genes, with σ^{70} predominating.]

B. Steps in RNA synthesis

The process of transcription of a typical gene of Escherichia coli (E. coli) can be divided into three phases: initiation, elongation, and termination. A transcription unit extends from the promoter to the termination region, and the initial product of transcription by *RNA pol* is termed the primary transcript.

1. **Initiation:** Transcription begins with the binding of the *RNA pol* holoenzyme to a region of the DNA known as the promoter, which is not transcribed. The prokaryotic promoter contains characteristic consensus sequences (Fig. 36.7). [Note: Consensus sequences are idealized sequences in which the base shown at each position is

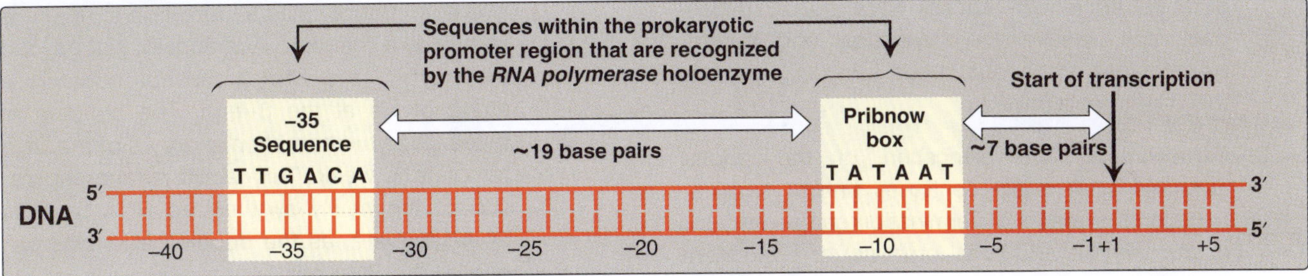

Figure 36.7 Structure of the prokaryotic promoter region. T = thymine; G = guanine; A = adenine; C = cytosine.

the base most frequently (but not necessarily always) encountered at that position.] Those that are recognized by prokaryotic *RNA pol* σ factors include the following.

a. **–35 Sequence:** A consensus sequence (5′-TTGACA-3′), centered about 35 bases to the left of the transcription start site (see Fig. 36.7), is the initial point of contact for the holoenzyme, and a closed complex is formed. [Note: By convention, the regulatory sequences that control transcription are designated by the 5′→3′ nucleotide sequence on the coding strand. A base in the promoter region is assigned a negative number if it occurs prior to (to the left of, toward the 5′-end of, or "upstream" of) the transcription start site. Therefore, the TTGACA sequence is centered at approximately base –35. The first base at the transcription start site is assigned a position of +1. There is no base designated "0".]

b. **Pribnow box:** The holoenzyme moves and covers a second consensus sequence (5′-TATAAT-3′), centered at about –10 (see Fig. 36.7), which is the site of melting (unwinding) of a short stretch (~14 base pairs) of DNA. This initial melting converts the closed initiation complex to an open complex known as a transcription bubble. [Note: A mutation in either the –10 or the –35 sequence can affect the transcription of the gene controlled by the mutant promoter.]

2. **Elongation:** Once the promoter has been recognized and bound by the holoenzyme, local unwinding of the DNA helix continues (Fig. 36.8), mediated by the *polymerase*. [Note: Unwinding generates supercoils in the DNA that can be relieved by *DNA topoisomerases* (see p. 600).] *RNA pol* begins to synthesize a transcript of the DNA sequence, and several short pieces of RNA are made and discarded. The elongation phase begins when the transcript (typically starting with a purine) exceeds 10 nucleotides in length. Sigma is then released, and the core enzyme is able to leave (clear) the promoter and move along the template strand in a processive manner, serving as its own sliding clamp. During transcription, a short DNA–RNA hybrid helix is formed (see Fig. 36.8). Like *DNA pol*, *RNA pol* uses nucleoside triphosphates as substrates and releases pyrophosphate each time a nucleoside monophosphate is added to the growing chain. As with replication, transcription is always in the 5′→3′ direction. In contrast to *DNA pol*, *RNA pol* does not require a primer and does not have a *3′→5′ exonuclease* domain for proofreading. [Note: Misincorporation

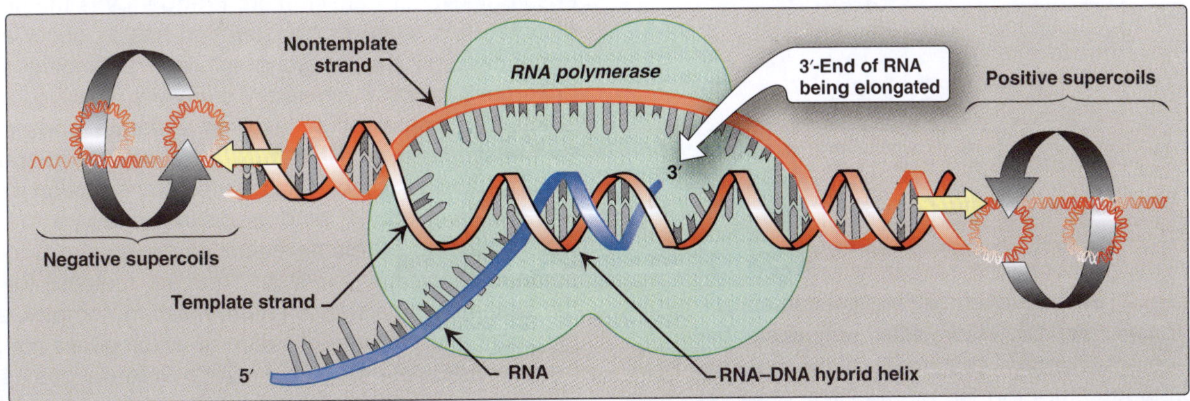

Figure 36.8 Local unwinding of DNA by *RNA polymerase* and formation of an open initiation complex (transcription bubble).

of a ribonucleotide causes *RNA pol* to pause, backtrack, cleave the transcript, and restart. Nonetheless, transcription has a higher error rate than does replication.]

3. **Termination:** The elongation of the single-stranded RNA chain continues until a termination signal is reached. Termination can be intrinsic (occur without additional proteins) or dependent upon the participation of a protein known as the ρ (rho) factor.

   a. **ρ-Independent:** Seen with most prokaryotic genes, this requires that a sequence in the DNA template generates a sequence in the nascent (newly made) RNA that is self-complementary (Fig. 36.9). This allows the RNA to fold back on itself, forming a GC-rich stem (stabilized by hydrogen bonds) plus a loop. This structure is known as a "hairpin." Additionally, just beyond the hairpin, the RNA transcript contains a string of Us at the 3′-end. The bonding of these Us to the complementary As of the DNA template is weak. This facilitates the separation of the newly synthesized RNA from its DNA template, as the double helix "zips up" behind the *RNA pol*.

   b. **ρ-Dependent:** This requires the participation of the additional protein **rho, which is a hexameric *ATPase* with *helicase* activity**. Rho binds a C-rich rho utilization (rut) site near the 5′-end of the nascent RNA and, using its *ATPase* activity, moves along the RNA until it reaches the *RNA pol* paused at the termination site. The ATP-dependent *helicase* activity of rho separates the RNA–DNA hybrid helix, causing the release of the RNA.

4. **Antibiotics:** Some antibiotics prevent bacterial cell growth by inhibiting RNA synthesis. For example**, rifampin (rifampicin) inhibits transcription initiation** by binding to the β subunit of prokaryotic *RNA pol* and preventing chain growth beyond three nucleotides (Fig. 36.10). Rifampin is important in the treatment of tuberculosis. **Dactinomycin (actinomycin D)** was the first antibiotic to find therapeutic application in tumor chemotherapy. It **inserts (intercalates) between the DNA bases and inhibits transcription initiation and elongation.**

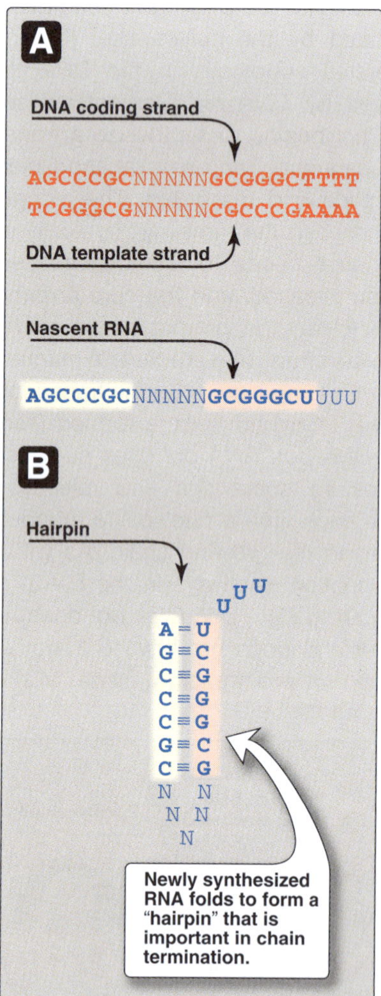

Figure 36.9 Rho-independent termination of prokaryotic transcription. **A.** DNA template sequence generates a self-complementary sequence in the nascent RNA. **B.** Hairpin structure formed by the RNA. N represents a noncomplementary base; A = adenine, T = thymine; G = guanine; C = cytosine; U = uracil.

IV. EUKARYOTIC GENE TRANSCRIPTION

The transcription of eukaryotic genes is a far more complicated process than transcription in prokaryotes. **Fidelity** is the site at which transcription starts. This is designated as +1. The core-promoter elements (see Fig. 36.12) ensure fidelity of transcription. **Frequency** is the number of times this event of transcription has to occur and is dependent on the transcription factors binding to the proximal and distal regulatory elements (see Fig. 36.12). Eukaryotic transcription involves separate *polymerases* for the synthesis of rRNA, tRNA, and mRNA. In addition, a large number of proteins called transcription factors (TF) are involved. TF bind to distinct sites on the DNA within the core promoter region, close (proximal) to it, or some distance away (distal). They are required for both the assembly of a transcription initiation complex at the promoter and the determination of which genes are to be transcribed. [Note: Each eukaryotic *RNA pol* has its own promoters and TF that bind core promoter sequences.] For TF to recognize and bind to their specific DNA sequences, the chromatin structure in that region must

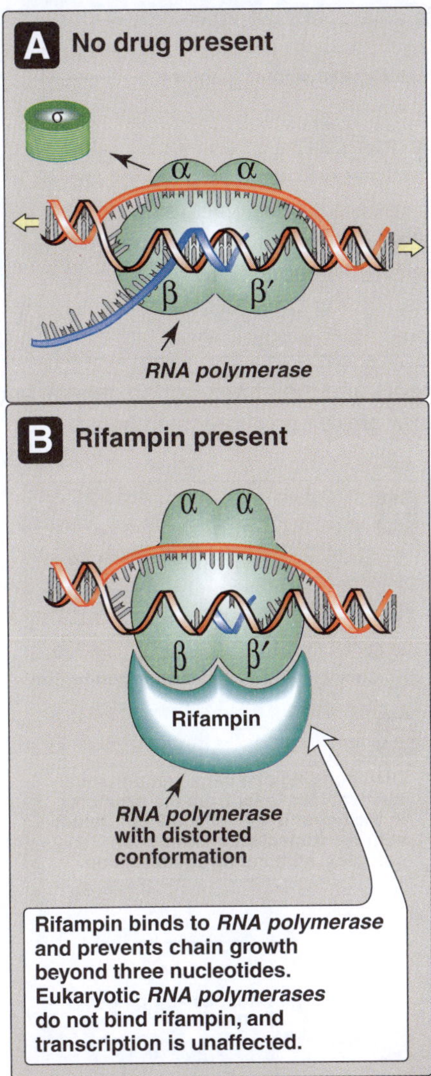

Figure 36.10 Inhibition of prokaryotic *RNA polymerase* by rifampin (rifampicin).

be decondensed (relaxed) to allow access to the DNA. The role of transcription in the regulation of gene expression is discussed in Chapter 38.

A. Chromatin structure and gene expression

The association of DNA with histones to form nucleosomes (see p. 608) affects the ability of the transcription machinery to access the DNA to be transcribed. Most actively transcribed genes are found in a relatively decondensed form of chromatin called euchromatin, whereas most inactive segments of DNA are found in highly condensed heterochromatin. The interconversion of these forms is called chromatin remodeling. A major component of chromatin remodeling is the covalent modification of histones (for example, the acetylation of lysine residues at the amino terminus of histone proteins), as shown in Figure 36.11. Acetylation, mediated **by *histone***

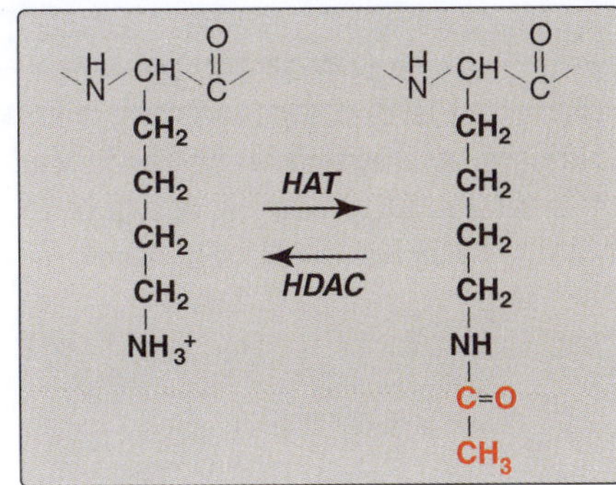

Figure 36.11 Acetylation/deacetylation of a lysine residue in a histone. Acetyl coenzyme A provides the acetyl group. *HAT = histone acetyltransferase*; *HDAC = histone deacetylase*.

***acetyltransferases* (*HAT*)**, eliminates the positive charge on the lysine, thereby decreasing the interaction of the histone with the negatively charged DNA. Removal of the acetyl group by ***histone deacetylases* (*HDAC*)** restores the positive charge and fosters stronger interactions between histones and DNA. [Note: The ATP-dependent repositioning of nucleosomes is also required to access DNA.]

B. Nuclear RNA polymerases

There are three distinct types of *RNA pol* in the nucleus of eukaryotic cells. All are large enzymes with multiple subunits. Each type of *RNA pol* recognizes particular genes. [Note: Mitochondria contain a single *RNA pol* that resembles the bacterial enzyme.]

1. **RNA polymerase I:** This enzyme synthesizes the precursor of the 28S, 18S, and 5.8S rRNA in the nucleolus.

2. **RNA polymerase II:** This enzyme synthesizes the nuclear precursors of mRNA that are processed and then translated to proteins. *RNA pol II* also synthesizes certain small ncRNA, such as snoRNA, snRNA, and miRNA.

   a. **Promoters for RNA polymerase II:** In some genes transcribed by *RNA pol II*, a sequence of nucleotides (TATAAA) that is nearly identical to that of the Pribnow box (see p. 623) is found centered ~25 nucleotides upstream of the transcription start site. This core promoter consensus sequence is called the **TATA, or Hogness, box**. In the majority of genes, however, no TATA box is present. Instead, different core promoter elements such as **Inr (initiator)** or **DPE (downstream promoter element)** are present (Fig. 36.12). [Note: No one consensus sequence is found in all core promoters.] Because these sequences are on the same molecule of DNA

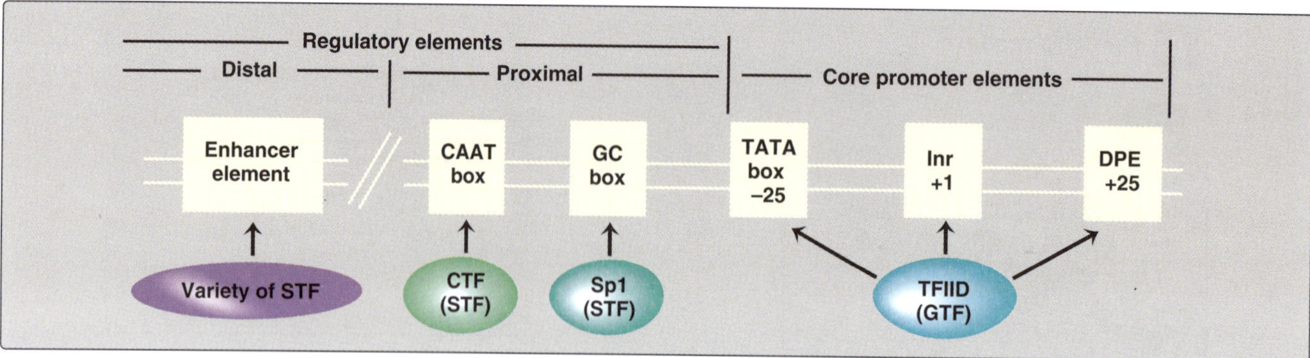

Figure 36.12 Eukaryotic gene cis-acting promoter and regulatory elements and their trans-acting general and specific transcription factors (GTF and STF, respectively). Inr = initiator; DPE = downstream promoter element.

as the gene being transcribed, they are cis-acting. The sequences serve as binding sites for proteins known as general transcription factors (GTF), which in turn interact with each other and with *RNA pol II*.

b. **General transcription factors:** GTF are the minimal requirements for recognition of the promoter, recruitment of *RNA pol II* to the promoter, formation of the preinitiation complex, and initiation of transcription at a basal level (Fig. 36.13A). GTF are encoded by different genes, synthesized in the cytosol, and diffuse (transit) to their sites of action, and so are trans-acting. [Note: In contrast to the prokaryotic holoenzyme, eukaryotic *RNA pol II* does not itself recognize and bind the promoter. Instead, TFIID, a GTF containing TATA-binding protein and TATA-associated factors, recognizes and binds the TATA box (and other core promoter elements). TFIIF, another GTF, brings the *polymerase* to the promoter. The *helicase* activity of TFIIH melts the DNA, and its *kinase* activity phosphorylates *polymerase*, allowing it to clear the promoter.]

c. **Regulatory elements and transcriptional activators:** Additional consensus sequences lie upstream of the core promoter (see Fig. 36.12). Those close to the core promoter (within ~200 nucleotides) are the proximal regulatory elements, such as the **CAAT and GC boxes.** Those farther away are the distal regulatory elements such as enhancers (see d. below). Proteins known as transcriptional activators or **specific transcription factors (STF)** bind these regulatory elements. STF bind to promoter proximal elements to regulate the frequency of transcription initiation and to distal elements to mediate the response to signals such as hormones (see p. 660) and regulate which genes are expressed at a given point in time.

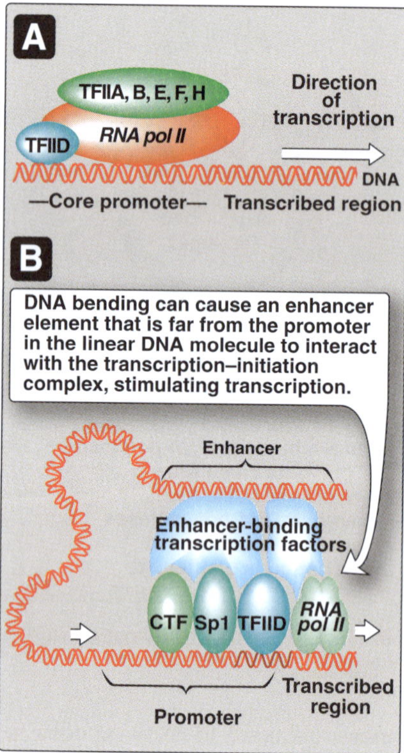

Figure 36.13 **A.** Association of the general transcription factors (TFII) and *RNA polymerase II (RNA pol II)* at the core promoter. [Note: The Roman numeral II denotes a TF for *RNA pol II*.] **B.** Enhancer stimulation of transcription. CTF = CAAT box transcription factor; Sp1 = specificity factor-1.

A typical protein-coding eukaryotic gene has binding sites for many such factors. STF have two binding domains. One is a DNA-binding domain, the other is a transcription activation domain that recruits the GTF to the core promoter as well as coactivator proteins such as the ***HAT* enzymes involved in chromatin modification.** [Note: Mediator, a multisubunit coactivator of *RNA pol II*–catalyzed transcription, binds the *polymerase*, the GTF, and the STF and regulates transcription initiation.]

Transcriptional activators bind DNA through a variety of motifs, such as the helix-loop-helix, zinc finger, and leucine zipper (see pp. 654–655).

d. **Role of enhancers:** Enhancers are special DNA sequences that increase the rate of initiation of transcription by *RNA pol II*. Enhancers are typically on the same chromosome as the gene whose transcription they stimulate (Fig. 36.13B). However, they can 1) be located upstream (to the 5′-side) or downstream (to the 3′-side) of the transcription start site, 2) be close to or thousands of base pairs away from the promoter (Fig. 36.14), and 3) occur on either strand of the DNA. Enhancers contain DNA sequences called response elements that bind STF. By bending or looping the DNA, STF can interact with other TF bound to a promoter and with *RNA pol II*, thereby stimulating transcription (see Fig. 36.13B). Mediator also binds enhancers. [Note: Although silencers are similar to enhancers in that they also can act over long distances, they reduce gene expression.]

e. **RNA polymerase II inhibitor:** α-Amanitin, a potent toxin produced by the poisonous mushroom Amanita phalloides (sometimes called the "death cap"), binds *RNA pol II* tightly and slows its movement, thereby inhibiting mRNA synthesis.

3. **RNA polymerase III:** This enzyme synthesizes tRNA, 5S rRNA, and some snRNA and snoRNA.

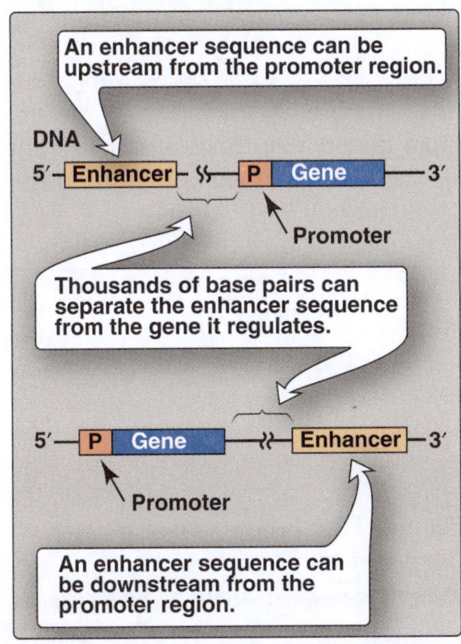

Figure 36.14 Some possible locations of enhancer sequences.

V. POSTTRANSCRIPTIONAL MODIFICATION OF RNA

A primary transcript is the initial, linear, RNA copy of a transcription unit (the segment of DNA between specific initiation and termination sequences). The primary transcripts of both prokaryotic and eukaryotic tRNA and rRNA are posttranscriptionally modified by cleavage of the original transcripts by *ribonucleases*. tRNA are further modified to help give each species its unique identity. In contrast, prokaryotic mRNA is generally identical to its primary transcript, whereas eukaryotic mRNA is extensively modified both co- and posttranscriptionally. Further, in eukaryotes, mRNA packaging factors called **TREX (transcription export)** are responsible for directing the mRNA out of the nucleus into cytoplasm through nuclear pores.

A. Ribosomal RNA

rRNA of both prokaryotic and eukaryotic cells are generated from long precursor molecules called pre-rRNA. The 23S, 16S, and 5S rRNA of prokaryotes are produced from a single pre-rRNA molecule, as are the 28S, 18S, and 5.8S rRNA of eukaryotes (Fig. 36.15). [Note: Eukaryotic 5S rRNA is synthesized by *RNA pol III* and modified separately.] The pre-rRNA are cleaved by *ribonucleases* to yield intermediate-sized pieces of rRNA, which are further processed (trimmed by *exonucleases* and modified at some bases and riboses) to produce the required RNA species. [Note: In eukaryotes, rRNA genes are found in long, tandem arrays. rRNA synthesis and processing occur in the nucleolus, with base and sugar modifications facilitated by snoRNA.]

B. Transfer RNA

Both eukaryotic and prokaryotic tRNA are also made from longer precursor molecules that must be

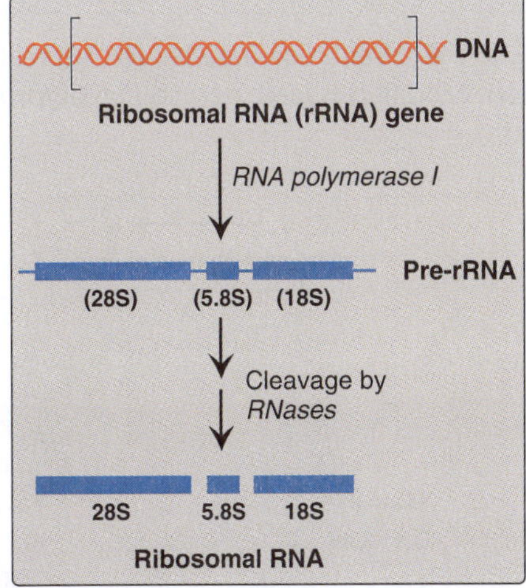

Figure 36.15 Posttranscriptional processing of eukaryotic ribosomal RNA by *ribonucleases* (*RNases*). S = Svedberg unit.

modified (Fig. 36.16). Sequences at both ends of the molecule are removed, and, if present, an intron is removed from the anticodon loop by *nucleases*. Other posttranscriptional modifications include addition of a –CCA sequence by *nucleotidyltransferase* to the 3′-terminal end of tRNA and modification of bases at specific positions to produce the unusual bases characteristic of tRNA.

C. Eukaryotic messenger RNA

The collection of all the primary transcripts synthesized in the nucleus by *RNA pol II* is known as heterogeneous nuclear RNA (hnRNA). The pre-mRNA components of hnRNA undergo extensive co- and posttranscriptional modification in the nucleus and become mature mRNA. These modifications usually include the following. [Note: *Pol II* itself recruits the proteins required for the modifications.]

1. **Addition of a 5′-cap:** This is the first of the processing reactions for pre-mRNA (Fig. 36.17). The cap is a 7-methylguanosine attached to the 5′-terminal end of the mRNA through an unusual 5′→5′-triphosphate linkage that is resistant to most *nucleases*. Creation of the cap requires removal of the γ phosphoryl group from the 5′-triphosphate of the pre-mRNA, followed by addition of guanosine monophosphate (from guanosine triphosphate) by the nuclear enzyme *guanylyltransferase*. Methylation of this terminal guanine occurs in the cytosol and is catalyzed by *guanine-7-methyltransferase*. S-Adenosylmethionine is the source of the

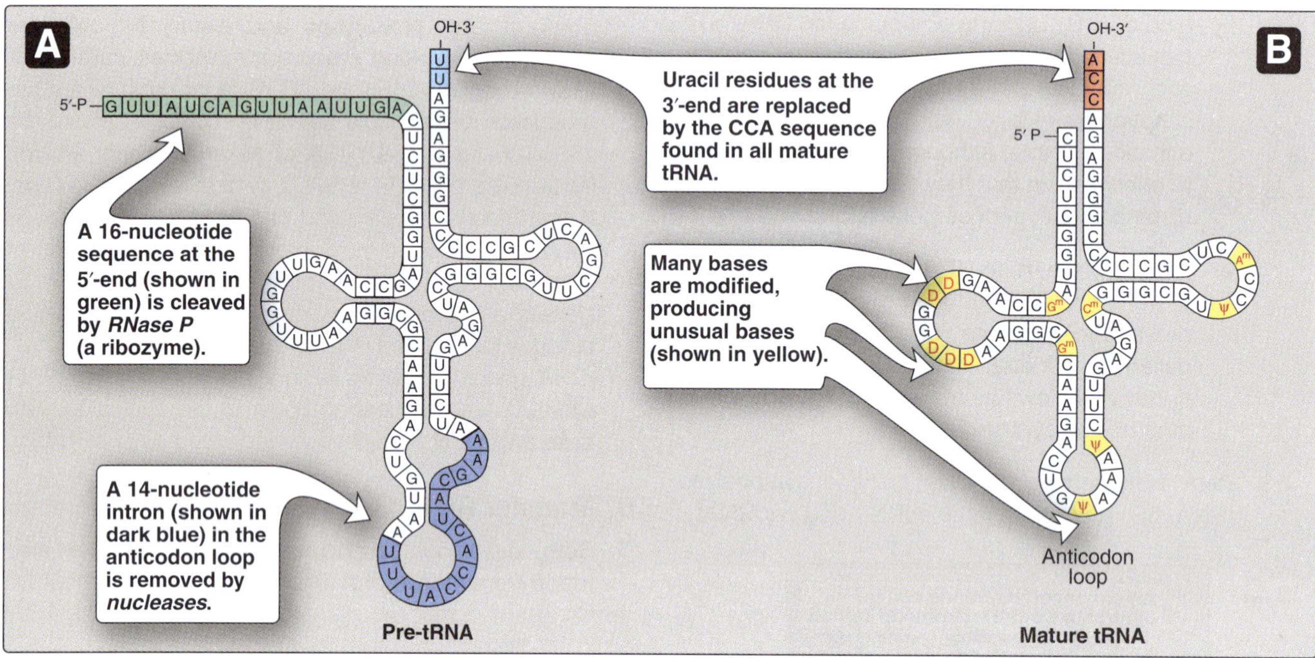

Figure 36.16 **A.** Precursor transfer RNA (pre-tRNA) transcript. **B.** Mature (functional) tRNA after posttranscriptional modification. Modified bases include D (dihydrouracil), ψ (pseudouracil), and ^{m}, which means that the base has been methylated.

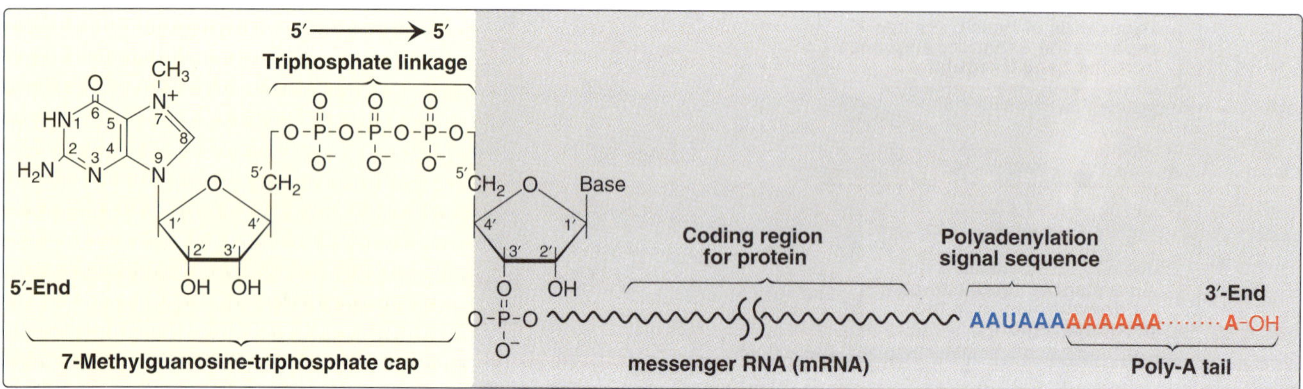

Figure 36.17 Posttranscriptional modification of mRNA showing the 7-methylguanosine cap and polyadenylate (poly-A) tail.

methyl group (see p. 399). Additional methylation steps may occur. The addition of this 7-methylguanosine cap helps stabilize the mRNA and permits efficient initiation of translation (see p. 642).

2. **Addition of a 3′-poly-A tail:** Most eukaryotic mRNA (with several exceptions, including those for the histones) have a chain of 40–250 adenylates (adenosine monophosphates) attached to the 3′-end (see Fig. 36.17). This poly-A tail is not transcribed from the DNA but rather is added by the nuclear enzyme, *polyadenylate polymerase*, using ATP as the substrate. The pre-mRNA is cleaved downstream of a consensus sequence, called the polyadenylation signal sequence (AAUAAA), found near the 3′-end of the RNA, and the poly-A tail is added to the new 3′-end. Tailing terminates eukaryotic transcription. Tails help stabilize the mRNA, facilitate its exit from the nucleus, and aid in translation. After the mRNA enters the cytosol, the poly-A tail is gradually shortened.

3. **Splicing:** Maturation of eukaryotic mRNA usually involves removal from the primary transcript of RNA sequences (introns or intervening sequences) that do not code for protein. The remaining coding (expressed) sequences, the exons, are joined together to form the mature mRNA. The process of removing introns and joining exons is called splicing. The molecular complex that accomplishes these tasks is known as the spliceosome. **Spliceosome** consists of primary transcript plus snRNA like U1, U2, U4, U5, U6 and approximately 60 other proteins. A few eukaryotic primary transcripts contain no introns (for example, those from histone genes). Others contain a few introns, whereas some, such as the primary transcripts for the α chains of collagen, contain >50 introns that must be removed.

   a. **Role of small nuclear RNA:** In association with multiple proteins, uracil-rich snRNA form five small nuclear ribonucleoprotein particles (snRNP, or "snurp") designated as U1, U2, U4, U5, and U6 that mediate splicing. They facilitate the removal of introns by forming base pairs with the consensus sequences at each end of the intron (Fig. 36.18). [Note: In systemic lupus erythematosus, an autoimmune disease, patients produce antibodies against their own nuclear proteins such as snRNP.]

   b. **Mechanism:** The binding of snRNP brings the sequences of neighboring exons into the correct alignment for splicing, allowing two transesterification reactions (catalyzed by the RNA of U2, U5, and U6) to occur. The 2′-OH group of an adenine nucleotide (known as the branch site A) in the intron attacks the phosphate at the 5′-end of the intron (splice-donor site), forming an unusual 2′→5′-phosphodiester

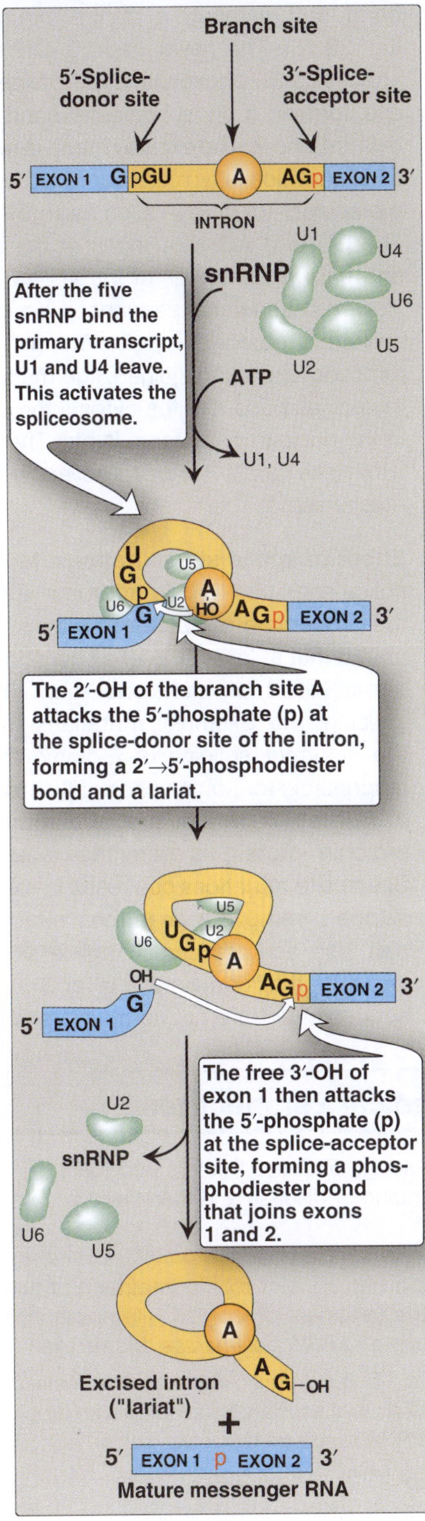

Figure 36.18 Splicing. [Note: U1 binds the 5′-donor site, and U2 binds the branch A and the 3′-acceptor site. Addition of U4–U6 completes the complex.] snRNP = small nuclear ribonucleoprotein particle.

bond and creating a "lariat" structure (see Fig. 36.18). The newly freed 3′-OH of exon 1 attacks the 5′-phosphate at the splice acceptor site, forming a phosphodiester bond that joins exons 1 and 2. The excised intron is released as a **lariat,** which is typically degraded but may be a precursor for ncRNA such as snoRNA. [Note: The GU and AG sequences at the beginning and end, respectively, of introns are invariant. However, additional sequences are critical for splice-site recognition.] After introns have been removed and exons joined, the mature mRNA molecules pass into the cytosol through pores in the nuclear membrane. [Note: The introns in tRNA (see Fig. 36.16) are removed by a different mechanism.]

c. **Effect of splice site mutations:** Mutations at splice sites can lead to improper splicing and the production of aberrant proteins. It is estimated that at least 20% of all genetic diseases are a result of mutations that affect RNA splicing. For example, mutations that cause the incorrect splicing of β-globin mRNA are responsible for some cases of **β-thalassemia**, a disease in which the production of the β-globin protein is defective (see p. 110). Splice site mutations can result in exons being skipped (removed) or introns retained. They can also activate cryptic splice sites, which are sites that contain the 5′ or 3′ consensus sequence but are not normally used.

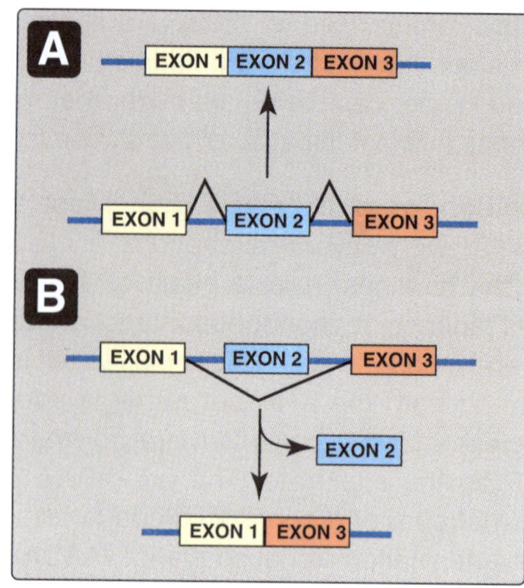

Figure 36.19 Alternative splicing patterns in eukaryotic messenger RNA (mRNA). The removal (skipping) of exon 2 from the mRNA in panel **B** results in a protein product that is different than the one made from the mRNA in panel **A**.

4. **Alternative splicing:** The pre-mRNA molecules from >90% of human genes can be spliced in alternative ways in different tissues. Because this produces multiple variations of the mRNA and, therefore, of its protein product (Fig. 36.19), it is a mechanism for producing a large, diverse set of proteins from a limited set of genes. For example, the mRNA for tropomyosin (TM), an actin filament–binding protein of the cytoskeleton (and of the contractile apparatus in muscle cells), undergoes extensive tissue-specific alternative splicing with production of **multiple isoforms of the TM protein**.

VI. Medical Perspectives

The knowledge of RNA structure and function, both in eukaryotes as well as prokaryotes has led to many advances in understanding of disease and therapeutics.

Drugs

Rifampicin has a major role in treatment of tuberculosis, being active against **Mycobacterium tuberculosis.** It is also used in the treatment of leprosy. Rifampicin inhibits transcription initiation by binding to the beta subunit of prokaryotic *DNA-dependent RNA polymerase*. Mammalian cell *RNA polymerase* is not affected by Rifampicin.

Actinomycin-D is a drug isolated from the bacterium **Streptomyces**. It works by insertion of phenoxazone rings between G-C bp of DNA as well as RNA. It interferes with *topoisomerase* as well as *RNA polymerase* activity. This hampers DNA replication as well as RNA synthesis by blocking chain elongation. This drug is used as a strong antibacterial as well as an anticancer drug.

Poisoning

Some mushrooms are known to be very poisonous as they have a toxin called **alpha-Amanitin** which inactivates *RNA polymerase II*.

Genetic Diseases

Errors in RNA synthesis or errors in posttranscriptional modification of RNA are the cause of many genetic diseases. For example, incorrect splicing of beta-globin mRNA causes a form of **beta thalassemia**. Other examples are splicing errors due to splice-site mutations leading to **spinal muscular dystrophy** and **Duchenne muscular dystrophy.**

Systemic lupus erythematosus is a disease caused by auto-antibodies against own nuclear proteins such as snRNP (which are involved in splicing).

VII. Chapter Summary

- Three major types of RNA participate in the process of protein synthesis: **ribosomal RNA** (**rRNA**), **transfer RNA** (**tRNA**), and **messenger RNA** (**mRNA**), as shown in Figure 36.20. They are unbranched polymers of nucleotides but differ from DNA by containing **ribose** instead of deoxyribose and **uracil** instead of thymine. **rRNA** is a component of the **ribosomes**. **tRNA** serves as an **adaptor** molecule that carries a specific amino acid to the site of protein synthesis. **mRNA** (coding RNA) carries genetic information from DNA for use in protein synthesis.
- The process of RNA synthesis is called **transcription**, and the substrates are **ribonucleoside triphosphates**. The enzyme that synthesizes RNA is ***RNA polymerase*** (***RNA pol***). In **prokaryotic** cells, the **core enzyme** has five subunits (2 α, 1 β, 1 β′, and 1 Ω) and possesses ***5′→3′ polymerase* activity** needed for transcription. The core enzyme requires an additional subunit, **sigma (σ) factor**, to recognize the nucleotide sequence (**promoter** region) at the beginning of the DNA to be transcribed. This region contains **consensus sequences** that are highly conserved and include the **–10 Pribnow box** and the **–35 sequence**. Another protein, **rho (ρ)**, is required for **termination** of transcription of some genes.
- There are three distinct types of *RNA pol* in the nucleus of **eukaryotic** cells. **RNA pol I** synthesizes the precursor of rRNA in the nucleolus. In the nucleoplasm, ***RNA pol II*** synthesizes the precursors for mRNA and some noncoding RNA, and ***RNA pol III*** synthesizes the precursors of tRNA and 5S rRNA. In both prokaryotes and eukaryotes, *RNA pol* does not require a primer. Proofreading involves the *polymerase* backtracking and cleaving the transcript.
- Core **promoters** for genes transcribed by ***RNA pol II*** contain **cis-acting** consensus sequences, such as the **TATA (Hogness) box**, which serve as binding sites for **trans-acting general transcription factors**. Upstream of these are **proximal** regulatory elements, such as the CAAT and GC boxes, and **distal** regulatory elements, such as **enhancers**. **Specific transcription factors** (transcriptional activators) and **Mediator complex** bind these elements and regulate the frequency of transcription initiation, the response to signals such as hormones, and which genes are expressed at any given time.
- Eukaryotic transcription requires that the **chromatin** be relaxed (decondensed) in a process known as **chromatin remodeling**.
- **A primary transcript** is a linear copy of a **transcription unit**, the segment of DNA between specific initiation and termination sequences. The primary transcripts of both prokaryotic and eukaryotic tRNA and rRNA are **posttranscriptionally modified**. The rRNA are synthesized from long precursor molecules called **pre-rRNA**. These precursors are cleaved and trimmed by *ribonucleases*, producing the three largest rRNA, and bases and sugars are modified. Eukaryotic 5S rRNA is synthesized by *RNA pol III* and is modified separately.
- Prokaryotic and eukaryotic tRNA are also made from longer precursor molecules (**pre-tRNA**). If present, an intron is removed by *nucleases*, and both ends of the molecule are trimmed by *ribonucleases*. A **3′-CCA** sequence is added, and bases at specific positions are modified.
- Prokaryotic mRNA is generally identical to its primary transcript, whereas eukaryotic **pre-mRNA** is extensively modified co- and posttranscriptionally. For example, a **7-methylguanosine cap** is attached to the 5′-end of the mRNA through a 5′→5′ linkage. A **long poly-A tail**, not transcribed from the DNA, is attached by *polyadenylate polymerase* to the 3′-end of most mRNA. Most eukaryotic mRNA also contains **intervening sequences** (**introns**) that must be removed for the mRNA to be functional. Their removal, as well as the joining of **expressed sequences** (**exons**), requires a **spliceosome** composed of **small nuclear ribonucleoprotein particles** (**"snurps"**) that mediate the process of **splicing**.
- Eukaryotic mRNA is **monocistronic**, containing information from just one gene, whereas prokaryotic mRNA is **polycistronic**.

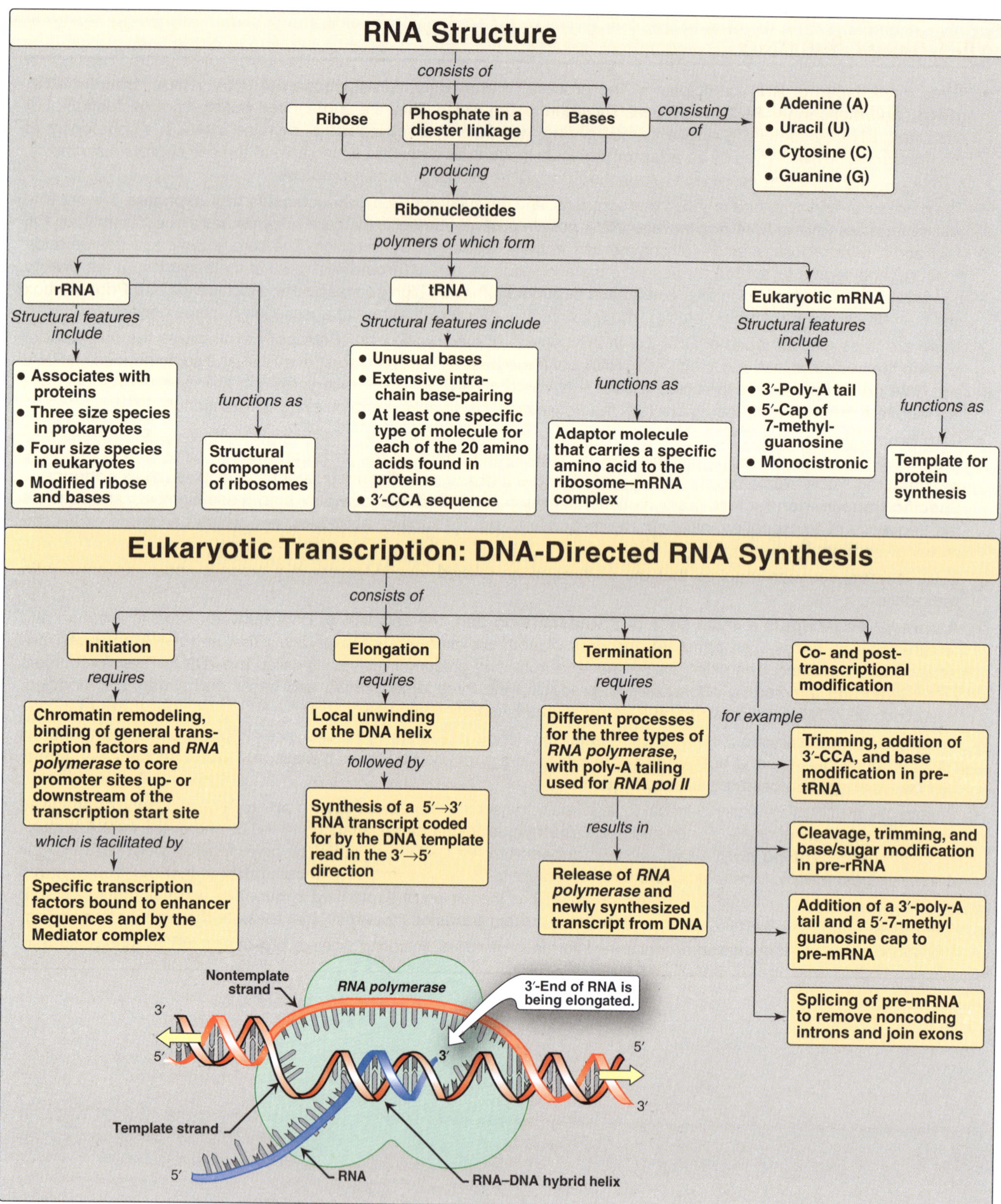

Figure 36.20 Key concept map for RNA structure and synthesis. rRNA = ribosomal RNA; tRNA = transfer RNA; mRNA = messenger RNA.

Study Questions

Short-Answer Questions

36.1. Enumerate the posttranscriptional modifications and describe splicing.
36.2. Discuss the types of RNA and their function.
36.3. Explain RNA editing with an example.
36.4. Discuss catalytic functions of RNA.

Long-Answer Questions

36.1. Discuss the process of prokaryotic gene transcription in detail. Explain with examples how this knowledge has been used for therapeutics.
36.2. Explain how prokaryotic and eukaryotic gene expression differs from each other while retaining some basic similarities. (The answer to this question will require inputs from chapter 38 also)

Multiple-Choice Questions

Choose the ONE best answer.

36.1. Which among the following is NOT a function of the poly A tail of the eukaryotic mRNA?
A. It protects the 3' end of the mRNA from degradation
B. It promotes the recognition of the mRNA during the initiation of translation.
C. It gets translated to a poly-lysine chain during protein synthesis, protecting the protein from carboxypeptidase mediated degradation.
D. It facilitates the export of mRNA from the nucleus to the cytoplasm for translation.

Correct answer = C.

36.2. Which among the following protein factors identify the promoter of protein-coding genes in eukaryotes?
A. Pribnow box
B. Rho
C. Sigma
D. TFIID

Correct answer = D.

36.3. Which of the following RNA polymerases is involved in the transcription of large eukaryotic rRNA genes, such as 18S rRNA and 28S rRNA?
A. RNA polymerase I
B. RNA polymerase III
C. RNA polymerase II
D. Mitochondrial RNA polymerase

Correct answer = A.

36.4. Which among the following processes can be utilized to generate multiple transcripts from the same primary transcript?
I. 5' capping II. 3' polyadenylation III. RNA editing IV. Splicing
A. I, II, III and IV
B. II, III, and IV only
C. III and IV only
D. I and IV only

Correct answer = B.

36.5. Lipid transport from the liver is mediated by apoB-100 and that from the intestine is mediated by apoB-48. However, both the proteins are coded by the same gene in the liver and the intestine. The nomenclature of the proteins derives from the fact that the intestinal protein is 48% length of the liver protein which is considered 100% long. Which of the following mechanisms is responsible for the difference in the lengths of the proteins?
A. RNA editing
B. Alternative polyadenylation
C. Differential promoter usage
D. Alternative splicing

Correct answer = A.

36.6. Statement 1: Dactinomycin is used as a chemotherapeutic agent in the treatment of various cancers.
Statement 2: Dactinomycin intercalates into DNA double helix inhibiting DNA replication, transcription remaining unaffected.
A. Both the Statements are correct, and Statement 2 is the correct explanation for Statement 1
B. Both the Statements are correct, but Statement 2 is NOT the correct explanation for Statement 1
C. Statement 1 is correct, but Statement 2 is NOT correct
D. Statement 1 is NOT correct, but Statement 2 is correct

Correct answer = C.

36.7. Which among the following statements about enhancers is correct?
A. They are invariably located within 100 bp of the gene promoter.
B. They are composed of a specific sequence of nucleotides.
C. They are always located 5' (i.e., upstream) to the promoter.
D. They interact with specific transcription factors and alter the structure of the DNA.

Correct answer = D.

36.8. Which of the following represents the correct sequence of events during the initiation of transcription?
A. Closed complex -> Open complex -> Synthesis of small fragments of RNA -> Escape from the promoter

B. Closed complex -> Synthesis of small fragments of RNA -> Open complex -> Escape from the promoter
C. Open complex -> Closed complex -> Escape from the promoter -> Synthesis of small fragments of RNA
D. Closed complex -> Escape from the promoter -> Open complex -> Synthesis of small fragments of RNA

Correct answer = A.

36.9. Which of the following is true about transfer-RNA (t-RNA)?

A. They are the most abundant RNA species in a cell.
B. The three dimensional structure of a tRNA is described as the clover-leaf model.
C. There are multiple tRNAs, each unique to a specific triplet codon.
D. They are rich in modified and unusual nitrogen bases.

Correct answer = D.

36.10. Which among the following is the RNA product of the DNA template sequence, 5′-GATCTAC-3′?

A. 5′-GAUGAUC-3′
B. 5′-CATCTAG-3′
C. 5′-GATCTAC-3′
D. 5′-CUAGAUG-3′

Correct answer = A.

Multiple-Choice Questions (NExT/NEET PG)

Choose the ONE best answer.

36.1. An 8-month-old male with severe anemia is found to have β-thalassemia. Genetic analysis shows that one of his β-globin genes has a mutation that creates a new splice-acceptor site 19 nucleotides upstream of the normal splice-acceptor site of the first intron. Which of the following best describes the new messenger RNA molecule that can be produced from this mutant gene?

A. Exon 1 will be too short.
B. Exon 1 will be too long.
C. Exon 2 will be too short.
D. Exon 2 will be too long.

Correct answer = D.

36.2. A 4-year-old child who easily tires and has trouble walking is diagnosed with Duchenne muscular dystrophy, an X-linked recessive disorder. Genetic analysis shows that the patient's gene for the muscle protein dystrophin contains a mutation in its promoter region. Of the choices listed, which would be the most likely effect of this mutation?

A. Initiation of dystrophin transcription will be defective.
B. Termination of dystrophin transcription will be defective.
C. Capping of dystrophin messenger RNA will be defective.
D. Splicing of dystrophin messenger RNA will be defective.

Correct answer = A.

36.3. A 28-year-old male, previously treated for tuberculosis (TB), presented with a recurrent cough, weight loss, and night sweats. Despite adherence to a rifampicin-based regimen, his symptoms persisted, and investigations confirmed the persistence of *Mycobacterium tuberculosis*. A genetic alteration in which of the following protein subunits/domains in the bacteria could have led to this persistence?

A. 3' to 5' exonuclease domain of DNA polymerase I
B. σ subunit of RNA polymerase
C. β subunit of RNA polymerase
D. Polyadenylate polymerase

Correct answer = C.

36.4. A 30-year-old woman arrived at the emergency department with acute onset of nausea, vomiting, and abdominal cramps, following a meal that included wild mushrooms (*Amanita phalloides*) she had foraged earlier that day. Within 48 hours, she developed signs of liver failure, including jaundice and altered mental status. Which of the following statements is NOT correct in this condition?

A. The clinical manifestations here are a result of a fungal infection of the liver.
B. Hepatic protein synthesis is severely compromised.
C. Transcription of several mRNAs in the liver is affected.
D. RNA polymerase I is not inhibited.

Correct answer = B.

36.5. AAUAAA is the consensus sequence for polyadenylation in the eukaryotic messenger RNA (mRNA). What will be the consequence of a gene variant that disrupts this sequence?

A. The length of the mRNA will be increased.
B. The length of the mRNA will be shortened.
C. The length of the mRNA will remain the same but that of the poly A tail will be decreased.
D. The length of the mRNA and the poly A tail will remain the same.

Correct answer = A.

Protein Synthesis

37

Learning Objectives

At the end of this chapter, the student will be able to demonstrate an understanding of:

- Genetic code
- The process of protein synthesis
- Mutations
- Protein targeting
- Posttranslational modifications

I. OVERVIEW

Genetic information, stored in the chromosomes and transmitted to daughter cells through DNA replication, is expressed through transcription to RNA and, in the case of messenger RNA (mRNA), subsequent translation into proteins (polypeptides) as shown in Figure 37.1. [Note: The proteome is the complete set of proteins expressed in a cell.] The process of protein synthesis is called translation because the "language" of the nucleotide sequence on the mRNA is translated into the language of an amino acid sequence. Translation requires a genetic code, through which the information contained in the nucleotide sequence is expressed to produce a specific amino acid sequence. Any alteration in the nucleotide sequence may result in an incorrect amino acid being inserted into the protein, potentially causing disease or even death of the organism. Newly made immature (nascent) proteins undergo a number of processes to achieve their functional form. They must fold properly, and misfolding can result in aggregation or degradation of the protein. Many proteins are covalently modified to alter their activities. Lastly, proteins are targeted to their final intra- or extracellular destinations by signals present in the proteins themselves.

II. THE GENETIC CODE

The genetic code is a "dictionary" that identifies the correspondence between a sequence of nucleotide bases and a sequence of amino acids. Each individual "word" in the code is composed of three nucleotide bases. These genetic words are called codons.

A. Codons

Codons are presented in the mRNA language of adenine (A), guanine (G), cytosine (C), and uracil (U). Their nucleotide sequences are always written from the 5′-end to the 3′-end. The four nucleotide bases are used to produce the three-base codons. Therefore, 64 different combinations of bases exist, taken three at a time (**a triplet code**), as shown in the table in Figure 37.2.

1. **How to translate a codon:** This table can be used to translate any codon and, thus, to determine which amino acids are coded for by an mRNA sequence. For example, the codon AUG codes for methionine ([Met], see Fig. 37.2). [Note: AUG is the initiation (start) codon for translation.] Sixty-one of the 64 codons code for the 20 standard amino acids.

2. **Termination codons:** Three of the codons, UAA, UAG, and UGA, do not code for amino acids but, rather, are termination (also called **stop, or nonsense) codons**. When one of these codons appears in an mRNA sequence, synthesis of the polypeptide coded for by that mRNA stops.

B. Characteristics

Usage of the genetic code is remarkably consistent throughout all living organisms. It is assumed that once the standard genetic code evolved in primitive organisms, any mutation (a permanent change in DNA sequence) that altered its meaning would have caused the alteration of most, if not all, protein

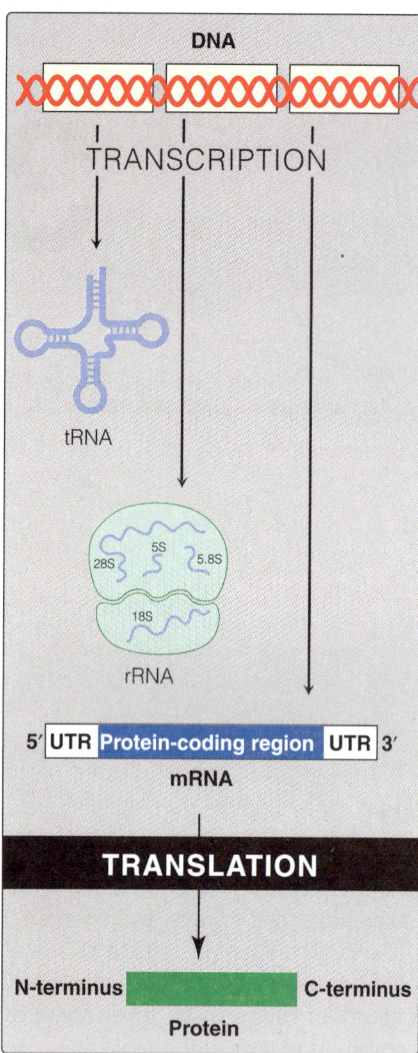

Figure 37.1 Protein synthesis or translation. tRNA = transfer RNA; rRNA = ribosomal RNA; mRNA = messenger RNA; UTR = untranslated region.

sequences, resulting in lethality. Characteristics of the genetic code include the following.

1. **Specificity:** The genetic code is **specific (unambiguous),** because a particular codon always codes for the same amino acid.

2. **Universality:** The genetic code is virtually universal insofar as its specificity has been conserved from very early stages of evolution, with only slight differences in the manner in which the code is translated. [Note: An exception occurs in mitochondria, in which a few codons have meanings different than those shown in Figure 37.2. For example, UGA codes for tryptophan (Trp).]

3. **Degeneracy:** The genetic code is **degenerate (sometimes called redundant**). Although each codon corresponds to a single amino acid, **a given amino acid may have more than one triplet coding for it**. For example, arginine (Arg) is specified by six different codons (see Fig. 37.2). Only Met and Trp have just one coding triplet.

4. **Nonoverlapping and commaless:** The genetic code is nonoverlapping and commaless, meaning that the code is read from a fixed starting point as a continuous sequence of bases, taken three at a time without any punctuation between codons. For example, AGCUGGAUACAU is read as AGC UGG AUA CAU.

C. Consequences of altering the nucleotide sequence

1. Changing a single nucleotide base (a point mutation) in the coding region of an mRNA can lead to any one of three results (Fig. 37.3). [Note: Sickle cell anemia is also caused by a point mutation, see Chapter 8, p. 107.]

   a. **Silent mutation:** The codon containing the changed base **may code for the same amino acid**. For example, if the serine (Ser) codon UCA is changed at the third base and becomes UCU, it still codes for Ser. This is termed a silent mutation.

   b. **Missense mutation:** The codon containing the changed base may code for a different amino acid. For example, if the Ser codon UCA is changed at the first base and becomes CCA, it will **code for a different amino acid** (in this case, proline [Pro]). This is termed a missense mutation.

   c. **Nonsense mutation:** The codon containing the changed base may become a termination codon. For example, if the Ser codon UCA is changed at the second base and becomes UAA, the new codon causes premature termination of translation at that point and the production of a shortened (truncated) protein. This is termed a nonsense mutation. [Note: The nonsense-mediated degradation pathway can degrade mRNA containing premature stops.]

2. **Other mutations:** These can alter the amount or structure of the protein produced by translation.

   a. **Trinucleotide repeat expansion:** Occasionally, a sequence of three bases that is repeated in tandem will become amplified in number so that too many copies of the triplet occur. If this happens within the coding region of a gene, the protein will contain many extra copies of one amino acid. For example, expansion of the

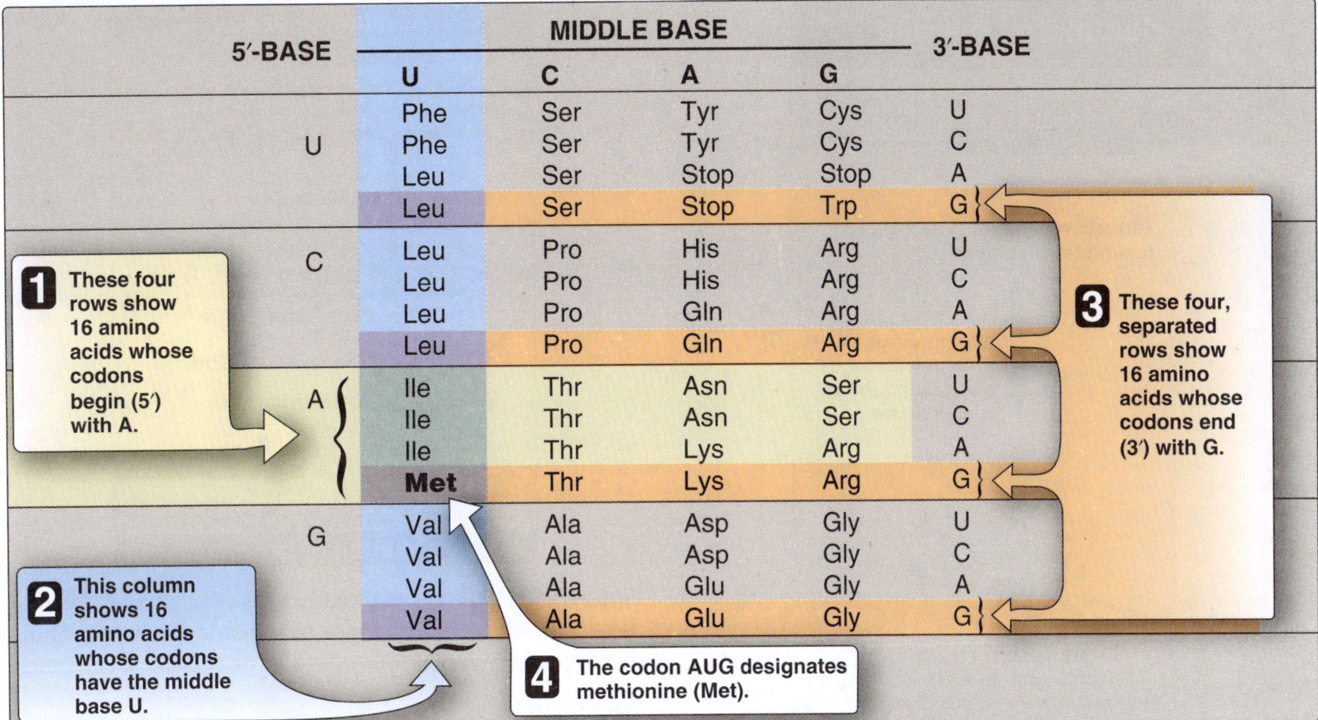

Figure 37.2 Use of the genetic code table to translate the codon AUG. A = adenine; G = guanine; C = cytosine; U = uracil. The three-letter abbreviations for many common amino acids are shown as examples.

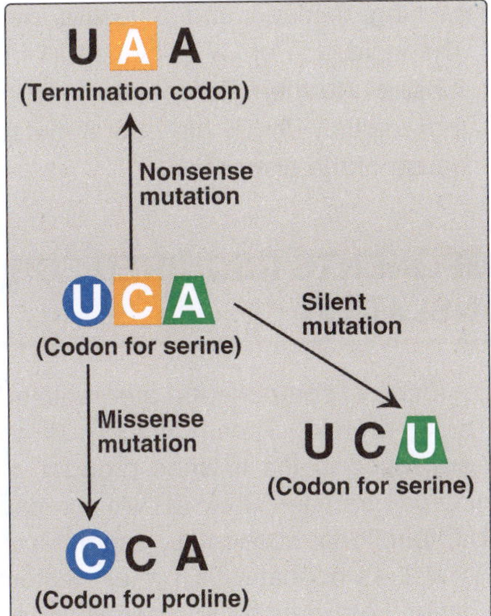

Figure 37.3 Possible effects of changing a single nucleotide base in the coding region of a messenger RNA. A = adenine; C = cytosine; U = uracil.

CAG codon in exon 1 of the gene for huntingtin protein leads to the insertion of many extra glutamine residues in the protein, causing the neurodegenerative disorder Huntington disease (Fig. 37.4). The additional glutamines result in an abnormally long protein that is cleaved, producing toxic fragments that aggregate in neurons. If the trinucleotide repeat expansion occurs in an untranslated region (UTR) of a gene, the result can be a decrease in the amount of protein produced, as seen in **fragile X syndrome and myotonic dystrophy**. Over 20 triplet expansion diseases are known. [Note: In fragile X syndrome, the most common cause of intellectual disability in males, the expansion results in gene silencing through DNA hypermethylation (see p. 664).]

b. **Splice site mutations:** Mutations at splice sites (see p. 630) can alter the way in which introns are removed from pre-mRNA molecules, producing aberrant proteins. [Note: In **myotonic dystrophy**, a muscle disorder, gene silencing is the result of splicing alterations due to triplet expansion.]

c. **Frameshift mutations:** If one or two nucleotides are either deleted from or added to the coding region of an mRNA, a frameshift mutation occurs, altering the reading frame. This can result in a product with a radically different amino acid sequence or a truncated product due to the eventual creation of a termination codon (Fig. 37.5). If three nucleotides are added, a new amino acid is added to the peptide. If three are deleted, an amino acid is lost. Loss of three nucleotides maintains the reading frame but can result in serious

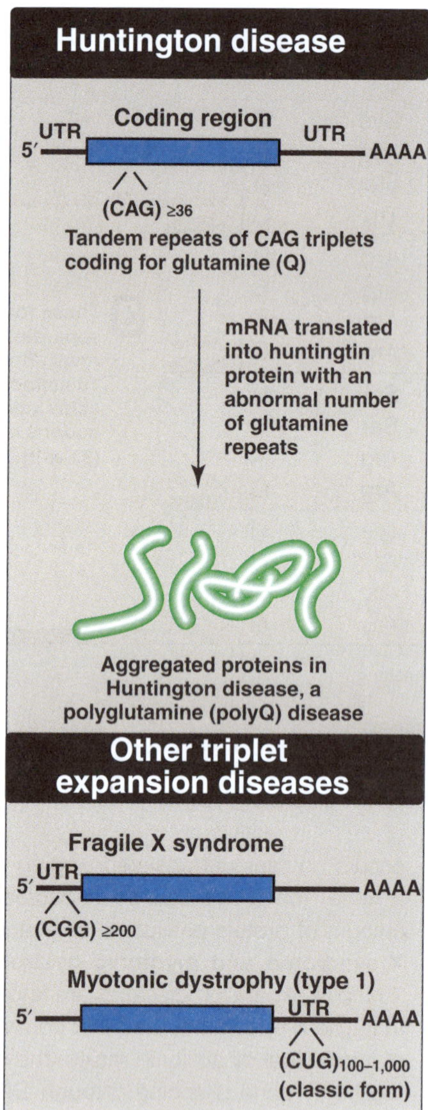

Figure 37.4 Tandem triplet repeats in messenger RNA (mRNA) causing Huntington disease and other triplet expansion diseases. [Note: In unaffected individuals, the number of repeats in the huntingtin protein is <27; in fragile X mental retardation protein, it is 5–44; and in *myotonic dystrophy protein kinase*, it is 5–34.] UTR = untranslated region; A = adenine; C = cytosine; G = guanine; U = uracil; Q = single-letter abbreviation for glutamine.

pathology. For example**, cystic fibrosis (CF)**, a chronic, progressive, inherited disease that primarily affects the pulmonary and digestive systems, is most commonly caused by deletion of three nucleotides from the coding region of a gene, resulting in the loss of phenylalanine (Phe, or F) at the 508th position (ΔF508) in the **CF transmembrane conductance regulator (CFTR)** protein encoded by that gene. This ΔF508 mutation prevents normal folding of CFTR, leading to its destruction by

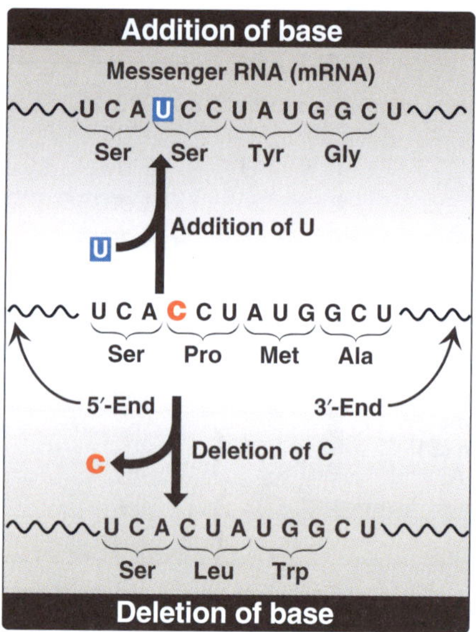

Figure 37.5 Frameshift mutations as a result of addition or deletion of a base can cause an alteration in the reading frame of mRNA. A = adenine; C = cytosine; G = guanine; U = uracil.

the proteasome. CFTR normally functions as a chloride channel in epithelial cells, and its loss results in the production of thick, sticky secretions in the lungs and pancreas, leading to lung damage and digestive deficiencies. The incidence of CF is highest (1 in 3,300) in those of Northern European origin. In >70% of individuals with CF, the ΔF508 mutation is the cause of the disease.

III. COMPONENTS REQUIRED FOR TRANSLATION

A large number of components are required for the synthesis of a protein. These include all the amino acids that are found in the finished product, the mRNA to be translated, transfer RNA (tRNA) for each of the amino acids, functional ribosomes, energy sources, and enzymes as well as noncatalytic protein factors needed for the initiation, elongation, and termination steps of polypeptide chain synthesis.

A. Amino acids

All the amino acids that eventually appear in the finished protein must be present at the time of protein synthesis. If one amino acid is missing, translation stops at the codon specifying that amino acid. [Note: This demonstrates the importance of having all the essential amino acids (see p. 397) in sufficient quantities in the diet to insure continued protein synthesis.]

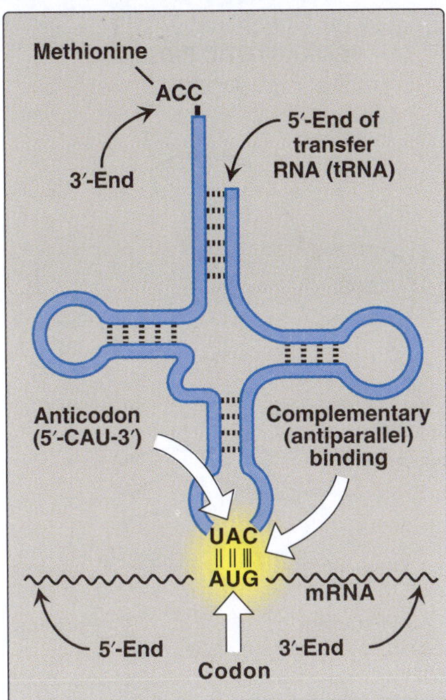

Figure 37.6 Complementary, antiparallel binding of the anticodon for methionyl-tRNA (CAU) to the messenger RNA (mRNA) codon for methionine (AUG), the initiation codon for translation.

B. Transfer RNA

At least one specific type of tRNA is required for each amino acid. In humans, there are at least 50 species of tRNA, whereas bacteria contain at least 30 species. Because there are only 20 different amino acids commonly carried by tRNA, some amino acids have more than one specific tRNA molecule. This is particularly true of those amino acids that are coded for by several codons.

1. **Amino acid attachment site:** Each tRNA molecule has an attachment site for a specific (cognate) amino acid at its 3′-end (Fig. 37.6). The carboxyl group of the amino acid is in an ester linkage with the 3′-hydroxyl of the ribose portion of the A nucleotide in the –CCA sequence at the 3′-end of the tRNA. [Note: A tRNA with a covalently attached (activated) amino acid is charged. Without an attached amino acid, it is uncharged.]
2. **Anticodon: Each tRNA molecule also contains a three-base nucleotide sequence, the anticodon, which pairs with a specific codon on the mRNA** (see Fig. 37.6). This codon specifies the insertion into the growing polypeptide chain of the amino acid carried by that tRNA.

C. Aminoacyl-tRNA synthetases

This family of 20 different enzymes is required for attachment of amino acids to their corresponding tRNA. Each member of this family recognizes a specific amino acid and all the tRNA that correspond to that amino acid (isoaccepting tRNA, up to five per amino acid). *Aminoacyl-tRNA synthetases* catalyze a two-step reaction that results in the covalent attachment of the α-carboxyl group of an amino acid to the A in the –CCA sequence at the 3′-end of its corresponding tRNA. The overall reaction requires ATP, which is cleaved to adenosine monophosphate and inorganic pyrophosphate (PP_i), as shown in Figure 37.7. The extreme specificity of the *synthetases* in recognizing both the amino acid and its cognate tRNA contributes to the high fidelity of translation of the genetic message. In addition to their synthetic activity, the ***aminoacyl-tRNA synthetases* have a proofreading, or editing activity** that can remove an incorrect amino acid from the enzyme or the tRNA molecule.

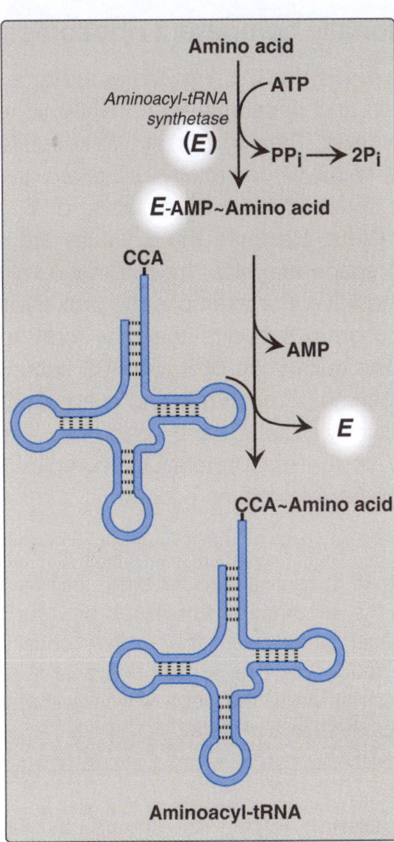

Figure 37.7 Attachment of a specific amino acid to its corresponding transfer RNA (tRNA) by an *aminoacyl-tRNA synthetase*. PP_i = pyrophosphate; P_i = inorganic phosphate; A = adenine; C = cytosine; AMP = adenosine monophosphate; ~ = high-energy bond.

D. Messenger RNA

The specific mRNA required as a template for the synthesis of the desired polypeptide must be present. [Note: In eukaryotes, mRNA is circularized for use in translation.]

E. Functionally competent ribosomes

As shown in Figure 37.8, ribosomes are large complexes of protein and ribosomal RNA (rRNA), in which rRNA predominates. They consist of two subunits (one large and one small) whose relative sizes are given in terms of their sedimentation coefficients, or **S (Svedberg)** values. [Note: Because the S values are determined by both shape and size, their numeric values are not strictly additive. For example, the prokaryotic 50S and 30S ribosomal subunits together form a 70S ribosome. The eukaryotic 60S and 40S subunits form an 80S ribosome.] Prokaryotic and eukaryotic ribosomes are similar in structure and serve the same function, namely, as the macromolecular complexes in which the synthesis of proteins occurs.

> The small ribosomal subunit binds mRNA and determines the accuracy of translation by insuring correct base-pairing between the mRNA codon and the tRNA anticodon. The large ribosomal subunit catalyzes formation of the peptide bonds that link amino acid residues in a protein.

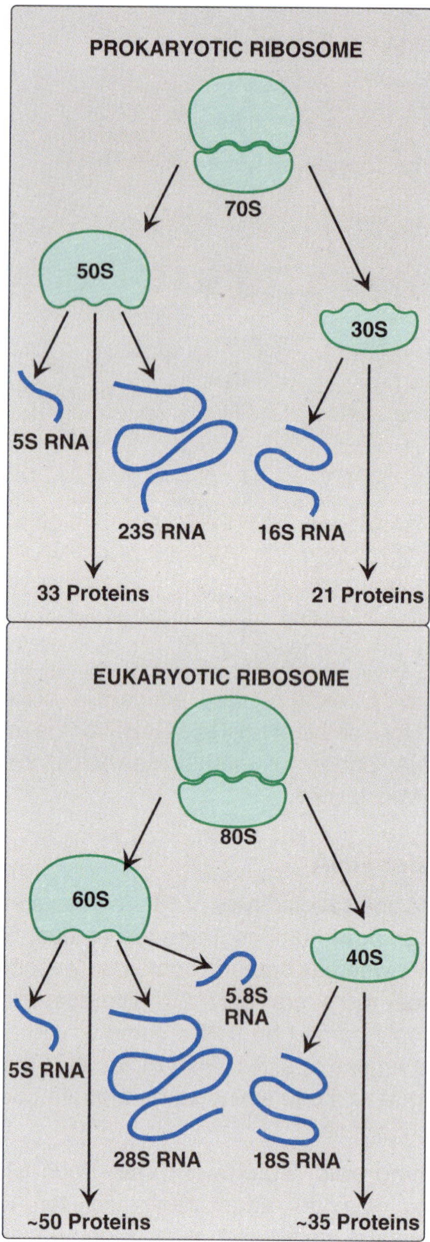

Figure 37.8 Ribosomal composition. [Note: The number of proteins in the eukaryotic ribosomal subunits varies somewhat from species to species.] S = Svedberg unit.

1. **Ribosomal RNA:** As shown in Figure 37.8, prokaryotic ribosomes contain three size species of rRNA, whereas eukaryotic ribosomes contain four (see Fig. 37.8). The rRNA are generated from a single pre-rRNA by the action of *ribonucleases*, and some bases and riboses are modified.

2. **Ribosomal proteins:** Ribosomal proteins are present in greater numbers in eukaryotic ribosomes than in prokaryotic ribosomes. These proteins play a variety of roles in the structure and function of the ribosome and its interactions with other components of the translation system.

3. **A, P, and E sites:** The ribosome has three binding sites for tRNA molecules: the A, P, and E sites, each of which extends over both subunits. Together, they cover three neighboring codons. During translation, the A site binds an incoming aminoacyl-tRNA as directed by the codon currently occupying this site. This codon specifies the next amino acid to be added to the growing peptide chain. The P site is occupied by peptidyl-tRNA. This tRNA carries the chain of amino acids that has already been synthesized. The E site is occupied by the empty tRNA as it is about to exit the ribosome. (See Fig. 37.13 for an illustration of the role of the A, P, and E sites in translation.)

4. **Cellular location:** In eukaryotic cells, the ribosomes either are free in the cytosol or are in close association with the endoplasmic reticulum (which is then known as the rough endoplasmic reticulum, or RER). **RER-associated ribosomes** are responsible for synthesizing proteins (including glycoproteins; see p. 269) that are to be exported from the cell, incorporated into membranes, or imported into lysosomes (see p. 272 for an overview of the latter process). **Cytosolic ribosomes** synthesize proteins required in the cytosol itself or destined for the nucleus, mitochondria, or peroxisomes. [Note: Mitochondria contain their own ribosomes (55S) and their own unique, circular DNA. Most mitochondrial proteins, however, are

encoded by nuclear DNA, synthesized completely in the cytosol, and then targeted to mitochondria.]

F. Protein factors

Initiation, elongation, and termination (or, release) factors are required for polypeptide synthesis. Some of these protein factors perform a catalytic function, whereas others appear to stabilize the synthetic machinery. [Note: A number of the factors are small, cytosolic G proteins and thus are active when bound to guanosine triphosphate (GTP) and inactive when bound to guanosine diphosphate (GDP). See p. 189 for a discussion of the membrane-associated G proteins.]

G. Energy sources

Cleavage of four high-energy bonds (see p. 163) is required for the addition of one amino acid to the growing polypeptide chain: two from ATP in the *aminoacyl-tRNA synthetase* reaction, one in the removal of PP_i and one in the subsequent hydrolysis of the PP_i, to two P_i by *pyrophosphatase*, and two from GTP, one for binding the aminoacyl-tRNA to the A site and one for the translocation step (see Fig. 37.13, p. 645). [Note: Additional ATP and GTP molecules are required for initiation in eukaryotes, whereas an additional GTP molecule is required for termination in both eukaryotes and prokaryotes.] Translation, then, is a major consumer of energy.

IV. CODON RECOGNITION BY TRANSFER RNA

Correct pairing of the codon in the mRNA with the anticodon of the tRNA is essential for accurate translation (see Fig. 37.6). Most tRNA (isoaccepting tRNA) recognize more than one codon for a given amino acid.

A. Antiparallel binding between codon and anticodon

Binding of the tRNA anticodon to the mRNA codon follows the rules of complementary and antiparallel binding, that is, the mRNA codon is read 5′→3′ by an anticodon pairing in the opposite (3′→5′) orientation (Fig. 37.9). [Note: Nucleotide sequences are always written in the 5′ to 3′ direction unless otherwise noted. Two nucleotide sequences orient in an antiparallel manner.]

B. Wobble hypothesis

The mechanism by which a tRNA can recognize more than one codon for a specific amino acid is described by the wobble hypothesis, which states that codon–anticodon pairing follows the traditional Watson-Crick rules (G pairs with C and A pairs with U) for the first two bases of the codon but can be less stringent for the last base. The base at the 5′-end of the anticodon (the first base of the anticodon) is not as spatially defined as the other two bases. Movement of that first base allows nontraditional base-pairing with the 3′-base of the codon (the last base of the codon). This movement is called wobble and allows a single tRNA to recognize more than one codon. Examples of these flexible pairings are shown in Figure 37.9. The result of wobble is that 61 tRNA species are not required to read the 61 codons that code for amino acids.

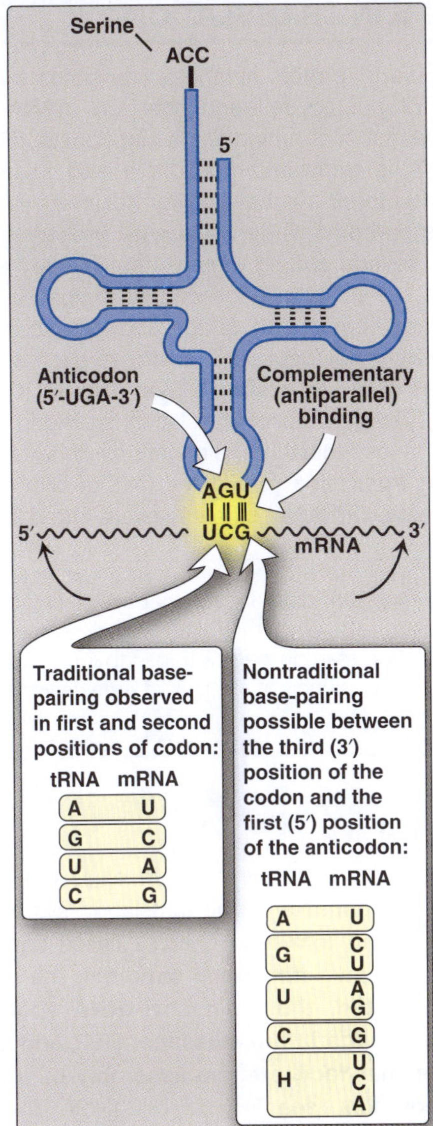

Figure 37.9 Wobble: Nontraditional base-pairing between the 5′-nucleotide (first nucleotide) of the anticodon and the 3′-nucleotide (last nucleotide) of the codon. Hypoxanthine (H) is the product of adenine deamination and the base in the nucleotide inosine monophosphate (IMP). A = adenine; G = guanine; C = cytosine; U = uracil; tRNA = transfer RNA; mRNA = messenger RNA.

V. STEPS IN TRANSLATION

The process of protein synthesis translates the 3-letter alphabet of nucleotide sequences on mRNA into the 20-letter alphabet of amino acids that constitute proteins. The mRNA is translated from its 5′-end to its 3′-end, producing a protein synthesized from its amino (N)-terminal end to its carboxyl (C)-terminal end. Prokaryotic mRNA often has several coding regions (that is, they **are polycistronic**; see p. 621). Each coding region has its own initiation and termination codon and produces a separate species of polypeptide. In contrast, each **eukaryotic mRNA has only one coding region (that is, it is monocistronic)**. The process of translation is divided into three separate steps: initiation, elongation, and termination. Eukaryotic translation resembles that of prokaryotes in most aspects. Individual differences are noted in the text.

> One important difference is that translation and transcription are temporally linked in prokaryotes, with translation starting before transcription is completed as a consequence of the lack of a nuclear membrane in prokaryotes.

A. Initiation

Initiation of protein synthesis involves the assembly of the components of the translation system before peptide-bond formation occurs. These components include the **two ribosomal subunits, the mRNA to be translated, the aminoacyl-tRNA specified by the first codon in the message, GTP, and initiation factors that facilitate the assembly of this initiation complex** (see Fig. 37.13). [Note: In prokaryotes, three initiation factors are known (IF-1, IF-2, and IF-3), whereas in eukaryotes, there are many (designated eIF to indicate eukaryotic origin). Eukaryotes also require ATP for initiation.] The following are two mechanisms by which the ribosome recognizes the nucleotide sequence (AUG) that initiates translation.

1. **Shine-Dalgarno sequence:** In Escherichia coli (E. coli), a purine-rich sequence of nucleotide bases, known as the Shine-Dalgarno (SD) sequence, is located six to ten bases upstream of the initiating AUG codon on the mRNA molecule (that is, near its 5′-end). The 16S rRNA component of the small (30S) ribosomal subunit has a nucleotide sequence near its 3′-end that is complementary to all or part of the SD sequence. Therefore, the 5′-end of the mRNA and the 3′-end of the 16S rRNA can form complementary base pairs, facilitating the positioning of the 30S subunit on the mRNA in close proximity to the initiating AUG codon (Fig. 37.10).

2. **5′-Cap:** Eukaryotic mRNA does not have SD sequences. In eukaryotes, the small (40S) ribosomal subunit (aided by members of the eIF-4 family of proteins) binds close to the cap structure at the 5′-end of the mRNA and moves 5′→3′ along the mRNA until it encounters the initiator AUG. This scanning process requires ATP. Cap-independent initiation can occur if the 40S subunit binds to an internal ribosome entry site close to the start codon. [Note: Interactions between the cap-binding eIF-4 proteins and the poly-A tail–binding proteins on eukaryotic mRNA mediate circularization of the mRNA and likely prevent the use of incompletely processed mRNA in translation.]

3. **Initiation codon:** The initiating AUG is recognized by a special initiator tRNA ($tRNA_i$). Recognition is facilitated by IF-2-GTP in prokaryotes and eIF-2-GTP (plus additional eIF) in eukaryotes. The charged $tRNA_i$ is the only tRNA recognized by (e)IF-2 and the only tRNA to go directly to the P site on the small subunit. [Note: Base modifications distinguish $tRNA_i$ from the tRNA used for internal AUG codons.] In bacteria and mitochondria, $tRNA_i$ carries an N-formylated methionine (fMet), as shown in Figure 37.11. After Met is attached to $tRNA_i$, the formyl group is added by the enzyme *transformylase*, which uses N^{10}-formyl

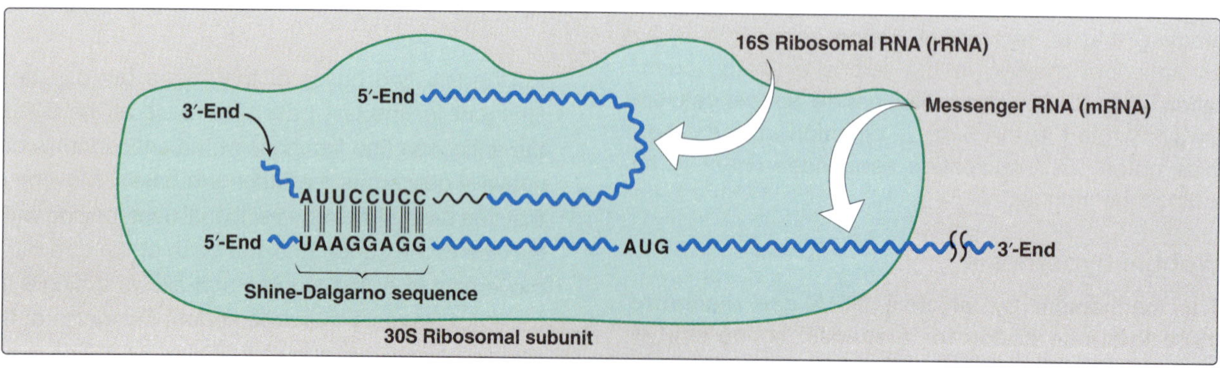

Figure 37.10 Complementary binding between prokaryotic mRNA Shine-Dalgarno sequence and 16S rRNA. S = Svedberg unit.

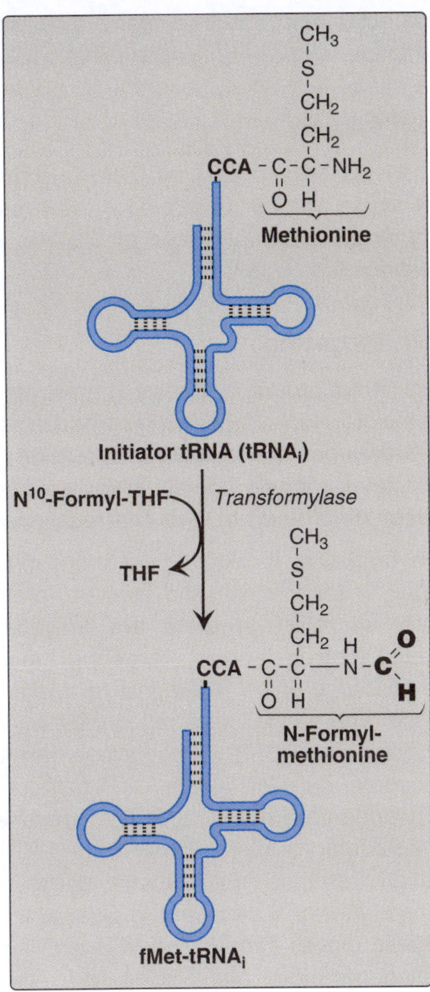

Figure 37.11 Generation of the initiator N-formylmethionyl-transfer RNA (fMet-tRNA$_i$). THF = tetrahydrofolate; C = cytosine; A = adenine.

tetrahydrofolate (see p. 403) as the carbon donor. In eukaryotes, tRNA$_i$ carries a Met that is not formylated. In both prokaryotic and eukaryotic cells, this N-terminal Met is usually removed before translation is completed. The large ribosomal subunit then joins the complex, and a functional ribosome is formed with the charged tRNA$_i$ in the P site. The A site is empty. [Note: Specific (e)IF function as anti-association factors and prevent premature addition of the large subunit.] The GTP on (e)IF-2 gets hydrolyzed to GDP. In eukaryotes, the guanine nucleotide exchange factor eIF-2B facilitates the reactivation of eIF-2-GDP through replacement of GDP by GTP.

B. Elongation

Elongation of the polypeptide involves the addition of amino acids to the carboxyl end of the growing chain. Delivery of the aminoacyl-tRNA whose codon appears next on the mRNA template in the ribosomal A site (a process known as decoding) is facilitated in E. coli by elongation factors EF-Tu-GTP and EF-Ts and requires GTP hydrolysis. [Note: In eukaryotes, comparable elongation factors are EF-1α-GTP and EF-1βγ. Both EF-Ts and EF-1βγ function in guanine nucleotide exchange.] Peptide-bond formation between the α-carboxyl group of the amino acid in the P site and the α-amino group of the amino acid in the A site is catalyzed by ***peptidyltransferase,*** an activity intrinsic to an rRNA of the large subunit (Fig. 37.12). [Note: Because this rRNA catalyzes the reaction, it is a ribozyme (see p. 130).] After the peptide bond has been formed, the peptide on the tRNA at the P site is transferred to the amino acid on the tRNA at the A site, a process known as transpeptidation. The ribosome then advances three nucleotides toward the 3′-end of the mRNA. This process is known as translocation and, in prokaryotes, requires the participation of EF-G-GTP (eukaryotes use EF-2-GTP) and GTP hydrolysis. Translocation causes movement of the uncharged tRNA from the P to the E site for release and movement of the peptidyl-tRNA from the A to the P site. The process is repeated until a termination codon is encountered. [Note: Because of the length of most mRNA, more than one ribosome at a time can translate a message. Such a complex of one mRNA and a number of ribosomes is called a polysome, or polyribosome.]

C. Termination

Termination occurs when one of the three termination codons moves into the A site. These codons are recognized in E. coli by release factors: RF-1, which recognizes UAA and UAG, and RF-2, which recognizes UGA and UAA. The binding of these release factors results in hydrolysis of the bond linking the peptide to the tRNA at the P site, causing the nascent protein to be released from the ribosome. A third release factor, RF-3-GTP, then causes the release of RF-1 or RF-2 as GTP is hydrolyzed (see Fig. 37.13). [Note: Eukaryotes have a single release factor, eRF, which recognizes all three termination codons. A second factor, eRF-3, functions like the prokaryotic RF-3. See Figure 37.14 for a summary of the factors used in translation.] The steps in prokaryotic protein synthesis, as well as some antibiotic inhibitors of the process, are summarized in Figure 37.13. The newly synthesized polypeptide may undergo further modification as described below, and the ribosomal subunits, mRNA, tRNA, and protein factors can be recycled and used to synthesize another polypeptide. [Note: In prokaryotes, ribosome recycling factors mediate separation of the subunits. In eukaryotes, eRF and ATP hydrolysis are required.]

D. Translation regulation

Gene expression is most commonly regulated at the transcriptional level, but translation may also be

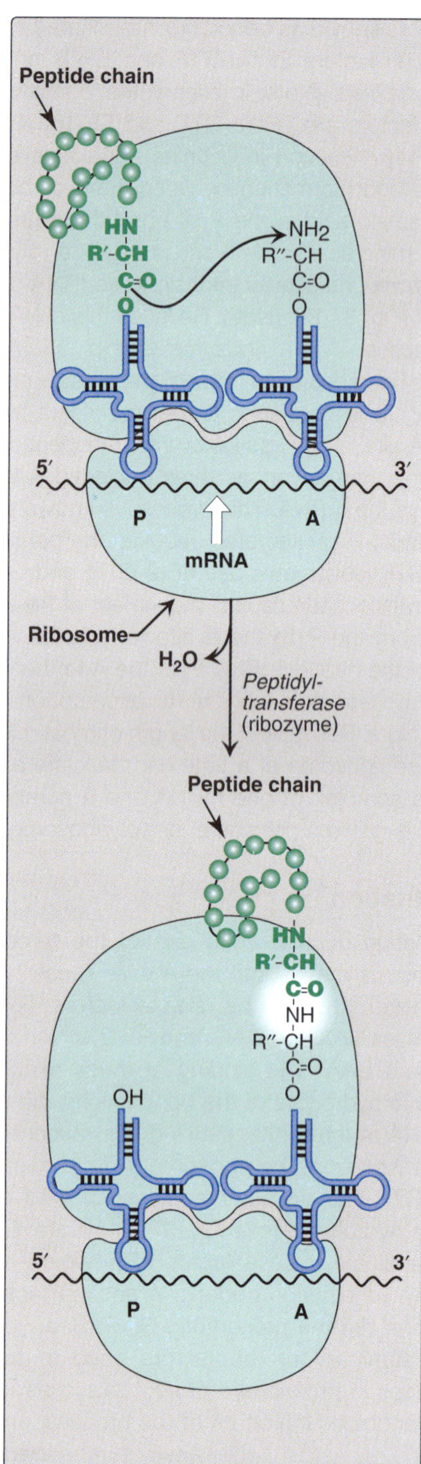

Figure 37.12 Formation of a peptide bond. Peptide-bond formation results in transfer of the peptide on the transfer RNA (tRNA) in the P site to the amino acid on the tRNA in the A site (transpeptidation). mRNA = messenger RNA; R′, R″ = different amino acid side chains.

regulated. An important mechanism by which this is achieved in eukaryotes is by covalent modification of eIF-2: Phosphorylated eIF-2 is inactive (see p. 664). In both eukaryotes and prokaryotes, regulation can also be achieved through proteins that bind mRNA and inhibit its use by blocking translation.

E. Protein folding

Proteins must fold to assume their functional, native state. Folding can be spontaneous (as a result of the primary structure) or facilitated by proteins known as **chaperones** (see p. 89).

F. Protein targeting

Although most protein synthesis in eukaryotes is initiated in the cytoplasm, many proteins perform their functions within subcellular organelles or outside of the cell. Such proteins normally contain amino acid sequences that direct the proteins to their final locations.

Protein targeting is required for many proteins like secretory, mitochondrial and nuclear proteins. For example, **secreted proteins are targeted during synthesis (cotranslational targeting)** to the RER by the presence of an N-terminal hydrophobic signal sequence. The sequence is recognized by the signal recognition particle (SRP), a ribonucleoprotein that binds the ribosome, halts elongation, and delivers the ribosome–peptide complex to an RER membrane channel (the translocon) via interaction with the SRP receptor. Translation resumes, the protein enters the RER lumen, and its signal sequence is cleaved (Fig. 37.15). The protein moves through the RER and the Golgi, is processed, packaged into vesicles, and secreted.

In contrast to secretory proteins, **the nuclear and mitochondrial proteins are synthesized in free cytoplasmic ribosomes and protein targeting here is not a cotranslational event.** Proteins targeted after synthesis (posttranslational) include nuclear proteins that contain an internal, short, basic nuclear localization signal rich in lysine and arginine which marks these proteins to pass through the nuclear pores; mitochondrial matrix proteins that contain an N-terminal, amphipathic, α-helical mitochondrial entry sequence; and peroxisomal proteins that contain a C-terminal tripeptide signal. 95% of mitochondrial proteins are synthesized outside the mitochondria and must then first be targeted to the outer mitochondrial membrane with an amino terminal sequence of 20–70 amino acids, rich in serine and threonine. Several proteolytic events occur to facilitate entry of these proteins into the mitochondrial inner membrane and matrix.

VI. CO- AND POSTTRANSLATIONAL MODIFICATIONS

Many polypeptides are covalently modified, either while they are still attached to the ribosome (cotranslational)

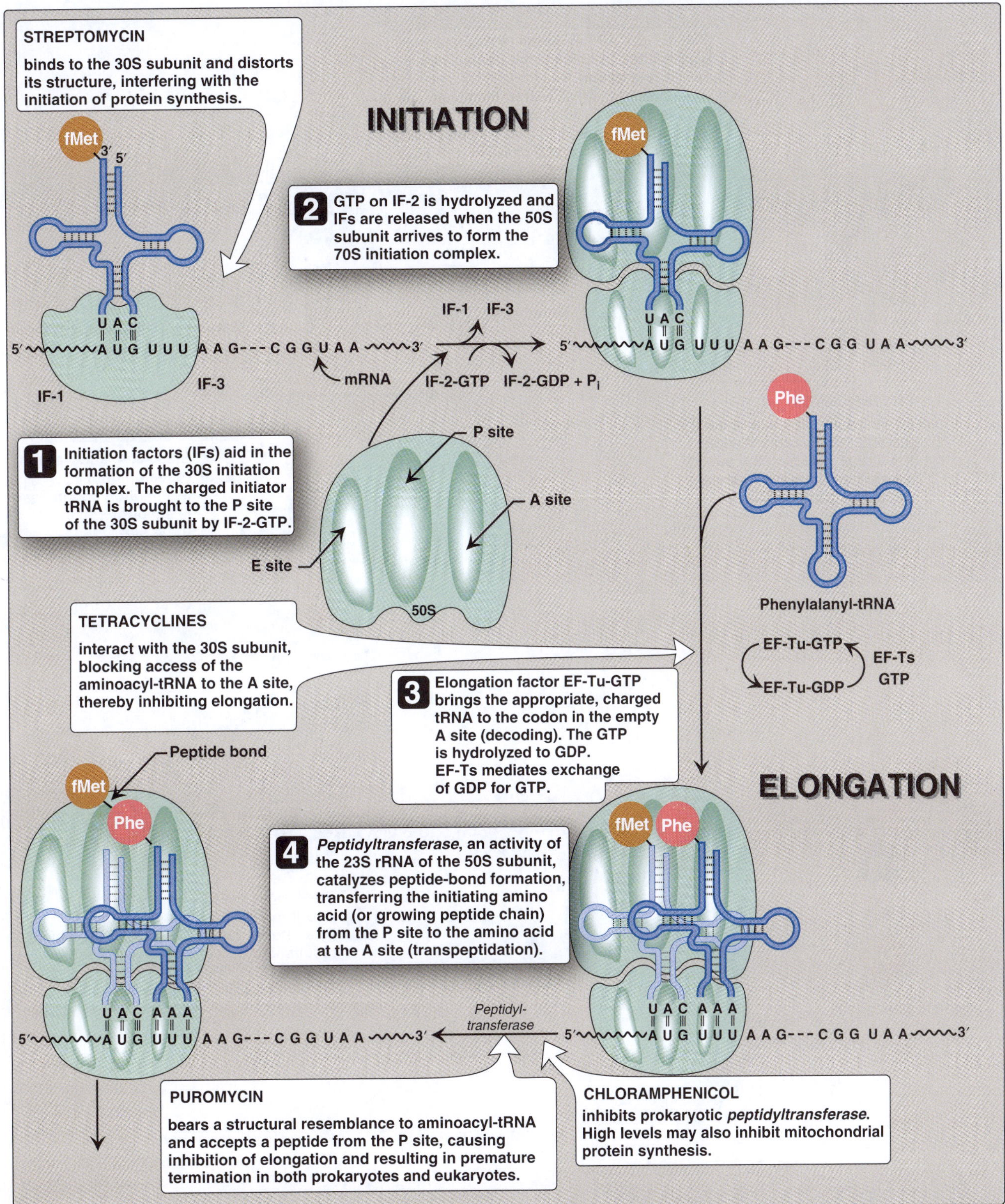

Figure 37.13 (continued on next page) **Steps in prokaryotic protein synthesis (translation), and their inhibition by antibiotics.** [Note: EF-Ts is a guanine nucleotide exchange factor. It facilitates the removal of guanosine diphosphate (GDP) from EF-Tu, allowing its replacement by guanosine triphosphate (GTP). The eukaryotic equivalent is EF-1$\beta\gamma$.] fMet = formylated methionine; S = Svedberg unit; Phe = phenylalanine; Lys = lysine; Arg = arginine; tRNA = transfer RNA; mRNA = messenger RNA.

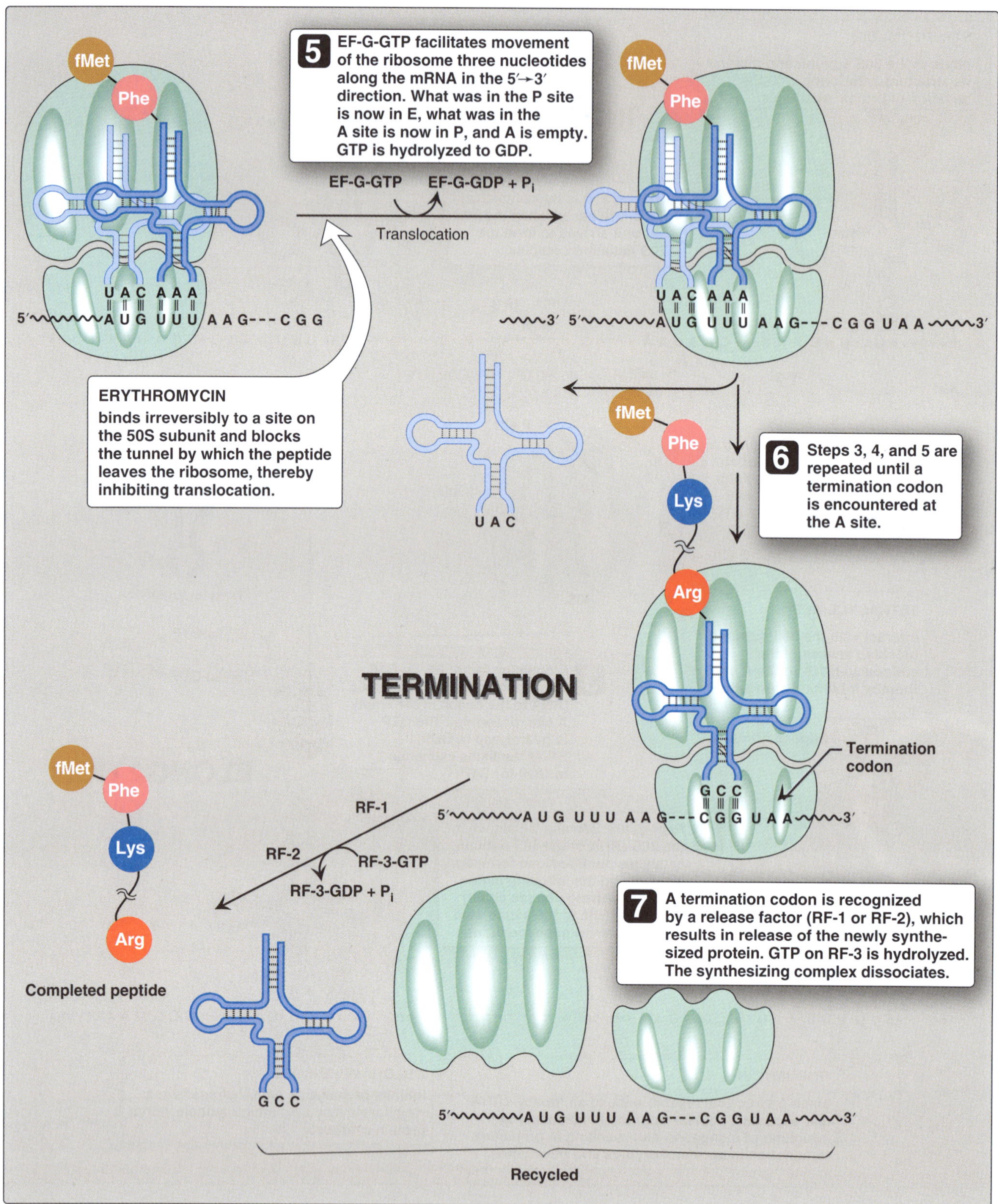

Figure 37.13 (Continued from previous page) [Note: **In eukaryotes, diphtheria toxin inactivates EF-2,** thereby inhibiting the translocation phase of elongation. **Ricin, a toxin from castor beans, removes a specific A from the 28S ribosomal RNA (rRNA)** in the large subunit of eukaryotic ribosomes, thereby inhibiting ribosomal function.]

Cell	Factor	Function
Initiation		
P E	IF-2-GTP eIF-2-GTP	Bring charged initiating tRNA to P site
P E	IF-3 eIF-3	Prevent association of subunits
Elongation		
P E	EF-Tu-GTP EF1α-GTP	Bring all other charged tRNA to A site
P E	EF-Ts EF-1βγ	Guanine nucleotide exchange factors
P E	EF-G-GTP EF-2-GTP	Translocation
Termination		
P E	RF-1, 2 eRF	Recognize stop codons
P E	RF-3-GTP eRF-3-GTP	Release of other RF

Figure 37.14 Protein factors in the three stages of translation. **P** = prokaryotes; **E**= eukaryotes; tRNA = transfer RNA; IF = initiation factor; EF = elongation factor; RF = release factor; GTP = guanosine triphosphate.

or after their synthesis has been completed (posttranslational). These modifications may include removal of part of the translated sequence or the covalent addition of one or more chemical groups required for protein activity.

A. Trimming

Many proteins destined for secretion are initially made as large, precursor molecules that are not functionally active. Portions of the protein must be removed by specialized *endoproteases*, resulting in the release of an active molecule. The cellular site of the cleavage reaction depends on the protein to be modified. Some precursor proteins are cleaved in the RER or the Golgi; others are cleaved in developing secretory vesicles (for example, insulin; see Fig. 27.4, p. 439); and still others, such as collagen (see p. 120), are cleaved after secretion.

B. Covalent attachments

Protein function can be affected by the covalent attachment of a variety of chemical groups (Fig. 37.16). Examples include the following.

1. **Phosphorylation:** Phosphorylation occurs on the hydroxyl groups of serine, threonine, or, less frequently, tyrosine residues in a protein. It is catalyzed by one of a family of ***protein kinases*** and may be reversed by the action of ***protein phosphatases***. The phosphorylation may increase or decrease the functional activity of the protein. Several examples of phosphorylation reactions have been previously discussed (for example, see Chapter 16, p. 231, for the regulation of glycogen synthesis and degradation).

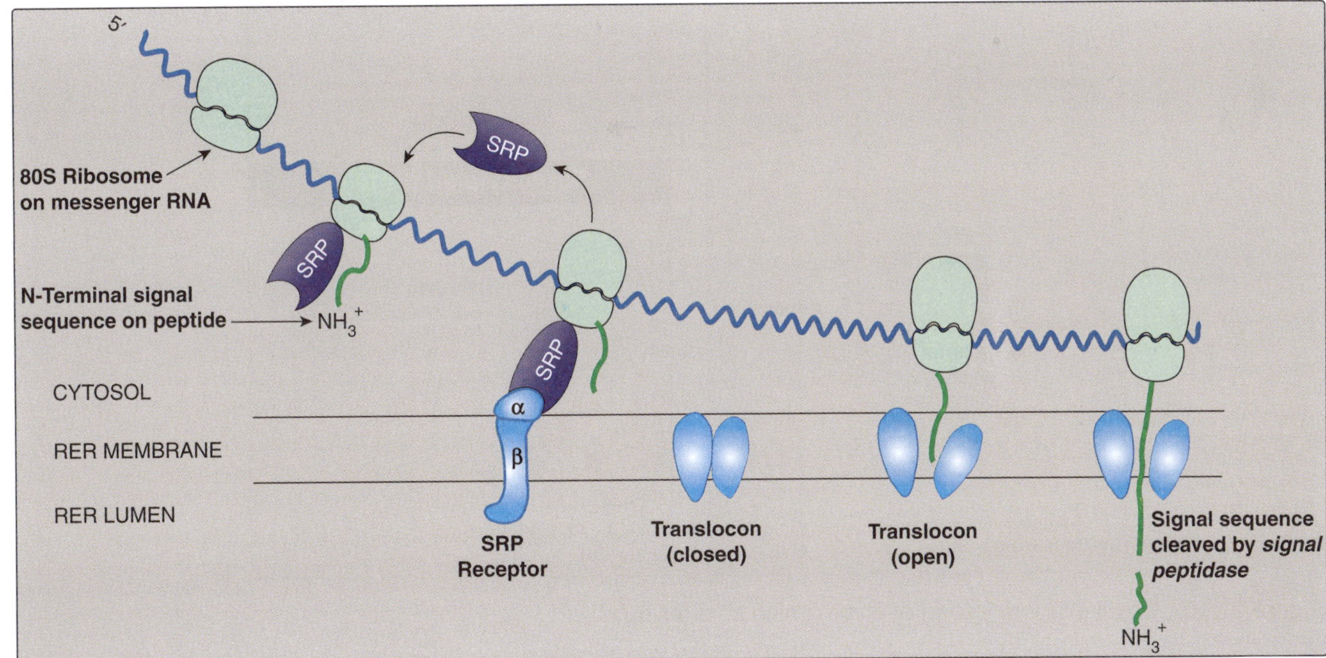

Figure 37.15 Cotranslational targeting of proteins to the rough endoplasmic reticulum (RER). SRP = signal recognition particle.

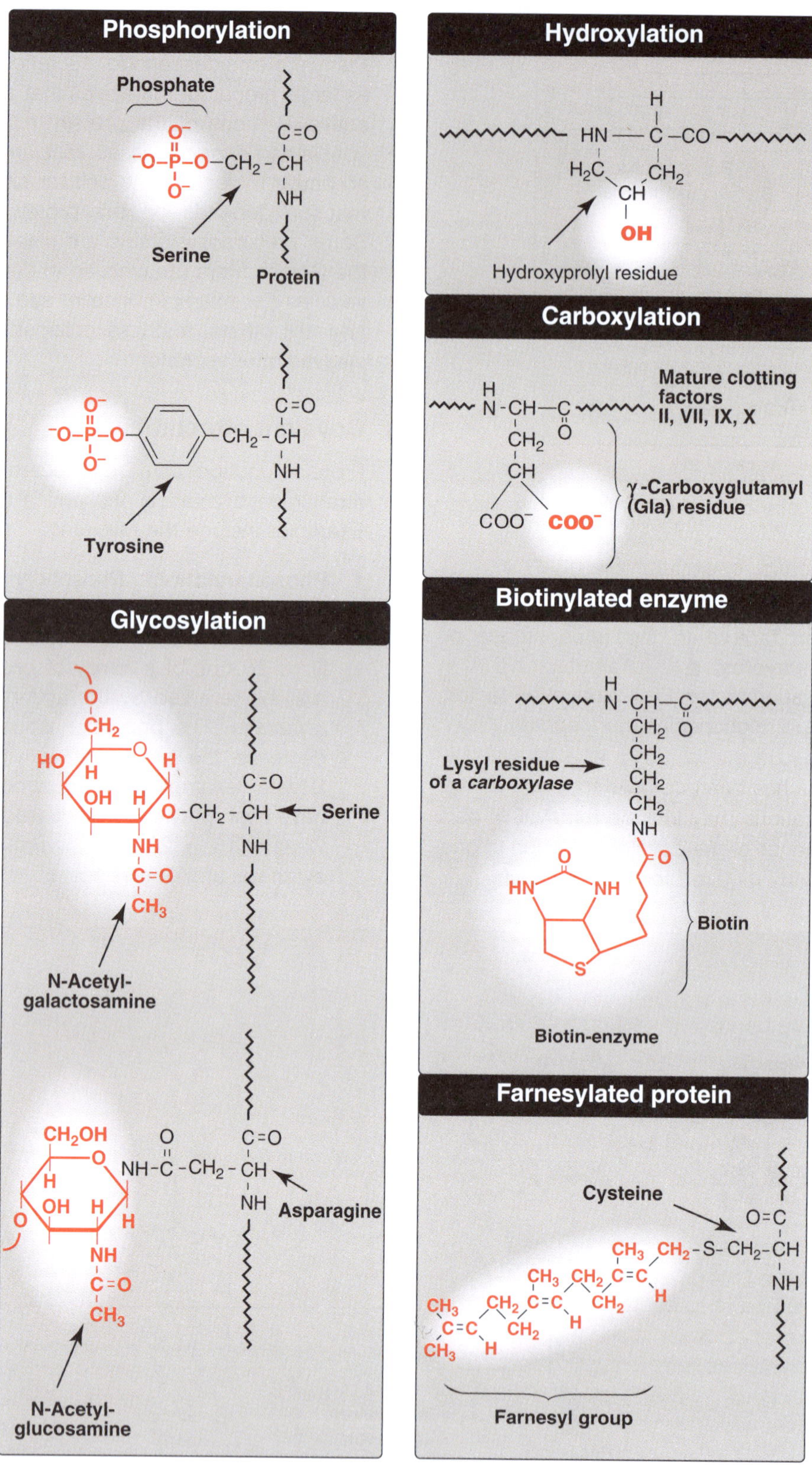

Figure 37.16 Covalent modification of some amino acid residues.

2. **Glycosylation:** Many of the proteins that are destined to become part of a membrane or to be secreted from a cell have carbohydrate chains added en bloc to the amide nitrogen of an asparagine (N-linked) or built sequentially on the hydroxyl groups of a serine, threonine, or hydroxylysine (O-linked). N-glycosylation occurs in the RER and O-glycosylation in the Golgi. (The process of producing such glycoproteins was

discussed on pp. 268–269.) N-glycosylated *acid hydrolases* are targeted to the matrix of lysosomes by the phosphorylation of mannose residues at carbon 6 (see p. 272).

3. **Hydroxylation:** Proline and lysine residues of the α chains of collagen are extensively hydroxylated by vitamin C–dependent *hydroxylases* in the RER (see p. 121).

4. **Other covalent modifications:** These may be required for the functional activity of a protein. For example, additional carboxyl groups can be added to glutamate residues by vitamin K–dependent carboxylation (see p. 544). The resulting γ-carboxyglutamate (Gla) residues are essential for the activity of several of the blood-clotting proteins. Biotin is covalently bound to the ε-amino groups of lysine residues of biotin-dependent enzymes that catalyze carboxylation reactions such as *pyruvate carboxylase* (see Fig. 15.3, p. 217). Attachment of lipids, such as farnesyl groups, can help anchor proteins to membranes (see p. 344). Many eukaryotic proteins are cotranslationally acetylated at the N-end. [Note: Reversible acetylation of histone proteins influences gene expression (see pp. 664–665).]

Chaperones are proteins which play a role in assembly and proper folding of synthesized proteins. This is important for the proteins to now become biologically active. Chaperones are present in various parts of cell like cytoplasm and endoplasmic reticulum. Chaperones usually exhibit *ATPase* activity and the binding of ATP and ADP has an important role in effective folding of the protein. They belong to a large family of proteins called heat shock proteins. **Heat shock proteins** are also called stress proteins since they get activated in presence of cellular stress like exposure to radiations, free radicals and toxins.

C. Protein degradation

Proteins that are defective (for example, misfolded) or destined for rapid turnover are often marked for destruction by ubiquitination, the covalent attachment of chains of a small, highly conserved protein called ubiquitin (see Fig. 24.3, p. 379). Proteins marked in this way are rapidly degraded by the proteasome, which is a macromolecular, ATP-dependent, proteolytic system located in the cytosol. For example, misfolding of the CFTR protein (see p. 638) results in its proteasomal degradation. [Note: If folding is impeded, unfolded proteins accumulate in the RER causing stress that triggers the unfolded protein response, in which the expression of chaperones is increased; global translation is decreased by eIF-2 phosphorylation; and the unfolded proteins are sent to the cytosol, ubiquitinated, and degraded in the proteasome by a process called ER-associated degradation.]

VII. Medical Perspectives

Protein synthesis inhibitors are used as drugs against bacterium. However, some toxins also inhibit eukaryotic protein synthesis and cause human disease. Further, some genetic diseases, diseases of posttranslational modifications and protein folding errors are also clinically relevant.

Drugs

Streptomycin is bactericidal against dividing Mycobacterium tuberculosis as it inhibits its protein synthesis by binding to the 30S mycobacterial ribosome

Tetracycline inhibits the initiation process of protein synthesis in bacteria by binding to the 30S subunit and further inhibiting the binding of the aminoacyl tRNA.

Chloramphenicol inhibits the process of elongation during protein synthesis by inhibiting the *peptidyl transferase* activity.

Erythromycin inhibits translocation by binding to the 50S subunit of the ribosome.

Toxins

Diphtheria is a serious infection caused by Corynebacterium diphtheriae. It causes difficulty in breathing, painful swallowing and may progress to serious complications. C. diphtheriae produces a protein toxin which binds to the EF-2 in eukaryotes and therefore stops translocation. **Diphtheria vaccination is provided to all children under the National Immunization Programme of India.**

Ricin, a toxic lectin naturally found in castor beans, prevents aminoacyl tRNA from binding to the A site. This inhibits protein synthesis.

Diseases

Alzheimer's disease, a very common form of dementia, may be caused by abnormal folding of proteins, unassisted by chaperones. It is characterized by senile neuritic plaques, neurofibrillary tangles and cerebral amyloid deposits.

VIII. Chapter Summary

- **Codons** are composed of three nucleotide bases presented in the messenger RNA (mRNA) language of **adenine** (**A**), **guanine** (**G**), **cytosine** (**C**), and **uracil** (**U**). They are always written **5′→3′**. Of the 64 possible three-base combinations, 61 code for the 20 standard amino acids and 3 signal termination of protein synthesis (**translation**).
- Altering the nucleotide sequence in a codon can cause **silent mutations** (the altered codon codes for the original amino acid), **missense mutations** (the altered codon codes for a different amino acid), or **nonsense mutations** (the altered codon is a termination codon).
- Characteristics of the genetic code include **specificity**, **universality**, and **degeneracy**, and it is **nonoverlapping** and **commaless** (Fig. 37.17).
- Requirements for protein synthesis include all the **amino acids** that eventually appear in the finished protein; at least one specific type of **transfer RNA** (**tRNA**) for each amino acid; one ***aminoacyl-tRNA synthetase*** for each amino acid; the **mRNA** coding for the protein to be synthesized; fully competent **ribosomes** (70S in prokaryotes, 80S in eukaryotes); **protein factors** needed for initiation, elongation, and termination of protein synthesis; and **ATP** and **guanosine triphosphate** (**GTP**) as energy sources.
- tRNA has an attachment site for a specific amino acid at its 3′-end and an anticodon region that can recognize the codon specifying the amino acid the tRNA is carrying.
- Ribosomes are large complexes of **protein** and **ribosomal RNA** (**rRNA**). They consist of **two subunits**, 30S and 50S in prokaryotes and 40S and 60S in eukaryotes. Each ribosome has three binding sites for tRNA molecules: the A, P, and E sites that cover three neighboring codons. The **A site** binds an **incoming aminoacyl-tRNA**, the **P site** is occupied by **peptidyl-tRNA**, and the **E site** is occupied by the **empty tRNA** as it is about to exit the ribosome. Recognition of an mRNA codon is accomplished by the tRNA **anticodon**, which binds to the codon following the rules of **complementarity** and **antiparallel** binding.
- The **wobble hypothesis** states that the first (5′) base of the anticodon is not as spatially defined as the other two bases. Movement of that first base allows nontraditional base-pairing with the last (3′) base of the codon, thus allowing a single tRNA to recognize more than one codon for a specific amino acid.
- For **initiation** of protein synthesis, the components of the translation system are assembled, and mRNA associates with the small ribosomal subunit. The process requires **initiation factors** (**IF**). In **prokaryotes**, a purine-rich region of the mRNA (the **Shine-Dalgarno sequence**) base-pairs with a complementary sequence on 16S rRNA, resulting in the positioning of the small subunit on the mRNA so that translation can begin. The **5′-cap** (bound by proteins of the eIF-4 family) on **eukaryotic mRNA** is used to position the small subunit on the mRNA. The **initiation codon** is **AUG**, and **N-formylmethionine** is the initiating amino acid in prokaryotes, whereas **methionine** is used in eukaryotes. The charged initiating tRNA ($tRNA_i$) is brought to the P site by **(e)IF-2**.
- In **elongation**, the polypeptide chain is lengthened by the addition of amino acids to the carboxyl end of its growing chain. The process requires **elongation factors** that facilitate the binding of the aminoacyl-tRNA to the A site as well as the movement of the ribosome along the mRNA. The formation of the peptide bond is catalyzed by ***peptidyltransferase***, which is an activity intrinsic to the rRNA of the large subunit and, therefore, is a **ribozyme**. Following peptide-bond formation, the ribosome advances along the mRNA in the **5′→3′ direction** to the next codon (**translocation**). Because of the length of most mRNA, more than one ribosome at a time can translate a message, forming a **polysome**.
- **Termination** begins when one of the three termination codons moves into the A site. These codons are recognized by **release factors**. The newly synthesized protein is released from the ribosomal complex, and the ribosome is dissociated from the mRNA.
- **Initiation**, **elongation**, and **termination** are driven by the hydrolysis of GTP. Initiation in eukaryotes also requires ATP for **scanning**.
- Numerous **antibiotics** interfere with the process of protein synthesis. Many polypeptide chains are covalently modified during or after translation. Such modifications include amino acid **removal**; **phosphorylation,** which may activate or inactivate the protein; **glycosylation**, which plays a role in **protein targeting**; and **hydroxylation** such as that seen in collagen.
- Protein targeting can be either **cotranslational** (as with secreted proteins) or **posttranslational** (as with mitochondrial matrix proteins).
- Proteins must **fold** to achieve their functional form. Folding can be spontaneous or facilitated by **chaperones**. Proteins that are defective (for example, misfolded) or destined for rapid turnover are marked for destruction by the attachment of chains of a small, highly conserved protein called **ubiquitin**. Ubiquitinated proteins are rapidly degraded by a cytosolic complex known as the **proteasome**.

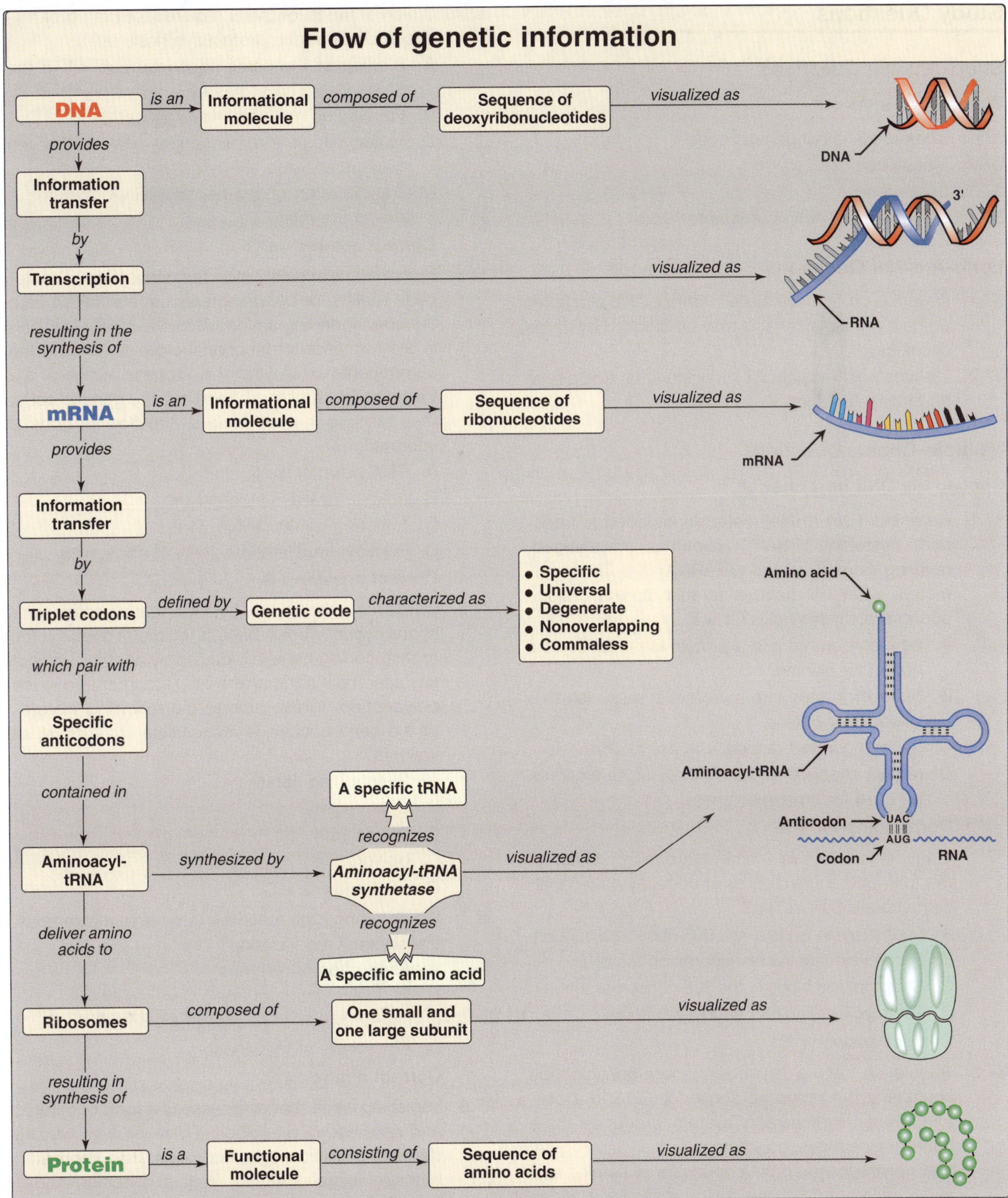

Figure 37.17 Key concept map for protein synthesis. mRNA = messenger RNA; tRNA = transfer RNA; A = adenine; G = guanine; C = cytosine; U = uracil.

Study Questions

Short-Answer Questions

Write short notes on:

37.1. Characteristics of genetic code
37.2. Mutations
37.3. Chaperones
37.4. Mechanism of action of streptomycin

Long-Answer Questions

37.1. Describe protein synthesis in detail. Give examples on how protein synthesis may be blocked by drugs or toxins.

37.2. Discuss posttranslational modifications. Add a note on protein targeting.

Multiple-Choice Questions

Choose the ONE best answer.

37.1. A transfer RNA (tRNA) molecule intended to transport cysteine ($tRNA^{cys}$) becomes mischarged, carrying alanine instead (ala-$tRNA^{cys}$). If left uncorrected, what will happen to this alanine residue during protein synthesis? It will:

A. be incorporated into a protein in response to a codon for alanine.
B. be incorporated into a protein in response to a codon for cysteine.
C. be incorporated randomly at any codon.
D. remain attached to the tRNA because it cannot be used for protein synthesis.

Correct answer = B.

37.2. Many antimicrobials inhibit translation. Which of the following antimicrobials is correctly paired with its mechanism of action?

A. Erythromycin binds to the 60S ribosomal subunit.
B. Puromycin inactivates elongation factor-2.
C. Streptomycin binds to the 30S ribosomal subunit.
D. Tetracyclines inhibit peptidyltransferase.

Correct answer = C.

37.3. Translation of a synthetic polyribonucleotide containing the repeating sequence CAA in a cell-free protein-synthesizing system produces three homopolypeptides: polyglutamine, polyasparagine, and polythreonine. If the codons for glutamine and asparagine are CAA and AAC, respectively, which of the following triplets is the codon for threonine?

A. AAC
B. ACA
C. CAA
D. CAC

Correct answer = B.

37.4. Which of the following is required for both prokaryotic and eukaryotic protein synthesis?

A. Binding of the small ribosomal subunit to the Shine-Dalgarno sequence
B. Formylated methionyl-transfer (t)RNA
C. Movement of the messenger RNA out of the nucleus and into the cytoplasm
D. Translocation of the peptidyl-tRNA from the A site to the P site

Correct answer = D.

37.5. Protein translation involves the conversion of nucleotide triplets, or codons present in the mRNA, into the corresponding amino acids, ultimately resulting in the formation of functional proteins. Among the components involved in this process, which of the following is capable of "reading" the genetic code, thus forming the keystone of the flow of genetic information?

A. RNA polymerase II
B. Aminoacyl tRNA synthetases
C. Anticodon of the tRNA
D. Peptidyl transferase activity of ribosomes

Correct answer = B.

37.6. One among the usually asymptomatic genetic variations occurs when a base of the triplet codon in the protein coding gene is replaced by another, without any alteration in the amino acid incorporated in the polypeptide. Which among the following features of the genetic code is responsible for such silent variants?

A. Overlapping nature
B. Unambiguity
C. Degeneracy
D. Universality

Correct answer = C.

37.7. Which among the following is an irreversible post-translational modification?

A. Acetylation of histones
B. Phosphorylation of HMG CoA reductase
C. Glycosylation of alkaline phosphatase
D. Proteolysis of trypsinogen

Correct answer = D.

37.8. Vanishing white matter disease is a form of chronic and progressive neurological disease with varying degrees of clinical manifestations. The implicated pathogenic gene variant in this condition codes for eIF2B which is required for the activation eIF2. Which among the following biological processes is maximally affected in this condition?

A. DNA replication
B. mRNA synthesis
C. rRNA synthesis
D. Protein synthesis

Correct answer = D.

37.9. Statement 1: It is important to consume all the essential amino acids in adequate quantities in the diet.
Statement 2: Protein synthesis halts if the charged tRNA carrying a specific amino acid is absent at the time of translation.

A. Both the Statements are correct, and Statement 2 is the correct explanation for Statement 1
B. Both the Statements are correct, but Statement 2 is NOT the correct explanation for Statement 1
C. Statement 1 is correct, but Statement 2 is NOT correct
D. Statement 1 is NOT correct, but Statement 2 is correct

Correct answer = A.

37.10. Which of the following is an exception to post-translational modifications?

A. De-amination of glutamine to glutamate
B. γ-carboxylation of glutamate to γ-carboxyglutamate
C. Hydroxylation of lysine to hydroxylysine
D. Phosphorylation of serine to phosphoserine

Correct answer = A.

Multiple-Choice Questions (NExT/NEET PG)

37.1. A 20-year-old man with a microcytic anemia is found to have an abnormal form of β-globin (Hemoglobin Constant Spring) that is 172 amino acids long, rather than the 141 found in the normal protein. Which of the following point mutations is consistent with this abnormality?

A. CGA → UGA
B. UAA → UAG
C. GCA → GAA
D. UAA → CAA

Correct answer = D.

37.2. A pharmaceutical company is studying a new antibiotic that inhibits bacterial protein synthesis. When this antibiotic is added to an in vitro protein synthesis system that is translating the messenger RNA sequence AUGUUUUUUUAG, the only product formed is the dipeptide fMet-Phe. What step in protein synthesis is most likely inhibited by the antibiotic?

A. Initiation
B. Binding of a charged transfer RNA to the ribosomal A site
C. Peptidyltransferase activity
D. Ribosomal translocation

Correct answer = D.

37.3. α1-Antitrypsin (AAT) deficiency can result in emphysema, a lung pathology, because the action of elastase, a serine protease, is unopposed. Deficiency of AAT in the lungs is the consequence of impaired secretion from the liver, the site of its synthesis. Proteins such as AAT that are destined to be secreted are best characterized by which of the following statements?

A. Their synthesis is initiated on the smooth endoplasmic reticulum.
B. They contain a mannose 6-phosphate targeting signal.
C. They are produced from translation products that have an N-terminal hydrophobic signal sequence.
D. They contain no sugars with O-glycosidic linkages because their synthesis does not involve the Golgi.

Correct answer = C.

37.4. In a patient with cystic fibrosis (CF) caused by the ΔF508 mutation, the mutant CF transmembrane conductance regulator (CFTR) protein folds incorrectly. The patient's cells modify this abnormal protein by attaching ubiquitin molecules to it. What is the fate of this modified CFTR protein?

A. It performs its normal function because the ubiquitin largely corrects for the effect of the mutation.
B. It is degraded by the proteasome.
C. It is placed into storage vesicles.
D. It is secreted from the cell.

Correct answer = B.

37.5. In 1978, Bulgarian dissident Georgi Markov was fatally poisoned in London in a notorious Cold War incident. While waiting for a bus on Waterloo Bridge, he was jabbed in the thigh with a modified umbrella that injected a tiny pellet containing ricin. Despite seeking medical attention, Markov's condition rapidly deteriorated, leading to his death 4 days later. What is the mechanism of delayed worsening of symptoms in this poisoning?

A. Ricin inhibits replication of the DNA, hence only dividing cells are affected.
B. Ricin selectively inhibits transcription of certain mRNAs; hence majority of transcription continues.
C. Ricin inhibits protein translation; hence the existing proteins are unaffected.
D. Ricin is metabolized to a more potent toxin in the body; this bioactivation has a latent period.

Correct answer = C.

Regulation of Gene Expression

38

Learning Objectives

At the end of this chapter, the student will be able to:

- Describe basic concepts of prokaryotic gene expression like lac operon and molecular switch
- Describe the special features of eukaryotic gene expression
- Explain DNA motifs
- Describe how knowledge of gene expression can be used for medical care

I. OVERVIEW

Gene expression refers to the multistep process that ultimately results in the production of a functional gene product, either ribonucleic acid (RNA) or protein. The first step in gene expression, the use of deoxyribonucleic acid (DNA) for the synthesis of RNA (transcription), is the primary site of regulation in both prokaryotes and eukaryotes. In eukaryotes, however, gene expression also involves extensive posttranscriptional and posttranslational processes as well as actions that influence access to particular regions of the DNA. Each of these steps can be regulated to provide additional control over the kinds and amounts of functional products that are produced.

All somatic cells have same genes with the exception of amplified and rearranged genes. Organisms adapt to environmental (cellular and external) changes by altering gene expression. Cells do not express all their genes. Also, not all genes are tightly regulated. For example, genes described as constitutive encode products required for basic cellular functions and so are expressed at essentially a constant level. They are also known as **"housekeeping"** genes. **Regulated genes, however, are expressed only under certain conditions**. They may be expressed in all cells or in only a subset of cells, for example, hepatocytes. This ability to express certain genes in particular cells in eukaryotes is called **tissue-specific expression**. The ability to regulate gene expression (that is, to determine if, how much, and when particular gene products will be made) gives the cell control over structure and function. It is the basis for cellular differentiation, morphogenesis, and adaptability of any organism. Control of gene expression is best understood in prokaryotes, but many themes are repeated in eukaryotes. Figure 38.1 shows some of the sites where gene expression can be controlled.

We can understand some key concepts of gene expression by first understanding the **response of gene to a signal** (type A, B, and C responses)**, lac operon system** in prokaryotes (an example of de-repression), **tryptophan operon** in prokaryotes (example of control by attenuation) and the prokaryotic **molecular switch** (genes regulating each other's expression).

Type A response to a gene signal implies that there will be increased expression of a gene only till the signal is present. Once the signal is removed, expression of the gene will come back to basal levels. Prokaryotic response to nutrients and eukaryotic response to hormones are examples of type A response. **In type B** response, increased amount of gene expression is transient even in the continued presence of regulatory signal. Some developmental gene expressions follow type B responses. **Type C** response, like response to Cro protein, is the increased expression of gene response that persists indefinitely even after termination of the regulatory signal.

II. REGULATORY SEQUENCES AND MOLECULES

Regulation of transcription, the initial step in all gene expression, is controlled by regulatory sequences of DNA that are usually embedded in the noncoding regions of the genome. The interaction between these DNA sequences

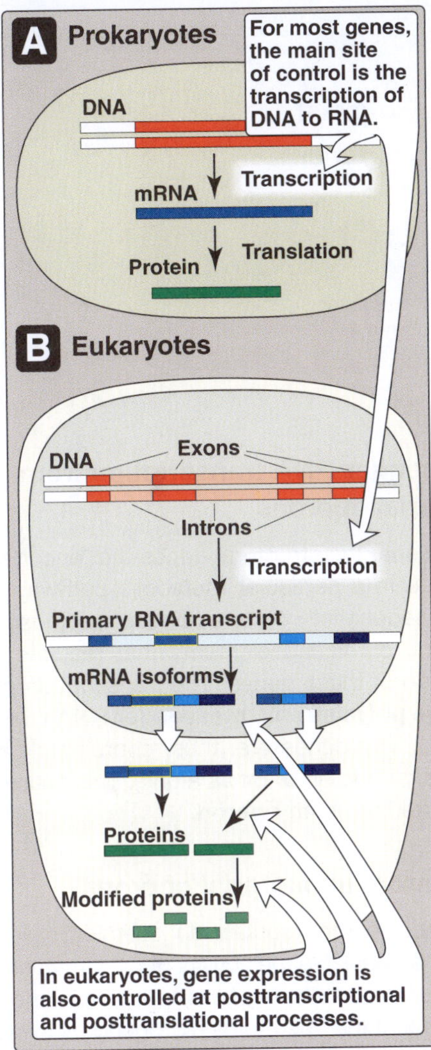

Figure 38.1 Control of gene expression. mRNA = messenger RNA.

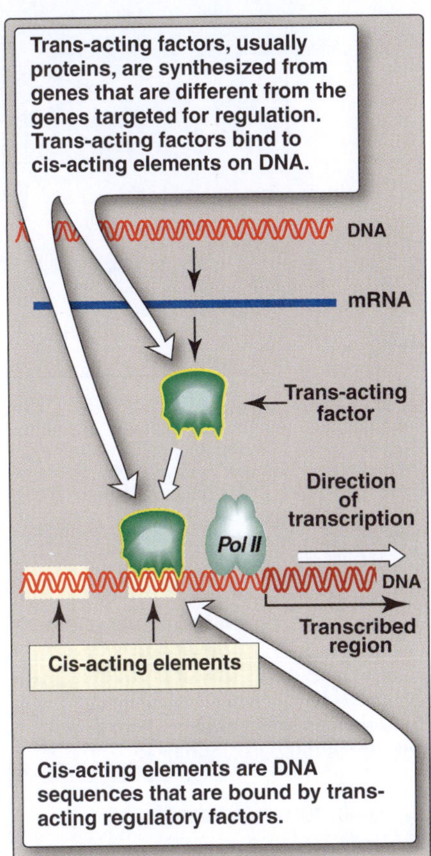

Figure 38.2 Cis-acting elements and trans-acting factors. mRNA = messenger RNA; *Pol II* = *RNA polymerase II*.

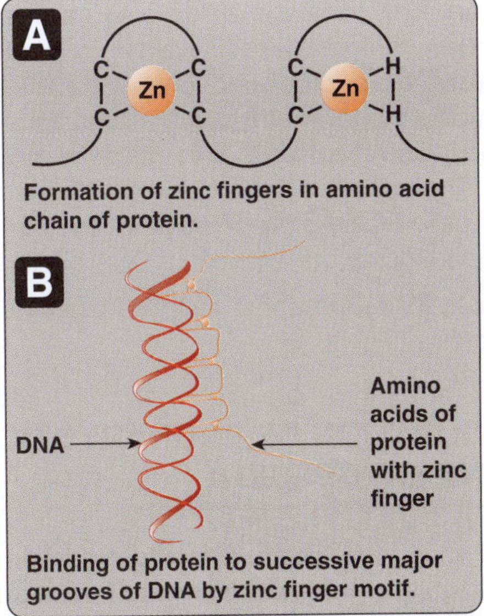

Figure 38.3 Zinc finger binding motif. **A.** Formation of zinc fingers in amino acid chain of protein. **B.** Binding of protein to successive major grooves of DNA by zinc finger motif.

and **regulatory molecules, such as transcription factors**, can induce or repress the transcriptional machinery, influencing the kinds and amounts of products that are produced. The regulatory DNA sequences are called cis-acting because they influence expression of genes on the same chromosome as the regulatory sequence (see p. 626). The regulatory molecules are called trans-acting because they can diffuse (transit) through the cell from their site of synthesis to their DNA-binding sites (Fig. 38.2). For example, a protein transcription factor (a trans-acting molecule) that regulates a gene on chromosome 6 might itself have been produced from a gene on chromosome 11. The binding of proteins to DNA is through **structural motifs such as the zinc finger** (Fig. 38.3**), leucine zipper** (Fig. 38.4**), or helix-turn-helix** (Fig. 38.5**)** in the protein.

Motifs mediate the binding of regulatory proteins to DNA. This binding to DNA is of high affinity, involving small regions of protein and maintained by H-bonds, ionic interactions and van der Waals forces. **Helix-turn-helix** binding can be understood by the example of cro protein binding to DNA. Cro has three antiparallel beta sheets and three alpha helices. Cro binds to DNA through alpha helix

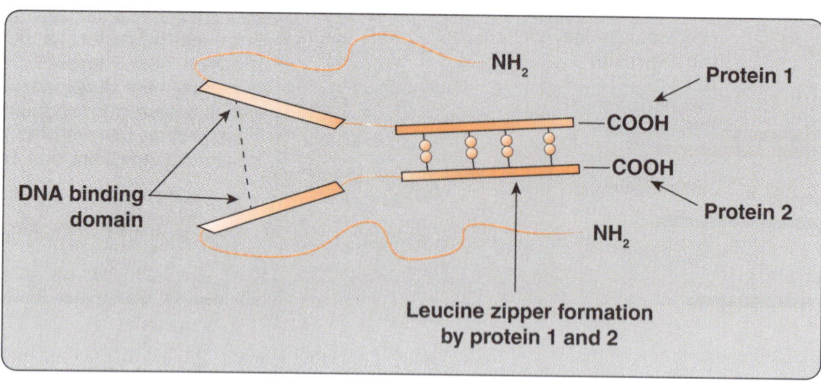

Figure 38.4 Leucine zipper motif.

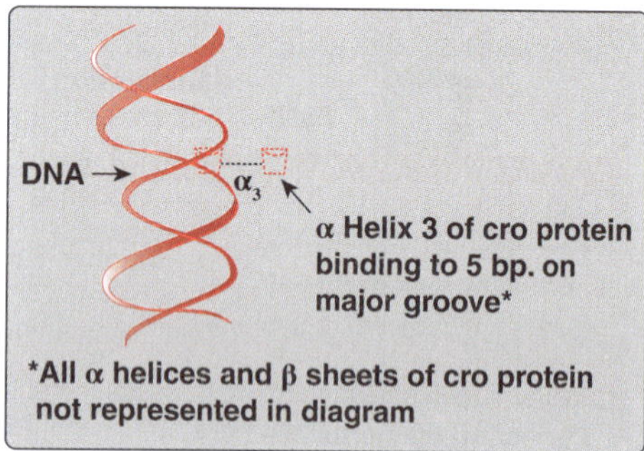

Figure 38.5 Helix-turn-helix motif.

3 binding to 5 bp on major groove (Fig. 38.5). An example of **zinc finger binding** is the binding of TF111A (a positive regulator of 5S RNA gene transcription) to DNA. There is a zinc bound to cysteine and/or histidine 12–13 amino acids apart. The presence of this zinc causes 'fingers" to be made in the protein chain which now binds to 5 bp in successive major grooves of DNA (Figs 38.3A and B). In the **leucine zipper motif,** proteins form an alpha helix with a periodic repeat of leucine residues at every 7th position. This makes a leucine line on one side of the protein. These leucines on one side of the protein "zip" with another such protein to form a DNA-binding region.

An example of leucine zipper motif is Fos and Jun regulatory proteins (Fig. 38.4).

III. REGULATION OF PROKARYOTIC GENE EXPRESSION

In prokaryotes such as the bacterium Escherichia coli (E. coli), regulation of gene expression occurs primarily at the level of transcription and, in general, is mediated by the binding of trans-acting proteins to cis-acting regulatory elements on their single DNA molecule (chromosome). [Note: Regulating the first step in the expression of a gene is an efficient approach, insofar as energy is not wasted making unneeded gene products.] Transcriptional control in prokaryotes can involve the initiation or premature termination of transcription.

A. Messenger RNA transcription from bacterial operons

In bacteria, the structural genes that encode proteins involved in a particular metabolic pathway are often found sequentially grouped on the chromosome along with the cis-acting elements that regulate the transcription of these genes. The transcription product is a single **polycistronic messenger RNA** ([mRNA], see p. 621). The genes are, thus, coordinately regulated (that is, turned on or off as a unit). This entire package is referred to as an **operon.**

B. Operators in bacterial operons

Bacterial operons contain an operator, a segment of DNA that regulates the activity of the structural genes of the operon by reversibly binding a protein known as the repressor. If the operator is not bound by the repressor, *RNA polymerase* (*RNA pol*) binds the promoter, passes over the operator, and reaches the protein-coding genes that it transcribes to mRNA. If the repressor is bound to the operator, the *polymerase* is blocked and does not produce mRNA. As long as the repressor is bound to the operator, no mRNA (and, therefore, no proteins) are made. However, when an inducer molecule is present, it binds to the repressor, causing the repressor to change shape so that it no longer binds the operator. When this happens, *RNA pol* can initiate transcription. One of the best-understood examples is the inducible lactose (lac) operon of E. coli that illustrates both positive and negative regulation (Fig. 38.6).

C. Lactose operon

The lac operon contains the genes that code for three proteins involved in the catabolism of the disaccharide lactose: the lacZ gene codes for β-galactosidase, which hydrolyzes lactose to galactose and glucose; the lacY gene codes for a *permease*, which facilitates the movement of lactose into the cell; and the lacA gene codes for *thiogalactoside transacetylase*, which acetylates lactose. [Note: The physiologic

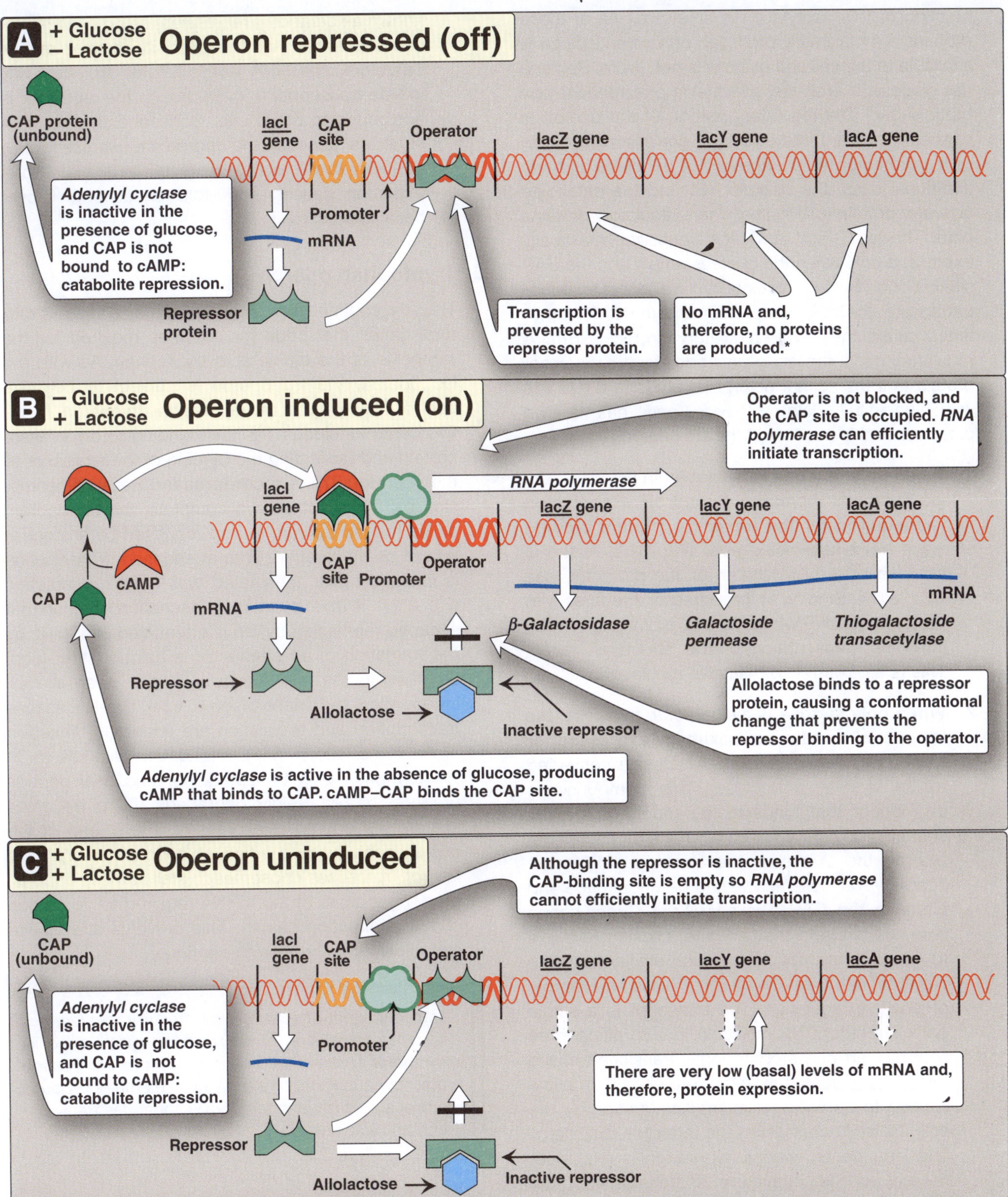

Figure 38.6 The lactose operon of Escherichia coli in the presence of **A.** only glucose, **B.** only lactose, and **C.** both sugars. *[Note: Even when the operon has been turned off, the repressor transiently dissociates from the operator at a slow rate, allowing a very low level of expression. The synthesis of a few molecules of *permease* (and β-*galactosidase*) allows the organism to respond rapidly should glucose become unavailable.] CAP = catabolite activator protein; cAMP = cyclic adenosine monophosphate; mRNA = messenger RNA.

function of this acetylation is unknown.] All of these proteins are maximally produced only when lactose is available to the cell and glucose is not. [Note: Bacteria use glucose, if available, as a fuel in preference to any other sugar.] The regulatory portion of the operon is upstream of the three structural genes and consists of the promoter region where *RNA pol* binds and two additional sites, the operator (O) and the catabolite activator protein (CAP) sites, where regulatory proteins bind. The lacZ, lacY, and lacA genes are maximally expressed only when the O site is empty and the CAP site is bound by a complex of cyclic adenosine monophosphate ([cAMP], see p. 190) and the CAP, sometimes called the cAMP regulatory protein (CRP). A regulatory gene, the lacI gene, codes for the repressor protein (a trans-acting factor) that binds to the O site with high affinity. [Note: The lacI gene has its own promoter and is not part of the lac operon.]

1. **When only glucose is available:** In this case, the lac operon is repressed (turned off). **Repression is mediated by the repressor protein binding via a helix-turn-helix motif** (Fig. 38.5) to the O site, which is downstream of the promoter (see Fig. 38.6A). Binding of the repressor interferes with the binding of *RNA pol* to the promoter, thereby inhibiting transcription of the structural genes. This is an example of negative regulation.

2. **When only lactose is available:** In this case, the lac operon is **induced (maximally expressed, or turned on)**. A small amount of lactose is converted to an isomer, allolactose. This compound is an inducer that binds to the repressor protein, changing its conformation so that it can no longer bind to the O site. In the absence of glucose, *adenylyl cyclase* is active, and cAMP is made and binds to the CAP. The cAMP–CAP trans-acting complex binds to the CAP site, causing *RNA pol* to initiate transcription with high efficiency at the promoter site (see Fig. 38.6B). This is an example of positive regulation. The transcript is a single polycistronic mRNA molecule that contains three sets of start and stop codons. Translation of the mRNA produces the three proteins that allow lactose to be used for energy production by the cell. [Note: In contrast to the inducible lacZ, lacY, and lacA genes, whose expression is regulated, the lacI gene is constitutive. Its gene product, the repressor protein, is always made and is active unless the inducer is present.]

3. **When both glucose and lactose are available:** In this case, **the lac operon is uninduced, and transcription is negligible, even if lactose is present at a high concentration**. *Adenylyl cyclase* is inhibited in the presence of glucose (a process known as catabolite repression) so no cAMP–CAP complex forms, and the CAP site remains empty. Therefore, the *RNA pol* is unable to effectively initiate transcription, even though the repressor is not bound to the O site. Consequently, the three structural genes of the operon are expressed only at a **very low (basal) level** (see Fig. 38.6C). [Note: Induction causes a 50-fold enhancement over basal expression.]

D. Tryptophan operon

The tryptophan (trp) operon contains five structural genes that code for enzymes required for the synthesis of the amino acid tryptophan. As with the lac operon, the trp operon is subject to negative control. However, for the repressible trp operon, negative control includes Trp itself binding to a repressor protein and facilitating the binding of the repressor to the operator: **Trp is a corepressor.** Because repression by Trp is not always complete, the trp operon, unlike the lac operon, is also regulated by a process known as attenuation. With attenuation, transcription is initiated but is terminated well before completion (Fig. 38.7). If Trp is plentiful, transcription initiation that escaped repression by Trp is attenuated (stopped) by the formation of an attenuator, a hairpin (stem-loop) structure in the mRNA similar to that seen in rho-independent termination (see p. 624). [Note: Because transcription and translation are temporally linked in prokaryotes (see p. 642), attenuation also results in the formation of a truncated, nonfunctional peptide product that is rapidly degraded.] If Trp becomes scarce, the operon is expressed. The 5′-end of the mRNA contains two adjacent codons for Trp. The lack of Trp causes ribosomes to stall at these codons, covering regions of the mRNA required for formation of the attenuation hairpin. This prevents attenuation and allows transcription to continue.

> Transcriptional attenuation can occur in prokaryotes because translation of an mRNA begins before its synthesis is complete. This does not occur in eukaryotes because the presence of a membrane-bound nucleus spatially and temporally separates transcription and translation.

E. The genetic switch of bacteriophage lambda

The **genetic switch of bacteriophage lambda** provides an example of **two genes regulating each other's expression through protein-DNA interactions**. A bacteriophage infects a bacterium and incorporates its genomic material inside the bacterial genome. Now, if there is presence of nutrition

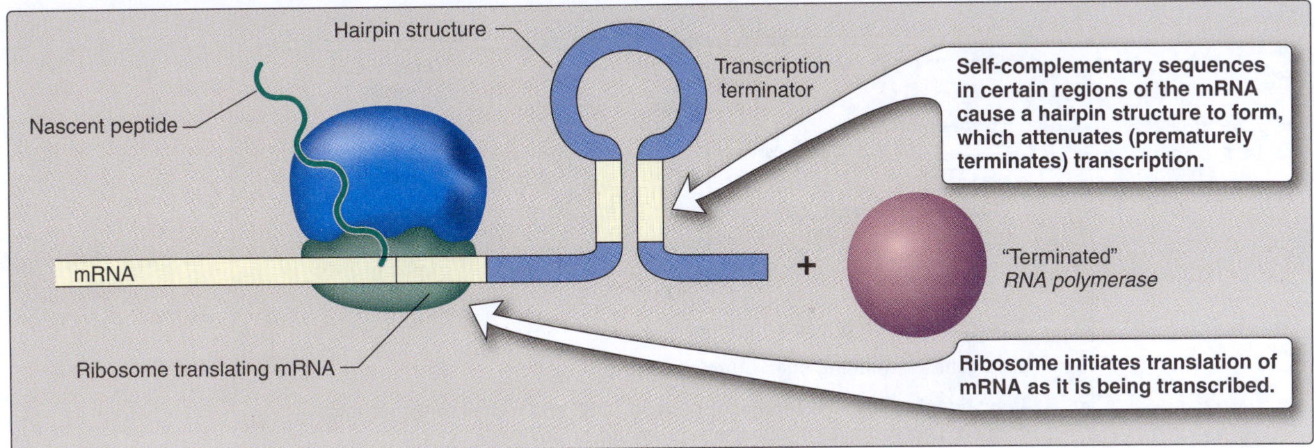

Figure 38.7 Attenuation of transcription of the trp operon when tryptophan is plentiful. mRNA = messenger RNA.

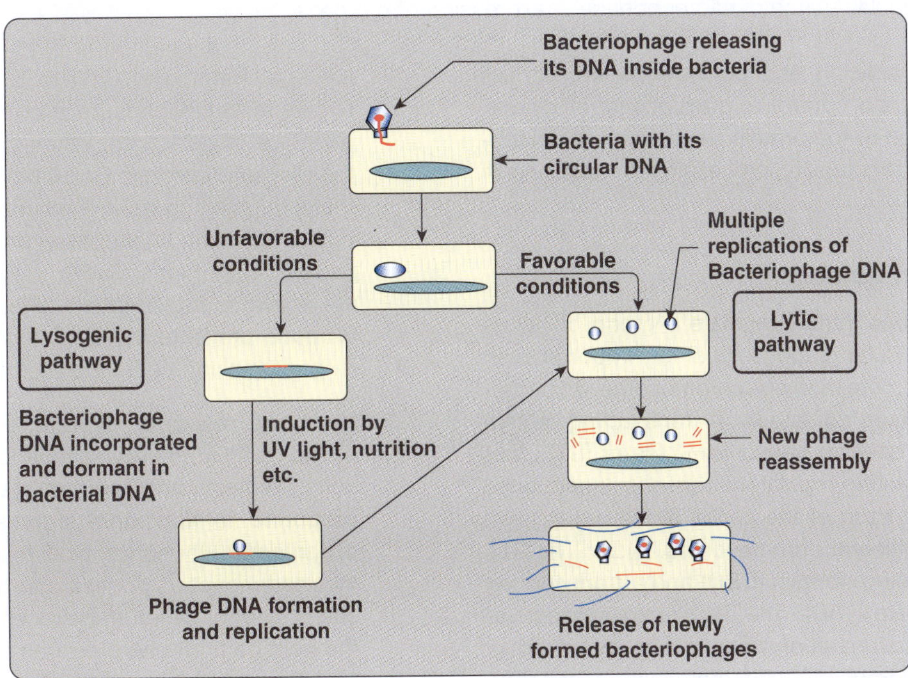

Figure 38.8 Molecular switch of bacteriophage λ: lytic and lysogenic phase.

and sunlight, it can go into the lytic phase, replicate inside the bacteria and release its copies by lysing the bacteria (Fig. 38.8). For this, it needs expression of Cro gene (Fig. 38.9).

The Cro protein will also suppress the repressor gene by protein-DNA interactions. The lytic phase is a type C response which is irreversible. In case there is non-availability of nutrients and sunlight, the bacteriophage will infect the bacteria but lie dormant. The dormancy will be supported by activation of repressor gene and suppression of Cro gene by protein-DNA interactions. In this lysogenic phase, repressor protein is made, which binds to the Cro promoter to suppress synthesis of Cro protein. Lysogenic phase is reversible and can proceed to lytic phase in favorable conditions.

IV. REGULATION OF EUKARYOTIC GENE EXPRESSION

The higher degree of complexity of eukaryotic genomes, as well as the presence of a nuclear membrane, necessitates a wider range of regulatory processes. As with the prokaryotes, transcription is the primary site of regulation. Again, the theme of trans-acting factors binding to cis-acting elements is seen. Operons, however, are not found in eukaryotes, which must use alternate strategies to solve the problem of how to coordinately regulate all the genes required for a specific response. In eukaryotes, gene expression is also regulated at multiple levels other than transcription. For example, the major modes of posttranscriptional regulation at the mRNA level are

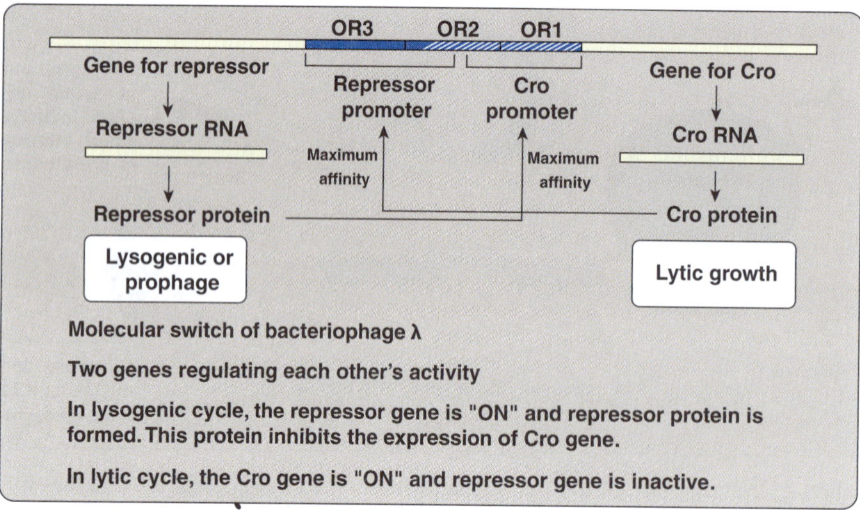

Figure 38.9 Molecular switch of bacteriophage λ: expression of cro gene.

alternative mRNA splicing and polyadenylation, control of mRNA stability, and control of translational efficiency. Additional regulation at the protein level occurs by mechanisms that modulate stability, processing, or targeting of the protein.

A. Coordinate regulation

The need to coordinately regulate a group of genes to cause a particular response is of key importance in organisms with more than one chromosome. An underlying theme occurs repeatedly: A trans-acting protein functions as a specific transcription factor (STF) that binds to a cis-acting regulatory consensus sequence (see p. 599) on each of the genes in the group even if they are on different chromosomes. [Note: The STF has a DNA-binding domain (DBD) and a transcription activation domain (TAD). The TAD recruits coactivators, such as *histone acetyltransferases* (see p. 624), and the general transcription factors (see p. 626) that, along with *RNA pol*, are required for formation of the transcription initiation complex at the promoter. Although the TAD recruits a variety of proteins, the specific effect of any one of them is dependent upon the protein composition of the complex. This is known as combinatorial control.] Examples of coordinate regulation in eukaryotes include the galactose circuit and the hormone response system.

1. **Galactose circuit:** This regulatory scheme allows for the use of galactose when glucose is not available. In yeast, a unicellular organism, the genes required to metabolize galactose are on different chromosomes. Coordinated expression is mediated by the protein Gal4 (Gal = galactose), a STF that binds to a short regulatory DNA sequence upstream of each of the genes. The sequence is called the upstream activating sequence Gal (UAS_{Gal}). Binding of Gal4 to UAS_{Gal} through zinc fingers in its DBD occurs in both the absence and presence of galactose. When the sugar is absent, the regulatory protein Gal80 binds Gal4 at its TAD, thereby inhibiting gene transcription (Fig. 38.10A). When present, galactose activates the Gal3 protein. Gal3 binds Gal80, thereby allowing Gal4 to activate transcription (Fig. 38.10B). [Note: Glucose prevents the use of galactose by inhibiting expression of Gal4 protein.]

2. **Hormone response system:** Hormone response elements (HRE) are DNA sequences that bind trans-acting proteins and **regulate gene expression in response to hormonal signals in multicellular organisms.** Hormones bind to either intracellular (nuclear) receptors (for example, steroid hormones; see p. 360) or cell-surface receptors (for example, the peptide hormone glucagon; see p. 444).

   a. **Intracellular receptors:** Members of the nuclear receptor superfamily, which includes the steroid hormone (glucocorticoids, mineralocorticoids, androgens, and estrogens), vitamin D, retinoic acid, and thyroid hormone receptors, function as STF. In addition to domains for DNA-binding and transcriptional activation, these receptors also contain a ligand-binding domain. For example, the steroid hormone cortisol (a glucocorticoid) binds intracellular receptors at the ligand-binding domain (Fig. 38.11). Binding causes a conformational change in the receptor that activates it. The receptor–hormone complex enters the nucleus, dimerizes, and binds via a zinc finger motif to DNA at a regulatory element, the glucocorticoid response element (GRE) that is an example of a **hormone response element** (HRE). Binding

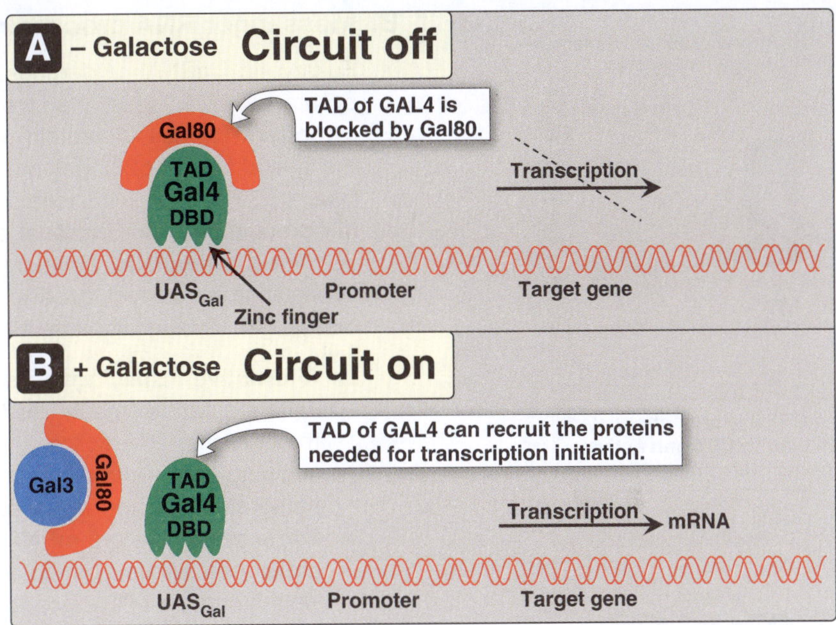

Figure 38.10 Regulation of galactose (gal) circuit in yeast in the **A.** absence and **B.** presence of galactose. [Note: Target genes, whether on the same or a different chromosome, each have an upstream activating sequence galactose (UAS$_{Gal}$).] TAD = transcription activation domain; DBD = DNA-binding domain; mRNA = messenger RNA.

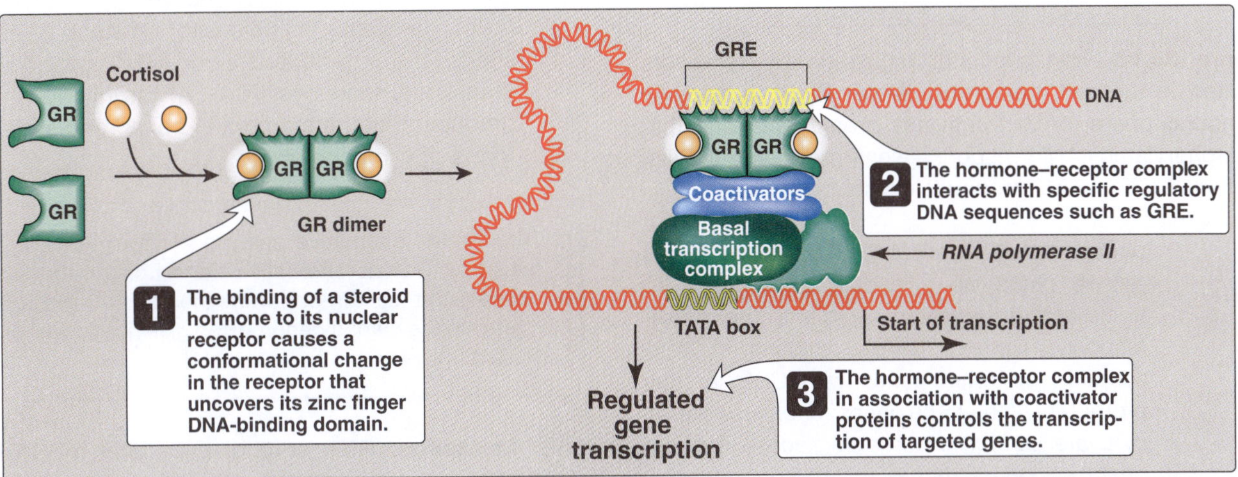

Figure 38.11 Transcriptional regulation by intracellular steroid hormone receptors. GRE = glucocorticoid response element; GR = glucocorticoid receptor.

allows recruitment of coactivators to the TAD and results in expression of cortisol-responsive genes, each of which is under the control of its own GRE. Binding of the receptor–hormone complex to the GRE allows coordinate expression of a group of target genes, even though these genes are on different chromosomes. The GRE can be located upstream or downstream of the genes it regulates and at great distances from them. The GRE, then, can function as a true enhancer (see p. 627). [Note: If associated with repressors, hormone–receptor complexes inhibit transcription.]

b. Cell-surface receptors: These receptors include those for insulin, epinephrine, and glucagon. Glucagon, for example, is a peptide hormone that binds its G protein–coupled plasma membrane receptor on glucagon-responsive cells. This extracellular signal is then transduced to intracellular cAMP, a second messenger (Fig. 38.12; also see Fig. 13.7 on p. 190), which can affect protein expression (and activity) through *protein kinase A*–mediated phosphorylation. In response to a rise in cAMP, a trans-acting factor (cAMP response element–binding [CREB] protein) is

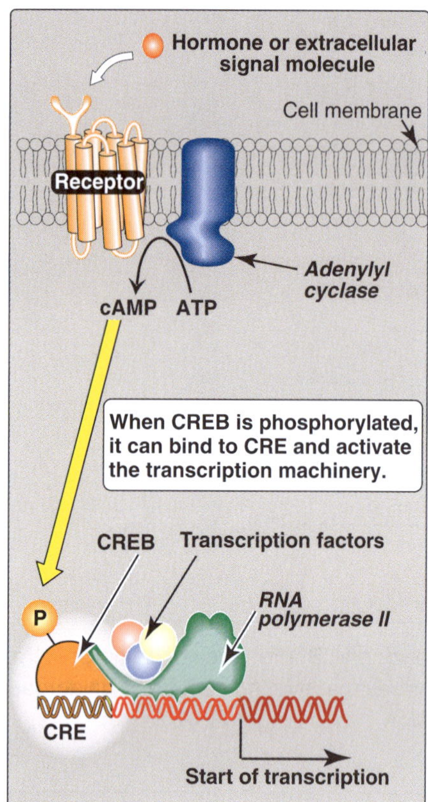

Figure 38.12 Transcriptional regulation by receptors located in the cell membrane. [Note: Cyclic adenosine monophosphate (cAMP) activates *protein kinase A* that phosphorylates cAMP response element–binding (CREB) protein.] CRE = cAMP response element.

phosphorylated and activated. Active CREB protein binds via a leucine zipper motif to a cis-acting regulatory element, the cAMP response element (CRE), resulting in transcription of target genes with CRE in their promoters. [Note: The genes for *phosphoenolpyruvate carboxykinase* and *glucose 6-phosphatase*, key enzymes of gluconeogenesis (see p. 215), are examples of genes upregulated by the cAMP/CRE/CREB system.]

B. Messenger RNA processing and use

Eukaryotic mRNA undergoes several processing events before it is exported from the nucleus to the cytoplasm for use in protein synthesis. Capping at the 5′-end (see p. 628), polyadenylation at the 3′-end (see p. 629), and splicing (see p. 629) are essential for the production of a functional eukaryotic messenger from most pre-mRNA. Variations in splicing and polyadenylation can affect gene expression. In addition, messenger stability also affects gene expression.

1. **Alternative splicing:** Tissue-specific protein isoforms can be made from the same pre-mRNA through alternative splicing, which can involve exon skipping (loss), intron retention, and use of alternative splice-donor or -acceptor sites (Fig. 38.13). For example, the pre-mRNA for tropomyosin (TM) undergoes tissue-specific alternative splicing to yield a number of TM isoforms (see p. 630). [Note: Over 90% of all human genes undergo alternative splicing.]

2. **Alternative polyadenylation:** Some pre-mRNA transcripts have more than one site for cleavage and polyadenylation. Alternative polyadenylation (APA) generates mRNA with different 3′-ends, altering the untranslated region (UTR) or the coding (translated) sequence. [Note: APA is involved in the production of the membrane-bound and secreted forms of immunoglobulin M.]

> The use of alternative splicing and polyadenylation sites, as well as alternative transcription start sites explains, at least in part, how the ~20,000 to 25,000 genes in the human genome can give rise to well over 100,000 proteins.

3. **Messenger RNA editing:** Even after mRNA has been fully processed, it may undergo an additional posttranscriptional modification in which a base in the mRNA is altered. This is known as **RNA**

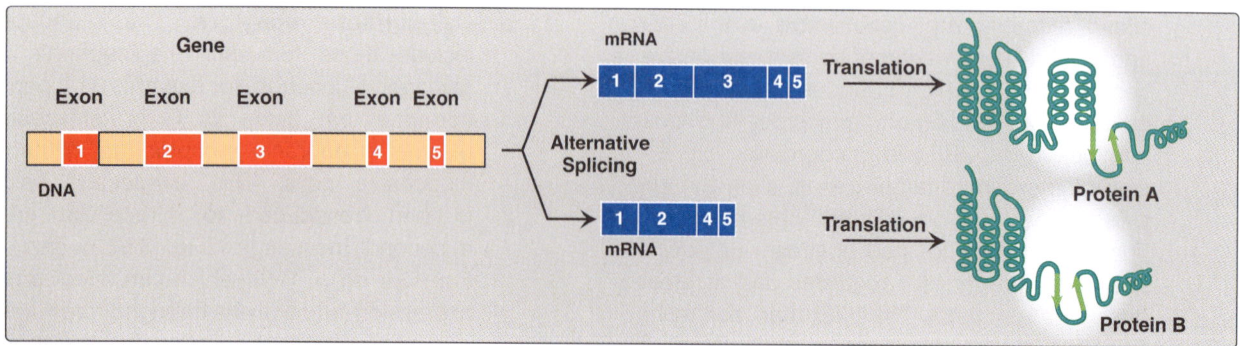

Figure 38.13 Tissue-specific alternative splicing produces different proteins, or isoforms, from a single gene. mRNA = messenger RNA.

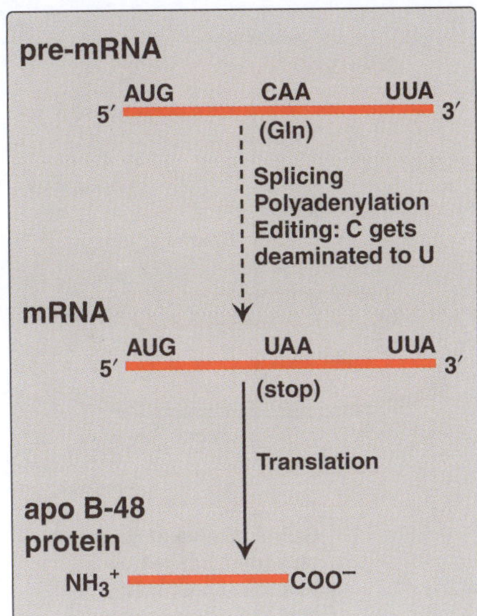

Figure 38.14 Editing of apolipoprotein (apo) B pre-mRNA in the intestine and generation of the apo B-48 protein needed for chylomicron synthesis. Gln = glutamine; mRNA = messenger RNA; A = adenine; C = cytosine; G = guanine; U = uracil.

editing. An important example in humans occurs with the transcript for apolipoprotein (apo) B, an essential component of chylomicrons (see p. 350) and very-low-density lipoproteins ([VLDL], see p. 352). Apo B mRNA is made in the liver and the small intestine. However, in the intestine only, the cytosine (C) base in the CAA codon for glutamine is enzymatically deaminated to uracil (U), changing the sense codon to the nonsense or stop codon UAA, as shown in Figure 38.14. This results in a shorter protein (apo B-48, representing 48% of the message) being made in the intestine (and incorporated into chylomicrons) than is made in the liver (apo B-100, full-length, incorporated into VLDL).

4. **Messenger RNA stability:** How long an mRNA remains in the cytosol before it is degraded influences how much protein product can be produced from it. Regulation of iron metabolism and the gene-silencing process of RNA interference (RNAi) illustrate the importance of mRNA stability in the regulation of gene expression.

   a. **Iron metabolism:** Transferrin (Tf) is a plasma protein that transports iron. Tf binds to cell-surface receptors (transferrin receptors [TfR]) that get internalized and provide cells, such as erythroblasts, with iron. The mRNA for the TfR has several cis-acting iron-responsive elements (IRE) in its 3′-UTR. IRE have a short stem-loop structure that can be bound by trans-acting iron regulatory proteins (IRP), as shown in Figure 38.15. When the iron concentration in the cell is low, the IRP bind to the 3′-IRE and stabilize the mRNA for TfR, allowing TfR synthesis. When intracellular iron levels are high, the IRP dissociate. The lack of IRP bound to the mRNA hastens its destruction, resulting in decreased TfR synthesis. [Note: The mRNA for ferritin, an intracellular protein of iron storage, has a single IRE in its 5′-UTR. When iron levels in the cell are low, IRP bind the 5′-IRE and prevent the use of the mRNA, and less ferritin is made. When iron accumulates in the cell, the IRP dissociate, allowing synthesis of ferritin molecules to store the excess iron. *Aminolevulinic acid synthase*

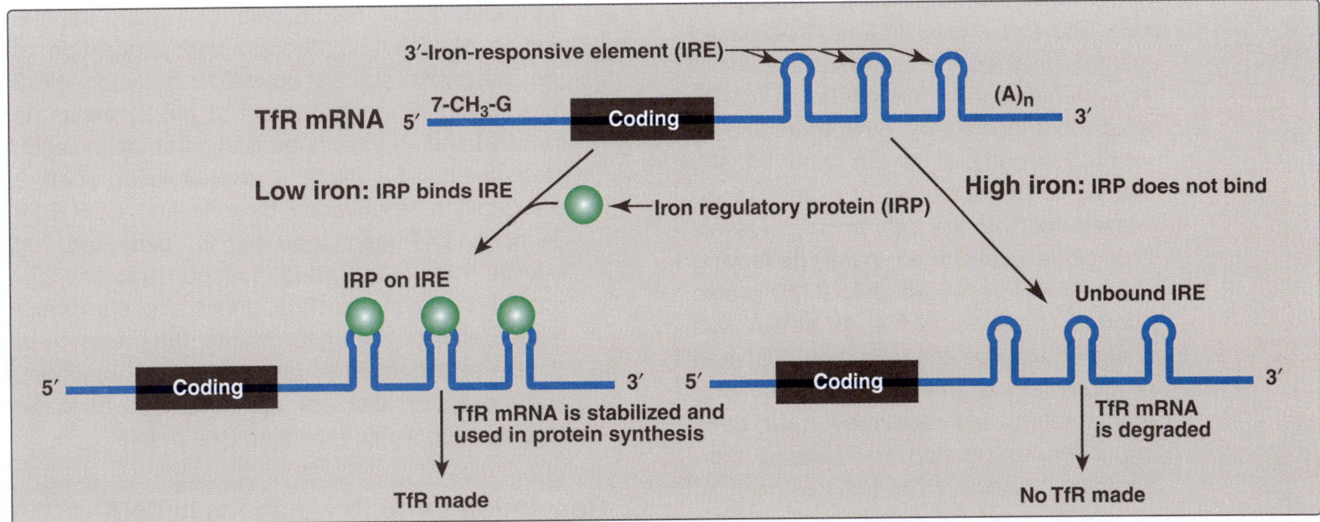

Figure 38.15 Regulation of transferrin receptor (TfR) synthesis. [Note: The IRE are located in the 3′-UTR (untranslated region) of TfR messenger RNA (mRNA).] $7\text{-}CH_3\text{-}G$ = 7-methylguanosine cap; $(A)_n$ = polyadenylate tail.

2, the regulated enzyme of heme synthesis (see p. 416) in erythroblasts, also contains a 5′-IRE.] (See Chapter 33 for a discussion of iron metabolism.)

b. **RNA interference: RNAi is a mechanism of gene silencing through decreased expression of mRNA, either by repression of translation or by increased degradation**. It plays a key role in such fundamental processes as cell proliferation, differentiation, and apoptosis. RNAi is mediated by short (~22 nucleotides), noncoding RNA called microRNA (miRNA). The miRNA arise from far longer, genomically encoded nuclear transcripts, primary miRNA (pri-miRNA), that are partially processed in the nucleus to pre-miRNA by an *endonuclease* (*Drosha*) then transported to the cytoplasm. There, an *endonuclease* (*Dicer*) completes the processing and generates short, double-stranded miRNA. A single strand (the guide or antisense strand) of the miRNA associates with a cytosolic protein complex known as the RNA-induced silencing complex (RISC). The guide strand hybridizes with a complementary sequence in the 3′-UTR of a full-length target mRNA, bringing RISC to the mRNA. This can result in repression of translation of the mRNA or its degradation by an *endonuclease* (*Argonaute/Ago/Slicer*) of the RISC. The extent of complementarity appears to be the determining factor (Fig. 38.16). RNAi can also be triggered by the introduction of exogenous double-stranded short interfering RNA (siRNA) into a cell, a process that has enormous therapeutic potential.

i. **RNA interference–based therapeutics:** The first clinical trial of RNAi-based therapy involved the neovascular form of age-related macular degeneration (AMD), which is triggered by overproduction of vascular endothelial growth factor (VEGF), leading to the sprouting of excess blood vessels behind the retina. The vessels leak, clouding and often entirely destroying vision (therefore, neovascular AMD is also referred to as wet AMD). An siRNA was designed to target the mRNA of VEGF and promote its degradation. Although considerable effort and resources have been expended to develop RNAi-based therapeutics, especially for the treatment of cancer, no products have gone from trials to the market. The research applications of RNAi, however, have grown rapidly.

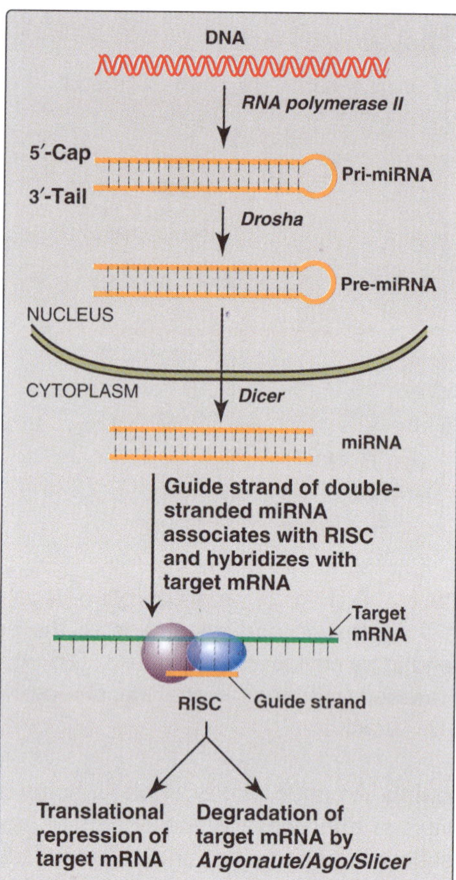

Figure 38.16 Biogenesis and actions of microRNA (miRNA). [Note: The extent of complementarity between the target messenger RNA (mRNA) and the miRNA determines the final outcome, with perfect complementarity resulting in mRNA degradation.] Pri = primary; RISC = RNA-induced silencing complex.

5. **Messenger RNA translation:** Regulation of gene expression can also occur at the level of mRNA translation. One mechanism by which translation is regulated is through phosphorylation of the eukaryotic translation initiation factor, eIF-2 (Fig. 38.17). Phosphorylation of eIF-2 inhibits its function and so inhibits translation at the initiation step (see p. 644). [Note: Phosphorylation of eIF-2 prevents its reactivation by inhibiting GDP-GTP exchange.] Phosphorylation is catalyzed by *kinases* that are activated in response to environmental conditions, such as amino acid starvation, heme deficiency in erythroblasts, the presence of double-stranded RNA (signaling viral infection), and the accumulation of misfolded proteins in the rough endoplasmic reticulum (see p. 649).

C. Regulation through variations in DNA

Gene expression in eukaryotes is also influenced by the accessibility of DNA to the transcriptional

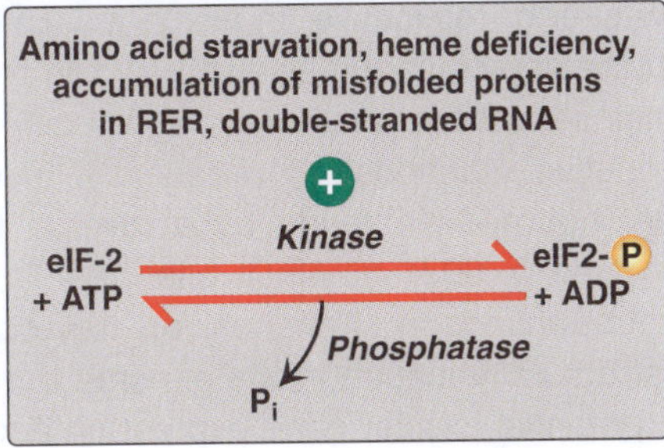

Figure 38.17 Regulation of translation initiation in eukaryotes by phosphorylation of eukaryotic translation initiation factor, eIF-2. RER = rough endoplasmic reticulum; ADP = adenosine diphosphate; P_i = inorganic phosphate; P = phosphate.

apparatus, the amount of DNA, and the arrangement of DNA. [Note: Localized transitions between the B and Z forms of DNA (see p. 598) can also affect gene expression.]

1. **Access to DNA:** In eukaryotes, DNA is found complexed with histone and nonhistone proteins to form chromatin (see p. 608). Transcriptionally active, decondensed chromatin (euchromatin) differs from the more condensed, inactive form (heterochromatin) in a number of ways. Active chromatin contains histone proteins that have been covalently modified at their amino terminal ends by reversible methylation, acetylation, or phosphorylation (see p. 625 for a discussion of histone acetylation/deacetylation by *histone acetyltransferase* and *histone deacetylase*). Such modifications decrease the positive charge of these basic proteins, thereby decreasing the strength of their association with negatively charged DNA. This relaxes the nucleosome (see p. 609), allowing transcription factors access to specific regions on the DNA. Nucleosomes can also be repositioned, an ATP-requiring process that is part of chromatin remodeling. Another difference between transcriptionally active and inactive chromatin is the extent of methylation of cytosine bases in CG-rich regions (CpG islands) in the promoter region of many genes. Methylation is by *methyltransferases* that use S-adenosylmethionine as the methyl donor (Fig. 38.18). Transcriptionally active genes are less methylated (hypomethylated) than their inactive counterparts, suggesting that DNA hypermethylation silences gene expression. Modification of histones and methylation of DNA are epigenetic in that they are heritable changes in DNA that alter gene expression without altering the base sequence.

2. **Amount of DNA:** A change up or down in the number of copies of a gene can affect the amount of gene product produced. An increase in copy number **(gene amplification)** has contributed to increased genomic complexity and is still a normal developmental process, e.g., ribosomal protein genes amplify as a new cell is being made. Gene amplification is also seen in some diseases and in response to particular chemotherapeutic drugs such as methotrexate, an inhibitor of the enzyme *dihydrofolate reductase* (*DHFR*), required for the synthesis of thymidine triphosphate (TTP) in the pyrimidine biosynthetic pathway (see p. 587). TTP is essential for DNA synthesis. Gene amplification results in an increase in the number of *DHFR* genes and resistance to the drug, allowing TTP to be made.

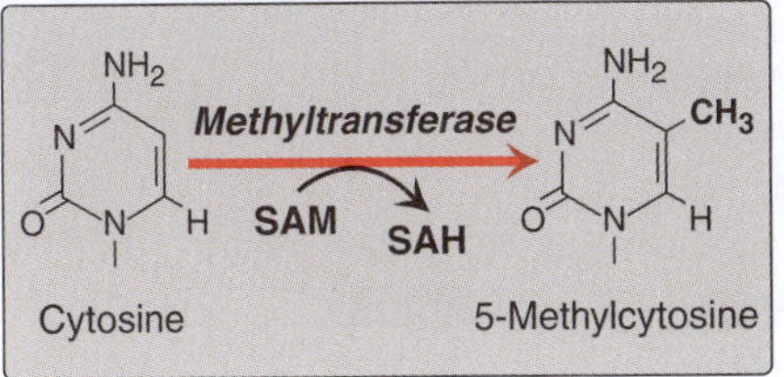

Figure 38.18 The methylation of cytosine in eukaryotic DNA. SAM = S-adenosylmethionine; SAH = S-adenosylhomocysteine

3. **Arrangement of DNA: The process by which immunoglobulins (antibodies) are produced by B lymphocytes involves permanent rearrangements of the DNA in these cells.** The immunoglobulins (for example, IgG) consist of two light and two heavy chains, with each chain containing regions of variable and constant amino acid sequence. The variable region is the result of somatic recombination of segments within both the light- and the heavy-chain genes. During B-lymphocyte development, single variable (V), diversity (D), and joining (J) gene segments are brought together through gene rearrangement to form a unique variable region (Fig. 38.19). This process allows the generation of 10^9–10^{11} different immunoglobulins from a single gene, providing the diversity needed for the recognition of an enormous number of antigens. [Note: Pathologic DNA rearrangement is seen with translocation, a process by which two different chromosomes exchange DNA segments.]

4. **Mobile DNA elements: Transposons (Tn) are mobile segments of DNA that move in an essentially random manner from one site to another on the same or a different chromosome.** Movement is mediated by *transposase*, an enzyme encoded by the Tn itself. Movement can be direct, in which *transposase* cuts out and then inserts the Tn at a

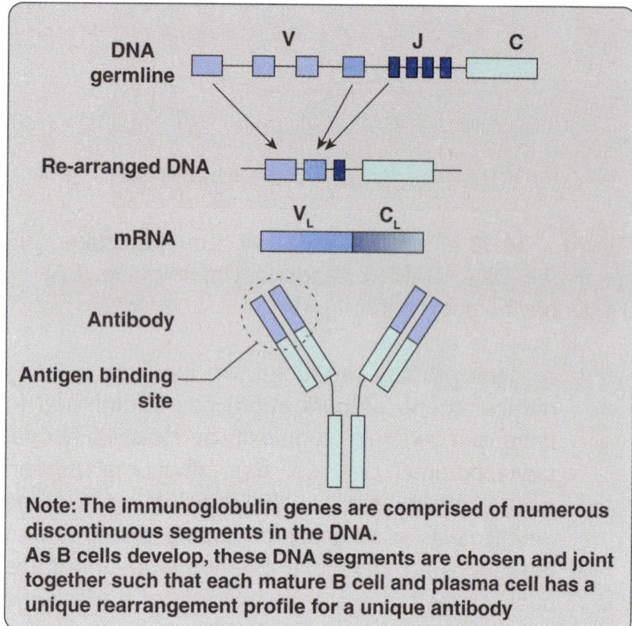

Figure 38.19 Immunoglobulin gene rearrangement in antibody-producing cell.

new site, or replicative, in which the Tn is copied and the copy inserted elsewhere while the original remains in place. In eukaryotes, including humans, replicative transposition frequently involves an RNA intermediate made by a *reverse transcriptase* (see p. 608), in which case the Tn is called a retrotransposon. Transposition has contributed to structural variation in the genome but also has the potential to alter gene expression and even to cause disease. Tn comprise ~50% of the human genome, with retrotransposons accounting for 90% of Tn. Although the vast majority of these retrotransposons have lost the ability to move, some are still active. Their transposition is thought to be the basis for some rare cases of hemophilia A and Duchenne muscular dystrophy. [Note: The growing problem of antibiotic-resistant bacteria is a consequence, at least in part, of the exchange of plasmids among bacterial cells. If the plasmids contain Tn-carrying antibiotic resistance genes, the recipient bacteria gain resistance to one or more antimicrobial drugs.]

Additional interesting features about eukaryotic gene expression

- In eukaryotes, there can be further binding of multiple proteins which lead to extra DNA bends and approximations that exponentially increase rate of transcription. This formation of **enhanceosomes** helps the eukaryotes to deal with an urgent and excessive need like increased transcription of human beta interferon gene in response to human immunodeficiency virus (HIV) infection.
- **Locus control regions (LCRs)** are segments of DNA that controls gene clusters which may be near or far apart from each other (but on the same chromosome). This control is a level above regulated expression of genes and may be related to chromatin structure. For example, an LCR controls expression of globin gene family over a large region of DNA. **Insulators** are segments of DNA beyond which the influence of LCR will not extend.
- **Gene switching** is a phenomenon in which transcription and translation of a given gene stops and a related one is turned on. This is seen in development of hemoglobin from embryonic to adult life and in immunoglobin response to an antigen (Chapter 43).
- Eukaryotes have tissues and therefore their differentiation requires some specific processes. **Tissue-specific expression** implies the production of particular products in particular tissues only, e.g., production of insulin from pancreas and immunoglobulins from B-lymphocytes. Some ways in which tissue-specific expression can be achieved include enhancer or enhancer-like elements which can act in a tissue-specific manner to regulate gene expression.
- **Genomic imprinting** is the inheritance of some genes in a silent state from one parent (imprinted by methylation which inhibits its expression), while the other parental gene is expressed. Further, eukaryotes also have the ability of **an SOS response** which is the coordinated regulation of unrelated genes having a common goal. For example, when there is severe DNA damage, the common goal of limiting damage and preventing transmission to daughter cells leads to coordination of DNA-repairing genes and cell division repression genes. **Gene loss** may also occur in eukaryotes. It is the loss of genetic material from cells, e.g., RBC which are without nucleus.
- The meiotically and mitotically heritable changes in gene expression which are not coded for in the DNA, are another challenge in understanding eukaryote gene expression. These **epigenetic modifications** have been targeted in drug therapies.
- **Gene silencing** is also one of the epigenetic process of gene regulation. It is the switching off a gene by mechanisms other than genetic modifications. Common ways are transcriptional gene silencing by histone modifications, DNA methylations and silencing by interfering RNA. This can be developmental as a part of tissue-specific expression and could also be a protection against transposons or viruses.

V. Medical Perspectives

- A resistance to methotrexate (MTX), an anticancer drug is often seen in patients receiving MTX for a long time. This is because of **gene amplification** of *dihydrofolate reductase (DHFR)* gene leading to increased number of *DHFR* genes, causing increased production of *DHFR*. Therefore, the clinician often needs to increase dosage of MTX in such patients
- As the individual grows from the embryonic stage to an adult, his hemoglobin changes to a different type. This is achieved by **gene switching**. Embryonic hemoglobin has two alpha and two epsilon chains. This is replaced by fetal Hb at 6 months of life which has two alpha and two gamma chains. After birth, two alpha and two beta chains (HbA1) or two alpha and two delta chains are synthesized (HbA2). Defects in gene switching can give rise to severe hemoglobinopathies.
- Gene switching is also responsible for class switching of immunoglobulins during transition from primary to secondary immune response. During **primary response IgM** is produced, but in **secondary response IgG** with same antigen specificity is synthesized.
- Therefore, to assess a recent infection, IgM antibodies are tested, and IgG antibodies are tested to assess an infection which happened some time ago.
- The inhibition of specific epigenetic enzymes can reverse the incorporation of an epigenetic mutation making these enzymes attractive targets for cancer therapy. Drugs which inhibit *DNA dimethyltransferase 1* (Azacytidine and Decitabine) are approved for treatment of adult myeloid leukemia (AML), chronic myelogenous leukemia (CML), and myelodysplastic syndrome (MDS). Vorinostat is a *histone deacetylase* inhibitor used to treat cutaneous T-cell lymphoma (CTCL).
- **Prader-Willi syndrome (PWS)** and its sister syndrome, **Angelman syndrome (AS),** are classical examples of genomic imprinting in humans. Both syndromes are associated with defects in chromosome 15. Paternal inheritance of a deletion of this region is associated with PWS (characterized by hypotonia, obesity, and hypogonadism), while maternal inheritance of the same deletion is associated with AS (characterized by epilepsy, tremors, and a perpetually smiling facial expression).

VI. Chapter Summary

- **Gene expression** results in the production of a functional gene product (either RNA or protein) through the processes of transcription and translation (Fig. 38.20).
- **Genes** can be either **constitutive** (always expressed, housekeeping genes) or **regulated** (expressed only under certain conditions in all cells or in a subset of cells). The ability to appropriately **induce** (**positively regulate**) or **repress** (**negatively regulate**) genes is essential in all organisms. Regulation of gene expression occurs primarily at **transcription** in both prokaryotes and eukaryotes and is mediated through **trans-acting proteins** binding to **cis-acting regulatory DNA elements.** I
- n **eukaryotes**, regulation also occurs through DNA **modifications** and through **posttranscriptional** and **posttranslational processing**. In **prokaryotes**, such as Escherichia coli, the coordinate regulation of genes whose protein products are required for a particular process is achieved through **operons** (groups of genes sequentially arranged on the chromosome along with the regulatory elements that determine their transcription).
- The **lac operon** contains the Z, Y, and A structural genes, the protein products of which are needed for the catabolism of lactose. It is subject to negative and positive regulation. When **glucose** is available, the operon is repressed by the binding of the **repressor protein** (the product of the lacI gene) to the **operator**, thus preventing transcription. When only **lactose** is present, the operon is induced by an isomer of lactose (**allolactose**) that binds the repressor protein, preventing it from binding to the operator. In addition, **cyclic adenosine monophosphate** (**cAMP**) binds the **catabolite activator protein** (**CAP**), and the complex binds the DNA at the **CAP site**. This increases **promoter** efficiency and results in the expression of the structural genes through the production of a **polycistronic messenger RNA** (**mRNA**). When both **glucose and lactose** are present, glucose prevents formation of cAMP, and transcription of these genes is negligible.
- The **trp operon** contains genes needed for the synthesis of tryptophan (Trp), and, like the lac operon, it is regulated by **negative control**. Unlike the lac operon, it is also regulated by **attenuation**, in which mRNA synthesis that escaped repression by Trp is terminated before completion. Transcription of **ribosomal RNA** and **transfer RNA** is selectively inhibited in prokaryotes by the **stringent response** to amino acid starvation.
- **Translation** is also a site of prokaryotic gene regulation: Excess ribosomal proteins bind the **Shine-Dalgarno sequence** on their own polycistronic mRNA, preventing ribosomes from binding.
- Gene regulation is more complex in **eukaryotes**. Operons are not present, but coordinate regulation of the transcription of genes located on different chromosomes can be achieved through the binding of trans-acting proteins to cis-acting elements as seen in the **galactose circuit** in unicellular yeast.

- In multicellular organisms, **hormones** can cause coordinated regulation, either through the binding of the **hormone receptor–hormone complex** to the DNA (as with steroid hormones) or through the binding of a protein that is activated in response to a **second messenger** (as with glucagon). In each case, binding to DNA is mediated through structural motifs such as the **zinc finger**. **Co-** and **posttranscriptional regulation** is also seen in eukaryotes and includes **alternative mRNA splicing** and **polyadenylation**, mRNA **editing**, and variations in mRNA **stability** as seen with **transferrin receptor** synthesis and with **RNA interference**. Regulation at the **translational level** can be caused by the **phosphorylation** and inhibition of **eukaryotic initiation factor-2**.
- Gene expression in eukaryotes is also influenced by **accessibility** of DNA to the transcriptional apparatus (as seen with **epigenetic** changes to histone proteins), the **amount** of DNA, and the **arrangement** of the DNA.

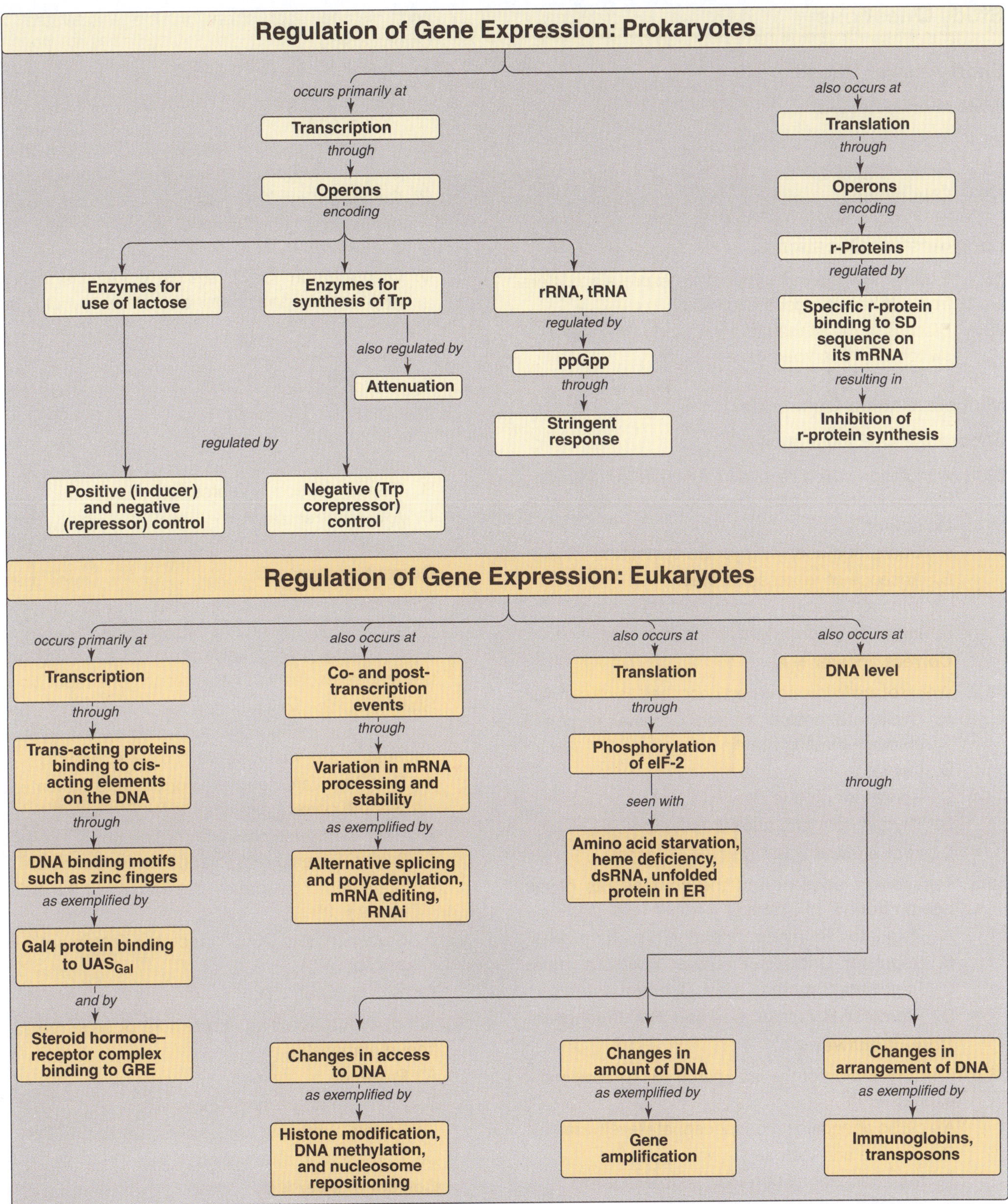

Figure 38.20 Summary of key concepts for the regulation of gene expression. Trp = tryptophan; rRNA, tRNA, mRNA = ribosomal, transfer, and messenger RNA, respectively; ppGpp; guanosine tetraphosphate; r-protein = ribosomal protein; SD = Shine-Dalgarno; Gal = galactose; UAS = upstream activating sequence; GRE = glucocorticoid response element; RNAi = RNA interference; eIF = eukaryotic initiation factor; ds = double stranded; ER = endoplasmic reticulum.

Study Questions

Short-Answer Questions

Write short notes on:

38.1. Tissue-specific expression
38.2. Gene amplification
38.3. DNA motifs

Long-Answer Questions

38.1. Explain how the eukaryotic gene expression differs from prokaryotic gene expression.
38.2. Discuss in detail the lac operon and the molecular switch of prokaryotic genes.

Multiple-Choice Questions

Choose the ONE best answer.

38.1. A mutation which results in which of the following is most likely to cause a reduced expression of the lac operon?
A. Decreased adenylyl cyclase synthesis
B. Absence of repressor protein
C. Impaired binding of operator to repressor
D. Impaired glucose uptake into the cell
Correct answer = A.

38.2. Which of the following is best described as cis-acting?
A. Cyclic adenosine monophosphate response element–binding protein
B. Operator
C. Repressor protein
D. Thyroid hormone nuclear receptor
Correct answer = B.

38.3. The correct sequence of the components of lac operon from 5' to 3' end is:
A. Regulator, Promoter, Operator, Structural gene
B. Regulator, Operator, Promoter, Structural gene
C. Regulator, Structural gene, Operator, Promoter
D. Promoter, Regulator, Operator, Structural gene
Correct answer = A.

38.4. The ZYA region of the lac operon will be maximally expressed if:
A. cyclic adenosine monophosphate levels are low.
B. glucose and lactose are both available.
C. the attenuation stem-loop is able to form.
D. the CAP site is occupied.
Correct answer = D.

38.5. X chromosome inactivation is a process by which one of two X chromosomes in mammalian females is condensed and inactivated to prevent overexpression of X-linked genes. What would most likely be true about the degree of DNA methylation and histone acetylation on the inactivated X chromosome?
A. Cytosines in CpG islands would be hypermethylated, and histone proteins would be deacetylated.
B. Cytosines in CpG islands would be hypomethylated, and histone proteins would be acetylated.
C. Cytosines in CpG islands would be hypermethylated, and histone proteins would be acetylated.
D. Cytosines in CpG islands would be hypomethylated, and histone proteins would be deacetylated.
Correct answer = A.

38.6. Regulation of an enzyme by which among the following mechanisms results in an alteration in its concentration?
A. Feedback inhibition
B. Gene induction or repression
C. Allosteric regulation
D. Reversible covalent modification
Correct answer = B.

38.7. Statement 1: Transcription attenuation, where the presence of an attenuator sequence in the form of a hairpin loop prematurely terminates transcription, is a feature exclusive to eukaryotic gene regulation.
Statement 2: Transcription and translation are spatially and temporally separated in eukaryotes because of the well-defined nuclear structure.
A. Both the Statements are correct, and Statement 2 is the correct explanation for Statement 1
B. Both the Statements are correct, but Statement 2 is NOT the correct explanation for Statement 1
C. Statement 1 is correct, but Statement 2 is NOT correct
D. Statement 1 is NOT correct, but Statement 2 is correct
Correct answer = D.

38.8. Which of the following is a feature of gene regulation in humans?
A. Presence of operons
B. Alternate polyadenylation
C. Protein synthesis using polycistronic mRNAs
D. Transcription attenuation
Correct answer = B.

38.9. Vorinostat, an inhibitor of histone deacetylases, is used in the management of certain hematological malignancies. What is the most likely effect of this drug on gene expression in the cells that are exposed to it?
A. Gene expression is increased due to increased access for transcriptional machinery

B. Gene expression is decreased due to decreased access for transcriptional machinery
C. Gene expression is unaltered
D. Gene expression is increased independent of the access of transcriptional machinery

Correct answer = A.

38.10. Despite the same structure of the antigen binding site and most of the constant region, and a similar capacity for binding to the same antigen, the B-cell receptor from a particular B-cell remains membrane bound, while the antibody produced is secreted. Which among the following is responsible for this difference?

A. mRNA editing
B. Gene silencing by miRNAs
C. Epigenetic gene regulation
D. Differential RNA processing

Correct answer = D.

Multiple-Choice Questions (NExT/NEET PG)

Choose the ONE best answer.

38.1. Various elements of the lac operon are: i–Promoter site, ii–Operator site, iii–Regulator gene, iv–Structural gene.
Functions of various elements of the lac operon are: a–Binding site for repressor molecule, b–Binding site for RNA polymerase, c–Codes for repressor molecules, d–Codes for the enzyme protein.
Which of the following is matched correctly?

A. i – b, ii – a, iii – c, iv – d
B. i – b, ii – c, iii – a, iv – d
C. i – b, ii – a, iii – d, iv – c
D. i – d, ii – c, iii – a, iv – b

Correct answer = A.

38.2. Which of the following is most likely to be true in hemochromatosis, a disease of iron accumulation?

A. The messenger RNA for the transferrin receptor is stabilized by the binding of iron regulatory proteins to its 3′-iron-responsive elements.
B. The messenger RNA for the transferrin receptor is not bound by iron regulatory proteins and is degraded.
C. The messenger RNA for ferritin is not bound by iron regulatory proteins at its 5′-iron-responsive element and is not translated.
D. The messenger RNA for ferritin is bound by iron regulatory proteins and is not translated.

Correct answer = B.

38.3. Patients with estrogen receptor–positive (hormone responsive) breast cancer may be treated with the drug tamoxifen, which binds the estrogen nuclear receptor without activating it. Which of the following is the most logical outcome of tamoxifen use?

A. Increased histone acetylation of estrogen-responsive genes
B. Increased growth of estrogen receptor–positive breast cancer cells
C. Inhibition of the estrogen operon
D. Inhibition of transcription of estrogen-responsive genes

Correct answer = D.

38.4. Fragile X syndrome is a disorder caused by a loss-of-function mutation in the FMR1 gene leading to reduced or absent levels of fragile X messenger ribonucleoprotein (FMRP). In majority of the cases, this loss of function is due to an unstable expansion of a CGG trinucleotide repeat in the 5' untranslated region and adjacent regions of the gene. Which among the following leads to the clinical manifestations in the presence of an increase in the number of the repeats?

A. Presence of the CGG repeats leads to alteration in the reading frame during translation.
B. CGG repeats get translated to arginine, and this disrupts the net charge in the protein.
C. Methylation of cytosines at the CGG repeats in the promoter leads to epigenetic silencing of the gene.
D. CGG repeats decrease the rate of ribosome binding to the FMRP mRNA, thus reducing the rate of protein synthesis.

Correct answer = C.

38.5. Givosiran is a small interfering RNA (siRNA) targeted against ALA synthase-1 (ALAS1) which is the rate limiting enzyme of heme synthesis. It has been approved for the treatment of acute hepatic porphyrias. Which of the following best describes mechanism of the action of this drug?

A. Givosiran decreases the ALAS1 mRNA levels by binding to it and targeting its degradation.
B. Givosiran decreases the ALAS1 protein synthesis by interfering with the ribosomal binding to the mRNA.
C. The drug disrupts the binding of charged tRNAs to the ALAS1 mRNA during translation.
D. Givosiran edits the ALAS1 mRNA to insert a premature stop codon during translation.

Correct answer = A.

Biotechnology and Human Disease

39

Learning Objectives

At the end of this chapter, the student will be able to:

- Describe the various techniques used in recombinant DNA technology
- Describe the medical uses of genetic methods

I. OVERVIEW

In the past, efforts to understand genes and their expression have been confounded by the immense size and complexity of human deoxyribonucleic acid (DNA). The human genome contains ~3 billion (10^9) base pairs (bp) that encode 20,000–25,000 protein-coding genes located on 23 chromosomes in the haploid genome. It is now possible to determine the nucleotide sequence of long stretches of DNA, and the entire human genome has been sequenced. This effort (called the **Human Genome Project** and completed in 2003) was made possible by several tools that have already contributed to our understanding of many genetic diseases (Fig. 39.1). These include 1) the discovery of ***restriction endonucleases*** that permit the cleavage of huge DNA molecules into defined fragments, 2) the development of cloning techniques that provide a mechanism for amplification of specific nucleotide sequences, and 3) the ability to synthesize specific probes, which has allowed the identification and manipulation of nucleotide sequences of interest. These and other experimental approaches have permitted the identification of both normal and mutant nucleotide sequences in DNA. This knowledge has led to the development of methods for the diagnosis of genetic diseases and some successes in the treatment of patients by gene therapy. [Note: The genomes of several viruses, prokaryotes, and nonhuman eukaryotes have also been sequenced.]

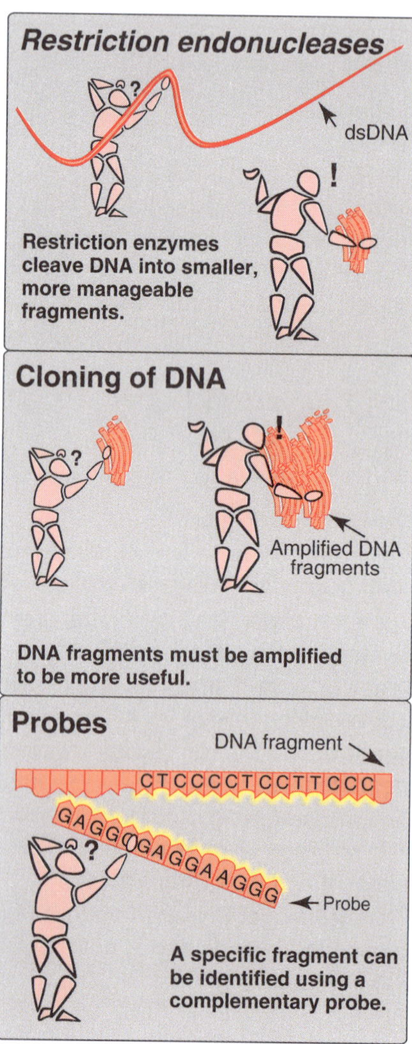

Figure 39.1 Three tools that facilitate analysis of human DNA. dsDNA = double-stranded DNA.

II. RESTRICTION ENDONUCLEASES

One of the major obstacles to molecular analysis of genomic DNA is the immense size of the molecules involved. The discovery of a special group of bacterial enzymes, called *restriction endonucleases* (restriction enzymes), which cleave double-stranded DNA (dsDNA) into smaller, more manageable fragments, opened the way for DNA analysis. Because each enzyme cleaves dsDNA at a specific nucleotide sequence (restriction site), restriction enzymes are used experimentally to obtain precisely defined DNA segments called restriction fragments.

A. Specificity

***Restriction endonucleases* recognize short stretches of dsDNA (4–8 bp) that contain specific nucleotide sequences**. These sequences, which differ for each restriction enzyme, are palindromes, that is, they exhibit twofold rotational symmetry (Fig. 39.2). This means that, within a short region of the dsDNA, the nucleotide sequence on the two strands is identical if each is read in the 5′→3′ direction. Therefore, if you turn the page upside down (that is, rotate it 180° around its axis of symmetry) the sequence remains the same.

> In bacteria, *restriction endonucleases* limit (restrict) the expression of nonbacterial (foreign) DNA through cleavage. Bacterial DNA is protected from cleavage by methylation of adenine at the restriction site. a

B. Nomenclature

A restriction enzyme is named according to the organism from which it was isolated. The first letter of the name is from the genus of the bacterium. The next two letters are from the name of the species. An additional letter indicates the type or strain (as needed), and a number (Roman numeral) is appended to indicate the order in which the enzyme was discovered in that particular organism. For example, *Hae*III is the third *restriction endonuclease* isolated from the bacterium Haemophilus aegyptius.

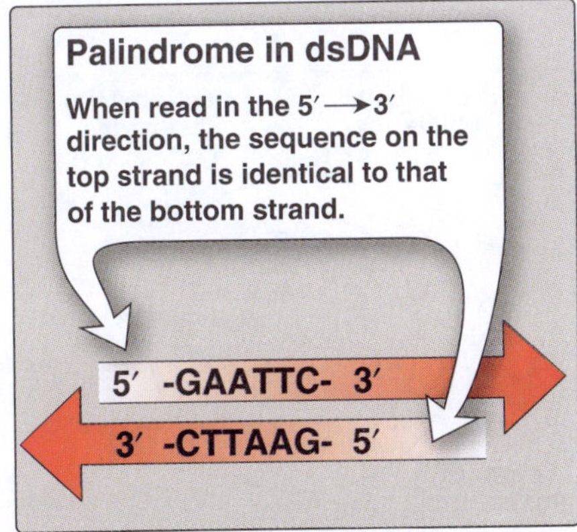

Figure 39.2 Recognition sequence of *restriction endonuclease Eco*RI shows twofold rotational symmetry. dsDNA = double-stranded DNA; A = adenine; C = cytosine; G = guanine; T = thymine.

C. Sticky and blunt ends

Restriction enzymes cleave dsDNA so as to produce a 3′-hydroxyl group on one end and a 5′-phosphate group on the other. Some *restriction endonucleases*, such as ***TaqI*, form staggered cuts** that produce sticky or cohesive ends (that is, the resulting DNA fragments have single-stranded regions that are complementary to each other), as shown in Figure 39.3. Other *restriction endonucleases*, such as **HaeIII, produce fragments that have blunt ends** that are entirely double stranded and, therefore, do not form hydrogen bonds with each other. Using the enzyme *DNA ligase* (see p. 602), sticky ends of a DNA fragment of interest can be covalently joined with other DNA fragments that have sticky ends produced by cleavage with the same *restriction endonuclease* (Fig. 39.4). [Note: A *ligase* encoded by bacteriophage T4 can covalently join blunt-ended fragments.]

D. Restriction sites

A DNA sequence that is recognized and cut by a restriction enzyme is called a **restriction site.** ***Restriction endonucleases*** cleave dsDNA into fragments of different sizes depending upon the size of the sequence recognized. For example, an enzyme that recognizes a specific 4-bp sequence produces many cuts in the DNA molecule, one every 4^4 bp.

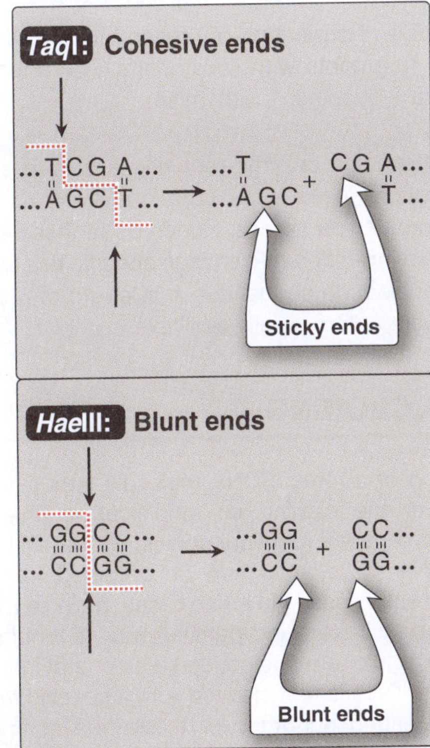

Figure 39.3 Specificity of *TaqI* and *HaeIII restriction endonucleases*. A = adenine; C = cytosine; G = guanine; T = thymine.

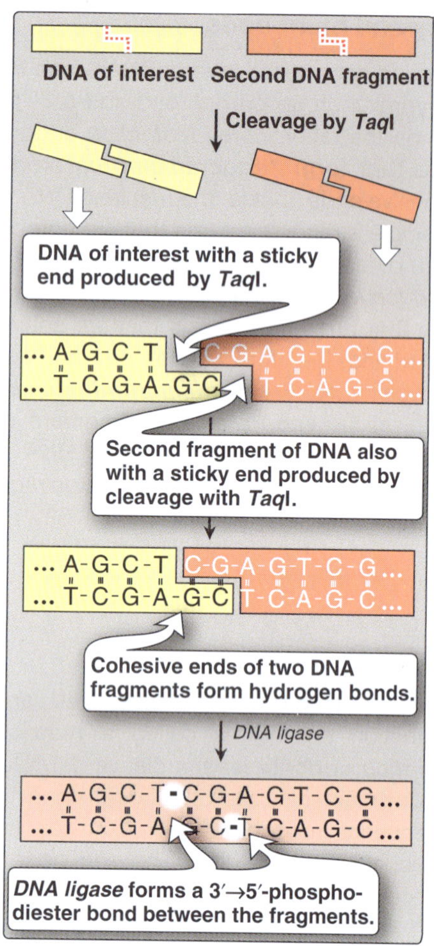

Figure 39.4 Formation of recombinant DNA from restriction fragments with sticky ends. A = adenine; C = cytosine; G = guanine; T = thymine.

In contrast, an enzyme requiring a unique sequence of 6 bp produces fewer cuts (one every 4^6 bp) and, therefore, longer pieces. Hundreds of these enzymes, each having different cleavage specificities (varying in both nucleotide sequences and length of recognition sites), are commercially available.

III. DNA CLONING

Introduction of a foreign DNA molecule into a replicating cell permits the cloning or, amplification (that is, the production of many identical copies) of that DNA. [Note: Human DNA for cloning can be obtained from blood, saliva, and solid tissue.] In some cases, a single DNA fragment can be isolated and purified prior to cloning. More commonly, to clone a nucleotide sequence of interest, the total cellular DNA is first cleaved with a specific restriction enzyme, creating hundreds of thousands of fragments. Each of the resulting DNA fragments is joined to a DNA vector molecule (referred to as a cloning vector) to form a hybrid, or recombinant, DNA molecule. Each recombinant molecule carries its inserted DNA fragment into a single host cell (for example, a bacterium), where it is replicated. [Note: The process of introducing foreign DNA into a cell is called transformation for bacteria and yeast and transfection for higher eukaryotes.] As the host cell multiplies, it forms a clone in which every bacterium contains copies of the same inserted DNA fragment, hence the name "cloning." The cloned DNA can be released from its vector by cleavage (using the appropriate *restriction endonuclease*) and isolated. By this mechanism, many identical copies of the DNA of interest can be produced. [Note: An alternative to amplification by biologic cloning, the polymerase chain reaction (PCR), is described on p. 686.]

A. Vectors

A vector is a molecule of DNA to which the fragment of DNA to be cloned is joined. Essential properties of a vector include the 1) **capacity for autonomous replication within a host cell**, 2) presence of at least one specific nucleotide sequence **recognized by a *restriction endonuclease***, and 3) presence of **at least one gene (such as an antibiotic resistance gene**) that confers the ability to select for the vector. Commonly used vectors include **plasmids and viruses**.

1. **Prokaryotic plasmids:** Prokaryotic organisms typically contain single, large, circular chromosomes. In addition, most species of bacteria also normally contain small, circular, extrachromosomal DNA molecules called plasmids (Fig. 39.5). Plasmid DNA undergoes replication that may or may not be synchronized to chromosomal division. Plasmids may carry genes that convey antibiotic resistance to the host bacterium and may facilitate the transfer of genetic information from one bacterium to another. They can be readily isolated from bacterial cells, their circular DNA cleaved at specific sites by ***restriction endonucleases***, and up to 15 kb (kilobases) of foreign DNA (cut with the same restriction enzyme) inserted. The recombinant plasmid vector can be introduced into a

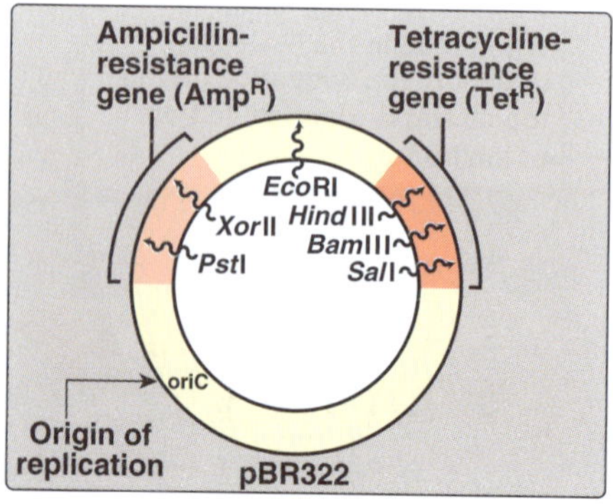

Figure 39.5 A partial map of pBR322 indicating the positions of its antibiotic resistance genes and 6 of the >40 unique sites recognized by specific *restriction endonucleases*. p = plasmid.

bacterium, producing large numbers of copies of the plasmid. The bacteria are grown in the presence of antibiotics, thus selecting for cells containing the hybrid plasmids, which provide antibiotic resistance (Fig. 39.6). **Artificial plasmids** are routinely constructed. An example is the classic pBR322 (see Fig. 39.5), which contains an origin of replication, two antibiotic resistance genes, and >40 unique restriction sites. Use of plasmids is limited by the size of the DNA that can be inserted.

2. **Other vectors:** The development of improved vectors that can more efficiently accommodate larger DNA segments, or express the passenger genes in different cell types, has aided molecular genetics research and therapeutics. In addition to the prokaryotic plasmids described above, naturally occurring viruses that infect bacteria (bacteriophage λ, for example) or mammalian cells (retroviruses, for example), as well as artificial constructs such as **cosmids** and **bacterial or yeast artificial chromosomes** (BAC or YAC, respectively), are currently used as cloning vectors. [Note: BAC and YAC can accept DNA inserts of 100–300 kb and 250–1,000 kb, respectively.]

Expression vector is the vector in which the protein can be actually synthesized, e.g., bacteria. The expression vector should have the transcriptional and translational machinery. The protein

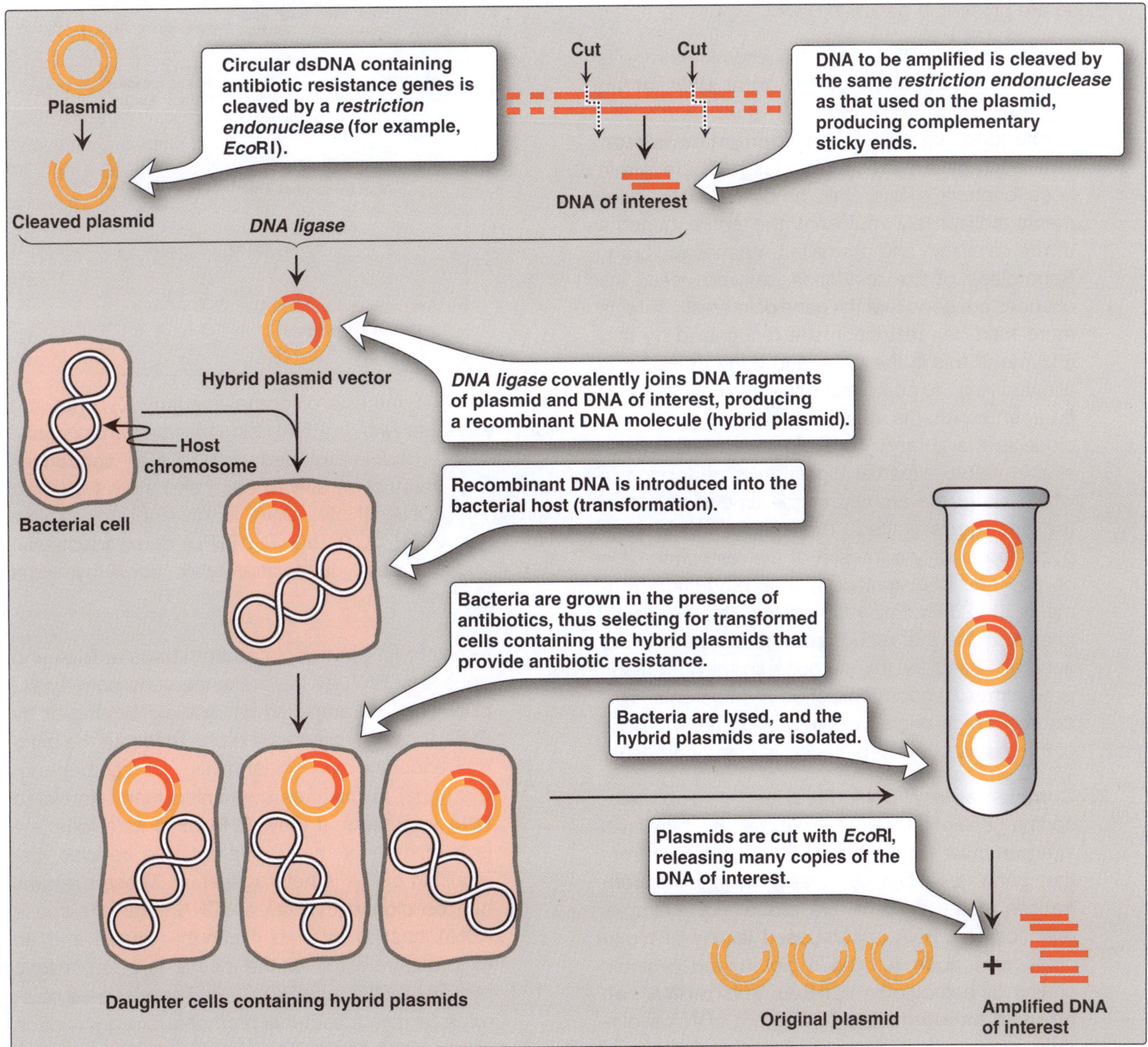

Figure 39.6 Summary of biologic gene cloning. [Note: Transformation is inefficient in that only a small percentage of the cells will contain the recombinant plasmid.] dsDNA = double-stranded DNA.

should be coded by the gene introduced in the vector by recombinant DNA technology.

B. DNA libraries

A DNA library is a collection of cloned restriction fragments of the DNA of an organism. Two kinds of libraries are commonly used: genomic libraries and complementary DNA (cDNA) libraries. **Genomic libraries ideally contain a copy of every DNA nucleotide sequence in the genome. In contrast, cDNA libraries contain those DNA sequences that only appear as processed messenger RNA (mRNA) molecules**, and these differ according to cell type and environmental conditions. [Note: cDNA lacks introns and the control regions of the genes, whereas these are present in genomic DNA.]

1. **Genomic DNA libraries:** A genomic library is created by **digestion of the total DNA of an organism with a *restriction endonuclease* and subsequent ligation to an appropriate vector**. The recombinant DNA molecules replicate within host bacteria. Thus, the amplified DNA fragments collectively represent the entire genome of the organism and are called a genomic library. Regardless of the restriction enzyme used, the chances are good that the gene of interest contains more than one restriction site recognized by that enzyme. If this is the case, and if the digestion is allowed to go to completion, the gene of interest is fragmented (that is, it is not contained in any one clone in the library). To avoid this usually undesirable result, a partial digestion is performed in which either the amount or the time of action of the enzyme is limited. This results in cleavage occurring at only a fraction of the restriction sites on any one DNA molecule, thus producing fragments of ~20 kb. Enzymes that cut very frequently (that is, those that recognize 4-bp sequences) are generally used for this purpose so that the result is an almost random collection of fragments. This insures a high degree of probability that the gene of interest is contained, intact, in some fragment.

2. **Complementary DNA libraries:** If a protein-coding gene of interest is expressed at a high level in a particular tissue, the mRNA transcribed from that gene is likely also present at high concentrations in the cells of that tissue. For example, reticulocyte mRNA is composed largely of molecules that code for the α-globin and β-globin chains of hemoglobin A (HbA). This **mRNA can be used as a template to make a cDNA molecule using the enzyme *reverse transcriptase*** (Fig. 39.7). Therefore, the resulting cDNA is a double-stranded copy of mRNA. [Note: The template mRNA is isolated from transfer RNA and ribosomal RNA by the presence of its poly-A tail.] cDNA can be amplified by biologic cloning or by PCR. It can be used as a probe to locate the gene that encodes the original mRNA (or fragments of the gene) in mixtures containing many unrelated DNA fragments. If the mRNA used as a template is a mixture of many different size species, the resulting cDNA is heterogeneous. These mixtures can be cloned to form a cDNA library. Because cDNA has no introns, it can be cloned into an expression vector for the synthesis of eukaryotic proteins by bacteria (Fig. 39.8). These special plasmids contain a bacterial promoter for transcription of the cDNA and a Shine-Dalgarno (SD) sequence (see p. 642) that allows the bacterial ribosome to initiate translation of the resulting mRNA molecule.

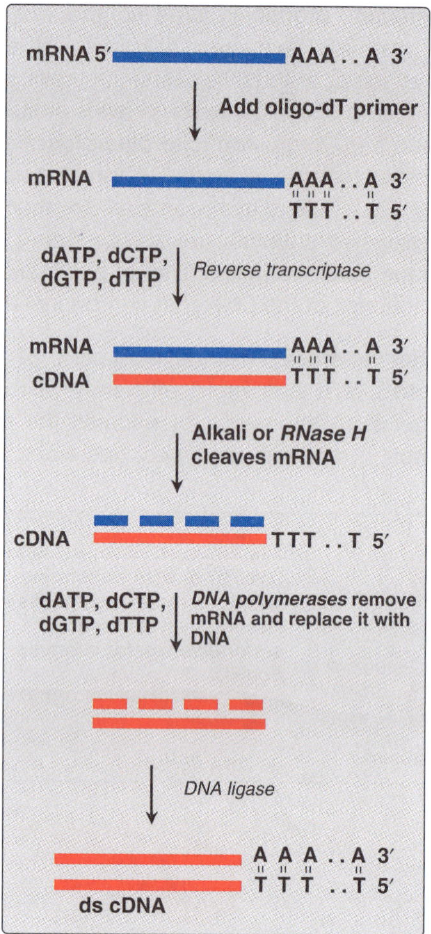

Figure 39.7 Synthesis of complementary DNA (cDNA) from messenger RNA (mRNA) using *reverse transcriptase*. Ligation of double-stranded (ds) DNA sequences containing a restriction site to each end allows biologic cloning of cDNA. [Note: DNA is resistant to alkaline hydrolysis.] dATP, dCTP, dGTP, dTTP = deoxyadenosine, deoxycytidine, deoxyguanosine, and deoxythymidine triphosphates.

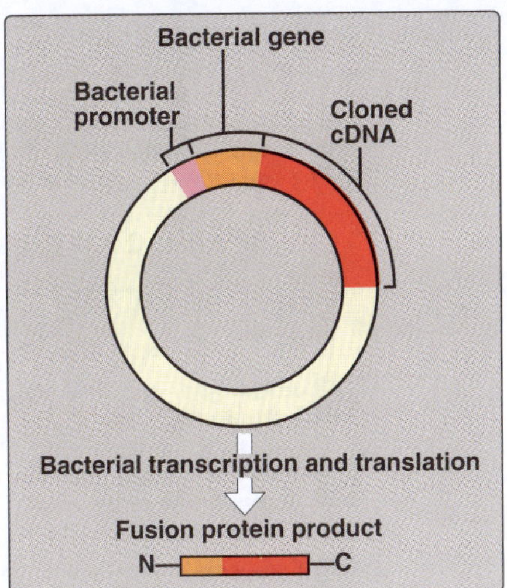

Figure 39.8 An expression vector. The product is a fusion protein that contains just some amino acids of the bacterial protein ■ and all the amino acids of the complementary DNA (cDNA)-encoded protein ■. [Note: Proteins are written from the amino (N)-terminus to the carboxy (C)-terminus.]

The cDNA is inserted downstream of the promoter and within a gene for a protein that is expressed in the bacterium (for example, lacZ; see p. 656), such that the mRNA produced contains an SD sequence, a few codons for the bacterial protein, and all the codons for the eukaryotic protein. This allows for more efficient expression and results in the production of a fusion protein. [Note: **Therapeutic human insulin is made in bacteria through this technology**. However, the extensive co- and posttranslational modifications required for most **other human proteins (for example, blood clotting factors) necessitates the use of eukaryotic, even mammalian, hosts.**]

C. Sequencing cloned DNA fragments

The base sequence of DNA fragments that have been cloned can be determined. The original procedure for this purpose was the **Sanger dideoxy chain termination method** illustrated in Figure 39.9. In this method, the single-stranded DNA (ssDNA) to be sequenced is used as the template for DNA synthesis by *DNA polymerase* (*DNA pol*). A radiolabeled primer complementary to the 3′-end of the target DNA is added, along with the four deoxyribonucleoside triphosphates (dNTP). The sample is divided into four reaction tubes, and a small amount of one of the four dideoxyribonucleoside triphosphates (ddNTP) is added to each tube. Because it contains no 3′-hydroxyl group, incorporation of a ddNMP terminates elongation at that point. The products of this reaction, then, consist of a mixture of DNA strands of different lengths, each terminating at a specific base. Separation of the various DNA products by size in an electric field using polyacrylamide gel electrophoresis, followed by autoradiography, yields a pattern of bands from which the DNA base sequence can be read. [Note: The shorter the fragment, the farther it travels on the gel, with the shortest fragment representing that which was made first (that is, the 5′-end).] In place of a labeled primer, a mixture of the four ddNTP linked to different fluorescent dyes and in a single reaction tube is now commonly used. The mixture is separated by capillary electrophoresis, the fluorescent labels are detected, and a color readout of the sequence is generated (Fig. 39.10). [Note: The Human Genome Project used variations of this technique to sequence the human genome.] Advances in sequencing technology, so-called **next generation, or high-throughput sequencing**, now allow the rapid sequencing of an entire genome with increased fidelity and decreased cost through the simultaneous (parallel) sequencing of many DNA pieces. [Note: Sequencing of the exome, that portion of the genome that encodes proteins, is now possible.]

IV. PROBES

Cleavage of large DNA molecules by restriction enzymes produces an enormous array of fragments. How can the DNA sequence of interest be picked out of such a mixture? The answer lies in the use of a probe, a short piece of ssDNA or RNA, labeled with a radioisotope, such as ^{32}P, or with a nonradioactive molecule, such as biotin or a fluorescent dye. **The sequence of a probe is complementary to a sequence in the DNA of interest, called the target DNA**. Probes are used to identify which band on a gel or which clone in a library contains the target DNA, a process called screening.

A. Hybridization to DNA

The **utility of probes hinges on the process of hybridization (or annealing)** in which a probe containing a complementary sequence binds a single-stranded sequence of a target DNA. ssDNA, produced by alkaline denaturation of dsDNA, is first bound to a solid support, such as a nitrocellulose membrane. The immobilized DNA strands are prevented from self-annealing but are available for hybridization to the **exogenous, radiolabeled, single-stranded probe**. The extent of hybridization is measured by the retention of radioactivity on the membrane. Excess probe molecules that do not hybridize are removed by washing the membrane.

B. Synthetic oligonucleotide probes

If the sequence of all or part of the target DNA is known, short, single-stranded oligonucleotide probes can be synthesized that are complementary to a small region of the gene of interest. If the sequence of the gene is unknown, the amino acid sequence of the protein, the final gene product, may be used to construct a nucleic acid probe using the genetic code as a guide. Because of the degeneracy of the genetic code (see p. 636), it is necessary to synthesize several oligonucleotides. [Note: Oligonucleotides can be used to detect single-base changes in the sequence to which they are complementary. In contrast, cDNA probes contain many thousands of bases, and their binding to a target DNA with a single-base change is unaffected.]

1. **Detecting the β^S-globin mutation:** A synthetic allele-specific oligonucleotide (ASO) probe can be used to detect the presence of the sickle cell

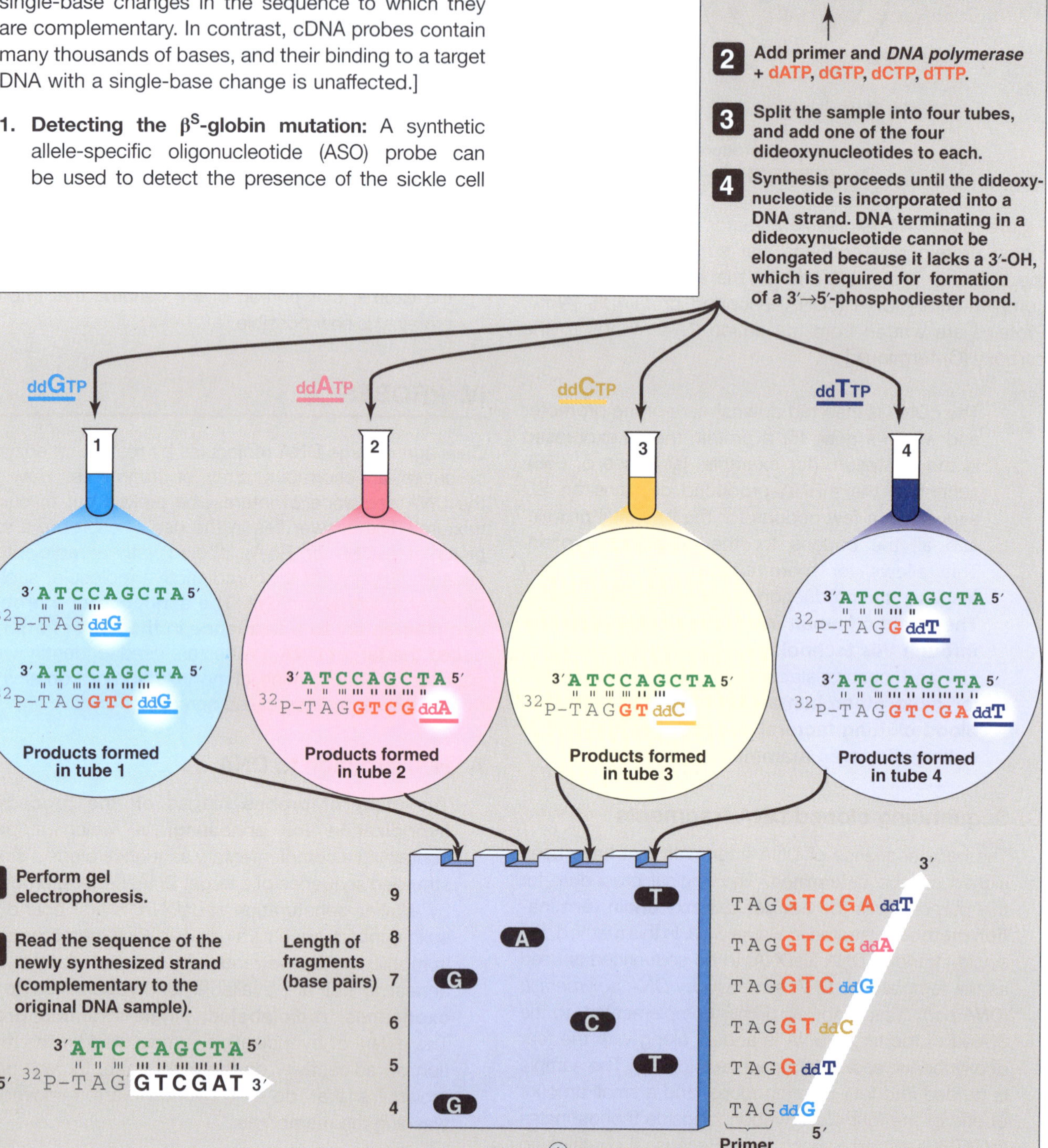

Figure 39.9 DNA sequencing by the Sanger dideoxy method. [Note: The original method utilized a radiolabeled primer. Fluorescent dye-labeled dideoxyribonucleoside triphosphates are now commonly used.] A = adenine; C = cytosine; G = guanine; T = thymine; d = deoxy; dd = dideoxy.

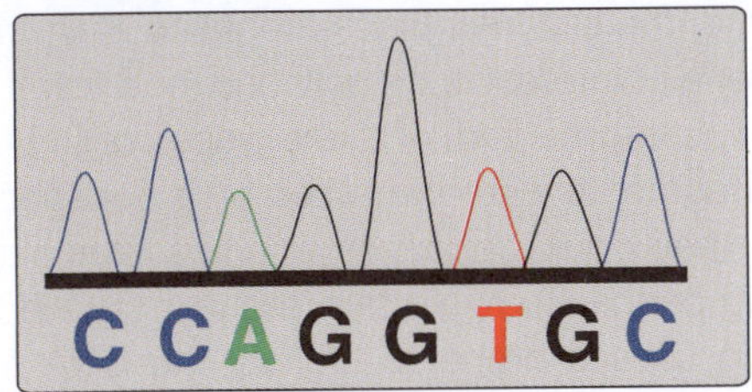

Figure 39.10 Color readout of a DNA sequence.

mutation in the β-globin gene (Fig. 39.11). DNA, isolated from white blood cells (WBC) and amplified, is denatured and applied to a membrane. A radiolabeled oligonucleotide probe, complementary to the point mutation (GAG → GTG, glutamate → valine) at codon 6 in patients with the β^S gene, is applied to the membrane. DNA isolated from a heterozygous individual (sickle cell trait) or a homozygous patient (sickle cell anemia) contains a sequence that is complementary to the probe and a double-stranded hybrid form can be detected. In contrast, DNA obtained from normal individuals is not complementary at this position and, therefore, does not form a hybrid (see Fig. 39.11). Use of a pair of such ASO probes (one specific for the normal allele and one specific for the mutant allele) allows all three possible genotypes (homozygous normal, heterozygous, and homozygous mutant) to be distinguished (Fig. 39.12). [Note: ASO probes are useful only if the mutation and its location are known.]

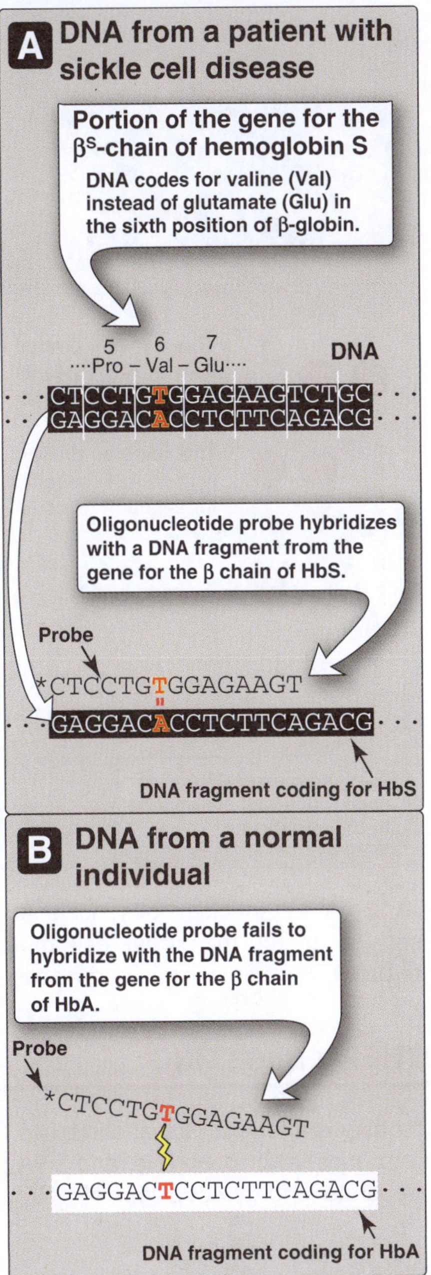

Figure 39.11 Allele-specific oligonucleotide probe detects hemoglobin (Hb) S allele. [Note: The * indicates ^{32}P radiolabel.]

C. Biotinylated probes

Because the disposal of radioactive waste is becoming increasingly expensive, nonradiolabeled probes have been developed. One of the most successful is based on the **vitamin biotin (see p. 534), which can be chemically linked to the nucleotides used to synthesize the probe.** Biotin was chosen because it binds very tenaciously to avidin, a readily available protein contained in chicken egg whites. Avidin can be attached to a fluorescent dye detectable optically with great sensitivity. Thus, a DNA fragment (displayed, for example, by gel electrophoresis) that hybridizes with the biotinylated probe can be made visible by immersing the gel in a solution of dye-coupled avidin. After washing away the excess avidin, the DNA fragment that binds the probe is fluorescent. [Note: Labeled probes can allow detection and localization of DNA or RNA sequences in cell or tissue preparations, a process called ***in situ* hybridization (ISH**). **If the probe is fluorescent (F), the technique is called FISH.]**

D. Antibodies

If no amino acid sequence information is available to guide the synthesis of a probe for direct detection of the DNA of interest, a gene can be identified indirectly by cloning cDNA in an expression vector that allows the cloned cDNA to be transcribed and translated. A labeled antibody is used to identify which bacterial colony produces the protein and, therefore, contains the cDNA of interest.

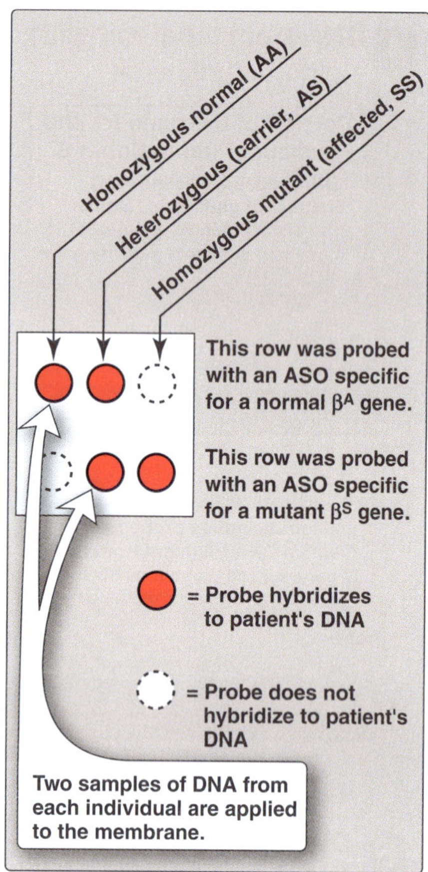

Figure 39.12 Allele-specific oligonucleotide (ASO) probes used to detect the sickle cell mutation and differentiate between sickle cell trait and disease.

V. SOUTHERN BLOTTING

Southern blotting is a technique that combines the use of restriction enzymes, electrophoresis, and DNA probes to generate, separate, and detect pieces of DNA.

A. Procedure

This method, **named after its inventor, Edward Southern**, involves the following steps (Fig. 39.13). First, DNA is extracted from cells, for example, a patient's WBC. Second, the DNA is cleaved into many fragments using a restriction enzyme. Third, the resulting fragments (all of which are negatively charged) are separated on the basis of size by electrophoresis. [Note: Because the large fragments move more slowly than the smaller ones, the lengths of the fragments, usually expressed as the number of base pairs, can be calculated from comparison of the position of the band relative to standard fragments of known size.] The DNA fragments in the gel are denatured and transferred (blotted) to a nitrocellulose membrane for analysis. If the original DNA represents the individual's entire genome, the enzymic digest contains ≥10^6 fragments. The gene of interest is on only one (or a few if the gene itself was fragmented) of these pieces of DNA. If all the DNA fragments were visualized by a nonspecific technique, they would appear as an unresolved blur of overlapping bands. To avoid this, the last step in Southern blotting uses a probe to identify the DNA fragments of interest. The patterns observed on Southern blot analysis depend both on the specific *restriction endonuclease* and on the probe used to visualize the restriction fragments. [Note: Variants of the Southern blot have been facetiously named northern if RNA is being studied (see p. 689) and western if protein is being studied, neither of which relates to anyone's name or to points of the compass.]

B. Mutation detection

Southern blotting can detect DNA mutations such as large insertions or deletions, trinucleotide repeat expansions, and rearrangements of nucleotides. It can also detect point mutations (replacement of one nucleotide by another; see p. 636) that cause the loss or gain of restriction sites. Such mutations cause the pattern of bands to differ from those seen with a normal gene. Longer fragments are generated if a restriction site is lost. For example, in Figure 39.13, person 2 lacks a restriction site present in person 1. Alternatively, the point mutation may create a new cleavage site with the production of shorter fragments. [Note: Most sequence differences at restriction sites are harmless variations in the DNA.]

VI. RESTRICTION FRAGMENT LENGTH POLYMORPHISM

It has been estimated that the genomes of any two unrelated people are 99.5% identical. With 6 billion bp in the diploid human genome, that represents variation in ~30 million bp. These genome variations are the result of mutations that lead to polymorphisms. A polymorphism is a change in genotype that can result in no change in phenotype or a change in phenotype that is harmless, causes increased susceptibility to a disease, or, rarely, causes the disease. It is traditionally defined as a sequence variation at a given locus (allele) in >1% of a population. Polymorphisms primarily occur in the 98% of the genome that does not encode proteins (that is, in introns and intergenic regions). A restriction fragment length polymorphism (RFLP) is a genetic variant that can be observed by cleaving the DNA into fragments (restriction fragments) with a *restriction endonuclease*. The length of the restriction fragments is altered if the variant alters the DNA so as to create or abolish a restriction site. RFLP can be used to detect human genetic variations, for example, in prospective parents or in fetal tissue.

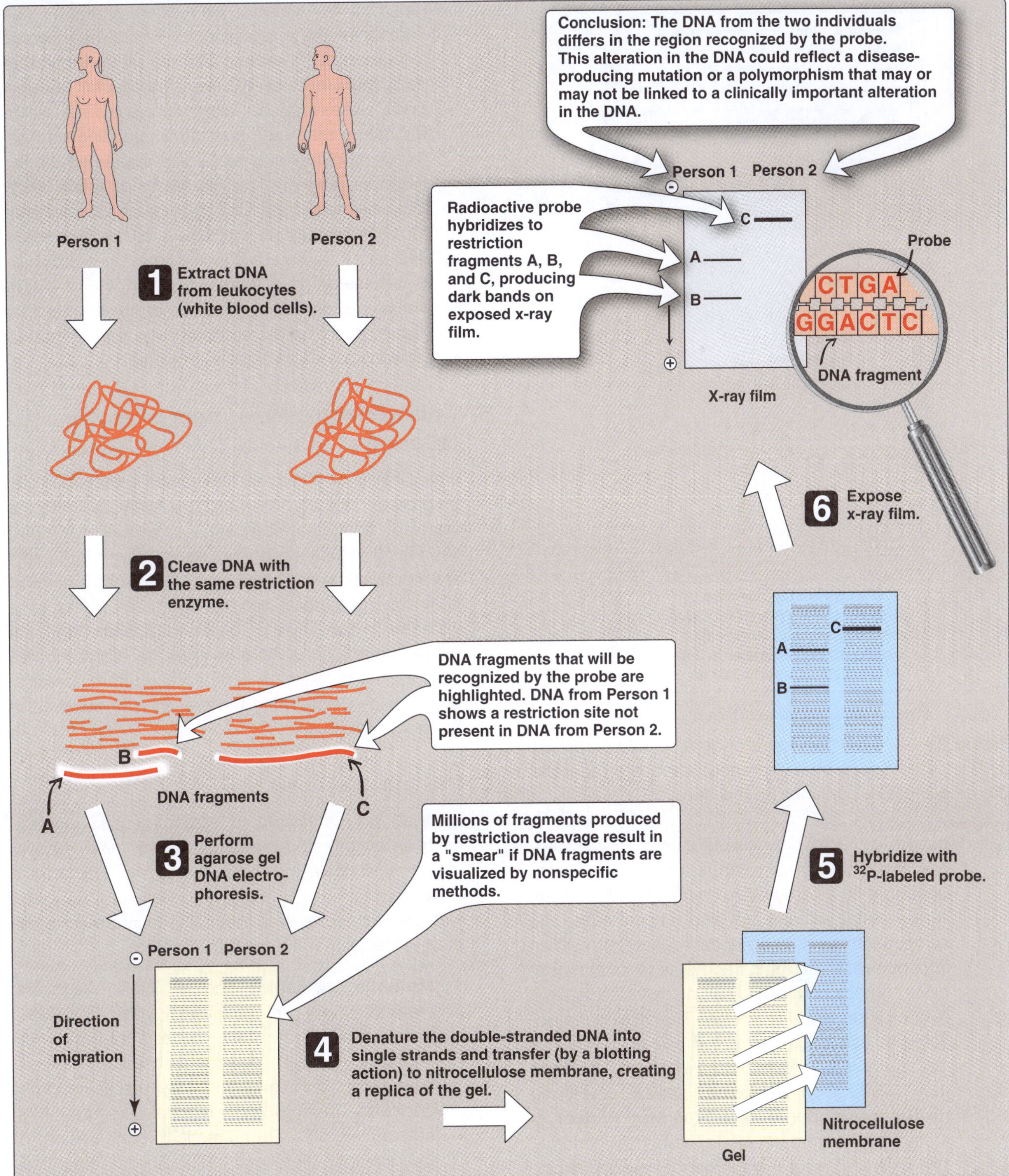

Figure 39.13 Southern blotting procedure. [Note: Nonradiolabeled probes are now commonly used.]

A. DNA variations resulting in RFLP

Two types of DNA variations commonly result in RFLP: single-base changes in the DNA sequence and tandem repeats of DNA sequences.

1. **Single-base changes:** About 90% of human genome variation comes in the form of single nucleotide polymorphisms (SNPs, pronounced “snips”), that is, variations that involve just one

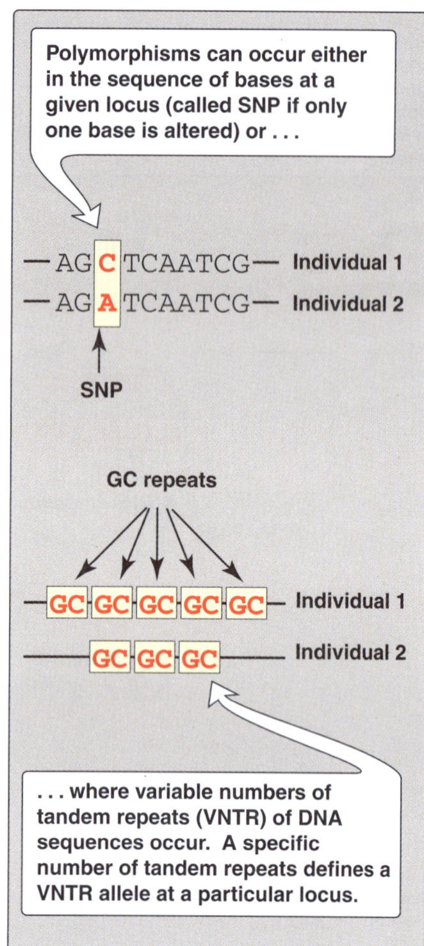

Figure 39.14 Common forms of genetic polymorphism. SNP = single nucleotide polymorphism; A = adenine; C = cytosine; G = guanine; T = thymine.

base (Fig. 39.14). The substitution of one nucleotide at a restriction site can render the site unrecognizable by a particular *restriction endonuclease*. A new restriction site can also be created by the same mechanism. In either case, cleavage with an *endonuclease* results in fragments of lengths that differ from the normal and can be detected by DNA hybridization (see Fig. 39.13). The altered restriction site can be either at the site of a disease-causing mutation (rare) or at a site some distance from the mutation. [Note: The HapMap, developed by **The International Haplotype Map Project**, is a catalog of common SNP in the human genome. The data are being used in genome-wide association studies (GWAS) to identify those alleles that affect health and disease.]

2. **Tandem repeats:** Polymorphisms in chromosomal DNA can also arise from the presence of a **variable number of tandem repeats (VNTR),** as shown in Figure 39.15. These are short sequences of DNA at scattered locations in the genome, repeated in tandem (one after another). The number of these repeat units varies from person to person but is **unique for any given individual and, therefore, serves as a molecular "fingerprint."** Cleavage by restriction enzymes yields fragments that vary in length depending on how many repeated segments are contained in the fragment (see Fig. 39.15). Many different VNTR loci have been identified and are extremely useful for DNA fingerprint analysis, such as in forensic and paternity cases. It is important to emphasize that these polymorphisms, whether SNP or VNTR, are simply markers, which, in most cases, have no known effect on the structure, function, or rate of production of any particular protein.

B. Tracing chromosomes from parent to offspring

If the DNA of an individual has gained a restriction site by base substitution, then enzymic cleavage yields at least one additional fragment. Conversely, if a mutation results in loss of a restriction site, fewer fragments are produced by enzymic cleavage. An individual who is heterozygous for a polymorphism has a sequence variation in the DNA of one chromosome and not in the homologous chromosome. In such individuals, each chromosome can be traced from parent to offspring by determining the presence or absence of the polymorphism.

C. Prenatal diagnosis

Families with a history of severe genetic disease, such as an affected previous child or near relative, may wish to determine the presence of the disorder in a developing fetus. Prenatal diagnosis, in association with genetic counseling, allows for an informed reproductive decision if the fetus is affected.

1. **Methods available:** The available diagnostic methods vary in sensitivity and specificity. Visualization of the fetus, for example, by ultrasound or fiberoptic devices (fetoscopy), is useful only if the genetic abnormality results in gross anatomic defects (for example, neural tube defects [NTD]). The chemical composition of the amniotic fluid can also provide diagnostic clues. For example, the presence of **high levels of α-fetoprotein is associated with NTD.** Fetal cells obtained from amniotic fluid or from **biopsy of the chorionic villi can be used for karyotyping,** which assesses the morphology of metaphase chromosomes. Staining and cell sorting techniques permit the rapid identification **of trisomies and translocations** that produce an extra

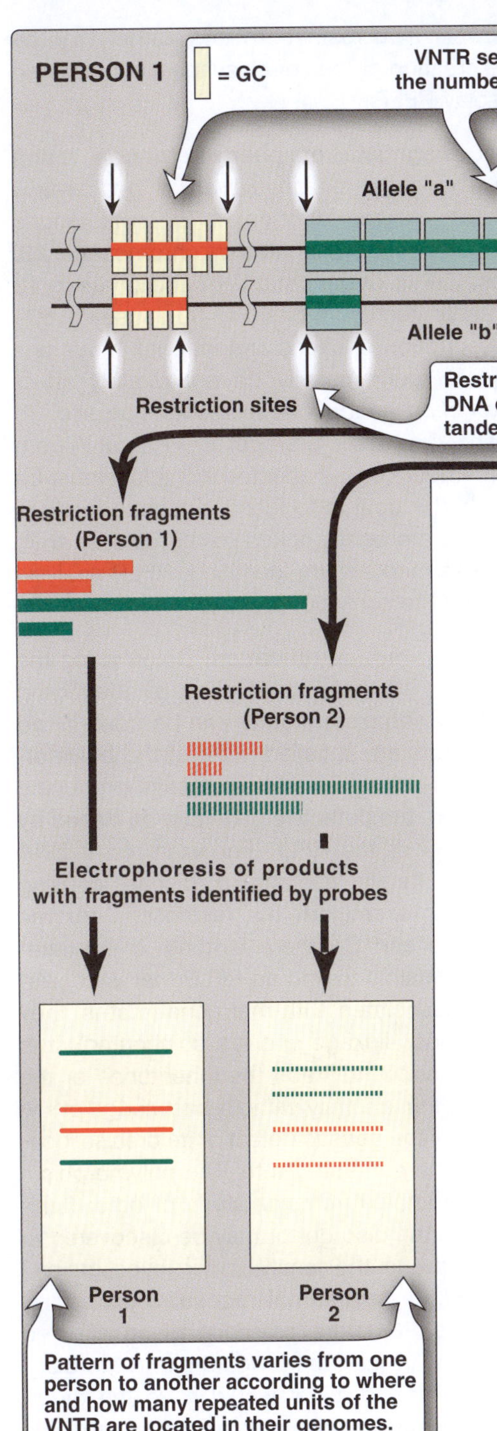

Figure 39.15 Restriction fragment length polymorphism of variable number tandem repeats (VNTR). For each person, a pair of homologous chromosomes is shown.

chromosome or chromosomes of abnormal lengths. However, molecular analysis of fetal DNA provides the most detailed genetic picture.

2. **DNA sources:** DNA may be obtained from blood cells, amniotic fluid, or chorionic villi (Fig. 39.16). For amniotic fluid, it was formerly necessary to grow cells in culture for 2–3 weeks in order to have sufficient DNA for analysis. The ability to amplify DNA by PCR has dramatically shortened the time needed for a DNA analysis.

3. **Direct diagnosis of sickle cell anemia using RFLP:** The genetic disorders of Hb are the most common genetic diseases in humans. In the case of sickle cell anemia (Fig. 39.17), the point mutation that gives rise to the disease (see p. 107) is actually one and the same mutation that gives rise to the polymorphism. However, direct detection by RFLP of diseases that result from point mutations is limited to only a few genetic diseases.

   a. **Early diagnostic efforts:** In the past, prenatal diagnosis of sickle cell anemia involved the determination of the amount and kinds of Hb synthesized in the nucleated red cells obtained from fetal blood. However, the invasive procedures to obtain fetal blood have a high mortality rate (~5%), and analysis cannot be carried out until late in the second trimester of pregnancy when HbA (and its HbS variant) begins to be produced.

   b. **RFLP analysis:** In sickle cell anemia, the sequence alteration caused by the point mutation abolishes the recognition site of the ***restriction endonuclease Mst*II: CCTNAGG** (where N is any nucleotide; see Fig. 39.17). Thus, the A-to-T mutation in codon 6 of the β^S-globin gene eliminates a cleavage site for the enzyme. Normal DNA digested with *Mst*II yields a 1.15-kb fragment, whereas a 1.35-kb fragment is generated from the β^S gene as a result of the loss of one *Mst*II cleavage site. Diagnostic techniques that allow **analysis of fetal DNA from amniotic cells or chorionic villus sampling** rather than fetal blood have proved valuable because they provide

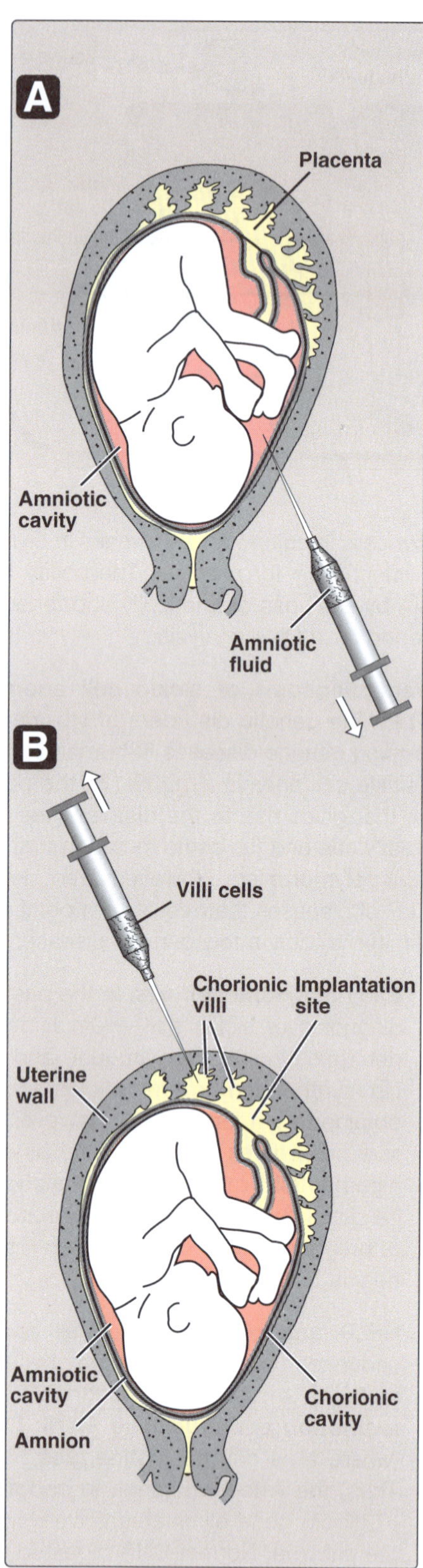

Figure 39.16 Sampling of fetal cells. **A.** Amniotic fluid. **B.** Chorionic villus.

safe, early detection of sickle cell anemia as well as other genetic diseases. [Note: Genetic disorders caused by insertions or deletions between two restriction sites, rather than by the creation or loss of cleavage sites, will also display RFLP.]

4. **Indirect diagnosis of phenylketonuria using RFLP:** The gene for *phenylalanine hydroxylase* (*PAH*), the enzyme deficient in phenylketonuria ([PKU], see p. 406), is located on chromosome 12. It spans ~90 kb of genomic DNA and contains 13 exons separated by introns (Fig. 39.18; see p. 629 for a description of exons and introns). Mutations in the *PAH* gene usually do not directly affect any *restriction endonuclease* recognition site. To establish a diagnostic protocol for PKU, DNA from family members of the affected individual must be analyzed. The goal is to identify genetic markers (RFLP) that are tightly linked to the disease trait. Once these markers are identified, RFLP analysis can be used to carry out prenatal diagnosis.

   a. **Mutant gene identification:** Determining the presence of the mutant gene by identifying the polymorphism marker can be done if two conditions are satisfied. First, if the polymorphism is closely linked to a disease-producing mutation, the defective gene can be traced by detection of the RFLP. For example, if DNA from a family carrying a disease-causing gene is examined by restriction enzyme cleavage and Southern blotting, it is sometimes possible to find an RFLP that is consistently associated with that gene (that is, they show close linkage and are coinherited). It is then possible to trace the inheritance of the gene within a family without knowledge of the nature of the genetic defect or its precise location in the genome. [Note: The polymorphism may be known from the study of other families with the disorder or may be discovered to be unique in the family under investigation.] Second, for **autosomal-recessive disorders, such as PKU,** the presence of an affected individual in the family would aid in the diagnosis. This individual would have the mutation present on both chromosomes, allowing identification of the RFLP associated with the genetic disorder.

   b. **RFLP analysis:** The presence of abnormal genes for *PAH* can be shown using DNA polymorphisms as markers to distinguish between normal and mutant genes. For example, Figure 39.19 shows a typical pattern obtained when DNA from members of an affected family is cleaved with an appropriate restriction enzyme and subjected to electrophoresis. The vertical arrows represent the cleavage sites

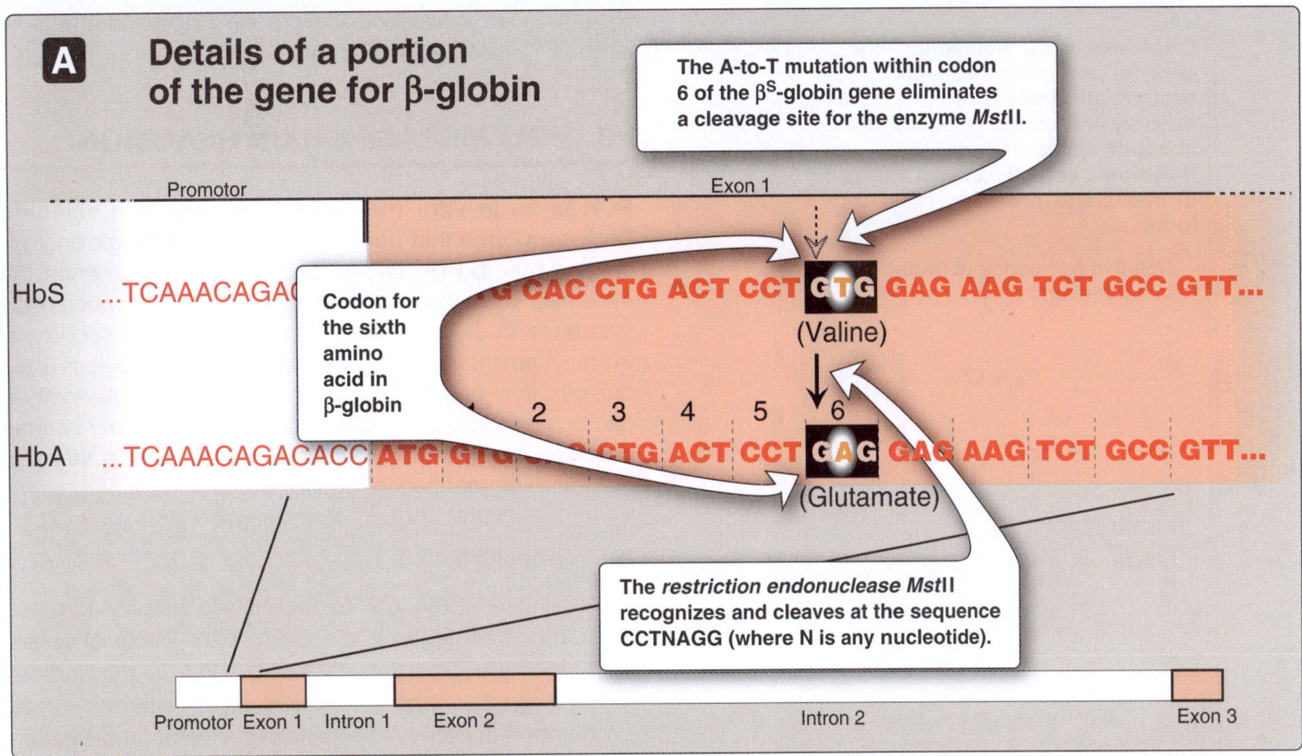

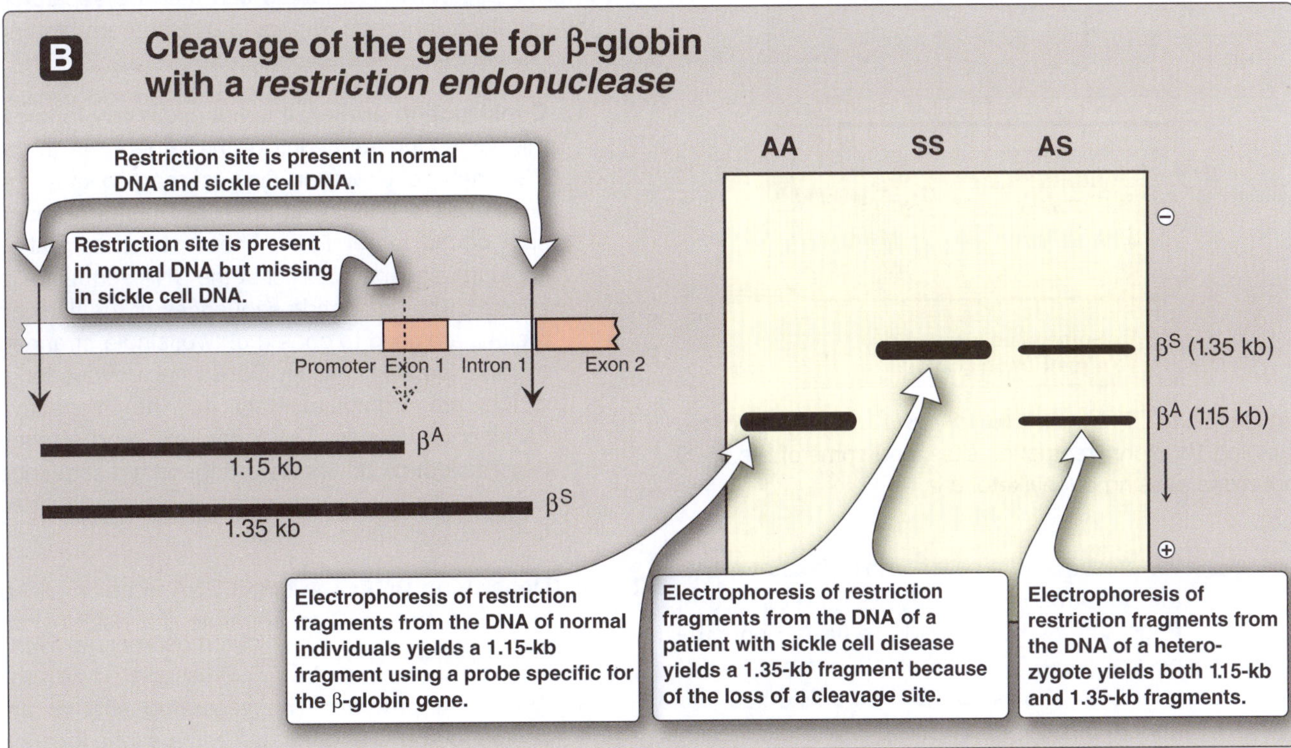

Figure 39.17 Detection of β^S-globin mutation. kb = kilobase (1 kb = 1,000 base pairs in double-stranded DNA); Hb = hemoglobin.

for the restriction enzyme used. The presence of a polymorphic site creates fragment "b" in the autoradiogram (after hybridization with a labeled *PAH*-cDNA probe), whereas the absence of this site yields only fragment "a."

Note that subject II-2 demonstrates that the polymorphism, as shown by the presence of fragment "b," is associated with the mutant gene. Therefore, in this particular family, the appearance of fragment "b" corresponds to

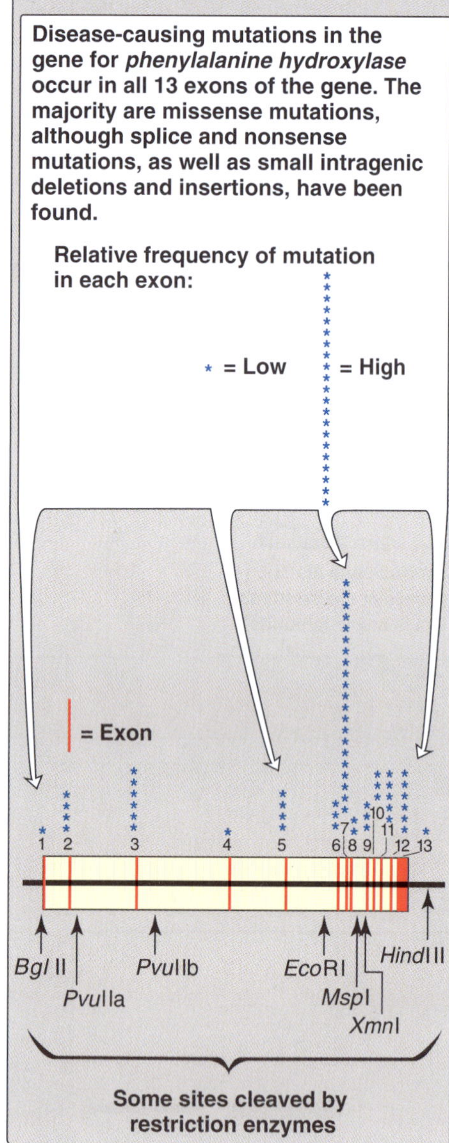

Figure 39.18 The gene for *phenylalanine hydroxylase* showing 13 exons, restriction sites, and some of the >500 mutations causing phenylketonuria.

the presence of a polymorphic site that marks the abnormal gene for *PAH*. The absence of fragment "b" corresponds to having only the normal gene. In Figure 39.19, examination of fetal DNA shows that the fetus inherited two abnormal genes from the parents and, therefore, has PKU.

c. **Value of DNA testing:** DNA-based testing is useful not only in determining if an unborn fetus is affected by PKU but also in detecting unaffected carriers of the mutated gene to aid in family planning. [Note: **PKU is treatable by dietary restriction of phenylalanine.** Early diagnosis and treatment are essential in preventing severe neurologic damage in affected individuals.]

VII. POLYMERASE CHAIN REACTION

PCR is an in vitro method for amplifying a selected DNA sequence that does not rely on the biologic (in vivo) cloning method described on p. 674. PCR permits the synthesis of millions of copies of a specific nucleotide sequence in a few hours. It can amplify the sequence, even when the targeted sequence makes up less than one part in a million of the total initial sample. The method can be used to amplify DNA sequences from any source, including viral, bacterial, plant, or animal. The steps in PCR are summarized in Figures 39.20 and 39.21.

A. Procedure

PCR uses *DNA pol* to repetitively amplify targeted portions of genomic or cDNA. Each cycle of amplification doubles the amount of DNA in the sample, leading to an exponential increase (2^n, where n = cycle number) in DNA with repeated cycles of amplification. The amplified DNA products can then be separated by gel electrophoresis, detected by Southern blotting and hybridization, and sequenced.

1. **Constructing primer:** It is not necessary to know the nucleotide sequence of the target DNA in the PCR method. However, it is necessary to know the nucleotide sequence of short segments on each side of the target DNA. These stretches, called flanking sequences, bracket the DNA sequence of interest. The nucleotide sequences of the flanking regions are used to construct two, single-stranded oligonucleotides, usually 20–35 nucleotides long, which are complementary to the respective flanking sequences. The 3′-hydroxyl end of each oligonucleotide points toward the target sequence (see Fig. 39.20). These synthetic oligonucleotides function as primers in PCR.

2. **Denaturing DNA:** The target DNA to be amplified is heated to ~95°C to separate the dsDNA into single strands.

3. **Annealing primers:** The separated strands are cooled to ~50°C and the two primers (one for each strand) anneal to a complementary sequence on the ssDNA.

4. **Extending primers:** *DNA pol* and dNTP (in excess) are added to the mixture (~72°C) to initiate the synthesis of two new strands complementary to the original DNA strands. *DNA pol* adds nucleotides to the 3′-hydroxyl end of the primer, and strand growth extends in the 5′→3′direction

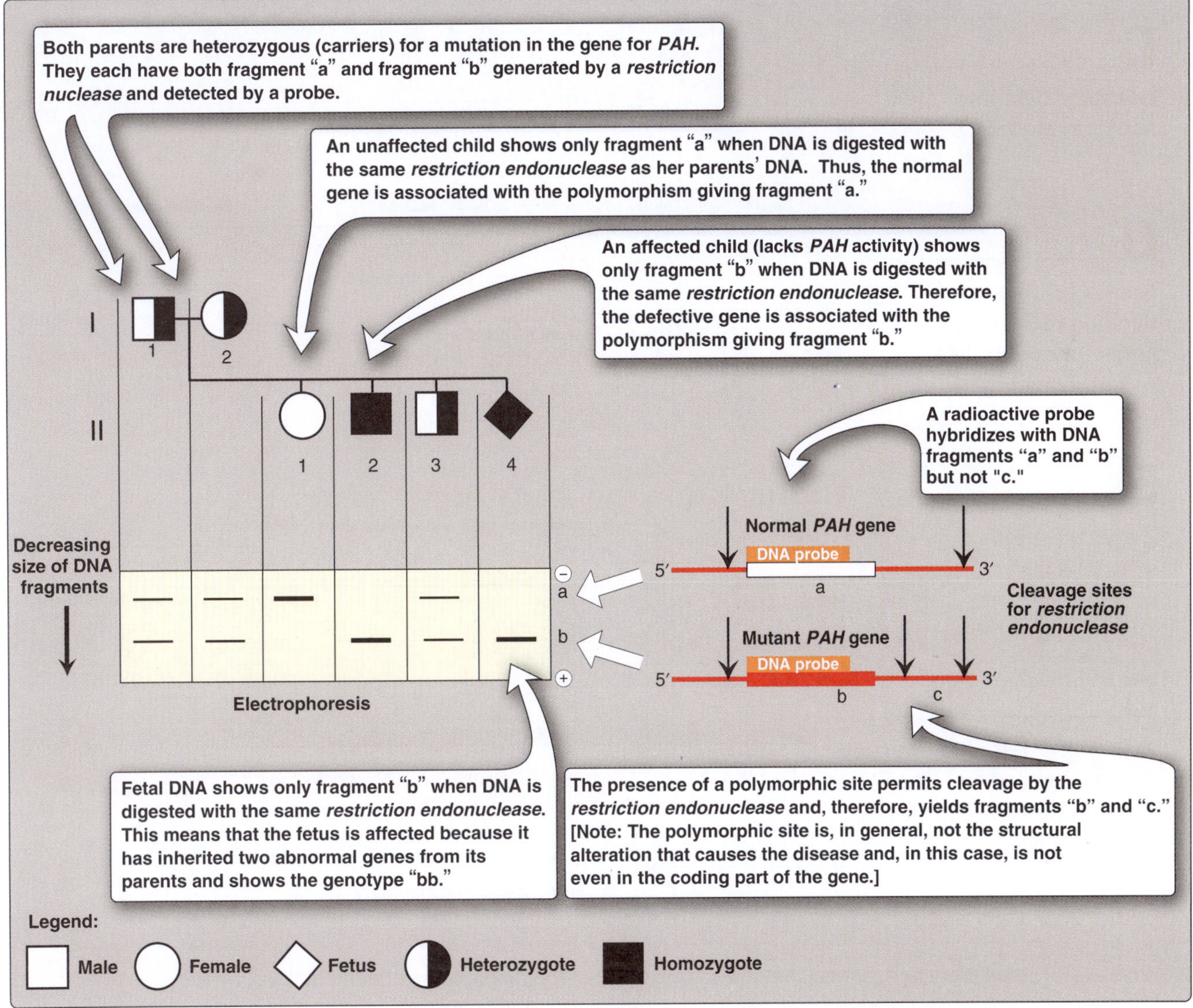

Figure 39.19 Analysis of restriction fragment length polymorphism in a family with a child affected by phenylketonuria (PKU), an autosomal-recessive disease. The molecular defect in the gene for *phenylalanine hydroxylase* (*PAH*) in the family is not known. The family wanted to know if the current pregnancy would be affected by PKU.

across the target DNA, making complementary copies of the target. [Note: PCR products can be several thousand base pairs long.] At the completion of one cycle of replication, the reaction mixture is heated again to separate the strands (of which there are now four). Each strand binds a complementary primer, and the step of primer extension is repeated. By using a heat-stable *DNA pol* (for example, *Taq* from the bacterium **Thermus aquaticus** that normally lives at high temperatures), the *polymerase* is not denatured and, therefore, does not have to be added at each successive cycle. However, *Taq* lacks proofreading activity. Typically, 20–30 cycles are run during this process, amplifying the DNA by a million-fold (2^{20}) to a billion-fold (2^{30}). [Note: Each extension product includes a sequence at its 5′-end that is complementary to the primer (see Fig. 39.20). Thus, each newly synthesized strand can act as a template for the successive cycles (see Fig. 39.21). This leads to an **exponential increase in the amount of target DNA with each cycle, hence, the name "polymerase chain reaction**."] Probes can be made during PCR by adding labeled nucleotides to the last few cycles.

B. Advantages

The major advantages of PCR over biologic cloning as a mechanism for amplifying a specific DNA sequence are **sensitivity and speed**. DNA sequences present in only trace amounts can be amplified to become the predominant sequence. PCR is so sensitive that DNA

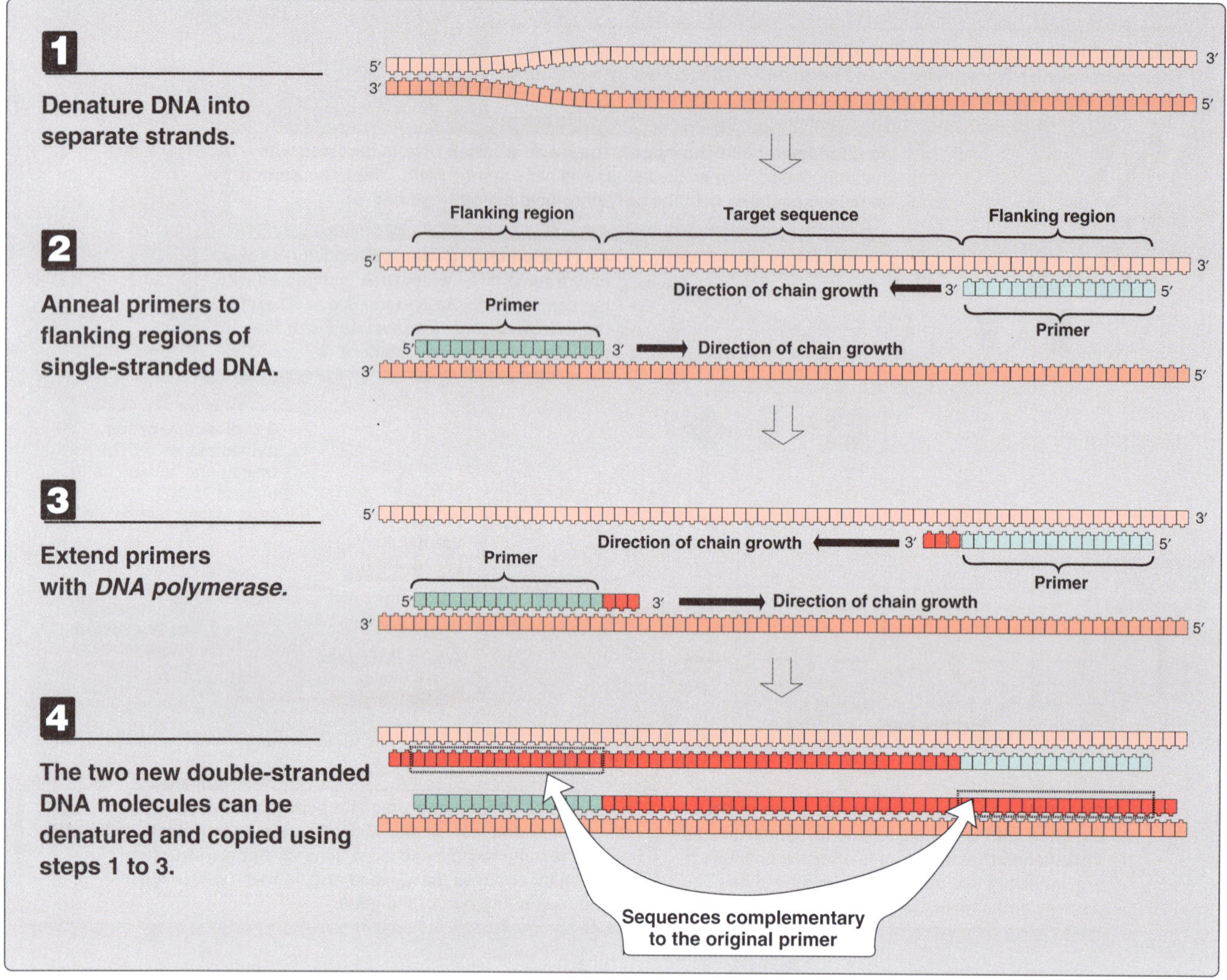

Figure 39.20 Steps (denature, anneal, extend) in one cycle of the polymerase chain reaction.

sequences present in an individual cell can be amplified and studied. Isolating and amplifying a specific DNA sequence by PCR is faster and less technically difficult than traditional cloning methods using recombinant DNA techniques.

C. Applications

PCR has become a very common tool in research, forensics, and clinical diagnostics.

1. **Comparison of a normal gene to its mutant form:** PCR allows the synthesis of mutant DNA in sufficient quantities for a sequencing protocol without laborious biologic cloning of the DNA.

2. **Forensic analysis of DNA samples:** DNA fingerprinting by means of PCR has revolutionized the analysis of evidence from crime scenes. DNA isolated from a single human hair, a tiny spot of blood, or a sample of semen is sufficient to determine whether the sample comes from a specific individual. The DNA markers analyzed for such fingerprinting are most commonly a type of polymorphism known as short tandem repeats. These are very similar to the VNTR described previously (see p. 682) but are smaller in size. [Note: Paternity testing uses the same techniques.]

3. **Detection of low-abundance nucleic acid sequences:** Viruses that have a long latency period, such as **human immunodeficiency virus (HIV),** are difficult to detect at the early stage of infection using conventional methods. PCR offers a rapid and sensitive method for detecting viral DNA sequences even when only a small proportion of cells harbors the virus. [Note: Quantitative PCR (qPCR), also known as real-time PCR, allows quantification of the amount (copy number) of the target nucleic acid after each cycle of amplification (that is, in real time) rather than at the end

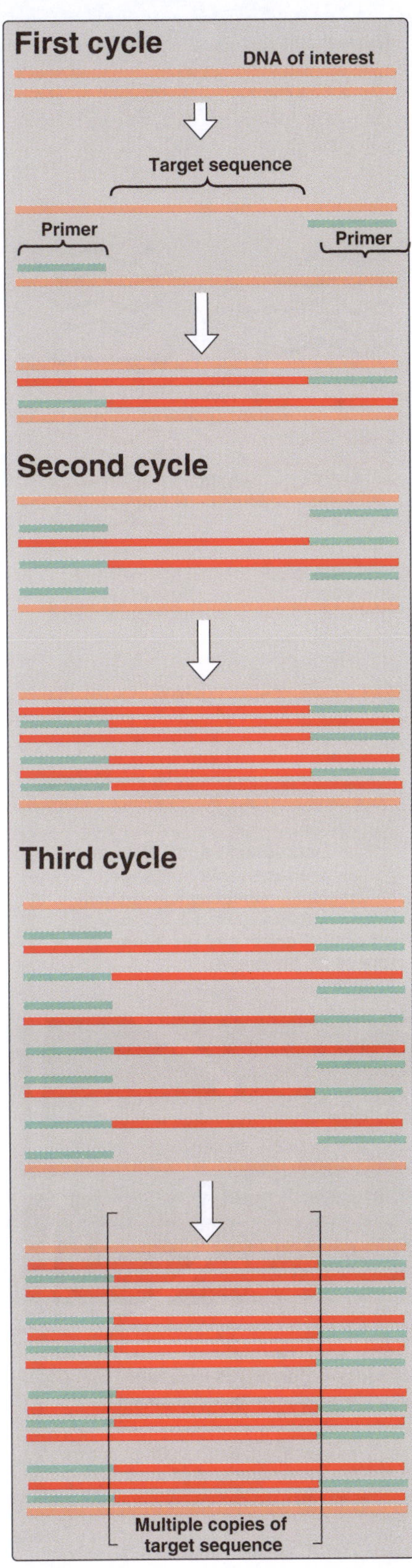

Figure 39.21 Multiple cycles of the polymerase chain reaction.

and is useful in determining viral load (the amount of virus).]

4. **Prenatal diagnosis and carrier detection of cystic fibrosis: Cystic fibrosis** is an autosomal-recessive genetic disease resulting from mutations in the gene for the cystic fibrosis transmembrane conductance regulator (CFTR) protein. The most common mutation is a three-base deletion that results in the loss of a phenylalanine residue from the CFTR protein (see p. 638). Because the mutant allele is three bases shorter than the normal allele, it is possible to distinguish them from each other by the size of the PCR products obtained by amplifying that portion of the DNA. Figure 39.22 illustrates how the results of such a PCR test can distinguish between homozygous normal, heterozygous (carriers), and homozygous mutant (affected) individuals.

> The simultaneous amplification of multiple regions of a target DNA using multiple primer pairs is known as multiplex PCR. It allows detection of the loss of ≥1 exon in a gene with many exons such as the gene for CFTR, which has 27 exons.

VIII. GENE EXPRESSION ANALYSIS

The tools of biotechnology not only allow the study of gene structure, but also provide ways of analyzing the mRNA and protein products of gene expression.

A. Determining messenger RNA levels

mRNA levels are usually determined by the hybridization of labeled probes to either mRNA itself or to cDNA produced from mRNA. [Note: Amplification by PCR of cDNA made from mRNA by retroviral *reverse transcriptase* (*RT*) is referred to as RT-PCR.]

1. **Northern blots:** Northern blots are similar to Southern blots (see Fig. 39.13), except that the sample contains a **mixture of mRNA molecules that are separated by electrophoresis,** then transferred to a membrane and hybridized with a radiolabeled probe. The bands obtained by autoradiography give a measure of the amount and size of the mRNA molecules in the sample.

2. **Microarrays:** DNA microarrays contain thousands of immobilized ssDNA sequences organized in an area no larger than a microscope slide. These microarrays are used to analyze a sample for the presence of gene variations or mutations **(genotyping)** or to determine the patterns of mRNA production **(gene expression analysis**), analyzing thousands of genes at the same

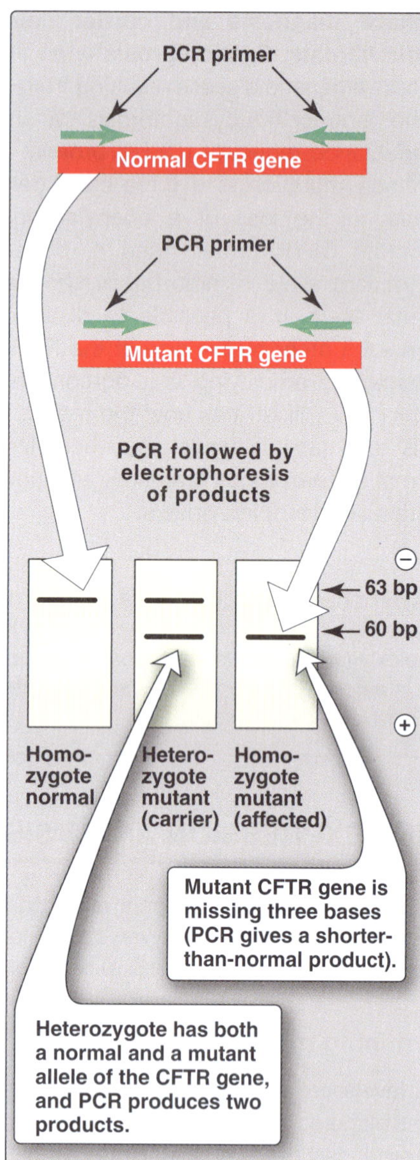

Figure 39.22 Genetic testing for cystic fibrosis (CF) using the polymerase chain reaction (PCR). [Note: CF is also diagnosed using allele-specific oligonucleotide analysis (see p. 677).] CFTR = cystic fibrosis transmembrane conductance regulator; bp = base pairs.

time. For genotyping analysis, the sample is from genomic DNA. For expression analysis, the population of mRNA molecules from a particular cell type is converted to cDNA and labeled with a fluorescent tag (Fig. 39.23). This mixture is then exposed to a gene (or, DNA) chip, which is a glass slide or membrane containing thousands of tiny spots of DNA, each corresponding to a different gene. The amount of fluorescence bound to each spot is a measure of the amount of that particular mRNA in the sample. DNA microarrays are used to determine the differing patterns of gene expression in two different types of cell (for example,

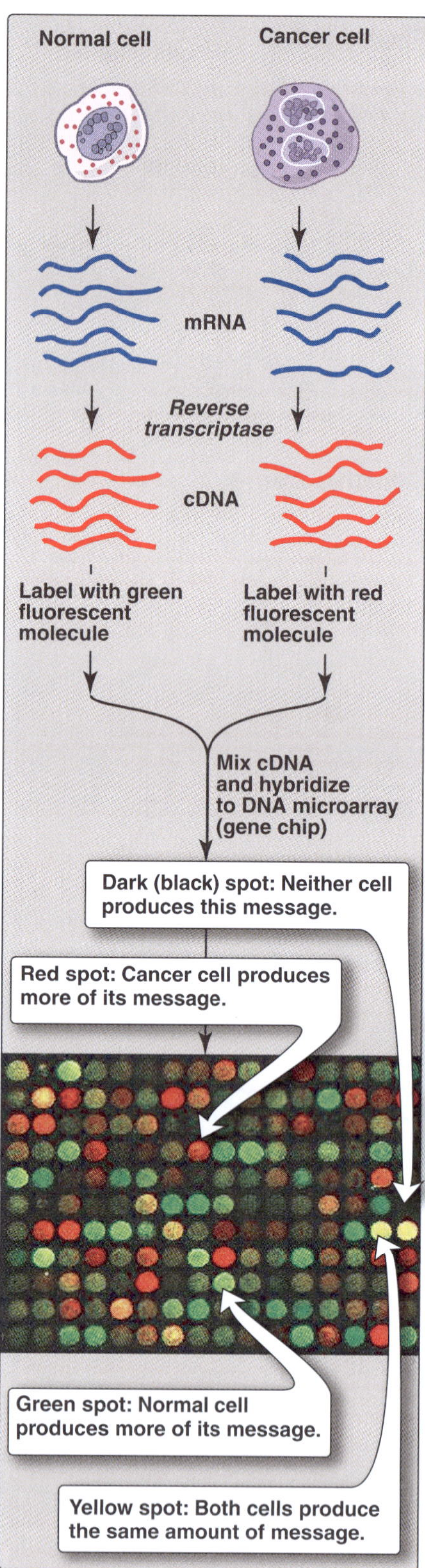

Figure 39.23 Microarray analysis of gene expression using DNA (gene) chips. [Note: Protein chips are also used.] mRNA = messenger RNA; cDNA = complementary DNA.

normal and cancer cells; see Fig. 39.23). They can also be used to subclassify cancers, such as breast cancer, to optimize treatment. [Note:

Microarrays involving proteins and the antibodies or other proteins that recognize them are being used to identify biomarkers to aid in the diagnosis, prognosis, and treatment of disease based on a patient's protein expression profile. Protein (and DNA) microarrays are important tools in the development of personalized (precision) medicine in which the treatment and/or prevention strategies consider the genetic, environmental, and lifestyle variations among individuals.]

B. Protein analysis

The kinds and amounts of proteins in cells do not always directly correspond to the amounts of mRNA present. Some mRNA are translated more efficiently than others, and some proteins undergo posttranslational modification. When analyzing the abundance and interactions of a large number of proteins, automated methods involving a variety of techniques, such as **mass spectrometry and two-dimensional electrophoresis**, are used. When investigating one, or a limited number of proteins, labeled antibodies (Ab) are used to detect and quantify specific proteins and to determine posttranslational modifications.

1. **Enzyme-linked immunosorbent assays:** These assays (known as ELISA) are performed in the wells of a microtiter dish. The antigen (protein) is bound to the plastic of the dish. The probe used consists of an Ab specific for the protein (such as troponin, see p. 142) to be measured. The Ab is covalently bound to an enzyme, which will produce a colored product when exposed to its substrate. The amount of color produced is proportional to the amount of Ab present and, indirectly, to the amount of protein in a test sample.

2. **Western blots:** Western blots (also called immunoblots) are similar to Southern blots, except that it **is protein molecules in the sample that are separated by electrophoresis** and blotted (transferred) to a membrane. The probe is a labeled Ab, which produces a band at the location of its antigen.

3. **Detecting exposure to human immunodeficiency virus: ELISA and western blots** are commonly used to detect exposure to HIV by measuring the amount of anti-HIV Ab present in a patient's blood sample. ELISA are used as the primary screening tool because they are very sensitive. Because these assays sometimes give false positives, however, western blots, which are more specific, are often used as a confirmatory test (Fig. 39.24). [Note: ELISA and western blots can only detect HIV exposure after anti-HIV Ab appear in the bloodstream. PCR-based testing for HIV is more useful in the first few months after exposure.]

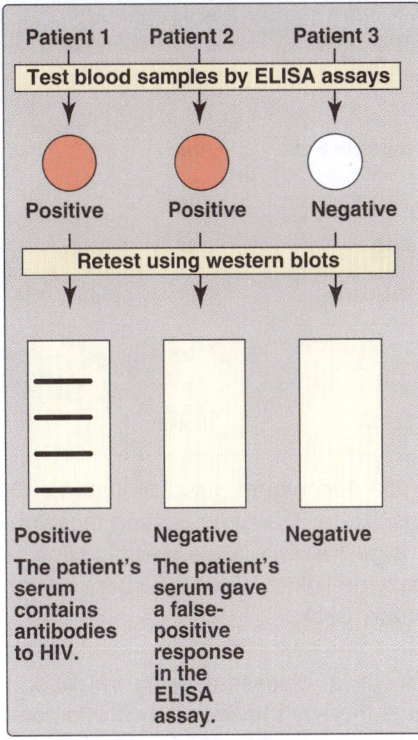

Figure 39.24 Testing for human immunodeficiency virus (HIV) exposure by enzyme-linked immunosorbent assays (ELISA) and western blots.

C. Proteomics

The study of the proteome, or all the proteins expressed by a genome, including their relative abundance, distribution, posttranslational modifications, functions, and interactions with other macromolecules, is known as proteomics. The 20,000–25,000 protein-coding genes of the human genome translate into well over 100,000 proteins when posttranscriptional and posttranslational modifications are considered. Although a genome remains essentially unchanged, the amounts and types of proteins in any particular cell change dramatically as genes are turned on and off. [Note: Proteomics (and genomics) required the parallel development of bioinformatics, the computer-based organization, storage, and analysis of biologic data.] Figure 39.25 compares some of the analytic techniques discussed in this chapter.

IX. GENE THERAPY

The goal of gene therapy is to treat disease through delivery of the normal, cloned DNA for a gene into the somatic cells of a patient who has a defect in that gene

TECHNIQUE	SAMPLE ANALYZED
Southern blot	DNA
Northern blot	RNA
Western blot	Protein
ASO	DNA
Microarray	cDNA or genomic DNA
	Protein
ELISA	Protein

Figure 39.25 Techniques used to analyze DNA, RNA, and proteins. [Note: The three blotting techniques involve the use of a gel.] ASO = allele-specific oligonucleotides. ELISA = enzyme-linked immunosorbent assay; cDNA = complementary DNA.

as a result of a disease-causing mutation. Because somatic gene therapy changes only the targeted somatic cells, the change is not passed on to the next generation. [Note: In germline gene therapy, the germ cells are modified, and so the change is passed on. A long-standing moratorium on germline gene therapy is in effect worldwide.] There are two types of gene transfer: 1) ex vivo, in which cells from the patient are removed, transduced, and returned, and 2) in vivo, in which the cells are directly transduced. Both types require use of a viral vector to deliver the DNA. Challenges of gene therapy include development of vectors, achievement of long-lived expression, and prevention of side effects such as an immune response. The first successful gene therapy involved two patients with **severe combined immunodeficiency disease (SCID)** caused by mutations to the gene for *adenosine deaminase* (*ADA*, see p. 585). It utilized mature T lymphocytes transduced ex vivo with a retroviral vector (Fig. 39.26). [Note: Human *ADA* cDNA is now used.] Since 1990, only a small number of patients (with a variety of disorders, such as **hemophilia, cancers, and certain types of blindness)** have been treated with gene therapy, with varying degrees of success.

Gene editing, as opposed to gene addition, allows a mutated gene to be repaired. Combinations of DNA-binding molecules (proteins or RNA) and *endonucleases* are used to identify and cleave the mutated sequence. Cleavage activates homologous recombination repair of dsDNA breaks (see p. 610) that integrates DNA containing the correct sequence into the gene. [Note: An *endonuclease* guided to a specific DNA sequence by a custom-designed RNA has been used in gene editing in human cell lines. The technique is based on (and named for) the prokaryotic CRISPR-*Cas9*

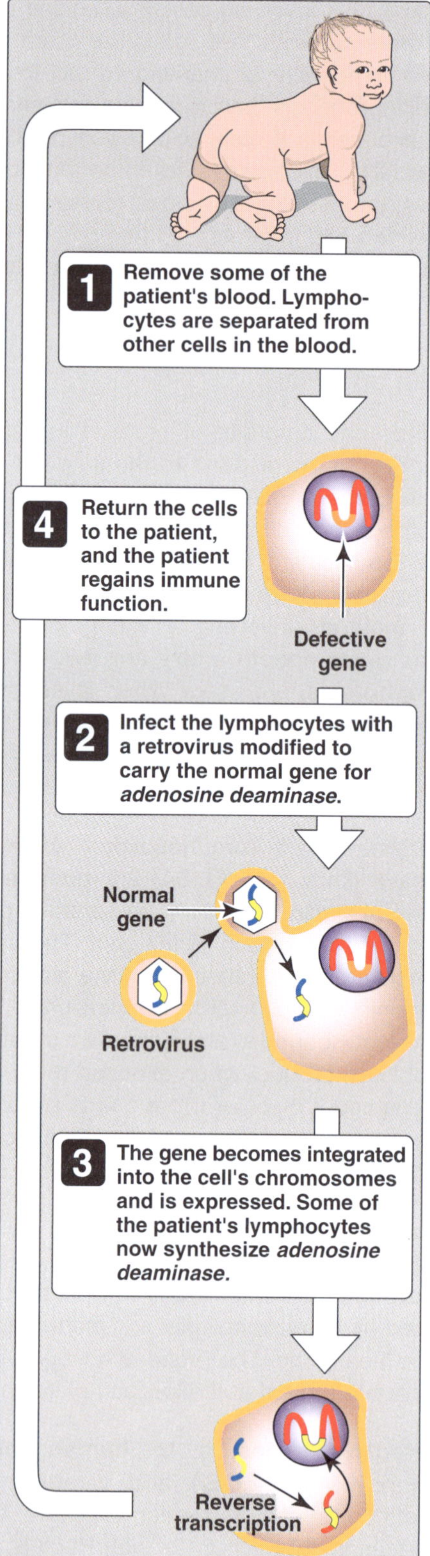

Figure 39.26 Gene therapy for severe combined immunodeficiency disease caused by *adenosine deaminase* deficiency. [Note: Bone marrow stem cells and a modified retroviral vector are now used.]

(clustered regularly interspaced short palindromic repeats **[CRISPR]**-associated protein) system that identifies and cleaves foreign DNA in bacterial cells. CRISPR is currently used in the laboratory but not in the clinic.]

Gene gun method is an innovative technique in genetic engineering, primarily in the research stage. In this method, nanoparticles coated with desired genes are propelled by Helium gas onto target cells, typically bacteria, with the aim of achieving transfection in laboratory settings.

X. TRANSGENIC ANIMALS

Transgenic animals can be produced by injecting a cloned foreign gene (a transgene) into a fertilized egg. If the gene randomly and stably integrates into a chromosome, it will be present in the germline of the resulting animal and can be passed from generation to generation. A giant mouse called "Supermouse" was produced in this way by injecting the gene for rat growth hormone into a fertilized mouse egg. [Note: Transgenic animals have been designed that produce therapeutic human proteins in their milk, a process called "**pharming.**" **Antithrombin,** an anticlotting protein, was produced by transgenic goats and approved for clinical use in 2009.] If the functional transgene undergoes targeted (not random) insertion, a knockin (KI) mouse that expresses the gene is created. Targeted insertion of a nonfunctional version of the transgene creates a **knockout (KO) mouse** that does not express the gene. Such genetically engineered animals can serve as models for the study of a corresponding human disease.

XI. Medical Perspectives

Recombinant DNA technology is used to make biomolecules for treatment, diagnosis of disease, pedigree analysis, gene therapy, forensic medicine and medical research.

- ***Making biomolecules for treatment***: Currently these techniques are being used to make in large quantities of insulin (human insulin), blood clotting factors VIII and IX and interferon for treatments, growth hormone, albumin, monoclonal antibodies and thrombolytic agents. Previously these biomolecules were taken from animals, other tissues or blood products and had the limitations of quantity production and transmission of infections.
- ***Diagnosis of disease***: Polymerase chain reaction (PCR) and restriction fragment length polymorphism (RFLP) are used for diagnosis of diseases such as sickle cell anemia, thalassemias and Duchenne muscular dystrophy.

  Infective diseases like HIV and tuberculosis can be diagnosed by PCR and specific probes, particularly in the window period.

  In the year 2020, the COVID-19 was declared a pandemic and the diagnosis was done by real-time PCR against specific genes of the SARS-CoV2 (severe acute respiratory syndrome-coronavirus 2).

  Genetic methods are also used in diagnosis of cancers and tissue typing for transplants.
- ***Genetic counseling*** is an important part of prenatal diagnosis of heritable diseases and provides the patient information about current pregnancies as well as aids planning for future pregnancies. This is especially important for families who have thalassemic traits, phenylketonuria, cystic fibrosis, Duchenne muscular dystrophy, etc.
- ***Gene therapy*** is the utilization of genetic engineering for treatment of disease. The vision is to correct the basic genetic abnormality by replacement of mutant gene by normal one or correction of mutation. **Gene therapy is still in the research stages, with limited success achieved thus far**.

  Gene augmentation, which is only partially successful, involves introduction of foreign genetic material into cell to compensate for defective product of mutant gene. The method of introduction in humans is by recombinant retroviruses, adenoviruses or plasmid liposome complexes. **Target cells so far used for gene augmentation are bone marrow cells, fibroblasts, tumor cells and epithelial cells of respiratory tract.**

  For example, *adenosine deaminase* (ADA) deficiency leads to lack of adequately functioning white blood cells and immune deficiency. A viral vector was used to incorporate ADA gene from a twin to his brother who suffered from severe ADA deficiency. Levels of ADA in recipient increased transiently.
- ***In forensic medicine***, genetic techniques are used for crime detection (sample could be hair, semen, etc.). For parenthood disputes and other issues, techniques used could be DNA fingerprinting, restriction map analysis, PCR, RFLP and use of specific probes.
- ***In medical research***, study of normal genes and those associated with disease is done by genetic methods. Gene for polycystic kidney disease and Duchenne muscular dystrophy is localized on chromosome by RFLP and chromosome walking. Fluorescence in situ hybridization (FISH) is used to localize gene for prolactin, insulin, and growth hormone.
- **Stem cells have an ability to divide for an unlimited period**. They can give rise to specialized cell types. This property is known as developmental plasticity.

  Embryonic stem cells and cells of umbilical cord **show developmental plasticity.** Stem cells may be **embryonic** (capable of differentiation) or **adult type** (limited capacity of differentiation). Stem cells can be **totipotent** (capable of producing all cell types of organism), **pluripotent** (capable of generating cells of three germ layers), **multipotent** (capable of generating only closely related cell types) and **unipotent** (generating only one cell type).

There have been trials in the use of stem cells in the treatment of stroke, brain injury, parkinsonism, myocardial infarction and spinal cord injury.

XII. Chapter Summary

- ***Restriction endonucleases*** are bacterial enzymes that cleave double-stranded DNA (dsDNA) into smaller fragments. *Restriction endonucleases* are also called **molecular scissors**. Each enzyme cleaves at a specific 4–8 base-pair sequence (a restriction site), producing DNA segments called **restriction fragments**. The sequences that are recognized are **palindromic**.
- Restriction enzymes form either **staggered cuts** (**sticky ends**) or **blunt-end cuts** on the DNA. Bacterial ***DNA ligases*** can join two DNA fragments from different sources if they have been cut by the same *restriction endonuclease*. This hybrid combination of two fragments is called a **recombinant DNA molecule**.

- **Recombinant DNA** is the altered DNA that results from the insertion of a sequence of deoxynucleotides not previously present into an existing molecule of DNA by enzymatic or chemical means. It is also called **chimeric** DNA if DNA is from two different sources.
- Amplification of the DNA may be done by PCR (in vitro) or cloning (in vivo). Introduction of a foreign DNA molecule into a replicating cell permits the **amplification** (production of many copies) of the DNA, a process called **cloning.**
- A **vector** is a molecule of DNA to which the fragment of DNA to be cloned is joined. Vectors must be capable of **autonomous replication** within the host cell, must contain at least one specific nucleotide sequence recognized by a *restriction endonuclease*, and must carry at least one gene that confers the ability to select for the vector such as an **antibiotic resistance gene**.
- Prokaryotic organisms normally contain small, circular, extrachromosomal DNA molecules called **plasmids** that can serve as vectors. They can be readily isolated from the bacterium (or artificially constructed), joined with the DNA of interest, and reintroduced into the bacterium, which will replicate, thus making multiple copies of the **hybrid** plasmid.
- A **DNA library** is a collection of cloned restriction fragments of the DNA of an organism.
- A **genomic library** is a collection of fragments of dsDNA obtained by digestion of the total DNA of the organism with a *restriction endonuclease* and subsequent ligation to an appropriate vector. It ideally contains a copy of every DNA nucleotide sequence in the genome. In contrast, **complementary DNA (cDNA) libraries** contain only those DNA sequences that are complementary to **processed messenger RNA** (mRNA) molecules present in a cell and differ according to cell type and environmental conditions. Because cDNA has no introns, it can be cloned into an **expression vector** for the synthesis of human proteins by bacteria or eukaryotes. Cloned, then purified, fragments of DNA can be sequenced, for example, using the **Sanger dideoxy chain termination method**.
- Probes are molecules used to detect the presence of a specific fragment of DNA or RNA or protein. Probes can be synthetic cDNA oligonucleotides (DNA probes), RNA oligonucleotides (RNA probes) and antibody probes. They can hybridize with specific DNA or RNA or cause an antigen-antibody bonding. For visualization, the probe may be radiolabeled, fluorescent or enzyme tagged.
- Probes are used to search samples for infectious agents, identify target in plaque or colony hybridization and blotting techniques as well as search libraries for specific genes.
- **Southern blotting** is a technique that can be used to detect specific sequences present in DNA. The DNA is cleaved using a *restriction endonuclease*, after which the pieces are separated by **gel electrophoresis** and are denatured and transferred (blotted) to a **nitrocellulose membrane** for analysis. The fragment of interest is detected using a probe. The human genome contains many thousands of **polymorphisms** (DNA sequence variations at a given locus). Polymorphisms can arise from single-base changes and from tandem repeats. A polymorphism can serve as a genetic marker that can be followed through families.
- A **restriction fragment length polymorphism (RFLP)** is a genetic variant that can be observed by cleaving the DNA into restriction fragments using a restriction enzyme. A base substitution in one or more nucleotides at a restriction site can render the site unrecognizable by a particular *restriction endonuclease*. A new restriction site also can be created by the same mechanism. In either case, cleavage with the *endonuclease* results in fragments of lengths differing from the normal that can be detected by **hybridization** with a probe. **RFLP analysis** can be used to diagnose genetic diseases early in the gestation of a fetus.
- The **polymerase chain reaction (PCR)**, another method for **amplifying** a selected DNA sequence, does not rely on the biologic cloning method. PCR permits the synthesis of millions of copies of a specific nucleotide sequence in a few hours. It can amplify the sequence, even when the targeted sequence makes up less than one part in a million of the total initial sample. The method can be used to amplify DNA sequences from any source. Applications of the PCR technique include 1) efficient comparison of a normal gene with a mutant form of the gene, 2) forensic analysis of DNA samples, 3) detection of low-abundance nucleic acid sequences, and 4) prenatal diagnosis and carrier detection (for example, of cystic fibrosis).
- The products of gene expression (mRNA and proteins) can be measured by techniques such as **northern blots**, which are like Southern blots except that the sample contains a mixture of **mRNA** molecules that are separated by electrophoresis, then hybridized to a radiolabeled probe; **microarrays** are used to determine the differing patterns of gene expression in two different types of cells (for example, normal and cancer cells); **enzyme-linked immunosorbent assays (ELISA)**; and **western blots (immunoblots)** are used to detect specific proteins.
- **Proteomics** is the study of all the proteins expressed by a genome. The goal of **gene therapy** is the insertion of a normal cloned gene to replace a defective gene in a **somatic cell**, whereas the goal of **gene editing** is the repair of a mutated gene. Insertion of a foreign gene (transgene) into the germline of an animal creates a **transgenic animal** that can produce therapeutic proteins or serve as gene **knockin** or **knockout** models for human diseases.

Study Questions

Short-Answer Questions

Write short notes on:

39.1. Restriction endonucleases
39.2. Vectors
39.3. DNA libraries
39.4. Probes

Long-Answer Questions

39.1. List the various techniques used in recombinant DNA technology. Describe any two in detail.
39.2. Discuss in detail the various areas in which medical diagnosis, treatment and research are utilizing genetic methodologies.

Multiple-Choice Questions

Choose the ONE best answer.

39.1. HindIII is a restriction endonuclease. Which of the following is a palindromic sequence that is most likely to be the recognition sequence for this enzyme?
A. AAGAAG
B. AAGAGA
C. AAGCTT
D. AAGGAA
Correct answer = C.

39.2. A physician would like to determine the global patterns of gene expression in two different types of tumor cells in order to develop the most appropriate form of chemotherapy for each patient. Which of the following techniques would be most appropriate for this purpose?
A. Enzyme-linked immunosorbent assay
B. Microarray
C. Northern blot
D. Southern blot
Correct answer = B.

39.3. Which of the following is an advantage of real-time PCR over conventional PCR?
A. It can amplify longer DNA fragments
B. It does not require thermal cycling
C. It can quantify the amount of DNA accurately
D. It requires fewer cycles to amplify DNA
Correct answer = C.

39.4. Which of the following techniques has restriction digestion of DNA as their component?
A. Gene cloning
B. Western blotting
C. Microarray for gene expression
D. Polymerase chain reaction
Correct answer = A.

39.5. Which among the following statements regarding polymerase chain reaction is true?
A. PCR requires an RNA primer which is usually ≥ 18 nucleotides long.
B. The entire sequence to be amplified must be known for the PCR to be set-up.
C. PCR using Taq polymerase is accurate because of the proof-reading activity of the enzyme.
D. PCR can be modified to quantify the amount of DNA or RNA present in the sample.
Correct answer = D.

39.6. Which of the following methods is NOT correctly matched against its application?
A. Analysis of short tandem repeats – DNA profiling in forensics
B. Sanger sequencing – detection of known mutations in a gene of interest
C. Western blotting – to study the expression of all the proteins in the cell simultaneously
D. CRISPR – Gene editing to correct disease causing gene variants
Correct answer = C.

39.7. Which of the following techniques can be used to study a well characterized triplet repeat disorder in the DNA in a given DNA sample from a patient?
I. Polymerase chain reaction
II. Northern blotting
III. Sanger sequencing
IV. Southern blotting
V. Microarray

A. I, III, and IV only
B. I, III, and V only
C. I, II, and IV only
D. IV only
Correct answer = A.

39.8. The diagram given below shows the results of PCR-RFLP in a family with sickle cell anemia, where PCR has been performed using primers flanking the MstII restriction site, followed by restriction digestion of the amplified DNA using MstII. Which among the following represents an individual who is homozygous for HbS? *Note that the restriction site is disrupted in the presence of HbS allele.*

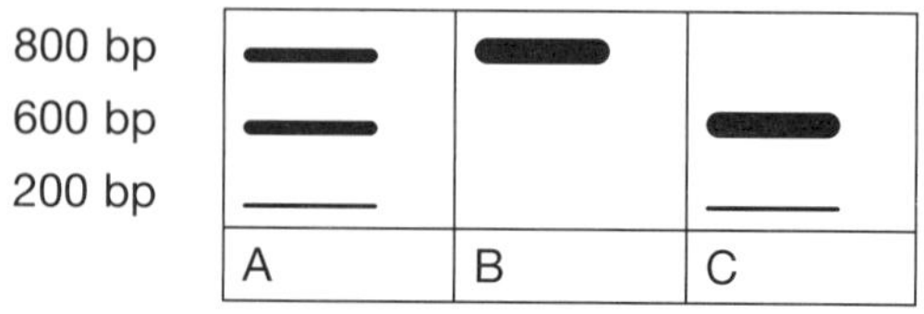

800 bp
600 bp
200 bp

A B C

A. Individual A
B. Individual B
C. Individual C
D. Individuals A and C

Correct answer = B.

39.9. All the following represent the genetic heterogeneity between two humans, EXCEPT:
A. Variable number of tandem repeats
B. Restriction fragment length polymorphisms
C. Short tandem repeats
D. Ultra-conserved elements

Correct answer = D.

39.10. Which among the following is the best technique to study the duplication of a chromosome (e.g., trisomy 21)?
A. Fluorescent in situ hybridization
B. Sanger sequencing
C. Western blotting
D. Polymerase chain reaction

Correct answer = A.

Multiple-Choice Questions (NExT/NEET PG)

Choose the ONE best answer.

39.1. An Ashkenazi Jewish couple has their 6-month-old son evaluated for listlessness, poor head control, and a fixed gaze. Tay-Sachs disease, an autosomal-recessive disease of lipid degradation, is diagnosed. The couple also has a daughter. The family's pedigree is shown below, along with Southern blots of a restriction fragment length polymorphism very closely linked to the gene for hexosaminidase A, which is defective in Tay-Sachs disease. Which of the statements below is most accurate with respect to the daughter?

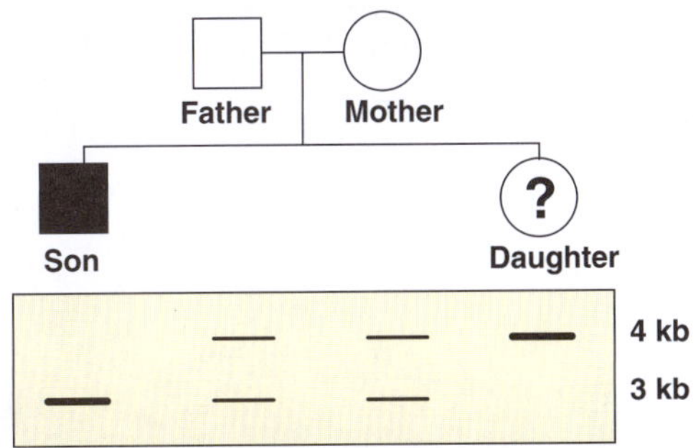

A. She has a 50% chance of having Tay-Sachs disease.
B. She has Tay-Sachs disease.
C. She is a carrier for Tay-Sachs disease.
D. She is homozygous normal.

Correct answer = D.

39.2. A 2-week-old infant is clinically suspected to have a urea cycle defect. Enzymic analysis showed no activity for ornithine transcarbamoylase (OTC), an enzyme of the cycle. Molecular analysis revealed the presence of a gene variant in the patient that was not found in the reference genome of healthy individuals. Which of the techniques listed below was most likely used to analyze OTC gene in this patient?
A. Sanger sequencing
B. Northern blot
C. Microarray
D. Western blot

Correct answer = A.

39.3. The image below shows the restriction digestion of a plasmid of ~8000 bp length by multiple restriction enzymes.

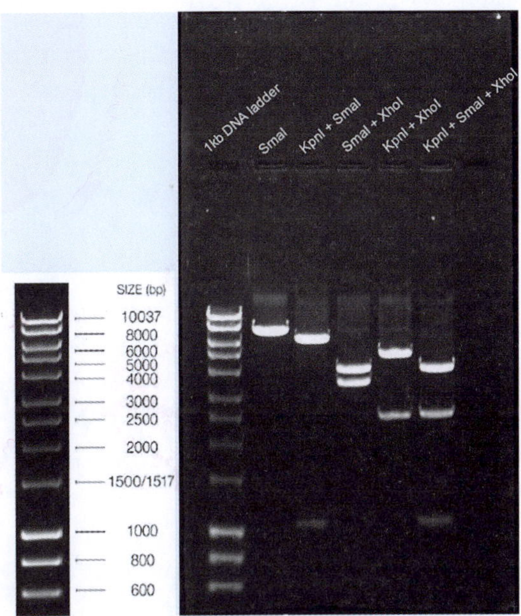

Photo courtesy: *Dr. Ashikh Seethy*, All India Institute of Medical Sciences, Guwahati, India.

Which among the following statements is true about this plasmid?
A. The plasmid has two restriction sites for SmaI
B. KpnI restriction site is located between SmaI and XhoI restriction sites
C. XhoI restriction site is located between SmaI and KpnI restriction sites
D. The SmaI and the XhoI restriction sites are the same

Correct answer = B.

39.4. Assuming that the pedigree shown below of a family where the son is affected with an autosomal recessive disorder, what is the probability that the daughter is a carrier?

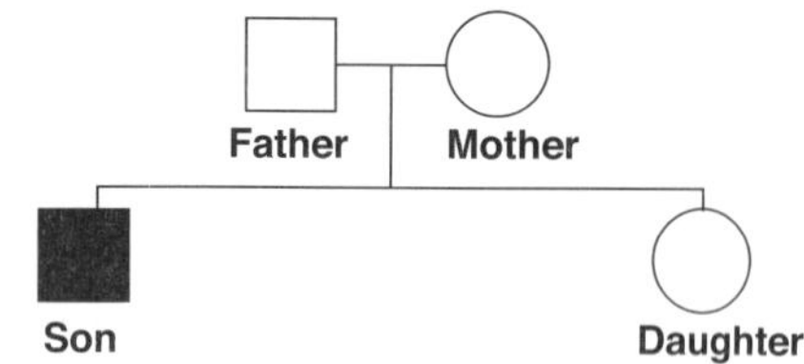

A. 1/2
B. 2/3
C. 1/3
D. 1/4

Correct answer = B.

39.5. In which of the following clinical situations is the use of next generation sequencing for disease diagnosis best suited?

A. The clinical features suggest a single gene disorder and the gene has only a few exons that are short in length.
B. The parents of the patient are heterozygotes for β-thalassemia, and the patient now has presented with severe anemia.
C. The patient has presented with a clinical syndrome where the cause could lie in one among a dozen genes that are known to be altered.
D. The genetic variant in both the parents are known, and now the fetus is to be tested for the same variant.

Correct answer = C.

Appendix
Unit VIII: Storage and Expression of Genetic Information

MUTATIONS AND OTHER COMMON ABNORMALITIES

Even as India controls communicable diseases, a demographic shift to non-communicable diseases as well as genetic diseases is seen. Most of such genetic disorders present as spontaneous abortions, bad obstetric history, mental retardations and malformations. Since India has a high birth rate, the burden of such diseases is a lot even though the incidence is less (*approx. 2% malformations in births). For example, about 20,000 cases of Down syndrome are born every year in India. Cases of beta-thalassemia major are about 1 in 3,000 births, but that leads to about 10,000 cases per year in India.

The amino acid disorders which we refer to as inborn errors of metabolism are also genetic diseases. Infants with mental retardation have been screened in large studies in India and have shown that the common disorders include tyrosinemia, alkaptonuria, maple syrup urine disease and phenylketonuria. Other genetic abnormalities seen are mucopolysaccharidosis, lysosomal diseases, Wilson disease, glycogen storage diseases and galactosemia.

Prenatal diagnosis, premarriage counseling and genetic counseling especially in patients with family history becomes very important as genetic diseases do not have a cure as of now and only supportive treatment can be given.

Universal prenatal genetic screening has been advised for Down syndrome, beta-thalassemia and neural tube defects in India. Newborn screening for amino acid and metabolic disorders is currently advised in symptomatic infants.

The primary healthcare physician should know enough about genetic diseases to suspect the disorders and refer to pediatric and genetic centers for further follow up.

Table 1 Some common genetic disorders in India and their diagnosis

Common Disorders	Symptoms	Genetic Cause	Diagnosis
Thalassemias	Weakness, paleness, slow growth or even mortality if defect is severe	Abnormal form or insufficient amount of Hb due to gene mutations and deletions in one or more of alpha chain or beta chains of hemoglobin (see chapter 8 and appendix of Unit II for details)	Peripheral smear showing microcytic hypochromic anemia, Hb electrophoresis showing abnormal hemoglobins, PCR-RFLP for mutations
Hemophilia	Bleeding for a long time after injury	Hemophilia A is due to mutation in the gene for factor VIII leading to deficiency of this clotting factor. Factor IX genetic mutations lead to hemophilia B	Complete blood count, activated partial thromboplastin time, prothrombin time and fibrinogen
Sickle cell anemia	Dizziness, fatigue, joint pains, shortness of breath	Glutamic acid at the sixth position on the β-chain is substituted by valine due to a point mutation	Peripheral smear shows sickle-shaped red blood cells, electrophoresis and PCR-RFLP used to diagnose

(Continued on next page)

*Exact epidemiological data of genetic disorders is not available in India as the research has been mainly focused in cities and there is a lack of pan India data.

Table 1 (Continued from previous page)

Common Disorders	Symptoms	Genetic Cause	Diagnosis
Down syndrome	Flat facial features, short neck, bulging tongue, eyes that slant upwards, mental retardation	Trisomy of chromosome 21	Prenatal quadruple test for maternal screening and karyotype for confirmation; after birth, infant has typical phenotype and also karyotyping can be done
Neural tube defects (NTDs)	Defects in brain, spine or spinal cord. Symptoms such as paralysis, loss of bowel and bladder control. Prenatal diagnosis is usually a finding on routine ultrasounds	Multiple genes and multiple environmental factors	Prenatal amniocentesis and ultrasound to see spinal defect, maternal serum alpha-fetoprotein levels increased. Patients who survive after being born with NTD can be detected by typical features, CT scans, ultrasound, etc.
Musculoskeletal disorders like Duchenne muscular dystrophy	Progressive skeletal and heart muscle weakness and loss ultimately leading to paralysis	X-linked recessive, deletions of exons in Dystrophin gene leading to a defective dystrophin	Extremely high *creatine kinase* levels, muscle biopsy

PCR = polymerase chain reaction; RFLP = restriction fragment length polymorphism

Clinical Case Questions

Chapter 34 Nucleotide Metabolism

34.1. A 45-year-old male presented to the orthopedic OPD with history of pain and swelling in big toe. On questioning, he gave history of occasional such pain particularly after a party the night before. He is a social drinker. On examination, the big toe and ankle looked swollen, red and tender. His blood tests showed uric acid levels of 8.6 mg/dl.

A. What is the suspected diagnosis and treatment of the disease?

Answer—A: The patient appears to suffer from acute gout. Normally uric acid is excreted through kidneys, but excess intake of purines may cause the deposition of uric acid crystals in joints causing redness, pain and swelling. Alcohol and a non-vegetarian diet are proven to increase uric acid levels as it has a high level of purines. The treatment is behavioral modifications, anti-inflammatory medication and allopurinol. Allopurinol inhibits conversion of xanthine to uric acid by inhibiting xanthine oxidase.

Chapter 35 DNA Structure, Replication, and Repair

35.1. A 30-year-old lady complains of high-grade fever and burning sensation while passing urine. The doctor sends a requisition for urine routine and microscopy tests.

The results show a field full of pus cells and urine albumin positive.

The doctor further orders a culture sensitivity of the urine. Culture grows E. coli which is sensitive to norfloxacin.

A. What is the mechanism of action of norfloxacin and will the patient's enzymes be targeted if this drug is given?

Answer—A: The drug norfloxacin is bactericidal as it inhibits bacterial DNA gyrase. It has no effect on human DNA topoisomerase or other DNA-associated enzymes and hence is considered safe to administer. The patient may therefore be started on norfloxacin in recommended doses.

35.2. A patient suffering from breast cancer is being given the drug doxorubicin. She has started having hair loss, vomiting, rashes and ulcers in mouth.

A. What is the reason for these side effects, and can they be avoided?

Answer—A: Cancerous cells are caused by defective cell replication resulting in a cluster of abnormal cells. Doxorubicin acts by inhibiting the action of DNA topoisomerase. This stops the progress of DNA replication. However, since the drug cannot differentiate between normal replicating cells and abnormal replicating cells, the side effects in cancer chemotherapy are common. Currently, only symptomatic and suppressive treatments can be given in such cases.

Chapter 36 RNA Structure, Synthesis, and Processing

36.1. A 54-year-old male patient came to hospital with complaint of cough for a month and sometimes blood-stained sputum. He also had fever, especially at night, and loss of weight. The doctor ordered blood tests, chest X-ray and sputum for acid-fast bacilli. The X-ray results showed classical upper lobe infiltrates with cavities consistent with a tuberculosis chest infection. Report for acid-fast bacilli was positive.

The doctor started the patient on five drugs which also included rifampicin. The patient is skeptical about taking rifampicin as he gives service in a leprosy home where rifampicin is given to leprosy patients.

A. What is the mechanism of action of rifampicin?
B. Is this drug effective in tuberculosis and leprosy both?

> **Answer:** Rifampicin inhibits transcription initiation by binding to the beta subunit of prokaryotic DNA-dependent RNA polymerase. Rifampicin is further able to cross blood-brain barrier and is also effective for intracellular bacteria as it crosses the cell membrane because of its lipid solubility.
>
> Rifampicin is exceptionally bactericidal for Mycobacterium leprae as well as effective against M. tuberculosis. Hence it is used in treatment of both tuberculosis and leprosy.

36.2. An 8-year-old child with pale complexion and excessive tiredness presented in the pediatric OPD. The doctor ordered a complete blood count which showed severe anemia with Hb 6.4g%. He then ordered an Hb electrophoresis. Hb electrophoresis showed HbA to be 84% (normal 95%–98%) Hb A2 to be 5% (normal 1.5%–3.5%) and Hb F to be 3% (normal 0.8%–2%). The doctor further ordered DNA mutational analysis by molecular techniques. The genetic lab gave a diagnosis of β-thalassemia minor.

A. What is the genetic error in β-thalassemia?
B. The doctor also requested for blood samples of the child's parents. What is the rationale behind this?

> **Answer:** β-thalassemia can be caused by mutations that affect globin gene expression including transcription, posttranscriptional modifications, translation or posttranslational modifications. The most common defects are mutations that cause derangement in splicing of the mRNA precursor.
>
> Since thalassemias are inherited disorders, it is important to know the variants that the parents carry. This will be useful for their own health, understanding the genetic traits that the family carries, and prenatal diagnosis of disease in case they plan for another child.

Chapter 37 Protein Synthesis

37.1. A 6-year-old boy was brought to the pediatric OPD by his mother. She said that he has stopped playing and complains of weakness and muscle pain. The mother gave history of her brother dying of similar complaints when he was 18 years old. On examination, lower limb muscle wasting was seen. Abnormal gait and difficulty in getting up from the ground was observed. The doctor ordered a blood test in which creatine kinase was 1,060 mg/dl. The doctor suspected Duchenne muscular dystrophy (DMD).

A. What is the cause of DMD?

> **Answer—A:** Duchenne muscular dystrophy occurs because of a defective dystrophin protein. Normally, dystrophin helps in anchoring the actin in the cytoskeleton to the extracellular matrix. In this X-linked disorder, exonal deletions or mutation in the gene cause formation of an abnormal dystrophin protein. Only males are affected.

B. How can the diagnosis be confirmed?

> **Answer—B:** The diagnosis is highly suspected if creatine kinase is extremely high. Muscle biopsy shows a deficient dystrophin. Further, confirmation is by multiplex PCR for genetic diagnosis of dystrophin gene abnormalities.

37.2. A 35-year-old male presented to the medical OPD with severe throat pain and difficulty in swallowing. On examination, a pharyngeal pseudomembrane and extensive exudate was seen. The physician made a diagnosis of diphtheria which was further confirmed by the laboratory.

A. What is the toxin produced by Corynebacterium diphtheriae?

B. What is the mechanism of action of this toxin?

Answers—A and B: Pathogenic strains of C. diphtheriae produce Diphtheria protein toxin which binds to the EF-2 in eukaryotes and therefore stops translocation.

Chapter 38 Regulation of Gene Expression

38.1. A 65-year-old patient was brought to the medical OPD by his wife. She complained that he keeps forgetting everything, particularly recent events. He tends to forget his way back to home when he goes out for walks. On examination, clinical signs and symptoms were unremarkable. Routine laboratory tests were normal. Brain imaging showed posteriorly predominant cortical and hippocampal atrophy suggestive of Alzheimer's disease.

A. What is the cause of Alzheimer's disease?

Answer—A: Alzheimer's disease is a gradual and progressive neurodegenerative disease characterized by dementia. It has a polygenic cause involving many genes like amyloid precursor protein, presenilin 1 and 2 and Apo E4. It is a protein conformational disease involving aggregation of neurofibrillary tangles and plaques.

Chapter 39 Biotechnology and Human Disease

39.1. A 6-year-old child with anemia presented to the pediatric OPD with failure to thrive. He also complained of breathlessness, back pain and fatigue. His blood was sent for normal hemogram and peripheral smear. The Hb was 7 g% and peripheral smear showed sickle-shaped red blood cells.

A. What further tests would you like to do? What is the methodology of these tests?

Answer—A: Sickle-shaped cells on peripheral smear are suggestive of sickle cell anemia. Hemoglobin electrophoresis can be done to check for hemoglobin S (HbS). For genetic diagnosis, DNA amplified by PCR and restriction fragment length polymorphism (RFLP) will identify the single nucleotide substitution (A to G) at the sixth position of beta-globin chain of hemoglobin. This replaces the amino acid glutamate with the less polar valine.

39.2. A patient presents to the medical OPD with complaints of fever and productive cough for past 1 month. Since his fever was mild and mostly in the evenings he was not very worried about it and put it down to tiredness. He got worried when he saw blood-stained sputum yesterday. His blood tests showed a high erythrocyte sedimentation rate (ESR) and lymphocytosis. His blood culture has been sent for tuberculosis, but report will come in 3 weeks. His sputum for acid-fast bacillus (AFB) is negative.

A. What other test can you do for this patient to affirm or rule out tuberculosis?

Answer—A: India has a high prevalence of tuberculosis, so index of suspicion should be high. It is not always possible to find acid-fast bacilli in sputum, even though the patient may have pulmonary tuberculosis. In this case, PCR for tuberculosis bacilli in sputum is a good choice as probes are made specific to genes in Mycobacterium tuberculosis and will amplify genetic material of bacilli if present. Nucleic acid amplification tests can detect M. tuberculosis ribosomal RNA or DNA directly from sputum samples.

Advanced Cases/Self-Directed Learning

VIII.1. A 6-year-old boy, born of second-degree consanguineous marriage, presented with extensive freckling over photo-exposed areas of the body, including the face, hands, and upper chest. Additionally, he had pigmented nodulo-ulcerative growths over the nose and around the eyes, which exhibited features of basal cell carcinoma and basi-squamous carcinoma on histopathological examination. The child also had marked conjunctival congestion and photophobia.

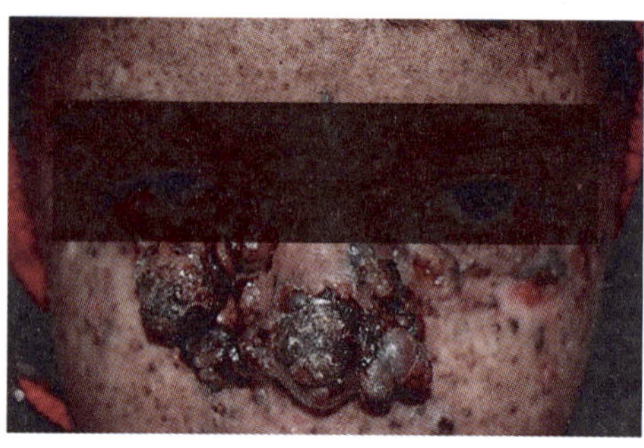

Photo courtesy: *Dr. Neha Taneja* and *Dr. Ananya Sharma*, All India Institute of Medical Sciences, New Delhi.

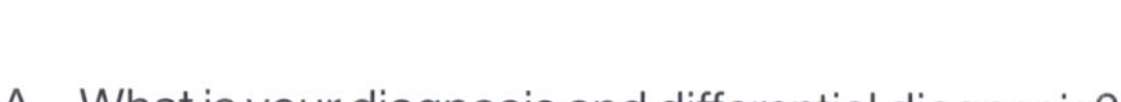

A. What is your diagnosis and differential diagnosis?

Answer—A: This is likely a case of Xeroderma pigmentosum. Other photosensitive disorders of childhood, such as Cockayne Syndrome and Bloom Syndrome, should be ruled out.

B. What is the pathogenic mutation and defect?

Answer—B: Xeroderma pigmentosum is due to a defect in DNA repair mechanisms and the mutation may be confirmed by genetic analysis of the genes involved in nucleotide excision repair of DNA.

C. What is the most important intervention for prevention of development of cancers in these patients?

Answer—C: Strict physical photoprotection and use of sunscreens and goggles that block visible light.

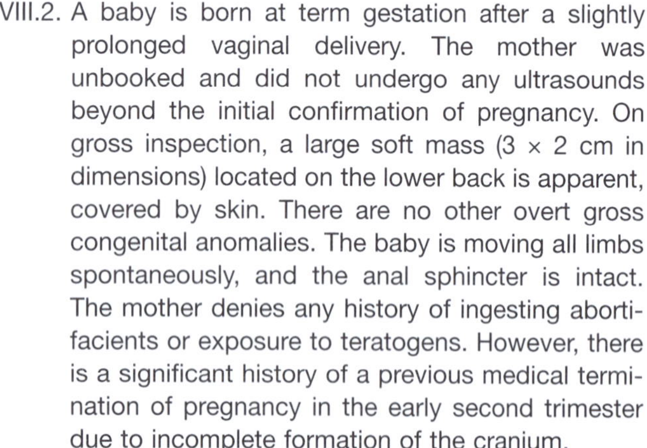

VIII.2. A baby is born at term gestation after a slightly prolonged vaginal delivery. The mother was unbooked and did not undergo any ultrasounds beyond the initial confirmation of pregnancy. On gross inspection, a large soft mass (3 × 2 cm in dimensions) located on the lower back is apparent, covered by skin. There are no other overt gross congenital anomalies. The baby is moving all limbs spontaneously, and the anal sphincter is intact. The mother denies any history of ingesting abortifacients or exposure to teratogens. However, there is a significant history of a previous medical termination of pregnancy in the early second trimester due to incomplete formation of the cranium.

A. What is the clinical phenotype and probable diagnosis?

Answer—A: It is likely a neural tube defect, most probably a meningocele, as it is covered by skin and there is no evidence of paraparesis.

B. What do you think was the indication for the previous medical termination of pregnancy based on the description given by the mother?

Answer—B: Anencephaly, another form of a neural tube defect.

C. What advice should the mother have been given in the previous pregnancy to reduce the likelihood of a recurrence of a neural tube defect?

Answer—C: Preconception folic acid (5 mg/day) that should be started at least 8–12 weeks before pregnancy and continued for at least 3 months into the first trimester.

VIII.3. A six-year-old boy born to a consanguineous couple presents with a history of delayed acquisition of gross motor milestones, including sitting, standing, walking, running, and climbing stairs. He was never able to jump. Over the past six months, the parents have noticed a gradual, painless decline in his ability to get up from the squatting position and negotiate stairs. However, the other developmental domains (fine motor, social-emotional, language, and cognitive) are all age-appropriate and have not displayed any regression. There is no history of any preceding febrile illness, loss of consciousness, or seizures. The boy's 18-year-old maternal uncle had a similar history and has been non-ambulatory for almost a decade.

On general physical examination, there is hypertrophy of both calves. Neurological examination reveals decreased power, normal tone and normal deep tendon reflexes. The cardiovascular system appears normal.

A. What is the probable diagnosis?

Answer—A: Dystrophinopathy results from a decreased amount and function of dystrophin, a membrane-associated protein present in the sarcolemma. Based on the early onset of symptoms, regression, and severity of impaired motor function, this case is most likely Duchenne muscular dystrophy.

B. What is the calf abnormality characterized by hypertrophy despite muscle weakness?

Answer—B: Pseudohypertrophy, caused by the deposition of fat and abnormal connective tissue.

C. Which biochemical and genetic tests will help you in establishing a diagnosis?

Answer—C: Serum creatine phosphokinase levels will be highly elevated, typically ranging from 15,000 to 20,000 U/L in Duchenne muscular dystrophy (DMD). Exon deletions of the DMD gene (located on Xp21) can be detected by multiplex polymerase chain reaction (PCR) in most cases.

VIII.4. An 18-month-old boy born out of a consanguineous marriage is brought to the OPD with parental concerns regarding easy bruising since a few months. These started following minor falls that occur once the child learned how to walk and has started roaming around the house. There have been two episodes of painful swelling of both knees associated with bluish discoloration. His 5-year-old sister is healthy. However, the mother is extremely apprehensive because his elder brother had similar complaints and expired when he was a toddler, following what she has heard described as a 'bleed in the brain' after a minor injury sustained while playing at home.

A. What is the probable diagnosis?

Answer—A: A heritable coagulopathy, most probably hemophilia since the inheritance is X-linked recessive.

B. What is the underlying pathophysiology?

Answer—B: Hemophilia A and B are caused by significant deficiencies of Factor VIII and Factor IX, respectively.

C. What is the mainstay of management?

Answer—C: Replacement therapy with concentrates of Factors VIII or IX depending upon the type of hemophilia.

VIII.5. A 26-year-old female patient with mild intermittent bronchial asthma presents to the emergency department with breathlessness since one day. She was apparently well until 3 days ago when she developed high-grade continuous fever associated with a dry cough. The cough worsened over the next 2 days with only transient relief from medications. She has developed progressive breathlessness, which worsens when lying down. The patient had a recent history of international travel and visiting crowded public places.

On examination: Dyspneic, restless

Pulse: 110/min, BP: 130/70 mmHg

Resp rate: 26/min with use of accessory muscles

O_2 saturation was 70% which increased to 88% on using oxygen mask

Arterial blood gas analysis showed Type 1 respiratory failure

Chest: Polyphonic wheeze

Chest X-ray shows multiple bilateral opacities

A. What is the likely diagnosis?

Answer—A: Viral pneumonia, likely COVID-19, is suspected, especially given the patient's history of recent international travel and visiting crowded public places.

B. What investigation will help us clinch the diagnosis?

Answer—B: RT-PCR for COVID-19 should be performed.

C. Which inflammatory markers are likely to be elevated?

Answer—C: CRP, IL-6, fibrinogen and ESR are likely to be elevated.

Unit IX

Immunology

Competencies Addressed

- **BI2.6** Discuss use of enzymes in laboratory investigations (enzyme-based assays)
- **BI10.3** Describe the cellular and humoral components of the immune system and describe the types and structure of antibody
- **BI10.4** Describe and discuss innate and adaptive immune responses, self/non-self recognition and the central role of T-helper cells in immune responses
- **BI10.5** Describe antigens and concepts involved in vaccine development

Unit IX: Overview

Immunology deals with protection of the self (the body) from invading organisms and environment. The immune system is able to distinguish self from nonself and respond only to nonself molecules. This lack of immunological response to self molecules is known as **immunological tolerance**. The immunological identity of self is defined in utero. Natural or innate immunity is what we are born with, and acquired (or adaptive) immunity is what we develop after birth by exposure to immunogens (Chapter 42).

Cellular immunity is mediated by T cells which recognize and kill pathogens at cellular level, and **humoral immunity** involves the production of soluble and secreted antibodies (e.g., IgM and IgG) by the B cells and plasma cells. Components of both cellular immunity and humoral immunity communicate with and assist each other (Chapter 43).

Antigen exposure leads to the formation of antibodies which in turn neutralize the antigens (Chapter 41). Some organs like **thymus, spleen, lymph nodes, and bone marrow**, and some specialized immunological cells like **T cells, B cells and natural killer cells** are more involved in immunological process (Chapter 40). These cells not only deal with the invaders but also remember them (memory cells) so that next-time recognition and response can be faster.

Allergy and hypersensitivity are exaggerated and inappropriate immune responses to an antigen, e.g., asthmatic response to a normal pollen (Chapter 43). **Immunodeficiency** is an inadequate response to an antigen, e.g., acquired immunodeficiency syndrome **(AIDS)** subsequent to HIV infection or severe combined immune deficiency **(SCID)**. Autoimmunity means an immunological attack against one's own cells; it is seen in many diseases like rheumatoid arthritis, systemic lupus erythematosus and type 1 diabetes mellitus.

The primary healthcare physician must be aware that immunity may be natural, acquired by exposure to infections or acquired by vaccination. In India, children are given **vaccines**, such as BCG, DPT, OPV and hepatitis B, under the **Universal Immunization Programme (UIP).** Immunization can be active when vaccines are given or passive when preformed immunoglobulins are given (Chapter 44). In blood transfusions, antigen-antibody typing of blood donor and recipient should be followed stringently to avoid reactions and even death (Chapter 43). The knowledge of immunology is also important for success of organ transplantation. The immune system is also therapeutically exploited to eliminate cancer cells. This field is known as cancer immunotherapy.

Introduction to Immune System and Immune Cells 40

Learning Objectives

At the end of this chapter, the student will be able to:

- Describe the organs and cells of the immune system
- Explain primary and secondary immunodeficiencies

I. OVERVIEW OF IMMUNE SYSTEM

Immune system is a complex system that has been organized to protect the body from other organisms and foreign substances. It comprises of several organs and cells which form a dynamic network that recognizes and eliminates the pathogens. The immune system is capable of distinguishing different types of pathogens by recognizing their molecular patterns and chemical differences. Human pathogens are mainly viruses (causing AIDS, influenza, etc.), bacteria (causing tuberculosis, tetanus), fungi (causing candidiasis, ringworm) and parasites (causing malaria). Exposure to pathogens allows immune system to respond with two types of responses, effector response and memory response. Pathogens are eliminated through effector response, whereas memory response protects the host from infection by same pathogen on subsequent occasions. Also, the memory response of the immune system has led to the development of vaccines against different types of pathogens. Immunity can be innate which provides first and second line of defense against pathogens or adaptive which provides third line of defense against pathogens. Innate immunity involves phagocytosis, receptor-mediated endocytosis and pinocytosis and is mediated through interferons and complements. Adaptive immunity which is more specific acts after few days of exposure to antigens and is mediated through B cells and T cells, and antibodies. Several organs and cells are responsible for the development of immune responses. Among the lymphoid organs, thymus, bone marrow, lymph nodes and spleen play a major role in the formation, differentiation and maturation of immune cells like lymphocytes and phagocytes. Any dysfunction of the immune system fails to protect the host against infections and can lead to debilitating disorders like allergies, asthma, autoimmune and immunodeficiency disorders.

II. FUNCTIONS OF THE IMMUNE SYSTEM

The main function of the immune system is to protect the host against infections and foreign substances. It eliminates pathogens through phagocytes and lymphocytes. Phagocytes like monocytes and macrophages internalize pathogens and degrade them. Lymphocytes degrade pathogens through B cells which produce antibodies or T cells which destroy the infected cells. Another function of the immune system is to discriminate between the body's own tissue (*self*) and foreign material (*non-self*). Immune system is also needed for the maintenance of gut flora in a manner that is beneficial to the body.

III. ORGANS OF THE IMMUNE SYSTEM AND THEIR ROLE IN IMMUNE FUNCTION

The organs of the immune system are broadly classified as primary and secondary lymphoid organs based on their functions.

Primary (central) lymphoid organs are the thymus and the bone marrow which are involved in the maturation of lymphocytes. ***Secondary (peripheral) lymphoid organs*** are lymph nodes, spleen and gut-associated lymphoid tissue (GALT) which is a part of mucosa associated lymphoid tissue (MALT), which allow interaction of matured lymphocytes with the antigen. Tertiary lymphoid organs are abnormal lymphoid structures which develop in some peripheral organs because of chronic inflammation.

A. Thymus

It is a flat organ situated above heart and comprises of cortex and medulla (Fig. 40.1). Cortex is in the

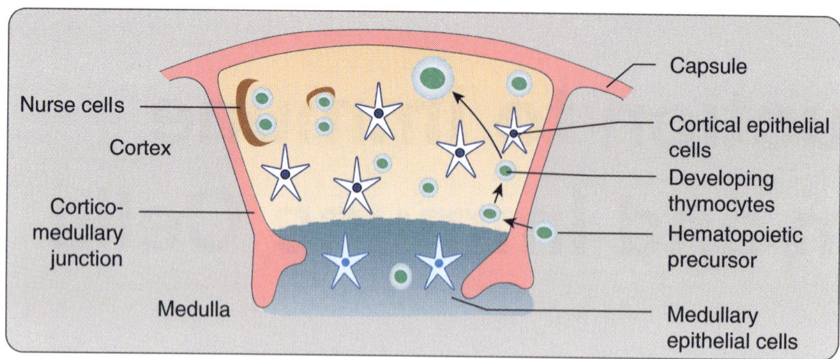

Figure 40.1 Diagrammatic representation of a cross-section of portion of thymus.

outer layer and contains immature T cells known as thymocytes. These T-cell precursors originate in the bone marrow, from where they migrate to the thymic cortex for maturation. Maturation is facilitated by the presence of cortical and medullary thymic epithelial cells. Some thymic epithelial cells in the outer cortex, called nurse cells have long membranous extensions. As thymocytes mature, they migrate to medulla and then leave the thymus. The main function of thymus is development and maturation of T cells. As age advances, the structure and function of thymus begin to decline. Thymus undergoes atrophy after puberty accompanied by reduction in the generation of T cells. Immune dysfunction in elderly people has been attributed to atrophy of thymus. **DiGeorge syndrome** which occurs due to congenital defect in the development of thymus, due to a deletion in the long arm of chromosome 22 is associated with a reduction in T cells and absence of cell-mediated immunity.

B. Bone marrow

Bone marrow comprises of hematopoietic stem cells and progenitor cells in a stroma composed of adipocytes, macrophages, mesenchymal cells, sympathetic neurons, and extracellular matrix. It is the site where hematopoietic cells are generated and distributed to other cells of the body. In humans, bone marrow is involved in proliferation and development of B cells and secretion of cytokines. Differentiation of B cells is achieved due to reticular cells which are located near the endosteum and central sinus.

C. Lymph nodes

They are bean-shaped structures, which are involved in neutralizing the antigens present in the lymphatic fluid draining into the lymph node. It consists of an outer layer (cortex) which contains B lymphocytes, macrophages, and dendritic cells, middle layer (paracortex) containing T lymphocytes and an inner layer (medulla) containing plasma cells which secrete antibodies (Fig. 40.2). Afferent lymphatic vessels enter the capsule at multiple sites and drain lymph into the subcapsular sinus. As lymph having antigen from tissues passes through the node, antigen is trapped by phagocytic cells and dendritic cells. Professional **antigen-presenting cells** (APCs - dendritic cells, B

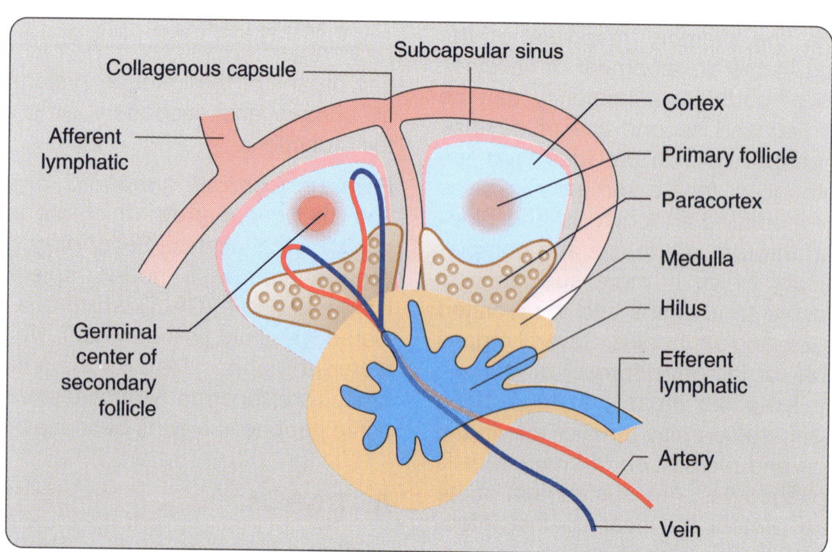

Figure 40.2 Diagrammatic representation of a cross-section of a portion of lymph node.

cells and macrophages) present these trapped antigens in association with major histocompatibility complex (MHC) molecules leading to the activation of T cells, which in turn activate B cells. Lymphoid follicles are of two types—primary follicles and secondary follicles. Primary follicles are aggregates of B-cells, while secondary follicles are characterized by the presence of a specialized feature known as germinal center. Germinal center comprises of activated and proliferating B cells, along with plasma cells. Thus, secondary follicles are generated only after exposure to the immunogen. During infection, the blood-borne lymphocytes migrate to lymph nodes in response to pathogens and proliferate further resulting in lymph node enlargement.

D. Spleen

It is an ovoid-shaped organ situated in left side of abdominal cavity. **Spleen plays a key role in trapping blood-borne antigens and removal of old and defective red blood cells.** It consists of two types of compartments—red pulp and white pulp (Fig. 40.3). **Red pulp** has venous sinuses which contain macrophages, lymphocytes, granulocytes, erythrocytes and platelets. Old and defective red blood cells are destroyed and removed in the red pulp by macrophages. **White pulp** located along the branches of the splenic artery consists of cells involved in the adaptive immune response, and is separated from the red pulp by a marginal zone. Immediately close to the arterial branches is the periarteriolar lymphatic sheath (PALS), which is rich in T cells. B cell rich zone is present between the PALS and the marginal zone. The marginal zone which surrounds the white pulp contains B cells, macrophages, and dendritic cells. Activation of B cells and T cells and presentation of antigen to T cells occur in periarteriolar lymphoid sheath. Surgical removal of spleen due to tumors or coagulation disorders increases the risk of bacterial infections and children are at higher risk compared to adults.

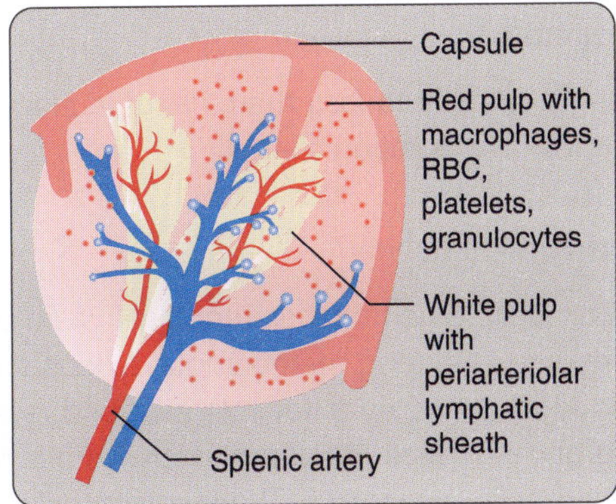

Figure 40.3 Diagrammatic representation of a cross-section of a portion of spleen.

IV. CELLS OF IMMUNE SYSTEM AND THEIR ROLE IN IMMUNE FUNCTION

A. B Lymphocytes (B cells/CD19/CD20 cells)

B cells develop in the bone marrow from hematopoietic stem cells (Fig. 40.4) and constitute about 20% of lymphocytes. Their major function is to **produce antibodies. Most of the B cells in circulation express immunoglobulins M and D (IgM and IgD), whereas B cells in tissues express IgG, IgE, and IgA.** Exposure of naïve B cells to foreign antigens leads to binding of antigens to antibodies resulting in maturation and differentiation of B cells to plasma cells and memory cells. Plasma cells produce antibodies and memory cells allow a rapid response if the same infection is encountered again. Human B cells can be identified mainly by markers like CD19, CD20 and CD21.

B. T Lymphocytes (T cells/CD3 cells)

T cells develop from hematopoietic stem cells in the bone marrow and mature in thymus (Fig. 40.4). T cells have different abilities to recognize antigens and are varied in their function. **APCs process peptide antigens and present them to T cells.**

T cells can be classified into "killer" or cytotoxic T cells ($CD8^+$ T cells), T-helper cells ($CD4^+$ T cells), and regulatory T cells.

1. **Cytotoxic T cells** destroy the infected cells and protect the body from certain bacteria, fungus and viruses that have the ability to survive and reproduce within the body's own cells. They also provide antitumor immunity.
2. **Helper T cells** assist B cells to produce antibodies and assist cytotoxic T cells in their attack on infected or transformed cells. Th1 cells secrete interleukin-2 (IL-2), interferon gamma, tumor necrosis factor-alpha (TNF-α), etc. which stimulate macrophages, natural killer (NK) cells and cytotoxic T cells. Th2 cells secrete IL-4, IL-5, IL-10, and IL-13 which stimulate B cells and synthesis of antibodies. Helper T cell subtypes include T_H-17 cells and Follicular Helper T cells (T_{FH} cells). T_H-17 cells produce IL-17 and are involved in antifungal and antibacterial immunity. T_{FH} cells assist B cells in the formation of germinal center.
3. **Regulatory T cells** are considered as negative regulators of the immune system as they suppress immune responses. It is essential for the optimal utilization of our immune system once the infection is cured. They secrete IL-10 and transforming

growth factor-beta (TGF-β) which suppress the function of T cell subtypes and dendritic cells. Suboptimal functioning of regulatory T cell ($T_{reg\ cells}$) can result in autoimmunity.

"Cluster of differentiation" (CD) antigens

They are molecules present on the surface of lymphocytes, which can be detected using monoclonal antibodies. The CD nomenclature was proposed and established in the 1st International Workshop and Conference on Human Leukocyte Antigens in 1982. Since then, more than 400 molecules have been designated. CD markers are used to identify lineage of cells. Various immune cells express different CD markers during different stages of development, which is used for their identification.

C. Natural killer cells (NK cells)

NK cells recognize and kill tumor cells and virus-infected cells. NK cells are derived from the bone marrow (Fig. 40.4) and are present in relatively low numbers in the human peripheral blood. NK cells constantly monitor other cells. The killing activity of NK cells depends on the balance of signals from activating receptors and inhibitory receptors on the NK cell surface. Healthy cells express MHC I receptors, which are recognized by inhibitory receptors on NK cells and prevent their killing. **Cancer cells and infected cells often lose their MHC I, leaving them vulnerable to NK cell killing**. They are particularly important in the defense against herpesvirus (herpes simplex virus [HSV], Epstein-Barr virus [EBV], chickenpox). NK cells destroy the targeted cells through antibody-dependent cell-mediated cytotoxicity (ADCC, see Chapter 43).

D. Neutrophils

Neutrophils (polymorphonuclear leukocytes/granulocytes) are the most abundant type of white blood cells. They develop from hematopoietic stem cells in the bone marrow. Their major role is to **ingest bacteria or fungi and kill them**. Neutrophils also employ both oxygen-dependent and oxygen-independent pathways to generate antimicrobial substances. Neutrophils play a larger role than macrophages in killing ingested microorganisms. The primary granules contain ***acid hydrolase*, lysozyme and *myeloperoxidase*,** and secondary granules contain lactoferrin and lysozyme. Antigen-antibody complexes (immune complexes) activate neutrophils releasing its contents causing destruction of pathogens. Neutrophils are responsible for the formation of "pus," and increase in the number of neutrophils in circulation is used as an indicator of acute bacterial infection.

Pus is composed primarily of live and dead neutrophils. The presence of pus generally reflects an acute inflammatory process. It may be localized, such as abscess, or disseminated through tissue planes. Infections which induce pus formation are called pyogenic infections.

E. Monocytes and macrophages

Both monocytes and macrophages are considered as mononuclear phagocytes in blood and tissues, respectively. Monocytes migrate to tissues and differentiate into macrophages. Macrophages are characterized by increase in size, increase in the number of organelles, and increase in the ability to phagocytose microorganisms. Macrophages are activated mainly by mediators of inflammation and cytokines.

Antigen-presenting cells (APCs)

APCs are leukocytes which are involved in the activation of T cells. Dendritic cells, macrophages and B cells are the major APCs. They express class II MHC molecules and present antigens to helper T cells.

T cells cannot be activated unless the antigen is presented along with MHC molecules by APCs. This is known as MHC restriction of T cells.

F. Eosinophils

Eosinophils constitute around 0%–6% of leukocytes. They contain cytoplasmic granules, which are rich in major basic protein and act as potent toxin for parasites. The eosinophils have Fc receptors for IgE and IgG. They can bind to the antibody-coated parasite and destroy it by antibody-dependent cell-mediated cytotoxicity (ADCC).

G. Basophils

Basophils constitute less than 1% of leukocytes and play a role in mediating allergy.

H. Dendritic cells

Dendritic cells arise from hematopoietic stem cells and are covered with long membranous extensions. They act as APCs and play a key role as a bridge between innate and adaptive immune systems. Four types of dendritic cells are known—Langerhans cells, inflammatory dendritic cells, classical (or conventional) dendritic cells, and plasmacytoid dendritic cells.

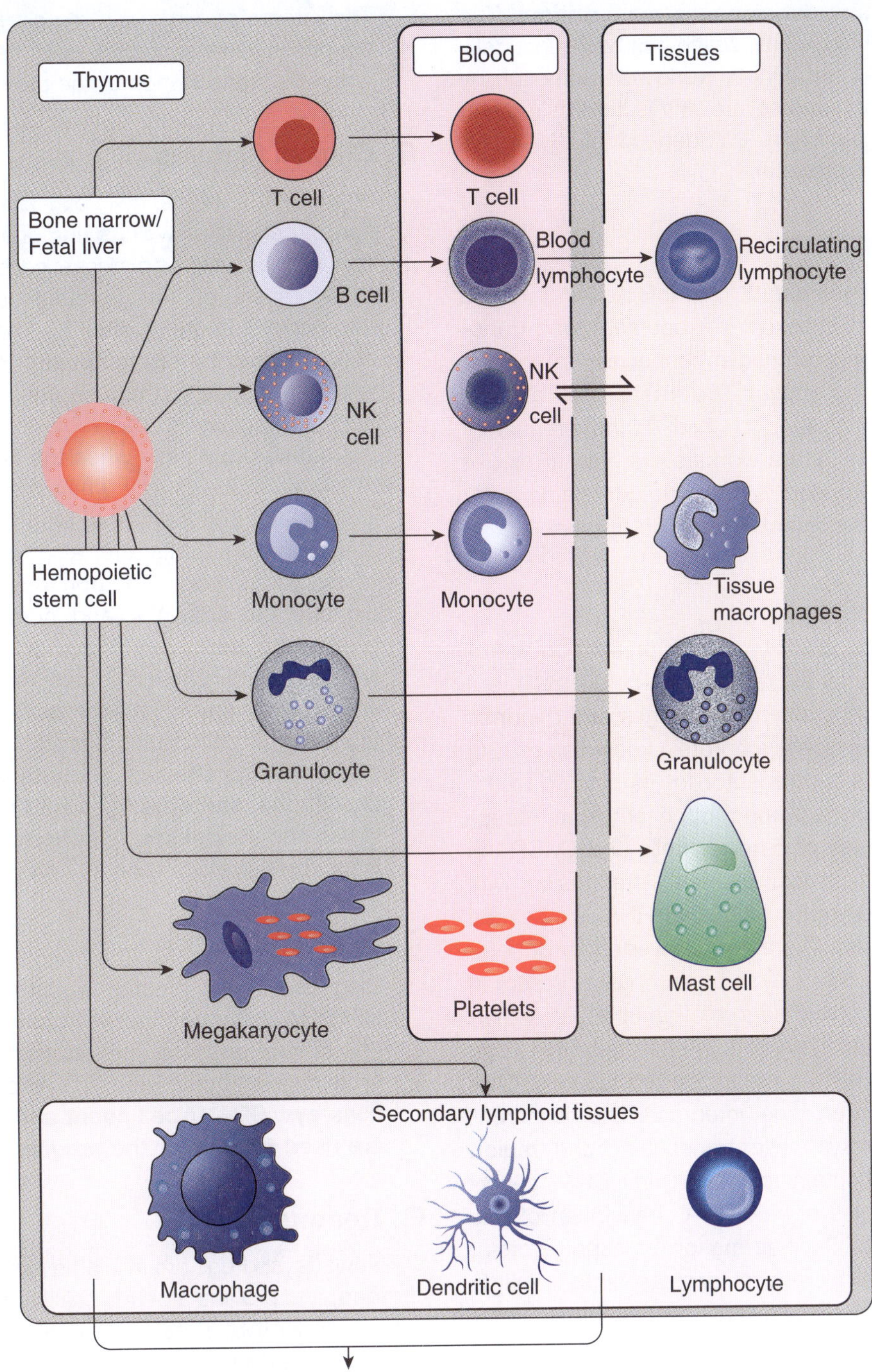

Figure 40.4 Key concepts of cells of the immune system.

V. COMPLEMENT SYSTEM

Complements are plasma proteins which assist the immune cells in eliminating the pathogens. They are synthesized in liver. They eliminate pathogens by cytolysis, inflammation, and by facilitating phagocytosis. Complements are activated either by classical pathway or by alternate pathway. These pathways are described in detail in chapter 48.

VI. HUMAN IMMUNODEFICIENCY VIRUS (HIV) AND ACQUIRED IMMUNODEFICIENCY SYNDROME (AIDS)

AIDS, which is the result of advanced stage disease due to HIV, is one of the common causes of secondary immune deficiency (Note: The most common cause

of secondary immune deficiency is malnutrition). The causative agent in AIDS is HIV, which has two genetically distinct forms—HIV-1 and HIV-2; worldwide, infection by HIV-1 is more common and severe. HIV is a retrovirus and replicates with the help of an RNA-dependent DNA polymerase (reverse transcriptase).

A. Transmission

The disease is transmitted from infected individuals through unprotected sexual intercourse, blood transfusion, and sharing of needles (especially in context of intravenous drug abuse). The disease can also be transmitted from mother to child during pregnancy and lactation. Healthcare workers are also at risk of infection when exposed to the virus-containing body fluids via mucosal contact, needle prick injury, etc.

B. Pathogenesis

The virus attaches to CD4 receptor present on leukocytes (predominantly expressed by T-helper cells, but also found in monocytes/macrophages and dendritic cells) via a viral surface glycoprotein known as gp120. gp120 also needs a coreceptor for its binding to the cells; this role is facilitated by chemokine receptors—CCR5 in case of T-helper cells and CXCR4 in case of monocytes/macrophages. After fusion with the host cell membrane, the viral RNA is released into the cytoplasm, where it is converted to double-stranded DNA by the reverse transcriptase present in the virus. This involves the formation of an RNA-DNA hybrid intermediate. The double-stranded DNA (also known as proviral DNA) integrates into the host DNA with the aid of an enzyme, integrase. The host transcription machinery, while transcribing the region of the genome containing the proviral DNA, generates multiple copies of viral RNA. Translation of this viral RNA results in the formation of a polyprotein, which is cleaved by a viral *protease* into multiple viral proteins including reverse transcriptase, integrase and *protease*, as well as several viral proteins that are required for the packaging of the virus. These proteins along with the viral RNA are packaged into new virions by incorporating the host cell membrane; the virions are released usually by cell lysis, resulting in the death of the host cells. Macrophages act as reservoirs of the virus during HIV infection.

The integrated virus can remain latent for prolonged periods within the infected cells; cell death ensues only when the transcription of the proviral DNA is active.

C. Clinical features

Since $CD4^+$ T-helper cells play a key role in the activation of cellular as well as humoral immunity, a reduction in the number of these cells leads to a drastic deterioration of the immune function leading to secondary immune deficiency. The full-blown disease characterized by **opportunistic infections** by otherwise innocuous pathogens (i.e., organisms that are non-pathogenic in case of an immunocompetent individual) occur only when the $CD4^+$ T cell numbers reach a nadir—usually when the $CD4^+$ T cell count is less than 200 cells/µl of blood. Apart from T cells, other immune cells are also affected. Production of cytokines from macrophages can lead to polyclonal B cell activation and increased immunoglobin levels, and malignant B cell proliferation leading to lymphomas. However, T cell mediated B cell activation is severely compromised due to reduction in the $CD4^+$ T cell count. Immune-surveillance against cancer is also affected, and **viral-mediated cancers like Kaposi sarcoma** (mediated by a type of herpesvirus) and **Hodgkin's lymphoma (due to Epstein-Barr virus**) are common in individuals with AIDS.

D. Diagnosis

Diagnosis of HIV infection is made by the detection of **HIV RNA using polymerase chain reaction**, or detection of **viral proteins (e.g., p24** which is a viral capsid protein) or antibodies against viral proteins using immunoassays. **$CD4^+$ T cell count and HIV viral RNA load are used to monitor the progression of disease.**

E. Treatment

Nowadays, drugs are available to reduce the viral RNA load, and provide a better quality of life in people living with HIV/AIDS. These drugs are used in combination, and target various steps in the viral replication cycle like membrane fusion, reverse transcription, integration into the host genome, cleavage of polyprotein, etc. Vaccine development for prevention of HIV infection still remains a challenge, primarily due to the high rate of mutations in the virus, attributed to a highly error-prone DNA synthesis by the reverse transcriptase.

VII. Medical Perspectives

Immunodeficiency Disorders

Immunodeficiency disorders occur due to a defect in immune system. It may be due to defect in the development of B cells and T cells, defect in the signaling required for activation of these cells or absence of complements or due to defect in effector functions of the immune system. It can be of two types—primary immunodeficiency and secondary immunodeficiency.

CD4/CD8 count ratio can be used as an index to measure the ratio of T-helper cells and cytotoxic T cells. the normal ratio is 2:1 and it is altered in certain immunodeficiency disorders.

Primary Immunodeficiency

It is due to immune dysfunction caused by genetic defects in the components of innate and adaptive immunity. It can affect both lymphoid and myeloid cells resulting in the loss of function in T cells/B cells and phagocytes. The affected individuals are susceptible to infections and mortality is higher at younger age. Examples for primary immunodeficiency disorders are **severe combined immunodeficiency (SCID), ataxia telangiectasia**, and Chediak-Higashi syndrome.

SCID is characterized by dysfunction of T cell and B cell. Deficiency of *adenosine deaminase* enzyme is one of the common causes of SCID. Recurrent infections, chronic diarrhea, pneumonia, and lesions of oral cavity are the most common clinical features. Reduction in the number of lymphocytes in blood is a feature of SCID.

Secondary Immunodeficiency

They are more common than primary immunodeficiency disorders and usually develop later in life. They occur due to acquired causes like HIV/AIDS and in patients who are on immunosuppressants, (e.g., glucocorticoids), and in recipients of radiotherapy. Diabetes mellitus can also cause immune dysfunction later in life.

VIII. Chapter Summary

- Immune system is an organized network in the body which protects the host from foreign particles like toxins and pathogens. The organs of immune system are classified into primary and secondary organs.
- Primary lymphoid organs are thymus and bone marrow which are involved in the maturation of lymphocytes.
- Secondary lymphoid organs are lymph nodes, spleen, and lymphatic tissues associated with the skin and the mucosal layers which allow interaction of matured lymphocytes with the antigen.
- The major immune cells are B cells, T cells, neutrophils, NK cells and APCs like macrophages and dendritic cells. B cells produce antibodies which eliminate pathogens by activating complement system and other effector mechanisms. Macrophages eliminate pathogens by phagocytosis and presenting them to T cells. Cytotoxic T cells destroy the infected cells by releasing cytokines, which are involved in apoptosis of infected cells (Fig. 40.4).
- Immunodeficiency disorders occur due to malfunction of the immune system. They may be primary if the defect is due to genetic causes or secondary if they are due to acquired causes.

Study Questions

Short-Answer Questions

40.1. What are antigen-presenting cells? Give two examples.

40.2. List various types of T cells and mention their functions.

40.3. Mention the clinical applications of CD4/CD8 counts.

40.4. Enumerate the types of lymphoid organs with suitable examples.

Long-Answer Questions

40.1. Describe various types of immune cells and their functions.

40.2. Describe the pathogenesis, clinical features and diagnosis of AIDS.

Multiple-Choice Questions

Choose the ONE best answer.

40.1. CD4/CD8 ratio is reduced in:
A. AIDS
B. Hashimoto's thyroiditis
C. Type 1 diabetes mellitus
D. Systemic lupus erythematosus

Correct answer = A.

40.2. Immune response is suppressed by:
A. Natural killer cells
B. Helper T cells
C. Regulatory T cells
D. B cells

Correct answer = C.

40.3. Which of the following cells is considered as an indicator of inflammation?

A. Neutrophil
B. Basophil
C. Eosinophil
D. Erythrocyte

Correct answer = A.

40.4. Which of the following have no role in immune system?

A. Thymus and bone marrow
B. Lymph nodes and spleen
C. GALT and MALT
D. Liver and kidneys

Correct answer = D.

40.5. Which of the following is not true regarding lymph node?

A. Lymph nodes have three layers.
B. Outer layer (cortex) contains B lymphocytes.
C. Middle layer (paracortex) contains T lymphocytes.
D. Inner layer (medulla) contains follicles which produce antibodies.

Correct answer = D.

40.6. What is not true of neutrophils?

A. Develop from progenitor cells in the bone marrow.
B. Major role is to kill bacteria and fungi.
C. Employ both oxygen dependent and oxygen independent pathways.
D. Pus cells are primarily composed of live and dead neutrophils.

Correct answer = A.

40.7. Which of the following is true for B cells?

A. B cells develop in the bone marrow and produce antibodies.
B. B cells in circulation express IgM and IgD.
C. B cells in tissues express IgG, IgE, and IgA.
D. All the above are true for B cells.

Correct answer = D.

40.8. Which of the following is not true?

A. Thymus produces mature T cells and their precursors.
B. Bone marrow have both the hematopoietic stem cells and progenitor cells.
C. Lymph nodes are bean-shaped.
D. Lymph nodes have both B and T Lymphocytes.

Correct answer = A.

40.9. For the normal functioning of the immune system the count of $CD4^+$ cells has to be:

A. At least 100 cells/µl of blood.
B. At least 200 cells/µl of blood.
C. At least 400 cells/µl of blood.
D. At least 600 cells/µl of blood.

Correct answer = D.

40.10. Which of the following is true about passive immunity?

A. The immunity causes reactions.
B. It develops immediately.
C. It lasts for few weeks or months.
D. All of the above.

Correct answer = C.

Multiple-Choice Questions (NExT/NEET PG)

Choose the ONE best answer.

40.1. Which of the following cells are involved in antigen presentation?

A. Eosinophils
B. Macrophages
C. T cells
D. Natural killer cells

Correct answer = B.

40.2. All of the following are causes of secondary immunodeficiency except:

A. Glucocorticoids
B. Radiation
C. HIV infection
D. Adenosine deaminase deficiency

Correct answer = D.

40.3. Which of the following immunity is obtained during a life time?

A. Innate immunity
B. Active immunity
C. Passive immunity
D. Both B and C

Correct answer = D.

40.4. Cells involved in innate immunity are:

A. Phagocytes
B. Macrophages
C. Natural killing cells
D. All of the above

Correct answer = D.

40.5. Name the nerve stimulator which causes inflammatory response.

A. Prostaglandins
B. Histamine
C. Bradykinin
D. Insulin

Correct answer = C.

Antigen and Antibody

41

Learning Objectives

At the end of this chapter, the student will be able to:

- Explain the types, structure, and function of immunoglobulins
- Classify and describe antigens
- Discuss the medical relevance of antigens and antibodies

I. OVERVIEW

Antigens and antibodies are important components of the immune system. When a foreign antigen such as microorganisms or toxins enter the body, our immune system responds by activation of B lymphocytes to secrete antibodies. Antigens bind to antibodies through epitopes which are immunologically active regions of the antigen. Antibodies which are constituents of B cells are membrane bound and those produced by plasma cells are secreted. Interaction of antigen with membrane-bound antibodies leads to proliferation of B cells and secreted antibodies mark the antigen for destruction. Most antigens contain different epitopes and a single B cell will produce monoclonal antibodies only, against one epitope of the antigen. Since many B cells are activated in response to multiple epitopes in the antigen, B cell responses to natural infections are usually polyclonal. The antigenic specificity of an antibody is determined by the organization and rearrangement of immunoglobulin genes during B cell maturation. Antigens and antibodies can be detected by several types of immunoassays which are useful for the detection of infections, and diagnosis of hormonal imbalance and cancers. Antibodies formed against self-antigens lead to autoimmune disease.

II. ANTIGENS

Antigens can be defined as molecules that contain distinct sites (epitopes) which are recognized by and interact with various components of the immune system, like B-cell and T-cell receptors.

A. Characteristics and classification of antigen

Antigens can be foreign substances or body's own proteins like autoantigens. They exhibit immunogenicity (induction of immune response) and immunological reactivity (react with antibodies or T cells). Antigens are classified based on immunogenicity, origin of antigen, source of antigen and on their ability to induce antibody formation (Fig. 41.1).

1. **Based on immunogenicity:** Antigens can be classified as (i) complete antigen/immunogen which are capable of inducing antibody formation or T-cell response and (ii) incomplete antigen/hapten which cannot induce the formation of antibodies or T-cell response.

2. **Based on the source of antigen:** Antigens can be exogenous antigens which enter the body through inhalation or ingestion. These antigens are presented to T helper cells through antigen-presenting cells (APCs) using class II major histocompatibility complex (MHC) molecules. Endogenous antigens are generated in the cell from the endogenous proteins. The abnormal peptides derived from viral infected cells and cancer cells are presented by MHC-I molecules to cytotoxic T-cells.

   Endogenous antigens can further be classified as:

   a. **Xenoantigens:** Antigens are from different species, like bacteria or viruses.

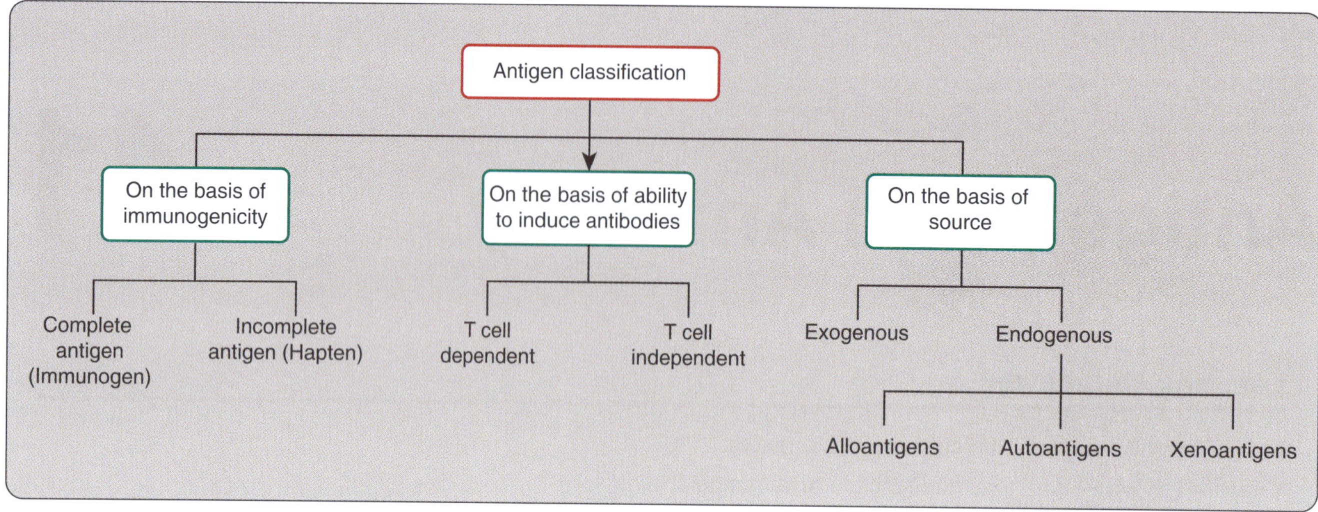

Figure 41.1 Classification of antigens.

b. **Alloantigens:** Antigens are from same species but of different individuals, like blood group antigens.

c. **Autoantigens:** Antigens from same individual, like lens protein or tumor cells.

Definitions

Immunogen	Immunogen is a substance that produces immune response
Antigen	Antigen refers to any molecule that can bind to the B cell receptors, or the T cell receptors when complexed in context of MHC complex.
Hapten	Haptens are low-molecular weight substances which cannot induce specific immune response by themselves, though they can bind to the receptors in the immune cells. They can be made immunogenic by combining with a suitable carrier.
Adjuvant	Adjuvants are substances which increase the immunogenicity of an antigen

3. **Based on the ability to induce antibodies:** Antigens can be

a. **T cell-dependent antigens:** B cells cannot respond to these antigens without costimulation by T cells. These are mostly protein antigens and are processed by APCs to be presented to T cells, which are then activated. Activated T-helper cells further lead to B cell activation and antibody production. Activated B and T cells also lead to formation of memory cells and immunological memory. They are rapidly cleared from the body and do not cause tolerance.

b. **T cell-independent antigens:** Antigens directly stimulate B cells to produce antibodies. These antigens are not processed by APCs and do not produce immunological memory. They remain in the body for long time and cause tolerance, e.g., bacterial lipopolysaccharides.

B. Determinants of immunogenicity

These are the properties which make a substance immunogenic.

- *Size*: Generally, larger molecules are more antigenic, and smaller molecules are less antigenic. Larger molecules are more immunogenic as they are readily phagocytosed and processed as compared to smaller molecules.
- *Nature of the molecule*: Proteins are more antigenic compared to carbohydrates/lipids/nucleic acids.
- *Foreignness*: Antigens from other individuals of the same species are less antigenic than those from other species.
- *Heterogeneity*: Heteropolymers are more antigenic compared to homopolymers.
- *Immunogen dosage and route of administration*: Optimal dosage and parenteral administration increase antigenicity.
- *Host genetic makeup*

III. IMMUNOGLOBULINS

Immunoglobulins, also known as antibodies, are glycoprotein molecules produced by plasma cells. They

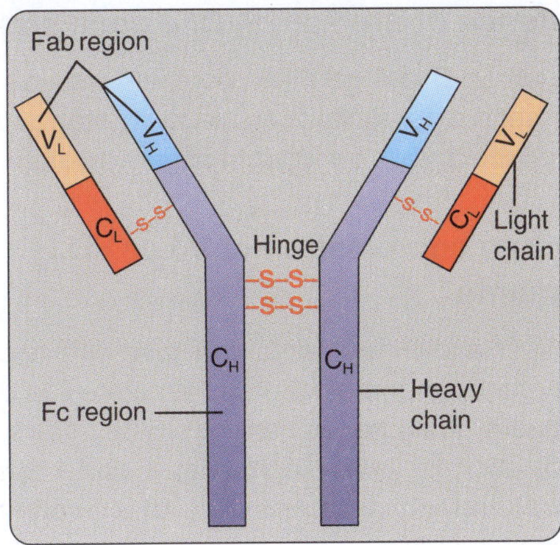

Figure 41.2 Structure of immunoglobulin. V_L = variable region of light chain; V_H = variable region of heavy chain; C_L = constant region of light chain; C_H = constant region of heavy chain.

participate in the humoral immune response by acting against specific antigens.

A. Structure of immunoglobulins

Immunoglobulins have four polypeptide chains (2 heavy chains and 2 light chains) linked with each other by disulfide bonds (Fig. 41.2). Heavy chains are designated as gamma (γ), alpha (α), mu (μ), delta (δ) and epsilon (ε). Light chains are designated as kappa (κ) and lambda (λ). A single antibody molecule will have only one type of light chain and one type of heavy chain. Immunoglobulins consists of:

1. **Variable region:** It is present in both the heavy and the light chains. The amino acid sequence in the variable region is highly different for antibodies that arise from different B cell clones. It provides specificity for antibodies to bind with antigens.

2. **Constant region:** For a single type of heavy chain or light chain, the constant region has the same amino acid sequences. It defines how a specific antibody will contribute to immune response.

3. **Hinge region:** It provides flexibility and facilitates binding of antibody with antigen.

4. **Fc region:** It comprises only of the heavy chains. It interacts with cell surface receptors called Fc receptors and some proteins of the complement system. (Fc stands for "fragment, crystallizable", because it crystallizes during cold storage.)

5. **Fab region:** It contains antigen-binding site and is composed of the variable regions of both heavy and light chains.

Types of Immunoglobulins (Isotype/Allotype)	Functions of Immunoglobulins
IgG (IgG1, IgG2, IgG3, IgG4)	Crosses placenta and provide immunity to fetus Complement activation, agglutination Responsible for Rh isoimmunization Enables secondary immune response
IgM	Enables primary immune response Complement activation, agglutination
IgA (IgA1, IgA2)	Secretory antibodies Agglutination
IgD	Facilitates immune response from B cells
IgE	Mediates allergy and anaphylaxis Causes mast cell degranulation and release of histamine

Figure 41.3 Classes of immunoglobulins and their functions.

B. Classification of immunoglobulins

Immunoglobulins can be classified into IgG, IgM, IgA, IgD and IgE based on type of heavy chain found in the molecule. Each class of immunoglobulins and their functions are given in Figure 41.3.

C. Immunoglobulin genes

The heavy and light chains of immunoglobulins are coded by immunoglobulin genes located on different chromosomes. In humans, the gene for heavy chain is located on chromosome 14, kappa (κ) light chain gene is on chromosome 2 and lambda (λ) light chain gene is on chromosome 22. The light chain genes contain V (variable region), J (joining region) and C (constant region) gene segments and the heavy chain gene contains V, D (diversity region), J and C gene segments. The VJ gene segment in light chain and VDJ gene segments in heavy chain code for variable regions in the light and heavy chains, respectively. The C gene segment codes for constant region in both the chains.

1. **Light chain gene segments:** In lambda chain, there are 30 V segments, 4 J segments and 7 C segments. In kappa chain, there are 40 V segments, 5 J segments and 1 C segment.

2. **Heavy chain gene segments:** In heavy chain, there are 39 V segments, 23 D segments, 6 J segments and series of C segments.

D. Immunoglobulin (Ig) gene rearrangement

The gene segments of immunoglobulin gene undergo rearrangement during maturation of B cells to form functional Ig gene. Rearrangement of variable region gene occurs in heavy chain followed by light chains. This results in the formation of immunocompetent B cells which can produce antibodies. Rearrangement of constant region genes in heavy chain results in class switching of immunoglobulins during primary and secondary immune responses.

1. **Variable-region gene rearrangement:** Rearrangement of V and J segments facilitates the expression of kappa and lambda light chains. After V-J joining, the rearranged light chain gene contains leader exon, introns, combined VJ segment and constant region. This then undergo transcription, posttranscriptional modifications like splicing and polyadenylation, and subsequent translation to form functional light chain. In heavy chain DNA, D and J segments join forming DJ segment which joins with V segment forming VDJ segments. The rearranged heavy chain DNA contains leader exon, introns, VDJ segment and series of constant region gene segments. The heavy chain gene transcribed along with Cm and Cd by *RNA polymerase* undergo splicing, polyadenylation and further translation to form m and d heavy chains which allow B cells to produce IgM and IgD antibodies. The variable-region gene rearrangement is facilitated by the recombination signal sequences (RSSs) surrounding VDJ segments, and proteins RAG-1/RAG-2 (recombination-activating genes) which allow joining of VJ and VDJ segments in light and heavy chains, respectively.
2. **Constant-region gene rearrangement (Class switching):** After antigenic stimulation of B cells, the expression of IgM changes to other isotypes (apart from IgM). This is known as class switching and involves rearrangement of heavy-chain DNA where the $V_HD_HJ_H$ unit can combine with any C_H gene segment. It involves switch regions (short repeats of GAGCT and TGGGG) located upstream of C segment, cytokines like IL-4 and an enzyme known as *activation-induced cytidine deaminase (AID)*. The details of class switching and its significance are described in the Chapter 43.

IV. ANTIBODY DIVERSITY

It is the ability of the immune system to produce antibodies against wide range of antigens. There are multiple ways through which antibody diversity can be generated.

A. Existence of multiple V, D, J gene segments

The human B cell germline contains multiple V, D, J gene segments, which can undergo rearrangements to produce a wide variety of antigen-binding regions.

B. Random combination of VJ or VDJ segments

The V, D and J segments can join with each other in any random combination to generate the diversity. The heavy chain genes can form 5,382 (39 × 23 × 6) combination of gene segments, λ chain forms 120 (30 × 4) and κ chain forms 200 (40 × 5) combination of gene segments.

C. Junctional flexibility

It refers to the flexibility of the region where VDJ segments join. During VDJ rearrangement, recombination signal sequences (RSSs) join accurately and coding sequences join without much accuracy. This leads to variation of amino acid sequences in the hypervariable region in immunoglobulins which causes increased antibody diversity.

D. Addition of P-nucleotides and N-nucleotides

During immunoglobulin gene rearrangement, enzymes coded by recombination-activating genes (RAG)-1/2 cleave DNA and generate hairpin-like structure. *Endonucleases* cleave hairpin, and complementary nucleotides are added generating palindromic sequences (P-nucleotides) at coding joint. *Terminal deoxynucleotidyl transferase (TdT)* catalyze the addition of nucleotides (N-nucleotides) at the site of V (D) J joining forming short amino acid sequences at variable region.

E. Somatic hypermutation

It is a process in which the point mutations in variable region gene segments alter the specificity of antibodies. It is called as hypermutation, since the frequency of mutation is 100,000 folds higher than the spontaneous mutation of other genes. Somatic hypermutation allows B cells to generate antibodies which can bind to pathogens in more precise manner. This process involves nucleotide substitution throughout the VJ or VDJ segments.

F. Association of heavy chains and light chains

Due to V(D)J rearrangement, 5,382 heavy chain genes and 320 light chain genes are generated. These heavy chains and light chains are associated randomly and generates increased antibody diversity.

V. ANTIGEN-ANTIBODY INTERACTION

It is the interaction between the epitope of the antigen and variable region of the antibodies, known as paratopes. Antigens are associated with the antibodies through hydrogen bonds, hydrophobic interactions, or van der Waals interactions. Various immunological assays used for diagnostic purposes are based on Ag-Ab interactions. (Epitope, also known as antigenic determinant, is the site of an antigen molecule that interacts with a specific site in the variable region of an antibody molecule, known as a paratope.)

A. Precipitation reactions

Interaction of antibodies with soluble antigens form precipitins. Precipitation reactions form the basis for immunodiffusion and immunoelectrophoresis.

1. **Immunodiffusion:** Here, antigens and antibodies diffuse toward each other in agar and form precipitate on interaction. Visible precipitation occurs in region where antigen and antibody concentrations are equal.
2. **Immunoelectrophoresis:** It is based on double immunodiffusion. Electrophoresis is done to separate the antigen mixture by charge. Troughs are cut into the agar gel parallel to the direction of the electric field, and antiserum is added to the troughs. Diffusion of antibody and antigen towards each other produce lines of precipitation where they meet. In laboratory, it is used for the diagnosis of conditions like amyloidosis and in infections like echinococcosis.

B. Agglutination reactions

Interaction of antibodies with particulate antigens form agglutinins. Excess antibodies inhibit agglutination reactions and this effect is known as prozone effect. Agglutination reactions are used for blood typing and in diagnosis of bacterial infections.

C. Immunoassays

Immunoassay measures the presence or concentration of analytes in a solution through the use of an antibody or an antigen. If radiolabeled antigens are used, it is known as radioimmunoassay. If enzymes are used, it is known as enzyme-linked immunosorbent assay (ELISA).

1. **Enzyme-linked immunosorbent assay:** In ELISA, the antigens are allowed to interact with antibody labeled to an enzyme. On addition of substrate, a colored product is formed. The enzymes commonly used are *alkaline phosphatase* and *horseradish peroxidase*. ELISA are of three types—indirect ELISA, competitive ELISA and sandwich ELISA. It is mainly used for the detection of HIV, HBsAg, hormones and tumor markers.

a. Types of ELISA

i. **Direct ELISA:** In direct ELISA, antigen (target) is immobilized to the plate (solid surface). An enzyme-labeled antibody against the antigen is added, which binds to the antigen. On addition of substrate, the enzyme linked to the antibody reacts with it to produce a visible signal that can be measured. The amount of signal is proportional to the amount of antigen (Fig. 41.4). *Horseradish peroxidase (HRP)* enzymes are commonly used to label antibodies, which acts on substrate such as tetramethylbenzidine (TMB) to produce color.

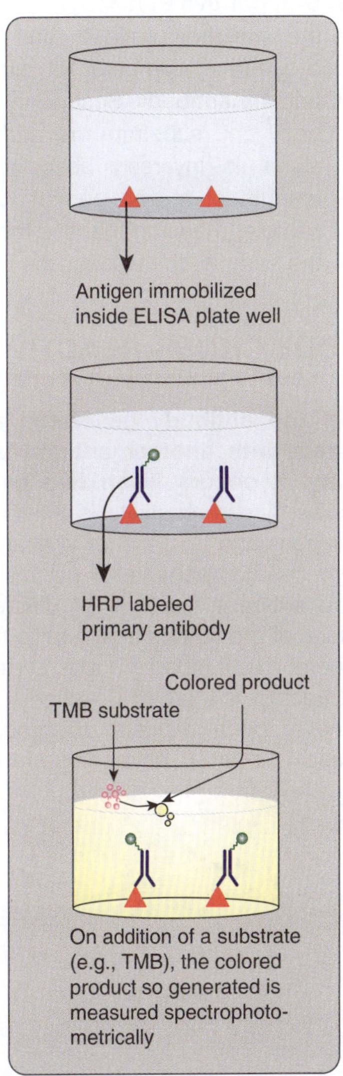

Figure 41.4 Direct ELISA. TMB = 3,3′,5,5′-Tetramethylbenzidine, HRP = Horseradish peroxidase.

ii. **Indirect ELISA:** In indirect ELISA, both a primary antibody (against the antigen, unlabeled) and a secondary antibody (against the primary antibody, labeled with enzyme) are used. The primary antibody binds to the antigen on the plate, and then the enzyme-labeled secondary antibody binds to the primary antibody. Addition of substrate produces a signal due to enzymatic reaction that can be measured.

iii. **Sandwich ELISA:** In sandwich ELISA, the antibody, known as capture antibody, is immobilized to the plate. Antigen of interest binds to the capture antibody. Then enzyme-labeled detection antibody is added. On addition of substrate, the enzyme reacts with its substrate to produce a visible signal that can be measured.

iv. **Competitive ELISA:** In competitive ELISA, the inhibitor antigen and the antigen of interest compete for binding to the primary antibody. Secondary labeled antibody and substrate are added. The final signal is inversely associated with the amount of the antigen of interest in the sample, meaning that the more antigen in the sample, the weaker the final signal.

VI. ANTIBODY CROSS-REACTIVITY

The ability of the antibody generated against one antigen to react with another antigen is known as cross-reactivity. It occurs when two antigens have similar epitopes. Cross-reactivity is exhibited by polysaccharide antigens and bacterial or viral antigens which are identical to cell components of normal host. A and B blood group antigens have minor differences in the terminal residues of the sugars attached to these surface proteins. In a person with blood group A, serum antibodies are produced against the group B antigen, although the person has never been exposed to group B antigen (and vice versa). The anti-B antibodies are produced by exposure to cross-reacting microbial antigens present on common intestinal bacteria.

Antibodies, produced against M protein of Streptococcus after throat infection are thought to cross-react with heart tissue, causing heart damage in rheumatic fever.

VII. FUNCTIONS OF ANTIBODIES

A. Opsonization

It is a process which uses molecules known as opsonins to mark the foreign antigen for eliminating them by phagocytosis (Fig. 41.5). It is mediated through Fc receptors which are present on phagocytic cells that bind to constant region of immunoglobulins. After phagocytosis, the foreign pathogens are destroyed through enzymatic digestion or by oxidative damage.

B. Complement activation

Complements are group of plasma proteins which can perforate cell membranes and are involved in direct or indirect elimination of pathogens, acting in conjunction with innate or adaptive immune systems. Binding of complements to pathogen-antibody complex destroys the pathogens through opsonization, perforation of the cell membrane of pathogens or by depositing the immune complexes in spleen for further clearance.

C. Neutralization

When a pathogen or toxin enters the body, antibodies are secreted which bind to pathogen or toxin and neutralize it (Fig. 41.5). Diphtheria antitoxin is a neutralizing antibody which neutralizes the diphtheria toxin.

D. Antibody-dependent cell-mediated cytotoxicity (ADCC)

Pathogens are coated by antibodies and killed by effector cells having receptors for Fc region of these antibodies.

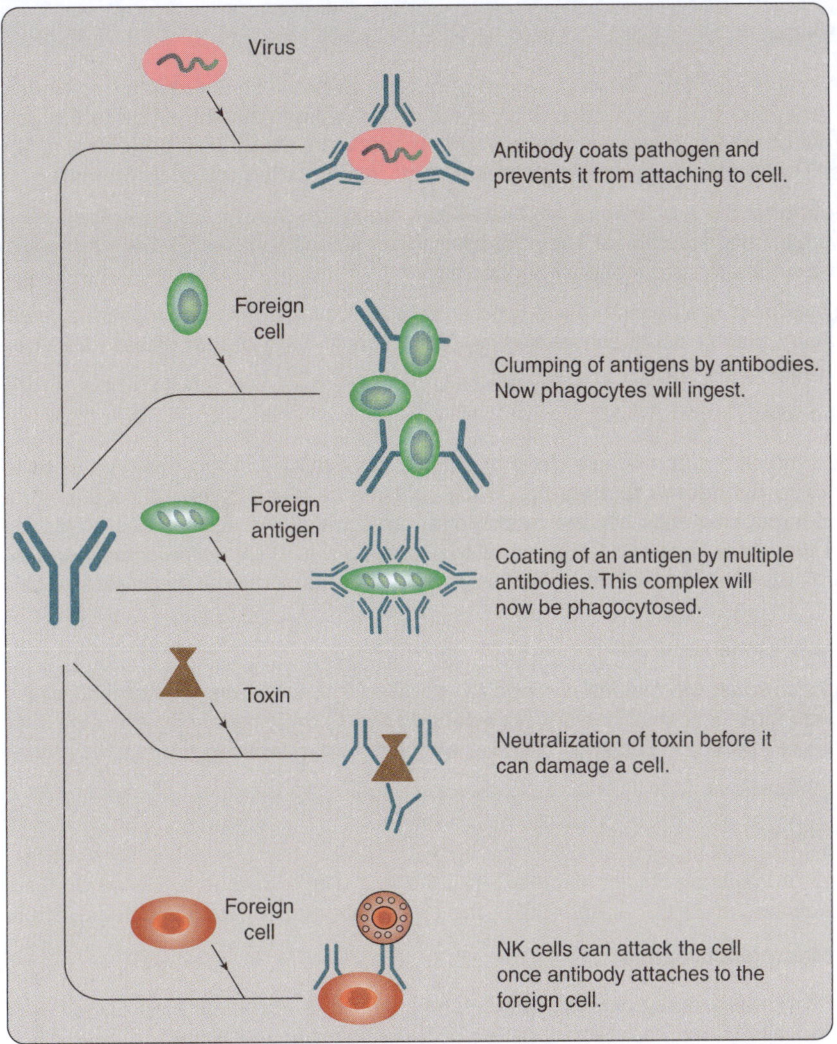

Figure 41.5 Key concepts of in vivo antigen-antibody interactions.

VIII. Medical Perspectives

A. Hypergammaglobulinemia

It is characterized by the presence of excess of immunoglobulin in blood. It occurs in infections, chronic liver disease, autoimmune disorders like rheumatoid arthritis, and in multiple myeloma. It is detected by the presence of increased gamma band in serum protein electrophoresis.

B. Autoimmune Disorders

These disorders occur due to the abnormal response of immune system against the host components. It is due to antibodies or T cells causing damage to host cell/organs/antigens. The possible mechanisms of autoimmunity include release of hidden antigens, cross-reacting antibodies or altered expression of MHC molecule. Autoimmune disorders may be organ specific like **type 1 diabetes mellitus** (antibodies, as well as T cells against pancreatic beta cells), **myasthenia gravis** (antibodies against acetylcholine receptors), **pernicious anemia** (antibodies against intrinsic factor or parietal cells of stomach) or it may be organ non-specific like **systemic lupus erythematosus (SLE), rheumatoid arthritis and multiple sclerosis** in which antibodies affect multiple organs.

1. **Systemic lupus erythematosus:** SLE is an autoimmune disease in which organs and tissues are damaged by autoantibodies and immune complexes. Most of the patients are women of childbearing age. Clinical features include fever, weakness, rashes and a characteristic butterfly rash on cheeks. Antinuclear and anti-ds-DNA antibodies are commonly detected.

C. Polyclonal Antibodies

These are antibodies synthesized from different clones of plasma cells, which are capable of binding to multiple epitopes of same or different antigens. They are produced by injecting the antigen several times to the animal which forms high titers of antibodies against the same antigen. Compared to monoclonal antibodies, polyclonal antibodies are inexpensive and easy to produce. The disadvantages are high cross-reactivity and batch-to-batch variability.

1. **Diagnostic applications:** Since polyclonal antibodies can recognize several epitopes, they are used in the diagnostic assays and hemagglutination reactions. They are used in immunoassays like ELISA, Western blotting, immunohistochemistry and flow cytometry as secondary antibodies.
2. **Therapeutic applications:** The plasma of a person who has recovered from an infectious illness contains antibodies to the infectious agent. Such plasma can be transfused into an ill victim of that disease as a treatment. This is known as **convalescent plasma therapy**.

D. Monoclonal Antibodies

These are antibodies synthesized from single clone of plasma cells which bind to one type of epitope. Monoclonal antibodies are synthesized by **hybridoma technology**. In this, plasma cells and myeloma cells are fused to generate hybrid cells (hybridoma) and hybridoma cells secrete monoclonal antibodies. The advantages of monoclonal antibodies are: high specificity since they bind to specific epitope of the antigen, minimal cross-reactivity and better quantification of proteins. Its disadvantages—the production is expensive and time consuming which requires cell culture and purification techniques.

1. **Applications of monoclonal antibodies**
   a. ***Diagnostic applications***: Monoclonal antibodies are used in several diagnostic techniques like ELISA, Western blotting and immunohistochemistry as primary antibody due to their specificity. It is also useful for detecting pregnancy, diagnosis of infections, detection of tumor antigens and estimating drug levels in the blood.
   b. ***Therapeutic applications***: Monoclonal antibodies are used in the treatment of cancer and rheumatoid arthritis.

E. Bence Jones Proteins

These are light chains of immunoglobulins excreted in urine in multiple myeloma. Bence Jones proteins are detected by heat test. They precipitate at 40–60°C. After 60°C, the precipitate disappears. On cooling, the precipitate reappears.

F. Waldenström's Macroglobulinemia

It is a malignancy of lymphoplasmacytoid cells that secrete IgM. It is associated with lymphadenopathy and hepatosplenomegaly, and the major clinical manifestations are of hyperviscosity syndrome.

IX. Chapter Summary

- Antigens which induce immune response can be a foreign substance or body's own proteins like autoantigens. Antigens are classified based on immunogenicity, source of antigen and on their ability to induce antibody formation. Immunogenicity of a substance depends on its size, nature, processing and presentation of an antigen, host genetic makeup, foreignness, and heterogeneity.
- Antibodies, produced by plasma cells, can coat the pathogens and prevent their attachment to cell or cause their clumping or make them vulnerable to phagocytosis or destruction by NK cells. Antibodies can also neutralize toxins before they can damage the cell.
- Immunoglobulins are of five types—IgG, IgA, IgM, IgD, and IgE. IgM is involved in primary immune response and IgG is involved in secondary immune response. The heavy and light chains of immunoglobulins are coded by immunoglobulin genes located on different chromosomes. The gene segments of immunoglobulin gene undergo rearrangement during maturation of B cells to form functional Ig gene.
- Antibody diversity is facilitated by the presence of multiple V, D, J segments, their random combination, junctional flexibility, addition of P and N nucleotides and somatic hypermutation.
- Antigens are associated with the antibodies through hydrogen bonds, hydrophobic interactions or van der Waals interactions. Various immunological assays used for diagnostic purposes are based on Ag-Ab interactions.
- Autoimmune diseases occur due to reaction of antibodies and T cells against the host antigens, cells or organs. Monoclonal antibodies are used in the diagnosis and treatment of cancers and viral infections.

Study Questions

Short-Answer Questions

41.1. Enumerate various mechanisms through which antibodies destroy the pathogens.
41.2. What are Bence Jones proteins? Mention their properties and clinical significance.
41.3. Mention the types and applications of ELISA.
41.4. What are the mechanisms of antibody diversity?
41.5. What is antibody cross-reactivity?

Long-Answer Questions

41.1. Classify antigens with suitable examples. Explain the factors affecting the immunogenicity of a substance.
41.2. Explain the structure of immunoglobulin with suitable illustrations. List various types of immunoglobulins and mention their functions.

Multiple-Choice Questions

Choose the ONE best answer.

41.1. Which of the following enzymes is used in ELISA?
A. Alkaline phosphatase
B. Aspartate transaminase
C. Alanine transaminase
D. Acid phosphatase
Correct answer = A.

41.2. Monoclonal antibodies are synthesized using:
A. DNA fingerprinting
B. Immunoelectrophoresis
C. Hybridoma technology
D. Fluorescence in situ hybridization
Correct answer = C.

41.3. Immunoglobulin which enables primary immune response is:
A. IgG
B. IgM
C. IgE
D. IgA
Correct answer = B.

41.4. Antigens are classified based upon:
A. Immunogenicity only.
B. Source of antigen only.
C. T cell dependency only.
D. All of the above.
Correct answer = D.

41.5. What is not true of the following?
A. Antigen can bind to either B or T cell receptors.
B. Hapten can produce the immune response by themselves.
C. Adjuvant increase the immunogenicity of the antigen.
D. Immunogen produce the immune response.
Correct answer = B.

41.6. Which best determine immunogenicity of antigen?
A. Size, chemical nature, foreign nature, and heterogeneity of antigen but not the genetic makeup of the host.
B. Size, chemical nature, foreign nature, and heterogeneity of antigen along with the genetic makeup of the host.
C. Size, chemical nature, foreign nature, and homogeneity of antigen along with the genetic makeup of the host.
D. Chemical nature, foreign nature, and heterogeneity of antigen along with the genetic makeup of the host. Immunogenicity is independent of the antigen size.
Correct answer = B.

41.7. Which of the following is true?
A. Variable, constant, Fab and not Fc regions are present on both the light and heavy chains of immunoglobulins.
B. Variable, constant, Fc and Fab regions are present on both the light and heavy chains of immunoglobulins.
C. Variable, Fc, and Fab but not the constant regions are present on both the light and heavy chains of immunoglobulins.
D. Variable, constant, Fc and not Fab regions are present on both the light and heavy chains of immunoglobulins.
Correct answer = A.

41.8. Human B cells have multisegments of:
A. V gene only.
B. V and D genes
C. D and J genes.
D. V, D, and J genes.
Correct answer = D.

41.9. Prozone effect means:
A. Formation of precipitins due to interaction of antigen with antibodies.
B. Double immunodiffusion
C. Inhibition of agglutination by excess of antibodies.
D. None of the above.
Correct answer = C.

41.10. Following is true of Bence Jones proteins:
A. Present in blood but not excreted in urine.
B. Get excreted in urine and are heat resistant.
C. Get excreted in urine and get precipitated at 40–60°C and precipitates are heat resistant.

D. Get excreted in urine precipitates formed between 40–60°C, disappear after 60°C.

Correct answer = D.

Multiple-Choice Questions (NExT/NEET PG)

Choose the ONE best answer.

41.1. Blood group antigens are examples of:

A. Alloantigens
B. Xenoantigens
C. Heterophile antigens
D. Autoantigens

Correct answer = A.

41.2. Which of the following is not true about polyclonal antibodies?

A. Bind to several epitopes
B. Therapeutic applications are limited
C. Highly specific
D. Cross reactivity is more

Correct answer = C.

41.3. True of light chain gene segments:

A. In lambda chain there are 30 V, 4 J, and 7 C segments.
B. In lambda chain there are 40 V, 5 J, and 1 C segments.
C. In lambda chain there are 30 V, 4 J, 23 D, and 7 C segments.
D. In lambda chain there are 20 V, 2 J, and 1 C segments.

Correct answer = A.

41.4. Which is true for human?

A. Gene for heavy chain of immunoglobulins is located on chromosome 14, kappa and lambda regions of light chain on chromosome 2 and 22 respectively.
B. Gene for heavy chain of immunoglobulins is located on chromosome 14, kappa and lambda regions of light chain on chromosome 22 and 2 respectively.
C. Gene for heavy chain of immunoglobulins is located on chromosome 24, kappa and lambda regions of light chain on chromosome 2 and 22 respectively.
D. Gene for heavy chain of immunoglobulins is located on chromosome 24, kappa and lambda regions of light chain on chromosome 22 and 2 respectively.

Correct answer = A.

41.5. Class switching is a function of:

A. Kappa region of light chain.
B. Lambda region of light chain.
C. Heavy chain.
D. None of the above.

Correct answer = C.

Innate and Adaptive Immunity

42

Learning Objectives

At the end of this chapter, the student will be able to:

- Explain the functions and mechanisms of innate immunity
- Discuss the mediators of adaptive immunity
- Explain the concept of self and non-self

I. OVERVIEW OF INNATE AND ADAPTIVE IMMUNITY

Innate immunity is a type of immunity which is involved in the prevention of the pathogen from entering the body before the occurrence of infection. It provides the first and second lines of defense mechanism (physical and chemical barriers being the first line and non-specific mechanisms such as phagocytes constitute the second line), and it eliminates the pathogen within few minutes to few hours after entering the body. It is mediated by the anatomical barriers and phagocytes. Even though it is nonspecific, innate immunity is considered as natural since it is present from birth. In contrast to innate immunity, **adaptive immunity, also known as acquired immunity** is developed after the host is exposed to foreign antigen. It is considered as the third line of defense mechanism in eliminating the pathogens. It is mediated by lymphocytes.

Most foreign antigens present at least few structural patterns that allow the body to differentiate between self and non-self. These are called pathogen-associated molecular patterns or PAMPs. They are recognized by pattern recognition receptors (PRRs). Many cells, like leukocytes, contain certain PRRs such as Toll-like receptors (TLRs), C-type lectin receptors. Further, the recognition is helped by unique specific T cell receptors (TCRs) and B cell receptors. Several such molecules help in identifying self from non-self in the body.

II. FUNCTIONS OF INNATE IMMUNE SYSTEM

Innate immune system forms various barriers in the body, both physical and chemical which prevent the entry of pathogens. It allows the immune cells to organize around the pathogens and eliminate them from tissues and organs through mediators like cytokines or through activation of complement proteins. It also presents the foreign antigen to immune cells and activates the adaptive immune system leading to destruction of the harmful microorganisms. Some components of innate immunity, such as skin, epithelium and complement are always active, while others like neutrophils and macrophages provide immediate protection if first lines of defense are breached.

A. Barriers of innate immune system

The host is protected from entry of pathogens through physical and chemical barriers.

1. **Physical barriers:** The physical barriers include skin**, mucosal lining of gastrointestinal tract and respiratory tract**, eyelashes and other body **hairs**. **Epidermis** (outer layer of the skin) contains keratinocytes, which are capable of producing peptides and cytokines which destroy the microorganisms. Sebaceous glands secrete fatty acids and create an acidic environment inhibiting the growth of pathogens. Desquamation of epithelial layer of skin, presence of sweat, peristalsis in gastrointestinal tract and **ciliary movements of respiratory tract** remove pathogens and prevent them from entering the body. Body fluids like **tears,** secreted by lacrimal glands located under the upper eyelid wash out pathogens entering the eyes. **Saliva** secreted from salivary glands washes over teeth and mouth and traps the infectious agents and removes them. Flushing mechanism of urine prevents colonization of pathogens.

The blood-brain barrier prevents the entry of pathogens into central nervous system from blood. It is formed from the endothelial cells of the capillary wall and astrocytes surrounding the endothelial cells. Defect in blood-brain barrier in trauma, neurological disorders and epilepsy allows pathogens to enter brain resulting in brain infection.

> **Barriers of innate immunity**
>
> **Anatomical barriers:** Skin and mucous membranes
>
> **Physiological barriers:** Temperature, chemical mediators (interferons, lysozymes, etc.), low pH
>
> **Phagocytic barriers:** Immune cells (monocytes, macrophages)
>
> **Inflammatory barriers:** Sequence of events in response to invading pathogens or tissue injury which leads to accumulation of fluid and phagocytic cells for clearing the pathogen.

2. **Chemical barriers:** The presence of chemicals like enzymes and proteins in the body fluids play a role in the removal of pathogens. **Lysozyme**, a hydrolytic enzyme which is present in sweat, tears and saliva can degrade the pathogens by cleaving their cell wall. Gastric juice contains **hydrochloric acid (HCl)**, mucus and enzymes which kills most of the microorganisms entering through food. Bacterial flora present in the gut also prevents the growth of harmful pathogens. The presence of fatty acids, bile acids and lactoferrin are essential for the maintenance of normal gut flora and for preventing the entry of pathogens through mucosa.

   Any breach in the physical barriers due to trauma, animal bite or insect bite allows pathogens to enter the body, leading to infections which can spread to other parts of the body if not treated properly. Certain bacteria like Helicobacter pylori neutralize the HCl secreted by stomach and form colonies in stomach resulting in gastritis and gastric ulcer.

B. Immune cells

Immune cells form the second line of defense of innate immunity. It They consist mainly of phagocytes comprising neutrophils, monocytes and macrophages (see Chapter 40). Phagocytes eliminate the pathogens through phagocytosis.

C. Phagocytosis

It refers to the ingestion of a pathogen by phagocytic cells. Here the phagocytic cell utilizes the protrusions of its cell membrane known as pseudopodia to engulf the bacteria and other pathogens (Fig. 42.1).

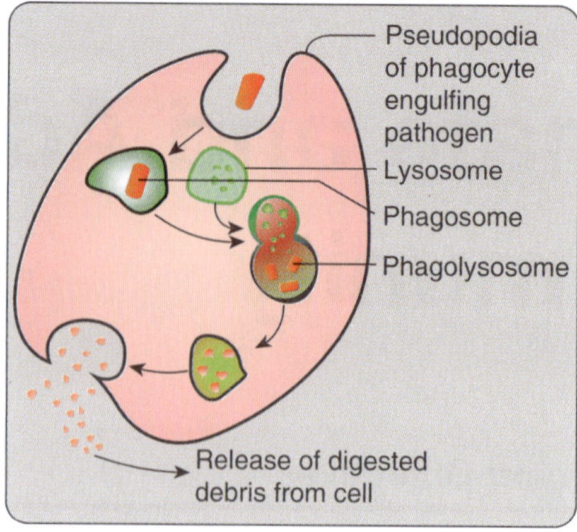

Figure 42.1 Mechanism of phagocytosis.

1. **Steps of phagocytosis**

   a. **Migration and chemotaxis of phagocytes:** It refers to the movement of phagocytes from bloodstream to the site of infection characterized by the presence of damaged tissues and pathogens. **Neutrophils** are recruited first at the site of infection, and monocytes are converted to **macrophages** as the infection progresses. Chemotaxis is influenced by variety of inflammatory molecules like cytokines and cell adhesion molecules. Phagocytes bind to the endothelial cells and migrate to the site of infection through junctions in between the endothelial cells. After crossing the endothelium, phagocytes move along the concentration gradient of chemokines, towards the site of inflammation.

   b. **Attachment to the surface of pathogens:** Phagocytes after chemotaxis, attach to the surface of pathogens through **toll-like receptors (TLRs) and/or opsonins**.

      i. **Toll-like receptors:** TLRs are protein receptors, which are present on the membranes of phagocytes, antigen-presenting cells (APCs) and lymphocytes. TLRs usually attach to the lipopolysaccharides or peptidoglycans in the cell wall of bacteria and nucleic acids of viruses leading to its activation. After activation, TLRs activate various signaling pathways leading to upregulation or downregulation of genes resulting in the production of cytokines, which initiate inflammatory responses causing either phagocytosis or lysis of the pathogen.

      ii. **Opsonins:** Opsonins are molecules which mark the pathogens for phagocytosis. Opsonins include circulating proteins like collectins, antibodies, and complements. The process of opsonization has been described in the Chapter 43.

c. **Formation of phagosome:** After phagocytes are attached to pathogens, the pseudopodia of phagocytes engulf the pathogen and it is internalized resulting in the formation of phagosome.

d. **Formation of phagolysosome:** Phagosomes combine with the lysosomes inside the cell leading to the formation of phagolysosome, which is acidic in nature and it results in the activation of enzymes involved in the degradation of pathogens.

e. **Destruction of pathogen:** Destruction of pathogens occurs through oxygen-dependent or oxygen-independent mechanisms. Oxygen-dependent mechanism involves activation of enzymes like *NADPH oxidase*, *myeloperoxidase* and generation of reactive oxygen species (like superoxide and hydroxyl radicals) and reactive nitrogen species (like NO and peroxynitrite). The oxygen-independent mechanism includes release of enzymes and proteins like lysozyme, lactoferrin and defensins.

2. **Evasion of phagocytosis by pathogens:** Many pathogens can evade phagocytosis with the help of structural components or through other mechanisms. Microbes like Staphylococcus and Listeria can kill phagocytes by secreting a protein known as leukocidin or cytolysins, which make transmembrane pores, respectively and Streptococcus inhibits its attachment to TLRs through capsules and M proteins.

III. SOLUBLE COMPONENTS OF INNATE IMMUNITY

Apart from phagocytes, there are certain soluble factors which can mediate innate immunity. These soluble factors are proteins which are either expressed on innate immune cells or present in circulation. Some components like interferons (IFNs) and defensins are synthesized at the site where tissue is damaged and infected, whereas other components like complements and acute phase proteins like C-reactive protein (CRP) are generated at distant sites and reach the site of infection through blood.

A. Acute phase proteins

Acute phase proteins are plasma proteins whose concentration increases or decreases during acute phase of illness. **CRP** is one of the acute phase proteins which mediates innate immunity. It belongs to family of pentraxins. It is synthesized in liver and induced by **cytokines like tumor necrosis factor-alpha (TNF-α), interleukin-1 and -6.** CRP binds to polysaccharides or phosphorylcholine on the surface of pathogens and promotes phagocytosis through complement activation.

B. Interferons

These are glycoproteins secreted by various immune cells in response to infection by virus or tumors. They are also released by virus-infected cells which induce an antiviral response in other cells in the locality. These proteins were named as interferons as they interfere with the replication of virus. They play a role in the regulation of immune response. There are three main types of IFNs.

1. **Interferon-α (IFN-α):** It is secreted by macrophages. It plays a role in innate immunity. The main function of IFN-α is to increase presentation of antigens and the expression of major histocompatibility complex (MHC) molecules.

2. **Interferon-β (IFN-β):** It is secreted by fibroblasts. Along with IFN-α, it mediates innate immunity. It is involved in the expression of class I MHC molecules. Interferon alpha and beta responses also help in mitigation of viral infections by inhibition of viral transcription and translation, and destruction of viral RNAs.

3. **Interferon-γ (IFN-γ):** It is secreted by T helper cells/natural killer (NK) cells. It is mainly involved in adaptive immune response, where it differentiates T helper cells and activates NK cells. It is responsible for delayed-type hypersensitivity reactions. IFN-γ increases the influx of macrophages into the graft and mediates graft rejection.

C. Cytokines

Cytokines are proteins which play a role in the development and regulation of immune response by influencing the interaction between inflammatory cells, lymphoid cells and hematopoietic cells. They are secreted by white blood cells. It includes chemokines that induce migration of leukocytes to the site of infection and interleukins (ILs) which assist the communication between leukocytes, tumor necrosis factor (TNF) and IFNs. They are secreted under the influence of various stimuli like stress and inflammation. Cytokines like IL-1, IL-6, IL-12 and TNF-α which are involved in innate immunity are secreted mainly by macrophages, and cytokines like IL-2, IL-4 and IL-5 which are involved in adaptive immunity are secreted by T cells (T_H1 cells secrete IL-2 and IFN-γ, T_H2 cells secrete IL-4 and IL-5).

1. **Mechanism of action:** Cytokines bind to the receptors of target cells and activate/inhibit the signal transduction pathways leading to the alteration in the expression of various genes of the target cells. Cytokines exert autocrine, paracrine or endocrine effect depending on the site of target cells. The actions of cytokines are either synergistic or antagonistic with other cytokines.

2. **Functions of interleukins:**
   - IL-1 and TNF-α are involved in inducing acute phase proteins and mediating inflammation.
   - IL-6 is involved in the proliferation of B cells and secretion of antibodies.
   - IL-2 induces proliferation of T cells and promotes activation of NK cells.
   - IL-5 is involved in the generation and activation of eosinophils.
   - IL-10 is an anti-inflammatory cytokine, and its levels are reduced during inflammation.

IV. ADAPTIVE IMMUNITY

Adaptive immunity develops during the lifetime of the host. It differs from innate immunity as it is highly specific and associated with the development of immunological memory. Immunological memory is the ability of immune system to respond rapidly to pathogens that have been encountered previously either through infection or vaccination.

A. Types of adaptive immunity

Adaptive immunity can be classified as **active and passive immunity** which are acquired either naturally or artificially. **Active immunity is termed as natural if it is acquired through contact with the pathogen (infections) and artificial if it is acquired through vaccines** which are developed from dead or attenuated pathogens or their components. **Passive immunity is obtained through natural route by transfer of antibodies from mother to fetus through placenta (e.g., IgG) and IgA by breast milk or through artificial route by administration of gamma globulins.** Adaptive immunity can also be classified as humoral immunity if it is mediated by B cells through production of antibodies or cellular immunity if it is mediated by T cells.

B. Mediators of adaptive immunity

Adaptive immunity is mediated by B cells, T cells, antibodies and APCs.

1. **Antigen-presenting cells:** Cells which trap antigens, process them and present them to T cells are known as APCs. They are involved in the storage and transport of antigens to peripheral lymphoid tissues. These cells also provide costimulatory signals to T cells leading to their activation. B cells, dendritic cells and macrophages are the major APCs which can activate a naïve T cell. APCs are classified into two types—professional and non-professional APCs. Dendritic cells, macrophages and B cells are professional APCs. Among the **professional APCs, dendritic cells and B cells constitutively express class II MHC,** while resting macrophages express them only minimally. Activated macrophages express high levels of major histocompatibility complex II (MHC II) molecules. Only dendritic cells constitutively express costimulatory molecules. B cells and macrophages have to be activated for expression of costimulatory molecules. Because of this, dendritic cells are considered the best APCs. **Non-professional APCs (e.g., fibroblasts, glial cells, pancreatic beta cells, etc.) have to be induced to express class II MHC molecules**. They are involved in antigen presentation for short duration. Dendritic cells are characterized by the presence of long protrusions known as dendrites and are distributed in circulation, skin, thymus and lymphatic vessels. Dendritic cells and macrophages ingest the pathogen, degrade it, transport it to lymph node and present it to T cells. B cells take up the antigen by endocytosis and present it to T cells.

2. **Antibodies:** Antibodies can degrade pathogens through opsonization, neutralization and activation of complements (see Chapter 41).

3. **T cell:** T cells provide the cell-mediated immunity. They are described in detail in Chapter 43.

V. MAJOR HISTOCOMPATIBILITY COMPLEX

Major histocompatibility complex is a complex of genes that codes for proteins on the cell surfaces that help the T cells to identify the pathogens. The TCR recognizes the antigens presented with MHC and activates the T cell resulting ultimately in the elimination of the pathogen. In humans it is found on chromosome 6 and is known as human leukocyte antigen (HLA) complex. It participates in cellular and humoral immunity and is responsible for differentiating self from non-self antigens. There are three types of MHC molecules.

A. Class I MHC molecules

These are **expressed on all nucleated cells** and are involved in the presentation of endogenous antigens to cytotoxic T cells. They contain an alpha (α) chain which is linked noncovalently to β_2 microglobulin, and this facilitates the expression of class I MHC molecules on the cell membrane. The α chain is connected to the cell membrane and contains three domains (α1, α2 and α3). The α3 segment serves as binding site for CD8 molecule of cytotoxic T cells. The peptide-binding groove is formed between α1 and α2 domains and it binds to peptide with 8–10 amino acids.

B. Class II MHC molecules

These are expressed on APCs and are involved in presentation of exogenous antigen to helper T cells. It contains alpha and beta chains which anchor the MHC to cell membrane. Unlike class I MHC molecule, class II MHC molecules lacks β_2 microglobulin. The peptide binding groove is formed between α1 and β1 domains and it binds to peptide with 15–18 amino acids.

C. Class III MHC molecules

They are not related to class I and II MHC molecules and are not involved in antigen presentation. Class III MHC genes code for components of complements and heat shock proteins and are involved in inflammation.

D. Regulation of MHC expression

Expression of class I and class II MHC molecules are regulated by certain transcription factors and cytokines. Among cytokines, IFN-α and TNF-α upregulate the expression of class I MHC molecules. IFN-γ increases the expression of class II MHC molecules by activating the transcriptional factors.

Reduced expression of class I MHC on the surface of cells leads to the evasion of immune response by certain viruses. Hepatitis B virus, adenovirus and cytomegalovirus reduce the expression of class I MHC molecules by binding to β_2 microglobulin, reducing the availability of transcription factors and the components required for assembly of class I MHC molecules.

MHC restriction

It is the property of T cells to recognize the antigens presented only by self MHC molecules. CD8$^+$ T cells are class I MHC restricted, while CD4$^+$ T cells are class II MHC restricted.

VI. ANTIGEN PROCESSING

Endogenous and exogenous antigens are presented to the T-cells using different pathways.

A. Cytosolic pathway

This pathway is used for the processing endogenous antigens (e.g., peptides derived from viral proteins and cancer antigens) and then they are presented on the cell membrane with class I MHC molecule.

In cytosolic pathway, endogenous antigens meant for degradation are attached to ubiquitin. The **ubiquitin-peptide complex** is then degraded to smaller peptides by an ATP-dependent cellular machinery known as proteasome. Proteasome is induced by IFN-γ and TNF-α. The degraded smaller peptides are transported to rough endoplasmic reticulum (RER) through a carrier known as transporter associated with antigen processing (TAP). TAP is a heterodimer, which is attached to RER. TAP contains ATP binding unit, and energy from ATP hydrolysis is used for the transport of peptide to RER. In RER, peptides interact with class I MHC. The interaction of MHC-I molecule with beta-2 microglobulin, as well as initial folding of MHC-I molecules are facilitated by chaperones like calnexin and calreticulin. The class I MHC-peptide complex is transported from RER to cell surface through Golgi apparatus.

[Note: **Chaperones** are proteins that facilitate folding and assembly of macromolecules like proteins].

B. Endocytic pathway

This pathway is used for the processing exogenous antigens (e.g., bacteria, virus, chemicals, etc.) and then they are presented on the cell membrane with class II MHC molecule.

In endocytic pathway, the exogenous antigens are trapped by APCs and internalized by either phagocytosis or endocytosis. The antigens which are internalized enter endosomes and move to lysosomes with gradual reduction in pH. In lysosomes, antigens are degraded by hydrolytic enzymes like ***proteases***, ***nucleases***, ***lipases*** **and** ***glycosidases***. Peptide antigens bind with class II MHC molecules. Class II MHC molecules, after their synthesis in the rough endoplasmic reticulum, contain a protein known as invariant chain which binds to peptide-binding cleft and prevent the binding of endogenous antigen. Invariant chain allows class II MHC molecules to enter endocytic compartment where the former gets degraded to form CLIP (class II-associated invariant chain peptide) fragments. The CLIP present in the peptide binding cleft is exchanged with the exogenous antigenic peptide, and class II MHC-peptide is transported to surface of cell membrane (Fig. 42.2).

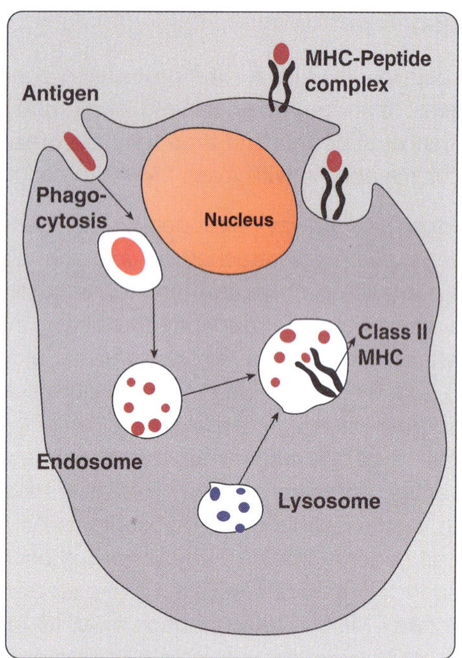

Figure 42.2 Exogenous antigen processing.

VII. ANTIGEN PRESENTATION

It is the process through which the processed antigen is presented to T cells in the form of short peptide fragments. The endogenous and exogenous antigenic peptides, which are presented on the cell surface with class I and class II MHC molecules, respectively binds to T cells through TCRs leading to activation of T cells. Exogenous antigens are presented to T helper cells and endogenous antigens are presented to cytotoxic T cells. Besides TCR-MHC-Peptide interaction, costimulation and cytokine signaling are also required for T cell activation.

VIII. Medical Perspectives

Inflammation

It is a response induced by tissue injury or infection due to breach in the anatomical barriers by pathogens, or other foreign molecules. Inflammation is needed to destroy the pathogen, to prevent the spreading of infection and to repair the damaged tissue. **Acute inflammation** may occur due to tissue damage and indicates early infection. **Chronic inflammation** is due to persistence of the instigating cause of inflammation and is associated with increased morbidity and mortality. Inflammation is characterized by swelling (tumor), redness (rubor), heat (calor), pain (dolor) and functio laesa (loss of function). Inflammatory response can occur in three stages. In the first stage, tissue injury leads to vasodilation mediated by histamine, kinins, leukotrienes and cytokines. Vasodilation leads to increase in blood flow which increases the warmness (heat) and redness in tissues. Increase in vascular permeability increases leakage of fluid from capillaries and its accumulation causing edema and swelling. In the second stage of inflammation, phagocytes stick to the vessel walls (margination), traverse through the fenestrations in the endothelial cells, and migrate to the site of infection/injury and facilitate phagocytosis. In the third stage of inflammation, tissue repair occurs by replacing the dead/damaged cells.

[Note: Pus is due to accumulation of dead cells and body fluids. Abscess is due to formation of pus in the cavity after tissue damage].

[Note: Chronic inflammatory disorders include connective tissue disorders like rheumatoid arthritis, and disorders like **atherosclerosi**s. **Rheumatoid arthritis** (RA) is an immune-mediated multisystem inflammatory disease that predominantly affects the synovial joints. Rheumatoid factors are antibodies against self IgG. The IgG-Rheumatoid factor complex deposit in joints, where they activate the complement cascade, leading to chronic inflammation. The disease can lead to joint destruction, deformity, and disability, and may also present with extra-articular manifestations. Aspirin is used in the treatment of inflammatory disorders as it reduces the synthesis of inflammatory mediators like prostaglandins by inhibiting the enzyme *cyclooxygenase*].

Fever

It is a systemic response to infections characterized by high body temperature. Cytokines released during inflammation increase the body temperature by elevating the hypothalamic temperature set point. The benefits of fever include abrogation of optimal temperature for the growth of pathogens, increasing the rate of activity of metabolic enzymes, and increase in the immune response. However, extreme elevations in body temperature can lead to enzyme denaturation and hazardous effects.

C-reactive protein

CRP is an acute phase protein released in the early phases of inflammation against a wide range of acute and chronic inflammatory conditions. CRP binds to the phosphocholine expressed on surface of dead or dying cells and some bacteria. It activates the complement system, increases release of IL-6 and cytokines and promotes phagocytosis by macrophages.

Therapeutic applications of interferons

IFNs are glycoproteins released from virus-infected cells. They confer antiviral protection on neighboring cells and increase the activity of cytotoxic T cells against tumors. IFN-α is used in the treatment of melanoma, hepatitis, lymphoma and leukemia. IFN-β is used in the treatment of multiple sclerosis and IFN-γ is used in the treatment of chronic granulomatous disease and osteopetrosis.

Disease related to cytokines

Bacterial septic shock is caused by the endotoxins present in the cell wall of bacteria. The endotoxins stimulate the release of IL-1 and TNF-α, which cause early sepsis and IL-6 and IL-8, which are associated with late sepsis.

IX. Chapter Summary

- When a pathogen enters the body, two arms of immune system—innate and adaptive immune systems act in eliminating it from the body.
- Innate immune system forms the first and second lines of defense, and adaptive immune system forms the third line of defense mechanism. Innate immunity components include anatomical barriers like skin and mucosa of GIT and respiratory tract (first line of defense) and phagocytic cells like macrophages and monocytes (second line of defense). Phagocytosis is the process of ingestion of pathogens by phagocytes. Innate immunity is induced by several soluble factors like IFNs, cytokines and acute phase proteins.
- Adaptive immunity develops after exposure to the antigen and consists of active and passive immunity which may be induced by either natural or artificial methods. The mediators of adaptive immunity include B cells, T cells, APCs and antibodies. APCs present the antigens to T cells with the help of MHC molecules. MHC molecules are classified into class I and class II which are involved in the presentation of endogenous and exogenous antigens, respectively
- Endogenous antigens are processed through cytosolic pathway and are presented to cytotoxic T cells (Fig. 42.3).
- Exogenous antigens are presented through endocytic pathway and are presented to helper T cells. Any breach in innate immunity is associated with inflammation characterized by swelling, pain, redness and increased heat.
- Viral infections are associated with reduced expression of class I MHC molecules leading to evasion of virus from the immune system.

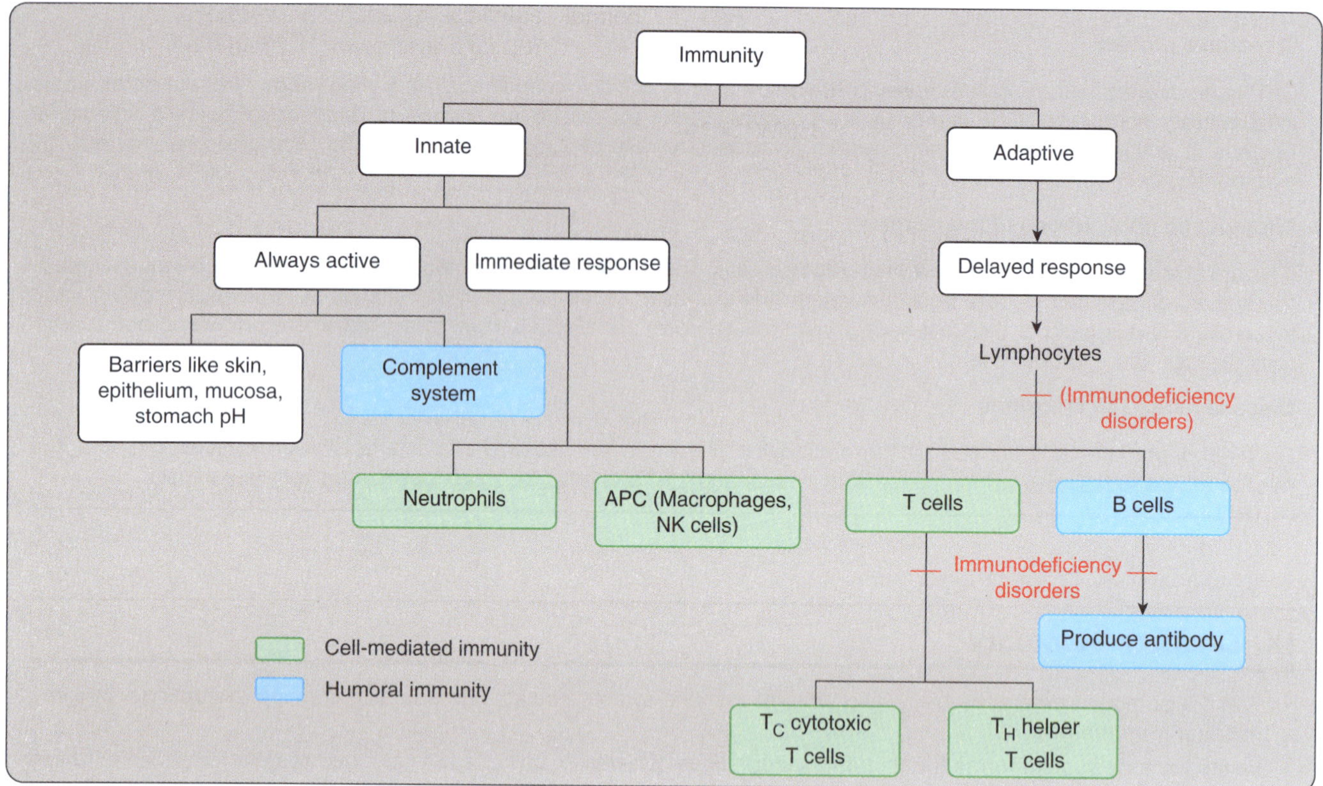

Figure 42.3 Key concept map of innate and adaptive immunity.

Study Questions

Short-Answer Questions

42.1. What are the physical barriers of innate immunity?

42.2. Mention the functions of antigen-presenting cells.

42.3. What are the features of adaptive immunity?

Long-Answer Questions

42.1. Describe the processing and presentation of antigens.

42.2. Explain the types, structure and functions of MHC molecules.

Multiple-Choice Questions

Choose the ONE best answer.

42.1. Which of the following is not a professional antigen presenting cell?

A. Pancreatic beta cells
B. Dendritic cells
C. Macrophages
D. B-cells

Correct answer = A.

42.2. Sebaceous glands inhibit the entry of pathogens by secreting:

A. Amino acids
B. Lactic acid
C. Fatty acids
D. Nucleic acids

Correct answer = C.

42.3. What is not true of innate immunity?

A. It is introduced by physical and chemical barriers.
B. It prevents the entry of pathogen in the body.
C. It ends up activating the adaptive immune system.
D. Neutrophils and macrophages have no role to play in innate immunity.

Correct answer = D.

42.4. Barriers of innate immunity are:

A. Anatomical, physiological, phagocytic, and inflammatory.
B. Anatomical, physiological, phagocytic but not inflammatory.
C. Anatomical, physiological, inflammatory but not phagocytic.
D. Anatomical and physiological only. Neither phagocytic nor inflammatory.

Correct answer = A.

42.5. What is not true of chemical barriers of innate immunity?

A. Various enzymes and proteins of body fluids play a role.
B. Lysozyme present in blood degrade the pathogen.

C. Gastric juice plays a role.
D. Lactoferrin and bile salts are also involved.

Correct answer = B.

42.6. Immune cells comprise:
A. T, B, and NK cells only.
B. T, B, NK cells, and neutrophils but not monocytes.
C. T, B, and NK cells, monocytes/macrophages but not neutrophils.
D. T, B, NK cells, neutrophils, and monocytes/macrophages.

Correct answer = D.

42.7. Toll-like receptors are present on:
A. Membranes of phagocytes, APCs, and lymphocytes.
B. Membranes of phagocytes and lymphocytes.
C. Membranes of phagocytes, APCs only.
D. Membranes of lymphocytes only.

Correct answer = A.

42.8. Which of the following describes the destruction of a pathogen (phagocytosis) by an oxygen-dependent mechanism?
A. Activation of NADPH oxidase only.
B. Activation of NADPH oxidase and myeloperoxidase.
C. Generation of reactive oxygen species only without activation of any oxidase or peroxidase enzymes.
D. Activation of NADPH oxidase and myeloperoxidase with generation of reactive oxygen species.

Correct answer = D.

42.9. What is true about secretion of interferons?
A. INF-α, INF-β, and INF-Υ get secreted by macrophages, fibroblasts, and helper cells respectively.
B. INF-α, INF-β, and INF-Υ get secreted by macrophages, helper cells, and fibroblasts respectively.
C. INF-α, INF-β, and INF-Υ get secreted by fibroblasts, macrophages, and helper cells respectively.
D. INF-α, INF-β, and INF-Υ get secreted by fibroblasts, macrophages, and helper cells respectively.

Correct answer = A.

42.10. All of the following are signs of inflammation except:
A. Rubor
B. Calor
C. Pallor
D. Dolor

Correct answer = C.

Multiple-Choice Questions (NExT/NEET PG)

Choose the ONE best answer.

42.1. Which of the following is an acute phase protein?
A. Tamm-Horsfall protein
B. C-reactive protein
C. Bence Jones protein
D. Fatty acid binding protein

Correct answer = B.

42.2. All of the following are proinflammatory cytokines except:
A. IL-10
B. IL-1
C. IL-2
D. IL-17

Correct answer = A.

42.3. What is the main function of interlukin-6?
A. Mediating inflammation
B. Proliferation of B cells and secretion of antibodies
C. Activation of NK cells
D. Generation and activation of eosinophils

Correct answer = B.

42.4. What is true of major histocompatibility complex?
A. Class I, II, and III MCH molecules are all involved in presentation of antigen.
B. Class I and II but not class III MCH molecules are all involved in presentation of antigen.
C. Class III MCH genes do not code for heat shock proteins.
D. Class III MCH do not have a role in inflammation.

Correct answer = B.

42.5. What is not true of C-reactive protein?
A. Released in acute and chronic inflammatory conditions.
B. Binds to phosphocholine expressed on surface of dead/dying cells and some bacteria.
C. Enhances the release of IL-6 and cytokines.
D. Induces the production of heat shock proteins.

Correct answer = D.

Humoral and Cellular Immunity

43

Learning Objectives

At the end of this chapter, the student will be able to:

- Describe the cellular and humoral components of the immune system
- Discuss the central role of T helper cells in immune response

I. OVERVIEW

Humoral immunity is the ability of the immune system to eliminate pathogens by antibodies. Antibodies bind to the antigens of the pathogen and remove them by either opsonization-mediated phagocytosis or by degradation of the cells by activation of complements. Hence, humoral immunity is also known as antibody-mediated immunity. Humoral immunity is assisted by T helper cells and cytokines which play a crucial role in the production of plasma cells from B cells. In contrast to humoral immunity, **cellular immunity is involved in the elimination of virus infected cells and cancer cells.** Cellular immunity is mediated by T cells which recognize the antigens of the pathogen presented on the surface of antigen-presenting cells. The cytokines secreted from T helper cells help in the activation of cytotoxic T cells and lysis of the infected cells.

II. T CELLS (T LYMPHOCYTES)

A. Types of T cells

1. **T helper cells (Th cells):** Th cells play a role in the immune response by activating other immune cells. They are also known as **CD4 cells**, as they express CD4 glycoprotein on their surface. Th cells are involved in the maturation of B cells to plasma cells and memory cells. They are also responsible for activation of cytotoxic T cells and macrophages. There are two types of Th cells. Th1 cells assist in phagocytosis and secrete cytokines like IL-2 and IFN-γ. Th2 cells activate B cells and secrete cytokines like IL-4, IL-10, and IL-5.
2. **Cytotoxic T cells (Tc cells):** They are also known as **CD8 cells**, as they express CD8 glycoprotein on their surface. Their main role is to destroy tumor cells and cells infected by virus. They are also involved in transplant rejection. They are activated in lymph nodes by antigen-presenting cells. They bind with antigenic peptides presented by class I MHC molecules and destroy their target cells by releasing perforin which creates pores in the membranes of infected cells and granzymes which cause apoptosis of tumor cells and viral infected cells.
3. **Memory T cells:** They are termed as memory cells as they provide memory to immune system for the infections acquired in the past, in a manner similar to memory cells of B lymphocytes. These cells persist for longer time after the infection is controlled. Re-exposure to same pathogens allows these cells to activate the immune system quickly and eliminate them. These cells may express either CD4 or CD8 and consist of central and effector memory cells.
4. **Regulatory T cells (Treg cells):** They were previously known as suppressor T cells as they suppress the cell-mediated immunity after the elimination of foreign antigen and maintain immune tolerance. They also suppress autoreactive T cells that had escaped the negative selection in thymus (explained later in this chapter) and prevent the occurrence of autoimmune disorders.

B. T cell receptor (TCR)

TCR is a protein, present on the surface of T cells and involved in the recognition of MHC-peptide complex by T cells. It is a heterodimer and consists of two types of chains—α and β (95% of T cells) or γ and δ (5% of T cells). The αβ TCR is highly specific and plays a role in adaptive immunity, whereas γδ TCR is mainly involved in innate immunity. TCR is membrane bound, and its interaction with antigen is weak compared to antibodies. Binding of MHC-peptide complex to TCR activates T cells through several signal transduction molecules.

1. **Structure of TCR:** αβ TCR consists of two chains, alpha and beta linked by disulfide bridges. Both the chains have a ***variable region***, responsible for the specific antigen recognition by the TCR and ***constant region*** which is involved in the attachment of TCR to the membrane. Below the constant region there is a connecting sequence, which connects two chains through disulfide linkage. Both the chains are attached to the plasma membrane through transmembrane region containing positively-charged amino acids. Each chain contains a cytoplasmic tail at C-terminal end. TCR is associated with another membrane protein known as CD3, forming a complex (TCR-CD3 complex). CD3 is essential for the expression of TCR on the membrane and subsequent signal transduction (Fig. 43.1). TCR-CD3 complex is involved in the recognition of antigen-MHC complexes, and the interaction between TCR and MHC-peptide complex is enhanced by the presence of cell adhesion molecules and coreceptors like CD2, CD4, CD8 and CD28.

2. **TCR gene rearrangement:** The genes of TCR are formed by rearrangement of V, D and J segments similar to immunoglobulins. The α and γ chains are formed by rearrangement of V and J segments, and β and δ chains are formed by the rearrangement of V, D and J segments. V-J and V-D-J joining during TCR gene rearrangement is catalyzed by recombination-activated genes (RAG-1 and RAG-2) which is expressed on the early thymocytes.

C. T cell development

Thymus plays an important role in the development, differentiation, and positive and negative selection of T cells. T cell development is predominant during fetal life and continues till puberty. T cell production declines after puberty owing to shrinkage of thymus. The hematopoietic stem cells in the bone marrow form lymphoid precursors which migrate to thymus and develop into T cells. In thymus, T cells are activated after exposure to

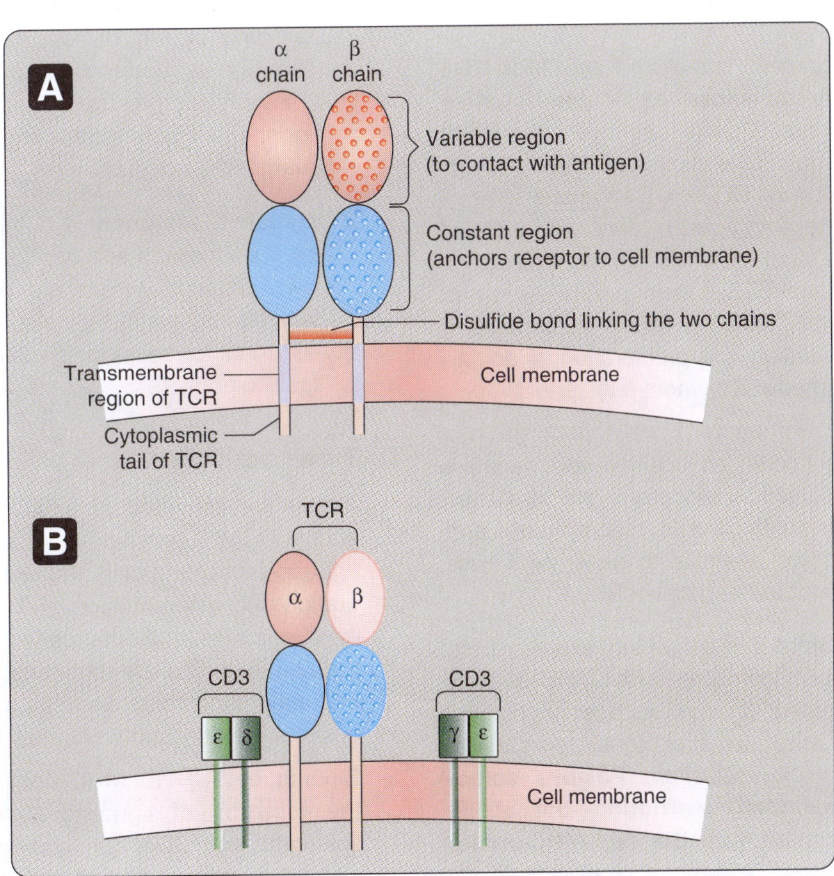

Figure 43.1 Structure of **A.** T-Cell receptor. **B.** Complex of TCR with two CD3.

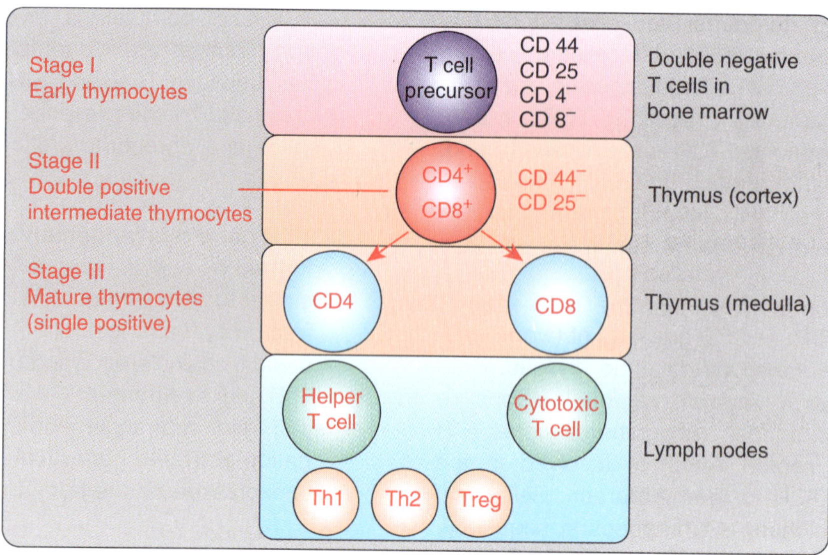

Figure 43.2 T cell development. Note that self-reacting T cells are eliminated in thymus or periphery.

antigens and are associated with several markers like CD2, CD4, CD8 and CD3 (Fig. 43.2).

The development of T cell precursors into mature T cells involves alterations in:

- TCR gene
- expression of TCR protein, and
- production of CD3 complex and coreceptors like CD4, CD8

During early development, immature T cells **lack CD4 and CD8 and they are known as double negative (DN) cells**. In thymus, double negative cells (early thymocytes) undergo proliferation and this is associated with expression of CD25, CD44 and TCR α, β and γ chain gene rearrangement. More than 90% of DN cells mature to form αβ T cells. During this stage, pre T-cell receptor (pre-TCR) is formed. After β chain is rearranged, CD4 and CD8 coreceptors are expressed resulting in the formation and proliferation of double positive cells (intermediate thymocytes).

During maturation, immature T cells undergo two types of selection known as positive and negative selection. These selection processes are mediated by thymic cells like dendritic cells, macrophages and epithelial cells and are essential to generate T cells which are MHC-restricted and self-tolerant.

1. **Positive selection:** T cells which express both $CD4^+$ and $CD8^+$ combine with MHC-peptide complex presented on the surface of cortical epithelial cells and get the signal for survival resulting in positive selection. **Positive selection determines MHC restriction—the ability of the TCR to bind with the self MHC molecules**. Those T cells which cannot bind to MHC, or those that bind to the MHC with a high affinity are eliminated by apoptosis. Only T cells with intermediate affinity binding to MHCs are selected for further development. Helper T cells (CD4 cells) are formed in response to positive selection on class I MHC, and cytotoxic T cells (CD8 cells) are formed due to positive selection of class II MHC. It was found by experimental studies that signals received through CD4 suppress CD8 gene resulting in the development of helper T cells and signals received through CD8 suppress CD4 gene resulting in the development of cytotoxic T cells. The T cells expressing only CD4 or CD8 are mature thymocytes.
2. **Negative selection:** It occurs due to recognition by the T cells of self-peptides derived from host's own proteins, which are expressed by medullary thymic epithelial cells. Negative selection is essential for developing tolerance to self-antigens (self-tolerance).

D. T cell activation

T cells are activated when TCR-CD3 complex interacts with MHC-peptide complex on the surface of antigen-presenting cells. This interaction allows proliferation and differentiation of T cells to memory cells or effector cells. Some genes (e.g., c-Fos, c-Myc, IL-2,3,6 and IFN) are expressed within few hours of interaction and some genes (e.g., cell adhesion molecules) are expressed within few days.

Binding of TCR to MHC-peptide complex initiates the assembly of signaling complex which includes lipid rafts and CD4/CD8 coreceptors. This signaling complex phosphorylates the tyrosine residues of ITAMs (immunoreceptor tyrosine-based activation motifs) leading to the activation of transcription factors involved in the gene expression, activation and differentiation of T cells.

Activation of T cells is also facilitated by the presence of costimulatory signals like CD28 on T cells and B7 molecules on the antigen-presenting cells. In the absence of costimulatory signals, T cells recognize peptide-MHC molecules but will not proliferate resulting in *clonal anergy*.

1. **Immunological synapse: It is a highly organized structure formed between T cells and antigen-presenting cells.** It consists of a central region, which contains TCR-CD3 complexes interacting with the MHC-peptide complex, interaction of costimulatory molecules (e.g., CD28 with B7 on the APC), interaction of CD4 or CD8 with the MHC, and an outer region, which contains adhesion molecules like lymphocyte function-associated antigen 1 (LFA-1). It is crucial for the activation of T cells.

2. **T cell differentiation to memory T cells and effector T cells:** Exposure to foreign antigens activates T cells resulting in secretion of IL-2. IL-2 binds to its receptor causing proliferation and differentiation of activated T cells generating effector T cells and memory T cells. Effector T cells are involved in secretion of cytokines, and activation of B cells and cytotoxic T cells. Memory T cells are derived from both naïve T and effector T cells after activation by antigens and generate secondary response against similar antigens.

III. B CELLS (B LYMPHOCYTES)

A. Features

The features of B cells are described in the Chapter 40. The maturation of B cells occurs in **bone marrow**, where it acquires B cell receptors (BCRs). After maturation, B cells migrate to lymphatic organs/periphery and activates to form plasma cell/memory cell after exposure to B cell antigens.

B. Functions of B cells

B cells are mainly involved in humoral immunity, a component of adaptive immunity. They secrete antibodies and eliminate the antigens by opsonization, complement activation or antibody-dependent cell-mediated cytotoxicity (ADCC). B cells are antigen-presenting cells and are involved in the presentation of antigens to helper T cells. They also secrete cytokines like IL-2, IL-4, IL-6, TNF-α, IFN-γ, etc.

C. B cell receptor

BCR consists of immunoglobulin molecule attached to the membrane and a heterodimer Ig-α and Ig-β. The tail of Ig-α and Ig-β in the cytoplasm contains immunoreceptor tyrosine-based activation motifs (ITAMs). The immunoglobulin molecules in the extracellular region recognize and bind to antigens. Cross-linking of membrane Ig molecules by antigen cause the interactions of cytoplasmic tails of Ig-α and Ig-β, producing intracellular signal. The tyrosine residues in ITAM are phosphorylated and allow binding of signaling molecules which initiate signal transduction after binding with antigen (Fig. 43.3).

D. B cell development

The development of B cells begins in the bone marrow. **B cell is said to be matured if it expresses IgD and IgM on its membrane.** After maturation, B cells move to periphery and convert to plasma cells or memory cells after exposure to antigens. Immature B cells that bind to self-antigens undergo apoptosis. About 10% of B cells reach maturity and exit the bone marrow.

The stromal cells of bone marrow send signals to lymphoid precursor cells which lead to combination of D-J segment on the H chain chromosome to become early pro-B cells. Pre-B cells are formed by rearrangement of V segment of heavy chain genes. Light chain rearrangement leads to formation of immature B cells, which express membrane IgM (mIgM) on the cell surface. Immature B cells leave the bone marrow. They are not functional yet, and exposure to antigen induces death or anergy. Transformation into mature B cells requires co-expression of IgD and IgM on the cell surface. These mature B cells undergo activation and clonal selection forming plasma cells and memory B cells (Fig. 43.4).

1. **Regulation of B cell development:** B cell development is regulated by cell adhesion molecules

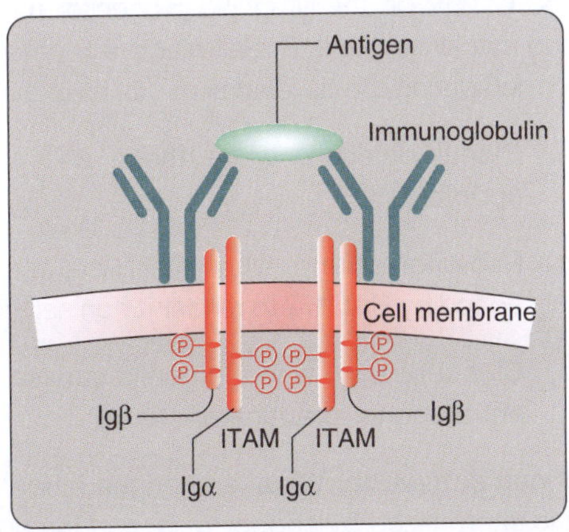

Figure 43.3 Cross-linked B cell receptors after binding antigen.

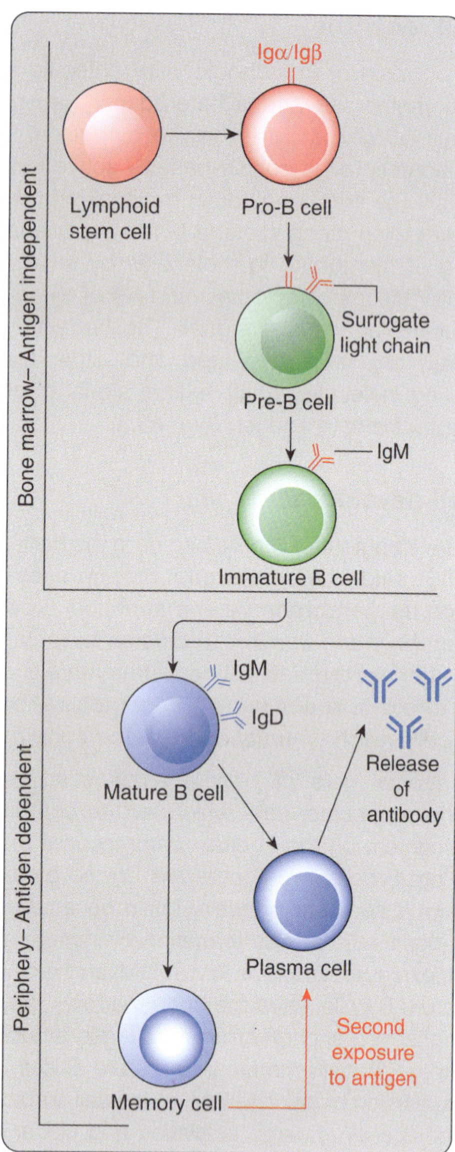

Figure 43.4 B cell development.

like stem cell factor on the stromal cell membrane and CD117 on the lymphoid progenitor membrane, and cytokines like IL-7. Self-reactive B cells that are formed during the development can have three fates:

a. Clonal deletion, when these cells undergo apoptosis.

b. Receptor editing, when the B cell receptor is edited by RAG1/2 to render it non self-reactive.

c. Clonal anergy, when the cells survive, but do not respond to antigenic stimuli.

2. **B cell activation:** When an antigen binds to BCR it is internalized by endocytosis, and the antigen is processed and presented by class II MHC molecules on the surface of B cells resulting in the activation of B cells. It requires helper T cells for its activation. Binding of MHC-peptide complex to Th cells activate it releasing cytokines which induce proliferation and differentiation of B cells to plasma cells or memory cells.

   a. **Memory cells:** These cells are formed from activated B cells and contribute to the immunological memory. These cells persist for a longer time and are found in abundance in spleen and epithelium. Memory cells are generated during primary immune response after infection or vaccination. Memory cells are identified by the presence of cell surface marker, CD27.

   b. **Plasma cells:** Plasma cells are formed from activated B cells. B cells undergo differentiation to form plasmablast, then proplasmacyte and finally plasma cells. Plasma cells secrete antibodies, and these antibodies reach the site of antigen through blood or lymphatic system and destroy them by neutralization and other processes. Plasma cells are identified by the presence of cell surface markers like CD27, CD138 and CD78.

IV. PRIMARY IMMUNE RESPONSE

It is the response of the immune system when it comes in contact with the antigen for the first time. It occurs mainly in **lymph node and spleen** and results from the **activation of naïve B or naïve T cells.** In this type of response, the immune system is capable of recognizing the antigen and produces antibodies and/or activated T-cells against it. The primary response is influenced by various factors like adjuvants, nature of antigen and route of antigen administration. Here, immunity is achieved after few days of encountering antigen, known as lag phase. Antibodies are produced after 4–5 days and reach peak after 7–10 days. **IgM** is the predominant antibody produced in primary response, but its affinity towards antigens is low (Fig. 43.5).

V. SECONDARY IMMUNE RESPONSE

It is the response of the immune system when it comes in contact with the antigen for the second or subsequent times. It occurs mainly in bone marrow followed by lymph node and spleen and results from the activation of memory cells produced during the primary immune response. In secondary response, immunity is established quickly compared to primary response. Antibodies are produced within 1–2 days and reach peak at 3–5 days. **IgG** is the predominant antibody produced in secondary response (Fig. 43.5). The antibodies have high affinity for antigens, and their levels remain higher for longer duration.

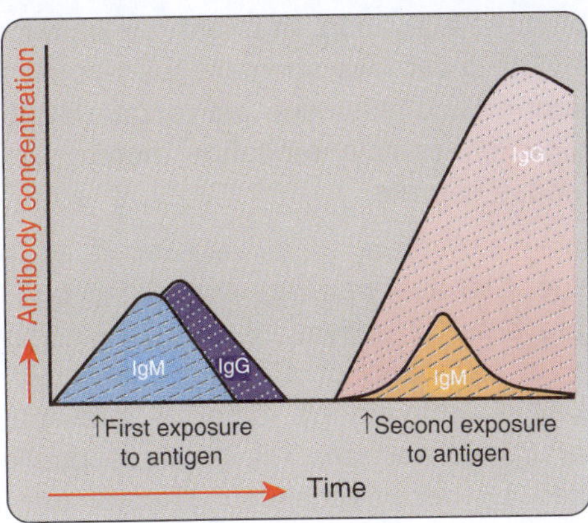

Figure 43.5 Primary and secondary humoral response to an antigen.

VI. CLASS SWITCHING OF IMMUNOGLOBULINS (ISOTYPE SWITCHING)

It is a process through which immunoglobulin production from B cells switches from one heavy chain type to another. During primary immune response IgM is predominantly produced, and in secondary immune response IgM is changed to IgG type. During class switching, the constant region of heavy chains of immunoglobulins is changed, but the variable region does not alter. Hence the antigen specificity is not affected, but the effector activity of the antibodies varies resulting in rapid removal of pathogens during secondary immune response. Class switching is facilitated by interaction between CD40 on B cell and CD40 ligand on helper T cells, and the presence of cytokines like IL-4, IL-5, IFN-γ and TGF-β.

During the course of immune response against an antigen, there is an increase in the affinity of antibodies towards the antigen. This is known as **affinity maturation** and occurs due to somatic hypermutation in immunoglobulin genes of B cells in germinal centers.

VII. COMPLEMENT

These are proteins/glycoproteins which complement the action of antibody. They are synthesized in liver, tissue macrophages, circulating monocytes and epithelial cells of gut. They are named by numbers (C1-C9) or by upper case letters (factor D) or followed by lower case letters (2a, 2b, etc. denoting peptide fragments). Those with enzymatic activity may also be denoted by a line over the complex (e.g., $\overline{C4b2a}$).

A. Complement functions

Complements, when activated, eliminate the pathogens through antibody-dependent or antibody-independent destruction, through phagocytosis, by activation of inflammation and by facilitating antigen-antibody complex clearance by hepatocytes and splenic cells.

B. Pathways of complement activation

Binding to immune complexes, pathogen-associated molecular patterns (PAMPS), or surface of pathogens activates complement proteins either by classical or alternate pathway or lectin pathway and ends with formation of membrane-attack complex (MAC).

1. **Classical pathway:** It involves components C1-C4 and is **activated by IgM and IgG.** The Fc portion of IgM undergoes conformational changes after binding with antigen and activates C1q. C1q activates C1s which cleaves C4 to C4b and C2 to C2a and C2b. C2a combines with C4b forming C3 convertase (C4b2a). C3 convertase hydrolyzes C3 to C3a and C3b. C3b combines with C3 convertase forming C5 convertase (C4b2a3b) which cleaves C5 to C5a and C5b. C5b combines with C6 to C9 forming MAC (Fig. 43.6).

2. **Alternate pathway:** It differs from classical pathway as it does not require antibodies. Since antibodies are not involved, it acts like innate immune defense mechanism against pathogens. In this pathway, complements are activated by C3, factors B and D, and properdin. It is initiated by cell wall components of pathogens and tumor cells. In alternate pathway, C3 is spontaneously, but slowly, converted into C3a and C3b. C3b binds to the surface of pathogens and even the host cells. However, high levels of sialic acid present on the surface of human cells inactivate C3b. On the

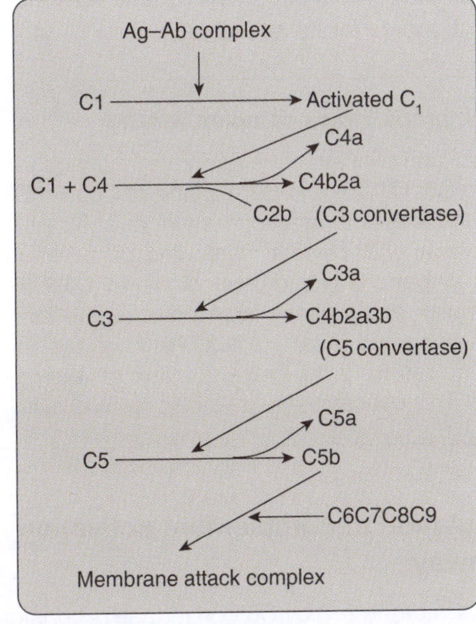

Figure 43.6 Classical complement pathway.

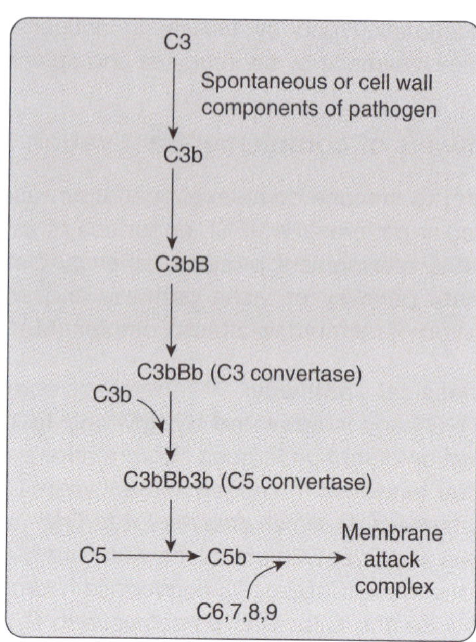

Figure 43.7 Alternate pathway.

surface of pathogens, the molecule remains active, to continue the alternate complement pathway. C3b is cleaved further by factor D to form C3 convertase (C3bBb). C3 convertase hydrolyzes C3 to C3b and forms C5 convertase (C3bBb3b). C5 convertase cleaves C5 to C5a and C5b. C5b combines with C6 to C9 forming MAC (Fig. 43.7).

3. **Lectin pathway:** Lectins are proteins that recognize and bind to specific carbohydrate targets. Binding of mannose-binding lectin (MBL) to mannose residues on glycoproteins or carbohydrates on the surface of microorganisms such as Salmonella, Listeria, and Neisseria activates this pathway. The lectin pathway also does not require antibodies for its activation.

Membrane attack complex (MAC)

It is a complex structure formed by interaction of C5b with C6, C7, C8 and C9. Binding of C5b with C6 and C7 induces conformational changes in C8 which then interacts with plasma membrane of the pathogens. C9, a perforin-like protein binds to C5bC6-C7-C8 complex and induces large pores in the membrane of pathogens, causing efflux of electrolytes and small molecules from the cell and influx of water into the cell. This causes osmotic imbalance and cell lysis.

C. Regulation of complement activation pathway

The step catalyzed by C3 convertase is considered as the rate-limiting step in all the three pathways. During complement activation, the host cell damage is less since C3b undergoes spontaneous hydrolysis. Also, regulator of complement activation (RCA) protein family which contain repeating motifs regulate C3 convertase enzyme.

Complement activation is also regulated by other proteins. Classical pathway is inhibited by C1 inhibitor (prevents C1 activation) and C4 binding protein (prevents formation of C3 convertase). Alternative pathway is inhibited by factor H (prevents formation of C3 convertase). All the three pathways are inhibited by decay-accelerating factor (dissociation of C3 convertase) and complement receptor type 1 (prevents formation of C3 convertase). MAC is inhibited by S protein which prevents insertion of MAC into cell membrane.

D. Biological effects of complement activation

Activation of complement enhances the defense mechanism against the pathogens. MAC mediates cell lysis; C3a, C4a and C5a mediate inflammation; C3b and C4b mediate opsonization and C3b is involved in the neutralization of virus (along with MAC) and clearance of antigen-antibody complexes. The complement system also causes neutralization of viruses by formation of larger viral aggregates, which reduces the net number of infectious viral particles. The binding of complement and/or antibody to a viral particle also blocks attachment to susceptible host cells.

E. Evasion of complement-mediated defense mechanism by microorganisms

Gram-negative bacteria (e.g., Escherichia coli, Pseudomonas, etc.) and Gram-positive bacteria (e.g., Streptococcus) prevent insertion of MAC into cell membrane through lipopolysaccharides and peptidoglycans, respectively in their cell wall. *Elastase* inactivates factors C3a and C5a. Several proteins in virus and fungi mimic complement regulatory proteins and inhibit the activation pathways.

VIII. OPSONIZATION

It is a process which uses **molecules known as *opsonins* to mark the foreign antigen for an immune response and eliminate them by phagocytosis**. Opsonins can be antibodies like IgG, complement components like C3b and C4b, and proteins in circulation like pentraxins and collectins. Opsonins mark the apoptotic cells and pathogens for clearance by phagocytes and activates complement pathways.

Opsonization is mediated through Fc receptors and complement receptors which are present on phagocytic cells and bind to constant region (Fc) of immunoglobulins or complement components, respectively. Binding of Fc with Fc receptors, and C3b/C4b to complement receptors activate signal transduction that facilitates phagocytosis of immune complexes. After phagocytosis, the pathogens are destroyed through enzymatic digestion or by oxidative damage.

Graft-versus-host (GVH) reaction

It is an immune reaction occurring in immunosuppressed subjects who have undergone transplantation. GVH reaction develops when immune cells from the donor attack the recipient host tissue. It usually occurs when bone marrow is transplanted to patients with leukemia or immunodeficiency disorders. Since donor and recipient are not genetically identical, donor's immune cells which are present in the graft identify the non-self MHC molecules in the recipient's organs and attack the host. The host cannot mount an immune response against donor's graft cells due to immunocompromised state (which preexists or is medically induced) before the transplantation.

The T cells from donors attack the host since the host tissue act as foreign antigen against graft cells. The T cells secrete proinflammatory cytokines like IFN-γ and TNF-α which induces inflammation in host tissue leading to rejection of graft. GVH reaction may be acute if it occurs within 100 days of transplantation or it may be chronic if it occurs after 100 days. It is characterized by clinical symptoms like diarrhea, skin lesions, jaundice and splenomegaly.

IX. ANTIBODY-DEPENDENT CELL-MEDIATED CYTOTOXICITY (ADCC)

It is a mechanism through which cytotoxic cells bind to target cells (e.g., pathogens) through antibodies and destroy them. ADCC is mediated by macrophages, NK cells, monocytes, neutrophils and eosinophils. These cells have Fc receptors, which bind to Fc region of immunoglobulins. Cytotoxic cells attach to target cells through antibodies and release their contents like lytic enzymes, perforin and tumor necrosis factor resulting in lysis of target cells.

X. HYPERSENSITIVITY REACTIONS

Hypersensitivity reactions are inappropriate immune reactions to an antigen resulting in extensive tissue damage in the host. These reactions occur during humoral or cell-mediated immune response. If the symptoms manifest early (within minutes/hours) the reactions are known as **immediate hypersensitivity**, and if it manifests late after few days, they are known as **delayed-type hypersensitivity**. There are four types of hypersensitivity reactions.

A. Type 1 (IgE-mediated) hypersensitivity

It is induced by allergens like plant pollens, food items (nuts, seafood, milk, eggs, etc.), insect venom and drugs like penicillin and salicylates. Allergen exposure stimulates B cells to secrete IgE. IgE binds to Fc receptors on mast cells and basophils. Subsequent exposure of the same allergen induces cross-linking of IgE causing degranulation of mast cells/basophils releasing histamine, cytokines, prostaglandins and leukotrienes. These mediators increase vascular permeability, vasodilation and smooth muscle contraction resulting in rhinitis, asthma, anaphylactic shock and atopic dermatitis.

B. Type II (antibody-mediated cytotoxic) hypersensitivity

It involves destruction of cells through antibodies. Antibodies can destroy the foreign antigens through activation of complements and creation of pores or through ADCC.

1. **Transfusion reactions:** These reactions occur when individual with one type of blood group is transfused with blood of an other group. For example, if a person with blood group A is given blood of group B, the anti-B antibodies (IgM) in recipient (having group A blood) bind to transfused blood cells and destroy them. The clinical features include fever, chills, backache, vascular collapse and detection of hemoglobin in urine.

   ABO blood group showing distribution of antigens and antibodies is given in blue box below.

Blood Group	Antigens on RBC	Serum Antibodies
A	A	B
B	B	A
AB	A+B	None
O	None	A+B

   When packed red cells are transfused, individuals with type O RhD negative blood are called universal donors. On the other hand, individuals with type AB RhD positive blood are called universal recipients. (These are relevant only in emergency situations when appropriately cross-matched blood is not available.)

2. **Hemolytic disease of the newborn (Erythroblastosis fetalis):** Here maternal antibodies (IgG) cross placenta and destroy fetal red blood cells due to their specificity for fetal blood group antigens. Erythroblastosis fetalis occurs commonly when mother is Rh^- and fetus is Rh^+.

The first child is not affected since mother is not exposed to fetal RBCs. During delivery, placenta is separated, and fetal cells enter maternal circulation and activate Rh-specific B cells producing Rh-specific plasma cells and memory cells. In the subsequent pregnancy, Rh antibodies are produced from activated plasma cells, which cross the placenta and destroy fetal RBCs. This condition can be treated by administering anti-Rh antibody to the mother at 28th week of gestation and within 72 hours of delivery.

C. Type III (immune complex-mediated) hypersensitivity

This type of reaction occurs due to excessive formation of immune complexes. These complexes are deposited on the walls of blood vessel, basement membrane of kidney and synovial membrane of the joints. These complexes recruit inflammatory cells (neutrophils) and activate the complement, which causes degranulation and releases their contents causing tissue damage. Immune complex deposition plays a role in the pathogenesis of rheumatoid disease, systemic lupus erythematosus, drug allergies and infections like meningitis and hepatitis.

D. Type IV (delayed-type) hypersensitivity

It is a localized inflammatory reaction induced by secretion of cytokines by activated helper T cells on contact with pathogens. The common pathogens involved are *Mycobacterium tuberculosis*, *Candida albicans*, Herpes simplex virus and measles virus. It has two phases—sensitization phase and effector phase. In sensitization phase, helper T cells (Th) proliferate and differentiate to Th1 cells after contact with antigens. In the effector phase, Th1 cells secrete cytokines like IL-3, IFN-γ and TNF-α which recruit macrophages. These macrophages are activated and present the antigens to T cells initiating delayed-type hypersensitivity reactions.

1. **Tuberculin test:** Type IV hypersensitivity is detected with skin test. Tuberculin test is done to detect the exposure to M. tuberculosis. In this test, the purified protein derivative (PPD) antigen derived from the bacteria is injected subcutaneously and the inflammatory reaction is observed after 48–72 hours at the site of injection. The presence of swollen, firm lesion indicates previous exposure to the bacteria. The positive reaction indicates the presence of Th1 cells which are specific to the bacterial antigen.

XI. Medical Perspectives

Hypersensitivity reactions to drugs are **unpredictable adverse drug reactions**. They manifest either as immediate reactions (within 1–6 h following drug intake) or delayed reactions (several hours to days later). It is very important to perform skin tests for sensitization to drugs known to cause such reactions such as beta-lactams (penicillin, oxacillin), immunoglobulins, contrast dyes, local anesthetics, etc.

Some people are allergic to certain foods. Symptoms of a **food allergy** can occur within minutes to hours of ingestion. They may include hives, itching, swelling of the lips, face, tongue and throat, or other parts of the body, wheezing, nasal congestion or trouble breathing. Some patients may also have abdominal pain, diarrhea, nausea, or vomiting. While any food can cause an adverse reaction, allergy to eggs, peanuts, shellfish and soy are common.

The household dust may contain pet hair, mold or pollen spores, or dust mite (tiny creatures frequently found in furniture, carpets, and bedding), all of which are common allergens. These allergens can cause an allergic reaction.

Latex, a component of gloves, is also a known allergen. To avoid this allergy, latex-free nitrile gloves are now commonly used. Some people can develop allergy to cloth material, known as textile contact dermatitis. It is rarely due to fiber itself and more commonly due to chemical, such as dye, which are used to treat cloth fibers.

Plasma Cell Disorders

These occur due to abnormal proliferation of B cells or plasma cell, forming a clone, which secrete excess of monoclonal antibodies or their fragments (light chain or heavy chain). These antibodies or their fragments are excreted in urine in the absence of renal dysfunction (overflow proteinuria). They account for 10%–15% of deaths from hematological malignancies. It includes monoclonal gammopathy, also known as paraproteinemia or dysproteinemia. **Monoclonal gammopathy** consists of several disorders like multiple myeloma, Waldenström macroglobulinemia, heavy chain disease and primary amyloidosis.

Multiple myeloma: It accounts for 1% of cancer deaths and is characterized by monoclonal proliferation of plasma cells with osseous and extraosseous manifestations. Vertebral column is commonly affected followed by ribs and skull. Presence of lytic lesions in long bones and punched-out lesions in skull are the common in multiple myeloma. The clinical features include bone pain and pathological fractures (extensive destruction of bones by malignant plasma cells), recurrent infections (suppression of humoral immunity) and renal failure (excretion of light chain causes tubular damage). It is diagnosed by the presence of M band (most commonly IgG or IgA) in the gamma region in serum protein electrophoretogram and excretion of Bence Jones proteins in urine. Diagnosis is confirmed by bone marrow examination which shows increase in plasma cells. β_2 microglobulin and *lactate dehydrogenase (LDH)* are used as prognostic markers in multiple myeloma.

XII. Chapter Summary

Humoral and cellular immune systems are involved in the destruction of pathogens mediated by B and T cells, respectively.

Development and differentiation of T cells occur in thymus. Immature T cells are converted to mature T cells by changes in (i) TCR gene, (ii) expression of TCR protein and (iii) production of CD3 complex and coreceptors CD4 and CD8.

Positive selection of T cells in thymus ensures self MHC restriction and negative selection ensures T cells which react with self-antigens undergo apoptosis. The development of B cells occurs in the bone marrow and involves rearrangement of V-D-J genes.

Mature B cells in the periphery express IgM and IgD on their surface and convert plasma cell or memory cell after exposure to antigens. Immature B cells that bind to self-antigens undergo clonal deletion, receptor editing, or clonal anergy. IgM is involved in primary immune response and IgG is involved in secondary immune response.

Antibodies cause lysis of infected cells either through activation of complement or through antibody-dependent cell-mediated cytotoxicity (ADCC) (Fig. 43.8). Activated Th cells play a central role in both humoral and cellular immunity.

Complement are activated either by classical, alternative or lectin pathway. All the three pathways form MAC which creates pores, produces osmotic instability and causes cell lysis. Abnormal proliferation of plasma cells results in plasma cell disorders like multiple myeloma.

Hypersensitivity reactions can be of type 1 (IgE mediated), type 2 (complement activation and ADCC by antibodies), type 3 (immune complex), type 4 (cellular, delayed).

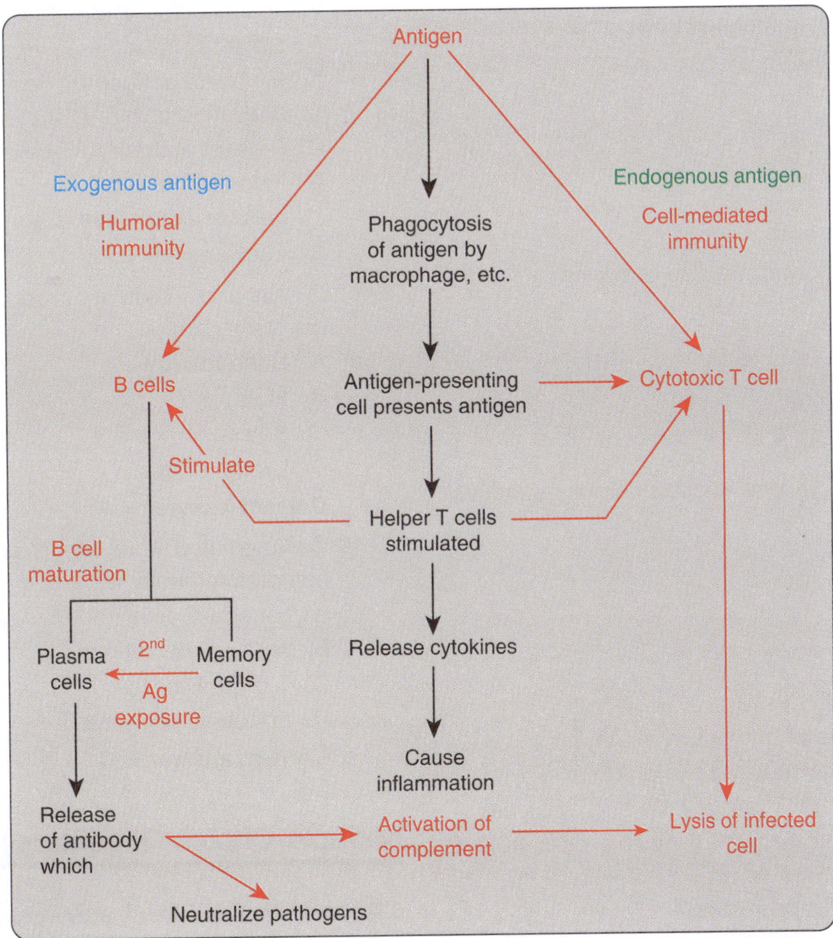

Figure 43.8 Key concept map of humoral and cell-mediated immunity.

Study Questions

Short-Answer Questions

Write short notes on:

43.1. Delayed-type hypersensitivity reaction with example.
43.2. Plasma cell disorders.
43.3. Differences between humoral and cell-mediated immunity.
43.4. How cytotoxic T cells destroy pathogens?
43.5. Structure of B cell receptor.

Long-Answer Questions

43.1. Explain the types of complement activation pathways and their regulation.
43.2. Describe T cell development emphasizing on the need for positive and negative selection.

Multiple-Choice Questions

Choose the ONE best answer.

43.1. The predominant immunoglobulin in the secondary immune response is:
A. IgM
B. IgG
C. IgA
D. IgE
Correct answer = B.

43.2. Which of the following acts as opsonins?
A. C1b
B. C2a
C. C2b
D. C3b
Correct answer = D.

43.3. Immunological synapse is essential for the activation of:
A. B cells
B. Neutrophils
C. T cells
D. Eosinophils
Correct answer = C.

43.4. Which is not true of humoral immunity?
A. It is antibody-mediated immunity.
B. It is assisted by T helper cells and cytokines.
C. It involves production of plasma cells from B cells.
D. It eliminates cancer cells.
Correct answer = D.

43.5. What is not true of T helper cells (Th cells)?
A. Also called CD4 or CD8 cells
B. Involved in the maturation of B cells to plasma cells
C. Cause activation of cytotoxic T cells and macrophages
D. Secrete cytokines
Correct answer = A.

43.6. What best describes regulatory T cells?
A. Also called suppressor T cells.
B. Suppress cell-mediated immunity and autoreactive T cells.
C. Prevent occurrence of autoimmune disorders.
D. All the above.
Correct answer = D.

43.7. What is not true of T cell development?
A. Thymus plays an important role.
B. Development of T cell precursors into mature T cells does not involve any alteration in TCR gene.
C. During early development, immature T cells lack CD4 and CD8.
D. PreT-cell receptor is also formed.
Correct answer = B.

43.8. What is not true of B cells?
A. Secrete antibodies.
B. Eliminate antigen by opsonization, complement activation or by ADCC.
C. Present antigen to T cells.
D. Involved in providing both the humoral and cellular immunities.
Correct answer = D.

43.9. B-cells and T-cells are produced by stem cells that are formed in:
A. Bone marrow
B. Liver
C. Spleen
D. Lymph nodes
Correct answer = A.

43.10. Cell-mediated immunity is carried by _____ while humoral immunity is carried mainly by _____.
A. B cells, T cells
B. Antibodies, antigens
C. T cells, B cells
D. Epitopes, antigens
Correct answer = C.

Multiple-Choice Questions (NExT/NEET PG)

Choose the ONE best answer.

43.1. The type of hypersensitivity reaction mediated by IgE is:
A. Anaphylactic response
B. Cytotoxic mediated response
C. Immune-complex mediated response

D. Arthus reaction

Correct answer = A.

43.2. B cell activation requires:

A. Memory T cells
B. Cytotoxic T cells
C. Helper T cells
D. Regulatory T cells

Correct answer = C.

43.3. The antigenic determinant is called:

A. Epitope
B. Epitome
C. Microbiome
D. Metagenome

Correct answer = A.

43.4. Which segment of chromosome 2 encodes for the V region of the antibody?

A. A segment
B. C segment
C. V segment
D. J segment

Correct answer = D.

43.5. Which of the following is not true for blood transfusion?

A. Nature of blood group is based upon the nature of antigen.
B. A person having blood group AB has no anti A and anti B antibodies.
C. A person having blood group O has neither antigen A or B.
D. Person with blood group AB and O are called universal donor and recipient respectively.

Correct answer = D.

Vaccines

44

Learning Objectives

At the end of this chapter, the student will be able to:

- Explain active and passive immunity
- Explain the concepts involved in vaccine development

I. OVERVIEW

Adaptive immunity can be induced either actively or passively. Passive immunity is induced by administration of preformed antibodies. It could be attained naturally when a mother transfers the antibodies to fetus/infant via placenta or breast milk. Injection of preformed antibodies against a specific disease induces artificial passive immunity. In active immunity, body's own system is involved in producing the antibodies to fight a given infection. Active natural immunity is brought by clinical or subclinical infections. Administration of vaccines induces active artificial immunity.

Vaccine is the suspension of a weakened or killed microorganism or subunit of a microorganism which evokes an immunological response without causing the disease. Vaccines are administered to not only combat prevailing disease but also to protect the body from subsequent infections. Vaccines are classified into six classes depending upon the nature of the pathogenic antigen and the technology used in its preparation. The antigen can either be the whole organism, its subunit, purified macromolecule or DNA/gene responsible for the production of the antigen. The immunity induced by various vaccines varies from few months (cholera vaccine) to entire life (measles vaccine).

II. IMMUNIZATION

Immunization is the process of inducing the body to develop resistance against a specific disease. Immunization can either be passive or active.

A. Passive immunization

Passive immunization deals with protection of an individual without involving his or her own immune system. The immunity is obtained by providing antibodies from another person or animal who has already been suffering from the same disease. Passive immunization is done under the following conditions:

- In immunodeficient patients
- In individuals who have already been exposed to a toxin or venom or a disease, e.g., measles

Passive immunization can either be acquired naturally or artificially.

1. **Passive natural immunization:** It is acquired from the mother before birth by the transfer of IgG antibodies via placenta and after birth by the transfer of IgA antibodies through breastfeeding.

2. **Passive artificial immunization:** It provides quick protection against acute infections (e.g., measles, tetanus, hepatitis A, hepatitis B, rabies, chickenpox, etc.) by injecting human or animal serum having already developed antibodies against a specific disease from which we want protection. Use of humanized monoclonal antibodies and antitoxins are also being made use of to provide passive artificial immunity.

B. Active immunization

In active immunization, **host immune system is induced to develop immunity**. The immunity against a specific disease is obtained by activating the immune system of an individual with administration of either the attenuated/inactivated form of the specific disease-producing organism or the toxoid responsible for producing that specific disease, or components of the agent causing infection (e.g., proteins, polysaccharides, or nucleic acids). Active immunization is done to provide **long-term protection by inducing immunological memory**. In active immunization, administered antigens cause proliferation of specific T and B lymphocytes resulting in the generation of specific memory cells. Active immunization is done under the following conditions:

- To attain eradication or containment of a disease
- To protect persons traveling to regions where certain communicable diseases are endemic (e.g., yellow fever).
- To protect armed forces against biological warfare
- To prevent high-risk groups against diseases for which no treatment is available (such as rabies vaccine in animal handlers)

III. VACCINES

Vaccines act by inducing immunity against a specific disease-causing agent. Vaccines can either be given orally, administered by injection or applied locally on mucosal layers. Vaccines can be administered subcutaneously (e.g., measles), intramuscularly (e.g., injectable polio vaccine [IPV]), orally (e.g., typhoid vaccine), intradermally (e.g., BCG vaccine) and intranasally (e.g., influenza vaccine), etc.

In brief, vaccines are used to stimulate immune system to fight disease-causing agents and give protection from subsequent infections by creating specific memory cells. The period for which most of the common vaccines retain their potency of being effective in inducing an immune response is summarized in Figure 44.1.

Immunity Maintenance Period	Vaccine
6 months in children 2 years in adults	Cholera vaccine
2 years	TAB vaccine (typhoid-paratyphoid A and B vaccines)
3–5 years	Pneumococcal vaccine, Pertussis vaccine
5–10 years	Tuberculosis-BCG (bacillus Calmette–Guérin) vaccine, diphtheria vaccine, *Haemophilus influenzae* type-B vaccine
10–15 years	Mumps vaccine, varicella vaccine, yellow fever vaccine, tetanus toxoid
>15 years	Polio vaccine, hepatitis B vaccine, rubella vaccine
Lifelong	Measles vaccine

Figure 44.1 Immunity maintenance period of vaccines.

A. Brief history of vaccine development

First attempt in the development of vaccines was made by giving children dried crusts of smallpox lesions from affected individuals to prevent the occurrence of the disease in the healthy children. However, the first vaccine against smallpox was formally developed by Edward Jenner in 1796 by using cowpox virus (vaccinia). Since then, significant developments in the fields of vaccinology and immunology have taken place which has led to the successful prevention or management of many diseases. The list of vaccines given as part of National Immunization Schedule in India is given in Figure 44.2. ***Through vaccination, smallpox has been successfully eradicated worldwide,*** **and polio cases have declined by 99%.** Vaccines have also been developed for typhoid fever, cholera, plague, tuberculosis, pneumococcal infection, yellow fever, some types of encephalitis and typhus.

B. Types of vaccines

To induce an immune response in the host, vaccines can either be prepared from the whole microorganism in its live, inactivated or dead form; part or subunit of the microorganism; purified products, obtained from the microorganisms which could either be proteins, peptides, polysaccharides, toxoids or the genetic material responsible for the production of the immune response. An ideal vaccine should have the following properties:

- Produce sufficient immunity which lasts for a longer period and provides long-term memory
- Should be safe and have high degree of efficiency
- Should have no or minimum requirement for the booster doses
- Should be relatively cheap, easy to store and administer

Depending upon the source used for their production, various developed vaccines can be categorized in the following types:

- Whole microorganism vaccines
- Purified pathogenic macromolecule vaccines/ subunit vaccines
- Synthetic polypeptide vaccines
- DNA vaccines
- Recombinant vector vaccines
- Multivalent vaccines

1. **Whole microorganism vaccines:** Most of the commonly used whole microorganism vaccines have been created from the pathogenic bacteria or viruses by attenuating or inactivating these microorganisms. Attenuation is done by growing

the microorganisms under abnormal conditions for a prolonged period of time, while inactivation of microorganisms is done by either heat or chemical treatment. The attenuated or inactivated microorganisms retain their ability of inducing an immune response while losing their pathogenicity. These modified bacteria/viruses can induce immunity without causing the disease. Depending upon the modifications introduced, whole microorganism vaccines have further been divided into following two categories.

a. **Live, attenuated vaccines:** These are formed from weakened microorganisms. Since they are very close to the natural pathogens, they end up creating a very strong and long-lasting immunity. Lifetime protection against a specific disease can be obtained by just one or two doses. Examples of attenuated vaccines having:

   i. **Bacterial origin:** Tuberculosis (BCG vaccine), cholera vaccine and typhoid vaccine.

   ii. **Viral origin:** Measles vaccine, mumps vaccine, rubella vaccine, polio vaccine (Sabin), rotavirus vaccine, varicella zoster (chickenpox) vaccine and yellow fever vaccine.

b. **Inactivated/killed vaccines:** Vaccines using inactivated/killed microorganisms do not provide very strong immunity. During inactivation, it is absolutely essential to maintain active conformation of the epitopes. Heat treatment is usually not preferred as it causes extensive denaturation of proteins, hence, formaldehyde or alkylating agents are often used to chemically inactivate the pathogen. Several doses of inactivated vaccine are required to maintain a desired level of immunity. Examples of inactivated vaccines having:

   i. **Bacterial origin:** Anthrax vaccine, cholera vaccine, pertussis vaccine, typhoid vaccine and plague vaccine.

   ii. **Viral origin:** Hepatitis A vaccine, influenza vaccine, polio (Salk) vaccine, rabies vaccine.

2. **Purified pathogenic macromolecule vaccines/ subunit vaccines:** Vaccines prepared from attenuated or inactivated whole microorganisms many a times cause complications/symptoms identical to the natural disease as sometimes the treated microorganisms revert back to their original form or if the inactivation remains inadequate, leaving live organisms in the preparation. Vaccine-related encephalitis symptoms have been reported. Many vaccine-related complications can be avoided by preparing vaccines using purified macromolecules from the pathogens. Three different types of vaccines prepared from purified macromolecules (inactivated exotoxins [toxoids], capsular polysaccharides and recombinant antigens) are currently being used.

   a. **Toxoids:** Symptoms of many diseases like diphtheria and tetanus are produced by the exotoxins produced by the pathogenic microorganisms. Purified exotoxins obtained from the pathogenic microorganisms are inactivated and converted to their respective toxoid by treatment with formaldehyde. It is essential to obtain detoxification of the exotoxins without inactivating the binding capacity of their specific epitopes. Sufficient quantities of the exotoxins are obtained by cloning the exotoxin gene and getting it expressed in a rapidly dividing host cell. Examples include diphtheria, tetanus and anthrax toxoid vaccines.

   b. **Capsular polysaccharides:** Vaccines using the purified capsular polysaccharides are based upon the principle that the hydrophilic nature of the capsular polysaccharides prevents the phagocytosis of the pathogens, and binding of their specific antibodies to these polysaccharides significantly enhances their rate of phagocytosis. Purified capsular polysaccharides have been used to prepare vaccines against *Streptococcus pneumoniae, Neisseria meningitidis* and *H. influenzae* type b (Hib) which are responsible for causing pneumococcal pneumonia, bacterial meningitis in all age groups and bacterial meningitis in children of less than 5 years of age, respectively.

   c. **Vaccines from pathogenic bacterial proteins produced by recombinant DNA technology:** Genes coding for surface proteins responsible for inducing immune response have been cloned and got successfully expressed in suitable host cells to produce large quantities of immunogenic proteins which are used to produce vaccines. Hepatitis B vaccine, prepared by recombinant DNA technology, was the first such vaccine to be approved for use in humans.

3. **Synthetic peptide vaccines:** Despite a lot of initial excitement, vaccines from synthetic peptides have not picked the desired momentum. Main drawback of peptides, when compared to proteins, is that they are much less effective to induce the humoral

and cellular immunological responses. Conjugates and adjuvants have been used with peptides to increase their protective immunity; however, such attempts have not yet been very successful.

4. **DNA vaccines:** In DNA vaccines, the gene coding for the antigenic protein of the pathogenic organism is directly inserted into the host cell via plasmid DNA. Perhaps by getting integrated into the host chromosomes, or by remaining as episomal DNA, the antigenic protein producing gene gets expressed, resulting in the production of a specific antigenic protein which in turn induces the body to give both the humoral and cell-mediated immune responses. DNA vaccines make the body's own cells as vaccine-making factories. DNA vaccines have following distinct advantages over the other forms of vaccines.
   - The antigenic protein is expressed in the host in their natural physiological environment without undergoing any denaturation or any other structural modification.
   - The immune response produced is an exact replica of the one produced by the active pathogen.
   - DNA vaccines induce the production of both the humoral and cell-mediated immune responses.
   - The immunity induced by DNA vaccines stays for a prolonged period as large number of memory cells gets generated.
   - Same plasmid can be used to insert DNA coding for many proteins, resulting in simultaneous immunization against many diseases.
   - DNA vaccines are relatively easy to produce at much lower costs.
   - Requires no refrigeration, storage or transportation.
   - Significant improvement in the delivery system can be made by coating with gold beads and using gene gun for their delivery.

   After successful trials in animal model systems, human trials using DNA vaccines are in progress for influenza, herpes, malaria and HIV. Only disadvantage of the DNA vaccines being that these can only be prepared when the antigen is a protein. These cannot be prepared when the antigenic macromolecules are polysaccharides in nature.

5. **Recombinant vector vaccines:** In principle, they are similar to the DNA vaccines with the sole exception that an attenuated virus or bacteria is used as a vector to transfer the antigen-producing microbial gene to the host cells. Various microorganisms (vaccinia virus, canarypox virus, attenuated poliovirus, BCG strain of Mycobacterium bovis, etc.) have been used for the production of vector vaccines. Usually, those microorganisms are used as vectors for vaccine production which have large genome (having hundreds of genes) because many genes can easily be inserted into them without affecting their ability of infection and replication in the host. These vaccines produce specific cytotoxic T cells, good immunity and long-lasting memory.

6. **Multivalent vaccines:** Multivalent vaccines are being developed to significantly increase the immunogenicity of the polypeptide, protein or recombinant DNA vaccines. It is achieved by synthesizing a polypeptide which has both the immunodominant B-cell and T-cell epitopes. Delivery system plays an important role in the development of multivalent vaccines having multiple copies of a given peptide or a mixture of peptides. Following delivery systems have been developed to deliver the multivalent vaccines to the immune system of the host.

   a. **Solid matrix:** First, monoclonal antibodies are attached to the solid matrix and then, the antibodies are saturated with the antigens. The antigens could be peptides, proteins, immunodominant epitopes for T and B cells, etc. The solid matrix-antibody-antigen (SMAA) complexes thus produced are capable of inducing much vigorous humoral and cell-mediated responses resulting in the increased immunogenicity and much higher rate of phagocytosis.

   b. **Liposomes:** Liposomes are also being used for the production of multivalent vaccines. Detergents are used to incorporate protein antigens into liposomes.

      Protein-detergent mixture is suspended in phospholipid media under conditions where removal of detergent results in liposome formation. Proteins in liposomes are oriented in a way that their hydrophobic residues are at the center, while their hydrophilic residues are towards the outer side facing the aqueous environment. Examples include influenza and HPV vaccines.

IV. VACCINATION

Vaccination is the process of inducing immunity in the host against specific diseases by the administration of the disease-specific antigenic material. Following immunization scheme is routinely used.

A. Primary vaccination

1. **One-dose vaccines:** BCG vaccine, mumps vaccine, rubella vaccine and yellow fever vaccine.
2. **Multiple-dose vaccines:** Polio vaccine, DPT vaccine, hepatitis B vaccine.

B. Booster vaccination

For some vaccines (DT and MMR vaccines), immunity levels start falling after some time and they need booster doses of vaccines to maintain constant levels of the specific antibodies.

C. Role of adjuvants in vaccination

Adjuvants are used to enhance the immune response of some vaccines, e.g., inactivated or subunit vaccines. Alum (aluminum salt) is the most commonly used adjuvant. Two adjuvants (MF59 & ASO4) have been specifically developed for human vaccines. Adjuvants, act by enhancing Th2 cell response with slow release of the antigen.

D. Immunization schedule from birth to adulthood

National Immunization Schedule (NIS) in India for Infants, Children and Pregnant Women is given in Figure 44.2.

Category	Vaccine	Administration Time
Pregnant women		
	TT-1	Early in pregnancy
	TT-2	4 weeks after TT-1
	TT-Booster	If received TT dose within last 3 years
Infants		
	BCG	At birth or at the earliest till 1 year of age
	Hepatitis-B – Birth dose	At birth or at the earliest
	OPV-0	At birth or at the earliest
	OPV-1, 2, and 3	At 6, 10, and 14 weeks
	Rotavirus vaccine	At 6, 10, and 14 weeks
	IPV (Inactivated Polio Vaccine)	At 6 and 14 weeks
	Pentavalent* 1, 2, and 3	At 6, 10, and 14 weeks
	Measles - 1st dose	9–12 completed months
	Japanese encephalitis† - 1st dose	9–12 completed months
	Vitamin - 1st dose	With measles at 9 completed months
Children		
	DPT booster-1	16–24 months
	Measles 2nd dose	16–24 months
	OPV booster	16–24 months
	Japanese encephalitis - 2nd dose	16–24 months
	Vitamin-A 2nd to 9th dose	16 months, then one dose after every 6 months till 5 years
	DPT booster-2	5–6 years
	TT	10 and 16 years

*Pentavalent vaccine contains a combination of DPT, Hepatitis-B & Hib vaccines.

†Japanese encephalitis (JE) vaccine was introduced after the campaign in selected epidemic districts of UP, West Bengal, Karnataka, Assam and Bihar.

Note: Interval between two doses of Pentavalent, oral polio vaccine (OPV) & hepatitis vaccine should not be less than 4 weeks.

Figure 44.2 National Immunization Schedule in India.

V. Medical Perspectives

Immunization is a key component of primary health care. Vaccines are one of the most successful health interventions which have significantly reduced incidence of infectious diseases and adverse health consequences. **India's National Immunization Schedule (NIS)** has been designed to protect children since birth, and at the ages when they are vulnerable to specific vaccine-preventable diseases.

India has successfully eradicated smallpox (1977), poliomyelitis (2014) and maternal and neonatal tetanus (2015) by use of vaccines.

If a child has received initial doses but is brought late for a subsequent dose of multi-dose vaccines, the next dose should be given, and the remaining doses should be completed at recommended interval as per the schedule.

Besides the vaccines covered under the NIS, several other vaccines, such as against typhoid and **human papillomavirus (HPV)** are available in the market.

Cervical cancer is caused by persistent infection with certain types of HPV. An effective and safe vaccine against HPV types 16 and 18, the two cancer-causing types of HPV, are available. Vaccination for girls before the age of 15 is recommended.

During international travel, certain vaccines are recommended for protection against diseases endemic to the country of origin or of destination. Besides protecting the traveler, they also prevent disease spread between countries. Some countries may also ask for proof of vaccination for travelers wishing to enter or exit the country. These include vaccines against meningococcal disease, tick-borne encephalitis and yellow fever.

People who are immunocompromised due to diseases (HIV, cancer, etc.) or immunosuppressive medical treatment have an increased risk of infection. Influenza and pneumococcal vaccines are given to these patients to provide optimal protection. Also, live vaccines are contraindicated in people who are severely immunocompromised as vaccine-related disease can be caused by unchecked infection.

People who work in certain jobs may be exposed to high-risk diseases at work and should be vaccinated against such diseases. **Healthcare workers should be vaccinated against Hepatitis B** and influenza. People who handle animals should be vaccinated against rabies.

Individual vaccination against a disease-causing pathogen can protect even those individuals in a community who have not been vaccinated against that pathogen, a phenomenon known as "herd immunity." This is only achieved when a very high proportion of community members are vaccinated, so it is difficult for the disease-causing organism, which spread by person to person transmission, to spread to unprotected persons.

Maternally-acquired antibodies are critical to protect infants during initial months of life and immunization of pregnant women is important not only to protect mothers from infection, but also to provide natural passive immunity to infants.

World Immunization Week is celebrated by the WHO in the last week of April (24 to 30 April) to promote the use of vaccines to protect people of all ages against diseases.

VI. Chapter Summary

- Development of resistance against specific diseases is immunization, which can be active as well as passive. Transfer of preformed antibodies, either naturally or artificially, is known as passive immunization. It provides immediate but short-lived protection. Active immunization induces the host immune system. This induction can occur naturally after exposure to pathogens or can be done artificially by using vaccines (Fig. 44.3).
- Vaccines can consist of whole microorganisms, their purified pathogenic macromolecules, synthetic polypeptides or DNA. Recombinant vector vaccines and subunit/multivalent vaccines are also available. Whole microorganisms can be live, attenuated or inactivated, killed. Live vaccines have the advantage of creating a very strong and long-lasting immunity. Whole organisms or their toxins or components can be inactivated by heat or chemicals to be used as vaccines. Some vaccines provide sufficient immunity with a single dose, while others require multiple booster doses. Adjuvants, such as alum salt, are also used to enhance the efficacy of vaccines.
- National Immunization Schedule of India provides vaccines for common diseases with huge burden for children as well as pregnant women.

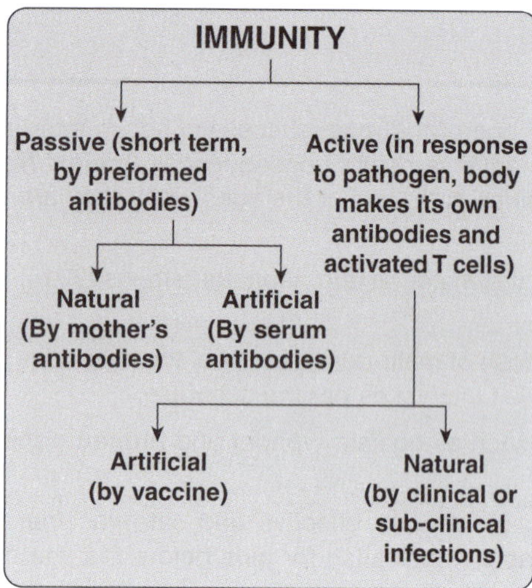

Figure 44.3 Key concept map of vaccines.

Study Questions

Short-Answer Questions

Write short notes on:

44.1. Various routes usually employed to deliver vaccines. Give one example of vaccine delivered by each such route.

44.2. Subunit vaccines. Discuss their immunological significance.

44.3. Role of adjuvants in vaccines.

44.4. Passive immunization.

Long-Answer Questions

44.1. What are attenuated and inactivated vaccines? How they are produced? Which of the two is better and why? Give at least two examples of viral and bacterial origin vaccines of each type.

44.2. What are DNA vaccines? Give their advantages and disadvantages over traditional vaccines.

Multiple-Choice Questions

Choose the ONE best answer.

44.1. Transplacental transfer of maternal IgG antibodies induces:
- A. Active natural immunity
- B. Active artificial immunity
- C. Passive natural immunity
- D. Passive artificial immunity

Correct answer = C.

44.2. Immunity of DPT vaccine lasts for:
- A. Few months only
- B. Less than 2 years
- C. 5–10 years
- D. More than 15 years

Correct answer = C.

44.3. Attenuation of microorganism for vaccine production is done by:
- A. Heating
- B. Treating with formaldehyde solution
- C. Treating with alkylating agents
- D. Growing under unfavorable conditions

Correct answer = D.

44.4. Which of the following is NOT a type of vaccine categorized based on the source used for their production?
- A. Weakened/killed microorganism or its subunit vaccine
- B. Purified pathogenic macromolecule vaccines/subunit vaccine
- C. DNA/RNA vaccines
- D. Recombinant vector vaccines
- E. Lipid vaccines

Correct answer = E.

44.5. Vaccine which is not of bacterial origin:
- A. Cholera vaccine
- B. Typhoid vaccine
- C. Plague vaccine
- D. Influenza vaccine

Correct answer = D.

44.6. What is not true for DNA vaccines?
- A. Response produced is exact replica of the one produced by active pathogen.
- B. Induce the production of both the humoral and cell-mediated immune response.
- C. Immunity produced is of prolonged duration.
- D. Only a small number of memory cells are generated.

Correct answer = D.

44.7. Vaccine which is not of viral origin:
- A. Polio
- B. Rabies
- C. Pertussis
- D. Hepatitis

Correct answer = C.

44.8. What is not true of multivalent vaccines?
- A. Increase the immunogenicity of subunit and purified pathogenic macromolecular vaccine.
- B. It synthesizes a polypeptide having immunodominant epitopes of both the B and T cells.
- C. Delivery systems are used for its delivery.
- D. Liposomes cannot be used for their production.

Correct answer = D.

44.9. First and second doses of Measles vaccines are given between:
- A. 6–8 weeks and 8–12 months respectively.
- B. 1–4 and 8–12 months respectively.
- C. 4–6 and 10–12 months respectively.
- D. 9–12 and 16–24 months respectively.

Correct answer = D.

44.10. Why are the vaccines produced by genetic engineering considered safe?
- A. They have antigen only from the coat of the pathogen.
- B. They are attenuated form of pathogen.
- C. They have killed form of pathogens.
- D. They have all the genes of the pathogen.

Correct answer = A.

Multiple-Choice Questions (NExT/NEET PG)

44.1. Which of the following is a subunit vaccine?
- A. Hepatitis B
- B. Measles
- C. Tetanus
- D. Diphtheria

Correct answer = A.

44.2. Which of the following vaccines is given immediately after the birth of the child?
- A. Anthrax
- B. Hepatitis B
- C. Hepatitis A
- D. Influenza

Correct answer = B.

44.3. What is not true about the use of adjuvants in vaccination?
- A. Enhance the immune response of some vaccines.
- B. Alum is the most commonly used adjuvant.
- C. MF59 and AS04 adjuvant are specific for human vaccines.
- D. Adjuvant act by enhancing Tc 2 cell response.

Correct answer = D.

44.4. For the Cholera vaccine the immunity maintenance period in adults is:
- A. 2 years.
- B. 6 years.
- C. 8 years.
- D. 12 years.

Correct answer = A.

44.5. ATS inoculation soon after injury that brings contact with the soiled objects gives protection against the attack of:
- A. Poliomyelitis
- B. AIDS
- C. Tetanus
- D. Influenza

Correct answer = C.

Appendix
Unit IX: Immunology

COMMON DISORDERS IN IMMUNOLOGY

An overview of common immunological disorders and their diagnosis is given in Table 1 below.

Table 1 Common immunological disorders and their diagnosis

Type of Disease	Name of Disease	Abnormality	Diagnosis
Immunodeficiency disorders	*Severe combined immunodeficiency disorder (SCID)*	Defect in adenosine deaminase	Low CD4 and CD8 counts
	Acquired immunodeficiency syndrome (AIDS)	HIV infection	Antibodies against HIV p24 antigen Detection of HIV-RNA
	Agammaglobulinemia	Defect in development and maturation of B cells	Low IgG levels in blood
Hypersensitivity	*Anaphylactic reaction (allergy)*	Cross-linking of IgE causes degranulation of mast cells	Elevated eosinophil count, IgE levels
	Hemolytic disease of the newborn	Destruction of fetal RBC by maternal antibodies	Increased reticulocyte count Positive direct Coombs' test
	Systemic lupus erythematosus	Immune complex–mediated tissue damage	Elevated antinuclear antibodies
	Contact dermatitis	Delayed-type hypersensitivity reaction	Positive skin test for allergens
Plasma cell disorders	*Multiple myeloma*	Monoclonal proliferation of plasma cells	Presence of M band in protein electrophoresis Presence of Bence Jones proteins in urine Increased plasma cell number in bone marrow biopsy
Paraproteinemia	*Waldenström macroglobulinemia*	Excessive proliferation of B lymphocyte clones	Elevated IgM in blood

Clinical Case Questions

Chapter 40 Introduction to Immune System and Immune Cells

40.1. A 6-year-old boy presents with fever for 2 months, cough for a month, decreased appetite and weight loss. His parents were diagnosed with AIDS 7 years earlier, but the child has never been taken to the hospital. The child is febrile, has decreased oxygen saturation (SpO_2 = 91% in room air) and tachycardia. General physical examination reveals dehydration, severe acute malnutrition, clinical pallor, extensive oral candidiasis and generalized lymphadenopathy. The chest displays clinical signs of bilateral pneumonia. There is moderate firm hepatosplenomegaly. A clinical diagnosis of probable pediatric HIV infection with opportunistic infections is made, and appropriate investigations are ordered.

A. How can pediatric HIV infection be suspected and confirmed in this child?

Answer—A: ELISA for antibodies against HIV can screen for HIV infection, but genetic testing is required for confirmation. During the window period (about 30 days to 3 months from the infection), antibodies may not yet be detectable, so ELISA for antibodies may be negative even in the presence of HIV infection. For confirmation during the window period and also for confirmation of ELISA results, genetic testing by PCR and Western blot for HIV RNA or HIV antigen (viral proteins like p24 capsid protein) respectively is advised. In this case, the child is likely to have IgG antibody to HIV.

B. How will you gauge the severity of immune suppression?

Answer—B: The severity of immune suppression is measured by the CD4% in children under 5 years and the CD4 cell values (cells/mm^3) in children above 5 years, comparing them with age-dependent cut-offs. CD4% is the percentage of white blood cells that are CD4 cells. In adults, the normal CD4 count is between 500 to 1000 cells/mm^3. A CD4 count less than 200 cells/mm^3 and/or opportunistic infections is a cause of concern as it implies immunosuppression and signifies the onset of Acquired Immunodeficiency Syndrome (AIDS).

40.2. A 57-year-old woman presented with chief complaints of fever of 103°F, nonproductive cough and dyspnea which has progressed over 1 week. She has anorexia and lost 12 kg over the last 3 months. She complained of several bouts of loose motions in last 3 months. She also told that she used heroin and cocaine intravenously few years ago. On examination, she was pale and in acute respiratory distress. Pelvic exam was normal except for vaginal candidiasis. Sputum examination was positive for Pneumocystis carinii.

A. What is the probable diagnosis?

Answer—A: History of intravenous drug abuse, significant weight loss over a short period of time and infections (diarrhea, vaginal candidiasis and pneumonia) are suggestive of AIDS. Pneumocystis carinii pneumonia (PCP) is a life-threatening lung infection that can affect people with weakened immune systems, such as those with AIDS.

B. Which investigations will you perform to confirm the diagnosis?

Answer—B: Diagnosis of HIV infection is made by the detection of antibodies against viral proteins using immunoassays and confirmed by detection of viral proteins (e.g., p24) or HIV RNA using polymerase chain reaction.

40.3. A 3-month-old baby was brought to pediatric emergency clinic with history of repeated fever, developmental delay, palpitations, and cyanosis. Physical examination revealed cleft palate, and echocardiography revealed patent ductus arteriosus.

A. What is the probable diagnosis? Mention the defect in this condition.

Answer—A: The probable diagnosis is DiGeorge syndrome. It is characterized by recurrent infections, congenital heart disease, and cleft palate. DiGeorge syndrome is typically due to the deletion of genes in the chromosome 22 at location 22q11.2.

Chapter 41 Antigen and Antibody

41.1. A 50-year-old farmer presented with the complaints of easy fatigue, weakness, weight loss and bleeding from nose and gums. His blood report showed elevated lactate dehydrogenase, ESR, uric acid and low hemoglobin. Blood IgM levels were increased.

A. What is the possible diagnosis? What is the basis of the symptoms?

Answer—A: The probable diagnosis is Waldenström macroglobulinemia. It is characterized by an excessive proliferation of B lymphocyte clones. It is due to mutation in gene MYD88. Environmental factors like exposure to farming, pesticides and wood dust increase the risk of the disease. The symptoms are due to increase in IgM monoclonal protein, which increases the viscosity of the blood by forming aggregates to each other.

41.2. A 35-year-old female presented to immunology clinic with fever, swelling and pain in the joints of the wrist and fingers, and rashes on her face. Physical examination revealed enlarged lymph nodes. Her blood report showed leukopenia, low platelets and low red blood cells. Antinuclear antibodies were found to be elevated.

A. What is the probable diagnosis? What are the mechanisms of autoimmunity?

Answer—A: The probable diagnosis is systemic lupus erythematosus (SLE). In SLE, the body's immune system produces antibodies against body's proteins in the cell nucleus. The possible mechanisms of autoimmunity include release of hidden antigens, cross-reacting antibodies or altered expression of MHC molecule.

Chapter 42 Innate and Adaptive Immunity

42.1. A 45-year-old female presents with complaints of joint pain and stiffness, predominantly in her hands and wrists, which has been progressively worsening over the past 6 months. She also reports morning stiffness lasting for more than an hour. On examination, there is tenderness, swelling and warmth noted in multiple joints, including the proximal interphalangeal (PIP) joints and wrists. Additionally, she exhibits ulnar deviation of the fingers and swan-neck deformity. She denies any recent trauma or fever. X-rays reveal findings such as periarticular osteopenia, joint space narrowing, erosions are seen. Further investigations show elevated levels of rheumatoid factor and anti-CCP antibodies.

A. What is the diagnosis and what type of hypersensitivity is this disease an example of? What is the basis of treatment using Aspirin?

Answer—A: This is a case of rheumatoid arthritis and a Type 3 hypersensitivity. Rheumatoid arthritis is a chronic inflammatory condition characterized by arthritis of peripheral joints. It is the most common form of chronic inflammatory arthritis. Rheumatoid factors are antibodies against self IgG. The IgG-Rheumatoid factor complex deposit in joints, where they activate the complement cascade, resulting in a type III hypersensitivity reaction, leading to chronic inflammation. Increase in vascular permeability increases leakage of fluid from capillaries and its accumulation causing edema and swelling. Aspirin is used in the treatment of rheumatoid arthritis as it reduces the synthesis of inflammatory mediators like prostaglandins by inhibiting the enzyme cyclooxygenase.

B. Patient was also started on low dose Methotrexate therapy. However, she developed anemia. Discuss the possible cause.

Answer—B: Folate deficiency is often seen when patients are put on Methotrexate. This leads to anemia and it occurs because Methotrexate is antagonistic to folate. Methotrexate inhibits the enzyme dihydrofolate reductase thereby decreasing the pool of reduced folates. Folate supplementation is required when a person is taking Methotrexate.

Chapter 43 Humoral and Cellular Immunity

43.1. A 15-year-old girl presents to the emergency department with sudden onset shortness of breath, wheezing and chest tightness. She reports a history of intermittent episodes of cough and dyspnea, usually triggered by exposure to cold air or allergens. She denies any recent respiratory infections. On examination, she is tachypneic with bilateral expiratory wheezes and decreased air entry. Her oxygen saturation is 88% on room air. Chest X-ray is normal. Pulmonary function testing (PFT) shows a decrease in FEV1/FVC and reversibility of FEV1 after bronchodilator administration.

A. What is the diagnosis and type of hypersensitivity?

Answer—A: The diagnosis is bronchial asthma and this is a Type 1 hypersensitivity.

B. What types of white blood cells are involved, and what type of arachidonic acid derivative mediates this response?

Answer—B: The cells involved are eosinophils, and leukotrienes (derived from arachidonic acid) mediate this response through the lipoxygenase pathway.

43.2. The 20-year-old male, diagnosed with acute myeloid leukemia (AML), underwent a hematopoietic stem cell transplant from his sister. After 30 days, he developed a maculopapular rash starting on his face and spreading to the rest of his body. He also developed watery, blood-streaked diarrhea and painful oral ulcers. The patient has been on oral immunosuppression with cyclosporine and steroids.

Investigations:

LFT: AST/ALT 234/214 IU

KFT: Urea/creatinine—54/1.2 mg/dl

CBC: Hb/TLC/platelets—9 g/4800 cumm/90,000

It is given that the Hb and platelet counts are below normal ranges, while the TLC is on the lower side of normal.

A. What is this reaction called?

Answer—A: The reaction is Graft-versus-Host (GVH) reaction. It is an immune reaction occurring in subjects who have undergone transplantation.

B. Explain the basis of this reaction.

Answer—B: GVH reaction develops when immune cells from the donor attack the recipient host tissue. It occurs when graft contains immunocompetent cells and donor T cells mount an immune response against the host.

43.3. A 72-year-old male patient presented with joint pain, fever, loss of appetite and difficulty in the movement of limbs. Past history revealed recurrent fractures at various parts of the body. His serum urea and creatinine levels were elevated. Radiological investigations showed lytic lesions in tibia and radius bones. Serum electrophoresis showed a band in the gamma region.

A. What is the probable diagnosis?

Answer—A: The probable diagnosis is multiple myeloma.

B. What is the basis of his clinical features?

Answer—B: The clinical features include bone pain and pathological fractures (extensive destruction of bones by malignant plasma cells), recurrent infections (suppression of humoral immunity) and renal failure (excretion of light chain causes tubular damage).

C. How is the diagnosis confirmed in this condition?

Answer—C: It is diagnosed by the presence of M band in the gamma region in the protein electrophoretogram and presence of increased plasma cells in bone marrow biopsy.

Chapter 44 Vaccines

44.1. A middle-aged pregnant lady living in a remote area visited a hospital and was concerned about the health of her unborn baby. Her husband had poliomyelitis when he was 4 years old as his parents never took him to the health center for immunization. She was worried if her child might also have poliomyelitis.

A. What advice would you give to the pregnant lady?

Answer—A: She should be advised to undergo regular antenatal checkup so that the status of her health as well as that of her baby could be monitored. Sensitize her about the importance of immunization during pregnancy and of the newborn. She should also be told that Polio is not a hereditary disease and therefore cannot be transmitted directly from a father to his child. Polio is an infection that can very well be prevented by vaccine. On 27 March 2014, the World Health Organization (WHO) declared India a polio-free country.

B. Explain the immunization schedule that she should follow during pregnancy and also, the National Immunization Schedule for her baby after birth.

Answer—B: Refer to Figure 44.2.

C. Name the polio vaccines, their types and protection period provided by them.

Answer—C:

S. No.	Name of Polio Vaccines	Type of Vaccine	Period of Protection
1.	Oral polio vaccine (OPV) (Sabin)	Live attenuated	>15 years
2.	Inactivated polio vaccine (IPV) (Salk)	Inactivated	>15 years

Unit X

Special Topics

Competencies Addressed

- **BI5.5** Interpret laboratory results of analytes associated with metabolism of proteins
- **BI6.13** Describe the functions of the kidney, liver, thyroid and adrenal glands
- **BI6.14** Describe the tests that are commonly done in clinical practice to assess the functions of these organs (kidney, liver, thyroid and adrenal glands)
- **BI6.15** Describe the abnormalities of kidney, liver, thyroid and adrenal glands
- **BI7.5** Describe the role of xenobiotics in disease
- **BI7.6** Describe the antioxidant defense systems in the body
- **BI7.7** Describe the role of oxidative stress in the pathogenesis of conditions such as cancer, complications of diabetes mellitus and atherosclerosis
- **BI10.1** Describe the cancer initiation, promotion oncogenes and oncogene activation. Also focus on p53 and apoptosis
- **BI10.2** Describe various biochemical tumor markers and the biochemical basis of cancer therapy

Unit X: Overview

Uncontrolled growth of cells, due to mutations in oncogenes, tumor suppressor genes or genes controlling apoptosis leads to tumors, either benign or malignant. Cancer cells differ from normal cells in metabolic, morphological, and many other aspects. Proto-oncogenes present in the normal tissues can be mutated to oncogenes or the function of tumor suppressor genes can be lost, leading to accumulation of mutations and subsequently, tumors. Genetic, environmental, physical, chemical or biological factors can predispose to cancer (Chapter 45).

Human body is exposed to a number of foreign compounds, such as drugs, industrial chemicals, food additives, environmental pollutants, which can be harmful and therefore, need to be chemically altered to an excretable form. Such compounds, known as xenobiotics, are metabolized by specific pathways. Sometimes, compounds which are produced by metabolism of xenobiotics are more harmful than the original compound (Chapter 46).

Oxidative stress can cause damage to deoxyribonucleic acid (DNA), protein and lipids. Antioxidants, which include enzymes and vitamins, can counter the effects of oxidants. Oxidative stress leads to various conditions like cancer formation, atherosclerosis and complications of diabetes mellitus. Toxicity due to oxidative stress is also used therapeutically, as in radiation treatment for cancer (Chapter 47).

Specific organs perform specific functions. Evaluation of a specific analyte associated with a particular organ can help in diagnosing or monitoring a disease. Functioning of an organ can be evaluated by a group of investigations, known as organ function tests. A normal test does not rule out organ dysfunction, but an abnormal test almost always signifies underlying disease or dysfunction. These tests may involve assessment of metabolites processed by that organ or markers associated with injury to that organ. A group of analytes can also indicate the location of injury, such as BUN-creatinine ratio and fractional excretion of sodium for differentiating prerenal azotemia from intrarenal azotemia (Chapter 48).

A primary healthcare physician must have the knowledge of common causes of oxidative stress and its effects on health. Knowledge of cancer-causing factors is essential to promote reduction of exposure to such factors and for screening in predisposed population, which can reduce risk of cancer and its associated morbidity. Tumor markers are commonly used to detect or monitor progression or effectiveness of the therapy. Drugs which are metabolized through cytochrome P450 system can worsen the manifestations of porphyrias and this must be considered while prescribing such drugs. Also, the differences in xenobiotic metabolism can alter a person's response to a drug or its adverse effects. While evaluating a clinical condition, assessment needs to be made keeping in mind the affected organs and metabolic pathways to make a correct diagnosis. For example, in case of hyperbilirubinemia, excessive hemolysis results in prehepatic jaundice (increased unconjugated bilirubin), while defective excretion leads to posthepatic jaundice (increased conjugated bilirubin).

Cancer

45

Learning Objectives

At the end of this chapter, the student will be able to:

- Describe the differences between normal and cancer cells
- Explain the characteristics of oncogenes and tumor suppressor genes
- Describe apoptosis
- Discuss tumor markers with examples

I. OVERVIEW

For normal growth and damaged cell replacement, cells divide in a highly controlled way regulated by specific gene products. Uncontrolled growth leads to tumors, which could either remain localized (benign tumors) or their cells could travel to different organs/tissues to cause malignancy. Uncontrolled growth and malignancy occur as a result of mutations in cancer-forming oncogenes and/or tumor suppressor genes. Cancer-forming oncogenes are normally present in the body as inactive proto-oncogenes, whose conversion to oncogenes is also regulated by specific genes. Bioregulators which control the transformation of a normal cell into a cancerous cell bring about this change by acting either as growth factors or their receptors, stimulators of oncogenes or inhibitors of tumor suppressor genes. Specific substances either produced by their respective tumors or by the body in response to the presence of specific tumors are used as biomarkers to screen people for the presence of specific cancers or to monitor the tumor progression or the effectiveness of their therapeutic process.

II. INTRODUCTION

Uncontrolled growth of cells in tissues forms tumors. **Tumors can be either benign or malignant. Malignant tumors are designated as cancer.** The term "cancer" is derived from the Latin word "cancrum," meaning "crab" and the cancer cells migrate to other tissues of the body like the feet of the crab. Fatality rate of cancer patients is second only to the patients suffering from cardiovascular diseases. Approximately 25% of all deaths in the world are due to cancer. In developed countries also, more than 20% of all deaths have been attributed to cancer. At the molecular level, cancer comprises a group of approximately 200 different diseases.

A. Important properties of malignant tumors

- Cell **growth and multiplication is very rapid due to loss of contact inhibition** and cells become insensitive to anti-growth factors.
- Cells **invade various tissues/organs of the body**. Cells of the benign tumors never undergo metastasis or spread to other tissues or organs of the body.
- Cells stimulate **angiogenesis**.
- Cells become immortal due to **inhibition of apoptosis**.
- Cells undergo **dedifferentiation**, whereby they regress from a specialized function to a simpler state reminiscent of stem cells. They may also stop producing certain specialized proteins produced by specialized cells.

B. Differences between normal and cancer cells

Cancer cells differ from the normal cells in the following processes:

1. **Cell cycle time:** Normally the cell cycle consisting of G1, S, G2, and M phases is of 18–24 hours (Fig. 45.1). Although the cell cycle time stays the same in both normal and cancer cells, yet, **in tumor cells the percentage of the cells in the dividing phase increases by 2–5 times**.

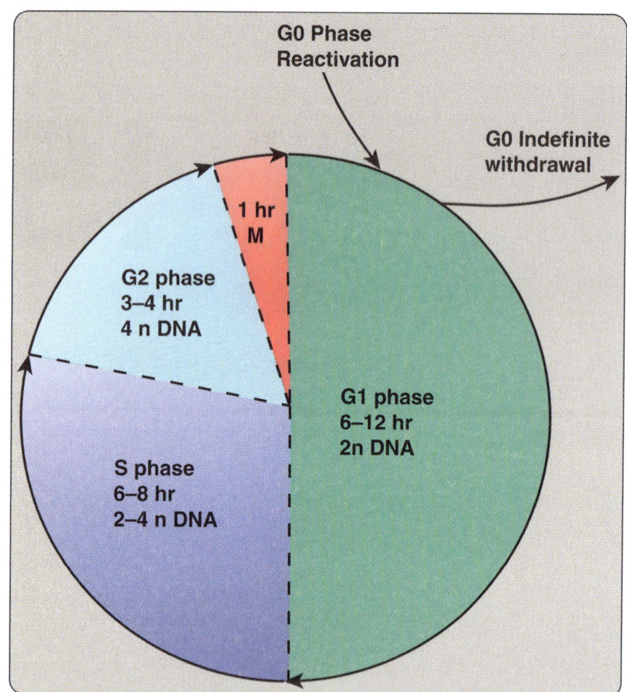

Figure 45.1 Cell cycle - Interphase is divided into G1, S, and G2 phases. One cycle is separated from the other by Mitosis (M). Cells may withdraw from the G0 or reenter from it.

This difference helps us to develop cytotoxic drugs which can selectively kill the dividing cells without having any effect on the resting cells.

2. **Cell proliferation:** Proliferation coefficient is the ratio of the cells in the dividing phase to the ones in the resting phase.

   Doubling time in tumor cells varies from 10–450 days. With a doubling time of 100 days, malignant tumors may achieve a size of 1 cm in 8–10 years.

3. **Morphological changes:** A normal cell becomes malignant after undergoing transformation. Malignant cells are morphologically different from normal cells. These cells are **characterized by a large irregular nucleus, prominent nucleoli and scarce cytoplasm**. Normal cells occur in monolayer, but the malignant cells have **multilayered structure**.

4. **Contact inhibition:** Contact inhibition is a property of normal cells only. When one set of normal multiplying cells comes in contact with another set of normal multiplying cells, further multiplication stops due to contact inhibition. For example, when a deep cut in the skin occurs, cells on both sides of the cut start multiplying to fill the gap. Further multiplication stops by contact inhibition, when the gap gets filled and multiplying cells from both sides come in contact. However, **contact inhibition is lost in cancer cells**.

5. **Tissue adherence:** Normal cells adhere to other tissue cells with the help of a protein called vinculin. During malignant transformation of cells, vinculin becomes inactive due to phosphorylation by *tyrosine kinase*. Inactivation of vinculin **decreases the adhesion ability of cancer cells resulting in facilitation of metastasis**.

6. **Role of sialylation:** A characteristic property of the cancer cells is that their surface is highly negatively charged due to its higher content of N-acetylneuraminic acid (NANA). **Presence of negative charges on the surface of cancer cells prevents their adhesion.** Role of sialidase and its inhibition in malignancy, opens the door for developing effective diagnostic and therapeutic strategies of future.

7. **Metabolic alterations:** Significant alterations in metabolic pathways occur as normal cells get transformed into cancer cells. Cancer cells metabolize sugars mainly by anaerobic glycolysis by eliminating the role of enzymes of citric acid cycle and mitochondrial electron transport chain. Production of several enzymes and proteins can also be altered.

8. **Immortal nature of the cancer cells:** Cancer becomes immortal due to inactivation of apoptosis (programmed cell death).

9. **Metastasis:** Collagenase and stromelysin released by cancer cells help them **pass through membrane barriers leading to migration**. Migration usually takes place via blood vessels or lymphatics. Normal cells stay localized due to the presence of adhesion molecules. A significant decrease in important adhesion molecules, such as E-cadherin, facilitates the migration of cancer cells. Cancer cells also have increased activity of matrix metalloproteinases. These enzymes help in migration of cancer cells by degrading the basement membrane proteins.

10. **Angiogenesis in cancer:** Tumor cells, due to their rapid growth, need extensive blood supply. Hypoxia is an important stimulant for **neoangiogenesis** (formation of new blood vessels). Tumor cells secrete angiogenic substances, such as vascular endothelial growth factor (VEGF), which help them meet their demand for blood supply.

    VEGF, acting through *tyrosine kinase* receptors, present on the membranes of endothelial cells of blood vessels and lymphatic system, stimulates the multiplication of endothelial cells via specific

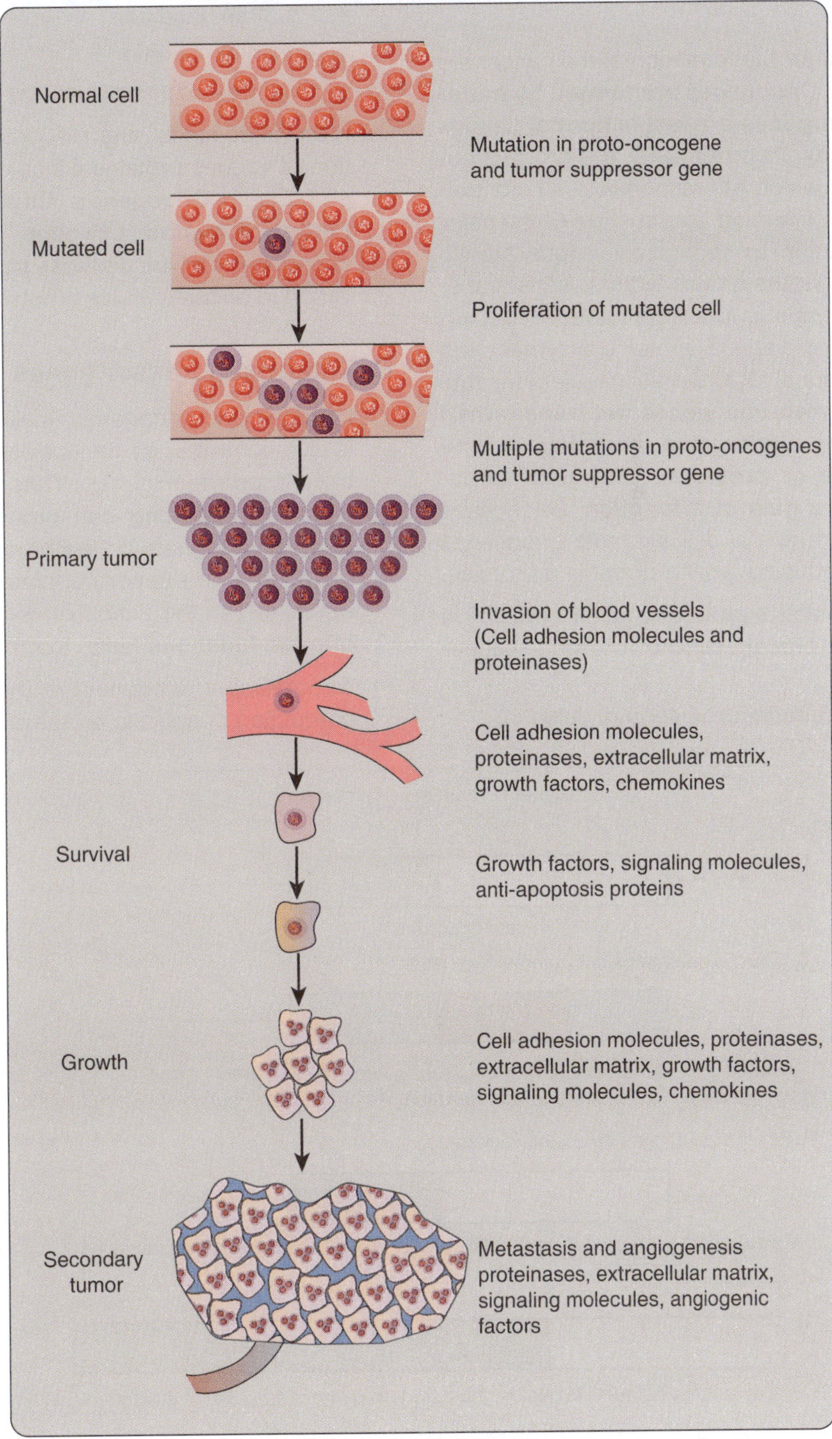

Figure 45.2 Development of cancer by gene mutations, proliferation and metastasis.

signal transduction pathways. VEGF is upregulated by hypoxia-inducible factor-1 (HIF-1), whose concentration increases manifold during hypoxia. Understanding the mechanism of tumor angiogenesis assumes special significance because inhibiting angiogenesis can arrest tumor formation and its growth.

III. MOLECULAR BIOLOGY OF CANCER CELLS

Uncontrolled growth in cancer cells is due to mutations in the genes that regulate cell cycle and proliferation, differentiation, and survival of cells (Fig. 45.2).

A. Oncogenes

Genes responsible for the formation of the cancer are called oncogenes. **Oncogenes are formed by mutation of proto-oncogenes present in normal tissues of the human body**. Approximately 100 proto-oncogenes are known which are responsible for controlling the growth, proliferation, and survival of the cells. These genes function by producing various growth factors, receptors for the growth factors, signal transduction proteins, various *kinases* involved in activation/deactivation of various proteins/enzymes and different transcriptional factors which regulate the rate of transcription. **Proto-oncogenes get transformed to oncogenes by undergoing mutation in one allele of its gene or promoter**. Oncogene formation is a result of a **gain of a function**. Oncogenes are produced from normal cellular proto-oncogenes responsible for producing growth-promoting proteins.

Conversion of an inactive proto-oncogene to an active oncogene can be brought by the following mechanisms (Fig. 45.3):

- Mutations (point mutation, insertion, deletion)
- Promoter insertion
- Enhancer insertion
- Chromosomal translocation

Proto-oncogenes are responsible for the production of various proteins which control various diverse processes. **Oncogenes thus produced can be further activated by gene amplification**. Proto-oncogenes and the diseases produced after their activation to oncogenes are enlisted below in Figure 45.4.

B. Tumor suppressor genes

In addition to oncogenes, malignant tumor formation is also controlled by the activity of tumor suppressor/onco-suppressor genes. **These genes are responsible for inhibiting cell proliferation in response to various signals.** In the process of transformation, mutation in the tumor suppressor gene makes them insensitive to the action of various signals. The **loss of gene functions** leads to malignancy.

Tumor suppressor genes are involved in controlling the production of specific proteins which perform various

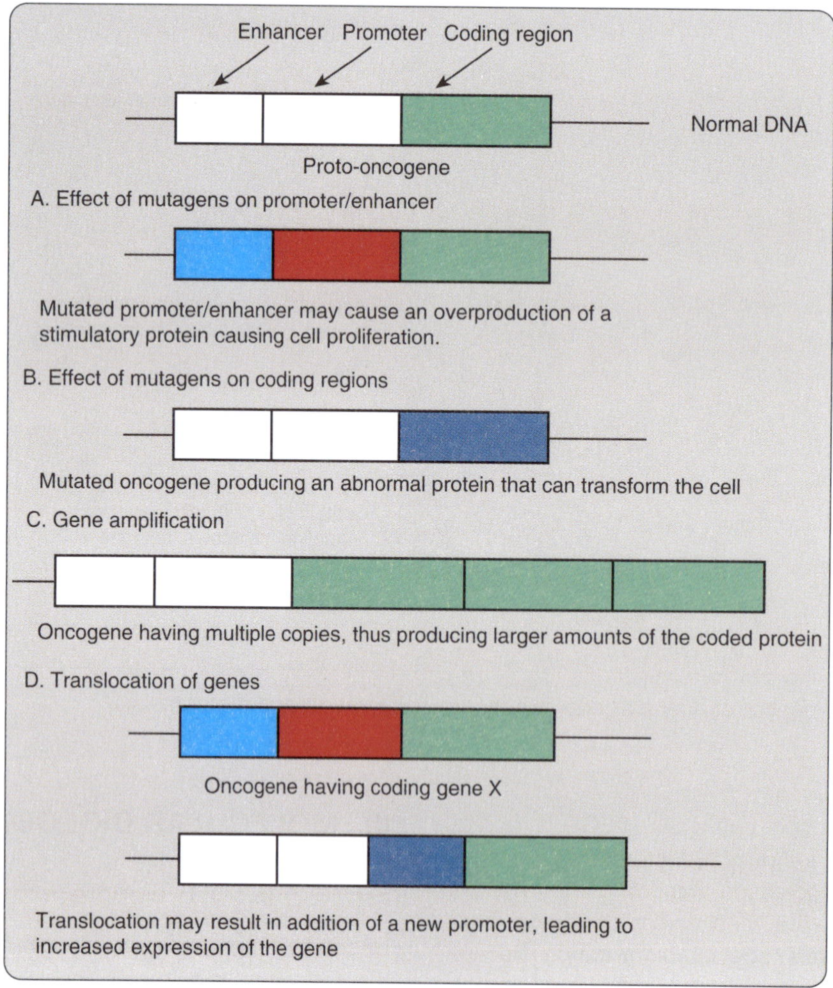

Figure 45.3 Mutagenic agents damaging chromosomal DNA.

Proto-Oncogene and its Activation Mechanism(s)	Process Influenced by Proteins Produced by Oncogenes	Oncogene-Associated Cancers
Cyclin D—Translocation Amplification	Regulation of cell cycle	Breast, liver and esophageal cancers
CDK4—Point mutation, Amplification	Regulation of cell cycle	Sarcoma and melanoma
ras—Point mutation	Signal transduction	Lung, colon, thyroid, and pancreatic cancers
akt2—Amplification	Signal transduction	Ovarian cancer
src—Overexpression	Signal transduction	Breast cancer
raf—Overexpression	Signal transduction	Myeloid leukemia
sis—Overexpression	Growth factors	Osteosarcoma
int-2—Amplification	Growth factors	Breast and bladder cancers
erb-B1—Overexpression	Growth factor receptors	Lung carcinoma
erb-B2—Amplification	Growth factor receptors	Breast, ovary, lung and stomach cancers
PDGFR—Overexpression	Growth factor receptors	Myelomonocytic leukemia
SMO—Amplification	Growth factor receptors	Basal cell sarcoma
Hox11—Translocation	Transcription factors	T cell leukemia
Myc—Translocation Amplification	Transcription factors	Burkitt's lymphoma, Lung carcinoma
Fos, Jun—Phosphorylation	Transcription factors	Breast cancer
Bcl-2—Translocation	Apoptosis regulators	Follicular B cell lymphoma
RARα—Translocation	Hormone receptors	Acute promyelocytic leukemia

Note: In this figure, name of proto-oncogene is written in small letters and italics (*ras, myc*) and its protein product capital, non-italics (Myc).

Figure 45.4 Oncogene-associated cancers produced in human beings.

functions. Some of the tumor suppressor genes and the processes by which they control the production of specific cancers are enlisted below:

1. p53, by regulating cell cycle and apoptosis, controls the production of many cancers.
2. p16, which regulated cell cycle, is associated with lung and pancreatic cancers.
3. BRCA 1, which is important for DNA repair, is associated with breast cancer.
4. Transforming growth factor beta (TGF-β), which regulates receptor functions, is associated with colon cancer.
5. Dysfunction of E-cadherin, which mediates cell adhesion, is associated with stomach cancer.

   It is important to note that the production of cancer is due to the overall effect of the activity of oncogenes and inactivation of tumor suppressor genes (Fig. 45.5).

IV. ETIOLOGY OF CARCINOGENESIS

Origin of all cancer cells is **multifactorial**. Formation of malignant tumors is net manifestation of genetic, environmental, physical, chemical or biological factors. These cancer-causing factors can be classified into predisposing factors and mutagens. All cancer cells originate from one mutated cell which by multiplication produces a tumor.

A. Predisposing factors

1. **Age:** Occurrence of cancer is mainly associated with old people (>50 years of age). However, cancer has also been known to occur in children, viz. retinoblastomas, neuroblastomas, Wilms' tumor, etc.
2. **Heredity:** Inherited high-risk factors (precancerous) play an important part at least in some types of cancers. For example, in some families, many women develop breast cancer and/or

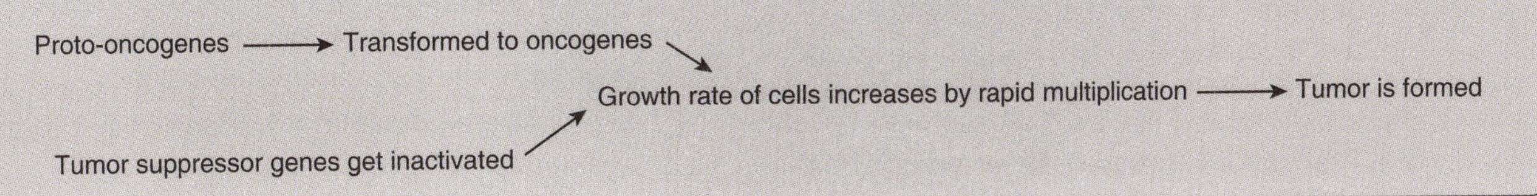

Figure 45.5 Mechanism of tumor formation.

ovarian cancer, often at a younger than the usual age for these cancers. This is known as Hereditary Breast and Ovarian Cancer syndrome (HBOC), caused by an inherited mutation in either the BRCA1 or BRCA2 gene.

Hereditary nonpolyposis colorectal cancer (HNPCC), adenomatous polyposis of colon (APC), Li-Fraumeni syndrome (p53 mutations) and retinoblastoma (RB-1 mutations) also show hereditary risk.

3. **Environmental factors:** In approximately 80% of human cancer cases, various environmental factors discussed below play a significant role in the development of cancer.

   a. **Lifestyle**

      i. **Cigarette smoking:** Aromatic hydrocarbons, nicotine, carbon monoxide, and nitrogen dioxide present in the cigarette smoke are responsible for developing lung cancer. Smoking 10 cigarettes per day increases the chance of occurrence of lung cancer by at least 15 times. The activity of glutathione S-transferase (GST) is responsible for detoxification of various chemicals present in smoke. Persons deficient in this enzyme are much more susceptible to the occurrence of cancer.

      ii. **Tobacco chewing:** Oral cancers in human beings have been attributed to tobacco chewing. In India, 20% of oral cancers are due to tobacco chewing.

      iii. **Alcohol consumption:** Increases the chance of occurrence of oral, pharyngeal, esophageal and liver cancers.

      iv. **Dietary:** Certain fungus, such as Aspergillus flavus, which grows on rice, wheat and other grains under humid conditions, produce aflatoxin, a potential hepatocarcinogen. Fungus growing on the cattle fodder enters the human system with the intake of cow's milk. The consumption of infected grains can lead to formation of liver cancer.

      v. **Occupational:** Exposure to asbestos, benzene, naphthylamines, beryllium, etc. predisposes the human body to the formation of various types of cancers.

      vi. **Therapeutic drugs:** Use of some therapeutic drugs lead to the formation of liver and gastrointestinal (GI) tract cancers.

B. Mutagens

Any substance which by damaging DNA causes mutation is called a **mutagen**. Substances which by enhancing the rate of mutation convert a normal cell to a cancerous cell are called **carcinogens. All carcinogenic substances are mutagenic.** Mutagens can be physical, chemical or biological/viral in nature.

1. **Physical/radiation energy:** Any radiation rich in energy can act as a carcinogen.

   Radiation energy can cause mutation by damaging the DNA by the following mechanisms.
   - Causing single-stranded breaks in the DNA
   - Causing double-stranded breaks in the DNA
   - Deletion of purine or pyrimidine bases of DNA
   - Formation of pyrimidine dimers
   - Cross-linkage of DNA strands

   a. **Visible light:** It is mostly non carcinogenic; however, it **can become carcinogenic in the presence of photosensitive agents** which generate reactive oxygen species.

   b. **UV light:** It is an important carcinogenic agent. Sunlight is its main source. Increase in the exposure time and decrease in the quantity of melanin, increase the chances of occurrence of skin cancer. Exposure to UV light causes cancer primarily by forming pyrimidine dimers and secondarily by immunosuppression.

   c. **X-rays and γ-rays:** They can also induce mutations by radiation energy with the formation of reactive oxygen species (ROS)/free radicals, e.g., OH^-, superoxides, etc.

      Particulate radiations by α-particles and neutrons are highly carcinogenic. Occurrence of higher incidence of leukemias in Japan after atomic attack, higher incidence of lung cancer in miners working in radioactive ore mines, and development of thyroid cancer in kids after exposure of their necks to therapeutic radiations provided experimental evidences that ionizing radiations can lead to malignancy of specific organs/tissues.

2. **Chemical carcinogens:** It has been established that more than 75% of the human cancers are primarily due to their exposure to chemicals. Well-documented groups of chemicals, with specific examples, are enlisted below:
   - Polycyclic aromatic hydrocarbons (benzopyrene, dimethylbenzanthracene)
   - Aromatic amines (2-acetylaminofluorene, N-methly-4-aminoazobenzene/MAB)

- Nitrosamines (dimethylnitrosamine, diethylnitrosamine)
- Therapeutic drugs (alkylating agents, e.g., cyclophosphamide, diethylstilbestrol)
- Naturally occurring compounds (dactinomycin, aflatoxin B_1)

a. **Mechanism of action of chemical carcinogens:** Chemical carcinogens cause mutagenesis by **either directly acting upon their targets** (DNA, RNA, proteins of expression) as carcinogens **or these compounds act as procarcinogen and have to be converted to active carcinogens by the enzymes present in the human body**. Carcinogens act like "electrophiles," i.e., deficient in electrons, and they act by binding to the "nucleophiles" i.e., electron-rich nitrogen and oxygen atoms present in bases of DNA, RNA or amino acids of expression proteins.

i. **Mechanism of procarcinogen activation:** Enzymes of cytochrome P450 system present in the endoplasmic reticulum (ER) of the liver are primarily responsible in converting an inactive procarcinogen to an active carcinogen.

ii. **Mechanism of action of carcinogens:** Most of the chemical carcinogens act by covalently binding to either the atoms present in the bases of DNA/RNA molecules or to their phosphodiester bonds. The adduct product thus formed causes mutation by wrongly pairing with the base of the complementary strand of the DNA during replication. Figure 45.7 demonstrates that how binding of dimethylnitrosamine to DNA results in damaging the DNA structure. **Normally guanine pairs with cytosine; however, methylguanine pairs with thymine during replication resulting in mutation**.

In another mechanism, compounds like ethidium bromide and proflavines can form adducts with DNA bases, leading

Ames Test

How is it made sure that a drug or a product (e.g., cosmetics, preservatives) intended for human use is not mutagenic, and does not predispose to cancer? To test this possibility, drugs or similar products are subjected to a versatile testing process before marketing—though it dates back to 1970s, Ames test still finds its use (Fig. 45.6). This test uses a strain of Salmonella typhimurium which is defective in histidine synthesis, due to mutations in genes involved in histidine biosynthesis. Since the bacteria require histidine for growth, they do not multiply and form colonies when grown on growth media deficient in histidine. If the bacteria form colonies in the presence of the chemical being tested, this indicates that the chemical has caused mutations in the previously mutated genes involved in histidine synthesis, reverting their function to normal. Thus, presence of bacterial colonies denotes the potential of the chemical to cause mutations. A modified version of the test uses a rat-liver extract which is added to the chemical. This modification allows the detection of a chemical which can cause mutations as a result of bioactivation.

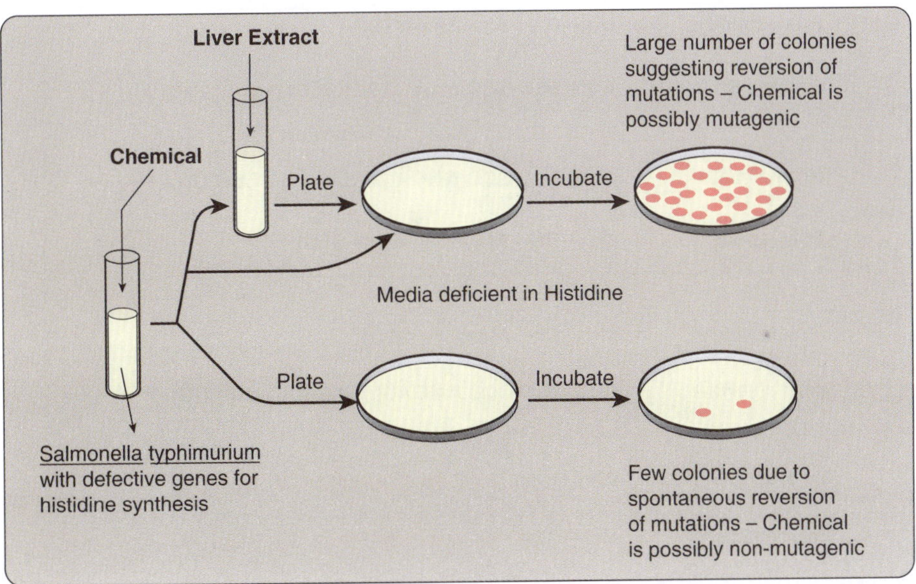

Figure 45.6 Ames test.

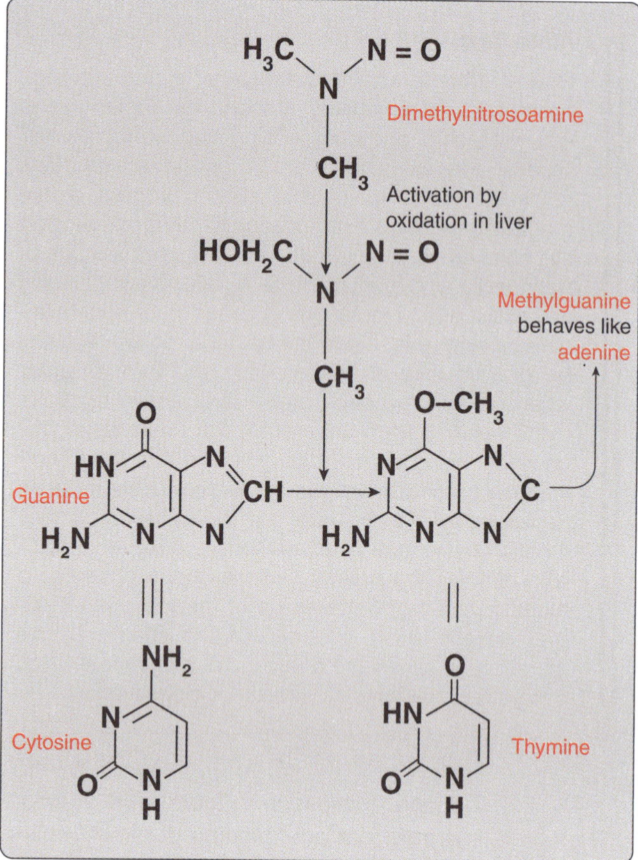

Figure 45.7 Mechanism of mutation caused by nitrosamines.

to mutations. This adduct formation can result in either the addition or deletion of a base in a DNA strand, which can cause frameshift mutations (Fig. 45.8).

3. **Biological carcinogens:** Approximately 20% of human cancers are caused by viral genome which could either be DNA or RNA. Some of the important DNA and RNA viruses causing cancer in human beings are enlisted below along with the type of cancer caused by them.

   a. **DNA viruses:** DNA viruses mainly causes tumors in animals. Following are the DNA viruses which causes tumors in human beings:

      i. **Human DNA viruses**

      - **Epstein-Barr virus (EBV)**: It belongs to the herpes family and produces:
        - **Burkitt's lymphoma,** a tumor of B lymphocytes. The disease is endemic in Africa.
        - **Nasopharyngeal carcinoma,** which is endemic in China.
      - **Hepatitis B virus (HBV):** It is responsible for causing liver cancer.
      - **Human papillomavirus (HPV):** It causes the following types of cancers:
        - **Multiple warts** (benign squamous cell papilloma) which in 30% cases become malignant. HBV types 1, 2, 4 and 7 are responsible for causing them.
        - **Cervical cancers:** In 90% of cases, HBV types 16 and 18 are responsible for causing them.

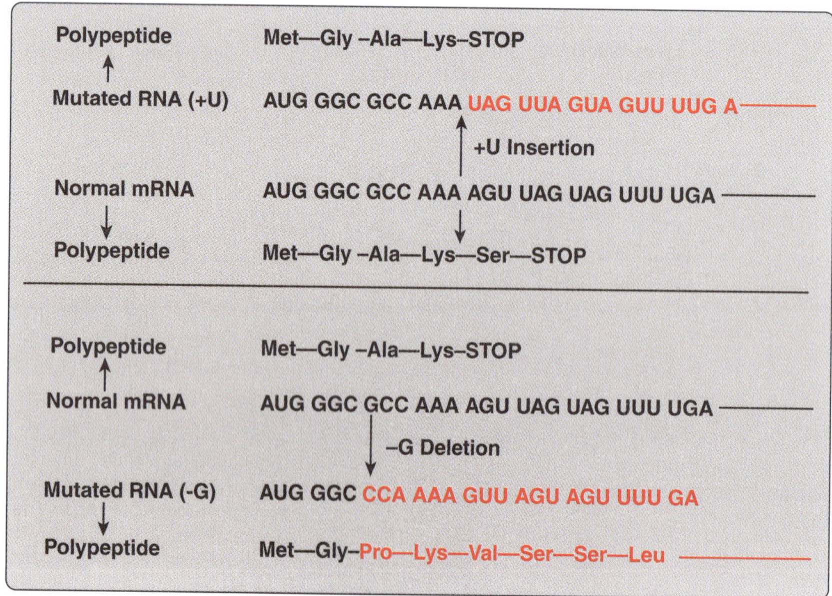

Figure 45.8 Mechanism of frameshift mutation caused by addition or deletion of a base.

> **Examples of specific proteins produced by DNA viruses**
>
> T antigens—Produced by polyomavirus
>
> EIA & EIB—Produced by adenoviruses
>
> E6 & E7—Produced by papillomaviruses

ii. **Mechanism of action of DNA viruses:** Viral DNA gets integrated in the host genome without completing its replicative cycle. The viral genes which have been transcribed earlier produce specific proteins which act on the host genome. Proteins produced by the viral genome modify the regulatory checks and balances of the host genes resulting into their transformation and occurrence of malignancy.

b. **RNA viruses:** All RNA viruses are retroviruses. Their RNA first gets converted to DNA by reverse transcriptase before insertion into the host genome. Examples include:

i. **Human T-cell leukemia virus (HTLV):** It causes leukemia/lymphoma in human beings, which is endemic in Japan. Viral "tat" gene produces proteins which are responsible for causing transformation. Recently, a "Tax" protein produced by HTLV has been shown to cause blood cancer leukemia in human beings by reacting with the Tax BP2 protein produced by the human genome.

ii. **Rous sarcoma virus:** It is responsible for causing tumors in animals. The viral genome contains four genes which are responsible for causing transformation in host. The four genes with the functions of the proteins produced by them are given below:

- Gag—produces specific viral antigens
- Pol—produces reverse transcriptase
- Env—produces protein for viral envelope
- Src—produces sarcoma

c. **Products of the oncogenes:** Oncogenes produce following types of proteins which are responsible for causing transformation. Approximately 50% of the produced proteins are *tyrosine kinases*.

i. **Tyrosine kinases:** These regulate the activity of the target proteins by covalently modifying them by phosphorylating their tyrosine residues. The target proteins can either be of receptors or present in the cytoplasm.

- **Receptor proteins**:
  - Epidermal growth factor (EGF) coded by ErbB
  - Platelet-derived growth factor (PDGF) coded by Neu
  - Colony-stimulating factor-1 (CSF-1) coded by c-fms
- **Cytoplasmic kinases:** pp60 coded by src

ii. **Guanosine triphosphate (GTP)-binding proteins** coded by Ras oncogenes

iii. **Nuclear proteins** coded by myc, fos and myb oncogenes

d. **Anticarcinogenic substances occurring in nature:** Natural products having anticancer potentials either **act as nonspecific cancer inhibitors due to their antioxidant properties or act as specific inhibitors of a particular tumor only.** Their antioxidant properties help them to act as scavengers of ROS which are highly carcinogenic. Other anticancer natural products act as inhibitors of some cancers, but their mechanisms of action are not known.

i. **General anticarcinogenics which act as antioxidant:**

- β-Carotene
- Vitamin A
- Vitamin C
- Vitamin E
- Lycopene

ii. **Specific anticarcinogenic:**

- Quercetin specifically inhibits prostate cancer
- Sinigrin specifically inhibits cancer of pancreas

Anti-mutagens can impede the conversion of a pro-mutagen into a mutagen, deactivate the mutagen, or hinder the interaction between the mutagen and DNA.

V. ROLE OF GROWTH FACTORS AND THEIR RECEPTORS IN CANCER

Growth factors are either polypeptide or protein in nature. They are mitogenic and may be involved in initiating cellular differentiation, remodeling and migration, and play a major role in the differentiation of stem cells.

Oncogenes code for the production of growth factors and their receptors. Some growth factors act as growth inhibitors, e.g., TGF-β. Growth factors work by binding with their specific receptors to initiate various signal transduction pathways. Receptors are usually transmembrane proteins, with both external and cytoplasmic domains. Cytoplasmic domains have *kinase* activity.

Both the growth factors and their receptors can become oncogenic under the influence of specific mutations, and the **increased concentration of growth factors and/or their receptors leads to the proliferation of the cells resulting in the formation of malignant tumors.**

Classic example is that of PDGF which binds to its receptor and through phospholipase-inositol triphosphate-diacylglycerol system, activates cellular MYC and FOS genes, which in turn stimulates mitosis.

Few important growth factors with their function are mentioned below:

- EGF: Stimulates growth of epidermal and epithelial cells
- Erythropoietin (EPO): Regulates early erythropoietic cell development
- Fibroblastic growth factors (FGFs): Promote cell proliferation
- Interleukins: Act by influencing the activity of immune system
- Nerve growth factor (NGF): Acts on neurons
- PDGF: Stimulates growth of mesenchymal and glial cells
- Transforming growth factor-α (TGF-α): Acts like EGF

A. Mechanism of action of growth factors

Growth factors behave like hormones and may act by either of the following ways:

1. **Endocrine:** Their site of action is different from the site of their production. They are carried to the site of action from their site of production via blood.
2. **Autocrine:** Their sites of action and production are same.
3. **Paracrine:** They act on cells in the vicinity of site of secretion.

VI. CANCER AND APOPTOSIS

Apoptosis occurs normally during development and aging to maintain cell populations in tissues. Apoptosis also occurs as a defense mechanism against the damaged tissues. Such damage could have been caused by radiation, toxins, free radicals, inactive growth factors, hormones, T cells of the immune system, etc. Apoptosis involves the participation of a cascade system of various enzymes and proteins. **Apoptosis is a programmed cell death** and differs from necrosis which is not genetically controlled and unlike apoptosis its products are not automatically removed and cause local inflammation.

A. Mechanism of apoptosis

Apoptosis is very tightly regulated and consists of extrinsic and intrinsic pathways. Apoptosis occurs in two phases, i.e., initiation and execution.

1. Phases of apoptosis

a. **Initiation phase:** The **extrinsic pathway** starts with binding of the ligands (tumor necrosis factor-α [TNF-α] and Fas) to the death receptors present on the cell surface. Death receptors are proteins, and some of these can also bind to death domain of adapter proteins (FAS-associated protein with death domain [FADD]). Intracellular events (e.g., oxygen loss, damaged DNA, and various mutations) activate the **intrinsic pathway**, which cause changes in the inner mitochondrial membrane.

b. **Execution phase:** It involves action of cysteine *protease* called **caspases** which act by breaking the peptide bond between aspartate and cysteine amino acid residues of proteins. Caspases are present in the form of large molecular weight inactive procaspases.

2. Apoptosis pathways:
Apoptosis occurs by two pathways.

a. **Death receptor pathway (external pathway):** Major steps are:

- Binding of the ligands (TNF-α and Fas) to transmembrane death receptors present on the cell surface.
- The adapter (FADD) can also bind to death receptors.
- Binding of the above molecules forms a complex which converts procaspase-8 to caspase-8.
- Caspase-8 acts like an initiator and converts procaspase-3 to caspase-3.
- Caspase-3 acts like an effector after getting activated by various molecules.
- Caspase-3 causes cell death by digesting various proteins responsible for maintaining the structure and functions of the cell.

i. **Regulation of death receptor pathway:** Regulation of this pathway is done by the following proteins:

- FLIP (FLICE [FADD-like IL-1β-converting enzyme]-inhibitory protein) inhibits the activation of caspase-8.

- IAPs (inhibitors of apoptosis) inhibit the activation of caspase-3.
- SMAC (second mitochondria-derived activator of caspase) inhibits the action of FLIP and IAPs.

b. **Mitochondrial pathway (internal pathway):** Major steps are:
- Damaged DNA, inactive DNA repair enzymes, ROS, radiation, etc., causes pore in the outer mitochondrial membrane resulting in the release of cytochrome c into the cytoplasm.
- Apaf-1 reacts with cytochrome c in the cytoplasm and forms an apoptosome complex and the energy for complex formation is provided by ATP.
- Apoptosome activates procaspase 9 and causes its dimerization to caspase-9.
- Caspase-9 activates caspase-3.
- Caspase-3 causes the death of the cell by digesting important structural and functional molecules.
- **Caspase-9 act as an initiator of internal pathway, whereas caspase-8 is the initiator of external pathway**.

i. **Regulation of the mitochondrial pathway:** It occurs by the following ways:
- Activation of *p53* stimulates BAX (proapoptotic factor) which decreases mitochondrial membrane potential resulting in increase in its porosity.
- Bcl-2 is anti-apoptotic which opposes the above action.

ii. **Regulation of pro- and anti-apoptotic pathways:** Proteins of Bcl-2 family are divided into two groups, i.e., pro- and anti-apoptotic.
- Pro-apoptotic family members are further subdivided into two subclasses, i.e., channel formers (Bax, Bak & BK) and BH3 group (Bad, Bid & Bod/Bim).
- Anti-apoptotic family members are Bcl-2, Bcl-x and Bcl-w.

The balance between the pro- and anti-apoptotic regulators determine whether a cell is going to commit suicide or not. Pro-apoptotic channel formers stimulate the formation of channels in the mitochondria, while the anti-apoptotic family after getting inserted into the mitochondrial membrane inhibits the process of channel formation. Anti-apoptotic members also by binding to cytoplasmic Apaf inhibit the formation of apoptosome complex. The mechanism of apoptosis is summarized in Figure 45.9.

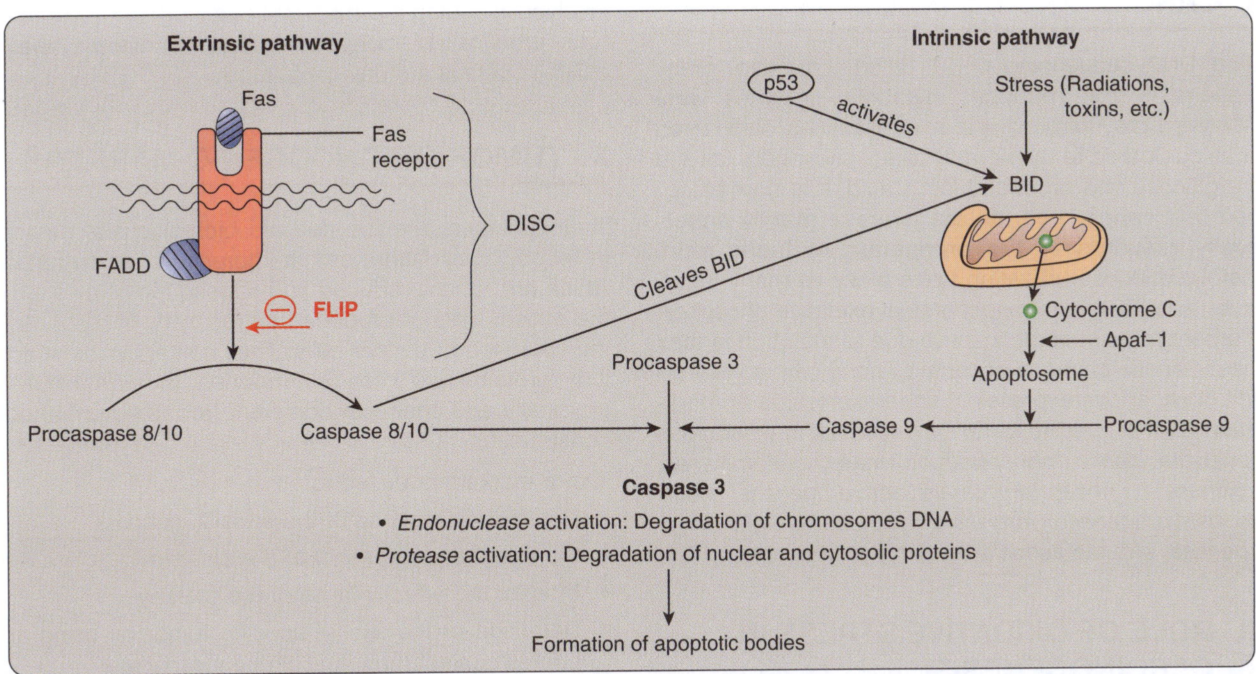

Figure 45.9 Mechanism of apoptosis. Apaf1 = Apoptotic *protease* activating factor 1; DISC = Death inducing signaling complex; FADD = FAS–associated death domain; BID = BH3 interacting-domain death agonist; FLIP = FLICE inhibitory protein.

p53

The p53 gene is a tumor suppressor gene, located on chromosome 17. **p53 plays a critical role both in the G1/S check point (in which cells arrest prior to DNA replication) and in the G2/M check point (in which arrest occurs before mitosis).** In the cell, p53 protein binds DNA and stimulates another gene to produce a protein called p21. When p21 is complexed with cell division-stimulating protein (cdk2), the cell cannot pass through to the next stage of cell division.

Mutant p53 is unable to bind DNA effectively, and as a consequence the p21 protein is not made. Thus cells divide uncontrollably, and form tumors. Mutations in p53 are found in most tumor types. However, when one copy of mutated p53 is inherited, person is at a very high susceptibility for development of several independent tumors in different tissues in early adulthood. This rare condition is known as **Li-Fraumeni syndrome**.

B. Cancer cells are anti-apoptotic

Cancer cells are immortal because they have developed mechanisms to inhibit apoptosis. The normal balance between pro- and anti-apoptotic systems is disturbed as the former gets inhibited and the latter gets overexpressed. **The most commonly mutated gene in various cancers is p53, which causes upregulation of the antiapoptotic factors while downregulating the proapoptotic factors**.

VII. CANCER AND METABOLISM

During DNA sequencing of 21 different tumors, many mutations in the important metabolic enzymes were observed. Otto Warburg and his colleagues discovered that as compared to the normal cells, cancer cells utilized more glucose and produced lactic acid. It is now established that **cancer cell utilizes glucose mainly anaerobically resulting in the production of lactic acid.** Anaerobic glucose utilization takes place as many tricarboxylic acid (TCA) and mitochondrial oxidative phosphorylation enzymes become inactive due to mutation in their genes. Genetic changes occurring during carcinogenesis result in either overexpression, underexpression or modification of the specific gene of metabolism. Alterations in *pyruvate kinase, phosphofructokinase, glutaminases, hexokinase II, phosphoglycerate dehydrogenase, isocitrate dehydrogenase, fatty acid synthase, lactate dehydrogenase*, etc. are associated with many cancers.

VIII. ROLE OF OXIDATIVE STRESS IN CARCINOGENESIS

Oxidative stress is defined as an imbalance between production and destruction of various ROS, free radicals and their reactive metabolites. The oxidative stress causing damage to DNA by various reactive oxygen species (ROS) and reactive nitrogen species (RNS) has been found to play an important role in the initiation and progression of the process of carcinogenesis. Damage to the skin by various environmental factors, damage to the internal organs and the aging process all have been directly or indirectly linked to the action of ROS and RNS resulting from the oxidative stress. Under normal conditions in living cells, excessive ROS and RNS are destroyed by various antioxidant mechanisms present in the body. (See Chapter 47 for details of oxidative stress and antioxidant mechanisms.) In normal cells, low-level concentrations of ROS and RNS are required for signal transduction before they get destroyed by various antioxidant mechanisms prevalent in the body. High concentrations of ROS and RNS species help the cancer cells to maintain their high metabolic rate which is essential to provide required energy for cell proliferation. Monitoring the concentration of ROS and RNS and developing various ways to regulate their concentration assume clinical significance to not only study the progression of the malignant tumors but also to develop various diagnostic and therapeutic strategies.

IX. ROLE OF miRNAs IN CANCER

Micro RNAs (miRNAs) are single-stranded RNA molecules which play an important role in carcinogenesis by controlling the transcription and/or translation of some of the important proteins involved in tumor formation or the process of apoptosis. Some miRNAs act as tumor suppressor by hydrolyzing specific mRNA species involved in coding the important carcinogenic or antiapoptotic proteins. In many cancerous conditions, specific miRNA species are downregulated.

X. TUMOR OR ONCOGENIC MARKERS

Tumor markers are the specific biomolecules, detected in blood, body fluids, cell membranes or in cytoplasm, which are either produced and subsequently released by the cancer cells or produced by the body in response to the presence of cancer cells. They can either be specific ions, proteins, enzymes or hormones. Tumor markers are physically and chemically different from the normal body constituents.

Tumor markers are used to:

- Detect the presence of the specific cancers
- Monitor the progression of the disease
- Monitor the effectiveness of the therapy
- Detection of the site of specific tumor by using antibodies against their specific tumor markers

In blood and body fluids, tumor markers can be detected by radioimmunoassays and enzyme immunoassays. Using the immunohistochemical analyses, cellular skeletons, lymphocyte surface antigens, cytoplasmic markers, oncogene products and cell adhesion molecules can be detected in tissue biopsies.

Please see Table 1 of Appendix Unit X for a list of tumor markers (p. 826).

XI. CANCER TREATMENT

Therapy used for the management of cancer depends upon the type of cancer and the class it belongs to.

Surgery and radiotherapy are the most commonly employed tools that are effective in destroying the cancer cells. These are very effective against solid cancers.

Chemotherapy is most effective in the management of leukemias, advanced lymphomas, choriocarcinomas and some other soft tissue malignancies. Chemotherapy is often supplemented by radiotherapy or surgery of the primary cancers. Surgery can also be followed by chemotherapy or radiotherapy to manage residual tumors. Chemotherapy acts by inhibiting cell division and their multiplication. It is most effective where the doubling time of the tumor is very high. Its effectiveness is inversely proportional to the size of the tumor. Cytotoxic drugs arrest the activity of both the cancer cells and the normal human cells in the dividing phase. Cell destruction follows the first-order kinetics, i.e., drug dose-dependent destruction is proportional to the percentage of cell and not on their number. Based upon their chemical structure and the mechanism of action, various anticancer drugs can be classified into 11 classes as shown in Figure 45.10.

Cancer immunotherapy uses substances made by the body or in a laboratory to boost the immune system and help the body find and destroy cancer cells. It can be used alone or along with other cancer treatments.

The different types of immunotherapies include the following:

- Monoclonal antibodies
- T-cell therapy
- Cancer vaccines

Targeted therapy for cancer is a form of molecular targeting that selectively acts on cancer cells, distinct from hormonal therapy or chemotherapy. It uses monoclonal antibodies and is an evolving method of cancer treatment.

S. No.	Class of Drug	Class Member	Specific Member	Mechanism of Action: Enzyme Activity/Conversion Inhibited
1.	Vitamin analog	Folic acid analog	Methotrexate	*Dihydrofolate reductase* Inhibitor
2.	Base analog	Purine analog	6-Mercaptopurine	IMP → AMP
		Guanine analog	Thioguanine	Inhibit conversion of IMP → AMP
		Uracil/thymine analog	5-fluorouracil	*Thymidylate synthase*, dUMP → dTMP
3.	Amino acid analog	Glutamine analog	Azaserine	Synthesis of glutamine
4.	Enzyme		*L-asparaginase*	
5.	Antibiotics		Mitomycin C	
		Adriamycin	Actinomycin D	Prevent transcription by binding to guanine
		Adriamycin	Anthracyclines	Stimulate *topoisomerase*-mediated breaks in DNA
		Camptothecin		Stimulate *topoisomerase I*-mediated breaks in DNA
6.	Alkylating agents		Cyclophosphamide	Inhibit DNA strand separation
7.	Alkaloids		Vincristine	Inhibit cytoskeleton assembly and spindle movement
			Vinblastine	Inhibit cytoskeleton assembly and spindle movement
8.	Glycoside	Etoposide	Podophyllotoxin	Stimulate *topoisomerase II* mediated breaks in DNA
9.	Organometallic compounds	Platinum compounds	Cisplatin	Form intrastrand DNA adducts
10.	Small molecule inhibitor	Imatinib		*Tyrosine kinase*
11.	Monoclonal antibody	Trastuzumab		Binds to epidermal growth factor receptor (EGRF) in cancer cells, inhibiting downstream signaling

Figure 45.10 Anticancer drugs.

CAR T-cells

These are laboratory modified T cells, in which the gene for a special receptor that binds to a certain protein on the patient's cancer cells is added to the T cells. The special receptor is called a chimeric antigen receptor (CAR). As different types of cancers have different antigens, each kind of CAR T cell therapy is made to fight a specific kind of cancer antigen.

An ideal tumor marker

An ideal tumor marker must satisfy the following criteria:

- Method of its detection should be simple and cost-effective.
- It should have high tumor specificity and sensitivity.
- It should be detectable in early stages to provide a lead-time over clinical diagnosis.
- It should have short half-life to allow frequent serial monitoring of the marker levels.
- The levels of the marker should correlate reliably with the tumor size/volume.

  It is important to note that some of the biomolecules detected in cancer patients can also be detected in persons under non-cancerous conditions also. For example,

  - Prostate specific antigen (PSA) is a glycoprotein present in patients having cancer of prostate gland. PSA has also been shown to be present in persons having prostatitis or benign prostatic hyperplasia (BPH).
  - Carcinoembryonic antigen (CEA), has not only been found in patients having various types of cancers but in persons having ulcers of colon, liver cirrhosis and people smoking more than 5–10 cigarettes per day.

It thus becomes important that when specific biomolecules are used as biomarkers to detect the presence of cancer, results obtained must be confirmed by other tests such as biopsies and imaging techniques (Appendix Unit X: Tables 1 and 2).

XII. Medical Perspectives

Cancer is the second leading cause of death globally, causing about 10 million deaths annually. Lung, prostate, colorectal, stomach and liver cancer are the most common types of cancer in men, while breast, colorectal, lung, cervical and thyroid cancer are the most common among women.

30–50% of cancers can be prevented by avoiding risk factors, early detection of cancer and management of patients who develop cancer. Many cancers have a high chance of cure if diagnosed early and treated adequately.

Tobacco use is the most important risk factor for cancer, accounting for approximately 22% of cancer deaths. Besides tobacco, being overweight or obese, having unhealthy diet with low fruit and vegetable intake, lack of physical activity, alcohol use, exposure to radiations and pollution are important preventable risk factors.

Vaccination and screening programs are effective interventions to reduce the global burden of specific types of cancer, such as against human papillomavirus (HPV) that causes cervical cancer and against the hepatitis B virus (HBV) that causes liver cancer.

Cervical cancer can be prevented by screening women through population-based programs. Screening aims to detect the disease at the precancer stage, using various approaches such as visual inspection with acetic acid or Lugol's iodine, Papanicolaou test (Pap test or Pap smear) and HPV DNA testing. In many of the developed countries, the annual incidence and mortality from this cancer have reduced by 50–70% since the introduction of screening programs.

Patients with advanced cancers may suffer a lot due to cancer progression and also due to co-existing diseases, and side effects of cancer treatment, which are physical, psychosocial and emotional. These lead to poor quality of life, and palliative care is provided to such patients to improve the quality of life through the prevention and relief of suffering by early identification and impeccable assessment and treatment of pain and other problems—physical, psychosocial and spiritual.

Cancer cachexia is a progressive condition resulting from a negative protein-energy balance, marked by the loss of lean body mass, including skeletal muscle and adipose tissue. It is accompanied by a systemic inflammatory response, increased proteolysis, lipolysis, and autophagy. Adequate nutritional supplementation and supportive therapy are essential components of management.

XIII. Chapter Summary

- The term "cancer" refers to uncontrolled growth of cells leading to malignancy.
- Cancers of lung, stomach, colon, liver, breast, and prostate account for approximately 20% of all deaths occurring the world over.
- Cancer-producing genes are called oncogenes. Oncogenes are produced from proto-oncogenes.
- Proto-oncogenes code for proteins which stimulate cell division, inhibit cell differentiation, and inhibit apoptosis (programmed cell death). Proto-oncogene to oncogene conversion, i.e., gene transformation is brought about by amplification, translocation, and action of various mutagenic agents.
- Tumor suppressor genes are responsible for inhibiting cell proliferation in response to various signals.
- Mutagens can either be physical, chemical, biological (DNA and RNA viruses) or radiation/ionization based.
- Hereditary and lifestyle (cigarette smoking, tobacco chewing alcohol consumption, etc.) and occupation act as important predisposing factors.
- The occurrence of malignancy is a result of the stimulation of oncogenes and the inhibition of tumor suppressor genes, as well as the dysregulation of genes controlling apoptosis (Fig. 45.11).
- Specific biomolecules, detected in blood, body fluids, cell membranes or in cytoplasm, which are either produced and subsequently released by the cancer cells or produced by the body in response to the presence of cancer cell are called as tumor markers.
- Tumor markers are used to either detect the presence of specific cancers, their progression or the effectiveness of the therapy.

Therapy for the management of cancer can either be surgery, radiation, chemotherapy or immunotherapy.

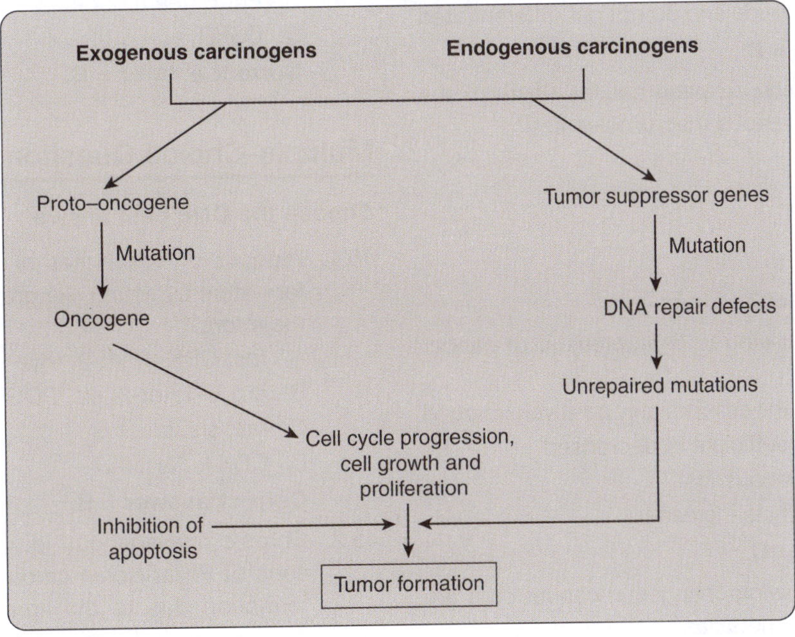

Figure 45.11 Key concept map of carcinogenesis.

Study Questions

Short-Answer Questions

Briefly describe the following.

45.1. Metabolic alterations during cancer formation.

45.2. How cancer cells differ from normal cells?

45.3. Significance of miRNA in cancer management.

45.4. Conversion of a proto-oncogene to an oncogene.

Long-Answer Questions

45.1. Discuss the mechanism in detail by which various growth factors act to induce carcinogenesis.

45.2. Giving specific examples, discuss the role of viruses in cancer formation.

45.3. Discuss the important properties which an ideal tumor should possess. Describe the significance of the use of tumor markers in cancer management giving limitations if any.

Multiple-Choice Questions

Choose the ONE best answer.

45.1. Cancer is mainly caused by:
A. Uncontrolled mitosis
B. Uncontrolled meiosis
C. Cell rupture
D. Loss of immunity
Correct answer = A.

45.2. Cancer-causing oncogenes are not expressed because of the presence of:
A. Proto-oncogenes
B. Tumor promoters
C. Tumor suppressor genes
D. Jumping genes/transposons
Correct answer = C.

45.3. Basic difference between normal cells and cancer cells is that:
A. Cancer cells divide continuously
B. Normal cells are bigger than cancer cells
C. Normal cells are immortal, but cancer cells are mortal
D. Cancer cells divide and do not get differentiated
Correct answer = D.

45.4. In apoptosis, which caspase causes death of the cell by breaking various macromolecules?
A. Caspase-9
B. Caspase-8
C. Caspase-6
D. Caspase-3
Correct answer = D.

45.5. Which of the following is characteristic of cancer cells?
A. The surface of the cells is highly positively charged
B. Proliferation coefficient is decreased
C. Contact inhibition is lost
D. Adhesion ability is increased
Correct answer = C.

45.6. Which of the following changes in cancer cells are helpful to them in metastasis?
A. Activation of vinculin that decreases the adhesion ability leading to metastasis
B. Release of collagenase and stromelysin that helps in passing through membrane barriers
C. Decreased secretion of vascular endothelial growth factor (VEGF) that increases angiogenesis
D. Metabolism of sugars mainly by aerobic glycolysis
Correct answer = B.

45.7. Which of the following is not an anticarcinogenic antioxidant?
A. β-Carotene
B. Vitamin C
C. Vitamin E
D. Vitamin K
Correct answer = D.

45.8. What is the primary cause of oncogene formation?
A. Loss of a function
B. Gain of a function
C. Mutations in tumor suppressor genes
D. Chromosomal translocation
Correct answer = B.

45.9. Which of the following is NOT a mechanism of oncogene activation?
A. Point mutation
B. Chromosomal translocation
C. Gene silencing
D. Gene amplification
Correct answer = C.

45.10. Which hormone receptor is targeted in hormone-sensitive breast cancer?
A. HER2
B. Estrogen receptor (ER)
C. Androgen receptor
D. EGFR
Correct answer = B.

Multiple-Choice Questions (NExT/NEET PG)

Choose the ONE best answer.

45.1. Product of which set of genes induces cancer formation by acting like growth factors and/or their receptors?
A. ras, CDK-4, Int-2, Myc
B. erb B-1, int-2, sis, PDGFR
C. ras, sis, src, fos
D. CDK4, Cyclin D
Correct answer = B.

45.2. Chronic myelogenous leukemia, due to the presence of Philadelphia chromosome, leads to transformation due to the creation of a factor which is nonregulated and aberrant. This factor is best described as which of the following?
A. Transcription factor
B. Growth factor receptor
C. Growth factor
D. Tyrosine kinase
Correct answer = D.

45.3. In a patient diagnosed with a melanoma, molecular analysis indicated that the tumor had sustained a loss of p16(INK4) activity (inhibitor of cyclin-dependent kinase 4). Such a gene would be best classified as which of the following?

A. Dominant oncogene
B. Tumor suppressor
C. Proapoptotic factor
D. Antiapoptotic factor
E. Growth factor

Correct answer = **B**.

45.4. Based on statements given below, choose the correct option.

Statement 1: Mutation of p53 can lead to various tumors.

Statement 2: Protein called p53 prevents the cells from moving to next stage of cell division.

A. Both the Statements are correct, and Statement 2 is the correct explanation for Statement 1
B. Both the Statements are correct, but Statement 2 is NOT the correct explanation for Statement 1
C. Statement 1 is correct, but Statement 2 is NOT correct
D. Statement 1 is NOT correct, but Statement 2 is correct

Correct answer = **C**.

45.5. A physician in a rural clinic sees a child with swelling of the jaw, loosening of the teeth, and swollen lymph nodes. Karyotype analysis of blood cells shows a translocation between chromosomes 8 and 14. It is diagnosed to be Burkitt lymphoma. This rapidly growing tumor most likely due to which of the following?

A. EBV activation
B. Bcr-abl activation
C. BCl-2 activation
D. Constitutive myc expression

Correct answer = **D**.

Xenobiotics

46

Learning Objectives

At the end of this chapter, the student will be able to:

- Describe the role of xenobiotics in disease
- Discuss the various phases of biotransformation reactions of xenobiotics
- Explain the metabolism of ethanol

I. OVERVIEW

The human body is exposed to a large number of chemical substances that are extrinsic or foreign to the normal chemical environment and metabolism of our cells. Examples of such chemical entities include food additives and preservatives, pollutants, environmental toxins, pharmacological agents, dyes, detergents, synthetic polymers, cosmetics, biocides (like insecticides, herbicides, antimicrobials), etc. **"Xenobiotic," the collective term used to denote such exogenous molecules** is derived from the Greek words *xenos* meaning stranger and *biotic* meaning related to life. Any chemical entity is toxic when the cells are exposed to them in supraphysiological quantities and/or duration and/or physical states. **The human body metabolizes xenobiotics to render them nontoxic and prevent their accumulation.** This is achieved by a series of enzymatic reactions termed as biotransformation. Hepatocytes play a central role in the process, though a number of these enzymes are present in most of the tissues. A sound knowledge of the metabolism of xenobiotics is important in understanding pharmacology, toxicology, cancer biology and the advancing field of pharmacogenomics (explained later in Section VI. Pharmacogenomics and personalized medicine, p. 787). This chapter covers various aspects of xenobiotic metabolism—the enzymatic reactions involved, effects of biological variations in the activity of these enzymes, rationale for the clinical management of certain drug toxicities, and other related clinical aspects.

II. METABOLISM OF XENOBIOTICS INVOLVES DISTINCT PHASES

Most hydrophilic molecules cannot cross the lipid barrier created by the plasma membrane, unless there are specific carriers that are present to transport them into the cell. Many hydrophilic molecules are removed from the body as such (without undergoing any metabolism) in urine. On the contrary, lipophilic molecules cannot be excreted in a similar way. Further, they have a propensity to cross the membranes, to enter the cell and mediate toxic effects, and to accumulate in lipid-rich tissue sites like organelle membranes and adipocytes. Thus, **for elimination from the body, lipophilic molecules are first rendered hydrophilic, and are then transported out of the body through urine or bile (and ultimately in feces)** (Fig. 46.1). Thus, the purpose of xenobiotic metabolism is to convert such lipophilic molecules into a hydrophilic form that can be excreted.

Xenobiotic metabolism involves multiple steps. **In phase 1 reactions, lipophilic molecules are transformed into relatively hydrophilic molecules by exposing or generating a polar group**; some xenobiotics are excreted soon after the phase 1 reactions. However, many xenobiotics undergo the subsequent **phase 2 reactions, where the products of phase 1 reactions (or sometimes, the xenobiotics as such) are conjugated with molecules (like glutathione, glucuronic acid, etc.) making them further hydrophilic**. This facilitates their transport outside the cells via various transporter proteins. In

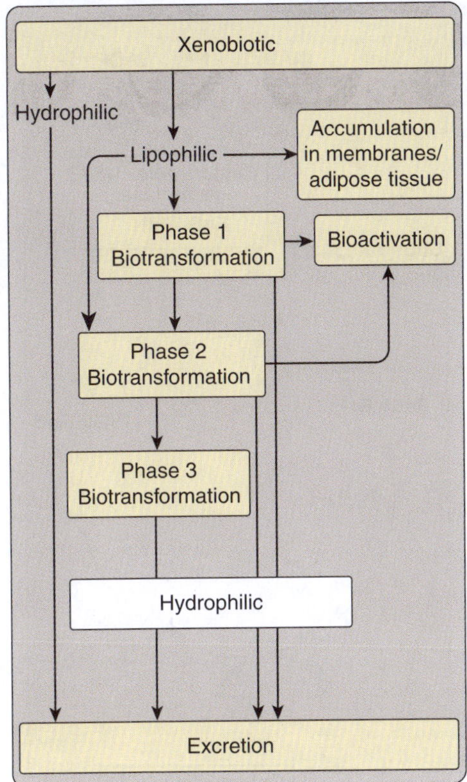

Figure 46.1 Fate of xenobiotics in the body.

certain cases, there can be additional chemical modifications, like further conjugation reactions, of the products of phase 2 reactions, before extruding the metabolites out of the cell—these are termed as **phase 3 reactions**. The phase 1, 2 and 3 reactions together form what is also termed as biotransformation. If the xenobiotic is a drug, the study of its metabolism and excretion is the part of a field known as pharmacokinetics.

In case of orally administered drugs, certain amount of the drug absorbed from the intestinal lumen is inactivated by the enzymes present in the small intestinal epithelial cells. The remaining unaltered drug reaches the liver via the portal circulation, where the hepatic xenobiotic metabolizing enzymes act on some or most of the drug. Thus, only a portion of the orally administered drug reaches the systemic circulation and is available to the target tissues due to the biotransformation in intestines and liver. This is termed as "**first-pass effect**." More of the drug gets metabolized during the subsequent "passes" through the liver, until it is completely metabolized and eliminated. If the first-pass effect significantly reduces the availability of a drug to the target organs, routes other than oral administration are considered. This is known as parenteral (*para*: away, *enteral*: intestines) administration, and examples include intravenous, intramuscular, and subcutaneous administration of drugs.

Although the intended purpose of xenobiotic metabolism is the inactivation and elimination of these molecules, sometimes biotransformation results in the activation of the molecule. This process, termed as bioactivation, has clinical implications. **An inactive or minimally active form of a drug—known as prodrug—can undergo biotransformation to generate the active pharmacological agent.** For example, codeine is a drug which is used in the management of pain. Codeine undergoes phase 1 biotransformation reaction to form morphine, which binds to its receptors to bring about this pharmacological effect; codeine as such cannot bind to these receptors with high affinity. Bioactivation is also responsible for activation of procarcinogens. The reactive molecules that are generated by this process form adducts with DNA so as to introduce mutations, leading to carcinogenesis (see Chapter 47, Section III.A. DNA damage, p. 795). This is exemplified by benzo[*a*]pyrene, a polyaromatic hydrocarbon (PAH), present in tobacco smoke. Benzo[*a*]pyrene, a procarcinogen, undergoes a series of phase 1 biotransformation reactions to form benzo[*a*]pyrene-diol-epoxide derivative which forms adducts with deoxyguanine in DNA, leading to mutations and an increased risk of lung cancer.

Apart from their actions on xenobiotics, the enzymes and reactions involved in biotransformation are also involved in metabolism of certain endogenous molecules. Bilirubin is conjugated with UDP-glucuronic acid to form hydrophilic bilirubin monoglucuronide and diglucuronide (conjugated bilirubin) to facilitate its excretion. This is similar to certain conjugation reaction involved in phase 2 of biotransformation. The first step in conversion of cholesterol to bile acids for excretion is its oxidation mediated by the enzyme *7-α-hydroxylase*. This enzyme belongs to a family of *hydroxylases* involved phase 1 reactions (see Chapter 23, Section V.B, p. 346). A similar enzyme, *1-α-hydroxylase*, is involved in activation of vitamin D to form 1,25-dihydroxycholecalciferol or calcitriol. The first step in the ω-oxidation of fatty acids (see Chapter 21, Section IV.C. Peroxisomal α-oxidation, p. 313) is also catalyzed by a member of the same group of enzymes.

> Microsomes are fragments of endoplasmic reticulum that are produced during high-speed centrifugation after the disruption of the cells. Undoubtedly, they are not present in living cells. Hence, it is to be understood that all the enzymes that are mentioned to be microsomal are actually a part of the endoplasmic reticulum in the living cell.

III. PHASE 1 REACTIONS RENDER THE XENOBIOTICS MORE HYDROPHILIC

Phase 1 biotransformation reactions either expose polar groups in the xenobiotics, or add new ones to them, so as to make the xenobiotics more hydrophilic. The products that are sufficiently hydrophilic are excreted, but more often the phase 1 reactions generate intermediates that are ready to undergo subsequent phase 2 reactions. Phase 1 reactions include hydroxylation, dehydrogenation, hydrolysis, etc. The enzymes catalyzing phase 1 reactions are localized mostly, but not exclusively, to the microsomes.

Among the phase 1 reactions, the most prominent ones are hydroxylation reactions, and the most studied enzymes involved in hydroxylation reactions belong to the cytochrome P450 superfamily.

A. Cytochrome P450 (CYP) superfamily

Cytochrome P450 (CYP) superfamily of enzymes are hemoproteins which are present in all the nucleated cells of the body. The highest concentrations are found in liver, followed by intestines and kidneys. **"P450" refers to the wavelength of 450 nm at which CYP enzymes absorb maximum light <u>in vitro</u>, when the heme (Fe^{2+}) of these enzymes are bound to carbon monoxide.** There are at least 57 functional CYP genes in humans and they are responsible for the metabolism of 50% to 70% drugs that are currently in use, apart from their role in metabolism of other xenobiotics.

The cytochrome P450 enzymes bring about their action by hydroxylation reactions. The general reaction catalyzed is:

$$AH + O_2 + NADPH + H^+ \rightarrow AOH + H_2O + NADP^+$$

The oxygen molecule to be incorporated to the substrate is activated by its binding to the Fe^{2+} of the heme prosthetic group of the CYP (Fig. 46.2). One of the oxygen atoms is added to the substrate (AH) to form AOH; hence the reaction is called hydroxylation. The other oxygen atom gets protonated by H^+ associated with NADPH to be released as water. Two electrons that are required for the reaction are also derived from one molecule of NADPH. **Since only one oxygen atom is added to the substrate, the enzymes are also known as *monooxygenases*.** As the oxygen atoms have two different fates—getting incorporated into the substrate and getting protonated to form water—the enzymes are also known as "**mixed-function oxidases.**"

> A superfamily of genes/proteins is a set of large number of genes/proteins that have been formed evolutionarily as a result of gene duplications. They usually share a common protein domain but have otherwise diverged to such an extent that their functions are nonoverlapping.

Since there are large number of enzymes in the cytochrome P450 superfamily, a systemic nomenclature is in place to unambiguously name the individual

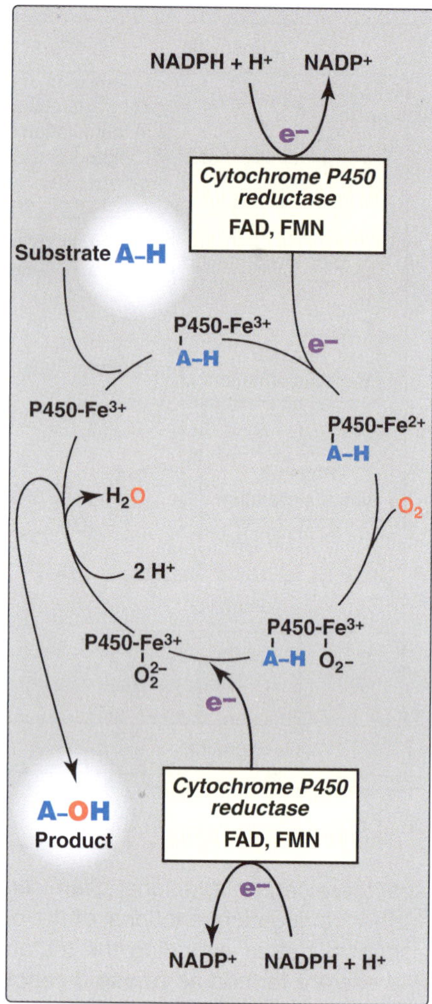

Figure 46.2 Reaction mechanism of cytochrome P450 *monooxygenases*.

members (and their genes), and to identify their similarity with each other. For example, let us consider the phase 1 biotransformation reactions of benzo[*a*]pyrene and codeine mentioned in the Section II. Metabolism of xenobiotics involves distinct phases. Benzo[*a*]pyrene is oxidized to benzo[*a*]pyrene-diol-epoxide derivative by a cytochrome P450 enzyme known as CYP1A1; codeine is converted to morphine by another cytochrome P450 enzyme, CYP2D6. Thus, the enzyme names have a root "CYP" followed by three more characters—a number, followed by an alphabet and another number (Fig. 46.3). The first character denotes "family"; if the amino acid sequence of two CYP enzymes are >40% similar, they are included in the same family. The second character represents "sub-family," where the members share >55% amino acid sequence similarity. The final letter represents the individual enzymes. Thus, CYP2D6 is the sixth member of family 2 and subfamily 2D. The genes coding for the cytochrome P450 enzymes are named after the respective enzymes, but with the

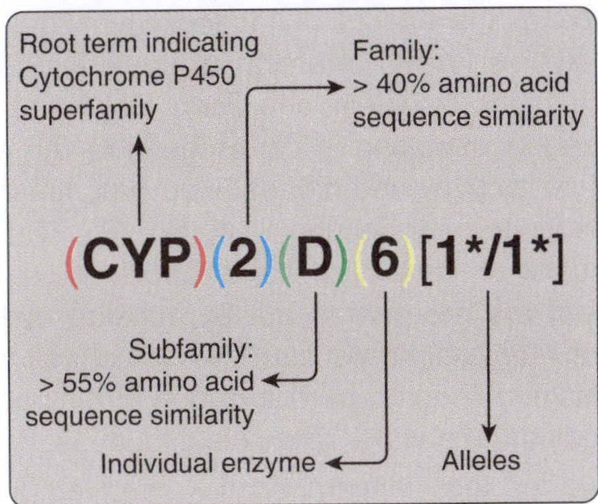

Figure 46.3 Nomenclature of cytochrome P450 enzymes.

names in italics—e.g., CYP2D6 enzyme is coded by *CYP2D6* gene.

Many of the CYP genes show polymorphisms (see Chapter 39, Section VI. Restriction fragment length polymorphism, p. 680), i.e., a particular CYP gene with one or more nucleotide variants can exist in the population. For example, most of the population have "normal" or "wild-type" *CYP2D6*; their CYP2D6 enzyme will have "normal" activity. About 10% of the individuals have nucleotide variants of *CYP2D6* that result in a decreased enzyme activity. In these individuals, administration of codeine for pain relief will show suboptimal effects. They are termed as **poor metabolizers.** Interestingly some allelic variants lead to increased CYP enzyme activity. For example, individuals with *CYP2D6*2/*2* metabolize codeine to morphine rapidly and show increased pain relief with codeine administration. They are termed as **ultra-rapid metabolizers**. Whether poor metabolism or ultra-rapid metabolism is of any advantage or disadvantage depends heavily on the clinical context. In case of a drug with severe adverse effects, poor metabolism leads to persistence of the drug in the body, increasing the risk of adverse effects. In another scenario, increased activity of CYP1A1 leads to increased conversion of benzo[*a*]pyrene to mutagenic intermediates. 10% of the population with such allelic variants are at a higher risk of lung cancer when exposed to cigarette smoke.

Though the number of CYP genes are only close to 57, the CYP enzymes are involved in the phase 1 biotransformation of thousands of molecules. Thus, **these enzymes have a broad substrate specificity and most of them catalyze the metabolism of more than one xenobiotic. Despite the fact that the basic reaction catalyzed by CYP is monooxygenation, they can mediate a large number of reaction types.** Revisiting the previously cited example, CYP2D6-mediated monooxygenation of codeine (which is chemically methylmorphine) results in its O-demethylation to form morphine:

$$AO\text{-}CH_3 + O_2 + NADPH + H^+ \rightarrow AOH + HCHO + H_2O + NADP^+$$

Other reaction types catalyzed by cytochrome P450 enzymes include N-demethylation, deamination, S-oxidation, N-oxidation, dehalogenation, epoxide formation, etc. Though most of the xenobiotics are inactivated by these processes, some of others undergo bioactivation, examples of which has already been mentioned.

Of the 57 CYP genes, 50 encode CYP enzymes that are localized to the endoplasmic reticulum (or microsomes), and these are primarily involved in xenobiotic metabolism. The other seven genes code for CYP enzymes that are localized to the mitochondria, and these are expressed by tissues like adrenal cortex, testes and ovaries, placenta, liver and kidney. Their reaction mechanism is similar to their microsomal counterparts—they catalyze hydroxylation reactions—but they require an iron-sulfur protein known as adrenodoxin for the electron transport from NADPH to the Fe^{2+} of heme in the CYP. Their substrate specificity is also limited. **Some of the reactions catalyzed by the mitochondrial CYP enzymes include:**

- Steroidogenesis (see Chapter 23, Section VII.A, p. 359) to form hormones like estrogens and progesterone (in ovaries and placenta), testosterone (in testes), and glucocorticoids and mineralocorticoids (in adrenal cortex).
- Most of the steps in conversion of cholesterol to bile acids (remember that the rate-limiting step is catalyzed by *7-α-hydroxylase* which is a microsomal enzyme—see Chapter 23, Section V.B, p. 346).
- Hydroxylation of vitamin D in the liver by *vitamin D 25-hydroxylase* to form 25-hydroxy-cholecalciferol.

B. Cytochrome P450 enzymes are responsible for certain drug-drug interactions

Enzyme induction refers to the increased synthesis of an enzyme, in response to presence of another molecule known as enzyme inducer. This occurs mostly at the DNA level, where the enzyme inducer leads to increased transcription of the gene coding for the enzyme, leading to increased mRNA and hence, protein levels. The opposite happens in enzyme repression. **Many drugs (and other xenobiotics) act as inducers or repressors of CYP enzymes, thus determining the level and thus, overall activity of these enzymes.** If a particular drug (or xenobiotic), say "X," can induce or repress a CYP enzyme which is involved in the metabolism of another drug, say "Y," coadministration of drugs "X" and "Y" to a patient may result in untoward effects due to increased or decreased metabolism of "Y." This can lead to

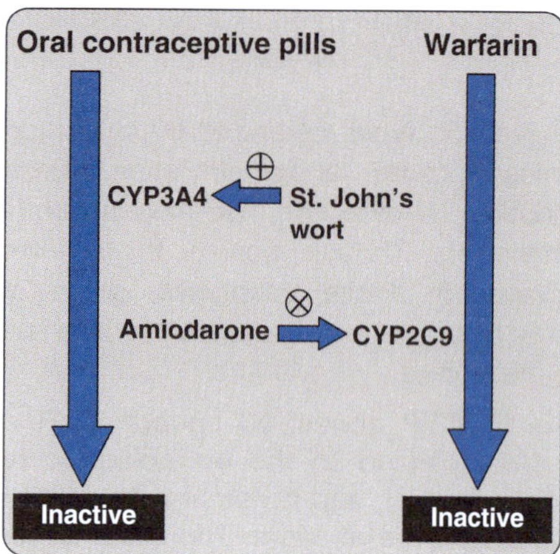

Figure 46.4 Examples of induction and repression of CYP enzymes.

suboptimal pharmacological effects or toxicity, as the case may be. Two such scenarios are explained below:

- Drug components in many oral contraceptive pills (OCPs) are metabolized by the enzyme CYP3A4. St. John's wort, a common herbal preparation used in alternative medicine induces this enzyme. **If females who are on oral contraceptives consume St. John's wort, the metabolism of the OCPs is accelerated, which might result in contraceptive failure and unwanted pregnancies** (Fig. 46.4).
- Warfarin is an oral anticoagulant used in the treatment and prevention of thromboembolism. It is metabolized by CYP2C9, which is inhibited (or repressed) by amiodarone, a drug used in the treatment of arrythmias (Fig. 46.4). Since some arrythmias can predispose to thromboembolism, warfarin is often coadministered with antiarrhythmic drugs. But **patients who are on amiodarone should be carefully monitored if warfarin is coadministered, as amiodarone inhibits the metabolism of warfarin leading to increased chances of bleeding tendencies in these patients**.

The above examples also give an insight into the importance of asking the patient for a history of drug intake, while prescribing a new medication.

C. Other phase 1 reactions

Apart from the microsomal cytochrome P450 *monooxygenases*, phase 1 reactions are also catalyzed by certain other enzymes.

- Flavin-containing *monooxygenases* (FMOs), like CYP enzymes, are microsomal enzymes with highest tissue expression in liver. They catalyze NADPH- and O_2-dependent hydroxylation reactions. Unlike CYP enzymes, they are not subjected to xenobiotic-mediated enzyme induction or repression.
- One of the effects of monooxygenation by CYP enzymes is the formation of epoxides. Epoxides are highly reactive and form DNA and protein adducts, resulting in carcinogenesis. An example of benzo[*a*]pyrene-diol-epoxide has already been mentioned in Section III.A, p. 782. Thus, it is crucial to catabolize these epoxides, and these reactions are carried out by *epoxide hydrolases*. *Epoxide hydrolases* can be microsomal or cytoplasmic. They convert epoxides to dihyrodiols, by addition of a water molecule (Fig. 46.5).
- Two of the *dehydrogenases* involved in phase 1 biotransformation reactions are explained in Section III.D, p. 784.

D. Ethanol is metabolized by three distinct enzymatic reactions related to phase 1 biotransformation

Small amount of ethanol is produced in the body as a result of metabolism and bacterial fermentation in the intestines. Whether present in the body as a result of metabolism or because of ingestion, ethanol is first oxidized to acetaldehyde, and then to acetate. Acetate is converted to acetyl CoA by the enzyme acyl CoA synthetase, which is then oxidized in the citric acid cycle (Fig. 46.6). The first step—conversion of ethanol to acetaldehyde—can be catalyzed by three different enzymes, each present in a different subcellular location (Fig. 46.6):

1. **The major enzyme involved in the catabolism of ethanol to acetaldehyde is *alcohol dehydrogenase*.** It is a cytoplasmic NAD$^+$-dependent enzyme, predominantly present in liver.

2. **Microsomal ethanol oxidizing system (MEOS) is predominated by CYP2E1.** This enzyme is induced by ethanol. Since CYP2E1 is also involved in activation of some procarcinogens in diet and tobacco smoke, increased consumption of alcohol can predispose to cancer. Oxidation of ethanol by MEOS also produces reactive oxygen species which result in hepatocellular injury (see Chapter 47, Section III. Oxidative stress and biological effects of cellular oxidants, p. 795).

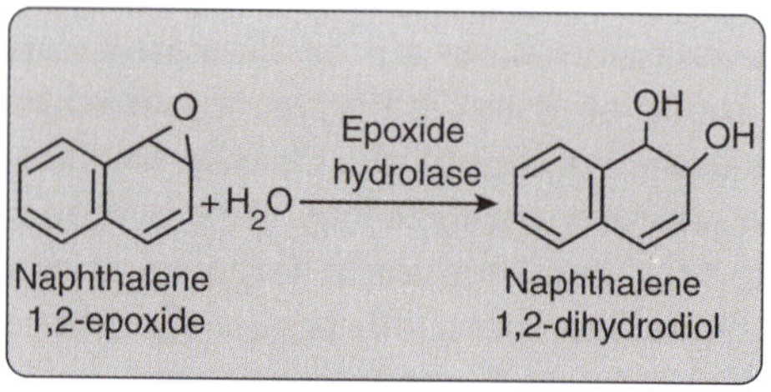

Naphthalene 1,2-epoxide + H_2O —Epoxide hydrolase→ Naphthalene 1,2-dihydrodiol

Figure 46.5 Example of reaction catalyzed by *epoxide hydrolases*.

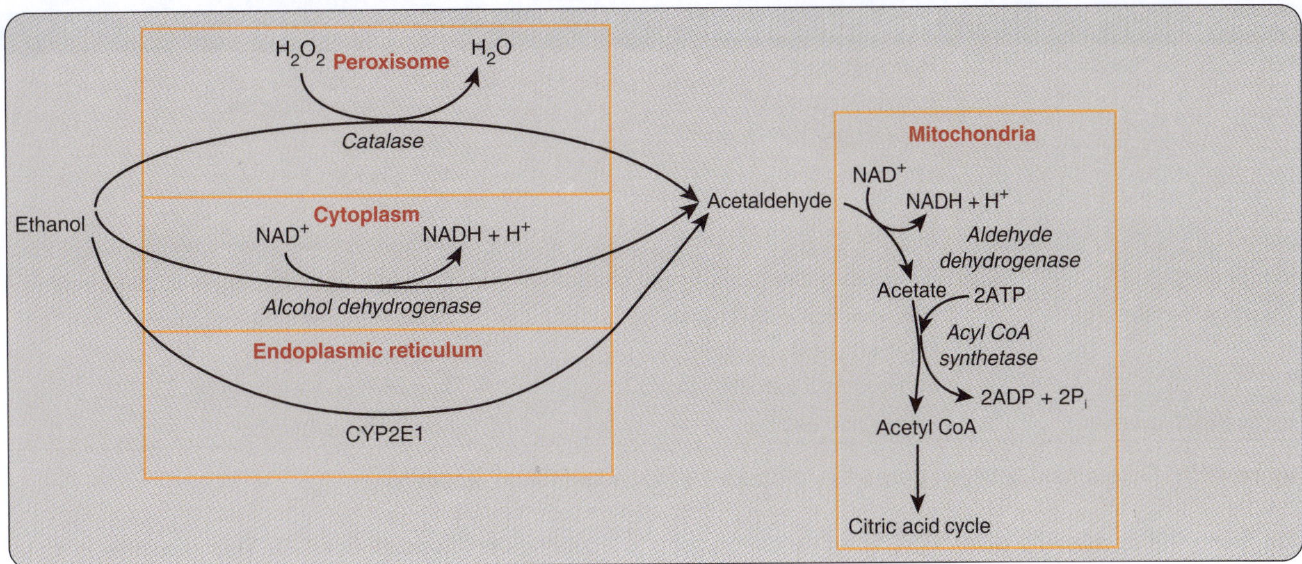

Figure 46.6 Metabolism of ethanol.

3. **Peroxisomal oxidation of ethanol to acetaldehyde mediated by catalase** is a minor pathway, accounting for less than 5% of ethanol catabolism.

Acetaldehyde generated as a result of the above reactions forms adducts with numerous cellular proteins, resulting in cellular dysfunction. Toxicity caused by acetaldehyde is one of the reasons for alcoholic hepatitis and cirrhosis in chronic alcohol consumption. Elevated acetaldehyde levels also cause cutaneous vasodilation and increased skin temperature, facial flushing, nausea, headache and dryness of mouth—the classic manifestations of a "hangover." Acetaldehyde is converted to acetate by a mitochondrial NAD^+ dependent enzyme—*aldehyde dehydrogenase (ALDH)* (Fig. 46.6). Like CYP enzymes, *ALDH* also has polymorphic variants. Majority of Asians have an enzyme variant with low activity. This predisposes these individuals to long-term adverse effects of acetaldehyde. Pharmacological inhibition of *ALDH* by disulfiram is used in the treatment of alcohol use disorder. **In a patient who is on disulfiram, intake of ethanol leads to accumulation of acetaldehyde and immediate development of symptoms of hangover. So, disulfiram is used to ensure abstinence from alcohol.**

Methanol, the lesser homologue of ethanol, is present in antifreeze, industrial solvents and adulterated alcoholic beverages. Methanol is toxic at a relatively low plasma concentration. **Methanol is metabolized by *alcohol dehydrogenase* and *ALDH* into formaldehyde and formic acid, respectively. Formic acid is responsible for the manifestations of methanol poisoning. It leads to metabolic acidosis and death.**

Treatment of methanol poisoning involves inhibition of *alcohol dehydrogenase*, so that methanol is not metabolized into formic acid. This can be achieved by **fomepizole**, a competitive inhibitor of *alcohol dehydrogenase*. **Ethanol also can be used as a competitive inhibitor**, since the affinity of ethanol for *alcohol dehydrogenase* is 10–20 times higher than that of methanol. Unmetabolized methanol is excreted via lungs.

IV. PHASE 2 BIOTRANSFORMATION REACTIONS INVOLVE CONJUGATION

Xenobiotics that are metabolized by phase 1 reactions to products hydrophilic enough for excretion are removed from the body in urine and/or bile. However, **many xenobiotics undergo the subsequent phase 2 reaction before excretion.** Indeed, one of the purposes of phase 1 reactions is to generate reactive groups that can engage in the phase 2 reactions. **Common to all the phase 2 reactions is the conjugation of an endogenous molecule to the substrate, to generate a hydrophilic product.** The molecules added to the substrates can be glucuronic acid (glucuronidation), glutathione, acetate (acetylation), methyl group (methylation), sulfate (sulfation), and amino acids. The enzymes catalyzing glucuronidation are present in the lumen of endoplasmic reticulum; most other enzymes are cytoplasmic.

Some xenobiotics directly enter phase 2 without undergoing phase 1 reactions. Like in case of phase 1 reactions, phase 2 reactions also sometimes lead to bioactivation. An example for the previous two scenarios is the phase 2 biotransformation of morphine. Morphine, without undergoing phase 1 biotransformation, is conjugated with UDP-glucuronic acid to form morphine-6-glucuronide; morphine-6-glucuronide is much more potent

Property	Phase 1 Reactions	Phase 2 Reactions
Nature of the reaction	Hydroxylation Epoxide hydrolysis Dehydrogenation	Glucuronidation Glutathione conjugation Acetylation Methylation Sulfation Conjugation with amino acids
Mechanism	Exposure/generation of polar groups	Conjugation with endogenous molecules
Effect of the reaction	Mild increase in hydrophilicity Inactivation of xenobiotic Sometimes, bioactivation	Huge increase in hydrophilicity Facilitates excretion Sometimes, bioactivation
Intracellular location	Mostly microsomal	Mostly cytoplasmic

Figure 46.7 Differences between phase 1 and phase 2 biotransformation reactions.

than morphine in relieving pain. Morphine-6-glucuronide is water soluble and is excreted in urine.

The differences between phase 1 and phase 2 biotransformation reactions are given in Figure 46.7.

A. Glucuronidation

UDP-glucuronic acid acts as the donor for glucuronidation reactions, the most common phase 2 reaction. The reactions are catalyzed by microsomal enzymes known as *UDP-glucuronosyltransferases (UGTs)*. There are two families of UGTs—UGT1 and UGT2. Apart from the conjugation of xenobiotics, endogenous molecules like bilirubin and steroid hormones also undergo glucuronidation, making them water soluble before excretion.

UGT1A1, which is involved in glucuronidation of bilirubin is also involved in metabolism of certain drugs. 4%–16% of the population has a low UGT1A1 enzyme activity (i.e., 10%–33% of the normal enzyme activity) due to a promoter mutation or a missense mutation of the *UGT1A1* gene, and this condition is termed as **Gilbert syndrome**. The condition as such is harmless and presents as unconjugated hyperbilirubinemia. But when certain drugs (e.g., irinotecan, an anticancer drug) that are metabolized with the help of UGT1A1-mediated glucuronidation are administered to individuals with Gilbert syndrome, inadvertent drug side effects may be encountered.

Certain drugs like barbiturates can induce UGTs leading to drug-drug interactions.

B. Glutathione conjugation

Glutathione (GSH) is a tripeptide thiol with the chemical name γ-glutamyl-cysteinyl-glycine. Many properties of glutathione are conferred to it by the highly reactive thiol (-SH) group present in one of its constituent amino acids—cysteine; hence the abbreviation GSH. **Xenobiotics form a thioether (-S-) linkage with glutathione (Fig. 46.8). This reaction is catalyzed by enzymes known as *glutathione-S-transferases (GSTs)*.** Cytoplasmic GSTs are important in the xenobiotic metabolism, and are present in higher concentrations in liver. Due to their high concentrations, they also bind nonspecifically (and without any catalytic effect) to molecules like bilirubin and steroid molecules to retain these molecules in liver cytoplasm for their metabolism by other enzymes like *UGTs*. Previously, *GSTs* were known as ligandins.

Glutathione-conjugated xenobiotics usually undergo phase 3 biotransformation (see Section V. Phase 3 biotransformation reactions and excretion of metabolized xenobiotics, p. 787) reactions before excretion.

C. N-Acetylation

The donor for N-acetylation reactions is acetyl CoA. The reactions are catalyzed by *N-acetyltransferases (NATs)*. Polymorphic variants of NATs exist, and this is relevant in the metabolism of certain drugs. An example is **isoniazid (INH) which is used in the treatment of tuberculosis. Depending on the genotype, individuals can be "fast acetylators," "normal acetylators," or "slow acetylators." Fast acetylators rapidly metabolize the drug, resulting in inadequate antitubercular effects. Slow acetylators are susceptible to INH toxicity, which manifests as peripheral neuropathy** (due to pyridoxine deficiency—see Chapter 32, Section V.A. Clinical indications for pyridoxine, p. 531) and liver damage.

D. Other phase 2 biotransformation reactions

The other common phase 2 biotransformation reactions are listed below:

- **Methylation** reactions or addition of methyl groups to the substrates are catalyzed by various *methyltransferases*. Depending on the atom receiving

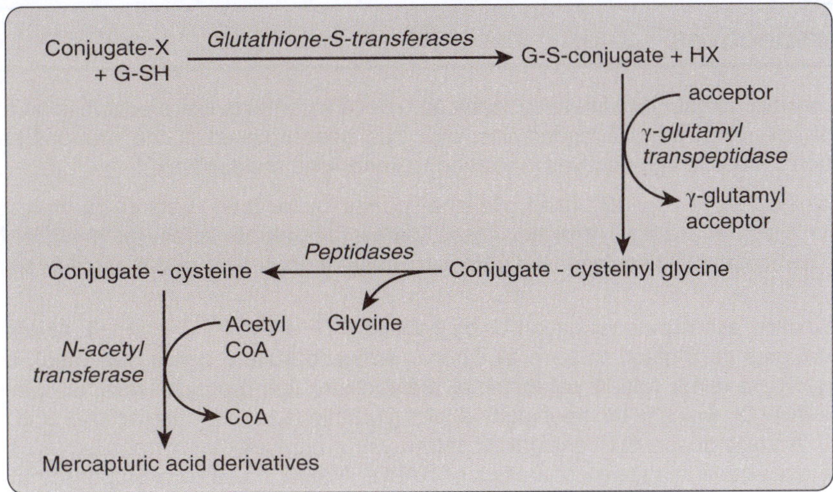

Figure 46.8 Conjugation of xenobiotic (conjugate-x) with glutathione (GSH) and phase 3 biotransformation.

the methyl group, these enzymes can be categorized into N-, O- or *S-methyltransferases*. S-adenosylmethionine (SAM) acts as the donor of the methyl group.

- **Sulfation** is catalyzed by *sulfotransferases* and result in the formation of highly water-soluble sulfate esters that can be excreted in urine and bile. The donor of the sulfate group is 3'-phosphoadenosine-5'-phosphosulfate (PAPS) (Fig. 22.6, p. 325), also known as "active sulfate." Apart from xenobiotic metabolism, PAPS also play a role in the formation of sulfated glycolipids and proteoglycans (see Chapter 19, Section III.E. Sulfate group addition, p. 266).
- **Xenobiotics can also be conjugated with amino acids like glycine and glutamate.** Sodium benzoate and potassium benzoate are commonly used food preservatives. Benzoate is converted to benzoyl CoA which reacts with glycine to form hippuric acid. The latter is excreted in urine.

V. PHASE 3 BIOTRANSFORMATION REACTIONS AND EXCRETION OF METABOLIZED XENOBIOTICS

Sometimes, a phase 3 biotransformation precedes the excretion of the products of phase 2 reactions. A typical example is the metabolism of products formed after conjugation with glutathione. The fate of a glutathione-conjugated xenobiotic is shown in Fig. 46.8. The γ-glutamyl group of glutathione gets transferred to other acceptor molecules by the enzyme, *γ-glutamyl transpeptidase (GGT)* (see Chapter 48, Section III.B. Biochemical markers of cholestasis, p. 807). *Peptidases* remove the glycine residue to form cysteine conjugate of the xenobiotic, which is then acetylated to form mercapturic acid derivatives that are excreted in urine or bile.

There are two major protein families involved in the efflux of the metabolized xenobiotics into the biliary canaliculi and/or tubular lumen of the nephrons. They are solute carrier (SLC) transporter proteins and multidrug resistance-associated proteins (MRPs). The latter belong to the ATP-binding cassette (ABC) transporter family (see Chapter 23, Section VI.E, p. 358).

VI. PHARMACOGENOMICS AND PERSONALIZED MEDICINE

Throughout this chapter, we have seen that polymorphic variants exist for genes that encode various drug metabolizing enzymes like *CYPs*, *glucuronosyltransferases*, *NATs*, etc. The response generated to various pharmacological agents thus depend on the genotype of the individual. While those individuals with wild-type alleles follow the expected course of drug action, this is not the case with poor metabolizers and ultra-rapid metabolizers. They are prone to experience drug toxicity or suboptimal drug effects as the case may be. If we have a prior knowledge of the genotype and gene expression patterns of an individual, the person can receive tailored medications according to the functioning of the drug-metabolizing enzymes. Doses can be adjusted, and drugs causing untoward effects can be avoided. **This concept is known as personalized medicine, and the study of the role of genome in response to drugs is known as pharmacogenomics.**

VII. Medical Perspectives

The importance of xenobiotic metabolism in various clinically relevant contexts like predisposition to cancers, drug-drug interactions, methanol poisoning, pharmacogenomics, etc., has been covered in the previous sections. This section describes some additional medical perspectives in relation to xenobiotic metabolism.

Paracetamol or acetaminophen is a widely used over-the-counter medication used in the management of fever and pain. The safe dose in adults is 3 g/day. At higher doses, paracetamol leads to severe hepatotoxicity and hepatocellular damage. Indeed, paracetamol poisoning is the common cause of acute liver failure and liver transplantation in many countries.

- Paracetamol is detoxified either by conjugation or by hepatic CYP enzymes (Fig. 46.9). At pharmacological doses, 90% of paracetamol gets conjugated to form glucuronic acid and sulfate derivatives which are excreted in urine; CYP enzymes play only a minor role in paracetamol metabolism; this produces minimal quantities of N-acetyl-p-benzoquinoneimine (NAPQI) which is then conjugated with glutathione to form mercapturic acid derivatives which are excreted (Fig. 46.8). At toxic doses, the conjugation pathways for paracetamol metabolism are overwhelmed and the CYP enzymes take over, generating large quantities of NAPQI. NAPQI has a high propensity to bind to -SH groups. Binding rapidly to the -SH groups of glutathione, NAPQI depletes the intracellular glutathione stores. Following this, NAPQI irreversibly binds various intracellular proteins with -SH groups, forming NAPQI-protein adducts. A widespread cellular dysfunction, especially mitochondrial dysfunction, ensues and leads to hepatocellular injury.

  The specific antidote in paracetamol poisoning is N-acetylcysteine. A thiol (-SH) containing compound, N-acetylcysteine acts by at least two distinct mechanisms—(1) by replenishing the intracellular glutathione stores and (2) by directly binding to NAPQI, impeding the binding of NAPQI to cellular proteins.
- We have seen that some xenobiotics are excreted after conjugation with amino acids—an example of benzoate (indeed benzoyl CoA) conjugating with glycine to form hippuric acid was given in Section IV.D, p. 787. The same principle applied in a reverse direction is used in the treatment of certain disorders. In urea cycle defects leading to hyperammonemia, excretion of nitrogen in the form of urea is not possible. As an alternative mechanism for nitrogen excretion, sodium benzoate can be administered in these individuals. Benzoate, after getting converted to benzoyl CoA, combines with glycine to form hippuric acid, which is excreted. Since glycine contains one nitrogen atom, this whole process aids in removal of one nitrogen from the body, per sodium benzoate molecule administered.
- Heme is a negative regulator of the enzyme *ALA synthase 1*, the rate-limiting enzyme of heme synthesis. Cytochrome P450 enzymes are hemoproteins. When these enzymes are induced leading to more enzyme production, the intracellular level of heme is reduced (as heme gets incorporated into the newly synthesized enzymes). Thus, the inhibition for heme synthesis is removed during cytochrome P450 induction resulting in increased formation of precursors of heme. In porphyrias, where the functioning of various enzymes of heme synthesis is diminished, these intermediates accumulate in increasing amounts, precipitating the disease manifestations. This is the rationale for avoidance for alcohol and CYP enzyme inducers in patients with porphyria.

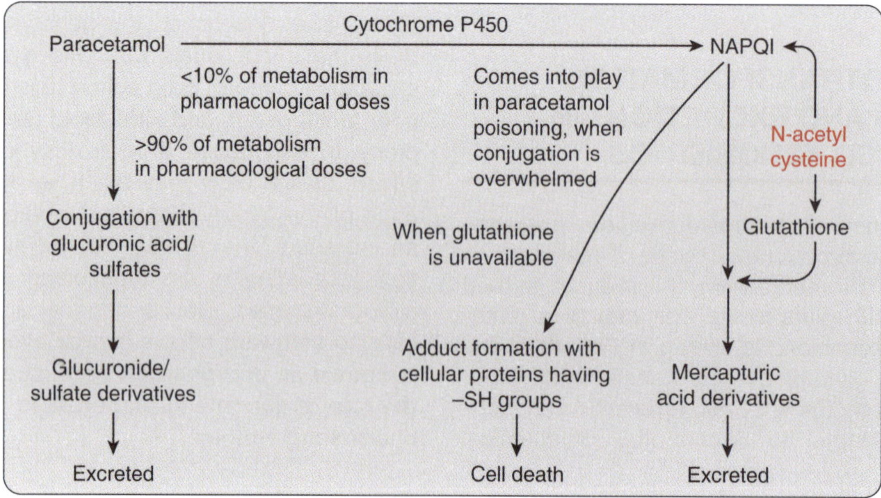

Figure 46.9 Paracetamol poisoning. N-acetylcysteine is used as the antidote.

VIII. Chapter Summary

- Xenobiotics are chemical substances that are extrinsic or foreign to the normal chemical environment and metabolism of human cells.
- Human body metabolizes xenobiotics to render them nontoxic and prevents their accumulation. Hepatocytes play a central role in this biotransformation, though a number of these enzymes are present in most of the tissues.
- In phase 1 reactions, lipophilic molecules are transformed into relatively hydrophilic molecules by exposing or generating a polar group.
- The most prominent phase 1 reactions are the hydroxylation reactions, and the most studied enzymes involved in hydroxylation reactions belong to the cytochrome P450 superfamily. They are also involved in the metabolism of endogenous molecules like cholesterol and bile acids.
- Apart from cytochrome P450 superfamily, phase 1 reactions are also mediated by *flavin-containing monooxygenases* (FMOs), *epoxide hydrolases*, and *dehydrogenases*.
- Many xenobiotics undergo the subsequent phase 2 reactions, where the products of phase 1 reactions are conjugated with endogenous molecules to make them further hydrophilic, and to facilitate their transport outside the cells via various transporter proteins.
- Phase 2 reactions involve conjugations. The molecules added to the substrates can be glucuronic acid (glucuronidation), glutathione, acetate (acetylation), methyl group (methylation), sulfate (sulfation), and amino acids (Fig. 46.10).
- Some xenobiotics undergo a phase 3 reaction also before excretion.
- Sometimes, biotransformation results in the activation of the molecule—this process is termed as bioactivation. Activated precursor molecules include prodrugs and procarcinogens. Apart from their actions on xenobiotics, the enzymes and reactions involved in biotransformation are also involved in metabolism of certain endogenous molecules.
- Polymorphic variants exist for genes that encode various drug metabolizing enzymes like *CYPs*, *glucuronosyltransferases*, *N-acetyltransferases (NATs)*, etc. and these polymorphisms are important in drug administration and disease susceptibility.
- The study of the role of genome in response to drugs is known as pharmacogenomics.
- Knowledge of xenobiotic metabolism is also important in management of conditions like methanol poisoning, paracetamol toxicity, etc. and in screening of chemicals for mutagenesis.

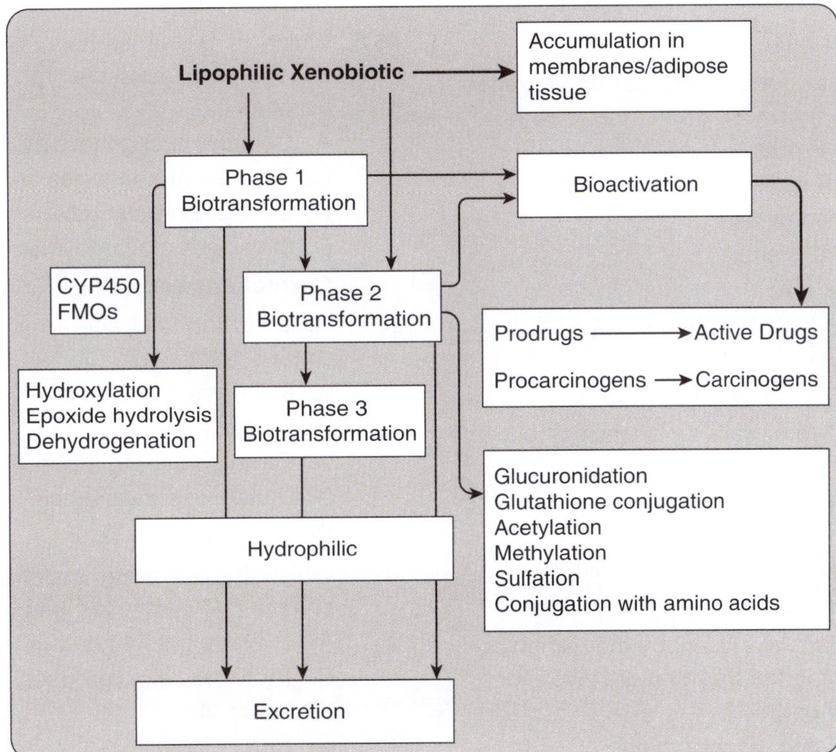

Figure 46.10 Key concept map of xenobiotic metabolism.

Study Questions

Short-Answer Questions

46.1. Explain the reason for the higher risk of cancer in individuals who consume ethanol.

46.2. Explain the role of cytochrome P450 in drug interactions.

46.3. Explain the concept of pharmacogenomics.

Long-Answer Questions

46.1. Describe the phases of xenobiotic metabolism with details of glucuronidation.

46.2. In the context of alcohol metabolism, answer the following:

A. What are the biochemical pathways by which ethanol is metabolized?

B. Name two situations where pharmacological inhibition of an enzyme involved in alcohol metabolism is useful.

Multiple-Choice Questions

Choose the ONE best answer.

46.1. Which of the following is NOT a phase I reaction of metabolism of xenobiotics?

A. Oxidation
B. Reduction
C. Hydroxylation
D. Conjugation

Correct answer = D.

46.2. The main enzyme for the metabolism of xenobiotics is:

A. Cytochrome P 450
B. Glutathione S-transferase
C. Glutathione isomerase
D. Glucuronyl transferase

Correct answer = A.

46.3. What is NOT CORRECT about xenobiotics and their metabolism?

A. Xenobiotics are lipophilic.
B. By phase I hydroxylation, xenobiotics become hydrophilic.
C. Phase II reactions have no effect on their hydrophilicity.
D. Hydrophilic xenobiotics gets excreted.

Correct answer = C.

46.4. For the metabolism of ethanol by the human body enzymes present in the following are involved:

A. Peroxisomes only.
B. Peroxisomes and cytoplasm only.
C. Peroxisomes, endoplasmic reticulum, and cytoplasm only.
D. Peroxisomes, endoplasmic reticulum, mitochondria, and cytoplasm.

Correct answer = D.

46.5. What is incorrect about phase II reactions?

A. It involves conjugation of an endogenous molecule to the substrate/xenobiotic to form a hydrophilic product.
B. Conjugation can occur with glucuronic acid, glutathione, acetate, methyl group, sulfate, and amino acids.
C. Conjugation enzymes are mostly present in cytosol and endoplasmic reticulum.
D. Epoxide hydrolysis takes place in phase II reactions.

Correct answer = D.

46.6. Which of the following is metabolized like xenobiotic?

A. Myoglobin
B. Hemoglobin
C. Bilirubin
D. Biliverdin

Correct answer = C.

46.7. Primary site for the metabolism of xenobiotics and drugs is:

A. Kidneys
B. Liver
C. Pancreas
D. Lungs

Correct answer = B.

46.8. Which of the following is a characteristic of the endogenous molecule to be transferred to a xenobiotic?

A. Complex endogenous molecules get transferred.
B. Small polar molecules are attached.
C. Strong non polar groups are attached.
D. Molecules are large in size.

Correct answer = B.

46.9. Which of the following enzymes is not involved in phase I reactions?

A. Monoamine oxidase
B. Esterases
C. Flavin containing monooxygenases
D. Glutathione transferase

Correct answer = D.

46.10. Which the following statements regarding cytochrome P450 enzymes is true?

A. They are present only in the liver.
B. They are dioxygenases with high substrate specificity.
C. They are present in mitochondria and microsomes.
D. They act exclusively on xenobiotics only.

Correct answer = C.

Multiple-Choice Questions (NExT/NEET PG)

Choose the ONE best answer.

46.1. The example of conversion of an active to more active metabolite, i.e., bioactivation is:

A. Codeine to morphine
B. Digitoxin to digoxin
C. Iproniazid to isoniazid
D. Phenacetin to paracetamol

Correct answer = A.

46.2. Which of the following is NOT an example of phase 2 biotransformation reaction?

A. Acetylation
B. Glucuronidation
C. Sulfation
D. Hydroxylation

Correct answer = D.

46.3. A patient on warfarin was also started on amiodarone for the management of arrythmias. On the next follow-up, his prothrombin time and international normalized ratio (INR) were found to be increased. Which of the following is the most likely reason for this finding?

A. Amiodarone increases absorption of warfarin from the gastrointestinal tract.
B. Amiodarone leads to a decrease in the metabolism of warfarin.
C. Amiodarone has an activity similar to that of warfarin.
D. Amiodarone inhibits absorption of vitamin K from the intestines.

Correct answer = B.

46.4. Which statement about glutathione-S-transferases is INCORRECT?

A. They are present in high concentrations in liver.
B. They bind to bilirubin and steroids to conjugate these molecules with glutathione.
C. They form thioether linkages between glutathione and the substrate.
D. They are involved in phase 2 biotransformation reactions.

Correct answer = B.

46.5. Which of the following amino acids is involved in the detoxification of benzoates?

A. Methionine
B. Glycine
C. Glutamate
D. Cysteine

Correct answer = B.

Free Radicals, Oxidative Stress, and Antioxidant Defense Mechanism

47

Learning Objectives

At the end of this chapter, the student will be able to:

- Describe the beneficial and deleterious effects of free radicals
- Describe various antioxidant defense systems in the body
- Explain the role of free radicals in the pathogenesis of cancer, diabetes mellitus, and atherosclerosis

I. OVERVIEW

Free radicals refer to those chemical species with an unpaired electron in its outer orbital. They accept or donate the unpaired electron to attain stability and are hence highly reactive. Various oxygen containing free radicals and other unstable highly reactive chemical entities possessing oxygen, constitute what is termed as reactive oxygen species (ROS). Similar molecular entities which contain a nitrogen atom are termed as reactive nitrogen species (RNS). Collectively known as cellular oxidant species, both ROS and RNS are produced in our body during the course of normal cellular metabolism, and their production is enhanced by certain chemical, biological and environmental stimuli. Due to the inherent nature of these entities to react with other cellular molecules so as to achieve stability, ROS and RNS invariably bring about alterations in the structures of macromolecules like nucleic acids, proteins and lipids. Thus, they can have far-reaching effects on the homeostasis of the cell. Understandably, various defense mechanisms exist to counter the effects of these oxidants, so as to prevent the alterations they induce in the cell (Fig. 47.1). This chapter describes the generation and the effects of ROS and RNS in a mammalian cell, various antioxidant defense systems that act to balance these effects, and the mechanisms by which the imbalance between the oxidant and antioxidant systems initiate or drive the progression of physiological and pathological states like aging, cancer, diabetes mellitus, and atherosclerosis.

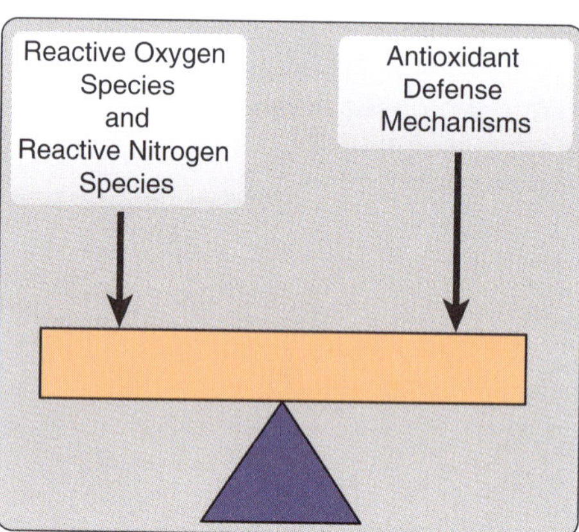

Figure 47.1 Redox homeostasis of the cell is maintained by a fine balance between the oxidant and the antioxidant levels.

II. GENERATION OF CELLULAR OXIDANT SPECIES

Cellular oxidant species comprises of ROS and RNS. **The prominent ROS in the cell are superoxide radical, hydrogen peroxide, hydroxyl radical (Fig. 47.2), and hypochlorite anion, while the major RNS are peroxynitrite, nitrous acid, and nitric oxide.** Among the ROS, hydroxyl radical is the most potent oxidant. As evident from their chemical structure, not all the oxidant species contain an unpaired electron in the outer orbital. However, even those ROS and RNS that are not free radicals can lead to

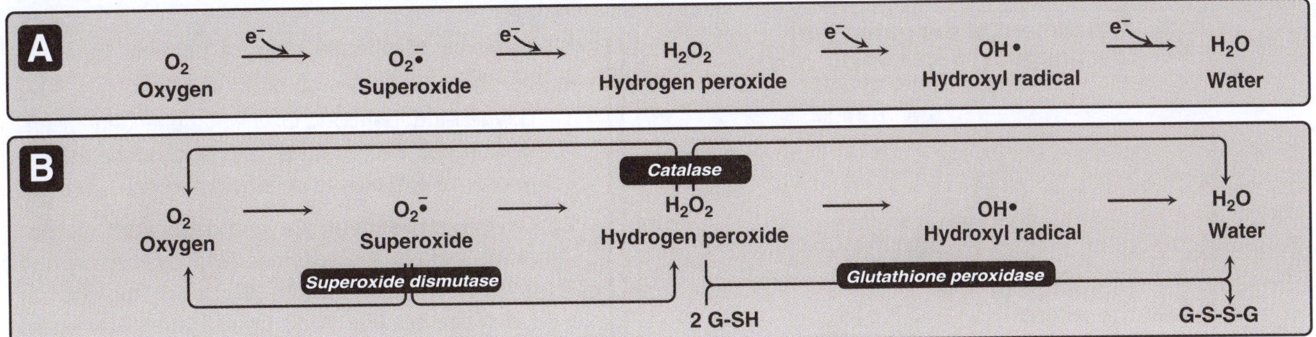

Figure 47.2 **A.** Formation of reactive intermediates from oxygen. e^- = electrons. **B.** Actions of antioxidant enzymes. G-SH = reduced glutathione; G-S-S-G = oxidized glutathione.

generation of free radicals, and subsequently their deleterious consequences. Multiple mechanisms that occur in the cell can lead to the formation of oxidant species.

A. Biological oxidation

Biological oxidation is a major source of free radical generation. The final phase of oxidation of cellular metabolites occurs in the electron transport chain (ETC) located in the inner mitochondrial membrane. Here, NADH and $FADH_2$ donate their electrons to the complexes I and II, respectively, to regenerate NAD^+ and FAD. These electrons are transported along complex III to complex IV, where they are accepted by oxygen, the final electron acceptor, to get converted into water (see Fig. 11.8). The efficiency of this system is not 100%, and electron leak occurs at different complexes of the ETC, more so in case of aged mitochondria. In the normal scenario, a single molecule of oxygen accepts 4 electrons and 4 protons within the complex IV to generate two molecules of water.

$$O_2 + 4H^+ + 4e^- \rightarrow 2H_2O$$

But in the milieu of mitochondrial electron leakage, the oxygen molecule tends to accept 1, 2, or 3 electrons, thus getting partially oxidized to superoxide, peroxide or hydroxyl ions, respectively. This electron leakage occurs in all nucleated cells, and is a major reason for cellular aging.

Certain *oxidoreductase* enzyme reactions can also generate oxidant species, for instance—those catalyzed by *PGH_2 synthase* (see Chapter 22, Section VIII.A. Prostaglandin and thromboxane synthesis, p. 333) and *lipoxygenase* (see Chapter 22, Section VIII.C. Leukotriene synthesis, p. 336) generate hydroperoxides, and *xanthine oxidase* (see Fig. 34.15) and very long chain *fatty acyl CoA dehydrogenase* which generate H_2O_2 (see Chapter 21, Section IV.B.8. Peroxisomal β-oxidation, p. 313).

Very-long-chain acyl-CoA + FAD → Very-long-chain trans-Δ^2-enoyl-CoA + $FADH_2$

$$FADH_2 + O_2 \rightarrow FAD + H_2O_2$$

B. Cellular inflammation

Activated leukocytes (e.g., activated macrophages and neutrophils) generate oxidants during the course of cellular inflammation (Fig. 47.3). Presence of pathogens leads to activation of phagocytosis (*phago*: eat, *cyte*: cell) by these cells, and one mechanism by which the destruction of the pathogens is brought about is by oxidant-dependent killing. **The inflammatory cells undergo respiratory burst, a metabolic reaction which leads to generation of abundant quantities of superoxide ions.** (The rapid consumption of O_2 that accompanies formation of superoxide is referred to as the **respiratory burst**). This reaction is catalyzed by a multisubunit enzyme, NADPH oxidase.

$$NADPH + 2O_2 \rightarrow NADP^+ + 2O_2^{\cdot -} + H^+$$

The components of the enzyme are assembled to form the active multisubunit within the phagolysosomal membrane, exposing the engulfed pathogens to oxidants. Superoxide spontaneously, as well as enzymatically, gets converted to hydrogen peroxide. **Hydrogen peroxide as such is a weak ROS, but it can be converted to hydroxyl ion, a highly potent ROS** (The formation of hydroxyl ion is described in the Section C). **Hydrogen peroxide can also react with chloride in presence of the enzyme *myeloperoxidase*, which is present in azurophilic granules, to form other ROS like hypochlorite anion.**

$$Cl^- + H_2O_2 \rightarrow OCl^- + H_2O$$

Presence of pathogens also leads to elevated levels of the enzyme—*inducible nitric oxide synthase* (iNOS) (see Fig. 18.6) in the phagocytic cells, resulting in production of significant amounts of nitric oxide (NO). NO can react with superoxide to form peroxynitrite ($ONOO^-$). The ROS and RNS thus generated lead to damage of cellular components like nucleic acids, proteins and lipids of the pathogen, leading to cell lysis. These mechanisms, originally intended to eliminate

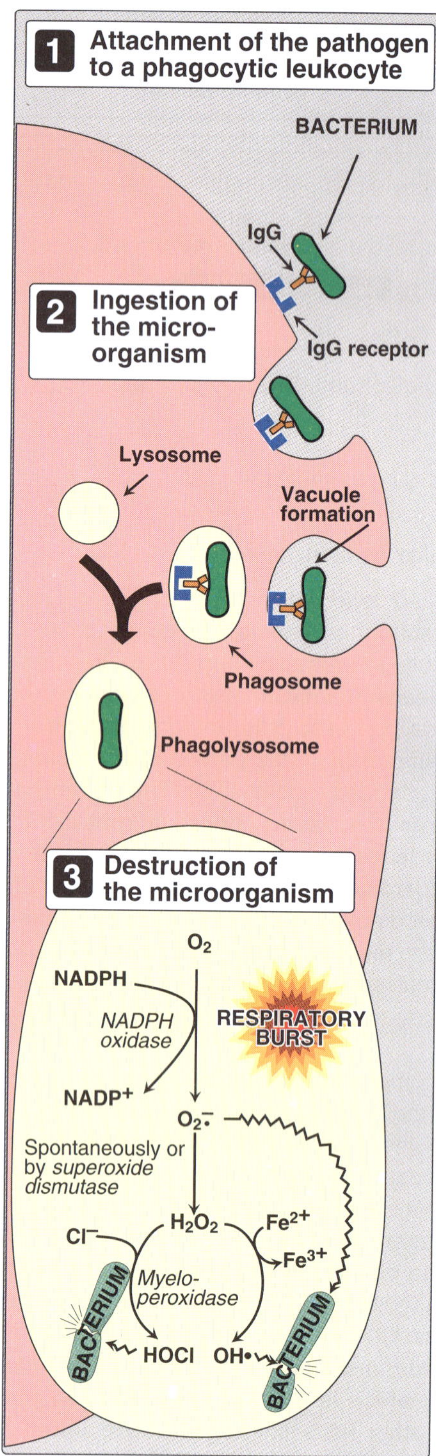

Figure 47.3 Phagocytosis and the oxygen (O_2)-dependent pathway of microbial killing. IgG = immunoglobulin G; NADP(H) = nicotinamide adenine dinucleotide phosphate; $O_2^{\bar{\bullet}}$ = superoxide; H_2O_2 = hydrogen peroxide; HOCl = hypochlorous acid; OH• = hydroxyl radical.

pathogens, also come into play in chronic inflammation, where persistent activation of the phagocytic cells by various immune stimuli leads to continuous generation of oxidant species, ultimately resulting in tissue damage.

Mechanisms of phagocytosis of bacteria by white blood cells

1. **Oxygen-independent:** Oxygen-independent mechanisms use pH changes in phagolysosomes and lysosomal enzymes to destroy pathogens.
2. **Oxygen-dependent:** Oxygen-dependent mechanisms include the enzymes *NADPH oxidase* and *myeloperoxidase* (*MPO*) that work together in killing bacteria (Fig. 47.3). Overall, the *MPO* system is the most potent of the bactericidal mechanisms.

Deficiencies in *MPO* do not confer increased susceptibility to infection because peroxide from *NADPH oxidase* is bactericidal. However, rare genetic deficiencies in *NADPH oxidase* cause chronic granulomatous disease (CGD) characterized by severe, persistent infections and the formation of granulomas (nodular areas of inflammation) that sequester the bacteria that were not destroyed.

C. Fenton reaction and Haber-Weiss reaction

Transition metals like iron and copper can generate free radicals by Fenton reaction.

$$H_2O_2 + Fe^{2+} \rightarrow Fe^{3+} + OH^{\cdot} + OH^-$$

Since most of the intracellular free iron is available as the ferric form (Fe^{3+}), precursor reactions are required to generate the ferrous form required for Fenton reaction.

$$Fe^{3+} + O_2^{\cdot -} \rightarrow Fe^{2+} + O_2$$

The overall process can be summed up as the formation of hydroxyl radical from peroxide and superoxide, the so-called Haber-Weiss reaction:

$$O_2^{\cdot -} + H_2O_2 \rightarrow OH^{\cdot} + O_2 + OH^-$$

Since the oxidant species thus formed can damage proteins containing iron or copper complexes to release further free iron or copper intracellularly, the deleterious effect of transition metal-induced oxidative stress is much higher. **Higher level of oxidative toxicity mediated by iron can even lead to a specific form of programmed cell death known as ferroptosis** (*ferro*: iron, *ptosis*: falling). Hence, in the human body, transition metals like iron and copper are mostly complexed with storage or transport proteins (like ferritin, ceruloplasmin and transferrin) to limit the availability of the free intracellular forms, thus to minimize their toxic effects.

D. Other sources of cellular oxidant species

Ultraviolet rays, especially those in the UV-A spectrum (315–400 nm) which are not absorbed by the ozone layer, can excite the outer orbital electrons of water and oxygen molecules within the skin cells, leading onto the formation of superoxide and hydroxyl free radicals. A similar effect is also seen with ionizing radiations, including **X-rays and gamma rays**.

Free radicals can also be formed from the **metabolism of exogenous compounds like drugs and chemicals**. This is exemplified by the generation of $CCl_3^{\cdot}$ from carbon tetrachloride, and similar free radicals from other halogenated hydrocarbons. **Cigarette tar** is a source of chemicals that can accelerate generation of free radicals like hydroxyl radical.

III. OXIDATIVE STRESS AND BIOLOGICAL EFFECTS OF CELLULAR OXIDANTS

Certain cellular states lead to a situation where the generation of the cellular oxidant species is at such extents that the counterregulatory antioxidant defense mechanisms are overwhelmed. Thus, the levels of the oxidants remain persistently high in such situations, leading to aberrations from steady state and ultimately, cell toxicity. This condition is known as **oxidative stress**. Cellular oxidants bring about their detrimental effects by multiple mechanisms.

A. DNA damage

Oxidant species can cause cell injury causing multiple insults to the deoxyribonucleic acid (DNA) including **double-strand breaks, single-strand breaks, and formation of DNA adducts**. DNA adducts refer to the DNA which is covalently bonded to a chemical species; they can be hazardous because certain oxidized bases and nucleotide adducts are identified different from their parent bases by the DNA polymerases. This leads to introduction of mutations in the daughter strand during replication. For example, hydroxyl radical can lead to oxidation of guanine to 8-oxoguanine. Unlike guanine which forms a Watson-Crick base pair with cytosine, 8-oxoguanine base pairs with adenine. This leads to introduction of a transversion mutation (cytosine to adenine) in the daughter strand (Fig. 47.4). Normally such base perturbations are countered by the DNA repair systems, but in presence of free radicals, the DNA repair mechanisms are overwhelmed, and alterations in DNA can lead to arrest of cell cycle and cessation of cell division (this is termed “cellular senescence”).

B. Protein damage and unfolded protein response

Oxidant-induced damage affects protein function by **modification of side chain of amino acids leading to alteration in the structure and function**. Oxidation of cysteine residues involved in disulfide bond formation can affect folding of proteins into their native functional forms. Oxidation of critical residues involved in catalysis and ligand binding affects enzyme and receptor functions, respectively. For instance, oxidant chemicals present in tobacco smoke lead to oxidation of methionine residues of proteins to form methionine sulfoxide. Formation of sulfoxide at critical methionine residues (positions 351 or 358) of α1-antiproteinase, a protein involved in limiting the activity of proteinases, nullifies its activity, leading to degradation of proteins like elastin. Elasticity of the lung, when compromised leads to diseases like emphysema.

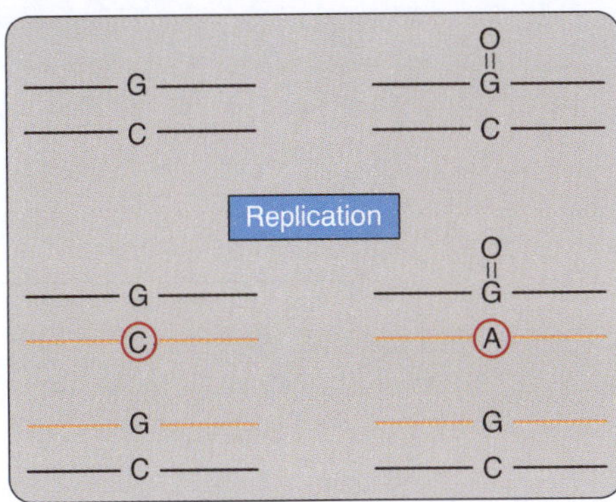

Figure 47.4 Oxidation of guanine by hydroxyl radical forms 8-oxoguanine which base pairs with adenine, during replication. Note the transversion from cytosine to adenine (red circles) in the newly formed daughter strand (yellow).

Oxidants also cause accelerated proteolysis and also, generation of protein adducts that can lead to abnormal protein folding. When such unfolded proteins accumulate, the cell resorts to a salvage mechanism known as “**unfolded protein response**” (UPR) (see Chapter 37, Section VI.C. Protein degradation, p. 649). If the extent of damage is beyond the limits of UPR to contain, apoptotic pathways are activated, resulting in cell death.

C. Lipid peroxidation

ROS and RNS can react with lipid components of the biological membranes (plasma membrane and membranes of organelles) to cause widespread membrane damage. The usual targets are the polyunsaturated fatty acids (PUFAs) present in the membrane phospholipids.

Lipid peroxidation happens in three distinct phases (Fig. 47.5).

1. **Initiation:** Free radical-mediated lipid peroxidation is initiated when the allylic hydrogen of the methylene group is removed from the PUFA by an oxidant species, for instance hydroxyl radical, to

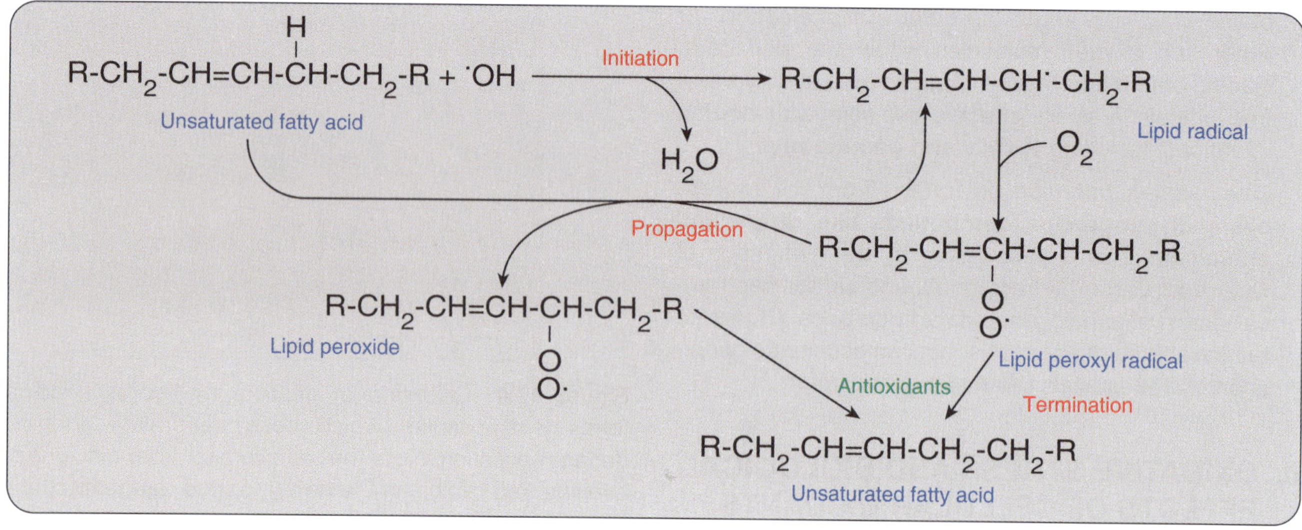

Figure 47.5 Lipid peroxidation proceeds through initiation and propagation steps. Note that a chain reaction is initiated at the propagation step, leading to an amplified generation of lipid peroxyl radical and lipid peroxides. This continues till these products are quenched by some cellular antioxidant which acts on the lipids.

generate a lipid radical; the hydroxyl group itself is converted to water.

2. **Propagation:** The lipid radical generated reacts with a molecular oxygen to form a lipid peroxyl radical, which can react with more PUFAs, generating lipid peroxides, and regenerating lipid radicals. **The regeneration of lipid radicals in this step initiates a chain reaction, highly amplifying the production of lipid peroxides**. Lipid peroxides can also lead to the formation of alkoxyl and peroxyl radicals, further contributing to the propagation of the chain reaction.

$$ROOH + Fe^{2+} \rightarrow RO^{\cdot} + Fe^{3+} + OH^-$$

$$ROOH + Fe^{3+} \rightarrow ROO^{\cdot} + Fe^{2+} + H^+$$

3. **Termination:** The chain reaction continues to produce lipid peroxides, till the antioxidant defense mechanisms in the cell take over and quench the lipid radicals and eliminate the lipid peroxides (see Section IV for the details).

D. Physiological functions of cellular oxidants

It should be emphasized that though oxidative stress is deleterious, **oxidant species mediate critical cellular functions when their levels are within the physiological limits**. We have already seen that ROS and RNS have a crucial role in the defense against pathogens, when phagocytosed pathogens are neutralized by activated macrophages and neutrophils; ROS and RNS are also important in mediating inflammation. Nitric oxide has diverse functions as potent vasodilator, signaling molecule, inhibitor of platelet aggregation, and inflammatory mediator (see Chapter 18, Section IV.E. Nitric oxide synthesis, p. 252). ROS also has a role in the control of ventilation by carotid bodies where they are involved in oxygen sensing. It has also been seen that various proteins involved in gene expression and signal transduction (like transcription factors, receptors, *protein kinases*, and *phosphatases*) when modified by the cellular oxidant species or their downstream products, fine-regulate their activity, thus mediating important physiological functions.

> Membrane lipid peroxidation leads to alteration in the transmembrane potential, disruption of organellar membranes leading to dysfunction of organelles, and induction of apoptosis. Products of lipid peroxidation can also form DNA adducts, leading to generation of mutations. Oxidant-mediated changes can also occur in lipid components apart from PUFA. Cholesterol can undergo oxidation to keto-, hydroxy-, and hydroperoxy-cholesterol. Free radical-mediated oxidation of low-density lipoprotein (LDL) to form oxidized LDL is a major mechanism for the pathogenesis of atherosclerosis (see Section VI.C).

IV. ANTIOXIDANT DEFENSE MECHANISMS

Because the cellular oxidants lead to unwanted effects when their levels are unchecked, various mechanisms exist in the cell to contain their levels. Though some oxidants, like superoxide radical, to certain extent get spontaneously converted to less potent forms, well-defined antioxidant mechanisms are present in the cell to actively quench the effect of oxidants. The antioxidant defense mechanisms in the body can be categorized according to the levels at which they act. The first line of defense is provided by the preventive antioxidants, which abrogate the formation of oxidants (e.g., metal-binding proteins like transferrin), while the second line of defense is maintained by chain-breaking antioxidants, which nullify the effect of oxidants by terminating the chain initiation or chain propagation steps. Certain antioxidants act preferentially or exclusively in the aqueous phase, while others are predominantly lipophilic. The latter, for obvious reasons, are important in keeping lipid peroxidation at bay. A third level of defense also exists, comprised by various cellular mechanisms involved in the repair or clearing of the macromolecules that had undergone oxidant-mediated damage. This is exemplified by DNA repair enzymes (see Chapter 35, Section VI. DNA repair, p. 610) and unfolded protein response (see Chapter 37, Section VI.C. Protein degradation, p. 649).

Free radical scavengers are antioxidants that eliminate free radicals produced during mitochondrial respiratory reactions of aerobic cellular metabolism. They typically accomplish this by enhancing the levels or synthesis of GSH and by inhibiting peroxidation.

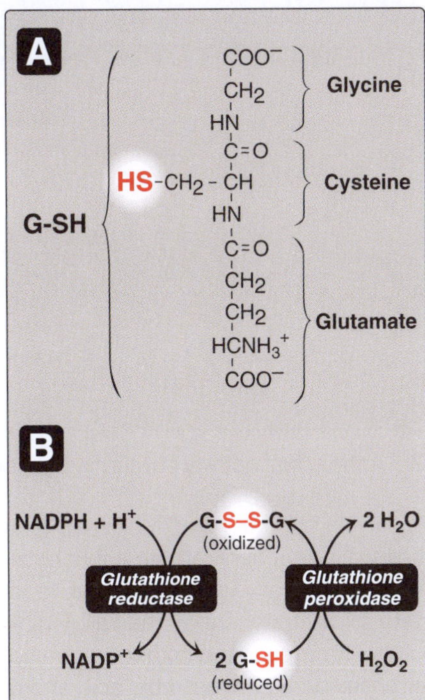

Figure 47.6 **A.** Structure of reduced glutathione (G-SH). [Note: Glutamate is linked to cysteine through a γ-carboxyl, rather than an α-carboxyl.] **B.** The roles of G-SH and reduced nicotinamide adenine dinucleotide phosphate (NADPH) in the reduction of hydrogen peroxide (H_2O_2) to water. G-S-S-G = oxidized glutathione.

A. Enzymes with antioxidant effects

Various cellular enzymes can limit the generation of potent ROS from milder forms, acting as preventive antioxidants. **Catalase is present in membrane-bound organelles called peroxisomes which are the sites of generation of H_2O_2 during oxidation of very long-chain fatty acids** (see Section II.D). The reaction catalyzed is:

$$2H_2O_2 \rightarrow 2H_2O + O_2$$

Localization of *catalase* at the site of ROS generation helps in efficiently negating the generation of other free radicals from H_2O_2. Peroxides can also be acted upon by various *peroxidases* in the cell, to form hydroxides, which do not possess oxidant effects.

$$ROOR' + AH_2 \rightarrow ROH + R'OH + A$$

A prominent antioxidant enzyme in most cell types is *glutathione peroxidase* (see Fig. 47.2), which **utilizes glutathione** (GSH) to reduce peroxides (Fig. 47.6). This enzyme **requires selenium** and catalyzes the following reaction, in accordance with the general equation mentioned above.

$$ROOR' + 2GSH \rightarrow ROH + R'OH + GSSG$$

Reduced glutathione (GSH) needs to be regenerated from the oxidized form (GSSG) for further catalysis by *glutathione peroxidase*. This regeneration is catalyzed by *glutathione reductase*, an NADPH-dependent flavoenzyme.

$$GSSG + NADPH + H^+ \rightarrow 2GSH + NADP^+$$

Glutathione, with the aid of these enzymes, thus plays a crucial role in the cellular antioxidant defense. It also has to be noted that adequate levels of NADPH are required for the normal maintenance of this system, and conditions in which NADPH levels are depleted predispose to oxidant injury.

Dismutation refers to the generation of two molecules—one with a higher and the other with a lower oxidation state, from a compound of intermediate oxidation state (or vice versa). Different isoenzymes of ***superoxide dismutase*** (SOD) catalyze the dismutation of superoxide to hydrogen peroxide, the latter being further acted upon by *catalases* and *peroxidases*.

$$O_2^{\cdot -} + 2H^+ \rightarrow H_2O_2$$

Mutations in Cu/Zn-dependent SOD is seen in a large number of familial cases of amyotrophic lateral sclerosis (ALS), also known as Lou Gehrig's disease (after the American baseball player who succumbed to the disease). One postulated mechanism for the nerve damage seen in this condition is the oxidant-mediated damage due to suboptimal functioning of SOD.

B. Antioxidant role of vitamins

Vitamin E (tocopherol), the most potent lipophilic antioxidant, plays a crucial role in termination of lipid peroxidation, and is the major chain-breaking antioxidant.

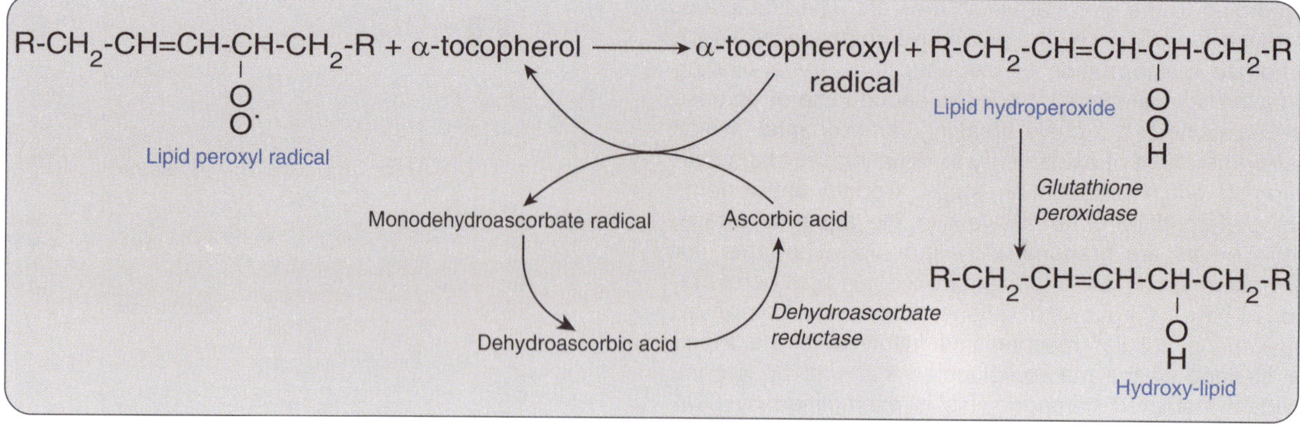

Figure 47.7 Antioxidant activity of vitamin E (α-tocopherol) and vitamin C (ascorbic acid) in termination of lipid peroxidation.

The reaction of α-tocopherol with lipid peroxides forms relatively inert hydroxy-fatty acids, the former getting converted to α-tocopheroxyl radical (Fig. 47.7). The α-tocopherol is regenerated from α-tocopheroxyl radical by other antioxidants like ascorbic acid and retinol.

Vitamin C (ascorbic acid) exerts a major antioxidant activity by reduction of α-tocopheroxyl radical to regenerate vitamin E (Fig. 47.7). Monodehydroascorbate radical formed in this step is a stable radical and can be reduced to dehydroascorbic acid enzymatically or nonenzymatically. Thus, vitamin C can have a multipronged effect in cellular antioxidant defense. Dehydroascorbic acid generated requires reducing agents like glutathione for its enzymatic reduction (by *dehydroascorbate reductase*) to form ascorbic acid again. Ascorbate can also directly act on ROS like H_2O_2 and hydroxyl radical, and on RNS.

C. Other antioxidants

Apart from vitamins C and E, the other dietary antioxidants include **β-carotene**—the precursor molecule of vitamin A, **lycopene**—the carotenoid found in tomatoes, **selenium** (mostly because of its cofactor activity for *glutathione peroxidase*) and **polyphenols** (like resveratrol which is present in red wine, and catechin present in green tea).

Two major antioxidants in plasma are bilirubin and uric acid. Bilirubin can scavenge oxidant species in solution as well as membranes. In humans, almost half of the antioxidant activity in plasma is contributed by uric acid. Indeed, our serum uric acid levels are at values near the saturation and is much higher than the other primates. This is a probable adaptation to counter the lack of production of vitamin C by human body in contrast to the other primates. The higher levels of uric acid, preventing oxidative damage, could be a reason for increased human longevity, compared to the primate counterparts.

D. Antioxidants in excess can act as prooxidants

Because of the apparent benefits of antioxidants by reducing the damage caused by oxidant species, various clinical trials explored the efficacy of supplementation of antioxidants to prevent disease generation or progression. But the results of these studies were counterintuitive, leading to the proposition that—in excess, antioxidants can have a prooxidant effect. For instance, supplementation of vitamin E can lead to elevated levels of α-tocopheroxyl radical, which can oxidize various lipid components of the cell, by persisting in the cell for longer durations. Also, excess removal of oxidants from the cell can compromise their function because of the disruption of the physiological functions like cell signaling, which are mediated by oxidants.

V. ESTIMATION OF CELLULAR OXIDANT SPECIES

Because of the detrimental effects of oxidative stress on biological systems, it is essential to assess the levels of cellular oxidant species and/or their downstream products, to study the effects of various interventions that are generated to counter them.

- The classical technique to estimate lipid peroxidation relies on the formation of products like malondialdehyde (MDA) ($O=CH-CH_2-CH=O$) as a result of decomposition of lipid peroxides. MDA reacts with thiobarbituric acid (TBA) to form an adduct (thiobarbituric acid reactive substances—TBARS), which can be measured colorimetrically.
- Peroxides can be estimated by their ability to oxidize Fe^{2+} to Fe^{3+}. In a sulfuric acid solution, the Fe^{3+} formed, complexes with xylenol orange (XO) present in the reagent to form a purple product which can be quantified colorimetrically.

- The ratio of GSH to GSSG (reduced glutathione/oxidized glutathione) is also another crucial indicator of the cellular redox status.
- Cellular oxidant levels can also be estimated with the help of various dyes that fluoresce in their oxidized form. In this case, a higher fluorescence level indicates the presence of oxidative stress.

VI. Medical Perspectives

Oxidative stress can affect all the cell types in the body and can lead to numerous disease states. Indeed, current studies suggest that almost all diseases have some contribution by oxidant species in their development and/or progression. This section describes some well-established biological and pathological conditions associated with oxidative stress.

A. Aging

Free radical theory of aging implicates free radicals as the cause of aging. As the cell or the organism ages, the more its constituents get exposed to oxidative stress. This results in perturbations in the homeostasis due to the effect of free radicals on DNA, lipids and proteins. Mutations also lead to cellular dysfunction and induction of apoptosis. All these, when seen at the level of organism, are contributing factors for aging. Wrinkled skin in aged individuals is a direct evidence of free radicals on aging. This occurs due to decreased synthesis of collagen and elastin, due to the free radical-mediated disruption of protein synthesis.

B. Cancer

A number of conditions associated with oxidative stress, like long-term exposure to ultraviolet radiation and chronic inflammation, are predisposing conditions to the development of cancer. In carcinogenesis, a mechanism similar to aging operates; but the cells that have accumulated the mutations because of oxidative stress, escape senescence and apoptosis, and continue to proliferate. This leads to generation of additional mutations and elusion from the cell cycle regulatory mechanisms, leading to abnormal, unregulated proliferation—and thus cancer.

Interestingly, radiation therapy used in treatment of certain malignancies, renders its effect on cancer cells due to toxicity generated by free radicals. The ionizing radiation used in radiation therapy leads to generation of oxygen-derived free radicals (especially hydroxyl radical) in the tumor environment, by ionization of intracellular molecules like oxygen and water. This results in the death of cancer cells, mostly due to extensive membrane and DNA damage.

C. Atherosclerosis

Oxidation of lipids and proteins present in the LDL leads to the generation of oxidized LDL (ox-LDL). Recall that LDL is removed from circulation by peripheral cells including macrophages, by the LDL receptor, which is highly regulated (see Chapter 23.VI.D.2. Endocytosed cholesterol and cholesterol homeostasis, p. 354). On the other hand, ox-LDL is taken up by the macrophages present in the tunica intima of the blood vessels, via an unregulated scavenger receptor (SR-B1), leading to accumulation of increased amounts of cholesterol within these cells. Formation of the so called "foam cells" leads to the development of atherosclerotic plaque in the blood vessels (see Fig. 23.22). Ox-LDL is also directly toxic to the endothelial cells, accelerating the generation of plaque. One of the mechanisms by which high-density lipoprotein (HDL) protects against atherosclerosis is due to the presence of the enzyme *paraoxonase-1*, which is an antioxidant that protects against lipid peroxidation of LDL.

Oxidant-mediated injury is also implicated in paradoxical tissue damage to brain and myocardium following stroke or myocardial infarction, respectively, when circulation is re-established to these tissues following a period of ischemia. This condition, which leads to significant organ damage, is termed ischemia-reperfusion injury.

D. Diabetes mellitus

Destruction of pancreatic islets in type 1 diabetes mellitus, which is an autoimmune condition, is partly due to the leukocyte-mediated generation of free radicals, which accompanies inflammatory infiltration of these cells. Oxidative damage is also implicated in the pathogenesis of complications of type 2 diabetes mellitus. Persistent hyperglycemia leads to entry of glucose into non-insulin dependent tissues like nervous tissues, retina and nephrons. In such tissues, glucose is converted to sorbitol by *aldose reductase* (see Fig. 17.4), which is an NADPH-dependent reaction. This leads to depletion of NADPH in individuals with type 2 diabetes mellitus. As described previously (Section IV.A), NADPH plays a central role in antioxidant defense, since it is required for maintaining adequate levels of reduced glutathione (GSH). The resultant oxidative stress due to reduced levels of GSH in neurons, retina, and nephrons is one of the mechanisms that leads to complications of diabetes mellitus like neuropathy, retinopathy, and nephropathy.

E. Glucose-6-phosphate dehydrogenase (G6PD) deficiency

The destruction of erythrocytes due to oxidative stress, leading to hemolysis, has been described in detail in Chapter 18, Section.V. G6PD deficiency, p. 253.

VII. Chapter Summary

- Free radicals have an unpaired electron in their outer orbital and are highly reactive.
- The prominent reactive oxygen species in the cell are superoxide radical, hydrogen peroxide, hydroxyl radical, and hypochlorite anion.
- The major reactive nitrogen species are peroxynitrite, nitrous acid, and nitric oxide.
- Oxidative stress refers to the condition when the levels of the oxidants remain persistently high. This can cause damage to DNA, protein and lipids. DNA damage leads to mutations, protein damage leads to proteolysis, protein dysfunction and unfolded protein response, and lipid damage leads to formation of chain reactions where lipid peroxides are formed.
- ROS and RNS also serve important physiological functions like signal transduction and antimicrobial defense.
- Antioxidants that counter the effects of ROS and RNS can be broadly classified into preventive and chain-breaking antioxidants. Antioxidants include various antioxidant enzymes, vitamins C and E, certain other dietary molecules, bilirubin and uric acid. Vitamin E is the most potent lipophilic antioxidant and is a chain-breaking antioxidant.
- In excess quantities, antioxidants can show prooxidant effects.
- Oxidative stress leads to various conditions like aging, cancer formation, atherosclerosis, and complications of diabetes mellitus (Fig. 47.8).

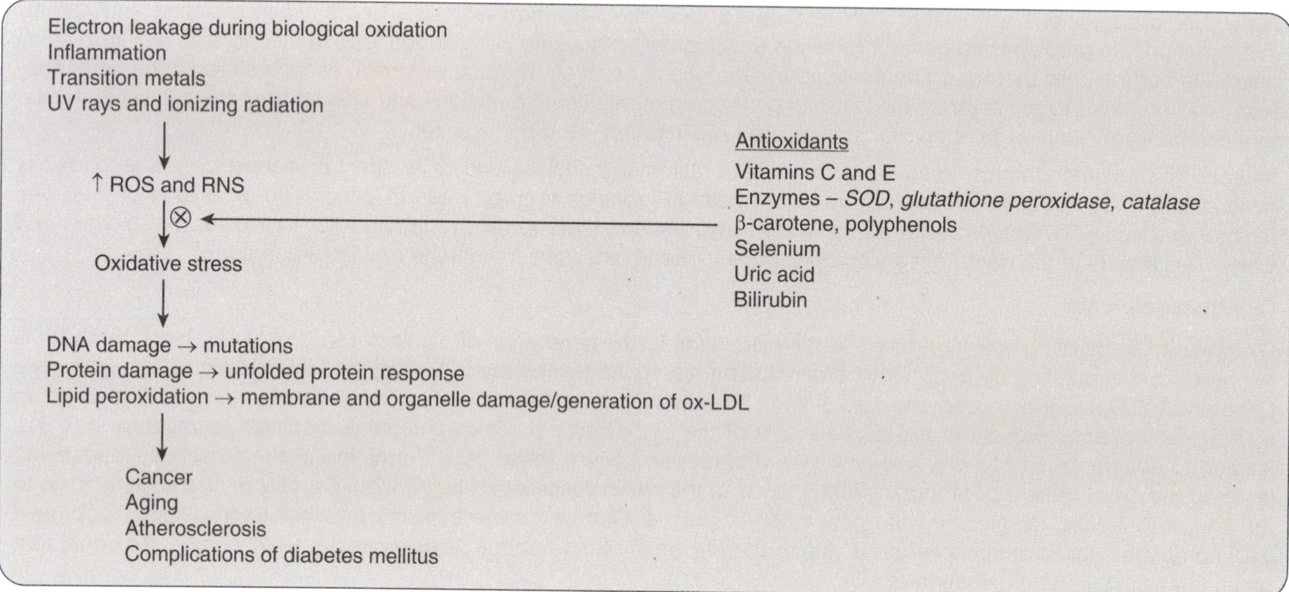

Figure 47.8 Key concept map of antioxidant defense mechanisms and oxidative stress. UV = ultraviolet; ROS = reactive oxygen species; RNS = reactive nitrogen species; SOD = *superoxide dismutase*; DNA = deoxyribonucleic acid; ox-LDL = oxidized low-density lipoprotein.

Study Questions

Short-Answer Questions

47.1. Which is the most potent lipophilic antioxidant? How does it exert its effect?

47.2. Describe the principles of various techniques that are used to assess oxidative stress.

47.3. Free radicals and reactive oxygen species are not the same. Explain why?

47.4. Classify antioxidants with suitable examples.

Long-Answer Questions

47.1. Describe the cellular processes where oxidant species are generated.

47.2. Describe the deleterious effects of free radicals on cells.

Multiple-Choice Questions

Choose the ONE best answer.

47.1. Which of the following enzymes, when defective in phagocytes, leads to immune deficiency?

A. Glutathione reductase

B. Glutathione peroxidase
C. NADPH oxidase
D. Superoxide dismutase

Correct answer = C.

47.2. Which of the following elements possesses antioxidant activity?
A. Selenium
B. Iron
C. Copper
D. Calcium

Correct answer = A.

47.3. Which of the following is a marker enzyme for peroxisomes?
A. Prostaglandin H_2 synthase
B. Catalase
C. Lipoxygenase
D. NADPH oxidase

Correct answer = B.

47.4. Which of the following is the characteristic of free radicals?
A. Have paired electrons in the outer shell.
B. They accept or donate an unpaired electron to attain stability.
C. They accept or donate the paired electron to attain stability.
D. Free radicals are oxygen containing only.

Correct answer = B.

47.5. Cellular oxidant species are:
A. Oxygen containing free radicals and highly reactive oxygen containing entities only.
B. Highly reactive oxygen and nitrogen containing entities only.
C. Highly reactive entities having nitrogen only.
D. Oxygen containing free radicals and highly reactive species containing oxygen or both oxygen and nitrogen.

Correct answer = D.

47.6. Which of the following is NOT a part of reactive oxygen species (ROS)?
A. Superoxide
B. Hydrogen peroxide
C. Nitrous acid
D. Hydroxyl radical

Correct answer = C.

47.7. Molecular oxygen is converted to water in mitochondria. The process requires that a single molecule of O_2 accepts:
A. 4 electrons and 4 protons to generate one molecule of water.
B. 4 electrons and 4 protons to generate two molecules of water.
C. 2 electrons and 2 protons to generate one molecule of water.
D. 2 electrons and 2 protons to generate two molecules of water.

Correct answer = B.

47.8. For oxygen-dependent phagocytosis by activated macrophages and neutrophils, the following enzyme(s) are involved:
A. Superoxide dismutase and NADPH oxidase only.
B. NADPH oxidase and myeloperoxidase only.
C. Superoxide dismutase, myeloperoxidase and NADPH oxidase.
D. Superoxide dismutase, myeloperoxidase, nitric oxide synthase and NADPH oxidase.

Correct answer = D.

47.9. Which of the following is true for the humans?
A. Iron and copper are present mostly bound to storage or transport proteins to minimize their toxic effect due to formation of free radicals by their free forms.
B. Iron and calcium are present mostly bound to storage or transport proteins to minimize their toxic effect due to formation of free radicals by their free forms.
C. Iron and magnesium are present mostly bound to storage or transport proteins to minimize their toxic effect due to formation of free radicals by their free forms.
D. Calcium and magnesium are present mostly bound to storage or transport proteins to minimize their toxic effect due to formation of free radicals by their free forms.

Correct answer = A.

47.10. Oxidative stress may damage the DNA by causing:
A. Single or double stranded beaks.
B. Formation of adducts, single stranded breaks only.
C. Single or double stranded breaks, formation of adducts.
D. Formation of adducts only.

Correct answer = C.

Multiple-Choice Questions (NExT/NEET PG)

Choose the ONE best answer.

47.1. In sickle cell anemia patients, intravascular hemolysis of the defective RBCs leads to release of free heme into circulation. These patients also experience vaso-occlusion leading to damage of peripheral organs. Quenching of which of the following molecules by free heme leads to vasoconstriction?
A. Hydrogen peroxide

B. Nitric oxide
C. Glutathione
D. Hydroxyl radical

Correct answer = B.

47.2. All the following enzymes catalyze reactions that generate oxidant species, EXCEPT:

A. Very long-chain fatty acyl CoA dehydrogenase
B. Xanthine oxidase
C. Glutathione reductase
D. Lipoxygenase

Correct answer = C.

47.3. Which of the following elements is involved to handle oxidative stress?

A. Calcium
B. Iron
C. Potassium
D. Selenium

Correct answer = D.

47.4. What is the role of ascorbic acid (vitamin C) in the antioxidant activity of vitamin E (α-tocopherol)?

A. Ascorbic acid forms lipid peroxides
B. Ascorbic acid reduces α-tocopheroxyl radical to regenerate vitamin E
C. Ascorbic acid forms dehydroascorbic acid
D. Ascorbic acid increases ROS production

Correct answer = B.

47.5. Which of the following is incorrect about antioxidant enzymes?

A. Catalase is present in peroxisomes and degrades H_2O_2 to H_2O and O_2.
B. Glutathione peroxidase reduces peroxides particularly hydrogen peroxide (H_2O_2) and lipid peroxides.
C. Glutathione reductase, a selenium containing enzyme, plays an important role in cellular antioxidant defense mechanism.
D. Superoxide dismutase catalyzes the conversion of the superoxide radical ($O_2^{\cdot -}$) into hydrogen peroxide (H_2O_2) and oxygen (O_2).

Correct answer = C.

48

Organ Function Tests

Learning Objectives

At the end of this chapter, the student will be able to:

- Describe the functions of the kidney, liver, thyroid and adrenal glands, and discuss the tests that are commonly done in clinical practice to assess the functions of these organs
- Describe the diseases associated with abnormal function of kidney, liver, thyroid, and adrenal glands

I. OVERVIEW

Since diseases are associated with dysfunction at the cellular level, and an alteration in the cellular homeostasis results in changes in the chemical environment of the cell which is afflicted, analysis of chemical parameters can aid in the diagnosis of diseases.

The term **"Organ Function Tests" refer to a battery of clinical investigations, biochemical and otherwise (e.g., radiological investigations) that are employed to identify the abnormal functioning of the organ under question**. A normal test result does not always rule out the possibility of organ dysfunction, whereas an abnormal test result almost always result from a deranged physiology. Thus, paradoxically, organ function tests are most useful in a scenario of organ dysfunction. This chapter will cover the common "Organ Function Tests" performed in a clinical chemistry laboratory, viz. liver function tests, renal function tests, thyroid function tests, and tests for adrenal dysfunction. Before the description on the tests performed to assess each of these organs, a summary of the physiological functions of these organs are given, so as to provide a perspective to the tests involved to identify the organ dysfunction.

One of the effects of injury to cells in any tissue is the release of the contents of the cells into the extracellular milieu, some of which invariably enter the plasma. Minute quantities of such molecules may be present in plasma as a result of the usual wear and tear of the tissues. However, in cases of tissue injury, their levels will be significantly elevated. Thus, in clinical scenarios where an organ injury is suspected, a clue to the diagnosis can be made by estimating the analytes that originate from the organ involved. This forms the basis of biochemical investigations in numerous conditions of tissue injury—e.g., elevation of cardiac troponins (cTnt and cTnI) and CK-MB isoenzyme in myocardial infarction (see Chapter 10, Section.IX.C.2. Diagnosis of myocardial infarction, p. 142), and elevation of *lactate dehydrogenase* (LDH) in serum in hemolysis (because of release of this enzyme from RBCs into the blood).

It may be noted that the enzymes that are present in plasma are classified into functional and nonfunctional plasma enzymes. As the name suggests, the former have some specific biological function in the plasma and are present in free form or are attached to the endothelial cells exposed to blood. Their substrates are present in the plasma. Examples include ***lipoprotein lipase****, enzymes involved in blood coagulation (the* ***clotting factors****) and proteins of complement system. Since they perform vital biological functions in the blood, a deficiency or low levels of these enzymes in blood results in diseases like familial chylomicronemia in lipoprotein lipase deficiency, and coagulation disorders in clotting factor deficiencies.* ***Nonfunctional plasma enzymes do not have their substrates present in plasma*** *and thus, they do not serve any physiological function in plasma. They are actually tissue-localized enzymes, with biological actions that are specific to the tissue(s) in which they are located. They enter the plasma as the result of tissue turnover or disease, and thus their levels are elevated in case of tissue injury. As mentioned previously, they are diagnostically significant, and elevated levels of these enzymes in plasma are suggestive of some pathology. For example,* ***amylase and lipase*** *(plasma levels are elevated in pancreatitis), LDH, and* ***CK-MB****.*

However, there is a catch—more than one tissue can release the same nonfunctional plasma enzyme or similar analytes into the plasma following tissue injury. For instance, damage to either salivary glands or pancreas can lead to elevation in plasma levels of amylase, as the enzyme is present in both these tissues. Thus, it is important to consider the clinical features, and/or employ a panel of analytes to diagnose the pathology under consideration.

II. FUNCTIONS OF LIVER

The liver is the largest solid organ in the human body, and it acts as the hub of metabolism. It plays a crucial role in the metabolism of carbohydrates, amino acids, lipids, nucleotides and various micronutrients. Since liver directly receives venous outflow from the intestines via the hepatic portal system, many drugs and toxins (xenobiotics) that enter the body via oral route are metabolized here (first-pass metabolism or first-pass effect). Liver is also involved in metabolism of systemically administered drugs, and endogenous molecules with a potential harmful effect. The liver is also involved in the synthesis of a large number of physiologically important molecules. It has been observed that the liver can regenerate to its original size even after removal of two-thirds of the original tissue mass, and this forms the basis of living-donor liver transplantation.

Hepatocytes, the parenchymal cells of liver accounts for nearly 60% of total cells in the liver. The functions of liver can be broadly categorized into metabolic functions, synthetic and storage functions, and detoxification and excretory functions.

A. Metabolic functions

Metabolic functions of liver are individually detailed in various chapters throughout this book. To summarize, these include:

1. **Carbohydrate metabolism:** Glycogenesis and glycogenolysis, gluconeogenesis (the other major site being renal medulla), galactose metabolism and fructose metabolism.

2. **Lipid metabolism:** Synthesis of lipoproteins, phospholipids and cholesterol, conversion of cholesterol into bile acids (which forms bile salts that are important in absorption of dietary lipids), removal of intermediate-density lipoprotein (IDL), low-density lipoprotein (LDL), high-density lipoprotein (HDL), and chylomicron remnants from circulation, ketone body synthesis, and fatty acid synthesis and oxidation.

3. **Protein metabolism:** Transamination, oxidative deamination, urea formation, and gluconeogenesis and ketogenesis from R groups of catabolized amino acids.

4. **Nucleotide metabolism:** The liver is the primary site for purine synthesis and uric acid formation.

5. **Vitamin and micronutrient metabolism:** Regulation of iron absorption via hepcidin synthesis, incorporation of copper into ceruloplasmin for transport, synthesis of 25-hydroxycholecalciferol (Fig. 32.23), and isomerization of retinol (part of Wald's visual cycle – See Chapter 32, Section XI.E.1. Visual cycle, p. 538). The liver is also the site of vitamin K cycle (Fig. 32.26).

B. Synthetic and storage functions

Apart from the molecules synthesized as a part of the intermediary metabolism mentioned in the previous section, liver is also involved in the synthesis of a variety of substances. The liver also acts as the site of storage of many of these.

1. **All the major serum proteins except the immunoglobulins are synthesized in the liver.** These include albumin (which is important in maintaining the plasma volume, and also in transport of molecules), other transport molecules like transferrin, transthyretin, ceruloplasmin and hormone-binding globulins, lipoproteins, proteins of the complement system, and acute phase reactants (see X. Medical Perspectives) like C-reactive protein.

2. **Hepatocytes are the major sites of synthesis of all the clotting factors except calcium and factor VIII.** The latter, along with von Willebrand factor are produced by hepatic sinusoidal endothelial cells. The liver is also the site of the microsomal enzyme *vitamin K epoxide reductase* and the vitamin K cycle, which is involved in the γ-carboxylation of glutamic acid residues of factors II, VII, IX and X, which is necessary for the functioning of these coagulation factors.

3. **The liver is the storage site for fat soluble vitamins,** viz. vitamins A, D, E and K. Of these, vitamin A is stored after esterification into retinyl palmitate in specialized cells known as hepatic stellate cells or Ito cells. The liver also stores vitamin B_{12}, the only water-soluble vitamin that is stored in the body. The liver stores carbohydrates in the form of glycogen, and also stores minerals like iron and copper.

C. Detoxification and excretory functions

Exogenous compounds like drugs and chemicals known as xenobiotics are acted upon by cellular enzymes. At least 50% of drugs administered to humans and numerous other chemicals are acted upon by the microsomal cytochrome **P450 *monooxygenase* system** (see Chapter 46, Section III.A. The cytochrome P450 (CYP) superfamily, p. 782), which in most of the cases is involved in inactivation of the compound, in what is known as the phase 1 reactions of xenobiotic metabolism. Phase 2 involves conjugation of the products of phase 1 reactions with molecules (e.g., glucuronic acid) to render them hydrophilic and promote their excretion. The liver plays a central role in both phase 1 and phase 2 reactions of xenobiotic metabolism.

The liver is also involved in **detoxification of endogenous molecules like ammonia to urea** via urea cycle. Hepatocytes are also involved in the **conjugation of bilirubin** with UDP-glucuronic acid to form bilirubin mono- and diglucuronides (conjugated bilirubin), which are then excreted in the bile.

The compounds that are detoxified by liver are excreted through bile or in urine via the kidneys. The liver is also involved in removal of cholesterol from the body, either by conversion of cholesterol into bile acids or by facilitating cholesterol excretion as such into the bile. Bile also acts as the portal for removal of other endogenous molecules including certain hormones.

D. Miscellaneous functions

1. **Kupfer cells** are tissue macrophages localized to liver; they are involved in phagocytosis of pathogenic microorganisms and removal of senescent red blood cells (RBCs).
2. The liver also acts as a site of hematopoiesis during intrauterine life.
3. The liver also forms a component of the renin-angiotensin-aldosterone system (RAAS) as angiotensinogen is synthesized by the hepatocytes (see Fig. 5.8).

III. LIVER FUNCTION TESTS

Diseases of liver account for more than 2 million deaths worldwide—with the majority occurring as a complication of **cirrhosis** (fibrosis of liver accompanied by nodular regeneration), and the rest are mostly due to **viral hepatitis and hepatocellular carcinoma**. Liver diseases account for 3.5% of the global deaths and liver is the second most commonly transplanted solid organ (most common being the kidney). Further, the number of individuals with liver dysfunction are on the rise due to the increasing prevalence of **alcohol use** (leading to fatty liver, alcoholic hepatitis, and cirrhosis) and metabolic diseases like **diabetes mellitus and obesity** which can cause conditions like fatty liver (nonalcoholic fatty liver disease or NAFLD) and hepatitis (steatohepatitis), eventually progressing to cirrhosis. Thus, it is pertinent to diagnose and monitor liver diseases in a prompt manner, and biochemical investigations are crucial in achieving this. Further, **a number of drugs can lead to hepatotoxicity**, for example, antituberculosis drugs and drugs used in the treatment of human immunodeficiency virus (HIV)/acquired immunodeficiency syndrome (AIDS).

Liver has a considerable functional and compensatory reserve; hence when the liver is diseased, not all of its functions may be impaired. Thus, biochemical assessment of liver function (or dysfunction) involves a number of analytes. In the subsequent discussion, these analytes are classified according the dysfunctions of liver that they are suggestive of.

A. Biochemical markers of liver injury

Hepatocellular damage is classically defined by the elevated levels of two enzymes in the plasma, viz. ***alanine aminotransferase (ALT)*** **and** ***aspartate aminotransferase (AST).*** The older names of these enzymes *serum glutamate pyruvate transaminase (SGPT)* and *serum glutamate oxaloacetate transaminase (SGOT)* are rarely used in clinical practice nowadays. The reactions catalyzed by these enzymes are shown in Fig. 24.8; note that both these enzymes use pyridoxal phosphate (PLP, the active form of vitamin B_6) as the coenzyme.

ALT is cytoplasmic and is mostly expressed by the liver, with some amounts present in the kidneys, skeletal muscles, and other tissues; *AST* exists as cytoplasmic and mitochondrial isoenzymes, with very high contents in liver, myocardium, skeletal muscles and kidneys, and lower amounts in pancreas and RBCs. Thus, ***ALT* is the more "liver-specific"** among these transaminases. In most of the liver diseases, the level of *ALT* is higher than that of *AST* mostly because of a short half-life of *AST* (18 hours) compared to that of *ALT* (36 hours). Exceptions are alcoholic hepatitis, cirrhosis, ischemic liver injury, and hepatocellular carcinoma, where AST/ALT ratio (known as De Ritis ratio) is usually >1. This may be because of mitochondrial damage in these conditions leading to the leakage of the mitochondrial *AST* into circulation.

The common causes of hepatocellular damage are listed in (Fig. 48.1). The etiology of hepatocellular damage can be identified based on the extent of elevation of the transaminases from the upper reference limit (URL). The highest elevations are seen in ischemic liver injury, followed by hepatitis due to toxicity (e.g., paracetamol overdose or α-amanitin, see Chapter 36, Section IV.B.2.e. RNA polymerase II inhibitor, p. 627) and viruses (hepatitis A, B and C). Alcoholic hepatitis, though associated with extensive liver injury only causes moderate elevation of transaminases because of low levels of vitamin B_6 in alcohol use disorder (because of low dietary intake). It may be recalled that a PLP is the coenzyme for *ALT* and *AST*.

A serial estimation of ***transaminases*** is more useful than a single value in monitoring the progress of the

ALT and AST >15 times URL (severe elevation) or >10,000 IU/l (massive elevation)
Acute viral hepatitis A, B and C
Acetaminophen (paracetamol) toxicity
Ischemic liver injury
ALT and AST 5–15 times URL (moderate elevation)
Alcoholic hepatitis
Autoimmune hepatitis
ALT and AST 2–5 times URL (mild elevation)
Carcinoma of liver (hepatocellular carcinoma or metastasis)
Cirrhosis
Nonalcoholic fatty liver disease (NAFLD)
Hemochromatosis (see Chapter 33, Section.III.B.5, p. 558 and Chapter 47, Clinical Case Question 47.1, p. 831)
Wilson's disease (see Chapter 33, Section.III.A.2, p. 556)
α1-antitrypsin deficiency (see Chapter 9, Section III.B.2, p. 125)
ALT and AST up to 2 times URL (borderline elevation)
Chronic viral hepatitis B and C

Figure 48.1 Common causes of hepatocellular damage.

disease. A sudden massive drop in the level of *transaminases* is suggestive of impending liver failure as the capacity of the whatever remaining hepatocytes to synthesize *ALT* and *AST* is severely compromised. A persistent low level of *transaminases* can be seen in vitamin B_6 deficiency.

Apart from the *transaminases*, activity of two other enzymes can also be estimated to assess hepatocellular damage, though they are not a part of the routine liver test panel.

- ***Glutamate dehydrogenase (GDH)*** (Fig. 24.11) is a mitochondrial enzyme with higher concentration in the centrilobular areas of the liver (Fig. 48.2). Since these areas are far from the portal triad, they receive relatively less blood supply and are thus

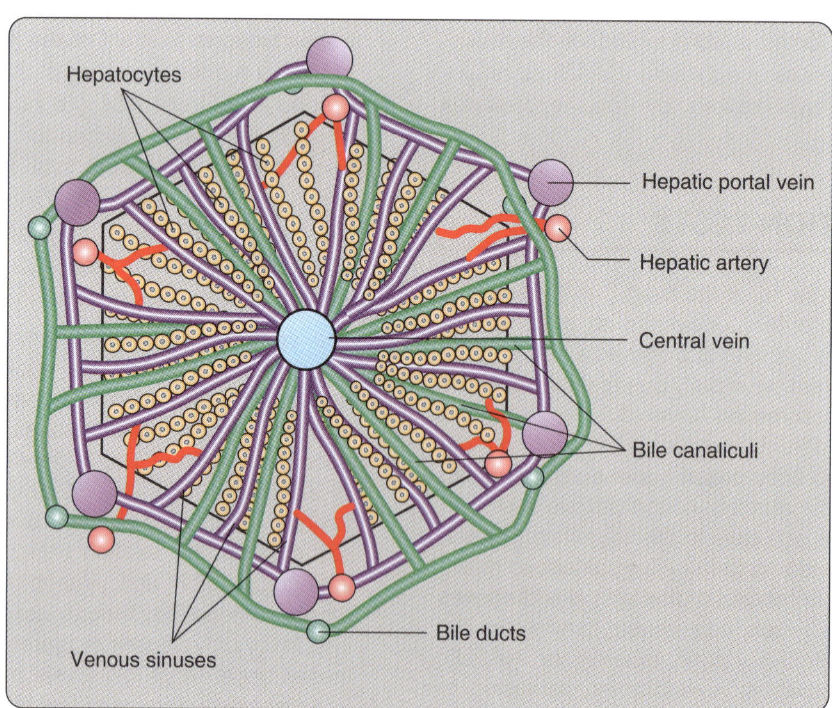

Figure 48.2 Microscopic structure of liver lobule. *Glutamate dehydrogenase* (GDH) is distributed in the centrilobular areas, Transaminases in the periportal region, and glutathione S-transferase (GST) is uniformly distributed. Note the close relation between the hepatocytes and the bile canaliculi.

prone to ischemic injury—*GDH* level in plasma is thus elevated in ischemic liver injury.

- ***Glutathione S-transferases (GSTs)*** are involved in the detoxification of xenobiotics by conjugation with glutathione. They also bind to bilirubin once the latter is delivered to the hepatocytes, and facilitate the conjugation reaction of bilirubin with UDP-glucuronic acid by *UDP-glucuronyl transferase A1 (UGT1A1)*. *GSTs* are distributed uniformly throughout the hepatic lobules, and elevations in plasma are seen in all types of hepatocellular injury.

B. Biochemical markers of cholestasis

Cholestasis (*chole*: bile) occurs when there is an **obstruction to bile outflow**. This can occur in two broad scenarios, viz. intrahepatic obstruction and extrahepatic obstruction. Intrahepatic biliary obstruction occurs commonly in the background of an associated hepatocellular injury, where the inflammation associated with the hepatitis leads to obstruction of the biliary canaliculi. Thus, it is seen in **alcoholic, viral and toxic hepatitis**. It can also occur in certain autoimmune afflictions where small bile ducts are injured, e.g., primary biliary cholangitis (previously known as primary biliary cirrhosis). Extrahepatic biliary obstruction occurs due to obstruction of the larger bile ducts, outside the liver, in causing what is termed as obstructive jaundice or post-hepatic jaundice. Causes include stones, tumors or parasitic infections in the bile duct, carcinoma head of pancreas and malignancy around ampulla of Vater (periampullary carcinoma). The biochemical markers of cholestasis include serum bilirubin and three nonfunctional plasma enzymes, viz. ***alkaline phosphatase*** **(ALP), γ-glutamyl transpeptidase (GGT) and 5'-nucleotidase (5'-NT).**

1. **Bilirubin:** "Total bilirubin" is composed of two fractions, viz. "direct bilirubin" and "indirect bilirubin." **"Direct bilirubin" and "indirect bilirubin" are considered to be representative of the conjugated bilirubin and unconjugated bilirubin**, respectively, that is present in circulation. Since unconjugated bilirubin is converted to conjugated bilirubin within the hepatocytes, this process is impaired in hepatocellular damage. Moreover, the intrahepatic biliary obstruction associated hepatocellular injury prevents excretion of conjugated bilirubin; this conjugated bilirubin regurgitates back into the circulation. Thus, in intrahepatic cholestasis, both direct and indirect bilirubin are elevated in the serum. On the contrary, extrahepatic cholestasis is characterized by a predominant elevation of direct bilirubin only.

2. **ALP:** ***Alkaline phosphatase*** is involved in hydrolysis of organic phosphate esters, but its exact

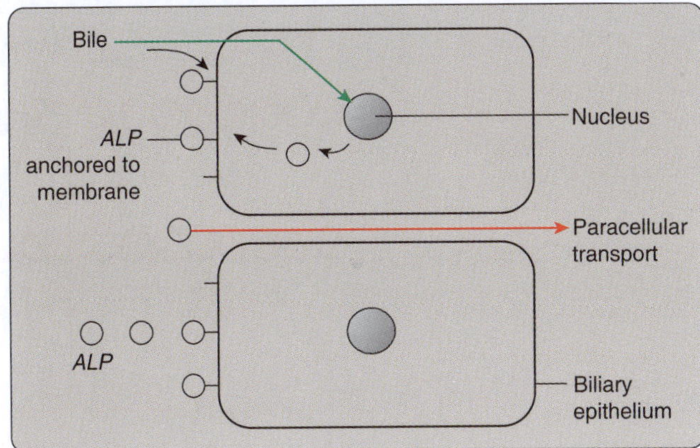

Figure 48.3 Location of *alkaline phosphatase (ALP)* in the lining cells of the bile ducts. In cholestasis, detergent action of bile acids releases the enzyme from the membrane, which enters the circulation via a paracellular pathway. Bile also enhances the enzyme production.

physiological function is not known. It has **multiple isoenzymes and isoforms** with a widespread tissue distribution, with highest concentrations in bone, intestine, liver, and placenta. In the liver, the enzyme is present on the sinusoidal surface of the hepatocytes and also in the cells lining the bile canaliculi and the larger bile ducts (Fig. 48.3), anchored on the outer leaflet of the plasma membrane. Thus, ALP is an **"ectoenzyme." ALP is physiologically elevated in children (because of high osteoblastic activity) and in pregnancy (due to the placental isoenzyme).** In **cholestasis, elevation of ALP can be due to two mechanisms**, both attributed to bile salts, the excretion of which is impaired—(1) bile salts induce the production of ALP by acting at the deoxyribonucleic acid (DNA) level, (2) bile salts, owing to their detergent action, release ALP from the canalicular membrane; the released ALP enters the circulation via a paracellular mechanism (Fig. 48.3). The reference interval of ALP in adults is 33–96 U/l. Levels >4 times the URL is seen in extrahepatic obstruction and primary biliary cholangitis, with values sometimes reaching 10–12 times the URL. In the other common causes of intrahepatic obstruction and parenchymal liver diseases, the levels are usually 2–4 times the URL.

3. **GGT:** γ-glutamyl transpeptidase is involved in the synthesis of glutathione, and also in the absorption of neutral amino acids in the kidneys, hepatobiliary system, and intestines via a pathway known as **Meister's cycle** (Fig. 48.4). The enzyme has a predominant membrane-bound fraction and a minor microsomal fraction. The membrane-bound fraction is elevated in biliary obstruction. Interestingly, *GGT* is also elevated in alcohol

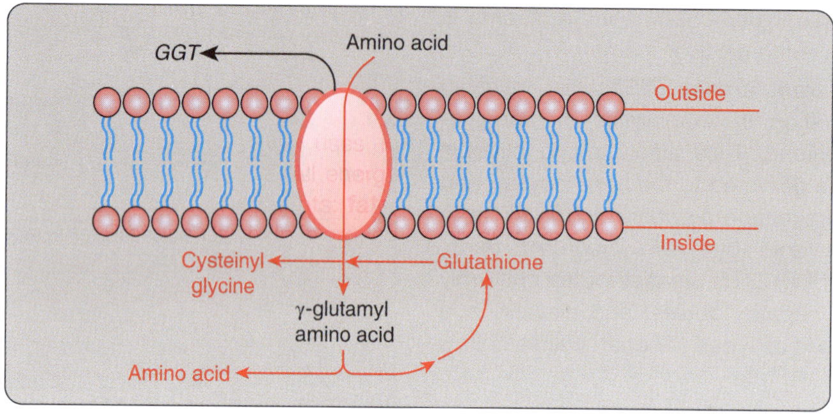

Figure 48.4 Meister's cycle. GGT = γ-glutamyl transpeptidase.

abuse and alcoholic liver disease. This is probably due an enzyme induction of the microsomal *GGT* by alcohol, or due to increased leakage of *GGT* from hepatocytes into the circulation by alcohol-mediated hepatocyte injury, or both.

4. **5'-NT:** ***5'-nucleotidase*** **plays a role in the catabolism of purines** (Fig. 34.15). Like ALP, it is an ectoenzyme present on the sinusoidal surface of the hepatocytes and also in the cells lining the bile canaliculi. In intra- and extrahepatic cholestases, the levels are increased in the circulation because of the release of the enzyme from the membrane by the bile salts. However, unlike *ALP* which is increased in children and in bone diseases, *5'-NT* is relatively more specific to biliary obstruction.

C. Biochemical markers of synthetic functions of liver

Although a large number of molecules are synthesized by the liver, only two of these are routinely assessed to evaluate the synthetic functions of liver, viz. albumin and prothrombin (and other clotting factors), and the assays used are estimation of serum albumin, and prothrombin time (PT) and international normalized ratio (INR), respectively.

1. **Albumin:** Albumin is exclusively synthesized by the liver. A decrease in serum albumin levels (termed as hypoalbuminemia) occurs commonly due to loss of albumin in urine (e.g., in nephrotic syndrome), and decreased production of albumin by the liver (seen in hepatic dysfunction and in severe malnutrition). **Albumin has a half-life of ~14 days, hence a reduction in serum albumin levels is not manifested in acute liver diseases.** Thus, the **diagnostic utility of albumin is mostly in chronic liver diseases**. Indeed, low albumin levels (<2.8 g/dl) is associated with a reduction in survival in patients with cirrhosis.

2. **PT and INR:** The liver plays two important roles in the context of coagulation factors, viz. synthesis of coagulation factors, and vitamin K-dependent γ-carboxylation of glutamyl residues of factors II (prothrombin), VII, IX and X which is required for their physiological function. Thus, **the synthesis and/or functioning of prothrombin is impaired** in **liver dysfunction and biliary obstruction,** as the latter leads to impaired absorption of fats and fat-soluble vitamins (including vitamin K) from the intestines (Fig. 20.7). **The levels and/or functioning of clotting factors (I, II, V, VII and X) are assessed by PT.** PT evaluates the extrinsic and common pathways of coagulation; it is estimated by calculating the time required for plasma (which is anticoagulated by Ca^{2+} depletion using citrate) to clot, after addition of Ca^{2+} and tissue factor or tissue thromboplastin (factor III) (Fig. 48.5). Reference interval is 11–13 seconds. **In order to standardize the estimation of PT across laboratories, another entity, INR was introduced.**

$$INR = \left[\frac{PT\ of\ Test}{PT\ of\ Control}\right]^{ISI}$$

where,

PT of Test is the PT for the patient whose sample is being tested,

PT of Control is the mean PT of ≥30 healthy individuals, and

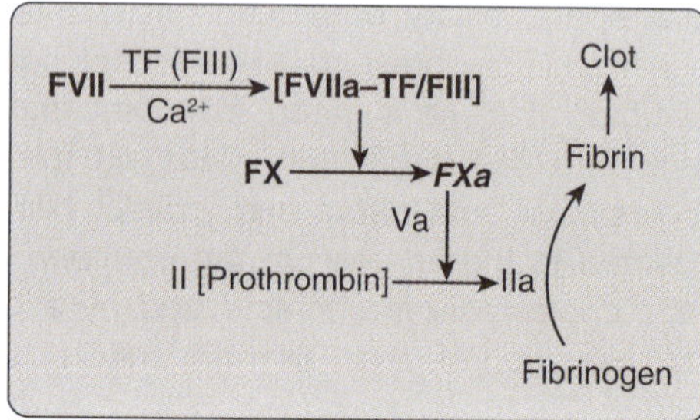

Figure 48.5 Estimation of prothrombin time (PT). Time for fibrin clot after addition of factor III and Ca^{2+} is determined.

ISI stands for International Sensitivity Index, which is a value assigned based on the thromboplastin reagent in comparison to an international reference reagent.

Note: Calculation of INR is primarily used for monitoring patients who are receiving warfarin as a part of anticoagulation therapy (see Fig. 32.26).

Unlike albumin, half-life of prothrombin is short (60 hours). Thus, PT is deranged in both acute as well as chronic liver diseases.

D. Markers of viral hepatitis A to E

Confirmatory tests for viral hepatitis due to hepatitis A, B, C, D and E viruses are based either on the detection of viral proteins/nucleic acids in circulation or on the detection of immune response, i.e., generation of immunoglobulins (Igs or antibodies) against the viral components. The nucleic acids are detected after amplification using polymerase chain reaction. Immune response to any infection involves two phases—a primary immune response characterized by the generation of IgM, which is of a short duration, and a secondary immune response which is sustained and is characterized by the formation of IgG. Half-life of IgG (2–3 weeks) is much more prolonged than that of IgM (5–6 days). IgM is a marker of acute infection, and IgG is suggestive of chronic infection (or recovery from the illness). The specific IgM and IgG against viral proteins are detected using immunological techniques like enzyme-linked immunosorbent assay (ELISA). A detailed discussion of the viral markers is beyond the scope of this book; some of the features of viral hepatitis due to hepatitis viruses A to E and the tests performed to confirm these are summarized in Figure 48.6.

E. Other Biochemical tests in liver diseases

In liver failure, the ability of liver to convert ammonia to urea is impaired, leading to a **reduction in urea levels** in circulation and **elevation in plasma ammonia**. (*Note: A similar finding is seen in urea cycle disorders as well. see Chapter 24, Section VI.C. Hyperammonemia, p. 390.*) Similarly, liver failure is also associated with a decrease in the serum uric acid levels.

Serum proteins are constituted by albumin and globulins. Since **albumin** is produced exclusively by the liver, and immunoglobulins, one of the major globulins, are produced by plasma cells (which originate from B lymphocytes), albumin: globulin (known as A:G ratio) is altered in chronic liver diseases. Reference interval for serum albumin is 3.5–5.5 g/dl and for total serum globulin is 2.0–3.5 g/dl, and for A:G ratio is 1.5–2.5:1. The **A:G ratio lowering or even reversal is seen especially in chronic viral hepatitis** where albumin synthesis is low, and there **is an increased production of IgG against the viral proteins.**

Hemochromatosis (see Chapter 33, Section.III.B.5, p. 558 and Chapter 47, Clinical Case Question 47.1, p. 831) is associated with iron overload and an increase in serum iron and transferrin saturation. Wilson's disease (see Chapter 33, Section.III.A.2, p. 556) is confirmed by an increase in the copper content in liver and **increased urinary copper excretion**. Autoantibodies can be detected using ELISA for **autoimmune hepatitis.** In conditions like **Wilson's disease and hereditary hemochromatosis**, genetic testing for detection of mutations in ***ATP7B*** **gene in the former, and *HFE* and other associated genes** in the latter, is the mainstay in diagnosis.

If biochemical tests are insufficient to pinpoint a diagnosis, liver biopsy, and histopathological examination

Name of the Virus	Genetic Material	Transmission	Chronic Infection	Diagnostic Tests
HAV	RNA	Feco-oral	No	Anti-HAV IgM ELISA
HBV	DNA	Bloodborne or due to contact with infected body fluids or sexual contact	Yes	ELISA for HBV surface antigen (HBsAg) and for IgM against HBV core antigen (anti-HBc IgM) PCR for HBV DNA
HCV	RNA			PCR for HCV RNA ELISA for Anti-HCV
*HDV**	RNA			PCR for HDV RNA ELISA for Anti-HDV *Markers of HBV infection
HEV	RNA	Feco-oral	Rare	ELISA for Anti-HEV IgG and IgM

*HDV infection can occur only in a setting of HBV infection

Figure 48.6 Viral hepatitis due to hepatitis virus A to E. HAV - HEV = hepatitis (A, B, C, D, E) virus; PCR = polymerase chain reaction; ELISA = enzyme-linked immunosorbent assay.

of the tissue obtained may be warranted. (A summary on the approach to diagnose a liver disease using the tests discussed above is given in Figure 48.17.)

IV. FUNCTIONS OF THE KIDNEYS

The kidneys are paired organs located in the retroperitoneum, the blood-flow to which is nearly 25% of the cardiac output. The functional unit of the kidney is the nephron, with each kidney having 0.6–1.5 million nephrons. The glomeruli filter ~125 ml of blood per minute while most of the glomerular filtrate is reabsorbed by the tubules, ultimately to generate 1 ml of urine per minute. Production of urine, albeit an important one, is just one among the multitude of functions that are carried out by the kidneys. Functions of the kidneys can be broadly classified into the following.

A. Excretory functions

The fluid that enters the proximal tubule following glomerular filtration is an ultrafiltrate of the plasma. The **glomerular filtration barrier** ensures the selection of molecules that are filtered based on their size and charge. The molecular weight cutoff for glomerular filtration is around 70 kDa as the molecules above this size are not filtered. Thus, the glomerular filtrate is composed of water, small molecules (like glucose, amino acids, urea, creatinine and uric acid), electrolytes (like Na^+, K^+, Cl^-, bicarbonate and inorganic phosphate), and low molecular weight proteins. Large proteins and cellular components of the blood are not filtered, and presence of these in urine suggests a loss of integrity of the filtration barrier, or addition of these to the urine at the postrenal level (e.g., due to injury to urethra by a calculus, or bleeding from a tumor in the urinary bladder, or a lower urinary tract infection).

The filtration barrier has a negative charge due to the presence of heparan sulfate proteoglycans, hence, negatively charged molecules have a further low molecular weight cutoff. This is the reason why albumin with a molecular weight of 66.5 kDa is not filtered by the glomeruli. However, in **nephrotic syndrome**, this charge barrier is lost, and albumin is lost from the circulation into the urine, leading to a decrease in serum albumin levels.

Apart from glomerular filtration, some of the molecules can also enter the tubular fluid via secretion at the level of proximal and distal tubules; some molecules leave the tubular fluid by **tubular reabsorption**. The kidneys help in excretion of a large number of molecules including metabolic end products like urea, uric acid and creatinine, surplus unwanted molecules like electrolytes, many of the drugs and other xenobiotics, enzymes (e.g., *amylase*), and short peptides.

B. Homeostatic functions

The kidneys, apart from the facilitation of the removal of many molecules, regulate the levels of many molecules in the body. This is achieved by a fine control of glomerular filtration as well as the secretion and reabsorption at the level of tubules. The kidneys play a key role in the maintenance of **water and electrolyte balance** and this involves many hormones that act on the nephrons, like antidiuretic **hormone (ADH), aldosterone, atrial natriuretic peptide (ANP) and B-type natriuretic peptide (BNP)** (see Chapter 5). Thus, electrolyte imbalance is a feature of a large number of kidney diseases. By maintaining water and electrolyte balance, the kidneys ultimately regulate blood volume, and hence blood pressure.

The kidneys also regulate the blood pH by controlling bicarbonate generation, bicarbonate reabsorption, free H^+ excretion, and combination of H^+ in the tubular fluid with phosphate and ammonia to cause further H^+ excretion (see Chapter 4, Section II.C, p. 27). They are thus the main line of defense against acidosis and alkalosis.

Apart from the liver, renal medulla is also a site for gluconeogenesis.

C. Endocrine functions

Three hormones, their functions widely different from each other are produced by the kidneys.

1. The juxtaglomerular cells on the walls of the afferent arteriole is a part of the juxtaglomerular apparatus (see Fig. 5.8). Macula densa acts as a **sensor of the Na^+ levels** in the tubular fluid. When the Na^+ levels are low, signaling from the macula densa leads to the production of renin from the juxtaglomerular cells. This ultimately leads to the production of angiotensin II and aldosterone.

2. The interstitial cells in the peritubular capillary bed of the kidneys produce **erythropoietin (*erythro*: red, *poiesis:* production) in response to hypoxia**. Erythropoietin acts on the stem cells in the bone marrow to increase the RBC production. Erythropoietin was found to be used by certain athletes for performance enhancement; the use for this purpose is banned by various regulatory authorities. Impaired erythropoietin production in chronic kidney disease (CKD) leads to anemia.

3. The **final step in the activation of vitamin D**, the formation of 1,25-dihydroxycholecalciferol (calcitriol) from 25-hydroxycholecalciferol is mediated by the microsomal enzyme *1-α-hydroxylase* (Fig. 32.23). The enzyme is induced by high level of parathyroid hormones and low levels of phosphates. The kidneys also play a role in the inactivation of

25-hydroxycholecalciferol to 24,25-dihydroxycholecalciferol by *24-hydroxylase*, when *1-α-hydroxylase* is inhibited. CKD is associated with impaired calcium and phosphate homeostasis, and bone and parathyroid abnormalities.

V. RENAL FUNCTION TESTS

Because of the diverse functions of the kidney, renal diseases can present with a varied number and type of clinical manifestations. Further, a large number of renal diseases are asymptomatic till they have reached an advanced stage. In such scenarios, biochemical investigations to assess renal function (or rather, dysfunction) is important in diagnosis, severity assessment, and monitoring of the conditions. Renal tests also provide clues for initiation of renal replacement therapy in the form of dialysis or renal transplantation. Assessment of renal function is also important before the administration of nephrotoxic drugs that can exacerbate preexisting kidney disease. In an era where lifestyle diseases are rampant, renal tests are also important in detecting renal complications of hypertension and diabetes mellitus. Small vessel injury in these conditions (microangiopathy) can affect the glomeruli, leading to nephropathy, which can progress to CKD. Preemptive testing is required to forestall such life-threatening complications.

It may be noted that factors such as exercise and fever can result in the temporary excretion of excess protein in urine by nephrons. This is due to increased renal blood flow and is termed functional proteinuria. Additionally, temporary functional proteinuria may occur due to dehydration.

The assessment of renal function involves biochemical tests to assess the glomerular function and integrity, and tests to identify the underlying cause of the renal disease (e.g., urine analysis, urinary protein estimation, etc.).

A. Assessment of glomerular function

Glomerular filtration rate (GFR) is the volume of fluid filtered by all the glomeruli of both the kidneys into their Bowman's capsules in unit time. GFR which depends on the age, sex and size of the body of the individual, is ~130 ml/min in males and ~120 ml/min in females. Thus, **roughly 180 l of blood is filtered by the kidneys every day, to produce ~1.5 l of urine**. GFR is an indicator of the number and functioning of the nephrons. **A reduction in GFR** is seen in glomerular diseases (e.g., glomerulonephritis, diabetic nephropathy, and hypertensive nephropathy), in conditions associated with a **reduction in the renal perfusion** (e.g., hypovolemia, heart failure, renal artery stenosis), and due to an increased back pressure on the glomeruli (this is seen in urinary tract obstruction). GFR can be reduced in **acute kidney injury** (AKI) as well as CKD (by definition, CKD is a renal abnormality that persists for >3 months).

GFR cannot be measured directly. It is estimated with the help of a concept which is termed as clearance, which is the amount of blood from which a substance is eliminated by both the kidneys in unit time. For a substance that is freely filtered by the glomeruli, which does not have tubular reabsorption or tubular secretion or metabolism in the kidney, the amount filtered will be equal to the amount excreted in the urine. Thus,

GFR (ml/min) × Concentration of the substance in plasma, P (mg/dl) = Concentration of the substance in urine, U (mg/dl) × Urine flow rate, V (ml/min).

Hence, **GFR (ml/min) or clearance** = **(U x V)/P**

The ideal substance for estimation of clearance is inulin, a polymer of fructose, as it freely filtered by the glomeruli, and does not have tubular reabsorption or tubular secretion or metabolism in the kidney. However, inulin has to be administered exogenously, which is time-consuming, and the assays for inulin are not very perfect. Hence, endogenous molecules (those that are synthesized in the body) which can be assayed easily are used as surrogates for inulin clearance/for calculation of GFR. Three commonly used analytes are:

1. **Creatinine:** Creatinine is the product of dehydration and cyclization of creatine (see Fig. 26.19), which takes place in the muscles (hence the name; *crea*: flesh). A fixed amount of creatinine is produced and released into the circulation every day, depending on the muscle mass of the individual. Reference interval of serum creatinine is **0.5–0.9 mg/dl in females and 0.6–1.2 mg/dl in males**; the higher value in males is due to the increased muscle mass. Diet can also affect serum creatinine levels, if the sample has been collected after consumption of meat.

   Creatinine is freely filtered by the glomeruli, and does not undergo tubular reabsorption, however, a certain amount (10%–40%) of creatinine enters the tubular fluid because of tubular secretion. Due to this, creatinine clearance overestimates the GFR. Also, in chronic renal diseases, the kidneys compensate for nephron loss by hyperfiltration in the existing nephrons. Because of both these reasons, the **changes in serum creatinine and creatinine clearance set in only after the disease has been progressed significantly**. This has to be kept in mind while interpreting serum creatinine levels in patients with risk factors for CKDs.

2. **Cystatin C:** Cystatin C is a low molecular weight protein (12.8 kDa), a *cysteine protease* inhibitor, that is synthesized by all the nucleated cells in the body. It is released into the circulation at a relatively constant rate, that unlike creatinine is independent

of muscle mass and diet. It is freely filtered by the glomeruli and there is no tubular secretion or reabsorption; cystatin C is indeed broken down within the renal tubules and the levels in urine are very low in normal individuals. The **concentration of cystatin C in serum reflects the glomerular function and is inversely related to the GFR**; eGFR calculations based on serum cystatin C levels are also available. A caveat regarding cystatin C is that its production is altered in thyroid diseases, diabetes mellitus, and obesity.

Estimated GFR (eGFR)

Three parameters are required for calculating the creatinine clearance applying the formula for clearance given above (U × V/P), including V, the urine volume (which necessitates the collection of urine voided during a 24-hour period), and U, the concentration of creatinine in that 24-hour urine sample. Since collection, transport, etc. of the urine sample is time-consuming and cumbersome, other formulas exist, for calculation of GFR based on the estimation of serum creatinine only. The GFR thus calculated is termed eGFR or estimated GFR. Several formulas exist for the calculation of eGFR. One of them, the Modification of Diet in Renal Diseases (MDRD) formula is given below as an example:

eGFR (ml/min per 1.73 m^2) = 186.3 × serum creatinine$^{-1.154}$ × age$^{-0.203}$ × (0.742, if female) × (1.21, if black). Note that the formula considers age, sex, and race, as well as the body size (i.e., per 1.73 m^2) for calculation of the GFR.

3. **Urea:** Urea, the end product of catabolism of amino group of amino acids, is produced in the liver from ammonia via the urea cycle. More than 90% of the urea produced is excreted by the kidney. Urea is freely filtered by the glomeruli, and there is no tubular secretion or active tubular reabsorption; however, there is a significant passive diffusion of urea from the tubules to the renal interstitium which is determined by the urine flow rate. (*Note: Urea also plays a physiological function in reabsorption of water by maintaining the medullary concentration gradient*). The passive transport of urea into the interstitium is mediated by urea transporters, which are activated by the action of ADH.

   The reference interval of serum urea is **15–40 mg/dl**; this varies with age and is slightly lower in females. An elevation in the serum urea is termed as **azotemia**; when azotemia is associated with clinical manifestations and fluid, electrolyte and/or metabolic abnormalities, the clinical syndrome is called **uremia.** Even though serum urea level varies inversely with GFR, estimation of serum urea is not a very useful surrogate for GFR estimation. This is because of the following reasons:

   a. Urea levels are **affected by protein** (and amino acid) catabolism. Urea levels are higher after a diet rich in proteins, and also in severe malnutrition and other chronic diseases associated with a protein breakdown (e.g., loss of muscle mass).

   b. Because of the passive reabsorption, **urea clearance underestimates the GFR.** Since the passive reabsorption is high in situations where the urine flow rate is low, in diseases where urine outflow is obstructed (postrenal azotemia, see Section V.B and Fig. 48.7), urea levels are disproportionately high in the circulation.

   However, urea has an important role in identifying the etiology of the renal dysfunction (see Section V.B).

In many countries including the United States, serum urea is denoted by a related term called **Blood Urea Nitrogen or BUN**, since historic assays for urea were based on the measurement of nitrogen in blood. Molar mass of nitrogen is 14 g/mol and molar mass of urea is 60 g/mol. 1 molecule of urea (NH_2-CO-NH_2) contains 2 atoms of nitrogen. Thus, ratio of mass of nitrogen to mass of urea is 28/60 which is 0.47. Hence,

1 mg/dl of serum urea ≡ 0.47 mg/dl of BUN, and conversely, 1 mg/dl of BUN ≡ 2.14 mg/dl of serum urea.

Reference interval for BUN is 7–20 mg/dl (This corresponds to serum urea of ~ 15–43 mg/dl).

B. Tests to identify the etiology of renal diseases

Renal diseases are broadly classified into AKI and CKD. **AKI is characterized by a sudden deterioration of renal function (in hours to days, usually <48 hours)** and elevation of serum urea and creatinine (azotemia), the cause of which can be at the prerenal, intrarenal or postrenal levels. Prerenal azotemia stems usually from a reduction in the glomerular perfusion; intrarenal azotemia occurs due to an intrinsic damage to the kidneys; and postrenal azotemia is due to an obstruction to the flow of urine. The causes are listed in Figure 48.7. The **CKD is an insidious process (the damage persisting for >3 months)** with varying degrees of reduction in the GFR and varying severity of albuminuria (excretion of albumin in the

Classification	Level of Defect	Causes
Acute Kidney Injury (AKI) *Disease evolves in <48 hours*	Prerenal azotemia	Decreased renal perfusion seen in: • Hypovolemia • Heart failure
	Intrarenal azotemia	Intrinsic disorders of kidney leading to renal parenchymal damage: • Ischemia • Infections • Nephrotoxic drugs and other agents
	Postrenal azotemia	Sudden obstruction to the urine outflow: • Stones in renal pelvis, ureters or bladder • Benign enlargement of prostate • Carcinoma of prostate or bladder • Abdominal or pelvic tumors causing external obstruction of ureters • Stricture/stenosis of ureters
Subacute kidney injury or acute kidney diseases and disorders (AKD) *Disease evolves in 48 hours to 3 months*	Etiology overlaps with AKI and CKD	
Chronic kidney disease (CKD) *Disease persists for >3 months*	Prerenal diseases	Low circulating blood volume leading to decreased renal perfusion—seen in: • Heart failure • Cirrhosis Decreased renal perfusion due to atherosclerosis or vessel abnormalities: • Renal artery stenosis
	Intrinsic renal disease	Chronic disorders of kidney leading to renal parenchymal damage: • Hypertension • Diabetes mellitus • Autoimmune diseases • Polycystic kidney disease • Glomerulonephritis
	Postrenal obstruction	Persistent obstruction to the urine outflow: • Benign enlargement of prostate • Carcinoma of prostate or bladder • Abdominal or pelvic tumors causing external obstruction of ureters • Stricture/stenosis of ureters

Figure 48.7 Classification and causes of kidney disease.

urine). As expected, CKD is associated with widespread systemic abnormalities, and when progressed mandates renal replacement therapies. Common causes of CKD include diabetes mellitus and hypertension. Other causes are listed in Figure 48.7.

Some of the **biochemical investigations to identify the etiology of AKI and/or CKD** are detailed below:

1. **Proteinuria:** In healthy individuals, the daily excretion of proteins in the urine is <150 mg, and most of this is contributed by uromodulin or Tamm-Horsfall protein which is derived from the loop of Henle, and other small proteins that are filtered by the glomeruli. **Proteinuria is the excretion of >150 mg of protein in the urine in 24 hours**. This is classified into:

   a. **Glomerular proteinuria:** Seen in glomerular diseases due to **a loss of the integrity of the filtration barrier** (e.g., nephrotic syndrome, diabetic nephropathy, glomerulonephritis). The major protein lost via urine is albumin. Other evidences of glomerular involvement include presence of RBCs in the urine (in inflammatory conditions, i.e., glomerulonephritis), decreased serum albumin levels (in nephrotic syndrome), etc.

b. **Tubular proteinuria:** Small proteins that are filtered by the glomeruli are generally reabsorbed by the tubules. However, this is impaired in tubular dysfunction, and then these small proteins start appearing in the urine, e.g., β2-microglobulin. **Increased levels of β2-microglobulin** in urine is seen in renal diseases that affect the tubules, e.g., due to nephrotoxic drugs.

c. **Overflow proteinuria:** This occurs due to an increase in the production or release of a protein, which then gets filtered in increased amounts by the glomeruli, resulting in their increased presence in urine. This is seen in **multiple myeloma**, a malignant proliferation of plasma cells, which results in increased production of immunoglobins. The low molecular weight immunoglobin light chains (molecular weight: 25 kDa) appear in the urine, which are known as Bence-Jones proteins. Overflow proteinuria is also seen in muscle damage leading to myoglobin release and myoglobinuria.

Glomerular proteinuria is also classified according to the amount of albumin lost in urine. Loss of >3.5 g of albumin in the urine over 24 hours characterizes nephrotic range proteinuria. In early stages of renal involvement in diabetes mellitus and hypertension, the albumin loss in urine is minimal, and this requires special tests for detection.

2. **Albumin excretion rate (AER): Not more than 30 mg of albumin is excreted in the urine in healthy individuals in 24 hours.** AER of >300 mg/day is seen in individuals with nephropathy due to diabetes mellitus and hypertension, but by this time the renal diseases can go irreversible despite the control of the glycemic levels and the blood pressure. Thus, to prevent renal complications, albumin excretion in the urine from **30–300 mg/day** has to be detected. However, routine urine tests employ a device known as dipstick which cannot detect such low levels of albumin. [*Note: A dipstick is a reagent strip with reagents for one or more analytes; it can be dipped in the collected urine sample. Proteins, or other analytes as the case may be, react with the reagents in the dipstick to give a change in color which can be compared with a standard color chart to obtain an approximate quantification of the analyte* (Fig. 48.8).]

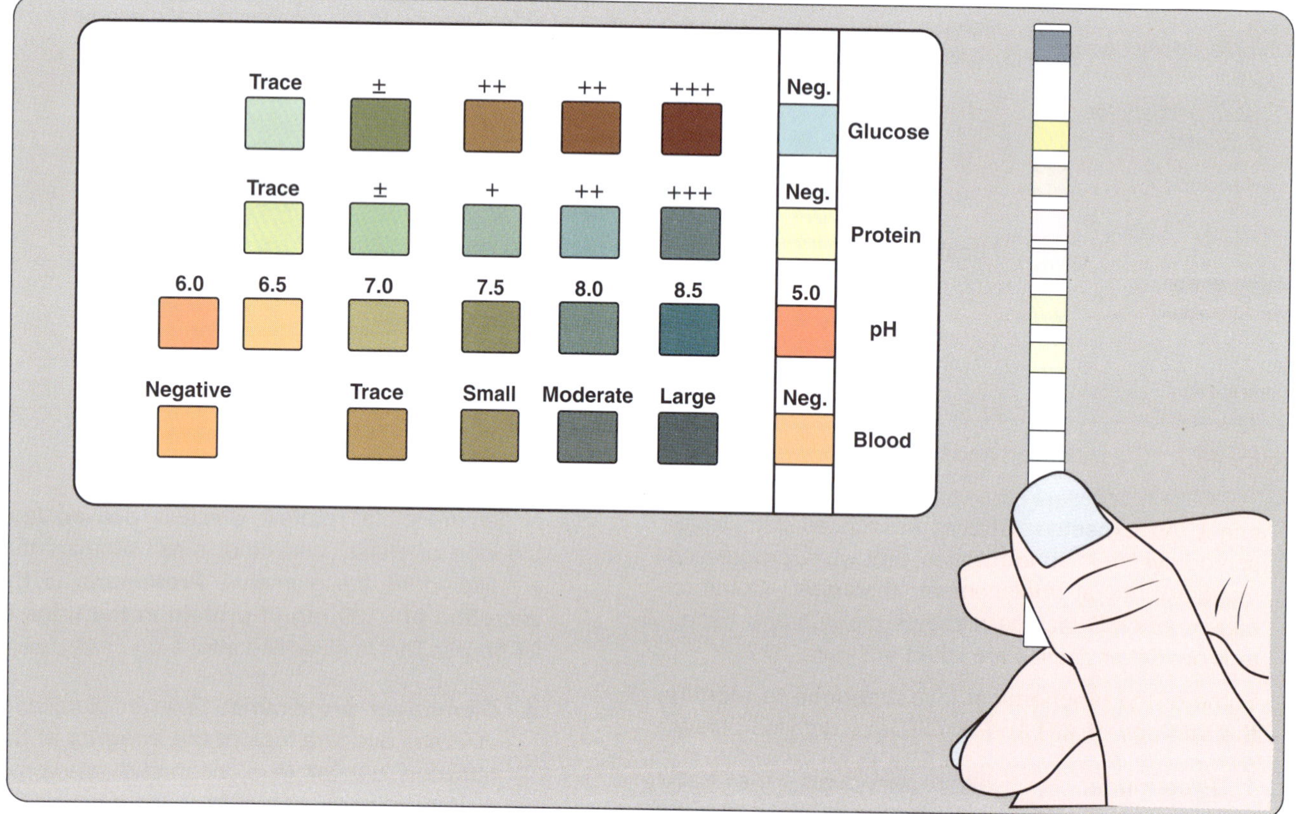

Figure 48.8 Urine dipstick.

Albumin excretion rate of 30–300 mg/day, detected using immunological techniques, is termed as **microalbuminuria**. This acts as a predictor for future renal disease, and also future cardiovascular morbidity and mortality (as the diseases involve micro- and macroangiopathy). Since a 24-hour urine collection for AER is tedious, albumin excretion can also be expressed as a ratio to urine creatinine levels (to normalize the variations in the urine volume). Thus, in a spot urine sample, **Albumin-Creatinine Ratio (ACR)** can be estimated and a value of 30–300 mg/g has been found to be approximately equivalent to AER of **30–300 mg/day.**

AER is also used to classify CKD, along with GFR. This is shown in Figure 48.9.

3. **Urine analysis:** This usually employs urine dipstick and urine microscopy. Some common analytes tested, besides protein and albumin, are:

    a. **RBCs and hemoglobin:** Presence of heme in the urine (hemoglobinuria) is detected by utilizing the peroxidase activity of heme, which changes the color of a chromogen. Both free heme and heme present in RBCs can give a positive test. Thus, this can be seen in **hemoglobinuria and myoglobinuria**. Detection of RBCs is done by microscopy—their presence suggests glomerular damage (glomerulonephritis), tumors (e.g., renal cell carcinoma, carcinoma of urinary bladder etc.), injury to the urinary tract (e.g., stones) or infections.

    b. **Leukocyte esterase and WBCs:** Presence of leukocyte esterase released by neutrophils and macrophages is suggestive of **urinary tract infection**. WBCs can be detected using microscopy.

Parameter	Range	Grade	Severity
Glomerular filtration rate (ml/min/1.73 m²)	≥90	G1	Normal
	60–89	G2	Mild decrease
	45–59	G3a	Moderate decrease
	30–44	G3b	
	15–29	G4	Severe decrease
	<15	G5	Renal failure
Urine albumin creatinine ratio (mg/g)	<30	A1	Normal to mild increase
	30–300	A2	Moderate increase
	>300	A3	Severe increase

Note: Both GFR and ACR are used to predict the prognosis of CKD, e.g. both G3a with A3 and G4 with A1 will have an increased risk of progression of CKD.

Figure 48.9 Classification of chronic kidney disease (CKD).

c. Presence of **glucose in urine** is seen in uncontrolled diabetes mellitus (when plasma glucose levels are >180 mg/dl). It can also be seen in patients receiving sodium-glucose cotransporter-2 (SGLT2) inhibitors and in individuals with renal glucosuria (see Chapter 5, Sections VIII, p. 51)

d. **Specific gravity:** Also termed as relative density, this is the density of a liquid in comparison to the density of water. The reference interval for urine specific gravity is **1.012–1.024.** It is a measure of the concentrating ability of the renal tubules. A low and fixed specific gravity (termed as isosthenuria) is seen in CKD and in diabetes insipidus. A high specific gravity is commonly seen in dehydration, and syndrome of inappropriate antidiuretic hormone secretion (SIADH; due to increased concentration of urine due to less water content) and in proteinuria and diabetes mellitus (due to increased presence of solutes in the urine) (see Chapter 5, Section II.D, p. 41)

4. **Prerenal azotemia** can be differentiated from intrarenal azotemia by biochemical analysis. Two tests employed for this are:

    a. **BUN-serum creatinine ratio:** The glomerular filtration of both urea and creatinine is decreased due to hypoperfusion in prerenal azotemia. However, urea is reabsorbed in increased amounts due to the activation of urea transporter by increased ADH levels in hypovolemia. Thus, the ratio between urea and creatinine is increased in prerenal azotemia. A BUN/serum creatinine of >20:1 is suggestive of prerenal azotemia, while the ratio is 10–15:1 in intrarenal azotemia.

    b. The term **Fractional Excretion of sodium (or FENa)** refers to the percentage of sodium that is filtered by the kidneys, which is excreted in the urine. The kidneys normally reabsorb most of the sodium that is filtered; in hypovolemia and reduced glomerular perfusion, body tries to conserve sodium, and the FENa is decreased. FENa is given by the formula:

$$FENa = \frac{Urine\ Sodium/Serum\ Sodium}{Urine\ Creatinine/Serum\ Creatinine} \times 100$$

Reference interval is 1–2%. In conditions of prerenal azotemia, the FENa is <1%.

C. Other tests in kidney diseases

A number of other biochemical tests are useful in renal diseases. Most of the cases of AKI and advanced CKD are associated with electrolyte and acid-base abnormalities, which can be diagnosed by estimation of serum Na^+ and K^+, and arterial blood gas analysis, respectively. Bone abnormalities in kidney disease requires estimation of **serum Ca^{2+}, phosphate, parathyroid hormone (PTH), vitamin D levels, and *alkaline phosphatase*,** for assessment. Urinary tract infections are detected by **urine culture and antibiotic sensitivity testing**. Some infections like tuberculosis might require molecular diagnostic testing involving nucleic acid amplification. Calculi in kidneys and urinary tract are mostly diagnosed by imaging studies. **Etiology of calculi** (e.g., uric acid calculi, and cystine stones—see Fig. 24.6), are diagnosed by biochemical tests and microscopy. Most of the postrenal causes of AKI and CKD are diagnosed by **imaging** studies. In prostate cancer, serum levels of a tumor marker known as **PSA (prostate-specific antigen, a *serine protease* present in prostatic tissue)** are elevated. If a diagnosis cannot be made using noninvasive modalities, a renal biopsy may be required for confirmation of the diagnosis.

VI. FUNCTIONS OF THYROID HORMONES

The thyroid gland produces three hormones, viz. **thyroxine (T_4) and triiodothyronine (T_3) by the thyroid follicular cells, and calcitonin (by the parafollicular cells or C cells).** The latter is involved in calcium homeostasis (see Fig. 32.23), where it regulates the calcium resorption in the bones by cells known as osteoclasts; calcitonin ultimately decreases the plasma calcium concentration, especially in children, where osteoclastic activity is high because of rapid bone remodeling. In adults, physiological levels of calcitonin do not have a marked effect on plasma calcium levels; however, pharmacological doses of calcitonin are used in the management of hypercalcemia.

The production of T_4 and T_3 has been described in detail in Chapter 33, Section.IV.A, p. 560. T_3 and T_4 secretion are controlled primarily by thyroid-stimulating hormone (TSH or thyrotropin) from the anterior pituitary, the release of which is in turn regulated by thyrotropin-releasing hormone (TRH) from hypothalamus. **TRH is a tripeptide amide (pyroglutamyl-histidyl-proline-amide) synthesized in the supraoptic and paraventricular nuclei** and is stored in the median eminence of hypothalamus. It reaches the anterior pituitary by hypothalamic-hypophyseal portal system, where it binds to TRH receptors of the **TSH-secreting cells**. A signaling cascade following the binding of TRH to its receptor leads to release of TSH, which binds to the TSH receptor in the thyroid gland (Fig. 48.10). TSH has multiple effects on the thyroid gland wherein it increases the iodide trapping by activating the transporter involved in this process (Na^+/I^- symporter), it increases by iodination of tyrosine, enhances the proteolysis of thyroglobulin to release the thyroid hormones

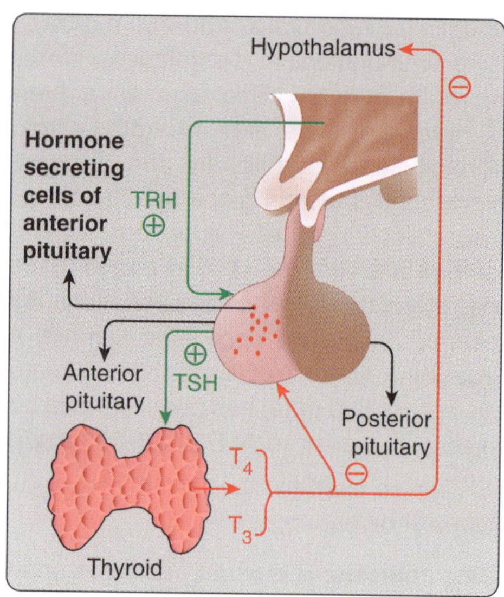

Figure 48.10 Regulation of thyroid hormone secretion. TSH = thyroid-stimulating hormone; TRH = thyrotropin-releasing hormone.

into the circulation, and increases the number of follicular cells.

T_3 is the potent form (four times more potent than T_4), but **>80% of the hormone secreted from the thyroid gland is T_4**. In the circulation, T_3 and T_4 are bound to proteins like albumin, transthyretin (transports thyroid hormones and retinol-binding protein), and thyroid-binding globulin. **Only the protein unbound forms (known as free T_3 and free T_4, i.e., fT_3 and fT_4) are available for hormonal activity** and these constitute only a minute fraction of the total T_3 and T_4; **99.98% of T_4 and 99.8% of T_3 in the circulation is protein bound**. At the tissue site of action, T_3 and T_4 cross the plasma membrane to enter the cytoplasm, where the enzymes 5'-deiodinases convert fT_4 to fT_3, the active form. fT_3 binds to the thyroid hormone receptors. **Thyroid hormone receptor (TR)** is present in the DNA at specific hormone receptor elements, along with another receptor known as retinoid-X-receptor (RXR). Binding of fT_3 to the receptor complex increases gene transcription, and hence the synthesis of a large number of proteins, which ultimately bring about the effector functions (Fig. 48.11). TSH receptor antibodies are a common cause of Graves disease.

The actions of thyroid hormones are diverse and numerous. Some of these are listed below:

1. **Effects on general metabolism:** Thyroid hormones enhance the number and the functioning of mitochondria. Thus, they stimulate the ATP production, and overall metabolism. Thyroid hormones are important in maintaining the **basal metabolic rate (BMR)** (see Chapter 31, Section III.B.1, p. 503). Thyroid hormones also stimulate the activity of ion channels including the *Na^+-K^+ ATPase*. Since this pump utilizes ATP, and thyroid hormones increase the ATP production in the mitochondria, there is an overall increased recycling of

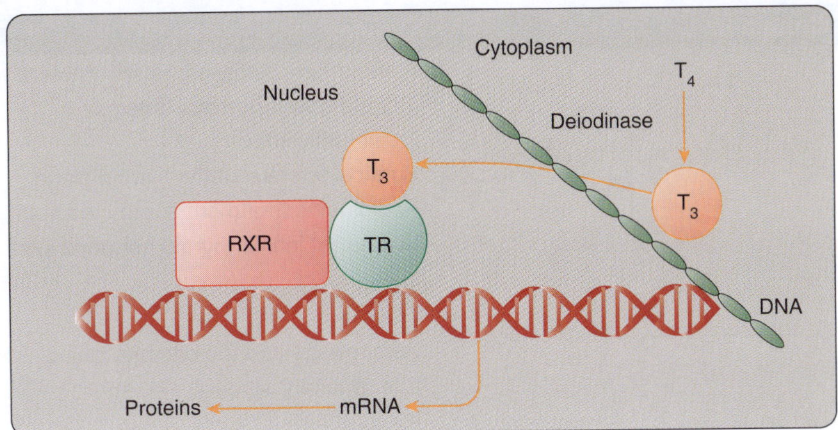

Figure 48.11 Mechanism of action of thyroid hormones. TR = thyroid hormone receptor; RXR = retinoid-X-receptor.

ATP. Since ATP production is only ~30% efficient and the rest of the energy during the process is dissipated mostly as heat, thyroid hormones play an important role in thermogenesis and thus, maintenance of body temperature. Thyroxine also acts as an uncoupler, and also increases the levels of uncoupling protein, UCP1, in the brown adipose tissue, contributing further to thermogenesis (see Chapter 11, Section VI.A.2.c and Fig. 11.16, p. 170).

2. **Effects on carbohydrate, lipid, and protein metabolism:** Thyroid hormones stimulate carbohydrate metabolism by promoting insulin secretion and uptake of glucose into tissues, and by stimulating glycolysis as well as gluconeogenesis. Lipid catabolism is also enhanced—thyroid hormones stimulate lipolysis in adipose tissue, and β-oxidation of fatty acids. By increasing the expression of LDL-receptor, thyroid hormones decrease the cholesterol levels in the circulation. Thyroid hormones also promote protein catabolism and muscle protein breakdown.
3. **Effects on cardiovascular system:** Thyroid hormones have a direct stimulatory effect on cardiac activity. Also, they increase expression of catecholamine receptors, enhancing the action of epinephrine and norepinephrine on the heart. Thyroid hormones increase heart rate and cardiac output, thus increasing the oxygen supply to the peripheral tissues to promote the intermediary metabolism.
4. **Effect on nervous system:** Thyroid hormones are essential for the normal development of nervous system. Hypothyroidism during intrauterine life and early childhood can lead to intellectual disability. Thyroid hormones increase the activity of neuronal synapses, decrease sleep, and increase skeletal muscle activity and intestinal motility.

The feedback regulation of thyroid secretion is shown in Fig. 48.10. The secretion of TSH is inhibited directly by the thyroid hormones by their action on the anterior pituitary, as well as indirectly by their effect on TRH release. Even small changes in thyroid hormone levels can affect the TSH levels to a larger extent. This is important while assessing the thyroid function using hormone assays.

VII. THYROID FUNCTION TESTS

Hypothyroidism refers to a reduction in the thyroid hormone levels in the circulation. Based on the organ affected, hypothyroidism can be **primary (defect lies in the thyroid gland), secondary (defect in the pituitary), or tertiary (defect in the hypothalamus)**. The causes are listed in Figure 48.12. Clinical features of hypothyroidism include lethargy, weight gain, **increased sleep, exhaustion, cold intolerance, constipation, decreased heart rate, abnormal tendon reflexes, delayed growth and development (in children)**, etc. Severe hypothyroidism can even result in a comatose state.

Causes of **hyperthyroidism**, i.e., increased circulating levels of thyroid hormones, are also given in Figure 48.12. Hyperthyroidism mostly occurs due to a defect at the level of thyroid gland, or due to exogenous intake of iodine or thyroid hormones. Clinical features are mostly diametrically opposite to that of hypothyroidism and **include restlessness, weight loss, sleeplessness, anxiety, heat intolerance, diarrhea, tachycardia, and increased tendon reflexes**. Persistent hyperthyroidism can lead to arrythmias and heart failure.

The diagnosis of both hypo- and hyperthyroidism is confirmed by hormone assays. The following are some considerations in the biochemical diagnosis of thyroid diseases:

1. Thyroid function tests classically comprise of estimations of serum levels of **TSH, fT_3, and fT_4.** These are estimated using special **immunoassays.**
2. Although the major fractions of T_3 and T_4 are in the protein-bound form, the biologically active form is the **free hormones** (denoted by fT_3 and fT_4). Also, numerous factors like illnesses, medications, and

	Condition	Causes
Hypothyroidism	Primary (defect in the thyroid gland)	Autoimmune diseases: Hashimoto's thyroiditis Thyroid developmental defects Iodine deficiency Surgical removal of the thyroid gland Radioiodine treatment (for thyroid ablation) Drugs that inhibit thyroid hormone synthesis
	Secondary (defect in the pituitary)	Tumor Trauma Hemorrhage into the pituitary gland After pituitary surgery
	Tertiary (defect in the hypothalamus)	Rare
Hyperthyroidism	Primary (defect in the thyroid gland)	Graves' disease (autoimmune disease with presence of thyroid-stimulating immunoglobulins or TSH receptor antibodies) Thyroid tumors (adenoma) Goiter (hyperthyroidism in a setting of goiter is known as toxic nodular goiter)
	Secondary (defect in the pituitary)	TSH-secreting adenoma
	Other causes	Initial phases of thyroid gland destruction leading to secretion of stored hormones Intake of thyroid hormones Iodine-induced hyperthyroidism

Figure 48.12 Causes of hypothyroidism and hyperthyroidism.

genetic factors can influence the protein binding of thyroid hormones. Hence, fT_3 and fT_4 estimation is preferred over the assay of total T_3 and total T_4.

3. The most important test for diagnosing thyroid dysfunction is **serum TSH**. A normal TSH virtually rules out thyroid hormone abnormalities. Thus, the diagnostic algorithms for workup of hypo- and hyperthyroidism start with estimation of serum TSH.
4. Since most of the thyroid hormone abnormalities can be detected using fT_4 alone, fT_3 estimation has only a very limited role among the thyroid function tests. One particular scenario where fT_3 is useful is in T_3-toxicosis—a condition where TSH is low and fT4 is normal; it is seen in **toxic nodular goiter and Graves' disease**, due to increased conversion of T_4 to T_3.
5. In **primary hypothyroidism**, fT_4 is reduced and as the result of removal of the feedback inhibition on pituitary, TSH is elevated. In secondary hypothyroidism, TSH is reduced as a result of pituitary defects, leading to decreased stimulation of thyroid gland and hence decreased fT_4 (Fig. 48.10). A similar finding is seen in hypothalamic disorders (tertiary hypothyroidism) also, but this is rare. If suspected, a reduction in TRH levels points to tertiary hypothyroidism.
6. If TSH is high, but fT_4 is normal, the condition is termed as **subclinical hypothyroidism**. Most of such patients are asymptomatic; however, a large number of them progress to overt hypothyroidism. Such patients should be followed up with estimations of TSH and fT_4, and treatment should be started, like in hypothyroidism, if TSH levels are too high.
7. In **hyperthyroidism** due to a TSH-secreting tumor of pituitary, TSH and fT_4 are elevated. In hyperthyroidism arising to due thyroid or exogenous causes, fT_4 is elevated but TSH is suppressed by the feedback inhibition (Fig. 48.10). One exception is T_3-toxicosis (see point 4). The other exception is subclinical hyperthyroidism where TSH is low, but both fT_4 as well as fT_3 is normal. Such patients should be followed up, as they can develop frank hyperthyroidism; treatment, as in hyperthyroidism, should be initiated if TSH values are too low.
8. Figure 48.13 gives a flowchart for the diagnosis of hypo- and hyperthyroidism based on the thyroid function tests.

Some other investigations are important in identifying the etiology of the thyroid diseases. **Graves' disease** is an autoimmune disease with autoantibodies that stimulate the TSH receptors in the thyroid gland, as well as autoimmune reactions involving the periorbital tissue. The latter leads to a typical clinical feature of Graves' disease—exophthalmos (Fig. 33.16). Apart from **thyroid function tests, TSH receptor-binding immunoglobulins** can also be assayed for diagnosis of Graves' disease. Another autoimmune disease is **Hashimoto's thyroiditis,** characterized by lymphocytic infiltration of the thyroid, and

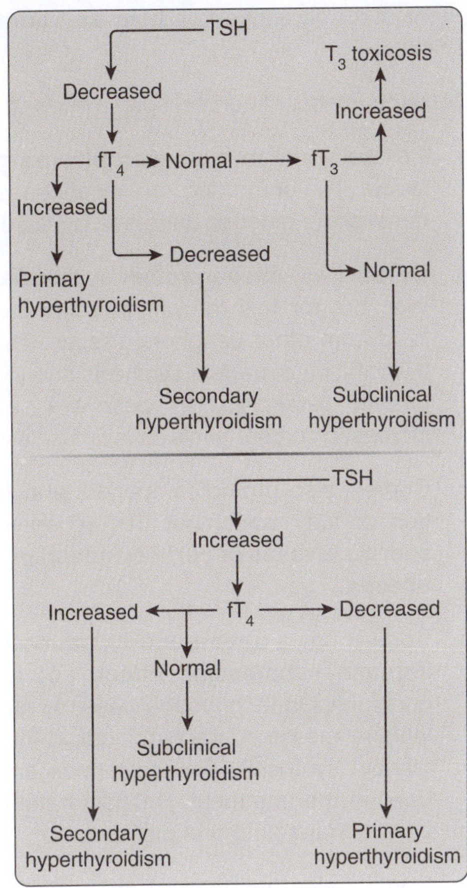

Figure 48.13 Diagnosis of thyroid hormone abnormalities. TSH = thyroid-stimulating hormone.

this eventually leads to hypothyroidism. **Serum anti-TPO (thyroid peroxidase) antibodies** are elevated in Hashimoto's thyroiditis. **Serum calcitonin** is used as a tumor marker in medullary carcinoma of thyroid, which is a tumor that arises from the parafollicular cells (or C cells); monitoring calcitonin levels after treatment is important to detect the recurrence of the tumor. Similarly, **serum thyroglobulin** levels are used in following up patients with thyroid malignancies who underwent thyroidectomies, for detection of residual disease and tumor recurrence.

VIII. FUNCTIONS OF THE ADRENAL GLAND

The adrenal gland is made of two embryologically and functionally distinct, yet anatomically proximal regions—adrenal medulla and adrenal cortex.

A. Adrenal medulla

The adrenal medulla forms one-fifth of the gland and is closely associated with the sympathetic nervous system. It receives nerve innervation from the centers in brainstem and spinal cord, and in response to stimuli **produces catecholamine hormones—norepinephrine and epinephrine**. The synthesis of catecholamines from tyrosine and their degradation is detailed in Chapter 26, Section III.A, p. 423. **Note that the degradation products of epinephrine and norepinephrine are metanephrine/normetanephrine and vanillylmandelic acid (VMA)**. Apart from the adrenal medulla, norepinephrine is also synthesized by the sympathetic nerve endings. Epinephrine and norepinephrine mediate their effects by binding to cell surface receptors known as adrenergic receptors (α_1, α_2, β_1, β_2)—these are G-protein coupled receptors, and bring about their actions with the help of second messengers including, but not limited to, cyclic adenosine monophosphate (cAMP). Norepinephrine binds mostly to the α-adrenergic receptor, with minor binding to the β receptors; epinephrine binds to both the receptors with almost equal affinity. These **receptors have a widespread tissue distribution**, hence the diverse actions of these hormones. For example:

- **α_1 receptor** is present on vascular smooth muscles stimulation of which leads to vasoconstriction (and increase in blood pressure).
- **α_2 receptor** is present in nerve terminals, where it inhibits release of norepinephrine.
- **β_1 receptor** is present in the myocardium, and stimulation leads to increased heart rate and myocardial contractility (thus increasing the cardiac output and blood pressure). In adipose tissue, activation of this receptor leads to activation of *hormone-sensitive lipase* and lipolysis.
- **β_2 receptor** is present in bronchial smooth muscle and activation of this receptor leads to bronchodilation. The receptor also leads to activation of glycogenolysis in the liver.

To summarize, action of epinephrine and norepinephrine prepares the body for a "fight or flight" reaction by increasing the blood supply to the muscles, increasing energy production by glycogenolysis and lipolysis, and so on. **Many drugs acting on the adrenergic receptors are in clinical use—bronchodilators** used in patients with bronchial asthma activate β_2 receptor, β_1 receptor blockers are used in the management of hypertension, and nasal decongestants activate α_1 receptor. **Adrenaline (epinephrine)** as such is used as a lifesaving drug in cardiac arrest, and in anaphylaxis (which is a severe form of allergic reaction characterized by vasodilation and low blood pressure, and severe bronchoconstriction).

B. Adrenal cortex

The adrenal cortex forms four-fifths of the adrenal gland and is made of three histologically distinct zones—zona glomerulosa (which secretes mineralocorticoids, mainly aldosterone), zona fasciculata (which secretes glucocorticoids, principally cortisol), and zona reticularis (which produces androgens and some amount of estrogen).

All these hormones are synthesized from cholesterol and have a steroid ring in their structure. The synthesis of these hormones is given in Chapter 23, Section VII.A, p. 359 and their regulation in Chapter 23, Section VII.B, p. 359.

In brief, **cortisol production and androgen production are regulated by adrenocorticotropic hormone (ACTH) from the anterior pituitary**. The ACTH secretion is controlled by **corticotropin-releasing hormone (CRH)** from the paraventricular nucleus of the hypothalamus. ACTH acts on the cell membrane G-protein coupled receptors of the adrenocortical cells, and activates the enzymes involved in cortisol synthesis (Fig. 48.14). Physical and mental stress and tissue damage also leads to CRH release and cortisol synthesis—cortisol is hence considered as the **"stress hormone."** Cortisol has an inhibitory effect on ACTH and CRH formation in pituitary and hypothalamus, respectively.

ACTH also stimulates the initial steps of mineralocorticoid production. Otherwise, mineralocorticoid synthesis and secretion are increased by angiotensin II which is a part of the **RAAS**, and also by increased potassium levels in the extracellular fluid (ECF). Aldosterone increases Na^+ reabsorption and K^+ excretion, thus maintaining ECF volume and ECF Na^+ levels, leading to decreased renin levels, thus negatively regulating its own secretion.

Since the hormones from adrenal cortex are lipophilic, they are transported in the circulation in a protein-bound form—90%–95% of cortisol in the circulation is bound to cortisol-binding globulin and albumin; 60% of aldosterone also binds to plasma proteins. Free form is usually detected in hormone assays. Many of the adrenocortical hormones also show a **diurnal variation** in the secretion, e.g., cortisol levels are the highest during the morning hours, soon after waking up.

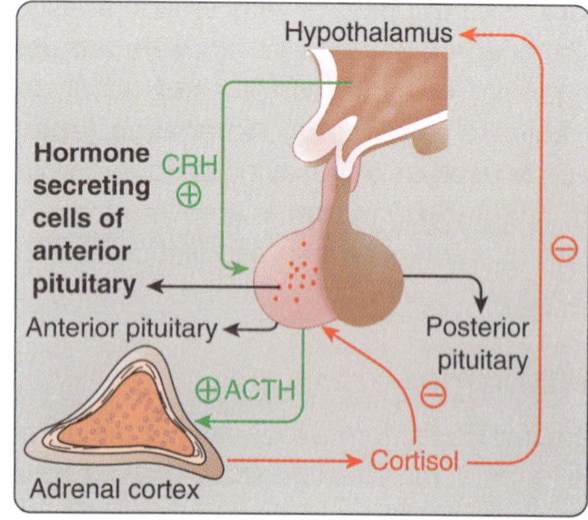

Figure 48.14 Regulation of cortisol secretion. CRH = corticotropin-releasing hormone; ACTH = adrenocorticotropic hormone.

Functions of the adrenocortical hormones are summarized below:

1. **Cortisol:** Effects of cortisol span a huge spectrum:

    a. Cortisol leads to **increased plasma glucose levels** by promoting gluconeogenesis and diminishing glucose utilization by tissues.

    b. **Cortisol increases protein synthesis** by the liver, but for this the amino acids are mobilized from other peripheral tissues. This leads to inhibition of protein synthesis and increased protein catabolism in peripheral tissues—increased cortisol levels cause muscle wasting.

    c. Cortisol also promotes lipolysis and β oxidation of fatty acids, but this action is tissue specific. **Excess of cortisol indeed promotes obesity**.

    d. Cortisol has a **potent anti-inflammatory and immune suppression effect**. For example, cortisol inhibits *phospholipase* A_2 and thus inhibits release of arachidonic acid for prostaglandin synthesis. Corticosteroids are hence used in the treatment of inflammatory conditions and autoimmune diseases.

    e. Cortisol also has a **weak mineralocorticoid effect** as it can bind weakly to the mineralocorticoid receptor.

2. **Aldosterone: Aldosterone increases the reabsorption of sodium and secretion of potassium by the renal tubules**. By reabsorbing sodium, aldosterone maintains the ECF volume. Detailed mechanism of the action of aldosterone on the renal tubules is given in Chapter 5, Section III.C, p. 44. Aldosterone is also involved in renal regulation of pH, as it promotes secretion of H^+ into the tubular fluid (see Chapter 4, Table 2).

3. Adrenal androgens, the most important being dehydroepiandrosterone have only a weak effect as such, are probably involved in the development of male sex organs. Peripheral conversion of adrenal androgens to testosterone is responsible for most of the effects of these hormones.

IX. TESTS FOR ADRENAL FUNCTION

The most common disorder of the adrenal medulla is the presence of a relatively rare catecholamine-secreting tumor, known as **pheochromocytoma** (the tumor when stained with chromium salts gives a dark color, hence the name – *phio*: dark, *chroma*: color, *cytoma*: tumor). 10% of these tumors are malignant. It is clinically characterized by the features of sympathetic stimulation—palpitation, sweating, anxiety, headache, weight loss, hypertension and arrythmias. The diagnosis is made by **estimating metanephrine in plasma**, and 24-hour urine excretion of metanephrine and **VMA**. Localization of the tumor is done by imaging studies.

The rest of this section will focus on the disorders of adrenal cortex, and the tests used in their diagnosis.

A. Tests for Cushing's syndrome

Cushing's syndrome is characterized by an excess of cortisol in the circulation. The clinical manifestations include weight gain and obesity, lethargy, muscle weakness, edema, hypertension, hyperglycemia, hypernatremia and hypokalemia. The causes can be exogenous **corticosteroid administration** (e.g., in patients with autoimmune diseases), ACTH secretion by a pituitary adenoma (this is termed **Cushing's disease**), ectopic ACTH production by tumors outside the pituitary (e.g., **small cell lung carcinoma**), and tumors of the adrenal cortex (**adenoma and carcinoma**). Establishing the diagnosis and the underlying etiology is critical in the management of Cushing's syndrome.

An increased salivary cortisol levels or 24-hour urine free cortisol levels is suggestive of Cushing's syndrome. Diagnosis can also be established by performing a low-dose (1 mg) **dexamethasone suppression test**. Dexamethasone is a highly potent synthetic corticosteroid; 1 mg of this drug is administered at midnight and serum cortisol is measured at 8 AM the next morning. In normal individuals, dexamethasone suppresses the hypothalamic CRH and pituitary ACTH production, resulting in decreased cortisol levels. This is not seen in Cushing's syndrome at low doses of dexamethasone.

The first step in establishing the cause of Cushing's syndrome is **the estimation of plasma ACTH.** In adrenal tumors, ACTH will be decreased due to the feedback inhibition by high cortisol levels. A high ACTH is seen in pituitary tumors and in ectopic ACTH production. The differentiation between these is achieved by a high-dose (8 mg) **dexamethasone suppression test.** The principle of the test is that in excess production of ACTH by the pituitary, a high dose of dexamethasone suppresses the ACTH, and thus cortisol levels; this negative feedback is not seen with ectopic ACTH-producing tumors. Thus, after a midnight administration of 8 mg dexamethasone, if plasma cortisol is suppressed, the diagnosis of Cushing's disease is highly likely.

The tests in Cushing's syndrome are summarized in a flowchart in Fig. 48.15.

B. Tests for adrenal insufficiency

Adrenal insufficiency is characterized by a decrease in the cortisol and aldosterone levels in the circulation. The clinical manifestations include weakness and fatigue, weight loss, dehydration, hypotension, hypoglycemia, hyponatremia and hyperkalemia. The causes can be **primary adrenal insufficiency or Addison's disease** (defect lies in the adrenal glands; causes include autoimmunity, tuberculosis, hemorrhage, metastasis of tumors, etc.), **secondary adrenal insufficiency** (seen in pituitary diseases) and **tertiary adrenal insufficiency** (seen in hypothalamic diseases, and in high-dose glucocorticoid therapy—this suppresses CRH production in hypothalamus, and a sudden drug withdrawal leads to adrenal insufficiency). Laboratory diagnosis involves establishing the diagnosis of adrenal insufficiency and establishing the cause.

Diagnosis of adrenal insufficiency is made by the demonstration of **low circulating cortisol levels**—this can be done by estimating serum cortisol levels in the early morning (6 AM) or by estimating morning salivary cortisol levels. If cortisol levels are normal, a short **ACTH stimulation test** will help in the diagnosis. This involved administration of 250 µg of cosyntropin (synthetic ACTH, with 24 amino acids, which is shorter than ACTH which has 39 amino acids) and monitoring

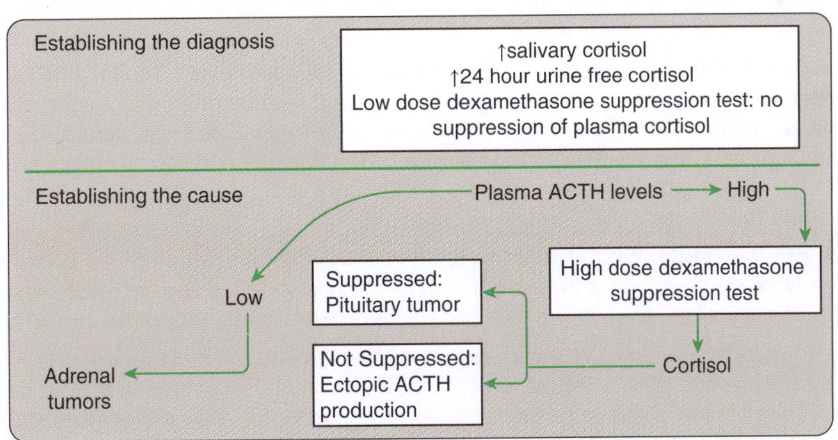

Figure 48.15 Diagnosis of Cushing's syndrome. ACTH = adrenocorticotropic hormone.

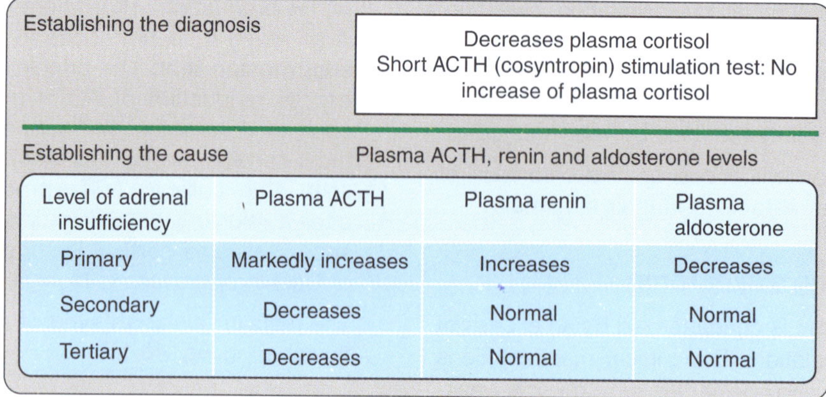

Establishing the diagnosis

Decreases plasma cortisol
Short ACTH (cosyntropin) stimulation test: No increase of plasma cortisol

Establishing the cause — Plasma ACTH, renin and aldosterone levels

Level of adrenal insufficiency	Plasma ACTH	Plasma renin	Plasma aldosterone
Primary	Markedly increases	Increases	Decreases
Secondary	Decreases	Normal	Normal
Tertiary	Decreases	Normal	Normal

Figure 48.16 Diagnosis of adrenal insufficiency. ACTH = adrenocorticotropic hormone.

the changes in serum cortisol levels before, and at 60 minutes of administration. A failure of the serum cortisol to increase after cosyntropin administration is suggestive of adrenal insufficiency.

The cause of adrenal insufficiency can be identified by measuring **plasma ACTH, renin, and aldosterone levels**. In **Addison's disease**, ACTH and renin levels are increased due to absence of negative feedback by cortisol and aldosterone, and aldosterone levels are decreased. Since the aldosterone production is more dependent on RAAS than ACTH, in secondary and tertiary adrenal insufficiency, renin and aldosterone levels are normal despite a decrease in the ACTH level. From a treatment point of view, distinction between secondary and tertiary adrenal insufficiency is not important.

The tests in adrenal insufficiency are summarized in a flowchart in Fig. 48.16.

C. Tests for mineralocorticoid excess

Increased plasma aldosterone levels lead to hypernatremia and hypokalemia. In primary hyperaldosteronism, the level of defect is in the adrenals. This is seen in **Conn's syndrome** (aldosterone-secreting tumor of adrenal cortex) or in **adrenal hyperplasia**. In such cases, serum aldosterone will be high, and renin levels will be low, due to feedback inhibition. In secondary hyperaldosteronism, the defect is in the renal level (e.g., decreased renal blood flow leading to increased renin levels, or a renin-producing tumor); both renin and aldosterone levels will be elevated in such conditions.

X. Medical Perspectives

This section describes some additional medical perspectives in relation to the organ functions and tests described in this chapter.

Acute phase reactants: These are the serum proteins the levels of which increase or decrease by 25% in presence of acute or chronic inflammation. They are produced by the liver, and their synthesis is altered in response to various molecules like cytokines that are associated with inflammation. The proteins whose levels are increased in inflammation are termed positive acute phase reactants, and examples are C-reactive protein, ferritin, complement proteins, ceruloplasmin, etc. Negative acute phase proteins have their serum concentrations reduced during inflammation and include albumin, transthyretin, retinol-binding protein, and transferrin.

We have seen that in alcoholic hepatitis, serum *AST* levels are higher than *ALT* levels. Thus, an ***AST/ALT* >2 is suggestive of alcohol use disorder**. It was also explained that ***GGT*** is elevated because of alcohol intake, due to induction of the microsomal isoenzyme and alcohol-induced leakage of the enzyme from the cell. Thus, serum *AST/ALT* and *GGT* act as markers of alcohol use disorder. Other markers include **serum ethanol levels** and serum **carbohydrate-deficient transferrin** (CDT) levels. Transferrin is normally glycosylated by addition of sialic acid residues, and this process is impaired in chronic alcoholics, leading to increase in CDT levels.

5'-deiodinases which are involved in peripheral conversion of T_4 to the active form, i.e., T_3 are selenium dependent enzymes. The activity of the enzymes is thus impaired in selenium deficiency, leading to hypothyroidism. *5'-deiodinase* is also a target of certain drugs that are used in the treatment of hyperthyroidism, e.g., **propranolol and propylthiouracil.**

During a nuclear fallout, one of the radioactive isotopes that is released into the atmosphere is I^{131}. Uptake of this radioactive isotope by thyroid tissue can lead to the increased concentration of this isotope in thyroid gland, leading to future thyroid malignancies. Intake of normal iodine (in the form of potassium iodide) can prevent the accumulation of radioactive iodine in the gland, and this forms the **basis of iodide supplementation during a nuclear fallout.**

XI. Chapter Summary

- Organ tests are performed to detect an organ dysfunction (rather than organ function). In most scenarios, more than one biochemical test is required for this purpose to be achieved. Functions, disorders, and tests for four organs—liver, kidney, thyroid, and adrenals, were discussed in this chapter.
- The liver has important functions in metabolism of macro- and micromolecules. It also synthesizes a large number of plasma proteins including albumin and clotting factors and is involved in detoxification of xenobiotics and endogenous molecules. Liver tests include those for hepatocellular injury—the *transaminases*, enzyme and non-enzyme markers of cholestasis, biomarkers for synthetic functions of liver, markers of viral hepatitis, etc. (Fig. 48.17).
- Functions of the kidneys can include excretion of metabolic waste products, homeostasis of water, electrolytes and pH, and production of hormones like renin, erythropoietin, and calcitriol. Glomerular function is assessed by indirect markers of glomerular filtration rate (GFR) like serum cystatin and serum creatinine. Formulas are available for calculation of estimated GFR using these parameters. Prerenal and intrarenal azotemias can be distinguished by fractional excretion of sodium (FENa) and blood urea nitrogen (BUN)-serum creatine ratio. Other tests for identifying the etiology of renal diseases include detection of proteinuria, albumin excretion rate, and urine analysis for detection of other abnormal analytes.
- Thyroid hormones play a key role in metabolism and general functioning of the body. Thyroid-stimulating hormone (TSH) regulates the thyroid hormone production. Hyperthyroidism and hypothyroidism can be diagnosed by thyroid function tests, of which TSH is the most important.
- Adrenal medulla produces catecholamine hormones—their level is elevated in pheochromocytoma. Adrenal cortex produces cortisol, aldosterone, and androgens. Hyperfunctioning of the cortex can be detected by suppression tests, while adrenal insufficiency can be diagnosed by stimulation tests.

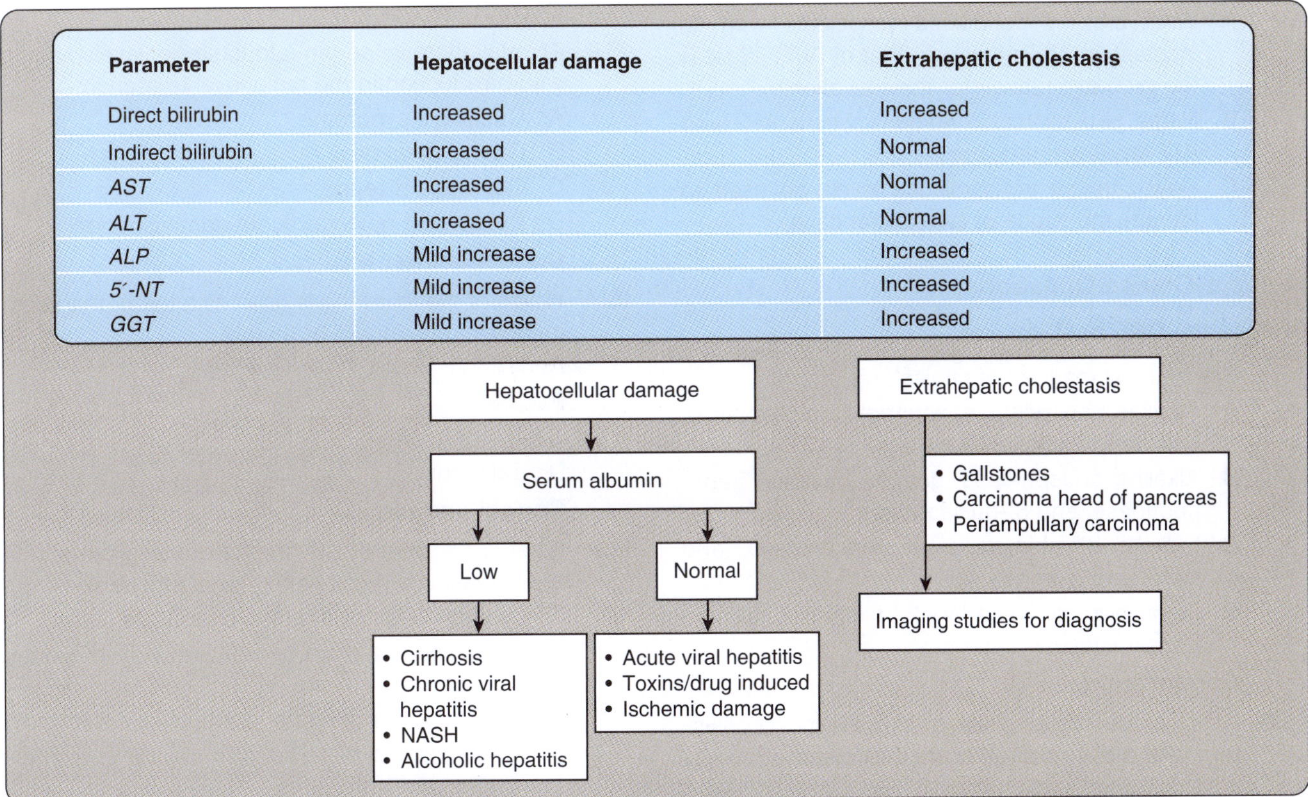

Parameter	Hepatocellular damage	Extrahepatic cholestasis
Direct bilirubin	Increased	Increased
Indirect bilirubin	Increased	Normal
AST	Increased	Normal
ALT	Increased	Normal
ALP	Mild increase	Increased
5′-NT	Mild increase	Increased
GGT	Mild increase	Increased

Figure 48.17 Key concept map – diagnosis of liver diseases. *AST = aspartate aminotransferase; ALT = alanine aminotransferase; ALP = alkaline phosphatase; 5′-NT = 5′-nucleotidase; GGT = γ-glutamyl transpeptidase*; NASH = nonalcoholic steatohepatitis.

Study Questions

Short-Answer Questions

48.1. What is the mechanism(s) for elevation of serum alkaline phosphatase in cholestasis?

48.2. What is the clinical significance of estimating albumin excretion rate?

48.3. What is cosyntropin? What is the clinical utility of this molecule?

48.4. How can thyroid function tests suggest a diagnosis of subclinical hyperthyroidism?

Long-Answer Questions

48.1. With regard to serum cystatin C as a marker for renal function, answer the following:
   A. What is the physiological function of cystatin C?
   B. Why is it considered a better marker of renal function compared to serum creatinine?
   C. Name two non-renal conditions associated with elevation of serum cystatin C.

48.2. Answer the following questions regarding the utility of serum transaminases in liver dysfunction:
   A. What are the conditions associated with an increase in AST relative to that of ALT? What is the probable reason for this elevation?
   B. Name two clinical conditions where AST and ALT levels are reduced.
   C. How can the transaminase levels be used to identify the cause of liver dysfunction?

Multiple-Choice Questions

Choose the ONE best answer.

48.1. Biochemical markers of cholestasis are:
   A. Alkaline phosphatase, aspartate transaminase, and alanine transaminase
   B. Alkaline phosphatase, gamma glutamyl transpeptidase, and 5'-nucleotidase
   C. Indirect bilirubin, alkaline phosphatase, and lactate dehydrogenase
   D. Direct bilirubin, aspartate transaminase, alanine transaminase

   Correct answer = B.

48.2. Which of the following scenarios is the most consistent with a diagnosis of acute liver disease?
   A. Decreased serum albumin, normal prothrombin time
   B. Decreased serum albumin, increased prothrombin time
   C. Normal serum albumin, increased prothrombin time
   D. Increased serum albumin, normal prothrombin time

   Correct answer = C.

48.3. Persistent low levels of transaminase can be seen due to deficiency of:
   A. Vitamin B_6
   B. Vitamin B_1
   C. Vitamin K
   D. Vitamin B_{12}

   Correct answer = A.

48.4. Which of the following is NOT a feature of pre-renal azotemia?
   A. Low blood pressure
   B. Increased BUN/serum creatinine
   C. Increased FENa
   D. Increased serum creatinine

   Correct answer = C.

48.5. Which of the following conditions is associated with overflow proteinuria?
   A. Diabetic nephropathy
   B. Nephrotoxic drugs
   C. Glomerulonephritis
   D. Multiple myeloma

   Correct answer = D.

48.6. An elevation in serum aldosterone levels, along with a reduction in the renin level is seen in:
   A. Cushing's syndrome
   B. Conn's syndrome
   C. Cushing's disease
   D. Exogenous glucocorticoid administration

   Correct answer = B.

48.7. Which of the following is used as a tumor marker in medullary carcinoma of thyroid?
   A. fT_4
   B. TSH
   C. Thyroglobulin
   D. Calcitonin

   Correct answer = D.

48.8. What is the primary advantage of using cystatin C over creatinine in estimating renal function?
   A. Cystatin C is not affected by muscle mass.
   B. Creatinine is a more sensitive marker of kidney function.
   C. Cystatin C is not reabsorbed unlike creatinine.
   D. Creatinine is a more specific marker of glomerular filtration rate.

   Correct answer = A.

48.9. Which of the following equations is commonly used to estimate the glomerular filtration rate (GFR) based on serum creatinine levels, age, sex, and race?
   A. Modification of Diet in Renal Disease (MDRD) equation

B. Cockcroft-Gault equation
C. Schwartz equation
D. CKD-EPI equation
Correct answer = A.

48.10. The most sensitive marker for diagnosing thyroid dysfunction is:
A. TT3
B. fT3
C. fT4
D. TSH
Correct answer = D.

Multiple-Choice Questions (NExT/NEET PG)

Choose the ONE best answer.

48.1. The current gold standard test for the diagnosis of pheochromocytoma is:
A. 24-Hour urine metanephrines and catecholamines
B. Plasma-free metanephrines
C. Plasma catecholamines during a provocative test
D. MRI of the adrenal glands
Correct answer = B.

48.2. A 44-year-old male presents with recurrent episodes of jaundice. Laboratory investigations reveal mildly elevated bilirubin level which is predominantly unconjugated and normal serum ALT, AST, and ALP levels. The most likely diagnosis is:
A. Chronic alcoholic liver disease
B. Primary biliary cirrhosis
C. Gilbert's syndrome
D. Crigler-Najjar syndrome
Correct answer = C.

48.3. A 37-year-old female presents with symptoms of hyperthyroidism. Thyroid function tests show elevated serum fT3 and fT4 levels and suppressed TSH level. What additional test can help to differentiate Graves' disease from other causes?
A. Thyroid ultrasound
B. Thyroid peroxidase antibody (TPOAb) assay
C. Thyroid uptake and scan
D. Thyroid stimulating immunoglobulin (TSI) assay
Correct answer = D.

48.4. Estimation of serum T3, T4, TSH is based on which technique?
A. Reflectometry
B. potentiometry
C. immunoassay
D. nephelometry
Correct answer = C.

48.5. Which of the following tests is indicative of chronic kidney disease (CKD) rather than acute kidney injury (AKI)?
A. Urinary protein-to-creatinine ratio > 3.5 g/g
B. Increased urinary sodium concentration
C. Urine osmolality > 500 mOsm/kg
D. Normal-sized kidneys on renal ultrasound
Correct answer = A.

Appendix
Unit X: Special Topics

CLINICAL RELEVANCE FOR PRIMARY HEALTHCARE PHYSICIANS

Pharmaceutical agents constitute the major xenobiotic group to which humans are exposed. Besides phase 1 and phase 2 transformation, active membrane transporters that function to shuttle drugs and other xenobiotics across cellular membranes also play an important role in xenobiotic metabolism. Drugs that diffuse passively across the apical membranes of intestinal epithelial cells can be extruded back into the intestinal lumen by transporters or vice versa. Apart from human enzymes, gut bacterial community (microbiome) also metabolize xenobiotics and can significantly modulate drug bioavailability for orally administered drugs.

Drugs, such as rifampicin, which induce cytochrome P450 (CYP450), can alter the metabolism, and therefore, the pharmacological effect or adverse effect of coadministered drugs. Adverse reactions of several drugs, such as doxorubicin and cisplatin (anticancer drugs) and azathioprine (antiretroviral) are mediated though oxidative stress. Increased hemolysis due to antimalarial drugs (quinine, chloroquine and primaquine), seen in G6PD deficiency, is also mediated through oxidative stress (see Chapter 18, V. G6PD deficiency, p. 253). Free radicals also have several beneficial roles when maintained at low concentrations. Phagocytes synthesize free radicals for destruction of invading pathogens. Defective nicotinamide adenine dinucleotide phosphate (NADPH) oxidase system leads to inability to produce O_2^-, causing granulomatous disease. Free radicals, such as nitric oxide (NO), are also involved in a number of cellular signaling pathways.

In India, there is a National Programme for Prevention and Control of Cancers, Diabetes, Cardiovascular Diseases and Stroke (NPCDCS) which focuses on health promotion through behavior change, opportunistic screening at all levels in the healthcare delivery system and management of chronic non-communicable diseases, especially cancer, diabetes, cardiovascular diseases (CVDs) and stroke. Also, National Cancer Registry Programme (NCRP) is being run by the Indian Council of Medical Research, with the objectives of obtaining an overview of patterns of cancer in different parts of the country and to calculate estimates of cancer incidence.

When prescribing chemotherapy to cancer patients, the probability of hyperuricemia should be considered and appropriate preventive steps should to be taken (see Chapter 34, V.C: Diseases associated with purine degradation, p. 586). An upcoming treatment for cancer is chimeric antigen receptor (CAR) T-cell therapy, in which a patient's T cells are modified to express a receptor against certain protein on the patient's cancer cells. Such CAR T-cells can attack cancer cells and destroy them.

Table 1 Classes of tumor markers with associated cancer(s)

Class of Tumor	Specific Markers	Associated Cancer(s)
Tumor-associated antigen	Carcinoembryonic antigen (CEA)	Carcinomas of GIT and embryonic tissue
	A-fetoprotein (AFP)	Liver carcinoma
	Tissue polypeptide antigen (TPA)	Breast, prostate
	Tennessee antigen (Tenagen)	Lung, stomach, pancreas
Carbohydrate antigen (CA)	CA 125	Ovary
	CA 15-3	Breast carcinoma
	CA 19-9	Pancreas
	CA 72-4 (TAG-72)	GIT, ovary, breast
	CA 50	GIT, lung
	Mammary serum antigen (MSA)	Breast

(Continued on next page)

Table 1 (Continued from previous page)

Class of Tumor	Specific Markers	Associated Cancer(s)
	MAM-6	Breast
	Mucin-like carcinoma-associated antigen (MCA)	Breast
	MAP	Breast
Pregnancy-associated antigen	β-hCG	Testes and ovaries, germ cell tumors
	SP-1	Human placenta
	Human placenta lactogen (HPL)	Human placenta
	α2 Placenta associated globin	Human placenta
Enzymes	*Prostatic acid phosphatase (PAP)*, Prostate specific antigen (PSA)	Prostate gland
	Neuron-specific enolase (NSE)	Neuroendocrine tumor
	Total LDH	Germ cells, breast, lung, leukemias
	LDH1	Germ cell, ovary
	LDH 2,3,4	Leukemia, lymphomas
	LDH 5	Liver, breast, prostate
Isoenzymes	*Alkaline phosphatase (ALP)*	
	Liver	Liver
	Bone	Bone
	Regan	Lung, ovary, breast, uterus
	Nagao	Pancreas, bile duct
	Creatine kinase	
	CK-BB	Adenocarcinoma, prostate carcinoma
	α1-antitrypsin	Germ cell tumors of Testes and ovaries
Hormones	ACTH	Lung, pancreas
	Catecholamine carcinoids	Pheochromocytoma
	Insulin	Insulinoma
	Glucagon	Glucagonoma
	Gastrin	Gastrinoma
	Serotonin	Carcinoid
	Calcitonin	Carcinoma of thyroid and lung
Serum proteins	Bence-Jones proteins	Multiple myeloma
	Immunoglobulins	Macroglobulinemia, multiple myeloma

GIT = gastrointestinal tract; β-hCG = beta-human chorionic gonadotropin; LDH = *lactate dehydrogenase*; ACTH = adrenocorticotropic hormone.

Table 2 Frequently used biomarkers with values indicative of malignancy

Tumor Marker	Values Indicating Malignancy
Alpha fetoprotein (AFP)	>1,000 ng/l
Carcinoembryonic antigen (CEA)	>10 ng/ml
Beta hCG	>10,000 IU/l
Cancer antigen 125 (CA-125)	>35 U/ml
Prostate specific antigen (PSA)	>10 ng/l

Beta hCG = beta-human chorionic gonadotropin.

Table 3 Interpretation of liver function test (LFT)

Biochemical Findings	Probable Diagnosis	Further Investigations
Total bilirubin: ↑↑ Direct bilirubin, *AST*, *ALT* and *ALP*: within reference limits *LDH*: ↑	Hemolytic anemia	Hemoglobin Peripheral smear examination
Total bilirubin: ↑↑ Direct bilirubin: ↑ *AST* and *ALT*: ↑↑↑ *ALP*: ↑ PT and INR: ↑ Serum Albumin: within reference limits	Acute viral hepatitis	Viral markers
Total bilirubin: ↑↑ Direct bilirubin: ↑ *AST* and *ALT*: ↑ *ALP*: ↑ PT and INR: ↑↑ Serum Albumin: ↓↓	Chronic liver disease	Viral markers *GGT*: for alcoholic liver disease Imaging studies
Total bilirubin: ↑↑ Direct bilirubin: ↑↑ *AST* and *ALT*: within reference limits *ALP*: ↑↑↑ PT and INR: ↑↑ Serum Albumin: within reference limits	Extra-hepatic obstruction	*GGT* and *5′-NT* Imaging studies

AST = *aspartate aminotransferase*; *ALT* = *alanine aminotransferase*; *ALP* = *alkaline phosphatase*; *LDH* = *lactate dehydrogenase*; PT = prothrombin time; INR = international normalized ratio; *GGT* = *gamma-glutamyl transpeptidase*; *5′-NT* = *5′-nucleotidase*.

Clinical Case Questions

Chapter 45 Cancer

45.1. A 62-year-old CEO reported in the urology OPD with complaints of increased frequency of micturition, weak urinary stream, occasional episodes of hematuria, constant back pain and bouts of constipation. A digital rectal examination showed that the patient had enlarged prostate gland with several palpably discrete nodules.

A. Which laboratory investigation should be done to rule out carcinoma of the prostate in this patient?

Answer—A: Laboratory investigation: Prostate-specific antigen (PSA)—levels above 10 ng/ml (reference range 0.0–4.0 ng/ml) is suggestive of carcinoma of prostate.

Ultrasonography-guided biopsy and histological evaluation is recommended for confirmation and staging of carcinoma.

Radiological investigations: MRI studies, chest, abdominal, and pelvic CT scans can be done to evaluate the extent of disease and detect metastatic lesions.

B. What is the possible cause for the complaint of constant back pain in this patient?

Answer—B: Most common sites for the metastatic spread of prostate cancer are the hip, spine, and pelvic bones by direct invasion or by traveling through blood or lymphatic system. Thus, the patient may present with back pain, weak bones with increased risk for fractures, numbness and weakness from spinal cord compression, stiffness or pain in the hip, thighs, or back and high blood calcium levels.

45.2. A 32-year-old woman reported in the dermatology OPD with induration and asymmetry of her breasts. On physical examination, her left breast was found to be smaller and edematous. Erythema, induration and peau d'orange were observed. Left axillary lymph nodes were palpable. She had no personal or family history of breast or ovarian cancer and was referred to the surgery department for evaluation.

A. Which laboratory investigations can be done in suspected cases of carcinoma breast?

Answer—A: A deep skin biopsy and molecular analysis for estrogen receptors (ERs), progesterone receptors (PRs) and human epidermal growth factor receptor 2 (HER2) can be done to confirm the diagnosis of carcinoma breast.

B. What are the common treatment modalities used?

Answer—B: Mastectomy to remove cancerous tissue and the draining lymph nodes, radiotherapy to kill or shrink cancerous cells, chemotherapy to kill fast-growing cancer cells and targeted therapy against HER2/neu protein in case of HER2-positive breast cancer are the common treatment modalities.

Chapter 46 Xenobiotics

46.1. A 45-year-old male was brought to the emergency with complaints of nausea, vomiting and abdominal pain. Physical examination showed tenderness in the right hypochondrium. Further questioning revealed that he had consumed at least 20 tablets of paracetamol (650 mg), about 24 hours before, because he was emotionally disturbed after an argument with his wife.

Laboratory investigations showed highly elevated aspartate transaminase (AST) and alanine transaminase (ALT) levels, and an elevated prothrombin time. The treatment protocol for management for paracetamol poisoning was initiated, which included administration of N-acetylcysteine.

A. How is paracetamol normally metabolized by the body?

Answer—A: Paracetamol is detoxified in the liver, either by conjugation or by CYP enzymes. At pharmacological doses, 90% of paracetamol gets conjugated to form glucuronic acid and sulfate derivatives which are excreted in urine; CYP enzymes play a minor role and produce minimal quantities of N-acetyl-p-benzoquinone imine (NAPQI) which is then conjugated with glutathione to form mercapturic acid derivatives which are excreted.

B. What is the reason for highly elevated AST and ALT levels and elevated prothrombin time in this patient?

Answer—B: At toxic doses, paracetamol is metabolized by CYP enzymes, generating large quantities of NAPQI. Binding rapidly to the -SH groups of glutathione, NAPQI depletes the intracellular glutathione stores. Following this, NAPQI irreversibly binds various intracellular proteins with -SH groups, forming NAPQI-protein adducts. Widespread cellular dysfunction, especially mitochondrial dysfunction, that follows causes hepatocellular injury, releasing the intracellular enzymes into circulation, causing marked elevations in AST and ALT levels. Liver dysfunction also impairs synthesis and γ-carboxylation of clotting factors, leading to elevated prothrombin time.

C. What is the rationale for the administration of N-acetylcysteine in this patient?

Answer—C: N-acetylcysteine has free -SH groups and acts by two distinct mechanisms—it replenishes the intracellular glutathione stores, facilitating excretion of NAPQI; it directly binds to NAPQI, impeding the binding of NAPQI to cellular proteins.

46.2. A 25-year-old male was brought to the emergency department with complaints of visual disturbances, headache and nausea. He was otherwise healthy but gave a history of consumption of locally made alcohol the previous night. On examination, his respiratory rate was 30/min (normal 12–16/min), and his vision was decreased. Arterial blood gas analysis and serum electrolyte estimations revealed high anion-gap metabolic acidosis. A provisional diagnosis of methanol poisoning was made, and treatment was initiated.

A. How is methanol metabolized by the body?

Answer—A: Methanol is metabolized by alcohol dehydrogenase in the cytoplasm to formaldehyde. Formaldehyde is converted to formic acid by aldehyde dehydrogenase. Some amount of unmetabolized methanol is excreted as such in breath.

B. What is the reason for high-anion gap metabolic acidosis?

Answer—B: Formic acid which is formed as a result of methanol metabolism is a nonvolatile acid; accumulation of formic acid reduces blood pH and causes metabolic acidosis. Anion gap is due to increase in the levels of unmeasured anions during routine arterial blood gas and serum electrolyte analysis. These include anions other than chloride and bicarbonate. Presence of methanol and formic acid in this case causes elevation of anion gap.

C. Which are the antidotes used in methanol poisoning?

Answer—C: Treatment of methanol poisoning involves inhibition of alcohol dehydrogenase, so that methanol is not metabolized into formic acid. This can be achieved by fomepizole, a competitive inhibitor of alcohol dehydrogenase. Ethanol also can be used as a competitive inhibitor, since the affinity of ethanol for alcohol dehydrogenase is 10–20 times higher than that of methanol. Unmetabolized methanol is excreted via lungs.

Chapter 47 Free Radicals, Oxidative Stress and Antioxidant Defense Mechanism

47.1. A 50-year-old male visited his family doctor with complaints of fatigue and joint pains for more than a year. He gave no history of smoking or alcohol consumption. His uncle died of liver cancer about 10 years ago. The physician noted grayish pigmentation of the skin. Ultrasound examination of the abdomen revealed hepatomegaly (enlarged size of the liver) with irregular margins. His liver function tests were mildly deranged; plasma iron: 50 µmol/l (Reference: 7–25 µmol/l) and serum ferritin: 3200 µg/l (Reference: 29–248 µg/l in males) were elevated. A diagnosis of hemochromatosis was suspected, which was confirmed by genetic testing for mutations in *HFE* gene. Since the condition is associated with increased body iron levels, iron chelation therapy with deferoxamine was initiated.

A. Hemochromatosis can clinically present with manifestations of diabetes mellitus, due to destruction of pancreatic tissue. What could be the mechanism for the tissue destruction?

Answer—A: Hemochromatosis is a disorder of iron metabolism, where intracellular level of iron is increased, due to defects in the regulation of its absorption. The iron that is accumulated in the initial stages of hemochromatosis is complexed with ferritin, but as the level of iron increases further, free iron starts to accumulate. As described in Section II.C, free iron leads to formation of oxidant species, which leads to oxidative stress and tissue damage.

B. What could be the possible reason of the family history of liver cancer in this case?

Answer—B: The effect of oxidant species on DNA leading to development of mutations (see Sections III.A and VI.B) predisposes individuals with hemochromatosis to malignancy. Primary hemochromatosis is an inherited disease, and hence the positive family history.

47.2. Paraganglioma refers to a rare group of tumors originating from neuroendocrine cells. In patients with familial paraganglioma, a frequently mutated set of genes encodes for the subunits of *succinate dehydrogenase* (SDH) enzyme.

A. What is the function of SDH?

Answer—A: SDH is located in the inner mitochondrial membrane and is involved in both citric acid cycle and electron transport system (ETS). The enzyme catalyzes the conversion of succinate to fumarate (see Fig. 14.5). It also forms the complex II of ETS where $FADH_2$ donates its electrons to get oxidized to FAD.

B. What is the mechanism by which the above mutations lead to tumor formation?

Answer—B: Mutations in subunits of SDH lead to electron leakage from this complex, leading to partial oxidation of oxygen to generate ROS (see Section II.A). Again, DNA damage due to ROS predisposes to carcinogenesis.

Chapter 48 Organ Function Tests

48.1. A 21-year-old student was admitted to the hospital after he developed yellowish discoloration of the skin and sclera and low-grade fever. He reported that he had nausea and vomiting for the past 10 days and a decreased appetite. On examination, he had fever of 102°F. Abdominal examination revealed tenderness in the right hypochondrium. The urine was high colored. The total serum bilirubin was 7.2 mg/dl, direct bilirubin 3.0 mg/dl, AST 1,723 IU/l and ALT 2,675 IU/l. He denied history of sexual contact, recent surgery, blood transfusion, and injections.

A. What is the most likely diagnosis?

Answer—A: Acute viral hepatitis, most likely hepatitis A. Clues to diagnosis include presence of fever which suggests infection, and presence of pain in the hypochondrium which suggests inflammation of the liver. Elevation of both direct and indirect bilirubin suggests hepatocellular jaundice, and severe elevation of transaminases is seen in hepatocellular injury. Absence of sexual contact, recent surgery, blood transfusion, and injections may point to feco-oral transmission—that is hepatitis A and E, but hepatitis E is most commonly seen in pregnant females.

B. What could be the serum ALP level in this patient? Why?

Answer—B: Elevated, but only 2–4 times the upper reference limit. The inflammation associated with hepatitis leads to intrahepatic cholestasis, and an elevation in the serum ALP levels.

C. What is the reason behind elevated bilirubin and transaminases?

Answer—C: The liver dysfunction in hepatitis impairs the conjugation of bilirubin, leading to increased unconjugated bilirubin in the circulation. Increase in conjugated bilirubin is due to intrahepatic cholestasis. Transaminases are elevated due to their release from the hepatocytes as a result of hepatocellular injury.

D. How is the diagnosis confirmed?

Answer—D: By performing an ELISA for serum anti-hepatitis A virus (HAV) IgM. Elevated levels suggest a recent hepatitis A infection.

48.2. A 32-year-old female was at her doctor's office with the chief complaints of fatigue, and weight gain despite a poor appetite. She attributed this to her inability to keep up with her normal routine workouts due to loss of energy and overall sleepiness even throughout the day. She also had complaints of feeling cold all the time and difficulty swallowing and said that her throat feels swollen.

The doctor noticed that she looked depressed and had a hard time concentrating while questions were being asked. Examination revealed a pulse rate of 58 beats per minute and coarse dry skin. Her tendon reflexes showed a markedly delayed relaxation phase. Palpation of the neck revealed a non-tender, irregular, firm swelling front of neck, which moves downwards with swallowing.

Laboratory investigations revealed serum TSH: 23 mIU/l (Reference limits: 0.35–6.20 mIU/l), serum free T4: 6.44 pmol/l (reference limits: 9.00–23.12 pmol/l), elevated levels of TPO antibodies and serum total cholesterol of 255 mg/dl.

A. What is the most probable diagnosis?

Answer—A: Primary hypothyroidism, most likely Hashimoto's thyroiditis. The symptoms like fatigue, weight gain, cold intolerance, and drowsiness, as well as signs like bradycardia, delayed tendon reflexes, and neck swelling are indicative of thyroid disease and hypothyroidism. A low fT_4 and a high TSH further confirms primary hypothyroidism. Elevated anti-TPO antibodies suggest Hashimoto's thyroiditis.

B. What is the advantage of estimating free T_4 over total T_4 in a patient with thyroid disease?

Answer—B: Free T_4, rather than bound form is the one that enters the cells to get converted into the active T_3. Also, the levels of bound forms are altered by the levels of the proteins to which T_4 binds, which is affected by illnesses and medications.

C. If you are given a choice of performing a single test to assess the thyroid function, which test will you prefer? Why?

Answer—C: TSH. A normal TSH level rules out thyroid hormone abnormality. TSH levels are highly sensitive to small alterations in T_3 and T_4 and is the first to change in thyroid diseases. Thus, performing TSH alone as the first step in diagnosis is much cost effective.

D. What is the reason for elevated serum total cholesterol in this patient?

Answer—D: Expression of LDL-receptors requires the action of thyroid hormones. In hypothyroidism, a reduction in the number of LDL-receptors leads to decreased clearance of cholesterol from the circulation.

E. Name two common autoimmune diseases affecting the thyroid gland. How do they differ in the clinical presentation?

Answer—E: Graves' disease presents as hyperthyroidism. Hashimoto's thyroiditis during the initial phase (when there is associated thyroid follicle damage) also can cause hyperthyroidism, but hypothyroidism sets in as the disease progresses.

48.3. A term newborn born to an unbooked primigravida is noted to have rapidly deepening jaundice at around 12 hours of life. Although the vitals are stable, the newborn is found to have pallor and hepatosplenomegaly. Phototherapy is initiated immediately, and investigations are ordered. The mother's blood group is found to be A negative, while the baby is A positive.

(**Note:** *Clinical Case 48.3 requires knowledge from Chapters 43 and 48)*

A. What is the clinical phenotype and etiological diagnosis?

Answer—A: Pathological jaundice (onset within 24 hours of birth) due to Rh incompatibility.

B. What abnormalities will you find in the liver function tests?

Answer—B: The total serum bilirubin and direct fraction will be elevated and demonstrate a rapid rise (>5 mg/dl in 24 hours or >0.5 mg/dl/h). Serum liver enzymes (SSGOT, SGPT) and total proteins (albumin and globulin) will be within normal limits.

C. What precautions should have been taken during the current pregnancy to have prevented this condition?

Answer—C: Administration of two doses of anti-D Immunoglobulin at 28 weeks and 34 weeks of gestation.

48.4. A 4-year-old boy presents with gradually increasing facial puffiness, abdominal girth and swelling of both legs and his scrotum over the past week. This is the third time this has happened. The first episode occurred when he was two years old, which resolved spontaneously within a month. The child was taken to a hospital when it occurred the second time at the age of 3.5 years. The family was told that he had a renal disorder and was started on medication for around 2 months, after which the symptoms abated. The treating clinician orders some investigations based on the primary diagnosis that he is considering.

A. What does the clinical phenotype suggest? What is the triad of features that should be present?

Answer—A: Nephrotic syndrome. This is characterized by massive proteinuria, hypoalbuminemia and generalized edema.

B. Which investigations will confirm the clinical diagnosis?

Answer—B: Proteinuria > 1 $g/m^2/d$ (3+ or 4+ on bedside dipstick urinary examination), serum albumin levels < 2.5 g/dL and plasma total cholesterol levels > 250 mg/dl.

C. What is the pathophysiology of the massive proteinuria in minimal change nephrotic syndrome?

Answer—C: It is still not completely clear. The favored hypothesis is possible immune-mediated obliteration of the epithelial foot processes and basement membrane of the glomerular podocytes. Alteration in the normal charge barrier to macromolecules like albumin leads to the characteristic selective proteinuria.

48.5. A 56-year-old female, an office receptionist, visits the medical OPD with complaints of fatigue and mild discomfort in right upper quadrant of abdomen. The discomfort worsens after a meal but is not accompanied by any nausea, vomiting or change in bowel habits. There is no yellowish discoloration of eyes or urine. The lady has Type 2 diabetes and hypertension since the past 6 years and is on medication for the same. She leads a sedentary lifestyle with minimal physical activity, does not smoke and drinks alcohol only on special occasions.

On examination: Liver is palpable 2 cm below right costal margin (span 15 cm)

No icterus or signs of chronic liver disease

Investigations:

Liver function tests show mildly elevated AST and ALT

Fasting blood glucose 130 mg/dl and HbA1c 7%

Lipid profile: Triglycerides are elevated and low HDL cholesterol

Ultrasound abdomen: Increased echogenicity of the liver consistent with hepatic steatosis

Transient elastography (FibroScan) indicates no significant fibrosis

A. What is the probable diagnosis?

Answer—A: NAFLD (nonalcoholic fatty liver disease) in view of enlarged liver, asymptomatic elevation of liver enzymes and hepatic steatosis.

B. What are the common risk factors of the condition?

Answer—B: Diabetes mellitus type 2, hypertension, obesity and hypertriglyceridemia are the common risk factors.

C. What is the hallmark of NAFLD?

Answer—C: Intrahepatic steatosis is a hallmark of NAFLD.

Figure Sources

Figure 7.12: Modified from Garrett RH, Grisham CM. *Biochemistry*. Philadelphia, PA: Saunders College Publishing, 1995. Figure 6.36, p. 193.

Figure 7.13 (panel C, top): Horizon Symposium on Protein Folding and Disease, Italy. Stuttgart, Germany: Nature Publishing Group, Macmillan Publishers, 2002.

Figure 8.1A: Illustration: Irving Geis. Rights owned by Howard Hughes Medical Institute. Not to be used without permission.

Figure 8.21: Photo courtesy of Photodyne Incorporated, Hartland, WI.

Figure 9.3: Electron micrograph of collagen: Natural Toxin Research Center. Texas A&M University Kingsville. Collagen molecule modified from Mathews CK, van Holde KE, Ahern KG. *Biochemistry*, 3rd ed. Boston, MA: Addison Wesley Longman, Inc., 2000:175. Figure 6.13.

Figure 9.4: Modified from Yurchenco PD, Birk DE, Mecham RP, eds. *Extracellular Matrix Assembly and Structure*. San Diego, CA: Academic Press, 1994.

Figure 9.8: From Buhler K. Images in clinical medicine. *N Engl J Med.* 1995;332(24):1611.

Figure 9.10: Photo from Web site Derma.de.

Figure 9.11: From Jorde LB, Carey JC, Bamshad MJ, et al. *Medical Genetics*, 2nd ed. Mosby, St. Louis, MO, 1999. http://medgen.genetics.utah.edu/index.htm

Chapter 9, Question 9.3: From Berge LN, Marton V, Tranebjaerg L, et al. Prenatal diagnosis of osteogenesis imperfecta. *Acta Obstet Gynecol Scand.* 1995;74(4):321–323.

Figure 18.11: The Crookston Collection, University of Toronto.

Figure 22.13: *Urbana Atlas of Pathology*, University of Illinois College of Medicine at Urbana-Champaign. Image number 26.

Figure 22.19: Interactive Case Study Companion to Robbins Pathologic Basis of Disease.

Figure 23.12: Custom Medical Stock Photo.

Figure 25.21: Success in MRCO path: http://www.mrcophth.com/iriscases/albinism.html.

Figure 25.23A: Rubin E, Farber JL. *Pathology*, 2nd ed. Philadelphia, PA: J. B. Lippincott, 1994:244. Figure 6-30.

Figure 25.23B: Gilbert-Barness E, Barness L. *Metabolic Disease*. Natick, MA: Eaton Publishing, 2000:42. Figure 15.

Figure 26.6: Rich MW. Porphyria cutanea tarda. *Postgrad Med.* 1999;105:208–214.

Figure 26.7: Department of Dermatology, University of Pittsburgh. http://www.upmc.edu/dermatology/MedStudentInfo/introLecture/enlarged/vespct.htm.

Figure 26.11: Custom Medical Stock Photo, Inc.

Figure 26.14: Phototake.

Figure 27.2: Childs G. http://www.cytochemistry.net/.

Figure 27.13: Modified from Cryer PE, Fisher JN, Shamoon H. Hypoglycemia . *Diabetes Care.* 1994;17:734–753.

Figure 28.5: Phototake.

Figure 30.5: Corbis.

Figure 30.6: Gibson W, et al. Hypoglycemia. *J Clin Endocrinol Metab.* 2004;89(10):4821.

Figure 31.22 A, B: Centers for Disease Control and Prevention Public Health Image Library, Atlanta, GA.

Figure 32.4: Matthews JH. Queen's University Department of Medicine, Division of Hematology/Oncology, Kingston, Canada.

Figure 34.16: Wuthrich DA, Lebowitz AL. Tophaceous gout. Images in clinical medicine. *N Engl J Med.* 1995;332:646.

Figure 34.18: WebMD Inc. http://www.samed.com/sam/forms/index.htm.

Online Content, Chapter: Blood Clotting, Figure 10: Foerster J, Lee G, Lukens J, et al. *Wintrobe's Clinical Hematology*, 10th ed. Philadelphia, PA: Lippincott Williams & Wilkins, 1998.

Online Content, Chapter: Blood Clotting, Figure 20: Cohen BJ, Taylor JJ. *Memmler's the Human Body in Health and Disease*, 10th ed. Baltimore, MD: Lippincott Williams & Wilkins, 2005.

Online Content, Clinical Cases (Integrative Cases - Case 1), p. 3: Gold DH, Weingeist TA. *Color Atlas of the Eye in Systemic Disease*. Baltimore, MD: Lippincott Williams & Wilkins, 2001.

Online Content, Clinical Cases (Focused Cases - Case 2), p. 20: Goodheart HP. *Goodheart's Photographs of Common Skin Disorders*, 2nd ed. Philadelphia, PA: Lippincott Williams & Wilkins, 2003.

Online Content, Clinical Cases (Focused Cases - Case 7), p. 24: McConnell TH. *The Nature of Disease Pathology for the Health Professions*. Philadelphia, PA: Lippincott Williams & Wilkins, 2007.

Index

Note: Followed by numbers, italic *t* and *f* denotes Table and Figure, respectively

A

B

C

D

E

M

N

O

P

R

S

T